PATTERSON'S
GERMAN-ENGLISH DICTIONARY
FOR CHEMISTS

PATTERSON'S
GERMAN-ENGLISH DICTIONARY FOR CHEMISTS

by
AUSTIN M. PATTERSON

Revised by
James C. Cox

Edited by
George E. Condoyannis

FOURTH EDITION

JOHN WILEY & SONS, INC.

New York / Chichester / Brisbane / Toronto / Singapore

Library of Congress Cataloging in Publication Data:

Patterson, Austin M. (Austin McDowell), 1876–1956.
 Patterson's German-English dictionary for chemists / revised by James C. Cox; edited by George E. Condoyannis.—4th ed.
  p. cm.
 Rev. ed. of: A German-English dictionary for chemists. 3rd ed. 1950.
 Includes bibliographical references.
 ISBN 0-471-66991-1
 1. Chemistry—Dictionaries—German. 2. German language—Dictionaries—English. I. Cox, James C. II. Condoyannis, George E. (George Edward), 1915– . III. Patterson, Austin M. (Austin McDowell), 1876–1956. A German-English dictionary for chemists. IV. Title.
QD5.P3 1991
540′.3—dc20

90-45365
CIP

To
WILLIAM ALBERT NOYES
1857–1941

EDITOR'S PREFACE TO THE FOURTH EDITION

I was a young instructor in German and other modern languages when lexicography was thrust upon me, along with the harmless drudgery attributed to it by its classic practitioner, Samuel Johnson. I was not altogether unqualified for such a pursuit. Although I was born in New York, my first language was German. My intimate familiarity with the language evidently impressed a colleague who recommended me to a publisher with whom he was attempting to compile a new up-to-date German-English dictionary because of their utter disillusionment with existing works of this sort. The project eventually collapsed of its own weight. We especially lacked experience in limiting its parameters, but even in its failure it turned me into a seasoned lexicographer.

Meanwhile I had joined the staff of the Massachusetts Institute of Technology, where I taught science and engineering students to read technical literature in German as well as French and Russian. Ultimately this led to my being asked to revise the Patterson dictionary. I thought the job would be relatively easy until I started working through the letter A. I must have gone through the first quarter of this letter ten times or more before I had a format that satisfied me. Meanwhile Dr. Cox was brought into the project. Since he is an expert in the field of chemistry, which, after all, is not my real forte, I was now able to concentrate fully on the linguistic aspect of the undertaking. Eventually I came up with a satisfactory system of notation and arrangement, and I accept full responsibility for it, including all my own innovations.

But there was another factor I had to deal with. The third edition of Patterson appeared in 1950, but much of its content harked back to an earlier day, in many ways a simpler and less threatening era when environmental conditions were a matter of much less concern. So while Dr. Cox concentrated on bringing the chemistry and its nomenclature up to date, I concentrated on bringing the vocabulary of the fourth edition up to the eighties and nineties of the twentieth century and to try to anticipate the needs of readers in the twenty-first. We are living in an era preoccupied with the environment, with toxic waste, with new and old eco-systems, with the effect of food and its processing on our health. It is also an age of space travel, of nuclear power and its effect on our lives, of solar energy and the search for nuclear fusion. The German terms for environmental protection, toxic waste, space shuttles, microwave-safe utensils and food additives may be more likely to turn up in popular periodicals than in technical chemical journals, yet they could hardly be omitted from a modern edition of Patterson.

From its first edition on, this dictionary has never limited itself to chemistry in the narrow sense; it has always included basic German vocabulary as well as that of allied fields such as biology and medicine. And there are more such fields today than ever before, not the least of them being biochemistry and food processing. I had to maintain a balance between concentration on highly technical terms in these fields and a more generalized terminology; the former would have doubled or tripled the size of the dictionary.

I not only read popular science articles in German, I also made frequent visits to Germany and Austria, where I walked through department stores and supermarkets reading labels, read catalogs and publicity brochures, and engaged relatives and friends in conversations on environmental matters. After Chernobyl I did not even have to bring up the topics myself.

I gladly accept full responsibility for the inclusion of the above-described vocabulary and for eliminating some outdated terms, old-fashioned spellings, Nazi-era terms, and unsuccessful attempts at Germanization of foreign words such as *Alm* for *Aluminium*.

Since I grew up in the United States, my English is naturally American and this is reflected in my choice of English translations, spelling, and idioms. And although I am fully aware of how languages change, I am not yet ready to accept as standard usage most of the illiteracies now so prevalent in popular speech, and even in print, on radio, and on television. (For example, I still regard *media*, *bacteria*, *criteria*, *phenomena*, and *crises* as plurals.)

I am also a firm believer in the importance of basic grammar and its standard terminology. There is no better way of defining the difference in function of a word than in terms of whether it is a noun, verb, adjective, and so on. There is no better way of defining the uses of a verb than by defining it as transitive, intransitive, and so on. We hardly need to be reminded that the world has gone through vast changes since 1950, changes that have affected not only the field of chemistry, but all natural sciences, changes in both the English and German languages, and changes in lexicography. I have endeavored to bring the Patterson dictionary into line with the most recent trends and practices of the field. But I have also gone beyond that and introduced some notations and devices of my own. I can only hope that all my efforts have resulted in a useful, up-to-date Fourth Edition of *Patterson's German-English Dictionary for Chemists*.

GEORGE E. CONDOYANNIS

New York, N.Y.
September 1991

EDITOR'S BIBLIOGRAPHY: FOURTH EDITION

1. Lutz Mackensen: *Deutsches Wörterbuch, vierte verbesserte und erweiterte Auflage,* 1962, Pfahl-Verlag, Baden-Baden.
2. *Wörterbuch der Deutschen Gegenwartssprache,* Akademie-Verlag, Berlin (DDR), 1977 (6 vols.).
3. DUDEN, *Das große Wörterbuch der Deutschen Sprache in sechs Bänden* Bibliographisches Institut, Mannheim, Dudenverlag, 1977.
4. *Oxford-Harrap Standard German-English Dictionary* Clarendon Press, Oxford, England, 1974 to date. (First 3 vols. A–R, 4th vol. S–Z not yet available in May 1991).
5. *Der Neue Muret-Sanders/Langenscheidts Enzyklopädisches Wörterbuch der Englischen und Deutschen Sprache:* Teil II Deutsch-Englisch 2 Bände, 1974.
6. Neville, Johnston and Boyd, *A New German-English Dictionary for Chemists,* Princeton & Toronto, Van Nostrand, 1964.

PREFACE TO THE FOURTH EDITION

At a dinner meeting of the Division of Chemical Education of the American Chemical Society in 1946, I happened to be seated next to Dr. Austin M. Patterson, the author of previous editions of *A German-English Dictionary for Chemists,* with whom I carried on a lively discussion, during which he commented to me that some day his work in the previous editions would have to be revised.

From that time on, I began the practice of keeping a notebook containing all the German words I encountered in papers that I translated and abstracted as an abstractor for *Chemical Abstracts.* This is a practice that I continued for nearly 40 years through the thousands of chemical papers in German that I abstracted for *Chemical Abstracts.*

At a later date I was assigned for a 10-year period to the translation and abstraction of all papers in *Acta Chemica Scandinavica* appearing in German and French.

In these endeavors, I was greatly encouraged by Dr. Evan J. Crane, at that time the Editor of *Chemical Abstracts,* and by my associates at the University of Delaware, especially Dr. William A. Mosher, Head of the Department of Chemistry, Dr. Glenn S. Skinner, Professor of Organic Chemistry, Dr. Elizabeth Dyer, Professor of Biological Chemistry, and Dr. Quaesita C. Drake, Professor of History of Chemistry.

After several decades my notebook grew to be voluminous, and I wrote to John Wiley & Sons, Inc. in New York to inquire whether the new words I had collected, as well as the additional meanings of words in Patterson's Third Edition, might be of interest. Shortly thereafter I signed a contract to undertake a revision to be known as the Fourth Edition of *A German-English Dictionary for Chemists.* The present work is the result.

Dr. Patterson's plan for the Third Edition has generally been followed. All parts of speech are indicated by the customary marks. Some terms distinguished by the label "Old Chem." have been discontinued to make room for newer terms, while most terms having the same (or practically the same) spelling in German and English have been omitted.

The vocabulary in this edition contains an estimated total of 65,000 terms, primarily for chemists and chemical engineers but useful, as well, in other fields of science and technology. The most recent nomenclature is followed.

I am grateful to Dr. Philip C. Manor, Senior Editor, John Wiley & Sons, Inc. and his staff for continued helpfulness, kindness, and cooperation. I am also grateful to my parents, James C. and Maggie L. Cox (now deceased), both of whom were college German teachers, from whose tongue I first heard the German language spoken, and to my wife, Alma T. Cox, who has been a constant help.

As in all past editions, corrections and suggestions are welcomed.

JAMES C. COX, JR.

Plainview, Texas
September 1991

PREFACE TO THE THIRD EDITION

Although this book is designed primarily for chemists and chemical engineers it contains a large number of terms from related fields of science and hence has proved useful also to physicists, biologists, geologists, etc. It was the first technical dictionary, as far as the author knows, to include a vocabulary of *general* words (see the Introduction). Special attention has been paid to prefixes, suffixes, and abbreviations. All the parts of speech are indicated by the customary marks. Labels are freely used to indicate the field in which a German word has a particular English meaning. Many obsolete or antiquated chemical terms are defined as a help to reading the older literature; where they might be confused with modern meanings they are distinguished by the label *"Old Chem."* or in some other way. (It should be remembered that some terms that have been replaced in chemistry are still in use in pharmacy or industry.) Words that have the same, or practically the same, spelling in German and English have in general been omitted in order to economize space.

The Introduction furnishes hints for using the book and devotes special attention to the peculiarities of German chemical nomenclature.

The vocabulary in this edition contains an estimated total of 59,000 terms, some of the larger additions being in the fields of chemical technology, electronics, and warfare. Additional meanings are given for many terms that were listed in the earlier editions. The latest decisions in nomenclature are followed; in particular, the Stock valence names for inorganic compounds are freely given, as well as the older names.

The German-English vocabularies chiefly consulted are as follows: *General:* Duden, *Orthographisches Wörterbuch;* Flügel-Schmidt-Tanger, *Deutsch-Englisches Wörterbuch;* Hebert-Hirsch, *New German-English Dictionary;* Muret-Sanders, *Encyclopädisches Deutsch-Englisches Wörterbuch. Technical:* Allied Control Commission for Germany, *Glossary of Technical Terms of the Oil Industry;* Artschwager, *Dictionary of Biological Equivalents;* Bading, *Wörterbuch der Landwirtschaft;* Cissarz-Jones, *German-English Geological Terminology;* De Vries, *German-English Science Dictionary;* Eger, *Technologisches Wörterbuch;* Freeman, *Das Englische Fachwort;* Freeman, *Deutsch-Englisches Spezial-Wörterbuch für das Maschinenwesen;* Freeman, *Elektrotechnisches Englisch;* Freeman, *Fachwörterbuch der Metallurgie;* Harras, *Technologisches Wörterbuch;* Hinnerichs, *Deutsch-Englisches Fachwörterbuch;* Hoyer-Kreuter-

Schlomann, *Technologisches Wörterbuch* (6th edition); I. G. Farbenindustrie, *Fachwörterbuch für die Farbstoffe und Textilhilfsmittel verbrauchenden Industrien;* King-Fromherz, *German-English Chemical Terminology;* König-Bohrisch, *Warenlexikon für den Verkehr mit Drogen und Chemikalien;* Lang-Meyers, *German-English Medical Dictionary;* Lingen, *Deutsch-English Fachwörterbuch Papier und Druck;* Mayer, *Chemisches Fachwörterbuch;* Partridge, *Dictionary of Bacteriological Equivalents;* Pietzsch, *Fremdsprachliches Optisches Wörterbuch;* Regen-Regen, *German-English Dictionary for Electronics Engineers and Physicists;* Ruis, *Bergbau Fachwörterbuch;* v. Seebach, *Fachwörterbuch für Bergbautechnik und Bergbauwirtschaft;* Singer, *German-English Dictionary of Metallurgy;* Slater, *Pitman's Technical Dictionary of Engineering and Industrial Science;* U.S. Air Force *German-English Technical Dictionary* (ed. Leidecker); Webel, *German-English Technical and Scientific Dictionary;* Wittfoht, *Kunststofftechnisches Wörterbuch;* and various glossaries. Textbooks, journals, and dealers' catalogs have also been examined for words.

The author is grateful to the many friends and correspondents who have cooperated in improving the dictionary. He is specially indebted to Prof. William T. Hall, Prof. C. W. Foulk, Prof. Paul Rothemund and Miss Janet D. Scott for assistance in proofreading and for suggestions; to Dr. E. J. Crane and the *Chemical Abstracts* corps for general help; to Prof. H. V. Knorr for coöperation on terms in atomic structure; and to Mr. T. E. R. Singer for aid in procuring German books after World War II. Others who deserve special mention are: Dr. Nicolas W. Baklanoff, Dr. D. D. Berolzheimer, Dr. C. C. Davis, Dr. Francis C. Frary, Miss Mildred W. Grafflin and associates, Dr. Geo. C. O. Haas, and Prof. Kurt F. Leidecker.

As in the past, corrections and suggestions for additions are welcomed.

Austin M. Patterson

Xenia, Ohio
February, 1950

OLD STYLE
GERMAN LETTERS

With Their Roman Equivalents

Although for the last seventy years most scientific literature has been printed in the roman alphabet (the one also used in English), the old-style alphabet usually called *Fraktur* or *Gotisch* is given here because it was still occasionally used until about 1950.

𝔄 α	A a	𝔒 o	O o
𝔄̈ ä	Ä ä	𝔒̈ ö	Ö ö
𝔅 b	B b	𝔓 p	P p
ℭ c	C c	𝔔 q	Q q
𝔇 d	D d	𝔑 r	R r
𝔈 e	E e	𝔖 ſ s*	S s
𝔉 f	F f	𝔗 t	T t
𝔊 g	G g	𝔘 u	U u
𝔥 h	H h	𝔘̈ ü	Ü ü
𝔍 i	I i	𝔙 v	V v
𝔍 j	J j	𝔚 w	W w
𝔎 f	K k	𝔛 x	X x
𝔏 l	L l	𝔜 y	Y y
𝔐 m	M m	𝔷 z	Z z
𝔑 n	N n	ß	ß

*The second form of the small letter occurs in final position only, not merely in whole words, but also in their component syllables.

THE GREEK ALPHABET

A α	alpha		Ξ ξ	xi
B β	beta		O o	omicron
Γ γ	gamma		Π π	pi
Δ δ	delta		P ϱ	rho
E ε	epsilon		Σ σ	sigma
Z ζ	zeta		ς	*final* sigma
H η	eta		T τ	tau
Θ θ	theta		Y υ	upsilon
I ι	iota		Φ φ	phi
K ϰ	kappa		X χ	chi
Λ λ	lambda		Ψ ψ	psi
M μ	mu		Ω ω	omega
N ν	nu			

ABBREVIATIONS

a	adjective.	comb.	combination, combined
abbv	abbreviation	compar	comparative
accdg	according	compd(s).	compound(s)*
accus	accusative	Cosmet.	Cosmetics
adj	adjective	Cryst.	Crystallography, Crystals
adv	adverb, adverbial	Cybern.	Cybernetics
Aero.	Aeronautics	dat	dative
affirm	affirmative	def art	definite article
Am.	American	demons	demonstrative
amt	amount	deriv	derived, derivative†
Anal.	Analysis	d.i.o.	dative indirect object
Anat.	Anatomy	Dterg.	Detergents
ans	answer	Dye.	Dyes, Dyeing
art	article	Econ.	Economics
Astron.	Astronomy	EG	East German(y)
Atom.	Atomic Physics	e.g., e.g.	for example
aux.	auxiliary	Elec.	Electricity, Electric
Bact.	Bacteria, Bacteriology	Electron.	Electronics
Biochem.	Biochemistry	emphat.	emphatic
Biol.	Biology	endg(s)	ending(s)
Blch.	Bleach(ing)	Engg.	Engineering
bldg	building	Entom.	Entomology
Bot.	Botany	equiv	equivalent
Brew.	Brewing	erron	erroneous(ly)
Brit.	British	esp	especially
cap	capital(ized)	Eur.	European
Ceram.	Ceramics	exc	except
cf, Cf	compare	exclam	exclamation
Chem.	Chemistry, Chemical	Explo.	Explosives
Cmptrs.	Computers	expr	expression(s)
cnjc.	conjunction	f	feminine
Com.	Commerce		

*Refers to compound words rather than chemical compounds.
†Refers to derived words rather than chemical derivatives.

fig	figurative(ly)	*m*	masculine
Fin.	Finance, financial	*Mach.*	Machinery
Flot.	Flotation	*Magn.*	Magnetics, Magnets
Foun.	Founding, Foundry	*masc*	masculine
fpl	feminine plural	*Math.*	Mathematics
Frac.	Fraction(s)	*Meas.*	Measurements
fut	future (tense)	*Med.*	Medical, Medicine
genit	genitive	*Meteor.*	Meteorology
genl	general(ly)	*Metll.*	Metallurgy
Geog.	Geography, geographical	*Mfg.*	Manufacturing
		Micros.	Microscopy
Geol.	Geology, geological	*Mil.*	Military
Ger	German(y)	*Min.*	Mineral(s), Mining
gram	grammar, grammatical	*mpl*	masculine plural
		n	neuter
(h)	perfect tenses formed with aux v **haben**[*]	*Nat. Sci.*	Natural Science(s)
		neg	negative
		Nomencl.	Nomenclature
(h or s)	aux v **haben** or **sein**[*]	*nomin*	nominative
		npl	neuter plural
(h/s)	aux v **haben** & **sein**	*Nucl*	Nuclear (Physics)
Hist.	History, historical	*num*	numeral, numerical
hyph	hyphenated	*obj*	object
imperat	imperative	*obsol*	obsolete
impers, imps	impersonal	*obsolsc*	obsolescent
incl	including	*oft*	often
indef	indefinite	*opp*	opposite
indir quot	indirect quotation	*Opt.*	Optics
Indus.	Industry, industrial	*orig*	original
inf	infinitive	*Org*	Organic (Chemistry)
inorg, Inorg.	Inorganic (Chemistry)	*p.a*	participial adjective
		p. adv	participial adverb
insep	inseparable	*Parlam.*	Parliamentary
Instrum.	Instrumentation	*Petrog.*	Petrography
interrog	interrogative	*Petrol.*	Petroleum
intj	interjection	*pfx*	prefix
Lat.	Latin	*Pharm.*	Pharmacy, Pharmaceutics
lit	literal(ly)		
Lthr.	Leather	*Philos.*	Philosophy

[*]See Editor's Introduction **10.5**

Phot.	*Photography*	*specif*	*specifical(ly)*
phr	*phrase(s)*	*Spect.*	*Spectrography*
Phys	*Physics*	*Stat.*	*Statistics*
Physiol.	*Physiology.*	*sthg*	*something*
pl	*plural*	*subord*	*subordinate,*
Polit.	*Politics, political*		*subordinating*
possess	*possessive*	*superl*	*superlative*
pp	*past participle*	*syll*	*syllable*
pred	*predicate*	*Synth.*	*Synthetics*
prep	*preposition*	*Techn.*	*Technology*
pres	*present (tense)*	*Termin.*	*Terminology*
pres p	*present participle*	*Tex.*	*Textiles*
Print.	*Print(ing)*	*Tgph.*	*Telegraph*
pron	*pronoun*	*Thmodyn.*	*Thermodynamics*
psn, *psn*	*person(al)*	*Tlph.*	*Telephone*
Pyro.	*Pyrotechnics*	*TN*	*Trade Name*
Quant.	*Quantity*	*transl., Transl.*	*translated,*
quot	*quotation*		*Translation*
recip	*reciprocal*	*Transp.*	*Transportation*
reg	*regular*	*unemph*	*unemphatic*
relat	*relative*	*usu*	*usual(ly)*
resp	*respective(ly)*	*v*	*verb*
rflx	*reflexive*	*var*	*variation*
(s)	perfect tenses	*v aux.*	*auxiliary verb*
	formed with *aux v*	*Vet.*	*Veterinary*
	sein	*vi*	*intransitive verb*
sbdy	somebody	*vr*	*reflexive verb*
sbjc	*subjunctive*	*vt*	*transitive verb*
Sci.	*Science, scientific*	*vt/i*	*transitive and*
sep	*separable*		*intransitive verb*
sfx	*suffix*	*WG*	*West German(y)*
sg	*singular*	*Zool.*	*Zoology*
spec	*special*		

Abbreviations

EDITOR'S INTRODUCTION ON FORM, ARRANGEMENT AND WISE USE OF THE DICTIONARY

A two-language dictionary is usually designed for the specific needs of readers fluent in one but not necessarily both of the languages it deals with. This one is tailored to the requirements of English-speaking readers of German technical literature in chemistry and allied fields. It assumes that their knowledge of English will aid them in choosing from among the given English translations the one that best fits the context.

SPECIAL NOTE: The examples of entries quoted in this introduction are simplified in order to focus attention on the particular point being illustrated. In most cases three consecutive dots, . . . , show where portions of the actual entry have been omitted. When necessary, phrases or sentences are included to illustrate a point. These are preceded by a special symbol, >, used only in this introduction, to indicate that they are not part of any entry.

1 GENERAL WORDS. From the very beginning the Patterson dictionaries have always included a richer general vocabulary than many pocket dictionaries, and for two reasons: first, to save the user the trouble of looking up the more common German words in a separate dictionary, and second, because many general words have one or more technical meanings. In a general work these meanings are often either absent or buried among other senses. In this dictionary the special chemical meaning is given priority by being listed first or indicated in other ways, such as by the abbreviation (*Chem.*).

2 SELF-EVIDENT WORDS. This edition continues the Patterson tradition of omitting from the vocabulary so-called "cognates," words common to English and German, even with a slight variation in spelling, which are so readily recognizable that they are not likely to be looked up. Exceptions are:

a Cognates that also have other meanings: **China I** *n* China. **II** *f* (=CHINARINDE) cinchona, Peruvian bark.

b Different meanings: **Advokat** *m* (-en, -en) lawyer.

And a special warning is in order here: Many names of chemical compounds with standardized international endings like **-at, -it, -id, -an, -en, -in, -on,** have English equivalents ending with a final -e (sulfate, chloride, ketene, etc). When the German names do have a final **-e,** they are *plurals* (**Sulfate** = sulfates; etc). A final **-en** may be either a chemical ending or a *dative plural* ending: **Terpen** means "terpene;" **Terpenen** means not "terpenene," but "terpenes." Another warning: if the "obvious" translation does not fit, look up the word in spite of its similarity to English!

3 SPELLING

3.1 Standard German spelling is used, but variant spellings still occur in scientific literature. Warnings regarding these variations appear at the beginning of the listings for the letters **C, I, J, K, T, Z.** The user is advised to read these carefully.

3.2 Umlaut: Ordinarily umlaut has no effect on alphabetical order:

> **Bezug bezüglich Bezugnahme**
> **Anfang Anfänger anfänglich anfangs.**

But if two consecutive entries differ only in that one has an umlaut, the non-umlaut word comes first:

> **fallen** before **fällen, zahlen** before **zählen, Losung** before **Lösung, Stuck** before **Stück.**

Although **Ae, Oe, Ue** are sometimes substituted for **Ä, Ö, Ü,** especially when using a typewriter, this dictionary uses only the umlaut letters.

3.3 The Disappearing e:

a When derivative words are formed from nouns ending in **-el** or verbs ending in **-eln** by means of a suffix beginning with a vowel, notably **-ung** and **-ig,** the **e** before the **l** is practically silent in normal German pronunciation, so there is a strong tendency to drop the **e** in the written word as well:

> **Handel, handeln** → **Handlung** action (*instead of* "**Handelung**")
> **entwickeln** → **Entwicklung** development (*instead of* "**Entwickelung**")

The more elevated or technical the style of the context, the more the **e** is likely to be preserved:

> **Nickel** → **vernickeln** → **Vernickelung** or **Vernicklung** nickel-plating
> **Kugel m** ball, sphere → **kugelig** or **kuglig** spherical, globular; rolled up

Adjectives ending in **-el** regularly drop the **e** when a grammatical ending (always beginning with e) is added:

> **übel:** >**üble Zustände** bad conditions
> **edel:** >**ein edles Metall** a noble metal

The short stems of these adjectives make recognition difficult, so the dictionary lists them in both forms (with and without the disappearing **e**):

> **edel a 1** noble (*incl* metals) . . . *Cf* EDLE (**edel** is the "endingless" form).
> **edle a** noble, etc = EDEL (**edle** consists of stem **edl-** plus ending **-e.**)

Sometimes one of two **e**'s is dropped:

> **dunkel** *a* "dark, dim, obscure" drops the **e** in the same way as **übel** and **edel;** with the ending **-en,** either the **e** before or the one after the **l** may disappear:
>
> "**dunkelen**" → **dunklen** or **dunkeln**   "**übelen**" → **üblen** or **übeln**

b The same often applies to adjectives ending in **-er:**

 sauer *a* (with endings **saur** . . .) sour, acid . . .

But the dictionary also lists:

 saur (+ *adj endgs*) sour, etc. *Cf* SAUER

In older chemical nomenclature compound adjectives ending in **-sauer** designate a salt of the corresponding acid (see the full version of the entry **sauer**):

 Essig *m* (-e) vinegar; (*in compds oft*) acetic, aceto-

 essig·sauer *a* acetate of, acetic . . . :e—e **Tonerde** aluminum acetate (e—e stands for **essigsaure**)

The reader should always be prepared to find the **e** missing when endings are added to

 sauer: saure, saurer, saures, etc.

And as with **dunkel** above, one of two **e**'s may disappear: "**saueren**" → **sauren** or **sauern.**

c When an **e** is optional, the dictionary shows it in parentheses whenever alphabetical order permits:

 Arz(e)nei *f* (-en) medicine **Dunk(e)lungs·bad** *n* (*Dye.*) saddening liquor

Otherwise each form is listed separately, the second in a shortened form with a cross-reference to the first:

 kugelig *a* spherical, globular; rolled-up **kuglig** *a* spherical, etc = KUGELIG

 Teuerung *f* (-en) general price rise *cf* TEUER **Teurung** *f* (en) price rise = TEUERUNG

Cross-references (anything printed in SMALL CAPITALS) can be very helpful. Do not ignore them!

3.4 Triple Consonants

When compounding of nouns and adjectives results in a triple consonant, such a combination of letters is always simplified to a double consonant:

 Zell·leib *m* "cell body" is spelled **Zelleib**

 Kenn·nummer *f* "reference number" is spelled **Kennummer**

 schnell·laufend *a* "fast-running, high-speed" is spelled **schnellaufend**

In order to maintain strict alphabetical order, these compound words must be listed separately in conformity with their shortened spelling and will therefore not be found among the paragraphed **Zell·**, **Kenn·** or **schnell·** compounds, although you may find a cross-reference there to the two-consonant spelling. Moreover, when it is necessary to separate such a word at the end of a line, it reverts to the three-consonant spelling: **Zell-leib,** but it must still be looked up as **Zelleib.** The three-to-two letter simplification also applies to a few separable-prefix verbs:

 volladen (= voll·laden), *pres* **vollädt** or **lädt** . . . **voll,** *pp* **vollgeladen**

Since double **s** in final position is always spelled **ß** no shortening of the spelling is necessary in words like **Fluß·säure** and **Schluß·stein**. *See* **3.5.**

3.5 ss, ß: Standard German spelling includes one letter not used in English: the so-called double s symbol **ß**, called **Eszet** in German. Both **ss** and **ß** represent a *voiceless* s-sound where the letter *s* alone would represent a *voiced* s (like English *z*); **wessen** vs **Wesen, Maße** vs **Hase**. **ss** is used between two short vowels (**Masse, Kessel**), otherwise **ß** is used, *i.e.* between a long and a short vowel (**Maße**), in final position after either a long or a short vowel (**Maß, muß**) and before a consonant (**mißt**). When a word is printed entirely in capitals, **ss** is used in all situations: MASSE for either **Masse** or **Maße**. This is important for the dictionary user because all cross-references are printed in small capitals. No distinction is made between **ß** and **ss** in alphabetizing: **Faß** . . . **Fassade** . . . **faßbar** . . . **fassen** . . .

Another warning: If you come across a word with **ss** before an ending (**des Fasses, die Fässer**), be prepared to find it in the dictionary with a final **ß**: **Faß** n (Fässer). And still another: In Swiss publications **ß** is often avoided and replaced by **ss**.

3.6 Capitalization. The fact that all German nouns and words used as nouns are capitalized causes some problems for lexicographers:

a When identical words differ only as regards capitalization, the small-letter spelling precedes: **essen** before **Essen, laden** before **Laden, öde** before **Öde.** This can be important because of the way entries with a common beginning are often paragraphed. If you are looking for **Angst,** you may find it in the paragraph beginning with **angst;** the same applies to **fett** and **Fett** and similar pairs. If the word you want is not at the beginning of a line or paragraph, look inside. All key words are printed in **boldface** type.

b Compound nouns and adjectives are listed in continuous paragraphs. Since nouns are capitalized and adjectives (as well as other parts of speech) are not, the differences are indicated as follows:

> **Voll·mehl** n whole grain flour. **-milch** f whole milk. **v–mundig** a (*Beer,* etc),
> full-bodied. **V–mundigkeit** f body, fullness.

> (all words in this paragraph begin with **voll: -milch** stands for **Vollmilch,**
> **v–mundig** stands for **vollmundig, V–mundigkeit** for **Vollmundigkeit.**
> *See also* **5.3.**

4. KNOWLEDGE OF GRAMMAR: The dictionary necessarily assumes that you are familiar with basic German grammar and its terminology. All words are tagged as to part of speech: noun (by gender: *m*asculine, *n*euter, *f*eminine or *pl*ural), *pro*noun, *a*djective, *art*icle, *v*erb, *adv*erb, conjunction (*cnjc*), *prep*osition, interjection (*intj*), *num*eral and a few others.

5 SPACE-SAVING DEVICES:

The ideal dictionary would start each entry on a new line. Unfortunately such an arrangement would consume so much space as to make the price prohibitive. Saving space thus becomes the lexicographer's overriding concern.

5.1 The best space-saving device is to **paragraph consecutive entries** that have a common initial element:

   Abseite° *f* . . . **abseitg** *a* . . . **abseits 1** *adv* . . . **2** *prep* . . .
   ab·teilen *vt* . . . **Abteilung** *f* . . . **abteilungs·weise** *adv* . . .

So if the entry you are looking for is not at the beginning of a line or paragraph, look for it further inside the paragraph. All key words are printed in boldface type.

5.2 More Details on Paragraphing:

Note the following arrangement:

   konsistent *a* solid, compact, firm; consistent. **Konsistenz** *f* (-en) solidity,
      etc. **··messer** *m* consistometer

This involves two space-saving devices:

a The abbreviation "etc" indicates that the remaining meanings of **Konsistenz** can be derived from the adjective **konsistent,** namely: compactness, firmness; consistency.

b The notation **··messer** indicates a compound noun whose first component is the immediately preceding entry, namely the noun **Konsistenz; ··messer** therefore stands for **Konsistenzmesser.**

c Another example:

   Person *f* (-en) person . . . **Personal** *n* personnel. **p··los** *a* unmanned

The notation **p··** refers to the immediately preceding entry in the paragraph, hence **p··los** stands for **personallos.** *See also* **18.3, 18.9a.**

5.3 Paragraphed Compounds:

By far the most frequent application of paragraphing to save space occurs in listing compound words (usually nouns and adjectives):

   Ader·haut *f* choroid. **-häutchen** *n* chorion. **-hautspanner** *m* ciliary muscle.
      -holz *n* wood cut with the grain.

Each hyphen in this paragraph stands for the word **Ader: Aderhäutchen,** etc.

Strict alphabetical order requires that the **Ader·** compounds be interrupted after **-holz** (= **Aderholz**) to list the adjectives **aderig, äderig** *a* . . . after which the **Ader·** compounds resume in a new paragraph:

   Ader·knoten, -knopf *m* varicose vein. **-laß** *m* bloodletting

So if you are looking for the entry **Aderlaß,** you would have to look in the next **Ader·** paragraph, and for **Adernetz** in a later paragraph after further interruptions. If a compound noun or adjective or even a verb is not found in the first paragraph of such compounds, look in a later one.

5.4 The first (and sometimes only) series of **compounds** is often **appended** to a plain noun entry:

> **Achsel** *f* (-n) shoulder . . . **-bein** *n* shoulder blade. **-drüse** *f* axillary gland. **-grube, -höhle** *f* armpit, axilla. **a–ständig** *a* axillary
> *Cf* **5.2** and **18.3.**

5.5 Unlisted Compounds:

Obviously not all possible compound nouns and adjectives and all prefixed verbs can be listed. Writers are at liberty to make up new ones at will, and any new invention, device, technique or process has to be given a new name, which will more than likely be a new compound noun. The reader can resort to several expedients to track down the meaning of an unlisted compound. The simplest and most obvious method is to break the word down into its components: **Öl-schwemme** splits up into **Öl** "oil" and **Schwemme** *f;* the latter entry has a division **2** with the translation "glut"; this adds up to the translation "oil glut."

5.6 The dictionary also provides some **aids in implementing this process:** Translations may be inserted after the first component at the head of a paragraph of compounds:

> **Ätz·** etching; caustic corrosive: **-farbe** *f* etching ink; (*Tex.*) discharge paste . . .

Although a large number of compounds follows, there is always the possibility that the one you are looking for is not listed. At least in that case the translations listed for **Ätz·** will help in interpreting any additional compounds.

5.7 In some entries a separate division marked "(*in compds.*)" is included to provide translations of the word when it occurs as a prefix or first component of a compound:

> **außen 1** *adv, pred a* outside: **nach a.** to(ward) the outside, out(ward) . . . **2** (*in compds*) outer, external, exterior, outside, outdoor; foreign: **Außen·antenne** *f* outside (outdoor, external) antenna . . .

The translations given under **2** will also help to interpret any **außen-**compounds not explicitly listed in the dictionary.

5.8 But what if there are **no specific translations** listed for the first component of a series of compound words? As a practical example, assume that you want to look up the word **Angriffsziel.** First, it helps to know that compounding often involves a connective **s, n** or **en** added to the first component. With a little effort you can find an entry **Angriff** and a paragraph beginning with **Angriffs;** but no specific translations for this first component. Second, scan the **Angriffs·** paragraph for the common element that will give you the basic meaning or meanings of **Angriffs·** . Then check these with the entry **Angriff,** where you will find confirmation that all these words deal with "attack, assault, offensive(s)" or some kind of aggression. If you do not already know the meaning of **Ziel,** look it up (goal, aim, destination, etc). Putting the two concepts together, you get "goal or aim of the attack," and with a little ingenuity you can interpret this as "objective." By the same process **Angriffsvorbereitungen** comes out literally as "attack preparations," or in more typical English, "preparations for the attack."

5.9 Some entries are listed only as **first components** of compounds, e.g.:

Einphasen· (*in compds*) single-phase

Although no actual compounds are listed, this entry will help to interpret compounds like **Einphasenstrom**"single-phase current." *See also* **18.2.**

Further space-saving devices are mentioned throughout this introduction.

6. Form and Arrangement of Entries
6.1 English Equivalents and Translations:

People who have learned to use their native language know that it is a rare word that has only one meaning. When a second language enters the linguistic picture, the problem is compounded by the sometimes unnerving realization that each meaning of an English word will have a different translation in any other language. Of course, the converse is also true. The majority of the German words listed in this dictionary—other than very specific scientific terms—have a variety of meanings, each of which can be expected to have a different English translation or set of translations. Simply listing all these translations, or even a selection of them, after the German word only confuses or repels the user. The translations must be systematically organized by grouping and defining areas of meaning. This can be a simple matter in most entries but in others it can become a highly complex undertaking.

6.2 Selection of Translations

First of all, the translations given in this dictionary for each German entry are not all possible ones. They are usually chosen to give an overall representative picture of the range or ranges of meaning of the German word. There is no guarantee that the translations given will always result in a perfectly polished English rendition of the passage in question. That can only be achieved by applying one's own command of good English style to the task.

6.3 For most entries the job of **organizing the translations** can be accomplished by using commas to separate more or less interchangeable equivalents, while semicolons are used to separate more distinct ranges of meaning:

ab·dünsten *vt* evaporate, vaporize; graduate (brine); steam, stew

Particularly sharp distinctions are best made clear by using numbered divisions, and in more complex entries there may be a whole array of such numbered divisions, within which there may be translations separated by commas and semicolons. Sometimes even the Arabic-numbered divisions are not sufficient to cope with the complexities of an entry. Roman numerals are used mainly in verb entries to separate transitive, intransitive, reflexive, and auxiliary uses, each of which may have several areas of meaning grouped in Arabic-numbered divisions. Even capital letters may be necessary to designate larger categories:

an·hängen A (*reg*) **I** *vt* (*oft* + **an**) **1** hang on, couple on, hitch on (to). **2** append, add, attach (to). **3** (*Tlph*): **den Hörer a.** hang up. **4** (*d.i.o*) impose, inflict (on), burden, saddle (with); palm off (on). **II. sich a. 5** (**an** or *dat*) hang on, cling; attach oneself (to), force oneself (on). **B** (*reg* or *) **III** *vi* (*usu* + *dat*) **6** be attached, cling (to); **es hängt ihnen an** they are burdened with it. **7** be an adherent (*or* a follower) of. **8** (*Tlph*.) hang up.

In noun entries Roman numerals serve to show that the noun has two or more distinct ranges of meaning that may go back to separate origins, so that it is actually a matter of homonyms:

Ton *m* **I** (-e) clay. **II** (⁻e) **1** tone . . . ; sound, note; audio . . .

Often the homonyms even have different genders:

Messer I *m* (-) meter, measuring device. **II** *n* (-) knife . . .

Roman numerals may also be necessary for words that are both adverbs and conjunctions, prepositions and prefixes, and many other such combinations. Every numbered division of an entry is to be regarded as a distinct unit. Any abbreviation or illustrative phrase that appears in it applies to that division only.

6.4 Total or near-total equivalents:

Occasionally there will be a German word that coincides with an English one in all or almost all meanings:

Grad *m* (-e) degree (*all meanings*); rank

Magazin *n* (-e) magazine (*all meanings*)

6.5 Phrases and Idiomatic Expressions:

a Some words are frequently used in common phrases or expressions, and no bilingual dictionary can claim to do these words justice without including such phrases and their translations, which are rarely literal:

Zug *m* (⁻e) . . . **9** tendency, trend: **der Z. der Zeit** the trend of the times

Note that the key word is quoted by using only its initial. If the word has to be used in the quoted phrase with an ending, the ending is added after a dash:

absehbar *a* . . . foreseeable: **in a—er Zeit** in the foreseeable future (**a—er** = **absehbarer**)

Zug *m* (⁻e) . . . **4** breath, puff; draft, sip **—in vollen Z⁻en genießen** enjoy fully

Note first of all how the umlaut is indicated in the phrase: (Z⁻en = Zügen). Note also that in this instance the phrase is introduced by a dash instead of a colon. This is done when the phrase is highly idiomatic and the translation is far from literal.

b A phrase may also serve to clarify an ambiguous English translation:

> **Sache** *f* (-n) . . . 3 cause: **gemeinsame S. machen** make common cause . . .

(The phrase makes it clear which meaning of "cause" is intended.)

> *See also* **18.4** and **18.11f, g.**

c Some words are used mainly in phrases:

> **Anspruch°** *m* claim, demand: **A. haben (auf)** have a claim (on), have a right (to) . . . ; **in A. nehmen** take up, occupy (time, space), make demands on, keep busy

Here the first phrase is fairly literal and is introduced by a colon. The second phrase is more idiomatic. When two or more phrases are included in a division, the more literal ones come first, the most idiomatic last. All phrases after the first are separated by semicolons.

d Some words are used only in one or more standard phrases; in that case no general translation is given:

> **Betracht** *m:* **in B. kommen (ziehen)** come (take) into consideration

Or a word may have many meanings, one of which occurs only in a phrase. The entry **Mittel** *n* (-) has six divisions with various meanings, and then:

> **7: sich ins M. legen** mediate, intercede

6.6 The same notation is used when a **phrase** is **quoted within a paragraph** of compounds:

> **Kern·macher** *m* coremaker. **-macherei** *f* (*Foun.*) core-making (shop) . . .
>   **-seife** *f* curd soap: **abgesetzte K.** curd-settled soap

Since all of the compounds in the paragraph begin with **Kern·** , the **K.** in the phrase stands for **Kernseife.**

6.7 Space limitations prohibit listing all of the many slightly **idiomatic contexts** in which some words are used. In that case you can use a combination of imagination and ingenuity. In the entry **Hintergrund** *m* "background," the frequent phrase **in den H. treten** is not explicitly included, but a literal translation yields the wording "step into the background." It is not difficult to reword this into the more familiar English phrase "recede into the background." Similar considerations hold when **Vordergrund** (foreground) and **Ruhestand** (retirement) are used with **treten:** the literal translations with "step" can easily be reworded as "come to the fore" and "go into retirement."

7 PARTS OF SPEECH

7.1 Nouns: Plurals

One of the chief innovations of this edition is to include plurals of nouns. This is done to aid the reader in finding the correct entry. If you try to look up the word **Anlässe** and find only **Anlaß**, the notation (..lässe), specifying the plural, will assure you that you have found the right entry. See item **j.** Plurals are given in standard notation, using a hyphen to represent the noun in its original singular form:

a No change: **Kasten** *m* (-)

b Umlaut but no ending: **Vogel** *m* (:) (*pl* is **Vögel**)

c Ending -e: **Jahr** *n* (-e) (*pl* **Jahre**)

d Umlaut and -e: **Baum** *m* (:e) (*pl* **Bäume**)

e Ending -er with umlaut where possible, *i.e.* on **a**, **o** or **u**:

    Huhn *n* (:er) (*pl* **Hühner**)   **Mann** *m* (:er) (*pl* **Männer**)

    Kind *n* (-er) (*pl* **Kinder**)   **Haus** *n* (:er) (*pl* **Häuser**)

f Ending -se: **Ergebnis** *n* (-se) (*pl* **Ergebnisse**)

If the plural of any of the preceding nouns appears with an -n added to the plural ending, you have come across a dative plural:

    zu den Vögeln; vor Jahren; unter den Bäumen; mit den Kindern

g Ending -n, -en, or -nen, no umlaut:

    Blume *f* (-n) (*pl* **Blumen**)   **Nadel** *f* (-n) (*pl* **Nadeln**)

    Bett *n* (-en) (*pl* **Betten**)   **Ärztin** *f* (-nen) (*pl* **Ärztinnen**)

h Ending -s, no umlaut; only on nouns of foreign origin:

    Auto *n* (-s) (*pl* **Autos**)   **Scheck** *m* (-s) (*pl* **Schecks**) (*French* cheque)

i Some nouns of Latin origin form their plural with a Germanized ending:

    Material *n* (-ien) (*pl* **Materialien**) (*Latin pl* materialia)

But the majority of nouns of Latin, Greek or Italian origin substitute either their original plural ending or a new German one for the singular ending. A special notation is used to indicate this:

    Studium *n* (..dien) (*pl* **Studien**)   **Museum** *n* (..seen) (*pl* **Museen**)

    Drama *n* (..men) (*pl* **Dramen**)   **Lexikon** *n* (..ika *or* ..iken) (*pl* **Lexika** *or* **Lexiken**)

    Konto *n* (-s, ..ti *or* ..ten) (*pl* **Kontos, Konti** *or* **Konten**) (*Ital.* conto, *pl* conti)

Since many such nouns have a singular identical to an English word, they are not actually listed in the dictionary, but their plural may be there as a separate entry. This is especially necessary where the shortness of the stem of such nouns makes their plural difficult to recognize:

    Viren viruses: *pl of* **Virus**   **Seren** sera, serums: *pl of* **Serum**

    Museen museums: *pl of* Museum

j Nouns ending in **ß**, whether of native or foreign origin, require a special notation:

    Kompaß *m* (..asse) (ship's) compass   **Fluß** *m* (Flüsse) . . . river . . .

    See also **3.5**

k Nouns listed with two endings in parentheses (with a comma between) are usually so-called n-types or masculine weak nouns:

    Biologe *m* (-en, -en) biologist   **Photograph** *m* (-en, -en) photographer

The first of the two endings is attached to the noun in all remaining *singular* case forms (genitive, dative, and accusative: **des Biologen, dem Biologen, den Biologen**), while the second ending is the one used for all plural case forms: **die, der, den Biologen**.

l If no plural or reference to where a plural can be found is given for a simple (uncompounded) noun, it usually means the noun is never or rarely used in the plural:

> **Milch** *f* milk . . . , **Wärme** *f* heat, warmth . . . and the names of chemical elements like **Bor** *n* boron, **Natrium** *n* sodium, **Phosphor** *m* phosphorus, etc

m Some nouns are always plural:

> **Ferien** *pl* vacation   **Leute** *pl* people   **Kosten** *pl* cost(s), expenses

n Plurals are not listed for derivative (*usu* prefixed) nouns:

> **Abseite°** *f* . . . (*incl Tex.*) wrong side . . .

unless the corresponding noun does not exist or is not listed:

> **Abartung** *f* (-en) (There is no noun "**Artung.**")
>
> **Ankunft** *f* (ː̈e) (There is no noun "**Kunft.**")
>
> **Anlaß** *m* (..lässe) (There is no noun "**Laß.**")
>
> **Auseinandersetzung** *f* (-en) (The noun **Setzung** is not listed.)

The small raised circle ° marking the entry **Abseite** is a signal that this word is a prefixed derivative. A look at surrounding entries reveals that the prefix is **Ab-**; the plural will be found under **Seite.** *See also* **18.14.**

o Plurals are not listed for compound nouns:

> **Abwasser·anlage** *f* sewage disposal plant   (For *pl* see **Anlage.**)
>
> **Arbeits·markt** *m* labor market   (For *pl* see **Markt.**)

But the last component may be a word used mainly in the plural and be so listed: **Apotheker·waren** *fpl* pharmaceuticals, drugs

p Sometimes the plural of one noun will look like the singular of another:

> **Zweck** *m* (-e) . . . purpose . . . *cf* ZWECKE
>
> **Zwecke** *f* (-en) . . . tack . . . *cf* ZWECK

Usually cross-references are included in each entry so that you can check both entries to make sure you have the right word.

7.2 Adjectives used as nouns, including compounds, are marked (*adj endgs*):

> **Gerade** *f* (*adj endgs*) straight line
>
> **Verwandte** *m, f* (*adj endgs*) relative, family member

7.3 Verbal Nouns

The user is expected to know that nouns can be derived from verbs in the following three ways:

a By simply capitalizing the infinitive and using it as a neuter noun; the English translation usually ends in -ing:

> **schmelzen*** . . . melt . . . **das Schmelzen** melting >**beim Sch.** in (on, when) melting.

Such nouns are not listed unless they have some additional meaning not derivable from the verb infinitive by using the ending -ing:

> **an·faulen** *vi* (s) begin to rot . . . **Anfaulen** *n* incipient rot

an·halten* I *vt* **1** stop, bring to a halt, detain . . . **3 (zu)** encourage, urge (to).
    II *vi* **4** stop, come to a halt, pause. **5 (um)** apply (for). **6** last, keep up, continue.
    Anhalten *n* **1** stop, halt: **zum A. bringen** bring to a halt. **2** detention . . .
    3 arrangement . . . **4** application. **5** continuance, persistence . . .

Note that such nouns are appended to the verb entry, and remember that in addition to the translations given, these nouns can always have as well the translations derived from the infinitive by using the -ing ending. For **Anhalten** these would be "stopping, halting, detaining, encouraging, urging, pausing, applying, lasting, keeping up, continuing." The context must determine whether the translation will be one of these or one of those listed in the entry **Anhalten** itself.

Sometimes space is saved by not explicitly listing the verbal noun but using the capitalized initial in a phrase included in the verb entry:

    an·sieden* *vt* boil . . . —**A. und Abdunkeln** stuffing and saddening

The A. stands for **Ansieden.**
b Nouns are also formed from verbs—especially prefixed verbs—by changing the final **-en** of the infinitive to **-ung** and capitalizing the resulting word, which is always a feminine noun:

    trennen *vt* separate . . . → **Trennung** *f* (-en) separation

    an·wenden* *vt* use, . . . apply → **Anwendung** *f* (-en) use, application . . .

Verbs ending in **-eln** or **-ern** require some spelling adjustments:

    entwickeln *vt* . . . develop → **Entwicklung** *f* (-en) development (*See also* **3.3.**)

    wandern *vi* . . . wander, migrate → **Wanderung** *f* (-en) wandering, migration

The meaning of these **-ung** nouns often coincides or overlaps with that of the capitalized infinitive: **die Wanderung** and **das Wandern** can both mean "wandering *or* migration." By using your knowledge of English with a dash of ingenuity you can usually make up your own translations for these **-ung** nouns, and this is exactly what the dictionary relies on you to do:

    nach·eichen *vt* recalibrate, restandardize; realine. **Nacheichung°** *f* recalibration, etc.

It should not be difficult to fill in the "etc" by deriving "restandardization; realinement" from "restandardize; realine."

The verb entry **ab·brennen** has many divisions with a rich variety of meanings, of which a selection is given here:

    ab·brennen* I *vt, vi*(s) **1** burn off (away, down); deflagrate. **II** *vi* **2** fire off (gun, etc). **3** calcine; temper. **4** dip, pickle. **5** cauterize; sear off, singe. **6** (*Ceram.*) finish firing . . .

A few lines down the page the noun entry **Abbrennung** is presented as follows:

    Abbrennung° *f* burning off (away, down); deflagration, etc *cf* ABBRENNEN

The "etc" and the "*cf* ABBRENNEN" direct you to look at the entry **ab·brennen** and, according to context, convert any relevant meanings of the verb to use as nouns: firing off; calcination; tempering; dipping, pickling; cauterization; searing off; singeing; completion of firing.

The same procedure can be applied even when the **-ung** noun is not listed at all:

> **ab·isolieren** *vt* isolate, insulate; seal (surface coating)

For **Abisolierung**, not listed, the translations would be "isolation, insulation; sealing."

c There is a third way to form verbal nouns, usually from a strong (irregular) past tense, but occasionally also from a present-tense stem:

> **Gabe** *f* giving, gift (*from* geben*, gab)
>
> **Stand** *m* stand, status (*from* stehen*, stand)
>
> **Wuchs** *m* growth, stature (*from* wachsen*, wuchs)
>
> **Zwang** *m* force, coercion (*from* zwingen*, zwang)
>
> **Schlag** *m* blow, stroke (*from* schlagen*)

Prefixed nouns are especially frequent: **Abgabe, Zustand, Nachwuchs, Zuschlag.** Some occur only with prefixes:

> **Annahme** *f* (-n) acceptance; assumption (*from* an·nehmen,* nahm an)
>
> **Anlaß** *m* (..lässe) cause, occasion (*from* an·lassen*)

Verbs whose past tense has the vowel **o** have corresponding verbal nouns with a **u:**

> **Fluß** *m* (Flüsse) flow, river, etc (from fließen,* floß); *also* **Abfluß, Einfluß,** etc
>
> **Zug** *m* (Züge) pull, draw, etc (*from* ziehen*, zog); *also* **Abzug, Zuzug,** etc

Such nouns are listed as complete separate entries, but it is important to know that they are verbal nouns when this term is used in entries like:

> **unter I** *prep* . . . 4 (*esp* + *verbal nouns*) with, accompanied by: **u. Bezugnahme auf** with reference to

7.4 Nouns are also **derived from adjectives** by means of the suffixes **·heit** and **·keit**, which produce feminine nouns. The second of these is a variation of the first and is added mainly to adjectives ending in **-bar, -ig, -lich** and **-sam:**

> **frei** *a* free . . . → **Freiheit** *f* (-en) freedom, liberty
>
> **fruchtbar** *a* fruitful, fertile: **Fruchtbarkeit** *f* fruitfulness, fertility

Sometimes **-ig-** is inserted before adding **-keit: Festigkeit** *f* "solidity, strength" is derived from **fest** *a* solid, firm . . . (There is no adjective **"festig."**) All adjectives ending in **-haft** also insert **-ig-:**

> **mangelhaft** *a* defective, flawed → **Mangelhaftigkeit** *f* defectiveness

As with the nouns ending in **-ung,** space limitations make it impossible to list all nouns ending in **-heit** or **-keit,** so here too the reader will have to derive the translation from the adjectives:

biegsam *a* flexible, pliable; supple; pliant; docile. **Biegsamkeit** *f* flexibility, etc

For the "etc" you can easily make up "pliability, suppleness, pliancy, and docility."

8 ADJECTIVES are listed without endings. You are expected to know enough about adjective endings or their absence to look up the words concerned.

8.1 Comparatives and Superlatives: Irregular or umlauted ones are given in adjective entries:

> **groß** *a* (¨-er,..ößt-): (*compar* größer, *superl* größt-; the superlative always has an ending or appears in the phrase **am größten.**)

Highly irregular comparatives like **besser** and **lieber** are, of course, listed separately. You should also know enough about adjectives to be wary of the ending **-er**, which does not always mean the form is a comparative:

> >**ein lauter Schrei** a loud cry: look up **laut.**
> >**Es wird immer lauter.** It keeps getting louder: look up **laut.**
> >**lauter Unsinn** nothing but nonsense: look up **lauter.** (If you look up **laut,** you will see *cf* LAUTER to direct you to the right entry.)
> >**ein lauterer Ton** a pure tone: look up **lauter.**
> >**Soviel ist sicher . . .** This much is certain . . . : look up **sicher.**
> >**ein sicherer Schluß** a sure (*or* safe) conclusion: look up **sicher.**
> >**ein sichererer Schluß** a safer conclusion: look up **sicher.**

8.2 Since **superlatives** almost always have adjective endings, they are often listed with an **-e** or at least with some indication that there will always be an ending:

> **beste** *a* (*superl of* GUT) . . .   **kleinst . .**   **groß** *a* (¨-er, ..ößt-)

A few adverbial superlatives have no ending and are listed separately:

> **äußerst . . .** *adv* extremely, highly   **höchst . . .** *adv* highly . . .
> **möglichst . . .2** *adv* . . . : **m. bald** as soon as possible . . .

8.3 Adjectives in -e: Some adjectives end in the letter **e,** which also serves as the ending **-e;** others have an optional final **-e:**

> **leise**   **spröde**   **mild(e)**

When such adjectives have to add other grammatical endings, the final **-e** is made part of the ending: >**ein spröder Stoff** a brittle substance

8.4 No endings: Some adjectives never have endings and are so marked:

> **allerlei** *a* (*no endgs*) all kinds of   >**a. Versuche** all kinds of experiments
> **lauter . . . II** *a* (*no endgs*) nothing but, sheer . . .   >**l. Unsinn** sheer (*or* nothing but) nonsense

This includes adjectives formed from geographical names by means of a suffix **er:**

> >**diese Stuttgarter Fabriken** these Stuttgart factoreis

8.5 adjectives can be formed from **family names** with a suffix **-sch** plus grammatical endings. The best English equivalent is a possessive form with 's:

>**die Avogadrosche Zahl** Avogadro's number

9 ADVERBS

Since in German nearly any adjective may be used adverbially simply by using it without an ending, you are ordinarily left to make up these adverbial meanings for yourself by using the endings -ly or -ally in English:

Das ist leicht getan. That is easily done.

Translations made up in any other way are listed in a separate division labeled *adv, adv oft* or *as adv:*

laut *a . . . 1* loud. **2** noisy . . . **4** *adv oft* aloud

This arrangement conveys the information that the translation "aloud" is another possibility in addition to the regularly derived (but not explicitly listed) "loudly" and "noisily."

augenblicklich 1 *a* immediate, instant; momentary. **2** *adv oft* at the moment

As an adverb, this word can mean "at the moment," but don't overlook the other possible translations: "immediately, instantly; momentarily."

10 VERBS

10.1 All **verbs** listed in the dictionary are marked with one or more of the following **abbreviations:**

vt transitive verb (used with an accusative-case direct object) (*See also* **10.2.**)

vi intransitive verb (used without a direct object) (*See also* **10.5, 10.6, 10.7.**)

vt/i transitive and intransitive with the same translation(s)

vr reflexive verb (used with an accusative-case reflexive pronoun, *usu* **sich**).
   In this dictionary the abbreviation *vr* is used for verbs that are always reflexive: **an·schicken** *vr:* **sich a. (zu)** get ready (to) . . .
   Others have the appropriate division identified by the word **sich:**

zeigen 1 *vt* show, indicate . . . **2** *vi* **(auf)** point (to). **3: sich z.** appear, show up. See **10.3, 10.4, 10.6.**

v aux auxiliary verb

   a haben, sein used with past participle to form perfect tenses

   b werden used in its present tense with an infinitive to form a future tense and used in any tense with a past participle to form passive tenses

   c modal auxiliary (**dürfen, können,** etc) used with an infinitive. *See also* **10.8.**

These categories are usually listed in the order given here (*vt, vi, vr, v aux*), but they are sometimes grouped together if any of them have a common translation: **ab·wechseln** *vt, vi* & **sich a.** alternate

10.2 *vt:* Transitive verbs frequently used with both direct and indirect objects have subdivisions marked *d.i.o* (dative indirect object):

vor·stellen I *vt* **1** (*d.i.o*) introduce, present (to); point out (to) . . .

or more explicitly with both direct and indirect objects in generic form:

zu·muten *vt* (*d.i.o*): **einem etwas z.** attribute sthg to sbdy, expect sthg of sbdy   (**einem** is the dative case of **man**.)

10.3 *vt* + **sich:** Transitive verbs can also be used with a reflexive indirect object in the dative case, indicated by the abbreviation (+*dat rflx*) or by the generic objects **sich etwas:**

an·sehen I *vt* look at . . . **5** (*dat rflx*): **sich etwas a.** have (*or* take) a look at sthg . . .   (= have oneself a look . . .)

vor·stellen I *vt* . . . **3: sich etwas v.** imagine sthg, conceive of sthg . . . (*lit.* present sthg to oneself)

If a reflexive pronoun (*usu* **sich**) is used as an indirect object (dative case), the meaning you are looking for will be listed under *vt,* not under **sich** + initial of the verb. It would also help you to study the entry **sich.**

10.4 *vr:* Reflexive verbs can also be used reciprocally. Usually such uses are given in a division marked (*recip*):

treffen* . . . **IV: sich t. 8** (+ **mit**) meet, have a meeting (with). **9: es trifft sich** it turns out, it happens. **V: sich t.** (*recip*) meet (each other)

But occasionally a verb is used only reciprocally:

ab·kompensieren *v recip* (+ **sich, einander**) compensate each other, cancel out

10.5 *vi:* Intransitive verbs that form their perfect tenses with the auxiliary verb **sein** are marked (s). Those that form their perfects with **haben** are not marked unless there are also some meanings for which **sein** is used:

auf·stehen* 1 *vi*(h) be (*or* stand) open. **2** *vi*(s) get up, rise (up), stand up; rebel

Verbs marked (h *or* s) refer to regional differences in usage. Some verbs, notably **hängen, liegen, sitzen, stehen,** and sometimes their prefixed derivatives, are regularly used with the auxiliary **haben** in the northern part of the German-speaking area and with **sein** in the southern part. Verbs marked (h, s) leave it up to you to identify which is which if you happen to come across the verb in a perfect tense:

sprühen . . . **III** *vi (h. s)* . . . **5** flow, gush; fly (*as:* sparks). **6** scintillate, sparkle, glitter. **7** sprinkle, drizzle

The auxiliary **sein** is more likely to be encountered where there is a suggestion of motion toward a destination (divisions **5** and **7**), like water drizzling or sparks flying onto a surface.

10.6 *vr:* Theoretically, almost any transitive verb can be converted into a reflexive verb by simply using it with an accusative-case reflexive pronoun (**sich** oneself). But while this is easily and frequently done in German, the literal translation is rarely possible: If **beherrschen** can mean "control," then **sich b.** can be interpreted as "control oneself," and in this case it turns out to be the correct translation. In this dictionary such literal translations of the reflexive are usually not listed and you are left to make them up yourself:

    schonen *vt* spare, treat gently; take care of

If you should come across a sentence like >**Sie sollten sich schonen,** you can safely translate it: They ought to spare themselves, *or* treat themselves gently, *or* take care of themselves.

Far more often, however, the translation of a reflexive verb will be a phrase:

    ab·tragen* . . . **II** *vt* & **sich a.** wear out. **III** *vi* & **sich a.** stop bearing fruit.

Note that the first of these translations (wear out) can be interpreted either transitively (to wear sthg out) or intransitively (sthg wears out). *As a translation of a German reflexive, only the intransitive interpretation is valid.* The second meaning (stop bearing fruit) can only be intransitive, even in English.

A one-word translation of a reflexive is even more likely to have an intransitive meaning. The verbs **bewegen** and **rühren** can both be translated (among many other ways) as "move", either as *vt* or as **vr: sich b.** or **sich r.** Be sure you fully understand that in their transitive meaning they imply moving something, while as **sich bewegen** and **sich rühren** they imply that something moves.

10.7 *vi:* Verbs with **objects in the genitive or dative** case are technically intransitive. In this dictionary the case is specified by the abbreviations (*genit*) or (*dat*) at the beginning of the division concerned:

    bedürfen* *vi* (*genit*) need, require
        >**Das bedarf keiner Erklärung.** That needs no explanation.
    helfen* . . . (*dat*) help, aid, assist; be of use (to)
        >**Wer hilft diesen Leuten?** Who will help these people?

For verbs used with prepositions, *see* **12.1.**

10.8 *v aux:* **a** In addition to the modal auxiliaries, several other verbs can be used with an infinitive without **zu.** The main ones are **helfen, hören, sehen** and especially **lassen,** which has a whole variety of meanings in such combinations. (*See* the entry **lassen*.**)

b But many more verbs can be used in combination with an infinitive preceded by **zu:**

    > **Sie versuchten es zu wiederholen.**   They tried to repeat it.
    > **Jemand hatte vergessen, das Licht auszulöschen.** Someone had forgotten to put out the light.

Such verbs are sometimes marked *v aux* (e.g. **vermögen**) and where necessary are accompanied by the notation (+ **zu** + *inf*). The translation is usually literal:

    gedenken *vi* . . . **2** (+ **zu** + *inf*) intend (to)

10.9 Regular and Stem-Changing (or Weak and Strong) Verbs

a Principal parts (*past tense stem, past participle*) are listed in addition to the *infinitive* for all unprefixed stem-changing verbs as part of the entry. The third person singular of the present tense is appended to these if it involves an additional stem change:

gehen* (ging, gegangen) **bringen*** (brachte, gebracht)

fahren* (fuhr, gefahren; fährt) **nehmen*** (nahm, genommen; nimmt)

b Principal parts are listed for prefixed stem-changing verbs only if there is no corresponding unprefixed verb:

gewinnen* (gewann, gewonnen) **verlieren*** (verlor, verloren)

c The asterisk * is used to mark all stem-changing verbs, prefixed or unprefixed. Since principal parts are not normally included with prefixed verbs, the asterisk serves as a signal that the stem-changed forms can be found under the unprefixed verb (*see also* **18.12**):

d Some verbs are stem-changing only in some meanings and regular in others; in that case only the stem-changing division is marked with the asterisk and, if the verb is unprefixed, includes the principal parts:

weichen I (*reg*) *vt, vi* (s) soak, steep, soften. **II*** (wich, gewichen) *vi* (s) (*usu* + *dat, oft* + **vor**) yield, give way (to) . . .

The abbreviation (*reg*) warns you that there is also a stem-changing division. Occasionally the difference is made more striking by also including the regular principal parts:

hängen I (hängte, gehängt) **1** *vt* hang (up), suspend; (*Tex.*) age (dyed material). **2: sich h. (an)** hang on, cling (to). **II*** (hing, gehangen) *vi* (h *or* s). **3** hang, be suspended . . .

e The abbreviation (*reg*) is also used if the verb can be regular or stem-changing, without regard to meaning:

wenden* (*reg or* wandte, gewandt) *vt* & **sich w.** turn . . . (*Reg forms:* wendete, gewendet)

f The abbreviation (*reg*) is also used as a warning against mistaking some verbs as stem-changing when they are not:

beratschlagen *vi* (*reg*) confer, deliberate

You should be aware that many verbs beginning with one of the inseparable prefixes, especially **be..**, **er..** and **ver..**, are often made up from nouns or adjectives:

Ursache *f* cause (*noun*) → **verursachen** *vt* cause (*verb*)

übrig *a* left over, superfluous → **erübrigen 1** *vt* save, put aside; spare. **2: sich e.** be(come) superfluous

The noun may begin with a prefix that would be separable as part of a verb, but if there is an inseparable prefix before the separable one, the verb is always inseparable:

Einfluß° *m* influence (*noun*) → **beeinflussen** *vt* influence (*verb*)

If the noun ends in a stem that is identical to the stem of a stem-changing verb, it is easy to mistake a derivative verb for a stem-changing one:

Ratschlag *m* advice → **beratschlagen** *vi* (*reg*) confer, deliberate

   (*NOT* based on **schlagen***: *past:* beratschlagte, *pp:* beratschlagt)

Auftrag *m* assignment, etc → **beauftragen** *vt* entrust . . . ; authorize . . .

   (*NOT based on* **tragen***: *past:* beauftragte, *pp:* beauftragt)

10.10 Alphabetical listing of stem-changed verb forms:

Principal parts of stem-changing verbs are listed not only in verb entries but also individually; subjunctives are listed only if they are not easily derived from the past tense:

ging went: *past of* GEHEN   **nimmt** takes: *pres of* NEHMEN

gegangen gone: *pp of* GEHEN   **stünde** would stand: *sbjc of* STEHEN

For reasons of space conservation such entries are incomplete in that they give only one of several possible translations. If the one given does not fit the context, look up the verb entry (GEHEN, NEHMEN, etc) and adapt the suitable translation by changing it to a past, past participle, etc. Moreover, the translation given is not necessarily the most frequent one; the choice is often based on the availability of space, since such entries aim to occupy no more than one line of print. There are other space-saving restrictions on these listings as well:

a Past Participles are listed only if their stem vowel is different from that of the infinitive:

listed: **gegossen**. . . : *pp of* GIESSEN   **gebracht**. . . : *pp of* BRINGEN

not listed: **gegeben**. . . : *pp of* GEBEN   **geschlafen**. . . : *pp of* SCHLAFEN

If you find that a past participle ending in **-en**, like **gegeben** or **geschlafen**, is not listed, simply drop off the **ge-** prefix and look up the resulting infinitive.

b If you know your grammar, you can usually trace a prefixed verb form to its infinitive:

To identify **verschob**, you would have to detach the prefix, look up **schob**, which is listed as the past tense of **schieben**; then reattach the prefix and look up **verschieben**.

To identify **verschoben**, you would first have to check your context to verify whether this is a past participle or the plural of a past tense. In the latter case you would then proceed as for **verschob**. If it is a past participle, detach the prefix and replace it with **ge-**. Actually either procedure will lead you to the verb **schieben** and from that to **verschieben**.

But to facilitate the process, this dictionary lists specific present and past tenses and also participles of prefixed verbs, with certain restrictions:

Present and past tenses are given in their singular forms only and for verbs with inseparable prefixes only, so you would find **verschob** but not **vorschob** or **schob vor.**

Past participles of stem-changing verbs are given with all inseparable but only the most frequent separable prefixes, so you would find **verschoben** and **vorgeschoben** but not **stattgefunden** (See **c** below.):

    verschob(en) shifted: *past* (*& pp*) *of* VERSCHIEBEN

    vorgeschoben pushed forward: *pp of* VORSCHIEBEN

To identify **stattgefunden,** look up **gefunden,** which leads you to **finden,** and then look up **statt·finden.**

To identify **vorschob** or **schob vor,** look up **schob,** which leads you to **schieben,** and by attaching the prefix, to **vorschieben.**

Any past participle ending in **-en** must be part of a stem-changing verb, but only those with a stem vowel different from that of the infinitive are listed, so you will find **verschoben** and **vorgeschoben** (**o** is different from the **ie** of **ver·schieben** or **vor·schieben**) but you will not find **abgegeben** or **aufgetragen.** So if you look for these last two and find them not listed, simply drop out the **-ge-** and you have the infinitives **auf·geben** and **auf·tragen.**

If you come across **vergeben** or **ertragen** as past participles, you will find them as infinitives. (Such past participles are identical to their infinitives.)

To identify forms like **einzuhalten, zurückzugewinnen** or **vorzustellen,** drop out **-zu-** and look up **ein·halten, zurück·gewinnen** or **vor·stellen.**

c The listing of these verb forms for stem-changing verbs with separable prefixes is restricted to the more frequent prefixes, and here too only one of the verb's translations can be given for each form; for the others you will have to look up the infinitive:

    aufgestiegen ascended: *pp of* AUFSTEIGEN

For other meanings and translations of **aufgestiegen,** consult the entry AUFSTEIGEN. For verbs with less frequent prefixes, these present, past and past participle forms are not listed, e.g. for **statt·finden** and for many verbs in which the infinitive of one verb serves as a separable prefix on another verb: **hängen·bleiben, stecken·lassen,** etc.

d While each prefixed verb entry is normally started on a new line, those with longer prefixes like **zusammen** or **auseinander,** require a space-saving device: they are paragraphed, often interspersed with noun entries:

    Zusammen·prall° *m* clash, collision, impact. **z—·prallen** *vi*(s) (**mit**) collide, clash (with)

    auseinander *adv & sep pfx* apart, in two, to pieces: **·brechen*** *vt, vi*(s) break apart. **·breiten** *vt* spread out, unfold . . . **·setzen I** *vt* 1 (*d.i.o*) explain (to). **II: sich a.** (**mit**) 2 study closely, analyze. 3 confront . . . **A—setzung** *f* (-en) 1 explanation. 2 analysis. 3 argument, confrontation . . .

Note how the raised dot before a verb infinitive (**·setzen**) designates a separable verb with the prefix that heads the paragraph, in this case **auseinander.**

e Past participle, past-tense and present-tense forms with longer prefixes are also paragraphed:

> **zurück/gewonnen** regained: *pp of* ZURÜCKGEWINNEN. **-geworfen** rejected: *pp of* ZURÜCKWERFEN. **-gezogen** secluded . . . : *pp of* ZURÜCKZIEHEN. **Z— gezogenheit** *f* seclusion . . .

The slash mark / serves as a warning that this is a paragraph of past participles and not one of verbs like **zurück·gewinnen.** *See* 18.13c.

10.11 Dual Prefixes: You should also know that the prefixes **durch, über, um, unter** and in a few verbs also **wider** and **wieder,** are sometimes separable and sometimes inseparable. This dictionary uses a special format to deal with these verbs:

a If the verb is *always separable,* the listing is identical to that of separable verbs:

> **um·leiten** *vt* divert, detour; reroute

b When the verb is *always inseparable,* it is listed with the abbreviation (*insep*) immediately after the infinitive:

> **umgrenzen** (*insep*) *vt* **1** limit, bound. **2** enclose . . .

c When the verb can be *both separable and inseparable,* the following notation is used:

> **um/gehen* I** (*sep*) *vi*(s) go round, circulate . . . **II** (*insep*) *vt* get around, circumvent, bypass . . .

d Occasionally a prefixed verb can be *either separable or inseparable* with the same meaning:

> **wider/spiegeln** (*sep or insep*) **1** *vt* reflect, mirror. **2: sich w.** be reflected, be mirrored

See 18.13b.

10.12 Principal parts (past tense, past participle, present tense) **of dual-prefix verbs** (*see* 10.11) are listed subject to certain restrictions:

a Present and past tense forms are listed only for inseparable verbs, but a reminder is added that the separable prefix may be found attached to such a verb form if it occurs at the end of a dependent clause:

> **umlief** circled: *past of* UMLAUFEN **II** (*insep*) & *oft* **I** (*sep*)
> **übertritt** violates, etc: *pres of* ÜBERTRETEN **II** (*insep*) & *oft* **I** (*sep*)

Note that the one translation given in each entry is valid only for the *inseparable verb;* for other meanings of the inseparable verb and *all meanings of the separable verb,* look up the infinitive entries **um/laufen** or **über/treten.**

b Past participles of separable verbs will include the prefix **ge-:**

> **durchgebrochen** broken through: *pp of* DURCHBRECHEN (*sep*)

c Past participles of inseparable verbs have no **ge-** prefix:

durchbrochen 1 broken through: *pp of* DURCHBRECHEN (*insep*). **2** *p.a* open-work(ed), filigreed, perforated
Re *p.a, see* **11.2.**

10.13 The separable prefix **wieder** is often attached to a verb that already has another separable prefix. When the verb is separated, the two prefixes appear as two separate words:
>**Die Stadt baute das abgebrannte Gebäude wieder auf.**
The city built the burned-down building up again.
The verb must be looked up under **wieder·auf·bauen.**

10.14 Almost **any** kind of **word can become a separable prefix** if it is habitually used with one or more verbs in a meaning perceived to be idiomatic. Even verb infinitives can be used in this capacity, as in:
    kennen·lernen . . . *vt* meet, become acquainted (*or* familiar) with, get to
       know . . . (*lit* learn to know)
But not every word used with a verb in such a combination is treated as a prefix, i.e. is attached to the verb when the verb comes last in a sentence or clause. Sometimes a distinction is made between a literal combination and a figurative one:
    kalt stellen (2 words) put on ice, refrigerate
    (**kalt** is a separate word; the combination **kalt stellen** is listed in the dictionary as an idiom in the entry **kalt**)
    kalt·stellen (one word) shelve, render harmless ("put on ice" figuratively!)
    (**kalt** used as a separate prefix; **kalt·stellen** is listed among the **kalt** compounds)
For the reader it will be difficult to tell from some contexts which one of these two has been encountered:
>**Man stellt das Gemisch auf zwei Stunden kalt.**
The mixture is refrigerated for two hours.
>**Man stellte den Plan auf einige Zeit kalt.**
The plan was shelved for some time.
The only solution is to look in both places, particularly since German writers are not always certain as to whether they should write the expression as one word or two, especially where an infinitive or a past participle is involved:
>**das Gemisch wird zwei Stunden lang kalt gestellt.** (refrigerated: 2 words)
>**Man sollte diesen Plan kaltstellen.** (shelve: one word)
This is the way the words would appear after being carefully edited according to the rules, but in the absence of an editor the spellings might be less "correct": **kaltgestellt** and **kalt stellen.**
The final authority in such cases is a reference work called **Duden:** *Recht-*

schreibung.† But this is itself based on prevailing usage among writers; **Duden** does not dictate correct spelling; it merely tells what most writers are doing, and this may change from one edition to the next. Dictionaries, on the other hand, are much less frequently revised: the preceding edition of this one dates back to 1950.

While **zurück, zusammen** and **zurecht** are considered separable prefixes, **zugute, zugrunde, zunichte** and **zutage** are, at this writing, not. Fortunately the word order is the same, and the last three occur only in combination with a specific number of verbs, which the dictionary lists in paragraphed form:

> **zugrunde** *adv* **1: z. gehen** (*oft* + **an**) be destroyed, be ruined (by); die (of).
>
> **2: z. legen** (*d.i.o*) use as a basis (for) . . .

Derivative nouns are written as one word: **Zugrundelegung** *f,* and so are participial adjectives like **zugrundeliegend** (*see* **11**).

One more example: in the entry **klein** *a* . . . "small, little" there is an idiomatic expression **k. drehen** "turn down (gas, radio)." If you come across a sentence like:

> **Die Flamme muß klein gedreht werden.** (The flame must be turned down.)

you would be wise to look up **klein** rather than **drehen;** in case **"kleingedreht"** were erroneously spelled as one word, you might be tempted to look for **kleindrehen,** but you would not find it. What really matters now is knowing what to do next: look up **klein!**

11 PARTICIPLES AND PARTICIPIAL ADJECTIVES

A participle is defined as a verb form that can be used as an adjective:

> the spoken language   the shining sun

Participles usually derive their meanings from their parent verbs but many of them tend to expand their range of meaning as adjectives, taking on meanings not covered by the verb. And like most other adjectives, they can also be used adverbially. Such participles are referred to in this dictionary as *participial adjectives* (*p.a.*) and when used adverbially, as *participial adverbs* (*p.adv*).

11.1 Present Participles are formed by adding a **-d** to the infinitive:

> **stehend**   **ausblühend**   **zunehmend**
>
> standing   efflorescent(ly)   increasing(ly)

Because they are so easily formed and readily recognized, and also so easily translated by using the -ing ending in English, the dictionary saves space by not listing them unless they have acquired meanings not directly derivable from the parent verb or translatable with the usual -ing ending, as is true of **ausblühend**

†Published by Dudenredaktion, Wiesbaden, Federal Republic of Germany.

"efflorescent" above. The reader is also expected to recognize one of these forms if it is used as an adverb (no ending):

>**Das wird zunehmend beobachtet.**   That is increasingly observed.

As an adjective, a participle will have regular adjective endings:

>**die zunehmenden Schwierigkeiten**   the increasing difficulties

Present participles with special adjective meanings are appended to their parent verb entries:

> **aus·harren** *vi* persist, persevere. **ausharrend** *p.a* persistent
> (**ausharrend** is listed chiefly because the translation "persistent" does not end in -ing.)
> **aus·gleichen*** 1 *vt* & **sich a.** level (even, balance) out . . . **ausgleichend** *p.a:*
> **a—e Farbe** complementary color
> **genügen** *vi* suffice; be enough . . . **genügend** *p.a* sufficient

11.2 Past participles

a The formation of these is a part of elementary German grammar that every reader must be familiar with. Here we are concerned mainly with the use of past participles as adjectives and adverbs:

> >**eine sehr gefragte Ware**   a much demanded article, an article much in demand

> >**Die Ware ist jetzt sehr gefragt.**   The article is in great demand now.

The word **gefragt** is used here in a meaning not directly derivable from the parent verb **fragen.** It is therefore listed as a separate entry with a cross-reference to **fragen,** and the verb entry **fragen** also has a cross-reference to **gefragt.**

In case you recognize **gefragt** as a form of **fragen** and look it up there, the *cf* (cross-reference) will direct you to **gefragt** *p.a* as a separate entry giving only this special meaning. If you do not look up **fragen** but go directly to the entry **gefragt,** the *cf* FRAGEN will remind you that **gefragt** could also, in a different context and with no ending, have its literal meaning "asked."

b Since past participles of stem-changing verbs are listed for grammatical reasons anyway, additional meanings not derivable from the parent verb are then added in a separate division marked *p.a:*

> **abgeschieden I** separated, etc: *pp of* ABSCHEIDEN. **II** *p.a* 1 secluded, isolated.
> 2 departed, deceased. **Abgeschiedenheit** *f* seclusion, isolation

c Past participles of regular (non-stem-changing) verbs are not listed unless they have special meanings as adjectives or other parts of speech:

> **abgerechnet** *prep, adv* (+ *accus*) excluding, deducting, except for:
> **Verluste a.** excluding losses; *see also* ABRECHNEN

d Past participles without a **ge-** prefix are appended to the parent verb entry if this is alphabetically feasible; otherwise they are listed separately:

verbürgen *vt* guarantee; confirm. **verbürgt** *p.a* (*oft*) authentic

verbreiten I *vt* spread, disseminate. **2** circulate, broadcast . . .

verbreitet *a* **1: stark (weithin, überall) v.** widespread, widely used (*or* known). **2** *adv* (*in weather reports*) in many areas. *Cf* VERBREITEN

> (**verbreitet** could not be appended to the entry **verbreiten** because of an intervening entry **verbreitern**)

verbunden I combined, etc: *pp of* VERBINDEN. **II** *p.a* **1** obliged. **2: v. sein mit** involve. **3** (*Tlph*): **falsch v. (sein)** (have the) wrong number

12 PREPOSITIONS occur in phrases with noun or pronoun objects whose grammatical case (genitive, dative, accusative) is governed by the individual preposition. They are often identical in form to separable prefixes (**an, auf, vor,** etc) but should not be confused with them:

> \>**Sie kommen durch das Labor.** They are coming through the lab.
>
> (**durch** is a preposition locked in the phrase **durch das Labor.** The verb is **kommen.**)
>
> \>**Sie kommen heute durch.** They are coming through today.
>
> (**durch** is a separable prefix; the verb is **durch·kommen.**)
>
> \>**Das kommt auf das Wetter an.** That depends on the weather.
>
> (**auf** is a preposition locked in the phrase **auf das Wetter; an** is a separable prefix; the verb is **an·kommen.**)

12.1 Prepositions are listed in the dictionary with their most typical and frequent **meanings,** but they can almost always have additional English equivalents that are valid only in combination with certain verbs:

von does not normally mean "on," but that is its English equivalent in:

ab·hängen* . . . **3** *vi* (+**von**) depend (on)

It can not be said that **von** *means* "on"; it can only be said that in idiomatic combinations with this verb **von** corresponds to "on." The dictionary includes such prepositions in parentheses with verb entries, where space permits, as an aid to the user. Remember that these are prepositions that occur in phrases. They are not separable prefixes, not part of the verb.

So also in

bemühen . . . **II: sich b.** exert oneself; (*esp* + **um**) make efforts (on behalf of, for, to get)

denken* . . . **2** *vi* (**an**) think (of)

An English preposition can also correspond to a German grammatical case:

gehören *vi* **1** (*dat*) belong (to) . . .

German verbs used with prepositional phrases may have an English translation involving a direct object and no preposition. The verb **antworten** will serve as an illustration of this and the preceding possibilities:

antworten *vi* (*dat* or **auf**) reply, respond (to); answer
Note how this applies to specific sentences:
>**Sie antworten dem Volkszähler.** (*dative*)
  They answer the census taker. (*direct object, no preposition*)
  They reply (*or* respond) to the census taker. (*preposition* "to")
>**Sie antworten auf seine Fragen.** (*prep* **auf** + *accus case*)
  They answer his questions. (*direct obj, no prep*)
  They reply (*or* respond) to his questions. (*prep* "to")

12.2 Prepositions that follow their object are an item that should be part of
the reader's knowledge of German grammar, but usually an example is included
in the entry:

wegen 1 *prep* (*usu* + *genit*) on account of, because of: **w. des Wetters, des
  Wetters w.** because of the weather
nach I *prep* (*dat*) **1** after . . . **4: n., je n.** according to, by: . . . **unseren Untersuchungen n.** according to our investigations . . .

13 CONJUNCTIONS usually cause little trouble, but the dictionary does identify
some conjunctions as subordinating (*subord cnjc*). This is especially important
where the same word can be both a conjunction and some other part of speech. In
context the subordinating conjunction can be recognized by the fact that it is
followed by dependent word order (finite verb last):

da I *adv* there, that's where . . . **II** *adv* & *adv cnjc* **4** then, and then . . .
  IV *subord cnjc* since, inasmuch as, considering that
>**Da entstehen manchmal gewisse Schwierigkeiten.** (*adv*)
  That is where certain difficulties sometimes arise.
>**Da manchmal gewisse Schwierigkeiten entstehen** . . .(*subord cnjc*)
  Since certain difficulties sometimes arise . . .

während 1 *prep* (+ *genit*) during, in the course of. **2** *subord cnjc* while,
  whereas . . .
>**während des Tages** during the day
>**Während diese Vorbereitungen getroffen werden,** . . .
  While these preparations are being made . . .

14 ABBREVIATIONS used in German are listed alphabetically, as if the letters
comprising them spelled a word. An abbreviation is often converted from singular to plural by doubling its final letter, so to find **Lsgg.** (Lösungen), look up **Lsg.**
(Lösung). If there is a corresponding English abbreviation, it is given as a
translation with the full word following it in parentheses:

Lsg. *abbv* (Lösung) sol. (solution)

15 PREFIXES AND SUFFIXES are often listed as separate entries, usually with a
raised dot ·:

15.1 a ur · *pfx* **1** original, primitive, primeval, elemental, proto-. **2** great-:
Urenkel greatgrandchildren. **3** genuine; standard, master. **4**
extremely *cf* URALT

b Inseparable verb prefixes, which are never used as independent
words, are listed with two dots after them instead of the raised dot: **be..,
ent.., er.., miß.., ver.., zer..**). Since these prefixes have no actual translations, the dictionary can only attempt to define their effect on the meanings of the verbs they are attached to:

ent.. *insep pfx: implies* **1** (*deprivation, removal, undoing*) de-; dis-; un- *cf*
ENTAKTIVIEREN, ENTDECKEN). **2** (*origin, incipient action*) *cf* ENTFLAMMEN,
ENTSTEHEN. **3** *esp vi*(s) (*escape*) *cf* ENTGEHEN

c Since the vast majority of prefixes can also occur as independent words,
or at least coincide in form with independent words, they are most likely to
be listed inside a paragraph beginning with the independent word:

ab I *adv* off, away, down–**ab und zu** now and then, back and forth *cf* AUF. **2**
prep: **ab Hamburg** from (*or* leaving) Hamburg; **ab 10. Juli** beginning
July 10. **3** *sep pfx* off, away, down; *also implies quick or brief treatment, cf* ABBACKEN **2**, *or cessation of activity, cf* ABBLEICHEN *vi*

Haupt 1 *n* (ꞋꞋer) head, chief. **2** (*as pfx*) main, chief, principal, most important:
-abteilung *f* main (*or* principal) division

halb I *a* half. **II** *adv oft* halfway, by halves, partly, almost; (*Time*) e.g.: **h.
elf** half past ten. **III** *pfx* half-, semi-, hemi-, demi-, medium. **IV** *sfx* half:
viereinhalb four and a half . . .

15.2 Suffixes

While some suffixes can exist as independent words (*cf* **halb** *in* **15.1** *above*),
most of them are listed as separate entries:

·bar *sfx* (*attached to verb stems*) -able, -ible: **ausdehnbar** *a* expandable, expansible *cf* HALTBAR, LÖSBAR, etc.

Haft I *m* (-e *or* -en) **1** . . . **II** *f* . . . **·haft** *sfx* (*forms adj from nouns*) e.g.
schmerzhaft painful . . .

·halben *sfx* on behalf (*or* account) of: **ihrethalben** on their account

16 TYPE, PUNCTUATION AND SIGNS

16.1 Type

a Regular roman type is used for English translations, some abbreviations
(sthg, sbdy, etc), German plural endings (**7.1**), comparatives and superlatives
of adjectives (**8.1**), principal parts of verbs (**10.9**) and for the full words equivalent to abbreviations in both languages (**14**):

u.ä. *abbv* (und ähnliches) and the like   **Nr.** *abbv* (Nummer) no. (number)

b Boldface roman type is used for all German text except as specified in **a**

above. **Boldface** is also used for Arabic and Roman division numbers and for letters used to introduce divisions and subdivisions. (*See* **17**.)

c *Italics* are used for most abbreviations (*genl, cnjc; Chem., Biol.*), and in explanatory matter, as in the model entry **·bar** quoted under **15.2**.

d SMALL CAPITALS are used for all cross-references. WORDS in small capitals are to be interpreted as cross-references even if not preceded by the abbreviation *cf*:

> **gab** gave: *pp of* GEBEN   **hieher** *adv* here, this way = HIERHER

Warning: Cross-references can be very helpful in finding the information you are looking for. DON'T ignore them!

17 NUMBER AND LETTER SYMBOLS IN BOLDFACE

17.1 Arabic Numbers are used to mark off the main divisions of an entry.

17.2 Roman Numerals are used when even larger divisions are needed, as is often the case with verb entries to designate *vt, vi* and *sich* (*vr*), when one or more of these break down into Arabic-numbered divisions.

17.3 Letter symbols are used to mark off divisions or subdivisions where numbers alone don't suffice. Most letter symbols are boldface capitals (**A, B, C,** etc) (*See* model entry **anhängen** under **6.3**). But small boldface letters may sometimes be used to mark off different meanings of a phrase quoted within a numbered division or to clarify finer distinctions in meanings of compound words:

> **schießen*** . . . **2** *vi* (s) shoot, dash, dart . . . : **in die Höhe sch. (a)** shoot up, grow tall, **(b)** go flying up
>
> **Sau·bohne** *f* feed bean, (*specif*): **(a)** vetch, **(b)** soybean

17.4 When you come across a **succession of numbers** in an entry, e.g.: *cf* SEIN **IA1,2. III** *adv* **2** . . ., look for a period. The number *after* the period starts the next numbered division (in this case **III**); the numbers *before* the period are part of the cross-references at the end of the preceding divison, in this case: *see Divisions* **IA1** *and* **2** *of the entry* SEIN.

18 PUNCTUATION AND SIMILAR SIGNS AND SYMBOLS

Since there is no standard order in which punctuation marks are to be listed, they are arranged here in order of importance and frequency. An alphabetical index is given herewith to help you find the information you need:

asterisk *	**18.12**	hyphen -	**18.7**
colon :	**18.6**	parentheses ( )	**18.11**
comma ,	**18.4**	period .	**18.1 a, b**
dash —	**18.8, 18.9**	plus sign +	**18.10**
degree mark °	**18.14**	raised dot ·	**18.2, 18.3, 18.9**
dots in succession	**18.1 c, d, e**	semicolon ;	**18.5**
equal sign =	**18.15**	slash mark /	**18.13**

18.1 a The period . is used to mark the end of a numbered division, but no period is placed at the end of an entry. The period is also used for capitalized abbreviations denoting fields of study: (*Chem., Dterg., Lthr., Zool.*)

b The period is also used after the initial(s) that represent the key word when it is quoted within an entry:

> **ab·tragen*** . . . **III: sich a.**   **Anfang** *m* . . . beginning . . . **von A. an** from the very beginning.

c Successive dots (two or three) are placed at the end of a word to indicate that it is incomplete, which usually means that it always has a grammatical ending (**kleinst..**)

d Two successive dots are placed before a plural ending for a noun entry if the ending replaces a singular ending: **Studium** *n* (..ien)

e Two dots are placed after inseparable verb prefixes listed as separate entries (**be..**) (*See also* **15.1b.**)

18.2 The raised dot · is a special symbol used as a means of focusing attention on various kinds of word compounding.

a The raised dot is used to mark compound nouns, adjectives, adverbs and a few other compounds:

> **Kali·bergwerk** *n* potash mine. **·blau** *n* Prussian blue
> (Only the first compound noun in a paragraph has the raised dot.)
> **Kanella·rinde** *f* canella bark
> (Even if a compound word does not initiate a paragraph of compounds, it is shown with the raised dot to identify it as a compound.)
> **Einphasen·** single-phase (Here the raised dot indicates that the entry is to be interpreted as a first component of possible but not specifically listed compounds like **Einphasen·motor** *m* single-phase motor.)

b The raised dot also marks prefixes and suffixes:

> **ur·** *pfx* **1** original, primitive . . .
> **·bar** *sfx* (*attached to verb stems*) -able, -ible . . .

c The raised dot in a verb entry always marks a separable verb:

> **an·kommen***   **kalt·stellen**   **zusammen·setzen**

In a paragraph of compounds, an initial raised dot marks a separable verb whose prefix is the key word that heads the paragraph:

> **Zusammen·halt** *m* . . . **z–haltbar** *a* . . .**·halten I** *vt* . . .
> (compound noun)   (compd adj)   (separable verb)
>   (**·halten** stands for **zusammen·halten**)

For combinations of the raised dot with a hyphen, *see* **18.3**
For combinations of the raised dot with a dash, *see* **18.9**.

18.3 The raised dot followed by a hyphen ·-

This combination is used to append compound words to the end of an ordinary simple word entry:

> **Abdampf°** *m* exhaust (*or* waste) steam . . . **·-apparat** *m* evaporating apparatus, evaporator. **-druckregler** *m* exhaust steam pressure regulator . . . (**·-apparat** stands for **Abdampf·apparat, -druckregler** for **Abdampf·druckregler**) *see* 5.4

18.4 The comma , is used mainly to separate more or less synonymous or interchangeable translations of the key word; often a second translation helps to clarify a possibly ambiguous first one:

> **Majorität** *f* majority, age of consent   **lügen*** . . . *vi* lie, tell a lie

(The words "majority" and "lie" are ambiguous without the second translation.)

18.5 The semicolon ; separates groups of synonymous translations from other such groups:

> **an·leiten** *vt* direct, lead; instruct, train; guide

> There are three areas of meaning here, separated by the semicolons. The first two have two translations each; the third has one. (*See also* 6.5c.)

18.6 The colon :

a The colon is used between Arabic or Roman division numbers and the word **sich** when it marks the reflexive use of a verb:

> **ab·wickeln I** *vt* . . . **II: sich a.** . . . .

or between the abbreviation *vr* and the word **sich**:

> **an·schicken** *vr:* **sich a.** . . . .

b The colon is used in verb-form entries:

> **gegessen** eaten: **pp of** ESSEN

c The colon is used when an illustrative or idiomatic phrase is given in any division of an entry:

> **absehbar** *a* foreseeable, within sight: **in a—er Zeit** in the foreseeable future

> **aus·gehen*** *vi* (s) . . . **5** run out . . . , fail, give out . . . (*oft + dat of psn*): **ihnen geht etwas aus** they run out of . . . sthg

> **bestellt** *p.a* (*usu pred*) **1: b. sein** have an appointment. **2: es ist gut (schlecht) b. (mit, um)** things are in a good (bad) way (with, for)

> **brauchen I** *vt* **1** need. **2** (*of time*): **sie b. einen Tag** it takes them a day . . . *See also* 6.5.

d The colon is used to introduce compound words when specific translations are given for the first component:

> **Anfangs·** initial, starting: **-bad** *n* starting bath. **-bahn** *f* initial orbit (*or* path) . . .

Cyan *n* cyanogen; (*in compds oft*) cyanic, cyano-, cyanide: **-alkali** *n* alkali cyanide. **-alkyl** *n* alkyl cyanide

18.7 The hyphen -

In addition to its customary uses, the hyphen also has some specialized applications:

a It is used in the listing of paragraphed compound nouns and adjectives as shown in **18.5d** *above and in* **5.3, 5.4** *and* **5.6.**

b The hyphen is used, with and without umlaut, to show noun plurals, as explained in **7.1**, and to show comparatives and superlatives of adjectives, as explained in **8.1.**

c It may appear at the end of a word or prefix, or at the beginning of a suffix, especially in English, to indicate that this element occurs only in unhyphenated compounds: *See* **Cyan** *in* **18.6d** above, with special attention to the translation "cyano-"; note "half-", etc in the entry **halb III** in **15.1c** and "-able" in **·bar, 15.2.**

d When words are separated at the end of a line, it is a rule in German spelling to break **ck** as **k—k: Zuk·ker = Zucker.**

18.8 The dash —

a The dash is used to introduce an idiomatic expression (*see* **6.5**) quoted in an entry, especially if the translation of the phrase is far from literal and the English equivalent of the key word does not appear in the translation:

chemisch *a* chemical—**ch. reinigen** dry-clean . . .

b The dash is also used after the initial letter of a quoted word when an ending is added. The entry **chemisch** (see **a** above) also lists the expression **ch—e Reinigung** "dry cleaning, dry cleaner's"; **ch—e** stands for **chemische.**

c The dash is used to mark the change from capital to small letter or vice versa in a paragraph of compound words:

Arbeits·markt *m* labor market. **a—mäßig** *a, adv* as regards work. **A— methode** *f* working method, procedure . . .

auseinander *adv & sep pfx* apart, in two, to pieces: **·brechen*** *vt, vi*(s) break apart. **·breiten** *vt* spread out, unfold . . . **A—setzung** *f* (-en) **1** explanation. **2** analysis. **3** argument . . .

18.9 The dash followed by a raised dot —·

a This notation is used to mark a change from capital to small letter or vice versa when compound words are appended to a simple key word:

primär *a* primary. **P—·akt** *m* primary act (**P—·akt = Primärakt**)

Prisma *n* (..men) prism. **p—·ähnlich** *a* prismatoid (**p—·ähnlich = prismaähnlich**)

b The dash plus raised dot is also used to mark a separable verb after a prefixed noun in a paragraph headed by a prefix or a word that can be used as a separable prefix:

Zusammen·prall *m* collision . . . **z——·prallen** *vi* (s) collide

(**z——·prallen** is to be interpreted as a separable verb **zusammen·prallen.**)

18.10 The plus sign + can be interpreted as "with" and is often used with the prepositions or grammatical cases that occur in combination with verbs (*see* **12.1** *and* **10.7**), and sometimes with nouns:

bemühen . . . **II: sich b. 3** exert oneself; (*esp* + **um**) make efforts (on behalf of, for, to get). **4** (+ *adv of place*) take the trouble to go (*or* come) to . . .

gedenken* *vi* . . . **2** (+ **zu** + *inf*) intend (to)

Kenntnis *f* (-se) . . . (+ *genit*) knowledge (of), acquaintance, familiarity (with)

weichen . . . **II*** *vi*(s) **8** (*usu* + *dat, oft* + **vor**) yield, give way (to); (+ **von**) move, budge (from)

zugrunde *adv* **1: z. gehen** (*oft* + **an**) be destroyed, be ruined (by)

18.11 Parentheses are used in fairly obvious ways, but the following details may help to clarify a few points:

a Parentheses are put around nongrammatical abbreviations (*oft*), (*usu*), (*genl*), (*specif*); abbreviations and sometimes full words that designate fields of study (*Math.*), (*Mach.*), (*Metll.*), (*Metals*); plurals of nouns (¨e), (..ien); comparatives and superlatives of adjectives and adverbs (¨er, ¨ste); principal parts of verbs: **treffen*** (traf, getroffen; trifft) and the fully spelled-out equivalents of abbreviations in both languages: **Nr.** *abbv* (Nummer) no. (number). *See* **16.1a.**

b Parentheses are often put around corresponding items in German and English, such as the prepositions and grammatical cases indicated in **18.10.** They also save space in listing individual principal parts of verbs as separate entries:

verlieh(en) imparted, etc: *past* (& *pp*) *of* VERLEIHEN

(**verlieh** is the *past,* **verliehen** the *pp*)

Similarly in the entry:

Betracht *m:* **in B. kommen (ziehen)** come (take) into consideration

(**in B. kommen** corresponds to "come into consideration", **in B. ziehen** corresponds to "take into consideration.")

In the entry **kein** the words **"keine(r) (keines) von beiden** neither of the two" stand for **"keine, keiner** or **keines von beiden** neither of the two."

c Parentheses are also used to mark alternate translations in a space-saving format. When one alternate is given, the word *or* is included:

Kern·frucht *f* stone (*or* pit) fruit (This obviously stands for "stone fruit, pit fruit.")

ab·sehen* I *vt* . . . **4** foresee, predict: **nicht abzusehen** impossible to predict (*or* to tell)

When two or more alternates are given, they are set in parentheses without *or:*

Kennummer *f* . . . identification (index, license, reference) number

(Each of the words in parentheses can be substituted for the word "identification" to specify the kind of number involved here.)

d Parentheses also mark alternate items often involving different meanings:

Kern·macherei *f* (*Foun.*) core-making (shop) (= core making *or* core-making shop)

kein *a* no, not a(ny) (= not a *or* not any)

Keto·verbindung *f* keto(nic) compound (= keto compound *or* ketonic compound)

Vermoderung *f* rot(ting) (= rot *or* rotting)

e Parentheses may also set off optional items:

Kern·mehl *n* prime (quality) flour (= prime flour *or* prime-quality flour)

verlöten *vt* solder (together) (= solder *or* solder together)

dämpfen *vt* . . . **2** damp(en) (= damp *or* dampen)

Genick *n* (-e) (back of the) neck (= neck *or* back of the neck)

German letters in parentheses are always optional:

bitumin(is)ieren *vt* bituminize (The German verb is either **bituminieren** or **bituminisieren.**)

See also **3.3** *for the* "disappearing e."

f Parentheses may set off items that are not part of the English translations but are included by way of explanation. Usually this involves clarification of a potentially ambiguous English translation (*See also* **18.4**):

klein . . .*a* . . . **k. drehen** (or **stellen**) turn down (gas, radio)

"Turn down" can have several meanings in English: adding "(gas, radio)" makes it clear what "turn down" means here.

kern·fein *a* outer, peripheral, valence (electrons)

(The translations are valid only with reference to electrons.)

veröden *vt* . . . **2** (*Med.*) deaden (nerves)

("Deaden" is valid for **veröden** only in this specific medical sense.)

verreiben *vt* . . . **2** . . . rub out (spots) (No other connotation of "rub out" is valid.)

versäumen *vt* **1** miss, lose; waste (time) (The translation "waste" is valid for **versäumen** only with reference to time.)

vor·binden* *vt* tie on (an apron, etc) (In conformity with the meaning of the German prefix **vor,** the verb can refer only to tying on a garment worn in front.)

g Parenthetical explanations are sometimes needed to clarify ambiguous English words used as translations:

liegen* . . . *vi* . . . (*lit* = be in a reclining position) . . .

Magen *m* (⸚) **1** stomach (*as internal organ only*), maw, gizzard . . .

Magazin *v* (-e) magazine (*all meanings*)
   (*See also* **6.5b.**)

18.12 The asterisk * is used to mark a stem-changing verb:

stehen* (stand, gestanden) . . . *vi* (h *or* s) . . . stand

verstehen* *vt/i* . . . understand

auf·stehen* 1 *vi* (h) be (*or* stand) open. **2** *vi* (s) get up, rise (up), stand up . . .

weichen I (*reg*) *vt, vi* (h) . . . **II*** (wich, gewichen) *vi* (s) . . . (*See* **10.5, 10.9d**)

18.13 The slash mark /

g The slash is used when the suffix **·chen** is added after a single **s:**

Gläs/chen *n* little glass, vial   (derived from **Glas** *n* glass, etc)

Bläs/chen *n* small blister, vesicle   (derived from **Blase** *f* bubble, etc)

This is done to show that the resulting **sch** is not the same as the **sch** in words like **schon** or **Fisch,** and to help you identify the suffix **·chen.**

b The slash is used in place of the raised dot to show that a prefixed verb is both separable and inseparable:

um/gehen* I (*sep*) **1** *vi* (s) go round, circulate . . . **II** (*insep*) *vt* get around, circumvent, bypass

c The slash also replaces the raised dot in the initial entry of a paragraph of past participles of separable verbs:

zurück/gewonnen regained: *pp of* ZURÜCKGEWINNEN. **-geworfen** rejected: *pp of* ZURÜCKWERFEN . . .

d The slash is used to save space in alternate wordings:

rückwärts *adv* . . . **2** (in the) rear, in back: **nach/von r.** toward/from the rear

18.14 The equal sign =:

The use of this sign is ordinarily self-explanatory, but here are some examples:

liegen* . . . *vi* (h *or* s) **1** lie (*lit* = be in a reclining position) . . .

bemüht *p.a:* **b. sein (um)** = **sich bemühen 3 (+ um)**

Means: **bemüht sein** is synonymous with **sich bemühen** in division **3** of the entry BEMÜHEN.

Metallack *m* metal lacquer = **Metall·lack**

Means this is one of those compound nouns in which a triple consonant has been reduced to a double one, as explained in **3.4.**

Hammel *m* (- *or* ⸚) **1** sheep, ram, wether. **2** = ⸚**fleisch** *n* mutton

Means that the word **Hammel** may also be used as a synonym for **Hammelfleisch** "mutton."

The following examples show how the equal sign may be used for cross-references:

hintan· *sep pfx:* . . . ·**setzen** *vt* put last; neglect; ignore . . . ·**stehen*** *vi* . . . ·**stellen** *vt* put last, etc = ·**SETZEN**

(For the rest of the meanings of **hintan·stellen,** look under **hintan·setzen.** The words are synonymous.)

hieher *adv* here, this way = HIERHER

(The word **hieher** is a variant form of the word **hierher.**)

mussieren *vi* effervesce, etc = MOUSSIEREN

(Another variant spelling; usually the spelling that comes first alphabetically is the one that gives the full list of meanings.)

Haide *f* (-n) heath = HEIDE (Actually **Heide** is the more frequent spelling.)

kuglig *a* spherical = KUGELIG (A case of the disappearing e; *see* **3.3.**)

ONE FINAL WARNING: Don't be hasty in picking a translation out of a longer entry. Read, or at least scan, the whole entry. The English word you need may well be the last one.

Besides, a tremendous amount of time and effort went into devising all the details of notation used in the entries. Learn how to use them to your advantage. With a good knowledge of basic German grammar and a little imagination, you can use this dictionary to help you read and understand German scientific and technical writings. The more you consult this introduction, the more experienced you will become in making wise use of the entries. And perhaps you will also gain some insight into the problems and difficulties a lexicographer has to cope with in making up a bilingual dictionary.

GEORGE E. CONDOYANNIS

New York, N.Y.
September 1991

PATTERSON'S
GERMAN-ENGLISH DICTIONARY
FOR CHEMISTS

A

a. *abbv* **1** (an, am) on. **2** (anno) in the year. **3** (asymmetrisch) asymmetric(al). **4** (aus) from, of

A *abbv* **1** (*Elec.*) ampere(s). **2** (Arbeit) work. **3** (*Chem.*) Argon. **4** (Atomgewicht) at. wt.

A. *abbv* **1** Alkohol. **2** (Annalen) Annals, Journal. **3** (Anno) in the year

Ä. *abbv* (Äther) ether

Aal *m* (-e) eel

a.a.O. *abbv* **1** (an anderen Orten) elsewhere. **2** (am angeführten *or* angegebenen Orte) in the place cited, loc. cit.

Aas *n* (-e) **1** carrion, carcass. **2** scrapings (from hides). **3** bait. **aasen 1** *vt* flesh, scrape (hides). **2** *vi* graze; (**mit**) squander

Aas-geruch *m* carrion odor. **-käfer** *m* carrion beetle. **-schmiere** *f* (*Lthr.*) flesh-side dubbing. **-seite** *f* flesh side

ab 1 *adv* off, away, down—**ab und zu** now and then, back and forth *cf* AUF. **2** *prep:* **ab Hamburg** from (*or* leaving) Hamburg; **ab 10. Juli** beginning July 10. **3** *sep pfx* off, away, down; *also implies quick or brief treatment, cf* ABBACKEN **2,** *or cessation of activity, cf* ABBLEICHEN *vi*

ab-aasen *vt* (*Lthr.*) flesh

Abaka *m* abacá, Manila hemp

abänderlich *a* modifiable, variable, alterable. **Abänderlichkeit** *f* modifiability, variability, etc

ab-ändern *vt* modify, change, alter; amend. **Abänderung°** *f* modification, change, alteration; variation, variant; amendment **Abänderungs-patent** *n* reissue patent. **-spielraum** *m* variation tolerance

ab-arbeiten 1 *vt* work off; rough out; tire

out. **2: sich a.** tire oneself out

Abart° *f* variety, modification, variant. **ab-arten** *vi*(s) vary; deviate; degenerate. **abartig** *a* anomalous, irregular; variant. **Abartung** *f* (-en) deviation, etc.

ab-ästen *vt* prune, trim (trees)

ab-atmen, ab-ätmen *vt* glow (a cupel), cupel; anneal; desiccate

ab-ätzen *vt* corrode (off), eat away. **abätzend** *p.a* corrosive

Abb. *abbv* (Abbildung) illus., fig.

ab-backen* *vt* **1** bake dry, overbake. **2** fry quickly, brown

ab-balgen *vt* flay, skin

Abbau *m* (-e) **1** breakdown, decomposition, degradation. **2** analysis. **3** demolition, dismantling. **4** mining; exploitation. **5** abolition, termination, ending. **6** reduction. **7** dismissal. **8** exhaustion (of mines). **9** (*Physiol.*) catabolism

abbaubar *a* degradable, decomposable: **biologisch a.** biodegradable

Abbau-bereich *m* breakdown range. **-dichte** *f* degradation density

ab-bauen I *vt* **1** break down, decompose, degrade. **2** analyze. **3** demolish, dismantle. **4** mine; exploit. **5** abolish; do away with; end, terminate. **6** reduce, cut down (on). **7** dismiss. **8** exhaust (mines). **9** (*Physiol.*) catabolize. **II: sich a. 10** break down, decompose, decay. **11** be abolished; end, terminate. **12** be reduced (exhausted, catabolized). **abbauend** *p.a* (*Physiol.*) catabolic

Abbau-erscheinung *f* degenerative change. **a-fähig** *a* decomposable, (bio)degradable;

1

(*Min.*) minable, workable. **A–ferment** *n* digestive enzyme. **-grad** *m* degree of disintegration. **-krankheit** *f* (*Agric.*) virus disease. **-produkt** *n* 1 (*Chem.*) decomposition (*or* degradation) product. 2 (*Biol.*) waste product. **-schritt** *m* cleavage step. **-stoff** *m* decomposition (*or* degradation) product. **-stufe** *f* stage of degradation; (*Biochem.*) catabolic stage. **-testung** *f* assay of degradation. **-verfahren** *n* degradation process. **a–verhindernd** *a* disintegration-inhibiting, stabilizing

ab·beißen* *vt* bite off; nip off, break off

Abbeiz·druck *m* (*Tex.*) discharge printing

ab·beizen *vt* remove (*or* treat) with corrosives, scour, cauterize; (*Metll.*) dip, pickle; (*Tex.*) discharge; (*Lthr.*) dress; (*Metals*) deoxidize. **abbeizend** *p.a* corrosive. **Ab·beizer** *m* (-) (paint) remover; corrosive. **Abbeiz·mittel** *n* corrosive, pickling agent; paint remover. **Abbelzung** *f* scouring, etc *cf* ABBEIZEN. **Abbeiz·verfahren** *n* cloth printing by color discharge

ab·bekommen* *vt* get (spots, etc) off, remove

ab·bersten* *vi*(s) burst off, crack off

ab·berufen* *vt* call away; recall. **Ab·berufung°** *f* recall; decease, demise

ab·bestellen *vt* cancel (an order for); countermand. **Abbestellung°** *f* cancellation; countermanding

ab·beugen *vt usu* = ABBIEGEN *vt;* (*Opt.*) scatter, diffract; (*Bot.*) take cuttings from

ab·bewirken *vt* give rise to, bring about, effect

ab·bezahlen *vt* pay off (in installments *or* with a final payment)

ab·biegen* **I** *vt* 1 bend (off), snap off. 2 take off in layers. 3 deflect; diffract. **II** *vi*(s) turn off, branch off, deviate. **Abbie·gung°** *f* deflection; diffraction; deviation

Abbild° *n* copy, image; portrait. **ab·bilden** 1 *vt* picture, portray; illustrate; represent. **2: sich a.** form an image. **abbildend** *p.a* (*Opt.*) image-forming. **Ab·bildung°** *f* illustration, figure; representation; picture, portrait

Abbildungs·fehler *m* optical aberration. **-gleichung** *f* imaging equation. **-güte** *f* definition, quality (of the image).

-kristallisation *f* mimetic crystallization. **-vermögen** *n* (*Micros.*) resolving power

ab·bimsen *vt* rub (off) with pumice; (*Lthr.*) buff, friz

Abbinde·geschwindigkeit *f* rate of setting

ab·binden* **I** *vt* 1 untie, unbind, loosen. 2 tie off. 3 castrate. 4 remove (by chemical combination). 5 frame. 6 (*Yarn*) skein. 7 (*Plastics*) cure. **II** *vi*(s) (*Cement,* etc) set, harden

Abbinde·vorgang *m* curing (*or* setting) process. **-zeit** *f* curing (*or* setting) time

Abblase·hahn *m* blowoff valve

ab·blasen* **I** *vt* 1 blow off (*or* away); blast (castings). 2 distill off. 3 call off, cancel. **II** *vi* (*Gases*) expand

Abblase·ventil *n* blowoff valve

ab·blassen *vi*(s) fade, lose color, pale

ab·blättern 1 *vt* strip of leaves; peel off; descale. 2 *vi & sich a.* scale off, exfoliate, shed leaves; effloresce. **Abblätterung°** *f* scaling off, exfoliation

Abblätterungs·mittel *n* exfoliative

ab·bläuen 1 *vt* blue (fabrics). 2 *vi* lose its blue color

ab·bleichen 1 *vt* bleach (out). 2 *vi* (h) finish bleaching. 3 *vi*(s)* fade, pale

ab·blenden 1 *vt/i* screen off, shade; dim (lights). 2 *vt* (*Phot.*) stop down; black out

Abblend·schalter *m* (headlight) dimmer (switch)

ab·blicken *vi* (*Metll.*) grow dull after the flash or "blick", tarnish

ab·blitzen *vi*(s) miss fire; (h) (*of lightning*) cease flashing

ab·blühen *vi*(s) wither; (h) cease blooming

ab·borken *vt* peel the bark off, (de)bark

ab·böschen *vt* slope, incline; bevel

Abbrand° *m* 1 burnt (*or* roasted) ore; slag, (forging) waste. 2 loss by burning. 3 burn (of a cigar); calcination, combustion, burn-up. **-·geschwindigkeit** *f* rate of combustion

ab·brauchen *vt* use up, wear out

ab·brauen 1 *vt* brew (thoroughly). 2 *vi* finish brewing

ab·braunen *vi* lose its brown color

ab·bräunen 1 *vt* brown; dye brown. 2 *vi* lose its brown color

ab·brausen 1 *vt* rinse (off); drench; spray. 2 *vi* (h) cease roaring; cease effervescing. 3 *vi*(s) depart with a roar

ab·brechen* I *vt* 1 break off. 2 tear down, demolish. 3 break up (*or* down), disassemble, take down (*or* apart). 4 bring to an end, terminate. II *vi* (h) 5 stop (short), end (suddenly). III *vi*(s) break off, snap off

ab·breiten *vt* stretch out, flatten

ab·bremsen *vt* brake, retard, decelerate

abbrennbar *a* combustible, deflagrable. **Abbrennbarkeit** *f* combustibility, deflagrability

Abbrenn·bürste *f* contact-breaking brush

ab·brennen* I *vt, vi*(s) 1 burn off (away, down); deflagrate. II *vt* 2 fire off (gun, etc). 3 calcine; temper. 4 dip, pickle. 5 cauterize; sear off, singe. 6 (*Ceram.*) finish firing. III *vi*(s) 7 miss fire. 8 lose everything in a fire. **Abbrenner°** *m* deflagrator

Abbrenn·glocke *f* deflagrating jar. **-löffel** *m* deflagrating spoon. **-mittel** *n* deflagrating agent, deflagrator

Abbrennung° *f* burning off (away, down); deflagration, etc *cf* ABBRENNEN

ab·bringen* *vt* (**von**) 1 remove (from), get (sthg) off. 2 turn, divert ‛from). 3 dissuade (from), talk (out of)

ab·bröckeln 1 *vt, vi*(s) break off in bits, chip off. 2 *vi*(s) crumble away; drop off

Abbruch° *m* 1 break-off, breaking off. 2 tearing down, demolition. 3 break-up, breakdown. 4 dismantling, disassembly. 5 breach. 6 fragment. 7 discontinuance, cessation, termination. 8 harm, damage: **A. tun** (+ *dat*) do harm (to), impair; **zum A. gereichen** be detrimental. *Cf* ABBRECHEN. **-·effekt** *m* break-off effect

abbrüchig *a* 1 crumbly, brittle. 2 detrimental

Abbruch·linie *f* line of cleavage. - **reaktion** *f* chain-terminating reaction, interfering (competing) reaction

abbruchs·reif *a* ready for demolition (termination, etc) *cf* ABBRUCH. **A—substanz** *f* (*Biochem.*) terminating reagent

ab·brühen *vt* scald, parboil. **Abbrüh·kessel** *m* scalding kettle

ab·buchen *vt* deduct. **Abbuchung°** *f* deduction

ab·bürsten *vt* brush off

ab·dachen *vt* 1 slope, incline; bevel. 2 unroof. **abdachig** *a* sloping, inclined. **Abdachung** *f* (-en) 1 slope, incline; escarpment. 2 unroofing

ab·dämmen *vt* dam up (*or* off); restrain. **Abdämmung°** *f* damming up (*or* off); restraint; embankment

Abdampf° *m* exhaust (*or* waste) steam— **mit A. geheizter Vorwärmer** exhaust-feed heater. **-·apparat** *m* evaporating apparatus, evaporator. **-druckregler** *m* exhaust steam pressure regulator

ab·dampfen 1 *vt/i* evaporate (off), vaporize, boil down. 2 *vi*(s) steam away, take off. **Abdampfen** *n* (removal by) evaporation

ab·dämpfen *vt* 1 evaporate, volatilize. 2 steam, scald, stew. 3 quench (charcoal). 4 damp(en), tone down

Abdampf·gefäß *n* evaporating vessel. **-heizung** *f* waste-steam heating. **-kapelle** *f* evaporating capsule. **-kasserolle** *f* evaporating pan. **-kessel** *m* evaporating kettle. **-kondenserwasserabschneider** *m* exhaust steam separator. **-leitung** *f* exhaust steam line. **-maschine** *f* 1 evaporator. 2 (*Ceram.*) = **-ofen** *m* slip kiln. **-rückstand** *m* residue on evaporation. **-sammelstück** *n* exhaust steam collector. **-schale** *f* evaporating dish. **-speicher** *m* evaporating storer. **-stutzen** *m* exhaust hood

Abdampfung *f* (-en) evaporation, volatilization

Abdampfungs· = ABDAMPF·: **-vorrichtung** *f* evaporating device, evaporator

ab·danken 1 *vt* dismiss, retire. 2 *vi* resign, retire; abdicate. **Abdankung** *f* (-en) dismissal; retirement; resignation; abdication

ab·darren *vt* dry, kiln-dry, desiccate; cure (by drying); (*Metll.*) liquate

Abdarr·temperatur *f* (*Brew.*) finishing temperature

Abdeck·blech *n* cover plate

ab·decken *vt* 1 uncover, unroof; skin, flay. 2 cover up, shield; screen (off) (rays). 3 cover, meet (requirements). 4 (*Sugar*) wash. 5 (*Etchings*) stop out. **Abdecker°** *m* animal skinner, flayer. **Abdecke-**

rei *f* (-en) skinner's (yard; trade). **Abdecker(ei)·fett** *n* fat from skinning
Abdeck·papier *n* carpet felt paper. **-platte** *f* cover (*or* access) plate
Abdeckung° *f* uncovering, etc *cf* ABDECKEN
Abdeck·vorrichtung *f* covering device
ab·deichen *vt* dam (off), dike (off)
ab·dekantieren *vt* distill off, remove
ab·dichten *vt* seal, lute, calk, pack; waterproof; insulate. **Abdichtung°** *f* 1 sealing, luting; calking; waterproofing. 2 (air-, water-)tightness. 3 seal, closure; gasket; washer. 4 electrolytic reaction on cell walls
ab·dicken *vt* boil down, evaporate (down), thicken (by boiling), inspissate. **Abdickung** *f* boiling down, etc
ab·dissoziieren *vt/i* dissociate (off)
ab·dorren *vi*(s) wither and dry up
ab·dörren *vt* dry (up), desiccate; (*Metll.*) roast, sweat (ores)
Abdörr·ofen *m* refining furnace. **-stein** *m* lead ore containing silver or copper
Abdraht *m* turnings
ab·drängen *vt* push aside, displace
Abdreh·diamant *m* abrasive diamond
ab·drehen 1 *vt* turn off (gas, etc); turn (on a lathe); twist off; wring (neck); unscrew. 2 *vi* & **sich a.** turn away, veer off
Abdreh·maschine *f* lathe; (*Ceram.*) finishing machine. **-späne** *mpl* turnings. **-spindel** *f* (*Ceram.*) finishing machine. **-stahl** *m* turning tool
Abdrehung° *f* torsion; turning, etc *cf* ABDREHEN
Abd·Rk. *abbv* (Abderhaldensche Reaktion) Abderhalden reaction
ab·drosseln *vt* throttle, choke (off), shut off
Abdruck *m* (-e *or* ⁼e) 1 printing; copying. 2 impression; copy, proof. 3 reprint; (off-)print; cast. 4 (*Micros.*) replica
ab·drucken *vt* 1 (re)print; copy. 2 mold (by pressure)
ab·drücken I *vt* 1 make an impression of. 2 press off, force off; wring. 3 release (spring). 4 fire (gun); squeeze (trigger). II *vi* pull the trigger, fire
Abdruck·film *m* replica film. **-gips** *m* impression plaster. **-häutchen** *n* replica film. **-lauge** *f* release liquor. **-masse** *f* molding material. **-präparat** *n* replica specimen. **-recht** *n* right of reproduction, copyright. **-schraube** *f*. set screw
Abdrucks·recht *n* copyright = ABDRUCKRECHT
Abdruck·technik *f* replication technique
ab·duften *vi*(s) lose its aroma, fade; get lost
ab·dunkeln 1 *vt* darken, black out; (*Colors*) sadden *cf* ANSIEDEN. 2 *vi*(s) darken, deepen
Abdunklungs·mittel *n* (*Dyes*) saddening agent
ab·dunsten *vi*(s) evaporate, vaporize
ab·dünsten *vt* evaporate, vaporize; graduate (brine); steam, stew
Abdunstung, Abdünstung *f* (-en) evaporation
Abdünstungs·haus *n* (*Salt*) graduation house
abduzieren *vt* (*Physiol.*) abduct
Abelmosch(us) *m* abelmosk, musk
Abel·prüfer *m* Abel tester. **-test** *m* Abel test
abend *adv,* **Abend** *m* (-e) evening, night: **heute a.** tonight, this evening. **--brot** *n,* **-essen** *n* supper, evening meal. **-land** *n* western country; West, Occident. **-länder** *m* Westerner, Occidental. **a–ländisch** *a* western, occidental. **abendlich** *a* (of the) evening. **abends** *adv* in the evening; evenings. **abend·wärts** *adv* westward
Abenteuer *n* (-) adventure
aber 1 *cnjc* but, however. 2 *adv* ever; now. 3: **Aber** *n:* **ohne Wenn und A.** without ifs or buts
Aber·glaube *m* superstition. **abergläubisch** *a* superstitious
aberkannt denied: *pp* of ABERKENNEN
ab·erkennen* *vt:* **einem etwas a.** deny sbdy sthg, deprive sbdy of sthg. **Aberkennung°** *f* denial, deprivation
abermalig *a* repeated; further, new. **abermals** *adv* again: **aber und a.** again and again
ab·ernten *vt* reap, harvest; reap the crops off (the fields)
ab·erregen *vt* deactivate, deenergize
Aberwitz *m* madness, absurdity. **aberwitzig** *a* mad, absurd

ab·essen* 1 *vt* (*usu* + **von**) eat . . . off. 2 *vi* finish eating

Abessinien *n* Abyssinia

ab·fachen *vt* divide (off), partition (off), compartmentalize

ab·fädeln *vt* unstring, unthread

abfahrbar *a* removable

ab·fahren* I *vt* 1 remove; cart (*or* carry) away. 2 patrol (in a vehicle). 3 wear down (*or* out) (tires, etc). II *vi*(s) 4 depart, leave. 5 be snubbed. **Abfahrt°** *f* departure

Abfall° *m* 1 falling off. 2 decrease, drop, diminution, deterioration, decline. 3 slope, declivity. 4 desertion, defection. 5 (*oft pl* **Abfälle**) waste, trash, refuse, scrap(s), garbage, cuttings. 6 (+ **gegen, neben**) unfavorable contrast (with)

Abfall· *usu* waste: **-baumwolle** *f* cotton waste. **-beseitigung** *f* waste removal (*or* disposal). **-brennstoff** *m* refuse fuel. **-eimer** *m* trash (*or* garbage) can. **-eisen** *n* scrap iron

Abfallauge *f* spent lye = **Abfall·lauge**

ab·fallen* *vi*(s) 1 (*oft* + **von**) fall off, drop (off). 2 decrease, diminish; decline, deteriorate. 3 be left over (*usu* as waste). 4 lose weight (intensity, brilliance, altitude, etc). 5 slope (away). 6 desert, defect. 7 be rebuffed. 8 (+ **gegen, neben**) contrast unfavorably (with). **abfallend** *p.a oft* = ABFÄLLIG

Abfall·erzeugnis *n* waste product

Abfälle·verwertung *f* utilization of waste (products), waste recycling

Abfall· *usu* waste: **-fett** *n* waste fat (*or* grease). **-grenze** *f* critical limit. **-gut** *n* (recoverable) waste material

abfällig *a* 1 decreasing, diminishing; declining; decaying; falling off. 2 sloping. 3 (*Bot.*) deciduous. 4 adverse, unfavorable; derogatory

Abfall· *usu* waste: **-kohle** *f* waste coal (*or* charcoal), coal dust. **-koks** *m* coke breeze, refuse coke. **-kurve** *f* decay curve (of isotopes). **A.-Labor** *n* waste-sampling laboratory. **-lauge** *f* spent lye = ABFALLAUGE. **-marke** *f* (*Dye.*) off shade. **-produkt** *n* waste product; by-product. **-rohr** *n* waste pipe; draft (*or* suction) tube. **-säckchen** *n* (air) sickness bag. **-säure** *f* waste (*or*

spent) acid. **-stoff** *m* waste material (*or* matter); (*specif*) sewage. **-tonne** *f* trash barrel. **-topf** *m* waste jar. **-vernichtung** *f* waste destruction. **-vernichtungsanlage** *f* (*genl*) waste disposal unit; (*specif*) incinerator. **-verwertung** *f* waste salvage; waste- (*or* by-)product utilization. **-ware** *f* rejected merchandise, rejects. **-wärme** *f* waste heat. **-wasser** *n* (*genl*) waste water; (*specif*) sewage. **-wirtschaft** *f* waste recycling. **-zeit** *f* drop-out (*or* decay) period

ab·falzen *vt* trim; (*Skins*) scrape, skive

ab·fangen* *vt* 1 catch, capture (*incl Nucl. Phys.*), intercept; secure, take (up), absorb, seize; fix; trap (an intermediate reaction). 2 overtake. 3 straighten out; pull (a car) out of a skid, pull (a plane) out of a dive. 4 forestall; thwart. 5 support, prop up. **Abfänger°** *m* interceptor, scavenger; trapping agent

Abfang·radius *m* pull-out radius. **-schritt** *m* capture step. **-verfahren** *n* interception process

ab·färben I *vt* 1 dye (thoroughly). 2 finish dyeing. II *vi* 3 lose color, fade, run, bleed. 4 (+ **auf**) rub off (on), impart its color (to); stain; (*Min.*) give a colored streak (to)

abfärbig *a* discolorable, apt to rub off (its color)

ab·fasern *vi* & **sich a.** lose fibers, unravel, fray

ab·fassen *vt* 1 write, draw up; word, formulate. 2 take hold of, seize; arrest. 3 bend (iron). **Abfasser** *m* (-) writer. **Abfassung°** *f* 1 writing, wording, version. 2 drawing up, formulation. 3 seizure; arrest

ab·federn I *vt* 1 spring, fit with springs. 2 (*Explo.*) cushion. 3 defeather. II *vi* shed feathers, molt

Abfege·mittel *n* detergent

ab·fegen *vt* sweep (away *or* off); clean, depurate. **abfegend** *p.a* detergent. **Abfegung** *f* depuration, cleaning; sweeping

ab·feilen *vt* file off; polish. **Abfeilicht** *m* filings

ab·feilschen *vt:* **einem etwas a.** bargain sthg from sbdy

ab·feimen *vt* skim off *cf* ABGEFEIMT

ab·fertigen *vt* 1 send off, dispatch. 2 serve; attend to, deal with. 3 finish (off). 4 rebuff. **Abfertigung°** *f* sending off, dispatch(ing); service; attention; clearance; rebuff

Abfertigungs·schein *m* (customs) declaration; clearance, permit. **-stelle** *f* (*Com.*) dispatching office

Abfett° *n* (fat) skimmings; drain-oil scourings. **ab·fetten** *vt* skim fat (*or* grease) off, degrease

ab·feuern *vt* fire (off), discharge

Abfeuer·vorrichtung *f* firing device

ab·fiedeln *vt* (*Metll.*) skim off (dross)

ab·filtern, ab·filtrieren *vt* filter off. **Abfiltrierung°** *f* filtering off, filtration

ab·finden* 1 *vt* pay (off), compensate. 2: **sich a.** (**mit**) settle, reach an agreement (with); resign oneself (to); put up (with). **Abfindung** *f* (-en) compensation, indemnification; agreement, settlement

ab·flachen *vt* flatten, level off; bevel *cf* ABGEFLACHT. **Abflachung** *f* (-en) 1 flattening, leveling; bevel(ing); rounding off (of a curve). 2 flat spot

ab·flammen *vt* 1 singe, flame. 2 (*Hides*) grease, tallow. 3 gas

ab·flauen 1 *vt* wash, scour; buddle (ore). 2 *vi*(s) diminish, let up, subside

ab·fleischen *vt* flesh, scrape (hides)

ab·fliegen* 1 *vt, vi*(s) fly off, fly away. 2 *vt* patrol by air. 3 *vi*(s) take off; drop off

ab·fließen* *vi*(s) flow (*or* run) off; drain (off); discharge

ab·fluchten *vt* line out, mark out, align. **Abflucht·linie** *f* alignment

Abflug° *m* 1 take-off, flight *cf* ABFLIEGEN. 2 matter carried by air currents

Abfluß° *m* 1 flowing off; efflux, run-off, flow-off; discharge. 2 outlet, drain. 3 effluent, waste water. *Cf* ABFLIESSEN. **-geschwindigkeit** *f* efflux velocity. **-graben** *m* culvert; drainage ditch. **-hahn** *m* discharge cock (*or* tap). **-kühler** *m* efflux condenser. **-leitung** *f* drain (*or* run-off) pipe. **a—los** *a* having no outlet: **a—e Wanne** isolated basin. **A—menge** *f* amount of discharge (run-off, effluent). **-rinne** *f* outlet trough, discharge (*or* run-off) gutter. **-rohr** *n* outlet, drainage tube,

waste (*or* drain) pipe. **-ventil** *n* discharge (*or* run-off) valve. **-wasser** *n* waste water

Abfolge° *f* 1 sequence, order: **A. der Zeit** chronological sequence. 2 succession. 3 (*Geol.*) (magmatic) origin

ab·fordern *vt* demand; recall (a person)

ab·formen *vt* form, shape, mold; copy, model. **Abform·masse** *f* molding compound

ab·forsten *vt* deforest, clear; cut down

Abfrage·anlage *f* (*Mach.*) interrogator

ab·fragen *vt* 1 question, interrogate. 2 get information from. 3 (*Instrum.*) read; (*Cmptrs.*) access

ab·fräsen *vt* mill off (*or* away)

ab·fressen* *vt* eat off (*or* away), eat up, consume; corrode (off), remove with caustics

ab·frieren* *vi*(s) freeze (off); be frostbitten, be nipped by frost

ab·frischen *vt* (*Brew.*) change (steep water)

Abfuhr° *f* 1 carting (*or* carrying) away; outlet. 2 transport(ation), conveyance; removal. 3 rebuff, defeat. **--dünger** *m* night soil

ab·führen I *vt* 1 lead away, carry off, conduct away. 2 lead off, exhaust (waste steam), dissipate (heat). 3 draw (wire). 4 pay (out), credit. 5 (*Med.*) purge, evacuate. **II** *vi* 6 have a laxative effect. 7 move one's bowels. **abführend** *p.a* laxative, purgative; efferent, excretory; **a—es Brausepulver** Seidlitz powder

Abführ·gang *m* (*Physiol.*) efferent duct. **-mittel** *n* laxative, purgative. **-pille** *f* laxative pill. **-salz** *n* laxative salt

Abführung° *f* leading away, etc *cf* ABFÜHREN; outlet, discharge

Abführungs·gang, -kanal *m* excretory (*or* efferent) duct. **-mittel** *n* laxative. **-rohr** *n* discharge tube (*or* pipe)

Abfüll·apparat *m* emptying apparatus, etc *cf* ABFÜLLEN; (*Brew.*) racker. **-bütte** *f* (*Brew.*) racking square

ab·füllen *vt* 1 fill (a container). 2 pour, decant, draw (off), drain off, rack (liquid)— **Wein (auf Flaschen) a.** bottle wine; **Bier in Dosen a.** can beer. 3 empty (a container). 4 skim off (dross)

Abfüll·flasche *f* receiving bottle. **-maschine** *f* bottling machine. **-pipette** *f* delivery pipette. **-schlauch** *m* siphon tube; (*Brew.*) racking hose. **-trichter** *m* filling funnel

Abfüllung° *f* filling, pouring, etc *cf* ABFÜLLEN

Abfüll·vorrichtung *f* emptying device

ab·furchen *vt* furrow; divide. **Abfurchung** *f* (-en) division, segmentation

ab·futtern, ab·füttern *vt* line, case; feed.

Abfütterung° *f* lining; feeding

Abgabe° *f* 1 delivery; yield; giving off, release, output, emission. 2 (*Gases*) evolution, escape. 3 dissipation; (*Electron.*) donation. 4 firing (of a shot). 5 sale. 6 surrender, submission, handing in (of tickets, accounts, etc). 7 tax, duty. 8 (*Com.*) draft; payment

abgabe(n)·frei *a* tax-free, duty-free. **-pflichtig** *a* taxable, dutiable

Abgabe·rohr *n*, **-röhre** *f* delivery (outlet, escape, emission) tube

Abgang° *m* 1 leaving, departure, exit. 2: **in A. geraten** fall into disuse. 3 successful completion (of secondary school). 4 outlet; flux, passing, discharge; effluent. 5 escape, leakage; miscarriage. 6 loss, decrease (in value). 7 (*Com.*) sale; market. 8 *pl:* **Abgänge** waste, scraps, tailings, etc

abgängig *a* 1 waste, scrap. 2 missing. 3 salable, selling well. 4 deteriorating, declining; worn out. 5 departing. 6 sloping

Abgang·papier *n* waste paper

Abgangs·dampf *m* exhaust steam

Abgängsel *n* (-) (piece of) refuse, scrap

Abgangs·rohr *n* waste pipe. **-säure** *f* waste (*or* spent) acid. **-zeugnis** *n* secondary school diploma

ab·gären* *vi* 1 ferment completely. 2 cease fermenting

Abgas° *n* waste (exhaust, flue) gas, gaseous emission. **··prüfegerät** *n* exhaust analyzer. **-spelcherofen** *m* regenerative furnace recuperator. **-staub** *m* exit-gas dust. **-vorwärmer** *m* waste gas preheater, heat economizer

ab·gautschen *vt* (*Paper*) couch

ab·geben* I *vt* 1 give off, emit. 2 give up, hand in (*or* over), surrender. 3 lose; yield. 4 give (opinion, etc), make (offer, etc). 5 serve as, act as, make (a good chemist,

etc). **II: sich a.** (**mit**) concern oneself (with)

abgebissen bitten off: *pp of* ABBEISSEN

abgeblichen bleached, etc: *pp of* ABBLEICHEN

abgebogen bent, etc: *pp of* ABBIEGEN

abgeborsten burst off: *pp of* ABBERSTEN

abgebracht removed, etc: *pp of* ABBRINGEN

abgebrannt I burned down, etc: *pp of* ABBRENNEN. II *p.a* 1 (*Metals*) refined. 2 penniless, broke

abgebrochen 1 broken off: *pp of* ABBRECHEN. 2 *p.a* abrupt. **Abgebrochenheit** *f* abruptness

abgebrüht *p.a* case-hardened; callous; *see also* ABBRÜHEN

abgebunden untied, etc: *pp of* ABBINDEN

abgedrechselt *p.a* rounded off; stilted, affected

abgedroschen *p.a* trite, hackneyed

abgefeimt *p.a* crafty, cunning; *see also* ABFEIMEN

abgeflacht *p.a* flat; oblate; shallow; *see also* ABFLACHEN

abgeflogen taken off, etc: *pp of* ABFLIEGEN

abgeflossen flowed off *pp of* ABFLIESSEN

abgefressen corroded: *pp of* ABFRESSEN

abgefroren frostbitten: *pp of* ABFRIEREN

abgefunden paid off, etc: *pp of* ABFINDEN

abgegangen left, etc: *pp of* ABGEHEN

abgegessen eaten off: *pp of* ABESSEN

abgeglichen equalized: *pp of* ABGLEICHEN

abgeglitten slid off: *pp of* ABGLEITEN

abgegolten settled: *pp of* ABGELTEN

abgegoren fermented: *pp of* ABGÄREN

abgegossen decanted: *pp of* ABGIESSEN

abgegrenzt *p.a* discrete; *see* ABGRENZEN

abgegriffen 1 worn out, etc: *pp of* ABGREIFEN. 2 *p.a* well-worn, hackneyed

abgehaart *p.a* hairless; *see also* ABHAAREN

abgehackt *p.a* staccato; *see also* ABHACKEN

ab·gehen* I *vt* 1 patrol; pace off. II *vi*(s) 2 go away, leave. 3 be sent, be delivered— **a. lassen** send off. 4 (*usu* + **von**) deviate, depart (from), go astray. 5 come off (*or* out). 6 (*Med.*) be discharged. 7 sell, be sold. 8 be deducted. 9 come (*or* turn) out (well, badly). 10 be lost; depreciate. 11 (+ *dat*) be lacking (to). **abgehend** *p.a* (*oft*) outgoing

abgehoben lifted off: *pp of* ABHEBEN

abgeholfen helped: *pp of* ABHELFEN

abgekämpft *p.a* worn out, exhausted

Abgeklärtheit *f* 1 clarity. 2 tranquillity; serenity; poise. *Cf* ABKLÄREN

abgeklungen died out *pp of* ABKLINGEN

abgekürzt *p.a* abridged, concise; *see also* ABKÜRZEN

abgelaufen *p.a* run-down; expired, terminated; *see also* ABLAUFEN

abgelebt *p.a* obsolete; decrepit. **Abgelebtheit** *f* obsoleteness; decrepitude

abgelegen *p.a* 1 ripe, mature. 2 remote, isolated. *See also* ABLIEGEN. **Abgelegenheit** *f* 1 ripeness, maturity. 2 remoteness, isolation

ab·gelten* *vt* settle, pay in full; compensate. **Abgeltung** *f* (-en) settlement, full payment; compensation

abgemessen *p.a* 1 measured, precise. 2 appropriate. *See also* ABMESSEN. **Abgemessenheit** *f* 1 exactness, precision. 2 appropriateness

abgeneigt *p.a* disinclined, reluctant, averse. **Abgeneigtheit** *f* disinclination, reluctance, aversion *cf* ABNEIGUNG

abgenommen taken off *pp of* ABNEHMEN

Abgeordnete *m, f* (*adj endgs*) deputy, delegate

abgeplattet *p.a* flattened, oblate; (*Nucleus*) deformed; *see also* ABPLATTEN

ab·gerben *vt* tan (thoroughly)

abgerechnet *prep, adv* (+ *accus*) excluding, deducting, except for: **Verluste a.** excluding losses; *see also* ABRECHNEN

abgerieben rubbed off: *pp of* ABREIBEN

abgerissen 1 torn off: *pp of* ABREISSEN. 2 *p.a* abrupt, disconnected; incoherent

abgeronnen run off: *pp of* ABRINNEN

abgerundet *p.a* 1 rounded off; fire-polished (glass edges). 2 round (numbers). 3 dished

abgerungen wrested: *pp of* ABRINGEN

abgesagt *p.a* declared, sworn (enemy); *see also* ABSAGEN

Abgesandte *m, f* (*adj endgs*) envoy, delegate *cf* ABSENDEN

Abgeschäumte(s) *n* (*adj endgs*) skimmings, scum *cf* ABSCHÄUMEN

abgeschieden I separated, etc: *pp of* ABSCHEIDEN. II *p.a* 1 secluded, isolated.

2 departed, deceased. **Abgeschiedenheit** *f* seclusion, isolation

Abgeschlagenheit *f* (*Econ.*) general decline *cf* ABSCHLAGEN

abgeschliffen ground off: *pp of* ABSCHLEIFEN. **Abgeschliffenheit** *f* polish, refinement

abgeschlossen 1 closed off, etc: *pp of* ABSCHLIESSEN. 2 *p.a* closed, sealed; self-contained. **Abgeschlossenheit** *f* 1 seclusion, isolation. 2 exclusiveness. 3 closure. 4 completeness (in itself)

abgeschmackt *a* tasteless, insipid; absurd. **Abgeschmacktheit** *f* poor taste, tastelessness; absurdity

abgeschmolzen melted off: *pp of* ABSCHMELZEN

abgeschnitten cut off: *pp of* ABSCHNEIDEN

abgeschoben pushed off: *pp of* ABSCHIEBEN

abgeschoren shorn off: *pp of* ABSCHEREN

abgeschossen fired *pp of* ABSCHIESSEN

abgeschrieben copied: *pp of* ABSCHREIBEN

abgeschritten paced off: *pp of* ABSCHREITEN

Abgeschwindigkeit *f* rate of discharge

abgeschwollen subsided: *pp of* ABSCHWELLEN

abgeschwommen floated away: *pp of* ABSCHWIMMEN

abgeschwungen centrifuged: *pp of* ABSCHWINGEN

abgesehen *p.a* 1 (**auf**) intended (for), aimed (at). 2 (**von**) apart (from), disregarding: **a. davon, daß** apart from the fact that. *Cf* ABSEHEN

abgesessen settled: *pp of* ABSITZEN

abgesogen suctioned: *pp of* ABSAUGEN

abgesotten boiled off: *pp of* ABSIEDEN

abgespannt *p.a* unstrung, unnerved; exhausted; *see also* ABSPANNEN. **Abgespanntheit** *f* exhaustion; fatigue

abgesprochen denied, etc: *pp of* ABSPRECHEN

abgesprungen jumped off: *pp of* ABSPRINGEN

abgestanden *p.a* stale, flat, dead; *see also* ABSTEHEN

abgestiegen descended: *pp of* ABSTEIGEN

abgestochen tapped, etc: *pp of* ABSTECHEN

abgestorben died, etc: *pp of* ABSTERBEN

abgestrichen scraped: *pp of* ABSTREICHEN

abgestuft *p.a* differential (centrifugation); *see also* ABSTUFEN

abgestumpft *p.a* blunt, dull; blasé; *see also* ABSTUMPFEN. **Abgestumpftheit** *f* bluntness, dullness; (mental) apathy

abgesunken sunk down: *pp of* ABSINKEN

abgetan put away: *pp of* ABTUN

abgetragen *p.a* threadbare, shabby; *see also* ABTRAGEN

abgetrieben expelled, etc: *pp of* ABTREIBEN

abgetroffen dripped down: *pp of* ABTRIEFEN

abgewandt averted: *pp of* ABWENDEN

abgewichen deviated: *pp of* ABWEICHEN II

abgewiesen rejected: *pp of* ABWEISEN

ab·gewinnen* *vt* (*d.i.o*) win away, get, wrest, extract (from)

abgewogen 1 weighed, etc: *pp of* ABWÄGEN & ABWIEGEN. 2 *p.a* organized (movement); (carefully) considered

ab·gewöhnen *vt* 1 (**von**) wean (from), break (sbdy) of the habit (of). **2: sich etwas a.** give up (the habit of) doing sthg

abgewonnen won away: *pp of* ABGEWINNEN

abgeworfen cast off: *pp of* ABWERFEN

abgewunden unwound: *pp of* ABWINDEN

abgezirkelt *p.a* (elaborately) precise *cf* ABZIRKELN

abgezogen drawn, etc: *pp of* ABZIEHEN

ab·gießen* *vt* pour off, decant; drain; (*Foun.*) cast; (*Metll.*) pour

ab·giften *vt* detoxify

Abglanz *m* reflection

ab·glasen *vt* (*Lthr.*) glaze

ab·glätten *vt* smooth (off), polish (off)

Abgleich *m* (-e) balance. **ab·gleichen*** *vt* 1 make equal, equalize, even up. 2 level, smooth; balance (accounts). 3 (*Print.*) justify (lines). **Abgleichung°** *f* equalization, evening up; leveling; balance; (*Print.*) justification

ab·gleiten* *vi*(s) slide off, slip off; glance off; decline. **Abgleitung°** *f* slip; decline

ab·gliedern 1 *vt* dismember. 2 *vi* segment off, form lateral offshoots. **Abgliederung°** *f* dismemberment; lateral offshoot, outgrowth

ab·glühen 1 *vt* (*usu Metal*) heat red-hot; anneal, temper; (*Wine*) mull. 2 *vi*(s) & **sich**

a. cease glowing, cool down

ab·graben* *vt* dig away; provide with ditches; drain; cut off (water supply)— **einem das Wasser a.** cut the ground from under sbdy

ab·grasen *vt* graze (on); exploit; (**nach**) scour (an area) (for)

ab·graten *vt* trim, burr, remove the flash from. **Abgrat·schrott** *m* trimmings

ab·greifen* *vt* 1 wear out. 2 grope. 3 measure off; span (distance). 4 (*Elec.*) tap (a circuit). *Cf* ABGEGRIFFEN

abgrenzbar *a* limitable, definable

ab·grenzen *vt* (de)limit; mark off; define *cf* ABGEGRENZT. **Abgrenzung** *f* (-en) (de)limitation; demarcation; definition; boundary

Abgriff° *m* measuring off; spanning; (*Elec.*) tap *cf* ABGREIFEN

Abgrund° *m* abyss, chasm. **abgründig** *a* unfathomable; (*adv oft*) immeasurably. **abgrund·tief** *a* abysmally deep, unfathomable

Abguß° *m* 1 pouring off, decantation. 2 casting, cast. 3 drain; sink. *Cf* ABGIESSEN. **-kasten** *m* trap, drain

abh. *abbv* (abhängig) dependent

Abh. *abbv* (Abhandlung) treatise, etc; (*in Journals, oft*) transactions

ab·haaren 1 *vt* remove hair from, depilate. 2 *vi* lose hair, molt. *Cf* ABGEHAART

ab·haben* *vt*: **etwas a. von . . .** get one's share of . . .

ab·hacken *vt* chop off (*or* down), cut down *cf* ABGEHACKT

ab·haken *vt* 1 unhook. 2 mark off, check off

ab·halten* *vt* 1 hold off, keep off (*or* away), ward off. 2 keep, prevent (from doing). 3 detain, hinder. 4 withstand. 5 hold (a meeting, etc). **Abhaltung°** *f* 1 previous engagement. 2 keeping, prevention. 3 detention. 4 withstanding. 5 holding (of meetings, etc)

ab·handeln *vt* 1 treat (of), discuss, debate (on). 2 settle, bargain (for), bargain down (price)

abhanden kommen* *vi*(s) be lost, be mislaid

Abhandlung° *f* treatise, paper, article, dissertation; discussion; (*pl oft*) transactions (of a society)

Abhang° *m* slope, declivity; hillside
ab·hängen* 1 *vt* take down; disconnect. 2 *vt/i* hang up (the phone). 3 *vi* (+ **von**) depend (on), be a function (of); hang down (from)
abhängig *a* sloping, inclined; (*usu* + **von**) dependent, depending (on). **Abhängige** *m, f* (*adj endgs*) (*incl Math.*) dependent, variable. **Abhängigkeit** *f* (-en) 1 dependence. 2 slope. 3: **in A. von** depending on, as a function of
Abhängigkeits·verhältnis *n* dependent relationship, state of dependence
ab·hären 1 *vt* unhair. 2 *vi* lose hair. = ABHAAREN
ab·härten *vt* harden; (*Iron, Steel*) temper; (*usu* + **gegen**) toughen (one's body), inure (to). **Abhärtung** *f* hardening; tempering; toughening
ab·harzen *vt* take resin from, tap for resin
ab·hauen* 1 *vt* hew, chop down; chop off. 2 *vi* leave; stop work
ab·häuten *vt* skin, peel; (*Metll.*) skim, scum
abhebbar *a* removable
ab·heben* 1 *vt/i* lift off, take off. 2 *vt* remove, skim off; siphon off; collect, withdraw (dividends). 3: **sich a. (von)** contrast (with), stand out (against). **Abhebung°** *f* removal; withdrawal; skimming off; siphoning off
ab·hebern *vt* siphon off
ab·heften *vt* 1 unstaple, unstitch. 2 staple up, stitch up (in a separate batch)
ab·heilen *vi*(s) heal up
ab·heizen *vt* cool down
ab·helfen* *vi* (*dat*) help, remedy, correct
ab·hellen* 1 *vt* clarify, clear up; (*Wine*) fine. 2: **sich a.** (become) clear
ab·hetzen 1 *vt* overwork, tire out; harass. 2: **sich a.** wear oneself out
Abhilfe° *f* remedy, corrective measure
Abhitze *f* waste heat; heat loss. **-·kanal** *m* waste-heat flue. **-kessel** *m* waste-heat boiler. **-ofen** *m* waste-heat oven. **-verwertung** *f* waste-heat utilization
ab·hobeln *vt* plane (off), smooth
abhold *pred. a* (*dat*) unfavorable, averse (to)
ab·holen *vt* fetch, call for, pick up
Abhol·fach *n* private post-office box

ab·holzen *vt* deforest, clear (of timber)
Abhör·apparat *m* (*Med.*) stethoscope; (*Electron.*) monitor(ing device)
ab·horchen *vt* (*Med.*) sound, tap, auscultate; (*incl Electron.*) find out by eavesdropping, monitor. **Abhorchen** *n* sounding, auscultation; eavesdropping, monitoring
ab·hören *vt* 1 hear, listen to; examine (orally); eavesdrop on; (*Electron.*) monitor, tap. 2 (*dat*) find out (from) by eavesdropping (*or* monitoring)
Abhub *m* scum, waste, dross
ab·hülsen *vt* shell, hull, husk, peel
ab·hungern *vr:* **sich a.** starve oneself
Abies-öl *n* fir oil, abies oil
Abietin·säure *f* abietic acid. **-äthylester** *m* ethyl abietate
ab·irren *vi*(s) deviate, stray. **abirrend** *p.a* aberrant, etc. **Abirrung°** *f* deviation, aberration
ab·isolieren *vt* isolate, insulate; seal (surface coating)
Abitur *n* (-e) comprehensive final examinations (for graduation from secondary school)
Abk. *abbv* (Abkürzung) abbr(ev).
ab·kalken *vt* unlime, delime
ab·kälten *vt* cool down
ab·kämmen *vt* card (wool)
ab·kanten *vt* bevel; trim; fold (metal)
ab·kappen *vt* decapitate, cut the top off
ab·kapseln *vt* encyst, encapsulate
ab·karten *vt* prearrange, plot; trump up
ab·kaufen *vt* (*usu* + *d.i.o*) buy (from)
Abkehr *f* withdrawal, alienation; renunciation. **ab·kehren** 1 *vt* sweep (off); turn away, avert. 2 *vi*(s) (*Min.*) stop working. 3: **sich a. (von)** turn away (from)
ab·keltern *vt* press (wine)
ab·ketten *vt* unchain
ab·kippen *vt* dump (load of gravel, etc)
ab·klappen *vt* fold down, swing down (*or* out)
ab·klären I *vt* 1 clarify, clear; fine. 2 (according to method): filter, decant, elutriate, etc. 3 (*Sugar*) decolor. 4 (*Colors*) brighten; (*Dye*) strip off, boil off. II: **sich a.** 5 clear (up)
Abklär·flasche *f* decanting bottle (*or*

flask). **-topf** *m* decanting jar
Abklärung° *f* clarification, clearing, etc *cf*
ABKLÄREN
Abklärungs- *oft* = ABKLÄR-: **-methode** *f*
method of clarification. **-mittel** *n* clarifying agent
Abklatsch *m* (-e) print, copy; (brush) proof;
(stereotyped) plate; poor imitation. **ab-
klatschen** *vt* 1 print (off), strike (off)
(proofs). 2 cast (pattern, etc). 3 stereotype;
imprint, copy; make a poor copy of.
4 (*Chem.*) transfer (wet material into a
vessel). **Abklatsch-präparat** *n* impression preparation
ab-klemmen *vt* pinch (*or* squeeze) off
Abkling-dauer *f* decay time
ab-klingen* *vi*(s) die away (*or* out), fade
away (*or* out), damp; lose radioactivity,
decay. **Abkling-konstante** *f* decay constant. **Abkling(ungs)-kurve** *f* radioactive decay curve, extinction curve
Abkling-wert *m* decay value. **-zeit** *f* decay
time
ab-klopfen *vt* beat (dust out of); knock
... off; scale off; (*Med.*) percuss. auscultate
ab-klören *vt* boil out (for redyeing)
ab-knallen *vt* fire (a gun); detonate. 2 *vi*(s)
explode, go off
ab-kneifen* *vt* pinch (*or* nip) off
ab-knicken *vt* snap (*or* break) off; bend at a
sharp angle
ab-knistern *vt/i* decrepitate
abkochbar *a* decoctible. **abkoch-echt** *a*
boilproof, fast to boiling. **ab-kochen** *vt*
boil (thoroughly); decoct; (*Tex.*) scour, degum. **Abkoch-mittel** *n* decoction medium. **Abkochung** *f* (-en) decoction; (thorough) boiling
Abkomme *m* (-n,-n) descendant
ab-kommen* *vi*(s) 1 depart, deviate. 2 get
away. 3 be spared. 4 be descended. 5 go out
of style (*or* use). **Abkommen** *n* (-)
1 agreement, accord; settlement. 2: **im A.
sein** be going out of style (*or* use)
Abkommin *f* (-nen) descendent
abkömmlich *a* free, off (work)
Abkömmling *m* (-e) descendant; derivative
ab-kompensieren *v recip* (+ **sich, ein-**

ander) compensate each other, cancel out
ab-köpfen *vt* decapitate
ab-koppein *vt* uncouple, unfasten
Abkratz-eisen *n* scraper, **ab-kratzen** *vt*
scrape (off)
ab-kreiden *vi* (*Paint, etc*) chalk
ab-krücken *vt* rake off, rabble
Abkühl-apparat *m* cooler, refrigerator
ab-kühlen 1 *vt, vi*(s) & **sich a.** cool (down *or*
off). 2 *vt* chill, refrigerate; anneal.
Abkühler° *m* cooler
Abkühl-faß *n* cooler, cooling vat; (*Glass*)
annealing oven. **-gefäß** *n* cooling vessel.
-mittel *n* cooling agent, refrigerant,
coolant
Abkühlung° *f* cooling, refrigeration; chilling; annealing
Abkühlungs- *oft* = ABKÜHL- **cooling:
-dauer** *f* duration of cooling. **-pause** *f*
cooling-off period. **-raum** *m* refrigeration
chamber. **-spannung** *f* cooling strain
Abkühl-verlust *m* loss on cooling
Abkunft *f* (-e) origin, descent, extraction
ab-kuppeln *vt* uncouple, disconnect; disengage (clutch)
ab-kürzen *vt* shorten, abbreviate; truncate
cf ABGEKÜRZT. **Abkürzung°** *f* abbreviation; shortened version
Ablade-gewicht *n* shipping weight
ab-laden* *vt* unload, discharge. **Ablade-
platz** *m* 1 unloading area. 2 dumping
ground. **Ablader** *m* (-) unloader; ste-
vedore, longshoreman
Ablade-schein *m* discharge (*or* unloading)
certificate. **-stelle** *f* 1 unloading area.
2 dumping ground
Abladung° *f* unloading, discharge
Ablage° *f* depositing; checking (for storage); storage place, depot; checkroom; filing cabinet
ab-lagern *vt* deposit; store; dump, unload.
2 *vt/i* age, mellow, mature, season. 3 *vi* (h
or s) deteriorate (in storage). **4: sich a.**
settle, be deposited. **Ablagerung°** *f*
1 deposit(ing); storage; sediment(ation);
sedimentary layer; dumping. 2 aging,
maturing, mellowing, seasoning. 3 deterioration (in storage). **Ablagerungs-
menge** *f* amount of deposit (etc.)
Ablaktation *f* weaning, ablactation; in-

arching. **ablaktieren** *vt* wean, ablactate; inarch. **Ablaktierung** *f* = ABLAKTATION

ab·längen *vt* cut to length; cut into lengths

Ablaß° *m* (. . . lässe) 1 drainage, discharge. 2 cessation. 3 discount, deduction. 4 drain, outlet, vent

ab·lassen* I *vt* 1 let off; draw (*or* siphon) off. 2 discharge, drain, decant; blow off, run off. 3 (*Steel*) anneal. 4 (*Soap*) cleanse. 5 dispatch, send off. 6 sell. 7 give a discount of . . . II *vi* 8 (*Colors*) fade, come off. 9 (+ **von**) leave off, cease, give up

Ablaß·graben *m* drainage ditch. **-hahn** *m* delivery (discharge, drain) cock. **-kante** *f* feather edge. **-leitung** *f* delivery (discharge, drain) pipe. **-öffnung** *f* outlet, vent. **-rohr** *n*, **-röhre** *f* outlet tube (*or* pipe), drain pipe. **-schieber** *m* drain sluice valve. **-schraube** *f* drain plug. **-ventil** *n* outlet (*or* drain) valve

Ablast° *f* failure (of an organ) to develop

Ablations·eigenschaften *fpl* ablation properties. **-system** *n* ablator system

Ablauf° *m* 1 run(ning) (off); discharge; effluent; filtrate; (*Pipette*) delivery. 2 outlet, drain, sink. 3 slope. 4 (*Time*) lapse, expiration. 5 course (of a reaction, etc); process. 6 (*Sugar*) run-off syrup, molasses. **-brett** *n* drain board

ab·laufen* *vi*(s) 1 run off (*or* down); drain (off), empty. 2 run (its course). 3 go, turn out (well, badly). 4 expire, run out. 5: a. lassen let run off; discharge, launch; rebuff. **Ablaufen·lassen** *n* running off, discharge, launching; rebuff

Ablauf·flüssigkeit *f* discharge (*or* drainage) liquid, waste effluent. **-hahn** *m* discharge (*or* drain) cock. **-öl** *n* expressed oil, run oil. **-rinne** *f* run-off (*or* drainage) channel, gutter. **-rohr** *n* **-röhre** *f* outlet (*or* discharge) pipe (*or* tube). **-streifen** *m* (*Coating*) drainage mark. **-stutzen** *m* run-off (*or* drainage) nozzle. **-trichter** *m* (inverted) draining (*or* condensing) funnel (to carry off vapors). **-ventil** *n* drain cock, outlet valve. **-wasser** *n* run-off water, effluent

Ablauge° *f* waste lye (*or* liquor); washing out. **ab·laugen** *vt* lixiviate, wash the lye out of; steep in lye

ab·läutern *vt* 1 clarify, purify, refine. 2 strain, filter. 3 draw off. 4 (*Ore*) wash. **Abläuterung** *f* clarification, purification, etc. **Abläuterungs·vorrichtung** *f* clarifying device

Ableben *n* death, demise, decease

ab·lecken 1 *vt* lick off. 2 *vi*(s) drip (trickle, leak) down

ab·ledern *vt* buff; skin

ab·legen I *vt* 1 lay aside, discard. 2 take off (*esp* outer clothing). 3 deposit. 4 take (exam, proof, oath, etc), make (vow); give, submit (proof, accounting, etc). 5 unwind, unroll. 6 (*Bot.*) layer. II *vi* take off one's wraps. **Ableger** *m* (-) offshoot; (*Bot.*) layer, slip, cutting

ab·lehnen *vt* decline, reject, refuse, turn down. **ablehnend** *p.a* negative; deprecatory. **Ablehnung** *f* (-en) rejection, refusal

ab·lehren *vt* 1 gage, true up. 2 dress, trim

ableitbar *a* derivable, etc *cf* ABLEITEN

ab·leiten I *vt* 1 derive; deduce. 2 divert, turn away, conduct away. 3 draw off, drain off. 4 (*Elec.*) ground, shunt off. 5 (*Math.*) differentiate. II: **sich a.** (+ **von**) be derived (from). **Ableiter°** *m* (*Elec.*) 1 conductor. 2 spark-gap (*or* lightning) arrester. **Ableit(e)·rohr** *n* delivery tube (*or* pipe). **Ableiter·stange** *f* conducting bar

Ableitung° *f* 1 derivation; derivative; deduction; diversion, deflection, turning away; removal by conduction; drainage. 2 (*Elec.*) leakage, stray current; branch circuit; grounding; shunt(ing). 3 (*Math.*) differential coefficient. *Cf* ABLEITEN

Ableitungs·gewebe *n* conducting tissue. **-graben** *m* drainage ditch; (open) sewer. **-kanal** *m* drainage channel; sewer. **-mittel** *n* derivative, revulsive. **-rinne** *f* delivery (*or* drainage) channel (duct, gutter). **-rohr** *n*. **-röhre** *f* drain pipe. **-wort** *n* derivative

ablenkbar *a* divertible, distractable

ab·lenken I *vt* 1 divert, deflect; distract. 2 draw off. 3 (*Light*) diffract. 4 fault. II *vi* 5 turn aside, deviate. 6 digress, change the subject. III: **sich a.** relax, find distraction. **Ablenker°** *m* deflector

Ablenk·fehler *m* deflection aberration.

-feld *n* deflecting field. **-platte** *f* deflecting plate. **-registrierung** *f* deflector registration

Ablenkung *f* (-en) diversion; deflection, distraction; diffraction; deviation, digression; relaxation *cf* ABLENKEN

Ablenkungs·kraft *f* deflecting force. **-vervielfacher** *m* deflection multiplier

Ablenk(ungs)·winkel *m* angle of deflection (*or* diffraction)

ablesbar *a* readable, discernible

Ablese· (*Instrum.*) reading: **-einrichtung** *f* reading device. **-fehler** *m* error in reading. **-fernrohr** *n* reading telescope. **-genauigkeit** *f* accuracy of reading. **-lupe** *f* reading lens

ab·lesen* *vt* 1 read off (*esp* on a scale); tell (from appearance); infer. 2 pick (up *or* off), gather

Ablese· (*Instrum.*) reading: **-schärfe** *f* sharpness of reading. **-vorrichtung** *f* reading device. **-wert** *m* (meter, dial) reading, value (read from a scale)

Ablesung *f* (-en) 1 (meter, dial) reading; reading off. 2 picking, gathering; crop, harvest

Ablesungs·fehler *m* error in reading. **-gennauigkeit** *f* accuracy of reading. **-verfeinerung** *f* refinement of reading

ab·lichten *vt* 1 (*Dye.*) lighten, reduce; shade. 2 (*Phot.*) photostat. **Ablichtung°** *f* lightening, reducing; shading; photostat(ing)

ab·liefern *vt* deliver. **Ablieferung°** *f* delivery

ab·liegen* *vi* (h *or* s) 1 lie at a distance, be remote. 2 ripen, mature. 3 deteriorate from nonuse. **abliegend** *p.a* remote *cf* ABGELEGEN

ablösbar *a* detachable, removable, separable; redeemable *cf* ABLÖSEN

ablöschbar *a* slakable (lime); temperable; quenchable

ab·löschen *vt* 1 quench, extinguish, put out (fire). 2 dry up, slake (lime). 3 (*Steel*) chill, temper. 4 blot (ink); erase. **Ablösch·flüssigkeit** *f* quenching (*or* tempering) liquid

Ablöse·arbeit *f* (+ **von**) energy needed (to). **-energie** *f* bonding energy

ab·lösen I *vt* 1 loosen, detach; split off;

separate, remove (cautiously). 2 release. 3 relieve (on duty). 4 replace, supersede. 5 pay off, redeem. 6 discharge (obligation). 7 (*Biochem.*) abstract. **II: sich a.** 8 come loose (*or* off); dissolve (out). 9 be relieved. **III. sich** *or* **einander a.** alternate, spell each other. **ablösend** *p.a* (*oft*) 1 alternate. 2 oncoming (work shift). 3 (*Med.*) resolvent

ablöslich *a* detachable, removable, separable. **Ablöslichkeit** *f* detachability, etc.

Ablösung° *f* 1 loosening, detachment; splitting off; separation; (cautious) removal. 2 relief, replacement. 3 redemption. 4 alternation, quid pro quo. 5 (*Biochem. oft*) abstraction. *Cf* ABLÖSEN

Ablösungs·energie *f* separation energy. **-fläche** *f* cleavage surface; (*Geol.*) jointing plane, joint. **-mittel** *n* (*Molding*) release agent. **-richtung** *f* (*Cryst.*) cleavage plane. **-system** *n* relay system

ab·löten *vt* unsolder

Abluft° *f* outgoing (*or* exhaust) air, waste (*or* foul) air. **-kanal** *m*, **-lutte** *f* exhaust air duct

ab·lüften *vt* air, expose to air

Abluft·rohr *n* air outlet tube (*or* pipe)

ab·lutieren *vt* unlute

ab·machen *vt* 1 remove, detach, undo. 2 close (account). 3 settle, agree on. 4 stipulate. **Abmachung** *f* (-en) 1 removal, detachment. 2 settlement, agreement. 3 stipulation

ab·magern *vi* (s) lose weight, become emaciated. **Abmagerung** *f* weight loss, reducing; emaciation. **Abmagerungs·kur** *f* (weight) reducing program, diet

ab·mähen *vt* mow off; mow down

ab·mahnen *vt* (**von**) warn (against)

ab·maischen *vt/i* finish mashing. **Abmaisch·temperatur** *f* final mashing temperature

ab·malen 1 *vt* paint (a copy of). 2: **sich a.** be reflected

ab·marken *vt* mark off, mark out

Abmaß *n* 1 off size. 2 deviation (from correct measurement); (*Mach.*) tolerance

ab·matten *vt* dull (a surface finish); deaden (gold); fatigue, exhaust

ab·meißeln *vt* chisel off

abmeßbar *a* measurable
ab·messen* *vt* measure off; estimate, gage *cf* ABGEMESSEN. **Abmessung°** *f* measurement, dimension; estimate
ab·mildern *vt* moderate, alleviate
ab·mindern *vt* lessen, diminish, reduce. **Abminderung°** *f* diminution, reduction
ab·mischen *vt* mix, blend. **Abmischung°** *f* blend(ing), mix(ing); mixture; admixture
ab·montieren *vt* dismantle; disconnect
ab·mühen *vr:* **sich a.** toil, labor
ab·mustern *vt* (*Tex.*) pattern; (+ **gegen, mit**) compare (a color with a standard)
ab·nagen *vt* gnaw (off); erode
Abnahme *f* (-n) 1 removal; amputation. 2 decrease, decline: **in A. kommen** decline. 3 (*Biochem. oft*) depression, decay. 4 acceptance (of ordered goods). 5 purchase, taking delivery. 6 audit, examination, review. *Cf* ABNEHMEN. **··prüfung** *f* acceptance test. **-schein** *m* receipt. **-versuch** *m* acceptance test
ab·narben *vt* buff, scrape (hides)
abnehmbar *a* removable, detachable
ab·nehmen* **I** *vt* 1 take (off, away, down), remove; amputate. 2 (*Elec.*) collect, pick up (current). 3 (*d.i.o*) relieve (of); exact (from); charge (for). 4 audit, inspect (for acceptance). 5 accept, take delivery of; buy; order. 6 (*oft* + **daß**) believe (someone's assertion that . . .). **II** *vi* 7 decrease, decline, diminish, wane. 8 lose weight. 9 decay. **Abnehmen** *n:* **im A.** on the wane (*or* decline). **Abnehmer** *m* (-) 1 buyer, customer, consumer, subscriber. 2 (*Elec.*) (current) collector; point contact. 3 (*Tex. Mach.*) doffer
Abneigung° *f* reluctance, aversion, disinclination; deviation; slope *cf* ABGENEIGT
abnorm *a* abnormal, anomalous. **Abnormität** *f* (-en) abnormality, anomaly
ab·nötigen *vt* (*d.i.o*) extort (wring, exact, force) (from)
ab·nutschen *vt* suction-filter, aspirate; (*Dye.*) hydroextract
abnutzbar *a* consumable, liable to wear. **Abnutzbarkeit** *f* liability to wear, wearing quality
ab·nutzen, ab·nützen 1 *vt* use up, wear out; abrade. **2: sich a.** wear out (*or* off); fray; get used up. **Abnutzung, Abnützung** *f*

wear and tear, attrition, depreciation; abrasion
abnutzungs·beständig *a* wear-resistant. **A—erscheinung** *f* symptom of attrition, sign of wear. **-festigkeit** *f* resistance to wear; abrasion resistance. **-geschwindigkeit** *f* rate of wear. **-größe** *f* coefficient of wear. **-prüfer** *m*, **-prüfmaschine** *f* wear tester. **-satz** *m* degree of deterioration. **-widerstand** *m* resistance to wear; abrasion resistance
Abo. *abbv* (Abonnement) subscription
Aböl° *n* waste oil. **ab·ölen** *vt* 1 oil off; remove oil from. 2 oil thoroughly
Abonnement *n* (-s) subscription; season ticket. **Abonnent** *m* (-en, -en), **Abonnentin** *f* (-nen) subscriber; season-ticket holder. **abonnieren** *vt/i* subscribe (to); have (*or* get) a season ticket (to)
ab·ordnen *vt* delegate *cf* ABGEORNETE. **Abordnung°** *f* delegation, deputation
Abort *m* 1° (*deriv of* Ort) toilet, lavatory. 2 (= ABORTUS) (-e) abortion; miscarriage
abortieren *vt* abort. **Abortiv·mittel** *n* abortive agent, abortifacient
Abort·öl *n* lavatory oil
Abortus *m* (-,-) abortion; miscarriage. **··bazillus** *m* Bang's bacillus
ab·oxidieren *vt* oxidize off
ab·packen *vt* 1 unload. 2 pack up (in batches)
ab·palen *vt* shell
ab·pälen *vt* unhair, depilate (hides)
ab·passen *vt* 1 measure, fit, adjust. 2 watch for, wait for, head off
ab·pellen *vt* peel, skin
Abperl·effekt *m* water-repellent effect
ab·pflöcken *vt* mark off with pegs
ab·pflücken *vt* pick, pluck; gather
ab·plaggen *vt* cut (sod)
ab·platten 1 *vt* flatten, level, make oblate. **2: sich a.** become flattened (leveled, faceted). *Cf* ABGEPLATTET. **Abplattung** *f* (-en) 1 flattening, leveling. 2 flatness, oblateness
ab·platzen *vi*(s) break off (explosively); crack (snap, chip, spall, split) off
Abprall° *m* rebound, recoil, ricochet, bounce; reflection. **ab·prallen** *vi*(s) (*oft* + **an**) rebound, recoil, ricochet (off); reverberate; bounce (*or* glance) off; be re-

flected. **Abprallungs·winkel** m angle of rebound (or reflection)

ab·pressen vt 1 press (or squeeze) off, express; press apart. 2 squeeze in a press. 3 (d.i.o) extort (from). **Abpreß·maschine** f press

Ab·produkt° n waste product; by-product

ab·puffern vt buffer off, prevent by buffering

ab·pumpen vt pump off, drain by pumping

Abputz m plaster(ing), (re)coating. **ab·putzen** vt 1 plaster, coat. 2 clean off, wipe off, cleanse. 3 trim (off)

Abquetsch·druck m squeezing pressure. **-effekt** m (Tex.) mangle expression; (degree of) hydroextraction

ab·quetschen vt squeeze off (or out); (Tex.) mangle, hydroextract

Abquetsch·fläche f (Engg.) mating surface. 2 = **-grat** m flash (of a mold)

Abquick·beutel m filter bag

ab·quicken vt purify; refine (with mercury); separate (gold, silver) from amalgam

ab·radieren vt erase (off), scratch out

ab·rahmen vt skim (the cream off). **Abrahmer** m (-) skimmer

ab·rammen vt ram, trap

Abrasivität f abradability

ab·raspeln vt rasp off, file (smooth)

ab·raten* vt (usu + d.i.o) dissuade (from), advise against (doing)

ab·rauchen vi(s) evaporate in fumes, fume (off), vaporize

ab·räuchern vt fumigate thoroughly

Abrauch·raum m evaporating chamber. **-schale** f evaporating dish

Abraum° m rubbish, trash; rubble; (Min.) overburden

ab·räumen vt clear away, remove; clear (table, etc). **Abräumen** n (Brew.) clearing the kiln

Abraum·halde f slag heap. **-salz** n abraum (or rubble) salt, saline deposit; laxative salt. **-sprengung** f stripping with explosives. **-stoff** m waste

Abräumungs·schlag m final cut(ting)

ab·rechen vt rake off

ab·rechnen I vt 1 deduct; discount. 2 settle, reckon up (accounts). 3 (d.i.o) debit, credit (to a psn's account). II vi 4 settle accounts. Cf ABGERECHNET. **Abrechnung°** f deduc-

tion; discount; settlement; statement (of account)—**auf A.** on account; **in A. bringen** = ABRECHNEN I

Abrede° f agreement, understanding; denial—**in A. stellen** deny, dispute.

ab·reden 1 vt (+ **von**) dissuade (from doing), talk out of (doing). 2 vt/i arrange; reach an agreement (on)

ab·regnen 1 vt spray from an aircraft. 2: **sich a.** stop raining

Abreibe·festigkeit f resistance to abrasion

ab·reiben* vt rub off; scrape, scour, rub smooth; grate; grind (colors); wear down; rub down. **Abreibung°** f rubbing (off); rubdown; abrasion; attrition

Abreicherung f (-en) depletion

Abreise° f departure. **ab·reisen** vi(s) depart

Abreiß·bogenspektrum n intermittent arc spectrum

ab·reißen* I vt 1 tear off, rip off, detach. 2 pull (or tear) down, demolish. 3 draw, sketch. II vi(s) 4 tear (off), break (off). 5 stop short, be interrupted, come to an end. Cf ABGERISSEN

Abreiß·feder f antagonistic spring. **-funken** m (contact-)breaking spark. **-hebel** m interrupter. **-sand** m abrasive sand. **-zünder** m friction igniter; make-and-break igniter

ab·resten vt (Brew.) decant

Abrichte·lauge f (Soap) weak caustic liquor

ab·richten vt 1 fit (out), adjust, true. 2 level; plane, dress; (Mach.) rough-machine. 3 train, break in. 4 (Soap) strengthen. **Abrichter°** m trainer. **Abrichtung°** f fitting, adjustment, truing; training, breaking-in; rough-machining; (Soap) strengthening, strong change

Abrieb m (-e) 1 attrition, abrasion. 2 abraded material; rubbings, grindings, dust. **a—·fest** a abrasion-resistant. **A—·festigkeit** f abrasion resistance; abradability; attrition index. **-prüfmaschine** f abrasion tester

Abriebs·zahl f index of abrasion

Abrieb·widerstand m abrasion resistance

ab·rieseln vi(s) trickle (or drizzle) down

Abrin n abrine, N-methyltryptophane

ab·rinden vt peel, strip, bark, decorticate

ab·ringen* vt (d.i.o) wrest, wring (from)

ab·rinnen* vi(s) run (or flow) down, run off

Abriß° m 1 outline; sketch, design, summary. 2 coupon. **··erregung** f break excitation (or response)

Abrohr° n outlet (or discharge) pipe

ab·rollen 1 vt, vi(s) & **sich a.** roll off, roll by, unroll. 2 vi(s) & **sich a.** develop, unfold. **Abroller** m (-) roller

ab·röschen vt air (paper)

ab·rosten vi(s) rust off (or away), corrode (away)

ab·rösten vt roast (thoroughly); heat (for sublimation)

Abrostung f rusting (off), corrosion

Abröstung f thorough roasting

ab·röten 1 vt dye red. 2 vi lose its red color

ab·rücken 1 vt push away, displace; shift downward. 2 vt, vi(s) (oft + **von**) move away (from). 3 vi(s) (+ **von**) disassociate oneself (from); march away (from)

Abruf° m 1 announcement, proclamation. 2 (Com.) recall: **auf A.** subject to recall, on call

ab·rufen* vt 1 call away. 2 recall, call back, call in. 3 call off (or out), announce. 4 (Cmptrs.) call up, access, retrieve. **Abrufung** f (-en) recall, call-back; announcement; access(ing); retrieval

ab·rühren vt stir, beat (thoroughly)

ab·runden vt 1 round (off), finish off. 2 (Perfume) blend. Cf ABGERUNDET. **Abrundung°** f rounding (off); roundness, curvature. **Abrundungs·mittel** n (Perfume) blending agent

abruß·echt a fast to rubbing, abrasionfast. **ab·rußen** vi (Colors) rub off, crock, lose color

ab·rüsten 1 vt dismantle, disassemble, take down (or apart). 2 vi disarm. **Abrüstung°** f dismantling, disassembly; disarmament

Abrutsch m (-e) sheer drop, precipice; downhill glide. **ab·rutschen** 1 vt wear off (or out). 2 vi(s) slip, slide, glide, plunge (off or down)

ab·rütteln vt shake off

abs. abbv (absolut) absolute

Abs. abbv (Absender) sender; (on Mail) from

ab·sacken I vt 1 sack, bag, pack in bags.

2 unload. II vi(s) 3 sink, drop. 4 break down, sag. 5 shrink. III: **sich a.** 6 become encysted, sacculate. **Absackung** f (-en) bagging, packing in bags; sinking, dropping; breakdown; sagging; encystment, sacculation

Absage° f cancellation, refusal, rejection; renunciation. **ab·sagen** I vt 1 cancel, call off; decline. II vi 2 decline, cancel an engagement, send an excuse. 3 (dat) renounce, repudiate. Cf ABGESAGT

ab·sägen vt saw off

ab·saigern vt liquate, etc = ABSEIGERN

ab·salzen vt salt (thoroughly). **Absalzen** n (Soap) partial salting out

ab·sättigen vt saturate; neutralize. **Absättigung°** f saturation, neutralization. **Absättigungs·versuch** m saturation test

Absatz° m 1 deposit, sediment; precipitate. 2 pause, break: **ohne A.** without a break. 3 interval, intermission; period: **in Absätzen** intermittently. 4 paragraph, section. 5 ledge, setback, offset; (stair) landing. 6 heel (of a shoe). 7 sale, market: **A. finden** find a market, sell well. **--bassin** n settling tank. **-gebiet** n (Com.) market, selling area. **-gestein** n sedimentary (or stratified) rock

absätzig a interrupted, intermittent; inferior, faulty (wool); (Min.): **a. sein** break off, peter out

Absatz·teich m settling pond. **a—weise** adv interruptedly, intermittently; fractionally; gradually. **A—zeichen** n paragraph mark

ab·säuern vt acidify, sour

Absaug·anlage f suction extractor; induced-draft installation; exhaust unit

Absaug(e)· suction: **-flasche** f suction extractor. **-leitung** f suction piping (line, duct), vacuum line; exhaust

ab·saugen* vt 1 suck off, suction-filter, draw off by suction; exhaust, aspirate; vacuum(-clean). 2 squeegee. 3 (Dye.) hydroextract

Absaug·entfeuchter m vacuum desiccator

Absauge·öffnung f suction outlet. **-pumpe** f suction (or vacuum) pump

Absauger m (-) suction fan, (dust) exhauster; (Dye.) hydroextractor; (Paper) suction box

Absaug(e)·trockenofen *m* vacuum drying oven. **-vorrichtung** *f* suction apparatus

Absaug·finger *m* suction tube. (*For other* **Absaug·** *compds see also* ABSAUGE·)

Absaugung *f* suction, suction filtration; exhaust; vacuum line. **Absaugungs·anlage** *f* suction installation

Abschabe·mittel *n* abrasive agent

ab·schaben 1 *vt* scrape, shave, pare (off); clean; abrade. 2: **sich a.** wear out, become threadbare (*or* shabby). **Abschabsel** *n* (-) (*usu pl*) scrapings, shavings, parings. **Abschabung** *f*(-en) scraping (off), cleaning, abrasion

ab·schaffen *vt* abolish, do away with, discontinue; give up. **Abschaffung** *f* (-en) abolition, discontinuance, giving up

ab·schälen 1 *vt* peel (off); shell, pare, bark, decorticate. 2: **sich a.** peel (*or* scale) off, come off

Abschäler *m* (-) knife edge (of a separation nozzle system). **--weite** *f* distance between knife edge and curved wall

ab·schalten *vt* 1 disconnect, switch off, turn off. 2 shut (*or* close) down. 3 control. 4 tune out. **Abschalter°** *m* turn-off switch; *pl oft* controls

Abschalt·stab *m* control rod. **-system** *n* disconnection system

Abschaltung° *f* disconnection, switching off, turning off; control; tuning out *cf* ABSCHALTEN

Abschälung *f* (-en) peeling off, shelling, etc *cf* ABSCHÄLEN

ab·schärfen *vt* 1 taper, sharpen to a point. 2 bevel, chamfer; round off (corners). 3 pare

ab·schatten 1 *vt* shade (off); silhouette, outline. 2: **sich a.** cast its shadow

ab·schattieren *vt/i* shade off. **Abschattierung°** *f* shading off; nuance

Abschattung° *f* shading (off); silhouette, outline; screen, shade

abschätz·bar *a* appraisable

ab·schätzen *vt* 1 estimate, value, appraise. 2 tax. 3 disparage, belittle. **Abschätzer** *m* (-) appraiser. **abschätzig** *a* disparaging, derogatory, contemptuous. **Abschätzung°** *f* estimate, (e)valuation; appraisal

Abschaum° *m* skimmings, scum; dross, dregs. **ab·schäumen** *vt* skim (off), remove scum from. **Abschäumer** *m* (-) skimmer, skimming device. **Abschaum·löffel** *m* skimmer, skimming ladle. **Abschäumung°** *f* skimming (off), removal of scum

abscheldbar *a* separable

Abscheide·gefäß *n* separating vessel. **-geschwindigkeit** *f* rate of separation (from solution). **-kammer** *f* separating (*or* condensation) chamber

ab·scheiden* 1 *vt & sich a.* separate (out); precipitate; deposit. 2 *vt* eliminate, disengage; secrete; refine (metals). 3: **sich a.** be eliminated; be disengaged; be secreted; be refined; seclude oneself. 4 *vi*(s) die, depart. *cf* ABGESCHIEDEN. **Abscheiden** *n* separation; precipitation; departure, decease. **abscheidend** *p.a* displaced, liberated.

Abscheider *m* (-) separator, precipitator; deflector; trap; refiner; settling vessel, settler

Abscheidung° *f* separation; precipitation, deposit, deposition, settling; sediment; elimination; secretion; refining

Abscheidungs·ausbeute *f* deposition yield. **-form** *f* type (*or* pattern) of deposition, separation form. **-mittel** *n* means of separation, separating agent; precipitant. **-polarisation** *f* deposition polarization. **-produkt** *n* secretion. **-spannung** *f* deposition potential. **-verfahren** *n* separation process. **-vorrichtung** *f* separator

ab·schelfe(r)n *vt, vi & sich a.* scale (off)

ab·scheren* *vt* shear (off), cut (off); clip, crop

Abscher·festigkeit *f* shearing strength. **-fläche** *f* shear plane

Abscherung *f* (-en) shear (strength); shearing (off)

Abscherungs·beanspruchung *f* shearing strain. **-festigkeit** *f* shearing strength

Abscheu *m* abhorrence, aversion, abomination; disgust

ab·scheuern 1 *vt* scour (off), rub (off), cleanse. 2 *vt & sich a.* wear thin, wear away. **Abscheuerung** *f* attrition, wear(ing away); scouring, etc

abscheulich *a* detestable, abominable, atrocious. **Abscheulichkeit** *f*(-en) abomination; atrociousness, detestability

ab·schichten *vt* separate into layers

ab·schicken *vt* send (off), forward, mail. **Abschickung°** *f* sending (off), forwarding, mailing

ab·schieben* 1 *vt* push (away), shove off; shed, get rid of; expel, deport. 2 *vi*(s) shove off, clear out. **Abschiebung°** *f* expulsion, deportation

Abschied *m* (-e) 1 departure, leave(-taking), farewell: **A. nehmen** take leave, say goodbye. 2 dismissal, discharge. 3 retirement, resignation. 4 certificate of leaving

ab·schiefern *vt, vi* & **sich a.** flake (*or* scale) off; exfoliate

ab·schießen* I *vt* 1 shoot down. 2 shoot off, discharge, fire (guns). 3 sink (mines). II *vi*(s) 4 shoot (*or* go shooting) down. 5 rush off (*or* away). 6 fade, lose its color

ab·schilfern *vt* & **sich a.** scale off, peel off

ab·schirmen *vt* screen (off), shield, protect, cover

Abschirm·faktor *m* shielding factor. **-konstante** *f* screening constant. **-korrektur** *f* screening correction. **-strom** *m* shielding current. **-zahl** *f* screening number

Abschirmung *f* (-en) screening, shielding—**negative A.** deshielding

Abschirmungs· (*oft* = ABSCHIRM·) **-glied** *n* degree (*or* coefficient) of shielding (*or* screening). **-zone** *f* shielding zone

ab·schlachten *vt* butcher, slaughter

ab·schlacken *vt* remove slag from

Abschlack-öffnung *f* cinder notch (*or* tap)

Abschlag° *m* 1 cutting (down), felling. 2 fragments, chips, etc. 3 rebound. 4 refusal. 5 outlet, drain; waste (*or* drainage) water; overflow. 6 partition; compartment. 7 abatement, decline, let-up. 8 (*Com.*) drop in price; reduction, discount; deduction; payment on account, installment: **auf A.** on account, in installments

ab·schlagen* I *vt* 1 knock (beat, chop) off; chop down; repulse. 2 lower in value, bring down the price of. 3 refuse (requests, etc.). 4 dismantle, take down (*or* apart). 5 drain off (water). 6 partition (off). 7 stamp, impress; (*Foun.*) stop (molten metal); (*Pharm.*) sieve; (*Tex.*) back off,

slough. II *vi* (h *or* s) 8 rebound, bounce. 9 fail, go wrong. 10 abate, let up; drop (in place). III: **sich a.** 11 deposit, settle

abschlägig *a* 1 negative, unfavorable (answer). 2 brittle

Abschlamm *m* sludge; sludge removal

abschlämmbar *a* separable (*or* removable) by washing out

ab·schlammen, ab·schlämmen *vt* wash, elutriate, separate by washing; decant; clear of mud. **Abschlämmer** *m* (-) thickener. **Abschlämmung** *f* washing; decantation

ab·schleifen* 1 *vt* grind off, abrade; polish (off), smooth; sharpen, whet. 2: **sich a.** wear (off); wear smooth

Abschleif·festigkeit *f*, **-härte** *f* abrasion resistance

Abschleifsel *n* (*usu pl*) grindings, dust

Abschleifung° *f* grinding (off), abrasion; polishing; sharpening; attrition

Abschleif·versuch *m* abrasion test

ab·schleimen *vt* deslime; (*Sugar*) clarify

ab·schlemmen *vt* elutriate, separate by washing; decant

ab·schleppen *vt* tow (*or* drag) away

Abschleuder·maschine *f* centrifuge; hydroextractor

ab·schleudern *vt* centrifuge; hydroextract (yarn); fling (throw, cast) off; eject

abschließbar *a* closable; salable

ab·schließen* I *vt* 1 close (off), shut off; seal off; occlude. II *vt/i* 2 conclude, finish, end, complete. 3 lock up, lock (the doors). III *vi* 4 (**mit**) reach an agreement, settle accounts (with). **abschließend** 1 *p.a* conclusive, final. 2 *adv* in conclusion. **Abschließung°** *f* closing (off), shutting (off); locking up, sealing off; occlusion

Abschluß° *m* 1 closing, conclusion, termination, close. 2 exclusion. 3 agreement, deal, contract. 4 balancing, rounding off. 5 closing device, shut-off, cut-off; seal, sealing off

Abschluß·bericht *m* final report. **-blatt** *n* balance sheet. **-blech** *n* separation plate. **-blende** *f* front diaphragm (of camera). **-dichtung** *f* seal. **-gewebe** *n* protective tissue. **-hahn** *m* stopcock. **-kappe** *f* cap, bonnet. **-konferenz** *f* final (*or* closing) conference. **-rechnung** *f* final account (*or* bill). **-schieber** *m* slide valve. **ventil** *n* shut-off valve. **-vorrichtung** *f* closing device, shut-off

ab·schmecken *vt* season to taste. **abschmeckend** *p.a* unpleasant-tasting, unsavory, insipid

Abschmelz·draht *m* fuse wire, fusible wire, safety fuse

ab·schmelzen* 1 *vt* melt (off), fuse, separate by melting; seal (off). 2 *vi*(s) melt off

Abschmelz·konstante *f* fusion coefficient. **-schweißung** *f* flash welding. **-sicherung** *f* (*Elec.*) safety fuse (*or* cutout), electric fuse

Abschmelzung *f* melting (off), etc. *cf* AB- SCHMELZEN; (*Metll.*) liquation

ab·schmieren 1 *vt* grease (thoroughly), lubricate. 2 *vi* smear, slur (*as:* colors)

ab·schminken *vt, vi & sich a.* remove make-up

ab·schmirgeln *vt* grind (*or* polish) with emery (*or* pumice)

ab·schmutzen *vt* rub off, soil, stain, blot

Abschmutz·makulatur *f,* **-papier** *n* offset paper

Abschn. *abbv* (Abschnitt) sec. (section)

ab·schnallen *vt* unbuckle

Abschneide·apparat *m* (*Mach.*) cutter. **-frequenz** *f* cut-off frequency

ab·schneiden* I *vt* 1 cut (off); detach; end, clip, cut short; (*Math.*) truncate. II *vi* 2 end, give out. 3 do, perform (well, badly). 4 be (*or* take) a short cut. III: **sich a.** 5 (**gegen**) stand out (against). **Abschneiden** *n* cutting (off); trimming; performance. **Abschneider°** *m* cutter, cutting device

Abschneide·radius *m* cut-off radius

Abschnitt° *m* 1 section; segment, portion, division, part, area, sector, phase; paragraph, chapter; partition. 2 coupon, stub. 3 period (of time). 4 intercept. 5 (*esp pl*) cuttings, trimmings, chips, scraps, shreds

Abschnittling *m* (-e) cutting

Abschnitzel° *n* shred, chip, paring

ab·schnüren *vt* untie; tie off, ligature; constrict, throttle; measure, mark off (with string); rope off. **Abschnürung** *f* (-en) constriction; tying off; ligature, etc.

ab·schöpfen *vt* skim (off); ladle out

ab·schrägen *vt* slope, slant, bevel, taper.

Abschrägung *f* (-en) sloping; slope, bevel

ab·schrauben *vt* screw off, unscrew

Abschreck·bad *n* (*Metll.*) quenching bath. **-bottich** *m* (*Metll.*) quenching tank

ab·schrecken *vt* 1 chill, cool; (*Metll.*) quench (steel). 2 (*Water*) render lukewarm. 3 scare off, deter. **abschreckend** *p.a* deterrent

Abschreck·grube *f* quenching tank. **-guß** *m* chill (*or* quenching) cast. **-härtung** *f* quench hardening. **-mittel** *n* 1 deterrent. 2 cooling agent; (*Metll.*) quenching medium. **-temperatur** *f* (*Steel*) quenching temperature

Abschreckung *f* (-en) 1 chilling, quenching. 2 deterrence, deterrent. *Cf* AB- SCHRECKEN. **Abschreckungs·mittel** *n* deterrent; means of repulsion

Abschreck·zeit *f* chilling (*or* quenching) time

ab·schreiben* I *vt* 1 copy, transcribe. 2 revoke, cancel; write off, depreciate. 3 deduct. II *vi* 4 (*dat*) write (to) canceling an engagement (an order, etc). **Abschreibung** *f* (-en) copy(ing), transcription; cancelation; depreciation, amortization; deduction

ab·schreiten* 1 *vt* pace off. 2 *vi*(s) stride away

Abschrift° *f* copy, transcript; duplicate. **abschriftlich** *a* copied; *adv* as a copy, in duplicate

ab·schülfern *vi*(s) scale off, flake off

ab·schuppen 1 *vt* scale (fish). 2 *vi & sich a.* scale off, flake off

ab·schürfen *vt* (*oft* + *dat rflx:* sich) scrape, skin, excoriate (e.g. knuckles)

Abschuß° *m* discharge, firing (of guns); shooting down *cf* ABSCHIESSEN

abschüssig *a* sloping; steep, precipitous. **Abschüssigkeit** *f* slant, steep slope; steepness, precipitousness

ab·schütteln *vt* shake off (down, out), cast off; get rid of

ab·schütten *vt* pour off, pour out

ab·schützen 1 *vt* shield off, screen off; drain (pond). 2 *vt/i* (*Metll.*) cut off (the blast)

ab·schwächen I *vt* 1 weaken, subdue, diminish, reduce, attenuate. 2 (*Lthr.*) soften, mellow. 3 (*Phot.*) clear, reduce. II *vi*(s) & **sich a.** 4 weaken, subside, level off. **Ab-schwächer** *m* (-) reducer, reducing agent; (*Phot.*) clearing agent. **Abschwächung** *f* (-en) weakening, diminution, reduction, attenuation, mellowing, clearing

Abschwächungs·faktor *m* attenuation factor. **-lösung** *f*(*Phot.*) clearing solution. **-mittel** *n* thinner, diluent, reducer

abschwärzen 1 *vt* blacken. 2 *vi* lose its color, come off black

ab·schwefeln *vt* 1 desulfurize. 2 impregnate with sulfur. 3 coke (coal). 4 (*Ores*) calcine. **Abschwef(e)lung** *f* desulfurization; coking; calcination

ab·schweifen I *vt* 1 (*Tex.*) wash, steep (yarn); ungum (silk). 2 *vi*(s) deviate, stray, digress. **Abschweifung** *f* (-en) deviation, digression

ab·schwelen *vt* calcine, roast; vacuum-distill

ab·schwellen* *vi*(s) 1 subside, decrease, wane; (*Med.: Swelling*) shrink, detumesce

ab·schwemmen *vt* wash (away); rinse; erode; cleanse, scour; elutriate

ab·schwenken 1 *vt* rinse (off); wash (away). 2 *vi*(s) turn off (*or* aside); deviate

ab·schwimmen* 1 *vt* swim (a distance); separate by float-and-sink method. 2 *vi*(s) float (*or* swim) away; clear out

ab·schwingen* I *vt* 1 centrifuge, hydroextract. 2 fan, winnow; sift. 3 shake off. II *vi*(s) & **sich a.** (**von**) swing off

ab·schwirren 1 *vt* centrifuge. 2 *vi*(s) whiz away

Abschwitzel *n* (-) chip, paring

ab·schwitzen *vt* 1 sweat off (weight). 2 (*Lthr.*) sweat; (*Hides*) depilate

Abscisse *f* (-n) abscissa

ab·sedimentieren *vt* deposit as sediment

absehbar *a* 1 foreseeable: **in a—er Zeit** in the foreseeable future. 2 conceivable

ab·sehen* I *vt* 1 (fore)see: **nicht abzusehen** impossible to foresee, not in sight. II *vt* (*d.i.o*) 2 learn by watching. 3 copy (from). 4 tell by looking at. *Cf* ABGESEHEN. III *vi* (**von**) 6 refrain (from); disregard. 7 copy (from)

Abseide° *f* floss silk

ab·seifen I *vt* 1 wash with soap. 2 rinse the soap off. 3 test (welds) with soap solution.

II *vt* & **sich a.** soap down, lather up

ab·seigern *vt* 1 (*Metll.*) liquate (out); separate by fusion. 2 (*Min., Engg.*) plumb

Abseih·beutel *m* filtering (*or* straining) bag. **-tuch** *n* filtering (*or* straining) cloth

ab·seihen *vt* filter (off), strain (off), percolate; decant. **Abseihung** *f* (-en) filtration, etc

ab·sein* *vi* 1 be (broken, torn) off. 2 be at a distance. 3 be worn out. *Cf* AB

Abseite° *f* 1 (*incl Tex.*) wrong (*or* reverse) side. **Abseiten·stoff** *m* (*Tex.*) reversible fabric

abseitig *a* 1 remote, out-of-the-way. 2 strange, out-of-the-ordinary. 3 irrelevant (remark). **abseits** 1 *adv* aloof, aside; remote(ly); off (to one side). 2 *prep* (*genit*) off, apart from

ab·senden* *vt* send off, forward, mail, ship. **Absender°** *m* sender, shipper *cf* ABS. **Absendung°** *f* sending, forwarding, mailing, shipping

ab·sengen *vt* singe (off)

ab·senken 1 *vt* lower; (*Min.*) sink (shaft); (*Bot.*) plant (cuttings), layer (plants). 2: **sich a.** slope downward. **Absenker** *m* (-) layer, sprout; descendant. **Absenkung°** *f* lowering; sinking; planting; layering; (*Geol.*) downthrow

Absetz·anlage *f* sedimentation plant

absetzbar *a* removable, subject to dismissal; salable, marketable; precipitable. **Absetzbarkeit** *f* removability, liability to dismissal; salability; precipitability

Absetz·bassin *n*, **-becken** *n* settling tank (*or* basin). **-behälter** *m* settling vessel, settler; straining chamber; thickener; draining tank. **-bereich** *m* sedimentation area. **-bottich** *m* settling tub (*or* vat); sedimentation tank

ab·setzen I *vt* 1 deposit, precipitate. 2 put (*or* set) down, drop. 3 remove, take off, strike (out). 4 dismiss, depose; wean. 5 sell, market; dispose of; deduct. 6 set off, contrast; set in type; offset. 7 lay off, plot; indent; start a new line. 8 (*Cheese*) pitch. II *vi* 9 stop, end; break off; pause. III: **sich a.** 10 settle (out); deposit; subside. 11 withdraw, leave. 12 stand out. **Absetzen** *n* depositing, etc.—**ohne A.** without a break. **Absetzen·lassen** *n* deposition, sedimentation, settling. **Ab-**

setzer *m* (-) settler, settling agent
Absetz·geschwindigkeit *f* rate of sedimentation. **-teich** *m* settling pond
Absetzung *f* (-en) 1 deposition, sedimentation, precipitation; settling. 2 removal, dismissal. 3 deduction
Absetz·verhinderung *f* inhibition of sedimentation. **-verhütungsmittel** *n* anti-settling agent. **-zeit** *f* settling time; (*Cheese*) pitching point. **-zisterne** *f* settling cistern
ab·sichern *vt* protect, secure, safeguard. **Absicherung°** *f* protection, safeguard-(ing)
Absicht *f* (-en) intention, purpose, aim: **die A. haben** (**zu . . .**) have the intention (to), intend (to); **mit A.** on purpose, intentionally; **ohne A.** unintentionally
ab·sichten *vt* sift out, separate; bolt (flour)
absichtlich *a* intentional
ab·sickern *vi*(s) seep (*or* trickle) down
ab·sieben *vt* sieve, screen, sift
Absieb·kalk *m* lime residue (from sieves)
ab·sieden* 1 *vt* & *vi*(s) boil (off). 2 *vt* decoct, extract by boiling
ab·sinken* *vi*(s) sink down (*or* away); fall off, drop. **Absinken** *n* drop, decline
ab·siphonieren *vt* siphon off
Absitz·behälter *m* settling vessel. **-bütte** *f* (*Brew.*) settling tub. **-dauer** *f* settling (*or* precipitation) time
ab·sitzen* I *vt* 1 serve (time, term). II *vi* (h *or* s) 2 deposit, settle: **a. lassen** let settle. 3 dismount
absol. *abbv* (absolut) absolute, abs.
absolut *a* absolute; (*Oils*) essential. **A—·betrag** *m* absolute value; total amount. **-schwarz** *n* (*Dye.*) absolute black. **a—trocken** *a* absolutely dry, bone dry. **A—trockengewicht** *n* absolutely dry weight
absolvieren *vt* absolve; finish, complete; pass (a test)
absonderlich *a* strange, peculiar: *adv oft* specially, particularly. **Absonderlichkeit** *f* (-en) peculiarity, idiosyncrasy
ab·sondern *vt* 1 separate, detach, isolate; insulate. 2 abstract; extract; segregate. 3 (*Physiol.*) secrete, excrete. **absondernd** *p.a* (*Physiol.*) secretory, excretory. **Absonderung°** *f* separation, isolation; abstraction, extraction; segregation; secretion, excretion; (*Geol.*) jointing

Absonderungs·drüse *f* secretory gland. **-flüssigkeit** *f* (liquid) secretion. **-stoff** *m* secreted material. **-vermögen** *n* (*Physiol.*) secretory power. **-vorgang** *m* process of secretion (*or* segregation). **-werkzeug** *n* secretory organ
Absorbens *n* (..entien) absorbent
absorbierbar *a* absorbable. **Absorbierbarkeit** *f* absorbability
absorbieren *vt* absorb; (*Chem.*) occlude. **absorbierend** *p.a* absorbent, absorptive: **a—es Mittel** absorbent. **Absorbierung** *f* absorption
Absorptiometer° *n* absorption meter. **absorptiometrisch** *a* absorptiometric
Absorptions· absorption, absorptive: **-anlage** *f* absorption unit (*or* equipment). **a—erzeugend** *a* absorbefacient, absorption-producing. **-fähig** *a* absorptive, absorbent. **A—fähigkeit** *f* absorptive capacity. **-gewebe** *n* absorbent tissue. **-kante** *f* absorption limit. **-kohle** *f* absorptive charcoal. **-kontinuum** *n* absorptive band. **-kristallspektrum** *n* crystal absorption spectrum. **-küvette** *f* absorption bulb (*or* cell). **-messer** *m* absorptiometer. **-mittel** *n* absorbent. **-öl** *n* absorbent oil. **-röntgenspektrum** *n* X-ray absorption spectrum. **-sprung** *m* absorption jump. **-streifen** *m* absorption band. **-vermögen** *n* absorptive power, absorbency. **-wärme** *f* heat of absorption
abspaltbar *a* capable of being split off, cleavable, separable. **Abspaltbarkeit** *f* cleavability
ab·spalten 1 *vt*, *vi*(s) & **sich a.** split off, cleave off, spall, separate; dissociate. 2 *vt* liberate; eliminate; crack (chemically). **Abspaltung°** *f* splitting off, cleavage; spallation, separation; dissociation; liberation; elimination; fission
Abspaltungs·arbeit *f* work (*or* energy) of fission (cleavage, separation, etc.). **-kurve** *f* liberation curve. **-reaktion** *f* abstraction reaction. **-regel** *f* rule of elimination. **richtung** *f* direction of elimination. **-wahrscheinlichkeit** *f* abstraction probability
ab·spanen, ab·spänen *vt* clear of shavings; clean with shavings (*or* steel wool); wean (a pig)
ab·spannen *vt* 1 relax (the tension on), re-

lease (from tension) (a spring, etc). **2** (allow to) expand. **3** cut off (steam). **4** unhitch, uncouple. **5** brace, anchor. **6** strip (a mold). *cf* ABGESPANNT. **Abspanner** *m* (-) step-down transformer

Abspann·klemme *f* anchor clamp

Abspannung° *f* relaxation (of tension), release; expansion; uncoupling; fatigue, lassitude, exhaustion; bracing, anchoring *cf* ABSPANNEN

ab·spateln *vt* scrape off (with a spatula)

abspenstig *pred a:* **a. machen** (*d.i.o*) lure away, entice, estrange, alienate (from)

absperrbar *a* capable of being shut off, etc *cf* ABSPERREN

ab·sperren *vt* **1** shut off, seal off, cut off; confine. **2** stop, block. **3** laminate (wood)

Absperr·flüssigkeit *f* confining (*or* sealing) liquid. **-glied** *n* shut-off. **-hahn** *m* stopcock. **-schieber** *m* slide (*or* sluice, *or* gate) valve

Absperrung° *f* **1** shutting off, sealing off; confinement; stopping, stoppage, blocking. **2** barrier, obstruction

Absperr·ventil *n* cut-off valve, stop (*or* check) valve. **-vorrichtung** *f* shut-off (*or* isolating) device

ab·spiegeln *vt* reflect, mirror. **Abspiegelung** *f* (-en) reflection

ab·spielen 1 *vt* play (through); play at sight; play back. **2: sich a.** take place, come off, be done

Abspiel·nadel *f* phonograph needle

ab·splittern *vt, vi*(s) & **sich a.** splinter off, split off

Absprache° *f* agreement, arrangement

ab·sprechen* *vt* **1** *d.i.o.* deny. **2** discuss; arrange, agree on. **absprechend** *p.a* unfavorable, adverse

ab·spreizen *vt* brace, prop, support

ab·sprengen I *vt* **1** break off, blast off; drive off. **2** water, sprinkle. **II** *vi*(s) **3** gallop off. **Absprenger** *m* (-) cutter (for glass tubes)

ab·springen* *vi*(s) **1** jump (leap, spring) off (away, down). **2** fly off, come loose; burst off, crack off; rebound; (*Wood*) warp; (*Iron*) be red-short. **3** break away, desert; digress (suddenly)

ab·spritzen 1 *vt* wash off; hose down, spray. **2** *vi*(s) splash, spatter, spray (off)

ab·sprudeln *vi*(s) bubble away

Absprung° *m* jump, dive, descent; take-off; rebound; offshoot; (*Phys.*) reflection

Abspül·bad *n* rinsing bath

ab·spulen *vt* unwind, unreel

ab·spülen *vt* wash (off), rinse (off, away). **Abspülung°** *f* washing (off), etc

Abst.*abbv* (ABSTAND) distance, etc.

ab·stammen *vi*(s) descend; come, be derived. **Abstämmling** *m* (-e) descendant. **Abstammung** *f* (-en) origin, descent, derivation

Abstammungs·beziehung *f* zoological cognateness. **-lehre** *f* theory of descent

Abstand° *m* **1** distance (*incl* interatomic); interval; spacing: **A. halten** keep one's distance, keep away. **2** difference, disparity. **3: A. nehmen von** desist (*or* refrain) from

Abstand·halter *m* (*Mach.*) spacer

abständig *a* deteriorating; flat, stale

Abstands·differenz *f* difference between distances. **-geld** *n* compensation; forfeit money. **-halter** *m* distance (*or* interval) marker. **-lage** *f* distance between positions. **-messer** *m* position finder. **-quadrat** *n* square distance. **-quadratgesetz** *n* inverse square law. **-ring** *m* spacer. **-scheibe** *f* appropriately spaced winding. **-summe** *f* (amount of) compensation. **-vektor** *m* distance vector

ab·statten *vt* render, give; pay (a visit)

abstäub·echt *a* non-crocking (colors)

ab·stauben, ab·stäuben 1 *vt* dust (off), brush off. **2** *vi* (*of colors*) crock. **Abstauber** *m* (-) duster

Abstech·eisen *n* edging tool

ab·stechen* I *vt* **1** tap, draw, drain (off). **2** mark off (with pinpricks, by scribing); etch (a picture); cut (*or* trim) off. **3** stab, stick, slaughter. **4** defeat. **5** put the safety catch on (a gun). **II** *vi* **6** (+ **von, gegen**) contrast (with), stand out (against). **ab·stechend** *p.a* contrasting. **Abstecher** *m* (-) side trip; (*Metll.*) tapper

ab·stecken *vt* **1** mark out (with pins, pegs). **2** unfasten, undo

Absteck·fähnchen *n* surveyor's flag. **-leine** *f* tracing cord. **-pfahl** *m* marker, stake, post. **-pflock** *m* (marking) peg

ab·stehen* *vi* (h *or* s) **1** (*oft* + **von**) stand apart (*or* off), stand at a distance, stay away, be distant (from), stand clear (of)— **a. lassen** let stand. **2** protrude, jut out. **3** desist, refrain (from). **4** spoil, go bad (flat, stale), decay, decompose; die off, perish. *Cf* ABGESTANDEN. **abstehend** *p.a* distant; spreading, protruding; deteriorating

Absteh·ofen *m* (*Metll.*) holding furnace

ab·steifen *vt* stiffen; thicken; reinforce

ab·steigen* *vi*(s) descend; alight, get off

abstellbar *a* that can be turned (*or* switched) off (disconnected, removed); remediable, redressable *cf* ABSTELLEN

Abstell·brett *n* storage shelf

ab·stellen I *vt* **1** put (*or* set) down. **2** stow, put up (aside, away); park (a vehicle). **3** turn (cut, shut, switch) off. **4** disconnect, stop. **5** remedy, redress, rectify. **6** (**auf**) aim (at), focus (on). **II**: **sich a. 7** stand away (aside, at a distance)

Abstell·hahn *m* stopcock. **-hebel** *m* cut-off (shut-off, stop) lever

Abstellung° *f* setting down; putting aside; parking; turning off; disconnection; remedy, redress, rectification; stop motion *cf* ABSTELLEN

Abstell·vorrichtung *f* cut-off device

ab·stempeln *vt* stamp

ab·sterben* *vi*(s) **1** die (off, out). **2** fade, wither, decay; grow numb; (*of lime*) airslack; (*Sugar*) change to opaque crystalline mass

abstergieren *vt* cleanse, purge

Abstich° *m* tapping, drawing; tapped metal, run-off. **··loch** *n*, **·öffnung** *f* tap hole. **-pfanne** *f* (*Metll.*) tap ladle. **-schlacke** *f* flush slag

Abstieg *m* (-e) descent; decline; downward path

ab·stielen *vt* take stems off (fruit)

ab·stillen *vt* wean

ab·stimmen I *vt* **1** (*oft* + **auf**) tune, attune (to); relate (to). **2** (*usu* + **auf**) coordinate, synchronize (with), balance off (against). **3** lower the pitch of. **4** formulate (mixtures). **II** *vi* **5** vote. **Abstimmer** *m* (-) tuner

Abstimm· tuning: **-knopf** *m* tuning knob. **-schärfe** *f* tuning sharpness, selectivity. **-vorrichtung** *f* tuning device, tuner

Abstimmung° *f* **1** tuning, attuning. **2** coordination, synchronization. **3** balancing. **4** formulation (of mixtures). **5** checking, control. **6** vote, voting

ab·stoppen 1 *vt/i* stop, halt. **2** *vt* time (with a stopwatch)

ab·stöpseln *vt* **1** stop(per), cork. **2** uncork

Abstoß° *m* rebound; push-off. **ab·stoßen*** *vt* **1** repel, push off, rub (escape, grind, knock) off; sleek off (fat). **2** (*Med.*) reject. **3** (*Lthr.*) buff, degrain. **4** dispose of. **abstoßend** *p.a* repellent, repulsive. **Abstoß·fett** *n* currier's table grease. **Abstoßung** *f* (-en) **1** repulsion, removal by abrasion; (*Fat*) sleeking off. **2** (*Med.*) rejection (e.g. of transplanted organs). **3** (*Lthr.*) buffing, degraining. **4** disposal

Abstoßungs·kraft *f* repulsion, repulsive force. **-sphäre** *f* repulsive envelope

abs. tr. *abbv* (absolut trocken) abso. dry

abstrahieren *vt* abstract

Abstrahl° *m* reflected ray. **ab·strahlen 1** *vt* reflect; radiate; sand-blast. **2** *vi*(s) be reflected, radiate. **Abstrahlung°** *f* reflection, radiation

Abstraktions·fähigkeit *f* abstractability

Abstrebe·kraft *f* centrifugal force

Abstreich·eisen *n* scraper; skimmer

ab·streichen* I *vt* **1** scrape (skim, wipe) off; strike out. **2** sweep (with a searchlight). **3** deduct. **II** *vi*(s) **4** fly, slip, steal away. **Abstreicher** *m* (-) scraper, skimmer

Abstreich·löffel *m* skimming ladle. **-messer** *n* scraping knife, (*Tex.*) doctor (blade)

ab·streifen I *vt* **1** strip (off); slip off; cast off, shed. **2** patrol. **II** *vi*(s) **3** deviate, stray. **Abstreifer** *m* (-) (*Metll.*) stripper; scraper, scraping knife; skimmer. **Ab·streif·messer** *n* scraping knife; (*Tex.*) doctor (blade)

ab·streiten* *vt* deny; (*d.i.o*) dispute, contest (sbdy's . . .)

Abstrich° *m* **1** scrapings, skimmings; dross, scum, (*specif*: skim of arsenates, etc, on molten lead). **2** downstroke. **3** deduction; (amount of) cutback. **4** (*Med.*) smear. *Cf* ABSTREICHEN. **~blei** *n* lead skim, crude litharge

ab·stricken 1 *vt* knit. **2** *vt/i* finish knitting

ab·strömen *vi*(s) flow off (*or* away); stream out; (*Gas*) escape

ab·stufen 1 *vt* grade, graduate; arrange in steps (*or* terraces); shade off. **2: sich a.** be graduated, be arranged in steps; grade (off), shade off. *Cf* ABGESTUFT. **Abstufung** *f* (-en) grad(u)ation; arrangement in steps; shading

ab·stumpfen 1 *vt* neutralize, saturate (acids); truncate; blunt, dull. **2: sich a.** become blunt (*or* dull). *Cf* ABGESTUMPFT. **Abstumpfung** *f* (-en) neutralization, saturation; truncation, etc. **Abstumpfungs·fläche** *f* (*Cryst.*) truncating face

Absturz° *m* fall, plunge; crash; precipice. **··bauwerk** *n* drop structure

ab·stürzen I *vt* **1** throw down; dump. **2** take the lid off. **II** *vi*(s) fall (down, off), plunge; crash; drop away (steeply)

ab·stutzen *vt* clip, dock, trim

ab·stützen *vt* support, prop up, shore up

ab·sublimieren *vt* remove by sublimation

ab·suchen *vt* search, examine, comb (an area); pick off (*or* up)

Absud° *m* decoction, extract; (*Dye.*) iron (*or* chrome) mordant

Absüß·bottich *m* edulcorating vat

ab·süßen *vt* **1** sweeten. **2** desugar(ize). **3** purify by washing, edulcorate; wash (out) (a precipitate)

Absüß·kessel *m* edulcorating vessel. **-spindel** *f* (*Sugar*) sweet-water spindle

Absüßung *f* sweetening; desugaring; purification by washing; edulcoration; washing out

Abszisse *f* (-n) abscissa

Abt. *abbv* (ABTEILUNG) div., sec., dept.

ab·tasten *vt* feel (out); probe; make contact with; select; (*Phys., Electron.*) scan; (*Med.*) palpate. **Abtaster°** *m* scanner, pickup unit. **Abtastung** *f* (-en) scan(ning); palpation

Abtast·kopf *m* scanning head. **-szintigraphie** *f* scanning

ab·tauen *vt, vi*(s) (*Ice*) melt off, thaw off

Abteil° *n* compartment

ab·teilen *vt* **1** separate, divide (off). **2** graduate. **3** classify. **Abteilung°** *f* department; division, section; separation; graduation; classification; compartment. **abteilungs·weise** *adv* part by part, piece by piece

ab·teufen *vt* sink (a shaft *or* well); deepen

Abtitrierungs·punkt *m* end point (of titration)

ab·tönen *vt* tone down; shade (off). **Abtön·farbe** *f* tinting color. **Abtönung°** *f* shading; shade, tint; tone

ab·toppen *vt* defront (coal tar benzene)

ab·töten *vt* destroy, kill (off); exterminate

Abtrag *m* (⁼e) **1** removed material. **2** excavation. **3** payment. **4** damage, disadvantage. **ab·tragen* I** *vt* **1** carry (*or* clear) away; remove. **2** level; demolish; pull down. **3** pay off. **4** lay off, plot. **5** erode; denude. **II** *vt & sich a.* wear out *cf* ABGETRAGEN. **III** *vi & sich a.* stop bearing fruit

abträglich *a* injurious, harmful

Abtrags·rate *f* **1** rate of erosion. **2** sputtering rate. **-stufe** *f* step in the erosion process; stage of wear and tear

Abtragung *f* (-en) **1** clearing away, removal. **2** leveling, demolition. **3** paying off, payment; liquidation (of debt). **4** wear, deterioration; erosion; denudation. **Abtragungs·gebiet** *n*, **-raum** *m* area of erosion, denuded area

Abtränk·brühe *f* (*Lthr.*) tanning liquor

ab·tränken *vt* saturate, impregnate (with tanning liquor); (*Lthr.*) fill (the pit) with tanning liquor

Abtransport *m* (-e) carting away, removal by vehicle. **ab·transportieren** *vt* cart (carry, transport) away (*or* off)

Abtreib(e)·apparat *m* still, distilling apparatus. **-herd** *m* refining hearth. **-hütte** *f* refinery. **-kapelle** *f* refining cupel. **-kolonne** *f* separating (*or* distilling) column. **-mittel** *n* expulsive agent, purge; abortifacient

ab·treiben* I *vt* **1** drive off (out, down), expel. **2** separate (by distillation); refine (by cupelation), cupel; cream (butter). **3** clear (woods). **4** (*Med.*) abort. **II** *vi*(s) **5** go off course. **abtreibend** *p.a* (*Med.*) abortive, abortifacient

Abtreibe·ofen *m* cupel furnace

Abtreiber° *m* dephlegmator

Abtreib· *see also* ABTREIB(E)·: **-gas** *n* expelled gas. **-rad** *n* secondary drive wheel

Abtreibs·alter *n* age (of wood) when cut

Abtreib·säule *f* stripping (*or* distilling) column. **-schmelze** *f* refining melt

Abtreibung *f* (-en) **1** driving out, expulsion. **2** separation; refinement; cupelation. **3** clearing. **4** abortion. *Cf* ABTREIBEN

Abtreib·welle *f* driven (*or* secondary) shaft

Abtrenn·arbeit *f* work of separation

abtrennbar *a* separable, detachable

ab·trennen 1 *vt* separate; dissociate; sever, detach; disconnect. **2: sich a.** separate, come off. **Abtrenner** *m* (-) separator; disconnector. **Abtrennung°** *f* separation; dissociation; severing, detachment; disconnection

Abtrennungs·arbeit *f* work of separation

ab·treten* I *vt* **1** tread (off, down), wear out (by walking). **2** pace off. **3** (*d.i.o*) transfer, convey, assign, yield, cede (to). **II** *vi*(s) **4** retire, resign, withdraw; yield. **5** alight; stop (off). **abtretend** *p.a* (*oft*) outgoing. **Abtreter** *m* (-) (foot) scraper, mat; transferrer. **Abtretung** *f* (-en) **1** treading (off, down); wearing out. **2** pacing off. **3** transfer, conveyance; assignment; cession. **4** retirement

Abtrieb° *m* **1** driving off, expulsion. **2** distillation. **3** slack (of a cable). **4** felling; felled timber; clearing. **5** negative buoyancy

Abtriebs·ertrag *m* yield of timber. **-fläche** *f* cleared area. **-säule** *f* distilling (stripping, separating) column. **-schlag** *m* recent cutting (of trees)

ab·triefen* *vi*(s) drip (*or* trickle) down

Abtrift° *f* drift, leeway

Abtritt° *m* **1** retirement, withdrawal. **2** toilet

Abtritts·dünger *m* cesspool fertilizer. **-grube** *f* cesspool

ab·trocknen *vt, vi*(s) & **sich a.** dry off

Abtropf·bank *f* drain(er), (drip-) draining rack. **-blech** *n* drip pan. **-brett** *n* drainboard

ab·tröpfein *vi*(s) ooze, seep; (*Water*) drip

ab·tropfen *vi*(s) drip (off), drain, trickle. **Abtropfer** *m* (-) drainer

Abtropf· drain(ing): **-gefäß** *n* drainer. **-gestell** *n* draining rack. **-gewicht** *n* drained weight. **-kasten** *m* (*Paper*) drain chest; (*Metll.*) drain table. **-pfanne** *f* drain (*or* drip) pan; (*Tinplating*) list pot. **-schale** *f* draining dish (*or* basin), perforated drainer

ab·trüben *vt* dull, darken, sadden (colors). **Abtrübungs·farbstoff** *m* saddening dye

ab·tupfen *vt* dab (dry)

Abut·säure *f* abutic acid

ab·wägen* *vt* weigh (off, out); ponder; level, balance *cf* ABGEWOGEN. **Abwägung°** *f* weighing (off, out); pondering; balancing—**bei A. (aller Umstände)** on balance, (on weighing all circumstances)

ab·wällen *vt* simmer; blanch; bring to a boil

ab·walzen *vt* roll; compact with a roller

ab·wälzen *vt* **1** roll away. **2** (+ **auf**) shift (blame) (to)

abwandelbar *a* modifiable

ab·wandeln 1 *vt* & **sich a.** change, alter. **2** *vt* modify, run through changes. **3: sich a.** vary. **Abwandler°** *m* modifier

ab·wandern *vi*(s) wander off, (e)migrate. **Abwanderung°** *f* wandering (off), (e)migration

Abwandlung° *f* modification, change, variation. **Abwandlungs·produkt** *n* derivative

Abwärme *f* waste heat. **ab·wärmen** *vt* **1** heat, warm. **2** let cool down

ab·warten *vt* await, wait for; wait out; (+ **es**) wait and see. **abwartend** *p.a:* **a—e Haltung** wait-and-see attitude; **sich a. verhalten** take a wait-and-see attitude

abwärts *adv* downward, down(hill)

Abwärts·bewegung *f* downward motion, decline. **-gasen** *n* downrun (of water gas). **a—gerichtet** *a* inverted, directed downward. **A—geschwindigkeit** *f* downspeed, downward velocity. **-transformator** *m*, **-umkehrer** *m* (*Elec.*) stepdown transformer

abwaschbar *a* washable

ab·waschen* *vt* wash (off, out), rinse, cleanse. **Abwasch·mittel** *n* cleanser, cleansing agent. **Abwaschung°** *f* washing, rinsing, cleansing; bath; ablution, sponge bath

Abwasser° *n* aqueous emission; waste water (*or* liquor); liquid sewage. **··abfluß** *m* waste water run-off (*or* effluent). **-anlage** *f*, **-betrieb** *m* sewage disposal plant. **a—biologisch** *a* sewage-biological. **A—klärtrichter** *m* save-all tray. **-kontrolle** *f* sewage (*or* waste-water) control. **-leitung** *f* drain

ab·wässern *vt* 1 wash. 2 free from water, drain

Abwasser·reinigung *f* waste-water purification, sewage treatment. **-reinigungsanlage** *f* sewage (*or* waste-water) treatment plant (*or* unit). **-rohr** *n* drain (*or* waste) pipe. **-technik** *f* sewage (*or* waste) engineering

ab·wechseln 1 *vt, vi* & **sich a.** alternate. 2 *vt* (inter)change. **abwechselnd** 1 *p.a* alternating, alternate; intermittent; variable. 2 *adv* (*oft*) by turns. **Abwechs(e)lung** *f* (-en) alternation; (inter)change; variation; variety: **zur A.** for a change. **abwechslungs·reich** *a* varied, diversified

Abweg° *m* byway, wrong way (*or* road)— **auf A—e führen** lead astray; **auf A—e geraten** go astray. **abwegig** *a* mistaken, misguided, aberrant. **abwegs** *adv* out of the way, off the (beaten) track

Abwehr *f* warding off; defense; resistance. **ab·wehren** *vt* ward off, repulse, repel. **abwehrend** *p.a* defensive

Abwehr·ferment *n* protective (*or* defensive) ferment (*or* enzyme); antibody. **-maßnahme** *f* defensive measure. **-mittel** *n* preventive, prophylactic. **-stoff** *m* defensive (*or* protective) substance, repellent. **-zellsystem** *n* protective cell system

ab·weichen I (*reg*) *vt, vi*(h) soak (off), soften (by steeping), macerate. **II** *vi**(s) deviate, diverge; differ, vary. **abweichend** *p.a* deviating, divergent, irregular. **Abweicher** *m* (-) variant. **Abweichung** *f* (-en) 1 deviation, divergence; variation; declination (of the needle); (*Opt.*) aberration; anomaly. 2 soaking, softening, maceration. **Abweichungs·quadrat** *n:* **mittleres A.** mean square deviation

ab·weisen* *vt* rebuff, send away; turn away, reject, repel. **abweisend** *p.a* repellent; unfavorable, adverse, negative: **sich a. verhalten** take a negative attitude. **Abweisung°** *f* rebuff(ing); rejection, repulsion

ab·weißen 1 *vt* whiten. 2 *vi* lose whiteness

ab·welken 1 *vt* (*Lthr.*) dry partially, sammy. 2 *vi*(s) wilt, wither. **Abwelk·maschine** *f* (*Lthr.*) sammying machine

abwendbar *a* preventable, avertable. **ab·**

·wenden* 1 *vt* & **sich a.** turn away. 2 *vt* avert. **Abwendung°** *f* prevention; turning away

ab·werfen* *vt* 1 throw off (out, down), shed; cast off, discard; drop. 2 emit, give off, discharge. 3 break down (rock); run off (slag); (*Tinning*) melt off (the list). 4 (*Com.*) yield

Abwerf·ofen *m* refining furnace. **-pfanne** *f* refining pan; (*Tinning*) list pot

Abwerg *m* waste tow (*or* oakum)

ab·werten *vt* 1 devalue, devaluate; depreciate. 2 estimate, appraise. **Abwertung°** *f* devaluation; depreciation; appraisal

abwesend *a* 1 absent. 2 absent-minded, distracted. **Abwesenheit** *f* 1 absence. 2 distraction, reverie

ab·wickeln I *vt* 1 unwind, unreel. 2 wind up; carry out, perform; settle; develop; (*Geom.*) rectify. **II: sich a.** 3 develop (its course), go (smoothly). **Abwickel·spule** *f* feed reel. **Abwick(e)lung°** *f* settlement, completion; development; (*Geom.*) rectification; unwinding

Abwiege·maschine *f* weighing machine

ab·wiegen* *vt* weigh, etc = ABWÄGEN

Abwind° *m* down current, downdraft

ab·winden* *vt* & **sich a.** unwind, reel off

ab·winkeln *vt* bend. **Abwink(e)lung** *f* bending

ab·wirtschaften 1 *vt* ruin, mismanage; (*Agric.*) exhaust (soil). 2 *vi* & **sich a.** go to ruin

ab·wischen *vt* wipe (off)

ab·wittern *vi*(s) weather away, erode

Abwitterungs·halde *f* weathered slope

ab·wracken *vt* wreck, scrap

Abwrack·metall *n* scrap metal

Abwurf° *m* 1 throwing (*or* casting) off, shedding, discarding, dropping. 2 castoff, drop (*esp* from a plane); dumping. 3 refuse, rubbish. 4 coarse plaster, roughcast. 5 (*Com.*) yield, profit

ab·würgen *vt* strangle; choke

abyssisch *a* abysmal; abyssal

abzahlbar *a* payable in installments

abzählbar *a* enumerable, countable

Abzähl·bedingung *f* condition for enumeration

ab·zahlen *vt* pay off; pay in installments; pay on account

ab·zählen *vt* count (off *or* out), enumerate

Abzahlung° f payment on account (or in installments); paying off—**auf A.** on the installment plan, on time

Abzählung° f enumeration

Abzapf·arbeit f (Metll.) drawing off, tapping

ab·zapfen vt draw off, tap; bleed, drain

ab·zehren vt, vi(s) & **sich a.** waste (away); corrode, **abzehrend** p.a wasting, consumptive (disease). **Abzehrung°** f wasting (away), corrosion; (Med.) consumption

Abzeichen° n mark, sign; badge, insignia. **ab·zeichnen** 1 vt draw, copy, trace; mark off. **2: sich a.** stand out, attract attention

ab·zentrifugieren vt centrifuge off; separate by centrifuging

Abzieh·apparat m distilling apparatus; proof press. **-bad** n (Dye.) stripping bath

abziehbar a deductible, subtractable

Abzieh·bild n transfer (picture), decal(comania). **-blase** f retort, still, alembic

ab·ziehen* I vt 1 draw (off, away), pull off; peel (off), remove, withdraw. 2 tap, drain; siphon off; rack (beer), decant (liquids); bottle, distil: **abgezogene Wasser** distilled liquors. 3 distract, divert. 4 skim. 5 sharpen, hone; plane; buff, polish. 6 subtract, deduct. 7 transfer (designs); copy; (Phot.) print. 8 fire (gun). 9 degum (silk); print. (Dye.) strip, boil off (color); (Lthr.) strip, clear; string (beans). 10 check, test (new clock). II vi(s) 11 leave, depart. 12 (of gases) escape. **Abzieher** m (-) proof puller; printer; abductor (muscle)

Abzieh·farbe f transfer color. **-flotte** f (Dye.) stripping bath. **-halle** f (Brew.) racking room. **-kolben** m still; condenser. **-mittel** n (Dye.) stripping agent. **-muskel** m abductor (muscle). **-papier** n pottery tissue; proof paper; transfer paper. **-presse** f proof press. **-stein** m whetstone

Abziehung° f removal, withdrawal; drainage; sharpening; subtraction, deduction; transfer, etc. Cf ABZIEHEN, oft = ABZUG 2

Abzieh·zahl f subtrahend. **-zünder** m friction igniter

ab·zielen vt/i (+ **auf**) aim (at)

ab·zirkeln vt define (precisely), pin down cf ABGEZIRKELT. **Abzirk(e)lung** f precise definition (or measurement); preciseness

Abzug° m 1 hood, fume cupboard. 2 scum (esp on molten lead), dross, sharp slag. 3 outlet, drain; gutter, sewer; vent. 4 deduction: **in A. bringen** deduct; **in A. kommen** be deducted. 5 tax (withheld). 6 proof (sheet), copy; (Phot.) print. 7 trigger. 8 departure, withdrawal; retreat; retirement. 9 outflow, discharge, drainage; escape (of gases). Cf ABZIEHEN

abzüglich adv & prep (oft + genit) less, minus, deducting

Abzugs·blei n lead obtained from dross. **-bogen** m (Print.) proof (sheet); (Explo.) firing spring. **-dampf** m drawn-off vapor; (specif) exhaust steam. **-dose** f junction box. **-fadenkraft** f strength of finished thread. **-gas** n chimney (or flue) gas; waste gas. **-graben** m culvert, drainage ditch; (open) sewer. **-haube** f fume hood. **-kanal** m sewer, drain; discharge duct. **-klappe** f trap. **-kupfer** n copper obtained from dross. **-öffnung** f outlet, vent. **-papier** n duplicating (proofing, printout) paper. **-raum** m hood; room with hoods. **-rohr** n, **röhre** f waste (drain, outlet) pipe (or tube). **-schacht** m hood flue, ventilating shaft. **-schalter** m firing connector. **-schrank** m hood, fume cupboard. **-ventil** n outlet valve. **-vorrichtung** f escape device; ventilator; constant-level apparatus (for water baths). **-wärme** f waste heat. **-werke** npl (Metll.) scum, dross

ab·zwacken vt (oft + d.i.o) grab, extort (from)

Abzweig° m branch, turn-off. **-dose** f (Elec.) junction box

ab·zweigen 1 vt trim (off); divide; divert. 2 vi(s) branch off. 3 vi(h) & **sich a.** branch, divide, fork; branch off

Abzweig·leitung f branch line. **-muffe** f (cable) connecting box, branch T. **-reaktanzspule** f feeder reactor. **-stromkreis** m (Elec.) derived (or branch) circuit, shunt

Abzweigung f (-en) division, diversion; branch(ing) (off); junction; (Elec.) shunt, branch line

Abzweig(ungs)·verhältnis n branching ratio. **-widerstand** m shunt resistance

ab·zwicken vt pinch (or nip) off

ab·zwingen* vt (d.i.o) extort (from)

ab·zwirnen vt unwind (thread); **A. über**

Kopf upstroke twisting
Ac . . . : see also **Ak . . . & Az . . .**
Acaj(o)u m (-s) cashew (tree)
Acceleren pl accelerators
accessorisch a accessory
Accise f (-n) excise
Aceconit·säure f aceconitic acid
Acenapht(h)indan n benzindacine
Acet· acet(o), acetic cf AZET·
Acetaldehyd·ammoniak n acetaldehyde ammonia, 1-aminoethanol
Acetaldol n aldol; 2-hydroxybutanol
acetalisieren vt/i form an acetal. **Acetalisierung** f acetal formation
acetamino· acetamino-, acetamido-
Acet·anhydrid n acetic anhydride
Acetat n (-e) acetate; (Tex.) (acetate) rayon. **··film** m cellulose acetate film. **-kunstseide** f acetate silk, rayon. **-seide** f acetate rayon. **-spinnfaser** f acetate (or rayon) spin fiber. **-(zell)wolle** f spun acetate rayon
acetenyl· ethynyl
Acetessig·äther, -ester m acetoacetic ester, ethyl acetoacetate. **-säure** f acetoacetic acid, diacetic acid. **-säureanilid** n acetoacetanilide. **-säureäthylester** m acetoacetic (or diacetic) ester
Acetimido·äthylester·hydrochlorid n ethyl imidoacetate hydrochloride
Acetin n acetin, glyceryl acetate. **··blau** n (Dyes) acetin blue
Aceto·ameisensäure·ester m acetonyl formate
Acetol n (-e) acetol (hydroxyacetone)
Aceton·chloroform n chloroform, α-α-α-trichloro-t-butyl alcohol. **-diäthylmercaptol** n acetone ethylthioacetyl
Acetonitril n (-e) acetonitrile, methyl cyanide
Aceton·säure f acetonic (α-hydroxyisobutyric) acid. **-trockenpulver** n acetone powder
Aceto·persäure f per(oxy)acetic acid. **-veratron** n acetoveratrone (3, 4-dimethoxy-acetophenone)
acetoxylieren vt acetoxylate
Acet·persäure f per(oxy) acetic acid. **-phenetedin** n phenacetin
Acetrizoe·säure f acetrizoic acid
Acet·säure f acetic acid

Acetum n (..ta) vinegar—**A. pyrolignosum** acetic acid
Acetur·säure f acetylglycine
Acetyl·bernsteinsäureester m ethyl acetylsuccinate. **-cellulose** f cellulose acetate. **-chlorür** n acetyl chloride
Acetylen n acetylene. **··bindung** f acetylene linkage, triple bond. **-dicarbonsäure** f acetylenedicarbolic acid. **A.-Dissousgas** n dissolved acetylene. **A—gebläse** n oxyacetylene torch
Acetylenid n acetylide
Acetylen·kalkschlamm m carbide lime sludge. **-kupfer** n copper acetylide. **-ruß** m acetylene black. **-sauerstoffbrenner** m oxyacetylene blowpipe. **-sauerstoffschweißung** f oxyacetylene welding. **-schnittbrennerei** f acetylene cutting. **-silber** n silver acetylide
acetylenyl· ethynyl-
Acetyl·essigsäure f acetoacetic acid. **-harnstoff** m acetylurea
Acetylid n (-e) acetylide
acetyliden· vinylidene-
acetylieren vt acetylate. **Acetylier·kolben** m acetylating flask. **Acetylierung** f acetylation
Acetylierungs·kolben m acetylation flask. **-mittel** n acetylating agent
Acetyl·kohlenwasserstoff m acetylenic hydrocarbon. **-muraminsäure** f (AMS) acetylmuramic acid (AMA). **-säure** f acetic acid. **-schwefelsäure** f acetylsulfuric acid. **-senföl** n acetyl mustard oil. **-sulfoessigsäure** f acetylsulfoacetic acid. **-zahl** f acetyl number
Achat m (-e) agate. **a—ähnlich, -artig** a agate-like, agatine. **-farben** a agate-colored. **A—glätte** f flint glazing. **a—haltig** a agate-containing, agatiferous. **A—mörser** m agate mortar. **-porzellan** n agate ware. **-reibschale** f agate mortar. **-schellack** m shellac substitute. **-schneide** f agate knife-edge. **tulpe** f agate
Achema abbv (Ausstellung für chemisches Apparatewesen) Exhibition of Chemical Apparatus
Achillen·öl n achillea oil
Achmit m (Min.) acmite
Achromasie f (Opt.) achromatism
Achromat m (-en, -en) achromat(ic lens).

achromatisieren *vt* achromatize

Achs·büchse *f* axle box, journal box.

-druck *m* axle load (*or* pressure)

Achse *f* (-n) **1** (*Math.*) axis. **2** (*Mech.*) axle, shaft

Achsel *f* (-n) **1** shoulder; (*Physiol.*) axilla. **2** (*Bot.*) axil. **·-bein** *n* shoulder blade. **-drüse** *f* axillary gland. **-grube, -höhle** *f* armpit, axilla, **a—ständig** *a* axillary

Achsen·abschnitt *m* **1** (*Math.*) axial intercept. **2** (*Cryst.*) axial index. **-abstand** *m* interaxial distance. **a—bürtig** *a* axial. **A—ebene** *f* (*Opt.*) axial plane. **a—förmig** *a* axial. **A—kreuz** *n* system of coordinates; (*Opt.*) cross hairs. **-lager** *n* axle bearing. **-linie** *f* axis. **a—los** *a* anaxial, having no axis. **A—neigung** *f* inclination of the axis. **-öl** *n* axle oil. **a—parallel** *a* concentric, with parallel axes. **A—schlauch** *m* axial tube. **-schmiere** *f* axle grease. **-schnitt** *m* axial intercept. **-schraube** *f* (*Mach.*) axle nut. **-skelett** *n* chordoskeleton. **-strang** *m* dorsal chord, notochord. **a—symmetrisch** *a* axially symmetric(al). **A—verhältnis** *n* axial ratio. **-winkel** *m* (*Cryst.*) interaxial angle

Achs·fett *n* axle grease

achsial, achsig *a* axial. **·achsig** *sfx* e.g. **zweiachsig** two-axle(d)

Achs·lager *n* axle bearing, journal box. **-linie** *f* axis. **a—parallel** *a* concentric; with parallel axis. **-recht** *a* axial. **A—richtung** *f* axial direction

acht I *num* **1** eight: **zu a.** in a group of eight. **2: a. Tage** a week. **3** (*as pfx, oft*) oct(o)-. II *sep pfx* care, attention *cf* ACHT 1, ACHTGEBEN

Acht *f* **1** attention, care: **A. geben** (*or* **haben**) **auf** pay attention to = ACHTGEBEN, ACHTHABEN; **außer A. lassen** disregard; **sich in A.** (*or* **acht**) **nehmen** take care, be careful. **2** ban, outlawry. **3** (-en) (figure) eight

achtbar *a* respectable

acht·basisch *a* octabasic

achte *a* eighth

Acht·eck *n* octagon. **a—eckig** *a* octagonal

Achtel *n* (-) **1** (*Frac.*) eighth. **2** (*as pfx in old Chem. Nomencl.*) octo-, octa-, e.g. **Achtelkohleneisen** octoferric carbide. **·-kreis** *m* (*Geom.*) oxtant. **-schlag** *m* 45° angle

achten 1 *vt* regard, respect, esteem. **2** *vi* (**auf**) pay attention (to)—**darauf a., daß** ... take care that ..., see to it that ...

ächten *vt* outlaw, ban

achtens *adv* in the eighth place, eighthly

achtens·wert, -würdig *a* worthy of esteem, estimable

Achter *m* (-), **-figur** *f* (figure) eight. **-gruppe** *f* octet (of electrons)

achterlei *a* (*no endgs*) eight kinds of

Achter·periode *f* period of the (periodic) table. **-ring** *m* eight-membered ring. **-schale** *f* shell of eight electrons, octet shell

achtfach, achtfältig *a* eightfold

Acht·flach *m* octahedron. **a—flächig** *a* octahedral. **A—flächner** *m* octahedron. **-füßler** *m* octopod

acht·geben* *vi* (**auf**) take care (of), be careful (of); look out (for); pay attention (to)

acht·gliedrig *a* eight-membered. **-kantig** *a* octagonal

acht·haben* *vi* (**auf**) pay attention (to)

achtlos *a* careless, inattentive, heedless. **Achtlosigkeit** *f* (-en) carelessness, inattention, heedlessness

acht·mal *adv* eight times. **-malig** *a* eight-time, (repeated) eight times

Acht·ring *m* eight-membered ring

achtsam *a* heedful, attentive. **Achtsamkeit** *f* (-en) heedfulness, attentiveness; attention

acht·seitig *a* eight-sided; eight-page. **-stündig** *a* eight-hour. **-tägig** *a* eight-day, week-long

Achtung *f* attention, notice; esteem, respect; (*exclam*) watch out!, take care!

Ächtung *f* (-en) outlawing, ban

acht·wertig *a* octavalent. **Acht·wertigkeit** *f* octavalence

achtzehn *num* eighteen. **Achtzehner·periode** *f* period of eighteen. **-schale** *f* shell of eighteen (electrons)

achtzehnte *a* eighteenth

achtzig *num* eighty. **achtzigste** *a* eightieth

ächzen *vi* groan

acidifizieren *vt* acidify

Acidität *f* acidity

Acido·ion *n* complex (*or* acido) ion

Acidolyse *f* (-n) acidolysis

Acido·salz *n* complex (*specif.* acido-) salt.
-**verbindung** *f* acido compound
acidylieren *vt* acylate, acetylate
Acker *m* (⁻) (cultivated) field; farm; acre.
-**bau** *m* farming, agriculture
Ackerbau· agricultural: -**kunde** *f*, -**lehre** *f* agronomy
Acker·beere *f* dewberry (*esp Rubus caesius*), bramble. -**boden** *m* (arable *or* surface) soil. -**bohne** *f* field (*or* broad) bean (*Vicia fabia*). -**distel** *f* Canada thistle (*Cirsium arvense*). -**erde** *f* (arable) soil. -**günsel** *m* ground pine (*Ayuga chamaepitys*). -**kamille** *f* wild camomile. -**krume** *f* (arable) soil, topsoil. -**land** *n* arable land. -**mennig** *m* agrimony. -**minze** *f* corn mint (*Mentha arvensis*)
ackern *vt/i* till (the soil); plow
Acker·schädling *m* agricultural pest. -**schleppe** *f* drag, harrow. -**senf** *m* charlock, wild mustard. -**vieh** *n* livestock, farm animal(s). -**walze** *f* roller. -**weide** *f* green pasture. -**wirtschaft** *f* agriculture. -**zeug** *n* agricultural equipment
Acon . . . *See also* AKON . . .
Aconit·säure *f* aconitic (achilleric, equisetic, citridic) acid
Acon·säure *f* aconic acid
Acr . . . *See also* AKR . . .
Acrit *n* DL-mannitol
Acro·reiz *m* formative stimulus
Acryl·faser *f* acrylic fiber. -**harz** *n* acrylic resin. -**nitril** *n* acrylonitrile. -**säure** *f* acrylic acid.
Acrylsäure·amid *n* acrylamide. -**ester** *m* acrylic ester, ethyl acrylite. -**methyle·ster** *m* methyl acrylate
Actihiazin·säure *f* actihiazic acid, 4-thiazolidone-2-caproic acid
Actiniden·reihe *f* actinide series
activieren *vt* activate
acyclisch *a* acyclic(al)
Acylamidophosphorsäure·ester *m:* NA. N-acylphosphoramidate
acylieren *vt* acylate. **Acylierung** *f* acylation
Acylphosphorsäure·ester *m* acyl phosphate
ad *prep* (*Latin*) (up) to; re(garding)
a.d. *abbv* (an der, an dem, etc) on (the), at

(the), to (the) *cf* AN
Adamantyl·essigsäure *f* adamantene-1-acetic acid
Adamin *n* (*Min.*) adamite
Adams·apfel *m* **1** Adam's apple. **2** shaddock; grapefruit. -**nadel** *f* yucca, Adam's needle
Adaptations·fähigkeit *f* adaptability
adaptieren *vt* adapt. **adaptiv** *a* adaptive
adäquat *a* adequate
Addend *n* (-en *or* -a) addendum
addierbar *a* additive, that can be added
addieren *vt/i* add. **addierend** *p.a* additive; *adv* oft by addition. **Addier·maschine** *f* adding machine. **Addierung** *f* (-en) addition
additionell *a* additional
additions·fähig *a* capable of addition, additive. **A—fähigkeit** *f* capacity for addition, additive power. **a—freudig** *a* ready to undergo additions, having good addition properties. **A—freudigkeit** *f* readiness to undergo additions. -**maschine** *f* adding machine. -**vermögen** *n* additive power
additiv *a*, **Additiv** *n* (-e) additive. **Additivität** *f* (-en) additivity
Adduktierungs·verfahren *n* addition method
Adduktoren·eindruck *m* adductor impression, scar
Adel *m* nobility, noble birth, aristocracy. **adelig** *a* noble, aristocratic. **adeln** *vt* ennoble; dignify; confer noble rank on
adenotrop *a* adenotropic
Adenyl·säure *f* adenylic acid
Ader *f* (-n) **1** vein (*in various senses*); blood vessel, artery: **zur A. lassen** bleed. **2** (*Wood*) grain. **3** (*Wire*) strand; (*Cable*) core
ader·blätterig *a* inophyllous
Äderchen *n* (-) little vein, veinlet
Ader·haut *f* choroid (membrane). -**häutchen** *n* chorion. -**hautspanner** *m* ciliary (tensor) muscle. -**holz** *n* wood cut with the grain
aderig, äderig *a* **1** veiny, veined; venous, vascular. **2** streaked. **3** (*Wire*) strand(ed): **mehraderiges Kabel** multiple-strand cable

Ader·knoten, -kropf *m* varicose vein. **-laß** *m* venesection, bloodletting

Adermin *n* adermine, pyroxidine (vitamin B6)

adern, ädern *vt* vein; grain, marble

Ader·netz *n* venous network. **-netzschlag-ader** *f* choroid artery. **-presse** *f* tourniquet. **a—reich** *a* richly veined, full of veins. **-rippig** *a* (*Bot.*) nerved. **A—schlag** *m* pulse (beat), pulsation. **-schnur** *f* flexible cable; extension cord

Aderung, Äderung *f* (-en) veining; marbling, grain; venation

Ader·verkalkung *f* arteriosclerosis. **-wasser** *n* blood serum, lymph

Adhärenz *f* (-en) adherence. **adhärieren** *vi* (**an**) adhere (to). **Adhäsion** *f* (-en) adhesion

Adhäsions·eigenschaft *f* adhesive property. **-fähigkeit** *f* adhesiveness, adhesive capacity. **-kraft** *f* adhesive power; force of adhesion. **-masse** *f* adhesive (substance). **-prüfer** *m* adhesion tester. **-spannung** *f* adhesive stress

adhäsiv *a*, **Adhäsiv** *n* (-e) adhesive

Adiabate *f* (-n) adiabatic curve

adiatherman *a* a(dia)thermanous

Adika·fett *n* Adica fat

adipid *a* fat, fatty. **adipidieren** *vt* grease. **Adipinat** *n* (-e) adip(in)ate. **Adipin·-säure** *f* adipic acid

Adipocire *n* (-n) adipocere

Adipomal·säure *f* adipomalic acid (α-hydroxyl-α-methylglutaric)

adipös *a* adipose

Adipo·weinsäure *f* adipotartaric (*or* dihydroxyadipic) acid

Adjustage *f* (-n) adjustment. **adjustieren** *vt* adjust, regulate. **Adjustierung** *f* (-en) adjustment

Adjuvans *n* (..antien) adjuvant. **-wirkung** *f* adjuvant activity

Adler·fett *n* (-) eagle. **-stein** *m* (*Min.*) modular goethite, eaglestone. **-vitriol** *n* Salzburg vitriol (mixed $CuSO_4$, $FeSO_4$)

adlig *a* noble, aristocratic = ADELIG

Admiralitäts·metall *n* admiralty brass (Cu 70, Zn 29, Sn 1)

Admolekül *n* adsorbed molecule

Adonit *n* adonitol

adoptieren *vt* adopt. **adoptiv** *a* adoptive, adopted

adoucieren *vt* sweeten; wash, edulcorate; temper, anneal; (*Iron*) decarbonize

Adoucier·gefäß *n* annealing pot (*or* vessel). **-ofen** *m* tempering (*or* annealing) furnace

Adr. *abbv* (Adresse) address

adrenergisch *a* adrenergic

Adressant *m* (-en, -en) writer; sender, consignor; drawer (of a note). **Adressat** *m* (-en, -en) addressee, consignee; drawee (of a note)

Adreß·buch *n* directory. **-büro** *n* registry

Adresse *f* (-n) (postal) address. **addressieren** *vt* address

Adria·bindung *f* (*Tex.*) diagonal rib

adrig, ädrig *a* veiny, etc = ADERIG

Adsorbat *n* (-e), **Adsorbend(um)** *n* (..den) adsorbed substance, adsorbate

adsorbierbar *a* adsorbable. **Adsorbierbarkeit** *f* adsorbability

adsorbieren *vt* adsorb. **adsorbierend** *p.a* adsorbent

Adsorpt *n* (-e) adsorbate, material adsorbed

Adsorptions·analyse *f* adsorption analysis, chromatography. **a—fähig** *a* adsorptive. **A—fähigkeit** *f* adsorptiveness, adsorptive capacity. **-haut** *f*, **-häutchen** *n* adsorbed (*or* adsorption) film. **-koeffizient** *m* solubility coefficient. **-kohle** *f* activated charcoal. **-kraft** *f* adsorptive power. **-lage** *f* adsorption site. **-mittel** *n* adsorbent. **-wärme** *f* heat of adsorption

Adstringens *n* (..genzien) astringent. **Adstringenz** *f* astringency. **adstringierend** *p.a* astringent

Adteil° *m* absorbed component

Adular *m* (-e) adularia, moonstone

Adurol *n* (*Phot.*) (chloroquinol) developer

ad. us. propr. *abbv* (*Pharm.*) (ad usum proprium) (*Lat.*) for personal use

adventiv *a* adventitious

Advokat *m* (-en, -en) lawyer

Ae. *abbv* (Äther) ether

Ae . . . *oft* used in place of **Ä . . .**

Å.E. *abbv* (Ångströmeinheit) Ångstrom unit, Å

AeDTE *abbv* (Aethylen·diamin·tetra·essigsäure) ethylenediaminetetraacetic acid, EDTA

AEF abbv (Ausschuß für Einheiten und Formeln) Committee on Units and Formulas

A-Effekt m abbv (alternierender Effekt) alternating effect

Å-Einheit f Ångström Unit

aelotrop a aelotropic, double-refracting

Aeq. abbv (Aequivalent) equivalent

Aequations-stellung f equal longitudinal division

a/erob a aerobic. **A/erob** n (-en) (Bact.) aerobe. **A/erobier** m (usu pl) aerobic bacteria

A/erobiont m (-en, -en) aerobe. **a/erobiontisch** a aerobic. **A-erobiose** f (-n) aerobiosis

a/erobisch a aerobic

a/erogen a aerogenic. **A--gas** n aerogene gas

a/erophag a aerophagic

A/erosol-behälter m aerosol container. **-sprühdose** f aerosol spray can. **-treibmittel** n aerosol propellant

a/erostatisch a aerostatic

aeth. abbv (ätherisch) ethereal

Aetz . . . = ÄTZ . . .

a.f. abbv (aschefrei) ash-free

Affäre f (-n) affair, matter

Affe m (-n, -n) ape, monkey

Affekt m (-e) emotion, ardor, passion

affektieren vt affect, put on, feign. **Affektiertheit** f (-en) affectation

Affektions-preis m fancy price

affektiv a emotional; affective. **affektlos** a dispassionate, apathetic. **Affektlosigkeit** f dispassionateness, apathy

äffen vt ape, mimic; mock

Affen·brotbaum m baobab tree

Affiche f (-n) poster

afficieren vt affect

affiliiert p.a affiliated

affin a affine. **Affinade** f (-n) washed raw sugar. **affin·ähnlich** a similar in affinity. **Affination** f (-en) refining (esp of gold with H_2SO_4; also of sugar)

affinieren vt refine, affine. **Affinierung** f (-en) refining, affination

Affinität f (-en) (+ zu) affinity (for, toward); free energy

Affinitäts·einheit f unit of affinity. **-konstante** f affinity (or dissociation) constant (of acid or base)

Affinivalenz f (-en) atomicity, valence

affizierbar a impressionable, easily affected (or moved). **affizieren** vt affect; move (emotionally)

a.f.S. abbv (auf folgender Seite) on the following page

After m (-) 1 anus, rectum. 2 bottom. 3 waste (matter), refuse; excrement. 4 residuary ore

After· pseudo-, false; secondary, neo-; after-, back-, hind-; anal, rectal: **-bildung** f malformation; secondary growth. **-blatt**, **-blättchen** n (Bot.) stipule. **-darm** m (Anat.) rectum. **-erz** n residuary ore. **-geburt** f afterbirth. **-gewebe** n heteroplastic tissue. **-holz** n deadwood. **-kegel** m (Geom.) conoid. **-kohle** f slack (coal), coal dust, cinders. **-kristall** m pseudomorph. **-kugel** f (Geom.) spheroid. **-mehl** n coarse flour, seconds; pollard. **-moos** n algae, seaweed. **-schörl** m axinite. **-seite** f (Lthr.) bellies. **-silber** n silver containing dross. **-weisheit** f sophistry, pseudowisdom

a.G. abbv (auf Gegenseitigkeit) mutual

A.G., A.-G. abbv 1 (Atomgewicht) atomic weight. 2 (Aktiengesellschaft) joint-stock company, corporation; (US) Inc., (Brit.) Ltd. 3 (Amtsgericht) local court

Agaricin n, **--säure** f agari(ci)c acid

Agarizin n (Pharm.) agaricin(e)

Agat m (-e) agate

Agathalin n agathalene (1,3,5-trimethyl-naphthaline)

Agathen·disäure f agathenedioic acid

Agave·faser f agave fiber, sisal. **-hanf** m agave (or aloe) hemp

ÄGDÄ abbv EGDE, ethyline glycol dimethyl ether

Agene-Verfahren n agene (flour-bleaching) process

Agens n (Agenzien) agent. **Agent·schaft** f, **Agentur** f (-en) agency

Agenzien agents: pl of AGENS

Agfa abbv (Aktiengesellschaft für Anilinfarbenfabrikation) Aniline Dye Mfg. Co., Inc.

Agglomerat n (-e) 1 agglomerate. 2 sinter

cake. **Agglomerier·anlage** *f* sintering plant. **agglomerieren** *vt/i* agglomerate. **Agglomerierung** *f* (-en), **Agglomerisation** *f* (-en) agglomeration

Agglutinations·fähigkeit *f* agglutinability

agglutinieren *vi* agglutinate

Aggregat *n* (-e) aggregate; outfit; (*Mach.*) set, unit, plant; (*Elec.*) (power) generator. **--zustand** *m* state of aggregation, state (of matter), physical state. **aggregieren** *vt/i* aggregate

aggressiv *a* aggressive, offensive; corrosive. **A—·beständigkeit** *f* resistance to chemical attack (*or* corrosion). **Aggressivität** *f* aggressiveness

Ägide *f* (-n) egis

agieren *vi* act

Agio *n* (-s) premium (on money exchange), agio

Agitakel *m* (-) stirrer, mixer

agitatorisch *a* agitative. **agitieren** *vt* agitate, stir up. **Agitiertheit** *f* (state of) agitation

Agon *n* (-e) co-ferment

Agraffe *f* (-n) clasp

Agrar· agrarian: **-gemeinschaft** *f* agrarian (*or* farming) community. **Agrarier** *m* (-), **agrarisch** *a* agrarian

Agrochemikalien *pl* agricultural chemicals (*usu* pesticides)

Agronom *m* (-en, -en) agronomist. **Agronomie** *f* agronomy. **agronomisch** *a* agronomic(al)

Agrume *f* (-n) citrus (fruit). **Agrumen·öl** *n* agrumen oil

Ag(t)·stein *m* amber; **schwarzer A.** jet

Ägypten *n* Egypt. **Ägypter** *m*, **ägyptisch** *a* Egyptian

Ah *abbv* (Amperestunde) amp(ere) hour

Ahl·beere *f* (black) currant

Ahle *f* (-n) 1 awl, punch, broach. 2 (*Print.*) point. 3 whetstone

Ahn *m* (-en, -en) ancestor; grandfather

Ahne *f* (-n) 1 chaff. 2 grandmother

ähneln *vi* (*dat*) be similar (to), resemble

ahnen 1 *vt* surmise, suspect. 2 *vt/i* (*dat*): **ihm ahnt . . .** he has a premonition (*or* a foreboding) (of, that . . .); **a. lassen** foreshadow, forbode

Ahnen·keim *m*, **-keimplasma** *n* germ plasma. **-reihe** *f* line of ancestors. **-tafel** *f* genealogical table

ähnl. *abbv* (ähnlich) sim., like

ähnlich *a* 1 (*dat*) similar (to), analogous (to), resembling, (a)like: **ä. sehen** (*dat*) look (*or* be) like, resemble; **u. ä—es** and the like; **ä. wie . . .** just as, the same as . . .2 (*as sfx*) -like, -oid. **Ähnlichkeit** *f* (-en) similarity, resemblance, likeness; analogy; similitude

Ähnlichkeits·betrachtung *f* consideration of similitude. **-prinzip** *n* principle of similarity

Ahnung *f* (-en) 1 idea, notion. 2 presentiment, foreboding. 3 anticipation

ahnungs·los *a* unsuspecting. **-schwer, -voll** *a* ominous; full of foreboding

Ahorn *m* (-e) maple. **--holz** *n* maple (wood). **-melasse** *f* maple syrup. **-saft** *m* maple sap. **-säure** *f* aceric acid. **-zucker** *m* maple sugar

Ähre *f* (-n) (*Bot.*) ear, spike

Ähren·früchte *fpl* grains, cereals. **-gras** *n* cereal (grass). **-lese** *f* gleaning. **-spitze** *f* (*Bot.*) awn, beard, glume

ährig *a* eared, spiked

Aich·amt *n* ga(u)ging office. = Eichamt. **aichen** *vt* ga(u)ge, calibrate = eichen

Aich·metall *n* Aich('s) metal (Cu-Zn-Fe alloy)

A.I.F. *abbv* (Arbeitsgemeinschaft Industrieller Forschungsgemeinschaften) Union of Industrial Research Associations

Ajoran·öl, Ajowan·öl *n* ajowan oil

Akajou *m* acajou (cashew *or* mahogany tree); cashew nut. **··gummi** *n* cashew (*or* mahogany) gum. **-holz** *n* mahogany. **-öl** *n* cashew nut oil

Akaroid·harz *n* acaricide =

Akaschu *m* cashew, etc = akajou

Akazie *f* (-n) acacia, common locust tree

Akazien·gummi *n* acacia gum, gum arabic. **-rinde** *f* wattle bark

Akazin(·gummi) *n* gum arabic

Akelei *f* (-en) (*Bot.*) columbine, aquilegia

Akkaroid·harz *n* accaroid resin

akklimatisieren *vt* acclimatize

Akkommodations·breite *f* amplitude, extent of accommodation. **-leistung** *f* adaptation capacity

akkommodieren *vt* accommodate, adapt

Akkord *m* (-e) 1 chord. 2 agreement, settlement, accord. 3 contract—**im A. arbeiten** do piecework. **--arbeit** *f* piecework, job work. **-satz** *m* piecework rate

Akkordeon·balg *m* accordeon bellows

akkordieren *vi* agree, settle, come to terms

Akku *m* (-s) storage battery. **--bohrer** *m* battery-powered drill. **Akkumulator** *m* (-en) storage battery, accumulator. **Akku(mulator)·säure** *f* battery acid, electrolyte. **-zelle** *f* storage battery cell

akkumulieren *vt* accumulate

akkurat *a* accurate, exact. **Akkuratesse** *f* (-n) accuracy, exactness

Akme *f* acme; (*Med.*) crisis. **--gelb** *n* Acme yellow, chrysoidine

A-Kohle *f* activated charcoal (*or* carbon)

Ako . . . *See also* Aco . . .

Akon *m/n* kapok

Akonin·säure *f* aconinic acid. **Akonit** *m* (-en, -en) aconite. **Akonitin·säure** *f* aconitic acid ($C_6 H_6 O_4$); (*more recently:*) aconitinic acid ($C_{31} H_{35} NO_{10}$)

Akonit·knollen *mpl* aconite (root). **-säure** *f* aconitic acid

Akon·säure *f* aconic acid

Akr . . . *See also* ACR . . .

akrib *a* meticulous. **Akribie** *f* meticulousness. **akribisch** *a* meticulous

Akridin·gelb *n* acridine yellow. **-säure** *f* acrid(in)ic acid

Akryl· acrylic: **-nitril** *n* acrylonitrile. **-säure** *f* acrylic acid

Akt *m* (-e) act, action; deed

Akt . . . *See also* ACT . . .

akt. *abbv* (aktiv) active

Akte *f* (-n) 1 act, deed. 2 document, file, record: **zu den A—n nehmen** (put on) file

Akten·bündel *m* file. **-deckel** *m* (filing) folder. **a—mäßig** *a* documentary. **A—papier** *n* deed (*or* record) paper. **-schrank** *m* filing cabinet. **-stück** *n* (filed) document, file

Aktie *f* (-n) share (of stock)

Aktien·besitzer *m* stockholder, shareholder. **-gesellschaft** *f* (joint stock) company *cf* A.G. II. **-markt** *m* stock market

Aktion *f* (-en) action; lawsuit; campaign

Aktionär *m* (-e) stockholder, shareholder

aktiv *a* active; activated. **Aktiv** *n* (-en) detail, task force. **Aktiva** *npl* assets

Aktivator *m* (-en) activator, promoter, activating agent. **a—frei** *a* unpromoted

Aktiv·bestand *m* assets; effective strength

Aktiven *npl* assets; task forces

Aktiv·gehalt *m* active content(s)

aktivieren *vt* activate

Aktivierungs·energie *f* energy of activation. **-mittel** *n* activating agent, activator. **-wärme** *f* heat of activation. **-zahl** *f* activation number

Aktivin *n* choramine-T.

Aktivität *f* (-en) activity

Aktiv·kieselsäure *f* silica gel. **-kohle** *f* activated charcoal (*or* carbon)

Aktivum *n* (..va *or* ..ven) asset

aktuell *a* 1 timely, present (-day), current; up-to-date. 2 actual

Akustik *f* acoustics. **akustisch** *a* acoustic

akut *a* acute

akzedieren *vi* (*dat.*) accede (to); join

Akzelerator *m* (-en) accelerator

Akzent *m* (-e) accent. **akzentuieren** *vt* accentuate

Akzept *n* (-e) (*Com.*) acceptance. **Akzeptant** *m* (-en, -en), **Akzeptor** *m* (-en) acceptor

Akzessorie *f* (-n), **akzessorisch** *a* accessory

akzidentell *a* accidental

Akzise *f* (-n) excise (tax)

Alabaster·gips *m* high-grade plaster of Paris. **-glas** *n* (devitrified opaque) alabaster glass. **alabastern** *a* (made of) alabaster, alabaster-white

Alan *n* alane, aluminum hydride. **Alanat** *n* alanate, tetrahydroaluminate

Alant *m* (-e) elecampane. **--beere** *f* black currant. **-kampher** *m* helenin. **-öl** *n* elecampane oil; alantin; inulin. **-wurzel** *f* (*Pharm.*) inula, elecampane root

alarmieren 1 *vt* alarm. 2 *vi* sound the alarm

Alaun *m* (-e) alum; **A. von Rocca** roche (*or* Roman) alum

alaun·artig *a* alumic, aluminous. **A—bad** *n* alum bath. **-beize** *f* (*Dye.*) aluminous mordant; alum bath; (*Lthr.*) aluming, alum steep. **-brühe** *f* (*Lthr.*) alum pick (*or* set-up)

alaunen *vt* alum, treat with alum

Alaun·erde *f* alumina. **a—erdehaltig** *a* aluminiferous. **A—erdesulfat** *n* aluminum sulfate. **-erz** *n* alunite. **-faß** *n* (*Lthr.*) alum vat. **-festigkeit** *f* (*Paper*) alum resistance. **a—förmig** *a* aluminiform. **-gar** *a* dressed with alum, alumed, tawed. **A—gerber** *m* tawer. **-gerberei** *f* tawing; tawery. **-gerbung** *f* alum tawing. **-gips** *m* artificial marble (made with alum and gypsum). **a—haltig** *a* aluminous, aluminiferous. **A—hütte** *f* alum works

alaunieren *vt* alum, treat with alum

alaunig *a* aluminous

Alaun· alum **-karmin** *m* (*Micros.*) alum carmine. **-kies** *m* aluminous pyrites. **-mehl, pulver** *n* powdered (*or* precipitated) alum. **a—sauer** *a* aluminate of . . . **—a—es Salz** aluminate. **A—schiefer** *m*, **-schiefererz** *n* alum slate (*or* shale). **-seife** *f* aluminous soap. **-sieden** *n* alum boiling (*or* making). **-siederei** *f* alum works. **-spat, -stein** *m* alum stone, alunite

alaunt *pres:* alums & *pp:* alumed *of* ALAUNEN

Alaunung *f* aluming

Albertol *n* modified phenol-formaldehyde resin

albumin·artig *a* albuminoid. **A—gehalt** *m* albumin content

Albuminimeter *m or n* albuminometer

albuminisieren *vt* albuminize

Albumin·kupfer *n* copper albuminate. **-papier** *n* albumin(ized) paper. **-stoff** *m* albuminous substance, protein. **-substanz** *f* albuminoid substance. **-urie** *f* albuminuria

Alchemie *f* alchemy. **alchemisch** *a* alchemical. **alchemistisch** *a* alchemistic(al)

Alchimie *f* alchemy

aldehydisch *a* aldehydic

aldehydo· formyl-

Aldehyd·oxim *n* adloxime. **-säure** *f* aldehydic (*or* aldonic) acid

Aldon·säure *f* aldonic acid

Alektoron·säure *f* alectoronic acid

Alembroth·salz *n* alembroth (salt) $(2NH_4Cl \cdot Hg\ Cl_2 \cdot H_2O)$

alepisch *a:* **a—er Gallapfel** Aleppo gall

Aleuritin·säure *f* aleuritic acid

Aleuron·korn *n* (*Bot.*) aleurone grain

Alfa *f* esparto. **~gras** *n* alfalfa grass, lucerne

Alfenid *n* (-e) nickel-copper-zinc alloy

Alfol *n* aluminum foil

Algarot(t)·pulver *n* algaroth, powder of Algaroth (SbOCl)

Alge *f* (-n) alga, seaweed

algebraisch *a* algebraic

Algen·flora *f* algal fleece. **-kohle** *f* boghead coal. **-pilz** *m* algal fungus, phycomycete. **-säure** *f* algic acid, norgine

Algol·farbe *f*, **-farbstoff** *m* algol dye

Alhidade *f* (-n) (*Opt.*) alidade

aliphatisch *a* aliphatic. **Aliphatoretin** *n* aliphatic resin

Alit(h) *n* (-e) (*Cement*) alite

alitieren *vt* aluminize, calorize (steel)

Alizarin·altrot *n* alizarin old red, Turkey red. **-farbe** *f* alizarin color (*or* dye). **-gelb** *n* alizarin yellow. **-sulfosäure** *f* alizarinsulfonic acid

Alk. *abbv* (Alkohol) alcohol

alkal. *abbv* (alkalisch) alkaline

Alkaleszenz *f* alkalinity

Alkali·alkoholat *n* alkali (metal) alkoxide. **-amid** *n* alkali metal amide. **a—arm** *a* poor in alkali, low-alkali. **-artig** *a* alkaloid(al). **-bildend** *a* alkaligenous, alkali-forming. **A—bindemittel** *n* alkali binding agent. **a—echt** *a* alkali-fast, alkali-resistant. **A—echtfarbe** *f* alkalifast color (*or* dye). **a—empfindlich** *a* alkali-sensitive (or -labile)

Alkalien alkalies: *pl of* **Alkali** *n*

alkali· alkali(ne): **-enthaltend** *a* containing alkali, alkaline. **-fest** *a* alkali-proof. **-führend** *a* alkaloidal. **A—gestein** *n* alkaline rock. **-halogenid** *n* alkali halide. **-kalkgestein** *n* calc-alkali rock. **-lauge** *f* alkali solution, lye. **-messer** *m* alkalimeter. **-messung** *f* alkalimetry

alkalinisch *a* alkaline. **Alkalinität** *f* alkalinity

Alkalisator *m* (-en) alkalizer

alkalisch *a* alkaline: **a—e Luft** alkaline air (*Priestley's* name for ammonia); **a. machen** alkalize, render alkaline; **a. reagierend** giving an alkaline reaction; **a. wirken** react alkaline

Alkali·schmelze f alkali fusion; alkali melt
alkalisierbar a alkalizable. **alkalisieren** vt alkalize
alkali·stabil a stable to alkali. **A—stabilität** f stability to alkali
Alkalität f alkalinity
alkaloid·artig a, **alkaloidisch** a alkaloidal
Alkaloid·lösung f alkaloidal solution. **-vergiftung** f alkaloid poisoning
Alkalose f (Med.) alkalosis
Alkamin n (-e) alcamine, amino alcohol
Alkan n alkane, paraffin
Alkanna·rot n alkanna red, anchusin, alkanet. **-tinktur** f alkanet extract, alkanin. **-wurzel** f alkanet (or henna) root
Alkarsin n cacodyl oxide
Alkazid·verfahren n process for extracting sulfur from gases
Alkein n dialkylamino ethanol ester
Alken n (-e) alkene, olefin
Alkin n (-e) alkyne. **--säure** f alkynoic acid
Alkogel n alcogel
alkoh. abbv (alkoholisch) alcoholic
Alkohol m (-e) alcohol, (specif) ethyl alcohol: **fester A.** solid alcohol, metaldehyde fuel.
alkohol·artig a alcoholic, alcohol-like. **A—artigkeit** f alcoholicity. **-blutprobe** f blood test for alcohol. **a—fest** alcohol-proof. **-frei** a nonalcoholic. **-haltig** a alcoholic, containing alcohol
alkoholisch a alcoholic
alkoholisch·wässerig a aqueous-alcoholic
alkoholisierbar a alcoholizable
alkoholisieren vt alcoholize
alkohol·löslich a alcohol-soluble, soluble in alcohol. **A—lösung** f alcohol(ic) solution. **-messer** m alcoholometer. **a—reich** high (or rich) in alcohol, highly alcoholic. **A—säure** f alcohol acid. **-spiegel** m alcohol level (or content). **a—unlöslich** a insoluble in alcohol. **A—vergällung** f denaturing of alcohol. **-vergärung** f alcohol(ic) fermentation
Alkoholyse f alcoholysis
Alkoven m (-) alcove, recess
alkoxyalkylieren vt convert into an alkoxy compound by alkylation
alkoxylieren vt alkoxylate
Alkyd·harz n alkyd resin. **-lack** m alkyd resin varnish

Alkyl·harnstoff m: **N-A.** N-alkylurea. **-harz** n alkyd resin
alkylieren vt alkylate. **Alkylierung** f alkylation. **Alkylierungs·mittel** n alkylating agent
Alkyl·rest m alkyl residue (or group). **-sulfosäure** f alkyl sufonic acid
all a, pron all; whole; every: **alle** all of them, all people, everyone; **alle beide** both (of them); **alle Tage** every day; **alle zwei (drei, vier) Tage** every other (third, fourth) day cf ALLES
All n universe, space
allabendlich a (occurring) every evening
allantoisch a allantoic. **Allanto(is)·säure** f allantoic acid, allantoin
Allantoxan·säure f allantoxanic acid
Allantur·säure f allanturic acid
all·bekannt a universally known; notorious
All·buch n encyclopedia
alledem: bei (or **mit**) **a.** for all that, in spite of that; **bei** (or **mit**) **a., daß** . . . in spite of the fact that . . . , even though . . .
Allee f (-n) avenue, boulevard
allein 1 pred a alone, single. 2 adv only, merely. 3 cnjc but
Allein·besitz m sole ownership; exclusive property. **-gerbstoff** m self-tannin. **-gerbung** f self-tanning. **-handel** m monopoly. **-hersteller** m sole producer
alleinig 1 a sole, only, exclusive. 2 adv alone
Allein·inhaber m sole owner. **-sein** n solitariness, isolation, aloneness. **a—stehend** p.a 1 isolated; insulated. 2 single; unmarried; separate, detached. 3 independent. **A—vertrieb** m monopoly, sole distributorship
allel a allelic. **Allel** n (-e), **Allelogen** n (-e) allelic gene. **allelomorph** a allelomorphic, allelic
allelotrop a allelotropic. **Allelotropie** f allelotropism
allemal adv always, every time—**a. wenn** whenever; **ein für a.** once and for all
Allen n (-e) allene
allenfalls adv 1 at all events, in any case. 2 at most. 3 possibly
allent·halben adv everywhere
aller·best a best (of all), very best. **-dings** adv to be sure, of course. **-erste** a very

first. **-feinste** *a* very fine(st). **-größte** *a* very greatest (largest, biggest). **-hand** *a* (*no endgs*) of all kinds, all sorts of

Allerheiligen·holz *n* logwood

aller·höchste *a* very highest, utmost. **-jüngste** *a* very (*or* most) recent. **-kleinste** *a* minute

allerlei *a* (*no endgs*) all kinds of. **Allerlei** *n* variety, mixed bag. **--gewürz** *n* 1 mixed spice. 2 allspice

aller· very, (most) of all: **-letzt** *a* very last, ultimate; very recent. **-meist** *a* most (of all), principal. **-mindeste** *a* least (of all). **-nächst** 1 *a* nearest (of all). 2 *adv* close (by). **-neuest** *a* very latest. **-orten, -orts** *adv* everywhere. **-seits** *adv* on all sides. **-wärts** *adv* everywhere. **-wenigste** *a* least (of all). **-wesentlichste** *a* most essential. **-wichtigste** *a* most important, paramount

alles *n* (*adv endgs*) all, everything; anything: **vor allem** above all, first of all; **a. in allem** all in all; **a. eher** (*or* **andere**) **als** anything but *cf* ALL *a*. **--fressend** *a* omnivorous. **A—fresser** *m* mixed-food eater, omnivore. **-kleber** *m* all-purpose adhesive. **-schneidemaschine** *f* universal cutting machine

alle·zeit *adv* at all times, always; (at) any time

all·fällig *a* possible, any . . . that may occur; *adv* possibly, if the occasion should arise

allg. *abbv* (allgemein) gen(l). (generally)

Allgebrauch· *pfx* all-purpose

allgemach *adv* gradually, by degrees

allgemein *a* general, common; universal; (*Med.*) constitutional; epidemic: **im a—en** in general, generally; **A—es** *n* general information; **a. machen** generalize

allgemein·gültig *a* universally (*or* generally) valid. **A—gültigkeit** *f* universal (*or* general) validity. **-gut** *n* common property

Allgemeinheit *f* (-en) generality; universality

Allgemein·kost *f* normal diet. **-kosten** *pl* general expenses. **-krankheit** *f* (*Med.*) systemic disorder. **-reaktion** *f* general reaction. **a—verständlich** *a* generally (*or* universally) understandable, popular. **A—wirkung** *f* general effect

All·gewalt *f* omnipotence. **all·gewaltig** *a* omnipotent

All·heilmittel *n* universal remedy, panacea

Allheit *f* (-en) totality, universality

Allianz *f* (-en) alliance. **alliieren** *vt & sich a.* ally (oneself). **Alliierte** *m, f* (*adj endgs*) ally

all·jährig, -jährlich *a* yearly, annual. **-mächtig** *a* almighty, omnipotent. **-mählich** *a* gradual; *adv oft* by degrees. **-monatlich** *a, adv* monthly, every month. **-nächtlich** *a, adv* nightly, every night

allochthon *a* allochthonous

allogam *a* allogametic

Allomerie *f* allomerism. **allomerisch** *a* allomeric. **Allomerismus** *m* allomerism

allomorph *a* allomorphic. **Allomorphismus** *m* allomorphism

Allonge *f* (-n) 1 adapter. 2 (*Zinc*) prolong, extension. 3 condenser. 4 flyleaf

Allopathie *f* homeopathy. **allopathisch** *a* homeopathic

Allophan·säure *f* allophanic acid

Alloschleim·säure *f* allomucic acid

Allotellur·säure *f* allotelluric acid

allotrop *a* allotropic. **Allotropie** *f*, **Allotropismus** *m* allotropy, allotropism

Alloxur·basen *fpl* alloxuric (*or* purine) bases. **-körper** *m* alloxuric substance, purine

Allozimt·säure *f* allocinnamic acid

allseitig *a* universal; versatile; *adv oft* on all sides, in all directions. **Allseitigkeit** *f* universality; versatility

Allseit·lage *f* ability to present any surface. **a—wendig** *a* multilateral

Allstrom·motor *m* universal (*or* AC-DC) motor

all·stündlich *a* hourly; *adv oft* every hour

Alltag *m* workday; everyday world; daily routine. **alltäglich** *a* 1 daily; (*Med.*) quotidian. 2 commonplace, everyday, routine

all·umfassend *a* all-embracing, universal. **-verbreitet** *a* universal; widespread. **-wissend** *a* all-knowing, omniscient. **-wöchentlich** *a* weekly; *adv oft* every week

Allylen *n* propyne, methyl acetylene

Allylin *n* allyl ether of glycerol

Allyl·jodid *n* allyl iodide. **-rhodanid** *n* allyl thiocyanate. **-senföl** *n* allyl mustard oil, allyl isothyanate. **a—ständig** *a* allylic. **A—verbindung** f allyl(ic) compound

All·zeichner m pantograph. **-zeugung** f pangenesis

allzu adv far too, entirely too; oft as pfx, e.g.: **--weit** much too far

Allzweck(s)·dose f all-purpose container (esp for refrigerators). **-werkstoff** m all-purpose material

Alm f (-en) alpine pasture

Almanach m (-e) almanac

Aloe f aloes. **--auszug** m extract of aloes. **a—haltig** a aloetic. **A—harz** n aloe resin. **-holz** n agalloch. **-mittel** n (Med.) aloetic. **-rot** n aloe red. **-saft** m aloe juice, aloes. **-säure** f aloetic acid. **-tinktur** f tincture of aloes

Aloetin·säure f aloetic acid. **aloetisch** a aloetic

Alpak(k)a n 1 (Metll.) German (or nickel) silver. 2 (Tex.) alpaca

Alpen·mehl n lycopodium powder. **-milch** f alpine milk. **-ranken** fpl (Bot.) bittersweet. **-rose** f rhododendron

alphabetisch a alphabetic(al)

Alpha·gras n alfa grass, esparto. **-milchsäure** f (α-) lactic acid. **-strahl** m alpha ray. **-strahler** m alpha (ray) emitter. **-teilchen** n alpha particle. **-welle** f alpha rhythm. **-zerfall** m alpha decay

Alphina·teilchen n alphina particle (2 protons + 1 neutron)

Alp·käse m Alpine cheese. **-ranken** fpl (Bot.) bittersweet

Alraun m (-e), **Alraune** f (-n) mandrake. **Alraun·wurzel** f mandragora

als cnjc 1 as, like; (esp + past) when; (+ sbjc) as if; (after compar) than cf ANDERS 2. 2 to be: **man empfindet es a. lästig** it is felt to be bothersome cf ERWEISEN 2, HERAUSSTELLEN 2. 3 (after neg) but: **nichts a. Gas** nothing but gas. 4: **als ob** as if. Cf SOWOHL

als·bald adv at once. **-dann** adv then; thereupon

Alse f (-n) shad

also 1 adv & cnjc accordingly, therefore, so; thus; then; that is. 2 intj well

alt a old, aged; ancient cf ÄLTER, ÄLTESTE— **Alter Mann** (Min.) backfill

alt·angesehen a old and respected, vener-

able. **-angesessen** a long-established. **-backen** a stale. **-bekannt** a long-known, well-known. **-bewährt** a long-standing; tried and tested

Alte (adj endgs) m old man; f old woman

alt·ehrwürdig a time-honored, venerable. **A—eisen** n old (or scrap) iron

Alter n age; old age

älter a older; elder(ly) cf ALT

Alterantia npl alteratives

Alteration f (-en) 1 alteration. 2 excitement; consternation. **alterieren** 1 vt alter; disturb, upset. 2: **sich a.** be disturbed, be(come) upset

altern 1 vt age. 2 vi (h/s) age, grow old; deteriorate, decline. **Altern** n aging, senescence

Alternanz f (-en) alternation. **alternieren** vi alternate. **alternierend** p.a alternate, alternating

alters: von (or **seit**) **a. her** from olden times, since time immemorial

Alters· (old) age, senile: **-abstufung** f age gradation. **a—bedingt** a caused by age, age-induced. **A—bestimmung** f (usu Geol.) age determination. **-entartung** f senile deterioration. **-erscheinung** f old-age (or senile) phenomenon, symptom of old age. **-folge** f (order of succession by) seniority; chronological series. **-genosse** m contemporary, person of the same age. **-grenze** f age limit. **-heilkunde** f geriatrics. **-rang** m seniority, priority by age. **a—schwach** a decrepit, age-weakened. **A—schwäche** f decrepitude, weakness of old age. **-stufe** f age level. **-tod** m death of old age. **-veränderung** f senile change, change due to age. **-vorsprung** m head start in age, seniority

Altertum n (..tümer) antiquity; ancient times. **altertümlich** a ancient; old-time; adv in an ancient (or old-time) manner

Altertums·forscher m archeologist. **-forschung** f, **-kunde** f, **-wissenschaft** f archeology

Alterung f aging cf ALTERN

Alterungs·beschleunigung f acceleration of aging. **a—beständig** a age-resistant. **A—erscheinung** f symptom of aging. **-härtung** f age-hardening. **-prognose** f prediction of aging. **-schutzmittel** n age

resister, antiager; (*specif*) anti-oxident.
-verfahren *n* aging procedure (*or* technique). **-vorgang** *m* aging process
älteste *a* oldest, eldest *cf* ALT
alt-gedient *a* veteran. **A—gesell(e)** *m* foreman. **-gestein** *n* country rock. **a—gläubig** *a* orthodox. **A—gold** *n* old gold.
-grad *m* (*Math.*) old (*or* conventional) degree, 1/360 circle. **-gummi** *n* scrap (*or* used) rubber
Althee *f* (-n) **1** marsh mallow (*Althaea officinalis*). **2** (*Pharm.*) = **-wurzel** *f* althaea
alt-hergebracht, -herkömmlich *a* traditional, customary. **A—holozän** *n* Early Holocene. **-holz** *n* mature timber. **a—klug** *a* precocious. **A—kupfer** *n* old (*or* scrap) copper
ältlich *a* elderly, oldish
Alt-malz *n* stored malt. **-material** *n* scrap. **-metall** *n* old (*or* scrap) metal. **a—modisch** *a* old-fashioned. **A—öl** *n* used oil. **-papier** *n* old (used, waste) paper. **-rot** *n* (*Dye.*) Old Red, Turkey Red
Altrot-grundierung *f* (*Dye.*) old red ground. **-verfahren** *n* old red process, Turkey-red dyeing
Alt-schadenwasser *n* (*Pharm.*) yellow mercurial lotion. **-stoff** *m* scrap, waste (products). **a—vulkanisch** *a* old-volcanic. **A—waren** *fpl* junk, scrap goods. **-weibersommer** *m* Indian summer. **-wolle** *f* reclaimed wool. **-zeit** *f* primeval times
Alu· *pfx, short for* Aluminium *cf* ALU-RAHMEN
Aludel-ofen *m* aludel furnace
Alumen *n* alum
alumetieren *vt* aluminum-plate (by spraying)
Aluminat *n* (-e) aluminate. **-lauge** *f* aluminate liquor, (sodium) aluminate solution; (*Paper*) green liquor
alumin(is)ieren *vt* aluminize
Aluminium *n* aluminum, *Brit.* aluminium. **-blech** *n* sheet aluminum; aluminum sheet. **-bor** *n* aluminum boride. **-eisen** *n* ferroaluminum. **-farbe** *f* aluminum paint. **-fluorwasserstoffsäure** *f* fluoaluminic acid, hydrogen hexafluoroaluminate. **-grieß** *m* (coarse) aluminum powder; aluminum shot. **-guß** *m* alumi-

num casting(s), cast aluminum. **a—haltig** *a* containing aluminum. **A—hütte** *f* aluminum reduction plant. **-hydrat** *n* aluminum hydroxide. **-lack** *m* aluminum lake. **-oxid** *n* alumina. **-oxidhydrat** *n* aluminum hydroxide
Alumino-thermie *f* aluminothermic process. **a—thermisch** *a* aluminothermic
Aluminoxid *n* aluminum oxide
Alum(in)o-silikat *n* aluminosilicate
Alundum-ziegel *m* alundum brick
alunitisieren *vt* alunitize
Alu-Rahmen *m* aluminum frame *cf* ALU·
am *contrac* = **an dem** *cf* AN; (+ *superl*): *e.g.*
am längsten (the) longest (*adv or pred a*)
amalgam-bildend *a* amalgam-forming.
amalgamierbar *a* amalgamable. **amagamieren** *vt* & **sich a.** amalgamate
Amalgamier-pfanne *f* amalgamating pan. **-salz** *n* amalgam salt
Amalgamierung *f* (-en) amalgamation
Amalgam(ier)-verfahren *n* amalgamation process. **Amalgamier-werk** *n* amalgamating mill
Amalin-säure *f* amalic acid
Amarant *m* (-e) amaranth. **-holz** *n* purpleheart, purplewood
Amause *f* (-n) **1** imitation jewel. **2** enamel
Amazonen-stein *m* Amazon stone, amazonite
Amber *m* amber: **grauer A.** ambergris
amber-artig *a* amber (-like). **A—fett** *n*, **-harz** *n* = **Amberine** *f* ambrain
Amber-öl *n* oil of amber. **-stoff** *m* ambrain
Amboß *m* (..bosse) anvil; (*Anat.*) incus
Ambra *m, f, n* amber: **gelber A.** yellow (*or* ordinary) amber; **grauer A.** ambergris; **flüssiger A.** liquidambar. **-baum** *m* amber tree. **-fett** *n* ambrain. **-holz** *n* yellow sandalwood. **-öl** *n* oil of amber. **-staude** *f* (*Bot.*) amber plant
Ambrette-moschus *m* ambrette musk. **-öl** *n* ambrette (*or* amberseed) oil
Ambrettol(in)-säure *f* ambrettolic acid
ambulant *a* **1** ambulant, ambulatory. **2** itinerant. **3** outpatient; *adv oft* as an outpatient. **Ambulanz** *f* (-en) **1** outpatient department. **2** ambulance. **ambulatorisch** *a* ambulatory
aMe *abbv* (atomische Masseneinheit) amu (atomic mass unit)

Ameise f (-n) ant
Ameisen· form-, formic: **-aldehyd** m formaldehyde. **-amyläther, -amylester** m amyl formate. **-äther** m, **-naphta** n ethyl formate. **-persäure** f per(oxy)formic acid. **a—sauer** a formate of: **a—es Natrium** sodium formate; **a—es Salz** formate. **A—säure** f formic acid
Ameisensäure·nitril n formonitrile, HCN. **-salz** n formic salt, formate.
Ameisen·sulfonsäure f sulfoformic acid. **-weinäther** m ethyl formate
Amel·korn n emmer. **-mehl** n starch
Amerikaner m (-), **Amerikanerin** f (-nen), **amerikanisch** a American—**a— er Nußbaum** black walnut (tree)
Ametall° n nonmetal
Amiant(h) m amianthus (silky asbestos). **a—·artig** a amianthine
Amid·gruppe f amido group cf AMIDO·
amidieren vt amidate, convert to an amide. **Amidierung** f (-en) amidation
Amido· (as ligand) amido-; (otherwise) amino (obsol for org compds): **-azobenzol** n amidoazobenzene. **-benzol** n amidobenzene. **-essigsäure** f amidoacetic acid. **-kohlensäure** f amidocarbonic acid, carbamic acid. **-säure** f (inorg) amido acid; (org, obsol) amino acid. **-schwefelsäure, sulfo(n)säure** f amidosulfuric acid, sulfamic acid. **-toluol** n amidotoluene, toluidine
Amid·säure f amide acid; (esp in compds) -amic acid. **-stickstoff** m amide nitrogen
amikroskopisch a amicroscopic, submicroscopic
amin·artig a amine-like
aminieren vt aminate, convert into an amine. **Aminierung** f (-en) amination
Amino·benzol n aminobenzene, aniline. **-essigsäure** f α-A. aminoacetic acid, glycine, **-kohlensäure** f carbamic acid. **-phyllin** n theophylline
Aminoplast m (-en, -en) aminoplastic, aminoresin. **-harz** n amino resin. **-kunststoff** m aminoplastic
Aminoquecksilber·chlorid n mercuriammonium chloride, NH_2HgCl. **-chlorür** n aminomercurous (or mercurous ammonium) chloride, NH_2Hg_2Cl
Amino·säure f amino acid. **-säurerest** f amino acid residue. **-verbindung** f amino compound
Amin·oxid n amine oxide
Amino·zucker m amino sugar
Amin·säure f amino acid; amic acid. **-schwarz** n amine black
Ammen·generation f asexual generation. **-zeugung** f asexual reproduction
Ammer f (-n) (Zool.) yellowhammer
Ammiak n ammine
Ammin·verbindung f ammine compound
Ammon m ammonia; (in compds, oft) ammonium: **oxalsaurer A.** ammonium oxalate. **-alaun** m ammonium (or ammonium) alum. **-azotat** n ammonium nitrate. **-formiat** n ammonium formate
Ammoniacum n gum ammoniac
Ammoniak n 1 ammonia; (in old names of salts) ammonium: **salzsaures A.** ammonium chloride. 2 ammoniac. **-alaun** m ammonium (or ammonia) alum
ammoniakalisch a ammoniacal; **a—e Lösung** liquid ammonia
ammoniak·arm a low-ammonia
Ammoniakat n (-e) 1 ammoniate. 2 ammonia addition compound, (specif) ammine
Ammoniak· (oft = AMMONIUM·) **-bestimmung** f ammonia determination. **a— bindend** a combining with ammonia. **A—flüssigkeit** f ammonia water, aqueous ammonia, (Indus.) ammoniacal liquor. **-gummi** n gum ammoniac. **a— haltig** a containing ammonia, ammoniacal. **A—harz** n gum ammoniac. **-kupferchlorür** n ammoniacal cuprous chloride. **-laugung** f ammonia leaching. **-leitung** f 1 conduction of ammonia. 2 ammonia pipe(line). **a—reich** a high-ammonia, rich in ammonium. **A—rest** m ammonia residue (amidogen, NH_2). **-salpeter** m ammonium nitrate. **-soda** f ammonia (or Solvay) soda. **-sodaverfahren** n ammonia-soda (or Solvay) process. **-wäsche** f ammonia scrubbing; ammonia recovery plant. **-wäscher** m ammonia scrubber. **-wasser** n ammonia water; (Indus.) ammonia(cal) liquor. **-weinstein** m potassium ammonium tartrate. **-zusatz** m addition (or admixture) of ammonia
ammonisieren vt ammoniate
Ammonium n ammonium; (Pharm.): **A. bromatum (chloratum, jodatum)** am-

monium bromide (chloride, iodide)
Ammonium·ferrisulfat *n* ferric ammonium sulfate. **-ferrosulfat** *n* ferrous ammonium sulfate. **-platinchlorid** *n* ammonium chloroplatinate. **-rest** *m* ammonium radical. **-salpeter** *m* ammonium nitrate. **-seife** *f* ammonia soap. **-sulfhydrat** *n* ammonium hydrosulfide. **-zinnchlorid** *n* ammonium chlorstannate, pink salt
Ammon· (*in compds, oft* = AMMONIUM·) **-karbonat** *n* ammonium carbonate. **-pulver** *n* explosive ammonium powder (made by evaporating NH_4NO_3 solution with powdered charcoal). **-salpetersprengstoff** *m* ammonium nitrate explosive. **-sulfatsalpeter** *n* mixture of ammonium sulfate and nitrate
Amnesie *f* amnesia
Amnion·flüssigkeit *f* amniotic fluid. **-säure** *f* amniotic acid, allantoin. **-wasser** *n* amniotic fluid
Amöbe *f* (-n) ameba, amoeba
amorph, amorphisch *a* amorphous
amortisieren *vt* amortize
am. P. *abbv* (amerikanisches Patent) Am. Pat. (American patent)
Amp. *abbv* (Ampère) amp (ampere)
Ampel *f* (-n) (signal) light (*or* lamp); (*Pharm.*) ampoule
Ampere·meter *m* or *n* ammeter. **-zahl** *f* amperage
Amperometrie *f* amperometry, dead-stop technique. **amperometrisch** *a* amperometric
Ampfer *m* (*Bot.*) sorrel, dock (*Rumex*)
Amphibie *f* (-n) (*Zool.*) amphibian. **amphibisch** *a* amphibian, amphibious
amphichro(i)tisch *a* amphichro(t)ic
Amphiole *f* (-n) ampoule
amphiprotid *a* amphiprotic
Ampho·gerbstoff *m* amphoteric tannin
Ampholyt *m* (-en, -e *or* -en) ampholyte
amphoter(isch) *a* amphoteric
Ampulle *f* (-n) (*Pharm.*) ampoule; (*Anat.*) ampulla. **Ampullen·fabrikation** *f* manufacture of ampoules. **a—förmig** *a* ampulliform, ampoule-shaped
amputieren *vt* amputate
Amsel *f* (-n) blackbird
Amt *n* (-̈er) **1** office (*both* place *and* position): **ein A. bekleiden** hold (an) office.

2 (government) department, board, council, court. **3** post, position, (official) appointment. **4** employment. **5** duty, function
amtieren *vi* hold office; officiate. **amtierend** *p.a* (*oft*) acting, pro tem(pore)
amtlich *a* official
Amt·mann *m* district administrator
Amts·alter *n* seniority. **-beförderung** *f* promotion to a higher office. **-befugnis** *f*, **-bereich** *m* jurisdiction. **-bericht** *m* official report. **-blatt** *n* official gazette. **-buch** *n* official register. **-genosse** *m* colleague in office. **-geschäfte** *npl* official business (*or* duties). **a—mäßig** *a* official. **A—schimmel** *m* bureaucratic excess(es), red tape: **den A. reiten** reel out the red tape. **-vertreter** *m* deputy. **-verwalter** *m* official in charge, administrator
Amyl·alkohol *m* amyl alcohol
Amylium nitrosum *n* (*Pharm.*) amyl nitrite
Amyl·jodid *n* amyl iodide
amyloklastisch *a* amyloclastic
Amylolyse *f* (-n) amylolysis
Amylo·verfahren *n* (*Alcohol*) amylo process (of saccharification)
Amyl·oxyhydrat *n* amyl alcohol
Amylum *n* starch
an I *prep* **1** at, by, along, against, to. **2** in, on. **3** near, next to. **4** about. **5** with respect to. **6** (in the way) of: **Vorrat an Kohle** supply (in the way) of coal, coal supply. **7** (*Transl. oft varies with verb:* **man denke an** . . . think of . . .).**—an (und für) sich** in itself, in themselves, per se, intrinsically *cf* AM, BIS, ENTLANG, VORBEI. **II** *adv* on, onward, along, up: **von nun an** from now on. **III** *sep pfx* (*attached to verbs*) on, onward, along, up; (*also indicates arrival, starting*) *cf* ANKOMMEN, ANSEHEN, etc
·an *sfx* (*Org. Chem.*) **1** (*saturated hydrocarbons and heterocyclic parent compds.*) -ane: **Methan** methane. **2** (*in other cases*) -an: **Pyran** pyran; **Urethan** urethan(e)
Anacardiaceen·harz *n* Anacardiaceae resin
Anacard(in)säure *f* anacardic acid
anaerob *a* anaerobic. **Anaeroben** *pl* anaerobes, anaerobic bacteria

Anaeroben·kolben *m* anaerobic flask. **-kultur** *f*, **-züchtung** *f* culture of anaerobes

anaerobisch *a* anaerobic

anal. *abbv* (analytisch) anal. (analytic)

Analepticum *n* (..ica), **analeptisch** *a* analeptic, restorative

Analgeticum *n* (..ica) analgesic

analog *a* analogous; **A—es gilt für** . . . the same applies to (*or* for) . . . ; **das eine dem andern a. setzen** draw an analogy between the one and the other

Analogie *f* (-n) analogy: **A. des B mit C** analogy between B and C

Analogie·fall *m* analogous case. **-rechner** *m* analog computer. **—schluß** *m* analogy

Analogon *n* (..ga) analog(ue)

Analog·rechenglied *n* analog computing element

Analphabet *m* (-en, -en) illiterate (person)

Analysator *m* (-en) analyzer. **A.-Nicol** *n* analyzing Nicol (prism)

Analyse *f* (-n) analysis

Analysen· analytical **-befund** *m* analytical finding (*or* result). **-einwa(a)ge** *f* weighed sample for analysis. **a—fein** *a* finely powdered for analysis. **-fertig** *a* ready for analysis. **A—formel** *f* empirical formula. **-gang** *m* course (*or* process) of analysis. **-probe** *f* sample for analysis. **a—rein** *a* analytically pure, (of) analytical grade. **A—substanz** *f* substance to be analyzed. **-waage** *f* analytical balance. **-zeit** *f* analysis time

analysierbar *a* analyzable

analysieren *vt* analyze

Analytik *f* analytics, analytical methods. **Analytiker** *m* (-) analyst. **analytisch** *a* analytic(al)

Anämie *f* anemia

Ananas *f* (-se) **1** pineapple. **2** large strawberry. **-·äther** *m* pineapple essence (ethyl butylate). **-erdbeere** *f* large strawberry. **-essenz** *f* = ÄTHER. **-frucht** *f* (whole) pineapple. **-öl** *n* banana oil (amyl acetate). **-saft** *m* pineapple juice

anaphylaktisch *a* anaphylactic. **Anaphylaxe, Anaphylaxie** *f* anaphylaxis.

an·arbeiten *1* *vt* join (on), attach (to). **2** *vi* (**gegen**) work (against), counteract

an·arten *1* *vt* assimilate. **2** *vi(s)* & **sich a.** assimilate, become assimilated; (+ **an**)

become natural (to) *cf* ANGEARTET

Anästhesie *f* anesthesia. **anästhesieren** *vt* anesthetize. **Anästhesin** *n* ethyl-p-amino-benzoate. **Anästhetikum** *n* (..ika *or* ..iken) anesthetic. **anästhetisch** *a*, **a—es Mittel** anesthetic

Anatas *m* (-e) anatase, octahedrite

an·atmen *vt* breathe (up)on

Anatom *m* (-en, -en) anatomist. **Anatomie** *f* (-n) anatomy. **anatomieren** *vt* dissect, anatomize. **Anatomiker** *m* (-) anatomist. **anatomisch** *a* anatomical

an·ätzen *vt* **1** cauterize. **2** begin to etch (*or* corrode), etch (*or* corrode) slightly. **3** (+ **an**) etch (on, onto). **Anätzung°** *f* cauterization

an·backen* **1** *vt* start to bake (*or* fry), bake (*or* fry) slightly. **2** *vi(s)* (**an**) bake on (to); stick, get stuck (to); cake (on)

an·bahnen **1** *vt* pave the way for, prepare, open up. **2: sich a.** start (to develop), open up. **Anbahnung** *f* (-en) preparation, initiation, first step, opening up

an·ballen *vr*: **sich a.** clot, cake, pack

Anbau° *m* **1** cultivation, culture. **2** addition, annex, new wing. **3** colony. **anbaubar** *a* **1** capable of cultivation. **2** adaptable to extensions (*or* additions). **an·bauen** *I* *vt* **1** cultivate; start to raise (*or* grow). **2** (+ **an**) build (*or* add) on (to). **II** *vi* add on, extend. **III**: **sich a.** settle (down). **Anbauer** *m* (-) cultivator, grower; colonist, settler

Anbau·fähigkeit *n* possibility of cultivation. **-gebiet** *n* area for (new) cultivation

anbaulich *a* **1** capable of cultivation. **2** adaptable to extensions (*or* additions)

Anbau·stoffwechsel *m* anabolism. **-teil** *m* add-on part (*or* unit). **-versuch** *m* (*Agric.*) field trial. **-wert** *m* cultivation value. **a—würdig** *a* worthy of cultivation

Anbeginn *m* beginning *cf* ANFANG, BEGINN

an·behalten* *vt* keep on (wearing)

anbei *adv* enclosed (herewith)

an·beißen* **1** *vt* bite into. **2** *vi* bite, nibble, take the bait

an·beizen *vt* mordant

an·belangen *vt* concern: **was das anbelangt** as far as that is concerned

an·bequemen: sich a. (*dat*) adapt oneself (to)

an·beraumen *vt* appoint, set (date, etc)

anberegt *a* aforementioned

Anbetracht *m* consideration, view: **in A. dessen, (daß)** in view of (the fact) that . . .

an·betreffen* *vt* concern, regard: **was das anbetrifft** as regards that, as for that

an·biegen* *vt* (**an**) attach (to)

an·bieten* 1 *vt* offer. 2: **sich a.** offer oneself; volunteer

Anbildung° *f* new growth

an·binden* 1 *vt* (+ **an**) tie, bind (on, to) *cf* ANGEBUNDEN. 2 *vi* (**mit**) get involved (with)

Anbinde·zettel *m* tag

an·blaken *vt* blacken, smoke(-stain)

an·blasen* *vt* 1 blow (in a blast furnace); blow on (*or* at); seal on (with the blast); blow up, fan (fire). 2 intone (wind instrument). 3 announce, sound (with horns, trumpets). *Cf* ANGEBLASEN

an·bläuen* *vt* blue, tinge with blue

Anblick° *m* sight, aspect, view. **an·blicken** *vt* look at, glance at

an·bluten *vi* (*Colors*) bleed, run

Anbohr·apparat *m* boring apparatus, drilling rig. **an·bohren** *vt* 1 bore, drill (into); tap; perforate. 2 scuttle (ship)

an·bräunen *vt* brown

an·brechen* 1 *vt* break into; start, open. 2 *vi*(s) begin; appear; dawn, break (*as:* day); begin to spoil. **anbrechend** *p.a:* **bei a—em Tag** at daybreak

an·brennen* 1 *vt* set fire to, light; scorch; calcine. 2 *vi*(s) catch fire, begin to burn; burn, get scorched; calcine

anbringbar *a* salable, marketable. **an·bringen*** *vt* 1 bring in; place, put (in, on, down); mount, install, attach. 2 dispose of, find a place for; sell, market. 3 lodge, present (complaints, etc). 4 accuse, report (to authorities). *Cf* ANGEBRACHT. **Anbringe·ort** *m* suitable (*or* appropriate) place

Anbruch° *m* 1 beginning, opening; break (of day)—**A. der Nacht** nightfall. 2 crack, fracture. 3 (*Min.*) first ore; freshly opened lode. 4 decay; decayed wood. **anbrüchig** *a* decaying, rotten, moldy. **Anbrüchigkeit** *f* putrescence, rottenness, moldiness

an·brühen *vt* scald; infuse, steep (in hot water)

Anchovis *f* (-) anchovy

Anchusa·säure *f* anchusic acid (anchusin)

Anciennität *f* seniority

Andacht *f* (-en) devotion. **andächtig** *a* devout, devoted

an·dämmen *vt* dam up

an·dampfen *vt/i* deposit by evaporation

Anda·öl *n* anda oil

an·dauen *vt* predigest; partly digest

an·dauern *vi* last, continue, persist. **andauernd** *p.a* lasting, continual, steady

Andenken *n* (-) memento, souvenir; memory: **zum A.** in memory, as a souvenir

ander . . . *a, pron* other (one), different (one); (*with indef pron*) else: **ein a—er** another (one); **etwas a—es** something else; **nichts a—es (als)** nothing else, (nothing but); **unter a—em** among other things; **einmal über das a—e** over and over, repeatedly; **eins ins a—e gerechnet** counting in the one with the other; **am a—(e)n Tag** next day; **zum a—en** for another thing *cf* ANDERS, ALLES

änderbar *a* alterable, changeable. **Änderbarkeit** *f* alterability, changeability

ander(e)n·falls *adv* otherwise, or else. **-orts** *adv* elsewhere. **-teils** *adv* on the other hand

andermal: ein a. some other time

andern 1 = anderen, *cf* ANDER. 2 (*in compds*) = ANDER(E)N·

ändern 1 *vt* change, alter. 2: **sich ä.** change

anders 1 = anderes *cf* ANDER. 2 *adv* otherwise, different(ly), (in) a different (*or* another) way; (*with indef pron*) else: **jemand a.** someone else; **nicht a.** in no other way; **a. gesagt** in other words; **a. laufen** run a different course; **a. lauten** have a different wording; **a. werden** change; **sich a. besinnen** change one's mind; **wenn a.** if . . . indeed, provided; **wenn a. nicht** unless; **a. als hier** differently from here; **a. als wir es tun** not the way we do it, in a way different from ours

anders·artig *a* different, of a different kind. **-denkend, -gesinnt** *a* of a different opinion; dissenting, dissident. **-wertig** *a* of a different valence. **-wie** *adv* (in) some other way. **-wo** *adv* elsewhere, somewhere else. **-woher** *adv* from somewhere else. **-wohin** *adv* to some other place

anderthalb 1 *num* one and a half. 2 *pfx* sesqui: **-basisch** *a* sesquibasic. **-fach** *a* one and a half times, sesqui-: **a—faches Chlorid (Oxid)** sesquichloride (sesquioxide). **-(fach)kohlensauer** a sesquicarbonate of . . .

Änderung f(-en) alteration, change, variation

Änderungs·geschwindigkeit f rate of change

ander·wärtig a other, further. **-wärts** adv elsewhere. **-weitig** a, adv other(wise), in another way (or place), elsewhere

an·deuten vt indicate, suggest; imply; hint (at). **Andeutung°** f indication, suggestion; implication; hint. **andeutungs·weise** a, adv by way of suggestion (implication, a hint)

an·dicken vt/i begin to thicken (stiffen, set)

Andorn m, **··kraut** n (Bot.) horehound

an·dorren vi(s) dry on (and stick to)

Andrang m crowd; rush; (Med.) congestion. **an·drängen** (oft + **an**) 1 vt push, thrust (against). 2 vi press forward, crowd in; impinge (on, against). 3: **sich a.** press close (to)

andre = **andere** cf ANDER . . .

an·drehen vt turn on; twist on

Andreh·kurbel f (starting) crank

andrer, andres = **anderer, anderes** cf ANDER . . .

an·dringen* vi(s) press forward, crowd in; importune, threaten

Andruck° m pressure; (Print.) proof; (Space Techn.) inertial force. **an·drücken** (oft + **an**) 1 vt press on (to); damage by pressure. 2: **sich a.** press close (to), nestle (against)

an·dunkeln vi darken, (begin to) grow dark

an·dunsten, an·dünsten vi(s) be deposited by evaporation

an·eifern vt urge on

an·eignen vt: **sich etwas a.** acquire (appropriate, assimilate) sthg (for oneself). **Aneignung** f(-en) acquisition; appropriation; assimilation

aneinander adv & sep pfx on (at, to, against) each other (or one another) cf AN; together: **·fügen** vt join. **-geflanscht** p.a joined by flanges. **·grenzen** vi adjoin, come into (or be in) contact. **·ketten** vt chain together, link. **A—kettung** f linkage. **a—·lagern** vt stack together. **A—**

lagerung° f juxtaposition, stacking. **a—·liegen** vi lie side by side, be in juxtaposition. **·prallen** vi(s) collide. **·reihen** vt line up, align; arrange in a row; string together. **·schließen*** vt connect, join. **·schmelzen*** vt & **sich a.** fuse (melt, weld) together. **·schmiegen: sich a.** press close together, cling together. **·setzen** vt put side by side; bring into contact. **·stellen** vt place together (or side by side)

an·ekeln vt disgust, nauseate

anelastisch a inelastic

Anelland n (-en) anellated ring

anellieren vt anellate, join by "orthofusion" (e.g. two rings in naphthalene); **anelliertes Derivat** fused-ring derivative. **Anellierung** f (-en) anellation, (ring) fusion

Anemonin·säure f anemoninic acid

Anemon·kampfer m anemonin. **-säure** f anemonic acid

Anemostat m (-e) diffuser

an·erbieten* 1 vt offer, tender. 2: **sich a.** volunteer, offer (to do)

anerboten offered, etc: pp of ANERBIETEN

anerk. abbv = **anerkannt** recognized: pp of ANERKENNEN. **anerkanntermaßen** adv admittedly

anerkennbar a recognizable. **an·erkennen*** vt recognize; acknowledge; admit. **anerkennens·wert** a worthy of recognition. **Anerkennung°** f recognition, acknowledgment; appreciation. **anerkennungs·wert** a worthy of recognition

Aneth n dill. **Anethol** n anethole, anise camphor

Aneurin n thiamine

an·fächeln vt fan

an·fachen vt fan, blow (into a flame); stir up, arouse; incite; kindle. **Anfachung** f fanning; arousing; incitement; kindling.

Anfachungs·stadium n initial stage

an·fahren* I vt 1 bring (carry, drive) up (a load, etc). 2 run into; head for. 3 start (running) (a machine). 4 attack, tear into. II vi(s) 5 arrive, drive up. 6 start up. 7 (Min.) go down. **Anfahrt°** f 1 arrival, approach. 2 trip, ride. 3 start

Anfall° m 1 amount formed (or collected); yield; usu pl: **Anfälle** revenue. 2 (incl Med.) attack, fit. 3 (Min.) prop, stay. **an·fallen*** I vt 1 attack. II vi(s) 2 accrue, ac-

cumulate. 3 be obtained, be yielded; result; fall due. 4 be necessary (*or* involved). 5 (*Dye.*) go on. **anfallend** *p.d* 1 accruing, resulting. 2 aggressive
anfällig *a* 1 susceptible, disease-prone. 2: **a. werden** become available. **Anfälligkeit** *f* susceptibility
Anfang° *m* beginning, start; origin: **gleich zu A.** right at the beginning; **von A. an** from the (very) beginning; **es steckt in den A⸚en** it is in its initial stage. **anfangen*** *vt/i* 1 begin, start. 2 do, go about.
Anfänger *m* (-) beginner. **anfänglich** *a* initial, original; *adv oft* = **anfangs** *adv* at first, originally, in the beginning
Anfangs- initial, starting: **-bad** *n* starting both. **-bahn** *f* initial orbit (*or* path). **-buchstabe** *m* initial (letter). **-glied** *n* first member (of a series). **-gründe** *mpl* elements, rudiments. **-punkt** *m* starting point, origin. **-stadium** *n* initial stage. **-teil** *m* initial part; base; lamella. **-wertaufgabe** *f* initial-value problem. **-zustand** *m* initial state (*or* condition)
anfärbbar *a* colorable, dyeable. **Anfärbbarkeit** *f* coloring (*or* dyeing) capacity
Anfärbe-methode *f* colorimetric method
an-färben *vt* color, dye, stain; tinge. **Anfärbe-vermögen** *n* coloring (*or* dyeing) power. **Anfärbung** *f* (-en) coloring, dyeing, staining
an-fassen 1 *vt* grasp, take hold of; tackle, go about. 2: **sich** (e.g. **weich**) **a.** feel (soft, etc) to the touch
an-faulen *vi*(s) begin to rot *cf* ANGEFAULT. **Anfaulen** *n* incipient rot. **anfaulend** *p.a* rotting, putrescent
anfechtbar *a* contestable, disputable
an-fechten* *vt* 1 contest, dispute; impugn. 2 trouble, concern. 3 tempt. **Anfechtung** *f* (-en) disputation; impugnment; temptation
an-fertigen *vt* make, manufacture, produce; prepare. **Anfertigung°** *f* making, manufacture, production; preparation
an-fesseln *vt* (*oft* + **an**) chain, shackle (to)
an-fetten *vt* oil, lubricate; (mix with) grease
an-feuchten *vt* moisten, dampen. **Anfeuchter** *m* (-) moistener; (*Brew.*) sparger. **Anfeuchtung** *f* (-en) moistening, dampening

an-feuern *vt* 1 light (an oven, etc.), fire. 2 prime (a rocket). 3 animate, arouse, excite. **Anfeuerung°** *f* lighting, firing; priming; animation, arousal, excitation
an-firnissen *vt* varnish, give a coat of varnish
an-flammen *vt* ignite, inflame; rouse
an-flechten* *vt* (**an**) weave (into, onto)
an-fliegen* **I** *vt* 1 fly to, land at (a place). 2 come over (*as:* a mood). **II** *vi*(s) 3 come flying, arrive by air. 4 (*Dye.*) rush on; effloresce. 5 (*dat*) come over. *Cf* ANGEFLOGEN
Anflug° *m* 1 efflorescence, incrustation, bloom, film. 2 tinge; trace, hint, suggestion. 3 approach (flight), arrival (by air). 4 young growth (from windborne seeds)
Anfluß° *m* alluvium; afflux, onflow
an-fluten *vi*(s) flow in, come flooding in
an-fordern *vt* demand, require, order; request. **Anforderung°** *f* demand, requirement, order; request
Anfr. *abbv* = **Anfrage°** *f* inquiry, question. **an-fragen** *vi* inquire, ask
Anfraß° *m* corrosion, pitting; erosion
an-fressen* *vt* corrode; eat into, attack, gnaw at; erode *cf* ANGEFRESSEN. **anfressend** *p.a* (*oft*) corrosive. **Anfressung** *f* (-en) corrosion; erosion; attack (by vermin, etc) *cf* PUNKTFÖRMIG. **anfressungs-beständig** *a* corrosion- (*or* erosion-) resistant
an-frieren* *vi*(s) (**an**) freeze on (*or* fast) (to)
an-frischen *vt* freshen (up); reduce (oxides); refine (metals); varnish (paintings). **Anfrischer** *m* (-) refiner (of metals). **Anfrisch-herd, -ofen** *m* refining furnace. **Anfrischung** *f* freshening up; reduction; refining; varnishing. **Anfrischungs-ofen** *m* refining furnace
Anfuge° *f* enclosure; insertion. **an-fugen** *vt* (**an**) fit, join (onto)
an-fügen *vt* (**an**) join, attach, add (to). **Anfügung°** *f* joining; attachment, addition
an-fühlen 1 *vt* feel, touch, handle. 2: **sich a.** feel (to the touch)
Anfuhr *f* bringing, carting, delivery
anführbar *a* quotable, adducible
an-führen *vt* 1 lead, guide; train. 2 quote, cite; adduce. 3 cheat, swindle. **Anführer°** *m* 1 leader; instructor. 2 ringleader. 3 swindler. **Anführung°** *f* 1 leadership.

2 quotation, citation. 3 cheating. **Anführungs·zeichen** *npl* quotation marks
an·füllen 1 *vt & sich a.* fill (up). 2 prime, charge; cram, stuff. **Anfüllung°** *f* filling (up); priming; charging; cramming, stuffing

ang. *abbv* (angewandt) applied

Ang. *abbv* (Angebot) offer; (*Com.*) supply

Angabe° *f* 1 statement, specification; report(ing); indication: **ohne A. von ...** without stating (*or* specifying) ... ; **unter A. von ...** stating, specifying ... 2 *pl* **Angaben** *oft* information, data, particulars; instructions, directions. *Cf* ANGEBEN. **Angaben·verarbeitung** *f* data processing

angängig, angänglich *a* feasible; permissible

Angärung° *f* prefermentation

angeartet *p.a* innate, inborn. *See also* ANARTEN

Angeb. *abbv* (Angebot) offer; (*Com.*) supply

angebbar *a* specifiable; assignable *cf* ANGEBEN

an·geben* I *vt* 1 state, specify. 2 report; indicate. 3 set, fix. 4 denounce. II *vi* boast. **Angeber°** *m* 1 stater. 2 denunciator, informer. 3 braggart. **angeberisch** *a* denunciatory; boasting

angebissen bitten, etc: *pp of* ANBEISSEN

angeblasen 1 blown: *pp of* ANBLASEN. 2 *p.a:* **wie a. kommen** come out of the blue

angeblich *a* alleged; supposed, ostensible; nominal; *as adv usu* = **angeblicher·maßen** *adv* allegedly; as stated

angebogen attached: *pp of* ANBIEGEN

angeboren *p.a* inborn, innate, congenital

Angebot° *n* offer, bid; quotation—**A. und Nachfrage** supply and demand

angeboten offered: *pp of* ANBIETEN

angebracht *p.a* fitting, appropriate, suitable; timely. *See also* ANBRINGEN

angebrannt scorched: *pp of* ANBRENNEN

angebrochen 1 opened, etc: *pp of* ANBRECHEN. 2 *p.a* partially spoiled; broken open

angebunden 1 tied: *pp of* ANBINDEN. 2 *p.a* busy, tied up; tied down. 3 *pred a:* **kurz a. sein** (**mit**) be abrupt (with)

an·gedeihen* *vt:* **einem etwas a. lassen** grant a person sthg, bestow sthg on a psn

Angedenken *n* remembrance, recollection

angedrungen importuned: *pp of* ANDRINGEN

angefault *p.a* rotten, decayed. *See also* ANFAULEN

angeflanscht *p.a* flange-mounted

angeflochten woven: *pp of* ANFLECHTEN

angeflogen 1 *pp of* ANFLIEGEN: **a. kommen** come on suddenly (*or* without effort), be windborne. 2 *p.a* incrusted; spontaneous (growth), windborne

angefochten disputed: *pp of* ANFECHTEN

angefroren frozen on: *pp of* ANFRIEREN

angegangen 1 concerned, etc: *pp of* ANGEHEN. 2 *p.a* (*Food, etc*) slightly rotten, tainted, turned

angeglichen adapted: *pp of* ANGLEICHEN

angeglommen glowed: *pp of* ANGLIMMEN

angeglüht *p.a* red-hot *cf* ANGLÜHEN

angegossen 1 cast on: *pp of* ANGIESSEN. 2 *p.a:* **es sitzt wie a.** it fits like a glove

angegriffen 1 attacked: *pp of* ANGREIFEN. 2 *p.a* (*usu*) exhausted, worn out. **Angegriffenheit** *f* exhaustion; fatigue

Angehänge *n* (-) pendant; appendage

Angehäufe *n* conglomerate; aggregate, heap

an·gehen* I *vt* 1 attack, tackle; take hold of; take up (the fight with), combat; approach (a psn). 2 (+ *indef pron*) concern: **es geht ihn nichts an** it doesn't concern him. II *vi*(s) 3 start (up), begin; go on, light up (*as:* fire, light). 4 go on (*as:* glove); be absorbed (*as:* dyes). 5 rise. 6 turn, go bad (*as:* meat). 7 be feasible, be possible, do: **es geht an** it will do well, it will be all right. *Cf* ANGEGANGEN. **angehend** *p.a* aspiring; incipient. **angehends** *adv* at first, in the beginning

angehoben raised: *pp of* ANHEBEN

an·gehören *vi* (*dat*) belong (to), be part (of); be related (to). **angehörig** *a* (*dat*) belonging, related (to). **Angehörige** *m, f* (*adj endgs*) member; relative; adherent; inhabitant

Angeklagte *m, f* (*adj endgs*) accused, defendant

angeklungen been heard: *pp of* ANKLINGEN

angekränkelt *p.a* sickly, unwholesome, diseased; tainted

angekrochen crept: *pp of* ANKRIECHEN

Angel *f* (-n) **1** hinge; pivot. **2** fishing rod, fishhook

an·gelangen *vi*(s) arrive

Angeld° *n* down payment

angelegen *p.a* important, of consequence—**wir lassen es uns a. sein** we take an active interest in it, we make it our concern *cf* ANLIEGEN. **Angelegenheit** *f* (-en) affair, matter; concern, business. **angelegentlich** *a* pressing, urgent; earnest, serious—**sich a. bemühen** do one's utmost

angelegt *p.a & pp:* **er hat es a. (auf)** he has it in (for); **es ist darauf a. (zu)** it is intended (to), it is so arranged (that)

angel·förmig *a* hooked, barbed. **A—gerät** *n* fishing tackle. **-haken** *m* fishhook

Angelica·lacton *n* angelica lactone. **-säure** *f* angelic acid

Angel·kreis *m* polar circle. **-leine** *f* fishline

angeln 1 *vt* hook, trap, fish. **2** *vi* (*oft* + **nach**) fish, angle (for)

an·geloben *vt* vow

Angel·punkt *m* pivotal point. **-rute** *f* fishing rod

angelsächsisch *a* Anglo-Saxon

Angel·schnur *f* fishline. **-stern** *m* pole star

angem. *abbv* (angemeldet) (*Pat.*) applied for

angemessen 1 measured: *pp of* ANMESSEN. **2** *p.a* (*usu* + *dat*) appropriate, suitable (to, for), adequate; in conformity (with). **Angemessenheit** *f* appropriateness, suitability; adequacy

angenähert *p.a* approximate *cf* ANNÄHERN

angenehm *a* pleasant, agreeable

angenommen 1 taken on: *pp of* ANNEHMEN. **2** *p.a* assumed (name, conclusion), adopted (child). **3** (+ *clause*) **a., es ist wahr** assuming it is true

angepaßt *p.a* appropriate. *See also* ANPASSEN. **Angepaßtheit** *f* (-en) adaptation, adjustment

angepriesen praised: *pp of* ANPREISEN

angequollen sprouted: *pp of* ANQUELLEN

Anger *m* (-) pasture; grassy lane

angerannt run into: *pp of* ANRENNEN

angeraucht *p.a* smoke-blackened *cf* ANRAUCHEN

angeregt *p.a* animated, lively *see also* ANREGEN

angerieben rubbed, etc: *pp of* ANREIBEN

angerissen 1 torn, etc: *pp of* ANREISSEN. **2** *p.a* opened, half-begun

angerochen sniffed at: *pp of* ANRIECHEN

angeschieden deposited: *pp of* ANSCHEIDEN

angeschienen shone on: *pp of* ANSCHEINEN

angeschimmelt *p.a* slightly moldy *cf* ANSCHIMMELN

angeschliffen ground *pp of* ANSCHLEIFEN

angeschlossen 1 attached, etc: *pp of* ANSCHLIESSEN. **2** *p.a* associated, affiliated; (*Radio, TV*) hooked up

angeschmolzen fused on: *pp of* ANSCHMELZEN *vi*

angeschnitten cut into: *pp of* ANSCHNEIDEN

angeschoben 1 pushed etc: *pp of* ANSCHIEBEN. **2** *p.a* batch-baked (bread)

angeschossen shot at: *pp. of* ANSCHIESSEN

angeschrieben 1 written, etc: *pp of* ANSCHREIBEN. **2** *p.a:* **ist (gut, schlecht) a. bei** . . . has a (good, bad) reputation with . . .

angeschwemmt *p.a* alluvial; settled. *See also* ANSCHWEMMEN

angeschwollen swollen: *pp of* ANSCHWELLEN

angeschwungen started swinging: *pp of* ANSCHWINGEN

angesehen 1 looked at: *pp of* ANSEHEN. **2** *p.a* respected, esteemed, (highly) regarded

angesessen 1 clung: *pp of* ANSITZEN. **2** *p.a* settled (as a resident)

Angesicht° *n* **1** face: **von A. zu A.** face to face. **2** brow. **3:** **imA.** (+ *genit*) in the presence (of). **angesichts** *prep* (*genit*) in view (of); in the presence (of); within sight (of)

angesogen absorbed: *pp of* ANSAUGEN

angesotten boiled: *pp of* ANSIEDEN

angespannt *p.a* taut, tense; strained; utmost, rapt (attention) *cf* ANSPANNEN

angesponnen plotted: *pp of* ANSPINNEN

angesprochen addressed: *pp of* ANSPRECHEN

angesprungen reacted *pp of* ANSPRINGEN

angestammt *p.a* hereditary

angestanden delayed, etc: *pp of* ANSTEHEN

angestaut *p.a* pent-up; *see also* ANSTAUEN

Angestellte *m, f (adj endgs)* (salaried) employee (in private sector) *cf* ANSTELLEN 3

angestiegen risen: *pp of* ANSTEIGEN

angestochen 1 pierced: *pp of* ANSTECHEN. **2** *p.a* worm-eaten

angestorben devolved: *pp of* ANSTERBEN

angestoßen 1 collided, etc: *pp of* ANSTOSSEN. **2** *p.a* damaged; bruised (fruit); chipped (glass)

angestrengt *p.a* strenuous, arduous *see also* ANSTRENGEN

angestrichen painted: *pp of* ANSTREICHEN

angetan 1 put on, etc: *pp of* ANTUN —**das hat es ihnen a.** that has captivated them. **2** *p.a* dressed up; (+ **von**) impressed (by); **danach a. zu** likely to

angetrieben driven: *pp of* ANTREIBEN

angetroffen found, etc: *pp of* ANTREFFEN

angew. *abbv* (angewandt) applied

angewachsen 1 grown on: *pp of* ANWACHSEN. **2** *p.a* rooted (to the spot), (*Bot.*) adnate

angewandt 1 used: *pp of* ANWENDEN. **2** *p.a:* **a—e Wissenschaften** applied sciences

angewiesen 1 instructed: *pp of* ANWEISEN. **2** *p.a:* **a. sein (auf)** be dependent (on)

angewöhnen 1 *vt:* **einem etwas a.** get sbdy accustomed to sthg; (+ *dat rflx*): **sich etwas a.** get accustomed (*or* used) to sthg. **Angewohnheit°** *f* habit. **Angewöhnung** *f* habituation

angeworben solicited: *pp of* ANWERBEN

angeworfen thrown, etc: *pp of* ANWERFEN

angewurzelt *p.a* rooted (to the) spot *cf* ANWURZELN

angezeigt *p.a* advisable, (*esp Med.*) indicated. *See also* ANZEIGEN

angezogen 1 attracted, etc: *pp of* ANZIEHEN. **2** *p.a* slack (malt)

an·gießen* *vt* **1** pour on, pour in. **2** water, sprinkle; baste. **3** (*Metll.*) (**an**) cast on (to), cast integral (with). **4** (*Ceram.*) match, conform (colors). *Cf* ANGEGOSSEN

an·gilben *vi(s)* (turn) yellow

an·glänzen *vt* shine (up)on

an·gleichen* 1 *vt* (*d.i.o*) & **sich a.** (*dat*) adapt, adjust, assimilate (to). **2** *vt* bring closer (to), (try to) match (with). **3: sich a.** come closer (to). **Angleichung°** *f* adapta-

tion, adjustment, assimilation

an·gliedern 1 *vt* (*usu* + **an**) incorporate (into); affiliate, link (with); annex (to). **2: sich a.** (**an**) be incorporated (into), be affiliated (with); be annexed (to). **Angliederung°** *f* incorporation, affiliation, linkage; annexation

an·glimmen* 1 *vt* cause to smolder. **2** *vi(s)* begin to glow (*or* smolder); ignite slowly

an·glühen 1 *vt* heat to glowing, heat red hot; illuminate with a glow; mull (wine). **2** *vi(s)* start to glow *cf* ANGEGLÜHT

angreifbar *a* attackable, open to attack, assailable, vulnerable; corrodible. **Angreifbarkeit** *f* vulnerability (to attack), assailability; corrodibility

an·greifen* I *vt* **1** attack, assault, storm; contest; tackle. **2** go about (a matter). **3** touch; affect. **4** strain, wear out, exhaust. **5** corrode. **6** break (*or* dip) into (supplies). **7** embezzle. **II** *vi* **8** attack—**mit a.** lend a hand. **III: sich a. 9** (**rauh, glatt**) feel (rough, smooth). **angreifend** *p.a* aggressive, offensive; corrosive; exhausting. **Angreifer** *m* (-) **1** attacker, aggressor, assailant. **2** grip, holder. **angreiferisch** *a* aggressive

an·grenzen *vi* (**an**) border (on), adjoin. **angrenzend** *p.a* bordering, adjoining, neighboring. **Angrenzer** *m* (-) neighbor

Angriff° *m* attack, assault; offensive—**in A. nehmen** undertake, tackle (a project); **zum A. vorgehen** take the offensive

Angriffs·einrichtung *f* offensive set-up. **-fähigkeit** *f* offensive capability. **-geist** *m* aggressiveness, militancy. **-geschwindigkeit** *f* rate of attack. **-linie** *f* line of attack. **a—lustig** *a* aggressive. **A—mittel** *n* attacking agent; corrosive. **-punkt** *m*, **-stelle** *f* point of attack (*or* application). **-stellung** *f* aggressive (*or* offensive) stance (*or* posture). **a—tüchtig** *a* pathogenic, virulent. **-weise** *adv* aggressively; by way of attack. **A—weise** *f* mode of attack

angst *pred a:* **ihnen ist a. vor** . . . they are afraid of . . . ; **ihnen wird a.** they are getting frightened. **Angst** *f* (ˉe) fear; anxiety: **A. haben vor** . . . be afraid of . . . ; **A. bekommen** get frightened; **A. machen** (*dat*) frighten (a psn)

angst·erfüllt *a* fearful. **A—gefühl** *n* feeling of anxiety, trepidation

ängstigen 1 *vt* frighten, alarm; worry. **2: sich ä.** be frightened (alarmed; worried)

ängstlich *a* afraid, alarmed; anxious, worried; timid, nervous. **Ängstlichkeit** *f* fear, alarm, anxiety, worry; timidity, nervousness

Ångström·einheit *f* Ångström unit

Angst·schweiß *m* cold sweat. **-zustand** *m* state of anxiety

Anguß° *m* 1 pouring on; watering, sprinkling, basting; casting on, integral casting. 2 (*Metll., Print.*) waste material on casting, (*specif*) deadhead, feedhead, sprue, lug, flash. 3 (*Ceram.*) coating, engobe. *Cf* ANGIESSEN. **--farbe** *f* (*Ceram.*) colored coating clay, engobe color. **-ver·teiler** *m* (*Molding*) runner

anh. *abbv* (anhydrisch) anhydrous

Anh. *abbv* (Anhang) app. (appendix)

an·haben* *vt* 1 have on, be wearing. 2 (*d.i.o*): **das kann ihnen nichts a.** that can not harm them

an·hacken *vt* hack into, damage by hacking

an·haften *vi* (*usu* + *dat*) adhere, stick, cling (to); be inherent (in). **Anhaften** *n* adhesion. **anhaftend** *p.a* adhesive; clinging; inherent

an·häkeln *vt* (*oft* + **an**) crochet on, hook on (to)

an·haken I *vt* 1 hook (on), fasten with a hook. 2 check off, mark off (on a list). II: **sich a.** hook on

Anhalt° *m*1 point of support, foothold; clue; (supporting) evidence. 2 stop, halt; stopover

an·halten* I *vt* 1 stop, bring to a halt, detain; hold (back), check. 2 (**an**) hold up (to), hold (against). 3 (**zu**) encourage, urge (to). II *vi* 4 stop, come to a halt, pause. 5 (**um**) apply (for). 6 last, keep up, continue. **Anhalten** *n* 1 stop, halt: **zum A. bringen** bring to a halt. 2 detention, retention, checking, holding (back). 3 encouragement, urging. 4 application. 5 continuance, persistence. **anhaltend** *p.a* lasting, continuous, ceaseless

Anhalte·potential *n* stopping potential.

-punkt *m*, **-stelle** *f* 1 (*Transp.*) stop. 2 point of support, etc = ANHALTSPUNKT. 3 fixed point

Anhalts·punkt *m* point of support; clue, lead, something to go on; evidence, proof; reference point; (*Mach.*) fulcrum

Anhaltungs·mittel *n* astringent

anhand, an Hand *adv* (+ **von**), *prep* (*genit*) with the aid (of), on the basis (of)

Anhang° *m* 1 appendix, addendum, supplement. 2 following, (circle of) followers, adherents. 3 family, relatives; dependents

Anhänge·adresse *f* address tag

an·hängen A (*reg*) I *vt* (*oft* + **an**) 1 hang on, couple on, hitch on (to). 2 append, add, attach (to). 3 (*Tlph.*): **den Hörer a.** hang up. 4 (*d.i.o*) impose, inflict (on), burden, saddle (with); palm off (on). II: **sich a.** 5 (**an** or *dat*) hang on, cling; attach oneself (to), force oneself (on). B (*reg or **) III *vi* (*usu* + *dat*) 6 be attached, cling (to); **es hängt ihnen an** they are burdened with it. 7 be an adherent (or a follower) of. 8 (*Tlph.*) hang up. **anhängend** *p.a* hanging; appended, attached. **Anhänger** *m* (-) 1 trailer. 2 pendant; tag; loop (on a coat, etc). 3 adherent, follower. 4 (*Bot.*) scion, graft. **Anhängerschaft** *f* following

Anhänge·wagen *m* trailer. **-zettel** *m* tag

anhängig *a* 1 attached, adherent, annexed; dependent. 2 pending—**a. machen** file (in court). 3 (*dat*) devoted (to)

anhänglich *a* devoted, faithful; (emotionally) attached. **Anhänglichkeit** *f* devotion, faithfulness, attachment

Anhängsel *n* (-) pendant; appendage, adjunct, attachment

Anhangs·gebilde *n* appendage. **-kraft** *f* adhesion, adhesive force. **a—weise** *adv* as a supplement, as an appendix

anharmonisch *a* anharmonic. **Anharmonizität** *f* anharmonicity. **Anharmonizitäts·glied** *n* anharmonicity term

Anhauch *m* breath; tinge. **an·hauchen** *vt* breathe on, tinge

an·hauen* *vt* chop (or cut) into, start cutting (mowing, chopping)

an·häufeln *vt* ridge, pile up earth around (plants)

an·häufen 1 *vt* & **sich a.** pile up, accumu-

late, aggregate, agglomerate. **2** *vt* amass.
Anhäufer *m* (-) accumulator. **Anhäufung** *f* (-en) pile-up, accumulation, aggregation, agglomeration
an·heben* **1** *vt* raise, lift, jack up. **2** *vt/i* start, begin
an·heften *vt* attach, fasten; *(specif)* staple (stitch, clip, tack) on; put up. **Anheftung** *f* attachment, fastening, stapling, etc
Anheftungs·linie *f* line of junction. **-punkt** *m*, **-stelle** *f* point of attachment
an·heilen **1** *vt* cause to heal up. **2** *vi*(s) heal up; heal together—**es heilt wieder an** it grows back on
anheim·fallen* *vi*(s) *(dat)* fall prey (to); devolve, fall (on)
anheim·geben* **1** *vt* (*d.i.o*) leave (to), leave in the hands (of). **2: sich a.** (*dat*) entrust oneself (to)
anheim·stellen* *vt* (*d.i.o*) leave, entrust (to)
anheischig: sich a. machen (**zu**) offer (to), obligate oneself (to)
an·heizen **1** *vt* light, fire, start (a fire in) (stove, boiler); kindle, ignite. **2** *vi* light the fire, turn on the heat. **Anheiz·periode, -zeit** *f* heating-up period, warm-up time
Anhieb *m:* **auf (den ersten) A.** at the first stroke (*or* attempt), right off
Anhöhe° *f* hill, elevation
an·holen *vt* fetch, get; (*Rope*) draw tight, haul up
an·hören **1** *vt* (*oft + dat rfix*) hear, listen to; (*d.i.o*) tell (*or* detect) by listening (to)—**das läßt sich a.** that's worth listening to. **2: sich** (e.g. **gut**) **a.** sound (good)
Anhub *m* lift. **Anhubs·moment** *n* upstroke, initial stroke
anhyd. *abbv* (anhydrisch) anhydrous
Anhydrämie *f* anhydremia
anhydrieren **1** *vt* dehydrate, make anhydrous. **2: sich a.** become anhydrous. **anhydrisch** *a* anhydrous. **anhydrisieren** *v* dehydrate, etc = ANHYDRIEREN. **Anhydrierungs·mittel** *n* dehydrating agent
Anhydro·säure *f* anhydro acid (e.g. pyrosulfuric)
Anhydroxid *n* (-e) anhydride, anhydrous oxide

Anil *n* indigo
Anilin· aniline: **-blau** *n* aniline blue. **-chlorhydrat** *n* aniline hydrochloride. **-einbadschwarz** *n* one-dip aniline black. **-klotzschwarz** *n* stop-padded aniline black. **-öl** *n* aniline oil, technical aniline. **-salz** *n* aniline salt (*or* hydrochloride). **-sulfosäure** *f* aniline sulfonic (*or* aminobenzenesulfonic) acid
Anilismus *m* anilism, anilinism
Anil·säure *f* anilic acid
Animalien *npl* animal bodies; animal food. **animalisch** *a* animal, beastly. **animalisieren** *vt* (*Tex.*) animalize
Animé·gummi *n*, **-harz** *n* (gum) animé
animieren *vt* animate, enliven; encourage. **animiert** *p.a* animated, lively
Animosität *f* (-en) animosity
an·impfen *vt* (*Crystals*) seed
Animus *m* animus; intention; idea, inkling
Anionen·austauscher *m* anion exchanger
anionotrop *a* anionotropic. **Anionotropie** *f* anionotropy
Anis *m* anise, aniseed. **a—·ähnlich** *a* anise-like. **A—branntwein** *m* anisette. **-geist** *m* (*Pharm.*) spirit of anise; anisette. **-kampfer** *m* anise camphor, anethole. **-likör** *m* anisette. **-öl** *n* anise (*or* aniseed) oil
anisotrop *a* anisotropic
Anis·saat *f*, **-same(n)** *m* aniseed, anise seed. **-säure** *f* anisic acid. **-wasser** *n* anisette. **-zucker** *m* sugared anise
an·jagen *vt* drive on (to greater speed)
anjetzt *adv* now
an·kalken *vt* limewash, whitewash
an·kämpfen *vi* (**gegen**) struggle (against)
Ankauf° *m* purchase, acquisition; buying (up). **an·kaufen** **1** *vt* buy (up). **2: sich a.** buy land (and settle on it). **Ankaufs·preis** *m* purchase price—**zum A.** at cost
Anker *m* (-) anchor; (*Elec.*) armature, rotor. **-·geld** *n* anchorage (fee) **-grund** *m* anchorage (area). **-kern** *m* armature
ankern **1** *vt* anchor, moor. **2** *vi* drop anchor; ride at anchor
Anker·platte *f* anchor plate. **-platz** *m* anchorage. **-spannung** *f* armature voltage. **-spule** *f* armature coil. **-stelle** *f* anchorage. **-strom** *m* armature current.

-welle *f* armature shaft. **-wicklung** *f* armature winding

an·ketten *vt* chain (up); tie up

an·kitten *vt* cement (on); putty (on)

Anklage° *f* accusation, charge; indictment; impeachment. **an·klagen** *vt* accuse, charge; indict; impeach

an·klammern 1 *vt* (**an**) clip, clamp, pin, fasten (on, onto). **2: sich a.** (**an**) cling (to), take hold (of)

Anklang° *m* 1 accord, harmony. 2 (**an**) echo, reminiscence (of). **3: A. finden** (**bei**) appeal (to), find favor (with), be(come) popular (with)

an·kleben 1 *vt* (*oft* + **an** or + *dat*) stick (paste, glue) on (to); agglutinate. 2 *vi* (h) (*dat*) adhere, cling (to). 3 *vi*(s) (**an**) stick, get stuck (to). **4: sich a.** (**an**) cling (to), **anklebend** *p.a* adhesive, agglutinative. **Ankleber** *m* (-) bonding agent, agglutinant

an·kleiden *vt* & **sich a.** dress

an·kleistern *vt* (**an**) paste on, stick on (to)

an·klemmen *vt* (**an**) clamp on, clip on (to)

an·klingen* *vi* be heard, be audible, be detectable; (**an**) be reminiscent (of)

an·klopfen 1 *vt* (*Med.*) percuss, sound. 2 *vi* (*oft* + **bei**) knock (at sbdy's door)

an·knacksen *vt* crack, make a crack in

an·knipsen *vt* snap (switch, turn) on

an·knöpfen *vt* (**an**) button on (to)

an·knüpfen 1 *vt* start, open, enter into; (*oft* + **an**) tie on, link, connect (to)—**wieder a.** resume. 2 *vi* (**an**) tie in (with), refer (to); (**mit**) make contact (with). **Anknüpfung** *f* (-en) starting, opening; connection, tie-in; reference; contact. **An·knüpfungs·punkt** *m* starting point; point of contact; point of reference, tie-in

an·kochen *vt* bring to a boil, parboil

an·kohlen *vt* char (partially)

an·kommen* *vi*(s) 1 arrive; approach, come in. 2 (**gegen**) cope (with), hold one's own (against). 3 (**bei**) find a job (with); find favor (with); win (the) approval (of)—**übel a.** (**bei**) be badly received (by). 4 (*dat*) get at (a psn). 5 (*impers,* + **auf**) depend (on), be a matter (of); be important, matter: **es kommt darauf an, ob** . . . it depends on (*or* it is a matter of) whether . . . ; **darauf kommt es** (**nicht**)

an that is (not) what matters; **wir lassen es darauf a.** we will risk it. 6 (+ *dat* or *accus, &* + *adv,* e.g. **hart**) affect, strike (a psn) (severely). 7 (*accus*) come over, overcome, overtake

ankömmlich *a* accessible

Ankömmling *m* (-e) newcomer, new arrival; (*Com.*) new article (*or* product)

an·koppeln *vt* (*oft* + **an**) leash; couple on (to)

an·körnen *vt* 1 mark with a center punch. 2 lure, bait

an·kreiden *vt* mark with chalk; chalk up

an·kreuzen *vt* mark with a cross; check off

an·kriechen* 1 *vt* creep up on, come over. 2 *vi*(s) creep, crawl (up)

Ankultur° *f* young culture

an·künd(ig)en *vt* announce, give notice; herald; usher in; declare. **Ankündiger** *m* (-) announcer. **Ankündigung°** *f* announcement, notice; declaration; prospectus

Ankunft *f* (..künfte) arrival

Ankunfts·ort *m* place of arrival. **-zeit** *f* time of arrival

an·kuppeln *vt* (*oft* + **an**) couple on (to); leash

Anküpung *f* onset of vatting

an·kurbeln *vt* crank up, start up; get . . . moving, stimulate, give a boost (to). **Ankurbeln** *n,* **Ankurbelung** *f* cranking up, pump priming, boost(ing), stimulation

Anl. *abbv* (Anlage) unit; equipment; design

an·lächeln *vt* smile at (*or* upon); attract, look good to

Anlage° *f* 1 enclosure—**in der A. senden wir** enclosed we send. 2 laying out; layout, plan, arrangement; design; draft, outline; planning stage, initial stage. 3 park, grounds, (public) garden. 4 installation; unit; equipment; plant, works. 5 (*oft* + **zu**) inclination, predisposition, tendency (to); talent, gift, aptitude (for); rudiment, hereditary factor, germ. 6 investment. **--kosten** *pl* installation expenses

an·lagern 1 *vt* accumulate, deposit; (*Geol.*) accrete; (*Chem.*) add on, take up, combine (with), absorb. **2: sich a.** accumulate, be deposited; (*Chem.*) be added, be taken up, combine (with), be absorbed. **Anlage-rung°** f accumulation, deposit; addition, taking up, combination, absorption; accretion

Anlagerungs-erzeugnis n addition product. **-komplex** m addition compound

Anlage-substanz f basic substance. **-teil** m basic (piece of) equipment. **-teilchen** n fundamental particle. **-typus** m (Biol.) genotype

an·landen 1 vt bring on land, put ashore. 2 vi(s) land, disembark. **Anlandung°** f putting ashore; landing; alluvial deposit

an·langen I vt 1 concern, regard: **das a—d, was das anlangt** as regards that, as for that. 2 touch; take hold of. II vi(s) 3 (usu + **an**) arrive at, reach

Anlaß m (..lässe) 1 cause, reason, provocation: **aus welchem A.** for what reason. 2 occasion: **aus A. des** on the occasion of. 3 opportunity. 4 (Metll.) temper. **5: auf A.** (+ genit or von) by order (of), on instructions (from). **--dauer** f duration of tempering (or drawing)

an·lassen* I vt 1 start, turn on (motor, etc). 2 keep on (coat, etc); leave on (light, etc), leave running (motor, etc). 3 (Metll.) temper, draw; **blau a.** heat-tint, blue; (Glass) anneal. 4 (e.g. **hart**) criticize, speak to (harshly). II: **sich a.** 5 start; (+ **zu**) get ready (for): **sich gut a.** start well, look promising, do (or come along) well. **Anlassen** n (Metll.) tempering; annealing. **Anlasser** m (-) starter, starting device; control(ler), regulator

Anlaß- tempering; starting:-**farbe** f (Steel) tempering color, oxidation tint; (Lthr.) graining color. **-härte** f tempering hardness. **-härtung** f temper hardening. **-hebel** m starting lever (or handle). **-kolben** m trigger

anläßlich prep (genit) on the occasion (of)

Anlaß-magnet m starting magnet. **-mittel** n (Metll.) tempering medium. **-ofen** m tempering oven, annealing furnace. **-öl** n tempering oil, drawing oil. **-wert** m (Phys.) yield value. **-wirkung** f tempering (drawing, annealing) effect. **-zeit** f tempering time; induction period

Anlauf° m 1 start (incl motor, etc); running start; starting (or take-off) run; onset.

2 attempt, try. 3 attack, assault. 4 slope, incline. 5 (Metll.) tempering color; tarnish; film (on metal, glass); mist, cloud, steam, condensation (on glass). 6 (Mach.) tappet, catch, shoulder. 7 rising, swelling

an·laufen* I vt 1 make for, head for; call (at a port). II vi(s) 2 start, start running; run up, arrive on the run; (+ **zu**) run (for), make a run (or a running start) (for)— **den Motor a. lassen** start the motor. 3 (**gegen**) run (against, into), rush (at); collide (with), oppose, fight. 4 accumulate, mount up; (**auf**) run, amount (to). 5 rise, slope up (gently). 6 (Water) rise, swell. 7 take on a coating (or coloration); get steamed up, mist, cloud (up); tarnish, oxidize; turn moldy; (Dyed Goods) recolor; turn (a color): **blau a.** turn blue, take on a blue coloration; **angelaufene Fenster** steamed (-up) windows. 8 (Nucl. Phys.) go critical. 9 (**bei**) offend; **übel a.** (**bei**) meet with a rebuff (from)

Anlauf-farbe f (Metll.) tempering color; tarnish. **-hilfe** f tempering agent. **-kurve** f filling curve (of a tank, pipe). **-reaktion** f tarnishing reaction. **-spannung** f initial voltage; reaction voltage; collision voltage

Anlaufs·rechnung f preliminary calculation

Anlauf·strecke f starting (reaction, collision) distance. **-temperatur** f tempering temperature. **-zeit** f filling time (of a pipe, tank); (Mach.) starting time; reaction time

Anlege·apparat m feed mechanism

an·legen I vt 1 (oft + **an**) A (genl) put on, put (up) against, lay on, apply (to); put on (clothes). B (specif): **Feuer a.** set (a) fire; **Gewehr a. (auf)** aim a gun (at); **Hand a.** lend a hand; **Holz (Kohle) a.** put wood (coal) on the fire; **Vorrat a.** lay in a supply. 2 lay out, plan, design; make up, draw up; prepare, start; organize, arrange; build, install cf ANGELEGT. 3 invest; (**für**) spend (for, on). II vi 4 land, call (at a port). III:**sich a.** 5 be deposited. 6 (**mit**) get involved, get into a dispute (with)

Anlege-öl n gold size. **-pyrometer** n contact pyrometer

Anleger m (-) investor

Anlege·thermoelement *n* surface thermo-couple

Anlegung *f* (-en) 1 putting on, laying on, application. 2 laying out, planning, design. 3 preparation, starting. 4 organization, arrangement. 5 construction, installation. 6 investment; spending. 7 landing. *Cf* ANLEGEN

Anlegungs·periode *f* formative period, planning stage

Anlehen *n* (-) loan

an·lehnen (*usu* + **an**) 1 *vt* & **sich a.** lean (on, against). 2 *vt* leave ajar. 3: **sich a,** follow, imitate; stick to (a model). **Anlehne·punkt** *m* point of support. **Anlehnung** *f* (-en) 1 protection, support; dependence, reliance. 2 imitation: **in** (*or* **unter**) **A. an** . . . in imitation of . . . , following the model of . . .

Anleihe *f* (-n) loan, borrowing

an·leiten *vt* direct, lead; instruct, train; guide. **Anleiter°** *m* instructor, trainer; guide; director. **Anleitung°** *f* instruction, training; direction; guidance; *pl usu* directions, instructions; (*as title, oft*) manual

an·lernen *vt* 1 train. 2 learn, acquire: **angelernte Künste** learned (*or* acquired) skills. **Anlernling** *m* (-e) trainee

an·liefern *vt* deliver; furnish, supply. **Anlieferung°** *f* delivery, supply. **Anlieferungs·zustand** *m* condition as delivered

an·liegen* *vi* 1 fit. 2 (*oft* + **an**) border (on), adjoin. 3 (*dat*) be of concern (to) *cf* ANGELEGEN. **Anliegen** *n* (-) 1 proximity, closeness. 2 request; desire, wish; concern, affair: **ein A. haben** (**an**) have a request (to make of), ask a favor (of). **anliegend** *p.a* 1 fitting; **eng a.** close-fitting, snug. 2 adjoining, adjacent. 3 enclosed (in a letter). **Anlieger** *m* (-) adjoining resident, neighbor

an·locken *vt* lure, entice, attract. **anlockend** *p.a* alluring, attractive. **Anlockkung** *f* (-en) allurement, enticement, attraction. **Anlockungs·mittel** *n* lure, bait

an·lösen *vt* begin to dissolve, treat superficially with solvent. **Anlösung°** *f* partial solution

an·löten *vt* (**an**) solder on (to). **Anlötung** *f* soldering (on)

an·lügen *vt* lie to, tell a lie to

Anm. *abbv* 1 (Ammerkung) note. 2 (Anmelder) patent applicant

an·machen *vt* 1 (**an**) fasten, attach (to). 2 put on (light), light (lamp, stove, fire), kindle (fire); turn on (radio, TV). 3 prepare, make, mix (salad, concrete, etc); adulterate (wine); temper (colors); slack, slake (lime). **Anmach(e)·wasser** *n* (*Cement*) mixing (*or* slaking) water, water of plasticity

an·mahnen *vt* 1 issue a warning about, send a reminder about (debts, etc). 2 urge, admonish. **Anmahnung°** *f* reminder, warning

an·malen *vt* paint (on)

Anmarsch° *m* 1 approach, approaching march: **im A. sein** be approaching. 2 way (*or* distance) to go. **an·marschieren** *vi* (s) march in (*or* up), approach on the march. **Anmarsch·weg** *m* way (*or* distance) to go

an·maßen 1 *vt*: **sich etwas a.** claim sth for oneself, take sth over (without legal authority). 2: **sich a.** (**zu** + *inf*) claim, presume (to). **anmaßend,** *p.a,* **anmaßlich** *a* arrogant, presumptuous; overbearing. **Anmaßung** *f* (-en) arrogance, arrogant thing to do (to say, etc); presumption; overbearing behavior. **anmaßungs·voll** *a* arrogant, presumptuous; overbearing

an·melden I *vt* 1 announce. 2 register, make known, put on record. 3 register (a psn with the authorities, for vital statistics). II: **sich a.** 4 announce one's presence. 5 (*oft* + **zu**) register, apply (for). 6 register (one's residence with the authorities). **Anmelder** *m* (-) patent applicant

Anmelde·schein *m* registration form. **-stelle** *f* registrar's (*or* registration) office. **-tag** *m* registration day; (patent) application date

Anmeldung° *f* 1 announcement; registration, registry; application. 2 registration (*or* registrar's) office. **Anmeldungs·formular** *n* registration (*or* application) form

an·mengen *vt* mix, blend, temper, dilute

an·merken *vt* 1 (*oft* + *d.i.o*) notice (sth about a psn or thing); (also + **an**) tell, detect (by appearance, etc): **sie lassen sich nichts a.** they don't show any (tell-tale) signs. **2** note, remark. **3** make a note of, take down. **Anmerker** m (-) observer; annotator. **Anmerkung** f (-en) note, footnote, annotation; remark, observation. **anmerkungs·wert** a noteworthy

an·messen* vt **1** (d.i.o) measure a psn for (a garment, etc), cut . . . to fit. **2** (Phys., Astron.) measure the distance to (a heavenly body). Cf ANGEMESSEN

Anmisch·behälter m mixing vessel. **an·mischen** vt mix, blend. **Anmisch·gefäß** n mixing vessel

an·montieren vt mount, attach, fasten (on)

anmoorig a boggy, peaty

an·mustern vt/i hire (on), sign (on)

Anmut f charm, grace, **an·muten** vt (+ adv) affect, strike (in a certain way)

Ann. abbv (Annalen, usu Annalen der Chemie) Annals (or Jnl.) of Chemistry

an·nagen vt nibble at, gnaw at

an·nähen vt sew on

an·nähern I vt **1** (oft + d.i.o) adapt (to), bring closer (to). **II: sich a. 2** (dat) approach, come closer to. **3** (recip) approach (each other), come closer together. Cf ANGENÄHERT. **annähernd** p.a approximate; (as adv) approximately, nearly: **nicht a.** not nearly. **Annäherung** f (-en) approach; approximation; (re)conciliation, rapprochement

Annäherungs·grad m degree of approximation. **-kraft** f centripetal force. **-linie** f line of approach, asymptote. **-punkt** m point of approximation. **-schalter** m proximity switch. **-versuch** m attempt at reconciliation (or reapprochement), approach. **-weg** m line of approach. **a—weise** adv approximately, by way of approximation. **A—wert** m approximation, approximate value

Annahme f (-n) **1** acceptance; approval; adoption, taking in. **2** taking on; assumption, hypothesis, supposition. Cf ANNEHMEN. **-stelle** f receiving office. **-verweigerung** f non-acceptance, refusal to accept. **-verzug** m delay in acceptance

Annalen pl annals, records, journal

Annalin n (Paint, Paper) annaline (precipitated $CaSO_4$)

an·nässen vt moisten, dampen

annehmbar a **1** acceptable, admissible. **2** plausible, conceivable. **3** agreeable, pleasant; tolerable. **Annehmbarkeit** f acceptability, admissibility. **2** plausibility, conceivability. **3** agreeableness, pleasantness; tolerability

an·nehmen* vt **1** accept. **2** approve (of). **3** adopt; pass (bill, motion). **4** admit (applicants); receive (guests). **5** take (on), assume (new form, etc). **6** assume, suppose. **II: sich a. 7** (genit) take an interest in (psn, cause), look into (a matter), embrace, espouse (a cause). Cf ANGENOMMEN, ANNAHME. **annehmens·wert** a acceptable, admissible; plausible; agreeable. **Annehmer** m (-) accepter, receiver, recipient

Annehmlichkeit f (-en) convenience; pl oft amenities

an·neigen (an) 1 vt incline (toward). **2: sich a.** lean, incline (toward); converge (with). **anneigend** p.a convergent. **Anneigung°** f convergence

annektieren vt annex; commandeer. **Annektierung** f (-en) annexation; commandeering

annelieren vt fuse, condense (rings); annelate. **Annelierung** f (-en) annelation

an·netzen vt moisten, dampen

Annexion f (-en) annexation

an·nieten vt (an) rivet on (to)

anno adv in the year—**a. dazumal** a long time ago, away back then

Annonce f (-n) ad(vertisement). **annoncieren** vt/i advertise, announce

Annuität f (-en) annuity

annullieren vt annual, nullify, cancel. **Annullierung** f (-en) annulment, nullification, cancelation

Anobie f (-n) anobiid

an·öden vt **1** bore, tire. **2** molest, annoy

Anoden·batterie f anode (or B) battery. **-belastung** f anode loading; anode dissipation. **-dichte** f anodic density. **-dunkelraum** m anode dark space. **-fallschwingungen** fpl decreasing amplitude of oscillations at the anode. **-glimmlicht** n anode glow. **-kreis** m plate circuit. **-legierung** f anode alloy. **-raum** m anodic region. **-schlamm** m anode slime (or

mud). **-schutznetz** *n* anodic screening grid. **-spannung** *f* anode potential. **-strahl** *m* positive (*or* canal) ray
anodisch *a* anodic. **anodisieren** *vt* anodize
Anol *n* anol, propylenephenol
an·ölen *vt* oil, coat with oil, lubricate
anomal, anomalisch *a* anomalous. **Anomalität** *f* (-en), **Anomalie** *f* (-n) anomaly
anomer *a* anomeric
Anon *n* anon, cyclohexane
anonym *a* anonymous. **Anonymität** *f* anonymity
an·ordnen *vt* 1 order, prescribe; regulate. 2 arrange, organize, set up. **Anordner** *m* (-) director; arranger, organizer. **Anordnung°** *f* regulation; order, prescription; arrangement, organization, set-up: **eine A. treffen** issue an order; **A—en treffen** make arrangements
anorg. *abbv* (anorganisch) inorg.
Anorganiker *m* (-) inorganic chemist **Anorganikum** *n* (..ka *or* ..ken) inorganic chemical. **anorganisch** *a* inorganic
anormal *a* abnormal
anorthotyp *a* (*Cryst.*) monoclinic
an·oxidieren *vt* oxidize mildly (*or* partly), begin to oxidize
an·packen *vt* seize, grasp, take hold of. 2 grip, overcome, overpower. 3 treat (harshly). 4 tackle (a problem)
an·pappen 1 *vt* paste, stick (on). 2 *vi*(s) get stuck, stick
anpaßbar *a* adaptable, adjustable
an·passen 1 *vt* (*d.i.o*) fit, tailor (to); adapt, adjust (to). 2: **sich a.** (*dat*) adapt (oneself), adjust (to). *Cf* ANGEPASST. **Anpassung** *f* (-en) fitting, tailoring; adaptation, adjustment
Anpassungs·erscheinung *f* adaptation phenomenon, mimicry. **a—fähig** *a* adaptable. **A—fähigkeit** *f* adaptability. **-gepräge** *n* adaptation feature. **-mensch** *m* adaptable person; opportunist. **-vermögen** *n* adaptability. **-wechsel** *m* change in adaptation
an·pasten *vt* make into a paste, prepare (ointment)
an·peilen *vt* get a bearing on; aim at. **Anpeilung** *f* (-en) direction finding
an·pfählen *vt* prop up; fasten to a stake
an·pflanzen *vt* plant (freshly), start to grow

(a crop); lay out (a garden). **Anpflanzung°** *f* 1 planting, growing. 2 plantation
an·pflocken, an·pflöcken *vt* tether to a post
an·pfropfen *vt* (an) graft (on, onto)
an·pinseln *vt* paint, brush some paint on
an·pochen *vi* knock (at the door)
an·polymerisieren *vt* prepolymerize. **Anpolymerisierung°** *f* prepolymerization
Anprall *m* (-e) impact, collision. **an·prallen** *vi*(s) (an, gegen) strike (on, against), collide (with). **Anprall·punkt** *m* point of impact. **-winkel** *m* angle of impact
an·preisen* *vt* praise; recommend. **Anpreisung** *f* (-en) praise; recommendation; sales pitch
Anpreß·druck *m* contact pressure
an·pressen *vt* (an) press (on, onto, against)
Anprobe° *f* fitting. **an·proben, an·probieren** *vt* try on (garment)
an·quellen* *vi*(s) sprout
an·quicken *vt* amalgamate. **Anquick·faß** *n* amalgamating tub (*or* vat). **-silber** *n* silver amalgam. **Anquickung** *f* (-en) amalgamation
an·quirlen *vt* blend in (with a whisk)
Anrat° *m* (piece of) advice. **an·raten*** *vt* (*d.i.o*) advise (to use), recommend (to)
an·rauchen *vt* 1 light, start to smoke (cigars, etc); break in (pipe). 2 blow smoke at; smoke (up), blacken (darken, color) with smoke. *Cf* ANGERAUCHT
an·räuchern *vt* 1 smoke (meat) slightly. 2 fumigate (plants). 3 perfume with incense
an·rechnen *vt* (*d.i.o*) 1 charge (a price). 2 count (to sbdy's credit); allow (as credit): **einem etwas zum Verdienst a.** consider sthg a credit to sbdy; **wir rechnen es uns zur Ehre an** we consider it an honor. 3 take into account. **Anrechnung°** *f* charging; counting (*or* allowing) as credit, allowance—**in A. bringen** = ANRECHNEN 2
Anrecht° *n* (usu + **auf**) claim, right, title (to)
Anrede ° *f* (form of) address. **an·reden** *vt* address, speak to, start a conversation with
an·regen I *vt* 1 bring up for consideration, suggest. 2 prompt, inspire (to do); stimu-

late, excite; activate. **II** *vi* **3** be stimulating, act as a stimulant. *Cf* ANGEREGT.

anregend *p.a* stimulating, provocative: **a—es Mittel** stimulant, excitant: **a. wirken** act as a stimulant. **Anreger** *m* (-) instigator, initiator, moving spirit; stimulator. **Anregung** *f* (-en) suggestion, inspiration, prompting; stimulation, stimulant, stimulus; excitation, activation: **in A. bringen** = ANREGEN

Anregungs· excitation: **-arbeit, -energie** *f* excitation energy. **-grenze** **f** excitation limit, **-mittel** *n* stimulant, excitant. **-spannung** *f* excitation potential. **stärke** *f* strength of excitation, work function. **-stufe** *f* excitation level. **-zustand** *m* state of excitation; high-energy state

an·reiben* **1** (**an**) rub (on, against); apply; paste on. **2** grind colors—**mit Wasser a.** dilute

an·reichern **I** *vt* **1** enrich, fortify: **mit Fluor angereicheries Wasser** fluorine-enriched (*or* fluoridated) water. **2** accumulate, raise its content of. **3** concentrate. **4** yield in substantial amount. **II: sich a. 5** be(come) enriched. **6** accumulate. **Anreicherung** *f* enrichment, fortification; accumulation, increase of content; concentration; substantial yield. **Anreicherungs·kaskade** *f* enrichment cascade

Anreich·lech *m* enriched matte

an·reihen **I** *vt* **1** (*d.i.o*) add (to a series). **2** baste (on). **3** line up, align, arrange in a series; string (beads). **II: sich a. 4** (*dat*) be added (to), follow; join in; get on line, line up—**sich würdig a.** be a worthy successor. **Anreihung** *f* (-en) series, sequence

an·reißen* *vt* **1** tear a little way into, make a small tear in. **2** tear (*or* break) open; break into; break out, start using; start (outboard motor) *cf* ANGERISSEN. **3** pluck (strings). **4** pull in (customers), pressure (people) to buy. **5** tense, pull in (arms, legs). **6** scribe, mark, trace a line on. **7** broach, bring up (a subject). **anreiße·risch** *a* high-pressure (selling)

Anreiz° *m* stimulus; incentive, inducement; temptation. **an·reizen** *vt* **1** stimulate; titillate. **2** (**zu**) induce, encourage,

spur on (to). **3** (*Elec.*) energize, excite.

anreizend *p.a* attractive, provocative.

Anreizung° *f* stimulation; titillation; inducement, encouragement; excitation

an·rennen* **I** *vt* **1** run into, rush (at); storm. **2** (*dat rflx*): **sich** (e.g. **den Kopf**) **a.** bump (one's head). **III** *vi*(s) **3** run up, arrive on the run. **4** (**an, gegen**) run (into, against), run headlong (against)

Anrichte *f* (-n) sideboard, serving table; cupboard. **an·richten** **1** prepare, serve (food); mix (paints); dress (ore). **2** cause, perpetrate, do (damage, etc). **3** aim at, sight. **Anrichter°** *m* preparer; (ore) dresser; assayer. **Anrichtung** *f* preparation, serving, mixing, dressing; causing, perpetration

an·riechen* *vt* **1** sniff at. **2** (*d.i.o*) tell by smelling, smell (sthg on sbdy)

Anriß° *m* **1** initial tear (*or* break); surface crack. **2** scribe, mark. **3** outline

an·ritzen *vt* scratch (the surface of); scarify

an·rollen *vt, vi*(s) roll in; taxi (an airplane)

an·rosten *vi*(s) get somewhat rusty, start rusting; (*oft* + **an**) rust on (to), get stuck by rusting. **Anrostung** *f* slight rusting, incipient rust; rusting on

an·rüchig *a* disreputable. **Anrüchigkeit** *f* disrepute

an·rücken **1** *vt* (*oft* + **an**) move up, bring near (to). **2** *vi*(s) move in, come closer, approach

Anruf° *m* **1** (warning) call, signal; appeal. **2** (phone) call. **an·rufen*** **1** *vt* call (to); challenge. **2** call (up), ring up, phone.**3** call for, appeal to; (*oft* + **als, um**) call on, appeal to (as a witness, for aid). **Anrufung** *f* calling; telephoning; appeal(ing)

an·rühren *vt* **1** touch, handle. **2** stir, mix, grind; temper. **3** touch on (a subject). **4** stir, move (emotionally)

Anrühr·gefäß *n* mixing vessel; (*Flot.*) conditioner. **-wasser** *n* mixing water

an·rußen *vt* coat with soot, smoke (up)

ans *contrac* = **an das** *cf* AN *prep*

Ansaat° *f* sowing, seeding. **Ansaats·stärke** *f* extent of seeding

an·säen *vt* sow, seed

Ansage° *f* announcement, notification, notice. **an·sagen** **I** *vt* **1** announce, present, introduce *cf* KAMPF. **2** dictate (letters).

3 tell; betray (secrets). **II** *vi* **4** act as announcer. **III: sich a. 5** announce oneself (*or* one's will); show up. **Ansager** *m* (-), **Ansagerin** *f* (-nen) announcer; emcee

an·salzen *vt* salt lightly

an·sammeln *vt* & *sich a.* gather, collect, accumulate. **Ansammlung°** *f* gathering, collection, accumulation; crowd.

Ansammlungs·apparat *m* condenser

ansässig *a* resident: **a. sein** be a resident, reside; **a werden, sich a. machen** take up (permanent) residence, settle; **die A—en** the permanent residents, the local inhabitants. **Ansässigkeit** *f* residency, resident status

Ansatz° *m* **1** (*Chem.*) deposit, sediment; accretion; arrangement, set-up (for an experiment); batch; starting materials. **2** (*Yeast*) start. **3** (*Geol.*) alluvial deposit. **4** (*Dye.*) standard. **5** (*Bot.*) stipula, germ, spore. **6** extension (piece), added piece. **7** layer, coating, crust, incrustation. **8** tendency, disposition. **9** attempt, first step; starting position; rudiment; initial (*or* incipient) stage, first sign: **A. machen (zu)** take the first step (toward), get into position (to, for), make an attempt (at); **gute A̤—e zeigen** show good promise; **in den ersten A̤—en stecken-bleiben** get stuck in the initial stages. **10** (*Physiol.*) base, root, starting point (of leg, etc); (*Muscles*) insertion, attachment: **Haar·ansatz** hairline. **11** projecting part, projection, lug, flange; shoulder, appendage. **12** (*Fin.*) appropriation, budgeted amount; quotation; charge, rate; estimate, evaluation: **in A. bringen** estimate, evaluate; **außer A. bleiben** be left out of consideration. **13** (*Math.*) statement (of the problem); trial solution; member (of a series). *Cf* ANSETZEN

Ansatz·bad *n* (*Dye.*) first (initial, starting) bath. **-bildung** *f* (formation of a) coating (*or* deposit). **-bottich** *m* preparing vessel. **a—frei** *a* deposit-free, unincrusted. **A—gefäß** *n* addition tube; preparing vessel. **-lösung** *f* starting solution. **-punkt** *m* **1** starting point. **2** point of attachment (*or* insertion). **3** boring point (for an oil well). **-rohr** *n* **1** extension tube (*or* pipe), attached (*or* inserted) tube (*or* pipe). **2** con-

necting tube; side arm; nozzle. **3** (*Anat.*) supraglottal passage. **schraube** *f* set screw. **-stelle** *f* point of attachment. **-stück** *n* extension piece, attachment, attached part. **-stutzen** *m* nozzle attachment; extra attachment

an·säuern *vt* acidify, sour; (*Baking*) leaven, add yeast (to). **Ansäuerung** *f* acidification; leavening

Ansauge·gerät *n* suction device; aspirator. **-hub** *m* suction stroke

an·saugen* **I** *vt* **1** suck in, suck up; draw in; aspirate. **2** suck at (for siphoning). **3** (*Chem.*) adsorb; absorb. **4** attract, lure. **II** *vi* **5** suck, take the breast. **6: Pumpe a. lassen** prime the pump. **III: sich a. 7** adhere (*or* fasten itself) by suction

Ansauge·rohr *n* suction tube (*or* pipe), induction pipe

Ansaug· (= ANSAUGE·) **-heber** *m* siphon

Ansaugung *f* suction, intake; aspiration; (forced) induction; adsorption, absorption *cf* ANSAUGEN

Ansaug·vorrichtung *f* suction device

an·schaffen *vt* acquire, obtain; procure, order; (*oft* + *dat rflx*): **sich etwas a.** get oneself sthg. **Anschaffung** *f* (-en) acquisition; procurement

Anschaffungs·kosten *pl* cost of acquisition. **-preis** *m* acquisition (*or* purchase) price. **-wert** *m* acquisition value

an·schalen *vt* peel partly; start peeling

an·schalten *vt* turn on, switch on. **2** (*an*) connect (to)

an·schärfen *vt* sharpen; strengthen (a bath)

an·schauen *vt* **1** look at; regard, contemplate. **2** (+ *dat rflx*): **sich etwas a.** have (*or* take) a look at sthg. **anschauend** *p.a* (*oft*) intuitive; contemplative

anschaulich *a* **1** clear, vivid. **2** visual, graphic; descriptive. **3** intuitive. **Anschaulichkeit** *f* clarity, vividness; visual quality; descriptiveness; intuitiveness

Anschauung *f* (-en) **1** view, opinion. **2** (*oft* + **von** *or in compds*) idea, conception (of). **3** (empirical) observation, (visual) experience, perception: **zur A. bringen** make visually perceptible. **4** intuition. **5** contemplation

Anschauungs·begriff *m* intuitive idea.

-bild *n* mental image; visual image. **-erkenntnis** *f* intuitive knowledge. **-material** *n* illustrative material. **-unterricht** *m* visual instruction; object lesson. **-weise** *f* perception, way of looking at things

an·scheiden* *vt* deposit. **Anscheidung°** *f* deposit(ing)

Anschein *m* (outward) appearance: **allem A. nach** to all appearances, in all likelihood; **sich den A. geben** put on an appearance, pretend; **es hat den A.** it looks; **dem A. nach** apparently. **an·scheinen*** *vi* shine (up)on. **anscheinend** *p.a*, **anscheinlich** *a* (*both usu adv*) apparent(ly), seeming(ly)

an·schichten *vt* stack in layers; stratify

an·schicken *vr:* **sich a.** (**zu**) get ready, get set, prepare (to, for), be about (to)

an·schieben* *vt* 1 push to start, give a push. 2 (**an**) push up (against) *cf* ANGESCHOBEN. **Anschieber** *m* (-) extension (piece), table leaf; (*Baking*) kissing crust; (*Brew.*) workman. **Anschiebsel** *n* (-) piece added on, extension

an·schienen *vt* 1 fasten with metal bands. 2 put in splints

an·schießen* I *vt* 1 shoot at, snipe at; wound (with a shot). 2 shoot off; test-fire (a gun). II *vi*(s) 3 dash in, rush up. 4 crystallize, shoot into crystals: **Kristalle schießen an** crystals form. **Anschießgefäß** *n* crystallizing vessel, crystallizer. **Anschießung** *f* crystallization

an·schimmeln *vi*(s) start getting moldy *cf* ANGESCHIMMELT

an·schirren *vt* harness, hitch up

Anschlag° *m* 1 touch; tap, strike; striking, impact. 2 posted notice, bulletin: **durch A.** by posting on the bulletin board. 3 plot, scheme, conspiracy. 4 attack, assault; attempt (on sbdy's) life. 5 estimate, assessment—**in A. bringen** take into account, make allowances for. 6 firing (of a gun); aiming (*or* firing) position. 7 (*Mach.*) stopping device, e.g.: trip, dog, check (on a vise), (door) stop. 8 barking (of dogs). **··brett** *n* bulletin board

an·schlagen* I *vt* 1 strike, hit; touch, tag; strike (a tone, pose, tempo, etc). 2 nail (up, on); post, put up (on a bulletin board).

3 tap (a keg). 4 aim (guns). 5 estimate, rate, value. II *vi* (h) 6 (*oft* + **bei**) have the desired effect (on), take effect, work (out); **gut a.** (*oft*) do well. 7 start sounding, ring, strike; start singing (talking, barking). III *vi*(s) 8 (**an**) bump (into), run, strike (against). IV: **sich a.** 9 bump into sthg

anschlägig *a* inventive, resourceful; skillful, deft. **Anschlägigkeit** *f* inventiveness, resourcefulness; skill, deftness

Anschlag·wert *m* estimated value; bodybuilding value (of food). **-zettel** *m* bulletin, poster

an·schlammen 1 *vt* suspend; mix into a paste; deposit mud on, cover with mud; silt up; elutriate. 2: **sich a.** get silted up

an·schleifen* *vt* 1 grind (an edge on), sharpen (slightly); start grinding (*or* cutting); (+ *d.i.o*) grind (on, onto). 2 drag in, drag along

an·schließen* I *vt* (**an**) 1 chain up, make fast (to,) fasten (onto). 2 connect (to), hook up with); attach, add (to). II *vi* 3 (**an**) adjoin; follow. 4 (*oft* + **eng**) fit (closely). 5 make connections. III: **sich a.** 6 (**an**) border (on), adjoin; follow. 7 (*dat*) make contact (with); join; fall in (with); concur (in), agree (with). *Cf* ANGESCHLOSSEN. **anschließend** *p.a* adjacent, adjoining; immediately following, subsequent; connecting; *adv oft* thereafter, then, followed by. *Cf* ANSCHLUSS

Anschliff° *m* polished surface (*or* section); polish. **··winkel** *m* grinding angle; included angle

Anschluß° *m* 1 (train, bus, phone, electric) connection; (telephone) extension; (radio, TV) hook-up—**mit elektrischem A.** to plug in. 2 (**an**) contact (with); affiliation (with); acquaintance. 3 (**an**) union (with). 4: **im A. an** following (up on), subsequent to; with reference to. *Cf* ANSCHLIESSEN. **··dose** *f* junction box; wall socket. **-gerät** *n* plug-in accessory. **-kasten** *m* junction box; wall socket. **-klemme** *f* terminal clamp, binding post. **-leiter** *m* lead terminal. **-leitung** *f* branch line (pipe, wire, etc); extension (*or* plug-in) cord. **-rohr** *n* connecting (*or* branch) pipe (*or* tube). **-schlauch** *m* connecting hose. **-schnur** *f* (*Elec.*) extension (*or* plug-in) cord. **-stück**

n connecting piece, gooseneck, nipple. **-wert** *m* connecting load. **-zelle** *f* (*Bot.*) hypophysis

an·schmauchen *vt* smoke up, soot up

an·schmecken *vt* sample, take a taste of

an·schmelzen (an) 1 *vt* fuse on (to). 2 *vi*(s)* melt on, get fused on (to). **Anschmelz·herd** *m* smelting furnace

an·schmieden *vt* (**an**) weld on (to); chain (to)

an·schmiegen 1 *vt* & **sich a. (an)** snuggle up, press close (to). 2: **sich a. (an)** hug. **anschmiegsam** *a* pliable, tractable; adaptable; affectionate. **Anschmieg·samkeit** *f* pliability, tractability; adaptability; affection

an·schmieren I *vt* 1 smear, daub. 2 cheat, swindle. II: **sich a.** get stained (*or* smeared)

an·schmutzen 1 *vt* soil, dirty, stain. 2: **sich a.** get dirty (soiled, stained). **Anschmutzung** *f* (-en) soil(ing), stain(ing)

an·schnallen 1 *vt* (**an**) buckle on (to). 2: **sich a.** fasten one's seat belt. **Anschnall·gurt** *m* safety (*or* seat) belt

an·schneiden* *vt* 1 cut into, cut a piece of (*or* off); start cutting (*or* mowing). 2 broach, bring up (a subject). 3 notch. 4 (*Metll.*) gate. 5 get a bearing on; sight; aim at, focus

Anschnitt° *m* 1 first cut (*or* slice). 2 cut side; cutting. 3 cut, slice: **im A. verkaufen** sell by the slice. 4 notch. 5 (*Metll.*) gating

an·schoppen 1 *vt* choke, engorge, congest. 2: **sich a.** become engorged (*or* congested)

Anschove *f* (-n), **Anschovis** *f* (-) anchovy

anschraubbar *a* attachable with screws, (made) to screw on. **an·schrauben** *vt* (**an**) screw (*or* bolt) on (to)

Anschreibe·buch *n* notebook, memo book; credit (sales) record; score (book)

an·schreiben* I *vt* 1 (*oft* + **an**) write (*or* put) down (on)—**was steht angeschrieben?** what does it say (on the sign)? 2 (*d.i.o*) charge (to sbdy's account): **a. lassen** buy (*or* take) on credit, have . . . charged to one's account. 3 write, apply to . . . by letter. 4 (*Math.*) escribe (circles). II *vi* 5 keep score. *Cf* ANGESCHRIEBEN. **Anschreiben** *n* (-) covering (*or* accompanying) letter. **Anschreibe·tafel** *f* scoreboard

Anschrift° *f* (postal) address

an·schuldigen *vt* accuse, charge. **Anschuldigung** *f* (-en) accusation, charge

an·schüren *vt* fan (fire); stir up, foment

Anschuß° *m* 1 first shot (fired). 2 crystallization; crop of crystals. **--gefäß** *n* crystallizing vessel, crystallizer

an·schütten 1 *vt* pour (*or* spill) liquid on. 2 fill up, put a filling in. 3 store up (grain)

an·schwängern *vt* impregnate, saturate. **Anschwängerung** *f* impregnation, saturation

Anschwänz·apparat *m* sprinkler, sprayer. **an·schwänzen** *vt* sprinkle, sparge

an·schwärzen *vt* blacken, stain black; slander

an·schwefeln *vt* treat (*or* fumigate) with sulfur

an·schweißen *vt* (**an**) weld on (to)

an·schwellen 1 *vt* swell, blow up, inflate. 2 *vi*(s)* swell (up), get swollen; grow, increase. **Anschwellung°** *f* (slight) swelling, intumescence; increase

an·schwemmen *vt* wash up, deposit, float . . . ashore *cf* ANGESCHWEMMT. **Anschwemm·filter** *n* settling (*or* alluvial) filter. **Anschwemmung** *f* (-en) alluvium, alluvial deposit

an·schwingen* *vi* start swinging (*or* oscillating)

Anschwöde·brei *m* (*Lthr.*) liming paste. **an·schwöden** *vt* paint (flesh side of hides) with lime

Anschwung° *m* initial impetus, push-off

an·sehen* I *vt* 1 look at; regard, judge. 2 respect, have regard for; (+ **nicht**, *oft*) disregard. 3 (**für, als**) regard (as), take (for). 4 (*d.i.o*) tell (*or* see) by looking at, recognize as. 5 (*dat rflx*): **sich etwas a.** have (*or* take) a look at sthg. 6 (*usu*): **mit a.** put up (with), tolerate. II: **sich a.** 7 (*oft* + **wie**) appear, look (like). *Cf* ANGESEHEN. **Ansehen** *n* 1 looking, view. 2 sight, appearance: **vom A. kennen** know by sight; **sich das A. geben** put on the appearance, make oneself appear; **allem A. nach** to all appearances. 3 (high) regard, esteem; distinction; repute, reputation—**ohne A. der Person** without respect of person

ansehnlich *a* 1 appreciable, considerable.

2 respectable-looking. 3 good-looking, handsome. **Ansehnlichkeit** *f* appreciability; respectable appearance; good looks

Ansehung *f* 1: **in A.** (+ *genit*) in consideration (of), with respect (to). 2: **ohne A. der Person** without respect of person

ansetzbar *a* attachable; applicable; preparable, mixable; crystallizable, capable of being deposited

an·setzen I *vt* 1 put up, set up; put in place (*or* position); lay on, apply; put (*or* raise) to one's lips. 2 put to work, use; put on the trail of. 3 (*oft* + **an**) sew on, add on, attach (to). 4 mix, prepare; set up, arrange; state (problem); charge (furnace). 5 sprout, grow (leaves, etc), get covered with, put on a coat of (rust, mold) 6 set a time for, schedule. 7 estimate, assess; reckon; set, quote (price); charge (at a rate). II *vi* 8 sprout. 9 start, set in; (+ **gut**) be off to a good start; set well; (+ **zu**) get set (to, for). 10 put on weight. 11 (*Phys.*) act (*as:* a force). III: **sich a.** 12 deposit; accumulate, accrete; form. IV *vi & sich a.* 13 stick (to pan or container). 14 effloresce; crystallize. *Cf* ANSATZ. **Ansetz·stelle** *f* point of attachment. **Ansetzung** *f* 1 putting up, setting up; application. 2 addition, attachment. 3 preparation. 4 depositing; accretion, accumulation. 5 setting a time, scheduling. 6 quotation (of a price). 7 estimate; assessment. 8 growing, sprouting. **Ansetzungs·stelle** *f* point of attachment

Ansicht *f* (-en) 1 view, opinion. 2 inspection: **zur A.** for inspection, on approval. **ansichtig** *a:* **a. werden** (*genit*) catch sight (of) **Ansichts·sache** *f* matter of opinion. **-tafel** *f* table, tabulation, summary

Ansiedelei *f* (-en) settlement. **an·siedeln** 1 *vt* settle (people in a place). 2: **sich a.** settle (down), take up permanent residence

an·sieden* *vt* boil; blanch; scorify; mordant (by boiling)—**A. und Abdunkeln** stuffing and saddening *cf* ANSUD

Ansiedler *m* (-) settler, colonist. **An·siedlung** *f* (-en) settling; settlement, colony *cf* ANSIEDELN

Ansiedung *f* boiling; blanching; scorifica-

tion; mordanting *cf* ANSIEDEN

Ansinnen *n* (-) (unreasonable) demand *cf* STELLEN 8

an·sintern *vi*(s) sinter, form a sinter; (*Geol.*) form stalactites

an·sitzen* *vi* (h *or* s) cling, adhere; fit (snugly) *cf* ANGESESSEN

ansonsten *a* apart from that, otherwise; or else

an·spannen I *vt* 1 harness, hitch up. 2 pull taut, tense; (over)tax, strain; exert (force). 3 (**zu**) harness, call on (sbdy to . . .). II: **sich a.** exert (*or* strain) oneself. *Cf* ANGESPANNT. **Anspanner** *m* (-) tensor muscle. **Anspannung** *f* (-en) harnessing; tension (*incl Mach.*), strain, stress; straining; exertion

an·spielen 1 *vt/i* start playing. 2 *vi* (**auf**) allude (to), hint (at). **Anspielung** *f* (-en) allusion, hint

an·spinnen* I *vt* 1 join on (threads). 2 scheme, plot; hatch (plots). 3 start, get (sthg) underway. II: **sich a.** 4 spin its cocoon. 5 get started (*or* underway), develop

an·spitzen *vt* 1 point, sharpen (to a point). 2 goad, prod

Ansporn *m* spur, prod; stimulus, incentive. **an·spornen** *vt* spur (on), prod, goad; stimulate

Anspr. *abbv* (Anspruch) claim, etc

Ansprache° *f* address, talk, speech; conversation; someone to talk to *cf* ANSPRECHEN

ansprechbar *a* regardable *cf* ANSPRE-CHEN 3

an·sprechen* I *vt* 1 address, start a conversation with, accost. 2 (*oft* + **um**) turn to, apply to (for) *cf* ANGESPROCHEN. 3 (**als**, **für**) designate (as), declare (to be); regard (as): **es ist als ernst anzusprechen** it is to be regarded as serious. 4 please, appeal to: **a—d eingeteilt** pleasingly arranged. 5 claim. II *vi* 6 (**bei**) be pleasing, be appealing (to). 7 (**auf**) (*incl Mach.*) respond (to). *Cf* ANSPRACHE

Ansprech·geschwindigkeit *f* speed of response, response time. **-grenze** *f* limit of response, effective limit. **-zeit** *f* response (*or* reaction) time

an·sprengen I *vt* 1 sprinkle. 2 (*Min.*) blow up, blast. II *vi*(s) come galloping up

an·springen* I *vi*(s) 1 (*incl Mach.*) respond,

react, start running. 2 bound along. **II** *vt* jump up on; leap at, attack; strike, overcome

an·spritzen *vt* sprinkle, spatter, splash (on)

Anspruch° *m* claim, demand: **A. haben (auf)** have a claim (on), have a right (to) *cf* ERHEBEN 1; **in A. nehmen** take up, occupy (time, space), make demands on, keep busy

anspruchslos *a* modest, unassuming, unpretentious. **Anspruchslosigkeit** *f* modesty, unpretentiousness. **anspruchsvoll** *a* demanding, fastidious, pretentious

an·spulen *vt* reel on, put on a reel

an·spülen *vt* wash up, wash ashore; deposit, accrete. **Anspülung** *f* (-en) washing up; (alluvial) deposit; accretion

an·stacheln *vt* spur (on), goad, prod; incite

an·stählen *vt* coat (*or* point) with steel

Anstalt *f* (-en) 1 institution, establishment. **2: A—en machen** (*or* **treffen**) make a move (to), make preparations (*or* arrangements) (to)

Anstand° *m* 1 decency; good manners; dignity. 2 hesitation; (*usu pl*) objections. **A. nehmen** hesitate, take offense. 3 *oft pl:* **Anstände** trouble, difficulties. **anständig** *a* 1 decent, proper, respectable. 2 considerable. **Anständigkeit** *f* decency, propriety, respectability

anstands·halber *adv* for the sake of decency (*or* propriety). **-los** *a* unhesitating; *adv oft* without hesitation (*or* objection), freely

an·stapeln *vt* pile up, stack up

anstatt 1 *prep* (*genit*) instead (of), in place (of). 2 *cnjc* (+ **zu**) instead of (..ing)

an·stäuben *vt* dust, sprinkle with powder

an·stauen *vt* dam up, impound; pile up, build up *cf* ANGESTAUT

an·staunen *vt* marvel at, gaze at in wonder. **anstaunens·wert, -würdig** *a* marvelous, astonishing

Anstauung° *f* damming up, impounding; build-up, pile-up *cf* ANSTAUEN

an·stechen* *vt* 1 pierce, puncture, prick, penetrate. 2 tap (keg). 3 stitch on. 4 turn on (gas). *Cf* ANGESTOCHEN

an·stecken I *vt* 1 pin on (badge); put on (ring). 2 tap (keg). 3 set fire to; light, kindle. 4 infect, contaminate. **II** *vi* be con-

tagious, be infectious. **III: sich a. (bei)** catch a disease (from), be infected (by). **ansteckend** *p.a* infectious, contagious. **Ansteckung** *f* (-en) infection, contamination

Ansteckungs·gefahr *f* danger of infection (*or* contagion). **-kraft** *f* infectiousness, virulence. **-quelle** *f* source of infection. **-stoff** *m* infectious matter, virus. **-vermögen** *n* contagiousness, infectiousness; pathogenic potency

an·stehen* *vi* 1 stand in line, line up. 2 (*dat*) be becoming (to), befit, suit. 3 (**zu**) hesitate (to), 4 be pending; be scheduled, be set; (+ **zu**) be up, be due (for). 5 be delayed; (+ **lassen**) delay, put off, postpone. 6 (*Geol.*) crop out. 7 (**auf**) be dependent (on). **Anstehen** *n* (*Geol.*) outcropping (rock). **anstehend** *p.a* (*Geol.*) outcropping, naturally occurring; **a—es Gestein** bedrock

an·steifen *vt* starch, stiffen; prop out

an·steigen* *vi*(s) rise, climb, ascend; increase *cf* ANSTIEG

anstellbar *a* 1 adjustable. 2 employable

Anstell·bottich *m* (*Brew.*) starting tub

anstelle *prep* (*genit*), *adv* (**von**) instead, in place (of)

an·stellen I *vt* 1 set, put (in position); add on (to a series); (**an**) put up, lean (against). 2 turn on (motor, etc); adjust. 3 employ, hire *cf* ANGESTELLTE; put to work; requisition (for a job). 4 make, perform, do, undertake, carry out; try (*Transl. varies with noun object*) *e.g.*: **Versuche a.** make attempts, do (*or* undertake) experiments. 5 start, initiate, institute;—(*Brew.*): **die Würze a.** pitch the wort. 7 (e.g. **gut, schlecht**) go about (a matter) (well, badly). **II: sich a.** 8 get on line, line up. 9 behave, act, carry on. **Ansteller** *m* (-) employer (in private sector). 2 perpetrator; initiator

Anstell·hefe *f* (*Brew.*) pitching yeast

anstellig *a* skillful, handy, clever. **Anstelligkeit** *f* skill, handiness, cleverness

Anstell·temperatur *f* (*Brew.*) starting temperature

Anstellung *f* (-en) **I** 1 putting (up), adding (on). 2 turning on; adjustment. 3 employment, hiring; requisitioning. 4 performance, carrying out; attempt. **5** initiation, institution (of proceedings). **6** perpetration, commission; indulgence. *Cf* AN-STELLEN 1–5. **II** position, job

Anstell·winkel *m* angle of incidence.

-würze *f* (*Brew.*) pitching wort

an·stemmen *vt* & **sich a.** (*oft* + **gegen**) brace oneself (against)

an·sterben* *vi*(s) (*dat*) devolve (on)

ansteuerbar *a* approachable. **an·steuern** *vt* steer toward, head for, make for; approach. **Ansteuer·schaltung** *f* control switch. **Ansteuerung** *f* approach, entrance

Anstich° *m* **1** tapping (of a keg); freshly tapped beer (*or* wine): **im A.** on tap. **2** puncture; wormhole (in fruit). **3** groundbreaking; first spadeful of earth. *Cf* ANSTECHEN

an·sticken *vt* embroider on

Anstieg *m* (-e) rise, climb, ascent; (upward) slope; increase *cf* ANSTEIGEN

an·stiften *vt* perpetrate, commit; incite, instigate. **Anstifter°** *m* perpetrator, person responsible; instigator. **Anstiftung°** *f* perpetration, commission; inciting, instigation

an·stimmen *vt* strike up, intone; start

an·stocken *vi*(s) turn moldy

Anstoß° *m* **1** impact, collision. **2** impetus. **3** kickoff. **4** offense: **A. nehmen** (**an**) take offense (at). **5** hitch, obstacle: **ohne A.** without a hitch, without hesitation (*or* faltering); **Stein des A—es** stumbling block. **6** (butt) joint; (*Tex.*) overlap

an·stoßen* **I** *vt* **1** give a push (to), set in motion (with a push); push off, kick off. **2** prod, nudge. **3** bump into, run against. **4** crush. **5** join; butt-joint. **II** *vi* (h) **6** falter, hesitate (in one's speech); **mit der Zunge a.** lisp. **III** *vi*(s) **7** (**bei**) be offensive (to), offend. **8** (**an**) bump (into sthg), collide (with sthg): **mit dem Fuß a.** bump one's foot, run one's foot against sthg. **9** (**an**) border (on), butt up (against); abut, adjoin. *Cf* ANGESTOSSEN, ANSTOSS. **Anstößer** *m* (-) neighbor (on adjoining property)

anstößig *a* offensive, improper. **Anstößigkeit** *f* (-en) offensiveness, impropriety

an·strahlen *vt* illuminate, light up; spotlight, floodlight; irradiate; beam at (*or* on). **Anstrahlung°** *f* illumination; irradiation. **Anstrahl·winkel** *m* angle of irradiation

Anstrebe·kraft *f* centripetal force

an·streben 1 *vt* strive for (*or* to reach); aim at (*or* for); hope to gain, aspire to. **2** *vi* strive (upward); tower; (**gegen**) resist, oppose

an·strecken *vt* stretch (taut)

Anstreiche *f* (-n) paint. **an·streichen*** *vt* **1** paint; brush. **2** mark, check off; underline. **3** strike (matches). **Anstreicher** *m* (-) (house) painter

an·strengen 1 *vt* strain; overstrain, tire, tax; initiate, institute (proceedings). **2** *vi* be tiring (taxing, exhausting). **3: sich a.** exert oneself. *Cf* ANGESTRENGT. **an·strengend** *p.a* (*oft*) strenuous, arduous. **Anstrengung** *f* (-en) exertion, effort; strain; initiation (of proceedings)

Anstrich° *m* **1** painting (of a wall, etc). **2** (coat of) paint. **3** (outer) appearance, aspect; veneer; air; tinge. **-bindemittel** *n* paint binder. **-farbe** *f* color of (the) paint, painting color. **a—fertig** *a* ready-mixed (paint). **A—film** *m* coat(ing) of paint. **-masse** *f* coating compound, filming. **-mittelharz** *n* surface coating resin. **-stoff** *m* painting compound (= paint, varnish, lacquer). **-untergrund** *m* undercoating, primer

Anström·bedingungen *fpl* turbulent flow conditions

an·strömen 1 *vt* flow toward; wash up, deposit. **2** *vi*(s) stream in, crowd in. **Anströmung°** *f* incident flow

an·stücke(l)n *vt* piece on, patch on; append. **Anstück(l)ung** *f* (-en) added piece, patch; addendum; piecing (*or* patching) on

Ansturm° *m* attack, assault, onslaught; run, rush. **an·stürmen** *vi*(s) (**auf, gegen**) storm, attack, charge, rush (at); resist, fight back (against); storm in, rush in. **anstürmend** *p.a* (*oft*) onrushing

an·stützen 1 *vt* prop up. **2: sich a.** (**an, gegen**) lean (on, against)

an·suchen *vi*: **bei einem um etwas a.** apply to sbdy for sthg, request sthg from sbdy. **Ansuchen** n (-) request; application; petition. **Ansucher°** m applicant, petitioner

Ansud m (-e) **1** blanching, partial boiling, scorification, mordanting. **2** boiled-up liquid cf ANSIEDEN

an·summen vt & **sich a.** add up

an·süßen vt sweeten slightly

antaphrodisisch a anaphrodisiac

Antarktis f Antarctic(a). **antarktisch** a Antarctic

an·tasten vt **1** touch; touch on. **2** question. **3** violate

antedatieren vt antedate

an·teeren vt tar

an·teigen vt make into a paste (or dough); premix. **Anteig·mittel** n pasting agent

Anteil m **1** share, part, portion. **2** interest, sympathy, concern: **A. nehmen (an)** take an interest (in); sympathize (with). **--haber** m shareholder; participant. **an·teilig** a shared; proportionate, pro rata **anteil·los** a indifferent, unconcerned. **-mäßig** a proportionate, pro rata; shared. **A—nahme** f (an) interest, participation (in), concern (for). **A—schein** m, **Anteil(s)·verschreibung** f share, stock certificate

an·tempern vt (Glass) anneal

Anthelminticum n (..tica) anthelmintic, vermifuge

Anthocyan n anthocyanin

Anthrac . . . see also ANTHRAZ . . .

Anthracen n anthracene

Anthrachinolin n anthraquinoline. **Anthrachinon** n anthraquinone. **Anthrachinonazin** n anthraquinonazine. **anthrachinoid** a anthraquinoid

Anthranil·säure f anthranilic acid

Anthraphenon n anthraphenone (9-benzoylanthracene)

Anthrazen n anthracene

Anthrazit n anthracite. **a—·artig** a anthracitic. **A—bildung** f anthracitization. **anthrazitisch** a anthracitic. **Anthrazit·kohle** f anthracite

Anthrazyl n anthracyl

Anthroe·säure f anthroic acid

Anti·babypille f contraceptive pill. **a—bindend** a anti-bonding. **A—biotikum**

n (..tika) antibiotic. **-blockmittel** n antiblocking agent

anticipieren vt anticipate

an·ticken vt **1** touch lightly. **2** start (a clock)

Anti·dot n (-e), **-doton** n (..ta) antidote. **-fäulnis** f anti-fouling (or anti-rotting) agent. **-febrin** n antifebrin, acetanilide. **-fermentativ** n enzyme-inhibiting agent. **-flammittel** n fire-retarding agent. **-hydroticum** n (..tica) antiperspirant

antik a ancient; antique; **a—er Purpur** Tyrian purple

Antikatalysator m (-en) anticatalyst

Antikathoden·leuchte f anticathode luminescence

Antike f (-n) antiquity; antique cf ANTIK

anti·ketogen a antiketogenic. **A—klinale** f anticline. **-klopfmittel** n antiknock (agent). **-klumpmittel** n anticoagulant. **-körper** m antibody. **-krackmittel** n anticracking agent

Antilyssa·serum n antirabic serum

Antimon n antimony. **·-arsen** n antimony arsenide, (Min.) allemontite. **-arsenfahlerz** n tetrahedrite. **a—artig** a antimonal, antimony-like. **A—blei** n antimonial lead, lead containing antimony, lead-antimony alloy

Antimonblei·blende f boulangerite. **-kupferblende** f bournonite. **-spat** m bindheimite

Antimon·blende f kermesite. **-blüte** f antimony bloom, valentinite (Sb_2O_3). **-butter** f butter of antimony ($SbCl_3$). **-chlorid** n antimony pentachloride. **-chlorür** n antimony trichloride. **-fahlerz** n tetrahedrite. **-glanz** m antimony glance, antimonite, stibnite. **-goldschwefel** m golden antimony sulfide (Sb_2S_5). **-halogen** n antimony halide. **a—haltig** a antimonial, containing antimony

Antimoniat n (-e) antimonate

antimonig a antimonious: **a—e Säure antimonious acid. ·-sauer** a antimonite of. **A—säureanhydrid** n antimony trioxide

Antimonin n antimony lactate. **antimonisch** a antimonic, antimonial. **Antimonium** n antimony

Antimon·jodür n antimony triiodide. **-ka-**

liumtartrat *n* tartar emetic, antimonyl potassium tartrate. **-kermes** *m* kermes mineral. **-kupferglanz** *m* antimonial copper glance (bournonite *or* chalcostibite). **-nickel** *n* breithauptite. **-nickelglanz** *m* ullmannite. **-ocker** *m* antimony ocher (cervantite *or* stibiconite), (hydrated) oxide of antimony. **-oxid** *n* antimony trioxide. **-oxychlorür** *n* antimony oxychloride (SbOCl). **-phyllit** *n* valentinite. **a—reich** *a* rich in antimony, high-antimony. **A—safran** *m* antimonial saffron, crocus of antimony. **-salz** *n* antimony salt (*specif* $SbF_3 \cdot (NH_4)_2SO_4$). **a—sauer** *a* antimonate of. **A—säure** *f* antimonic acid. **-säureanhydrid** *n* antimonic anhydride (Sb_2O_5). **-silber** *n* antimonial silver, dyscrasite (silver antimonide). **-silberblende** *f* pyrargyrite. **-silberglanz** *m* stephanite. **-spat** *m* valentinite. **-spiegel** *m* antimony mirror. **-sulfid** *n* antimony pentasulfide. **-sulfür** *n* antimony trisulfide. **-wasserstoff** *m* antimony hydride, stibine. **-weiß** *n* antimony white (Sb_2O_3). **-zinnober** *m* kermes mineral

Anti·neuralgicum *n* (..gica) (*Med.*) antineuralgic (remedy). **-oxidationsmittel** *n*, **-oxydans** *n* (..danten), **-oxygen** *n* antioxidant, antioxydizing agent. **a—phlogistisch** *a* antiphlogistic. **-podisch** *a* antipodal, opposite

an·tippen *vt* touch lightly; touch on, hint at

anti·pyretisch *a* antipyretic

Antiquar *m* (-e) 1 second-hand bookseller. 2 antique dealer. **Antiquariat** *n* (-e) 1 second-hand bookstore; antique shop. **antiquarisch** *a* second-hand; antiquarian **antiquiert** *a* antiquated. **Antiquität** *f* (-en) 1 antiquity. 2 antique

Anti·scabiosum *n* scabies remedy. **-schaummittel** *n* anti-foam agent

Antiseptikum *n* (..ika), **antiseptisch** *a* antiseptic

antiskorbutisch *a* antiscorbutic

Antistatikum *n* (..ika) antistatic agent. **antistatisch** *a* antistatic

Anti·stellung *f* (*Oximes*) anti configuration

antistokes'sche Linie (*Phys.*) anti-Stokes line

Anti·teilchen *n* anti-particle

Antithese *f* (-n) antithesis

Anti·weinsäure *f* mesotartaric acid

antizipieren *vt* anticipate

Antlitz *n* (-e) face, countenance

an·tönen I *vt* 1 (*Dye.*) stain, tint. 2 hint at. II *vi*(s) begin to sound

Antonius·feuer *n* St. Anthony's fire, erysipelas

Antons·kraut *n* (*Bot.*) willow herb

Anzonit *n* (*Min.*) antozonite, fluorite

Antrag *m* (..t⸚e) 1 request, application, petition. 2 proposal, proposition, offer. 3 (*Parlam.*): **einen A. stellen** make a motion. **an·tragen*** 1 *vt* & **sich a.** offer. 2 *vt* apply; bring, carry in. **Antrag·steller** *m* proposer, mover

an·transportieren *vt* bring in (by vehicle)

an·träufeln *vt* dribble on, drip on

Antrazen *n* anthracene. **-öl** *n* anthracene oil

an·treffen* *vt* 1 find, run across. 2 affect, concern. **antreffend** *p.a* concerned

an·treiben* I *vt* 1 drive (on), propel, impel, urge on. 2 force (plants). II *vi* (h) (*Bot.*) put out shoots, sprout. III *vi*(s) drift ashore. **Antreiber** *m* (-) (slave) driver; instigator. **Antreibung** *f* driving (on), urging; instigation; forcing (of plants) *cf* ANTRIEB

an·treten* I *vt* 1 enter upon, start (on); accede to; take over. 2 start (by stepping on a pedal); apply (brakes). 3 tread down. 4 approach; confront. II *vi*(s) 5 appear, show up; report (for duty); fall in; line up. 6 participate; (**gegen**) compete (with). 7 (**an**) be added (to), join

Antrieb° *m* 1 (*incl Mach.*) drive. 2 (inner) drive; motivation: **aus eigenem A.** of one's own accord. 3 stimulus, incentive; inducement

Antriebs·kette *f* drive chain. **-kraft** *f* driving force, motive power. **-leistung** *f* driving force, power. **-maschine** *f* (driving) engine. **-mittel** *n* 1 means of propulsion. 2 incentive. **-rad** *n* drive wheel, driver. **-scheibe** *f* drive pulley. **-spindel** *f* drive shaft. **-strahl** *m* propulsion jet. **-stoff** *m* propellant, fuel. **-vorrichtung** *f* drive mechanism. **-welle** *f* drive shaft

Antritt° *m* 1 start, beginning. 2 assumption, takeover (of), accession (to), entry (upon). **3** first (*or* bottom) step. **4** acceleration, spurt, sprint. *Cf* ANTRETEN

Antritts-geld *n* admission (*or* entry) fee. **-rede** *f* inaugural address. **-vorlesung** *f* inaugural lecture

an-trocknen *vi*(s) **1** begin to dry, dry partly. **2** (*usu* + **an**) dry on (to), dry and stick (to). **Antrocknen** *n* partial drying

an-tun* *vt* **1** (*d.i.o*) do (harm, violence) (to), inflict (on); show (kindness) (to). **2** put on (garment). *Cf* ANGETAN

Antwerpen *n:* **A—er Blau** Antwerp blue

Antwort *f* (-en) answer, reply, response. **antworten** *vi* (*dat* or **auf**) reply, respond to; answer. **Antworter** *m* (-) respondent, person answering. **antwortlich** *prep* (*genit*) in reply (to)

an-vertrauen *vt* & **sich a.** (*d.i.o*) entrust (oneself) (to), confide (in). **Anvertrauen** *n* trust; commitment. **anvertraut** *p.a* (*oft*) held in trust

an-verwandeln *vt* (*dat rflx*): **sich Ideen a.** assimilate ideas

anverwandt *a* related. **Anverwandte** *m, f* (*adj endgs*) relative, related person

an-visieren *vt* aim at; sight; take bearings on (*or* of)

Anvulkanisation *f* (-en) prevulcanization, pre-cure. **an-vulkanisieren** *vt* prevulcanize, pre-cure, scorch (rubber)

Anw. *abbv* (Anwendung) application, use

Anwachs *m* **1** growth, increase. **2** swelling. **3** accumulation, accrual, accretion. **4** accrued interest. **5** (*Geol.*) alluvion

an-wachsen* *vi*(s) **1** grow on; take root, be rooted. **2** grow, increase, swell, rise. **3** accumulate. **Anwachsen** *n* growth, increase, rise; accumulation *cf* ANGEWACHSEN, ANWUCHS

Anwachs-kurve *f* growth curve

an-wählen *vt* (*Tlph*) dial (e.g. a data bank)

Anwalt *m* (-e *or* ..wⁱe) **1** lawyer, attorney; (legal) representative. **2** advocate

an-wandeln 1 *vt* (*imps*) come over, befall. **2** *vi*(s) approach slowly. **Anwandlung** *f* (-en) fit, seizure; inclination; whim, mood

an-wärmen *vt* warm up, warm slightly; preheat; (*Glass*) anneal. **Anwärmer** *m* (-) preheater, (heat) economizer. **Anwärmzeit** *f* heating (*or* warm-up) time

Anwärter *m* (-) candidate, aspirant

an-wässern *vt* moisten (with water), irrigate

an-wehen *vt* (*imps*) blow on; come over

an-weichen *vt* **1** soften slightly. **2** soak slightly

an-weisen* *vt* **1** (*usu* + **zu** + *inf*) instruct, direct, order (to). **2** instruct, train; break in. **3** (*d.i.o*) assign (to). **Anweiser** *m* (-), **Anweiserin** *f* (-nen) **1** instructor. **2** usher

an-weißen *vt* whiten, whitewash

Anweisung *f* (-en) **1** (*genl*) instruction(s). **2** (*specif*) order(s); training; directions *cf* GEBRAUCHSANWEISUNG. **3** assignment (to a job, etc). **4** (*Fin.*) remittance; draft, money order

anwendbar *a* usable, applicable. **Anwendbarkeit** *f* usability, applicability

an-wenden* *vt* use, (+ **auf**, *usu*) apply (to) *cf* ANGEW., ANGEWANDT. **Anwendung** *f* (-en) use, application: **A. finden** be used, be applied; **in A. bringen** put to use, apply; **zur A. kommen** be put to use, be applied

Anwendungs-bereich *m* **-gebiet** *n* field (*or* range) of application, scope. **-möglichkeit** *f* possible application, applicability; opportunity to apply. **-technik** *f* technique of use; method of application. **-techniker** *m* expert in the field. **a—technisch** *a* (of) technical service . . . **A—weise** *f* mode of application. **-zweck** *m* application

an-werben* *vt* solicit, hire, recruit; **sich a. lassen** enlist

an-werfen* **I** *vt* **1** throw (*or* fling) at. **2** roughcast (a wall); (+ **an**) throw on (mortar). **3** start, crank up, throw into gear; turn on. **II** *vi* have the first throw

Anwesen *n* (-) estate, piece of (real) property; premises; farm

anwesend *a* present, attending. **Anwesenheit** *f* presence, attendance

an-widern 1 *vt* disgust, turn . . . off. **2** *vi* arouse disgust

an-wirbeln 1 *vt* fasten with a turnbolt. **2** *vi*(s) arrive in a whirl; (**gegen**) whirl, spin (against)

an-wittern 1 *vt* sniff out. **2** *vi* begin to weather; effloresce

an-wohnen *vi* (*dat*) live near; attend

Anwuchs° *m* **1** rooting, taking root.

2 young forest growth; forest preserve.
3 growth, increase 4 marine fouling (of
ships). *Cf* ANWACHSEN

Anwurf° *m* 1 roughcast, plaster(ing).
2 first draft. 3 (*Motors*) priming, starting,
cranking up. 3 deposit, alluvium. 4 slan-
derous remark; false accusation. *Cf* AN-
WERFEN. **~kurbel** *f* starting crank

an·würgen *vt* grip, crimp

an·wurzeln *vi*(s) take root *cf*ANGEWURZELT

Anz. *abbv* 1 (Anzahl) number. 2 (Anzeige)
ad(vertisement). 3 (Anzeiger) advertiser

Anzahl° *f* number

an·zahlen *vt* pay on account, pay as a deposit; pay down. **Anzahlung°** *f* payment on
account; deposit; down payment

an·zapfen *vt* tap (a keg, wires)

Anzapf·ventil *n* bleeder valve

An·zeichen° *n* sign, indication; mark.
an·zeichnen *vt* 1 (**an**) draw (a picture) on.
2 mark, indicate

Anzeige *f*(-n) 1 ad(vertisement), announcement; notice, notification. 2 complaint,
accusation, charge: **zur A. bringen** report, file a charge of. 3 sign, indication;
(instrument) reading; indicator. **~appa-
rat** *m*, **-gerät** *n* indicating instrument,
indicator. **-mittel** *n* (*Chem.*) indicator

an·zeigen I *vt* 1 indicate, mark; show, be a
sign of; (*Instrum., oft*) read, register.
2 report, announce; (*d.i.o*) inform, notify
(sbdy) of. 3 report (to the authorities), file
charges against. **II: sich a.** 4 show up,
appear, be indicated. 5 announce oneself.
anzeigend *p.a* indicative. **Anzeiger** *m* (-)
1 indicator, gage; registering instrument.
2 index. 3 (*Math.*) exponent. 4 advertiser
(*also* in newspaper names). 5 informant

Anzeige·röhre *f* visual-indicator tube;
electric eye. **-tafel** *f* indicator board. **-vor-
richtung** *f* indicator, registering device

Anzeigung *f* indication, registering; reporting, filing of charges; announcement; advertisement *cf* ANZEIGEN

an·zementieren *vt* (**an**) cement on (to)

Anzettel *m* (-) (*Tex.*) warp. **an·zetteln** *vt*
1 (*Tex.*) warp; frame. 2 hatch, contrive
(plots); instigate. **Anzettler** *m* (-) instigator, schemer. **Anzettlung** *f* warping;
(plot) hatching, scheming; instigation

an·ziehen* I *vt* 1 pull on, put on (clothes);

dress (a psn). 2 (*incl Phys.*) draw, attract;
absorb. 3 pull up, draw up; adduct (muscle); pull (a door) to; pull taut, tighten;
drive (a screw). 4 raise, grow, breed, cultivate. 5 cite, quote. **II** *vi* (h) 6 start to pull,
pull, pull out, start moving. 7 rise, increase, advance. 8 turn cold. 9 harden,
set; bite, grip, take hold; tighten up; take
effect. 10 be attractive. 11 attract moisture. **III** *vi*(s) 12 come marching up (*or*
in); move in; approach. 13 move, change
residence. **IV: sich a.** dress, get dressed
Cf ANGEZOGEN. **anziehend** *p.a* adducent
(muscle); attractive, appealing; astringent. **Anziehung°** *f* 1 attraction, appeal.
2 gravitation, pull; adduction; tension.
3 rise, increase. 4 absorption. 5 adhesion.
Cf ANZUG

Anziehungs·bereich *m*. **-gebiet** *n* range of
attraction (*or* appeal); gravitational field.
-kraft *f* 1 force (*or* power) of attraction,
attractive force. 2 drawing power, appeal.
-punkt *m* center of attraction; drawing
card. **-wirkung** *f* attractive effect

an·zinnen *vt* tin, coat with tin

Anzucht *f* 1 raising, growing, cultivation,
breeding. 2 nursery

an·zuckern *vt* sugar, sprinkle with sugar

Anzug° *m* 1 suit (of clothes). 2 approach,
advance: **im A. sein** be approaching, be
afoot. 3 tightening. 4 taper. 5 first move.
Cf ANZIEHEN 1, 3, 9, 13, IV

anzüglich *a* insinuating, personal. **Anzüg-
lichkeit** *f* (-en) insinuation, personal
remark

Anzugs·kraft *f* starting power, tractive
force. **-moment** *n* initial torque. **-tag** *m*
starting (*or* first) day (on a job). **-ver-
mögen** *n* (*Cars*) getaway

an·zünden *vt* light, kindle, ignite, set afire.
Anzünder° *m* (cigarette) lighter; igniter.
Anzündung° *f* lighting, kindling, igni-
tion

an·zweifeln *vt* doubt, call into question

Aorta *f* (..ten) aorta. **Aorten·ast** *m* branch
of the aorta. **-kammer** *f* left ventricle.
-wand *f* wall of the aorta

Aouara·öl *n* African palm oil

Ap. *abbv* (Apotheker) druggist, pharmacist

A.P. *abbv* (amerikanisches Patent) Amer.
pat.

apart a striking, distinctive; special; adv oft apart, aside

Apathie f(-n) apathy. **apathisch** a apathetic

Apatit·kristall m crystal of apatite, phosphorite

aperiodisch a aperiodic

Apertur·begrenzung f aperture limitation

Apfel m (··) apple. **-äther** m ethyl malate; malic ether; apple essence. **-branntwein** m apple brandy, applejack. **-brei** m applesauce. **-dorn** m crabapple tree

Äpfel·eisenextrakt m (Pharm.) ferrated extract of apples

Apfel·frucht f(Bot.) pome. **-kernöl** n apple seed oil, amyl valerate. **-most** m sweet cider. **-mus** n applesauce. **-öl** n apple oil, malic ether. **-saft** m apple juice. **-salbe** f pomatum. **a—sauer** a malic, malate of. **A—säure** f malic acid. **-schnaps** m apple brandy, applejack. **-schorf** m apple scab. **-sine** f(sweet) orange

Apfelsinenschalen·effekt m, **-erscheinung** f orange-peel effect (of coatings); orange peeling. **-öl** n orange peel oil; essence of orange oil, oil of sweet orange

Apfel·trester m pomace. **-wein** m apple wine; apple cider

Aphel n (-e), **Aphelium** n (..lien) aphelion, electron position farthest from the nucleus

Aphrodin n yohimbine

Apiol n apiol, parsley camphor

Apion·säure f apionic acid

Apocampher·säure f apocamphoric acid

Apochinen n apoquinene

Apochinin n (-e) apoquinine

apochromatisch a apochromatic

Apogluzin·säure f apoglucic acid

Apohydrochinin n (-e) apohydroquinine

Apokaffein n (-e) apocaffeine

Apokampher·säure f apocamphoric acid

Apokodein n (-e) apocodeine

apolar a nonpolar

A-Polymerisation f addition polymerization

aposteriorisch a empirical, a posteriori

Apotheke f (-n) (prescription) pharmacy; medicine chest (or kit). **Apotheker** m (-) apothecary, pharmacist

Apotheker·buch n pharmacopoeia; dispensatory. **-farbe** f drug (or pharmacist's) color. **-gewicht** n apothecary (or troy) weight. **-kunst** f (science of) pharmacy. **-ordnung** f dispensatory. **-verein** m pharmaceutical society, pharmacists' association. **-waage** f apothecary (or pharmacist's) scale(s). **-waren** fpl pharmaceuticals, drugs. **-warenhändler** m wholesale druggist. **-wesen** n pharmaceutical field (or business). **-wissenschaft** f pharmacology

App. abbv (Apparat) apparatus

Apparat m (-e) 1 apparatus; device, appliance; gadget, paraphernalia. 2 (tele)phone: am A. on the line, speaking. 3 oft = PHOTOAPPARAT, RASIERAPPARAT, etc

Apparate·bau m apparatus (appliance, instrument) construction. **-brett** n instrument panel

Apparaten·kunde f knowledge of apparatus

Apparate·schnur f appliance electric cord. **-teil** m apparatus (or appliance) part. **-tisch** m instrument table. **-wesen** n appliance (apparatus, instrument) field

Apparat-färberei f machine dyeing

apparativ a, adv as regards apparatus, by means of apparatus

Apparat·satz m set of apparatus

Apparatur f (-en) apparatus; (set of) equipment, machinery, gadgetry

Appell m (-e) appeal; roll call, inspection. **appellieren** vi (an) appeal (to)

Appetit m appetite. **appetitlich** a appetizing

appetitlos a lacking appetite, anorectic: a. sein have no appetite. **Appetitlosigkeit** f lack of appetite, anorexia

applizieren vt apply, administer; appliqué. **Applizierung** f(-en) application

apportieren vt/i retrieve

Appret n (Tex.) size, sizing; stiffening; finish. **appretieren** vt size, stiffen; finish, dress. **Appretier·masse** f finishing material, size. **Appret·kocher** m (Tex.) finish boiler. **Appretur** f (-en) finishing; dressing; finish, size; (Lthr.) seasoning

appretur·echt a finish-resistant, unaffected by dressing. **A—körper** m filler, loading (agent). **-leim** m dressing size.

-**masse** *f* dressing, sizing, size. -**mittel** *n* finishing agent, sizing agent; dressing (material). -**öl** *n* dressing oil, finishing oil. -**präparat** *n* dressing. -**pulver** *n* dressing powder. -**seife** *f* dressing soap. -**verfahren** *n* finishing process. -**zeit** *n* finishing time

approbieren *vt* approve; certify

approximativ *a* approximate

Aprikose *f* (-n) apricot

Aprikosen·kernöl *n* apricot kernel oil. -**pfirsich** *m* nectarine

apriorisch *a* nonempirical, deductive

aprotid *a* aprotic

aptieren *vt* adapt, adjust; get (sthg) ready

apyrisch *a* apyrous, noncombustible, fireproof

Aquametrie *f* determination of aqueous content

Aquarell *n* (-e) water color (painting). -**farbe** *f* water color. **aquarellieren** *vt/i* paint in water colors

Äquation *f* (-en) equation

Äquator *m* equator. **äquatorial** *a* equatorial

Aquavit *m* aqua vitae, brandy, liquor

äqui· *pfx* equi- e.g. **äquimolekular**

Äquinoktium *n* (..tien) equinox

äquivalent *a*, **Äquivalent** *n* (-e) equivalent. -**leitfähigkeit** *f* equivalent conductance

Äquivalenz *f* (-en) equivalence. -**punkt** *m* (*Analysis*) end point. -**verbot** *n* exclusion principle

aquotisieren *vt* aquate. **Aquotisierung** *f* (-en) aquation

Aquo·verbindung *f* aquo compound

Ar *n* (-) are, 100 sq m

ar. *abbv* (aromatisch) aromatic

Ära *f* (Ären) era

Araber *m* (-), **Araberin** *f* (-nen) Arab. **Arabien** *n* Arabia

Arabin·gummi *n* gum arabic. -**säure** *f* arabi(ni)c acid

arabisch *a* Arabic, Arabian: **a—es Gummi** gum arabic; **a—er Balsam** balsam of Mecca, balm of Gilead

Arabit *n* arabitol

Arabon·säure *f* arabonic acid

Arachidon·säure *f* arachidonic acid

Arachin·säure *f* arach(id)ic acid

Arachis·öl *n* arachis oil, peanut oil

Aräo·messer *m*, **Aräometer** *n* (-) hydrometer, areometer. **Aräometrie** *f* hydrometry, areometry. **aräometrisch** *a* hydrometric, areometric

Arb. *abbv* (Arbeit) work

Arbeit *f* (-en) 1 (*genl, incl Phys. Mach.*) work: **in A. haben** be working on, have in preparation; **in A. nehmen** start work on; **in A. geben** have work started on. 2 toil, labor (*incl Polit.*); effort, trouble. 3 employment, job. 4 piece of work; job, project, task, assignment, chore. 5 (piece of) needlework, sewing. 6 written test (*or* examination); paper, essay, composition; report; dissertation. 7 study, investigation, research project. 8 workmanship. 9 fermentation. 10 operation, activity; procedure

arbeiten I *vt* 1 work on; make, produce, fashion. **II** *vi* 2 work, do work; toil, labor. 3 (**an**) work (on), study, prepare. 4 ferment, be in ferment; be active; rage; (*Dough*) rise. 5 (*Mach.*) run, operate. 6 yield a profit. **III: sich a.** 7 (e.g. **zu Tode**) work oneself (to death). 8 work one's way (to, toward, through). **arbeitend** *p.a* working, operating, active

Arbeiter *m* (-) worker, workman; employee; laborer. -**ausstand** *m* strike, walkout

Arbeiterin *f* (-nen) worker, employee

Arbeiterschaft *f* (-en) 1 (all the) workers, employees; working personnel. 2 working class, labor

Arbeiter·verein *m* workers' organization (*or* association); union

Arbeit·geber *m* employer. -**nehmer** *m* employee

arbeitsam *a* 1 industrious, hard-working. 2 work-filled (life, days). **Arbeitsamkeit** *f* industriousness

Arbeits·änderung *f* change in work. -**anweisung** *f* working (*or* operating) instructions. -**äquivalent** *n* mechanical equivalent. -**aufwand** *m* expenditure of work, effort (involved). -**ausschuß** *m* working (*or* executive) committee. -**band** *n* assembly line. -**bedarf** *m* (*Engg.*) energy requirement(s). -**bedingungen** *fpl* working (*or* operating) conditions.

-belastung *f* work load, operating load. **-bezeichnung** *f* work designation. **-druck** *m* working pressure; pressure of work. **-einheit** *f* 1 unit of work, erg. 2 work(ing) unit. **-einsparung** *f* saving of labor. **-einstellung** *f* 1 work stoppage. 2 attitude toward one's work. **a-fähig** *a* fit for work, able-bodied. 2 operational, in operating condition. **-fähigkeit** *f* 1 fitness for work, working capacity. 2 operational condition. **-feld** *n* field of work, sphere of activity. **-festigkeit** *f* fatigue strength, strength limit. **-fläche** *f* bearing surface; working surface. **-fuge** *f* expansion (*or* construction) joint. **-gang** *m* 1 step (*or* stage) of work, operation; working procedure (process, cycle). 2 course of manufacture. 3 walkway, gangway. **a-geil** *a*, **A-geile** *m,f* (*adj endgs*) workaholic. **-gemeinschaft** *f* 1 working team, study group, seminar. 2 syndicate, alliance, association. 3 employer-employee partnership. **-größe** *f* quantity of work. **-gruppe** *f* team, task force. **-gut** *n* working material. **-hub** *m* working stroke, power stroke. **-kammer** *f* workroom; laboratory. **-körper** *m* working medium. **-kraft** *f* 1 capacity to work; effort. 2 worker. 3 *pl* **-kräfte** workers, personnel; labor(force), manpower. **-kreis** *m* 1 working team, study group. 2 field of work. **-lager** *n* labor camp. **-leistung** *f* amount of work performed; output; efficiency. **-lohn** *m* wages

arbeitslos *a* unemployed, out of work. **Arbeitslosigkeit** *f* unemployment

Arbeits·markt *m* labor market. **a-mäßig** *a*, *adv* as regards work. **A-methode** *f* working method, procedure. **-okular** *n* working eyepiece. **-ordnung** *f* 1 organization of work. 2 work regulations

arbeit·sparend *a* labor-saving

Arbeits·platz *m* 1 workplace. 2 (*esp in pl*) job(s). **-prozeß** *m* 1 working procedure. 2 employment (*or* labor) set-up (of a country). **-raum** *m* workroom, (work)shop; laboratory. **-schritt** *m* operation, step (in a working procedure). **-schutz** *m* 1 industrial safety, safety in the workplace. 2 safety device (*or* appliance). **-sitzung** *f* business meeting. **-spiel** *n* time cycle (of

operations). **-stätte** *f* workplace; workroom, (work)shop; laboratory. **-stelle** *f* 1 (available) job. 2 place of employment. 3 workplace, shop. **-strom** *m* (*Elec.*) open-circuit current; working (*or* operating) current. **-stube** *f* workroom, (work)shop; study, office. **-stufe** *f* stage of work, operating stage. **-stunde** *f* working hour; man hour. **-tag** *m* working day, workday. **a-täglich** *a*,*adv* per working day. **A-takt** *m* working tempo (*or* rate); operational cycle. **a-teilig** *a* based on a division of labor. **A-teilung** *f* division of labor. **-temperatur** *f* working temperature. **-tisch** *m* work table, workbench. **-tüchtigkeit** *f* working capacity (*or* efficiency). **-tür** *f* (*Metll.*) charging door. **-verfahren** *n* working procedure (*or* method). **-vermittlung** *f* employment office (*or* bureau). **-vermögen** *n* 1 working capacity. 2 (*Phys.*) energy, capacity for work. **-verrichtung** *f* work performance. **-vorgabe** *f* (suggested) specifications (*or* plan of work). **-vorgang** *m* operating procedure. **-vorhaben** *n* operating plan. **-walze f** (*Metll.*) work roll. **-weise** *f* working method, way of working, (operating) procedure; the way it works. **-wert** *m* value in terms of work, mechanical equivalent. **-wut** *f* passion for work. **-zeit** *f* 1 working hours. 2 operating time (for a particular job *or* operation). 3 time for work. 4 time per piece of work. **-zeug** *n* tools, work(ing) equipment. **-zimmer** *n* workroom, (work)shop; study, office

Arbuse *f* (-n) watermelon; arbutus

Arbutus·fruchtöl *n* arbutus oil

Arch. *abbv* (Archiv, Archive) archive(s)

archaisch *a* archaic, ancient

archäisch *a* (*Geol.*) Archaen, azoic

Archäolog *m* (-en, -en) archeologist. **archäologisch** *a* archeological

architektonisch *a* architectural

Archiv *n* (-e) 1 archives, records, files; (*also used in names of periodicals*). 2 records office. **-tinte** *f* record(ing) ink

Arcus·cosinus *m* (*Math.*) inverse cosine, arc cos, $\cos^{-1}$. **-sinus** *m* inverse sine, arc sin, $\sin^{-1}$

Areal *n* (-e) grounds; area. **-verschiebung** *f* shift in location

Areca·nuß, Areka·nuß f areca (or betel) nut. **-nußöl** n areca (nut) oil (or fat)
arg a 1 bad, wrong: **im A—en liegen** be in a bad way. 2 evil; mean, nasty. 3 severe, grave. 4 inveterate. Cf ÄRGER a, ÄRGSTE.
Arg n malice; harm; guile
Argand·brenner m Argand burner
Argentin·druck m metal printing
Argentinien n Argentina
ärger a (compar of ARG) worse
Ärger m anger, aggravation; annoyance, trouble. **ärgerlich** a 1 angry; annoyed. 2 annoying, aggravating. **ärgern** 1 vt anger, make angry, aggravate. **2: sich ä.** get angry, be annoyed. **Ärgernis** f (-se) annoyance; offense
Arg·list f guile, craft, cunning; malice. **a-listig** a guileful, crafty, cunning, malicious
arglos a guileless, innocent. **Arglosigkeit** f guilelessness, innocence
Argon·hochstrombogen m high-current argon arc
ärgste a (superl of ARG) worst
Argument n (-e) (incl Math.) argument. 2 (Math.) independent variable; polar angle. **argumentieren** vi argue
Argwohn m suspicion. **argwöhnen** vt suspect. **argwöhnisch** a suspicious
a. Rh. abbv (am Rhein) on the Rhine
arithmetisch a arithmetic(al): **a—e Reihe** arithmetic progression
Arkatom·schweißen n atomic hydrogen welding
Arktis f, **arktisch** a Arctic
arm 1 a poor; low(-grade); deficient; weak; (Gas) lean. 2 sfx poor in... , low- cf ASCHEARM
Arm m (-e) 1 arm: **unter die A—e greifen** (+dat) come to the aid (of). 2 branch. 3 (cross)beam; shaft
Armatur f (-en) armature; fitting, mounting; accessory; reinforcement (of concrete)
Armaturen·brett n instrument board (or panel). **-front** f dashboard
Arm·band n bracelet. **-banduhr** f wristwatch. **-binde** f sling. **-blei** n refined (silver-free) lead. **-bruch** m fracture of the arm

Armee f (-n) army
Ärmel m (-) sleeve. **--kanal** m English Channel
armenisch a: **a—e Erde** Armenian bole
ärmer a poorer cf ARM a
Arm·erz n low-grade ore. **-füß(l)er** m brachiopod. **-gas** n lean gas
armieren vt arm, equip; armor, reinforce. **Armierung** f (-en) arming, armament, equipment; armor(ing), reinforcement
Arm·leuchter m candelabrum; (Bot.) stonewort, chara
ärmlich a poor(-looking), shabby; miserable; scanty, meager
Arm·oxide npl (Lead) skimmings
arm·selig a pitiful, miserable, paltry. **Armseligkeit** f pitifulness, etc
ärmste a poorest cf ARM a
arm·treiben* vt (Metll.) concentrate
Armut f 1 povery, indigence. 2 poor quality, low grade
Arnika·blüten fpl arnica flowers, (Pharm.) arnica. **-wurzel** f arnica root (or rhizome)
Arom(a) n (..men) aroma; essence
Aroma·rückgewinnung f volatile recovery. **-stoff** m aromatic substance
Aromaten mpl aromatic substances, aromatics. **aromatisch** a aromatic: **a—es Mittel** (Pharm.) aromatic (substance). **aromatisieren** vt aromatize; perfume. **Aromatisierung** f aromatization. **Aromatizität** f aromaticity
Aron m arum. **--gewächse** npl Araceae
Arons·stab m arum (Arum maculatum). **-stärke** f arum starch
Aron·stab m arum. **-wurzel** f arum root
arretieren vt arrest; impound; stop; lock. **Arretierung** f (-en) arrest; seizure; stop; (Mach.) detent
Arrhenius·darstellung f Arrhenius diagram
arrodieren vt erode. **Arrosion** f (-en) erosion
Arrow·mehl n arrowroot flour, (Pharm.) arrowroot, maranta
Arsanil·säure f arsanilic acid
Arsen n arsenic. **--abstrich** m arsenic skimmings. **-antimon** n antimony arsenide, (Min.) allemontite. **-antimonnickelglanz** m corynite, arsonian ull-

mannite, -äscher m (Lthr.) arsenic lime liquor. -blende f arsenic blend —gelbe A. orpiment; rote A. realgar. -blüte f arsenic bloom, arsenolite, As_2O_3. -chlorid n arsenic trichloride. -dampf m arsenic(al) vapor. -eisen n iron (or ferric) arsenide, arsenical iron, (Min.) löllingite. -eisensinter m petticite. -fahlerz n tennantite. a–führend a arseniferous. A–gelb n orpiment. -glas n arsenic glass, vitreous As_2O_3. -halogen n arsenic halide. a–haltig a arsenical, containing arsenic

Arseniat n (-e) arsenate

arsenig a arsenious. --sauer a arsenite of. A–säure f arsenious acid. -säuresalz n arsenite

Arsenik m,n arsenic —gelber A. orpiment; roter A. realgar; weißer A. white arsenic, arsenolite

Arsenikal·fahlerz n tennantite, gray copper ore

Arsenikalien npl arsenicals, arsenical compounds (or preparations). arsenikalisch a arsenical

Arsenikal·kies m löllingite, leucopyrite, arsenopyrite, mispickel

Arsenik· see also ARSEN·: -antimon n antimony arsenide, allemontite, (Min.) -blei n lead arsenite. -bleispat m (Min.) mimetite. -blumen fpl, -blüte f arsenic bloom, arsenolite, As_2O_3. -fahlerz n tennantite. -gegengift n arsenic antidote. -glas n arsenic glass, vitreous As_2O_3—gelbes A. orpiment; rotes A. realgar. -hütte f arsenic works. -kalk m arsenolite. -kobalt, -kobaltkies m cobalt arsenide, (Min.) skutterydite. -kupfer n copper arsenide. -öl n: ätzendes A. caustic oil of arsenic, $AsCl_3$. -säure f arsenic acid; unvollkommene A. arsenious acid. -silber n arsenical silver. -sinter m scorodite. -spießglanz m allemontite. -vitriol n arsenic sulfate

Arsen·jodid n arsenic triiodide. -jodür n arsenic diiodide. -kies m arsenical pyrites, (arsenopyrites or löllingite). -kobalt m cobalt arsenide. -kupfer n copper arsenide, (Min.) domeykite. -mehl n arsenic(al) dust. -metall n metallic arsenic, metal arsenide. -nickel n nickel arsenide, (Min.) niccolite. -nickelglanz m gersdorffite. -nickelkies m niccolite, cloanthite

Arsenobenzol n arsenobenzene

Arsenopyrit m arsenical pyrites

Arsen·oxid n arsenic trioxide. -polybasit n pearceite. -präparat n arsenical (preparation). -probe f arsenic test (or sample). -rohr n, -röhre f arsenic tube. -rot n red arsenic, realgar (As_2S_3). -rotgültigerz n proustite. -rubin m realgar, ruby arsenic (AsS). a–sauer a arsenate of. A–säure f arsenic acid. -schwarz n black (or beta-) arsenic. -silber n silver arsenide. -silberblende f proustite. -sulfid n arsenic pentasulfide. -sulfür n arsenic trisulfide

Arsenür n (-ous) arsenide

Arsen·wasserstoff m arsenic hydride (arsine, AsH_3). -zink n zinc arsenide

Arsinig·säure f arsinic acid

Arsin·säure f arsonic acid

arsonieren vt arsonate, introduce the arsonic group (into)

Arson·säure f arsonic acid

Art f (-en) 1 way, manner: auf diese A. (und Weise) in this manner, this way; in der A., daß. . . in such a way that . . . ; nach A. (der, des) in the manner (of). 2 way to act, manners. 3A sort, kind, type, nature; (Biol.) species: eine A. (+noun) a kind (or sort) of . . . ; aus der A. schlagen go one's own way, break out of the mold, degenerate. B (in compds., usu generic): Pecharten kinds (or types) of pitch(es)

Art. abbv (Artikel) art. (article)

artbar a arable

art·eigen a characteristic of the species, true to type. Art-eigenschaft f species characteristic

arten I vt 1 form, fashion. 2 cultivate (land). II vi(s) 3 (+nach) turn out (like), take (after). 4: gut a. thrive. cf GEARTET

Arten·wandel m mutation

Arterie f artery. arteriell a arterial

Arterien· arterial:-druck m arterial (or blood) pressure. -kammer f left ventricle. -verkalkung f arteriosclerosis

arteriös *a* arterial

artesisch *a* artesian

art·fremd *a* alien, foreign (to the species); extraneous. **-gemäß** *a* true to type.

Art·genosse *m*, **-genossin** *f* member of the same species; fellow man (woman). **art·gleich** *a* of the same species, (*Biol.*) co-specific, identic. **Art·gleichheit** *f* identity (of species)

artig *a* good, well-behaved; well-mannered, polite; nice; *adv oft* well

·artig *a sfx* -like, -ous, -ic, -style *cf* **asbest·artig, asphalt·artig, neu·artig**

Artikel *m* (-) article, item; entry; (*Tex.*) style. **a·weise** *adv* item by item

artikulieren *vt* articulate

Artischocke *f* (-n) artichoke

Art·kreuzung *f* (*Biol.*) crossing of species

artlich *a* qualitative

Artung *f* quality, nature; inclination

Art·verband *m* group(ing) of species.

art·verwandt *a* naturally related, kindred. **Art·wärme** *f* specific heat

arylieren *vt* arylate

Arz(e)nei *f* (-en) medicine, drug. **·bereiter** *m* pharmacist. **-bereitung** *f* preparation of medicines, pharmacy. **-buch** *n* pharmacopeia; dispensatory. **-essig** *m* medicated vinegar. **-formel** *f* medical formula. **-gabe** *f* dose (of medicine). **-geruch** *m* medicinal odor. **-gewicht** *n* officinal weight. **-handel** *m* drug trade. **-händler** *m* druggist. **a–kräftig** *a* medicinal; curative, therapeutic. **A–kugel** *f* (*Pharm.*) bolus. **-kunde, -kunst** *f* pharmacology, pharmaceutics. **-lehre** *f* pharmacology

arzneilich *a* medicinal, pharmaceutical

Arznei·maß *n* officinal measure. **-mittel** *n* medicine, remedy

Arzneimittel·kunde *f,* **-lehre** *f* pharmacology. **-sucht** *f* drug addiction. **-träger** *m* (*Pharm.*) carrier, excipient

Arznei·seife *f* medicated soap. **-stoff** *m* medicinal substance, pharmaceutical. **-trank** *m* (medicinal) potion. **-waren·kunde** *f* pharmacology. **-wesen** *n* field of pharmaceuticals. **-wissenschaft** *f* pharmacology, medicinal science. **-zettel** *m* medical prescription

Arzt *m* (-e) physician, doctor. **Ärzteschaft** *f* (-en) **1** medical profession (group, circle).

2 = **Ärzte·verein** *m* medical society.

Ärztin *f* (-nen) physician, doctor

ärztl. *abbv* (ärztlich) med. (medical)

ärztlich *a* medical; *adv oft* by a doctor

As *n* (-se) ace

as· *abbv* (asymmetrisch) asym. (asymmetrical)

A.S. *abbv* (Ampère-Stunde) amp/hr (ampere hour)

Asant *m:* **stinkender A.** asafetida; **wohlriechender A.** gum benzoin. **--öl** *n* asafetida oil

Asaron·säure *f* asaronic acid

Asarum·kampfer *m* asarum camphor, asaron

Asbest *m* asbestos: **schillernder A.** chrysotile asbestos. **a—·artig** *a* asbestoid, asbestine, asbestiform, amianthine. **A–aufschlämmung** *f* asbestos suspension. **-drahtnetz** *n* asbestos wire gauze. **-flocken** *fpl* asbestos flocks (*or* wool). **-kordel** *f* asbestos string

Asbestolith *n* short-fibered asbestos. **Asbestose** *f* (-n) asbestosis

Asbest·schale *f* asbestos dish. **-schicht** *f* asbestos layer; (*Filtration*) asbestos mat. **-schnur** *f* asbestos string

Asbolan *m* (*Min.*) asbolite

Aschanti·nuß *f* peanut

Asch·becher *m* ashtray. **-blau** *n* zaffer. **-blei** *n* native bismuth. **a–bleich** *a* ashen, pale

Asche *f* ash, ashes, cinders: **gebundene A.** inherent ash (in coal)

Äsche *f* (-n) **1** (*Zool.*) grayling. **2** (*Bot.*) ash (tree) = ESCHE

asche·arm *a* low-ash. **-bildend** *a* ash-forming. **-frei** *a* ash-free. **A–gehalt** *m* ash content

Aschel, Äschel *n* (*Metll.*) sullage, ash from cobalt ores; ash spot (in iron, steel)

Asch·eimer *m* ashcan, garbage can

aschen *vt* ash

aschen·arm *a* low-ash. **A–becher** *m* ashtray. **-eimer** *m* ashcan, garbage can. **-ermittlung** *f* determination of ash content. **-fall** *m* ash pit, ash pan. **a–fleckig** *a* (*Metll.*) sullage-specked. **-frei** *a* ash-free, ashless. **A–gehalt** *m* ash content. **-halde** *f* ash (*or* slag) heap; (refuse) dump. **-kasten** *m* ash box (*or* bin). **-kübel** *m* ashcan.

-lauge f lye from ashes. **-ofen** m calcining oven. **-raum** m ash pit. **-salz** n potash, K_2CO_3. **-schmelzpunkt** m ash fusion temperature. **-schwefel** m sulfur in the ash(es). **-tonne** f ashcan, ash barrel. **-trecker** m (Min.) tourmaline. **-verflüssigung** f fusion of the ash(es), clinkering. **-vulkan** m cinder core. **-zieher** m (Min.) tourmaline

Ascher m (-) 1 ashtray. 2 = ÄSCHER

Äscher m (-) 1 (Lthr.) tanner's pit, lime pit. 2 (Soap) ash cistern. 3 (Masonry) slaked lime, 4 (Ceram.) tin ashes. 5 (Glass) buck ashes, lye ashes. **-brühe** f (Lthr.) lime liquor

asche-reich a rich (or high) in ash, high-ash

Äscher- (Lthr.): **-faß** n liming tub. **-flüssigkeit** f lime liquor. **-grube** f lime pit. **-kalk** m lime

äschern vt ash; (Lthr.) lime, slacken (hides); treat with lye

Äscher-ofen m (Ceram.) frit kiln

Äscherung f ashing; liming; treatment with lye cf ÄSCHERN

Asche-schichtenkurve f (Coal) instantaneous ash curve

Asch-fall m ash pit, cinder pit. **a--farben** a ash-colored, ashy. **-grau** a ash-gray

aschig, äschig a ashy

aschist a (Petrog.) aschistic

Asch-lauch m shallot, eschalot. **-raum** m ash box (bin, pit)

Ascorbin-säure f ascorbic acid

Asebotosid n (-e) phlorrhizin

äsen vi graze, browse

Asepsis, Aseptik f asepsis

Asien n Asia

Asparagin-säure f aspartic acid

Aspe f (-n) aspen (tree)

asphalt-artig a asphalt-like, asphaltic, bituminous. **A--beton** m asphalt(ic) concrete. **-gestein** n native asphalt. **a--haltig** a asphaltic, containing asphalt

asphaltieren vt asphalt

Asphalt-kitt m asphalt cement (or mastic). **-lack** m asphalt varnish. **-mastix** m mastic asphalt. **-pech** n bituminous pitch. **-pflaster** n asphalt pavement. **a--reich** a high-asphalt **A--stein** m native (or crude) asphalt, asphalt stone, mineral pitch.

-stoff m asphaltene. **-verfahren** n asphalt (or bitumen) process

asphärisch a aspherical

Aspirateur m (-e) aspirator. **aspirieren** 1 vt aspirate. 2 vi (auf) aspire (to)

asporagen a non-sporing

aß ate: past of ESSEN

Assam-kautschuk m Assam rubber

Assanierung f (-en) sanitation; purification

Assekuranz f (-en) insurance. **Assekurat** m (-en, -en) insured (person). **assekurieren** vt insure

Assel f (-n) isopod; wood louse

Assimilat m (-en, -en) assimilate(d material). **assimilatorisch** a assimilatory, assimilative

assimilierbar a assimilable. **Assimilierbarkeit** f assimilability. **assimilieren** vt assimilate

Assistent m (-en, -en), **Assistentin** f (-nen) assistant. **Assistenz** f assistance. **-arzt** m resident (physician)

Associé m (-s) partner

assortieren vt assort, sort

assouplieren vt render pliable (or supple); half-boil (silk)

Assoziations-faser f associative fiber. **-grad** m degree of association

assoziieren 1 vt associate. 2: sich a. (mit) associate, enter into a partnership (with)

Ast m (=e) 1 branch, bough, limb. 2 leg (of a curve). 3 knot (in wood). 4 (Geol.) spur

Astat n astatine. **astatisieren** vt astatize, render astatic. **Astatium** n astatine

ästen 1 vt prune, trim the branches of. 2 vi put out branches

Asterin-säure f asertic acid

ast-frei a free of branches, branchless

ästhetisch a esthetic, aesthetic

astig, ästig a branched, branchy; knotted, gnarled

Ast-loch n knot hole. **a--los** a branchless. **-rein** a clear(ed) of branches

Astronom m (-en, -en) astronomer. **Astronomie** f astronomy

Ästung f (-en) pruning

Ast-werk n branches (of a tree)

ASU abbv (Ammoniumsulfat) ammonium sulfate

Äsung f grazing, browsing (of cattle)

Asyl *n* (-e) asylum, sanctuary

asynchron *a* asynchronous

aszendent *a* (*Math.*) primary. **Aszendent** *m* (-en, -en) (*Geol.*) hypogene; ascendant

At. *abbv* **1** (Atom) atom. **2** (Atmosphäre) atmosphere

-at *sfx* (*in names of salts*) -ate

Ata, ata *abbv* (Atmosphären absolut) atmospheres absolute (kg/cm^2)

Atem *m* breath; breathing, respiration; spirit: **A. holen** catch one's breath

atembar *a* respirable, breathable

Atem- breathing, respiratory: **-beklemmung** *f* shortness of breath. **-einsatz** *m* respirator canister. **-gerät** *n* respirator. **-höhle** *f* air chamber, respiratory cavity. **-maske** *f* gas mask, respirator. **-not** *f* shortness of breath. **-pause** *f* breathing space. **a–raubend** *a* breath-taking. **A–rohr** *n*, **-röhre** *f* breathing (*or* respiratory) tube. **-weg** *m* respiratory passage. **-zug** *m* breath

At.-Gew. *abbv* (Atomgewicht) at. wt. (atomic weight)

äth. *abbv* (ätherisch) ethereal

Äthal *n* ethal, cetyl alcohol

Äthan *n* ethane

Äthanal *n* ethanal, acetaldehyde

Äthanol-säure *f* glycollic (*or* hydroxyacetic) acid

Äthanoyl *n* ethanoyl, acetyl

Äthan-säure *f* ethanoic (*or* acetic) acid

Äthen *n* ethene. **Äthenyl** *n* ethenyl

Äther *m* ether

äther. *abbv* (ätherisch) ethereal

äther-ähnlich *a* ethereal, ether-like. **Ä–art** *f* type of ether, *pl* oft ethers. **ä–artig** *a* ethereal. **Ä–auszug** *m* ether(eal) extract. **-bildung** *f* etherification. **ä–förmig** *a* etheriform. **-haltig** *a* ethereal, containing ether

ätherifizieren *vt* etherify. **Ätherifizierung** *f* etherification

ätherisch *a* ethereal; volatile, essential (oil) **-riechend** *a* ethereal-smelling

ätherisierbar *a* etherizable. **ätherisieren** *vt* etherize. **Ätherisierung** *f* etherization

äther-löslich *a* ether-soluble

äthern *vt* etherize

Äther-schwefelsäure *f* ethylsulfuric acid, ethyl hydrogen sulfate. **-schwingung** *f* ethereal vibarion. **ä–unlöslich** *a* ether-

insoluble, **Ä–weingeist** *m* (*Pharm.*) spirit of ether (solution of ether in alcohol). **-zahl** *f* ether index

Äthin *n* ethine, ethyne, acetylene. **Äthinilierung** *f* (-en) addition of acetylene

Äthion-säure *f* ethionic acid

Äthyl *n* (-e) ethyl. **Äthylat** *n* (-e) ethylate

Äthyl-äther *m* ethyl ether

Äthylen *n* (-e) ethylene. **-bindung** *f* ethylene linkage, double band

Äthyliden *n* ethylidene. **-milchsäure** *f* ethylidene lactic acid

äthylieren *vt* ethylate. **Äthylierung** *f* (-en) ethylation

Äthyl-senföl *n* ethyl mustard oil (C_2H_5NCS). **-sulfhydrat** *n* ethyl hydrogen sulfide. **-wasserstoff** *m* ethyl hydride. **-zinnsäure** *f* ethylstannic acid

Aticon-säure *f* aticonic acid

Ätiologie *f* etiology

Atlas *m* (-se) **1** atlas. **2** satin. **3** atlas bone. **-appretur** *f* satin finish. **a–artig** *a* satiny, satined. **A–band 1** *m* volume of an atlas. **2** *n* satin ribbon. **-erz** *n* fibrous malachite. **-gips** *m* fibrous gypsum. **-glanz** *m* satin luster. **a–glänzened** *a* satiny, satin-lustered. **A–holz** *n* satinwood. **-kies** *m* fibrous malachite. **-papier** *n* glazed paper. **-spat, -stein** *m* satin spar (fibrous calcium carbonate). **-vitriol** *n* white vitriol, (hydrous) zinc sulfate

Atm. *abbv* (Atmosphären) atm. (atmospheres)

atmen *vt/i* breathe, inhale, respire

Atmolyse *f* atmolysis, gas diffusion

Atmosphäre *f* (-n) atmosphere

Atmosphären-druck *m* atmospheric pressure. **-überdruck** *m* atmospheric excess pressure, gauge pressure

Atmosphärilien *pl* substances in the atmosphere; atmospheric influences

atmosphärisch *a* atmospheric

Atmung *f* breathing, respiration

Atmungs- respiratory: **-apparat** *m* breathing apparatus, respirator. **-einrichtung** *f* provision for respiration. **-gerät** *n* respirator. **-größe** *f* respiratory quotient. **-kohlensäure** *f* respiratory carbon dioxide. **-weg** *m* respiratory passage. **-werkzeug** *n* respiratory apparatus (*or* organs)

Atom-abstand *m* interatomic distance.

-anordnung f atomic arrangement (or configuration)

atomar a atomic

Atom- atomic **-bau** m atomic structure. **-begriff** m concept of the atom. **-bindung** f atomic linkage (or bond), covalent bond. **a–bindend** a: **-e Kraft** = **A–bindungskraft** f, **-bindungsvermögen** n atomic combining power, valence. **-brenner** m atomic reactor. **-formfaktor** m atomic scattering factor. **-gramm** n gram atom. **a–haltig** a atomic. **A–hülle** f atomic envelope (or sheath) (planetary electrons surrounding the nucleus)

atomig a atomic. **Atomigkeit** f atomicity

Atom-innere n (adj endgs) interior of the atom

atomisch a atomic

atomisieren vt atomize

atomistisch a atomic

Atomizität f atomicity

Atom- atomic: **-kern** m atomic nucleus. **-kraftanlage** f atomic pile. **-lage** f atomic layer, atomic position

atomlich a atomic

Atom- atomic: **-meiler** m atomic pile (or reactor). **-müll** m atomic (or nuclear) waste. **-ordnung** f atomic arrangement (or configuration). **-punktfolge** f atomic series. **-rand** m atomic surface. **-ring** m ring of atoms, cyclic compound. **-rumpf** m atomic residue (or core) (after removal of outer electrons). **-säule** f atomic pile. **-strahl** m atomic beam. **-tafel** f atomic (or periodic) table. **a–theoretisch** a as regards atomic theory. **A–verband** m union of atoms, atomic union. **-verbindung** f non-ionic compound. **-verkettung** f atomic linkage. **-verschiebung** f atomic displacement, rearrangement of the atoms (in the molecule). **-versuch** m atomic test. **-wärme** f atomic heat. **-werk** n atomic energy plant. **-zahl** f 1 atomic number. 2 number of atoms. **-zeichen** n atomic symbol. **-zertrümmerung** f atom smashing, atomic fission

atoxisch a atoxic, non-toxic

Atractyl-säure f atractylic acid

Atramentierung f phosphate rustproofing (of steel). **Atrament-stein** m inkstone, native copperas

Atranorin-säure f atranorinic acid

Atranor-säure f atranoric acid

Atrar-säure f atraric acid

atro. abbv (absoluttrocken) bone-dry

Atrolaktin-säure f atrolactic acid

Atropa-säure f atropic acid

Atrop-isomerie f molecular dissymetry due to restricted rotation

attenuieren vt attenuate

Attest n (-e) certificate. **attestieren** vt certify; attest

Attich m (-e) dwarf elder, danewort

Attrappe f (-n) dummy, mock-up; sham

Atü, abbv (Atmosphärenüberdruck) atmospheric excess pressure

atypisch a atypical

Ätz- caustic, corrosive: **-alkali** n caustic alkali. **ä–alkalisch** a caustic alkaline. **Ä–ammon** n (Pharm.) ammonia solution. **-ammoniak** n caustic ammonia (ammonium hydroxide), ammonia water. **-artikel** m (Tex.) discharge style. **-bad** n etching bath

ätzbar a etchable; corrodible; corrosive; (Tex.) dischargeable. **Ätzbarkeit** f etchability, etc

Ätz-baryt m caustic baryta, barium hydroxide. **-beizdruck** m (Tex.) discharge printing. **-beize** f (Tex.) chemical discharge, discharge mordant. **ä–beständig** a caustic-resisting; (Tex.) discharge-resisting. **Ä–bild** n etch pattern, etched picture (or figure). **-boden** m (Tex.) discharge ground. **-druck** m discharge printing; engraving; etching

Ätze f (-n) etching fluid (esp aqua fortis); (Tex.) discharge

atzen vt feed cf ÄTZEN

ätzen vt etch, corrode; cauterize; eat, bite; (Tex.) discharge cf ATZEN. **ätzend** p.a caustic, corrosive

Ätz- etching; caustic, corrosive: **-farbe** f etching ink; (Tex.) discharge paste. **-figur** f (Engg.) etched figure; (Cryst.) etching figure. **-flüssigkeit** f caustic liquid; etching fluid. **-folie** f etched foil. **-gift** n caustic poison. **-grübchen** n etched pit. **-grund** m etching (or Tex.: discharge) ground; (Etching) protective coating, resist. **-kali** n caustic potash. **-kalischmelze** f caustic potash fusion. **-kalk** m caustic lime: **gebrannter Ä.** quicklime, unslaked lime; **gelöschter Ä.**

slaked lime. **-kalklösung** *f* limewater. **-kraft** *f* causticity, corrosiveness, corrosive power. **-kunst** *f* art of etching. **-lack** *m* discharge lake. **-lage** *f* 1 corrosive layer. 2 caustic condition. **-lauge** *f* caustic lye (*or* liquor). **-lösung** *f* caustic (*or* etching) solution. **-magnesia** *f* caustic magnesia. **-mittel** *n* corrosive (caustic, etching) agent; etching medium; (*Tex.*) discharging agent. **natron** *n* caustic soda. **-papp** *m* (*Tex.*) resist. **-paste** *f* corrosive (*or* etching) paste; (*Dye.*) discharge paste

ätzpolieren *vt* etch-polish, polish with etching liquid

Ätz· etching; caustic, corrosive: **-probe** *f* etching test (*or* sample). **-salz** *n* corrosive salt. **-schliff** *m* ground section (of metal) for etching. **-silber** *n* lunar caustic (silver nitrate). **-stein** *m* caustic potash. **-stoff** *m* caustic, corrosive (material). **-strontian** *m* caustic strontia (strontium oxide *or* hydroxide). **-sublimat** *n* corrosive sublimate (mercuric chloride). **ä–technisch** *a*, *adv* regarding (*or* by) etching (techniques). **Ä–tinte** *f* etching ink

Atzung *f* feeding, feed, food *cf* ÄTZUNG

Ätzung *f* (-en) etching; corrosion, cauterization

Ätz·verfahren *n* caustic (corrosive, etching) process. **-verhalten** *n* behavior to etching. **-wasser** *n* etching fluid (*esp* aqua fortis); mordant. **-weiß** *n* (*Tex.*) white discharge. **-wirkung** *f* corrosive action; (*Tex.*) action of the discharge

a.u.a. *abbv* (auch unter andern) also among others

auch *adv* & *cnjc* 1 also, too, likewise, as well; **a. nicht** not . . . either, neither *cf* SONDERN, SOWOHL. 2 even: **a. das nicht** not even that; **a. wenn** even if. 3 (*after interrog cnjc*) .. ever: **was ... a.** whatever *cf* IMMER. 4 (*oft rhetorical, intensifies verb, no transl*): **kann man es a. glauben?** can you (really) believe it?; **wer weiß das a.?** who knows that (anyway)?; **so klein es a. sei** small as it may be *cf* NOCH

Audienz *f* (-en) audience (with a V.I.P.)

Auditorium *n* (..rien) 1 auditorium. 2 audience, listeners

Aue *f* (-n) meadow(land), pasture

Auer· Welsbach: **-brenner** *m* Welsbach (gas) burner. **-metall** *n* Welsbach metal (pyrophoric cerium alloy). **-strumpf** *m* gas mantle

auf I *prep* (*dat, accus*) 1 on (top of) upon, onto; in, at, to *cf* BIS. 2 in answer to; **a. das hin** on the strength of that; **a. einmal** all at once, suddenly. 3 (*Time*): **a. einige Zeit** for some time (to come); **a. unbestimmte Zeit** for an indefinite period of time; **von heute a. morgen** (a) from today till tomorrow, (b) overnight. 4: **a. . . . zu** toward *cf* ZU V 26. 5: **wieviel Mark a. den Dollar?** how marks per (*or* to the) dollar? *cf* FALLEN 4, KOMMEN 3 H. 6: **aufs** (+ *superl*): **aufs beste versuchen** try one's level best; **aufs eifrigste** in a most eager manner. 7 (*idiomatic accdg to verb*) *cf* ANTWORTEN, FREUEN 3, PASSEN, VERTRAUEN VORBEREITEN. II *adv* 8 up (*esp* = on one's feet), upward. 9 (*Door, etc*) open; (*Gas, etc*) (turned) on. 10: **a. und ab, a. und nieder** up and down, back and forth *cf* VON 5. 11: **a. daß** so that. III *sep pfx* 12 up (*various meanings*); on; up against; open; apart

auf·achten *vi* pay attention, watch out

auf·ackern *vt* plow up

auf·arbeiten I *vt/i* catch up (on) (unfinished work). II *vt* 2 use up, wear out. 3 work up, process (raw material). 4 refurbish, renovate. 5 pry open. III: **sich a.** 6 wear oneself out. 7 work one's way up. **Aufarbeitung** *f* using up, wearing out; processing, refurbishing, renovation; prying open. **Aufarbeitungs·anteil** *m* stage in processing

auf·ästen *vt* prune (trees)

auf·atmen *vi* draw a breath, breathe freely (*or* with relief)

auf·ätzen *vt* 1 etch on. 2 open (*or* skin) with caustic or corrosive. 3 force-feed

auf·backen* *vt* warm up (in the oven *or* by frying)

auf·ballen *vt* bale, lump together

Aufbau *m* (-e *or* -ten) 1 construction, buildup; erecting, putting up; assembly, mounting. 2 reconstruction. 3 (*Chem.*) synthesis. 4 composition, make-up organization. 5 structure. 6 superstructure; (*esp. Cars*) body. 7 (*Spect.*) aufbau (of energy levels). *Cf* AUFBAUEN

auf·bauchen *vt, vi* & **sich a.** swell up, puff up

Aufbau·einheit *f* standard (machine-tool) unit

auf·bauen I *vt* 1 build (up), construct, erect. 2 rebuild, reconstruct. 3 put together, assemble, arrange, compose, synthesize. 4 (**auf**) set up, erect (on); base (on). **II** *vi* 5 (**auf**) be based (on), rise up (on). **III: sich a.** 6 be composed. 7 stand (erect), take a (standing) position; line up. 8 rise up, tower. **aufbauend** *p.a* constructive; synthetic; based (on)

Aufbau·gerät *n* (kit) component, part

auf·bäumen 1 *vt* (*Weaving*) beam, wind up. 2: **sich a.** rear up; rebel

Aufbau·prinzip *n* construction principle (governing atomic structure). **-produkt** *n* synthetic product

auf·bauschen 1 *vt, vi* & **sich a.** swell (up), puff up. 2 *vt* inflate, exaggerate

Aufbau·stoff *m* 1 building material. 2 compounding ingredient. 3 material for synthesis. 4 (*Dterg.*) builder

Auf·bauten superstructure(s): *pl of* AUFBAU 6

auf·bearbeiten *vt* process. **Aufbearbeitung°** *f* processing

auf·begehren *vi* 1 protest, rebel. 2 effervesce

auf·beißen* *vt* bite open

auf·beizen *vt* 1 freshen with mordant. 2 (*Furniture*) restain. 3 (*Med.*) open with caustic

auf·bekommen* *vt* 1 get as an assignment. 2 get (a door) open. 3 manage to eat up

Aufbenzolung *f* (-en) rebenzolation, aromatization (of coal gas)

aufbereitbar *a* easily prepared (*or* processed). **Aufbereitbarkeit** *f* ease of preparation, etc

auf·bereiten *vt* 1 prepare (for use). 2 (*Metll.*) dress, wash; separate (ore). 3 refine, purify (water). 4 process (statistics, evidence). **Aufbereiter** *m* (-) processor, processing technologist. **Aufbereitung** *f* (en) preparation, dressing, washing; separation, refinement; purification; processing

Aufbereitungs·anlage *f* preparing (dressing, processing) plant; screening room.

-grad *m* degree of preparation (separation, refinement, processing); state of subdivision. **-verlust** *m* loss in preparation (processing, etc); ore-dressing loss

auf·bersten* *vi*(s) burst (*or* crack) open

auf·bessern *vt* improve, renovate; raise (pay). **Aufbesserung** *f* (-en) improvement, renovation; (pay) raise; raising

auf·beulen *vt* buckle (out of shape)

auf·bewahren *vt* keep, store, preserve. **Aufbewahrung** *f* (safe)keeping, storage, preservation

Aufbewahrungs·ort *m* 1 place where sthg is stored. 2 = **-raum** *m* storage place, storehouse, depository

auf·biegen* *vt* bend up; bend open; turn up

auf·bieten*vt* muster up; exert, use; call upon (to help, etc). **Aufbietung** *f* (-en) mustering up, exertion; call, appeal

auf·binden* *vt* 1 undo, untie. 2 tie up, fasten; tuck up; bind (books). 3 (**auf**) tie on (to). 4 (*d.i.o.*) dupe (sbdy) into believing (sthg)

auf·blähen *vt* 1 swell (up), inflate, bloat. 2: **sich a.** swell up, become inflated; strut, boast. **Aufblähung** *f* (-en) swelling (up), inflation, bloating, strutting, boasting; (*Med.*) flatulence

auf·blasen* 1 *vt* blow up, inflate; puff out; exaggerate. 2: **sich a.** show off, put on airs

auf·blättern 1 *vt* open, unfold; leaf through. 2 *vi* (*Min., Med.*) exfoliate. 3: **sich a.** unfold, open up. **Aufblätterung** *f* (-en) opening (up); unfolding; exfoliation

auf·bleiben* *vi*(s) 1 stay up (late). 2 stay open

Aufblick° *m* upward glance; (*Metll.*) gleam, shine (after cupelation); (*Assaying*) fulguration, blick. **auf·blicken** *vi* look (*or* glance) up(ward); (*Metll.*) gleam, shine; (*Assaying*) fulgurate, give the "blick", brighten

auf·blitzen *vi*(s) flash (up) (like lightning)

auf·blühen *vi*(s) bloom, blossom (out); effloresce, form a flower-like mass; thrive. **Aufblüh·zeit** *f* efflorescence, blossoming time

auf·bohren *vt* bore (*or* drill) open

auf·brauchen *vt* use up, exhaust. **Aufbrauchs·frist** *f* (*Food*) expiration date;

(*on Packages*) use before . . .

auf·brausen *vi* (h/s) effervesce, bubble up, ferment; flare up, surge. **aufbrausend** *p.a* effervescent

auf·brechen* I *vt* 1 break open; rip up, eviscerate. 2 break up; plow. II *vi*(s) 3 break (*or* burst) open; blossom. 4 break up, disintegrate. 5 show up, emerge. 6 leave, start out

auf·breiten *vt* spread out; lay out, display. **Aufbreitungs·effekt** *m* spreading effect

auf·brennen* I *vt* 1 burn up (fuel). 2: **Zeichen a.** (*dat*) brand a mark (on). 3 open with caustic. 4 refine (metal). 5 sulfur (wine). II *vi*(s) 6 burn up. 7 flare up; start to burn

auf·bringen* *vt* 1 raise, gather (in), rustle up (funds); muster up, drum up. 2 bring up (*or* out), introduce (a novelty), start (rumors). 3 anger, provoke *cf* AUFGE-BRACHT. 4 get (a door) open. 5 put up, set up; put on, apply. 6 capture (ship)

auf·brodeln *vi*(s) bubble up, boil up

Aufbruch° *m* 1 breaking up, break-up; opening up. 2 plowed land. 3 start, departure. 4 emergence: **im A. sein** be emerging, be awakening. 5 crevice; fracture. 6 (animal) entrails

auf·brühen *vt* boil up, brew

auf·bürden *vt* (d.i.o) burden (sbdy) with

auf·dämmern *vi*(s) 1 dawn; (+ **in** + *dat*) dawn on (a psn). 2 begin to appear

auf·dampfen 1 *vt* (auf) steam (on, onto), deposit by vaporization, distill (on), vaporize (on). 2 *vi*(s) evaporate, rise in a vapor

auf·dämpfen *vt* steam; press up, iron up

Aufdampf·schicht *f* evaporated film. **-temperatur** *f* deposition temperature (in the evaporation process)

auf·darren *vt* kiln-dry, dessicate

auf·decken *vt* 1 uncover, open, expose. 2 reveal, disclose. **Aufdeckung°** *f* uncovering, exposure; revelation, disclosure

auf·docken *vt* roll up; bundle; shock (grain)

auf·dornen *vt* drift, ream (out) (holes); enlarge, expand (pipe)

auf·dörren *vt* dry, desiccate

auf·drängen* I *vt* 1 force open. 2 (d.i.o) force (sthg on sbdy), make inescapably

clear (to). II: **sich a.** (*dat*) 3 force oneself (on). 4 suggest itself forcibly, become inescapably clear (to)

auf·drehen I *vt* 1 turn on (light, radio). 2 wind up (clock). 3 roll up (hair). 4 rev up (engine). 5 unscrew, unravel; screw open. 6 (*Ceram.*) throw. II *vi* 7 increase one's speed

aufdringlich *a* 1 obtrusive, pushy, insistent. 2 (*Odor*) pungent. **Aufdringlich·keit** *f* 1 obtrusiveness, pushiness, insistence. 2 pungency

auf·dröseln *vt* unravel, unwind

Aufdruck° *m* printing, printed words; imprint, stamp; overprint. **auf·drucken** *vt* (**auf**) print, imprint (on): **aufgedruckter Stromkreis** *m* printed circuit

auf·drücken *vt* 1 press (*or* push) open. 2 (*d.i.o*) press, impress, stamp (on, onto)

auf·duften *vi*(s) give off an odor (*or* a fragrance)

auf·dunsten 1 *vt* reheat, steam (food). 2 *vi*(s) evaporate, rise as a vapor

auf·dünsten *vt* 1 reheat, steam (food). 2 evaporate, vaporize

aufeinander *adv & sep pfx* on (onto, to, against) each other (*or* one another). **A–·folge** *f* succession, series. **a–·folgen** *vi* follow (upon) (*or* succeed) each other. **·folgend** *p.a* consecutive, successive. **·häufen** *vt* heap up, pile up. **·legen** *vt* super(im)pose, lay on top of each other. **A–legen** *n* (*Math.*) superposition. **a–·passen** *vi* fit, match. **A–prall** *m* collision, clash. **a–·prallen** *vi*(s) collide, clash. **A–stoß** *m* collision, clash, impact. **a–·stoßen*** *vi*(s) collide, clash

Auf·elektron° *n* outer (*or* valence) electron

Aufenthalt *m* (-e) 1 (length of) stay (in a place). 2 stop, stopover, layover. 3 delay. 4 residence, abode, whereabouts. **Aufenthalts·dauer, -zeit** *f* residence time

auf·erlegen *vt* (d.i.o) (super)impose (on)

auf·erziehen* *vt* rear, bring up; train. **auferzogen** *pp*

auf·essen* *vt* eat up, consume

auf·fahren* I *vt* 1 bring up, bring on; marshal, line up (data, arguments). 2 (*Min.*) drive (gallery). 3 tear up (road). II *vi*(s) 4 start up, be startled; jump up, flare up. 5 ascend, ride up. 6 pull up (in a vehicle).

7 tailgate; (**auf**) run (into). **Auffahren** *n* (*oft*) tailgating; rear-end collision. **auffahrend** *p.a* irascible, quick-tempered. **Auffahrt**° *f* ascent; driving up (in a vehicle); driveway, approach road; up ramp
auf·fallen* *vi*(s) **1** attract attention, be striking. **2** (**auf**) land, come down (on), strike; fall (on). **auffallend** *p.a* striking, conspicuous; incident (light). **auffallenderweise** *adv* strikingly
auffällig *a* striking, conspicuous; incident (light). **Auffälligkeit** *f* conspicuousness
Auffall·winkel *m* angle of incidence
auf·falten *vt* **1** fold up(ward). **2** unfold
Auffang·behälter *m* collecting vessel, receiver
Auffange·gefäß *n* collecting vessel, receiver. **-glas** *n* (*Opt*.) object glass (*or* lens), objective
auf·fangen* *vt* **1** catch; collect, pick up; capture; intercept. **2** deflect, ward off
Auffange·platte *f* collector plate
Auffänger *m* (-) collector, receiver; catcher. **--strom** *m* interceptor current. **-system** *n* capture system
Auffange·gefäß *n* collecting vessel, receiver. **-glas** *n* (*Opt*.) object glass (*or* lens), objective. **-kolben** *m* receiving flask. **-pfanne** *f* collecting (*or* receiving) pan. **-querschnitt** *m* (*Nucl. Phys.*) capture cross-section. **-rinne** *f* (*Furnaces*) collecting launder. **-rohr** *n*, **-röhre** *f* collecting tube (*or* cylinder). **-schale** *f* collecting dish; drip pan; oil collector
auf·färben *vt* redye, dye over again; touch up the dye on. **Auffärber**° *m* job (*or* touch-up) dyer
auf·fasern *vt* separate into fibers, unravel
auf·fassen *vt* **1** grasp, understand. **2** (*oft* + **als**) interpret (as). **Auffassung**° *f* **1** grasp, comprehension. **2** view, opinion; conception; interpretation
auf·feuchten *vt* wet, moisten (again)
auffindbar *a* discoverable, traceable
auf·finden* *vt* find, discover; detect, trace; locate. **Auffindung** *f* (-en) discovery, detection, location
auf·flackern *vi*(s) flare up; deflagrate
auf·flammen *vi*(s) flare up, flame up, blaze up, burst into flame; deflagrate
auf·fliegen* *vi*(s) **1** fly up, take off (in

flight), rise. **2** blow up, explode. **3** be exposed. **4** break up, collapse, fail. **5** fly open
auf·fließen* *vi*(s) flow up; (**auf**) flow on(to)
Aufflug° *m* **1** upward flight; take-off; rise, ascent. **2** blow-up. **3** exposure. **4** collapse, failure. *Cf* AUFFLIEGEN
Auffluß° *m* flow, upflow *cf* AUFFLIESSEN
auf·fordern *vt* ask, invite; call upon; urge, encourage; challenge. **Aufforderung**° *f* request; invitation; encouragement; challenge
auf·fressen* *vt* eat up, devour; corrode
auf·frischen *vt* **1** freshen up, refresh, revive. **2** brush up, touch up; restore, regenerate. **3** (*Brew.*) change (steep water). **II** *vi*(h) be refreshing. **III** *vi*(s) (*esp Meteor.*) swell (*as:* wind). **Auffrischung** *f* freshening (up), refreshment, revival; brush-up, touch-up; restoration, regeneration; changing (of steep water). **Auffrischungs·mittel** *n* freshening agent, regenerative; (catalytic) activator
aufführbar *a* performable, presentable; quotable
auf·führen **I** *vt* **1** perform, put on. **2** bring up, present; cite, quote, list, adduce, mention. **3** put up, erect, raise. **II:** sich a. behave, conduct onself. **Aufführung**° *f* performance; presentation; citing, quotation, mention(ing); erection, construction
auf·füllen **I** *vt* **1** fill (up); refill, replenish; bottle (wine). **2** mix with broth. **II** *vt/i* dish up (food). **Auffüllung**° *f* filling (up); refilling, replenishment; dishing up
Aufg. *abbv* (AUFGABE) assignment, etc
Aufgabe° *f* **1** assignment, task, job (to do); function. **2** problem (to solve); lesson, homework. **3** (*usu* + *genit* **or von**) giving up, relinquishment, abandonment (of), quitting; closing (of a shop); resignation, retirement (from). **4** mailing; handing in; checking (of baggage, etc); placing (of an order). **5** (*Mach.*) feed(ing) (of work to a machine). *Cf* AUFGEBEN
Aufgabe·bunker *m* feed hopper. **-drehzahl** *f* input rpm. **-gut** *n* material for processing. **-leistung** *f* throughput
Aufgaben·bereich *m* scope (*or* range) of one's duties. **-stellung** *f* statement (*or* definition) of the problem; setting of the task

Aufgabe· mailing, checking; feed(ing): **-schein** *m* receipt. **-station** *f* point of origin. **-stempel** *m* (receipt) stamp. **-trichter** *m* feed funnel. **-vorrichtung** *f* feeding apparatus, feeder; charging device. **-zeit** *f* feed(ing) time

Aufgang° *m* **1** entrance ramp (*or* stairway); way up; ascent. **2** (*Sun, Moon*) rise, rising. **3** opening (of season). **4** upstroke (of piston). **5** thaw; break-up; disappearance; (**in**) absorption (in an effort). *Cf* AUFGEHEN

auf·gären* *vi*(s) ferment (and rise); effervesce

auf·geben* *vt* **1** assign, give (to do, to solve). **2** give up, relinquish, abandon, quit, resign, retire from; close (shop). **3** mail, check, hand in; place (order, ad). **4** feed, charge (into furnace). **5** serve, dish up (food). *Cf* AUFGABE. **Aufgeber** *m* (-) mailer, sender, consignor

aufgebissen bitten open: *pp of* AUFBEISSEN

aufgeblieben stayed up: *pp of* AUFBLEIBEN

aufgebogen bent up: *pp of* AUFBIEGEN

aufgeborsten burst open: *pp of* AUFBERSTEN

Aufgebot° *n* **1** notice; summons. **2** input; mass, body (of people). **3** exertion (of strength)

aufgeboten mustered up: *pp of* AUFBIETEN

aufgebracht 1 raised, etc: *pp of* AUFBRINGEN. **2** *p.a* angry, furious

aufgebrannt burned up, refined: *pp of* AUFBRENNEN

aufgebrochen broken up: *pp of* AUFBRECHEN

aufgebunden tied up: *pp of* AUFBINDEN

aufgedunsen *a* swollen, bloated

aufgeflogen flown up: *pp of* AUFFLIEGEN

aufgeflossen flowed up: *pp of* AUFFLIESSEN

aufgefunden found: *pp of* AUFFINDEN

aufgegangen risen: *pp of* AUFGEHEN

aufgegessen eaten up: *pp of* AUFESSEN

aufgeglommen lighted (*or* flickered) up: *pp of* AUFGLIMMEN

aufgegoren fermented: *pp of* AUFGÄREN

aufgegossen brewed: *pp of* AUFGIESSEN

aufgegriffen grabbed: *pp of* AUFGREIFEN

auf·gehen* *vi*(s) **1** rise (*as:* sun, dough); swell (*as:* lime); spread out, get fat. **2** sprout, spring up, come up; (a)rise (within one, *as:* feelings, ideas). **3** open, come open, burst (open). **4** be revealed;

(*oft* + *dat*) become clear (to). **5** (*Math.*) come out even; (+**in**) divide evenly (into). **6** (**in**) be absorbed (in), merge (into), be(come) devoted (to). **7** be used up, be spent (*as:* money). **8** (*Dye.*) go up, be absorbed. **9** go up (in smoke, flames)

aufgehoben 1 picked up; kept: *pp of* AUFHEBEN. **2** *p.a* taken care of, safe

Aufgeld° *n* **1** surcharge, extra charge. **2** earnest money; margin

aufgelegen lain, etc: *pp of* AUFLIEGEN

aufgelegt *pred a* disposed, inclined, in a . . . mood. *See also* AUFLEGEN

aufgelockert *p.a* **1** loose; informal, relaxed. **2** scattered (clouds). *See also* AUFLOCKERN

aufgenommen absorbed, picked up: *pp of* AUFNEHMEN

aufgequollen soaked: *pp of* AUFQUELLEN

aufgerichtet *p.a* erect, upright *cf* AUFRICHTEN

aufgerieben rubbed: *pp of* AUFREIBEN

aufgerissen 1 torn open: *pp of* AUFREISSEN. **2** *p.a* open-minded, receptive

aufgeschliffen ground, etc: *pp of* AUFSCHLEIFEN

aufgeschlossen 1 revealed, etc: *pp of* AUFSCHLIESSEN. **2** *p.a* open-minded, receptive

aufgeschmolzen melted down: *pp of* AUFSCHMELZEN

aufgeschnitten cut open (*or* up): *pp of* AUFSCHNEIDEN

aufgeschoben postponed: *pp of* AUFSCHIEBEN

aufgeschossen shot up, risen: *pp of* AUFSCHIESSEN

aufgeschrieben written down: *pp of* AUFSCHREIBEN

aufgeschwollen swollen (up): *pp of* AUFSCHWELLEN

aufgeschwungen risen, soared: *pp of* AUFSCHWINGEN

aufgesogen absorbed: *pp of* AUFSAUGEN

aufgesotten boiled: *pp of* AUFSIEDEN

aufgesprossen sprouted, germinated: *pp of* AUFSPRIESSEN

aufgesprungen cracked, jumped up: *pp of* AUFSPRINGEN

aufgestanden got(ten) (*or* stood) up: *pp of* AUFSTEHEN

aufgestiegen ascended: *pp of* AUFSTEIGEN

aufgestochen punctured: *pp of* AUF-STECHEN

aufgestrichen spread (on): *pp of* AUF-STREICHEN

aufgetan opened (up): *pp of* AUFTUN

aufgetrieben buoyed up: *pp of* AUFTREIBEN

aufgetroffen struck: *pp of* AUFTREFFEN

aufgewandt expended: *pp of* AUFWENDEN

aufgeweckt *p.a* alert, quick-witted. *See also* AUFWECKEN

aufgewiesen shown: *pp of* AUFWEISEN

aufgewogen balanced: *pp of* AUFWIEGEN

aufgeworfen thrown up: *pp of* AUFWERFEN

aufgewunden wound up: *pp of* AUFWINDEN

aufgezogen raised, etc: *pp of* AUFZIEHEN

aufgezwungen forced: *pp of* AUFZWINGEN

auf·gichten *vt* charge (furnace)

auf·gießen* *vt* 1 brew (tea, etc). 2 pour on, add (liquid) *cf* AUFGUSS

auf·gischen *vi* foam up; spray up; ferment

Aufglasur·farbe *f* overglaze (*or* on-glaze) color

auf·gliedern *vt* break down, classify, organize. **Aufgliederung°** *f* classification, breakdown

auf·glimmen* *vi/*(s) light up, flicker up

auf·glühen *vi*(s) blaze up, start to glow

auf·graben* *vt* dig up

auf·greifen* *vt* pick up, grab, snatch

aufgrund *adv & prep* (= **auf Grund**) (+ *genit* or **von**) on the basis, by reason (of)

Aufguß° *m* infusion, brew *cf* AUFGIESSEN. **-apparat** *m* sparger. **-gefäß** *n* infusion vessel, digester. **-probe** *f* pour test. **-tierchen** *npl* Infusoria. **-verfahren** *n* infusion process; mashing process

auf·haben* I *vt* 1 have on (hat). 2 have (door) open. 3 have as an assignment. II *vi* (*Shops*) be open

auf·hacken *vt* hack up; hack open

auf·haken 1 *vt* unhook. 2: **sich a.** come unhooked

Aufhaldung *f* piling up, dumping

Aufhalt° *m* stop; check(ing) (of motion). **auf·halten*** I *vt* 1 stop, detain, keep, hold up (*or* back); check (motion). 2 hold open; keep open. II *vi* 3 (**mit**) stop, cease (doing). III:**sich a.** 4 stay, be keeping oneself. 5 (**mit**) take time, bother (with), linger (over). 6 (**über**) find fault (with), criticize. **Aufhalter** *m* (-) stop(ping) device, check; brake. **Aufhaltung°** *f* delay

auf·hängen *vt* hang (up), suspend. **Aufhänger** *m* (-) hanger; loop; peg. **Aufhängung** *f*(-en) suspension; mounting; hanging (up)

auf·haspeln *vt* 1 (**auf**) reel (*or* wind) on (to). 2 hoist up

auf·häufen *vt & sich a.* heap up, pile up, accumulate. **Aufhäufung°** *f* heaping (up), accumulation

auf·heben* I *vt* 1 pick up, lift (up), raise. 2 keep, store, save *cf* AUFGEHOBEN 2. 3 end, adjourn; abolish, do away with; annul, repeal. II *vt & sich a.* (*recip*) 4 neutralize, balance, offset, cancel (each other); (*esp Colors*) compensate (each other), be complementary. III: **sich a.** 5 rise, get up, lift oneself (up). 6 disappear, dissolve, be resolved

Aufhebung *f* (-en) 1 picking up, lifting, raising. 2 keeping, storing, saving. 3 ending, adjournment; abolition; annulment, repeal. 4 neutralization, balancing, offsetting, cancellation. 5 rise. 6 dissolution, disappearance. *cf* AUFHEBEN

auf·heitern *vt & sich a.* brighten up, clear up, cheer up. **Aufheiterung** *f* (-en) brightening, clearing, cheering (up); (*Meteor. oft*) bright period

auf·heizen *vt & sich a.* heat up. **Aufheizer°** *m* heater; stoker, fireman

auf·hellen 1 *vt & sich a.* brighten (up), clear (up), lighten (in color). 2 *vt* illuminate; clarify, elucidate. 3: **sich a.** be illuminated (clarified, elucidated). **Aufheller** *m* (-): **optischer A.** optical bleach. **Aufhellung** *f*(-en) brightening, clearing, lightening (up); illumination; clarification, elucidation

Aufhellungs·lage *f* (*Min.*) position of maximum illumination. **-mittel** *n* clearing (*or* brightening) agent

Aufhol·bedarf *m* need to catch up

auf·holen 1 *vt/i* catch up (on: arrears); catch up to, overtake; make up (for: delay). 2 *vt* haul up, hoist up, raise

auf·horchen *vi* prick up one's ears, sit up and take notice

auf·hören *vi* (**mit**) stop, cease (doing)

auf·keimen *vi*(s) sprout, bud, germinate

auf·kippen *vt* tilt up, tip up *cf* AUFKLAPPEN

auf·kitten *vt* (**auf**) cement on (to)

aufklappbar *a* hinged, folding, tip-up.

auf·klappen vt 1 open, unfold, unclasp (book, hinged lid, penknife, etc). 2 fold up, close (by folding upward, e.g. oven door); tip up; raise (collar). **Aufklapp·sitz** m tip-up seat

auf·klaren vi (Meteor.) clear (up)

auf·klären I vt 1 clear up, clarify. 2 enlighten. **II: sich a.** 3 clear up, brighten. **Aufklärung** f (-en) clearing up, clarification; enlightenment, rationalism (incl 18th Cent. Philos.)

auf·kleben vt (auf) paste on (to). **Aufkleber** m (-) sticker

auf·kleistern vt (auf) paste on (to)

auf·klinken vt unlatch

auf·klopfen 1 vt break (or crack) open. 2 vi (auf) rap, tap (on)

auf·kochen I vt 1 boil, bring to a (quick) boil. 2 warm up, reboil. 3 prime (boilers). **II** vi 4 boil up, come to a boil. 5 (für) cook a meal (for). **Aufkoch·gefäß** n boiler; reboiler.

auf·kohlen vt (Metll.) (re)carburize, cement. **Aufkohlungs·mittel** n (re)carburizing agent

auf·kommen* vi(s) 1 come up, arise, spring up. 2 come into use (into fashion). 3 rally, recover. 4 catch up, gain. 5 (für) accept responsibility (for), meet (expenses, etc). 6 (gegen) stand up, assert oneself (against). 7 come in (as: earnings). **Aufkommen** n rise, emergence; recovery

Aufkosten pl overhead (expenses)

auf·kratzen vt 1 scratch (up), scrape (up). 2 scratch open. 3 rake, stir (a fire). 4 dress, nap (fabrics); card (wool)

auf·kräusen vi (Brew.) form a curly head

auf·kühlen vt coal; (Brew.) aerate

auf·kündigen vt give notice of terminating; terminate, cancel, revoke, annul

auf·kupfern vt copper, coat with copper

Aufl. abbv (Auflage) ed. (edition)

Auflade·geschwindigkeit f rate of loading, (Elec.) rate of charging

auf·laden* vt 1 load (goods) on (vehicles); (+ dat rflx):**sich etwas a.** burden oneself with sthg. 2 load up (vehicles). 3 (Elec.) charge. 4 (Mach.) boost (engine). **Aufladung°** f 1 loading, boosting, (super)charging. 2 load; (Elec.) charge (build-up); (Explo.) top charge

Auflage° f 1 (Books) edition; printing; (Newspaper) circulation. 2 line (of goods). 3 covering, coating, lining, (seat) pad, overlay. 4 order, directive. 5 stipulation, condition. 6 production quota (or norm). 7 rest, support (for arm, tool, gun). 8 filter medium

Auflage·dicke f coating thickness. **-fläche** f bearing surface. **-gewicht** n 1 coating weight. 2 vertical pressure (of tone arm)

Auflagen·dicke f thickness of layers

Auflager° n 1 (Engg.) support, bearing, seat. 2 (Metll.) feeder, feeding mechanism

auf·lagern I vt 1 store up. 2 support, mount (on bearings). **II** vt, vi(s) 3 (Geol.) aggregate, form a layer (of). **III** vi(s) 4 (dat) be super(im)posed (on). **Auflagerung°** f storage; mounting (on bearings); aggregation, layering; super(im)position, stratification

Auflagerungs·schicht f stratified layer

auf·lassen* vt 1 leave open. 2 let . . . get up, let rise, release (balloon, etc). 3 give up, close (down) (an enterprise). 4 convey, transfer (property). **Auflassung** f (-en) 1 abandonment, closing (down). 2 conveyance, transfer

Auflauf° m 1 gathering crowd; mob, riot. 2 accumulation (of funds, debts). 3 (Conveyor Belts) upward run. 4 gangplank, ramp. 5 pudding, soufflé (baked in a mold)

auf·laufen* vi(s) 1 (oft + auf) run aground; run into (or against); (Trailers) overrun. 2 sprout, spring up; rise (as: plants, dough, flood). 3 accumulate, accrue, mount up; swell. 4 move up front. 5 (+ auf) run up; wind up (on). **Aufläufer°** m mechanical stoker, feeder, feeding mechanism

auf·leben vi(s) revive, come to life again

auf·legen I vt 1 (oft + auf) put, lay, load (on), apply (to); hang up (phone). 2 (d.i.o) impose (on). 3 print, publish (book); start (new line of goods). 4 issue (publication, securities). 5 lay out, display. 6 lay up (ship, vehicle). **II** vi 7 hang up (the phone). 8 lay down one's cards. **III: sich a.** 9 (oft + auf) rest one's elbows (on). Cf AUFGELEGT

Aufleger m (-) (furnace) stoker, feeder

Auflege·schuß m (Min.) mud cap(ping)

Auflegung f (-en) laying on, application; imposition; issue, issuance; display cf AUFLEGEN

auf·lehnen 1 vt & **sich a.** (auf) lean on, rest (one's elbows) on. **2: sich a.** (gegen) rebel (against). **Auflehnung** f (-en) rebellion

auf·leimen vt glue on

auf·lesen* vt pick up, gather

auf·leuchten vi(s) light up; (dat) dawn (on)

Auflicht° n reflected light. **--mikroskop** n reflecting microscope

auf·liefern vt mail, send

auf·liegen* I vi **1** (usu + **auf**) lie, rest (on). **2** (dat) weigh (on). **3** be out of service. **4** be on display, be available. **II: sich a. 5** get bed sores. **aufliegend** p.a available, on display

auf·listen vt list, tabulate

auf·lockern 1 vt & **sich a.** loosen (up), relax; scatter, disperse. **2** vt relieve (monotony). **Auflockerung°** f loosening, relaxation; dispersion; relief

Auflockerungs·mittel n loosening agent. **-sprengung** f (Explo.) breaking-in shot

auf·lodern vi(s) blaze up, flare up

auflösbar a soluble, resolvable. **Auflösbarkeit** f solubility, resolvability

auf·lösen I vt & **sich a. 1** dissolve; decompose; disperse. **II** vt **2** loosen, undo, untie, unravel. **3** (re)solve (problem, dissonance). **4** give up, break up (household). **III: sich a. 5** loosen up, come undone (untied, unraveled). **6** be resolved. **7** deliquesce. **auflösend** p.a solvent, dissolvent. **Auflöser°** m dissolver, dissolving agent (or apparatus); (Paper, etc) mixer

auflöslich a soluble. **Auflöslichkeit** f solubility

Auflösung° f **1** (esp Chem.) solution. **2** dissolution, decomposition, dispersion; analysis (opp. of synthesis). **3** loosening (up), undoing, untying, unraveling. **4** resolution. **5** (Malt) mellowness, friability. Cf AUFLÖSEN

auflösungs·fähig a soluble, capable of solution. **A–gefäß** n dissolving vessel, dissolver. **-geschwindigkeit** f velocity of (re)solution, decomposition, etc. **-grenze** f (Micros.) limit of resolution. **-kraft** f dissolving power; resolution power. **-mittel** n solvent. **-nafta, -naphta** f or n solvent

naptha. **-pfanne** f (Sugar) blow-up pan, clarifier. **-vermögen** n solvent power; (Micros.) resolving power. **-wärme** f heat of solution. **-zeit** f resolving time (of instruments)

auf·löten vt **1** unsolder. **2** (auf) solder on(to)

auf·machen I vt **1** open; launch, start; set up, establish. **2** undo, untie. **3** put up (signs). **4** make up, dress up; package; glamorize. **5** play up (news item). **6** make up, draw up (bill, etc). **7** get up (steam). **II** vi **8** open (up), open the door. **III: sich a. 9** (oft + **als**) dress up (as). **10** start out, start on one's way. **11** (**zu** + inf) get set (to), proceed (to). **12** (Wind) blow up, start blowing

Aufmach·material n (Biol.) mounting material

Aufmachung f (-en) **1** packaging, get-up, make-up; window dressing: **in großer A. bringen** package (or play up) in a big way, glamorize. **2** statement (of account)

auf·maischen vt mash up; remash

auf·merken 1 vt note (down). **2** vi (auf) pay attention (to)

aufmerksam a attentive: **einen a. machen** (auf) call sbdy's attention (to). **Aufmerksamkeit** f (-en) attentiveness; attention; token of esteem

auf·mischen vt mix up (a preparation)

auf·montieren vt (usu Mach.) (auf) mount (on)

auf·muntern vt cheer up; encourage

Aufnahme f (-n) **1** reception: **A. finden** meet with a (good) reception, be accepted, be adopted. **2** reception room. **3** admission (to an institution); acceptance. **4** absorption, assimilation, intake. **5** taking up, start (of work), inauguration—**in A. kommen** come into use (or fashion). **6** recording (in documents), on film, tape, disc); surveying, mapping; plotting (of curves). **7** (written, disc) record; (tape, disc) recording; photograph. **8** (Mach.) adapter; (tool) carrier; mounting, installation. Cf AUFNEHMEN

Aufnahme·apparat m **1** camera. **2** recording equipment (or device). **a–fähig** a **1** receptive. **2** (incl. Phys.) absorbent, absorptive. **3** (Chem.) absorbable. **A–fähigkeit** f **1** receptiveness, receptivity.

2 absorbing capacity. 3 (*Phys.*) capacity; (*Magn.*) susceptibility. 4 (*Chem.*) absorption power. 5 radar pick-up performance. **-gerät** *n* 1 camera. 2 recording equipment. **-geschwindigkeit** *f* rate of absorption. **-kolben** *m* absorption flask. **-kopf** *m* recording head. **-pipette** *f* pipette calibrated to take up a definite volume *cf* AUSFLUSSPIPETTE. **-prüfung** *f* entrance examination. **-raum** *m* (film, recording) studio. **-vermögen** *n* absorptive power. **-willigkeit** *f* consumer acceptance

Aufnahms· = AUFNAHME.

aufnehmbar *a* available; absorbable; assimilable

auf·nehmen* *vt* 1 pick up (objects, scent, ideas). 2 receive, take in (people, ideas, news). 3 absorb, assimilate. 4 admit (as member, etc), accept. 5 include, adopt (in programs, publications). 6 hire. 7 hold, accommodate. 8 borrow (money); raise (loan). 9 take up (work, relations), pick up (interrupted work); enter into (relations). **10: es a. mit** compete with, be a match for. 11 take down, record; photograph; survey, map. *Cf* AUFNAHME. **aufnehmend** *p.a* absorptive. **Aufnehmer** *m* (-) receiver; adapter; taker-up; (tape) detector head

auf·nötigen *vt* (*d.i.o*) force (on)

auf·opfern *vt* & **sich a.** sacrifice (oneself). **aufopfernd** *p.a* selfless, loyal

auf·oxidieren *vt* oxidize (to a higher state of oxidation)

auf·passen 1 *vt* fit on. 2 *vi* (*oft* + **auf**) pay attention (to), take care (of)

auf·peitschen *vt* whip up, excite

auf·perlen *vi*(s) bubble (up)

auf·pfropfen *vt* (*Bot.*) graft on (to)

auf·platzen *vi*(s) burst (break, split) open, rupture

auf·prägen 1 *vt* (**auf**) stamp, impress (on). **2:sich a.** (*dat*) leave its mark (on)

Aufprall *m* impact, collision; impingement. **auf·prallen** *vi*(s) (**auf**) collide (with), strike (against)

auf·pressen *vt* (*d.i.o*) (im)press (on)

auf·pumpen *vt* pump up, inflate

Aufpunkt° *m* test point; peak (of curve)

Aufputz *m* attire, finery. **auf·putzen** *vt* 1 clean up. 2 decorate, deck out, dress up

auf·quellen* I *vt* 1 soak, steep. 2 parboil. II *vi*(s) 3 swell up, bloat. 4 rise, well up

auf·raffen I *vt* 1 pick up; rake up, collect. II: **sich a.** 2 get up (on one's feet). 3 (*oft* + **zu**) pull oneself together (to do . . .). 4 (**aus**) pull oneself (out of)

auf·ragen *vi* rise, tower (up)

auf·rahmen 1 *vt* stretch on a frame (*or* tenterhooks) *cf* RAHMEN. 2 *vi* (*Milk*) (form) cream *cf* RAHM

auf·rauhen *vt* roughen (up); nap, teazel (cloth), card (wool)

auf·räumen I *vt* 1 tidy up, clean up. 2 clear (*or* rake) away. II *vi* 3 clean up. 4 take its toll. 5 (**mit**) do away (with), sweep away. **Aufräumung°** *f* tidying (up), clean-up, clearance, clearing (away)

auf·rechnen *vt* 1 count up, reckon up. 2 (*d.i.o*) charge (extra) (for); charge, chalk up (to). 3 (**gegen**) balance off (against). **Aufrechnung°** *f* reckoning up; extra charge

aufrecht *a* upright, erect; honorable; courageous. **aufrecht·erhalten*** *vt* maintain, uphold. **Aufrechterhaltung** *f* maintenance, upholding

auf·regen *vt* excite; upset, agitate, irritate. **Aufregung°** *f* excitement; agitation, irritation; confusion. **Aufregungs·mittel** *n* excitant, irritant

auf·reiben* *vt* 1 rub (sore), chafe. 2 grate (up). 3 ream. 4 wear out, wear down. 5 wipe out. **Aufreibung°** *f* rubbing, chafing; grating; reaming; exhaustion; attrition; annihilation

auf·reihen I *vt* 1 string (beads). 2 set up in a row. II *vt* & **sich a.** 3 line up

auf·reißen* I *vt*, *vi*(s) 1 tear (rip, split, crack) open, open wide; break up. II *vt* 2 tear (up), rip (up); lacerate. 3 pull up sharply. 4 plot, sketch, draw; outline *cf* AUFRISS. III: **sich a.** 5 jump up. *Cf* AUFGERISSEN

auf·richten I *vt* 1 raise, pick up; help to one's feet; set up; straighten up. 2 straighten out; (set) right, level off. 3 cheer up, encourage, give a boost (to). 4 erect, put up. 5 fix up, repair. II: **sich a.** 6 rise, get up (and stand erect); sit up; straighten up, draw oneself up. *Cf* AUFGERICHTET

aufrichtig *a* sincere, honest; upright.
Aufrichtigkeit *f* sincerity, honesty; uprightness
Aufrichtung° *f* 1 raising, erection, putting up, setting up; straightening up (*or* out). 2 boost, encouragement. *Cf* AUFRICHTEN
Aufriß° *m* 1 design, sketch; (*specif*) elevation, vertical projection. 2 outline. *Cf* AUFREISSEN
auf·rollen *vt* 1 unroll, unwind; reveal. 2 roll up, wind up. 3 bring up (subject)
auf·rücken *vi*(s) move up, advance, rise
Aufruf° *m* call, appeal, proclamation; calling of one's name. **auf·rufen*** *vt* call, call on; call in, recall
Aufruhr *m* uproar, commotion; turmoil; rebellion. **auf·rühren** *vt* stir up, agitate. **Aufrührer** *m* (-) rebel, agitator. **aufrührerisch** *a* rebellious
auf·runden *vt* round off (a number)
auf·rüsten *vt/i* arm. **Aufrüstung°** *f* armament; arms build-up
auf·rütteln *vt* shake up, arouse. **aufrüttelnd** *p.a* rousing, nerve-shaking
aufs *contrac* = **auf das** *cf* AUF 1,6; (+*superl*): e.g. **a. schnellste** in the fastest way possible
Aufsage° *f* notice; dismissal; cancelation.
auf·sagen I *vt* 1 recite. 2 (*d.i.o*) cancel, terminate, break off. II *vi* (*dat*) give notice; dismiss. **Aufsagen** *n* recitation. **Aufsagung** *f* (-en) cancelation, termination
Aufsalzung *f* salinization
auf·sammeln *vt* pick up, gather up; store up
aufsässig *a* rebellious; hostile
Aufsättigung *f* saturation
Aufsatz° *m* 1 superstructure, crown, top (piece); dome; piece set on top. 2 (*Mach.*) fixture, attachment. 3 gunsight. 4 (written) composition; (technical) article, paper, thesis. 5 (*Dye.*) topping. 6 (*Tex.*) cover print. *Cf* AUFSETZEN. **-farbe** *f* (*Dye.*) topping color. **-farbstaff** *m* topping dye. **-schlüssel** *m* socket wrench
auf·säuern *vt* acidify (again)
aufsaugbar *a* absorbable. **Aufsaugbarkeit** *f* absorbability
Aufsauge·fähigkeit *f* absorptivity. **-flüssigkeit** *f* absorption liquid
auf·saugen* *vt* suck up, soak up, absorb;

aspirate. **aufsaugend** *a* absorbent: **a–es Mittel** absorbent. **Aufsauger°** *m* absorber. **Aufsauge·ton** *m* absorbent clay.
Aufsaugung *f* absorption, suction
Aufsaugungs·fähigkeit *f* absorptive ability (*or* capacity). **-verfahren** *n* absorption process. **-vermögen** *n* absorptive power. **-wärme** *f* heat of absorption
auf·schalten: sich a. (*Radar*) lock on
auf·schärfen *vt* sharpen; (*Dye.*) strengthen
auf·schaukeln *vt* amplify, boost (waves)
auf·schäumen *vi* foam (up), froth, effervesce
auf·schichten *vt* pile up, stack; arrange in layers, stratify. **Aufschichtung°** *f* piling up; stratification
auf·schieben* *vt* 1 put off, postpone. 2 push open
auf·schießen* *vi*(s) shoot up, rise, sprout
Aufschlag° *m* 1 impact; crash; bounce, bound (of ball). 2 (price, tax) increase; surcharge, extra charge. 3 lapel; cuff; facing; (*Tex.*) warp. 4 (*Forest*) young growth. 5 (*Med.*) compress, application
Aufschlage·buch *n* reference book
auf·schlagen* I *vt* 1 break (crack, split) open. 2 open (book, lid, eyes); unbung (keg); handle (hides). 3 turn up, raise (collar, curtain, eyes), roll up (sleeves). 4 put up, set up (structure). 5 take up (residence). 6 raise, increase (price); (+**auf**) add (to). 7 put on, apply (poultice). II *vi*(h) 8 (*oft* + **auf**) strike, beat (on). 9 (*Prices*) rise, increase. III: *vi*(s) 10 (*oft* + **auf, gegen**) strike, hit, crash, impinge (on, against). 11 shoot up, rise; blaze up. 12 spring (*or* snap) open
Aufschlag·gerät *n* blender, mixer. **-zünder** *m* percussion fuse; impact detonator
auf·schlämmen *vt* suspend, make into a paste; reduce to slime. **Aufschlämmung** *f* 1 suspension, flotation; reduction to paste (*or* slime). 2 paste, slime; sludge. **Aufschlämm·verfahren** *n* flotation process
auf·schleifen* *vt* (**auf**) grind, engrave (on)
auf·schlemmen *vt* suspend, etc = AUFSCHLÄMMEN
Aufschließ·arbeit *f* decomposition operation

aufschließbar *a* decomposable etc *cf* AUF-SCHLIESSEN

auf·schließen* I *vt* 1 decompose, break down; convert to nonrefractory form (e.g. by fusion with alkali carbonates); hydrolyze (starch); solubilize; pulp (paper); crush (ore). 2 open, unlock; reveal. 3 make accessible (*or* available). 4 open (land) to development. *Cf* AUFGESCHLOSSEN. II *vi*(s) close ranks. *Cf* AUFSCHLUSS

Aufschließ·machine *f* ore crusher. **-mischung** *f* decomposition mixture

Aufschließung *f* (-en) decomposition, break-down, etc *cf* AUFSCHLIESSEN 1, AUFSCHLUSS

Aufschließungs·vermögen *n* decomposing power

auf·schlitzen *vt* slit open; slash, rip

Aufschluß° *m* 1 decomposition, break-down, etc *cf* AUFSCHLIESSEN 1. 2 opening (up); development; revelation. 3 information, explanation, elucidation. 4 (*Geol.*) exposure, (natural) section. **·behälter** *m*, **-gefäß** *n* (*Chem.*) digester

auf·schlüsseln *vt* codify, classify; decode

Aufschluß·grad *m* degree of decomposition. **-mischung** *f* decomposition mixture. **-mittel** *n* decomposing agent. **-ofen** *m* sintering furnace. **a—reich** *a* instructive, informative. **A—verfahren** *n* decomposition (*or* disintegration) process. **-verhältnis** *n* (*Geol.*) condition of exposure

auf·schmelzen* 1 *vt* melt down (open, apart); smelt; (**auf**) fuse on (to). 2 *vi*(s) melt (away); thaw. **Aufschmelzen** *n* (*Metll.*) flow-melting, (*esp Tinplate*) thermal reflowing. **Aufschmelzung** *f* melting, fusion

auf·schneiden* I *vt* 1 cut open, lance; dissect. 2 cut up, slice up. II *vi* 3 brag, boast

Aufschnitt° *m* sliced meat, cold cuts. **·maschine** *f* slicing machine

auf·schöpfen *vt* scoop up, dip up

aufschraubbar *a* unscrewable, screw-off; screw-on (lid, etc)

auf·schrauben I *vt* 1 unscrew, screw off (*or* open). 2 (*oft* + **auf**) screw on (to). 3 screw tighter, tighten; turn up (gas, radio). 4 intensify; force up (prices, statistics). II: **sich a.** 5 come unscrewed

auf·schrecken 1 *vt* startle, rouse. 2 *vi* (s) be startled, start up (in fright)

auf·schreiben* 1 write (*or* take) down; (+ *dat reflx*):**sich etwas a.** make a note of sthg. 2 book (at police station). 3 prescribe, write a prescription for. 4 (+ **lassen**) have ... charged to one's account

Aufschrift° *f* 1 inscription, caption, title; (text of a) sign. 2 label. 3 address

auf·schrumpfen *vt* (*usu* + **auf**) shrink on (to)

Aufschub *m* delay, postponement; extension

auf·schüren *vt* stir up, poke, stoke (fire)

auf·schürfen *vt* & **sich a.** graze; abrade (skin)

Aufschütt·dichte *f* apparent density

auf·schütteln *vt* shake up

auf·schütten *vt* 1 pour, dump (on); (*Geol.*) deposit. 2 heap (up), pile (up). 3 spread, scatter; spread material for (roads, dams). 4 lay up, store up

Aufschütt·faß *n* (*Dye.*) settling vat. **-trichter** *m* feed hopper

Aufschüttlung° *f* 1 pouring, dumping (on) etc. *cf* AUFSCHÜTTEN. 2 dike, embankment

auf·schwefeln *vt* treat with sulfur

Aufschweiß° *m* weld, welded seam. **auf·schweißen** *vt* 1 cut open with a welding torch. 2 (*usu* + **auf**) weld (on, onto). 3 hard-face. **Aufschweiß·legierung** *f* hard-facing alloy

auf·schwellen 1 *vt* swell, blow up, inflate. 2 *vi**(s) swell up, tumefy

auf·schwemmen I *vt* 1 deposit, wash up (silt, etc). 2 suspend. 3 soak. 4 bloat, swell, fatten. II *vi* 5 be fattening, cause bloating. **Aufschwemmung** *f* (-en) deposit(ing); suspension; soaking; swelling, bloating

auf·schwingen* *vr*: **sich a.** rise, soar, surge (up). **Aufschwung°** *m* upswing; (up)-surge, rise; impetus, boost; (*Econ.*) boom

auf·sehen* *vi* look up. **Aufsehen** *n* sensation. **a—erregend** *a* sensational. **Aufseher** *m* (-), **Aufseherin** *f* (-nen) overseer, guard; supervisor, foreman

auf·setzen I *vt* 1 put on; put ... on the fire (to cook). 2 sew on; appliqué. 3 put down, set (foot). 4 attach (on top); build on, mount. **5** serve (food). **6** set up, put in an upright position. **7** draw up (document). **8** (*Dye.*) top; (*Tex.*) cover-print. **II** *vt/i* **9** touch down, land. **III: sich a. 10** sit up. *Cf* AUFSATZ

Aufsetz·gewicht *n* fractional weight. **-kasten** *m* (*Dye.*) jig, jigger

Aufsicht° *f* **1** supervision, surveillance; monitoring, proctoring (duty). **2** supervisor, monitor, proctor, person in charge. **3** top view; (*Engg. Drawing*) plan. **4: in der A.** in reflected light. **5** (*Dye.*) underhand appearance, undertone

Aufsichts·farbe *f* reflected (*or* surface) color. **-rat** *m* board of directors

auf·sieden* **1** *vt* bring to a brief boil; reboil; boil out; blanch (silver). **2** *vi*(s) come to a (brief) boil. **Aufsieden** *n* boiling up; ebullition

aufsitzend *p.a* perched (*or* mounted) on top

auf·spalten 1 *vt* & **sich a.** split (up, open); open up (rings); depolymerize. **2** *vt* cleave; resolve (isomers). **3** *vi*(s) (*Biol.*) segregate. **Aufspaltung°** *f* splitting up, cleavage; opening up; depolymerization; resolution, segregation

auf·spannen *vt* **1** stretch (out); open (umbrella). **2** spread; mount (map on board). **Aufspanner** *m* (-) step-up transformer

auf·speichern 1 *vt* store, lay up; amass; bottle up. **2** *vt* & **sich a.** accumulate. **Aufspeicherung** *f* (-en) storage, accumulation

auf·sperren *vt* (throw) open; spread apart; unlock

auf·splittern *vt, vi*(s), **sich a.** split up, splinter, fragment

auf·sprengen *vt* **1** break (force, blast) open, blow up. **2** (**auf**) sprinkle (on)

auf·sprießen* *vi*(s) spring up, sprout, germinate

auf·springen* *vi*(s) **1** jump (leap, spring) up; (**auf**) jump (on). **2** bounce. **3** spring (fly, split, burst) open. **4** crack, chap. **aufspringend** *p.a* (*Bot.*) dehiscent

auf·spritzen *vt* & *vi*(s) spray, squirt, spurt (up)

auf·sprudeln *vi*(s) boil up, bubble up, effervesce

auf·sprühen 1 *vt* spray (on). **2** *vi*(s) spray up

auf·spunden, auf·spünden *vt* unbung, tap

auf·spüren *vt* track down, ferret out, unearth

Aufst. *abbv* (Aufstellung) formulation, graph

auf·stampfen 1 *vt* trample; ram. **2** *vi* stamp (one's foot)

auf·stapeln *vt* stack up, pile up; store

auf·stäuben *vt* spray (dust, atomize) on

auf·stechen* *vt* **1** prick open, puncture; lance. **2** spade up. **3** spear, impale. **4** fix, retouch. **5** (+**auf**) pin on (to)

aufsteckbar *a* attachable, slip-on, pin-up

Aufsteck·blende *f* (*Opt.*) slip-on diaphragm; lens cover

auf·stecken I *vt* **1** pin (stick, put) up. **2** put on, stick on, slip on; attach. **3** gain, earn. **II** *vt/i* give up

Aufsteck·glas *n* (*Opt.*) slip-on lens

auf·stehen* **1** *vi*(h) be (*or* stand) open. **2** *vi*(s) get up, rise (up), stand up; rebel

auf·steigen* *vi*(s) **1** (**auf**) get on (vehicle, horse), climb on. **2** ascend, rise; climb up. **3** fly up, take off. **4** come up (*as:* storm), arise, be aroused. **5** advance, move up, be promoted. **aufsteigend** *p.a* ascendant

auf·stellen I *vt* **1** (*genl*) put up, erect; (*specif*) assemble, install, mount. **2** (*genl*) set (up), establish; (*specif*) formulate, make up, draw up. **II: sich a. 3** take a (standing) position, line up. **4** stand up (*or* on end). **Aufstellung°** *f* putting up, erection; assembly; installation, mounting; setting up, set-up; establishment; formulation, making up, drawing up; statement; curve; graph; tabulation

auf·stickern *vt* (*Metll.*) nitride

Aufstieg *m* (-e) rise, ascent, climb; advancement, promotion *cf* AUFSTEIGEN

auf·stöbern *vt* scare up; unearth, dig up

auf·stocken 1 *vt* add a story to (a bldg.). **2** *vt/i* (*Econ. & fig*) increase; accumulate; beef up

Aufstoß° *m* impact. **auf·stoßen*** **1** *vt* push (*or* kick) open; strike, put down (on the ground). **2** *vi*(h) belch, eruct. **3** *vi*(h *or* s) ferment again; (*of wine*) turn acid; (*dat*) (*Foods*) repeat; **sauer a.** cause heartburn (*or* hyperacidity). **4** *vi*(s) (*dat*) come to (one's) attention, strike

auf·strahlen *vi* (begin to) shine, beam, fluoresce. **Aufstrahler** *m* (-), **Aufstrahl·**

stoff *m* fluorescent substance
auf·streichen* *vt* spread, apply, brush on
auf·streuen *vt* (**auf**) sprinkle, dust (on).
Aufstreu·pulver *n* sprinkling powder
Aufstrich° *m* 1 spreading on, application, brushing on. 2 finish, (*specif*) paint, varnish, lacquer, stain. 3 coat(ing). 4 (sandwich) spread. 5 upstroke. 6 auction
Aufstrom° *m* updraft, upcurrent. **-vergaser** *m* updraft carburetor
auf·stützen *vt* prop up; support; rest, lean
auf·suchen *vt* look up, seek out; pick up
auf·summen, auf·summieren *vt & sich s.* add up
auf·tauchen *vi*(s) emerge; appear; arise.
Auftauchen *n* emergence; appearance; rise
auf·tauen *vt/i* thaw (out), defrost. **Auftau·punkt** *m* thawing point
auf·teilen *vt* divide (up); distribute; parcel out. **Aufteilung°** *f* division; distribution; parceling out
Auftrag *m* (. .t⁼e) 1 assignment, mission. 2 instructions, orders—**im A.** (+*genit* or **von**) by order of, on behalf of. 3 (*Com.*) order; contract: **in A. geben** contract for, order. 3 (*Engg.*) embankment, fill. 4 application (of paint, etc); layer, coat(ing)
auf·tragen* **I** *vt* 1 serve (food). 2 put on, lay on, apply (paint, etc); (*Welding*) deposit. 3 (*d.i.o*) order, ask (to do); entrust (*or* charge) with. 4 wear out (clothes). 5 plot, lay out, mark off. 6 charge (furnace). **II** *vi* 7 (*Clothes*) be bulky
auftrag·schweißen *vt/i* deposition-weld
auf·träufeln 1 *vt* apply in drops. 2 *vi*(s) drip, fall in drops
auf·treffen* *vi*(s) (**auf**) strike (against), impinge (on), impact. **Auftreffen** *n* (*Opt.*) incidence. **auftreffend** *p.a* (*Light*) incident
Auftreff·punkt *m*, **-stelle** *f* point of impact (*or* incidence). **-winkel** *m* angle of incidence (*or* impact)
auf·treiben* **I** *vt* 1 rouse (up), force (to get) up; buoy up. 2 blow (whirl, raise) up (dust). 3 swell, distend, inflate; cause (bread) to rise; blow (up) (glass). 4 open out (holes). 5 drive on (barrel hoops). 6 bring to market. 7 find, get hold of. **II** *vt/i* 8 sublime. **III** *vi*(s) 9 swell up, dis-

tend, expand; rise (as; dough). 10 shoot (rise, fly) up. 11 run aground. *cf* AUFTRIEB. **Auftreiber°** *m* (*Glassblowing*) flanger, reamer. **Auftreibung** *f* (-en) blowing (up); distension, inflation; finding, obtaining; sublimation; expansion
auf·trennen 1 *vt* open up, rip up; undo. 2:**sich a.** come undone
auf·treten* *vi*(s) 1 enter, appear. 2 arise, occur. 3 (**gegen**) take a stand (against), oppose. 4 (*usu* + *adv*) act, behave, proceed. 5 tread, put one's foot down. *Cf* AUFTRITT. **Auftreten** *n* 1 entrance, appearance. 2 rise, occurrence. 3 (**gegen**) stand (against), opposition (to). 4 action, behavior; attitude
Auftrieb° *m* 1 buoyancy, uplift, upward thrust. 2 impetus, lift. 3 plankton. **-mittel** *n* buoying agent; swelling agent
Auftriebs·kraft *f* buoyant (*or* lifting) force
Auftritt° *m* entrance, appearance; scene; tread
auf·trocknen 1 *vt* wipe up. 2 *vt, vi*(s) dry up
auf·tropfen 1 *vt* apply drop by drop. 2 *vi*(s) drip (on)
auf·tun* *vt & sich a.* open (up)
auf·türmen *vt & sich a.* pile up, heap up, tower
auf·vulkanisieren *vt* (**auf**) vulcanize (on)
auf·wachen *vi*(s) wake up, awaken
auf·wachsen* *vi*(s) grow up, arise. **Aufwachsung** *f* (-en) growth; filament growth. **Aufwachs·verfahren** *n* filament growth method (of decomposing volatile metallic iodides on hot filament)
auf·wallen *vi*(h *or* s) boil (bubble, billow) up; seethe; rise (suddenly). **Aufwallung°** *f* bubbling, ebullition; effervescence; billowing; (sudden) rise; (*Metll.*) wildness
Aufwand *m* expenditure, expense; effort; extravagance
auf·wärmen *vt* warm up, reheat; rehash
aufwärts *adv & sep pfx, prep* (*accus*) up, upward. **A--bewegung** *f* upward motion, improvement. **-gasen** *n* up run (of water gas). **-transformator** *m* (*Elec.*) step-up transformer
auf·waschen* *vt/i* wash up, clean up, cleanse
auf·wecken *vt* wake up, awaken, arouse *cf* AUFGEWECKT

auf·weichen 1 *vt, vi*(s) soften (up), soak. **2** *vt* (*Metll.*) temper (colors). **aufweichend** *p.a* emollient

aufweisbar *a* demonstrable

auf·weisen* *vt* show, present, produce

auf·weiten *vt* widen, stretch, expand

auf·wenden* *vt* expend, put to use, employ, devote. **aufwendig** *a* extravagant; elaborate, sophisticated; demanding. **Aufwendung°** *f* expenditure, expense

auf·werfen* I *vt* **1** throw up (dam, wall), pile up (earth). **2** toss, raise, turn up, curl (up). **3** throw open. **4** dig up. **5** throw on (coal). **6** bring up (topic). **II: sich a. 7** be arrogant. **8** (**als**) pose, set oneself up (as)

auf·werten *vt* upgrade; raise in value, appreciate

auf·wickeln I *vt* **1** wind (up), wind on. **2** unwrap, unwind. **II: sich a. 3** come unwound. **Aufwickel·spule** *f* take-up reel

auf·wiegen* *vt* compensate, balance, offset

auf·winden* *vt* wind up, wind on; haul up

auf·wirbeln 1 *vt, vi*(s) whirl up. **2** *vt* fluidize

auf·wischen *vt* wipe (up), mop (up)

Aufwuchs° *m* growth; young growth

auf·wühlen *vt* root up; churn up, stir up

auf·zahlen *vt* pay down

auf·zählen *vt* count out (*or* up), enumerate, itemize. **Aufzählung°** *f* enumeration, itemization. **Aufzählungs·reihe** *f* series of summations; frequency distribution

auf·zehren I *vt* **1** absorb. **2** eat up, consume; use up. **3** sap (one's) strength. **II: sich a. 4** be absorbed (consumed, used up). **5** wear oneself out. **Aufzehr.grad** *m* degree of absorption. **Aufzehrung** *f* absorption; consumption

auf·zeichnen *vt* sketch; write down, record, register; (video)tape; plot, trace. **Aufzeichner°** *m* recorder; plotter, tracer. **Aufzeichnung°** *f* **1** sketching; note-taking, recording, registration; plotting, tracing. **2** written record, log; *esp pl* notes

auf·zeigen *vt* show, demonstrate, point out

auf·ziehen* I *vt* **1** pull up, hoist; raise; suck up. **2** pull open (*or* apart); pull out (stops). **3** stretch, put on (e.g. string on violin); mount, paste on. **4** raise, bring up (child), cultivate (plants) *cf* AUFZUCHT. **5** wind up (spring). **6** start, plan, organize, arrange. **7** tease. **8** agitate (yeast). **9** (*Med.*) draw (injection). **II** *vi*(s) **10** pull up, arrive (*usu* in a group), march up, appear, show up. **11** come up, arise (*as:* storm). **12** (*Dye.*) go on, be absorbed, become attached. **III: sich a. 13** be self-winding. *Cf* AUFZUG

Aufzieh·geschwindigkeit *f* (*Dye.*) absorption rate. **-krücke** *f* (*Brew.*) rouser. **-vermögen** *n* (*Dye.*) absorptive power, substantivity

auf·zischen *vi*(s) rise (*or* start) with a hiss

Aufzucht° *f* breeding, raising, cultivation

Aufzug° *m* **1** arrival; marching up; parade procession. **2** elevator; crane, hoist; pump. **3** attire. **4** act (of a play). *Cf* AUFZIEHEN **1, 10**

auf·zwängen *vt*, **auf·zwingen* 1** *vt* (*d.i.o*) force, impose (on). **2: sich a.** force oneself to get up; (*dat*) force itself (on)

Aug·apfel *m* eyeball

Augapfel·bindhaut *f* ocular conjunctiva. **-gefäßhaut** *f* chorioid (membrane). **-haut** *f* sclerotic coat, sclera

Aug·bolzen *m* eyebolt

Auge *n* (-n) **1** eye: **im A. haben** have one's eye on; **ins A. fassen** keep one's eyes on; **vor A–n haben** be aware of. **2** eyelet; lug; bubble (in liquid, beads); hole (in bread, cheese); luster (on fabric); knot (in wood); (*Bot. oft*) bud

äugeln *vt* **1** ogle. **2** bud, graft

Augen· eye, visual, optic, ocular: **-achat** *m* (*Min.*) cat's eye. **-ader** *f* ophthalmic vein. **-aderhaut** *f* chorioid. **-arzt** *m* oculist, opthalmologist. **-bindehaut** *f* ocular conjunctiva

Augen·blick *m* movement, minute, instant. **augenblicklich 1** *a* immediate, instant; momentary. **2** *adv oft* at the moment. **Augenblicks·bild** *n* snapshot

Augen· eye, visual, optic(al), ocular: **-bogen** *m* iris. **-bolzen** *m* eyebolt. **-braue** *f* eyebrow. **-deckel** *m* eyelid. **-drüse** *f* lacrimal gland. **a–fällig** *a* conspicuous, evident, obvious. **A–flüssigkeit** *f:* **glasartige A.** vitreous humor; **wasserartige A.** aqueous humor. **-gefäßhaut** *f* chorioid. **-heilkunde** *f* ophthalmology. **-höhle** *f* eye socket, orbital cavity

Augenhöhlen·(*in compds*) orbital

Augen· eye, visual, optic(al), ocular: **-hornhaut** *f* cornea. **-kammerwasser** *n*

aqueous humor. **-licht** *n* eyesight. **-linse** *f* crystalline lens; eye lens, eyepiece. **-loch** *n* 1 pupil. 2 orbital cavity. **-marmor** *m* eye-spotted marble. **-maß** *n* measure by eye: **gutes A. haben** have a sure eye. **-merk** *n* attention. **-nichts** *n* nihil album (zinc oxide). **-pfropfen** *m* bud-grafting. **-pinsel** *m* (fine camel's hair) eye brush. **-punkt** *m* point of sight, visual point. **-ring** *m* iris

Augen·schein *m* 1 appearance(s). 2 inspection, examination—**in A. nehmen** inspect, examine. **augenscheinlich** *a* apparent; evident. **Augenscheinlichkeit** *f* apparent evidence; obviousness

Augen·schirm *m* eyeshade. **-schutzbrille** *f*, **schützer** *m* protective (*or* safety) goggles, safety glasses. **-schwarz** *n* retinal pigment. **-stein** *m* 1 white vitriol, zinc sulfate. 2 lacrimal calculus. **-stern** *m* pupil (of the eye). **-wasser** *n* eye lotion. **-weiß** *n* white of the eye, sclera, scleroritic. **-windung** *f* frontal convolution. **-wulst** *m,f* optical swelling. **-wurz, -wurzel** *f* dandelion, wood anemone, mountain parsley. **-zahn** *m* eyetooth. **-zeuge** *m* eyewitness. **-zirkel** *m* iris

augit·artig, augit·haltig, augitisch *a* augitic, (*Cryst.*) monoclinic. **Augit·spat** *m* (*Min.*) (*genl*) silicate, (*specif*) augite, acmite, amphibole, wollastonite

Aureole *f* (-n) aureole; arc flame

Auri- auric, (trivalent) gold: **-chlorid** *n* auric chloride. **-chlorwasserstoffsäure** *f* chloroauric acid. **-cyanid** *n* auric cyanide. **-cyanwasserstoffsäure** *f* cyanauric acid. **-pigment** *n* orpiment. **-rhodanwasserstoffsäure** *f* aurothiocyanic acid, thiocyanatoauric acid

Auro- aurous, (univalent)gold:**-chlorid** n aurous chloride. **-chlorwasserstoffsäure** *f* chloroaurous acid (HAuCl₂). **-cyanid** *n* aurous cyanide. **-cyanwasserstoffsäure** *f* aurocyanic acid. **-rhodanwasserstoffsäure** *f* aurothiocyanic acid

aus I *prep* (*dat*) 1 out(of), from. 2 (made) of. 3 because of. **II** *adv & sep pfx* 4 out. 5 over, done, through, finished. 6 (*Clothes, Gas, Light, Radio, etc*) off. 7: **auf etwas a. sein** be out to get sthg; **ein und aus** in and out;

von hier a. (starting) from here

aus·arbeiten 1 *vt* work out (in detail), elaborate, develop; formulate. 2 *vi* cease (*or* finish) working. 3: **sich a.** get exercise (*or* a workout); work itself out, develop. **Ausarbeitung** *f* (-en) working out, elaboration, development; formulation; exercise, workout

aus·arten *vi*(s) deteriorate, degenerate; run wild

aus·äthern *vt* extract with ether

aus·atmen 1 *vt/i* breathe (out), exhale. 2 *vi* breathe one's last

aus·ätzen *vt* 1 cauterize. 2 destroy with caustics. 3 etch (out). 4 discharge (colors)

aus·balancieren *vt* balance out, counterbalance

Ausball·masse *f* (*Lthr.*) bottom filler

Ausbau° *m* 1 removal; dismantling. 2 expansion, (full) development, elaboration; finishing. 3 rebuilding, reconstruction, remodeling. 4 outbuilding. *Cf* AUSBAUEN

aus·bauchen 1 *vt & sich a.* bulge (belly, swell, flare) (out). 2 *vt* emboss

aus·bauen *vt* 1 remove, take out, dismantle. 2 expand, develop fully; elaborate; finish (the interior of). 3 (**als, zu**) rebuild, remodel (as, into)

aus·bedingen* *vt* (+ *dat rflx:*sich) stipulate, reserve. **ausbedingen** *pp*

aus·beißen* *vi* (*Geol.*) crop out *cf* AUSGE-BISSEN

aus·beizen *vt* cauterize

aus·bessern *vt* repair, mend, correct. **Ausbesserungs·arbeiten** *fpl* repair work

aus·beulen *vt* 1 swell (bag, round) out. 2 smooth out, flatten (dents, bumps)

Ausbeute *f* yield, output; profit; result(s). **--erhöhung** *f* increase in the yield. **aus·beuten** *vt* make (full) use of, expoit; work (mines). **Ausbeutung** *f* (-en) exploitation

aus·biegen* **I** *vt* 1 bend outward. 2 straighten. **II** *vi*(s) (*dat*) dodge, avoid

aus·bilden I *vt* 1 educate. 2 train. 3 develop. 4 arrange, devise, form. **II: sich a.** 5 get an eduction; get (full) training; specialize. 6 form, develop. **Ausbildung°** *f* education; training; development; arrangement, design

Ausbiß° *m* (*Min.*) outcrop

aus·bitten* *vt* (+ *dat rflx:* **sich**) request, demand

Ausblase·dampf *m* exhaust steam. **-hahn** *m* blow-off (*or* drain) cock

aus·blasen* *vt* 1 blow out (flame; smoke; pipe). 2 blow off, exhaust (steam). 3 shut down (blast furnace). *Cf* AUSGEBLASEN

Ausblase·ventil *n* blow-off valve

aus·blassen *vi*(s) (grow) pale, fade (out)

aus·bleiben* *vi*(s) 1 stay out (*or* away), be absent. 2 fail to appear, fail to occur, be lacking. **Ausbleiben** *n* absence; nonappearance, nonoccurrence

aus·bleichen 1 *vt* bleach out. 2 *vi*(s)* fade, pale

aus·bleien *vt* line with lead

aus·blenden *vt* (*Opt.*) stop out (*or* down); shield (from rays); scatter (rays); fade out. **Ausblendung** *f* (-en) stopping out (*or* down); shielding; scattering; fading out, fade-out

Ausblick° *m* view; outlook, prospect

aus·blühen *vi* fade; go by, cease blooming; effloresce. **ausblühend** *p.a* efflorescent. **Ausblühung** *f* (-en) fading; bloom(ing); efflorescence

aus·bluten *vi*(s) 1 stop (*or* finish) bleeding; bleed to death. 2 (*Dye.*) bleed, mark off

aus·bohren *vt* bore (drill, ream) out. **Ausbohrung°** *f* boring, drilling; bore

Ausbrand° *m* 1 burning out, total combustion. 2 degree of oxidation. 3 annealing; fritting. **--versuch** *m* (*Metll.*) scaling test

aus·braten* 1 *vt, vi*(s) roast through; roast out. 2 *vt* heat; render, fry lean. 3 *vi*(s) (*Fat*) be fried (*or* roasted) out

aus·brechen* I *vt, vi*(s) 1 break out. II *vt* 2 knock out (teeth); knock, cut (holes in walls). 3 break through (walls). 4 break off, prune, lop off. 5 bore out; drive (tunnel, shaft); clear (furnace). 6 (*Min.*) work (a lode); quarry (stone). 7 throw up, regurgitate. III *vi*(s) 8 skid. 9 erupt. 10 (**in**) break (into: anger, etc). *Cf* AUSBRUCH

Ausbreite·maß *n* degree of spread (*or* slump)

aus·breiten I *vt & sich a.* 1 spread (out), unfold, expand, extend; open (arms). II *vt* 2 lay out, display. 3 flatten. 4 diffuse.

5 floor (grain). III: *sich a.* 6 spread oneself (out), sprawl. 7 (+ **über,** *oft*) talk extensively (about), expatiate (on). 8 (*Concrete*) slump

Ausbreite·probe *f* flattening (hammering; flow) test. **-tisch** *m* flow table

Ausbreitung *f* (-en) 1 spread, spreading (out), unfolding, expansion, extension. 2 laying out, display. 3 flattening. 4 diffusion. 5 (*Grain*) flooring

Ausbreitungs·ebene *f* plane of (wave) propagation. **-koeffizient** *m* diffusion coefficient. **-kugel** *f:* Ewaldsche A. (*Cryst.*) sphere of reflection. **-widerstandsverfahren** *n* increased resistance method

aus·brennen* I *vt* 1 scorch, parch. 2 burn out (weeds, vermin). 3 cauterize. 4 anneal (metals); frit (glass). II *vi*(s) 5 burn down, go out (*as:* candles). 6 be burned out, be gutted. 7 (*Explo.*) blow out

aus·bringen* *vt* 1 extract, produce, yield. 2 hatch (eggs); bear (young). 3 remove, get (sthg) out (*or* off). 4 empty (glass). 5 spread (rumor)

aus·bröckeln *vi*(s) flake, crumble

Ausbruch° *m* 1 wine from the ripest grapes. 2 escape, breakout. 3 outbreak: **zum A. kommen** break out. 4 outburst; eruption. 5 excavation. *Cf* AUSBRECHEN

aus·brühen *vt* parboil, brew; scald

aus·brüten *vt* hatch (out); inoculate

aus·buchten 1 *vt* cut curves into, scallop, indent. 2 *vi & sich a.* curve outward, project, bulge. **Ausbuchtung** *f* (-en) 1 indentation. 2 outward curve, bay, projecting part. 3 turnout (for parking)

Ausbund *m* model, paragon

Ausbutterungs·grad *m* churning efficiency

aus·dampfen 1 *vt* evaporate; emit as steam (*or* as a vapor). 2 *vi*(h) cease streaming. 3 *vi*(s) come out as steam

aus·dämpfen *vt* clean out with steam; smoke out; brew; (let) evaporate

Ausdampfung *f* evaporation

Ausdauer *f* endurance, perseverance. **ausdauernd** *a* enduring, persevering; hardy; (*esp Bot.*) perennial

ausdehnbar *a* expansible, extensible;

ductile. **Ausdehnbarkeit** f expansibility, extensibility, ductility

aus·dehnen vt & sich a. expand, spread; stretch, extend; distend, dilate cf AUSGEDEHNT. **Ausdehnung°** f 1 spread, expansion; extension; distension, dilation. 2 expanse, spaciousness, extent. 3 (Math.) dimension

Ausdehnungs·arbeit f work done in expanding. **-fuge** f expansion joint. **-koeffizient** m coefficient of expansion. **-kraft** f expansive force. **-messer** m dilatometer, expansometer. **-vermögen** n expansibility; extensibility; dilatability. **-zahl** f expansion coefficient

ausdenkbar a conceivable, imaginable. **aus·denken*** vt (usu + dat rflx: sich) think up (or out), imagine, conceive (of); contrive, devise, invent: **nicht auszudenken** beyond conception, unimaginable

aus·deuten vt interpret, explain

aus·dorren, aus·dörren vt, vi(s) dry up, dry out, wither, desiccate, parch; season (timber)

aus·drehen vt turn off (or out); strip (screws)

Ausdruck m (..d⁻e) expression: **zum A. bringen** express; **zum A. kommen** be expressed cf AUSDRÜCKEN

aus·drucken vt finish printing, print off (book); print out (in full)

aus·drücken I vt 1 express. 2 press out, squeeze out; stub out, snuff out. **II: sich a.** express onself, be expressed. **Ausdrücker** m (-) ejector. **ausdrücklich** a express, explicit

ausdruckslos a expressionless, inexpressive; adv oft without expression. **Ausdruckslosigkeit** f absence of expression, inexpressiveness

Ausdrucks·mittel n means of expression. **a—voll** a expressive. **a—weise** f manner of expression, way of expressing onself

aus·dünnen vt & sich a. thin out

Ausdunst° m exhalation, etc = AUSDUNSTUNG

ausdünstbar a evaporable. **Ausdünstbarkeit** f evaporability

aus·dunsten, aus·dünsten 1 vt exhale, give off, exude (vapors, fumes). 2 vi evap-

orate; exhale; perspire, transpire. **Ausdunstung, Ausdünstung** f (-en) evaporation, exhalation; perspiration, transpiration; emanation, (exhaled) vapor

Ausdünstungs·apparat m evaporating apparatus. **-messer** m atmometer, evaporimeter

aus·egalisieren vt equalize, level up

auseinander adv & sep pfx apart, in two, to pieces: **·brechen** vt, vi(s) break apart. **·breiten** vt spread out, unfold. **·bringen*** vt separate, get (things) apart (or separated). **·entwickeln: sich a.** develop in different directions. **·fahren*** vi(s) separate (suddenly); scatter, disperse, diverge. **·fallen*** vi(s) fall apart, fall to pieces, crumble. **·falten** vt unfold, spread out. **·fliegen*** vi(s) fly apart, separate (quickly); disperse. **·gehen*** vi(s) 1 separate, part. 2 diverge, differ. 3 come apart; break up, dissolve. 4 disperse. 5 spread out; expand. **·gehend** p.a (oft) divergent. **·halten*** vt keep apart; tell apart, distinguish. **·legen** vt (d.i.o) explain (to). **·liegen*** vi lie (or be located) (a distance) apart. **·nehmen*** vt take apart, dismantle. **·reißen*** vt, vi(s) tear (or break) apart, separate. **·rücken** vt, vi(s) move (or shift) apart. **·setzen** I vt 1 (d.i.o) explain (to). **II: sich a.** (mit) 2 study closely, analyze (a topic). 3 confront. 4 (also + über) have an argument (or a clarification of positions), exchange views (with sbdy, on a topic). 5 come to an agreement. **A—setzung** f (-en) 1 explanation. 2 analysis. 3 argument, confrontation, dispute; exchange of views. 4 agreement. **a—ziehen*** 1 vt pull apart, stretch out. 2 vi(s) separate, move away from each other

aus·entwickeln vt & sich a. develop fully

auserkoren p.a chosen, selected

auserlesen p.a select, choice, excellent

aus·ersehen*, aus·erwählen vt choose, select; designate

aus·exponieren vt expose fully

aus·fahren* I vt 1 take out for a ride (or drive). 2 deliver (by vehicle). 3 make full use of (one's driving skill); (Mach.) run at full capacity. 4 raise, lower, slide over, extend (machine parts). 5 run (or wear) ruts into (roads) *cf* AUSGEFAHREN. **6** hold (a race). **II** *vt, vi*(s) **7** keep on the outside lane of (curves). **III** *vi*(s) **8** depart, pull (run, go, drive) out. **9** go out for a drive (*or* ride).

ausfahrend *p.a* (*oft*) erratic, angular (movements). **Ausfahrt°** *f* **1** departure. **2** drive, ride, trip. **3** exit (for vehicles)

Ausfall° *m* **1** precipitation, precipitate. **2** falling out (*or* off), loss, drop, lapse; deficiency, deficit. **3** absence; absentee, casualty. **4** failure, breakdown; stoppage; (cardiac) arrest. **5** sally, sortie. **6** (violent) outburst, attack. **7** (*Elec.*) shock. **8** (*Nucl. Phys.*) fallout. *Cf* AUSFALLEN

Ausfäll-apparat *m* precipitating apparatus

ausfällbar *a* precipitable

Ausfall-eisen *n* off-grade iron

aus-fallen* *vi*(s) **1** precipitate (out), be deposited. **2** drop out, fall out. **3** not take place, not appear; be absent, be missing; be canceled. **4** fail, break down. **5** come out, turn out (well, badly). **6** stick out, jut out. *Cf* AUSFALL, AUSFALLEND, AUSGEFALLEN

aus-fällen *vt* precipitate (out)

ausfallend *p.a* **1** precipitated, deposited, **2** absent, missing, lost, canceled. *Cf* AUSFALLEN

Ausfall-öffnung *f* discharge opening. **-schmelzung** *f* (*Metll.*) off-heat

Ausfalls-erscheinung *f* deficiency symptom

Ausfall-straße *f* exit road; arterial road

Ausfällung° *f* precipitation, depositing

Ausfall-winkel *m* angle of reflection (*or* emergence)

aus-färben I *vt* **1** give the last dye to (material); dye (completely). **2** decolorize, extract the color from. **3** exhaust (dye bath). **II** *vi* **4** lose color

Ausfärb-vorrichtung *f* color extractor

aus-fasern *vt, vi*(s) & **sich a.** unravel, fray

aus-faulen *vi*(s) rot (out); (*Sewage*) digest

aus-feilen *vt* file out, file down; polish

aus-fertigen *vt* make out, draw up, issue.

Ausfertigung *f* (-en) **1** making out, drawing up, issuing (of documents). **2** copy, draft—**in zweifacher A.** in duplicate; **in dreifacher A.** in triplicate

aus-fetten *vt* **1** de-fat, degrease. **2** scour (wool). **3** grease (a pan)

aus-feuern *vt* burn out (casks); warm (rooms)

aus-filtern *vt* filter out

ausfindig machen *vt* find (out), discover

aus-fleischen *vt* flesh (hides)

aus-flicken *vt* patch up

aus-fließen* *vi*(s) flow out, run out; discharge; emanate; spill all its contents

aus-flocken *vt, vi*(s) separate in flakes (*or* flocks), flocculate

Ausflucht° *f* excuse, evasion; subterfuge

Ausflug° *m* **1** excursion, outing. **2** (*Birds*) flight

Ausfluß° *m* **1** outflow, efflux. **2** effluent; discharge; emanation; secretion. **3** outlet, mouth. **4** result. **--geschwindigkeit** *f* outflow (*or* efflux) velocity, rate of discharge. **-hahn** *m* outflow (*or* discharge) cock. **-loch** *n*, **-öffnung** *f* discharge opening, outlet, exit. **-pipette** *f* pipette calibrated to deliver a definite volume *cf* AUFNAHMEPIPETTE. **-rohr** *n*, **-röhre** *f* outlet (*or* discharge) pipe. **-zeit** *f* outflow (*or* discharge) time; delivery time (of pipettes)

aus-folgen *vt* hand out, issue

aus-formen *vt* form, shape, mold

aus-forschen *vt* **1** question, sound out. **2** spy out, ferret out. **3** locate

aus-fragen *vt* question, interrogate

aus-fransen *vt, vi*(s) fray (out)

aus-fräsen *vt* mill out; countersink, recess

aus-fressen* *vt* corrode (out), eat out (*or* away)

aus-frieren* I *vt* **1** freeze (up); concentrate by freezing. **2** separate by freezing; freeze out (gas residues). **II** *vi*(h) **3** cease freezing. **III** *vi*(s) **4** be killed by frost. **5** freeze (through, up, solid). **5** (*Plants*) loosen up with frost. **Ausfrier-tasche** *f* freezing (*or* liquid air) trap

Ausfuhr *f* (-en) export, exportation

ausführbar *a* **1** feasible, practicable. **2** exportable. *Cf* AUSFÜHREN 3,4

Ausfuhr-bewilligung *f* export license

aus-führen *vt* **1** lead out, take out. **2** display, show off. **3** export. **4** carry out, (put into) effect, execute; commit, perpetrate.

5 make, do, produce; build, construct. 6 work out (in detail), elaborate. 7 explain in detail. 8 purge; evacuate, excrete; secrete. 9 (*Metll.*) assay. **ausführend** *p.a* (*oft*) 1 executive. 2 excretory, secretory. **Ausführer°** *m* exporter

Ausführ·gang *m* excretory duct

ausführlich *a* detailed; *adv* in detail. **Ausführlichkeit** *f* (fullness of) detail

Ausführung° *f* 1 exportation. 2 carrying out, effectuation, execution, implementation; commission, perpetration. 3 production; completion, accomplishment, achievement; construction. 4 working out, elaboration. 5 workmanship, quality. 6 model, style, finish. 7 evacuation. 8 *pl, oft* details; remarks, comments. *Cf* AUSFÜHREN

Ausführungs·form *f* working form, model. **-gang** *m* excretory duct. **-mittel** *n* purgative. **-rohr** *n*, **-röhre** *f* discharge tube (*or* pipe), outlet

Ausfuhr·zoll *m* export duty

aus·füllen *vt* 1 fill (in, up, out); stuff, pad. 2 occupy fully, absorb, engross. 3 satisfy fully, fulfil

Ausfüll·masse, Ausfüllungs·masse *f* filling, stuffing, packing. **Ausfüll·stoff** *m* packing (*or* filling) material

aus·futtern, aus·füttern *vt* 1 line. 2 feed. 3 stuff. **Ausfütterung°** *f* lining, liner; feeding; stuffing

ausg. *abbv* (ausgegeben) issued

Ausg. *abbv* (Ausgabe) edition, issue, no.

Ausgabe° *f* 1 issuing, distribution. 2 pickup (window, counter). 3 edition, issue. 4 *usu pl* expense(s), expenditure(s). *Cf* AUSGEBEN

Ausgang° *m* 1 start, starting point, origin. 2 going out, outing. 3 time off, day off. 4 *pl* outgoing mail (*or* goods). 5 exit, way out; outlet. 6 end, close (of a passage, period); ending. 7 outcome. 8 (*Elec.*) output. *Cf* AUSGEHEN

ausgangs *prep* (*genit*) at the end of

Ausgangs· starting, initial; outlet: **-annahme** *f* initial assumption. **-basis** *f* starting point. **-gestein** *n* parent rock. **-material** *n* raw material. **-produkt** *n* primary product. **-punkt** *m* starting point, origin, point of departure. **-rohr** *n*, **-röhre** *f* out-

let (waste, eduction) pipe (*or* tube). **-spannung** *f* (*Elec.*) output voltage. **-sprache** *f* initial (*or* original) language (in translating). **-stellung** *f* initial position. **-stoff** *m* initial (starting, primary, parent, raw) material. **-straße** *f* exit road. **-zoll** *m* export duty

aus·garen *vt* (*Metll.*) kill (steel); carbonize (coke)

aus·gären* I *vt* 1 throw off (by fermentation). II *vi*(h/s) 2 finish fermenting; rise (from fermentation). 3 mature fully

Ausgar·zeit *f* (*Steel*) quiescent period

aus·gasen *vt* fumigate

Ausgas·kolonne *f* gas discharge stack

aus·geben* I *vt* 1 give (*or* hand) out, distribute; deal (out); issue, emit. 2 (*als*) pass off (as). 3 yield. II *vt/i* 4 spend (money). III *vi* 5 be ample, go far enough. 6 (*Slaking Lime*) swell. III: **sich a.** 7 exhaust oneself. 8 (**als**) pass oneself off (as)

ausgebeten requested: *pp of* AUSBITTEN

ausgebissen *p.a* (*Geol.*) outcropping *cf* AUSBEISSEN

ausgeblasen *p.a* exhaust (steam); *see also* AUSBLASEN

ausgeblichen faded: *pp of* AUSBLEICHEN*

ausgeblieben been absent, stayed out: *pp of* AUSBLEIBEN

ausgebogen bent outward, straightened: *pp of* AUSBIEGEN

ausgebracht extracted: *pp of* AUSBRINGEN

ausgebrannt burned out, etc: *pp of* AUSBRENNEN

ausgebreitet *p.a* widespread, extensive; open *cf* AUSBREITEN

ausgebrochen broken out: *pp of* AUSBRECHEN

ausgedacht devised: *pp of* AUSDENKEN

ausgedehnt *p.a* spacious, extensive; *see also* AUSDEHNEN

ausgedient *p.a* worn-out, superannuated; retired

ausgefahren *p.a* beaten (path); *see also* AUSFAHREN

ausgefallen *p.a* strange, eccentric; *see also* AUSFALLEN

ausgeflossen flowed out, run out: *pp of* AUSFLIESSEN

ausgefroren frozen, etc: *pp of* AUSFRIEREN

ausgegangen 1 gone out, etc: *pp of* AUS-

GEHEN. **2** *p.a* gone, used up; sold out

ausgeglichen 1 balanced: *pp of* AUS-GLEICHEN. **2** *p.a* even, smooth; even-tempered, steady; harmonious. **Ausgeglichenheit** *f* evenness, smoothness; even temper, poise, harmony; compensated state

ausgeglitten slipped: *pp of* AUSGLEITEN

ausgegoren matured: *pp of* AUSGÄREN

ausgegossen poured out: *pp of* AUSGIESSEN

ausgehangen posted: *pp of* AUSHÄNGEN **II** *vi*

aus·gehen* *vi*(s) **1** go out (*incl* light, fire); (+ **lassen**) send out; (*Geol.*) crop out. **2** (**auf**) be out (for, after), be bent (on). **3** (**von**) start out, emanate, come (from); originate (from, with): **von da a–d** starting from there. **4** come out, end, close (well, badly); go: **leer a.** go empty-handed; **frei a.** go (scot-)free. **5** run out (*as:* supplies); fail, give out (*as:* breath); (*dat*): **ihnen geht etwas aus** they run out of (*or* they lose) sthg. **6** fade, run (*as:* colors); (*Tex.*) lose color; come out (*as:* hair). **7: es geht an ihnen aus** they bear the brunt of it. **8** (*Baking*) ferment. *Cf* AUSGEGANGEN. **ausgehend** *p.a* outgoing, ending, toward the end of. **Ausgeher** *m* (-), **Ausgeherin** *f* (-nen) messenger

ausgehoben dug up: *pp of* AUSHEBEN

ausgehöhlt *p.a* hollow *cf* AUSHÖLEN

ausgeholfen helped out: *pp of* AUSHELFEN

ausgekannt been informed *cf* AUSKENNEN

ausgeklügelt *p.a* ingenious; *see also* AUSKLÜGELN

ausgeklungen died away, ended: *pp of* AUSKLINGEN

ausgekocht *p.a* spent (hops); *see also* AUSKOCHEN

ausgelassen *p.a* exuberant, unrestrained; *see also* AUSLASSEN

ausgelastet *p.a* fully occupied, kept busy; challenged

ausgeleiert *p.a* worn(-out); hackneyed

ausgelegen been available (*or* on view) *pp of* AUSLIEGEN

ausgemacht *p.a* **1** sure, definite, certain. **2** absolute, out-and-out. *Cf* AUSMACHEN

ausgenommen 1 taken out, etc: *pp of* AUSNEHMEN. **2** *p.a* except(ing), except. **3** *cnjc* unless, except if

ausgeprägt *p.a* marked, distinctive, pronounced; *see also* AUSPRÄGEN

aus·gerben *vt* **1** (*Lthr.*) tan (fully). **2** (*Steel*) weld

ausgerieben rubbed out: *pp of* AUSREIBEN

ausgerissen torn up: *pp of* AUSREISSEN

ausgeronnen run out: *pp of* AUSRINNEN

ausgerungen wrung out: *pp of* AUSRINGEN

ausgesandt emitted: *pp of* AUSSENDEN

ausgeschieden precipitated: *pp of* AUSSCHEIDEN

ausgeschliffen ground, etc: *pp of* AUSSCHLEIFEN

ausgeschlossen 1 excluded: *pp of* AUSSCHLIESSEN. **2** *p.a* out of the question

ausgeschnitten cut out, clipped: *pp of* AUSSCHNEIDEN

ausgeschossen shot up, etc: *pp of* AUSSCHIESSEN

ausgeschrieben written out: *pp of* AUSSCHREIBEN

ausgeschwungen swung out: *pp of* AUSSCHWINGEN

ausgesogen drained, etc: *pp of* AUSSAUGEN

ausgesonnen devised: *pp of* AUSSINNEN

ausgesotten boiled out: *pp of* AUSSIEDEN

ausgesprochen 1 pronounced: *pp of* AUSSPRECHEN. **2** *p.a* pronounced, decided (e.g., effect)

ausgesprungen broken out: *pp of* AUSSPRINGEN

aus·gestalten *vt* **1** (**als**) work out (as); (**zu**) turn (into). **2** elaborate, develop. **3** arrange, organize. **4** fit out, equip. **Ausgestaltung°** *f* development; arrangement; equipment

ausgestanden endured: *pp of* AUSSTEHEN

ausgestiegen gotten off: *pp of* AUSSTEIGEN

ausgestochen dug up: *pp of* AUSSTECHEN

ausgestorben died out: *pp of* AUSSTERBEN

ausgestrichen struck out: *pp of* AUSSTREICHEN

ausgetrieben expelled: *pp of* AUSTREIBEN

ausgetrunken drunk up: *pp of* AUSTRINKEN

ausgewachsen *p.a* full-grown; *see also* AUSWACHSEN

ausgewichen dodged: *pp of* AUSWEICHEN **2**

ausgewiesen expelled: *pp of* AUSWEISEN

ausgewogen 1 weighed out, etc *pp of* AUSWÄGEN & AUSWIEGEN. **2** *p.a* balanced, harmonious. **Ausgewogenheit** *f* (-en) balance, harmony

ausgeworfen ejected: *pp of* AUSWERFEN

ausgewrungen wrung out; *pp of* AUSWRINGEN

ausgewunden wrung out; unscrewed: *pp of* AUSWINDEN

ausgezackt, ausgezahnt *p.a* jagged; serrated, indented, notched

ausgezeichnet *p.a* 1 excellent, outstanding. 2 distinct, well-defined; **a–er Punkt** (*Math.*) singular point, (*Opt.*) cardinal point. See also AUSZEICHNEN

ausgezogen 1 extracted, etc: *pp of* AUSZIEHEN. 2 *p.a* solid (line)

ausgiebig *a* ample, abundant; productive; extensive; thorough; (*Dye.*) strong. **Ausgiebigkeit** *f* abundance; productivity; extensiveness; thoroughness; (*Dye.*) yield, strength; (*Paint*) spreading (*or* covering) power

aus·gießen* *vt* 1 pour out, discharge; spread. 2 empty (a vessel). 3 fill up (a casting mold); stereotype. 4 put out (fire) with water. **Ausgießung** *f* (-en) effusion, outpouring, pouring out, discharging; spreading, etc

aus·gipsen *vt* fill with plaster

aus·gischen *vi* cease frothing

aus·glätten *vt* smooth out

Ausgleich *m* (-e) 1 agreement, settlement; conciliation, compromise. 2 leveling out, equalization; balance, compensation: **als** (*or* **zum**) **A. für** to balance, to make up for. 3 (*in compds*) See AUSGLEICH(S)·

aus·gleichen* 1 *vt & sich a.* level (even, balance) out. 2 adjust, settle, compensate (for), equalize. 3: **sich a.** be balanced *cf* AUSGEGLICHEN. **ausgleichend** *p.a:* **a–e Farbe** complementary color. **Ausgleicher** *m* (-) equalizer, compensator

Ausgleich(s)·behälter *m* surge tank. **-bunker** *m* surge bunker (*or* hopper). **-druck** *m* equalizing pressure. **-filter** *n* balanced filter; (*Phot.*) equalizing filter. **-getriebe** *n* differential (gear). **-gewicht** *n* counterweight. **-grube** *f* (*Metll.*) soaking pit. **-punkt** *m* equalization point. **-rechnung** *f* calculation to correct errors, (*specif*) method of least squares. **-spannung** *f* (*Elec.*) compensating voltage.

-vorrichtung *f* compensating device

Ausgleichung° *f* 1 equalization, etc = AUSGLEICH. 2 equilibrium

Ausgleichungs·strom *m*, **-strömung** *f* compensating (*or* equalizing) current

aus·gleiten* *vi*(s) slip (by accident)

aus·glühen I *vt* 1 (*Metll.*) anneal, calcine, roast; reheat. 2 (*Med.*) cauterize (wound). II *vt, vi*(s) burn out. III *vi*(h) cease glowing, cool down. **Ausglüh·topf** *m* (*Glass*) annealing pot. **Aus·glühung** *f* annealing, calcining, etc

aus·graben* *vt* dig up, dig out, excavate

ausgriefend *p.a* sweeping, wide-ranging

Ausguß° *m* 1 drain, outlet; sink. 2 spout; lip. 3 casting; stereotype. 4 pouring out, effusion. 5 (*Burettes*) delivery. 6 (*Metll.*) ingot. **--beton** *m* grouted concrete. **-leitung** *f* outlet (*or* discharge) piping. **-masse** *f* casting composition; filling compound. **-mörser** *m* lipped mortar. **-pfanne** *f* (ingot) mold. **-rinne** *f* pouring spout. **-rohr** *n*, **-röhre** *f* drain (waste, delivery) pipe. **-schnauze** *f* pouring spout; nozzle. **-ventil** *n* discharge (*or* escape) valve. **-wasser** *n* waste water

aus·halten* I *vt* 1 endure, stand. 2 (*Min.*) pick out, sort out. II *vi* 3 bear up, hold up (*or* out), last

aus·hämmern *vt* hammer out

aus·händigen *vt* (*d.i.o*) hand over, deliver

Aushang° *m* bulletin; bulletin board. **Aushänge·bogen** *m* proof sheet. **aus·hängen** I *vt* 1 hang out, put up, display. 2 unhinge. 3 pick up (telephone). II *vi** 4 be posted (on the bulletin board), be on display. III: **sich a.** (*Clothes*) smooth out by hanging

Aushänge·schild *n* 1 (shop *or* professional) sign(board), shingle. 2 false front

aus·harren *vi* persist, persevere. **ausharrend** *p.a* persistent

aus·härten 1 *vt* (age-)harden, (*Metll.*) temper. 2 *vi* (*Synth.*) cure. **Aushärtung** *f* age-hardening; tempering; curing. **Aushärtungs·katalysator** *m* curing agent

aus·harzen *vi* exude (resin); (*Lthr.*) spew

Aushauch· (*in compds*) exhalation, expiratory. **aus·hauchen** *vi* breathe out, exhale. **Aushauchung** *f* (-en) exhalation

Aushebe·arbeit *f* (*Metll.*) ladling out

aus·heben* *vt* 1 dig (up). 2 lift out; dislocate; select. 3 raid. 4 recruit, draft. 5 pump out, ladle out, empty
aus·hebern *vt* pump out, siphon out
aus·hecken *vt* (+ *dat rflx:***sich**) hatch (out)
aus·heilen 1 *vt* heal; cure. 2 *vi*(s) heal up. 3 *vi*(s) & **sich a.** be cured
aus·heizen *vt* heat thoroughly, anneal
aus·helfen* *vi* (*dat*) help out, assist
Aushieb° *m* 1 cutting (through). 2 assay
Aushilfe° *f* 1 help, assistance; makeshift, substitute, stand-by. 2 relief (person), temporary (employee). *Cf* AUSHELFEN
Aushilfs· temporary, relief, substitute: **-mittel** *n* expedient, makeshift, emergency measure. **a–weise** *adv* temporarily, as a makeshift, in an emergency
aus·höhlen *vt* hollow out, scoop out; excavate; undermine, erode. *cf* AUSGEHÖHLT. **Aushöhlung°** *f* hollowing (out) etc. 2 hollow, depression; cavity; excavation; erosion
Aushub *m* excavation; excavated material
aus·hülsen *vt* hull, shell, husk
aus·hungern *vt* starve (out)
aus·jäten *vt* weed (out)
aus·kalten *vt/i* cool (thoroughly), chill
aus·kehlen *vt* channel, groove, flute
aus·keilen *vi* & **sich a.** taper off, peter out
aus·keimen *vi* germinate; cease germinating
aus·kellen *vt* ladle out
aus·kennen* *vr:* **sich a.** know one's way (around)
aus·kippen *vt* dump out, pour out; empty
aus·kitten *vt* cement up, fill with cement
Ausklang° *m* closing, end
ausklappbar *a* folding, hinged
aus·kleiden *vt* 1 line, coat (internally). 2 undress, strip. **Auskleidung°** *f* lining
aus·klingen* *vi* 1(h) die away. 2(s) end
aus·klinken 1 *vt* release, disengage. 2: **sich a.** be released, separate
aus·klügeln *vt* contrive, devise (ingeniously) *cf* AUSGEKLÜGELT
aus·koagulieren *vt* coagulate
aus·kochen I *vt* 1 boil (down), boil the life out of. 2 clean (sterilize, purify) by boiling; scour, buck. 3 render, extract, decoct. 4 (*Min.*) burn off (defective blast charge). *Cf* AUSGEKOCHT. II *vi* 5 cease boiling.

6 boil down (to nothing). 7 (*Explo.*) blow out: **a–der Sprengschuß** blown-out shot. **Auskocher** *m* boiler; bucking keir
Auskohl·bad *n* carbonizing bath
aus·kohlen *vt* carbonize (wool, etc). **Auskohlung** *f* (-en) carbonization. **Auskohlungs·verfahren** *n* carbonizing process
Auskolkung *f* (-en) (*esp Geol.*) pothole, kettle; (*Rivers*) undermining erosion
aus·kommen* *vi*(s) 1 (*usu* + **mit**) get along (with); manage, make do (with). 2 be hatched. 3 escape. 4 (*Fire*) break out. **Auskommen** *n* (*usu*) livelihood. **auskömmlich** *a* adequate; comfortable
aus·kopieren *vt* copy (out), (*Phot.*) print out. **Auskopier·papier** *n* print-out paper
Auskoppel·prisma *n* diverging prism
aus·körnen *vt* pick the grains out of; thresh (grain); gin (cotton)
aus·kratzen *vt* scratch out; erase; scrape (out); (*Metll*). rake out, rabble out. **Auskratzung** *f* (-en) erasing; scraping, curettage
aus·kristallisieren *vt* & *vi*(s) crystallize (out)
aus·krücken *vt* rake out
aus·kühlen *vi*(s) cool off, cool (thoroughly)
Auskunft *f* (..k⁼e) 1 (item of) information. 2 = **Auskunfts·büro** *n* information bureau (desk, window, center)
aus·kuppeln *vt* disconnect, disengage; uncouple
ausl. *abbv* (ausländisch) for. (foreign)
aus·laden* 1 *vt* unload, discharge. 2 *vi* project, jut out. **ausladend** *p.a* projecting, jutting; bulging; spreading; extensive, broad, sweeping; overhanging. **Ausladung°** *f* unloading; discharge; projection, overhang; range, swing
Auslage° *f* 1 (window) display; show window. 2 (*usu pl*) outlay, expense(s)
aus·lagern *vt* 1 take out of storage. 2 caseharden (aluminum). 3 (*Brew.*) settle (beer)
Ausland *n* foreign country (*or* countries) – **im A., ins A., nach dem A.** abroad. **Ausländer** *m* (-), **Ausländerin** *f* (-nen) foreigner; alien. **ausländisch** *a*, **Auslands·** (*in compds*) foreign, international; alien; exotic

Auslaß *m* (..lässe) outlet, vent; exhaust, discharge. **aus·lassen* I** *vt* 1 let out, emit; vent, discharge; exhaust. 2 melt (down). 3 leave out, omit. 4 let go of; leave alone. **II** *vi* 5 give out, fail. **III: sich a.** 6 speak one's mind. *Cf* AUSGELASSEN. **Auslaß·ende** *n* (*Mach., Containers*) discharge end. **Auslassung** *f* (-en) 1 emission, discharge, exhaust. 2 melting. 3 omission.4 utterance

Auslauf° *m* 1 running out, outflow; leakage, leak. 2 outlet, drain; (*River*) mouth. 3 finishing run, runout, landing run; coasting to a stop. 4 start, starting out. 5 projection. 6 running space, space to run around. **~becher** *m* efflux cup; viscometer

aus·laufen* I *vt* 1 run the full course of. **II** *vi*(s) 2 run (flow, spill, leak) out. 3 blot, run, (*esp Colors*) bleed. 4 set out, depart. 5 run dry, be drained (*or* emptied). 6 run down, decelerate, coast to a stop. 7 end, be discontinued; run out, expire; dwindle off; wear out. **III: sich a.** 8 get exercise. **auslaufend** *p.a* (*oft*) outgoing

Ausläufer° *m* 1 outer edge, fringe; far reaches; spur, foothills. 2 branch, offshoot, runner; process, (*Bot.*) stolon. 3 (*Spect.*) attendant line, satellite. **~wurzelstock** *m* rhizome

Auslauf·flasche *f* overflow flask. **-hahn** *m* drain cock. **-pipette** *f* delivery pipette. **-probe** *f* (*Metll.*) pouring test. **-rohr** *n*, **-röhre** *f* outlet (delivery, discharge) tube (*or* pipe). **-spitze** *f* delivery tip (of a burette). **-zeit** *f* efflux time; (*Pipette*) delivery time

Auslauge· lixiviating, leaching, extraction: **-behälter** *m* lixiviating (leaching, extraction) tank. **-hülse** *f* extraction thimble. **-kasten** *m* leaching vat

aus·laugen *vt* leach (out), lixiviate, extract; wash out; steep (*or* wash) in lye; buck; (*Sugar*) diffuse. **Auslauger** *m* (-) leacher, extractor. **Auslaugerei** *f* (-en) leaching plant. **Auslauge·trichter** *m* extraction funnel. **Auslaugung** *f* (-en) leaching (out), lixiviation, extraction, etc

aus·leeren *vt* empty, evacuate; purge. **Ausleerungs·mittel** *n* (*Med.*) evacuant, purgative

aus·legen *vt* 1 put out, lay out. 2 exhibit, display. 3 (*usu* + **mit**) lay, pave, line, coat, cover (with); inlay, veneer. 4 interpret. 5 design. **Ausleger** *m* (-) (*Mach.*) arm, bracket, jib, (crane) beam; cantilever

Auslege·schrift *f* patent application open to inspection. **-tag** *m* (*Pat.*) public release date

Auslegung *f* (-en) 1 display, exhibition. 2 laying, paving, lining, covering; inlay. 3 interpretation. 4 design

Ausleih·bibliothek *f* lending library

aus·lenken *vt* deflect. **Auslenk·härte** *f* deflection hardness

aus·lernen *vi* finish learning; complete one's apprenticeship

Auslese *f* (-n) selection; elite, pick of the crop; choice (*or* vintage) wine. **~fähigkeit** *f* selectivity. **aus·lesen*** *vt* 1 select, pick (out); weed out. 2 read through. **Ausleser°** *m* selector, sorter, separator

aus·leuchten I *vt* 1 light, illuminate (fully); floodlight. 2 project (on a screen). **II** *vi* stop shining, go out. **Ausleuchtung** *f* (-en) 1 (complete) illumination, floodlighting. 2 projection. 3 extinction

aus·liefern *vt* 1 surrender, hand over; extradite. 2 distribute, deliver

aus·liegen* *vi* be on view, be available (for inspection, use or purchase)

aus·löffeln *vt* spoon out, ladle out

auslösbar *a* releasable, launchable; separable; redeemable *cf* AUSLÖSEN

aus·löschen *vt* 1 put out, extinguish. 2 wipe out, annihilate. **Auslöschung°** *f* extinguishment; annihilation

aus·lösen *vt* 1 set off, touch off, trigger, arouse; launch, release; disengage, uncouple. 2 dissolve out; take out, separate; shell (peas, etc). 3 liberate, emit, give off. 4 buy off, redeem. **Auslösung°** *f* release, triggering; disengagement; dissolving out; separation; liberation; emission; redemption

aus·lüften *vt* air, ventilate

aus·machen *vt* 1 turn off (light, radio); put out (fire). 2 settle, agree on, arrange. 3 get, find. 4 make (up), constitute; amount to, total. 5 (*usu* + **etwas, nichis,** etc) matter, make a difference. 6 make out, discern, sight; establish. 7 dig up;

shell, gut; remove. *Cf* AUSGEMACHT. **Aus-machung** *f* (-en) settlement, agreement, stipulation, arrangement

aus·mahlen *vt* grind up (thoroughly)

aus·malen *vt* 1 paint; color. 2 describe, depict. 3 (+ *dat reflx* **sich**) imagine

Ausmaß° *n* 1 extent, scale. 2(*usu pl*: **Ausmaße**) dimensions

aus·mauern *vt* line with masonry; brick up

aus·mergeln *vt* exhaust, emaciate

aus·merzen *vt* eliminate, eradicate

aus·messen* *vt* measure; gage

ausmittig *a* eccentric

aus·münden *vi* (in) empty, lead, open out (into), end (in). **Ausmündung°** *f* (river) mouth; outlet, orifice

aus·mustern *vt* muster out; reject; phase out

Ausnahme *f* (-n) 1 exception. 2 (*in compds* = AUSNAHMS·) exceptional, special, emergency: **-fall** *m* exceptional case

ausnahms·los *a, oft adv* without exception, unanimous(ly), indiscriminate(ly). **-weise** *adv* as (*or* by way of) an exception

aus·nehmen* I *vt* 1 take out; dig up. 2 clean (out); eviscerate. 3 raid (a place). 4 except, exclude *cf* AUSGENOMMEN. 5 make out, discern. II: **sich a.** (*usu + adv*) look, appear (good, bad, etc). **ausnehmend** *p.adv* exceptionally

ausnutzbar *a* utilizable; exploitable. **aus·nutzen, aus·nützen** *vt* makes use of, utilize; take advantage of, exploit. **Ausnutzung** *f* utilization, exploitation

aus·ölen *vt* oil thoroughly

aus·packen *vt/i* unpack; reveal (secrets)

aus·pendeln 1 *vt* measure with a pendulum. 2 *vi* cease swinging (*or* oscillating)

aus·pflanzen *vt* transplant

aus·photometrieren *vt* investigate (record, evaluate) photometrically

aus·pichen *vt* tar, coat with pitch

aus·pinseln *vt* swab, paint (all over)

aus·planimetrieren *vt* investigate with a planimeter

aus·platten *vt* flatten out; hammer flat; pave with slabs (*or* tiles)

aus·plätten *vt* iron out, press

aus·prägen 1 *vt* coin; stamp. 2: **sich a.** be stamped; take distinct shape. *Cf* AUS-

GEPRÄGT. **Ausprägung** *f* coinage; stamp, impress; precise expression

aus·pressen *vt* (ex)press, squeeze out; pressure-dry; wring out

aus·probieren *vt* try out, test (thoroughly)

Auspuff° *m* exhaust. **·hub** *m* exhaust stroke. **-topf** *m* muffler, silencer

aus·pumpen *vt* pump (out), evacuate

Ausputz *m* trimming(s), ornamentation; (*Tex.*) card waste. **aus·putzen** *vt* clean out (*or* up); trim, prune

aus·quetschen *vt* squeeze out, press; crush out; wring out

aus·radieren *vt* erase, rub out

aus·rangieren *vt* discard, cast off; scrap

aus·rasten 1 *vt* (*Mach.*) disengage. 2 *vi* rest

aus·rauchen 1 *vt* finish smoking, smoke all of. 2 *vi*(s) go stale, lose its taste (*or* aroma)

aus·räuchern *vt* fumigate, smoke out

aus·räumen *vt* 1 clear out, remove. 2 clean out, empty. 3 (*Metll.*) rake out. 4 eliminate. 5 (*Geol.*) wash away

aus·reagieren *vi* react fully; finish reacting

aus·rechnen *vt* figure out, calculate

Ausrede° *f* excuse

aus·reden 1 *vt* (*d.i.o*) talk out of (doing). 2 *vi* finish talking

aus·reiben* *vt* rub (out), wipe (out); scour

aus·reichen *vi* suffice, be enough; (+ **mit**, *oft*) manage, make do (with). **aus·reichend** *p.a* sufficient, adequate

aus·reichern *vt* enrich, strengthen

aus·reifen *vt, vi*(s) ripen, mature (fully)

aus·reinigen *vt* clean out (thoroughly); purge

aus·reißen* I *vt* 1 tear (rip, pull) out. 2 uproot. II *vi*(s) 3 tear (off, away), come off; get torn. 4 run away. **Ausreißer** *m* (-) runaway (*incl Chem.*); wild shot; aberration, fluke

aus·renken *vt* dislocate, sprain

aus·richten *vt* 1 align, straighten, line up. 2 direct, orient; bring into line. 3 convey, pass on (regards). 4 arrange (and pay for); hold, put on (a function). 5 accomplish.6 (*Min.*) open up (and develop) (a lode). **Ausrichtung°** *f* 1 alignment, straightening. 2 direction, orientation. 3 convey-ing (of regards). 4 arrangement.

5 accomplishment. 6 (opening up and) development

aus·ringen* 1 *vt* wring out. 2 *vi* finish the fight

aus·rinnen* 1 *vi*(s) run (trickle, leak) out

aus·roden *vt* root out, uproot; clear

aus·rollen 1 *vt, vi*(s) unroll, roll out. 2 *vt* pay out (cable, etc). 3 *vi*(h *or* s) roll to a stop. **Ausroll·grenze** *f* plastic limit

aus·rotten *vt* root out, eradicate, exterminate

aus·rücken 1 *vt* disengage, take out of gear. 2 *vt, vi*(s) move out. 3 *vi*(s) run away. **Ausrücker** *m* (-) disengaging gear

Ausruf° *m* exclamation; proclamation. **aus·rufen *** 1 *vt/i* call out, cry out, exclaim; proclaim. 2 *vt* offer for sale. **Ausrufung** *f* (-en) exclamation; proclamation; offering for sale

aus·ruhen *vi* & **sich a.** rest

aus·rühren *vt* precipitate

aus·runden *vt* & **sich a.** round out, hollow out

aus·rüsten *vt* & **sich a.** arm, provide, outfit, equip (oneself). **Ausrüstung°** *f* 1 arming, providing, outfitting. 2 armament(s); equipment

aus·rutschen *vi*(s) slip; go astray

Aussaat *f* 1 sowing; (crystal) inoculation. 2 seed. **aus·säen** *vt* sow; inoculate (with crystals); disseminate; (*Beer*) pitch

Aussage° *f* statement, assertion; declaration; testimony. **aus·sagen** *vt* state, assert, declare; testify

aus·saigern *vt* liquate = AUSSEIGERN

aussalzbar *a* capable of being salted out

aus·salzen *vt* salt out, separate by addition of salt; (*Soap*) grain. **Aussalzung** *f* (-en) salting out; graining

Aussatz *m* leprosy; (*Bot.*) scab

aus·säuern *vt* deacidify

aus·saugen* *vt* suck out, suck dry; drain, exhaust; bleed

Aussaug·pumpe *f* suction pump

aus·schachten *vt* excavate; sink (a well)

ausschaltbar *a* capable of being cut out (*or* switched off)

aus·schalten I *vt* 1 switch off; cut out. 2 disengage. 3 eliminate, exclude. II *vi*(s) 4 drop out. **Ausschalter°** *m* cut-out (switch); circuit breaker. **Ausschalt·stellung** *f*

"off" position. **Ausshaltung** *f* switching off; cutting out; disengagement; elimination, exclusion; dropping out

Ausschank *m* (..scḧe) 1 (retail) sale (of beverages). 2 tavern, bar

aus·schärfen *vt* 1 neutralize, deaden; dull (colors). 2 bevel. 3 (*Dye.*) sharpen, strengthen

Ausschau *f:* **A. halten nach** be on the lookout for. **aus·schauen** *vi* look, appear = AUSSEHEN 2

aus·schaufeln *vt* shovel (scoop, bail) out

aus·scheiden* I *vt* 1 separate (out), give out, expel; precipitate, deposit; discharge, secrete, excrete. 2 sort out, eliminate, exclude. II *vi*(s) 3 be eliminated, be excluded. 4 drop out, withdraw, retire. III: **sich a.** be precipitated, settle out. **Ausscheider** *m* (-) (*Med.*) carrier. **Ausscheidung°** *f* 1 separation. 2 deposit, precipitate. 3 discharge; secretion; excretion, excrement. 4 elimination, exclusion. 5 withdrawal; retirement

Ausscheidungs·härtung *f* (*Metll.*) precipitation hardening. **-mittel** *n* separating agent, precipitant. **-produkt** *n* separation product; by-product. **-punkt** *m* separation (*or* precipitation) point

aus·schenken *vt* pour, sell, dispense (beverages)

aus·schießen* I *vt* 1 shoot out. 2 discard, scrap. II *vi*(s) shoot up, sprout

aus·schlachten *vt* 1 cut up; cannibalize (for parts). 2 capitalize on, exploit

aus·schlacken *vt* clear of slag (*or* dross)

Ausschlag° *m* 1 rash, skin eruption. 2 exudation, efflorescence (on walls). 3 (*Lthr.*) bloom. 4 deflection (of needle), swing (of pendulum), tipping (of scales): **den A. geben** tip the scales, be the deciding factor. 5 (*Phys.*) amplitude (of vibrations). 6 (*Bot.*) sprout, shoot. 7 lining; trimming. 8 (*Brew.*) run of wort from the kettle. **--becken** *n* settling pond. **-bütte** *f* (*Brew.*) underback. **-eisen** *n* punch, piercing tool; tool for pounding iron

aus·schlagen* I *vt* 1 knock out. 2 beat out (fire). 3 pound, punch (holes). 4 beat, hammer flat, planish. 5 reject, decline. 6 line, trim, cover. 7 express (oils). 8 (*Min.*) pound out and separate (lodes). 9

(Brew.) turn out (wort) from the kettle. **II**
vi(h) **10** *(Clocks)* finish striking; *(Heart)*
beat its last. **11** lash (strike, kick) out. **III**
vi (h/s) **12** sprout, bud. **13** sweat, exude
moisture; *(Lthr.)* spew. **14**
(Scales) tip; *(Needle)* be deflected. **IV** *vi*(s)
15 (**zu**) turn out (to be)

ausschlag·gebend *a* decisive *cf* AUS-
SCHLAG 4. **A–winkel** *m* angle of deflection. **-würze** *f (Brew.)* finished wort

aus·schlämmen *vt* levigate; dredge; clear
of mud

aus·schleifen* **I** *vt* **1** whet, grind (out).
2 hollow-grind. **3** rebore, ream out. **4**
(Soap) grain. **II: sich a. 5** wear out (by
friction)

Ausschleuder·maschine *f* centrifuge.
aus·schleudern **I** *vt* **1** hurl out, fling out;
eject, expel. **2** emit. **3** centrifuge. **4** hydroextract (yarns). **II** *vi*(s) skid, swerve

aus·schließen* *vt* **1** lock out. **2** shut out, exclude, eliminate. **3** expel. *Cf* AUS-
GESCHLOSSEN. **ausschließlich 1** *a, adv*
exclusive(ly). **2** *prep (genit, dat)* excluding, exclusive of

Ausschließungs·prinzip *n* exclusion
principle

aus·schlüpfen *vi*(s) slip out, *(Zool. oft)* be
hatched

Ausschluß° *m* exclusion; expulsion: **unter
A. der Wärme** with exclusion of heat

aus·schmelzen* **1** *vt* melt (down), render; smelt; fuse; purify by melting;
(Metals) liquate. **2** *vi*(h) cease melting. **3**
vi(s) melt

aus·schmieren *vt* smear; lubricate, grease

aus·schneiden* *vt* **1** cut out, clip; excise,
prune. **2** punch (*or* stamp) out. **3** (sell at)
retail

Ausschnitt° *m* **1** cut-out, hole, notch; neck-
line. **2** clipping. **3** excerpt, extract; section, sector, part. **4** *(Phot.)* detail, trimmed negative

aus·schöpfen *vt* scoop (bail, ladle) out; exhaust. **Ausschöpf·kelle** *f* scoop, ladle

aus·schrauben *vt* unscrew

aus·schreiben* *vt* **1** write out, make out,
draw up. **2** copy out. **3** announce, advertise; offer (prizes, bids). **4** set (a date for)

Ausschreitung *f* (-en) excess, outrage

Ausschuß° *m* **1** committee. **2** rejects, scrap,

refuse; *(Paper)* broke. **3** (bullet) exit
wound (*or* hole). **-ware** *f (Com.)* rejects

aus·schütteln *vt* shake out; agitate; extract
by shaking with a solvent

aus·schütten *vt* pour (*or* dump) out; empty;
spill

aus·schwärmen *vt* **1** swarm out; *(fig)*
spread, sprawl. **2** cease swarming

aus·schwefeln *vt* fumigate with sulfur.
Ausschwefelung *f* **1** fumigation with
sulfur. **2** sulfur bloom

aus·schweifen 1 *vt* curve outward; cut in
an outward curve. **2** *vi*(s) go to excess; digress. **ausschweifend** *p.a* unbridled; extravagant. **Ausschweifung** *f* (-en) **1** outward curve (*or* bulge). **2** excess, extravagance. **3** digression

aus·schweißen 1 *vt* weld out, cleanse (iron)
by welding. **2** *vi* bleed

aus·schwemmen *vt* wash (rinse, flush) out

aus·schwenken *vt* **1** centrifuge, hydroextract. **2** swing out, swivel out. **3** rinse
out. **Ausschwenk·maschine** *f* centrifuge, hydroextractor

aus·schwingen* **1** *vt/i* swing out. **2** *vi* stop
oscillating, swing to a halt; *(Oscillations)*
die out, decay

Ausschwing·maschine *f* centrifuge.
-strom *m* decay current

aus·schwirren *vt* centrifuge, hydro-extract

aus·schwitzen *vt*, *vi*(s) sweat out, exude

aus·sedimentieren *vi* deposit sediment

aus·sehen* *vi* **1** (**nach**) look out, watch
(for). **2** (**wie, nach**) look (like); (**als
ob**) look (as if). **Aussehen** *n* look(s),
appearance

aus·seigern 1 *vt* liquate out, separate by
liquation. **2** *vi (Crystals)* segregate. **Aus-
seigerung** *f* (e)liquation, segregation

aus·seimen *vt* clarify (honey)

außen 1 *adv, pred a* outside: **nach a.**
to(ward) the outside, out(ward); **von a.**
from (the) outside. **2** *(in compds)* outer,
external, exterior, outside, outdoor; foreign: **A–·antenne** *f* outside (outdoor, external) antenna. **-anstrich** *m* exterior
paint(ing). **-bewitterung** *f* outdoor
exposure

aus·senden* *vt* send out, emit

Außen·druck *m* external pressure. **-drüse**
f external gland

Aussendung° *f* sending out, emission

Außen- *cf* AUSSEN 2: **-durchmesser** *m* outside diameter. **-elektron** *n* outer electron. **-handel** *m* foreign trade. **-haut** *f* outer skin. **-leitung** *f* outside line (*or* supply); (*Brew., Elec.*) outer circuit. **-luft** *f* external air, atmosphere. **-lunker** *m* (*Metll.*) surface cavity. **-ministerium** *n* foreign ministry, (*U.S.*) State Department. **a-mittig** *a* eccentric, off-center. **A-rinde** *f* (*Bot.*) outer bark, periderm. **-seite** *f* outer side (*or* surface), outside, exterior. **-taster** *m* external calipers. **-verwitterung** *f* outdoor exposure (*or* weathering). **-welt** *f* outside world. **-wulst** *m* external rim (*or* bead)

außer I *prep* (*dat*) **1** out of, outside (of), beyond. **2** except (for), in addition (to), other than, beside(s); (*after neg oft*) but: **a. sich** beside oneself. **II** *cnjc* **3:a.** (**wenn**) unless, except if; **a.daß** unless, except (that). **III** *in compds, cf* AUSSER·

äußer *a* (*always with endgs*) outer, outside, external; foreign *cf* ÄUSSERE, ÄUSSERST

außer·achsig *a* eccentric. **A-achtlassung** *f* disregard, ignoring *cf* ACHT *f* **1**. **a-amtlich** *a* unofficial. **A-betriebsetzung** *f* discontinuance, taking out of service, suspension, closing *cf* BETRIEB

außerdem *adv* in addition, besides, moreover

außer·dienstlich *a* off-duty, unofficial

Äußere *n* (*adj endgs*) **1** exterior, outside. **2** external appearance. **3** foreign affairs. *Cf* ÄUSSER

außer·gewöhnlich *a* unusual, extraordinary

außerhalb 1 *prep* (*genit*) outside (of) **2** *adv* (on the) outside

außer·irdisch *a* extraterrestrial. **A-kraftsetzung** *f* repeal, suspension, rescission

äußerlich *a* **1** outward, external. **2** superficial. **Äußerlichkeit** *f* (-en) (*oft pl*) outward appearance(s), externals; nonessentials, frills; formality; superficiality

außermittig *a* eccentric, off-center

äußern 1 *vt* express. **2: sich ä.** express oneself (*or* one's opinion); manifest itself, show up

außer·ordentlich *a* extraordinary, unusual

äußerst I *a* **1** remotest, outermost. **2** greatest, utmost, extreme. **3** latest, final. **4** worst. **II** *adv oft* extremely, highly. **Äußerste** n (*adj endgs*) absolute limit, extreme (measure). *Cf* ÄUSSER

außerstand(e) *pred a* (*usu* + **zu**) incapable (of), in no condition (*or* position) (to)

Äußerung *f* (-en) expression; remark, comment *cf* ÄUSSERN

außer·wesentlich *a* nonessential

aus·setzen I *vt* **1** put out, set out. **2** set free, abandon. **3** (*d.i.o*) expose, subject (to). **4** interrupt, suspend; postpone, defer. **5** *usu:* **etwas auszusetzen haben** (**an**) have objections (to), find fault (with), criticize. **6** (*usu Tex.*) line. **II** *vi* **7** pause, stall; skip a beat. **8** misfire. **9** (**mit**) stop working (on), suspend. **III: sich a. 10** (*dat*) expose (*or* subject) oneself (to). **aussetzend** *p.a* discontinuous, intermittent. **Aussetzung** *f* (-en) **1** abandonment. **2** exposure, subjection. **3** suspension; pause; deferment. **4** criticism, objection

Aussicht° *f* view; outlook, prospect—**das steht in A.** there is hope (or a good chance) of that; **in A. nehmen** consider, plan; **etwas in A. stellen** hold out the hope of sthg

aussichts·los *a* hopeless. **A-punkt** *m* lookout, vantage point. **a-reich** *a* promising

aus·sickern *vi*(s) trickle out, ooze out, percolate

aus·sieben *vt* sift (sort, screen, filter) (out)

aus·sieden* *vt* boil (out); blanch (silver)

aus·sinnen* *vt* devise, contrive, think up

aus·sintern *vi*(s) trickle out, ooze out

aus·soggen *vt/i* crystallize out *cf* SOGGEN

aus·söhnen 1 *vt* reconcile. **2: sich a.** be(come) reconciled

aus·solen *vt* leach

aus·sondern 1 *vt* sort out, sift out; separate; eliminate; secrete, excrete. **2: sich a.** be sorted out (sifted out, separated, etc). **aussondernd** *p.a* secretory, excretory. **Aussonderung°** *f* sorting out, sifting out; separation, etc

aus·sortieren *vt* sort (out), pick out

aus·spannen I *vt* **1** stretch (out), span,

spread (out); (*Tex. oft*) tenter. **2** release; unleash, unfasten; take out. **II** *vi* **3** relax, rest. **III: sich a. 4** extend, expand. **Ausspannung°** *f* relaxation

aus·sparen *vt* **1** save (up), reserve. **2** omit, avoid, bypass. **3** leave open (*or* blank). **Aussparung°** *f* **1** saving, reservation; omission. **2** recess, open space, blank (space)

aus·sperren *vt* lock out, exclude

Aussprache° *f* pronunciation; discussion. **-kreis** *m* discussion group

aus·sprechen *I* *vt* **1** pronounce. **2** express, state. **3** finish saying. **II** *vi* **4** finish speaking. **III: sich a. 5** express oneself (*or* one's opinion). **6** be expressed. **7** unburden oneself. **8** have a (frank) discussion. *Cf* AUSGESPROCHEN

aus·sprengen *vt* **1** blast out. **2** sprinkle (water). **3** spread (rumor)

aus·springen* *vi*(s) **1** snap out of place; chip out, break out; desert. **2** jut out, project. **ausspringend** *p.a* salient (angle)

aus·spritzen 1 *vt* spray (out), flush out, hose out; inject; syringe. **2** *vt, vi*(s)squirt out. **Ausspritzer** *m* (-) (*Enameling*) spit-out

Ausspruch° *m* saying, statement, remark; decision

aus·sprühen *vt, vi*(s) spray (out)

Aussprungs·winkel *m* angle of reflection

aus·spülen *vt* wash (flush, rinse) out; erode

aus·staffieren *vt* outfit, furnish, equip

aus·stampfen *vt* stamp out; (*Metll.*) ram, line

Ausstand° *m* **1** strike, walkout. **2** outstanding debt. **ausständig** *a* **1** striking, on strike. **2** outstanding, unpaid; lacking

aus·stanzen *vt* stamp, punch out

aus·statten *vt* **1** provide, furnish, equip; endow; vest. **2** decorate, dress up; package. **Ausstattung** *f* (-en) **1** furnishing, equipment, endowment. **2** packaging; make-up, format, decor

aus·stauben *vt* dust, shake the dust out of

aus·stechen* **1** spade up, dig up; poke out; punch out. **2** outdo, eliminate . . . as a rival

aus·stehen* *I* *vt* **1** endure, suffer. **II** *vi* **2** be on display. **3** be outstanding (pending,

overdue), *usu.*: **das steht noch aus** that has not come in yet; **a. haben** be expecting, have coming (to one). **4** (go on) strike.

ausstehend *p.a* outstanding, overdue, receivable

aus·steifen *vt* stiffen, brace, reinforce

aus·steigen* *vi*(s) (*oft* + **aus**) **1** climb out (of). **2** get off, get out of (a vehicle). **3** resign (from), quit. **4** be eliminated (from), drop out (of). **Aussteiger** *m* (-) dropout

aus·stellen *vt* **1** post, set (out), put out. **2** switch off. **3** display, exhibit. **4** issue, make out (a document). **5** object = AUSSETZEN **5. Aussteller** *m* (-) exhibitor; issuer, drawer. **Ausstellung°** *f* display, exhibit; exhibition, exposition, fair; issuing; objection

Ausstellungs·glas *n* display glass; specimen jar

aus·sterben* *vi*(s) die out, become extinct

Aussteuerung° *f* **1** equipment; supply (as of energy). **2** modulation, adjustment (of sound). **3** expiration of social insurance benefits

Ausstich° *m* choicest wine

Ausstieg *m* (-e) (*oft* + **aus**) **1** exit (from). **2** abandonment, giving up (of)

aus·stopfen *vt* stuff; fill up, plug up

aus·stöpseln *vt* unstopper, unplug

Ausstoß° *m* **1** output, production. **2** emission. **3** ejection. **4** (*Brew.*) tapping, broaching. **5** torpedo tube. **aus·stoßen*** *vt* **1** emit, discharge, eject, expel; extrude; eliminate, excrete. **2** poke out, knock out. **3** give out, utter (sounds). **4** turn out, produce. **Ausstoßer** *m* (-) ejector

Ausstoß·ladung *f* bursting charge. **-leistung** *f* extrusion output. **-produkt** *n* waste product. **-system** *n* cleansing system

Ausstoßung *f* (-en) ejection, emission, expulsion; extrusion; elimination

Ausstrahl *m* emission. **aus·strahlen** *I* *vt* **1** radiate, emit. **2** transmit, broadcast, air. **II** *vi*(h) **3** exert its influence. **III** *vi*(s) **4** (*von*) radiate, be emitted (from). **ausstrahlend** *p.a* radiant. **Ausstrahlung°** *f* **1** radiation, emission. **2** transmission, broadcast(ing), airing. **3** influence, effect

Ausstrahlungs·fläche *f* radiation surface. **-verlust** *m* radiation loss. **-vermögen** *n*

radiating power, emissivity

aus·strecken *vt* & **sich a.** stretch (out), extend

aus·streichen* I *vt* 1 strike out. 2 smooth out, iron out; level. 3 spread; smear, grease. II *vi* 4 (*Geol.*) crop out

aus·streuen *vt* scatter, disseminate; sprinkle

Ausstrich° *m* 1 (*Med.*) smear. 2 (*Geol.*) (line of) outcrop. 3 (*Min.*) stream tin. **--präparat** *n* smear preparation

aus·strömen 1 *vt* radiate, emit, pour out. 2 *vi*(s) stream out, flow out; escape; emanate, radiate. **Ausströmung°** *f* emission; outflow, efflux, effluence; escape; emanation, effusion

Ausströmungs·rohr *n* delivery pipe (*or* tube)

Ausstülpung *f* (-en) protrusion; eversion

aus·stürzen *vt* pour out, empty, discharge

aus·suchen *vt* 1 pick out, choose, select; sort. 2 search (an area). **Aussucher°** *m* sorter

aus·süßen *vt* wash (a precipitate); edulcorate; sweeten; (*Brew.*) sparge

Aussüß·glas *n* wash bottle (for precipitates). **-pumpe** *f* leaching pump. **-rohr** *n*, **-röhre** *f* washing tube. **-vorrichtung** *f* wash bottle (for precipitates); sparger

aus·tarieren *vt* tare, counterbalance

Austausch *m* exchange, interchange; barter; replacement, substitution; (*Biol.*) crossover. **--azidität** *f* exchange acidity

austauschbar *a* exchangeable, interchangeable

Austausch·boden *m* plate of fractionating column

aus·tauschen *vt* (*oft* + **gegen**) exchange (for), interchange; barter; replace, substitute; (*Biol.*) cross over. **Austausch·energie** *f* exchange energy. **Austauscher** *m* (-) exchanger

austausch·fähig *a* exchangeable. **A–harz** *n* ion-exchange resin. **-kombination** *f* recombination crossover. **-schicht** *f* interfacial layer. **a–sauer** *a* (rendered) acid by ion exchange. **A–stoff** *m* substitute. **-werkstoff** *m* substitute material. **-wert** *m* crossover value

aus·teeren *vt* tar

aus·teilen *vt* distribute, deal out; dish out

austenitisch *a* austenitic

Auster *f* (-n) oyster

aus·tilgen *vt* exterminate, eradicate

Austonen *n* change in (color) tone

Aust. P. *abbv* (Australisches Patent) Austral. pat.

Austrag *m* (..t⁻e) discharge; settlement, decision: **zum A. kommen** reach a settlement (*or* a decision); **zum A. bringen** settle, decide, arbitrate. **--brücke** *f* discharge weir. **aus·tragen*** I *vt* 1 deliver, distribute. 2 settle, decide. 3 play (game), hold (contest). 4 remove, discharge. 5 wear out. 6 bear (a child) to full term. II *vi* 7 cease bearing (fruit)

Australien *n* Australia. **australisch** *a* Australian

austreibbar *a* expellable

aus·treiben* 1 *vt* drive out, expel, eject; liberate. 2 *vt/i* sprout. **Austreibung** *f* (-en) expulsion; liberation

aus·treten* I *vt* 1 tread (out), trample (out). 2 (*Footwear*) wear out; break in. II *vi*(s) 3 step out, come out, emerge, issue; escape; extravasate, be secreted. 4 (*River*) overflow (its banks). 5 protrude. 6 withdraw, leave, resign. **austretend** *p.a* (*oft*) emergent

aus·trinken* *vt* drink up

Austritt° *m* 1 stepping out, emergence, issue; escape; extravasation, secretion. 2 overflow. 3 protrusion. 4 withdrawal, resignation. 5 exit, outlet, vent, exhaust; exit point. *Cf* AUSTRETEN

Austritts·arbeit *f* work function (of electrons). **-geschwindigkeit** *f* discharge rate, escape velocity. **-öffnung** *f* discharge opening; steam exhaust part; outlet, exit. **-winkel** *m* angle of emergence (reflection, refraction)

aus·trocknen 1 *vt, vi*(s) dry up, dry out, season. 2 *vt* desiccate parch, drain. 3 *vi*(s) become parched (*or* desiccated). **austrocknend** *p.a* desiccative: **a–es Mittel** drying agent. **Austrockner°** *m* desiccator, drier. **Austrocknungs·rinde** *f* crust due to drying

aus·tröpfeln, austropfen 1 *vi*(h) cease trickling (*or* dripping). 2 *vi*(s) trickle (*or* drip) out

aus·üben *vt* practise; exert, exercise; carry

out, execute. **ausübend,** *p.a* professional; executive

Ausverkauf° *m* (bargain) sale. **aus·verkaufen** *vt* sell out

Auswaage° *f* 1 weighing out. 2 weighed substance. 3 (*opp of* EINWAAGE) state of substance when weighed. 4 final precipitate. *Cf* AUSWÄGEN

aus·wachsen* **I** *vt* 1 outgrow. **II** *vi*(s) 2 sprout. 3 attain full growth. 4 (**in**) end (in). **III: sich a.** 5 be outgrown, disappear. 6 (**zu**) grow, develop (into). *Cf* AUSGEWACHSEN

auswägbar *a* weighable. **aus·wägen*** *vt* weigh out; balance; calibrate (by weighing) = AUSWIEGEN

Auswahl° *f* choice, selection. **aus·wählen** *vt* choose, select. **auswählend** *p.a* selective

Auswahl·prinzip *n* selection principle. **-regel** *f* (*Spect.*) selection rule

aus·walken *vt* 1 roll out. 2 (*Lthr.*) block

aus·walzen *vt* (*Metll.*) roll out

aus·wandern *vi*(s) emigrate; shift. **Auswanderung°** *f* emigration

aus·wärmen *vt* anneal; warm, heat; roast. **Auswärm(e)·ofen** *m* annealing furnace

auswärtig *a* outside, out-of-town; foreign. **auswärts** *adv & sep pfx* out(side); out of town; abroad; outward

aus·waschen* *vt* wash (out); clean, cleanse; scour, scrub; elutriate; erode. **Auswasch·flasche** *f* wash bottle. **Auswaschungs·verlust** *m* loss due to washing

aus·wässern 1 *vt* soak, steep. 2 *vi* be soaked

auswechselbar *a* exchangeable, interchangeable; replaceable. **aus·wechseln** *vt* (*oft* + **gegen**) exchange (for); interchange; replace. **Auswechs(e)lung°** *f* exchange, interchange; replacement

Ausweg° *m* way out, escape; outlet

aus·weichen 1 *vt* soak (off), soften (out). 2 *vi*(s)* (*usu* + *dat*) turn aside (from), dodge, avoid, evade. **ausweichend** *p.a* evasive. **Ausweichung** *f* (-en) 1 soaking. 2 avoidance, evasion

aus·weiden *vt* gut, eviscerate

Ausweis *m* (-e) identification (papers); (bank) statement. **aus·weisen* I** *vt* 1 expel. 2 (*als*) delcare (to be), identify (as).

5 show, proof of, demonstrate; show, indicate. 4 report. 5 reserve, earmark. **II: sich a.** 6 identify onself, prove one's identity. 7 furnish proof; (**als**) prove (to be). 8 become evident. **Ausweisung°** *f* expulsion

aus·weiten *vt & sich a.* stretch, widen, expand, extend

auswendig 1 *adv* by heart. 2 *a* external

aus·werfen* *vt* 1 cast, throw out, eject, discharge. 2 cough up, expectorate. 3 shovel out. 4 allocate, grant. *Cf* AUSWURF

aus·werten *vt* 1 evaluate, interpret. 2 make (practical) use of, utilize. 3 (*Math.*) plot, calculate **Auswert·gerät** *n* analyzer

aus·wettern *vt* weather (a storm)

aus·wickeln *vt* unwrap

aus·wiegen* *vt* weigh out, balance = AUSWÄGEN

aus·winden* *vt* wring out; unscrew

aus·wirken I *vt* 1 knead. 2 work, take (salt out of pans). **II: sich a.** 3 work up; (+ **zu,** *oft*) turn out (to be). 4 (**auf**) have an effect (on), affect. 5 (**in**) result (in). **Auswirkung°** *f* effect, consequence

aus·wischen *vt* wipe (out), wipe clean; erase

aus·wittern 1 *vt, vi*(s) weather, wear away; season; decompose, decay; effloresce; (*Min.*) bloom out. 2 *vt* get the scent of

aus·wringen* *vt* wring out

Auswuchs° *m* 1 excrescence, (morbid) growth; (*Med. oft*) tumor; (*Agric.*) sprouting in the ear. 2 aberration, excess, abuse. 3 (harmful) outgrowth. 4 premature germination

aus·wuchten *vt* (counter) balance, equilibrate

Auswurf° *m* 1 ejection, discharge. 2 ejected material; refuse, scum, dregs. 3 expectoration, sputum; droppings, excrement. **··leitung** *f* discharge piping. **-stoffe** *mpl* droppings, excretions. **auswurfs–befördernd** *a* (*Med.*) expectorant

aus·zahlen 1 *vt* pay out, pay off; cash (checks, etc). **2: sich a.** pay (off), be worthwhile

aus·zählen *vt* count out

Auszahlung° *f* payment, disbursement; cashing

aus·zehren *vt* exhaust, consume, waste away. **Auszehrung** *f* (*Med.*) consumption

aus·zeichnen I *vt* 1 decorate, honor.
2 distinguish. 3 mark, tag. II: **sich a.**
4 distinguish oneself, be distinguished.
Cf AUSGEZEICHNET. **Auszeichnung°** *f*
decoration, award; distinction, marking,
tagging; mark, tag

aus·zentrifugieren *vt* centrifuge (out)

ausziehbar *a* extractable, extensible; ductile; telescopic

aus·ziehen* I *vt* 1 pull out, extract; exhaust
(drugs). 2 quote. 3 extend, stretch; draw
(wire). 4 draw (a line) solid *cf* AUSGEZOGEN
2; trace, plot, ink in (lines). 5 bleach out
(colors). 6 eviscerate. 7 take off (clothes);
undress, strip (sbdy). II *vi*(s) 8 set (march,
move) out; leave, pull out; escape. III:
sich a. 9 undress, take off one's wraps. *Cf*
AUSZIEHEN

Auszieh·rohr *n*, **-röhre** *f* telescopic tube.
-tisch *m* extension table. **-tubus** *m* (*Micros.*) draw tube. **-tusch** *m*, **-tusche** *f*
drawing ink

Ausziehung° *f* 1 extraction. 2 quotation.
3 extension; (wire) drawing. 4 tracing,
plotting (of lines). 5 bleaching out. 6 evis-
ceration. 7 undressing. *Cf* AUSZIEHEN I,
AUSZUG

Auszubildende *m,f* (*adj endgs*) trainee

Auszug° *m* 1 departure, moving out. 2 quotation, excerpt. 3 abstract, digest. 4 extraction; extract, essence, tincture, infusion, decoction. 5 (*Mach.*) pullout, telescopic part, extension; (table) leaf. **-mehl**
n superfine flour

authentisch *a* authentic

Auto *n* (-s) automobile, car: **mit dem A.** by
car; **A. fahren** drive, go by car. **-bahn** *f*
(super)highway. **-benzin** *n* motor fuel,
gasoline. **-fahrer** *m* motorist, (car) driver

autogen *a* autogenic, autogenous. **A-·
schweißung** *f* autogenous welding

Auto·katalysator *m* autocatalyst. **-ka-
talyse** *f* autocatalysis. **-klav** *m* (-en) autoclave. **-lack** *m* automobile lacquer

Autolyse *f*(-n) autolysis. **autolysieren** *vt/i*
autolyze

Automat *m* (-en, -en) automatic machine
(*or* device); automat; vending (*or* slot)
machine

Automaten·stahl *m* machining steel

autonom *a* autonomous, self-governing

Auto·poliermittel *n* automobile polish

Autor *m* (-en) author

Autoren·recht *n* author's rights; copyright. **-referat** *n* author's report (*or*
abstract). **-register** *n* index of authors.
-tantiemen *fpl* author's royalties. **-ver-
zeichnis** *n* index of authors

Auto·reparaturwerkstatt *f* auto(mobile)
repair shop

Autorin *f*(-nen) author(ess)

autorisieren *vt* authorize. **Autorität** *f*(-en)
authority

Auto·schlauch *m* inner tube

Autotypie *f* half-tone engraving

autoxydabel *a* (*with endgs.* . **dable,** etc)
autoxidizable

auxochrom *a* auxochromic, auxochromous

Aventurin *n* aventurine (glass). **aven-
turisieren** *vt* aventurize

avinieren *vt* add alcohol to (wine)

Avidität *f* avidity

Avis *m or n* (-e) notification. **avisieren** *vt*
notify

Avivage *f* reviving, etc *cf* AVIVIEREN

avivier·echt *a* unaffected by brightening.
avivieren *vt* 1 revive, restore. 2 (*Dye.*)
brighten, clear, tone, top. 3 (*Silk*) scroop.
Avivier·mittel *n* brightening agent

ä.W. *abbv* (äußere Weite) outside diameter

Axe *f*(-n) axis

Axial·druck *m* axial thrust. **-schlag** *m*
(*Engg.*) end-play. **-schub** *m* axial thrust.
a–symmetrisch *a* axisymmetrical. **A–
zug** *m* axial tension

Axt *f*(-̈e) axe; hatchet

a.Z. *abbv* 1 (als Zugabe) as an extra. 2 (auf
Zeit) on time, on the installment plan

A.Z., AZ *abbv* (Azetylzahl) acetyl no.

Aze . . *see also* ACE . .

Azelaïn·säure *f* azelaic acid

Azelaon *n* (-en) cyclooctane

Azen *n* (-e) azene, aliphatic diazo compound

Azenaphthen *n* (-e) acenaphthene. **Aze-
naphthylen** *n* (-e) acenapthylene

Azenium *n* negatively charged nitrogen

azeotropisch *a* azeotropic, constant-
boiling

Azetal *n* (-e) acetal

Azetat *n* (-e) acetate

Azetessig·äther *m* acetoacetic ester, ethyl
acetoacetate

Blattern *fpl* variola, (*specif*) smallpox
blättern 1 *vt* pick leaves off. 2 *vi*(h) (**in**) leaf
(through). 3 *vi*(s) & **sich b.** flake off, exfoliate; (*Baking*) rise in flakes
Blätter·pilz, **-schwamm** *m* leaf fungus.
-serpentin *m* (*Min.*) antigorite. **-spat** *m*
foliaceous spar
Blatter·stein *m* variolite
Blätter·teig *m* flake pastry. **-tellur** *n* foliated tellurium, nagyagite. **-ton** *m* foliated clay. **-zeolith** *m* foliated zeolite,
(*Geol.*) heulandite
Blatt·farbstoff *m* leaf pigment. **-feder** *f*
leaf spring. **-federchen** *n* plumule = B–
KEIM. **-flechte** *f* leaf lichen. **-flecken-
krankheit** *f* (*Bot.*) leaf spot (disease). **b–
förmig** *a* leaf-shaped, foliate; laminated,
flaked. **B–gelb** *n* xanthophyll. **-gewebe**
n leaf tissue, mesophyll. **-gold** *n* leaf
gold, gold leaf (*or* foil). **-goldgrundöl** *n*
gold size. **-grün** *n* leaf green, chlorophyll.
-keim *n* plumule; (*Malt*) acrospire.
-kohle *f* foliated (*or* slate) coal. **-kupfer** *n*
sheet copper, copperfoil. **-lack** *m* shellac.
-laus *f* plant louse, aphis. **-mark** *n* (*Bot.*)
mesophyll. **-metall** *n* leaf metal, foil
blättrig *a* leafy, etc = BLÄTTERIG
Blatt·scheide *f* leaf sheath. **-schichtung** *f*
lamination. **-silber** *n* silver leaf. **-spreite**
f leaf blade, lamina. **-stahl** *m* sheet steel.
-stiel *m* leaf stalk, petiole. **-werk** *n* foliage. **-zinn** *n* tinfoil
blau *a* 1 blue: **b–e Erde** blue earth, (*specif*)
earthy vivianite; **b–e Säure** purple acid
(H_2SO_5N). 2 poached (fish). **Blau** *n* blue
(color) *cf* BERLINER. **--asche** *f* saunders
blue, blue verditer
Bläu·bad *n* bluing bath
Blau·beere *f* blueberry, huckleberry. **b–
blank** *a* polished blue. **B–bleierz** *n*
pyromorphite. **-bruch** *m* (*Iron*) blue-
shortedness. **b–brüchig** *a* blue-short,
brittle at blue heat. **B–carmin** *n* indigo
carmine. **-dämpfen** *n* (*Ceram.*) blue
smoking. **-druck** *m* blueprint(ing); (*Tex.*)
blue (*specif* indigo) print
Bläue *f* 1 blue(ness). 2 (laundry) bluing
Blaueisen·erde *f* earthy vivianite. **-erz** *n*,
-spat *m*, **-stein** *m* vivianite; **faseriger
B.** crocidolite

Bläuel *m* (-) 1 bluing bag. 2 = BLEUEL beetle. **bläueln** *vt* 1 blue. 2 = BLEUELN beetle
blau·empfindlich *a* sensitive to blue
blauen, **bläuen** 1 *vt* blue; dye blue; whiten
(with bluing). 2 *vi* turn blue
Blauerei *f* (-en) indigo dyehouse
Blau·erz *n* vivianite. **-farbe** *f* color (of)
blue, smalt. **-farbenglas** *n* smalt. **-fär-
ber** *m* dyer in blue. **-färbung** *f* blue col-
or(ation). **-fäule** *f* blue rot, sap rot, sap
stain (of wood). **-gas** *n* Blau oil gas. **b–
gesäuert** *a* 1 cyanide of. 2 impregnated
with prussic acid. **-glühend** *a* blue-hot.
B–glut *f* blue-white incandescence. **b–
grau** *a* blue-gray. **-grün** *a* blue-green. **B–
holz** *n* logwood, campeche wood. **-kohl** *m*
red cabbage. **-kreuz** *n* blue-cross poison
gas (diphenylchloroarsine or other ster-
nutative). **-küpe** *f* blue (*specif* indigo) vat.
-lack *m* blue lake
bläulich *a* bluish. **--grau** *a* bluish-gray
Blau·masse *f* (*Rayon*) supramonium solution. **-mühle** *f* smalt mill. **-ofen** *m* flowing (*or* shaft) furnace. **-öl** *n* blue oil; (*Org
Chem.*) cerulignol. **-papier** *n* blueprint
paper. **-pause** *f* blueprint tracing; blue
print. **-probe** *f* blue test. **b–rot** *a* blue-
red, violet. **B–salz** *n* potassium ferrocyanide. **-sand** *m* coarsest smalt. **b–
sauer** *a* cyanide of: **b–es Salz** cyanide;
gelbes b–es Kali potassium ferrocyanide; **rotes b–es Kali** potassium ferricyanide. **B–säure** *f* hydrocyanic acid
Blausäure·gas *n* hydrogen cyanide. **b–
haltig** *a* containing hydrocyanic acid. **B–
verbindung** *f* cyanide, (*specif*) hydrogen
cyanide compound. **-vergiftung** *f* hydrocyanic acid poisoning
Blau·schimmelkäse *m* blue-veined
cheese. **-schlamm** *m* (*Geol.*) blue mud.
-schörl *m* cyanite. **b–schwarz** *a* blue-
black. **B–spat** *m* lazulite. **-stein** *m* blue-
stone, blue vitriol; lazulite; (*Metll.*) blue
metal, blue matte. **-stich** *m* blue tinge (*or*
tint). **-stift** *m* blue pencil. **-stoff** *m*
cyanogen. **-sucht** *f* cyanosis
Bläuung *f* (-en) bluing. **Bläuungs·farb-
stoff** *m* bluing color (*or* dye)
blau·verschoben *a* displaced toward blue.
B–violett *n* blue-violet (color). **-vitriol** *n*

blue vitriol. **b—warm** *a* blue-hot. **B—wärme** *f* blue heat. **-wasser** *n* Schweitzer's reagent, cupric ammonium sulfate solution. **-wassergas** *n* blue (water) gas
Blech *n* (-e) 1 sheet metal, (*specif*) sheet iron, sheet steel–(**weißes**) B. tinplate, "tin"; **schwarzes B.** black iron plate. 2 sheet of metal (iron, tinplate, etc), plate; baking tin, cookie sheet. **-ausschnitt** *m*, **-ausstoß** *m* (*Metll.*) blank. **-büchse** *f*, **-dose** *f* tin (box, container); (tin) can (esp food). **-drucktinte** *f* tinplate ink. **-eisen** *n* sheet iron
blechern *a* (of) tin(plate); tinny
Blech·flasche *f* narrow-necked tin vessel. **-gefäß** *n* tin container (*or* vessel). **-glühofen** *m* plate-heating furnace, **-hafen** *m* tin pot (pan, can)
blechig *a* tinny
Blech·kanne *f* tin pot, tin can. **-kasten** *m* tin (*or* sheet-iron) box. **-lack** *m* tinplate (*or* sheet-metal) varnish. **-messing** *n* sheet brass. **-rohr** *n* sheet-iron tube. **-schere** *f* tin shears. **-schmied** *m* tinner, tinsmith; sheet-metal worker. **-stärke** *f* (metal) plate (*or* sheet) thickness. **-tafel** *f* iron sheet (*or* plate). **-trichter** *m* tinplate funnel. **-trommel** *f* tin(plate) drum. **-ware** *f* tinware, sheet-metal ware
Blei (-e) 1 *n* (*Metal*) lead; lead weight. 2 *m or n* pencil: **mit B. schreiben** write in pencil, = BLEISTIFT. **-abgang** *m* lead dross (*or* scoria). **-ablagerung** *f* lead deposit. **-abschirmung** *f* lead shield(ing). **-antimonerz** *n*, **-antimonglanz** *m* zinkenite. **-arbeit** *f* 1 (*Metll.*) lead smelting. 2 (*Indus.*) plumbing. **-arsenglanz** *m* sartorite. **-arsenik** *n* lead arsenate (*or* arsenide). **-arsenit** *n* (*Min.*) dufrenoysite. **-art** *f* type (*or* variety) of lead. **b–artig** *a* leadlike, leady, plumbeous. **B–asche** *f* lead dross (*or* ash); gray oxide film (on lead exposed to air). **-auskleidung** *f* lead lining. **-azetat** *n* lead acetate. **-bad** *n* molten lead bath. **-baum** *m* lead tree, arbor saturni
bleiben* (blieb, geblieben) *vi*(s) remain, stay, keep on being; (**bei**) stick (with); (+ *dat or pred a*) be left (to, for); (*oft with other verb as sep pfx*) *cf* STEHENBLEIBEN, etc. **bleibend** *p.a* lasting, permanent; du-

rable, stable; fast (colors). **bleiben·lassen*** *vt* leave undone (*or* alone), refrain from (doing)
Blei·bergwerk *n* lead mine. **-blech** *n* sheet lead. **-block** *m* 1 (*Explo.*) lead (*or* Trauzl) block. 2 (*Metll.*) lead pig. **-blüte** *f* mimetite. **-braun** *n* lead dioxide. **-büchse** *f* lead box (case, can, container)
bleich *a* pale, white
Bleich·bleach(ing): **-anlage**, **-anstalt** *f* bleaching plant, bleachery. **-artikel** *m* (*Tex.*) bleach style. **-bad** *n* bleaching bath (*or* liquor)
bleichbar *a* bleachable
Bleich·beize *f* bleaching mordant. **-chlor** *n* bleaching (*or* active) chlorine
Bleiche *f* (-n) 1 bleachery. 2 pallor
bleich·echt *a* bleach-fast, bleach-resistant
bleichen 1 *vt* bleach, whiten. 2 *vi*(s) (*oft**: blich, geblichen) face, lose color, turn white. **Bleicher** *m* (-) bleacher. **Bleich·erde** *f* bleaching (*or* fuller's) earth, absorbent clay. **Bleicherei** *f* bleachery, bleaching plant
Bleich· bleaching: **-essenz** *f* soda bleach liquor (sodium hypochlorite solution). **-fähigkeit** *f* bleachability. **-faß** *n* bleaching vat. **-fleck** *m* bleaching stain, bleach spot. **-flotte, -flüssigkeit** *f* bleaching liquor (*or* fluid). **-grad** *m* degree of bleaching (*or* of bleachability). **-gut** *n* material to be bleached. **-holländer** *m* bleaching machine; (*Paper*) poacher, poaching machine. **-kalk** *m* (calcium chloride) bleaching powder. **-kasten** *m* bleaching vat. **-kessel** *m* bleaching boiler (*or* keir). **-lauge** *f* bleaching lye (*or* liquor)
Blei·chlorid *n* lead chloride
Bleich·mittel *n* bleach, bleaching agent; decolorant. **-moos** *n* sphagnum moss. **-öl** *n* bleaching oil
Blei·chromat *n* lead chromate
Bleich· bleaching: **-salz** *n* bleaching salt. **-sand** *m* bleached sand. **-säure** *f* chloric acid. **-sellerie** *m* celery root. **-soda** *f* washing soda. **-sucht** *f* chlorosis. **-ton** *m* bleaching clay. **-vorbereitungsmittel** *n* pre-bleach. **-wasser** *n* bleaching liquor (*or* solution), e.g. Javelle water. **-wirkung** *f* bleaching action
Blei·dampf *m* lead vapor (*or* fumes).

-draht m lead wire. **b–empfindlich** a 1 sensitive to lead. 2 (*Motor Fuel*) lead-responsive. **B–empfindlichkeit** f 1 (*Motor Fuel*) response to tetraethyl lead. 2 (*Cars*) fouling tendency caused by lead deposits. **-eisenstein** m (*Metll.*) iron-lead matte

bleien vt lead, treat with (tetraethyl) lead

Blei·erde f earthy cerussite

bleiern a lead; leaden, dull, heavy

Blei·erz n lead ore. **-essig, -extrakt** m vinegar of lead, (aqueous solution of basic) lead acetate. **-fahlerz** n (*Min.*) bournonite. **-farbe** f lead color, lead paint. **b–farben, -farbig** a lead-colored. **B–feder** f lead pencil. **-folie** f lead foil. **b–frei** a lead-free, unleaded. **-führend** a lead-bearing, plumbiferous. **B–gang** m (*Min.*) lead vein. **b–gefüttert** a lead-lined. **B–gehalt** m lead content. **-gelb** n lead chromate; massicot; yellow lead monoxide. **-gerät** n lead implement (or apparatus). **-gewinnung** n lead extraction (or production). **-gießer** m plumber; lead founder. **-glanz** m lead glance, galena. **-glas** n lead glass; (*Min.*) anglesite. **-glasur** f lead glaze. **-glätte** f litharge. **-glimmer** m cerussite in micaceous form, plumbogummite. **-grube** f lead mine. **-gummi** 1 m lead-pencil eraser. 2 n (*Rubber*) lead rubber; (*Min.*) plumboresinite. **b–haltig** a lead-bearing, plumbiferous. **B–hornerz** n, **-hornspat** m phosgenite. **-hütte** f lead works. **-hydroaluminat** n (*Min.*) plumbogummite. **-ion** n lead ion

bleiisch a leady, lead-like

Blei·kalk m lead calx, lead oxide. **-kammerkristalle** mpl lead chamber crystals (nitrosulfuric acid). **-könig** m lead regulus. **-krankheit** f lead poisoning, plumbism. **-krätze** f lead waste, (*specif*) lead dross. **-kristall** m lead glass (or crystal). **-kugel** f lead ball (pellet, bullet). **-lasur** f linarite. **-leder** n (*Metll.*) silver with high lead content. **-lot** n 1 plummet; plumb bob. 2 lead solder. **-löter** m lead burner. **-lötung** f lead soldering. **-mangan** n lead manganese. **-mehl** n lead dust. **-mennige** f minium, red lead. **-mine** f pencil lead (replacement). **-molybdänspat** m (*Min.*) wulfenite. **-mulde**

f lead pig. **-mulm** m earthy galena. **-niere** f (*Min.*) bindheimite. **-öl** n lead acetate in oil of turpentine. **-oxid** n lead oxide, (*specif*) PbO**–rotes B.** minium, Pb_3O_4. **-oxidhydrat** n lead hydroxide. **-oxidsalz** n (bivalent-)lead salt. **-oxydul** n basic suboxide, Pb_2O; lead monoxide, PbO. **-oxyduloxid** n minium, Pb_3O_4. **-papier** n 1 lead paper (for hydrogen sulfide tests). 2 lead foil. **-peroxid** n lead dioxide, PbO_2. **-pflastersalbe** f (*Pharm.*) diachylon ointment. **-plattenprobe** f (*Explo.*) lead plate test. **-rauch** m lead fumes (or smoke). **b–reich** a rich in lead, high-lead. **B–rohr** n, **-röhre** f lead pipe. **-rot** n red lead, minium. **-safran** m orange lead. **-salbe** f lead ointment, cerate of lead acetate. **-salpeter** m lead nitrate. **-salz** n lead salt, (*specif*) lead acetate. **-sammler** m (*Elec.*) lead storage battery. **b–sauer** a plumbate of. **B–saum** m (*Physiol., Med.*) lead line, blue gum line in lead poisoning. **-säure** f plumbic acid. **-schale** f lead dish (or basin). **-schaum** m lead dross (or ash), litharge. **-scheelat** n lead tungstate; (*Min.*) stoltzite. **-schlange** f lead coil. **-schlick** m lead slime. **-schnee** m lead snow ($PbSO_4 \cdot Pb(OH)_2$). **-schwamm** m lead sponge. **-silberfahlerz** n (*Min.*) malinowskite. **-spat** m cerussite—**gelber B.** wulfenite; **roter B.** crocoisite. **-speise** f lead speiss. **-spiegel** m specular galena. **-stein** m lead matte. **-stift** m (lead) pencil. **-superoxid** n lead dioxide, PbO_2. **-teträäthyl** n tetraethyl lead. **-vitriol** n 1 lead vitriol (or sulfate). 2 = **-vitriolspat** m anglesite. **-wasser** n Goulard water. **-wasserstoff** m lead hydride. **-weiß** n white lead. **-weißsalbe** f lead carbonate ointment. **-wismutglanz** m (*Min.*) galenobismutite. **-wolframat** n lead tungstate, (*Min.*) stoltzite. **-wurz** f (*Bot.*) lead wort, plumbago. **-zinnlot** n lead-tin solder. **-zinnober** m red lead, minium. **-zucker** m sugar of lead, lead acetate

Blend·art f hybrid (species)

Blende f (**-n**) 1 (*Min.*) glance, blende; (*specif*) zinc blende (ZnS). 2 screen, blind, shield, stop, baffle; (*Phot.*) shutter, diaphragm; fade-in, fade-out; niche, blind window. **blenden** vt 1 blind, dazzle.

2 screen (from light). 3 dull, dim. 4 deceive. 5 blend. **blendend** *p.a* dazzling, radiant

Blenden·öffnung *f* (*Phot.*) diaphragm aperture

Blend·farbstoff *m* color for sighting

Bleuel *m* (-) (*Tex.*) beetle, beetling machine. **bleueln** *vt* beetle

blf. *abbv* (blätterförmig) in flakes, in leaflets

blg. Pat *abbv* (belgisches Patent) Belg. Pat.

Blick *m* (-e) 1 glance, look. 2 sight, eye(s). 3 view. 4 insight. 5 flash; (*Metll.*) flashing (as of molten silver), fulguration, "blick". **blicken** *vi* 1 glance, look, peek: **sich b. lassen** show up, put in an appearance. 2 (*Metll.*) flash, give the "blick"

Blick·feld *n* field of vision. **-gold** *n* refined gold still containing silver. **-punkt** *m* 1 point of view. 2 focal point; (*oft in titles, e.g.*) **B. Umwelt** Focus on the Environment. **-silber** *n* (*Metll.*) refined silver still containing impurities, lightened silver. **-winkel** *m* 1 (*Opt.*) angle of view, visual angle. 2 point of view

blieb *stayed: past of* BLEIBEN

blies blew: *past of* BLASEN

blind *a* blind; dull, tarnished, clouded; false; (*Tests*) blank. **B—·darm** *m* (*Anat.*) caecum, (*oft inaccurately*) appendix. **-darmentzündung** *f* appendicitis. **-darmleder** *n* goldbeater's skin. **-feuer** *n* blank fire. **-gänger** *m* (*Explo.*) dud; misfire

Blindheit *f* blindness

Blind·kohle *f* underburned charcoal. **-leistung** *f* (*Elec.*) reactive power. **-leitwert** *m* (*Elec.*) susceptance. **-licht** *n* flashlight

blindlings *adv* blindly, rashly

blind·machen *vt* blunt (glass)

Blind·probe *f* blank sample; blank test. **-schlauch** *m* flexible tube closed at one end. **-schleiche** *f* blindworm, slowworm. **-strom** *m* (*Elec.*) wattless (*or* reactive) current. **-versuch** *m* blank test. **-wert** *m* blank (*or* reagent) value. **-widerstand** *m* (*Elec.*) reactance

blinken 1 *vt/i* blink; flash; signal. 2 *vi* sparkle, shine, gleam. **Blinker** *m*, **Blink·licht** *n* blinker, flashing signal

blinzeln *vi* blink; wink; twinkle

Blitz *m* (-e) (flash of) lightning; (*Phot.*) flash. **-ableiter** *m* lightning rod. **blitzen** *vi* 1 flash, sparkle; gleam. 2: **es blitzt** there is (a flash of) lightning, lightning is flashing. 3 take flash pictures

Blitz·lampe *f* flashbulb. **-licht** *n* (*Phot.*) flash(light). **-röhre** *f* fulgurite; flashtube. **-schlag** *m* lightning stroke. **-strahl** *m* streak of lightning. **-würfel** *m* flashcube

Bllg. *abbv* (biologische Lösung) biol. sol.

Block *m* I (-̈e) block (of stone, etc); ingot, pig; bar, cake (of soap); (tree) log. II (-s) *or* (-) block (of houses). III (-̈e) *or* (-s) bloc, coalition; (writing) pad; book, block (of tickets, etc). **-blei** *n* pig lead. **Blöckchen** *n* (-) small ingot (*or* pig). **Block·eisen** *n* ingot iron (*or* steel). **blocken** *vt* block; (*Metll.*) bloom, cog

Block·färbung *f* (*Micros.*) staining *in toto*. **-form** *f* ingot mold. **-hahn** *m* stopcock. **-holz** *n* wood in logs

blockieren *vt* block, blockade

Block·kondensator *m* (*Elec.*) blocking (*or* stopping) condenser. **-kupfer** *n* ingot copper. **-lehm** *m* (*Geol.*) boulder clay. **-mühle** *f* block mill. **-pech** *n* pitch in lumps. **-polymerisation** *f* bulk polymerization. **-schnitt** *m* die, matrix. **-seigerung** *f* (*Metll.*) segregation in the ingot. **-walzwerk** *n* blooming mill. **-zinn** *n* block tin

blöd(e) *a* 1 stupid, foolish. 2 feeble-minded, imbecilic. 3 awkward, shy. **Blöd·sinn** *m* nonsense; foolishness; dementia

blondieren *vt/i* bleach (hair)

bloß I *a* 1 bare, naked. 2 nothing but, mere. 3 pure (e.g. coincidence). **II** *adv* oft only, merely, just. **III** *sep pfx cf* BLOSS·LEGEN etc

Blöße *f* (-n) 1 nakedness. 2 vulnerability, weak spot; loophole: **sich eine B. geben** lay oneself bare (to attack, etc), expose a weak spot. 3 pelt, unhaired hide

bloß·legen, bloß·stellen *vt* & **sich b.** lay (oneself) bare, expose (oneself) (to ridicule, etc)

Blt. *abbv* (Blättchen) leaflet(s), lamella(s)

blühen *vi* bloom, blossom; flourish

Blume *f* (-n) 1 flower, bloom. 2 (*Wine*) aroma, bouquet. 3 (*Beer*) head. 4 (*Dye.*) scum, flurry

Blumen·blatt *n* petal. **-blau** *n* anthocyanin. **-erde** *f* garden mold. **-gelb** *n* yellow flower pigment. **-kohl** *m* cauliflower. **-seite** *f* (*Lthr.*) hair side. **-staub** *m* pollen. **-tee** *m* imperial tea: **-topf** *m* flower pot. **-zucht** *f* flower growing, floriculture. **-zwiebel** *f* flower bulb

blumig *a* flowery, florid

Blunzen *f* (-) blood sausage = BLUTWURST

Blut *n* blood. **-achat** *m* blood agate. **-ader** *f* vein, blood vessel. **b–arm** *a* 1 anemic. 2 destitute. **B–armut** *f* anemia, blood deficiency. **-bahn** *f* bloodstream, **-bank** *f* blood bank. **b–bildend** *a* blood-forming. **B–blume** *f* arnica, bloodflower. **-druck** *m* blood pressure. **-druckmesser** *m* sphygmomanometer, blood pressure gage

Blüte *f* (-n) blossom, bloom, flower; (*fig*) climax, apex

blut·echt *a* non-bleeding (colors). **B–egel** *m* leech. **-eisenstein** *m* hematite. **-eiweiß** *n* blood (*or* serum) albumin

bluten *vi* bleed (*incl Colors*); shed blood; sweat, exude sap, run

Blüten· blossom, flower: **-blatt** *n* petal. **-farbstoff** *m* flower pigment. **-hüllenblatt** *n* sepal. **-kelch** *m* calyx. **-krone** *f* corolla. **-öl** *n* flower oil, attar. **-pflanze** *f* flowering plant, phanerogam. **-staub** *m* pollen. **-stengel** *m* peduncle

Blut·entziehungsmittel *n* hemagog

Blüten·wachs *n* flower wax. **-weiß** *n* calcium sulfate used in loading paper

Blut·erguß *m* hemorrhage, effusion of blood

Bluter·krankheit *f* hemophilia

Blut·erz *n* hematite

Blüte·zeit *f* flowering, blossom(ing) time; golden age, heyday

Blut·farbe *f* blood color (*or* pigment). **-farbstoff** *m* blood pigment, hemochrome—(**roter**) **B.** hemoglobin. **-faser·stoff** *m* fibrin. **-fleck(en)** *m* bloodstain. **-fluß** *m* hemorrhage, flow of blood. **b–flüssig** *a* hemorrhagic. **B–flüssigkeit** *f* blood plasma, hemolymph. **b–führend** *a* blood-carrying, blood-circulating. **B–gefäß** *n* blood vessel. **-gerinnung** *f* blood coagulation, clotting. **b–gierig** *a* bloodthirsty. **B–gift** *n* hematoxin. **-haargefäß** *n* capillary (blood vessel). **-hochdruck** *m*

high blood pressure. **-holz** *n* logwood. **-hund** *m* bloodhound

blutig *a* bloody, sanguinary; gory; (*Meat*) rare

Blut·klumpen *m* blood clot. **-kohle** *f* blood charcoal. **-konserve** *f* (unit of) stored blood. **-körperchen** *n* blood corpuscle. **-kraut** *n* bloodroot; blood-stanching herb. **-kreislauf** blood circulation, bloodstream. **-kruste** *f* scab. **-kuchen** *m* 1 blood clot. 2 placenta. **-lauf** *m* blood circulation. **-laugensalz** *n*: (**gelbes**) **B.** potassium ferrocyanide; **rotes B.** potassium ferricyanide. **b–leer** *a* anemic, bloodless. **-los** *a* bloodless. **B–mehl** *n* blood meal. **-mittel** *n* (*Med.*) blood tonic. **-plättchen** *n* blood platelet. **-probe** *f* 1 blood test. 2 blood sample. **b–reich** *a* blood-filled, plethoric. **B–reinigung** *f* blood cleansing. **b–rot** *a* blood-red, crimson. **B–ruhr** *f* bloody dysentery. **-scheibe** *f* blood corpuscle. **-schlag** *m* apoplexy. **-seuche** *f* anthrax. **-spendedienst** *m* blood-donor service. **-spender** *m* blood donor. **-stein** *m* bloodstone, hematite. **b–stillend** *a* blood-stanching, hemostatic, styptic. **B–stillstift** *m* styptic pencil. **-stillungsmittel** *n* hemostatic (*or* styptic) (agent). **-strom** *m* blood stream. **-sturz** *m* hemorrhage. **-sucht** *f* hemophilia. **-übertragung** *f* blood transfusion. **-umlauf** *m* blood circulation

Blutung *f* (-en) bleeding; hemorrhage

blut·unterlaufen *a* bloodshot. **B–untersuchung** *f* examination of blood, blood test. **-vergiftung** *f* blood poisoning. **-versorgung** *f* blood supply. **-wärme** *f* blood heat. **-wasser** *n* blood serum. **-wassergefäß** *n* lymphatic vessel. **-wolle** *f* blood (*or* plucked) wool. **-wurst** *f* blood sausage (*or* pudding). **-wurz(el)** *f* 1 (*genl*) bloodroot. 2 (*specif*) **a** a red-sapped plant; **b** blood-stanching plant. **-zentrifuge** *f* hematocrit. **-zuckerspiegel** *m* blood sugar level. **-zufuhr** *f* (*Med.*) blood supply

BM³ *abbv* (Betriebskubikmeter) actual cubic meters (as measured under operating conditions)

Bö *f* (-en) (*Meteor.*) gust; squall

Bobine *f* (-n) bobbin, spool, winding reel

Bock I *m* (⸚e) 1 (*Zool.*) buck, ram. 2 trestle,

jack, horse, stand, stool, (saw)buck. 3 blunder. 4 pigheaded person. **II** *n* bock beer. **-asche** *f* coal ashes. **-bier** *n* bock beer. **-holz** *n* guaiacum wood, lignum vitae

bockig *a* balky, stubborn, refractory

Bock-käfer *m* capricorn (*or* longhorn) beetle. **-leder** *n* goat leather, kid, buckskin. **-nuß** *f* butternut, souari nut. **-säure** *f* hircic acid. **-seife** *f* mountain (*or* rock) soap

böckseln *vi* (*Wine*) taste of H_2S

Bocks·horn *n* goat's horn; carob (pod). **-hornkraut** *n* fenugreek. **-hörnlein** *n* carob bean

Bock·sprung *m* caper; handspring; (*fig*) sudden change, quantum jump

Boden *m* (⁻) **1** (top)soil, earth, land. **2** ground. **3** floor. **4** bottom, base. **5** (*Distil.*) plate, tray. **6** attic, garret, loft. **7** (*Dye.*) blotch. *Cf* GRUND. **-analyse** *f* soil analysis. **-anspruch** *m* soil requirement. **-art** *f* (type of) soil. **-beize** *f* (*Paint*) floor stain. **-bestandteil** *m* soil constituent. **-chemie** *f* soil chemistry. **-decke** *f* soil cover(ing). **-elektrode** *f* hearth electrode. **-entleerer** *m* bottom discharge car (*or* truck). **-erschöpfung** *f* soil exhaustion. **-ertrag** *m* crop yield, harvest. **-fläche** *f* floor (*or* ground) space, area. **b—fremd** *a* allochthonous. **B—glocke** *f* (*Distil.*) bubble cap. **-güte** *f* soil quality. **-hals** *m* neck (*or* tubulation) at the bottom. **-hefe** *f* grounds, dregs, lees (of yeast). **-impfstoff** *m* soil inoculum. **-impfung** *f* soil inoculation. **-kappe** *f* (*Distil.*) bubble (*or* tray) cap. **-kolloid** *n* soil colloid. **-kolonne** *f* plate-type column. **-körper** *m* **1** soil substance. **2** substance at the bottom of a solution, (*specif*) solid (*or* crystalline) phase (in saturated solutions). **-kraft** *f* soil fertility. **-kunde** *f* soil science, pedology. **-kupfer** *n* (*Copper*) bottoms. **-lack** *m* floor varnish. **-leder** *n* sole leather. **-leim** *m* (*Soap*) bottoms. **b—los** *a* **1** bottomless, abysmal. **2** boundless, unmitigated. **3** outrageous. **B—mechanik** *f* soil mechanics. **-mehl** *n* fecula, starch. **-müdigkeit** *f* soil exhaustion. **-nähe** *f* (*Aero.*) ground (*or* sea) level. **-nährstoff** *m* soil nutrient. **-nutzung** *f* soil utiliza-

tion. **-öl** *n* floor oil. **-pappe** *f* (*Agric.*) mulch paper. **-personal** *n* ground crew. **-pflege** *f* soil preservation. **-probe** *f* **1** soil test. **2** soil sample. **-satz** *m* deposit, sediment; grounds, dregs; (*Salt*) bitterlings. **-satzgestein** *n* sedimentary rock. **-säure** *f* soil acid. **-schädling** *m* soil pest. **-schätze** *mpl* mineral resources. **-schicht** *f* bottom layer; (*Geol.*) bed, lowest stratum. **-schlamm** *m* sediment, sludge, settlings. **-see** *m* (*Geog.*) Lake of Constance. **b—ständig** *a* indigenous, native, soil-rooted; (*Geol.*) autochthonous. **B—stein** *m* bottom stone; bed (of furnace); bed stone (of a mill); (*Salt*) bittern. **-teig** *m* underdough. **-untersuchung** *f* soil investigation (*or* examination). **-ventil** *n* bottom valve. **-vergiftung** *f* soil poisoning (*or* contamination). **-wachs** *n* floor wax. **-wasser** *n* ground (*or* soil) water. **-zahl** *f* **1** base (number). **2** (*Distil.*) number of plates. **-ziegel** *m* paving tile. **-zustand** *m* soil condition

bog bent: *past of* BIEGEN

Bogen *m* (- *or*) ⁼ **1** curve, bend, arc (*incl* Math. & Elec.); detour, turn; circle. **2** slur, ligature. **3** arch. **4** bow. **5** sheet (of paper). *Cf* BAUSCH. **-flamme** *f* arc flame. **b—förmig** *a* bowed, curved, arched. **B—lampe** *f* arc lamp. **-leimung** *f* (*Paper*) tub sizing. **-licht** *n* arc light. **-linie** *f* curved line. **-maß** *n* (*Math.*) radian measure. **-rohr** *n*, **röhre** *f* bent (*or* curved) pipe (*or* tube). **-schweißung** *f* arc welding. **-skala** *f* curved scale. **-spektrum** *n* arc spectrum. **-stück** *n* curved piece; (*Brew.*) return bend. **b—weise** *adv* (*Paper*) by the sheet

Bohle *f* (-n) plank

Böhmen *n* Bohemia. **böhmisch** *a* Bohemian; **b—er Granat** (*Min.*) pyrope. **Böhmit** *m* (*Min.*) boehmite

Bohne *f* (-n) bean (*incl* coffee)

Bohnen·erz *n* pea ore (oolitic limonite). **-kaffee** *m* real (bean) coffee. **-keimling** *m* bean sprout. **-kraut** *n* summer savory. **-mehl** *n* bean meal (*or* flour). **-öl** *n* bean oil, (*specif*) soy bean oil. **-ranke** *f* beanstalk

Bohner·masse *f* floor-polishing wax.

bohnern *vt/i* wax, polish (the floor).
Bohner·wachs *n* floor wax
Bohn·erz *n* pea (*or* bean) ore, oolitic limonite
bohren 1 *vt/i* bore, drill (a hole). **2** *vt* drive (sthg into the ground). **3** *vi* penetrate, drill away; persist. **4: sich b.** bore (*or* drill) one's way. **Bohrer** *m* (-) drill (bit *or* machine), auger, gimlet; perforator
Bohr·fett *n* drilling grease (*or* paste). **-guß** *m* (easily drilled) malleable cast iron, drilling casting. **-gut** *n* borings. **-hammer** *m* (compressed) air hammer. **-kern** *m* drill(ed) core. **-loch** *n* drill hole, borehole; (drilled) well; worm hole. **-maschine** *f* (mechanical *or* electric) drill, drill press. **-mehl** *n* borings, bore dust. **-öl** *n* (*Mach.*) cutting (*or* drilling) oil. **-plattform** *f* drilling platform. **-probe** *f* **1** core sample. **2** drill test. **-rohr** *n* casing (pipe). **-rohrbenzin** *n* casing-head gasoline. **-schlamm, -schmand** *m* drilling slime (*or* sludge). **-späne** *mpl* drill shavings, drillings. **-stahl** *m* steel boring tool, drilling steel. **-stelle** *f* drilling site. **-stock** *m* boring stock, borer. **-turm** *m* oil derrick
Bohrung *f* (-en) **1** drilling, boring. **2** drill hole; (drilled) well. **3** bore, caliber
Bohr·versuch *m* experimental drilling; drill test. **-vorrichtung** *f* boring (*or* drilling) device; drilling jig. **-winde** *f* bit brace. **-wurm** *m* (*Zool.*) shipworm, marine borer
Boje *f* (-n) buoy
Bol *m* bole, bolus
Bolivien *n* Bolivia
Bolle *f* (-n) onion bulb. **Bollen·gewächs** *n* bulb plant
Boll·werk *n* bulwark
Bologneser *a* (*no endgs*) (of) Bologna: **B. Flasche** Bologna flask; **B. Spat** Bologna stone; **B. Leuchtstein** fluorescent barium sulfate
Bolus *m* (-) bole, bolus—**roter B.** reddle, red ocher; **weißer B.** kaolin. **-erde** *f* bole
Bolzen *m* (-) bolt; stud, pin, rivet. **-gelenk** *n* pin joint. **-lager** *n* trunnion bearing. **-mutter** *f* (bolt) nut. **-scheibe** *f* (bolt) washer

Bombage *f* swelling (of food cans)
Bombe *f* (-n) bomb; tank, (gas) cylinder
Bomben·ofen *m* bomb (*or* Carius tube) furnace. **-rohr** *n* bomb tube. **-sauerstoff** *m* tank oxygen. **b-sicher** *a* **1** bombproof. **2** dead certain. **B-stickstoff** *m* tank nitrogen. **-zielvorrichtung** *f* bombsight
Bomse *f* (-n) (*Ceram.*) support
Bon *m* (-s) coupon, voucher, ticket, claim check, receipt
Bonduc·nuß *f* bonduc nut, nicker nut
bonifizieren *vt* compensate, reimburse
Bonität *f* (-en) solvency, good quality
bonitieren *vt* value, appraise, assess
Boot *n* (-e) boat. **Boots·lack** *m* boat (*or* spar) varnish
Bor *n* boron; (*in compds oft*) boric, boro-: **-aluminium** *n* aluminum boride. **-ameisensäure** *f* boroformic acid
Borat·glas *n* borate glass
Bor·äthan *n* bor(o)ethane, diborane (B_2H_6). **-äthyl** *n* borethyl, triethylborine
Borax·blei *n* lead borate. **-eisen** *n* iron borate. **-honig** *m* -borax honey. **-kalk** *m* calcium borate. **-perle** *f* borax bead. **b-sauer** *a* borate of. **B-säure** *f* boric acid. **-see** *m* borax lake. **-spat** *m* boracite. **-weinstein** *m* potassium borotartrate, (*Pharm.*) tartarus boratus
Borazit *n* boracite
Bor·butan *n* borobutane, tetraborane (B_4H_{10}). **-chlorid** *n* boron trichloride
Bord I *n* (-e) shelf; edge. **II** *m* (-e) (ship)board
bordeaux·rot *a* Bordeaux red, claret. **B-brühe** *f* = **Bordelaiser Brühe** Bordeaux mixture
bördeln *vt* **1** border, edge, flange. **2** (*Metal*) bead
Bor·diamant *m* crystalline boron
Boretsch *m* borage
Bor·fluorid *n* borofluoride. **-fluorwasserstoff** *m* (hydro)fluoric acid (HBF_4). **-flußsäure** *f* fluoboric acid
Borg *m* borrowing: **auf B.** on credit
Bor·gehalt *m* boron content
borgen *vt* (*usu*) borrow, (*oft*) lend
Borid *n* (-e) boride. **borig** *a* borous. **Borium** *n* boron
Bor·jodid *n* boron iodide. **-kalk** *m* calcium

borate. **-karbid** n boron carbide
Borke f (-n) (tree) bark; rind, crust; scab.
  Borken·käfer m bark beetle
Bor·methyl n trimethylborine, boron
  methyl, $B(CH_3)_3$
Born m (-e) fount(ain); salt pit; brine
Bor·nitrid n boron nitride
Boro·fluorid n fluoroborate. **-kalzit** n
  borocalcite
Boronat n borohydride
Boro·wolframat n tungstoborate
Bor·puder m boric acid powder
Borratz, Borretsch m borage
Bor·salbe f (Pharm.) boric acid ointment.
  b—sauer a borate of. **B—säure** f boric
  acid
Borsäure·anhydrid n boric anhydride
  (B_2O_3). **-anomalie** f (Glass) boric oxide
  anomaly. **-weinstein** m potassium boro-
  tartrate = BORAXWEINSTEIN
Börse f (-n) 1 stock market (or exchange).
  2 purse
Borst m (-e) crack, fissure cf BORSTE
Bor·stahl m boron steel
Borste f (-n) bristle cf BORST
Bor·stickstoff m boron nitride
borstig a 1 bristly. 2 cracked
Borte f (-n) border, edging
Bor·verbindung f boron compound. **-was-
  ser** n boric acid solution. **-wasserstoff** m
  boron hydride. **-wolframsäure** f tung-
  stoboric (or borotungstic) acid
bös a bad = BÖSE. **-artig** a malicious,
  vicious; malignant; pernicious, harmful
Böschung f (-en) slope; embankment.
  Böschungs·winkel m slope angle
böse a 1 evil, bad, vicious. 2 harmful, noxious. 3 sore, aching. 4 angry. 5: **b—es
  Wesen** epilepsy; **b—es Wetter** (Min.)
  chokedamp
boshaft a malicious
bosseln vt emboss. **bossieren** vt 1 emboss.
  2 mold, shape. **Bossier·wachs** n molding
  wax
Boswellin·säure f boswellic acid
bot offered: past of BIETEN
Botanik f botany. **Botaniker** m (-) botanist
Bote m (-n,-n) messenger; harbinger.
  Botschaft f (-en) 1 message. 2 embassy
Böttcher m (-) cooper, barrelmaker

brach I a fallow, idle. II broke: past of
  BRECHEN
Brache f (-n) 1 fallow land. 2 fallow period
Brachy·achse f (Cryst.) brachyachsis
Brack n refuse, waste. **-gut** n refuse;
  damaged goods. **brackig, brackisch** a
  brackish. **Brack·wasser** n brackisch
  water
Bramme f (-n) slab (of iron), ingot bloom
Brammen·block m slab ingot. **-walzwerk**
  n (Metll.) slabbing mill
Branche f (-n) line of work, profession;
  trade, industry. **Branchen·kenntnis** f
  professional expertise
Brand m (-e) 1 burning, combustion. 2 fire:
  in B. on fire cf GERATEN. 3 baking,
  calcination, refining. 4 charge (of ovens,
  kilns) 5 firebrand. 6 brand (on cattle, etc.)
  7 burning thirst. 8 (Med.) gangrene;
  (Bot.) blight, smut. **-balsam** m ointment
  for burns. **-blase** f blister caused by burning. **-bombe** f incendiary bomb. **-eisen** n
  1 burnt iron. 2 branding iron. **-erz** n inflammable ore (e.g. bituminous shale,
  idrialite). **-fäule** f brown rot. **b—fest** a
  fireproof. **B—fläche** f burned surface.
  -fleck(en) m burn (mark). **-gelb** n reddish yellow. **-geschoß** n incendiary shell.
  -gold n refined gold. **-granate** f incendiary shell (or grenade). **-harz** n empyreumatic resin. **-herd** m source of fire;
  trouble spot
brandig a 1 (smelling, tasting) burnt.
  2 gangrenous. 3 blighted. Cf BRAND
Brand·kitt m fireproof cement (or lute).
  -legung f arson. **-loch** n fuse hole, vent.
  -lunte f slow match. **-mal** n (burn) scar;
  brand; stigma. **brandmarken** vt (pp
  gebrandmarkt) brand, stigmatize.
  Brand·messer m pyrometer. **-mittel** n
  burn (blight, gangrene) remedy; escharo-
  tic. **-neu** a brand-new. **B—öl** n empyreumatic oil. **-pilz** m smut fungus.
  -probe f fire test, fire assay. **b—rot** a fiery red. **B—salbe** f anti-
  burn ointment. **-satz** m (Mil.) incendiary
  charge. **-säure** f pyroligneous acid.
  -schaden m fire damage. **-schiefer** m bituminous shale (or schist), bone coal.
  -schutz m fire prevention (or protection).

-**schutztür** f fire safety door. **b–sicher** a fireproof. **B–silber** n refined silver. -**sohle** f insole. **b–stiftend** a incendiary. **B–stifter** m arsonist, -**stiftung** f arson. -**tür** f fireproof door

Brandung f (-en) surf; surge, roar

Brand·versicherung f fire insurance. -**wunde** f burn (wound). -**zeug** n incendiary charge. -**ziegel** m firebrick

Bränke f (-n) (*Brew.*) yeast tub

brannte burned: *past of* BRENNEN

Brannt·hefe f spent yeast. -**kalk** m caustic lime, quicklime

Branntwein m (*genl*) (distilled) spirits, liquor; (*specif*) brandy. -**blase** f liquor still. -**brenner** m distiller. -**brennerei** f 1 distillery. 2 distillation of spirits. -**essig** m brandy vinegar. -**geist** m alcohol. -**hefe** f alcohol ferment. -**mixtur** f (*Pharm.*) brandy mixture. -**prober** m alcoholometer. -**vergiftung** f alcohol(ic) poisoning. -**waage** f alcoholometer

branstig a smelling (*or* tasting) burnt

Brasil·holz n brazil wood

brasilianisch a Brazilian. **Brasilien** n Brazil

Brasilin·säure f brazilinic acid

Brasil·kopalinsäure f brazilcopalinic acid. -**kopalsäure** f brazilcopalic acid. -**säure** f brazilic acid

Brasse(n) m (..ssen) (*Zool.*) bream

Brassicasterin n brassicasterol

Brassidin·säure f brassidic acid. **Brassin·säure** f brassic acid

Brassyl·säure f brassylic acid

brät roasts, fries: *pres of* BRATEN

braten* (briet, gebraten; brät) vt/i 1 roast; bake; fry. 2 calcine; burn; scorch. **Braten** m (-) 1 roasting meat. 2 roast, roasted meat. -**fett** n drippings

Brat·frischarbeit f roasting and refining. -**pfanne** f frying pan. -**rost** n grill. -**spieß** n roasting spit. -**wurst** f (grilled) veal-and-pork sausage

Bräu m (-e) 1 beer; brew; malt liquor; quantity brewed at one time. 2 brewery. 3 beer tavern

Brau·bottich m brewing vat

Brauch m (⁻e) custom, usage

brauchbar a usable; useful. **Brauchbar-**

keit f usability; usefulness, utility

brauchen I vt 1 need. 2 (*of time*): **sie b. einen Tag dazu** it takes them a day to do it. 3 use. II v aux (**zu** + *inf*) need (to), have (to). *See also* GEBRAUCHEN

Brauchtum m (..t⁻er) custom, usage

Brauch·wasser n water for industrial use

Braue f (-n) (eye)brow

brauen vt/i brew. **Brauer** m (-) brewer. **Brauerei** f (-en) brewery; brewing

Brauer·hefe f brewer's yeast. -**lack** m brewer's varnish. -**pech** n brewer's pitch

Brau· brewing: -**gerste** f brewing barley. -**haus** n brewery. -**kessel** m (*Brew.*) kettle, copper. -**meister** m brewmaster

braun a, **Braun** n brown; **b—er Glaskopf** limonite *cf* BERLINER

Braun·algen fpl brown (*or* fission) algae. -**bleierz** n pyromorphite

Braune (*adj endgs*) 1 m bay (horse). 2 m,f brown-haired person. 3 m coffee with milk

Bräune f 1 brownness; (sun)tan. 2 (*Med.*) quinsy; angina; diphtheria. 3 (*Min.*) limonite, etc = BRAUNERZ

Brauneisen n, -**erz** n, -**mulm** m brown iron ore, limonite. -**ocker** m brown iron ocher. -**stein** m limonite = BRAUNSTEIN-(ERZ)

Braunelle f (-n) self-heal (*Prunella vulgaris*)

bräunen 1 vt dye brown; burnish; (*Cooking*) brown. 2 vt&vi(s) tan. 3: **sich b.** get a tan; turn brown

Braun·erz n limonite; vivianite; sphalerite. -**färbung** f brown color(ation). **b—gefärbt** a browncolored. -**gelb** a brownish yellow. **B–glühhitze** f dark red heat. **b–grün** a brownish green. **B–heil** n self-heal = BRAUNELLE. -**holz** n logwood, Brazil wood. -**holzpapier** n, (-**holzpappe** f) paper (cardboard) made from steamed mechanical woodpulp. -**kalk** m dolomite. -**kohl** m kale; savoy cabbage. -**kohle** f brown coal, lignite

braunkohlen· lignite, brown-coal: -**haltig** a ligniferous. **B–klein** n lignite breeze (*or* slack). -**schiefer** m lignite shale. -**schwelerei** f lignite coking (plant). -**schwelgas** n lignite distillation gas. -**teer** m lignite tar. -**teerpech** n lignite pitch

Braun·kräusen *fpl* (*Brew.*) fuzzy heads.
-lauge *f* brown liquor
bräunlich *a* brownish. **--gelb** *a* brownish yellow
Braun·manganerz *n* (*Min.*) manganite.
-rost *m* brown (leaf) rust. **b--rot** *a* brownish red. **B--rot** *n* brownish red; colcothar; Indian red. **b--rotgrau** *a* drab. **-rötlich** *a* reddish brown. **B--salz** *n* brown dye
Braunsche Röhre *f* cathode ray tube
Braun·schliff *m* (*Paper*) steamed mechanical woodpulp, brown woodpulp. **b--schwarz** *a* brown-black, dark brown
Braunschweig *n* Brunswick. **Braun-schweiger Grün** *n* Brunswick green
Braun·späne *mpl* logwood shavings. **-spat** *m* dolomite
Braunstein *m* pyrolusite, manganese dioxide—**roter B.** rhodochrosite; **schwarzer B.** hausmanite. **-blende** *f*, **-kies** *m* alabandite. **-kiesel** *m* rhodonite. **-rahm** *m*, **-schaum** *m* bog manganite, wad. **b--stichig** *a* brownish. **B--tran** *m* (*Lthr.*) blubber, thick cod oil
Bräunung *f* browning; burnishing; tanning; brown coloring *cf* BRÄUNEN
Brau·pfanne *f* (*Brew.*) kettle, copper
Brause *f* (-n) **1** shower (bath). **2** spray head. **3** effervescent soft drink, fizz. **4** (*in compds oft*) effervescent: **--bad** *n* shower (bath). **-limonade** *f* effervescent soft drink, fizz. **-lithiumcitrat** *n* effervescent lithium citrate. **-magnesia** *f* effervescent citrate of magnesia
brausen 1 *vt* spray, sprinkle **2** *vi* (h) roar, rage; fizz, effervesce. **3** *vi* (h) & **sich b.** (take a) shower. **4** *vi*(s) rush, dash. **brausend** *p.a* effervescent
Brause·pulver *n* effervescent powder— **englisches B.** Seidlitz powder. **-salz** *n* effervescent salt. **-sieb** *n* spray sieve. **-ton** *m* bituminous clay. **-vorrichtung** *f* spraying (*or* sprinkling) device. **-wasser** *n* soda water. **-wein** *m* sparkling wine
Braut *f* (-̈e) fiancée; bride
Brau·wasser *n* water for brewing; (mash) liquor. **-wesen** *n* brewing trade. **-zucker** *m* brewer's sugar
brav *a* good, well-behaved, fine; *adv oft* well
BRD *abbv* (Bundesrepublik Deutschland)

Federal Republic of Germany
Breccie *f* (-n) (*Geol.*) breccia. **Brecci-en·textur** *f* brecciated structure. **brecci-ös** *a* brecciated
Brech·anlage *f* crushing plant. **-arznei** *f* emetic. **-backen** *fpl* crusher (*or* compacter) jaws
brechbar *a* refrangible; breakable, brittle
Brech·bohne *f* snap bean. **-eisen** *n* crowbar
brechen* (brach, gebrochen; bricht) **I** *vt*, *vi*(s) & **sich b. 1** break. **2** (*Opt.*) refract. **II** *vt* **3** fold; crush, pulverize; bruise; split. **4** mine; quarry. **5** pick, pluck. **6** blend (colors). **7** boil off (silk). **8** vomit, cough up. **III** *vi* (h) **9** vomit, throw up. **10** (mit) break (relations) (with). **IV** *vi*(s) **11** snap, crack; crease, crinkle; collapse. **12** decompose. **13** (*Dye baths*) change. **V** (*reg:* brechte, gebrecht) beat, break (flax, hemp). *Cf* GEBROCHEN; *see also* GEBRECHEN. **bre-chend** *p.a* **1** refractive. **2: b. voll** full to bursting, jam-packed. **brechen·er-regend** *a* emetic. **Brecher** *m* (-) breaker (*incl* wave); crusher
Brech·gut *n* crushed material. **-haufen** *m* (*Brew.*) broken pieces. **-koks** *m* crushed coke. **-körner** *npl* castor beans. **-kraft** *f* refractive power. **-mittel** *n* emetic. **-neigung** *f* nausea. **-nuß** *f* nux vomica. **-pulver** *n* emetic powder. **-punkt** *m* point of refraction. **-reiz** *m* nausea. **-ruhr** *f* diarrhea with vomiting. **-sand** *m* crushed-stone sand. **-stange** *f* crowbar. **-stoff** *m* emetic
Brechung *f* breaking, fracture; (*Opt.*) refraction *cf* BRECHEN
Brechungs· (*Opt.*) refraction, refractive: **-ebene** *f* plane of refraction. **-exponent**, **-index** *m* refractive index. **-messer** *m* refractometer. **-verhältnis** *n* refractive index. **-vermögen** *n* refractivity. **-werk** *n* crushing mill. **-winkel** *m* angle of refraction. **-zahl** *f* refractive index
Brech·walzwerk *n* crushing mill (*or* rollers). **-wein** *m* (*Pharm.*) wine of antimony, antimonial wine. **-weinstein** *m* tartar emetic. **-werk** *n* crusher. **-wurz(el)** *f* ipecac
Bredtsche Regel Bredt's rule
Brei *m* (-e) **1** porridge, puree, mush; paste,

pulp; mash; slurry; magma; (*Ceram.*) slip.
2 (*in compds, e.g.*): **Apfelbrei** applesauce; **Haferbrei** (cooked) oatmeal; **Kartoffelbrei** mashed potatoes. **b—·artig, breiig** *a* mushy, pulpy; mashed, semifluid

breit *a* wide, broad *cf* BREITMACHEN; LANG 6; WEIT 1. **-blättrig** *a* broad-leaved. **Breite** *f* (-n) width, breadth; latitude. **breiten** *vt* & **sich b.** spread, extend. **Breiten·grad** *m* (degree of) latitude **breit·flanschig** *a* wide-flanged. **-gefächert** *a* wide-ranging. **B—leder** *n* sole leather. **breit·machen: sich b.** 1 spread out, take up room. 2 show off, make a display. **Breit·schlitzdüse** *f* (*Plastics*) fishtail extruding die. **b—spurig** *a* 1 broad-gauge. 2 pompous. **B—waschen** *n* (*Tex.*) open-width washing **Brei·umschlag** *m* poultice cataplasm. **b—weich** *a* pulpy-soft, pappy

Brekzie *f* (-n) (*Geol.*) breccia = BRECCIE

Bremer *a* (*no endgs*) (of) Bremen: **B. Blau** Bremen blue

Bremse *f* (-n) 1 brake. 2 horsefly. **bremsen** 1 *vt* brake; check, moderate, retard. 2 *vi* put on the brake(s); cut down, slow down. 3: **sich b.** control (*or* restrain) oneself

Brems·feld *n* (*Elec.*) retarding field. **-flüssigkeit** *f* brake fluid. **-futter** *n* brake lining. **-gitter** *n* (*Elec.*) suppressor (*or* cathode) grid. **-klotz** *m* brake shoe (*or* block). **-leistung** *f* braking (horse)power. **-mittel** *n* (*Flot.*) depressant. **-stoff** *m* (*Nucl.*) moderator. **-strahlen** *mpl*, **-strahlung** *f* rays (*or* radiation) caused by retardation of particles. **-substanz** *f* (*Nucl.*) moderator

Bremsung *f* (-en) braking, retardation, slowdown; moderation; braking effect *cf* BREMSEN

Brems·vermögen *n* braking (*or* retarding) power. **-wirkung** *f* braking (*or* retarding) effect

Brenke *f* (-n) (*Brew.*) yeast tub

Brenn· burning, of combustion, calorific; distilling; (*Opt.*) focal: **-achse** *f* focal axis. **-apparat** *m* still, distilling apparatus; branding machine. **-arbeit** *f* fire assaying. **-ätzverfahren** *n* pyrography

brennbar *a* combustible, (in)flammable: **b—e Luft** (*Old Chem.*) flammable air (= hydrogen). **Brennbarkeit** *f* combustibility

Brenn·berge *mpl* carboniferous shale. **-blase** *f* alembic, still. **-cylinder** *m* (*Pharm.*) moxa. **-dauer** *f* 1 burning time. **2** flash duration. **3** lighting hours, life (of lamps). *Cf* BRENNZEIT

Brenne *f* (-n) (*Metll.*) pickle, dip

Brenn·ebene *f* focal plane. **-eisen** *n* cautery; curling iron; branding iron. **-element** *n* fuel cell

brennen* (brannte, gebrannt; *sbjc* brennte) I *vt* 1 burn; (*Light, oft*) keep on, keep lit. 2 roast, calcine, anneal; fire, bake. 3 scald; singe; cauterize. 4 pickle (metal). 5 distill. 6 brand. 7 curl (hair). II *vi* 8 burn; sting, smart. 9 (*Sun*) beat down. 10 (*Light*) be on, be lit. 11 (zu) be burning (*or* itching) (to do). 12 (*imps*): **es brennt** there is a fire, something is burning. III: **sich b.** 13 burn oneself, get burned (stung, scalded). 14 be mistaken. *Cf* GEBRANNT. **Brennen** *n* combustion, burning, etc. **brennend** *p.a* ardent, fervent; urgent; caustic; parching. **Brenner** *m* (-) 1 (*Mach.* or *psn*) burner; combustor; time burner. 2 torch, blowpipe. 3 brickmaker. 4 distiller

Brennerei *f* (-en) 1 distillation; calcination; (lime, brick) burning. 2 distillery; brickworks, kiln; limekiln. **-betrieb** *m* distillery (operation). **-hefe** *f* distillery yeast. **-maische** *f* distillery mash. **-schlempe** *f* distillery vinasse

Brenner·kopf *m* burner tip. **-mündung** *f* burner orifice. **-öffnung** *f* burner port (of furnace). **-stein** *m* burner brick

Brennessel *f* (-n) (stinging) nettle

Brenn·farbe *f* (*Ceram.*) fired color. **b—fest** *a* nonflammable. **B—fläche** *f* 1 (*Chem.*) caustic surface. 2 (*Opt.*) focal plane. **-fleck** *m* focal spot. **-freudigkeit** *f* combustibility, flammability. **-gas** *n* fuel (*or* combustible) gas. **-gerste** *f* distilling barley. **-geschwindigkeit** *f* rate of combustion (*or* distillation). **-gestein** *n* combustible rock. **-glas** *n* burning glass, focusing glass. **-gut** *n* material to be burned (*or*

distilled). **-heim** n still head. **-herd** m
hearth, refining furnace. **-holz** n firewood. **-kammer** f combustion chamber;
firebox. **-kapsel** f (Ceram.) sagger; (Explo.) cap. **-kegel** m (Ceram.) pyrometric
cone. **-kessel** m retort. **-kolben** m distilling flask, alembic. **-kraftmaschine** f internal combustion engine. **-linie** f focal
line. **-linse** f burning lense, convex lens.
-luft f air for combustion; (Old Chem.) inflammable air (= hydrogen). **-material** n
fuel. **-mittel** n caustic, corrosive. **-nessel**
f nettle = BRENNESSEL. **-ofen** m burning
(or baking) oven, kiln; roasting (or calcining) furnace. **-öl** n fuel oil. **-palme** f jaggery palm. **-punkt** m focal point; focus;
burning (or fire) point. **-säure** f pickling
acid. **-schieferöl** n creosote. **-schnei-
den** n (Engg.) oxy-gas (or flame) cutting. **-schnitt** m (Distil.) cut, fraction.
-schwindung f (Ceram.) shrinkage in
firing. **-silber** n (copper-)silvering amalgam. **-spiegel** m concave mirror. **-spiri-
tus** m fuel alcohol. **-stahl** m blister steel.
-staub m powdered fuel, fuel dust. **-stoff**
m fuel, combustible
Brennstoff- fuel: **-aufwand** m fuel consumption. **-element** n fuel cell (or element) **-verbrauch** m fuel consumption.
-wert m fuel value. **-ziegel** m fuel
briquet
Brenn·strahl m 1 (Opt.) focal ray. 2 (Math.)
focal radius. **-stunde** f (Elec.) lighting
hour. **-suppe** f browned-flour soup
brennte would burn, etc: sbjc of BRENNEN
Brenn·versuch m burning test. **-wärme** f
heat of combustion. **-weite** f focal length.
-wert m fuel value. **-zeit** f 1 (Elec. Light)
lighting hours. 2 (Molding) cure time.
3 (Phot.) flash duration. 4 (Rockets, etc)
combustion time. **-ziegel** m firebrick;
fuel briquet. **-zone** f combustion zone.
-zünder m fuse (train). **-zylinder** m
(Pharm.) moxa
Brenz m (-e) 1 empyreuma, burnt smell.
2 pl inflammables, combustibles
Brenz· pyro: **-apfelsäure** f maleic acid.
-cain n (Pharm.) pyrocain. **-catechin** n
pyrocatechin, pyrocatechol
brenzeln vi have a burnt (or an empyreumatic) odor

brenzen vt pyrolyze
Brenz·essigäther m, **-essiggeist** m acetone. **-gallussäure** f pyrogallic acid,
pyrogallol. **-harz** n empyreumatic resin.
b–holzsauer a of pyroligneous acid. **B–
holzsäure** f pyroligneous acid. **-ka-
techin** n pyrocatechin, pyrocatechol.
-katechingerbstoff m pyrocatechol
tannin
brenzig, brenzlich a 1 smelling (or tasting) burnt. 2 empyreumatic; **b—e Säuren** pyro acids. 3 critical, ticklish
Brenz· pyro-: **-öl** n empyreumatic oil. **-säu-
re** f pyro acid. **-schleimsäure** f pyromucic (or furoic) acid. **-traubenalkohol**
m pyroracemic alcohol (hydroxyacetone).
-traubensäure f pyroracemic (or pyruvic) acid
Bresche f (-n) 1 breach—**eine B. schlagen**
clear the way. 2 (Geol.) breccia
Brett n (-er) 1 board, plank—**aus B—ern**
(oft) (made) of wood, wooden. 2 shelf. 3:
(schwarzes) B. bulletin board. 4 (pl oft)
skis. **brettern** a wooden, made of boards
Bretter· wooden, board(s): **-planke** f board
fence. **-wand** f wooden partition; board
fence
Brett·mühle f sawmill
Brezel f (-n) pretzel
Brezzie f (-n) (Geol.) breccia
bricht breaks: pres of BRECHEN
Brief m (-e) letter; book (of needles, etc).
-·bogen m (sheet of) letter paper. **Brief-
chen** n note. **Brief·kasten** m mailbox.
brieflich a written; adv by letter, in
writing
Brief·marke f postage stamp. **-papier** n
writing paper, stationery. **Briefschaften**
fpl correspondence; papers. **Brief-
·stempel** m postmark. **-tasche** f wallet,
billfold. **-träger** m, **-trägerin** f letter carrier. **-waage** f postal (or letter) scale.
-wechsel m correspondence
Bries n (-e), **Brieseldrüse** f thymus
(gland)
briet roasted, fried: past of BRATEN
Brikett n (-e) briquet. **-·bindemittel** n briquet binder. **brikettieren** vt make bri-
quets out of
Brikett·kohle f briquet (char)coal. **-pech** n
briqueting pitch

brillant *a* brilliant, dazzling. **Brillant** *m* (-en,-en) (polished) diamond. **-gelb** *n* brilliant yellow

Brille *f* (-n) 1 (pair of) eyeglasses. 2 gland (of a stuffing box). 3 toilet seat

Brillen· (*usu Zool.*) spectacled: **-bär** *m* spectacled bear. **-ofen** *m* spectacle furnace

Brinell·härte *f* Brinell hardness. **brinellieren** *vt* test with the Brinell machine. **Brinell·probe** *f* 1 Brinell test. 2 Brinell sample

bringen* (brachte, gebracht) I *vt* 1 bring, take (to a place or psn), get (sthg to a place). 2 put. 3 publish, print; list, include (in a list). 4 present, offer. 5 yield, earn; produce. 6 (+ **mit sich**) involve, lead to; **das bringt es mit sich, daß** that brings it about that... 7 (+ **hinter sich**) **a** get (sthg) done, **b** cover (a distance). 8 (**um**) deprive, rob, cheat (of) *cf* UMBRINGEN. 9 get done, achieve: **sie b. es noch weit** they will get far; **es dahin b., daß**... get things to the point where...; **er brachte es sogar zum Chef** he even got as far as being boss. II *vt/i:* **das bringt (es)** that does it, that hits the spot

brisant *a* (high-)explosive; shattering, disruptive; brisant. **Brisanz** *f* (-en) explosiveness, explosive effect; brisance, shattering power. **-geschoß** *n* high-explosive shell. **-sprengstoff** *m* high explosive, disruptive

Brise *f* (-n) breeze

Bristol·pappe *f* Bristol board

Britisch·gummi *n* British gum, dextrin

Bröckchen *n* (-) small piece, bit, crumb. **bröckelig** *a* crumbly, brittle, friable, fragile. **Bröckeligkeit** *f* brittleness, friability, fragility. **bröckeln** *vt, vi*(s) crumble, break up. **Bröckel·stärke** *f* lump starch

brocken *vt* crumble, break up. **Brocken** *m* (-) piece, morsel; lump; fragment; crumb. **-gestein** *n* (*Petrog.*) breccia. **-glas** *n* broken glass, cullet. **-stärke** *f* lump starch. **-stein** *m* (*Petrog.*) breccia. **-torf** *m* lump peat. **b–weise** *adv* piecemeal

bröcklig *a* crumbly, etc = BRÖCKELIG

Brodel *m* fumes, etc = BRODEM. **brodeln** *vi* bubble, seethe, boil; effervesce, ferment

Brodem *m* vapor, steam; fumes; exhala-

tion; heavy smell; (*Min.*) foul air, damp

Brokat *m* (-e) brocade. **-farbe** *f* brocade color (powdered bronze); brocade dye

Brom *n* bromine, (*in compds, oft*) bromo-, bromide of **-ammon(ium)** *n* ammonium bromide. **-äther** *m* ethyl bromide. **-äthylformin** *n* (*Pharm.*) bromaline. **-beere** *f* blackberry. **b–beerrot** *a* blackberry-red. **B–benzol** *n* bromobenzene. **-cyan** *n* cyanogen bromide, bromocyanogen. **-eisen** *n* iron bromide. **-fluor** *m* bromine fluoride. **-goldkalium** *n* potassium auribromide (*or* bromoaurate)

Bromid *n* (-e) bromide *cf* BROMÜR

bromierbar *a* capable of bromination.

bromieren *vt* brominate

bromig *a* bromous

brominieren *vt* brominate

Brom·ion *n* bromine ion. **Bromismus** *m* (*Med.*) bromism

Brom· bromine, bromide (of), bromo-: **-jod** *n* iodine bromide. **-kalium** *n* potassium bromide. **-kalk** *m,* **-kalzium** *n* calcium bromide. **-kampfer** *m* bromocamphor, (*Pharm.*) monobromated camphor. **-kohlenstoff** *m* carbon (tetra)bromide. **-körper** *m* (*Colloids*) bromine body (*or* ion). **-kresol** *n* bromocresol. **-lauge** *f* bromine lye (bromine passed into NaOH solution). **-lost** *m* (*Poison Gas*) 2,2'-dibromodiethyl sulfide. **-metall** *n* metallic bromide. **-methylat** *n* methobromide. **-natrium** *n* sodium bromide. **-natron** *n* sodium hypobromite. **-öldruck** *m* (*Phot.*) bromoil painting

Bromometrie *f* bromometry

Brom· bromine, bromide (of), bromo-: **-salz** *n* bromide; bromate. **b–sauer** *a* bromate of. **B–säure** *f* bromic acid. **-spat** *m* bromyrite, AgBr. **-stickstoff** *n* nitrogen bromide. **-toluol** *n* bromotoluene. **-übertrager** *m* bromine carrier

Bromür *n* (-e) (-ous) bromide *cf* BROMID

Brom·vergiftung *f* bromine (*or* bromide) poisoning. **-wasser** *n* bromine water

Bromwasserstoff *m* hydrogen bromide; hydrobromic acid. **b–sauer** *a* hydrobromide of. **B–säure** *f* hydrobromic acid

Brom·zahl *f* bromine number

Bronze *f* (-n) bronze. **b–artig** *a* bronzelike, bronzy. **B–blau** *n* bronze (*or* reflex)

blue). **-färben** n bronzing. **b–farbig** a bronze-colored. **B–gießerei** f bronze foundry; bronze founding. **-glanz** m bronze luster. **-guß** m bronze casting; cast bronze. **-lack** m bronze varnish; bronzing lacquer. **-tinktur** f bronzing liquid

bronzieren vt bronze

Bronzier-pulver n bronzing powder. **-salz** n bronzing salt, antimony chloride

bronzig a bronzy

Brosame m (-n,-n) or f (-n) crumb

Brös/chen n (-n) sweetbread

broschieren vt stitch, sew; bind in paper covers; brocade. **broschiert** p.a paperbound, in booklet (or brochure) form. **Broschüre** f (-n) brochure, booklet

Brot n (-e) (loaf of) bread. **--brei** m bread paste. **Brötchen** n (-) roll cf BELEGT

Brot-gärung f panary fermentation, leavening of bread. **-geschmack** m (Brew.) steam taste. **-herr** m employer. **-korn** n breadstuff. **-laib** m loaf of bread. **b–laibartig** a loaf-like. **-los** a 1 unemployed. 2 unprofitable. **B–maschine** f bread slicer. **-masse** f breadstuff. **-raffinade** f loaf sugar. **-rinde** f bread crust. **-studium** n bread-and-butter education. **-wurzel** f cassava; yam. **-zucker** m loaf sugar

Brownsche Bewegung f Brownian movement

Bruch I m (ᵉe) 1 break, breach, crack, fissure, cleft; (Geol.) fault. 2 break-up, breakage, rupture; (Med.) hernia. 3 crash, crack-up. 4 breach, violation. 5 quarry. 6 broken pieces, scraps; fragments. 7 trash, junk, shoddy goods. 8 crease, fold, wrinkle, crimp. 9 (Math.) fraction. 10 (Cheese) curd. 11: **zu B. gehen, in die B--e gehen** fail, break down. **II** m (ᵉe) or n (ᵉer) swamp, marsh

Bruch- breaking; fracture, fraction(al); scrap: **-beanspruchung** f breaking stress. **-belastung** f breaking load. **-bildung** f 1 (Min.) fracturing. 2 (Geol.) faulting. 3 (Med.) herniation. 4 (Yeast) flocculence. **-blei** n scrap lead. **-dehnung** f breaking elongation; (Paper) stretch. **-ebene** f plane of fracture. **-eisen** n scrap iron. **b–fest** a breakage-resistant, tena-

cious; fracture-proof. **B–festigkeit** f fracture resistance. **-fläche** f surface of fracture. **-gefüge** n fracture. **-gewicht** n fractional weight. **-glas** n broken glass, cullet. **-gramm** n fraction of a gram. **-grammgewicht** n fractional gram weight. **-grenze** f breaking point

bruchig a swampy, marshy cf BRUCH II

brüchig a 1 brittle, fragile, friable; (Tex.) tender, easily split; (Geol.) clastic, fragmental. 2 cracked, full of cracks. 3 shaky, weak, frail. 4 (Metll.: in compds) -short cf BLAUBRÜCHIG. **Brüchigkeit** f brittleness, fragility, etc. Cf BRUCH I

Bruch- breaking; fracture, fraction(al); scrap: **-kupfer** n scrap copper. **-last** f breaking load. **-metall** n broken (or scrap) metal. **-modul** n modulus of rupture. **-moment** n (Phys.) breaking moment. **-probe** f breaking (or breakdown) test. **-punkt** m breaking point. **-rechnung** f (Math.) (figuring with) fractions. **-reis** m broken rice. **-riß** m (Metll.) failure crack. **b–sicher** a breakproof, unbreakable. **B–silber** n broken (or scrap) silver. **-spannung** f breaking stress; tensile strength. **-stein** m quarry stone, rubble. **-stelle** f (point of) fracture. **-strich** m (Math.) fraction line (stroke, bar). **-stück** n fragment, piece, scrap; (pl oft) debris, scrap. **-teil** m fraction. **-verformung** f deformation on breaking. **-zahl** f fractional number

Brücke f (-n) bridge (incl Elec.); platform; (Anat.) pons; (Polymers) cross-link

Brücken-atom n bridge (or bridging) atom. **b–bildend** a bridge-forming, bridging. **B–bindung** f bridge bond(ing). **-glühzünder** m (Explo.) bridge-wire cap. **-ring** m bridged ring. **-sauerstoff** m bridging (or connecting) oxygen. **-schaltung** f (Elec.) bridge (connection). **-schlag** m bridging the gap. **-waage** f platform balance (or scales). **-widerstand** m (Elec.) bridge resistance. **-zoll** m bridge toll

Brüden m (-) 1 evaporating liquid; water vapor, steam; cloud of vapor (or steam). 2 = BRODEM fumes, etc. **--dampf** m vapor, steam (from an evaporating liquid)

Bruder *m* (-) brother
Brühe *f* (-n) 1 (*genl*) liquid left after cooking or processing: (*specif*) broth, stock; liquor, (*Lthr.*) juice, drench, ooze; (*Tobacco*) sauce. 2 soup; consommé, bouillon; brew, decoction. 3 dishwater, slop. **brühen** *vt* 1 scald, douse (*or* treat) with hot water. 2 brew, decoct. 3 scour, kier-boil. 4 soak (in hot water)
brüh·gar *a* (*Lthr.*) liquor-tanned. **-heiß** *a* scalding-hot. **B–kartoffeln** *fpl* potatoes boiled in meat broth. **-malz** *n* proteolytic malt. **-messer** *m* (*Lthr.*) barkometer. **-saft** *m* (*Sugar*) scalding juice. **-wasser** *n* 1 water for scalding. 2 stock (*or* liquor) after scalding. **-würfel** *m* bouillon cube. **-wurst** *f* sausage for boiling
brüllen *vt/i* bellow, roar, howl; (*Cows*) low, moo
brummen 1 *vt/i* mumble, grumble, growl. 2 *vi* hum, buzz; drone; throb. **Brummen** *n* growl; hum, buzz; drone; rumble
Brunelle *f* (-n) self-heal (Prunella vulgaris)
Brunft *f* (-e) estrus, heat, mating (*or* rutting) season; (*in compds*) estrual, rutting
Brünier·beize *f* (*Metll.*) bronzing pickle. **brünieren** 1 *vt* burnish, polish; brown, bronze. 2 *vi* turn brown. **Brünier·stein** *m* burnishing stone, bloodstone
Brunnen *m* (-) well; fountain, spring; (mineral) waters. **-flasche** *f* mineral-water (*or* siphon) bottle. **-kresse** *f* watercress. **-salz** *n* well salt; brine salt. **-wasser** *n* well (*or* spring) water
Brunst *f* (-e) 1 ardor, fervor. 2 = BRUNFT estrus, etc. **brünstig** *a* ardent, fervent; rutting, in heat
Brust *f* (-e) breast; chest, thorax; brisket; front (of a blast furnace). **--angst** *f* angina pectoris. **-arznei** *f* chest medicine, pectoral. **-beere** *f* jujube, zizyphus. **-bein** *n* breastbone, sternum. **-beindrüse** *f* thymus gland. **-bohrer** *m* breast drill. **-drüse** *f* mammary gland. **-fell** *n* pleura. **-fellentzündung** *f* pleurisy. **-fieber** *n* bronchitis. **-gang** *m* thoracic duct. **-kasten, -korb** *m* thorax. **-krampf** *m* asthma. **-milch** *f* breast milk, human milk; pectoral emulsion. **-pulver** *n* pectoral

(*specif* licorice) powder. **b–reinigend** *a* expectorant. **B–ton** *m* resonant voice. **-wirbel** *m* dorsal (*or* thoracic) vertebra
Brut *f* (-en) 1 brooding, sitting (on eggs); hatching. 2 brood, progeny. **--apparat** *m* incubator. **-ei** *n* hatching egg
brüten 1 *vt* hatch, plot, scheme, brew (up). 2 *vi* brood; sit (on eggs). **Brüter** *m* (-) breeder: (*Nucl.*) **schneller B.** fast breeder
Brut·gewinn *m* (*Nucl.*) breeding gain. **-hitze** *f* stifling heat. **-kasten** *m*, **Brut·** or **Brüt·ofen** *m* incubator. **Brut·reaktor** *m* breeder (reactor). **-schrank** *m* incubator. **-stätte** *f* breeding ground, hotbed. **-stoff** *m* fertile (*or* breeder) material
brutto *a* (no endgs) & *adv* gross, overall. **B--eigenschaft** *f* bulk property. **-formel** *f* empirical (*or* molecular) formula. **-gewicht** *n* gross weight. **-reaktion** *f* overall (*or* gross) reaction. **-sozialprodukt** *n* gross national product
Brut·wärme *f* blood heat; incubation heat
Bruzin *n* brucine. **Bruzit** *n* (*Min.*) brucite
BSB. *abbv* (biochemischer Sauerstoffbedarf) B.O.D. (biochemical oxygen demand)
bspw. *abbv* (beispielsweise) for example
B-Stoff *m* bromoacetone (tear gas)
BTÄ. *abbv* (Bleitetraäthyl) T.E.L. (tetraethyl lead)
Bub *m* (-en,-en) boy; apprentice
Bubonen·pest *f* bubonic plague
Bucco·blätter *npl* (*Pharm.*) buchu
Buch *n* 1 (-er) book: **zu B—e stehen** be on the books; **B. führen** (*or* **halten**) keep book(s); **zu B—e schlagen** cost, add up to (on the books). 2 (*pl* -) (*Paper Meas.*) quire. **--besprechung** *f* book review. **-binder** *m* bookbinder. **-binderpappe** *f* (book)binder's board. **-druck** *m* typography, printing. **-drucker** *m* (book) printer. **-druckerei** *f* printing plant (*or* shop)
Buchdrucker·farbe, -schwärze *f* printer's ink
Buchdruck·farbe *f* printing ink. **-firnis** *m* printer's varnish
Buche *f* (-n) beech (tree)
Buch·ecker *f* (-n) beechnut. **-eckernöl** *n*

beechnut oil. **-eichel** f = **Buchel** f (-n) beechnut

buchen I a (of) beech(wood). II vt 1 put (or enter) on the books, record. 2 book, reserve (seats, etc)

Buchen- beech: **-asche** f beech ashes. **-holz** n beech(wood). **-holzteer** m beech tar

Bücherei f (-en) library. **--verzeichnis** n library catalog

Bücher-schau f 1 book exhibition. 2 book review. **-schrank** m bookcase. **-verzeichnis** n book list (or catalog)

Buch-führer m, **-führerin** f bookkeeper, accountant. **-führung** f bookkeeping, accounting. **-gold** n leaf gold. **-halter** m, **-halterin** f bookkeeper. **-handel** m book trade. **-händler** m, **-händlerin** f bookdealer. **-handlung** f bookstore

Büchner-trichter m Buechner funnel

Buch-nuß f beechnut. **-(nuß)öl** n beech (nut) oil

Buch-prüfer m auditor

Buchs-baum m box tree. **-baumholz** n boxwood

Buchse f (-n) (Mach.) bushing; (Elec.) socket

Büchse f (-n) 1 (tin) can; box, canister. 2 rifle

Büchsen-fleisch n canned meat. **b-förmig** a box-shaped. **B--frucht** f canned fruit. **-gemüse** n canned vegetable(s). **-macher** m gunsmith

Buchsen-metall n bush (or bearing) metal

Büchsen-milch f canned milk. **-mühle** f barrel mill. **-öffner** m can opener. **-pulver** n gunpowder. **-schuß** m gunshot, rifle shot. **-stein** m iron pyrites

Buchstabe m (-n,-n) letter (of the alphabet), character, type. **Buchstaben-rechnung** f algebra. **buchstabieren** vt/i spell. **buchstäblich** a literal

Bucht f (-en) 1 bay, bight; safe harbor. 2 (Anat., Med.) fossa, sinus. 3 (animal) shed, pigsty

Buchu-blätter npl buchu (leaves)

Buch-umschlag m book jacket

Buchung f (-en) 1 entry (on the books), bookkeeping entry. 2 booking, reservation. 3 booking, recording (of income, etc); registration

Buchungs-fehler m booking (bookkeep-

ing, accounting) error cf BUCHUNG. **-maschine** f bookkeeping (or accounting) machine. **-wert** m book value

Buch-weizen m buckwheat

Buckel m (-) 1 (person's) back. 2 hump, hunched back. 3 hill, knoll; knob, boss

bücken vr: sich b. bend (down, over), lean; stoop

Bücking, Bückling m (-e) smoked herring

Bucko-, Bucku-blätter npl buchu (leaves)

Bude f (-n) booth; shack; dump, pad; shop; shed

Büffel m (-) buffalo. **--leder** n buff (leather). **-leim** m leather glue

Bug m (-e or -̈e) 1 bow (of ship); nose (of plane). 2 bend, crease. 3 (Meat.) hock; shoulder; chuck

Bügel m (-) 1 bow. 2 (coat) hanger; stirrup; temple (of eyeglasses). 3 handle, (carrying) strap; bail, yoke. 4 frame. 5 clamp, clip. **6: durchlöcherter B.** shelf in a pneumatic trough. **--brett** n ironing board. **b--fest** a fast to ironing. **B--eisen** n (pressing) iron. **-falte** f (pressed) crease. **b--faltenbeständig** a crease-resistant. **B--kaltsäge** f bow hacksaw

bügeln vt/i iron, press (clothes)

Bügel-säge f bow saw. **-schälchen** n dish, cup (of bent metal). **-tisch** m (folding) ironing board

bugsieren vt tow; steer, maneuver

Bühne f (-n) 1 platform, scaffold. 2 theater, stage

buk baked, etc: past of BACKEN I

büken vt boil (or steep) in lye = BEUCHEN

Bukett n (-e) bouquet

Bukko-, Bukku-blätter npl buchu leaves

Bulbe f (-n) (Bot.) bulb; pseudobulb

Bulgarien n Bulgaria

Bullrich-salz n (Pharm.) bicarbonate of soda

Buna m or n (Synth.) buna (rubber)

Bund I m (-̈e) 1 bond—**im B. mit** in league with. 2 relationship. 3 agreement, (com)pact. 4 organization, union, association, society, league. 5 alliance; (con)federation. 6 conspiracy. 7 federal government (German, Austrian, Swiss). 8 waistband; neckband, collar; hoop. II n (-e) bunch, bundle **--bolzen** m flange bolt

Bündel n (-) bundle, bunch; parcel, packet

(*incl* waves); pencil (of rays). **bündeln** *vt* bundle, bunch; focus, concentrate

Bundes· (German, Austrian, Swiss) Federal *cf* BUND 7: **-amt** *n* federal office (*or* agency). **-anstalt** *f* federal institute (*or* agency). **b–eigen** *a* federally owned (*or* operated). **B–ministerium** *n* Federal (cabinet) Ministry (*or* Department). **-rat** *m* 1 (*Germany, Austria*) Federal Council of Constituent States (Upper House). 2 (*Switzerland*) Federal Executive Council. **-republik** *f* Federal Republic *cf* BRD. **-tag** *m* Federal House of Representatives, Parliament, Lower House. **-wehr** *f* Federal Armed Forces

bündig *a* 1 concise, convincing; conclusive *cf* KURZ. 2 obligatory. 3 flush, level (floorboards, etc)

Bündnis *n* (-se) relationship; association, alliance; pact

Bund·rohr *n* flanged pipe. **-stahl** *m* faggot (*or* bundled) steel

Bunsen·brenner *m* Bunsen burner. **-flamme** *f* Bunsen (burner) flame. **-kohle** *f* carbon for the Bunsen cell. **-ventil** *n* Bunsen valve

bunt *a* 1 colorful, bright; gaudy, many-colored. 2 motley, spotted, speckled. 3 varied. 4 confused, tangled, jumbled. 5 lively, wild

Bunt·ätzdruck *m* (*Tex.*) colored discharge printing. **-ätze** *f* (*Dye.*) color discharge. **-ätzfarbe** *f* color discharge paste. **-bleiche** *f* bleaching of colored goods. **-bleierz** *n* pyromorphite. **-druck** *m* color print(ing). **-farbe** *f* 1 variety of color. 2 iron oxide pigment

Bunte·bürette *f* Bunte gas burette

bunt·färben *vt* color, dye, stain; mottle

Bunt·färberei *f* dyeing in different colors. **b–farbig** *a* many-colored, motley. **-fleckig** *a* speckled, dappled (in various colors). **B–gewebe** *n* colored fabric. **-glas** *n* stained glass. **-kupfererz** *n,* **-kupferkies** *m* bornite. **-metall** *n* nonferrous metal. **-öllack** *m* colored oil varnish. **-papier** *n* colored (speckled, mottled) paper. **-papierfarbe** *f* surface paper color. **-pappe** *f* colored cardboard. **-reserve** *f* (*Tex.*) color resist. **-sandstein** *m* colored sandstone. **b–scheckig** *a* mottled, many-

colored. **-schillernd** *a* iridescent, chatoyant. **B–schichte** *f* size for (*or* sizing of) colored goods. **-stift** *m* colored pencil. **-webartikel** *mpl* woven colored goods

Bürde *f* (-n) burden

Bureau *n* (-s) office, etc = BÜRO

Bürette *f* (-n) burette

Büretten·bürste *f* burette brush. **-flasche** *f* volumetric flask. **-gestell** *n* burette stand. **-klemme** *f* burette clamp. **-schwimmer** *m* burette float. **-trichter** *m* funnel for filling burettes

Burg *f* (-en) 1 castle, stronghold. 2 beaver lodge

Bürge *m* (-n,-n) guarantor; bondsman; security. **bürgen** *vi* (**für**) stand bail, vouch (for)

Bürger *m* (-) citizen; middle-class person; bourgeois. **-krieg** *m* civil war. **-kunde** *f* civics

bürgerlich *a* civil; middle-class; bourgeois; home-style (food)

Bürger·meister *m* mayor, **-recht** *n* civil right. **Bürgerschaft** *f* (-en) 1 citizens, citizenry. 2 city council. **Bürger·steig** *m* sidewalk. **Bürgertum** *n* middle class; bourgeoisie

Bürgschaft *f* (-en) security, guarantee; bail: **B. leisten** (**für**) stand security (*or* bail) (for), guarantee

Burgunder *m* (-) Burgundy (wine). **-harz**, **-pech** *n* Burgundy pitch

burnettisieren *vt* burnettize, impregnate with zinc chloride

Büro *n* (-s) office. **-maschine** *f* office (*or* business) machine. **-zeit** *f* office hours

Bursche *m* (-n,-n) young fellow; orderly; student fraternity member

Bürste *f* (-n) brush (*incl* Elec.). **bürsten** *vt/i* brush

Bürsten·entladung *f* (*Elec.*) brush discharge. **-feuer** *n* (*Elec.*) brush sparking

Bus *m* (-se) bus. **-bahnhof** *m* bus terminal

Busch *m* (-̈e) bush; tuft; woods; grove

Büschel *m* (-) bunch, cluster, tuft; pencil, sheaf (of lines, rays); (*Elec.*) brush. **-entladung** *f* (*Elec.*) brush discharge. **b–förmig** *a* bunched, clustered, tufted

Busch·holz *n* undergrowth, brushwood

Busen *m* (-) 1 bosom; heart. 2 gulf, bay

Bussard *m* (-e) buzzard

Buße *f* (-n) penance; penalty, fine. **büßen**
vt/i atone, pay the penalty (for), make
good

Busse buses: *pl of* Bus

Bussole *f* (-n) compass, theodolite; galvanometer

Büste *f* (-n) bust; dressmaker's dummy.
Büsten·former, -halter *m* brassiere

Butan *n* butane

Butt *m* (-e) flatfish, turbot, brill

Bütte *f* (-n) tub, vat, butt; back (basket);
tank

Bütten·färbung *f* (*Paper*) vat (*or* pulp)
coloring. **b—gefärbt** *a* (*Paper*) vat-
colored; unbleached. **B—leimung** *f* vat (*or*
pulp) sizing. **-papier** *n* vat (*or* handmade)
paper

Butter *f* butter; (*in compds oft*) butyric,
butyryl, butyro-: **b—·ähnlich, -artig** *a*
buttery, butyraceous. **B—amylester** *m*
amyl butyrate. **-äther** *m* butyric ether
(ethyl butyrate). **-baum** *m* shea tree.
-blume *f* buttercup. **-brot** *n* bread and
butter; open sandwich. **-farbe** *f* butter
color. **-faß** *n,* **-fertiger** *m* butter churn.
-fett *n* butterfat, butyrin. **-gärung** *f* butter fermentation. **-gelb** *n* butter yellow.
b—haltig *a* containing butter,

butyraceous

butterig *a* buttery

Butter·messer *m* butyrometer. **-milch** *f*
buttermilk

buttern 1 *vt* butter. 2 *vi* churn; yield butter

Butter·nuß *f* butternut. **-öl** *n* clarified butter. **-persäure** *f* per(oxy)butyric acid.
-prüfer *m* butter tester. **b—sauer** *a* butyrate of. **B—säure** *f* butyric acid

Buttersäure·äther *m* ethyl butyrate; *pl*
butyric esters. **-gärung** *f* butyric fermentation. **-pilz** *m* butyric acid bacteria

Butter·schmalz *n* clarified butter

butylieren *vt* butylate

Butyl·jodid *n* butyl iodide. **-kautschuk** *m*
butyl rubber

Butyra·persäure *f* per(oxy)butyric acid

Butzen *m* (-) 1 (*Min.*) ore pocket. 2 (*Glass,
Metal*) lump. 3 (apple) core. 4 (*Anat.*)
secretion

b.w. *abbv* (bitte wenden) please turn over

Bz. *abbv* (Benzol) benzene; commercial
benzol

bzgl. *abbv* (bezüglich) respecting, regarding

Bzl. *abbv* (Benzol) benzene, benzol

Bzn. *abbv* (Benzin) (motor) gasoline

bzw. *abbv* (BEZIEHUNGSWEISE) or, etc

C

Words not listed under C can be looked up under **K** (if the C is followed by **a, o, u, l** or **r**) or under **Z** (if the C is followed by **e, i, y, ä** or **ö**).

C *abbv* (Celsius) centigrade

C. *abbv* (Chemisches Zentralblatt) German Chemical Abstract Journal

ca. *abbv* (circa = ZIRKA) approx., about

Caban·holz *n* camwood

Cabesa·seide *f* cabeca, cabesse (silk)

Cabuja·faser *f* cabuya

Cachau, Cachou *n* catechu. **cachoutieren** *vt* dye with catechu (*or* cutch)

Cade·öl, Cadin·öl, Cadi·öl *n* oil of cade

cadmieren *vt* treat (*or* plate) with cadmium

Cadmium·gelb *n* cadmium yellow. **Cadmium-Normalelement** *n* standard Weston cadmium cell. **Cadmium·stab** *m* (*Nucl.*) cadmium rod. *See also* KADMIUM·

Caesium *n* cesium

Cainca·säure *f* cahinic acid. **-wurzel** *f* cahinca (root). **Caincin** *n* cahincin

Cajeput·geist *m* spirit of cajuput. **-öl** *n* cajuput oil

cal, Cal *abbv* (Calorie) gram calorie, kilogram calorie (*resp*)

Calain·säure *f* calaic acid

calcificieren *vt* calcify

calcinieren *vt* calcine. **Calcinier·herd** *m* calcining hearth

Calcium·gehalt *m* calcium content. **c—haltig** *a* containing calcium. **C—hydrat** *n* calcium hydroxide **-hydrür** *n* calcium hydride (CaH$_2$). **-iodid** *n* calcium iodide. **-metall** *n* metallic calcium. **-spiegel** *m* (*Biol.*) calcium level. **-sulfhydrat** *n* calcium hydrogen sulfide. **-sulfuret** *n* calcium sulfide. **-superoxid** *n* calcium peroxide

Caledonisch·braun *n* Caledonian brown

calibrieren *vt* calibrate

Calisaya·rinde *f* calisaya bark

Calit *n* calite, insulator ceramic

Calmus·öl *n* oil of calamus

Calorien·wert *m* caloric value

Calorimeter·bombe *f* bomb calorimeter

calorimetrieren *vt* measure with the calorimeter. **Calorimetrierung** *f* (-en) calorimetric measurement. **calorimetrisch** *a* calorimetric

calorisch *a* caloric, thermal

Calorstat *m* (-en,-en) thermostat

Calypsol *n* (*TN*) cold-resistant grease

Camba(l)holz *n* camwood

Cambogia·säure *f* gambogic acid

Camille *f* (-n) camomile

Campane *f* (-n) bell jar

Campeche·holz *n* campechy wood, logwood

Campfer *m* camphor; *for compds see* KAMPFER·

Camphan·säure *f* camphanic acid

Campher *m* camphor = CAMPFER, KAMPFER

camphotetisch *a* (*Org. Chem.*) camphothetic, isocampholytic

Canada·balsamöl *n* oil of Canada balsam (*or* turpentine). **canadisch** *a* Canadian

Canarien·farbe *f* canary yellow; canarin. **-grün** *n* canary green. **-vogel** *m* canary

cancerogen *a* cancerogenic, carcinogenic

Candara·wachs *n* carnauba wax

Candela *f* candle(power)

canneliert *a* channeled

Cannel·kohle *f* cannel coal

Cantharin·säure *f* cantharidic acid. **Canthar·säure** *f* cantharic acid

Caoutschuk *m* caoutchouc; *for compds cf* KAUTSCHUK·

capillar *a* capillary; *for compds cf* KAPILLAR

Capo·messer *m* meter for gases

Caprar·säure *f* capraric acid

Capri·blau *n* Capri blue

157

Caprin·säure f capric acid
Capron·aldehyd m hexanol. **-fett** n caproin
Capryl·säure f caprylic acid
Caragheen·moos n carageen moss, Irish moss
Caranna·gummi n, **-harz** n caranna (resin)
Carbamid·säure f carbamic acid
carbamin·sauer a carbamate of. **C–säure** f carbamic acid
Carbanil n phenyl isocyanate. **-·säure** f carbanillc acid
Carbäthoxyl n carbethoxyl
Carbazot·silizium n silicocarbonitride
Carbeniat n carbonion. **Carbenium** n carbonium
carbidisch a carbide
Carbid· carbide: **-kalk** m carbide lime. **-kohle** f carbide carbon. **-ofen** m carbide furnace. **-schlamm** m carbide sludge. **-sprit** m acetaldehyde
Carbinol n methyl alcohol
Carbochindolin n carboquindoline
Carbol n phenol, carbolic acid; (in compds) carbolic, carbolized, carbolated: **-kalk** m carbolated lime. **-öl** n methanol. **c–sauer** a phenolate (or phenoxide) of. **C–säure** f carbolic acid, phenol. **c–schwefelsauer** a (Pharm.) sulfocarbolate of. **C–schwefelsäure** f sulfocarbolic (or sulfophenic) acid, p-phenolsulfonic acid. **-seife** f carbolic soap. **-sulfosäure** f sulfocarbolic acid = **C–SCHWEFELSÄURE**. **-vergiftung** f carbolic acid poisoning. **-wasser** n aqueous solution of phenol. **-watte** f carbolized cotton wool
Carbon m carbon; (Geol.) Carboniferous. **-·amid** n carboxamide, carboxylic acid amide. **-ester** m carboxylic acid ester
Carbonat·härte f (Water) carbonate hardness. **-ion** n carbonate ion
Carbonatisierung f (Geol.) calcification, carbon(at)ation
Carbonat·puffer m carbonate buffer. **-wasser** n chalk (or carbonate) water
carbonieren vt carbonate, carbonize. **Carbonisation** f (-en) carbonization. **carbonisch** a carbonaceous; (Geol.) Carboniferous

Carbonisier· carbonizing: **-anlage, -anstalt** f carbonizing plant. **c–echt** a resistant to carbonizing
carbonisieren vt carbonize, carbonate; carburize
Carbonitrieren n (Metll.) carbonitriding
Carbon·säure f carboxylic acid; (rarely) carbonic acid. **-spat** m (Min.) apathic carbonate
Carbonyl·gruppe f carbonyl group. **carbonylieren** vt add CO to a triple bond. **Carbonyl·wasserstoff** m carbonyl hydride
Carborund n carborundum
Carboxyl·gruppe f carboxyl group
carbozyklisch a carbocyclic
Carbür n (-e) (-ous) carbide
Carburator m (-en), **Carburations·apparat** m carburetor
carburieren vt carburet, carburize. **Carburier·öl** n carbureting (or carburizing) oil. **Carburierung** f (-en) carbureting, carbur(iz)ation. **Carburierungs·stufe** f number of C–C bonds (indicating primary, secondary or tertiary carbon atoms)
cardieren vt (Tex.) card
Carlinin·säure f carlinic acid
Carmin·säure f carminic acid
Carnauba·säure f carnaubic acid. **-wachs** n carnauba wax
Carneol m (-e) carnelian
Carolin·säure f carolinic acid
Caro'sche Säure f Caro's acid
Carotin n carotene
cartesisch a Cartesian
Carthamin·säure f carthamic acid
Carub m (-en), **Carube** f (-n) carob
Cärul·, cärul· cerul...
casein·artig a casein-like, caseous. **C–farbstoff** m casein paint. **-natrium** n sodium caseinate
Cäsium n cesium. **-·alaun** m cesium alum. **c–bedeckt** a cesium-coated, cesiated
Cassawa·stärke f tapioca
Casselmann·grün n Casselmann's Green
Cassie f (-n) cassia. **-·blütenöl** n cassia flower oil. Cf **KASSIEN. Cassiöl** n oil of cassia
Cassius·purpur m purple of Cassius

Castilianer Seife *f* Castile soap
Castor-nuß *f* castor bean. **-öl** *n* castor oil
Cathartin-säure *f* cathartic acid
Cautschuk *m* caoutchouc = KAUTSCHUK
cbcm, ccm *abbv* (Cubic centimeter) cc
cd *abbv* (Candela) c.p. (candlepower)
Ceara-kautschuk *m* Ceara rubber
Ceder *f* (-n) cedar. **Cedern-öl** *n* cedar (wood) oil
Cedrat-öl *n* citron oil
Cedron-samen *m* cedron seed
Cedro-öl *n* citron oil
Ceiba-wolle *f* kapok
Celaster-öl *n* celastrus oil, stafftree oil
Celit *n*, **Celith** *m* (*Cement*) celite
Cellit *n* cellite, cellulose acetate
Cellon-lack *m* (*TN*) cellulose acetate liquor
cellulose-artig *a* cellulose-like, cellulosic. **C–gärung** *f* cellulose fermentation. **c–haltig** *a* containing cellulose. **C–lack** *m* cellulose lacquer. **-lackfarbe** *f* pigmented cellulose lacquer. **-schleim** *m* cellulose (*or* pulp) slime
Celsius-grad *m* degree Celsius (*or* centigrade)
Cer *n* cerium. **Cerat** *n* (-e) cerate
Cerealien *npl* cereals
Cerebron-säure *f* cerebronic acid
Cer-eisen *n* ferrocerium. **-eisenzünder** *m* "flint" lighter. **-erde** *f* cerium earth. **-erz** *n* cerium ore, *esp* cerite
Ceresin *n* paraffin wax
Ceri- cerium, ceric: **-chlorid** *n* ceric chloride
Cerin *n* cerotic acid. **--stein** *m* cerite
Cerit-erde *f* cerite earth
Cer-mischmetall *n* mixed (*or* misch) metal (mixture of rare-earth metals)
Cero- cerium, cerous: **-ion** *n* cerous ion
Cerotin *n* ceryl alcohol. **--säure** *f* cerotic acid
Cer-oxid *n* cerium oxide
Cetan-zahl *f* cetane number
Cetolein-säure *f* cetoleic acid
cet. par. *abbv* (*Lat:* ceteris paribus) other things being equal
Cetyl-säure *f* cetylic (*or* palmitic) acid
Ceylon-zimt *m* Ceylon cinnamon. **-öl** *n* oil of cinnamon
Chagrin *n* shagreen

Chair-leder *n* suede
Chalcedon *m*, **--stein** *m* chalcedony
chalkophil *a* chalcophilic, having a high affinity for sulfur
Chalzedon *m* chalcedony = CHALCEDON
Chamäleon *n* (-s) (*Zool. & Min.*) chameleon. **--lösung** *f* potassium permanganate solution
Chamois-leder *n* chamois (leather)
Chamotte *f* (-n) chamotte *cf* SCHAMOTTE
Champagner *m* champagne. **--weiße** *f* foaming pale beer
Champignon *m* (-s) (edible) mushroom
changeant *a* (*Tex.*) changeable, iridescent, shot: **ch. färben** shot-dye. **Ch--stoff** *m* shot fabric. **changierend** *a* shot, etc = CHANGEANT
Chappe-seide *f* spun silk
chaptalisieren *vt* treat (grape juice) with sugar (by Chaptal's method)
charakterisieren *vt* characterize. **Charakteristik** *f* (-en) characterization. **Charakteristikum** *n* (..ika) characteristic, trait. **charakteristisch** *a* (**für**) characteristic (of)
Charge *f* (-n) 1 (*genl*) load; batch, lot. 2 (*incl Metll., Nucl.*) charge. 3 (*Metll. oft*) heat; blow. **chargen-weise** *a*, *esp adv* by the load (*or* batch), charge by charge. **chargieren** *vt/i* charge, load (a furnace)
Charpie *f* lint
Chassis *n* (-) 1 underframe; (*Cars, Radio*) chassis. 2 (*Synth.*) frame, chase, bolster. 3 (*Tex.*) trough
Chaulmugra-öl *n* chaulmoogra oil. **-säure** *f* chaulmoogric acid
Chaussee *f* (-n) highway, road; avenue. **chaussieren** *vt* macadamize
Chebulin-säure *f* chebulic acid
Chef *m* (-s) chief, head, boss. **--chemiker** *m* chief chemist. **-exekutive** *m* (-n,-n) chief executive (officer)
Chefin *f* (-nen) chief, head, boss
Cheiranthus-säure *f* cheiranthic acid
Chelat-bildner *m* chelating agent
Chelidam-säure *f* chelidamic acid. **Chelidon-säure** *f* chelidonic acid. **chelieren** *vt* chelate
chem. *abbv* (chemisch) chemical
Chemiatrie *f* chemiatry, iatrochemistry

Chemie *f* chemistry. **Ch.-Aktien** *fpl* (*Fin.*) chemical stocks. **Chemie-anlage** *f* chemical plant. **-arbeiter** *m* (unskilled) chemical worker. **-facharbeiter** *m* skilled chemical process worker. **-faden** *m*, **-faser** *f* chemical (man-made, synthetic) fiber. **Ch.-Ingenieur** *m* chemical engineer. **Ch.-Ingenieurwesen** *n* chemical engineering. **Chemie-kupferseide** *f* cuprammonium rayon. **-laborant** *m* chemistry laboratory assistant. **-laborfachwerker** *m* trained chemical laboratory worker. **-spinnfaser** *f* chemical (*or* synthetic) fiber

Chemigraphie *f* (-n) photoengraving (shop). **Chemigraphen-lack** *m* chemigraphic varnish

Chemikalie *f* (-n) chemical (substance) **chemikalien-fest** *a* chemically resistant. **Ch.-handel** *m* trade in chemicals **chemikalisch** *a* chemical

Chemiker *m* (-), **Chemikerin** *f* (-nen) chemist

Chemiker-kolorist *m* color chemist. **-stelle** *f* position as a chemist

Chemilumineszenz *f* chemiluminescence **chemisch** *a* chemical—**ch—e Reinigung** dry cleaning, dry cleaner's; **ch. reinigen** dry-clean

Chemisch-blau *n* chemic blue (indigo extract). **-gelb** *n* patent (*or* Cassel) yellow (lead oxychloride). **-grün** *n* sap green. **ch.-physikalisch** *a* physicochemical. **Ch—rot** *n* Venetian red. **ch.-technisch** *a* chemicotechnical, technochemical. **ch.-technologisch** *a* (of) chemical engineering

chemisieren *vt* chemicalize, (over)treat with chemicals. **Chemisierung** *f* chemicalization, etc

Chemismus *m* (..ismen) chemism, chemical behavior, chemistry (of an organism)

chemo-taktisch *a* chemotactic(al). **Ch-technik** *f* technochemistry, application of technology in chemistry. **-techniker** *m*, **-technikerin** *f* chemically trained independent laboratory technician. **ch—technisch** *a* chemotechnical, technochemical. **Ch—therapie** *f* chemotherapy

Chem-Ztg. *abbv* (Chemiker-Zeitung) Chemists' Journal

Chemurgie *f* (-n) chemurgy

-chen *sfx* (*makes neuter diminutives out of nouns*): **Blättchen** *n* small (*or* little) leaf *cf* BLATT; *cf* TEIL, TEILCHEN

Chester-käse *m* Cheshire cheese

chevillieren *vt* polish

Chevrette *f* chevrette, kidskin

Chibou-harz *n* cachibou

Chiffre *f* (-n), **chiffrieren** *vt/i* code, cipher

chilenisch *a* Chilean

Chile-salpeter *m* Chile saltpeter ($NaNO_3$)

Chin- quin. *cf* CHININ, CHINON, etc

China **I** *n* (*Geog.*) China. **II** *f* (= CHINARINDE) cinchona, Peruvian bark. **--alkaloid** *n* cinchona alkaloid. **-basen** *fpl* cinchona (*or* quinine) bases. **-baum** *m* cinchona tree. **ch—baumartig** *a* cinchonaceous

Chin- quin-: **-acetophenon** *n* quinacetophenone. **-acridin** *n* quinacridine

China-eisenwein *m* (*Pharm.*) bitter wine of iron. **-gerbesäure** *f* quinotannic acid. **-gras** *n* China (*or* cambric) grass, ramie. **-holz** *n* cinchona wood. **-kunde** *f* sinology

Chinaldin *n* quinaldine. **--säure** *f* chinaldic acid

Chin-alizarin *n* quinalizarin. **-amin** *n* quinamine

China-öl *n* balsam of Peru. **-rinde** *f* cinchona (*or* Peruvian) bark. **-rindensäure** *f* quinic acid. **-rot** *n* cinchona red. **ch—sauer** *a* of quinic acid, quinate of. **Ch—säure** *f* quinic (*or* cinchonic) acid. **-seide** *f* China (*or* Chinese) silk. **-silber** *n* China silver (silver-plated nickel)

Chinäthylin *n* quinethyline

China-tinktur *f* tincture of cinchona. **-toxin** *n* cinchona toxin. **-wein** *m* quinine wine. **-wurzel** *f* chinaroot

Chin-azolin *n* quinazoline. **-azolon** *n* quinazolone. **-dolin** *n* quindoline

Chinen *n* quinene

chinesisch *a* Chinese—**ch—es Papier** India paper; **ch—e Tusche** India ink

Chinesisch-grün *n* Chinese green, lokac. **-rot** *n* Chinese red (red mercuric sulfide), vermillion

Chinetum *n* quinetum

Chin-hydrin *n* quinhydrine. **-hydron** *n* quinhydrone

Chinicin *n* quinicine. **Chinid** *n* (-e)

quinide. **Chinidin** *n* quinidine
chinieren *vt* (*Tex.*) cloud, weave (*or* print) chiné
Chinin *n* quinine—**Ch. und Ammoniak** ammoniated quinine
Chinindolin *n* quinindoline
Chinin·eisen *n*, **-eisencitrat** *n* citrate of iron and quinine. **-säure** *f* quininic acid. **-wein** *m* quinine wine
Chinisatin *n* quinisatin
Chinit *n* quinitol, quinite
Chinizarin *n* quinizarin
Chino· quino...
chinoid *a* quinoid(al). **Chinoidin** *n* quinoidine
Chinol *n* quinol
Chinolin *n* quinoline. **-blau** *n* quinoline blue
Chinolinium *n* quinolinium
Chinolin·säure *f* quinolinic acid
Chinolizin *n* quinolizine. **Chinolon** *n* quinolone. **Chinomethan** *n* quinomethane
Chinon *n* (-e) quinone
Chino·pyran *n* quinopyran. **-pyridin** *n* quinopyridine
Chinosol *n* (*Pharm.*) quinosol
Chinotin *n* quinotine
Chino·toxin *n* quinotoxin. **-tropin** *n* quinotropin
Chinova·bitter *n* quinova bitter, quinovin. **-säure** *f* quinovic (*or* chinovic) acid
Chinovin *n* quinovin. **Chinovose** *f* quinovose. **Chinoxalin** *n* quinoxaline
Chinoyl *n* quinoyl
Chinuclidin *n* quinuclidine
Chiretta *f* chirata, chiretta
Chirurg *m* (-en,-en) surgeon. **Chirurgie** *f* (-n) surgery. **chirurgisch** *a* surgical
Chlf. *abbv* chloroform
Chlor *n* chlorine; (*in compds*) chloro·, chloride (of): **Chlorcalcium** calcium chloride
chlor·ähnlich *a* chlorine-like, chlorinous. **Ch·alaun** *m* chloralum. **-alkalien** *npl* alkali-metal chlorides. **-allyl** *n* allyl chloride. **-ammonlauge** *f* ammonium chloride solution. **-arsenik** *n* arsenic chloride. **-arseniklösung** *f* (*Pharm.*) arsenious acid solution, hypochloric solution of arsenic. **-arsinkampfstoff** *m* chlorodiphenylarsine, adamsite. **ch·ar-**

tig *a* chlorine-like, chlorinous
Chlorat *n* (-e) chlorate. **·-ätze** *f* (*Dye.*) chlorate discharge. **-chlor** *n* chlorine in chlorate form
Chlor·äther *m* ethylene chloride. **-äthyl** *n* ethyl chloride, chloroethane. **-äthylen** *n* ethylene chloride. **-äthyliden** *n* ethylidene chloride
Chloration I *f* chlorination. **II** = **Chlorat·ion** *n* chlorate ion
Chlorat·sprengstoff *m* chlorate explosive
Chlor·ätze *f* (*Tex.*) chlorate discharge. **-azetyl** *n* acetyl chloride. **-azetylchlorid** *n* chloroacetyl chloride. **-benzol** *n* chlorobenzene. **-bestimmung** *f* chlorine determination. **-bleiche** *f* chlorine bleach(ing). **-bleichlauge** *f* chlorine bleach liquor, *esp* NaOCl solution. **-bleispat** *m* phosgenite. **-brom** *n* bromine chloride. **-bromsilber** *n* silver chlorobromide, (*Min.*) embolite. **-calciumrohr** *n*, **-calciumröhre** *f* calcium chloride tube. **-chrom** *n* chromium chloride, (*specif*) chromic chloride. **-chromsäure** *f* chlorochromic acid. **-cyan** *n* cyanogen chloride. **ch–echt** *a* chlorine-fast. **Ch·echtheit** *f* fastness to chlorine. **-eisen** *n* iron (*specif* ferric) chloride
Chloreisen·oxid *n* ferric chloride. **-oxydul** *n* ferrous chloride
chloren *vt* chlore, chlorinate; (*Blch.*) gas
Chlor·entwickler *m* chlorine generator. **-entwicklung** *f* evolution (*or* generation) of chlorine. **-entwicklungs·apparat** *m* chlorine generator. **-entzinnung** *f* detinning with chlorine. **-essigsäure** *f* chloroacetic acid. **ch–fest** *a* chlorine-resistant. **-frei** *a* chlorine-free. **Ch–gas** *n* chlorine gas. **-gehalt** *m* chlorine content. **-gold** *n* gold (*specif* auric) chloride. **-goldnatrium** *n* sodium chloroaurate. **ch–haltig** *a* containing chlorine. **Ch–hydrat** *n* hydrochloride; chlorine hydrate. **-hydrin** *n* chlorohydrin
Chlorid *n* (-e) (*usu* -ic) chloride. **·-chlor** *n* chlorine in chloride form. **ch–haltig** *a* containing chloride(s). **Ch–lauge** *f* chloride solution
chlorierbar *a* capable of chlorination
chlorieren *vt* chlorinate; (*Blch.*) chemick; (*Metll.*) chloridize. **Chlorierer** *m* (-)

chlorinator. **Chlorier·turm** *m* chlorinating tower. **Chlorierung** *f* chlorination, chloridizing
chlorierungs·fähig *a* chlorinatable, chloridizable. **Ch—mittel** *n* chlorinating (*or* chloridizing) agent
chlorig *a* chlorous: **ch—e Säure** chlorous acid. **··sauer** *a* of chlorous acid, chlorite of. **Ch—säure** *f* chlorous acid
Chlor·ion *n* chlorine ion
Chlorit *n* (-e) chlorite. **chloritisch** *a* chloritic. **chloritisieren** *vt* chloritize
Chlorit·schiefer *m* chlorite (*or* chloritic) schist. **-spat** *m* spathic (*or* foliated) chlorite
Chlor·jod *n* iodine (mono)chloride. **-kali** *n* 1 potassium hypochlorite. **2** = **-KALIUM**. **-kalilösung** *f* (*Pharm.*) solution of chlorinated potassa, Javelle water. **-kalium** *n* potassium chloride
Chlorkalk° *m* chloride of lime, bleaching powder. **··bad** *n* (*Blch.*) chemic vat. **-bleiche** *f* chemicking, bleaching with bleaching powder. **-flüssigkeit** *f* (*Pharm.*) solution of chlorinated lime, bleaching powder solution. **-kufe** *f* chemic vat. **-lauge** *f* bleaching powder liquor. **-lösung** *f* bleaching powder solution
Chlorkalzium° *n* calcium chloride. **··rohr** *n*, **-röhre** *f* calcium chloride tube
Chlor·kautschuk *m* chlorinated rubber. **-knallgas** *n* chlorine detonating gas (explosive mixture of chlorine and hydrogen)
Chlorkohlen·oxid *n* carbonyl chloride. **-oxidäther** *m* ethyl chloroformate. **-säure** *f* chlorocarbonic (*or* chloroformic) acid, ClCOOH. **-säureamid** *n* carbamyl chloride, H_2NCOCl. **-stoff** *m* carbon chloride: **(vierfacher) Ch.** carbon tetrachloride. **-stoffsäure** *f* chlorocarbonic acid. **-wasserstoff** *m* chlorinated hydrocarbon
Chlor· chloride (of): **-kupfer** *n* copper chloride. **-lauge** *f* chloride of soda = **-NATRON**. **-messer** *m* chlorometer. **-messung** *f* chlorometry. **-metall** *n* metallic chloride. **-metalloid** *n* metalloid chloride. **-methyl** *n* methyl chloride, chloromethane. **-methylat** *n* methochloride. **-methylen** *n* methylene chloride.

-natrium *n* sodium chloride. **-natron** *n* sodium hypochlorite. **-natronlösung** *f* (*Pharm.*) solution of chlorinated soda, Labarroque's solution
Chloro·benzal *n* benzal (*or* benzylidene) chloride. **-benzil** *n* dichlorobenzil, $C_6H_5COCCl_2C_6H_5$
chloroformieren *vt* chloroform
Chloro·gensäure *f* chlorogenic acid. **-goldsäure** *f* chloroauric acid, (*Com. oft*) "gold chloride". **-jodid** *n* chloroiodide, iodochloride. **ch—metrisch** *a* chlorometric. **Ch—pikrin** *n* chloropicrine, trichloronitromethane. **-platinsäure** *f* chloroplatinic acid
Chlorose *f* (*Med.*) chlorosis
Chlor· chlorine, chloro-, chloride (of): **-oxid** *n* chlorine oxide. **-phosphor** *m* phosphorus chloride. **-phosphorstickstoff** *m* phosphorus nitride dichloride. **-phosphoryl** *n* phosphorus oxychloride. **-pikrin** *n* chloropicrin. **-platin** *n* platinic chloride. **-platinsäure** *f* chloroplatinic acid. **-quecksilber** *n* mercury (*or* mercuric) chloride. **-räucherung** *f* chlorine fumigation. **ch—sauer** *a* chlorate of. **Ch—säure** *f* chloric acid. **-säureanhydrid** *n* chloric anhydride, chlorine pentoxide. **-schwefel** *m* sulfur monochloride. **-silber** *n* silver chloride. **-silberspat** *n* cerargyrite. **-silizium** *n* silicon tetrachloride. **-soda** *f* sodium hypochlorite. **-stickstoff** *m* nitrogen chloride. **-strom** *m* stream of chlorine. **-sulfonsäure** *f* chlorosulfonic acid, chlorosulfuric acid. **-toluol** *n* chlorotoluene. **-übertrager** *m* chlorine carrier
Chlorung *f* (-en) chlorination
Chlorür *n* (-ous) chloride: **Eisenchlorür** ferrous chloride
Chlor·vinyl *n* vinyl chloride
Chlorwasserstoff *m* hydrogen chloride; hydrochloric acid. **··äther** *m* ethyl chloride. **ch—sauer** *a* chloride (*or* hydrochloride) of. **Ch—säure** *f* hydrochloric acid
Chlor·wismut *n* bismuth trichloride
Chlorylen *n* trichloroethylene
Chlor· chloride (of): **-zink** *n* zinc chloride. **-zinn** *n* tin chloride. **-zinnbad** *n* tin chloride bath. **-zyan** *n* cyanogen chloride

Chm. *abbv* (Chemie) chem. (chemistry)
Cholämie *f* (*Med.*) cholemia
Cholan·säure *f* cholanic acid
Cholecampher·säure *f* colecamphoric acid (choloidanic acid)
Cholein·säure *f* choleic acid, choleinic acid
Cholesterin *n* (-e) cholesterol, cholesterin.
·-säure *f* cholesteric acid, cholesterol.
-spiegel *m* cholesterol level
Chol(in)·säure *f* cholic acid
Chondronin·säure *f* chondroninic acid.
Chondron·säure *f* chondronic acid
Chor *m* (-̈e) chorus, choir
Chordale *f* (*adj endgs*) radial axis
Chordaten *pl* (*Zool.*) chordata, chordates
Chorde *f* (-n) (*Anat.*) (spinal) cord; string, chord; sinew, tendon
Chordonier *mpl* (*Anat.*) chordata
Christ·dorn *m* holly, ilex
Christofle *n* (*TN*) nickel silver
Christophs·kraut *n,* **-wurz** *f* baneberry
Christ·palmöl *n* castor oil. **-wurz** *f* (*Bot.*) black hellebore; spring pheasant's eye
Chrom *n* chromium; chrome. **·-alaun** *m* chrome alum
Chromat *n* (-e): **saures Ch.** dichromate.
·-ätze *f* (*Dye.*) chromate discharge. **-gelb** *n* chrome yellow. **-grün** *n* chrome green.
chromatographieren *vt* chromatograph, subject to chromatographic adsorption
Chrom·azetat *n* chromium acetate. **-bad** *n* chrome bath. **-beize** *f* chrome mordant
chrombeizen *vt* chrome-mordant
chrom·beständig *a* chrome-resistant. **Ch·blei** *n* lead chromate. **-bleispat** *m* (*Min.*) crocoite. **-braun** *n* chrome brown.
-chlorid *n* chromic chloride. **-chromit** *n* (*Min.*) kaemmerite. **-chlorür** *n* chromous chloride. **-druck** *m* (*Tex.*) chrome printing. **ch–echt** *a* (*Dye.*) chrome-resistant (sodium or potassium chromate). **Ch–echtschwarz** *n* fast chrome black.
-eisen(erz) *n,* **-eisenstein** *m* chromite, chrome iron ore. **ch–enthaltend** *a* containing chromium. **Ch–entwicklungsfarbstoff** *m* chrome-developed dye. **-erz** *n* chromium ore. **-farbe** *f* chrome color.
-fluorid *n* chromic fluoride. **ch–gar** *a* chrome-tanned. **Ch–gehalt** *m* chromium content. **-gelatine** *f* chromatized (*or* bichromated) gelatin. **-gelb** *n* 1 chrome

yellow (lead chromate). 2 Cologne yellow (lead chromate and sulfate). **-gerberei, -gerbung** *f* chrome tanning. **-glimmer** *m* chrome mica. **-granat** *m* chrome red.
-grün *n* chrome green. **-guß** *m* cast chrome steel. **ch–haltig** *a* containing chromium, chromiferous. **Ch–hydrat, -hydroxid** *n* chromic hydroxide. **-hydroxydul** *n* chromous hydroxide
Chromi· chromic: **-chlorid** *n* chromic chloride. **-cyanwasserstoffsäure** *f* chromicyanic acid
Chromiak *n* chrome-ammine
Chromier·artikel *m* (*Tex.*) chrome style
chromierbar *a* chromable. **chromieren** *vt* chrome
Chromier·farbstoff *m,* **Chromierungsfarbe** *f* chrome dye
Chromi· chromic: **-hydroxid** *n* chromic hydroxide. **-oxid** *n* chromic oxide. **-rhodanwasserstoffsäure, -sulfocyansäure** *f* chromithiocyanic acid
Chrom·kali *n:* **rotes Ch:** potassium dichromate; **gelbes Ch.** potassium chromate.
-kaliumalaun *m* potassium chrome alum. **-karbid** *n* chromium carbide.
-lack *m* chrome lake. **-leder** *n* chrome leather. **-leim** *m* chrome gelatin (*or* glue).
-magnesit *n* chrome magnesite. **-metall** *n* chrome metal. **-natron** *n:* **rotes Ch.** sodium dichromate; **gelbes Ch.** sodium chromate. **-nickelstahl** *m* chrome-nickel steel
Chromo· chromous: **-chlorid** *n* chromous chloride
Chrom·ocker *m* chrome ocher
Chromo·hydroxid *n* chromous hydroxide.
-ion *n* chromous ion
Chromolyse *f* (-n) chrom(at)olysis
Chromo·oxid *n* chromous oxide
chromophor *a* chromophoric, chromophorous
Chromosom *n* (-en) chromosome
Chromotrop·säure *f* chromotropic acid
Chromo·verbindung *f* chromous compound
Chromoxid° *n* chromium (*or specif* chromic) oxide. **·-farbe** *f* chromic oxide color. **-grün** *n* chromic oxide green.
-hydrat *n* chromic hydroxide. **-natron** *n* sodium chromite. **-salz** *n* chromic salt

Chrom·oxychlorid *n* chromium oxychloride, chromyl chloride. **-oxydul** *n* chromous oxide

Chromoxydul·hydrat *n* chromous hydroxide. **-salz** *n* chromous salt. **-verbindung** *f* chromous compound

Chrom·rindleder *n* chromated neat's leather. **-rot** *n* chrome red. **-salpetersäure** *f* chromonitric acid. **-salz** *n* chromium salt, chromate. **ch–sauer** *a* chromate of: **ch—es Kali** potassium chromate; **doppeltchromsaures Kali, rotes ch—es Kali** potassium dichromate. **Ch–säure** *f* chromic acid

Chromsäure·anhydrid *n* chromic anhydride. **-salz** *n* salt of chromic acid, chromate

Chrom·schwarz *n* (*Dye.*) chrome black. **-schwefelsäure** *f* chromosulfuric acid. **silber** *n* silver chromate. **-stahl** *m* chrome steel. **-stein** *m* chromium brick. **-sud** *m* chrome mordant. **-sulfat** *n* chromium sulfate. **-sulfozyanid** *n* chromium thiocyanate. **-sulfür** *n* chromous sulfide. **-trioxid** *n* chromium trioxide. **-überzug** *m* chromium coating (*or* plating). **-wolframstahl** *m* chrome-tungsten steel. **-ziegel** *m* chrome brick. **-zyanid** *n* chromium cyanide

Chronik *f* (-en) chronicle. **chronisch** *a* chronic

Chrysalide *f* (-n) chrysalis

Chrysatropa·säure *f* chrysatropic acid

Chrysen·chinon *n* chrysenequinone

Chrysochinon *n* chrysoquinone. **Chrysokoll** *n* chrysocholla. **Chrysolith** *m* (*Min.*) chrysolite

Chrysophan·säure *f* chrysophanic acid

chrysotilitisch *a* (*Asbestos*) chrysotile

Chylus *m* chyle

Chymus *m* chyme

Ciba·blau *n* ciba blue. **-rot** *n* ciba red

Cibebe *f* (-n) cubeb, large raisin

Cichorie *f* (-n) chicory

Cider·branntwein *m* cider (*or* apple) brandy. **-essig** *m* cider vinegar. **-wein** *m* cider wine, fermented cider

Cie. *abbv* (Compagnie) Co. (company)

cimolisch *a* Cimolian: **c—e Erde** *f* Cimolean earth, cimolite

Cinchonin·säure *f* cinchoninic acid. **Cinchon·säure** *f* cinchonic acid

Cineol·säure *f* cineolic acid

cinnamal· *pfx* cinnamylidene

Cinnabarit *n* cinnabar, mercury ore

circa, cirka *adv* about, approximately

cirkulieren *vi* circulate

Cis·körper *m* cis compound. **-stellung** *f* cis position

Cistus·öl *n* cistus (*or* labdanum) oil

Citat *n* (-e) quotation = ZITAT. **citieren** *vt/i* cite, quote = ZITIEREN

Citrat *n* (-e) citrate = ZITRAT

Citrazin·säure *f* citrazinic acid

claircieren *vt* (*Sugar*) clear, clarify; purge

Claisen·kolben *m* Claisen flask (for vacuum distillation)

Cloake *f* (-n) cloaca; sewer, cesspool

Cloison *f* (-s) partition, septum

Cnidium·säure *f* cnidic acid

Cobyrin·säure *f* cobyric acid

Coca·blätter *npl* coca leaves. **-säure** *f* cocaic acid, alpha-truxillic acid. **-wein** *m* (*Pharm.*) wine of coca

Coccin·säure *f* coccinic acid

Cochenille *f* (-n) cochineal

Cochenillen·farbstoff *m* cochineal dye. **-schildlaus** *f* cochineal insect

Cochenille·säure *f* cochenillic acid

Cocon *m* (-s) cocoon

Cocos *f* (-) coconut = KOKOS

Cod·öl *n* cod-liver oil

Coffearin *n* caffearine

Coffein *n* caffeine

Coffolin *n* caffoline (1,3,6-trimethylallantoin)

Coffur·säure *f* coffuric acid

Cognak *m* (-e) cognac. **-öl** *n* oil of cognac

cohobieren *vt* cohobate

cokondensieren *vt* copolymerize

Colamin *n* ethanolamine

Cola·nuß *f* cola nut

Colatur *f* (-en) filtrate

Cölestin *m* (-e) (*Min.*) celestite, celestine

Coli·bakterium *n* bacterium coli

Coliform·gehalt *m* coliform count

Collatol·säure *f* collatolic acid

Collorescin *n* (*TN*) methyl cellulose

Colombo·wurzel *f* calumba (root)

Colorister *m* (-) dyer

color·tüchtig *a* capable of having color

Columba·säure *f* columbic acid (from ca-lumba)

Columb·eisen *n* (*Min.*) columbite

Columbia·kopalinsäure *f* columbiacopalinic acid. **-kopalsäure** *f* columbiacopalic acid

Columbo·wurzel *f* calumba (root)

Colza·öl *n* colza oil

Compakt·platte *f* compact disc

computer·gesteuert *a* computer-controlled. **computerisieren** *vt/i* computerize. **computer·unterstützt** *a* computer-assisted

conaxial *a* coaxial

conchieren *vt* shell (out)

Condurit *n* conduritol

conglobieren *vt* heap up

Congo·farbe *f* Congo color (*or* dye)

Conidien· (*Bot.*) conidial

Conifern·harz *n* fir resin

Coniin *n* conine, coniine

Conima·harz *n* conima (resin)

conphas *a* in phase

Conspersa·säure *f* conspersic acid

Constitual·kampf *m* environmental struggle

Continü·küpe *f* (*Dye.*) continuous vat

Copaiva·balsam *m* copaiba (balsam). **-öl** *n* oil of copaiba. **-säure** *f* copaivic acid

Copoly·addukt *n* addition copolymer. **-harnstoff** *m* urea copolymer. **-kondensat** *n* copolymer. **-kondensation** *f* condensation copolymerization

copolymer *a* copolymeric. **Copolymerisat** *n* (-e) copolymerized product

Cops·färberei *f*, **-färbung** *f* cop dyeing

Corduan·leder *n* cordovan (leather)

Cornicular·säure *f* cornicularic acid

Corozo·nuß *f* ivory nut

corr. *abbv* (corrigiert) corrected; proofread

Cörulignon *n* cerulignone

Cosekante *f* (-n) cosecant. **Cosinus** *m* (..nen) cosine

Costus·säure *f* costic (*or* costusic) acid

Cotangente *f* (-n) cotangent

Cotarn·säure *f* cotarnic acid

Coto·rinde *f* coto bark

Cotta *f* (*pl* Cotten) (*Metll.*) bloom, lump, coke

cottonisieren *vt* cottonize

Cotton·öl *n* cottonseed oil

Couch·tisch *m* coffee table

Couleur *f* (-en) 1 color. 2 caramel; burnt sugar; dark coarse smalt

Coulomb·feld *n* coulombic field

coupieren *vt* cut (short), etc = KUPIEREN

Coupüre *f* (-n) (*Tex.*) reduced print; reduction of print paste

couragiert *a* courageous

Covellin *n* (*Min.*) covellite

C-Polymerisation *f* condensation polymerization

crabb·echt *a* (*Tex.*) crabbing-resistant

Crack·benzin *n* cracked gasoline

cracken *vt* (*Petrol.*) crack. **Cracken** *n* I cracking. II crackene. **Cracken·chinon** *n* crackenequinone

Crack·kessel *m* cracking retort. **-prozeß** *m* cracking process

Crackung *f* (-en) cracking (of petroleum). **Crack·verfahren** *n* cracking process

Craig·verteilung *f* countercurrent distribution

creme *a* cream(-colored) = KREM. **Creme** *f* (-s) cream = KREM

Crêpe·kautschuk *m/n* crepe rubber

Cresyl·säure *f* cresylic acid, cresol

Cribia·teil *m* phloem element

Crocein·säure *f* croceic acid

Croton·alkohol *m* crotyl alcohol. **-harz** *n* croton resin. **-säure** *f* crotonic acid

Croupon *m* (-s) (*Lthr.*) crop, butt. **crouponieren** *vt* crop, round (hides)

C-Stoff *m* rocket fuel

Cubeben I *n* cubebene. II cubebs: *pl of* **Cubebe** *f*

Cubeben·öl *n* cubeb oil. **-pfeffer** *m* cubebs. **-säure** *f* cubebic acid

Cuite·seide *f* boiled-off silk

Cumalin *n* coumalin. **--säure** *f* coumalic acid

Cumar·aldehyd *m* coumaraldehyde (o-hydroxycinnamaldehyde)

Cumaril·säure *f* coumarilic acid

Cumarin·säure *f* coumarinic acid

Cumaron·harz *n* coumarone resin

Cumar·säure *f* coumaric acid

Cumin·öl *n* cumin (seed) oil. **-samen** *m* cumin seed. **-säure** *f* cum(in)ic acid

Cumo·chinol *n* cumoquinol. **-chinon** *n* cumoquinone

Cumol *n* (-e) cumene

Cuoxam *n* cuprammonium

Cupri· cupric: **-ammoniumsulfat** *n* cuprammonium (*or* tetra-ammine cupric) sulfate. **-azetat** *n* cupric acetate. **-verbindung** *f* cupric compound. **-weinsäure** *f* cupritartaric acid

Cupro· cuprous: **-asbolan** *n* (*Min.*) cuproasbolite. **-chlorid** *n* cuprous chloride. **-cyanür** *n* cuprous cyanide, cuprocyanide. **-faden** *m* cuprammonium rayon. **-faser** *f* cuprammonium rayon staple. **-mangan** *n* cupromanganese alloy. **-uranit** *n* (*Min.*) torbernite. **-verbindung** *f* cuprous compound

Cuproxam *n* cuprammonium

Curcuma *n* turmeric = KURKUMA

Cusk·hygrin *n* cuscohygrine

Cusko·rinde *f* cusco bark

Cutinin·säure *f* cutinic acid. **Cutin·säure** *f* cutic acid

Cüvette *f* (-n) cuvette, etc = KÜVETTE

Cyan *n* cyanogen; (*in compds of*) cyanic, cyano-, cyanide; **-alkali** *n* alkali cyanide. **-alkyl** *n* alkyl cyanide

Cyanamid·calcium *n* calcium cyanamide. **-natrium** *n* sodium cyanamide

Cyanamido·dikohlensäure *f* cyanamido-dicarboxylic acid. **-kohlensäure** *f* cyanamidocarboxylic (*or* cyanocarbamic) acid

Cyan·ammonium *n* ammonium cyanide. **-äther** *m* cyanic ester. **-äthyl** *n* ethyl cyanide. **-bad** *n* cyanide bath. **-barium** *n* barium cyanide. **-benzol** *n* cyano-benzene. **-bromid** *n* cyanogen bromide. **-calcium** *n* calcium cyanide. **-chlorid** *n* cyanogen chloride. **-doppelsalz** *n* double cyanide salt. **-eisen** *n* iron cyanide

Cyaneisen·kalium *n* potassium ferro-cyanide. **-verbindung** *f* iron-cyanogen compound, (*specif*) ferrocyanide

Cyan·essigsäure *f* cyanoacetic acid. **-gas** *n* cyanogen gas. **-gold** *n* gold cyanide. **-goldkalium** *n* potassium auricyanide (*or* aurocyanide). **c–haltig** *a* containing cyanogen (*or* a cyano group). **C–härtung** *f* (*Metll.*) cyanide hardening, cyaniding

Cyanid *n* (-ic) cyanide; *cf* CYANÜR. **c—frei** *a* cyanide-free. **cyanidieren** *vt* (*Metll.*)

cyanide. **Cyanid·laugerei** *f* (*Metll.*) cyanide (extraction) process

Cyanin·farbe *f* cyanin dye. **cyanisieren** *vt* cyanize. **Cyanit** *n* (-e) (*Min.*) kyanite

Cyan·jodid *n* cyanogen iodide. **-kali** *n* potassium cyanide. **c–kalisch** *a* (of) potassium cyanide. **C–kalium** *n* potassium cyanide. **-kobalt** *m* cobalt cyanide. **-kohlensäure** *f* cyanocarbonic acid. **-kupfer** *n* copper cyanide. **-laugerei, -laugung** *f* cyanidation, cyaniding. **-lösung** *f* cyanide solution. **-metall** *n* metallic cyanide. **-methyl** *n* methyl cyanide. **-natrium** *n* sodium cyanide. **-platin** *n* platinum cyanide. **-quecksilber** *n* mercuric cyanide

Cyano·eisensäure *f* hydrogen hexacyano-ferrate. **-ferrat** *n* (hexa)cyanoferrate, ferrocyanide

Cyanquecksilber·kalium *n* mercuric po-tassium cyanide. **-oxid** *n* mercuric cyanide. **-salz** *n* mercuric cyanide (salt)

Cyan·salz *n* cyanide, cyanogen salt. **-salz-badhärtung** *f* (*Metll.*) cyanide harden-ing, cyaniding. **c–sauer** *a* cyanate of. **C–säure** *f* cyanic acid; cyano acid. **-schlamm** *m* cyanide sludge. **-senföl** *n* cyanomustard oil. **-silber** *n* silver cyanide. **-stickstoff** *m* cyanonitride. **-stickstofftitan** *n* titanium nitride cyanide. **-toluol** *n* cyanotoluene

Cyanür *n* (-ous) cyanide

Cyanur· cyanuric, cyanuryl: **-chlorid** *n* cyanuric chloride. **-säure** *f* cyanuric acid

Cyan·verbindung *f* cyanogen compound, cyanide. **-vergiftung** *f* cyanide poison-ing. **-wasserstoff** *m* hydrogen cyanide; hydrocyanic acid

cyanwasserstoff·sauer *a* cyanide of. **C–säure** *f* hydrocyanic (*or* prussic) acid

Cyan·zink *n* zinc cyanide

Cyclisieren *n*, **Cyclisierung** *f* (-en) cycliz-ation. **cyklisch** *a* cyclical

Cyclit *n* (-e) cyclitol, polyhydroxycyclohex-ane

Cyklo·geraniumsäure *f* cyclogeranic acid. **-kautschuk** *m*/*n* cyclized rubber

Cylinder *m* (-) = ZYLINDER

Cymol *n* cymene

Cypern *n* (*Geog.*) Cyprus. **-holz** *n* wood of the Spanish elm

Cypresse *f* (-n), **Cypressen-baum** *m* cypress (tree). **-holz** *n* cypress (wood). **-nuß** *f* cypress cone. **-öl** *n* cypress oil

cyprisch *a* Cyprian, Cypriot; **c—es Vitriol** blue vitriol, copper sulfate

Cyste *f* (-n) (*Med.*) cyst

cysten-artig *a* cyst-like, cystic. **C—flüssig-keit** *f* cystic fluid. **c—förmig** *a* cysti-form

Cytidyl-säure *f* cytidylic acid

Cystinuriker *m* (-) victim of cystinuria

C.Z. *abbv* 1 (Cellulosezahl) cellulose number. 2 (Chemisches Zentralblatt) German Chemical Abstracts (Journal)

D

d. *abbv* **1** (der, die, das, etc) the, of the, to the, etc. **2** (dieser, etc) this, these, of this, etc

D. *abbv* **1** (Dichte) density. **2** (Deutsch...) Germ. (German). **3** (Durchmesser) diam.

da I *adv* **1** there, that's where: **da oben (unten, drüben)** up (down, over) there; **hier und da** here and there, now and then *cf* **4**; **da und da** in such and such a place. **2** here, present. **3** in that case, on that occasion. **II** *adv & adv cnjc* **4** then, and then, so then, at that point, that's when. **III** *adv cnjc* **5** (*esp after* **gerade** *or* **kaum** *in preceding clause*) (just, hardly...) when, that. **IV** *subord cnjc* **6** since, inasmuch as, considering that. **7** (*esp after time expr*) when. **V** *sep pfx* here, present; there; with one(self) *cf* DA-SEIN. **VI** *pfx* (*on prep*) prep + it, that, them *cf* DAFÜR, DAGEGEN etc., DAR II

D.A. *abbv* (Deutsches Arzneibuch) German Pharmacopoeia. **D.A.-B.** *abbv* (Deutsches Apothekerbuch) German Dispensatory

dabei (*meanings derived from* BEI *prep*) **I** *adv* **1** next to it (that, them) *cf* DA **VI;** nearby. **2** in doing so, in that connection, in this matter. **3: d. sein** be in on it, be involved; be present; be enclosed (*or* attached); see also **7. 4: nichts d.** nothing to it. **5** at the same time. **6** also, besides. **7: d. sein** (+ **zu** + *inf*) be (in the process of) ...ing. **II** *adv cnjc* and yet. **III** *sep pfx* there, present; by: **dabei·bleiben*** *vi*(s) **1** stick to it. **2** remain unchanged—**es bleibt d.** that is final, that is how matters stand. **d—·stehen*** *vi* stand by. *Cf* WOBEI

da·bleiben* *vi*(s) stay (there *or* here)

Dach *n* (⁼er) roof; dome (of a boiler)—**unter D. und Fach** safe and sound; **unter D. und Fach bringen** get (sthg) under shelter, get (sthg) finished. **·-bau** *m,* **-be-deckung** *f,* **-belag** *m* roofing. **-blech** *n* sheet metal for roofing. **-boden** *m* attic, garret, loft. **-decker** *m* roofer. **-deckerei, -deckung** *f* roofing

Dachel, Dächel *n* (-) (*Metll.*) lump, bloom

Dach·fenster *n* skylight; dormer window. **-filz** *m* roofing felt. **-first** *m* roof ridge. **-gesellschaft** *f* holding company. **-gestein** *n* (*Min.*) overlying rock

dachig *a* imbricate

Dach·kammer *f* attic, garret. **-kohle** *f* upper coal. **-organisation** *f* umbrella (*or* parent) organization. **-pappe** *f* roofing paper (*or* composition), tar paper. **-pfanne** *f* pantile

Dachs *m* (-e) badger

Dach·schiefer *m* roof slate. **-schindel** *f* (roof) shingle

Dächsel I *m* (-) dachshund. **II** *f* (-n) adze

Dachs·fett *n* badger fat. **-hund** *m* badger hound, dachshund

Dachstein *m* **1** roof tile (*or* slate). **2** (*Geol.*) bituminous shale. **3** (*Min.*) roof rock. **4** (*Geog.*) Dachstein (*mtn. in Austria*)

dachte 1 thought: *past of* DENKEN. **2** roofed: *past of* DACHEN

Dachung *f* (-en) roofing

Dach·wurz *f* common houseleek. **-ziegel** *m* roof(ing) tile. **-ziegelei** *f* tilery, tile kiln

dad. gek. *abbv* (dadurch gekennzeichnet) characterized by (the fact...)

dadurch *adv* (*meanings derived from* DURCH *prep*) **1** through there, through it (that, them). **2** thereby, by that means, by (doing) that. **3: d. ...** , **daß** by ...ing, by virtue of the fact that

dafür I *adv* (*meanings derived from* FÜR *prep*) **1** for it (that, them), therefor; in favor of it (that, them) *cf* DA **VI. 2** instead (of that), to make up for that. **3: d. ...** , **daß** for ...ing; to make up for the fact that. **II** *adv cnjc* (*usu* + **ja, doch**) after all. *Cf* DAFÜRKÖNNEN

Dafürhalten *n* opinion

dafür·können* *vt:* **sie können nichts dafür** they can't help it, they are not to blame for it

dagegen I *adv* (*meanings derived from* GEGEN *prep*) *cf* DA **VI: 1** against it (that, them). **2** compared to it (that, them), in comparison to. **3** in exchange (*or* return

168

for it (that, them). **II** *adv cnjc* on the other hand, however. **III** *sep pfx* (*implying meanings of* I 1,2): **dagegen·halten*** *vt* hold up to (*or* up against) it, compare; **d— ·stellen: sich d.** oppose, resist (it)

dagewesen 1 existed, etc: *pp of* DASEIN. 2 *p.a:* **noch nie d.** unprecedented

daheim *adv* at home

daher I *adv* from there, that's where ... from—**es kommt** (*or* **rührt**) **d., daß** it is due to the fact that *cf* HERRÜHREN. **II** *adv cnjc* therefore, hence, that's why. **III** *sep pfx* (*on verbs of motion, conveying*) here, to this place; (*usu as pp* + **kommen**) e.g. **dahergerannt kommen** come running up

daherum *adv* around here, (t)hereabouts

dahin I *adv* 1 there, to that place, in that direction. 2: **bis d.** up to here, up to there; till then, by then. 3 away, gone; dead, defunct. **II** *adv* (+ *certain verbs*) 4 (+ **gehören**) belong there, be one of them. 5 (+ *motion, conveying*) to that point, so far; (+ *daß*) to the point where. 6 (+ **gehen**): **es geht d., daß** (*or* **zu** + *inf*) it comes down to the fact that (or to ...ing), it aims at ...ing *cf* DAHINGEHEND. 7 (+ *thinking, interpreting*) to that effect; (+ *daß*) to the effect that. **III** *sep pfx* 8 (*in any of the uses under* II). 9 (*Motion*) along, on, by. 10 (*implies dying*) *cf* DAHINSCHEIDEN

dahin·gehen* *vi*(s) 1 go (*or* walk) along. 2 pass (on). 3 pass away (= die).

dahingehend *p.a* (*usu as adv*) to that effect; (+ *daß*) to the effect that

dahingestellt *p.a* undecided, uncertain: **d. bleiben** be left undecided, be a moot point

dahin·scheiden* *vi*(s) pass away (= die)

dahin·stehen* *vi* remain undecided

dahinten *adv* back there, in back

dahinter *adv & sep pfx* behind it (that, them); after it; aware of it *cf* DA **VI**, HINTER. **d—·kommen*** *vi*(s) discover the secret

damalig *a* former, of that time, of those days. **damals** *adv* then, at that time

damassé *a* (*no endgs*) damask(ed)

Damast *m* (-e) damask. **--stahl** *m* damask (*or* Damascus) steel. **-stoff** *m* (*Tex.*) damask

Damaszener *m* (-), *a* (*no endgs*) Damascene, (of) Damascus; damask.

damaszieren *vt* damask, damascene

Dambonit *n* dambonitol

Dame *f* (-n) 1 lady. 2 (*Cards, Chess*) queen. 3 (*Checkers*) king. 4 (game of) checkers

Damen·binde *f* sanitary napkin. **-brett** *n* checkerboard. **-spiel** *n* (game of) checkers

Dam·hirsch *m* fallow deer. **-hirschfell** *n* buckskin

damit I *adv* (*meanings derived from* MIT *prep*) 1 with it (that, them), therewith *cf* DA **VI.** 2 by that, thereby; (+ *daß, oft*) by ...ing. **II** *adv cnjc* thus, so, therefore. **III** *subord cnjc* so that, in order that

Damm *m* (-̈e) 1 dam. 2 dike, embankment. 3 pier, jetty, breakwater. 4 roadway. 5 (*Anat.*) perineum

Dammar·firnis *m* dammar varnish. **-harz** *n* dammar resin

Dämmaterial° *n* (= **Dämm·material**) insulating material

Dämmen *vt* 1 dam (up), hold back, curb. 2 dampen

Damm·erde *f* 1 mold, humus. 2 foundry sand

dämmern *vi* 1 drowse. 2 (*imps*) dawn; get dark. **Dämmerung** *f* (-en) half-light, twilight; dawn; dusk. **Dämmer·zustand** *m* semiconscious state

Damm·grube *f* foundry pit. **-platte** *f* acoustical tile. **-rutsch** *m* landslide. **-stein** *m* damstone (of a furnace)

Dämm·tafel *f* insulating tile. **Dämmung** *f* (-en) 1 damming (up), etc *cf* DÄMMEN. 2 insulation

Dampf *m* (-̈e) (water) vapor, steam; mist, fog; fume, smoke: **direkter D.** live steam; **indirekter D.** exhaust steam; **stickender D.** choke damp. **--antrieb** *m* steam drive

Dämpf·apparat *m* steaming apparatus

dampf·artig *a* vaporous. **D—artikel** *m* (*Tex.*) steam style. **-auflösung** *f* volatilization. **-bad** *n* steam bath; vapor bath. **-behälter** *m* steam vessel. **-betrieb** *m* steam operation (*or* drive); steam-driven plant (*or* unit). **-blase** *f* 1 bubble of vapor (*or* steam). 2 steam-heated still. **-blasen** *n* steam injection; distilling with steam.

-blasenbildung *f* steam bubble formation; (*Mach.*) vapor lock. -bleiche *f* steam bleaching. -bleicherei *f* steam bleaching; steam bleachery. -bügeleisen *n* steam iron. -chlor, -chloren *n* steam chemicking. -darre *f* steam kiln. d–dicht *a* steamtight, vaportight. D–dichte, -dichtigkeit *f* vapor density, density of steam. -dichtebestimmung *f* vapor density determination. -dichtung *f* steam packing. -druck *m* 1 vapor pressure; steam pressure. 2 steam printing. -drukkerei *f* steam (color) printing (plant) Dampfdruck-erniedrigung *f* vapor-pressure lowering. -messer *m* manometer; (steam) pressure gauge. -regler *m* steam pressure regulator. -zünder *m* vapor pressure igniter

Dampf-düse *f* steam nozzle. d–echt *a* steam-resistant. -empfindlich *a* susceptible to steaming

dampfen 1 *vt* smoke, puff (cigar, etc). 2 *vi* (h) steam, give off steam; smoke, fume. 3 *vi* (s) steam, go puffing (to a place)

dämpfen *vt* 1 steam (food, clothes, wood, etc); braise; stew. 2 damp(en), muffle, subdue, tone down; smother, put out (fire). 3 damp down, bank (blast furnace). 4 evaporate (fruit). 5 ease (pain)

Dampf-entöler *m* oil extractor. -entwässerer *m* steam desiccator (*or* drier). -entwickler *m* steam (*or* vapor) generator. -entwicklung *f* evolution of steam; steam generation. -entzug *m* (*Distil.*) steam stripping

Dampfer *m* (-) steamer, steamboat, steamship

Dämpfer *m* (-) 1 steamer, steam cooker. 2 damper, mute, muffle. 3 (*Radio*) baffle. 4 shock absorber

Dampf-erzeuger *m* steam generator, boiler. -erzeugung *f* steam generation; production of vapor. -farbe *f* (*Dye.*) steam color. -färberei *f* steam printing. -faß *n* steam drum

dampff. *abbv* (dampfförmig) vaporous, etc

dampf-förmig *a* vaporous; in the form of steam. D–gas *n* dry (*or* superheated) steam. -gebläse *n* steam blast; steam blower. -gummi *n* dextrin. -hahn *m*

steam cock. d–haltig *a* containing vapor (*or* steam); humid, moist. D–härtung *f* steam hardening (*or* curing). -heizrohr *n* steam (heating) pipe. -heizung *f* steam heat. -holzschliff *m* steamed mechanical wood pulp. -hülle *f* steam jacket; vaporous envelope

dampfig *a* vapory, steamy

Dampf-kanal *m* steam pipe; steam port. -kasten *m* steam chest; steaming chamber. -kessel *m* steam boiler

Dampfkessel-anlage *f* steam boiler unit; steam plant. -bekleidung *f* boiler casing. -blech *n* boiler plate. -kohle *f* steam (boiler) coal. -speisung *f* (steam) boiler feed

Dampfkoch-apparat *m*, -kessel *m*, -topf *m* steam cooker; sterilizer; pressure cooker

Dampf-kochung *f* steam cooking. -kolben *m* steam piston. -kraft *f* steam power. -kraftwerk *n* steam power plant. -kühlung *f* vapor cooling. -leitung *f* steam piping, steam pipe; vapor (*or* steam) conduction. -leitungsrohr *n* steam pipe. -mantel *m* steam jacket. -maschine *f* steam engine. -messer *m* steam gauge; manometer

Dämpf-mittel *n* neutralizer (for acids)

Dampf-mühle *f* steam mill. -nudeln *fpl* (sweet yeast) dumplings. -ofen *m* steam oven (*or* furnace). -orgel *f* calliope. -pfeife *f* steam whistle. -phase *f* vapor phase. -probe *f* steam test. -rohr *n*, röhre *f* steam pipe. -rohrleitung *f* steam piping. -schale *f* evaporating dish. -scheider *m* vapor separator (to catch liquid carried up in evaporation). -schiff *n* steamboat. -schiffahrt *f* steam(ship) navigation. -schlange *f* steam coil. -schmierung *f* steam lubrication. -schwarz *n* (*Dye.*) steam black. -schwelung *f* destructive distillation with steam. -spannung *f* vapor (*or* steam) tension. -spülung *f* (*Water gas*) steam purge. -strahl *m* steam jet. -strom *m* current of steam (*or* vapor). -topf *m* autoclave; steam digester (*or* sterilizer); pressure cooker. -trichter *m* steam funnel. -trockenapparat *m* steam drier. -trockenofen, -trockenschrank *m* steam drying oven. **-trockner** *m* steam drier. **-turbine** *f* steam turbine. **-überdruck** *m* excess vapor (*or* steam) pressure. **-überhitzer** *m* steam superheater

Dämpfung *f* (-en) steaming; braising; stewing; damp(en)ing, muffling, subduing; smothering, extinguishing; banking; evaporation *cf* DÄMPFEN

Dämpfungs-klappe *f* damper. **-vorrichtung** *f* damping device. **-zylinder** *m* dash pot

Dampf-ventil *n* steam valve. **-verbrauch** *m* steam consumption. **-vulkanisation** *f* (*Rubber*) steam cure. **-walze** *f* steamroller. **-wärme** *f* heat of vaporization; vaporizing temperature. **-wäscherei** *f* steam laundry. **-wassertopf** *m* steam trap. **-weg** *m* steam pipe (*or* port). **-zucker** *m* steam-refined sugar. **-zuführung, -zuleitung** *f* steam supply

danach *adv* (*meanings derived from* NACH *prep*) *cf* DA VI. 1 after it (that, them), thereafter, after(ward), later, then. 2 accordingly, according to that, by that. 3 (*varies according to verb* + **nach**): **d.** greifen reach for it (that, them); **d.** fragen ask (*or* about) it, etc; **d.** schmecken (reichen) taste (smell) of it, etc; **d.** aussehen look like it, etc

daneben I *adv* (*meanings derived from* NEBEN *prep*) 1 next to it (that, them), beside it (that, them) *cf* DA VI; alongside, nearby; next door. 2 off (*or* wide of) the mark. II *adv* & *adv cnjc* 3 besides (that), in addition (to that). 4 at the same time. 5 in comparison. III *sep pfx* (*with verbs of motion and aiming, implies missing the mark*): **daneben-gehen*** *vi*(s), **d—-treffen*** *vi* miss (the mark)

Däne *m* (-n,-n) Dane. **Dänemark** *n* Denmark

Daniellscher Hahn (*Oxy-hydrogen Burners, etc*) Daniell tap

dänisch *a* Danish

dank *prep* (*usu* + *dat*) thanks (to), owing (to)

Dank *m* thanks, gratitude; reward. **dankbar** *a* thankful, grateful; rewarding, worthwhile; durable

danken 1 *vt* (*d.i.o*) appreciate; owe; at-tribute. 2 *vi* (*dat*) thank. **dankend** *p.a* grateful; *as adv oft* with thanks. **dankenswert** *a* welcome, much appreciated

dann *adv* then: **d. und wann** now and then; **auch d. ...., wenn** even when (*or* if); **nur** (*or* **erst**) **d. ...., wenn** only if (*or* when), not ... until

Dän.P. *abbv* (dänisches Patent) Danish patent

D'Anvers-blau *n* Antwerp blue

D.Ap. *abbv* (Deutsches Apothekerbuch) German Dispensatory

Daphne-öl *n* mezereon oil

dar I *sep pfx cf* DARBIETEN, DARLEGEN, DARREICHEN, DARSTELLEN, DARTUN. II *pfx on preps* = DA VI, *cf* DARAN, DARIN etc

daran 1 *adv* (*meanings derived from* AN *prep*) on (at, in, by, along, against, to, next to) it (that, them), thereon, thereat, thereto *cf* DA VI, DAR II. 2 *sep pfx:* **daran-gehen*** *vi*(s), **d—-machen** *vr:* **sich d.** got to it, get to work (on it); **d—liegend** *p.a* adjacent, adjoining

darauf I *adv* (*meanings derived from* AUF *prep*) 1 on (onto, at, to, in, into) it (that, them), thereon, thereto: **oben d.** on top of it *cf* DA VI, *cf* ANKOMMEN 5. 2 then, thereafter, thereupon, on the basis of that. II *sep pfx* on it, upon it. **-folgend** *p.a* following, subsequent. **-hin** 1 *adv* (+ *question clause or* + **ob**) as to, to see (whether, etc). 2 *adv* & *adv cnjc* thereupon; on that basis; in answer to that. **D—lassen** *n* (*Brew.*) doubling

daraus *adv* (*meanings derived from* AUS *prep*) from (*or* out of) it (that, them), therefrom *cf* DA VI, DAR II

darben *vi* starve; suffer privation

dar-bieten* *vt* (*d.i.o*) offer, present (to)

darein *adv* & *sep pfx* in it, into it—**sich darein-mischen** *vr:* interfere

darf may, is permitted: *pres of* DÜRFEN

Dargebot *n* occurrence; supply

dargeboten offered: *pp of* DARBIETEN

darin *adv* (*meanings derived from* IN *prep*) in (*or* into) it (that, them), therein *cf* DA VI, DAR II

dar-legen *vt* explain, expound; state, define, set forth; demonstrate, present.

Darlegung f (-en) explanation; statement; demonstration, presentation

Darlehen n (-) (*Com.*) loan

Darm m (-e) gut, intestine—**blinder D.** caecum *cf* BLINDDARM; **dicker D.** colon; **dünner D.** small intestine; **gerader D.** rectum; **langer D.** ileum; **leerer D.** jejunum

Darm·bein n ilium. **-drüse** f intestinal gland. **-entleerung** f evacuation of the intestine, bowel movement. **-entzündung** f enteritis. **-fäule** f dysentery. **-fell** n peritoneum. **-gang** m intestinal tract. **-gicht** f ileus, colic. **-gift** n enterotoxin. **-haut** f intestinal membrane. **-katarrh** m enteritis. **-knochen** m ilium. **-kot** m feces. **-netz** n omentum. **-rohr** n intestinal tube. **-ruhr** f dysentery. **-saite** f catgut, gut string. **-schlauch** m gut. **-schnitt** m enterotomy. **-stein** m enterolith. **-stiel** m vitelline duct. **-system** n digestive tract. **-tiere** npl metazoa. **-verdauung** f intestinal digestion. **-verschluß** m intestinal obstruction. **-verstopfung** f (intestinal) constipation. **-wand** f intestinal wall. **-windung** f intestinal convolution. **-zotte** f intestinal villus

Darr·arbeit f (*Metll.*) liquation

Darraum m (= **Darr·raum**) kiln chamber. **-gewicht** n kiln-dry weight

Darr·boden m drying (*or* kiln) floor. **-brett** n drying board

Darre f (-n) 1 (kiln) drying, desiccation; kilning; (*Metll.*) liquation. 2 drying kiln, drying room, oast; (*Brew.*) malt kiln; (*Metll.*) liquation hearth. 3 atrophy, withering; consumption (in animals)

dar·reichen vt 1 offer. 2 administer (drugs, etc)

Darreife f (= **Darr·reife**) (*Brew.*) kiln fitness

darren vt 1 (kiln-) dry. 2 (*Metll.*) liquate; torrefy

Darr·fax m (*Brew.*) kilnman. **-fläche** f drying (*or* kiln) surface. **-gekrätz** n liquation (*or* copper) dross (*or* slag). **-gewicht** n kiln-dry weight. **-haus** n drying house, kiln. **-holz** n kiln-dried wood. **-kammer** f drying room (*or* chamber). **-kupfer** n liquated copper. **-malz** n kiln-dried (*or*

cured) malt. **-ofen** m drying oven; (*Brew.*) malt kiln; (*Metll.*) liquation hearth. **-raum** m: see DARRAUM. **-reife** f: see DARREIFE. **-schrank** m drying cabinet. **-staub** m malt dust. **-sucht** f atrophy; consumption = DARRE 3

Darrung f (-en) (kiln-) drying; (*Metll.*) liquation, torrefaction

Darst. abbv (Darstellung) prep. (preparation)

darstellbar a 1 preparable, producible, manufacturable. 2 representable, portrayable

dar·stellen I vt 1 (*incl Chem.*) prepare, produce, synthesize. 2 represent, portray, present. 3 describe, (*Math.*) construct. 4 play (*or* act) the part of. 5 constitute, be. II: **sich d.** present itself, appear. **darstellend** p.a representational; descriptive; performing (art, artist). **Darstellung** f (-en) 1 preparation, production, synthesis. 2 representation, portrayal, presentation. 3 description, (*Math.*) construction. 4 acting; interpretation; performance

Darstellungs·verfahren n process of preparation (representation, etc). **-weise** f manner (*or* method) of preparation (representation, etc). *cf* DARSTELLUNG

dar·tun* vt prove, verify, demonstrate

darüber I adv (*meanings derived from* ÜBER *prep*) *cf* DA VI, DAR II. 1 above (over, across) it (that, them). 2 (+ **hinaus**) beyond it (that, them). 3 more, higher, beyond (*or* above) that. 4 (*esp re time*) past it (*or* that), beyond (it *or* that). 5 meanwhile, in the process, while (one is) preoccupied with it. 6 about (*or* concerning) it (that, them). II sep pfx (*in meanings* 1 & 4 *above*)—**sich darüber·machen** get to work on it

darübergelagert p.a super(im)posed

darum I adv (*meanings derived from* UM *prep*) 1 around it (that, them), around there *cf* DA VI, DAR II. 2: **d. ...,** **weil** because; **d. ...,** **damit** (*or* **daß**) so that. II adv cnjc 3 for that reason, that's why, that is the reason (why). 4: **aber ... d.** but in spite of that. III sep pfx around it; above (*or* over) it

darunter I adv (*meanings derived from* UN-

TER *prep*) **1** under it (that, them), thereunder, underneath *cf* DA **VI,** DAR **II. 2** under, less than (that). **3** among them, including. **4** (*with verbs of understanding, imagining*) by it, by that *cf* VERSTEHEN, VORSTELLEN. **III** *sep pfx* underneath, **-liegend** *p.a* underlying, subjacent

das *n* **1** *def art* the. **2** *demons pron* that, this. **3** *relat pron* that, which, who

das. *abbv* (DASELBST) there, etc

DAS. *abbv* (Deutsche Auslegeschrift) Germ. pat. applic. (open to inspection)

da·sein* **1** be there, be present; be available. **2** exist. **Dasein** *n* existence, life

daselbst *adv* there, in that place, ibidem

dasjenige *n pron* (*usu + genit* or *+ relat pron*) that (of), the one (that, which)

daß *subord cnjc* **1** that, so that. **2** (*after* **da; dar·** *prep*) e.g.: **die Lösung liegt darin, daß...** the solution lies in ...ing (*or* in that ...) *cf* DA **VI,** DADURCH **2,** DAFÜR **3** etc

dasselbe *n pron* & *a* the same, it

da·stehen* *vi* stand (there), be (in a condition *or* position)

Datei *f* (-en) data bank

Daten *npl* **1** data. **2** *pl of* DATUM dates. **-verarbeitung** *f* data processing

datieren *vt/i* date

dato: bis d. to (this) date

Dattel *f* (-n) date. **-baum** *m,* **-palme** *f* date palm. **-pflaume** *f* persimmon, date plum. **-wein** *m* date wine. **-zucker** *m* date sugar

Datum *n* (Daten) (calendar) date; datum

Daube *f* (-n) (barrel) stave. **Dauben·holz** *n* barrel (*or* cooper's) wood

Dauer *f* **1** duration, period; life (of patents). **2** permanence: **(nicht) von D. sein** be of long (short) duration, (not) last long; **auf die D.** in the long run; **D. im Wechsel** stability in a changing environment. **3** (*Food*) (preparation *or* cooking) time

Dauer· permanent, continuous, lasting, perennial, chronic; **-ausscheider** *m* (*Biol.*) chronic carrier. **-beanspruchung** *f* continuous stress, fatigue stress. **-belastung** *f* steady (*or* permanent) load. **-beobachtung** *f* continuous observation. **-betrieb** *m* continuous operation. **-biegefestigkeit** *f* bending limit (*or* endurance); (*Plastics*) flex life. **-biegespan-**

nung *f* repeated flexural stress. **-bieg(e)versuch** *m* endurance bending test. **-bier** *n* lager beer. **-bleiche** *f* continuous bleaching. **-bruch** *m* fatigue fracture. **-butter** *f* canned butter. **-ei** *n* winter egg; resting zygote. **-entladung** *f* continuous discharge. **-farbe** *f* permanent (*or* lasting) color. **-festigkeit** *f* durability; (*Engg.*) fatigue strength, endurance. **-flamme** *f* pilot flame. **-fleisch** *n* preserved meat. **-form** *f* permanent form; permanent mold. **-frost** *m* permafrost. **-gebrauch** *m* continuous (*or* extended) use. **-gewächs** *n* perennial (plant). **-gewebe** *n* permanent tissue

dauerhaft *a* durable, lasting, permanent; fast (colors); touch (leather). **Dauer·haftigkeit** *f* durability, permanence

Dauer·haltbarkeit *f* (*Engg.*) fatigue endurance. **-hefe** *f* permanent (*or* dried) yeast. **-leistung** *f* continuous output. **-licht** *n* continuous illumination. **-magnet** *m* permanent magnet. **-milch** *f* sterilized (condensed, evaporated, powdered) milk

dauern *vi* last, go on; take (a period of time), take long. **dauernd** *p.a* lasting, permanent; continual; perennial

Dauer·pflanze *f* perennial (plant). **-probe** *f* endurance test. **-prüfmaschine** *f* endurance testing machine. **-prüfung** *f* **1** endurance test. **2** long-lasting test. **-riß** *m* fatigue crack. **-schlagfestigkeit** *f* impact endurance. **-schlagversuch** *m* repeated impact test. **-schwingbruch** *m* fatigue fracture. **-schwingversuch** *m* fatigue test. **-spore** *f* resting spore. **-standfestigkeit** *f* resistance to creep stress. **-standversuch** *m* creep test. **-strichlaser** *m* continuous-beam laser. **-strom** *m* continuous current. **-tauchversuch** *m* total immersion test. **-träger** *m* (*Biol.*) chronic carrier. **-veränderung** *f* permanent change. **-verhalten** *n* durability, permanence. **-versuch** *m* endurance test. **-waren** *fpl* preserved food(s), preserves. **-weide** *f* permanent pasture. **-welle** *f* permanent wave. **-wiese** *f* permanent meadow. **-wirkung** *f* lasting effect. **-wurst** *f* hard smoked sausage, hard salami. **-zelle** *f* resting spore. **-zug** *m* (*Rubber*) permanent set (on expansion). **-zustand** *m* steady state; permanent condition

Daumen *m* (-) thumb; (*Mach.*) tappet, finger, cam, dog. **-abdruck** *m* thumbprint. **-mutter** *f* (*Mach.*) thumb nut. **-rad** *n* cam wheel. **-schraube** *f* thumb screw. **-welle** *f* camshaft

Däumling *m* (-e) 1 thumb (of glove); rubber thumb(stall). 2 (*Mach.*) tappet, cam, lifter; wiper

Daune *f* (-n) down(y) feather; *pl oft* down. **daunen·dicht** *a* downproof

davon I *adv* (*depends on meanings of* VON; *cf* DA VI) 1 of it (that, them), thereof. 2 from it (that, them) therefrom: **weit d. entfernt** far from it. 3 by it (that, them), thereby. 4 (out) of it (that, them). 5 (out) of it (or that) (*i.e. made of some material*). 6 because of that. 7 away, gone. **II** *sep pfx* away: **davon·tragen*** *vt* carry away; come away with, gain (e.g. victory), sustain (loss)

davor I *adv* (*meanings derived from* VOR *prep, cf* DA **VI.**) 1 in front of it (that, them). 2 before (that). 3 from (or of) it (that, them). **II** *sep pfx* in front (of it), before it

dawider *adv & sep pfx* against (*or* contrary to) it (that, them) *cf* WIDER *prep*, DA VI

dazu I *adv* (*meanings derived from* ZU *prep, cf* DA **VI**) 1 to it (this, that, them), thereto—**d. kommt noch, daß...** in addition (there is the fact that)...; **es kam d., daß...** things got to the point where... *cf* KOMMEN. 2 for it, that, them. 3 in answer to that, with reference to that. 4 to go with it (or that). 5 for this (or that) purpose: **es ist d. da, ...zu** it exists for the purpose of ...ing. **II** *adv cnjc, usu:* **noch d.** besides, in addition. **III** *sep pfx* to it: **dazu·geben***, **·tun*** *vt* add (to it); **d—·gehören** *vi* belong to it, be part of it; **d—gehörig** *a* that goes with it, pertaining to it, accompanying it. **d—·kommen*** *vi*(s) arrive on the scene; be added (to it); **d—·lernen** *vt/i* expand one's knowledge (of)

dazwischen 1 *adv* between, among (them) *cf* DA **VI;** (in) between. 2 *sep pfx* between, inter- (*oft implies intervening, interfering*): **dazwischen·kommen*** *vi*(s) get between (them); get in the way, interfere; come (or belong) in between. **d—liegend** *p.a* lying between, intermediate. **d—**

·schalten, ·schieben*, **·stellen** *vt* put (or push) between (them), interpose, interpolate. **d—·treten*** *vi*(s) step in (between), intervene

DB *abbv* (Deutsche Bundesbahn) Ger. Fed. RR.

DBGM *abbv* (deutsches Bundesgebrauchsmuster) Fed. Germ. Registered Design

d.Bl. *abbv* (dieses Blattes) of this periodical

DBP *abbv* (Deutsche Bundespost) Ger. Fed. P.O. **DBP(a)** *abbv* (Deutsches Bundespatent/angemeldet) Ger. Fed. Pat. (applied for)

dch. *abbv* (DURCH) through, by

DD *abbv* (Dampfdichte) vapor density

D.D. *abbv* (Dichten) densities

DDR *abbv* (Deutsche Demokratische Republik) (East) Ger. Dem. Rep.

DE *abbv* 1 (Dielektrizitätskonstante) dielec. const. 2 (dieselelektrisch) dieselelec.

Debatte *f* (-n) debate: **zur D. stehen** be debatable

Debet *n* (-s) debit (side)

deca-,Deca- *cf* DEKA..

decarbonisieren *vt* decarbonize, decarburize

decarboxylieren *vt* decarboxylate

Dechema *abbv* (Deutsche Gesellschaft für chemisches Apparatewesen) German Society for Chemical Apparatus Technology

dechiffrieren *vt* decipher

Dechsel *f* (-n) adze

Deck· cover(ing), protective *cf* DECKEN: **-ablauf** *m* (*Sugar*) wash syrup. **-anstrich** *m* 1 top (*or* finishing) coat (of paint). 2 deck paint. **-appretur** *f* (*Tex.*) top finish. **-blatt** *n* overlay; fly leaf; (cigar) wrapper; (*Bot.*) bract

Decke *f* (-n) 1 cover(ing), coat(ing), wrapper, (outer) layer. 2 skin, hide, integument. 3 blanket, (bed)cover. 4 (table)cloth. 5 ceiling, top, roof. 6 (*Sugar*) cleansing, purging; fine liquor. 7 (*Brew.*) head

Deckel *m* (-) lid, cover (of containers, books); top, cap. **-abfall** *m* carding waste. **-kanne** *f* pot with a lid

deckeln *vt* provide with a cover, cover with a lid

Deck·email *n* enamel top coat(ing)

decken I *vt/i* 1 cover (*most meanings*). **II** *vt* 2 protect. 3 set (table). 4 (*Sugar*) cleanse, purge, fine, clay. **III: sich d.** (*recip*) coincide. **Decken** *n* covering, protection; (*Paints*) covering power, opacity; (*Sugar*) claying. **deckend** *p.a* (*Paints, usu*) opaque, of good covering power

Decken·erguß *m* alluvial soil. **-gewebe** *n* epitheleal tissue. **-lampe** *f* ceiling (*or* dome) light

Decker *m* (-) coverer, roofer; (*Tobacco*) wrapper. **-druck** *m* (*Tex.*) blotch print

deck·fähig *a* (*Paint*) opaque, of good covering power. **D–fähigkeit** *f* covering power, opacity. **-farbe** *f* final coat, body (opaque, coating) paint (*or* color). **d–farbig** *a* opaque, of good body. **D–firnis** *m* covering (*or* protective) varnish. **-frucht** *f* (*Agric.*) cover crop. **-gebirge** *n* (*Geol., Min.*) overburden. **-geflecht** *n* protective tissue. **-glas, gläs/chen** *n* cover glass. **-grün** *n* Paris green. **-hülle** *f* covering; concealing cover (*or* shell). **-hütchen** *n* cover cap; (*Explo.*) capsule

deckig *a* (*Biol.*) imbricate

Deck·kläre *f*, **-klärsel** *n* (*Sugar*) wash liquor (for cleansing crystals). **-kraft** *f* covering power, body (of pigments, dyes). **-lack** *m* coating lacquer (*or* varnish). **-mittel** *n* covering medium (*or* material). **-name** *m* trade name; code name; pseudonym, alias. **-papp** *m*, **pappe** *f* (*Tex.*) resist (paste). **-pappdruck** *m* resist printing. **-plättchen** *n* (*Micros.*) cover plate (*or* slide). **-schicht** *f* covering (*or* protective) layer; top coat (of varnish); (*Geol.*) overlying stratum. **-schleuder** *f* (*Sugar*) washing centrifuge

Decksel *n* (-) (*Sugar*) crystal-cleansing liquor

Deck·sirup *m* syrup obtained in sugar refining, treacle. **-span** *m* veneer. **-stopfen** *m* flanged (*or* rimmed) stopper

Deckung *f* (-en) **1** covering, cover(-up). **2** protection, shelter. **3** concealment, mask. **4** collateral, security. **5** (*Phot.*) density. **6** (*Geom.*) congruence—**zur D. bringen** cause to coincide

Deck·vermögen *n* covering (*or* hiding) power. **-weiß** *n* zinc (*or* opaque) white. **-weißfarbe** *f* (*Varnish*) white enamel. **-wort** *n* code word. **-zelle** *f* cortical cell. **-ziegel** *m* cover tile; coping brick

decolorieren *vt* decolorize

Decreusage *f* (*Silk*) determination of gum

Decyl·säure *f* decylic acid, decalic acid

deduzieren *vt* deduce

defekt *a* defective. **Defekt·leiter** *m* defect semiconductor

Defibreur *m* (-e) (*Paper*) pulp grinder. **defibriniert** *p.a* defibrinated

definieren *vt* define. **definiert** *p.a* (*oft*) definite

definitions·gemäß *adv* by definition; as defined

deflagrieren *vt* deflagrate

Deformations·schwingung *f* deformation vibration. **deformieren** *vt* deform

Defo·wert *m* defo (*or* plasticity) value

degallieren *vt* degall

Degea *abbv* (Deutsche Gasglühlicht-Auer-Gesellschaft) Ger. Welsbach Gas Lighting Soc.

Degebo *abbv* (Deutsche Gesellschaft für Bodenmechanik) Ger. Soil Mechanics Society

Degen *m* (-) sword

degenerieren *vi*(s) degenerate

deglutieren *vt* swallow

degommieren *vt* degum

degorgieren *vt* remove the sediment from

degradieren *vt* degrade

degra(i)ssieren *vt* degrease = ENTFETTEN

Degras *n* (*Lthr.*) degras, stuff, dubbing

degummier·echt *a* degumming-resistant. **degummieren** *vt* degum; malt; strip, boil off (silk)

Degussa *abbv* (Deutsche Gold- und Silberscheideanstalt) Ger. Gold & Silver Inst.

dehnbar *a* **1** stretchable, expandable, extendable. **2** flexible, elastic. **3** (*Metll.*) ductile, malleable. **4** (*Gas*) dilatable. **Dehnbarkeit** *f* stretchability, etc. **Dehnbarkeits·messer** *m* ductilometer, dilatometer

dehnen *vt* & **sich d.** stretch (*lit & fig*), expand, extend; dilate, distend; lengthen (a sound)

dehn·fähig *a* stretchable, etc = DEHNBAR. **D–kraft** *f* force (*or* power) of expansion

Dehnung *f* (-en) stretching; expansion, extension; dilation, distension; lengthening; diastole

Dehnungs·ermüdung *f* tensile fatigue. **d–**

fähig *a* expansible, extensible. **D—fuge** *f* expansion joint. **-kraft** *f* power of expansion = DEHNKRAFT. **-lehre** *f* strain gage. **-messer** *m* extensometer, dilatometer. **-meßstreifen** *m* strain gage. **-modul** *n* (Young's) modulus of elasticity. **-rest** *m* residual elongation. **-spannungskurve** *f* stress-strain curve. **-wärme** *f* heat of expansion. **-zahl** *f* coefficient of expansion

Dehydracet·säure *f* dehydroacetic acid

Dehydratation *f* (-en), **Dehydratisierung** *f* (-en) dehydration

dehydrieren *vt* dehydrogenate; dehydrate. **Dehydrierung** *f* (-en) dehydrogenation, dehydration. **Dehydrierungs·anlage** *f* dehydrogenation plant

Dehydro·benzol *n* benzyne. **-campher-säure** *f* dehydrocamphoric acid. **-china-cridon** *n* dehydroquinacridone. **-chinon** *n* dehydroquinone. **-schleimsäure** *f* dehydromucic acid

Deich *m* (-e) dike

Deil *m* (-e) bloom, lump, cake (of metal)

Deka *n* (-) decagram, 10 grams

Dekade *f* (-n) decade

Dekaden· decade, decadic, decimal: **-einheit** *f* decade unit

dekadisch *a* decadic; **d—e Logarithmen** common logarithms

Dekaeder *n* (-) decahedron

dekantieren *vt* decant. **Dekantier·topf** *m* (earthenware) decanting jar

Dekapier·bad *n* (*Metll.*) pickling bath. **de-kapieren** *vt* scour, descale, pickle (metals)

dekarbonisieren *vt* decarbonize, decarburize

dekatieren *vt* hot-press, decatize, pre-shrink by steaming. **Dekatur** *f* (-en) hot-pressing, pre-shrinking by steaming. **de-katur·echt** *a* hot-press-resistant

dekaustizieren *vt* decausticize

deklinieren *vi* (*Phys., Astron.*) deviate, decline

Dekokt *n* (-e) decoction

dekonzentriert *a* (Opt.) defocused

Dekorations·farbe *f* decorative color. **-lack** *m* decorative varnish. **-stoffe** *mpl* (*Tex.*) furnishings

Dekor·brand *m* (*Ceram.*) enamel fire. **de-korieren** *vt* decorate

dekrepitieren *vt* decrepitate

Dekyl *n* (-e) decyl

deliqueszieren *vi* deliquesce

Delle *f* (-n) dent; depression, hollow

Delphin *m* (-e) dolphin

dem *dat of* DAS, DER *as def art; demons a & pron, relat pron* (to) the, (to) that, (to) the one; (to) which, (to) whom—**wenn d. so ist** if that is so *cf* SEI

Demant·blende *f* eulytite. **-spat** *m* adamantine spar, corundum

dementsprechend 1 *a* corresponding, expected. **2** *pred a* as expected. **3** *adv* accordingly

demethylieren *vt* demethylate

dem·gegenüber *adv* as opposed to that, on the other hand. **-gemäß** *adv* accordingly

demjenigen *of* DERJENIGE, DASJENIGE (to) that, (to) the one, etc

dem·nach *adv* according to that, accordingly. **-nächst** *adv* soon

demolieren *vt* demolish

Demontage *f* (-n) dismantling, disassembly. **demontierbar** *a* dismountable, detachable, capable of disassembly. **de-montieren** *vt* dismantle, disassemble, take apart

demselben *a & pron, dat of* DASSELBE, DER-SELBE (to) the same

demulzieren *vt* demulsify, soften

Demut *f* humility. **demütig** *a* humble

demzufolge *adv* accordingly, consequently

den *def art, demons pron* **1** *accus of* DER the, that (one), him. **2** *dat of* DIE *pl* (to) the, (to) those

denaturieren *vt* denature, denaturize. **De-naturier(ungs)·mittel** *n* denaturant. **de-naturisieren** *vt* denaturize

Dendrachat, Dendriten·achat *m* dendritic agate

dendrisch, dendritisch *a* dendritic

denen *dat pl* **1** *demons a & pron* (to) those, (to) the ones. **2** *relat pron* (to) which, (to) whom

Denitrier·bad *n* denitrating bath. **de-nitrieren** *vt* denitrate. **Denitrier·mittel** *n* denitrating agent. **denitrifizieren** *vt* denitrify

denjenigen 1 *accus of* DERJENIGE the one, that. **2** *dat of* DIEJENIGEN *pl* (to) the ones, (to) those

Denk·ansatz *m* incipient idea. **-anstoß** *m* provocative idea, impulse to start a line of thought. **-art** *f* mentality, way of thinking **denkbar** *a* thinkable, conceivable; *adv oft* e.g.: **d. schlecht** as bad(ly) as one can imagine
denken* (dachte, gedacht) 1 *vt/i* think, imagine, reflect. 2 *vi* (**an**) think (of). 3 *vt* (+ *dat rflx* **sich**) imagine, conceive of. *Cf* GEDACHT, *see also* GEDENKEN, LASSEN
Denk·fabrik *f* think tank. **-mal** *n* (-mäler) monument, memorial. **-methode** *f* way of thinking. **-modell** *n* hypothetical model. **-münze** *f* commemorative coin. **-pause** *f* time out for thinking. **-schrift** *f* 1 memoir. 2 memorial, commemorative article. 3 memorandum. **-übung** *f* mental exercise. **d–würdig** *a* memorable, noteworthy. **D–würdigkeiten** *fpl* memorabilia
denn I *adv* 1 then. 2 *no specif transl* (*implies interest or impatience, esp in questions*). 3: **es sei d., daß** unless. II *cnjc* 4 for, because. 5 (*after compar*) than
dennoch *adv cnjc* yet, still, nevertheless
denselben *accus of* DERSELBE, *dat of* DIESELBEN *pl* (to) the same
Densität *f* (-en) density
Depesche *f* (-n) message, dispatch
Dephlegmator *m* (-en) fractionating column. **dephlegmieren** *vt* dephlegmate
Deplacement *n* (-s) displacement. **deplacieren** *vt* displace
Depolarisator *m* (-en) depolarizer. **depolarisieren** *vt* depolarize
Deponie *f* (-n) garbage dump, waste repository
deprimieren *vt* depress
deproteinisieren *vt* deproteinize
depur. *abbv* (*Lat.* depuratus, etc) purified
der I *def art* 1 *m sg nomin* the. 2 *f sg genit & dat* (of, to) the. 3 *pl genit* (of) the. II *demons a & pron* (*same cases as* I) that, those, the one(s). III *relat pron m sg nomin & f sg dat* who, to whom, that, (to) which
Der. *abbv* (Derivat) derivative
derart *adv* so (much), in such a way, (in) that way. **derartig** 1 *a* such. 2 *adv* so, etc = DERART
derb *a* 1 solid, robust, sturdy. 2 coarse. 3 rude, uncouth. 4 (*Min.*) massive

Derb·erz *n* massive ore. **-gehalt** *m* solid (*or* cubic) content, actual volume
Derbheit *f* (-en) 1 solidity, robustness, etc *cf* DERB 1–4. 2 uncouth remark
Derb·holz *n* log wood. **d–stückig** *a* lumpy, large-sized (ore). **-wandig** *a* stout-walled, thick-walled
dereinst *adv* some day; once. **dereinstig** *a* future
deren *f & pl genit* 1 *demons a* her, its, their, the latter's. 2 *demons pron* of it, of them, of those, of the latter. 3 *relat pron* whose, of which
derer *demons pron genit pl* (*usu* + *relat pron*) of those (who, which...), of the ones
dergestalt *adv* so (much), in such a way
dergl. *abbv* = **dergleichen** 1 *a* such. 2 *pron* the like. 3: **nichts d.** nothing of the kind
Derivat *n* (-e) derivative. **derivieren** *vt* derive
derjenige *m demons a & pron* (*usu* + *genit* or + *relat pron*) that (of...), the one (who, which ...). **derjenigen** *demons a & pron genit* 1 *f* (of) that, (of) the one. 2 *pl* (of) those, (of) the ones
derlei *a & pron* (*no endgs*) such (things)
dermaßen *adv* so (much), to such an extent
dermatisch *a* cutaneous
Derr·säure *f* derric acid
derselbe *m a & pron* the same, he, it. **derselben** *a & pron* 1 *f genit* (of) the same, (of) her, (of) it. 2 *pl genit* (of) the same, (of) them
derweil(en) 1 *adv* meanwhile. 2 *subord cnjc* while
derzeit *adv* 1 at this time, at present. 2 at that time, then. **derzeitig** I *a* 1 present. 2 then, of that time. II *adv* = DERZEIT
des *genit of* DER *m*, DAS *n* (of) the
des· *pfx* dis-, de- *cf* DESINFEKTION, DESHYDRIEREN
des. *abbv* (desmotrop) desmotropic
Desaggregation *f* (-en) 1 disaggregation, disintegration. 2 peptization, depolymerization, degradation
desaktivieren *vt* deactivate
desamidieren, desaminieren *vt* deaminate
desarsenisieren *vt* dearsenicate
Desensibilisator *m* (-en) desensitizer. **desensibilisieren** *vt* desensitize

desgl. *abbv* = **desgleichen 1** *adv* likewise, ditto. **2** *pron* the like

deshalb *adv cnjc* therefore, for that reason

deshydrieren *vt* dehydrate

Desinfektion *f* (-en) disinfection. **Desinfektions·mittel** *n* disinfectant. **Desinfektor** *m* (-en) disinfector; disinfectant. **desinfektorisch** *a* disinfectant

Desinfiziens *n* (..zienten) disinfectant. **desinfizieren** *vt* disinfect. **desinfizierend** *p.a* disinfectant

desintegrieren *vi* disintegrate

desjenigen *genit of* DERJENIGE, DASJENIGE of that, of the one

desmotrop *a* desmotropic. **Desmotropie** *f* desmotropy, desmotropism

Desodorans *n* (..ranten), **Desodorations·mittel** *n* deodorizer, deodorant. **desodor(is)ieren** *vt* deodorize. **Desodorisierungs·mittel** *n* deodorant

Desoxidations·mittel *n* deoxidizing agent. **-wirkung** *f* deoxidizing action (or effect)

desoxidieren *vt* deoxidize

desselben *genit of* DERSELBE, DASSELBE (of) the same

dessen *m,n genit* **1** *demons a* his, its, the latter's. **2** *demons pron* of him, of it, of that. **3** *relat pron* whose, of which. **··ungeachtet** *adv* nevertheless

Dessin *n* (-s) design, pattern. **··druck** *m* pattern printing. **dessinieren** *vt* design

dest., Dest. *abbv* (destilliert, Destillation) distilled, distillation

Destillat *n* (-e) distillate. **Destillateur** *m* (-e) distiller

Destillations· = DESTILLIER·: **-apparat** *m* distilling apparatus, still. **-aufsatz** *m* fractionating column, distillation head. **-benzin** *n* straight-run gasoline. **-gas** *n* distillation gas, coke-oven gas. **-gefäß** *n* distilling vessel. **-hut** *m* receiver. **-kokerei** *f* by-product coking; by-product coke plant. **-kolben** *m* distilling flask. **-ofen** *m* distilling furnace; by-product coke oven. **-rohr** *n*, **-röhre** *f* distillation tube (*or* column). **-tiegel** *m* distilling crucible. **-vorlage** *f* distillation receiver

Destillat·sammelgefäß *n* receiver, (*Petrol.*) rundown tank

Destillier· *see also* DESTILLATIONS·: **-appa-**

rat *m* distilling apparatus

destillierbar *a* distillable

Destillier· = DESTILLATIONS· distilling: **-betrieb** *m* distilling plant. **-blase** *f* distilling vessel, still. **-brücke** *f* distillation bridge

destillieren *vt/i* distill

Destillier· = DESTILLATIONS· distilling: **-glas** *n* (glass) distilling flask. **-haus** *n* still (house). **-helm** *m* still head. **-kolonne, -säule** *f* distilling (*or* fractionating) column. **-topf** *m* distilling pot. **-vorstoß** *m* receiver adapter

desto *adv* (+ *compar*) all the, so much the. **2** *correlative cnjc* e.g.: **je mehr, d. besser** the more, the better

Destricta·säure *f* destrictic acid. **Destrictin·säure** *f* destrictinic acid

des·wegen *adv cnjc* therefore, for that reason

deszendent *a* descendent; (*Geol.*) supergene

Detacheur *m* (-e) (*Dye.*) spot cleaner. **detachieren** *vt* cleanse of stains; detach. **Detachier·mittel** *n* spot remover. **Detachur** *f* (-en) spot removing

detonieren *vt* detonate: **d—de Strecke** (*Explo.*) detonating column; **d—der Sprengstoff** = **Detonier·sprengstoff** *m* detonator

Deul *m* (-e) (*Metll.*) bloom, (iron)ball, lump

deutbar *a* explainable, interpretable

deuten 1 *vt* interpret, explain. **2** *vi* (**auf**) point (to); (fore)bode. **3: sich d.** be interpreted

deuterieren *vt* deuterate

Deuterium·wasserstoff *m* hydrogen deuteride. **Deuter(ium)·oxid** *n* deuterium oxide

deutlich *a* clear, distinct. **Deutlichkeit** *f* clarity, distinctness; (*esp Opt.*) definition

deutsch *a, adv* German: **auf d.** in German; **d—e Folie** German (tin-alloy) foil; **d—es Essigverfahren** quick vinegar process; **d—es Geschirr** *n* (*Paper*) mill stamping. **Deutsch** *n* German (language). **Deutsche 1** *m,f* (*adj endgs*) German (person). **2** *n* German (language)

Deutschland *n* Germany: **Bun-**

desrepublik D. Federal Republic of Germany

Deutung f (-en) interpretation, explanation

Devarda·legierung f Devarda's alloy

Devise f (-n) 1 motto, slogan; trade mark. 2 pl foreign exchange (or currency)

devonisch a (Geol.) Devonian

Dewar·gefäß n, **Dewarsches Gefäß** Dewar vessel

Dextronen·säure f dextronic (or D-gluconic) acid

dezentrieren vt decenter

dezidiert a dedicated, exclusive

Dezi- deci-: **-gramm** n decigram

dezimal a decimal. **D—·bruch** m decimal fraction. **Dezimale** f (-n) decimal

Dezimal·komma n decimal point. **-stelle** f decimal place. **-waage** f decimal balance

DFB abbv (Druckfeuerbeständigkeit) (Ceram.) refractoriness under load

DFG abbv (Deutsche Forschungsgemeinschaft) German Research Council

d.G. abbv (durch Güte) by favor of

dgl. abbv = DERGLEICHEN, DESGLEICHEN the like, etc

DGMK abbv (Deutsche Gesellschaft für Mineralwissenschaft und Kohlechemie e.V.) Ger. Soc. for Petroleum Science and Coal Tar Chemistry

d.h. abbv (das heißt) i.e., that is

dH abbv (deutsche Härte) (Water) German degrees of water hardness

DHD-Anlage f abbv (Dehydrierungsanlage) dehydrogenation plant

d.i. abbv (das ist) i.e., that is

Dia n (-s) (Phot.: short for **Diapositiv**) slide, transparency. **--betrachter** m slide viewer

Diacetin n glycerol diacetate. **Diacet·säure** f acetoacetic acid

Diachylon·salbe f diachylon ointment

diadoch a isomorphous. **Diadochie** f (-n) isomorphism

Diagenese f (-n) diagenesis

Dialur·säure f dialuric acid

Dialysator m (-en) dialyzer. **Dialyse** f (-n) dialysis. **dialysieren** vt dialyze. **Dialysier·papier** n dialyzing paper. **dialysisch, dialytisch** a dialytic

Diamant m (-en,-en) diamond. **d—·artig, diamanten** a diamond(-like), adamantine

diamant·förmig a diamond-shaped, diamond-like. **D–glanz** m adamantine luster. **-kitt** m diamond cement, galbanum. **-mörser** m diamond mortar (small steel mortar). **-pulver** n diamond dust. **-schwarz** n diamond black. **-spat** m adamantine spar, corundum. **-stahl** m very hard steel, tool steel

diametral, diametrisch a diametric(al)

Diamido- diamino-, diamido- cf AMIDO-: **-benzol** n diaminobenzene. **-toluol** n diaminotoluene

Diamin·farbstoff m diamine dye

Dianen·baum m silver tree, Arbor dianae

Dianil·gelb n dianil yellow

diaphan a diaphanous

Diaphorese, Diaphoresie f (-n) diaphoresis. **diaphoretisch** a diaphoretic

Diaphragma n (..gmen) diaphragm

Diarrhöe f diarrhea

Diastase·wirkung f diastatic action

Diät f (-en) diet. **Diätetik** f dietetics

diatherm(an) a diathermic, diathermanous

Diatomee f (-n) (Biol.) diatom, pl diatoma. **--erde** f diatomaceous earth, kieselguhr. **-schlamm** m (Geol.) diatom ooze

Diazo·benzol n diazobenzene. **-körper** m diazo compound. **-salz** n diazo(nium) salt

diazotieren vt diazotize. **Diazotierungs·farbstoff** m diazotizing dye; diazotizable dye

Diazotypie f diazo copying process

Diazoxid n diazo oxide

dibenzoyliert a dibenzoyl-, having two benzoyl groups introduced

Dibrom·benzol n dibromobenzene. **-polysulfan** n polysulfur dibromide

Dicarbon·säure f dicarboxylic acid

Dichinol n diquinol cf CHINO...

Dichlor·äthylen n dichloroethylene. **-disulfan** n disulfur dichloride. **-methan** n dichloromethane. **-polysulfan** n polysulfur dichloride

dichroitisch a dichroic, dichroitic

dichrom·sauer a dichromate of. **D–säure** f dichromic acid

Dichro·salz *n* dichroic salt

dicht *a* 1 dense, thick. 2 (water)tight, leak-proof, sealed: **d. machen** seal; **d. halten** be leakproof. 3 tightly packed; close-textured, compact; impervious; (*Min.*) massive. 4 close (succession): (*usu adv*) (+ **an, bei, nach**) close to (at, behind); **d. vor** close (*or* shortly) before

dicht·brennen* *vt* (Ceram.) vitrify

Dichte *f* (-n) 1 density, thickness; (wa-ter)tightness; compactness; impervious-ness *cf* DICHT. 2 (*Paper*) bulking quality; (*Paint, etc*) body. **·bestimmung** *f* density determination. **-flasche** *f* specific-gravity bottle. **-messer** *m* density measuring instrument, densi(to)meter; (*spec-if*) hydrometer. **-messung** *f* density measurement, densi(to)metry

dichten I *vt* seal, make leakproof, lute, calk; condense, compact. II *vt/i* write (po-etry, literature)

Dichter *m* (-), **Dichterin** *f* (-nen) poet, writer

dicht· close, tight(ly), dense(ly): **-faserig** *a* close-fibered, close-grained. **-gedrängt** *a* tightly pressed, compact. **-gepackt** *a* tightly (*or* thickly) packed

Dichtheit *f* 1 density, etc = DICHTE 1. 2 con-sistency (of liquid)

Dichtigkeit *f* density, etc = DICHTE

Dichtigkeits·grad *m* degree of density (tightness, etc). **-messer** *m* densimeter

Dicht·polen *n* (*Copper*) (dense) poling. **-ring** *m* gasket. **d—schließend** *a* tightly closing, tight(-fitting) (lid, etc). **D—schweißung** *f* close weld(ing)

Dichtung *f* (-en) I 1 sealing, packing, lut-ing. 2 seal; washer; gasket; joint; septum. II poetry (and literature); poetic (*or* liter-ary) work

Dichtungs·fett *n* packing (*or* sealing) grease. **-masse** *f*, **-material** *n*, **-mittel** *n* sealing (calking, luting, packing) mate-rial (*or* agent). **-ring** *m* washer, spacer, gasket, packing ring. **-säule** *f* (*Distil.*) packed column. **-schnur** *f* sealing tape. **-stoff** *m* sealing material = -MASSE

dick *a* 1 thick, stout; fat. 2 swollen. 3 dense. 4 viscous. 5 (great) big, hefty, heavy. 6 *adv oft* in a big way

Dick·auszug *m* thickened (*or* inspissated)

extract. **-darm** *m* (*Anat.*) large intestine; colon

Dicke *f* (-n) I 1 thickness; (sheet metal) gauge. 2 stoutness, corpulence. 3 density. 4 bulk, large size. 5 viscosity, consistency. II mother (liquor), dregs

dicken *vt/i* thicken

Dicken·messer *m* thickness gauge, calipers

Dick·farbe *f* thick color (*or* paint). **d—flüssig** *a* viscous, viscid. **D—flüssigkeit** *f* viscosity; thick (*or* viscous) liquid. **d—grell** *a* dead-white (pig iron). **D—häuter** *m* pachyderm

dickig *a* consistent, firm, stiff. **Dickigkeit** *f* (-en) consistency, firmness, stiffness

Dick·lauge *f* concentrated lye

dicklich *a* thickish, stoutish, fattish

Dick·maische *f* (*Brew.*) thick mash, decoc-tion. **-milch** *f* curdled milk, curds. **-mittel** *n* thickener, thickening agent. **-öl** *n* thick oil, heavy oil, stand oil; viscous by-product. **-ölfirnis** *m* thick oil varnish. **d—ölig** *a* oily-viscous. **D—pfanne** *f* con-centration pan. **-saft** *m* juice concen-trated by evaporation; (*Sugar*) thick juice, syrup. **d—schalig** *a* thick-shelled, thick-skinned; husky (grain). **D—schlamm** *m* thick mud (*or* sludge), thick-ened slurry. **d—sirupartig** *a* thick and syrupy. **D—spülung** *f* fluid mud. **-stoff** *m* viscous material. **d—tafelig** *a* in thick plates. **D—teer** *m* thick tar, heavy tar. **-trübe** *f* (*Ores*) regenerated dense medi-um. **d—wandig** *a* thick-walled

Dicyan *n* dicyanogen, cyanogen gas

didodekaedrisch *a* (*Cryst.*) didodecahed-ral

die *f & pl* 1 *def art* the. 2 *demons a & pron* that (one), those, the one(s); she, her. 3 *relat pron* who(m), which, that

Dieb *m* (-e) thief. **·stahl** *m* theft, larceny

diejenige *demons a & pron* (*usu + genit or + relat pron*) 1 that (of...), the one (who, which...). 2 *pl* those (of...), the ones (who, which...)

Diele *f* (-n) 1 (floor)board, plank, deal. 2 floor(ing). 3 vestibule, entrance hall

Dielektrikum *n* (..ka *or* ..ken) dielectric. **Dielektrizitäts·konstante** *f* dielectric constant

Dien *n* (-e) diene
dienen *vi* (*dat* or **als, zu**) serve (as, for)
dienlich *a* (*oft* + *dat*) useful, of help (to); advisable
Dienometrie *f* determination of diene content (by Diels-Alder reaction). **dienophil** *a* dienophilic *cf* DIEN
Dienst *m* (-e) service; duty; employment, job, work: **außer D.** out of service, retired; **den D. versagen** break down
Diens·tag *m* Tuesday
Dienst·alter *n* seniority
dienstbar *a:* **sich etwas d. machen** make use of, harness sthg
dienst·bereit *a* ready for service. **D–herr** *m* employer. **-leistung** *f* service. **-leistungsbranche** *f* service industry
dienstlich *a* official; *adv oft* on official business, in one's official capacity
Dienst·lohn *m* wage(s). **-sache** *f* (matter of) official business. **-stelle** *f* department, administrative office. **-stunden** *fpl* office hours
dies *demons pron, oft a, n* this = DIESES
diesbez. *abbv* = **diesbezüglich** *a* relevant; *adv* in this regard
dieselbe *f*, **dieselben** *pl* 1 *a* & *pron* the same. 2 *pron: f* she, her, it; *pl* they, them
Diesel·motor *m* diesel engine. **-zahl** *f* diesel (index) number
diese(r), dieses *demons a* & *pron* (*adj endgs*) 1 *a* this, *pl* these. 2 *pron* this one, these; he him; she, her; it; they, them. 3 *a* & *pron* (*esp paired with forms of* JENER) the latter—**dies(es) und jenes** this and that
diesig *a* misty, hazy
dies·jährig *a* this year's. **-mal** *adv* this time. **-malig** *a* present, (of) this time. **-seits** *prep* (*genit*) (on) this side of
Dietrich *m* (-e) skeleton key, picklock
Differential·gleichung *f* differential equation. **-rechnung** *f* (differential) calculus. **-strom** *m* differential current
Differenz·bildung *f* (*Math.*) subtraction
Differenzial· = DIFFERENTIAL·
differenzieren 1 *vt/i* differentiate. 2: **sich d.** be differentiated. **differenziert** *p.a* (*oft*) discriminating, sophisticated
Differenz·strom *m* differential current
differieren *vi* differ

difform *a* deformed
diffundieren *vt/i* diffuse
Diffuseur *m* (-e) diffuser, diffusion cell
diffusions·fähig *a* diffusible. **D–fähigkeit** *f* diffusibility. **-geschwindigkeit** *f* velocity of diffusion. **-glühen** *n* (*Metll.*) diffusion annealing, homogenizing. **-hülse** *f* diffusion shell. **-luftpumpe** *f* diffusion pump. **-produkt** *n* diffusate. **-vermögen** *n* diffusibility. **-wärme** *f* heat of diffusion
Digallus·säure *f* digallic acid
digerieren *vt* digest
Digerier·flasche *f* digestion bottle. **-ofen** *m* digestion oven, digester
Digerierung *f* (-en) digestion
Digestions·salz *n* digestive salt (potassium chloride). **-kolben** *m* digestion flask
Digestiv·salz *n* digestive salt. **-mittel** *n* digestive
Digestor *m* (-en) digester. **Digestorium** *n* (..rien) hood, fume cupboard
Digital·technik *f* digital technology
Digitogen·säure *f* digitogenic acid. **Digito·säure** *f* digitoic acid
digonal *a* twofold, two-sided
digyn(isch) *a* (*Bot.*) digynous
Digyre *f* (-n) (*Cryst.*) diagonal rotary axis
Diharnstoff *m* diurea
dihetero·atomig *a* diheteroatomic
dihexaedrisch *a* (*Cryst.*) dihexahedral
dihydratisch *a* dihydric
Dijodid *n* (-e) diiodide; (*in compds*) diiodo-
Dika·brot *n* dica bread
Dikarbon·säure *f* dicarboxylic acid
diktieren *vt/i* dictate
Dilatanz *f* (-en) dilatance
Dilem·öl *n* dilem oil, Java (patchouli) oil
Dilettant *m* (-en,-en), **Dilettantin** *f* (-nen) dilettante, amateur
Dilitur·säure *f* dilituric (or nitrobarbituric) acid
Dille *f* (-n) 1 nozzle, socket. 2 dill
Dill·kraut *n* dill (weed). **-öl** *n* dill oil. **-samen** *m* dill seed
dimensionieren *vt* dimension
dimer *a* dimeric. **dimerisieren** *vt* dimerize
dimethyliert *a* dimethylated, dimethyl-
dimetrisch *a* (*Cryst.*) dimetric (tetragonal)
Dimilch·säure *f* dilactic (or dilactylic) acid
dimorph(isch) *a* dimorphous. **Dimorphie**

f, **Dimorphismus** *m* dimorphism
DIN *abbv* (Deutsche Industrienormalien) German Industrial Standards
Dinas·stein *m* Dinas (*or* silica) brick *cf* DINA·
Dina·ton *m* Dinas clay
Dinatrium *n* disodium
Dina·ziegel *m* Dinas (*or* silica) brick
Ding *n* (-e *or* -er) thing; object; matter, affair
dingen* *vt* (dingte, gedungen) hire
Ding·glas *n* (*Opt.*) objective
Dinicotin·säure *f* dinicotinic acid
dinitriert *a* having two nitrogen groups, dinitrated, dinitro-
Dinkel *m* spelt, German (bearded) wheat
Dino·kristall *m* (*Geol.*) phenocryst
diorit·haltig *a* containing diorite, dioritic
Diorsellin·säure *f* diorsellinic acid
Dioxid *n* (-e) dioxide
Dioxy· dioxy-, dihydroxy-: **-benzol** *n* dihydroxybenzene. **-chinolin** *n* dihydroxyquinoline. **-chinon** *n* dihydroxyquinone. **-naphtalin** *n* dihydroxynaphthalene. **-toluol** *n* dihydroxytoluene
diözisch *a* diæcious
diphasisch *a* diphase, diphasic
Dipheno·chinon *n* diphenoquinone
Diphen·säure *f* diphenic acid
Diphenyl·arsenchlorür *n* diphenylarsenious chloride, chlorodiphenylarsine. **-borchlorid** *n* diphenylboron chloride. **-schwarz** *n* diphenyl black
diphosphor·sauer *a* pyrophosphate of. **D–säure** *f* pyrophosphoric acid
Diphtherin·säure *f* diphtheric acid. **Diphtheritis** *f* diphtheria
dipl. *abbv* (diplomiert) certified. **Dipl.-Ing.** *abbv* (Diplom-Ingenieur) graduate (licensed, professional) engineer
Diplo·eder *m* (*Cryst.*) diplohedron, diploid. **-kokkus** *m* (..kken) diplococcus (*pl* ..cci)
Diplom *n* (-e) diploma, certificate; *in compds* = **diplomiert** *a* certified, holding a diploma, Bachelor of Science in ...
Dipol *m* (-e) dipole. **d—·frei** *a* nonpolar. **D–kraft** *f* dipole force. **-molekel** *n* polar molecule
Dippels·öl *n* Dippel('s) oil
Diptam *m* (*Bot.*) dittany, fraxinella
dipyramidal *a* bipyramidal

Direkt·aufnahme *f* live recording. **-dampf** *m* live steam. **-druck** *m* (*Tex.*) direct printing. **-farbe** *f* direct (cotton) dye. **d–färbend** *a* direct-coloring. **D–farbstoff** *m* direct dye, substantive dye
Direktion *f* (-en) management
Direkt·schrift *f* (*Instrum.*) direct recording. **d–ziehend** *a* (*Dye.*) direct
Dirhizon·säure *f* dirhizoic acid
Dirhodan *n* dithiocyanogen
Dirigent *m* (-en,-en) 1 director, manager, head. 2 conductor. **dirigieren** *vt/i* 1 direct, manage, rule. 2 conduct
Disauerstoff *m* dioxygen, molecular oxygen. **Disäure** *f* diacid
Dischwefel·chlorid *n* disulfur dichloride. **-säure** *f* disulfuric (*or* pyrosulfuric) acid
dischweflig *a* disulfurous, pyrosulfurous
diskontieren *vt* discount
diskontinuierlich *a* discontinuous
diskret *a* discreet; (*Sci.*) discrete, discontinuous
diskutieren *vt, vi* (*oft* + **über**) discuss
dispensieren *vt* dispense; exempt, excuse
Dispergator *m* (-en), **Dispergens** *n* (..entien) dispersion agent
dispergierbar *a* dispersible. **Dispergierbarkeit** *f* dispersibility
dispergieren *vt* disperse. **dispergierend** *p.a* dispersive
dispergier·fähig *a* dispersible. **D–mittel** *n* dispersing agent
Dispergierung *f* (-en) dispersion
Dispergierungs·mittel *n* dispersing agent. **-zahl** *f* dispersion number
Dispergier·vermögen *n* dispersing power
dispers *a* disperse(d)
Dispersions·farbe *f* emulsion paint. **-farbstoff** *m* disperse dye. **-grad** *m* degree of dispersion. **-mittel** *n* dispersion medium (*or* agent). **-vermögen** *n* dispersing power
Dispersität *f,* **Dispersitäts·grad** *m* degree of dispersion
Disponent *m* (-en,-en) managing clerk
disponibel *a* available
disponieren **I** *vt* 1 allot, set aside. **II** *vi* 2 make arrangements. 3 place orders. 4 (+ **über**) have ... at one's disposal; make free use (*or* disposition) of. **disponiert** *pred p.a* 1 in (good, bad) shape (*or* form).

2 (+ **für**) susceptible, inclined (to).
3 (+ **zu** + *inf*) disposed, inclined (to)
Disposition *f* (-en) **1** plan, arrangement.
2 disposition, mood, inclination; predisposition. **3** (**über**) disposal (of). **4** (descriptive) outline, prospectus
Disproportionierung *f* (-en) disproportionation
Diss. *abbv* **1** (Dissoziation) dissoc. **2** = **Dissertation** *f* (-en) dissertation, thesis
dissezieren *vt* dissect
dissimilieren 1 *vt* (*Biol.*) decompose, catabolize. **2** *vt/i* dissimilate
dissociieren *vt/i* dissociate
Dissous-gas *n* dissolved (acetylene) gas
Dissoz. *abbv* (Dissoziation) dissoc.
Dissoziations-arbeit *f* dissociation energy. **-grad** *m* degree of dissociation. **-grenze** *f* dissociation limit. **-vermögen** *n* dissociating power. **-wärme** *f* heat of dissociation. **-zustand** *m* dissociated state (*or* condition)
dissoziierbar *a* dissociable. **dissoziieren** *vt/i* dissociate
Distel *f* (-n) thistle
Disthen *m* (-e) disthene; (*Min.*) **blauer D.** kyanite
Distickstoff *m* dinitrogen, molecular nitrogen. **-oxid** *n* dinitrogen oxide, nitrous oxide
distillieren *vt* distill *cf* DESTILL...
Disulfamin-säure *f* disulfamic acid. **Disulfan** *n* dihydrogen disulfide. **Disulfat** *n* disulfate, pyrosulfate. **Disulfit** *n* (-e) disulfite, pyrosulfite, (*Com.*) metabisulfite. **Disulfo-säure** *f* disulfonic acid
Dita-rinde *f* dita rind
Dithiokohlen-säure *f* dithio(l)carbonic acid
dithionig *a* dithionous; hyposulfurous; thiosulfuric. **-sauer** *a* dithionite of
Dithion-säure *f* dithionic acid
Dithiophosphor-säure *f* dithiophosphoric acid
Diüberjod-säure *f* diperiodic acid
diuretisch *a* diuretic
Divaricatin-säure *f* divaricatinic acid. **Divaricat-säure** *f* divaricatic acid. **Divar-säure** *f* divaric acid
divergierend *p.a* diverging, divergent
divers *a* various, sundry; different

Diversche Flüssigkeit *f* Diver's solution
dividieren *vt/i* divide
Diwein-säure *f* ditartaric acid
Diwolfram-säure *f* ditungstic acid
Dizimt-säure *f* dicinnamic acid
Dizyan° *n* dicyanogen
d.J. *abbv* **1** (dieses Jahres) of this year. **2** (der Jüngere) jr. (junior)
DK *abbv* **1** (Dielektrizitätskonstante) dielectric constant. **2** (Dezimal-Klassifikation) UDC (Universal Decimal Classification)
DKT *abbv* (Dikaliumtartrat) potassium tartrate
d.M. *abbv* (dieses Monats) of this month
Dm, Dmr *abbv* (Durchmesser) diam. (diameter)
DNA *abbv* (Deutscher Normenausschuß) German Standards Committee
d.O. *abbv* (der (die, das) Obige) the above
Dobel *m* (-) peg = DÜBEL
Döbel *m* (-) **I** (*Zool.*) chub, fallfish; (*Bot.*) darnel. **II** peg = DÜBEL
doch I *adv cnjc* **1** yet, but. **2** (*follows verb*) since (+ *emphat verb*): **kommt es d. oft so vor** since it does often happen this way. **II** *adv* **3** still, nevertheless. **4** (*ans to neg statement or question*) on the contrary. **5** (*unemph, oft no specif transl*) **A** (+ *imper*) e.g. **man beachte d.** just notice. **B** (*anticipates agreement or affirm ans*) after all, of course. **6** (*emphat, intensifies verb*) e.g.: **es dissoziiert d.** it does dissociate
Docht *m* (-e) wick. **-garn** *n* wick yarn. **-kohle** *f* cored carbon
Docke *f* (-n) skein, hank, bundle; roller, mandrel; baluster; plug. **docken** *vt* **I** wind (into skeins). **II** dock (ships)
Dodekaeder *n* (-) dodecahedron. **dodekaedrisch** *a* dodecahedral
Dogge *f* (-n) mastiff; Great Dane
Dogger-erz *n* (*Min.*) dogger ore
Döglings-tran *m* doegling oil, arctic sperm oil
Dohle *f* (-n) **I** (*Zool.*) jackdaw. **II** (*Engg.*) drain, culvert = DOLE
Doisynol-säure *f* doisynolic acid
DOK *abbv* (Dinitro-Ortho-Kresol)
Doktorand *m* (-en,-en) doctoral (*or* PhD) candidate. **Doktor-arbeit** *f* doctor's (*or* PhD) thesis

dokumentieren vt document. **Dokumentier·tinte** f record ink

Dolde f (-n) (Bot.) umbel; (Hops) cone, strobile

dolden·blumig, -blütig a umbelliferous. **D–pflanze** f umbelliferous plant

Dole f (-n) drain, culvert

dol(l)ieren vt (Lthr.) pare, shave

Dolmetsch m (-e) interpreter. **dolmetschen** vt/i interpret, translate. **Dolmetscher** m (-), **Dolmetscherin** f (-nen) interpreter

Dolomiten pl (Geol.) Dolomites

dolomit·haltig, dolomitisch a dolomitic. **dolomitisieren** vt (Min.) dolomitize. **Dolomit·kalk** m dolomitic lime

Dom m (-e) 1 cupola, vault, dome. 2 head (of brewing still). 3 cathedral

Doma n (..men) (Min.) dome

Donau f Danube (River)

Donnansches Gleichgewicht Donnan equilibrium

Donner m thunder. **--keil** m 1 thunderbolt. 2 belemnite. **donnern** vt/i thunder; **es donnert** there is thunder. **Donners·tag** m Thursday

Donner·stein m thunderstone, belemnite. **-strahl** m lightning flash

Doppel n (-) duplicate, double

doppel· double, doubly; bi-, di-; see also Doppelt·: **-atomig** a diatomic. **D–ausschlag** m double deflection (of a needle). **d–basisch** a dibasic. **D–bier** n double beer. **-bild** n double image. **-bindung** f double bond. **-boden** m double bottom. **d–bodig** a double-bottomed; ambiguous. **D–bogen** m double bend; U–tube; double sheet (of paper). **d–borsauer** a biborate of, tetraborate of. **-brechend** a doubly refracting. **D–brechung** f double refraction. **d–breit** a (of) double-width. **D–brillantscharlach** m double brilliant scarlet. **-chlorid** n double chloride, bichloride, dichloride. **-chlorzinn** n stannic chloride, $SnCl_4$. **d–chromsauer** a bichromate of, dichromate of. **-deutig** a ambiguous. **-fädig** a double-threaded, bifilar. **-farbig** a dichroic, dichromatic. **D–farbigkeit** f dichroism. **-färbung** f (Micros.) double staining; counterstaining.

-fernrohrlampe f binocular magnifier. **-gänger** m double, alter ego. **-gas** n mixture of coal gas & water gas. **-gestaltung** f dimorphism. **-gitter** n (Elec.) double grid. **-glocke** f double bell jar. **-härten** n (Metll.) double quench. **d–hochrund** a biconvex. **-hohlrund** a biconcave. **D–jodquecksilber** n mercuric iodide. **-kessel** m double (or jacketed) boiler (or kettle). **d–kohlensauer** a bicarbonate of. **-lebig** a amphibious. **D–linie** f (Spect.) double line, doublet. **d–logarithmisch** a log-log. **D–lötung** f double seal (in Dewar flask). **-mantel** m (double-wall) jacket. **-mantelgefäß** n double-walled vessel. **-metall·** bimetallic..

doppeln vt double; duplicate; fold; twist **Doppel·ofen** m double furnace (or oven). **d–oxalsauer** a binoxalate of. **D–oxid** n dioxide; double oxide. **-punkt** m 1 (Print.) colon. 2 (Math.) double point. **d–polig** a bipolar. **D–salz** n double salt. **-schale** f double dish, Petri dish. **-schicht** f double layer (or shift). **d–schichtig** a two-layered. **-schlicht** a dead-smooth; superfine (file)

Doppelschwefel·eisen n iron disulfide. **d–sauer** a bisulfate of. **D–säure** f pyrosulfuric acid. **-zinn** n stannic sulfide **doppel·schwefligsauer** a bisulfite of. **-seitig** a bilateral; double-faced. **D–schweißstahl** m double-shear steel. **-silikat** n disilicate. **d–sinnig** a ambiguous. **D–spat** m Iceland spar. **d–spurig** a two-track, double-track. **-stark** a double-strength. **D–stock·, d–stöckig** a double-deck. **D–strahlenbundspektrometer** n double-beam spectrometer. **D–stück** n duplicate. **-sulfat** n double sulfate, bisulfate. **-sulfit** n double sulfite, bisulfite

doppelt 1 a double, twofold, twice as much; dual, duplex—**das D—e** twice as much. 2 adv doubly, twice (as). 3 (in compds)·bi(n)-, di-, see also DOPPEL·: **·-basisch** a bibasic, dibasic. **-brechend** a birefringent. **-hochrund** a convexo-concave, biconvex. **-hohl** a concavo-concave. **-kieselsauer** a bisilicate of. **-kleesauer** a binoxalate of. **-kohlensauer** a bicarbonate of

= DOPPELKOHLENSAUER: **d-es Salz** bicarbonate. **-normal** *a* double (*or* twice) normal

Doppelton·farbe *f* double-tone color

doppelt·selenigsauer *a* biselenite of. **-selensauer** *a* biselenate of. **-titansauer** *a* bititanate of. **-ungesättigt** *a* doubly unsaturated. **-vanadinsauer** *a* bivanadate of. **-weinsauer** *a* bitartrate (*or* acid tartrate) of. **-wirkend** *a* double-acting. **-wolframsauer** *a* bitungstate of

Doppel·umsetzung *f* double decomposition

Doppelung *f* (-en) doubling = DOPPLUNG

Doppel·verbindung *f* double compound. **-wägung** *f* double weighing. **d–wandig** *a* double-walled. **D–weghahn** *m* two-way cock. **d–wirkend** *a* double-acting; amphicroic, amphoteric. **D–wirkung** *f* double action. **-wurzel** *f* (*Math.*) double root. **-zentner** *m* double centner, 100 kilo(gram)s. **-zersetzung** *f* double decomposition. **-zünder** *m* combination fuse. **d–züngig** *f* two-faced, double-dealing

Döpper *m* (-) (*Mach.*) riveting die (*or* set)

Dopplung *f* (-en) doubling = DOPPELUNG

Dorf *n* (-̈er) village

Dorn *m* I (-en) thorn, spine. II (-e *or* -̈er) arbor, mandrel; pin; tongue; spike; (*Metll.:Copper*) slag. **d——artig** *a* thornlike, spiny. **D–busch** *m* thorn bush

dornen *vt* pierce; punch, drift; indent. **Dornen·stein** *m* (*Salt Mfg.*) thornstone. **Dörner·schlacke** *f* (*Metll.*) bulldog. **dornig** *a* thorny, spiny. **Dorn·stein** *m* thornstone = DORNENSTEIN

dorren *vi*(s) dry (up, out), wither

dörren *vt* dry (out), dehydrate, desiccate; cure (by dryng), scorch

Dörr·fleckenkrankheit *f* (wood) blight. **-fleisch** *n* dried meat; smoked bacon. **-gemüse** *n* dried vegetable(s)

Dorr·klassierer, -klassifikator *m* Dorr classifier

Dörr·obst *n* dried fruit. **-ofen** *m* drying oven

Dorschleber·tran *m,* **-öl** *n* cod-liver oil

dort *adv* there. **-hin** *adv* there, to that place. **dortig** *a* there, of (*or* in) that place

Dose *f* (-n) I can, tin; box, capsule; (*Elec.*) socket. II (*Med.*) dose = DOSIS

Dosen I cans, etc; *pl of* DOSE I. II doses: *pl of* DOSIS

Dosen·barometer *n* aneroid barometer. **-blech** *n* sheet "tin" for cans. **-fleisch** *n* canned meat. **-gemüse** *n* canned vegetable(s). **-konserven** *fpl* canned food. **-libelle** *f* circular spirit level. **-milch** *f* canned (evaporated) milk. **-öffner** *m* can opener

Dosier·anlage *f* dosing plant (*or* unit). **-automat** *m* automatic dosage dispenser

dosieren *vt* I dose, measure out; administer in doses; meter. 2 (*Wine*) sweeten, dose (champagne). **dosiert** *p.a* (*oft*) calculated, measured. **Dosierung** *f* (-en) 1 dosing, dosage; metering. 2 dose, measured amount. **Dosier·vorrichtung** *f* dosing device

Dosis *f* (*pl* Dosen) dose. **--leistung** *f* dose rate

Dossierung *f* (-en) slope

Dost *m* (-e) marjoram; origanum. **Dosten·öl** *n* origanum oil. **Dost·kraut** *n* origanum

dotieren *vt* fund, endow

Dotter *m,n* (-) (egg) yolk, vitellus; (*Bot.*) camelina, gold of pleasure, wild flax. **--bläs/chen** *n* yolk sac. **-blume** *f* buttercup; marsh marigold. **-gelb** *n* egg yolk; egg yellow. **-öl** *n* camelina (seed) oil. **-sack** *m* yolk sac

doublieren *vt* 1 (*Tex.*) double. 2 (*Metll.*) plate

Dozent *m* (-en,-en) university lecturer

D.P. *abbv* (deutsches Patent) German patent. **D.P.a.** *abbv* (deutsche Patentanmeldung) Ger. pat. application

D.Prior. *abbv* (deutsche Priorität) German priority

DR *abbv* (Deutsche Reichsbahn) German State Railroads (in DDR)

Drache *m* (-n,-n) dragon. **Drachen** *m* (-) kite

Drachen·blut *n,* **-gummi** *n* dragon's blood (*or* resin)

Dragée *f* (-n) *or* *n*(-s) 1 (sugarcoated) pill. 2 (piece of) candy. **dragieren** *vt* sugarcoat (pills, candy)

Dragon, Dragun *m,n*(-s) tarragon

Draht *m* (⁼e) wire, filament; (*Shoemaking*) (waxed) twist; (*Wood*) grain; (*in compds oft*) telegraphic, cable: **-anschrift** *f* cable (*or* telegraphic) address. **-asbestgewebe** *n* asbestos wire gauze. **-band** *n* metal tape. **-brief** *m* telegram. **-bund** *m* wire coil. **-bürste** *f* wire brush. **-dreieck** *n* wire triangle

drahten I *vt/i* wire, telegraph. **II** *a* (of) wire

Draht· wire: **-faser** *f* (*Paper*) wire mold. **-gaze** *f* wire gauze. **-geflecht** *n* wire netting. **-gewebe** *n* wire mesh. **-gewicht** *n* wire weight, balance rider. **-glas** *n* wire (*or* armored) glass. **-klemme** *f* wire clamp; (*Elec.*) binding post. **-korb** *m* wire basket. **-lack** *m* wire enamel. **-lehre** *f* wire gauge. **d–los** *a* wireless; (of) radio. **D–netz** *n* wire net(ting). **-prüfung** *f* wire testing. **-ring** *m* wire ring (*or* coil). **-rolle** *f* coil of wire. **-schleife** *f* wire loop. **-seil** *n* wire cable (rope, cord). **-seilbahn** *f* cable railway, funicular. **-sieb** *n* wire sieve (*or* strainer). **-stärke** *f* wire diameter. **-stift** *m* wire nail (tack, pin). **-walzwerk** *n* wire rolling mill. **-zange** *f* wire-cutting pliers. **-ziehen** *n* wire drawing. **-ziehstein** *m* wire-drawing die. **-zug** *m* wire-drawing; wire mill

drainieren *vt/i* drain

drall *a* 1 tight(ly twisted). 2 buxom. **Drall** *m* (-e) 1 twist; spin, rotation. 2 torque. 3 spiral thread; rifling

dran *adv & sep pfx* on it, etc = DARAN

Drän *m* (-e), **drähnen** *vt/i* drain

drang penetrated: *past of* DRINGEN

Drang *m* 1 urge, (inner) drive, impetus, impulse. 2 stress, pressure

drängen 1 *vt/i* press, push, drive. 2 *vi* be urgent; (+ **nach**) strive (for). 3: **sich d.** crowd, press, push (one's way). *Cf* GEDRÄNGT

dränieren *vt/i* drain

Drän·rohr *n*, **-röhre** *f* drain pipe

Dränung *f* (-en) draining, drainage

Draß *m* (oil) dregs

Drastikum *n* (..ika) powerful laxative.

drastisch *a* 1 drastic, rigorous. 2 vivid, graphic

drauf *adv & sep pfx* thereupon, etc = DARAUF. **Drauf·sicht** *f* top view, plan (view)

draus *adv & sep pfx* out of it, etc = DARAUS.

draußen *adv* outside, out: **dort d.** out there

drechseln 1 *vt* turn (out, on a lathe). 2 *vi* (**an**) work (on)

Dreck *m* 1 dirt, mud; muck; excrement. 2 dross. **--eisen** *n* pig iron

Dreh· rotary, rotation; revolving: **-achse** *f* axis of rotation; pivot, fulcrum; knife edge (of a balance). **-band** *n* spinning band. **-bank** *f* lathe

drehbar *a* turnable, rotary; rotating; twistable: **d.** (**eingesetzt** or **gelagert**) pivoted. **Drehbarkeit** *f* turnability; rotation, rotary capacity; versatility

Dreh·bewegung *f* rotary motion. **-bolzen** *m* privot (pin). **-bottich** *m* rotating tub. **-ebene** *f* plane of rotation (*or* revolution)

drehen 1 *vt & sich d.* turn, rotate. 2 *vt* twist; crank; roll; contrive. 3 *vi* turn, veer; (+ **an**) tamper (v.ith). 4: **sich d.** revolve, pivot, spin—**es dreht sich um...** it is a matter of... **drehend** *p.a* (*usu*) **sich d.** revolving, rotating, rotary; spinning

Dreher *m* (-) turner; rotator, rotor; crank

Dreh· rotary, rotation; revolving; pivoted, swivel: **-faß** *n* drum, tumbler. **-feder-waage** *f* torsion balance. **-federzahl** *f* torsion coefficient. **-feld** *n* (*Elec.*) rotating field. **-festigkeit** *f* torsional strength. **-gelenk** *n* swivel joint. **-geschwindigkeit** *f* speed (of rotation); rotary speed; revolutions per minute. **-impuls** *m* angular velocity. **-inversions-achse** *f* (*Cryst.*) rotary inversion axis. **-kalzinierofen** *m* revolving roaster. **-kocher** *m* rotary boiler. **-kolben** *m* rotary piston. **-kondensator** *m* revolving roaster. **-kraft** *f* torsional force, torque. **-kran** *m* (*Mach.*) rotary (*or* swivel) crane. **krankheit** *f* (*Vet.*) staggers. **-kreuz** *n* turnstile; rotating shutter; (*Brew.*) sparger. **-maschine** *f* lathe. **-moment** *n* moment of rotation, torque. **-ofen** *m* rotary furnace (*or* kiln). **-punkt** *m* pivot(al point), central point, fulcrum. **-richtung** *f* direction of rotation. **-rohrofen** *m* rotary furnace. **-schalter** *m* (*Elec.*) rotary switch. **-scheibe** *f* 1 turntable. 2 potter's wheel. 3 (telephone) dial. **-schieber** *m* rotary slide valve. **-sinn** *m* direction of rotation (*or* revolution). **-späne** *mpl* (lathe) turnings. **-spannung** *f* torsional strain. **-spiegelung** *f* (*Cryst.*) rotary reflection, combined rotation and reflection. **-spindel** *f* revolving (*or* live) spindle. **-sprenger** *m* rotary sprinkler. **-spulgalvanometer** *n* rotary coil galvanometer. **-stahl** *m* turning (*or* lathe) tool. **-strom** *m* rotary (*or* three-phase) current; alternating current. **-symmetrie** *f* rotational symmetry. **-teller** *m* turntable. **-trommel** *f* rotary drum

Drehung *f* (-en) rotation, turn(ing); torsion, twist(ing); revolution; spin

Drehungs- rotary, etc: see *also* DREH-: **-achse** *f* axis of rotation (*or* revolution). **-festigkeit** *f* resistance to torsion, torsional strength. **d-frei** *a* nonrotary, irrotational. **D-grad** *m* degree of rotation. **-kraft** *f* rotary power. **-schwingung** *f* torsional vibration. **-streuung** *f* rotary dispersion. **-vermögen** *n* rotary power. **-winkel** *m* angle of rotation

Dreh- rotary, rotation; revolving; pivoted, swivel: **-vektor** *m* rotation vector. **-vermögen** *n* rotary power. **-waage** *f* torsion balance. **-wert** *m* (*Chem.*) specific rotation. **-winkel** *m* angle of rotation. **-zahl** *f* number of revolutions; rpm. **-zähler** *m* revolution counter. **-zahlgeber** *m* tachometer, speedometer. **-zahlmesser** *m* revolution counter. **-zapfen** *m* pivot. **-zylinder** *m* revolving cylinder

drei *num* three; (*as pfx oft*) tri-, triple. **Drei** *f* (-en) (figure) three

drei-achsig *a* triaxial. **-atomig** *a* triatomic. **-basisch** *a* tribasic. **-basischphosphorsauer** *a* orthophosphate of. **D-bein** *n* tripod. **-blatt** *n* trefoil. **d-blätt(e)rig** *a* three-leaved. **D-brenner** *m* triple burner. **-eck** *n* triangle. **d-eckig** *a* triangular

Dreiecks-koordinaten *fpl* triangular coordinates. **-lehre** *f* trigonometry

dreier of three: *genit of* DREI. **Dreier** *m* (-) (figure, number) three. **-gemisch** *n* triple mixture. **-gruppe** *f* 3-group, group of three. **d-lei** *a* (*no endgs*) of three kinds, three kinds of. **D-stoß** *m* three-body (*or* triple) collision

dreifach *a* triple, threefold, treble; (+

Chem. compd) tri: **d-es Chlorjod** iodine trichloride. **D-bindung** *f* triple bond. **-essig** *m* triple vinegar. **-d-fädig** *a* three-ply (yarn). **-frei** *a* trivariant. **D-verbindung** *f* ternary compound

Dreifarben- three-color, trichrome: **-druck** *m* three-color print(ing)

drei-farbig *a* three-color(ed), trichromatic. **D-farbigkeit** *f* trichroism. **d-flächig** *a* three-faced, trihedral. **D-flächner** *m* (*Geom.*) trihedron. **d-flammig** *a* three-flame, triple (burner). **D-fuß** *m* tripod. **d-füßig** *a* three-footed, tripedal. **-gliedrig** *a* three-membered, three-part, (*Math.*) trinomial. **-gruppig** *a* three-group, of three groups. **D-gutscheidung** *f* three-product separation. **d-halsig** *a* three-necked. **D-halskolben** *m* three-necked flask

Dreiheit *f* (-en) triad; trinity

Drei-kantfeile *f* three-cornered (*or* triangular) file. **d-kantig** *a* triple-edged, triangular. **-kernig** *a* trinuclear

Dreikörper-, -apparat *m* triple-effect apparatus. **-verdampfer** *m* triple-effect evaporator

Dreileiter-system *n* (*Elec.*) three-wire system

dreimal *adv* three times. **dreimalig** *a* three(-time), threefold. **Dreimal-schmelzerei** *f* (*Metll.*) three-stage refining

drein *adv & sep pfx* in(to) it = DAREIN

Dreiphasen-, dreiphasig *a* three-phase

drei-polig *a* three-pole. **D-punkt-** three-point. **drei-quantig** *a* three-quantum, of three quanta. **D-ring** *m* three-membered ring. **-salz** *n* trisalt (neutral salt of a triacid); triple salt. **d-säurig** *a* triacid. **-schenkelig** *a* three-legged. **D-schenkelrohr** *n* three-way tube, T-tube. **d-schichtig** *a* three-layered. **-seitig** *a* three-sided, trilateral, triangular

dreißig *num* thirty. **dreißigste** *a* thirtieth

dreist *a* bold; brazen, arrogant

Dreistein-wurzel *f* feverroot (*Triosteum*)

drei-stellig *a* three-place (number)

Dreistoff- three-component, ternary: **-diagramm** *n* ternary (*or* triangular) diagram. **-legierung** *f* three-component (*or* ternary) alloy

Drei-strom *m* (*Elec.*) three-phase current

drei·stufig *a* three-step, three-stage. **-stün-dig** *a* three-hour. **D–system** *n* three-component system. **d–teilig** *a* three-part, tripartite. **-undeinachsig** *a* (*Cryst.*) monotrimetric (hexagonal). **-wandig** *a* three- (*or* triple-)walled

Dreiweg(e)· - three-way: **-hahn** *m* three-way cock. **-stück** *n* three-way piece (*or* tube)

drei·wertig *a* trivalent. **D–wertigkeit** *f* trivalence. **d–winkelig** *a* triangular. **D–zack** *m* trident. **d–zählig** *a* threefold, triple, ternary; (*Cryst.*) trigonal

dreizehn *num* thirteen. **dreizehnte** *a* thirteenth

Drell *m* (*Tex.*) drill(ing) = DRILLICH

Dresbacher·blau *n* Dresbach (Prussian) blue

dreschen* (drosch, gedroschen; drischt) *vt/i* thresh, thrash (out)

dressieren *vt* 1 dress (meat, fabrics); finish (metals). 2 train (animals)

drgl. *abbv* (dergleichen) the like

D.R.G.M *abbv* (Deutsches Reichs-Gebrauchsmuster) Ger. registered design

Drill·achse *f* (*Quantum Theory*) rotator. **-bohrer** *m* drill

drillen *vt/i* drill; twist (threads)

Drillich *m* (-e) (*Tex.*) drill(ing); ticking, denim

Drilling *m* (-e) triplet; (*Cryst.*) tripling; (*Mach.*) lantern (wheel). **·-salz** *n* triple salt. **Drillings·pumpe** *f* triplex pump

Drillung *f* (-en) drilling; torsion, twist(ing) *cf* DRILLEN. **Drillungs·kristall** *m* twister crystal

drin *adv* in it, etc = DARIN

Dr.-Ing. *abbv* (Doktor-Ingenieur) graduate engineer (with a doctorate in technology)

dringen* (drang, gedrungen) *vi* 1 (s) penetrate, force one's way. 2 (h) (**auf**) insist (on). 3 (*h or* s) (**in**) importune, urge. **dringend** *p.a* urgent. *Cf* GEDRUNGEN

dringlich *a* urgent, pressing. **Dringlichkeit** *f* urgency

drinnen *adv* inside, in: **da d.** in there

drischt threshes: *pres of* DRESCHEN

dritt *a* third—**zu d.** in threes, in a group of three

Drittel *n* (-) (one) third. **·-alkohol** *m* (*Bact.*) Ranvier's alcohol (one part alcohol to three parts water). **d–sauer** *a* tribasic. **D–silber** *n* tiers-argent, aluminical silver (one part silver to two of aluminum or nickel silver)

drittens *adv* thirdly, in the third place

dritt·letzt *a* third last, third from the end. **-rangig** *a* third-rate

droben *adv* (up) above = OBEN

Droge *f* (-n) drug

drogen·abhängig *a* drug-addicted. **D–abhängige** *m/f* (*adj endgs*) drug addict. **-händler** *m* drug dealer. **-handlung** *f* 1 drug dealing, drug business. 2 *oft* = DRO-GERIE. **-kunde** *f* pharmacology. **-waren** *fpl* drugs, pharmaceuticals

Drogerie *f* (-n) (non-prescription) drugstore (also selling cosmetics)

Drogett *n* (*Tex.*) drugget

Drogist *m* (-en,-en) druggist

drohen *vi* (*dat*) threaten, menace. **drohend** *p.a* (*oft*) impending, threatened

Drohne *f* (-n) drone. **dröhnen** *vi* drone, rumble, groan

Drohung *f* (-en) threat

drosch threshed: *past of* DRESCHEN

Drossel *f* (-n) 1 (*Zool.*) thrush. 2 (*Vet.*) windpipe. 3 (*Mach.*) throttle (valve), choke. **·-ader** *f* jugular vein. **-bein** *n* collar bone, clavicle. **-klappe** *f* throttle valve; damper

drosseln *vt* throttle, choke (off); reduce, cut down, cut short

Drossel·spule *f* (*Elec.*) choke coil, reactor. **-ventil** *n* throttle valve

DRP *abbv* (Deutsches Reichspatent) German Patent (pre-1945) *cf* DBP(A)

Dr. phil. *abbv* Ph.D. **Dr. rer. nat.** *abbv* (Doctor rerum naturalium) Doctor of Science

drüben *adv* over there, abroad *cf* HÜBEN

drüber *adv & sep pfx* over it, etc = DARÜ-BER, *cf* DRUNTER

Druck *m* (-e *or* ⸚e) 1 pressure: **im D.** under pressure, in a bind. 2 burden. 3 squeeze. 4 printing, impression: **in D. geben (gehen)** send (go) to press; **im D.** in (*or* on) the press. 5 print, type: **im D. erscheinen** appear in print. **·-abfall** *m* drop in pressure. **d–abhängig** *a* pressure-dependent. **D–abnahme** *f* drop in pressure. **-ansatz** *m* (*Tex.*) printing color (*or* paste). **-anstieg** *m* rise in pressure. **-apparat** *m* pressure apparatus. **-ausgleich** *m* equalization of pressure. **-behälter** *m* pressure reservoir. **-belastung** *f*: **unter D.** under pressure. **-beuche** *f* bucking under pressure. **-birne** *f* acid egg, globular cistern for raising liquids by gas pressure. **-blau** *n (Dye.)* printing blue. **-bombe** *f* pressure bomb. **-buchstabe** *m* type, printed letter. **-destillation** *f* pressure distillation. **d–dicht** *a* pressure-tight, pressurized. **D–einfluß** *m* influence of pressure. **-einheit** *f* unit of pressure. **-elektrizität** *f* piezoelectricity

drucken *vt/i* print; stamp

drücken I *vt/i* **1** press, squeeze, pinch. **II** *vi* **2** oppress, depress. **3** lower, force down. **drückend** *p.a* oppressive

Druck-entlastung *f* relief from pressure. **-entwicklung** *f* pressure build-up

Drucker *m* (-) printer

Drücker *m* (-) handle, knob; push button; trigger; *(Metll.)* pusher

Druckerei *f* (-en) printer's (establishment); printing plant; *(Tex.)* print works

Drucker-farbe *f* printer's ink

Druck-erhitzung *f* heating under pressure. **-erhöhung** *f* increase of pressure. **-erniedrigung** *f* decrease of pressure

Drucker-schwärze *f* printer's ink

Druck· pressure; printing: **-farbe** *f* printing color (*or* ink). **-fehler** *m* misprint, typographical error. **d–fest** *a* pressure-resistant. **D–festigkeit** *f* resistance to pressure; compressive strength. **-filter** *n* pressure filter. **-firnis** *m* printer's varnish. **-flasche** *f* pressure bottle, aspirator. **-flüssigkeit** *f* hydraulic medium. **-gärung** *f* pressure fermentation. **-gas** *n* compressed (*or* pressure) gas. **-gefälle** *n* pressure gradient (*or* drop). **-gefäß** *n* pressure vessel. **-gefäßversuch** *m* autoclave test. **-guß** *m* **1** *(Metll.)* pressure die-casting. **2** *(Plastics)* injection molding. **-heft** *n* (printed) pamphlet, brochure. **-höhe** *f (Mach.)* head. **-hydrierung** *f* hydrogenation under pressure. **-kammer** *f* pressure chamber. **-kapsel** *f* **1** load cell (*or* capsule). **2** flameproof casing. **-kattun** *m* printed calico. **-kessel** *m (Mach.)* pressure tank. **-kleister** *m* printing starch. **-knopf** *m* pushbutton. **-kocher** *m* pressure cooker (*or* boiler). **-kolben** *m* **1** pressure flask.

2 *(Mach.)* piston; plunger. **-kraft** *f* compressive force; crushing stress. **-kriechversuch** *m* compressive creep test. **-lack** *m* printer's lake. **-legung** *f* printing. **-leitung** *f* high-pressure piping. **-luft** *f* compressed air. **-lufthammer** *m* jackhammer, pneumatic (*or* air) hammer. **-luftpumpe** *f* **1** air compressor. **2** tire pump. **-maschine** *f* printing machine. **-messer** *m* pressure gauge, manometer. **-minderer** *m* pressure reducer. **-minderventil** *n* pressure-reducing valve. **-papier** *n* printing paper. **-platte** *f* **1** printing plate. **2** pressure plate. **-probe** *f* **1** printer's proof. **2** compression test (specimen); pressure test. **3** *(Barley)* squeezing test. **-prüfung** *f* pressure testing. **-pumpe** *f* pressure (*or* force) pump. **-regler** *m* pressure regulator. **-rohr** *n*, **-röhre** *f* pressure tube; force pipe. **-rückgang** *m* pressure drop. **-sache** *f* printed matter. **-schacht** *m (Metll.)* pressure pit. **-schalter** *m* pushbutton (*or* pressure) switch. **-schieferung** *f (Geol.)* rock cleavage; schistosity. **-schlauch** *m* pressure tubing (*or* hose). **-schmierung** *f* pressure lubrication; forced(-feed) lubrication. **-schraube** *f* pressure screw, setscrew; *(Aero.)* pusher propeller. **-schreiber** *m* pressure recorder, manograph. **-schrift** *f* **1** print (*or* block) letters. **2** printing. **3** (printed) publication. **-schwarz** *n (Dye.)* printing black. **-seite** *f* **1** printed page. **2** compression face. **-spannung** *f* compressive stress. **-steigerung** *f* rise in pressure. **-stempel** *m* **1** punch, stamp. **2** (plunger-)piston. **-stock** *m (Print.)* cut, electro(type). **-stollen** *m (Metll.)* pressure gallery. **-stutzen** *m* pressure connection (*or* joint). **-taste** *f* pushbutton, press-key. **d–unabhängig** *a* pressure-independent. **D–ventil** *n* pressure (*or* discharge) valve. **-veränderung** *f* change in pressure. **-verfahren** *n* printing process. **-vergasung** *f* gasification under pressure. **-verlust** *m* loss of pressure. **-verminderung** *f* decrease in pressure. **-versuch** *m* compression (pressure, crushing) test. **-waage** *f* pressure balance. **-walze** *f* **1** pressure roll(er) (*or* cylinder). **2** printing roller. **-wasser** *n* water under pressure; (*in compds*) hydraulic. **-welle** *f*

pressure wave. **-zeug** n printing cloth.
-zunahme f increase in pressure.
-zwiebel f pressure bulb

Druden·fuß m (Bot.) lycopodium, club
moss

drum adv & cnjc about it, etc; that is why,
etc = DARUM

drunter adv under it, etc = DARUNTER: **d.
und drüber** in a confused tangle

Drusch m (-e) **1** threshing. **2** threshed
grain

Drüs/chen n (-) glandule, small gland cf
DRÜSE

Druse f (-n) **1** druse, geode. **2** (Vet.) stran-
gles. **3** pl lees, dregs; husks. **4** (Med.) sulfur granules

Drüse f (-n) gland

drusen vi turn dreggy (or turbid)

Drüsen· gland(ular): **d—·artig** a glandular

Drusen·asche f calcined (wine) lees.
-branntwein m spirits distilled from fermented lees. **d—förmig** a drusy, drused

Drüsen·gewebe n glandular tissue.
-krankheit f glandular disease. **-kropf**
m goiter

Drusen·öl n grapeseed oil (from wine
dregs)

Drüsen·saft m glandular secretion

Drusen·schwarz n Frankfort black (from
dregs)

Drüsen·sekret n glandular secretion

drusig a drusy, drused cf DRUSE

drüsig a gland-like, glandular cf DRÜSE

d.s. abbv (das sind) these are

Dschungel m (-) jungle

Dschunke f (-n) junk (sailboat)

dsgl. abbv (desgleichen) likewise, ditto

ds. J. abbv (dieses Jahres) of this year

dt. abbv (deutsch) German

D.T.A. abbv (Differentialthermoanalyse)
differential thermal analysis

D.T.G.A. abbv (differentialthermogravi-
metrische Analyse) differential thermogravimetric analysis

dtsch. abbv (deutsch) German

Dtz., Dtzd. abbv (Dutzend) doz.

d.U. abbv (der Unterzeichnete) the undersigned

Dübel m (-) (wall) plug; peg, dowel

Dublette f (-n) doublet, duplicate

dublieren vt double; (Metll.) concentrate,
plate. **Dublier·stein** m concentrated
metal, regulus

ducken 1 vt humiliate; duck (one's head).
2: sich d. duck, crouch; submit

Ducker m (-) siphon = DÜKER

Duck·stein m calcareous tufa

duff a dull, dead

Duft m (-̈e) fragrance, aroma; vapor, mist.
düfteln vi have a slight aroma. **duften** vi
have a fragrance: **es duftet** there is a fragrance. **duftend** p.a fragrant, sweet-
smelling. **Duft·essig** m aromatic vinegar. **duftig** a filmy; misty; aromatic

duftlos a odorless, unscented. **Duft·
losigkeit** f odorlessness

Duft·öl n fragrant (or aromatic) oil. **d—
reich** a (richly) fragrant. **D—stoff** m perfume; aromatic principle; essential (or
aromatic) oil. **-wasser** n scented water

Dugong·öl n dugong oil

Düker m (-) siphon

Duktilität f ductility

Dulcit n dulcite, dulcitol

dumm a dumb, stupid, foolish, silly

dumpf a **1** musty, stuffy (air). **2** dull, hollow,
muffled (sound). **3** apathetic. **4** vague.
5 numb. **Dumpfheit** f mustiness, stuffiness, etc. **dumpfig** a musty, moldy

Düne f (-n) dune. **Dünen·sand** m dune
sand

Dung m manure, (natural) fertilizer, dung.
d—·artig a dung-like, stercoraceous

Dünge·gips m gypsum for manuring.
-harnstoff m fertilizer urea. **-jauche** f
dung water, liquid manure. **-kalk** m manuring lime. **-mittel** n fertilizer

düngen 1 vt/i manure, fertilize. **2** vt use as
fertilizer. **Dünge·pulver** n powdered
fertilizer

Dünger m (-) manure, fertilizer. **-bedarf**
m fertilizer requirement(s)

Dung·erde f mold, compost, humus

Dünger·fabrik f fertilizer factory. **d—
fordernd** a requiring fertilizer. **D—salz**
n fertilizer salt. **-streuer** m manure
spreader. **-wert** m manurial value

Dünge·salz n saline manure; fertilizer salt

Dung·fabrik f fertilizer factory. **-jauche** f
dung water, liquid manure. **-mittel** n fer-

tilizer. **-pulver** n powdered manure. **-salz** n saline manure; fertilizer salt

Düngung f (-en) manuring, fertilization

dunkel a dark, dim, obscure. **Dunkel** n dark(ness), obscurity. **d—·farbig** a darkcolored. **D–färbung** f darkening, dyeing in a dark color. **-feld** n (Opt.) dark field, dark ground. **-feldbeleuchtung** f darkfield illumination. **-glühhitze** f dull red heat (above 700°)

Dunkelheit f (-en) darkness, dimness; obscurity. **Dunkel·kammer** f darkroom; camera obscura. **dunkeln** 1 vt/i darken; deepen, sadden (colors). 2 vi get dark

Dunkel-öl n (Petrol.) black oil. **-raum** m dark space; darkroom

Dunkelrot·glut f dull red heat (above 700°). **-gültigerz** n dark red silver ore, pyrargyrite

Dunkel·stellung f (Opt.) dark position. **-strahlung** f invisible radiation. **-strom** m dark current. **-tran** m dark train oil, (specif) dark brown cod oil

Dunk(e)lungs·bad n (Dye.) saddening liquor

dünn a thin, sparse, dilute; rare. **D—·bier** n weak beer. **d–blättrig** a thin-leaved; laminated. **D–darm** m small intestine. **-darmentzündung** f enteritis

Dünne f thinness, sparseness, etc cf DÜNN. **dünnen** vt thin, dilute

dünn·flüssig a highly liquid (or fluid); watery. **D–flüssigkeit** f thinness, liquidity, low viscosity; thin liquid. **d–geschichtet** a thinly layered. **-häutig** a thin-skinned, filmy

Dünnheit f thinness; sparseness, diluteness, rarity

Dünn·maische f (Brew.) thin mash. **-saft** m (Sugar) thin juice. **-säure** f dilute(d) acid. **d–schalig** a thin-skinned, thin-shelled

dünn·schlagen* vt beat thin, beat out (flat)

dünn·schleifen* vt grind thin

Dünn·schliff m (Min.) thin section. **d–schuppig** a in thin scales. **D–stein** m (Metll.) thin mat(te); (Jewelry) table stone. **d–tafelig** a in thin plates. **D–teer** m thin tar, fluid tar

dünn·walzen vt roll thin, roll out (flat)

dünn·wandig a thin-walled

Dunst m (ː͏e) 1 vapor, steam; fume(s); mist, haze. 2 dust shot, bird shot. 3 medium grind (of meal). **·-abzug** m, **-abzugshaube** f hood (for fumes). **-abzugsrohr** n vent pipe. **d–artig** a vaporous. **D–bad** n vapor bath. **-bläs/chen** n vapor bubble

dunsten vi (give off) steam, exhale vapor, fume; vaporize

dünsten 1 vt stew, steam (food). 2 vi give off fumes (or vapors)

Dunst·essig m aromatic vinegar. **-fang** m hood. **d–förmig** a vaporous. **D–glocke** f smog blanket. **-hülle** f vaporous envelope, atmosphere

dunstig a steamy, vaporous; misty; hazy. **Dunstigkeit** f steaminess; mistiness; haze

Dunst·kreis m atmosphere; aura. **-loch** n vent, airhole. **d–los** a vaporless, fumeless. **D–messer** m atmometer. **-milch** f evaporated milk. **-mittel** n (Lthr.) mulling agent. **-obst** n stewed fruit. **-rohr** n ventilation pipe. **-vulkanisation** f (Rubber) vapor cure

Dünung f (-en) (ground) swell

duplizieren vt duplicate, double

Dural·blech n duralumin sheet

durch I prep (accus) 1 through. 2 (esp with passive) by. 3 by (means of): **d. Erhitzen, d. Erhitzung** by heating. 4 as a result of. 5 for, throughout (a period of time). 6 (Math.) over, divided by. 7 (Time) past II adv 8 through. 9: **ist d.** (oft) has passed, has gone by. 10 over, done, finished. 11 cut (or worn) through. 12 matured. III pfx (sep & insep) through, thoroughly, without interruption (oft + phr with prep DURCH)

durch·arbeiten I vt 1 work out, make a thorough study of. 2 knead thoroughly. 3 (Dye.) pole. II vt/i work through (the day, etc). III vt & sich d. work one's way through. **Durcharbeitung** f (-en) thorough study; elaboration

durch·ätzen vt eat through, corrode

durchaus adv 1 thoroughly, through and through. 2 absolutely, definitely

durch·backen* vt/i bake through (or thoroughly)

durch/beißen* 1 vt (sep & insep) 1 bite

through, bite in two. **2** (*sep*): **sich d.** struggle one's way through

durch/beizen *vt* (*sep & insep*) **1** corrode through (*or* thoroughly). **2** steep thoroughly

durch·belichten *vt* irradiate thoroughly

durch·beuteln *vt* bolt (flour)

durch·biegen* **1** *vt* bend, deflect; break (by bending). **2** *vi* & **sich d.** sag. **Durchbiegung°** *f* bending, flexure; sagging; deflection

Durchbiegungs·festigkeit *f* transverse bending strength. **-messer** *m* deflectometer

durch·bilden *vt* **1** train (*or* educate) thoroughly. **2** perfect, develop thoroughly. **3** design in detail

durchbiß bit through: *past of* DURCHBEISSEN (*insep*)

durchbissen bitten through: *pp of* DURCHBEISSEN (*insep*)

durch·bittern *vt* make thoroughly bitter

durch/blasen* (*sep & insep*) **1** *vt/i* blow through. **2** *vt* inflate

durch/blättern (*sep & insep*) *vt* **1** leaf (*or* page) through. **2** split into lamellas

durch/bläuen *vt* (*sep & insep*) blue thoroughly

durch·blicken *vi* **1** (*usu* + **durch**) look through. **2** become apparent; (+ **lassen**) let it be known

durchblies inflated, etc: *past of* DURCHBLASEN (*insep & oft sep*)

durchbluten (*insep*) *vt* supply with blood. **Durchblutung** *f* blood supply; circulation

durch·bohren *vt* (+ **durch**), **durchbohren** (*insep*) *vt* bore through; penetrate; pierce, perforate. **Durchbohrung°** *f* penetration, perforation

durchbrach broke through: *past of* DURCHBRECHEN (*insep & oft sep*)

durch/brechen* **1** (*sep & insep*) *vt*, *vi*(s) break through, break in two; pierce, perforate. **2** (*insep*) *vt* violate, infringe. *Cf* DURCHBROCHEN

durch·brennen* **I** *vt* **1** burn through, burn a hole in. **II** *vi*(s) **2** burn out, blow (*as:* a fuse), melt and break. **3** get red hot through and through. **4** (*usu Mach.*) race. **5** run away

durchbricht breaks through: *pres of* DURCHBRECHEN (*insep & oft sep*)

durch·bringen* **I** *vt* **1** bring through, pull (put, push) (sthg) through (e.g. a crisis); support. **2** squander. **II**: **sich d.** struggle through, make ends meet

durchbrochen **1** broken through: *pp of* DURCHBRECHEN (*insep*). **2** *p.a* openwork(ed), filigreed, perforated

Durchbruch° *m* **1** breakthrough; breakout, eruption: **zum D. kommen** break through (*or* out). **2** aperture; breach. **3** (*Brew.*) collapse (of the head)

durchdacht(e) thought out: *pp* (& *past*) *of* DURCHDENKEN (*insep*), (*oft past of sep*)

durch/dampfen (*sep & insep*) *vt* fill with vapor (*or* fumes)

durchdämpfen (*insep*) *vt* steam

durch·denken* (*sep & insep*) *vt* think through (out, over)

durch·diffundieren *vt*, *vi*(s) diffuse through

durchdrang penetrated: *past of* DURCHDRINGEN (*insep & oft sep*)

durchdringbar *a* penetrable, permeable. **Durchdringbarkeit** *f* penetrability, permeability

durch/dringen* **1** *vt* (*sep & insep*) penetrate, permeate. **2** *vi*(s) (*sep*, + **mit**) be successful (with). **durchdringlich** *a* penetrable, permeable. **Durchdringlichkeit** *f* penetrability, permeability. **Durchdringung** *f* (-en) penetration, permeation

Durchdringungs·fähigkeit *f* penetrating power. **-komplex** *m* convalent complex. **-vermögen** *n* penetrating power. **-zwilling** *m* (*Cryst.*) (inter)penetration twin

durch·drücken *vt* press through

durchdrungen penetrated, etc: *pp of* DURCHDRINGEN (*insep*). **Durchdrungenheit** *f* permeation

durchduften (*insep*) *vt* fill with fragrance

durch/dunsten, durch/dünsten 1 *vt* (*sep & insep*) permeate with vapor. **2** *vi*(s) (*sep*) be exhaled; pass through as a vapor

durcheilen (*insep*) *vt* rush through

durcheinander *adv* & *sep pfx* in(to) confusion (*or* disorder). **Durcheinander** *n* confusion, mix-up, tangle. **d—·schütteln** *vt* shake up (thoroughly)

durch/fahren* 1 (*sep*) *vi*(s) drive (go, run, move, travel) through (without stopping); travel all (day, night, etc). 2 (*insep*) *vt* pass through (*or* across), cross; negotiate, take (curves); (*fig oft*) run through. **Durchfahrt°** *f* thoroughfare, passage, transit **Durchfall°** *m* 1 screenings. 2 diarrhea. 3 failure. **durch·fallen*** *vi*(s) fall through, fail; (*Light*) be transmitted; (*Beer*) clear; (*of beer head*) fall back

durchfärbbar *a* penetrable with color. **Durchfärbe·mittel** *n* (*Dye.*) penetrating agent. **durch·färben** 1 *vt* (*Dye.*) penetrate, dye thoroughly. 2 *vi* come through, bleed. **Durchfärbe·vermögen** *n* (*Dye.*) penetrating power

durch·faulen *vi*(s) rot through

durchfeuchten (*insep*) *vt* soak through, saturate, moisten thoroughly

durch·feuern *vt* fire (*or* heat) thoroughly

durch·filtern, durch·filtrieren *vt* filter through

durch·finden* *vr:* **sich d.** find one's way through

durchfließen* (*insep*) *vt* flow through, diffuse: *past* **durchfloß**, *pp* **durchflossen**. **Durchfluß** *m* continuous flow, flowthrough; diffusion

Durchfluß·messer *m* flow meter. **-wanne** *f* (*Glass*) continuous tank. **-zeit** *f* (*Flow Systems*) residence time

durchforschen (*insep*) *vt* investigate (search, explore) thoroughly. **Durchforschung°** *f* thorough investigation (search, exploration)

durchfraß corroded: *past of* DURCHFRESSEN (*insep & oft sep*)

durch/fressen* (*sep & insep*) *vt & sich d.* eat through, corrode (through)

durch/frieren* (*sep & insep*) *vi*(s) freeze up, freeze through (*or* thoroughly)

durchfrißt corrodes: *pres of* DURCHFRESSEN (*insep & oft sep*)

durchfror(en) froze(n) through *past & pp of* DURCHFRIEREN *esp insep*

durchfuhr passed through, etc: *past of* DURCHFAHREN 2 (*esp insep*). **Durchfuhr°** *f* transportation (through an area), transit

durchführbar *a* feasible. **Durchführbarkeit** *f* feasibility. **durch·führen** *vt* lead through; carry out (*or* through);

implement, organize; (*Tinplate*) dip, wash. **Durchführung** *f* implementation; organization; workmanship. **Durchführungs·form** *f* working form, model

Durchgabe° *f* 1 filtration. 2 transmission. 3 broadcast

Durchgang° *m* 1 passage(way), duct. 2 transit; transition. 3 thoroughfare: **D. verboten** no thoroughfare, no entry. 4 runthrough, (re)check, replay. **durchgängig** *a* 1 universal. 2 open, permeable. 3 uninterrupted *cf* DURCHGEHEND II

Durchgangs·hahn *m* straight-through tap. **-stadium** *n* transitional stage. **-ventil** *n* straight-through valve. **-widerstand** *m* (*Elec.*) volume resistance

durch·gären* *vi*(s) ferment thoroughly

durch·geben* *vt* 1 transmit. 2 announce; broadcast. 3 filter, strain

durchgebissen bitten through: *pp of* DURCHBEISSEN (*sep*)

durchgebogen deflected: *pp of* DURCHBIEGEN

durchgebracht put through: *pp of* DURCHBRINGEN

durchgebrannt burned out: *pp of* DURCHBRENNEN

durchgebrochen broken through: *pp of* DURCHBRECHEN (*sep*)

durchgedacht thought out: *pp of* DURCHDENKEN (*sep*)

durchgedrungen penetrated: *pp of* DURCHDRINGEN (*sep*)

durchgefroren frozen up: *pp of* DURCHFRIEREN (*sep*)

durchgefunden found one's way through: *pp of* DURCHFINDEN

durchgegangen gone through: *pp of* DURCHGEHEN

durchgegoren thoroughly fermented: *pp of* DURCHGÄREN

durchgegossen poured through: *pp of* DURCHGIESSEN

durchgegriffen penetrated: *pp of* DURCHGREIFEN

durch·gehen* *vi*(s) 1 (*usu* + **durch**) go (pass, run) through; be transmitted. 2 pass, be accepted; (+ **lassen**) overlook, let pass. 3 bolt, run away; get out of control. 4 (*Fuse*) blow. 5 (*Nucl.*) melt down. 6 (+ *accus*) go through, suffer, experience. 7 (+

accus) run through, check (a list, etc.). **durchgehend I** *p.a* 1 pervading. 2 transmitted (light). 3 uninterrupted. 4 through (car, train). **II** *adv oft* = **durchgehends** *adv* all the way; without a break; all day

durchgekrochen crawled through: *pp of* DURCHKRIECHEN (*sep*)

durchgelesen read through: *pp of* DURCHLESEN

durchgenommen taken up: *pp of* DURCHNEHMEN

durch·gerben *vt* (*Lthr.*) tan thoroughly. **Durchgerbungs·zahl** *f* tanning number

durchgerieben chafed: *pp of* DURCHREIBEN

durchgerissen broken: *pp of* DURCHREISSEN

durchgeschienen shone through: *pp of* DURCHSCHEINEN

durchgeschmolzen fused: *pp of* DURCHSCHMELZEN

durchgeschnitten cut through: *pp of* DURCHSCHNEIDEN (*sep*)

durchgeschossen shot through: *pp of* DURCHSCHIESSEN (*sep*)

durchgeschrieben carbon-copied: *pp of* DURCHSCHREIBEN

Durchgeseihte(s) *n* (*adj endgs*) filtrate

durchgesogen sucked through: *pp of* DURCHSAUGEN (*sep*)

durchgesotten boiled thoroughly: *pp of* DURCHSIEDEN

durchgestanden withstood: *pp of* DURCHSTEHEN

durchgestochen pierced: *pp of* DURCHSTECHEN (*sep*)

durchgestrichen crossed out: *pp of* DURCHSTREICHEN (*sep*)

durchgetrieben driven through: *pp of* DURCHTREIBEN

durchgezogen moved through: *pp of* DURCHZIEHEN (*sep*)

durch·gießen* *vt* pour through; strain, filter

durch·glühen 1 *vt* heat red hot; anneal (thoroughly). 2 *vi*(s) burn out

durch·greifen* *vi* 1 take decisive action. 2 penetrate. **durchgreifend** *p.a* decisive, drastic, radical, thoroughgoing; penetrating

Durchgriff° *m* penetration; decisive action; (*Electron.*) inverse amplification

factor, penetration factor

Durchguß° *m* 1 filter, strainer. 2 filtration, pouring through. 3 gutter, sink

durch·halten *vi* hold out, hang on

Durchhang° *m* sag, dip; deflection

durch·härten *vt* (*Metll.*) depth-harden

durch/heizen (*sep & insep*), **durchhitzen** (*insep*) *vt* heat thoroughly, heat through

durchkälten (*insep*) *vt* chill thoroughly

durch/klären (*sep & insep*) *vt* clarify thoroughly

durch/kneten (*sep & insep*) *vt* knead thoroughly; (*Metll.*) rabble

durch·kochen *vt* boil thoroughly

Durchkohlung *f* thorough carbonization

Durch·kommen* *vi*(s) come (get, pull) through, succeed, manage

durch/kreuzen I (*sep*) 1 *vt* cross out. **II** (*insep*) **A** *vt* 2 cross, cut across, intersect. 3 frustrate, thwart. **B: sich d.** (*recip*) intersect (each other). **Durchkreuzungs·zwilling** *m* (*Cryst.*) (inter)penetraton twin

durch/kriechen* I (*sep*) *vi*(s) crawl through. **II** (*insep*) *vt*: **es durchkreuzt einen** it gradually comes over one. **durchkroch(en)** *past* (& *pp*) (*insep*)

durch/kühlen (*sep & insep*) *vt* cool thoroughly

Durchlaß *m* (..lässe) 1 passage, way through; opening, outlet. 2 gutter, culvert. 3 filter, sieve. 4 (*Elec.*) band pass. 5 transmission (of light). 6 permission to pass. **durch·lassen*** *vt* let through, let pass, be permeable to; filter, strain; transmit. **durchlässig** *a* permeable, leaky; translucent; diathermanous. **Durchlässigkeit** *f* permeability, leakiness; translucence; diathermancy; transmission (factor). **Durchlassung** *f* filtering, straining; transmission

Durchlauf° *m* 1 processing time, pass. 2 run-through, continuous flow. 3 passage. 4 colander. 5 diarrhea. **durch/laufen* I** (*sep*) *vi*(s) leak (drip, filter) through. **II** (*insep*) *vt* cover, run (a distance); go through (a course of study). **III** (*sep & insep*) run through. **durchlaufend** *p.a* continuous. **Durchläufer·mineral** *n* worldwide mineral

Durchlauf·messer *m* flow meter (for water, gas, etc). **-ofen** *m* tunnel kiln. **-öl** *n* flux oil (for asphalt)

durch·laugen *vt* lye (*or* lixiviate) thoroughly

durchläutern (*insep*) *vt* purify, refine thoroughly

durch·leiten *vt* lead (pass, conduct) through

durch·lesen* *vt* read through

durch/leuchten I (*sep*) *vi* 1 shine through. 2 become apparent. **II** (*insep*) *vt* 3 irradiate; (trans)illuminate; fluoroscope; scan; candle. 4 analyze. **durchleuchtend** *p.a* translucent, transparent. **Durchleuchtung°** *f* irradiation, (trans)illumination; fluoroscopy; scanning, candling

Durchlicht° *n* transmitted light

durchlief ran through, etc: *past of* DURCHLAUFEN (*insep & oft sep*)

durchlochen (*insep*) *vt* perforate, punch, pierce; core (fruit). **durchlöchern** *vt* (*insep*) perforate; punch (*or* shoot) full of holes

durch/lüften I (*sep*) *vt/i* air (*or* ventilate) thoroughly. **II** (*insep*) aerate. **durch·lüftungs·fähig** *a* capable of ventilation (*or* aeration)

durch·machen I *vt* 1 go through, finish (courses, etc). 2 live through, experience, undergo. **II** *vi* carry on without a break

durchmaß crossed, etc: *past of* DURCHMESSEN (*insep & oft sep*)

durch/messen* *vt* **I** (*sep*) measure completely. **II** (*insep*) travel through, cover (a distance); cross, traverse

Durchmesser° *m* diameter

durch/mischen (*sep & insep*) *vt* mix thoroughly, blend; intermix

durchmißt crosses, etc: *pres of* DURCHMESSEN (*insep & oft sep*)

durch/mustern (*sep & insep*) *vt* inspect, look through; scrutinize, examine closely; scan

durch/nässen I (*sep*) *vi* soak through. **II** (*insep*) *vt* soak (thoroughly), drench; steep

durch·nehmen* *vt* take up, study, run through

durchnetzen (*insep*) *vt* soak, etc = DURCHNÄSSEN **II**

durch·ölen *vt* oil thoroughly

durch·patentieren *vt* (*Wire*) patent by

passing through a furnace and then cooling

durch·pausen *vt* trace (with tracing paper)

durch·perlen *vi*(s) bubble through

durch·pressen *vt* press through; strain, filter

durchprüfen (*insep*) *vt* check (*or* test) thoroughly

durchqueren (*insep*) *vt* cross, traverse

durch·quetschen *vt* squeeze through; strain, filter

durch/räuchern (*sep & insep*) *vt* smoke, fumigate (thoroughly)

durch·rechnen *vt* calculate (in detail)

durch·reiben* *vt* rub through, chafe; fray; strain

durch·reißen* *vt & vi*(s) tear in two, break

durch·rieseln *vi*(s) trickle through

durch·rosten *vi*(s) rust through

durch·rösten *vt* roast thoroughly

durch·rühren *vt* stir thoroughly, agitate; strain

durchs *contrac* = **durch das** *cf* DURCH *prep*

durch/sättigen (*sep & insep*) *vt* saturate thoroughly

Durchsatz° *m* charge, throughput (of furnace, pipeline); fuel consumption; infiltration; diffusion. **--leistung** *f* throughput. **-ofen** *m* tunnel furnace

durch/säuern (*sep & insep*) **I** *vt* 1 acidify, make sour. 2 leaven thoroughly. **II** *vi*(s) become thoroughly leavened

durch/saugen* (*sep & insep*) *vt* suck through, force through by suction

durchschaubar *a* transparent, clear, comprehensible. **durchschauen** (*insep*) *vt* see through

durch·scheinen* *vi* shine through, become evident, show through. **durchscheinend** *p.a* translucent, transparent

durch/schießen* **I** (*sep*) 1 *vt & vi*(s) shoot through. 2 *vi*(s) dash through. **II** (*insep*) *vt* 3 shoot through; pass through (*as:* shots); flash through (one's mind). 4 interleave, slip-sheet. 5 interweave

Durchschlag° *m* 1 strainer, colander; filter. 2 (*Mach.*) punch, drift. 3 piercing, puncture. 4 (*Tunnels*) breakthrough. 5 (*Elec.*) flashthrough, flashover, blowout. 6 carbon copy. **--boden** *m* perforated bottom

durch/schlagen* **I** (*sep*) **A** *vt* **1** break in two, break apart. **2** strain, press through a strainer, filter. **B** *vi*(h) have a laxative effect. **C** *vi* (h/s) (+ **durch**) come (strike, leak) through, penetrate, pierce; blot (*as: paper*); (*Elec.*) flash through, blow out; (fig) take effect. **D: sich d.** manage, struggle one's way through. **II** (*insep*) crash through, penetrate, pierce. **durchschlagend** *p.a* decisive, conclusive, cogent; striking; effective. **Durchschläger** *m* (*Mach.*) punch

Durchschlag·festigkeit *f* (*Elec.*) dielectric (*or* disruptive) strength. **-papier** *n* carbon paper

Durchschlags·kraft *f* **1** penetrating power; striking power; (*esp fig*) impact. **2** effectiveness; conclusiveness. **3** (*Paper*) puncture resistance

Durchschlag·spannung *f* (*Elec.*) breakdown voltage

durchschlug penetrated, etc: *past of* DURCHSCHLAGEN (*insep & oft sep*)

durch·schlüpfen *vi*(s) slip through

durch·schmelzen* **1** *vt, vi*(s) melt through, fuse (thoroughly). **2** *vi*(s) (*Nucl.*) melt down. **Durchschmelzung** *f* (-en) **1** thorough melting, fusing. **2** (*Nucl.*) meltdown. **3** internal seal

durch·schmieren *vt* grease (*or* lubricate) thoroughly

durchschmolzen *p.a* fused, melted down

durch/schneiden* (*sep & insep*) *vt* cut (slice, plow) through; intersect, cross, cut across. **durchschnitt** *past* (*insep*). **Durchschnitt°** *m* average, mean, cross-section

durchschnitten cut through: *pp of* DURCHSCHNEIDEN (*insep*)

durchschnittlich *a* average; *adv* on the average

Durchschnitts·bestimmung *f* average determination. **-ertrag** *m* average yield (*or* profit). **-muster** *n*, **probe** *f* average sample. **-punkt** *m* point of intersection. **-temperatur** *f* mean temperature. **-wert** *m* average value. **-zahl** *f* (arithmetic) mean, average, mean

durch·schreiben* *vt* make a carbon (copy) of. **Durchschreib·papier** *n* carbon (*or* copying) paper

Durchschrift° *f* carpon copy

durchschoß, durchschossen interwove(n), etc: *past (& pp) of* DURCHSCHIESSEN (*insep & past oft sep*)

Durchschuß° *m* **1** shot that goes clear through. **2** penetration wound. **3** (*Tex.*) woof, weft. **4** (*Print.*) interlinear space

durch·schütteln *vt* shake up (thoroughly), agitate

durchschwängern (*insep*) *vt* impregnate thoroughly

durchschwefeln (*insep*) *vt* sulfurize thoroughly

durch·schwitzen *vt/i* sweat (seep, ooze) through

durch·sehen* **I** *vt* **1** look (*or* search) through (papers, etc). **2** run through (by reading), proofread, examine. **II** *vi* (+ **durch**) **3** see (*or* look) through. **4** grasp (the meaning of). **5** stick (out) through

durch·seihen *vt* filter, strain, percolate. **Durchseiher°** *m* filter, strainer, percolator

durch/setzen I (*sep*) **A** *vt* **1** sift, put through a sieve. **2** push through, achieve, carry out—**es d., daß ...** manage to bring it about that ... **B: sich d.** assert oneself; be successful, gain acceptance. **II** (*insep*) *vt* riddle, intersperse; intermingle. **durchsetzt** *p.a* (**mit**) shot through (with)

Durchsicht° *f* **1** perusal, inspection, examination. **2** clear view, vista. **3** undertone (of a color); transparency; transmitted light. **durchsichtig** *a* transparent, clear. **Durchsichtigkeit** *f* transparency

Durchsichts·bild *n* (*Phot.*) diapositive. **-farbe** *f* transmitted color; transparent color

durch·sickern *vi*(s) trickle (filter, seep, percolate, ooze) through

durch/sieben *vt* **I** (*sep*) sift, screen. **II** (*insep*) perforate, riddle with holes

durch·sieden* *vt* boil thoroughly

durch·sintern *vi*(s) trickle (*or* percolate) through

durchsog(en) sucked through: *past (& pp) of* DURCHSAUGEN (*insep & past oft sep*)

durch·spülen *vt* rinse through (*or* thoroughly); irrigate

durchstach pierced, etc: *past of* DURCHSTECHEN (*insep & oft sep*)

durch/stechen* **I** (*sep*) (+ **durch**) **1** *vt* stick (e.g. a needle) through. **2** *vi* stick (*or* protrude) through. **II** (*sep & insep*) *vt* pierce, stab; cut through; (*Metll.*) smelt. **Durchstich°** *m* piercing, cutting through; excavation. **durchsticht** pierces: *pres*, **durchstochen** pierced *pp of* DURCHSTECHEN (*insep*)

durch·stehen* *vt* withstand, weather

durchstieß pierced: *past of* DURCHSTOSSEN (*insep*)

durch/stoßen* **I** (*sep*) **1** *vt* push through; knock a hole through; wear through (*or* thin). **II** (*sep. vi*(s) & (*insep*) *vt* pierce, penetrate, break through

durch/strahlen **I** (*sep*) *vi* shine (*or* radiate) through. **II** (*insep*) *vt* irradiate, radiograph; penetrate with rays. **Durchstrahlung°** *f* irradiation, radiography; penetration with rays

durch/streichen* *vt* **I** (*sep*) **1** strike out, cross out. **2** strain, put through a strainer. **II** (*insep*) cut across, sweep through (an area); **d—de Linie** trajectory. **durchstrich(en)** *past* (& *pp*)

Durchstrom *m* (*Elec.*) polyphase current. **durch/strömen** (*sep*) *vi*(s) & (*insep*) *vt* flow (run, stream) through. **Durchströmung°** *f* flowing (*or* streaming) through, perfusion. **Durchströmungs·methode** *f* flow method

durchsuchen (*insep*) *vt* search thoroughly

durchsüßen (*insep*) *vt* sweeten thoroughly, edulcorate

durch·tränken (*insep*) *vt* saturate, impregnant, soak. **Durchtränkung** *f* saturation, etc

durch·treiben* *vt* drive (force, press) through; distill; strain. **Durchtreiber°** *m* (*Mach.*) drift (punch)

durchtrieben *a* cunning, crafty

Durchtritt° *m* passage (through); penetration

Durchvulkanisation° *f* thorough vulcanization

durch·wachsen* *vi*(s) (+ **durch**) grow through. **durchwachsen** (*insep*) *p.a* (*oft* + **von**) **1** marbled, streaky (meat, etc). **2** interspersed (with). **3** overgrown (with)

durch/wandern **I** (*sep*) *vi*(s) wander (travel, pass, migrate) through. **II** (*insep*) *vt* wander through; cross, cover (an area); diffuse through (a body)

durch/wärmen 1 *vt* (*sep & insep*) warm

thoroughly; warm moderately till temperature is equalized. **2** (*sep*) **sich d.** get thoroughly warmed up. **durchwärmig** *a* diathermic. **Durchwärmigkeit** *f* diatherma(n)cy. **Durchwärm·zeit** *f* time needed to equalize temperature on heating

durch·waschen* *vt* wash thoroughly

durch·wässern (*insep*) *vt* soak, drench; irrigate

Durchweg° *m* passage, way through. **durchweg(s)** *adv* consistently, without exception

durch/weichen* **I** (*sep*) *vi*(s) get soaked. **II** (*insep*) soak thoroughly, steep. **Durchweichungs·grube** *f* (*Metll.*) soaking pit

durch·werfen* *vt* sift, screen, bolt; traject

Durchwirbelung *f* agitation

Durchwurf° *m* **1** riddle, sieve. **2** mesh. **3** screenings. **4** (*Coke*) through-breeze

Durchzeichen·papier *n* tracing paper. **durch·zeichnen** *vt* trace

durch/ziehen* **I** (*sep*) **A** *vt* **1** pull (a needle) through. **2** cut (a ditch, canal) through. **3** ram (push, railroad) (e.g. a plan) through. **4** (*Metll.*) strip. **B** *vi*(s) **5** move through. **6** become seasoned; (+ **lassen**) let steep. **C: sich d.** **7** pass (*or* run) through, be pervasive. **II** (*insep*) *vt* **8** move (pass, march) through (an area). **9** pervade, permeate. **10** crisscross. **Durchziehen·lassen** *n* allowing (heat) penetration. **Durchzieh·glas** *n* (*Micros.*) slide

durchzog(en) pervaded, etc: *past* (& *pp*) *of* DURCHZIEHEN (*insep & past oft sep*)

Durchzug° *m* **1** (through) draft: **D. machen** create a draft, ventilate. **2** march (move, passage) through (an area). **3** (*Tex.*) drawing frame. **4** (*Engg.*) through girder. **--ofen** *m* tunnel furnace

dürfen* (durfte; *pres sg* darf) *v aux* (+ *inf*) **1** be allowed, be permitted (to); *pres oft* may; *neg oft* must not. **2** *sbjc*; **dürfte** (*oft*) might, could: **dürfte sein** might be, probably is. **3** (*verb of motion oft omitted*): **sie d. ins Labor** (**gehen**) they are allowed (to go) into the lab

dürftig *a* meager, scanty, poor, wretched; needy

Duro·chinon *n* duroquinone. **-hydrochinon** *n* durohydroquinone

Durol *n* durene

Duroplast m thermosetting plastic. **duroplastisch** a thermosetting

dürr a 1 dry, arid. 2 thin, lean. 3 unadorned. **Dürre** f (-n) 1 dryness, aridity; thinness, leanness; plainness, sobriety. 2 drought. **Dürr-erz** n 1 dry (low-grade, barren) ore. 2 lead-free silver

Durst m thirst—**D. haben** be thirsty. **dursten, dürsten** vt/i thirst, be thirsty. **durstig** a thirsty. **durst-stillend** a thirst-quenching

Dusche f (-n) 1 shower (bath). 2 spray. 3 (vaginal) douche. 4 douching syringe

Düse f (-n) 1 jet, nozzle; tuyere; blast pipe. 2 head (of burner). 3 (wire-drawing) die

Düsen•antrieb m jet propulsion. **-austritt** m orifice. **-flugzeug** n jet plane (or aircraft). **d–gefärbt** a spun-dyed, dope-dyed. **D–maschine** f jet plane. **-regler** m nozzle regulator. **-rohr** n, **-röhre** f blast pipe, jet tube. **-stock** m bustle (or blast) pipe

düster a dark, gloomy, somber; dim, shady

Düte f (-n) bag = TÜTE

Dutzend n (-e) dozen

Dyn n (Meas.) dyne. **--aktivität** f surface activity

Dynamik f (-en) dynamics. **dynamisch** a dynamic

dysprotid a proton-donating. **Dysprotid** n (-e) proton donor

dystektisch a dystectic

dz., Dz. abbv (Doppelzentner) 100 kilograms

dz(t). abbv (derzeit, derzeitig) at (or of) this time

E

E. *abbv* (Erstarrungspunkt) freezing point

Ebbe *f* (-n) ebb(tide); low ebb, decline

ebd. *abbv* (ebenda) ibidem

eben I *a* 1 flat, level. 2 even, smooth.
3 plane. **II** *adv* 4 just, exactly, precisely:
nicht e. not exactly. 5 just about, barely

Eben·baum *m* ebony tree

Eben·bild *n* image, likeness. **e–bürtig** *a*
(*oft* + *dat*) equal (in rank) (to). **-da, -dort**
adv in the same place, exactly there,
ibidem

Ebene *f* (-n) plain; plane; level. **ebenen** *vt*
level, etc = EBNEN

Ebenen·gruppe *f* (*Cryst.*) plane group.
Ebene·normale *f* (*Geom.*) perpendicular
to a plane. **Ebenen·winkel** *m* dihedral
angle

eben·erwähnt *a* just-mentioned. **-falls** *adv*
likewise, also; (+ **nicht**) not . . . either.
-flächig *a* plane, flat-surfaced

Ebenheit *f* flatness, evenness, smoothness

Eben·holz *n* ebony. **ebenieren** *vt* ebonize,
inlay with ebony

Eben·maß *n* harmony, symmetry. **-maßflä-
che** *f* plane of symmetry. **e—mäßig** *a*
symmetrical, proportionate. **-so** *adv* 1 the
same way, just as much. 2 (+ *a,adv* or *in
compds*) just (as). **-solche** *a, pron* (*adj endgs*) just such (a one), similar, the same
kind (of)

ebenso·oft *adv* just as often. **-sehr** *adv* just
as (much). **-viel** *a, pron* just as much.
-wenig *a, pron* just as little, no more.
-wohl *adv* just as well

Eber *m* (-) boar, hog. **--esche** *f* mountain
ash, service tree. **-raute** *f* southernwood,
abrotanum. **-wurz(el)** *f* carline thistle (*or*
root)

ebne *a* even, etc = EBEN I. **Ebne** *f* (-n) plain,
etc = EBENE. **ebnen** *vt* smooth; level, flatten; plane

ebullieren *vi* boil up, bubble; break out

Ecballium·säure *f* ecballic acid

Echappé-öl *n* recovered oil

Echse *f* (-n) saurian; lizard *cf* EIDECHSE

echt I *a* 1 genuine, real, true; pure. 2 (*Colors*) fast; (*Tex.*) fast-dyed, ingrain.
3 (*Math., Phys., oft*) proper. **II** *sfx* -fast,
-proof, -resistant

Echt· (*esp Tex. & Dye.*) fast: **-base** *f* fast-
color base. **-(baumwoll)blau** *n* fast (cotton) blue. **-färben** *n* fast-color dyeing.
e—farbig *a* fast-color dyed

Echtheit *f* genuineness; fastness *cf* ECHT
1, 2

Echtheits·grad *m* (*Dye.*) degree of fastness. **-prüfung** *f* (*Dye.*) fastness test

Echt·säurefuchsin *n* fast acid fuchsine.
-weiß *n* lead sulfate pigment

Eck *n* (-e *or* -en) 1 corner, angle. 2 (*Cryst.*)
summit. 3 (*in compds usu*) -angle, -gon:
Fünfeck pentagon. *cf* ECKE

Ecke *f* (-n) 1 corner, nook; angle. 2 edge
(plane angle); (*Cryst.*) summit; (*Engg.*)
quoin. *Cf* ECK

Eck·eisen *n* angle iron; corner plate

Ecker *f* (-n) acorn; beechnut. **--doppe** *f*
acorn cup

Eck·holz *n* squared timber, beam

eckig I *a* 1 angular, cornered, square; (*in
compds also*) -gonal: **sechs·eckig** hexagonal *cf* DREIECKIG

Eck·punkt *m* corner (point). **-stein** *m* cornerstone. **-zahn** *m* canine (tooth)

Ecrasement *n* crushing

Ecrü·seide *f* raw silk

edel *a* 1 noble (*incl* metals). 2 fine(-quality).
3 (*Chem.oft*) electropositive; inert (gases);
rare (earths, gases). 4 vital (organs). 5
precious (metal, stone); rich (vein of ore).
6 pure-bred. *Cf* EDLE

Edeleanu·verfahren *n* (*Petrol.*) edeleanui-
zation

Edel·erde *f* rare earth. **-erz** *n* rich ore (*usu*
containing precious metals). **-faser** *f* staple fiber (*or specif* rayon). **-galmei** *m*
smithsonite. **-gamander** *m* wall germander. **-gas** *n* inert (*or* rare) gas

Edelgas·hülle *f* inert gas shell (of electrons). **-rumpf** *m* inert gas core. **-schale** *f*

(completed) electron shell (of an inert gas)

Edel·glanz *m* lustrous finish. **-holz** *n* fine wood. **-kastanie** *f* Spanish chestnut. **-kohle** *f* low-ash coal. **-kunstharz** *n* fine plastic, casting resin. **-metall** *n* 1 noble metal. 2 precious metal. **-metallwaage** *f* bullion balance. **-mist** *m* barnyard manure. **e—mütig** *a* noble, magnanimous. **E—pelz** *m* genuine (*or* valuable) fur. **-pilzkäse** *m* blue-veined cheese. **-porzellan** *n* hard porcelain. **-reis** *n* (grafting) scion (*or* slip). **-rost** *m* patina. **-salze** *npl* abraum salts (overlying rock salt). **-schrott** *m* pure scrap (metal). **-spat** *m* adularia, moonstone. **-stahl** *m* high-quality alloy steel. **-stein** *m* jewel, gem, precious stone. **-tanne** *f* silver fir. **-zement** *m*/*n* barium (*or* strontium) cement

-eder *sfx* (*Geom.*) -hedron

Edersche Lösung Eder's (actinometry) solution

edieren *vt*/*i* edit; publish

edle *a* noble, etc = EDEL

edulkorieren *vt* edulcorate, wash

E.E. *abbv* (Entropieeinheit) entropy unit

Efeu *m* ivy. **efeu·umrankt** *a* ivy-entwined

Effekt *m* (-e) 1 effect. 2 result, yield. 3 *pl* (-en) (negotiable) securities. 4 (*in compds usu*) = EFFEKTIV 1. **effektiv** 1 *a* actual, real, net; effective. 2 *adv oft* in effect; (+ *neg, usu*) at all, whatever. **Effektiv·wert** *m* effective value. **Effekt·kohle** *f* flame carbon

Effervescens *n* (. . entien) (*Pharm.*) effervescent (remedy). **Efferveszenz** *f* effervescence. **efferveszieren** *vi* effervesce

effleurieren *vt* (*Lthr.*) buff

effloreszieren *vi* effloresce

Effluvien *npl* effluvia

effundieren *vt*/*i* effuse. **Effundierung** *f* (-en) effusion

Eg. *abbv* (Eisessig) glacial acetic acid

egal *a* 1 even, regular. 2 equal, alike: **es ist e.** (+ *dat*) it doesn't matter (to). 3 (+ *interrog*) e.g.: **e. was** no matter what. **egalisieren** *vt* 1 even (out), equal(ize). 2 level (*incl Dyes*), flatten. 3 dye evenly. **Egalisierer** *m* (-) equalizer; (*Dye.*) leveling agent

Egalisier·farbstoff *m* leveling dye. **-lack**

m (*Lthr.*) top coat(ing)

Egalisierungs·mittel *n* (*Tex.*, *Dye.*) leveling agent

Egalität *f* (-en) equality

Egel *m* (-) leech

Egge *f* (-n), **eggen** *vt*/*i* harrow

Egoutteur *m* (-e) (*Paper*) dandy roll

egrenieren *vt* gin (cotton). **Egrenier·maschine** *f* cotton gin

ehe *subord cnjc* before; **e. nicht** not until

Ehe *f* (-n) marriage, matrimony

ehedem *adv* (back) then, in former times

ehem. *abbv* (ehemalig, ehemals) former(ly)

ehemalig *a* former, past. **ehemals** *adv* formerly

eher *adv* earlier, sooner; rather; more easily; more likely; more nearly—**alles e. als** anything but

ehern *a* (of) brass, brazen; (of) bronze

ehest *a*; *adv* **am e—en** = **ehestens** *adv* (at the) earliest, soonest; most likely, most easily

ehmalig *a*, **ehmals** *adv* = EHEMALIG, EHEMALS

ehrbar *a* honorable, respectable; honest

Ehr·begierde *f* ambition

Ehre *f* (-n) honor, reputation; credit; self-respect. **ehren** *vt* honor, esteem, respect. **ehren·amtlich** *a* volunteer; *adv oft* as a volunteer. **ehrenhaft** *a* honorable, respectable, honest

Ehren· honor(able), honorary: **-mitglied** *n* honorary member. **-preis** 1 *m* prize, reward. 2 *m*/*n* (*Bot.*) speedwell, veronica. **-stelle** *f* honorary post. **e—voll** *a* honorable

Ehrgeiz *m* ambition. **ehrgeizig** *a* ambitious

ehrlich *a* honest; honorable

Ei *n* (-er) egg, ovum. **~albumin** *n* egg albumin, ovalbumin

Eibe *f* (-n), **Eiben·gewächse** *npl* (*Bot.*) yew(s)

Eibisch *m* (-e) hibiscus, marshmallow (plant). **~wurzel** *f* marshmallow root. **-zucker** *m* marshmallow (as a confection)

Eich· 1 gauging *cf* EICHEN *vt*. 2 oak *cf* EICHE I: **-amt** *n* gauging office, office of weights and measures. **-apfel** *m* oak gall, gall nut. **-baum** *m* oak tree

Eiche *f* (-n) **I** oak. **II** gauging = EICHUNG
Eichel *f* (-n) 1 acorn. 2 glans penis. **-becher**
m, **-doppe** *f* acorn cup. **e—förmig** *a*
acorn-shaped, glandiform. **E—(kern)öl**
n acorn oil. **-napf** *m* acorn cup. **-zucker** *m*
acorn sugar, quercitol
eichen I *a* oak(en). **II** *vt* gauge, calibrate,
standardize; test, adjust (weights, measures) *cf* GEEICHT
Eichen *n* I: **Ei/chen** (-) little egg, ovule *cf*
EI. **II** calibration, gauging *cf* EICHEN **II** *vt*
Eichen· oak: **-gallapfel** *m*, **-galle** *f* oak gall.
-gerbsäure *f*, **-gerbstoff** *m* quercitannic
acid, oak tannin. **-holz** *n* oak(wood).
-kern *m* acorn. **-kufe** *f* oak vat. **-lohe** *f*
(oak) tanbark. **-mehl** *n* powdered oak
bark. **-rinde** *f* oak bark. **-rin-
dengerbsäure** *f* quercitannic acid. **-sa-
men** *m* acorn
Eicher *m* (-) gauger
eich· gauging *cf* EICHEN *vt*; oak *cf* EICHE
I: **-fähig** *a* adjustable, capable of calibration. **E—gas** *n* standard gas. **-gerät** *n*
calibrating apparatus. **-holz** *n* oak(wood).
-horn, **-hörnchen** *n*, **-katze** *f* squirrel.
-kern *m* acorn. **-lohe** *f* (oak) tanbark.
-mehl *n* powdered oak bark. **-kurve** *f*
calibration curve. **-maß** *n* calibrating
standard (*or* instrument), gauge. **-mei-
ster** *m* gauger, adjuster. **-metall** *n* ster-
rometal. **-nagel** *m* gauge mark. **-punkt**
m reference point. **-schein** *m* certificate
of standardization
Eichung *f* (-en) calibration, etc *cf* EICHEN **II**
Eich·wert *m* calibration value (*or* standard)
Eid *m* (-e) oath
Eidechse *f* (-n) lizard
Eiderdaunen *pl* eiderdown
Eidg. *abbv* (Eidgenössisch) Swiss (Federal)
Eidgenosse *m* (-n, -n) confederate. **eidgenössisch** *a* 1 confederate, federal. 2 (*cap*)
Swiss (Federal). **Eidgenossenschaft°** *f*
federation; (*specif*) Swiss Federation
Ei-dotter *n* egg yolk. **-dotterfett** *n* lecithin
Eier eggs: *pl of* EI. **-albumin** *n* egg al-
bumin. **-brikett** *n* egg-shaped briquette;
(*Coal*) ovoid. **-drüse** *f* corpus luteum.
-frucht *f* eggplant. **-gang** *m* oviduct. **-öl** *n*
egg(-yolk) oil. **-pulver** *n* egg (*or* custard)
powder. **-schale** *f* eggshell. **-schalen-**

glanz *m* eggshell luster. **-stein** *m* oolite,
egg stone. **e—steinartig** *a* oolitic. **E—
stock** *m* ovary. **-weiß** *n* white of eggs
Eifer *m* zeal, eagerness, ardor. **eifern** *vi* agitate; be eager; strive
Eifer·sucht *f* jealousy. **e—süchtig** *a*
jealous
Ei·formbrikett *n* egg-shaped briquette.
ei-förmig *a* egg-shaped, oviform
eifrig *a* zealous, eager, ardent
eig. *abbv* 1 (eigentlich) actual(ly). 2 EIGEN
own, etc. **Eig.** *abbv* (Eigenschaft) property
Ei-gelb *n* egg yolk
eigen *a* 1 (one's) own, of one's own, individual. 2 private. 3 (*oft* + *dat*) peculiar (to),
characteristic (of). 4 unique. 5 odd,
strange. 6 (*as sfx, oft*) -owned: **bun-
des-eigen** (*WG*) federally owned; **volks-
·eigen** (*EG*) state-owned. **Eigen·** individual, intrinsic, specific, proper, self-,
auto-, idio-: **-anlage** *f* private installation. **-art** *f* peculiarity, individuality;
originality. **e—artig** *a* strange, peculiar,
singular. **E—artigkeit** *f* peculiarity,
strangeness. **-atmung** *f* autorespiration
(of yeast). **-belastung** *f* dead load, dead
weight. **-bewegung** *f* proper (*or* individual) motion, spontaneous movement.
-dämpfung *f* self-damping. **-drehim-
puls** *m* proper (*or* characteristic) angular
momentum. **-drehung** *f* proper rotation.
-farbe *f* proper (intrinsic, natural) color.
e—farbig *a* (self-)colored. **E—farbstoff**
m self-dye, self-color. **-form** *f* own peculiar form, characteristic form. **-frequenz**
f proper (characteristic, natural) frequency. **-funktion** *f* proper (*or* characteristic)
function; eigenfunction. **-gewicht** *n* own
(specific, proper) weight; light (*or* dead)
weight. **-halbleiter** *m* intrinsic semiconductor. **e—händig** *a* personal; (*adv oft*)
with one's own hand(s). **E—last** *f* dead
load, dead weight. **-leitfähigkeit** *f* auto-
conductivity. **e—mächtig** *a* arbitrary;
unauthorized, independent. **E—mittel** *n*
1 (*Med.*) specific remedy. 2 (*pl oft*) private
means. **-name** *m* proper name. **e—
nützig** *a* selfish, egoistic
eigens *adv* expressly
Eigenschaft *f* (-en) property, characteris-

tic; quality, distinctive feature; capacity

Eigenschafts·verkettung f correlation of properties. **-wort** n adjective

Eigen·schwingung f proper (natural, characteristic) vibration. **e—schwingungsfrei** a aperiodic, deadbeat. **-sicher** a (Elec.) intrinsically safe. **E—sinn** m stubbornness; wilfullness. **E—spannung** f **1** internal stress. **2** (Elec.) natural voltage. **-strahlung** f proper (or characteristic) radiation; fluorescent radiation

eigentlich 1 a actual, real; original; (+ dat) peculiar (to), characteristic (of). **2** adv actually, really; in actual fact, in reality; (+ interrog, oft) anyway

Eigentum n (. . -er) property, possession; ownership. **Eigentümer I** m (-) owner. **II** pl of EIGENTUM

eigentümlich a **1** strange, odd; remarkable. **2** (oft + dat) peculiar (to). **Eigentümlichkeit** f (-en) peculiarity, idiosyncrasy. **2** individuality, uniqueness

Eigentums· of property, proprietary; wholly owned

Eigen·vergiftung f autointoxication. **-vergrößerung** f actual size; primary magnification. **-viskosität** f specific viscosity. **-volumen** n specific (or individual) volume. **-wärme** f specific heat; body heat. **-wert** m proper (characteristic, inherent) value, eigenvalue. **-widerstand** m internal resistance. **-zündung** f self-ignition; compression ignition. **-zustand** m proper (or characteristic) state

eignen: sich e. (als, für, zu) be suitable (to, for) cf GEEIGNET. **Eignung** f (-en) aptitude, qualification; suitability

eigtl. abbv (eigentlich) actual, etc

Ei·klar n (-) white of egg

Eikosan n (e)icosane, eikosane

Eil· fast, express; special-delivery

Eile f hurry, haste; speed: **E. haben** be in a hurry, be urgent. **eilen 1** vi (s) & **sich e.** hurry, rush, hasten. **2** vi (h) be urgent—**damit eilt es nicht** there is no hurry about that. **eilends** adv quickly, speedily

eil·fertig a hasty, rash. **Eil·gut** n express (or special-delivery) freight

eilig a hurried; urgent—**es e. haben** be in a hurry. **eiligst** adv speedily

Eil·zug m fast (or express) train

Eimer m (-) bucket, pail—**im E.** in the can, taken care of. **-kettenbagger** m bucket chain dredge; bucket excavator

ein I (oft with adj endgs) **A** indef art a, an. **B** num, pron one; (as pfx oft) mono-, uni-, single cf EINER. **II** adv & sep pfx in, into, intro-: **ein und aus** in and out

ein·achsig a uniaxial. **-adrig** a single-strand

einander recip pron each other, one another; (oft combined with prep) cf ANEINANDER, ZUEINANDER, etc

ein·armig a one-armed

ein·äschern vt **1** incinerate, calcine, reduce to ashes. **2** cremate

einatembar a respirable

ein·atmen vt/i inhale. **Einatmung** f inhalation

ein·atomar, ein·atomig a monatomic

ein·ätzen vt (oft + in) corrode, etch (in, into)

ein·äugig a monocular; one-eyed; (fig) narrow-minded, one-track-minded. **Einäugigkeit** f narrow-mindedness, one-track mind

Einbad·, ein·badig a one-bath, one-dip. **Einbad·schwarz** n (Dye.) one-dip black. **-verfahren** n single-bath process

Einbahn·straße f one-way street

ein·balgen vt encyst

ein·balsamieren vt embalm

Einband° m (book) binding, cover

ein·basig, -basisch a monobasic

Einbau° m **1** installation; incorporation; built-in part, attachment. **2** (in compds) built-in; recessed. **-antenne** f built-in (or recessed) antenna. **ein·bauen** vt (oft + in + accus) **1** build in, build (into); install (in). **2** incorporate (in). Cf EINGEBAUT

einbau·fertig a ready to install. **E—spüle** f built-in (or recessed) dishwasher

ein·begreifen* vt (oft + mit) include (in, with); pp **einbegriffen**

ein·behalten* vt hold back, retain

ein·beizen vt **1** etch in. **2** pickle

ein·betonieren vt set in concrete

ein·betten vt embed, set, mount

ein·beziehen* vt involve, include. **Einbeziehung°** f: **unter E. von** involving

einbezogen included: pp of EINBEZIEHEN

ein·biegen* 1 vt & **sich e.** bend (inward, down). **2** vt deflect. **3** vi (s) turn (the corner); bend. **4: sich e.** sag. **Einbiegung°** *f* turn, bend; deflection; curvature

ein·bilden *vt* (+ *dat rflx* **sich**) **1** imagine. **2** (+ **etwas, viel**, etc, + **auf**) be (somewhat, highly) conceited (about) *cf* EINGEBILDET. **Einbildung°** *f* **1** (figment of one's) imagination, fantasy. **2** conceit. **Einbildungs·kraft** *f* (power of) imagination

Einbinde·linie *f* tie line (in phase diagrams)

ein·binden* *vt* **1** bandage; tie up, wrap up. **2** bind (books). **3** include

ein·blasen* *vt* (*oft* + **in**) blow in; blow (into), insufflate, inject (gas), bubble through

ein·blauen, ein·bläuen *vt* blue

ein·blenden 1 *vt/i* blend (in); fade in; insert. **2** *vt* (**auf**) focus, concentrate (rays) (on)

Einblick° *m* **1** view; insight. **2** (*Opt.*) eyepiece

Einbrand° *m* burn; (welding) penetration. **·tiefe** *f* depth of burn (*or* fusion)

ein·brechen* I *vt* **1** break down. **II** *vi* (*oft* + **in**) **2** (h *or* s) break in, invade. **3** (s) break through, collapse. **4** (s) set in, fall (*as*: darkness)

Einbrenn(e) *f* (. . nnen) browned flour, roux. **Einbrenn·emaille** *f* baking enamel

ein·brennen* I *vt* **1** brand. **2** brown (flour for a roux). **3** bake (enamel); harden, fix (colors); fumigate, sulfur (wine barrels); tallow (skins). **II: sich e.** (*dat*) leave its mark (on)

Einbrenn·farbe *f* ceramic color. **·firnis** *m* stoving varnish; baking enamel (*or* lacquer). **·harz** *n* stoving resin. **·kunst** *f* encaustic art. **·lack** *m* baking enamel (*or* lacquer). **·ofen** *m* (*Ceram.*) glost oven

ein·bringen* *vt* **1** (*oft* + **in**) bring in, deliver; introduce, insert, put (into). **2** harvest; capture. **3** earn, yield. **4** (*oft* + **wieder**) make up (for), recover

ein·brocken *vt* crumble (a solid into a liquid)—(**sich, einem**) **etwas e.** get (oneself, sbdy) into trouble

Einbruch° *m* **1** collapse, drop. **2** breakthrough, penetration, invasion. **3** onset. **4** break-in

ein·brühen *vt* steep in boiling water, scald

Einbuchtung *f* (-en) indentation, dent; (*Geol.*) bay

ein·bürgern 1 *vt* naturalize, adopt. **2: sich e.** be naturalized; settle; establish oneself; be adopted; gain currency

Einbuße° *f* (*oft* + **an**) loss (of); **E. tun** (*dat*) inflict loss (on). **ein·büßen** *vt/i* lose

ein·dämmen *vt* **1** embank (a river); dam up. **2** check, (bring under) control

Eindampf·apparat *m* evaporating apparatus

ein·dampfen I *vt* **1** (*also* **ein·dämpfen**) boil down, evaporate; concentrate by evaporation; inspissate. **II** *vi* (s) **2** boil down, get thick(ened). **3** come steaming in

Eindampf·gerät *n* evaporator. **·rückstand** *m* evaporation residue. **·schale** *f* evaporating dish (*or* pan)

Eindampfung *f* (-en) evaporation, etc *cf* EINDAMPFEN

ein·decken I *vt* **1** cover (plants, roof). **2** roof (a house). **3** (**mit**) supply (with). **II: sich e.** (**mit**) lay in a supply (of)

Eindecker *m* (-) monoplane

Eindeckung° *f* covering, etc *cf* EINDECKEN

ein·dellen *vt* dent

eindeutig *a* unequivocal, unmistakable, clear

ein·dichten *vt* condense

Eindick·anlage *f* concentrating plant. **ein·dickbar** *a* capable of thickening (*or* concentration). **Eindicke** *f* (-n) thickening, concentration; inspissation. **ein·dicken** *vt* thicken, concentrate; inspissate. **Eindicker** *m* (-) thickener. **Eindickung** *f* (-en) *f* **1** = EINDICKE. **2** livering (of paints); bodying (of oil). **Eindickungs·mittel** *n* thickening agent. **Eindick·zyklon** *m* cyclone thickener

ein·diffundieren *vt* diffuse in

ein·dimensional *a* one-dimensional

ein·dorren *vi* (s) dry up and shrink

ein·dosen *vt* can (food). **Eindosen** *n* canning

ein·drängen 1 *vi* (**auf**) crowd in (on). **2: sich e.** intrude; penetrate

Eindreh·schnecke *f* (*Mach.*) feed screw

ein·dringen* *vi* (s) **1** (**in**) penetrate, force one's way (into), invade. **2** (**auf**) assail, attack; crowd in (on); (**auf, zu** + *inf*) pressure (sbdy) to (do sthg). **eindringlich** *a* penetrating; forceful, convincing; urgent. **Eindring·tiefe** f depth of penetration

Eindruck m (.. d⁻e) impression; imprint; dent. **ein·drucken** vt (Tex.) block in, ground in; imprint, print in. **ein·drücken** I vt 1 (usu + in) press (in, into), impress (on). 2 dent, push in, crush. **II: sich e.** (in) press (in, into), make an impression (in, on), impress itself (on)

Eindruck·farbe f (Tex.) grounding in. -härte f indentation hardness. -messer m penetrometer

eindrucks·voll a impressive

ein·dunsten 1 vt (also **ein·dünsten**) evaporate (down), concentrate by evaporation; impregnate with vapor. 2 vi(s) evaporate (down), dry up; become concentrated (by evaporation). **Eindunstung** f (-en) evaporation; impregnation with vapor

ein·düsen vt inject (through a nozzle)

eine a, an; one: f of EIN I & EINER II

ein·ebnen vt level (off), even out

einein·deutig a one-to-one. -halb num one and a half. -wertig a uni-univalent

einem dat of EIN I, EINER II & MAN

einen I vt & sich e. unite. II accus of EIN I, EINER II & MAN

ein·engen vt concentrate; confine, restrict

einer I (of, to) a, an, one: genit or dat f of EIN I. II pron (some)one, somebody, a person. **Einer** m (-) unit, one-digit number

einerlei a (no endgs) & adv 1 indifferent, immaterial; (esp + dat) all the same (to): **e., ob** regardless of whether. 2 the same, of one kind. 3 monotonous. **Einerlei** n monotony

einerseits adv on the one hand; on one side

eines I of a, of an, of one: genit m & n of EIN I. II n pron one (thing)

einfach I a 1 simple, plain, easy; ordinary, regular; adv oft just. 2 single, one-way; adv oft once. 3 elementary (e.g., substances). 4 primary (colors). II pfx 5 (hyph., in old Chem. Nomencl.) proto-, mono-, -ous: **e.-Chlorjodid** iodine monochloride; **e.-Schwefeleisen** iron protosulfide, ferrous sulfide. 6 (Chem.: in compds with sfx **·sauer**) neutral: **einfach·kohlensauer** neutral carbonate of cf EINFACHSAUER

einfach·brechend a singly refracting. **E—bindung** f single bond. -chlorzinn n stannous chloride. **e—frei** a having one degree of freedom, univariant

Einfachheit f simplicity, singleness

einfach·ionisiert a simply ionized. **E—kiste** f (Tinplate) base box (22,557 m²). **e—sauer** a monohydrogen (salt of a polybasic acid): **e—es Orthophosphat** monohydrogen orthophosphate cf EINFACH. 6. **E—schicht** f monolayer. **e—schwefelsauer** a neutral sulfate of. -wirkend a single-acting, single-action

ein·fädeln I vt 1 thread (needle). 2 initiate; contrive. II vi thread the needle. III: sich e. (in) merge (into traffic)

Einfaden· single-filament. **einfadig** a unifilar, single-stranded(ed)

ein·fahren* I vt 1 bring in (harvest, etc) by vehicle. 2 retract. 3 break in (vehicles, animals). II vi(s) 4 arrive, pull in. 5 go down (into the mine). 6 bring in the harvest. III: sich e. become customary cf EINGEFAHREN. **Einfahrt°** f 1 (vehicle) entrance. 2 arrival (by vehicle). 3 entry, signal to enter

Einfall° m 1 idea. 2 invasion. 3 incidence (of light). 4 collapse. 5 decay. **ein·fallen*** vi(s) 1 collapse, cave in, sink. 2 (in) invade; break in, interrupt. 3 (Locks, etc) click, engage. 4 (Light) shine (in), be incident. 5 dip, slope downward. 6 join in. 7 set in, start. 8 (dat) occur (to a psn) (as: an idea, a memory). **einfallend** p.a incident (light)

Einfalls·lot n perpendicular, ordinate. **e—reich** a ingenious. **E—strahl** m incident ray. -winkel m angle of incidence

Einfalt f simplicity; artlessness; simple-mindedness. **einfältig** a simple; artless; simple-minded

Einfang m 1 enclosure. 2 (incl Nucl.) capture. **ein·fangen*** vt capture, catch, seize. **Einfang·querschnitt** m (Nucl) capture cross-section. **Einfangs·beute** f (Nucl.) capture yield (or efficiency). **Einfangung** f (-en) capture, seizure

ein·färben vt dye (thoroughly) (to hide stains or obliterate patterns); dye in the grain; ink. **Einfärbe·vermögen** n (Dye.) penetrating power. **einfarbig** a one-color, monochromatic. **Einfärbung°** f thorough dyeing, etc

ein·fassen vt 1 edge, border; trim; surround; enclose. 2 set (precious stones). **Einfassung°** f enclosure; border; setting

ein·fetten vt grease, oil, lubricate

ein·feuchten vt moisten, wet, dampfen

ein·finden* vr: **sich e.** appear, show up

Einfl. abbv (Einfluß) influence

einflammig a single-flame

ein·flechten* vt (oft + **in**) weave (in, into)

ein·fließen* vi(s) 1 flow in. 2 (+**lassen**) let flow in, insert, add; (usu + **daß**) hint at (the fact that)

ein·flößen vt (d.i.o) administer (to); instill (in); arouse (in)

Einfluß m 1 influence, effect. 2 influx, inflow. **e—·reich** a influential. **E—·rohr** n, **-röhre** f influx (or inlet) tube (or pipe)

ein·formen vt (Ceram.) mold on a jigger

einförmig a uniform; monotonous

ein·fressen* vr: **sich e.** (in) eat its way (into), corrode; penetrate (into)

ein·frieren* vt, vi(s) freeze (up), congeal

Einfrier·punkt m freezing point. **-temperatur** f (Glass) setting point

ein·fritten vt frit

ein·fügen 1 vt (**in**) fit in(to); insert, interpolate (in, into). 2: **sich e.** (**in**) adapt oneself (to), fit (oneself) in(to). **Einfügung°** f insertion, interpolation; adaptation

Einfuhr f (-en) 1 imports, importation. 2 (Med.) intake

ein·führen vt introduce, insert; import. **Einführung°** f introduction, insertion

Einfuhr·waren fpl imports, imported goods

ein·füllen vt (**in**) pour (into) (vessels), fill (vessels) with . . . , (Metll.) charge (a furnace) with . . .

Einfüll·hahn m feed cock. **-stoff** m filling, packing. **-trichter** m (feed) funnel, (feed, charging, loading) hopper. **-wägepipette** f weighing pipette

Eingabe° f 1 insertion; (Mach.) feeding; (Cmptrs.) input. 2 administration (of medication). 3 submission (of reports). 4 petition, applicaiton. 5 (Patents) amendment. Cf EINGEBEN

Eingang° m 1 entrance; entry: **E. finden** gain entry, win acceptance. 2 beginning.

3 pl receipts. **eingängig** a 1 (Mach.) single-thread. 2 clear, understandable. **eingangs** adv at the outset. **Eingangs·** (in compds) initial, inaugural; input

eingebaut 1 p.a: **e—e Verunreinigung** (Cryst.) lattice impurity. 2 built in, etc: pp of EINBAUEN

ein·geben* I vt (d.i.o) 1 administer (medication) (to). 2 feed, put (into machines, computers). 3 hand in, submit (reports, applications). II vt/i (usu + **zu** + inf) persuade, inspire (to . . .). III vi (**um**) apply (for)

eingebildet p.a conceited. See also EINBILDEN

eingebogen bent: pp of EINBIEGEN

eingeboren a inborn, innate; native

eingebracht brought in: pp of EINBRINGEN

eingebrannt branded: pp of EINBRENNEN

eingebrochen broken in: pp of EINBRECHEN

eingebunden bound: pp of EINBINDEN

Eingebung f (-en) inspiration

eingedenk adv, pred a (genit) mindful (of)

eingedorrt p.a dried; shrunken. See also EINDORREN

eingedrungen penetrated, assailed: pp of EINDRINGEN

eingefahren 1 p.a customary, familiar. 2 brought in, etc: pp of EINFAHREN

eingeflochten woven in: pp of EINFLECHTEN

eingeflossen flowed in: pp of EINFLIESSEN

eingefroren frozen up: pp of EINFRIEREN

eingefunden shown up: pp of EINFINDEN

eingegangen received, shrunken, etc: pp of EINGEHEN

eingegossen cast: pp of EINGIESSEN

eingegriffen intervened: pp of EINGREIFEN

ein·gehen* vi(s) 1 (usu + **in**, oft + accus) enter (into), (+ accus, oft) incur: (Chem.) **Verbindungen e.** enter into compounds. 2 (dat) dawn (on), be credible (understandable, welcome) to, come (easily, etc) (to): **es geht ihnen nicht ein** they can't grasp it. 3 come in, arrive, be received. 4 shrink, perish, die; close down; cease to exist. 6 (**auf**) agree, be receptive (to), go (into: a subject). **eingehend** p.a 1 detailed, thorough; (adv oft) in detail; in depth. 2 (Geom.) re-entrant

Eingemachte(s) *n* (*adj endgs*) preserves, canned food *cf* EINMACHEN

eingenommen 1 taken in, etc: *pp of* EINNEHMEN. **2** *p.a* (+ **für, gegen**) prejudiced (in favor of, against); (+ **von**) fascinated (with). **Eingenommenheit** *f* (-en) prejudice; fascination

eingerannt smashed: *pp of* EINRENNEN

eingerieben rubbed: *pp of* EINREIBEN

eingerissen torn down: *pp of* EINREISSEN

eingesalzen *p.a* salt-cured, corned *cf* EINSALZEN

eingeschliffen 1 ground (in): *pp of* EINSCHLEIFEN. **2** *p.a* habituated, reflex

eingeschlossen included: *pp of* EINSCHLIESSEN

eingeschmolzen melted down: *pp of* EINSCHMELZEN

eingeschnitten cut in: *pp of* EINSCHNEIDEN

eingeschoben inserted: *pp of* EINSCHIEBEN

eingeschrieben registered: *pp of* EINSCHREIBEN

eingeschritten intervened: *pp of* EINSCHREITEN

eingesogen absorbed: *pp of* EINSAUGEN

eingesotten boiled down: *pp of* EINSIEDEN

eingespielt *p.a* **1** broken in, practiced; routine. **2** tried and tested. **3** (+ **auf**) attuned to; (+ **aufeinander**) attuned to each other, coordinated. *See also* EINSPIELEN

eingesprengt *p.a* isolated; *see also* EINSPRENGEN

eingesprochen objected: *pp of* EINSPRECHEN

eingesprungen helped out; clicked: *pp of* EINSPRINGEN

eingestanden 1 admitted: *pp of* EINGESTEHEN. **2** vouched *pp of* EINSTEHEN

eingestiegen boarded: *pp of* EINSTEIGEN

eingestochen pierced: *pp of* EINSTECHEN

eingestrichen coated: *pp of* EINSTREICHEN

eingetrieben driven in: *pp of* EINTREIBEN

eingetroffen arrived: *pp of* EINTREFFEN

eingewandt objected: *pp of* EINWENDEN

Eingeweide *n* viscera, entrails; intestines; (*in compds*) visceral, intestinal

eingewiesen instructed: *pp of* EINWEISEN

eingewoben woven in: *pp of* EINWEBEN

eingewogen weighed in: *pp of* EINWÄGEN & EINWIEGEN

ein·gewöhnen *vr*: **sich e.** became acclimatized

eingeworfen interjected: *pp of* EINWERFEN

eingezogen conscripted: *pp of* EINZIEHEN

ein·gießen* *vt/i* pour; cast; infuse. **Eingießung** *f* (-en) pouring; casting; infusion, transfusion

eingipsen *vt* set in plaster (of Paris); put a plaster cast on; fasten with plaster; plaster up

eingliedern *vt* incorporate, integrate

ein·graben* *vt* bury; dig (in); engrave

ein·greifen* *vi* (*oft* + **in**) **1** lock, penetrate (into); (*Gears*) mesh (with), engage. **2** intervene, interfere (in). **3** act, have an effect (on) (*incl Chem.*). **Eingreifen** *n* mesh(ing), engagement; gearing; intervention, interference; (chemical) action. **eingreifend** *p.a* decisive, radical

ein·grenzen *vt* enclose; bound, limit

Eingriff° *m* **1** (*Mach.*) engagement: **im** (or **in**) **E.** in gear, in mesh. **2** interference, encroachment. **3** (*Med.*) operation, surgery. *Cf* EINGREIFEN

Einguß° *m* **1** pouring in, casting. **2** (*Metll.*) mold, ingot. **3** (*Foun.*) gate, runner. **4** infusion. *Cf* EINGIESSEN. **-trichter** *m* feeder funnel, sprue

Einhalt *m*: **E. gebieten** (*dat*) check, restrain; **E. tun** (*dat*) stop, halt. **ein·halten*** I *vi* **1** keep (to), maintain. **2** abide by, observe. **3** take (*or* tuck) in. II *vi* (**in, mit**) stop, cease (. . ing). **Einhaltung** *f* observance (of laws, etc)

ein·händigen *vt* hand over, deliver

ein·hängen I *vt* **1** hang, suspend; install; hinge. **2** hitch on, couple on. II *vt/i* (*Tlph.*) hang up

ein·hauen* *vt* chisel in, cut in; smash in

einhäusig *a* monoecious

Einhebel· single-top, single-level: **-mischer** *m* mixing faucet

einheimisch *a* native, indigenous; endemic

Einheit *f* (-en) unit, unity. **einheitlich** *a* **1** uniform, homogeneous; *adv oft* alike. **2** united, unified. **Einheitlichkeit** *f* uniformity, homogeneity

Einheits· unit, uniform, united: **-fläche** *f* (*Cryst.*) parametral plane. **-gewicht** *n* unit (standard, uniform) weight; specific gravity. **-gitter** n (Cryst.) unit lattice. **-kreis** m unit radius circle. **-masse** f unit mass. **-pol** m (Phys.) unit pole. **-preis** m unit (or uniform) price. **-preisgeschäft** n one-price store. **-satz** m (Com.) flat rate. **-volumen** n unit volume. **-wert** m unit value. **-zelle** f unit cell

einhellig a unanimous

einher adv & sep pfx along

ein·holen vt 1 make up for; catch up with, overtake. 2 haul in (or down). 3 shop for, buy: **e. gehen** go shopping

ein·hüllen vt wrap (up), envelop, sheathe; embed

ein·hülsen vt 1 encyst, encapsulate. 2 (Nucl.) can

einhundert num one hundred

einig pred a united (esp + rflx **sich**) in accord. **einige** a/pron some, a little, a few: **e. wenige** just a few cf EINIGES. **einigemal** adv a few times

einigen I vt & **sich e.** unite. II dat pl of EINIGE

einigermaßen adv to some extent, somewhat

einiges something, a few things: nsg of EINIGE

Einigkeit f unit, unanimity, union

Einigung f 1 agreement, settlement. 2 reconciliation. 3 unification. **Einigungs·kitt** m (adhesive) cement; putty

ein·impfen vt 1 inoculate. 2 instill, implant

einjährig a one year('s); one year old; annual

ein·kalken, ein·kälken vt (treat with) lime, soak in lime water

ein·kapseln 1 vt encapsulate, encyst, encase. 2: **sich e.** become encapsulated (encysted, encased)

Einkauf° m purchase; buying. **ein·kaufen** vt/i shop (for), buy: **e. gehen** go shopping. **Einkaufs·zentrum** n shopping center

einkeim·blättrig a monocotyledonous

ein·kellern vt store in the cellar; lay in

ein·kerben vt notch, indent; carve (in). **Einkerbung** f (-en) notch(ing), indentation

einkernig a mononuclear

ein·kitten vt cement in; putty in

ein·klammern vt 1 bracket, put in brackets (or parentheses). 2 clamp

Einklang° m harmony, accord

ein·kleiden vt clothe, dress

ein·klemmen vt catch, jam; clamp, wedge (in)

ein·kochen 1 vt boil, stew (esp for preserves). 2 vt, vi(s) boil down, evaporate, thicken (by boiling)

Einkommen n (-) income; revenue

Einkristall° m single crystal, monocrystal. **-faden** m single-crystal filament

ein·krücken vt crutch (soap)

Einkrustung f (-en) incrustation

Einkünfte fpl earnings; income revenue

ein·laben vt curdle, coagulate

ein·laden* 1 vt/i load (on, in). 2 vt invite

Einlage° f 1 insert, insertion; (specif): enclosure (in a letter), stiffener, pad(ding), arch support, insole, filler, temporary tooth filling, garnish, pieces of meat (in soup), steel reinforcement (in concrete), charge (in a furnace). 2 (bank) deposit; investment

ein·lagern I vt 1 store, stockpile. 2 deposit. 3 embed, intercalate. II: **sich e.** form a deposit (or an intercalation). **Einlagerungs·verbindung** f intercalation compound

Einlaß m (.. lässe) 1 inlet, intake. 2 admission, admittance. **-druck** m intake pressure. **ein·lassen*** I vt 1 admit, let in. 2 insert, set (in), embed, inlay. 3 wax (a floor). II: **sich e.** get involved

Einlaß· inlet, intake: **-farbe** f (Lthr.) graining color. **-grund** m (Paints) scaler. **-öffnung** f inlet, intake. **-rohr** n inlet tube. **-stück** n inserted piece, insert. **-ventil** n intake (or inlet) valve

Einlauf° m 1 influx. 2 intake, inlet. 3 arrival; receipt (of mail); incoming mail. 4 soup thickener. 5 enema. 6 (Mach.) break-in run. **-druck** m intake pressure. **e—echt** a unshrinkable; shrink-resistant

ein·laufen* I vi(s) 1 arrive, come in; be received. 2 run in, flow in. 3 shrink, contract. II: **sich e.** be broken in; start running smoothly. **einlaufend** p.a incoming

Einlauf·pipette f pipette calibrated to hold a given volume. **-schnecke** f feed(ing)

screw. **-trichter** *m* inflow funnel; feed hopper

ein·laugen *vt* soak in lye, buck

ein·legen *vt* 1 put (in), insert; enclose; inlay. 2 set (hair). 3 pickle, preserve, put up, can. 4 deposit (in bank). 5 lodge (complaints, etc). 6 (*Mach.*) engage, go into (gear)

ein·leiten *vt* 1 introduce. 2 start, open, initiate. 3 trigger. 4 conduct (*incl Elec.*); inject, feed, let flow (in). **Einleitung°** *f* 1 introduction. 2 opening, initiation. 3 preparation. 4 conduction; injection; feed, inflow. **Einleitungs·rohr** *n* inlet pipe, (gas) feed pipe

ein·lenken 1 *vt* guide, steer; (*Med.*) set (limbs). 2 *vi*(h) relent. 3 *vi*(s) turn

ein·leuchten *vi* (*dat*) be(come) clear (to), dawn (on). **einleuchtend** *p.a* clear, convincing

ein·liefern *vt* bring in, deliver

Einlinien·spektrum *n* spectrum of single lines

ein·lösen *vt* redeem, cash

ein·löten *vt* solder in

ein·machen *vt* 1 preserve, put up, can (food); pickle. 2 temper (lime). **Einmache·salz** *n* preserving salt. **Einmach·essig** *m* pickling vinegar

ein·mahlen *vt* grind in; intergrind

ein·maischen *vt* (*Brew.*) dough in

Einmaisch·temperatur *f* doughing-in temperature (of the mash). **-wasser** *n* mash liquor

einmal *adv* 1 once, one time; **noch e.** once more; **wieder e.** once again; **auf e. (a)** suddenly, **(b)** at once, at the same time. 2 some time, some day. 3 for one thing. 4: **nicht e.** not even. 5 *oft no transl* (*modifies meaning of neighboring words*): **nun e.** (*implies no alternative*): **das ist nun e. so** that's the way things are. *Cf* SCHON 7. **E—·eins** *n* multiplication table(s). **einmalig** *a* 1 single, solitary, lone. 2 onetime, single-occurrence. 3 unique. **Einmal·schmelzerei** *f* (*Iron*) single refining

ein·marinieren *vt* pickle, marinate

ein·mauern *vt* wall in, immure; embed

ein·mengen, ein·mischen 1 *vt* mix in, blend in. 2: **sich e.** interfere

ein·mitten *vt* center

Einmolekül° *n* single (giant) molecule

ein·motten *vt* mothball, put in mothballs.

ein·münden *vi* (**in**) 1 empty, discharge, disembogue (into). 2 lead, open (into). 3 join, inosculate

einmütig *a* unanimous. **Einmütigkeit** *f* unanimity

Einnahme *f* (-n) 1 consumption, intake (of food); taking (of medicine). 2 conquest. 3 *usu pl* receipts, earnings, income

ein·nehmen* *vt* 1 take in; earn, receive, collect; **mit e.** include. 2 consumer (food), take (medicine). 3 take on (a load of). 4 capture, conquer. 5 (*oft* + **für sich**) captivate; gain (*or* win) (sbdy's) favor; (**gegen**) alienate, prejudice (against) *cf* EINGENOMMEN. 6 take (up), occupy. **einnehmend** *p.a* winning, captivating

einnormal *a* (*Solutions*) normal, 1N

Einöde *f* (-n) desert, wilderness

ein·ölen *vt* oil, lubricate

ein·ordnen *vt* 1 arrange, classify; sort; fit in, put in place. 2: **sich e.** find (*or* take) one's place

ein·packen *vt/i* pack (up); wrap (up)

ein·passen *vt* & **sich e.** fit in, adapt (oneself)

Einphasen· (*in compds*), **einphasig** *a* single-phase, monophase

ein·pichen *vt* coat with pitch

ein·pilieren *vt* mill (soap)

ein·pökeln *vt* pickle, corn

einpolig *a* unipolar

ein·polymerisieren *vt* incorporate by polymerization

ein·prägen *vt* impress, imprint (on). **einprägsam** *a* impressive; easily remembered

ein·pressen *vt* press (squeeze, force) in. **Einpreß·mörtel** *m* grout

einprozentig *a* one-percent

ein·pudern *vt* (sprinkle with) powder, dust

ein·pumpen *vt* pump in

einquantig *a* one-quantum

Einquell·bottich *m* soaking tub, steeping vessel. **-quellen** *vi* steep, soak

Einquell·kufe *f* steeping tub. **-wasser** *n* steeping water

ein·rahmen *vt*, **Einrahmung** *f* (-en) frame

ein·rammen *vt* ram in

ein·räuchern *vt* fill with smoke. fumigate

ein·räumen *vt* 1 (**in**) move (sthg) (in, into), put . . . away (*or* in its place). 2 furnish. 3 (*d.i.o*) relinquish (to); concede, grant (to), let . . . have

ein·rechnen *vt* (*oft:* **mit e.**) count in, include

Einrede° *f* objection, contradiction, plea. **ein·reden** I *vt* (*d.i.o or dat rflx* **sich**) talk a psn (*or* talk oneself) into believing. II *vi* 2 (+ *dat* & + *zu* + *inf*) persuade (to do), talk (into doing). 3 (**auf**) talk persuasively (to)

ein·regeln, ein·regulieren *vt* adjust, regulate

ein·reiben* *vt* rub (down), smear; wax. **Einreibe·** (*see also* EINREIBUNGS·): **-salbe** *f* rub, rubbing ointment. **Einreibung°** *f* rubbing; liniment, embrocation. **Einreibungs·mittel** *n* (body) rub, liniment

ein·reichen *vt* hand in, present, submit

ein·reihen 1 *vt* (**in**) assign (to), incorporate (in, into); (**unter**) rank (among), file (under). 2: **sich e.** (**in** *or dat*) join (the ranks of)

einreihig *a* single-series, single-row

ein·reißen* I *vt* 1 tear (into), make a tear (in). 2 tear down, demolish. 3 (*dat rflx* **sich**) run (a splinter) into (one's hand). II *vi* (s) 4 get torn, tear (from the edge inward). 5 make inroads, gain ground. **Einreiß·festigkeit** *f* tearing strength

ein·rennen* *vt* 1 smash, run down. 2 melt down. **Einrennen** *n* melting down, first smelting

ein·richten I *vt* 1 adjust, arrange, regulate, organize. 2 furnish, equip. 3 set up, establish. 4 set (limbs). 5 reduce (mixed numbers). II: **sich e.** 6 furnish one's home. 7 settle, move in, establish oneself. 8 limit oneself, economize; manage. 9 (+ **auf**) prepare (for). **Einrichtung°** *f* 1 arrangement, organization, set-up; institution; fixture, device, equipment, installation; (set of) furniture. 2 furnishing, setting up, establishment; setting (of limbs). 3 *pl oft* furnishings, facilities

Einriß° *m* tear, rip (on the edge); crack, fissure

ein·rollen *vt* & **sich e.** roll up, curl up

ein·rosten *vi*(s) get rusty, get rust-covered

ein·rücken 1 *vt* insert, put in; (*Mach.*) engage, throw in (clutch). 2 *vi*(s) move (in), enter, report for duty

ein·rühren *vt* stir (in), mix; temper (mortar)

ein·rußen *vt* soot up, cover with soot

ein·rütteln *vt* compact by vibration, shake down

eins 1 *num* one. 2 *pron* = EINES. II one (thing). **Eins** *f* (-en) (figure) one

ein·sacken 1 *vt* bag, pack in sacks; encyst. 2 *vi* (s) sag; cave in. **Einsack·stelle** *f* sunk spot (in a casting)

ein·säen *vt* sow in (crystals into solution)

ein·salben *vt* rub with ointment; embalm

ein·salzen *vi* salt (down), corn, brine

einsam *a* solitary, alone, deserted; secluded

Einsattlung *f* (-en) dip, depression, saddle

Einsatz° *m* 1 insert, insertion, inset, inserted piece; (*specif*) charge (in furnaces); liner (in centrifugal drums); bed spring; filler, filling; tray. 2 case hardening. 3 onset, start. 4 stake(s), risk. 5 use, application. 6 service, duty. 7 action, engagement, effort; input. *Cf* EINSETZEN

einsatz·bereit *a* ready for use (service, action), operational. **-fähig** *a* (*Metll.*) casehardenable. **E—gewichte** *npl* (set of) nested weights. **-gut** *n* (*Metll.*) charging material. **-härtung** *f* case hardening. **-kessel** *mpl* set of (nested) kettles. **-korb** *m* test-tube basket (*or* case). **-mittel** *n* (*Metll.*) case-hardening compound, carburizer. **-pulver** *n* cementing powder. **-rohr** *n* inlet tube. **-stahl** *m* casehardened steel. **-stück** *n* insert, inserted piece. **-tür** *f* charging door

ein·säuern *vt* acidify; pickle; leaven; silo. **Einsäuerung** *f* acidification; pickling; leavening; silage. **Einsäuerungs·bad** *n* acid bath

Einsaug(e)· absorbent; lymphatic: **-mittel** *n* absorbent. **ein·saugen*** 1 *vt* suck (*or* drink) in, suck up, absorb; inhale. 2: **sich e.** be absorbed, soak in. **einsaugend** *p.a* absorbent, absorptive

Einsaug·mittel *n* absorbent. **-rohr** *n* suction tube

Einsaugung f absorption, suction, imbibation. **Einsaugungs·fähigkeit** f absorptivity

einsäurig a monoacid, monacid

ein·schalten I vt 1 turn on (light, radio, motor), cut (or connect) in (to a circuit). 2: . . . Gang e. shift into . . . gear. 3 interject, interpolate, throw in. 4 call in (usu as mediator). **II** vi turn on the power (or current). **III:sich e.** cut oneself in(to the proceedings), intervene, step in, get involved. **Einschalter°** m circuit closer, switch. **Einschaltung°** f turning on, connection; shift (into . . . gear); interjection, interpolation; calling in; intervention, involvement

ein·schärfen vt (d.i.o) impress (on)

ein·schätzen vt (usu + als) estimate, judge, rate, assess (as, at). **Einschätzung°** f estimate, estimation, judgment, rating, assessment

ein·schenken vt pour (liquid, beverage)

ein·schichten vt 1 layer, put (up) (or pack) in layers. 2 embed; stratify. **einschichtig** a one-layer, single(-layered)

ein·schieben* vt shove in, shove (bread) into the oven; insert, interpolate

einschl. abbv (einschließlich) incl.

ein·schlafen* vi(s) fall asleep, go to sleep

ein·schläfern vt/i make drowsy, put to sleep; narcotize. **einschläfernd** p.a soporific, narcotizing, hypnotic. **Einschläferungs·mittel, Einschlaf·mittel** n sleep-inducing (soporific, hypnotic) agent

Einschlag° m 1 impact, strike, hit. 2 wrapping. 3 tinge, touch, admixture. 4 (Tex.) weft; fold, tuck. 5 sulfur match (for casks). 6 handshake. **ein·schlagen*** I vt 1 pound (drive, dig) (sthg) in(to). 2 smash (break, bash) in. 3 wrap, cover. 4 fold over, tuck in. 5 take, choose, strike out on (path, direction). 6 dip (sheet metal). 7 sulfur (wine). **II** vi 8 (oft + auf) strike, hit (at). 9 (usu + bei) make a hit, be a success (with). 10 shake hands

einschlägig a relevant, pertinent, appropriate

Einschlag·lupe f folding magnifier. **-papier** n wrapping paper. **-punkt** m point of impact. **-tiefe** f impact depth

Einschlämmung° f sedimentation; (Bot.) watering; (Engg.) sluicing

ein·schlauchen vt pour in (through a hose)

ein·schleifen* vt grind (in); reseat (valves) cf EINGESCHLIFFEN

ein·schließen* vt 1 lock in (up, away), shut in, confine; trap. 2 enclose, surround. 3 include. 4 occlude (gas, etc). **einschließlich** adv & prep (genit) including, inclusive (of). **Einschließung°** f locking in, enclosing, surrounding; inclusion; occlusion

ein·schlucken vt swallow up, absorb

ein·schlüpfen vi (s) slip in, slide in; soak in

Einschluß° m 1 inclusion: **unter** (or **mit**) **E.** (+ **von** or + genit) including. 2 confinement. 3 enclosure. 4 occlusion. 5 encirclement. Cf EINSCHLIESSUNG. **-energie** f impact energy, energy input. **-lack** m (Micros.) varnish for cell making. **-mittel** n (Micros.) embedding (or mounting) medium. **-rohr** n, **-röhre** f seated tube. **-thermometer** n, m enclosed-scale thermometer. **-verbindung** f inclusion (occlusion, clathrate) compound

ein·schmalzen vt grease, lubricate, oil

ein·schmauchen vt smoke, fumigate

ein·schmelzen* 1 vt, vi(s) melt down, remelt; (+ **in**) fuse (with). 2 vt assimilate; seal in; (Med.) cavitate. 3 vi (s) be assimilated

Einschmelz·flasche f bottle closed by fusion. **-glas** n fusible glass (for sealing in platinum points, etc). **-legierung** f addition alloy. **-probe** f (Metll.) meltdown test, first test. **-rohr** n, **-röhre** f sealing tube, ampule. **-schlacke** f first slag. **-sirup** m remelt syrup (or liquor)

Einschmelzung f (-en) meltdown, melting down, remelting; fusion; assimilation; sealing in ; cavitation cf EINSCHMELZEN. **Einschmelzungs·glas** n fusible glass (for ampules, etc)

Einschmelz·zucker m remelt sugar

ein·schmieren vt smear; grease, lubricate

ein·schneiden* 1 vt cut in(to); notch, indent; cut up and put in. 2 vi cut in(to); be incisive. 3: **sich e.** cut (one's way) in(to). **einschneidend** p.a incisive; trenchant

Einschnitt° m 1 incision, section, cut. 2 indentation, notch; slot. 3 excavation. 4 incisive event, turning point

ein·schnüren *vt* tie up, lace up, constrict. **Einschnürung** *f* (-en) constriction; tying up; (*Metll.*) necking

ein·schränken *vt* limit, restrict, reduce

ein·schrauben *vt* screw in(to)

ein·schreiben* *vt* & **sich e.** register

ein·schreiten* *vi*(s) intervene, step in

ein·schrumpfen *vi* (s) shrink, shrivel up, contract

Einschuß° *m* 1 admixture, impurity. 2 (*Tex.*) weft

ein·schütten *vt* pour (in), put in, feed (in)

ein·schwärzen *vt* blacken; ink

ein·schwefeln *vt* sulfur, sulfurize

ein·schwöden *vt* lime (hides)

ein·sehen* I *vt* 1 look in(to), examine. 2 realize, recognize; (+ **warum**) see (why). II *vt/i* see in(to), observe. **Einsehen** *n* insight, realization—**ein E. haben** have consideration

ein·seifen *vt* soap up, lather

ein·seihen *vt/i* infiltrate

einseitig *a* one-sided, unilateral; unbalanced

ein·senken 1 *vt* sink, lower; plant. 2: **sich e.** sink in, subside; make an impression. **Einsenk·punkt** *m* (*Glass*) softening point

Einser *m* (-) (figure, number) one; (*Money*) single

ein·setzen I *vt* 1 put in, set in, insert; install; appoint; name; start. 2 establish, set up. 3 preserve (food). 4 risk, stake. 5 earmark, set aside. 6 use, put to use, employ, apply; call (*or* assign) to duty. 7 case-harden, carburize. II *vi* start, set in. III: **sich e.** make an effort, (+ **für**, *oft*) support, advocate. **Einsetzen** *n* (*Ceram.*, *oft*) charge. **Einsetzer** *m* (-) combustion boat

Einsetz·gewichte *npl* nest of weights. **-tür** *f* charging door

Einsicht *f* (-en) 1 view (of, into). 2 look, inspection, examination: **E. haben** (*or* **nehmen**) **in** inspect, examine. 3 insight, realization; understanding, discernment. **einsichtig** *a* discerning, judicious, insightful

ein·sickern *vi* (s) trickle in, soak in, infiltrate

ein·sieden* 1 *vt*, *vi*(s) boil down. 2 *vt* preserve

Einsonderungs·drüse *f* endocrine gland

ein·spannen *vt* 1 put in, insert. 2 clamp, grip, mount. 3 frame. 4 put to work, harness. **Einspann·länge** *f* free length of a specimen (between clamps)

ein·sparen *vt* save, economize (on); eliminate. **Einsparung** *f* (-en) saving, economy

ein·speicheln *vt* (in)salivate, mix with saliva

ein·sperren *vt* lock in (*or* up), shut in, confine

ein·spielen I *vt* 1 bring into play. 2 record. 3(+ **aufeinander**) balance; attune (to each other). 4: **seine Kosten e.** pay its way. II: **sich e.** 5 warm up, get into its stride. 6 (+ **aufeinander**) become attuned to each other. *Cf* EINGESPIELT

Einsprache° *f* objection, protest = EINSPRUCH. **ein·sprechen*** *vi* 1 (**gegen**) object (to), protest (against.) 2 (**für**) advocate

ein·sprengen *vt* 1 sprinkle. 2 intersperse, scatter, disseminate; interstratify. 3 break in (down, open), blast. *Cf* EINGESPRENGT. **Einsprengling** *m* (-e), **Einsprengsel** *n* (-) interspersed body (*or* substance), scattered remnant; xenocryst

ein·springen* *vi* (s) 1 help out; (+ **für**) take over, substitute (for). 2 (*locks*) click, snap. 3 shrink. 4 recede, be set back. 5 (*Angles*) reenter

Einspritz·anlage *f* injection unit (*or* system)

ein·spritzen *vt* inject, squirt in, syringe. **Einspritzer** *m* (-) injector, syringe

Einspritz·hahn *m* injection cock. **-rohr** *n* injection tube

Einspritzung *f* (-en) injection

Einspruch° *m* objection, protest. **Einspruchs·frist** *f* period for lodging objections

Einsprung° *m* shrinkage *cf* EINSPRINGEN

einst *adv* 1 formerly, at some time. 2 some day, sometime

ein·stampfen *vt* stamp, ram down, compress; mash, pulp; pulverize

Einstands·preis *m* cost price

ein·stanzen *vt* stamp, punch

ein·stäuben *vt* (spray with) dust, powder

ein·stechen* I *vt/i* **1** prick (a hole). **2** stick in. **II** *vi* **3** (**in**) penetrate (into), puncture, pierce. **4** (**auf**) stab (at)

ein·stecken *vt* **1** (put in one's) pocket. **2** put up with. **3** (**in**) stick in(to). **4** mail (letters). **5** lock up. **6** outdo

ein·stehen* *vi*(s) **1** (**für**) vouch, stand up, be responsible (for); guarantee. **2** start work

ein·steigen* *vi*(s) (**in**) get on, enter, board; join, go (*or* get) into

Einstell· set, adjusting, focusing

einstellbar *a* adjustable

ein·stellen I *vt* **1** put (up); park (vehicle); stow. **2** leave (sthg somewhere). **3** set, adjust, focus; (*Solutions, oft* + **auf**) standardize (against), make up (to a standard volume). **4** hire, employ. **5** stop, suspend, discontinue, terminate. **II** *vt, vi* (**auf**) **6** tune (in on). **III** *vi* **7** focus. **IV: sich e.** **8** appear, show up; arise; set in. **9** (+ **auf**) get set, adjust (to), adapt (*or* attune) oneself (to). **Einsteller** *m* (-) regulator; thermostat

Einstell·lupe *cf* EINSTELLUPE. **-marke** *f* reference mark. **-schraube** *f* adjusting (*or* set) screw

Einstellung° *f* **1** parking, leaving, putting (up). **2** setting, adjustment, focus(ing). **3** standardization; (*Chem.*) standard (for solutions). **4** hiring. **5** stoppage, suspension, discontinuance, termination. **6** tuning in. **7** (+ **zu**) attitude (toward), opinion (of). *Cf* EINSTELLEN

Einstellupe *f* focusing lens = EINSTELL·LUPE

Einstich° *m* puncture, injection. **-farbe** *f* (*Dye.*) fancy color

Einstieg *m* (-e) boarding; entrance; start

einstig *a* one-time, former *cf* EINST

ein·stimmen 1 *vt* tune. **2** *vi* join in, chime in

einstimmig *a* unanimous

Einstoff·system *n* one-component system

ein·stöpseln *vt* put in, stick in (cork, *Elec.* plug)

ein·stoßen* *vt* push in, smash in, bash in

ein·strahlen 1 *vt, vi* (s) beam, radiate. **2** *vt* irradiate. **Einstrahlung°** *f* absorbed radiation; absorption (of radiation); irradiation, isolation

ein·streichen* *vt* **1** coat, smear. **2** putty up,

fill in. **3** (+ **in**) file a notch (into). **4** pocket, reap

ein·streuen *vt* scatter, sprinkle; intersperse

Einstrich·mittel *n* coating agent

ein·strömen *vi* (s) stream in, flow in. **Einströmen** *n* influx

Einströmungs·rohr *n*, **-röhre** *f* inlet (*or* admission) tube (*or* pipe)

Einström·ventil *n* inlet valve

ein·stufen *vt* rate, grade; assign (to a class, etc)

einstufig *a* single-stage, simple

einstündig *a* one-hour, one hour long

Einsturz° *m* collapse; subsidence. **ein·stürzen 1** *vt* batter down, demolish. **2** *vi* (s) collapse, cave in

einstweilen *adv* meanwhile; for the time being. **einstweilig** *a* temporary, provisional

ein·sumpfen *vt* wet, soak

ein·süßen *vt* sweeten, edulcorate

eintägig *a* one-day, one-day-old; ephemeral

ein·tauchen 1 *vt, vi*(s) dip (in). **2** *vt* immerse

Eintauch·refraktometer *n* immersion refractometer. **-tiefe** *f* depth of immersion

Eintausch° *m*, **ein·tauschen** *vt* (**für, gegen**) exchange (for)

eintausend *num* one thousand

ein·teeren *vt* tar

ein·teigen *vt* cover with (*or* wrap in) dough; (*Brew.*) dough in (malt)

einteilbar *a* divisible, classifiable *cf* EINTEILEN

Einteilchen° *n* single particle

ein·teilen *vt* **1** divide. **2** classify. **3** arrange, organize, graduate. **4** (**zu**) assign (to)

einteilig *a* one-piece, unipartite

Einteilung° *f* division; classification; arrangement, organization *cf* EINTEILEN

eintönig *a* monotonous. **Eintönigkeit** *f* monotony

Eintracht *f* harmony, unity. **einträchtig** *a* harmonious

Eintrag *m* (.. t͞e) **1** entry, item (in a list). **2: E. tun** (*dat*) be detrimental (to). **3** (*Tex.*) weft, woof. **ein·tragen* I** *vt* **1** carry in(to), bring in. **2** yield, earn. **3** enter, record, register (on a list, etc); (*Math.*) plot. **II: sich e.** enter, register, sign (one's name), enroll. **einträglich** *a* profitable. **Eintra-**

gung f (-en) entry, item (on a list, etc); registration

ein·tränken vt steep, dip, soak, impregnate

ein·träufeln vt instill, infuse, put (drops) in

ein·treffen* vi(s) arrive; happen, come true

ein·treiben* vt drive in, gather in, collect

ein·treten* vi(s) 1 (**in**) enter (in, into); join. 2 set in, start. 3 arise; occur, take place. 4 (**für**) stand up (for)

ein·trichtern vt pour in with a funnel

Eintritt° m entry, admission (incl price), admittance; onset; (Opt.) incidence cf EINTRETEN

Eintritts·druck m, **-spannung** f pressure on admission. **-stelle** f point of entry. **-temperatur** f temperature on admission

ein·trocknen vi(s) dry (up), shrink in drying

ein·tröpfeln, ein·tropfen vt drop in; instill

ein-und-einachsig a (Cryst.) orthorhombic

ein·verleiben vt incorporate (into)

Einvernehmen n understanding, agreement

einverstanden pred a in agreement, agreed. **Einverständnis** n agreement, understanding

Einw. abbv 1 (Einwaage: next entry). 2 (Einwirkung) action, influence

Einwaage° f 1 weighing (in). 2 weighed sample; weight of the sample

Einwäge·löffel m weighing-in spoon

ein·wägen* vt weigh (and put) in; level. **Einwägung°** f weighing in; leveling; amount weighed.

ein·walken vt 1 full closely; force (oil, etc) by fulling. 2 vi & sich e. shrink by fulling

Einwand m (. . ẅ-e) objection

ein·wandern vi(s) 1 diffuse in. 2 immigrate. **Einwanderung°** f 1 diffusion. 2 immigration

einwand·frei a flawless, unobjectionable

einwärts adv & sep pfx inward

ein·waschen* vt wash

einwässerig a (Dye.) one-bath

ein·wässern vt soak, steep, lay in water

ein·weben* 1 vt weave in, interweave. 2: sich e. weave a cocoon

Einweg·dose f (Food) no-return (or non-returnable) can. **-flasche** f no-return bottle. **-gleichrichter** m half-wave rectifier

einweibig a monogynous

Einweich·bottich m steeping vat, soaking tub

ein·weichen vt 1 soak, steep; (Flax) ret. 2 digest, macerate

Einweich·flüssigkeit f soaking liquor. **-mittel** n steeping agent

ein·weisen* vt 1 commit, assign (to a place). 2 install (in office). 3 brief, instruct. 4 guide, direct

einwellig a (Elec.) single-phase

ein·wenden* vt bring up . . . as an objection; (+ **daß**) raise the objection (that). **Einwendung** f (-en) objection

ein·werfen* vt throw in, interject; mail

einwertig a monovalent, univalent. **Einwertigkeit** f monovalence, univalence

ein·wickeln vt wrap (up), enclose. **Einwickel·papier** n wrapping paper. **Einwicklung** f wrapping

ein·wiegen* vt 1 pack by weight. 2 = EINWÄGEN

ein·willigen vi (in) consent, agree (to)

ein·wirken 1 vt work (or weave) in. 2 vi (**auf**) act, have an effect (on). **Einwirkung°** f action, effect

Einwirkungs·dauer f duration of the action. **-produkt** n resultant product

Einwohner m (-) inhabitant, resident

Einwurf° m 1 interjection, remark. 2 (mailing, coin) slot. 3 insertion, throwing in. 4 charging, feeding. 5 hopper. Cf EINWERFEN. **-trichter** m feed hopper. **-zucker** m remelt charge

ein·zahlen vt pay in, deposit. **Einzahlung°** f deposit; payment

ein·zehren vi (s) diminish by evaporation. **Einzehrung** f loss by evaporation

ein·zeichnen vt draw (or sketch) in; note (or mark) down

Einzel· single, individual: **-antrieb** m individual drive. **-beobachtung** f single observation. **-beschickung** f single charge, (single) batch. **-bestimmung** f single determination. **-blech** n single tinplate. **-darstellung** f single presentation; monograph. **-erscheinung** f single (or isolated) phenomenon. **-fall** m individual (or particular) case. **-gänger** m individualist. **-handel** m retail

Einzelheit f (-en) detail, particular (point)

Einzeller m (-) unicellular organism, protozoon. **einzellig** *a* unicellular, single-cell(ed)

Einzel·lupe *f* single-lens (*or* simple) magnifier. **-messung** *f* single measurement. **-molekül** *n* single molecule

einzeln I *a* 1 single, individual; solitary— **das e—e** the details; **e—es** a few details. 2 *pl oft* occasional, isolated, a few scattered. **II** *adv oft* one by one—**im e—en** (**a**) in detail, (**b**) retail

Einzel· single, individual: **-potential** *n* single (electrode) potential. **-schrift** *f* monograph. **e—stehend** *a* isolated, solitary. **E—teil°** *n* component (part), detail. **-teilchen** *n* single particle. **-verfahren** *n* batch process. **-verkauf** *m* retail (sale). **-vorgang** *m* single (separate, individual) process (*or* reaction). **-wert** *m* single (*or* individual) value. **-wesen** *n* individual (person, organism)

ein·ziehen* **I** *vt* 1 conscript. 2 insert. 3 pull in, retract. 4 draw in, suck in, absorb; inhale. 5 call in (notes, etc); withdraw; confiscate; 6 eliminate; pull down. 7 reduce (metal sheets). 8 gather, collect. 9 indent, set back. **II** *vi(s)* 10 move (come, march) in. 11 soak (seep, sink) in. **III:** **sich e.** shrink, contract

Einziehungs·mittel *n* absorbent

einzig *a* only, sole; unique: **der** (**das**) **e—e** the only one (only thing). **--artig** *a* unique

ein·zuckern *vt* sugar; preserve (with sugar)

Einzug° *m* 1 arrival, entry; moving in: **E. halten in** move in(to). 2 indentation. 3 gathering, collecting. *Cf* EINZIEHEN 8, 9, 10

Einzugs·gebiet *n* catchment area. **-luft** *f* intake (*or* aspirated) air

Einzwecks·maschine *f* one-purpose machine

Ei·pulver *n* dried (*or* powdered) egg. **ei–rund** *a* oval

Eis *n* 1 ice; ice cream. 2 (*in compds oft*) glacial, freezing: **-abkühlung** *f* cooling (*or* refigeration) with ice. **e—ähnlich** *a* ice-like, glacial. **E—alaun** *m* rock alum. **-ansatz** *m* deposit (*or* layer) of ice. **e—artig** *a* ice-like, icy, glacial. **E—belag, -beleg** *m* coating of ice. **-bereitung** *f* ice-making. **-beutel** *m* icebag. **-bildung** *f* formation of

ice. **-blume** *f* 1 frost pattern. 2 (*Bot.*) ice plant

Eisblumen·bildung *f* (*Varnish*) reticulation. **-farbe** *f* (*Paint*) crackle finish. **-glas** *n* frosted glass. **-lack** *m* crystal lacquer

Ei·schale *f* eggshell

Eiseln *n*, **Eiselung** *f* (*Varnish*) crystallization

Eisen *n* 1 (*specif*) iron, (*oft*) steel; iron (*or* steel) instrument (*or* utensil), iron, horseshoe. 2 (*in compds oft*) ferro-, ferruginous: **-abbrand** *m* iron waste. **-abfälle** *mpl* scrap iron. **-abscheidung** *f* separation of iron. **e—ähnlich** *a* iron-like, ferruginous. **E—alaun** *m* iron alum, ferric potassium sulfate. **-amiant** *m* fibrous silica. **-andradit** *m* (*Min.*) skiargite. **-ammonalaun** *m* ammonium ferric alum. **-antimonerz** *n*, **-antimonglanz** *m* (*Min.*) berthierite. **e—arm** *a* low-iron, poor in iron; (*Med.*) iron-deficient. **E—arsenik** *n* iron arsenide, (*specif, Min.*) loellingite, leucopyrite. **e—artig** *a* iron-like, ferruginous, chalybeate. **E—arznei** *f* ferruginous remedy, medicine with iron. **-asbest** *m* (*Metll.*) fibrous silica. **-aufnahme** *f* absorption of iron. **-bahn** *f* railroad. **-bedarf** *m* iron requirements. **-beize** *f* (*Dye.*) iron mordant, copperas black. **-beizung** *f* iron pickling (*or* mordanting). **-beschwerung** *f* weighting with iron. **-beton** *m* reinforced concrete. **-bitterkalk** *m* ferruginous dolomite, ferromagnesian limestone

Eisenblau *n* vivianite. **--druck** *m* blueprint(ing). **-erde** *f* earthy vivianite. **-papier** *n* blueprint paper. **e—sauer** *a* ferrocyanide of. **E—säure** *f* ferrocyanic acid. **-spat** *m* vivianite

Eisen· iron, steel, ferrous, ferric: **-blech** *n* sheet (of) iron, iron plate. **-blende** *f* pitchblende. **-block** *m* ingot, bloom (*or* block) of iron. **-blumen** *fpl* iron flowers, ferric chloride. **-blüte** *f* flos ferri, coralloid aragonite. **-bor** *n* iron boride. **-braun** *n* iron (*or* mineral) brown. **-bromid** *n* iron (*or* ferric) bromide. **-bromür** *n* ferrous bromide. **-bromürbromid** *n* ferrosoferric bromide. **-brühe** *f* iron mordant (*or* liquor). **-cement** *m/n* ferro-concrete. **e—chamois** *a* (*no endgs*) iron-buff. **E—chinawein** *m* iron and quinine wine. **-chlorid** *n* iron (*or* ferric) chloride. **-chlorür** *n* ferrous chloride. **-chlorürchlorid** *n* ferrosoferric chloride. **-chlorwasserstoff** *m* ferrichloric acid. **-chrom** *n* ferrochrome; (*Min.*) chromic iron, chromite. **-chromschwarz** *n* iron-chrome black. **-chrysolith** *m* hyalosiderite; fayalite. **-cyanfarbe** *f* iron cyanogen pigment. **-cyanid** *n* iron (*or* ferric) cyanide. **-cyankalium** *n* potassium ferrocyanide. **-cyanür** *n* ferrous cyanide. **-cyanürcyanol** *n* Prussian blue. **-drahtnetz** *n* iron gauze. **e—empfindlich** *a* sensitive to iron. **E—erde** *f* ferruginous (*or* iron) earth; (*Ceram.*) hard stoneware—**blaue E.** earthy vivianite. **-erz** *n* iron ore. **-fänger** *m* iron trap. **e—farbig** *a* iron-colored. **E—feile** *f*, **-feilicht** *m*, **-feilspäne** *mpl* iron filings. **-feilstaub** *m* fine iron filings. **-feinschlacke** *f* iron refinery slag. **-firnis** *m* varnish for iron. **-flasche** *f* iron cylinder (for gases). **-fleck** *m* iron spot (*or* stain). **e—fleckig** *a* iron-stained. **E—flüssigkeit** *f* iron liquid (*or* liquor). **e—frei** *a* iron-free, nonferrous, **E—frischerei** *f* iron refining (*or* refinery); puddling. **-frischflammofen** *m* puddling furnace. **-frischschlacke** *f* finery cinders. **e—führend** *a* iron-bearing, ferriferous. **E—gallustinte** *f* iron gallate ink. **-gang** *m* iron lode, iron-ore vein. **-gans** *f* iron pig. **-gerbung** *f* iron tannage. **-gestein** *n* ironstone. **-gestell** *n* iron stand (*or* frame). **-gewinnung** *f* iron production. **-gießerei** *f* iron foundry; iron founding. **-gilbe** *f* yellow ocher. **-gitter** *n* iron grating. **-glanz** *m* iron glance, specular iron (form of hematite). **-glas** *n* (*Min.*) fayalite. **-glimmer** *m* micaceous iron ore (form of hematite). **-glimmerschiefer** *m* itabirite. **-granat** *m* iron garnet, almandite. **e—grau** *a* iron-gray. **E—graupen** *fpl* granular bog iron ore. **-grund** *m* iron liquor (*or* mordant). **-guß** *m* iron casting; cast iron. **e—haltig** *a* containing iron, ferriferous; ferruginous, chalybeate. **E—hammerschlag** *m* iron (hammer) scale. **e—hart** *a* hard as iron, iron-hard. **E—hart** *n* 1 ferriferous gold sand. 2 (*Bot.*) vervain. **-holz** *n* ironwood. **-hut** *m* 1 (*Bot.*)

aconite, monkshood. 2 (*Min.*) gossan. **-hütte** *f* iron works, forge **Eisenhütten· **(ferro)metallurgical: **-kunde** *f* iron metallurgy. **-leute** *pl of* **-mann** *m* (iron) metallurgist. **e—männisch** *a* ferrometallurgical. **E—werk** *n* iron (*or* steel) mill. **-wesen** *n* iron metallurgy **Eisen·** iron, steel; ferric, ferrous: **-hydroxid** *n* ferric hydroxide. **-hydroxydul** *n* ferrous hydroxide. **-jodid** *n* iron (*or* ferric) iodide. **-jodür** *n* ferrous iodide. **-jodürjodid** *n* ferrosoferric iodide. **-kalium** *n* potassium ferrate. **-kaliumalaun** *m* iron potassium alum, ferric potassium sulfate. **-kalk** *m* (*Min.*) siderite. **-karbid**, **-karburet** *n* iron carbide. **-kern** *m* iron core. **-kies** *m* iron pyrites, pyrite— **hexagonaler E.** pyrrhotite; **rhombischer E.** marcasite. **-kiesel** *m* ferruginous flint (*or* chert). **e—kiesführend** *a* pyritic. **E—kitt** *m* iron-rust cement. **-kobalterz** *n*, **-kobaltkies** *m* cobaltite (*or* smaltite) rich in iron. **-kohlenoxid** *n* iron carbonyl. **-kohlenstoff** *m* iron carbide. **-koks** *m* ferrocoke. **-korn** *n* (*Metll.*) ferrite grain. **-kraut** *n* vervain, verbena. **-lack** *m* iron varnish (*or* lake). **-lebererz** *n* hepatic iron ore. **-leder** *n* iron-tanned leather. **-lila** *n* alizarin lake. **e—magnetisch** *a* ferromagnetic. **E—mangan** *n* ferromanganese. **-manganerz** *n* manganiferous iron ore, ferriferous manganese ore. **-mann** *m* (*Min.*) scaly red hematite. **-mennige** *f* red ocher. **-mittel** *n* (*Med.*) ferruginous remedy, iron tonic. **-mohr** *m* æthiops martialis (powdered black Fe_3O_4); earthy magnetite. **-molybdän** *n* ferromolybdenum. **-mulm** *m* earthy iron ore. **-natronsalz** *n* iron sodium salt. **-nickel** *n* ferronickel. **-nikkelkies** *m* petlandite. **-niere** *f* eaglestone. **-ocker** *m* (iron) ocher **Eisenoxid°** *n* iron (*specif* ferric) oxide; (*with adj* **-sauer**, e.g.): **salpetersaures E.** ferric nitrate. **-gelb** *n* iron yellow. **-hydrat** *n* ferric hydroxide. **-oxydul** *n* ferrosoferric oxide. **e—reich** *a* rich in iron oxide. **E—rot** *n* red iron oxide. **-salz** *n* ferric salt. **-sulfat** *n* ferric sulfate **Eisenoxydul°** *n* ferrous oxide; (*with adj*

-sauer, e.g.): **salpetersaures E.** ferrous nitrate. **-hydrat** n ferrous hydroxide. **-oxid** n ferrosoferric oxide, magnetic iron oxide (Fe_3O_4). **-salz** n ferrous salt. **-sulfat** n ferrous sulfate. **-verbindung** f ferrous compound

Eisen· iron, steel; ferric, ferrous: **-pastille** f (Pharm.) reduced iron lozenge. **-pecherz** n (Min.) limonite, pitticite; triplite. **-phosphor** m iron phosphide. **-platte** f cast iron plate. **-portlandzement** m/n Portland blast-furnace cement. **-probe** f iron test; iron sample. **-quarz** n ferriferous quartz. **-quelle** f 1 source of iron. 2 chalybeate spring. **-rahm** m porous hematite (or limonite). **e—reich** a iron-rich, rich in iron. **E—reihe** f iron series. **-resin(it)** n (Min.) humboldtine. **-rhodanid** n ferrous thiocyanate, (Min.) rhodanite. **-rhodanür** n ferrous thiocyanate. **-rogenstein** m oolitic iron ore. **-rostwasser** n iron liquor (or mordant). **-rot** n colcothar, red iron oxide. **-safran** m saffron (or crocus) of Mars. **-salmiak** m (Pharm.) ammoniated iron, iron and ammonium chloride. **-sand** m ferruginous sand. **-sau** f pig of iron. **e—sauer** a ferrate of. **E—säuerling** m chalybeate water, water with iron salts. **-säure** f ferric acid. **-schaum** m 1 (Metll.) kish. 2 (Min.) porous hematite. **-schlacke** f iron slag (or dross). **-schmelze** f iron smelting (smeltery, founding) **-schmelzhütte** f iron foundry. **-schörl** m bog iron ore. **-schrot(t)** m scrap iron. **e—schüssig** a ferriferous, ferruginous. **E—schutz** m rustproofing

Eisenschutz·anstrich m anti-rust paint. **-farbe** f protective paint for iron. **-mittel** n anti-rust (or rustproofing) agent

Eisen· iron, steel, ferric, ferrous: **schwamm** m iron sponge, spongy iron. **-schwarz** n 1 black iron oxide. 2 lampblack; graphite, plumbago. 3 (Dye.) iron (or copperas) black; precipitated antimony. **-schwärze** f 1 graphite, ground black lead. 2 earthy magnetite. 3 (Dye.) iron black, iron liquor. 4 (Lthr.) currier's ink. **-selenür** n ferrous selenide. **-silberglanz** m sternbergite. **-silizium** n iron silicide. **-sinter** m iron dross (or

scale); pitticite. **-späne** mpl iron turnings (or borings). **-spat** m siderite, spathic iron ore. **-spiegel** m specular iron, hematite. **-staub** m iron dust. **-stein** m ironstone, iron ore **-spatiger E.** spathic iron ore, siderite. **-steinmark** n lithomarge (kaolinite, halloysite) containing iron, tetraolite. **-suchgerät** n iron detector. **-sulfat** n iron (specif ferrous) sulfate. **-sulfid** n iron (specif ferric) sulfide. **-sulfür** n ferrous sulfide. **-sumpferz** n bog iron ore. **-titan** n ferrotitanium; (Min.) ilmenite. **-ton** m clay ironstone; iron clay. **-tongranat** m almandite. **-trümmerlagerstätte** f clastic iron-ore deposit. **-vanadin** n ferrovanadium. **-violettholz** n granadilla wood, red ebony. **-vitriol** n vitriol, botryogen. **-wasser** n chalybeate water. **-weg** m magnetic path (or circuit). **-wein** m (Pharm.) iron wine. **-weinstein** m tartrated iron, iron and potassium tartrate. **-werk** n ironworks, iron mill. **-wolfram** m ferrotungsten. **-zeit** f iron age. **-zement** m/n reinforced concrete. **-zinkblende** f marmatite. **-zinkspat** m ferruginous calamine. **-zinnerz** n ferriferous cassiterite. **-zucker** m (Pharm.) saccharated ferric oxide. **-zuckersyrup** m (Pharm.) syrup of ferric oxide. **-zyan** n ferrocyanide; (in compds) ferrocyanic cf EISEN·CYAN . . .

eisern a (made of) iron, steel; cast-iron; ironclad; ferrous—(Min.) **e—er Hut** gossan

Eis·essig m glacial acetic acid. **-farbe** f (Dye.) ice color. **e—gekühlt** a ice-cooled. **E—glas** n frosted glass

eisig a icy. **Eisigkeit** f iciness

Eis·kasten m icebox. **-kraut** n ice plant. **·krem** f ice cream

eis·kühlen vt cool with ice, ice-cool. **Eis·kühlung** f ice cooling. **-meer** n: **nördliches E.** Arctic Ocean; **südliches E.** Antarctic Ocean. **-phosphorsäure** f glacial phosphoric acid. **-punkt** m freezing point (of water). **-regen** m sleet; hail. **-schale** f ice bowl. **-schrank** m icebox; refrigerator. **-spat** m glassy feldspar, sanidine, rhyacolite. **-stein** m cryolite. **-warndienst** m ice(berg) warning service. **-würfel** m ice cube

eitel *a* (*with endgs usu* **eitl** . .) 1 idle, vain. 2 pure, sheer, nothing but
Eiter *m* pus. **e—·ähnlich, -artig** *a* purulent. **E—bakterien** *npl* pyogenic bacteria. **e—befördernd** suppurative. **-bildend** *a* pyogenic. **E—bildung** *f* pus formation, suppuration, pyogenesis. **e—erzeugend** *a* pyogenic. **E—gang** *m* fistula
eiterig *a* purulent, festering. **eitern** *vi* suppurate, fester
Eiterungs·erreger *m* suppurative agent
Eiter·vergiftung *f* pyemia. **e—ziehend** *a* pyogenetic
eitle *a* vain, etc = EITEL
eitrig *a* purulent = EITERIG
Eiweiß I n egg white, albumen; albumin, protein. **II** (*in compds*) albumin(ous); protein: **-abbau** *m* proteolysis. **-abbauprodukt** *n* protein decomposition product. **e—ännlich** *a* albuminoid. **-arm** *a* albumin-poor, low-protein; (*Rubber*) deproteinized. **E—art** *f* variety of albumin (*or* protein). **e—artig** *a* albuminous, albuminoid. **E—bedarf** *m* albumin (*or* protein) requirement(s). **-drüse** *f* albuminous gland. **-faser** *f* protein fiber. **e—förmig** *a* albuminous. **E—gerüst** *n* protein framework (*or* structure). **e—haltig** *a* containing albumen (albumin, protein), albuminous. **E—harnen** *n* albuminuria. **-körper** *m* albuminous substance, protein, egg albumin. **-leim** *m* albumin glue; gluten protein. **-nahrung** *f* albuminous food. **-papier** *n* albuminized paper. **e—reich** *a* albumin-rich, protein-rich, high-protein. **-spaltend** *a* proteolytic. **E—spaltung** *f* protein cleavage, proteolysis) cleavage of albumin. **-sparmittel** *n* albumin (*or* protein) sparer. **-stoff** *m* albumen, albumin, protein, albumin substance. **-verbindung** *f* albuminous compound, protein; albuminite. **e—zersetzend** *a* proteolytic, protein-decomposing
Ei·zelle *f* egg cell, ovum, oocyte
EK, E.K. *abbv* 1 (elektromotorische Kraft) EMF. 2 (Entkeimung) sterilization
Ekel *m* disgust, loathing; nausea. **e—erregend, ekelhaft** *a* disgusting, nauseating. **ekeln** 1 *vt/i* (*accus, dat*) disgust, repel. **2: sich e.** (**vor**) be disgusted (by).

eklatant *a* sensational, striking
eklig *a* disgusting
ekliptisch *a* eclipsed (conformation)
Ekrü·seide *f* ecru (silk)
E-kupfer *n* = ELEKTROKUPFER
Ekzem *n* (-e) eczema
E.L. *abbv* (Einheitslack) unit lacquer
Elaidin·säure *f* elaidic acid
Elain·säure *f* oleic acid
Elaio-, Eläo- el(a)eo-, elaio-: **Eläolith** *m* eleolite, (*Geol.*) nepheline
Elaste *pl* elastomers
Elastik *n* (-s) elastic material. **elastisch** *a* elastic. **Elastizität** *f* elasticity
Elastizitäts·grenze *f* elastic limit. **-modul** *n* modulus of elasticity. **-verstärker** *m* (*Rubber*) elasticator. **-zahl** *f* modulus of elasticity
Elatin·säure *f* elatinic acid. **Elat·säure** *f* elatic acid
Elch *m* (-e) elk
Elefanten·laus *f* cashew nut. **-nuß** *f* ivory nut
elekt., Elekt. *abbv* (elektrisch, Elektrizität) elec.
Elekta·wolle *f* first-class wool
elektiv *a* selective, elective. **E—·nährboden** *m* (*Bact.*) selective medium
Elektrik *f* (science of) electricity. **Elektriker** *m* (-) electrician. **elektrisch** *a* electric(al)
elektrisierbar *a* electrifiable. **elektrisieren** *vt* electrify, charge. **Elektrisier·maschine** *f* electrostatic generator. **Elektrisierung** *f* (-en) electrification, charging
Elektrizitäts· electric(ity): **-druck** *m* voltage, potential. **-ladung** *f* electric charge. **-menge** *f* quantity of electricity. **-messer** *m* electrometer, electric meter. **-messung** *f* electrometry. **-werk** *n*, **-zentrale** *f* powerhouse, power generating station
Elektro· electro-, electric(al): **-chemie** *f* electrochemistry
Elektroden·abstand *m* distance between electrodes. **-kohle** *f* electrode carbon. **e—los** *a* electrodeless, without electrodes. **E—spalt** *m* spark gap
Elektro· electro-, electric(al): **-eisen** *n* electrical iron. **-flußstahl** *m* electric steel. **-herd** *m* electric stove (*or* range). **-herd-**

ofen *m* electric hearth furnace. **-hoch-ofen** *m* electric shaft furnace. **-inge-nieur** *m* electrical engineer. **-kochplatte** *f* electric hotplate. **-korund** *m* electro-corundum. **-kupfer** *n* electrical copper. **Elektrolyse** *f* (-n) electrolysis. **--bad** *n* electrolytic bath

Elektrolysen-endlauge *f* spent electrolyte. **-schlamm** *m* electrolytic slime

elektrolysieren *vt* electrolyze

Elektrolyt *m* (-en, -en) electrolyte. **--eisen** *n* electrolytic iron

Elektromotor° *m* electric motor. **elektro-motorisch** *a* electromotive *cf* EMK

Elektronen- electron(ic): **-abgabe** *f* electronic emission. **-anordnung** *f* electron configuration. **-aufnahme** *f* electron acceptance. **-bahn** *f* electronic orbit (*or* path). **-bremsung** *f* retardation of electrons. **-defektleitung** *f* electron-defect conduction. **-donator** *m* electron donor. **-drall** *m* electron spin. **-duett** *n* electron pair. **-einfang** *m* capture of electrons. **e—erzeugend** *a* electron-generating, electrogenic. **E—haftstelle** *f* electron trap. **-hülle** *f* electronic sheath (*or* envelope); electron shell. **-lehre** *f* electronics. **-optik** *f* electron optics. **-ort** *m* electron position (*or* locus). **-rechner** *m* electronic calculator. **-röhre** *f* electron (*or* vacuum) tube. **-rumpf** *m* electronic core. **-schleuder, -spritze** *f* electron gun, betatron. **-sprung** *m* electron jump. **-stauung** *f* electron accumulation. **-steuerung** *f* electronic control. **-stoß** *m* electronic collision (*or* impact). **-strahlen** *mpl* electron rays (radiation, beams). **-übergang** *m* electron transition (*or* migration). **-über-mikroskop** *n* electron microscope, super-microscope. **e—überschußleitend** *a* electron-excess semiconducting. **E—un-terhülle** *f* electron subshell. **e—unter-leitend** *a* electron-defect semiconducting. **E—verfielfacher** *m* electron multiplier. **-wucht** *f* electronic collision force

Elektronik *f* electronics. **Elektroniker** *m* (-) electronic technician (*or* expert). **elek-tronisch** *a* electronic

Elektro- electro, electric(al): **-ofen** *m* electric furnace. **e—optisch** *a* electro-optic(al).

Elektrophor *m* (-en) electrophorus. **Elek-tro-phorese** *f* electrophoresis. **elek-troplattieren** *vt* electroplate

Elektro-porzellan *n* electrical porcelain. **-positivität** *f* electropositive state (*or* quality). **-roheisen** *n* (*Iron*) electric pig. **-stahl** *m* electric(al) steel. **-technik** *f*, **e—technisch** *a* electrical engineering. **E—werk** *n* (electric) powerhouse

Element *n* (-e) 1 (*usu*) element. 2 (*Elec.*) cell

elementar *a* elementary. **E—-ladung** *f* elementary (*or* unit) charge. **-stoff** *m* elementary substance, element. **-quantum** *n* elemental quantum

Element-art *f* kind of element; type of cell

Elementar-teilchen *n* elementary (*or* subatomic) particle. **-zelle** *f* unit cell

Elementen-glas *n* (glass) battery jar. **-messung** *f* stoichiometry. **-verwand-lung** *f* transformation of elements, transmutation

Element- cell, battery: **-gefäß** *n*, **-glas** *n* battery jar. **-kohle** *f* cell carbon. **-schlamm** *m* battery mud

Elemi-harz *n* (gum) elemi. **-öl** *n* elemi oil. **-säure** *f* elemic acid

Elemol-säure *f* elemolic acid. **Elemon-säure** *f* elemonic acid

Elen *m/n* elk

elend *a* pitiful, miserable. **Elend** *n* misery **Elen-fett** *n* elk fat

elf *num* eleven. **Elf** *f* (-en) (number) eleven **Elfenbein°** *n* ivory. **e—ähnlich, -artig** *a* ivory-like, eburnean. **-farbig** *a* ivory colored. **E—karton** *m* (*Paper*) ivory board. **-substanz** *f* dentine. **-surrogat** *n* ivory substitute, artificial ivory. **e—weiß** *a* ivory-white

elfte *a* eleventh. **Elftel** *n* (-) (*Frac.*) eleventh

eliminieren *vt* eliminate

Elixier *n* (-e) 1 elixir. 2 sweetened alcoholic drink

Ellag(en)-(gerb)säure *f* ellagic acid

Ell-bogen *m* elbow

Elle *f* (-n) 1 (*Meas.*) yard(stick). 2 (*Anat.*) ulna

Ellen-bogen *m* elbow. **-waren** *fpl* dry goods

Eller *f* (-n) (*Bot.*) alder

Ellipsen-bahn *f* elliptic orbit

ellipsoid-förmig *a* ellipsoidal

elliptisch *a* elliptic(al)

Elmi- *pfx, short for* ELEKTRONMIKROSKOP-

Elms·feuer n St. Elmo's fire
Eloxal·verfahren n anodizing process (for aluminum). **eloxieren** vt anodize (metal); eloxate
Els·beere f chokecherry (tree). **Else** f (-n) 1 (*Bot.*) alder; chokecherry. 2 (*Fish*) shad.
Elsen·holz n alder (wood)
Elsholtzia·säure f elsholtzic acid
Elster f (-n) magpie
elterlich a parental. **Eltern** pl parents
Eluat n (-e) eluate, extract. **eluieren** vt eluate, extract, wash out
elutrieren vt elutriate, wash
Eluvial·boden m eluvial soil
Email n enamel. **e—·artig** a enamel-like. **E—belag** m enamel coating. **-draht** m enameled wire. **-fluß** m enamel flux. **-geschirr** n enamelware. **-glas** n enamel (*or* fusible) glass. **-lack** m enamel (varnish)
Emaille f enamel. **-lackfarbe** f enamel (paint). **emaillieren** vt enamel. **Email-lier·ofen** m enameling oven. **Email-lierung** f (-en) enameling.
Email·ofenlack m stoving enamel. **-ware(n)** f (pl) enamelware
Ematel·verfahren n (*Aluminum*) anodic oxidation process
Embelia·säure f embelic acid
Embolie f (-n) embolism
Embryonal·zustand m embryonic stage
Emetäthylin n emetethyline
emetisch a emetic
Emissions·fähigkeit f emissivity. **-lehre** f emission theory. **-verhältnis** n emissivity. **-vermögen** n emissive power
emittieren vt emit
EMK abbv (elektromotorische Kraft) EMF
emollieren vt soften
E-modul n = ELASTIZITÄTSMODUL
empfahl recommended: *past of* EMPFEHLEN
empfand felt: *past of* EMPFINDEN
Empfang m (. . änge) receipt; reception: **in E. nehmen** receive, accept. **empfangen*** 1 vt get, receive. 2 vt/i (*Physiol.*) conceive. **Empfänger** m (-) receiver (*incl Electron.*); recipient. **empfänglich** a (**für**) 1 receptive (to). 2 susceptible (to). **Empfänglichkeit** f receptivity; susceptibility **Empfängnis** f (-se) (*Physiol.*) conception. **-verhütung** f contraception
Empfangs· receiving: **-schein** m receipt

empfehlen (empfahl, empfohlen; empfiehlt) I vt 1 recommend. 2 entrust. II: **sich e.** 3 take leave; (*imps*): **es empfiehlt sich zu** . . . it is advisable to . . . **emp-fehlens·wert** a worthy of recommendation; advisable. **Empfehlung** f (-en) recommendation, reference
empfiehlt recommends: *pres of* EMPFEHLEN
empfindbar a sensible, perceptible. **Emp-findbarkeit** f sensibility, perceptibility
empfinden* 1 vt/i feel. 2 vt experience, have a sensation of; (+ **als**) feel (to be). **Empfinden** n feeling, sense. **emp-findlich** a 1 sensitive; (*esp Fabrics*) susceptible; (*oft in compds*) e.g.: **lichtemp-findlich** sensitive to light: **e. machen** sensitize. 2 delicate. 3 irritable. 4 sensible, perceptible. 5 considerable, appreciable; severe. **empfindlichen** vt sensitize. **Empfindlichkeit** f (-en) sensitivity, susceptibility, etc cf EMPFINDLICH. **Emp-findlich·machung**, **Empfindlichung** f sensitization. **Empfindung** f (-en) feeling, sensation. **Empfindungs·zelle** f sensory cell
empfing received: *past of* EMPFANGEN
empfohlen recommended: *pp of* EMP-FEHLEN
empfunden felt: *pp of* EMPFINDEN
empirisch a empirical
empor adv & sep pfx up(ward)
empören I vt 1 infuriate. II: **sich e.** 2 (**über**) become indignant (about). 3 (**gegen**) rebel (against)
empor·heben* vt lift up, raise. **e—·kommen** vi (s) come up, rise, advance. **·ragen** vi (h,s) (**über**) tower, loom, rise (above). **·schnellen** vi (s) rise (up) suddenly, skyrocket. **·steigen** vi (s) rise (up), advance; (+ *accus*) go up, climb. **·streben** vi (h,s) strive (upward); aspire, aim high
emprotid a proton-accepting. **Emprotid** n (-e) proton acceptor
Emscher·brunnen m (*Sewage*) Imhoff tank
emsig a diligent, industrious, hardworking
Emulgator m (-en) emulsifier. **emulgier-bar** a emulsifiable. **emulgieren** vt/i emulsify
Emulgier·fähigkeit f emulsifying capacity. **-mittel** n emulsifying agent
Emulgierung f (-en) emulsification.

Emulgierungs·mittel n emulsifying agent

emulsions·bildend a emulsion-forming.

emulsionieren vt/i emulsify, emulsionize

Emulsions·bildner m emulsifier. **-bildung** f formation of an emulsion. **e—fähig** a emulsifiable. **E—farbe** f distemper color, emulsion paint. **-kolloid** n emulsion sol, emulsion. **-polymerisat** n emulsion polymer

emulzieren vt emulsify, emulsionize

Enantiotropie f enantiotropy, enantiotropism

End· end, final, terminal: **-anzeiger** m indicator. **-ausbeute** f final yield. **-ausschalter** m limit switch. **-bahn** f final orbit (or path). **-darm** m (Anat.) rectum. **-destillat** n final distillate. **-dicke** f finished thickness. **-druck** m final pressure

Ende n (-n) 1 end, close, termination: **am E.** (a) over, finished, through, (b) = VIELLEICHT maybe, possibly; **letzten E—s** in the end, in the final analysis; **zu E.** over, finished; **zu E. kommen** (or **gehen**) end, come to an end; **zu E. bringen** (or **führen**) end, finish, bring to an end; **ein E. machen** (or **bereiten**) (+ dat) put an end (to); **ein E. machen** (a) finish (up), (b) (+ **mit**) end, finish. 2 butt (end). 3 (small) piece, distance

End·ecke f **-eck** n terminal angle, summit

endemisch a endemic

enden vi end, terminate

End· final, terminal, end: **-ergebnis** n final result (or product). **-ertrag** m final yield. **-erzeugnis** n final product. **-fläche** f terminal face. **-gas** n tail gas. **-geschwindigkeit** f terminal velocity. **-glied** n terminal (or end) member; final link. **-gruppe** f end group. **-gruppenbestimmung** f end group assay. **e—gültig** a (absolutely) final, finally valid; adv oft once and for all, definitely

endigen vi end, finish, terminate

Endivie f (-n) endive

End· end, final, terminal: **-kohlenstoff** m terminal carbon (atom). **-lage** f end (or final) position. **-lauge** f (residual, spent) liquor (or solution). **-laugenkalk** m (Agric.) product made by lime treatment of

KCl mother liquor

endlich 1 a final, finite, definite. 2 adv finally, at last; after all. **Endlichkeit** f limitation, finiteness. **endlos** a endless, interminable; infinite

End· final: **-lösung** f final solution. **-melasse** f blackstrap molasses. **-nutzung** f final yield

Endosmose f endosmosis. **endosmotisch** a endosmotic

Endothel n (-ien) endothelium

endotherm(isch) a endothermic

End· end, final, terminal: **-produkt** n end (or final) product. **-punkt** m end (point), final point. **-schalter** m limit switch. **-schwefelung** f (Sugar) final sulfitation. **-stadium** n final stage. **e—ständig** a terminal, end; (Bot.) apical. **E—stärke** f ultimate strength (or thickness). **-stellung** f end (or final) position. **-stück** n endpiece, terminal piece. **-summe** f grand total, bottom line. **-termin** m closing (or terminal) date, deadline

Endung f (-en) ending, termination

End·urteil n final decision (judgment, verdict). **-vergären** n complete fermentation. **-wert** m final value. **-ziel** n ultimate aim, final goal. **-ziffer** f final digit. **-zustand** m final state. **-zweck** m final aim, ultimate purpose

Energetik f energetics, science of energy. **energetisch** a energy(-related) cf ENERGISCH

Energie·abgabe f release of energy. **-bedarf** m energy demand (or requirement). **-berg** m energy barrier. **-betrag** m amount of energy. **-freigabe** f release of energy. **e—liefernd** a energy-producing. **E—prinzip** n law of conservation of energy. **-quantelung** f energy quantization. **e—reich** a energy-rich. **E—satz** m law of conservation of energy. **-schwelle**, **-sperre** f energy barrier. **-stufe** f energy level (or stage). **-versorgung** f energy supply. **-wirtschaft** f power industry. **-zufuhr** f energy supply. **-zustand** m energy state

energisch a energetic, vigorous cf ENERGETISCH

eng a 1 narrow, limited, restricted. 2 close; confining. 3 tight(-fitting)

Engagement *n* (-s) involvement. **engagieren** *vt* hire, engage. **2: sich e. (in)** engage, get involved (in). **engagiert** *p.a* (*oft*), committed, dedicated; (*as adv*) with dedication

eng·benachbart *a* closely adjacent

Enge *f* (-n) **1** narrowness, closeness, tightness. **2** confinement, restriction. **3** narrow passage. **4** predicament, bind

Engel *m* (-) angel. **--rot** *n* colcothar. **-süß** *n* (*Pharm.*) polypody root. **-wurz(el)** *f* (*Pharm.*) angelica

Enghals·flasche *f* narrow-necked bottle. **enghalsig** *a* narrow-necked

Engländer *m* (-) **1** Englishman. **2** monkey wrench

Engler·grad *m* Engler degree (*or* number)

englisch *a* English: **e—** colcothar; **e—e Schwefelsäure, e—es Vitriolöl** English sulfuric acid (*or* oil of vitriol) (ordinary lead-chamber acid); **e—e Erde** rotten stone; **e—es Gewürz, e—es Piment** pimento, allspice; **e—es Kollodium** flexible collodion; **e—e Küpe** (*Dye.*) bisulfite zinc-lime vat; **e—es Leder** *n* moleskin; **e—es Pflaster** *n* court plaster; **e—es Pulver** *n* powder of Algaroth (SbOCl); **e—es Salz** Epsom salt

Englisch·blau *n* royal blue. **-gelb** *n* patent yellow, lead oxychloride. **-leder** *n* moleskin; **-pflaster** *n* court plaster. **-rot** *n* colcothar. **-salz** *n* Epsom salt

eng·lochig *a* fine-pored, with small holes. **-maschig** *a* close-meshed, fine-meshed

Engobe *f* (-n) (*Ceram.*) engobe, slip. **engobieren** *vt* slip, coat with engobe

Eng·paß *m* (narrow) pass; bottleneck. **e— porig** *a* fine-pored, finely porous

en gros *adv* wholesale. **Engros·preis** *m* wholesale price

eng·wandig *a* narrow-bore

enkaustisch *a* encaustic

Enkel *m* (-) **I** grandson, grandchild. **II** ankle. **Enkelin** *f* (-nen) granddaughter

Enlevage *f* (-n) (*Dye.*) discharge. **--druck** *m* discharge printing

Enolisierung *f* (-en) enolization

enorm *m* enormous, tremendous

ent· *insep pfx: implies* **1** (*deprivation, removal, undoing*) de-; dis-, un- *cf* ENTAK-TIVIEREN, ENTDECKEN. **2** (*origin, incipient action*) *cf* ENTFLAMMEN, ENTSTEHEN. **3** *esp vi* (s) (*escape*) *cf* ENTGEHEN

entaktivieren *vt* deactivate, render inactive

entalkoholisieren *vt* dealcoholize

entalkylieren *vt* dealkylate

entamidieren *vt* deprive of amidogen, deamidize

entarretieren *vt* release, unlock

entarsenisieren *vt* dearsenicate

entarten 1 *vt* render degenerate, debase. **2** *vi* (s) degenerate. **entartet** *p.a* degenerate; abnormal. **Entartung** *f* (-en) degeneration. **Entartungs·reaktion** *f* degeneration reaction

Entaschung *f* ash removal

entasphaltisieren *vt* deasphalt(ize)

entband released, etc: *past of* ENTBINDEN

entbasen *vt* (*Soil.*) deprive of bases

entbasten *vt* degum (silk), decorticate. **Entbastungs·mittel** *n* degumming (*or* decorticating) agent

entbehren 1 *vt/i* do without. **2** *vt* dispense with. **3** *vi* (*imps*): **es entbehrt** (+ *genit*) there is a lack (of)

ent·binden* 1 *vt* disengage, liberate; evolve (gases); release. **2** *vi, vt* (*passive*) give birth (to). **Entbindung** *f* (-en) disengagement, liberation, evolution; release; delivery, (giving) birth, parturition

Entbindungs·anstalt *f* maternity hospital. **-flasche** *f* generating flask (*or* bottle). **-rohr** *n* delivery tube

entbittern *vt* remove the bitterness from

entblättern *vt* defoliate. **Entblätterungs· mittel** *n* defoliant

entbleien *vt* remove lead from

entblößen *vt* bare, uncover; (+ *genit*) deprive, strip (of)

entbrannt 1 *p.a* inflamed. **2** erupted: *pp of* ENTBRENNEN. **entbrannte** erupted: *past of* ENTBRENNEN

entbrausen *vi* (s) escape with effervescence

entbrennen* *vi* (s) break out; erupt *cf* ENT-BRANNT

entbromen *vt* debrominate

entbrühen *vt* (*Ores*) recover (the medium from the cleaned products)

entbunden released, etc: *p.p of* ENTBINDEN

entbutanisieren *vt* debutanize

entchloren *vt* dechlorinate
entdampfen 1 *vt* free from vapor. **2** *vi* (s) escape as vapor
entdecken *vt* **1** discover. **2** (*d.i.o*) reveal (to). **Entdecker** *m* (-) discoverer; detector. **Entdeckung** *f* (-en) discovery, revelation
entdenaturieren *vt* free from denaturants
entdolomitisieren *vt* dedolomitize
entdröhnen *vt* soundproof; absorb (sound). **Entdröhnungs‑mittel** *n* sound-absorbing medium
entdunsten 1 *vt* devaporize. **2** *vi* (s) evaporate
Ente *f* (-n) **1** (*Zool.*) duck. **2** duck-shaped vessel. **3** hoax, false report
entehren *vt* dishonor
enteisen *vt* defrost, de-ice
enteisenen *vt* deprive of iron, deferrize. **Enteisenung** *f* removal of iron, deferrization
Enteiser *m* (-) de-icer. **Enteisung** *f* de-icing, anti-icing
enteiweißen *vt* deproteinize
entemulgieren, entemulsionieren *vt* de-emulsify
Enten‑fett *n* duck fat. **-fuß** *m* (*Bot.*) duck's-foot, podophyllum
entfachen *vt* kindle; provoke, unleash
Entfall° *m* waste, scrap; loss; cancellation. **entfallen*** *vi* (s) **1** be dropped, be cancelled; be lost as scrap (*or* waste). **2** escape one's memory **3** (*dat*) slip out of one's hands. **4** (**auf**) come (to), fall (to) (one's share), be allotted (to)
entfalten I *vt* **1** unfold, unroll, spread out. **2** develop, cultivate, elaborate. **3** let loose, unleash (activity, energy); display, exhibit (talent, ability). **II: sich e. 4** unfold, open up. **5** develop, blossom out. **Entfaltung** *f* unfolding; spreading; development; cultivation; display
entfärbbar *a* decolorizable, bleachable. **entfärben 1** *vt* decolor(ize), bleach. **2: sich e.** fade, pale, discolor. **Entfärbung** *f* decolor(iz)ation, bleaching; fading, paling; paleness
Entfärbungs‑flüssigkeit *f* bleaching fluid. **-kohle** *f* decolorizing carbon. **-messer** *m* decolorimeter. **-mittel** *n* decolorizing agent, decolorant
entfasern *vt* defiber; string (beans)

entfernen 1 *vt* remove. **2: sich e.** withdraw, depart, move away. **entfernt** *p.a* distant, remote; *adv oft* far away—**weit davon e.** far (removed) from it; **nicht im e—esten** not in the least. **Entfernung** *f* (-en) distance, range; removal; departure
Entfernungs‑gesetz *n* (*Phys.*) inverse square law. **-kraft** *f* centrifugal force. **-messer** *m* range finder
entfesseln *vt* unchain, free, let loose
entfetten *vt* **1** remove grease (fat, oil) from; degrease; unoil; scour; (*Metll.*) unwax. **2** reduce (in weight). **Entfettung** *f* removal of grease, etc; weight-reducing
Entfettungs‑kur *f* (weight-)reducing program. **-mittel** *n* **1** scouring (*or* degreasing) agent, fat remover. **2** weight-reducing remedy
entfeuchten *vt* dehumidify, desiccate
entfiel was dropped: *past of* ENTFALLEN
entfilzen *vt* unfelt, remove the felt from
entflammbar *a* (in)flammable. **Entflammbarkeit** *f* (in)flammability
entflammen 1 *vt* inflame, kindle, (a)rouse. **2** *vi* (s) ignite, flash; break out. **3** *vi* (s) & **sich e.** be inflamed, be aroused. **Entflamm‑prüfung** *f* ignition (*or* flash) test. **Entflammung** *f* (-en) inflammation, ignition, flash; arousal
Entflammungs‑probe, -prüfung *f* ignition (*or* flash) test. **-punkt** *m* flash point. **-temperatur** *f* kindling temperature (*esp* of kerosene, etc), flash point
entflecken *vt* remove spots from, unspot. **Entfleckung** *f* spot removing. **Entfleckungs‑mittel** *n* spot remover
entfleischen *vt* flesh, remove the flesh from
entfließen* *vi* (s) (*dat*) flow, emanate (from)
entflocken *vt, vi* (s) deflocculate
entfloß, entflossen emanated: *past & pp of* ENTFLIESSEN
entfremden 1 *vt* (*usu* + *d.i.o*) alienate, estrange (from). **2: sich e.** (*dat*) become alienated (*or* estranged) from. **Entfremdung** *f* (-en) alienation, estrangement
entfrosten *vt* defrost
entführen *vt* carry off, abduct
entfuseln *vt* remove fuel oil from, rectify
entgalt compensated: *past of* ENTGELTEN

entgangen escaped, etc: *pp of* ENTGEHEN

entgasen *vt* degas(ify), free from gas; de-aerate; (*Coal*) carbonize, distill. **Entgasung** *f* (-en) degassing, degasification; de-aeration; dry distillation, carbonization

Entgasungs·gas *n* coke-oven gas. **-mittel** *n* degassing agent. **-wärme** *f* (*Coke*) heat of carbonization

entgegen 1 *prep & pred a* (*dat*) contrary to, against; toward. 2 *sep pfx* toward, to meet; against counter-: **e—·arbeiten** *vi* (*dat*) work against, counteract. **·gehen*** *vi*(s) head for, face; go to meet. **e—gesetzt** *p.a* opposite, contrary *cf* E—·SETZEN. **e—·halten*** *vt* (*d.i.o*) 1 oppose, say in rebuttal. 2 hold out (to), offer. 3 contrast (with). **·kommen*** *vi*(s) (*dat*) 1 come toward, (come to) meet. 2 cooperate with, accomodate. **E—kommen** *n* cooperation, courtesy. **e—kommend** *p.a* cooperative, accommodating. **e—·nehmen*** *vt* accept, receive. **·setzen** *vt & sich e.* (*dat*) contrast (with), oppose, counter *cf* E—GESETZT. **·stehen*** *vi* (*dat*) conflict (with), stand in the way (of), oppose; face. **·treten*** *vi*(s) (*dat*) 1 step up (to meet). 2 take steps (against), oppose. 3 face, confront. 4 deny (rumors). **·wirken** *vi* (*dat*) counteract

entgegnen *vt* reply, retort, say in rebuttal

entgehen* *vi*(s) (*dat*) escape; avoid

entgeisten *vt* dealcoholize, free from alcohol

entgelten* vt 1 compensate. 2 pay for

entgerben *vt* detan *cf* GERBEN

entgerbern *vt* wash, scour (wool)

entgiften *vt* detoxify, decontaminate. **Entgiftungs·mittel** *n*, **-stoff** *m* detoxifier, decontaminating agent

entgilt compensates: *pres of* ENTGELTEN

entging escaped: *past of* ENTGEHEN

entglasbar *a* devitrifiable. **entglasen** *vt* devitrify

entgleisen *vi*(s) derail; deviate, go wrong

entgolden *vt* extract the gold from; ungild

entgolten compensated: *pp of* ENTGELTEN

entgummieren *vt* degum

enth. *abbv* (enthaltend) containing

enthaaren *vt* unhair, depilate. **enthaarend** *p.a* depilatory. **Enthaarungs·mittel** *n* depilatory

enthalogenisieren *vt* dehalogenate, dehalinate

Enthalpie *f* (-n) enthalpy, heat content

enthalten* **I** 1 *vt* contain, include. 2: **sich e.** (*genit*) refrain, abstain (from). **II** *pred a* present (as a component). **enthaltsam** *a* abstemious, abstinent. **Enthaltsamkeit** *f* abstemiousness, abstinence

enthärten *vt* soften; (*Metll.*) anneal. **Enthärtungs·mittel** *n* (water) softener

entharzen *vt* remove the resin from, deresinify. **Entharzung** *f* deresinification

enthäuten *vt* skin, flay

entheben* *vt* (*genit*) relieve (of); dismiss (from)

enthefen *vt* remove sediment from (wine)

enthielt contained; *past of* ENTHALTEN

enthirnen *vt* decerebrate

enthob(en) relieved: *past* (*& pp*) *of* ENTHEBEN

entholzen *vt* delignify

enthüllen *vt* unveil; reveal, disclose

enthülsen *vt* husk, hull, shell, peel

enthydratisieren *vt* dehydrate

entionisieren *vt* de-ionize. **Entionisierung°** *f* de-ionization

entkalken *vt* decalcify, delime

Entkalkungs·flüssigkeit *f* decalcifying (*or* deliming) solution. **-mittel** *n* deliming agent

entkam escaped: *past of* ENTKOMMEN

entkampfern *vt* decamphorate

entkarboxylieren *vt* decarboxylate

entkeimen 1 *vt* remove the sprouts from; degerm(inate); disinfect, sterilize, pasteurize. 2 *vi* (s) germinate, sprout. **Entkeimung°** *f* removal of sprouts; degermination, disinfection, pasteurization; germination. **Entkeimungs·filtration** *f* sterile filtration

entkieseln *vt* desilicify

entkleiden *vt* undress, strip

entkletten *vt* free from burrs, (de)burr

entkoffeinieren *vt* decaffeinate

entkohlen *vt* decarbonize, decarburize

entkommen* *vi*(s) (*dat*) escape (from)

entkörnen *vt* thresh (grain); gin (cotton)

entkräft(ig)en *vt* weaken, debilitate; invalidate

entkrusten *vt* take the crust off; descale

entkupfern *vt* free from copper, decopper

entkuppeln vt uncouple; disengage, disconnect

entladen* I vt 1 unload; discharge. 2 (+ genit or + von) relieve (of), free (from). II: **sich e.** 3 discharge, go off, break (out). 4 (+ genit or + von) get rid (of); free oneself (from)

Entlade·spannung f (Elec.) discharge loading. **-strom** m discharge current

Entladung° f discharge; unloading

Entladungs·büschel m corona discharge. **-rohr** n, **-röhre** f (Elec.) Geissler (or discharge) tube. **-spannung** f discharge potential

entlang prep (dat, accus), adv, sep pfx along(side): **dem** (or **am**) **Fluß e.** along the river. **e—·bewegen** vr: **sich e.** move along

entlassen* vt release, dismiss, discharge

entlasten vt release, ease the strain on; reduce; exonerate; save (sbdy) trouble. **Entlastung°** f relief; reduction; exoneration. **Entlastungs·ventil** n relief (or bleeder) valve

entlauben vt defoliate, strip the leaves off

entlausen vt delouse

entledigen (+ genit) 1 vt release (from). 2: **sich e.** get rid (of); discharge (an obligation)

entleeren vt empty, evacuate, void, drain, deflate; discharge. **Entleerungs·kammer** f discharging chamber

entlegen a remote, out-of-the-way

entlehnen vt (d.i.o) borrow, derive, take (from)

entleimen vt degum; desize; unglue

entleuchten vt deprive of light, render nonluminous. 2 vi (s) (begin to) shine, radiate

entließ dismissed: past of ENTLASSEN

entlud discharged: past of ENTLADEN

entlüften vt 1 drain the air from, bleed, exhaust, de-aerate. 2 ventilate. **Entlüfter°** m exhauster, de-aerator; ventilator

Entlüftungs·loch n vent. **-rohr** n vacuum pipe (or tube); ventilating pipe. **-schacht** m ventilation shaft. **-ventil** n de-aerating valve, vent valve

entmagnetisieren vt demagnetize

entmangan(is)ieren vt demanganize. **Entmanganung** f demanganization

Entmessingung f debrassing

entmethylieren vt demethylate

Entmilzung f (-en) splenectomy

entmineralisieren vt demineralize

entmischen vt & **sich e.** separate (into its components), unmix; disintegrate, decompose, dissociate; demulsify; (Metll.) segregate. **Entmischung°** f separation, etc; (Metll.) coring

entmutigen vt discourage

entnahm inferred: past of ENTNEHMEN

Entnahme f taking; inferring cf ENTNEHMEN. **·rohr** n extraction (or sampling) tube. **-schnecke** f extraction screw

entnässen vt deprive of moisture, dry

entnebeln vt defog, free from fog (mist, fumes). **Entnebelung** f mist (fog, fume) dispersion

entnehmen* vt (d.i.o) take, draw (from, out of); gather, infer, conclude (from)

entnerven vt enervate

entnimmt infers: pres of ENTNEHMEN

entnommen inferred: pp of ENTNEHMEN

entölen vt de-oil, remove the oil from. **Entöler** m (-) de-oiler, oil separator

Entomologe m (-n, -n) entomologist

Entoplasma n (. . men) endoplasm(a)

entorientieren vt deorient

entparaffinieren vt dewax, deparaffin

entpechen vt remove the pitch from

entphenolen vt dephenolize, dephenolate

entphosphoren vt dephosphorize

Entpolymerisierung f (-en) depolymerization

entpressen vt (d.i.o) press (out of); extract (from)

entpropanisieren vt depropanize

entquellen* vi(s) (dat) stream, spring (from). **Entquellung** f (-en) streaming, springing; shrinkage (of gels); syneresis

Entquickungs·anlage f demercurizing plant

entquillt springs: pres of ENTQUELLEN

entquoll(en) sprang, (sprung), past (& pp) of ENTQUELLEN

entrahmen vt skim (the cream off)

entrann escaped: past of ENTRINNEN

enträtseln vt decipher, puzzle out, solve

entregen vt de-energize, deactivate

entreißen* vt (d.i.o) tear away, snatch (from)

entrichten *vt* submit, pay

entriechen* *vt* deodorize

entriegeln *vt* unbolt, unlock; release

entrieseln *vi*(s) trickle away

entrinden *vt* strip the bark (*or* rind) off, decorticate

entrinnen* *vi* (s) (*dat*) flow, run, escape (from)

entrippen vt (*Bot.*) unrib (leaves)

entriß, entrissen snatched: *past & pp of* ENTREISSEN

entroch(en) deodorized: *past (& pp) of* ENTRIECHEN

entronnen escaped: *pp of* ENTRINNEN

Entropie *f*(-n) entropy

entrosten *vt* clean the rust off, free from rust. **Entrostungs·mittel** *n* rust remover

entröten *vt* take the red color out of

entrüicheln *vt* deodorize

entsagen *vi* (*dat*) renounce, give up

entsalzen *vt* free from salt; demineralize

entsanden *vt* remove the sand from, desand

entsann remembered: *past of* ENTSINNEN

entsäuern *vt* deacidify; deoxidize; (*Carbonates*) decarbonize

entschädigen *vt* compensate, reimburse

entschälen *vt* shell, peel; (*Silk*) scour, degum

entschäumen 1 *vt* skim (the foam off), scum, despumate. 2 *vi*(s) foam up, effervesce. **Entschäumer** *m* (-) skimmer, scummer, foam remover; antifoam agent

entscheiden* *vt, vi & sich e.* decide *cf* ENTSCHIEDEN. **entscheidend** *p.a* decisive. **Entscheidung°** *f* decision

entscheinen* *vt* deluster, debloom. **Entscheinungs·mittel** *n* deblooming agent

entschied decided: *past of* ENTSCHEIDEN

entschieden 1 decided: *pp of* ENTSCHEIDEN. 2 *p.a* firm, determined; decisive; (*esp as adv*) decided(ly), definite(ly). **Entschiedenheit** *f* firmness, determination; decisiveness; definiteness

entschien(en) debloomed: *past (& pp) of* ENTSCHEINEN

entschlacken *vt* free from slag (*or* dross). **Entschlackung** *f* slag (*or* dross) removal

entschlammen, entschlämmen *vt* free from mud (*or* slime)

entschleimen *vt* free from slime

entschlichten *vt* (*Tex.*) remove the size from, desize, unsize

Entschlichtungs·bad *n* desizing bath. **-mittel** *n* desizing agent

entschlickern *vt* (*Metll.*) free from dross

entschließen*: sich e. decide, resolve, make up one's mind. **Entschließung** *f* (-en) (*esp Parlam.*) resolution

entschloß sich decided: *past of* ENTSCHLIESSEN. **entschlossen** 1 decided: *pp of* ENTSCHLIESSEN. 2 *p.a* resolute, resolved, determined. **Entschlossenheit** *f* resoluteness, determination.

Entschluß° *m* decision: **einen E. fassen** reach a decision

entschlüsseln *vt* decode, decipher

entschmelzen* *vi*(s) melt away. **entschmilzt** *pres.* **entschmolz(en)** *past (& pp)*

entschuldigen *vt & sich e.* excuse (oneself)

entschwand disappeared: *past of* ENTSCHWINDEN

entschwefeln *vt* desulfurize

Entschwef(e)lungs·anlage *f* desulfurization unit. **-mittel** *n* desulfurizing agent

entschweißen *vt* degrease, scour (wool)

entschwinden* *vi* (s) disappear, vanish, fade (*or* slip) away. **entschwunden** *pp*

entseifen *vt* rinse, free from soap

entsetzen I *vt* 1 horrify. 2 (*genit*) remove, dismiss (from). II: **sich e.** be horrified. **Entsetzen** *n* horror. **entsetzlich** *a* horrible, atrocious; awful

entseuchen *vt* disinfect, decontaminate

entsilbern *vt* desilver(ize)

entsilizieren *vt* desiliconize

entsinnen* *vr*: **sich e.** (*genit*) remember

entsintern *vt* desinter

entsonnen remembered: *pp of* ENTSINNEN

entsorgen *vt* dispose of. **Entsorgung** *f* (waste) disposal (*esp Nucl.*)

Entsorgungs·park *m*, **-zentrum** *n* nuclear waste disposal site

entspannen 1 *vt & sich e.* relax; (*Gas*) expand. 2 *vt* (*Glass*) anneal; (*Water*) reduce the surface tension of. 3: **sich e.** ease, find relaxation. **Entspannung°** *f* 1 relaxation; (*gas*) expansion; (*Glass*) annealing; (*Water*) lowering of surface tension. 2 (*Petrol.*) flash evaporation

Entspannungs·gas n let-down gas. **-temperatur** f (Glass) annealing temperature. **-turm** m flash tower. **-ventil** n relief valve

entspiegeln vt (Glass) eliminate reflections from

entspr. abbv (entsprechend) corresponding, etc

entsprach corresponded: past of ENTSPRECHEN

entsprang arose, etc: past of ENTSPRINGEN

entsprechen* vi (dat) correspond, conform (to), match; comply (with). **entsprechend 1** p.a corresponding, equivalent; appropriate, suitable; adequate. **2** prep (dat) according (to), in accord (with). **Entsprechung** f(-en) correspondence, conformity; compliance; accord(ance); correlation. **Entsprechungs·ziffer** f coefficient of correlation

entspricht corresponds: pres of ENTSPRECHEN

entsprießen* vi (s) (dat) sprout, spring (from)

entspringen* vi (s) **1** (dat) spring; arise; escape (from). **2** (River) have its source, rise

entsprochen corresponded: pp of ENTSPRECHEN

entsproß, entsprossen sprouted: past & pp of ENTSPRIESSEN

entsprungen sprung; risen, etc: pp of ENTSPRINGEN

entstählen vt (Metll.) soften; unsteel (lamina)

entstammen vi(s) (dat) originate (from)

entstand(en) arose (arisen) past (& pp) of ENTSTEHEN

entstänkern vt deodorize. **Entstänkerungs·patrone** f deodorizing cartridge

entstauben, entstäuben vt dust (off), dedust; remove the dust from. **Entstauber** m (-) dust remover (or arrester). **Entstaubung** f dust removal, dust exhaust

entstehen* vi (s) arise, originate, have its origin; be formed: **eben e—d** nascent, incipient. **Entstehung** f (-en) origin, rise, genesis, formation

Entstehungs·art f mode of origin, the way it arises (originates, is formed). **-lehre** f genetics. **-moment** n moment of formation. **-weise** f = -ART. **-zentrum** n center of origin (or formation). **-zustand** m nascent (or embryonic) state

entsteinen vt take the stones (pits, seeds) out of. **Entsteiner** m (-) stoner, seeder

entstellen vt disfigure, distort

entstielen vt remove the stalk (or stem) from

entstören vt (Electron.) free from static; debug. **Enstörer** m (-) static suppressor, debugger. **Entstörung°** f suppression of static; debugging

entströmen vi (s) (dat) stream, flow, issue, escape (from)

enttäuschen vt disappoint, disillusion

entteeren vt detar, free from tar. **Entteerer** m (-) tar extractor. **Entteerung** f tar removal (extraction, separation)

enttrüben vt uncloud, clear (up)

Entüberhitzer m (-) desuperheater

entvölkern vt depopulate

Entvulkanisation f devulcanization. **entvulkanisieren** vt devulcanize

entw. abbv (entweder) either

Entw. abbv (Entwicklung) devel., evol.

entwachsen I vt dewax. **II*** vi (s) (dat) grow out (of), outgrow

Entwaldung f (-en) deforestation

entwarf designed: past of ENTWERFEN

entwärmen vt deprive of heat

Entwässerer m dehydrator. **entwässern I** vt **1** free from water, drain. **2** dehydrate. **3** concentrate, rectify. **II** vi drain, discharge. **entwässert** p.a (oft) anhydrous. **Entwässerung** f (-en) drainage; dehydration; concentration; rectification; hydroextraction

Entwässerungs·mittel n dehydrating agent. **-rohr** n, **-röhre** f drain pipe

entweder cnjc: **e. . . . oder** either . . . or

entweichen* vi (s) (dat) escape, leak (from). **Entweichungs·ventil** n escape (or delivery) valve

entwerfen* vt sketch, design, outline

entwerten vt devalue, debase; cancel, invalidate

entwesen vt disinfest, free from vermin

entwich(en) escaped: past (& pp) of ENTWEICHEN

entwickelbar a developable

entwickeln 1 vt & **sich e.** develop (incl Phot.), evolve. **2** vt unfold, reveal, display; liberate, generate, produce. **ent-wickelnd** *p.a* (*oft*) nascent

Entwickler *m* (-) developer (*incl Phot. & Dye.*); evolver; (gas) generator. **-bottich** *m* developing tank. **-schale** *f* developing dish (*or* tray).

Entwicklung° *f* development; (*Phot. usu*) developing; evolution; unfolding, display; liberation, generation (of gases)

Entwickungs- development, developing; evolution; (gas) generating: **-bad** *n* developing bath. **-dose** *f* developing tank. **e—fähig** *a* capable of development (*or* evolution); viable; underdeveloped. **E—farbstoff** *m* developing dye. **-flüssigkeit** *f* developing fluid, developer. **-gefäß** *n* (gas) generating vessel; generator. **-geschichte** *f* history of the development (of . . .), evolutionary history. **-hilfe** *f* development(al) aid. **-kolben** *m* generating flask. **-land** *n* developing country. **-lehre** *f* theory of evolution. **-rohr** *n*, **-röhre** *f* delivery tube (*or* pipe). **-stadium** *n* developmental stage; stage of development. **-stand** *m* state of development. **e—träge** *a* slow in developing. **E—zustand** *m* state of development

entwirft designs: *pres of* ENTWERFEN

entwirren *vt* unravel, disentangle

entwöhnen 1 *vt & sich e.* (*genit*) rid (oneself) of the habit (of). 2 *vt* wean. **entwöhnt** *p.a* unaccustomed

entworfen designed: *pp of* ENTWERFEN

entwuchs outgrew: *past of* ENTWACHSEN II

Entwurf° *m* sketch, design, outline, plan

entwurzeln *vt* uproot, eradicate

entzerren *vt* eliminate distortion from, rectify, compensate

entziehen* (*dat*) 1 *vt* take (away), withdraw, revoke (from); deprive (sbdy) of; extract. 2: *sich e.* withdraw, hide, escape (from). **Entziehung°** *f* withdrawal, revocation, deprivation; extraction

entziffern *vt* decipher

entzinken *vt* dezinc, free from zinc

entzinnen *vt* detin, free from tin

entzischen *vi* (s) (*dat*) escape (from) with a hiss

entzog(en) withdrew (withdrawn): *past (& pp) of* ENTZIEHEN

entzücken *vt* delight. **entzückend** *p.a* delightful

entzuckern *vt* free from (*or* deprive of) sugar. **Entzuckerung** *f* extraction of sugar

Entzug° *m* withdrawal, etc = ENTZIEHUNG

entzündbar *a* ignitable; (in)flammable; inflamable

entzünden 1 *vt & sich e.* ignite. 2 *vt* kindle; inflame. 3: *sich e.* catch fire; be(come) inflamed; arise

entzundern *vt* free from scale, (de)scale

entzündlich *a* 1 inflammable 2 (*Med.*) inflammatory, with inflammation

Entzündung° *f* inflammation (*incl Med.*); ignition, kindling; (*Med.*) (*as last part of compd. noun, oft*) -itis *cf* NIERENENTZÜNDUNG

Entzündungs- ignition: **-gemisch** *n* ignition mixture. **-punkt** *m* kindling (igniting, burning) point. **-rohr** *n*, **-röhre** *f* ignition tube. **-temperatur** *f* ignition temperature

entzwei *adv & sep pfx* in two, apart. **entzweien** 1 *vt* separate, disunite. 2: *sich e.* fall out, become disunited; be at variance

entzweigen *vt* debranch (protein)

Enzian *m* gentian

Enzyklopädie *f* (-n) encyclopedia

Enzym-art *f* type of enzyme. **enzymatisch** *a* enzymatic.

Enzym-beize *f* (*Lthr.*) enzyme bate. **-entschlichtung** *f* enzyme desizing. **-wirkung** *f* enzyme action

Eocän, Eozän *n* (*Geol.*) Eocene. **Eolithenzeit** *f* (*Geol.*) Eolithic

E.P. *abbv* 1 (englisches Patent) English (*or* British) patent. 2 (Erstarrungspunkt) freezing (*or* solidification) point. 3 (Erweichungspunkt) softening point

Epidemie *f* (-n), **epidemisch** *a* epidemic

Epidotisierung *f* epidotization

Epihydrin-säure *f* epihydrinic acid

epimer *a* epimeric

Epithel *n* (-e) epithelium. **-körperchen** *n* epithelial gland. **-zelle** *f* epithelial cell

Epizucker-säure *f* episaccharic acid

Epoche *f* (-n) epoch, era. **e—machend** *a* epoch-making, epochal

Epoxyd-harz *n* epoxy resin

Eppich *m* celery

Eprouvette *f* (-n) test tube

EPS *abbv* (effektive Pferdestärke) actual horsepower

EPZ *abbv* (Eisenportlandzement) Portland blast-furnace cement

Equiset·säure *f* equisetic acid

er *pron* he, it (*referring to a masc. noun*)

er· *insep pfx* (*no specif transl*) *implies:* **1** (*successful completion*) *cf* ERBAUEN, ERZEUGEN. **2** (*incipient or spontaneous action*) *cf* ERBEBEN, ERSCHEINEN. **3** (*achievement of the condition specified by an adjective*) *cf* ERGÄNZEN (*from* GANZ), ERKÄLTEN (*from* KALT)

Er. *abbv* (Erstarrungspunkt) freezing (*or* solidification) point

erachten *vt* (**für, als**) regard (as), consider, deem. **Erachten** *n:* **meines E—s** in my opinion

erarbeiten *vt* **1** work out, compile. **2** (*usu* + *dat rflx* **sich**) acquire by work(ing)

Erb. heredit(ar)y, genetic *cf* ERBE

erbat requested: *past of* ERBITTEN

erbauen *vt* **1** erect, build. **2** edify

Erbe I *m* (-n, -n) heir. **II** *n* inheritance

erbeben *vi* (s) (start to) quake, tremble

Erb·einheit *f* (*Biol.*) genetic factor, gene

erben *vt* inherit

erbeten requested: *pp of* ERBITTEN

erbieten* *vr:* **sich e.** (**zu** + *inf*) offer (to + *v*)

Erbin *f* (-nen) heir(ess)

Erbin·erde *f* erbia, erbium oxide

erbitten* *vt* **1** solicit, request. **2: sich e. lassen** give in to persuasion

erbittern *vt* embitter. **erbittert** *p.a* (*oft*) stubborn, vehement

Erb·körperchen *n* chromosome. **-krankheit** *f* hereditary disease

erblasen* *vt* (*Metll.*) blast, blow, subject to the blast; produce in a blast furnace

erblassen *vi*(s) (turn) pale; fade, die

erblich *a* hereditary. **Erblichkeit** *f* heredity

erblicken *vt* **1** catch sight of, see. **2** recognize, detect

erblies blasted: *past of* ERBLASEN

Erblindung *f* going blind, loss of sight

erbohren *vt* obtain (e.g. oil) by drilling; tap (e.g. deposits) by boring; bore (a well)

erborgt *a* borrowed; artificial

erbot(en) offered: *past* (*& pp*) *of* ERBIETEN

erbötig *a* (**zu**) ready, willing (to, for)—**sich e. machen** (**zu**) offer (to + *v*)

erbrach broke open: *past of* ERBRECHEN

erbracht(e) produced: *pp* (*& past*) *of* ERBRINGEN

erbrechen* **1** *vt* break open. **2** *vt/i* vomit

erbreiten *vt* broaden

erbricht breaks open: *pres of* ERBRECHEN

erbringen* *vt* **1** bring (profit); yield, produce. **2** pay, meet (costs). **3** furnish (proof)

erbrochen broken open: *pp of* ERBRECHEN

Erbschaft *f* (-en) inheritance

Erbse *f* (-n) pea

Erbsen·baum *m* Siberian acacia. **-erz** *n* pea ore (limonite). **e—förmig** *a* pea-shaped, pisiform. **-groß** *a* pea-sized. **E—größe** *f* pea size. **e—grün** *a* pea-green. **E—mehl** *n* pea flour, pea meal. **-stein** *m* peastone, pisolite (granular limestone)

Erbs·kohle *f* pea coal

Erb·substanz *f* (*Biol.*) germ plasm

Erd· earth(y), ground, geo-, terrestrial, mineral *cf* ERDE: **-ableitung** *f* (*Elec.*) ground (connection). **-achse** *f* earth's axis

erdacht(e) devised: *pp* (*& past*) *of* ERDENKEN

Erdalkali *n* alkaline earth. **-halogen** *n* alkaline-earth halide. **-salze** *npl* alkaline-earth metal salts

erd·alkalisch *a* alkaline-earth

Erd· earth, etc *cf* ERDE: **-anziehung** *f* gravitational attraction. **-apfel** *m* potato. **-arbeiten** *fpl* excavation, digging. **-art** *f* type of earth (*or* soil). **e—artig** *a* earth-like, earthy. **E—asphalt** *m* earthy (*or* native) asphalt. **-bahn** *f* earth's orbit. **-ball** *m* terrestrial globe. **-beben** *n* earthquake. **-bebenkunde** *f* seismology. **-bebenwarte** *f* seismological station. **-beere** *f* strawberry. **-beschleunigung** *f* acceleration of gravity. **-beschreibung** *f* geography. **-birne** *f* Jerusalem artichoke. **-boden** *m* ground, soil. **-bohrer** *m* soil borer. **e—braun** *a* khaki(-colored). **E—brot** *n* (*Bot.*) sow bread. **-chlorid** *n* earth-metal chloride

Erde *f* (-n) earth (*incl* planet); ground, soil, dirt; floor, **e—·haltig** *a* containing earth, earthy

Erd·eichel *f* **1** peanut. **2** tuberous (*or* heath) pea

erden *vt* **1** (*Elec.*) ground. **2** treat with fuller's earth

Erd·enge *f* isthmus

erdenkbar *a* conceivable. **erdenken*** *vt* think up, devise, invent. **erdenklich** *a* conceivable

Erd- earth, ground, etc *cf* ERDE: **-erschütterung** *f* earth tremor. **-farbe** *f* earthy (*or* mineral) color. **e—farben**, **-farbig** *a* earth-colored. **-feucht** *a* earth-moist. **E—flachs** *m* amianthus. **-floh** *m* flea beetle. **-forscher** *m* geologist. **-gang** *m* (*Min.*) vein; tunnel, gallery. **-gas** *n* natural gas. **-gasbenzin** *n* natural gasoline. **-gasgashead** (*or* drip) gasoline. **-gasruß** *m* (natural) gas black. **-gelb** *n* yellow ocher. **e—gelb** *a* ocherous, ocher-yellow. **E—gemisch** *n* mixture of earths. **-geruch** *m* earthy odor. **-geschmack** *m* earthy taste, flavor of the soil. **-geschoß** *n* ground floor. **-glas** *n* selenite. **-grün** *n* mineral green, green verditer (basic $CuCO_3$); green earth (celadonite *or* glauconite). **-gürtel** *m* (terrestrial) zone. **e—haltig** *a* earthy, containing earth. **E—harz** *n* asphalt, bitumen; fossil resin—**gelbes E.** amber. **e—harzig**, **-harzartig**, **-harzhaltig** *a* asphaltic, bituminous

erdichten *vt* invent, fabricate (ideas, tales)

erdig *a* 1 earthy; earth-stained. 2 terrestrial. 3: **e—es Wasser** natural water containing calcium salts

Erd- earth(y); terrestrial; ground, soil; mineral: **-innere** *n* (*adj endgs*) interior of the earth. **-kabel** *n* underground cable. **-kalk** *m* marly limestone. **-kern** *m* earth's core. **-kobalt** *m* earthy cobalt, asbolite—**grüner E.** nickel ocher, annabergite; **roter E.** cobalt crust, earthy erythrite; **roter strahliger E.** cobalt bloom, erythrite; **schwarzer E.** asbolane, asbolite. **-kohle** *f* brown lignite, earthy coal. **-körper** *m* terrestrial body. **-krume** *f* topsoil, vegetable mold, black earth. **-kruste** *f* earth's crust. **-kugel** *f* terrestrial globe. **-kunde** *f* geography; geology. **-leiter** *m* (*Elec.*) ground wire. **-leitung** *f* underground line (wire, pipe). **e—magnetisch** *a* geomatic. **E—mandel** *f* chufa, ground almond; chufa tuber; tuberous pea. **-maus** *f* field mouse (*or* vole). **-mehl** *n* silicious earth = KIESELGUHR. **-meßkunst** *f* geodesy. **metall** *n* earth metal: **alkalisches E.** alkaline-earth metal. **-nuß** *f* 1 (*genl*)

groundnut. 2 (*specif*) peanut; pignut. **-nußöl** *n* peanut (*or* arachis) oil. **-nußsäure** *f* arachidic acid. **-oberfläche** *f* earth's surface. **-öl** *n* petroleum

Erdöl- petroluem: **-bearbeitung** *f* petroleum refining. **e—haltig** *a* containing petroleum, petroliferous. **E—industrie** *f* petroluem (*or* oil) industry. **-vorkommen** *n* petroleum deposit, oilfield

Erd-oxid *n* oxide of an earth metal. **-pech** *n* mineral pitch, asphalt, bitumen

erdpech-artig *a* asphaltic. **-haltig** *a* containing asphalt, asphaltic

Erd-probe *f* soil sample; soil test. **-rauch** *m* (*Bot.*) fumitory. **-reich** *n* 1 earth, ground, soil. 2 world, (whole) earth. **-rinde** *f* earth's crust. **-rohr** *n* underground pipe

erdrücken 1 crush to death. 2 oppress, overpower. **erdrückend** *p.a* (*oft*) oppressive

Erd- earth(y), terrestrial; ground, soil; mineral: **-rutsch** *m* landslide. **-salz** *n* rock salt. 2 saline efflorescence on the soil. **-säuren** *fpl* acids of earth elements (V, Nb, Ta, Pa). **-scheibe** *f* sow bread = ERDBROT. **-schellack** *m* acaroid resin. **-schicht** *f* layer of earth, stratum. **-schierling** *m* (*Herbs*) hemlock. **-schlacke** *f* earthy slag. **-schluß** *m* (*Elec.*) ground, current leakage. **-scholle** *f* clod; (*Geol.*) block; (*fig,* = ERDE) soil. **-schwamm** *m* mushroom. **-schwefel(samen)** *m* lycopodium, club moss. **-schwere** *f* force of gravity. **-seife** *f* mountain soap (type of clay). **-stein** *m* eaglestone, aetites. **-stoß** *m* earth tremor. **-strich** *m* zone, region. **-strom** *m* earth current. **-talg** *m* mineral tallow, hatchettite, ozocerite. **-talk** *m* earthy talc. **-teer** *m* mineral tar, bitumen, maltha. **-teil** *m* continent. **-trabant** *m* Earth satellite

erdulden *vt* suffer, endure, put up with

Erd-umkreisung *f* circling (of) the earth. **-umlaufbahn** *f* earth orbit

Erdung *f* (-en) (*Elec.*) ground(ing)

Erd-verbindung *f* (*Elec.*) ground connection. **e—verlegt** *a* underground, laid in the ground. **E—wachs** *n* ozocerite, mineral (*or* earth) wax. **-wärme** *f* heat of the earth. **-zeitalter** *n* geological era

ereignen *vr*: **sich e.** happen, occur. **Ereig-**

nis *n* (-se) event, occurrence

ereilen *vt* overtake

Eremakausie *f* eremacausis, slow combustion

Erf. *abbv* (Erfinder) inventor

erfahren* I *vt* 1 find out (about), learn. 2 experience, undergo; (*esp* + ·**ung** *noun*): **eine Verbesserung e.** undergo an improvement, improve. II *p.a* experienced, proficient. **Erfahrung** *f* (-en) 1 experience: **gute (schlechte) E—en machen (mit)** do well (badly), be satisfied (disappointed) (with). 2: **in E. bringen** find out (about) = ERFAHREN I

Erfahrungs· experience, empirical: -**beweis** *m* empirical proof. -**formel** *f* empirical formula. **e—gemäß** *a* by experience, empirical. **e—kunde, -lehre** *f* empiricism. **e—mäßig** *a* by experience, empirical. -**reich** *a* experienced. **E—wert** *m* empirical (*or* experimental) value

erfand invented: *past of* ERFINDEN

erfaßbar *a* graspable; attainable, obtainable (*esp* for statistical records). **erfassen** *vt* 1 take (get, catch) hold of. 2 catch and drag along. 3 seize, grip, overcome. 4 grasp, comprehend. 5 (*esp Stat.*) (seek out and) register; include, extend to. **Erfassung°** *f* seizure; grasp; registration, inventory, survey. **Erfassungs·grenze** *f* (*Reagents*) sensitivity (limit)

erfinden* *vt* invent. **Erfinder** *m* (-) inventor. **Erfinder·geist** *m* inventive spirit. **erfinderisch** *a* inventive. **Erfindung** *f* (-en) invention; fiction, fabrication. **e—gemäß** *a* (*Pat.*) according to the invention

Erfolg *m* (-e) success; result, consequence: **E. haben** meet with success, get results.

erfolgen *vi* (s) 1 take place, occur, be undertaken (made, carried out, given, received). 2 (+ **auf**) follow. *Cf* ERFOLGT

erfolg·los *a* unsuccessful. -**reich** *a* successful

erfolgt *p.a* completed: **nach e—er Verbrennung** after (the completed) combustion, after the combustion has taken place *cf* ERFOLGEN

erforderlich *a* necessary, required. **erfordern** *vt* require, demand. **Erfordernis** *n* (-se) requirement, demand

erforschen *vt* explore, investigate, examine; discover. **Erforscher°** *m* explorer; discoverer. **Erforschung°** *f* exploration, etc; discovery

erfragen 1 *vt* ascertain, find out. 2 *vt/i* inquire

erfreuen 1 *vt* delight, please. 2: **sich e. (an** or + *genit*) enjoy. **erfreulich** *a* pleasant. **erfreulicherweise** *adv* fortunately. **erfreut** *p.a* pleased, glad

erfrieren* 1 *vt* (*Metll.*) chill; (+ *dat rflx* **sich**) freeze, get (e.g., one's finger) frostbitten. 2 *vi* (s) freeze (to death); get frostbitten

erfrischen 1 *vt, vi,* **sich e.** refresh (oneself). 2 *vt* (*oft*) freshen, revive; refrigerate. **Erfrischung** *f* (-en) refreshment; refrigeration

erfror(en) froze(n): *past* (& *pp*) of ERFRIEREN

erfuhr found out: *past of* ERFAHREN

erfüllen *vt* fill, impregnate; fulfill

erfunden invented: *pp of* ERFINDEN

ergab yielded: *past of* ERGEBEN

ergangen fared, etc: *pp of* ERGEHEN

ergänzen I *vt* 1 supplement 2 complete, round out. 3 replace. 4 add, supply. II *vt* & *esp recip*: **sich** (*or* **einander**) **e.** complement (each other). **ergänzend** *p.a* supplementary; additional **Ergänzung** *f* (-en) 1 supplement 2 completion. 3 replacement. 4 addition; supply(ing). 5 complement

Ergänzungs·band *m,* -**buch** *n* supplement(ary volume). -**farbe** *f* complementary color. -(**nähr**)**stoff** *m* (food) supplement. -**werk** *n* supplement(ary volume *or* work). -**winkel** *m* complementary angle

Erg.-Bd. *abbv* (ERGÄNZUNGSBAND) supp. vol.

ergeben* I *vt* 1 yield, produce, result in. 2 prove, show. II: **sich e.** 3 arise, offer itself. 4 turn out (this *or* that way). 5 (*dat*) devote (*or* dedicate) oneself (to). 6 (*dat*) resign oneself, give in (to); become addicted (to). 7 (*dat or* + **in**) yield, surrender (to), acquiesce (in). 8 (*usu* + **aus**) result, follow, be clear (from), be proved (by). III *p.a.* (*usu pred*) devoted

Ergebnis *n* (-se) result; product; yield. **e—**

·los *a* unsuccessful, fruitless; futile

ergehen* I *vi* (s) **1** be issued. **2: etwas über sich e. lassen** submit to sthg. **3** (*imps + dat*): **es ist ihnen gut (schlecht) ergangen** they fared well (badly). **II: sich e. 4 (in)** indulge (in). **5 (über)** expatiate (on)

ergiebig *a* fruitful, productive; abundant. **Ergiebigkeit** *f* fruitfulness, productiveness; abundance, (high) yield; (*Lime*) volume yield; (*Colors*) tinctorial power

ergießen* **1** *vt* pour, discharge. **2: sich e.** pour, stream

ergibt yields, etc: *pres of* ERGEBEN

Ergine *pl* (*collective term for:*) vitamins, minerals and hormones

erging fared, etc: *past of* ERGEHEN

erglänzen *vi* (s) (start to) gleam (shine, glow)

erglühen *vi* (s) start to glow; be kindled

ergoß, ergossen poured: *past & pp of* ERGIESSEN

Ergosterin *n* ergosterol

ergreifen* *vt* **1** take hold of, seize, grasp; catch. **2** overcome. **3** affect, stir. **4** take (steps, initiative, etc); take up (profession). **ergriff(en)** *past* (*pp*)

ergründen *vt* fathom, probe into the depths of

ergrünen *vi* (s) (turn) green

Erguß° *m* effusion, discharge, outpouring. **·-gestein** *n* effusive rock

Ergw. *abbv* (ERGÄNZUNGSWERK) supp. vol.

erh. *abbv* (erhitzt) heated. **Erh.** *abbv* (Erhitzung) heating

erhaben *a* **1** raised, elevated. **2** convex; embossed, (in) relief. **3** lofty, sublime. **4 (über)** superior (to). **Erhabenheit** *f* (-en) elevation; convexity; sublimity; superiority

erhaltbar *a* preservable; obtainable

erhalten* I *vt* **1** keep, maintain, preserve, save. **2** support, keep (alive). **3** receive, be given. **4** take on (e.g., a new meaning). **5** obtain. **II: sich e. 6** be preserved, keep (oneself), endure, last; survive. **7** support oneself; (+ **mit, von,** *oft*) survive (on). **erhältlich** *a* available, obtainable. **Erhaltung** *f* conservation; preservation; maintenance; support *cf* ERHALTEN

Erhaltungs·gesetz *n* conservation law. **-mittel** *n* preservation; antiseptic; (means of) support. **-umsatz** *m* basal metabolism

erharten *vi* (s) harden, set

erhärten **1** *vt & vi*(s) harden, solidify. **2** *vt* confirm, verify. **Erhärtung** *f* hardening; confirmation

erheben* **1** *vt* raise; make, file (claims, complaints, etc). **2: sich e.** rise; arise; rebel. **erheblich** *a* considerable, appreciable. **Erhebung** *f* (-en) **1** elevation, high spot **2** raising, making, filing. **3** rebellion, revolt. **4** inquiry, investigation

erheischen *vt* demand, require

erheitern **1** *vt & sich e.* brighten (up), cheer (up). **2: sich e.** (*oft*) clear (up)

erhellen **1** *vt & sich e.* light (up), brighten (up); clear (up). **2** *vt* (*oft*) clarify. **3** *vi* (s) become clear (*or* evident). **Erhellung** *f* (-en) illumination

erhielt preserved: *past of* ERHALTEN

erhitzen **1** *vt* heat (up); excite, stimulate; pasteurize. **2: sich e.** get heated, get excited (*or* flushed). **erhitzend** *p.a* exciting, stimulating. **Erhitzer** *m* (-) heater; still. **erhitzt** *p.a* excited; flushed. **Erhitzung** *f* heating; excitement, stimulation; pasteurization

Erhitzungs·mikroskop *n* high-temperature microscope. **-rückstand** *m* residue on heating

erhoffen *vt* (*dat rflx* **sich**) hope for, hope to attain

erhob(en) raised: *past* (& *pp*) *of* ERHEBEN

erhöhen **1** *vt* raise, increase; promote. **2: sich e.** rise, increase. **erhöht** *p.a* (*oft*) higher, greater. **Erhöhung** *f* (-en) raising; rise; elevation; increase; promotion

erholen *vr*: **sich e.** recover, rest, relax. **Erholung** *f* (-en) recovery, convalescence; rest, relaxation. **Erhol·zeit** *f* recovery time

Erichsen·tiefung *f* (*Metll.* Erichsen cupping (test)

Erigenon·öl *n* oil of fleabane

erinnern (*usu* + **an**) **1** *vt* remind (of). **2** *vi* be reminiscent (of). **3: sich e.** remember. **Erinnerung** *f* (-en) memory; reminder— **in E. bringen** = ERINNERN **1, 2**

Eriochrom·schwarz *n* eriochrome black
Erk. *abbv* (Erkennung) detection, etc
erkalten *vi* (s) cool, grow cold
erkälten 1 *vt* cool, chill. 2: **sich e.** catch (a) cold. **erkältet** *p.a*: **e. sein** have a cold
Erkaltung *f* (spontaneous) cooling
Erkältung *f* (-en) 1 cooling, chilling. 2 (common) cold
erkannt(e) recognized: *pp* (& *past*) *of* ER-KENNEN
erkaufen *vt* buy (dearly), pay a price for
erkennbar *a* recognizable
erkennen* I *vt* 1 recognize; identify, detect; diagnose: (**sich**) **zu e. geben** reveal (one's identity). 2 recognize the true nature of. 3 realize. 4: **e. lassen** (*oft*) show, reveal. II *vi* (**auf**) hand down a judgment (of)
erkenntlich *a* 1 recognizable. 2 grateful: **sich e. zeigen** show one's gratitude
Erkenntnis I *f* (-se) realization, discovery; recognition; (full) knowledge, understanding. II *n* (-se) (legal) judgment
Erkennung *f* recognition; identification; detection; diagnosis; realization
Erkennungs·marke *f* identification tag. **-zeichen** *n* 1 sign of recognition. 2 identification mark (*or* sign)
erklärbar *a* explainable, explicable
erklären I *vt* 1 explain. 2 declare. 3 (**zu**) appoint (as). II: **sich e.** 4 (**aus**) be explained (by), be explainable: **daraus erklärt sich** . . . this explains . . . 5: **das erklärt sich von selbst** that is self-explanatory. 6 explain oneself; (*oft* + **für**) declare oneself (to be). **erklärend** *p.a* explanatory. **erklärlich** *a* explainable, understandable. **erklärt** *p.a* declared, avowed; professed. **Erklärung°** *f* explanation; declaration. **Erklärungs·versuch** *m* attempted explanation; explanatory experiment
erkranken *vi* (s) fall ill; (**an**) fall victim, succumb (to: an illness). **erkrankt** *p.a* sick, ill, diseased. **Erkrankung** *f* (-en) succumbing; illness, being ill. **Erkrankungs·fall** *m*: **im E.** in case of illness
erkundigen *vr*: **sich e.** (**nach, ob**) inquire, get information (about; as to whether). **Erkundigung** *f* (-en) inquiry: **E—en einholen** (*or* **einziehen**) make inquiries, gather information

Erl. *abbv* (Erläuterung) explanation
erlag succumbed: *past of* ERLIEGEN
erlangen *vt* gain, reach, attain
Erlanger·blau *n* Erlanger (Prussian) blue. **-leder** *n* glove kid (leather)
Erlaß *m* (. . lasse) 1 issuance. 2 decree.
erlassen* *vt* 1 issue (order, etc). 2 (*d.i.o*) exempt (from); absolve, relieve (of).
erläßlich *a* pardonable, dispensable
erlauben *vt* (*d.i.o*) allow, permit. **Erlaubnis** *f* (-se) permission; license, permit
erläutern *vt* explain, clarify
Erle *f* (-n) alder (tree)
erleben *vt* 1 (live to) see. 2 live through, experience, have (an experience.) **Erlebnis** *n* (-se) (personal) experience
erledigen *vt* 1 take care of, settle (a matter); finish off (an opponent). 2 vacate
erlegen I *vt* kill. II succumbed: *pp of* ERLIEGEN
·erlei *sfx* (*added to num* & *Quant.*) e.g.:
dreierlei three (different) kinds of;
vielerlei many (different) kinds of
erleichtern I *vt* 1 relieve. 2 (*d.i.o*) facilitate, make easy (for). 3 lighten (in weight). 4 unburden. II *vi* 5 provide relief. III: **sich e.** 6 unburden oneself. 7 relieve oneself. **Erleichterung** *f* (-en) relief; facilitation
erleiden* *vt* suffer, undergo
Erlen·baum *m* alder (tree). **-holz** *n* alder wood. **-kohle** *f* alder charcoal
Erlenmeyer·kolben *m* Erlenmeyer (conical) flask
erlernen *vt* learn; acquire (a skill)
erlesen *p.a* select, choice, exquisite
erleuchten *vt* & **sich e.** light (up), illuminate
erliegen* *vi*(s) break down; (*dat*) succumb (to)—**zum E. kommen** (**bringen**) come (bring) to a halt
erließ issued: *past of* ERLASSEN
erlischt goes out: *pres of* ERLÖSCHEN
erlitt(en) suffered: *pres* (& *pp*) *of* ERLEIDEN
erlogen *p.a* fabricated, false
Erlös *m* (-e) proceeds, profit(s)
erlöschen* (erlosch, erloschen; erlischt) *vi*(s) 1 expire, die out, become extinct; slake (*as*: lime). 2 (*also reg*: erlöschte, etc) go out (*as*: fire, light), be extinguished. **erloschen** *p.a* (*oft*) spent, dead, extinct

erlösen vt liberate; redeem; realize (profit)

ermächtigen vt empower, authorize

ermangeln vi (genit) lack. **Ermangelung** f: **in E.** (+ genit) for lack (of), in the absence (of)

ermaß judged: pres of ERMESSEN

ermäßigen vt moderate, reduce (esp prices)

ermatten vt, vi (s) weaken, tire, wear out

ermessen* vt judge, appreciate. **Ermessen** n judgment; discretion

ermißt judges: pres of ERMESSEN

ermitteln vt ascertain; find. **Ermittlung** f 1 ascertainment; determination: **E—en anstellen** make inquiries. 2 (esp as sfx) referral service

ermöglichen vt make possible, permit

ermüden vt, vi (s) tire, weary. **Ermüdung** f tiring, fatigue, exhaustion

Ermüdungs·bruch m fatigue fracture. **-erscheinung** f symptom of fatigue. **-festigkeit** f fatigue strength. **-gift** n fatigue toxin. **-grenze** f endurance (or fatigue) limit. **-kampfstoff** m (Mil.) harassing agent. **-prüfung** f fatigue test. **-schutzmittel** n fatigue inhibitor. **-stoff** m product of fatigue. **-versuch** m fatigue (or endurance) test

ermuntern vt arouse; encourage, urge

ermutigen vt encourage

ernähren 1 vt feed; nourish. 2: **sich e.** make a living; (**von**) live (on). **ernährend** p.a nutritive, nourishing. **Ernährung** f 1 nutrition, nourishment; alimentation. 2 food (supply). 3 livelihood, support. substance

Ernährungs·bedürfnis n food requirement. **-boden** m nutritive medium. **-flüssigkeit** f nutritive liquid. **-geschäft** n nutrition. **-güter** npl foodstuffs. **-kanal** m alimentary canal. **-krankheit** f nutritional disease. **-kunde**, **-lehre** f dietetics. **-stoff** m, **-substanz** f food substance, nutrient. **-vergiftung** f food poisoning, alimentary intoxication. **-wert** m nutritive (or nutritional) value

ernannt(e) appointed: pp (& past) of ERNENNEN

ernennen* vt name; appoint, nominate

erneue(r)n vt renew; repair; replace; restore; refresh (colors); regenerate (steam). **Erneuerung** f (-en) renewal, re-

pair. **erneut** p.a (oft) fresh; adv again

erniedrigen vt lower, decrease; degrade. **Erniedrigung** f (-en) lowering, decrease; degradation

ernst a serious, earnest. **Ernst** m seriousness: **im E.** seriously, in earnest; **es ist ihr** (or **ihnen**) **E.** they mean it seriously. **Ernst·fall** m emergency. **ernsthaft** a serious, earnest

Ernte f (-n) harvest, crop. yield. **-ertrag** m crop yield. **-maschine** f harvester

ernten vt/i harvest, reap

erobern vt conquer. **Eroberung** f (-en) conquest

erodieren vt erode

eröffnen vt/i & **sich e.** open (up), reveal (oneself). **eröffnend** p.a (Med.) aperient. **Eröffnung°** f opening; revelation

erörtern vt discuss, debate

erproben vt test. **Erprobung** f (-en) test(ing)

erquicken vt refresh, revive

errang achieved: past of ERRINGEN

erraten* vt guess (correctly). **errät** pres

erratisch a erratic

errechnen vt calculate, figure out

erregbar a excitable, irritable; sensitive

erregen vt 1 arouse, cause. 2 excite, stimulate; agitate, irritate. 3 energize. **erregend** p.a: **e—es Mittel** stimulant

Erreger I m(-)1 cause; (Med.) pathogen, bacillus, virus. 2 exciter, stimulant, agitator. 3 energizer II (in compds oft) exciting: **-feld** n exciting field. **-frequenz** f exciting frequency. **-salz** n exciting salt (e.g., NH_4Cl in Leclanché cell). **-spannung** f excitation voltage. **-strom** m (Elec.) exciting current

erregt p.a (oft) active cf ERREGEN. **Erregung** f (-en) arousal; excitation, stimulation, agitation, irritation; energizing. **Erregungs·flüssigkeit** f exciting fluid

erreichbar a attainable. **erreichen** vt reach, attain, achieve

erreifen vi (s) mature, ripen. **Erreifung** f maturation, ripening

erretten vt rescue, save

errichten vt erect, put up; establish

erriet guessed: past of ERRATEN

erringen* vt win, gain; acquire; achieve

errungen achieved: pp of ERRINGEN. **Er-**

rungenschaft *f* (-en) achievement; improvement; acquisition

ersah saw, etc: *past of* ERSEHEN

ersann devised: *past of* ERSINNEN

Ersatz I *m* replacement, substitute; restitution, compensation; reinforcement(s); extra(s), spare(s). **II** (*in compds, oft*) synthetic, artificial: **-brennstoff** *m* synthetic (*or* substitute) fuel. **-elektron** *n* equivalent (*or* replacing) electron. **-fähigkeit** *f* replacement capability, equivalency. **-gewicht** *n* equivalent (weight). **-kaffee** *m* coffee substitute. **-leim** *m* size (*or* glue) substitute; synthetic glue. **-leistung** *f* restitution, replacement (of defective goods); compensation; damage(s). **-menge** *f* equivalent (amount). **-mittel** *n* substitute, surrogate, (artificial) *or* synthetic) replacement. **-stoff** *m* substitute, synthetic (material). **-stromkreis** *m* equivalent circuit. **-stück** *n*, **-teil** *n/m* replacement (part), spare part. **-ware** *f* substitute, synthetic (merchandise). **-wert** *m* replacement value

ersaufen* *vi* (s) drown; be (or get) flooded; (*Lime*) overslake

ersäufen* *vt* drown

erschaffen* *vt* create

erschallen *vi* (s) resound, ring out

erscheinen* *vi* (s) appear, show up. **Erscheinung** *f* **I** (-en) 1 phenomenon; appearance: **in E. treten** appear. 2 sight; apparition. 3 publication (of sthg written). **II** (*as sfx oft*) symptom *cf* ERMÜDUNGSERSCHEINUNG

Erscheinungs·form *f* form (in which sthg appears or occurs), phase, state, manifestation. **-jahr** *n*, (**-tag**) *m* year (date) of publication (*or* of issue)

erschien(en) appeared: *past* (& *pp*) *of* ERSCHEINEN

erschlaffen *vi*(s) slacken, flag; tire

erschließen* *vt* 1 open (up). 2 reveal. 3 develop (an area). 4 (**aus**) infer (from). **Erschließung** *f* (-en) opening; revelation; development

erschloß, erschlossen revealed: *past & pp of* ERSCHLIESSEN

erschmelzen* *vt* melt, smelt. **erschmilzt** *pres.*, **erschmolz(en)** *past* (& *pp*)

erschöpfen 1 *vt* exhaust. 2: **sich e.** be ex-

hausted. **erschöpfend** *p.a* exhaustive. **Erschöpfung** *f* exhaustion

erschrak was frightened: *past of* ERSCHRECKEN *vi*

erschrecken I *vt* frighten, scare. **II*** *vi* (s) (erschrak, erschrocken; erschrickt) be frightened, get scared. **erschreckend** *p.a* frightening, frightful

erschuf created: *past of* ERSCHAFFEN

erschüttern 1 *vt* shake (up), upset, disrupt, shock. 2 *vi* shake, vibrate; (*Mach.*) chatter, tremble. **Erschütterung** *f* (-en) upset(ting), shock; vibration; concussion; convulsion; (seismic) tremor; (*fig*) shake-up, disruption

erschütterungs·fest *a* shockproof, vibration-proof. **-frei** *a* vibration-free, concussion-free. **E—gebiet** *n* area of seismic disturbance, tremor area. **-ladung** *f* (*Explo.*) cracking charge. **-zünder** *m* concussion fuse

erschwang afforded: *past of* ERSCHWINGEN

erschweren *vt* 1 make heavy; weight, load (silk); reduce (dyes). 2 make difficult, aggravate, complicate. **Erschwerung** *f* (-en) 1 weighing, loading; reduction; 2 aggravation, complication, obstacle, impediment

erschwingen* *vt* afford (to pay for). **erschwinglich** *a* affordable. **erschwungen** afforded: *pp of* ERSCHWINGEN

ersehen* *vt* (**aus**) see, gather (from)— **hieraus ist zu e.** from this it is evident

ersetzbar *a* replaceable

ersetzen *vt* replace, make good; serve as; compensate, make up (for). **Ersetzung** *f* (-en) replacement, restitution; compensation *cf* ERSATZ

ersichtlich *a* evident, apparent; obvious, visible

ersieht sees, etc: *pres of* ERSEHEN

ersinnen* *vt* devise, contrive; conceive

ersoff(en) drowned: *past* (& *pp*) *of* ERSAUFEN

ersonnen devised: *pp of* ERSINNEN

erspann spun: *past of* ERSPINNEN

ersparen *vt* (*d.i.o* or *dat rflx*: **sich**) save, spare (sbdy *or* oneself sthg). **Ersparnis** *f/n* (-se) (*usu pl*) saving(s), economy. **ersparnis·halber** *adv* for the sake of economy

erspinnen* *vt* spin (fibers); **ersponnen** *pp*

ersprießlich *a* fruitful, profitable

erst 1 *a* first: **der (die, das) E—e** the first (one), the best (one); **in e—er Linie** first of all, primarily: **fürs e—e** for one thing *cf* ERSTERE, ERSTENS. **2** *adv* first(ly); only, not until: **e., wenn es dissoziiert** only once it (*or* not until it) dissociates; **e. recht** all the more. **3** (*in compds oft*) first, prime, primary, initial *cf* ERSTKLASSIG, ERSTSTROM ect

erstarren *vi*(s) solidify; congeal; freeze; set (*as* : cement); coagulate, become petrified; become torpid. **estarrt** *p.a* (*oft*) petrified, rigid. **Erstarrung** *f*(-en) **1** solidarity, rigidity; paralysis. **2** solidification, congelation, coagulation, freezing **Erstarrungs·gefüge** *n* solidification structure. **-gestein** *n* igneous rock. **-kurve** *f* freezing-point curve. **-punkt** *m* freezing point; setting (solidification, coagulation) point. **-wärme** *f* heat of fusion

erstatten *vt* **1** repay. **2** submit (reports, etc)

erstaunen 1 *vt* astonish. **2** *vi* (s) be astonished. **Erstaunen** *n* astonishment—**in E. setzen** astonish. **erstaunlich** *a* astonishing

Erst·ausführung *f* prototype. **-ausscheidung** *f* first separation; first substance to separate out. **e—best** *a* first (one) available

erstellen *vt* **1** set up; put up. **2** make available, put out, produce. **3** draw up. **4** plot (a curve)

erstens *adv* in the first place, first of all

erstere *a* first; former *cf* LETZTERE

erst·erwähnt, -genannt *a* first-mentioned

ersticken 1 *vt, vi* (s) stifle, suffocate; **e— der Kampfstoff** (*Mil.*) lung irritant. **2** *vt* (*oft*) suppress. **Erstickung** *f* (-en) suffocation, asphyxia(tion)

erst·klassig *a* first-class, first-rate

Erstling *m* (-e) first-born. **Erstlings·werk** *n* first work (publication, opus)

Erst·luft *f* primary air (in combustion). **e— malig** *a* first-time. **-mals** *adv* for the first time. **E—milch** *f* colostrum

Erstp. *abbv* (Erstarrungspunkt) freezing point

erstrahlen *vi*(s) shine (forth), gleam; sparkle

Erst·produkt *n* **1** first product. **2** (*Sugar*) first (crop of) crystals. **e—rangig** *a* first-rank; top-priority

erstreben *vt* strive for, aspire to. **erstrebenswert** *a* worthwhile, worth striving for

erstrecken *vr*: **sich e.** extend, stretch

Erst·schlag *m* (*Mil.*) first strike. **-strom** *m* (*Elec.*) primary current

ersuchen *vt* request; (+ **um**) implore (for)

ertappen *vt* catch (by surprise)

erteilen *vt* (*d.i.o*) give, grant; hand out; assign (to); bestow (on). **Erteilungs·akten** *fpl* patent records

ertönen *vi*(s) ring out, resound

Ertrag *m* (. . äge) yield, crop; income; profit. **ertragen*** *vt* bear, endure, tolerate. **erträglich** *a* tolerable; passable

ertrag·los *a* unprofitable, unproductive. **-reich** *a* high-yield, profitable

Ertrags·ermittlung *f* ascertainment of yield. **e—fähig** *a* productive. **E—fähigkeit** *f* productivity. **-feststellung** *f* determination of the yield. **-vermögen** *n* productivity, producing power

ertrank drowned: *past of* ERTRINKEN

ertränken *vt* drown

ertrinken* *vi* (s) drown

ertrüben *vi* (s) turn cloudy, become turbid

ertrug endured: *past of* ERTRAGEN

ertrunken drowned: *past of* ERTRINKEN

erübrigen *vt* **1** save, make superfluous; (have . . . to) spare. **2**: **sich e.** be(come) superfluous

Eruca·säure *f* erucic acid

eruieren *vt* find out; locate

Eruptiv·gang *m* eruptive vein (*or* lode); dike. **-gestein** *n* eruptive (*or* igneous) rock

erw. *abbv* **1** (erwärmt) warmed. **2** (erwähnt) mentioned

erwachen *vi*(s) awaken, wake up

erwachsen *p.a* grown-up, adult. **Erwachsenen·bildung** *f* adult education

erwägen* *vt* weigh, consider, ponder. **Erwägung** *f* (-en) deliberation, consideration: **in E. ziehen** take into consideration; **E—en anstellen** deliberate (v)

erwählen *vt* choose; elect

erwähnen *vt* mention

erwarb acquired: *past of* ERWERBEN

erwärmen *vt* warm; heat (moderately). **Erwärmungs-kraft** *f* heating (*or* calorific) power. **-verlust** *m* heating (*or* thermal) loss

erwarten *vt* await, expect. **Erwartung** *f* (-en) expectation, expectancy

erwartungs-gemäß *adv* as expected. **E— wert** *m* expectation value

erwecken *vt* awaken, wake up; arouse

erwehren *vr*: **sich e.** (*genit*) refrain (from), avoid; resist, fend off

erweichen *vt, vi* (s) soften; soak; plasticize. **erweichend** *p.a* (*oft*) emollient. **Erweicher** *m* softener

Erweichungs-mittel *n* softening agent; (*Pharm.*) emollient, demulcent. **-punkt** *m* softening point. **-vermögen** *n* softening power; (*Coal*) fusibility

erweisen* I *vt* 1 prove, demonstrate. 2 show, display (gratitude, etc). 3 do (favors, etc). II: **sich e.** (*usu* + **als**) prove, turn out (to be)—**sich dankbar e.** show one's gratitude

erweitern *vt* & **sich e.** widen, expand, extend; enlarge, distend, dilate. **erweiterungs-fähig** *a* expandable, extendable, having add-on possibilities

Erwerb *m* 1 pay, earnings. 2 source of income. 3 acquisition. **erwerben*** *vt* earn; acquire; purchase. **erwerbs-tätig** *a* gainfully employed

erwidern 1 *vt* return, acknowledge. 2 *vt/i* (**auf**) reply, respond (to). **Erwiderung** *f* (-en) reply

erwies(en) proved: *past* (& *pp*) *of* ERWEISEN

erwirbt acquires: *pres of* ERWERBEN

erwog(en) weighed: *past* (& *pp*) *of* ERWÄGEN

erworben acquired: *pp of* ERWERBEN

erwünscht *p.a* desired; welcome

Erz *n* (-e) I ore. II bronze (*or* similar alloy). III (*in compds*) 1 ore. 2 bronze. 3 (*as pfx*) arch-, extremely: **-abfälle** *mpl* (ore) tailings. **-ader** *f* ore (*or* metal) vein

erzählen *vt/i* tell (a story), relate, report. **Erzählung** *f* (-en) story; account; report

Erz-arbeit *f* (*Tin*) concentrate smelting. **-arbeiter** *m* bronze (*or* metal) worker. **e—arm** *a* poor in metal, lean. **E— aufbereitung** *f* ore dressing. **-bergbau** *m* (ore) mining. **-brecher** *m*, **-brechma-**

schine *f* ore crusher. **e—bringend** *a* mineralizing. **E—bringer** *m* (*Geol.*) mineralizing solution. **-brocken** *m* lump of ore. **e—dumm** *a* arch-stupid, extremely stupid. **E—eisen** *n* mine iron (obtained from ores without fluxes)

erzeugen *vt* produce; manufacture; generate; beget. **Erzeugende** *f* (*adj endgs*) (*Math.*) generatrix. **Erzeuger** *m* (-) producer; manufacturer; generator; begetter. **Erzeugnis** *n* (-se) product; produce. **Erzeugung** *f* (-en) production; manufacturer; generation; begetting

Erzeugungs-kosten *pl* production costs. **-ort** *m* place of production (*or* manufacture)

Erz- ore; metal, bronze: **-fall** *m* (*Min.*) ore shoot. **-farbe** *f* bronze color. **-flöz** *n* ore seam. **-frischverfahren** *n* (*Metll.*) ore process; direct process. **e—führend** *a* ore-bearing. **E—gang** *m* ore vein, lode. **-gebirge** *n* (*Geog.*) Ore Mountains (in Saxony) **-gestein** *n* ore-bearing rock; native ore. **-gewinnung** *f* ore mining. **-gicht** *f* charge of ore. **-gießer** *m* bronze (*or* brass) founder. **-grünfrischen** *n* (*Iron*) refining with ore. **-graupe** *f* coarse grain of ore. **-grube** *f* mine, pit. **e— haltend, -haltig** *a* ore-bearing. **E— hütte** *f* smelter

erziehen* *vt* 1 bring up (and educate), raise (children); (*oft* + **zu**) raise (to be); train, educate (for, to be); discipline. 2 cultivate, grow (plants). **Erzieher** *m* (-), **Erzieherin** *f* (-nen) educator; trainer; tutor. **erzieherisch, erziehlich** *a* educative, educational, pedagogical; *adv oft* by education, by proper training. **Erziehung** *f* (-en) upbringing (and early education); raising, training; discipline. 2 good breeding, manners. 3 cultivation, growing (of plants)

Erziehungs-anstalt *f* 1 educational institution. 2 reform school. **-fach** *n* education (as a profession). **e—fähig** *a* educable, teachable. **E—kunde** *f* (field of) education, pedagogy. **-lehre** *f* pedagogy, theory of education. **-rat** *m* educational council, board of education. **-wesen** *n* (field of) education, pedagogy. **-wissenschaft** *f* science of education

erzielbar *a* achievable; attainable, obtainable

erzielen *vt* 1 achieve; attain, reach; obtain; make (profit), score (success). 2 strive for, aim at

erzittern *vi*(s) shake, tremble; quake; vibrate

Erz·kies *m* ore-bearing pyrites. **-klauber** *m* ore picker. **-klein** *n* ore fines. **-körper** *m* ore body. **-kunde** *f* (extractive) metallurgy, mineralogy, ore science. **-kundige** *m, f, (adj. ends)* mineralogist. **-lager** *n* bed of ore, ore deposit. **-lagerstätte** *f* ore deposit. **-lagerstättenkunde** *f* mineralogy of ore beds. **-laugerei** *f* ore leaching (plant). **-laugung** *f* ore leaching. **-metalle** *npl* ore metals, heavy metals. **-mittel** *n* ore (as opposed to gang). **-mühle** *f* ore mill. **-muster** *n* ore sample. **-mutter** *f* matrix. **-nest** *n* ore pocket. **-niere** *f* kidney-shaped ore. **-ofen** *m* ore (smelting) furnace

erzog educated: *past of* ERZIEHEN. **erzogen** 1 *pp of* ERZIEHEN. 2 *p.a:* **gut** (**schlecht**) **e.** well- (ill-)bred

Erz·probe *f* ore sample; ore assay(ing). **-röster** *m* ore burner, calciner. **-scheidekunst** *f* art of sorting ores. **-scheider** *m* ore separator. **-schicht** *f* 1 ore layer. 2 (daily) ore output. **-schlamm** *m* ore slime, ore slurry. **-schlauch** *m* ore pipe, ore chimney. **-stahl** *m* ore (*or* mine) steel *cf* ERZEISEN. **-staub** *m* ore dust. **-stück** *n* lump of ore. **-tasche** *f* ore pocket. **-trennung** *f* ore separation. **-tugend** *f* cardinal virtue

erzürnen 1 *vt* anger, infuriate. 2 *vi* (s) & **sich e.** be angered, become furious. **erzürnt** *p.a* angry, furious

Erz·verhüttung *f* ore reduction. **-wäsche** *f* ore washing (plant) **-wespe** *f* (*Zool.*) chalcid(ian)

erzwang forced: *past of* ERZWINGEN

erzwingbar *a* (en)forceable, compellable; obtainable by force

erzwingen* *vt* force, compel; (+ **von**, *oft*) obtain (*or* win) (from) by force; wrest, wring (from); enforce. **Erzwingung** *f* forcing; obtaining by force, wresting; enforcement

erzwungen forced: *pp of* ERZWINGEN. **erz-**

wungenermaßen *adv* by force, under duress

Erz·zement *m/n* iron ore cement. **-zerkleinerung** *f* ore crushing. **-ziegel** *m* ore briquette

es 1 *pron* (*usu*) it; (*oft no transl*): **er meint es gut** he means well. 2 (*expletive*) there (*or no transl*): **es war niemand da** there was nobody there, nobody was there; **es gibt** there is, there are

Esbachs·reagens *n* Esbach reagent

Eschappé·öl *n* recovered (*or* reprocessed) oil

Esche *f* (-n) ash (tree)

Eschel *m* (-) zaffer; finest grade of smalt; (*Foun.*) black spot

eschen *a* ashen, (made of) ash (wood)

Eschen·ahorn *m* box elder. **-holz** *n* ashwood. **-wurz** *f* (*Bot.*) fraxinella, dittany

Eschweger Seife *f* Eschwege soap

Esdragon *m* tarragon. **-öl** *n* oil of tarragon (*or* estragon)

e.s.E., E.S.E. *abbv* (elektrostatische Einheit) electrostatic unit

Esel *m* (-) 1 ass, donkey. 2 fool, (stupid) ass. 3 (*Mach.*) ram, monkey; (*Brew.*) scooper; (saw)horse; easel, frame. **-arbeit** *f* drudgery. **Eselei** *f* (-en) asininity; foolish thing to do

Esere·samen *m* Calabar bean

esocyclisch *a* endocyclic

Eskadron *f* (-en) squadron

Esparsette *f* (-n) (*Bot.*) sainfoin

Espe *f* (-n) aspen, poplar. **espen** *a* (of) aspen

Espen·holzschliff *m* aspen woodpulp. **-stoff** *m* (*Paper*) aspen pulp

eßbar *a* edible; **etwas E—es** something to eat. **Eßbarkeit** *f* edibility

Eß·begier *f* craving for food, appetite

Esse *f* (-n) 1 forge, hearth. 2 chimney; stack

Eß·eisen *n* (*Metll.*) tuyère, twyer

essen* (aß, gegessen; ißt) *vt/i* eat: (**etwas**) **zu e. geben** give something to eat. **Essen** *n* (-) 1 food; meal; dinner, supper. 2 (*Geog.*) (*industrial Ruhr Valley city, cf* ESSE)

Essen· forge; chimney: **-gas** *n* chimney (*or* flue) gas. **-klappe** *f* chimney damper

essentiell *a* essential. **Essenz** *f* (-en) essence

Esser *m* (-) eater: **starker (schwacher) E.** big (poor) eater. **eß-gierig** *a* ravenous **Essig** *m* (-e) vinegar; (*in compds oft*) acetic, aceto-: **-aal** *m* vinegar eel. **e—ähnlich** *a* acetic, vinegar-like. **E—älchen** *n* vinegar eel. **-aldehyd** *m* acetaldehyde. **e—artig** *a* acetic, vinegar-like. **E—äther** *m* ethyl acetate. **-bakterie** *f* acetic acid bacterium, acetobacter. **-baum** *m* elm-leaved (*or* tanner's) sumac; vinegar plant. **e—bereitung** *f* vinegar making. **-bildner** *m* acetifier. **-bildung** *f* acetification. **-brauerei** *f* vinegar works. **-essenz** *f* essence of vinegar (80% acetic acid solution). **-ester** *m* acetic ester, ethyl acetate. **-fabrik** *f* vinegar works. **-gärung** *f* acetic fermentation. **-geist** *m* acetone. **-gurke** *f* pickle, pickled cucumber. **-gut** *n* alcoholic liquor for the quick vinegar process. **e—haltig** *a* acetic, containing vinegar. **E—honig** *m* oxymel, honey and vinegar. **-kastanie** *f* Spanish chestnut. **-konserve** *f* pickle, pickled food. **-messer** *m* acetometer, vinegar tester, oxymeter. **-mutter** *f* mother of vinegar. **-naphta** *f* ethyl acetate. **-pilz** *m* mother of vinegar, vinegar plant. **-prober, -prüfer** *m* vinegar tester, acetometer. **-rose** *f* red (*or* French) rose. **e—salpetersauer** *a* mixed acetate-nitrate of. **E—salz** *n* acetate. **e—salzsauer** *a* mixed acetate-chloride of **essig·sauer** *a* acetate of, acetic; sour, vinegary *cf* SAUER: **e—es Ammoniak** *n* ammonium acetate; **e—e Eisenflüssigkeit** (*Pharm.*) solution of ferric acetate, liquor ferri acetati; **e—es Eisen(oxid)** ferric acetate; **e—es Eisenoxydul** ferrous acetate; **e—es Salz** acetate; **e—e Tonerde** aluminum acetate, (*Pharm.*) Burow's solution

Essigsäure° *f* acetic acid. **-amid** *n* acetamide. **-anhydrid** *n* acetic anhydride. **-(äthyl)äther, äthylester** *m* ethyl acetate. **-gärung** *f* acetic fermentation. **e—haltig** *a* containing acetic acid. **E—messer** *m* acetometer, vinegar tester. **-messung** *f* acetometry. **-rest** *m* acetyl group. **-salz** *n* acetate

Essig·schaum *m* flower of vinegar. **e—schwefelsauer** *a* mixed acetate-sulfate of. **E—siederei** *f* vinegar works. **-sirup**

m oxymel, honey and vinegar. **-sprit** *m* essence of vinegar, triple vinegar (10–12% acetic acid). **-ständer** *m* vinegar tun; graduator (in quick vinegar process). **-stich** *m* vinegar plant. **-stube** *f* vinegar room (warm room for quick vinegar process). **-zucker** *m* (*Pharm.*) oxysaccharum

Eß-kohle *f* forge coal, low-volatile bituminous (12–19% volatiles). **-löffel** *m* tablespoon. **-lust** *f* appetite, delight in eating. **eß-reif** *a* ripe enough to eat, mature. **Eß-stunde** *f* mealtime, dinner (*or* supper) hour. **-waren** *fpl* food (products), victuals

Ester *m* (-) ester; (*specif*) ethyl ester. **e—-ähnlich, -artig** *a* ester-like, in the form of an ester. **E—bildung** *f* ester formation, esterification. **-gummi** *n* ester gum. **-harz** *n* resin (*or* rosin) ester

esterifizieren *vt* esterify. **Esterifizierung** *f* (-en) esterification

Ester-lack *m* ester varnish. **-öl** *n* ester oil. **-pyrolysen** *fpl* pyrolyses of esters. **-säure** *f* ester acid. **-verseifung** *f* saponification of an ester. **-zahl** *f* ester number

Estragon *m* tarragon. **-essig** *m* tarragon vinegar. **-öl** *n* tarragon oil

Estrich *m* (-e) plaster (*or* composition) floor, plastered stone floor. **-gips** *m* estrich flooring plaster (anhydrous $CaSO_4$ similar to Keene's cement). **-stein** *m* estrich stone (paving brick)

Estrifikation *f* (-en) esterification. **estrifizieren** *vt* esterify

etablieren 1 *vt* establish. **2: sich e.** become established, establish oneself; settle. **Etablissement** *n* (s) establishment, (place of) business

Etage *f* (-n) floor, story (of a building); tier, deck; (*Geol.*) stage

Etagen·haus *n* apartment house. **-heizung** *f* single-story heating (unit). **-ofen** *m* multiple-deck oven, shelved oven *or* kiln). **-presse** *f* multilayer press. **-sieb** *n* multi-deck sieve. **-ventil** *n* multiple valve. **-wohnung** *f* whole-floor apartment

Etagere *f* (-n) stand (with shelves), set of shelves; shelf

Etalon·apparat *m* standard(ized) apparatus

Etappe f (-n) **1** stage, phase. **2** (rest) stop, stopping place. **3** (*Mil.*) rear area (behind the lines), communications zone. **4** day's march

etappen·weise adv by stages. **E—wesen** n system of communications

Etat m (-s) budget; military strength. **etatisieren** vt budget

Etat· (*see also* **Etats·**): **-jahr** n budgetary (*or* fiscal) year. **e—mäßig** a, adv **1** according to the budget, planned. **2** permanent, permanently employed

Etats·jahr n budgetary (*or* fiscal) year

Ethik f ethics. **Ethiker** m (-) ethical philosopher; ethicist. **ethisch** a ethical

ethnisch a ethnic

Etikett n (-e) label, tag. **Etikette** f **1** etiquette. **2** label, tag. **etikettieren** vt label, tag.

etliche I a & pron **1** pl some, a few, several; quite a few, a whole lot (of). **2** sg some (amount of), a little; quite a bit (of), quite a lot (of)

Etui n (-s) case (for cigarettes, glasses)

etw. abbv (etwas) sthg. (something)

etwa adv **1** about, approximately—**in e.** to some extent. **2** for example. **3** sometimes. **4** possibly, by any chance: **nicht e.** not by any chance; **nicht e., daß** (or **als ob**) . . . not that . . .

etwaig a any possible. **etwaigenfalls** adv possibly

etwas I indef pron **1** anything, something: **ohne e. zu sagen** without saying anything; **e. Neues** something new. **2** some, a bit: **e. von der Säure** some of the acid. **3**: **so e.** such a thing, anything like that; (+ **von**, *oft*) such. **II** a (*no endgs*) **4** some, a little: **mit e. Geduld** with a little patience. **IV** adv **5** somewhat, rather, a bit: **e. besser** a bit better. **Etwas** n **1** (e.g. **ein gewisses**) **E.** a certain something. **2** (little thing, trifle, insignificant thing

etwelche a, pron some, a few, several = ETLICHE

E.T.Z. abbv (Elektrotechnische Zeitschrift) Elec. Engg. Jnl.

Eukalyptus·öl n eucalyptus oil

Euchinin n euquinine

Euchlor, Euchlorin n euchlorine

Eudiometrie f (-en) eudiometry. **eu-**

diometrisch a eudiometric

Eugen·glanz m (*Min.*) polybasite

Euklas m (*Min.*) euclase

euklidisch a Euclidian

Eule f (-n) owl

Europa n Europe. **europäisch** a European

eustasisch a Eustachian

Eutektikum n (. . ika), **eutektisch** a eutectic. **eutektoidisch** a eutectoid

Euter n (-) udder

ev. abbv (eventuell) possibly, etc. **eV** abbv (Elektronenvolt) electron volt(s). **e.V.** abbv (eingetragener Verein) membership corporation

evakuieren vt evacuate. **Evakuierungs·kessel** m vacuum boiler (*or* pan)

evaporieren vt & vi (s) evaporate

event. abbv (eventuell) possibly

eventuell I a **1** possible, any . . . that may arise. **II** adv **2** possibly, perhaps; (*oft equiv to aux v* may): **es evaporiert e.** it may (possibly) evaporate. **3** if necessary. **4** in that case

Evernin·säure f everninic acid. **Evern·säure** f evernic acid

evomieren vt vomit

evtl. abbv (EVENTUELL) possibly, etc

EWG abbv (Europäische Wirtschaftsgemeinschaft) European Economoic Union (Common Market)

ewig a **1** eternal, everlasting, perpetual. **2** constant, endless. **3** adv oft in (*or* for) ages, forever. **Ewigkeit** f (-en) eternity, age(s); perpetuity

Exaktheit f exactness, precision, accuracy

Examen n (-) examination

exc . . . : *see also* **exz . . .**

excentrisch a eccentric cf EXZENTER etc

Exempel n (-) example: **ein E. statuieren** (**an**) make an example (of); **die Probe aufs E. machen** put it to the test

Exemplar n (-e) copy: sample; specimen

Exhaustor m (-en) exhauster, exhaust fan

Existenz f (-en) **1** existence, life. **2** livelihood, income. **e—·bedrohend** a life-threatening. **-fähig** a viable, capable of existing. **existieren** vi exist

exkl. abbv (exklusiv) exclusive(ly). **exklusive** adv excluding, exclusive of

Exkurs m (-e) digression

exogen a exogenous

exolieren vt anodize
exotherm(isch) a exothermic
expedieren vt expedite, rush; send (off), mail, ship. **Expedition** f (-en) 1 expedition. 2 shipping (or mailing) department
Experimentator m (-en) experimenter. **experimentell** a experimental. **Experimentier-anlage** f experimental equipment (or laboratory). **experimentieren** vi experiment. **experimentier-freudig** a eager to experiment. **Experimentierung** f experimentation
Experte m (-n, -n), **Expertin** f (-nen) expert. **Expertise** f 1 expertise. 2 expert's report
explizieren vt explain
explizit a explicit. **explizite** adv explicitly
explodierbar a explodable, explosive
explodieren vi (s) explode, blow up. **explodierend** p.a (oft) explosive. **explosibel** a explosive, explodable. **Explosivität** f explodability
explosions-artig a explosive. **E—druck** m explosion pressure. **e—fähig** a explosive. **E—fähigkeit** f explosiveness. **-gefahr** f danger of explosion. **-gemisch** n explosive mixture. **-grenze** f explosive limit. **-kraft** f explosive power (or force). **-motor** m internal combustion engine. **-raum** m explosion chamber. **e—sicher** a explosionproof. **E—stoß** m explosive impact, explosion impulse. **-vorgang** m explosive process. **-wärme** f heat of explosion
Explosiv-geschoß n explosive projectile. **-maschine** f internal combustion engine.

-stoff m explosive. **-stoffchemie** f chemistry of explosives
Exponat n (-e) exhibit
exponieren vt & **sich e.** (dat) expose (oneself) (to). **Exposition** f (-en) exposition; (Phot.) exposure
Exsikkator m (-en) exsiccator, desiccator
Exsudat n (-e) exudate
Extr. abbv (Extrakt) extract
extra 1 a (no endgs) extra, special. 2 adv extra, specially, separately; all the more; on purpose. 3 (in compds oft) extra, special, super: **E—-benzin** n supergasoline, (specif) high-octane gasoline
extrahierbar a extractable. **extrahieren** vt extract. **Extrahierung** f (-en) extraction
Extrakt-brühe f extract liquor
Extrakteur m (-e) extractor
Extraktions-hülse f extraction shell (or thimble)
Extrakt(iv)-stoff m extractive (substance or matter)
Extrakt-wolle f extract(ed) wool
extrapolatorisch a extrapolational; adv by extrapolation. **extrapolieren** vt extrapolate
extrem a, **Extrem** n (-e) extreme. **extremistisch** a extremist. **extrem-kurz** a ultrashort
exz . . . : see also **exc . . .**
Exzenter m (-) eccentric. **Exzentrizität** f (-en) excentricity. **exzentrisch** a eccentric
EZ. abbv (Esterzahl) ester no.

F

F. *abbv* **1** (fast) almost. **2** (fein) fine. **3** (fest) solid. **4** (folgend) fol., following. **5** (für) for

F. *abbv* **1** Fahrenheit. **2** (feinste Sorte) finest grade. **3** (Fusionspunkt) m.p., melting point

Fa *abbv* (Feines in der Aufgabe) fines in raw coal for redusting

f.a.B. *abbv* (frei an Bord) f.o.b., free on board

Fabel *f* (-n) fable; fiction; plot (of a story). **fabelhaft** *a* fabulous. **Fabel·lehre** *f* mythology

Fabrik *f* (-en) factory, works, (work)shop, mill. **-anlage** *f* manufacturing plant

Fabrikant *m* (-en, -en) manufacturer, maker

Fabrik·arbeit *f* factory work. **-arbeiter** *m*, **-arbeiterin** *f* factory worker

Fabrikat *n* (-e) **1** manufactured product (*or* article). **2** make, brand. **3** (*Tex.*) fabric

Fabrikation *f* (-en) manufacture, manufacturing, production; processing

fabrikations·echt *a* fast to processing. **F— fehler** *m* factory flaw, production defect. **-gang** *m* processing, manufacturing process. **-wasser** *n* water for manufacturing use, industrial water

fabrikatorisch *a* manufacturing, industrial

Fabrik· factory, manufacture(d): **-betrieb** *m* factory operation. **-erzeugnis** *n* manufactured product. **f—fertig** *a* **1** factory-built. **2** prefabricated. **F—gold** *n* strong gold leaf. **-leiter** *m* factory (*or* plant) manager. **-marke** *f* trade mark. **f— mäßig** *a* industrial, factory- (*or* machine-)made; *adv oft* by factory methods, by machine. **-neu** *a* brand-new. **F—öl** *n* oil for manufacturing use, factory-grade oil. **-schiff** *n* factory ship, (*specif*) fish-processing ship. **-ware** *f* manufactured (*or* factory-made) article(s). **-wäsche** *f* factory scouring (of wool). **-wesen** *n* fac-

tory system; manufacturing industry. **-zeichen** *n* trade mark

fabrizieren *vt* make, manufacture, produce

fäcal *a* fecal. **Fäcalien** *npl*, **Fäces** *pl* feces

Facette *f* (-) facet, edge. **facettiert** *a* faceted

Fach I *n* (-̈er) **1** compartment, pigeonhole *cf* DACH; chamber; cell; shelf. **2** subject (of study); field (of specialization), line (of work); department. **II** (*in compds*) expert, professional, specialized. **III** *sfx*: **·fach** -fold *cf* DREIFACH, EINFACH, ZWEIFACH

Fach·arbeiter *m* specialist, expert; skilled worker. **f—artig** *a* cellular. **F—arzt** *m* (*Med.*) specialist. **-ausbildung** *f* technical (specialized, professional) education. **-ausdruck** *m* technical term. **-bildung** *f* = -AUSBILDUNG. **-blatt** *n* trade (technical, professional) journal

fächeln 1 *vt* fan. **2** *vi* blow, waft

Fächer I *m* (-) (non-rotating) fan; flapper. **II** *pl of* FACH

Fächer·brenner *m* fantail burner. **f— förmig** *a* fan-shaped. **F—gerste** *f* bearded barley

fächerig *a* **1** fan-shaped. **2** divided into compartments; locular

Fächer·stein *m* (*Min.*) ripidolite

Fach·gebiet *n* field of work, (professional) specialty. **f—gemäß** *a* professional, workmanlike. **F—genosse** *m*, **-genossin** *f* professional colleague. **f—gerecht** *a* professional, workmanlike. **-hochschule** *f* technical institute, professional school. **-kenner** *m* expert, specialist, professional. **-kenntnis** *f* technical (*or* specialized) knowledge. **-kollege** *m* professional colleague. **-kraft** *f* skilled worker, specialist. **f—kundig** *a* professionally competent, expert; experienced. **-kundlich** *a* technical. **F—leute** *pl of* -MANN, -MÄNNIN experts, professionals

fachlich *a* professional, technical

Fach·literatur *f* technical (*or* trade) literature. **-mann** *m*, **-männin** *f (pl* -LEUTE) expert, specialist, professional. **f—männisch** *a* workmanlike, expert, professional; technical, specialized. **F—presse** *f* technical press. **f—mäßig** *a* professional. **F—ordnung** *f* classification. **-schrift(en)tum** *n* technical (*or* trade) literature. **-schule** *f* technical (*or* specialized) school. **-simpelei** *f* shop talk. **-sprache** *f* technical language (*or* jargon). **f—sprachlich** *a* technical; *esp adv* in technical language. **-technisch** *a* technical, specialized. **F—werk** *n* 1 framework, lattice(-work); half-timbering. 2 checkerwork. 3 truss, strut. **-wissenschaft** *f* specialized branch of science, specialty. **-wort** *n* technical word (*or* term). **-wörterbuch** *n* technical (*or* specialized) dictionary. **-zeitschrift** *f* trade (technical, specialized) journal

Facit *n* (-e) sum total, result

Fackel *f* (-n) torch. **-baum** *m* 1 marsh elder. 2 = **-föhre** *f* Scotch pine. (*Pinus silvestris*). **-glanz** *m* perfect clarity (of wine). **-kohle** *f* cannel coal. **-palme** *f* sago palm

Façon *f* (-s) fashion, manner = FASSON

Factis *n* Factice (*TN*), vulcanized oil

Fädchen *n* (-) filament, fiber, fine thread

fade *a* insipid, flat; boring; stale

Faden *m* **I** (⁝) 1 thread; string; filament, fiber; strand; suture: **Fäden ziehen** get stringy. 2 (*Wood*) grain. 3 *pl oft* ties, links. **II** (-) (*Meas.*) fathom. **f—artig** *a* threadlike, filamentous. **-dünn** *a* threadlike, thin as a thread. **F—elektrometer** *n* string electrometer. **-element** *n* fiber, filament. **F—förmig** *a* threadlike, filiform. **F—gerüsttheorie** *f* filar theory of protoplasmic structure. **-glas** *n* spun (fiber, filigree) glass. **-holz** *n* cordwood. **-kreuz** *n* (*Opt.*) cross hairs, reticule. **-länge** *f* (*Dye.*) length of fiber. **-mikrometer** *n* wire (*or* filar) micrometer. **-molekül** *n* filamentary molecule, molecular chain, linear macromolecule. **-netz** *n* (*Opt.*) cross-hairs, reticule. **-nudeln** *fpl* vermicelli. **-pilze** *mpl* filamentous fungi, Hyphomycetes. **f—scheinig** *a* threadbare. **F—spannung** *f* (*Elec.*) filament voltage. **-strich** *m* (*Opt.*)

cross-hair. **-strom** *m* (*Elec.*) filament current. **-wurm** *m* threadworm, nematode. **-zähler** *m* thread counter. **f—ziehend** *a* ropy, stringy

fädig *a* thready, stringy, filamentous

fähig **I** *a* 1 (*oft* + **zu**) capable (of), able (to). 2 talented. 3 qualified, fit. **II** *sfx* **-able**, **-ible**, **-ive**, capable (of) *cf* ANPASSUNGSFÄHIG, ERTRAGSFÄHIG. **Fähigkeit** *f* (-en) 1 ability, capability, capacity; qualification, fitness. 2 (*as sfx oft*) **-ability**, **-ivity** *cf* LEITFÄHIGKEIT

fahl **I** *a* 1 pale; wan; faded. 2 (*Color*) fallow (pale yellow); fawn (yellowish brown); dun (grayish brown); earth-colored; (*in compds*) pale, **-ish** *cf* FAHLBLAU

Fahl·band *n* (*Min.*) fahlband, band of metallic sulfides in rock. **f—blau** *a* pale (*or* grayish) blue, bluish. **F—erz** *n* (*genl*) fahlore; (*specif*): **dunkles F.** tetrahedrite, **lichtes F.** tennantite. **-leder** *n* russet leather (for uppers). **f—rot** *a* fawn, pale yellowish brown. **F—stein** *m* pale gray slate

fahnden *vi* (**nach**) search (for)

Fahne *f* (-n) 1 flag, banner, colors. 2 (*Chem.*) policeman (for transferring precipitates). 3 vane. 4 trail (of smoke). 5 (*TV*) streak, flare. 6 (*Print.*) galley proof. 7 bad breath

Fahnen·schraube *f* wing nut, wing screw

Fahr·bahn *f* roadway; channel, lane

fahrbar *a* 1 mobile, traveling; portable. 2 navigable, passable

Fahr·benzin *n* (standard) gasoline (for cars). **-damm** *m* roadway; (railroad) embankment. **-dynamik** *f* (*Cars*) driver dynamics

Fähre *f* (-n) ferry

fahren* (fuhr, gefahren; fährt) **I** *vt* 1 drive, haul, take (in a vehicle). 2 drive (a vehicle). 3 run, operate (machines). **II** *vi* (s) 4 (*of persons*) go, travel, ride (in a vehicle); drive (a car). 5 (*of vehicles, machines*) run. 6 start (up); clash, dart, flash. 7 (+ **in**, **aus**, *oft*) slip (into, out of). 8: **mit der Hand über etwas f.** run one's hand over sthg. **fahrend** *p.a* traveling, mobile; itinerant. **fahren·lassen*** *vt* drop, give up, abandon, neglect. **Fahrer** *m* (-) driver

Fahr·gast *m* passenger. **-geld** *n* fare.

-karte *f* (transportation) ticket. **f—läs-sig** *a* negligent. **F—plan** *m* timetable, schedule. **-rad** *n* bicycle. **-schein** *m* ticket. **-stuhl** *m* 1 elevator, hoist, lift. 2 wheelchair

Fahrt *f* (-en) 1 trip, journey, ride, drive. 2 travel, (forward) motion: **während der F.** while in motion. 3 way, distance (to go); "go" signal—**freie F.** clear road ahead, green light. 4 speed, velocity. 5: **in F.** in stride; in motion

fährt drives, etc: *pres of* FAHREN

Fährte *f* (-n) trail, track

Fahr·wasser *n* channel, groove. **-weg** *m* 1 way, distance (to go). 2 roadway. **-zeug** *n* vehicle

fäkal *a* fecal. **Fäkalien** *npl* feces. **Fäkal·stoff** *m* fecal substance

Faktis *n* Factice (*TN*), vulcanized oil

faktisch *a* actual, de facto; *adv oft* in reality

faktisieren *vt* sulfurize (oils)

Faktor *m* (-en) 1 (*incl Math.*) factor. 2 manager, agent

Faktoren·tabelle *f* table of factors. **-zerlegung** *f* factoring

Faktor·öl *n* sulfurized oil

Faktur(a) *f* (. . ren) invoice

Fäkulenz *f* feculence, sediment, dregs

Fakultät *f* (-en) 1 (academic) school, department; faculty. 2 (*Math.*) factorial. **fakultativ** *a* elective, optional

falb *a* 1 dun-colored. 2 pale, faded

Falke *m* (-n, -n) falcon

Fall *m* (⸚e) 1 fall, drop; collapse, downfall: **zu F. bringen** cause the downfall of. 2 case, instance: **auf alle F—e** in any case. 3 (*as sfx*) **·fall** case: **im Notfall** in case of emergency

Fäll·apparat *m* precipitation apparatus. **-bad** *n* precipitating (*or* settling) bath

fällbar *a* 1 precipitable. 2 fit to fell. **Fäll·barkeit** *f* 1 precipitability. 2 fitness for felling

Fall·beschleunigung *f* gravitational acceleration

Fäll·bottich *m* precipitating vat

Falle *f* (-n) 1 trap; bolt; valve. 2 *dat of* FALL

fallen* (fiel, gefallen; fällt) *vi* (s) 1 fall, drop; (*Chem.*) separate (out), be deposited. 2 drop off; (*Min.*) dip. 3 drop out, be dropped. 4 (**auf**) strike; fall (to one's lot);

be equivalent to: **wieviel Mark f. auf den Dollar?** how many marks are equivalent to a dollar? 5 (**an** + *psn*) be inherited (by). 6 (**unter**) come, be included (under). 7 (**in,** *oft*) lapse (into); (**in** + *color*) verge (on). 8 (*Words,* etc) be spoken, be mentioned. 9: **es fällt ihnen leicht (schwer)** it is easy (difficult) for them. *See also* GEFALLEN

fällen *vt* 1 precipitate. 2 fell, cut down. 3 pronounce (judgment, decision). 4 erect (a perpendicular)

fallen·lassen* *vt* drop, discard

Fäller *m* (-) precipitator

Fallfilm·verdampfer *m* falling film evaporator

Fäll·flüssigkeit *f* precipitating liquid

Fall·gesetz *n* law of gravity. **-hammer** *m* drop hammer; (*Explo.*) falling weight. **-hammerprobe** *f* (*Explo.*) falling-weight test. **-härte** *f* impact ball hardness. **-höhe** *f* 1 height of fall. 2 depth of sedimentation

fallieren *vi* fail, go bankrupt

fällig *a* due, payable; mature. **Fälligkeit** *f* (date of) maturity, due date

Fäll·kessel *m* precipitating vessel; (*Cellulose*) hydrolizer

Fall·kraut *n* (*Bot.*) arnica. **-maschine** *f* (*Phys.*) Atwood's machine

Fäll·methode *f* precipitation method. **-mittel** *n* precipitating agent, precipitant

Fall·pendel *n* (*Explo.*) friction pendulum. **-probe** *f* drop test; (*Coke*) shatter test

Fäll·produkt *n* precipitate

Fall·rohr *n* down pipe, waste pipe; downcomer

falls *subord cnjc* in case

Fall·schirm *m* parachute. **-silber** *n* deposited silver. **-strom** *m* downdraft. **-sucht** *f* epilepsy

fällt *pres of* FALLEN (falls) & FÄLLEN (precipitates)

Fall·tür *f* trap door

Fällung *f* (-en) precipitation; felling *cf* FÄLLEN

Fällungs·becherglas *n* precipitation beaker. **-flüssigkeit** *f* precipitating liquid. **-lösung** *f* mother liquor. **-mittel** *n* precipitating agent, precipitant. **-wärme** *f* heat of precipitation

Fall·weg *m* downward path. **f—weise** *a* occasional. **-werk** *n* drop machine; pile driver; stamp(ing machine). **-winkel** *m* angle of fall (*or* of inclination)

falsch *a* **1** false; counterfeit, artificial, pseudo-. **2** two-faced. **3** wrong, mistaken. **4** *adv oft* the wrong way; *esp* + *v*, *oft* mis-: **f. auslegen** misinterpret *cf* AUSLEGEN 4. **·-blau** *a* navy blue

fälschen *vt/i* falsify, counterfeit, forge; adulterate. **Fälscher** *m* (-) falsifier, counterfeiter, etc

Falschheit *f* (-en) falseness, duplicity; falsehood; deception

fälschlich *a*, *esp adv* false(ly), wrong(ly), etc = FALSCH. **fälschlicherweise** *adv* falsely; wrongly

Falsch·luft *f* excess (*or* infiltrated) air

Fälschung *f* (-en) falsification; counterfeit(ing), forgery; adulteration *cf* FÄLSCHEN

Fälschungs·mittel *n*, **-stoff** *m* adulterant

Falte *f* (-n) fold, crease; pleat; wrinkle, inner recess; (*Geol. oft*) flexure—**F—n werfen** pucker. **falten** *vt* fold, crease; pleat; wrinkle

Falten·blatt *n* folder, leaflet. **-filter** *n* folded (plaited, fluted) filter. **f—frei, -los** *a* creaseless, wrinkle-free. **F—punkt** *m* point of sharp change of direction, inflection point (of a curve). **-rohr** *n* flexible tube (*or* pipe). **-werfen** *n* puckering, creasing, crimping

Falter *m* (-) folder, creaser. **2** (*Zool.*) lepidoptera, butterfly, moth, **3** third stomach of ruminants

faltig *a* full of folds, creased, wrinkled, puckered; *adv oft* in folds (creases, wrinkles)

·faltig, ·fältig *sfx* ·fold, ·ple, -ply: **vierfältig** fourfold, quadruple, four-ply

Falt·probe *f* folding test. **-punkt** *m* plait point (of a phase diagram)

Faltung *f* (-en) **1** folding, creasing, etc *cf* FALTEN. **2** (*Geol.*) fold, plication. **3** (*Math.*) convolution. **Faltungs·punkt** *m* plait point

Falt·versuch *m* bending (*or* folding) test

Falz *m* (-e) **1** fold, crease. **2** joint, seam. **3** groove, flute, rabbet. **falzen** *vt* **1** fold,

crease. **2** seam. **3** groove, flute, rabbet. **4** (*Lthr.*) pave, shave

Falz·festigkeit *f* folding endurance; crease resistance. **-span** *m* shaving, skiving. **-ziegel** *m* interlocking tile

Fam. *abbv* = **Familie** *f* (-n) family

fand found: *past of* FINDEN

Fang *m* (⁻e) **1** catching, capture. **2** catch; yield; booty. **3** trap. **4** coup de grace. **5** *pl* claws, talons; clutches; fangs. **fangen*** (fing, gefangen; fängt) *vt* catch, capture; soften (hides) *cf* GEFANGEN. **Fänger** *m* (-) catcher; trap; (*Mach.*) catch, check; absorber

Fang·gitter *n* (*Elec.*) suppressor grid. **-mittel** *n* (*Elec.*) getter(ing agent). **-stoff** *m* **1** pulp, material from the save-all. **2** (*Elec.*) getter

fängt catches, etc: *pres of* FANGEN

Fantasie· (*Com.*) fancy: **-artikel** *mpl* fancy goods

Faraday·käfig *m* Faraday cage. **faradisch** *a* faradic

Farb· color, dye: **-abänderung** *f* change in color. **-abstreichrakel** *m/f* color doctor. **-abstufung** *f* color gradation.· **anstrich** *m* coat of color (*or* paint). **-auszugnegativ** *n* (*Phot.*) color-separation negative. **-bad** *n* dye bath. **-band** *n* typewriter ribbon

färbbar *a* colorable, stainable, readily dyed. **Färbbarkeit** *f* colorability, stainability

Farb· color, dye: **-becher** *m* dye (testing) beaker. **f—beständig** *a* color-fast, fadeproof. **F—bier** *n* dark beer for coloring. **-brühe** *f* dye liquor. **-deckschicht** *f* color coat (of paint). **-dia(positiv)** *n* color transparency (*or* slide). **-druck** *m* color print(ing)

Farbe *f* (-en) **1** color, pigment; dye, stain; paint. **2** (*Lthr.*) weak ooze (container). **3** (*Lead*) skimmings

Färbe· color(ing), stain(ing), dye(ing): **-artikel** *m* (*Tex.*) dyed style. **-bad** *n* dye bath. **-becher** *m* dye beaker (*or* pot). **-bier** *n* dark beer for coloring. **-bottich** *m* dye vat (*or* beck). **-brühe** *f* dyeing liquor. **-buch** *n* dye book

farb·echt *a* color-fast, color-proof; (*Phot.*)

orthochromatic. **F—echtheit** f color fastness, color stability

Färbe· color(ing), dye(ing): **-eigenschaft** f coloring (or dyeing) property. **f—fähig** a dyeable, stainable, colorable. **F—faß** n dye(ing) beck (vat, tub). **-flechte** f archil, dyer's mass. **-flotte** f dye liquor (or bath). **-flüssigkeit** f dyeing liquid, staining fluid. **-ginster** m (Bot.) dyer's broom. **-grad** m degree of coloring. **-gut** n goods to be dyed

farbe·haltend a color-fast, holding (its) color

Färbe· color(ing), dye(ing): **-holz** n dyewood. **-kessel** m dyeing kettle. **-kraft** f dyeing (or coloring) power. **-kraut** n dyer's weed—**gelbes F.** yellowweed. **-kufe**, **-küpe** f dye(ing) vat. **-lack** m lac dye

farbelos a colorless; achromatic

Färbe·malz n amber malt cf FARBMALZ. **-mittel** n coloring agent, pigment, dye

·farben sfx -colored cf BLEIFARBEN,etc

färben 1 vt color, dye, stain cf WOLLE. 2: **sich f.** be(come) colored, turn

Farben· color(s), dye, paint: **-abbeizmittel** n paint remover. **-abstreichmesser** n color doctor, knife for removing excess dye. **-abweichung** f chromatic aberration. **-änderung** f color change. **-auftrag** m coat of color. **-band** n spectrum. **-bild** n colored image, spectrum. **-bildung** f color formation, coloration. **f—blind** a colorblind. **F—bogen** m iris. **-brechung** f refraction of colors; color blending. **-brühe** f dye(ing) liquor. **-buchdruck** m color printing. **-chemie** f color (paint, dye) chemistry. **f—chemisch** a colorchemical, relating to dye (or paint) chemistry. **F—distel** f (Bot.) safflower. **-druck** m color print(ing). **f—empfindlich** a color-sensitive. **F—empfindung** f sensation (or perception) of color. **-erde** f colored earth. **-erscheinung** f 1 color phenomenon. 2 appearance of color. **f—erzeugend** a color-producing, chromogenic. **F—erzeuger** m chromogen. **-erzeugung** f production of colors. **-fabrik** f dye (or paint) factory. **-fabrikant** m dye (or paint) manufacturer. **-fehler** m color defect;

chromatic aberration. **-filter** n color filter (or screen). **f—froh** a colorful. **F—gebung** f coloring, coloration; color scheme. **-glanz** m brilliance of color. **-glas** n colored (or stained) glass. **-gleichheit** f equality of color. **-industrie** f color (dye, paint) industry. **-karte** f color chart. **-körper** m coloring body; coloring matter. **-körperchen** n pigment granule. **-lack** m lake. **-lehre** f science of color, chromatics. **-leiter** f color scale. **f—los** a colorless. **F—lösungsmittel** n color (or dye) solvent. **-malz** n amber malt cf FARBMALZ. **-maßstab**, - **meßapparat** m colorimeter. **-meßkunst**, **-messung** f chromatometry, colorimetry. **-mischung** f 1 color mixing (or blending). 2 mixture of colors. **-mühle** f color mill. **-ofen** m enameling furnace. **-probe** f color (or dye) test, dye trial. **-rand** m iris. **-reiber** m color grinder. **f—reich** a colorful, richly colored. **F—reihe** f color range. **f—rein** a 1 colorless. 2 pure in color. **-richtig** a orthochromatic. **F—ringe** mpl (Newton's) interference rings. **-saum** m color(ed) fringe. **-schiller** m play of colors, iridescence, schiller. **f—schillernd** a iridescent, chatoyant. **F—schimmer** m iridescence, shimmering color. **-schmelzofen** m enameling furnace. **-schwund** m color fading. **-sehen** n color vision. **-sinn** m color sense. **-skala** f color scale (or chart). **-spiel** n play of colors, iridescence. **-spritzpistole** f paint spray gun. **-steindruck** m chromolithography. **-stift** m colored pencil (or crayon). **-stoff** m dye, etc = FARBSTOFF. **-strahl** m colored ray. **-stufe** f color gradation, tint. **-stufenmesser** m tintometer. **-tafel** f color table (or chart). **-ton** m color tone, hue. **-übergang**, **-umschlag** m color change. **f—verändernd** a color-altering. **F—veränderung** f change in color; discoloration, **-verdünner** m color thinner. **-waren** fpl coloring materials = FARBWAREN. **-wechsel** m play (or change) of colors. **-wurzel** f madder. **-zerstreuung** f chromatic dispersion (or abberation)

Färbe·prozeß m dyeing (process), coloring process

Färber *m* (-) dyer. **-baum** *m* sumac (tree); Venetian sumac. **-blume** *f* woodwaxen

Farb·erde *f* (*Ceram.*) colored coating clay

Färber·distel *f* (*Bot.*) safflower; sawwort

Färberei *f* (-en) **1** dyeing. **2** dyer's trade. **3** dyer's, dyehouse

Färber·eiche *f* dyer's oak, quercitron

färberei·chemisch *a* dye-chemical

Färber·erde *f* Armenian bode. **-flechte** *f* archil, dyer's moss. **-flotte** *f* dye liquor (*or* bath). **-ginster** *m* dyer's broom. **-holz** *n* dyer's wood, dyewood

färberisch *a* (of) dyeing, tinctorial

Färber· dyer's: **-kraut** *n* dyer's weed. **-kamille** *f* yellow camomile. **-kreuzdorn** *m* (*Bot.*) dyer's buckthorn. **-lack** *m* lac dye. **-maulbeerbaum** *m* dyer's mulberry. **-meister** *m* master (*or* head) dyer. **-moos** *n* archil, dyer's moss. **-ochsenzunge** *f* dyer's alkanet. **-rinde** *f* quercitron bark. **-rot** *n* alizarin. **-röte** *f* madder. **-saflor** *m* safflower

Farb·erscheinung *f* **1** color phenomenon. **2** appearance of color

Färber·waid *m* dyer's woad. **-wau** *m* yellowweed. **-wurzel** *f* madder

Färbe·sieb *n* color sieve. **-stoff** *m* coloring matter, dye(stuff). **-trog** *m* staining trough (*or* box). **-vermögen** *n* coloring (*or* tinctorial) power, dyeing value

Farbe·waren *fpl* coloring materials = Farbwaren

Färbe·weise *f* dyeing method. **-wurzel** *f* madder. **-zwecke** *mpl* dyeing purposes

Färb·faß *n* dye(ing) vat

Farb· color, dye: **-fehler** *m* color defect; chromatic aberration. **f—fertig** *a* ready for dyeing (*or* coloring). **-film** *m* color film. **-filter** *n* color filter (*or* screen). **-flotte** *f* dye liquor (*or* bath). **-flüssigkeit** *f* (*Micros.*) staining solution. **-fülle** *f* (*Dye.*) body. **-gebung** *f* coloring; color scheme. **-gehalt** *m* color (*or* dye) content. **-glas** *n* color(ed) (*or* stained) glass. **-gut** *n* goods to be dyed. **-holz** *n* dyewood

farbig 1 *a* colored (*incl use as sfx, cf* BLEIFARBIG); stained; chromatic; colorful. **2** *adv, oft* in color. **Farbigkeit** *f* (depth of) color, colorfulness

Farb· color, dye: **-kissen** *n* ink pad.

-kochapparat *m* color boiler (*or* kettle). **-kopie** *f* color print, colored copy. **-körper** *m* coloring matter (*or* substance), pigment; chromosome. **f—korrigiert** *a* color-corrected, achromatized. **F—kraft** *f* coloring (*or* tinctorial) power, dyeing value. **f—kräftig** *a* **1** having strong coloring (*or* dyeing) power; (*Dye.*) productive. **2** strong in color. **-kreide** *f* colored chalk. **-kreisel** *m* color sensitometer. **-kreiselmethode** *f* color-disc method. **-küche** *f* color shop. **-kuchen** *m* dye (*or* color) cake. **-kunde** *f* color (*or* dye) technology

farbl. *abbv* (farblos) colorless

Farb·lack *m* color lake; lacquer

farblos *a* colorless; achromatic. **Farblosigkeit** *f* colorlessness; achromatism

Farb· color, dye: **-lösung** *f* **1** dyeing liquid. **2** (*Micros.*) staining solution. **-malz** *n* coloring malt, (roasted) amber malt. **-material** *n* dye(stuff), coloring (*or* dyeing) material. **f—messend** *a* colorimetric. **F—messer** *m* colorimeter. **-messung** *f* colorimetry. **f—metrisch** *a* colorimetric. **F—mine** *f* colored pencil lead; colored ink cartridge. **-mittel** *n* coloring agent. **-mühle** *f* color mill. **-muster** *n* color pattern; color sample. **-nuance** *f* tint, shade (of color) **-oxid** *n* colored oxide. **-pflanze** *f* dye plant. **-rakel** *m/f* color doctor. **-raster** *m* color screen. **-register** *n* color index. **f—richtig** *a* orthochromatic. **F—ruß** *m* carbon black. **-säure** *f* color acid. **-sehen** *n* color vision. **-stärke** *f* depth (*or* intensity) of color. **-stein** *m* dyestone. **-stift** *m* colored pencil

Farbstoff *m* dye(stuff); coloring matter; pigment, stain. **-aufnahme** *f* dye absorption **-base** *f* color base. **-bildung** *f* dye (*or* color) formation; (*Biol.*) chromogenesis. **f—erzeugend** *a* chromogenic. **-haltig** *a* pigment-bearing, containing dye. **F—lösung** *f* dye (*or* staining) solution

Farb· color, dye: **-stufe** *f* color gradation, tint. **-substanz** *f* coloring substance (*or* matter). **-tiefe** *f* depth of color. **-ton** *m* **1** color tone, tint, shade. **2** coloring clay. **-trennung** *f* color separation. **-treue** *f* color fidelity. **-trog** *m* color (*or* dye) trough (*or* vat). **-überzug** *m* coat of paint. **-umschlag** *m* color change

Färbung *f* (-en) **1** dyeing, coloring, staining. **2** coloration, pigmentation; hue, tinge. **3** dye

färbungs·fähig *a* capable of being dyed (colored, stained). **F—mittel** *n* coloring agent, dye, pigment, colorant. **-vermögen** *n* coloring (*or* tinctorial) power

Farb- color, dye: **-veränderung** *f* alteration of color. **-wandel** *m* color change. **-waren** *fpl* coloring (*or* dyeing) materials; dyes, paints, pigments. **-wechsel** *m* color change; play of colors. **-wiedergabe** *f* color reproduction. **-zelle** *f* (*Biol.*) pigment cell. **-zerstäuber** *m* color atomizer; paint spray(er). **-zusatz** *m* color additive (*or* admixture); addition of color

Farin *m* (-e), **Farinade** *f* (-n), **Farin·zucker** *m* brown (*or* moist) sugar, muscovado

Farn *m* (-e), **-kraut** *n* fern

farnkraut·artig *a* fern-like. **F—öl** *n* fern oil. **-wurzel** *f* (*Pharm.*) aspidium

Färse *f* (-n) heifer

Fasan *m* (-e) pheasant

Faschine *f* (-n) fascine, bundled brushwood

Fase *f* (-n) (*Mach.*) bevel, chamfer; thread

Fasel I *m* (-) male breeding animal; young (of animals), brood. **II** *f* (-n) kidney bean

fäseln *vt* unravel

Faser *f* (-n) **1** fiber; filament, thread, string; grain (of wood); (*Brew.*) pulp. **2** (*in compds oft*) fibrous, fibro: **f—·ähnlich** *a* fibroid. **F—alaun** *m* halotrichite. **f—artig** *a* fibery, fibrous, filamentary; grained. **F—asbest** *m* fibrous asbestos. **-asche** *f* fiber ash. **-bildung** *f* fiber formation, fibrillation. **-blende** *f* fibrous sphalerite

Fäserchen *n* (-) little fiber, fibril, filament

Faser·färbung *f* coloring (*or* coloration) of (the) fiber(s). **f—förmig** *a* fibrous, fibri(lli)form, filiform. **F—gehalt** *m* fiber content; (*Paper*) fiber yield. **-gewebe** *n* fibrous tissue. **-gips** *m* fibrous gypsum. **-glas** *n* fiberglass. **-haut** *f* (*Anat.*) fibrous membrane

faserig *a* fibrous, fibery, stringy; filamentous. **--kristallinisch** *a* fibrocrystalline

Faser· fiber, fibrous: **-kalk** *m* fibrous

calcite (*or* aragonite). **-kiesel** *m* fibrolite; fibrous (*or* radiated) quartz, sillimanite. **-kohle** *f* fibrous coal. **-kunde** *f* fiber technology. **f—los** *a* fiberless

fasern *vi* fray, shed fibers, lint; ravel (out); (*Paper*) flake, get mottled

Faser· fiber, fibrous: **-pappe** *f* fiberboard. **-pflanze** *f* fiber (*or* fibrous) plant; textile plant. **-platte** *f* fiberboard. **-quarz** *n* **1** fibrous quartz. **2** (*Min.*) fibrillite, fibrous sillimanite. **-richtung** *f* direction of the fiber (*or* grain). **-rohstoff** *m* fibrous raw material. **-salz** *n* fibrous salt. **-schutzmittel** *n* fiber-protecting agent. **-serpentin** *n* (*Min.*) chrysolite. **-stoff** *m* fibrin; fibrous material: **vegetalischer F.** vegetable fibrin (fiber, gluten, cellulose). **f—stoffartig, -stoffhaltig, -stoffig** *a* fibrinous. **F—tonerde** *f* fibrous synthetic alumina. **-torf** *m* fibrous peat

Faserung *f* (-en) fibering, fibrillation; texture; grain

faser·zellartig *a* fibrocellular. **F—zeolith** *m* natrolite

fasrig *a* fibery, fibrous, etc = FASERIG

Faß *n* (Fässer) barrel, keg; vat, tub; drum— **Bier vom F.** beer on tap, draft beer

Fassaden·farbe *f* outdoor paint, house paint

faßbar *a* comprehensible, graspable

Faß·bier *n* draft beer. **-bohrer** *m* gimlet.

Fäßchen *n* keg, small barrel; drum

Faß·daube *f* barrel stave

fassen I *vt* **1** grasp, grab, take hold of, seize; comprehend—**zu f. bekommen** get hold of. **2** catch, capture; arrest. **3** (*Mach.*) engage (gears, etc). **4** hold (a certain volume); contain, include. **5** set, mount. **6** barrel (wine). **7** word, formulate, express. **8** (*esp in idioms + noun*) conceive, form (ideas, opinions, plans) *cf* GEDANKE, MEINUNG, PLAN; reach (decisions) *cf* ENTSCHLUSS; take root *cf* WURZEL; gain (a foothold) *cf* FUSS; *cf* AUGE. **9** catch, hold, take, set. **10** (+ **an, auf, in, gegen**) reach (for), grasp (at). **III: sich f. 11** control (*or* get hold of) oneself. express oneself. *Cf* GEFASST

Faß· barrel, drum: **-färbung** *f* drum dyeing. **f—gar** *a* (*Lthr.*) drum-tanned. **F—**

gärung f cask fermentation. **-gärungs-system** n (Brew.) cleansing system. **-geläger** n cask deposit, bottoms. **-gerben** n drum tanning. **-geschmack** m taste of the barrel (cask, keg). **-hahn** m (barrel) tap, spigot. **-leimung** f (Paper) tub sizing

faßlich a comprehensible, conceivable

Faß·loch n bunghole

Fasson f (-s) fashion, style; way, manner; shape. **fassonieren** vt fashion; mold; form, shape

Faß·pech n 1 cooper's pitch. 2 barreled pitch. **-reif(en)** m barrel hoop. **-schmiere** f (Lthr.) drum stuffing. **-talg** m barreled tallow

Fassung f (-en) 1 grasping, etc; seizing, comprehension cf FASSEN 1. 2 frame, setting, mounting; (Elec.) socket. 3 wording, formulation; version; text. 4 (self-)control, composure. 5 conception, formation; reaching (of decisions), taking (root) etc cf FASSEN 8

Fassungs·kraft f (mental) capacity, comprehension. **-raum** m (volumetric) capacity. **-vermögen** n capacity (mental or volumetric); volume, content

Faß·waren fpl barreled goods, merchandise in barrels. **-zwickel** m (Brew.) try cock

fast adv almost, nearly

Fastage f (empty) barrels, kegs, vats

fasten vi fast. **Fasten** pl, **-zeit** f (time of) fasting, fast; (specif) Lent

faszinieren 1 vt fascinate. 2 vi be fascinating

fatal a disastrous, awful; awkward, embarrassing

fauchen vi spit (like a cat), hiss

faul a 1 rotten, spoiled, putrid, decayed. 2 foul. 3 bad, phony, shady. 4 lazy; slow. 5 (Metll.) brittle, short. **faulbar** a putrescible

Faul·baum m (Bot.) 1 alder buckthorn. 2 bird cherry. **-becken** n septic tank. **-brand** m 1 (Bot.) smut. 2 (Med.) moist gangrene. **-bruch** m (Metll.) shortness, brittleness. **f—brüchig** a (Metll.) short, brittle, **F—bütte** f fermenting trough; rotting vat

Fäule f putrefaction; rottenness; (dry) rot, mold, blight; (Min.) rotten lode

faulen vi (s) rot, spoil, go bad; putrefy, decay; (Indus.) ferment

fäulen vt rot, cause to putrefy; (Paper) ferment

faulend p.a putrescent; putrid, septic; moldy cf FAULEN

Faulenzer m (-) 1 idler. 2 book of tables. 3 (Brew.) Y connection

Faul·gas n sewer gas. **-grube** f fermenting pit

faulig a rotting, putrescent; putrefactive; moldy

Faul·kern m (Wool) heart rot

Fäulnis f rot, rottenness; putrefaction, decay; sepsis. **--alkaloid** n ptomaine. **-bak-terien** pl putrefactive bacteria. **-base** f ptomaine. **f—befördernd** a septic, putrefactive. **-beständig** a rotproof, decay-resistant; imputrescible. **-bewirkend** a septic, putrefactive. **F—erreger** m putrefactive agent. **-erscheinung** f symptom of putrefaction. **f—fähig** a putrefiable, putrescible. **F—gärung** f putrefactive fermentation. **-geruch** m putrid (or rotten) odor. **-gift** n septic poison. **f—hemmend**, **-hindernd** a antiseptic, antiputrefactive. **F—keim** m putrefactive germ. **-pilz** m rot fungus. **f—sicher** a imputrescible; anti-fouling. **-unfähig** a imputrescible, unputrefiable. **-verhindernd**, **-verhütend** a antiseptic, antiputrefactive; anti-fouling. **-widrig** a antiseptic. **-wirkend** a putrefactive, septic

Faul·raum m (Sewage) digestion tank. **-schlamm** m (Geol.) sapropel; (Sewage) activated sludge; (Metll.) debris. **-schlammgas** n sapropel gas. **-tier** n (Zool.) sloth

Faulung, Fäulung f putrefaction, rotting, decay

Faul·weizen m wheat smut. **-werden** n putrefaction, rotting, decay

Faum m (-̈e) foam, froth

Faust f (-̈e) fist; hand; grasp, grip.—**auf eigene F.** on one's own. **-formel** f rule of thumb, rough formula. **f—groß** a fist-size. **F—handschuh** m mitten. **-regel** f rule of thumb. **-säge** *f* hand saw. **-skizze** *f* rough sketch

Fayence·blau *n* china blue. **-druck** *m* (*Tex.*) faïence printing

Fäzes *pl* feces

Fazette *f* (-n) facet

Fazier *f* (-) facies, face, aspect

Fazit *n* sum total; upshot, net result, bottom line; **das F. ziehen** figure out the sum total (*or* the net result)

fbls. *abbv* (farblos) colorless

Fe. *abbv* (Feines im Entstauben) undersize in dedusted coal

fechten* (focht, gefochten; ficht) *vi* fight, fence

Feder *f* (-n) 1 feather, plume. 2 pen. 3 (*Mach.*) spring. 4 (*Wood*) tongue. **f— ähnlich** *a* spring-like, feathery. **F— alaun** *m* feather alum (halotrichite, alumogen *or* epsomite); amianthus. **f—artig** *a* spring-like; feathery. **F—barometer** *n* aneroid barometer. **-bart** *m* vane (*or* web) of a feather, vexillum. **-busch** *m* plume; tuft, crest

Federchen *n* (-) little feather (pen, spring) *cf* FEDER; (*Bot.*) plumule

Feder·druckmesser *m* spring manometer. **f—echt** *a* (*Tex.*) downproof. **F—erz** *n* feather ore, plumose silver ore, jamesonite. **-fahne** *f* 1 feather flag, trimmed feather. 2 vexillum, vane of a feather. **f— förmig** *a* feather-shaped, plumiform; pinnate. **-führend** *a* responsible, serving as the leader. **F—führung** *f* leadership; central handling. **-gips** *m* fibrous gypsum. **f—haar** *n* filamentous hair. **f—hart** *a* spring-hard. **F—härte** *f* spring hardness (*or* temper). **-harz** *n* (India) rubber, caoutchouc; elaterite

federig *a* 1 feathery, feathered (*esp in compds*); plumrose. 2 springy, elastic

Feder·kiel *m* quill. **-kleid** *n* plumage. **-klemme** *f* spring clip. **-kontakt** *m* spring contact. **-kraft** *f* spring tension; elasticity, resiliency; (spring) cushioning effect

federn I *vt* 1 spring, equip with springs. **II** *vi* 2 be resilient (springy, elastic), bounce. 3 lose feathers, malt. **federnd** *p.a* spring(y), resilient, elastic; sprung *cf* GEFEDERT

Feder·ring *m* spring (*or* lock) washer. **-salz** *n* feather salt, fibrous (*or* plumose) mineral. **-schloß** *n* spring lock. **-sieb** *n* (*Slurry*) Zimmer screen

Federung *f* (-en) 1 springs, spring suspension, springing. 2 springiness, resilience, elasticity

Feder·ventil *n* spring valve. **-vieh** *n* poultry. **-waage** *f* spring balance. **-wechsel** *m* molting. **-weiß** *n* amianthus; French chalk; powdered talc; fibrous gypsum; steatite. **-zange** *f* spring pincers. **-zeichnung** *f* pen-and-ink drawing. **-zug** *m* stroke of the pen

fedrig *a* feathery; elastic, etc = FEDERIG. **Fedrigkeit** *f* elasticity

Fege *f* (-n) sieve, screen, riddle

Fege·maschine *f* sweeper; winnower

fegen 1 *vt, vi* (s) sweep. 2 *vt* polish, scour; winnow

Fege·salpeter *m* saltpeter sweepings. **-sand** *m* scouring sand. **-schober** *m* (*Salt*) scum pan

Fegsel *n* (-) (*usu pl*) sweepings

Fehde *f* (-n) feud. **·-handschuh** *m* gauntlet: **den F. aufheben** (*fig*) accept the challenge

fehl 1: f. am Platz (or **Ort**) out of place **2** (*in compds & as sep pfx*) wrong, false, faulty; mis-. **Fehl: ohne F.** flawless, faultless

Fehl·anweisung *f* wrong indication, misdirection. **-anzeige** *f* 1 (instrument) misreading. 2 false report. 3 report of a shortage

fehlbar *a* fallible

Fehl·betrag *m* deficit, deficiency. **-bohrung** *f* (*Oil, Gas*) dry hole. **-druck** *m* misprint

fehlen *vi* 1 be missing (lacking, absent). **2** (+ *dat*) be wrong (with), ail (a psn.). **3** (*imps, usu* + *dat*): **das fehlt ihnen** they lack (*or* miss) that; **es fehlt ihnen an** . . . they lack . . . ; **es fehlte nicht an** . . . there was no lack of . . . **4** do wrong. **Fehlen** *n* absence; deficiency. **fehlend** *p.a* missing, lacking, absent

Fehler *m* (-) mistake, error; flaw, defect; fault. **·-ausgleichung** *f* equalization (*or* adjustment) of error(s). **f—behaftet** *a* erroneous, affected by error. **F—**

beseitigung f elimination of errors. **-berechnung** f calculation of error(s). **f—frei** a flawless, faultless, perfect. **F—grenze** f limit of error; tolerance
fehlerhaft a faulty; erroneous; defective
fehlerlos a flawless, faultless
Fehl·ernte f bad harvest
Fehler·quelle f source of error
Fehl·farbe f 1 discolored cigar wrapper. 2 (Tex.) off color. **-geburt** f miscarriage
fehl·gehen* vi (s) go wrong; go astray
fehl·geordnet a (Cryst.) disordered, dislocated. **F—gewächs** n crop failure. **-gewicht** n short weight
fehl·greifen* vi make a mistake, grab (or touch) the wrong thing
Fehl·griff m mistake, blunder; slip of the hand. **-guß** m (Metll.) faulty casting
Fehlingsche Lösung Fehling('s) solution
Fehl·kochung f defective boiling. **-konstruktion** f construction error; faulty structure, shoddy product
fehl·leiten vt mislead, misdirect, lead astray
Fehl·ordnung f disorder; (Cryst.) defect, dislocation. **-rechnung** f miscalculation
fehl·schießen* vi miss the mark
Fehl·schlag m failure. **fehl·schlagen*** vi fail
Fehl·schluß m false inference, wrong conclusion. **-schuß** m misfire; miss. **-zündung** f misfire
feien vt fortify, make invulnerable
Feier f (-n) celebration, festival. **feierlich** a festive; solemn, ceremonial
feiern 1 vt/i celebrate. 2 vt observe. 3 vi quit work; take time off from work; rest, not go to work
Feier·stunde f leisure hour. **-tag** m holiday
feig a cowardly, etc = FEIGE a
Feig·bohne f lupine, fig bean
feige a cowardly; rotten, underhanded
Feige f (-n) fig
Feigen·baum m fig tree. **-wachs** n gondang wax. **-wein** m fig wine
feil 1 a venal, mercenary. 2 a & sep pfx for sale: **feil·bieten** vt offer for sale
Feile f (-n) 1 file. 2 filings. 3: **letzte F.** finishing touches, final polish. **feilen** vt/i file; polish, put the finishing touches on

feilen·hart a (Metll.) file-hard. **F—härte** f file-hardness
feil·halten* vt offer for sale
Feil·hieb m file cut, notch
Feilicht n filings
feilschen vi bargain, haggle
Feilsel n (-) filings
Feil·späne mpl, **-staub** m filings. **-strich** m file cut
fein 1 a fine; thin, delicate; distinguished, refined; (in compds oft) finely cf FEIN-TEILIG, precision cf FEINMESSUNG, refining cf FEINPROZESS. 2 sep pfx fine; (also implies refining)
Fein· fine: **-abgleich** m fine balance. **-arbeit** f fine (or precision) work; (Metll.) (re)fining. **-aufspaltung** f fine separation. **-bau** m fine structure; atomic structure. **-berge** mpl (Ores) fine waste. **-bewegung** f (Opt.) slow motion, fine adjustment. **f—blasig** a fine-bubbled, fine-blistered. **F—blech** n thin(-gauge) sheet metal. **-brand** m highly rectified spirits
fein·brennen* vt refine. **Fein·brenner** m refiner. **-chemikalien** npl fine chemicals
feind pred a hostile. **Feind** m (-e) enemy. **feindlich** a 1 (of the) enemy. 2 hostile
Fein·druckmesser m micromanometer
Feindschaft f (-en) hostility; enmity
Feine f fineness; (Min.) fines
fein·einstellen vt/i micro-adjust, fine-tune. **Fein·einstellung** f fine (or micro-) adjustment; fine tuning. **-eisen** n fine iron; light section steel
feinen vt deoxidize; refine, purify
Fein·erz n fine ore. **-farbe** f (Dye.) pastel shade. **f—faserig** a fine-fibered, fine-grained
feinfein a very fine, superfine
Fein·feuer n refinery; running-out fire. **-frost** m, **-frostware** f deepfreeze food(s). **-gefüge** n fine structure, microstructure; fine texture. **-gehalt** m fineness (of metals); proportion of fine metal (in coins). **-gehaltsstempel** m hallmark. **f—gekühlt** a precision-cooled. **F—gemisch** n fine mixture; lean mixture (of fuel and air). **f—gepulvert** a finely powdered. **-geschichtet** a thin-layered, thin-

bedded. **-gespitzt** a sharp- (or fine-) pointed. **-geteilt** a finely divided (or graduated). **F—gewicht** n refined (or precision) weight. **-gold** n fine gold. **-grus** m (Min.) fines, smalls. **-gut** n fine material, fines

Feinheit f (-en) 1 fineness, thinness. 2 fine point, subtlety. 3 refinement, distinction **Feinheits·grad** f (degree of) fineness. **-grenze** f limit of fineness

Fein·kalk m ground quicklime. **-keramik** f fine ceramics, pottery; porcelain. **-kernseife** f curd toilet soap. **-kies** m fine ore (or gravel); fines. **-kohle** f fine coal, slack. **-korn** n fine grain. **-korneisen** n fine-grain(ed) iron. **-kornentwickler** m (Phot.) fine-grain developer. **f—körnig** a fine-grained. **F—kost** f, **-kostladen** m deli(catessen). **f—kristallinisch** a finely crystalline. **F—kupfer** n high-purity copper. **f—lockig** a finely perforated, fine-pored. **F—lunker** m pinhole

fein·machen vt refine. **fein·mahlen** vt grind fine, triturate, pulverize

Fein·mahlgut n material to be pulverized. **f—maschig** a fine-mesh(ed). **-mechanik** f precision engineering. **f—mehlig** a finely powdered. **F—messen** n precision measurement. **-messer** m micrometer; vernier

Feinmeß·diagramm n stress-strain diagram. **-gerät** n precision measuring instrument. **-lehre** f micrometer caliper (or gauge). **-schraube** f micrometer screw **Fein·messung** f precision measurement. **-metall** n fine metal. **-mühle** f 1 fine-grinding mill, pulverizer. 2 (Paper) wood-pulp refiner. **-ofen** m refining furnace. **-passung** f close (or tight) fit. **f—porig** a finely porous, fine-pored. **F—probe** f delicate test. **-prozeß** m refining process. **-puddeln** n (Metll.) refining, puddling **fein·pulvern** vt powder fine, pulverize. **fein·pulv(e)rig** a finely powdered, pulverulent

Fein·schlacke f refinery slag. **-schlag** m fines; stone chips

fein·schleifen* vt fine-grind, grind smooth. **Fein·schliff** m fine grind, fine-ground finish

fein·schmeckend a savory, fine-tasting. **F—schmecker** m gourmet, epicure. **-schnitt** m fine cut, thin slice. **-schraube** f fine-adjustment screw. **-schrot** m/n fine shot (groats, meal); (Brew.) finely crushed malt. **f—schuppig** a fine-scaled, fine-scaly. **F—schutt** m fine debris. **-sehrohr** n microscope. **-seife** f toilet soap. **-silber** n fine (or refined) silver. **f—sinnig** a sensitive. **F—skala** f fine scale, vernier. **-sprit** m (Alcohol) fine spirits

feinst a 1 (superl of FEIN) finest. 2 (esp in compds) very fine, superfine

Fein·stein m (Metll.) converter mat(te)

fein·stellen vt adjust finely, regulate

Fein·stellschraube f micrometer (or fine-adjustment) screw

feinst·gepulvert a very finely powdered. **F—kohle** f (coal) fines

Fein·strukter f fine structure. **f—strukturell** a fine-structure(d)

Feinst·sehrohr n ultramicroscope

fein·stückig a small-size

Fein·talg m refined tallow. **f—teilig** a finely divided. **F—treiben** n (Metll.) refining

Feinung f (-en) (re)fining. **Feinungs·schlacke** f refining (or final) slag

Fein·verschiebung f fine displacement. **f—verteilt** a finely divided. **F—waage** f precision balance. **-waschmittel** n light-duty detergent. **-werk** n (Instrum.) precision. **f—zellig** a fine-celled, finely cellular. **-zerkleinert** a finely divided (or ground). **-zerrieben** a finely ground. **F—zerstäuber** m fine sprayer, atomizer. **f—zerteilt** a finely divided. **F–zeug** n pulp; (Paper also) stuff. **-zeugholländer** m (Paper) beater, beating machine. **-zink** n high-grade zinc. **-zinn** n grain tin. **-zucker** m refined sugar

feist a fat, plump; rich (soil, food)

Felbel m/f shag (cloth or its nap)

Feld n (-er) 1 (genl) field. 2 (specif, oft) land, ground; open country; area, sphere, range; pane, panel; compartment; (Cmptrs) array—**zu F—e ziehen** move into battle; **Argumente ins F. führen** bring up arguments. **--bau** m agriculture, farming. **-beleuchtung** f (*Micros.*) ground illumination. **-brand** m (*Ceram.*) clamp burning

Feldchen n (-) little field, small area; (*Bot.*) areole

Feld·dichte f intensity of the field. **-dienst** m (military) field duty. **-elektron** n field electron

Felderung f (*Biol.*) tesselation

feld·frei a field-free. **F—frucht** f farm product, (*esp pl*) produce. **-gerät** n agricultural implements (*or* equipment). **-gleichung** f (*Math.*) field equation. **-kamille** f camomile. **-klee** m white clover; rabbit-foot clover. **-kümmel** m wild caraway; wild thyme. **-linien** fpl lines of force. **-messen** n land surveying. **-minze** f (*Bot.*) corn basil. **-ofen** m (*Ceram.*) clamp kiln. **-rauch** m, **-raute** f (*Bot.*) fumitory. **-salat** m (*Bot.*) lamb's lettuce. **-spat** m feldspar

feldspat·ähnlich, **-artig** a feldspathic. **F—gestein** n feldspathic rock. **f—haltig** a, **feldspatisch** a feldspathic. **Feldspat·steingut** n feldspathic earthenware

Feld·stärke f field intensity. **-stecher** m field glass(es). **-stein** m 1 fieldstone. 2 boulder. 3 boundary stone. 4 compact feldspar. **-strom** m (*Elec.*) field current. **-thymian** m wild thyme. **-versuch** m field experiment (*or* trial). **-winde** f (*Bot.*) small bindweed, field convolvulus. **-wirtschaft** f agriculture. **-ziegel** m (*Ceram.*) clamp(-burnt) brick. **-zug** m (*Mil.*) campaign

Felge f (-n) 1 (wheel) rim, 2 fallow land

Fell n (-e) skin, hide, pelt; (rough) sheet (of masticated rubber). **-abfälle** mpl skin cuttings, hide parings. **-handel** m trade in (raw) skins

Felleim m (= **Fell·leim**) hide glue

Fell·motte f clothes moth. **-schmitzer** m skin dyer. **-späne** mpl hide parings

Felpel m/f shag = FELBEL

Fels m (-en, -en) rock; cliff = FELSEN (*incl compds*) **-alaun** m rock alum. **-art** f type of rock

Felsen m (-) rock; cliff = FELS (*incl compds*) **-ader** f rock seam, dike. **-alaun** m rock alum. **-boden** m rock soil. **-gebirge** n Rocky Mountains. **-glas** n rock glass, re-

sistant glass for combustion tubes. **-glimmer** m mica. **-gras** n Iceland moss. **f—hart** a rock-hard. **F—öl** n rock oil, petroleum. **-salz** n saltpeter

Fels· rock, see also FELSEN·: **-gestein** n rock (material). **-glimmer** m mica

felsig a rocky. **felsisch** a (*Geol.*) felsic

Fels·öl n petroluem = FELSENÖL. **-partie** f rocky area, cliff. **-quarzit** m rock quartzite

femisch a femic

Fenchel m (-) fennel. **-holz** n sassafras wood. **-öl** n fennel oil. **-wasser** n (*Pharm.*) fennel water

Fenchocampher·säure f fenchocamphoric acid

Fenchol·säure f fencholic acid

Fenn n (-e) fen, marsh, bog

Fenster n (-) window. **-angel** f casement hinge. **-glimmer** m (*Min.*) muscovite. **-kitt** m glazing putty. **-leder** n chamois, window-cleaning leather. **f—los** a windowless. **F—scheibe** f window pane

Ferien pl (*usu* school) vacation

Ferment n (-e) enzyme; ferment(ing), fermentation. **fermentieren** vt/i ferment. **Fermentierung** f (-en) fermentation. **Ferment·wirkung** f ferment (*or* enzyme) action

fern 1 a distant, remote, far(away); long past; *adv oft* far. 2 (*in compds oft* long-distance, long-range; tele . . . ; remote(-control); television. 3 *sep pfx* away, at a distance; remote

Fern·ablesung f remote reading

Fernambuk(·holz) n Brazil wood.

Fern·anzeiger m remote indicator. **-aufnahme** f telephotograph(y). **-bedienung**, **-betätigung** f remote control. **-bild** n 1 television picture. 2 telephoto(graph)

fern·bleiben* vi(s) stay away

Ferne f (-n) distance; distant place(s) (past, future)

ferner (*compar of* FERN) 1 a further. 2 adv futher(more), besides; + v: **es wird auch f. so sein** it will continue to be so. **Ferner** m(-) glacier

Fern·gas n piped (*or* pipeline) gas. **f—gesteuert** a remote-controled. **F—glas** n telescope; binoculars

fern·halten* *vt &* **sich f.** keep away, keep at a distance

Fern·hörer *m* (*Tlph., Radio, TV*) receiver. **-leitung** *f* 1 long-distance (transmission) line (*or* pipeline). 2 (*Tlph.*) trunk line

fern·lenken *vt* remote-control. **Fernlenk-·körper** *m* guided missile

fern·liegen* *vi* be far, be remote; be far-fetched

Fern·meldenetz *n* (long-distance) communications network. **-meldewesen** *n* telecommunication. **-meldung** *f* (*Instrum.*) remote signal (*or* indication). **-messer** *m* range finder. **f—mündlich** *a* telephonic; (*esp as adv*) by telephone. **F—photographie** *f* telephotograph(y). **-rakete** *f* long-range rocket. **-rohr** *n* telescope. **-rohrlupe** *f* telescopic magnifier. **-ruf** *m* telephone call (*or* number). **-schalter** *m* remote control switch. **-schaltung** *f* remote control (connection). **-schreiben** *n* teletype message. **-schreiber** *m* teletype(writer), telex(machine)

Fernseh· television: **-apparat** *m* TV set (*or* receiver). **-bild** *n* TV picture. **-bildschirm** *m* TV screen

fern·sehen* *vi* watch TV. **Fernsehen** *n* television

Fern·seher *m* 1 TV set. 2 TV viewer. **-sicht** *f* view perspective **f—sichtig** *a* far-sighted

Fernseh·kolben *m*, **-rohr** *n*, **-röhre** *f* cathode ray tube, picture tube

Fernsprech·, Fernsprecher° *m* telephone

Fern·stecher *m* binoculars

fern·steuern *vt* remote-control. **Fern-·steuerung** *f* remote control

Fern·unterricht *m* correspondence school (course). **-verkehr** *m* long-distance transportation (*or* communication). **-waffe** *f* long-range weapon. **-wirkung** *f* remote effect **-zeichnung** *f* perspective drawing. **-zündung** *f* remote-control ignition

Ferri· ferric, ferri: **-acetat** *n* ferric acetate. **-ammonsulfat** *n* ammonium ferric sulfate. **-bromid** *n* ferric bromide. **-chlorid** *n* ferric chloride **-chlorwasserstoffsäure** *f* ferrichloric acid. **-cyan** *n* ferricyanogen. **-cyaneisen** *n* ferrous ferri-cyanide (Turnbull's blue). **-cyanid** *n* ferric cyanide, ferricyanide

Ferricyan· . . ferricyanide: **-kalium** *n* potassium ferricyanide

Ferri·cyanür *n* (-ous) ferricyanide

Ferricyan·verbindung *f* ferricyanide. **-wasserstoff** *m*, **-wasserstoffsäure** *f* ferricyanic acid

Ferrid·cyankalium *n* potassium ferricyanide

Ferri·eisencyanür *n*, **-ferrocyanid** *n* ferric ferrocyanide (Prussian blue)

Ferriferro· ferrosoferric: **-jodid** *n* ferrosoferric iodide

Ferri· ferric, ferri: **-hydroxid** *n* ferric hydroxide. **-ion** *n* ferric ion. **-jodat** *n* ferric iodate. **-jodid** *n* ferric iodide. **-kalium-sulfat** *n* ferric potassium sulfate. **-nitrat** *n* ferric nitrate. **-oxid** *n* ferric oxide. **-phosphat** *n* ferric phosphate. **-rhodanid** *n* ferric thiocyanate. **-salz** *n* ferric salt. **-sulfat** *n* ferric sulfate. **-sulfid** *n* ferric sulfide

Ferrit *m/n* ferrite, (*Chem. oft*) ferrate. **ferritisch** *a* ferritic

Ferri·verbindung *f* ferric compound. **-zyan** *n* ferricyanogen = FERRICYAN

Ferro· ferrous, ferro-, ferroso-: **-acetat** *n* ferrous acetate. **-ammonsulfat** *n* ammonium ferrous sulfate. **-bor** *n* ferroboron. **-bromid** *n* ferrous bromide. **-chlorid** *n* ferrous chloride. **-chrom** *n* ferrochrome, ferrochromium. **-cyan** *n* ferrocyanogen. **-cyaneisen** *n* ferric ferrocyanide (Prussian blue). **-cyanid** *n* ferrous cyanide, ferrocyanide

Ferrocyan· ferrocyanide: **-kalium** *n* potassium ferrocyanide. **-verbindung** *f* ferrocyanide. **-wasserstoff** *m*, **-wasserstoffsäure** *f* ferrocyanic acid

Ferro·elektrikum *n* ferroelectric substance

Ferroferri·cyanid *n* ferrous ferricyanide. **-oxid** *n* ferrosoferric oxide

Ferro·hydroxid *n* ferrous hydroxide. **-ion** *n* ferrous ion. **-jodid** *n* ferrous iodide. **-kaliumsulfat** *n* ferrous potassium sulfate. **-karbonat** *n* ferrous carbonate. **-legierung** *f* ferro-alloy. **f—magnetisch** *a* ferromagnetic. **F—magnetikum** *n* ferromagnetic (substance). **-mangan** *n* ferromanganese. **-molybdän** *n* ferromolybdenum. **-nitrat** *n* ferrous nitrate. **-oxalat**

n ferrous oxalate. **-oxid** *n* ferrous oxide. **-phosphat** *n* ferrous phosphate. **-phosphor** *m* ferrophosphorus. **-salz** *n* ferrous salt. **-silizium** *n* ferrosilicon. **-sulfat** *n* ferrous sulfate. **-sulfid** *n* ferrous sulfide. **-titan** *n* ferrotitanium. **-vanadin** *n* ferrovanadium. **-verbindung** *f* ferrous compound. **-wolfram** *m/n* ferrotungsten. **-zirkon** *n* ferrozirconium. **-zyan** *n* ferrocyanogen

Ferse *f* (-n) heel

fertig I *a* 1 ready, prepared. 2 ready-made. 3 finished, complete(d); *pred. oft* worn out, done in, ruined, bankrupt. 4 accomplished, skilled. **II** *as pfx oft* prefabricated, ready-to-use. **III** *adv oft* perfectly. **IV** *adv* (+ *v*) & *sep pfx* ready; finished; done; finish (doing sthg): **f. lesen (schreiben, polieren)** finish reading (writing, polishing) *cf* FERTIGBRINGEN, etc

Fertig·bauteil *m* prefabricated building component

fertig·bearbeiten *vt* finish machining

fertig·bekommen* *vt* get (sthg) done, etc = FERTIGBRINGEN

Fertig·beton *m* ready-mixed concrete

fertig·bringen* *vt* accomplish, finish, get (sthg) done; **es f., zu** (or **daß**) manage to, bring it about that . . .

fertigen *vt* turn out, produce, manufacture; draw up (a document)

Fertig·erzeugnis *n* finished (ready-made, prefabricated) product. **-form** *f* (*Glass*) finishing mold. **-frischen** *n* (*Metll.*) final purification. **-frischofen** *m* finishing furnace. **f—gekocht** *a* ready-cooked. **F—gerüst** *n* (*Metll.*) finishing mill. **-guß** *m* iron as cast. **-gut** *n* manufactured (or finished) goods. **-haus** *n* prefabricated house

Fertigkeit *f* readiness: skill, dexterity, talent; accomplishment; fluency

fertig·machen *vt* 1 get (sthg) ready. 2 finish, complete. 3 wear out, do in. 4 pitch (soap) **Fertig·macher** *m* finisher. **-mörtel** *m* ready-mixed mortar

Fertig·packung *f* pre-packaging. **-produkt** *n* finished product. **-raffination** *f* (*Lead*) refining. **-schlacke** *f* (*Metll.*) white slag. **-schneiden** *n* finishing cut.

-sinterung *f* (*Metll.*) final sintering

fertig·stellen *vt* finish, complete; get (sthg) ready. **Fertig·stellung** *f* production, finishing; completion

Fertig·teil *m/n* prefabricated part. **-teilbau** *m* prefabricated construction

Fertigung *f* (-en) production, manufacture; product; copy (of a document)

Fertigungs·straße *f* assembly line

Fertig·walzwerk *n* finishing mill. **-ware** *f* finished article (or merchandise)

fertig·werden* *vi* (s) (**mit**) deal, cope (with); manage

Fertilität *f* fertility

Ferul(a)·säure *f* ferulic acid

Fessel *f* (-n) chain, *pl oft* fetters. **fesseln** *vt* chain, fetter, bind; captivate

fest I *a* 1 solid, firm, hard. 2 tight. 3 fixed, rigid. 4 steady, permanent; durable. 5 binding, definite, set, firmly established. 6 secure, fortified. **II** *adv oft* closely, strongly **III** *sep pfx* solid, firm(ly), fast; tight(ly); definite(ly). **IV** *sfx* -proof, -resistant *cf* FEUERFEST, HITZEFEST

Fest *n* (-e) celebration; festival; feast

fest·backen *vi* (h *or* s) cake (*or* stick) together firmly; clinker, frit

Fest·bett *n* fixed bed

fest·binden* *vt* bind fast, fasten; unite. **f—·brennen*** *vi*(s) sinter; seize

Feste *f* (-n) 1 compact rock. 2 (*Min.*) pillar. 3 stronghold; strength

fest·fahren* 1 *vt, vi* (s) run aground. 2 *vi*(s) get stuck

fest·fressen*: **sich f.** seize, jam, stick; become fixed

fest·frieren* *vi*(s) freeze (*or* get frozen) (solid)

Fest·gehalt *m* solid content

fest·gipsen *vt* stick (together) with gypsum plaster. **f—·haften** *vi* stick fast, cling, adhere firmly

fest·halten* I *vt* 1 hold fast, hold on to, retain. 2 seize; keep, detain; arrest. 3 record (permanently) **II** *vi* (**an**) hold on, stick, cling (to). **III: sich f.** hold on tight, cling. **festhaltend** *p.a* adhesive; tenacious, retentive

festigen 1 *vt* establish firmly; strengthen, fortify; consolidate. 2: **sich f.** become firm (established, fortified)

Festigkeit *f* solidity, firmness; (*Metals*) tenacity; (*Mettl.*, *esp as sfx*) strength, resistance *cf* BIEGEFESTIGKEIT

Festigkeits-grenze *f* resistance limit, breaking point. **-lehre** *f* strength of materials. **-probe** *f*, **-prüfung** *f* strength test(ing). **-zahl** *f* coefficient of resistance

fest-keilen *vt* wedge tight. **f—·kitten** *vt* cement (*or* lute) tight. **·kleben** *vt*, *vi*(s) stick fast

Fest·körper *m* solid (body). **Festkörper·chemie** *f* solid-state chemistry

Fest·land *n* mainland; continent

fest·legen *vt* 1 set definitely, fix; define. 2 establish, lay down (firmly); bed, moor. **Fest·legung** *f* setting, fixing, fixation; definition; establishment; agreement, convention

festlich *a* festive; solemn, ceremonial

fest·liegend *p.a* fixed, settled

fest·machen *vt* 1 solidify; consolidate. 2 make fast, fasten, attach. 3 decide, agree on, fix

Fest·maß *n* solid (*or* cubic) measure. **-meter** *n/m* cubic meter. **-preis** *m* fixed price. **-punkt** *m* fixed point; fixed target

Fest·schrift *f* commemorative publication

fest·setzen 1 *vt* set, fix, determine, establish; schedule. 2: **sich f.** settle(down), get lodged. **Festsetzung** *f* setting, fixing. establishment; scheduling; settlement

Fest·sitz *m* tight (*or* snug) fit; force fit. **fest·sitzen*** *vi* sit tight, be stuck, be firmly lodged

fest·stehen* *vi* stand firm; be certain (definite, fixed, established). **feststehend** *p.a* certain, definite, etc

feststellbar *a* determinable, ascertainable

fest·stellen *vt* 1 determine, ascertain; discover, detect. 2 note, observe. 3 state, declare. **Feststellung** *f* determination, ascertainment; discovery; detection; observation; statement, declaration

Fest·stern *m* fixed star. **-stoff** *m* solid (matter)

Feststoff·gehalt *m* solid content. **-geschwindigkeit** *f* solid feed rate. **-katalysator** *m* fixed catalyst

Festungs·achat *m* fortification agate

fest·wachsen* *vi* (s) grow on, adhere, take root

fest·weich *a* semisolid

fest·werden* *vi*(s) solidify. **Fest·werden** *n* solidification. **-wert** *m* fixed value, constant; coefficient. **-zahl** *f* fixed number

fett *a* 1 fat. 2 fatty, aliphatic; adipose; greasy, oily. 3 thick, hefty; rich, hearty. 4 fertile. 5 profitable; prosperous. 6 (in) boldface (type). 7 (*Varnish*, *etc*) high-grade, long-oil **f—es Puddeln** (*Iron*) pig boiling. **Fett** *n* (-e) fat, grease. **·-abschneider** *m* grease separator. **f—ähnlich** *a* fat-like, fatty, lipoid. **F—aldehyd** *m* aliphatic aldehyde. **-ansatz** *m* fat(ty) deposit. **-appretur** *f* (*Lthr.*) oil dressing. **f—arm** *a* low-fat. **-aromatisch** *a* aliphatic-aromatic, alkyl-substituted aromatic. **-artig** *a* fat-like, fatty, lipoid, adipose, **F—bedarf** *m* fat requirement. **-bestandteil** *m* fatty constituent. **-bestimmung** *f* fat determination. **- bildung** *f* fat formation. **-brühe** *f* (*Lthr.*) fat liquor. **-büchse** *f* grease cup, lubricator. **f—dicht** *a* fat-tight, greaseproof. **F—druck** *m* boldface type

fett·drucken *vt* print in boldface

Fett·drüse *f* sebaceous gland

Fette *f* (-n) fatness, greasiness; corpulence

fett·echt *a* fat-insoluble

Fetteilchen° *n* fat particle = **Fett·teilchen**

fetten I *vt* 1 grease, oil, lubricate. 2 stuff, dub (leather); compound (oil). **II** *vi* give off grease, be greasy. **fettend** *p.a* greasing, grease-depositing

Fett·entziehung *f* fat extraction. **-farbe** *f* fat-soluble color, oil color. **-farbstoff** *m* color for fats; (*Micros.*) fat stain; (*Biochem.*) lipochrome, lutein. **-fleck** *m* grease spot. **f—frei** *a* fat-free. **-gar** *a* (*Lthr.*) oil-tanned. **F—garleder** *n* oil-tanned leather, chamois. **-gas** *n* oil gas. **f—gebunden** *a* attached to an aliphatic carbon atom. **-gedruckt** *a* (printed in) boldface. **F—gehalt** *m* fat (*or* grease) content. **-gerbung** *f* oil tannage (*or* dressing). **-geruch** *m* odor of grease. **-geschmack** *m* greasy taste. **-gewebe** *n* adipose tissue. **-glanz** *m* greasy luster. **f—glänzend** *a* greasy-lustered. **-haltig** *a* fatty, greasy, containing fat. **F—härtung** *f* hardening (*or* hydrogenation) of fat(s). **-harz** *n* oleoresin. **-harzreserve**

f (*Tex.*) fat resin, resist. **-haut** *f* (*Lthr.*) corium, derma, adipose membrane

Fettheit *f* **1** fatness, corpulence **2** fattiness, greasiness, oiliness. **3** richness. **4** fertility. *Cf* FETT

fettig *a* fatty, greasy, oil; adipose. **Fettigkeit** *f* (-en) fattiness, greasiness, oiliness; *pl oft* food fats, fatty (*or* fat-rich) foods

Fett·industrie *f* fat industry. **-kalk** *m* fat lime. **-käse** *m* cheese with 40% fat. **-kohle** *f* fat (*or* bituminous) coal, high-voltage coking coal. **-körper** *m* fatty (*or* greasy) matter, fatty (*or* aliphatic) compound. **-kraut** *n* (*Bot.*) butterwort, steepgrass. **-kügelchen** *n* fat globule

fett·lickern *vt* (*Lthr.*) fat-liquor

fett·lösend *a* fat-dissolving. **F—löser** *m* fat solvent, fat dissolver. **-löserseife** *f* fat-dissolving soap. **f—löslich** *a* soluble in fat. **F—lösungsmittel** *n* solvent for fats. **-magen** *m* abomasum, fourth stomach (of ruminants). **-nappe** *f* grease spot (in fabrics). **-öl** *n* fatty oil.

Fetton *m* fuller's earth, plastic clay (= **Fett·ton**)

Fett·presse *f* (*Mach.*) grease gun. **-puddeln** *n* (*Iron*) wet puddling, pig boiling. **-quarz** *n* greasy (lustered) quartz. **-radikal** *n* alkyl radical. **f—reich** *a* fat-rich, high-fat. **F—reif** *m* bloom (on chocolate). **-reihe** *f* aliphatic series. **-reservemittel** *n* fat (*or* wax) resist. **-rest** *m* fatty residue. **-salbe** *f* fatty ointment. **f—sauer** *a* of (a) fatty acid; sebacate of. **F—säure** *f* fatty (*or* aliphatic) acid; sebacic acid. **-schicht** *f* layer of fat (*or* grease). **-schliff** *m* greased ground joint. **-schmiere** *f* (*Lthr.*) fat liquor. **-schmierung** *f* grease lubrication. **-schwarz** *n* fat-soluble black. **-schweiß** *m* fatty sweat; yolk (of wool). **-seife** *f* fat (*or* lard) soap. **-sein** *n* (*Wine*) ropiness. **f—spaltend** *a* fat-splitting, lipolytic. **F—spaltung** *f* fat cleavage, lipolysis. **-spritze** *f* grease gun. **-stein** *m* eleolite. **-stift** *m* wax pencil (*or* crayon). **-stoff** *m* fat, fatty matter. **-stoffwechsel** *m* fat metabolism. **-sucht** *f* **1** fatty degeneration. **2** obesity

Fettung *f* greasing, oiling, etc *cf* FETTEN

Fettungs·mittel *n* (*Lthr.*) fat-liquoring agent

Fettusche *f* lithographic ink = **Fett·tusche**

Fett·verbindung *f* fatty (*or* aliphatic) compound. **-verseifung** *f* saponification of fat. **-verzinnung** *f* grease tinning. **-wachs** *n* adipocere. **f—wachsartig** *a* adipocerous. **F—wolle** *f* wool in the yolk, grease wool

Fetzen *m* (-) shred, scrap, torn piece; rag

feucht *a* **1** damp, moist, humid. **2** wet: **auf f—em Weg(e)** by wet process. **Feuchte** *f* (-n) dampness, moistness, humidity; wetness

Feuchte·gleichgewicht *n* equilibrium moisture content. **-kammer** *f* moist chamber. **-messer** *m* hygrometer, moisture tester; psychrometer

feuchten *vt* moisten, dampen, wet. **Feuchter** *m* (-) moistener, damper; wetter

Feucht·gewicht *n* moist weight; drained weight

Feuchtheit, Feuchtigkeit *f* **1** moistness, dampness; wetness. **2** moisture, humidity—**vor F. schützen!** keep dry

Feuchtigkeits·anzeiger *m* moisture indicator, hygrometer, hygroscope. **-aufnahme** *f* absorption of moisture. **f—fest** *a* moisture-resistant (*or* -proof). **-gehalt** *m* moisture content. **F—grad** *m* degree of moisture (*or* humidity). **-messer** *m* hygrometer, psychrometer. **-niederschlag** *m* moisture deposit. **-prüfer** *m* moisture tester. **f—sicher** *a* moisture-proof (*or* resistant). **F—zeiger** *m* hygrometer, hygroscope, moisture meter

Feucht·kammer *f* moist chamber. **-kugelthermometer** *n/m* wet-bulb thermometer. **-lagern** *n* damp storage

feuchtlich *a* slightly moist

Feucht·mahlung *f* wet grinding. **-messer** *m* hygrometer, moisture-tester; psychrometer. **-walze** *f* damp(en)ing roll. **f—warm** *a* moist and warm. **F—wasserfarbe** *f* water color

Feuer *n* (-) **1** fire; flame. **2** furnace, hearth, forge. **3** (signal) light, beacon

Feuer·anzünder *m* kindler. **f—beständig** *a* fire-resistant, fireproof; refractory. **F—beständigkeit** *f* resistance to fire; re-

fractoriness. **-bestattung** f cremation.
-beton m refractory concrete. **-blende** f
fire blende, pyrostilpnite. **-brücke** f fire
bridge. **-büchse** f firebox. **-darre** f drying
kiln. **f—fangend** a (in)flammable. **-far-
big** a flame-colored. **-fest** a fireproof; refractory: **f—er Ton** fireclay; **f—er Ziegel**
firebrick. **F—festigkeit** f fireproofness;
refractoriness. **-festton** m fireclay.
-fläche f heating surface. **f—flüssig** a
molten, fused. **F—führung** f firing. **-gas**
n flue gas, burnt gas. **-gefahr** f danger of
fire; fire hazard. **f—gefährlich** a dan-
gerously (in)flammable, fire-hazardous.
F—gefährlichkeit f (in)flammability,
fire-hazardousness. **-gradmesser** m
pyrometer. **-gradmessung** f pyrometry.
-grube f ash pit. **-haken** m fire hook,
poker. **-haut** f (fire-resistant) lining,
liner. **-holz** n firewood
feuerig a fiery, etc = FEURIG
Feuer-kammer f fire chamber, firebox.
-kitt m fireproof cement. **-krücke** f furnace rake, rabble. **-kunst** f pyrotechnics.
-leitung f priming; (Mil.) fire control. **f—
los** a fireless; lusterless. **F—löschap-
parat, -löscher** m fire extinguisher
Feuerlosigkeit f lack of brilliance (or
luster)
Feuer-luft f furnace gas. **-material** n fuel,
combustible. **-melder** m fire alarm.
-messer m pyrometer. **-messung** f
pyrometry
feuern I vt/i 1 light a fire (in), stoke (furnace, etc). 2 (Mil.) fire. II vt 3 burn, use as
fuel. 4 (Wine) sulfur. 5 fling. 6 fire (=
dismiss). III vi burn, glow, spark; be fiery
Feuer-politur f fire polish. **-porzellan** n
refractory porcelain. **-probe** f fire test;
crucial test. **-punkt** m fire point; (Opt.)
focus; (Min.) hearth. **-raum** m fireplace;
firebox; furnace, hearth. **-regen** m rain
of fire, fiery rain. **-rohr** n 1 fire tube,
flue. 2 firearm. **-rohrkessel** m fire-tube
boiler. **f—rot** a fiery red. **F—saft** m
(Iron) slag bath. **-schlauch** m fire hose.
-schutz m fire prevention (or protection);
fireproofing. **-schutzfarbe** f fire-
resistant paint. **-schutzmittel** n fireproofing agent. **-schwader** m fire damp.
-schwamm m tinder, punk

Feuers-gefahr f danger of fire
feuer-sicher a fireproof, fire-resistant.
F—stein m flint; firebrick **f—steinartig**
a flintlike, flinty. **F—steinknolle** f flint
nodule. **-stelle** f (burning area of the) furnace. **-strahl** m jet (or flash) of fire. **-ton**
m fireclay
Feuerung f 1 fire-kindling; furnace-
stoking, heating; firing. 2 fuel. 3 firebox,
hearth
Feuerungs-anlage f furnace hearth fireplace. **-bedarf** m fuel requirement(s).
-gewölbe n furnace arch. **-material,
-mittel** n fuel. **-öl** n fuel oil. **-raum** m firebox, furnace, hearth. **-rost** m firebox
grate. **f—technisch** a pyrotechnic
feuer-verbleit a hot-leaded. **-vergoldet** a
fire-gilt. **F—vergoldung** f fire gilding.
-versicherung f fire insurance. **-ver-
silberung** f fire silvering. **-verzinkung** f
hot-dip galvanizing. **-verzinnung** f hot
tin-coating. **-waffe** f firearm. **-wehr** f fire
department. **-werk** n fireworks. **-werker**
m pyrotechnist. **-werkerei** f, **werker-
kunst** f pyrotechnics. **f—widerstands-
fähig** a fire-resistant. **F—zange** f fire
tongs. **-zeug** n lighter. **-zeugbenzin** n
lighter fluid. **-ziegel** m firebrick. **-zug** m
fire tube, flue. 2 train of explosives
feurig a 1 fiery, red-hot, burning,
flaming—**f—er Schwaden** firedamp.
2 igneous. 3 (Wine) generous. 4 (Colors)
bright, loud. **--flüssig** a igneous. **-rot** a
fiery-red, red-hot
Feurung f firing, heating etc = FEUERUNG
ff abbv 1 (feinfein) superfine. 2 (feuerfest)
fireproof, refractory. 3 (und folgende) and
following, et seq.
FF., Ff. abbv (Fusionspunkte) melting
points
F.f. abbv (Fortsetzung folgt) (to be) continued
Fibel f (-n) primer; manual
Fiber f (-n) fiber. **--dichtung** f fiber packing. **-karton** m fiberboard
Fibranne f rayon staple
Fibrille f (-n) fibril. **Fibrillen-struktur** f
fibrillar structure
fibrin-haltig a containing fibrin, fibrinous.
fibrinös a fibrinous
fibrös a fibrous

ficellieren *vt* wire (bottles)
ficht fights: *pres of* FECHTEN
Fichte *f* (-n) spruce (*Picea excelsa*); (*oft inaccurately*) pine, fir
Fichten·harz *n* spruce resin, rosin. **-holzschliff** *m* mechanical woodpulp. **-holzteer** *m* spruce tar. **-nadelöl** *n* pine-needle oil. **-nadelteer** *m* pine-needle tar. **-pech** *n* spruce resin (*or* pitch). **-rinde** *f* spruce bark, (*esp Tanning*) fir bark. **-säure** *f* pinic acid. **-span** *m* spruce (*or* pine) chip (shaving, splint). **-sprossen** *fpl* (*Pharm.*) pine shoots. **-wolle** *f* pine wool. **-zellstoff** *m* (*Paper*) spruce pulp. **-zucker** *m* pinitol, pinite
F.i.D. *abbv* (Faden im Dampf) (*Thermometry*) thread in vapor
Fieber *n* fever, temperature. **--anfall** *m* attack of fever. **f—artig** *a* feverish, febrile. **F—arznei** *f* febrifuge, antipyretic, fever remedy. **f—fest** *a* immune to fever. **F—frost** *m* fever chills, ague
fieberhaft *a* feverish, fevered
Fieber-klee *m* buck bean (*Menyanthes*). **-kraut** *n* feverfew. **-mittel** *n* fever remedy, febrifuge, antipyretic
fiebern *vi* have a fever; be feverish; be delirious; (+ **nach**) crave (for)
Fieber·pulver *n* powdered fever remedy, (antimonic) (anti-)fever powder. **-rinde** *f* cinchona (*or* Peruvian) bark. **-thermometer** *n* fever (*or* clinical) thermometer. **F—vertreibend, -widrig** *a* antifebrile, fever-combating. **F—wurz(el)** *f* bitterwort, yellow gentian
fiebrig *a* feverish, febrile
Fieder *f* (-n) small feather; leaflet, pinnule. **f—förmig** *a* pinnate
fiel fell: *past of* FALLEN
Fig. *abbv* (Figur) fig. (figure)
Figuren·druck *m* (*Tex.*) topical (*or* figure) printing. **-druckartikel; ** *m* (*Tex.*) topical style
figurieren 1 *vt* figure. 2 *vi* appear
figürlich *a* figurative
fiktiv *a* fictitious
Filial· branch: **-anstalt** *f* branch establishment
Filiale *f* (-n) branch (establishment)
Filia·säure *f* filicic acid
Filicin·säure *f* filicinic acid

Filigran *n* (-e) filigree
Filix·säure *f* filicic acid
film·bildend *a* film-forming. **F—bildschicht** *f* (*Phot.*) film emulsion (layer, coat). **-druck** *m* (*Tex.*) screen printing. **-schicht** *f* film layer
Filter·abwurf *m* discharged filter cake. **-ablaß** *m* filter draw-off. **-anlage** *f* filtration plant
filterbar *a* filterable
Filter·beutel *m* filter bag. **-bewegung** *f* seepage flow. **-blatt** *n* filter (paper, leaf). **-dauer** *f* filtering time. **-element** *n* filter cell. **-fläche** *f* filter surface, filtering area. **-gehäuse** *n* filter case (casing, housing). **-gerät** *n* filtering apparatus (*or* device). **-gestell** *n* filter stand. **-gewebe** *n* filter cloth. **-gläs/chen** *n* small filter glass (*or* tube). **-kegel** *m* filter cone. **-kissen** *n* filter pad. **-kohle** *f* filter(ing) charcoal. **-körper** *m* filtering material. **-kuchen** *m* filter cake; (*Petrol.*) slack wax. **-lage** *f* filter layer (*or* bed). **-masse** *f* (*Brew.*) pulp, filter mass. **-material** *n* filtering material. **-mittel** *n* filtering aid (*or* medium). **-mundstück** *n* filter tip (on cigarettes)
filtern *vt/i* filter, strain
Filter·nutsche *f* suction filter. **-plättchen** *n* (small) filter plate. **-platte** *f* filter plate. **-pressentuch** *n* filterpress cloth. **-rand** *m* edge of the filter. **-rest** *m* residue on the filter, filtration residue. **-rohr** *n*, **-röhre** *f* filter tube (*or* pipe), screen pipe. **-rückstand** *m* filtration residue. **-schablone** *f* filter-cutting form (*or* template). **-schale** *f* filter(ing) dish. **-schicht** *f* filter bed (*or* layer). **-schoner** *m* filter cone. **-sieb** *n* filtering sieve (*or* strainer). **-stäbchen** *n* filter stick. **-stein** *m* filter(ing) stone. **-stoff** *m* filter material (*or* cloth). **-tiegel** *m* filter crucible. **-träger** *m* filter holder (*or* ring). **-trommel** *f* drum filter. **-tuch** *n* filter cloth. **-turm** *m* filter(ing) tower
Filterung *f* (-en) filtering, filtration
Filterungs·dauer *f* filtering time. **-hilfsmittel** *n* filtering aid
Filter·vorsatz *m* air filter (in a gas mask). **-wäger** *m* filter weigher. **-watte** *f* filter wadding. **-zelle** *f* filter cell (*or* unit)

Filtrat n (-e) filtrate
filtrations·fähig a capable of filtration, filterable. **F—kraft** f filtration power
Filtrat·leistung f filtrate output
Filtrier· filter(ing): **-aufsatz** m filtering attachment
filtrierbar a filterable. **Filtrierbarkeit** f filterability
Filtrier· filter(ing): **-beutel** m filtering bag, percolator. **-druck** m filtration pressure. **-einlage** f (charge of) filtering material
filtrieren vt filter, strain. **Filtrierer** m (-) filter, strainer; filtration plant
Filtrier· filter(ing), (in many compds = FILTER·): **-erde** f filtering earth. **-hut** m (paper) filtering funnel. **-korb** m filtering basket. **-papierstreifen** m strip of filter paper. **-sack** m filtering bag, bag filter. **-schale** f filtering dish (with perforated bottom). **-schwierigkeit** f difficulty in filtering. **-stativ** n filter stand. **-stutzen** m (cylindrical) filtering jar (with lip). **-tasse** f filtering cup (with perforated cover for a funnel)
Filtrierung f (-en) filtering, filtration
Filtrum n (- . . tren) filter
Filz m (-e) felt; felt hat; (Bot.) tomentum; (Metll.) slime ore. **filz·artig** a felt-like; tomentous
Filzbarkeit f felting property
filzen vt/i felt
filz·fähig a capable of felting. **F—fähigkeit** f felting property. **-falte** f (Paper) crease in the felt. **-gewebe** n felted tissue. **-haar** n felted hair
filzig a feltlike; tomentous
Filz·malz n felted malt. **-schlauch** m (tubular) felt jacket. **-stift** m felt-tip marker. **-tuch** n felt(ed) cloth, blanket. **-wirkung** f felting effect. **f—wollen** a (of or like) felted wool
Fimmel m (-) 1 (fimble) hemp. 2 (Min.) wedge. 3 (esp in compds) mania
finanzieren vt finance
finden* (fand, gefunden) 1 vt find, discover; meet with; get, win, gain: **sie f. es gut** they find it (or think it is) good. 2 vi & **sich f.** find one's way. 3: **sich f.** be found; occur, happen; turn out (well); (+ **in**) reconcile oneself (to)

findig a ingenious, resourceful. **Findigkeit** f ingenuity, resourcefulness
Findling m (-e) 1 foundling. 2 (Geol.) boulder, erratic block
fing caught: past of FANGEN
Finger·abdruck m fingerprint. **f—fertig** a dexterous, skilful. **F—futter** n rubber finger, finger stall. **-hut** m 1 thimble; 2 (Bot.) = **-hutkraut** n foxglove, digitalis
Fingerling m (-e) rubber finger, finger stall
Finger·nagel m fingernail. **-probe** f rule of thumb. **-spitze** f fingertip. **-spitzengefühl** n intuition; sensitivity. **-stein** m belemnite. **-zeig** m hint; tip, indication; warning
fingieren vt feign, fake; simulate. **fingiert** p.a fictitious, imaginary
Finne f (-n) 1 pustule, pimple. 2 bladder worm. 3 fin. 4 claw; peen (of a hammer). 5 stud; peg
finster a dark, gloomy; obscure, sinister. **Finsternis** f (-se) darkness; gloom; obscurity; (Astron.) eclipse
Fiole f (-n) vial, phial
Firma f (. . men) firm, company. **firmen·eigen** a company-owned
firn a (Wine) well-matured; last year's
Firn m (-e) glacier snow; granular snow. **·-blau** n glacier blue. **-eis** n glacier ice
Firnis m (-se) varnish; veneer. **f—·artig** a varnish-like. **F—ersatz** m varnish substitute. **-glasur** f spirit enamel **-papier** n glazed paper
firnissen vt varnish. **Firnis-Sumach** m poison sumac; varnish tree
First m (-e) 1 ridge. 2 top, summit. 3 (roof)ridge, (roof)top. 4 = **Firste** f (-n) (Min.) roof
Fisch m (-e) fish **·-augenstein** m apophyllite. **-bein** n whalebone. **-blase** f fish bladder; isinglass. **-düngemittel** n, **-dünger** m fish manure (or guano). **-eier** npl roes
fischen vt/i fish. **Fischerei** f (-en) fishery; fishing
Fischer·gelb n cobalt (or Aurelian) yellow
Fisch·fang m 1 fishing. 2 catch (of fish). **-gift** n fish poison
fischig a fishy, fish-like
Fisch·körner npl (Bot.) Indian berries.

-**lebertran** m fish liver oil. -**leim** m fish glue, isinglass; (*Pharm.*) ichthyocolla. -**leimgummi** n sarcocolla. -**mehl** n fish meal. -**milch** f milt, fish blubber. -**öl** n fish oil; ichthyol. -**otter** f (old world) otter. -**schwanzbrenner** m fishtail burner. -**speck** m fish blubber. -**tran** m fish oil, train oil. -**zucht** f fish culture, pisiculture

Fiset(t)·holz n (young) fustic

Fistel f (-n) fistula. --**holz** n young fustic. -**kassie** f purging cassia, cassia fistula. -**stimme** f falsetto (voice)

fistulös a fistular, fistulous

Fitsche f (-n) hinge; fishplate

Fitze f (-n) (*Tex.*) hank, skein

fix a 1 fixed, steady. 2 quick, nimble. 3 smart, clever. 4: **f. und fertig** all set and ready, (all) finished

Fixage f (-n) (*Phot.*) fixing, fixation. **Fixateur** m (-e) (fixative) sprayer

Fix·bleiche f bleaching with (bleaching) powder. -**färberei** f tipping (of plush), stain dyeing; fast dyeing

Fixier·bad a fixing bath

fixierbar a fixable. **Fixierbarkeit** f fixability

fixieren vt 1 (**an**) affix (on), attach (to). 2 fix (incl *Phot.*); set (incl wool, hair); arrange; lay down (rules, conditions). 3 harden. 4 mordant (silk). 5 stare at

Fixier·flüssigkeit f fixing liquid (or liquor). -**mittel** n fixing agent, fixative; hardening agent. -**natron** n sodium thiosulfate. -**salz** n (*Phot.*) fixing salt, fixer, "hypo" (sodium thiosulfate). -**ton** m fixing clay

Fixierung f (-en) attaching; fixing, fixation; setting; arranging; hardening cf FIXIEREN 1-3

Fixierungs·mittel n fixing (or hardening) agent

Fix·punkt m fixed point. -**stern** m fixed star. -**wert** m fixed value

fl. abbv 1 (flüssig) liquid. 2 (flüchtig) volatile

Fl. abbv (Flüssigkeit) liquid

flach a 1 flat; plain, level, even, smooth. 2 low; shallow. 3 gentle (slope). --**bodig** a flat-bottomed. **F—brenner** m flat-flame burner. -**draht** m flat wire. -**druck** m flat printing; lithoprinting

Fläche f (-n) flatness, level. 2 (flat) surface; plane; plain; sheet. 3 (flat) space, expanse. 4 area. 5 face, facet

Flach·eisen n flat bar-iron; flat steel

Flächen·anziehung f surface attraction, adhesion. -**berührung** f surface contact. -**dichte** f surface density. -**druck** m surface pressure; (*Tex.*) blotch print(ing). -**einheit** f unit of area (or surface). -**farbe** f surface color. **f—förmig** a flat, plane, sheet-like. **F—gewicht** n area weight. -**gitter** n layer lattice. -**größe** f area, amount of surface

flächenhaft a flat, plane, sheet-like

Flächen·helle, -**helligkeit** f surface luminosity. -**inhalt** m surface area. -**kathode** f flat (or plate) cathode. -**maß** n surface (or square) measure. -**messer** m planimeter. -**messung** f planimetry. -**raum** m surface area. **f—reich** a polyhedral, many-faced. **F—satz** m theorem of conservation of areas. -**trägheitsmoment** n angular impulse. -**umriß** m perimeter, outline. -**wert** m surface value. -**winkel** m plane angle. -**zahl** f 1 square number. 2 number of surfaces. **f—zentriert** a (*Cryst.*) face-centered

Flach·feile f flate file. **f—gewunden** a planispiral. **F—glas** n flat glass; plate glass

Flachheit f flatness; shallowness

-**flächig** sfx -faced, -hedral

Flach·land n flat land, level country. **f—·legen** vt lay (down) flat, flatten. **F—·müllerei** f (*Flour*) flat (or low) milling

Flächner m (-) polyhedron

Flach·profil n flat section. -**riemen** m flat belt. **f—rund** a round and flat-bottomed

Flachs m flax. --**bau** m flax growing

Flach·schliff m (sur)face grind(ing). -**schnitt** m horizontal section

flächse(r)n a flaxen

flachs·farben, -**farbig** a flax-colored. -**gelb** a flaxen. **F—gewebe** n linen (goods). -**leinwand** f flax linen. -**samen** m flaxseed, linseed. -**spinnerei** f flax spinning (mill). -**stein** m amianthus

Flachs·stab m flat bar. -**stahl** m flat steel (bar). -**stange** f flat bar

Flachs·wachs n flax wax

Flach·wasser n shallow water. -**zange** f

(pair of) flat-nose pliers

flackerig *a* flickering; flaring. **flackern** *vi* flicker; flare

Flacon *m/n* (-s) vial; (small) flask

Fladder·mine *f* land (*or* contact) mine

Fladen·brot *n* pita (bread). **-lava** *f* black lava

Flader *f* (-n) (*Wood*) mottle, speckle, vein. **fladerig** *a* mottled, speckled, veined

Flagge *f* (-n) flag

Flak *f* (-s) *abbv* (Flugabwehrkanone) AA gun, antiaircraft (gun). **-·granate** *f* AA (*or* antiaircraft) shell

Flakon *m/n* (-s) vial, etc = Flacon

flambieren *vt* flame, singe

Flame *m* (-n, -n) Fleming *cf* FLÄMISCH

Flamm· flame *cf* FLAMMEN·

flammbar *a* (in)flammable. **Flämmchen** *n* (-) little flame, flamelet. **Flamm·druck** *m* chiné printing

Flamme *f* (-n) flame. **flammen I** *vt* 1 flame(-treat). 2 singe, sear. 3 (*Tex.*) cloud (fabrics), water, moiré (silk). **II** *vi* blaze, flame; flare up; be aflame; flash

flammen·beständig *a* flame-resistant, flameproof. **F—blume** *f* (*Bot.*) phlox. **-bogen** *m* flaming (luminous, electric) arc. **f—farbig** *a* flame-colored. **F—färbung** *f* flame coloration. **-feuer** *n* blazing fire. **-fläche** *f* flame surface. **-frischarbeit** *f* firing in a reverberatory furnace. **-härten** *n* flame-hardening. **-höhe** *f* height of (the) flame. **-kern** *m* inner core (of a gas flame). **-lichtmesser** *m* flame photometer. **f—los** *a* flameless. **F—ofen** *m* flame (*or* reverberatory) furnace. **-opal** *m* fire opal. **-rohr** *n*, **-röhre** *f* fire-tube, flue. **-rückschlag** *m* backfiring, strike-back (of burner flame). **f—sicher** *a* flameproof. **F—strahl** *m* jet of flame. **-werfer** *m* flamethrower

flamm·farbig *a* flame-colored; rainbow-colored. **F—färbung** *f* flame coloration. **-gas** *n* flame gas

flammieren *vt* flame(-treat); singe; cloud (fabrics), water (silk)

flammig *a*1 flame-like, flaming. 2 (*Tex.*) clouded, shadowed; (*Silk*) watered, moiré—**f. meliert** jaspé. 3 (*Wood*) veined, grained

Flamm·kohle *f* flame- (*or* open-) burning

coal. **-ofen** *m* flame (*or* reverberatory) furnace. **-ofenfrischen** *n* reverberatory refining. **-ölbombe** *f* (flaming) oil bomb. **-punkt** *m* flash (ignition, fire) point. **-punktprüfer** *m* flashpoint tester. **-rohr** *n*, **-röhre** *f* fire tube, flue. **-rohrkessel** *m* fire-tube boiler. **-ruß** *m* furnace black. **-schutzmittel** *n* flameproofing agent. **-schwarz** *n* furnace black. **f—sicher** *a* flameproof. **F—sichermachen** *n* flameproofing. **-spritzen** *n* (*Ceram.*) flame spraying. **f—widrig** *a* flame-resistant

flandrisch *a* (of) Flanders, Flemish

Flanell *m* (-e) flannel

Flanke *f* (-n) flank, side. **flankieren** *vt* flank

Flan(t)sch *m* (-e), **Flan(t)sche** *f* (-n), **flan(t)schen** *vt* flange. **Flanschen·verbindung** *f* flange(d) joint (*or* union)

Flarakete *f* (*short for* Flugabwehrrakete) AA (*or* antiaircraft) rocket

Fläschchen *n* (-) small bottle (*or* flask), vial

Flasche *f* (-n) 1 bottle, flask; jar; (gas) cylinder: **Wein auf F—n ziehen** bottle wine. 2 (*Foun.*) casting box, flask. 3 (*Mach.*) (pulley) block

Flaschen·abteilung *f* (*Brew.*) bottling department. **-bier** *n* bottled beer. **f—förmig** *a* bottle- (*or* flask-)shaped. **F—gas** *n* cylinder gas, bottled gas. **-gestell** *n* bottle (*or* flask) rack. **-glas** *n* bottle glass. **f—grün** *a* bottle-green. **F—hals** *m* bottleneck (*lit* only). **-inhalt** *m* contents of the bottle (flask, cylinder). **-kappe**, **-kapsel** *f* bottle cap (*or* top). **-kürbis** *m* bottle gourd, calabash. **-öffner** *m* bottle opener. **-schild** *n* bottle (*or* flask) label. **-zug** *m* set of pulleys, block and tackle

Flaser *f* (-n) vein(ing), grain, streak; flaw. **flaserig** *a* veined, grained; streaky

flatterhaft *a* volatile; unstable; fickle

Flatter·mine *f* land (*or* contact) mine

flattern *vi* flutter, flap; wobble; be unstable

Flatter·ruß *m* (flaky) lampblack

Flattier·feuer *n* choked fire

flau *a* faint, feeble, weak; dull, slack, flat. **flauen** 1 *vt* buddle, rinse. 2 *vi* (s) slacken, weaken

Flaum *m* down, fuzz, fluff; nap, pile.

flaumig a downy, fluffy; frothy
Flaus(ch) m (-e) tuft; pad; nappy cloth
Flavaspid·säure f flavaspidic acid
Flavean·wasserstoff m cyanothioformamide
Flavian·säure f flavianic acid, Napthol Yellow 5
Flavin n (-e) flavine, lyochrome
fl. Best. abbv (flüchtige Bestandteile) volatile constituents
Flechse f (-n) tendon, sinew
Flechte f (-n) 1 braid, plait. 2 lichen. 3 herpes
flechten* (flocht, geflochten; flicht) vt/i weave, braid, plait, twine
Flechten·farbstoff m lichen dye. **-pilz** m lichen fungus. **-rot** n orcein. **-säure** f lichen(ic) (or fumaric) acid. **-stärkemehl** n lichenin, moss starch. **-stoffe** mpl lichens, lichen-like substances
Fleck m (-e or -en) 1 spot, stain, blot. 2 patch. 3 place, spot: **vom F. kommen** get ahead. **-analyse** f spot analysis
flecken 1 vt/i spot, stain, blot. 2 vt patch (up). 3 vi make headway
Flecken m (-) 1 (= FLECK) spot, patch, etc. 2 (as pl) measles. 3 place, town. **-bildung** f spotting, staining. **f—empfindlich** a susceptible to spotting (or staining), stain-prone. **F—entfernung** f spot removing. **-entfernungsmittel** n spot remover. **f—frei, -los** a spotless, stainless. **F—seife** f spot-cleaning soap, scouring soap. **-wasser** n spot-removing liquid, (specif) Javelle water
Fleck·fieber n spotted fever, (specif) typhus
fleckig a spotty, spotted, stained; flawed
Fleck·mittel n spot remover. **-schierling** m poison hemlock. **-seife** f stain-removing soap. **-stein** m scouring stone, stain-removing clay. **-storchschnabelwurzel** f (Pharm.) geranium, cranesbill. **-wasser** n liquid spot- (or stain-) remover
Fledermaus° f (Zool.) bat. **-brenner** m batswing burner
Flegel m (-) 1 flail. 2 lout, boor
flehen vi (um) implore, plead (for)
Fleisch n meat; flesh, pulp (of fruit); (in compds oft) muscle, sarco·: **-beschau** f

meat inspection. **-brühe** f meat broth, bouillon. **-eiweiß** n meat protein (or albumin)
fleischen vt flesh (hides). **Fleischer** m (-) butcher. **Fleischerei** f butcher shop, meat store. **Fleischer·talg** m unmelted tallow
Fleisch·farbe f flesh color. **f—farbig** a flesh-colored. **F—faser** f muscle fiber. **-fäulnis** f putrefaction of meat. **f—fressend** a meat-eating, carnivorous. **F—gift** n meat poison, ptomaine. **-gummi** n sarcocolla. **-hacker** m butcher. **-hackmaschine** f meat grinder. **-hauer** m butcher
fleischig a meaty; fleshy; pulpy cf FLEISCH
Fleisch·kloß m meatball. **-kohle** f animal charcoal. **-konserve** f canned (preserved, processed) meat. **-kost** f meat diet. **-leim** m, **-leimgummi** n sarcocolla
fleischlich a meat(y); carnal, fleshly
fleisch·los a meatless; fleshless. **F—mehl** n meat meal, tankage. **-milchsäure** f sarcolactic acid. **f—rot** a flesh-pink. **F—saft** m meat juice (or broth) **-seite** f flesh side (of hides). **-spalt** m (Lthr.) flesh split. **-tee** m beef tea (or broth). **-vergiftung** f meat poisoning. **-waren** fpl meat products. **-wasser** n meat broth. **-wolf** m meat grinder. **-zucker** m inositol, inosite
Fleiß m diligence, hard work, pains—**mit F., zu F.** on purpose. **fleißig** a industrious, diligent, hard-working
flicht weaves: pres of FLECHTEN
flicken vt mend, repair, patch (up)
Flick·gewebe n scar tissue. **-masse** f patching material. **-werk** n patchwork
Flieder m (-) (Bot.) lilac; elder
Fliege f (-n) (Zool.) fly. **fliegen*** (flog, geflogen) vt, vi(s) fly
Fliegen·draht m, **-fenster** n window screen. **-gift** n fly poison. **-gitter** n window screen. **-holz** n quassia wood. **-kobalt** m native arsenic. **-netz** n mosquito net. **-pilz, -schwamm** m fly agaric, fly amanita. **-stein** m native arsenic
Flieger m (-) flier, aviator. **-benzin** n aviation gasoline. **-bild** n aerial photograph. **-bombe** f aerial bomb
fliegerisch a flying, aerial, aeronautical
Flieger·schutz m air defense

Flieg·lack *m* airplane varnish (*or* dope)
fliehen* (floh, geflohen) **1** *vt/vi* (s) flee, escape. **2** *vt* shun, avoid
Flieh·kraft *f* centrifugal force
Fliese *f* (-n) tile; flagstone; paving brick.
fliesen *vt* tile; pave
Fließ *n* (-e) brook, creek; (*Metll.*) yield.
-·arbeit *f* assembly-line work. **-band** *n*
assembly line
fließbar *a* fluid. **Fließbarkeit** *f* fluidity.
Fließbarkeits·dauer *f* setting time (of
enamel)
Fließ·bereich *m* plastic range. **-betrieb** *m*
assembly-line operation. **-bett** *n* fluid
bed. **-bettkatalysator** *m* fluidized catalyst. **-bild** *n* flow diagram. **-druck** *m*
flow (*or* hydraulic) pressure. **-eigenschaft** *f* rheological property
fließen* (floß, geflossen) *vi*(s) **1** flow, run.
2 be soaking wet. **3** melt. **4** (*Time*) fly,
fleet. **fließend** *p.a* (*oft*) fluid, liquid; fluent; mobile—**f—e Hitze** melting heat;
f—e Bindung *f* floating linkage
fließ· flow: **-fähig** *a* fluid, flowable; fusible.
F—fertigung *f* assembly-line production. **-figur** *f* flow figure. **-geschwindigkeit** *f* flow rate (*or* velocity). **-glätte** *f*
wet litharge. **-grenze** *f* (*Metals*) flow limit, yield point. **-harz** *n* oleoresin, (*specif*)
turpentine. **-katalysator** *m* fluid catalyst. **-kohle** *f* liquid coal, colloidal suspension of coal in oil. **-koksverfahren** *n*
fluid coking. **-körper** *m* flowing substance, non-solid. **-kunde** *f* rheology.
-naht *f* welded seam. **-papier** *n* absorbent
(filter, blotting) paper. **-pressen** *n* extrusion molding. **-probe** *f* flow (*or* pour) test;
flow sample. **-punkt** *m* flow point; yield
point. **-sand** *m* quicksand. **-straße** *f* assembly
line. **-verhalten** *n* flow behavior (*or* characteristics). **-vermögen** *n* fluidity, mobility. **-verzug** *m* flow distortion. **-wasser** *n* **1**
running water. **2** (*Physiol.*) lymph
Flimmer *m* (-) glimmer; (false) glitter; tinsel, mica; (*Physiol.*) cilium, (*in compds,
oft*) ciliary. **flimmern** *vi* glimmer, shimmer, scintillate. **Flimmer·schein** *m*
sparkling luster.
flink *a* quick, fast; alert; nimble, deft

flinken *vi* shine, sparkle, glisten
Flinte *f* (-n) gun, (*specif*) shotgun
Flinten·lauf *m* gun barrel. **-schrot** *m* (gun)
shot
Flint·glas *n* (*Opt.*) flint glass. **-stein** *m*
flintstone
Flinz *m* (-e) siderite; dark shale, Helvetian
sand
Flitsch *m* (-e), **Flitsche** *f* (-n), **Flitschen·erz** *n* ore in glittering laminas.
Flitsch·holz *n* (*Wood*) flitch
Flitter *m* (-), *f* (-n) spangle, tinsel, glitter;
shining platelet, flake; small change.
-draht *m* tinsel wire. **-erz** *n* ore in glittering laminas. **-glas** *n* pounded glass for
frosting. **-gold** *n* Dutch metal, tombac,
brass foil
flittern *vi* glitter, glisten, sparkle
Flitter·sand *m* micaceous sand. **-silber** *n*
silver-colored tinsel
flitzen *vi*(s) flit, dart, scoot
Flitz·wagen *m* jeep
Fll. *abbv* (Flüssigkeiten) liquids, fluids
flocht wove: *past of* FLECHTEN
Flocke *f* (-n) flake; tuft, flock, floc(cule).
flocken 1 *vt, vi & sich f.* flake. **2** *vt* pull
(wool) into tufts. **3** *vi* flocculate, fuzz out
flocken·artig *a* flocculent, flake-like. **F—
bildner** *m* flocculant, flocculating agent.
-bildung *f* flocculation. **-bildungszerstörer** *m* deflocculant. **-erz** *n* mimetite.
-graphit *m* flake graphite. **-papier** *n*
flock paper. **-reaktion** *f* flocculation reaction. **f—rissig** *a* flaky. **F—salpeter** *m*
efflorescent saltpeter. **-stoff** *m* (*Tex.*) nap
cloth. **-wolle** *f* waste wool
flockig *a* flaky, flocculent. **-·käsig** *a* curdy
Flock·mittel *n* flocculant. **-seide** *f* floss
silk, silk waste
Flockung *f* (-en) flaking, flocculation.
Flockungs·mittel *n* flocculating agent
Flock·wolle *f* flock (*or* short) wool
flog flew: *past of* FLIEGEN
floh fled: *past of* FLIEHEN
Floh *m* (-̈e) flea. **f—·braun** *a* puce(-colored). **F—farbe** *f* puce (color). **-kraut** *f*
fleabane. **-krautöl** *n* pennyroyal oil. **-samen** *m* fleawort (*or* psyllium) seed
Flor *m* (-e) **1** bloom, blossom. **2** gauze. **3** nap,
pile. **-·decke** *f* (*Tex.*) pile

Florentiner Flasche Florentine receiver (for essential oils)
Florett·seide *f* floss silk
Flor·garn *n* (*Tex.*) lisle (thread)
Florida·(bleich)erde *f* Florida (bleaching) earth
Flor·stoff *m* gauze
floß flowed: *past of* FLIESSEN
Floß *n* 1 (*pl* Flossen) pig (iron): **blumiges F.** white pig with granular fracture. 2 (*pl* -e) raft, float, *cf* FLOSSE
Flosse *f* (-n) 1 fin; flipper. 2 float. 3 (*Metll.*) pig (of iron). *Cf* FLOSS
Floß·eisen *n* white pig iron
flößen I *vt/i* 1 float. II *vt* 2 pour; instill. 3 skim (milk). 4 refine (tin)
Flossen·bett *n* (*Metll.*) pig mold
flossig *a* finned
Floß·ofen *m* flowing furnace
Flotations·chemie *f* flotation chemistry. **f—fähig** *a* floatable. **F—schäumer** *m* flotation frother. **-verfahren** *n* flotation process
flotativ *adv* by flotation. **flotierbar** *a* floatable.
flotieren *vt* float (ores)
flott *a* 1 brisk (*esp* reactions), lively, fast, quick—**f. gehen** thrive. 2 smart, stylish. 3 floating; (*pred*) afloat
Flotte *f* (-n) 1 (dye, bleaching) liquor (*or* bath). 2 fleet, navy
Flotten· (*Dye.*) liquor: **-flüssigkeit** *f* dye liquor. **-gefäß** n color reservoir. **-länge** *f* volume of liquor. **-menge** *f* amount (*or* ratio) of liquor. **-stand** *m* height of the liquor. **-verhältnis** *n* liquor consistency; ratio of goods to liquor
flottieren 1 *vt/i* float. 2 *vi* fluctuate
flott·machen *vt* get afloat, get into running order
Flott·seide *f* untwisted silk. **-stahl** *m* ingot (*or* run) steel
Flötz, Flöz *n* (-e) layer, stratum, bed, seam.**-erz** *n* ore in beds. **-gestein** *n* stratified rock. **-kalk** *m* bedded limestone. **-vergasung** *f* underground gasification (of coal)
Fluat *n* (-e) fluosilicate. **fluatieren** *vt* treat (waterproof, weatherproof) with fluosilicate
Fluch *m* (-e), **fluchen** *vi* curse
Flucht *f* (-en) 1 flight, escape. drain. 2 row,

line; suite; alignment; flight (of stairs). 3 play, swing (of a door, etc) **-burg** *f* (place of) refuge, haven. **-ebene** *f* vanishing plane
fluchten *vt* align. 2 *vi* be in alignment
flüchten *vi* (s) flee, escape, take refuge
Flucht·holz *n* leveling rule, level
flüchtig *a* 1 volatile, ethereal: **f—e Salbe** volatile (ammonia) liniment; **f—es Laugensalz** ammonium carbonate. 2 fleeting, transient. 3 fleeing, fugitive. 4 hasty, hurried. 5 superficial, cursory, casual; flighty, 6 fragile, brittle
flüchtigen *vt* volatilize
Flüchtigkeit *f* (-en) volatility, transience; haste; superficiality, flightiness; fragility, brittleness; careless mistake *cf* FLÜCHTIG
Flucht·linie *f* vanishing line. **-linientafel** *f* nomogram, nomographic chart; alignment chart
Flüchtling *m* (-e) escapee, refugee
Flucht·punkt *m* vanishing point. **f—recht** *a* flush, in alignment
Flug *m* (-e) flight; flock—**im. F.** hastily, in an instant. **-abwehr** *f* antiaircraft (defense). **-asche** *f* flying ash(es), flue dust. **-bahn** *f*, flight path; trajectory, orbit. **-benzin** *n* aircraft gasoline. **-betrieb** *m* 1 air traffic. 2 flying (operations). **-blatt** *n* handbill, flier, leaflet
Flügel *m* (-) 1 wing. 2 vane, blade, fin. 3 (window) casement. 4 grand piano. **f—·förmig** *a* wing-shaped. **flügelig** *a* winged; bladed, vaned
Flügel· wing(ed), vane(d); fan: **-gebläse** *n* rotary fan, fan blower. **-gruppen** *fpl* branched groups. **-kühler** *m* fan cooler. **f—los** *a* wingless. **F—mine** *f* finned (*or* winged) bomb. **-mutter** *f* (*Mach.*) wing nut. **-pumpe** *f* vaned (*or* semi-rotary) pump. **-rad** *n* fan (vane, bladed) wheel; propeller; (turbine) rotor; (*Brew.*) flighter. **-schraube** *f* wing nut. **-tür** *f* double door; folding door(s)
flug·fähig *a* 1 volatile 2 light, easily flying (dust, etc). 3 airworthy. **F—feld** *n* flying field, airfield
flügge *a* full-fledged
Flug·hafen *m* airport. **-koks** *m* flue coke, coke dust. **-körper** *m* flying object; mis-

sile; space vehicle. **-kraft** f power of flight. **-linie** f 1 flight path, trajectory. 2 airline. **-mehl** n mill dust. **-plan** m flight (or airline) schedule. **-platz** m airfield, airport. **-post** f air mail. **-rost** m light film of rust

flugs adv quickly, instantly

Flug·sand m drifting (or wind-blown) sand. **-schrift** f handbill, flier, leaflet. **-staub** m airborne dust; flue dust; (Lead Metll.) fume. **-staubkammer** f dust (or condensing) chamber. **-technik** f aviation; aerodynamics. **-treibstoff** m aviation fuel. **-weg** m airway; flight path— **auf dem F.** by air. **-wesen** n aviation, aeronautics. **-zeug** n (air)plane, aircraft. **-zeuglack** m airplane varnish (or dope)

Fluh f (-en) rock cliff; concrete. **fluhen** vt cover with (or form of) concrete

fluidifizieren vt fluidize

Fluidität f fluidity. **Fluido·plaste** pl liquid plastic masses. **Fluidum** n (.. da) fluid; fluidity; aura

fluktuieren vi fluctuate. **fluktuös** a fluctuating

Flunder m(-) flounder

Fluo·borat n fluoroborate **-borsäure** f fluoboric acid. **-kieselsäure** f fluorosilicic acid

Fluor n fluorine; (in compds, oft) fluo(ro)-, fluoride: **-ammonium** n ammonium fluoride. **-anthenchinon** n fluoranthenequinone. **-arsen** n arsenic fluoride. **-benzol** n fluo(ro)benzene. **-bor** n boron trifluoride. **-borsäure** f fluoboric acid. **-calcium** n calcium fluoride. **-dioxid** n fluorine dioxide

Fluorescenz, Fluoreszenz f fluorescence. **f—·erregend** a fluorogenic. **F—farbe** f fluorescent dye (color, paint, ink). **·licht** f fluorescent light. **-messer** m fluorometer. **-schirm** m fluorescent screen

fluoreszieren vi fluoresce, be fluorescent. **fluoreszierend** p.a fluorescent

Fluor·gehalt m fluorine content. **f—haltig** a containing fluorine

Fluorid n (-e) (usu -ic) fluoride cf FLUORÜR

fluorieren vt/i fluorinate

Fluor· fluo(ro)-, fluor(ine), fluoride: **-jod** n fluorine iodide. **-kalium** n potassium fluoride. **-kalzium** n calcium fluoride.

-kiesel m silicon fluoride. **-kieselsäure** f fluosilicic acid. **-kohlenstoff** m carbon fluoride. **-metall** n metallic fluoride. **-natrium** n sodium fluoride. **-phosphat** n fluophosphate. **-salz** n fluoride. **-silikat** n, **-sillziumverbindung** f fluosilicate. **-stickstoff** m nitrogen trifluoride. **-streuköder** m fluorinated insecticide. **-tantalsäure** f fluotantalic acid. **-toluol** n fluo(ro)toluene

Fluorür n (-e) (-ous) fluoride cf FLUORID

Fluorwasserstoff° m hydrogen fluoride; hydrofluoric acid. **f—·sauer** a (hydro)fluoride of. **F—säure** f hydrofluoric acid

Fluor·zink n zinc fluoride. **-zinn** n tin fluoride

Fluo·sulfosäure f fluosulfonic acid

fluotieren vi form a fluosilicate

Flur I m (-e) vestibule, (entrance) hall. **II** f (-en) 1 (open) field. 2 (Tex.) pile

Fluse f (-n) end of a thread

Fluß m (Flüsse) 1 flow; fluency; flux: **im F.** in flux; **in F.** in motion, flowing, moving. 2 (Metll.) state of fusion; molten metal (or mass). 3 river, stream. 4 (Min.) fluorspar. 5 (Med.) dicharge; catarrh. 6 (Gems) paste. 7 (Soap) figging

flüss. abbv (flüssig) liq. (liquid)

Fluß· flow; river: **-bett** n river bed. **-bild** n flow sheet. **-bildung** f (Soap) figging. **-dichte** f flux density. **-eisen** n lowcarbon (mild, soft, ingot) steel (or iron). **-erde** f earthy fluorite **-fieber** n rheumatic fever. **-gebiet** n river basin. **-gold** n river (or stream) gold. **-harz** n (gum) animé

flüssig a liquid, fluid, fluent, free-flowing; adv oft in liquid form—**f. machen** liquefy; **f. werdend** liquescent; **f—es Chlorzink** (Pharm.) solution of zinc chloride; **f—er Extrakt** fluidextract. **Flüssig·gas** n liquefied (petroleum) gas. **Flüssigkeit** f (-en) 1 liquid, fluid; liquor; (Physiol.) humor. 2 liquidity, fluidity, fluency, free flow

Flüssigkeits· liquid, fluid, hydraulic, hydrostatic: **-bad** n liquid bath. **-dichtemesser** m hydrometer. **-druck** m hydrostatic (liquid, fluid) pressure. **-förderung** f transport of liquids. **-grad** m degree of fluidity, viscosity. **gradmesser** *m* viscosimeter. **-heber** *m* siphon; (*Elec.*) salt bridge. **-kurve** *f* liquidus curve (for phase systems). **-linse** *f* (*Micros.*) immersion objective. **-maß** *n* liquid measure. **-menge** *f* amount of liquid. **-messer** *m* liquid meter, hydrometer. **-oberfläche** *f* liquid surface. **-pressung** *f* liquid (or hydraulic) pressure. **-reibung** *f* fluid friction. **-säule** *f* column of liquid. **-spiegel** *m* surface of a liquid; fluid level. **-spindel** *f* hydrometer float. **-stand** *m* fluid level. **-strom** *m* liquid stream, current of liquid. **-zerstäuber** *m* atomizer

Flüssig·kristall *m* liquid crystal. **-kristallanzeige** *f* liquid crystal (or falling-leaf) display

flüssig·machen *vt* 1 liquefy. 2 convert into cash; make available. **Flüssigmachen** *n*, **-machung** *f* liquefaction; conversion into cash

flüssig·werden* *vi* (s) turn (or become) liquid, liquefy, liquesce, fuse. **Flüssigwerden** *n* liquefaction, liquescence, fusion. **f—werdend** *p.a* liquescent

Fluß·kalkstein *m* limestone (used as a) flux. **-kies** *m* river gravel (or shingle). **f—kieselsauer** *a* fluosilicate of. **F—kieselsäure** *f* fluosilicic acid. **-metall** *n* liquid metal. **-mittel** *n* flux, fluxing material; antirheumatic. **-ofen** *m* flowing furnace. **-pferd** *n* hippopotamus. **-pulver** *n* flux powder, powdered flux. **-punkt** *m* melting point. **-sand** *m* river sand. **f—sauer** *a* (hydro)fluoride of. **F—säure** *f* hydrofluoric acid. **-schmiedeeisen** *n* ingot iron

Flußspat° *m* fluorspar, fluorite. **-erde** *f* earthy fluorite. **-säure** *f* hydrofluoric acid

Fluß·stahl *m* ingot steel; medium carbon steel. **-stein** *m* compact fluorite. **-ton** *m* river clay. **-wasser** *n* river water. **f—wasserstoffsauer** *a* (hydro)fluoride of. **F—wasserstoffsäure** *f* hydrofluoric acid

flüstern *vt/i* whisper

Flut *f* (-en) flow, high tide; flood; *pl oft* waters. **fluten** 1 *vt/vi*(s) flood, stream. 2 *vt* set afloat. 3 *vi* (h) swell, surge. **Flut·licht**

n floodlight. **Flutung** *f* flooding

Flux·mittel *n* flux

flz., flztr. *abbv* (flächenzentriert) face-centered

focht fought: *past of* FECHTEN

Föderation *f* (-en) federation. **föderativ** *a* federal, federated

Föhn *m* warm dry south wind *cf* FÖN

Föhre *f* (-n) Scotch pine (or fir)

Fokus *m* (-se), **fokussieren** *vt/i* focus. **Fokus·tiefe** *f* focal depth

folg. *abbv* (folgend . . .) fol. (following)

Folge *f* (-n) 1 (dire) consequence: **etwas zur F. haben** result in sthg, lead to sthg. 2 sequence, series, succession; order. 3: **in der F.: a** subsequently, **b** = **für die F.** in (the) future. 4: **F. leisten** (+ *dat*) follow up (on); comply (with). 5 installment; number, issue (of a publication). **-erscheinung** *f* consequence, result; aftereffect

folgen 1 *vi*(s) (+ *dat* or + **auf**) follow; comply with, observe: **daraus folgt** . . . from this it follows . . . ; **auf A folgt B** after A comes B. 2 *vi* (h) (*dat*) obey. **folgend** *p.a* following—**im f—en** in what follows, in the following discussion

folgender·maßen, -weise *adv* as follows, in the following manner

folgen·los *a* inconsequential, without result(s). **-reich** *a* having many consequences. **-schwer** *a* of serious consequence; momentous, grave; far-reaching

Folge·produkt *n* secondary product; (*Nucl.*) daughter product. **-punkte** *mpl* (*Magn.*) poles, consequent points. **-reaktion** *f* consequent (or secondary) reaction. **f—recht, -richtig** *a* logical, consistent. **F—richtigkeit** *f* (logical) consistency

folgern *vt/i* conclude, infer. **Folgerung** *f* (-en) conclusion, inference, induction

folge·widrig *a* illogical, inconsistent. **F—wirkung** *f* consequence, resultant. **-zeit** *f* time (or period) following—**in der F.** subsequently. **-zeitschalter** *m* sequence timer

folglich *adv* consequently, therefore

folgsam *a* obedient

Folie *f* (-n) foil; film, metal leaf; silvering (on mirrors). **Folien·abdruckverfahren**

n (Micros.) replica technique. **foliieren** *vt* foliate, coat with foil; silver (mirrors); page (a book)

Follikel *m* (-) follicle

Folin·säure *f* folinic acid

Fol·säure *f* folic acid

Fön *m* (-e) (hot-air) hair drier *cf* FÖHN

Fond *m* (-s) **1** basis, foundation. **2** gravy base. **3** *(Dye.)* ground, bottom, **4** back-(ground). *Cf* FONDS. **-farbe** *f* ground color (or shade)

Fonds *m* (-) fund(s) *cf* FOND

fönen *vt* dry (usu hair) with hot air

Fonzier·maschine *f* *(Paper)* staining machine

forcieren *vt* force, strain, overtax; push, promote; speed up, step up

Forcier·krankheit *f* strain disease (of metals). **-probe** *f* accelerated test

Ford·becher *m* Ford viscometer

Förder·anlage *f* conveyor (system). **-band** *n* conveyor belt

Förderer *m* (-) **1** promoter, patron. **2** accelerator. **3** conveyor

Förder·gut *n* output, goods to be transported. **-höhe** *f* delivery (or pressure) head (of a pump). **-kohle** *f* run-of-the-mine coal

förderlich *a* (oft + *dat* or + **für**) useful, favorable (to, for)

Förder·menge *f* output. **-mittel** *n* means of transportation; conveyor

fordern *vt* **1** demand, require; claim; charge. **2** challenge. **3** summon

fördern *vt* **1** promote, foster; accelerate. **2** convey, haul, deliver, feed *cf* ZUTAGE. **3** mine, extract; put out

Förder·quantum *n* *(Min.)* output. **-schnecke** *f* screw conveyor

Forderung *f* (-en) demand, requirement; claim; challenge *cf* FORDERN

Förderung *f* (-en) promotion, fostering; conveying, hauling, delivery, feed(ing); mining, extraction; *(Min.)* output *cf* FÖRDERN. **Förderungs·mittel** *n* *(Pharm.)* adjuvant

Förder·wagen *m* *(Min.)* mine car, tram. **-werk** *n* conveyor; elevator. **-wirkung** *f* promoting effect

Forelle *f* (-n) trout

Forellen·eisen *n* mottled white pig iron

Form *f* (-en) **1** form, shape, design. **2** (casting) mold; *(Soap)* frame; *(Metll.)* tuyère. **3** (social) convention—**in aller F.** with all formalities. **-änderung** *f* change of form; deformation, strain

formänderungs·fähig *a* deformable; plastic, ductile. **F—fähigkeit** *f* deformability; plasticity, ductility. **-messer** *m* deformometer, strainometer

Form·art *f* form species. **-artikel** *m* molded article (or pl oft goods)

Format *n* (-e) format, size, shape; stature

Formation *f* (-en) formation; *(Geol.)* system and era. **Formations·kunde** *f* *(Geol.)* stratigraphy

formbar *a* moldable, workable, plastic. **Formbarkeit** *f* moldability, workability, plasticity

form·beständig *a* shape-retaining, dimensionally stable. **F—bildung** f structure; *(Biol.)* morphogeny. **-eisen** *n* structural iron (or steel)

Formel *f* (-n) formula. **-bild** *n* (structural) formula

Form·element *n* structural element

Formel·gewicht *n* formula weight. **-gleichung** *f* chemical equation. **-register** *n* formula index

formen *vt* form, shape, mold; *(Soap)* frame

Formen·einstreichmittel *n* mold lubricant, mold release agent *cf* FORM 2. **-lehre** *f* morphology. **-naht** *f* *(Glass)* seam. **-sand** *m* molding sand. **-trennmittel** *n* mold release agent *cf* FORM 2

Form·erde *f* molding (or modeling) clay

Formerei, Förmerei *f* (-en) **1** molding shop. **2** molding (operation), pattern molding

Former·pech *n* molder's pitch. **-stoff** *m* plastic (or molding) material

Form·faktor *m* *(Cryst.)* atomic scattering factor. **-festigkeit** *f* stability of form. **-gebung** *f* fashioning, shaping; shape, design. **f—gerecht** *a* true to form, undistorted. **F—gestaltung** *f* (industrial) design, styling. **f—getreu** *a* true to form. **F—gips** *m* (high-grade) plaster of Paris. **-guß** *m* shaped casting; die casting. **f—haltend** *a* shape-retaining. **F—heizung**

f (Rubber) mold cure. **-holz** *n* plastic wood
Formiat *n* (-e) formate
formieren *vt* shape, form (up)
Formier·säure *f (Elec.)* forming acid
·förmig *a sfx* -shaped, -form(ed), ·oid
Form·kasten *m (Foun.)* molding box, flask.
   f—konstant *a* constant in form (*or*
   shape). **F—körper** *m* molded body.
   -lehm *m* casting loam. **-lehre** *f* profile
   gauge
förmlich *a* 1 formal. 2 literal, virtual; *a &
   adv* downright
Formling *m* (-e) molded article; casting;
   briquet
formlos *a* formless, shapeless, amorphous
Form·maschine *f* molding (*or* shaping)
   machine. **-masse** *f* molding material (*or*
   composition)
Formopersäure *f* per(oxy)formic acid
Form·presse *f* molding press. **-preßholz** *n*
   molded plywood. **-puder** *m* molding
   powder. **-sand** *m* molding (*or* foundry)
   sand **-schwindung** *f* shrinkage. **-stahl** *m*
   structural steel. **-stein** *m* folded brick;
   (*Lead*) pipe stone. **-stück** *n* shaped part.
   -teil *m* molded part. **-ton** *m* molded clay.
   -trennmittel *n* mold release agent
Formular *n* (-e) form, blank (to fill out)
formulieren *vt* formulate. **Formulierung**
   f (-en) formulation
Formung *f* (-en) forming, formation, shaping, molding; (*Soap*) framing
Form·veränderung *f* change of form, modification; deformation. **-wechsel** *m*
   change of form
formylieren *vt* formylate. **Formyl·säure** *f*
   formic acid
Form·ziffer *f* form (*or* shape) factor
forsch *a* spirited, vigorous, dashing
forschen *vi* investigate, do research;
   (**nach**) search (for). **Forscher** *m* (-), **For-
   scherin** *f* (-nen) investigator, scientist,
   researcher. **Forschung** *f* (-en) investigation, research
Forschungs· research: **-anstalt** *f* research
   institute. **-gebiet** *n* field of research.
   -geist *m* spirit of research (*or* inquiry).
   -gemeinschaft *f* research association—
   Deutsche F. German Research Council.
   -rakete *f* explorer rocket. **-sonde** *f* re-

search probe. **-vereinigung** *f* research
   association
Forst *m* (-e/-en) (cultivated) forest. **För-
   ster** *m* (-) forester, forest ranger.
Forst·kunde *f* forestry. **-leute** *pl* for-
   esters *cf* FORSTMANN
forstlich *a* (of the) forest, (of) forestry
Forst·mann *m* (*pl* -leute) forester. **-wesen**
   n, **-wirtschaft** *f* forestry
fort *adv & sep pfx* 1 away, gone *cf* FORT-
   SCHAFFEN. 2 on(ward), forward, ahead,
   pro· (*denoting continuation or progress*) *cf*
   FORTSCHRITT, FORTSETZEN: **in einem f.**
   continuously, without a break; **und so f.**
   and so on (*or* forth). *Cf* SOFORT. **··an** *adv*
   (t)henceforth, from now (*or* then) on
Fortbestand *m* continuation, continued
   existence. **fort·bestehen*** *vi* continue (to
   exist), survive
fort·bewegen 1 *vt* move (away, ahead),
   budge. 2: **sich f.** move ahead (along, forward). **Fortbewegung°** *f* forward motion, locomotion, progression
Fortbildung *f* continuing education. **Fort-
   bildungs·gewebe** *f (Bot.)* cambium
fort·bringen* I *vt* 1 carry (*or* take) away,
   remove. 2 carry forward, advance; support, maintain. II: **sich f.** get ahead, advance (oneself); make one's way
Fortdauer *f* continuation, (continued) duration; continued existence; perpetua-
   tion. **fort·dauern** *vi* continue (to exist);
   last
fort·diffundieren *vt* diffuse away
Fortdruck° *m* separate print; run-on
fort·entwickeln *vt* & **sich f.** continue
   developing
fort·fahren* I *vt* 1 haul away. II *vi* (s) 2 go
   (ride, drive, travel) away, depart. 3 continue, go on
Fortfall *m* dropping (out), discontinuance,
   disappearance; cessation: **in F. kommen**
   cease to exist. **fort·fallen*** *vi*(s) drop out,
   disappear; cease to exist
fort·fließen* *vi*(s) flow away (*or* off), run
   off
fort·führen *vt* 1 lead (take, carry) away).
   2 continue, carry on
Fortgang° *m* 1 departure. 2 progress: **F.
   haben** make progress. **fortgegangen**

departed: *pp of* **fort·gehen*** *vi* (s) **1** go away, depart, leave. **2** go on, continue

fortgeglommen continued to glimmer: *pp of* FORTGLIMMEN

fortgehoben lifted off: pp of FORTHEBEN

fortgenommen taken away: *pp of* FORTNEHMEN

fortgeschritten 1 *p.a* late (hour). **2** advanced, etc: *pp of* FORTSCHREITEN

fortgesetzt *p.a* constant. *See also* FORTSETZEN

fortgezogen moved away, etc: *pp of* FORTZIEHEN

fort·glimmen* *vi* continue to glimmer (*or* smolder)

fort·heben* 1 *vt* lift off; remove, eliminate. **2**: sich f. withdraw, disappear

fort·kochen *vt/i* **1** boil away (*or* off). **2** boil on

fort·kommen* *vi* (s) **1** get away, escape, disappear. **2** get along, (make) progress; thrive. **Fortkommen** *n* **1** escape; disappearance. **2** progress. **3** livelihood

fort·lassen* *vt* omit; let (sthg, sbdy) leave. **Fortlassung** *f* (-en) omission

Fortlauf *m* continuity, continuation. **fort·laufen*** *vi* (s) **1** run on, continue. **2** run away, escape. **fortlaufend** *p.a* continuous, continual, constant

fort·leben *vi* live on, continue to exist, survive. **Fortleben** *n* continued existence, survival

fort·leiten *vt* carry off, conduct (away); transmit

Fortnahme *f* removal; seizure. **fort·nehmen*** *vt* take away; (+ *d.i.o*) take (from)

fort·oxidieren *vt/i* oxidize off

fort·pflanzen 1 *vt & sich f.* propagate, transmit. **2**: sich f. spread, be propagated, be transmitted, travel; reproduce. **Fortpflanzung°** *f* propagation, transmission; reproduction. **Fortpflanzungs·geschwindigkeit** *f* velocity (*or* rate) of propagation (*or* transmission)

fort·rücken *vt, vi*(s) **1** move away. **2** move on, advance

Forts. *abbv* (Fortsetzung) cont. (continuation, continued)

Fortsatz° *m* **1** continuation, prolongation, extension. **2** (*Anat.*) process, appendix.

3 (*Mach.*) catch, stop

fort·schaffen *vt* remove, get rid of

Fortschr. *abbv* (Fortschritt) progress

fort·schreiten* *vi* (s) progress, advance. **fortschreitend** *p.a* progressive. **Fortschritt°** *m* (*sg & pl*) progress: **F—e machen** make progress. **fortschrittlich** *a* progressive

fort·setzen *vt & sich f.* continue, carry on (with): **wird fortgesetzt** (to be) continued. **Fortsetzung** *f* continuation; installment (of a serialized article)—**F. folgt** to be continued; **F. auf Seite 10** continued on page 10

fort·spülen *vt* wash away, rinse away

fortwährend *a* constant, continuous, incessant

fort·wirken *vi* remain effective, live on

fort·ziehen* 1 *vt* drag along (*or* away). **2** *vi* (s) move away, move on

fossil *a* fossil(ized): **f—es Wachs** fossil wax, ozocerite. **Fossil** *n* (-ein) fossil. **fossil·führend, fossil(ien)·haltig** *a* fossiliferous, fossil-bearing

fötal *a* fetal

foto-, Foto *n*(-s) = PHOTO etc: **foto·chemisch** *a* photochemical

Foulard·färbung *f* pad, dyeing. **foulardisieren** *vt* (*Dye.*) (slop-)pad

Fourier·reihen *fpl* Fourier series

F.P. *abbv* **1** französisches Patent) French patent. **2** (Fusionspunkt, Fließpunkt) m.p. (melting point)

Fp. *abbv* **1** (Fusionspunkt) m.p. (melting point). **2** (Füllpulver) shell explosive— **Fp. 02** TNT

fr. *abbv* (frei) free. **Fr.** *abbv* **1** (Franken) franc(s). **2** (Frau) Mrs.

Fracht *f* (-en) freight, cargo; freightage, hauling charge. **~brief** *m* waybill, bill of lading. **-dampfer** *m* freighter, cargo ship. **f—frei** *a* freight-free, freightage paid. **F—geld** *n* freightage, freight (-hauling) charge(s). **-gut** *n* freight, cargo. **-unkosten** *pl* freight charges

Frage *f* (-n) **1** question: **außer F.** beyond question; **F—n stellen** ask questions; **einen Punkt in F. stellen** question a point; **in F. kommen** come into question (*or* consideration); **nicht in F. kommen**

be out of the question. **2** matter, affair.
-bogen *m* questionnaire

fragen 1 *vt/i* (*oft* + **nach**) ask, inquire
(about); (+ **um**) ask (for). **2: sich f.** wonder, be curious—**es fragt sich,
ob** . . . the question is whether . . . *Cf*
GEFRAGT

Frage-steller *m* questioner. **-stellung** *f*
1 question, problem. **2** formulation of the
question. **-zeichen** *n* question mark

fraglich *a* **1** questionable, doubtful. **2** *e.g.*:
der f—e Punkt the point concerned (*or*
in question)

frag-los *adv* unquestionably. **-würdig** *a*
questionable, doubtful

frakt. *abbv* (fraktioniert) fractionated

fraktionär *a* fractional

Fraktionier-apparat *m* fractionating apparatus. **-aufsatz** *m* fractionating attachment, distilling tube. **-boden** *m* fractionating plate (*or* tray). **-destillation** *f*
fractional distillation

fraktionieren *vt* fractionate. **fraktioniert**
p.a oft, e.g.: **f—e Destillation** fractional
distillation

Fraktionier- fractionating: **-kolben** *m*
fractionating flask. **-turm** *m* fractionating tower, rectifying column

Fraktionierung *f* (-en) fractionation

Fraktions-aufsatz *m* fractionating attachment (*or* column), distilling tube. **-hut** *m*
fractionating receiver. **-sammler** *m* fraction collector. **-schnitt** *m* fraction, cut.
-vorlage *f* fractionating receiver

Fraktur *f* (-en) fracture

Frambösie *f* (*Med.*) frambesia, yaws

Frankfurter-schwarz *n* Frankfort black

franko *adv* postpaid, free

Frankreich *n* France

Franse *f* (-n) fringe; frayed end

franz. *abbv* (französisch) Fr. (French)

Franz-band *n* calf binding; calf-bound
book. **-branntwein** *m* **1** French brandy.
2 (*Pharm.*) (cheap brandy used as) rubbing alcohol. **-gold** *n* (pale) French leaf
gold (alloyed with silver)

Franzose *m* (-n, -n) **1** Frenchman.
2 monkey wrench

Franzosen-harz *n* guaiacum. **-holz** *n*
guaiacum, lignum vitae

Französin *f* (-nen) French woman (lady,

girl). **französisch** *a* French—**f—e
Beeren** Avignon berries; **f—es Leder** *n*
glove leather

Franz-topas *m* smokey topaz

frappant *a* striking

Fräs- (*Mach.*) milling. **Fräse** *f* (-n) milling
machine (*or* cutter). **fräsen** *vt* (*Mach.*)
mill. **Fräser** *m* (-) milling machine
(operator)

Fräs-maschine *f* milling machine. **-saat-
maschine** *f* rotary seed spreader

fraß corroded: *past of* FRESSEN. **Fraß** *m* (-̈e)
1 (animal) feed. **2** corrosion. **3** insect or
rodent damage. **4** (*Med.*) caries

Fraß-gang *m* (larval *or* insect) gallery, tunnel. **-gift** *n* (stomach) insecticide; rodent
poison

Frau *f* (-en) **1** woman, **2** wife. **3** lady. **4** Mrs.,
Ms

Frauen-arzt *m* gynecologist. **-distel** *f*
Scotch thistle. **-eis** *n* (*Min.*) selenite.
-flachs *m* toad flax. **-glas** *n* (*Min.*)
selenite—**russisches F.** muscovite.
-haar *n* (*Bot.*) maidenhair

frauenhaft *a* womanly

Frauen-handschuh *m* (*Bot.*) columbine,
lady's glove. **-heilkunde** *f* gynecology.
-milch *f* human milk. **-rechte** *npl* wom-
en's rights. **-schuh** *m* (*Bot.*) ladies' slipper. **-spat** *m* (*Min.*) selenite. **-spiegel** *m*
Venus's looking glass

Fräulein *n* (-) **1** young lady. **2** single lady.
3 Miss. **4** (*Tlph.*) operator

fraulich *a* womanly

Fraunhofer(sche) Linien Fraunhofer
lines

frbl. *abbv* (farblos) colorless

Frdl. *abbv* (Friedländers Fortschritte der
Teerfarbenfabrikation) Friedlaender's
"Progress in the Production of Coal Tar
Dyes."

frech *a* impudent, fresh; brazen. **Frech-
heit** *f* (-en) **1** impudence, brazenness.
2 impudent (*or* brazen) thing to say (*or* do)

frei I *a* & *sep pfx* **1** free. **2** unbound, uncombined (*esp Chem.*). **3** clear (path); open
(space); blank; vacant, available. **4** open-
air, outdoor. **5** released, open (to the public); freed, liberated (*incl Chem.*). **6** (time)
off (from work). **II** *sfx* **-frei, -less, an-, non-**
cf ALKOHOLFREI, WASSERFREI

Frei·ballon *m* free balloon

Freiberger Aufschuß (*Anal.*) fusion with sulfur and alkali carbonate

frei·beweglich *a* freely moving, mobile. **F—brief** *m* license

frei·drehen *vt* (*Ceram.*) throw

Freie *n* (*adj ends*) open air

Frei·fallmischer *m* gravity mixer. **f—fließend** *a* free-flowing

Frei·gabe *f* release; decontrol. **frei·geben*** *vt* **1** set free, release, open (to the public); permit. **2** vacate; leave open. **3** give (time) off

freigebig *a* generous. **Freigebigkeit** *f* generosity

Frei·gebung *f* release. **f—geworden** *p.a* liberated *cf* FREIWERDEN

frei·halten* *vt* **1** reserve, keep open (clear, free, available). **2** treat (*usu* to drinks)

Freiharz·gehalt *n* free resin content. **-leim** *m* (*Paper*) acid size. **f—reich** *a* rich in free resin

Freiheit *f* (-en) freedom, liberty; exemption; (*Phase Systems*) degree of freedom; **–in F. setzen** set free, liberate. **Freiheits·grad** *m* degree of freedom

Frei·heizung *f* (*Rubber*) open (*or* steam) cure

frei·kommen* *vi* (s) get away; be released; be acquitted

Frei·lage *f* exposed site (*or* position). **-lager** *n* bonded warehouse. **-lagerprüfung** *f* (*Paint*) exposure (testing)

frei·lassen* *vt* set free, liberate; decontrol. **F—lassung** *f* release, liberation; decontrol. **-lauf** *m* freewheel(ing)

frei·legen *vt* expose

Frei·leitung *f* open-air wires (wiring, piping); (*Elec. oft*) overhead transmission line

freilich *adv* to be sure, of course; admittedly

freiliegend *a* exposed

Freilicht·, Freiluft· open-air. outdoor: **Freiluft·bewitterung** *f* open-air weathering

frei·machen 1 *vt* stamp, put postage on (mail). **2** *vi* take time off. **3: sich f. (von)** free oneself, disengage oneself (from); get rid (of); make oneself available

Frei·marke *f* postage stamp. **-mut** *m* frankness, candor. **-raum** *m* **1** open (*or* unused) space; (*Tanks*, etc) ullage. **2** (*fig*) room for action, leeway. **f—schaffend** *a* freelance. **F—sein** *n* free state, freedom

frei·setzen *vt* set free, liberate, release. **F—setzung** *f* liberation, release

frei·sinnig *a* liberal(-minded). **F—spiegel·stollen** *m* (*Metll.*) free-flow gallery

frei·sprechen* *vt* acquit, absolve, exonerate

Frei·staat *m* republic; free state. **-tag** *m* Friday

frei·werden* *vi* (s) be freed (liberated, released). **freiwerdend** *p.a* nascent

frei·willig *a* voluntary, spontaneous; gratuitous; *adv oft* of one's own free will. **F—willige** *m/f* (*adj engs*) volunteer. **-zeit** *f* leisure time, time off

fremd *a* **1** foreign. **2** strange. **3** extraneous. **4** other people's, someone else's. **··artig** *a* odd, peculiar, exotic. **F—atom** *n* heteroatom. **-befruchtung** *f* cross-fertilization. **-bestandteil** *m* foreign (*or* extraneous) ingredient (*or* constituent). **-bestäubung** *f* cross-pollination

Fremde 1 *m/f* (*adj ends*) stranger; foreigner; visitor, tourist. **2** *f* foreign land(s)**—in der F.** abroad

Fremd·einfluß *m* external effect, outside influence. **f—farbig** *a* artifically colored. **F—gas** *n* extraneous gas, gaseous impurity. **-körper** *m* extraneous (*or* foreign) body (*or* object). **-metall** *n* foreign metal. **f—sprachig, -sprachlich** *a* foreignlanguage. **F—stoff** *m* extraneous (*or* foreign) substance, impurity. **f—stoffig** *a* heterogeneous. **F—stoffigkeit** *f* heterogeneity. **-strom** *m* (*Elec.*) stray current. **-wasser** *n* added water. **-wort** *n* foreign word, word of foreign origin. **-zündung** *f* ignition by external means

Frequenz *f* (-en) **1** (*incl Phys.*) frequency. **2** attendance (= number of people). **3** traffic (density). **4** pulse rate; heartbeat. **-bereich** *m* frequency range. **-gesetz** *n* law of frequency. **-messer** *m* frequency (*or* wave) meter, ondometer. **f—moduliert** *a* frequency-modulated, FM. **-wert** *m* frequency (value)

fressen* (fraß, gefressen; frißt) **1** *vt* consume, devour. **2** *vt/i* eat; corrode. **3** *vi*

(*Mach.*) bind, jam, seize; **um sich f.** spread. **4: sich f. (in)** eat (into), corrode. **Fressen** *n* meal, feed, feast. **fressend** *p.a* corrosive

freß‧gierig *a* greedy, voracious. **F—zelle** *f* phagocyte

Freude *f* (-n) joy, pleasure, delight; happiness. **freudig 1** *a* joyful, glad. **2** *sfx*, e.g.: **sublimier‧freudig** having a tendency to sublime, readily subliming; **arbeits‧ ‧freudig** ready (*or* eager) to work, industrious. **Freudigkeit** *f* **1** joyfulness, gladness. **2** *as sfx* tendency to . . . , ease of . . .

freuen 1 *vt* gladden, delight—**es freut mich** I am happy. **II: sich f. 2 (über, an)** be glad (*or* happy) about, enjoy. **3 (auf)** look forward (to)

Freund *m* (-e), **Freundin** *f* (-nen) friend. **freundlich** *a* friendly, nice; (*esp as sfx*) favorable, beneficial *cf* UMWELTFREUND-LICH. **Freundschaft** *f* (-en) friendship; (circle of) friends

Friede(n) *m* peace, peacefulness, tranquillity. **Friedens‧korps** *n* Peace Corps

fried‧fertig *a* peaceable. **F—hof** *m* cemetery

friedlich *a* peaceful

frieren* (fror, gefroren) **1** *vt*: **es friert ihn** he is freezing. **2** *vi* (h,s) freeze (up). *Cf* GEFRIEREN. **Frieren** *n* chill, shivering, ague. **Frier‧punkt** *m* freezing point

Frigen *n* Freon, refrigerant

friktionieren *vt* **1** subject to friction. **2** glaze by friction (e.g., cellulose)

Friktions‧antrieb *m* (*Mach.*) friction drive. **-messer** *m* tribometer, friction meter. **-zünder** *m* friction igniter

frisch *a* **1** fresh: **von f—em** afresh. **2** new. **3** lively, cheerful. **4** cool. **5** green (hide). **6** (*Metll.*) refined

Frisch‧arbeit *f* fining (process). **f— backen** *a* fresh-baked, freshly baked. **-bereitet** *a* freshly prepared. **F—blei** *n* refined lead. **-dampf** *m* live steam

Frische *f* freshness, vitality; coolness

Frisch‧eisen *n* refined iron

frischen *vt* **1** (*Metll.*) fine, refine; revive (litharge); reduce (lead). **2** reclaim (rubber, oil). **Frischer** *m* (-) (re)finer. **Frischerei** *f* (-en) **1** refinery. **2** (re)fining. **-‧roheisen** *n* forge pig

Frisch‧erz *n* raw ore. **-esse** *f* refining furnace. **-feuer** *n* refining fire. **-feuereisen** *n* refined iron, charcoal hearth iron. **-feuerstahl** *m* reinforced steel. **-fleisch** *n* fresh meat. **f—gebacken** *a* freshly baked, oven-fresh; (*fig*) brand-new. **-gebrannt** *a* freshly burned. **-gefällt** *a* freshly precipitated. **-gelöscht** *a* freshly quenched (*or* slaked). **F—gewicht** *n* fresh (undried, green) weight. **-glätte** *f* litharge containing silver

Frischhalte‧beutel *m* plastic food-storage bag. **frisch‧halten*** *vt* keep fresh. **Frischhalte‧packung** *f* plastic food wrapping

Frisch‧ fresh(ly); refined, refining:- **haut** *f* green hide. **-herd** *m* refining furnace; forge hearth. **-lauge** *f* fresh lye (*or* liquor); (*Paper*) white liquor. **-lech** *m* crude mat(te)

Frischling *m* (-e) (*Metll.*) scoria; (*Zool.*) young boar

Frisch‧ fresh(ly); refined, refining: **-luft** *f* fresh air. **-ofen** *m* refining furnace. **-periode** *f* (*Metll.*) oxidizing period. **-prozeß** *m* (*Metll.*) (re)fining process. **-roheisen** *n* pig iron for refining. **-schlacken** *fpl* refinery cinders (*or* slag). **-schlamm** *m* raw (*or* fresh) sludge. **-stahl** *m* refinery steel, German steel. **-stück** *n* (*Copper*) liquation cake

Frischung *f* (-en) (re)fining, etc *cf* FRISCHEN

Frisch‧verfahren *n* (*Iron*) refining process. **-wasser** *n* fresh(ly drawn) water *cf* SÜSSWASSER. **-wirkung** *f* (*Metll.*) (re)fining

frißt corrodes: *pres of* FRESSEN

Frist *f* (-en) **1** (set) period of time (up to a deadline), time allowed. **2** time limit, deadline. **3** extension (of time allowed). **fristen** *vt* eke out (one's existence)

Fritte *f* (-n) frit. **-‧glasur** *f* fritted glaze. **fritten 1** *vt* frit. **2** *vt/i* sinter. **3** *vi* cohere. **Fritten‧porzellan** *n* soft (*or* frit) porcelain

Fritt‧ofen *m* frit kiln; (*Glass*) calcar. **-porzellan** *n* frit porcelain

Frittung *f* (-en) fritting, sintering

Frl. *abbv* (Fräulein) Miss, etc

froh *a* glad, joyous, happy. **fröhlich** *a* joyful, happy, cheerful

Froh·sinn *m* cheerfulness. **f—wüchsig** *a* fast-growing

fromm *a* religious; good-natured, docile; well-meant

Frontal·zusammenstoß *m* head-on collision

fror froze: *past of* FRIEREN

Frosch *m* (⁻e) frog; (*Mach.*) cam, dog, bar, arm. **~arten** *fpl* batrachians. **-laich** *m* frog spawn

Froschlaich·pflaster *n* (*Pharm.*) lead plaster. **-pilz** *m* (*Bot.*) leuconostoc

Frosch·löffel *m* (*Bot.*) water plantain (*Alisma*)

Frost *m* frost; chill(s); **5 Grad F.** 5 degrees below zero C. **f—·beständig** *a* frost-resistant. **F—beständigkeit** *f* resistance to frost.

frösteln 1 *vt* (*imps*): **es fröstelt sie** = **2** *vi*: **sie f.** they feel chilly

frost·empfindlich *a* sensitive to frost

frosten 1 *vt* deep-freeze. **2** *vi*: **es frostet** there is a frost. **Froster** *m* (-) freezing compartment, freezer (of a refrigerator)

Frost·grad *m* degree below 0°C. **f—hart** *a* **1** frost-resistant. **2** (*Bot.*) hardy

frostig *a* frosty, chilly, frigid

Frost·mischung *f* freezing mixture. **-mittel** *n* antifreeze. **-punkt** *m* freezing point. **-schaden** *m* frost damage. **-schnitt** *m* frost section. **-schutzmittel** *n* antifrost agent; antifreeze. **f—sicher** *a* frostproof

frottieren *vt* rub (down), massage

Frucht *f* (⁻e) **1** fruit. **2** produce; grain, crop. **3** product. **4** fetus; embryo; child. **~abtreibung** *f* abortion. **-abtreibungsmittel** *n* abortifacient. **f—artig** *a* fruit-like. **F—äther** *m* fruit essence, (*specif*) amyl acetate

fruchtbar *a* fruitful, fertile, productive

Frucht·blatt *n* carpel. **-brand** *m* ergot. **-branntwein** *m* fruit brandy. **f—bringend** *a* fruitful, productive

fruchten *vi* bear fruit, be of use: **das fruchtet nichts** that's of no use

Frucht·essenz *f* fruit essence (*or* extract). **-essig** *m* fruit vinegar. **-fleisch** *n* fruit pulp. **-folge** *f* (*Agric.*) crop rotation. **-gelee** *n* fruit jelly. **-halter** *m* uterus. **-haut** *f* seed coat. **-knoten** *m* (*Bot.*) ovary.

-konserve *f* fruit preserves. **f—los** *a* fruitless. **F—mark** *n* fruit pulp. **-saft** *m* fruit juice. **-säure** *f* fruit acid. **-schale** *f* (*Bot.*) pericarp. **-wand** *f* endocarp. **-wasser** *n* **1** amniotic fluid. **2** fruit juice. **-wasserhaut** *f* amnion. **-wechsel** *m* crop rotation. **-zucker** *m* fruit sugar, levulose

frug asked: (*unauthorized*) *past of* FRAGEN

früh *a, adv* (& *oft in compds*) early, premature; *adv oft* (in the) morning: **morgen f.** tomorrow morning. **Frühe** *f* morning. **früher** *a, adv* **1** earlier, sooner, **2** former(ly), (*esp as adv*) before; **es war f.** it used to be. **Früh·erkennung** *f* (*esp Med.*) early detection. **frühestens** *adv* at the earliest

Früh·holz *n* spring wood. **-jahr** *n* = **Frühling** *m* spring(time)

früh·reif *a* **1** early-ripe(ning). **2** precocious. **F—stück** *n* breakfast. **f—stücken** *vt/i* have (for) breakfast

früh·treiben* *vt* (*Agric.*) force

früh·zeitig *a* early; premature, precocious. **F—zündung** *f* premature ignition; pre-ignition; backfire

Fuchs *m* (⁻e) **1** fox; fox fur. **2** (*Horse*) sorrel, bay. **3** freshman. **4** flue. **5** (*Metll.*) tramp iron, (piece of) unmeltable iron; accidental impurity. **~brücke** *f* flue bridge. **-fett** *n* fox fat. **-gas** *n* flue gas

fuchsig *a* **1** fox-red, reddish brown. **2** furious

Füchsin *f* (-nen) vixen

fuchsin·schweflig *a*: **f—e Säure** Schiff's reagent. **F—tinktion** *f* fuchsin stain(ing)

Fuchs·kanal *m* flue. **f—rot** *a* fox-red, reddish brown

Fuchtel *f* (-) control, domination

Fuder *n* (-) **1** wagonload. **2** (large) wine measure

Fug *m*: **mit F. und Recht** with every right

Fugazität *f* fugacity

Fuge *f* (-n) **1** joint, seam. **2** crack + gap. **3** groove. **fugen** *vt* join(t), fit together, seal

fügen I *vt* **2** form, join, assemble. **2** shape (wood, etc) for jointing. **3** (**an**) join, attach (to, onto). **4** (**zu**) add (to). **5** (*usu* + **daß**) ordain, decree (that). **II**: **sich f. 6** (**in**) fit (into), find its place (in); submit (to). **7** (**an**) adjoin; follow. **8** (**zu**) fit in, go (with). **9** (*dat*) give in, adapt oneself (to). **10** (*oft* + **daß**) turn out (right), come about (that)

fugenlos *a* jointless, seamless

füglich *adv* **1** conveniently. **2** justifiably

fügsam *a* tractable, pliant

Fügung *f* (-en) dispensation (of fate)

fühlbar *a* palpable, noticeable; perceptible—**sich f. machen** make itself felt

fühlen *vt, vi,* **sich f.** feel. **Fühlen** *n* feeling, sensation. **Fühler** *m* (-) feeler, antenna, sensor, detector, probe. **Fühlung** *f* (-en) feeling, sensation; touch, contact: **F. nehmen** (**mit**) have (*or* make) contact (with).

Fühlung·nahme *f* (making) contact

fuhr drove, rode, etc: *past of* FAHREN

Fuhre *f* (-en) **1** (truck)load. **2** hauling, carting

führen I *vt/i* **1** lead. **II** *vt* **2** guide, steer; drive; bring (up). **3** carry, bear. **4** carry on, conduct, keep (up). **5** wield, handle, use; run (machines). **III: sich f.** behave, conduct oneself. **·führend** *p.a sfx* -bearing, ·iferous. **Führer** *m* (-), **Führerin** *f* (-nen) **1** leader. **2** guide; pilot. **3** driver. **Führerschaft** *f* leadership, direction. **Führer·schein** *m* driver's license. **Führung** *f* (-en) **1** leadership, management. **2** leading position, lead. **3** guidance. **4** guided tour. **5** guide channel (*or* mechanism). **6** carrying (on), conduct. **7** handling, use, operation. **8** behavior

Führungs·stange *f* guide rod. **-stift** *m* guide pin. **-system** *n* guide system

Fuhr·werk *n* horse-drawn vehicle, cart

Fulgen·säure *f* fulgenic acid

Füll· filling, bottling; feed; stuffing: **-apparat** *m* filling (*or* bottling) machine(ry). **-appretur** *f* (*Tex.*) filling (finish). **-dichte** *f* packing density

Fülle *f* (-n) **1** abundance. **2** fullness, richness; body, depth (*esp* of colors). **3** filling, stuffing. **4** corpulence

füllen I *vt* **1** fill. **2** run, pour (liquid or bulk material into vessels). **3** stuff. **4** load (paper). **5** plump (leather). **II: sich f.** fill (up). **Füllen** *n* **I** filling, pouring; stuffing, etc. **II** (-) foal, colt. **Füller** *m* (-) filler; loader; fountain pen

Fuller·erde *f* fuller's earth

Füll· filling, bottling, loading: **-feder** *f*

fountain pen. **-flüssigkeit** *f* filling liquid; (*Micros., etc*) immersion liquid. **-gerbung** *f* plumping tannage. **-gut** *n* packing, stuffing; contents. **-halter** *m* fountain pen. **-haus** *n* filling house (*or* room); (*Soap*) frame room **-horn** *n* cornucopia. **-kelle** *f* filling ladle. **-koks** *m* bed coke. **-körper** *m* filling (*or* packing) material, filler. **-körpersäule** *f* packed column (*or* tower). **-körperturm** *m* packed tower. **-kraft** *f* filling power. **-löffel** *m* cf FÜLLÖFFEL. **-masse** *f* filling material; (*Sugar*) fillmass, massecuite. **-material** *n*, **-mittel** *n* filler, filling (packing, loading) material; stuffing

Fullöffel *m* (= **Füll·löffel**) filling ladle

Füll· filling, bottling, loading: **-pulver** *n* (*Explo.*) filling powder—**F. 02** TNT. **-rohr** *n*, **-röhre** *f* filling tube, feed pipe, spout. **-säure** *f* battery acid, electrolyte

Füllsel *n* (-) filler; filling; stuffing; stopgap

Füll·stand *m* level (of contents in a container). **-stoff** *m* filler, filling (packing, loading) material; stuffing. **f—stoffarm** *a* lightly loaded, loosely filled. **-stoffreich** *a* fully packed, heavily loaded. **F—strich** *m* filling (*or* "full") mark. **-ton** *m* filling clay. **-trichter** *m* filling (*or* loading) funnel, hopper

Füllung *f* (-en) **1** filling, packing, stuffing. **2** bottling. **3** load, charge. **4** seal. **5** (door) panel

Füllungs·grad *m* degree of filling

fulminant *a* fulminating. **fulminieren** *vi* fulminate

Fulminur·säure *f* fulminuric acid

Fulvo·säure *f* fulvic acid

fum. *abbv* fumaroid

Fumar·säure *f* fumaric acid

Fund *m* (-e) discovery, find; object found

Fundament *n* (-e) foundation; base; (*Metll.*) bottom plate, soleplate

Fund·büro *n* lost-and-found office

fundieren *vt* **1** put on a sound basis; found, lay the foundation(s) of. **2** fund. **Fundierung** *f* (sound) establishment, founding; funding

fündig *a* rich, productive; ore-bearing—**f. werden** strike ore (deposits)

Fund·ort *m* finding place, locality (of a find). **-sache** *f* found article. **-stätte** *f* locality (of a find, discovery). **-stelle** *f* lost-and-found office. **-zettel** *m* (*Min.*) locality tag

fünf *num* five; *as pfx oft* penta-, quintuple. **Fünf** *f* (-en) (figure) five). **f—·atomig** *a* pentatomic. **-basisch** *a* pentabasic. **F—eck** *n* pentagon. **f—eckig** *a* pentagonal

Fünfer *m* (-) (number, figure) five

fünferlei *a* (*no endgs*) of five different kinds

fünffach *a* quintuple, fivefold, (*esp* + *Chem. compd*) penta: **f—es Chlorid** pentachloride; **f.-Chlorphosphor** phosphorous pentachloride

fünf·gliedrig *a* five-member(ed). **-mal** *adv* five times. **F—ring** *m* five-member(ed) ring. **f—seitig** *a* five-sided, pentahedral

fünft *a* fifth—**zu f.** in a group of five. **Fünftel** *f* (-) (*Frac.*) fifth. **fünftens** *adv* fifthly, in the fifth place

fünf·wertig *a* pentavalent, quinquevalent. **F—wertigkeit** *f* pentavalence, quinquevalence

fünfzehn *num* fifteen. **fünfzehnt** *a* fifteenth

fünfzig *num* fifty. **fünfzigst** *a* fiftieth

fungieren *vi* function, act, serve

Fungisterin *n* fungisterol

fungizid *a* fungicidal. **fungös** *a* fungous

Funk *m* radio. **-·anlage** *f* radio station. **-apparat** *m* radio (set). **-bericht** *m* radio report

Funke *m* (-ns, -n) spark = FUNKEN. **fun·kelig** *a* sparkling, glittering. **funkeln** *vi* sparkle, glitter

funkel·nagelneu *a* brand-new

funken 1 *vt/i* radio, transmit. 2 *vi* work; sparkle; click, work out right

Funken *m* (-) spark. **-·auswurf** *m* spark emission. **-bildung** *f* spark formation, sparking. **-entladung** *f* spark discharge. **f—frei** *a* sparkless; (*Elec.*) non-arcing. **F—garbe** *f* shower of sparks. **-geber** *m* spark coil, sparking device. **-holz** *n* touchstone. **-induktor** *m* spark (*or* inductor) coil). **f—los** *a* sparkless. **F—sammler** *m* spark condenser. **-schlagweite** *f* sparking distance. **-spannung** *f* spark voltage. **-spektrum** *n* spark spectrum **-spiel** *n* play of sparks. **-sprühen** *n* spark emission, scintillation. **f—**

sprühend *a* spark-emitting, scintillating. **F—station** *f* radio room **-strecke** *f* spark gap. **-überschlag** *m* arcing, flashover. **-zünder** *m* spark igniter; (*Explo.*) jump-spark cap. **-zündung** *f* spark ignition

Funker *m* (-) radio operator

Funk· radio: **-gerät** *n* radio (set), transmitter. **-meßgerät** *n* radar set. **-meßtechnik** *f* radar direction finding. **-netz** *n* radio network. **-technik** *f* radio technology

Funktion *f* (-en) 1 (*incl Math.*) function: **in F. treten** start to function (*or* operate); **in F. sein** be operating. 2 office, official capacity. **funktionell** *a* functional

Funktionen·tafel *f* table of functions

funktionieren *vi* work, function, operate

Funktions·erprobung *f* operational test(ing). **f—fähig** *a* operable, functional. **F—probe, -prüfung** *f* operational test. **f—sicher** *a* reliable (in operation). **F—störung** *f* (*Med.*) functional disorder, malfunction. **f—tüchtig** *a* operational, in working order

Funk·weg *m*: **auf dem F.** by radio. **-wesen** *n* (field of) radio (broadcasting, communications)

für *prep.* (*accus*) 1 (*usu*) for. 2: **Tag f. Tag** day after day; **Stück f. Stück** piece by piece. 3: **was f.** (**ein**) what kind of. 4: **f. sich** by itself, apart, separate; per se *cf* AN I. 5: **das hat etwas f. sich** there is something to be said for that. *Cf* HALTEN

Furche *f* (-n) furrow; groove channel; wrinkle. **furchen** *vt* furrow, groove; wrinkle. **Furchen·spatel** *f* grooved spatula. **furchig** *a* furrowed

Furcht *f* (*usu* + **vor**) fear (of). **furchtbar** *a* fearful; terrible, awful; terrific. **fürchten** *vt*, *vi*, **sich f.** fear, be afraid. **fürchterlich** *a* fearful; terrible. **furchtsam** *a* timid

Furchung *f* (-en) furrowing; segmentation; cleavage

Furil·säure *f* furilic acid

Furnier *n* (-e) veneer. **furnieren** *vt* veneer; inlay, incrust

Furnier·holz *m* veneer wood; plywood. **-platte** *f* plywood

Furon·säure *f* furonic acid

Furoylierung *f* furoylation

fürs *contrac* = **für das** *cf* FÜR

Fürsorge *f* 1 care; solicitude. 2 welfare (service); relief, assistance

Fürst *m* (-en, -en) (sovereign) prince. **fürstlich** *a* princely; lavish

Furt *f* (-en) ford

Fürwort° *n* pronoun

Fusel *m* (-) 1 fusel oil. 2 rotgut, bad liquor. **--branntwein** *m* liquor containing fusel oil. **f—frei** *a* free from fusel oil. **F—geruch** *m* odor of fusel oil. **f—haltig** *a* containing fusel oil

fuselig *a* 1 containing fusel oil. 2 intoxicated

Fusel-öl *n* fusel oil

Fusion *f* (-en) 1 (*usu, incl Nucl.*) fusion. 2 (*Com., Econ.*) merger. **fusionieren** *vt* fuse; amalgamate. **Fusions·punkt** *m* fusing (*or* melting) point

Fuß *m* (-̈e) 1 foot; footing, base: **zu F.** on foot; **auf gutem (gleichem, vertrautem) F.** on good (equal, intimate) terms, on a good (etc) footing: (**festen**) **F. fassen** gain a (firm) footing (*or* foothold). 2 (*in compds oft*) ped·, pedal: **-arzt** *m* chiropodist, podiatrist. **-blatt** *n* 1 sole (of the foot). 2 (*Bot.*) mayapple, duck's foot. **-boden** *m* floor; flooring

Fußboden· floor: **-belag** *m* floor covering. **-decklack** *m* floor varnish. **-platte** *f* floor tile. **-wachs** *n* floor wax *or* polish). **-ziegel** *m* floor tile; paving brick

Fuß·brett *n* pedal

Fussel *f* (-n) (piece of) lint

fußen *vi* (**auf**) rest, be based (on)

Fuß·gänger *m* pedestrian. **-gelenk** *n* ankle (joint). **-gestell** *n* pedestal, base. **-glätte** *f* black impure litharge. **-hebel** *m* pedal. **-mehl** *n* low-grade flour. **-nagel** *m* toenail. **-note** *f* footnote. **-pflege** *f* pedicure. **-pfund** *n* foot-pound. **-puder** *m* foot powder. **-punkt** *m* (*Math.*) foot; (*Astron.*) nadir. **-teppich** *m* carpet, rug. **-tritt** *m* 1 kick. 2 pedal. **-weg** *m* 1 foot path. 2 distance to walk. **-wippschalter** *m* foot (*or* pedal) switch. **-wurzel** *f* (*Anat.*) tarsus. **-zehe** *f* toe. **-zylinder** *m* round glass jar

Fustage *f* 1 empties (= containers, barrels, crates, etc). 2 charge for container(s)

Fustik *m*, **--holz** *n* (young) fustic

Futter *n* (-) 1 feed, fodder, (animal) food, 2 lining, coating, casing. 3 chuck (of a lathe). **Futteral** *n* (-e) case, sheath, bag (for eyeglasses, sliderule, knife, etc)

Futter· feed; lining: **-bohne** *f* feed bean. **-erbse** *f* field pea. **-gerste** *f* winter (*or* feed) barley. **-getreide** *n* feed grain(s). **-gewächse** *npl* feed (*or* forage) plants. **-gras** *n* green fodder. **-hefe** *f* feed yeast. **-kalk** *m* feed lime, calcium additive for feed (*usu* calcium acid phosphate). **-klee** *m* red clover. **-kohl** *m* feed kohlrabi. **-korn** *n* feed grain. **-kräuter** *npl* feed (*or* forage) plants. **-leder** *n* lining leather. **-masse** *f* (*Metll.*) lining material. **-mehl** *n* low-grade flour, screenings. **-mittel** *n* feed material; lining material

füttern *vi* 1 feed. 2 line, case, coat *cf* FUTTER

Futter· feed, lining: **-pflanze** *f* forage plant. **-rohr** *n* casing (of a bore-hole). **-rübe** *f* common turnip; feed beet, mangel(-wurzel). **-salz** *n* additive salt for feed (*usu* CaHPO$_4$). **-stein** *m* lining brick. **-stoff** *m* 1 feed(ing material). 2 lining (material). **-stroh** *n* forage (*or* feed) straw

Fütterung *f* (-en) 1 feeding (time); feed. 2 lining; casing, sheathing

Fütterungs·stoff *m* lining material **-versuch** *m* feeding experiment

Futter·verbrauch *m* feed (*or* food) consumption. **-wert** *m* feed (*or* forage) value. **-wicke** *f* common vetch. **-wurzel** *f* forage root

F.Z$_J$ *abbv* (Farbzahl gegen Jod) iodine number

G

G. *abbv* **1** (Gewicht) wt. (weight). **2** (Gesellschaft) soc., co. (society, company)
gab gave: *past of* GEBEN
Gaban·holz *n* camwood; (red) sandalwood
Gabe *f* (-n) **1** gift. **2** dose, dosage. **3** talent. **4** donation
Gabel *f* (-n) fork; crotch; (*Bot.*) tendril
gabel·artig, -förmig, gabelig *a* forked, bifurcated
Gabel·hubstapler *m* fork lift truck. **-klammer, -klemme** *f* pinch clamp (for ground joints). **-kneter** *m* (forked) mixer, kneader. **-knoten** *m* crotch, fork. **-mehl** *n* superfine flour.
gabeln 1 *vt* spear. **2: sich g.** fork, divide
Gabel·rohr *n*, **-röhre** *f* forked tube, Y-tube. **-stapler** *m* fork lift truck. **-teilung** *f* forking, bifurcation
Gabelung *f* (-en) fork; forking, bifurcation
Gadolin·erde *f* gadolinia, gadolinium oxide
gaffen *vi* gape, stare, gawk
Gagat *m* (-e), **-kohle** *f* (*Min.*) jet
gähnen *vi* yawn
Galakturon·säure *f* galacturonic acid
Galanga·öl *n* galanga oil, galingale oil. **-wurzel** *f* galingale (root)
Galban *n*, **-harz** *n* galbanum
Galenit *n* (*Min.*) galena, lead glance
Galgant *m* (-en, -en) galingale, = GALANGA
Gali(t)zen·stein *m* (white) vitriol, zinc sulfate; **blauer G.** blue vitriol
galizisch *a* Galician
gall·abtreibend *a* cholagog, bile-removing
Gallamin·blau *n* gallamine blue
Gall·apfel *m* gall(nut), nutgall
Galläpfel·abkochung *f* nutgall decoction. **-aufguß** *m* nutgall infusion. **-beize** *f* (*Dye.*) gall steeping. **-gerbsäure** *f* gallotannic acid. **g–sauer** *a* gallotannate of . . . **G–säure** *f* gallotannic (ordinary tannic) acid. **-tinktur** *f* tincture of gallnuts
Gallat *n* (-e) gallate
Gall·beize *f* (*Dye.*) gall steep

Galle *f* (-n) **1** (*genl.*) gall. **2** (*specif*) bile; nutgall, gallnut; flaw; bubble, blister; (*Glass*) sandiver; (*Min.*) nodule; (*Anat.*) gall bladder
Gall·eiche *f* gall oak
gallen *vt* **1** remove the gall from. **2** (*Dye.*) gall, steep with gallnuts. **gällen 1** *vt* = GALLEN. **2** *vi* (s) turn bitter
gallen·abführend *a* cholagog. **G–absonderung** *f* bile secretion. **g–artig** *a* biliary. **G–behälter** *m* gall bladder. **g–bitter** *a* bitter as gall. **G–bitter** *n* picromel. **-blase** *f* gall bladder. **-blasenstein** *m* gallstone. **-braun** *n* bilirubin. **-darm** *m* duodenum. **-farbstoff** *m* bile pigment. **-fett** *n* cholesterol. **-fettsäure** *f* bile acid. **-fistel** *f* biliary fistula. **-gang** *m* bile duct. **-gelb** *n* bilirubin. **-gerbstoff** *m* (nut)gall tannin (gallotannic acid). **-grün** *n* biliverdin. **-salz** *n* bile salt. **-säure** *f* bile acid. **-seife** *f* ox-gall soap. **-stein** *m* gallstone, biliary calculus. **-steinfett** *n* cholesterol. **-stoff** *m* bile substance (*or* constituent). **-süß** *n* picromel. **-talg** *m* cholesterol. **g–treibend** *a* cholagog. **G–wachs** *n* cholesterol. **-zucker** *m* picromel
Gallert *n* (-e) **1** gelatin, jelly. **2** colloid, gel. **3** glue
gallert·ähnlich, -artig *a* gelatinous, jelly-like; colloidal
Gallerte 1 *f* (-n) gelatin, etc. = GALLERT. **2** *pl of* GALLERT
Gallert·gewebe *f* gelatinous tissue
gallertig *a* gelatinous, jelly-like
Gallert·kapsel *f* gelatin capsule. **-masse** *f* gelatinous mass, jelly-like substance. **-moos** *n* Iceland moss. **-säure** *f* pectic acid. **-schicht** *f* gelatin(ous) layer. **-substanz** *f* colloid(al substance)
Gallet·seide *f* silk waste
Gall·gerbsäure *f* gallotannic acid
Galli· gallic (of trivalent gallium): **-chlorid** *n* gallic chloride
gallieren *vt* treat with gallnut decoction

gallig *a* biliary; bilious; bitter, ill-tempered
Galli·hydroxid *n* gallic hydroxide. **-oxid** *n* gallic oxide
Gallipot·harz *n* galipot
Galli·salz *n* gallic salt
gallisieren *vt* gallize (wine)
Gallitzen·stein *m* white vitriol = GALIT-ZENSTEIN
Gallium·chlorid *n* gallium (*specif* gallic) chloride. **-chlorür** *n* gallous chloride. **-oxydul** *n* gallous oxide
Gallo·chlorid *n* gallous chloride. **-oxid** *n* gallous oxide. **-verbindung** *f* gallous compound
Gall·seife *f* (ox-)gall soap. **-stoff** *m* bile substance, bilin. **-sucht** *f* jaundice
Gallus *m* nutgalls, gallnuts. **--aldehyd** *m* gallaldehyde. **-gerbsäure** *f* (gallo)tannic acid, tannin. **g--sauer** *a* gallate of. **G--säure** *f* gallic acid. **-säuregärung** *f* gallic fermentation. **-tinte** *f* nutgall ink
Galmei *m* (-e) calamine
galt was valid, etc: *past of* GELTEN
galvanisch *a* galvanic; electrolytic; electro-; **g. verchromen (verkupfern, versilbern)** electroplate with chromium (copper, silver), chromium-plate, (copperplate, silver-plate)
galvanisieren *vt* galvanize. **Galvanisierung** *f* (-en) galvanization, galvanizing.
Galvanismus *m* galvanism
Galvano·plastik *f* galvanoplastics. **-stegie** *f* electroplating, electrodeposition. **-technik** *f* electroplating (technology)
Gambe *f* (-n) (*Ceram.*) jamb
Gammagraphie *f* gamma-ray inspection
Gamma·säure *f* gamma acid. **-strahl** *m* gamma ray
Gamon *n* (-e) gamete hormone, gamone.
gang *pred a:* **g. und gäbe** the accepted thing, nothing unusual
Gang *m* (-e) **1** gait, walk; pace. **2** hall(way), passage; aisle; path; (*Physiol.*) duct, vas; (*Min.*) gallery; seam; vein. **3** errand; way to go; walk, trip. **4** motion, progress: **in G. bringen** (or **setzen**) set in motion, get (sthg) going; **in G. kommen** start moving; **im G—(e) sein** be in progress, be going on; **toter G.** lost motion, dead travel. **5** course (*incl* meals): **seinen G. gehen** run its course. **6** (*esp Mach.*) operation: **in**

G. sein be in operation (or motion), **in einem G.** in one operation (or motion). **7** (*Mach.*) gear (of autos); thread, pitch (of screws). **--art** *f* **1** gait; pace. **2** (*Geol.*) matrix, gangue
gangbar *a* **1** passable, negotiable (road). **2** pervious. **3** current, customary. **4** operational, working. **5** workable, feasible. **6** fast-selling, marketable. **7** (*Pharm.*) officinal
Gang·erz *n* vein ore. **-gestein** *n* (*Min.*) gang(ue). **-höhe** *f* pitch (of a screw)
gängig *a* **1** current. **2** marketable, salable
Gang·masse *f* (*Min.*) gang(ue)
Gangrän *f* (-e) gangrene
Gang·stein *m* (*Min.*) gang(ue). **-unterschied** *m* (*Elec.*) phase difference
Gans *f* (-e) **1** goose. **2** large lump (*esp* of salt). **3** (*Mettl.*) pig. **4** (*Min.*) hard rock
Gänse·blume *f* daisy. **-fett** *n* goose fat. **-füßchen** *npl* quotation marks. **-hals** *m* gooseneck. **-haut** *f* gooseflesh; goose pimples. **-kiel** *m* goose quill; quill pen. **-leberpastete** *f* paté de foie gras. **-marsch** *m* single file
Gänserich *m* (-e) **1** gander. **2** (*Bot.*) tansy
Gänse·schmalz *n* goose fat (or grease)
ganz I *a* **1** whole, complete, entire. **2** (*in time expr usu*) all, whole: **den g—en Tag** all day; **eine g—e Stunde** a whole hour. **3** *pl:* e.g. **g—e zehn Minuten** all of ten minutes. **4:** **im g—en** on the whole, all told *cf* GROSS **1. II** *adv* **5** (*usu*) wholly, completely, entirely, altogether. **6** all: **in g. Asien** in all (of) Asia; **g. naß** all wet. **7** (+ *Location*) e.g.: **g. oben** away (or all the way) at the top. **8** very; quite: **g. klein** very (or quite) small; **g. alt** quite (or really) old. **9** (+ *a* or *adv, oft*) e.g. **g. schön** fairly (or pretty) nice. **10:** **g. und gar** completely, absolutely. *Cf* GANZE(s)
Ganz, Gänz *f* (*Min.*) solid rock; (*Mettl.*) pig
Gänze *f* entirety
Ganze(s) *n* (*adj endgs*) whole (thing); total
Ganz·fabrikat *n* finished product. **-form** *f* pig mold, pig bed
Ganzheit *f* entirety; whole
Ganz·holländer *m* (*Paper*) beater, pulper. **-holz** *n* unhewn (round) timber. **g--jährig** *a* all year. **G--leder** *n* full leather (binding). **-leimen** *n* (*Paper*) hard-sizing

gänzlich a complete; whole, entire

Ganz·mahlen n (Paper) beating

Ganzmetall· all metal

Ganz·reifenregenerat n (Rubber) whole-tire reclaim. **-stoff** m finished product; (Paper) whole stuff, pulp. **-stoffholländer** m pulper = GANZHOLLÄNDER. **g-tägig** a, **-tags** adv all day, full time. **-zahlig** a integral, whole-numbered

Ganzzeug n finished product; (Paper) whole stuff, pulp. **-holländer** m (Paper) beater, pulper. **-kasten** m (Paper) stuff chest

gar I a 1 thoroughly cooked, done. 2 ready (for use, cultivation, etc). 3 (genl) finished, fully processed; (specif) (Lthr.) tanned, dressed; (Metll.) refined: **g. machen** finish, tan, dress, refine. II adv 4 even. 5 quite, very: **g. so kurz** so very short; **g. viel(e)** quite a lot. 6 (+neg) e.g.: **g. nicht** not at all, simply not; **g. kein Gas** no gas whatever (or at all). 7 possibly; (oft + neg) (not) by any chance. 8 (+zu) e.g.: **g. zu groß** entirely (far, much) too big. 9: **und (nun) g.** not to mention; more than ever. Cf. GANZ 10

Gär·ansteig m period of increasing enzyme activity

Garantie f (-n), **garantieren** vt, vi(+ **für**) guarantee

Garanzin n garancine

Gar·arbeit f refining

Garaus m : **den G. machen** (+ dat) kill, destroy, put an end to

gärbar a fermentable. **Gärbarkeit** f fermentability

Garbe f (-n) 1 sheaf, bundle. 2 (Metll.) fagot, pile. 3 (Bot.) caraway; yarrow.

gärben vt (Iron) pile and weld, refine

Gär·bottich m fermenting vat (or tub)

Gar·brand m (Ceram.) finishing burn.

gar·brennen* vt fire to maturity

Gar·brühe f finishing (or dressing) liquor

Garb·stahl, Gärb·stahl m shear steel, refined iron

Gär· fermentation: **-bütte** f fermenting vat (or tub). **-chemie** f zymurgy, fermentation chemistry. **-dauer** f duration of fermentation

Gardine f (-n) curtain

Gär·druck m fermentation pressure.

-dünger m fermented manure

Gare f (-n) 1 finished state, doneness; refined state; (Lthr.) dressed state. 2 (Agric.) readiness (of soil) for cultivation, good soil condition. 3 (Lthr.) dressing; dressing medium (or paste); bundle of 24 skins. 4 (Metll.) refining; refinery. 5 (Min.) cleavage plane (of granite, etc)

Gäre f fermentation; yeast; (Wine) bouquet

Gar·eisen n 1 refined iron. 2 trial rod (for testing melted copper)

garen vt 1 finish. 2 refine (metals); coke (coal). 3 dress (leather). 4 cook thoroughly (or till well done)

gären vi (lit*gor, gegoren; fig:reg.) ferment

Gar·erz n roasted ore

Gär·erzeugnis n fermentation product. **g-fähig** a fermentable. **G–fähigkeit** f fermentability

Gar·faß n dressing vat (or tub)

Gär·faß n fermenting cask, union. **-flüssigkeit** f fermentable liquid

gar·frischen vt refine thoroughly

Gär·führung f method of fermentation

Gar·gang m (Metll.) (normal) refining process

Gär·gas n fermentation gas. **-gefäß** n fermentation vessel

Gar·gekrätz n refinery slag

Gär·geschirr n fermenting vessel(s)

Garheit f finished (or Metll. refined) state

Gar·herd m refining hearth

Gär· fermenting: **-kammer** f fermenting room. **-kölbchen** n fermenting saccharimeter

gar·kochen vt cook (or boil) till done

Gär·kraft f fermenting power. **g–kräftig** a fast-fermenting (esp yeast)

Gar·krätze f refinery slag. **-kupfer** n refined copper. **-leder** n dressed leather

Gär·lehre f zymology. **-luft** f foul air, mephitis

gar·machen vt finish; dress; refine cf GAR 3

Gär·messer m zymometer. **-mittel** n ferment, enzyme

Garn n (-e) 1 yarn; thread; twine. 2 net; snare. **-druck** m (Tex.) yarn printing

Garnele f (-n) shrimp, prawn

Garn·färberei f yarn dyeing. **g–farbig** a yarn-dyed

garnieren vt garnish, trim; line

Garnitur *f* (-en) 1 garnish(ing), trim(ming); lining. 2 fitting(s), mounting(s); furniture. 3 set; outfit. 4 (first, second) rank, team, string

Garn·kötzer *m* (*Tex.*) cop. **-nummer** *f* yarn count. **-schlichtung** *f* yarn sizing. **-wickel** *m* wound package (of yarn)

Gar·ofen *m* refining furnace. **-probe** *f* refining assay

Gär· fermentation: **-probe** *f* fermentation sample (*or* test). **-raum** *m* fermentation chamber

Gar·rösten *n* (*Metll.*) finishing roast. **-schaum** *m* (*Iron*) kish. **-scheibe** *f* (*Copper*) disc of refined copper. **-schlacke** *f* refining slag, rich slag. **g–schmelzig** *a* : **g—es Eisen** low-silicon pig; open white pig. **G–span** *m* (*Copper*) coating of metal on the trial rod.

Gär·stoff *m* ferment

Gar·stück *n* lump of purified salt

Gär·tätigkeit *f* fermentative activity

Garten *m* (⸚) garden. **-bau** *m*, **-baukunst** *f* gardening, horticulture. **-distel** *f* artichoke. **-koralle** *f* garden pimento. **-kressenöl** *n* garden cress oil. **-kunst** *f* horticulture. **-raute** *f* garden (*or* common) rue

Gärtner *m* (-) gardener. **Gärtnerei** *f* (-en) 1 gardening. 2 truck garden (*or* farm). 3 plant (*or* tree) nursery. **gärtnerisch** *a* horticultural

gär·tüchtig *a* (*Yeast*) vital

Garung *f* (-en) finishing; refining; dressing; thorough cooking; coking (of coal) *cf* GAREN

Gärung *f* (-en) fermentation

Gärungs· (resulting from) fermentation: **-alkohol** *m* ethyl alcohol. **-amylalkohol** *m* fermentation amyl alcohol. **-buttersäure** *f* butyric acid. **-chemie** *f* fermentation chemistry, zymurgy. **g–erregend** *a* fermentative, zymogenic. **G–erreger** *m* ferment. **g–erzeugend** *a* fermentative, zymogenic. **-fähig** *a* fermentable. **G–fähigkeit** *f* fermentability. **-fuselöl** *n* fusel oil. **-gewerbe** *n* fermentation industry. **g–hemmend** *a* fermentation arresting, antifermentative. **-kölbchen** *n* fermentation tube. **-kraft** *f* fermentative power. **-küpe** *f* (*Dye.*) fermentation (*or*

steeping) vat. **-lehre** *f* zymology. **-messer** *m* zymometer. **-mikrobe** *f* fermentation organism. **-mittel** *n* fermenting agent, ferment. **-pilz** *m* fermentation fungus. **-probe** *f* fermentation test. **-stoff** *m* ferment. **-technik** *f* zymotechnology. **g–technisch** *a* zymotechnic(al). **-unfähig** *a* unfermentable. **G–verfahren** *n* fermentation process. **g–verhindernd** *a* antifermentative, antizymotic. **G–vermögen** *n* fermentative power. **-vorgang** *m* fermentation process. **g–widrig** *a* antizymotic, antifermentative

Garungs·zeit *f* (*Coke*) coking period

Gärungs·zeit *f* fermentation period

Gär· fermentation: **-verfahren** *n* fermentation process. **-vermögen** *n* fermenting capacity

Gar·waage *f* (*Salt*) brine gauge

Gär·wärme *f* heat of fermentation

Gar·zeit *f* (*Coke*) coking period

Gas *n* (-e) 1 gas. 2 gas(oline), *only in:* G. geben step on the gas; G. wegnehmen take one's foot off the gas (pedal) *cf* BENZIN

Gas·abführung *f* 1 removal of gas. 2 gas outlet. **-abgabe** *f* evolution (*or* escape) of gas. **-ableitung** *f* 1 removal (*or* drawing off) of gas. 2 gas outlet (pipe), gas delivery tube. **-abwehr** *f* (*Mil.*) defense against gas. **-abzug** *m* 1 drawing off of gas. 2 gas outlet; (*Blast Furnace*) downcomer. **g–ähnlich** *a* gaseous, gas-like. **-analytisch** *a* (of) gas analysis. **G–anfall** *m* amount of gas collected, gas yield. **-angriff** *m* gas attack. **-anlage** *f* gas plant, gas works. **-ansammlung** *f* accumulation of gas. **-anstalt** *f* gas works, gashouse. **-anstaltkoks** *m* gas(house) coke. **-anzeiger** *m* gas detector. **-anzug** *m* gasproof suit (*or* clothing). **-anzünder** *m* gas lighter (*or* igniter). **g–arm** *a* low-gas, (*Coal*) lean. **G–art** *f* type of gas. **g–artig** *a* gas-like, gaseous. **G–artigkeit** *f* gaseousness. **-aufnahme** *f*, **-aufsaugung** *f* gas absorption. **-aufzehrung** *f* gas consumption. **-ausbruch** *m* outburst of gas. **-austreibung** *f* gas expulsion, outgassing. **-austritt** *m* escape of gas. **g–behaftet** *a* gas-contaminated. **G–behälter** *m* gas container (holder, tank).

g–beheizt *a* gas-heated. **G–bekleidung** *f* gasproof clothing. **-beleuchtung** *f* gas illumination, gas lighting. **-benzin** *n* (*Petrol.*) light ends (butane and propane); bottle gas. **-bereitung** *f* gas manufacture, gas production. **-beschädigte** *m, f* (adj endgs) (poison) gas casualty. **-beschuß** *m* gas shellfire. **g–bildend** *a* gas-forming. **G–bildner** *m* gas former, gas producer. **-bildung** *f* gas formation, gas production. **g–bindend** *a* gas-fixing, gettering: **g–es Mittel** getter. **G–bindung** *f* gas fixation, (*specif*) gettering. **-bläs/chen** *n* small gas bubble. **-blase** *f* gas bubble; (*Metll.*) blowhole. **-bleiche** *f* gas bleaching; (*Paper*) patching. **-bohrloch** *n* gas well. **-bohrung** *f* 1 drilling for gas. 2 gas well. **-bombe** *f* 1 gas cylinder. 2 (*Mil.*) gas bomb. **-brand** *m* gas gangrene. **-brenner** *m* gas burner

gäschen *vi* foam, froth

gas·dicht *a* gas-tight. **G–dichte** *f* gas density. **-dichtigkeit** *f* gastightness, impermeability to gas. **-druck** *m* gas(eous) pressure. **-druckmesser** *m* manometer, pressure gauge. **-durchlässigkeit** *f* gas permeability. **-einschluß** *m* gas occlusion. **-einsteller** *m* gas regulator

gasen 1 *vt* gas. 2 *vi* give off gas *cf* GEGAST **Gas·entartung** *f* gas degeneration. **-entbindung** *f* generation of gas

Gasentbindungs·flasche *f* gas generator (flask). **-rohr** *n* gas delivery tube

Gas·entladung° *f* (*Elec.*) gas discharge. **-entladungslampe** *f* gas discharge lamp. **-entschwefelung** *f* desulfurization of gas. **-entweichung** *f* escape of gas. **-entwicklung** *f* evolution of gas. **-entwickler**, **-entwicklungsapparat** *m* gas generator. **-erdmine** *f* (*Mil.*) chemical land mine. **-erkennung** *f* gas detection (or recognition). **-erkennungsmittel** *n* gas detector. **g–erzeugend** *a* gas-producing. **G–erzeuger** *m* gas generator, gas producer. **-erzeugergas** *n* producer gas. **-erzeugung** *f* gas production. **-erzeugungsanlage** *f* gas plant, gas works; gas-producing unit

gasf. *abbv* (gasförmig) gaseous

Gas·fabrik *f* gas works. **-fach** *n* gas engineering. **-fahrt** *f* maximum yield operation of gas works. **-fang** *m* gas catcher (or collector). **g–fest** *a* gasproof, gas-tight. **G–festkörperchromatie** *f* gas-solid chromatography. **-feuerung** *f* 1 gas firing. 2 gas-fired furnace. **-film** *m* gaseous film. **-flammkohle** *f* open-burning (or high-volume) coal (35–40% volatile material). **-flammofen** *m* gas-fired reverberatory furnace. **-form** *f* gaseous state. **g–förmig** *a* gaseous. **G–förmigkeit** *f* gaseousness. **-frischen** *n* (*Metll.*) gas puddling. **g–führend** *a* gas-bearing, containing gas. **G–führungssystem** *n* gas inlet system. **-gebläse** *n* gas blast (apparatus). **g–gefeuert** *a* gas-fired, gas-heated. **G–gehalt** *m* gas content. **-gemenge**, **-gemisch** *n* gaseous mixture. **-geruch** *m* odor of gas. **-geschoß** *n* gas projectile (or shell). **g–geschützt** *a* gasproof(ed). **G–gesetz** *n* gas law. **-gestalt** *f* gaseous form. **g–getrieben** *a* gas-driven. **G–gewinnung** *f* gas production (or extraction). **-gips** *m* aerated gypsum plaster. **-glocke** *f* gas bell (or holder). **-glühlicht** *n* incandescent gas light. **-glühlichtkörper** *m* gas mantle. **-granate** *f* gas grenade (or shell). **-hahn** *m* gas tap (knob, cock). **-hälter** *m* gas container. **g–haltig** *a* containing gas. **G–handwerfer** *m* (*Mil.*) chemical mortar (or hand grenade). **-haut** *f* gas(eous) film. **-hebel** *m* gas throttle. **-hohlraum** *m* gas pocket. **-hülle** *f* gaseous envelope

gasieren *vt* (*Tex.*) gas, singe

gasig *a* gaseous, gassy

Gas·kalk *m* gas lime. **-kampf** *m* gas (or chemical) warfare. **-kampfflasche** *f* cloud gas attack cylinder. **-kampfstoff** *m* war gas. **-kette** *f* (*Elec.*) gas cell. **-kocher** *m* gas plate (or cooker). **-kohle** *f* (high-volatile) gas coal (28–35% volatile material), gas retort carbon. **-kokerei** *f* coke oven plant (for local gas). **-kraftmaschine** *f* gas engine, **g–krank** *a* gassed, gas-poisoned. **G–krieg** *m* gas (or chemical) warfare. **-leitung** *f* gas pipe(line); gas flue. **-leitungsröhre** *f* gas (conducting) pipe. **-luftgemisch** *n* gas-and-air mixture. **-maschine** *f* gas(-driven) engine.

-menge f amount of gas. **-messer** m gas meter. **-meßröhre** f, **-messungsrohr** n gas-measuring tube; eudiometer. **-mine** f 1 gas (or chemical) mine. 2 (gas) projector drum. **-mischung** f 1 mixing of gases. 2 gas(eous) mixture. **-mörser** m (Mil.) chemical mortar. **-motor** m gas motor (or engine); internal combustion engine. **-nitrieren** n (Metll.) gas nitriding. **-ofen** m 1 gas stove (oven, furnace). 2 gas-fired kiln. 3 coke oven. **-öl** n gas (or solar) oil

Gasol n liquefied petroleum gas

Gasolin n (-e) gasoline, (specif) light petroleum fraction boiling at 30–80°C

Gasometer m/n (-) gas tank (or container) **gasös** a gaseous

Gas·pressung f gas pressure. **-probe** f gas sample, gas test. **-prüfer** m gas tester, eudiometer. **-quelle** f gas well. **-raum** m gas space. **-raumvolumen** n volume of gas space; interstitial space. **g–reich** a rich in gas, high in gas content; (Coals) fat. **G–reiniger** m gas purifier. **-reinigungsmasse** f gas-purifying material. **-rest** m gas residue. **-rohr** n, **-röhre** f gas tube; gas pipe. **-ruß** m gas soot, gas black. **-sack** m gas bag. **-sammelrohr** n gas collecting tube. **-sammler** m gas collector, gas tank. **-sauger** m gas exhauster. **-schiefer** m cannel coal. **-schlauch** m gas tube. **-schmelzschweißung** f gas (or autogenous) welding. **-schutz** m 1 protection (or defense) against (poison) gas. 2 (in compds) anti-gas, gas-defense, e.g.: **-schutzlager** n anti-gas (or gas-defense) depot. **-schwaden** m (cloud of) gas fumes. **-schweißbrenner** m gas torch. **-schweißen** n gas welding

Gasse f (-n) 1 alley; (esp Austrian) street. 2 passage(way), path. 3 (Foun.) channel **gas·sicher** a gasproof. **G–silikat** n cellular silicate (buiding material). **-spannung** f gas pressure. **-spüren** n gas (leak) detection. **-spürstoff** m gas detection agent. **-stoffwechsel** m gaseous metabolism, gas exchange; respiration. **-strahl** m gas jet. **-strom** m stream of gas. **-strömung** f gas flow. **-strumpf** m gas mantle. **-sumpf** m dense low-flying gas cloud

Gast m (ꞈe) guest; stranger; extraneous (or foreign) object. **-arbeiter** m (imported) foreign worker

Gas·technik f gas engineering. **g–technisch** a (of) gas engineering. **G–teer** m gas tar. **-teilung** f T-tube, gas supply divider. **-trenner** m gas separator. **-trennung** f separation of gas(es). **-trockenapparat** m gas drier. **-uhr** f 1 gas meter. 2 gas circulating pump. **g–undurchlässig** a impermeable to gas. **G–ventil** n gas valve. **-verbrauch** m gas consumption. **-verdichter** m gas compressor. **-verflüssigung** f liquefaction of gas(es). **-vergiftung** f gas poisoning. **-verschluß** m gas seal. **-volumen** n gas volume. **g–volumetrisch** a gasometric. **G–vulkanisation** f (Rubber) gas cure. **-waage** f gas balance. **-wanne** f gas (or pneumatic) trough. **-waschaufsatz** m gas-washing attachment. **-wäsche** f gas scrubbing. **-waschflasche** f gas-washing bottle. **-wasser** n gas liquor. **-wechsel** m gas exchange, respiration; (Physiol.) gaseous metabolism. **-werfer** m gas (or chemical) projector. **-werk** n gas works. **-wolke** f gas cloud. **-zähler** m gas meter. **-zementieren** n gas cementation. **-zufuhr** f gas supply. **-zug** m gas flue, gas duct. **-zuleiter** m, **-zuleitungsrohr** n gas inlet (pipe or tube). **-zünder** m gas lighter. **-zustand** m gaseous condition (or state)

Gatsch m 1 slush. 2 (Petrol.) slack (or crude scale) wax

Gat(t) n (-e) orifice, mouth

Gatte m (-n, -n) husband

Gatter n (-) 1 grating, lattice; lattice(d) gate. 2 picket fence. 3 fenced enclosure. 4 (Cmptrs.) gate. 5 (Tex.) creel (of spinning frame). 6 frame saw. **gattern** vt 1 fence in. 2 refine (tin)

gattieren vt 1 (Metll., Tex.) mix (ores, fibers). 2 sort, classify. 3 (Metll.) make up the charge (or mixture) of. **Gattierung** f (-en) 1 mixing. 2 mixture (of ores). 3 sorting

Gattin f (-nen) wife

Gattung f (-en) kind, type, category, genre; model; (Nat. Sci.) genus

Gattungs·begriff *m* generic concept.
-name *m* generic name

Gau *m* (-e) (*usu* rural) district

Gauch·heil *n* (*Bot.*) 1 scarlet pimpernel.
2 self-heal

gaufrieren *vt* (*Tex., Paper*) gauffer, goffer

gaukelhaft *a* deceptive, illusory. **gaukeln**
vi flutter, waver, flicker. **Gaukler** *m* (-)
juggler, illusionist

Gaultheria-öl *n* oil of wintergreen

Gaumen *m* (-) (hard) palate. **--segel** *n*,
-vorhang *m* soft palate

gautschen *vt* (*Paper*) couch

Gay·erde *f*, **-salpeter** *m* native saltpeter
earth

Gaze *f* (-n) gauze, cheesecloth. **--papier** *n*
tissue paper

gbr. *abbv* 1 (gebräuchlich) usual, customary. 2 (gebraucht) used. 3 (gebrannt)
burned

GDCh *abbv* (Gesellschaft Deutscher Chemiker) Soc. of Ger. Chemists

GE *abbv* (Gewichtseinheit) unit of weight

Geäder *n* (system of) veins, blood vessels;
venation; graining. **geadert, geädert**
p.a veined, venose, grained, marbled

geartet *p.a* constituted, disposed *cf* ARTEN

geb. abbv 1 (gebildet) formed; educated.
2 (geboren) born; née. 3 (gebunden)
bound

Gebäck *n* baked goods, *esp* pastry, cookies

gebacken baked; fried: *pp of* BACKEN

Gebälk(e) *n* timberwork, beams, rafters

gebändert *a* banded

gebar gave birth: *past of* GEBÄREN

Gebärde *f* (-n) gesture

Gebaren *n* behavior, demeanor

gebären* *vt* (gebar, geboren; gebiert) bear,
give birth to, bring forth. **Gebären** *n*
childbirth. **Gebär·mutter** *f* womb, uterus; matrix

Gebäude *n* (-) building, structure, edifice;
mine. **--reiniger** *m* (-) building cleaner

gebd. *abbv* (gebunden) bound

Gebein *n* (e) bones; skeleton

geben* (gab, gegeben; gibt) **I** *vt* 1 (*usu*)
give; hand over; offer; sell. 2 put. 3 make,
lead to; yield. 4 play (the part of); perform. 5 (+ **auf**) attach importance (to).
6 (+ **von sich**) give off (*or* out), emit;

utter; *cf* VERSTEHEN, VERLIEREN. **II** *vt*
(*imps*): **es gibt** there is, there are, is (are)
available, exist(s) **—was gibt's?** what's
happening? **es gibt Fleisch** we are having meat (for dinner, etc.). **III: sich g.** 7
(+ *dat*) surrender, yield (to). 8 offer itself.
9 let up, subside; (*cloth*) stretch. 10 turn
out all right. 11 (+ *adv*) act, behave; pretend to be. 12 (+ **in**) resign oneself (to). *Cf*
GEGEBEN; *cf* ERKENNEN, ZUFRIEDEN. **Geber** *m* (-) giver, donor; (*Tgph., etc*) sender,
transmitter

gebeten asked: *pp of* BITTEN

gebiert gives birth: *pres of* GEBÄREN

Gebiet *n* (-e) area, field, territory

gebieten* (gebot, geboten) 1 *vt* order, command. 2 *vi* (*oft* + **über**) rule (over), have
command (of)

gebieterisch *a* imperious

Gebilde *n* (-) formation, structure, shape

Gebinde *n* (-) 1 bundle, pack; skein, hank;
truss. 2 range (of tiles). 3 barrel

Gebirge *n* (-) 1 mountain range, mountains. 2 (*Min.*) rock; gang(ue). **gebirgig** *a*
mountainous

Gebirgs·art *f* type of rock. **-bildung** *f*
mountainous (*or* rock) formation; oro-
genesis. **-kunde, -lehre** *f* orology. **-zug** *m*
mountain range

Gebiß° *n* 1 (set of) teeth, dentition. 2 denture

gebissen bitten: *pp of* BEISSEN

Gebläse *n* (-) 1 blower, fan; supercharger.
2 blast. 3 bellows. 4 blowtorch. **--kies** *m*
blasting sand (*or* grit). **-lampe** *f* blast
lamp. **-luft** *f* blast air, air blast. **-ma-
schine** *f* blast engine, blower. **-messer** *m*
blast meter. **-ofen** *m* blast furnace.
-röhre *f* blast pipe, tuyère. **-vorrichtung**
f blast (*or* blowing) apparatus. **-wind** *m*
blast

geblättert *p.a* foliate(d); lamellar *cf* BLÄT-
TERN

geblichen bleached: *pp of* BLEICHEN

geblieben stayed: *pp of* BLEIBEN

geblümt *a* flowered, floral

Geblüt *n* blood, heritage

gebogen bent, curved: *pp of* BIEGEN

geboren *p.a* born; *pp of* GEBÄREN

geborgen rescued: *pp of* BERGEN

geborsten burst: *pp of* BERSTEN

Gebot *n* (-e) 1 commandment; basic principle; requirement. 2 bid. 3: **einem zu G(— e) stehen** be at sbdy's service (*or* disposal)

geboten *pp of* BIETEN (offer) & GEBIETEN (command)

gebr. *abbv* 1 (gebräuchlich) customary, usual. 2 (gebraucht) used. 3 (gebrannt) burned

Gebr. *abbv* (Gebrüder) bros.

gebrach lacked: *past of* GEBRECHEN

gebracht brought: *pp of* BRINGEN

gebrannt 1 burned: *pp of* BRENNEN. 2 *p.a oft* burnt, calcined (e.g. lime), toasted, roasted, baked —**g—e Erde** terra cotta; **g—e Magnesia** magnesia usta (MgO); **g—er Gips** plaster of Paris; **g—es Wasser** distilled water (*or* liquor)

Gebräu *n* (-e) brew; gyle

Gebrauch *m* (-̈e) 1 use: **in G. nehmen** start using. 2 *pl* customs, habits. **gebrauchen** *vt* use —**nicht zu g.** useless, of no use *cf* GEBRAUCHT

gebräuchlich *a* usual, customary, current

Gebrauchs· use; usual: **-anweisung** *f* directions for use. **-dosis** *f* usual dose. **g—fähig** *a* usable, serviceable. **-fertig** *a* ready for use. **G—gegenstand** *m* utensil; article (of daily use). **-güter** *npl* commodities, consumer goods. **-last** *f* (*Mach.*) working load. **-muster** *n* sample for use (*or* for testing); (*Patents*) registered design, short-term patent. **-porzellan** *n* household china. **g—tüchtig** *a* sturdy, long-wearing. **G—vorschrift** *f* directions for use. **-ware** *f* utensils, articles for daily use. **-wasser** *n* tap water. **-wert** *m* utility value. **-zweck** *m* intended use

gebraucht 1 *pp of* BRAUCHEN (need) & GEBRAUCHEN (use). 2 *p.a* used, second-hand, (*oft in compds*): **G—ware** *f* used article(s), second-hand goods

gebrech *a* (*Min.*) brittle, friable. **Gebrech** *n* brittle (*or* crumbly) rock

gebrechen* *vi* (*imps*): **es gebricht (ihnen) an . . .** there is (they have) a lack of . . .

Gebrechen *n* (-) infirmity; ailment. **gebrechlich** *a* frail, fragile; infirm

gebremst *p.a* limited, reduced: **Waschmittel mit g—em Schaum** low-sudsing

detergent *cf* BREMSEN

gebricht lacks: *pres of* GEBRECHEN

gebrochen 1 *pp of* BRECHEN (break) & GEBRECHEN (lack). 2 *p.a* fractional; crooked, bent; ruined; subdued

Gebrüder *mpl* brothers

Gebrüll *n* roaring, bellowing, etc *cf* BRÜLLEN

Gebrumm *n* hum(ming); muttering, grumbling

Gebühr *f* (-en) 1 fee, charge, rate. 2: **nach G.** appropriately; **über G.** excessively. **gebühren** 1 *vi* (*imps, dat*): **es gebührt ihnen** they deserve it. 2: **sich g.** be fitting. **gebührend** *p.a* appropriate

Gebund *n* (-e) bundle, bunch, skein

gebunden I *pp of* BINDEN. II *p.a* 1 bound (*incl Chem.*). 2 fixed, definite. 3 obligated. 4 dependent. 5 (*Chem., Phys.*): sp. **g—e Wärme** latent heat. 6 (*Chem.*): **g. halten** hold in combination. **Gebundenheit** *f* restraint; restriction; obligation; dependence; latency

Geburt *f* (-en) birth. **gebürtig** *a* native (-born)

Geburts·helfer *m* obstetrician. **-helferin** *f* midwife. **-hilfe** *f* obstetrics, midwifery. **g—hilflich** *a* obstetric. **G—kunde, -lehre** *f* obstetrics. **-ort** *m* place of birth. **-tag** *m* birthday. **-wehen** *npl* labor (pains)

Gebüsch *n* bushes, shrubbery, thicket

gedacht I *pp of* DENKEN (think) & GEDENKEN (intend). II *p.a* 1 (*esp* + **für**) meant, intended (for). 2 imagined. 3 e.g. **g—es Element** the element in question, the aforementioned element. **gedachte** intended: *past of* GEDENKEN

Gedächtnis *n* 1 memory. 2 (*in compds, oft*) commemorative, memorial

Gedanke *m* (-ns,-n) thought, idea; intention —**kein G. (daran)** inconceivable, not by any means

Gedanken·austausch *m* exchange of ideas. **-folge** *f*, **-gang** *m* train of thought. **g—los** *a* thoughtless

gedanklich *a* intellectual, mental

Gedärm(e) *n* (..me) intestines, bowels

Gedeck *n* (-e) 1 covering. 2 cover, place (at table). 3 menu, full-course meal. **gedeckt** *p.a* (*Colors*) opaque. *See also* DECKEN

gedeihen* (gedieh, gediehen) *vi*(s) 1 thrive;

(*Slaking Lime*) swell. **2** (**zu**) mature, develop (to, into). **gedeihlich** *a* beneficial, favorable

gedenken* *vi* **1** (+ *genit*) remember; commemorate; mention. **2** (+ *zu* + *inf*) intend (to). **Gedenken** *n* memory, remembrance

Gedicht *n* (-e) poem

gediegen *a* **1** sound, solid, superior, well-made. **2** (*Min.*) native, pure. **Gediegenheit** *f* soundness, solidity, superiority, high quality; purity

gedieh(en) thrived: *past* (*&pp*) *of* GEDEIHEN

Gedränge *n* **1** (pushing) crowd. **2** jam, scramble, rush. **3** bind, fix. **gedrängt** *p.a.* **1** compact; concise. **2** packed, crowded: **g. voll** crowded to capacity. *Cf* DRÄNGEN

gedroschen threshed: *pp of* DRESCHEN

gedrückt *p.a* depressed; *see also* DRÜCKEN

gedrungen 1 penetrated: *pp of* DRINGEN. **2** *p.a* compact; squat

Geduld *f* patience. **geduldig** *a* patient

gedunsen *a* swollen, bloated

gedurft been suffered: *pp of* DÜRFEN

geeicht *p.a* (*usu* + **auf**) expert, well-versed (in) *cf* EICHEN

geeignet *p.a* suitable, suited, fit *cf* EIGNEN

geerbt inherited: *pp of* ERBEN

Geer-ofen *m* (*Rubber*) Geer-Evans (aging) oven

gef. *abbv* **1** (gefunden) found. **2** GEFÄLLIG(ST) kind(ly), etc

Gefahr *f* (-en) danger, risk: **in G. bringen** endanger; **G. laufen** run the risk; **auf die G. hin** . . . at the risk . . . **gefährden** *vt* endanger. **Gefahren·punkt** *m* danger point. **gefährlich** *a* dangerous. **gefahr·los** *a* safe, without risk

Gefährt *n* (-e) vehicle *cf* GEFÄHRTE

Gefährte *m* (-n, -n) **Gefährtin** *f* companion, associate *cf* GEFÄHRT

Gefäll·draht *m* potentiometer wire

Gefälle *n* (-) **1** slope, (down)grade, gradient. **2** descent, drop, fall —**durch natürliches G.** by gravity. **3** downward trend. **4** difference, disparity. **5** (*Rivers*) head, fall

gefallen I* *vi* (*dat or dat rflx*) be pleasing, appeal (to): **es gefällt ihnen** it appeals to them, they like it; **sich etwas g. lassen** put up with sthg; **sie g. sich in dieser Rolle** they enjoy playing this part. **II** *p.p of* FALLEN (fall) & GEFALLEN (appeal).

Gefallen 1 *m* (-) favor. **2** *n:* **G. finden (an)** take pleasure (in). **gefällig** *a* **1** obliging. **2** pleasant, pleasing. **3** *adv* (*in polite phr*) please, kindly: **zur g-en Beachtung** please note

gefällt 1 is pleasing: *pres of* GEFALLEN I. **2** precipitated: *pp of* FÄLLEN

Gefangene *m,f* (*adj endgs*) prisoner, captive. **Gefängnis** *n* (-se) prison; imprisonment

Gefäß *n* (-e) vessel (*incl Physiol.*), container; (*in compds oft*) vascular: **-barometer** *n* cistern barometer. **-bau** *m* vascular structure. **-einmündung** *f* anastomosis. **-haut** *f* vascular membrane

gefäßig *a* vascular

Gefäß·kunde *f* (art of) ceramics. **-lehre** *f* angiology. **-ofen** *m* retort (crucible, pot) furnace. **-pflanze** *f* vascular plant

gefaßt *p.a* **1** calm, self-possessed. **2**: **g. sein (auf)** be prepared (for). **3**: **sich g. machen (auf)** prepare (*or* brace) onself (for)

Gefäß·versuch *m* (*Agric.*) pot experiment. **-wand** *f* wall of the vessel

Gefecht *n* (-e) **1** fight, battle —**außer G.** out of action; **ins. G. führen** bring into play. **2** fencing match. **Gefecht·kopf** *m* warhead (of missiles, etc)

gefedert *p.a* elastic, springy; spring-loaded; sprung *cf* FEDERN

gefeit *a* (**gegen**) immune, invulnerable (to)

Gefieder *n* feathering, plumage. **gefiedert** *a* feathered, plumed

gefiel was pleasing: *past of* GEFALLEN I

gefl. *abbv* (gefällig) kind(ly)

Geflecht *n* (-e) network, netting, lattice(work); texture; plexus

geflissentlich *a* deliberate, studied, studious

geflochten woven: *pp of* FLECHTEN

geflogen flown: *pp of* FLIEGEN

geflohen fled, flown: *pp of* FLIEHEN

geflossen flowed: *pp of* FLIESSEN

Gefluder *n* (-) (*Min.*) launder, flume

Geflügel *n* poultry. **geflügelt** *a* winged, pinnate, alate —**g—e Worte** familiar quotations

gefochten fought: *pp of* FECHTEN

Gefolge *n* (-) **1** (VIP's) attendants. **2**: **im G.** as a consequence. **Gefolgschaft** *f* (-en) adherents; **G. leisten** (+ *dat*) show allegiance (to)

Gef. P. *abbv* (Gefrierpunkt) frz. pt.

gefragt *a* in demand, sought-after. *See also* FRAGEN

gefräßig *a* voracious

Gefrier·anlage *f* refrigerating plant. **-apparat** *m* freezing(-point) apparatus; freezer

gefrierbar *a* freezable, congealable. **Gefrierbarkeit** *f* freezability, congealability

Gefrier·durchschnitt *m* (*Micros.*) frozen section

gefrieren* *vi* (s) freeze, congeal. **Gefrierer** *m* (-) freezer, congealer

Gefrier· freezing, frozen, refrigeration: **-fach** *n* freezing compartment. **-fleisch** *n* frozen meat. **g–getrocknet** *p.a* freeze-dried. **G–lagerung** *f* frozen storage. **-probe** *f* frost test. **-punkt** *m* freezing point. **-salz** *n* freezing salt, (*specif*) ammonium nitrate. **-schiff** *n* refrigerator ship. **-schnitt** *m* frozen section. **-schrank** *m* freezer **-schutzmittel** *n* antifreeze (agent). **-trocknung** *f* freeze-drying. **-truhe** *f* freezer, freezing chest. **-wagen** *m* refrigerator car (*or* truck)

Gefrisch *n* regenerated material, (*specif*) reclaimed rubber

gefror froze: *past of* GEFRIEREN. **gefroren** frozen: *pp of* FRIEREN & GEFRIEREN. **Gefrorene(s)** *n* (*adj endgs*) ice cream

Gefüge *n* (-) 1 structure; framework; (inner) make-up; system. 2 (*Min.*) bed, stratum. **--art** *f* type of structure, etc. **-bestandteil** *m* structural (*or Coal:* banded) constituent. **-lehre** *f* science of structure, metallography. **g–los** *a* structureless, amorphous. **G–prüfung** *f* structure (*or* texture) test. **-spannung** *f* structural stress. **-struktur** *f* jointed structure

gefügig *a* 1 pliable, malleable; supple. 2 compliant, docile

Gefühl *n* (-e) feeling; emotion; sensitivity. **gefühllos** *a* insensitive, callous

Gefühls·nerv *m* sensory nerve

gefunden found: *pp of* FINDEN

geg. *abbv* (gegen) toward; against

gegangen gone: *pp of* GEHEN

gegast *p.a* gassed: **g—e Lauge** (*Paper*) gassed liquor (tower liquor treated with gas from the cookers) *cf* GASEN

gegeben I given: *pp of* GEBEN. II *p.a* 1 (*esp Math.*) given. 2 actual, present. 3 appropriate. **gegebenenfalls** *adv* possibly; if the need arises. **Gegebenheit** *f* (-en) 1 given (fact). 2 *usu pl* data; (given) conditions, realities

gegen I *prep* (*accus*) 1 against, contrary to. 2 to, toward. 3 compared to. 4 (in exchange) for. 5 (+ *quantity*) close to, almost; (+ *clock time, oft*) shortly before, at about. II *pfx* counter-, contra-, anti-, opposite, opposing.

Gegen· counter-, anti-: **-arznei** *f* antidote. **-bemerkung** *f* (opposing) reply, rejoinder. **-bewegung** *f* countermovement. **-beweis** *m* counterevidence. **-beziehung** *f*, **-bezug** *m* correlation. **-bild** *n* counterpart, antitype; inverted image

Gegend *f* (-en) region, area, vicinity, neighborhood

Gegendruck° *m* counterpressure, resistance

gegeneinander *adv & sep pfx* against (toward, compared to) each other; reciprocally, mutually

Gegen· counter, contra·, anti·: **g–elektromotorisch** *a* counterelectromotive. **G–email** *n* counterenamel. **-farbe** *f* complementary (*or* contrast) color. **-färbung** *f* contrast coloring (*or* staining). **-feuer** *n* backfire; counterfire. **-flüssigkeit** *f* backtitration solution. **-füßler** *mpl* antipodes. **-gewicht** *n* counterweight, counterbalance. **-gift** *n* antidote; antitoxin. **-grund** *m* counterargument, objection. **-induktivität** *f* mutual inductance. **-ion** *n* oppositely charged ion. **-klopfmittel** *n*, **-klopfstoff** *m* anti-knock agent. **-kopplung** *f* (*Elec.*) feedback. **-kraft** *f* counterforce, opposing force. **-lauf** *m* countercurrent, reverse motion. **g–läufig** *a* countercurrent; opposite, opposing; counterrotating; antidromic. **G–lötstelle** *f* cold junction (in thermocouples). **-maßnahme**, **-maßregel** *f* countermeasure. **g–mischbar** *a* consolute. **G–mittel** *n* antidote, remedy. **-mutter** *f* lock nut. **g–phasig** *a* opposite-phased. **G–polung** *f* reversal (*or* opposition) of polarity. **-pro-**

be *f* cross-check, control test. **-reaktion** *f* counterreaction, opposite reaction. **-reiz** *m* 1 counterirritation. 2 =**reizmittel** *n* counterirritant. **-reizung** *f* counterirritation. **-satz** *m* 1 opposite. 2 contrast. 3 antagonism, difference. **g–sätzlich** *a* opposite, opposing; contrasting; antagonistic. **G–sätzlichkeit** *f* oppositeness, contrariness. **-schein** *m* reflection; counterglow. **-schmelz** *m* counterenamel. **-seite** *f* opposite side; reverse; opponents. **g–seitig** *a* mutual, reciprocal; (*adv* + **sich,** *oft*) each other. **G–seitigkeit** *f* reciprocity —**auf G.** on a mutual basis. **-sinn** *m* contrary sense, opposite direction. **g–sinnig** *a* in the opposite sense (*or* direction); irrational. **G–sonne** *f* mock sun, parhelion, anthelion. **-sprechanlage** *f* intercom. **-spülung** *f* counterwashing, reverse circulation

Gegenstand° *m* 1 object. 2 article, item. 3 subject (*incl* school); topic. **gegenständig** *a* opposite, opposed. **gegenständlich** *a* objective; concrete

Gegenstands·glas *n* object glass; (*Opt.*) objective. **g–los** *a* 1 abstract. 2 groundless, baseless. 3 invalid; pointless

Gegen·stoff *m* antisubstance, antibody; antidote. **-strahl** *m* 1 reflected ray, reflection. 2 reverse jet. **-strom** *m,* **-strömung** *f* countercurrent, counterflow. **-stück** *n* counterpart; mate, match. **-takt·** (*in compds*) push-pull . . . **-teil** *n,* **g–teilig** *a* contrary, opposite, converse

gegenüber I *prep* (*dat*) (*oft follows its obj*) 1 opposite, facing; face to face with. 2 toward. 3 in relation to. 4 compared to. 5 in the face of. **II** *adv* opposite, on the other side. **III** *sep pfx* opposite, facing. **--liegend** *p.a* opposite

gegenüber·sehen* *vr:* **sich g.** (*dat*) find oneself facing (*or* confronted with)

gegenüber·stehen* *vi* (*dat*) face, be confronted with

gegenüber·stellen *vt* (*d.i.o*) confront, place face to face, juxtapose (with); compare (to). **Gegenüberstellung°** *f* confrontation; juxtaposition; comparison

Gegen·uhrzeigersinn *m* counterclockwise direction. **-versuch** *m* control experiment

Gegenwart *f* 1 present (time). 2 presence. **gegenwärtig** *I a* 1 present, current. 2 (*dat*) in one's memory. **II** *adv* oft at present

Gegen·wehr *f* defense, resistance. **-welle** *f* (*Engg.*) countershaft. **-wert** *m* equivalent (value). **g–wirkend** *a* counteractive, reactive. **G–wirkung** *f* countereffect, counteraction, reaction. **-zug** *m* countermove

gegessen eaten: *pp of* ESSEN
geglichen equaled: *pp of* GLEICHEN
gegliedert *p.a* articulate(d), jointed; distinct. *See also* GLIEDERN
geglitten glided: *pp of* GLEITEN
geglommen glowed: *pp of* GLIMMEN
Gegner *m* (-) opponent. **gegnerisch** *a* opposing
gegolten been valid, etc.: *pp of* GELTEN
gegoren fermented: *pp of* GÄREN
gegossen poured, cast: *pp of* GIESSEN
gegr. *abbv* (gegründet) founded
gegriffen grasped: *pp of* GREIFEN
geh. *abbv* (geheftet) stitched *cf* HEFTEN
Geh. *abbv* (Gehalt) content(s); salary
Gehaben *n* behavior, conduct
Gehalt I *m* (-e) 1 content: **G. an Fett** fat content. 2 substance, body; value. **II** *m/n* (-er) salary. **g–los** *a* insubstantial. **-reich** *a* substantial
Gehalts·bestimmung *f* determination of content, assay; (*esp Chem.*) analysis. **-rübe** *f* standard feed beet
gehalt·voll *a* substantial
Gehänge *n* (-) 1 pendant. 2 hanger, suspension attachment. 3 slope. **gehangen** hung: *pp of* HÄNGEN *vi.* **Gehänge·schutt** *m* rock debris
gehäuft *p.a* e.g.: **g—er Teelöffel** heaping teaspoon. *See also* HÄUFEN
Gehäuse *n* (-) case, casing, housing; shell; core. **--schnecke** *f* snail
geheftet *p.a* stitched, stapled (but not bound). *See also* HEFTEN
geheim *a* secret, concealed: **g—e Tinte** invisible ink. **G--mittel** *n* secret remedy; patent medicine. **Geheimnis** *n* (-se) secret, mystery. **geheimnisvoll** *a* mysterious; secretive
Geheim·schrift *f* code, cipher. **-tinte** *f* invisible ink

Geheiß *n* command, order, bidding

gehen* (ging, gegangen) *vi*(s) **I 1** (*usu*) go; (+ **an,** *oft*) go (at), get to work (on) —**vor sich g.** go on, proceed. **2** walk. **3** leave, depart. **4** (*esp Mach.*) run, work; be (turned) on; move. **5** sound; (*specif*) ring, click, open, close, etc. **6** reach. **7** (+ **nach**) face (east, west, etc). **II** (*imps*) **8** be possible, work. **9** (+ **auf**) be aimed (at). **10** (*dat.*): **es geht ihnen gut** they are (doing, faring) well. **11** (+ **um**): **es geht um** . . . it is a matter (*or* a risk) of . . . **gehen·lassen*** **I** *vt* **1** let go of, neglect. **II: sich g.** let oneself go, neglect oneself

geheuer *a:* **nicht g.** risky, weird, not reassuring

Gehilfe *m* (-n, -n) **Gehilfin** *f* (-nen) helper

Gehirn *n* (-e) brain; (*in compds oft*) cerebral: **-anhang** *m* pituitary body, hypophysis. **-entzündung** *f* encephalitis. **-erschütterung** *f* brain concussion. **-fett** *n* cerebrin. **-hautentzündung** *f* meningitis. **-rinde** *f* cerebral cortex. **-schädel** *m* cerebral cranium. **-schlag** *m* cerebral apoplexy, stroke. **-wäsche** *f* brainwashing. **-wassersucht** *f* hydrocephaly

gehoben 1 lifted: *pp of* HEBEN. **2** *p.a* high; solemn; animated

geholfen helped: *pp of* HELFEN

Gehölz *n* (-e) **1** forest grove, woods. **2** woodwork. **3** *pl* woody plants. **-pollen** *n* tree pollen

Gehör *n* **1** (sense of) hearing, ear. **2** audience, hearing; chance to be heard. **3** (*in compds oft*) audio-, auditory, acoustic

gehorchen *vi* (*dat*) obey; respond (to)

gehören I *vi* (+ *dat* or + *phr of place*) belong. **2** (+ **zu**) be a part (of), be one (of), be among; go (with): **dazu gehört auch** . . . among these is also . . . , that also includes . . . **3** e.g.: **dazu g. besondere Kenntnisse** that takes (*or* requires) special knowledge. **4** (+ *pp*) ought to be . . . . **II: sich g.** be proper, be fitting

gehörig *a* **1** belonging; relevant. **2** fitting, appropriate. **3** good and (proper), vigorous, emphatic, thorough. **4** *adv oft* with a vengeance

Gehör·lehre *f* acoustics

gehorsam *a* obedient. **Gehorsam** *m*, **Gehorsamkeit** *f* obedience

gehört *pp of* HÖREN (hear) & GEHÖREN (belong)

Gehör·wasser *n* fluid of the inner ear

Gehre *f* (-n) **1** wedge, gusset. **2** miter (joint), bevel. **Gehren** *m* (-) = GEHRE 2. **Gehrung** *f* (-en) = GEHRE 1. **Gehrungs·säge** *f* miter(-box) saw

Geier *m* (-) vulture

Geige *f* (-n) violin, fiddle

Geigen·harz *n* rosin, colophony. **-lack** *m* violin varnish

Geiger·zähler *m*, **-zählrohr** *n* Geiger counter

geil I *a* **1** lush, luxuriant. **2** lecherous, lustful; (*Zool.*) in heat. **3: g—es Fleisch** proud flesh. **II** *sfx* eager

geirrt erred: *pp of* IRREN

Geiserit *m* (*Min.*) geyserite

Geiß *f* (-en) she-goat; doe. **-bart** *m*, **-bartskraut** *n* (*Bot.*) goatsbeard. **-blatt** *n* honeysuckle. **-bock** *m* he-goat

Geißel *f* (-n) flagellum, cilium; whip, scourge. **-färbung** *f* (*Micros.*) flagellum staining. **-organismus** *m*, **-tierchen** *n*, **-träger** *m* (*Biol.*) flagellate

Geiß·fuß *m* **1** (*Bot.*) goutweed, herb Gerard. **2** (*Tools*) nail-puller; carver's V-tool; crowbar; dental forceps

Geißlersches Rohr *n*, **Geißler·röhre** *f* Geissler tube

Geiß·raute *f* (*Bot.*) goat's rue

Geist *m* (-er) **1** spirit. **2** mind, intellect. **3** ghost. **4** (*oft in compds*) spirit of, liqueur: **Bergtee·geist** spirit of wintergreen *cf* KIRSCHGEIST

geistes·abwesend *a* absent-minded. **G—blitz** *m* brainstorm. **-gegenwart** *f* presence of mind. **g—krank** *a* mentally ill. **G—wissenschaften** *fpl* humanities

geistig *a* **1** mental; intellectual; spiritual. **2** spirituous, alcoholic; strong (wine, etc); volatile. **Geistigkeit** *f* **1** intellectuality; spirituality. **2** spirituousness

geistlich *a* **1** religious; clerical. **2** spiritual

geist·los *a* spiritless, lifeless, dull. **-reich, -voll** *a* witty, spirited; brilliant, ingenious

Geiz *m* **1** avarice, stinginess. **2** (-e) (*Hort.*) shoot. **geizig** *a* miserly, stingy; (+ **nach**) greedy (for)

gek. *abbv* (gekennzeichnet) characterized

gekannt known: *pp of* KENNEN

gekernt *a* having a nucleus, nucleated *cf* KERN

geklommen climbed: *pp of* KLIMMEN

geklungen sounded: *pp of* KLINGEN

gekniet *p.a* knee-jointed, geniculate *cf* KNIE(E)N

gekniffen pinched: *pp of* KNEIFEN

Geknister *n* 1 decrepitation. 2 crackling (sound)

gekonnt 1 *p.a* skilful, masterly. 2 been able: *pp of* KÖNNEN

gekörn(el)t *a* granular. *See also* KÖRNE(L)N

Gekrätz *n* (metal) waste, dross, refuse. **-ofen** *m* (*Metll.*) almond furnace

Gekriech *n* (*Geol.*) creep

gekrochen crept: *pp of* KRIECHEN

Gekröse *n* mesentery; tripe; chitterlings; giblets. **Gekrös·gang** *m* pancreatic duct

gekünstelt *a* artificial

gel. *abbv* (gelöst) dissolved

Geläger *n* dregs, sediment

gelagert *p.a* 1 situated, placed **—anders g.** different; **in besonders g—en Fällen** in special cases. 2 (*esp Mach.*) supported, mounted. 3 (*Geol.*) bedded. *See also* LAGERN

gel·ähnlich *a* gel-like

Gelände *n* (-) terrain, territory; area; grounds

Geländer *n* (-) railing, banister

gelang succeeded: *past of* GELINGEN

gelangen I *vi*(s) 1 (+ in, an, auf, nach) get (to), come (to), arrive (at), reach. 2 (+ zu) attain; (+ zu + *verbal noun: usu* = passive): **zur Verteilung g.** (= **verteilt werden**) be distributed; **zum Abschluß g.** be concluded. II *pl past of* GELINGEN

gel·artig *a* gel-like, gelatinous

Gelaß *n* (..asse) room, chamber; space

gelassen *p.a* cool, calm; poised; serene. *See also* LASSEN

Gelatine *f* gelatin. **g—·artig** *a* gelatinous. **G—folie** *f* sheet gelatine; gelatine foil. **-leim** *m* gelatin glue. **-stichkultur** *f* gelatin stab culture

gelatinieren *vt/i* gelatinize, gel; (*Rubber*) set up. **Gelatinierung** *f* gelatinization, gelling; set-up. **Gelatinier·vermögen** *n* gelling power. **gelatinisieren** *vt/i* gelatinize. **gelatinös** *a* gelatinous

geläufig *a* 1 common, current. 2 (*oft + dat*) familiar (to). 3 fluent, easy; *adv oft* with facility. **Geläufigkeit** *f* 1 currency. 2 familiarity. 3 fluency, ease, facility

gelaunt *a:* **gut g.** in a good mood *cf* LAUNE

Geläut(e) *n* (..te) 1 ringing. 2 chimes; carillon

gelb *a* yellow: **g—e Erde** yellow ocher; **g—es Blutkraut** (*Bot.*) goldenseal; **g—es Blutlaugensalz** potassium ferrocyanide; **g—er Ingwer** turmeric; **g—e Queck·silbersalbe** (*Pharm.*) mercuric nitrate ointment; **g—e Rübe** carrot

Gelb *n* yellow. **-antimonerz** *n* cervantite. **-beeren** *fpl* yellow (*or* buckthorn) berries. **-beize** *f* (*Dye.*) buff liquor. **-beizen** *n* (*Metll.*) yellowing. **g—blausauer** *a* ferrocyanide of. **G—blausäure** *f* ferrocyanic acid. **-bleierz** *n* yellow lead ore, wulfenite. **g—braun, -bräunlich** *a* yellowish brown. **G—brenne** *f* pickle (for brass)

gelb·brennen* *vt* (*Metll.*) dip-pickle. **G—brennsäure** *f* pickling acid

Gelbe *n* (*adj endgs*) yellow, yolk (of eggs)

Gelbeisen·erz *n* yellow clay ironstone, xanthosiderite **—okriges G.** yellow ocher; copiapite. **-kies** *m* pyrite. **-stein** *m* yellow ironstone

gelbeln *vi* turn yellowish. **gelben** *vt/i* yellow

Gelb·erde *f* yellow ocher. **-erz** *n* yellow ore, (*specif*) chalcopyrite, limonite, sylvanite

gelb·färben *vt* color yellow

gelb·farbig *a* yellow-colored. **G—färbung** *f* 1 yellow coloration. 2 yellowing, yellow dyeing. **-fieber** *n* yellow fever. **g—gar** *a* (*Lthr.*) tanned

gelb·gießen* *vt* cast in brass. **Gelb·gießerei** *f* brass foundry

Gelb·glas *n* King's yellow. **-glut** *f* yellow heat (*appx* 1100°C). **g—grau** *a* yellowish gray. **-grün** *a* yellowish green. **G—guß** *m* yellow brass. **-harz** *n* yellow resin

Gelbheit *f* yellowness

Gelb·holz *n* (old) fustic; yellowwood

Gelb·bildung *f* gel formation

Gelb·ingwer *m* turmeric. **-kali** *n* potassium ferrocyanide. **-körper** *m* (*Med.*) corpus luteum. **-kraut** *n* (*Bot.*) yellowweed, dyer's rocket. **-kreuzstoff** *m* mustard gas, "yellow cross" blister gas.

-kupfer n brass; crude copper. **-kup-fererz** n chalcopyrite
gelblich a yellowish; (oft in compds) e.g. **--grün** yellowish green
Gelb·nickelkies m (Min.) millerite. **-pech** n yellow pitch. **-rübe** f carrot. **-scheibe** f (Phot.) yellow filter. **-stich** m yellow cast (or tinge). **g--stichig** a yellow-tinged. **G--stoff** m (poison) mustard gas. **-sucht** f jaundice, icterus. **g--süchtig** a jaundiced, icteric. **-weiß** a yellowish white, cream-colored. **G--wurz(el)** f yellowroot, turmeric
Geld n (-er) money. **--anlage** f investment. **-buße** f fine. **-frage** f matter of money. **-hilfe** f financial aid, money. **-mittel** npl (financial) means. **-münze** f coin. **-sache** f matter of money. **-schrank** m safe. **-strafe** f fine. **-stück** n coin. **-verschreibung** f promissory note. **-wesen** n 1 (field of) finance. 2 monetary system
Gele npl (of GEL) gels
Gelee n (-s) jelly
gelegen I lain, etc: pp of LIEGEN. See also GELEGT. II p.a 1 located, situated. 2 (oft + dat) convenient, opportune (for); adv oft at the right moment. **3: ihnen ist viel daran g.** it matters a lot to them. **Gelegenheit** f (-en) 1 opportunity, chance. 2 occasion. 3 facilities. 4 bargain. **gelegentlich** I a 1 chance, accidental. 2 incidental, casual. 3 occasional. II adv 4 occasionally, on occasion; when there is a chance. 5 sometime(s), now and then. III prep (genit) on the occasion of, in connection with
gelegt laid, put: pp of LEGEN. See also GELEGEN
gelehrig a teachable, quick to learn. **Gelehrsamkeit** f learning, scholarship, erudition. **gelehrt** p.a learned, educated, erudite. See also LEHREN. **Gelehrte** m, f (adj endgs) scholar, learned person
Geleise n (-) track; channel = GLEIS
geleiten vt escort, accompany cf LEITEN. **Geleit·wort** n preface, foreword
Gelenk n (-e) joint, articulation; hinge; (in compds oft) articular, articulated, synovial: **-band** n ligament. **-basalt** m flexible basalt. **-fahrzeug** n articulated vehicle. **-fläche** f articular surface. **-flüssig-keit** f synovial fluid

gelenkig a 1 jointed, articulate(d); hinged. 2 flexible, supple. 3 nimble, agile
Gelenk·quarz n flexible sandstone, itacol-umite. **-saft** m, **-schleim** m, **-schmiere** f (Anat.) synovia. **-schwamm** m white swelling. **-steifheit** f ankylosis. **-wasser** n synovial fluid
gelernt p.a skilled. See also LERNEN
Geleuchte n (-) miner's lamp
Gelf m silver-bearing pyrites. **--erz, -kup-fer** n chalcopyrite
gel·förmig a gel-like
geliehen lent: pp of LEIHEN
gelieren vt cause to gel, gelatinize. **Gelier·hilfe** f gelling aid. **Gelierungs·mittel** n gelling agent
gelind(e) a mild, gentle; soft; slight: **g. kochen** simmer
gelingen* (gelang, gelungen) vi(s) (oft + dat) 1 be successful, succeed; be possible: **der Versuch gelang (ihnen)** the experiment was successful (for them); **es gelang, zu zeigen . . .** it was possible to show . . .; **es gelang ihnen zu zeigen . . .** they succeeded in showing . . . 2 turn out (well, badly). Cf GELUNGEN
gelitten suffered: pp of LEIDEN
geloben vi vow, pledge, promise
gelobt pp of LOBEN (praise) & GELOBEN (vow)
gelogen lied: pp of LÜGEN
gelöscht extinguished, etc.: pp of LÖSCHEN: **g--er Kalk** slaked (or hydrated) lime
gelöst p.a relaxed, loose. See also LÖSEN. **Gelöste(s)** n (adj endgs) dissolved substance, solute. **Gelöst·sein** n dissolved state, (being in) solution
Gelsemin n gelsemine, gelsemium, jasmine. **--säure** f gelsemic acid. **-wurzel** f gelsemium root
gelt a dry, barren
Gelte f (-n) tub, bucket
gelten* (galt, gegolten; gilt) I vt/i 1 be worth, count (as, for), amount to. 2 (imps): **es gilt** it matters, it is a matter of. III vi 3 be valid, hold (true), apply, count; be in force; be final. 4 (+ als, für) be considered, be recognized (as). 5 (+ dat) be meant (for), be directed (to, toward). 6 be permissible. III: **g. lassen** recognize, accept (as valid). **geltend** p.a 1 valid, applicable; in force, current; accepted, recog-

nized. **2: g. machen** bring up, assert; (*esp* + **bei**) bring to bear, exert (on); (+ **daß**) make (*or* bring up) the point that . . . **3: sich g. machen** assert onself; make itself noticeable. **Geltendmachung** *f* assertion; exertion; enforcement

Gelt· barren: -tier *n* barren animal

Geltung *f* (-en) **1** worth, value, validity, applicability; currency; recognition; acceptance; respect, prestige *cf* GELTEN. **2: G. haben** be valid (applicable, current, recognized, accepted, permissible) = GELTEN. **3: zur G. bringen** bring out, set off (effectively); gain acceptance for. **4: zur G. kommen** come out, show up (effectively), become effective, gain recognition. **5: G. verschaffen** (+ *dat, incl* **sich**) gain recognition (*or* acceptance) (for).

Geltungs·bereich *m* range, scope (of validity, application, effectiveness)

gelungen 1 succeeded: *pp of* GELINGEN. **2** *p.a* successful; ingenious, great

Gelüst(e) *n* (..te) desire, craving

Gel·zustand *m* gel condition

gem. *abbv* **1** (gemein, gemeinsam) common. **2** (gemischt) mixed. **3** (gemahlen) ground, milled. **4** (gemäß) according to. **5** (gemessen) measured

gemächlich *a* leisurely, unhurried

gemacht *p.a* pretended, artificial. *See also* MACHEN

Gemahl *m* (-e) husband

gemahlen *a* ground, powdered *cf* MAHLEN

Gemahlin *f* (-nen) wife

gemahnen *vi* (**an**) be a reminder (*or* warning) (of). *see also* MAHNEN

Gemälde *n* (-) painting, picture, portrait. **··firnis** *m* picture (*or* painter's) varnish

gemäß I *a* suited, agreeable. II (*prep*) (*dat*) (*oft follows its obj*) according to, in accordance with. III *sfx* according to, in conformity with. **gemäßigt** *p.a* moderate, temperate. *See also* MÄSSIGEN

Gemäuer *n* (-) masonry, stonework; walls

gemein *a* **1** ordinary, common; mutual: **g. haben (mit)** have in common (with). **2** mean, nasty; vulgar. **3** (*in compds*) mutually; common, public, general. *Cf* GEMEINSAM

Gemeinde *f* (-n) community; municipality; congregation

Gemeinheit *f* (-en) commonness; vulgarity;

nasty thing to do (*or* say)

Gemein·kosten *pl* overhead (expenses). **g–nützig** *a* **1** general-purpose; public-welfare. **2** nonprofit. **3** cooperative. **4** charitable

gemeinsam *a* common, joint, mutual; *adv oft* together, in common: **g. haben** have in common *cf* GEMEIN 1. **Gemeinsamkeit** *f* **1** (*sg*) community, solidarity. **2** (*pl*) things in common

Gemeinschaft *f* (-en) community, association, fellowship, partnership. **gemeinschaftlich** *a* common, cooperative, joint. **Gemeinschafts·forschung** *f* cooperative research

gemein·verständlich *a* commonly (generally, universally) understood

Gemeng· (= GEMENGE·): **-anteil** *m* ingredient, constituent (of a mixture)

Gemenge *n* (-) **1** (*genl*) mixture; (*specif*) (*Glass*) frit, (*Petrog.*) conglomerate. **2** fray, melee

Gemenge· (= GEMENG·): **-asche** *f* (*Assaying*) test ash. **-gestein** *n* (*Geol.*) conglomerate. **-material** *n*, **-(roh)stoff** *m*, **-teil** *m* ingredient, constituent (of a mixture); (*Coal*) maceral

Gemengsel *n* (-) mixture; hodgepodge, mishmash

gemessen 1 measured: *pp of* MESSEN. **2** *p.a* deliberate; well-considered

gemieden avoided: *pp of* MEIDEN

Gemisch *n* (-e) mixture, mix; composition; alloy; conglomerate. **··bildner** *m* emulsifier. **-kraftstoff** *m* composite fuel. **-schmierung** *f* oil-in-gasoline lubrication

gemischt *p.a* composite, mixed. *See also* MISCHEN. **··basig** *a* (*Petrol.*) mixed-base, intermediate-base. **G–bett** *n* mixed bed. **g–körnig** *a* of various grain sizes. **G–kraftstoff** composite fuel. **-waren** *fpl* groceries; general merchandise

Gemme *f* (-n) gem; (*Biol.*) gemma

gemocht liked, etc: *pp of* MÖGEN

gemolken milked: *pp of* MELKEN

Gemse *f* (-n) chamois, alpine goat. **Gemsen·fett** *n* chamois fat. **Gems·leder** *n* chamois (leather)

Gemüll *n* rubbish

Gemüse *n* (-) vegetable(s)

gemußt had to: *pp of* MÜSSEN

gemustert *p.a* (*Fabric*) figured, patterned, print(ed). *See also* MUSTERN

Gemüt *n* (-er) **1** spirit, temperament, turn of mind. **2** feeling, (*pl oft*) people's feelings; heart. **3** soul, spirit (= person). **gemütlich** *a* **1** cozy, pleasant. **2** genial, good-natured. **3** leisurely. **Gemütlichkeit** *f* **1** coziness; congenial atmosphere. **2** geniality, good nature. **3** leisurely pace

gen. *abbv* (genannt) named, mentioned

Gen *n* (-e) gene

genannt named, etc: *pp of* NENNEN

genas recovered: *past of* GENESEN

genau *a* **1** exact, precise; *adv oft* right: **g. da** right there. **2** accurate. **3** close, thorough (study, etc); detailed; *adv oft* in detail. **4** meticulous, particular. **5** strict. **6** (+ **nehmen**): **sie nehmen es damit sehr g.** they are very exact (precise, thorough) about it; **Worte g. nehmen** take words literally *cf* GENAUGENOMMEN; **Genaueres** further details, more exact information. **Genauigkeit** *f* exactness, precision; accuracy; closeness, thoroughness, meticulousness; strictness

Genauigkeits·arbeit *f* precision work. **-grad** *m* degree of accuracy. **-grenze** *f* limit of accuracy

genaugenommen *adv* strictly speaking, technically

genauso *adv* (= **genau so**) just as; (*oft pfx to adj & adv*): **genausogut** just as good (*or* well)

genehm *a* (*usu pred* + *dat*) agreeable, convenient, acceptable (to, for)

genehmigen *vt* approve (of), agree (to); permit. **Genehmigung** *f* (-en) approval, permission, license

geneigt *p.a* **1** sloping. **2** willing. **3** (*esp* + *dat*) (favorably) disposed (to, toward). *See also* NEIGEN. **Geneigtheit** *f* inclination, willingness; (favorable) disposition

generalisieren *vt/i* generalize

General·direktor *m* president, general manager. **-nenner** *m* common denominator. **-register** *n* collective index. **-versammlung** *f* general meeting (*oft* of stockholders)

Generation *f* (-en) **1** generation. **2** (*Yeast*) stage

Generator·betrieb *m* producer operation. **-gas** *n* producer gas. **-kohle** *f* gas-

producer coal. **-teer** *m* producer-gas tar

generell *a* general

generisch *a* generic

Genese *f* (-n) (*Biol.*) genesis.

genesen* (genas, genesen) *vi* (s) recover

Genetik *f* genetics. **genetisch** *a* genetic

Genever *m* Holland gin

Genf *n* Geneva. **Genfer** *a* (*no endgs*) Geneva(n)

Gen·forschung *f* genetic research

genial *a* of genius, brilliant. **Genialität** *f* genius, brilliance

Genick *n* (-e) (back of the) neck

Genie *n* (-s) genius

genieren **1** *vt* bother, embarrass. **2: sich g.** be bothered, be embarrassed

genießbar *a* palatable, enjoyable. **genießen*** (genoß, genossen) enjoy, have (to eat, drink)

genommen taken: *pp of* NEHMEN cf GENAUGENOMMEN

genoß enjoyed: *past of* GENIESSEN

Genosse *m* (-n, -n) **1** companion, associate, partner; comrade. **2** (*oft in compds*) mate, fellow: **Schulgenosse** *m* schoolmate; **Amtsgenosse** fellow official

genossen enjoyed: *pp of* GENIESSEN

Genossenschaft *f* (-en) association, partnership, fellowship; society, company; (*Com. oft*) cooperative

Genossin *f* (-nen) companion, etc *cf* GENOSSE

Genre *m* (-s) genre, kind, category; (*Tex.*) style

Gen·technologie *f* genetic engineering

Gentiana·blau *n* gentian (*or* aniline) blue. **-violett** *n* gentian violet

Genua *n* (*Geog.*) Genoa. **genuesisch** *a* Genoese, of Genoa

genug *a* (*no endgs*) *adv* enough. **Genüge** *f* **1: zur G.** sufficiently, only too well. **2: G. tun** (*or* **leisten**) (*dat*) comply (with), satisfy, fulfil. **genügend** *p.a* enough, sufficient; satisfactory. **genügsam** *a* modest, easily satisfied. **genug·tun*** *vi* (*dat*) comply (with), satisfy. **Genugtuung** *f* satisfaction; redress

Genuß *m* (..üsse) **1** pleasure, enjoyment. **2** benefit, use. **3** consumption (of food, etc). **-·mittel** *n* condiment, stimulant; pleasure consumption item (any substance consumed for pleasure rather than nutrition).

g-süchtig *a* pleasure-seeking. **G-zweck** *m* purpose of enjoyable consumption, table purpose

Gen·veränderung *f* gene mutation

Geochemie *f* geochemistry

geodätisch *a* geodetic

Geolog(e) *m* (..gen, ..gen) geology

geordnet *p.a* tidy, orderly; systematic. *See also* ORDNEN

Georgien *n* (*Geog.*) Georgia, Gruzia (USSR)

Georgine *f* (-n) dahlia

Gepäck *n* baggage, luggage

gepfiffen whistled: *pp of* PFEIFEN

gepflegt *p.a* well-groomed, tidy *cf* PFLEGEN

Gepflogenheit *f* (-en) habit, custom, usage

Gepräge *n* (-) 1 stamp, imprint. 2 characteristic(s), special character

gepriesen praised: *pp of* PREISEN

gequollen swollen: *pp of* QUELLEN

gerad·, Gerad· straight, direct, right. *See also* GRAD·, GRAD·

gerade I *a* 1 straight, direct. 2 erect, upright. 3 even (number). 4 *adv* just, exactly, precisely: **g. recht** just right; **g. daneben** right next to it. 4 (*oft as pfx to advs*): *cf* GERADEAUS, GERADESO. **II** *sep pfx* straight: **gerade·stellen** set straight, straighten (out). **Gerade** *f* (*adj endgs*) straight line

gerade· (*see also* GERAD·) **-aus** *adv* 1 straight ahead. 2 straightforward(ly). **-faserig** *a* straight-fibered (*or* granted). **-heraus, -hin** *adv* bluntly, point-blank. **-kettig** *a* straight-chained. **-linig** *a* rectilinear, straight-line. **-so** *adv* (= GENAUSO) just as; (*oft as pfx to advs*): **geradesooft** just as often. **-wegs** *adv* straight, directly. **-zu** *adv* 1 blunt(ly). 2 straight, directly. 3 almost, just about. 4 simply, downright, really

gerad· (= GERADE·): **-sichtig** *a* direct-vision. **-wandig** *a* straight-walled (*or* sided). **-wertig** *a* even-valence. **-zahlig** *a* even-numbered

gerafft *p.a* concise, condensed. *See also* RAFFEN

Geranium·säure *f* geranic acid

gerann coagulated: *past of* GERINNEN

gerannt run: *pp of* RENNEN

gerät gets (into): *pres of* GERATEN

Gerät *n* (-e) device, instrument; apparatus; utensil, implement; appliance; (TV, radio) set; gadget; *pl oft* equipment, gear

Geräte·bau *m* instrument making, appliance manufacture. **-brett** *n* instrument panel. **-glas** *n* apparatus (*or* instrument) glass

geraten I* (geriet, geraten; gerät) *vi* (s) 1 turn out (right), come out (well). 2 (**an, auf, in,** etc + *accus*) get, come (to, into); fall (into); (*oft* + **ins** + *verbal noun*) e.g. **ins Sieden g.** start (*or* get to) boiling; **in Brand g.** catch fire. 3 (**nach**) take (after). **II** 4 recovered: *pp of* RATEN. 5 *p.a* advisable

Geräte·park *m* stock of apparatus (*or* instruments)

Geratewohl: aufs G. at random, haphazardly

Gerätschaft *f: usu pl* (-en) implements, utensils

geraum *a*: **g-e Zeit** a long time. **geräumig** *a* spacious, roomy

Geräusch *n* (-e) noise, sound; murmur; rattle. **-isolierung** *f* soundproofing. **g-los** *a* noiseless. **-sicher** *a* soundproof. **-voll** *a* noisy

Gerb· tanning (*See also* GERBE·): **-anlage** *f* tannery. **-auszug** *m* tanning extract

gerbbar *a* tannable

Gerb·beschleuniger *m* tanning accelerator. **-brühe** *f* tan(ning) liquor. **-mittel** *n* tanning material, tan

gerben *vt/i* 1 tan (hides). **—weiß g.** taw; **sämisch g.** chamois. 2 (*Grain*) hull. 3 (*Metals*) polish; (*Steel*) (re)fine, **Gerber** *m* (-) tanner, tawer; (chamois) leather dresser. **Gerber·baum** *m* tanner's sumac. **Gerberei** *f* (-en) tanning; tannery. **-abfälle** *f* tanner's waste

Gerber·fett *n* (*Lthr.*) dégras, stuff. **-hof** *m* tan(ning) yard. **-kalk** *m* slaked lime. **-lohe** *f* 1 tanbark. 2 tan liquor. **-rot** *n* phlobaphene. **-strauch** *m* 1 (*Rhus coriaria*) tanner's sumac. 2 (*Coriaria*) ink plant. **-wolle** *f* pelt (slipe, tanner's) wool

Gerbe· tanning (*See also* GERB·): **-theorie** *f* theory of tanning. **-vermögen** *n* tanning power

Gerb· tanning (*See also* GERBE·): **-extrakt** *m* tanning extract. **-grube** *f* tanning pit. **-leim** *m* tannic acid glue; hide glue (*or* sizing). **-lohe** *f* 1 tanbark. 2 tan liquor. **-lösung** *f* tanning (*or* tannic) acid solution. **-mittel** *n* tan, tanning material.

-pflanze f tanniferous plant. -rinde f tanbark. g—sauer a tannate of. G—säure f tannic acid. -säuremesser m tannometer, barkometer. -stahl m shear (polished, tilted) steel; burnisher. -stoff m tan, tanning agent; (specif) tannin, tannic acid

gerbstoff· tannin(g): -artig a tannin-like. G—auszug m tannin(g) extract. -bestimmung f determination of tanning matter. -gehalt m tannin content. g—haltig a tanniferous. G—rot n phlobaphene. -vorbeize f tanning mordant

Gerbung f (-en) tanning, tannage; tawing; chamoising; hulling, polishing; (re)fining cf GERBEN

Gerb·vermögen n tanning power. -wert m tanning value

gerecht I a 1 just, fair. 2: g. werden (dat) do justice (to); conform (to); comply (with). 3 skilled. II sfx in accordance with, true to: formgerecht true to form

gerechtfertigt justified: pp of RECHTFERTIGEN

Gerechtigkeit f justice, fairness

Gerede n talk; gossip, rumor

geregelt p.a orderly, regular cf REGELN

gereichen vi (e.g.): es gereicht ihnen zum Vorteil (zum Nachteil) it is (or it works) to their advantage (their disadvantage)

Gereiztheit f (state of) irritation cf REIZEN

gereuen vt (imps) cause regret (to)

Geriatrie f geriatrics

Gericht n (-e) 1 (law) court; judgment. 2 dish (= food); course (of a meal)

gerichtet p.a 1 directed, vector (e.g. quantity). 2 oriented, -minded: konservativ g. conservative-minded. Cf RICHTEN, ZIELGERICHTET

gerichtlich a judicial, (of the) court; legal; forensic; (adv oft) in court

Gerichts· court, judicial: -amt n court, tribunal. -arzt m medical examiner

Gerichtsbarkeit f jurisdiction

Gerichts·chemie f forensic chemistry. -chemiker m forensic chemist. -hof m law court. -weg m: auf dem G. by legal proceedings

gerieben rubbed: pp of REIBEN

geriet got (into, etc): past of GERATEN

gerillt a grooved, ribbed cf RILLE

gering a small, little (in quantity), (pl oft) few; slight, minor; low cf GERINGST.

geringer a (compar) less, inferior, etc.

Geringe(s) n (adj endgs) small quantity, slight amount, trifle

gering·fügig a trifling, trivial; adv oft slightly. -haltig a low-grade, lean, low-content. G-haltigkeit f low quality. (grade, content)

gering·schätzen vt depreciate, underrate

geringst a, superl of GERING least, etc: nicht im g—en not in the least

gering·wertig a low-value, low-grade, inferior

gerinnbar a coagulable. Gerinnbarkeit f coagulability

Gerinne n (-) 1 small stream. 2 channel; flume, sluice. gerinnen* vi (s) (gerann, geronnen) coagulate, curdle, congeal; gel; set (as: cement). Gerinnsel n (-) 1 coagulum, coagulated mass, curd; clot. 2 streamlet. Gerinn·stoff m coagulant, coagulator. Gerinnung f coagulation, curdling, congelation; setting (of cement)

gerinnungs·fähig a coagulable. G—fähigkeit f coagulability. -masse f coagulum; gel. -mittel n coagulant. -punkt m coagulation point. -zeit f coagulation (or setting) time

Gerippe n (-n) skeleton, framework; ribbing. gerippt p.a ribbed, fluted, finned; (Paper) laid cf RIPPE

gerissen 1 torn, broken: pp of REISSEN. 2 p.a sly, crafty

geritten ridden: pp of REITEN

Germ m, f yeast

German n germanium. Germani· germanic: -chlorid n germanic chloride.

Germanium· germanium, germanic, germanous: -chlorid n germanic chloride. -chlorür n germanous chloride. -fluorwasserstoffsäure f fluogermanic acid. -säure f germanic acid. -wasserstoff m germanium hydride

Germano· germanous: -sulfid n germanous sulfide

Germer m (-) (Bot.) (white) hellebore, veratrum

gern(e) adv 1 gladly, readily, willingly: es verbindet sich g. mit . . . it readily combines with . . . ; sie tun es g. they like to do it, they are glad to do it; **sie haben (mögen, wollen) es g.** they like it, they are fond of it; **sie mögen** (or **wollen**) **g.** (+ *v*) they like to . . . **2** usually

gerochen smelled: *pp of* REICHEN

Geröll(e) *n* gravel, pebbles; round stones; rubble; detritus, scree

geronnen *pp of* RINNEN (flow) & GERINNEN (coagulate)

Geron-säure *f* geronic acid

Gerste *f* (-n) barley

Gersten- barley: **-graupen** *fpl* pearl barley. **-korn** *n* barleycorn. **-samenöl** *n* barley (seed) oil. **-schleim** *m* barley water. **-schrotbeize** *f* barley dressing. **-stoff** *m* hordein. **-zucker** *m* barley sugar

Gerte *f* (-n) switch, rod

Geruch *m* (-e) odor, smell. **--befreiungsmittel** *n* deodorant. **-belästigung** *f* subjection to foul odors. **g-frei** *a* odorless, odor-free. **G--gras** *n* fragrant (*specif* vernal) grass

geruchlich *a, adv* regarding odors

geruch-los *a* **1** odorless: **g. machen** deodorize. **2** unable to smell, anosmic. **Geruchlosigkeit** *f* **1** odorlessness. **2** anosmia. **Geruchlos-machung** *f* deodorization

Geruch-messer *m* odorimeter. **g--reich** *a* fragrant, odorous

geruchs- smell, odor, olfactory: **-beseitigend** *a*: **g--es Mittel** deodorant, deodorizer. **G--organ** *n* organ of smell, olfactory organ. **-probe** *f* olfactory test. **-reiz** *m* olfactory stimulus (*or* irritant). **-sinn** *m* sense of smell

Geruch(s)-stoff *m* aromatic substance

Geruchs-verbesserer *m* deodorant, deodorizer

geruch-zerstörend *a*: **g--es Mittel** deodorant

Gerücht *n* (-e) rumor, hearsay

Gerümpel *n* junk, trash

gerungen wrestled: *pp of* RINGEN

Gerüst *n* (-e) **1** framework; scaffold(ing). **2** skeleton (*fig*); plan, outline; structure. **3** (*Chem.*) ring system. **4** (*Biol.*) stoma, reticulum. **--eiweiß** *n* scleroprotein, albuminoid. **-faser** *f* structural fiber. **-silikat** *n* three-dimensional network silicate. **-stoff** *m* (*Dterg.*) builder. **-werk** *n* framework, (*Anat.*) stroma

ges. *abbv* **1** (gesamt) total. **2** (gesättigt) saturated. **3** (gesetzlich) legal, by law. **Ges.** *abbv* (Gesellschaft) Co., Soc.

gesalzen *p.a* salted *cf* SALZEN

Ges.-Amt. *abbv* (Gesundheitsamt) Health Dept., Board of Health

gesamt *a & oft in compds* whole, entire, complete, total; general, overall. **G--alkali** *n* total alkali. **-analyse** *f* total analysis. **-ansicht** *f* general (*or* overall) view. **-ausgabe** *f* complete edition. **-bedarf** *m* total requirement. **-brechung** *f* (*Phys.*) total refraction. **g--deutsch** *a* all-German, concerning all of Germany (*or* all German-language areas). **G--echtheit** *f* all-round fastness (of colors, etc). **-eigenschaftsbild** *n* overall profile of properties. **-energie** *f* **1** total energy. **2** (*Thmodyn.*) internal energy. **-ergebnis** *n* overall result. **-ertrag** *m* total output (*or* yield)

Gesamtheit *f* (-en) whole, entirety, totality: **in seiner** (or **ihrer**) **G.** on the whole

Gesamt- total, general, overall: **-ladung** *f* total (*or* net) charge. **-lösliche(s)** *n* (*adj endgs*) total soluble matter. **-maß** *n* overall dimension(s). **-spannung** *f* (*Elec.*) total (*or* overall) voltage. **-summe** *f* (sum) total. **-überblick** *m*, **-übersicht** *f* general (*or* overall) view. **-verhalten** *n* general behavior. **-wert** *m* total (*or* overall) value

gesandt sent: *pp of* SENDEN. **Gesandte** *m, f* (*adj endgs*) diplomatic agent, envoy, ambassador

Gesäß *n* (-e) posterior; seat (of pants)

gesät sowed: *pp of* SÄEN

gesätt. *abbv* (gesättigt) saturated. *cf* SÄTTIGEN

Geschabsel *n* (-) scrapings

Geschäft *n* (-e) **1** (*genl.*) business; (*specif*) deal, firm, company. **2** store, shop. **3** *pl* affairs, duties. **geschäftig** *a* busy, active, bustling. **geschäftlich** *a* business; business-like; (*adv oft*) on business, as regards business

Geschäfts-betrieb *m* **1** business activity. **2** business enterprise. **-führer** *m* **1** (business) manager. **2** general secretary (of a society). **-schluß** *m* closing time. **-stelle** *f* office, place of business

geschah happened: *past of* GESCHEHEN

gescheckt *a* dappled, mottled

geschehen* (geschah, geschehen; geschieht) *vi* (s) happen, be done. **Geschehen** *n* (-), **Geschehnis** *n* (-se) happening, occurence, event

gescheit *a* intelligent, clever; sensible *cf* KLUG

Geschenk *n* (-e) gift, present *cf* SCHENKEN

Geschichte f (-n) 1 history. 2 story. 3 affair; mess. **geschichtlich** *a* historical

Geschick *n* (-e) 1 skill, ability. 2 fate, destiny. **Geschicklichkeit** *f* skill, cleverness

geschickt I *a* skilful, skilled, clever. II sent: *pp of* SCHICKEN

Geschiebe *n* 1 pushing, etc *cf* SCHIEBEN. 2. (glacial) drift; rubble, boulders

geschieden separated: *pp of* SCHEIDEN

geschieht happens: *pres of* GESCHEHEN

geschienen shone, seemed: *pp of* SCHEINEN

Geschirr *n* I (*sg only*) 1 (set of) dishes; china. 2 earthenware. 3 harness. 4 tools, gear, equipment; utensils. II (-e) 5 vessel, pot. 6 (*Lthr.*) vat. **~leder** *n* harness leather

geschlängelt *p.a* winding, convoluted. See also SCHLÄNGELN

Geschlecht *n* (-er) 1 (male, female) sex; gender. 2 family, stock. 3 race, species, genus (*incl Math.*). 4 generation. **geschlechtlich** *a* sexual; generic. **Geschlechtlichkeit** *f* sexuality. **geschlechtslos** *a* sexless, asexual, nonsexual

Geschlechts· sex(ual), generic: **-anziehung** *f* sexual affinity. **-art** *f* generic character, genus. **-drüse** f sex (*or* genital) gland, (*specif*) ovary, testicle, gonad. **-gen** *n* sex gene. **-glied** *n* sex organ. **-krankheit** f venereal disease. **-pflanze** f phanerogam. **-reife** f puberty. **g-reizend** *a* aphrodisiac, sexually stimulating. **G-teil** *m* sex (*or* genital) organ. **-trieb** *m* sex(ual) drive, sex. **-verkehr** m sexual relations, sex

geschlichen sneaked: *pp of* SCHLEICHEN

geschliffen ground: *pp of* SCHLEIFEN

geschlissen worn out: *pp of* SCHLEISSEN

geschloffen slipped: *pp of* SCHLIEFEN

geschlossen I closed: *pp of* SCHLIESSEN. II *p.a* 1 close(d). 2 private. 3 compact, self-contained. 4 complete, finished. 5 continuous, unbroken. 6 tightly knit, close-knit. 7 uniform, consistent. 8 inland. 9 *adv oft* unanimously, solidly, in a body. **Geschlossenheit** f 1 unanimity, solidarity. 2 compactness. 3 unity, completeness. 4 uniformity, consistency. 5 closure

geschlungen wound, etc: *pp of* SCHLINGEN

geschm. *abbv* (geschmolzen) molten, etc

Geschmack *m* (-̈e) taste (*all meanings*). **geschmacklich** *a* as regards taste, gustatory. **geschmacklos** *a* tasteless. **Geschmacklosigkeit** f tastelessness

Geschmacks·becher *m* taste bud. **-muster** *n* (registered) design. **-sinn** *m* sense of taste. **-stoff** *m* flavoring. **-verstärker** *m* taste enhancer (e.g. MSG). **g-verstärkt** *a* taste-enhanced

geschmackvoll *a* tasteful

Geschmeide f jewelry

geschmeidig *a* 1 supple; smooth; lithe. 2 pliant, pliable, flexible. 3 (*Metals*) soft, ductile, malleable. **Geschmeidigkeit** f suppleness, etc

Geschmiede·hitze f forging heat

geschmiedet *p.a*: **g—es Eisen** wrought iron. See also SCHMIEDEN

geschmissen thrown: *pp of* SCHMEISSEN

geschmolzen melted: *pp of* SCHMELZEN; *as p.a oft* molten

geschnitten cut: *pp of* SCHNEIDEN

geschnoben snorted: *pp of* SCHNAUBEN

geschoben shoved: *pp of* SCHIEBEN

Geschöpf *n* (-e) creature; figment

geschoren shorn; *pp of* SCHEREN

Geschoß *n* (..osse) 1 missile, projectile; bullet; shell. 2 (*Bot.*) shoot, sprout. 3 floor, story (of a building). **--bahn** f trajectory, path of a projectile

geschossen shot, etc.: *pp of* SCHIESSEN

Geschrei *n* screaming; cry; clamor, outcry

geschr. *abbv* = **geschrieben** written: *pp of* SCHREIBEN

geschrieen screamed: *pp of* SCHREIEN

geschritten strode: *pp of* SCHREITEN

geschult *a* schooled, trained, skilled

geschunden exploited, etc: *pp of* SCHINDEN

geschuppt *p.a* scaly. See also SCHUPPEN

Geschür *n* (*Metll.*) dross, scoria

geschüttet *p.a* bulk, heaped (grain, coal, etc). See also SCHÜTTEN

Geschütz *n* (-e) gun, cannon; (*esp pl*) artillery. **-bronze** f, **-guß** m, **-metall** n gun metal. **-rohr** n gun (or cannon) barrel

Geschw. abbv (Geschwindigkeit) speed, velocity

geschwefelt p.a: **g—er Alkohol** thiol, mercaptan: **g—er Äther** thioether, organic sulfide cf SCHWEFELN

geschweige denn not to mention, let alone

geschwiegen been silent, etc: pp of SCHWEIGEN

geschwind a fast, swift, speedy. **Geschwindigkeit** f (-en) speed, velocity **geschwindigkeits·bestimmung** a speed-determining. **G–grenze** f speed limit. **-konstante** f velocity constant. **-messer** m speed indicator; speedometer, tachometer; chronograph. **-quadrat** n: **mittleres G.** mean square velocity. **-regler** m speed governor. **-verteilung** f velocity distribution. **-zähler** m tachometer, speedometer

Geschwister pl brother(s) and sister(s), siblings

geschwollen swelled: pp of SCHWELLEN; as p.a oft swollen

geschwommen swum: pp of SCHWIMMEN

geschworen 1 festered: pp of SCHWÄREN. 2 sworn: pp of SCHWÖREN. **Geschworene** m, f (adj endgs) juror; pl oft jury

Geschwulst f (-e) 1 swelling. 2 tumor

geschwunden shrunk(en), dwindled pp of SCHWINDEN

geschwungen swung: pp of SCHWINGEN

Geschwür n (-e) ulcer, abscess, sore. **geschwürig** a ulcerated, abscessed

Geselle m (-n,-n) 1 journeyman. 2 mate, companion. 3 fellow. **gesellen** 1 vt (zu) join, unite (with). 2: **sich g.** (zu) associate (with), join. **gesellig** a social, sociable. **Gesellin** f (-nen) 1 journeywoman. 2 = GESELLE 2. 3 woman, person

Gesellschaft f (-en) 1 society (all meanings). 2 company (all meanings). 3 (Com. also) corporation; partnership. 4 group, crowd. 5 party (= gathering). **gesellschaftlich** a social

Gesellschafts· social: **-wissenschaft** f 1 sociology, 2 pl social sciences

Gesenk n (-e) 1 (forging) die; swage. 2 (Min.) blind shaft; staple pit; winze; sump. 3 (fishing) sinker. 4 slope; ravine. 5 (Bot.)

layer. **--schmieden** n drop forging. **-schmiedestück** n drop forging (= forged piece). **-stahl** m die steel

gesessen sat: pp of SITZEN

Gesetz n (-e) law (incl Sci.); statute, rule. **-entwurf** m (legislative) bill. **g–gebend** a legislative. **G–gebung** f legislation **gesetzlich** a lawful, legal; **g. geschützt** protected by law, (specif) patented, copyrighted, registered. **Gesetzlichkeit** f lawfulness, legality

gesetzmäßig a 1 lawful, legal. 2 regular, conforming to (natural) law (or to the rule). **Gesetzmäßigkeit** f lawfulness, legality; regularity

gesetzt 1 p.a sedate. 2 cnjc: **g. den Fall** = **gesetztenfalls** cnjc, adv assuming, supposing (that) cf SETZEN 9. **Gesetztheit** f sedateness

gesetzwidrig a illegal, unlawful

ges. gesch. abbv (gesetzlich geschützt) protected by law cf GESETZLICH

gesichert p.a secure. See also SICHERN

Gesicht n I (-er) face; look, appearance. II (no pl) (eye)sight; vision; **aus dem G. verlieren** lose sight of; **zu G. bekommen** lay eyes on. III (-e) vision, apparition

Gesichts· face, facial, vision, visual: **-achse** f visual axis. **-eindruck** m visual impression. **-feld** n field of vision. **-kreis** m 1 field of vision. 2 sight, view. 3 (Astron.) horizon. **-krem** f face cream. **-linie** f 1 facial line. 2 visual line. **-maske** f 1 face mask. 2 face guard. 3 face pack, facial. **-packung** f face pack. **-punkt** m 1 point of view. 2 (Opt.) visual point. **-rose** f facial erysipelas. **-sinn** m sense of sight. **-strahl** m visual ray. **-täuschung** f optical illusion. **-wasser** n face lotion. **-winkel** m 1 facial angle. 2 visual angle. 3 point of view. **-zug** m (facial) feature

Gesims n 1 molding, cornice. 2 mantelpiece. 3 (window) sill. 4 (incl Geol.) ledge

gesinnt a (oft + dat) oriented, disposed (toward); (esp in compds) -minded. **Gesinnung** f (-en) mind, attitude, convictions

Gesinter n (-) agglomerate

gesittet a well-bred, civilized

gesoffen drunk: pp of SAUFEN

gesogen sucked: *pp of* SAUGEN

gesondert *p.a* separate *cf* SONDERN *vt*

gesonnen 1 contemplated: *pp of* SINNEN. 2 *p.a* pred: **g. sein, zu** . . . be inclined to . . .

gesorgt *p.a* (**für**) provided (for): **es ist für solche Fälle g.** such cases are provided for *cf* SORGEN

gesotten boiled: *pp of* SIEDEN

gespalten *p.a* cleft *cf* SPALTEN

gespannt *p.a* taut; tense, strained; (+**auf**) curious (about), anxious (to see). *See also* SPANNEN. **Gespanntheit** *f* tautness; tension; anxiety

Gespenst *n* (-er) specter, ghost; phantom

Gesperr(e) *n* (..rre) 1 (spun) yarn. 2 tissue, web. 3 cocoon. **-faser** *f* textile fiber. **-pflanze** *f* textile (*or* fiber) plant

gespieen spewed *pp of* SPEIEN

gesplissen spliced: *pp of* SPLEISSEN

gesponnen spun: *pp of* SPINNEN

Gespräch *n* (-e) 1 conversation, talk. 2 (telephone) call. **Gesprächs·kreis** *m* discussion group

gesprochen spoken: *pp of* SPRECHEN

gesprossen sprouted: *pp of* SPRIESSEN

gesprungen sprung, etc: *pp of* SPRINGEN

gest. *abbv* (gestorben) died, deceased

gestachelt *p.a* thorny, spiny, prickly *cf* STACHELN

Gestade *n* (-) shore, bank, coast

Gestalt *f* (-en) shape, form, configuration; figure; build, stature, aspect. **-änderung** *f* change of shape (*or* form, etc)

gestalten 1 *vt* shape, mold, fashion; make, create, design, arrange. 2: **sich g.** take shape; present itself; (+ **als**) turn out (to be). **gestaltend** *p.a* creative, formative

Gestalt·festigkeit *f* stability of form. **-lehre** *f* morphology

gestaltlos *a* amorphous, shapeless, formless. **Gestaltlosigkeit** *f* amorphousness, etc

Gestaltung *f* (-en) form; formation, shaping; creation; design(ing), construction; arrangement; configuration

Gestalt·veränderung *f* change of shape (*or* form), transformation, metamorphosis

gestand congealed, etc: *past of* GESTEHEN

gestanden 1 *pp of* STEHEN (stand) &

GESTEHEN (congeal). 2 *p. adv:* **offen g.** admittedly, to be perfectly frank

Geständnis *n* (-se) confession, acknowledgment

Gestänge *n* (-) system of rods (levers, poles, bars); rod linkage

Gestank *m* (-e) stench, stink *cf* STINKEN

gestatten *vt/i* allow, permit, make possible. **gestattet** *p.a* permissible

Geste *f* (-n) gesture

gestehen* 1 *vt/i* confess, admit, acknowledge. 2 *vi*(s) coagulate, congeal, curdle. *Cf* GESTANDEN

Gestein° *n* stone, rock *cf* ANSTEHEND; (*in compds: see also* GESTEINS·): **-art** *f* rock type. **g–bildend** *a* rock-forming. **G–bohrer** *m* rock drill. **-glas** *n* (*Geol.*) (volcanic) glass. **-kunde, -lehre** *f* petrology

Gesteins· (*see also* GESTEIN·): **-bildung** *f* rock formation. **-gang** *m* (*Geol., Min.*) dyke. **-mantel** *m* (*Geol.*) lithosphere. **-rest** *m* (*Ceram.*) rock residue. **-schicht** *f* layer of rock (*or* stone). **-schutt** *m* rock debris, talus. **-sprengmittel** *n* (rock) blasting explosive. **-unterlage** *f* rocky subsoil

Gestell *n* (-e) 1 rack, frame, case. 2 stand, support, base, mount(ing), bed. 3 (*Blast Furnace*) hearth

gestellt *p.a* 1: **gut g.** well off. 2: **auf sich selbst g.** left to one's own devices. *Cf* STELLEN

gestern *adv* yesterday

gesternt *a* starred, starry *cf* STERN

Gestiebe *n* (-) dust, powder; wind-driven snow

gestiegen climbed, etc: *pp of* STEIGEN

Gestirn *n* (-e) star; constellation. **gestirnt** *a* starred, starry

gestoben scattered: *pp of* STIEBEN

Gestöber *n* (-) (snow, dust) flurry; squall

gestochen 1 pricked: *pp of* STECHEN. 2 *p. adv* (*Phot.*) **g. scharf** needle-sharp

gestockt *p.a* plump (hides). *See also* STOCKEN

gestohlen stolen: *pp of* STEHLEN

gestorben died: *pp of* STERBEN

Gesträuch *n* bushes, shrubbery; thicket

gestreckt *p.a* 1 linear (molecule). 2 180° (angle). 3 wrought (iron). *Cf* STRECKEN

gestrichelt *p.a* dotted, (*more accurately*)

dashed (line) *cf* STRICHELN

gestrichen I deleted: *pp of* STREICHEN. **II** *p.a*
2 level (cupful, etc): **g. voll** brimful. **3:
frisch g.!** wet paint!

gestrig *a* yesterday's *cf* GESTERN

gestritten disputed: *pp of* STREITEN

Gestrüpp *n* (-e) underbrush, thicket

Gestüb(b)e *n* (-) **1** (*Metll.*) brasque, furnace
lining (of coke breeze and loam). **2** (*Coal*)
culm, slack

gestuft *a* multi-stage; stepwise, graduated
cf STUFE

gestunken stunk: *pp of* STINKEN

gestürzt *p.a* concave, dished. *See also*
STÜRZEN

Gesuch *n* (-e) request, application; petition

gesucht *p.a* **1** sought-after, in demand.
2 (*esp in ads*) wanted. **3** studious; fastidious; far-fetched. *Cf* SUCHEN

gesund *a* healthy, sound, well. **G-·brun-
nen** *m* mineral spring(s), spa. **gesunden**
vi(s) recover. **Gesundheit** *f* health;
healthfulness. **gesundheitlich** *a* (of, as
regards) health; sanitary, hygienic

Gesundheits·amt *n* board of health. **g-
halber** *adv* for reasons of health. **G-
pflege** *f* health care, hygiene. **g-
schädigend, -schädlich** *a* health-
impairing, unhealthy, unhygienic. **G-
wesen** *n* public health, field of health
care. **-zustand** *m* state of (one's) health

Gesundung *f* (-en) recovery; convalescence

gesungen sung: *pp of* SINGEN

gesunken 1 sunk: *pp of* SINKEN. **2** *p.a*
lowered, reduced

Getäfel *n* (-) paneling, wainscotting *cf*
TÄFELN

getan 1 done: *pp of* TUN. **2** *p.a:* **damit ist es
(nicht) g.** that is (not) all, that does (not
do) it

getigert *a* striped, streaked

getränk *n* (-e) beverage, drink

getrauen: sich g. dare, venture

Getreide *n* (-) grain (crop). **-art** *f* type of
grain, cereal. **-boden** *m* granary. **-brand**
m (cereal) smut, uredo. **-branntwein** *m*
grain whisky. **-brennerei** *f* grain distillery. **-flocken** *fpl* grain flakes, (breakfast)
cereal. **-frucht** *f* cereal. **-grube** *f* grain
silo. **-kleie** *f* bran. **-korn** *n* grain (of
cereal, wheat, etc.). **-krebs** *m* grain wee-

vil. **-kümmel** *m* grain kümmel (liqueur).
-mehl *n* grain meal, flour, **-prober, -prü-
fer** *m* grain tester. **-schnaps** *m* grain liquor (*or* brandy). **-speicher** *m* granary,
grain silo. **-stein** *m* grain liquor scale.
-stroh *n* straw (from cereal plants)

getrennt *p.a* separate. *See also* TRENNEN

getreu *a* (*usu* + *dat*) faithful, true (to); *oft
as sfx:* **lebensgetreu** true to life. **ge-
treulich** *a, usu adv* loyal(ly), faithful(ly),
true

Getriebe *n* (-) (*Mach.*) transmission, drive;
gearing, gear train; works, mechanism.
2 bustle, activity

getrieben 1 driven: *pp of* TREIBEN. **2** *p.a:*
g-es Eisen wrought iron; **g-e Pflanzen**
hothouse plants

Getriebe·öl *n* transmission oil. **-rad** *n*
(transmission) gear. **-turbine** *f* geared
turbine

getroffen I 1 met, etc: *pp of* TREFFEN **2** *p.a*
appropriate, accurate (description, etc).
II dripped, etc: *pp of* TRIEFEN

getrogen deceived: *pp of* TRÜGEN

getrost *a* confident; *adv usu* without hesitation, without a qualm

getrübt *p.a* clouded, dull, tarnished, etc =
TRÜBE; *see also* TRÜBEN

getrunken drunk: *pp of* TRINKEN

gettern *vt/i* getter. **Getter·substanz** *f*
getter

geübt *p.a* practiced, experienced; skilled.
See also ÜBEN. **Geübte** *m, f* practiced (*or*
experienced) person. **Geübtheit** *f* experience; skill

geviert *a* square(d), quadratic. **Geviert** *n*
(-e) square (*incl Math.*): **drei Meter im
G.** three meters square; **eine Zahl ins G.
bringen** square a number

gew. *abbu* (gewöhnlich) usu., usual(ly)

Gew. *abbu* (Gewicht) wt. (weight), gravity.
Gew.-% *abbu* (Gewichtsprozent) pct. by
wt.

Gewächs *n* (-e) **1** plant, herb. **2** produce.
—**eigenes G.** homegrown. **3** vintage.
4 growth, swelling, tumor

gewachsen *p.a:* **der Aufgabe g.** equal to
the task *cf* WACHSEN

Gewächs·haus *n* greenhouse, conservatory. **-kunde, -lehre** *f* botany. **-reich** *n*
vegetable kingdom

gewahr *pred a:* **g. werden** (+ *genit* or *accus*) become aware (of), notice

gewahren *vt* perceive, notice

gewähren 1 *vt* grant, allow; provide. 2 *vi:* **g. lassen** let go (*or* have) one's own way

gewährleisten *vt* (*insep, pp* gewährleistet) guarantee, vouch for

gewahrt *pp of* WAHREN (keep) & GEWAHREN (notice)

Gewalt *f* (-en) 1 power. 2 authority. 3 control. 4 force, violence: **G. antun** (*dat*) do violence (to). 5 might, strength. **gewaltig** *a* 1 powerful, mighty. 2 enormous, tremendous; *adv oft* terribly. **gewaltsam** *a* violent; forcible, (*adv oft*) by force.

Gewaltsamkeit *f* violence, force

Gewand *n* (⁻er) garment; clothing, garb

gewandt 1 turned: *pp of* WENDEN. 2 *p.a* agile; adept, skilled. **Gewandtheit** *f* (-en) agility, adeptness, skill

gewann won, etc: *past of* GEWINNEN

gewärtig *a* 1 aware. 2 *pred:* **g. sein** = **gewärtigen** *vt* expect, realize; be prepared for

Gewässer *n* (-) body of water; (*pl oft*) waters. **-kunde** *f* hydrology. **-reinhaltung** *f* water conservation (*or* purity control). **-reinigung** *f* water purification

Gewebe *n* (-) 1 tissue (*incl Biol.*), web. 2 fabric, textile. 3 wire mesh (*or* netting). 4 (*in compds oft*) histo(logical): **g-ähnlich** *a* weblike, webbed. **G-draht** *m* gauze wire. **-farbstoff** *m* histohematin. **g-feindlich** *a* tissue-destroying. **G-flüssigkeit** *f* tissue fluid. **-industrie** *f* textile industry. **-kultur** *f* tissue culture. **-kunstleder** *n* leathercloth. **-lehre** *f* histology. **-züchtung** *f* tissue culture

Gewebs- = GEWEBE 4: **-brei** *m* (*Bact.*) tissue pulp. **-lehre** *f* histology. **-waren** *fpl* textile goods, textiles

geweckt *p.a* alert, wide-awake *cf* WECKEN.

Gewecktheit *f* alertness

Gewehr *n* rifle, gun; (small) arms. **-geschoß** *n* bullet. **-granate** *f* rifle grenade. **-munition** *f* small-arms ammunition. **-pulver** *n* rifle powder. **g-schußsicher** *a* bulletproof. **G-sprenggranate** *f* high-explosive rifle grenade

Geweih *n* (-e) (set of) antlers, horns

gewellt *p.a* wavy; rolling, undulating (terrain). *See also* WELLEN *vt*

Gewerbe *n* (-) trade; occupation, vocation, profession; (branch of) industry. **-ausstellung** *f* trade fair, industrial exposition. **-fleiß** *m* industrial (*or* manufacturing) activity. **-hygiene** *f* industrial hygiene. **-salz** *n* industrial(-grade) salt. **-schule** *f* (trade, vocational, technical) school. **-treibende** *m, f* (*adj endgs*) tradesperson, professional; artisan

gewerblich *a* trade, industrial, commercial

gewerbs·mäßig *a* professional. **-tätig** *a* professionally active

Gewerk *n* (-e) trade, craft; guild. **-genossenschaft** *f* trade (*or* labor) union

Gewerkschaft *f* (-en) 1 labor union. 2 mining company

Gewerk·verein *m* labor union

gewesen 1 been: *pp of* SEIN *v.* 2 *p.a* former, past

gewichen yielded: *pp of* WEICHEN II

Gewicht *n* (-e) weight: **spezifisches G.** specific weight (*or* gravity); **ins G. fallen** matter, count, carry weight; **G. legen (auf)** lay weight (on), attach importance (to)

gewichten *vt* (*Stat.*) weight

gewichtig *a* heavy, weighty

Gewichts· weight: **-abgang** *m* weight deficiency (*or* loss). **-abnahme** *f* weight loss. **-analyse** *f* gravimetric analysis. **g-analytisch** *a* gravimetric. **G-änderung** *f* change in weight. **-bestimmung** *f* determination of weight. **-bruch** *m* weight fraction. **-einheit** *f* unit of weight. **g-konstant** *a* of constant weight. **G-konstanz** *f* constant (*or* constancy of) weight. **-konzentration** *f* concentration by weight. **-last** *f* dead weight. **-leder** *n* heavy leather. **g-los** *a* weightless. **-mäßig** *a, adv* by weight. **G-menge** *f* quantity by weight. **-prozent** *n* percent(age) by weight. **g-prozentig** *a* weight-percent, percent-by-weight. **G-satz** *m* set of weights. **-strom** *m* mass flow. **-stück** *n* weight. **-teil** *m* part by weight. **-verlust** *m* weight loss, loss of (*or* in) weight. **-zunahme** *f* weight gain, increase in weight. **-zusammensetzung** *f* composition by weight

Gewichtung f (-en) (*Stat.*) weighing

gewiegt a crafty; experienced cf WIEGEN **II**

gewiesen shown, etc: *pp of* WEISEN

gewillt a willing; prepared

Gewinde n (-) **1** (screw) thread. **2** skein. **--bohrer** m (screw) tap. **-glas** n screw-cap glass (jar). **-rohr** n threaded pipe. **-schneidöl** n screw-cutting oil. **-steigung** f screw pitch. **-stift** m threaded pin

gewinkelt p.a angled; angular cf WINKEL

Gewinn m (-e) **1** profit. **2** gain. **3** proceeds, yield; winnings. **4** win; winning. **5** acquisition. **g--abwerfend** a profitable. **gewinnbar** a winnable; obtainable; extractable. **gewinn-bringend** a profitable

gewinnen* (gewann, gewonnen) **I** vt/i **1** win. **II** vt **2** gain. **3** obtain, acquire; extract (ores), recover (metals). **4** produce, prepare (chemicals). **5** get to, reach. **6** (für) win over (to, for). **7: es über sich g., zu** . . . bring oneself to . . . **III** vi **8** improve; (*usu* + **an**) gain. **gewinnend** p.a winning, engaging. **Gewinnung** f obtaining, acquisition; extraction; recovery; production, preparation

Gewinnungs-weise f manner (*or* way) of obtaining (acquiring, extracting, etc) cf GEWINNUNG

gewirbelt p.a vertebrate cf WIRBELN

Gewirr n (-e) confusion, chaos, tangle

gewiß a certain, sure; *adv. oft* one can be sure

Gewissen n (-) conscience. **gewissenhaft** a conscientious. **gewissenlos** a unscrupulous

gewissermaßen adv to a certain extent, in a way

Gewißheit f (-en) certainty; conviction

Gewitter n (-) (thunder)storm. **gewittrig** a stormy

gewoben woven: *pp of* WEBEN

gewogen 1 weighed: *pp of* WÄGEN & WIEGEN. **2** p.a (+dat) favorably inclined (to), fond (of)

gewöhnen vt & sich g. (an) get (sbdy) used (*or* accustomed) (to) cf GEWOHNT, GEWÖHNT. **Gewohnheit** f (-en) **1** habit. **2** custom. **gewöhnlich** a usual, ordinary, common. **gewöhnlicherweise** adv usually, etc

gewohnt I lived: *pp of* WOHNEN. **II** a **1** usual, habitual. **2** *pred* (+*accus*) = **gewöhnt** p.a (an) used, accustomed (to)

Gewöhnung f habituation, acclimatization; habit

Gewölbe n (-) **1** vault; dome; arch; crown (of furnace). **2** (storage) cellar. **gewölbt** p.a convex; dished. *See also* WÖLBEN

Gewölk n clouds

gewollt p.a intentional, deliberate cf WOLLEN

gewonnen won: *pp of* GEWINNEN

geworden become, etc: *pp of* WERDEN

geworfen thrown: *pp of* WERFEN

gewrungen wrung: *pp of* WRINGEN

Gew. T. abbv (Gewichtsteil) part by weight

gew. Temp. abbv (gewöhnliche Temperatur) genl. temp.

Gewühl n (-e) (milling) crowd; turmoil

gewunden 1 wound: *pp of* WINDEN. **2** p.a convoluted; spiral

gewürfelt p.a check(er)ed. *See also* WÜRFELN

Gewürm n worms; vermin. **gewürmelt** p.a vermiculated

Gewürz n (-e) spice; seasoning, condiment. **g--artig** a spicy, aromatic. **G--essig** m aromatic vinegar

gewürzhaft a spicy, aromatic. **Gewürzhaftigkeit** f spiciness, aromatic quality. **gewürzig** a spicy, aromatic

Gewürz-kraut n spice plant. **-nelke** f clove. **-nelkenöl** n oil of cloves. **-pflanze** f aromatic (*or* spice) plant. **-pulver** n (*Pharm.*) aromatic powder. **-stoff** m spice aromatic. **-strauch** m allspice

gewürzt p.a spiced, spicy, aromatic cf WÜRZEN

Gewürz-tinktur f aromatic tincture. **-wein** m spiced wine

gewußt known: *pp of* WISSEN

gez. abbv (gezeichnet) signed

gezackt p.a jagged. *See also* ZACKEN

Gezähe n (miner's) tools

gezahnt, gezähnt p.a toothed, serrate(d), indented cf ZAHNEN

Gezeiten fpl tide(s)

Gezeug n tools, gear, equipment

gezielt p.a purposeful, goal-oriented; precisely aimed, specific; adv oft for a definite purpose cf ZIELEN

geziemen *vi* (*dat*), **sich g.** be fitting (proper, appropriate). **geziemend** *p.a* fitting, appropriate. *See also* ZIEMEN

geziert *p.a* coy, affected. *See also* ZIEREN

gezinkt *p.a* galvanized; dovetailed *cf* ZINKEN

gezogen 1 pulled, etc: *pp of* ZIEHEN. **2** *p.a* (*Cars*) front-wheel-drive

Gezücht *n* brood; riffraff

gezweiteilt *a* bipartite, divided in two

gezwungen forced: *pp of* ZWINGEN. **gezwungenermaßen** *adv* perforce, of necessity

gg. *abbv* (gegen) against, opposite

ggf. *abbv* (gegebenenfalls) if necessary

Ggw. *abbv* (Gegenwart) presence

Ghedda·säure *f* gheddic acid

GHz *abbv* (Gigahertz) gigahertz

Giberellin·säure *f* gibberellic acid

gibt gives: *pres of* GEBEN —**es g.** there is (*or* are)

Gicht *f* (-en) **I** (*Metll.*) **1** (furnace) mouth, throat, top. **2** charge. **II** (*Med., Bot.*) gout

gicht·artig *a* gouty, arthritic. **-brüchig** *a* palsied. **G–gas** *n* blast-furnace gas, top gas. **-glocke** *f* (*Metll.*) stopper bell (of furnace). **-mittel** *n* gout remedy. **-rauch** *m* blast-furnace smoke. **-rose** *f* peony; rhododendron. **-rübe** *f* bryony. **-schwamm** *m* (*Blast Furnace*) top incrustation. **-staub** *m* blast-furnace flue dust

Gichtung *f* charging (of a furnace)

Gicht·verschluß *m* (*Blast Furnace*) bell and hopper. **g–widrig** *a* antiarthritic

Giemen *n* (*Med.*) dyspnea, râle

Gier *f* craving; greed. **gierig** *a* avid, greedy

Gieß·arbeit *f* casting; founding. **-bach** *m* torrent

gießbar *a* pourable; castable

Gieß·becken *n*, **-beckenknorpel** *m* arytenoid (cartilage). **-bett** *n* (*Metll.*) casting bed, pig bed. **-brei** *m* slurry

gießen* (goß, gegossen) **1** *vt/i* pour (*incl* rain); water (plants); cast (metal, etc), found, teem; mold. **2** (*Phot.*) coat (film). **Gießer** *m* (-) caster, founder; pitcher, pouring vessel; smelter

Gießerei *f* (-en) founding, casting. **2** = **-betrieb** *m* foundry, casting house. **-eisen** *n* foundry pig (iron). **-formkoks** *m* formed foundry coke. **-harz** *n* foundry resin.

-koks *m* foundry coke. **-ofen** *m* melting furnace. **-roheisen** *n* foundry pig (iron). **-schwarz** *n* foundry blacking. **-wesen** *n* founding

Gießer·schwärze *f* (*Foun.*) black wash

Gieß·erz *n* bronze. **g–fähig** *a* castable, pourable. **G–fähigkeit** *f* castability, pourability. **g–fertig** *a* ready for casting (*or* pouring). **G–folie** *f* cast film. **-form** *f* casting (*or* shell) mold. **-grube** *f* casting pit. **-harz** *n* 1 casting resin. 2 cast resin. **-haus** *n*, **-hütte** *f* foundry, casting house. **-kanne** *f* watering can. **-kasten** *m* casting mold. **-kelle** *f* casting ladle. **-koks** *m* foundry coke. **-kopf** *m* feeding head

Gießling *m* (-e) casting

Gieß·löffel *m* casting ladle. **-masse** *f* casting material, (*Ceram.*) castable. **-mutter** *f* matrix, mold. **-ofen** *m* founding furnace. **-pfanne** *f* casting ladle, **-probe** *f* pour(ing) test. **-sand** *m* molding sand. **-schlicker** *m* (*Ceram.*) casting slip. **-schnauze** *f* pouring spout. **-stein** *m* porous granite. **-tiegel** *m* melting pot. **-topf** *m* pot (*or* jar) with a pouring lip. **-trichter** *m* (pouring) gate, runner. **-waren** *fpl* cast-metal goods. **-zement** *m/n* quick-setting alumina cement

Gift *n* (-e) **1** poison, toxin; venom, virus. **2** spite. **g–·abtreibend** *a* antitoxic, antidotal. **-artig** *a* poisonous. **G–arznei** *f* antidote. **-äscher** *m* (*Lthr.*) arsenic lime. **-baum** *m* poison oak. **-blase** *f* venom sac. **-drüse** *f* poison gland. **-dunst** *m* poisonous (*or* toxic) vapor. **-erz** *n* arsenic ore. **-fang** *m* **1** poison tower (*or* fume trap), receiver (for sublimed As_2O_3). **2** poison (*or* venom) fang. **g–fest** *a* immune to poison; venomproof. **-frei** *a* poison-free, nonpoisonous, nontoxic. **G–gas** *n* poison (*or* toxic) gas. **-gruppe** *f* toxin group. **g–haltig** *a* toxic, containing poison. **G–heber** *m* poison siphon. **-hütte** *f* arsenic works (*or* plant)

giftig *a* **1** poisonous, toxic; venomous; virulent. **2** spiteful. **Giftigkeit** *f* poisonousness, toxicity, etc.

Gift·jasmin *m* Carolina jasmine. **-kies** *m* **1** arsenopyrite, arsenical pyrites; mispickel. **2** loellingite. **-kobalt** *m* native arsenic. **-kraut** *n* poisonous plant (*or* herb), (*specif*) wolfbane, monkshood, etc. **-krieg** *m* chemical (*or* poison-gas) warfare. **-kunde** *f* toxicology. **-lattich** *m* prickly (*or* hemlock) lettuce. **-lehre** *f* toxicology. **g–los** *a* nonpoisonous, nontoxic. **G–mehl** *n* arsenic dust, flowers of arsenic, powdered arsenic trioxide. **-mittel** *n* antidote. **-müll** *m* toxic waste. **-nebel** *m* (*Mil.*) toxic smoke. **-pille** *f* 1 poison pill. 2 antidotal pill. **-rauch** *m* sublimed arsenic trioxide; (*Mil.*) toxic smoke. **-regen** *m* aerial gas spray. **g–reich** *a* poisonous, etc. = GIFTIG. **G–schlange** *f* poisonous snake. **-schutz** *m* protection against poison(ing). **-schwamm** *m* toxic fungus, poisonous mushroom. **-sekret** *m* toxic secretion. **-stoff** *m* poisonous substance, toxin, toxic agent. **-sumach** *m* poison sumac; poison ivy. **g–trächtig** *a* toxic. **G–turm** *m* poison (*or* receiving) tower (for arsenic fumes) = GIFTFANG 1

Giftung *f* (-en) (*Biochem.*) toxicogenation **Gift-wende** *f* swallowwort. **g–widrig** *a* antitoxic, antidotal. **G–wirkung** *f* toxic effect. **-wurzel** *f* 1 poisonous root. 2 contrayerva; swallowwort

Gigant *m* (-en,-en) giant, colossus

Gilbe *f* (-n) 1 yellow(ish) color (*or* dye); yellowing agent; yellow ocher; dyer's rocket. 2 jaundice. **gilben** *vt, vi*(s) yellow. **gilbig** *a* ocherous

Gilb- yellow: **-kraut** *n* yellowweed, dyer's rocket. **-wurzel** *f* turmeric

gilt is valid, etc: *pres of* GELTEN

giltig *a* valid = GÜLTIG. **Giltigkeit** *f* validity

ging went: *past of* GEHEN

Ginster *m* (*Bot.*) broom, *Genista*

Gipfel *m* (-) top; peak, summit; apex. **-höhe** *f* peak (*or* summit) altitude; (*Aero.*) (absolute) ceiling. **-konferenz** *f* summit conference

gipfeln 1 *vt* prune, top. 2 *vi* culminate

Gipfel-punkt *m* high(est) point, culmination, peak, summit, top. **-trieb** *m* (*Bot.*) top (*or* terminal) shoot

Gipfelung *f* (-en) culmination

Gipfel-wert *m* top (*or* peak) value, maximum

Gips *m* (-e) plaster (of Paris), plaster cast; (*Min.*) gypsum, $CaSO_4$. **-abdruck, -abguß** *m* plaster cast. **-arbeiter** *m* plas-

terer. **g–artig** *a* gypseous, plastery. **G–brei** *m* plaster paste. **-brennen** *n* gypsum burning. **-brennerei** *f* 1 plaster kiln. 2 calcination of gypsum. **-brennofen** *m* plaster kiln

gipsen I *vt* plaster; put in a cast. II *a* plaster; gypseous. **Gipser** *m* (-) plasterer

Gips-erde *f* earthy gypsum; gypseous soil. **-form** *f* plaster (of Paris) mold. **-gu(h)r** *f* earthy gypsum. **-guß** *m* plaster cast(ing). **g–haltig** *a* gypsiferous, containing gypsum. **G–härte** *f* hardness due to gypsum. **-kalk** *m* plaster lime. **-kitt** *m* plaster cement. **-lösung** *f* calcium sulfate solution. **-mehl** *n* powdered plaster (*or* gypsum). **-mergel** *m* gypsum marl. **-mörtel** *m* plaster, stucco. **-niederschlag** *m* calcium sulfate precipitate. **-ofen** *m* plaster kiln. **-platte** *f* plasterboard. **-spat** *m* sparry gypsum, selenite. **-stein** *m* plaster stone, hard gypsum deposit. **-teer** *m* plaster-and-tar mixture. **-verband** *m* plaster cast

gischen *vi* foam, froth; spray; ferment. **Gischt** *m* (-e), *f* (-en) foam, froth; spray; fermentation

Gitter *n* (-) grating, grill; (*esp Cryst.*) lattice, trellis; screen; (iron) bars; (*esp Electron.*) grid; (*Math.*) graph. **-abteilung** *f* grid leak. **-abstand** *m* lattice constant (*or* distance). **-anordnung** *f* (crystal) lattice structure. **g–artig** *a* latticed, gridlike. **G–bau** *m* lattice structure. **-draht** *m* 1 wire netting. 2 (*Elektron.*) grid wire. **-ebene** *f* lattice plane. **-elektron** *n* lattice electron. **-energie** *f* (*Cryst.*) lattice energy. **-farbe** *f* grating color. **-fehlbaustelle** *f* crystal lattice defect position. **g–förmig** *a* latticed, gridlike; lattice-shaped. **-fremd** *a* foreign to the lattice. **G–konstante** *f* lattice (*or* grating) constant. **-leerstelle** *f* lattice vacancy. **-loch** *n* lattice hole (*or* void). **-lücke** *f* interstitial hole, lattice vacancy. **-masche** *f* grid mesh. **-platz** *m* lattice position. **-punkt** *m* lattice point. **-rost** 1 *n* grating, grill. 2 *m* (*Bot.*) cluster-cup (rust). **-spannung** *f* grid voltage. **-spektrograph** *m* grating spectrograph. **-stein** *m* (*Furnace*) checker brick. **-steuerung** *f* grid control. **-störstelle** *f* lattice defect. **-störung** *f* lattice distortion. **-strom** *m* grid current. **-ty-**

pus *m* lattice type. **-werk** *n* lattice work, grill work, checkerwork. **-widerstand** *m* grid resistance. **-zelle** *f* lattice cell. **-zusammenhang** *m* lattice structure
Gl. *abbv* (Gleichung) equation
Glabrat·säure *f* lecanoric acid
glacé·gar *a* (*Lthr.*) alum-tanned. **G–gerbung** *f* alum tanning. **-handschuh** *f* kid glove. **-karton** *m* glazed cardboard. **-leder** *n* glacé leather. **-papier** *n* glazed paper
glacieren *vt* 1 gloss, glaze, (*Tex.*) glacé, friction-calender. 2 frost, ice. 3 freeze
glandern *vt* (*Tex.*) calender
Glanz *m* 1 gloss, luster. 2 brightness, brilliance, radiance, glitter: **mit G.** brilliantly. 3 polish, shine. 4 (*Diamonds*) water. 5 (*Min., esp in compds*) glance, sulfide mineral. 6: **farbiger G.** dichroism. **--appretur** *f* (*Tex.*) glaze. **-arsenikkies** *m* arsenopyrite, loellingite. **-blech** *n* polished sheet metal. **-blende** *f* alabandite. **-braunkohle** *f* bright lignite. **-braunstein** *m* hausmannite. **-brenne** *f* (*Metll.*) burnishing bath, bright dip. **-bürste** *f* polishing brush. **-decklack** *m* gloss coating varnish
Glänze *f* (-n) polish; glaze, size
Glanz· gloss, glaze, polish; brilliant: **-effekt** *m* luster, gloss(y look). **-einfall** *m* (*Opt.*) glazing incidence. **-eisen** *n* silvery iron, (*Min.*) cohenite, schreibersite. **-eisenerz** *n* specular hematite. **-eisenstein** *m* limonite
glänzen 1 *vt* polish, burnish, glaze. 2 *vi* shine, glisten, glitter; be shiny (*or* brilliant). **glänzend** *p.a* shiny, lustrous, brilliant, glossy
Glanz· gloss, glaze, polish; brilliant: **-erz** *n* argentite, graphite. **-farbe** *f* 1 brilliant color. 2 glazing varnish. 3 gloss ink. **g–fein** *a* brilliant. **G–firnis** *m* glazing varnish. **-gold** *n* burnished gold; (*Ceram.*) brilliant gold; (imitation) gold foil. **-karton** *m* glazed cardboard. **-kattun** *m* glazed calico. **-kobalt** *m* cobalt glance, cobaltite, smaltite. **-kohle** *f* glance coal, vitrain. **-kohlenstoff** *m* lustrous carbon. **-kopf** *m* hematite. **-lack** *m* glazing (*or* brilliant) varnish. **-leder** *n* patent (*or*

glacé) leather. **-leinwand** *f* glazed linen, buckram
glanzlos *a* lusterless, lackluster, dull, dim, dead, mat. **Glanzlosigkeit** *f* dullness, lusterlessness, lack of luster
Glanz· gloss, etc: **-manganerz** *n* manganite. **-messer** *m* gloss meter. **-messing** *n* polished brass. **-metall** *n* speculum metal. **-mittel** *n*, **Glänzmittel** *n* lustering (*or* polishing) agent. **-öl** *n* gloss oil. **-papier** *n* glazed paper. **-pappe** *f* glazed cardboard. **-pasta**, **-paste** *f* polishing paste. **-platin** *n* burnished platinum
glanzpolieren *vt* burnish, polish (to a gloss)
Glanz·presse *f* (*Tex.*) calender. **g–reich** *a* lustrous, brilliant. **G–rot** *n* colcothar. **-ruß** *m* lampblack, lustrous soot
glanz·schleifen* *vt* polish, burnish, buff
Glanz·silber *n* polished silver, (*Min.*) silver glance, argentite. **-stahl** *m* polished steel. **-stärke** *f* gloss starch. **-stoff** *m* glazed (*or* glossy) fabric; artificial silk; (*TN*) cuprammonium rayon
glanz·stoßen* *vt* (*Lthr.*) glaze
Glanz·streifenkohle *f* clarain-vitroclarain. **-stück** *n* showpiece; highlight. **g–voll** *a* brilliant, splendid. **G–weiß** *n* brilliant white; talc. **-wichse** *f* polishing wax. **-winkel** *m* glancing angle, angle of grazing incidence
Glanz·wirkung *f* polishing (*or* glazing) action (*or* effect)
Glas *n* (¨er) 1 (*usu*) glass. 2 (*specif*) (glass) jar; lens, (*pl oft*) (eye)glasses. **--abfall** *m* glass waste, cullet. **-achat** *m* obsidian. **g–ähnlich**, **-artig** *a* glasslike, glassy, vitreous. **G–artigkeit** *f* glassiness, vitreousness. **-ätzkunst** *f* art of etching glass. **-ballon** *m* glass carboy (bulb, balloon, flask). **-becher** *m* glass beaker. **-bereitung** *f* glassmaking. **-biegen** *n* glass bending. **-bild** *n* transparency, slide. **-bildner** *m* glass former, vitrifier. **-bildung** *f* glass formation, vitrification. **-birne** *f* electric light bulb. **-blase** *f* bubble in glass. **-blasen** *n* glassblowing. **-bläser** *m* glassblower. **-bläserei** *f* 1 glassblowing. 2 glassworks
Glasbläser·lampe *f* glassblower's lamp.

-pfeife, -röhre *f* glassblower's pipe, glassmaker's blowpipe

Glas·brennen *n* annealing of glass. **-brocken** *mpl*, **-bruch** *m* broken glass; cullet

Gläs/chen *n* (-) little glass (*or* jar), vial

Glas·dicke *f* thickness of (the) glass. **-dose** *f* glass jar (*usu* pillbox-shaped). **-düse** *f* glass nozzle. **-elektrizität** *f* vitreous (positive) electricity

glasen *vt* glaze. **Glaser** *m* (-) glazier

Gläser glasses: *pl of* GLAS

Glas·erde *f* silicious earth

Glaserei *f* (-en) glazing; glazier's shop

Glaser·kitt *m* glazier's putty

gläsern *a* 1 (made of) glass. 2 (*Sci.*) vitreous. 3 glassy, glazed. 4 brittle. 5 crystal(-clear). **Gläsernheit** *f* vitreousness, glassiness; brittleness

Glas·erz *n* silver glass, argentite. **-fabrikation** *f* glassmaking, glass manufacture. **-faden** *m* glass thread. **-farbe** *f* glass color. **-färben** *n*, **-färbung** *f* glass staining. **g–farbig** *a* glassy, hyaline. **G–faser** *f* glass fiber; fiberglass; fiber optic

Glasfaser·kabel *n* fiber optic cable. **-schichtstoff** *m* glass fiber laminate. **g–verstärkt** *a* glass-fiber-reinforced

Glas·federmanometer *n* spoon gauge. **-fehler** *m* flaw in glass. **-feuchtigkeit** *f* vitreous humor. **-fluß** *m* 1 glass flux. 2 paste (for imitation gems). 3 enamel. **g–förmig** *a* glasslike, vitriform. **G–fritte** *f* glass frit. **-frittentiegel** *m* sintered glass crucible. **-galle** *f* glass gall, sandiver. **-gerät** *n* glass utensil. **-gespinst** *n* spun glass, glass thread. **-gewebe** *n* glass cloth. **-glanz** *m* frost(ing), pounded glass; (*Min.*) vitreous luster. **g–glänzend** *a* glassy, glass-lustered. **G–glocke** *f* 1 bell jar. 2 glass cover (globe, dome, bell). **g–grün** *a* bottle-green. **G–hafen** *m* glass(-melting) pot. **-hahn** *m* glass tap (*or* cock). **g–hart** *a* hard as glass; brittle. **G–härte** *f* (*Steel*) glass hardness; temper. **-härten** *n* glass tempering. **-haube** *f* glass cap (bell, cover). **-haus** *n* greenhouse. **-haut** *f* cellophane; vitreous film; glass (*or* hyaline) membrane. **g–hell** *a* clear as glass, crystal-clear. **G–herstellung** *f* glassmaking, glass manufacture. **-hütte** *f* glassworks

Glasierbarkeit *f* (*Ceram.*) glaze take

glasieren *vt* 1 glaze, enamel. 2 frost, ice. 3 vitrify. **Glasierung** *f* (-en) 1 glazing, enameling, etc. 2 glaze, enamel

glasig *a* 1 glassy, glazed. 2 brittle. 3 transparent; crystal(-clear). 4 hyaline; vitreous

Glas·isolator *m* glass insulator. **-kalk** *m* glass gall. **-kasten** *m* glass case. **-kattun** *m* glass cloth. **-keil** *m* glass prism. **-kirsche** *f* 1 glass ampule. 2 amarelle, translucent cherry. **-kitt** *m* glass cement (*or* putty); diamond cement. **g–klar** *a* transparent, crystal-clear. **G–klotz** *m* glass block. **-kolben** *m* glass flask. **-kopf** *m* (*Min.*): **brauner G.** limonite; **roter, eigentlicher G.** (fibrous *or* reniform) hematite, kidney ore; **schwarzer G.** psilomelane. **-körper** *m* vitreous humor. **-kraut** *n* (*Bot.*) pellitory; glasswort. **-kugel** *f* glass bulb (*or* ball). **-kügelchen** *n* glass bead; small glass bulb. **-kühlen** *n* annealing of glass. **-kühlofen** *m* glass-annealing furnace. **-lack** *m* glass varnish. **-lava** *f* volcanic glass; (*specif*) hyalite, obsidian. **-leinen** *n* spun glass, glass cloth. **-lot** *n* glass solder. **-macherpfeife** *f* (glass)blowing iron, blowpipe. **-macherseife** *f* glassmaker's soap, decolorizer. **-malerei** *f* 1 glass painting (*or* staining). 2 stained glass (window). **-malz** *n* brittle malt. **-masse** *f* vitreous mass, "metal" —**rohe G.** frit. **-mehl** *n* powdered glass, glass meal. **-messer** *n* glass knife (*or* cutter). **-nutsche** *f* (sintered) glass suction filter; glass Buchner funnel. **-ofen** *m* glass furnace (*or* tank). **-opal** *m* hyalite. **-papier** *n* (abrasive) glass paper. **-pech** *n* hard (*or* stone) pitch. **-perle** *f* glass bead. **-pinsel** *m* glass brush (for acids, etc). **-platte** *f* glass plate (*or* platter); glass slide. **-porzellan** *n* vitreous porcelain. **-pulver** *n* glass powder. **-quarz** *n* hyaline (*or* transparent) quartz. **-rohr** *n*, **-röhre** *f* glass tube. **-salz** *n* glass gall. **-sand** *m* glassmaking sand. **-satz** *m* glass batch (*or* composition). **-schale** *f* glass dish.

-schaum *m* glass gall, sandiver. **-schei-be** *f* pane (*or* layer) of glass. **-scherben** *fpl* (pieces of) broken glass, cullet. **-schicht** *f* layer of glass. **-schlacke** *f* glass gall. **-schlange** *f* glass spiral. **-schleifen** *n* glass grinding. **-schliff** *m*, **-schliffverbindung** *f* ground(-glass) joint. **-schmalz** *n* (*Bot.*) glasswort, salicornia. **-schmelz** *m* enamel. **-schmelze** *f* glass melt (*or* batch). **-schmelzofen** *m*, **schmelzwanne** *f* glass-melting furnace (*or* tank). **-schmutz** *m* glass gall. **-schneider** *m* glass cutter. **-schörl** *m* axinite. **-schreibstift** *m* glass-marking pencil. **-schreibtinte** *f* glass-marking ink. **-schüssel** *f* glass dish (*or* bowl). **-schweiß** *m* glass gall. **-seife** *f* glass(maker's) soap, MnO_2. **-sorte** *f* grade of glass. **-spiegel** *m* 1 glass mirror. 2 molten glass level (*or* surface). **-stab** *m*, **-stange** *f* glass rod. **-stapelfaser** *f* glass staple fiber. **-staub** *m* glass dust; powdered (*or* ground) glass. **-stein** *m* 1 axinite. 2 paste (for imitation gems). 3 glass brick. **-stöpsel** *m* glass stopper. **-stöpselflasche** *f* glass-stoppered bottle. **-tafel** *f* sheet of glass. **g—technisch** *a* (of) glass technology. **G—temperofen** *m* glass annealing furnace. **-tiegel** *m* glass crucible. **-topf** *m* glass-melting pot. **-träne** *f* glass tear, Prince Rupert's drop

Glasur *f* (-en) glaze, enamel; frosting, icing; gloss (*esp of Lthr.*) **-blau** *n* zaffer. **-brand** *m* (*Ceram.*) glaze baking, glost firing

glasuren *vt* glaze, enamel; frost, ice

Glasur-erz *n* potter's ore, alquifou, glazing galena. **-farbe** *f* overglaze color. **-fluß** *m* glazing flux. **-ofen** *m* glaze kiln, glost oven. **-schlicker** *m* stopped glaze. **-stein** *m* vitrified brick

glasurt glazed, etc: *pp of* GLASUREN

Glasur-ziegel *m* glazed brick

Glas-verbindung *f* glass joint (*or* connection). **-wanne** *f* glass trough. **-waren** *npl* glassware. **-watte** *f* glass wadding. **-wolle** *f* glass wool, spun glass. **-zustand** *m* vitreous state

glatt I *a* 1 (*usu*) smooth; flush, flat; straight. 2 slick, slippery; icy. 3 glib. 4 complete, out-and-out; sheer, pure; clean (break),

flat (denial, rejection); (*incl adv*) outright: **g—e drei Tage** all of (*or* no less than) three days; (*Math.*) **g. aufgehen** come out even. 5 clear, unequivocal; plain, simple. II *sep pfx* smooth, flat *cf* GLATTFEILEN, GLATTHÄMMERN

Glätt-anfrischen *n* reduction of litharge

glättbar *a* smoothable, etc *cf* GLÄTTEN

Glatt-brand *m* (*Ceram.*) glost (*or* glaze) burn. **g—brennen*** *vt* glost-fire. **G—brennofen** *m* glaze kiln

Glätte *f* 1 smoothness, flatness, straightness. 2 slipperiness, slickness; glibness. 3 completeness, purity, cleanness, evenness. 4 clarity, plainness, simplicity. 5 litharge. 6 slippery conditions. *Cf* GLATT

Glatt-eis *n* smooth (*or* glare) ice; icy conditions

glätten 1 *vt & sich g.* smooth (down, out), even out, level off; calm down. 2 *vt* polish, burnish; finish, put a smooth finish on; glaze, calender. 3: **sich g.** subside

glatterdings *adv* outright, straight, flatly

Glatt-färberei *f* plain dyeing. **g—feilen** *vt* file smooth. **G—feuer** *n* (*Ceram.*) sharp (*or* full) fire

Glätt-frischen *n* reduction of litharge

glatt-hämmern *vt* hammer flat. **Glattheit** *f* smoothness, etc = GLÄTTE

Glätt-maschine *f* (*Tex.*) calender

Glätt-masse *f* smoothing (*or* polishing) compound

Glätt-ofen *m* (*Ceram.*) glost (*or* finishing) kiln

glatt-randig *a* smooth-edge(d). **G—scherbe** *f* potsherd

glatt-streichen* *vt* smooth down (*or* out)

Glättung *f* smoothing; polishing, burnishing; finishing; glazing, calendering *cf* GLÄTTEN. **Glättungs-mittel** *n* smoothing (*or* polishing) agent

Glatt-walze *f* plain roll, smoothing roller **Glätt-walze** *f* glazing roll(er)

Glatt-walzwerk *n* plain roll crusher. **-wasser** *n* (*Brew.*) last run of wort

Glätt-werk *n* glazing machine

Glaube(n) *m* belief; faith. **glauben** *vt/i* believe, think: **sie g. es zu wissen** they believe (*or* think) they know it; **wer glaubt ihnen diese Behauptung?** who believes this claim of theirs?

Glauber·salz n 1 Glauber's salt (sodium sulfate decahydrate). 2 (*Min.*) mirabilite

glaubhaft a credible, plausible; convincing

gläubig a believing; trusting, credulous; faithful. **Gläubige** m, f (*adj endgs*) faithful believer. **Gläubiger** m (-) creditor

glaublich a: **es ist kaum g.** it is hardly believable (*or* credible)

glaubwürdig a authentic; reliable, trustworthy

glauch a 1 glaucous, sea-green; clear, bright. 2 (*Ore*) poor. **Glauch·erz** n poor (*or* low-grade) ore

glaukonitisch a glauconitic

Glaukophan·säure f glaucophanic acid

Glazial·zeit f glacial period

gleich I a 1 equal: **g—e Mengen** equal amounts; **zweimal zwei (ist) g. vier** two times two equals four. 2 same, (a)like: **zur g—en Zeit** at the same time; **g—e Ladungen** like charges; **sie sind einander g.** they are alike; (*also:* they are equal to each other *cf* 1); **es ist ihnen g.** it's all the same to them; **g. behandeln** treat alike (*or* the same way); **ganz g., wie (wer, wo)** no matter how (who, where). II adv oft immediately, right (away): **sie kommen g. wieder** they are coming back right away (*or* coming right back). III as pfx oft equi·, homo·, iso· cf GLEICHACHSIG, GLEICHARTIG, etc. IV sep pfx equal, etc: cf GLEICHMACHEN, GLEICH-STELLEN

gleich·achsig a equiaxial; coaxial. **-armig** a equal-armed. **-artig** a similar, of the same type; uniform, homogeneous. **G-artigkeit** f similarity, uniformity, homogeneity. **g—atomig** a equiatomic. **-bedeutend** a synonymous, equivalent; (*Math.*) homologous. **-berechtigt** a equally entitled, having equal rights. **G-berechtigung** f equal rights

gleich·bleiben* vi(s) remain equal (constant, invariable). **gleichbleibend** p.a constant, invariable

gleich·deutig a synonymous, equivalent; tautomeric. **G-deutigkeit** f synonymy, equivalence; tautomerism. **-druck** m constant pressure. **-drucklinie** f isobar

gleichen* (glich, geglichen) (*oft* + *dat*) 1 vi equal; resemble, be (a)like. 2 vt adapt,

make equal (to); equalize; smooth, level (out)

gleich·entfernt a equidistant

gleicher·gestalt adv likewise. **-maßen** adv equally, to the same extent. **-weise** adv (in) the same way

Gleich·fälligkeitstrennung f separation by sedimentation. **g—falls** adv likewise. **-farbig** a isochromatic, like-colored. **G-feld** n constant field. **-flügler** mpl (*Zool.*) Homoptera. **g—förmig** a regular, even; uniform, homogeneous; monotonous. **G-förmigkeit** f regularity, evenness; uniformity, homogeneity; monotony. **-gang** m synchronism. **g—gekörnt** a even-grained. **-gerichtet** p.a rectified cf GLEICHRICHTEN. **-gesinnt** a like-minded. **-gestaltet, -gestaltig** a homomorphic; (*Cryst.*) isomorphous. **G-gewicht** n equilibrium, balance: **sich (or einander) das G. halten** balance (out), cancel out

Gleichgewichts·bedingung f condition for equilibrium. **-gemisch** n equilibrium mixture. **-isometrie** f tautomerism. **-lage** f equilibrium position. **-lehre** f statics. **-störung** f disequilibrium, disturbance of equilibrium. **-verhältnis** n equilibrium ratio. **-verschiebung** f displacement (*or* shift) of equilibrium. **-zustand** m state of equilibrium

gleich·groß a of equal size, equally large. **-gültig** a indifferent; immaterial. **G-gültigkeit** f indifference; insignificance

Gleichheit f (-en) equality, likeness; similarity; identity

gleich·ionig a having a common ion. **G-klang** m unison, consonance. **-laufschwankung** f (*Electron.*) flutter, wow

gleich·kommen* vi(s) (*dat*) equal, match; be equivalent (to)

Gleich·lauf m synchronism; parallelism. **g—laufend, -läufig** a synchronous; parallel

gleich·machen vt (*d.i.o*) make equal (*or* equivalent) (to); equalize; level (out). **Gleichmachung** f equalization, leveling

Gleich·maß n regularity, evenness; uniformity; symmetry, proportion, harmony. **g—mäßig** a regular, even; uniform; symmetrical, proportionate, harmonious. **G-mäßigkeit** f = GLEICHMASS. **g—mittig** a

concentric. **-molar** *a* equimolar. **G–mut**
m equanimity, even temper. **g–mütig** *a*
even-tempered. **-namig** *a* like(-named);
having a common denominator

Gleichnis *n* (-se) simile; parable; image
gleich·phasig *a* cophasal, of like phase.
-prozentig *a* of equal percentage. **G–
raum** *m* constant volume

gleich·richten *vt* (*Elec.*) rectify; (*Radio,
TV*) detect, demodulate (waves, etc). **G–
richter** *m* rectifier; (radio, video) detector. **-richtung** *f* rectification; detection;
demodulation

gleichsam *adv* as it were, so to speak;
practically

gleich·schenk(e)lig *a* (*Geom.*) isosceles.
-schwer *a* 1 of equal weight. 2 equally
difficult. **-seitig** *a* 1 (*Geom.*) equilateral.
2 (*Tex.*) double-faced

gleich·setzen *vt* (*d.i.o*) equate (with);
equalize

gleich·sinnig *a* of (*or* in) the same sense (*or*
direction), equidirectional. **G–spannung**
f direct current voltage. **-span-
nungskompensator** *m* potentiometer

gleich·stellen *vt* (*d.i.o*) equate, put on a par
(with); coordinate, synchronize (with)

gleich·stoffig *a* homogeneous. **G–stoffigkeit** *f* homogeneity. **-strom** *m* (*Elec.*)
direct current; (*Mach.*) unidirectional (*or*
parallel) flow. **-stromleitung** *f* direct current line. **g–teilig** *a* homogeneous; of
equal parts, isomeric; (*Colloids*) monodisperse. **G–teilung** *f* equipartition;
isomerism

Gleichung *f* (-en) equation

gleich·verhaltend *a* of similar properties
(*or* behavior). **G–verteilung** *f* equal distribution (*or* division), equipartition.
-verteilungsgesetz *n* equipartition law.
g–viel *adv* 1 (*usu* + *wie, wo,* etc) no matter (how, where, etc). 2 (*erron for* **gleich
viel**) just as much (*or* many), an equal
amount (*or* number) of. **-weit** *adv:* **g. abstehen** be equidistant. **G–wert** *m* equivalent. **g–wertig** *a* of equal value (*or* valence), equivalent. **G–wertigkeit** *f*
equivalence. **g–winklig** *a* equiangular.
-wohl *adv, cnjc* nevertheless, yet. **-zeitig**
a 1 simultaneous, synchronous, iso-

chronous; *adv oft* at the same time. 2 contemporary. **-zellig** *a* like-celled, homo-
cellular. **-zusammengesetzt** *p.a* of like
composition

Gleis *n* (-e) track = GELEISE
gleißen *vi* shine, gleam, glisten
Gleit·bahn *f* slide; chute; guideway, slide-
way; (*Cryst.*) glide plane. **-bombe** *f*
glide(r) bomb. **-ebene** *f* glide (*or* slip)
plane. **-eigenschaft** *f* frictional property
gleiten* (glitt, geglitten) *vi*(s) glide, move,
run; slide; slip; skid; **g. lassen** (*oft*) pass,
run (a hand over sthg, etc). **Gleiter** *m* (-)
glider

gleit·fähig *a* capable of gliding (*or* sliding).
G–fähigkeit *f* gliding (*or* sliding) ability.
-fläche *f* glide (*or* slip) plane; (*Mach.*)
slide, sliding surface. **-flug** *m* (*Aero.*) gliding (flight). **-fleugzeug** *n* glider (plane).
g–frei *a* nongliding, nonslip(ping). **G–
kontakt** *m* slide contact. **-körper** *m* lubricant. **-lager** *n* slide (plain, friction)
bearing. **-linie** *f* (*Metll.*) slip band; (*Cryst.*)
glide line. **-maß** *n* modulus of elasticity in
shear. **-masse** *f* (*Pharm.*) lubricating
jelly. **-mittel** *n* lubricant. **-modul** *n* shear
modulus; transverse modulus (of elas-
ticity). **-pulver** *n* talc. **-reibung** *f* sliding
friction. **-richtung** *f* direction of sliding
(gliding, slip). **-ring** *m* slip ring. **-schiene**
f guide rail; slide bar. **-schutz** *m* anti-skid
(*or* -slip) device (*or* protection). **-sitz** *m*
(*Engg.*) sliding fit. **-skala** *f* sliding scale.
-spiegelebene *f* (*Cryst.*) glide plane

Gleitung *f* gliding, slip(ping), etc. *cf*
GLEITEN

Gleitungs·fläche *f* sliding surface, gliding
plane. **-zahl** *f* shear modulus

Gletscher *m* (-) glacier. **--boden** *m* glacial
soil. **-salz** *n* 1 Epsom salt, magnesium
sulfate. 2 (*Min.*) epsomite. **-schutt** *m* glacial drift (*or* debris)

glich equaled, etc: *past of* GLEICHEN
Glied *n* (-er) 1 member; limb; part (*esp* of
body); penis. 2 joint, articulation. 3 link.
4 generation. 5 (*Math.*) term. *Cf* REIHE
Glieder·füß(l)er *m* arthropod. **-gicht** *f* arthritic gout

·gliederig *sfx* -membered, -limbed
gliedern I *vt* 1 organize, arrange. 2 classify,

divide, break down. **II: sich g. (in)** divide,
be divided, break down (into), consist (of).
See also GEGLIEDERT

Glieder·tier n (*Zool.*) articulate

Gliederung f (-en) organization, arrangement; classification, breakdown; articulation

Glieder·zahl f number of members, etc *cf*
GLIED

Glied·maßen *npl* limbs, extremities.
-nummer f number of a member in a
series

·gliedrig *sfx* -membered, -limbed

Glied·wasser n synovial fluid

glimmen* (glomm, geglommen) *vi* glimmer, glow; smolder

Glimm·entladung f glow discharge

Glimmer m mica —**grüner G.** torbenite;
(*in compds oft*) = **g--artig** a micaceous.
G--blatt n mica sheet. **-blättchen** n mica
scale (*or* lamina). **-erde** f micaceous
earth. **g--führend, -haltig** a micaceous

glimmerig a micaceous

Glimmer·kalk m micaceous lime. **-kerze** f
mica spark plug

glimmern *vi* glimmer, gleam; shimmer

Glimmer· mica(ceous): **-plättchen** n
(small) mica plate. **-platte** f mica plate.
g--reich a mica-rich, strongly micaceous.
G--schiefer m mica(ceous) schist, mica
slate. **-ton** m micaceous clay, (*specif*) illite

Glimm·lampe f glow (*or* discharge) lamp;
(*Phot.*) ready light. **-licht** n glow-
discharge light. **-lichtrohr** n, **-licht-
röhre** f glow-discharge tube

glimmrig a micaceous = GLIMMERIG

Glimm·röhre f (vacuum) discharge tube

glimpflich a mild, lenient; *adv oft* lightly

glitschen *vi* (s) slip, slide. **glitsch(r)ig** a
slippery

glitt glided: *past of* GLEITEN

glitzern *vi* glitter, glisten

Gln. *abbv* (Gleichungen) equations

Globus m (-se) (*Geog.*) globe

Glöckchen n (-) little bell (*or* bell jar)

Glocke f (-n) 1 (*usu*) bell. 2 (*specif:* bell-
shaped objects) bell jar (*or* glass); bellflower; bubble cap; globe; bulb; (air
pump) receiver; (sink) plunger; (*also oft
in compds*) *cf* GICHTGLOCKE

Glocken·blume f bellflower, campanula.
-boden m bubble-cap tray (*or* plate).
-bronze f, **erz** n bell metal. **-filter** n bell-
jar filter. **-form** f 1 bell shape. 2 bell mold.
g--förmig a bell-shaped, bell-mouthed.
G--gießer m bell founder. **-gut** n bell metal. **-kurve** f bell-shaped curve. **-metall** n,
-speise f bell metal. **-trichter** m bell funnel (as on thistle tubes). **-verfahren** n
bell process (*or* method). **-wäscher** m
bell-type scrubber

glomm glimmered; *past of* GLIMMEN

Gloriosin n phytosterolin

Glossar n (-e) glossary, vocabulary

glotzen *vi* stare, gape; goggle

Glover·säure f Glover (tower) acid. **-turm**
m Glover tower

Glucin·erde f glucinia (beryllia)

Glück n 1 luck, (good) fortune: **G. haben** be
lucky; **zum G.** luckily, fortunately: **ein
G., daß . . .** a lucky thing that . . . ; **etwas auf gut G. tun** take a chance on
doing sthg. 2 success. 3 happiness

glucken *vi* 1 cluck. 2 sit (on eggs), brood

glücken *vi* (s) 1 (*usu* + *dat*) be successful,
succeed. 2 work out right, turn out well. =
GELINGEN

glücklich a lucky, fortunate; happy; *adv oft*
safely. **glücklicherweise** *adv* fortunately

Glücks·fall m lucky chance, stroke of luck

Gluco . . . *See also* GLUKO . . . , GLYKO . . .

Glucoron·säure f glucoronic acid

Glucosamin·säure f glucosaminic acid

Glüh·asche f glowing ashes, embers.
-aufschluß m decomposition by ignition.
-behandlung f (*Metll.*) annealing. **g--
beständig** a stable at red heat. **G--birne**
f incandescent bulb. **-draht** m 1 hot (*or*
glowing) wire, incandescent filament.
2 annealed wire

Glühe f (-n) 1 glow(ing). 2 (*Iron*) chafery

Glüh·eisen n glowing (*or* red-hot) iron. **g--
elektrisch** a thermionic. **G--elektron** n
thermal (*or* thermionic) electron. **-elek-
tronenabgabe** f thermionic (electron)
emission

glühen I *vt* 1 heat red-hot (*or* to glowing).
2 anneal. 3 ignite. 4 calcine. 5 mull
(wine). II *vi* 6 glow, be aglow, be red hot. 7

be incandescent. **8** burn. **glühend** *p.a*
red-hot, aglow, incandescent
Glüh·faden *m* incandescent filament.
-farbe *f* glowing color, temper color
glühfrischen *vt* anneal, malleablize
Glüh· (red-)hot, incandescent: **-gut** *n* material to be annealed. **-hitze** *f* red (incandescent, annealing) heat. **-kasten** *m* annealing box. **-kathode** *f* hot cathode. **-kiste** *f* annealing box. **-kolben** *m* retort. **-kopf** *m* (*Mach.*) hot bulb. **-körper** *m* incandescent body (*or specif* mantle, filament).
-lampe *f* incandescent lamp. **-licht** *n* incandescent light
Glühlicht·brenner *m* incandescent burner. **-körper, -strumpf** *m* incandescent mantle
Glüh· (red-)hot, incandescent; **-masse** *f* incandescent material (*or* body, *or specif* crude strontium oxide). **-ofen** *m* **1** annealing (*or* glowing) furnace. **2** (*Ceram.*) hardening-on kiln. **-phosphat** *n* calcined phosphate. **-ring** *m* ring support for ignitions. **-sand** *m* refractory sand. **-schälchen** *n* igniting dish, incinerating capsule. **-schale** *f* roasting dish; cupel. **-schiffchen** *n* combustion boat. **-span** *m* iron scale. **-stahl** *m* malleable cast iron. **-stoff** *m* incandescent material. **-strumpf** *m* incandescent mantle. **-temperatur** *f* annealing temperature. **-topf** *m* annealing pot (*or* box), cementing box
Glühung *f* (-en) heating to incandescence; annealing; ignition; calcining; mulling; malleablizing *cf* GLÜHEN
Glüh·verlust *m* loss on ignition. **-wachs** *n* gilder's wax. **-wein** *m* mulled wine. **-würm** *m* glowworm. **-zone** *f* zone of incandescence. **-zünder** *m* electric igniter. **zündholz** *n* fuse(e)
Gluk ... *See also* GLUC ... , GLYK ...
Glukon·säure, Glukose·säure *f* gluconic acid
Glukosid *n* (-e) glucoside, glycoside
Glukosurie *f* (-n) glucosuria, glycosuria
Glut *f* (-en) **1** fire, blaze. **2** *pl oft* live coals, embers. **3** glow, incandescence. **4** (intense) heat; (*Ceram.*) glow heat. **5** ardor
—**in G.** ablaze, aglow, incandescent; ardent

Glutamin·säure *f* glutamic acid
Glutar·säure *f* glutaric acid
Glut·asche *f* (glowing) embers. **g—flüssig** *a* molten, fused (by high heat). **G—hitze** *f* glowing heat
glutinös *a* glutinous
Glut·messer *m* pyrometer. **g–rot** *a* glowing red
Gluzinium *n* glucinum (beryllium)
Gluzin·säure *f* glucic acid
Glyc ... *See also* GLYK ... , GLYX ...
Glycerin·galle *f* glycerinated bile. **-leim** *m* glycerin jelly. **g–phosphorsauer** *a* glycerophosphate of. **G–phosphorsäure** *f* glycerophosphoric acid. **-säure** *f* glyceric acid
Glycin *n* **1** glycine *cf* GLYZINE. **2** = **-erde** *f* glucine (beryllia)
Glycium *n* glucinum (beryllium)
Glyk ... **1** glyc ... **2** (*referring to glucose*) gluc ...
Glykochol·säure *f* glycocholic acid. **Glykogallus·säure** *f* glycogallic acid. **Glykokoll** *n* glycocoll, glycine
Glykol·säure *f* glycolic acid. **-schwefelsäure** *f* glycosulfuric acid. **Glykolur·säure** *f* glycoluric acid
Glykolyse *f* glycolysis. **glykolitisch** *a* glycolitic
Glykosämie *f* glucemia
Glykose *f* glucose
Glykosid *n* (*genl*) glycoside; (*Glucose deriv.*) glucoside
Glykoson *n* glucosone
Glykosurie *f* glycosuria, glucosuria
Glykuron·säure *f* glucuronic acid
Glyoxal·säure, Glyoxyl·säure *f* glyoxalic acid
Glyz ... = GLYC ...
Glyzerid *n* (-e) glyceride
Glyzerin *n* glycerol, glycerin *cf* GLYCERIN·
Glyzerol *n* glycerol
Glyzid *n* (-e) glycide. **-säure** *f* glycidic acid
Glyzine *f* (-n) wisteria. **Glyzinen·blütenöl** *n* oil of wisteria
G.m.b.H. *abbv* (Gesellschaft mit beschränkter Haftung) Ltd., Inc., Corp.
Gmelinsches (or **Gmelins**) **Salz** Gmelin's salt, potassium ferricyanide
G-Modul *n* shear modulus
Gnade *f* mercy; grace; clemency; favor

Gnaden·kraut *n* hedge hyssop, gratiola.
-stoß *m* finishing blow, coup de grace
gnädig *a* merciful; gracious
Gneis *m* 1 (*Geol.*) gneiss. 2 (*Med.*) seborrhea. **gneisig** *a* gneissic
Gneist *m* (*Lthr.*) scud, scurf
gnomonisch *a* (*Cryst.*) gnomonic
Goa·pulver *n* Goa powder
Gold *n* gold; (*in compds, Chem. oft*) auric, aurous. **g--ähnlich** *a* gold-like. **G--ammoniak** *n* fulminating gold.
-anlegeöl *n* gold size. **-anstrich** *m* gilding, gilt paint. **g--artig** *a* gold-like, golden. **G--äther** *m* solution of gold chloride in ether. **-auflösung** *f* gold solution. **-bad** *n* (*Phot.*) gold (toning) bath. **-barren** *m* gold bar (*or* ingot). **-belag, -beleg** *m* gold coating (plating, foil). **-beryll** *m* chrysoberyl. **-blatt** *n* gold leaf (*or* foil). **-blick** *m* flash (*or* "blick") of gold. **-bromür** *n* aurous bromide. **-bronze** *f* gold bronze (paint). **-butt** *m*, **-butte** *f* (*Zool.*) plaice, flounder. **-chlorid** *n* gold (*specif* auric) chloride; (*as last part of compd*) tetrachloroaurate. **-chlorwasserstoffsäure** *f* chloroauric acid. **-cyanid** *n* gold (*specif* auric) cyanide. **-cyanür** *n* aurous cyanide. **-doublé** *n* rolled gold. **-draht** *m* gold wire. **-erde** *f* auriferous earth. **-farbe** *f* gold color (*or* paint). **g--farben, -farbig** *a* gold-colored, golden. **G--firnis** *m* gold (*or* gilding) varnish. **-fisch** *m* goldfish. **-flitter** *m* gilt tinsel, gold spangle. **-folie** *f* gold foil. **g--führend** *a* goldbearing, auriferous. **g--gelb** *a* golden yellow. **G--gewicht** *n* 1 weight in gold. 2 troy weight. **-glanz** *m* golden luster. **g--glänzend** *a* golden-shiny. **G--glätte** *f* litharge. **-glimmer** *m* yellow mica. **-grube** *f* gold mine. **-grund** *m*, **-grundöl** *n* gold size. **g--haltig** *a* auriferous, containing gold. **-hell** *a* golden-bright. **G--jodid** *n* gold (*specif* auric) iodide. **-jodür** *n* aurous iodide. **-käfer** *m* gold beetle. **g--käfergrün** *a* beetle-green. **G--kaliumbromür** *n* potassium aurobromide. **-kaliumcyanür** *n* potassium aurocyanide. **-kies** *m* gold-bearing pyrites (sand, gravel). **-klee** *m* yellow trefoil. **-klumpen** *m* gold nugget. **-könig** *m* gold regulus. **-korn** *n* grain of gold. **-krätze** *f* gold dross

(*or* scrap). **-kraut** *n* (*Bot.*) 1 common groundsel. 2 field scabious. 3 moneywort.
-kupfer *n* Mannheim (*or* imitation) gold, pinchbeck. **-lack** *m* 1 gold varnish. 2 (*Glass, Min.*) aventurine. 3 (*Bot.*) wallflower. **-legierung** *f* gold alloy. **-leim** *m* gold size. **-macher** *m* alchemist. **-macherei, -macherkunst** *f* alchemy. **-natriumchlorid** *n* sodium aurichloride. **-natriumchlorür** *n* sodium aurochloride. **-ocker** *m* finest yellow ocher. **-orange** *f* golden orange. **-oxid** *n* gold (*specif* auric) oxide
Goldoxid·ammoniak *n* fulminating gold. **-salz** *n* auric salt, aurate. **-verbindung** *f* auric compound
Gold· gold, auric, aurous: **-oxydul** *n* aurous oxide **-papier** *n* gold (*or* gilt) paper. **g--plattiert** *a* gold-plated. **G--plombe** *f* gold filling **-probe** *f* gold assay (test, sample). **-purpur** *m* purple of Cassius. **-quarz** *n* auriferous quartz. **-regen** *m* laburnum. **g--reich** *a* gold-rich, highly auriferous. **G--rot** *n* (*Metalworking*) rouge. **-rubinglas** *n* ruby glass with gold colorant. **-rute** *f* goldenrod. **-salpeter** *m* gold nitrate. **-salz** *n* gold salt, (*specif*) sodium tetrachloroaurate, $NaAuCl_4$. **g--sauer** *a* aurate of. **G--säure** *f* auric acid. **-schale** *f* gold dish (*or* cup); cupel. **-schaum** *m* (real *or* imitation) gold leaf; Dutch foil; (*Metll.*) gold crust. **-scheiden** *n* gold refining. **-scheidewasser** *n* aqua regia. **-scheidung** *f* gold refining. **-schlag** *m* gold leaf (*or* foil). **-schläger** *m* gold beater. **-schlägerhaut** *f* goldbeater's skin. **-schlich** *m* gold slimes (*or* schlich). **-schmied** *m* goldsmith. **-schnitt** *m* gilt edge. **-schwefel** *m* 1 gold sulfide. 2 antimony pentasulfide, Sb_2S_5. **-seife** *f* (*Min.*) gold-bearing alluvium. **-siegellack** *m* gold sealing wax; aventurine. **-silber** *n*, **-silberlegierung** *f* electrum. **-staub** *m* gold dust. **-stein** *m* goldbearing stone, goldstone, aventurine. 2 touchstone. 3 chrysolite. **-streichen** *n* gold (streak) test. **-stück** *n* gold piece (*or* coin). **-sulfür** *n* aurous sulfide. **-tellur** *n* gold telluride: (*Min.*) sylvanite. **-thioschwefelsäure** *f* aurothiosulfuric acid. **-tropfen** *mpl* (*Pharm.*) ethereal tincture of ferric

chloride. **-vitriol** *n* gold sulfate. **-waage** *f* gold balance (*or* scales). **-währung** *f* gold standard. **-zahl** *f* (*Colloids*) gold number. **-zyanid** *n* gold (*specif* auric) cyanide. **-zyanür** *n* aurous cyanide

Golf I *m* (-e) gulf. **II** *n* golf

Gondel *f* (-n) gondola; nacelle, car

gönnen 1 *vt* allow, grant, let . . . have. **2** (+ *dat reflx.* **sich**) permit (oneself sthg). **Gönner** *m* (-), **Gönnerin** *f* (-nen) patron, benefactor, backer

Gooch·tiegel *m*, **Gooch'scher Tiegel** Gooch crucible

gor fermented: *past of* GÄREN

goß poured: *past of* GIESSEN

Gosse *f* (-n) gutter, drain; hopper

Göthit *n* (*Min.*) goethite

Gott *m* (⁼er) God, god

Götter·baum *m* tree of heaven, Ailanthus. **-fabel, -lehre** *f* mythology. **-speise** *f* ambrosia. **-trank** *m* nectar

Gottes·gnadenkraut *f* hedge hyssop

Göttin *f* (-nen) goddess

Goudron *m* asphalt, bitumen, tar. **Goudronné·papier** *n* tar paper

gr. *abbv* **1** (Gramm) g. (gram). **2** (granuliert) gran(ulated). **3** (groß) lge. (large), gt. (great)

Grab *n* (⁼er) grave, tomb

graben* (grub, gegraben; gräbt) *vt/i* dig; engrave, carve

Graben *m* (⁼) ditch, trench

Gräber I *m* (-) digger. **II** *pl of* GRAB

Grab·legung *f* interment, burial. **-scheit** *n* spade, shovel. **-schrift** *f* epitaph. **-stichel** *m* graver, (en)graving tool, chisel

gräbt digs: *pres of* GRABEN

grad *adv* directly, straight = GERADE

Grad *m* (-e) degree (*all meanings*); rank

Grad· (*also* = GERADE·): **-abteilung** *f* **1** scale, graduation. **2** degree mark (on a scale). **3** map (*or* chart) with geographical coordinates. **g–aus** *adv* straight ahead = GERADEAUS. **G–bogen** *m* **1** graduated arc. **2** protractor. **-einteilung** *f* scale, graduation (in degrees). **g–faserig** *a* straight-fibered (*or* -grained). **G–flügler** *m* (*Zool.*) Orthoptera

Gradheit *f* straightness

grad·heraus *adv* frankly, straightforwardly

Gradientien·ofen *m* gradient furnace

gradieren *vt* **1** grade, graduate, calibrate. **2** (*Metll.*) refine, improve in quality. **3** (*Salt*) test (for density, with a hydrometer).

Gradier·haus *n* salt graduation house. **Gradierung** *f* grading, graduation, calibration; refining; testing for density

Gradier·waage *f* brine gauge. **-werk** *n* (*genl*) concentration plant; (*specif*) cooling tower, (salt) concentration plant

·gradig *sfx* -degree, -grade *cf* HOCHGRADIG. **·grädig** *sfx* (*Chem.*) -concentration *cf* HOCHGRÄDIG

Grädigkeit *f* (-en) density, concentration

grad·kettig *a* straight-chained. **G–kreis** *m* graduated circle (*or* arc). **-laufapparat** *m* gyroscope. **g–linig** *a* straight-line, (recti)linear. **G–messer** *m* **1** graduator. **2** degree-scaled measuring instrument. **3** criterion, yardstick. **-teiler** *m* vernier. **-teilung** *f* scale, graduation

graduell *a* gradual

graduieren *vt* graduate, divide into grades. **Graduierung** *f* (-en) graduation

Grad·verwandtschaft *f* graduated affinity, degree of affinity. **g–wegs** *adv* directly, straight = GERADEWEGS. **-weise** *a* gradual, *adv oft* by degrees. **-zahlig** *a* even-numbered

Graf *m* (-en, -en) count. **Gräfin** *f* (-nen) countess

grafisch *a* graphic

Grafit *m* graphite = GRAPHIT

Gram *m* grief. **grämen** *vt* & **sich g.** grieve

Gram·färbung *f* (*Bact.*) Gram stain(ing)

Gramm *n* (-e) gram. **·äquivalent** *n* gram equivalent. **-atom** *n* gram atom. **-bruchteil** *m* fraction of a gram

Grammen·flasche *f* gram (*or* specific gravity) bottle

Gramm·grad *m* gram degree. **-ion** *n* gram ion

Gram· (= GRAMM·): **-mol** *n* mole = **-molekül** *n* gram molecule. **-mol(ekül)volumen** *n* gram molecular volume

Gramsche Färbung (*Bact.*) Gram staining

Gran, Grän *m/n* (-e) (*Meas.*) grain

Granadil·holz *n* granadilla wood
Granalien *pl* granules; shot. **·gebläse** *n* (*Metll.*) shot blasting
Granat *m* (-e) **I** (*Min.*) garnet. **II** (*Zool.*) shrimp. *Cf* GRANATE. **·apfel** *m* pomegranate. **·apfelbaumalkaloid** *n* pomegranate alkaloid. **g–artig** *a* garnet-like. **G–baum** *m* pomegranate tree. **·dodekaeder** *n* rhombic dodecahedron
Granate I *f* (-n) 1 grenade, shell. 2 garnet. 3 pomegranate. **II** *pl of* GRANAT
Granat·füllung *f* shell filling. **·gestein** *n* garnet rock
Granatill·holz *n* granadilla wood. **·öl** *n* physic-nut oil. **·samen** *m* physic nut
Granato·eder *n* (-) rhombic dodecahron
Granat·rinde *f* pomegranate bark. **g–rot** *a* garnet(-red). **G–ton** *m* garnet shade. **·trichter** *m* shell hole (*or* crater). **·werfer** *m* trench mortar. **·wurzelrinde** *f* pomegranate root bark
Grand *m* (-e) 1 coarse sand, fine gravel. 2 wheat bran. 3 (water) trough. 4 (*Brew.*) underback. *Cf* GRANT
granieren *vt* granulate, grain
granit·ähnlich *a* granitoid, = **·artig** *a* granite-like, granitic. **G–email** *n* granite enamel
graniten *a* granite, granitic
granit·farbig *a* granite-colored. **G–felsen** *m* granitic rock. **g–förmig** *a* granitiform
granitisch *a* granitic
Granit·papier *n* jaspé (*or* mottled) paper
Granne *f* (-n) (*Bot.*) awn, arista, beard
Grant *m* 1 (*Brew.*) underback. 2 (*Engg.*) clay mixed with crushed quartz. *Cf* GRAND
Granulat *n* (-e) granular (*or* granulated) material. **granulieren** *vt/i* granulate. **Granulierung** *f* granulation
Grapen *m* (-) iron pot
Graphik *f* (-en) 1 graphics, graphic arts. 2 graph, diagram. 3 print, engraving. **Graphiker** *m* (-) graphic (*or* commercial) artist, engraver. **graphisch** *a* 1 graphic. 2 (of) printing: **g—e Farbe** printing ink
graphit·ähnlich, **·artig** *a* graphite-like, graphitoid. **G–bildung** *f* graphitization. **g–haltig** *a* graphitic

graphitieren *vt* graphitize, coat with graphite. **graphitisch** *a* graphitic. **graphitisieren** *vt* graphitize
Graphit·kohle *f* graphitic carbon. **·masse** *f* graphite paste. **·mine** *f* 1 (pencil) lead refill. 2 graphite mine. **g–reich** *a* rich in graphite, strongly graphitic. **G–rohrofen** *m* carbon tube furnace. **·säure** *f* graphitic acid. **·schmiere** *f*, **schmiermittel** *n* graphite grease (*or* lubricant). **·schwärze** *f* graphite blacking, blacklead wash. **·spitze** *f* (arc-lamp) carbon. **·stein** *m* graphite brick. **·stift** *m* 1 (lead) pencil. 2 pencil lead
Gras *n* (-̈er) grass. **g–artig** *a* grassy, gramineous. **Gräs/chen** *n* (-) blade of grass. **Gras–ebene** *f* grassy plain
grasen *vt/i* graze
gras·fressend *a* herbivorous, graminivorous. **·grün** *a* grassgreen. **G–halm** *m* blade of grass
grasig *a* grassy, grassgrown
Gras·leinen *m* grass cloth. **·narbe** *f* turf, sod. **·öl** *n* grass (*specif* citronella) oil
grassieren *vi* rage, be rampant, spread
gräßlich *a* awful, terrible
Gras·wurzel *f* 1 couch grass, (*Pharm.*) triticum. 2 grass root
Grat *m* (-e) 1 ridge. 2 sharp (*or* cutting) edge. 3 flash (on castings), burr. 4 (roof) groin
Gräte *f* (-n) (fish)bone
grat·frei *a* ridgeless, smooth
grau *a* gray; drab; dim, misty (past) —**g—e (Quecksilber-)Salbe** *f* mercurial ointment. **·blau** *a* gray-blue. **·braun** *a* gray-brown, dun. **G–braunstein** *m* (*Min.*) manganite. **·braunsteinerz** *n* pyrolusite. **·eisenerz** *n* marcasite
grauen I *vi* 1 (grow) gray. 2 dawn. **II** *vi* (*imps*): **ihnen graut vor** . . . they dread . . . 4: **sich g. (vor)** dread. **Grauen** *n* dread
Grau·erz *n* galena. **g–farben, ·farbig** *a* gray-colored. **·gelb** *a* grayish yellow. **G–glanzoxidbad** *n* bath for coloring metals with film of arsenic or antimony. **·golderz** *n* nagyagite. **·grün** *a* gray-green, **G–gültigerz** *n* tetrahedrite. **·guß** *m* gray cast iron; gray-iron casting. **·kalk** *m* 1

gray (*or* dolomitic) lime. **2** crude calcium acetate. **-keil** *m* (*Opt.*) gray (*or* neutral) wedge. **-keilbeleuchtung** *f* (*Phot.*) gray-scale exposure. **-kobalterz** *n* gray cobalt ore, jaipurite. **-kupfererz** *n* chalcocite; tennantite; tetrahedrite. **-leiter** *f* (*Opt.*) gray scale

gräulich *a* greyish *Cf* GREULICH

Grau·manganerz *n* manganite, gray manganese ore **—lichtes G.** polianite, pyrolusite. **g—meliert** *a* mottled (*or* mixed) gray, graying. **G—metall** *n* gray metal (*or specif* pewter)

Graumont·samen *m* pumpkin seed

Grau·pappe *f* (gray) cardboard, pasteboard

Graupe *f* (-n) **1** (grain of) peeled (hulled, pearl) barley. **2** grain, granule. **3** knot (in cotton)

Graupel *f* (-n) (*usu pl*) sleet, fine hail. **graupeln** *vi* (*imps*) sleet, hail lightly

Graupen·erz *n* granular ore. **-kobalt** *m* smaltite. **-schleim** *m* barley water. **-schörl** *m* aphrizite, black tourmaline. **-suppe** *f* barley soup

graupig *a* granular

grau·rötlich *a* reddish gray, roan

graus *a* horrible, gruesome. **Graus** *m* **1** horror. **2** rubble

grausam *a* cruel, atrocious. **Grausamkeit** *f* (-en) cruelty, atrocity

Grau·schimmel *m* gray mold. **-schleier** *m* (*Phot.*) gray (*or* chemical) fog. **g—schwarz** *a* grayish black. **G—spiegel** *m* gray spiegeleisen. **-spiegelglanzerz**, **-spiegelglaserz** *n* stibnite, gray antimony ore **—haarförmiges G.** jamesonite. **-spießglanz** *m* stibnite, antimony glance. **-stein** *m* (*Petrog.*) graystone. **g—violett** *a* gray-violet. **G—wacke** *f* (*Petrog.*) graywacke. **g—weiß** *a* gray-white

Graveur *m* (-e) engraver

Gravidität *f* (-en) pregnancy

gravieren *vt* engrave. **gravierend** *p.a* aggravating; serious, grave; decisive

Gravitations·kraft *f* gravitational force

Gravur *f* (-en), **Gravüre** *f* (-n) engraving

Grazie *f* (-n) grace, charm

grd. *abbv* (Grad) degree(s)

greifbar *a* **1** within reach, handy *cf* NAHE,

NÄHE. **2** available. **3** tangible. **4** obvious.

Greifbarkeit *f* handiness, easy reach; availability; tangibility; obviousness

greifen* (griff, gegriffen) **I** *vt* **1** grasp, seize, grab, take hold of **—es ist zu g.** it is tangible *cf* LUFT, PLATZ. **2** touch. **3** set, estimate (price, etc.). **II** *vi* **4** (*esp* + **nach**) reach (for), grasp (at); (+ **zu**, *oft*) resort (to). **5** (+ **in**, *oft*) interfere (with); mesh (with). **6** (of wheels, brakes, pliers) grip, bite, take hold. **7** (+ **an**, *oft*) touch, put one's hand to. **8** mat, cake. **9: um sich g.** spread. **Greifen** *n:* **zum G. nahe** within reach. **Greifer** *m* (-) **1** (*genl*) gripping device. **2** (*specif*) grip(per), grab, claw, hook; scoop

Greif·organ *n* prehensile organ. **-zange** *f* (pair of) gripping tongs. **-zirkel** *m* (pair of) calipers

greis *a* aged, hoary. **Greis** *m* (-e) old man

Greisen *m* (*Min.*) greisen (quartz-mica rock)

Greisen·alter *n* old age

Greisin *f* (-nen) old woman

grell *a* glaring, dazzling; sharp; loud, shrill **—g—es Roheisen** white pig iron

Gremium *n* (..mien) group, body; committee

Grenadill·holz *n* granadilla wood

Grenz· limit, limiting; boundary: **-alkohol** *m* limit alcohol, alkanol. **-bedingung** *f* limiting (*or* boundary) condition. **-begriff** *m* borderline concept. **-belastung** *f* limit load; (*Distil.*) flooding point. **-biegespannung** *f* ultimate flexural stress. **-dextrin** *n* limit dextrin. **-dichte** *f* limit(ing) density

Grenze *f* (-n) boundary; frontier; border(line). **2** limit, *pl oft* bounds: **sich in G—n halten** stay within bounds, be limited. **grenzen** *vi* (**an**) border (on). **grenzenlos** *a* boundless, unlimited; infinite

Grenz· *cf* GRENZE: **-fall** *m* limiting case; borderline case. **-fläche** *f* boundary surface, surface of contact, interface

Grenzflächen·chemie *f* surface chemistry. **-energie** *f* interfacial energy. **-erscheinung** *f* interfacial phenomenon. **-katalyse** *f* contact catalysis. **-spannung** *f* interfacial tension. **g—wirksam** *a* interface-active

Grenz‧ *cf* GRENZE: **-gebiet** *n* borderline area; neighboring field (of study). **-gesetz** *n* limit(ing) law. **-gewebe** *n* epidermis, integument. **-kohlenwasserstoff** *m* saturated hydrocarbon, (*specif*) alkane, paraffin. **-korn** *n* near-mesh material. **-kurve** *f* limit(ing) curve. **-last** *f* limit (*or* maximum) load. **-lehre** *f* limit gauge. **-leistung** *f* peak performance; maximum output. **-linie** *f* boundary line, borderline. **-mal** *n* boundary marker. **-nutzen** *m* marginal utility. **-scheide** *f* boundary line. **-scheidung** *f* demarcation. **-schicht** *f* boundary (*or* marginal) layer. **-schichtenenergie** *f* interfacial energy. **-spannung** *f* limiting stress. **-strahl** *m* borderline ray. **-stromdichte** *f* (*Elec.*) limit current density. **-struktur** *f* canonical form (of molecular structure). **g‧überschreitend** *a* cross-border. **G‧verbindung** *f* terminal (compound) member; saturated aliphatic compound. **-viskosität** *f* (*Polymers*) intrinsic viscosity. **-viskositätszahl** *f* limiting viscosity number. **-wellenlänge** *f* threshold wavelength. **-wert** *m* limit, limiting (*or* marginal) value; (*Biol.*) threshold value. **-winkel** *m* critical angle. **-zahl** *f* limit(ing) number (value, figure). **-zelle** *f* heterocyst, limiting (*or* boundary) cell. **-zustand** *m* extreme test condition; limiting state; borderline condition

Greuel *m* (-) horror, outrage, atrocity. **greulich** *a* horrible, outrageous, atrocious. *Cf* GRÄULICH

Greze *f*, **Grez‧seide** *f* raw silk

Gricken *m* (-) buckwheat

Griebe *f* (-n) (*usu pl*) **1** greaves, cracklings. **2** sifting (*or* sieve) residue. **Grieben** *m* (-), **Griebs** *m* (-e) **1** (seed) core. **2** throat

Griechenland *n* Greece. **griechisch** *a* Greek **—g—es Heu** fenugreek

gries(e)lig *a* gritty, gravelly; granulated

Grieß *m* **1** grit, coarse-ground (*or* granular) material; (*specif*) shot, (coal) breeze, coarse sand. **2** farina, semolina; grits. **3** (*Med.*) gravel. **4** (*TV*) snow, granulation effect. **‧‧bildung** *f* grit (*or* gravel) formation. **-holz** *n* (*Pharm.*) nephritic wood

grießig *a* gritty, gravelly; granulated

Grieß‧kohle *f* pea (*or* small) coal. **-mittel** *n* (*Med.*) remedy for gravel. **-stein** *m* **1** urinary calculus. **2** nephrite, jade. **-wurzel** *f* (*Pharm.*) pareira (brava)

griff grasped: *past of* GREIFEN

Griff *m* (-e) **1** grasp, grip, hold. **2** control; knack: **im G. haben** have control (*or* a good grasp) of, have the knack of; **in (den) G. bekommen** get control of, have the knack of. **3** movement: **mit ein paar G—en** with a few (deft) move(ment)s. **4** reach, grab: **einen G. tun (A)** (+ **nach**) reach for, make a grab for, **(B)** (+ **in**) reach, dig (into); **ein guter G.** a good move, a good choice (*or* buy). **5** handle, knob; lever; handhold; handrail. **6** (*Tex.*) feel, handle (of cloth)

griff‧bereit *a* ready for use, ready to use

Griffel *m* (-) stylus, slate pencil; (*Bot.*) style, pistil. **g—förmig** *a* styloid, styliform. **G—schiefer** *m* (*Min.*) pencil slate

griffig *a* **1** easily held (*or* handled), handy. **2** with a good grip, non-slip, non-skid. **3** (*Tex.*) with a good feel (*or* handle), slightly rough: **g. machen** scroop, roughen. **4** gritty, granular. **Griffigkeit** *f* **1** easy grip; handiness. **2** gripping power; (tire) traction; frictional adhesion. **3** (*Tex.*) good feel (*or* handle). **4** granularity

Griff‧knopf *m* knob. **-mutter** *f* (*Mach.*) knurled nut. **-stange** *f* handrail. **-stopfen** *m* stopper with a thumb piece. **-stück** *n* handle, grip

grignardieren *vt* grignardize, subject to the Grignard reaction. **Grignardierung** *f* (-en) grignardization

Grill *m* grill; (outdoor) barbecue. **‧‧kohle** *f* barbecue (*or* grill) charcoal. **-kohle-anzünder** *m* charcoal lighter

Grille *f* (-n) **1** (*Zool.*) cricket. **2** whim

Grimmdarm *m* (transverse) colon. **‧‧entzündung** *f* colitis

Grimmen *n* abdominal pains (*or* cramps), colic

Grind *m* (-e) scurf, scab, mange. **grindig** *a* scurfy, scabby, mangy. **Grind‧wurzel** *f* bitterdock root

Grinsen *vi* grin; (*Colors*) stare; (*Metals*) begin to melt

Grippe *f* (-n) **1** influenza, flu. **2** irrigation canal

grit‧haltig *a* gritty, grit-bearing

grob a (⸚er, ⸚st...) 1 coarse. 2 rough, crude, rude. 3 gross. **G—·ätzung** f coarse (or macro-)etching. **-blech** n (*Metal*) rough (or heavy) plate. **-bruch** m coarse-grained fracture. **g—dispers** a coarsely disperse

Gröbe f (-n) (degree of) coarseness

Grob·einstellung f coarse adjustment. **-eisen** n merchant iron

gröber a coarser, etc.: *compar of* GROB

Grob·erz n coarse ore. **g—fadig** a coarse-threaded. **-faserig** a coarse-fibered (or -grained). **G—feile** f coarse file. **-gefüge** n coarse (or macro-)structure. **g—gepulvert** a coarsely powdered. **G—gewicht** n gross weight. **g—grießig** a coarse-grained, coarsely granular. **G—gut** n coarse material

Grobheit f (-en) 1 coarseness, roughness, etc *cf* GROB. 2 abusive remark

Grob·kalk m coarse limestone. **-keramik** f earthenware, ordinary ceramic wear; heavy clayware. **g—keramisch** a (of) baked clay. **G—kies** m coarse gravel. **-kohle** f coarse (lump, open-burning) coal. **-korn** n coarse grain. **g—körnig** a coarse-grained. **-kristallin** a coarsely crystalline

gröblich a 1 rather coarse, roughish. 2 gross

Grob·lunker m cavity, pipe hole. **g—maschig** a coarse-mesh(ed). **-mechanisch** a marcromechanical, large-scale. **G—mikrinit** n (*Coal*) massive micrinite. **G—mörtel** m coarse mortar. **g—porig** a large-pored, coarsely porous. **G—sand** m coarse sand

grob·schleifen* vt rough-grind

Grob·schnitt m coarse cut; thick slice. **-schrot** m, n 1 coarse grits. 2 (*Brew.*) coarsely ground malt. **-schutt** m coarse debris (or refuse). **-sieb** n coarse (or wide-mesh) sieve. **-spiegel** m coarse spiegel iron

gröbste a roughest, etc: *superl of* GROB

grob·strahlig a coarsely radiated (or fibrous. **G—struktur** f coarse structure, macrostructure. **-waschmittel** n heavy-duty detergent. **-zerkleinerung** f coarse crushing (or pulverization). **-zuschlag** m (*concrete*) coarse aggregate

Groenhart·holz n greenheart wood

Grönland n Greenland

Groschen m (-) 1 (*Austria*) 1/100 SCHILLING. 2 (any small coin)

groß 1 a (größer, größt..) big, large, great; major; tall; capital (letter): **im g—en** on a large scale, in bulk, wholesale; **im g—en und ganzen** on the whole; **um ein g—es** by a large amount, greatly. 2 *pfx* large(-scale), major: **G—·anlage** a large-scale (or major) installation (plant, unit). **g—artig** a grand, magnificent. **G—betrieb** m 1 large-scale operation. 2 wholesale trade (or business). **g—blätterig** a large-leaved, coarsely foliated (or laminated). **G—britannien** n Great Britain. **g—britannisch** a British, of Great Britain. **-chemie** f chemical industry

Größe f (-n) 1 bigness, largeness, large size. 2 height. 3 size, extent; magnitude, value, quantity. 4 importance, significance. 5 greatness. 6 great person, leader, celebrity

Größen·lehre f mathematics; geometry. **-maß** n size, dimensions. **g—mäßig** a according to size (magnitude, quantity); quantitative. **G—nummer** f (numerical) size. **-ordnung** f 1 order of magnitude. 2 arrangement by size (quantity, magnitude. **g—ordungsmäßig** a in order of magnitude.

großen·teils *adv* largely, in large part

größer a bigger, etc: *compar of* GROSS

Groß·erzeugung f mass production. **-fabrikation** f large-scale manufacture. **g—fabrikatorisch** a large-scale, industrial. **-faserig** a large (or heavy-) fibered. **G—feuer** n conflagration; (*Ceram.*) full fire. **-feuerungsanlage** f large-scale furnace. **-filter** n coarse filter. **g—flächig** a widespread; *adv* over a wide area. **G—gewerbe** n large-scale (business) operation. **-handel** m wholesale (business). **-händler** m wholesaler. **g—herzig** a big-hearted, magnanimous. **G—hirn** n cerebrum. **-hirnrinde** f cerebral cortex. **-industrie** f large-scale industry. **g—jährig** a of age. **G—koks** m large-scale (or blast-furnace) coke. **-kraftwerk** n super-powerhouse. **-kreis** m great circle. **g—lückig** a coarsely porous, wide-meshed. **-mütig** a generous, magnanimous. **G—**

ofen *m* large furnace. **-serie** *f* large-scale assembly line. **-spore** *f* (*Biol.*) macrospore. **-stadt** *f* major city, metropolis. **g—stückig** *a* lumpy, in large pieces. **G–tat** *f* (major) achievement

größte *a* biggest, etc: *superl of* GROSS
Groß-technik *f* large-scale technology. **g—technisch** *a* major technological, (on a) large scale. **-teil** *m* majority, greater part; major part

größtenteils *adv* for the most part, mainly
größt-möglich *a* greatest (*or* largest) possible. **G–wert** *m* maximum value

Groß-verkauf *m* wholesale. **-versuch** *m* major (*or* large-scale) experiment. **-zahlforschung, -zahluntersuchung** *f* statistical research

groß-ziehen* *vt* raise, rear, bring up
groß-zügig *a* 1 large-scale. 2 generous
Grotte *f* (-n) grotto
grub dug, etc: *past of* GRABEN
Grübchen *n* (-) pit, fossule; dimple; (*Biol.*) lacuna. **--bildung** *f* pitting
Grube *f* (-n) pit (*incl Min., Med.*) mine; quarry; hole, cavity; fovea
grübeln *vi* ponder, brood
Gruben- mine, pit: **-abfall** *m* mine waste. **-bau** *m* mine digging, mining (operations). **-betrieb** *m* mining (operation). **-brand** *m* mine fire. **-dampf** *m* choke damp. **-feld** *n* coalfield. **g—feucht** *a* freshly mined, green. **G–feuchte, -feuchtigkeit** *f* pit-moisture (of coal, sand). **g—gar** *a* (*Lthr.*) pit-tanned. **G–gas** *n* mine gas, (*specif*) methane; fire damp. **-gas-anzeiger** *m* methanometer. **-gerbung** *f* pit tanning. **-gut** *n* mine output, minerals. **-kalk** *m* pit lime. **-klein** *n* smalls, slack. **-kohle** *f* (mineral) coal, pit coal (*or* charcoal). **-lampe** *f*, **-licht** *n* miner's (acetylene, safety) lamp. **-pulver** *n* miner's (blasting) powder. **-sand** *m* pit sand. **-staub** *m* mine dust. **-verkohlung** *f* pit burning (of charcoal). **-wasser** *n* mine water. **-wetter** *n* mine damp, (*specif*) fire damp

Grude *f* (-n) 1 hot ashes, embers. 2 coke-breeze stove. 3 = **Gruden-koks** *m* lignite coke breeze

grün *a*, **Grün** *n* green —**g—e Erde** glauconite = GRÜNERDE

Grün-ablauf *m* (*Sugar*) green syrup. **g—blau** *a* greenish blue. **G–bleierz** *n* green lead ore (pyromorphite, mimetite). **g—braun** *a* greenish brown

Grund *m* (⸚e) 1 ground, soil; land: **von G. auf** (*or* **aus**) from the ground up, thoroughly; **etwas in G. und Boden fahren** run sthg into the ground (= ruin it); **in G. und Boden** thoroughly, completely *cf* BODEN. 2 bottom, depths: **zu G—e gehen** go to the bottom (*or* to ruin) *cf* ZUGRUNDE; **einer Sache auf den G. gehen** get to the bottom of a matter. 3 valley. 4 foundation, basis: **auf G.** (+ **von** *or* + *genit*) on the basis (of), by reason (of) *cf* 5; = AUFGRUND; **im G—e (genommen)** basically. 5 reason, cause: **aus diesem G(—e)** for this reason; **das hat seinen guten G.** there is good reason for that. 6 (*in compds oft*) fundamental, basic, thorough: **-ablaß** *m* bottom outlet. **-anstrich** *m* (*Paint*) base (*or* priming) coat. **-bahn** *f* ground orbit. **-bau** *m* foundation. **-baustein** *m* basic building unit; (*Polymers*) monomer. **-bedingung** *f* fundamental condition. **-bedürfnis** *n* basic need. **-begriff** *m* basic concept. **-besitz** *m* real estate. **-besitzer** *m* landowner. **-bestandteil** *m* basic constituent; base (for ointments, etc). **-düngung** *f* soil fertilization. **-eigenschaft** *f* basic property (*or* characterstic). **-einheit** *f* basic (*or* fundamental) unit; base unit. **-email** *n* base (coat of) enamel

gründen I *vt* 1 found, establish, lay the foundation of. 2 (**auf**) base (on). 3 (*Paint*) prime, size = GRUNDIEREN. 4 groove. **II** *vi* (**auf, in**) & **III: sich g.** (**auf**) be based, rest (on). **Gründer** *m* (-) founder

grund- basic(ally), fundamental: **-falsch** *a* basically (*or* radically) false (*or* wrong). **G–farbe** *f* 1 (*Opt.*) primary color. 2 (*Paint*) priming coat, undercoat. 3 (*Dye.*) bottom color. **-färbung** *f* bottom dyeing. **-faser** *f* elementary fiber. **-feuchtigkeit** *f* soil moisture. **-firnis** *m* priming varnish. **-fläche** *f* 1 (surface) area. 2 (*Math.*) base, basal surface. 3 (*Mach.*) floor space (for a machine). **-flüssigkeit** *f* (*Colloids*) dispersion medium; suspending liquid. **-form** *f* basic (*or* primary) form. **-formel** *f* basic formula. **-gebirge** n primitive rock, bedrock. **-gedanke** m basic thought (or idea). **-gesetz** n basic (or fundamental) law, constitution. **-gestein** n bedrock. **-gewebe** n 1 (Bot.) primary tissue, stroma. 2 (Tex.) ground weave. **-gleichung** f basic equation. **-heil** n (Pharm.) mountain parsley; androseme

Grundier·anstrich m (Paint) priming coat, undercoat. **-bad** n (Dye.) bottom bath

grundieren vt 1 (Paint) prime, undercoat. 2 (Gilding) size. 3 (Paper) stain. 4 (Dye.) bottom, impregnate. 5 (Tex.) prepare

Grundier·farbe f 1 (Paint) primer, undercoat; flat color. 2 (Enamel) ground coat. 3 (Dye.) bottom color. **-farbstoff** m bottoming color (or dye). **-firnis** m priming varnish, filler. **-flotte** f bottoming bath. **-lack** m priming varnish, filler. **-mittel** n (Paint) filler, sealer. **-salz** n (Tex.) preparing salt (sodium stannate). **-schicht** f priming coat

Grundierung f (-en) 1 priming, undercoat(ing); sizing; staining; bottoming; preparing cf GRUNDIEREN. 2 (Paper) texture

Grundierungs·bad n (Dye.) impregnation bath. **-mittel** n (Paint) grounding, primer, surfacer

Grund·immunität f primary (or natural) immunity. **-irrtum** m basic error. **-kapital** n 1 capital (stock). 2 original capital. **-konstante** f fundamental constant. **-körper** m fundamental (or parent) substance. **-kraft** f primary (or primitive) force. **-kreis** m (Math.) base circle

gründl. abbv (gründlich) thorough(ly)

Grund·lack m (Paint) base coat (varnish, lacquer). **-ladung** f (Explo.) base charge. **-lage** f foundation, basis; base; (pl oft) fundamentals. **-lagenforschung** f basic research. **-lager** n main bearing. **g-legend** a fundamental, basic. **G–legung** f founding, (laying the) foundation. **-legierung** f base alloy. **-lehre** f fundamental doctrine

gründlich a thorough, painstaking. **Gründlichkeit** f thoroughness

Grund· fundamental, basic: **-linie** f 1 base (line). 2 pl oft main features, outline. **g–**

los a 1 groundless, baseless. 2 bottomless; muddy. **G–maß** n basic unit (of measurement). **-masse** f 1 (Petrog.) groundmass, matrix. 2 (Paint) filler, bottoming paste. 3 (Anat.) stroma. **-material** n base material; matrix. **-metall** n (Electroplating) basis metal. **-mischung** f base mix(ture); master batch. **-mol** n base mole. **-molekül** n base molecule, monomer. **-mörtel** m concrete. **-niveau** n ground level. **-planung** f basic planning. **-platte** f base (or bed) plate. **-preis** m basic (or list) price. **-problem** n basic problem. **-reaktion** f basic (or essential) reaction. **-recht** n basic right (or privilege). **-regel** f basic rule; axiom. **-riß** m 1 ground plan. 2 outline, sketch. **-satz** m principle. **g–sätzlich** a fundamental, basic; adv oft in (on, as a matter of) principle. **G–sauerteig** m sourdough, starting leaven. **-schicht** f 1 fundamental (or primary) layer. 2 ground course. 3 bedrock. **-schule** f elementary (or primary) school. **-schwingung** f fundamental vibration. **-seife** f soap base. **-sorte** f (Bot.) native variety. **-stein** m 1 foundation stone, cornerstone. 2 lower millstone. **-stock** m matrix; basic stock. **-stoff** m 1 element. 2 base, raw material. 3 (Paper) ground pulp. **-stoffindustrie** f basic industry. **-stoffwechsel** m basal metabolism. **-strich** m (Paint) 1 downstroke. 2 base (first, primary) coat. **-stück** n (piece of) real estate, plot of ground. **-stufe** f elementary level. **-substanz** f 1 element, basic substance. 2 (Anat.) ground substance, matrix. **-teil** m basic part, element. **-teilchen** n fundamental particle. **-ton** m 1 general tone, keynote. 2 fundamental (or primary) tone. **-umsatz** m (Biol.) basal metabolism

Gründung f (-en) 1 founding, foundation, establishment. 2 priming, sizing. 3 grooving

Grün·dünger m green (or vegetable) manure. **-düngung** f green manuring

Grund· basic(ally), fundamental; thoroughly: **-ursache** f basic (or primary) cause. **g–verschieden** a basically (or radically) different. **G–versorgung** f basic care. **-versuch** m basic experiment.

-wasser *n* (under)ground water. **-wasserspiegel** *m* water table. **-zahl** *f* 1 basic number. 2 cardinal number. 3 base, radix. **-zug** *m* 1 basic feature. 2 (*pl, esp in book titles*) fundamentals, outline. **-zustand** *m* (*Nucl.*) ground state (*or* level)

Grüne I *f* green(ness). **II** *n* (*adj endgs*) country(side)

Grüneisen-erde *f*, **-erz** *n*, **-stein** *m* green iron ore, dufrenite

grün-empfindlich *a* sensitive to green

grünen *vi* be (grow, turn) green; flourish. **grünend** *p.a* verdant

Grün-erde *f* green earth, glauconite, celadonite. **-erdelack** *m* green-earth lake. **-feuer** *n* green fire. **-fläche** *f*(green) lawn; park area. **-frucht** *f* green crop(s). **g-gelb** *a* greenish-yellow. **-gesalzen** *a* (*Lthr.*) green-salted, green-cured. **-grau** *a* green-gray. **G-holz** *n* greenheart (wood). **-kalk** *m* gas lime. **-kohl** *m* kale. **-kreuz** *n*, **-kreuzkampfstoff** *m* "green cross" (poison) choking gas. **-land** *n* grassland

grünlich *a* greenish (*oft in compds + colors*, e.g.:) **-blau** *a* greenish-blue

Grün-malz *n* green malt. **-mist** *m* green manure. **-öl** *n* green (*or* anthracene) oil. **-rost** *m* verdigris. **g-rostig** *a* aeruginous, verdigris-covered. **G-sand(stein)** *m* (*Geol.*) greensand. **g-schwarz** *a* greenish-black. **G-span** *m* verdigris, green copper tarnish, (*specif*) basic copper acetate

Grünspan-blumen *fpl* crystals of verdigris. **-essig, -geist, -spiritus** *m* spirit of verdigris = KUPFERGEIST

Grün-spat *m* diopside. **-star** *m* glaucoma. **-stein** *m* (*Geol.*) greenstone. **-stich** *m* greenish tinge. **g-stichig** *a* greenish, green-tinged. **G-stift** *m* green pencil. **-sucht** *f* chlorosis. **-zeug** *n* greens, (green) vegetables

Gruppe *f* (-n) group

Gruppen-erscheinung *f* group phenomenon. **-geschwindigkeit** *f* group velocity. **-reagens** *n* (+ **auf**) group reagent (for). **-schaltung** *n* (*Elec.*) series-parallel connection. **-silikat** *n* silicate with large discrete ions (*or* with a ring structure). **g-weise** *a, usu adv* by (*or* in) groups, groupwise

gruppieren *vt* group. **Gruppierung** *f* (-en) grouping

Grus *m* (-e) 1 rubble, gravel, grit; (*Geol.*) debris, detritus. 2 (*Coal, etc*) breeze, slack, fines; (*Anthracite*) culm. **-boden** *m* soil with rock debris. **-kakao** *m* cacao husks. **-kohle** *f* small coal, slack, breeze. **-koks** *m* coke fines (*or* breeze)

Gruß *m* (-e) greeting. **grüßen** *vt* greet

Grütz- gritty, pulpy, (*Med.*) sebaceous, atheromatous: **-beutel** *m* atheroma, sebaceous cyst. **-brei** *m* mush, porridge *cf* BREI

Grütze *f* 1 grits; porridge. 2 brains, sense

Grütz-schleim *m* gruel, watery porridge

G-säure *f* G-acid (2-naphthol-6,8-disulfonic)

GS-Verfahren *n, abbv* (Gleichstrom-Schwefelsäure . . .) direct current sulfuric acid process for anodizing aluminum

GT, Gtl. *abbv* (Gewichtsteil) part by weight

Gtt. *abbv* (Gutta, *Lat.*) drops

GTZ *abbv* (Gesellschaft für technische Zusammenarbeit) Society for Technological Cooperation

Guaj . . . guai . . . e.g. **Guajadol** *n* guaiadol

Guajak *m* guaiacum. **-blau** *n* guaiacum blue. **-harzsäure** *f* guaiaretic acid

Guajakin-säure *f* guaiacinic acid. **Guajakon-säure** *f* guaiaconic acid

Guajak-säure *f* guaiacic acid. **-seife** *f* guaiac soap

Guajaret-säure *f* guaiaretic acid

Guajen *n* guaiene. **-chinon** *n* guaienequinine

Guanyl-säure *f* guanylic acid

Guäthol *n* guaethol (*o*-orthoxyphenol)

gucken *vi* look, peep; *with sep pfxs* = SCHAUEN, SEHEN. **Guck-loch** *a* peep-hole, sight-hole

Guhr *f* (kiesel)guhr

Guignet-grün *n* Guignet's green

guillochieren *vt* guilloche

Gulden *m* (-) guilder, florin

güldisch *a* auriferous. **Güldisch(silber)** *n* auriferous (*or* doré) silver (alloy)

Gülle *f* liquid manure

Gulon-säure *f* gulonic acid

gültig *a* valid. **Gültigkeit** *f* validity

Gummi I *n* rubber; (*Bot., oft*) gum, resin.

II *m* (-s) (rubber) eraser; rubber band. **III** *m,n* (-s) condom. **-abfälle** *mpl* scrap rubber. **g-ähnlich** *a* rubbery, rubberlike; gum-like, gummy. **G-arabicum** *n* gum arabic. **-art** *f* type of rubber (*or* gum). **g-artig** *a* rubbery, etc = -ÄHNLICH. **G-artikel** *m* rubber article (*or pl oft* goods). **-auflösung** *f* rubber (*or* gum) solution. **-auskleidung** *f* rubber lining. **-ballon** *m* rubber balloon (*or* bulb). **-band** *n* rubber band; elastic. **-baum** *m* rubber (*or* gum) tree. **-blase** *f* rubber bulb. **-bleispat** *m* plumbogummite. **-chemiker** *m* rubber chemist. **-chromverfahren** *n* bichromated gum process. **-dichtung** *f* rubber gasket (seal, packing). **-druck** *m* offset printing

Gummidruck·ball *m* rubber (pressure) bulb. **-farbe** *f* offset printing ink. **-lack** *m* gum printing varnish

Gummi elasticum rubber

gummieren *vt* rubberize; gum (labels, etc); (*Gilding*) size. *Cf* GUMMIERT

Gummi·ersatz(stoff) *m* rubber substitute, (*specif*) factice

gummiert *p.a* rubber-like; rubberized

Gummi·erz *n* (*Min.*) gummite. **-fahne** *f* rubber flag, policeman (for precipitates). **-fichte** *f* balsam fir. **-form** *f* rubber mold

gummig *a* gummy, gummous

Gummi·gärung *f* mucous fermentation. **-gebläse** *n* rubber bellows. **-gutt** *n* gamboge. **g-haltig** *a* containing rubber (*or* gum), gummy. **G-harz** *n* gum (*or* rubber) resin. **g-harzig** *a* of a gum resin. **G-hütchen** *n* rubber cap. **-isolierung** *f* rubber insulation. **-kneter** *m* rubber masticator, dough mill. **-lack** *m* gum lac. **-lösung** *f* rubber (*or* gum) solution. **-lösungsbenzin** *n* (*Petrol.*) rubber solvent. **-milch** *f* latex. **-papier** *n* gummed paper. **-pflaster** *n* rubber paving. **-pfropf(en)** *m* rubber stopper. **-puffer** *m* rubber buffer. **-rahmen** *m* rubber gasket. **-riemen** *m* rubber belt (or strap). **-sauger** *m* rubber suction bulb. **-säure** *f* gummic (*or* arabic) acid. **-scheibe** *f* rubber disc. **-schlauch** *m* rubber hose (tube, tubing). **-schleim** *m* gum mucilage, (*Pharm.*) (acacia) mucilage. **-schwamm** *m* rubber sponge, sponge rubber. **-sirup**

m (*Pharm.*) syrup of acacia. **-spat** *m* plumbogummite. **-spund** *m* rubber bung. **-stein** *m* hyalite. **-stempel** *m* rubber stamp. **-stoff** *m* 1 gum substance, gummy matter. 2 rubber(ized) cloth. 3 (*Min.*) gummite. **-stopfen, -stöpsel** *m* rubber stopper. **-tragant** *m* (-en, -en) gum tragacanth. **-tuch** *n* rubberized cloth. **-überzug** *m* rubber coating (*or* film). **-verbindung** *f* rubber connection. **-verdickung** *f* gum thickening. **-verfestigungsharz** *n* rubber compounding resin. **-walze** *f* rubber roller. **-wasser** *n* gum water. **-wischer** *m* rubber wiper (*or* "policeman"). **-zahl** *f* (*Paper*) rubber number. **-zucker** *m* arabinase

gummös *a* gummous, gummy

Gundelrebe *f*, **Gundermann** *m* ground ivy

Gunst *f* favor: **zu G—en** (+ *von* or + *genit*) in favor (of). *See also* ZUGUNSTEN. **günstig** *a* favorable. **günstigste** *a* (*superl*) most favorable, optimum: **im g—en Fall** at best

Gur *f* (kiesel)guhr, diatomite

Gura·nuß *f* guru (*or* cola) nut

Gur·dynamit *m, n* guhr dynamite

Gurgel *f* (-n) throat. **-ader** *f* jugular vein. **-mittel** *n* (*Pharm.*) gargle

gurgeln *vi* gargle; gurgle

Gurgel·wasser *n* (*Pharm.*) gargle

Gurjunen *n* guriunene

Gurke *f* (-n) cucumber. **Gurken·kraut** *n* (*Bot.*) borage

Gurt *m* (-e) 1 strap, belt. 2 flange

Gürtel *m* (-) 1 belt; girdle. 2 zone. **-reifen** *m* radial tire. **-rose** *f* (*Med.*) shingles, herpes zoster

gürten 1 *vt* gird(le). 2: **sich g.** gird oneself

Gurt·förderband *n*, **-förderer** *m* conveyor belt

Guß *m* (Güsse) 1 (metal, plastic) casting; (type) font; cast iron. 2 pouring, casting (of molten metals, etc), founding (of type) **—aus einem G.** cast in one piece, **aus anderem G.** cast in a different mold. 3 spout, gate (of a mold). 4 gutter. 5 grain hopper. 6 (*Brew.*) mash liquor. 7 gush, jet, stream, dash (of liquid); (*Med.*) affusion. 8 (rain) shower, downpour. 9 icing, frosting. *Cf* GIESSEN. **-aluminium**

n cast aluminum. **-asphalt** *m* mastic asphalt

gußbar *a* pourable, castable = GIESSBAR

Guß-barren *m* ingot. **-beton** *m* cast (*or* poured) concrete. **-blase** *f* blowhole, casting flaw. **-blei** *n* cast lead. **-block** *m* ingot. **-bruch** *m* cast metal scrap. **-eisen** *n* cast iron; pig iron. **g-eisern** *a* cast-iron. **G-email** *n* cast iron enamel. **g-fähig** *a* castable. **G-fehler** *m* casting flaw. **-flasche** *f* molding flask. **-form** *f* casting mold; (*Soap*) frame. **-fuge** *f* casting burr (*or* seam). **-haut** *f* skin (of a casting). **-karbid** *n* cast carbide alloy. **-kasten** *m* casting (*or* molding) box. **-kokille** *f* cast iron mold. **-legierung** *f* casting alloy. **-loch** *n* gate (of a mold). **-masse** *f* pouring (*or* casting) compound. **-messing** *n* cast brass. **-metall** *n* cast (*or* casting) metal. **-modell** *n* casting model. **-mörtel** *m* concrete, cement. **-naht** *f* casting burr (*or* seam). **-pfanne** *f* (*Metall.*) casting ladle. **-riß** *m* casting crack. **-rohr** *n*, **-röhre** *f* cast iron pipe. **-schlicker** *m* (*Ceram.*) casting slip. **-schrott** *m* cast iron scrap. **-späne** *mpl* cast iron chips (*or* borings). **-stahl** *m* cast steel. **g-stählern** *a* cast-steel. **G-stein** *m* sink, drain. **-stück** *n* cast, casting. **-wachs** *n* casting wax. **-waren** *fpl* castings, cast goods, foundry goods. **-zink** *n* cast zinc

gut I *a* **1** good —**im g—en** amicably; **das hat g—e Zeit** there is ample time for that. **2** (*pred, oft*) a good thing; all right, OK —**es g. sein lassen** let it go at that. **II** *adv* well *cf* GEHEN 10 —**gut(e) zwei Stunden** a good two hours. **III** *pfx* **3** (*esp on p.a*) e.g. **gutschmeckend** good-tasting; *cf* GUTERHALTEN. **4** (*sep*) good; on credit; *cf* GUTHABEN, GUTHEISSEN. **Gut** *n* (ⁱ-er) **1** good: **G. und Böse** good and evil. **2** possessions, property *cf* HABE. **3** goods, merchandise; *esp pl* **Güter** freight; (*esp with adj of material*) ware: **irdenes G.** earthenware. **4** (working) material *cf* FARBGUT, SCHÜTTGUT; charge, feed. **5** estate, farm, piece of real estate, property

Gut-achten *n* expert opinion; estimate; certificate; report. **g-artig** *a* benign; mild, good-natured. **G-befund** *m* approval

Güte *f* **1** (good) quality, grade (of material); (sound) fidelity. **2** goodness, kindness. **-anforderung** *f* quality requirement. **-bestimmung** *f* quality determination (*or* regulation). **-eigenschaft** *f* quality characteristic. **-einteilung** *f* grading, classification by quality. **-grad** *m* grade, (degree of) quality; efficiency. **-klasse** *f* quality class, grade. **g-mäßig** *a* as regards quality. **G-messer** *m* eudiometer

Güter goods, merchandise, freight: *pl of* GUT *n*

Gut-erhalten *a* well-preserved, well-maintained

Güter-verkehr *m* freight traffic (*or* transportation). **-wagen** *m* (railroad) freight car. **-zug** *m* freight train

Güte-steigerung *f* improvement in quality. **-überwachung** *f* quality control. **-verhältnis** *n* efficiency. **-wert** *m*, **-zahl** *f* index of quality, quality factor. **-zeichen** *n* quality seal

Gut-gewicht *n* **1** fair weight. **2** weight allowance. **3** commercial moisture tolerance

gut-haben* *vt* have credit for. **Guthaben** *n* (-) credit, balance

gut-heißen* *vt* approve (of), sanction. **Gutheißung** *f* (-en) approval, sanction

gütig *a* kind(-hearted)

gut-machen *vt* make good, make up (for)

gut-sagen *vt*, *vi* (+ **für**) vouch (for), guarantee

Gut-schein *a* coupon, voucher; (claim) check. **g-schließend** *a* tight-fitting (stoppers, lids)

gut-schreiben* *vt:* credit (to a psn *or* account). **Gut-schrift** *f* (-en) credit

Gutti *n* gamboge

gut-unterrichtet *a* well-informed. **-willig** *a* willing. **-ziehend** *a* well-drawing, with a good draft

Gymnasium *n* (..ien) (9-year) secondary school (preparing for university)

Gynocardin-säure *f* gynocardinic acid. **Gynocard-säure** *f* gynocardic (*or* chaulmoogric) acid

Gyps *m* gypsum, etc = GIPS

Gyre *f* (-n) rotatory axis. **Gyroide** *f* (-n) (*Cryst.*) alternating axis

Gyrophor-säure *f* gyrophoric acid

H

h (*International symbol: Lat.* hora) hr(s)., o'clock

h. (*abbv*) **1** (hochschmelzend) high-melting. **2** (heiß) hot. **3** (hoch) high

H. *abbv* **1** (Härte) hardness. **2** (Höhe) ht. (height), alt. **3** (Haben) credit. **4** (in *Beilstein*, = Hauptwerk) main part

ha *abbv* (Hektar) hectare

Haag *m* (*Geog.*) the Hague

Haar *n* (-e) **1** (*usu, oft as pl*) hair. **2** fiber; filament. **3** (*Tex.*) nap, pile. **4** (*in compds, oft*) capillary, (*oft implies extreme fineness*): **-alaun** *m* capillary alum. **-ansatz** *m* (person's) hair line. **h–artig** *a* hairlike, capillary. **H–balg** *m* hair follicle. **-balgdrüse** *f* sebaceous gland. **-beize** *f* depilatory. **-bleiche** *f* hair bleach(ing). **-blutgefäß** *n* (*Anat.*) capillary. **-draht** *m* hair-thin wire (*or* filament), capillary wire. **-drüse** *f* sebaceous gland. **h–dünn** *a* hair-thin

haaren 1 *vt* unhair, depilate. **2** *vi* & **sich h.** lose (*or* shed) hair

Haares·breite *f* hair's breadth

Haar· hair, capillary: **-erz** *n* capillary ore. **-erzeugungsmittel** *n* hair restorer. **-farbe** *f* **1** hair color. **2** = **-färbemittel, -färbungsmittel** *n* hair dye. **-faser** *f* capillary filament (*or* fiber). **h–faserig** *a* filamentous, capillary. **H–feder** *f* **1** (*Mach.*) hairspring. **2** (*Zool.*) filoplume. **h–fein** *a* hair-thin, hairline; capillary. **h–förmig** *a* hair-shaped, capillary. **H–gefäß** *n* capillary vessel. **h–genau** *a* meticulous, accurate (*or* exact) to a hair; *adv. oft* to a hair, to a T

haarig *a* hairy; pilose; nappy, napped

Haar·kies *m* capillary pyrites, nickel sulfide; (*Min.*) millerite. **-klauberei** *f* hairsplitting. **h–klein** *a* minutely detailed, *adv oft* down to the last detail. **H–kraft** *f* capillarity. **-kupfer** *n* capillary copper. **-linie** *f* hairline. **-lockerungsmittel** *n* depilatory

haarlos *a* hairless, bald; (*Tex.*) napless.

H--machen *n* (*Lthr.*) unhairing

Haar·nadel *f* hairpin. **-pinsel** *m* camel's hair paintbrush. **-riß** *m* hairline crack; (*Ceram.*) craze. **h–rissig** *a* hairline-cracked, crazed —**h. werden** craze. **H–rohr, -röhrchen** *n* capillary tube. **-röhrchenkraft** *f* capillarity. **-röhre** *f* capillary tube. **-röhrenanziehung** *f* capillary attraction. **-sack** *m*, **-säckchen** *n* hair follicle. **-salbe** *f* hair ointment; pomade, pomatum. **-salz** *n* **1** hair salt, fibrous alunogen, halotrichite *or* epsomite. **2** magnesium sulfate efflorescence. **h–scharf** *a* razor-sharp, hairline; meticulous; *adv oft* to a hair, by a hair's breadth. **H–schärfe** *f* hairline accuracy. **-schneidemaschine** *f* hair clippers. **-schwefel** *m* capillary sulfur. **-schweif** *m* coma, tail (of a comet). **-seite** *f* (*Lthr.*) grain (*or* hair) side. **-silber** *n* capillary silver. **-spalte** *f* hairline crack. **-spalterei** *f* hairsplitting. **-sprung** *m* hairline crack. **-stern** *m* comet. **-strang** *m* (*Bot., Pharm.*) peucedanum; hair cord. **h–sträubend** *a* hair-raising, frightful. **H–trockner** *m* hair dryer. **-vertilgungsmittel** *n* dilatory. **-vitriol** *n* capillary epsomite. **-wachs** *n* pomade. **-waschmittel** *n* shampoo. **-wasser** *n* hair lotion (*or* tonic). **-wurzel** *f* hair root, fibrous root. **-zelle** *f* hair (*or* capillary) cell

Habe *f* property, possessions *cf* GUT *n*

haben* (hatte, gehabt; hat) **I** *vt* (*usu*) have, have got: **das hat etwas für** (*or* **auf) sich** that has (got) sthg to be said for it; **er hat es an der Leber** (*or* **mit der Leber zu tun)** he has (got) sthg wrong with his liver; *Cf* WOLLEN. **2** get: **man hat nichts (etwas, viel) davon** you get nothing (sthg, a lot) out of it; **das haben sie von ihren Eltern** they got (received, inherited) that from their parents. **3** (*Meas., etc*) measure, weigh, be (an amt. on a scale, etc): **es hat 15 Grad** it (*or* the temperature) is 15 degrees; **wir h.**

10 Uhr it is 10 o'clock; **wir h. den 10. Juni** it is the 10th of June. **4** (*spec expr*): **Hunger h.** (= **hungrig sein**) be hungry *cf* ANGST, DURST, EILE, EILIG, GERN, NÖTIG, RECHT, SCHLAF, UNRECHT; **was h. sie?** what is wrong (*or* the matter) with them?; **es hatte viel Schnee** there was a lot of snow; **welche Farbe hat der Stein?** what color is the stone? **5: zu h.: es ist hier (nicht) zu h.** it is (not) available here; **sie sind dafür (nicht) zu h.** they (don't) go for that. **II** *v aux* **6** (+ *pp*): **sie h. es gesehen** (a) they have seen it, (b) they saw it. **7** (+ *inf*): **sie h. es an der Wand hängen** they have it hanging on the wall; **sie h. gut reden** they may well talk, it is easy for them to talk. **8** (+ *inf with* **zu**): **sie h. noch einen Versuch zu machen** they have another experiment to do; **die Lösung hat ein paar Stunden zu kühlen** the solution has to (*or* is supposed to) cool for a couple of hours. **III: sich h.** act; carry on, make a fuss. **Haben** *n* credit (column) *cf* SOLL

Haber·verfahren *n* Haber process

habhaft *pred a:* **h. werden** (+ *genit*) catch, get hold of

Habicht *m* (-e) hawk. **Habichts·kraut** *n* hawkweed

Habilitations·schrift *f* qualifying thesis for university lecturers. **habilitieren** *vi* & **sich h.** qualify to lecture

habituell *a* habitual. **Habitus** *m* bearing; (*Cryst.*) habit

Hab·sucht *f* avarice, greed.

Haché *n* (-s) hash. **hachieren** *vt* **1** mince, chop up. **2** cross-hatch, shade

Hack·braten *m* meatloaf

Hacke *f* (-n) **I** hoe, pick, matlock; hatchet, ax. **II** heel. **hacken** *vt* chop, mince, hack; hoe; pick; peck. **Hacken** *m* (-) heel. **Hacker** *m* (-) chopper, hacker, hoer

Hack·fleisch *n* chopped (minced, ground) meat. **-früchte** *fpl* root (*or* tuber) crops. **-kultur** *f* truck gardening. **-messer** *n* chopping knife, chopper, cleaver. **-salz** *n* minced-meat salt. **-schnitzel** *npl* (*Lumber*) hogged chips; wood chips

Häcksel *n* chopped straw. **häckseln** *vt* chop

Hack·span *m* chip

Hader *m* **I** (-) quarrel, dispute. **II** (- *or* -n)

rag. **haderig** *a* (*Iron*) short, unweldable.

hadern *vi* quarrel, wrangle

Hadern· (*Paper:* rag): **-brei** *m* rag pulp. **-halbstoff** *m* rag half stuff. **-kocher** *m* rag boiler. **-lade** *f* rag chest. **-papier** *n* rag paper. **-schneider** *m* rag cutter (*or* chopper)

hadrig *a* (*Iron*) short, unweldable

Hafen *m* (-) **1** harbor, port; haven. **2** pot, crock; (*Glass*) (melting) pot. **--ofen** *n* (*Glass*) pot furnace. **-schiffahrt** *f* harbor shipping. **-zange** *f* crucible tongs

Hafer *m* oats. **-flocken** *fpl* oat flakes, rolled oats. **-gras** *n* wild oats. **-malz** *n* oat malt. **-mehl** *n* oatmeal. **-schleim** *m* oatmeal, oat gruel. **-stärke** *f* oat starch. **-wurz(el)** *f* salsify

Hafner *m* (-) **1** potter. **2** stove fitter. **-erz** *n* potter's ore, fine galena; alquifou

Haft I *m* (-e *or* -en) **1** adhesiveness, hold. **2** fastener, hook, clasp. **II** *f* arrest, custody; detention; (*in compds, oft*) liability, responsibility. **-haft** *sfx* (*forms adj from nouns*) e.g. **schmerzhaft** painful *cf* ·HAFTIGKEIT

haftbar *a* responsible, liable. **Haftbarkeit** *f* responsibility, liability

Haft·druck *m* solution (*or* adhesion) pressure

haften *vi* **1** (**für**) vouch, be liable (*or* responsible) (for) **2** = **haften·bleiben** *vi* (s) (**an**) stick, cling, adhere (to). **haftend** *p.a* liable, responsible

haft·fähig *a* adhesive. **H–fähigkeit** *f* adhesiveness, adhesion. **-festigkeit** *f* **1** adhesiveness, adhesion. **2** firmness of attachment (*or* bonding), tenacity. **3** (*Dyes*) substantivity. **-festigkeitsprobleme** *npl* bonding problems. **-fläche** *f* surface of adhesion. **h–freudig** *a* adherent; (*of tires*) road-holding. **H–glas** *n* contact lens

·haftigkeit (*sfx for nouns formed from adjs in* ·**haft**) e.g. **Schmerzhaftigkeit** painfulness *cf* ·HAFT

Haft·intensität *f* intensity (*or* degree) of adhesion; solution pressure; affinity. **-kleber** *m* contact (*or* pressure-sensitive) adhesive. **-kraft** *f* adhesion, adhesive force (*or* power). **-masse** *f*, **-mittel** *n* adhesive. **-pflicht** *f* liability. **h–pflichtig** *a* liable. **H–reibung** *f* static (*or* adhesive) friction. **-schicht** *f* adhesive layer **-sitz** *m* (*Mach.*) tight (*or* force) fit. **-spannung** *f* adhesive stress. **-stelle** *f* point of attachment

Haftung *f* (-en) 1 adhesion. 2 adsorption. 3 liability, responsibility *cf* GMBH

Haft-vermittler *m* bonding agent, adhesion promoter. **-vermögen** *n* adhesion, adhesive power. **-wasser** *n* 1 adhering water. 2 (*Soil*) film water. 3 (*Petrol.*) connate water

Hag *m* (-e) 1 hedge; fence. 2 hedged (*or* fenced) enclosure, *esp* pasture. 3 pile (of bricks). **-apfel** *m* crabapple

Hage·buche *f* hornbeam. **-butte** *f* rose hip (*or* haw), rosebush pseudocarp; dog rose. **-buttenöl** *n* rose-hip oil. **-dorn** *m* hawthorn

Hagel *m*, **hageln** (*Meteor.*) *vi* hail

hager *a* haggard, lean (*incl Soil*)

Hage·rose *f* dog rose

Hahn *m* (⁻e) 1 cock (*incl* rooster); stopcock; tap, faucet, spigot; valve

Hahnen·fett *n* stopcock (*or* tap) grease. **-fuß** *m* 1 (*Ceram.*) cockspur. 2 (*Bot.*) ranunculus, crowfoot. **-sporn** *m* 1 cockspur. 2 (*Bot.*) plectranthus; ergot

Hahn·fett *n* stopcock grease. **-fuß** *m* ranunculus. **-hülse** *f* stopcock shell (*or* barrel). **-kegel** *m*, **-kücken**, **-küken** *n* stopcock plug (*or* stopper). **-messer** *m* flowmeter. **-schlüssel** *m* stopcock (tap, faucet) wrench. **-sitz** *m* stopcock seat(ing). **-stöpsel** *m* stopcock stopper (*or* plug). **-trichter** *m* tap (*or* separatory) funnel, funnel with a stopcock

Hai *m* (-e) shark

Haide *f* (-n) heath = HEIDE

Haifisch *m* shark. **-lebertran** *m* shark-liver oil. **-leder** *n* sharkskin (leather). **-tran** *m* shark (liver) oil

Hain *m* (-e) grove. **-buche** *f* hornbeam. **-simse** *f* woodrush, glowworm grass

Häkchen *n* (-) 1 little hook, hooklet. 2 check mark (on lists, etc). 3 comma, apostrophe; *as pl oft* quotation marks

häkeln *vt/i* crochet; hook

haken 1 *vt* hook; plow. 2 *vi* get hooked; get caught; get stuck. **Haken** *m* (-) 1 hook; peg; clasp; (*Puddling*) rabble. 3 snag,

hitch, catch. 4 check mark (on lists, etc)

haken·förmig *a* hook-shaped, hook-like. **H-kreuz** *n* swastika. **-probe** *f* (*Sugar*) hook test. **-wurm** *m* hookworm

hakig *a* hooked; hackly (fracture)

H-aktiv *a* active-hydrogen

halb I *a* half. II *adv oft* halfway, by halves, partly, almost; (*Time*) e.g. **h. elf** half past ten. III *pfx* half-, semi-, hemi-, demi-, medium. IV *sfx* 1 half: **viereinhalb** four and a half. 2 (*in adv of location*) *cf* AUSSERHALB. 3 on account of *cf* DESHALB, WESHALB, ·HALBEN, ·HALBER

Halb·acetal *n* hemiacetal. **-alaun** *m* impure alum. **-aldehyd** *m* hemialdehyde, half aldehyde. **-anthrazit** *m* semi-anthracite. **-art** *f* subspecies. **-ätze** *f* (*Tex.*) half discharge. **h-automatisch** *a* semi-automatic. **-beruhigt** *a* (*Metll.*) half-killed, balanced. **-beständig** *a* metastable. **-beweglich** *a* semiportable. **H-bildung** *f* half-baked education, semi-literacy. **-bleiche** *f* half bleach; cream color. **-chlorschwefel** *m* sulfur monochloride. **-croupon** *m* (*Lthr.*) bend. **h-cylindrisch** *a* semicylindrical. **-dauernd** *a* semipermanent. **-deckend** *a* semiopaque. **H-decker**, **-deckflügler** *mpl* (*Zool.*) Hemiptera. **-dunkel** *n* semidarkness; twilight, dusk. **h-durchlässig** *a* semipermeable. **H-durchmesser** *m* radius, semidiameter. **h-durchscheinend** *a* semitranslucent. **-durchsichtig** *a* semitransparent. **-edel** *a* semiprecious. **H-edelstein** *m* semiprecious stone. **h-eirund** *a* semioval. **H-element** *n* (*Elec.*) half cell (*or* element)

halben *vt* halve, divide in two

·halben *sfx* on behalf (*or* account) of: **ihrethalben** on their account; *see also* ALLENTHALBEN

halber *prep* (+ *genit*) *& sfx* on account (*or* behalf) of, for the sake of: **der Vorsicht h.**, **vorsichtshalber** for the sake of caution

Halb·erzeugnis *n* semi-finished product, semimanufacture. **-ester** *m* half-ester, acid ester. **-fabrikat** *n*, **-fertigware** *f* semi-finished product (*or* goods) = HALBERZEUGNIS. **h-fest** *a* semisolid;

semifixed; semipermanent. **-fett** *a* 1
(*Coal*) semibituminous. 2 (*Oil Varnish*)
medium. **-flächig** *a* (Cryst.) hemihedral.
H–flächner *m* (*Cryst.*) hemihedron.
-flügler *mpl* Hemiptera. **h–flüssig** *a*
semiliquid, semifluid. **H–franz(band)**
m (volume bound in) half-calf (*or* half-
leather). **h–gefüllt** *a* half-filled.
-geleimt *a* half-sized. **H–gerbung** *f* partial tanning. **h–gesättigt** *a* half-
saturated. **-gesäuert** *a* semi-acidified.
-geschmolzen *a* half-melted, semifused.
H–glanz *m* semi-gloss. **-glanzkohle** *f*
clarain. **-gold** *n* imitation gold (e.g. sim-
ilor). **-gräser** *npl* (*Bot.*) rushes. **-gut** *n*
(*Metll.*) half-and-half tin-and-lead alloy.
h–hart *a* half-hard, medium. **H–harz** *n*
crude resin
Halbheit *f* (-en) halfway measure, half-
hearted effort
Halb·holländer *m* (*Paper*) washing engine. **-hydrat** *n* hemihydrate
halbieren *vt* halve, cut in two, bisect.
Halbierende *f* (*adj endgs*) bisector
Halb· half, semi-: **-insel** *f* peninsula. **h–**
jährig *a* six-month(-old). **-jährlich** *a*
semiannual. **-klassisch** *a* semiclassical.
H–kochen *n* partial boiling, parboiling.
-koks *m* semicoke. **-kokung** *f* semicoking, partial carbonization. **-kolloid** *n*
semicolloid. **-kreis** *m* semicircle. **h–**
kreisförmig, -kreisrund *a* semicircular. **H–kugel** *f* hemisphere. **h–**
kugelförmig *a* hemispheric(al).
-lasierend *a* (*Colors*) semitransparent.
H–lasurblei *n* caledonite. **-leder** *n* half-
leather (binding). **-leinen** *n*, **-leinwand** *f*
half-linen; half-cloth (binding). **h–**
leitend *a* semiconducting. **H–leiter** *m*
semiconductor. **-literkolben** *m* half-liter
flask. **-mahlen** *n* partial grinding, breaking in. **h–matt** *a* (*Paper*) semimat, semi-
gloss. **H–mattglanz** *m*, **-mattglasur** *f*
semimat, semigloss (glaze). **H–messer** *m*
radius. **-metall** *n* semimetal, metalloid.
-mikro . . . semimicro-. **h–mineralisch**
a (*Pigments*) metallo-organic. **-monat-**
lich *a* semimonthly. **H–mond** *m* half-
moon, crescent. **h–mondförmig** *a*
crescent-shaped. **H–nitril** *n* seminitrile.

-öl *n* (*Paint*) half-oil, thinned oil. **-por-**
zellan *n* semiporcelain. **-prisma** *n*
(*Cryst.*) hemiprism. **-reduktion** *f* semi-
reduction, partial reduction. **-reserve** *f*
(*Tex.*) half-resist. **h–rund** *a* half-round,
semicircular, hemispheric(al). **-sauer** *a*
semi-acid. **H–schatten** *m* half-shade,
(*Opt, esp in compds*) half-shadow; (*Astron.*) penumbra. **-schattenplatte** *f*
(*Opt.*) quarter-wave plate. **-schwe-**
feleisen *n* iron hemisulfide. **-schwe-**
felkupfer *n* cuprous sulfide. **-seide** *f*
half-silk, silk-and-cotton mixture. **-seite**
f half page, half sheet. **h–seitig** *a* half-
page; (*Med.*) unilateral; *adv oft* on one
side. **-selbsttätig** *a* semiautomatic.
-sicher *a* metastable. **H--silikastein** *m*
semisilica brick. **-stahl** *m* semisteel. **h–**
starr *a* semirigid. **H–stoff** *m* (*Paper*) half
stuff, rag pulp. **-stoffholländer** *m* (*Paper*) breaker. **-stoffwerk** *n* half-stuff
mill. **-stufenpotential** *n* half-wave potential. **h–stündig** *a* half-hour('s).
-stündlich *a* half-hourly, *adv oft* every
half hour. **-symmetrisch** *a* hemisym-
metrical. **-tägig** *a* half-day, half a day's;
part-time. **-teigig** *a* plastic. **-tief** *a* shallow, medium-deep. **H–ton** *m* (*incl Phot.*)
half-tone. **h–trocknend** *a* semi-drying.
H–verkokung *f* semicoking.
-vitriolblei *n* lanarkite, Pb_2SO_5. **h–**
walzig *a* hemicylindrical. **-warm** *a* 1
lukewarm. 2 (*Soap*) half-boiled. **H–**
waren *fpl* semifinished goods.
-wassergas *n* semiwater gas. **h–wegs**
adv halfway, more or less. **-weich**
a half-soft, partly soft. **H–weiß**
n (*Tex.*) half-bleach. **-wert** *m*
1 half value. 2 (*Stat.*) median.
-wertdruck *m* half-value pressure.
-wertsbreite *f* (*Spect.*) half-width.
-wertzeit *f* (*Nucl. Phys.*) half-life. **-wolle**
f (*Tex.*) half-wool, wool-and-cotton mixture. **-wollfarbstoff** *m* mixed-fiber (fabric) dye. **-wüchsige** *m, f* (*adj endgs*)
minor, juvenile. **-wüste** *f* semi-desert. **h–**
zahlig *a* half-integral. **H–zeit** *f* (*Nucl.
Phys.*) half-life. **-zellstoff** *m* 1 hemi-
cellulose. 2 (*Paper*) semi-chemical pulp,
half stuff. **-zeug** *n* 1 semi-finished prod-

uct. 2 (*Paper*) half stuff. **-zinn** *n* tin alloyed with lead, base tin. **-zirkel** *m* semicircle

Halde *f* (-n) **1** refuse heap, dump. **2** slope, hillside

Halden-erz *n* waste-heap ore. **-koks** *m* stock(piled) coke. **-schlacke** *f* dump slag

half helped: *past of* HELFEN

Halfa *f*, **-gras** *n* esparto (grass)

Hälfte *f* (-n) **1** half: **es zur H. tun** do half of it. **2** (*Lthr.*) side. **hälften** *vt* halve; bisect.

Hälft-flächner *m* (Cryst.) hemihedron

Hall *m* (-e) **1** (resonant, ringing) sound. **2** echo

Halle *f* (-n) **1** hall (= bldg. or room). **2** lobby, concourse. **3** (work)shop; shed, hangar

hallen *vi* ring, resound; echo

Hallen- indoor: **-bahn** *f* indoor track

Halm *m* (-e) **1** blade (of grass); stalk, stem, culm; (single) straw **—Getreide auf dem H.** standing grain (crop). **2** fuze. **-früchte** *fpl* cereals

Halo *m* (-s *or* -nen) **1** halo. **2** (*in compds, oft, Chem.*) salt, halo(geno): **-chemie** *f* halochemistry, chemistry of salts. **h—chemisch** *a* halochemical. **H—chromie** *f* halochromism

Halogen- halogen, halide, halo(geno)-: **-alkyl** *n* alkyl halide

Halogenid *n* (-e) halide

halogenierbar *a* halogenatable, capable of halogenation. **halogenieren** *vt* halogenate. **Halogenierung** *f* (-en) halogenation

Halogen- halogen(o-), halide: **-kohlenstoff** *m* carbon halide. **-metall** *n* metallic halide

Halogeno-salz *n* halo-salt

Halogen-säure *f* halogen acid. **-schwefel** *m* sulfur halide. **-silber** *n* silver halide. **h—substituiert** *a* halogen-substituted, halo-: **h—e Chinone** haloquinones. **H—übertrager** *m* halogen carrier

Halogenür *n* (-e) (-ous) halide

Halogenwasserstoff *m* hydrogen halide. **-säure** *f* hydrohalic acid. **-verbindung** *f* hydrogen halide

Haloid *n* (-e), **-salz** *n* halide, haloid salt. **-wasserstoff** *m* hydrogen halide

Hals *m* (-e) **1** neck (*incl fig*). **2** throat. **3** (*Sci.,*

fig) cervix; collar; stem (of thermometer); (*Lthr.*) shoulder. **-ader** *f* **1** jugular vein. **2** carotid artery. **-band** *n* neckband, neck ribbon, necklace; collar. **-blutader** *f* jugular vein. **-bräune** *f* throat infection; (*specif*) strep throat, diphtheria, croup. **h—brecherisch** *a* breakneck; daredevil. **H—drüse** *f* cervical gland. **-mandel** *f* tonsil. **-röhre** *f* trachea. **-schmerzen** *mpl* sore throat. **h—starrig** *a* obstinate, stiffnecked. **H—tuch** *n* neckerchief, scarf. **-weh** *n* sore throat. **-wirbel** *m* cervical vertebra

-halsig *a sfx* -necked *cf* LANGHALSIG

halt! stop!, halt!

Halt *m* (-e) **1** halt, stop: **H. gebieten** (+ *dat*) call a halt, put a stop (to); **(einen) H. machen** (make a) stop. **2** (foot)hold, footing, (point of) support; mainstay; stability; security

hält holds, etc.: *pres of* HALTEN

haltbar *a* **1** strong, sturdy, durable, wear-resistant; (*pred oft*) **h. sein (a)** (*Materials*) wear well, **(b)** (*Foods*) keep well; **h. machen** preserve (foods); (*on packages*) **h. bis 1. 6.** use before June 1; **begrenzt h., nicht gut h.** perishable. **2** stable, lasting, permanent; (*Colors*) fast: **h. machen** stabilize. **3** tenable, defensible. **haltbaren** *vt* (*pp* gehaltbart) preserve, conserve. **Haltbarkeit** *f* **1** durability, wearing quality, wear resistance; (*Foods*) keeping quality. **2** permanence; (*Colors*) fastness. **3** tenability. **Haltbarmachen** *n*, **Haltbarmachung** *f* **1** stabilization. **2** = **Haltbarung** *f* conservation, preservation

Halte-kontakt *m* (*Elec.*) holding contact. **-kraft** *f* **1** (*Phys.*) cohesion. **2** holding (keeping, wearing, lasting) power

halten* (hielt, gehalten; hält) **I** *vt, vi,* **sich h. 1** hold. **2** keep. **II** *vt* **3** maintain. **4** observe (rules, occasions) **—h., was man verspricht** keep one's promises. **5** give (lectures); have (meals); subscribe to (periodicals). **6** (+ **gut, schlecht**) treat (well, badly). **7** detain; stop, bring to a halt. **8** (+ **von**) e.g. **was h. sie davon?** what do they think (*or* what is their opinion) of it? **9: es h. (mit)** deal (with), handle; have an attitude (toward). **10: es h. (mit** + *psn*) take sides (with). **III** *vt* & **sich h.** (+ **für**) take (for), consider (oneself) (to be): **man hielt es für ein Element** they took it for (*or* considered it) an element. **IV** *vi* 11 stop, (come to a) halt. **12** (+ **an sich**) control oneself. **13: h. dafür, daß** . . . maintain, be of the opinion that . . . **14** (+ **auf**) set store (by); attach importance (to). **15** (+ **zu**) take sides (with). **V** *vi* & **sich h.** last, continue, wear (well), remain stable. **VI: sich h. 16** hold out, endure. **17** keep one's balance. **18** contain (restrain, control) oneself. **19** (+ **an**) stick to, go by (principles, etc)

Halte·punkt *m* **1** stop, stopping place. **2** halting point, point of arrest. **3** (*Phys., Metll.*) critical point. **4** (*Metll.*) point of (re)calescence. **5** break, arrest (in a curve)

Halter *m* (-) **1** holder, holding device, (*specif*) handle, clamp, stand, rack. **2** support, bracket. **3** fountain pen. **4** owner. **-arm** *m* supporting arm, bracket

Halte·ring *m* retaining ring

Halterung *f* (-en) holder, etc. = HALTER 1

Halte·stelle *f* (car, bus) stop; station. **-zeit** *f* pause, halt; stopping (retention, residence, dwell) time

haltig I *a* (*Ore*) high-grade. **II** *as sfx, also* **·hältig** containing, -bearing, -ferous: **goldhaltig** gold-containing, containing gold, gold-bearing, auriferous

haltlos *a* **1** unsteady, unstable; insecure; infirm. **2** helpless, lacking firm support. **3** unrestrained. **4** untenable; unfounded, baseless. **5** (*Paper*) tender. **Haltlosigkeit** *f* unsteadiness, instability, etc

halt·machen *vi* stop, (come to a) halt

Haltung *f* (-en) **1** posture, bearing. **2** position, pose, stance. **3** attitude. **4** composure, selfcontrol. **5** well-being. **6** keeping (of cattle, etc)

Halt·ware *f* **1** preserved (*or* canned) food; *pl oft* preserves. **2** (*Pharm.*) confection, electuary, syrup

Häm *n* (*Biol., Chem.*) haem, heme; (*as pfx* = **Blut·**, blood-) h(a)em-. **-alaun** *m* (*Micros.*) hemalum (stain).

Hämatin *n* hematin. **Hämatit** *n* hematite

Hamburger I *m* person from Hamburg, Hamburgian. **II** *a* (of) Hamburg: **H. Weiß**

Hamburg white. **hamburgisch** *a* Hamburgian, (of) Hamburg

Hämin *n* h(a)emin

Hammel *m* (- *or* ⸗) **1** sheep, ram, wether. **2** = **·-fleisch** *n* mutton. **-klauenöl** *n* sheepsfoot oil. **-talg** *m* mutton tallow

hämmerbar *a* malleable; forgeable *cf* HÄMMERN. **Hämmerbarkeit** *f* malleability, forgeability

Hammer·brecher *m* hammer mill. **-eisen** *n* wrought iron. **h–gar** *a* (*Copper*) tough (-pitch)

hämmern 1 *vt/i* hammer, pound. **2** *vt* forge

Hammer·schlacke *f* hammer (*or* forge) scale. **-schlag** *m* **1** hammer (*or* smithy) scale, tri-iron tetroxide. **2** hammer blow. **-werk** *n* hammer mill

hämo· *pfx* h(a)emo- *cf* HÄM. **H–-chininsäure** *f* hemoquinic acid. **-globinurie** *f* hemoglobinuria

Hämolyse *f* hemolysis. **hämolytisch** *a* hemolytic

Hämopyrrol *n* hemopyrrole

hämorrhagisch *a* hemorrhagic

Hamsterei *f* hoarding. **hamstern** *vt/i* hoard

Hand f (⸗e) hand: **von H. (hergestellt)** (produced) by hand; **bei der H. haben** have ready (*or* handy); **beide H⸗e voll zu tun haben** have one's hands full; **einem zur H. gehen** lend sbdy a helping hand; **an H.** (+ *genit* or + **von**) with the aid (of), by means (of), on the basis (of) = ANHAND; **es liegt auf der H.** it is obvious; **H. d(a)rauf!** it's a deal!; **aus freier H.** spontaneously; **das hat H. und Fuß** that has a firm foundation; that is something tangible; **in der H.** (or **in den H⸗en**) **haben** (*oft*) have control of; **von langer H. vorbereiten** prepare well in advance; **Ausgabe letzter H.** last edition supervised by the author; **die letzte H. an etwas legen** put the finishing touches on sthg; **unter der H.** secretly, surreptitiously; **von der H. weisen** brush aside, reject; **es ist zur H.** it is available; (*on messages:*) **zu H⸗en von** . . . attention . . .

Hand· hand, manual: **-arbeit** *f* handiwork; manual labor. **-aufgabe** *f* hand feed(ing).

-ausgabe f pocket edition. **-betrieb** m manual (or hand) operation. **-buch** n handbook, manual. **-druck** m (Tex.) block printing. **-einstellung** f hand (or manual) adjustment

Handel m 1 trade, business, commerce, dealings: **H. treiben** trade, do business, carry on commerce; **im H. (a)** commercially, **(b)** on the market; **in den H. bringen (kommen)** put on (come onto) the market. 2 (business) deal, transaction. 3 business (establishment), store, shop. 4 matter, affair. 5 (usu pl:⸚) quarrel, dispute

handeln I vi 1 act; (**gut, schlecht**) **an einem h.** treat sbdy (well, badly). 2 (oft + **mit**) do business, deal, trade (in). 3 (+ **von**) deal (with), be (about) (a topic). 4 (oft + **um**) bargain, negotiate (for). **II** vt (passive): **gehandelt werden** be traded, be bought and sold. **III: sich h.: es handelt sich um** . . . it is a question of . . .

Handels- commercial, trade: **-amt** n board of trade. **-analyse** f commercial analysis. **-artikel** m article of commerce, commodity. **-benzol** n commercial benzene, benzol. **-betrieb** m trading concern, commercial establishment. **-betriebslehre** f business administration. **-bezeichnung** f trade name. **-bilanz** f 1 balance of trade. 2 commercial balance sheet. **-blatt** n trade journal. **-chemiker** m commercial analytical chemist. **-dünger** m commercial fertilizer. **h-einig** pred a.: **h. werden (mit)** agree on terms (or on a deal) (with). **H-eisen** n merchant iron. **-form** f commercial form (grade, quality). **-freiheit** f freedom of trade. **h-gängig** a commercial, marketable. **H-gärtner** m truck farmer (or gardener). **-gesellschaft** f (trading, business) company. **-gewicht** n trade (or avoirdupois) weight. **-harz** n commercial rosin. **-harzleim** m commercial rosin size. **-kammer** f chamber of commerce. **-kautschuk** m commercial rubber. **-kupfer** n commercial copper. **-laboratorium** n commercial laboratory. **-leute** pl (esp of H–MANN) tradespeople. **-mann** m tradesman. **-marine** f merchant navy. **-marke** f trade mark, brand name. **-messe** f trade

fair. **-minister** m (cabinet) minister of commerce. **-name** m trade name. **-öl** n commercial oil. **-oxid** n commercial (specif antimonious) oxide. **-präparat** n commercial preparation. **-recht** n commercial law. **-silber** n commercial (impure) silver. **-sorte** f commercial variety (or grade). **h-üblich** a customary (usual, standard) commercial (or trade). **H-verein** m commercial association. **-verkehr** m trading, commercial traffic (or dealings). **-ware** f commercial product (or article), merchandise. **-wert** m commercial value. **-zeichen** n trade mark, commercial brand. **-zink** n commercial zinc, spelter. **-zweck** m commercial purpose. **-zweig** m branch of trade (or of commerce)

Hand-fertigkeit f manual skill (or dexterity). **h-fest** a 1 strong, solid, ironclad. 2 downright, outright. **H-feuerung** f hand firing. **-gebläse** n hand blowpipe. **-gebrauch** m 1 manual use. 2 everyday use. **h-gefertigt** a handmade. **H-gelenk** n wrist (joint) —**aus dem H.** easily. **h-gerecht** a handy. **-geregelt** a hand-operated, hand-controlled. **-greiflich** a 1 tangible; plain, obvious. 2 violent. **H-griff** m 1 handle, grip, hilt; lever. 2 (deft) maneuver (of the hand), flick of the wrist, pl oft manipulations, manual techniques —**einen H. tun** lend a hand, lift a finger. **-habe** f handle; (+ **für, gegen,** oft) way of dealing (with)

handhaben vt (pp gehandhabt) handle, manage, deal with; operate (machines); apply (methods, etc); administer (laws). **Handhabung** f handling, management; operation; application; administration. **handhabungs-sicher** a safe for handling

Hand-hebel m hand lever. **h-heiß** a lukewarm, warm to the touch. **H-kurbel** f (Math.) (manual) crank. **-langer** m 1 unskilled worker. 2 underling. 3 accomplice

Händler m (-) dealer, merchant; storekeeper

handlich a handy; manageable. **Handlichkeit** f handiness; manageability

Handlung f (-en) 1 business (establishment); shop, store. 2 act, thing to do; action; ritual; plot (of a story)

Handlungs·weise f way of dealing (acting, doing business); conduct

Hand·lupe f (hand) magnifying glass. **-muster** n hand specimen (or sample). **-probe** f 1 manual test. 2 hand sample

hand·puddeln vt/i puddle by hand

Hand· oft manual: **-rad** n handwheel; spinning wheel. **-regel** f rough rule. **-reibe** f hand grater. **-reichung** f assistance, helping hand. **-scheidung** f manual sorting. **-schrift** f 1 handwriting. 2 manuscript. **-schuh** m glove; mitten. **-schuhkasten** m glove compartment; (Nucl. Phys.) glove box. **-strich** m (Ceram.) hand molding; (as pfx) hand-molded. **-stück** n hand(-size) specimen. **-tuch** n towel. **-umdrehen** n: im H. in no time, quick as a flash. **-verkauf** m over-the-counter sale, (Pharm.) non-prescription sale. **-versuch** m small-scale experiment. **-voll** f (no pl) handful. **-waage** f hand balance. **-wagen** m hand cart. **h-warm** a lukewarm, warm to the touch. **H−wärme** f heat of the hand. **-werfer** m (Mil.) mortar. **-werk** n 1 handicraft, (skilled) trade. 2 hand (or manual) work. **-werker** m 1 handicraftsman, (skilled) craftsman. 2 workman, mechanic. **h-werksmäßig** a mechanical, workmanlike (but uninspired). **H−werkzeug** n tools, instruments (of a craftsman). **-wörterbuch** n concise (desk, handy reference) dictionary. **-wurzel** f carpus, wrist(joint)

Hanf m hemp. **--dichtung** f hemp packing. **-faden** m hemp fiber (or twine). **-korn** n hempseed. **h-korngroß** a hempseed-size(d). **H−nessel** f, **-nesselkraut** n hemp nettle. **-öl** n hempseed oil. **-samen** m, **-saat** f hempseed. **-säure** f linoleic acid. **-seil** n hemp rope

Hang m (⁻e) 1 slope, incline. 2 (usu + zu) tendency, inclination (to, toward)

Hänge f (-n) 1 drying room (house, rack), suspension rack; (Tex., Dye.) ager. 2 (in compds) hanging, suspension, suspended, overhead: **--bahn** f suspended monorail; cable tramway. **-farbe** f (Lthr.)

suspender. **-klemme** f suspension clamp

hängen I (hängte, gehängt) 1 vt hang(up), suspend; (Tex.) age (dyed material). **2: sich h. (an)** hang on, cling (to). II* (hing, gehangen) vi (h or s) 3 hang, be suspended; hang out (or around); (+ **voller,** oft) be loaded (with). 4 (+ **an**) stick, cling (to); be fond (of), be attached (to). 5 be caught, be hung up; be in trouble, (+ **bei,** oft) be in debt (to). 6 (oft + **nach**) lean (toward). 7 be pending, be undecided, hang fire. 8 (Metll.: in a furnace) scaffold.

hängen·bleiben* vi (s) 1 remain (or keep) hanging. 2 (an) stick, adhere, cling (to). 3 get stuck (hung up, caught). 4 jam (up), (Mach.) seize. 5 (Metll.) scaffold.

Hängenbleiben n (Metll.) scaffolding.

hängend p.a 1 hanging, suspended, pendant; sagging, pendulous. 2 (Mach. oft) inverted, underslung; overhead. **hängen·lassen*** vt 1 leave (hanging). 2 let hang (droop, sag)

Hänge·rahmen m suspension frame. **-tropfen** m hanging drop

Hansa·gelb n Hansa yellow

Hansel m (Brew.) return wort

Hantel f (-n) dumbbell. **h--förmig** a dumbbell-shaped. **H−modell** n dumbbell model

hantieren vi 1 be busy, busy oneself. 2 (mit) use, handle; (Mach.) run, operate. 3 (an) work (on), tinker, fiddle (with). **Hantierung** f (-en) 1 activity, work, pl oft doings. 2 handling, operation; tinkering. 3 trade, occupation

hapern vi (imps): **es hapert an** . . . there is a lack of . . .

Härchen n (-) (little) hair; cilium; (Bot., pl) villi

Harfe f (-n) 1 harp. 2 rack, frame

Häring m (-e) herring

Harke f (-n). **harken** vt/i rake

harmlos a harmless, innocent, unsuspecting

harmonieren vi harmonize. **Harmonik** f harmonics. **harmonisch** a harmonic(al), harmonious

Harn m urine. **--absatz** m urinary sediment. **-absonderung** f urinary secretion. **h-abtreibend** a diuretic. **H−analyse** f urinalysis. **-apparat** m urinary apparatus. **h–artig** *a* urine-like. **H–ausscheidung** *f* excretion of urine. **-benzoesäure** *f* hippuric acid. **-blase** *f* (urinary) bladder. **-blasengang** *m* urethra. **-blau** *n* (*Biochem.*) indican, uroxanthin

harnen *vi* urinate

harn·fähig *a* (*Biochem.*) capable of passing into the urine. **H–farbstoff** *m* urinary pigment. **-gang** *m* urinary duct, (*specif*) ureter. **-gärung** *f* fermentation of urine. **-geist** *m* ammonium chloride

harnig *a* uric; urinous

Harnisch *m* (-e) armor; (*incl Tex.*) harness; (*Geol.*) slickenside

Harn·kolloid *n* urinary colloid. **-kraut** *n* diuretic plant. **-lehre** *f* urinology. **-leiter** *m* ureter; catheter. **-messer** *m* urinometer. **-mittel** *n* diuretic. **-niederschlag** *m* urinary deposit. **-oxid** *n* xanthic acid, xanthine. **-phosphor** *m* urinary phosphorus. **-probe** *f* 1 test for urine. 2 urine sample. **-prüfung** *f* urine testing; urine analysis. **-röhre** *f* urethra. **-röhren·** (*in compds*) urethro-, urethral. **-rosa** *n* urorosein. **-ruhr** *f* diabetes. **-salz** *n* 1 urinary salt. 2 microcosmic salt. **-satz** *m* urinary deposit. **h–sauer** *a* urate of. **H–säure** *f* uric acid. **-stein** *m* urinary calculus. **-stoff** *m* urea. **-stoffharz** *n* urea resin. **h–treibend** *a* diuretic. **H–untersuchung** *f* urinalysis. **-vergiftung** *f* uremia. **-waage** *f* urinometer. **-weg** *m* urinary passage. **-zucker** *m* sugar in the urine, urinary glucose

harren *vi* (*usu* + *genit* or + **auf**) wait (for)

hart I *a* 1 hard, tough; firm; hard-boiled (egg). 2 harsh, difficult; rigorous. 3 (*Min.*) refractory. **II** *adv* 4 hard, (+ **an**, *oft*) hard (by), close (to). 5 firmly; harshly; rigorously

härtbar *a* 1 hardenable; temperable. 2 (*Plastics*) thermosetting. **Härtbarkeit** *f* hardenability; temperability; thermosetting ability

Hart·blei *n* hard (antimonial) lead. **-borst** *m*, **-borste** *f* (*Steel*) crack formed during hardening. **-brandstein**, **-brandziegel** *m* hard-burned brick. **-braunstein** *m* braunite.

hart·brennen* *vt* hard-burn

Härte *f* (-n) 1 hardness, toughness; firmness. 2 harshness, difficulty, rigor. 3 (*Steel*) temper. *Cf* HART. **-bad** *n* hardening bath; (*Metll.*) tempering bath. **h–beständig** *a* (*Tex.*) hard-water-resistant, unaffected by hard water. **H–bestimmung** *f* determination of hardness. **-bildner** *m* 1 (*Water*) hard-water salt. 2 (*Metll.*) hardening constituent. **-flüssigkeit** *f* tempering liquid. **-grad** *m* 1 (*genl*) degree of hardness. 2 (*Steel*) (degree of) temper

Hart·eisen *n* hard iron

Härte·kessel *m* autoclave. **-messer** *m* hardness gauge. **-messung** *f* measurement of hardness. **-mittel** *n* hardening (*or Metll.* tempering) agent; (*Phot.*) hardener

härten 1 *vt*, *vi*, **sich h.** harden. 2 *vt* (*Metll.*) temper; case-harden; (*Plastics*) cure

Härte·ofen *m* hardening (*or Metll.* tempering) furnace. **-öl** *n* (*Steel*) hardening oil. **-probe** *f* hardness test. **-prüfer** *m* hardness tester. **-prüfung** *f* hardness test(ing). **-pulver** *n* hardening (tempering *or esp Metll.* cementing) powder

Härter *m* (-) hardener

Härte·riß *m* (*Steel*) crack formed in hardening. **-salz** *n* (*genl*) 1 salt causing hardness. 2 (*Metll.*) carburizing salt. **-skala** *f* (*esp* **Mohssche**) **H.** (Mohs's) scale of hardness. **-stufe** *f* degree of hardness. **-temperatur** *f* hardening temperature. **-tiefe** *f* depth of hardening. **-trog** *m* hardening trough. **-verfahren** *n* hardening (*or Metll.* tempering) process (*or* procedure). **-verzug** *m* distortion due to hardening. **-wasser** *n* hardening (*or* tempering) water. **-wert** *m* hardness value. **-zahl** *f* hardness number

Hart·faser *f* hard-fiber. **-fett** *n* hard (*or* solid) fat. **-feuerporzellan** *n* hard porcelain. **-filter** *n* hard(ened) filter. **-floß** *n* spiegeleisen. **-flügler** *mpl* Coleoptera. **-gas** *n* solid carbon dioxide. **h–gebrannt** hard-burned: *pp* of HARTBRENNEN. **-gefroren** *a* frozen solid. **-gegossen** chill-cast: *pp* of HARTGIESSEN. **-gekocht**, **-gesotten** *p.a* hard-boiled. **H–gestein** *n* hard rock. **h–gewalzt** *a* hard-rolled.

H–gewebe n indurated fabric. **h–ge-zogen** hard-drawn: pp of HARTZIEHEN **hart·gießen*** vt chill-cast, case-harden **Hart·glas** n hard(ened) glass. **-glasbecher** m hard-glass beaker. **-gummi** n hard rubber, vulcanite, ebonite. **-guß** m chill casting; chilled cast iron; case-hardened casting. **-gußasphalt** m stone-filled asphalt. **-harz** n hard (or solid) resin. **-heu** n St. John's wort. **-holz** n hardwood; laminated wood. **-kautschuk** m hard rubber. **-kies** m, **-kobalterz** n, **-kobaltkies** m skutterudite. **-kupfer** n hard copper. **h–leibig** a constipated

Härtling m (Metll.) hard slag, salamander **Hart·lot** n brazing solder. **hart·löten** vt/i hard-solder, braze

Hart·machen n hardening. **-manganerz** n psilomelane; braunite. **-metall** n 1 hard metal, (specif) metal (hard, cemented) carbide, hard pewter. 2 (as pfx, oft) cemented carbide. **-metallegierung** f sintered carbide alloy. **h–näckig** a stubborn, persistent. **H–papier** n (genl) hard paper; (specif) kraft (bank, plastic-impregnated) paper. **-pappe** f fiberboard. **-pech** n hard (or dry) pitch. **-platte** f hardboard. **-porzellan** n hard(-fired) porcelain. **-post** f bank (or typewriter) paper. **-putzgips** m hard-burnt gypsum plaster; Keene's (or Parian) cement. **-riegel** m (Bot.) 1 privet. 2 cornel, dogwood. **-salz** n (impure) hard salt. **h–schalig** a hard-shelled. **H–spat** m andalusite. **-spiritus** m solid alcohol; metaldehyde. **-stahl** m hard steel. **-steingut** n (Ceram.) hard (feldspathic) white ware (earthenware, stoneware). **-trockenöl** n hard-drying oil

Härtung f (-en) hardening; tempering; case-hardening; curing; heat treatment cf HÄRTEN

härtungs·fähig a hardenable; temperable. **H–kessel** m hardening vessel, autoclave; (Oil) hydrogenator. **-kohle** f hardening carbon. **-mittel** n hardening agent. **-verfahren** n hardening process

Hart·verchromung f hard chromium plating. **-wachs** n hard wax. **-waren** fpl hardware. **-wasser** n hard water. **-weizen** m hard (or durum) wheat. **-werden** n hardening. **-wurst** f hard sausage (esp salami). **-zerkleinerung** f crushing of hard material

hart·ziehen* vt (Metll.) hard-draw **Hart·zinn** n hard pewter

Harz I n (-e) resin; rosin; gum. II m Harz Mountains

harz·ähnlich, -artig a resin-like, resinous. **H–alkohol** m resin alcohol. **-austauscher** m ion-exchange resin. **-baum** m resiniferous (or gum-yielding) tree, (specif) pitch pine. **-bildung** f resin formation; (Gasoline) gumming. **-brei** m (Plastics) resin magma. **-cerat** n (Pharm.) rosin cerate. **-elektrizität** f resinous (negative) electricity

harzen I vt 1 extract resin from. 2 resinate, brush with resin. II vi 3 give off resin. 4 get sticky (with resin), gum (up). 5 gather resin.

Harzer a (no endings) Harz(-Mountain) **Harz·essenz** f rosin spirit. **-ester** m resin ester, ester gum. **-esterlack** m gum lacquer. **-fett** n rosin grease. **-fichte** f pitch pine. **-firnis** m resin varnish. **-fleck** m (Paper) resin (or rosin) spot. **-fluß** m 1 resinous exudation, oleoresin. 2 resinosis. **h–flüssig** a gummy, viscous. **-förmig** a resinous, resiniform. **-frei** a rosin-free, nonresinous. **H–galle** f (Wood) resin gall. **-gang** m (Bot.) resin duct. **-gas** n resin gas. **h–gebend** a resin-yielding, resiniferous. **H–gebirge** n Harz Mountains. **h–gebunden** a resin-bonded. **H–gehalt** m resin (or rosin) content. **-geist** m resin spirit, pinolin. **h–geleimt** a (Paper) rosin-sized. **H–gerbung** f resin tanning. **-geruch** m resinous odor. **-glanz** m resinous luster. **h–haltig** a resiniferous, containing resin. **H–holz** n resinous wood

harzig a resinous: **h. machen** resinify **Harz·** resin(ous), rosin: **-karbollösung** f (Paper) solution of rosin in carbolic acid. **-kernseife** f rosin (curd) soap. **-kiefer** f pitch pine. **-kitt** m resinous cement. **-kocher** m (Paper) rosin boiler. **-kohle** f bituminous coal. **-körper** m resinous substance. **-lack** m resin varnish; res-

in(ous) lake. **-leim** m (*Paper*) rosin size.
-leimverseifung f (*Paper*) rosin-size
cutting. **h–liefernd** a resin-yielding. **H–
masse** f resinous mass (*or* composition);
(*Paper*) rosin size. **-milch** f suspension of
resin (*or* rosin). **-naphtha** f, n resin oil.
-öl n rosin oil. **-pech** n resin(ous) pitch,
(common) rosin. **h–reich** a rich in resin
(*or* rosin), highly resinous. **H–reserve** f
(*Tex.*) resin resist. **-röhre** f resin duct.
-saft m resinous juice (*or* sap). **-salbe** f
(*Pharm.*) rosin cerate. **-salz** n resinate.
h–sauer a resinate of. **H–säure** f resin
(*or specif* resinic) acid. **-seife** f resin (*or*
rosin) soap; resinate. **-sikkativ** n rosin
drier. **-spiritus** m rosin (*or* resin) spirit.
-spur f trace of resin (*or* rosin). **-stippe** f
(*Paper*) rosin speck (*or* spot). **-stoff** m resinoid, resinous substance. **-talgseife** f
tallow-rosin soap, yellow household soap.
-tanne f pitch fir (*or* pine). **-teer** m resinous tar. **-wasser** n resin water. **-zahl** f
resin (*or* rosin) number. **-zement** n resinous cement

Haschee n (-s) hash (without potatoes)
haschen I vt 1 catch. 2 chase. **II** vi 3 (**nach**)
snatch, grab (at); strive, fish (for). 4 take
hash(ish)

haschieren vt hash, mince, make hash of
(meat)

Haschisch n hashish, Indian hemp
Hase m (-n,-n) hare; (*fig*) (scared) rabbit
Hasel I m (-) (*Zool.*) dace. **II** f (-n) (*Bot.*)
hazel (bush). **-nuß** f hazelnut, filbert. **h–
nußbraun** a hazel(nut brown). **H–öl** n
hazelnut oil. **-wurz** f hazelwort, asarum,
ascarabacca. **-wurzöl** n oil of asarum
Europaeum

Hasen·klee m rabbit-foot clover; hare's-
foot trefoil; wood (*or* sheep's) sorrel. **-kohl**
m wood (*or* cuckoo) sorrel; nipplewort.
-maus f chinchilla. **-ohr, öhrlein** n (*Bot.*)
hare's ear. **-schwanz** m hare's-tail grass
Haspe f (-n) 1 hasp, hinged clasp. 2 staple
Haspel f (-n) 1 hasp. 2 hank, skein, bobbin.
4 winder, winch. 5 (*Lthr.*) paddle wheel.
-äscher m (*Lthr.*) paddle lime. **-farbe** f
(Lthr.) paddle liquor

haspeln vt/i reel, wind (on, off); hoist
Haß m hate, hatred. **hassen** vt/i hate
häßlich a ugly; nasty, spiteful

Hast f haste. **hastig** a hasty
hat has: *pres of* HABEN. **hatte** had: *past of*
HABEN. **hätte** *sbjc of* HABEN 1 would have.
2 (*in if-clause & in indir quot, usu*)
had

Haube f (-n) 1 hood. 2 dome, cupola,
cowl(ing). 3 cap. 4 helmet. 5 hair drier.
6 crest, tuft (on birds). 7 second stomach
(of ruminants). **Hauben·lerche** f (*Zool.*)
crested lark; (*Ceram.*) cone protector

Haubitze f (-n) howitzer. **Haubitz·zünder**
m howitzer fuse

Hauch m (-e) 1 breath. 2 breath of air. 3
aura, atmosphere. 4 whiff. 5 trace, hint,
tinge. 6 thin layer, film, bloom (on fruit).
7 haze, condensed moisture. **h–·dünn** a
paper-thin, filmy. **hauchen** vt/i breathe

Haue f (-n) hoe, mattock, pick
hauen (*pp* gehaut *or* gehauen) **I** vt 1 beat;
hit, strike, punch; pound. 2 smash. 3 chop,
cut (*incl* trees). 4 bang, slam. 5 fling.
6 carve, sculpt. 7 hoe. **II** vi strike (out), hit.
Hauer m (-) 1 chopper, (tree-)cutter;
miner, pickman. 2 wine grower. 3 boar.
4 boar's tusk. 5 (*Tools*) punch. 6 cleaver;
hunting knife

häufbar a (*Stat.*) cumulative cf HÄUFEN
Häufchen n (-) little pile (*or* heap); clump,
cluster

Haufe(n) m (..ens, ..en) 1 pile, heap, stack
—**über den Haufen werfen** (*or* **rennen**) upset, knock over. 2 mass, (whole)
lot. 3 bunch, clump, cluster, batch. 4
(*Brew.*) couch, floor, piece. **häufen** 1 vt &
sich h. pile up, accumulate. 2 vt heap up,
amass. 3: **sich h.** increase (in frequency);
spread. *Cf* GEHÄUFT

Haufen·führen n (*Brew.*) couching, flooring. **-röstung** f (*Metll.*) heap roasting.
-verkohlung f charcoal burning in long
rectangular piles. **h–weise** a, adv in piles
(heaps, bunches, droves, large quantity);
frequent(ly); abundant(ly). **H–wolke** f
cumulus cloud

häufig a frequent; widespread; adv oft often. **Häufigkeit** f frequency, frequent occurrence. **Häufigkeits·kurve** f frequency curve

Häufung f (-en) 1 piling (*or* heaping) up,
amassment; accumulation —**in der H.** in
the aggregate, in quantity. 2 increase.

3 frequent occurrence. *Cf* HÄUFEN

Hauf·werk *n* (*Min.*) crude ore, run of the mine; (*Geol.*) (heap of) debris; (*Concrete, etc*) aggregate

Hau·hechel *f* (*Bot.*) restharrow

Haupt 1 *n* (-er) head; chief. **2** (*as pfx*) main, chief, principal, most important: **-abteilung** *f* main (*or* principal) division. **-achse** *f* main (*or* principal axis. **-agens** *n* principal agent. **-amt** *n* main office. **-anteil** *m* main portion. **-bedingung** *f* chief condition. **-bedürfnis** *n* chief requirement. **-bestandteil** *m* chief constituent. **-bindung** *f* chief (*or* principal) bond. **-brennpunkt** *m* main focus. **-erzeugnis** *n* chief product. **-fach** *n* major (subject); chief line of work. **-farbe** *f* primary (*or* principal) color. **-form** *f* principal form; master mold. **-gärung** *f* principal (*or* primary) fermentation. **-gesetz** *n* main (chief, fundamental) law. **-gitter** *n* (*Cryst.*) parent lattice. **-kette** *f* principal chain. **-klasse** *f* chief class. **-lauf** *m* (*Distil.*) main fraction. **-leitung** *f* (*incl Elec.*) main (pipe, cable). **-linie** *f* main (*or* chief) line. **-mann** *m* (*pl* -leute) captain; chieftain. **-menge** *f* principal amount, main mass, bulk. **-merkmal** *n* main feature, chief characterstic. **-nährstoff** *m* chief nutrient (*or* food). **-nenner** *m* (lowest) common denominator. **-ölbad** *n* (*Dye.*) white steeping. **-patent** *n* master patent **-pflaster** *n* (*Pharm.*) opium plaster. **-platz** *m* main square; most important place, center. **-post** *f*, **postamt** *n* main (*or* general) post office. **-punkt** *m* **1** main point. **2** (*Opt.*) cardinal point. **-quantenzahl** *f* principal (*or* total) quantum number. **-quartier** *n* headquarters. **-quelle** *f* main source. **-register** *n* general index. **-reihe** *f* principal series. **-rohr** *n*, **-röhre** *f* main (pipe *or* tube). **-rolle** *f* leading (*or* major) part (*or* role) *cf* ROLLE. **-sache** *f* most important thing, main issue, essential point **—in der H.** in the main, chiefly. **h–sächlich** *a* principal, essential; *adv usu* mainly, chiefly. **H–satz** *m* **1** basic law, proposition, main theorem. **2** main (*or* independent) clause. **-schalter** *m* (*Elec.*) main (*or* master) switch. **-schlagader** *f* main artery; aorta.

-schluß *m* (*Elec.*) main circuit; (*as pfx*) series-wound. **-schlüssel** *m* master key. **-schnitt** *m* principal section. **-schwingung** *f* principal vibration. **-serie** *f* principal series. **-sicherung** *f* (Elec.) main fuse (*or* cut-out). **-spirale**, **-spule** *f* primary coil. **-stadt** *f* capital (city). **-strom** *m* (*Elec.*) primary current. **-stück** *n*, **-teil** *m* main piece (part, component), principal part. **-typus** *m* principal type. **-unterschied** *m* chief difference. **-ursache** *f* main cause (*or* reason). **-valenz** *f* primary valence (in Werner theory). **-verkehrsstunden** *fpl* rush hours. **-versammlung** *f* general meeting; main meeting. **-verwendungsgebiet** *n* chief field of use. **-vorkommen** *n* chief occurrence. **-wert** *m* chief value. **-wirkung** *f* chief effect (*or* action). **-wissenschaft** *f* principal (*or* fundamental) science. **-würze** *f* (*Brew.*) first wort. **-zahl** *f* cardinal number. **-zug** *m* main feature. **-zweck** *m* main purpose

Haus *n* (-er) **1** house, home; building; (*esp as sfx*) shop: **außer H.** out, away from home; **Verkauf außer H.** sold to take out; **nach H—e** home (*as destination*); **zu H—e** at home; **von H. aus** by nature, originally, to begin with; **H—er bauen auf . . .** rely on, trust . . . **2** (snail) shell. **3** (*as pfx oft*) domestic: **-arznei** *f* home (*or* domestic) remedy. **-aufgabe** *f* homework; home assignment. **-besitzer** *m* home owner; landlord. **-bockkäfer** *m* longhorn beetle. **-brand** *m* domestic (*or* home-heating) fuel. **-brennöl** *n* home-heating oil

hausen *vi* **1** live, subsist. **2** carry on extravagantly. **3** rage, play havoc

Hausen *m* (-) beluga. **-blase** *f*, **-glas** *n* ichthyocol, fish glue, isinglass

Haus·gebrauch *m* home (*or* domestic) use. **-geflügel** *n* domestic fowl(s). **h–gemacht** *a* home-made. **H–gerät** *n* home furnishings, household utensils. **-halt** *m* **1** household. **2** house(keeping). **3** budget. **4** (*Med.*) metabolism. **5** (*in compds, oft*) economy: **Wasserhaushalt** water economy (of an area)

haus·halten* *vi* (*oft* + **mit**) be economical (with), husband (e.g. one's resources). **Haus·hälter** *m* housekeeper; manager.

h–hälterisch *a* economical, thrifty
Haushalts·farbe *f* household paint (color, dye). **-seife** *f* household soap
häuslich *a* 1 domestic. 2 economical, thrifty
Hausmacher· (*in compds*) home-made; homespun
Haus·mittel *n* household (*or* home) remedy. **-müll** *m* household refuse, garbage. **-schwamm** *m* dry rot. **-seife** *f* household (*or* home-made) soap
Hau·stein *m* ashlar, freestone
Haus·tier *n* domestic animal. **-tür** *f* front (*or* street) door. **-wanze** *f* bedbug. **-wäsche** *f* domestic wash(ing). **-wirtschaft** *f* housekeeping, home management (*or* economics). h–**wirtschaftlich** *a* domestic, household. **H–wurz** *f* houseleek
Haut *f* (-̈e) skin; hide; membrane; film, pellicle; bloom; crust; coating — (*Physiol.*) **durchsichtige H.** cornea; **harte H.** sclerotic coat; (*in compds, oft*) cutaneous, membranous: **-arzt** *m* dermatologist. h–**ätzend** *a* vesicant, skin-corroding. **H–ausschlag** *m* skin eruption. **-bildung** *f* skin (*or* film) formation; (*Paint*) skinning. **-blöße** *f* pelt. **-bräune** *f* 1 suntan. 2 (*Med.*) croup
Häutchen *n* (-) membrane, pellicle, film
Haut·creme *f* skin cream. **-drüse** *f* cutaneous gland
häuten 1 *vt* skin, flay. 2: **sich h.** shed its skin, molt; peel
Haut·entgiftungsmittel *n* skin decontaminant. **-erkrankung** *f* skin disease. **-farbe** *f* skin color. **-farbstoff** *m* skin pigment. **-gift** *n* (*Mil.*) skin poison, vesicant
häutig *a* membranous, cutaneous; covered with (a) skin; (*esp in compds*) skinned: **dünnhäutig** thin-skinned (*lit.*)
Haut· skin, cutaneous, membranes: **-krankheit** *f* skin disease. **-krebs** *m* skin cancer, epithelioma. **-krem** *f* skin cream. **-lehre** *f* dermatology **-leim** *m* hide glue (*or* size). h–**los** *a* skinless, dermal. **H–mittel** *n* skin preparation. **-öl** *n* skin oil (*or* lotion). **-pflege** *f* skin care. **-pulver** *n* hide powder. **-reaktion** *f* skin (*or* cutaneous) reaction. **-reibung** *f* skin friction. h–**reinigend** *a* skin-cleansing. **H–reiz** *m*,

-reizung *f* skin irritation. **-salbe** *f* skin ointment. **-schicht** *f* (*Physiol.*) dermal layer; (*Bot.*) periderm. **-substanz** *f* hide substance. **-talg** *m* sebaceous matter. **-tang** *m* edible dulse (*or* seaweed).
Häutung *f* (-en) skinning, flaying; shedding of the skin, molting, peeling, desquamation *cf* HÄUTEN
Haut·verhinderungsmittel *n* (*Paint*) anti-skinning agent. **-wirkung** *f* (*Elec.*) skin effect. **-wolle** *f* skin wool, pulled wool
H.D. *abbv* (Hochdruck) high pressure
heb·ärztlich *a* obstetric(al)
Hebe· lifting, raising, hoisting: **-arm**, **-baum** *m* lifting lever, crowbar. **-bock** *m* (lifting) jack. **-daumen** *m* cam, tappet. **-eisen** *n* crowbar. **-kraft** *f* lifting power, leverage
Hebel (-) lever. **-beziehung** *f* (*Phase Diagrams*) lever rule. **-kraft** *f* leverage. **-schalter** *m* (*Elec.*) knife (*or* lever) switch. **-waage** *f* beam balance. **-werk** *n* system of levers. **-wirkung** *f* lever action, leverage
Hebe·magnet *m* lifting (*or* loading) magnet
heben*** (hob, gehoben) **I** *vt* 1 raise, lift, elevate. 2 lift off, remove. 3 dig up. 4 improve, enhance, promote. **II: sich h.** 5 rise, go up. 6 improve, be enhanced, be promoted. 7 (*recip*) cancel out. *Cf* GEHOBEN
Hebe·pumpe *f* lift(ing) pump. **-punkt** *m* fulcrum
Heber *m* (-) 1 siphon. 2 pipette, dropper; syringe. 3 lift(er), lever; (*Mach.*) jack. 4 (*Anat.*) levator (muscle). 5 (*Elec.*) salt bridge. **-barometer** *n* siphon barometer. **-haarrohr** *n* capillary siphon
hebern *vt* siphon; pipette
Hebe·rohr *n* siphon (tube, pipe)
Heber·pumpe *f* siphon pump. **-rohr** *n* siphon (tube, pipe). **-säuremesser** *m* syringe hydrometer for acids. **-schraube** *f* jack screw. **-schreiber** *m* siphon recorder
Hebe·vorrichtung *f* lift(er), lifting device; lever. **-werk**, **-zeug** *n* lifting (*or* hoisting) device, e.g.: hoist, block and tackle, winch, jack, etc
Hebling *m* (-e) (*Mach.*) tappet
Hebung *f* (-en) 1 raising, lifting, etc *cf*

HEBEN; improvement, enhancement, promotion. 2 (*Geol.*) elevation

hecheln vt (*Tex.*) hackle, comb

Hecht m (-e) 1 (*Zool.*) pike. 2 (*Lthr.*) back, butt

Heck n (-e) stern, rear (end), tail cf HECKE. **-antrieb** m rear-engine drive

Hecke f (-n) I 1 (*Bot.*) hedge. 2 (*Zool.*) brood, litter; breeding (time, place). II pl of HECK

Hecken·winde f large bindweed

heck·getrieben a rear-engine driven

Hede f tow; oakum

Hederich m hedge mustard, charlock

Heer n (-e) army; host, horde

Heeres·gerät n army equipment (or matériel)

Heer·straße f (military) highway

Hefe f (-n) 1 yeast; (*Brew.*, *oft*) barm. 2 dregs, sediment; scum. **--apparat** m (*Brew.*) yeast propagator. **-bier** n rest (or barm) beer. **-block** m block of yeast. **-brühe** f liquid yeast. **-decke** f yeast head. **-ernte** f crop of yeast. **-form** f yeast mold (or press). **-gabe** f (*Brew.*) quantity for pitching. **-geben** n (*Brew.*) pitching, adding (of) yeast to the wort. **-gut** n leaven. **-keim** m yeast germ. **-mehl** n yeast powder

hefen·ähnlich a yeasty, yeast-like

Hefe·nährmittel n yeast nutrient. **-nahrung** f yeast food

Hefen·art f type of yeast. **h--artig** a yeasty, yeast-like. **H--artigkeit** f yeastiness. **-keim** m yeast germ. **-pflanze** f, **-pilz** m yeast plant (or fungus). **-pulver** n yeast powder. **h--trüb** a yeasty, muddy (from yeast)

Hefe· (of) yeast: **-nukleinsäure** f ribonucleic acid, nucleic acid of yeast. **-pilz** m yeast fungus (or plant), **-preßsaft** m yeast press juice. **-rasse** f type of yeast. **-reinkultur, -reinzucht** f pure yeast culture. **-satz** m yeast sediment. **-stück** n sponge dough, moistened yeast cake. **-teig** m yeast dough. **-trieb** m (*Brew.*) yeast head. **-vermehrung** f yeast multiplication (or growth). **-zucht, -züchtung** f yeast culture

hefig a yeasty; dreggy cf HEFE

Hefner·kerze f Hefner candle

Heft n (-e) 1 handle, grip; reins. 2 notebook;

(stitched or stapled) book(let), pamphlet; issue, number (of a periodical); part (of a serially published book)

heften 1 vt fasten; stick; stitch, baste; staple cf GEHEFTET. **2: sich h.** cling

heftig a violent, severe, heavy. **Heftigkeit** f violence, severity, force

Heft·klammer f 1 paper clip. 2 staple. **-maschine** f 1 stapler. 2 (book) stitching machine. **-pflaster** n adhesive plaster (or tape). **-schweißen** n stitch-welding. **h--weise** adv in parts, serially cf HEFT 2. **H--zwecke** f thumb tack

hegen vt 1 preserve, care for, look after. 2 harbor feelings of . . . ; hold, have (opinions, emotions)

Hehl n: **kein H. machen (von, aus)** make no secret (of); **ohne H.** without secrecy, frankly. **hehlen** 1 vt/i conceal, abet (a crime). 2 receive stolen goods, fence

hehr a sublime, magnificent

Heide 1 f (-n) heath; heather. 2 m (-n,-n) heathen; pagan. **h--ähnlich** a heathlike, ericaceous. **H--korn** n buckwheat. **-kraut** n heather

Heidel·beere f (European) blueberry

Heide·sandboden m sandy heath soil. **-torf** m (heath) peat

Heidin f (-nen) heathen; pagan

heikel a delicate; sensitive, touchy; fussy

heil a 1 whole, undamaged; intact, (safe and) sound. 2 cured, well: **h. werden** be cured, get well, be mended; **h. machen** cure, mend

Heil n 1 welfare, well-being. 2 salvation. 3 (good) fortune. 4 (in compds, usu) healing, medi(ci)nal, curative: **--anstalt** f sanatorium, hospital

heilbar a curable

heil·bringend a beneficial, salutary. **H--brunnen** m mineral well (or spring)

Heilbutt m (-e) halibut

heilen 1 vt cure. 2 vt,vi(s) heal (up)

heil·fähig a curable. **H--formel** f (*Med.*) prescription

heilig a holy, sacred

Heiligen·harz n guaiacum (resin). **-holz** n holy (or guaiacum) wood (lignum vitae). **-kraut** n lavender cotton. **-stein** m (*Pharm.*) lapis divinus, cuprum aluminatum

Heil· healing, curative, therapeutic; medicinal: **-kraft** f curative (or therapeutic) power. **h–kräftig** a curative, therapeutic. **H–kraut** n medicinal herb, officinal plant. **-kunde, -kunst** f art of healing, medicine, therapeutics. **-mittel** n remedy, therapeutic. **-mittellehre** f pharmacology **-pflanze** f medicinal (or officinal) plant. **-pflaster** n healing (or medicated) plaster. **-quelle** f medicinal (or mineral) spring. **-salbe** f medicinal (or therapeutic) ointment

heilsam a wholesome, beneficial; therapeutic

Heil·serum n antitoxic serum. **-stoff** m remedy, therapeutic. **-stofflehre** f pharmacology, pharmacy

Heilung f healing, cure. **Heilungs·vorgang** m healing process

Heil·verfahren n therapy, method of healing. **-wasser** n medicinal water. **-wert** m therapeutic value. **-wesen** n (field of) medicine, therapeutics. **-wirkung** f curative (or therapeutic) effect. **-wissenschaft** f science of medicine

heim adv & sep pfx home(ward). **Heim** n (-e) home

Heimat f home(land), native country (town, etc); habitat. **--vertriebene** m, f (adj end-gs) refugee (from former Ger. territories)

Heimchen n (house) cricket

heimisch a 1 domestic. 2 native, indigenous. 3 acclimatized. 4 adv, pred, oft at home

Heim·lauf mj return course (or stroke)

heimlich a secret; hidden; stealthy. **Heimlichkeit** f (-en) secrecy; stealth; secret

heim·suchen vt strike, afflict

Heirat f (-en) marriage, **heiraten** vt marry

heischen 1 vt demand. 2 vi (nach) ask (for)

heiser a hoarse

heiß a hot, torrid; ardent. **H–·abschneider** m hot catchpot. **-bearbeitung** f hot-processing. **-blasegas** n hot blast gas. **-blasen** n (Water gas) (air) blow. **h–blütig** a hot-blooded. **-brüchig** a (Iron) hot-short

heiß·brühen vt scald

Heiß·dampf m superheated steam

Heiße f (Metll.) charge

heißen* (hieß, geheißen) **I** vt 1 mean —**das heißt** that is, i.e. cf SOVIEL. 2 call (by a name). **II** vi 3 be called, be named: **wie h. sie?** what are they called?, what is their name? 4: **es heißt** (a) they say, it is said; (b) it says (e.g. on this page); (c) (+ inf, e.g) **nun heißt es langsam vorgehen** now it is a matter of proceeding slowly. **III** v aux (+ inf) (pp heißen) order, ask (to do)

heiß·erblasen a (Iron) hot-blast. **-gar** a (Metll.) too hot; (Iron) kishy. **H–gasschweißen** n hot-gas welding. **h–gesiegelt** p.a heat-sealed. **-gradig** a (Ores) difficult to fuse. **H–kühlung** f cooling by evaporation (with liquid, esp glycol). **-lauf** m (Metll.) hot run

heiß·laufen* vi (s) (Mach.) overheat, run hot

Heiß·leiter m hot conductor, thermistor. **-lösen** n hot dissolving. **-lötstelle** f hot-solder junction. **-luftbad** n hot-air bath. **-lufttrocknung** f hot-air drying. **-prägen** n heat embossing

heiß·pressen vt hot-press

Heiß·siegelklebstoff m heat-sealing adhesive

heiß·siegeln vt heat-seal

Heiß·sprühverfahren n hot(-air) spray drying. **-trub** m (Beer) hot break. **h–umstritten** a hotly disputed. **-verformbar** a malleable (or moldable) when hot. **H–verlösung** f thermal dissolution. **-verschweißen** n (Plastics) heat sealing. **-wasserbehälter** m hot-water tank. **-wasserheizung** f hot-water heat(ing). **-werden** n heating (up), becoming hot. **-wind** m (Metll.) hot blast. **-windofen** m hot-blast furnace

Heister m (-), f (-n) sapling

·heit sfx (usu forms nouns from adj.) **frei** free: **Freiheit** freedom, liberty

heiter a serene; cheerful; clear, fair (weather)

Heiz· heat(ing): **-anlage** f heat(ing) (in a bldg, etc); heating unit (or plant). **-apparat** m heating apparatus. **-bad** n heating bath. **-band** n heating tape (or band)

heizbar a heatable

heiz·beständig a heat-resistant. **H–**

dampf *m* (heating) steam. **-draht** *m* heating wire, hot filament

Heize *f* (*Steel*) charge of pig (in smelting)

Heiz-effekt *m* heating (*or* calorific) effect. **-einrichtung** *f* heating system (*or* installation)

heizen 1 *vt/i* heat, light a fire (in). **2** *vt* stoke. **Heizer** *m* (-) **1** fireman, stoker, furnace operator. **2** heater. **--stand** *m* firing floor

Heiz- heat(ing): **-faden** *m* heating filament. **-fähigkeit** *f* heating capacity. **-fläche** *f* heating surface. **-gas** *n* fuel (*or* heating) gas. **-haube** *f* heating mantle. **-gerät** *n* **1** heating appliance, heater. **2** radiator. **-kammer** *f* heating chamber; firebox. **-kessel** *m* (heating) kettle, cauldron. **2** boiler. **-kissen** *n* heating pad. **-körper** *m* **1** heating element. **2** heating appliance, heater. **3** radiator. **-kraft** *f* heating (*or* calorific) power. **-kraftwerk** *n* heating (power) plant (*or* unit). **-kranz** *m* ring burner. **-lampe** *f* heat(ing) lamp. **-loch** *n* fire door, stoke hole. **-luftmotor** *m* hot-air motor. **-mantel** *m* heating jacket (*or* mantle). **-material, -mittel** *n* fuel. **-ofen** *m* heating furnace (oven, stove). **-öl** *n* fuel (*or* heating) oil. **-platte** *f* hotplate. **-raum** *m* **1** combustion chamber, firebox. **2** furnace (*or* boiler) room. **3** fireplace. **-rohr** *n*, **-röhre** *f* **1** heating pipe (*or* tube). **2** fire tube. **-röhrenkessel** *m* fire-tube boiler. **-schlange** *f* heating coil. **-schrank** *m* heating chamber, oven. **-spirale, -spule** *f* heating coil. **-stoff** *m* fuel **-streifen** *m* heating tape. **-strom** *m* (*Elec.*) heating current. **-tisch** *m* (*Micros.*) heating (*or* hot) stage

Heizung *f* (-en) **1** heat, (central) heating system. **2** radiator; heater, heating unit. **3** (*Engg., Metll.*) firing

Heizungs-anlage *f* heating plant. **-vorrichtung** *f* heating appliance (*or* device)

Heiz- heat(ing): **-verlust** *m* loss of heat. **-vermögen** *n* heating capacity. **-vorgang** *m* heating process. **-vorrichtung** *f* heating device (*or* appliance). **-walze** *f* heated drying roll. **-wert** *m* heating (*or* calorific) value: **oberer (unterer) H.** gross (net) calorific value. **-wertmesser** *m* calorimeter. **-wicklung** *f* heating coil. **-wider-**

stand *m* heating resistance. **-wirkung** *f* heating effect. **-wirkungsgrad** *m* heating efficiency. **-zeit** *f* heating period. **-zug** *m* heating flue. **-zweck** *m* (*usu pl*) heating purpose(s)

Hektan *n* hectane

Hektar *n* (-e) hectare (10000m^2 = 2.47 acres)

Hektograph *m* (-en, -en) (spirit) duplicator

Held *m* (-en, -en) hero

helfen* (half, geholfen; hilft) *vi* (*dat*) help, aid, assist; be of use (to). **Helfer** *m* (-), **Helferin** *f* (-nen) helper, aid(e), assistant

Helio-echtrot *n* (*Dyes*) sunfast (*or* heliofast) red. **-farbstoff** *m* helio dye

Heliogen-farbstoff *m* phthalocyanin

Heliotropin *n* piperonal

hell *a* **1** bright, light (*opp of* dark). **2** clear, lucid. **3** loud, resounding. **4** clever. **5** great. **6** utter, sheer. **--blau** *a* light blue. **-braun** *a* light brown. **-brennend** *a* (*Ceram.*) light-burning

Helle *f* **1** brightness, light(ness); luminosity. **2** clarity

hellen *vt* & *sich h.* lighten; brighten; clarify

helleuchtend *a* brightly luminous

hell-farbig *a* light-colored. **H--feldbeleuchtung** *f* bright-field illumination. **h--gelb** *a* light yellow. **-getönt** *a* light-colored. **-grau** *a* light gray. **-grün** *a* light green

hellicht *a:* **am h--en Tag** in broad daylight

Helligkeit *f* brightness; light(ness); clearness; luminosity, light intensity. **Helligkeits-wert** *m* luminosity value

Hell-kammer *f* camera lucida. **h--matt** *a* semidull, semimat. **-rot** *a* light red. **-rotglühend** *a* bright-red-hot. **H--rotglut** *f* bright red heat (*appx* 950°C). **h--rotwarm** *a* bright-red-hot. **H--tran** *m* clear fish oil. **h--weiß** *a* clear (*or* bright) white

Helm *m* (-e) **1** helmet. **2** helm. **3** (still) head; hood. **4** cupola, dome. **--kolben** *m* distilling flask. **-rohr** *n*, **-schnabel** *m* beak, nose (of a still)

Hemd *n* (-en) **1** shirt. **2** (*Blast Furnace*) shell

Hemellit-säure *f* hemellitic acid

Hemi/eder *n* hemihedron, hemihedral crystal. **Hemiedrie** *f* hemihedrism. **hemiedrisch** *a* hemihedral
Hemimellit·säure *f* hemimellitic acid
Hemipin·säure *f* hemipinic acid
hemitrop *a* (*Cryst.*) hemitrope, twinned
Hemme *f* (-n) **1** hindrance, restraint, obstruction. **2** brake. **hemmen** *vt* **1** stop, arrest. **2** retard, brake; hinder; obstruct, impede. **3** inhibit. **Hemmnis** *n* (-se) hindrance, obstacle, impediment
Hemm·schuh *m* **1** brake shoe. **2** hindrance, obstacle. **-stoff** *m* inhibitor
Hemmung *f* (-en) **1** stopping, arresting, etc *cf* HEMMEN. **2** hindrance; obstruction; impediment; inhibition
Hemmungs·körper *m* inhibitor, decelerator. **h–los** *a* uninhibited, uncontrolled
Heneikosan *n* heneicosane
Hengst *m* (-e) stallion
Henkel *m* (-) handle. **--schale** *f* casserole
Henne *f* (-n) hen
her *adv & sep pfx* (*Usage varies as to separate or attached spelling with verbs*) **I** here, to this place, in this direction (i.e. toward the speaker) **1** (*implies bringing*): **Wasser h.!** bring some water! **2** (*implies surrender*) **h. damit!** hand it over! **3: es ist drei Tage h., daß** . . . it has been three days since (*or* that) . . . **4: das ist** (*or* **damit ist es**) **nicht weit h.** that does not amount to much. *Cf* HIN. **II 5** (*moving*) along: **hinter (neben, vor) ihnen hergehen** walk along behind (beside, in front of) them. **III** from (*esp in comb. with* **wo, da, dort,** *oft sfx*): **wo fließt es h.?, woher fließt es?** where does it flow from?; **von dort h.** from there. *Cf* DAHER, WOHER; BISHER, HIE(R)HER, JEHER, WEITHER. **IV** (*pfx on other sep pfxs*): **here-in!** come in! *Cf* HERAB, HERAN, HERVOR etc
herab *adv & sep pfx* down(ward). **·drücken** *vt* press down, depress; force down; reduce. **·fließen*** *vi* flow (*or* run) down. **-hängend** *p.a* hanging (down), dangling, pendant. **·lassen* 1** *vt* let down, lower. **2: sich h.** condescend. **·mindern** *vt* reduce, dminish. **·rieseln** *vi* trickle down. **·setzen** *vt* **1** lower, reduce. **2** disparage, downgrade. **H–setzung** *f* lowering, reduction; disparagement, downgrading. **h–**

·sinken* *vi* (s) drop (*or* sink) down. **·steigen*** *vi* (s) step (*or* climb) down, descend. **-transformieren** *vt* (*Elec.*) step down. **·tropfen** *vi* drop (*or* drip) down
Heraldik *f* heraldry
heran *adv & sep pfx* near, close; up (to): **·bilden** *vt* bring up, educate, train. **·gehen*** *vi*(s) (*usu* + **an**) get close to, approach; start on, tackle (a job). **·kommen*** *vi*(s) come up (to), approach; reach. **·machen: sich h. (an)** get to work (on); come up (to). **·nahen** *vi*(s) (**an**) approach: **·reichen** *vi* (**an**) come (*or* reach) up (to). **·wachsen*** *vi*(s) grow up — **die H—den** the rising generation, the young people. **·ziehen* I** *vt* **1** pull up (close). **2** call (*or* draw) upon (for service), make use of; cite, quote. **3** consider. **4** bring up, train. **II** *vi*(s) approach
heraus *adv & sep pfx* out, forth; published, known: **zur Tür h.** out the door. **h–·arbeiten 1** *vt* carve; work out; bring out, show. **2: sich h.** work one's way out. **·bekommen*** *vt* **1** get (sthg) out, remove; elicit. **2** find out; figure out; solve. **3** get change. **·bilden 1** *vt* develop, elaborate; create. **2: sich h.** arise, evolve. **·bringen* 1** bring out. **2** issue, publish. **3** get (sthg) out, remove. **4** elicit; find out; reveal. **5** produce (sound); say (words). **·destillieren** *vt* distill out. **·drehen** *vt* unscrew. **·dringen*** *vi*(s) break out. **·finden* 1** *vt* find out. **2** *vi & sich h.* find one's way out. **·fordern** *vt* challenge, provoke. **H–gabe** *f* **1** editing, editorship; issuance, publication. **2** handing over, deliverance. **h–·geben* I** *vt* **1** hand out. **2** hand over, deliver. **3** issue, publish. **4** edit. **II** *vt/i* give change. **H–geber** *m* editor; publisher. **h–·greifen*** *vt* pick out, single out; cite. **·haben*** *vt* e.g. **sie haben es heraus** (*1*) they have got(ten) it out; (*2*) they have found (figured, puzzled) it out. **·heben* I** *vt* **1** lift out. **2** distinguish, emphasize. **II: sich h.** stand out. **3** (*Math.*) be eliminated, cancel out. **·holen** *vt* (*usu* + **aus**)

get (take, draw) out (of), extract (from). **·kommen*** *vi*(s) **1** come out, get out; appear, be published, be issued. **2: es kommt aufs gleiche heraus** it all amounts to the same thing. **·laugen** *vt* leach out. **·lösen** *vt* dissolve out; pick out. **-nehmbar** *a* removable. **·nehmen*** *vt* take out, remove **—sich Freiheiten (sich zuviel) h.** take (too many) liberties. **·oxidieren** *vt* oxidize off. **·pipettieren** *vt* pipette out. **·pressen*** *vt* press (squeeze, wring) out. **·ragen** *vi* (**aus, über**) project, protrude, stand out (from, above). **-ragend** *p.a* outstanding. **·schlagen*** (*usu* + **aus**) **I** *vt* **1** knock out (of), strike (sparks) (from). **2** carve out (of). **3** gain (from). **II** *vi*(s) leap out, burst out. **·schleudern** *vt* fling out, centrifuge (out), eject. **·schneiden*** *vt* cut out, excise. **·schwimmen*** *vi*(s) float (*or* swim) out, separate by flotation. **·sichten** *vt* screen (*or* separate) out. **·spritzen** *vi*(s) spurt out. **-spülen** *vt* rinse out, wash out. **·stecken** *vt* & *vi** stick out. **·stehen*** *vi* stand out, protrude. **·stellen 1** *vt* emphasize, publicize. **2: sich h.** (*oft* + **als**) be revealed (as), turn out, prove (to be). **·strecken** *vt* stretch out, stick out. **·treiben*** *vt* drive out, expel. **·treten*** *vi*(s) step out; come out, emerge; protrude. **·waschen*** *vt* wash (leach, rinse) out. **·ziehen* 1** *vt* pull out, take out, extract. **2** *vi*(s) move out

herb *a* bitter, harsh; austere; tart (taste); dry (wine)

herbei *adv* & *sep pfx* here, in, up, over (*implying approaching motion*): **·eilen** *vi*(s) hurry over, come rushing in. **·führen** *vt* bring about. **·holen** *vt* fetch (over), go (and) get. **·schaffen*** *vt* bring over; get (hold of), procure; raise (money), produce (evidence)

Herberge *f* (-n) inn; shelter

Herbheit *f* bitterness, etc *cf* HERB

Herbiegeversuch: Hin- und H. reverse bending test

herblich *a* sourish, subacid; somewhat harsh

Herbst *m* (-e) autumn, fall. **--rose** *f* hollyhock. **-zeitlose** *f* meadow saffron

Herd *m* (-e) **1** (cook)stove. **2** hearth (*incl*

Metll.), (*specif*) furnace bottom. **3** fireside; home. **4** focus, center (*esp Med.*). **5** (*Ores*) buddle. **6** (*in compds, Med.*) focal. *Cf* HERDE. **--asche** *f* hearth ashes. **-auskleidung** *f* (*Metll.*) hearth lining. **-blei** *n* furnace lead. **-brücke** *f* fire bridge

Herde I *f* (-n) herd, flock; crowd. **II** *pl of* HERD

Herden·vieh *n* drove cattle; gregarious animals; herd of cattle

Herd·erkrankung *f* focal infection. **-flammofen** *m* open-hearth (*or* reverberatory) furnace. **-flotation** *f* (*Ores*) table flotation, agglomerate tabling. **-formerei** *f* open sand casting

Herdfrisch·arbeit *f* (*Metll.*) refinery (*specif* open-hearth) furnace. **-eisen** *n* refinery process iron

Herd·frischen *n* open-hearth refining

Herdfrisch·prozeß *m* refinery process. **-schlacke** *f* refinery slag. **-stahl** *m* refined steel. **-verfahren** *n* refinery process

Herd·futter *n* hearth lining (*or* fettling). **-gewölbe** *n* furnace arch. **-glas** *n* glass running onto the hearth. **-guß** *m* open sand casting. **-ofen** *m* hearth kiln (*or* furnace). **-raum** *m* heating chamber. **-rost** *n* fire grate. **-schlacke** *f* (*Metll.*) hearth slag, clinker(s). **-sohle** *f* hearth bottom (*or* bed). **-stahl** *m* refined steel

herein *adv* & *sep pfx* in: **zur Tür h.** in (through) the door; **h!** come in! **von draußen h.** (in) from the outside. **h-·brechen*** *vi*(s) **1** set in; break (*as:* storm); fall (*as:* night). **2** (+ **über**) befall, overtake (*as:* disaster). **·fallen*** *vi*(s) **1** fall in. **2** (*esp* + **auf**) be taken in (by), fall (for, e.g. a trick). **·ziehen* 1** *vt* pull (draw, drag) in. **2** *vi*(s) come in, move in

her·führen 1 *vt/i* lead here (*or* along). **2** *vt* refresh (yeast)

herg. *abbv* (hergestellt) produced, made

Hergang *n* course (of events), how it happened; incident; circumstances

her·geben* *vt* hand (over); give up, yield

hergebracht 1 brought here: *pp of* **her·bringen. 2** *p.a* traditional, conventional

her·gehen* *vi*(s) **1** walk (along); (+ **hinter** + *dat*) follow along behind. **2** (*imps*): **es**

geht her (+*adv*) things are happening (*or* going along) . . .

hergeholt 1 fetched, brought: *pp of* **her·holen.** 2 *p.a*: **weit h.** far-fetched

Hering *m* (-e) herring. **Herings·lake** *f* herring pickle. **-öl** *n*, **-tran** *m* herring oil

her·kommen* *vi*(s) 1 come here; get here. 2 (*in comb. with* **wo**) come from *cf* WOHER. 3 (*usu* + **von**) originate, be derived (from). **Herkommen** *n* custom, usage; tradition. **herkömmlich** *a* long-established, conventional, traditional. **Herkunft** *f* origin, derivation; descent, extraction

her·leiten I *vt* 1 bring (in), conduct (*esp* by pipe). 2 derive; deduce, infer. II: **sich h.** be derived. **Herleitung°** *f* derivation; deduction

Hermelin *n*, *m* (-e) ermine

Hermes·finger *m* (*Pharm*.) hermodactyl

hermetisch *a* hermetic, airtight

Hermitesch *a* (*Math*.) Hermitian

hernach *adv* afterward, later

hernieder *adv* & *sep pfx* down(ward)

Herr *m* (-n, -en) 1 (gentle)man. 2 sir, Mr. (*omit in English when used with other titles in German, e.g.* **H. Doktor**): **die H—en Mitarbeiter** the co-workers. 3 master, boss; lord

Herren·pilz *m* edible mushroom. **-tier** *n* (*Zool*.) primate

her·richten *vi* 1 fix up, spruce up. 2 prepare, arrange. 3 renovate. 4 (*Wood*) season

herrlich *a* magnificent, splendid, glorious. **Herrlichkeit** *f* (-en) magnificence, splendor, glory; *pl usu* wonderful things

Herrschaft *f* (-en) 1 rule, authority, control; predominance. 2 *pl:* **die H—en** (ladies and) gentlemen

herrschen *vi* 1 rule, govern; reign. 2 prevail

her·rühren *vi* (**von**) be traceable, be due (to), come, result (from)

her·sagen *vt* recite, repeat

her·schaffen *vt* bring here, procure, produce

her·schieben* *vt* shove along, shove over here

Herst. *abbv* (Herstellung) (*Chem*.) preparation

her·stammen *vi* (usu + **von**) come, be derived (from)

herstellbar *a* producible, manufacturable

her·stellen I *vt* produce, prepare, manufacture, make; **wieder h.** restore, renovate **—hergestellt sein** have recovered (from an illness). 2: **sich h.** arise, be produced. **Hersteller** *m* (-) producer, manufacturer; maker; (*Film*) production manager. **Herstellung°** *f* production, preparation, manufacture; recovery

Herstellungs·mittel *n* restorative. **-verfahren** *n* method of production; manufacturing process

Hertz *n* hertz, cycle (per second)

herüber *adv* & *sep pfx* over (here), across

herum 1 *adv* over, past. 2 *adv* & *sep pfx* (*oft* + **um**) around, about: **um Mittag h.** around (*or* at about) noon. **h--drehen** *vt* & **sich h.** turn around; (**um**) revolve (about, around). **·kommen*** *vi*(s) (*oft* + **um**) get (*or* come) around, circumvent. **·liegen*** *vi* lie around. **·spritzen** *vt* squirt (*or* spray) around (*or* about)

herunter *adv* & *sep pfx* down, off: **die Treppe h.** down the stairs. **h--bringen*** *vt* 1 bring down, take down, get down. 2 run down, ruin. **·drücken** *vt* press (*or* force) down, depress; oppress. **-gekommen** *p.a* run-down, dilapidated, seedy. *See also:* **·kommen*** *vi*(s) 1 come down, get down. 2 run down, deteriorate. **·lassen*** *vt* let down, lower; (*Brew*.) strike out (wort). **·rieseln** *vi*(s) trickle down. **·schwemmen** *vt* wash down. **·setzen** *vt* lower, reduce; disparage. **·spielen** *vt* play down. **·trocknen** *vt* dry down, reduce by drying. **·tropfen** *vi*(s) drip down. **·wirtschaften** *vt* ruin. **·ziehen*** *vt* pull (*or* drag) down; (*Tex*.) strip (down)

hervor 1 *adv* out from: **hinter dem Schirm h.** out from behind the screen. 2 *sep pfx* forth, forward, out: **·brechen*** *vi*(s) burst (*or* pop) out, break forth; gush; erupt; appear (suddenly). **·bringen*** *vt* 1 bring (get, take) (sthg) out. 2 bring forth, produce. 3 give rise to. 4 say, utter. **H–bringung** *f* (-en) product. **h--gehen*** *vi*(s) 1 come forth (*or* out), arise. 2: **daraus geht hervor, daß** . . . from this it is evident, that . . . **·heben*** 1 *vt* emphasize, stress; set off. 2: **sich h.** stand out. **H–hebung** *f* emphasis, accentuation. **h--ragen** *vi* 1 protrude. 2 rise

(above). **3** stand out (from, above).
-ragend *p.a* outstanding. **·rufen*** *vt* **1** call
out. **2** call forth, evoke. **3** cause. **4** (*Phot.*)
develop. **·stechen*** *vi* stand out, be conspicuous. **-stechend** p.a conspicuous.
-stehend *p.a* protruding; prominent.
·treten* *vi*(s) **1** step out (*or* forward). **2**
come out, appear. **3** stand out; bulge, protrude. **-tretend** *p.a* prominent, salient.
·tun*: sich h. distinguish oneself

Herz *n* (-ens, -en) heart; center, core; (*in
compds oft*) cardiac: **--arznei** *f* cardiac
remedy, **-bahn** *f* cardioid path (*or* orbit).
-beutel *m* pericardium; (*in compds*) per-
icardial. **-bräune** *f* angina pectoris. **-ent-
zündung** *f* (myo)carditis. **-fäule** *f* (*Wood*)
heart rot. **-fell** *n* pericardium. **h--förmig**
a heart-shaped, cordate, cordiform. **H--
fraktion** *f* (*Petrol. Distil.*) middle fraction, heart cut. **-gekröse** *n* mesocardium.
-gespann *n* (*Bot.*) motherwort. **-gift** *n*
cardiotoxin

herzhaft *a* hearty, courageous

Herz·haut *f*: **äußere H.** pericardium: **innere H.** endocardium. **-holz** *n* heartwood

herzig *a* cute, dear, sweet

Herz· heart, cardiac: **-infarkt** *m* cardiac infarction. **-kammer** *f* chamber of the
heart, *specif* ventricle. **-klappe** *f* cardiac
valve. **-klopfen** *n* heart palpitation(s).
-kurve *f* (*Math*) cardioid. **-leiden** *n* heart
ailment (*or* disease). **h--leidend** *a* cardiac, suffering from a heart ailment

herzlich *a* hearty, cordial, sincere; *adv oft*
very, quite; **h. wenig** precious little

Herz·mittel *n* cardiac remedy

Herzog *m* (-e) duke. **Herzogin** *f* (-nen)
duchess

Herz· heart, cardiac: **-pulver** *n* cardiac
powder. **-reiz** *m* cardiac stimulant (*or* irritant). **-sack** *m* pericardium. **-schlag** *m* **1**
heartbeat. **2** heart attack. **-stillstand** *m*
cardiac arrest. **-stück** *n* (*usu + genit*)
heart (and soul) (of)

herzu *sep pfx, e.g.:* **herzu·kommen*** *vi*(s)
arrive on the scene (and join those here)

Herz· heart, cardiac: **-verpflanzung** *f*
heart transplant. **-vorhof** *m*, **-vorkam-
mer** *f* auricle. **-wasser** *n* pericardiacal
fluid

Hesperitin·säure *f* hesperitinic acid

Hessen·fliege *f* Hessian fly. **hessisch** *a*

Hessian. **Hessisch·purpur** *n* Hessian
purple

Hessonit *m* (*Min.*) essonite

hetero·atomig *a* heteratomic. **-cyclisch** *a*
heterocyclic

heterogen *a* heterogeneous. **Hetero-
genität** *f* heterogeneity

Heteroklin *m* (*Min.*) braunite

heterolog *a* heterologous

Heterolyse *f* (-n) heterolysis

Hetero·polysäure *f* heteropoly acid. **-ring-
bildung** *f* ring formation containing unlike atoms. **-zimtsäure** *f* heterocinnamic
acid. **h--zygotisch** *a* heterozygous.
-zyklisch *a* heterocyclic. **H--zyklus** *m*
heterocycle

Heu *n* hay

heuer *adv* this year

heulen *vi* howl, roar; cry

Heu·pferd *n* grasshopper

heurig *a* this year's: **der H—e** this year's
(new) wine

Heu·schnupfen *m* hay fever. **-schrecke** *f*
grasshopper, locust. **-schreckenbaum** *m*
locust tree

heute *adv* today —**h. abend** tonight; **h.
morgen** this morning; **h. nacht** (**a**)
tonight, (**b**) last night (after midnight);
noch h. no later than today. **heutig** *a* today's, of today, present-day. **heutzutage**
adv nowadays

Hevea·baum *m* hevea tree, rubber tree

Hexa·eder *n* hexahedron. **h--edrisch** *a*
hexahedral. **H--edrit** *n* hexahedrite.
-gyre *f* (*Cryst.*) hexagonal rotary axis. **h--
gyrisch** *a* (*Cryst.*) having a sixfold rotation axis. **-gyroidisch** *a* (*Cryst.*) having a
hexad alternating axis. **-kontan** *n* hexa-
contane. **-kosan** *n* hexacosane

Hexan *n* hexane

Hexanaphten *n* hexanaphthene (cyclohexane)

Hexan·säure *f* hexanoic acid

Hexa·vanadinsäure *f* hexavanadic acid.
h--zyklisch *a* hexacyclic

Hexe *f* (-n) witch; hag *cf* HEXEN

Hexen *n* hexene *cf* HEXE

Hexen· witch's: **-kraut** *n* mandragora,
mandrake; enchanter's nightshade.
-mehl, **-pulver** *n* lycopodium (powder),
witch meal. **-meister** *m* sorcerer, wizard.

-schuß *m* (*Med.*) lumbago
Hexit *n* hexitol
Hexogen *n* trimethylene trinitramine
Hexol·salz *n* "hexol" complex salt
Hexon·säure *f* hexonic acid
Hexylen *n* hexene
HF *abbv* (Hochfrequenz) high frequency
HGB *abbv* (Handelsgesetzbuch) commercial law code
hie *adv* 1 here, = HIER: **h. und da** (a) here and there, (b) now and then. 2 (*oft in compds + prep*) *cf* HIEBEI, HIEMIT, etc, *cf* HIEHER
hieb struck, etc: *obsolsc past of* HAUEN
Hieb *m* (-e) 1 blow, stroke; cut. 2 (tree-)cutting. 3 dig, insinuation
hiebei *adv* in doing this, etc = HIERBEI
hieher *adv* here, this way = HIERHER
hielt held, etc: *past of* HALTEN
hiemit *adv* with this, etc = HIERMIT
hier *adv* here; at this point; in this matter; (*in compds & prep, oft*) this. **-an** *adv* in (on, at, by) this, herein, hereon, hereat, hereby *cf* AN, DARAN. **-auf** *adv* after this, hereupon; on this, hereon *cf* AUF, DARAUF. **-aus** *adv* from (*or* out of) this, herefrom *cf* AUS, DARAUS. **-bei** *adv* in doing this, in this case (process, connection), on this occasion, at this time *cf* BEI, DABEI. **-durch** *adv* through here; by this means, by doing this, hereby *cf* DURCH, DADURCH. **-für** *adv* for this *cf* FÜR, DAFÜR. **-gegen** *adv* against this. **-her** *adv & sep pfx* here, to this place (*always implies approaching motion*). **-herum** *adv* 1 around this way. 2 around here, hereabouts. **-hin** *adv* here **—h. und dorthin** this way and that. **-in** *adv* in this matter, herein *cf* IN, DARIN. **-mit** *adv* with this, herewith, hereby *cf* MIT, DAMIT. **-nach** *adv* after this, according to this *cf* NACH, DANACH. **-orts** *adv* hereabouts, in this place. **H—sein** *n* presence. **h—über** *adv* (over, above, across) here; about this (matter) *cf* ÜBER, DARÜBER. **-um** *adv* around this, around here *cf* UM, DARUM. **-unter** *adv* under here, under this; among these; by this *cf* UNTER, DARUNTER. **-von** *adv* of this, hereof; from this, herefrom; by this, hereby; about this *cf* VON, DAVON. **-zu** *adv* for this (purpose); to this, hereto, toward this; on this point (*or* matter); to go with this *cf* ZU, DAZU.

-zulande *adv* in this country (*or* area)
hiesig *a* of this place, local
hieß (was) called, etc: *past of* HEISSEN
hievon *adv* of this, etc = HIERVON
hiezu *adv* for this, etc = HIERZU
Hilfe *f* (-n) 1 help, aid, assistance; relief; *pl* aids **—zu H. nehmen** make use of. 2 helper, assistant; *pl oft* help. **-leistung** *f* aid, assistance
hilf·los *a* helpless. **-reich** *a* helpful
Hilfs· assistant, auxiliary: **-apparat** *m* accessory, stand-by (apparatus). **-arbeiter** *m*, **-arbeiterin** *f* (temporary, unskilled) helper, assistant, *pl oft* help. **h—bedürftig** *a* needing help. **H—behälter** *m* auxiliary container; reserve tank. **-beize** *f* (*Dye.*) auxiliary mordant. **-bereitschaft** *f* helpfulness. **-brennstoff** *m* auxiliary fuel. **-buch** *n* handbook, manual. **-dienst** *m* auxiliary (*or* emergency) service. **-dünger** *m* auxiliary fertilizer. **-geld** *n* financial aid, subsidy. **-größe** *f* (*Math.*) auxiliary (quantity). **-kraft** *f* 1 helper, assistant; *pl* help. 2 temporary employee. 3 (*Mach.*) servo power. **-leitung** *f* auxiliary (pipe, electric) line (*or* circuit). **-maßnahme** *f* remedial (*or* emergency) measure. **-mittel** *n* aid, expedient, resource; *esp pl* means; (*Med.*) adjuvant. **-quelle** *f* source of aid, resource, expedient. **-stativ** *n* auxiliary stand (*or* support). **-stoff** *m* auxiliary material, aid. **-strom** *m* (*Elec.*) auxiliary current. **-teilung** *f* auxiliary graduation. **-vorrichtung** *f* auxiliary device. **-werkzeug** *n* auxiliary tool (*or* organ). **-wissenschaft** *f* auxiliary (*or* ancillary) science

hilft helps, aids: *pres of* HELFEN
Him·beere *f* raspberry
Himbeer·essig *m* raspberry vinegar. **-limonade** *f* raspberry soda (pop, ade). **-spat** *m* rhodochrosite
Himmel *m* (-) sky; heaven; canopy. **h—an** *adv* skyward. **-blau** *a* sky-blue, azure, cerulian. **-hoch** *a* sky-high: **h. loben** praise to the skies; **h. überlegen** vastly superior
Himmels· heavenly, celestial: **-gegend** *f* 1 region of the sky. 2 compass point (*or* direction). **-körper** *m* heavenly (*or* celestial) body. **-kunde** *f* astronomy. **-licht** *n*

celestial light. **-richtung** *f* 1 (compass) direction. 2 cardinal point. **-schlüssel** *m* primrose. **-strich** *m* zone, region

himmel·weit *a* enormous, immense

himmlisch *a* celestial, heavenly

hin I *adv* 1 there, to that place, that way, in that direction (*implies motion away from the speaker, oft reinforcing preps*): **gegen die Tür h., zur Tür hin** toward the door; **bis zum Labor h.** up to (*or* as far as) the lab. 2: **h. und her** back and forth, going and coming, up and down: **h. und her gehen, h. und hergehen** go back and forth. 3: **h. wie her** neither here nor there, all the same. 4: **h. und wieder** now and again. 5: **h. und zurück** there and back; round trip. 6: **auf . . . h.** (a) in response to, (b) as to, for (*implies testing, investigation*) *cf* GEFAHR, (c) (+*period of time*) for (the duration of). 7 along (*oft in comb. with preps, implying extent of motion*): **zwischen den Linien h.** (moving) along between the lines; **am Rand h.** along the edge; **über die ganze Fläche h.** (extending) over the entire surface. 8: **vor sich h.** (talking) to oneself. 9 gone, lost; ruined, done for. **II** *sep pfx* there, etc (*meanings same as adv* 1, 6, 7); *also:* 10 down *cf* HINSCHREIBEN, HINSETZEN, etc. 11 (*in comb. with wo*) where(to): **wo fließt es h.?, wohin fließt es?** where does it flow to? 12 (*implies death, destruction*) *cf* **hinrichten. III** (*prefixed to other advs or sep pfxs to specify direction away from the speaker; also occur as preps following nouns*) *cf* HINAUF, HINUNTER, etc

hinab 1 *prep* (+*accus*): **den Weg h.** down the path. 2 *adv & sep pfx* down(ward): **·fallen** *vi*(s) fall down, drop. **·wandern** *vi*(s) migrate down(ward)

hinan *adv, sep pfx, prep* (+*accus*) up(ward)

hinauf 1 *prep* (+*accus*): **den Fluß h.** up the river. 2 *adv & sep pfx* up(ward): **·arbeiten: sich h.** work one's way up. **-gesetzt:** h—er Index superscript *cf* HINAUFSETZEN. **·kommen*** *vi*(s) come up, move up(ward). **·setzen I** *vt* 1 raise, increase. 2 superscribe, write as a superscript. **II** *vt & sich* h. move up, seat (oneself) up higher. **·transformieren** *vt* (*Elec.*) step up. **·wandern** *vi*(s) migrate up(ward). **·ziehen*** 1 *vt & sich* h. pull

(oneself) up. 2 *vi*(s) move up(ward) (*or* upstairs); extend up(ward)

hinaus *adv & sep pfx* 1 (*usu*) out: **zur Tür h.** out the door. 2 *e.g.:* **nach der Straße h.** facing the street. 3: **auf (soviel Zeit) h.** for (so much time) to come. 4: **über (die Grenze, die Zeit, das Alter) h.** beyond (the limit, the time, the age); **über (den Betrag) h.** over (and above) (the amount). 5 (*in comb. with wo, sollen*): **wo soll das noch hinaus (führen)?** where (*or* what) will this lead to? 6 (*pfx only*) (*implies delay, postponement*) *cf* HINAUS·SCHIEBEN, **·ZIEHEN: ·gehen*** *vi*(s) 1 go out. 2 lead out; face (a direction, etc). 3 (+**über**) go beyond, exceed, surpass. **·kommen*** *vi*(s) 1 come out, get out(side). 2 + (**über**) get beyond. 3 (+**auf**) lead (to), end up (as) —**es kommt auf eins (dasselbe, das gleiche) hinaus** it comes to the same thing. **·laufen*** *vi*(s) 1 run out. 2 (**auf**) aim (at), amount (to); lead (to), end up (as) *cf* HINAUSKOMMEN 3. **·ragen** *vi* stick out (*or* up). **·schieben*** *vt* 1 push out. 2 delay, postpone. 3 **·waschen*** *vt* wash out. **·ziehen*** **I** *vt* 1 pull (haul, draw) out. 2 extend. 3 delay, postpone. **II** *vi*(s) move out. **III: sich h.** drag on, take a long time

Hinblick *m:* **im** (*or* **in**) **H. auf** in view of, with regard to

hin·bringen* *vt* take there; spend, pass (time)

hinderlich *a* troublesome, cumbersome, impeding; *pred:* **h. sein** (*oft* + *dat*) be a hindrance (to), get in the way (of). **hindern** *vt/i* (+*zu*) e.g. **nichts hindert sie (daran), es zu tun** nothing keeps (hinders, prevents) them from doing it; **das hindert (sie) nicht** that is no hindrance (to them), that does not get in the(ir) way. **Hindernis** *n* (-se) hindrance, obstacle. **Hinderung** *f* (-en) hindrance. **Hinderungs·grund** *m* obstacle.

hin·deuten *vi* (+*auf*) point (to)

hindurch 1 *prep* (+*accus of time*) e.g.: **das (ganze) Jahr h.** throughout the year. 2 *adv & sep pfx* (*oft* + *durch*) e.g.: **durch die Membran h.** (right) through the membrane **·dringen*** *vi*(s) (+*durch*) penetrate, permeate (through). **·fallen*** *vi*(s) fall (pass, drop) through. **·gehen*** *vi*(s) (*usu* + *durch*) go (walk, pass, fit) through, traverse; experience. **·leiten** *vt* lead (*or* conduct) through. **·saugen*** *vt* suck (draw, suction) through. **·sehen*** *vt/i* see (*or* look) through. **·treten*** *vi*(s) step (*or* pass) through, penetrate. **·ziehen*** (*oft* + **durch**) 1 *vt* pull (*or* run) (sthg) through. 2 *vi*(s) move, (*or* make one's way) through. 3: **sich h.** run through, pervade

hinein *adv & sep pfx* (*usu* + **in**, *reinforcing the meaning of* **in**) (right, all the way) in, into, inside: **·bringen*** *vt* bring (take, put, get, introduce) (into). **·diffundieren** *vt, vi*(s) diffuse (into). **·dringen*** *vi*(s) penetrate (into). **·düsen** *vt* inject (into). **·gehen*** *vi*(s) go (pass, fit) in(to). **·geraten*** *vi*(s) get caught, trapped, involved), stray in(to). **·greifen*** *vi* reach in(to). **·kommen*** *vi*(s) come, get, stray in(to). **·mischen** 1 *vt* mix in (with). 2: **sich h.** interfere (in, with). **·ziehen*** 1 *vt* pull (draw, haul) in(to); involve (in). 2 *vi*(s) move in(to), make one's way in(to)

hin·fahren* I *vt* 1 drive (take, haul) there. **II** *vi*(s) 2 go (ride, drive) there. 3 (+*mit*) run (e.g. one's hand) over. 4 pass away, depart

hinfällig *a* 1 frail, infirm. 2 invalid, baseless. 3 deciduous. **Hinfälligkeit** *f* frailty, invalidity

hinfort *adv* henceforth

hin·führen *vi* lead (to)

hing hung: *past of* HÄNGEN *vi*

Hingabe *f* devotion, dedication; sacrifice; surrender

hin·geben* I *vt* 1 hand, pass (to). 2 give away, give up; sacrifice. **II: sich h.** (*dat*) devote oneself (to); indulge (in); abandon oneself (to); surrender (to). **hingebend** *p.a* devoted. **Hingebung** *f* devotion, dedication; sacrifice; surrender. **hingebungs·voll** *a* devoted. **hingegeben** *p.a* (+*dat*) devoted (to); absorbed, lost (in)

hingegen *adv* on the other hand; however

hin·gehen* *vi*(s) 1 go (there). 2 (*of time*) pass, go by. 3: **es mag h.** it can pass (*or* be overlooked); **es h. lassen** let it pass, overlook it. *Cf* HIN 1

hin·halten* *vt* 1 hold out (to), offer. 2 put off (to gain time), keep waiting; stave off. 3 maintain

hinken *vi* limp, be lame

hin·kommen* *vi*(s) 1 get there —**wo ist es hingekommen?** where did it disappear to?; **wo kämen wir da hin?** where would that take us? 2 (+*mit*) make do (with), get along (on). 3 be (*or* go far) enough. 4 work out all right. 5 be correct

hinlänglich *a* sufficient, adequate

hin·nehmen* *vt* 1 accept, take. 2 put up with. 3 fascinate

hin·neigen (+*zu*) 1 *vt* bend, incline (sthg) (toward). 2 *vi* lean, be inclined, gravitate (to, toward). 3: **sich h.** bend, lean (over) (toward)

Hinreaktion *f* forward reaction

hin·reichen 1 *vt* hold out (to), hand over (to). 2 *vi* suffice, be sufficient, go far enough. **hinreichend** *p.a* sufficient, adequate

hin·richten *vt* 1 execute, put to death. 2 (+**auf**) direct (sthg) (toward)

hin·schreiben* *vt* write (*or* put) down

hin·setzen 1 *vt* put (*or* set) down. 2: **sich h.** sit down, take a seat

Hinsicht° *f* respect, regard: **in H. auf** = **hinsichtlich** *prep* (+*genit*) with respect (*or* regard) to

hin·stellen 1 *vt* put (down), place; set up. 2 *vt & sich h.** (+*als*) make (oneself) appear (as), represent (oneself) (as). 3: **sich h.** (go *or* come and) stand, take a (standing) position, place oneself

hintan· *sep pfx:* **·halten*** *vt* hold back, check. **·setzen** *vt* put last; neglect, ignore. **·stehen*** *vi* take second (*or* last) place. **·stellen** *vt* put last, etc = **·SETZEN**

hinten *adv* (in *or* at the) back, in (*or* at) the rear, at the end; **h. im Buch** at the back of the book; **h. im Hof** back in the yard; **ganz h.** all the way (in) back; **h. und vorn** back and front, (*fig.*) in every way; **nach h.** (to the) back, backward; **von h.** from behind, from the back. **·-an** *adv & sep pfx* at the back (*or* rear). **·drein** *adv* later, afterward, after that. **·(he)rum** *adv* around the back way, in a roundabout way; on the quiet, under the counter. **·über** *adv & sep pfx* over backward

hinter 1 *prep* behind, in back of: **h. dem Schirm** (**hervor**) (out from) behind the screen; **h. etwas kommen (a)** find out (what's behind) sthg, (**b**) get the knack of sthg; **h. ihnen her** after them, in pursuit of them *cf* HER; **etwas h. sich haben** be done (*or* be through) with sthg; **etwas h. sich bringen** get sthg over and done (with); **sich h. etwas (her)machen** get to work on sthg. **2** *a* back, rear, hind, posterior. **3** *insep pfx* (*oft implies underhand or secret actions*) *cf* HINTERBRINGEN

Hinter· back, rear, hind, posterior: **-ansicht** *f* rear view. **-asien** *n* Eastern Asia. **-bein** *n* hind leg

hinterbleiben* *vi*(*s*) (*insep*) remain (*or* be left) behind; survive. **Hinterbliebene** *m,f* (*adj endgs*) survivor

hinterbringen* *vt* (*insep*) (*d.i.o*) secretly inform (of)

hinterdrehen *vt* (*insep*) (*Mach.*) back off, relieve

hintereinander *adv & sep pfx* one behind (*or* after) the other; in a row, in succession, without a break; (*Elec.*) in series; (*Mach.*) in tandem. **--weg** *adv* in rapid succession

Hinter·gedanke *m* ulterior motive

hintergehen* *vt* (*insep*) deceive; circumvent

Hinter·grund *m* background. **h–gründig** *a* cryptic, inscrutable; subtle. **-halt** *m* in the back. **H–halt** *m* ambush. **h–hältig** *a* underhand(ed). **H–haupt** *n* back of the head, occiput

Hinterhaupts· (*in compds*) occipital

hinterher *adv & sep pfx* **1** (along) behind; in pursuit. **2** after(ward); later

hinterkleiden *vt* (*insep*) coat the back of, back

Hinter·kopf *m* back of the head, occiput

hinterlassen* *vt* (*insep*) leave, bequeath. **Hinterlassenschaft** *f* (-en) bequest, estate; heritage

hinterlegen *vt* (*insep*) **1** deposit. **2** leave in safekeeping; file; check. **3** (+*mit*) back (with)

Hinter·leib *m* **1** hindquarters. **2** (*Entom.*) abdomen; metasoma. **h–listig** *a* deceitful, underhand

hinterschneiden* *vt* (*insep*) (*Plastics*) undercut

Hinter·schnitt *m* (*Plastics*) undercut. **-seite** *f* reverse side. **-sinn** *m* deeper (hidden) meaning

hinterst *a* (*superl*) hindmost

Hinter·treffen *n* disadvantage: **ins H. geraten** be put at a disadvantage, fall behind

hintertreiben* *vt* (*insep*) obstruct, frustrate

hinüber *adv & sep pfx* over (there), across: **·destillieren** *vt* distill over. **·greifen*** *vi* (**in**) reach (*or* extend) over (into). **·reißen*** *vt* (*Distil.*) carry over

hin- und her· *sep pfx* back and forth: **hin- und her·bewegen: sich . . .** move back and forth, reciprocate; oscillate. **Hin- und Herbewegung** *f* back-and-forth (*or* reciprocating) motion, oscillation. **hin- und hergehend** *p.a* reciprocating, oscillating

hinunter 1 *prep* (+*accus*) down: **den Berg (den Fluß) h.** down the mountain (the river), downhill (downstream) **2** *adv & sep pfx* down(ward): **·setzen** *vt* put down (*or* below); write as a subscript: **hinuntergesetzter Index** subscript

hinweg *adv & sep pfx* away; **über . . . h.** over (and beyond), past; **sie sind darüber h.** they have got over it: **·denken*** *vt* think of as absent (*or* nonexistent). **·gehen*** *vi*(*s*) (**über**) pass over; ignore, overlook. **·lesen*** *vi* (**über**) overlook in reading. **·sehen*** *vi* (**über**) look over; overlook, ignore. **·setzen*: sich h.** (**über**) overlook, disregard

Hinweg° *m* (the) way there *Cf* RÜCKWEG

Hinweis *m* (-e) hint; reference. **hin·weisen* 1** *vt:* **einen auf etwas h.** call someone's attention to sthg. **2** *vi* (**auf**) hint (at); refer (to); point (to); point out

hinwiederum *adv* on the other hand; in turn

hin·welken *vi*(*s*) wither away. **·werfen*** *vt* **1** throw down (*or* away). **2** jot down. **3** drop (*esp* remarks). **·wirken** *vi* (**auf**) aim (at), seek to achieve. **·ziehen* I** *vt* **1** delay. **2** (**zu**) attract (to). **II** *vi*(*s*) **3** move, go (there). **4** move along, drift. **III: sich h.** extend, stretch; drag on, take time; (*oft* + **über**) extend (over), last (some time). **·zielen** *vi* (**auf**) aim (at), be intended (for)

hinzu I *adv* in addition, besides: **h. kommt noch, daß . . .** in addition . . . *Cf* HINZUKOMMEN. **II** *sep pfx* **1** there, on the scene (*implies arrival to join those present*). **2** (*implies addition*): **·addieren** *vt* add on. **·fügen** *vt* add, append. **H–fügung** *f* addition. **h–·geben*** *vt* add. **·gießen*** *vt* pour in (*or* on), add. **·kommen*** *vi*(s) **1** be added **—es kommt noch hinzu, daß** . . . in addition (there is the fact that) . . . **2** come and join . . . , arrive (on the scene.) **-kommend** *p.a* additional. **·nehmen*** *vt* take in addition, incorporate. **·setzen** *vt* add. **·träufeln** *vt* add in drops, add drops of. **·treten*** *vi*(s) **1** be added. **2** arrive. = **·KOMMEN**. **H–tritt** *m* addition; appearance, accession. **h–·tropfen** *vt* add in drops = **·TRÄUFELN**. **·tun*** *vt* add. **·zählen** *vt* count in, include. **·ziehen*** *vt* **1** call in, consult. **2** add

hippur·sauer *a* hippurate of. **H–säure** *f* hippuric acid

Hirn *n* (-e) brain = **GEHIRN**; (*in compds oft*) cerebral: **·abstastung** *f* brain (*or* CAT) scan. **-anhang** *m*, **-anhangsdrüse** *f* hypophysis, pituitary gland. **-fett** *n* cerebrin. **-haut** *f* meninx. **-hautentzündung** *f* meningitis. **-holz** *n* crosscut timber. **-mark** *n* medullary brain matter. **-rinde** *f* cerebral cortex. **-säure** *f* cerebric acid. **-schädel** *m*, **-schale** *f* cranium, skull. **-schnitt** *m* **1** brain section. **2** (*Wood*) crosscut

Hirsch *m* (-e) deer, stag. **-·dorn** *m* (*Bot.*) buckthorn. **-horn** *n* staghorn, hartshorn

Hirschhorn· hartshorn: **-geist** *m* spirits of hartshorn, (aqua ammonia). **-öl** *n* hartshorn oil. **-salz** *n* salt of hartshorn (commercial ammonium carbonate). **-schwarz** *n* hartshorn black (fine boneblack). **-spiritus** *m* spirits of hartshorn

Hirsch·klee *m* **1** hemp agrimony. **2** yellow sweet clover. **-leder** *n* deerskin, buckskin. **-talg** *m* deer tallow (*or* fat). **-zunge** *f* (*Bot.*) hart's tongue

Hirse *f* millet. **-·erz** *n* oolitic hematite. **-fieber** *n* miliary fever, miliaria. **-korn** *n* millet seed

Hirsen· = **HIRSE·** **·-eisenstein** *m* = **Hirse(n)·stein** *m* oolitic hematite

Hirt *m* (-en, -en) herder, herdsman, shepherd

Hirten·täschchen *n*, **-tasche** *f*, **-täschlein** *n* (*Bot.*) shepherd's purse

Histo·chemie *f* histochemistry. **-genese** *f* histogenesis

histologisch *a* histological

historisch *a* historic(al)

Hitz·draht *m* hot wire

Hitze *f* heat; passion. **·-altern** *n* heat aging. **-ausgleich** *m* heat balance. **h–beständig** *a* heat-resistant, thermostable; (*Colors*) fast to ironing; (*Brick*) refractory. **H–einheit** *f* heat (*or* thermal) unit. **-einwirkung** *f* action (*or* influence) of heat. **h–empfindlich** *a* heat-sensitive. **H–erzeugung** *f* heat generation. **h–fest** *a* heat-resistant. **H–gegenwert** *m* heat equivalent. **H–gehärtet** *a* (*Plastics*) thermoset. **H–grad** *m* degree of heat. **h–härtbar, -härtend** *a* thermosetting. **H–härtung** *f* thermosetting. **-messer** *m* pyrometer. **-messung** *f* pyrometry.

hitz·empfindlich *a* heat-sensitive

Hitze·probe *f* heat test. **-schild** *n* heat shield. **-strahlung** *f* heat radiation. **-welle** *f* heat wave. **-wirkung** *f* action of heat, heat (*or* thermal) effect. **-wirkungsgrad** *m* thermal efficiency

hitzig *a* **1** hot, heated. **2** (*Med.*) acute. **3** hot-tempered, passionate. **4** inflammatory

HK *abbv* (Hefnerkerze) Hefner candle

hl *abbv* (Hektoliter) hectoliter

Hl. *abbv* (Halbleder) half-leather. **Hlw.** *abbv* (Halbleinwand) half-cloth

hob lifted, raised: *past of* **HEBEN**

Hobbock *m* (-s) (sealed) tin (shipping) container

Hobel *m* (-), **hobeln** *vt/i* plane (smooth). **Hobel·späne** *mpl* shavings

hoch (höher, höchst; *Cf* **HOHE**) **1** *pred a*, *adv* high, tall; great(ly). **2** (*Colors*) intense, bright, deep. **3** superior, superscript (number). **II** *adv* highly. **III** *adv* & *sep pfx* up. **Hoch** *n* (-s) (*Meteor.*) high

hoch·aktiv *a* highly active. **-angeregt** *a* highly excited

hoch·arbeiten: sich h. work one's way up

hoch· high-, highly, elevated; relief: **-atom-ar, -atomig** *a* high-atomic. **H–ätzung** *f* relief engraving. **-bau** *m* **1** building construction. **2** above-ground building (*or* structure). **3** (*Min.*) overground operations. **-bauklinker** *m* engineering brick.

-behälter *m* overhead (*or* elevated) container (tank, bunker). **h–beschleunigt** *a* highly accelerated. **-best.** *abbv* = **-beständig** *a* highly stable (*or* resistant). **H–betrieb** *m* **1** intense activity. **2** rush (hour); busy season. **3** peak operation. **-bild** *n* relief (sculpture). **h–bildsam** *a* highly plastic. **-blau** *a* bright blue, azure. **-brisant** *a* high-explosive. **H–burg** *f* citadel. **h–dichtig** *a* high-density. **-dispers** *a* highly dispersed. **H–druck** *m* **1** high pressure. **2** relief printing

Hochdruck· high-pressure: **-dampf** *m* high-pressure steam. **-gebiet** *n* (*Meteor.*) high-pressure area; anticyclone. **-sauerstoff** *m* compressed oxygen

Hoch· high-, highly: **-ebene** *f* plateau. **-email** *n* embossed enamel. **h–empfindlich** *a* highly sensitive. **H–empfindlichkeit** *f* extreme sensitivity. **h–erhitzt** *a* highly heated. **-evakuiert** *a* highly evacuated. **H–explosivstoff** *m* high explosive. **h–farbig** *a* bright-colored. **-fein** *a* superfine; choice, select. **-fest** *a* (*Steel*) high-strength, high-tensile. **-feuerfest** *a* highly fireproof (*or* refractory). **-flüchtig** *a* highly volatile. **-frequent**, **-frequenzmäßig** *a* high-frequency. **-gekohlt** *a* (*Metll.*) high-carbon. **-geladen** *a* (*Elec.*) highly charged. **-gespannt** *a* high-tension; high-pressure; highly stressed; highly compressed; highly superheated; (*fig*) far-reaching. **H–glanz** *m* high polish, brilliance

Hochglanz·blech *n* mirror-finish sheet (metal). **-kalander** *m* supercalender. **-lack** *m* high-gloss varnish (*or* lacquer). **-papier** *n* high-gloss (*or* enameled) paper. **h–poliert** *a* highly polished, burnished. **H–politur** *f* high-luster polish

hoch·gliedrig *a* highly articulated, many-membered. **H–glühen** *n* (*Metll.*) grain-coarsening annealing. **h–gradig** *a* high-grade, high-degree; *adv oft* to a high degree, highly. **-grädig a** (*Chem.*) high-concentration. **-grün** *a* bright-green

hoch·halten* *vt* **1** hold up (high). **2** uphold. **3** honor, esteem

hoch·haltig *a* high-content, high-grade, rich

hoch·heben* *vt* raise up, lift up

hoch·herzig *a* high-minded, generous. **-hydraulisch** *a* (*Lime*) eminently hydraulic. **-kant** *adv* **1** upright; on edge; on end. **2** (*Metll.*) **h. gegossen** top-poured. **H–kante** *f* edge

hoch·klappen *vt* fold (turn, tip) up(ward), raise

hoch·klassig *a* high-class, high-grade. **-komprimiert** *a* highly compressed. **-konz.** *abbv* = **-konzentriert** *a* highly concentrated, high-concentration. **-legiert** *a* high-alloy. **H–leistungs·** (*in compds*) (*usu Mach.*) high-capacity, heavy duty. **h–lichtstark** *a* (*Phot.*) high-speed, ultra-speed. **-molekular** *a* high-molecular(-weight); macromolecular. **H–moortorf** *m* hill peat. **-müllerei** *f* (*Flour*) high milling. **h–mütig** *a* arrogant; proud

Hochofen° *m* blast furnace. **··gasteer** *n* furnace-gas tar. **-schlacke** *f* blast-furnace slag. **-zement** *m* Portland blast-furnace cement.

hoch· high-, highly: **··orange** *a* bright (*or* deep) orange. **-phosphorhaltig** *a* high-phosphorus. **-polymer** *a* high-polymer, highly polymerized. **-prozentig** *a* high-percentage. **-quantig** *a* high-quantum-number. **H–quarz** *n* high-temperature quartz, β-quartz. **h–rein** *a* high-purity. **-rot** *a* bright (*or* deep) red. **-rund** *a* convex. **H–scharlach** *m* cochineal scarlet

hoch·schlagen* **1** *vt* turn (fold, tip) up, raise. **2** *vi*(s) rise, well up, run high

hoch·schlagfest *a* high-impact-strength, highly shock-resistant. **-schmelzend** *a* high-melting. **-schnell** *a* high-speed. **H–schule** *f* university; university-level professional school: **technische H.** institute of technology. **h–selig** *a* late, deceased. **-siedend** *a* high-boiling. **-siliziert** *a* high-silicon. **H–sommer** *m* midsummer. **-spannung** *f* high tension (*or* voltage)

Hochspannungs·erzeuger *m* (*Elec.*) high-voltage generator. **-leitung** *f* high-tension line (wire, circuit). **h–sicher** *a* high-tension-proof. **H–strom** *m* high-tension current

höchst **1** *a* (*superl of* HOCH) highest, tallest; greatest; topmost; utmost; maximum — **es ist h–e Zeit** it is high time. **2** *adv* highly, extremely —**aufs h—e** highly, to the highest degree. **3** as pfx = **1** & **2**. Cf HOHE

Hoch·stand m high level; high point, maximum height; high-water mark

Höchst·belastung f maximum (or peak) load. **-besetzungszahl** f maximum number of electrons in the shell. **-betrag** m maximum (amount). **-druck** m maximum (or highest) pressure

hochstehend a: **h—e Zahl** superior number, superscript

hoch·steigen* vi(s) rise(up), increase

höchstens adv at most, at best

Höchst·fall m maximum case. **-frequenz** f very high frequency, VHF. **-geschwindigkeit** f maximum speed; speed limit. **-last** f maximum load. **-leistung** f maximum performance (or output). **-menge** f maximum. **h—möglich** a highest (or highly) possible. **-prozentig** a maximum percent(age)

Hoch·strom m (Elec.) heavy current

höchst·siedend a highest-boiling. **H—spannung** f maximum tension; extra high tension, EHT. **-temperatur** f maximum temperature. **h—wahrscheinlich** a most probable; adv oft in all probability. **H—wert** m maximum value. **-wertigkeit** f maximum valence. **-zahl** f highest (or maximum) number

hoch·temperaturbeständig a high-temperature-resistant. **-tonerdhaltig** a of high alumina content, highly aluminous. **H—tour** f (usu pl) high speed, high rpm: **auf H—en** at high speed, in high gear. **h—tourig** a high-speed. **h—trabend** a pompous, bombastic

hoch·treiben* vt drive up; work (or refine) highly

Hoch·vakuum n high vacuum. **h—vergärend** a (Yeast) high-attenuating; top-fermenting. **H—vergärung** f (Brew.) high attenuation (of wort); top fermentation. **h—viskös** a highly viscous. **-voltig** a high-voltage. **-voluminös** a highly voluminous. **H—wald** m timber forest. **-wärmeverkohlung** f high-temperature carbonization. **h—weiß** a intensely white. **-wertig** a high-valence; high-quality. **-wirksam** a highly active (or effective). **H—zahl** f (Math.) exponent. **-zeilen-**

fernsehen n high-definition television

Hoch·zeit f **1** wedding. **2** = **Hoch-Zeit** f peak period; zenith, height

hoch·ziehen* vt pull up; erect, raise

Höcker m (-) hump, hummock; bump, knob; cam. **höckerig** a humpbacked; bumpy, knobby

Hode m, f -n), **Hoden** m, **Hoden·drüse** f testicle. **Hoden·sack** m scrotum

Hof m (-̈e) **1** court, yard, patio. **2** farm(yard). **3** halo; aureole, corona; areola

hoffen vt/i (oft + **auf**) hope (for) —**h. lassen, daß** . . . give rise to the hope that . . . **hoffentlich** adv hopefully, let us hope (that). **Hoffnung** f (-en) hope

höflich a courteous, polite

Hof·rahm m farm cream

hohe a high, tall, etc = HOCH (which changes to **hoh** before adj endgs.)

Höhe f (-n) **1** height, altitude; high level — **in die H.** up(ward) = HOCH as sep pfx; **in der H.** up high, in the air. **2** hill, highland(s); elevation. **3** peak, top, summit. **4** level: **auf gleicher H.** on the same level, level (with), abreast. **5** (Geog.): **auf der H. von** in the latitude of. **6** magnitude, amount; extent. **7** (Water) head. **8** (Color) depth, intensity

Hoheit f (-en) majesty, nobility; sovereignty

Höhen·gas n (Aero.) high-altitude gas. **-krankheit** f altitude sickness. **-lage** f altitude elevation. **-messer** m altimeter; hypsometer. **-schichtung** f (Sedimentation) vertical distribution. **-sonne** f **1** mountain sun. **2** (orig TN) sun lamp, ultraviolet lamp. **-strahl** m cosmic ray. **-strahlung** f cosmic radiation. **h—symmetrisch** a (Cryst.) symmetrical about the vertical axis

Höhe·punkt m **1** high point. **2** peak; climax

höher a (compar of **hoch, hohe**) higher, taller, greater, more, etc: **h. bezahlen** pay more; **h—e Schule** (university) preparatory school. **—gliedrig** a higher-membered. **-quantig** a of higher quantum number. **-siedend** a higher-boiling. **-wertig** a higher-valent

hohl a hollow; concave. **H—·blockstein** m hollow block (or brick). **-drüse** f (Anat.) follicle.

Höhle f (-n) **1** cave. **2** hollow pit; burrow.

3 (incl Anat.) cavity, socket. **höhlen** vt hollow, excavate. **höhlen·artig** a cavelike, concave

hohl· hollow, concave: **-erhaben** a concavoconvex, **-geschliffen** a hollow-ground. **H–gitter** n concave grating. **-glas** n hollow (-blown) glass(ware). **-granate** f 1 hollow shell. 2 constant-temperature bath. **-guß** m hollow casting, cored work, **-hand** f palm (or hollow) of the hand. **-kehle** f groove, flute; fillet

höhlig a full of cavities; cavernous; honeycombed; (glass) blebby

Hohl·körper m hollow body. **-kugel** f hollow sphere. **-ladung** f (Explo.) hollow charge. **-leiter** m waveguide. **-linse** f hollow lens. **-mantel** m (Mach.) jacket. **-maß** n capacity measure, dry measure. **-perle** f hollow (glass) bead. **-prisma** n hollow prism. **-profil** n hollow section. **-raum** m hollow space, cavity

Hohlraum·bildung f cavitation. **-gehalt** m void volume. **-resonator** m cavity resonator. **-schießen** n cushioned blasting. **-strahlung** f cavity (or black-body) radiation. **-volumen** n pore space

hohl·rund a concave. **H–schicht** f air space. **-schliff** m hollow grinding. **-sog·bildung** f cavitation. **-spat** m andalusite, chiastolite. **-spiegel** m concave mirror. **-tier** n cœlenterate

Höhlung f (-en) cave, hollow; pit, excavation

Hohl·vene f (Anat.) vena cava. **-walze** f holly roll(er) (or cylinder). **-welle** f (Mach.) hollow shaft. **-zahn** m (Bot.) hemp nettle. **-ziegel** m hollow brick (or tile). **-zirkel** m inside calipers

H-Öl n hydrazine hydrate

hold pred a (dat) kind (to), fond (of)

Holder m (Bot.) elder = HOLUNDER

holen vt 1 fetch, (go and) get; call. 2 (+dat rflx sich) pick up, catch (a disease)

holl· abbv (holländisch) Dutch

Holländer m (-) 1 (Paper, etc) hollander, beating engine. 2 Dutchman. **holländern** vt beat (in a hollander). **Holländer·weiß** n Dutch white. **holländisch** a Dutch **—h—es Geschirr** (Paper) hollander; **h—es Öl, h—e Flüssigkeit** ethylene chloride

Hölle f (-n) hell

Höllen·öl n 1 curcas oil. 2 castor oil. 3 low-grade olive oil. **-stein** m lunar caustic; lapis infernalis

Holl. P. abbv (holländisches Patent) Dutch patent

Hollunder m elder, lilac = HOLUNDER

Holm m (-e) 1 bar, crossbeam, (fence) rail; spar. 2 handle. 3 island; hill. **-firnis** m spar varnish

Holo/eder n (Cryst.) holohedron. **Holo·edrie** f holohedrism. **holoedrisch** a holohedral

holokristallin a holocrystalline

holp(e)rig a rough, uneven; bumpy. **holpern** vi jolt, bump; stumble

Holunder m elder (tree); (spanischer, türkischer) H. lilac. **-beere** f elderberry. **-blüte** f elder blossom. **-wein** m elderberry wine

Holz n (-er) 1 wood; timber. 2 piece of wood: beam, bat, stick, etc. 3 (small) forest; thicket; trees. **-abfall** m, **-abfälle** mpl wood waste, scrap wood. **h–ähnlich** a woodlike, ligneous. **H–alkohol** m wood alcohol. **-amiant** m ligneous asbestos. **-apfel** m crabapple. **-apfelwein** m crabapple cider. **h–artig** a woody, ligneous, xyloid. **H–asbest** m ligneous asbestos. **-asche** f wood ashes. **-aschenlauge** f wood ash lye. **-äther** m methyl ether; **essigsaurer H.** methyl acetate. **-bearbeitung** f woodworking. **-beize** f wood stain. **-beizen** n wood staining. **-bindemittel** n adhesive for wood. **-blau** n logwood (blue). **-bohrer** m 1 (Tool) gimlet, auger, wood drill. 2 (Zool.) wood borer. **-bottich** m wooden vat (tub, tank). **-branntwein** m wood (or methyl) alcohol. **-brei** m woodpulp. **-büchse** f wooden box (or case). **-cassie** f cassia bark. **-chemie** f wood chemistry

Hölzchen n wood splint(er); (match)stick; match

Holz·dämpfer m (Paper) wood digester. **-deckel** m wooden lid (or cover). **-destillation** f (dry) distillation of wood. **-destillierungsanlage** f wood distilling plant

hölzern a wooden; ligneous

Holzessig m 1 acetic acid (made from:) 2 pyroligneous acid. **-geist** m wood alcohol. **-säure** f 1 wood vinegar. 2 =

HOLZESSIG. **h—sauer** *a* pyrolignite of — **h—es Eisen** iron liquor

holz·farbig *a* wood-colored: **H—farbstoff** *m* wood dye. **-faser** *f* 1 wood fiber. 2 lignin, etc = HOLZFASERSTOFF. 3 grain of the wood

Holzfaser·masse *f* woodpulp. **-papier** *n* woodpulp paper. **-platte** *f* (wood) fiberboard. **-stoff** *m* 1 wood fiber. 2 woodpulp. 3 lignin; cellulose; lignocellulose

Holz·fäule, -fäulnis *f* wood rot, dry rot. **-feuer** *n* wood fire. **-firnis** *m* wood varnish. **h—frei** *a* (*Paper*) wood-free. **H—fuß** *m* wooden foot (base, stand). **-gas** *n* wood gas. **-gattung** *f* type (variety, species) of wood. **-geist** *m* wood alcohol, methanol. **h—geistig** *a* containing wood alcohol. **H—gerbstoff** *m* wood tannin. **-gestell** *n* wooden stand (or frame). **-gewächs** *n* woody (or ligneous) plant. **-gewebe** *n* woody (or ligneous) tissue, xylem. **-gewerbe** *n* wood trade, lumber (or timber) industry. **-griff** *m* wooden handle. **-grundierfarbe** *f* (*Paint*) wood primer. **-gummi** *n* wood gum, xylan. **-handel** *m* wood (lumber, timber) trade (or business). **-harz** *n* wood resin (or rosin)

holzig *a* woody; stringy

Holz·kalk *m* pyrolignite of lime, crude calcium acetate. **-karton** *m* (wood) fiberboard, wood-pulp board. **-kassie** *f* coarse cassia bark, cassia lignea. **-kasten** *m* wooden box (case, vat). **-kern** *m* heartwood. **-kirsche** *f* wild cherry. **-kistchen** *n* (small) wooden box. **-kiste** *f* wooden box (case, crate). **-kitt** *m* plastic wood; painter's putty. **-klotz** *m* wood(en) block. **-kocher** *m* (*Paper*) wood digester (or cooker). **-kohle** *f* (wood) charcoal

Holzkohlen· charcoal: **-brennerei** *f* charcoal burning (plant). **-brennofen** *m* charcoal furnace. **-eisen** *n* charcoal (or Norway) iron. **-feuer** *n* charcoal fire. **-klein** *n* fine charcoal. **-lösche** *f* charcoal dust. **-meiler** *m* charcoal mound (or pile). **-ofen** *m* charcoal oven (or stove). **-pulver** *n* powdered charcoal. **-roheisen** *n* charcoal pig iron

Holz· wood(en); ligneous: **-konservierung** *f* wood preservation. **-konservierungs-mittel** *n* wood preservative. **-körper** *m* woody tissue, xylem. **-kugel** *f* wooden ball. **-kupfer(erz)** *n* (fibrous) olivinite. **-lack** *m* 1 (*Paint*) wood lacquer (varnish, finish). 2 stick lac. **-leim** *m* wood glue. **-mangold** *m* (*Bot.*) wintergreen, pyrola. **-masse** *f* wood paste; woodpulp. **-mehl** *n* wood flour (meal, dust, powder); sawdust. **-melasse** *f* wood molasses. **-naphta** *n* wood naptha, crude wood alcohol. **-öl** *n* wood oil, (*specif*) tung oil. **-öllack** *m* wood oil varnish. **-papier** *n* wood(pulp) paper. **-pappe** *f* (*Paper*) woodpulp board. **-pech** *n* wood pitch. **-pflanze** *f* woody plant. **-pulver** *n* wood powder, sawdust. **-reif(en)** *m* wooden hoop (or ring). **-rot** *n* redwood extract, hematoxylin. **-ruß** *m* wood soot. **-sägemehl** *n* sawdust. **h—sauer** *a* pyrolignite of. **H—säure** *f* pyroligneous acid. **-scheit** *n* log, piece of wood **-schleifer** *m* (*Paper*) woodpulp grinder. **-schliff** *m* (*Paper*) groundwood, (mechanical) woodpulp. **-schnitt** *m* woodcut. **-schutzmittel** *n* wood preservative. **-schwamm** *m* dry-rot fungus. **-schwarz** *n* wood black. **-span** *m* wood shaving (or chip). **-spanplatte** *f* chipboard. **-spiritus** *m* wood alcohol, methanol. **-stativ** *n* wooden stand (or easel). **-stein** *m* petrified wood. **-stich** *m* wood engraving, woodcut. **-stoff** *m* 1 chemical woodpulp. 2 (wood) cellulose. 3 lignin

holzstoff·haltig *a* containing woodpulp. **H—masse** *f* woodpulp

holzstoff·frei *a* (*Paper*) free from woodpulp

Holz· wood(en); ligneous: **-substanz** *f* wood substance, lignocellulose. **-tafel** *f* wooden sign(board). **-tee** *m* (*Pharm.*) wood tea. **-teer** *m* wood tar. **-teil** *m* (*Bot.*) xylem. **-terpentinöl** *n* (steam-distilled) wood turpentine. **-trank** *m* (*Pharm.*) wood drink. **-tränkung** *f* wood pickling. **-trocknung** *f* wood drying

Holzung *f* (-en) 1 small wood(s), grove. 2 tree-chopping, woodcutting

Holz· wood(en); ligneous: **-vergasung** *f* wood distillation. **-verkohlung** *f* 1 carbonization of wood. 2 destructive distillation of wood. 3 charcoal burning. **-verschlag** *m* 1 wooden partition. 2 wooden crate. **-verzuckerung** *f* saccharification (or hydrolysis) of wood (cellulose). **-watte, -wolle** *f* excelsior, wood wool. **-zellstoff** *m* woodpulp, wood cellulose, lignocellulose. **h–zerstörend** *a* wood-destroying. **H–zeug** *n* woodpulp. **-zimt** *m* cassia bark. **-zinn(erz)** *n* wood tin, fibrous cassiterite. **-zucker** *m* wood sugar, D-xylose; carbohydrate mixture from the hydrolysis of wood

Homo-brenzcatechin *n* homopyrocatechol. **-camphersäure** *f* homocamphoric acid. **h–chrom** *a* homochromous, of uniform color; (*Dye.*) monogenetic. **-edrisch** *a* homohedral. **H–gallusaldehyd** *m* homogallaldehyde

homogen *a* homogeneous. **homogenisieren** *vt* homogenize. **Homogen(e)ität** *f* homogeneity. **Homogen-kohle** *f* solid carbon

Homo-kaffeesäure *f* homocaffeic acid. **-kampfersäure** *f* homocamphoric acid

homolog *a* homologous. **Homolog** *n* (-e) homologue. **Homologie** *f* homology

Homolyse *f* homolysis. **homolytisch** *a* homolytic

homöomorph *a* homeomorphous. **Homöomorphie** *f* homeomorphism

homöopolar *a* homeopolar

Homo-phtalsäure *f* homophtalic acid. **-pinocamphersäure** *f* homopinocamphoric acid. **-polymer(e)** *n* homopolymer. **-veratrumaldehyd** *m* homoveratraldehyde. **-veratrumsäure** *f* homoveratric acid. **h–zygotisch** *a* homozygous. **-zyklisch** *a* homocyclic

Honig *m* honey. **h–ähnlich, -artig** *a* honey-like, melleous. **H–biene** *f* honeybee. **-essig** *m* oxymel. **h–farbig** *a* honey-colored. **-gelb** *a* honey-yellow. **H–geruch** *m* odor of honey. **-geschmack** *m* taste of honey. **-gras** *n* (*Bot.*) soft grass, holcus. **h–haltig** *a* honeyed, containing honey. **H–harnruhr** *f* diabetes mellitus. **-klee** *m* sweet clover, melilot. **-saft** *m* nectar. **-säure** *f* mellitic acid; (*Pharm.*) oxymel. **-scheibe** *f* honeycomb. **-seim** *m* (liquid) honey. **-stein** *m* honeystone, mellite. **-steinsäure** *f* mellitic acid. **-strauch** *m* honeyflower. **h–süß** *a* honey-sweet. **H–tau** *m* honeydew. **-trank** *m* mead. **-wabe** *f* honeycomb. **-wabenbau** *m* honeycomb (construction). **-wasser** *n* hydromel. **-wein** *m* mead, honey wine. **-zucker** *m* levulose

Honorar *n* (-e) honorarium, fee. **hono-** rieren *vt* pay (a fee) for; repay

hopfen *vt* hop, add hops to (beer)

Hopfen *m* hop(s). **--abkochung** *f* hop(s) concoction. **h–ähnlich, -artig** *a* hopslike. **H–aufguß** *m* infusion of hops. **-bau** *m* hop culture (*or* cultivation). **-baum** *m* hop tree. **-bitter** *n* hop bitters, lupulin. **-bittersäure** *f* hop bitter acid. **-bitterstoff** *m* hop bitters, lupulin. **-darre** *f*, **-darrofen** *m* hop kiln. **-drüse** *f* (*Bot.*) lupulinic gland. **-gerbstoff** *m* hop tannin. **-glattwasser** *n* hop sparge. **-harz** *n* hop resin. **-klee** *n* black medic, shamrock. **-mehl** *n* hop dust, lupulin. **-mehltau** *m* hop blight. **-öl** *n* hop oil — **spanisches H.** origanum oil. **-staub** *m* hop dust, lupulin. **-stopfen** *n* (*Brew.*) dry hopping. **-treber** *pl* spent hops. **-trieb** *m* (*Brew.*) frothy head, first stage of fermentation

Hör-apparat *m* 1 hearing aid. 2 sound locator

hörbar *a* audible. **Hörbarkeit** *f* audibility

horchen *vi* (+**auf**) listen (to); eavesdrop (on)

Horch-gerät *n* 1 listening device. 2 sound locator (*or* detector)

Horde *f* (-n) 1 horde, gang; tribe. 2 (*Brew.*) kiln floor. 3 (storage) hurdle, (drying) rack. 4 lattice

Horden-kontakt *m* grid-type catalyst. **-trockner** *m* rack (*or* shelf) drier. **-wäscher** *m* hurdle-type scrubber

hören 1 *vt* hear, listen (to) —**das läßt sich h.** that sounds good. 2 *vt/i* e.g.: **sie h. (Chemie) bei X** they attend (chemistry) lectures by X. 3 *vi* (**auf**) listen (to). *Cf* GEHÖREN. **Hörer** *m* (-) 1 earphone; (telephone) receiver. 2 & **Hörerin** *f* (-nen) student, listener (*esp* at lectures), *pl oft* = **Hörerschaft** *f* (listening) audience

Hör-frequenz *f* audio frequency. **-gerät** *n* 1 listening device. 2 sound detector

Horizont *m* (-e) horizon. **Horizontale** *f* (*adj endgs*) horizontal (line). **horizontieren** *vt* level (instruments)

Hormon-absonderung *f* hormone secretion. **h–artig** *a* hormonal, hormone-like. **H–drüse** *f* hormonal (*or* endocrine) gland. **h–haltig** *a* hormone-containing

Horn-abfall *m* horn waste (*or* chippings). **h–ähnlich, -artig** *a* horny, horn-like. **H–**

blatt *n* 1 horn blade (for cleaning mortars). 2 horn sheet (*or* lamina). 3 (*Bot.*) hornwort. **-blei** *n* 1 fused lead chloride. 2 = **-bleierz** *n* (*Min.*) phosgenite. **-blendeschiefer** *m* hornblende schist

Hörnchen *n* (-) 1 little horn. 2 (*Med., Zool.*) corniculum. 3 crescent; elbow (*fig.*). 4 cone

Horn·chlorsilber *n* horn silver, cerargyrite

hörnern *a* (of) horn, keratinous; horny

Hör·nerv *m* auditory nerve

Horn·erz *n* horn silver, cerargyrite. **-gebilde** *n* horn, horny process. **-gewebe** *n* horny tissue. **-gummi** *n* hard rubber. **-haut** *f* 1 horny skin, callus; dead skin. 2 cornea, (*in compds, usu*) corneal

hornig *a* horny, corneous; keratinous, keratose

hornisieren *vt* 1 hornify. 2 vulcanize

Hornisse *f* (-n) hornet

Horn·klee *m* crowtoe, bird's-foot trefoil. **-kobalt** *m* asbolite. **-löffel** *m* horn spoon. **-mohn** *m* horn (*or* sea) poppy. **-öl** *n* horn oil. **-quecksilber** *n* horn quicksilver, native mercurous chloride. **-schicht** *f* horny layer; epidermis. **-silber** *n* horn silver, e.g. silver chloride; (*Min.*) cerargyrite. **-spatel** *m* horn spatula. **-stein** *m* hornstone, chert, silex. **-stoff** *m*, **-substanz** *f* horn(y substance *or* matter), keratin

Hör·rohr *n* stethoscope. **-saal** *m* lecture room (*or* hall), auditorium. **-stein** *m* ear stone, otolith

Hortensien·blau *n* Prussian blue

Hosen·rohr *n* siphon-pipe (of a hot-blast oven)

HOZ *abbv* (Hochofenzement) Portland blast-furnace cement

Hptw. *abbv* (Hauptwerk) main part (e.g. of *Beilstein*)

hrsg. *abbv* (herausgegeben) published, issued. **Hrsg.** *abbv* (Herausgeber) publisher, editor

H-Säure *f* H-acid, 1-amino-8-naphthol-3,6-disulfonic acid

h.s.l. *abbv* (heiß sehr löslich) very soluble hot

H-Strahlen *mpl* H-rays, proton particles

Hub *m* (-̈e) 1 lifting. 2 lift (*incl Mach.*, e.g.

of a pump). 3 (*Mach.*) travel, stroke (of pistons); throw (of cranks). **--bewegung** *f* lifting motion; movement of the stroke

hüben *adv* on this side: **h. und** (*or* **wie**) **drüben** on this side and the other, on both sides

Hub·flug *m* buoyant flight. **-karre** *f*, **-karren** *m* lift truck. **-kolbenmotor** *m* reciprocating piston engine. **-kraft** *f* 1 lifting power. 2 buoyancy. **-platte** *f* pallet. **-rad** *n* 1 lifting wheel. 2 (*Sugar*) beet wheel. **-raum** *m* (*Mach.*) piston displacement; stroke volume; (*esp Cars*) cubic (cylinder) capacity

hübsch *a* 1 pretty, nice-looking. 2 nice. 3 tidy, considerable (sum, etc). 4 *adv oft* nice and . . . , properly, nicely

Hub·schaufel *f* lifter, lifting flight. **-schnecke, -schraube** *f* lifting screw (*or* worm). **-schrauber** *m* helicopter. **-stapler** *m* fork-lift (truck). **-untersatz** *m* pallet. **-vermögen** *n* lifting power. **-werk** *n* hoisting gear. **-zahl** *f* number of strokes per minute

Huf *m* (-e) hoof. **--eisen** *n* horseshoe

hufeisen·förmig *a* horseshoe-shaped. **H--fuß** *m* horseshoe base

Huf·fett *n* neat's foot oil; hoof ointment. **-lattich** *m* coltsfoot. **-nagel** *m* horseshoe nail. **-schlag** *m* 1 hoof beat. 2 horseshoeing. **-schmied** *m* blacksmith

Hüft·bein *n* hip bone

Hüfte *f* (-n) hip; haunch

Hüft·gelenk *n* hip joint

Huf·tier *n* hoofed animal, ungulate

Hüft·loch *n* (*Anat.*) obturator foramen. **-nerv** *m* sciatic nerve

Huf·träger *m* hoofed animal, ungulate

Hügel *m* (-) 1 hill, elevation. 2 mound. 3 protuberance, knob

Huhn *n* (-̈er) chicken. **Hühnchen** *n* (-) chick, pullet

Hühner·auge *n* corn (on the foot). **-biß** *m* (*Bot.*) chickweed. **-eiweiß** *n* egg white, egg albumin. **h--eiweißartig** *a* albuminous. **H--fett** *n* chicken fat. **-kot, -mist** *m* chicken droppings

huldigen *vi* (*dat*) pay homage (to); subscribe (to), advocate (a view); indulge (in), be addicted (to) (a practice)

hülfe would help: *sbjc of* HELFEN

Hülfe f help, aid = HILFE

Hülle f (-n) wrapping, wrapper, cover(ing); case, casing; jacket; sheath, envelope; shell (*incl Electron.*); wrap, (*incl pl*) clothing —**in** (or **die**) **H. und Fülle** in great abundance. **hüllen** vt wrap; veil, cloak

Hüllen·elektron n orbital (or shell) electron. **-stoff** m cover (or wrapping) material, (*specif*) balloon fabric

Hüll·rohr n encasing tube, jacket

Hülse f (-n) case, sheath; shell; (*esp Bot.*) husk, hull, pod; (*esp Mach.*) sleeve, collar; bushing, socket; (*Tex.*) empty bobbin; (*Nucl.*) can

hülsen·artig a 1 sheath-like, etc, cf HÜLSE. 2 leguminous. **H–baum** m 1 holly. 2 locust (or carob) tree. **-frucht** f legume; pulse. **h–fruchtartig, -früchtig** a leguminous. **H–früchtler** mpl Leguminosae, leguminous plants. **-gewächs** n legume, leguminous plant. **-mutter** f (*Mach.*) sleeve nut. **-pflanze** f leguminous plant. **-schliff** m socket joint. **-träger** m legume

human a humane

Humat n (-e) humate

Humifizierung f (*Soil*) humification

Humin·säure f humic acid. **-stoff** m., **-substanz** f humic (or humous) substance

Hummel f (-n) bumblebee

Hummer m (-) lobster

humos a humous

humpelig a limping. **humpeln** vi (h/s) hobble, limp

Humulin·säure f humulic acid

Humulo·chinon n humuloquinine

Humus·bildung f humus formation. **-boden** m humic (or humous) soil. **-decke** f covering of humus. **-erde** f humus, humic (or humous) soil. **-kohle** f humic coal. **h–reich** a richly humous, rich in humus. **H–säure** f humic acid. **-schicht** f layer of humus. **-stoff** m humous substance

Hund m (-e) 1 dog; hound. 2 (*Min.*) mine car, trolley. 3 (*Engg.*) rammer. 4 (*Metll.*) guard

Hunde·blume f dandelion. **-futter** n dog food

hundert num, **Hundert** n (-e) (one or a) hundred —**vom H.** percent. **hunderterlei** 1 a (*no endgs*) a hundred kinds of. 2

pron a hundred (different) things

Hundertel n (-) hundredth = HUNDERTSTEL

hundert·fach a hundredfold; adv oft a hundred times. **-gradig** a centigrade. **-mal** adv a hundred times. **-prozentig** a hundred-percent. **H–satz** m percentage

hundertste num a hundredth. **Hunderstel** n (-) (*Frac.*) (one) hundredth

hundert·teilig a centesimal, centigrade

Hunde·zahnspat m dogtooth spar (type of calcite)

Hündin f (-nen) (female) dog, bitch

Hunds·blume f dandelion. **-gift** n dogbane. **-gras** n couch grass, cocksfoot. **-kamille** f (dog's) camomile, mayweed. **-kohl** m dog's mercury; dogbane. **-wut** f rabies. **-zahn** m 1 canine tooth. 2 (*Bot.*) dog's tooth. **-zahnspat** m dogtooth spar (type of calcite). **-zunge** f, **-zungenkraut** n (*Bot.*) hound's tongue

Hunger m hunger: (*in compds oft*) starvation; **H. haben** be hungry. **-korn** n ergot. **hungern** vi hunger, be (or go) hungry. **Hungers·not** f famine

Hunger·stein m salt-pan scale (calcium & sodium sulfates). **-stoffwechsel** m starvation metabolism. **-tod** m (death from) starvation

hungrig a hungry

Hupe f (-n) horn. **hupen** vi sound the horn

hupfen, hüpfen vi (h/s) hop, jump

Hürde f (-n) 1 fence; pen, (sheep)fold. 2 hurdle. 3 (storage or drying) rack. **Hürden·trockner** m drying rack; hot-air drier (for vegetables, etc)

hurtig a quick, swift; nimble, agile

husten vi, **Husten** m cough. **-mittel** n cough remedy

Hut I m (¨e) 1 hat. 2 (*fig, incl Bot., Mach.*) cap, lid; (*Bot. also*) pileus. 3 (*Geol., Min.*) gossan. 4 scum (of wine). 5 layer of spent tanbark. **II** f 6 guard: **auf der H.** on one's guard. 7 care, protection. 8 pasture (land); herd (of cattle)

Hütchen n (-) small (or little) hat (cap, lid); capsule; (*Bot.*) pileus

hüten I vt 1 watch (over), guard. 2: **das Bett (das Haus) h.** stay in bed (stay at home). **II: sich h.** watch out, be careful, be on one's guard. **Hüter** m (-) guard, protector; keeper; attendant; herder

Hut·filz *m* hatter's felt. **-lack** *m* hat varnish. **-mutter** *f* (*Mach.*) cap nut. **-pilz** *m* pileate fungus, mushroom

Hütte *f* (-n) **1** hut, shack. **2** metallurgical plant; foundry, mill; glassworks

Hütten· metallurgical, foundry, smelting: **-abfall, -after** *m* metallurgical waste. **-arbeit** *f* foundry work. **-bims** *m* foamed slag. **-blei** *n* smelter lead. **-glas** *n* (*Glass*) pot metal. **-herr** *m* smeltery owner. **-kalk** *m* blast-furnace slag (used as fertilizer). **-katze** *f* lead colic. **-kokerei** *f* coke oven plant (at an iron foundry). **-koks** *m* metallurgical coke. **-kunde** *f* metallurgy. **-kundige** *m,f* (*adj endgs*) metallurgist. **-kupfer** *n* smelter (*or* pig) copper. **-leute** metallurgists, etc: *pl of* **-mann** *m* metallurgist, smelter, founder. **h–männisch, -mäßig** *a* metallurgical. **H–mehl** *n* white arsenic powder. **-nichts** *n* white tutty (impure zinc oxide). **-probierkunst** *f* metallurgical assaying. **-prüfer** *m* metallurgical assayer. **-rauch** *m* **1** flue dust, furnace (*or* smelter) smoke. **2** white arsenic. **3** (*Glass*) bloom. **-reise** *f* (*Metll.*) (furnace) campaign. **-rohkupfer** *n* crude (smeltery) copper. **-sand** *m* granulated blast-furnace slag. **-speise** *f* ore to be smelted. **-stein** *m* slag brick. **-technik** *f* (extractive) metallurgy. **-trichter** *m* conical funnel. **-werk** *n* metallurgical plant, iron (*or* steel) mill, foundry. **-wesen** *n* smelting, metallurgy. **-zement** *m/n* slag cement. **-zink** *n* smelter zinc, spelter. **-zinn** *n* (*Metll.*) grain tin

Hutze *f* (-n) scoop; hood. **hutzen·artig** *a* hood-like

Hutz·zinn *n* cap (*or* Malacca) tin. **-zucker** *m* loaf sugar

h.w.l. *abbv* (heiß wenig löslich) not very soluble when hot

HWS *abbv* (Halbwertsschicht) (*X-Rays*) half-value thickness

HWZ *abbv* (Halbwertzeit) half-life

Hyacinth·granat *m* cinnamon garnet, (h)essonite

Hyaluron·säure *f* hyaluronic acid

Hyäne *f* (-n) hyena. **Hyän·säure** *f* hyenic acid

Hyazinth *m* (h)essonite **—grönländischer H.** eudialyte

Hyazinthe *f* (-n) hyacinth. **Hyazinthen·öl** *n* hyacinth oil

Hyazinth·granat *m* cinnamon garnet, (h)essonite)

hybrid *a* **1** hybrid. **2** arrogant. **H–·rohöl** *n* mixed-base crude oil. **hybridisieren** *vt* hybridize

Hydantoin·säure *f* hydantoic acid

Hydnocarp(us)·säure *f* hydnocarpic acid

Hydracyl·säure *f* hydracylic (*or* 3-hydroxy-propanoic) acid

Hydrast·säure *f* hydrastic acid

Hydrat *n* (-e) hydrate. **hydrat.** *abbv* (hydrat(is)iert) hydrated

Hydratase *f* hydrase

Hydratation *f* (-en) hydration. **Hydratations·wärme** *f* heat of hydration

Hydrat·bildung *f* hydration, hydrate formation. **-cellulose** *f* regenerated cellulose. **h–haltig** *a* hydrated

hydratieren *vt* hydrate = HYDRATISIEREN

Hydrations·wärme *f* heat of hydration

hydratisch *a* hydrated. **hydratisieren 1** *vt* hydrate. **2: sich h.** be(come) hydrated. **Hydratisierungs·grad** *m* degree of hydration. **Hydrat·kalk** *m* hydrated lime

Hydratropa·aldehyd *m* hydratropaldehyde. **-alkohol** *m* hydratropic alcohol. **-säure** *f* hydratropic (*or* phenylpropionic) acid

Hydrat·wasser *n* water of hydration. **h–wasserhaltig** *a* containing water of hydration. **H–zellulose** *f* regenerated cellulose. **-zustand** *m* state of hydration

Hydraulik *f* hydraulics. **hydraulisch** *a* hydraulic. **Hydraulit** *m* hydraulic binder

Hydrazo·benzol *n* hydrazobenzene. **-körper** *m*, **-verbindung** *f* hydrazo compound

Hydrid *n* (-e) (-ic) hydride *cf* HYDRÜR

Hydrier·analage *f* hydrogenation plant. **hydrieren** *vt* hydrogenate; hydrate

Hydrier·pech *n* synthetic pitch. **-rückstand** *m* hydrogenation residue

Hydrierung *f* (-en) hydrogenation; hydration. **Hydrier·werk** *n* hydrogenation plant

Hydrindan *n* perhydroindane

hydro·aromatisch *a* hydroaromatic, alicyclic. **H–bromsäure** *f* hydrobromic acid. **-chinon** n hydroquinone. **-chino-xalin** n hydroquinoxaline. **-chlorkau-tschuk** m hydrochlorinated rubber. **-chlorsäure** f hydrochloric acid. **-crackreaktion** f hydrocracking (reaction). **-cumaron** n hydroc(o)umarone. **-cumarsäure** f hydroc(o)umaric acid. **-cyansäure** f hydrocyanic acid. **-disulfid** n dihydrogen disulfide. **-finierung** f hydrofining

hydrogenieren vt hydrogenate, hydrogenize. **Hydrogenisation** f hydrogenation.

hydrogenisieren vt hydrogenate, hydrogenize

Hydrogen-schwefel m hydrogen sulfide

hydrohalogenieren vt hydrohalogenate

Hydro-jodid n hydriodide. **-jodsäure** f hydriodic acid. **-kaffeesäure** f hydrocaffeic acid. **-karbür** n hydrocarbon. **-kautschuk** m hydrogenated rubber. **-kette** f (Elec.) hydrocell, hydroelement

Hydrolin n methylcyclohexanol

Hydrolyse f (-n) hydrolysis. **hydrolysierbar** a hydrolyzable. **hydrolysieren** vt/i hydrolyze. **Hydrolisierung** f (-en) hydrolysis, hydrolyzation. **Hydrolysierungs-fähigkeit** f hydrolyzing capacity, hydrolyzability. **Hydrolisier-zahl** f hydrolyzation number. **hydrolytisch** a hydrolytic **—h. gutes Glas** neutral (or high chemical durability) glass

Hydro-peroxid n hydrogen peroxide hydroperoxide. **-persulfid** n hydrogen persulfide

hydrophil a hydrophilic

hydrophob a hydrophobic. **Hydrophobie** f 1 (Med.) hydrophobia, rabies. 2 hydrophobic (or water-repellent) property. **hydrophobieren** vt make water-repellent. **hydrophobiert** p.a water-repellent

Hydro-phthalsäure f hydrophthalic acid. **-polysulfid** n hydrogen polysulfide. **h-schweflig** a hydrosulfurous. **H-silikat** n hydrated silicate, silicate hydrate

Hydrosulfit-ätze f hydrosulfite discharge. **-küpe** f hydrosulfite vat

Hydro-thionsäure f hydrosulfuric acid, hydrogen sulfide

hydrotrop a hydrotropic

Hydro-verbindung f hydro compound

Hydroxam-säure f hydroxamic (or hydro-

xylamidosulfuric) acid

Hydroxid n (-e) (usu higher-valent) hydroxide cf OXYD. **Hydroxydul** n (lower-valent) hydroxide

Hydroxyl-aminsulfonsäure f hydroxylamidosulfuric acid. **h-haltig** a containing hydroxyl

hydroxylieren a hydroxylate

Hydro-zellulose f hydrocellulose. **-zimtsäure** f hydrocinnamic acid

Hydrür n (-e) (lower-valent or -ous) hydride

Hygienik f hygiene, hygienics

Hygroskopizität f hygroscopicity

hylotrop a hylotropic

Hyperämie f hyperemia

Hyperbel f (-n) hyperbole. **--bahn** f hyperbolic path (or orbit)

Hyper-chlorat n perchlorate. **-chlorid** n perchloride. **-chlorsäure** f perchloric acid. **h-eutektisch** a hypereutectic. **-eutektoidisch** a hypereutectoid. **-fein** a hyperfine. **H-glukämie** f hyperglycemia. **-jodat** n periodate. **-jodsäure** f periodic acid. **h-mangansauer** a permanganate of. **H-mangansäure** f permanganic acid. **-oxid** n (hy)peroxide

hypersensibilisieren vt hypersensitize

Hyper-vitaminose f hypervitaminosis

Hypo-chloritlauge f hypochlorite liquor (or solution). **h--eutektisch** a hypoeutectic

Hypogäa-säure f hypogeic acid

Hypo-hirnsäure f hypocerebric acid. **-jodit** n hypoiodite. **h-phosphorig** a hypophosphorous

Hypophyse f (-n) hypophysis, pituitary gland

hypo-salpetrig a hyponitrous. **H-sulfit** n hyposulfite, dithionite

Hypothek f (-en) mortgage; burden

Hypothese f (-n) hypothesis. **hypothetisch** a hypothetical

Hypovitaminose f hypovitaminosis, vitamin deficiency

Hysterese f hysteresis

Hysterie f hysteria, hysterics

Hz abbv (Hertz) hertz, cycle (per second)

HZ abv 1 (Hydrolisierzahl) hydrolization number. 2 (Hertzzahl) Hertz unit, cycle per second

I

NOTE: Except in the word **Ion** ion, initial **I** before vowels is replaced by **J** in cognate words, e.g. **Jod** iodine.

i. *abbv* **1** (in, im) in (the). **2** (ist) is. **3** (imido) imino, imido

i.a. *abbv* (im allgemeinen) in general

i.A. *abbv* (im Auftrag) on behalf of; (*with signatures*) by (order of), per

i. allg. *abbv* (im allgemeinen) in general

Iatrochemie *f* iatrochemistry

i.b. *abbv* (im besonderen) in particular

i.B. *abbv* (+**auf**) **1** (in Bezug) with respect (to). **2** (in Berechnung) calculated on the basis (of)

ich *pron* I. **Ich** *n* (*oft in compds*) ego

Ichthyol *n* ichthyol, ichthammol, ammonium ichthosulfonate. **--seife** *f* ichthyol soap

·id *sfx* (*Chem.*) -ide *cf* -IDE

iD *abbv* **1** (im Durchschnitt) on the average. **2** (im Dampf) in steam, in the vapor state. **3** (im Dunkeln) in darkness

·ide *sfx, pl* (*Chem.*) -ides *cf* ·ID

Ideal·lösung *f* ideal solution

Idee *f* (-n) idea, concept. **ideell** *a* ideal; idealistic. **Ideen·verbindung** *f* association of ideas

identifizierbar *a* identifiable. **identifizieren** *vt* identify. **Identifizierung** *f* (-en) identification. **identisch** *a* identical. **Identität** *f* (-en) identity. **Identitäts·periode** *f* (*Cryst.*) identity period, repeat distance

Idiosynkrasie *f* (-n) idiosyncrasy; (*Med.*) allergy

idioten·sicher *a* foolproof

Idit *n* iditol

Idon·säure *f* idonic acid

Idozucker·säure *f* idosaccharic acid

Idrilin *n* hepatic cinnabar

Idris·öl *n* lemongrass oil

IE *abbv* **1** (Immunisierungseinheit) immunization unit. **2** (Insulineinheit) insulin

unit. **3** (Internationale Einheit) international unit

I-echt *a* (*Dye.*) of the best fastness

·ig *sfx* (*Chem.*) -ous *cf* CHLORIG

IG *abbv* (Interessengemeinschaft) trust, combine

Igel *m* (-) **1** (*Zool.*) hedgehog, porcupine. **2** (*Agric.*) harrow. **--weizen** *m* German bearded wheat, spelt

Ignatius·bohne *f* St. Ignatius bean

ihm *pron* **1** (*dat of* ER) (to, for) him (*or* it). **2** (*dat of* ES) (to, for) it

ihn *pron* (*accus of* ER) him, it

ihnen *pron* (*dat of* SIE *pl*) (to, for) them. **Ihnen** *pron* (*dat of* SIE) (to, for) you

ihr *pron* **1** (*dat of* SIE) (to, for) her, it. **2** (*possess, with endgs*) her, its, their. **Ihr** (*possess, with endgs*) your

ihrer **1** pron (*genit of* SIE) (of) her, (of) them. **2** (*form of* IHR 2). **Ihrer 1** *pron* (*genit of* SIE) (of) you. **2** (*form of* IHR)

ihrerseits *adv* on her (its, their) part, in her (its, their) turn

ihresgleichen *pron* her (*or* their) like (*or* kind), one like her (*or* it), ones like them

ihresteils *adv* on her (its, their) part

ihrethalben *adv* on her (its, their) account (*or* behalf)

ihrig *pron: der* (**das, die**) **i—e** hers, theirs, its own. **Ihrig** *pron* yours

i.J. *abbv* (im Jahre) in the year

IK, I.K. *abbv.* (Immunkörper) immune body

Ikat·verfahren *n* (*Tex.*) Ikat process

Ikoka·öl *n* icoca oil

Ikosaeder *n* (-) icosahedron. **Ikosite·traeder** *n* (-) icositetrahedron

i.L. *abbv.* (im Liter) per liter

illitisch *a* illitic

illuminieren *vt* illuminate

illusionistisch, illusorisch *a* illusory

illustrieren vt illustrate. **Illustrierte** f (adj endgs) pictorial magazine (or newspaper)
Iltis m (-se) European polecat, fitchet
im (contrac) = IN DEM in (the) cf IN, DEM
imaginär a imaginary
im allg(em). abbv (im allgemeinen) in general
Imber m ginger = INGWER
imbibieren vt imbibe
Imbiß m (..isse) snack, bite to eat; lunch
Imhoff-Brunnen m Imhoff tank
Imidichlorid n imido-chloride
Imido- 1 (in compds of imide nature) imido-.
2 (otherwise) imino-. **:-äther** m imido ester. **-säure** f imido acid. **-sulfonsäure** f imidosulfonic acid. **-thioäther** m imido thioester
imitieren vt/i imitate
Imker m (-) beekeeper. **Imkerei** f (-en) 1 beekeeping. 2 apiary
Immatrikulation f (-en) matriculation, enrollment (in a university)
Imme f (-n) (honey) bee
Immedial-farbe f immedial dye (or color).
-schwarz n (Dye.) immedial (or sulfur) black
Immediat-analyse f proximate analysis
immer adv 1 always, all the time. 2 (esp after question words) -ever: **was (wer, wo, wie) i. es sei** (or **es i. sei**) whatever (whoever, wherever, however) it may be. 3 (+wenn) e.g.: **i. wenn es regnet** whenever (or every time) it rains. 4 -ever: **für** (or **auf**) **i.** forever; **i. und ewig** forever, constantly. 5 (+compar) e.g.: **i. schneller** ever faster, faster and faster; **i. weiter** on and on; **i. wieder** again and again cf MEHR. 6 (+num) e.g.: **i. drei** three at a time; **i. der dritte** every third one. 7: **i. noch, noch i.** still. 8: **i. geradeaus (rechts, links) gehen** keep going straight ahead (to the right, to the left)
immer-fort adv all the time, constantly.
-grün a evergreen. **I-grün** n evergreen; periwinkle, myrtle. **i–hin** adv 1 in any case; even so. 2 at least. **-während** a everlasting, constant. **-zu** adv constantly, always; (+v, oft) keep (doing)
Immobilien npl real estate, real property
immun a immune. **immunisieren** vt im-

munize. **Immunität** f (-en) immunity.
Immun-körper m antibody. **Immuno-chemie** f immunochemistry. **Immun-stoff** m antibody
Impedanz f (-en) impedance
Impf- vaccination, inoculation. **impfbar** a inoculable. **impfen** vt vaccinate, inoculate; graft (plants); seed (crystals)
Impf-keim m (Cryst.) (extraneous) nucleus. **-kristall** m seed crystal
Impfling m (-e) 1 person (to be) vaccinated. 2 seed crystal
Impf-material n (Cryst.) seed material.
-stoff m vaccine
Impfung f (-en) vaccination, inoculation; seeding
Impf-versuch m inoculation experiment
Implantat n (-e) implant, implantation.
implantieren vt implant
implodieren vi(s) implode
imponieren vi (dat) impose. **imponierend, imposant** a impressive, imposing
Impotenz f impotence
Imprägnations-kupfererz n porphyry copper ore
Imprägnier-bad n impregnating bath. **imprägnierbar** a impregnable. **imprägnieren** vt impregnate
Imprägnier-lack m impregnation lacquer.
-mittel n impregnating agent. **-öl** n impregnating oil
Imprägnierung f (-en) 1 impregnation. 2 impregnating agent, coating, waterproofing
Imprägnierungs-mittel n impregnating (or waterproofing) agent. **-öl** n impregnating oil
Imprägnier-verfahren n impregnation process
improvisieren vt/i improvise
Impuls m (-e) impulse; impetus; momentum; (Elec. oft) pulse. **--erhaltung** f conservation of momentum. **-former** m impulse shaper. **-gabe** f impulse giving, (im)pulsing. **-höhe** f pulse height. **-höhenanalysator** m impulse level analyzer. **-momentquantenzahl** f spin (or angular momentum) quantum number. **-quantelung** f quantization of momen-

tum. **-quantenzahl** *f* spin quantum number. **-regler** *m* (*Elec.*) energy regulator. **-satz** *m* momentum principle, law of conservation of momentum. **-teilung** *f* (*Elec.*) pulse division. **-übertragung** *f* transfer of momentum. **-verzögerung** *f* impulse delay

imstande: i. sein (**zu** + *inf*) be capable (of . .ing), be in a position (to)

in *prep* **1** (*dat*) in, at. **2** (*accus*) to, into

·in *sfx* (*Chem.*) **1** (*Nitrogen compds & minerals*) ·ine. **2** (*triple-bound compds*) ·yne. **3** (*others*) ·in

inaktiv *a* inactive. **inaktivieren** *vt* inactivate. **Inaktivierung** *f* inactivation. **Inaktivität** *f* inactivity

Inangriffnahme *f* tackling, starting, launching (of a project, etc) *cf* ANGRIFF

Inanspruchnahme *f* (+*genit*) **1** stress, strain (on). **2** utilization (of), recourse (to). **3** demands (on). *Cf* ANSPRUCH

Inaug. Diss. *abbv* (Inauguraldissertation) doctoral dissertation, Ph.D. thesis

Inbegriff *m* embodiment, epitome, essence. **inbegriffen** *a* included *cf* EINBEGREIFEN

Inberührungkommen *n* (coming into) contact *cf* BERÜHRUNG

Inbetrachtziehung *f* (taking into) consideration *Cf* BETRACHT

inbetreff *prep* (+*genit*) regarding

Inbetriebnahme *f*, **Inbetriebsetzung** *f* putting into operation, inauguration, opening *cf* BETRIEB

inbezug (+**auf**) with regard (to) (*unauthorized spelling of* **in Bezug**) *cf* BEZUG

Inbringer *m* (-) conveyor

Inbrunst *f* ardor, fervor, **inbrünstig** *a* ardent, fervent

incarnat *a* flesh-colored

inchromieren *vt* (*Metll.*) chromize

Indanthren *n* (-e) indanthrene

indem *cnjc* **1** as, while. **2** (*esp* + *man*) by: **i. man es mit Säure behandelt** by treating it with acid

Inden *n* (-e) indene

Inder *m* (-), **Inderin** *f* (-nen) (Asian) Indian

indes(sen) I *adv* **1** however. **2** meanwhile. **II** *cnjc* while, whereas

Index·strich *m* index mark (*or* line). **-zahl** *f*

index number. **-zeiger** *m* indicator needle, pointer

India·faser *f* Mexican fiber

Indian *m* (-e) (*Zool.*) turkey

Indianer *m* (-), **Indianerin** *f* (-nen), **indianisch** *a* (American) Indian. **Indianisch·rot** *n* Indian red

Indien *n* India

indifferent *a* indifferent; (*Chem. oft*) neutral, inert. **Indifferenz** *f* (-en) indifference; (*Chem. oft*) neutrality, inertness

Indig *m* indigo (plant), anil

indigen *a* indigenous

Indig·kraut *n* indigo plant

indigo·artig *a* indigoid. **I–auszug** *m* indigo extract. **-blau** *n* indigo blue, indigotin. **-blauschwefelsäure** *f* indigosulfuric (*or* indigosulfonic) acid. **-druck** *m* indigo print(ing). **-farbe** *f* indigo color (*or* dye). **i–farbig** *a* indigo(-colored). **I–farbstoff** *m* indigo blue, indigotin. **-küpe** *f* indigo (*or* blue) vat (solution of indigo white). **-leim** *m* indigo gelatin (*or* gluten). **i–liefernd** *a* yielding indigo. **I–lösung** *f* indigo solution. **-messung** *f* indigometry. **-purpur** *m* indigo purple. **-rot** *n* indigo red, indirubin. **-salz** *n* indigo salt, indigotate. **-säure** *f* indigotic acid. **-schwefelsäure** *f* indigosulfuric (*or* indigosulfonic) acid. **-solfarbstoff** *m* (*Dye.*) indigosol. **-stein** *m* indicolite (blue tourmaline). **-stoff** *m* indigo blue, indigotin. **-sulfosäure** *f* indigosulfonic acid. **-weiß** *n* indigo white

Indig·rot *n* indigo red, indirubin. **-stein** *m* indicolite (blue tourmaline). **-weiß** *n* indigo white

Indi·ion *n* indium ion

Indikator *m* (-en) indicator; (isotopic) tracer

Indioxid° *n* indium oxide (In_2O_3)

indisch *a* Indian (of India) **—I—er Balsam** balsam of Peru; **i—e Bohne** St. Ignatius bean; **i—e Feige** prickly pear; **i—er Flachs** jute; **i—es Grasöl** palmarosa oil; **i—es Rohr** *n* rattan; **i—er Safran** turmeric. **I—·gelb** *n* Indian yellow, potassium hexanitracobaltate. **-hanftinktur** *f* tincture of Indian hemp. **-rot** *n* Indian red

Indi(um)·sulfid *n* indium (sesqui)sulfide.

Indium·sulfür *n* indium monosulfide

individuell *a*, **Individuum** *n* (.. duen) individual

Indiz *n* (-ien) indication, item of evidence

Indizes indices; indexes: *pl of* **Index** *m*

Indizien indications, evidence: *pl of* INDIZ, INDIZIUM *n*. **·-beweis** *m* circumstantial evidence. **i–los** *a* without evidence

indizieren *vt* 1 indicate. 2 label (isotopes)

Indizium *n* (..zien) indication, item of evidence

Indochinolin *n* indoquinoline

Indol·körper *m* indole substance, member of the indole group

Indoxazen *n* 1,2-benzisoxazole

Indoxyl·säure *f* indoxylic acid

induktions·frei *a* non-inductive. **I–härten** *n* (*Metll.*) induction hardening. **-ofen** *m* induction furnace. **-rolle, -spule** *f* induction coil. **-vermögen** *n* inductive capacity. **-wirkung** *f* inductive effect

Induktivität *f* (*Elec.*) inductance. **Induktorium** *n* (..ien) (*Elec.*) induction coil

indurieren *vt* indurate, harden

Industrie *f* (-n) industry; factory, manufacturing plant. **·-abfallstoff** *m* industrial waste material. **-abwasser** *n* industrial waste water (*or* liquor). **-erzeugnis** *n* industrial product. **-gerät** *n* industrial apparatus (device, equipment, gadget). **-güte** *f* industrial grade. **-lack** *m* industrial paint (finish, lacquer)

industriell *a* industrial. **Industrielle** *m,f* (*adj endgs*) industrialist, industrial manufacturer

Industrie·norm *f* industrial standard (*or* norm). **-ofen** *m* industrial furnace. **-zweig** *m* branch of industry

induzieren *vt* induce. **Induzierung** *f* (-en) induction; inductance

inegal *a* unequal

ineinander *adv & sep pfx* in(to) each other; *pfx oft* inter. —**i. übergehen** merge, blend, intermingle. **i—·fließen*** *vi*(s) flow (*or* run) into each other, interflow. **-fließend** *p.a* confluent. **-fügen** *vt* join, fit into each other. **·greifen*** *vi* mesh, interlock, engage (each other, *as:* gears). **·passen** *vi* fit together (*or* into each other),

nest. **·schieben*** *vt & sich* i. telescope. **·stecken** *vt* insert into each other. **·wachsen*** *vi* (s) merge, grow together. **·weben*** *vt* interweave

Inempfangnahme *f* reception, receipt *Cf* EMPFANG

Inertie *f* inertia, inertness. **inert·machen** *vt* render inert, deactivate

Infarkt *m* (-e) (*Med.*) infarction

Infektions·erreger *m* cause of infection, (*specif*) virus. **-herd** *m* focus of infection. **-krankheit** *f* infectious disease. **-quelle** *f* source of infection. **-träger** *m* carrier (of infection). **i–widrig** *a* infection-resistant

infektiös *a* infectious

Infekt·reaktion *f* reaction to infection

infiltrieren *vt/i* infiltrate

infizieren *vt* infect; (Cryst.) inoculate

Influenz *f* (-en) (*Elec.*) influence, electrostatic induction. **influenzieren** *vt* influence, affect; (*Elec.*) induce

infolge *prep* (+*genit*), *adv* (+**von** +*dat*) as a result (of), owing (to). **infolgedessen** *adv* consequently

Informatik *f* computer science

Information *f* (-en) (item of) information; (*pl usu*) information. **informieren** 1 *vt/i* inform. 2 *vi* give information. 3: **sich** i. get (*or* gather) information. **Informations·schrift** *f* informative leaflet (*or* note)

infra·rot *a* infrared. **Infraschall** *m* infrasound; (*in compds*) infrasonic, subsonic

Infundier·apparat *m* infusion apparatus. **-büchse** *f* infusion vessel, digester

infundieren *vt* infuse

Infusions·tierchen *npl* infusoria. **-verfahren** *n* (*Beer*) infusion mashing

Infusorien·erde *f* infusorial earth, kieselguhr

Ing. *abbv* (Ingenieur) engineer

Ingang·kommen *n* starting, initiation. **-setzung** *f* setting in motion, launching, putting into operation *Cf* GANG 4

Ingber *m* ginger = INGWER

Ingebrauchnahme *f* start of use (*or* utilization) *cf* GEBRAUCH

Ingenieur *m* (-e) engineer. **I.-Chemiker** *m* chemical engineer, industiral chemist.

Ingenieur·dienst *m* engineering ser-

vice. **-hochbau** *m* structural engineering. **-wesen** *n* engineering
ingeniös *a* ingenious
ingleichen *adv* equally, just as
Ingot·stahl *m* ingot steel
Ingrediens *n* (..nzien), **Ingredienz** *f* (-en) ingredient
Inguß° *m* ingot. (mold)
Ingwer *m* ginger; **gelber I.** turmeric. **-bier** *n* ginger beer. **-wurzel** *f* ginger root
Inh. *abbv* 1 (Inhalt) content(s). 2 (Inhaber) proprietor
Inhaber *m* (-) owner, proprietor
inhalieren *vt/i* inhale
Inhalt *m* (-e) 1 content(s). 2 area; volume; capacity. **inhaltlich** *a* in (*or* as to *or* with regard to) content
Inhalts·angabe *f* (table of) contents; summary, abstract. **-auszug** *m* abstract (of an article, etc). **i–leer, -los** *a* empty, hollow; meaningless. **-reich** *a* rich (in content), meaningful. **I–verzeichis** *n* table of contents, index
inhärent *a* inherent
inhibieren *vt* inhibit
inhomogen *a* inhomogeneous. **Inhomogenität** *f* (-en) inhomogeneity
Initiale *f* (-n) 1 initial. 2 (*Explo.*) primer, priming, charge
Initial·explosivstoff, -sprengstoff *m* (*Explo.*) detonator, primer. **-zünder** *m* (*Explo.*) primer. **-zündsatz** *m* (*Explo.*) priming composition. **-zündung** *f* primer, priming charge
initiieren *vt* initiate; (*Explo.*) explode, detonate
injizieren *vt* inject
inkarnat *a* incarnadine, flesh-colored. **Inkarnat** *n* flesh color. **-klee** *m* crimson clover
inkl. *abbv* (inklusive) incl. (including): **bis i.** up to and including
Inklination *f* (-en) inclination; (*Magn., Instrum.*) dip (of the needle). **Inklinations·nadel** *f* (*Magn.*) dip needle
Inkohärenz *f* incoherence
Inkohlung *f* coalification; carbonization
Inkohlungs·grad *m* degree of carbonization; (*Coal*) rank. **-sprung** *m* coalification jump
inkonsequent *a* inconsistent. **Inkonse-**

quenz *f* (-en) inconsistency
Inkrafttreten *n* taking effect, becoming effective: **vor I. dieser Ordnung** before this regulation became (*or* becomes) effective *cf* KRAFT
Inkreis° *m* inscribed circle
Inkret *n* (-e) incretion, endocrine secretion; hormone. **inkretorisch** *a* endocrine
inkrustieren 1 *vt* incrust. 2 *vi* become incrusted. **Inkrustierung** *f* (-en) incrustation
Inland *n* 1 inland, interior (of the country). 2 home country, homeland; (*in compds oft*) domestic: **im In- und Ausland** at home and abroad. **Inländer** *m* (-) native; inlander. **inländisch** *a* 1 inland. 2 native. 3 domestic; homemade
Inlett *n* (-e) (*Tex.*) ticking
inliegend *a* enclosed; *adv oft* herewith
Inlösunggehen *n* (act of) going into solution *cf* LÖSUNG 1
inmitten *prep* (+*genit*) in the midst (of), amid
inne·haben* *vt* 1 hold (an office). 2 own; occupy (premises)
inne·halten* 1 *vt* observe (rules, customs). 2 *vi* pause
innen 1 *adv* (on the) inside, within: **von i.** from the inside, from within; **nach i.** toward the inside, inward. 2 *pfx* inside, interior, inner, internal, domestic. **I—·ansicht** *f* interior view. **-aufnahme** *f* interior (*or* indoor) photograph(y). **-auskleidung** *f* (inner) lining. **-druck** *m* internal pressure. **-durchmesser** *m* inside diameter. **-feuerung** *f* internal furnace. **-fläche** *f* inner surface. **i—gewickelt** *a* internally wound. **I—gewinde** *n* internal (*or* female) thread. **-haut** *f* internal membrane, inner skin. **-heizung** *f* 1 interior heating. 2 internal firing. **-hütchen** *n* (*Explo.*) inner capsule. **-konus** *m* 1 inner cone. 2 inside taper. **-lack** *m* interior varnish. **-leben** *n* inner life; internal activity. **-leitung** *f* interior wiring (*or* piping); inner circuit. **i—liegend** *a* interior, inside. **I—lunker** *m* (*Metll.*) internal blow-hole, shrinkage cavity. **-maß** *n* inside measure(ment). **-ministerium** *n* (cabinet) ministry of the interior. **-mischer** *m* internal mixer. **-raum** *m* inte-

rior (space); (*pl oft*) rooms. **rinde** *f* inner rind (bark, crust); secondary cortex. **-sechskantschraube** *f* allen screw. **-seite** *f* inner side, inside. **-spannung** *f* internal stress (*or* tension). **-taster** *m* inside calipers. **-tränkung** *f* soaking in, penetration. **-wand, -wandung** *f* inner wall. **-weite** *f* inside width (*or* diameter). **-widerstand** *m* (*Elec.*) internal resistance. **-winkel** *m* interior angle. **i‑zentriert** *a* (*Cryst.*) body-centered

inner *a & pfx* inner, inside, internal, interior; domestic; (*as pfx also*) intra‑. **-atomar** *a* intra-atomic. **-betrieblich** *a* in-house, in-plant, internal

Innere *n* (*adj endgs*) interior, inside; core, heart

innerhalb *prep* (*genit, dat*), *adv* (+**von**) inside of, within

innerlich *a* inner, internal; inward: **i. ausgeglichen** internally compensated

inner‑molekular *a* intramolecular. **-nuklear** *a* intranuclear. **-sekretorisch** *a* internal-secretion

innerst *a* inmost, innermost. **Innerste** *n* (*adj endgs*) innermost part, core, heart

innert *prep* (*genit, dat*) inside of, within

inne‑werden* *vi*(s) (*genit*) become aware (of). **‑wohnen** *vi* (*dat*) be inherent (in). **-wohnend** *p.a* inherent

innig *a* 1 sincere. 2 intimate. **Innigkeit** *f* sincerity; intimacy

inokulieren *vt* inoculate

Inosilikat° *n* silicate with a chain structure

Inosin‑säure *f* inosic acid

in praxi *adv* in practice

ins (*contrac*) = IN DAS in(to) (the) *cf* IN, DAS

Insasse *m* (-n, -n) occupant, inmate; passenger

insb. *abbv* = **insbesondere** *adv* especially, in particular

Inschrift° *f* inscription

insekten‑fressend *a* insectivorous. **I‑kunde, -lehre** *f* entomology. **-mittel** *n* insect repellent. **-stich** *m* insect bite. **i‑tötend** *a* insecticidal. **I‑töter** *m*, **-vernichtungsmittel** *n*, **-vertilgungsmittel** *n* insecticide. **i‑vertreibend** *a* insect-repellent. **I‑vertreibungsmittel** *n* insectifuge, insect repellent. **-wachs** *n* insect wax

Insektizid *n* (-e) insecticide

Insel *f* (-n) island, isle; **I—n im Film** spots on the film. **-‑silikat** *n* orthosilicate

Inselt *n* tallow; suet

Inserat *n* (-e) (newspaper, magazine) ad(vertisement); announcement, notice. **inserieren** 1 *vt* insert; advertise. 2 *vi* advertise, put in an ad

insgesamt *adv* altogether; as a whole; (*esp* + *number*) a total of

insofern 1 *adv* in this regard, to this extent; (+**als** *in next clause*) to the extent that, insofar as. 2 *subord cnjc* if, to the extent that

insolieren *vt* insolate

insoweit 1 *adv* to this extent; (+**als**) insofar as. 2 *cnjc* to the extent that. = INSOFERN

Inspizient *m* (-en, -en) inspector. **inspizieren** *vt/i* inspect

instabil *a* unstable. **Instabilität** *f* (-en) instability

installieren 1 *vt* install. 2: **sich i.** take(up) occupancy (residence, office, etc)

instand *adv* (usu + *v*): **i. halten** maintain (in good condition); **i. setzen** put in good condition, fix up, repair, recondition; (*oft* + **zu** + *inf*) put in a position, enable (to). **I—‑haltung** *f* upkeep, maintenance

inständig *a* urgent. **Inständigkeit** *f* urgency

Instand‑setzung *f* repair, reconditioning; enablement *cf* INSTAND

instationär *a* non-stationary

instruieren *vt* instruct

Instrumentarium *n* tools, instruments, apparatus. **Instrumenten‑kunde** *f* instrumentation, instrument engineering

Integral‑rechnung *f* integral calculus. **Integrations‑weg** *m* (*Math.*) path of integration. **integrierbar** *a* integrable. **integrieren** *vt/i* integrate

Intensität *f* (-en) intensity. **intensiv** *a* intensive, intense. **Intensivität** *f* intensiveness. **intensivieren** *vt* intensify

interessant *a* interesting. **Interesse** *n* (-n) (*oft* + **an**) interest (in). **Interessen‑gemeinschaft** *f* 1 community of interest. 2 (*Com.*) combine, trust. **Interessent** *m* (-en, -en) person interested; prospect(ive customer). **interessieren** 1 *vt* interest;

(*esp* + **für, an**) get (sbdy) interested (in).
2 *vt/i* be interesting, be of interest (to). **3:**
sich i. (**für, an**) be interested (in), (+**an,
oft**) be in the market (for)

Interesterifizierung *f* interesterification

Interferenz·bild *n* (*Phys., Opt.*) interference pattern (*or* figure). **-streifen** *m*
intereference fringe. **interferieren** *vi*
(*Phys., Opt.*) interfere

Interferrikum *n* (..ka *or* ..ken) (*Magn.*) air
gap

Interhalogen·verbindung *f* interhalogen
compound

interimistisch *a* provisional, interim; *adv*
oft in the interim

inter·ionisch *a* interionic. **-kristallin** *a* intercrystalline. **-mediär** *a* intermediary,
intermediate. **-metallisch** *a* inter-
metallic. **-mittierend** *a* intermittent

intern *a* internal

interpolieren *vt/i* interpolate

interponieren *vt* interpose, insert,
interpolate

interpunktieren *vt* punctuate. **Inter-
punktion** *f* punctuation

intervenieren *vi* intervene

Interzonen· interzonal (= between East
and West Germany)

intim *a* intimate; personal

intra·atomar *a* intra-atomic. **-kutan** *a* in-
tracutaneous. **-perlitisch** *a* intrapearli-
tic. **-venös** *a* intravenous

Intrusiv·gestein *n* intrusive rock

Intumeszenz *f* (-en) intumescence

Invar·stahl *m* invar steel

Inventar *n*(-e) inventory, stock; furnishings, fixtures, equipment. **Inventur** *f*
(-en) stocktaking, inventory

Inversions·achse *f* (*Cryst.*) inversion axis.
-geschwindigkeit *f* rate of inversion

invertieren *vt* invert. **Invertierung** *f* (-en)
inversion

Invert·seife *f* invert soap, cationic detergent. **-zucker** *m* invert sugar

investieren *vt* invest. **Investition** *f* (-en)
investment

inwärts *adv* inward

inwendig *adv* inwardly, on the inside

inwiefern *adv* to what extent, in what
respect

inwieweit *adv* to what extent

inwohnend *a* inherent = INNEWOHNEND

inzidieren *vt* incise. **Inzision** *f* (-en)
incision

Inzucht *f* in-breeding

inzwischen *adv* meanwhile, during that
time; by now

Iod *n* iodine. *cf* JOD

Ion *n* (-en) ion. **ional** *a* ionic

Ionen· ion(ic): **-abstand** *m* interionic distance. **-art** *f* type of ion. **-austausch** *m*
ion exchange. **-beschuß** *m* ion bombardment. **-beweglichkeit** *f* ionic mobility.
-beziehung *f* ion association. **i–bildend**
a ion-forming. **I–bildung** *f* ion formation. **-bindung** *f* ionic bond. **-dichte** *f*
ionic density. **-falle** *f* ion trap. **-farbe** *f*
ionic color. **-gehalt** *m* ionic content. **-ge-
schwindigkeit** *f* ionic velocity. **-gewicht**
n ionic weight. **-gitter** *n* ionic lattice.
-gleichung *f* ionic equation. **-leit-
fähigkeit** *f* ionic conductance (*or* conductivity). **-leitungsstrom m** ionic conduction current. **-paar** *n* ion pair.
-polarisation *f* atomic polarization.
-quelle *f* ion source. **-reibung** *f* ionic friction. **-reihe** *f* ionic series. **-rumpf** *m* ionic
core. **-spaltung** *f* ionic cleavage, ionization. **-spreizung** *f* ion spread (of flocculation ratios). **-stärke** *f* ionic strength.
-stärkemesser *m* ionometer. **-stoß** *m*
ionic collision. **-trennungsarbeit** *f* energy of ionic dissociation. **-verbindung** *f*
ionic compound. **-wanderung** *f* ionic migration. **-wolke** *f* ionic cloud (*or* atmosphere). **-zustand** *m* ionic state

Ionigkeit *f* ionic character

Ionisations·kammer *f* ionization chamber. **-manometer, -vakuummeter** *n* ionization gauge

ionisch *a* ionic

ionisierbar *a* ionizable. **ionisieren** *vt/i* io-
nize. **Ionisierung** *f* (-en) ionization

Ionisierungs·arbeit *f* work of ionization.
-grenze *f* ionization limit. **-mittel** *n* ionizing agent. **-spannung** *f* ionic interference; disturbance of ionization

ionogen *a* ionic; electrovalent

Ionon *n* ionone

ionometrisch *a* ionometric. **Ionosphäre** *f* ionosphere. **Ionosphären·schall** *m* ionospheric resonance

I.P. *abbv* (italienisches Patent) Ital. pat.

Ipecacuanha·säure *f* ipecacuanhic acid

I.P.S. *abbv* (indizierte Pferdestärke) indicated HP

Ipuroi·säure *f* ipurotic acid

irden *a* earthen. **I—·gut** *n*, **-ware** *f* earthenware

irdisch *a* earthly, worldly, mundane; mortal, perishable

Ire *m* (-n, -n) Irishman

Iren *n* (*Min.*) irene

irgend *adv* possibly, in any way, at all: **wenn i. möglich** if at all possible —**i. etwas** (*1*) something (or other), (*2*) anything (at all; **i. jemand** (*1*) someone (or other), (*2*) anyone (at all); **i. so ein** (*1*) some kind of, (*2*) any kind of. **-·ein** *a* (*endgs like* **ein**) **1** some . . . (or other). **2** any . . . (at all). **-eine(r)** *pron* (*adj endgs*) **1** someone (or other), (some) one (or another). **2** anyone, any one (at all). **-eins** *n pron* **1** some particular one. **2** any one (at all). **-etwas** *pron* (*unauthorized spelling of* irgend etwas *cf* IRGEND). **-wann** *adv* **1** sometime (or other). **2** (at) any time, anytime (at all). **-was** *pron* something; anything = IRGEND ETWAS. **-welche** *a* some (kind of); any (kind of). **-wer** *pron* someone; anyone = IRGEND JEMAND. **-wie** *adv* **1** somehow (or other). **2** (in) any way (at all). **-wo** *adv* **1** somewhere (or other). **2** anywhere (at all). **-woher** *adv* **1** from somewhere (or other). **2** from anywhere (at all). **-wohin** *adv* **1** (to) somewhere (or other). **2** anywhere, to anyplace (at all)

iridisieren 1 *vt* iridize. **2** *vi* iridesce = IRISIEREN

Iridium·gold *n* iridium aurate

Irin *f* (-nen) Irishwoman *cf* IRE. **irisch** *a* Irish

Irisation *f* (-en) irisation, iridescence

Iris·blende *f* iris diaphragm. **-druck** *m* (*Tex.*) rainbow printing, iris print. **-fond** *m* (*Dye.*) rainbow ground. **-glas** *n* iridescent glass

irisieren 1 *vt* iridize, irisate. **2** *vi* be iridescent, iridesce. **irisierend** *p.a.* iridescent.

Irisierung *f* iridescence

Iris·öl *n* iris (*or* orris) oil. **-plastik** *f* (*Med.*) coreoplasty. **-wurzel** *f* (*Pharm.*) orris root

Irland *n* Ireland. **Irländer** *m* (-) Irishman. **Irländerin** *f* (-nen) Irishwoman. **irländisch** *a* Irish

Iron *n* (*Chem.*) irone

Ironie *f* (-n) irony. **ironisch** *a* ironic(al)

irr *a* insane; confused = IRRE. **I—·beere** *f* (*Bot.*) deadly nightshade

irre I *a* **1** mad, insane. **2** confused, muddled: **i. werden** (**an**) become uncertain (*or* confused) (about), lose confidence (in). **3** stray, lost. **II** *sep pfx* astray, mis-. **Irre I** *m*, *f* (*adj endgs*) mad (*or* insane) person, lunatic. **II** *f* wrong way (*or* track): **in die I. führen** lead astray, mislead

irreal *a* unreal

irreduzibel *a* irreducible

irre·fahren* *vi*(s) go astray, lose one's way. **·führen** *vt* lead astray, mislead; deceive. **-führend** *p.a* misleading; deceptive. **I—führung** *f* deception, attempt to mislead. **i—·gehen*** *vi*(s) go astray, go wrong; lose one's way

Irregularität *f* (-en) irregularity

irre·leiten *vt* misguide, lead astray. **-machen** *vt* confuse, disconcert

irren 1 *vi*(h) be wrong, be mistaken, err. **2** *vi*(s) wander, roam; stray. **3: sich i.** (*oft* + **in**) be mistaken, be wrong (in, about); misjudge

Irren·anstalt *f* mental hospital. **-arzt** *m* mental specialist, doctor in a mental hospital

Irre·sein *n* insanity, dementia

Irr·fahrt *f* wandering, odyssey. **-gang** *m* aberration; *pl* **-gänge** *oft* = **-garten** *m* maze, labyrinth

irrig *a* erroneous, wrong, mistaken. **irrigerweise** *adv* erroneously

irrigieren *vt* irrigate

Irr·lehre *f* false teaching, heresy. **-licht** *n* will o' the wisp. **-sinn** *m* insanity, madness. **i—sinnig** *a* insane, mad. **I—sinnige** *m*, *f* (*adj endgs*) insane person, lunatic. **-strom** *m* (*Elec.*) stray current

Irrtum *m* (-er) error, mistake. **irrtümlich** *a* erroneous, mistaken

Irrung *f* (-en) aberration

Irr·weg *m* wrong road (*or* track) **—auf I—en** (*oft*) astray

isabell(en)·farbig *a* cream-colored

Isacon·säure *f* isaconic acid

Isäthion·säure *f* isethionic acid

Isatropa·säure *f* isatropic acid

Isatin·säure *f* isatic acid

ischämisch *a* ischemic, locally anemic

Ischias *m* sciatica. **··nerv** *m* sciatic nerve

Island *n* Iceland. **isländisch** *a* Icelandic: **i—er** (*or* **Isländer**) **Doppelspat** = **Islandspat** *m* Iceland spar

isobar *a* isobaric. **Isobar** *n* (-e) (*Nucl.*), **Isobare** *f* (-n) (*Meteor.*) isobar. **Isobaren·kern** *m* isobaric nucleus. **isobarisch** *a* isobaric

isodispers *a* (*Suspensions*) monodisperse

Isodurol *n* isodurene

isoelektrisch *a* isoelectric

Isoferula·säure *f* isoferulic acid

Isohydre *f* (-n) isohydrore

Isohypse *f* (-n) contour line

Isolation *f* (-en) 1 (*genl*) isolation. 2 (*Elec.*) insulation. **Isolator** *m* (-en) (*Elec.*) insulator

Isolier·anstrich *m* insulating coat (of paint, etc). **-band** *n* insulating tape

isolierbar *a* capable of being isolated (*or* insulated). **isolieren** *vt* 1 isolate. 2 (*Elec.*) insulate. *Cf* ISOLIERT

Isolier·fähigkeit *f* insulating ability (property, capacity). **-farbe** *f* insulating paint. **-fehler** *m* defect in insulation. **-firnis** *m* insulating varnish. **-flasche** *f* thermos bottle, vacuum flask. **-flüssigkeit** *f* insulating liquid. **-gewebe** *n* insulating fabric. **-hülle** *f* insulating sheath (*or* covering). **-kitt** *m* insulating cement. **-lack** *m* insulating varnish. **-masse** *f*, **-material** *n* insulating material. **-mittel** *n* insulating agent, insulator. **-öl** *n* insulating oil. **-pappe** *f* insulating (card)board. **-rohr** *n*, **-röhre** *f* insulating tube (pipe, conduit). **-schicht** *f* insulating layer. **-schlauch** *m* (flexible) insulating tube. **-stein** *m* insulating brick. **-stoff** *m* insulating material (*or* substance), insulator, insulation

isoliert *p.a.*: **i—e Färbung** (*Micros.*) differential staining: **i—er Riechstaff**

(*Perfumes*) isolate; *cf* ISOLIEREN. **Isolierung** *f* (-en) isolation; insulation

Isolier·vermögen *n* insulating power. **-werkstoff** *m* insulating plastic. **-zustand** *m* state of insulation (*or* isolation)

isolog *a* isologous

isomer *a* isomeric. **Isomere** *n* (*adj endgs*) isomer. **Isomerie** *f* isomerism. **isomerisieren** *vt/i* isomerize. **Isomerisierung** *f* (-en) isomerization

isomorph *a* isomorphous. **Isomorphie** *f* isomorphism

Isoöl·säure *f* isooleic acid

Isop *m* hyssop. **··öl** *n* hyssop oil

Isopurpurin·säure *f* isopurpuric acid

isoster *a* isoelectronic

isotherm *a* isothermal. **Isotherme** *f* (-n) isotherm

Isotonie *f* isotonicity. **isotonisch** *a* isotonic

isotop *a* isotopic. **Isotop** *n* (-e) isotope

Isotopen·gemisch *n* isotopic mixture. **-gewicht** *n* isotope mass. **-indikator** *m* isotopic tracer. **-trennung** *f* isotopic separation. **-verhältnis** *n* isotopic ratio

Isotopie *f* isotopy, isotopism. **--effekt** *m* isotope effect. **-verschiebung** *f* isotopic shift

isotrop *a* isotropic. **Isotropie** *f* isotropy, isotropism

isotyp *a* isostructural. **Isotypie** *f* isostructural relations

isozyklisch *a* isocyclic

ißt eats: *pres of* ESSEN

ist *is*: *pres of* SEIN; (+*pp oft*) has: **i. gegangen** has gone

Ist· actual, effective: **Ist-Betrag** *m* actual amount

Istle *f* (-n) tampico fiber tree

Ist-Maß *n* actual size (*or* dimensions). **-Menge** *f* actual amount

I-Stoß *m* butt joint

Istrien *n* (*Geog.*) Istria

Ist·wert *m* actual value

·it *sfx* 1 (*usu, esp Chem. & Min.*) ·ite *cf* SULFIT, BAUXIT. 2 (*org hydroxylic compds*) ·itol *cf* MANNIT. (*Caution: Ger.* ·ite *is the plural of* ·it.)

i.T. *abbv* (in der Trockenmasse) in dry matter, on a dry-weight basis

Itacon·säure *f* itaconic acid

Italien *n* Italy. **italienisch** *a* Italian
Itamal·säure *f* hydrosuccinic acid
iterieren *vt/i* iterate
i.V. *abbv* 1 (im Vakuum) in vacuo. 2 (in Vertretung) by, per (*with signatures*)
Iva-kraut *n* musk milfoil. **-öl** *n* iva (*or*
musk milfoil) oil
i.W. *abbv* 1 (innere Weite) inside width (*or*
  diameter). 2 (in Worten) in words
i.W.v. *abbv* (im Wert von) in the value of,
  amounting to
Izod·wert *m* (*Metll.*) izod value

J

NOTE: In cognate words, initial **J** usually replaces **I** when followed by a vowel, as in **Jod,** iodine. Occasionally **J** will be found in older publications replacing **I** even before consonants: **Jllit** = **Illit** n illite. *Cf* NOTE under letter **I.**

J (*Chem. symbol*) JOD iodine, I
J. *abbv* 1 (*Jahr*) yr. (year). 2 Jahresbericht der Chemie
ja *adv* 1 yes. 2 (*interrog*) all right?, will you? 3 in fact, indeed. 4 (*emphat*) surely: **damit es ja (nicht) zerschmilzt** so it will be sure (not) to melt. 5 after all, of course (*or no transl., implies* "as we all know"): **Wasser gefriert ja bei 0°C** water does (as we all know) freeze at 0°C; **es kommt ja oft vor** it does, after all, happen often; **man kann es ja versuchen, aber** . . . you can try it, of course, but . . . 6 (*in exclam*) e.g.: **das ist ja unglaublich!** why, that's incredible!
Jacke *f* (-n) jacket
Jagara-zucker *m* jaggery (sugar)
Jagd *f* (-en) 1 hunt; hunting season (party, ground) —**auf die J. gehen** go hunting. 2 (*oft* + *nach*) chase, pursuit, search —**J. machen auf** pursue, hunt down. 3 game (animals). **--flieger** *m* fighter pilot. **-flugzeug** *n* fighter (*or* pursuit) plane
jagen I *vt* 1 hunt (game). 2 chase, pursue. 3 rout, drive (out). 4 drive, thrust (knife, etc); shoot (bullets). **II** *vi* (h) 5 go hunting. 6 (**auf**) hunt. 7 (**nach**) chase (after), pursue. **III** *vi*(s) dash, rush, race
Jäger *m* (-) 1 hunter. 2 rifleman. 3 (*Aero.*) fighter (*or* pursuit) plane
jäh *a* 1 sudden. 2 steep, precipitous. **jählings** *adv* suddenly, precipitously; headlong
Jahr *n* (-e) year. **--buch** *n* yearbook, annual; almanac
jahre-lang 1 *a* e.g.: **nach j—er Arbeit** after years of work. 2 *adv* for (many) years
Jahres- annual, year('s): **-ausbeute** *f* annual yield. **-beginn** *m* beginning of the

year. **-bericht** *m* annual report. **-erzeugung** *f* annual production (*or* yield). **-frist** *f* a year's time, space of a year. **-hauptversammlung** *f* annual general meeting. **-inhaltsverzeichnis** *n* annual index. **-lauf** *m* course of the year. **-schluß** *m* end of the year. **-schrift** *f* annual (publication). **-tag** *m* anniversary. **-treffen** *n* annual meeting (*or* convention). **-übersicht** *f* annual review. **-wechsel** *m*, **-wende** *f* turn of the year, new year. **-zahl** *f* year, date (of an event). **-zeit** *f* season. **j–zeitlich** *a* seasonal. **J–zuwachs** *m* annual growth
Jahrg. *abbv* (Jahrgang) age group, year
Jahr-gang *m* 1 age group: **der J. 1990** all persons born in 1990. 2 vintage. 3 year, one year's issues (of a publication). **-hundert** *n* (-e) century. **-hundertfeier** *n* centenary, hundredth anniversary
-jährig *sfx* e.g.: **fünfjährig** five-year('s) (war, duration); five-year-old, five years old
jährlich *a* yearly, annual
Jahr-millionen *pl* millions of years. **-tausend** *n* millennium, a thousand years. **-zehnt** *n* (-e) decade. **j–zeitlich** *a* seasonal
Jakobs-kraut *n* groundsel
Jalape *f* jalap. **Jalapen-harz** *n* jalap resin, jalapin. **-wurzel** *f* jalap root. **Jalapin-säure** *f* jalapic acid
Jalousie *f* (-n) Venetian blind
Jamaika-Pfeffer *m* pimento, allspice
Jammer *m* 1 misery. 2 lamentation. 3 pity, shame. **jämmerlich** *a* miserable, pitiful
jammern 1 *vt* move to pity. 2 *vi* lament, moan, wail
Jam-pflanze *f,* **Jam(s)-wurzel** *f* yam

Jänner m (*Austrian*) = **Januar** m January
Janus·grün n Janus green
Japan·holz n sapanwood
japanieren vt japan
japanisch a Japanese; **j—e Erde** catechu; **j—es Wachs** Japan wax
Japan·kampfer m Japan camphor. **-lack** m japan, Japan lacquer (*or* varnish). **-leder** n japanned (*or* patent) leather. **-leim** m agar-agar. **-papier** n rice paper; Japan paper (*or* vellum). **-säure** f japonic acid. **-schwarz** n Japan black. **-talg** m, **-wachs** n Japan wax
Japhis m jasper
Jararabad·harz n hyderabad (aloe) resin
jarowisieren vt vernalize
Jasmin m (-e) jasmin(e) **—gemeiner J.** syringa. **-·blütenöl** n oil of jasmin petals. **-öl** n jasmine oil
Jaspachat m jasper agate
jaspiert a marbled, jaspé
Jaspis m (-se) jasper. **-·gut** n jasperware. **-porzellan** n jasperated china. **-steingut** n jasperware
Jaspopal m jasper opal
jäten vt/i weed (out)
Jato abbv (Jahrestonnen) tons per year
Jatrochemie f iatrochemistry
Jauche f (-n) liquid manure; (*Med.*) ichor
jauchzen vi rejoice, jubilate
javanisch a Javanese; **j—es Wachs** gondang (*or* Java) wax
Java·zimt m Java cinammon
Javelle'sche Lauge f Javelle water
jawohl adv yes indeed
Jb. abbv (Jahrbuch) yearbook, etc
Jber. abbv (Jahresbericht) annual report
je I adv 1 ever **—seit** (*or* **von**) **je** since time immemorial; **wie eh und je** as (it) always (has been); **je und je** now and then. **2: je nach** according to cf NACH; **je nachdem** (1) (*as adv*) as the case may be, depending, (2) (+*cnjc, e.g.* **ob, wie,** *or* +*clause*) according to (whether, how, etc). **3** each, at a time: **je zwei Elektronen** two electrons each (*or* at a time); **je drei und drei** three at a time, by (*or* in) threes; **je das vierte** every fourth one. **4** e.g.: **10 Mark je Kilo** ten marks a (*or* per) kilogram. **II** cnjc (+*compar*) e.g.: **je länger je lieber** the

longer the better cf DESTO, UM 12
jede, jedem, jeden Cf JEDER
jedenfalls 1 adv in any case, anyway. **2** adv/cjnc at least
jeder Ia 1 each, every. **2** any. **II** pron (*adj endgs*) **3** each (*or* every) one, everyone. **4** any one, anyone
jederlei a (*no endgs*) every (*or* any) kind of
jedermann pron everyone, everybody
jederzeit adv (at) any time; any minute
jedes cf JEDER
jedesmal adv every time: **j., wenn es regnet** every time it rains. **jedesmalig** a actual(ly existing), in any particular case
jedoch adv/cnjc however, but
jegliche(r) 1 a any, all; each, every. **2** pron (*adj endgs*) each one, everyone
jeher: von j. since time immemorial, all along
jemals adv ever, = JE 1
jemand indef pron, sg only (*oft with adj endgs*) **1** someone, somebody, a person. **2** anyone, anybody. **-j. anders** someone (*or* anyone) else, **j. Fremdes** someone (*or* anyone) strange, some (*or* any) stranger cf IRGEND
Jenaer a (*no endgs*) (of) Jena: **J. Glas** Jena glass. **jenaisch** a = JENAER
jene, jenem, jenen cf JENER
Jenenser a (*no endgs*), **jenensisch** a (of) Jena
jener 1 a that, pl those. **2** pron (*adj endgs*) that one, pl those; (+*relat pron oft*) the one(s): **diese(r) . . . jene(r)** (*oft*) the latter . . . the former. **jenes** cf JENER
jenseitig a **1** other, opposite, of the other side. **2** other(-worldly), of the world beyond. **jenseits 1** adv on the other side; (*esp* + **von**) usu = **2** prep (+*genit*) beyond, on the other side of. **Jenseits** n the world beyond
Jerez·wein m sherry
Jesuiten·rinde f Jesuit (*or* cinchona) bark. **-tee** m Mexican tea
jett·schwarz a jet-black
jetzig a present, present-day. **jetzt** adv now **Jetzt·wert** m present value. **-zeit** f present (time)
jeweilig a particular, of (*or* at) that time, then-existing (-present, -valid); as adv =

jeweils *adv* 1 in each case, every time. 2 at that (particular) time. 3 at a time, each = JE 3

Jg *abbv* JAHRGANG year, etc

Jo . . . (*in Chem. names oft*) io . . . *cf* JONAN, etc

Joch *n* (-e) 1 yoke —**sich ein J. aufbürden** saddle oneself with a burden. 2 (mountain) pass, saddle, col. 3 (*in compds oft, Anat.*) zygo(matic), jugal. **··baum** *m* hornbean (tree). **-bein** *n* zygomat(ic bone). **-hefepilz** *m* zygosaccharomyces

Jod *n* iodine; (*in compds oft*) iodide, (*as substituent, usu.*) iodo-: **-alkyl** *n* alkyl iodide. **-ammon(ium)** *n* ammonium iodide. **-amyl** *n* amyl iodide. **-amylum** *n* starch iodide. **-arsen(ik)** *n* arsenic iodide. **j-artig** *a* iodine-like

Jodat *n* (-e) iodate

Jod-äther *m* iodoether. **-äthyl** *n* ethyl iodide

Jodation *f* (-en) iodination. **Jodat·ion** *n* iodate ion

jodatum *a* (*Pharm.*) iodide

Jod·azid *n* iodine azide. **-baryum** *n* barium iodide. **-benzin** *n* iodized benzine. **-benzoesäure** *f* iodobenzoic acid. **-benzol** *n* iodobenzene. **-blei** *n* lead iodide. **-bromchlorsilber** *n* (*Min.*) iodobromite. **-bromid** *n* iodine bromide. **-bromür** *n* iodine monobromide. **-chinolin** *n* iodoquinoline. **-chlor** *n* iodine chloride. **-chlorid** *n* iodine trichloride. **-chlorür** *n* iodine monochloride. **-cyan** *n* cyanogen iodide. **-dampf** *m* iodine vapor. **-dioxid** *n* iodine dioxide. **-eisen** *n* iron iodide. **-eiweiß** *n* (*Pharm.*) iodated albumen. **-essigester** *m* iodoacetic ester. **-farbzahl** *f* iodine color value. **-gehalt** *m* iodine content. **-gelb** *n* iodine yellow (pigment). **-gold** *n* gold iodide. **-gorgosäure** *f* iodogorgoic acid, diiodotyrosene. **-grün** *n* iodine green. **j-haltig** *a* iodiferous, containing iodine. **J-hydrat** *n* hydriodide, iodine hydrate. **-hydrin** *n* iodohydrin

Jodid *n* (-e) (-ic) iodide. **··chlorid** *n* iodide chloride. **-ion** *n* iodide ion

jodierbar *a* iodizable, capable of iodination. **jodieren** *vt* iodinate; iodize, iodate

jodig *a* iodous

Jodi·jodat *n* iodine iodate

Jodimetrie *f* iodimetry. **jodimetrisch** *a* iodimetric

Jodin·rot *n* iodine scarlet (red HgI_2)

Jod·ion *n* iodine ion (I^+)

Jodit *n* (-e) iodite, (*Min.*) iodyrite (AgI)

Jodi·verbindung *f* iodic compound

jodiziert *a* (*Pharm.*) iodized

Jod·jodat *n* iodine iodate. **-jodkalium-lösung** *f* potassium polyiodide solution (iodine in aqueous KI). **-kadmium** *n* cadmium iodide. **-kali(um)** *n* potassium iodide. **-kaliumstärkepapier** *n* starch iodide paper. **-kalzium** *n* calcium iodide. **-kasein** *n* iodinated casein. **-kohlenstoff** *m* carbon (tetra)iodide. **-kupfer** *n* copper iodide. **-lebertran** *m* iodated codliver oil. **-lithium** *n* lithium iodide. **-magnesium** *n* magnesium iodide. **-mangan** *n* manganese iodide. **-menge** *f* amount of iodine. **-messung** *f* iodometry. **-metall** *n* metallic iodide. **-methyl** *n* methyl iodide. **-methylat** *n* methiodide. **-mittel** *n* iodine (-containing) remedy. **-natrium** *n* sodium iodide. **-natron** *n* sodium hypoiodite. **j-normal** *a* normal to iodine

Jodo- iodo-, iodoxy- (IO_2): **-benzoesäure** *f* iodoxybenzoic acid. **-benzol** *n* iodoxybenzene. **-form** *n* iodoform

Jodol *n* (-e) iodol, tetra-iodopyrrole

Jodometrie *f* iodometry. **jodometrisch** *a* iodometric

Jodo·naphthalin *n* iodoxynaphthalene

Jodonium *n* iodonium

Jodoso·benzol *n* iodosobenzene (C_6H_5IO)

Jodoxy·naphtochinon *n* hydroxyiodonaphthoquinone ($C_{10}H_4O_2(OH)I$)

Jod· iodine, iodide, iodo-: **-papier** *n* iodized paper. **-pentoxid** *n* iodine pentoxide. **-phosphonium** *n* phosphonium iodide. **-phosphor** *m* phosphorus iodide. **-phtalsäure** *f* iodophthalic acid. **-quecksilber** *n* = **-quecksilberoxid** *n* (*old Termin.*) mercuric iodide. **-quecksilberoxydul** *n* (*old Termin.*) mercurous iodide. **-quelle** *f* 1 source of iodine. 2 mineral spring containing iodine. **-radium** *n* radium iodide. **-salbe** *f* iodine ointment. **-salz** *n* iodine salt, iodide. **j-sauer** *a* iodate of. **J-säure** *f* iodic acid. **-schwefel** *m* sulfur iodide. **-serum** *n* iodized serum. **-silber** *n* silver iodide, (*Min.*) iodyrite. **-silizium** *n* silicon iodide. **-stärke** f starch iodide. **-stick-stoff** m nitrogen iodide. **-thymol** n (*Pharm.*) thymol iodide. **-tinktur** f tincture of iodine. **-toluol** n iodotoluene. **-überträger** m iodine carrier

Jodür n (-e) (-ous) iodide

Jod·vergiftung f iodine poisoning, iodism. **-wasser** n iodine water. **-wasserstoff** m hydrogen iodide

Jodwasserstoff·äther m ethyl iodide. **j—sauer** a (hydr)iodide of. **J—säure** f hydriodic acid

Jod·wismut n bismuth iodide

Jodyrit n iodyrite

Jod·zahl f iodine number (*or* value). **-zimtsäure** f iodocinnamic acid. **-zink** n zinc iodide. **-zinkstärkepapier** n zinc iodine starch paper. **-zinn** n tin iodide. **-zinnober** m red mercuric oxide. **-zyan** n cyanogen iodide

Joghurt m yogurt

Johannis·beere f currant. **-beerwein** m currant wine. **-brot** n St. John's bread, carob (*or* locust) bean. **-brotkernmehl** n carob (*or* locust) bean flour. **-kraut** n St.-John's-wort. **-krautöl** n hypericon oil. **-wurzel** f male (*or* shield) fern, aspidium

Johimbe·rinde f yohimbé bark

Jon (*obsol*) = **Ion** (*also in compds*)

Jonan n ionane. **Jonen** n ionene. **Jonon** n ionone. **·-säure** f iononic acid

Jonosphäre f ionosphere

Jonquillen·öl n jonquil oil

Josen n iosene

Jota n (-s) iota, jot, tiny bit

J-Säure f J-acid, 2-amino-5-naphthol-7-sulfonic acid

Jubel m jubilation. **·-band** m jubilee (*or* anniversary) volume. **-feier** f, **-fest** n = **Jubiläum** n (..läen) jubilee

Juchten m/n, **·-leder** n Russia(n) leather. **-öl** n birch tar oil. **-rot** n Janus red. **j—rot** a Russian-leather-red

jucken vt/i itch, be itching (for, with, to do . . .). **Jucken** n itching, pruritus

Juck·stoff m itching powder (*or* agent), (*esp. Mil.*) utricant, nettle gas

Jude m (-n, -n) Jew; pl (*Metll.*) refinery scraps

Juden·dornbeere f jujube, zizyphus. **-harz, -pech** n asphalt. **-pilz, -schwamm**

m boletus satanas (*or* luridus)

Jüdin f (-nen) Jewish woman. **jüdisch** a Jewish

Jugend f youth; young people. **j—·kräftig** a youthfully vigorous. **jugendlich** a youthful; juvenile (under 18)

Jugoslawien n Yugoslavia

Julei, Juli m July

jung a (¨er, ¨st . . .) young, youthful; new, fresh, recent; overpoled (copper)

Jung·bier n new beer

Junge I m (-n, -n) boy. II n (*adj endgs*) young animal, cub, puppy, kitten, etc

jungen vi give birth (to young ones)

jünger a younger (*compar of* JUNG).

Jünger m (-) disciple

Jungfer f (-n) 1: alte J. old maid. 2 (*Metll.*) iron ladle. 3 (*Engg.*) rammer

Jungfern· maiden, virgin; native: **-blei** n first lead from the furnace. **-erde** f virgin soil. **-geburt** f parthenogenesis. **-glas** n selenite, transparent gypsum; mica. **-metall** n native (*or* virgin) metal. **-öl** n virgin oil. **-pergament** n vellum. **-quecksilber** n native mercury. **-schwefel** m virgin (*or* native) sulfur. **-wachs** n virgin wax. **-wein** m wine from unpressed grapes. **-zeugung** f parthenogenesis

Jungfrau f virgin; (*Geog.*) Jungrau (*mtn. in Alps*). **jungfräulich** a virgin

Jung·geselle m bachelor, single man. **-gesellin** f single woman

Jung·holz n sapwood

Jüngling m (-e) youth, young man

jüngst· I a (*superl of* JUNG) 1 youngest. 2 latest, most recent. 3 last. II adv recently, lately

Jung·tertiär n (*Geol.*) later Tertiary, Neocene, Neogene. **j—vulkanisch** a (*Geol.*) young volcanic. **-zeitlich** a recent

Juni, Juno m June

Jura m (*Geol. & Geog.*) Jura, (*Geol.*) Jurassic. II npl Law (as subject of study). **·-kalk** m Jurassic limestone. **jurassisch** a Jurassic. **Jura·zeitalter** n Jurassic period

just adv just; **j. nicht** not exactly

justierbar a adjustable. **Justier·einrichtung** f adjusting device. **justieren** vt adjust, set; justify (type)

Justier- adjusting: **-schraube** *f* adjusting (*or* set) screw. **-tisch** *m* adjusting table

Justierung *f* (-en) adjustment; justification (of type)

Justier·waage *f* adjusting balance

Justiz *f* administration of justice; (*in compds usu*) law, legal; judicial: **-wesen** *n* law, legal (*or* judicial) system

Jute·faden *m* jute fiber (*or* twine). **-faser** *f* jute fiber. **-leinwand** *f* burlap, gunny (cloth). **-stoff** *m* jute cloth (*or* material)

Juwel *n* (-en) jewel, *pl oft* jewelry. **Juwelier** *m* (-e) jeweler. **-arbeit(en)** *f(pl)* jewelry. **-borax** *m* octahedral borax, sodium tetraborate pentahydrate (used as soldering flux)

Jux *m* joke; fun

J.Z. *abbv* (Jodzahl) iodine number

K

(*Cf* note under **C**)

k *abbv* 1 (Karat) K. **2** (Kilo) kg

k. *abbv* (kalt) cold

K *abbv* (Kathode) cathode

K. *abbv* (elektrische Dissoziationskonstante) electric dissociation constant

Kabel *n* (-) **1** (*genl*) cable. **2** (extension) cord; wire. **··ausgußmasse** *f* cable compound

Kabeljau *m* (-e *or* -s) cod(fish). **··(leber)tran** *m* cod-liver oil

Kabel·lack *m* cable dope. **-litze** *f* cable strand. **-masse** *f* (plastic) sheathing compound. **-mischung** *f* (rubber) cable stock. **-öl** *n* cable oil. **-schmiere** *f* cable compound. **-schuh** *m* cable clip (*or* eye). **-teer** *m* cable tar. **-vergußmasse** *f* cable compound. **-wachs** *n* cable wax

Kabine *f* (-n) cabin, cab; booth, cubicle; (cable *or* aerial tramway) car

Kabinett *n* (-e) **1** small room, study. **2** (display) booth, small exhibition hall. **3** toilet. **4** (*DDR*) study and guidance center. **5** (*Polit.*) cabinet

Kabis *m* cabbage

Kabliau *m* cod(fish) = KABELJAU

Kachel *f* (-n) glazed (*or* Dutch) tile. **··ofen** *m* tiled stove

Kachu *n* catechu

Kadam·öl *n* cadam oil

Kadaver *m* (-) (dead) body, corpse; carcass

Kaddich, Kaddig *m* cade, Spanish juniper. **··öl** *n* = **Kade-öl, Kadi(n)·öl** *n* oil of cade

kadmieren *vt* cadmium-plate; treat with cadmium, cadmiate

Kadmium *n* cadmium. **··beize** *f* (*Dye.*) cadmium mordant. **-chromgelb** *n* cadmium chrome yellow. **-gehalt** *m* cadmium content. **-gelb** *n* cadmium yellow. **k–haltig** *a* containing cadmium, cadmiferous. **K–jodid** *n* cadmium iodide. **-oxyhydrat** *n* cadmium hydroxide. **k–plattiert** *a* cadmium-plated. **K–rhodanid** *n* cadmium thiocyanate. *See also* CADMIUM

Käfer *m* (-) beetle, bug; coleopter(on). **k–·artig** *a* coleopterous

Kaff *n* chaff, rubbish

Kaffal·säure *f* caffalic (*or* coffalic) acid

Kaffee *m* coffee. **k–··ähnlich** *a* coffee-like. **K–baum** *m* coffee tree. **-bohne** *f* coffee bean. **k–braun** *a* coffee-brown (*or* -colored). **K–ersatz(stoff)** *m* coffee substitute. **-gerbsäure** *f* caffetannic acid. **-kanne** *f* coffee pot. **-öl** *n* coffee oil. **-satz** *m* coffee grounds. **-säure** *f* coffeic acid. **-surrogat** *n* coffee substitute

Kaffein *n* caffeine

Käfig *m* (-e) cage. **··effekt** *m* cage effect. **-wicklungsmotor** *m* squirrel-cage motor

Kahinka·wurzel *f* cahinca root

kahl I *a* **1** bald; hairless; (*Tex.*) napless. **2** bare; denuded; unadorned. **3** leafless; defoliated. **4** desolate, barren, bleak, empty. **5: k. gehen** (*Metll.*) need no flux. II *sep pfx* bald, bare, etc. **Kahlheit** *f* baldness; bareness; defoliation; barrenness, etc.

Kahm *m* mold(y film) (*esp* on wine); scum. **kahmen** *vi* get moldy, form a mold

Kahm·haut *f* pellicle of mold, moldy film. **-hefe** *f* mycoderma

kahmig *a* moldy, stale; (*Wine*) ropy

Kahm·pilz *m* mold (fungus); mycoderm; mother of vinegar

Kahn *m* (⁀e) **1** (row)boat, canoe. **2** barge, lighter. **k–··förmig** *a* boat-shaped, scaphoid, navicular

Kai *m* (-e *or* -s) quay, wharf; waterside street

Kaiman *m* (-e) cayman (alligator)

Kaïnka·wurzel *f* cahinca root

Kaiser *m* (-) emperor; (*in compds, oft*) imperial: **··blau** *n* smalt. **-gelb** *n* mineral yellow. **-grün** *n* imperial (*or* Paris) green

kaiserlich *a* imperial

Kaiser·öl *n* Kaiser oil, kerosene. **-reich** *n*

371

empire. **-rot** n colcothar, imperial red (dye). **-schnitt** m (*Med.*) Caesarian section. **-wasser** n aqua regia. **-wurz** f masterwort

Kajaput·ol, Kajeput·öl n cajuput oil

Kakadu m (-s) cockatoo

Kakao m cocoa; cacao. **··baum** m cacao (tree). **-bohne** f cacao (*or* cocoa) bean. **-butter** f cocoa butter. **-likör** m crème de cacao. **-masse** f cocoa paste. **-öl** n cacao oil. **-pulver** n cocoa (powder)

Kakaorin n cacaorine

Kakao·rot n cocoa red

Kakerlak m (-en, -en) 1 (cock)roach. 2 albino; (*specif*) albino rabbit

Kakodyl n cacodyl. **··säure** f cacodylic acid. **-wasserstoff** m cacodyl hydride

Kakothelin n cacotheline

Kakoxen n cacoxene, cacoxenite

Kakteen cacti: *pl of* **Kaktee** f & **Kaktus** m cactus

Kal. *abbv* (Kalorie) Cal., kilo(gram) calorie

Kalabar·bohne f Calabar bean

Kalamin m calamine. **Kalaminthe** f (-n) (*Bot.*) 1 calamint. 2 wild basil

Kalander m (-) (*Tex.*) calender, glazing machine *cf* KALENDER. **kalandern, kalandrieren** vt calender, glaze

Kalb n (⁻er) (*Zool.*) calf; (*Meat*) veal. **Kalbe** f (-n) heifer

Kälber·lab n, **-magen** m rennet

Kalb·fell n calfskin. **-fleisch** n veal

Kalbin f (-nen) heifer

Kalb·lackleder n patent calf (leather). **-leder** n calfskin, calf (leather)

Kalbs·magen m rennet; calf's stomach. **-pergament** n vellum

Kalb·velour n (*Lthr.*) suede calf

Kalci . . . *See* **Calci..** or **Kalzi..**

Kaldaunen fpl 1 tripe. 2 (beef) innards

Kalender m (-) 1 calendar; almanac *cf* KALENDER. 2 omasum

Kaleszenz f calescence

kalfatern vt calk

Kali n (caustic) potash; potassium hydroxide; (*in old names of compds oft* = KALIUM). **··alaun** m potash (*or* potassium) alum, kalinite. **-ammonsalpeter** m potassium-chloride-ammonium-nitrate mixture. **-apparat** m potash apparatus

Kaliatur·holz n caliatour wood, sandalwood

Kaliber m (-) 1 caliber, gauge; bore. 2 (*Metll.*) groove, pass (of rollers). 3 inside calipers

Kali·bergwerk n potash mine. **-blau** n Prussian blue. **-bleiglas** n potash lead glass

kalibrieren vt 1 calibrate. 2 (*Metll.*) groove (rolls). **Kalibrier·pipette** f calibrating pipette. **Kalibrierung** f (-en) calibration; grooving

·kalibrig *sfx:* e.g. **groß·kalibrig** large-caliber(ed)

kalibrisch a calibrated, in calibration

Kali·dünger m potash fertilizer. **-düngesalz** n potassium salt fertilizer. **-eisenalaun** m potash iron alum, halotrichite. **-eisencyanür** n potassium ferrocyanide. **-endlauge** f (*Paper*) black liquor. **-feldspat** m potash feldspar, orthoclase. **-form** f potash mold

kalifornisch a California(n)

kali·frei a potash-free. **K–glas** n potash glass. **-glimmer** m potash mica, muscovite. **k–haltig** a containing potash. **K–hydrat** n potassium hydroxide. **-hydratlösung** f (*oft*) potash lye. **-kalk** m potash lime

Kaliko n (-s) calico; (*Books*) cloth binding. **··druck** m calico printing

Kali· potash: **-kugel** f potash bulb. **-lauge** f potash lye, caustic potash solution. **-magnesia** f potash magnesia, potassium magnesium sulfate. **-metall** n metallic potassium. **-natron** n: **weinsaures K.** potassium sodium tartrate. **-nephelin** n (*Min.*) kaliophilite. **-olivenölseife** f potash olive oil soap. **-pflanze** f glasswort, alkaline plant; saltwort. **-reihe** f (*Petrog.*) potash series. **-rohsalz** n crude potassium salt. **-salpeter** m saltpeter, potassium nitrate, niter. **-salz** n potassium salt. **-salzbergwerk** n potash mine. **-salzlager** n potash salt deposit (*or* bed). **-schmelze** f caustic potash fusion (*or* melt). **-schmierseife** f potash soft soap. **-schwefelleber** f (*Pharm.*) sulfurated potash, liver of sulfur (impure potassium sulfide). **-seife** f potash (*or* soft) soap.

-siederei *f* potash works. -sulfat *n* potassium sulfate. -tonerde *f* potassium aluminate

Kalium *n* potassium; (*Pharm.*): **K. bromatum** potassium bromide; **K. chloratum** potassium chloride; **K. chloricum** p. chlorate; **K. jodatum** p. iodide; **K. nitricum** p. nitrate; **K. nitrosum** p. nitrite. **--alaun** *m* potassium alum. -azetat *n* potassium acetate. -borfluorid *n* potassium fluoborate. -brechweinstein *m* tartar emetic, potassium antimonyl tartrate. -bromid *n* potassium bromide. -chlorid *n* potassium chloride. -chromalaun *m* potassium chromium alum. -eisencyanid *n* potassium ferricyanide. -eisencyanür *n* potassium ferrocyanide. -ferrisulfat *n* potassium ferric sulfate. -ferrocyanat, -ferrocyanür *n* potassium ferrocyanide. -gehalt *m* potassium content. -goldbromür *n* potassium bromoaurite. -goldchlorid *n* potassium chloroaurate. -goldcyanid *n* potassium auricyanide. -halogen *n* potassium halide. k–haltig *a* potassium-containing. **K–hydrat** *n* potassium hydroxide. -hydrid *n* potassium hydride. -hydroxid *n* potassium hydroxide. -hydrür *n* potassium subhydride (K_2H or K_4H_2). -iridichlorid *n* potassium chloroiridate. -jodat *n* potassium iodate. -jodid *n* potassium iodide. -jodidstärkepapier *n* potassium iodide starch paper. -kobaltnitrit *n* potassium cobaltinitrite. -kohlenoxid *n* potassium carboxide (or hexacarbonyl). -nitrat *n* potassium nitrate. -oxid *n* potassium oxide. -oxidhydrat *n* potassium hydroxide. -permanganat *n* potassium permanganate. -phosphat *n* potassium phosphate: **primäres K.** p. dihydrogen phosphate; **sekundäres K.** dipotassium hydrogen phosphate

Kaliumplatin-chlorid *n* potassium chloroplatinate. -chlorür *n* potassium chloroplatinite. -cyanid *n* potassium cyanoplatinate. -cyanür *n* potassium cyanoplatinite

Kalium-quecksilberjodid *n* potassium mercuric iodide. -rhodanat, -rhodanid

n potassium thiocyanate. -seife *f* potassium soap. -silbercyanid *n* potassium dicyanoargentate. -sulfat *n* potassium sulfate. -sulfhydrat *n* potassium bisulfide, p. hydrogen sulfide. -sulfozyanat, -sulfozyanid *n* potassium thiocyanate. -verbindung *f* potassium compound. -wasserstoff *m* potassium hydride. -wolframat *n* potassium tungstate. -zelle *f* (*Elec.*) potassium cell. -zyanid *n* potassium cyanide

Kali· potash: -verbindung *f* potash compound. -wasserglas *n* potash waterglass, potassium silicate. -werk *n* potash plant (or works)

Kalk *m* (-e) **1** (*genl*) lime. **2** (*specif*) lime mortar; lime fertilizer; whitewash; (Med.) limestone, chalk; (*Med., & Chem. in old names of compds. oft* = CALCIUM): **ätzender K.** caustic lime; **gebrannter K.** quicklime; **gelöschter K.** slaked lime. **--ablagerung** *f* lime (calcareous, limestone) deposit. -abscheidung *f* lime scum; separation of lime. -algen *fpl* calcareous algae. -alkalität *f* lime alkalinity. -ammonsalpeter *m* calcium ammonium nitrate, lime chalk. -anlagerung *f* (*Physiol.*) calcification. -ansammlung *f* accumulation of lime. -ansatz *m* lime deposit (or crust). -anstrich *m* whitewash. -anwurf *m* plaster. k–arm *a* lime-deficient. **K–arsen** *n* calcium arsenide; calcium arsenate insecticide. k–artig *a* calcareous. **K–äscher** *m* **1** (*Soap*) lixiviated ashes, soap white. **2** (*Lthr.*) lime pit. -asphalt *m* asphalt with limestone aggregate. -ausscheidung *f* separation (or deposit) of lime. -back *m* (*Sugar*) lime back. -bad *n* lime bath. -baryt *m* (*Min.*) calcareous barite. -bedürfnis *n* lime requirement. -behandlung *f* lime treatment, liming. -beize *f* (*Dye.*) lime mordant. k–beständig *a* lime-resistant. **K–beständigkeit** *f* resistance to lime. -bestimmung *f* determination of lime. -beton *m* lime concrete. -beuche *f* (*Bleaching*) lime boil. -bewurf *m* coat of plaster; lime plaster(ing). -bindevermögen *n* (*Dterg.*) calcium chelating power. -blau *n* blue verditer (basic copper carbonate pigment), azurite blue. **-boden** *m* limy (*or* calcareous) soil. **-borat** *n* calcium borate. **-brei** *m* lime paste (putty, wash), cream of lime. **-brennen** *n* lime burning. **-brennerei** *f* 1 lime burning. 2 = **-brennofen** *m* limekiln. **-bruch** *m* limestone quarry. **-brühe** *f* lime liquor; milk of lime; whitewash. **-chromgelb** *n* calcium chrome yellow. **-düngung** *f* lime fertilizing, soil liming

Kälke *f* (-n) (*Lthr.*) lime pit; lime

kalk·echt *a* lime-resistant, fast to lime. **K-echtheit** *f* fastness to lime. **-einlagerung** *f* calcareous deposit

Kalkeisen·augit *m* (*Min.*) hedenbergite. **-granat** *m* (*Min.*) andradite, lime-iron garnet. **-olivin** *n* (*Min.*) iron monticellite. **-stein** *m* ferruginous (*or* iron-bearing) limestone; impure limonite

Kalk·elend *n* 1 (*Wine*) chalkiness. 2 (*Blast Furnace*) lime set. **k-empfindlich** *a* lime-sensitive. **K-empfindlichkeit** *f* sensitivity to lime

kalken, kälken *vt* 1 treat (*or* fertilize) with lime. 2 whitewash

Kalk·entartung *f* calcareous degeneration. **-erde** *f* lime, calcium oxide, calcareous earth, **k-erdig** *a* calcareous

kalkern *vt* (treat with) lime

Kalk· lime, calcium: **-erz** *n* limy ore. **-fällung** *f* precipitation of lime. **-farbe** *f* lime color (*or* paint); lime-resistant pigment. **-faß** *n* (*Lthr.*) lime vat (*or* pit). **-fehler** *m* (*Cement*) lime error. **-feldspat** *m* lime feldspar, (*Min.*) anorthite. **k-frei** *a* lime-free. **-führend** *a* lime-bearing, calciferous, calcareous. **K-gebirge** *n* limestone mountains. **-gehalt** *m* lime content. **-glas** *n* lime glass. **-glimmer** *m* (*Min.*) margarite. **-granat** *m* lime garnet, grossularite; essonite; andradite. **-grieß** *m* milk-of-lime grit. **-guß** *m* grout, thin lime mortar

kalkhaft *a* limy, chalky, calcareous

kalk·haltig *a* calcareous, lime-bearing. **K-haltigkeit** *f* calcareousness. **-harmotom** *n* (*Min.*) stilbite. **-harnstoff** *m* urea-chalk (*or* -limestone) fertilizer. **-härte** *f* (*Water*) calcium hardness. **-harz** *n* hard-

ened rosin. **k-hold** *a* (*Plants*) lime-loving. **K-hütte** *f* lime kiln. **-hydrat** *n* hydrated lime, calcium hydroxide. **-hydrosilikat** *n* hydrated calcium silicate

kalkieren *vt* trace

Kalkier·leinwand *f* tracing cloth. **-papier** *n* tracing paper

kalkig *a* limy, chalky, calcareous

Kalk· lime, calcium, calcareous: **-kitt** *m* lime (*or* calcareous) cement. **-körperchen** *n* (*Physiol.*) calcareous body (*or* granule). **-lager** *n* lime(stone) deposit; lime stores. **-lauge** *f* lime lye (*or* liquor); calcium hydroxide solution. **-licht** *n* limelight, calcium light. **k-liebend** *a* (*Plants*) lime-loving, calciphilous. **K-löschen** *n* lime slaking (*or* hydrating). **-löschtrommel** *f* lime-slaking drum. **-lösung** *f* lime (*or* calcium hydroxide) solution. **-malerei** *f* fresco (*or* wall) painting. **-manganspat** *m* (*Min.*) manganocalcite. **-mangel** *m* lime deficiency, lack of lime. **-mehl** *n* (air-slaked) lime powder. **k-meidend** *a* lime-avoiding, calcifugous. **K-mergel** *m* lime (*or* calcareous) marl. **-messer** *m* calcimeter (to determine CO_2 in limestone, etc). **-metall** *n* calcium. **-milch** *f* 1 milk of lime. 2 whitewash. **-mörtel** *m* lime mortar. **-muschelbreccie** *f* coquinoid (*or* shelly) limestone. **-natronfeldspat** *m* (*Min.*) plagioclase. **-natronglas** *n* soda-lime glass. **-niederschlag** *m* lime precipitate. **-ofen** *m* limekiln. **-oligoklas** *m* (*Min.*) labradorite. **-olivin** *n* (*Min.*) calcio-olivine. **-pflanze** *f* calcareous plant. **-pulver** *n* lime powder, powered lime. **-rahm** *m* cream of lime. **k-reich** *a* lime-rich; (*Water*) hard. **K-saccharat** *n* calcium sucrate. **-salpeter** *m* calcium nitrate. **-salz** *n* lime (calcareous, calcium) salt. **-sand** *m* lime (*or* calcareous) sand. **-sandstein** *m* 1 (Geol.) calcareous sandstone. 2 (*Ceram.*) = **-sandziegel** *m* sand-lime (*or* calcium silicate) brick. **-sauerbad** *n* (*Bleaching*) lime sour. **-schatten** *m* (*Lthr.*) lime speck (*or* blast). **-schaum** *m* lime froth (*or* scum). **k-scheu** *a* (*Plants*) lime-shunning, calcifugous. **K-schicht** *f* lime(stone) (*or* chalk) layer. **-schiefer** *m* calcareous slate (or schist). **-schlacke** f lime slag. **-schlamm** m lime sludge (mud, putty). **-schwarz** n lime black. **-schwefelbrühe** f lime sulfur, calcium polysulfide solution. **-schwefelleber** f (Pharm.) sulfurated lime, lime hepar. **-seife** f lime (or calcium) soap: **unlösliche K.** solid lime-soap scum. **-silo** n (Metll.) limestone bin. **-silt** m calcareous mud. **-sinter** m calcareous sinter, (Min.) travertine. **-spalt** m (Lthr.) limed split. **-spat** m calc-spar, (Min.) calcite. **-stein** m limestone

Kalkstein- limestone: **-ablagerung** f limestone deposit. **-gut** n (Ceram.) calcareous earthenware. **-mehl** n ground limestone. **-zuschlag** m (Metll.) limestone flux

Kalk-stickstoff m calcium cyanamide, "lime nitrogen". **-sulfat** n sulfate of lime, calcium sulfate. **-talkspat** m dolomite. **-teig** m lime putty. **-tiegel** m lime crucible. **-tropfstein** m calcareous dripstone. **-tuff** m calcareous tufa, (Min.) travertine. **-tünche** f lime whitewash

Kalkül m or n (-e) 1 calculation: **ins K. ziehen** include in the calculations. 2 (Math.) calculus. **kalkulatorisch** a calculatory, of calculation; adv by calculation, mathematically. **kalkulieren** vt/i calculate

Kalk-untergrund m (Dye) lime ground. **-uranglimmer** m (Min.) autunite, metazeunerite. **-uranit** n (Min.) autunite. **-verbindung** f lime (or calcium) compound. **-verseifung** f saponification with lime. **-wasser** n limewater. **-weinstein** m calcareous tartar. **-werk** n lime works. **-zementmörtel** m lime-cement mortar. **-zuckerreinigung** f purification of sugar by means of lime. **-zufuhr** f addition (or supply) or lime. **-zuschlag** m lime(stone) flux

Kallait n (Min.) turquoise

Kallochrom n (Min.) crocoite

kallös a callous

Kalmie f (-n) mountain laurel

Kalmus m (-se) calamus, myrtle flag. **-öl** n calamus oil. **-wurzel** f calamus root, flagroot

Kaloreszenz f calorescence

Kalorien-gehalt m calorie content

Kalorifer m (-en) radiator; warm air furnace

kalorimetrisch a calorimetric: **k—e Bombe** bomb calorimeter

kalorisch a caloric, thermal. **kalorisieren** vt (Metll.) calorize

kaloritieren vt (Metll.) calorize, aluminize

Kalotte f (-n) 1 segment of a sphere. 2 skull cap. 3 (Med.) calotte. 4 (Metll.) cup, spherical indentation. **Kalotten-modell** n (Stuart) atomic model

kalt 1 a (-er, -este) cold; cool: **ihnen ist k., sie haben k.** they feel (or are) cold; **k. essen** have a cold meal; **es überlief sie k.** they got cold chills; **k—er Satz** (Brew.) cold-water extract of malt; **auf k—em Wege** (a) without heat, (b) without attracting attention, (c) not strictly according to the rules; **k. stellen** put on ice, refrigerate Cf KALT-STELLEN. 2 sep pfx cold, cool (also implies being unexcited, indifferent, inoperative, dead)

Kalt-absetzen n cold sedimentation. **-aushärtung** f (Metll.) cold tempering, age hardening at room temperature. **-bad** n cold bath. **-bearbeitung** f cold working. **-biegeprobe** f cold bending test. **k-bildsam** a plastic, ductile (when cold). **-bläsig** a (Metll.) refractory. **K—bleiche** f cold bleaching. **-blüter** m cold-blooded animal. **k-blütig** a cold-blooded. **K-bruch** m cold-shortness, cold brittleness. **k-brüchig** a cold-short. **K-brüchigkeit** f cold-shortness

Kälte f 1 cold (=low temperature). 2 coldness, frostiness, chilliness. 3 coolness. 4 frigidity. **--anlage** f refrigerating plant. **-anwendung** f use of cold (temperatures), refrigeration. **-bad** n cooling bath. **k-beständig** a frost-resistant, antifreezing. **K-beständigkeit** f resistance to cold. **-chemie** f cryochemistry, low-temperature chemistry. **-einheit** f unit of cold, frigorie. **k-empfindlich** a sensitive to cold. **K-empfindung** f sensation of cold. **k-erzeugend** a refrigerating, cryogenic, frigontic. **K-erzeuger** m refrigerator, freezing apparatus. **-erzeugung** f refrigeration, cooling. **k-fest** a cold-resisting. **K–festigkeit** f resistance to cold. **-grad** m degree(s) below 0°C. **-industrie** f refrigeration industry. **-leistung** f refrigeration capacity. **-maschine** f refrigerating machine. **-mischung** f freezing mixture. **-mittel** n refrigerating agent, refrigerant.

kälten vt chill, refrigerate

Kälte·probe f cold test. **-punkt** m (Oils) solidifying point

kalt·erblasen* vt (Iron) cold-blast

Kälte·regler m cryostat. **-schutzmittel** n antifreeze. **-sprung** m cold-weather crack, crack caused by the cold. **-starre** f stiffness due to cold. **-stoff** m freezing mixture. **-technik** f refrigeration. **-techniker** m refrigeration engineer. **-thermometer** n low-temperature thermometer. **-träger** m cooling (or refrigeration) medium, coolant, refrigerant. **-trub** m cold sludge. **-trübung** f (Beer) chill haze. **-verfahren** n cold process; cold storage

Kalt·färben n cold dyeing. **-formen** n (Metll.) cold working. **-gebläse** n cold blast. **k–gegossen** p.a cold-cast cf KALTGIESSEN. **-gelegen** p.a. out of operation, closed down cf KALTLIEGEN. **-gelegt** p.a shut-down, (Furnace) blown-out cf KALTLEGEN. **-gezogen** p.a cold-drawn cf KALTZIEHEN. **·gießen*** vt cast cold. **K–glasur** f cold glaze. **k–gründig** a cold-soil. **K–guß** m spoiled casting. **k–hämmerbar** a cold-malleable. **·hämmern** vt cold-forge. **·härten** vt (Metll.) strain-harden. **-härtend** p.a cold-setting. **K–hefe** f wild yeast. **-kathodenröhre** f cold-cathode tube. **-klebemasse** f cold adhesive. **-lack** m cold varnish. **k—·lagern** vt put (or keep) in cold storage. **K–lagerung** f cold storage. **k—·lassen** vt leave unaffected, fail to affect. **·legen** vt shut down, (Furnace) blow out cf KALTGELEGT. **K–leim** m cold glue, wood cement. **-leiter** m low-temperature conductor (whose resistance increases with temperature). **k—·liegen*** vi be out of operation, lie idle cf KALTGELEGEN. **·lösen** vt dissolve (in the cold). **·löslich** a cold-soluble. **K–löt(e)stelle** f (Thermocouple) cold junction.**-luft** f cold air. **-luftgas** n carburetted water gas. **k—·machen** vt put out of

commission, eliminate, liquidate. **K–nachverdichten** n (Metll.) cold-working after heat treatment. **-nachwalzwerk** n (Metll.) skin (or temper) pass mill. **k—·nieten** vt cold-rivet. **·recken** vt (Metll.) cold-strain. **k–rissig** a (Metll.) cold-short. **K–schliff** m (Paper) cold-ground pulp. **k—·schmieden** vt cold-hammer, hammer-harden. **K–schmiere** f (Lthr.) cold stuffing. **-sprödigkeit** f cold-shortness. **k—·stellen** vt shelve (fig); render harmless cf KALT 1. **·verarbeiten** vt (Metll.) cold-work. **K–verbindung** f (Thermocouple) cold junction. **-verfestigung** f cold hardening. **k—·verformen** vt cold-work, cold-mold. **K–vergütung** f natural aging (at ordinary temperatures). **k–vulkanisiert** a cold-cured, cold-vulcanized. **K–walze** f cold roll. **k—·walzen** vt cold-roll. **K–walzwerk** n cold reduction mill

Kaltwasser·bad n cold-water bath. **-farbe** f cold-water paint (color, dye). **-probe** f cold-water test. **-strom** m cold-water stream (or current)

kalt·werden* vi(s) get cold, cool. **K–wind** m cold blast. **k—·ziehen*** vt (Metll.) cold-draw

kalz. abbv (kalziniert) calcined

kalzi· See also CALCI.

Kalzinator m (-en) calciner. **kalzinieren** vt calcine. **Kalzinier·ofen** m calcining furnace

Kalzit n (-e) calcite

Kalzium n calcium

kam came: past of KOMMEN

Kamba(l)·holz n camwood

Kambium·schicht f cambium zone

Kambodscha n Cambodia, Kampuchea. **··lack** m Cambodia lac

kambrisch a, **Kambrium** n (Geol.) Cambrian

käme would come, etc: sbjc of KOMMEN

Kamee f (-n) cameo

Kamel n (-e) camel. **··garn** n mohair. **-haarpinsel** m camel's-hair brush

Kamelie f (-n) camellia

Kamera f (-s) camera

Kamerad m (-en, -en) comrade, friend; companion

Kamerun n (Geog.) Cameroons

Kamfer *m* camphor = KAMPFER
Kam·holz *n* camwood
kamig *a* moldy
Kamille *f* (-n) camomile. **Kamillen·öl** *n* camomile oil
Kamin *m* (-e) 1 fireplace. 2 chimney (*incl Geol.*), smokestack. **··brenner** *m* burner with a chimney. **-gas** *n* flue gas. **-haube** *f* chimney hood. **-ruß** *m* chimney soot
Kamm *m* (ːe) 1 comb (*incl* rooster's). 2 (*Tex.*) card. 3 (mountain) ridge; crest. 4 cog. 5 (*Meat*) neck, chine. **··abfall** *m* combings
kämmen I *vt* 1 comb. 2 (*Wool*) card. **II** *vi* mesh (*as:* cogs, gears). **III: sich k.** comb one's hair
Kammer *f* (-n) (*genl*) chamber; (*specif*) small room, closet; compartment. **··gitterstein** *m* (*Furnaces*) checker. **-kristall** *m* chamber crystal. **-ofen** *m* chamber oven; compartment kiln; (*Coke*) by-product oven. **-prozeß** *m* chamber process. **-säure** *f* chamber acid. **-schlamm** *m* chamber sludge. **-sohle** *f* chamber floor. **-tuch** *n* cambric. **-wasser** *n* (*Anat.*) aqueous humor. **-zerfaserer** *m* drum shredder
Kamm·fett *n* horse grease. **k–förmig** *a* comb-shaped. **K–garn** *n* combed yarn, (*specif*) worsted. **-gras** *n* (*Bot.*) dog's-tail. **-kies** *m* coxcomb pyrites, marcasite. **-rad** *n* cogwheel
Kammuschel *f* (=**Kamm·muschel**) (*Zool.*) scallop
Kamm·wolle *f* worsted. **-zug** *m* (*Tex.*) woolen slubbing (*or* top); cotton sliver
Kamomille *f* (-n) camomile
Kampagne *f* (-n) campaign; working season
Kampane *f* (-n) bell jar
Kampesche·holz *n* logwood, campeachy wood
Kampf *m* (ːe) battle, fight; struggle, contest: **den K. ansagen** (+*dat*) declare war (on)
Kampfen *n* camphene
kämpfen *vi* fight, battle; struggle
Kampfer *m* camphor
Kämpfer *m* (-) fighter, warrior, combatant
kampfer·artig *a* camphor-like, camphoraceous. **K–baum** *m* camphor tree. **-chinon** *n* camphorquinone. **-essig** *m* camphorated vinegar. **-geist** *m* spirit of camphor. **k–haltig** *a* camphorated, containing camphor
Kämpferid *n* kaempferide
kampferieren, kampfern *vt* camphorate
Kampfer·öl *n* camphor(ated) oil
Kämpferol *n* kaempferol
Kampfer·phoron *n* camphorphorone. **-säure** *f* camphoric acid. **-säuresalz** *n* camphorate. **-spiritus** *m* spirit of camphor. **-wein** *m* camphorated wine
Kampf·gas *n* war gas, poison gas
Kampfin *n* camphine
Kampfor *n* camphor = KAMPFER
Kampf·platz *m* field of battle, arena. **-stoff** *m* (*genl.*) substance used in warfare, (*specif*) poison gas
Kampfstoff· poison gas, chemical (warfare): **-beschluß** *m* chemical shell barrage. **-bombe** *f* chemical (*or* poison-gas) bomb. **-gehalt** *m* poison-gas content. **-schwaden** *m* (*usu pl*) poison-gas fumes. **-zerstäuber** *m* chemical spray apparatus
Kampf·wagen *m* (*Mil.*) armored car, tank
Kamphen *n* camphene
Kampher, Kamphor *m* camphor. *See* KAMPFER
Kanada·balsam *m* Canada balsam
kanadisch *a* Canadian
Kanal *m* (..näle) 1 canal, channel (*incl TV*), (*Radio*) band. 2 sewer, flue. **··bildung** *f* channeling. **-gas** *n* sewer gas
Kanalisation *f* (-en) 1 canalization. 2 sewerage, sewer system
Kanalisations·rohr *n*, **-röhre** *f* sewer pipe, drain. **-system** *n* sewerage (*or* drainage) system
Kanal·jauche *f* (liquid) sewage. **-ofen** *m* tunnel kiln. **-rohr** *n* sewer pipe, drain. **-ruß** *m* channel black. **-schleuse** *f* canal lock. **-strahl** *m* canal ray, positive ray. **-trockner** *m* tunnel drying oven. **-wasser** *n* sewage
Kanarien· canary: **··farbe** *f* canary color. **k–gelb** *a* canary-yellow. **K–hirse** *f* canary seed. **-vogel** *m* canary
kanarisch: die k—en Inseln Canary Islands
Kandel·kohle *f* cannel coal
kandeln 1 *vt* channel, flute, groove. **2** *vi* flow, run
Kandel·zucker *m* candy sugar, rock candy

kandieren vt 1 candy (fruit, etc.) 2 crystallize. 3 (*Seed*) coat (*or* treat) with copper sulfate

Kandis m sugar candy. **·-störzel** m mother liquor from sugar candy. **-zucker** m sugar candy

Kanditen pl candied fruit

Kaneel m (-e) cinnamon: **brauner K.** Ceylon cinnamon; **weißer K.** cannella, clove cassia. **·-bruch** m broken cinnamon. **-granat, -stein** m cinnamon stone, essonite

Kanella·rinde f canella bark

Kanevas m (-se) (embroidery) canvas

Känguruh n (-s) kangaroo

Kanin n (-e) rabbit fur. **Kaninchen** n (-) rabbit

kann can, is able to: *pres of* KÖNNEN

Kännchen n (-) little jug (pot, can)

Kanne f (-n) can (for liquids); (coffee, tea) pot; pitcher, jug

Kännel m (-) drainpipe

kannelieren vt flute, groove, channel

Kannel·kohle, Kännel·kohle f cannel coal

Kannen·zinn n pewter

kannte knew: *past of* KENNEN

Kanone f (-n) cannon, gun

Kanonen·bronze f, **-gut** n, **-metall** n gunmetal. **-ofen** m 1 round iron stove. 2 bomb (*or* sealed-tube) furnace. **-pulver** n cannon powder. **-rohr** n. **-röhre** f 1 gun barrel. 2 bomb-furnace tube

kanonisch a canonical

Känozoikum n (*Geol.*) Cenozoic

Kante f (-n) edge, border; (*Tex. oft*) edging; (*Spect.*) band head —**auf die hohe K. legen** put by, save (for later). **kanten** vt 1 stand on edge, tilt, cant. 2 edge, border. 3 square (stone). 4 chamfer (*or* bevel) the edges of

Kantel n (-) square ruler

Kanten·kante f (*Spect.*) head of heads. **-mörser** m angle shot mortar. **-schema** n (*Spect.*) Deslandres table. **-spannung** f edge stress. **k–weise** adv edgewise, on edge

Kantharide f (-n) cantharides, Spanish fly

Kant·holz n squared timber

kantig a 1 square: **k. behauen** square-cut. 2 squared. 3 angular. 4 (*as sfx*) -edged, -cornered

Kantille f (-n) gold (*or* silver) purl

Kantine f (-n) canteen

Kanüle f (-n) cannula, tubule

Kanya·butter f canya butter

Kanzel f (-n) pulpit; (university) chair; cockpit; (gun) turret

Kanzlei f (-en) (law, government, records) office, chancery. **·-papier** n legal paper (for records), foolscap. **-tinte** f record ink

kaolinisieren vt kaolinize. **Kaolinisierung** f kaolinization. **Kaolin·lager** n kaolin deposit

Kap n (-s) (*Geog.*) cape

Kap. abbv (Kapitel) Chap. (chapter)

Kapaun m (-e) capon

Kapazitanz f (*Elec.*) capacitance

Kapazität f (-en) capacity; authority (on a subject). **kapazitäts·arm** a low-capacity. **kapazitiv** a (*Elec.*) capacit(at)ive

Kapelle f (-n) 1 (*Sci.*) cupel; sand bath; evaporating (*or* subliming) dish; capsule; priming cap. 2 chapel. 3 (*Music*) band

Kapellen·abfall m loss in cupellation. **-asche** f bone ash. **-form** f cupel mold. **-gold** n fine gold. **-kläre** f cupel dust, bone ash. **-ofen** m cupeling (assay, sublimation) furnace, cupola. **-probe** f cupel test, cupellation. **-pulver** n cupel dust, bone ash. **-raub** m loss in cupellation. **-silber** n fine silver. **-träger** m cupel holder. **-zug** m loss in cupellation

kapellieren vt cupel

Kaper f (-n) (*Bot.*) caper

Kap·granat m Cape ruby (pyrope). **-gummi** n Cape gum

kapillar a capillary. **K–·affinität** f capillary affinity. **k–aktiv** a surface-active, active in lowering surface tension. **K–anziehung** f capillary attraction. **-aszension** f capillary rise. **-druck** m capillary pressure

Kapillare f (-n) capillary (tube, blood vessel)

kapillar·elektrisch a electrocapillary. **K–flasche** f. **-fläschchen** n (small) capillary flask. **k–förmig** a capillary, capillaceous. **K–gang** m capillary passage. **-gefäß** n capillary vessel. **k–inaktiv** a surface-inactive, not lowering surface tension

kapillarisieren vt investigate by capillary

analysis. **Kapillarität** *f* capillarity. **Ka-pillaritäts·anziehung** *f* capillary attraction

Kapillar·kraft *f* capillary force. **-kreislauf** *m* capillary circulation. **-rohr** *n*, **-röhrchen** *n*, **-röhre** *f* capillary tube. **-spannung** *f* capillary tension. **-strömung** *f* capillary flow. **-versuch** *m* capillary test (*or* experiment). **-wirkung** *f* capillary action

Kapitel *n* (-) chapter

Kappe *f* (-n) cap; hood, cowl; top, dome

kappeln *vt* top

kappen *vt* 1 cut off the top of, top (trees); cut. 2 caponize, castrate. 3 cap, tip

Kappen·flasche *f* bottle with a cap. **k–förmig** *a* cap- (hood-, dome-)shaped. **K–pfeffer** *m* Guinea pepper. **-quarz** *n* cap quartz

Kapp·hahn *m* capon

Kaprin·säure *f* capric acid

Kapron·säure *f* caproic acid

Kapsel *f* (-n) 1 capsule; case, box. 2 bubble cap; priming cap. 3 (*Foun.*) chill, mold. 4 (*Ceram.*) saggar. **k–artig** *a* capsular. **K–blitz** *m* (*Phot.*) flash (capsule, bulb). **-boden** *m* bubble tray (*or* plate). **-färbung** *f* (*Micros.*) capsule staining. **k–förmig** *a* capsular, capsule-shaped. **K–frucht** *f* (*Bot.*) capsule. **-guß** *m* casting in chills. **-lack** *m* capping (*or* capsule) lacquer. **-mutter** *f* cap(ped) nut

kapseln *vt* encase, enclose; cap; jacket, can

Kapsel·ofen *m* (*Ceram.*) saggar kiln. **-scherben** *fpl* broken saggar scrap. **-ton** *m* saggar clay

Kapselung *f* encasing; capping etc *cf* KAPSELN

kaputt *a* ruined, out of commission, done for

Kapuze *f* (-n) hood, cowl

Kapuziner·kresse *f* Indian hood, nasturtium

Kap·wein *m* Cape wine. **-wolle** *f* Cape wool

Karabiner·haken *m* snap hook, spring catch

Karageen·moos *n* carrageen, Irish moss

karaibisch *a* Caribbean

Karamel *n* caramel, burnt sugar. **karamel(is)ieren** *vt* caramelize. **Ka-**ramelle *f* (-n) caramel (candy).

Karamel·malz *n* crystal malt. **-zucker** *m* caramel (sugar)

Karap·öl *n* carap(a) oil

Karat *m* (-) carat. **·-gewicht** *n* troy weight. **-gold** *n* alloyed gold

karatieren *vt* alloy (gold or silver)

·karatig *sfx* -carat (gold, etc)

Karb.., karb.. *See also* CARB.., CARB..

Karbe *f* caraway

Karben *n* carbene

Karbid·hartmetall *n* sintered carbide

Karbol *n* phenol; (*Com.*) carbolic acid; (*as pfx*) carbolic, carbolated, carbolized

Karbo·nitrieren *n* (*Metll.*) carbonitriding. **-zink** *n* zinc carbonate (rust preventative)

Kardamom·öl *n* cardamom oil

Kardan·gelenk *n* universal (*or* Cardan) joint. **kardanisch** *a* universal, Cardan. **Kardan·welle** *f* Cardan shaft

karden *vt* (*Tex.*) card, tease(l)

Karden·band *n* (*Tex.*) sliver. **-distel** *f* teasel

kardieren *vt* (*Tex.*) card, tease(l)

kardinal·rot *a* cardinal(-red)

Kardioide *f* (-n) (*Geom.*) cardioid

Kardobenedikten·kraut *n* (*Bot.*) blessed thistle

Karfunkel *m* (-) carbuncle. **·-stein** *m* garnet, almandine

karg *a* (゠er, ゠ste) meager, scanty; poor, barren, sterile (soil) —**k. bemessen** highly limited; **k. sein** (**mit**) = **kargen** *vi* (**mit**) be sparing (*or* stingy) (with). **Kargheit** *f* meagerness, scantiness; barrenness, sterility. **kärglich** *a* meager, scanty. **Kärglichkeit** *f* meagerness, etc = KARGHEIT

karibisch *a* Caribbean

kariert *p.a.* checkered

Karite·öl *n* karite oil, shea butter

Karm·alaun *m* (*Micros.*) carmelum

Karmeliter·geist *m* (*Pharm.*) carmelite water

karmesin *a* crimson. **K–·beeren** *fpl* kermes (berries). **k–farbig, -rot** *a* crimson (colored)

Karmin *n* carmine. **·-blau** *n* indigo carmine. **-farbe** *f* carmine (color). **k–farben** *a* carmine(-colored). **K–lack** *m* carmine lake, cochineal

Karminochinon *n* carminoquinone

Karmin·rot *n* carmine (red). **-säure** *f* carminic acid. **-spat** *m* (*Min.*) carminite. **-stoff** *m* carminic acid.

karmoisin *a* crimson = KARMESIN

Karnallit *m* (*Min.*) carnallite

Karnauba·säure *f* carnaubic acid. **-wachs** *n* Carnauba wax, Brazilian wax

Karneol *m* (-e) carnelian

Kärnten *n* Carinthia. **kärtnerisch** *a* Carinthian

Karo *n* (-s) (*esp Tex.*) check, square

Karobe *f* (-n) carob (bean)

Karolus·zelle *f* (*Opt.*) Kerr cell

Karosserie *f* (-n) body (of a vehicle)

Karotide *f* (-n) carotid (artery)

Karotin *n* caroteen

Karotte *f* (-n) (small, short) carrot. **Karotten·öl** *n* carrot oil

Karpfen *m* (-) (*Zool.*) carp

Karraghen·moos *n* carrageen, Irish moss

Karre *f* (-n), **Karren** *m* (-) (push)cart, (hand) truck; dolly; jalopy. **Karren** *pl* (*Geol.*) channels, flutings

Karriere *f* (-n) career: K. machen have a successful career

karriert *p.a* checkered

Karst *m* (-e) 1 mattock, (two-pronged) hoe. 2 (*Geol.*) karst, limestone

Karstenit *n* (*Min.*) anhydrite

Karst·höhle *f* limestone cavern

Kartäuser·likör *m* chartreuse (liqueur). **-pulver** *n* Carthusian powder, antimonous sulfide. **-tee** *m* Mexican tea

Karte *f* (-n) (*usu*) card; map, chart; ticket; menu

Kartei *f* (-en) card index (file, catalog). **-karte** *f* index card. **-karton** *m* index (card)board. **-wesen** *n* card indexing

Kartell *n* (-e) cartel, trust

Karten·blatt *n* (playing) card; (map) sheet. **-papier** *n* cardboard; map paper. **-pappe** *f* index (card)board

kartesisch *a* Cartesian

Karthäuser· Carthusian, etc = KARTÄUSER·

kartieren *vt* 1 map, chart. 2 (card-) index, file

Kartoffel *f* (-n) potato. **-beizung** *f* potato disinfection. **-brei** *m* mashed potatoes.

-erntemaschine *f* potato digger. **-fäule** *f* potato rot. **-fuselöl** *n* potato fusel oil. **-käfer** *m* Colorado potato beetle, potato bug. **-kraut** *n* potato vine(s); potato tops. **-krebs** *m* (*Agric.*) potato wart. **-mehl** *n* potato flour. **-mus** *n* mashed potatoes. **-nährboden** *m* potato culture medium. **-sirup** *m* glucose syrup from potato starch. **-stärke** *f* potato starch. **-walzmehl** *n* roller-mill potato flour. **-zucker** *m* potato sugar, potato starch glucose

Karton *m* (-s) 1 cardboard. 2 cardboard box, carton. **Kartonage** *f* (-n) 1 cardboard cover (*or* box). 2 *pl* cardboard products. 3 (*Books*) board binding. **Kartonagen·leim** *m* boxmaker's glue. **Kartonage·pappe** *f* box cardboard. **kartonieren** *vt* 1 pack in cardboard. 2 (*Books*) bind in boards. **Karton·papier** *n* paperboard

Karthotek *f* card index (file, catalog)

Kartusche *f* (-n) 1 cartridge. 2 cartouche. **Kartusch·papier** *n* cartridge paper

Karube *f* (-n) carob (bean)

Karyokinese *f* (-n) (*Biol.*) karyokinesis, mitosis

Karzinom *n* (-e) (*Med.*) carcinoma. **karzinomatös** *a* carcinomatous

KAS. *abbv* (Kalkammonsalpeter) lime ammonium nitrate, nitro-chalk (fertilizer)

kaschieren *vt* 1 cover(up), conceal. 2 (*Tex.*) line, back; bond (together). 3 (*Paper*) coat, glue-back; laminate, double — **kaschierter Sack** multi-ply bag. **Kaschier·papier** *n* lining paper. **Kaschierung** *f* (-en) covering, concealment; lining, etc

Kaschmir 1 *n* (*Geog.*) Kashmir. 2 *m* (-e) (*Tex.*) cashmere

Kaschu *n* catechu, cutch. **kaschutieren** *vt* dye with catechu (*or* cutch)

Käse *m* (-) cheese. **k-artig** *a* cheesy, caseous. **K-bildung** *f* cheese formation, caseation. **-bruch** *m* (cheese) curd. **-firnis** *m* (*Med.*) vernix caseosa. **k-förmig** *a* cheese-shaped, flat cylindrical

Kasein *n* casein. **-farbstoff** *m* casein dye (*or* paint). **-seide** *f* casein silk (*or* rayon)

Käse· cheese: **-krebs** *m* cheese rind rot. **-lab** *n* rennet. **-leim** *m* casein glue. **-loch** *n* cheese eye. **-made** *f* cheese maggot.

-**milbe** *f* cheese mite

käsen *vi* make cheese; (*of milk*) curdle

Käse-öl *n* cheese oil. -**pappel** *f* (*Bot.*) wild mallow

Käserei *f* (-en) 1 cheese-making. 2 cheese dairy

Käse- cheese: -**reifung** *f* cheese ripening (*or* curing). **k-sauer** *a* lactate of. **K-säure** *f* lactic acid. -**stoff** *m* casein. -**wachs** *n* cheese wax. -**wasser** *n* whey

käsig *a* cheesy

Kasimir *m* (*Tex.*) cassimere

Kaskaden·schaltung *f* (*Elec.*) cascade connection. -**schauer** *m* (*Nucl. Phys.*) cascade shower

Kaskarillen·rinde *f* cascarilla bark

kaspisch *a* Caspian

Kassa *f* (..ssen) cashier's desk = KASSE

Kassaun-zucker *m* cassonade, raw sugar

Kassave *f* (-n), **Kassawa** *f* (-s) cassava. -**mehl** *n* cassava starch (*or* meal)

Kasse *f* (-n) 1 till; cash box; treasury. 2 cash (on hand). 3: **die K. führen** keep accounts. 4 cashier's desk; box (*or* ticket) office; teller's window. 5 fund. *Cf* KRANKENKASSE, SPARKASSE

Kasseler *a* (*no endgs*) (of) Cassel: **K. Erde** Cassel Earth. -**braun** *n* Cassel brown. -**gelb** *n* Cassel yellow. -**grün** *n* Cassel (*or* manganese) green

Kasserolle *f* (-n) casserole

Kassette *f* (-n) 1 (*genl*) cassette (*incl.* tape), case, box. 2 (*specif*) cash (*or* strong) box, jewel case, plate (slide, film) holder, sagger

Kassia, Kassie *f* (..ien) cassia; (*in compds*) = **Kassien·** cassia: -**blüten** *fpl* cassia buds. -**öl** *n* cassia oil. -**rinde** *f* cassia bark

kassieren I *vt* 1 cash in, collect. 2 charge. 3 take (in, over); pocket. 4 cancel, annul, withdraw. II *vi* collect payment. **Kassierer** *m* (-), **Kassiererin** *f* (-nen) cashier; teller; ticket agent; treasurer

Kassiopeium *n* lutecium, cassiopeium

Kassius·purpur *m* purple of Cassius

Kaßler = KASSELER: **K. Braun** Cassel brown

Kassonade *f* (-n) (*Sugar*) cassonade, cask sugar

Kastanie *f* (-n) chestnut

Kastanien·auszug *m* chestnut extract. **k-**

braun *a* chestnut(-brown). **K-holz** *n* chestnut (wood)

Kasten *m* (-) 1 box, chest, case; crate. 2 drawer. 3 cabinet, cupboard, closet. 4 (car) body. 5 frame. 6 vat, beck, tank. 7 (*Furnace*) crucible. 8 (*Foun.*) flask. 9 (*Ceram.*) sagger. 10 (*Jewelry*) collet. -**band** *n* trough conveyor. -**blau** *n* pencil blue (indigo blue produced by vat dyeing). -**mälzerei** *f* box (*or* compartment) malting. -**wagen** *m* 1 moving (*or* delivery) van. 2 (railroad) box car

Kastor-öl *n* castor oil

kastrieren *vt* castrate

kat.., Kat.. *See also* CAT.., CAT..

Katabolismus *m* catabolism

katalanisch *a* Catalan, Catalonian

Katalase *f* (-n) catalase

Katalog *m* (-e), **katalogisieren** *vt* catalog

Katalysator *m* (-en) catalyst; (*Cars*) catalytic converter. **Katalysator(en)·gift** *n* catalyst poison. **katalysatorisch** *a* catalytic. **Katalyse** *f* (-n) catalysis. **katalysieren** *vt* catalyze. **katalytisch** *a* catalytic

Kataphorese *f* (-n) cataphoresis, electrophoresis

katarrhalisch *a* catarrhal

Katechin *n* catechol. -**gerbstoff** *m* catechol tan. -**säure** *f* catechuic acid (catechol)

Katechu *n* catechu. -**braun** *n* catechu brown, cutch (color). -**gerbsäure** *f* catechutannic acid. -**säure** *f* catechuic acid, catechol

Kater *m* (-) 1 (tom)cat. 2 hangover

Kateschu *n* catechu = KATECHU

kathartisch *a* cathartic

Kathete *f* (-n) (*Geom.*) cathetus, arm (of a right triangle)

Kathetometer *m* or *n* (-) cathetometer

Kathoden·dichte *f* cathode density. -**dunkelraum** *m* cathode dark space. -**fläche** *f* cathode surface. -**licht** *n* cathode glow. -**niederschlag** *m* cathode deposit. -**raum** *m* cathodic space. -**röhre** *f* cathode-ray tube. -**schutz** *m* cathodic protection. -**strahl** *m* cathode ray. -**strahlenbündel** *n* cathode beam. -**strahlröhre** *f* cathode-ray tube. -**strahlung** *f* cathode radiation. -**strom** *m* cathode current. -**zerstäu-**

bung f cathodic sputtering
kathodisch a cathodic. **kathodoche-
misch** a cathodochemical
Katholyt m (-e or -en) catholyte
Kat·ion n cation. **kation(en)·aktiv** a
cation-active, cationic (surface-active
agents). **Kationen·austausch, -um-
tausch** m cation exchange. **kationoid** a
cationoid, electrophilic
Kattun m (-e) calico, cotton cloth. **--druck**
m calico printing. **-druckerei** f calico
printing (factory)
kattunen a (of) calico, cotton
Kattun·fabrik f calico factory. **-färberei** f
calico dyeing (factory). **-presse** f calico
press
Kätzchen n (-) 1 kitten. 2 catkin (of pussy
willow)
Katze f (-n) 1 (Zool.) cat. 2 (Paper) thread,
string (in pulp). 3 (Mach.) trolley
Katzen·auge n 1 (Zool. & fig, incl jewel)
cat's eye. 2 (glass) bull's eye. 3 reflector.
-augenharz n dammar resin. **-baldrian**
m (common) valerian. **-darm** m catgut.
-glimmer m, **-gold** n yellow mica; pyrite.
-kraut n cat thyme. **-minze** f catnip.
-minzöl n catnip oil. **-silber** n colorless
(or argentine) mica. **-sprung** m stone's
throw
kau· chewing, masticatory
kaubar a chewable, masticable
kauen vt/i chew, masticate
Kauf m (ẅe) purchase; buy **—zu K. stehen**
be for sale; **in K. nehmen** put up with.
--blei n merchant (or commercial) lead
kaufen 1 vt buy, purchase. 2 vi shop.
Käufer m (-) buyer, purchaser; customer
Kauf·gut n merchandise. **-handel** m 1 business, trading. 2 transaction, deal. **-haus** n
department store. **-kraft** f purchasing
power. **-laden** m store, shop. **-leute:** business people: pl of KAUFMANN
käuflich a 1 purchasable, commercially
available **—k. erwerben** acquire by purchase. 2 venal
Kauf·mann m 1 businessman; merchant,
dealer. 2 clerk. 3 grocer. Cf -LEUTE. **-män-
nisch** a commercial, business. **K-
muster** n commerical sample. **-zink** n
commercial zinc
Kau·gummi n chewing gum

Kaukasien n Caucasus. **kaukasisch** a
Caucasian
kaum adv hardly, scarcely, barely
Kau·magen m gizzard. **-mittel** n (Med.)
masticatory. **-pfeffer** m betel
Kaupren n cauprene
Kau·probe f mastication test
Kauri·harz n kauri (resin). **-nuß** kauri nut.
-öl n kauri oil
Kausal·nexus m casual relation(ship)
kaustifizieren vt causticize. **Kaustifizie-
rung** f (-en) causticizing. **kaustisch** a
caustic. **kaustizieren** vt causticize.
Kaustizität f causticity
Kau·tabak m chewing tobacco
Kautel f (-en) 1 (incl Med.) precaution. 2
proviso
Kauter m (-) cauterizing iron; cautery. **kau-
terisieren** vt cauterize. **Kauterium** n
(..ien) cauterizing agent; caustic. 2
cauterizing iron
kautschen vt (Paper) couch
Kautschin n caoutchene (dipentene)
Kautschucin n caoutchoucin, rubber oil
Kautschuk m or n caoutchouc, (pure India)
rubber: **geräucherter K.** smoked sheet;
(for compds. see also GUMMI·). **--ballon** m
rubber-balloon (or bulb). **-baum** m rubber tree. **-block** m cake of rubber. **k-
elastisch** a elastomeric. **K–feigenbaum**
m rubber plant. **-firnis** m rubber varnish.
-füllung f rubber filling (or filler).
-gewebe n rubber fabric. **-gift** n rubber
poison. **-kitt** m rubber cement. **-kuchen**
m cake of rubber. **-lack** m rubber varnish,
rubberized paint. **-lösung** f rubber solution. **-masse** f rubber paste. **-milch** f rubber latex. **-pflanze** f rubber(-yielding)
plant. **-rohr** n, **-röhre** f rubber tube. **-saft**
m rubber latex. **-schlauch** m rubber hose
(or tube). **-schnur** f rubber cord. **-stopfen**
m rubber stopper (or bung). **-surrogat** n
rubber substitute. **-waren** fpl rubber
goods
kautschutieren vt 1 rubberize. 2 make out
of rubber
Kauz m (ẅe) 1 tawny owl. 2 oddball
Kaverne f (-n) cavity, cavitation
Kawa·pfeffer m kava pepper. **-säure** f ka-
vaic (or kawaic) acid
kcal. abbv kilogram calorie

keck *a* bold; pert

Keder·band *n* sealing (*or* weather) strip

Kefir *m* kefir, fermented milk drink. **··knollen**, *mpl*, **-körner** *npl* kefir grains

Kegel *m* (-) **1** (Geom. & fig) cone: **über den K. werfen** cone (in sampling); **abgestumpfter K.** truncated cone, frustum. **2** bevel, taper. **3** bowling pin. **4** (*in compds oft*) conical: **··bahn** *f* bowling alley. **-brecher** *m* conical mill (*or* crusher). **-brenner** *m* cone burner. **-fallpunkt** *m* (*Ceram.*) pyrometric cone equivalent. **-feder** *f* volute spring. **-fläche** *f* conical surface. **-flasche** *f* conical (e.g. Erlenmeyer) flask. **k–förmig** *a* cone-shaped, conical; tapered. **K–getriebe** *n* bevel gearing. **-glas** *n* conical glass (*or* flask). **-hahn** *m* stopcock (with a tapered plug)

kegelig *a* conical

Kegel· cone, conical: **-lehre** *f* theory of conics. **-mühle** *f* cone mill. **-rad** *n* bevel wheel (*or* gear). **-schaft** *m* taper shank. **-schmelzpunkt** *m* (*Ceram.*) pyrometric cone equivalent. **-schnitt** *m* conic section. **-stift** *m* taper pin. **-stumpf** *m* truncated cone, frustum. **-ventil** *n* conical valve. **-verfahren** *n* (sampling) coning and quartering

keglig *a* conical = KEGELIG

Kehl· *cf* KEHLE: **-ader** *f* jugular vein. **-dekkel** *m* epiglottis. **-drüse** *f* thryoid gland

Kehle *f* (-n) **1** throat, larynx. **2** channel, flute, groove. **kehlen** *vt* flute, groove

Kehl·entzündung *f* laryngitis. **-kopf** *m* larynx; (*in compds, oft*): **Kehlkopf· laryngeal**

Kehr·bild *n* inverted image; (*Phot.*) negative

Kehre *f* (-n) turn, curve, bend; (reversing) loop

kehren **1** *vt/i* sweep (with a broom). **2** *vt* turn. **3** *vi*(h) turn around, reverse direction. **4** *vi*(s) return. **5: sich k.** turn (around). **6: sich nicht k. (an)** not care (about), pay no attention (to)

Kehricht *m* *or* *n* refuse, sweepings; garbage, trash. **··ofen** *m* incinerator

Kehr·maschine *f* (*Mach.*) sweeper. **-salpeter** *m* saltpeter sweepings. **-salz** *n* salt sweepings. **-seite** *f* reverse (side); wrong side

kehrt·machen *vi* turn back, about-face

Kehr·wert *m* reciprocal (value). **-wolle** *f* wool sweepings

Keil *m* (-e) **1** (*genl*) wedge. **2** (*Mach.*) key, cotter. **3** (*Tex.*) gore, gusset. **k–ähnlich**, **-artig** *a* wedge-like, cuneiform, sphenoid. **K–bein** *n* sphenoid (*or* cuneiform) bone. **-draht** *m* wedge wire. **-eckenschliff** *m* (*Glass*) bevel grinding

keilen *vt* wedge; key

Keil·nut(e) *f* (*Mach.*) keyway; wedge-shaped groove. **-paar** *n* pair of wedges, double wedge. **-riemen** *m* V-belt. **-riemenscheibe** *f* V-grooved pulley. **-verbindung** *f* keyed joint

Keim *m* (-e) **1** germ: embryo; seed, bud: **im K––(e) ersticken** nip in the bud. **2** (*Cryst.*) nucleus. **3** sprout, shoot. **4** (*in compds*). germ, germinal, germination; embryonic: **-apparat** *m* germinating apparatus

keimbar *a* germinable

Keim·bildner *m* nucleating agent. **-bildung** *f* nucleation, formulation of nuclei. **-blatt** *n* **1** (*Bot.*) cotyledon. **2** (*Biol.*) germinal layer. **k–blätterig** *a* cotyledonous. **-blattlos** *a* acotyledonous. **K–dauer** *f* germination time. **k–dicht** *a* germproof. **K–drüse** *f* (*genl*) generative gland; (*specif*) gonad. **-drüsenhormon** *n* sex hormone

keimen *vi* germinate

Keim·entwicklung *f* embryo (*or* germ) development. **k–fähig** *a* germinable, viable, capable of germinating. **K–fähigkeit** *f* germinating power, viability. **k–frei** *a* germ-free, sterile; free from nuclei. **K–freiheit** *f* freedom from germs (*or* nuclei), sterility. **-gift** *n* **1** poison from germs, bacterial toxin. **2** poison for germs. **k–haltig** *a* containing germs (*or* nuclei). **K–haut** *f* blastoderm. **-hülle** *f* (*Bot.*) pericarp. **-kern** *m* (*Biol.*) germinal nucleus. **-kraft** *f* germinating power. **-kristall** *m* seed crystal, crystal nucleus

Keimling *m* (-e) germ, embryo; seed crytal; seedling; sprout, shoot

keim·los *a* germless; free from nuclei. **K–plasma** *n* germ plasm(a). **-probe** *f*, **-prüfung** *f* germination test. **-ruhezeit** *f* dormant phase (of cells). **-stoff** *m*, **-substanz**

f germinal matter, blastema. **k–tötend** *a* germicidal: **k—es Mittel** germicide. **K–töter** *m* germ killer, germicide; sterilizer. **-träger** *m* germ carrier. **-trommel** *f* germinating drum. **k–unfähig** *a* incapable of germinating

Keimung *f* (-en) germination

Keimungs·dauer *f* duration of germination. **-energie** *f* germinative energy. **-fähigkeit** *f* germinating power

Keim·versuch *m* germination test. **k–voll** *a* full of germs (*or* nuclei). **K–weiß** *n* endosperm. **-wirkung** *f* germ action; (*Cryst.*) action of nuclei. **-würzelchen** *n* rootlet. **-zahl**, **-zählung** *f* bacterial count. **-zelle** *f* germ cell

kein *a* no, not . . . a(ny): **k—e zehn** not even ten. **keine(r)** *pron* (*adj endgs*) no one, not . . . anyone; none — **keine(r) (keines) von beiden** neither of the two

K-Einfang *m* (*Nucl. Phys.*) K-capture

keinerlei *a* (*no endgs*) no (kind of): **auf k. Weise** in no way

keines·falls *adv* in no case, not in any case. **-wegs** *adv* in no way, not in any way; by no means; not in the least

kein·mal *adv* not once, never

keins *n pron* none, not . . . any

kein·seitig *a* neutral. **Keinseitigkeit** *f* neutrality

·keit *sfx* (*usu forms nouns from adjs*) **haltbar** *a* durable: **Haltbarkeit** *f* durability *cf* ·HEIT

Keks *m* (-e) cookie, cracker

Kelch *m* (-e) **1** (*Bot.*) calyx, (flower) cup. **2** (*Zool.*) infundibulum. **3** cup, chalice. **4** (stem) glass, goblet. **·-blatt** *n* sepal. **k–förmig** *a* cup-shaped. **K–glas** *n* (cup-shaped) test glass. **-mensur** *f* measuring cup

Kelen *n* (*Pharm.*) kelene, ethyl chloride

Kelle *f* (-n) **1** ladle; scoop. **2** trowel. **kellen** *vt* ladle, scoop

Keller *m* (-) cellar. **Kellerei** *f* (-en) wine cellar (space)

Keller·geschoß *n* basement. **-hals** *m* (*Bot.*) daphne mezereon

Kelter *f* (-n) winepress. **keltern** *vt/vi* press (grapes). **Kelter·wein** *m* wine from pressed grapes

Keltium *n* celtium

Kemiri·öl *n* cemiri oil

Kenn· characteristic; identification *cf* KENNEN

kennbar *a* recognizable, identifiable

Kenn·daten *npl* characteristic data; constants, coefficients

Kennel·kohle *f* cannel coal

kennen* (kannte, gekannt; *sbjc* kennte) *vt* know, be acquainted (*or* familiar) with. **kennen·lehren** *vt* inform about. **kennen·lernen** *vt* get to know, become acquainted with. **Kenner** *m* (-) connoisseur, expert; one who knows

Kenn· characteristic, identification: **-größe** *f* characteristic magnitude. **-karte** *f* identification card. **-linie** *f* characteristic line (*or* curve); (*Electron.*) characteristic (of a tube). **-linienknick** *m* break (*or* inflection point) in a curve (*or* graph). **-marke** *f* identification mark (token, stub). **-merkmal** *n* characteristic (feature); criterion. **-nummer** *f: see* KENNNUMMER

kennte would know, etc: *sbjc of* KENNEN

kenntlich *a* recognizable, identifiable; discernible: **k. machen** identify, mark; **sich k. machen** make itself known, show up. **Kenntlichmachung** *f* (-en) identification, marking

Kenntnis *f* (-se) **I** *sg* **1** (+*genit*) knowledge (of); acquaintance, familiarity (with): (*in Book Titles*) **zur K. . . . .** contribution to the knowledge of . . . ; **K. haben (von)** have knowledge (of), be informed (about); **K. bekommen** (*or* **erhalten**) **(von)** gain knowledge (of), find out (about); **das entzieht sich ihrer K.** that escapes their knowledge. **2** (+ *genit*) insight (into). **3** information: **einen über etwas in K. setzen, einem etwas zur K. bringen** inform someone about sthg. **II** *pl* knowledge; education; experience

kenntnis·los *a* uninformed, ignorant. **K–nahme** *f* taking note of *cf* KENNTNIS II: **zur gefälligen K.** for your information, please take note. **k–reich** *a* knowledgeable, well-informed

Kennummer *f* (= **Kenn·nummer**) identification (index, license, reference) number

Kennung *f* (-en) **1** identification signal (*or*

lights). 2 landmark

Kenn·wert *m* characteristic value; parameter; measurable characteristic. **-zahl** *f* identification number, etc = KENNNUMMER. **-zeichen** *n* 1 characteristic, (distinctive) feature; symptom. 2 identification mark (letter, number); (*specif*) license (*or* registration) number

kennzeichnen *vt* (kennzeichete, gekennzeichnet) (*oft* + **als**) identify, mark, tag, characterize (as); be characteristic (of). **kennzeichnend** *p.a* 1 distinctive, distinguishing. 2 (+**für**) characteristic, typical (of). **Kennzeichnung** *f* (-en) marking, tagging, etc; characterization. **Kennzeichnungs·farbe** *f* sighting color

Kenn·ziffer *f* 1 identification (index, registration, license, code) number. 2 (*Math.*) characteristic (of a logarithm)

kentern *vi* 1 (h) change direction, turn, shift. 2 (s) overturn, capsize

Kephalin *n* kephalin, cephalin

Keramik *f* (-en) 1 ceramics. 2 piece of pottery, ceramic (article). **Keramiker** *m* (-) potter, ceramicist. **keramisch** *a* ceramic

Kerargyrit, Kerat *n* cerargyrite

Keratom *n* (-e) keratoma

Kerb *m* (-e), **Kerbe** *f* (-n) notch, nick; indentation; groove, slot; cut, incision

Kerbel *m* (-) (*Bot.*) chervil. **-rübe** *f* bulbous chervil

Kerb·empfindlichkeit *f* notch sensitivity

kerben *vt* notch, indent; knurl, mill; cut into

Kerbe·zähigkeit *f* impact strength

Kerb·festigkeit *f* impact resistance

kerbig *a* notched, indented; jagged; knurled, milled

Kerb·karte *f* notched data card. **-presse** *f* notching (*or* crimping) press. **-schlag** *m* impact

Kerbschlag·festigkeit *f* impact strength. **-probe** *f*, **-versuch** *m* notched-bar impact test. **-zähigkeit** *f* (notched-bar) impact strength

Kerf *m* (-e) insect. **-kunde** *f* entomology

Kermes *m* (-) kermes (=a scarlet dye); (*Min.*) kermesite. **-baum** *m* kermes oak (tree). **-beere** *f* kermes berry. **-eiche** *f* kermes oak. **-farbstoff** *m* kermes dye, kermic acid. **-rot** *n* scarlet. **-säure** *f* kermic

acid. **-schildlaus** *f* kermes insect

Kern *m* (-e) 1 (*esp Phys. & Biol.*) nucleus, (*in compds, usu*) nuclear. 2 (*esp Bot.*) kernel, pit, stone. 3 (*esp fig*) core, heart, essence, crux. 4 (*Soap*) curd. 5 *pl* (*Oils*) neats. **-·abstand** *m* (inter)nuclear distance. **-anregung** *f* nuclear excitation. **-antrieb** *m* nuclear propulsion. **-aufbau** *m* 1 nuclear structure (*or* synthesis). 2 core configuration (of a reactor). **-baustein** *m* nucleon. **-beschuß** *m* nuclear bombardment. **-bewegung** *f* nuclear motion. **-bildung** *f* nucleation. **-bindemittel** *n* (*Metll.*) core binder. **-blatt** *n* (*Lthr.*) butt. **-bohren** *n* core drilling. **-brennstoff** *m* nuclear fuel. **-chemie** *f* nuclear chemistry. **k-chloriert** *a* chlorinated in the nucleus. **-echt** *a* true to type. **K-eisen** *n* 1 mottled white pig iron. 2 (*Elec.*) core iron. **-eiweißkörper** *m*, **-eiweißstoff** *m* nucleoprotein. **-emulsion** *f* nuclear emulsion.

kernen 1 *vt* take the core (pit, etc) out of *cf* KERN. 2 *vi* (*Milk*) turn buttery

Kern· core, kernel, nuclear: **-energie** *f* nuclear energy. **-explosion** *f* nuclear explosion. **-farbe** *f*, **färbemittel** *n*, **-farbstoff** *m* nuclear stain. **-faser** *f* nuclear fiber (*or* filament). **k-faul** *a* rotten at the core (*or* heart). **K-fäule** *f* (*usu Wood*) stem (*or* heart) rot. **k-fern** *a* outer, peripheral, valence (electrons). **K-forschung** *f* nuclear research. **-frage** *f* pivotal question, central issue. **k-fremd** *a* extranuclear. **-frisch** *a* thoroughly fresh, fresh to the core. **K-frucht** *f* stone (*or* pit) fruit; pomaceous fruit. **-fusion** *f* nuclear fusion. **k-gebunden** *a* united to the nucleus; (*Electrons*) orbital, planetary. **K-gedanke** *m* central idea. **-gehalt** *m* nuclear content (*or* concentration). **-gehäuse** *n* core (of fruit). **k-gesteuert** *a* nuclear-controlled. **-gesund** *a* fundamentally sound (*or* healthy). **K-guß** *m* (*Foun.*) cored work. **k-haltig** *a* nucleated, containing a nucleus (kernel, core). **K-haus** *n* core (of fruit). **-haut** *f* epidermis, tegumen. **-hefe** *f* seed yeast. **-höhle** *f* (*Biol.*) nuclear vacuole. **-holz** *n* heartwood

kernig *a* 1 seed-filled, acinaceous. 2 granu-

lar. 3 solid, robust. 4 coarse; blunt, earthy.
5 powerful, strong. 6 pithy. 7 (*Lthr.*) full,
compact. 8 (*Liquors*) full. 9 (*as sfx oft*) -nu-
clear. **Kernigkeit** *f* solidity, robustness;
coarseness, bluntness; power, strength;
pithiness; (*Lthr.*) fulness, body
Kern· core, kernel, nuclear: **-isomer** *n* nuclear (*or* ring) isomer; isobaric isotope.
-isomerie *f* nuclear isomerism. **-ket-
tenreaktion** *f* nuclear chain reaction.
-körper *m*, **-körperchen** *n* (*Biol.*) nu-
cleolus. **-kraftwerk** *n* nuclear power
plant. **-ladung** *f* nuclear charge.
-ladungszahl *f* nuclear-charge number;
atomic number. **-leder** *n* butt (*or* band)
leather. **-linse** *f* nuclear lens. **k–los** *a*
without a nucleus, anucleate; coreless.
K–macherei *f* (*Foun.*) core-making
(shop). **k–magnetisch** *a* nuclear magnetic. **K–mehl** *n* prime (quality) flour.
-membran *f* nuclear membrane. **-milch** *f*
buttermilk. **-moment** *n* nuclear moment.
k–nah *a* close to the nucleus; inner (electrons). **K–obst** *n* pomaceous fruit. **-öl** *n*
kernel oil, (*specific*) palm kernel oil. 2
(*Foun.*) core oil. **-physik** *f* nuclear physics. **-physiker** *m* nuclear physicist.
-polymerie *f* nuclear polymerization;
polynuclearity (of coordination compounds). **-probe** *f* (*Geol.*) core sample.
-prozeß *m* nuclear reaction. **-pulver** *n*
coated smokeless powder. **-punkt** *m* nucleus; essential point. **-quadrupolkopp-
lung** *f* nuclear quadrupole coupling.
-raum *m* heartland; central area. **-reak-
tion** *f* nuclear reaction. **-reaktor** *m* nuclear reactor. **k–reaktorbestrahlt** *a*
reactor-radiated. **-reich** *a* nucleated. **K–
resonanz** *f* nuclear resonance. **-rück-
feinen** *n* (*Metll.*) core refining. **-saft** *m*
(*Biol.*) nuclear fluid, enchylema. **-salz** *n*
rock salt. **-sand** *m* core sand. **-schale** *f*
nuclear shell. **-schatten** *m* complete
shadow, umbra. **-schleife** *f* chromosome.
-schliff *m* cone joint; convex part of a
ground glass joint. **-schmelzen** *n* core
meltdown. **-schwarz** *n* carbon black
(from fruit kernels); (*Micros.*) nucleus
black. **-seife** *f* curd soap: **K. auf Leim-
niederschlag** curd-settled soap; **K. auf
Unterlage** completedly salted-out curd

soap; **abgesetzte K.** neat soap.
-spaltung *f* nuclear fission. **-spin** *m* nuclear spin. **-stäbchen** *n* chromosome.
-stoff *m* nuclear substance; nuclein.
-stoß *m* nuclear impact. **-strahler** *m* radioactive isotope. **-strahlung** *f* nuclear
radiation. **-streuung** *f* nuclear scattering. **-stück** *n* 1 heart, core; crucial feature. 2 (*Lthr.*) butt. **-synthese** *f* nuclear
synthesis. **-teilung** *f* nuclear (sub)divi-
sion; segmentation. **-theorie** *f* nuclear
theory. **-tinktion** *f* nuclear staining. **k–
trocken** *a* thoroughly dry. **K–
umwandlung** *f* nuclear transformation
(*or* transmutation). **-verbindungsachse**
f internuclear axis. **-verknüpfung** *f* linkage to a nucleus. **-verschmelzung** *f* nuclear fusion. **-waffe** *f* nuclear weapon.
-wechselwirkung *f* nuclear interaction.
-werkstoff *m* core material. **-wolle** *f*
prime wool. **-wuchs** *m* seedling. **-zahl** *f*
number of nuclei. **-zelle** *f* nuclear cell,
cytoplast. **-zerfall** *m* nuclear disintegration. **-zerplatzen** *n* bursting of the nucleus. **-zone** *f* 1 (*genl*) central (*or* core) area. 2
(*Biol.*) nuclear zone. 3 (*Metll.*) core
Kerosin *n* (-e) kerosene
Kerze *f* (-n) 1 candle. 2 (*Mach.*) spark plug. 3
(*Geol.*) pillar
Kerzen·docht *m* candle wick. **-filter** *m* candle filter, filter candle. **k–gerade** *a*
ramrod-straight, upright. **K–kohle** *f*
cannel coal. **-masse** *f* candle composition.
-nuß *f* candlenut. **-schein** *m* candlelight.
-stärke *f* candle power. **-zündholz** *n*
taper. **-zündung** *f* spark plug ignition
Kessel *m* (-) 1 kettle; cauldron. 2 boiler. 3
tank, reservoir; vat, (*Tex.*) kier. 4 (*Geog.*)
basin, valley, hollow. 5 (*Geol.*) caldera,
crater. 6 (animal's) burrow. **-·asche** *f* potash (potassium carbonate). **-betrieb** *m*
boiler operation. **-blech** *n* boiler plate.
-dampf *m* boiler (*or* live) steam. **-druck**
m boiler pressure. **-färberei** *f* kettle dyeing. **k–förmig** *a* kettle-shaped; basin-
shaped. **K–gas** *n* boiler flue gas. **-haus** *n*
boiler house. **-kochung** *f* (*Tex.*) kier boiling. **-kohle** *f* boiler (*or* steam) coal
kesseln *vt* (*Explo.*) chamber
Kessel·niederschlag *m* boiler scale. **-ofen**
m crucible furnace. **-platte** *f* boiler plate.

-**rohr** n boiler tube. -**schmied** m boilermaker, coppersmith. -**speisewasser** n boiler feed water, -**speisung** f boiler feed(ing). -**stein** m 1 boiler scale (sediment, deposit, incrustation). 2 compass brick

Kesselstein· boiler scale: -**ablagerung** f boiler scale deposit. -**beseitigung** f boiler scale removal. -**bilder, -bildner** m boiler scale former. -**bildung** f boiler scale formation. -**gegenmittel** n boiler compound, disincrustant. -**kruste** f boiler scale incrustation. -**lösemittel** n boiler scale solvent. -**mittel** n boiler compound, anti-incrustant. -**schicht** f layer of boiler scale. -**verhütung** f boiler scale prevention. -**verhütungsmittel** n boiler compound, anti-incrustant

Kessel·wagen m tank car; tank truck. -**wand** f boiler wall. -**wasser** n boiler water. -**zug** m boiler flue

Keten n(-e) ketene

Ketimid n (-e) ketimine

Ketipin·säure f ketipinic acid

Keton n (-e) ketone. **Ketonimid** n ketimine. **ketonisch** a ketonic

Keton·säure f keto(nic) acid. -**spaltung** f ketonic cleavage. -**zucker** m ketonic sugar

Keto·säure f ketonic acid. -**verbindung** f keto(nic) compound

Kett·baum m (Tex.) warp beam. -**druck** m (Tex.) warp print(ing)

Kette f (-n) 1 (genl.) chain; necklace; series. 2 (Tex.) warp. 3 (Elec., Chem.) cell; circuit; network. **ketten** vt chain, link

Ketten· chain: -**abbrecher** m chainreaction terminator. -**abbruch** m chain (-reaction) termination. -**antrieb** m chain drive. **k–artig** a chain-like. **K–bruch** m (Math.) continued fraction. -**druck** m warp printing. -**färben** n, -**färberei** f warp dyeing. -**fläche** f (Geom.) catenoid. **k–förmig** a chain-like, in the form of a chain. **K–glied** n link of a chain. -**isomerie** f chain isomerism. -**kokken** fpl streptococci. -**länge** f chain length. -**linie** f catenary (curve). -**molekül** n chain molecule. -**polymer** n chain (or linear) polymer. -**rad** n sprocket wheel. -**reaktion** f chain reaction. -**sili**

kat n chain-structured silicate. -**starter** m chain initiator. -**träger** m chain propagator. -**trieb** m chain drive. -**übertrager** m chain-transfer agent. -**übertragung** f (Kinetics) chain transfer. -**wachstum** n chain growth

Kett·streifen m (Tex.) warp stripe

keuchen vi gasp, pant, puff. **Keuch·husten** m whooping cough

Keule f (-n) 1 club, bat; pestle. 2 leg (of meat)

kfm. abbv (kaufmännisch) commercial

KG abbv (Kommanditgesellschaft) limited partnership. **KGaA** abbv (Kommanditgesellschaft auf Aktien) partnership limited by stock shares

kgl. abbv (königlich) royal

Khaki·farbe f khaki color. **k–farben** a khaki(-colored). **K–stoff** m khaki (cloth)

KHz abbv (Kilohertz) kc (kilocycle/s)

Kicher f (-n), **··erbse** f chick pea

Kid·leder n kid leather. -**öl** n kid oil

Kiebitz m (-e) (Zool.) lapwing, peewit

Kiefer I m (-) jaw(bone), maxilla; (in compds usu) maxillary. II f(-n) pine (tree), esp Scotch pine

Kiefer(n)· -**harz** n pine resin. -**holz** n pine(wood). -**nadel** f pine needle. -**nadelöl** n pine-needle oil. -**öl** n pine oil. -**sperre** f lockjaw. -**teer** m pine tar. -**zapfen** m pine cone

Kiel m (-e) 1 keel; carina. 2 quill. **··raum** m hold (of a ship)

Kieme f (-n) gill, branchia. **Kiemen·** (in compds) gill, branchial

Kien m resinous pine(wood). **··apfel** m pine cone. -**baum** m pine tree. -**harz** n pine resin. -**holz** n resinous pine (wood)

kienig a piny, resinous

Kien·öl n pine oil, oil of turpentine. -**pech** n pine pitch. -**ruß** m pine soot (or black), lampblack. -**stock** m (Metll.) carcass. -**teer** m pine tar

Kies m (-e) 1 pyrites. 2 gravel, pebbles, grit. **··abbrand** m burnt (or roasted) pyrites, purple ore, pyrites cinders. **k–ähnlich, -artig** a pyritous, gravelly, gritty. **K–beton** m gravel concrete. -**boden** m gravelly soil. -**brenner** m pyrites burner

Kiesel m (-) 1 flint. 2 silica, silex; (in compds usu) silicon, siliceous. 3 pebble. **··al**

gen *fpl* diatoms. **k–artig** *a* flinty, siliceous. **K–brunnen** *m* siliceous spring. **-cerit** *n* cerite. **-chlorid** *n* silicon tetrachloride. **-effekt** *m* (*Lthr.*) pebble effect. **-erde** *f* silica, flinty earth

Kieselerde-gel *n* silica gel. **k–haltig** *a* siliceous, siliciferous. **K–hydrat** *n* hydrated silica

Kieselfluor· ..fluosilicate: **-baryum** *n* barium fluosilicate. **-blei** *n* lead fluosilicate

Kiesel-fluorid *n* silicon tetrafluoride

Kieselfluor-kalium *n* potassium fluosilicate. **-natrium** *n* sodium fluosilicate. **-salz** *n* fluosilicate. **-säure** *f* fluosilicic acid. **-verbindung** *f* fluosilicate. **-wasserstoff** *m*, **-wasserstoffsäure** *f* fluosilicic acid

Kiesel-fluß *m* siliceous flux. **-flußsäure** *f* fluosilicic acid. **-gallerte** *f* silica gel. **-galmei** *m* siliceous calamine (hydrous zinc silicate); (*Min.*) hemimorphite. **-gel** *n* silica gel. **-gestein** *n* siliceous rock. **-gips** *m* vulpinite (anhydride). **-glas** *n* 1 flint glass, silica glass. 2 (*Min.*) siliceous calamine. **-gur** *f* kieselguhr, siliceous (diatomaceous, infusorial) earth. **-gurstein** *m* kieselguhr brick. **k–haltig** *a* siliceous. **-hart** *a* flint-hard. **K–härte** *f* hardness of flint. **-hydrat** *n* hydrated silica

kieselig *a* siliceous, flinty; pebbly, gravelly

Kiesel-kalk *m* siliceous limestone

Kieselkalk· (*Min.*) ilvaite. **k–haltig** *a* silicicalcareous. **K–spat** *m* (*Min.*) wollastonite. **-stein** *m* flinty (*or* cherty) limestone

Kiesel-knolle *f* flint nodule. **-körper** *m* siliceous body. **-kreide** *f* 1 siliceous chalk. 2 finely divided chalk. **-kristall** *m* crystal of silica, rock crystal. **-kupfer**, **-malachit** *n* (*Min.*) chrysocolla. **-mangan** *n* (*Min.*) rhodonite. **-mehl** *n* fossil flour, infusorial earth, kieselguhr. **-metall** *n* silicon. **-papier** *n* flint paper. **-pulver** *n* pebble powder. **k–reich** *a* siliceous; flinty, pebbly. **K–sandstein** *m* siliceous (*or* quartz) sandstone. **k–sauer** *a* silicate of. **K–säure** *f* silicic acid —**pyrogene K.** fumed silica

Kieselsäure-anhydrid *n* silicic anhydride, silicon dioxide, silica. **-düngung** *f* silica

(*or* silicate) fertilization. **-gallerte** *f*, **-gel** *n* silica gel. **-gehalt** *m* silica (*or* silicic acid) content. **-glas** *n* vitreous silica. **k–haltig** *a* siliceous, containing silicic acid. **-reich** *a* rich in silicic acid. **K–salz** *n* vitreous silica. **-seife** *f* silicated soap. **-verbindung** *f* silicic acid compound

Kiesel-schiefer *m* siliceous schist. **-schwamm** *m* siliceous sponge. **-sinter** *m* siliceous sinter. **-spat** *m* (*Min.*) albite. **-stein** *m* 1 pebble, flint(stone), gravelstone. 2 silica brick. **-steinlehm** *m* boulder clay. **-stoff** *m* silicon. **-ton** *m* clay slate, (*Min.*) silicide. **-tuff** *m* siliceous sinter. **-verbindung** *f* silicate. **-wasserstoff** *m* hydrogen silicide. **-wasserstoffsäure** *f* hydrosilicic acid. **-wismut(h)** *m* or *n* bismuth silicate. **-wolframsäure** *f* silicotungstic (*or* tungstosilicic) acid. **-zinkerz** *n*, **-zinkspat** *m* siliceous calamine; (*Min.*) hemimorphite, willemite. **-zuschlag** *m* siliceous flux

Kies-filter *n* gravel filter. **-grube** *f* gravel pit

kiesig *a* 1 gravelly, gritty. 2 pyritic

Kies-lager *n* 1 gravel deposit. 2 pyrites deposit. **-lagerstätte** *f* pyrites (*or* sulfide) deposit. **-ofen** *m* pyrites burner (*or* kiln). **-sand** *m* gravelly (*or* gritty) sand

Kiff *m* tanbark

Kilbrickenit *n* (*Min.*) geocronite, kilbrickenite

Kilo *n* (-) kilo (gram). **-gramm** *n* kilogram. **-hertz** *n* kilocycle per second. **-kalorie** *f* kilocalorie. **-meter** *n* kilometer. **k–meterweit** *adv* for (many) kilometers. **K–pond** *n* kilogram force. **-wattstunde** *f* kilowatt hour

Kimme *f* (-n) 1 notch; (*specif*) (rear) sight (of rifles). 2 edge, border. 3 chine, croze (of barrels)

Kind *n* (-er) child *cf* VON 5

Kinder-ernährung *f* child nutrition, baby feeding. **k–feindlich** *a* harmful to children. **-freundlich** *a* beneficial to children. **K–krankheiten** *fpl* 1 childhood diseases. 2 (*usu fig*) teething troubles, (*Mach.*) bugs. **-lähmung** *f* infantile paralysis, poliomyelitis. **k–los** *a* childless. **K–mehl** *n* powdered baby food.

-**nährmittel** n, -**nahrung** f baby food. -**pech** n meconium. -**pocken** fpl chicken pox. -**puder** m baby powder. -**pulver** n (Pharm.) compound powder of rhubarb. -**schleim** m, -**schmiere** f vernix casesa. -**schuhe** mpl baby shoes, (fig) infancy, childhood. -**seife** f baby soap. -**zahn** m baby tooth

Kindes·pech n meconium. -**wasser** n amniotic fluid

Kindheit f childhood, infancy

Kinds·pech n meconium

Kine·film m cinefilm, movie film

kinematographisch a cinematographic

Kinetik f kinetics. **kinetisch** a kinetic

Kinn n (-e) chin. -**backe** f, -**backen** m jowl. -**backendrüse** f maxillary gland. -**lade** f (lower) jaw, jawbone

Kino n (-s) 1 movie theater. 2 (Chem.) kino. -**baumöl** n kino oil. -**gerbsäure** f kinotannic acid. -**harz** n kino resin. -**leinwand** f movie screen

kippbar a tipping, tilting, tiltable. **Kipp–bewegung** f tilting (or tipping) motion

Kippe f (-n) 1: **auf der K. stehen** be tottering on the brink, be about to topple over; be uncertain. 2 gold balance. 3 (mining) dump

kippelig a unsteady, shaky, uncertain

kippen I vt 1 tip, tilt. 2 dump; pour. II vi(s) tip (or topple) over, fall. **Kipper** m (-) 1 tipple, tipping device. 2 dump truck (or car)

Kipp·generator m (Electron.) sweep generator. -**gerät** n 1 (Mach.) tipping (or tilting) device. 2 (Electron.) relaxation oscillator, timebase. -**karren** m tipping (or dump) cart. -**kessel** m tilting basin, swivel pan

kipplig a unsteady, shaky, uncertain

Kipp·ofen m tilting furnace

Kippsche(r) Apparat m Kipp('s) apparatus

Kipp·schwingung f (Electron., Spect.) wagging vibration. **k–sicher** a stable, steady, topple-proof. **K–spiegel** m tilting (or oscillating) mirror. -**vorrichtung** f tipping (or dumping) device. -**wagen** m dump car (or truck)

Kipp-Wipp-Tisch m (Min.) tilting table

Kips n (-e) (usu pl) kips, kipkins. -**garnituren** fpl kip shoulders and bellies. -**leder** n kip leather

kirnen vt 1 churn. 2 shuck, shell (peas)

kirre a tame

Kirsch m cherry brandy. -**baum** m cherry tree. -**branntwein** m cherry brandy. **k–braun** a cherry-brown

Kirsche f (-n) cherry; small bulb, pellet, berry

Kirschen·gummi n cherry gum

kirsch·farben, -farbig a cherry-colored. **K–geist** m kirsch, cherry brandy. -**glut** f cherry-red heat. -**gummi, -harz** n cherry gum. -**kern** m cherry stone (or pit). -**kernöl** n cherry kernel oil. -**lorbeer** m cherry laurel. -**lorbeer(blätter)öl** n cherry laurel oil. **k–rot** a cherry-red. **K–rotglut** f cherry-red heat (about 850°C) **k–rotwarm** a cherry-red-hot. **K–saft** m cherry juice. -**wasser** n kirsch, cherry liqueur (or brandy). -**wurzelkraut** n athamanta

Kissen n (-) cushion; pillow; bolster, pad

Kistchen n (-) small chest (box, case)

Kiste f (-n) crate, packing case, box, chest

Kisten·zucker m muscovado, raw sugar

Kitt m (-e) 1 putty, (sealing) lute, filler. 2 (adhesive) cement, mastic. 3 nonsense

kitt·artig a putty-like, cement-like

kittbar a lutable; cementable cf KITT, KITTEN

Kittel m (-) white (work) coat, lab coat, smock

kitten vt 1 putty, lute, seal. 2 cement

Kitt·erde f luting (or cementing) clay. -**harz** n propolis, bee glue. -**masse** f 1 cementing composition. 2 (Anat.) intercellular cement substance. -**messer** n putty knife. -**öl** n putty oil. -**substanz** f cement (substance)

Kitz n (-e), **Kitze** f (-n) Zool.) kid, fawn

Kitzel m (-) 1 itch. 2 tickle, tickling sensation. 3 titillation; urge, desire. **kitzeln** vt tickle

Kjeldahl·kolben m Kjeldahl flask

kkal. abbv (kleine Kalorie) small calorie

kl. abbv 1 (klein) small. 2 (kaum löslich) scarcely soluble

Kl. abbv (Klasse) class

Kladde f (-n) 1 notebook, pad. 2 rough draft. 3 daybook

Kladstein m place brick (not fully burned)

klaffen vi gape, yawn, be wide open; be split

kläffen vi bark; clamor; complain, gripe

Klafter f (-n) cord (of wood); fathom

Klage f (-n) 1 complaint, grievance. 2 (legal) action, suit. 3 lamentation, cry of woe. **klagen I** vt/i 1 complain (about). 2 take legal action (against), sue. **II** vi lament. **Kläger** m (-) plaintiff

Klage·schrift f writ

kläglich m pitiful

klamm a 1 clammy. 2 (frozen) stiff, numb. 3 short (of cash), hard up

Klammer f (-n) 1 clamp; (paper) clip; staple; clothespin; vise. 2 parenthesis — **eckige K—n** (square) brackets. **-·ausdruck** m parenthetical expression. **klammern 1** vt clamp, clip, staple; fasten. **2: sich k. (an)** cling (to)

Klampe f (-n) cleat, chock; (U-shaped) staple

klang sounded, rang: past of KLINGEN

Klang m (-e) 1 sound, ring, tone; clang. 2 (good) reputation. **··farbe** f tone color, timbre. **-lehre** f acoustics. **k–los** a soundless, mute. **K–regler** m tone control. **-zinn** n fine (or sonorous) tin

Klapp· folding, collapsible; hinged

klappbar a folding, collapsible; hinged

Klapp·deckel m hinged cover (or lid)

Klappe f (-n) 1 (hinged) door (cover, lid), drop lid; trap; trapdoor. 2 value; shutter; damper. 3 flap, dropleaf. 4 flyswatter. **klappen I** vt 1 fold up, turn up; fold down, drop. 2 catch. **II** vi 3 open, close (with a click); clack, clop. 4 (+**an**) knock (on). 5 (usu + **mit**) flap. 6 be successful, work (out right)

Klappen· valvular: **-ventil** n clack (flap, check) valve

Klapper f (-n) rattle, clapper. **klappern** vi 1 rattle, clatter; click clap; chatter. 2 shiver

Klapper·nuß f bladdernut. **-rose** f corn poppy. **-schlange** f rattlesnake. **-stein** m rattle stone, eaglestone, nodular clay ironstone

Klapp·messer n (folding) pocket knife.

-sitz m drop seat, hinged seat

klar I a 1 clear, obvious **–sich im k–en sein (über)** be sure, be aware (of); **ins k–e kommen (über)** get a clear idea (of). 2 conscious, sober. 3 ready; (pred oft) in order. 4 (Dye): **k. ausziehen** exhaust completely. **II** sep pfx clear

Klar n (-) white (of egg)

Klär· clearing, clarifying; settling: **-anlage** f clarifying plant; (Sewage) purification plant. **-apparat** m setting (or clarifying) apparatus. **-bad** n clearing bath. **-bassin** n settling tank. **-becken** n settling basin; filter bed. **-behälter** m settling tank (or vessel)

Klar·beschichtung f clear coating. **k–blickend** a clear-sighted

Klär·bottich settling vat; clearing tub

Klare n (adj endgs) clear (part of) liquid

Kläre f (-n) 1 clarity. 2 clarified liquid. 3 (Sugar) claire, fine liquor. 4 coal dust. 5 boneash. **klären 1** vt clarify, clear (up); purify, defecate. **2: sich k.** clear up, become clear. **Klären** n clarification, etc.

Klär·faß n cleaning cask (or tub), clarifier; (Brew.) cleansing cask; (Vinegar) fining tun. **-filter** n (clarifying) filter. **-flasche** f decanting bottle (or flask)

Klar·gas n clear gas

Klär·gas n digester (or sewage) gas. **-gefäß** n (Sugar) clarifier. **-gelatine** f isinglass. **-grube** f settling pit

Klarheit f clarity, clearness; purity

klarieren vt clear (incl Sugar), purge

Klär·kasten m (Paper) settler. **-kessel** m (esp Sugar) clarifier

klar·kochen vt boil till clear

Klar·kohle f small coal, slack

klar·kommen* vi(s) 1 (+**mit**) cope (with), manage; get along (with). 2 (**von**) get clear (of)

Klar·lack m clear varnish (or lacquer)

klar·legen vt clarify, explain

klar·löslich a dissolving to give a clear solution

Klär·lösung f clearing solution

klar·machen vt 1 make clear, explain. 2 (+dat rflx): **sich etwas k.** get sthg clear in ones' mind; **sich k., daß . . .** realize, understand that . . . 3 chop (wood). 4 clear, get (or make) ready

Klär·mittel n clarifying agent; (Beer) fining. **-pfanne** f (Sugar) clearing (or defecating) pan, clarifier

Klar·saft m (Sugar) clarified juice

Klär·schlamm m sewage sludge

klar·schleifen* vt grind smooth

Klärsel n (Sugar) claircé, clear(ing) liquor

Klarsicht· transparent: **-salbe** f anti-dim compound

Klär·staub m boneash

klar·stellen vt clear up, make clear

Klär·teich m water purification pond

Klärung f (-en) clarification; purification; filtration; (Sugar) defecation

Klärungs·faß n clearing cask, etc = KLÄRFASS. **-mittel** n clarifying agent. **-salz** n clearing salt

klar·werden* 1 vt (dat) become clear (to). 2: **sich k.** (über) get (sthg) clear in one's mind

Klär·werk n (Water) purification plant; (Sewage) treatment plant. **-werktechniker** m sewage treatment engineer. **-zentrifuge** f centrifugal clarifier

Klasse f (-n) 1 class; grade, rank. 2 classroom. **Klassen·ordnung** f classification

klassieren vt classify (esp ore); grade, sort. **Klassierer** m (-) classifier. **Klassierung** f (-en) classification

Klassifikation f (-en) classification. **Klassifikator** m (-en) classifier. **klassifizieren** vt classify

Klassiker m (-) classic. **klassisch** a classic(al)

klastisch a (Geol.) clastic

Klatsch m (-e) 1 handclap; smack, slap; splash. 2 gossip. **klatschen** vt/i 1 slap, smack, bang. 2 clap (hands), applaud. 3 splash. 4 throw. 5 tattle; gossip

Klatsch·präparat n smear preparation (between cover glasses which are then slid apart); dab impression. **-rose** f corn poppy

Klaube·band n picking belt. **klauben** vt gather, pick (out); sort (out)

Klaue f (-n) claw, paw, clutch. 2 hoof. 3 (Mach.) jaw. 4 scrawl. **klauen** vt/i 1 claw, clutch. 2 swipe, crib

Klauen·fett n neat's foot oil. **-mehl** n hoof meal. **-öl** n neat's foot oil. **-seuche** f: **Maul- und K.** hoof and mouth disease

Klause f (-n) cell, chamber; den

Klausel f (-n) clause; stipulation

Klaviatur f (-en) keyboard

Klavier n (-e) piano. **-draht** m piano wire

kleb· adhesive: **-abweisend** a nonsticking. **K–äther** m collodion. **-band** n adhesive tape

Klebe f (-n) 1 adhesive. 2 (Bot.) dodder. **-band** n adhesive tape; splicing tape. **-fähigkeit** f adhesiveness. **k–frei** a nonsticky. **K–korn** n (Biol.) microsome. **-masse** f, **-material** n adhesive. **-mittel** n adhesive (agent), agglutinant; (specif) glue, cement, paste, etc

kleben 1 vt stick, paste, glue, cement. 2 vi (usu + an) stick, adhere; cling (to); be adhesive (or sticky). **kleben·bleiben*** vi(s) stick, get stuck, cling. **klebend** I p.a 1 adherent, adhesive, (ag)glutinative. 2 sticky. II sticking, etc: pres p of KLEBEN

Klebe·pflaster n adhesive plaster

Kleber m (-) 1 gluten. 2 adhesive, etc = KLEBSTOFF. 3 sticker, paster, gluer. **-brot** n gluten bread. **-fiber** f (Flour) fiber, gluten thread. **-grieß** m gluten grits. **k–haltig** a glutinous, containing gluten

kleberig a sticky, etc = KLEBRIG

Kleber·kristallkorn n aleurone grain. **-leim** m gluten glue. **-mehl** n aleurone (grains), gluten flour

Klebe·rolle f roll of adhesive (or gummed) tape

Kleber·protein n gluten protein. **k–reich** a rich in gluten. **K–schicht** f (Bot.) aleurone layer

Klebe· adhesive (See also KLEB·): **-trocknen** n (Lthr.) pasting

kleb· adhesive (See also KLEBE·): **-fähig** a adhesive. **K–fähigkeit** f adhesiveness. **k–frei** a non-sticky, free from tackiness. **K–kitt** m adhesive cement. **-kraft** f adhesive power. **-kraut** n goosegrass, cleavers. **-lack** m adhesive lacquer. **-masse** f, **-material** n adhesive (material). **-mittel** n adhesive (agent), agglutinant; binder; (specif) glue, cement, gum, paste, etc. **-pflaster** n adhesive plaster

klebrig a sticky, tacky; adhesive; gummy; viscid; glutinuos. **Klebrigkeit** f stickiness, tackiness, etc. **Klebrigmacher** m (-) tackifier

Kleb· adhesive (*See also* KLEBE·): **-sand** *m* (*Ceram.*) loam sand; luting (*or* daubing) sand. **-schiefer** *m* (*Petrog.*) adhesive slate (*or* shale). **-schmalz** *m* adhering dirt. **-stelle** *f* splice (in tape). **-stoff** *m* (*genl.*) adhesive, binder; (*specif*) glue, gum, cement, paste, etc. **-streifen** *m* adhesive tape (*or* strip); gummed tape: **durchsichtiger K.** transparent (*or* Scotch) tape. **-vermögen** *n* adhesiveness. **-zettel** *m* gummed (*or* adhesive) label

klecken *vi* 1 (h) make spots (stains, blots). 2 (h) come along nicely. 3 (s) drip, spill

kleckern I *vt* 1 spill, drop (paint on floor, etc). **II** *vi* 2 make a mess, make spots, drip. 3 progress in fits and starts

Klecks *m* (-e) spot, stain (ink) blot; blotch. **klecksen** 1 *vt* spill, drop 2 *vt/i* daub, splash. 3 *vi* (h) make ink spots, blot. 4 *vi*(s) spill, drip (down)

Klee *m* clover, trefoil **—ewiger** (*or* **Luzerner**) **K.** alfalfa, lucern; **spanischer** (*or* **türkischer**) **K.** sainfoin. **·-baum** *m* hop tree. **-blatt** *n* clover (*or* trefoil) leaf. **-futter, -heu** *n* clover hay. **k—rot** *a* clover-red, purplish-red. **K—salz** *n* salts of sorrel, potassium hydrogen oxalate. **-samenöl** *n* cloverseed oil. **k—sauer** *a* oxalate of. **K—säure** *f* oxalic acid. **-seide** *f* clover dodder

Klei *m* clay (soil) *cf* KLEIE

Klei·absudbad *n* (*Dye.*) bran decoction *cf* KLEIE

Kleid *n* (-er) 1 (woman's) dress; (*sg, esp fig*) garment, garb. 2 (*pl only*) clothes, clothing. **kleiden I** *vt* 1 dress, clothe. 2 become, look good on. **II: sich k.** dress, get dressed, clothe oneself. **Kleider** *npl* 1 dresses; garments, garb. 2 clothes, clothing. *Cf* KLEID

Kleider·bad *n* dip dry-cleaning. **-bürste** *f* clothes brush. **-färberei** *f* 1 garment dyeing. 2 dyer's (shop). **-motte** *f* clothes moth. **-stoff** *m* garment (*or* clothing) material

Kleidung *f* (-en) clothing, clothes; garb, dress. **Kleidungs·stück** *n* article of clothing, garment

Kleie *f* bran

kleien·artig *a* bran-like, furfuraceous. **K—bad** *n*, **-beize** *f* bran drench (*or* steep).

-brot *n* bran bread. **-essig** *m* bran vinegar. **k—förmig** *a* bran-like, branny. **K—mehl** *n* pollard, bran with some flour content **-sucht** *f* (*Med.*) pityriasis. **-wasser** *n* bran water

Klei·erde *f* clay earth

kleiig *a* 1 clayey *cf* KLEI. 2 branny *cf* KLEIE

klein *a* small, little; short, slight, minor; trifling, petty; humble, meek: **bei k—em** little by little; **k—es Geld** small change; **im k—en** (a) on a small scale, (b) in detail; **von k. auf** since childhood *cf* VON 5; **k. drehen** (*or* **stellen**) turn down (gas, radio) *cf* KLEINE, KLEINST, WENIG, WINZIG. **Klein** *n* 1 (*Min.*) smalls. 2 giblets

Klein·asien *n* (*Geog.*) Asia Minor. **-bahn** *f* light railway. **-bessemerei** *f* small Bessemer plant. 2 small-scale Bessemerizing. **-betrieb** *m* small-scale operation (plant, enterprise, process). **-bildkamera** *f* miniature camera. **-bildphotographie** *f* microphotography. **-birne** *f* (*Metll.*) small converter. **-eisenwaren** *fpl* small ironware, hardware

kleinen *vt* pulverize, comminute

kleiner smaller, less, shorter, etc: *compar of* KLEIN

kleinern *vt* make smaller, (*esp Math.*) reduce

Klein·färber *m* clothes dyer. **-filter** *n* fine filter, microfilter. **-gärmethode** *f* microfermentation method. **-gebäck** *n* pastry, small baked goods. **-gefüge** *n* fine structure, microstructure. **-geld** *n* (small) change. **k—gepulvert** *a* finely powdered. **-gliedrig** *a* small-membered. **K—handel** *m* retail (trade)

Kleinheit *f* smallness, small size

Klein·hirn *n* cerebellum. **-hirnrinde** *f* cerebellar cortex. **-holz** *n* matchwood

Kleinigkeit *f* (-en) small matter, trifle; (little) bit; bite (to eat)

Klein·kohle *f* small (buckwheat, pea) coal; slack. **-koks** *m* small-sized coke, nut coke. **k—körnig** *a* small-grained, fine-grained. **k—krystallinisch** *a* finely crystalline, in small crystals. **K—küche** *f* kitchenette; portable electric stove. **-lader** *m* (*Elec.*) trickle charger. **-lebewesen** *n* microorganism

kleinlich *a* petty, small-minded
Klein·lichtbildkunst *f* microphotography.
k–lückig *a* finely porous, close-grained.
-maschig *a* fine-meshed. **K–messer** *m*
micrometer. **-mühle** *f* pulverizing mill
Kleinod *n* (-ien) jewel, gem
Klein·röhre *f* little (*or* miniature) tube. **k–
schuppig** *a* with fine scales, finely scaly.
K–spore *f* microspore
kleinst.. 1 *a* smallest, least, etc *cf* KLEIN —
bis ins k–e down to the last detail; **k–e
Lebewesen** microorganisms. 2 (*also as
pfx*) minimum
klein·städtisch *a* small-town, provincial.
K–steller *m* bypass (in Bunsen burner)
Kleinst·lebewesen *n* microorganism. **k–
möglich** *a* smallest possible
klein·stoßen* *vt* pound to pieces (in a
mortar)
klein·stückig *a* small-sized, in small pieces
Kleinst·wert *m* lowest (*or* minimum) value
Klein·verkauf *m* retail (sale). **-versuch** *m*
small-scale experiment (*or* test).
-winkelstreuung *f* small-angle scattering
Kleister *m* (-) (starch *or* flour) paste, size.
-bildung *f* formation of paste.
kleisterig *a* pasty, sticky. **kleistern** *vt/i*
paste, size
Kleister·pinsel *m* paste brush. **-tiegel** *m*
paste pot. **-zähigkeit** *f* viscidity (*or* glu-
tinousness) of starch
kleistrig *a* pasty, sticky = KLEISTERIG
Klemm·backe *f* (*Mach.*) jaw, clamp. **-brett**
n (*Elec.*) terminal board
Klemme *f* (-n) 1 clamp, clip; bobby pin;
(*Med.*) hemostat. 2 (*Elec.*) terminal. 3
(pair of) tongs, nippers, forceps. 4 bind,
predicament. **klemmen** 1 *vt* wedge, jam,
clamp; (+*dat rflx*): **sich den Finger k.**
jam (*or* pinch) one's finger. 2 *vi* be stuck,
be jammed. **3: sich k.** wedge oneself, get
wedged (*or* jammed); (+**hinter**) get at (a
job), get after (sbdy to do . . .)
Klemmen·spannung *f* (*Elec.*) terminal
voltage. **-träger** *m* terminal block
Klemmer *m* (-) 1 pinchcock. 2 nose glasses
Klemm·leiste *f* (*Elec.*) terminal block (*or*
strip). **-schraube** *f* clamp (*or* set) screw;
(*Elec.*) binding post (*or* screw). **-streifen**

m (*Elec.*) terminal strip
Klemmutter *f* (=**Klemm·mutter**) lock nut
Klempner *m* (-) tinsmith; plumber. **-ware**
f tinware, tinplate
Klette *f* (-n) bur(r), burdock
Kletten·bombe *f* (*Mil.*) adhesive bomb.
-erz *n* bur(r) ore. **-öl** *n* colza oil. **-wurzel** *f*
burdock root, lappa. **-wurzelöl** *n* burdock
root oil
klettern *vi*(s) climb
Kletter·pflanze *f* (*Bot.*) climbing plant,
creeper. **-prinzip** *n* cascade principle.
-verdampfapparat, -verdampfer *m*
climbing film evaporator
Klima *n* (-s *or* -te) climate. **-anlage** *f* air-
conditioning plant (*or* unit), air conditioner. **-kammer** *f* refrigerating
chamber
klimatisch *a* climatic
klimatisieren *vt* air condition. **Klimati-
sierung** *f* (-en) air conditioning
klimmen* (klomm, geklommen) *vi*(s) climb
klimpern *vi* tinkle, plink (*esp* on piano);
jingle
Klinge *f* (-n) blade; sword
Klingel *f* (-n) (jingle) bell. **-draht** *m* bell
wire. **klingeln** *vi* ring; jingle: **es klingelt**
the bell is ringing
Klingel·schnur *f* bell cord
klingen* (klang, geklungen) *vi* ring, sound
Kling·glas *n* flint glass. **-stein** *m*
clinkstone, phonolite
Klinik *f* (-en) clinic. **klinisch** *a* clinical
Klinke *f* (-n) 1 (*Mach.*) pawl, catch; (*Tlph.*)
jack. 2 latch; door handle
Klinker *m* (-) clinker, hard-burnt brick.
-pflaster *n* brick paving. **-ziegel** *m*
vitrified brick
Klink·werk *n* ratchet
Klino·achse *f* (*Cryst.*) clino axis. **-edrit** *n*
(*Min.*) clinohedrite. **-klas** *m* clinoclase,
clinoclasite. **-meter** *n* clinometer
Klippe *f* (-n) 1 cliff; rock, reef. 2 obstacle
Klipp·fisch *m* dried cod, stockfish, clipfish
klirren *vi* rattle, jingle, clink
Klischee *n* (-s) (printing) block, plate,
cliché. **klischieren** *vt* make a block of,
stereotype, electrotype
Klistier *n* (-e) enema, clyster
klitschig *a* soggy, doughy

Kloake f (-n) 1 sewer; cesspool; 2 cloaca
Kloaken·gas n sewer gas. **-wasser** n sewage
Kloben m (-) 1 pulley block. 2 vise. 3 log. 4 hook. **klobig** a clumsy, heavy
klomm climbed: past of KLIMMEN
Klon m (-e) clone. **klonisch** a (Med.) clonic
Klopf·bremse f antiknock agent
klopfen 1 vt knock, pound; beat (rugs, etc); break (stones); scale (boilers). 2 vi knock; beat (as: heart)
Klopf·feind m antiknock (agent). **k–fest** a knockproof, antiknock. **K–festigkeit** f resistance to knocking, (anti)knock rating. **-festigkeitswert** m antiknock value. **k–frei** a knockless, no-knock. **K–gegenmittel** n antiknock agent. **-grenze** f knocking (or detonation) limit; point of incipient knocking. **-holz** n mallet. **-neigung** f knocking tendency. **-sieb** n tapping (or shaking) sieve. **-wert** m antiknock value, octane rating
Klöppel m (-) 1 (Bells) clapper, (Elec.) hammer. 2 (Tex.) (lace) bobbin. 3 ball (of b. and socket). 4 tongue (of t. & groove). **--pfanne** f socket (of ball & s.)
Klosett·papier n toilet paper
Kloß m (⁻e) 1 lump, clod. 2 dumpling; meatball
Kloster n (-) cloister, monastery; convent
Klotz m (⁻e) 1 log; chunk, block (of wood). 2 clod (incl psn). **--artikel** m (Tex.) padded style. **-bad** n (Tex.) pad bath, padding liquor. **-druck** m slop-pad printing
klotzen vt (Tex.) pad, slop-pad
Klotz·farbe f (Dye.) padding color. **-fär·bung** f padded dyeing, slop-pad. **-flotte** f padding liquor
klotzig a 1 clumsy, crude. 2 (Wood) knotty, cross-grained. 3 adv awfully
Klotz· (Tex., Dye.): **-reserveartikel** m padded resist style. **-schwarz** n slop-pad black
Klotzung f (Tex.) pad, (slop-)padding
KLT abbv (kritische Lösungstemperatur) critical solution temperature
Kluft f (⁻e) 1 crevice, fissure, cleft; gap, chasm. 2 (Geol.) joint, fault. 3 (Metll.) (pair of) tongs. **klüften** vt cleave, split. **klüftig** a 1 split, cleft; fissured. 2 jointed, faulted. 3 easily cracking. **Klüftung** f

cleavage, splitting, segmentation
klug a clever, smart, intelligent **—k. werden (aus)** make sense (of). **klügeln** vi 1 split hairs. 2 (+an, über) brood (about), turn sthg over in one's mind.
Klugheit f cleverness, intelligence; wisdom; good sense
Klümpchen n (-) little lump; nodule.
Klumpen m (-) 1 lump, chunk, clod; nugget; clot. 2 (Metll.) ingot; bloom. 3 clump, cluster. **--bildung** f coagulation, formation of lumps, etc. **klumpig** a lumpy; clotted; clodlike
Kluppe f (-n) 1 clamp, clip; (pair of) tongs, pincers, calipers. 2 diestock, threading die. **Klupp·zange** f (pair of) forceps
Klystier n (-e) (Med.) enema, clyster
k.M. abbv (kommenden Monats) of the coming month
KMR abbv (Kernmagnetische Resonanz) nuclear magnetic resonance
Knabbel·kohle f (Coal) cobbles. **-koks** m crushed (or nut) coke
knabbern 1 vt nibble. 2 vi (an) gnaw (at)
knacken 1 vt crack; crush. 2 vi (h) crack, snap, click; crackle, crepitate. 3 vi(s) snap, crack (=break). **knackig** a 1 effective. 2 clean-cut, sharp; cute
Knagge f (-n) 1 knot (in wood). 2 cleat, bracket. 3 (Mach.) cam, log, tappet
Knall m (-e) 1 explosive sound (or report), detonation; pop, crack, bang cf URKNALL; (thunder) clap; sonic boom. 2 crash, collapse. 3: **Kn. und Fall** suddenly, without warning. **--aufsatz** m (Explo.) initiator. **-blei** n fulminating lead
knallen 1 vt slam, put down with a bang; shoot down. 2 vi (h) detonate, fulminate; pop, bang; go off. 3 vi(s) crash
Knall· detonating, fulminating; (Chem.) fulminate: **-erbse** f (Fireworks) (toy) torpedo. **k–fähig** a detonating, explosive. **K–flamme** f oxyhydrogen flame. **-gas** n detonating gas, (specif) explosive mixture of hydrogen and oxygen; (as pfx) oxyhydrogen, e.g.: **-gasgebläse** f oxyhydrogen blowpipe. **-gebläse** n oxyhydrogen blast. **-glas** n Rupert's drop, anaclastic glass; candle bomb. **-gold** n gold fulminate, fulminating gold. **-korken** m explosive cap. **-körper** m de-

tonator. **-kraft** *f* explosive force (*or* power). **-luft** *f* = KNALLUFT. **-pulver** *n* detonating powder. **-pyrometer** *n* explosion pyrometer. **-quecksilber** *n* mercury fulminate. **k-rot** *a* fire-engine red. **K-salpeter** *m* ammonium nitrate. **-satz** *m* detonating composition. **k-sauer** *a* fulminate of. **K-säure** *f* fulminic acid. **-silber** *n* silver fulminate

Knalluft *f* (=**Knall·luft**) detonating gas, explosive air

Knall·welle *f* explosive (*or* shock) wave. **-zucker** *m* (*Explo.*) nitrosaccharose. **-zünder** *m* detonator. **-zündmittel** *n* (detonating) priming object

knapp *a* 1 meager, scanty. 2 scarce, in short supply: **k. sein mit** be short of. 3 (too) tight, (e.g., shoes). 4 concise, terse. 5 bare, barely sufficient; *adv* (*esp* + *num*) barely, just under. 6 *adv* close (to, by)

Knappe *m* (-n, -n) miner

Knappheit *f* (-en) 1 meagerness, scantiness. 2 scarcity, shortage. 3 tight fit. 4 conciseness, terseness. 5 bare sufficiency. 6 closeness. *Cf* KNAPP

Knarre *f* (-n) 1 ratchet. 2 rattle. **knarren** *vi* creak, rasp, grate

Knast *m* (-e) 1 knot (in wood); snag. 2 jail

knattern *vi* (*refers to sounds*) 1 (*Motorcycle*) putter, sputter. 2 (*Machine gun*) rat-tat-tat. 3 flap (in wind). 4 crackle (in fire)

Knäuel *m* (-) 1 ball (of yarn, etc); tangled knot; crowd, agglomeration. **-binse** *f* (*Bot.*) common rush. **-drüse** *f* sweat gland. **k-förmig** *a* tangled; convoluted, coiled; globular, ball-shaped. **K-gras** *n* orchard grass

knäueln *vt* wind into a ball; agglomerate; tangle, snarl

Knauf *m* (-e) knob

Knäulchen *n* (-) little ball, etc *cf* KNÄUEL

knautschen 1 *vt* crumple. 2 *vi* wrinkle. **Knautsch·leder** *n* crush (*or* upholstery) leather

Knebel *m* (-) 1 stick; crossbar; carrying handle. 2 (*Bot.*) slip. 3 (*Zool.*) mastax. **-gelenk** *n* toggle joint

Knecht *m* (-e) 1 farmhand; slave. 2 trestle, sawhorse. 3 clamp, vise; jack

kneifen* (kniff, gekniffen) **I** *vt* 1 pinch, nip; press, squeeze (together); cause (gas-

tric) pain. 2 *vi* back down, retreat. **Kneifer** *m* (-) nose glasses

Kneif·zange *f* (pair of) nippers, pincers

knetbar *a* kneadable, plastic. **Knet-einrichtung** *f* kneading device

kneten *vt* knead; mold; masticate, mill; (*Ceram.*) pug. **Kneter** *m* (-) kneader

Knet·gummi *n* art gum. **-holz** *n* plastic wood. **-legierung** *f* wrought alloy. **-maschine** *f*, **-mühle** *f* **-werk** *n* kneading machine, malaxatory, masticator; (*Ceram.*) pug mill

Knick *m* (-e) 1 sharp bend, kink; break (in a curve). 2 fold, crease. 3 dent, crack. **knicken** 1 *vt* bend sharply (to the point of breaking), kink; break; fold, crease; crack. 2 *vi*(s) buckle, give way; break, snap

Knick·festigkeit *f* breaking strength; (*Metals*) buckling strength. **-last** *f* buckling load. **-punkt** *m* 1 break (in a curve). 2 buckling point, point of inversion. **-schwingung** *f* bending vibration. **-spannung** *f* breaking (*or* buckling) stress. **-stelle** *f* break, knee (in a curve)

Knickung *f* (-en) sharp bend(ing), kinking; breaking; folding, creasing; cracking; buckling; snapping; break, rupture *cf* KNICKEN

Knick·versuch *m* buckling test

Knie *n* (- *or* -e) knee —**etwas übers K. brechen** rush (*or* hurry) a matter. 2 (sharp) bend, angle, elbow. **knieen** *vi* (h *or* s), **sich k.** kneel. *Cf* GEKNIET

Knie·gelenk *n* knee joint; (*Mach. oft*) elbow joint. **-hebel** *m* toggle. **-kehle** *f* popliteal space, back of the knee

knien *vi* (h *or* s), **sich k.** kneel *cf* GEKNIET

Knie·rohr *n*, **-röhre** *f* elbow (pipe), bent tube (*or* pipe). **-scheibe** *f* kneecap, patella. **-stück** *n* (*Mach.*) elbow, angle (piece)

kniff pinched, etc: *past of* KNEIFEN

Kniff *m* (-e) 1 crease. 2 pinch. 3 trick, artifice. **knifflig** *a* tricky

knipsen **I** *vt* snap (*incl. Phot.*). 2 punch (holes). 3 flick, flip. **II** *vi* (*genl*) make a clicking sound; (*specif*) turn (flick, flip) a switch, snap (one's fingers), take a picture

knirschen *vi* grind, grate, crunch: **mit den**

Zähnen k. grind (or gnash) one's teeth.
Knirsch·pulver n coarse gunpowder
Knister·gold n Dutch metal (or gold), tinsel. **-kohle** f coal altered by igneous intrusions
knistern vi crackle, rustle; creak; (Chem., Elec.) decrepitate
Knister·salz n decrepitating salt
Knitter m (-) crease. **k–·arm** a crumple- (or crease-)resistant. **-fest, -frei** a creaseproof, non-creasing. **K–festigkeit** f crease resistance. **-gold** n tinsel, gold foil
knitterig a creased, crumpled. **knittern** vt/i crumple, crease
Knitter·papier n crinkled paper. **-zahl** f fold- (or crease-)test value
knittrig a creased, crumpled = KNITTERIG
Knoblauch m garlic. **k–·artig** a alliaceous, garlicky. **K–erz** n (Min.) scorodite. **-gamander** m water germander. **-öl** n garlic oil. **-zehe** f garlic clove
Knöchel m (-) ankle(bone), malleolus; knuckle
Knochen m (-) bone; (in compds, oft) osseous, osteo-: **-abfall** m bone waste. **k–ähnlich, -artig** a bony, osseous, osteoid. **K–asche** f bone ash. **-band** n ligament. **k–bildend** a bone-forming, osteoplastic. **K–bildung** f bone formation, ossification. **-bruch** m bone fracture. **-dämpfapparat** m bone steamer. **-dünger** m bone manure. **-dungmehl** n bone meal. **k–dürr** a raw-boned, haggard. **K–erde** f bone ash. **-fett** n bone fat (or grease). **-gallerte** f ossein, bone gelatin. **-gelenk** n joint. **-gerüst** n skeleton. **-gewebe** n bone (or bony) tissue. **k–hart** a bone-hard. **K–haut** f periosteum. **-kohle** f bone charcoal. **-kohleglühofen** m boneblack furnace. **-krebs** m bone cancer. **-lehre** f osteology. **-leim** m bone glue, osteocolla. **-mark** n bone marrow. **-masse** f osseous material. **-mehl** n bone meal (or dust). **-mühle** f bone mill. **-öl** n bone (neat's foot, Dippel) oil. **-porzellan** n bone china. **-säure** f (ortho)phosphoric acid. **-schrot** m or n crushed bone. **-schwarz** n boneblack. **-teeröl** n bone oil. **k–trocken** a bone-dry. **K–zelle** f bone cell, osteocyte

knöchern a bony, osseous; angular.
knochig a bony, osseous; scrawny
Knödel m (-) dumpling
Knöllchen n (-) nodule, small bulb (or knot), tubercle cf KNOLLE. **-·bakterie** f nodule bacterium, rhizobium
Knolle f (-n), **Knollen** m (-) 1 lump, clod. 2 bulb; tuber, corn. 3 knob, nodule
Knollen· bulbous, tuberous, nodular: **-blätterpilz** m (Bot.) death cup, amanita. **-fäule** f tuber (esp potato) blight
knollig a lumpy, in lumps; knobby; bulbous, nodular, tuberous
Knopf m (-̈e) button; stud; knob; head (of a pin). **Knöpfchen** n (-) little button (stud, knob)
Knopf·deckel m knobbed cover. **-druck** m press(ing) (of) a button: **auf einen K.** at (or with) the press of a button
knöpfen vt button (up)
knopf·förmig a button-shaped; knobbed. **K–loch** n buttonhole
Knopper f (-n) gallnut. **-·eiche** f valonia oak, aegilops
Knorpel m (-) cartilage, gristle. **k–·ähnlich, -artig** a cartilaginous, gristly. **K–gewebe** n cartilaginous tissue. **-haut** f (Anat.) perichondrium
knorpelig a cartilaginous, gristly
Knorpel·leim m chondrin. **-tang** m (Bot.) carrageen, sphaerococcus; (Pharm.) chondrus
knorplig a cartilaginous, gristly = KNORPELIG
Knorren m (-) 1 knot, gnarl (in wood). 2 snag, stump (esp of a branch). 3 knob, knobby protuberance. **knorrig** a knotty, gnarled; knobby; rude
Knöspchen n (-) budlet, gemmule cf KNOSPE
Knospe f (-n) bud. **knospen** vi bud, sprout
knoten 1 vt knot, tie a knot in, tie together. 2 vi & **sich k.** knot, get knotted. **Knoten** m (-) 1 knot. 2 node, knob, tubercle; ganglion. 3 crux, nub. 4 (in compds, oft) central: **-amt** n central office; telephone exchange. **-ebene** f nodal plane. **-fläche** f (Math.) nodal surface. **-kalk** m (Geol.) slate (or schist) containing spots of limestone. **-linie** f nodal line. **-punkt** m 1 node, nodal point. 2 junction, center. 3

ganglion. **-schiefer** *m* slate with spots of limestone. **-wurz** *f* figwort. **-zahl** *f* 1 number of nodes. 2 nodal number

Knöterich *m* (-e) knotgrass, polygonum

knotig *a* 1 knotty; knobby: 2 nodular, nodose; tuberous. 3 rude

knüllen *vt, vi,* **sich k.** crumple, crease.

Knüller *m* (-) smash hit, sensation; bargain

knüpfen I *vt* 1 knot, tie together; (*esp* + **an**) tie, attach (to). 2 weave, form. **II: sich k.** (**an**) attach itself, be attached (to), be associated (with)

Knüppel *m* (-) 1 club, cudgel; billet 2 (*Aero.*) control stick. **-holz** *n* logs. **-schaltung** *f* stick shift

knurren 1 *vt/i* growl, grumble. 2 *vi* rumble

knusp(e)rig *a* crisp, crunchy; appetizing

Koacervierung *f* coacervation

Koagel *n* (-e) coagel, caogulum

Koagulat *n* (-e), **-masse** *f* coagulum, clot

koagulierbar *a* coagulable. **koagulieren** *vt/i* coagulate. **Koagulierung** *f* (-en) coagulation

Koagulierungs-bad *n* coagulating bath. **k-beständig** *a* coagulation-resistant. **-empfindlich** *a* liable to coagulation. **-fähig** *a* coagulable. **K-fähigkeit** *f* coagulability. **-flüssigkeit** *f* coagulating liquid. **-mittel** *n* coagulating agent, coagulant. **-schutzwirkung** *f* anticoagulating effect. **-wärme** *f* heat of coagulation

koaleszieren *vi* coalesce

koal(is)ieren *vi* form a coalition, unite, merge

koaxial *a* coaxial

koazervieren *vt* coacervate

Kobalt *m* cobalt. **-am(m)in** *n* cobaltammin. **-arsenikkies** *m* glaucadot, glaucodate

Kobaltat *n* (-e) cobaltate

Kobalt-beize *f* (*Dye.*) cobalt mordant. **-beschlag** *m* cobalt bloom, erythrite. **-blau** *n* cobalt (*or* Thénard's) blue, smalt. **-bleierz** *n,* **-bleiglanz** *m* (*Min.*) clausthalite. **-blume, -blüte** *f* cobalt bloom, erythrite. **-chlorür** *n* cobaltous chloride. **-erde** *f* (*Min.*) cobalt crust. **-fahlerz** *n* cobaltian tetrahedrite. **-farbe** *f* cobalt pigment (*or* blue), smalt, powder blue. **-flasche** *f* cobalt glass bottle.

-gehalt *m* cobalt content. **-glanz** *m* cobalt glance, cobaltite. **-graupen** *fpl* amorphous gray cobalt. **-grün** *n* Rinmann's (Turkish, cobalt) green. **k-haltig** *a* cobaltiferous

Kobalti· (*in compds*) cobaltic, cobalti·

Kobaltiak *n* cobaltammine

Kobalti· cobalt(ic): **-chlorid** *n* cobaltic chloride. **-cyankalium** *n* potassium cobalticyanide. **-cyanwasserstoff** *m,* **-cyanwasserstoffsäure** *f* cobalticyanic acid

kobaltieren *vt* treat with cobalt; convert into a cobalt derivative

Kobalti·kaliumnitrit *n* potassium cobaltinitrite

Kobaltin *n* (*Min.*) cobaltite

Kobalti· cobalt(ic): **-oxid** *n* cobaltic oxide. **-sulfid** *n* cobaltic sulfide

Kobalt·kies *m* cobalt pyrites, linnaeite. **-manganerz** *n* asbolite, cobaltian wad (manganese dioxide). **-nickelkies** *m* linnaeite, siegenite. **-nickelpyrit** *m* bravoite

Kobalto· cobaltous, cobalto·: **-chlorid** *n* cobaltous chloride. **-cyanwasserstoffsäure** *f* cobaltocyanic acid. **-nitrat** *n* cobaltous nitrate. **-oxid** *n* cobaltous oxide. **-sulfat** *n* cobaltous sulfate. **-sulfid** *n* cobaltous sulfide

Kobalt·oxid *n* cobalt(ic) oxide. **-oxidhydrat** *n* cobaltic hydrate. **-oxidoxydul** *n* cobaltocobaltic oxide. **-oxidverbindung** *f* cobaltic compound. **-oxydul** *n* cobaltous oxide. **-oxyduloxid** *n* cobaltocobaltic oxide, China blue, tricobalt tetroxide. **-oxydulverbindung** *f* cobaltous compound

Kobalto·zyankalium *n* potassium cobaltocyanide

Kobalt·rosa *n* cobalt red. **-salz** *n* cobalt salt. **-spat** *m* sphaerocobaltite, cobaltocalcite. **-speise** *f* cobalt speiss. **-trockner** *m* cobalt drier. **-violett** *n* cobalt violet, cobalt arsenate (*or* phosphate) pigment. **-vitriol** *n* 1 cobalt vitriol, cobalt(ous) sulfate. 2 (*Min.*) bicberite

Kober *m* (-) basket (with a lid)

Koch 1 *m* (∸e) cook, chef. 2 *n* porridge. 3 (*in compds oft*) cooking, boiling: **-abschnitt** *m* period of cooking (*or* boiling). **-apparat** *m* cooking apparatus. **-becher** *m* beaker.

k–beständig *a* **1** stable on boiling (*or* cooking). **2** (*Dyes*) fast to boiling. **K–brenner** *m* cooking (*or* boiling) burner. **-buch** *n* cookbook. **-dauer** *f* duration of boiling. **k–echt** *a* fast to boiling (in water). **K–echtheit** *f* fastness to boiling **kochen 1** *vt/i* boil, cook —**leicht (leise, gelinde) k.** simmer. **2** *vi* seethe, heave. **Kochen** *n* boiling, ebullition; cooking, cookery

kochend·heiß *a* boiling (hot); just below boiling

Kocher *m* (-) (*Mach.*) cooker; electric kettle; hotplate; (*Paper*) digester. **··ausbeute** *f* cooker yield, yield on cooking (*or* boiling). **-lauge** *f* cooker (*or* cooking) liquor

koch·fertig *a* ready-to-cook. **-fest** *a* boil-proof, fast to boiling. **K–flasche** *f* boiling flask. **-fleck** *m* (*Tex.*) kier mark (*or* stain). **-flüssigkeit** *f* cooking liquor; liquid for boiling. **-gefäß** *n* boiling (*or* cooking) vessel, boiler; (*Paper*, *Tex.*) bucking kier. **-gerät** *n* **1** cooker. **2** = **-geschirr** *n* cooking (*or* boiling) utensil(s). **k–heiß** *a* boiling hot. **K–hitze** *f* boiling heat; cooking temperature

Köchin *f* (-nen) cook

Koch·käse *m* processed curd cheese. **-kessel** *m* kettle; cauldron; (*Paper*) pulp boiler (*or* digester). **-kläre** *f* (*Sugar*) filtered liquor (*or* syrup). **-kölbchen** *n* small boiling flask; fractional distillation flask. **-kolben** *m* boiling flask. **-kunst** *f* cookery, culinary art. **-lauge** *f* boiling lye. **-periode** *f* boiling period, boil. **-platte** *f* hotplate. **-probe** *f* boiling test. **-prozeß** *m* boiling (cooking, evaporating) process. **-puddeln** *n* slag (*or* pig) boiling. **-punkt** *m* boiling point. **-punktmesser** *m* (*genl*) boiling point apparatus; (*specif*) ebullioscope. **-rohr** *n*, **-röhre** *f* boiling (*or* distilling) tube

Kochsalz *n* common (*or* table) salt. **··bad** *n* salt (*or* brine) bath. **-flamme** *f* sodium chloride flame. **-gehalt** *m* salt content, salinity. **k–haltig** *a* salted, saline, salt-containing. **K–lauge** *f* (common salt) brine. **-lösung** *f* (common) salt solution, brine solution: **oxidische K.** salt solution containing hydrogen peroxide. **k–sauer**

a (hydro)chloride of. **K–säure** *f* hydrochloric acid

Koch·stadium *n* boiling (*or* cooking) stage. **-topf** *m* cooking pot (*or* vessel), saucepan; boiler, cooker

Kochung *f* (-en) boiling, cooking

Koch·vergütung *f* (*Metll.*) artificial aging (at 100°C). **-verlust** *m* loss on boiling, loss by evaporation. **-wasser** *n* (leftover) boiling water. **-zeit** *f* boiling (*or* cooking) time. **-zucker** *m* brown sugar; powdered sugar; sugar for cooking

Kockels·beere *f*, **-korn** *n* Indian berry (*or* cockle)

Kodäthylin *n* (-e) codethyline

Kode/in *n* codeine

Köder *m* (-) bait, lure; decoy

Kod·öl *n* cod(-liver) oil; rosin oil

Kodzu·faser *f* kodzu (*or* tapa) fiber

Koeff. *abbv* = **Koeffizient** *m* (-en, -en) coefficient

Koenzym° *n* coenzyme

koerzibel *a* coercible. **Koerziv·kraft** *f* coercivity, coercive force

koexistieren *vi* coexist

Koffe/in *n* caffeine

Koffer *m* (-) **1** trunk, (suit)case, traveling (bag); box. **2** (*in compds, oft*) portable (in a case): **-apparat** *m* portable machine (typewriter, recorder, etc). **-lack** *m* trunk varnish (*or* lacquer). **-leder** *n* luggage hide. **-pappe** *f* (*Paper*) panel board. **-raum** *m* trunk (of a car)

Kognak *m* cognac, brandy. **··öl** *n* oil of cognac (*or* wine), oenanthic ether

kohärent *a* coherent. **Kohärenz** *f* coherence. **kohärieren** *vi* cohere. **Kohäsion** *f* cohesion

Kohäsions·druck *m* cohesion pressure. **-kraft** *f*, **-vermögen** *n* cohesive force (*or* power)

kohäsiv *a* cohesive

Kohl 1 *m* (-e) cabbage. **2** (*in compds*) *oft* = **Kohle(n)** coal, carbon

Kohle *f* (-n) **1** coal. **2** charcoal. **3** (*esp Elec.*) carbon. **k–beheizt** *a* coal-heated, coal-fired. **K–bogen** *m* (*Elec.*) carbon arc. **-bürste** *f* (*Elec.*) carbon brush. **-chemie** *f* coal (tar) chemistry. **-druck** *m* (*Phot.*) carbon print. **-elektrode** *f* carbon electrode. **-fadenlampe** f carbon filament lamp. **-feuerung** f 1 heating with coal. 2 coal furnace. **k–frei** a carbon-free. **-führend** a coal-bearing; (esp Geol.) carboniferous. **K–gehalt** m carbon content. **-hülltechnik** f carbon coating technique. **-hydrat** n carbohydrate. **-hydrierung** f hydrogenation of coal. **-klemme** f (Elec.) carbon terminal. **-korn** n carbon granule. **-lampe** f carbon lamp. **-lichtbogen** m (Elec.) carbon arc

kohlen I vt 1 char, carbonize; carburize. **II** vi 2 char, get charred. 3 burn improperly, smoke (as: a wick). 4 take on a load of coal

Kohlen- coal; carbon: **-abbrand** m (Elec.) carbon burnoff. **-ablagerung** f carbon deposition, carbonization. **k–ähnlich**, **-artig** a coal-like, carbonaceous. **K–asche** f coal ash(es). **-aufbereitung** f coal dressing. **-becken** n 1 coal pan, brazier. 2 (Geol.) coal basin. **-benzin** n benzene, benzol. **-bergbau** m coal mining. **-bergwerk** n coal mine. **-beule** f (Med.) carbuncle. **-bleispat** m (Min.) cerussite, white lead ore. **-bleivitriolspat** m (Min.) lanarkite. **-blende** f anthracite. **-bogenlampe** f carbon arc lamp. **-bohrer** m charcoal drill. **-braun** n Cologne brown. **-brennen** n charcoal (or coal) burning. **-brenner** m charcoal burner. **-brennerei** f 1 charcoal burning. 2 charcoal kiln (or works). **-dampf** m gases from burning coal, coal gas(es). **-dioxid** n carbon dioxide. **-disulfid** n carbon disulfide. **-dithiolsäure** f dithiolcarbonic acid. **-drucksäule** f carbon pile **-dunst** m carbon monoxide fumes, vapor from coal(s). **-eisen** n iron carbide. **-eisenstein** m blackboard ironstone, carbonaceous iron carbonate. **-entgasung** f coal carbonization (or distillation) **-faden** m carbon filament. **-filter** n charcoal (or boneblack) filter. **-flöz** n coal seam (bed, layer), vein of coal. **-fördermaschine** f coal conveyor. **-förderung** f coal extraction; coal output. **-futter** n carbonaceous lining. **-gas** n coal gas; carbon monoxide. **-gestäube** n coal dust. **-gestein** n carbonaceous rock. **-gestübbe** n coal slack, coal and clay dust. **-grieß** m coal slack,

coal (or carbon) granules. **-grube** f coal mine. **-grus** m slack coal. **k–haltig** a carbonaceous, carboniferous. **K–haufen** m coal heap. **-hydrat** n carbohydrate. **-hydratabbau** m degradation (or catabolism) of carbohydrates. **-hydrierung** f hydrogenation of coal (or carbon)

Kohlenkalk m carboniferous limestone. **--spat** m anthraconite. **-stein** m carboniferous limestone

Kohlen-karbonit n (Expl., TN) kohlencarbonite. **-kasten** m coal bin. **-klein** n small coal. **-klemme** f (Elec.) carbon terminal. **-korn** n carbon granule. **-licht** n carbon light. **-lösche** f coal dust (or slack). **-mehl** n coal dust, powdered coal. **-meiler** m charcoal pile. **-monoxid** n carbon monoxide. **-mulm** m coal dust (or slack); charcoal dust. **-ölsäure** f carbolic acid. **-oxid** n carbon monoxide

Kohlenoxid-eisen n iron carbonyl. **-hämoglobin** n compound of carbon monoxide and hemoglobin. **-kalium** n potassium hexacarbonyl. **-knallgas** n explosive mixture of carbon monoxide and oxygen. **-nickel** n nickel carbonyl. **k--reich** a rich in carbon monoxide. **K–vergiftung** f carbon monoxide poisoning

Kohlen-oxychlorid n carbonyl chloride. **-oxysulfid.** n carbon oxysulfide. **-papier** n carbon paper. **-preßstein** m coal briquet. **-pulver** n coal (or charcoal) powder, powdered coal (or charcoal). **-puppe** f (Elec.) carbon rod. **-rückstand** m carbon residue. **-sandstein** m carboniferous sandstone. **k–sauer** a carbonic, carbonate of; **k–es Gas** carbon dioxide. **K–säure** f carbonic acid; carbon dioxide

Kohlensäure-anhydrid n carbonic anhydride, carbon dioxide. **-ausscheidung** f (Physiol.) carbon dioxide excretion. **-bestimmer** m apparatus for determining carbon dioxide. **-brot** n aerated bread. **-chlorid** n carbonyl choride. **-eis** n dry ice, solid carbon dioxide. **-entwicklung** f evolution of carbon dioxide. **-ester** m carbonic ester. **-flasche** f carbon dioxide cylinder. **-gas** n carbonic acid gas, carbon dioxide. **-gehalt** m carbon dioxide content. **k–haltig** a carbonated, containing carbonic acid. **K—messer** *m* carbonometer, anthracometer, instrument for measuring carbon dioxide. **-salz** *n* carbonate. **-schnee** *m* carbon dioxide snow, dry ice. **-verflüssigung** *f* liquefaction of carbon dioxide. **-verlust** *m* loss (*or* escape) of carbon dioxide. **-wäsche** *f* carbon dioxide washing (unit). **-wasser** *n* carbonated water, soda water

Kohlen·schiefer *m* bituminous (*or* coal-burning) shale. **-schlacke** *f* cinder, clinker. **-schlichte** *f* (*Foun.*) black wash. **k—schwarz** *a* coal-black. **K—schwarz** *n* carbon (charcoal, coal) black. **-schwefelwasserstoffsäure** *f* trithiocarbonic acid, H_2CS_2. **-spat** *m* anthraconite, whewellite. **-spitze** *f* carbon point. **-stab** *m* carbon rod. **-staub** *m* coal (*or* charcoal) dust. **-stein** *m* carboniferous limestone. **-stickstoff** *m* cyanogen. **-stift** *m* carbon rod. **-stoff** *m* carbon

kohlenstoff· carbon: **-arm** *a* poor in carbon, low-carbon. **K—art** *f* variety of carbon. **-ausscheidung** *f* separation of carbon. **-bestimmung** *f* carbon determination. **-bindung** *f* carbon linkage. **-eisen** *n* 1 iron carbide. 2 (*Min.*) cohenite. **-entziehung** *f* removal of carbon, decarbonization, decarburization. **k—frei** *see* KOHLENSTOFFREI. **K—gehalt** *m* carbon content. **-gehaltverhältnis** *n* (*Petrol*) carbon ratio. **-gerüst** *n* carbon skeleton. **k—haltig** *a* carboniferous, carbonaceous, carbon-containing. **K—hydrat** *n* carbohydrate. **-kalium** *n* potassium carbide. **-kern** *m* carbon nucleus. **-kette** *f* carbon chain. **-legierung** *f* carbon alloy. **-magnesium** *n* magnesium carbide. **-metall** *n* (metallic) carbide. **-oxychlorid** *n* carbonyl chloride, phosgene

kohlenstoffrei *a* carbon-free, noncarbonaceous

kohlenstoff· carbon: **-reich** *a* rich in carbon, high-carbon. **K—ring** *m* carbon ring, ring of carbon atoms. **-silicium** *n* carbon silicide, silicon carbide. **-skelett** *n* carbon skeleton. **-stahl** *m* carbon steel. **-stein** *m* carbon (*or* graphite) brick, fire brick. **-stickstofftitan** *m* titanium carbonitride. **-sulfid** *n* carbon disulfide. **-zahl** *f* number of carbon atoms

Kohlen· carbon, charcoal, coal: **-suboxid** *n* carbon suboxide. **-sulfid** *n* carbon disulfide. **-sulfidsalz** *n* thiocarbonate. **-technik** *f* coal technology. **-teer** *m* coal tar. **-teerfarbe** *f* coal-tar dye (*or* color). **-teerseife** *f* coal-tar soap. **-tiegel** *m* carbon (*or* charcoal). crucible; crucible with a carbonaceous lining. **-typ** *m* 1 type of coal. 2 (*Petrology*) bonded constituent (of coal). **-verbrauch** *m* coal consumption. **-verflüssigung** *f* liquefaction of coal. **-vergasung** *f* coal distillation (*or* gasification). **-verschwelung** *f* low-temperature carbonization of coal. **-vitriolbleispat** *m* (*Min.*) lanarkite. **-vorkommen** *n* occurrence of coal, coal deposit. **-wäsche** *f* coal washing plant. **-wassergas** *n* coal water gas

Kohlenwasserstoff *m* hydrocarbon. **-gas** *n* hydrocarbon gas **—leichtes K.** methane; **schweres K.** ethylene. **-gemisch** *n* hydrocarbon mixture. **k—haltig** *a* containing hydrocarbons, hydrocarbonaceous. **-löslich** *a* soluble in hydrocarbons. **K—öl** *n* hydrocarbon oil. **-verbindung** *f* hydrocarbon

Kohlenwertstoff *m* coal (by-)product. **-betrieb** *m*, **-gewinnungsanalge** *f* coal by-product plant

Kohlen·widerstand *m* carbon resistor. **-zeche** *f* coal mine. **-ziegel** *m* coal briquet

Kohle·papier *n* carbon paper

Köhler *m* (-) charcoal burner. **Köhlerei** *f* (-en) charcoal burning; charcoal plant

Kohle· coal, charcoal, carbon: **-rohr** *n*, **-röhre** *f* carbon tube. **-rohofen** *m* carbon tube furnace. **-schwelung** *f* coal distillation. **-stab** *m* carbon rod. **-stahl** *m* carbon steel. **-staub** *m* coal (*or* charcoal) dust. **-vergasung** *f* gasification of coal. **k—verstärkt** *a* carbon-strengthened

Kohl·hernie *f* (*Bot.*) clubroot. **-holz** *n* wood for charcoal

kohlig *a* coal-like, coal-bearing, carbonaceous

Kohl· cabbage; coal: **-palme** *f* cabbage palm. **-palmöl** *n* cabbage palm oil. **-rabi** *m* kohlrabi, cabbage turnip. **-raps** *m* rape(seed). **-rübe** *f* rutabaga. **-saat** *f* colza, rapeseed. **-saatöl** *n* colza (*or* rapeseed) oil. **-salat** *m* cole slaw, cabbage salad. **k—**

schwarz *a* coal-black, jet-black. **K–schwarz** *n* charcoal (*or* coal) black. **-sprossen** *fpl* Brussels sprouts

Kohlung *f* (-en) carbonization; carburization *cf* KOHLEN

Kohlungs·mittel *n* carbonizing (*or* carburizing) agent. **-stahl** *m* cementation steel

Kohlweißling *m* (-e) cabbage butterfly

kohobieren *vt* cohobate, distill repeatedly

Kohune(n)·nuß *f* cohune nut. **-(nuß)öl** *n* cohune oil

Koinzidenz *f* (-en) coincidence. **koinzidieren** *vi* coincide

Koji-säure *f* kojic acid

Kok *m* coke

Koka *f* coca. **--blätter** *npl* coca leaves

Kokain *n* cocaine. **--vergiftung** *f* cocainism, cocaine poisoning

Kokarden·erz *n* cocade (*or* ring) ore

Koke *m or f* coke. **koken** *vi* produce coke. **Koker** *m* (-) 1 coking plant. 2 (*Ceram.*) sagger

Kokerei *f* (-en) 1 coking. 2 coking plant. **·-gas** *n* coke-oven gas. **-ofen** *m* coke oven. **-teer** *m* coke(-oven) tar

Kokes· = KOKS**·**.

Kokille *f* (-n) ingot mold; (*Foun.*) chill (mold). **Kokillen·guß** *m* chill casting

Kokke *f* (-n) coccus

Kokkels·korn *n* cocculus indicus, Indian cockle

Kokken cocci: *pl of* KOKKE *f* & KOKKUS *m*. **·-ketten** *fpl* streptococci. **Kokkus** *m* (..kken) coccus

Kokon *m* (-s) cocoon. **--faser** *f* cocoon fiber (*or* thread)

Kokos· coco(nut): **-baum** *m* coco(nut) palm (*or* tree). **-butter** *f* coconut butter. **-faser** *f* coconut fiber, coir. **-fett** *n* coconut oil. **-garn** *n* coir yarn. **-nuß** *f* coconut

Kokosnuß·bast *m* coconut bast, coco fiber. **-fett** *n*, **-öl** *n*, **-talg** *m* coconut oil

Kokos·öl *n* coconut oil. **-palme** *f* coco palm. **-seife** *f* coconut oil soap. **-talg** *m* coconut oil. **-talgsäure** *f* cocinic acid (cocinin)

Koks *m* coke. **--abfall** *m* refuse coke. **k–ähnlich** *a* coke-like. **K–asche** *f* coke ash(es). **-ausbeute** *f*, **-ausbringen** *n* coke yield. **k–beheizt** *a* coke-fired, coke-heated. **K–bereitung** *f* coke making (*or* production), coking. **k–bildend** *a* coke-forming; carbonizable. **K–brennen** *n* coke burning, coking. **-erzeugung** *f* coke production. **-fahrt** *f* gas plant operation for maximum coke yield. **-filter** *n* coke filter. **-gas** *n* coke-oven gas; blue (water) gas. **-gicht** *f* coke charge. **-grus** *m* coke breeze, small coke. **-heizung** *f* heating with coke. **-hochofen** *m* coke blast furnace. **-klein** *n* small coke, coke breeze (*or* dust). **-kohle** *f* coking coal. **-lösche** *f* coke dust (*or* breeze). **-mehl** *n* coke dust, powdered coke. **-ofen** *m* coke oven. **-ofenkoks** *m* coke-oven coke. **-ofenteer** *m* coke-oven tar. **-roheisen** *n* coke pig iron. **-schicht** *f* layer of coke. **-staub** *m* coke dust. **-stein** *m* coke brick. **-teer** *m* coke(-oven) tar

Kokung *f* (-en) coking

Kola·nuß *f* cola nut

Kolatorium *n* (..ien) colator(ium), strainer

Kölbchen *n* (-) small flask

Kolben *m* (-) 1 (*Mach.*) piston; plunger, ram; bit (of soldering irons). 2 flask; globe; (*Elec.*) bulb. 3 (*Bot.*) spadix, spike; (corn) cob. 4 (rifle) butt. 5 (*Metll.*) bloom. **k–förmig** *a* bulb-shaped; club-shaped, clavate. **K–hals** *m* neck of a flask. **-hub** *m* piston stroke. **-manometer** *n* piston gauge. **-pumpe** *f* reciprocating pump. **-schimmel** *m* club mold. **-stange** *f* piston rod. **-stoß** *m* piston stroke. **-träger** *m* flask stand (*or* support). **-wicke** *f* groundnut, peanut

kolbig *a* knobby, knotty; club-shaped, nodular

Koli·bazillus *m* colon bacillus

Kolibri *m* (-s) humming bird

kolieren *vt* filter, strain, percolate

Kolier·rahmen *m* filtering frame. **-tuch** *n* filter(ing) cloth

Koli·gruppe *f* colon bacillus group. **-reinkultur** *f* colon bacillus pure culture

Kolkothar *m* colcothar, Venetian red

Kollagen *n* collagen

Kolleg *n* (-e) lecture course. **Kollege** *m* (-n, -n) **Kollegin** *f* (-nen) colleague, fellow worker (*or* employee). **Kollegium** *n* (..gien) 1 committee, council, board. 2 (teaching) faculty

Kollektor *m* (-en) (*Elec.*) commutator

Koller·gang *m* **-mühle** *f* edge mill (*or* runner)

kollern 1 *vt* grind (in an edge mill). **2** *vi* rumble, roll; gobble (like a turkey)

Kolli 1 *n* (-s) parcel, package, bale. **2** *pl of* KOLLO

kollidieren 1 *vi*(s) collide. **2** *vi*(h) conflict

Kollidin *n* (-e) collidine

kolligativ *a* colligative

Kollo *n* (-s *or* Kolli) piece of freight, parcel, package, bale

kollodisieren *vt* collodionize

Kollodium *n* collodion. **··deckfarbe** *f* nitrocellulose finish. **-faden** *m* collodion filament. **-fasermasse** *f* collodion cotton. **-haut** *f*, **-häutchen** *n* collodion film. **-lack** *m* nitrocellulose lacquer. **-seide** *f* collodion silk (nitrate rayon). **-überzug** *m* coat (*or* skin) of collodion. **-verfahren** *n* collodion process. **-wolle** *f* collodion cotton, soluble guncotton, pyroxilin

kolloid(al) *a* colloid(al)

Kolloidal·lösung *f* colloidal solution. **-stoff** *m*, **-substanz** *f* colloidal substance, colloid. **-zustand** *m* colloidal state

kolloid·arm *a* poor in colloids. **K–chemie** *f* colloid chemistry. **-chemiker** *m* colloid chemist. **k–chemisch** *a* colloidochemical. **-dispers** *a* colloidally disperse

Kolloidik *f* colloid science

Kolloidität *f* colloidal quality

Kolloid·lösung *f* colloid(al) solution. **-mühle** *f* colloid mill. **-stoff** *m*, **-substanz** *f* colloidal substance, colloid. **-teilchen** *n* colloid(al) particle. **-zustand** *m* colloid(al) state

Köln *n* Cologne. **Kölner** *a* (*no endgs*) (of) Cologne: **K. Braun** Cologne (Cassel, Vandyke) brown; **K. Erde** Cologne earth

kölnisch *a* (of) Cologne: **k—es Wasser** (eau de) Cologne. **K–braun** *n* Cologne (Cassel, Vandyke) brown. **-wasser** *n* (eau de) Cologne

Kolombo·wurzel *f* calumba (root)

Kolonial·macht *f* colonial power. **-waren** *fpl* **1** groceries. **2** colonial products. **-warengeschäft** *n* grocery store. **-zucker** *m* muscovado; cane sugar

Kolonie *f* (-n) colony. **kolonisieren** *vt* colonize

Kolonne *f* (-n) column; gang

Kolonnen·apparat *m* column apparatus. **-füllmaterial** *n* packing for fractionating columns. **-turm** *m* fractionating column (*or* tower); absorption tower

Kolophan *n* colophane

Kolophon·eisenerz *n* pitticite

Kolophonium *n* (*Chem.*) colophony, rosin. **··ester** *m* rosin ester, ester gum. **-lack** *m* rosin varnish. **-seife** *f* rosin soap

Kolophon·säure *f* colophonic acid

Koloquinte *f* (-n) colocynth. **Koloquinten·öl** *n* colocynth oil. **Koloquintin** *n* colocynthin

kolorieren *vt* color. **Kolorimetrie** *f* colorimetry. **kolorimetrisch** *a* colorimetric. **Kolorit** *n* coloring, coloration

Kolostral·milch *f* colostrum

Kolumbo·wurzel *f* calumba (root)

Kolumne *f* (-n) (*Print*) column

Kolza·öl *n* colza (*or* rape) oil

Koman·säure *f* comanic acid

Kombinations·bleiche *f* combination bleach. **-druck** *m* combination printing. **-färbung** *f* combination dyeing; (*Micros.*) differential staining. **-folie** *f* laminated film. **-gerbung** *f* combination tanning. **-schwarz** *n* (*Dye.*) combination black

Kombinatorik *f* (*Math.*) combinatory analysis

kombinieren 1 *vt/i* combine. **2** *vi* surmise. **Kombinier·farbstoff** *m* compound dye (stuff)

Kombo·fett *n* combo fat

Komen·säure *f* comenic acid

komisch *a* comic(al); strange, odd

Komitee *n* (-s) committee

Komma *n* (-s) **1** comma. **2** (*Math.*) (decimal) point: **drei K. fünf** three point five

Kommandit·gesellschaft *f* limited partnership: **K. auf Aktien** stock company with general partners

Kommando· command, control, signal: **-zentrale** *f* control center

kommen* (kam, gekommen) *vi*(s) **1** (*usu*) come. **A** (+*pp*) e.g.: **gelaufen k.** come running. **B** (+*inf*) e.g.: **etwas ansehen k.** come to look at sthg. **C** (+*auf, zu*) come (upon), arrive (at), think (of): **darauf k., daß** ... discover that ... **D** (+*dat*) come, occur (to); come over (sbdy) (*as:* desire); confront, approach: **sie k. dem Laien mit**

zuvielen **Sonderausdrücken** they confront the layman with too many specialized expressions. **E: es kommt zu . . .** it comes to (the point) of: **es kam zu Gewittern (zu Argumenten)** storms (arguments) came up (*or* arose). **F: k. lassen** send for. **2** get (to a place, into a situation); (**+zu** + *inf*) get (to do). **3 G** (*oft* + **an, auf, in, zu**) go, be assigned (to): **das kommt an die Wand** that goes on the wall. **H** (**+auf**) go (to), be allotted (to), be inherited (by): **auf jeden k. zwei** each one gets two; **wieviele Mark k. auf einen Dollar?** how many marks (are there) to a dollar? **4** happen, come about. **5** (**+von**) originate, be derived (from), be due (to): **das kommt davon, daß . . .** that is because . . . ; **daher kommt es, daß . . .** that is why . . . **6** (**+zu**) attain *cf* SICH **I** 1. **7** (**+um**) lose, be deprived of *cf* UMKOMMEN. **8** (**+hinter**) discover the secret (of). **9** (**auf** + *amt of money*) cost, come to. **10** (**+dazu**): **es kommt dazu, daß . . .** it comes to the point where . . . ; **man kommt nicht dazu (es zu tun)** you don't get a chance (to do it) (*or* get round to doing it) *cf* DAZU. **11 A** (**+***phr with nouns*) *See* main entries, e.g. AUSDRUCK (*for* **zum Ausdruck k.**). **B** (**+in, ins** + *verbal noun*) start: **in Fluß k.** start to flow; **ins Sieden k.** start (*or* get to) boiling. **kommend** *p.a* coming, future; next; following, subsequent

Kommentar *n* (-e) commentary. **kommentieren** *vt* comment on, annotate

kommerziell *a* commercial

Kommissar *m* (-e) **1** commissioner. **2** deputy, acting official. **kommissarisch** *a* temporary, provisional, acting, deputy

Kommode *f* (-n) dresser, cabinet

kommunizieren *vi* communicate

kommutieren *vt* commute; (*Elec.*) commutate

Komödie *f* (-n) **1** comedy. **2** act, put-on

kompakt *a* compact; sturdy, rugged

Kompaß *m* (..asse) (ship's) compass

Kompendium *n* (..ien) handbook, manual, compendium

Kompensations·farbe *f* complementary color

Kompensator *m* (-en) **1** compensator. **2**

(*Elec.*) potentiometer. **kompensatorisch** *a* compensatory. **kompensieren** *vt/i* compensate

kompetent *a* **1** authorized, responsible. **2** competent. **Kompetenz** *f* (-en) **1** jurisdiction. **2** competence, expertise

Komplementär·farbe *f* complementary color

Komplement·bildung, -bindung, -fixierung *f* (*Bact.*) complement fixation. **-winkel** *m* complementary angle

Kompletin *n* (*Biol.*) nutrilite

komplett *a*, **komplettieren** *vt* complete

komplex *a* complex, complicated —**k. gebunden** coordinated. **K—bildner** *m* complexing (sequestering, chelating) agent. **-bildung** *f* complex formation, sequestration, chelation. **-ion** *n* complex ion

Komplexität *f* (-en) complexity

komplexometrisch *a* complexometric

kompliziert *a* complicated. **Kompliziertheit** *f* (-en) complication, complexity

Komponente *f* (-n) component

Komposition *f* (-en) (*usu*) composition; blend; arrangement; (*Dye.*) tin composition (solution of tin in aqua regia); hard lead. **Kompositions·metall** *n* composition metal (a copper alloy)

Kompost·dünger *m* compost. **kompostierbar** *a* compost(able). **kompostieren** *vt/i* compost. **Kompostierungs·werk** *n* composting plant

Kompoundierung *f* (-en) (*Elec.*) compounding. **Kompound·öl** *n* (*Petrol.*) compounded oil

kompressibel *a* compressible. **Kompressibilität** *f* compressibility

Kompressions·wärme *f* heat of compression. **-welle** *f* compression wave

Kompressor·öl *n* compressor oil

komprimierbar *a* compressible. **Komprimierbarkeit** *f* compressibility. **komprimieren** *vt* compress

kompromißlos *a* uncompromising. **kompromittieren** *vt* compromise; discredit

konchieren *vt* conche, treat by tumbling

Konchylien *fpl* shells, shellfish

Kondens·apparat *m* condensing apparatus, condenser. **Kondensat** *n* (-e) condensate

Kondensations·apparat *m* condensing apparatus, condenser. **-druck** *m* condensation pressure. **-ergebnis** *n* condensation product. **-gefäß** *n* condensation vessel, receiver. **-harz** *n* condensation (*or* synthetic) resin. **-kammer** *f* condensing chamber. **-kern** *m* condensation nucleus. **-mittel** *n* condensing agent. **-punkt** *m* condensation point. **-raum** *m* condensation space, condensing chamber. **-rohr** *n,* **-röhre** *f* condensation tube; adapter. **-turm** *m* condensing (*or* cooling) tower. **-verlust** *m* loss from condensation. **-vorrichtung** *f* condensation tube; adapter. **-wärme** *f* heat of condensation. **-wasser** *n* condensation water

Kondensator *m* (-en) condenser; (*Opt.*) condensing lens

kondensierbar *a* condensable. **Kondensierbarkeit** *f* condensability. **kondensieren** *vt/i* condense. **Kondensierung** *f* (-en) condensation

Kondens·milch *f* evaporated milk

Kondensor *m* (-en) condenser; (*Opt.*) condensing lens

Kondens·rohr *n* condenser tube. **-topf** *m* condensing pot; steam trap. **-wasser** *n* condensed (condensing, condenser) water, condensate

Konditionier·anstalt *f* (*Tex.*) conditioning plant. **konditionieren** *vt* condition; climatize

Konditor *m* (-en) confectioner. **Konditorei** *f* (-en) cake (*or* confectionery) shop; café. **Konditor·farbe** *f* confectionary color

Konduktometer *n* conductivity meter. **konduktometrisch** *a* conductometric

Konfekt *n* candy

Konfektion *f* 1 ready-made clothing (manufacture). 2 assembly-line manufacture. **konfektionieren** *vt/i* manufacture (by assembly line); compound

Konferenz *f* (-en) conference, meeting. **konferieren** *vi* confer

Konfigurations·beweis *m* proof of configuration

konfiszieren *vt* confiscate

Konfitüre *f* (-n) fruit marmalade (*or* jam)

konform *a* 1 in agreement, conforming: **k. gehen** agree. 2 (+*dat*) in conformity (with). 3 (*Math.*) conformal

Konformations·analyse *f* conformational analysis. **konformativ** *a* conformational

konfus *a* confused

kongelieren *vi*(s) congeal

konglomeratisch *a* conglomerated

Kongo·rot *n* Congo red

kongruent *a* congruent. **kongruieren** *vi* be congruent. **Kongruenz** *f* (-en) congruence

Konidie *f* (-n) conidium

Konifere *f* (-n) conifer. **Koniferen·holz** *n* conifer wood, softwood

König *m* (-e) king; (*Metll.*) regulus. **Königin** *f* (-nen) queen. **-metall** *n* queen's metal, antimony and tin-bearing metal. **königlich** *a* royal, regal

König·reich *n* kingdom, realm. **-salbe** *f* (*Pharm.*) rosin cerate, basilicon ointment

Königs·blau *n* royal blue, cobalt blue, smalt. **-chinarinde** *f* calisaya bark, (yellow) cinchona bark. **-farbe** *f* royal blue. **-format** *n* king size. **-gelb** *n* 1 royal yellow, (synthetic) arsenic sesquisulfide. 2 chrome yellow. 3 massicot. **-grün** *n* Paris green. **-holz** *n* kingwood, violet wood. **-kerze** *f* (*Bot.*) Aaron's rod, common mullein. **-krankheit** *f* scrofula, king's evil. **-palme** *f* royal palm. **-pulver** *n* perfuming powder. **-salbe** *f* basilicon, resin ointment. **-säure** *f,* **-wasser** *n* aqua regia

Koniin *n* conine, coniine

konisch *a* conic(al), coniform; beveled. **Konizität** *f* conicity; (angle of) taper

Konjugations·linie *f* (*Phase Diagrams*) tie line. **konjugiert** *p.a* (*Chem.*) conjugated; (*Math.*) conjugate

Konjunktur *f* (-en) 1 economic (*or* market) situation. 2 boom —**K. haben** be at one's peak. **konjunturell** *a* boom, boom-time. **Konjunktur·tafel** *f* market chart

Konkav *a* concave. **K–gitter** *n* concave grating. **Konkavität** *f* concavity. **Konkav·spiegel** *m* concave mirror

konkordant *a* (*Geol.*) conformable

Konkrement *n* (-e) (*Med.*) concrement, concretion

konkret *a* concrete. **Konkret** *n* (*Perfumery*) concrete

Konkurrent *m* (-en, -en) competitor. **Konkurrenz** *f* (-en) competition. **k–los** *a* without competition, unrivaled; matchless. **konkurrieren** *vi* compete. **konkurrierend** *p.a* 1 competitive. 2 (*Math.*) correlating

Konkurs *m* (-e) bankruptcy: **K. machen** go bankrupt

können* (konnte; *pres sg* kann) **I** *v aux* (+*inf*) **1** be able (to), know how (to); *pres usu* can, *past & sbjc usu* could; **hätte tun k.** could have done; (*inf of verbs of motion oft omitted*): **sie k. nicht hin** (= **hinkommen**) they can't get there. **2** be possible, *pres* may, *past* might *cf* KONNTE. **II** *vt/i* 3 be able to do, know (*esp* a language) *cf* GEKONNT. **III** *vt e.g.* **sie k. nichts für . . .** they are not to blame for . . . *cf* DAFÜRKÖNNEN. **Können** *n* ability, skill. **Könner** *m* (-) expert

Konnossement *n* (-e) bill of lading

konnte could, was able: *past of* KÖNNEN: **es k. durchgeführt werden** it was possible to carry it out. **könnte** could, would be able: *sbjc of* KÖNNEN

Konode *f* (-n) conode, (*Phase Diagrams*) tie line

konphas *a* cophasal, in phase

konsequent *a* 1 consistent, determined, unswerving. **2** (*Geog.*) consequent (rivers). **Konsequenz** *f* (-en) 1 consistency, consistent action. **2** conclusion, inference. **3** (unpleasant) consequence

Konservator *m* (-en) curator; taxidermist

Konserve *f* (-n) 1 preserved (*or* canned) food, preserves. **2** stored blood. **3** (food) can (*or* jar); (frozen food) package. **4** blood bank jar. **5** (*Pharm.*) electuary. **6** recording, recorded image, replay: **Vorlesung aus der K.** recorded lecture

Konserven·büchse, -dose *f* (food) can, tin. **-büchsenlack, -dosenlack** *m* lacquer for food cans (*or* tins). **-fabrik** *f* (food) cannery. **-glas** *n* (preserved food) jar, canning jar. **-öffner** *m* can opener. **-ring** *m* canning (jar) rubber (*or* ring)

konservieren *vt* 1 preserve; retain, keep. **2** can, tap, bottle (food). **3** record (on tape, etc.). **konservierend** *p.a* (disease-)preventive. **Konservierung** *f* preservation, retention; canning; recording, taping

Konservierungs·firnis *m* preserving varnish. **-mittel** *n* preservative. **-verfahren**

n preserving process

konsistent *a* solid, compact, firm, consistent. **Konsistenz** *f* (-en) solidity, compactness, firmness; consistency; body (of paint, etc)

Konsistenz·messer *m* consistometer, viscosimeter. **-probe** *f* consistency test

Konsole *f* (-n) bracket; console, base

konsolidieren *vt* consolidate

Konsortium *n* (..tien) consortium, syndicate

konspergieren *vt* sprinkle, dust; coat (pills) with a conspergent. **Konspergier·mittel** *n* conspergent, dusting (sprinkling, coating) agent

konst., Konst. *abbv* 1 = **konstant** *a* constant; **k—e Härte** permanent hardness. **2** = **Konstante** *f* (-n) constant

konstant·(er)halten *vt* keep constant, maintain. **K—(er)haltung** *f* keeping constant, maintenance

Konstanz *f* constancy; (*Biol.*) constance

konstatieren *vt* note; state; confirm

Konstellation *f* (-eh) 1 situation, circumstances. **2** (*Astron., Chem., Med.*) constellation

Konstit. *abbv* (Konstitution) constitution. **konstituieren** *vt* constitute

Konstitutions·formel *f* structural (*or* constitutional) formula. **-wasser** *n* water of constitution

konstruieren *vt* 1 construct. **2** construe. **3** invent, fabricate. **4** design. **konstruiert** *p.a* artificial, fictitious. **konstruktiv** *a* 1 constructional, structural. **2** constructive

konsultieren *vt* consult

Konsum *m* 1 consumption. **2** (*pl* -e) consumer's cooperative, co-op store. **Konsument** *m* (-en, -en) consumer. **Konsum·genossenschaft** *f* consumers' cooperative, co-op store. **konsumieren** *vt* consume. **Konsum·verein** *m* = -GENOSSENSCHAFT

kontagiös *a* contagious. **Kontagium** *n* (..gien) contagion

Kontakt *m* (-e) 1 contact. **2** catalyst, contact mass. **-·abdruck** *m* contact print(ing). **-belastung** *f* catalyst throughput (*or* charge). **-bolzen** *m* contact bolt (pin, stud.) **-draht** *m* contact wire. **-druck** *m* contact print(ing). **-fläche** *f* contact surface. **-folie** *f* contact paper. **-geber** *m* contactor. **-gift** *n* 1 catalyst poison. 2 contact poison (*esp* insecticide). **-hersteller** *m* contact maker. **-infektion** *f* contagion. **-katalyse** *f* contact catalysis. **-klebstoff** *m* contact adhesive. **-knopf** *m* push button. **-körper** *m* contact body (*or* substance); catalyst. **-leistung** *f* space-time yield (of catalysts). **-linse** *f* contact lens. **-masse** *f* contact mass, catalyst. **-mittel** *n* contact agent, catalyst. **-oberfläche** *f* contact surface. **-ofen** *m* (*Catalysis*) reactor. **-raum** *m* contact (*or* catalyst) space. **-ruß** *m* contact black. **-schale** *f* contact lens. **-schnur** *f* (*Elec.*) extension cord. **-schraube** *f* contact screw. **-stoff** *m*, **-substanz** *f* contact substance, catalyst. **-verfahren** *n* contact (*or* catalytic) process. **-wirkung** *f* catalytic action (*or* effect)

Konter·bande *f* contraband (goods). **-mutter** *f* (*Mach.*) lock nut

kontern *vt/i* counter, contradict

Konter·spülung *f* counterwashing; reverse circulation

Kontinental·sockel *m* continental shelf. **-verschiebung** *f* continental drift

Kontinue·betrieb *m* continuous operation. **-küpe** *f* (*Dye.*) continuous vat. **-verfahren** *n* continuous process

kontinuierlich *a* continuous. **Kontinuität** *f* (-en) continuity

Konto *n* (-s, ..ti *or* ..ten) account. **Kontor** *n* (-e) (business) office; branch. **Kontorist** *m* (-en, -en) office clerk

kontrahieren 1 *vt* & *sich* k. contract. 2 *vt* make (contractual agreements). 3 *vi* enter into a contract

kontrainsulär *a* opposing the action of insulin

kontrapolarisieren *vt* counterpolarize

konträr *a* contrary, opposite

Konstrast·farbe *f* contrast color. **-färbung** *f* contrast coloring. **-füllung** *f* (*X-rays*) opaque meal

kontrastieren *vi* contrast

Kontrast·mahlzeit *f* (*X-rays*) opaque meal. **K–reich** *a* rich in contrast, high-contrast. **K–wirkung** *f* contrast effect

Kontravalenz° *f* contravalence

Kontroll· control, check: **·ablesung** *f* (*In-*

strum.) control (*or* check) reading

Kontrollampe *f* (=**Kontroll·lampe**) pilot (*or* indicator) light

Kontroll·bestimmung *f* control determination

Kontrolle *f* (-n) 1 check(ing), inspection, monitoring. 2 checkpoint. 3 control, supervision

Kontroll·gerät *n* control device, monitor. **-hahn** *m* test (*or* regulating) stop cock

kontrollierbar *a* controllable; verifiable

kontrollieren *vt* 1 check (on), inspect, monitor. 2 *vt* control; supervise

Kontroll·kolben *m* control flask. **-lampe** = KONTROLLAMPE. **-marke** *f* inspector's seal (*or* mark). **-messung** *f* control (*or* check) measurement. **-muster** *n* control (*or* check) sample. **-probe** *f* control sample (*or* test). **-punkt** *m* checkpoint. **-streifen** *m* control chart. **-versuch** *m* control test (*or* experiment)

Kontur *f* (-en) contour, outline. **Konturen·schwarz** *n* (*Dye.*) black for outlines

Konus *m* (-se) cone. **·hahn** *m* conical (*or* taper) cock (*or* plug). **-lehre** *f* (*Math.*) theory of conics. **-mischer** *m* cone mixer. **-rohr** *n* tapered tube. **-spülung** *f* conical rinsing

Konvektions·strom *m* convection current

konventionell *a* conventional

Konvergenz *f* (-en) convergence. **··stelle** *f* convergence limit. **konvergieren** *vi* converge. **konvergierend** *p.a* convergent

Konversions·artikel *m* (*Tex.*) conversion style. **-farbe** *f* conversion color. **-salpeter** *m* converted saltpeter (from $NaNO_2$ + KCl)

Konverter·birne *f* (*Metll.*) converter. **-futter** *n* converter lining. **-stahl** *m* converter (*or* Bessemer) steel. **-werk** *n* converter (*or* Bessemer) mill

Konvert·gas *n* converted gas

konvertibel, konvertierbar *n* convertible

konvertieren *vt* convert. **Konvertierung** *f* (-en) conversion

Konvexität *f* convexity

konz. *abbv* (konzentriert) conc(entrated). **Konz.** *abbv* (Konzentration) conc(entration).

Konzentrat *n* (-e) concentrate

Konzentrations·änderung f change in concentration. **-apparat** m concentrating apparatus, concentrator. **-gefälle** n concentration gradient. **-kette** f (Electrochem.) concentration cell. **-spannung** f (Elec.) concentration potential. **-stein** m (Copper) white metal. **-verhältnis** n ratio of concentration

Konzentrat·trübe f (Metll.) concentrate pulp

konzentrieren vt & sich k. concentrate. **Konzentrierung** f (-en) concentration cf KONZ.

konzentrisch a concentric

Konzept n (-e) 1 first draft, rough copy; notes. 2 plan(s). 3: **aus dem K. bringen** confuse, throw off the track. **-papier** n scratch (or scrap) paper

Konzession f (-en) concession, license

konzipieren 1 vt/i conceive. 2 vt draft, outline

Koordinaten·anfangspunkt m origin of coordinates. **-netz** n coordinate system. **-papier** n coordinate paper. **-zahl** f coordination number (or index)

Koordinations·gitter n coordination lattice. **-lehre** f coordination theory. **-verbindung** f coordination compound. **-wert** m coordination value. **-zahl** f coordination number

koordinativ a coordinate. **koordinieren** coordinate. **Koordinierung** f (-en) coordination

Kopaiva·öl n oil of copaiva. **-säure** f copaivic acid

Kopal·baum m copal tree. **-brecher** m copal disintegrator. **-firnis** m copal varnish. **-gummi** n copal gum. **-harz** n copal (resin)

Kopalin·säure f copalinic acid

Kopalke·rinde f copalche (or copalchi) bark

Kopal·lack m copal varnish. **-luftlack** m airproof copal varnish. **-mattlack** m dull copal varnish. **-öl** n copal oil. **-säure** f copalic acid

Köper m (-) (Tex.) (cotton) twill, drill. **köpern** vt twill

Kopf m (-̈e) head; (fig oft) top, crown — (Metll.) **verlorener K.** blind riser, dead end; **je** (or **pro**) **K.** per capita; **auf den K.**

stellen upset, turn upside down. **-·decke** f scalp. **-drüse** f cephalic gland. **-düngung** f (Fertilizer) top dressing

köpfen vt 1 behead; truncate; cut the top off. 2 vi form a head

Kopf·füß(l)er m cephalopod. **-haut** f scalp. **-hörer** m headphone(s)

·köpfig sfx -headed; (with numbers oft) -person

Kopf·kissen n pillow. **k–lastig** a top-heavy, nose-heavy. **-los** a 1 headless, acephalous. 2 rash, panic-stricken. **K–produkt** n (Distl.) forerunnings, first fraction. **-rechnen** n mental arithmetic. **-ring** m (Zinc) nozzle block. **-rose** f erysipelas. **-salat** m (head) lettuce. **-schimmel** m black (or bread) mold. **-schmerzen** mpl headache. **-schraube** f cap screw (or bolt). **-schuppen** fpl dandruff. **-spalt** m (Lthr.) cheek split. **-steuer** f poll tax. **-titel** m main heading (headline, caption). **-waschmittel** n shampoo. **-waschpulver** n (powdered) hair wash. **-wassersucht** f hydrocephalus. **-weh** n headache. **-zahl** f number of persons. **-zünder** m (Explo.) nose fuse

Kopie f (-n) (usu) copy; (Phot.) (contact) print

Kopier·druckfarbe f copying ink

kopieren vt copy; (Phot.) print

kopier·fähig a copiable. **K–farbe** f copying ink. **-leinwand** f tracing (or copying) cloth. **-mine** f copying lead (for pencils). **-papier** n copying (or duplicating) paper. **-presse** f copying press. **-stift** m copying pencil. **-tinte** f copying ink

Koppel I f (-n) 1 paddock, enclosed pasture. 2 leash; leashed animals. 3 coupler; connecting rod. **II** n (-) (waist) belt. **koppeln** vt 1 tie together; leash; yoke. 2 couple, link up. 3 hyphenate

köppeln vi (Brew.) froth in afterfermentation

Koppel·wirtschaft f rotation of crops

Kopplung f (-en) coupling; linking, linkage

Kopra·fett, -öl n copra (or coconut) oil

Koprolith m coprolite

Koprosterin n coprosterol

Kops·färbung f cop dyeing

kopulieren 1 vt graft. 2 vi unite, copulate

kor. abbv (korrigiert) corrected

Koralle f (-n) coral
korallen·artig a coral-like, coralloid. **K–baum** m coral tree. **-baumöl** n polypody oil. **-erz** n (Min.) coral ore, hepatic cinnabar. **-kalk** m (Geol.) coral rag. **-rot** n coral red. **-wurzel** f (Pharm.) polypody root
Korb m (-̇e) basket, hamper; (bee)hive; creel; (Min.) cage. **--blütler** m composite (flower). **-flasche** f carboy, demijohn. **-warenlack** m basketmaker's varnish. **-weide** f osier
Kordel f (-n) cord, twine. **Kordelei** f cordage. **Kordel·griff** m milled (or knurled) knob. **kordeln, kordieren** vt knurl
Kordit n cordite
Kordofan·gummi n Kordofan gum
Korduan n Cordovan (leather)
Koriander-(samen)öl n coriander (seed) oil
Korinthe f (-n) currant
Kork m cork (as material). **k--ähnlich, -artig** a corky, cork-like, suberose. **K–asbest** m mountain (or rock) cork; cork fossil. **-baum** m cork tree. **-bohrer** m cork borer. **-dichtung** f cork packing. **-eiche** f cork oak (or tree)
korken vt cork; uncork
Korken m (-) cork, stopper. **-zieher** m corkscrew
korkig a corky, cork-like
Kork·isolation f cork insulation. **-kohle** f cork charcoal, burnt cork. **-mehl** n cork powder (dust, meal). **-messer** n cork knife. **-pfropfen** m cork (stopper). **-presse** f cork press. **-säure** f suberic acid. **-schrot** m or n cork meal, granulated cork. **-schwarz** n cork black. **-schwimmer** m cork float **-spund** m cork (stopper). **-staub** m cork dust (or powder). **-stein** m cork board; cork brick. **-stoff** m suberin. **-stopfen, -stöpsel** m cork stopper. **-substanz** f suberin. **-teppich** m linoleum. **-verbindung** f cork connection. **-wachssäure** f (Org. Chem.) ceric acid. **-zieher** m corkscrew
Korn I n 1 grain (in most senses, incl Lthr. Tex., Phot., etc). 2 (-̇er) grain (of sand, etc), granule; kernel, seed; particle. 3 (Agric.) grain, cereal (refers to prevailing local crop, in Germany & Austria usu rye or wheat) —türkisches K. (Indian) corn, maize (K. alone does not refer to this). 4 standard, fineness (of metal); (Assaying) button. 5 (-e) gunsight, bead. **II** m grain liquor, schnaps (from local **Korn** in 3)
Korn·alkohol m grain alcohol. **-art** f type (or variety) of grain. **k–artig** a grainlike, granular. **K–bau** m grain growing. **-bereich** m particle size range. **-bildung** f grain formation; granulation (esp of Sugar); crystallization. **k–blau** a cornflower-blue. **K–blei** n grain lead (for assaying). **-blume** f cornflower, bluebottle. **-blumenblau** n cornflower blue. **-brand** m grain blight (or smut). **-branntwein** m grain liquor (or spirits), schnaps. **-brennerei** f grain distillery (or distilling). **-bürste** f (Assaying) button brush
Körnchen n (-) granule, little kernel, seed, etc
Korn·dichte f closeness of grain. **-eisen** n iron of granular fracture
Kornel·kirsche f, **Kornelle** f (-n) cornelian cherry
körneln vt granulate cf GEKÖRNELT. **Körnelung** f granulation
körnen vt granulate; (Lthr.) grain cf GEKÖRN(EL)T
Körner I m (-) center punch. **II** pl of **KORN** n; kermes; shot. **--asant** m asafetida in grains. **-form** f granular form (or shape). **-frucht** f cereal, grain. **-futter** n grain (feed). **-lack** m seed lac. **-leder** n grained leather. **k–reich** a grainy, full of grain. **K–spitze** f (Mach.) lathe center. **-zinn** n grain (or granular) tin
Körne·zähler m grain counter
Korn·feinheit f fineness of grain; particle size. **k–förmig** a granular, grain-shaped, in the form of grains. **K–gefüge** n grain structure. **-grenze** f grain boundary. **-grenzenkorrosion** f intergranular corrosion. **-grenzenriß** m intercrystalline crack. **-größe** f grain size. **-größenverteilung** f particle size distribution
Körn·haus n (Explo.) corning (or granulating) house
körnig a granular, granulated; grainy; gritty; grained; showing a grain; (Malt) brittle; (as sfx) -grained. **Körnigkeit** f

granularity, graininess. **Körnigkeits·messer** *m* granulometer
körnigkristallinisch *a* granular-crystalline
Korn·käfer *m* grain weevil. **-kammer** *f* granary. **-klasse** *f* grain size category (*or* class). **-kluft** *f* assayer's tongs. **-kochen** *n* (*Sugar*) boiling to grain. **-kupfer** *n* granulated copper. **-leder** *n* grain leather. **-mutter** *f* ergot. **-prüfer** *m* grain tester. **-puddeln** *n* puddling of fine-grained iron. **-pulver** *n* granular powder. **-schnaps** *m* grain liquor, whisky. **-sieb** *n* granulating sieve; grain sieve. **-spanne** *f* particle size range. **-stahl** *m* granulated steel. **-struktur** *f* grain (*or* granular) structure
Kornung *f* (-en) grain, granulation
Körnung *f* granulation, graining; pitting
Körnungs·aufbau *m* particle size distributing; grading. **-bereich** *m* particle size range
Korn·verfeinerung *f* grain refining. **-vergröberung, -vergrößerung** *f* coarsening (enlargment) of grain. **-waage** *f* 1 grain scales. 2 button (*or* assay) balance. **-wachstum** *n* grain (*or* crystalline) growth. **-wurm** *m* grain weevil. **-zange** *f* assayer's tongs. **-zinn** *n* (*Metll.*) grain tin. **-zucker** *m* granulated sugar. **-zusammensetzung** *f* grain composition
Körper *m* (-) (*genl, incl Chem., Phys., Anat.*) body; (*Chem.*) substance, compound; (*Algebra*) (number) field, domain; (*in compds oft*) bodily, corpor(e)al, physical, solid. **··abtastung** *f* body scan. **-bau** *m* body (*or* bodily) structure, build. **-beschaffenheit** *f* constitution
Körperchen *n* (-) corpuscle, particle
Körper· body, corpor(e)al, physical: **-farbe** *f* body color, pigment(ation). **-flüssigkeit** *f* body fluid. **-fülle** *f* corpulence. **-gehalt** *m* solid content (of lacquers, etc). **-gewicht** *n* body weight. **-größe** *f* stature. **-gruppe** *f* group of substances, compound group. **-klasse** *f* class of substances (*or* of compounds)
körperlich *a* bodily; corpor(e)al; physical; somatic; material; solid
körper· body, corpor(e)al, physical: **-los** *a* incorporeal, non-material. **K−maß** *n* 1

cubic measure. 2 (*usu pl*) body measurement(s). **-masse** *f* body weight. **-oberfläche** *f* body surface. **-pflege** *f* 1 personal hygiene. 2 physical culture. **-pflegemittel** *n* cosmetic, toilet article. **-schaden** *m* bodily harm; physical defect (*or* disability)
Körperschaft *f* (-en) corporation, (corporate) body
Körper· body, corpor(e)al, physical: **-stoff** *m* (organic) matter. **-teil** *m* part of the body. **-teilchen** *n* particle, (*specif*) molecule. **-verbindung** *f* (*Elec.*) ground (connection). **-verletzung** *f* physical injury. **-wärme** *f* body heat. **-zusammensetzung** *f* composition of the body
Korpuskel *n* (-n) corpuscle
korr. *abbv* (korrigiert) corrected
Korrasion *f* (-en) (*Geol.*) wind erosion, corrasion
Korrektions·mittel, Korrektiv·mittel *n* corrective (agent), corrigent. **Korrektor** *m* (-en) proofreader. **Korrektur** *f* (-en) 1 correction. 2 (printer's) proof: **K. lesen** read proof. **··abzug, -bogen** *m* proof sheet
Korrelat *n* (-e) correlative. **korrelieren** *vt* correlate
Korrespondenz·prinzip *n* (*Spect.*) correspondence principle
korrigieren *vt* correct; soften (water)
korrodierbar *a* corrodible. **Korrodierbarkeit** *f* corrodibility. **korrodieren** *vt* corrode. **korrodierend** *p.a* (*oft*) corrosive
korrosions·beständig *a* corrosion-resistant. **K−beständigkeit** *f* corrosion resistance. **-bildner** *m* corrosive agent. **-element** *n* corrosion cell. **k−empfindlich** *a* sensitive to corrosion. **K−ermüdung** *f* corrosion fatigue. **k−fest** *a* corrosion-resistant. **K−festigkeit** *f* corrosion resistance. **k−fördernd** *a* corrosion-promoting. **-frei** *a* corrosion-free, non-corroding. **-hemmend** *a* corrosion-inhibiting. **K−mittel** *n* corrosive. **-prüfung** *f* corrosion testing. **-schutz** *m* protection against corrosion. **k−schützend** *a* corrosion-resistant, anti-corrosive. **K−wirkung** *f* corrosive effect
Korund *m* (-e) corundum
Koschenille *f* (-n) cochineal

Kosekans *m* (-), **Kosekante** *f* (-n) cosecant
Kosinus *m* (-) cosine
Kosmetik *f* 1 cosmetics, beauty culture. 2 care, cleaning, grooming; (*esp in compds.*): **Linoleumkosmetik** linoleum care (*or* cleaning). **innere K.** (wholesome) diet. **Kosmetiker** *m* (-), **Kosmetikerin** *f* (-nen) cosmetician, beautician. **Kosmetikum** *n* (..ika) cosmetic, beauty aid. **kosmetisch** *a* 1 cosmetic. 2 well-groomed, cultivated; affected
kosmisch *a* cosmic
Koso·blüten *fpl* (*Pharm.*) cusso flowers
Kost *f* diet; food, fare; board *Cf* KOSTEN
kostbar *a* costly expensive; valuable, precious. **Kostbarkeit** *f* (-en) 1 great value, (*or* cost), preciousness. 2 valuable (*or* precious) object, treasure
kosten *vt* I taste; get a taste of, experience. II cost, require, take (money, effort, etc)
Kosten *pl* cost, expense(s): **auf K. von..** at the expense of . . . **--anschlag** *m* cost estimate. **-aufwand** *m* expenditure, expense. **k–frei, -los** *a* free (of charge)
köstlich *a* 1 delicious. 2 exquisite. 3 delightful
kostspielig *a* expensive
Kot *m* 1 feces, excrement. 2 mud, sludge. **··abgang** *m* defecation, bowel movement
Kotangens *m* (-), **Kotangente** *f* (-n) cotangent
kot·artig *a* fecal, excremental. **-ausführend, -ausleerend** *a* laxative, purgative, cathartic. **K–ausleerung** *f* fecal evacuation. **-bad** *n* dung bath. **-beize** *f* dung bate
Kote *f* (-n) elevation
Kötel *n* (-) dung, dropping(s)
Kotelett *n* (-en) (*Meat*) chop, cutlet
koten *vt/i* defecate. **Kot·geruch** *m* fecal odor. **kotig** *a* 1 fecal, stercoraceous. 2 muddy
kotonisieren *vt* 1 cottonize. 2 degum (silk)
Koto·rinde *f* coto bark
Kot·porphyrin *n* coproporphyrin. **-stein** *m* fecal concretion
kottonisieren *vt* cottonize
Kötzer *m* (-) (*Tex.*) cop
Kovalenz° *f* covalence
Kovolumen° *n* covolume
kp. *abbv* (Kilopond) kilogram force, kgf

Kp. *abbv* (Kochpunkt) boiling point
KP. *abbv* (Kugelpackung) (*Cryst.*) close packing
kpl. *abbv* (komplett) complete, full
Krabbe *f* (-n) crab; shrimp, prawn
krabbeln 1 *vt* tickle; itch. 2 *vi*(s) crawl
krabben *vt* (*Tex.*) crab
Krach *m* (-e) 1 noise; crash, bang. 2 quarrel, row. **krachen** *vi*(h/s) crack, creak; crash, bang
Krach·griff *m* (*Tex.*) scroop
krächzen *vi* croak
Krack·benzin *n* cracked gasoline
kracken *vt* (*Petrol.*) crack
Krack·gas *n* gas from cracking process. **-prozeß** *m* cracking (process)
Krackung *f* cracking
Krack·verfahren *n* cracking process
Krad *n* (-er) (*short for* KRAFTRAD) motorcycle
kraft *prep* (*genit*) by virtue (of)
Kraft *f* (-e) 1 force, power: **in K.** in force; **in K. treten** come into force, become effective; **in K. setzen** put in force; **außer K.** invalid, no longer in effect; **außer K. setzen** make invalid, repeal; **außer K. treten** go out of force, become invalid, be repealed. 2 strength: **bei K--en sein** have one's full strength, be in good health; **zu K--en kommen** gain (one's) full strength. 3 worker, employee; *pl off* troops, forces; (*esp in compds*): **Laborkräfte gesucht** lab assistants (*or* lab help) wanted. **··anlage** *f* power plant (*or* station). **-antrieb** *m* (*Mach.*) power drive. **-aufwand** *m* expenditure of energy (effort, power, force). **-äußerung** *f* manifestation of force (power, energy). **-bedarf** *m* power requirements (*or* demand). **k–betätigt** *a* power-driven. **K–brühe** *f* meat (*usu* beef) broth
Kräfte·bedarf *m* labor (*or* personnel) requirements
Kraft·eck *n* polygon of forces
kräfte·frei *a* force-free. **K–gleichgewicht** *n* equilibrium of forces
Kraft·einheit *f* unit of force
Kräfte· (*Phys.*): **-paar** *n* force couple. **-parallelogramm** *n* parallelogram of forces
Kraft·erzeugung *f* power production
Kräfte· (*Phys.*): **-spiel** *n* interplay of forces.

-**vieleck** n polygon of forces. -**zerlegung** f resolution of forces. -**zusammensetzung** f composition of forces

Kraft·fahrer m motorist, (car) driver. -**fahrzeug** n motor vehicle. -**feld** n (*Phys.*) force field. -**futter(mittel)** n concentrated feed. -**gas** n power (fuel, producer) gas. -**gaserzeuger** m gas producer

kräftig a strong, sturdy; powerful, firm; energetic, vigorous; substantial; (*Hides*) plump; (*Dyes*) full, heavy. **kräftigen** 1 vt strengthen, invigorate. 2: **sich k.** gain strength, grow strong

Kraft·konstante f (*Spect.*) force constant. -**lastwagen** m motor truck. -**lehre** f dynamics. -**leitung** f (*Elec.*) power line. -**linie** f line of force. **k–los** a powerless, feeble; (nutritionally) poor; invalid. **K–maschine** f engine, motor. -**mehl** n starch, amylum. -**messer** m dynamometer. -**mittel** n 1 forceful means. 2 powerful remedy, tonic. -**(pack)papier** n kraft paper, heavy wrapping paper. -**papierstoff** m kraft pulp. -**probe** f test of strength. -**quelle** f power source. -**rad** n motorcycle. -**rohr** n, -**röhre** f tube of force. -**sammler** m storage battery, accumulator. -**sitz** m (*Mach.*) force fit. -**speicher** m storage battery, accumulator. -**spiritus** m fuel alcohol. -**stoff** m 1 motor fuel, (*specif*) gasoline, diesel oil. 2 (*Paper*) kraft pulp. -**stoffwirtschaft** f fuel economy. -**strom** m electric power, power current. -**übersetzung** f mechanical advantage (of levers). -**übertragung** f power transmission. -**verbrauch** m power consumption. -**verkehr** m motor traffic. -**verstärker** m power amplifier. **k–voll** a powerful, strong, vigorous. **K–wagen** m motor vehicle, automobile. -**wechsel** m energy exchange. -**werk** n power station. -**wirkung** f (inter)action of forces. **k–wirtschaftlich** a power-saving, energy-conserving. **K–zentrale** f (central) power station

Kragen m (-) collar; neck; flange. -**·kolben** m flanged flask

Krähe f (-n) crow. **Krähen·auge** n nux vomica

Krähl m (-e) rake, (*Metll.*) rabble. -**·ofen** m rabble furnace

Krain n (*Geog.*) Carniola

Kralle f (-n) claw, talon; pl oft clutches

Kram m junk, claptrap

Krampe f (-n) (U-shaped) staple; cramp iron

Krampf m (¨e) cramp, spasm, convulsion. -**·ader** f varicose vein. **k–ähnlich**, -**artig** a cramplike, spasmodic, convulsive. **K–arznei** f antipasmodic (medication)

krampfen 1 vt dig (fingers into). 2: **sich k.** clench, clutch (at); go into cramps (*or* spasms)

Krampf·gift n convulsive poison

krampfhaft a convulsive, spasmodic; desperate

Krampf·mittel n anticonvulsive, antispasmodic (agent). **k–stillend** a sedative, antispasmodic. **K–wurzel** f valerian root

Kran m (-e) (*Mach.*) 1 crane, hoist. 2 faucet.

kranen vt hoist (*or* lift) with a crane

Kranewit·beere f juniper berry

Kranich m (-e) (*Zool.*) crane

krank a (¨er, ¨ste) sick, ill; ailing; injured (limb, organ). **Kranke** m, f (*adj endgs*) patient, invalid. **Kränke** f sickness, illness; epilepsy

kränkeln vi be ailing, be in poor health

kranken vi (**an**) be afflicted (with), suffer (from)

kränken 1 vt hurt, offend. 2: **sich k.** feel hurt; grieve. *Cf* KRÄNKEND

Kranken·anstalt f hospital. -**·bett** n sickbed

kränkend p.a hurtful, offensive cf KRÄNKEN

Kranken·haus n hospital. -**kasse** f health insurance (plan). -**kost** f sick (*or* invalid's) diet. -**pflege** f nursing. -**pfleger** m, -**pflegerin** f, -**schwester** f nurse. -**wagen** m ambulance. -**zimmer** n sick room

krankhaft a diseased, morbid, pathological

Krankheit f (-en) sickness, illness; disease

Krankheits·anfall m attack of illness. **k–erregend** a disease-causing, pathogenic. **K–erreger** m disease-causing (*or* pathogenic) agent, pathogen; cause of illness. -**erscheinung** f pathological phenomenon. **k–erzeugend** a disease-producing, pathogenic. -**halber** adv on account of illness. **K–keim** m disease germ. -**kunde,**

-lehre f pathology. **-stoff** m morbid (or disease-producing) matter. **-urlaub** m sick leave. **-verlauf, -vorgang** m course of a disease. **-widerstand** m resistance to disease. **k–widerständig** a disease-resistant

kränklich a sickly, in poor health

Kränkung f (-en) offense, insult

Kranz m (ᵉe) wreath; ring; rim (esp of wheels); circle (also = group); (Astron., Opt.) corona; lid (of a furnace); (Anat. in compds) coronal, coronary. **--arterie** f coronary artery. **-brenner** m ring (or crown) burner

Krapp m madder. **--artikel** m (Tex.) madder style. **-bad** n madder bath

krappen vt (Dye.) madder

Krapp-farbe f madder color (or dye). **-färben** n madder dyeing. **-farbstoff** m alizarin. **-gelb** n madder (or alizarin) yellow, xanthin. **-lack** m madder lake. **-rot** n alizarin, madder red

kraß a crass, gross; blatant; blunt

Krater-bildung f crater formation; pitting

Kratte(n), Krätten m (..tten) basket

Kratz m (-e) scratch. **-band** n scraper conveyor. **-beere** f blackberry

Krätz-blei n slag lead

Kratz-bürste f 1 wire (scraping) brush. 2 spitfire, bristly person

Kratze f (-n) scraper; (Metll.) rake, rabble; (Tex.) card. 2 scrapings

Krätze f (-n) 1 waste metal, metal cuttings (sweepings, scrapings), dross. 2 (Med.) itch, scabies

Kratz-eisen n scraping iron, scraper

kratzen 1 vt/i scratch, scrape. 2 vt card, tease (wool); (Metll.) rake, rabble; bother, annoy; irritate; tickle (also = please). 3 vi taste harsh. 4: **sich k.** scratch oneself. **kratzend** p.a harsh, irritating

Kratzer m (-) 1 scratch (wound). 2 scraper; rake, (Metll.) rabble

Krätzer m (-) 1 scraper; rake, (Metll.) rabble. 2 Tyrolean red wine; harsh (sour, tart) wine

kratz-fest a scratch-resistant. **K–festigkeit** f scratch resistance

Krätz-frischen n (Metll.) refining of waste

kratzig a scratchy, irritating; irritable

krätzig a itchy; scabby, scabious; (Glass) fibrous

Krätz-kupfer n copper from waste. **-messing** n brass cuttings. **-wurzel** f white hellebore (root)

kraus a 1 curly, curled; wavy. 2 wrinkled, creased. 3 confused; tangled; intricate, convoluted

Krause-gummi n crepe rubber

kräuseln 1 vt & **sich k.** curl; wrinkle, crinkle; ripple; pucker. 2 vt mill (coins)

Kräusel-welle f ripple

Krause-minze f curled mint; spearmint. **-minzöl** n spearmint oil

krausen vt & **sich k.** wrinkle, crinkle, crumple, crisp

Kräusen fpl (Brew.) (rocky) head

Kraus-gummi n crepe rubber. **-putz** m rough plaster (or cast). **-tabak** m shag

Kraut n (ᵉer) 1 herb. 2 greens, tops (of root vegetables & tubers) -ins **K. schießen** run to leaves, (fig) run wild. 3 cabbage; sauerkraut. **k--artig** a herblike, herbaceous

Kräuter- herb(al), aromatic: **-essig** m aromatic vinegar. **k–fressend** a herbivorous. **K–mittel** n herb (or vegetable) remedy. **-pollen** m herbaceous (or nontree) pollen. **-salbe** f herbal ointment. **-salz** n vegetable salt. **-tee** m herb tea. **-wein** m herb-seasoned (or medicated) wine. **-zucker** m (Pharm.) conserve, confection

Kraut-salat m cabbage salad, cole slaw

Kreatin n creatin(e)

Kreatur f (-en) creature

Krebs m (-e) 1 crustacean, crayfish, crab. 2 (Med.) cancer. 3 canker, wart. 4 grain, hard particle (in clay). 5 (Ore) knot

krebs-artig a crab-like, crustaceous; cancerous; cankerous. **K–behandlung** f cancer treatment. **k–erregend** a cancer-causing, carcinogenic. **K–erreger** m carcinogen, cause of cancer. **k–erzeugend** a carcinogenic. **K–geschwulst** f cancerous tumor, carcinoma. **-gewebe** n cancerous tissue

krebsig a cancerous; cankerous

Krebs-kohlenwasserstoff m carcinogenic hydrocarbon. **-milch** f (Med.) cancer juice. **-tier** *n* crustacean. **k–verdächtig** *a* suspected of being cancerous (*or* carcinogenic)

Kredit *m* (-e) credit; loan. **k–-fähig** *a* credit-worthy; solvent. **Kreditiv** *n* (-e) 1 credentials. 2 letter of credit

Kreide *f* (-n) 1 chalk. 2 (*Geol.*) Cretaceous. **k–-artig** *a* chalky, cretaceous. **K–bad** *n* chalk bath. **-boden** *m* chalky soil. **-flotte** *f* chalk liquor. **-grund** *m* chalk ground. **-gur** *f* agaric mineral (earthy calcium carbonate), lithomarge. **k–haltig** *a* containing chalk, cretaceous. **K–mehl** *n* powdered chalk

kreiden 1 *vt* (coat with) chalk. 2 *vi* (*Paint*) become chalky

Kreide·papier *n* enameled (*or* art) paper. **-paste** *f* chalk-and-glue cement. **-pulver** *n* powdered chalk, chalk dust. **-stein** *m* chalkstone. **-stift** *m* stick of chalk; crayon. **-strich** *m* chalk line. **k–weiß** *a* chalk-white. **K–weiß** *n* whiting. **-zeichnung** *f* chalk (*or* crayon) drawing

kreidig *a* chalky, cretaceous

Kreis *m* (-e) 1 circle, circuit; sphere, range. 2 administrative district, (*appx*) county. **--abschnitt** *m* segment (of a circle). **k–artig** *a* circular. **K–ausschnitt** *m* sector. **-bahn** *f* orbit, circular path. **-bewegung** *f* circular motion, rotation. **-blattschreiber** *m* circular chart recorder. **-blende** *f* (*Phot.*) iris diaphragm. **-bogen** *m* 1 arc of a circle. 2 circular arch. **-büschel** *n* (*Math.*) pencil of circles

kreischen *vt/i* scream, screech, squeal. **Kreischen** *n* screaming, etc *cf* KREIS/CHEN

Kreis/chen *n* (-) little circle *cf* KREIS

kreischend *p.a* shrill, strident *cf* KREISCHEN

Kreisel *m* (-) (spinning) top; gyroscope. **--bewegung** *f* gyroscopic motion, gyration. **-brecher** *m* rotary crusher. **-elektron** *n* spinning electron. **k–förmig** *a* top-shaped, turbinate. **K–gang** *m* circular motion, rotation. **-gebläse** *n* turboblower. **-kompaß** *m* gyrocompass. **-mischer** *m* rotary mixer. **-molekül** *n* spinning molecule

kreiseln *vi* 1 spin (like a top), rotate (with

precession). 2 spin a (toy) top

Kreisel·pumpe *f* rotary (*or* centrifugal) pump. **-rad** *n* rotor, impeller (of turbines, pumps). **-verdichter** *m* turbocompressor

kreisen *vi* 1 circle, orbit, revolve. 2 spin, rotate. 3 circulate

Kreis·fläche *a* circular area. **k–förmig** *a* circular, round. **K–frequenz** *f* angular frequency (*or* velocity). **-gang** *m* revolution, circular motion. **-grad** *m* angular degree. **-kolben** *m* rotary piston. **-lauf** *m* cycle; circuit; circulation; rotation. **-laufanregungsmittel** *n* analeptic. **k–laufend** *a* circulatory

Kreislauf·gas *n* recycled gas. **-mahlung** *f* closed-circuit grinding. **k–müde** *a* suffering from poor circulation

Kreis·linie *f* arc, circular line; circumference. **-messer** *n* circular knife. **-ordnung** *f* district regulation. **-prozeß** *m* cyclic process, cycle. **-pumpe** *f* circulating pump. **-ring** *m* annulus. **k–rund** *a* circular. **K–säge** *f* circular saw. **-schicht** *f* 1 circular layer. 2 annual ring (of a tree). **k–ständig** *a* cyclic. **K–umfang** *m* circumference, periphery. **-umlauf** *m* circulation, cycle. **-verfahren** *n* cyclic process. **-zylinder** *m* circular cylinder

Krem *f* (-e *or* -s) cream = CREME. **--farbe** *f* cream color

Kremnitzer Weiß *n* Kremnitz white

Krempe *f* (-n) brim; flange

Krempel *m* (-) 1 (*Tex.*) card; carding machine. 2 junk, stuff = KRAM. **krempeln** *vt* 1 (*Tex.*) card. 2 roll (*or* turn) up (sleeves, etc)

Kremser Weiß *n* Kremnitz white

Kren *m* horseradish

Kreosot·lack *m* creosote. **-öl** *n* creosot oil

krepieren I *vt* annoy. II *vi*(s) 1 burst, explode. 2 die, croak

krepitieren *vi* crepitate, crackle

kreponieren *vt* crepe

Krepon·stoff *m* crepon (material)

Krepp *m* (-e) crepe. **Kreppapier** *n* crepe paper. **kreppen** *vt* crepe, crinkle

Krepp·flüssigkeit *f* creping liquor. **-kautschuk** *m* crepe rubber. **-papier** = KREPPAPIER

Kresol *n* cresol. **--harz** *n* cresylic resin.

-natron *n* sodium cresylate. **-säure** *f* cresylic acid, cresol. **-seife** *f* cresol soap

Kresorcin *n* cresorcinol

Kresotin·säure *f* cresot(in)ic acid

kreß *a*, **Kreß** *n* orange (color)

Kresse *f* (-n) cress

Kresyl·harz *n* cresylic resin. **-säure** *f* cresylic acid

kretazisch *a* (*Geol.*) cretaceous

kreuz *adv:* **k. und quer** in all directions, crisscross. **Kreuz** *n* (-e) 1 cross; crossing; **übers K. legen** lay crosswise. 2 affliction, cross to bear. 3 small of the back, lower back

Kreuz· cross; lower back: **-beere** *f* buckthorn (berry). **-befruchtung** *f* cross fertilization. **-bein** *n* (*Anat.*) sacrum. **-bestäubung** *f* cross-pollination. **-blech** *n* heavy-gauge tinplate (over 0.35 mm thick). **-blume** *f* milkwort. **-blütler** *m* (*Bot.*) crucifer. **-dorn** *m* buckthorn

kreuzen 1 *vt* cross; cross-breed. 2 *vi*(h) & **sich k.** cross (paths), intersect; clash. 3 *vi*(s) cruise; tack. **Kreuzer** *m* (-) 1 cruiser. 2 (former small coin)

kreuz· cross; lower back: **-förmig** *a* cross-shaped, cruciform. **K–gelenk** *n* universal joint. **k–geschichtet** *a* cross-bedded, cross-layered. **K–gitter** *n* cross grating; surface lattice. **-hahn** *m* four-way stopcock. **-kopf** *m* (*Mach.*) crosshead. **-kraut** *n* groundsel. **-kümmel** *m* cumin. **-schichtung** *f* (*Geol.*) cross-bedding. **-schlagmühle** *f* hammer mill. **-schlitz** *m* crossed slot(s); (*Screws*) Phillips head. **-schmerzen** *mpl* lower back pain(s), lumbago. **k–schraffiert** *a* cross-hatched. **K–spule** *f* 1 (*Tex.*) cross-wound bobbin, cone. 2 (*Elec.*) cross coil. **-stein** *m* cross stone, (*Min.*) chiastolite, harmotome, staurolite. **-stück** *n* (four-way) cross piece. **-tisch** *m* (*Micros.*) mechanical stage

Kreuzung *f* (-en) 1 crossing, intersection. 2 hybridization; hybrid, cross-breed. **Kreuzungs·punkt** *m* (point of) intersection

Kreuz·verweisung *f* cross-reference. **-weg** *m* crossroad(s). **k–weise** *a* crosswise, transverse. **K–winkel** *m* T-square. **-zug** *m* crusade

kribbeln *vt/i:* **es kribbelt 1** it is swarming (*or* crawling) (with insects, etc). 2 (*dat or accus*) make itch (*or* tingle)

kriechen* (kroch, gekrochen) *vi*(s) creep, crawl; grovel

kriech·fest *a* creep-resistant. **K–festigkeit** *f* creep resistance (*or* strength). **-grenze** *f* 1 creep limit. 2 = **-grenzenspannung** *f* limiting creep stress. **-pflanze** *f* creeping plant, creeper. **-spannung** *f* (*Elec.*) leakage (*or* tracking) voltage. **-strom** *m* (*Elec.*) leakage (*or* tracking) current. **-tier** *n* reptile

Krieg *m* (-e) war, warfare

kriegen 1 *vt* get = BEKOMMEN. 2 *vi* make war

Krieg·führung *f* warfare

Kriegs·ersatzstoff *m* wartime substitute. **-gerät** *n* (piece of) war equipment. **-kunde** *f* military science. **-marine** *f* navy. **-mittel** *n* warfare resource

krimpen *vt* & *vi*(s) (*Tex.*) shrink

krimp·fähig *a* shrinkable. **K–fähigkeit** *f* shrinking property; (*Wool*) felting property. **k–frei** *a* nonshrinking

Kringel *m* (-) ring, curl(icue); cracknel

Krippe *f* (-n) crib; manger, feed trough

Krise *f* (-n) crisis. **krisenhaft** *a* critical

krispelig *a* (*Lthr.*) grained, pebbled. **krispeln** *vt* (*Lthr.*) grain; pebble, bruise (flesh side); board (grain side)

krist. *abbv* (kristallisiert, kristallinisch); **Krist.** *abbv* (Kristallisation, Kristallographie) cryst. (crystalline, etc)

Kristall 1 *m* (-e) (*Cryst.*) crystal. 2 *n* crystal (glass, ware). **-abscheidung** *f* separation of crystals. **-achse** *f* crystal(lographic) axis

Kristallack *m* (=**Kristall·lack**) crystal lacquer

kristall·ähnlich *a* crystal-like, crystalloid. **K–anschuß** *m* crop of crystals; crystalline growth. **k–artig** *a* crystal-like, crystalline. **K–aufbau** *m* crystal structure. **-ausscheidung** *f* separation of crystals. **-bau** *m* crystal structure. **-baufehler** *m* defect in crystal structure. **-baustein** *m* unit of crystal structure, crystal unit. **-benzol** *n* benzene of crystallization. **-bereich** *m* crystal domain. **-bildung** *f* crystal formation, crystallization; (*Sugar*) granulation. **-bildungs-punkt** *m* (*Sugar*) granulating pitch. **-brei** *m* (moist) crystal mass, crystal sludge (mush, slurry). **-chemie** *f* crystal chemistry

Kriställchen *n* little crystal

Kristall·chloroform *n* chloroform of crystallization. **-druse** *f* crystal druse (*or* cluster). **-ebene** *f* crystal plane (*or* face). **-ecke** *f* (solid) crystal angle. **k–elektrisch** *a* piezoelectric

kristallen *a* (of) crystal, crystalline. **Kristaller** *m* (-) crystallizer

Kristall·fabrik *f* glassworks, glass factory. **-fehler** *m* lattice defect. **-feuchtigkeit** *f* crystalline (*or* vitreous) humor. **-fläche** *f* crystal face. **k–förmig** *a* crystalline, crystalloid. **K–gestalt** *f* crystal form. **-gitter** *n* crystal lattice (*or* grating). **-glasur** *f* (*Ceram.*) crystalline glaze. **-grenze** *f* crystal boundary. **k–haltig** *a* crystalliferous, containing crystals. **K–haufwerk** *n* crystal aggregate. **-haut** *f* crystalline crust. **-häutchen** *n* crystalline film. **k–hell** *a* crystal-clear, transparent

Kristallier·schale *f* crystallizing dish

kristallin(isch) *a* crystalline. **Kristallinität** *f* (-en) crystallinity

Kristallinse *f* (=**Kristall·linse**) (*Anat.*) crystalline lens

Kristallisations·bassin *n* crystallizing basin (tank, cistern). **-differentiation** *f* fractional crystallization. **k–fähig** *a* crystallizable. **K–fähigkeit** *f* crystallizability. **k–freudig** *a* readily crystallizable. **K–gefäß** *n* crystallizing vessel, crystallizer **-kern** *m*, **-korn** *n* crystal nucleus. **-schale** *f* crystallizing dish **-vermögen** *n* crystallizing power. **-wärme** *f* heat of crystallization. **-wasser** *n* water of crystallization.

Kristallisator *m* (-en) 1 crystallizer; (for paraffin) chiller. 2 mineralizer

kristallisch *a* crystalline

kristallisierbar *a* crystallizable. **Kristallisierbarkeit** *f* crystallizability

Kristallisier·behälter *m* crystallizing receptacle (*or* tank)

kristallisieren *vt/i* crystallize

Kristallisier·gefäß *n* crystallizing vessel,

crystallizer. **-schale** *f* crystallizing dish

Kristallit *m* (-e) crystallite. **Kristalliten·gefüge** *n* crystallite structure

Kristall·kante *f* crystal edge. **-keim** *m* seed crystal. **-kern** *m* nucleus of crystallization. **-korn** *n* crystal grain. **-kunde** *f* crystallography. **k–kundlich** *a* crystallographic. **K–lack** *m* = KRISTALLACK. **-mehl** *n* crystal powder (*or* sand). **-messung** *f* crystallometry. **-nädelchen** *n* crystal needle, acicular crystal. **-oberfläche** *f* crystal surface

kristallogen(isch) *a* crystallogenic

Kristallograph *m* (-en, -en) crystallographer. **Kristallographie** *f* crystallography. **kristallographisch** *a* crystallographic

Kristall·öl *n* white spirit(s)

Kristallose *f* saccharin

Kristall·pulver *n* crystal powder. **-quarz** *n* quartz crystal. **-richtröhre** *f* (*Elec.*) crystal diode. **-säure** *f* crystalline fuming sulfuric acid (containing 40–60% free SO_3). **-schnitt** *m* crystal section (*or* cut). **-schwingung** *f* crystal vibration. **-soda** *f* crystal(lized) soda ($Na_2CO_3 \cdot 10H_2O$). **-stein** *m* rock crystal. **-verstärker** *m* transistor. **-wachstum** *n* crystal growth. **-waren** *fpl* crystal (glass)ware. **-wasser** *n* water of crystallization. **k–wasserfrei** *a* anhydrous. **-wasserhaltig** *a* containing crystal water. **K–winkel** *m* crystal angle. **-ziehen** *n* crystal growing. **-zucker** *m* granulated sugar, refined sugar in crystals

Kristfm. *abbv* (Kristallform) crystal(line) form

Kriterium *n* (..rien) criterion

Kritik *f* (-en) 1 criticism. 2 critique, review. 3 the critics. **Kritiker** *m* (-) critic. **kritisch** *a* critical. **kritisieren** *vt/i* criticize; review

kritteln *vi* carp, criticize

kritzeln *vt/i* scribble; scratch

Kroatien *n* Croatia. **kroatisch** *a* Croatian

kroch crept, crawled: *past of* KRIECHEN

Krokodil·leder *n* crocodile skin (*or* leather)

Krokon·säure *f* croconic acid

Krokus·papier *n* crocus paper

Kron *n* crown glass

Krone f (-n) crown; corona; crest. **krönen** vt crown

kronen·artig a crown-like, coronal; (A nat.) coronary. **K–aufsatz** m 1 column (of a still). 2 crown top (of a burner). **-brenner** m crown (or ring) burner. **-gold** n crown (18-carat) gold. **-mutter** f castle (or castellated) nut

Kron·glas n crown glass. **-kümmel** m cumin

Krons·beere f cranberry

Krönung f (-en) coronation

Kropf m (⁼e) 1 crop, craw (of birds, etc.) 2 enlargement, swelling; (Med.) goiter. 3 (Bot.) club root, crown gall (& other plant diseases). 4 (Paper) breasting. **--drüse** f thyroid gland

kröpfen I vt 1 bend (at right angles). 2 crank, offset. 3 cram, stuff. **II** vi fill one's craw

Kropf·rohr n, **-röhre** f bent tube (or pipe). **-wurz** f figwort. **-wurzel** f polypody root. **-zylinder** m hydrometer jar (with wide upper part)

kröseln vt 1 groove, croze. 2 trim, crumble off (glass edges)

Kröte f (-n) toad

Kroton·öl n croton oil. **-säure** f crotonic acid

Krücke f (-n) 1 crutch. 2 crook, bend. 3 rake, rabble, scraper. 4 crowbar

Krug m (⁼e) 1 pitcher, jug. 2 tavern

Kruke f (-n) stone (or earthenware) jug

krüllen vt 1 shell, hull (legumes.) 2 gather in folds, crumple. **Krüll·tabak** m shag

Krümchen n (-) (little) crumb, bit; particle

Krume f (-n) 1 crumb. 2 (Agric.) topsoil; (vegetable) mold. **krümelig** a crumbly; friable. **krümeln** vt/i crumble. **Krü-mel·torf** m lump peat. **Krümelung** f crumbling. **Krümel·zucker** m dextroglucose, dextrose

krumm (⁼er, ⁼ste) a crooked, bent, curved. **K--achse** f crank. **k–blätterig** a in curved (or bent) leaflets; curvifoliate. **K–darm** m ileum

krümmen vt & sich k. bend, curve; warp. **Krümmer** m (-) 1 curved piece. 2 (pipe) elbow, knee. 3 (exhaust) manifold. **--ver-bindung** f elbow joint

krumm·faserig a cross-grained. **K–holz** n

1 crooked (or curved) piece of wood. 2 knee piece, compass timber. 3 gambrel. 4 = **-holzbaum** m, **-holzkiefer** f knee pine. **-holzöl** n templin oil. **k–linig** a cur-vilinear, curved

Krümmung f (-en) 1 bending, curving, winding. 2 curvature. 3 bend, curve

Krümmungs·halbmesser m radius of cur-vature. **-mittelpunkt** m center of curvature

Krumm·werden n warping, twisting, buckling. **-zirkel** m calipers

krumös a friable, crumbly

krumpeln, krümpeln vt/i crumple, crinkle

krumpf shrink(ing), shrinkage: **-echt** a shrinkproof, nonshrink(able)

krumpfen vt, vi(s) shrink

krumpf·frei a nonshrinking. **K–maß** n, **-wert** m (amount of) shrinkage

Krupp m croup

Krüppel m (-) cripple, disabled person. **krüppelhaft, krüpp(e)lig** a crippled, disabled; deformed; dwarfed

Krustazee f (-n) crustacean

Kruste f (-n) crust, incrustation; scab, scale

krusten·artig a crust-like, crusty. **-bil-dend** a crust-forming, incrusting. **K–bildung** f crust formation, incrustation. **-tier** n crustacean

krustieren vt incrust

krustig a crusty, (in)crusted; scabby

Kryo·chemie f cryochemistry. **k–hydra-tisch** a cryohydric

Kryolith m cryolite

Kryophor m cryophorus

Kryoskopie f cryosocpy. **kryoskopisch** a cryoscopic

Krypto·game f cryptogam

Krystall m (-e) crystal = KRISTALL

krz. abbv (kreuz) cross(wise), transverse

k.s.l. abbv (kalt sehr löslich) very soluble cold

K.S.P. abbv (Kegelschmelzpunkt) P.C.E. (pyrometric cone equivalent)

k.T. abbv (konkrete Tatbestände) concrete facts

kub. abbv (kubisch) cu. (cubic)

Kuba·holz n Cuba wood (fustic). **-zucker** m Cuban sugar

Kubebe f (-n) cubeb

Kubeben·öl n cubeb oil. **-pfeffer** m cubeb (or Java) pepper
Kübel m (-) 1 pail; bucket. 2 vat, tub. 3 (Glass) blank, parison (mold)
Kuben cubes: pl of KUBUS
kub. Gew. abbv (kubisches Gewicht) sp. gr. (specific gravity), weight per cubic meter in tons
kubieren vt cube; calculate the volume of
Kubik·gehalt m cubic content, volume. **-inhalt** m cubic content. **-maß** n cubic measure. **-wurzel** f cube root. **-zahl** f cube (number)
kubisch a cubic(al). **-(raum)zentriert** a (Cryst.) cubic body-centered
Kübler·ware f buckets, tubs, etc Cf KÜBEL
Kubus m (pl Kuben) cube
Küche f (-n) 1 kitchen. 2 cooking, cuisine. 3: **kalte** (**warme**) **K.** cold (hot) meal(s)
Kuchen m (-) cake. **k-·bildend** a cakeforming, caking. **K-erz** n cake ore. **-form** f 1 form (or shape) of a cake. 2 cake form (or mold). **k-·förmig** a cake-shaped, cakelike
Küchen· kitchen: **-garten** m vegetable garden. **-gerät** n kitchen utensil. **-gewächs** n kitchen (or cooking) herb; vegetable. **-salz** n common salt. **-schabe** f (cock)roach. **-schelle** f (Bot.) pasque flower
Kücken n (-) chick, etc = KÜKEN
Kuckuck m (-e) cuckoo
Kufe f (-n) 1 tub, vat; tun; beck; tank; measure (of beer). 2 runner (on sleds, etc); skid rocker
Küfer m (-) cooper; cellarman
Kugel f (-n) 1 sphere, globe. 2 ball, bulb; (Anat.) head (of bone in ball joints). 3 bullet, pellet; shot. **-·abschnitt** m (Geom.) spherical segment. **k-ähnlich** a spheroid(al), ball-like. **-apparat** m bulb apparatus (e.g. potash bulbs). **k-artig** a spherical, globular, spheroid(al). **K-ausschnitt** m (Geom.) spherical sector. **-bakterien** fpl cocci, spherical bacteria
Kügelchen n (-) little sphere (globe, etc); pellet, bead, globule cf KUGEL
kugel·dicht a bulletproof. **K-diorit** m globular diorite. **-dreieck** n spherical triangle. **-druckprobe** f ball pressure test (for hardness), Brinell test. **-eisenstein**

m spherodiserite. **-fallmethode** f fallingball method. **-fallprobe** f dropping-ball test. **k-fest** a bulletproof. **K-fläche** f spherical surface (or area). **-flansch** m spherical joint. **-flasche** f balloon (spherical, round-bottomed) flask. **k-förmig** a spherical, globular. **K-funktion** f spherical harmonic. **-gelenk** n ball-and-socket joint. **-gestalt** f spherical form (or shape). **-graphit** m spheroidal (or nodular) graphite. **-graphitgußeisen** n ductile iron, spheroidal graphite, cast iron. **-griff** m ball handle. **-grünspan** m basic copper acetate. **-hahn** m globe (-shaped) stopcock. **-hahnpipette** f, **-hahnstechheber** m pipette with globe stopcock. **-hefe** f spherical yeast
kugelig a spherical, globular; rolled-up
Kugel· sphere, spherical, etc: **-kocher** m spherical boiler. **-kohle** f coal pebbles. **-körperchen** n (spherical) particle, corpuscle; (Cryst.) globulite. **-kreisel** m spherical top. **-kühler** m ball (or spherical) condenser. **-kupplung** f ball-and-socket joint. **-lack** m lac (or lake) in balls. **-lager** n ball bearing. **-lampe** f (Elec.) spherical bulb. **-linse** f spherical lens. **-mikrophon** n ball-stage (or omnidirectional) microphone. **-mischbrenner** m (Gas) burner with a spherical mixing chamber. **-mühle** f ball mill. **-mutter** f (Mach.) ball nut
kugeln 1 vt, vi(s) & **sich k.** roll. 2 vt & **sich k.** roll up (into a ball)
Kugel· spherical, etc: **-nickel** m or n nickel pellets. **-packung** f (Cryst.) packing of spheres, close packing. **-probe** f ball test. **-probehahn** m ball gauge cock. **-rohr** n, **-röhre** f bulb tube. **-rührer** m bulb (or ball) stirrer. **-schale** f 1 spherical shell. 2 ball cup. 3 (Mach.) spherical seating (of journal bearings). **-schaltung** f ball gear shift. **-schieber** m (Mach.) spherical stride. **-schliff** m spherical ground joint, ground ball-and-socket joint. **-schreiber** m ball-point pen. **k-sicher** a bulletproof. **K-spiegel** m spherical mirror. **-stopfen** m bulb stopper. **-strahlen** n shot-blasting (or peening). **-symmetrie** f spherical symmetry. **k-symmetrisch** a spherosymmetrical. **K-tee** m gunpowder tea.

-**tisch** (*Micros.*) ball stage. -**ventil** *n* ball valve. -**vorlage** *f* spherical (*or* globular) receiver. -**welle** *f* spherical wave. -**winkel** *m* spherical angle

kuglig *a* spherical, etc = KUGELIG

Kuh *f* (¨e) cow. -**butter** *f* butter from cow's milk, dairy butter. -**dünger** *m* cow manure (as fertilizer). -**euter** *n* 1 cow udder. 2 Bredt adapter for vacuum distillation. -**haut** *f* cowhide. -**kot** *m* cow manure

kuhkoten *vt* (*pp* gekuhkotet) fertilize with cow manure. **Kuhkot·salz** *n* (*Dye.*) dung salt

kühl *a* cool. **K--anlage** *f* cooling (*or* refrigerating) plant (*or* unit). -**apparat** *m* cooling apparatus, cooler, refrigerator; condenser. -**bottich** *m* cooling tub

Kühle *f* 1 cool(ness). 2 cooling tub. **kühlen** 1 *vt* & *vi*(s) cool (off). 2 *vt* chill; refrigerate; (*Glass*) anneal. **Kühler** *m* (-) cooler; (*Chem.*) condenser; (car) radiator

Kühler·gestell *n* condenser stand (*or* support). -**halter** *m* condenser holder. -**klammer** *f* condenser clamp. -**mantel** *m* condenser jacket. -**retorte** *f* condenser retort. -**stativ** *n* condenser stand (*or* support)

Kühl·falle *f* cold trap. -**faß** *n* cooling (vat, vessel, tub). -**fläche** *f* cooling surface. -**flüssigkeit** *f* cooling liquid, coolant. -**gefäß** *n* cooling vessel (*or* tank); cooler, refrigerator; condenser. -**geläger** *n* 1 sediment from cooling. 2 (*Brew.*) wort sediment, dregs. -**haus** *n* cold-storage plant. -**hauslager** *n* cold-storage depot. -**kammer** *f* cooling chamber. -**kanal** *m* cooling channel (*or* tunnel). 2 (*Glass*) lehr, annealing furnace. -**kasten** *m* cooling box, cooler. -**kette** *f* cold chain, chain of refrigeration processes. -**kokille** *f* cooling mold. -**lager** *n* cold-storage depot. -**luft** *f* cooling air. -**maische** *f* (*Sugar*) cooling crystallizer

Kuhlmann·grün *n* Kuhlmann's green, copper oxide-chloride pigment

Kühl·mantel *m* cooling jacket (*or* mantle). -**maschine** *f* cooling machine, refrigerator. -**mittel** *n* cooling agent, coolant, refrigerant. -**ofen** *m* annealing oven, (*Glass*) lehr. -**öl** *n* cooling oil (*Mach.*) cutting oil; (*Metll.*) quenching oil. -**pfanne** *f*

1 cooling pan. 2 (*Glass*) lehr pan. -**presse** *f* cold press. -**raum** *m* 1 cooling chamber, chill room. 2 cold storage area. 3 refrigerating space (of a refrigerator). -**rippe** *f* cooling fin. -**riß** *m* cooling crack; (*Ceram.*) dunt. -**rohr** *n*, -**röhre** *f* cooling tube (*or* pipe); condenser tube. -**salz** *n* freezing salt (ammonium nitrate). -**schiff** *n* 1 refrigerator ship. 2 cooling pan; (*Brew.*) cooling back. -**schlange**, -**schnecke** *f* condenser (*or* cooling) coil, distilling worm; (car) radiator coil. -**schrank** *m* refrigerator. -**spannung** *f* cooling stress. -**spirale** *f* cooling spiral, spiral condenser. -**tasche** *f* cooling chamber. -**temperatur** *f* annealing temperature. -**truhe** *f* freezer, freezing cabinet. -**tunnel** *m* annealing oven; (*Glass*) lehr. -**turm** *m* cooling tower

Kühlung *f* (-en) 1 cooling, refrigeration. 2 cooling device; cooling system. 3 cool(er) air (*or* weather), coolness

Kühlungs·grad *m* rate (*or* degree) of cooling

Kühl·verfahren *n* cooling (*or* refrigerating) process. -**vorrichtung** *f* cooling (*or* refrigerating) device. -**wagen** *m* refrigerator car (*or* truck). -**wasser** *n* cooling water; (*Pharm.*) lead water. -**werk** *n* cooling (*or* refrigerating) plant (installation, unit). -**wirkung** *f* cooling effect. -**zement** *m* bauxite waste cement. -**zylinder** *m* cooling cylinder, condenser

Kuh·milch *f* cow's milk. -**mist** *m* cow manure

kuhmisten *vt* (*pp* gekuhmistet) (*Lthr.*) treat with cow manure

kühn *a* bold, daring. **Kühnheit** *f* boldness, daring

Kuhpocken *pl* cowpox. -**gift** *n* vaccine virus. -**impfung** *f* vaccination

kuh·warm *a* cowwarm (milk)

Kuka·brühe *f* (*short for* Kupfer·Kalk·) Bordeaux mixture

Küken *n* (-) 1 chicken. 2 stopcock (plug). -**hahn** *m* stopcock

Kukuruz *m* American corn, maize

Külbchen *n* (*Glass*) parison, lump, ball. **Külbel·form** *f* parison mold

Kuli *m* (-s) 1 (*short for* KUGELSCHREIBER)

ball-point pen. **2** coolie
Kulilawan·öl *n* culilawan oil
kulinarisch *a* culinary
Kulisse *f* (-n) (sliding) stage scene(ry)
Kulm 1 *m* (-e) knoll. **2** *n* (*Geol.*) culm
kulminieren *vi* culminate
Kulm·kalk *m* carboniferous limestone
kultivierbar *a* cultivatable, tillable.
   kultivieren *vt* cultivate, till. **Kultivie-**
   rung *f* (-en) cultivation
Kultur *f* (-en) culture; cultivation; civilization. **kulturell** *a* cultural
Kultur·flasche *f* culture flask. **-geräte** *npl*
   agricultural implements. **-geschichte** *f*
   history of civilization. **-hefe** *f* culture
   yeast. **-kolben** *m* culture flask. **-medien**
   npl culture media. **-pflanze** *f* cultivated
   plant. **-röhre** *f* culture tube. **-schale** *f* culture dish. **-staat** *m* civilized state (*or* nation). **-stufe** *f* stage of civilization. **-ver-**
   anstaltung *f* cultural event. **-versuch** *m*
   cultivation experiment
Kumarin·säure *f* cumarinic acid
Kumin·säure *f* cum(in)ic acid
Kümmel *m* **1** caraway —**römischer K.**
   cumin. **2** = **·branntwein, -likör** *m* kümmel (liquor). **-öl** *n* caraway oil; cumin oil;
   carvone. **-schnaps** *m* kümmel (liquor)
Kümmer·form *f* degenerate (*or* subnormal) form *cf* KÜMMERN
kümmerlich *a* **1** stunted, undersized. **2**
   pitiful, miserable. **3** paltry, meager
kümmern 1 *vt* bother, concern, mean (sthg)
   to. **2** *vi* become stunted; live miserably. **3:**
   sich k. (um) care (for), take care (of); concern oneself (with, about); pay attention
   (to)
Kummet·leder *n* (*Lthr.*) collar back
Kumol *n* cumene
Kumpe *f* (-n) (*Dye.*) basin, vessel
kümpeln *vt* **1** dish, cup. **2** flange. **Kumpen**
   m (-), **Kumpf** *m* (⁓e) dish, basin; trough
kumulieren *vt/i* accumulate
Kumys *m* kumiss, mare's nest liquor
kund *pred a & sep pfx* known
kündbar *a* recallable, redeemable,
   terminable
Kunde I *m* (-n, -n) customer, client. **II** *f* **1**
   (piece of) news, announcement. **2** knowledge. **3: K. geben (von)** bear witness (to).

4 (*as sfx*) study, science (of), **·ology** *cf*
   ERDKUNDE, TIERKUNDE. **künden** *vi* (**von**)
   bear witness (to)
kund·geben* 1 *vt* make known, announce.
   2: sich k. appear, manifest itself.
   Kundgebung *f* (-en) **1** announcement,
   declaration. **2** demonstration
kundig *a* (*oft* + *genit*) informed (about), expert (in), well-versed (in)
kündigen 1 *vt/i* give notice. **2** *vt* dismiss;
   give notice of termination (*or* cancellation) of. **Kündigung** *f* (-en) notice (of termination); dismissal
Kundin *f* (-nen) customer, client
kund·machen *vt* announce, etc =
   KUNDGEBEN
Kundschaft *f* (-en) **1** clientele. **2** customer,
   client. **3** reconnaissance; (search for)
   information
kund·tun* *vt* announce, etc = KUNDGEBEN
kund·werden* *vi*(s) become known, be
   announced
·kunft *f* (⁓e) (*noun corresp. to verb* KOMMEN;
   occurs only with pfxs) *cf* ANKUNFT, AUS-
   KUNFT, ZUKUNFT, etc
künftig *a* future, coming, next, prospec-
   tive; *adv* = **·hin** *adv* in (the) future, from
   now on
Kunst *f* (⁓e) **1** art. **2** skill. **3** artifice, trick-
   (ery); *pl oft* tricks, wiles. **4** (*as sfx*) art of ..
   e.g. **Baukunst** architecture. **5** (*as pfx in*
   compds oft) artificial, synthetic; technical: **-asphalt** *m* pitch. **-ausdruck** *m* technical term; art(istic) term. **-baumwolle** *f*
   artificial cotton. **-benzin** *n* synthetic gasoline. **-bronze** *f* art bronze. **-butter** *f*
   (oleo) margarine. **-darm** *m* synthetic sausage casing. **-druck** *m* art print(ing).
   -druckpapier *n* art paper. **-dünger** *m*
   (synthetic) fertilizer. **-eis** *n* artificial ice
   -erzeugnis *n* **1** artificial (*or* synthetic)
   product. **2** work of art (*or esp* handicraft).
   -faser *f* synthetic fiber. **-fehler** *m:*
   ärztlicher K. medical malpractice. **k–**
   fertig *a* (technically *or* artistically)
   skilled. **K–fertigkeit** *f* technical (*or* artistic) skill. **-fett** *n* synthetic fat (*or*
   grease). **-feuer, -feuerwerk** *n* fireworks.
   -feuerwerkerei *f* (artistic) pyrotechnics.
   -fleiß *m* industry, industriousness.

-**gärtnerei** f 1 flower nursery. 2 horticulture. -**gewerbe** n commerical (or applied) art; arts and crafts, handicrafts. -**gewerbschule** f school of commercial arts (or of arts and crafts). -**glanz** m artificial luster. -**griff** m artifice, device, trick. -**gummi** n synthetic rubber; artificial gum. -**guß** m art casting. -**harz** n synthetic resin, plastic

Kunstharz·bindung f plastic bond. -**kleber** m synthetic resin adhesive. -**leim** m synthetic resin glue. -**preßmasse** f, -**preßstoff** m plastic (or synthetic resin) molding composition. **k—verleimt** a resin-bonded

Kunst·hefe f artificial yeast. -**heilmittel** n synthetic remedy. -**holz** n artificial (or laminated) wood. -**honig** m artificial honey. -**horn** n artificial horn, esp casein plastic. -**kautschuk** m or n synthetic rubber. -**keramik** f art pottery. -**kohle** f artificial coal (or carbon). -**korund** m artificial corundum, fused alumina. -**leder** n synthetic (or man-made) leather. -**lehre** f theory of art

Künstler m (-) artist. -**farbe** f artist's color. **Künstlerin** f (-nen) artist. **künstlerisch** a artistic cf GEKÜNSTELT

künstlich a artificial, synthetic, false

Kunst·licht n artificial light. -**masse** f artificial composition, synthetic plastic. -**mist** m compost. -**mittel** n 1 artificial means. 2 artistic means. -**öl** n synthetic oil. -**papier** n art (or enameled) paper. -**porzellan** n art porcelain, china. -**produkt** n 1 artificial (or synthetic) product. 2 work of art (or esp handicraft). -**rahm** m synthetic cream. -**schmalz** n artificial (or compound) lard. -**seide** f rayon, artificial silk. -**seide(n)faden** m rayon thread. -**seidensamt** m rayon velvet. -**seideverarbeitung** f rayon processing. -**spinnfaser** f synthetic textile fiber, (specif) staple fiber (or rayon). -**spinnstoff** m artificial spinning material, reclaimed material. -**sprache** f 1 synthetic language. 2 technical (or specialized) language. -**stein** m artificial (or reconstructed) stone, earthenware. -**stoff** m artificial (or synthetic) material, (specif) plastic

Kunststoffaser f (=**Kunststoff·faser**) synthetic (or man-made) fiber

Kunststoff·chemie f chemistry of synthetics (esp of plastic). -**industrie** f plastics industry

Kunststofffläche f (=**Kunststoff·fläche**) plastic surface

Kunststoff·latex m polymer latex. -**masse** f plastic base, synthetic composition

Kunst·stück n trick, feat. **k—voll** a artistic; artful, skilful. **K—werk** n work of art. -**wolle** f artificial (or reclaimed) wool, shoddy. -**wort** n 1 coined word. 2 technical term

kunterbunt a 1 many-colored; adv in many colors. 2 varied, diversified. 3 confused, jumbled, topsy-turvy; adv in rich (or colorful) variety

Kuoxam n, **··zellulose** f cuprammonium solution

Küpe f (-n) (dyeing) vat

kupellieren vt cupel. **Kupellier·ofen** m cupel furnace. **Kupellierung** f (-en) cupellation

küpen vt vat-dye

Küpen· (dyeing) vat: -**alkali** n vat alkali. -**ansatz** m setting of the vat. -**artikel** m (Tex.) vat style. -**bad** n dye liquor. -**blau** n vat (or indigo) blue. -**buntätze** f colored vat discharge. -**färberei** f vat dyeing. **k—farbig** a vat-colored, vat-dyed. **K—farbstoff** m vat dye. -**färbung** f vat dyeing. -**flotte** f vat liquor. -**geruch** m vat odor. -**schlamm** m vat sediment

Kupfer n copper. -**abfall** n waste copper, copper scrap. -**alaun** m (Pharm.) cuprum aluminatum, lapis divinus. -**ammoniaklösung** f cuprammonium solution. -**ammonium** n cuprammonium, Schweitzer's reagent, tetrammine cupric sulfate. -**antimonglanz** m chalcostibnite, wolfsbergite. **k—artig** a copper-like, coppery, cupreous. **K—asche** f copper scale. -**azetat** n copper acetate. -**azetylen** n copper acetylide. -**bad** n copper bath. -**barre** f copper bar (or ingot). -**beize** f copper mordant. -**blatt** n copper foil. -**blau** n blue verditer, azurite. -**blech** n sheet copper, copper foil (or sheet). -**blei** n copper-lead alloy

Kupferblei· (Min.): -**glanz** m cuproplum-

bite, alisonite. **-spat** *m* linarite —
prismatischer K. caledonite. **-vitriol** *n*
linarite
Kupfer· copper: **-blende** *f* tennantite.
-blüte *f* copper bloom, plush copper ore,
chalcotrichite. **k–braun** *a* copper-brown.
K–braun *n* tile ore, earthy ferruginous
cuprite = KUPFERROT. **-bromid** *n* copper
(*or specif* cupric) bromide. **-bromür** *n*
cuprous bromide. **-brühe** *f* (*Agric.*) copper spray. **-chlorid** *n* copper (*or specif*
cupric) chloride. **-chlorür** *n* cuprous
chloride. **-cyanid** *n* copper (*or specif*
cupric) cyanide. **-cyanür** *n* cuprous
cyanide. **-dorn** *m* slag from liquated copper. **-draht** *m* copper wire. **-drahtnetz** *n*
copper gauze. **-drehspäne** *mpl* copper
turnings. **-druck** *m* copperplate (printing). **-druckfarbe** *f* copperplate ink,
(*specif*) Frankfort black. **-druckschwarz**
n copperplate black. **-eisenvitriol** *n* copper iron sulfate, pisanite. **-elektrode** *f*
copper electrode. **k–empfindlich** *a*
copper-sensitive. **K–erz** *n* copper ore.
-fahlerz *n* gray copper ore, tetrahedrite.
-farbe *f* copper color; copper pigment. **k–**
farben, -farbig *a* copper-colored. **K–**
federerz *n* plush copper ore,
chalcotrichite. **-feilicht** *n*, **-feilspäne**
mpl copper filings. **-folie** *f* copper foil.
-formiat *n* copper formate. **k–führend** *a*
copper-bearing, cupriferous. **K–gare** *f*,
-garmachen *n* copper refining. **-gefäß** *n*
copper vessel. **-gehalt** *m* copper content.
-geist *m* spirit of verdigris. **-gerät** *n* copper utensil(s), *pl oft* copperware. **-gewebe**
n copper gauze. **-gewinnung** *f* extraction
of copper. **-gießerei** *f* 1 copper foundry. 2
copper founding. **-glanz** *m* 1 copper
glance, chalcocite —**blauer K.** digenite.
2 coppery luster. **-glas(erz)** *n* chalcocite.
-glimmer *m* chalcophyllite, micaceous
copper. **-glühspan** *m* copper scale. **-gold**
n gold-like copper alloy, e.g.: similar,
Mannheim gold. **-grün** *n* copper green,
verdigris, (*Min.*) chrysocolla. **-guß** *m* copper casting; cast copper. **k–haltig** *a* containing copper, cupriferous, cupreous. **K–**
hammerschlag *m* copper scale.
-hochofen *m* copper blast furnace. **-hüt-**
te *f* copper smeltery. **-hydrat, -hydroxid**

n cupric hydroxide. **-hydroxydul** *n*
cuprous hydroxide
kupferig *a* coppery, cupreous, copper-like
Kupfer· copper: **-indig(o)** *m or n* indigo copper, covellite. **-jodid** *n* copper (*or specif*
cupric) iodide. **-jodür** *n* cuprous iodide.
-kalk *m* 1 copper calx (*or* oxide). 2 (*Agric.*)
= **-kalkbrühe** *f* Bordeaux mixture.
-kapsel *f* copper capsule (*or* cup); copper
priming cap. **-kies** *m* copper pyrites,
chalcopyrite —**bunter K.** bornite. **k–**
kieshaltig *a* containing copper pyrites.
K–kohle *f* coppered carbon. **-könig** *m*
copper regulus. **-kunstseide** *f* cuprammonium rayon. **-lackdraht** *m* enameled
copper wire. **-lasur** *f* azurite. **-lauge** *f* copper solution. **-laugung** *f* copper leaching.
-lebererz *n* liver ore, cuprite. **-legierung**
f copper alloy. **-lösung** *f* copper solution.
-lot *n* copper solder. **-mangan** *n*
cupromanganese. **-manganerz** *n*
cuprian wad (a variety of manganese dioxide). **-massel** *f* pig copper. **-metall** *n*
metallic copper. **-münze** *f* copper coin
kupfern I *vt* treat (plate, sheathe) with copper. **II** *a* (of) copper; coppery, copper-
colored
Kupfer·nickel *m or n* copper nickel, niccolite —**weißer K.** cloanthite. **-nieder-**
schlag *m* copper precipitate (*or* deposit).
-ocker *m* tile ore, cuprite
Kupferoxid *n* copper oxide; (*in compds*
usu) cupric: **··ammoniak** *n* ammoniacal
copper oxide; (*as pfx usu*) cuprammonium, e.g.: **-ammoniakzellulose** *f*
cuprammonium rayon. **-hydrat** *n* cupric
hydroxide. **-salz** *n* cupric salt
Kupferoxydul *n* cuprous oxide; (*in compds*
usu) cuprous: **-hydrat** *n* cuprous hydroxide. **-salz** *n* cuprous salt
Kupfer·pecherz *n* 1 copper pitch ore,
melaconite. 2 chrysocolla. **-platte** *f* copper plate; copperplate. **k–plattiert** *a*
copper-plated. **K–pol** *m* (*Elec.*) copper
(i.e. positive) pole. **-präparat** *n* copper
preparation. **-probe** *f* copper assay, test
for copper. **-raffination** *f* copper refining.
-rauch *m* 1 copper smoke (*or* fumes);
white vitriol. 2 copperas. **-regen** *m* copper rain. **k–reich** *a* rich in copper, high-
copper. **K–reinigung** *f* copper refining.

-reserve f (*Tex.*) copper resist. **-rhodanat, -rhodanid** n copper (*or specif* cupric) thiocyanate. **-rhodanür** n cuprous thiocyanate. **-rohr** n **-röhre** f copper tube (*or pipe*). **-rohstein** m raw copper matte

Kupferron n cupferron

Kupfer- copper: **-rost** m copper rust, verdigris. **k--rot** a copper-red, coppercolored. **K--rot** n 1 copper oxide; (*Min.*) cuprite. 2 copperas. 3 copper red (color). **-rubin** m copper ruby glass. **-salpeter** n copper nitrate. **-salz** n copper salt. **-samterz** n velvet copper ore; (*Min.*) cyanotrichite. **-säure** f cupric acid. **-schachtofen** m copper shaft furnace. **-schalentest** m (*Gasoline*) copper dish test (for gumming). **-schaum** m (*Metll.*) copper scum; (*Min.*) tyrolite. **-scheibe** f copper disk. **-schiefer** m copper-bearing schist (*or shale*). **-schlacke** f copper slag. **-schlag** m copper scale (*or dross*). **-schlange** f copper coil. **-schnitzel** npl copper turnings. **-schwamm** m copper sponge; (*Min.*) tyrolite. **-schwärze** f black copper; (*Min.*) melaconite, tenorite. **-schwefel** m copper sulfide. **-seide** f cuprammonium rayon. **-silberglanz** m stromeyerite. **-sinter** m 1 copper pyrites. 2 copper scale. **-smaragd** m emerald copper; (*Min.*) dioptase. **-spat** m malachite. **-spinnfaser** f cuprammonium rayon staple. **-spiritus** m spirit of verdigris. **-stahl** m copper steel. **-stechen** n copper engraving. **-stein** m copper matte **--blasiger K.** blister copper, pimple metal; **weißer K.** white metal. **-stich** m copper engraving (*or print*). **-streckseide** f Lilienfeld (*or* stretched cuprammonium) rayon. **-sulfat** n copper (*or cupric*) sulfate. **sulfid** n copper (*or cupric*) sulfide. **-sulfocyanür** n cuprous thiocyanate. **-sulfür** n cuprous sulfide. **-überzug** m copper coating (covering, plating). **-uranglimmer** m (*Min.*) torbenite. **-uranit** n (*Min.*) 1 torbenite. 2 metazeunerite. **-verbindung** f copper compound. **-vergiftung** f copper poisoning. **-verhüttung** f copper smelting. **-vitriol** n blue vitriol, copper sulfate. **-wasser** n (*Min.*) copper (*or cement*) water; (*Tanning*) green copperas, ferrous

sulfate. **-wasserstoff** m copper hydride. **-werk** n 1 copper works (*or plant*). 2 copper plate (*or plating*). **-wismuterz** n (*Min.*) wittichenite; klapproth(ol)ite. **-wismutglanz** m (*Min.*) emplectite. **-zahl** f copper number (*or value*). **-ziegelerz** n tile ore, cuprite. **-zuschlag** m copper flux. **-zyanid** n cupric cyanide. **-zyanür** n cuprous cyanide

kupfrig a coppery, cupreous = KUPFERIG

kupieren vt cut, clip; dilute; reduce (dyes)

Kupol-(hoch)ofen m cupola (blast) furnace

Kupon n (-s) coupon; (*Tex.*) remnant

Kuppe f(-n) (round) top, dome, summit; tip; (round) head (of screws, etc); meniscus

Kuppel f (-n) cupola, dome; arch (of a furnace). **k--artig** a dome-like, domeshaped, arched. **kuppeln 1** vt couple, connect; (*Mach.*) engage, put in gear; (*Dye.*) develop. **2** vi shift gears. **Kuppel-ofen** m cupola furnace. **Kuppler** m (-) coupler; (*Dye.*) developer. **Kupplung** f(-en) coupling, connection; clutch (of a car); (*Dye.*) developing

Kupplungs-bad n (*Dye.*) developing bath. **-farbstoff** m coupling dye. **-färbung** f coupled dyeing. **-flotte** f(*Dye.*) developing liquor. **-verfahren** n coupling process

Kupri- cupri(c) = CUPRI-

Kuprit n (-e) cuprite

Kupro- cupro(us) = CUPRO-

Küpung f(-en) vat dyeing

Kupüre f(-n) (*Tex.*) reduction

Kur f (-en) cure, treatment. **--anstalt** f sanatorium

Kurbel f(-n) crank; crank (*or winch*) handle. **--gehäuse** n crankcase. **kurbeln** vt/i crank, wind, turn

Kurbel-pumpe f reciprocating pump. **-welle** f crankshaft. **-zapfen** m crankpin

Kürbis m (-se) squash, pumpkin, gourd, etc. **--baum** m calabash tree. **-gewächse** npl cucurbitaceae. **-kernöl** n calabash oil. **-korn** n, **-samen** m pumpkin (squash, gourd) seed

kurieren vt cure

Kurkuma f (..men) turmeric, curcuma. **--gelb** n curcumin. **-papier** n turmeric paper. **-säure** f curcumic acid. **-wurzel** f turmeric

Kur·ort *m* spa, health resort. **-pfuscherei** *f* (*Med.*) quackery

Kurs *m* (-e) **1** course *cf* KURSUS; policy. **2** circulation: **in K. (außer K.) setzen** put in (take out of) circulation. **3** rate of (monetary) exchange; stock quotation, price; **etwas steht hoch im K. (a)** (*lit*) the price (rate, quotation) of sthg is high, **(b)** (*fig*) sthg is highly esteemed (*or* is in great demand)

Kursant *m* (-en, -en) student in a course

Kürschner *m* (-) furrier. **Kürschnerei** *f* (-en) furriery. **Kürschner·waren** *fpl* furs, skins

Kurse courses, etc: *pl of* KURS & KURSUS

kursieren *vi* circulate, be current

kursiv *a* (*Print*) italic; *adv usu* in italics. **K–·schrift** *f* italics

Kurs·stativ *n* (*Micros.*) class microscope

Kursus *m* (Kurse) course (of study) *cf* KURS

Kurs·wert *m* market (*or* exchange) value

Kurve *f* (-n) curve

Kurven·ast *m* branch of a curve. **-bild** *n*, **-blatt** *n*, **-darstellung** *f* curve diagram, graph, graphic representation. **-eck** *n* angle in a curve. **-gipfel** *m* apex of a curve. **-lineal** *n* curved rule(r); French curve. **k–mäßig** *a* diagrammatic, *adv oft* as a curve. **-reich** *a* winding, curvaceous. **K–schar** *f* family (group, system) of curves. **-schiefe** *f* (*Math.*) curve skewness, allokurtosis. **-schreiber** *m* curve tracer. **-zacke** *f* jag (*or* sharp change in direction) of a curve. **-zeichnen** *n* curve plotting. **-zug** *m* **1** series of curves. **2** graph; trend line

kurz (ër, ëste) *a* **1** short, brief; curt, abrupt — (*Dye.*) **k—e Flotte** concentrated dye liquor. **2** *adv oft* in short, to put it briefly; a short time (before, etc), a short way (beyond, etc) **–vor –em** recently. **K–·arbeiter** *m* short-term (*or* temporary) worker. **k–armig** *a* short-armed; short-beam (balance). **K–beize** *f* (*Agric.*) short disinfection treatment. **k–brennweitig** *a* (*Opt.*) short-focus. **-brüchig** *a* brittle. **-dauernd** *a* short-term, short-lived, brief, transient

Kürze *f* shortness; brevity, short duration; conciseness **—in K.** shortly, in a short time

Kürzel *n* (-) short(ened) form, abbreviation, contraction; (*specif*) logo, acronym, etc; (*Radio*) call sign

kürzen *vt* shorten; cut (down), reduce (*incl* fractions)

kurzerhand *adv* without hesitation (*or* delay)

kurz·faserig *a* short-fiber(ed). **-flammig** *a* (*Coal*) short-flaming. **-fristig** *a* short-term, short-lived; *adv* on short notice. **-gefaßt** *a* concise. **-geschlossen** *a* (*Elec.*) short-circuited. **-haarig** *a* short-haired; (*Wool*) short-staple. **-halsig** *a* short-necked. **-kettig** *a* short-chain. **-lebig** *a* short-lived

kürzlich *adv* recently, a short time ago

kurz·ohrig *a* short-eared. **-periodisch** *a* short-period; high-frequency. **K–prüfung** *f* short (*or* accelerated) test. **-referat** *n* brief report, summary, abstract. **k–säulig** *a* short-columnar

kurz·schließen* *vt* short-circuit

Kurzschliff·halskolben *m* flask with a short ground neck

Kurzschluß *m* short circuit. **-ankermotor** *m* squirrel-cage motor. **-ofen** *m* short circuit furnace. **k–sicher** *a* short-circuit-proof, non-shorting. **K–strom** *m* short-circuit current

Kurz·schrift *f* shorthand, stenography. **k–sichtig** *a* near-sighted; shortsighted. **-skalig** *a* short-scale(d). **K–stäbchen** *n* bacillus. **k–stapelig** *a* short-staple(d) (fibers). **-um** *adv* in short, in brief

Kürzung *f* (-en) **1** shortening. **2** cut(-back), reduction (in). **3** short cut, shorter procedure. *Cf* KÜRZEN

Kurz·versuch *m* short (*or* quick) test. **-waren** *fpl* notions; (small) hardware. **k–weg** *adv* **1** curtly, flatly. **2** on the spur of the moment; without a moment's hesitation. **K–wegdestillation** *f* molecular distillation. **-welle** *f* short wave. **-wellen·= k–wellig** *a* short-wave. **K–wolle** *f* short wool fibers, nails. **-wort** *n* abbreviated (*or* telescoped) word, contraction; acronym. **-zeichen** *n* symbol. **k–zeitig** *a* brief, short-time. **K–zeitphotographie** *f* high-speed photography

küssen *vt/i* kiss

Kussin *n* kosin

Kusso *m* (*Pharm., Bot.*) cusso
Küste *f* (-n) coast, shore. **Küsten·strich** *m* coastline
Kustos *m* (..oden) custodian, keeper, curator
Kusu·fett *n* cusu fat
kutan *a* cutaneous. **K—·probe** *f* skin test
Kutera·gummi *n* kutira gum
kutinisieren *vt* cutinize
Kuti·reaktion *f* cutaneous reaction
Kutte *f* (-n) cowl; cowled cloak
Kuvertüre *f* (-n) (chocolate) coating
Küvette *f* (-n) 1 bulb. 2 cuvette; cell; dish, bowl, basin; dental flask; (*Phot.*) filter trough. 3 cap (of a watch). 4 rain water drain. **Küvetten·verschluß** *m* (*Opt.*) cell cover
Kux *m* (-e) mining share

KW *abbv* (Kohlenwasserstoff) hydrocarbon
K-wert *m* (*Plastics*) K-value
k.w.l. *abbv* (kalt wenig löslich) not very soluble when cold
Kwst. *abbv* (Kilowattstunde) kw-hour
KW-Stoff *m* (Kohlenwasserstoff) hydrocarbon
Ky . . . *variation of* **Cy** *or* **Zy** *in words of Greek origin*
Kyanidin *n* cyanidine, kyanidine. **kyanisieren** *vt* kyanize
Kyaphenin *n* cyaphenine, kyaphenine
Kynuren·säure *f* cynurenic acid. **Kynur·säure** *f* cynuric acid
Kyste *f* (-n) cyst *cf* ZYSTE
KZ *abbv* 1 (Kernzahl) number of crystal nuclei. 2 (Koordinationszahl) coordination number

L

l. *abbv* **1** (löslich) sol(uble). **2** (lies) read. **3** liter. **4** (links) left. **5** (linksdrehend) levorotatory

L. *abbv* lira, *pl* lire

..l. *abbv* (..lich) e.g. **lösl.** (löslich) sol(uble)

Laab, Lab *n* (-e) rennet

Labdan·gummi, -harz *n* la(b)danum

Lab·drüse *f* peptic gland

laben 1 *vt* curdle (with rennet). **2** *vi* curdle, turn sour. **3** *vt & sich l.* refresh (oneself)

Laberdan *m* (-e) salt cod(fish)

Lab·essenz *f* rennet extract. **-fähigkeit** *f* rennetability. **-ferment** *n* rennet ferment, rennin; chymosin, chymase

labil *a* labile, unstable, changeable. **Labilität** *f* (-en) lability, instability

Lab·käse *m* rennet cheese. **-kraut** *n* (*Bot.*) galium, bedstraw. **-magen** *m* rennet stomach, abomasum

Labor *n* (-e) (*short for* Laboratorium) lab. **Laborant** *m* (-en, -en), **Laborantin** *f* (-nen) lab(oratory) assistant. **Laboratorium** *n* (..ien) laboratory

Laboratoriums·tisch *m* laboratory table (*or* bench). **-versuch** *m* laboratory experiment. **-zwecke** *mpl* laboratory purposes

laborieren *vi* **1** work in a laboratory. **2** (+**an**) labor (at, on, over)

Labor·kittel *m* lab coat. **-koffer** *m* portable laboratory case. **l–mäßig** *a* laboratory(-like); (*adv*) in the lab, under laboratory conditions. **L–ofen** *m* laboratory furnace

Lab·probe *f* curd test. **-pulver** *n* rennet powder

Labrador·feldspat *m* labradorite. **labradorisieren** *vi* iridesce. **Labrador·stein** *m* labradorite

Lab·saft *m* gastric juice. **-stärke** *f* rennet strength

Lache *f* (-n) **1** puddle, pool. **2** laugh(ter). **3** cut, incision; blaze, boundary mark(er)

lächeln *vi* smile

lachen *vi* laugh

lächerlich *a* laughable, ridiculous

Lach·gas *n* laughing gas, nitrous oxide

Lachs *m* (-e) salmon. **--farbe** *f* salmon (color). **l–farben, -farbig** *a* salmon-colored. **L–öl** *n* salmon oil. **l–rot** *a* salmon-pink. **L–ton** *m* salmon color (*or* shade). **-tran** *m* salmon oil

Lacht *m* (finery) slag

Lack *m* (-e) lacquer, varnish, enamel, japan; (*Dyes*) lake; lac; fingernail polish. **--abbeizmittel** *n* varnish (*or* paint) remover. **-anstrich** *m* coat of lacquer (paint, varnish). **-arbeit** *f* lacquered work. **l–artig** *a* lake-colored; varnishlike, (*Bot.*) vernicose. **L–auflösungsmittel** *n* varnish remover. **-aufstrich** *m* coat of lacquer (paint, varnish). **-beize** *f* varnish remover. **-benzin** *n* (*Paints*) mineral spirits, white spirit. **-bildner** *m* (*Dye.*) lake former. **-bildung** *f* lake (varnish, lacquer) formation. **-bindemittel** *n* lacquer binding medium. **-draht** *m* lacquered (*or* enameled) wire

Lacke *f* (-n) puddle, pool = LACHE 1

lacken *vt* lacquer, etc; polish; = LACKIEREN

Lack·entferner *m* paint (*or* varnish) remover. **-farbe** *f* **1** lake. **2** lac dye. **3** colored lacquer (paint, varnish). **l–farbig** *a* lakecolored, laked. **L–farbstoff** *m* **1** (*Dyes*) lake dye(stuff). **2** varnish stain. **-firnis** *m* lac (*or* lake) varnish, lacquer; boiled oil (for paints). **-firnislack** *m* lac lake. **-fläche** *f* varnish surface; lacquer coat. **-harz** *n* paint and varnish resin (*or* gum), gum lac, copal. **-haut** *f* film of lacquer (*or* varnish)

lackier·echt *a* fast to overvarnishing, nonbleeding. **lackieren** *vt* **1** lacquer, varnish, enamel, japan, paint; polish (fingernails). **2** cheat

lackier·fähig *a* lacquerable, varnishable. **L–grün** *n* Paris green. **-ofen** *m* japanning stove

Lackierung *f* (-en) **1** lacquering, varnishing, etc *cf* LACKIEREN. **2** dope (paint)

Lack·industrie *f* paint and varnish industry. **l–isoliert** a enamel-insulated. **L–kunstharz** n synthetic resin (for paint or lacquer). **-lack** m lac lake, lac dye. **-lasurfarbe** f transparent colored varnish. **-laus** f lac insect. **-lausfarbstoff** m lac dye. **-leder** n patent leather, japanned (enameled, varnished) leather. **-leinöl** n linseed oil (for paint or varnish). **-lösungsmittel** n lacquer solvent

Lackmus m or n, **-farbstoff** m litmus. **-flechte** f litmus lichen, archil. **-lösung** f litmus solution. **-papier** n litmus paper. **l–sauer** a acid to litmus. **L–tinktur** f litmus tincture.

Lack·oberleder n patent upper leather. **-papier** n lacquered (or varnished) paper. **-pigment** n lake pigment. **-platte** f enameled hardboard. **-rohstoff** m raw material for varnish or lacquer. **l–sauer** a laccaate of. **L–säure** f laccaic acid. **-schicht** f coat of lacquer (or varnish). **-schuhe** mpl patent leather shoes. **-sieder** m varnish boiler. **-siegel** n wax seal. **-spalt** m (Lthr.) enameled (or patent) split. **-sud** m varnish boiling. **-tuch** n oilcloth. **-überzug** m coat of lacquer (varnish, paint). **-verdünner** m lacquer (or varnish) thinner. **-waren** fpl lacquered (or japanned) goods

Lacmus m or n litmus; = LACKMUS

Lactarin·säure f lactarinic acid

Lactar·säure f lactaric (or stearic acid)

Lacton·bindung f lactonic linkage. **-säure** f lactonic (or D-galactonic) acid

Ladan n **-gummi**, **-harz** n la(b)danum

Ladanum·öl n oil of la(b)danum

Lade f (-n) 1 drawer. 2 (Foun.) flask. 3 lathe, batten (of a loom). Cf LADEN

Lade·baum m derrick, loading crane. **-maß** n loading gauge

laden* (lud, geladen; lädt or ladet) **I** vt **1** load, (Elec.) charge; impose —**etwas auf sich l.** burden oneself with sthg, incur sthg. **2** invite. **3** summon, subpoena. **II** vi (take on a) load

Laden I m (-) store, shop; business (enterprise). **2** (usu pl) shutters, blinds. Cf LADE. **II** n loading, charging; inviting; summoning cf LADEN

Ladenburg·kolben m Ladenburg flask

Laden· store: **-inhaber** m storekeeper,

store owner. **-preis** m retail price. **-tisch** m counter. **-zentrum** n shopping center

Lader m (-) **1** loader. **2** charger, booster

Lade·raum m **1** loading space; (ship's) hold. **2** (Guns) (exploding) chamber. **-spannung** f (Elec.) charging voltage. **-strom** m (Elec.) charging current. **-widerstand** m (Elec.) charging resistance

lädieren vt damage, injure

lädt loads, etc: pres of LADEN

Ladung f (-en) **1** load; cargo; loading. **2** (esp Elec.) charge; charging. **3** summons

Ladungs·austausch m charge transfer. **-dichte** f charge density. **-einheit** f (unit of) charge. **-flasche** f Leyden jar. **-gemisch** n charge mixture, mixed charge. **-häufungsregel** f adjacent charge rule. **-sinn** m (Elec.) sign of a charge. **-träger** m charge carrier. **-vermögen** n electrostatic capacity. **-verteilung** f charge (or load) distribution. **-werfer** m (Mil.) projector, launcher. **-wolke** f charge cloud

lag lay, etc: past of LIEGEN

Lage f (-n) **1** position. **2** location. **3** situation —**L. der Dinge** state of affairs; **in der L., etwas zu tun** able (or in a position) to do sthg. **4** state, condition. **5** layer, stratum; coat(ing); ply; (Geol.) bed, deposit; (Tex.) nap. **6** (Mil.) salvo, round. **7** (Paper) quire. **8** (Aero.) altitude

lage·fest a stable (in a position). **L–festigkeit** f stability (of position)

Lägel m or n (-) keg, barrel

lagen·förmig a layered, bedded, banded. **L–haftung** f ply adhesion. **-holz** n plywood. **-textur** f (Geol.) banded structure. **l–weise** adv in layers

Lager n (-) **1** bed; place to sleep, couch. **2** (esp. Geol.) layer, stratum; deposit. **3** sediment. **4** (Mach.) bearing. **5** (pl also -) storage place, warehouse; stores, stock, supplies. **6** stand (for barrels); (Engg.) support. **7** (Biol., Bot.) thallus. **8** camp. **-art** f (Ores) gangue (worthless syngenetic minerals accompanying ore in sedimentary deposits). **-aufbau** m stockpiling. **-aufnahme** f inventory. **-bestand** m stock, inventory. **l–beständig** a stable in storage. **L–beständigkeit** f stability in storage; storage (or shelf) life. **-bier** n lager (beer). **-bock** *m* (*Mach.*) pedestal, bracket. **-buchse, -büchse** *f* (*Mach.*) bushing. **-butter** *f* cold-storage butter. **l–echt** *a* stable in storage. **L–fähigkeit** *f* storability, stability in storage; shelf (*or* storage) life; storage capacity. **-faß** *n* storage barrel (*or* vat). **l–fest** *a* storable, stable in storage. **L–festigkeit** *f* storability, stability in storage. **l–förmig** *a* stratified, in beds (layers, strata). **L–gang** *m* (*Geol.*) bedded vein, sheet, sill. **-gestein** *n* (*Ores*) gangue (enclosed pieces of older rock in sedimentary deposits). **-hartblei** *n* bearing-metal containing antimony and lead. **-haus** *n* storehouse, warehouse. **-keller** *m* storage cellar. **-legierung** *f* bearing alloy. **-metall** *n* bearing (*or* bushing) metal

lagern I *vt* 1 store, keep; age (e.g. beer). 2 lay (*or* put) (down), rest; bed, support. **II** *vi* 3 camp. 4 lie (down), rest. 5 brood. 6 age, season (in storage). 7 be stored; be deposited. **III: sich l.** lie (*or* settle) down; be laid flat. *Cf* GELAGERT

Lager-pflanze *f* thallophyte. **-platz** *m* 1 storage place, depot. 2 place to rest. 3 camp site. **-prozeß** *m* storage process. **-raum** *m* storeroom, storage space. **-reibung** *f* (*Mach.*) bearing friction. **l–reif** *a* 1 storage-ripened. 2 aged (beer). **L–reife** *f* maturity (after storage). **-schale** *f* (*Mach.*) bushing. **l–stabil** *a* stable in storage. **L–stätte** *f* 1 (*esp Geol.*) bed, deposit; layer, stratum. 2 resting place

Lagerstätten-körper *m* ore body. **-kunde, -lehre** *f* science (*or* geology) of mineral deposits, economic geology

Lagerung *f* (-en) 1 storage; aging. 2 layering, stratification. 3 arrangement. 4 (*Mach.*) bearing (arrangement). 5 support, seat. 6 grain (of a stone). 7 (*Cryst.*) orientation. 8 encampment. *Cf* LAGER, LAGERN

Lagerungs-isometrie *f* structural isomerism

Lager-wein *m* aged (*or* storage-seasoned) wine. **-weißmetall** *n* white bearing metal, babbitt metal

lage-weise *adv* in layers, in strata

lahm *a* lame; numb, paralyzed: **l. gehen (a)** limp, **(b)** slacken. **lahmen** *vi* 1 limp; be

lame. 2 slacken, flag. **lähmen** *vt* paralyze, numb; cripple. **Lahmheit** *f* lameness. **lahm·legen** *vt* paralyze; cripple. **lahm·liegen*** *vi* (h *or* s) be (*or* lie) paralyzed. **Lähmung** *f* (-en) paralysis; numbness

Lahn I *f* (*Geog.*) Lahn (River). **II** *m* (-e) flattened wire; tinsel yarn. **--gold** *n* tinsel, brass foil

Laich *m* (-e), **laichen** *vi* spawn

Laie *m* (-n, -n) layman, layperson; amateur

Lake *f* (-n) brine, pickle

Laken *n* (-) (linen) sheet; bath towel; sail

Lakmus *m* litmus = LACKMUS

Lakritze *f* (-n) licorice. **Lakritzen·saft** *m* extract of licorice

Laktat *n* (-e) lactate

Lakton *n* (-e) lactone

Lamberts·nuß *f* filbert, hazelnut

Lamelle *f* (-n) lamella; lamina; plate, fin, vane; (*Bot.*) gill; (*Mach.*) clutch plate

lamellen·förmig *a* lamelliform. **L–struktur** *f* lamellar structure

lamellieren *vt* laminate. **Lamellierung** *f* (-en) lamination

Lametta *n* (silver) tinsel

lamilar *a* laminar, laminated

laminieren *vt* laminate; (*Spinning*) draw

Lamm *n* (ẅer) lamb. **Lämmer·wolke** *f* cirrus cloud. **-wolle** = **Lamm·wolle** *f* lamb's wool

Lämpchen *n* (-) small (*or* little) lamp

Lampe *f* (-n) lamp; (light)bulb; (blow)torch

Lampen·arbeit *f* (*Glass*) blast-lamp work. **-brennstunde** *f* lamp hour. **-docht** *m* lamp wick. **-faden** *m* lighting filament. **-fassung** *f* lamp socket. **-ruß** *m* lampblack. **-schirm** *m* lampshade. **-schwarz** *n* lampblack

Land *n* (*usu* ẅer, *oft* -e) 1 land; country (*incl* countryside). 2 (federal) state, province. 3 (*in compds oft*) rural, farm; agricultural; countrywide: **-bau** *m* agriculture, farming. **-butter** *f* farm (*or* country) butter

Lande· (*in compds: Aero.*) landing: **-bahn** *f* landing strip

landen *vt, vi* (s) land

Land·enge *f* isthmus

Landes· national, state, provincial; native, domestic; agricultural: **-angehörigkeit** *f* nationality. **-aufnahme** *f* topographic survey. **-erzeugnis** n native (home, domestic; agricultural) product. **-sprache** f vernacular. **l–üblich** a customary (in the land)

Land·gut n (country) estate. **-karte** f map. **-krankheit** f countrywide (epidemic, endemic) disease. **-kreis** m rural district. **-kundig** a well-informed about the country. **-läufig** a (locally) customary. **L–leute** country people: pl of LANDMANN

ländlich a rural

Land·mann m farmer, country person cf LANDLEUTE. **-maschine** f farm machine. **-messer** m surveyor. **-polizei** f rural police

Landschaft f (-en) 1 landscape. 2 (rural) area, region

Land·see m lake (in the country). **-seuche** f epidemic

Lands·leute pl of **-mann** m, **-männin** f fellow countryman, person from the same district (province, state, country)

Land·stadt f country town. **-straße** f highway, road. **-strich** m region, district. **-tag** m state (or provincial) legislature

Landung f (-en) landing

landw. abbv (landwirtschaftlich) agric.

Land·wirt m farmer. **-wirtschaft** f agriculture. **l–wirtschaftlich** a agricultural

lang (länger, längst..) I a long; **auf l–e Sicht** long-term, for a long term. 2 tall (person). 3 thin, dilute(d): **l–e Flotte** dilute dye liquor. 4 (Wines) ropy. II adv 5 long, for a long time cf LANGE. 6 e.g. **l. hinfallen** fall down full length. 6: **l. und breit** at great length; **über kurz oder l.** sooner or later. III prep & adv, e.g.: **den** (or **am**) **Ufer l.** along the shore. IV prep (follows its object) e.g.: **zwei Tage l.** for two days; (oft as sfx): **tagelang** (lasting) for days Cf ZEITLANG. Cf LÄNGST. **--angehalten, -anhaltend** a prolonged, protracted. **-brennweitig** a long-focus. **-dauernd** a long-lasting, lengthy

lange adv e.g.: **l. warten** wait long, wait for a long time; **das ist schon l. her** that was a long time ago; **wir wissen das schon seit l—m** we have known that for a long time; **l. nicht so gut** not nearly as good cf LANG, LÄNGST, SO 16

Länge f (-n) 1 length: **in L. und Breite** at

great length; **der L. nach** lengthwise; **auf die L.** for any length of time; **in die L. ziehen** (or **dehnen**) drag out; **sich in die L. ziehen** drag on and on. 2 longitude. 3 (person's) height

lang·eiförmig a of long oval shape

längelang adv (at) full length (= LANG 6)

langen I vt 1 take, get (sthg from a place). II vi 2 (+**nach, in**) reach (for, into). 3 reach, extend (far enough); suffice, be enough. 4 (+**mit**) get along, make out (with an amount). Cf GELANGEN

längen vt lengthen; dilute

Längen· longitudinal, linear: **-ausdehnung** f linear expansion. **-bruch** m longitudinal fracture. **-durchschnitt** m longitudinal section. **-einheit** f unit of length. **-grad** m degree of longitude. **-maß** n linear measure. **schnitt** m longitudinal cut (or section)

länger a longer: compar of LANG; **l—e Zeit** a fairly long time

Langerhanssche(n) Inseln (Anat.) islands of Langerhans

länger·wellig a of greater wavelength

lang·faserig a long-fiber(ed), long-staple(d). **-flammig** a long-flame, long-flaming. **L–flammrohr** n long flame tube. **l–flurig** a (Tex.) long-haired, long-piled. **-fristig** a long-time, long-term. **-gärend** a long (-rising) (dough). **-gestielt** a long-handle(d); long-stem (or -stalked). **-gestreckt** a extended, elongated, stretched-out; adv oft lengthwise. **-gezogen** a long-drawn-out; lanky. **-haarig** a long-hair(ed); flossy (silk); long-staple (wool). **-halsig** a long-necked. **-j.** abbv = **-jährig** a long-term, of many years. **-kettig** a long-chain. **-lebig** a long-lived

länglich a 1 longish, rather long. 2 elongated. 3 oblong; oval

Lang·loch n elongated hole, slot. **-lochung** f elongated perforation, slotting. **l–rund** a oval, elliptical

längs 1 prep (+genit) along. 2 adv lengthwise

Längs· longitudinal: **-achse** f longitudinal axis

langsam a slow, adv slowly, gradually. **L–binder** m slow-setting cement. **-gang** m slow speed, low gear

Langsamkeit f slowness

langsam·laufend a slow-running, (s)low-speed. **-wüchsig** a slow-growing

Längs·bruch m longitudinal fracture

Lang·schliff m long-fibered mechanical pulp. **-schlitzbrenner** m (*Gas*) ribbon burner

Längs·dehnung f longitudinal expansion. **-durchschnitt** m longitudinal section. **-ebene** f longitudinal plane

Lang·sein n (*Wines*) ropiness

Längs·faser f longitudinal fiber

Lang·sieb n (*Paper*) endless wire, Fourdrinier web. **-siebmaschine** f Fourdrinier machine

längs·laufend a longitudinal, lengthwise. **L–richtung** f longitudinal direction (*or* axis) **—in der L.** lengthwise. **-schnitt** m longitudinal section. **-schwingung** f vibration. **l–seit(s)** prep (+genit), adv alongside. **L–streifen** m longitudinal stripe (*or* stria)

längst I a longest: superl of LANG. II adv 1 long, for a long time, long ago (*with past tenses in English*): **das weiß man (schon) l.** that has long been known, that has been known for a long time, that was known long ago. 2 e.g.: **das ist l. gut genug** that is good enough by far. 3 (+neg) e.g.: **l. nicht genug** not nearly enough, far from being enough; **das ist l. nicht mehr der Fall** that has not been the case for a long time. Cf LÄNGSTENS

Lang·stäbchen n (*Bact.*) bacillus. **l–stapelig** a long-staple(d)

längstens adv at most, at the longest

lang·stielig a long-stem (-stalked, -handled)

längst/lebig a longest-lived, of longest life

lang·strahlig a (*Cryst.*) radiating-acicular

Langstrecken· long-distance, long-range

Längs·welle f longitudinal wave

Längung f (-en) lengthening, elongation; dilution

Languste f (-n) spiny lobster, crayfish

lang·weilig a boring

Lang·welle f (*Phys.*) long wave (*esp* 100-1500 kc). **l–wellig** a long-wave, of long wavelength. **-wierig** a 1 protracted, tedious. 2 (*Med.*) chronic

Langzeit·erfolg m long-term success. **-verhalten** n long-term behavior

Lanthan n lanthanum. **--salz** n lanthanum salt. **-schwefelsäure** f lanthanum hydrogen sulfate

Lanze f (-n) lance, spear **—eine L. brechen** (*or* **einlegen**) (**für**) stand up (for). **lanzen·förmig** a lance-shaped, lanceolate

Lanzette f (-n) lancet

Lapacho·säure f lapachoic acid, lapachol

Lapis·druck m (*Tex.*) lapis style

Läppchen n (-) lap, lobe, lobule; small rag, piece of cloth cf LAPPEN

Lappen m (-) 1 rag, (dust, wash, polishing) cloth. 2 patch. 3 (*incl Anat.*) flap (of skin, etc), lobe, lap

läppen vt (*esp Mach.*) lap. **Läppen·mittel** n lapping abrasive

Lappen·rüssler m weevil

lappig a 1 flabby; limp. 2 paltry

Läppulver n (=**Läpp·pulver**) lapping abrasive

Lärche f (-n) (*Bot.*) larch, tamarack cf LERCHE

Lärchen·baum m larch (tree). **-(baum)-harz** n larch resin, Venetian turpentine. **-holzöl** n larchwood oil. **-pech** n larch pitch. **-pilz, -schwamm** m purging agaric. **-stoff** m coniferin. **-terpentin** n larch (*or* Venetian) turpentine

Lard·öl n lard oil

Laricin·säure f laricinic acid, maltol

Larixin·säure f larixinic (*or* laricic) acid

Lärm m noise, din, commotion. **--bekämpfung** f noise abatement. **lärmen** vi make noise, clamor. **lärmend** p.a, **lärmig** a noisy

Lärm·schutzeinrichtung f noise reduction unit. **-schwerhörigkeit** f noise-induced hearing loss

Larve f (-n) 1 larva. 2 mask. **larvizid** a larvicidal

las read: past of LESEN

lasch a slack; limp; weak; tasteless

Lasche f (-n) 1 (*Mach.*) butt strap; shackle plate; (rail) joint plate, fishplate. 2 flap, tongue (*esp* of shoe), gusset. 3 groove

Laser·strahl m laser beam

lasieren vt glaze. **lasierend** p.a (*oft*) transparent. **Lasier·farbe** f glaze (color)

Läsion f (-en) lesion

laß a (*with endgs:* **lasse,** etc) lazy, sluggish

lassen* (ließ, gelassen; läßt) **I** vt **1** leave (in a place, in a condition): **weit hinter sich l.** leave far behind; **unvollendet l.** leave unfinished. **2** let (come, go, get): **Luft aus dem Reifen l.** let air out of the tire. **3** let have, allow; **einem Zeit l.** let sbdy have time; **sich Zeit l.** take one's time. **4** desist, refrain (from); stop (doing), cease, give up: **das Rauchen l.** give up smoking. **5** lose (e.g. one's life). **6** give credit for: **das muß man ihnen l.** you have to give them credit for. **II** vi **7** (+**von**) give up: **vom Alkohol l.** give up alcohol. **III** v aux (pp lassen) (+inf) **8** let, allow, permit; cause, induce (to do), make, have (do): **sie l. es auftauen** they let it thaw, they allow (cause, induce) it to thaw; **wir ließen sie es noch einmal versuchen** we let (or we had) them try it again; **das läßt sie uns fremdartig erscheinen** that makes them look strange to us cf HOFFEN. **9** (oft used to turn vi into vt, e.g.): **den Motor anlaufen l.** start the motor cf LAUFEN, LAUFENLASSEN. **10** have (sthg done): **sie haben sich die Zähne putzen l.** they had their teeth cleaned. *Cf* BLEIBENLASSEN, FALLENLASSEN, GELTEN, HÖREN, RUFEN, SEHEN. **IV** v aux: **sich l.** (+inf) **11** e.g. **es läßt sich machen** it can be done, it is possible to do it; (transl oft not lit) e.g. **es läßt sich denken** it is conceivable. **12** let oneself be (+pp): **sie l. sich nicht abschrecken** they won't (let themselves) be deterred. **V: sich l.: sich zu l. wissen** manage to control oneself. *Cf* GELASSEN

lässig a casual, nonchalant. **Lässigkeit** f casualness, nonchalance

läßlich a casual, lenient; pardonable

läßt leaves, lets, etc: pres of LASSEN

Last f (-en) **1** load, burden: **zur L. fallen** (+dat) become a burden (to). **2** weight. **3** freight, cargo; ballast. **4** debt, debit, charge: **es geht zu L—en des Kunden, es wird dem Kunden zu L—en gebucht** it is charged to the customer, it is at the customer's expense. **5: einem etwas zur L. legen** blame sbdy for sthg. **~auto** n (motor) truck

lasten vi (auf) **1** rest, weigh heavily (on); brood (over). **2** e.g.: **auf dem Haus l. Schulden** the house is burdened with debts

Lasten·aufzug m freight elevator. **l—frei** a debt-free, unencumbered

Laster I m (-) (motor) truck. **II** n (-) vice

lästig a troublesome, annoying: **l. fallen** (dat) be annoying (to), pester

Last·kraftwagen m (motor) truck. **-schrift** f debit item. **-wagen** m (motor) truck; freight car. **-zug** m **1** freight train. **2** truck and trailer

Lasur I f (-en) **1** glaze; glazing. **2** azure, azurite. **II** m (-e) ultramarine, lapis lazuli. **~blau** n ultramarine. **-farbe** f **1** glazing color, varnish. **2** azure; ultramarine. **l—farben** a azure, sky-blue; ultramarine

Lasurit n azurite; lapis lazuli

Lasur·lack m (transparent) varnish. **-schleiflack** m transparent flatting varnish. **-spat** m lazulite. **-stein** m lapis lazuli **—unechter L.** chessylite

Latein n Latin (language). **lateinisch** a Latin; roman (type)

Latenz f latency. **~zeit** f latent period

lateritisch a (Soil) lateric

Laterne f (-n) lantern; street light, lamppost

Laternen·bild n lantern slide. **-ofen** m lantern-shaped oven. **-pfahl** m lamppost

latex·führend a latex-bearing, laticiferous

Latsche f (-n), **Latschen·kiefer** f dwarf pine, knee pine. **-öl** n templin oil

latschig a slipshod

Latte f (-n) lath, slat; picket; crossbar

Latten·kiste f crate. **-verschlag** m crate; wooden shed; lattice partition. **-zaun** m picket fence

Lattich m (-e) lettuce

Latwerge f (-n) electuary

Latz m (-e) bib, flap

lau a lukewarm; tepid; mild, balmy

Laub n foliage, leaves. **l—artig** a foliaceous. **L—baum** m deciduous tree. **-blatt** n deciduous leaf

Laube f (-n) arbor, bower; arcade

Laub· foliage, leaf: **-erde** f leaf (or vegetable) mold. **-grün** n leaf (or chrome) green. **-holz** n **1** deciduous trees. **2** hardwood, leaf wood. **-holzkohle** f hardwood charcoal. **-säge** f coping saw, jigsaw. **l– verlierend** a deciduous. **L–wald** m deciduous forest. **l–wechselnd** a deciduous **Lauch** m (-e) leek. **l–·artig** a leek-like. **-grün** a leek-green

laudieren vt gloss (cloth)

Laue·bild n Laue photograph. **-fleck, -punkt** m Laue spot

Lauer I f: **auf der L.** on the lookout, (lying) in wait. **II** m (-) 1 wine from the second pressing. 2 (Distil.) low wine, singlings

lauern vi (+auf) lie in wait (for)

Lauf m (-e) 1 run —**im vollen L.** in full swing. 2 (foot) race. 3 (esp Mach.) running, motion, operation. 4 flow, current; circulation. 5 track, path, orbit; trajectory. 6 course: **seinen L. nehmen** take its course; **im L. der Zeit** in the course of time; **im L. des Tages** during the day; **freien L. lassen** (+dat) give free rein (to). 7 (gun) barrel. **-·bahn** f 1 career. 2 (running) track. 3 orbit

laufen* (lief, gelaufen; läuft) **I** vi (s) 1 (usu) run. 2 walk (=go on foot). 3 sail. 4 melt. 5 leak, drip. 6 go (well, badly), take its course. **II** vt (in combinations like:) **Schlittschuh l.** skate; **Schi l.** ski. **laufend** p.a 1 constant, ongoing. 2 current (year, etc). 3 consecutive (numbers). 4: **l— es Bad** (Dye.) standing bath. 5: **auf dem l—en bleiben** keep abreast of developments. Cf BAND II. **laufen·lassen*** vt let go (free), let take its course cf LASSEN 9

Läufer m (-) runner (incl Tex., Min., Mach.); slider, traveler; rotor. **-·gewicht** n sliding weight. **-mühle** f edge mill

Lauf·feldröhre f (Elec.) traveling wave tube. **-fläche** f bearing surface; (tire) tread; journal. **-getriebe** n running gear, mechanism. **-gewicht** n sliding weight. **-glasur** f (Ceram.) flow glaze. **-gummi** n rubber tire tread. **-katze** f trolley; crane carriage. **-kran** m traveling crane. **-nummer** f serial number. **-paß** m dismissal, walking papers. **-pfanne** f (Sugar) cooler. **-rad** n (genl) running wheel; (specif) landing wheel, trailing wheel, pony wheel, rotor. **-richtung** f running direction. **-rolle** f (Mach.) roller, caster; (trolley) wheel. **-schlacke** f running slag.

-steg m catwalk, runway, gangway

läuft runs: pres of LAUFEN

Lauf·term m variable term; (Spect.) current term. **-treppe** f escalator. **-werk** n (genl) mechanism, running gear; (specif) clockwork, record changer, film feed, truck (on railroad car). **-zahl** f variable number, running (or continuous) variable. **-zeit** f running (playing, transit, transmission) time; term (of validity); cycle; (Electron., oft) time delay. **-zeit- schwingung** f electron oscillation

laug·artig a lye-like, alkaline. **L–asche** f leached (or alkaline) ashes

laugbar a leachable

Lauge f (-n) 1 lye, alkaline (or caustic) solution. 2 liquor, solution. 3 (Bleaching) buck. 4 brine. 5 electrolyte. 6 (esp Metals) leach. 7 wash, suds. **-·anlage** f leaching plant. **-behälter** m leaching vat (tub, tank). **l–beständig** a alkali-resistant. **L–bottich** m leaching vat (tub, tank). **l– empfindlich** a sensitive to alkali. **-fähig** a leachable. **L–mittel** n leaching agent

laugen vt leach, lixiviate; steep in lye; buck

Laugen· lye, alkali(ne), caustic: **-akku(mulator)** m alkaline battery. **l– artig** a lye-like, alkaline. **L–asche** f leached ashes. **-bad** n lye bath, liquor. **l– beständig** a alkali-resistant. **L–bottich** m lye (or liquor) vat (tank, tub); tub (of washing machines). **l–echt** a fast to (caustic) lye. **L–eiweiß** n alkali albumin(ate). **-essenz** f caustic soda (or potash) solution. **-faß** n lye (or leaching) vat (or tub). **l–fest** a lyeproof, alkaliresistant. **L–flüssigkeit** f liquor

laugenhaft a lye-like, alkaline

Laugen· lye, alkali(ne), caustic: **-lösung** f leaching solution. **-messer** m alkalimeter. **-probe** f lye (or liquor) test (or sample). **-rückstand** m leaching residue, **-salz** n alkaline (or lixivial) salt —**bernsteinsaures L.** potassium succinate; **feuerbeständiges L.** fixed alkali; **flüchtiges L.** sal volatile (commercial ammonium carbonate); **mineralisches L.** sodium carbonate; **vegetabilisches L.** potassium carbonate. **l–salzig** a alkaline, lixivial. **L–turm** m lye (or liquor) tower. **-waage** f lye (or liquor) hydrometer; (*Salt*) brine gauge. **-wasser** *n* weak lye (*or* liquor), aqueous alkaline solution. **-zusatz** *m* addition of lye (*or* liquor)

Laugerei *f* (-en) lixiviation, leaching; leaching plant

Lauge·rückstand *m* leaching (lixiviation, extraction) residue

Laug·flüssigkeit *f* (*Metals, etc*) liquor

laugieren *n* leach; (*Tex.*) kier-boil

laugig *a* lye-like, alkaline, caustic

Laug·rückstand *m* leaching (lixiviation, extraction) residue

Laugung *f* (-en) leaching, lixiviation; backing; (*Tex.*) kier-boiling *cf* LAUGEN

laugungs·fähig *a* leachable. **L—mittel** *n* leaching agent

Laune *f* (-n) **1** mood. **2** whim, caprice. *Cf* GELAUNT. **launenhaft** *a* moody, capricious. **launig** *a* witty, clever. **launisch** *a* moody, capricious

Laurin·aldehyd *m* lauraldehyde. **-fett** *n* laurin. **-säure** *f* lauric acid

Laus *f* (÷e) **1** louse. **2** flaw; (*specif*): knot (in wood), spot (on dyed fabric)

lauschen *vi* (*dat*) **1** listen (to). **2** eavesdrop (on)

Läuse·körner *npl* stavesacre (*or* sabadilla) seed(s); cocculus indicus. **-kraut** *n* lousewort

lausen *vt* (de)louse

Läuse·pulver *n* insect (*specif* delousing) powder. **-rittersporn** *m* stavesacre. **-samen** *m* stavesacre seed(s) = LÄU-SEKÖRNER

laut I *a* (-er, -est) **1** loud. **2** noisy. **3**: **l. werden** become (*or* be made) known (*or* public); be voiced; spread (*as:* rumors). **4** *adv oft* aloud. *Cf* LAUTER. **II** *prep* (*genit, dat*) according to (reports, etc)

Laut *m* (-e) sound *cf* LAUTE

lautbar *a*: **l. werden** become known (*or* public)

Laute *f* (-n) **1** lute. **2** (*Dye.*) crutch. *Cf* LAUT

lauten *vi* **1** sound (e.g. favorable). **2** e.g.: **die Antwort lautet** the answer reads (says, goes, is to the effect). **3** (+**auf**): **die Papiere l. auf den Namen** . . . the papers are made out in the name of . . . ; **das Verdikt lautet auf** . . . the verdict is . . .

läuten *vt/i* ring; **es läutet** the bell is ringing

lauter I *a* **1** pure; clear. **2** honest, genuine. **II** *a* (*no endgs,* + *noun*) nothing but, only, sheer; (+*pl oft*) a whole lot of. **III** louder: *compar of* LAUT: **l. stellen** turn up (the volume of)

Läuter *m* (-) **I** ringing device, alarm *cf* LÄUTEN. **II** (*Distil.*) low wine, singlings. **III** (*in compds oft*) purifying, clarifying, refining *cf* LÄUTERN, LÄUTEROFEN. **-batterie** *f* (*Brew.*) underlet

Läuter·beize *f* (*Turkey Red Dyeing*) white liquor

Läuter·boden *m* (*Brew.*) perforated false bottom of: **-bottich** *m* (*Brew.*) clarifying (*or* straining) vat, lauter tub

Läuterer *m* (-) purifier, clarifier

Läuter·feuer *n* refining fire. **-flasche** *f* wash(ing) bottle

Lauterkeit *f* **1** purity; clarity. **2** genuineness

läutern *vt* purify, clarify, clear; refine; rectify; wash; strain; (*Lthr.*) drum

Läuter·ofen *m* refining furnace (*or* kiln). **-pfanne** *f* clearing pan, clarifier. **-trommel** *f* washing drum

Läuterung *f* (-en) purifying, etc *cf* LÄUTERN

Läuterungs·bad *n* clearing bath. **-kessel** *m* melting pot, refining boiler, clarifier. **-mittel** *n* purifying agent

Lautheit *f* loudness, noisiness *cf* LAUT I

lautlos *a* soundless, silent

Laut·schutzwand *f* noise-abating wall, baffle. **-sprecher** *m* loudspeaker. **l—stark** *a* vociferous. **L—stärke** *f* loudness, sound volume. **-stärkeregler** *m* volume control. **-verstärker** *m* (sound) amplifier. **-verstärkung** *f* amplification (of volume)

lau·warm *a* lukewarm, tepid *cf* LAU

Laven *lavas: pl of* Lava *f*

Lavendel *m* lavender. **-blüte** *f* lavender blossom. **l—farben, -farbig** *a* lavender (-colored). **L—öl** *n* lavender oil. **-wasser** *n* lavender water

Lavez·stein *m* potstone

Lävo· levo: **-glukosan** *n* levoglucosan. **l—gyr** *a* levorotatory

Lavör *m* (-e) washer

Lävulin *n* levulin. **-säure** *f* levulinic acid.

Lävulose *f* levulose, D-fructose

Lawine f (-n) avalanche

Laxans n (..nzien) laxative

Laxheit f laxity

laxieren vt purge, treat with a laxative.

laxierend $p.a$ laxative

Laxier·mittel n laxative, aperient. **-salz** n laxative salt: **englisches L.** Epsom salt(s)

Lazarett n (-e) (*Mil.*) (field) hospital

Lazulith n (*Min.*) lazulite; cordierite

Lazur 1 f glaze, etc. 2 m lapis lazuli. = LASUR

L.B. *abbv* Landolt-Börnstein (tables)

leben 1 vt/i (*oft + von*) live (on, by, off). 2 vi be alive. **Leben** n (-) life. **lebend** $p.a$ living, (a)live; **l—er Kalk** quicklime. **Lebende** m,f (*adj endgs*) living person (*or* being). **lebendig** a living, (a)live, active; lively; **l—er Kalk** quicklime; **l—e Kraft** vis viva, kinetic energy. **Lebendigkeit** f liveliness, vitality

Lebens· life, vital: **-alter** n age; stage of life. **-art** f 1 way of living, lifestyle. 2 manners, (good) breeding. **-baum** m arbor vitae, white cedar; tree of life. **-bedarf** m necessities of life. **-bedingungen** *fpl* living (*or* survival) conditions. **-bedürfnisse** *npl* necessities of life. **-dauer** f life(time), lifespan; durability. **-fähig** a viable, capable of living. **L—fähigkeit** f viability. **-funktion** f vital function. **-gefahr** f danger (to life), mortal danger **—in L.** (*oft*) critically ill (*or* injured). **l—gefährlich** a life-threatening, life-endangering; critical. **L—gemeinschaft** f life partnership; symbiosis. **-gewohnheit** f living habit. **-größe** f life size. **-haltung** f standard of living. **-holz** n lignum vitae. **-kosten** pl cost of living. **-kraft** f vital force, vitality. **-kunde** f biology. **l—länglich** a lifelong, lifetime; *adv* oft for life. **L—lauf** m career; curriculum vitae, life history. **-lehre** f biology. **-luft** f (*Old Chem.*) vital air, oxygen. **-luftmesser** m eudiometer. **-mittel** n (*usu pl*) food; groceries

Lebensmittel· food: **-chemie** f food chemistry. **-farbe** f food coloring. **-geschäft** n food store (*or* market). **-gesetzgebung** f food legislation. **-gewerbe** n food trade (*or* industry). **-technik** f food technology. **-überwachung** f food monitoring. **-untersuchung** f 1 food inspection. 2 food research

lebens· life, vital: **-notwendig** a vital. **L—ordnung** f way of life; (prescribed) diet, regimen. **-prozeß** m vital process. **-raum** m 1 living space. 2 habitat. **-saft** m vital fluid; (*Bot.*) latex. **-substanz** f vital substance. **-tätigkeit** f vital activity. **-träger** m (*Biol.*) biophore. **-unterhalt** m living, livelihood. **-verhältnis** n vital function. **-versicherung** f life insurance. **-vorgang** m vital process. **-wandel** m (way of) life, lifestyle. **-wasser** n aqua vitae, liquor, *esp* brandy. **-weise** f way of life, lifestyle. **l—wichtig** a vital, essential; *pred oft* a matter of life and death. **L—zeichen** n sign of life. **-zeit** f lifetime **—auf L.** for life

Leber f (-n) liver, (*in compds oft*) hepatic: **-beschwerden** *fpl* liver trouble (*or* ailment). **-blende** f (*Min.*) voltzite, (reniform) sphalerite, zinc blende; (*Chem.*) zinc sulfide. **-blümchen** n, **-blume** f hepatica, liverwort. **l—braun** a liver-colored, liver-brown. **L—brei** m liver pulp. **-drüse** f hepatic gland. **-eisenerz** n pyrrhotite. **-erz** n (*Min.*) curite; hepatic ore, (*specif*) hepatic (*or* liver-brown) cinnabar. **-farbe** f liver color. **l—farben** a liver-colored. **L—galle** f hepatic bile. **-käse** m fried liver loaf. **-kies** m pyrrhotite. **-klette** f agrimony. **-kraut** n liverwort, hepatica. **-mittel** n liver (*or* hepatic) remedy. **-moos** n liverwort, hepatica. **-nucleinsäure** f nucleic acid of liver. **-opal** m (*Min.*) menilite, brown opaque opal. **-schwefel** m liver of sulfur, hepar. **-stärke** f glycogen. **-stein** m 1 (*Min.*) hepatite. 2 (*Med.*) gallstone, biliary calculus. **-tran** m fish-liver oil, (*specif*) codliver oil. **-wurst** f liverwurst, liver sausage

Lebe·wesen n living being, organism. **-wohl** n farewell

lebhaft a lively; vivid; active, busy; strong. **Lebhaftigkeit** f liveliness, vividness; activity, animation, bustle

Leb·kuchen m spice cake; gingerbread

leblos a lifeless, dead; dull, flat. **Leblo-**

sigkeit *f* lifelessness; dullness

Lebzeiten *fpl:* **zu L.** (+**von** or +*genit*) during the lifetime (of)

Lecanor·säure *f* lecanoric acid

Lech *m* (*Metll.*) matte, regulus, lech

leck *a* leaky, leaking. **Leck** *n* (-e) 1 leak. 2 (*Elec., Nucl.*) = **Leckage** *f* (-n) leakage

Leck·dampf *m* leaking steam. **l–dicht** *a* leakproof

lecken I 1 *vt/i* lick, lap. 2 *vi* (**an, über**) lick (at), sweep, play (over) (*as:* flames, gases). II *vi* leak, drip, trickle

lecker *a* delicious, tasty; charming. **L– ·bissen** *m* tidbit, delicacy. **Leckerei** *f* (-en) (piece of) candy; delicacy

Leck·salz *n* block salt (for cattle). **-scha-den** *m* leak damage. **-stein** *m* lickstone; dripstone. **-strom** *m* leakage current. **-suche** *f* leak detection. **-verlust** *m* leakage loss

ledeburitisch *a* (*Steel*) ledeburitic

Leden *n* ledene

Leder *n* (-) leather. **--abfall** *m, usu pl:* **-ab-fälle** waste (*or* scrap) leather, leather cuttings (*or* trimmings). **-abschnitt** *m* leather cutting. **l–ähnlich, -artig** *a* leathery, leather-like. **L–band I** *m* 1 leather binding. 2 leather-bound volume. **II** *n* leather strap. **l–beschlagen** *a* leather-covered (-lined, -trimmed). **-braun** *a* leather-colored, leather-brown. **L–etui** *n* leather case. **-farbe** *f* leather color, buff. **L–far-ben, -farbig** *a* leather-colored, buff. **L–fett** *n* leather grease, dubbing. **l–gelb** *a* buff. **L–gummi** *n* (India) rubber. **-han-del** *m* leather trade. **-harz** *n* (India) rubber. **-haut** *f* corium, cutis, dermis; sclera. **-kitt** *m* leather cement. **-kohle** *f* leather charcoal, carbonized leather. **-lack** *m* leather varnish. **-leim** *m* hide glue, glue from leather waste. **-mehl** *n* leather powder (*or* meal)

ledern I *a* leather(n), leathery; dull, dry. II *vt* polish (with a chamois cloth)

Leder·öl *n* leather oil. **-pappe** *f* leather board. **-pflegemittel** *n* leather conditioner. **-riemen** *m* leather belt (*or* strap). **-scheibe** *f* leather disc; leather washer. **-schmiere** *f*, **-schmiermittel** *n* dubbing, leather polish. **-schnur** *f* leather cord (*or* string). **-schwärze** *f* leather blackening,

currier's black. **-seite** *f* flesh side. **-tang** *m* seaweed. **-tuch** *n* leathercloth, leatherette. **-wichse** *f* leather polish. **-zucker** *m:* **brauner L.** licorice paste; **schwarzer L.** licorice (candy); **weißer L.** marshmallow paste

ledig *a* 1 (+*genit*) free (of, from), devoid (of). 2 vacant, unoccupied. 3 single, unmarried. **lediglich** *adv* only, solely, purely

Ledum·campher, -kampfer *m* camphor of ledum, ledol

leer *a* empty, vacant; blank; unoccupied; unloaded: **l. ausgehen** go empty-handed; **l. laufen** (*Mach.*) (run) idle, run light *cf* LEERLAUFEN

Leer·band *n* blank tape. **-darm** *m* jejunum

Leere *f* 1 vacuum; vacuity. 2 = **Leere** *n* (*adj endgs*) emptiness, void

leeren 1 *vt* empty, evacuate. 2: **sich l.** (become) empty

Leer·faß *n* (*Paper*) emptying vat. **-gang** *m* (*Mach.*) 1 neutral (gear). 2 lost motion. 3 backlash. **-gewicht** *n* unloaded (empty, light) weight; tare. **-gut** *n* containers (barrels, crates, etc), empties. **-lauf** *m* I (*Mach.*) 1 idling. 2 neutral gear. II wasted motion; idleness

leer·laufen* *vi* (s) run dry *cf* LEER, LEERLAUF

Leer·raum *m* empty space, vacuum. **-stelle** *f* 1 empty space (*or* spot), void. 2 (*Cryst.*) vacant lattice position, Schottky defect

Leerung *f* (-en) 1 emptying, evacuation *cf* LEEREN. 2 postal collection

Leer·versuch *m* blank experiment; no-load test

Lefze *f* (-n) (overhanging) lip (of animals)

Leg. *abbv* (Legierung) alloy

Lege·batterie *f* hen battery

legen I *vt* 1 lay (down), put (in a lying position). 2 deposit (*incl* money). 3 set (*incl* hair); arrange, lay out. 4 plant. 5 (*Math.*) draw, construct. II *vi* lay (eggs). III: **sich l.** 6 lie down (=get into a lying position). 7 subside, let up. 8 (+**auf**, *oft*) concentrate (on), specialize (in). 9 (+**auf, über, um**) settle, descend (on, over, about). *Cf* GELEGT, GELEGEN

Legg. *abbv* (Legierungen) alloys

legierbar *a* alloyable. **Legierbarkeit** *f* alloyability. **legieren** vt 1 alloy (metals). 2 thicken (soup, sauce). **Legierung** f (-en) alloy

Legierungs·bestandteil m alloy constituent. **-rohmassel** f alloy ingot (or pig). **-stahl** m alloy steel. **-zusatz** m alloy additive

Legitimation f(-en) 1 authentication. 2 authorization. 3 identification (papers), credentials

Leguminose f (-n) legume, pod plant

Lehm m (-e) 1 clay. 2 loam. 3 mud. **l-·artig** a clayey; loamy. **L-erde** f rich clay. **-formguß** m loam casting. **-gelb** n clay yellow. **-grube** f clay pit. **-guß** m loam casting. **l-haltig** a loamy; clayey, argillaceous

lehmig a loamy; clayey; muddy

Lehm·kalk m argillaceous limestone. **-kern** m loam core. **-kitt** m clay (or loam) lute. **-mergel** m clay (or loamy) marl. **-mörtel** m (fire)clay mortar. **-schmierung** f clay (or loam) luting. **-stein** m unburnt brick, adobe. **-ziegel** m clay brick; unburnt brick; adobe

Lehne f (-n) 1 back (rest, support). 2 arm (rest). 3 slope, incline. **lehnen** vt & sich l. lean

Lehner·seide f nitrate (or Chardonnet) rayon

Lehr· teaching, educational: **-anstalt** f educational institution. **-b.** (abbv) = **-buch** n textbook. **-buchweisheit** f book learning, textbook wisdom

Lehre f(-n) 1 teaching(s), doctrine. 2 (esp in compds) science, study; theory: **Wärmelehre** science (study, theory) of heat cf KUNDE 5. 3 lesson, warning; (piece of) advice. 4 conclusion. 5 apprenticeship. 6 (Engg.) gauge, calipers; jig, pattern, template

lehren vt/i teach, instruct; show cf GELEHRT

lehren·haltig a true to gauge. **L-toleranz** f gauge tolerance

Lehrer m (-) **Lehrerin** f (-nen) teacher, instructor

Lehr·fach n 1 (teacher's) field of instruction. 2 teaching profession. **-gang** m 1 course (of study). 2 (training) course. **-kurs(us)** m course of study

Lehrling m (-e) apprentice

Lehr·mittel n instructional medium, teaching material. **-plan** m curriculum; syllabus. **l-reich** a instructive. **L-ring** m ring gauge. **-satz** m theorem, proposition. **-stelle** f teaching position. **-stuhl** m (academic) chair

Leib m (-er) 1 body. 2 belly, abdomen. 3 womb. 4 waist. 5 one's self (or person): **im L(—e)** within one(self); **am L—e** personally; **auf den L. rücken** (+dat) attack. **-binde** f sash; waistband; abdominal binder

Leibes·frucht f fetus, embryo. **-kräfte** fpl: **aus L—n** with all one's might

leiblich a 1 bodily, physical; somatic —das **L—e** (one's) bodily needs. 2 real, blood-related. 3 (one's) own; adv oft personally

Leich·dorn m corn (on the toe)

Leiche f (-n) corpse, (dead) body; cadaver

Leichen·alkaloid n, **-base** f putrefactive alkaloid (or base), ptomaine. **-fett** n adipocere. **-gift** n cadaveric poison, ptomaine. **-öffnung** f autopsy. **-schau** f (coroner's) inquest, post-mortem. **-verbrennung** f cremation. **-wachs** n adipocere

Leichnam m (-e) corpse, (dead) body

leicht a 1 light (weight). 2 slight; mild. 3 easy; casual; adv or pfx readily (esp + adj in ..bar or ..lich): **l. erklärbar** easily (or readily) explained, easy to explain; **l. löslich** readily soluble cf SCHWER. **-aschig** a light-ash (coal). **L-benzin** n light gasoline; light petroleum (b.p. 60–110°C). **-benzol** n light benzol. **-beton** m lightweight concrete. **l-beweglich** a readily mobile (esp liquids), easily movable. **L-beweglichkeit** f easy mobility. **-brand** m slightly burned material. **-brennstoff** m light fuel. **l-entzündlich** a readily flammable

Leichter m (-) lighter (=ship)

leichter·siedend a lower-boiling

leicht·fallen* vi (s) (+dat) be easy (for), come easily (to)

leicht· easily, readily; light(ly): **-flüchtig** a readily volatile. **-flüssig** a easily liquefied, readily fusible; thinly liquid; mobile. **L-flüssigkeit** f easy liquefiability, ready fusibility; (high) fluidity, mobility. **-frucht** f light grain (e.g. oats, emmer). **l-gläubig** a credulous, gullible

Leichtigkeit *f* lightness; ease, facility

Leicht- light(ly); easy, easily: **-kalkbeton** *m* lightweight silicate concrete. **-kornbeton** *m* lightweight aggregate concrete. **-kristall** *m* (*Glass*) light crystal. l–l. *abbv cf* L–LÖSLICH. **L–legierung** *f* light alloy. l–löslich *a* readily soluble. **L–metall** *n* light metal. **-öl** *n* light oil. l–schmelzbar *a* easily fusible. **-schmelzend, -schmelzlich** *a* low-melting, **-siedend** *a* low-boiling. **L–spat** *m* ground gypsum pigment. **-stein** *m* lightweight brick (*or* block). **-stoff** *m* lightweight material. l–verbrennlich *a* readily combustible, free-burning. **-zersetzlich** *a* easily decomposed. **L–zuschlagstoff** *m* lightweight aggregate (material)

leid *pred a* **1: es tut ihnen l.** they regret it, they feel sorry (for it). **2: es ist (es wird) ihnen l.** they are (getting) tired of it. **Leid** *n* **1** suffering, pain. **2** grief, sorrow **—L. tragen (um)** mourn (for). **3** harm. **leiden*** (litt, gelitten) **1** *vt* tolerate; (+**mögen** or **können**, *oft*) like. **2** *vt/i* (*oft* + *an*) suffer (from), undergo. **3** *vi* (+**unter**) be damaged (by). **Leiden** *n* (-) **1** ailment, illness. **2** suffering, torment. **leidend** *p.a* **1** ailing, (chronically) ill. **2** (*also as sfx*) suffering, afflicted

Leidener Flasche Leyden jar

Leidenschaft *f* (-en) passion

leider *adv* unfortunately, we are sorry to say

leidig *a* troublesome, annoying

leidlich *a* **1** tolerable. **2** fair (health); *adv oft* fairly well

leihen* (lieh, geliehen) **1** lend. **2** (+*dat rflx*): **sich etwas l. (bei, von)** borrow sthg (from)

Leim *m* (-e) glue; (*Paper*) size: **pflanzlicher L.** vegetable glue, gluten; **farbloser L., weißer L.** gelatin; **aus dem L. gehen** come apart. **-anstrich** *m* coat of glue (*or* size). l–artig *a* gluey, glutinous; gelatinous. **L–aufstrich** *m* coat of glue (*or* size). **-bottich** *m* (*Tex.*) sizing beck. **-druck** *m* size printing; collotype

leimen *vt* glue; (*Paper*) size. **leimend** *p.a* adhesive, agglutinative **—ganz l.** (*Paper*) hard-size

Leim- glue, size: **-fabrik** *f* glue factory.

-farbe *f* (*Paints*) distemper; (*Paper*) size color. **-faß** *n* glue vat; (*Paper*) size tank. **-festigkeit** *f* **1** bonding strength. **2** (*Paper*) imperviousness (*or* resistance) due to sizing. **-flüssigkeit** *f* size. **-galle** *f* (*Paper*) dark precipitate from rosin size. **-gallerte** *f* gelatin jelly. l–gebend *a* glue-yielding; gelatinous; collagenic. **L–gewebe** *n* collagen; (*Bot.*) collenchyma. **-glanz** *m* (*Lthr.*) size. **-gut** *n* glue stock. **-harz** *n* resin adhesive

leimig *a* gluey, glutinous, adhesive

leim- glue, size: **-kaschiert** *a* a glue-backed. **L–kitt** *m* joiner's cement (*or* glue), putty. **-kraut** *n* (*Bot.*) campion, flybane. **-küche** *f* glue cooking. **-kufe** *f* (*Tex.*) sizing beck. **-leder** *n* leather scrap (for glue-making). **-lösung** *f* glue solution. **-niederschlag** *m* (*Soap*) nigre, salt lye solution. **-pflaster** *n* adhesive plaster. **-pinsel** *m* glue brush. **-prober** *m* glue tester. **-pulver** *n* glue powder, powdered glue. **-rohstoff** *m* glue(-making) stock. **-seife** *f* filled (*or* cold) soap (with glycerine and lye left in). **-sieder** *m* glue boiler. **-siederniederschlag** *m* (*Soap*) nigre, soap solution in salt lye forming as an underlayer in curd soap making. **-stoff** *m* sizing material: gluten. **-substanz** *f* gelatinous substance. **-süß** *n* glycine, glycocoll. **-tiegel, -topf** *m* glue pot. **-trog** *m* (*Paper*) sizing vat.

Leimung *f* **1** gluing. **2** (*Paper*) sizing: **L. in der Masse** engine sizing. **3** (*Paint*) distempering

Leim-wasser *n* size; glue water. **-werk** *n* glue factory. **-zucker** *m* glycine, glycocoll

Lein *m* (-e) flax; linseed. **--dotter** *f* (*Bot.*) gold of pleasure, dodder (seed). **-dotteröl** *n* cameline oil

Leine *f* (-n) line, rope, cord; leash

leinen *a* (of) linen. **Leinen** *n* (-) linen; (bookbinding) cloth. **--band** *1 m* cloth-bound volume. **2** *n* tape. **-faden** *m* linen fiber (*or* thread). **-stoff** *m* linen cloth

Lein-firnis *m* linseed varnish (thickened linseed oil). **-kraut** *n* toadflax. **-kuchen** *m* oil cake, linseed cake. **-mehl** *n* linseed meal. **-öl** *n* linseed oil

Leinöl-fettsäure *f* fatty acid of linseed oil. **-firnis** *m* linseed oil varnish, boiled lin-

seed oil. **-harzlack** *m* linseed oil resin varnish. **-kitt** *m* linseed oil putty. **-kuchen** *m* linseed oil cake. **l–sauer** *a* linoleate of. **L–säure** *f* linoleic acid. **-schlichte** *f* linseed-oil size. **-trockenprozeß** *m* drying of linseed oil
Lein·pflanze *f* flax (plant). **-saat** *f*, **-samen** *m* linseed, flaxseed
Leinsamen·abkochung *f* decoction of linseed. **-mehl** *n* flaxseed (*or* linseed) meal. **-öl** *n* linseed oil. **-schleim** *m* linseed mucilage
Lein·schmalz *m* hydrogenated linseed oil. **-tuch** *n* 1 linen (cloth). 2 (bed)sheet. **-wand** *f* linen; canvas; screen
Leiocom, Leiogomm *n* (-e) leicome, dextrin
Leipziger Gelb *n* chrome yellow, lead chromate pigment
leise *a* 1 soft, faint (sound); quiet. 2 mild, gentle; slight. 3: **l—r stellen** turn down (the volume of)
Leiste *f* (-n) 1 lath, slat, strip (of wood); molding *cf* LEISTEN. 2 (*Tex.*) selvage. 3 (*Anat.*) groin
leisten *vt* 1 accomplish; do, carry out, perform (work, etc). 2 make (payments, restitution, etc); lend, give, provide (help, services, etc); offer (resistance) *cf* FOLGE 4. 3 (+*dat rflx*): **sich etwas l.** (a) afford sthg, (b) indulge in sthg
Leisten I *m* (-) (shoemaker's) last. **II** *pl of* LEISTE lath, etc
Leisten·bruch *m* inguinal hernia. **l–förmig** *a* striplike, lath-shaped
Leistung *f* (-en) 1 accomplishment, achievement. 2 performance, execution. 3 making, lending, giving, providing, offering *cf* LEISTEN. 2. 4 payment; (performance of a) service. 5 energy; capacity, power; productivity. 6 output, production. 7 piece of work; workmanship
Leistungs·abgabe *f* (*Mach.*) output. **-bedarf** *m* power requirements. **l–fähig** *a* efficient; high-performance; fit, capable; serviceable. **L–fähigkeit** *f* efficiency; adequacy; fitness, capacity; serviceability. **-faktor** *m* (*Elec.*) power factor. **l–los** *a* (*Elec.*) wattless, powerless. **L—messer** *m* power meter, (*Elec.*) watt meter. **-regler** *m* (*Elec.*) energy regulator. **-stufe** *f* level

of performance (*or* efficiency, etc). **-transistor** *m* power transistor. **l–unfähig** *a* inefficient; unfit, incapable, unserviceable. **L–zeiger** *m* (*Elec.*) wattmeter
Leit·artikel *m* editorial; feature article. **-bahn** *f* transit path. **-bahnwelle** *f* transit path oscillation. **-bild** *n* ideal; example, model. **-blech** *n* guide (*or* baffle) plate. **-bündel** *n* (*Bot.*) vascular bundle
leiten *vt* 1 conduct. 2 lead, guide; steer, channel, route. 3 direct, manage, run. *Cf* GELEITEN
Leiter I *m* (-) 1 (*Phys.*) conductor. 2 leader, guide. 3 director, manager, head. **II** *f* (-n) ladder. **·-bahn** *f* conducting path. **Leiterin** *f* (-nen) director, manager, head
Leiter·kreis *m* (*Elec.*) circuit. **-tafel** *f* nomogram
Leitf. *abbv* (Leitfähigkeit) conductivity
Leit·faden *m* 1 guide, manual. 2 central theme. **l–fähig** *a* conductive. **L–fähigkeit** *f* conductivity, conductance
Leitfähigkeits·gefäß *n* conductivity cell. **-koeffizient** *m* conductance ratio, conductivity coefficient. **-messung** *f* conductivity measurement. **-wasser** *n* conductivity water
Leit·flamme *f* pilot flame. **-fossil** *n* zone fossil. **-gedanke** *m* main theme (*or* idea). **-gewebe** *n* vascular tissue. **-isotop** *n* radioactive tracer. **-linie** *f* 1 guide line. 2 (*Math.*) directrix. **-probe** *f* standard sample. **-rohr** *n*, **-röhre** *f* conducting tube (*or* pipe); conduit pipe; hose, feeder; delivery tube (for gases). **-salz** *n* conducting salt. **-satz** *m* guiding principle. **-schicht** *f* (*Geol.*) key bed, marker bed. **-stelle** *f* central (*or* director's) office; control station. **-stern** *m* lodestar, guiding star; polestar. **-strahl** *m* 1 (*Math.*) radius vector. 2 (*Aero.*) guide (*or* pilot) beam. **-technik** *f* basic technology
Leitung *f* (-en) 1 conduction, transmission. 2 (*Water, Gas, etc*) pipe(s), piping; (pipe)-line; main; duct; tubing. 3 (*Elec.*) wire(s), wiring; line; circuit. 4 (*Telph.*) line. 5 leading, leadership, guidance; channeling, routing. 6 management, direction, administration, supervision. *Cf* LEITEN
Leitungs·draht *m* conducting wire. **-elektron** *n* conduction electron. **l–fähig** *a*

conductive. **L–fähigkeit** f conductivity.
-gas n gas from the pipe(s). **-hahn** m gas
tap; water tap (or faucet). **-kraft** f conducting power. **-messer** m conductome-
ter. **-netz** n 1 water supply (system). 2 gas
pipe system. 3 electric wiring (system).
-rohr n, **röhre** f (genl.) supply tube (or
pipe); (specif) water pipe, gas pipe.
-schnur f electric cord. **-spannung** f
(Elec.) line voltage. **-system** n 1 (water,
gas) supply system, system of pipes. 2
(electric) wiring system. 3 (Biol.) vascular system. **-vermögen** n conducting
power, conductivity. **-wärme** f heat of
conduction. **-wasser** n tap water. **widerstand** m (Elec.) resistance, (specif) line
resistance. **-zelle** f conductivity cell

Leit·vermögen n conducting power, conductivity, conductance. **-werk** n computer control section. **-wert** m (Elec.) conductance, admittance, susceptance. **-zahl**
f 1 (coefficient of) conductivity. 2 code (or
index) number

Lekanor·säure f lecanoric acid

Lektion f (-en) 1 lesson. 2 (one-hour) class

Lektüre f (-n) 1 reading. 2 reading matter

Lemnische Erde f Lemnian earth

Lemongras·öl n lemon-grass oil

Lende f (-n) loin(s). **Lenden·** lumbar

lenkbar a guidable, guided; steerable;
tractable

lenken 1 vt steer; guide, direct, control. 2
vt/i lead (the way). **Lenk·rad** n steering
wheel. **lenksam** a tractable, manageable. **Lenkung** f (-en) steering, guidance,
direction, control; (Mach.) steering gear

Lenzin n ground gypsum, annaline

Leonecopalin·säure f leonecopalinic acid.
Leonecopal·säure f leonecopalic acid

Lepargyl·säure f azelaic acid

Lepidokrokit m (-e) (Min.) lepidocrocite

Lepra f leprosy

Lerche f (-n) lark cf LÄRCHE

lernen 1 vt learn. 2 vi study, do one's les-
sons. Cf GELERNT, KENNENLERNEN. **Ler-
nende** m,f (adj endgs) learner

Les·art f (variant) reading, version

lesbar a readable. **Lesbarkeit** f
readability

Lese f (-n) 1 picking, gathering; vintage. 2
selection

Lese· reading; picking: **-band** n picking
belt. **-buch** n reader (=book). **-gerät** n
microfilm viewer. **-glas** n reading lens.
-holz n gathered wood. **-lampe** f reading
lamp. **-lupe** f reading glass (or lens)

lesen* (las, gelesen; liest) 1 vt/i read. 2 vt
pick, gather; cull, sort. **Leser** m (-),
Leserin f (-nen) reader

leserlich a legible

Lese·stein m 1 bog iron ore (limonite). 2
residual rock, boulder. **-stoff** m reading
matter

letal a lethal

Letten m (-) (red) potter's clay

Letter f (-n) letter; type. **Lettern·gut, -me-
tall** n type metal

lettig a clayey, loamy

lettisch a Lettish, Latvian. **Lettland** n
Latvia

letzen vt refresh

letzt a 1 last, final: **zum l—en Mal** for the
last time, **bis ins l—e** to the last detail;
bis zum l—en to the bitter end cf ENDE 1,
HAND. 2 latest; most recent: **in l—er Zeit**
in recent times. 3 ultimate, extreme. 4
worst, poorest. Cf LETZTERE. **Letzt: zu guter L.** to top it all. **letzt·angeführt** a last-
quoted, last-mentioned. **-endlich** a final,
ultimate; (adv oft) in the final analysis

letzte(n)mal: 1: das letztemal the last
time. **2: zum letztenmal** for the last time

letztens adv 1 lastly, finally. 2 recently

letztere a latter

letztesmal adv last time

letzt·genannt a last-named, last-mentioned. **-hin** adv recently, lately. **-jährig** a
last year's

letztlich adv 1 finally. 2 in the end

letzt·malig a final; adv last, for the last
time

Leuchämie f leukemia

Leucht· light, illuminating, luminous:
-bakterien npl photogenic (or luminous)
bacteria. **-beugung** f diffraction of light.
-bombe f flash bomb, flare. **-brenner** m
illuminating burner. **-dauer** f duration of
illumination. **-dichte** f brightness, specific luminous intensity. **-draht** m lighting filament

Leuchte f (-n) 1 light; lamp, beacon. 2 luminary. **Leucht·elektron** n optical elec-

tron, photo-electron. **leuchten** *vi* 1 shine; glow; sparkle; luminesce. 2 (*oft* + *dat* or + **in**) shine a light (on, into); light the way (for). **Leuchten** *n* (*oft*) luminescense. **leuchtend** *p.a* (*oft*) bright, luminous. **Leuchter** *m* (-) 1 light source. 2 candlestick; chandelier

Leucht· illuminating, luminous: **-erscheinung** *f* luminous phenomenon. **-faden** *m* lamp filament, luminous filament. **l–fähig** *a* capable of luminescence. **L–farbe** *f* luminous paint (*or* color); phosphor. **-gas** *n* illuminating gas. **-gasanstalt** *f* illuminating gas plant, gas works. **-gebilde** *n* luminous pattern. **-geschoß** *n* (*Mil.*) star shell. **-granate** *f* pyrotechnic shell. **-granatwerfer** *m* pyrotechnic projector. **-hülle** *f* luminous envelope; photosphere. **-käfer** *m* firefly, glowworm. **-kraft** *f* illuminating power, luminosity

Leuchtkraft·bestimmung, -messung *f* photometry. **-verstärker** *m* phosphorogen

Leucht· illuminating, luminous: **-kugel** *f* signal flare; Very light. **-masse** *f* luminescent mass. **-material** *n* luminescent material, phosphor. **-mittel** *n* illuminant; pyrotechnic device. **-munition** *f* illuminating (flare, tracer) ammunition. **-öl** *n* illuminating oil, (*specif*) kerosene. **-patrone** *f* signal (*or* tracer) cartridge, Very cartridge. **-petroleum** *n* kerosene. **-pistole** *f* signal (*or* Very) pistol. **-puder** *m* luminous powder. **-punkt** *m* flash point. **-quarz** *n* glow crystal. **-rakete** *f* signal rocket, flare. **-raketengestell** *n* pyrotechnics projector. **-rohr** n, **-röhre** *f* illuminating (luminescent, gas discharge) tube; fluorescent tube; neon tube. **-satz** *m* (*Explo.*) tracer (*or* flare) composition. **-schirm** *m* luminescent (*or* fluorescent) screen. **-spur** *f* light trail; (*Mil.*) tracer trajectory

Leuchtspur· tracer: **-geschoß** *n* tracer projectile. **-satz** *m* tracer composition

Leucht· illuminating, luminescent: **-stab** *m* flashlight. **-stärke** *f* candlepower, brightness. **-stein** *m* phosphorescent stone. **-stoff** *m* luminous substance; luminescent (*or* phosphorescent) material.

l–stoffbeschichtet *a* coated with luminescent material. **L–stofflampe, -stoffröhre** *f* fluorescent lamp (*or* tube). **-strahl** *m* light ray. **-stromröhre** *f* neon tube. **-turm** *m* lighthouse. **-vermögen** *n* illuminating power, luminosity. **-wert** *m* illuminating value. **-wurm** *m* glowworm. **-zeichen** *n* light (*or* flare) signal

Leucin·säure *f* leucinic acid

leucit·führend *a* leucite-bearing, leucitic

leugnen *vt* deny, disclaim; contradict

Leukämie *f* leukemia

Leuko·base *f* leuco base. **-körper** *m* leuco compound. **leukokrat** *a* (*Petrog.*) leucocratic. **Leukolyse** *f* leucolysis, leucocytolysis

Leukon·säure *f* leuconic acid

Leukophan *m* leucophane. **Leukoplast** *m* 1 (clothless) adhesive tape. 2 (*Bot.*) leucoplast. **Leuko·verbindung** *f* leuco compound. **Leukozyt** *m* (-en, -en) leucocyte, white corpuscle. **Leukozyten·zahl** *f* leucocyte count. **Leukozytose** *f* leucocytosis

Leumund *m* (-e) reputation

Leuna·salpeter *m* Leuna saltpeter (ammonium sulfate nitrate). **-werk** *f* Leuna (fixed nitrogen) plant

Leute *pl* people; employees; *in compds or as sfx, oft serves as pl of* **-mann** *or* **-männin** *cf* LANDSLEUTE

leutselig *a* 1 affable. 2 condescending

Leuzin *n* leucin

Leuzit *m* (*Min.*) leucite

levantieren *vt* (*Lthr.*) grain, levant

levantisch *a* Levantine

levigieren *vt* levigate

Levkoje *f* (-n) (*Bot.*) stock, gillyflower

lexikalisch *a* lexicographic

Lexikon *n* (..ka *or* ..ken) encyclopedia, lexicon

Leyd(e)ner *a* (*no endgs*) (of) Leyden: L. **Blau** Leyden (*or* cobalt) blue; L. **Flasche** Leyden jar

Lezithin *n* lecithin

lfd. *abbv* (laufend) current; consecutive. **lfd. Nr.** *abbv* (laufende Nummer) serial (*or* running) number

Lfg. *abbv* (Lieferung) issue, number, part (of a publication)

lg. *abbv* (lang) long

Lg. *abbv* ligroin

Libelle f (-n) 1 dragon fly. 2 (spirit) level
licht a 1 light (opp. of dark); bright, shining;
lucid. 2 thin, sparse; clear; wide, open
(mesh, etc). 3: l—e weite inside diameter;
l—e Höhe headroom. 4 (Colors; oft as pfx)
pale, light (-colored): **lichtgrün** light (or
pale) green
Licht n (-er) 1 light, daylight: L. machen
put on the light; bei L. besehen on closer
examination. 2 candle. 3 luminary, genius. **--abschluß** m exclusion of light. **-al-
terung** f light aging. **-art** f type of light.
-äther m luminiferous ether. **-aus-
schluß** m exclusion of light. **-bedarf** m,
-bedürfnis n light requirement(s). l—
beständig a photostable; (Colors) light-
resistant. **L—bild** n 1 photograph. 2 (lantern) slide. l—**bildlich** a photographic. **L—
bildung** f production of light, photogenesis. **-bildzeichnung** f photogrammetry. l—**blau** a light blue. **L—blende** f
light shield (or stop), diaphragm. **-blitz** m
flash of light. **-blitzentladung** f flash
discharge. **-bogen** m luminous (or electric) arc
Lichtbogen- arc: **-bildung** f arc formation,
arcing. **-festigkeit** f arc resistance. **-ofen**
m arc furnace. **-schweißung** f arc welding. **-zündung** f arc ignition
licht-braun a light brown. **-brechend** a refracting, refractive. **L—brechung** f refraction of light. **-brechungsvermögen**
n refractivity, refractive power; refractive index. **L—bündel, -büschel** n beam
(or pencil) of light. l—**chemisch** a photochemical. **-dicht** a lighttight, light-
proof. **L—druck** m 1 (Phys.) pressure of
light. 2 (Print.) photochemical printing,
phototype, photogravure, collotype, he-
liotype, photolithography. **-druckgela-
tine** f photoengraving gelatine. l—**durchlässig** a light-transmitting, translucent.
L—durchlässigkeit f light transmission,
translucence. l—**durchsichtig** a transparent. **-echt** a lightfast; fadeproof, non-
fading. **L—echtheit** f fastness to light.
-effekt m luminous effect. **-eindruck** m
luminous impression. **-einstrahlung** f irradiation (by light). **-einwirkung** f action
(or effect) of light. l—**elektrisch** a photoelectric. **L—elektron** n photoelectron.

l—**empfindlich** a light-sensitive, photo-
sensitive. **L—empfindlichkeit** f sensitivity to light, photosensitivity
lichten 1 vt clear (esp forest); thin out. 2:
sich l. thin out, grow sparse; clear up,
brighten (up)
licht-entwickelnd a light-emitting, photogenic. **L—entwicklung** f evolution (or
emission) of light, photogenesis
Lichter-fabrik f candle factory. l—**loh** a
blazing, ablaze: l. brennen be ablaze
Licht-erscheinung f luminous phenomenon; optical phenomenon; luminosity. l—
erzeugend a light-producing, photogenic. **L—faden** m light filament.
-farbe f color of light. l—**farben** a light-
colored. **L—farbendruck** m photo-
mechanical color printing. l—**farbig** a
light-colored. **L—filter** n light filter.
-form f candle mold. **-fortpflanzung** f
transmission of light. l—**freundlich** a
light-transmitting; (Biol.) photophilic.
-gebend a light-giving, luminous. **-gelb**
a light yellow. **L—genuß** m (Biol.) utilization of light, relation of light absorption
to light intensity. **-geschwindigkeit** f velocity (or speed) of light. **-gießen** n candle
molding. **-glanz** m brilliance, brightness.
l—**grau** a light gray. **-grün** a light green.
L—güte f quality of light. **-haut** f (Phys.)
cathode glow. **-hof** m halo
Lichthof- halo: **-bildung** f halation. l—**frei**
a halo-free, free from halation. **-schutz-
mittel** n anti-halation agent.
-schutzschicht f anti-halation layer. l—
sicher a halo-proof, halo-free
Licht-hülle f luminous envelope. **-impuls**
m light pulse. **-jahr** n light year. **-ka-
talysator** m photocatalyst. **-kegel** m
cone of light, luminous cone. **-kohle** f
light carbon (for arc lamps). **-körper** m
luminous body. l—**korpuskel** n light corpuscle, photon. **-kranz** m (Astron.) corona. **-kreis** m circle of light, halo. **-kunde**
f, **-lehre** f optics. **-leimdruck** m collotype.
-leitung f (Elec.) lighting circuit. **-loch** n
opening for light; pupil (of the eye). l—**los**
a lightless, dark. **L—marke** f light spot.
-markengalvanometer n spot galvanometer. **-maschine** f (Elec.) dynamo,
generator. **-maß** n clearance. **-mast** m

lamppost. **-menge** f quantity of light. **-messer** m photometer. **-meßkunst, -messung** f photometry. **-mikroskop** n optical microscope. **-mühle** f radiometer. **-nelke** f (Bot.) campion, ragged robin. **-netz** n lighting system (or network). **-öl** n kerosene. **-pause** f blueprint, photocopy(ing). **-pauspapier** n photocopy (or blueprint) paper. **-pausverfahren** n photocopying (or blueprinting) process. **punkt** m luminous (or bright) spot, focus. **-punktabmesser** m flying-spot scanner. **-quant(um)** n light quantum, photon. **-quelle** f light source. **-rakete** f rocket flare. **l–reflektierend** a light-reflecting. **L–reiz** m light stimulus. **-reklame** f illuminated advertising (sign). **-riß** m crack due to light. **-schalter** m (Elec.) light switch. **-schein** m glow (or gleam) of light. **l–scheu** a light-shunning, shady; photophobic. **L–schirm** m (lamp)shade screen. **l–schluckend** a light-absorbing, optically absorptive. **L–schreiber** m photographic recorder. **-schutz** m protection from light, hood, shade. **l–schwach** a (Opt.) small-intensity, faint. **L–schwingung** f light vibration. **-spiegler** m (light) reflector. **-spiel** n motion picture, movie. **-sprecher** m optophone. **-spur** f light track (trace, trail). **l–stark** a (Opt.) strong-intensity, optically powerful. **L–stärke** f intensity (of light), candlepower; (Lenses) speed. **-stärkemesser** m photometer. **-stärkemessung** f photometry. **-steindruck** m photolithograph(y). **-strahl** m light ray, luminous ray, ray of light. **-strahlung** f light radiation. **-strom** m luminous flux; (Elec.) lighting current. **-talg** m candle tallow. **-täuschung** f optical illusion. **l–undurchlässig** a opaque. **-unecht** a non-lightfast, non-lightproof, fading, fugitive. **-unempfindlich** a insensitive to light

Lichtung f (-en) **1** clearing, thinning (out); brightening cf LICHTEN. **2.** (forest) clearing; opening; (Biol.) lumen.

licht·voll a luminous; lucid. **L–weg** m path of light. **-weite** f inside diameter; clearance. **-welle** f light wave. **-werfer** m (optical) projector. **-wirkung** f action of

light; luminous effect. **-zeichen** n light signal, signal light; (specif) traffic light. **-zeiger** m optical pointer. **-zelle** f **1** photoelectric cell. **2** (Biol.) visual cell. **-zerstreuung** f diffusion of light. **-ziehen** n candle dipping. **-zufuhr** f admission of light

licken vt polish, burnish

Licker·bad n (Lthr.) fat-lickering bath. **-material** n fat liquor

lickern vt fat-liquor

Licker·öl n leather oil. **-verfahren** n fatliquoring process

Lid n (-er) (eye)lid. **-bindehaut** f (Anat.) conjunctiva

lidern vt **1** obturate (guns). **2** (Engg.) pack with leather, gasket. **Liderung** f (-en) **1** obturation. **2** gasket; packing

Lid·schatten m eye shadow

lieb a **1** dear, beloved —**l. haben** be fond of; **es ist ihnen l., daß..** they are glad that.. **2** hearty, cordial. **3** nice, kind; lovable. **Liebe** f (-n) love. **lieben** vt/i love; like, be fond of. **liebens·wert, -würdig** a amiable, kind

lieber a **1** (compar of LIEB) dearer; nicer, kinder, etc.; adv oft better: **sie wiederholen es l.** they had better repeat it. **2** adv (compar of GERN) rather, like . . . better; **l. haben** prefer cf MÖGEN

lieb·haben* vt be fond of. **Lieb·haber** m (-) lover; enthusiast, fan. **Liebhaberei** f (-en) hobby

Liebig·kühler m, **Liebigscher Kühler** Liebig condenser

lieblich a charming, delightful, pleasant

Lieblings· favorite

liebste a **1** (superl of LIEB) dearest, nicest, kindest, etc. **2** adv **am l–n** (superl of GERN): **am l–n haben** like most (or best of all); **am l–n tun** like to do best cf MÖGEN

Lieb·stock, -stöckel m (Bot.) lovage. **Liebstöckel·rüßler** m weevil, alfalfa snout beetle. **Liebstock·wurzelöl** n oil of lovage

Lied n (-er) song

liederlich a **1** slovenly, disorderly. **2** dissolute

lief ran, etc: past of LAUFEN

Lieferant m (-en, -en) supplier. **lieferbar** a

deliverable, available. **Lieferer** *m* (-) supplier, delivery person. **Liefer·frist** *f* delivery deadline

liefern *vt* 1 deliver. 2 yield, produce. 3 supply, provide. **Lieferung** *f* (-en) 1 delivery; supply. 2 installment (of a serially published book). **lieferungs·weise** *adv* serially, in installments. **Liefer·wagen** *m* delivery van

liegen* (lag, gelegen) *vi* (h *or* s) 1 lie (*lit* = be in a reclining position); be, be situated, be located: **die Schwierigkeit liegt darin, daß . . .** the difficulty lies in the fact that . . . ; **die Sache liegt kompliziert** the matter is complicated. 2 stay, not budge; be stuck. 3 (*esp* + *auf*) rest, weigh, lie heavy (on). 4 (+*dat):* **das liegt ihnen (gut)** that suits them. 5 (+**an**): **das liegt am Wetter** that is due to the weather. 6 (+**an, daran** & +*dat*): **es liegt ihnen nichts am Ruhm** fame means nothing to them; **ihnen liegt daran, das zu tun** it means much to them to do that. 7 (+**bei**): **die Entscheidung liegt bei ihnen** the decision is up to them (*or* is their affair). 8: **l. haben** have in stock. 9: **l. lassen** leave (where it is) *cf* LIEGENLASSEN. *Cf* GELEGEN, GELEGT, LEGEN, HAND, NAME. **liegen·bleiben*** *vi* (s) 1 remain lying; stay put. 2 be stuck; break down. 3 be left (behind), be forgotten. 4 be left unfinished. 5 be left unsold. 6 be postponed. **liegend** *p.a* (*oft*) inclined; horizontal; located. **liegen·lassen*** *vt* 1 forget, leave (behind). 2 leave where it is. 3 leave undone. 4 bypass; **links l.** ignore

lieh lent, etc: *past of* LEIHEN

Liesegangsche Schichtungen *pl* Liesegang rings

ließ let, left, etc: *past of* LASSEN

liest reads: *pres of* LESEN

Ligand *n* (-en) ligand, attached atom or group

lignin·haltig *a* containing lignin. **-reich** *a* rich in lignin

lignit·artig *a* lignitic

Lignocerin·alkohol *m* lignoceric alcohol

Likariol *n* licareol (l-linalool)

Likör *m* (-e) liqueur. **l-·artig** *a* liqueur-like. **L-wein** *m* liqueur wine

lila *a* (*no endgs*) lilac(-colored). **--blau** *a* lilac blue. **L-farbe** *f* lilac color

Lilak *m* (-e) (*Bot.*) lilac

Lila·öl *n* lilac oil

Lilie *f* (-n) lily

Lilien·gewächse *npl* (*Bot.*) Liliaceae. **-öl** *n* lily oil

Lima·bohne *f* Lima bean. **-holz** *n* Lima wood

Limbus *m* (..bi) 1 (*Bot., Meas. Instr.*) limb. 2 limbo

Limes *m* (-) 1 (*Math.*) limit. 2 boundary palisade

Limette *f* (-n) lime (=fruit). **Limetten·saft** *m* lime juice. **Limett·öl** *n* lime oil

Limno·technik *f* limnological technology

Limonade *f* (-n) fruit soda, soft drink

Limone *f* (-n) citron; **saure** (or **eigentliche**) L. lemon; **süße** L. lime

Limonen *n* (-e) limonene

Limon·grasöl *n* lemon grass oil. **-öl** *n* lemon oil

Linaloë· linaloa

lind *a* 1 gentle, mild; soft. 2 (*Silk*) scoured. 3 yellowish green

Linde *f* (-n) linden (tree), lime (tree)

Linden·blüte *f* linden blossom. **-gewächse** *npl* (*Bot.*) Tiliaceae. **-holz** *n* linden wood, basswood. **-kohle** *f* linden charcoal

lindern *vt* relieve, soothe, alleviate. **lindernd** *p.a* (*oft*) palliative, lenitive. **Linderung** *f* relief, alleviation. **Linderungs·mittel** *n* palliative, lenitive

Lind·leder *n* chamois leather

Lineal *n* (-e) (*Meas.*) rule(r), straightedge. **linealisch, linear** *a* linear, lineal. **Linear·maß** *n* linear measure

Linie *f* (-n) line —**in erster L.** primarily; **in absteigender L.** in descending order

Linien·anordnung *f* arrangement of lines. **-folge** *f* (*Spect.*) series of lines. **l-förmig** *a* linear. **L-gebilde** *n* system of lines. **-grenze** *f* (*Spect.*) convergence limit. **-kern** *m* nucleus of a line. **-mitte** *f* middle of a line. **-netz** *n* network of lines; graph paper. **-paar** *n* pair of lines. **-papier** *n* lined (*or* ruled) paper. **-raster** *m* ruled (*or* line) screen. **l-reich** *a* well-lined. **L-spannung** *f* (*Elec.*) line voltage. **-spek-**

trum n line spectrum. **-umkehr** f (*Spect.*) line reversal. **-verschiebung** f displacement of lines. **-zahl** f number of lines. **-zug** m plotted line (*or* curve)

lin(i)ieren vt rule (with lines). **Lin(i)ier-·farbe** f ruling ink

·linig sfx -lined, -linear

link a left, left-hand; (*Tex.*) **l—e Seite** wrong (*or* reverse) side. **Linke** f (*adj endgs*) left hand. **linkisch** a awkward, clumsy

links adv & prep left, to (*or* on) the left; (*as pfx oft*) levo.., left-hand, left-wing: **·-dr.** abbv = **-drehend** a levorotatory, counterclockwise, left-hand. **L—drehung** f levorotation, left-handed polarization; counterclockwise motion; left turn. **-fruchtzucker** m levulose. D-fructose. **l—gängig** a left-turning, counterclockwise; left-hand (screw). **L—gewinde** n left-hand thread. **l—herum** adv around to the left. **L—lauf** m counterclockwise motion, motion to the left. **l—laufend, -läufig** a running (to the) left, counterclockwise. **L—milchsäure** f levolactic acid. **-polarisation** f levopolarization. **-quarz** n left-handed quartz. **-säure** f levo acid. **-schraube** f left-hand screw. **l—seitig** a left-side(d); left-bank (of rivers). **L—weinsäure** f levotartaric acid. **l—wendig** a left-handed

Linnen n linen

Linolen·säure f linolenic acid

Linol·säure f linoleic acid

Linse f (-n) 1 lens. 2 lentil. 3 bob (of pendulum)

linsen· lenticular: **-ähnlich, -artig** a lenticular, lens-shaped. **L—erz** n oölitic limonite. **-glas** n lens. **l—groß** a lentil-sized. **L—kupfer** n lirconite. **-satz** m system of lenses

lip(o)id·löslich a lipide-soluble

Lipolyse f (-n) lipolysis. **lypolytisch** a lipolytic

Lipon·säure f liponic acid

lipophil a lipophilic

Lippe f (-n) lip; (*Sci.*) labium, labellum

Lippen· lip, labial: **-stift** m lipstick

Liquat·salz n aluminum salt solution used as disinfectant or astringent

liqueszieren vi liquefy, melt

liquidieren vt liquidate; vi charge (for)

lisieren vt (*Lthr*) roll (splits)

Lissabon n Lisbon

Lisseuse f (-n) (*Tex.*) backwasher. **lissieren** vt backwash

List f (-en) 1 trick, artifice. 2 cunning, craft(iness). *Cf* LISTE

Liste f (-n) list, roll, roster *cf* LIST

Listen·preis m list price

listig a tricky, crafty *cf* LIST

Lit. abbv (Literatur) literature

Litauen n Lithuania. **Litauer, litauisch** a Lithuanian

literarisch a literary

Literatur·angaben fpl references to the literature, bibliography

Liter·gewicht n density (as weight per liter). **-inhalt** m capacity in liters. **-meßkolben** m liter measuring flask

..lith sfx (*Min.*) ..lite

Lithargyrum n (..ren) litharge

Lithion n lithia. **·-glimmer** m lithia mica. **-hydrat** n lithium hydroxide

Lithium·alanat n lithium aluminum hydroxide. **-jodid** n lithium iodide. **-metall** n metallic lithium. **-salz** n lithium salt

Lithographen·firnis m lithographic varnish. **-tinte** f, **Lithographier·tinte** f lithographic ink. **lithographisch** a lithographic

lithophil a (*Petrog.*) lithophilic

litt suffered, etc: *past of* LEIDEN

Litze f (-n) 1 braid, cord. 2 (*Elec.*) stranded wire; (extension) cord. 3 strand (of rope). 4 (*Tex.*) heddle. **Litzen·draht** m stranded wire

Lixiv n (-e) lixivium, lye

Lizenz f (-en) license. **·-gebühr** f (patent) royalty. **-nehmer** m licensee

l.J. abbv (laufenden Jahres) of the current year

Lkw abbv (Lastkraftwagen) motor truck

l.l., l. lösl. abbv (leicht löslich) readily soluble

l.M. abbv (laufenden Monats) of the current month

Loangocopal(in)·säure f loangocopal(in)ic acid

Lob n (word of) praise

Lobelle *f* (-n) lobelia

loben *vt* praise, commend *cf* GELOBEN. **lobend** *p.a* (*oft*) laudatory. **löblich** *a* praiseworthy

Lob·rede *f* eulogy

Loch *n* (ᵉer) hole, opening; cavity; pore; eye. **--blech** *n* perforated plate. **-blende** *f* perforated screen (*or* diaphragm). **-eisen** *n* (hole) punch

lochen *vt* punch (a hole into); puncture; perforate. **Locher** *m* (-) punch(er), perforator

löcherig *a* full of holes, perforated; porous; honeycombed

Löcher·spektrum *n* hole spectrum. **-stelle** *f* gap, hole, open space

Loch·fraß *m* pitting (corrosion). **-karte** *f* punched card. **-kultur** *f* (*Bact.*) stab culture. **-scheibe** *f* perforated disc. **-stein** *m* perforated brick

Lochung *f* (-en) perforation, (hole) punching

Loch·weite *f* inside diameter, width (of a hole). **-zirkel** *m* internal calipers

locken *vt* 1 lure, entice. 2 curl (hair)

Locken·wickler *m* hair curler

locker *a* 1 loose, lax. 2 porous, spongy, fluffy. **locker·lassen*** *vi* yield, give in. **Lockerheit** *f* looseness, laxity; porosity, sponginess. **lockern** *vt* & **sich l.** loosen, relax. **lockernd** *p.a* (*Electron.*) antibonding. **Locker·stelle** *f* (*Cryst.*) loose spot. **Lockerung** *f* (-en) loosening, relaxation

Lock·flamme *f* pilot flame (*or* light). **-mittel** *n* bait. **-ruf** *m* birdcall; siren song

Lode *f* (-n) (*Bot.*) sprig, shoot

Loden *m* (*Tex.*) loden, fulled woolen cloth

Loder·asche *f* light ash, flyash. **lodern** *vi* flame, blaze, leap

Löffel *m* (-) spoon, ladle, scoop; shovel; bailer; (*Surg.*) currette. **--bagger** *m* power shovel, excavator. **l--förmig** *a* spoon-shaped, spatular. **L--kraut** *n* scurvy grass. **-krautöl** *n* oil of spoonwort

löffeln *vt/i* spoon, ladle, scoop, bail

Löffel·stiel *m* spoon handle. **l--weise** *adv* by the spoonful

log lied: *past of* LÜGEN

Logarithmen·tafel *f* logarithm table. **logarithmieren** *vt/i* take the logarithm (of).

logarithmisch *a* logarithmic. **Logarithmus** *m* (..men) logarithm

Logik *f* logic. **logisch** *a* logical

loh *a* blazing, fiery

Loh·bad *n* vegetable tanning bath. **-beize** *f* 1 tan liquor. 2 tanning process. **-brühe** *f* bark liquor, ooze. **-brühleder** *n* ooze leather

Lohe *f* (-n) 1 tanbark; tan liquor. 2 blaze, fire, flames

Loh·eiche *f* tanbark oak

lohen 1 *vt* tan (hides), steep in tan liquor. 2 *vi* blaze, burn

Loh·erde *f* tan earth. **-falzspäne** *mpl* vegetable shavings. **-farbe** *f* tan (color). **l--farben, -farbig** *a* tan(-colored), tawny. **L--faß** *n* tanning vat. **l--gar** *a* (vegetable *or* bark) tanned. **L--gare** *f* (bark) tanning; dressing. **-gerber** *m* (bark) tanner. **-gerberei** *f* bark tanning; tannery. **-gerbung** *f* (bark *or* vegetable) tanning. **-grube** *f* tan pit. **-kuchen** *m* tan cake, tan brick. **-messer** *m* barkometer

Lohn *m* (ᵉe) 1 wage(s), pay. 2 reward. **lohnen** 1 *vt* reward, pay. 2 *vi* & **sich l.** pay, be worthwhile. **löhnen** *vt* pay wages to. **lohnend** *p.a* (*oft*) profitable

Lohn·färber *m* job dyer

Lohnung, Löhnung *f* pay, payment

Loh·probe *f* (*Lthr.*) bark test. **-prober, -prüfer** *m* barkometer. **-pulver** *n* tan powder. **-rinde** *f* tanbark, oakbark. **-schicht** *f* bark layer

Lohschmidtsche Zahl Lohschmidt number

Loipon·säure *f* loiponic acid

Lok *f* (-s) locomotive, engine

Lokal *n* (-e) (*genl*) place, premises; (*specif*) inn, restaurant, tavern, store, meeting place. **--element** *n* (*Elec.*) local cell. **-farbe** *f* natural (*or* true) color. **-geschmack** *m* local taste

lokalisieren *vt* localize. **Lokalisierung** *f* (-en) localization

Lokal·wirkung *f* local effect

Lolch *m* (-e) darnel, rye grass

Lorbeer *m* (-en) (*Bot.*) laurel, bay. **--blatt** *n* bay leaf, laurel leaf. **-blätteröl** *n* oil of laurel leaves. **-kampfer** *m* laurin. **-öl** *n* laurel oil, bay oil. **-ölsäure** *f* lauric acid. **-rose** *f* oleander. **-spiritus** *m* bay rum.

-talg m laurel tallow

Lore f (-n) **1** truck. **2** dump car; gondola car

los I pred a & adv **1** loose cf LOSE. **2: etwas ist l.** sthg is happening (or is up); (oft + mit) sthg is wrong (with). **3: l. sein** (+accus or genit) be rid (of), have lost. **4: etwas l. haben** have sthg to one's credit. **II l.!** let's go!, get going! **III** sep pfx **5** loose (implies release, liberation) cf LOSLASSEN. **6** un-: **los-binden** untie. **7** rid (also implies loss, cf **3**) cf LOSWERDEN. **8** off, away (implies starting out, firing) cf LOSGEHEN. **9** (+**auf**) (implies heading for or toward, oft to attack): **los-fahren** rush (at). **IV** sfx -less cf FARBLOS, GERUCHLOS, MACHTLOS

Los n (-e) lot; (lottery) ticket; prize; fate

Lös. abbv (Lösung) solution

Losantin n bleaching powder

lösbar a soluble; separable; (re)solvable.

Lösbarkeit f solubility; separability; (re)solvability

los·binden* vt untie, unbind

los·blättern vi exfoliate, defoliate

los--brechen* vt, vi(s) break loose (off, away, out)

Lösch·arbeit f **1** (Metll.) charcoal process. **2** fire fighting. **3** unloading. **-blatt** n blotter, **-dienst** m fire fighting service

Lösche f (-n) **1** charcoal dust, slack, culm; cinder, clinker. **2** quenching tub (or trough). **3** quench(ing material). **--beton** m cinder concrete

Lösche-Mühle f (-n) Loesche mill

löschen vt **1** extinguish, put out. **2** slake, quench cf GELÖSCHT. **3** cancel, erase, eradicate, blot (out). **4** unload. **Löscher** m (-) extinguisher, quencher; slaker, hydrator; blotter; unloader

Lösch·flüssigkeit f extinguisher fluid. **-gerät** n **1** fire extinguisher. **2** (recording) tape eraser. **-grube** f (lime) slaking pit. **-kalk** m slaked lime. **-karton** m thick blotting paper. **-kohle** f quenched charcoal

Loschmidtsche Zahl f Loschmidt's (=Avogadro's) number

Lösch·mittel n extinguishing (or quenching) agent. **-papier** n blotting paper. **-spannung** f (Elec.) extinction potential. **-spule** f blow-out coil. **-turm** m quenching tower

Löschung f (-en) extinguishing; slaking, quenching; cancelation, erasing, eradication; blotting (out); unloading cf LÖSCHEN

Lösch·wasser n quenching (or tempering) water

los·drehen 1 vt twist off, unscrew. **2: sich l.** come loose (or unscrewed)

los·drücken 1 vt press off, detach. **2** vi fire, pull the trigger

lose a loose cf LOS **1** & **3**

Löse· dissolving: **-gefäß** n dissolving vessel. **-geschwindigkeit** f rate of dissolving. **-gut** n material to be dissolved. **-kraft** f solvent power. **l--kräftig** a strongly solvent. **L--lauge** f solvent liquor. **-mittel** n solvent

lösen I vt & **sich l. 1** dissolve (esp Chem.). **2** separate. **3** loosen (up), relax. **II** vt **4** undo. **5** release. **6** solve, **7** annul, cancel. **8** fire (a shot). **9** buy (esp tickets). **III: sich l. 10** come loose (or undone). **11** move away, detach oneself; be released. **12** resolve itself, be solved. **13** be fired, go off (as: shots). Cf GELÖST. **lösend** p.a (oft) solvent; expectorant; laxative. **Löser** m (-) dissolver, solvent; solubilizer; solver; (Mech.) release

Löse·verlust m loss of solution. **-vermögen** n dissolving power. **-vorgang** m dissolving process, solvent action. **-wirkung** f dissolving action, solvent effect

los·gehen* vi (s) **1** leave, go away. **2** come loose. **3** (+**auf**) go (at, up, to), head (for); rush (at), attack. **4** go off (as: guns, etc). **5** start, cut loose

losgelöst p.a separate, free. See also LOSLÖSEN

los·kitten vt unglue, etc cf KITTEN

los·kommen* vi (s) get away; (+**auf**) come (at)

los·kuppeln vt uncouple, detach, disconnect, disengage

lösl. abbv (löslich) sol. (soluble). **Lösl.** abbv (Löslichkeit) solubility

los·lassen* vt **1** let go of; let loose, release; free. **2** fire off

löslich a soluble cf LÖSLICHMACHEN. **Löslichk.** abbv = **Löslichkeit** f solubility

Löslichkeits·bestimmung f solubility determination. **-druck** m solubility pressure. **-grenze** *f* limit of solubility. **-pro-dukt** *n* solubility product. **-steigerung** *f* increase in solubility. **-unterschied** *m* difference in solubility. **-veränderung** *f* change in solubility. **-verhältnis** *n* solubility relation. **-verminderung** *f* decrease in solubility. **-zahl** *f* solubility value

löslich·machen *vt* solubilize, render soluble. **Löslich·machung** *f* solubilization

loslösbar *a* separable, detachable

los·lösen 1 *vt* separate, detach; release, set free *cf* LOSGELÖST. **2: sich l.** come loose, separate

los·löten *vt* unsolder

los·machen I *vt* **1** set free, release. **2** detach; loosen; undo, unfasten; disengage. **II** *vi* **3** set sail. **4** get going. **III: sich l.** get away, free (*or* disengage) oneself

los·platzen *vi* (s) go off, explode

los·reißen* **1** *vt* tear off (away, loose). **2: sich l.** break loose; tear oneself away

Löß *m* loess

los·sagen: sich l. (von) renounce, disown

Löß·boden *m* loess soil

Los·scheibe *f* (*Mach.*) loose pulley

los·schrauben *vt* unscrew, screw off

los·sprechen* *vt* release, free; absolve; acquit

los·sprengen *vt* blast loose (off, away)

Lost *n* mustard gas

los·trennen *vt* separate, detach; undo

Lost·vergiftete *m,f* (*adj endgs*) mustard gas casualty

Losung *f* (-en) **1** watchword, motto. **2** (animal) droppings

Lösung *f* (-en) **1** solution (*incl Chem.*); dissolving: **in L. gehen** go into solution; **in L. bringen** dissolve. **2** dissolution. **3** annulment, cancelation. **4** resolution (of a problem, etc)

Lösungs·behälter *m* solution (storage) tank. **-benzin** *n* benzine, mineral (*or* white) spirits, petroleum distillate. **-benzol** *n* solvent benzene (*or* naphtha). **-dichte** *f* concentration. **-druck** *m* Nernst solution pressure; (*Electrochem.*) solution pressure. **-erscheinung** *f* phenomenon of solution. **-extraktion** *f* solvent extraction. **l–fähig** *a* capable of solution. **L–fähigkeit** *f* dissolving capacity.

-flüssigkeit *f* solvent (liquid). **-geschwindigkeit** *f* rate of solution. **-koeffizient** *m* coefficient of solution. **-kohle** *f* (*Metll.*) cementing carbon

Lösungsm. *abbv* = **Lösungs·mittel** *n* solvent

Lösungsmittel·beständigkeit *f* stability toward solvents. **-dampf** *m* solvent vapor. **l–echt** *a* solvent-resistant. **-frei** *a* solvent-free. **L–gemisch** *n* solvent mixture. **-rückgewinnung** *f* solvent recovery

Lösungs·mittler *m* solubilizer, solution aid. **-stärke** *f* strength of solution. **-tendenz** *f* tendency to dissolve. **-theorie** *f* theory of solution. **-vermittler** *m* solution aid, solubilizer, hydrotrope. **-vermögen** *n* dissolving power. **-vorgang** *m* dissolving process. **-vorschrift** *f* directions for dissolving. **-wärme** *f* heat of solution. **-wasser** *n* solvent water. **-wirkung** *f* solvent effect

los·werden* *vi* (s) (+*accus*) get rid of

los·wickeln *vt* unwind, unwrap, uncoil

los·ziehen* *vi* (s) **1** start out; (+**auf**) head (for). **2** (+**gegen**) inveigh (against)

Lot *n* (-e) **1** solder. **2** plumb (bob): **im L. (a)** in plumb, (**b**) in order; **ins L. bringen** set right. **3** sounding lead. **4** perpendicular. **5** (*Meas.*) 10 grams; (*Silver*): **16 L.** = 1000 parts fine (=pure)

Löt·apparat *m* soldering equipment

lötbar *a* solderable

Löt·blei *n* lead solder. **-brenner** *m* soldering burner, gas blowpipe. **-draht** *m* soldering wire

Löte *f* solder; soldering

Löt·eisen *n* soldering iron

loten 1 *vt/i* plumb, sound, take soundings (of). **2** *vi:* **in die Tiefe l.** investigate in depth

löten *vt* solder, braze; seal (glass); agglutinate

Löt·fett *n* soldering paste (*or* flux). **-flamme** *f* blowpipe flame. **-fluß** *m* soldering flux. **-flüssigkeit** *f* soldering fluid. **-fuge** *f* soldered (*or* brazed) joint. **-glas** *n* solder glass

Lothringen *n* (*Geog.*) Lorraine

lötig *a* **1** weighing 10 grams. **2** (*Ores*) containing 10 g silver per hundredweight. **3**

(*Silver*) fine, pure; (*as sfx on numbers*) . . . 16ths fine: **13lötiges Silber** 13/16 fine silver; **16lötiges Silber** 100% (*or* 1000 parts) fine silver. **Lötigkeit** *f* (*Silver*) fineness

Löt·kolben *m* soldering iron (*or* gun). **-lampe** *f* soldering lamp (*or* torch), blowtorch

Lot·leine *f* plumb line, sounding line

Löt·masse *f* soldering compound (*or* composition). **-metall** *n* soldering metal. **-mittel** *n* solder. **-paste** *f* soldering paste. **-probe** *f* blowpipe test

lot·recht *a* perpendicular, vertical

Löt·rohr *n*, **röhre** *f* blowpipe

Lot·röhre *f* sounding tube

Lötrohr·flamme *f* blowpipe flame. **-fluß** *m* blowpipe flux. **-gebläse** *n* blast lamp, blow lamp. **-kapelle** *f* blowpipe cupel. **-kohle** *f* blowpipe (test) charcoal. **-lampe** *f* blowpipe lamp. **-probe** *f* blowpipe test. **-probierkunde** *f* blowpipe analysis. **-prüfgerätschaften** *pl* blowpipe testing apparatus. **-reagens** *n* blowpipe reagent. **-versuch** *m* blowpipe experiment (*or* test)

Löt·salmiak *m* sal ammoniac for soldering. **-salz** *n* soldering salt, ammonium tetrachlorozincate. **-säure** *f* soldering acid

Lot·schnur *f* plumb line

Lotse *m* (-n, -n), **lotsen** *vt/i* pilot

Löt·stelle *f* 1 solder(ed) joint. 2 junction (of a thermocouple)

Lotte *f* (-n) (*Iron*) bloom, ball

Lötung *f* (-en) soldering, brazing, sealing; agglutination, adhesion *cf* LÖTEN

Löt·versuch *m* blowpipe (*or* soldering) test. **-wasser** *n* soldering fluid, (*usu*) zinc chloride solution. **-zinn** *n* soldering tin

Löwe *m* (-n, -n) lion

Löwen·maul *n* (*Bot.*) snapdragon; toadflax. **-zahn** *m* dandelion. **-zahnbitter** *n* taraxacin

Löwin *f* (-nen) lioness

LP-Beton *m abbv* (Luftporenbeton) air-entrained concrete

Lsg. *abbv* (Lösung) sol. (solution). **Lsgg.** *abbv* (Lösungen) sols. (solutions). **Lsgm.** *abbv* (Lösungsmittel) solvent

lt. *abbv* (laut) according to

Luch *f* (-̈e) or *n* (-e) marsh, swamp

Luchs *m* (-e) lynx

Lucke *f* (-n) (*Metll.*) blister

Lücke *f* (-n) gap; interstice; loophole; vacancy, deficiency

lücken·bildend *a* gap-forming. **L—bindung** *f* unsaturated linkage, multiple bond. **-büßer** *m* stopgap

lückenhaft *a* incomplete, sketchy; defective

lücken·los *a* gapless, complete; uninterrupted; nonporous; airtight. **L—parenchym** *n* spongy parenchyma. **-satz** *m* vacancy principle

luckig *a* (*Metll.*) porous, honeycombed

lückig *a* 1 incomplete; defective. 2 (*Metll.*) porous, honeycombed

lud loaded, etc.: *past of* LADEN

Luder *n* (-) 1 carrion, carcass. 2 slob; bitch

Ludolfsche Zahl *f* the number π (pi)

Luft *f* (-̈e) 1 air, atmosphere; open air, (*esp pl*) sky: **an der L.** in (the) air; **in die L. blasen** (or **sprengen**), **in die L. gehen** blow up, explode; **aus der L. greifen** invent, make up. 2 breath. 3 breeze, draft

Luft·ablaß *m* escape of air; air outlet. **-absäuger** *m* (air) exhauster, extractor. **-abscheider** *m* air separator, deaerator. **-abschluß** *m* exclusion of air; air trap. **-abschreckung** *f* (*Metll.*) air quenching. **-abwehr** *f* air defense, antiaircraft. **-abzug** *m* air outflow; air vent; air exhauster. **l—ähnlich** *a* gaseous, aeriform. **L—anfeuchter** *m* air moistener, humidifier. **-angriff** *m* air attack, air raid. **l—artig** *a* gaseous, aeriform. **L—aufbereitung** *f* air conditioning. **-aufnahme** *f* 1 air absorption. 2 aerial photograph. **-aufnehmer** *m* air receiver. **-auftrieb** *m* air buoyancy; distension. **-ausdehnungsmaschine** *f* hot-air engine. **-auslaß** *m* air outlet (vent, exhaust). **-ausschluß** *m* exclusion of air. **-bad** *n* 1 air bath. 2 outdoor bathing area. **-bakterium** *n* airborne bacterium. **-befeuchter** *m* air humidifier. **-befeuchtung** *f* air moistening. **-behälter** *m* air holder (tank, reservoir, chamber, sac). **-beschaffenheit** *f* condition (*or* composition) of the air. **l—beständig** *a* stable in air. **L—beständigkeit** *f* stability in air, resistance to air. **-bestandteil** *m* constituent of (the) air. **-bild** *n* 1 aerial (*or* air) photograph. **2** (*Opt.*) aerial image; vision, phantasm. **-bläs/chen** *n* small air bubble, blister. **-blase** *f* air bubble (pocket, bladder). **-bleiche** *f* open-air bleaching. **-bombe** *f* aerial bomb. **-bremse** *f* air brake. **-brennstoffgemisch** *n* fuel-air mixture. **-brücke** *f* airlift. **l–dicht** *a* airtight, hermetic. **L–dichte** *f* air (*or* atmospheric) density. **-dichtemesser** *m* densimeter. **-dichtheit** *f* airtightness. **-dichtigkeit** *f* air (*or* atmospheric) density. **l–dichtschließend** *a* hermetic. **L–druck** *m* air pressure, atmospheric pressure

Luftdruck·anzeiger *m* air pressure gauge (*or* indicator). **-messer** *m* barometer, manometer. **-pumpe** *f* pneumatic pump, air compressor. **-tendenz** *f* barometric tendency. **-unterschied** *m* difference in air pressure

luft·durchlässig *a* air-permeable, porous. **L–durchlässigkeit** *f* permeability by air. **-durchlässigkeitsprüfer** *m* (*Paper*) densometer. **-düse** *f* air jet; air nozzle. **l–echt** *a* resistant to atmospheric influence, air-stable. **L–einlaß** *m* air inlet (*or* intake). **-einschluß** *m* air inclusion (*or* occlusion). **-eintritt** *m* air inlet (*or* intake). **-einwirkung** *f* action (*or* effect) of air. **-eisenstein** *m* iron meteorite, (*Min.*) aerosiderite. **-elektrizität** *f* atmospheric electricity. **l–empfindlich** *a* sensitive to air

lüften 1 *vt/i* air, ventilate; aerate. **2** *vt* raise, lift; reveal; (*Mach.*) release, clear. **3: sich l.** rise; be revealed

luft·entzündlich *a* flammable in contact with air, pyrophoric

Lüfter *m* (-) ventilator, fan

Luft·erhärtung *f* air hardening. **-erhitzer** *m* air heater. **-erneuerung** *f* air replacement, ventilation. **-erscheinung** *f* atmospheric phenomenon. **-erscheinungslehre** *f* meteorology. **-fahrt** *f* aviation, aeronautics; air travel. **-fahrzeug** *n* aircraft. **-falle** *f* air trap. **-feuchtigkeit** *f* atmospheric moisture. **-feuchtigkeitsmesser** *m* hygrometer. **-filter** *n* air filter. **-förderung** *f* 1 air displacement. **2** air transportation. **l–förmig** *a* gaseous,

aeriform. **-frei** *a* air-free. **L–gang** *m* air duct. **-gas** *n* producer gas, air gas. **-gaskrieg** *m* chemical air war(fare). **l–gefüllt** *a* air-filled. **L–gehalt** *m* air content. **l–gekühlt** *a* air-cooled. **L–geräusche** *npl* static, atmospherics. **l–gesichtet** *a* airfloated. **-getrocknet** *a* air-dried. **L–gewehr** *n* airgun. **-güte** *f* air quality. **-gütemesser** *m* eudiometer. **-hafen** *m* airport. **-hahn** *m* air valve. **l–haltig** *a* containing air. **l–härtend** *a*, **L–härtung** *f* air-hardening. **-heer** *m* air lift, air lever. **-hefe** *f* aerial yeast. **-heizung** *f* hot-air heating, **-hülle** *f* atmosphere

luftig *a* 1 airy; aerial. **2** light(weight). **3** flighty

Luft·inhalt *m* air content. **-ion** *n* atmospheric ion. **-kalk** *m* air-hardening (*or* non-hydraulic) lime. **-kammer** *f* air chamber. **-kampf** *m* aerial combat, air battle. **-kampfstoff** *m* (*Mil.*) volatile chemical agent. **-kanal** *m* air duct. **-kessel** *m* air chamber (*or* reservoir). **-kissen** *n* air cushion. **-klimatisierung** *f* air conditioning. **-kohlensäure** *f* atmospheric carbon dioxide. **-kompressor** *m* air compressor. **-konditionierung** *f* air conditioning. **-kraftmaschine** *f* hot-air engine. **-kreis** *m* atmosphere. **-kühler** *m* air cooler. **-kühlung** *f* air cooling **-kunstseide** *f* aerated rayon. **-lack** *m* airproof (*or* air-drying) varnish. **-lauf** *m* air passage; air flow. **l–leer** *a* evacuated, vacuous; (*Tires*) airless: **l—er Raum** vacuum. **L–leere** *f* vacuum, absence of air. **-leeremesser** *m* vacuometer, vacuum gauge. **l–leitend** *a* air-conducting. **L–leitung** *f* 1 air piping, air (pipe)line. **2** (*Elec.*) overhead wire(s), aerial transmission line. **-loch** *n* air hole, vent. **-lücke** *f* air space (*or* gap). **-malz** *n* air-dried malt. **-mangel** *m* lack of air. **-mantel** *m* air mantle (*or* jacket). **-menge** *f* quantity of air. **-messer** *m* aerometer. **-meßnetz** *n* air-analysis network. **-messung** *f* aerometry. **-mörtel** *m* non-hydraulic mortar; (air-hardened) lime mortar. **-nitrit** *n* nitrite from atmospheric nitrogen. **-oxidation** *f* atmospheric oxidation. **-patentieren** *n* (*Wire*) patenting with cooling in air. **-pflanze** *f* airplant, epiphyte. **-pinsel** *m* air brush. **l—porenbildend** *a* air-entraining. **L—porenbildner** *m* air-entraining agent. **-post** *f* air mail. **-presser** *m* air compressor. **-pressung** *f* 1 air compression. 2 air pressure. **-prüfer** *m* air tester. **-pumpe** *f* air (*or* pneumatic) pump; tire pump. **-raum** *m* air space; atmosphere. **-reifen** *m* pneumatic tire. **l—reinigend** *a* air-purifying. **L—reiniger** *m* air purifier. **-rohr** *n* air pipe (tube, duct). **-röhre** *f* 1 air pipe (tube, duct). 2 windpipe, trachea. **-röhren-** tracheal: **-röhrenentzündung** *f* tracheitis. **-rückstand** *m* air residue, residual air. **-salpetersäure** *f* nitric acid from atmospheric nitrogen. **-sättigung** *f* saturation of the air. **-sauerstoff** *m* atmospheric oxygen. **-saugapparat, -sauger** *m*, **-saugpumpe** *f* aspirator, exhauster. **-säule** *f* air column. **-säure** *f* carbonic acid. **l—scheu** *a* (*Bact.*) anaerobic. **L—schicht** *f* layer (*or* film) of air. **-schiff** *n* airship; dirigible; blimp. **-schiffer** *m* aeronaut. **-schlag** *m* (*Pyro.*) petard. **-schlauch** *m* 1 air hose. 2 inner tube. **-schleuse** *f* air lock. **-schöpfen** *n* respiration; catching one's breath. **-schraube** *f* airscrew, propeller. **-schutz** *m* air(-raid) defense. **-schutzraum** *m* air-raid shelter. **-schwärmer** *m* (*Pyro.*) serpent. **-seide** *f* air silk, hollow rayon. **-seilbahn** *f* aerial (cable) tramway. **-sog** *m* air suction, vacuum. **-spalt** *m* (*Elec.*) air gap. **-spiegelung** *f* mirage. **-stählung** *f* air hardening. **-stickstoff** *m* atmospheric nitrogen. **-störung** *f* atmospheric disturbance. **-stoß** *m* air blast. **-strahl** *m* air jet; air blast. **-strecke** *f* 1 (*Elec.*) air gap. 2 (*Aero.*) airway. **-strom** *m*, **-strömung** *f* air stream, air current. **-stützpunkt** *m* air base. **-trennung** *f* separation of air. **l—trocken** *a* air-dry, air-dried. **L—-trockengewicht** *n* air-dry weight. **-trocknung** *f* air drying. **-überschuß** *m* excess of air. **l—undurchlässig** *a* air-impermeable. **L—undurchlässigkeit** *f* impermeability to air

Lüftung *f* (-en) airing, ventilation; aeration; raising, lifting; revelation; release. *cf* LÜFTEN

Luft·ventil *n* air valve. **-verdichter** *m* air condenser (*or* compressor). **-verdichtung**

f air condensation (*or* compression). **l—verdorben** *a* air-damaged, weathered. **L—verdrängung** *f* air displacement. **l—verdünnt** *a* partially air-exhausted. **L—verdünnung** *f* rarefaction of air; partial vacuum. **-verdünnungsmesser** *m* vacuum gauge. **-verflüssigung** *f* liquefaction of air. **-verflüssigungsanlage** *f* air-liquefying unit; liquid-air plant. **-vergütung** *f* (*Metll.*) heat treatment with air. **-verkehr** *m* air transporation (*or* traffic). **-verpester** *m* (-) air polluter (*or* pollutants). **-verunreinigung** *f* air pollution. **-volumen** *n* volume of air. **-vorwärmung** *f* preheating of air. **-vulkanisation** *f* (*Rubber*) open cure. **-waage** *f* aerometer. **-waffe** *f* air force. **-wärme** *f* air temperature. **-wechsel** *m* ventilation. **-weg** *m* (*Aero.*) airway —**auf dem L.** by air. 2 (*Anat.*) air passage, *pl* of respiratory tract. **-wehr** *f* air defense. **-weiche** *f* (*Malt*) air rest. **-welle** *f* air wave. **-wichte** *f* air density. **-widerstand** *m* air (*or* atmospheric) resistance, aerodynamic drag. **-winde** *f* air winch. **-wirkung** *f* action of the air, atmospheric effect. **-wurzel** *f* (*Bot.*) aerial root. **-zerstäubung** *f* atomization of air. **-ziegel** *m* air-dried brick. **-zufluß** *m* air influx. **-zufuhr** *f* air supply. **-zuführung** *f* air supply (duct). **-zug** *m* 1 draft (of air). 2 air draft. 3 (*Mach.*) air flue, (*Metll.*) air uptake. **-zünder** *m* pyrophorus, spontaneously igniting substance. **-zurichtung** *f* air conditioning. **-zustand** *m* atmospheric state (*or* condition). **-zutritt** *m* access (*or* admission) of air. **-zwischenkühler** *m* air intercooler. **-zwischenraum** *m* air gap (*or* interstice).

Lug *m:* **L. und Trug** lying and deception.

Lüge *f* (-n) lie, falsehood

lugen *vi* look (peek, peer) out

lügen* (log, gelogen) *vt/i* lie, tell lies.

Lügner *m* (-) liar

Luke *f* (-n) hatch; skylight

Lumie *f* (-n) sweet lemon, Citrus lumia

Luminal *n* phenobarbitol

Lumineszenz *f* (-en) luminescence. **lumineszieren** *vi* luminesce

Lumpen *m* (-) rag. **·-auskohlung** *f* carbonization of rags. **-bleiche** *f* rag bleaching. **-brei** *m* (*Paper*) first stuff, (first) rag pulp. **-butte, -bütte** f (*Paper*) rag tub. **-erz** n (*Min.*) impure stibnite. **-kocher** m rag boiler. **-papier** n rag paper. **-reißer** m (*Paper*) rag cutter, devil. **-stoff** m (*Paper*) rag pulp. **-wolf** m rag-tearing machine, devil. **-wolle** f shoddy. **-zucker** m coarse loaf sugar

Lunge f (-n) lung(s): **eiserne L.** iron lung

Lungen- lung, pulmonary: **-entzündung** f pneumonia. **-flügel** m (one) lung. **-gift** n lung poison (*or* irritant). **l–krank** a tubercular. **L–kraut** n lungwort. **-leiden** n pulmonary ailment. **l–schädigend** a harmful to the lungs. **L–schützer** m respirator. **-stein** m pulmonary concretion. **-sucht** f tuberculosis. **l–süchtig** a tubercular

Lunker m (-) cavity, shrinkage hole; pipe (in ingots). **-bildung** f (*Metll.*) cavitation, piping. **l–frei** a cavity-free; pipeless. **L–hohlraum** m cavity; pipe

lunkerig a cavitated, honeycombed; piped. **lunkerlos** a cavity-free; pipeless. **lunkern** vi develop cavities; pipe. **Lunkerung** f (-en) **1** cavity. **2** cavitation, piping. **3** (cavity-forming) shrinkage. **lunkrig** a cavitated; piped. *Cf* LUNKER

Lunte f (-n) **1** fuse, slow match. **2** (*Tex.*) rove, roving, slubbing. **Lunten-gewehr** n musket

Lupe f (-n) magnifying glass —**unter die L. nehmen** (*fig*) examine closely

Lupen-linse f magnifying lens. **l–rein** a flawless

Luppe f (-n) (*Metll.*) bloom, lump, puddle(d) ball, ingot. **Luppen-eisen** n ball (*or* puddled) iron. **-frischfeuer** n bloomery fire. **-frischhütte** f bloomery. **-stahl** m bloom steel, steel in blooms

Lupulin-säure f lupul(in)ic acid

Lurch m (-e) (*Zool.*) amphibian

Lust f (⁀e) **1** desire, inclination. **2** pleasure, joy

Lüster m (-) **1** luster. **2** (*Tex.*) gloss. **3** chandelier. **-farbe** f (*Ceram.*) luster color. **-glanz** m luster, gloss. **-glasur** f (*Ceram.*) lustrous glaze. **-stoff** m luster (*or* glossy) fabric

Lust-gas n laughing gas, nitrous oxide

lustig a **1** merry, jolly, cheerful. **2** amusing, funny —**sich l. machen (über)** make fun (of)

lustlos a joyless; listless *cf* LUST

lüstrieren vt (*Tex.*) luster, glaze (silk)

Lust-seuche f syphilis. **-spiel** n comedy

Luteo-kobaltchlorid n luteocobaltic chloride. **-säure** f luteoic acid. **-verbindung** f luteo compound

Lutidin-säure f lutidinic acid

lutieren vt lute

lutschen vt/i suck. **Lutscher** m (-), **Lutsch-propfen** m nipple, pacifier. **-stange** f lollipop

Lutte f (-n) (*Min.*) air vent (tube, pipe)

Lutter m (-) (*Distil.*) singlings, low wine. **luttern** vt/i draw off (mash) for the first time, distil low wine. **Lutter-prober** m low wine tester

Lüttich n (*Geog.*) Liège

Lux-boje f flare buoy

Luxus m luxury. **-artikel** m luxury item (*or* article). **-konsumption** f (*Biol.*) excess food consumption. **-leder** n de luxe (*or* luxury) leather. **-papier** n de luxe (*or* fancy) paper

Luzern n (*Geog.*) Lucerne. **Luzerne** f (-n), **Luzerner-klee** m (*Bot.*) alfalfa, lucern(e)

L.V. St. abbv (Landwirtschaftliche Versuchsstation) agricultural experiment station

l.W. abbv (lichte Weite) inside diam., bore

Lw., Lwd. abbv (Leinwand) linen, cloth (binding)

lydischer Stein Lydian stone, touchstone

Lymph-bildung f lymph formation. **-drüse** f lymphatic gland

Lymphe f (-n) **1** lymph. **2** vaccine

Lymph-gefäß n lymphatic vessel (*or* duct). **-körperchen** n lymph corpuscle, lymphocyte

Lyo-chrom n lyochrome. **l–phil** a lyophile, lyophilic. **-phob** a lyophobic. **-trop** a lyotropic

Lysalbin-säure f lysalbinic acid

Lyse f (-n) lysis

Lyserg-säure f lysergic acid

lytisch a lytic

M

m. *abbv* **1** (männlich) masc(uline). **2** (mit) with. **3** (mein) my. **4** (Minute) min(utes). **5** (Meter) meter(s)

M. *abbv* **1** (Masse) mass. **2** (Monat) mo(nth). **3** (Monatshefte für Chemie) Chemical Monthly. **4** (Mittelsorte) med(ium) gr(ade). **5** (Mark) mark(s). **6** (Modell) model. **7** (Molekulargewicht) mol. wt.

macerieren *vt* macerate. **Macerier·gefäß** *n* macerator

Mach·art *f* make, brand; style, design

machbar *a* feasible. **Machbarkeits·studie** *f* feasibility study

Mache *f* **1** sham, pretense. **2: in der M. haben** be at work on

machen I *vt* **1** (*usu*) make *cf* GEMACHT. **2** do; perform; manage. **3** make out (*or* up), write (out). **4** cause, give *cf* SCHAFFEN. **5** play (the part of). **6** (+**zu**) turn (into). **7:es macht etwas/nichts** it matters/ doesn't matter. **8** (*in many idiomatic expr. with nouns or adj; see under noun or adj: cf* BEGRIFF, MOBIL, MUNTER, NOTIZ, etc). II *vi* **9** carry on, take care of things. **10** (+**in**) deal (in). **11** (+*pred a*) e.g.: **diese Speisen m. dick** these foods make one fat (*or* are fattening). **12** (+**daß**) see (to it) (that). III: **sich m.** **13** get along, do (well). **14** (+**an, hinter**) get to work (on) *cf* WEG

Machenschaften *pl* machinations

Macher *m* (-) **1** maker, producer. **2** perpetrator

Machsche Zahl Mach number

Macht (⁀e) **1** power. **2** force, might. **mächtig** *a* **1** powerful, mighty. **2** (+*genit*) in control (*or* command) (of). **3** enormous, tremendous —**m. groß** huge, great big. **Mächtigkeit** *f* powerfulness; (*esp Geol.*) thickness. **machtlos** *a* powerless

·machung *sfx* (*forms nouns based on compound verbs ending in* ·MACHEN) *cf* BEKANNTMACHEN

Mach·werk *n* (shoddy) piece of work, botched-up job

Macis *m* mace. **··öl** *n* oil of mace

Madar·wurzel *f* mudar root

Mädchen *n* (-) girl

Made *f* (-n) **1** maggot, worm. **2** set screw. **Maden·loch** *n* wormhole. **-schraube** *f* set screw

madig *a* maggoty, worm-eaten

mafisch *a* (*Geol.*) mafic

Mafura·talg *m* mafura tallow

mag like(s); may: *pres of* MÖGEN

Magazin *n* (-e) magazine (*all meanings*)

Magdala·rot *n* Magdala red

Magen *m* (⁀) **1** stomach (as internal organ only), maw, gizzard. **2** (*in compds oft*) gastric, gastro-: **··arznei** *f* stomach remedy, stomachic. **-bitter** *m* stomach bitters, gastric cordial. **-brei** *m* chyme. **-brennen** *n* heartburn, pyrosis

Magendarm· gastrointestinal

Magen·drüse *f* gastric (*or* peptic) gland. **-entzündung** *f* gastritis. **-flüssigkeit** *f* gastric juice. **-gift** *n* gastric poison. **-haut** *f* lining of the stomach. **-inhalt** *m* contents of the stomach. **-krampf** *m* gastric (*or* stomach) cramp. **-krankheit** *f* stomach (*or* gastric) ailment. **-lab** *n* rennet. **-labdrüse** *f* gastric gland. **-lipase** *f* gastric lipase. **-mittel** *n* stomach remedy, stomachic. **-mund** *m* cardia. **-pförtner** *m* pylorus. **-saft** *m* gastric juice. **-säure** *f* stomach acid(ity). **-schleimhaut** *f* gastric mucosa. **-schlund** *m* esophagus. **-schmerzen** *mpl* gastric pain(s), stomach ache. **-schwäche** *f* dyspepsia. **m-stärkend** *a*, **M-stärkungsmittel** *n* stomachic. **-stein** *m* gastric concretion (*or* calculus). **-tropfen** *mpl* stomach drops. **-verdauung** *f* gastric digestion. **-verstimmung** *f* upset stomach. **-wand** *f* wall of the stomach. **-weh** *n* stomach ache. **-zwölffingerdarm** *m* gastroduodenum

mager *a* **1** lean (meat, person; coal). **2** low-fat, low-oil. **3** meager, scanty; poor (soil). **4** *specif:* **m—er Kalk** lime difficult to

451

slake; **m—es Öl** n mineral oil; **m—er Stoff** (*Paper*) short pulp
Mager·beton m lean-mix concrete. **-erz** n lean ore. **-kalk** m lean lime
Magerkeit f leanness; low fat (*or* oil) content; meagerness; poor quality. **Mager·keits·anzeiger** m (*Soils*) indicator of lime exhaustion
Mager·kohle f lean coal, (*specif*) semi-anthracite coal (containing 10–12% volatile matter). **-milch** f skim milk
magern vt make (or vi become) lean; thin, dilute; (*Ceram.*) shorten, grog
Mager·ton m lean clay
Magerungs·mittel n (*genl*) diluent; (*Ceram.*) grog
Magie f magic. **magisch** a magic(al)
Magister m (-) master's degree
magmatisch a magmatic
magn. abbv (magnetisch) magnetic
Magnesia·beize f (*Dye.*) magneisa mordant. **-glimmer** m magnesium mica, (*Min.*) biotite, phlogophite. **m—haltig** a magnesian, containing magnesia. **M—härte** f (*Water*) magnesium hardness. **-hydrat** n magnesium hydroxide. **-kalk** m magnesian lime(stone). **-milch** f milk of magnesia. **-mörtel** m Sorel cement. **-salpeter** m magnesium nitrate. **-salz** n magnesium salt. **-verbindung** f magnesia (*or* magnesium) compound
Magnesit·mehl n ground magnesite. **-spat** m magnesite. **-stein, -ziegel** m magnesite brick
Magnesium·blitz m, **-blitzlicht** n magnesium flash(light). **-branntkalk** m magnesium quicklime. **m—haltig** a containing magnesium. **M—licht** n (*Phot.*) magnesium (flash)light. **-löschkalk** m hydrated magnesium lime. **-mergel** m magnesian agricultural lime (with more than 15% $MgCO_3$). **m—organisch** a organomagnesium, magnesium-organic. **M—oxyhydrat** n magnesium hydroxide. **-rhodanid** n magnesium thiocyanate
Magnet·abschneider m magnetic separator. **-aufnahme** f magnetic recording. **-band** n magnetic (recording) tape. **-eisen** n magnet(ic) iron. **-eisenerz** n, **-eisenstein** m magnetic iron ore. **m—elek-**

trisch a magneto-electric. **M—feld** n magnetic field. **-feldröhre** f magnetron
magnet·haltig, magnetisch a magnetic
magnetisierbar a magnetizable. **Magnetisierbarkeit** f magnetizability. **magnetisieren** vt magnetize. **magnetisier·fähig** a magnetizable. **Magnetisierung** f magnetizing, magnetization
Magnetisierungs·fähigkeit f magnetizability. **-strom** m magnetizing current
Magnet·kern m magnet core. **-kies** m pyrrhotite, magnetic pyrites. **-kraft** f magnetic force. **-messer** m magnetometer. **-nadel** f magnetic needle
Magneto·chemie f magnetochemistry. **m—motorisch** a magnetomotive. **M—phon** n (-e) magnetic (tape) recorder. **-rotationsspektrum** n magnetic rotation spectrum
Magnet·schale f magnetic shell. **-scheider** m magnetic separator. **-speicher** m magnetic memory (bank). **-spule** f field coil, solenoid. **-stab** m bar magnet. **-stahl** m magnet steel. **-stein** m magnetite, lodestone. **-tonband** n magnetic recording tape. **-tonfilm** m video tape. **-trommelspeicher** m (*Cmptrs.*) magnetic drum memory. **-zünder** m magneto. **-zündung** f magnetic ignition
Magnolie f (-n) magnolia
Magsamen m poppy seed
Mahagoni n mahogany. **m—·braun** a mahogany-brown. **M—gummi** n mahogany (*or* cashew) gum. **-holzöl** n mahogany wood oil. **-nuß** f cashew nut
Mahd f 1 mowing. 2 mowed crop. 3 mowing time
mähen vt/i mow
Mahl n (-e) meal; dinner
Mahl·anlage f grinding (*or* milling) plant. **-ausbeute** f (*Flour*) extraction rate
mahlbar a millable, grindable
mahlen vt/i grind, mill; pulverize; (*Paper*) beat. *See also* GEMAHLEN
Mahl· grinding, milling: **-feinheit** f fineness of grinding. **-gang** m 1 grinding operation. 2 mill course, set of millstones. **-gut** n grist, mill feed, material to be ground (*or, Paper*: beaten). **-holländer** m (*Paper*) beating engine, beater. **-korn** n

grist, grain to be ground. **-körper** *mpl* grinding media. **-rückstand** *m* grinding residue. **-stein** *m* millstone

Mahlung *f* (-en) grinding, milling *cf* MAHLEN

Mahlungs·grad *m* degree of grinding (*or* fineness); (*Paper*) degree of beating, pulp freeness

Mahl·werk *n* grinding (*or* milling) plant. **-wirkung** *f* grinding (*or* pulverizing) effect. **-zahn** *m* molar (tooth)

Mahlzeit *f* (-en) meal

Mähne *f* (-n) mane

mahnen 1 *vt* (*esp* + **an**) remind (of); demand payment (of debts). 2 *vt/i* warn, admonish. **Mahner** *m* (-) reminder

Mähre *f* (-n) mare

Mähren *n* Moravia

Mai *m* May. **--baum** *m* birch. **-blume** *f* lily of the valley. **-fisch** *m* shad. **-glöckchen** *n* lily of the valley. **-käfer** *m* June bug. **-kraut** *n* woodruff

Mais *m* (Indian) corn, maize. **--brand** *m* corn smut. **-branntwein** *m* corn whisky

Maisch *m* (-e) mash = MAISCHE. **--apparat** *m* mash(ing) machine. **-bottich** *m* mash(ing) tub (*or* tun)

Maische *f* (-n) 1 (*Brew.*) mash. 2 (*Wine*) crushed grapes, must. 3 (*Sugar*) pulp, crystallizer. 4 (*Distil.*): (**weingare**) **M.** wash; **zweite M.** aftermash

maischen *vt* mash

Maische·pumpe *f* (*Brew.*) mash pump

Maisch· (*Brew.*) mash: **-gitter** *n* stirrer, rake. **-gut** *n* mash. **-kessel** *m* mash copper (*or* kettle). **-kühler** *m* mash cooler. **-pfanne** *f* mash copper (*or* pan). **-prozeß** *m* mashing process

Maischung *f* (-en) mashing; mash

Maisch· (*Brew.*) mash: **-ventil** *n* grains valve, trap. **-wasser** *n* mash liquor. **-würze** *f* mash wort, grain wash

Mais· corn, maize: **-flocken** *fpl* corn flakes. **-geist** *m* corn whisky. **-gelb** *n* corn yellow, maize (*as a color*). **-kleber** *m* zein, maize gluten (meal). **-kolben** *m* corncob. **-korn** *n* corn (*or* maize) kernel (*or* grain). **-krankheit** *f* pellagra. **-mehl** *n* corn meal, maize flour. **-öl** *n* corn (*or* maize) oil. **-pistille** *npl* cornsilk; (*Pharm.*) zea.

-schrot *m* crushed corn (*or* maize). **-sirup** *m* corn syrup. **-spiritus** *m* corn whisky. **-stärke** *f* cornstarch. **-stroh** *n* corn straw, maize fodder. **-zucker** *m* corn sugar

Maizena *n* cornstarch

Majole *f* (-n) ampoule

Majoran *m* marjoram. **--öl** *n* marjoram oil

majorenn *a* of age

Majorität *f* majority, age of consent

Majuskel *f* (-n) capital (letter)

makadamisieren *vt* macadamize. **Makadam·straße** *f* macadam(ized) road

Makassar·öl *n* macassar oil

Makel *m* (-) defect, flaw, blemish. **makelhaft** *a* defective, flawed. **makellos** *a* flawless

mäkeln *vi* (+**an**) find fault (with)

Maker *m* (-) sledgehammer, maul

Makkaroni *pl* mac(c)aroni

Makler *m* (-) broker

Mäkler *m* (-) fault-finder

Mako *m* *or* *n* Egyptian cotton

Makrele *f* (-n) mackerel

Makro·achse *f* (*Cryst.*) macroaxis. **-analyse** *f* macroanalysis. **-elementaranalyse** *f* elementary macroanalysis. **m—gravimetrisch** *a* macrogravimetric. **M—ion** *n* macro-ion

Makrolar *n* (-e) macromolecular (compound)

Makro·mechanik *f* macromechanics. **-molekül** *n* macromolecule

Makrone *f* (-n) macaroon

makroskopisch *a* macroscopic

Makro·untersuchung *f* macroanalysis

Makulatur *f* 1 (*Print.*) waste sheets; waste paper. 2 nonsense. **makulieren** *vt* (re)pulp

mal *adv* 1 (*also as sfx*) time(s): **2 m. 10** two times ten; **dreimal** three times; **das erstemal** the first time; **zum letztenmal** for the last time. 2 once, etc = EINMAL.

Mal *n* (-e) **I** time: **dieses M.** this time; **das letzte M.** the last time; **zum letzten M.** for the last time. **II** (-̈er) mark; birthmark; stigma; scar; monument

Malachit·grün *n* malachite green. **-grün·nährboden** *m* malachite green nutrient medium. **-kiesel** *m* chrysocolla

malaiisch *a* Malayan *cf* MALAYSISCH

Malakka·nuß f Malacca (or marking) nut
Malakolith m (-e) malacolite, salite
malaxieren vt malaxate
Malaye m (-n, -n) Malayan. **malaysisch** a
Malaysian
Malein·aldehyd m maleic aldehyde.
-**amidsäure** f maleamic acid. -**anilsäure**
f maleanilic acid
Maleinat·harz n maleic resin
Maleinimid n (-e) maleimide
Malein·säure f maleic acid
malen vt/i paint (esp art works). **Maler** m
(-) painter (esp artist). **Malerei** f (-en) (artistic) painting
Maler·email n painter's enamel. -**farbe** f
painter's (or artist's) color. -**firnis** m
painter's varnish. -**gold** n painter's gold,
ormolu. -**grundierung** f painter's
priming
malerisch a picturesque
Maler· painter's: -**kolik**, -**krankheit** f
painter's colic, plumbism. -**leim** m painter's size. -**leinwand** f painter's canvas.
-**stift** m artist's pencil. -**tuch** n canvas
Mal·farbe f painter's color
·malig sfx -time(s), -fold: **dreimalige
Anfrage** three-time inquiry cf MAL 2
mal·nehmen* vt multiply
Malon·ester m malonic ester, ethyl malonate. **m–sauer** a malonate of. **M–säure** f
malonic acid. -**säureäthylester** m ethyl
malonate
Malten n (-e) maltene
Malter m or n (-) 1 grain measure (appx 8
bushels). 2 mortar
Malve f (-n) (Bot.) mallow. **Malven·farbe** f,
m–farbig a mauve. **M–kraut** n mallow
Malwe f (-n) (Bot.) mallow
Malz n malt. -**aufguß** m malt infusion,
wort. -**auszug** m malt extract; wort. -**bottich** m malt vat (or tub). -**darre** f malt
kiln. -**eiweiß** n diastase, malt protein
malzen, mälzen vt malt. **Malzer, Mälzer**
m (-) maltster, maltman. **Mälzerei** f (-en)
1 malting. 2 malt house.
Malz· malt: -**essig** m malt vinegar. -**fabrik**
f malt house. -**fabrikation** f malt manufacture. -**gerste** f malting barley.
-**häufen** n couching. -**haus** n malt house.
-**keim** m malt sprout; pl oft culm.
-**maische** f malt mash. -**mehl** n malt

flour. -**milch** f malted milk. -**probe** f malt
test(ing). -**quetsche** f malt press. -**schrot**
n crushed malt, malt grist. -**stärke** f malt
starch. -**surrogat** n malt substitute. -**tenne** f malt floor. -**treber** pl spent malt,
brewer's grains, malt husks
Malzung, Mälzung f malting. **Mäl-
zungs·schwund** m malting loss
Malz·wendeapparat, -**wender** m malt
turner. -**zucker** m malt sugar, maltose
Mammut n (-e) mammoth. -**bronze** f delta
metal
mammutieren vt fill (casks)
Mammut·pumpe f air lift
man pron one, a person; we; they, people:
wie sagt m. . . ? how does one (or do you)
say . . ?; **m. sagt** they say, people say, it is
said; **m. erhitzt die Lösung** we heat the
solution; (in instructions): **m. erhitze die
Lösung** heat the solution
m.A.n. abbv (meiner Ansicht nach) in my
opinion
Manager·krankheit f executive stress
syndrome
manch(e) a, pron 1 sg many a. 2 pl some:
manche (Leute) sind dagegen some
(people) are against it
manchenorts adv in many a place; in some
places
mancherlei 1 a (no endgs) various, many
kinds of. 2 pron many (or all sorts of)
things
mancherorts adv in many a place; in some
places
manchmal adv sometimes
Mandarin·druck m (Tex.) mandarining
Mandarine f (-n) tangerine, mandarin
orange
Mandarinen·blätteröl n mandarin leaf
oil. -**öl** n tangerine oil. -**schalenöl** n oil of
mandarin peel
Mandarin·gelb n mandarin yellow
Mandel f (-n) 1 almond. 2 tonsil. 3 (Geol.)
geode. 4 shock (of grain). **m–artig** a
amygdaline, almond(-like). **M–blüte** f almond blossom. -**bräune** f tonsillitis,
quinsy. -**drüse** f tonsil. -**entzündung** f
tonsillitis. **m–förmig** a almond-shaped,
amygdaloid. **M–gummi** n almond gum.
-**kern** m almond kernel. -**kleie** f almond
bran; ground almond cake. -**milch** f al-

mond milk, emulsion of almond. **-nitril** *n* mandelonitrile. **-öl** *n* almond oil. **-säure** *f* mandelic acid. **-säureamid** *n* mandelamide. **-säurenitril** *n* mandelonitrile. **-seife** *f* almond soap. **-stein** *m* amygdaloid (rock). 2 (*Med.*) tonsillar concretion. **m–steinartig** *a* amygdaloidal. **M–stoff** *m* amygdalin. **-storax** *m* amydaloid storax

Mandschurei *f* Manchuria

Mangan *n* manganese. **··alaun** *m* manganese alum. **-amphibol** *m* (*Min.*) manganese amphibol. **m–arm** *a* low-manganese. **M–beize** *f* (*Dye.*) manganese mordant. **-bister** *m* manganese brown. **-blende** *f* alabandite. **-braun** *n* manganese brown (hydrated manganese oxide pigment). **-chlorid** *n* manganic chloride. **-chlorür** *n* manganous chloride. **-dioxid** *n* manganese dioxide. **-eisen** *n* ferromanganese. **-eisenstein** *m* (*Min.*) triplite. **-epidot** *m* (*Min.*) piedmontite. **-erz** *n* manganese ore **—graues M.** manganite, pyrolusite; **schwarzes M.** hausmannite. **-gehalt** *m* manganese content. **-glanz** *m* (*Min.*) alabandite. **-granat** *m* manganese garnet, (*Min.*) spessartite. **-grün n** manganese green. **m–haltig** *a* manganiferous, containing manganese. **M–hartstahl** *m* austenitic manganese steel. **-hydroxid** *n* manganic hydroxide. **-hydroxydul** *n* manganous hydroxide. **-hyperoxid** *n* manganese peroxide

Mangani· manganic, mangani-: **-chlorid** *n* manganic chloride. **-cyanwasserstoffsäure** *f* manganicyanic acid

manganig *a* manganous. **··sauer** *a* manganite of. **M–säure** *f* manganous acid

Mangani· manganic: **··hydroxid** *n* manganic hydroxide. **-phosphat** *n* manganic phosphate. **-salz** *n* manganic salt. **-sulfat** *n* manganic sulfate. **-verbindung** *f* manganic compound

Mangan·jodür *n* manganous iodide. **-karbid** *n* manganese carbide. **-kies** *m* (*Min.*) hauerite. **-kiesel** *m* (*Min.*) rhodonite. **-kupfer** *n* cupromanganese. **-kupfererz**, **-kupferoxid** *n* (*Min.*) crednerite. **-legierung** *f* manganese alloy. **-menge** *f* amount of manganese

Mangano· manganous, mangano-: **-azetat** *n* manganous acetate. **-chlorid** *n* manganous chloride. **-cyanwasserstoffsäure** *f* manganocyanic acid. **-dolomit** *n* (*Min.*) kutnahorite. **-ferrit** *n* (*Min.*) jacobsite. **-ferrum** *n* ferromanganese. **-hydroxid** *n* manganous hydroxide. **-ion** *n* manganous ion. **-karbonat** *n* manganous carbonate. **-melan** *m* psilomelane. **m–metrisch** *a* manganometric, by titration with permanganate. **M–oxid** *n* manganous oxide. **-phosphat** *n* manganous phosphate. **-salz** *n* manganous salt. **-sulfat** *n* manganous sulfate. **-verbindung** *f* manganous compound

Mangan·oxid *n* manganese (*specif*) manganic) oxide **—rotes M.** manganomanganic oxide, trimanganese tetroxide

Manganoxid·hydrat *n* manganic hydroxide. **-oxydul** *n* mangano-manganic oxide, manganese trioxide. **-salz** *n* manganic salt. **-verbindung** *f* manganic compound

Mangan·oxydul *n* manganous oxide— **schwefelsaures M.** manganous sulfate

Manganoxydul·hydrat *n* manganous hydroxide. **-oxid** *n* mangano-manganic oxide, trimanganese tetroxide. **-salz** *n* manganous salt. **-verbindung** *f* manganous compound

Mangan·pecherz *n* (*Min.*) triplite. **-peroxid** *n* manganese peroxide (*or* dioxide). **m–reich** *a* high-manganese, rich in manganese. **M–reserve** *f* (*Tex.*) manganese resist. **-salz** *n* manganese salt. **m–sauer** *a* manganate of. **M–säure** *f* manganic acid. **-säureanhydrid** *n* manganic anhydride. **-schaum** *m* (*Min.*) bog manganese, wad. **-schlamm** *m* Weldon process mud. **-schwamm** *m* (*Min.*) wad, bog manganese. **-schwarz** *n* manganese black. **-schwärze** *f* (*Min.*) pyrolusite. **-siliziumstahl** *m* silicomanganese steel. **-spat** *m* (*Min.*) rhodochrosite. **-stahl** *m* manganese steel. **-sulfat** *n* manganese sulfate. **-sulfür** *n* manganese sulfide. **-superoxid** *n* manganese dioxide. **-verbindung** *f* manganese compound. **-violett** *n* manganese violet (*or* phosphate pigment). **-vitriol** *n* manganese vitriol (*or* sulfate)

Mangel I *m* (⸚) **1** lack, deficiency; shortage: **aus M. an** for lack of; **M. leiden** suffer want. **2** defect, flaw. **II** *f* (-n) (*Mach.*) mangle

mangelhaft *a* defective, deficient, imperfect. **Mangelhaftigkeit** *f* defectiveness, deficiency, poor quality

Mangel·krankheit *f* deficiency disease. **-leiter** *m* defect (*or* p-type) semiconductor

mangeln I *vi* lack, be lacking; **es mangelt an** there is a lack of; (+*dat.*) e.g.: **es mangelt ihnen an Wärme** they lack (*or* they are short of) heat. **II** *vt* mangle (laundry).

mangelnd *p.a* lacking, deficient, (a) lack of

mangels *prep.* (*genit*) for lack (of)

Mang·futter *n* mixed grain

Mangle·baum *m* mangrove

Mangold *m* (-e) feed beet, mangelwurzel

Manie *f* (n) mania

Manier *f* (-en) manner; way to act; mannerism

Manila·hanf *m* Manila hemp

manipulieren *vt/i* manipulate

Manikettinuß·öl *n* maniketti nut oil

Manko *n* (-s) deficit, shortage; defect

Mann *m* (⸚er) man; husband *cf* LEUTE

Manna·stoff, -zucker *m* manna (sugar), manitol

Männchen *n* (-) **1** little man. **2** male (animal)

mannig·fach, -faltig *a* many and various, many different. **Mannigfaltigkeit** *f* (-en) variety, diversity, multiplicity; (*Math.*) manifold

Mannit *n* (-e) mannitol

männlich *a* male; masculine; manly. **Männlichkeit** *f* maleness; masculinity; manliness

Mann·loch *n* manhole

Mannoheptit *n* (-e) mannoheptitol

Mannon·säure *f* mannonic acid

Mannschaft *f* (-en) team; crew; staff; troop

Mannuron·säure *f* mannuronic acid

manometrisch *a* manometric

Manöver *n* (-) maneuver, trick. **-pulver** *n* (*Explo.*) blank-fire powder

Manschette *f* (-n) **1** cuff, *pl. oft* handcuffs. **2** (*Mach.*) collar, sleeve; circular gasket; piston packing

Mantel *m* (⸚) **1** (over)coat, cloak; mantle.

2 casting mold. **3** (*Mach.*) jacket, sheath, shell, case, casing. **4** (*Bot.*) aril. **5** (*Geom.*) lateral area; sheet. **-fläche** *f* (*Geom.*) generated surface. **-futter** *n* jacket (shell, case) lining. **-geschoß** *n* jacketed bullet, metallic cartridge. **-konus** *m* casing cone. **-kühlung** *f* jacket cooling. **-linie** *f* (*Geom.*) director line, directrix. **-platte** *f* liner plate. **-riß** *m* surface crack. **-rohr** *n* casing tube, jacket pipe. **-schicht** *f* coating, protective layer. **-sprengstoff** *m* sheathed explosive. **-tier** *n* (*Zool.*) tunicate

Mantisse *f* (-n) mantissa

manuell *a* manual; *adv oft* by hand

Manufaktur *f* (-en) **1** hand work. **2** handicraft factory (*or* shop). **3** *pl oft* piece goods, yard goods

Mappe *f* (-n) briefcase, portfolio

Maranta·stärke *f* arrowroot

Märchen *n* (-) fairy tale. **märchenhaft** *a* fairy-tale, fabulous, fantastic

Marder *m* (-) (*Zool.*) marten; (*in compds oft*) thief

Margarete *f* (-n) daisy

Margarine·käse *m* margarine cheese

margarin·sauer *a* margarate of. **M−säure** *f* margaric acid

Margerite *f* (-n) daisy

Marien·bad *n* water bath. **-distel** *f* milk thistle. **-glas** *n* (*Min.*) **1** mica, muscovite. **2** selenite. **-käfer** *m* ladybug. **-korn** *n* milk-thistle seed

Marine *f* (-n) navy. **-blau** *n* navy blue. **-legierung** *f* admiralty alloy (*or* brass). **-leim** *m* marine glue. **-öl** *n* marine oil

marinieren *vt* marinate

Mark I *n* **1** marrow; pulp, pith; medulla; (*as pfx oft*) medullary. **2** core, essence. **3** (fruit *or* tomato) paste. **II** *f* (-) mark (monetary unit). *Cf* MARKE

markant *a* striking, outstanding

mark·artig *a* pithy; marrow-like, medullar(y), myeloid

Markasit *m* (-e) marcasite. **-glanz** *m* tetradamite

Mark·bildung *f* (*Physiol.*) myelinization

Marke *f* (-n) **1** mark; marker. **2** stamp (*incl* postage). **3** token, coupon, chit; chip; tag. **4** trademark; brand, make. **5** (quality) grade. *Cf* MARK

Marken·artikel m brand-name product. **-butter** f high-grade butter. **-öl** n brand-name oil. **-schutz** m trademark protection. **-werkstoff** m brand-name material. **-zeichen** n brand logo (or symbol), trademark

Markette f virgin wax (cake)

Mark·fäule f (Wood) heart rot. **-flüssigkeit** f spinal fluid. **-holz** n pithy wood

markieren I vt 1 mark. 2 indicate, outline. 3 pretend (to be). II: **sich m.** be marked, stand out. **Markierung** f (-en) mark, marking, marker

markig a 1 marrowy, medullar(y). 2 pithy

Markise f (-n) awning. **Markisen·stoff** m canvas

Mark·knopf m (Anat.) medulla oblongata. **-lager** n medullary layer. **-masse** f medullary substance. **-nagelung** f (Med.) medullary nailing. **-riß** m (Wood) heart rot. **-saft** m medullary sap. **-scheide** f 1 (Min.) boundary line. 2 (Anat.) medullary (or myelin) sheath. **-scheider** m mine surveyor. **-stammkohl** m marrow-stem kale. **-stein** m milestone. **stoff** m medullary substance. **-strahl** m medullary ray. **-substanz** f medullary substance

Markt m (-̈e) market. **markten** vi buy and sell; bargain

markt·fähig a marketable. **-gängig** a marketable; market (practices, customs, etc.). **M–ordnung** f 1 market(ing) regulation(s). 2 (common) market organization. **-preis** m market price. **m–reif** a ready to put on the market, marketable. **M–reife** f marketability. **-tasche** f shopping bag. **-wert** m market value. **-wirtschaft** f market economy. **-zettel** m market report

Mark·zelle f medullary cell

Marmelade f (-n) marmalade, jam; marmalade tree

Marmelos·beere, -frucht f marmelos berry or fruit

Marmite f (-n) kettle, digester

Marmor m (-e) marble. **m–ähnlich, -artig** a marble-like, marmoreal. **M–bruch** m 1 marble quarry. 2 marble chips (or fragments). **-gips** m Keene's cement (doubly burnt gypsum with alum)

marmoriert p.a marbled, veined, mottled

Marmor·kalk m lime from marble. **-kiesel**

m hornstone. **-lack** m marble varnish, **-mehl** n marble dust. **marmorn** a (made of) marble. **Marmor–papier** n marbled paper. **-weiß** n whiting (pigment)

marode pred a worn out, exhausted

Marokain–papier n morocco paper. **marokkanisch** a Moroccan. **Marokko** n Morocco. **–gummi** n Morocco (or Berber) gum

Marone f (-n) chestnut

Maroquin m or n morocco (leather)

Marsch I m (-̈e) march. II f (-en) marsh. **–boden** m marshy soil. **-flugkörper** m cruise missile

Marseiller Seife Marseilles soap

Mars·gelb n Mars yellow

Marsh(i)sche Probe Marsh('s) test

martensitisch a (Metll.) martensitic

Martin· (Iron) Martin, open-hearth: **-flußeisen** n open-hearth iron. **-flußstahl** m open-hearth steel. **-ofen** m Martin (or open-hearth) furnace. **-ofenschlacke** f open-hearth slag. **-roheisen** n open-hearth pig iron. **-stahl** m open-hearth (or Martin) steel. **-stahlofen** m open-hearth steel furnace. **-verfahren** n Siemens-Martin (or open-hearth) process

Martitisierung f (Min.) martitization

Martius·gelb n Martius yellow

März m March. **Märzen·bier** n bock beer

Masche f (-n) 1 mesh, loophole. 2 stitch; loop; run (in cloth). 3 (perfect) gimmick, trick

Maschen·draht m screen wire. **m–fest** a runproof (hosiery). **M–größe** f size of the mesh. **-sieb** n screen (or wire-mesh) strainer. **-weite** f size of the mesh, **-werk** n network. **-zahl** f mesh (number)

maschig a mesh(ed), netted, reticulated

Maschine f (-n) 1 machine; engine. 2 machinery. 3 (Aero.) plane. 4 motorcycle. 5 typewriter **—M. schreiben** type. **m–geschrieben** a typewritten. **maschinell** a mechanical; adv oft by machine

Maschinen·bau m machine construction; mechanical engineering. **-bauer** m machine builder. **-element** n machine (or engine) part. **-fett** n lubricating grease. **-gas** n engine (or power) gas. **m–gefertigt** a machine-made. **-geschrieben** a typewritten. **M–gewehr** *n* machine gun. **-glanz** *m*, **-glätte** *f* machine finish. **-guß** *m* machine casting. **-kunde** *f* 1 mechanical engineering. 2 practical mechanics. **-lack** *m* machine (*or* engine) varnish. **-lager** *n* (*Mach.*) bearing. **-lageröl** *n* bearing oil. **-lehre** *f* 1 mechanical engineering. 2 practical mechanics. **-leistung** *f* machine output (performance, rating). **m–mäßig** *a* machine-like, mechanical; *adv* oft by machine. **M–mischung** *f* mechanical mixing. **-öl** *n* machine oil. **-papier** *n* 1 machine-made paper. 2 typewriter paper. **-raum** *m* 1 machine shop. 2 engine room. **-schaden** *m* engine trouble; mechanical breakdown. **-schlosser** *m* machine fitter. **-schmiere** *f* lubricating grease. **-schmierung** *f* machine (*or* engine) lubrication. **-schreiber** *m*, **-schreiberin** *f* typist. **-schrift** *f* typewriting; typescript. **m–schriftlich** *a* typewritten. **M–sieb** *n* wire cloth. **-sprache** *f* machine (*or* computer) language. **-strom** *m* (*Elec.*) generator current. **m–technisch** *a* mechanical. **-torf** *m* machine-cut peat. **-werk** *n* machinery. **-werkstatt** *f* machine shop. **-wesen** *n* mechanical engineering. **-zeitalter** *n* machine age

maschine·schreiben* *vi* type

Maschinist *m* (-en, -en) machinist

Maser I *f* (-n) 1 mark, spot, speck(le) *cf* MASERN. 2 (*Wood*) grain; knot. **II** *m* (-) (*Phys.*) maser. **-holz** *n* curly-grained wood. **maserig** *a* grained, veined; speckled, mottled. **masern** *vt* spot, speckle; mottle; grain, vein. **Masern** *fpl* measles *cf* MASER I. **Maser·papier** *n* speckled (*or* grained) paper. **Maserung** *f* (-en) 1 spotting, etc *cf* MASERN. 2 grain (of wood)

Maske *f* (-n) mask. **maskieren** *vt* mask

maß measured: *past of* MESSEN

Maß I *n* (-e) 1 (*genl*) measure. 2 measurement, dimension; size. 3 standard(s), yardstick, scale; gauge; titer. 4 extent, degree: **in großem M(—e)** to a great extent (*or* degree), in great measure; **ein gewisses M. an** . . . a certain degree of . . . ; **in dem M., wie** . . . to the extent that . . . 5 (*pl usu* -en) moderation, proportion; bounds, limits: **mit M. (und Ziel)** in moderation; **M. halten** be moderate, stay within bounds; **über die M–en** beyond all bounds (*or* measure). **II** *f* (-) liter (of beer). **-abweichung** *f* dimensional deviation. **-analyse** *f* volumetric analysis. **m–analytisch** *a* volumetric. **M–änderung** *f* change in dimensions (measurements, size). **-band** *n* measuring tape. **-beständigkeit** *f* dimensional stability. **-bürette** *f* measuring burette

Masse *f* (-n) 1 (*usu, incl Phys*) mass; bulk. 2 large quantity, lot, great deal: **eine M. (von) Verbindungen** a lot of compounds. 3: **die M.** (+*von or genit*) the majority (of). 4 matter, substance; dough, batter; composition, paste, pulp; batch (of matter, etc); stock; (*Foun.*) dry sand. 5 crowd. 6 (bankrupt's) assets, estate. 7 (*Elec.*) ground. **-bolzen** *m* (*Elec.*) grounding pin. **-filter** *n* pulp filter. **m–frei** *a* (*Elec.*) ungrounded. **M–herstellung** *f* mass production

Maß·einheit *f* unit of measurement

Masse·kern *m* powdered iron core. **-klemme** *f* (*Elec.*) ground terminal

Massel *f* (*Metll.*) pig; slab, bloom, ingot. **-bett** *n* pig bed. **-eisen** *n* pig iron. **-graben** *m* (*Metll.*) sow. **-trog** *m* ingot (*or* pig) mold

Massen· mass, bulk, large-scale, wholesale: **-analyse** *f* mass analysis. **-anziehung** *f* gravitation. **-bewegung** *f* mass movement (*or* motion). **-defekt** *m* mass defect. **-dichte** *f* mass density. **-einheit** *f* unit of mass (*incl* atomic). **-erzeugung** *f*, **-fabrikation** *f* mass production. **-gestein** *n* massive (*or* unstratified) rock. **-guß** *m* dry-sand (*or* duplicate) casting. **-güter** *npl* bulk goods; mass-produced goods

massenhaft *a* mass(ive), large-scale; *adv* oft on a large scale, in masses; (*as Quant.*, *oft*) masses of . . .

Massen· mass, bulk, large-scale, wholesale: **-herstellung** *f* mass production. **-messung** *f* measurement of mass. **-mittelpunkt** *m* center of mass. **-moment** *n* moment of inertia. **-platte** *f* battery plate. **-produktion** *f* mass production. **-punkt** *m* mass point, center of mass (*or* gravity). **-rübe** *f* low-grade feed beet.

-schwächung f (X-rays) mass absorption. **-schwerpunkt** m center of gravity. **-schwund** m disappearance of mass. **-stahl** m low-carbon steel. **-strahler** m mass radiator. **-teilchen** n (Phys.) mass particle, corpuscle. **-trägheit** f inertia. **-übergang** m mass transfer. **-verhältnis** n mass ratio, relative mass. **-verlust** m loss of mass. **-ware** f mass-produced article(s). **m—weise** a massive, wholesale, in masses; adv oft in large numbers. **M— widerstand** m mass(ive) resistance, mass inertia. **-wirkung** f 1 (Chem., Phys.) mass action. 2 mass appeal, effect on the masses. **-wirkungsgesetz** n law of mass action. **-wirkungskonstante** f equilibrium constant. **-zahl** f mass number. **-zentrum** n center of mass. **-zucht** f (Agric.) quantity growing. **-zuwachs** m increase in mass, mass accretion

Masse-schlamm m (Ceram.) body slip. **-schlicker** m (Ceram.) body paste. **-verbindung** f (Elec.) ground connection. **-zusatz** m (Paper) beater additive

Maß- measuring, size, standard: **-flasche** f measuring flask. **-flüssigkeit** f standard solution. **-formel** f (Pharm.) standard formula. **-gabe** f: **nach M.** (+genit) according to; **mit der M., daß** ... with the proviso that ... **m—gebend** a 1 pace setting, tone-setting, standard; pred oft: **m. sein** set the tone (or the standard), be the criterion. 2 authoritative; decisive. 3 leading (persons). **-geblich** a 1 substantial, considerable. 2 leading (persons). **M—gefäß** n measuring vessel. **m—gerecht** a true to size, accurate. **M—glas** n measuring glass; volumetric container. **m—haltig** a precise, true to size. **M— haltigkeit** f precision, dimensional accuracy. **-holder** m common (or field) maple

massig a massive; adv oft masses of ...

mäßig I a 1 moderate. 2 mediocre. **II** sfx: a -like, -wise; adv according to, as regards (or simply forms adjectives from nouns): **herbstm.** autumn-like, autumnal; **größenm.** according to size, as regards size, size-wise. **mäßigen** 1 vt moderate, temper. 2: **sich m.** control oneself; (become) moderate. **Mäßigkeit** f moderation; temperance; mediocrity. **Mäßigung** f (-en) moderation; self-control

Massikot n (-e) massicot, unfused lead monoxide

massiv a 1 massive; severe; adv oft on a huge scale. 2 solid, pure, unalloyed (gold, etc). **Massiv** n (-e) (Geol.) massif. **-holz** n solid timber

Maß-kontrolle n gauging. **-kunde** f metrology, study of measurements. **m—los** a boundless, unrestrained; adv oft inordinately **m—mäßig** a measurementwise, as regards measurement(s). **M— nahme** f (corrective, remedial) measure, step: **M—n treffen (gegen)** take steps (against). **-regel** f regulation, rule; (remedial) measure. **-röhre** f measuring tube, graduated tube, burette. **-stab** m yardstick; standard; scale. **m—stäblich** a to scale: **m. genaue Modelle** exact-scale models. **-stabsgerecht** a true to scale. **M—system** n system of measurement(s). **-teil** m part by measure. **-teilung** f graduation. **-toleranz** f dimensional tolerance. **m—voll** a moderate. **M—zeichnung** f scale drawing

Mast I m (-e or -en) mast. **II** f 1 fattening (of animals). 2 fattening feed, esp beech nuts. **-darm** m (Anat.) rectum

mästen 1 vt feed, fatten. 2: **sich m.** gorge oneself

Mast-fähigkeit f fattening capacity. **-hahn** m capon. **-kalbsbraten** m roast milk-fed veal

Mastix m mastic (=resin); gum mastic; asphalt mastic. **-branntwein** m mastic (=liquor). **-firnis** m mastic varnish. **-harz** n mastic (resin). **-kitt** m mastic (cement). **-lack** m mastic varnish. **-lösung** f mastic solution. **-öl** n gum mastic oil

mastizieren vt masticate

Mästung f (-en) 1 fattening. 2 fattening feed

Mast-vieh n 1 fattened cattle. 2 cattle for fattening

Masut n mazout, heavy fuel oil

Material n (-ien) 1 material; equipment. 2 evidence; (+an) body: **ein großes M. an Daten** a large body of data. **-behälter** m receptacle for material(s).

-eigenschaft f property of materials. **-ermüdung** f material fatigue. **-förderung** f transport (or output) of material. **-kennwert** m physical property. **-prüfung** f materials testing. **-prüfungsamt** n materials testing bureau. **-überhitzung** f overheating of materials. **-untersuchung** f investigation (or inspection) of materials

Materie f (-n) 1 (Phys.) matter. 2 material. 3 subject matter

materiell a material; materialistic; financial

Materien·ofen m (Glass) calcar

materieren vi suppurate

Materie·strahl m matter (or de Broglie) wave. **-teilchen** n particle of matter. **-welle** f matter (or de Broglie) wave

Mathematiker m (-) mathematician

Matratze f (-n) mattress

Matricaria·campher m L-camphor

Matrikel f (-n) (student, personnel) roll

matrisieren vt damp (paper)

Matrize f (-n) matrix; (female) mold, die; stencil; (Phot.) master negative

Matrizen·ofen m die-heating furnace. **-pappe** f (Paper) matrix board

Matsch m mud, mire; sludge; slush

matt a 1 weak, exhausted. 2 dull, lusterless; mat, flat; frosted (glass); (Metll.) tarnished, pasty. 3 subdued; faint, dim. 4 lame, ineffective, limp. 5 checkmated —m. setzen checkmate. **M-·anstrich** m flat paint. **-appretur** f dull finish. **-beize** f pickle for giving a dull surface. **-blau** n flat (dull, pale) blue, bleu céleste. **-blech** n pickled sheet metal; (Metll.) terneplate. **-brenne** f, **-brennen** n dull pickling. **-dekatur** f (Tex.) dull finish

Matte f (-n) 1 mat. 2 (alpine) pasture, meadow

Matt·eisen n white pig iron. **-email** n mat (or flat) enamel

Matten·stoff m matting

Matt·farbe f deadening (or mat) color. **m-gelb** a pale (or flat) yellow, cream (-colored). **-geschliffen** a ground, frosted. **M-glanz** m dull luster (or finish). **-glas** n ground (or frosted) glass. **-glasur** f (Ceram.) mat glaze. **-gold** n dead gold. **m-grün** a pale (or flat) green

Mattheit f dullness, flatness; paleness; lack of luster; opacity; faintness, dimness cf MATT

mattierbar a (Metals) tarnishable

mattieren vt dull, deluster, deaden; make flat (dull, etc); frost (glass); tarnish. **Mattierung** f (-en) 1 dulling, delustering, deadening; frosting; tarnishing. 2 matting. 3 mat (dull, flat) finish. **Mattierungs·mittel** n delustering (or dulling) agent

Mattigkeit f weakness; exhaustion; fatigue

Mattine f (-n) furniture lacquer

Matt· dull, flat, mat: **-kohle** f dull coal. **-kunstseide** f delustered rayon. **-lack** m flat (dull, mat) varnish (or lacquer). **m-rosa** a dull pink. **-rot** a dull red. **M-scheibe** f 1 (Opt.) ground glass plate (or screen). 2 (Phot.) focusing screen. 3 frosted window pane. 4 TV screen

matt·schleifen* vt frost, grind (glass)

Matt· dull, flat, mat: **-schliff** m 1 mat grinding, frosting. 2 mat-ground finish. **m-schwarz** a dull (or flat) black. **M-sein** n dullness, flatness; dimness. **-vergoldung** f dead-gilding. **m-weiß** a dull (or dead) white. **M-weißblech** n terneplate. **-zurichtung** f mat (or dull) finish

Matur n, **Matura** f secondary school final comprehensive examination(s) = ABITUR

Matze f (-n), **Matzen** n (-) matzoh

Mauer f (-n) (external) wall. **··fraß** m wall decay (because of crystallization of salts). **-gelb** n yellow badigeon. **-kalk** m mortar, masonry lime. **-klinker** m highstrength brick. **-kraut** n wall pellitory

mauern 1 vt build (of bricks). 2 vi stonewall

Mauer·pfeffer m (Bot.) stonecrop, sedum. **-salpeter** m, **-salz** n wall saltpeter, calcium nitrate efflorescence. **-stein** m building stone, brick (other than burnt clay)

Mauerung f 1 (wall) building; stonewalling cf MAUERN. 2 = MAUERWERK n masonry, walls, brickwork

Mauer·zement m masonry cement. **-ziegel** m (building or wall) brick

mauken 1 vt digest with water. 2 vi (Ceram.) weather, age. **Mauken** n (Ceram.) fermenting, aging

Maul *n* (-̈er) mouth (of animals): **M.- und Klauenseuche** *f* foot and mouth disease. **-beere** *f* mulberry. **-esel** *m* mule. **-klemme, -sperre** *f* lockjaw, tetanus. **-tier** *n* (*Zool.*) mule. **-wurf** *m* (*Zool.*) mole

Maurer *m* (-) mason, bricklayer

maurisch *a* Moorish

Maus *f* (-̈e) mouse

Mäuse·dreck *m* mouse droppings. **-fraß** *m* (plant) damage done by mice. **-gift** *n* 1 mouse (*or* rat) poison, rodenticide. 2 ratsbane, (white) arsenic

Mauser *f* (*Zool.*) molt(ing). **mausern** *vi* & **sich m.** 1 molt. 2 develop, blossom out

maus·fahl, -farben, -farbig, -grau *a* mouse-gray (*or* -colored), mousy

m.a.W. *abbv* (mit anderen Worten) in other words

maximal *a* maximum. **M·-geschwindigkeit** *f* maximum velocity

Maxi·valenz *f* maximum valence

Mazeral *n* (-e) (*Coal*) maceral

mazerieren *vt* macerate. **Mazerier·gefäß** *n* macerator

m.b.H. *abbv* Ltd., Inc. *Cf* G.M.B.H.

MC *abbv* (Methylcellulose) methyl cellulose

MD *abbv* (Mitteldruck) medium pressure

m.E. *abbv* (meines Erachtens) in my opinion

Me. *abbv* (Methanol) methanol

ME *abbv* 1 (Masseeinheit) (atomic) mass unit. 2 (Mache-Einheit) Mache unit. 3 (Mäuseeinheit) (*Biol.*) mouse unit

Mechanik *f* (-en) 1 mechanics. 2 mechanism. **Mechaniker** *m* (-) mechanician; machine operator, mechanic. **mechanisch** *a* mechanical. **mechanisieren** *vt* mechanize. **Mechanisierung** *f* mechanization. **Mechanismus** *m* (..smen) mechanism

Mecon·säure *f* meconic acid

med. *abbv* (medizinisch) medical, medicinal

Medaille *f* (-n) medal

Median·papier *n* medium paper

Medien media *pl of* MEDIUM

Medikament *n* (-e) medicine, medication

Mediterran·fieber *n* Mediterranean (*or* undulant) fever

Medizin *f* (-en) medicine. **Mediziner** *m* (-),

Medizinerin *f* (-nen) medic; medical student. **mediziniert** *a* medicated. **medizinisch** *a* medical, medicinal; **m.-chemisch** *a* medico-chemical

Meduse *f* (-n) medusa, jellyfish

Meer *n* (-e) sea, ocean. **-ablagerung** *f* marine deposit. **m·-bewohnend** *a* marine. **-blau** *a* sea blue, greenish blue. **M·-busen** *m* (*Geog.*) gulf. **-enge** *f* strait(s), channel

Meeres· sea, marine: **-bildung** *f* marine formation. **-boden** *m* sea bottom, ocean floor. **-chemie** *f* marine chemistry. **-einzeller** *m* marine protozoon. **-fläche** *f* sea area (of the globe); surface of the sea. **-forschung** *f* oceanography. **-grund** *m* sea bottom. **-höhe** *f* sea level; height above sea level. **-kunde** *f* oceanography. **-küste** *f* seacoast. **-leuchten** *n* marine phosphorescence. **-niveau** *n* sea level. **-oberfläche** *f* surface of the sea. **-sand** *m* sea sand. **-spiegel** *m* sea level. **-strand** *m* ocean beach, seashore. **-strömung** *f* ocean current

Meer· sea, marine: **-farbe** *f* sea green (*or* color). **-gras** *n* seaweed, kelp. **-kohl** *m* sea kale. **-kraut** *n* glasswort, saltwort. **-leuchten** *n* marine phosphorescence. **-rettich** *m* horseradish. **-salz** *n* sea salt. **-schaum** *m* 1 meerschaum, sepiolite. 2 sea foam. **-schlamm** *m* sea ooze. **-schwein** *n* porpoise. **-schweinchen** *n* guinea pig. **-schweinöl** *n*, **-schweintran** *m* porpoise oil. **-seide** *f* sea silk, seaweed fiber. **-tang** *m* seatang, coarse seaweed. **-torf** *m* sea peat. **-wachs** *n* (*Petrol.*) sea wax. **-wasser** *n* sea water. **-wasserentsalzung** *f* sea water desalin(iz)ation. **m·-wasserecht** *a* fast to sea water, seawater-resistant. **M·-zwiebel** *f* sea onion, squill

Megahertz *n* megacycle(s) per second

Meger·kraut *n* yellow bedstraw

Mehl *n* (-e) flour; meal; powder, dust. **m·-artig** *a* mealy, farinaceous. **M·-beere** *f* haw, hawthorn berry. **-beutel** *m* 1 flour sifter. 2 flour bag. **-fälschung** *f* adulteration of flour. **m·-fein** *a* fine as flour. **M·-früchte** *fpl* cereals. **-gips** *m* (*Min.*) earthy gypsum. **m·-haltig** *a* containing flour, farinaceous

mehlig *a* mealy, floury, farinaceous.
Mehligkeit *f* mealiness, flouriness
Mehl· flour, meal: **··kleister** *m* flour
paste. **-körper** *m* endosperm (of grains).
-kreide *f* earthy calcite, kieselguhr.
-prüfer *m* flour tester, aleurometer.
-pulver *n* meal powder; (*Pyro.*) finely
ground gunpowder. **-sand** *m* fine sand,
rock flour. **-schwitze** *f* roux. **-speise** *f*
farinaceous food; dish (*esp* dessert) made
with flour; cereal. **-staub** *m* flour-mill
dust. **-stoff** *m* farinaceous substance.
-tau *m* mildew. **-taupilz** *m* mildew
(fungus). **-zucker** *m* powdered sugar
mehr 1 *adv & pron* more; **immer m.** more
and more; **nicht m.** no more, no longer,
not . . . anymore; **nie m.** never again;
nicht m. lange not much longer. 2 (*as pfx
oft*) poly-, multi-. **Mehr** *n* 1 (*usu* + **an**)
increase (in, of), increased amount (of);
surplus, excess (of). 2 majority
Mehr· poly-, multi-; many; increased: **··ar-
beit** *f* extra (*or* increased) work; overtime
work. **m–atomig** *a* polyatomic. **M–
aufwand** *m* (+**an**) increased effort, input, expenditure (in the way of). **M–
badverfahren** *n* multiple-bath process.
m–bändig *a* multi-volume. **-basig** *a*
polybasic. **M–basigkeit** *f* polybasicity.
m–basisch *a* polybasic. **M–belastung** *f*
overload, additional (extra, increased, excess) load. **-besitz, -bestand** *m* surplus,
oversupply. **-betrag** *m* excess, increased
amount. **m–deutig** *a* ambiguous. **M–
deutigkeit** *f* ambiguity. **m–
dimensional** *a* multi-dimensional. **M–
drehung** *f* multirotation
mehren *vt & sich m.* increase, multiply
mehrere *a pl* several
mehrerlei 1 *a* (*no endgs*) several different.
2 *pron* various things
mehrfach *a* 1 multiple; (*Elec. oft*) multi-
plex. 2 repeated. 3 *adv* several times, repeatedly. **M–·bindung** *f* multiple bond.
-färbung *f* (*Micros.*) multiple staining.
-verdampfer *m* multiple-effect evaporator. **-wechselrichter** *m* (*Elec.*) multi-
vibrator. **-zünder** *m* multiple (*or* combination) fuse
mehr·fädig *a* multifil(ament). **-fältig** *a*
multiple = MEHRFACH. **M–farbendruck**

m multicolor printing. **m–farbig** *a* multicolored, polychromatic. **M–farbigkeit** *f*
polychromatism, (*Art*) polychromy. **m–
flammig** *a* multiple-flame (*or* -burner).
M–gehalt *m* excess content. **-gewicht** *n*
excess weight, overweight. **m–gliedrig** *a*
multiple-membered
Mehrheit *f* (-en) majority
mehr· multi-, poly-, many-; increased: **-her-
dig** *a* multiple-hearth. **-jährig** *a* several
years (of); perennial. **M–kanal·** multi-
channel. **m–kantig** *a* many-sided. **m–
kernig** *a* polynuclear. **M–körperpro-
blem** *n* multibody problem. **-kristall** *m*
polycrystal. **m–lagig** *a* multiple-layer.
M–leiterkabel *n* multi-core cable. **-lini-
enspektrum** *n* many-line spectrum. **m–
malig** 1 *a* repeated. 2 *adv* = **-mals** *adv*
repeatedly, several times. **M–nährstoff-
Dünger** *m* multinutrient fertilizer. **M–
phasen·** (*Elec.*) polyphase. **-phasenra-
kete** *f* multi-stage rocket. **m–phasig** *a*
polyphase, multiphase. **-polig** *a* multipolar. **-säurig** *a* poly-acid. **M–schichten-
glas** *n* laminated glass. **m–schichtig** *a*
multilayered. **M–schritt·** multi-stage.
m–seitig *a* many-sided, polygonal.
-stellig *a* multi-digit, several-place
(number); multi-positioned
Mehrstoff·katalysator *m* mixed catalyst.
-polymerisation *f* copolymerization.
-system *n* multi-component system
Mehr·strahlbrenner *m* multijet burner.
-stufenrakete *f* multi-stage rocket. **m–
stufig** *a* multi-stage. **-stündig** *a* several
hours (of). **-tägig** *a* several days (of).
-teilig *a* multiple-part, multi-component
Mehrung *f* increase, multiplication
Mehr·wegbehälter *m* returnable container. **m–wellig** *a* multiwave, (*Elec.*)
polyphase. **M–wert** *m* increase in value;
added (*or* surplus) value. **m–wertig** *a*
multivalent; (*Acids*) polybasic; (*Alcohols*)
polyhydric. **-wöchig** *a* several weeks (of).
M–zahl *f* 1 plural. 2 majority. **m–zählig**
a many-membered; (*Electrolytes*) multivalent; (*Complex Compds.*) multidentate.
-zellig *a* multicellular. **M–zünder** *m*
multiple fuse. **M–zweckmaschine** *f*
multipurpose machine
meiden* (mied, gemieden) *vt* avoid. **Mei-**

dung f avoidance

Meierei f (-en) dairy farm

Meile f (-n) mile. **Meilen‧stein** m milestone

Meiler m (-) **1** atomic pile. **2** charcoal (burner's) pile. **3** clamp (of bricks)

mein possess my, mine

meinen vt/i **1** think, express (or have) the opinion. **2** mean, intend. **3** say, remark

meine(n), meiner, meines gram var of MEIN. **meinerseits** adv for my part. **meinet‧halben, -wegen** adv on my account; for all I care

Meinung f (-en) opinion: **meiner M. nach** in my opinion; **eine M. fassen** form an opinion

Meinungs‧umfrage f opinion poll. **-verschiedenheit** f difference of opinion

Meiran m (-e) marjoram. **-‧öl** n marjoram oil

Meisch m (-e) mash, etc = MAISCHE

Meise f (-n) titmouse

Meißel m (-) chisel, cutting tool. **m‧förmig** a chisel-shaped. **meißeln** vt/i chisel, carve. **Meißel‧stahl** m chisel steel

meist 1 a (with endgs), adv: **am m—en** most; (oft as pfx): **meist‧begünstigt** most-favored; **M—betrag** m maximum amount. **2** adv usually, for the most part. **meistens** adv mostly, most of the time. **meisten‧teils** adv mostly, for the most part

Meister m (-) master; master craftsman; foreman; champion. **meisterhaft, meisterlich** a masterful, masterly. **meistern** vt master. **Meisterschaft** f (-en) **1** rank of master (craftsman). **2** mastery. **3** championship (contest)

Meister‧stück n masterpiece; master stroke. **-wurz(el)** f masterwort (or its rhizome)

Méker‧brenner m Méker burner

Mekonin‧säure f meconinic acid

Mekon‧säure f meconic acid

Melangallus‧säure f melanogallic (or metagallic) acid

Melange f (-n) mixture **-‧garn** n multicolorfiber yarn (made of colored fibers blended before spinning). **melangieren** vt mix

Melan‧glanz m (Min.) stephanite. **-glimmer** m (Min.) (ferro)stilpnomelane; cronstedtite

Melanochroit n (Min.) phoenicochroite

Melasse f (-n) molasses. **-‧maische** f molasses mash

Melassen‧säure f melassic acid. **-spiritus** m molasses spirit.

Melasse‧schlempe f molasses residue. **-schnitzel** npl molasses pulp. **-spiritus, sprit** m molasses spirit

Melde‧ registration, reporting; message, signal: **-amt** n registration office; vital statistics center

melden I vt report, announce. **II: sich m. 1** report. **2** announce one's presence (or coming), be heard from. **3** register, sign up. **4** answer (esp Tlph.). **5** offer answers (services, etc)

melde‧pflichtig a **1** reportable (illness), that must be reported. **2** (incl Mil.) required to register (or report). **Melder** m (-) **1** messenger. **2** (Mach.) signal, indicator, alarm. **Melde‧stelle** f registration (reporting, vital statistics) center. **-tafel** f indicator board

Meldola‧blau n Meldola blue

Meldung f (-en) **1** message, report, news item. **2** announcement. **3** offer to answer (participate, etc); application, entry, bid. **4** signal, indication, alarm

melieren vt mix, blend; mottle

Melier‧farbe f (Paper) mottling color. **-fasern** fpl mottling fibers. **-papier** n mottled paper

meliert p.a mixed, blended; mottled; sparkled; multicolored; graying, mixed-gray (hair)

Meliloten‧klee m melilot, sweet clover. **Melilot‧säure** f melilotic acid

meliorieren vt ameliorate

Melis m coarse loaf sugar

Melisse f (-n) melissa, balm (mint). **Melissen‧geist** m Carmelite water, spirit of melissa. **-kraut** n melissa, balm (mint). **-‧öl** n melissa oil, balm mint oil. **-spiritus** m, **-wasser** n Carmelite water

Melissin‧säure f melissic acid

Melis‧zucker m coarse loaf sugar

melken* (molk, gemolken) vt/i milk (cows, etc). **Melkerei** f (-en) dairy farm

Mellit(h)‧säure f mellitic acid

Mellon n Pharaoh's serpents (residue from pyrolysis of mercuric thiocyanide). **-‧was-**

serstoffsäure *f* hydromellonic acid

Melone *f* (-n) melon. **-öl** *n* melon (seed) oil

Meltau *m* mildew = MEHLTAU

Membran(e) *f* (..anen) membrane, diaphragm

Membran·gleichgewicht *n* membrane (*or* Donnan) equilibrium. **-hydrolyse** *f* membrane hydrolysis. **-pumpe** *f* diaphragm pump. **-ventil** *n* diaphragm valve

Mendelejew *German spelling of* Mendeleyev

Mendelsch *a* Mendelian, Mendel's

mengbar *a* miscible, mixable

Menge *f* (-n) 1 amount, quantity. 2 (*esp* + *noun*) lot, abundance, large quantity (of). 3 crowd. 4 (*Math.*) aggregate, set. **-einheit** *f* unit of quantity. **mengen** 1 *vt & sich m.* mix, blend. 2: *sich m.* (*usu* + *in*) meddle, interfere (in, with)

Mengen·anteil *m* constituent amount, proportion. **-bestimmung** *f* quantitative determination. **-lehre** *f* (*Math.*) theory of sets. **m-mäßig** *a* quantitative. **M-messer** *m* flow meter; volumeter. **-regelung** *f* quantity control. **-untersuchung** *f* quantitative examination. **-verhältnis** *n* quantitative relation, (relative) proportion, composition (ratio)

Meng· mixed, mixing: **-futter** *n* mixed food (*or* feed), mash. **-gestein** *n* (*Min.*) conglomerate. **-kapsel** *f* mixing capsule. **-korn** *n* mixed grain (*esp* wheat and rye). **-saat** *f* mixed seed. **-spatel** *m* mixing spatula. **-teil** *m* ingredient (of a mixture)

Mengung *f* (-en) mixing, blending; mixture, blend; hybridization. **Mengungs·verhältnis** *n* proportion of ingredients

Menhaden·tran *m* menhaden oil, American fish oil

Meniskus *m* (..sken) meniscus. **menisken·förmig** *a* meniscal, meniscoid

Mennige *f* red lead, minium. **-ersatz** *m* red-lead substitute. **-kitt** *m* red-lead cement

mennigen *vt* red-lead, paint with minium

mennig·farbig, -rot *a* minium-colored

Mensch *m* (-en, -en) 1 man (as a species), human being. 2 person, *pl usu* people. 3: **kein M.** nobody

Menschen· human, anthropoid: **-affe** *m* anthropoid ape. **m-ähnlich** *a* human (-like), anthropoid. **M-alter** *n* generation. **-blut** *n* human blood. **-feind** *m* misanthropist. **-fett** *n* human fat. **-fresser** *m* cannibal. **-freund** *m* humanitarian. **-geschlecht** *n* mankind, the human race. **-kenner** *m* judge of people. **-körper** *m* human body. **-kunde, -lehre** *f* anthropology. **-pocken** *fpl* smallpox. **-verstand** *m* common sense

Menschheit *f* man(kind), human race

menschlich *a* human; humane. **Menschlichkeit** *f* humanity; humaneness

Mensur *f* (-en) 1 measure. 2 = **-glas** *n* (glass) measuring vessel, graduated cylinder

Mentha·kampfer *m* mint camphor, menthol

Mepasin *n* (*TN*) mixture of C_{14}-C_{17} paraffin hydrocarbons

mercerisier·echt *a* fast to mercerizing. **mercerisieren** *vt* mercerize

Mercerisier·hilfsmittel *n* mercerizing aid. **-lauge** *f* mercerizing liquor

Mercur ... *See* MERKUR ...

Mergel *m* (-) marl. **-art** *f* variety of marl. **m-artig** *a* marly. **M-boden** *m* marly soil. **-düngung** *f* (*Agric.*) marling. **-erde** *f* earthy marl. **m-haltig** *a* marly, containing marl

mergelig *a* marly

Mergel·kalk *m* marly limestone

mergeln *vt* marl, treat with marl

Mergel·sandstein *m* marly sandstone. **-schiefer** *m* slaty marl. **-ton** *m* marly clay, argillaceous marl

Mergelung *f* marling, treatment with marl

Mer·gewicht *n* mer (*or* monomer) weight

Merino·rot *n* (*Dye.*) Turkey red

Meritektikum *n* (..iken) meritectic

merkbar *a* 1 perceptible, noticeable. 2: (**leicht, schwer**) **m.** (easy, hard) to remember

Merk·blatt *n* information leaflet, instructions. **-buch** *n* notebook, memo book

merken I *vt* 1 notice, perceive, note. 2 (+ *dat rflx: sich*) remember, commit to memory. 3 (+ *lassen*) show signs of, let . . . be detected. **II** *vi* (+ *auf*) pay attention (to)

merkens·wert *a* noteworthy

Merk·farbe *f* indicator color

merklich *a* **1** noticeable, perceptible. **2** marked. **3** considerable

Merk·mal *n* **1** sign, mark, indication. **2** feature, characteristic; (*Genetics*) character. **3** criterion. **-malsträger** *m* (*Biol.*) gene. **-tinte** *f* marking ink

Merkur *m* mercury. **--blende** *f* cinnabar. **-chlorid** *n* mercury (*or specif* mercuric) chloride. **-chlorür** *n* mercurous chloride. **-gelb** *n* mercury yellow. **-glanz** *m* selerian metacinnabar. **m--haltig** *a* mercurial, containing mercury. **M--hornerz** *n* (*Min.*) horn quicksilver, calomel

Merkuri· mercuri-, mercuric

Merkurialien *npl* (*Pharm.*) mercurials

Merkurial·mittel *n* (*Pharm.*) mercurial

Merkuri·ammoniumchlorid *n* mercuriammonium chloride. **-azetat** *n* mercuric acetate. **-chlorid** *n* mercuric chloride. **-cyanid** *n* mercuric chloride. **-cyanwasserstoffsäure** *f* mercuricyanic acid

merkurieren *vt* mercurize, combine with mercury. **Merkurierung** *f* (-en) mercurization

Merkuri·jodid *n* mercuric iodide. **-nitrat** *n* mercuric nitrate. **-oxid** *n* mercuric oxide. **-rhodanid** *n* mercuric thiocyanate. **-salz** *n* mercuric salt. **-sulfat** *n* mercuric sulfate. **-sulfid** *n* mercuric sulfide. **-sulfozyanid** *n* mercuric thiocyanate. **-verbindung** *f* mercuric compound

Merkur·jodid *n* mercury iodide

Merkuro· mercurous: **--azetat** *n* mercurous acetate. **-chlorid** *n* mercurous chloride. **-chrom** *n* (*Pharm.*) mercurochrome. **-jodid** *n* mercurous iodide. **-nitrat** *n* mercurous nitrate. **-oxid** *n* mercurous oxide. **-salz** *n* mercurous salt. **-sulfat** *n* mercurous sulfate. **-sulfid** *n* mercurous sulfide. **-verbindung** *f* mercurous compound

Merkur·oxid *n* mercury (*or specif* mercuric) oxide. **-silber** *n* silver amalgam. **-sulfid** *n* mercury (*or specif* mercuric) sulfide

merk·würdig *a* **1** remarkable, noteworthy. **2** strange, odd. **-würdigerweise** *adv* oddly enough. **M--zeichen** *n* mark, sign; marker. **-zettel** *m* memorandum, memo sheet (*or* slip)

Mer·mol *n* (*Polymers*) mer mole

Mero·chinen *n* meroquinene. **-tropie** *f* merotropism, merotropy

Mersolat *n* (*TN*) sodium alkanesulfate detergent

merzerisieren *vt* mercerize

mesomer *a* mesomeric. **Mesomerie** *f* mesomerism, resonance

Mesonen·zerfall *m* meson decay

Mesothor *n* mesothorium

Meso·weinsäure *f* mesotartaric acid

Mesoxal·säure *f* mesoxalic acid

mesozoisch *a* Mesozoic

Meß· measuring: **--analyse** *f* volumetric analysis. **-anordnung** *f* measuring arrangement (*or* device). **-apparat** *m* measuring apparatus, meter. **-apparatöl** *n* meter oil. **-band** *n* measuring tape, tape measure

meßbar *a* measurable

Meß· measuring: **-becher** *m* measuring cup. **-behälter** *m* measuring vessel (*or* tank). **-bereich** *m* range of measurement. **-brücke** *f* (*Elec.*) (Wheatstone) bridge. **-dose** *f* pressure gauge. **-draht** *m* slide wire (of Wheatstone bridge)

Messe *f* (-n) **1** (trade) fair. **2** mess(hall). **3** (religious) mass

Meß·einheit *f* unit of measure(ment). **-einteilung** *f* graduation

messen* (maß, gemessen; mißt) **1** *vt/i* measure; survey; gauge. **2: sich m.** (+mit) compare, compete (with); measure up (to). *Cf* GEMESSEN

Messer I *m* (-) measurer; meter, gauge. **II** (-) knife, cutter; doctor, wiper

Messer·brecher *m* knife crusher

Meß·ergebnis *n* result of measurement

Messer·klinge *f* knife blade. **-schalter** *m* (*Elec.*) knife switch. **m--scharf** *a* sharp as a knife. **M--schmied** *m* cutler. **-schneide** *f* knife edge. **-spitze** *f* **1** knife point; tip of the knife. **2** (*Meas.*) pinch (e.g. of salt). **-stahl** *m* knife (*or* cutlery) steel

meß· measuring: **-fähig** *a* measurable. **M--fehler** *m* error in measurement. **-flansch** *m* orifice plate (of a flowmeter). **-flasche** *f* measuring (*or* graduated) flask. **-form** *f* (*Glass*) parison. **-flüssigkeit** *f* **1** measuring liquid. **2** titrating solution. **m--geblich** *a* significant. **M--gefäß** *n* measuring vessel, measure. **-ge-**

nauigkeit f accuracy (or precision) of measurement. **-gerät** n measuring device (or instrument); meter; range finder. **-gerätetisch** m instrument table. **-glas** n measuring (or gauge) glass, graduated cylinder. **-grenze** f limit of measurement. **-heber** m measuring pipette. **-holz** n 1 dimension lumber. 2 measuring stick

Messing n (-e) brass. **--abstrich, -abzug** m brass scum. **m--artig** a brassy, brass-like. **M--artigkeit** f brassiness. **-blatt** n brass foil. **-blech** n 1 sheet brass. 2 brass sheet. **-blüte** f aurichalcite. **-brennen** n brass making. **-draht** m brass wire

messingen a (made of) brass; brassy

messing- brass: **-farben, -farbig** a brass-colored. **M--fassung** f brass mounting (or casing). **-folie** f brass foil. **-gewicht** n brass weight. **-gießer** m brass founder. **-gießerei** f brass foundry; brass founding. **-guß** m cast brass; brass casting(s). **-hartlot** n brass solder = -LOT. **-lack** m brass lacquer. **-lot** n brazing spelter, brass solder. **-netz** n brass netting (or gauze). **-rohr** n, **-röhre** f brass tube (or pipe). **-schlaglot** n brass solder, hard solder. **-späne** mpl brass shavings (or turnings). **-stange** f brass rod

Meß- measuring: **-instrument** n measuring instrument, meter. **-keil** m measuring wedge (inside calipers for tubing, etc). **-kelch** m measuring cup. **-kolben** m measuring flask. **-kopf** m measuring head. **-kunde** f metrology; surveying. **-kurve** f calibration curve. **-länge** f gauge length. **-lösung** f (Anal.) standard solution. **-lupe** f (Opt.) magnifying lens. **-mikroskop** n measuring microscope. **-mittel** n means of measuring; measuring instrument. **-okular** n (Opt.) micrometer eyepiece. **-pipette** f graduated (or volumetric) pipette. **-pumpe** f metering pump. **-punkt** m experimental value; point on a graph. **-rand** m orifice plate (of a flowmeter). **-reihe** f series of measurements. **-rohr** n, **-röhre** f measuring tube, burette. **-röhrenklemme** f burette clamp. **-schaufel** f measuring scoop. **-schreiber** m recorder, recording unit. **-serie** f series of measurements. **-sonde** f measuring probe. **-stab** m 1 measuring

rod. 2 liquor gauge (for casks). **-strecke** f measured distance (or trajectory). **-streifen** m (Instrum.) strip chart. **-strom** m signal current. **-technik** f applied metrology, techniques of measurement. **m--technisch** a metrological, concerning techniques of measurement. **M--trichter** m measuring funnel, metered hopper. **-trommel** f graduated drum. **-uhr** f dial gauge; measuring clock

Messung f (-en) measuring; measurement; metering; mensuration

Messungs-ergebnis n result of measurement

Meß- measuring: **-verfahren** n method (or process) of measuring. **-vorrichtung** f measuring device. **-wandler** m metering transformer. **-werk** n measuring mechanism. **-wert** m measured value. **-wesen** n mensuration; measuring instrumentation. **-winkel** m measuring (or reference) angle. **-zahl** f index (number); measured (or measuring) value. **-zeug** n measuring instruments. **-zylinder** m graduated cylinder

Met m mead

meta-antimonig a meta-antimonious. **M--antimonsäure** f meta-antimonic acid. **m--arsenig** a meta-arsenious. **M--arsensäure** f meta-arsenic acid. **-biose** f metabiosis. **-bisulfit** n metabisulfite, pyrosulfite. **-bleisäure** f metaplumbic acid

metabol a metabolic. **Metabolie** f (Physiol.) metabolism; (Biol.) polymorphism. **Metabolismus** m metabolism

Meta-borsäure f metaboric acid. **-cer** n metacerium. **-cymol** n metacymene. **-eisenoxid** n metaferric oxide. **-ferrihydrat** n metaferric hydroxide. **-gallussäure** f metagallic acid. **-kieselsäure** f metasilicic acid. **-kohlensäure** f metacarbonic (or ordinary carbonic) acid

Metall n (-e) metal; (in compds oft) metallic: **-abfall** m waste (or scrap) metal. **-ablagerung** f metal-bearing deposit

Metallack° m metal lacquer = **Metall-lack**

Metall-ader f metallic (or metal-bearing) vein. **m--ähnlich** a metallic, metalline. **M--amid** n metallic amide. **-anstrich** m metal paint. **m--arm** a of low metal con-

tent. **-artig** *a* metallic, metalline. **M–artigkeit** *f* metallicity, metallic nature. **-asche** *f* metallic ash. **-auftrag** *m* metallic coating (*or* plating). **-auskleidung** *f* metal coating (covering, lining). **-azid** *n* metallic azide. **-bad** *n* metallic bath. **-bau** *m* metal construction. **-bearbeitung** *f* metalworking. **-beize** *f* metallic mordant. **-beschlag** *m* metallic coating (*or* sheathing). **-beschreibung** *f* metallography. **-beschwerung** *f* (*Tex.*) loading with metallic salts. **-bindung** *f* metallic bond. **-blatt** *n* (thin) sheet of metal. **-blättchen** *n* metal foil (*or* leaf). **-blech** *n* metal sheet; sheet metal. **-carbid** *n* metallic carbide. **-chemie** *f* chemistry of metals. **-chlorid** *n* metallic chloride. **-chlorwasserstoffsäure** *f* acid containing chlorine, hydrogen and a metal. **-cyanid** *n* metallic cyanide. **-dampf** *m* metallic vapor. **-draht** *m* metal(lic) wire **Metallegierung°** *f* alloy = **Metall-legierung**
Metall-eigenschaft *f* metallic property. **-einkristall** *m* metal single crystal. **-emaillelack** *m* metal enamel varnish. **m–empfindlich** *a* sensitive to metal **metallen** *a* metallic
Metall-erz *n* metallic ore. **-faden** *m* metal filament (*or* thread). **-fadenlampe** *f* metal filament lamp. **-farbe** *f* metallic (*or* bronze) color. **m–farbig** *a* metal-colored. **M–farblack** *m* metallic paint. **-färbung** *f* coloring of metal, metallochrome; metallic coloring. **-fassung** *f* metal casing (*or* mounting). **-fläche** *f* metallic surface. **-folie** *f* metal foil. **m–förmig** *a* metalliform. **-frei** *a* nonmetallic, metal-free. **-führend** *a* metal-bearing, metalliferous. **M–gefäß** *n* metal vessel. **-gehalt** *m* metal(lic) content. **-gekrätz** *n* waste metal, metal scrapings. **-geld** *n* metallic currency, coin(s), specie. **-gemisch** *n* metallic mixture, alloy. **-gewebe** *n* wire gauze (*or* cloth). **-gewinnung** *f* extraction of metals, metallurgy. **-gießerei** *f* foundry; metal founding. **-gift** *n* metallic poison. **-gitter** *n* metal grating (*or* grid). **-glanz** *m* metallic luster. **m–glänzend** *a* having a metallic luster. **M–glas** *n* enamel. **-gold** *n* Dutch gold (*or* foil). **-guß** *m* metal

founding (*or* casting); nonferrous casting. **m–haltig** *a* containing metal, metalliferous. **M–hütte** *f* (nonferrous) smelter. **-hüttenkunde** *f* (nonferrous) metallurgy (*or* smelting). **-hydroxid** *n* metallic hydroxide
metallieren *vt* metallize
Metall-ion *n* metal(lic) ion
metallisch *a* metallic
metallisieren *vt* metallize. **Metallisierung** *f* metallization
Metallität *f* metallic nature
Metall-kalk *m* metallic oxide, metal calx. **-karbid** *n* metallic carbide. **-karbonyl** *n* metal carbonyl. **-keramik** *f* powder metallurgy. **-kitt** *m* metal cement. **-komplexsalz** *n* complex metallic salt. **-komplexsäure** *f* complex metallic acid. **-könig** *m* metallic regulus. **-korn** *n* 1 grain of metal. 2 *esp pl* granulated metal. **-krätze** *f* waste metal, metal filings. **-kühler** *m* metallic cooler (*or* condenser). **-kunde** *f* metallography; metallurgy. **-lack** *m* **-legierung** *f*, **-lösung** *f* See METALLACK, METALLEGIERUNG, METALLÖSUNG. **-membran** *f* metal diaphragm. **-meldegerät** *n* metal detector. **-mikroskop** *n* metallurgical microscope. **-mikroskopie** *f* microscopic metallurgy. **-mischung** *f* 1 metallic mixture, alloy. 2 alloying, mixing of metals. **-mischungswaage** *f* alloy balance. **-mohr** *m* metallic moiré. **-mutter** *f* matrix (*of* ores), (*Min.*) rider. **-niederschlag** *m* metallic precipitate (*or* deposit). **-niet** *n* metal rivet
Metallo-chemie *f* chemistry of metals. **-chromie** *f* metallochromy
Metallograph *m* (-en, -en) metallographer. **Metallographie** *f* metallography. **metallographisch** *a* metallographic
metall-organisch *a* organometallic
Metallösung° *f* metallic solution = **Metall-lösung**
Metall-oxid *n* metallic oxide. **-oxyhydrat** *n* metallic hydroxide. **-papier** *n* metallic paper. **-poliermittel** *n* metal polish. **-probe** *f* test for metal, assay. **-pulver** *n* metal powder. **-putzmittel** *n* metal polish. **-rohr** *n*, **-röhre** *f* metal tube (*or* pipe). **-rückstand** *m* metallic residue. **-safran**

m saffron (*or* crocus) of antimony. **-salz** *n* metallic salt. **m–salzbeständig** *a* resistant to metallic salt. **M–salzlösung** *f* metallic salt solution. **-säure** *f* metallic acid. **-schlacke** *f* metal slag. **-schlauch** *m* (flexible) metal tube (tubing, hose). **-schliff** *m* 1 ground metal section (*or* surface). 2 metal filings. **-schmelze** *f* 1 fused (*or* molten) metal. 2 metal fusion (*or* melting). **-schmelzofen** *m* nonferrous melting furnace. **-schwamm** *m* metallic sponge. **-seife** *f* metal soap, soap for metals. **-silber** *n* imitation silver foil. **-späne** *mpl* metal shavings (turnings, chips). **-spiegel** *m* metal(lic) mirror. **-spitze** *f* metal(lic) tip. **-spritzen(verfahren)** *n* metal spraying (process), metallization. **-staub** *m* metal(lic) dust. **-sulfid** *n* metallic sulfide. **-technik** *f* metal technology. **-teil** *m* metal(lic) part. **-teilchen** *n* metallic particle. **-tuch** *n* wire cloth. **-überziehung** *f* coating (*or* plating) with metal. **m–überzogen** *a* metal-plated (-coated, -sheathed). **M–überzug** *m* metallic coating (plating, sheathing, wash)

Metallurg *m* (-en, -en) metallurgist

metall·verarbeitend *a* metalworking. **M–verarbeitung** *f* metalworking, metal processing. **-verbindung** *f* metallic compound. **-vergiftung** *f* metal(lic) poisoning. **-verlust** *m* loss of metal. **-versetzung** *f* alloying; alloy. **-ware** *f* metalware, hardware. **-weiß** *n* zinc sulfide pigment. **-wolle** *f* metal wool. **-zaponlack** *m* lacquer for metals, metal japan

metamer *a* metameric. **Metamerie** *f* metamerism

metamorph *a* metamorphic. **Metamorphie** *f* metamorphism. **Metamorphose** *f* (-n) metamorphosis

Metanil-gelb *n* metanil(ine) yellow. **-säure** *f* metanilic acid

metantimonig *a* metantimonious. **Metantimon·säure** *f* metantimonic acid

meta·phosphorig *a* metaphosphorous. **-phosphorsauer** *a* metaphosphate of. **M–phosphorsäure** *f* metaphosphoric acid. **-säure** *f* meta acid. **m–somatisch** *a* metasomatic. **-ständig** *a* in the meta position. **M–stase** *f* metastasis.

-stellung *f* meta position. **-styrol** *n* polystyrene. **-vanadinsäure** *f* metavanadic acid. **-weinsäure** *f* metatartaric acid. **-wolframsäure** *f* metatungstic acid. **-xylol** *n* metaxylene, **m–zentrisch** *a* metacentric. **M–zinnober** *m* metacinnabar. **-zirkonsäure** *f* metazirconic acid. **-zuckersäure** *f* metasaccharic acid

Meteor·eisen *n* meteoric iron

meteorisch *a* meteoric

Meteorologe *m* (-n, -n) meteorologist. **meteorologisch** *a* meteorological

Meteor·schweif *m* meteor trail. **-stahl** *m* meteoric steel. **-staub** *m* meteoric dust. **-stein** *m* meteoric stone, aerolite

Meter·kerze *f* (*Opt.*) meter candle. **-kilogramm** *n* kilogram-meter. **m–lang** *a* meter-long, (*fig.*) yard-long; *pred.* a meter long. **M–maß** *n* 1 metric measure. 2 meter rule (stick, measure). **-tonne** *f* metric ton. **-zentner** *m* 100 kilograms, metric hundredweight

Meth. *abbv* (Methode) method

Methacryl·säure *f* methacrylic acid

Methämoglobin *n* methemoglobin

Methan *n* methane. **-kohlenwasserstoff** *m* paraffin hydrocarbon, alkane. **-säure** *f* methanoic (*or* formic) acid

Methazon·säure *f* methazonic acid

Methion·säure *f* methionic acid

Methode *f* (-n) method. **Methodik** *f* (-en) 1 methodology. 2 method, methodical procedure. **methodisch** *a* methodical; as regards methods, methodological

Methoxid° *n* methoxide

Methoxyl·gruppe *f* methoxy(l) group. **methoxylieren** *vt* methoxylate

Methyl·alkohol *m* methyl alcohol, methanol. **m–alkoholisch** *a* methanolic. **M–arsonsäure** *f* methylarsonic (*or* methanearsonic) acid. **-äther** *m* methyl ether. **-blau** *n* methyl blue. **-chlorid, -chlorür** *n* methyl chloride

Methylen·blau *n* methylene blue. **-gruppe** *f* methylene group. **-jodid** *n* methylene iodide

Methyl·gallusäthersäure *f* O-methylgallic acid, methyl ether of gallic acid. **-grün** *n* methyl green. **-hydrür** *n* methane

methylierbar *a* methylatable. **methyl(is)ieren** *vt* methylate

Methyl·jodid n methyl iodide. **-kautschuk** m methyl rubber

Methylo·stärke f methyl starch

Methyl·oxyhydrat n methanol. **-rot** n methyl red. **-schwefelsäure** f methyl hydrogen sulfate. **-senföl** n methyl isothiocyanate. **-verbindung** f methyl compound. **-wasserstoff** m methane. **-zahl** f methyl number (or value). **-zinnsäure** f methylstannic acid

Metier n (-s) profession, trade, line (of work)

Metochinon n metoquinone

metrisch a metric

Mett·wurst f hard smoked pork or beef sausage

Metze f (-n) (pre-metric dry Meas., appx) 3 quarts

Metzel·suppe f (liver) sausage broth

Metzger m (-) butcher

mexikanisch a Mexican

m.G. abbv (mit Goldschnitt) with gilt edges

M.G. abbv 1 (Molekulargewicht) molecular wt. 2 (Massenwirkungsgesetz) law of mass action

MHz abbv (Megahertz) megacycles per sec.

Micellen·aufbau m micellar structure

mich pron me; myself

mied avoided: past of MEIDEN

Miene f (-n) mien, face; (facial) expression, look

Mies·muschel f (edible) mussel

Miete f (-n) 1 (tenant's) rent; **zur M.** as a tenant. 2 (Agric.) (hay)stack; pit,clamp, **mieten** vt/i rent (as a tenant). **Mieter** m (-) tenant

Migräne f migraine

Mikanit n fabricated mica

Mikro·analyse f microanalysis. **-analytiker** m microanalyst. **-arbeit** f micro work. **-aufnahme** f microphotograph, photomicrograph. **-becher** m micro beaker

Mikroben·forschung f microbe research. **-tätigkeit** f microbe activity

Mikro·bestimmung f microdetermination

mikrobiell a microbial, microbe

Mikro·bild n micrograph. **-biologie** f microbiology

mikrobisch a microbial

Mikro·brenner m micro burner. **-chemie** f microchemistry. **-chemiker** m microchemist. **m−chemisch** a microchemical. **M−chirurgie** f microsurgery. **-dampfbad** n micro steam bath. **-druckmesser** m micromanometer. **-elementaranalyse** f elementary microanalysis. **-gasbrenner** m micro gas burner

Mikrographie f micrography. **mikrographisch** a micrographic

Mikro·härteprüfer m micro hardness tester. **m−kanonisch** a microcanonical

Mikroklas m (Min.) anorthoclase

Mikroklin n (Min.) microcline

mikro·kosmisch a microcosmic. **-kristallin(isch)** a microcrystalline. **M−lebewesen** n microorganism

Mikrolith m (-en, -en) microlite

Mikro·lunker m micropore, microcavity. **-maßstab** m microscale. **-mechanik** f micromechanics. **-meßuhr** f dial gauge. **-meter** n micrometer. **-meteruhr** f dial gauge. **m−metrisch** a micrometric. **M−methode** f micro method. **-nutsche** f micro suction filter, micro Büchner funnel. **m−porös** a microporous. **M−probe** f 1 micro test. 2 micro sample. **-röhre** f microtube. **-röntgenbild** n microradiograph. **-schnellbestimmung** f rapid microdetermination

mikroskopieren vt examine under a microscope

Mikroskop·linse f microscope lens. **-stativ** n microscope stage

Mikro·spatel m microspatula. **-stativ** n micro stand. **-strahl** m microbeam (of X-rays). **-tiegel** m microcrucible. **-tomschnitt** m microtome section. **-trichter** m micro funnel. **-verbrennung** f microcombustion. **-waage** f microbalance. **-welle** f microwave

Mikrowellen·herd, -ofen m microwave oven. **m−sicher** a microwave-safe. **M−technik** f microwave technology

mikrurgisch a micrurgic

Milbe f (-n) mite, tick. **Milben·bekämpfungsmittel** n miticide, tick killer

Milch f 1 milk 2 milt (of fish). 3 (in compds, oft) lactic, lacteal, lacto-, galacto-: **m−abscheidend, -absondernd** a lactating, milk-secreting. **M−absonderung** f lactation, milk secretion. **m−abtreibend** a antigalactic. **M–achat** *m* milk-white agate. **-ader** *f* lacteal vessel, milk vein. **m–ähnlich, -artig** *a* milk-like, milky, lacteal. **M–bestandteil** *m* constituent of milk. **-bier** *n* kumiss. **-bildung** *f* lactation. **-drüse** *f* mammary (*or* lacteal) gland. **-eiweiß** *n* milk protein

milchen I *v* 1 *vi* give (*or* yield) milk, be lactescent. 2 *vt* emulsify. **II** *a* milky.

milchend *p.a* milk-giving, lactescent

Milch-erzeugnis *n* 1 milk production. 2 dairy product. **-fälschung** *f* adulteration of milk. **-farbe** *f* milk color, milk(y) white. **m–farbig** *a* milk-white. **M–fett** *n* butterfat. **-flasche** *f* milk bottle. **m–fördernd** *a* promoting milk secretion. **-führend** *a* lactiferous. **M–gang** *m* lactiferous duct. **-gärung** *f* milk fermentation. **-gefäß** *n* lacteal vessel. **-gerinnung** *f* curdling of milk. **-gewinnung** *f* milk production. **-glanz** *m* (*Lthr.*) milk finish. **-glas** *n* 1 milk (*or* opal) glass; frosted glass. 2 (drinking) glass for milk. 3 breast glass. **m–glasartig** *a* opalescent, like frosted glass. **M–grad** *m* milk degree (on a galactometer). **-güte** *f* quality of the milk. **m–haltig** *a* lactiferous, containing milk. **M–harnen** *n,* **-harnfluß** *m* chyluria

milchig *a* milky, lacteal; galactic; emulsified. **--trüb** *a* milky, turbid

Milch-kanal *m* lactiferous duct. **-konserve** *f* preserved (*incl* canned, evaporated, powdered) milk. **-kügelchen** *n* milk globule, fat globule in milk. **-kuh** *f* milch (milk, dairy) cow. **-lab** *n* rennet. **-leistung** *f* milk yield. **-messer** *m* (ga)lactometer

Milchner *m* (-) (*Fish*) milter

Milch-porzellan *n* milk glass, glass porcelain. **-prober, -prüfer** *m* milk tester, lactometer. **-prüfung** *f* milk testing. **-pulver** *n* powdered milk. **-quarz** *n* milky quartz. **-rahm** *m* cream. **-rohr** *n,* **-röhre** *f* latex tube; lactiferous duct. **-saft** *m* (*Bot.*) latex; (*Physiol.*) chyle

milchsaft-führend *a* lactiferous; (*Physiol.*) chyliferous. **milch-saftig** *a* chylous

Milchsaft-schlauch *m* (*Bot.*) latex tube

milch-sauer *a* lactate of. **M–säure** *f* lactic acid

Milchsäure-ferment *n* lactic ferment. **-gä-**

rung *f* lactic fermentation. **-pilz** *m* lactic-acid bacillus. **-salz** *n* lactate. **-stäbchen** *n* lactic-acid bacillus

Milch-schicht *f* (*Phot.*) emulsion. **-schleuder** *f* cream separator. **-straße** *f* Milky Way. **m–treibend** *a* milk-yielding, lactiferous. **M–untersuchung** *f* milk inspection (*or* analysis). **-verfälschung** *f* adulteration of milk. **m–vermehrend** *a* promoting milk secretion. **-vertreibend** *a* antigalactic. **M–vieh** *n* dairy cattle. **-waage** *f* lactometer. **-wein** *m* kumiss. **m–weiß** *a* milk-white. **M–wirtschaft** *f* dairy; dairy farming. **m–wirtschaftlich** *a* dairy(ing). **M–wurz** *f* milkwort. **-z.** *abbv* (Milchzucker) lactose. **-zelle** *f* (*Bot.*) latex cell. **-zentrifuge** *f* cream separator. **-zersetzung** *f* decomposition of milk. **-zucker** *m* lactose, milk sugar

Milchzucker-agar *m* lactose agar. **-säure** *f* saccharolactic (*or* mucic) acid

mild(e) *a* mild, gentle; subdued; benevolent; (*Metals*) ductile —**m. gesagt** to put it mildly. **Milde** *f* mildness, gentleness; benevolence

mildern *vt* temper, mitigate; alleviate; (*Chem.*) correct (acidity). **mildernd** *p.a* 1: **m–e Umstände** mitigating circumstances. 2 (*Chem.*) corrective. 3 (*Pharm.*) mitigant, lenitive. **Milderung** *f* (-en) tempering, mitigation; alleviation; (*Chem.*) correction. **Milderungs-mittel** *n* mitigant, lenitive; demulcent; (*Chem.*) corrective

Mild-glanzerz *n* (*Min.*) polybasite. **m–hart** *a* moderately hard

Milieu *n* (-s) environment. **m--bedingt** *a* environmental

Militär *n* 1 army; armed forces. 2 (*in compds*) military . . . : **-zweck** *m* military purpose

Milliarde *f* (-n) billion, 10^9 *cf* BILLION

Millionstel *n* (-) (*Frac.*) millionth

Millonsche Base *f* Millon's base

Milo-korn *n* milo, sorghum

Milori-blau *n* milori (*or* Prussian) blue

Milz *f* spleen; (*in compds oft*) splenic: **-ader** *f* splenic vein. **-brand** *m* anthrax. **-brandgift** *n* anthrax virus. **-drüse** *f* spleen. **-fieber** *n* anthrax. **m–krank** *a* splenetic

Mimose *f* (-n) mimosa; delicate plant

Mimosen·gummi *n* (fine) gum arabic

min. *abbv* (minimum, minimal)

mind. *abbv* (mindestens) at least

minder 1 *a* (*also as pfx*) minor, lesser, lower, inferior. **2** *a, adv* (*compar*) less. **M—·betrag** *m* deficiency, deficit. **-druck** *m* lower (*or* diminished) pressure. **-ertrag** *m* lower (*or* decreased) yield. **-gehalt** *m* lower content, deficiency in content. **-gewicht** *n* underweight; short weight. **m—haltig** *a* inferior, low-grade, low-content

Minderheit *f* (-en) minority

minder·jährig *a* under-age, minor. **-mäßig** *a* undersize

mindern *vt* & *sich* m. lessen, diminish, decrease. **Minderung** *f* (-en) diminution, decrease

Minder·wert *m* inferior (*or* reduced) value. **m—wertig** *a* low(er)-valent; inferior, low-grade. **M—wertigkeit** *f* lower valance; inferiority, poor quality. **-zahl** *f* minority

mindest 1 *a* least, slightest, lowest; **nicht im m—en** not in the least; **zum m—en** at (the) least. **2** (*as pfx*) minimum: **M—betrag** *m* minimum amount. **-druck** *m* minimum pressure

mindestens *adv* at least

Mindest·erfordernis *m* minimum requirement. **-maß** *n* minimum. **-nenner** *m* least common denominator. **-wert** *m* minimum (*or* lowest) value

Mine *f* (-n) **1** (*Min., Explo.*) mine —**alle M—n springen lassen** spare no effort. **2** refill, cartridge (for ball-point pen); (refill) lead (for mechanical pencil)

Minen·bombe *f* land mine; high-explosive bomb. **-gas** *n* mine gas. **-granate** *f* mine grenade. **-körper** *m* shell body. **-pulver** *n* blasting powder. **-werfer** *m* mine thrower, trench mortar

Mineral·ablagerung *f* mineral deposit; deposition of minerals. **-analyse** *f* mineral analysis. **-bad** *n* mineral bath; spa. **-beize** *f* (*Dye.*) mineral mordant. **-bestand** *m* mineral composition. **-bestandteil** *m* mineral constituent. **-bildner** *m* mineral former, mineralizer. **-bildung** *f* mineral formation, mineralization. **-bister** *m* or *n* (*Paint*) manganese brown. **-blau** *n* mineral blue. **-brunnen** *m* mineral spring (*or* well). **-chemie** *f* mineral (*or* inorganic)

chemistry. **-dünger** *m* mineral fertilizer. **-farbe** *f* mineral color (*or* dye). **-faser** *f* mineral fiber. **-fett** *n* mineral fat (*or* grease); petrolatum, petroleum jelly; (*Min.*) ozocerite. **-führung** *f* paragenesis. **-gang** *m* mineral vein. **-gehalt** *m* mineral content. **-gelb** *n* mineral yellow. **-gemisch** *n* mineral mixture. **-gerbung** *f* mineral tanning. **-grün** *n* mineral green

Mineralien minerals: *pl* of **Mineral** *n*. **-·kunde** *f* mineralogy

Mineralisator *m* (-en) mineralizer

mineralisch *a* mineral; **m—es Leichenwachs** ozocerite; **m—er Spiritus** ethyl alcohol from acetylene

mineralisieren *vt* mineralize

Mineral·kautschuk *m* mineral rubber. **-kermes** *m* kermes mineral. **-kunde** *f* mineralogy. **-laugensalz** *n* sodium carbonate. **-lehre** *f* mineralogy. **-mohr** *m* ethiops mineral (amorphous HgS)

Mineraloge *m* (-n, -n) mineralogist. **Mineralogie** *f* mineralogy. **mineralogisch** *a* mineralogical

Mineral·öl *n* mineral oil. **-ölfirnis** *m* mineral-oil varnish. **-ölraffinat** *n* refined mineral oil. **-pech** *n* mineral pitch, asphalt. **-quelle** *f* mineral spring (*or* source). **-reich** *n* mineral kingdom. **-rot** *n* cinnabar. **-salz** *n* mineral salt. **-säure** *f* mineral acid. **-schmiermittel** *n* mineral lubricant. **-schmieröl** *n* mineral lubricating oil. **-schwarz** *n* mineral black, ground graphite (*or* slate). **-seife** *f* mineral soap. **-stoff** *m* mineral (matter). **-stoffgehalt** *m* mineral content. **-stoffwechsel** *m* mineral metabolism. **-talg** *m* mineral tallow, hatchettite, ozocerite. **-teer** *m* mineral tar, maltha. **-trennung** *f* separation of minerals. **-turpeth** *m* turpeth mineral (basic mercuric sulfate). **-vorkommen** *n* mineral deposit. **-wachs** *n* mineral wax, ozocerite. **-wasser** *n* mineral water. **-weiß** *n* mineral white. **-wolle** *f* mineral (*or* slag) wool

Minette·erz *n* minette ore

minieren *vt* mine

Minimal· minimal, minimum, lowest: **-betrag** *m* minimum (*or* lowest) amount. **-gehalt** *m* minimum content

Minimi·formel *f* empirical formula

Minister *m* (-) cabinet minister (*or* officer).

Ministerium *n* (..ien) cabinet ministry (*or U.S.* department)

Mini·valenz *f* minimum valence

minorenn *a* under-age, minor

Minorität *f* (-en) minority

Minuskel *f* (-n) small letter

Minus·leitung *f* (*Elec.*) negative lead. **-pol** *m* negative (*or* minus) pole. **-zeichen** *n* minus sign

minuten·lang *a* lasting a minute, lasting minutes; *adv* for a minute, for minutes. **-weise** *a*, *adv* by the minute. **M—zeiger** *m* minute hand

·minütig *sfx* -minute: **zehnminütige Pause** ten-minute break

minutlich *a* per minute

minuziös *a* detailed, precise; minute

Minze *f* (-n) mint

Mio *abbv* (Million/en) million(s)

Miocän, Miozän *n* (*Geol.*) Miocene

Mipolid *n* (-e) mixed polyamide

mir *pron* (*dat*) (to) me, myself

Mirabelle *f* (-n) yellow plum

Mirban·essenz *f* essence of mirbane, nitrobenzene. **-öl** *n* oil of mirbane, nitrobenzene

Misch· mixed, mixing: **-anlage** *f* mixing plant. **-apparat** *m* mixing apparatus, mixer

mischbar *a* mixable, miscible. **Mischbarkeit** *f* mixability, miscibility

Misch· mixed, mixing: **-behälter** *m* mixing vessel, mixer. **-bestand** *m* (*Agric.*) mixed crop (*or* growth). **-bett** *n* mixed bed. **-bettaustauscher** *m* mixed-bed exchanger. **-binder** *m* (*Cement*) mixed binder (of ground slag and hydrated lime). **-bleichen** *n* mixed bleaching. **-bottich** *m* mixing tub (*or* vat). **-brot** *n* (mixed) wheat-and-rye bread. **-dünger** *m* mixed (*or* compound) fertilizer, compost. **-einrichtung** *f* mixing device. **-element** *n* (isotopically) mixed element, mixture of isotopes

mischen I *vt* **1** mix, blend; alloy. **2** (+**in,** *oft*) add (to), mix in (with). **II: sich m. 3** mix, blend; (*esp* + **unter**) mingle (with); (+**in,** *oft*) blend in (with). **4** (+**in**) meddle (in); join (in). **Mischer** *m* (-) **1** mixer, blender; agitator; compounding apparatus. **2** mixing vessel. **--eisen** *n* (*Metll.*) mixer metal

Misch· mixed, mixing: **-erz** *n* mixed ore. **-ester** *m* mixed ester. **-fähigkeit** *f* mixing capability, miscibility. **-fällung** *f* coprecipitate, co-precipitation. **-farbe** *f* mixed color; secondary color. **m—farben, -farbig** *a* of mixed color. **M—farbstoff** *m* mixed coloring matter (*or* dye); component dye. **-färbung** *f* (*Micros.*) mixed staining; (*Dye.*) combination dyeing (*or* shade). **-flasche** *f* mixing bottle. **-form** *f* hybrid. **-frucht** *f* hybrid fruit. **-futter** *n* mixed feed. **-garn** *n* mixed (*or* blended) yarn. **-gärung** *f* mixed fermentation. **-gas** *n* mixed gas; (*specif*) semiwater gas. **-gefäß** *n* mixing vessel. **-gerbung** *f* combination tanning. **-getränk** *n* mixed drink. **-gewebe** *n* (*Tex.*) mixed fabric, union. **-glas** *n* mixing glass. **-gut** *n* **1** mixed material. **2** material to be mixed. **-hahn** *m* mixing faucet (tap, cock). **-harz** *n* mixed resin. **-hefe** *f* composite yeast. **-holländer** *m* mixing beater (*or* engine). **-infektion** *f* mixed infection. **-kammer** *f* mixing chamber. **-katalysator** *m* mixed catalyst, promoter. **-kegel** *m* mixing cone (of a burner). **-kessel** *m* mixing kettle. **-klebstoff** *m* two-component adhesive. **-komponente** *f* component of a mixture. **-korn** *n* mixed grain, (*specif*) wheat and rye. **-kraftstoff** *m* mixed fuel. **-kristall** *m* mixed crystal, crystalline solid solution

Mischling *m* (-e) hybrid, crossbreed; halfbreed

Misch· mixed, mixing: **-masch** *n* hodgepodge, mishmash. **-maschine** *f* mixing machine, mixer; (*Soap*) crutcher. **-material** *n* (*Rubber*) filler. **-metall** *n* mixed (*or* misch) metal, alloy with cerium or lanthanum. **-molekül** *n* mixed molecule. **-mühle** *f* mixing (*or* incorporating) mill. **-öl** *n* mixed oil. **-oxid** *n* mixed oxide (*specif, Metll.*, of lead and zinc). **-phase** *f* mixed phase. **-pipette** *f* mixing pipette. **-polymer** *n* mixed polymer, copolymer. **-polymerisat** *n* mixed polymerizate, copolymer. **-polymerisatharz** *n* copolymer resin. **-polymerisation** *f* copolymerization, cross-polymerization. **-potential** *n* mixed potential. **-probe** *f* mixed melting point test. **-pult** *n* (*Electron.*) mixer, mixing panel. **-raum** *m* mixing space (area,

chamber). **-rohr** *n*, **-röhre** *f* mixing tube.
-rührwerk *n* mixing (*or* stirring) apparatus. **-saft** *m* (*Sugar*) mixed juice. **-salz** *n* mixed salt. **-säure** *f* 1 (*genl*) mixed acid. 2 (*specif*) nitric and sulfuric acid mixture. 3 chamber crystals, nitrosyl hydrogen sulfate. **-schmelzpunkt** *m* mixed melting point. **-schnecke** *f* screw mixer. **-spektrum** *n* mixed spectrum. **-strom** *m* (*Elec.*) undulatory current. **-teil** *m* ingredient. **-trommel** *f* mixing drum. **-typus** *m* mixed (*or* hybrid) type

Mischung *f* (-en) 1 mixing. 2 mixture, blend

Mischungs-bestandteil *m* component (*or* ingredient) of a mixture. **m—fähig** *a* miscible. **M—formel** *f* mixing formula. **-gewicht** *n* combining weight. **-kohle** *f* (*Soda*) mixing coal. **-lücke** *f* miscibility gap. **-probe** *f* 1 mixing test. 2 test (*or* sample) of a mixture. **-rechnung** *f* (*Math.*) (rule of) alligation. **-regel** *f* rule (*or* law) of mixtures. **-verhältnis** *n* 1 mixing ratio (*or* proportion). 2 *pl* mixing conditions. **-wärme** *f* heat of mixing. **-zusatz** *m* addition to a mixture; filler

Misch-ventil *n* mixing (*or* mixer) valve. **-vorgang** *m* mixing process. **-walzwerk** *n* set of mixing rollers. **-werk** *n* mixing mill, mixer. **-zement** *m* or *n* mixed cement. **-zylinder** *m* stoppered mixing cylinder

Mispel *f* (-n) medlar

miß.. *insep pfx* (*with genl neg meaning*) mis-, dis-, de-, false, fail(ure) to . . .

mißachten *vt* disregard, ignore; neglect; despise. **Mißachtung** *f* disregard; neglect; contempt

mißarten *vi* (s) degenerate

Mißbildung° *f* deformity; malformation

mißbilligen *vt* disapprove of

Mißbrauch° *m,* **mißbrauchen** *vt* misuse, abuse

mißdeuten *vt* misinterpret

missen *vt* do without; miss

Mißerfolg° *m* failure

Mißernte° *f* crop failure; bad harvest

mißfallen* *vi* (*dat*) be displeasing (to), displease. **Mißfallen** *n* displeasure, disapproval. **mißfällig** *a* 1 disapproving, disparaging. 2 disagreeable

mißfarben, mißfarbig *a* discolored. **Mißfärbung·** *f* discoloration

mißfiel displeased: *past of* MISSFALLEN

mißförmig *a* misshapen

Mißgeburt° *f* 1 miscarriage. 2 monstrosity

Mißgeschick° *n* misfortune

Mißgestalt° *f* 1 deformity. 2 misshapenness. 3 miscreation. **mißgestaltet** *a* deformed, misshapen

miß·glücken *vi* (s) (*oft* + *dat*) fail. **mißglückt** *p.a* unsuccessful, abortive

Mißgriff° *m* mistake, blunder. **m—-sicher** *a* foolproof

mißhandeln *vt* mistreat. **Mißhandlung** *f* mistreatment

mißhellig *a* discordant. **Mißhelligkeiten** *fpl* disagreements

Mißklang° *m* dissonance

Mißkredit° *m:* **in M. bringen** discredit; **in M. geraten** fall into disrepute, be discredited

mißlang failed: *past of* MISSLINGEN

mißlich *a* troublesome; precarious

mißlingen* (mißlang, mißlungen) *vi* (s) (*oft* + *dat*) fail, be unsuccessful (for), miscarry. **mißlungen** *p.a* unsuccessful, abortive

Mißpickel *m* mispickel, arsenopyrite

mißraten 1* *vi* (s) fail, turn out wrong. 2 *p.a* unsuccessful; wayward, gone wrong. **mißriet** *past*

Mißstand° *m* bad state of affairs; nuisance; abuse; grievance

mißt measures: *pres of* MESSEN

Mißton° *m* dissonance

mißtrauen *vi* (*dat*), **Mißtrauen** *n* distrust. **mißtrauisch** *a* distrustful, suspicious

Mißverhältnis° *n* disproportion, disparity; incongruity; asymmetry

mißverstand(en) misunderstood: *past* (& *pp*) *of* MISSVERSTEHEN. **Mißverständnis** *f* (-se) misunderstanding. **mißverstehen*** *vt* misunderstand

Mißwachs *m* crop failure, bad harvest

Mißwirtschaft *f* mismanagement

Mist *m* 1 manure. 2 manure heap. 3 trash, rubbish. **--beet** *n* hotbed. **-beize** *f* 1 (*Lthr.*) dung bate. 2 (*Dye.*) gray steeping

Mistel *f* (-n) mistletoe

misten 1 *vt/i* clean (the manure out of) 2 *vt* (fertilize with) manure. 3 *vi* (*of animals*) drop manure

Mist·haufen *m* manure (*or* dung) heap. **-jauche** *f* liquid manure. **-pulver** *n* powdered manure, poudrette

mit I *prep* (*dat*) **1** with. **2** by (means of) —**m. Bleistift (Tinte)** in pencil (ink). **3** (*Age*): **m. 20 Jahren** at (the age of) twenty. **4** at, as of: **m. erstem Juni** as of June first. **5** of, about: **die Sache m. . . .** the matter of . . . ; **wie ist es m. . . . ?** how about . . . ? **6** (+*parts of body, no equiv. word*): **m. dem Finger zeigen** point one's finger. **7: sie haben es m. . . .** they have trouble with . . . , they are obsessed with . . . *cf* ALLEDEM. **II** *adv* **8** also, too, as well: **das gehört m. dazu** that is part of it too. **9** join in: **das trägt m. dazu bei** that joins in contributing to it. **10** be included: **das ist m. ein Grund** that is included among (*or* is one of) the reasons. **III** *pfx* co-, joint-, fellow-. **IV** *sep pfx* along, with one *cf* MITNEHMEN; join in, co- *cf* MITARBEITEN; include *cf* MITBEHANDELN

Mitarb. *abbv* (Mitarbeiter) co-worker

Mitarbeit° *f* collaboration. **mit·arbeiten** *vi* collaborate, join in working. **Mitarbeiter°** *m* co-worker, fellow worker, collaborator

mit·behandeln *vt* include in the treatment

mit·bekommen* *vt* realize, understand

Mitbenutzung° *f* joint use

mitberücksichtigen *vt* consider also, include in consideration

mitbewegt *a* (*of the atomic nucleus*) in relative motion, moving in unison. **Mitbewegung°** *f* associated (joint, ideomotor, synkinetic) motion; (*of atomic nuclei*) relative motion

Mitbewerber° *m* competitor

Mitdasein° *n* coexistence

miteinander *adv* together, with each other

miteinbegriffen, miteingeschlossen *p.a* included

mit·fällen *vt* coprecipitate. **Mitfällung°** *f* coprecipitation

mitführbar *a* entrainable. **mit·führen** *vt* **1** carry along, (*Chem.*) entrain. **2** comanage. **Mitführung°** *f* carrying along, entrainment; convection; comanagement

Mitgefühl° *n* sympathy

mitgegangen gone along: *pp of* MITGEHEN

mit·gehen* *vi* (s) go along

mitgeklungen resonated: *pp of* MIT-KLINGEN

mitgenommen taken along *pp of* MITNEHMEN

mitgerechnet *p.a* counting in *cf* MITRECHNEN

mitgerissen coprecipitated *pp of* MITREISSEN

mitgeschrieben taken down *pp of* MIT-SCHREIBEN

mitgeschwungen covibrated *pp of* MIT-SCHWINGEN

mitgezogen pulled along *pp of* MITZIEHEN

Mitgift *f* (-en) contribution; dowry

Mitglied° *n* member. **Mitgliedschaft** *f* membership

mit·halten* *vi* (mit) keep up (with), keep abreast (of)

Mithilfe° *f* assistance, cooperation

mithin *adv-cnjc* consequently, accordingly

mit·hören *vt* monitor

Mitis·grün *n* mitis (Paris, emerald) green

mit·klingen* *vi* resonate

mit·kommen* *vi* (s) come along; keep pace

Mitkopplung° *f* (*Elec.*) positive feedback

Mitleid *n* compassion, sympathy

Mitleidenschaft *f*: **in M. ziehen** involve, affect

mit·machen *vt/i* participate (in), join in

Mitmensch° *m* fellow human being

Mitnahme *f* taking along, (*Chem.*) entrainment: **unter M. von . . .** taking along, carrying away (with it), entraining. **--möbel** *n* portable (collapsible) furniture. **mit·nehmen*** *vt* **1** take: along (*or* with one); take out (food); (*Chem.*) entrain. **2** give a ride, carry, take (in a vehicle). **3** take in, observe; learn. **4** affect deeply; wear out, exhaust. **Mitnehmer** *m* (-) **1** (*Mach.*) carrier, carrying element (*or* plate); catch, driver. **2** (*in compds oft*) driving, carrying

Mitose *f* (-n) mitosis

mit·rechnen 1 *vt* count in, include in the calculation. **2** *vi* count, be included (in the calculation); join in the calculation. *Cf* MITGERECHNET

mit·reißen* *vt* **1** sweep (*or* drag) along. **2** (*Chem.*) carry down, coprecipitate; (*Distil.*) carry over. **2** carry away, rouse, stir

mitsamt *prep* (*dat*) together with (all)

mit·schleppen *vt* 1 drag (lug, tow) along. 2 (*Chem.*) carry down, coprecipitate; (*Distil.*) carry over

mit·schreiben* *vt/i* take notes (*or* dictation) (on), take down

mit·schwingen* covibrate, resonate. **mitschwingend** *p.a* (*oft*) sympathetic

mit·spielen I *vt/i* join in playing. **II** *vi* 1 be involved, play a part. 2 (+*dat* & +*arg*) wreak havoc (on), do in

mitt. *abbv* (mittels) by means of

Mitt. *abbv* (MITTEILUNG) message, etc

Mittag° *m* midday, noon. **mittag(s)** *adv* (at) noon

Mitte *f* (-n) middle, center; midpoint, middle ground; midst: **M. Juni** in the middle of June; **in der M. der Bahn** halfway down the track (*or* orbit)

mitteilbar *a* communicable, transmittable

mit·teilen I *vt* (*d.i.o*) 1 give (message), inform (of), tell (about). 2 impart, communicate, transmit (to). **II: sich m.** (*dat*) 3 impart itself, communicate itself (to), be transmitted (to), spread (to, among). 4 communicate (with), unburden oneself (to), confide (in). **mitteilsam** *a* communicative, talkative. **Mitteilung°** *f* 1 informing, telling; imparting, communication, transmission; confiding. 2 message, (item of) information, news item; communication

Mittel I *n* (-) 1 means, way, measure. 2 remedy. 3 tool, method; *pl oft* resources. 4 *pl* financial means, funds, capital. 5 (*Math.*) mean (value); average. 6 medium. **7: sich ins M. legen** mediate, intercede. **II** (*as pfx*) middle, central, intermediate; average, mean: **-alter** *n* Middle Ages. **m-alterlich** *a* medieval. **M-amerika** *n* Central America. **M-asien** *n* Central Asia

mittelbar *a* indirect, mediate

Mittel· middle, central; intermediate; average, mean; mes(o)-: **-bauchgegend** *f* mesogastric region. **-benzin** *n* medium-heavy gasoline. **-bereich** *m* medium range. **-binder** *m* medium-setting cement. **-blech** *n* (*Metll.*) medium plate. **-darm** *m* mesenteron, middle intestine. **m-dick** *a* medium-thick. **M-ding** *n* intermediate, cross (between . . .). **-druck** *m* medium pressure. **-eck** *n,* **-ecke** *f* (*Cryst.*) lateral summit. **-ernte** *f* average harvest. **-erz** *n* medium-value ore. **-europa** *n* Central Europe. **m-europäisch** *a* Central-European. **M-farbe** *f* intermediate color; secondary color. **-fehler** *m* average (*or* mean) error. **m-fein** *a* medium-fine. **M-fell** *n* mediastinum. **-fleisch** *n* perineum. **m-flüssig** *a* semifluid. **M-fuß** *m* metatarsus. **-gehirn** *n* mid-brain, mesencephalon. **-geschwindigkeit** *f* average speed; mean velocity. **-glied** *n* intermediate member; middle term; (*Anat.*) middle phalanx (*or* joint). **m-groß** *a* middle-sized. **M-größe** *f* medium size. **m-gut** *a* medium-quality. **M-gut** *n* middlings. **-hand** *f* metacarpus. **m-hart** *a* medium-hard. **M-hartpech** *n* medium-hard pitch. **-haut** *f* middle layer (*or* membrane), tunica media. **-hirn** *n* mid-brain, mesencephalon. **-keim** *m* (*Zool.*) mesoblast. **-klasse** *f* 1 medium grade (*or* quality). 2 intermediate class. 3 middle class. **m-körnig** *a* medium-grain(ed). **M-kraft** *f* resultant (force). **-lage** *f* middle (*or* intermediate) position. **m-ländisch** *a* Mediterranean. **-lang** *a* middle-length. **M-lauf** *m* (*Distil.*) middle fraction. **-lauge** *f* medium-strength lye. **-linie** *f* 1 center line. 2 median; bisector; axis. 3 (*Phys.*) neutral zone. **m-los** *a* without (financial) means, impecunious. **-mäßig** *a* average, medium(-quality), mediocre. **M-mäßigkeit** *f* mediocrity. **-meerfieber** *n* Mediterranean fever. **-ohr** *n* middle ear. **-öl** *n* medium (-heavy) oil. **-ost(en)** *m* Middle East. **-produkt** *n* intermediate product. **-punkt** *m* center, midpoint. **-punktsabstand** *m* distance between centers. **-rinde** *f* (*Bot.*) primary cortex

mittels *prep* (*genit*) by means of

Mittel·saft *m* (*Sugar*) middle sirup. **-salz** *n* neutral salt. **-schicht** *f* middle (*or* intermediate) layer. **-schule** *f* non-prep secondary school. **m-schwer** *a* medium-heavy, medium-weight. **M-senkrechte** *f* (*adj endgs*) mean perpendicular, perpendicular bisector. **m-siedend** *a* intermediate-boiling

Mittels·leute *pl of* **-mann** *m* middleman, agent, go-between

mittelst 1 *a* (*superl*) mid(dle)most, most central. **2** *oft erron. used for* MITTELS

Mittel· middle, central; medium, mean, average, intermediate: **-stadt** *f* medium-sized city. **-stand** *m* middle class. **m–ständig, -ständisch** *a* **1** middle-sized, medium-sized, intermediate. **2** middle, central. **3** (*Bot.*) perigynous. **-stark** *a* medium-strength. **M–steinzeit** *f* Mesolithic (*or* Medium Stone) Age. **-stellung** *f* intermediate (*or* meta) position. **-straße** *f* middle road (*or* course). **-strecke** *f* medium distance (*or* range). **-stück** *n* middle part (portion, section). **-stufe** *f* intermediate stage (phase, level, grade). **-teil** *m* middle part, midsection. **-temperatur-teer** *m* medium-temperature tar. **-wand** *f* **1** central wall; partition. **2** (*Paper*) midfeather. **3** (*Anat.*) mediastinum **-weg** *m* middle way (road, course); mean, medium; compromise. **m–weich** *a* medium-soft. **M–welle** *f* **1** (*Mach.*) countershaft. **2** (*Radio*) medium wave(length). **-wert** *m* **1** mean (*or* average) vaue; mean, median, average **2.** (*Sugar*) middle juice. **-zahl** *f* mean. **-zeug** *n* (*Paper*) middle (*or* second) stuff

mitten I *vt* center. **II** *adv* in the middle (*or* midst): midway: **m. in (an, auf)** (right) in(to) the middle of; **m. unter** (right) among; **m. aus** from the (very) middle of; **m. auf dem Weg** halfway there. **--drin** *adv* right in the middle (of it). **-durch** *adv* right through (the middle of it *or* them). **-richtig** *a* centered

Mitternacht° *f* midnight

mittig *a* (con)centric; central; **m—e Last** axial load

mittler *a* (*compar*) **1** middle, central. **2** medium, mean, average; medium-sized, medium-grade. **3** standard. **Mittler** *m* (-) mediator; intermediary; middleman. **mittlerweile** *adv* meanwhile. **Mittlung** *f* averaging, approximation to the mean

mit·tragen* *vt* carry along, join (*or* share) in carrying

Mittwoch *m* Wednesday

mitunter *adv* occasionally, now and then

mitverantwortlich *a:* **m. sein** share the responsibility

Mitverwendung° *f* concomitant (coincident, shared) use

Mitwelt *f* contemporary world, (one's) generation (*or* contemporaries)

mit·wirken *vi* (*usu* + **an, bei, in**) participate, cooperate, collaborate, play a part (in). **Mitwirkung** *f* participation, cooperation, collaboration

Mitwissen *n* sharing in the knowledge (secret, plot), knowledge, cognizance: **ohne ihr M.** without her knowledge. **Mitwisser** *m* (-), **Mitwisserin** *f* (-nen) person sharing the knowledge (*or* secret); coplotter, accessory. **Mitwisserschaft** *f* sharing in the knowledge (secret, plot), cognizance; complicity

mit·ziehen* 1 *vt* pull (draw, drag, sweep) along (*or* with it). **2** *vi* (s) go (move, tag) along, join in; cooperate

mixen *vt* mix. **Mixerei** *f* (-en) mixing; mix(ture); cocktail mix

Mix·rezept *n* mixed-drink recipe

Mixtion *f* (-en) gold size. **Mixtur** *f* (-en) mixture

Mizelle *f* (-n) micelle. **Mizellen·aufbau** *m* micellar structure

mk. *abbv* (mikroskopisch) microscopic

Mk. *abbv* (Marke, Markierung) mark, marking

M.K. *abbv* (Meterkerze) meter candle

mkr. *abbv* (mikroskopisch) microscopic

Möbel *n* (-) (piece of) furniture. **--lack** *m* furniture varnish, cabinet lacquer. **-leder** *n* upholstery (*or* furniture) leather. **-politur** *f* furniture polish. **-stoff** *m* upholestery fabric. **-stück** *n* piece of furniture. **-wagen** *m* moving van

mobil *a* **1** mobile, movable. **2** lively, spry, active. **3** ready for action **—m. machen =** **mobilisieren** *vt* mobilize. **Mobilität** *f* mobility. **Mobil·machung** *f* mobilization

möblieren *vt* furnish

mochte liked; might, etc: *past of* MÖGEN

möchte would like, etc: *sbjc of* MÖGEN

Mock *m*, **--stahl** *m* semisteel (made by direct refining of cast iron)

Mode *f* (-n) **1** fashion, style. **2** custom; fad. **3** (*in compds oft*) fashionable: **-farbe** *f* fashionable color. **-gewürz** n allspice, pimento

Model m (-) **1** module; modulus. **2** mold. **3** (*Tex., Print.*) block

Modell n (-e) model; (*Foun.*) pattern. **Modellack** (=**Modell·lack**) m model (*or* pattern) varnish. **Modell·anlage** f pilot plant. **-herstellung** f pattern making; production of models

modellieren vt model, mold

Modellier·holz n wooden modeling tool. **-masse** f (plastic) modeling composition. **-ton** m modeling clay. **-wachs** n modeling wax

modellig a model-like; elegant, chic

Modell·lack cf MODELLACK. **m–mäßig** a model-like; adv by means of (according to, in terms of) a model. **M–sand** m molding (*or* facing) sand. **-schreinerei** f pattern maker's (carpentry) shop. **-stoff** m modeling (*or* molding) material. **-treue** f truth to scale, faithfulness, accuracy (of a model). **-versuch** m pilot experiment, experiment using a model

modeln vt model, mold; modulate. **Modelung** f (-en) modeling, molding; modulation

Moder m **1** decay, rot(tenness); mold(iness), mustiness. **2** mud, mire. **-duft** m musty (*or* moldy) smell, odor of decay. **-erde** f mold, humus. **-fäule** f soft rot. **-fleck** m mold (*or* mildew) stain. **m–fleckig** a (*Tex.*) mildewed. **-geruch** m musty (*or* moldy) smell, odor of decay

moderhaft a musty, moldy; rotten

moderieren vt/i moderate

moderig a musty, moldy; rotten

modern I (*accent on first syll.*) vi molder, rot, decay. **II** (*accent on second syll.*) a modern

Moder·pilz m mold (*or* saprophytic) fungus. **-stein** m rottenstone. **-stoff** m (*genl*) decayed material, (*specif*) humus

Mode·stoff m (*Tex.*) fancy (*or* stylish) material. **-ton** m fashionable shade (*or* tint). **-waren** fpl fashion goods; fancy articles. **-wort** n vogue word

Modi modes, etc pl of MODUS

modifizieren vt modify. **Modifizierung** f (-en) modification

modrig a musty, etc = MODERIG

Modul m (-e) modulus; module. **modulieren** vt/i modulate

Modus m (Modi) mode, modus, means; rule

Mofa n (-s) (*short for* Motorfahrrad) motor bike

mögen* (mochte; *pres sg* mag) **I** vt (*pp* gemocht) like: **sie m. es (gern)** they like it; **sie m. es lieber (es am liebsten)** they like it better (like it best). **II** v aux (+inf) (*pp* mögen) **1** (*oft* + **gern**) like, care: **sie m. es (gern) tun** they like to do it; **sie mochten nichts davon hören** they didn't care to hear about it. **2** sbjc: **möchte** (*oft*) would like: **wir möchten darauf hinweisen, daß . . .** we would like to point out that . . . ; **sie möchten, daß wir . . .** they would like us to . . . **3** (*pres & möge*) may, can; (*past & sbjc*) might, could: **das mag sein** that may be; **m. sie recht behalten** may they prove to be right; **wie man es auch (or immer) nennen mag** whatever you may call it; **es mochte (es möchte) später gewesen sein** it might (*or* could) have been later; **das möchte schwer zu beweisen sein** that might be hard to prove; **es mag diesmal hingehen** let it pass this time. **4** (+**leiden**) like, care for (=**I**): **sie haben es nie leiden m.** they have never cared for it cf LEIDEN. **5** (+*implied verb of motion*) e.g.: **sie m. nicht mit** they don't care to go (*or* come) along

mögl. abbv (möglich) possible

möglich a **1** (*oft* + *dat*) possible, feasible (for): **etwas für m. halten** consider sthg. possible; **so rasch wie nur m.** as quickly as at all possible; **eine m—e Gefahr** a possible (*or* potential) danger; **alles M—e** everything possible cf 2; **im Rahmen des M—en** within the limits of what is possible. **2:** **alles m—e** all sorts of things; **alle m—en Probleme** all sorts of problems. **möglichenfalls** adv **1** if possible. **2** = **möglicherweise** adv possibly. **Möglichkeit** f (-en) **1** possibility, feasibility: **im Bereich der M.** within the range of possibility; **nach M.** as far as possible. **2** (*esp pl*) potentialities, eventualities. **3** opportunity. **möglichst** a **1** (*superl*) most

possible; *oft as noun, e.g.:* **sein M—es
tun** do all one can, do one's utmost. **2** *adv*
(+*a, adv*) *e.g.:* **der m. große Vorteil** the
greatest possible advantage; **m. bald** as
soon as possible. **3** *adv, e.g.:* **m. gleich
jetzt** right now if possible
Mohär *m* mohair
Mohn *m* poppy (flower, plant); poppy seeds.
-gewächse *npl* (*Bot*) Papaveraceae.
-kapsel *f,* **-kopf** *m* poppy capsule (*or*
head). **-öl** *n* poppy-seed oil. **m—rot** *a*
poppy-red. **M—saft** *m* poppy juice, opium.
-samen *m* poppy seed. **-säure** *f* meconic
acid. **-stoff** *m* narcotine. **-stroh** *n* poppy
straw (residue from extraction of poppy
heads and stems)
Mohr *m* **I** (*Old Chem.*) ethiops, (platinum
or metallic) black: **mineralischer M.**
ethiops mineral = MINERALMOHR. **II** (-e)
(*Tex.*) moiré, moreen, watered fabric
Möhre *f* (-n) carrot. **möhrenfarbig** *a*
carrot-colored. **M—farbstoff** *m* carotene
Mohren·hirse *f* sorghum, Indian millet
Möhren·öl *n* carrot oil
Mohr·rübe *f* carrot. **-rübenfarbstoff** *m*
carotene
Mohrsches Salz Mohr's salt (ferrous am-
monium sulfate hexahydrate)
Moiré·appretur *f* moiré finish. **moirieren**
vt moiré, water (fabrics)
mol. *abbv* (molekular) molecular
Mol *n* (-e) mol(e); (*as pfx*) molar. **Molalität** *f*
molality
Molar·gewicht *n* molar weight. **-volumen**
n molar volume. **-wärme** *f* molar heat
Molch *m* (-e) salamandrid, newt
Mol·dispersion *f* molar dispersivity
Molekül *n* (-e) molecule
Molekular· molecular: **-abstoßung** *f*
molecular repulsion. **m—aktiv** *a* surface-
active molecular. **M—anordnung** *f* molecular arrangement. **-anziehung** *f* molecular attraction. **-aufbau** *m* molecular
structure. **-bahnfunktion** *f* molecular
orbital. **-bewegung** *f* molecular motion
(*or* movement). **-brechung** *f,* **-bre-
chungsvermögen** *n* molecular refraction. **m—dispers** *a* molecularly disperse.
M—drehung *f* molecular rotation.
-druck *m* molecular pressure. **-erschei-
nung** *f* molecular phenomenon. **-formel** *f*

molecular formula. **-gewicht** *n* molecular weight: **Gewichtsmittel des M—s**
weight-average molecular weight;
Zahlenmittel des M—s number-
average molecular weight. **-gewichtsbe-
stimmung** *f* determination of molecular
weight. **-größe** *f* molecular size (magnitude, weight). **-kraft** *f* molecular force.
-reibung *f* molecular friction. **-sieb** *n*
molecular sieve, microfilter. **-störung** *f*
molecular disturbance. **-strahl** *m* molecular beam (*or* ray). **-verbindung** *f* molecular compound. **-verhältnis** *n* molecular proportion. **-wärme** *f* molecular heat.
-wirkung *f* molecular action (*or* effect).
m—zertrümmernd *a* molecular-distinte-
grating. **M—zustand** *m* molecular state
(*or* condition), molecularity
Molekül·bahnfunktion *f* molecular orbital. **-bau** *m* molecular structure. **-emis-
sionskontinuum** *n* continuous molecular emission spectrum. **m—formselektiv**
a molecular type-selective. **M—gasion** *n*
molecular gas ion. **-gewicht** *n* molecular
weight. **-gitter** *n* molecular lattice.
-haufen *m* molecular cluster. **-ionisa-
tionskontinuum** *n* continuous molecular ionization spectrum. **-masse** *f* mol(e).
-schicht *f* molecular layer. **-sieb** *n* molecular sieve, microfilter. **m—spektroskopisch** *a* molecular-spectroscopic; *adv*
by molecular spectroscopy. **M—spektrum** *n* molecular spectrum. **-strahl** *m*
molecular ray (*or* beam). **-teil** *m* molecular part. **-umlagerung** *f* molecular rearrangement. **-verbindung** *f* molecular
compound. **-vergrößerung** *f* molecular
enlargement. **m—zertrümmend** *a*
molecule-disintegrating (*or* -smashing).
M—zertrümmerung *f* molecule smashing, destruction of the molecule
Molen·bruch *m* mole fraction. **-verhältnis**
n mole (*or* molar) ratio
Moler·erde *f* moler earth
Molgew. *abbv* (Molgewicht) molar wt. **Mol.
-Gew.** *abbv* (Molekulargewicht) molec.
wt.
Mol·gewicht *n* molar weight, gram-
molecular weight. **-größe** *f* molar magnitude. **-ion** *n* mole ion
Molisierung *f* (-en) molization, recombina-

tion (of ions to form molecules)

molk milked: *past of* MELKEN

Molke f whey

Molken·butter f whey butter. **-eiweiß** n whey protein. **-farbstoff** m whey pigment. **-käse** m whey cheese. **-säure** f lactic acid. **-wesen** n dairying. **-wirtschaft** f dairy(ing)

Molkerei f (-en), **--wirtschaft** f dairy(ing)

molkig a wheyey

Mol·konzentration f molar concentration

Möller m (-) burden (of blast furnace). **--ausbringen** n burden yield. **möllern** (*Mettl.*) 1 vt charge, burden (the blast furnace). 3 vi mix the ore and fluxes (for the charge), mix the charge. **Möllerung** f (-en) 1 charging; mixing of the charge. 2 charge, burden

mollig a 1 cozy, warm. 2 plump

Mol.-Refr. abbv (MOLREFRAKTION)

Mol·refraktion f molar refraction (or refractivity). **-verhältnis** n molar ratio. **-volumen** n molar volume. **-wärme** f molar heat

Molybdän n molybdenum. **--blau** n molybdenum blue (or oxide). **-blei(erz)** n, **-bleispat** m wulfenite. **-eisen** n ferromolybdenum. **-erz** n molybdenum ore. **-glanz** m (*Min.*) molybdenite. **m--haltig** a molybdeniferous, containing molybdenum. **M--kies** m (*Min.*) molybdenite. **-ocker** m (*Min.*) molybdic ocher, (ferro)molybdite. **-oxid** n molybdenum oxide (higher than Mo_2O_3). **-oxydul** n molybdous oxide (MoO or Mo_2O_2). **m--sauer** a molybdate of. **M--säure** f molybdic acid. **-stahl** m molybdenum steel

Mol·zahl f number of moles. **-zustand** m molecular state

Moment I m (-e) moment, minute (appx): **im M.** at the moment. **II** n (-e) 1 (*Phys.*) moment; torque. 2 factor; impetus

momentan a 1 momentary; temporary; adv oft at (or for) the moment. 2 (*Phys.*) instantaneous. 3 sudden, abrupt. **M--geschwindigkeit** f instantaneous velocity. **-wert** m instantaneous value

Moment·aufnahme f **-bild** n (*Phot.*) instantaneous exposure (or photograph), snapshot. **-klemme** f instantaneous clamp. **-wert** m magneton. **-zünder** m instantaneous fuse, detonator. **-zündpille** f flash pellet primer. **-zündung** f instantaneous ignition (or firing)

Mon. abbv 1 (Monat) mo. (month). 2 (Monographie) monograph

monalkyliert a mono-alkylated

Monat m (-e) month. **monatelang** a months of . . . ; adv for (several) months. **monatlich** a monthly; adv oft by the month, every month

Monats· month's, monthly: **-bericht** m monthly report. **-fluß** m menses, menstruation. **-h.** abbv = **-heft** n monthly (journal): **M—e für Chemie** Chemical Monthly. **-schrift** f monthly publication

Mönch m (-e) 1 monk. 2 die, punch, stamp. 3 ridge tile

Mönchs·kappe f (*Bot.*) monkshood, aconite. **-kolben** m plunger, ram. **-pfeffer** m (*Bot.*) agnus castus, monk's pepper

Mond m (-e) moon; satellite; (in compds oft) lunar

Mondamin n corn starch, maize flour

Mond·bahn f lunar orbit

Möndchen n (-) 1 little moon; lune; lunula. 2 meniscus

mond· moon, lunar: **-förmig** a moon-shaped, crescent-shaped; lunate; (*Bot.*) lunulate. **M--gas** n Mond gas. **-gaserzeuger** m Mond gas producer. **-glas** n crown glass. **-hof** m lunar halo, ring around the moon. **-licht** n moonlight. **-milch** f agaric mineral, lublinite (soft mixture of calcite and water). **-sichel** f lunar crescent. **-stein** m (*Min.*) moonstone, selenite

Monier·eisen n concrete reinforcing bar(s)

monieren vt/i find fault (with), complain (about), deplore

Mono·acetin n monoacetin, glycerol monoacetate. **m--atomar** a monatomic. **-basisch** a monobasic. **M--brom** n atomic bromine. **-carbonsäure** f monocarboxylic acid. **-chlor** n atomic chlorine. **-chloressigsäure** f monochloroacetic acid. **-chlorhydrat** n monohydrochloride. **m--chrom** a monochromic, monochrome. **-chromatisch** a monochromatic. **-cyclisch** a monocyclic. **M—derivat** n mono derivative. **m--dimetrisch** a (*Cryst.*) tetragonal. **M--fil**

m or *n* monofilament

Monographie *f* (-n) monograph

mono·heteroatomig *a* monoheteratomic. **-hydratisch** *a* monohydric. **M–kalium·monopotassium. m–klin(isch)** *a* monoclinic

monomer *a* monomeric. **Monomer** *n* (-e) monomer

Mono·natrium· monosodium . . . **-oxy·**monohydroxy . . . , monoxy . . . **m–phasisch** *a* monophase, single-phase

Monopol *n* (-e) monopoly

Mono·sauerstoff *m* atomic oxygen. **-schicht** *f* monolayer. **-sulfosäure** *f* monosulfonic acid

monoton *a* monotonous. **Monotonie** *f* monotony

Monotropie *f* monotropy

monovariant *a* univariant

Mono·wasserstoff *m* atomic hydrogen. **m–zyklisch** *a* monocyclic

Monsun *m* (-e) monsoon

Montag° *m* Monday

Montage *f* (-n) 1 (*Mach., Engg.*) installation, construction; assembly. 2 (*Phot.*) montage. **-bahn, -straße** *f* assembly line. **-trupp** *m* construction gang. **-werk** *n* assembly plant

Montan· mountain; mining; montanic: **-al·kohol** *m* montanic alcohol. **-geologie** *f* mining geology. **-industrie** *f* mining industry; coal and steel industry. **-salpeter** *m* ammonium sulfate-nitrate (fertilizer). **-säure** *f* montanic acid. **-wachs** *n* montan wax

Monteur *m* (-e) fitter, installer; assembler; installation (*or* assembly) engineer (*or* mechanic)

montieren *vt* install, erect; assemble. **Montierung** *f* (-en) installation, etc

Mont·milch *f* mountain milk, lublinite = Monomilch

Montpellier·gelb *n* Montpellier yellow (lead oxychloride)

Moor *n* (-e) 1 moor, bog, marsh. 2 peat. **-bad** *n* mud bath. **-boden** *m* marshy soil. **-erde** *f* marshy (boggy, peaty) soil. **-grund** *m* marshland, bog. **-heide** *f* marshy heath

moorig *a* boggy, marshy; peaty

Moor·kohle *f* moor (*or* black subbitumin-ous) coal. **-lauge** *f* marsh lye, bog earth extract. **-mergel** *m* (peat) bog marl. **-salz** *n* marsh salt, salt from a peat bog. **-wasser** *n* marsh (bog, swamp) water

Moos *n* 1 (-e) moss. 2 (Möser) moor, bog. **-achat** *m* moss agate. **m–artig** *a* mossy. **M–beere** *f* cranberry. **-grün** *n* moss green. **-gummi** *n* sponge (*or* cellular) rubber

moosig *a* mossy

Moos· moss; moor: **-pflanze** *f* bryophyte. **-pulver** *n* lycopodium powder. **-stärke** *f* lichen starch, (iso)lichenin. **-tierchen** *n* (*Zool.*) bryozoan, polyzoan. **-torf** *m* peat (turf), surface (*or* sphagnum) peat. **-wuchs** *m* mossy growth

Mops *m* (ᵉe) 1 partially decarbonated limestone. 2 pug (dog)

Moral *f* (-en) morals; morale, moral

Moräne *f* (n) moraine. **Moränen·schutt** *m* moraine (*or* glacial) deposit

Morast *m* (-e *or* ᵉe) morass, bog. **-erz** *n* bog ore, limonite

Morcatragant *m* (-e) yellow tragacanth

Morbillen *pl* measles

Morchel *f* (-n) morel, (edible fungus)

Mord *m* (-e) murder

mordanzieren *vt* mordant

morden *vt/i* murder. **Mörder** *m* (-) murderer. **mörderisch** *a* 1 murderous. 2 = **mörderlich** *a* horrible, awful

morgen *adv* 1 tomorrow: **m. früh** tomorrow morning. 2 morning: **heute m.** this morning. **Morgen I** *m* (-) 1 morning. 2 east, Orient. 3 *appx* acre. **II** *n* tomorrow, future. **-land** *n* East, Orient. **m–ländisch** *a* Eastern, Oriental. **-rot** *n* rosy dawn

morgens *adv* in the morning, mornings

Moringa·gerbsäure, Morin·gerbsäure *f* moringatannic acid, maclurin

Morphium *n* morphine

morphotropisch *a* morphotropic

morsch *a* 1 decayed, rotten. 2 crumbling, decrepit. **morschen** *vi* rot, decay. **Morschheit** *f* decay, rottenness; decrepitude

Morse, Morse·alphabet *n* Morse code

Morselle *f* (-n) spiced almond lozenge

morsen *vt/i* send (in) Morse code

Mörser *m* (-) mortar (vessel & cannon) cf Mörtel. **-keule** *f* pestle

mörsern *vt/i* grind (in a mortar)
Mörser·mühle *f* mortar mill. **-probe** *f* mortar test. **-stössel** *m* pestle
Morse·schrift *f* Morse code
Mörtel *m* (building) mortar *cf* MÖRSER.
--fuge *f* mortar joint. **-gips** *m* retarded gypsum plaster. **-kelle** *f* trowel. **-schlamm** *m* thin mortar, grout. **-struktur** *f* (*Petrog.*) brecciated texture. **-wäsche** *f* thin mortar, grout
mosaik·artig *a* mosaic-like, tesselated; *adv* *oft* like a mosaic
mosaisch *a* Mosaic, of Moses
Moschus *m* musk. **-drüse** *f* musk gland. **-hirsch** *m* musk deer. **-korn** *n* musk seed, ambrette. **-körneröl** *n* musk seed oil, oil of ambrette seed. **-kraut** *n* cat thyme, moschatel. **-öl** *n* oil of sumbul. **-tier** *n* musk deer. **-wurzel** *f* musk root, root of *Ferrula sumbul*
Mosel 1 *f* Moselle River. **2** *m* = **Moseler** *m*, **Moselwein** *m* Moselle wine
Moskito *m* (-s) mosquito
Moskovade *f* (-n) muscovado, raw cane sugar
Most *m* (-e) **1** must; (unfermented) pressed fruit (*or* grape) juice. **2** cider. **3** new wine. **4** fruit wine. **--messer** *m* must gauge
Mostrich *m* (-e) mustard
Most·sirup *m* arrope, boiled-down must. **-waage** *f* must gauge (*or* hydrometer). **-wein** *m* lightly fermented fruit juice
Motiv *n* (-e) motive. **motivieren** *vt* motivate, account (*or* give the reason) for. **motiviert** *p.a* (*oft*) accounted for, logical. **Motivierung** *f* (-en) motivation; grounds, reason(s)
Moto. *abbv* (Monatstonnen) (metric) tons per month
Motor·anlasser *m* (engine) starter. **-antrieb** *m* motor (*or* power) drive: **mit M.** motor-driven, motorized. **-benzin** *n* gasoline
Motoren·benzin *n* gasoline. **-öl** *n* motor oil. **-raum** *m* engine room. **-schmieröl** *n* lubricating oil
motor·getrieben *a* motor-driven, motorized, powered
motorisch *a* **1** (*Mach.*) motorized, powered. **2** (*Physiol.*) motor, kinetic
motorisieren *vt* motorize

Motor·kanone *f* (*Mil.*) automatic cannon. **-leistung** *f* motor (*or* engine) output. **m-los** *a* motorless. **M--luftpumpe** *f* motor-driven air pump. **-rad** *n* motorcycle. **-roller** *m* motor scooter. **-strom** *m* (*Elec.*) motor current. **-treibmittel** *n*, **-treibstoff** *m* motor fuel. **-wagen** *m* motor(ized) vehicle (*or* car)
Motte *f* (-n) moth
motten·moth: **-echt** *a* mothproof. **M--fraß** *m* moth damage. **-kugel** *f* mothball. **-pulver** *n* moth powder. **-schutz** *m* mothproofing. **-schutzmittel** *n* mothproofing agent, moth repellent. **m--sicher** *a* mothproof
Motze *f* (-n) (*Glass*) marver
Mouliné *m* (-s), **--garn** *n* multicolored twist yarn. **moulinieren** *vt* throw, twist (*esp* silk.) **Moulinier·seide** *f* thrown silk
moussieren *vi* effervesce, sparkle, bubble. **moussierend** *p.a* effervescent
Movragenin·säure *f* mowrageninic acid. **Movragen·säure** *f* mowragenic acid. **Movra·säure** *f* mowric acid
Möwe *f* (-n) seagull
Mrd. *abbv* (Milliarde/n) billion(s)
Mrz. *abbv* (März) Mar. (March)
m.s. *abbv* (man sehe) see
ms- *abbv pfx* (meso-) **meso-**
m-ständig *a* meta, in the meta position
Mücke *f* (-n) gnat; midge; (*Opt.*) floater
Mucoinosit *n* mucoinositol
Mucon·säure *f* muconic acid
Mudar·gummi *n* mudar rubber
Mudde *f* (-n) mud(bank)
müde *a* tired, weary, fatigued. **Müdigkeit** *f* weariness, fatigue
Muff *m* (-e) **1** (fur) muff. **2** musty (*or* moldy) smell; mustiness
Muffe *f* (-n) (*Mach.*) sleeve, collar, socket (joint); coupling box; boss (for clamps)
Muffel I *f* (-n) (*Metll.*) muffle. **II** *m* (-) **1** (*Zool.*) muzzle. **2** grumbler; arch-conservative. **--farbe** *f* (*Ceram.*) muffle (*or* burnt-in) color. **-ofen** *m* muffle furnace. **-röstofen** *m* muffle roaster. **-verfahren** *n* retort process (for producing zinc)
Muffen·rohr *n* socket pipe. **-verbindung** *f* sleeve (*or* socket) joint, bell-and-spigot joint
muffig *a* musty, moldy

Mufflerie f (-n) 1 pottery, ceramic plant. 2 (*Metll.*) muffle, retort

Muga·seide f muga (silk)

Mühe f (-n) trouble, (strenuous) effort: **der M. wert** worth the trouble, worthwhile; **sich M. geben** (*or* **machen**) take pains, make an effort; **mit M. und Not** with tremendous difficulty; just barely. **mühelos** a effortless. **mühen** vr: **sich m.** take pains, struggle. **mühevoll** a laborious. **Mühe·waltung** f (taking the) trouble, effort

Mühle f (-n) mill; grinder

Mühlen·staub m mill dust

Mühl·rad n mill wheel. **-stein** m millstone

Mühsal f (-e) trouble, hardship. **mühsam** a arduous, wearisome; adv oft with a great effort

mul abbv (mit dem Uhrzeiger laufend) clockwise

Mulde f (-n) 1 trough; tray; tub; bucket; bowl. 2 mold; (*Metll.*) pig (mold). 3 hollow, valley, basin; (*Geol.*) syncline

mulden·artig a trough- (basin-, bowl-) shaped; (*Geol.*) synclinal. **M−band** n troughed belt conveyor, bucket conveyor. **-blei** n pig lead. **-gurt** m troughed (conveyor) belt. **-heizung** f (*Rubber*) cure in molds. **-mischer** m trough mixer. **-trockner** m trough dryer

mulinieren vt throw (*or* twist) (silk). **Mulinier·seide** f throw silk

Mull m (*Tex.*) (thin) muslin, mull; lint

Müll m 1 garbage, refuse, waste. 2 dust, dry mold. **-berg** m garbage (*or* refuse) heap. **-deponente, -deponie** f garbage dump. **-eimer** m garbage can (pail, bucket)

Müller m (-) miller

Mull·erde f humus, (leaf) mold; garbage compost

Müllerei f (-en) 1 (flour) milling. 2 flour mill

Müller·gaze f (fine) bolting cloth. **-tuch** n bolting cloth

Müllerin f (-nen) miller; miller's wife

Müll·feuerung f refuse furnace. **-halde** f garbage dump. **-kippe** f garbage dump truck

Mull·krapp m mull madder

Müll·schaufel, -schippe f dustpan. **-schlucker** m garbage disposal (chute).

-tonne f garbage can. **-verbrennung** f (garbage) incineration. **-verbrennungs-anlage** f incinerator

Mulm m 1 rotten powdered wood. 2 earth (*or* ore) dust. 3 weathered stone. 4 mold, humus 5 moldiness, decay. **mulmen** 1 vt pulverize. 2 vi (s) crumble, turn to dust (*or* powder). **mulmig** a 1 powdery. 2 rotten, moldy. 3 worm-eaten. 4 friable. 5 dubious, fishy: risky. 6 uneasy; queasy

multifil a multi-filament, multi-stranded. **Multifil** n (-e) multifilament

Multiplikator m (-en) multiplier. **multiplizieren** vt/i multiply. **Multiplizität** f multiplicity. **Multiplum** n (..pla) multiple

Multipoly·addukt n addition copolymer

multrig a moldy, dusty; decayed

Mumie f (-n) mummy. **Mumien·bildung** f mummification. **mumifizieren** vt & vi(s) mummify

Mumme f mum, strong beer

mummen vt wrap

München n Munich. **Münchner** 1 a (no endgs) (of) Munich. 2 m (-), **Münchnerin** f (-nen) Municher, inhabitant of Munich

Mund m (ẅ-) mouth; muzzle, orifice, aperture, stoma, vent. **-art** f dialect, local idiom. **-drüse** f oral (*or* buccal) gland

munden vi (dat) taste good (to) —**sich etwas m. lassen** relish (the taste of) sthg

münden vi (usu + in, auf) 1 discharge, empty (into). 2 open, run (into), lead (to, into)

mund·gerecht a bite-size, digestible; (oft + dat) tasty, palatable (to). **M−höhle** f oral cavity

mundig a pleasant-tasting (wine)

mündig a of age. **Mündigkeit** f age of consent, majority

Mund·klemme f lockjaw. **-leim** m gum to be moistened by mouth

mündlich a oral, verbal; adv oft by (word of) mouth, personally

Mund· mouth, oral: **-loch** n mouth, orifice; (*Explo.*) fuse hole. **-ring** m die-holder. **-ringhalter** m bolster, closing ring. **-schicht** f chemical layer (in a gas mask). **-schleim** m oral mucus. **-schleimhaut** f mucous membrane of the mouth. **-schwamm** m sponge filter (for force masks). **-speichel** *m* (oral) saliva. **-speicheldrüse** *f* salivary gland. **-sperre** *f* inability to close one's mouth, malocclusion of the jaws. **-spülglas** *n* bathroom glass (for toothbrushing, mouth-washing, etc)

M- und S-Reifen *m* (=Matsch- und Schneereifen) snow tire

Mund·stück *n* mouthpiece, nozzle; (extrusion) die. **-tuch** *n* (table) napkin

Mündung *f* (-en) mouth (also of rivers); beginning (of a street or road at an intersection); muzzle (of a gun); orifice, aperture

Mündungs·energie *f* muzzle energy. **-feuer** *n* muzzle flash. **m–feuerfrei** *a* flashless (powder). **M–geschwindigkeit** *f* muzzle velocity. **-weite** *f* (muzzle) bore, caliber

Mund·verdauung *f* oral digestion. **-wasser** *n* mouthwash

Munition *f* (-en) ammunition, *pl oft* munitions

munter *a* 1 awake; **m. machen, m. werden** awaken. 2 lively, cheerful; spry. 3 well, in good health. **Munterkeit** *f* liveliness, cheerfulness; spryness; good health

Muntz·metall *n* Muntz metal (copper-zinc alloy)

Münz·anstalt *f* mint (for currency). **-beschickung** *f* alloyage

Münze *f* (-n) 1 coin; token; medal. **—bare M.** hard cash; **für bare M. nehmen** take at face value. 2 (currency) mint. 3 (*Bot.*) mint. **münzen** *vt* 1 coin. 2: **gemünzt auf . . .** aimed at, meant for (*as:* remarks)

Münz·einwurf *m* coin slot

münzen·ähnlich *a* coin-like. **-förmig** *a* coin-shaped, nummular

Münz·erz *n* money (*or* mintage) ore. **-fernsprecher** *m* coin telephone, pay phone. **-gehalt** *m* fineness (of coins). **-gold** *n* mint (*or* standard) gold. **-kenner** *m* numismatist. **-legierung** *f* coinage alloy. **-metall** *n* coinage metal (*or* alloy). **-prägung** *f* coinage. **-probe** *f* assay. **-wesen** *n* coinage, minting. **-zusatz** *m* coinage alloy

mürbe *a* 1 brittle, crumbly, friable. 2 tender (food), (well-)done. 3 short (steel, dough). 4 exhausted, worn-down, de-

feated: **m. machen (bekommen, kriegen)** wear down, soften up; **m. werden** break down

Mürbe·machemittel *n* softening (*or* shortening) agent. **-teig** *m* short(-cake) dough

Mürbheit, Mürbigkeit *f* brittleness, crumbliness; friability; tenderness, shortness; worn-down condition *cf* MÜRBE

Murbruch° *m*, **Mure** *f* (-n), **Murgang°** *m* (*Geol.*) mudflow

muriatisch *a* muriatic, hydrochloric

murmeln *vt/i* murmur, mumble

Murmel·tier *n* (*Zool.*) marmot

Mus *n* (-e) pulp, mash; mush, puree; (fruit) jam *oft* = BREI

Musa·faser *f* Manila hemp

mus·artig *a* pulpy, mushy *cf* MUS

Muschel *f* (-n) 1 seashell, conch. 2 bivalve mollusk; *specif* mussel. 3 (*Anat.*) external ear. 4 (*Telph.*) earpiece. 5 wash basin, sink. 6 toilet bowl. 7 (*Metll.*) nozzle. **m–förmig** *a* shell-shaped. **M–gold** *n* ormolu

muschelig *a* conchoidal, shelly, shell-like

Muschel·kalk *m* 1 (*Geol.*) shelly triassic limestone. 2 shell lime. **-kalkstein** *m* shell limestone. **-linie** *f* (*Math.*) conchoid. **-marmor** *m* shell marble, lumachel. **-mergel** *m* shell marl. **-sand** *m* shell sand. **-schale** *f* (mussel) shell. **-schieber** *m* slide valve. **-seide** *f* byssus silk. **-tier** *n* shellfish, mollusk

Muscovaden·zucker *m* muscovado

Museen museums: *pl of* **Museum**

musikalisch *a* musical. **Musiker** *m* (-) musician

Musiv·arbeit *f* inlaid (*or* mosaic) work. **-gold** *n* mosaic gold, ormolu; stannic sulfide. **-silber** *n* mosaic silver

Muskat *m* (-e) nutmeg. **--balsam** *m* nutmeg butter. **-birne** *f* musk pear. **-blüte** *f* mace. **-blütenol** *n* mace oil

Muskate *f* (-n) nutmeg

Muskateller *m* muscatel (wine); muscat grape

Muskaten·blüte *f* mace

Muskat·fett *n* nutmeg butter. **-nuß** *f* nutmeg. **-nußöl, -öl** *n* nutmeg (*or* myristica) oil

Muskel *m* (-n) muscle; (*in compds. oft*) muscular, myo-: **-adenylsäure** *f* muscle adenylic acid. **-eiweiß** *n* myosin. **-empfin-**

dung *f* muscular sensation. **-farbstoff** *m* muscle pigment. **-faser** *f* muscle fiber. **-fleisch** *n* muscular substance. **-gewebe** *n* muscle tissue. **-gift** *n* muscle poison. **-kater** *m* muscular strain, charleyhorse. **-lehre** *f* myology. **-masse** *f* muscular mass (*or* substance). **-saft** *m* muscle fluid. **-stoff** *m* sacrosine. **-zucker** *m* inositol

Muskovaden-zucker *m* muscuvado

Muskovit, Muscowit *m* (*Min.*) muscovite

muskulös *a* muscular

Müsli *n* oat flakes with sugar, milk and fruit

muß must, have to, has to: *pres of* MÜSSEN

Muß *n* must (=sthg that has to be done)

Muße *f* leisure: **mit M.** at (one's) leisure

Musselin *m* (-e) muslin

müssen* (mußte; *pres sg* muß) *v aux* 1 (+*inf; pp* müssen) have to, be obliged to, be bound to, (*pres oft*) must: **es muß erhitzt werden** it must (*or* has to) be heated; **es mußte endlich mißlingen** it had to (*or* was bound to) fail in the end; **das muß nicht bedeuten . . .** that does not have to (*or* that need not) mean . . . ; **es müßte zunehmen** it would have to increase; **es muß sich verändert haben** it must have changed. 2 (*inf, esp verb of motion, left implied; pp* müssen *or* gemußt): **sie mußten ins Krankenhaus** they had to go to the hospital

mussieren *vi* effervesce, etc = MOUSSIEREN

müßig *a* 1 idle; inactive; unemployed. 2 useless, pointless

mußte had to, etc: *past of* MÜSSEN

Muster *n* (-) 1 model, example. 2 sample, specimen. 3. pattern, design. **-beispiel** *n* model, perfect example; paragon. **-betrieb** *m* model operation (enterprise, etc); pilot plant. **-exemplar** *n* model, perfect example; standard; type specimen, prototype. **-färbung** *f* sample dyeing. **-flasche** *f* sample bottle. **m–gültig** *a* exemplary = MUSTERHAFT

musterhaft *a* exemplary, model; standard; perfect, classic. **Musterhaftigkeit** *f* exemplariness; perfection

Muster-hahn *m* sampling tap, bleeder valve. **-karte** *f* pattern (*or* sample) card. **m–mäßig** *a* standard, (proto)typical

mustern *vt* 1 scrutinize, inspect. 2 pattern,

make a design on (cloth, etc) *cf* GEMUSTERT; emboss (paper); (*Dye.*) bring to shade

Muster-nehmer *m* sampler. **-probe** *f* sample, specimen. **-sammlung** *f* specimen collection. **-schutz** *m* 1 design patent (*or* copyright). 2 registered design. 3 registration of designs. **-stück** *n* 1 test specimen, sample. 2 model, etc = MUSTERBEISPEIL. **m–treu** *a* (*Tex.*) true-shade

Musterung *f* (-en) 1 scrutiny, inspection; physical (examination, for draftees). 2 patterning, figuring; embossing; bringing to shade. 3 pattern, design. *Cf* MUSTERN

Muster-vorlage *f* prototype (*or* *Dye.*: shade) to be matched. **-zeichnung** *f* design; designing. **-zieher** *m* sampler, sampling device

Mut *m* courage; spirit(s); mood

Mutant *m* (-en, -en), **Mutante** *f* (-n) mutant

mutarotieren *vt* mutarotate

muten *vi* claim a mining concession. **Muter** *m* (-) claimant for a mining concession

mutieren *vi* 1 mutate. 2 (of voice) change, break

mutig *a* courageous, spirited

mutmaßen *vt* (*pp* gemutmaßt) suppose, conjecture. **mutmaßlich** *a* presumable. **Mutmaßung** *f* (-en) supposition, conjecture

Mutter *f* 1 (-̈) mother; matrix. (*in compds oft*) maternal, parent; uterine. 2 (-n) (*Mach.*) nut, female screw. **–boden** *m* 1 (*Biol.*) parent tissue. 2 (*Agric.*) native soil; topsoil. **-erde** *f* 1 garden mold. 2 native soil. **-essig** *m* mother of vinegar. **-faß** *n* (*Vinegar*) mother vat. **-fieber** *n* puerperal fever. **-form** *f* 1 (*Ceram.*) master mold. 2 (*Biol.*) parent form. **-gesellschaft** *f* parent company. **-gestein** *n* 1 (*Min.*) gangue, matrix. 2 (*Petrol.*) parent (*or* source) rock. 3 (*Geol.*) bedrock. **-gewebe** *n* mother tissue, matrix; (*Anat.*) uterine tissue. **-gewinde** *n* (*Mach.*) female thread. **-gummi, -harz** *n* galbanum, gum resin. **-hefe** *f* mother (*or* parent) yeast. **-korn** *n* ergot

Mutterkorn-pilz *m* ergot fungus. **-säure** *f* ergotic acid. **-vergiftung** *f* ergotism

Mutter- mother, maternal: **-kraut** *n* feverfew. **-kuchen** *m* placenta. **-kümmel** *m* cumin. **-lauge** *f* mother liquor, parent wash. **-leib** *m* womb, uterus

mütterlich *a* motherly, maternal

mutter- mother, maternal; parent; uterine: **-los** *a* motherless. **M–mal** *n* birthmark, mole. **-nelke** *f* mother clove **-pech** *n* (*Med.*) meconium. **-pflanze** *f* parent plant. **-schaf** *n* ewe. **-scheibe** *f* (*Mach.*) nut washer. **-scheide** *f* vagina. **-schlüssel** *m* nut wrench. **-sicherung** *f* (*Mach.*) nut lock. **-sprache** *f* mother tongue, native language. **-stoff** *m*, **-substanz** *f* parent substance; matrix. **-verschluß** *m* (*Mach.*) nut lock, lock nut. **-zelle** *f* mother (*or* parent) cell. **-zimt** *m* cassia

Mutung *f* (-en) claim for a mining concession

Mut-wille *m* wilfulness. **mutwillig** *a* wilful

Mütze *f* (-n) 1 cap. 2 (*Bot.*) calyptra. 3 (*Zool.*) honeycomb stomach (of ruminants). **mützenförmig** *a* cap-shaped

MV. *abbv* (Massenverhältnis) proportion (*or* ratio) of constituents

mval *abbv* (Milliäquivalent) milli-equivalent

m.W. *abbv* (meines Wissens) as far as I know

MWG *abbv* (Massenwirkungsgesetz) law of mass action

Mycel *n* (-ien) mycelium = MYZEL etc. **--pilze** *mpl* (*Bot.*) Hyphomycetes, Moniliales

Mycophenol-säure *f* mycophenolic acid

mydriatisch *a* mydriatic

Mykologie *f* mycology

Mykose, Mykosis *f* (..osen) mycosis

Mykosterin *n* mycosterol

mykotisch *a* (*Med.*) mycotic

Myon *n* (-en) muon, mu-meson

Myrica- (or **Myrika-**) **fett, -wachs** *n* myrtle wax

Myristicin-säure *f* myristicic acid

Myristin-säure *f* myristic acid

Myron-säure *f* myronic acid

Myrosin-behälter *m* myrosin cell

Myrrhe *f* (-n), **Myrrhen-harz** *n* myrrh. **-öl** *n* oil of myrrh

Myrte *f* (-n) myrtle

Myrten-gewächse *npl* (*Bot.*) Myrtaceae, myrtles. **m–grün** *a* myrtle green. **M–öl** *n* myrtle oil. **-säure** *f* myrtenic acid. **-wachs** *n* myrtle wax

mystifizieren *vt* mystify; mislead

mystisch *a* mystic(al)

Myzel(ium) *n* (..lien), **Myzele** *f* (-n) mycelium

N

n *abbv* (Nutzeffekt) efficiency

n. *abbv* **1** (NACH) after, etc. **2** (NETTO) net. **3** (neu) new. **4** (neutral; normal). **5** (NÖRD-LICH) northern, etc

N. *abbv* **1** (Nachmittag) aft. (afternoon). **2** (Nacht, nachts) (at) night. **3** (Nord, Norden) N, No. (North)

n.A. *abbv* (neue Auflage *or* Ausgabe) new printing (*or* edition)

Nabe *f* (-n) (*Mach.*) hub, nave; boss

Nabel *m* (-) **1** navel, umbilicus. **2** (*Bot.*) hilum. **3** (*Tech.*) eye, focus.**-kraut** *n* (*Bot.*) navelwort. **-punkt** *m* (*Math.*) umbilic. **-schnur** *f* umbilical cord. **-schwein** *n* peccary

nach I *prep* (*dat*) **1** after: **erst n. dem Erhitzen** not until after heating; **n. einer Stunde** after one hour, one hour ·later, (with)in an hour; **Viertel n. acht** a quarter after (*or* past) eight. **2** (*esp* + *Geog. name*) to. **3** toward, -ward (*oft* + **hin** *or* **zu**): **n. außen hin** (*or* **zu**) toward the outside, outward; **n. oben** upward, upstairs. **4: n., je. n.** according to, by: (**je**) **n. dem Wetter** according to the weather; **unseren Untersuchungen n.** according to our investigations; **es geht n. dem Gewicht** it goes by weight. **5** for: **n. etwas suchen** (**streben, rufen**) look (strive, call) for sthg. **6** about: **n. etwas fragen** ask about sthg.; **die Frage n. der Ursache** the question regarding (*or* as to) the cause. **II** *adv*: **n. und n.** little by little, gradually; **n. wie vor** just as before. **III** *sep pfx* **7** after (*implies following*) *cf* NACHGEHEN, NACHFOLGEN. **8** (*implies revising, correcting, verifying*): *cf* NACHMESSEN, NACHPRÜFEN. **9** (*implies imitation, copying, repeating*) *cf* NACHAHMEN, NACHDRUCKEN. **10** (*implies investigation, study*) *cf* NACHFORSCHEN. **11** (*implies yielding, slackening*) *cf* NACHGEBEN, NACHLASSEN. *Cf* NACHDEM

nach·ahmen *vt* imitate, copy, counterfeit *cf* NACHGEAHMT. **Nachahmung** *f* (-en) im-itation, copy(ing), counterfeit(ing)

Nacharbeit° *f* **1** make-up work. **2** copy, imitation. **3** finishing, finishing touches; touch-up work; refinishing, retouching. **4** maintenance (work). **nach·arbeiten I** *vt* **1** make up (for). **2** copy, imitate; follow. **3** finish, touch up; refinish, retouch. **4** maintain. **II** *vi* do extra work

nach·arten *vi*(s) (*dat*) take after, resemble

nach·äschern *vt* ash; (*Lthr.*) lime (after unhairing)

Nachavivage° f (*Tex.*) after-brightening

Nachbar *m* (-n, -n) neighbor; (*in compds oft*) neighboring, vicinal, adjoining, adjacent: **-atom** *n* neighboring atom. **-gebiet** *n* neighboring (*or* adjoining) territory; related subject. **Nachbarin** *f* (-nen) neighbor. **Nachbar·linie** *f* adjacent (adjoining, neighboring) line

Nachbarschaft *f* (-en) neighborhood. **Nachbarschafts·bild** *n* (*Cryst.*) coordination diagram

Nachbar·stellung *f* neighboring (*or* adjacent) position

nach·bearbeiten *vt* rework, (re)finish, retouch, dress; edit, revise. **Nachbearbeitung** *f* reworking, (re)finishing, etc; editing, revision

nach·behandeln *vt* aftertreat, give follow-up treatment. **Nachbehandlung°** *f* aftertreatment, follow-up treatment

Nachbehandlungs·bad *n* aftertreating bath. **-farbstoff** *m* dye requiring aftertreatment

nach·beizen *vt* **1** (*Dye.*) redye. **2** (*Tex.*) sadden. **3** (*Lthr.*) drench

Nachbeschickung° *f* (*Metll.*) aftercharge; aftercharging

nach·bessern *vt* improve on, touch up, retouch; repair; strengthen (solutions); replant. **Nachbesserung°** *f* improvement, touch-up, retouching, etc.

Nachbild° *n* **1** copy, replica. **2** after-image. **nach·bilden** *vt* **1** copy, reproduce, duplicate; recreate. **2** counterfeit. **3** make a model of. **Nachbildung°** f 1 copying, reproducing, etc, 2 copy, reproduction, duplicate; counterfeit; model

Nachblasen n (Metll.) afterblow

nach·bläuen vi(s) turn bluish (subsequently)

Nachbleiche° f (Tex.) final bleach. **nach·bleichen** vt rebleach, afterbleach

nach·bohren vt rebore

Nachbrand° m second burning (baking, fire)

nach·brauen vt rebrew

nach·brennen* 1 vt re-burn, afterburn; reroast. 2 vi smolder; hang fire. **Nachbrenner°** m hangfire shell; afterburner

nach·chromieren vt (Dye.) afterchrome, back-chrome. **Nachchromierung°** f afterchroming. **Nachchromierungs·farbstoff** m afterchroming (or afterchromed) dye

Nachdampf° m (Min.) afterdamp, choke damp. **nach·dampfen** vt resteam

nach·datieren vt post-date

nach·decken vt (Dye.) fill up, top

nachdem I cnjc 1 after. 2 since, (inasmuch) as. **II** adv: je n. depending (on circumstances); (usu + interrog cnjc, e.g.) je n., wie ... depending on (or according to) how ...

nach·denken* vi (+ über) think over, think (about), reflect, meditate (on)

Nachdruck° m 1 emphasis; vigor: mit N. emphatically, vigorously. 2 reprint(ing). **nachdrücklich** a emphatic; vigorous. **Nachdruck·recht** n copyright

nach·dunkeln 1 vt darken, (Dye.) sadden. 2 vi(s) darken, turn darker (later)

nach·eichen vt recalibrate, restandardize; realine. **Nacheichung°** f recalibration, etc

nach·eifern vi (dat) vie (with), emulate

nach·eilen vi(s) (dat) 1 hurry, rush (after, to catch up to). 2 (Elec.) lag (in phase). 3 (Metll.: of rollers) creep. **Nacheil·winkel** m angle of lag

nacheinander adv 1 one after the other, one by one. 2 in succession

Nachen m (-) boat, skiff

nach·erhitzen vt reheat; (Metll.) temper

Nachf. abbv (Nachfolger) successor(s)

nach·fallen* vi(s) fall (after); cave in

nach·färben vt redye, redip; counterstain

Nachfederung f hysteresis

nach·fetten vt (Lthr.) fat-liquor, stuff

Nachfilter° n second filter. **nach·filtern** vt refilter

nach·fixieren vt afterfix

Nachflammer m (-) (Mil.) flareback

nach·fleischen vt (Lthr.) reflesh

nach·fließen* vi(s) (dat) flow after, flow again

Nachfolge° f 1 succession (in office); right to take over a position: die N. antreten take over the position. 2: N. finden be imitated. 3 sequence. **--kern** m product nucleus. **nach·folgen** vi(s) (oft + dat) 1 follow (after), pursue. 2 succeed (in office). 3 come (or happen) later. 4 emulate. **nachfolgend** p.a 1 following, subsequent: im n—en in the following (paragraph, section, etc), below, hereinafter. 2 succeeding, incoming (office holder). 3 consequent, resulting. **Nachfolger** m (-), **Nachfolgerin** f (-nen) successor; follower

nach·formen vt copy; (Plastics) postform

nach·forschen vt, vi (dat) investigate; search. **Nachforscher°** m investigator. **Nachforschung°** f investigation, (re)search

Nachfrage° f 1 inquiry. 2 (com.) demand Cf ANGEBOT. **nach·fragen** vi inquire

Nachfraktionierturm° m afterfractionating tower. **Nachfraktionierung** f afterfractionation

nach·führen vt follow along (or up) with. **Nachführ·motor** m slaving torque motor. **Nachführung°** f 1 (sidereal) tracking. 2 (Mach.) drive

nachfüllbar a refillable. **nach·füllen** vt 1 refill. 2 e.g. Benzin (in den Tank) n. put more gasoline in (the tank), (fill up the tank with gas). **Nachfüll·flasche** f refillable (or returnable) bottle. **-packung** f returnable container, refill package

Nachgare° f refinished state; refinishing

nach·gären* vi(s) undergo afterfermentation, ferment again. **Nachgär·faß** n (Brew.) cleansing cask. **Nachgärung°** f afterfermentation, secondary fermentation

Nachgärungs·faß n (Brew.) cleansing vat. **-hefe** f afterfermentation yeast

Nachgasen *n* subsequent gassing

nachgeahmt *p.a* imitation; counterfeit(ed). *See also* NACHAHMEN

nach·geben* I *vt* **1** give more of. **2** overlook, let pass. **3: einer Sache nichts n. an . . .** be in no way inferior to a thing in the way of . . . **II** *vi* **4** give (way), buckle. **5** (*dat*) give in, yield (to)

nachgebrannt reburned, etc: *pp of* NACHBRENNEN

Nachgeburt° *f* afterbirth. **Nachgeburts·** (*in compds*) placental

nachgedacht thought, *pp of* NACHDENKEN

nachgeflossen flowed after: *pp of* NACHFLIESSEN

nachgegangen followed: *pp of* NACHGEHEN

nachgegoren afterfermented: *pp of* NACHGÄREN

nachgegossen poured more, etc: *pp of* NACHGIESSEN

nach·gehen* *vi*(s) (*dat*) **1** go after, follow. **2** follow up on, look into. **3** (*of timepieces*) be slow; lose (time). **4** engage (*or* indulge) in, pursue (an activity). **5** (continue to) affect

nachgeholfen helped along: *pp of* NACHHELFEN

nachgeklungen resounded: *pp of* NACHKLINGEN

nachgelassen *p.a* (*oft*) posthumous. *cf* NACHLASSEN

nachgemacht *p.a* imitation, artificial; counterfeit. *cf* NACHMACHEN

nachgerade *adv* **1** gradually. **2** finally, by this time. **3** really, downright

Nachgerbe·brühe *f* retanning liquor. **nach·gerben** *vt* retan. **Nachgerb·öl** *n* retanning oil. **Nachgerbung°** *f* retanning

Nachgeruch° *m* aftersmell, lingering odor

nachgesandt forwarded: *pp of* NACHSENDEN

Nachgeschmack *m* aftertaste

nachgeschliffen reground: *pp of* NACHSCHLEIFEN

nachgeschrieben taken down: *pp of* NACHSCHREIBEN

nachgestanden followed etc: *pp of* NACHSTEHEN

nachgewiesen detected: *pp of* NACHWEISEN

nachgewogen reweighed: *pp of* NACHWIEGEN

nachgezogen dragged: *pp of* NACHZIEHEN

nachgiebig *a* yielding, soft, unresisting; pliable; lenient. **Nachgiebigkeit** *f* softness, weak (*or* lack of) resistance; pliability; leniency

nach·gießen* I *vt* **1** pour some more. **2** make a cast of, cast from. **II** *vi* pour another (glass, cup, etc)

nach·gilben *vi*(s) turn yellow (later)

nach·glätten *vt* smooth (polish, etc) again *cf* GLÄTTEN

nach·glühen 1 *vt* reheat, reanneal. **2** *vi* continue glowing, have an afterglow, phosophresce

nach·grübeln *vi* (**über**) ponder (about, over)

nach·grünen *vi* (s) turn green (later)

Nachguß° *m* **1** second (*or* subsequent) cast(ing). **2** second pouring; refill(ing). **3** (*Brew.*) aftermash, second wort; sparging (water)

Nachhall° *m* reverberation, echo. **nach·hallen** *vi* reverberate, echo

nach·halten* 1 *vt* make up, hold (*e.g.* a meeting) later. **2** *vi* last, persist. **nachhaltig** *a* lasting, persistent; *adv oft* for a long time afterward. **Nachhaltigkeit** *f* lasting quality, durability, persistence

nach·härten *vt* harden subsequently (*or* later)

nach·heizen *vt* reheat, heat some more

nach·helfen* *vi* (*dat*) **1** help along, assist; push, boost. **2** coach, tutor

nachher *adv* later, afterward, after that. **nachherig** *a* later, subsequent

Nachhilfe° *f* (extra) help, assistance; coaching, tutoring

nach·hinken *vi*(s) limp (*or* lag) along behind

Nachhirn° *n* afterbrain, metencephalon

nach·hitzen *vt* reheat; (*Metll.*) temper

nach·holen *vt* **1** make up (for), recover, catch up on. **2** pick up later

Nachhut° *f* rear guard

Nachimpfung° *f* reinoculation

Nachinfektion° *f* reinfection

Nachklang° *m* echo, resonance; aftereffect

nach·klingen* *vi* resound, echo; leave a lingering effect

nach·kochen 1 *vt* boil again. **2** *vt/i* continue to boil

Nachkomme m (-n, -n) descendant.
nach·kommen* vi(s) (dat) 1 follow, come
later. 2 keep up (with). 3 fulfil, satisfy;
comply (with); observe, keep (promises).
Nachkommenschaft f progeny; posterity. **Nachkömmling** m (-e) 1 descendant.
2 latecomer; late-born child
Nachkondensator° m secondary condenser
Nachkriegs· post-war: **-zeit** f post-war era
nach·krispeln vt regrain (leather)
nach·kühlen vt recool, aftercool. **Nach-
kühler°** m aftercooler, recooler.
Nachkühlung° f aftercooling, recooling
nach·kupfern vt aftercopper, aftertreat
with copper. **Nachkupferung°** f after-
coppering
Nachkupplung° f aftercoupling
nach·laden* vt reload, recharge; load (or
charge) later. **Nachladung°** f subsequent
charge (or load), reload, recharge
Nachlaß m (.. lässe) 1 bequest; estate,
property left after death; posthumous
works (of a writer, etc). 2 price reduction,
discount. 3 allowance. 4 remission, abatement, let-up. **nach·lassen* I** vt 1 temper,
anneal (glass, metals). 2 let in more (water, steam, etc). 3 let (gases) expand. 4
relax (one's hold on), loosen, slacken. 5
remit, cancel; allow (someone a remission, a discount). 6 leave (behind) (esp after death), bequeath. **II** vi 7 let up, subside; give in; give up. 8 fall off, decline. 9
(+ **zu** + inf) cease (doing)
nachlässig a careless, negligent; indifferent. **Nachlässigkeit** f carelessness,
etc
Nachlauf m 1 (Distil.) last runnings,
faints. 2 (Engg.) caster (angle) (of wheels)
nach·laufen* vi(s) (dat) run after
Nachlauf·flotte f (Dye.) feeding liquor
nach·leimen vt reglue, resize
Nachlese f 1 gleaning. 2 later selection,
posthumous collection (of works). **nach·
lesen*** vt/i 1 glean. 2 read up on, look up;
reread
nach·leuchten vi keep on shining, phosphoresce; luminesce after excitation.
Nachleuchten n persistence of luminescence (after excitation). **nachleuchtend**
p.a persistent (luminescence), afterglow-

ing. **Nachleucht·pigment** n afterglow
(or luminescent) pigment. **-wirkung** f afterglow (or luminiscent) effect
nach·liefern vt supply (or deliver) later, deliver more of; supplement. **Nachliefe-
rung°** f later (or repeat) delivery;
supplement(ing)
Nachluft f additional air
nachm. abbv (nachmittags) PM
nach·machen vt 1 (usu.): **einem etwas n.**
imitate (or copy) sthg someone is doing;
counterfeit. 2 make up (later). Cf
NACHGEMACHT. **Nachmachung°** f imitation, copy(ing), etc
Nachmaische f aftermash
nachmalig a subsequent. **nachmals** adv
subsequently, afterward
nach·mattieren vt deluster, deaden (subsequently)
nach·messen* vt/i remeasure, check (the
measurements), control (the size)
Nachmittag° m afternoon. **nachmittags**
adv afternoons, in the afternoon
Nachmühlen·öl n inferior olive oil (from
second pressing)
Nachmuster·färben n dyeing to pattern
Nachnahme f (-n) C.O.D. (shipment)
Nachname° m family name, last name
nach·nuancieren vt shade off (later)
Nachöl° n inferior (olive) oil (from second
pressing)
Nachoxidation° f afteroxidation
nach·polieren vt repolish, polish up again
Nachpresse° f re-pressing machine; (Paper) press rolls. **nach·pressen** vt re-press
Nachprodukt° n afterproduct, by-product.
--füllmasse f (Sugar) low-grade massecuite. **-zucker** m low-grade sugar
nachprüfbar a verifiable. **nach·prüfen** vt
verify (re)check (out); reexamine, retest.
Nachprüfung° f verification, (re)-
check(ing); reexamination, retesting
nach·putzen vt repolish, clean further,
touch up
Nachreaktor° m postreactor, second
reactor
nach·rechnen vt/i recalculate, check (the
calculation). **Nachrechnung°** f recalculation, check(ing)
Nachrede° f 1 detraction; slander. 2 epi-
logue

nach·regeln *vt* readjust, rearrange

Nachreife *f* afterripening, ripening in storage. **nach·reifen** *vi*(s) afterripen, ripen after picking (*or* in storage). **Nachreifung** *f* afterripening, etc

Nachricht *f* (-en) message, news (item), *pl usu* news; (*in compds oft*) communication(s)

nach·richten *vt* readjust, reset, realine

Nachrichten·dienst *m* news service. **-mittel** *n* means of communication. **-satellit** *m* communications satellite. **-sendung** *f* news broadcast. **-technik** *f* communications engineering. **-wesen** *n* (field of) communications

Nachruf° *m* obituary (notice, speech)

nach·sacken *vi*(s) settle, sink, sag

nach·salzen *vt* resalt, add more salt to

Nachsatz° *m* 1 addition, addendum; postscript. 2 (*Dye.*) additional feed(ing). **-lösung** *f* (*Dye.*) feed liquor

nach·schalten *vt* connect (*or* switch in) beyond (*or* as a later stage)

nach·schärfen *vt* 1 resharpen, regrind. 2 restrengthen (a solution)

Nachschlage·buch *n* reference book. **nach·schlagen*** *vt/i* look (it) up (in), consult, refer (to). **Nachschlage·werk** *n* reference work

nach·schleifen* *vt* regrind, resharpen

Nachschlüssel *m* passkey, duplicate key

nach·schmecken *vi* leave an aftertaste

nach·schmieren *vt* relubricate

nach·schreiben* 1 *vt* take notes on, take down (a lecture); copy. 2 *vi* take notes, take dictation. **Nachschrift°** *f* 1 transcript, notes. 2 postscript, addendum

Nachschub *m* 1 supplies; reinforcements. 2 (*Baking*) new batch

Nachschuß° *m* 1 backshot, second shot. 2 additional payment

nach·schütten *vt* put (*or* pour) on more

Nachschwaden *mpl* afterdamp; choke damp; shotfiring fumes

nach·schwärzen *vt* blacken (leather) by topping with iron salts

nach·schwefeln *vt* (*Tex.*) stove afterward

nach·schwelen *vi* continue to smolder

nach·sehen* I *vt* 1 look over, check. 2 look up (in a reference work). 3: **einem etwas**

n. overlook, condone sthg someone does. II *vi* 4 go and see, check. 5 (*dat*) follow with one's eyes

nach·seigern *vt* (*Metll.*) liquate again

nach·senden* *vt* forward (mail); send later. **Nachsendung°** *f* forwarding; forwarded mail; delayed shipment

nach·setzen I *vt* 1 put (*or* place) after. 2 add (*or* feed in) more; refill (*or* recharge) with. II *vi*(s) (+ *dat*) pursue. **Nachsetz·löffel** *m* adding spoon (*or* ladle) (for ores, etc)

Nachsicht° *f* 1 indulgence, leniency: **N. haben (mit)** be indulgent (*or* lenient) (with). 2 hindsight *cf* VORSICHT

nach·spannen *vt* retighten, readjust; posttension (concrete)

Nachspeise° *f* dessert

nach·speisen *vt* replenish, re-feed

Nachspiel° *n* sequel, postlude

nach·spülen I *vt* 1 rerinse; rinse out. 2 wash down (food, pills). II *vi* 3 rinse one's mouth. 4 flush the toilet

nächst 1 *a* (*superl of* NAH, NAHE) next, nearest—**in n—er Zeit** in the near future; **der n—e beste** *cf* NÄCHSTBEST; **fürs n—e** for the time being *cf* NÄHE. 2: **am n—en** *pred a, adv* (*superl of* NAH, NAHE) nearest, closest. 3 *prep* (*dat*) next to. **-best** *a* first available, first that comes along. **Nächste** *m* (*adj endgs*) neighbor, fellow human being

Nachstech·bier *n* feed beer. **nach·stechen*** *vt* (*Brew.*) top up, refeed

nach·stehen* *vi* (h *or* s) (*dat*) 1 come after, follow. 2 be inferior (*or* second) (to). **nachstehend** *p.a* 1 inferior. 2 following; *adv* hereinafter: **im n—en** in the following paragraph (pages, lines, etc)

nachstellbar *a* (re)adjustable. **nach·stellen** I *vt* 1 (re)adjust, reset. 2 set back (a timepiece). 3 put (*or* place) after. II *vi* (h *or* s) (+ *dat*) pursue, persecute. **Nachstellung°** *f* (re)adjustment, resetting; postposition; pursuit; persecution. **Nachstell·vorrichtung** *f* adjusting device

nächstemal: das n. (the) next time

nächstens *adv* soon; shortly; next time

nächstesmal *adv* next time

nächst·folgend *a* next (in order), following. **-höher** *a* next higher. **-jährig** *a* next year's. **-liegend** *a* closest, nearest (at hand); most obvious. **-näher** *a* next near(by). **-niedrig(er)** *a* next lower

nach·strahlen *vi* phosphoresce, continue radiating. **Nachstrahler°**, **Nachstrahl·stoff** *m* phosphorescent substance, phosphor. **Nachstrahlung°** *f* phosphorescence

nach·streben *vi* (*dat*) strive (for), aspire (to)

nach·strömen *vi*(s) (*dat*) stream after, follow up (behind a moving body)

nach·suchen I *vt* **1** look for, search. **2** look up (words, etc). **3** inquire into, investigate. **II** *vi* (+ **um**) apply (for)

Nachsud *m* afterboiling, second evaporation; second brewing; (*Dye.*) second mordant

Nacht *f* (⁼e) night; (*fig*) darkness, blackout

nach·tanken *vt/i* refuel

nach·tannieren *vt* backtan (hides)

Nacht·blau *n* night blue. **-blindheit** *f* night blindness, nyctalopia. **-creme** *f* night cream

Nachteil° *m* disadvantage. **nachteilig** *a* unfavorable, disadvantageous

Nacht·falter *m* moth, miller. **-grün** *n* night green

Nachtigall *f* (-en) nightingale

Nachtisch° *m* dessert

Nacht·kerze *f* **1** night light. **2** (*Bot.*) evening primrose. **n–leuchtend** *a* (*Biol.*) noctilucent, bioluminescent

nächtlich *a* night(ly), nocturnal, nighttime

Nacht·mahl *n* supper, evening meal

nach·tönen 1 *vt* retint, touch up (in color); (*Dye.*) aftertone, backtone (to a tinted white). **2** *vi* resound, continue to sound

Nachtrag *m* (..äge) addendum, supplement. **nach·tragen*** *vt* **1** add (in writing), put down (later). **2: einem etwas n.** (a) carry sthg up to sbdy; (b) hold sthg (as a grudge) against sbdy. **nachtragend** *p.a* vindictive. **nachträglich** *a* supplementary, additional; later, subsequent

Nachtreter *m* (-) follower, adherent

Nachtripper° *m* (*Med.*) gleet

nach·trocknen *vt* redry, dry some more

nachts *adv* at night

Nachtschatten *m* nightshade. **--gewächse** *npl* Solanaceae

Nacht·schicht *f* night shift. **-violenöl** *n* rocket oil

Nachverbrennung° *f* aftercombustion

nach·verdampfen *vt* re-evaporate

Nachvergasung° *f* aftergasification

Nachverseifung° *f* aftersaponification

Nachvulkanisation *f* aftervulcanization, overcure (of rubber)

Nachw. *abbv* (NACHWEIS) proof, etc

nach·wachsen* *vi*(s) grow back, grow in again—**die n—de Generation** the new generation *cf* NACHWUCHS

nach·walzen *vt* **1** (*genl*) reroll. **2** (*Steel*) planish, cold-roll

nach·wärmen *vt* reheat, afterheat. **Nachwärme·ofen** *m* reheating furnace

nach·waschen* *vt* rewash, wash (after treatment)

Nachweiche *f* (*Brew.*) couching. **nach·weichen** *vt* couch

Nachweis *m* (-e) **1** detection, (confimatory) test, identification; (*Med.*) indication. **2** proof, evidence, demonstration. **3** certificate, voucher. **4** (*esp at end of compds*) list, index; referral service. **nachweisbar** *a* **1** detectable, testable, identifiable. **2** provable, demonstrable. **Nachweis·büro** *n* information office, referral service. **nach·weisen*** *vt* **1** detect, identify, confirm by test. **2** (*oft* + *d.i.o.*) prove, demonstrate, furnish (*or* find) proof (*or* evidence) (of, against). **3** (+ *d.i.o*) direct (*or* refer) (sbdy) to, inform of. **nachweislich** *a* **1** detectable, etc. **2** provable, etc. = NACHWEISBAR. **Nachweis·mittel** *n* means of detection (*or* proof); indicator. **Nachweisung°** *f* **1** detection, etc. **2** proof, etc. **3** referral. *Cf* NACHWEISEN. **Nachweisungs·mittel** *n* means of detection (*or* proof), indicator

Nachwelt *f* posterity

Nachwert *m* later (*or* future) value

nach·wiegen* *vt* reweigh, check the weight of

nach·wirken *vi* have a lasting effect (*or* an aftereffect); continue having its effect. **Nachwirkung°** *f* **1** lasting (*or* lingering) effect. **2** aftereffect. **3** subsequent (*or* delayed) effect. **Nachwirk·zeit** *f* aftereffect period, (*specif*) decay period

Nach·wuchs *m* 1 second growth, after-growth. 2 new (*or* younger) generation

Nachwürze *f* second wort

nach·zahlen *vt* pay later (*or* retroactively), pay in addition

nach·zählen *vt* recount, count over again, check the count

Nachzahlung° *f* later (additional, retroactive) pay(ment)

Nachzählung° *f* recount

nach·ziehen* **I** *vt* 1 drag (*or* pull) along behind. 2 tighten up (screws), straighten up. 3 redraw (wire). 4 (re)trace, go over (lines); touch up. 5 raise (*or* grow) more (plants, animals). 6 train (a successor). *Cf* ZIEHEN 11 (+ NACH SICH). **II** *vi*(s) 7 (+ *dat*) follow, move to join. 8 follow suit. 9 make the next move

Nachzucht *f* progeny

Nachzügler° *m* straggler, latecomer

Nachzündung° *f* (*Mach.*) retarded ignition

Nacken *m* (-) (nape of the) neck; (*in compds*) cervical: **-drüse** *f* cervical gland

nackt *a* naked, bare; plain; (*in compds oft*) gymno-: **-früchtig** *a* gymnocarpous. **Nacktheit** *f* nakedness, etc. **Nackt·keimer** *m* acotyledon. **-samer** *m* gymnosperm. **n–samig** *a* gymnospermous. **N–schnecke** *f* (*Zool.*) slug

Nadel *f* (-n) 1 pin. 2 needle; stylus. 3 (*in compds oft*) acicular: **n–artig** *a* needlelike, acicular. **N–ausschlag** *m* deflection of the needle. **-baum** *m* conifer(ous tree). **-blatt** *n* acicular leaf, needle

Nädelchen *n* (-) little needle (*or* pin)

Nadel·eisenerz *n*, **-eisenstein** *m* needle iron ore, göthite in acicular crystals. **-erz** *n* needle ore, aikinite in acicular crystals. **-faser** *f* acicular fiber. **n–förmig** *a* needle-shaped, acicular. **N–holz** *n* 1 conifers (*esp* pines and firs). 2 softwood

Nadelholz·kohle *f* softwood charcoal. **-teer** *m* pine tar. **-zellstoff** *m* softwood cellulose, conifer pulp

nadelig *a* needly, acicular

Nadel·lager *n* needle bearing. **-öhr** *n* eye of the needle. **-punktierung** *f* acupuncture. **-spitze** *f* pin point; needle point. **-stein** *m* needlestone, natrolite. **-stich** *m* pinprick;

pinhole. **-ventil** *n* needle valve. **-zeolith** *m* needle zeolite, natrolite. **-zinnerz** *n* (needle-shaped *or* acicular) cassiterite

Nafta *f* *or* *n* naphtha. **Naftalin** *n* naphthalene

Nagel *m* (-) nail; spike; peg; claw: **an den N. hängen** give up, quit. **-bohrer** *m* gimlet

Nägelchen *n* (-) little nail; brad; tack

Nägelein *n* (-) 1 little nail; brad; tack. 2 carnation, clove. **-öl** *n* oil of cloves. **-pfeffer** *m* allspice, pimento. **-rinde** *f*, **-zimt** *m* clove bark, clove cinnamon

Nagel·feile *f* fingernail file. **-härter** *m*, **-härtewasser** *n* (*Cosmet.*) nail-hardening solution. **-haut** *f* cuticle. **-lack** *m* nail polish. **-lackentferner** *m*, **-lackentfernungsmittel** *n* nail-polish remover

nageln 1 *vt* nail; (*Med.*) pin (bones). 2 *vi* (of engines) knock

nagel·neu *a* brand-new. **N–pflege** *f* manicure. **-politur** *f* nail polish. **-probe** *f* nail test. **-schere** *f* nail scissors. **-zange** *f* 1 nail clippers. 2 (nail) pincers

nagen *vt/i* gnaw; erode, corrode. **nagend** *p.a* (*oft*) nagging (pain, etc). **Nager** *m* (-), **Nage·tier** *n* rodent

Näglein *n* (-) 1 little nail; brad; tack. 2 carnation; clove. = NÄGELEIN

nah 1 *a* near, etc = NAHE. 2 *pfx* near(by); local; short-range

Näh· sewing *cf* NÄHEN, NÄHFADEN

Näh·aufnahme *f* close-up (view)

nah(e) **I** *a* (näher, nächst ..) 1 near, close; nearby, neighboring: **in naher Zukunft** in the near future. 2 approaching, imminent. **II** *adv* & *pred a* (*oft* + *dat*) (näher, am nächsten) near(by), closely, close (to): **aus nah und fern** from near and far; **n. verwandt** closely related; **n. an** (*or* **bei**) near, close to; **n. daran**, . . . **zu tun** close to doing, about to do . . . ; **ein n. liegendes** (*or* **gelegenes**) **Gebäude** a nearby building; **von nahem** at close range; **greifbar n.** within easy reach. **III** *prep* & *pred a* (*dat*) near, close (to): **n. der Ecke** near (*or* close to) the corner; **dem Tode n.** near death; **dem Zergehen n.** close to (*or* on the verge of) melting. **IV** *sep pfx* (*usu fig*) *cf* NAHELEGEN etc

Nähe *f* (-n) nearness, closeness, proximity;

neighborhood, vicinity; presence: **hier in der N.** here in this neighborhood, near here; **in unserer N.** in our presence, near us; **aus der N. ansehen** look at closely (*or* close-up); **aus nächster N.** closely, at close range; **in der N.** (+ *genit or* + **von**) near; **ganz in der N.** right nearby; **in die N. kommen** come near; **in greifbarer N.** within easy reach

nahebei *adv* nearby

nahe·gehen* *vi* (*d.i.o*) affect (closely)

nahe·kommen* *vi*(s) (*dat*) come close (to), approach, approximate

nahe·legen *vt* (*d.i.o*) urge (upon), strongly recommend (to); make plain (to)

nahe·liegen* *vi* (h *or* s) be obvious (evident, likely), suggest itself. **naheliegend** *p.a* obvious, evident, likely

nahen *vi*(s) & **sich n.** approach

nähen *vt/i* sew, stitch, suture; lace (a belt)

näher I *a* (*compar of* NAHE) **1** nearer, closer, *adv oft* more closely; **sich etwas n. ansehen** take a closer look at sthg; **n. treten** come in, step right up. **2** further, more detailed, *adv oft* in greater detail: **n—e Angaben** further particulars, more detailed specifications; **die n—en Umstände** the full particulars. **II** *sep pfx* (*usu fig*) *cf* NÄHERBRINGEN, etc

Näher *m* (-) sewer (=one who sews)

näher·bringen* *vt* (*d.i.o*) make clearer (to)

Nähere(s) *n* (*adj endgs*) further details, particulars: **N—s folgt** details follow

näher·kommen* *vi*(s) (*dat*) approach (more closely); gain a better understanding (of)

Näherin *f* (-nen) seamstress

näher·liegen* *vi* (h *or* s) be more obvious (*or* likely)

nähern 1 *vt* bring up (*or* close) to. **2: sich n.** (*dat*) come up (to), approach, approximate; (*recip*) come closer together; approximate each other

näher·treten* *vi*(s) (*dat*) become more familiar (with)

Näherung *f* (-en) approach; approximation: **in erster N.** to a first approximation

Näherungs·formel *f* approximation formula. **-lösung** *f* approximate solution. **-verfahren** *n* method of approximation.

n—weise *adv* by (way of) approximation. **N—wert** *m* approximate value

nahe·stehen* *vi* (h *or* s) (*dat*) be close, be closely related (to). **nahestehend** *p.a* close, closely related

nahezu *adv* almost, nearly

Näh· sewing: **-faden** *m* (sewing) thread. **-garn** *n* sewing yarn

nahm took, etc: *past of* NEHMEN

Näh·maschine *f* sewing machine

·nahme *f noun corresponding to verb* NEH-MEN; *occurs only with pfxs or in compds, cf* ABNAHME, INANSPRUCHNAHME

nähme would take, etc: *sbjc of* NEHMEN

Näh·nadel *f* sewing needle

Nah·ordnungserscheinung *f* short-range phenomenon. **-ost** *m* Near East

Nähr· nutrient, nutritive: **-agar** *m* nutrient agar. **-äquivalent** *n* nutritive equivalent. **-blatt** *n* (*Bot.*) storage leaf. **-boden** *m* **1** fertile soil, breeding ground. **2** (*Biol.*) culture (*or* nutrient) medium. **3** matrix, substrate (for fungi). **-bodenflasche** *f* culture flask. **-bouillon, -brühe** *f* nutrient broth. **-creme** *f* nutrient (skin-)cream

nähren I *vt* **1** nourish, feed; nurse, breast-feed. **2** feed, fuel (a fire). **3** nurture, foster (a hope, etc). **II** *vi* be nourishing (*or* nutritious). **III: sich n.** eat; (+ **von**) live (on); (+ **von**) earn a living (from, by). **nährend** *p.a* (*oft*) nutritious

Nähr· nutrient, nutritive; alimentary: **-flüssigkeit** *f* nutrient fluid. **-gang** *m* alimentary canal; nutrient duct. **-gelatine** *f* nutrient gelatine. **-gehalt** *m* nutritive content. **-gewebe** *n* nutrient tissue, endosperm

nahrhaft *a* **1** nourishing, nutritious: **n. kochen** cook nourishing food. **2** lucrative (trade). **3** productive (soil). **Nahrhaftigkeit** *f* nutritiousness; lucrativeness; productivity

Nähr· nutrient, nutritive: **-hefe** *f* nutrient yeast. **-kasein** *n* edible casein. **-kraft** *f* nutritive power. **n—kräftig** *a* nutritious. **N—lösung** *f* nutrient solution. **-medium** *n* nutrient medium. **-mittel** *npl* **1** (*genl*) food(stuffs). **2** (*specif*) (prepared) cereals, pasta. **-präparat** *n* food supplement. **-saft** *m* nutrient juice (*or* sap); chyle.

-salz *n* nutrient salt. **-stand** *m* agriculture. **-stoff** *m* nutritive substance, food, nutrient

nährstoff·arm *a* poor in nutrients, nutritionally deficient. **N—gehalt** *m* food (*or* nutrient) content. **n—reich** *a* nutritionally rich (food); fertile (soil). **-schonend** *a* nutrition-guarding (packaging)

Nähr·substanz *f* nutritive (*or* nutrient) substance. **-substrat** *n* nutrient base (*or* medium)

Nahrung *f* 1 food, nourishment, nutrition, diet. 2 livelihood. 3 fuel; impetus. 4 (*Lthr.*) tawing paste

Nahrungs·aufnahme *f* food intake. **-bedarf** *m* food requirements. **-brei** *m* chyme. **-dotter** *m* food (*or* nutritive) yolk, deutoplasm. **-flüssigkeit** *f* nutritive liquid; chyle. **-kanal** *m* alimentary canal. **-kette** *f* food chain. **-milch** *f* (*Physiol.*) chyle

Nahrungsmittel° *n* (*sg & pl*) food; (*pl*) foodstuffs, provisions. **-chemie** *f* food chemistry. **-chemiker** *m* food chemist. **-fälschung** *f* food adulteration. **-industrie** *f* food industry. **-kunde** *f* (science of) nutrition. **-schaden** *m* bodily damage caused by contaminated food. **-technik** *f* food technology. **-verfälschung** *f* food adulteration

Nahrungs·pflanze *f* food plant. **-rohr** *n*, **-röhre** *f* alimentary canal. **-saft** *m* 1 nutrient juice. 2 (*specif: Med.*) chyle; (*Bot.*) sap. **-stoff** *m* nutritive substance (*or* matter). **-störung** *f* nutritional disturbance. **-teilchen** *n* nutritive element (*or* particle). **-vergiftung** *f* food poisoning. **-vorrat** *m* food supply (*or* reserves). **-wert** *m* nutritional value. **-zufuhr** *f* food intake

Nähr·vorrat *m* reserve food, food in stock. **-wasser** *n* nutrient solution. **-wert** *m* nutritive (*or* food) value

Naht *f* (-̈e) seam; joint; weld; suture. **n—los** *a* seamless; smooth. **N—schweißen** *n* seam welding

Näh·wachs *n* sewing wax

Nah·werfer *m* short-range flame thrower. **-ziel** *n* immediate (*or* short-term) goal

Name *m* (-ns, -n) name; reputation: **nur dem N—n nach** in (*or* by) name only; **mit N—n X** by the name of X; **in je-** mandes **N—n handeln** act in someone's name (*or* on someone's behalf)

Namen- (*compds*) See NAMEN(S)·

namenlos *a* 1 nameless. 2 unutterable

namens 1 *adv* by the name of. 2 *prep* (*genit*) in the name of, on behalf of

Namen(s)·gebung *f* naming; nomenclature. **-liste** *f* list of names, roster, roll. **-reaktion** *f* name reaction. **-register** *n* index of names, author index. **-schild** *n*, **-tafel** *f* name plate. **-verzeichnis** *n* index of names, author index

namentlich 1 *a & adv* by name. 2 *adv* especially

namhaft *a* 1 considerable. 2 renowned, well-known. 3: **n. machen** name, specify

nämlich I *a* same. II. *adv* namely, that is. III *adv-cjnc* for: **es war n. schon zu spät** for it was already too late; **weil es n. schmilzt** because, as a matter of fact, it melts

Nancy·säure *f* lactic acid

nannte named, etc.: *past of* NENNEN

Nanziger Säure lactic acid

Napf *m* (-̈e) bowl, dish, basin; pan; cake form; (*Bot.*) cup(ule). **Näpfchen** *n* (-) little bowl (dish, etc); (*Mettl.*) cup; blank (for blasting caps). **-kobalt** *m* flaky metallic arsenic. **napf·förmig** *a* bowl-shaped, dished

Napht-, Naphta *f* naphtha = NAPHTH(A)

Naphtha *f or n* naphtha, petroleum, oil

Naphthacen *n* (-e) naphthacene. **-chinon** *n* naphthacenequinone

Naphtha·chinon *n* naphthaquinone. **-feld** *n* oil field

Naphthaldehyd° *m* naphthaldehyde

Naphthalin *n* naphthalene. **-derivat** *n* naphthalene derivative. **-farbe** *f* naphthalene dye. **-gelb** *n* naphthalene yellow. **-öl** *n* naphthalene oil. **-reihe** *f* naphthalene series. **-rosa** *n* naphthalene pink (Magdala red). **-rot** *n* naphthalene (*or* Magdala) red. **-säure** *f* naphthalenic acid. **-sulfo(n)säure** *f* naphthalenesulfonic acid

Naphthal·säure *f* naphthalic acid

Naphthanthrachinon *n* naphthanthraquinone

Naphtha·pech *n* naphtha pitch. **-quelle** *f* oil well

Naphthazen n (-e) naphthacene
Naphthazin n (-e) naphthazine
Naphthen n (-e) naphthene. **Naphthenat** n (-e) naphthenate. **naphthenisch** a naphthenic
Naphthen·kohlenwasserstoff m naphthenic hydrocarbon, naphthene. **-säure** f naphthenic acid. **-seife** f naphthenic soap. **-trockenstoff** m (Paint) naphthenate drier
Naphthidin n (-e) naphthidine
Naphthionat n (-e) naphthionate. **Naphthion·säure** f naphthionic acid
Naphtho·brenzcatechin n naphthopyrocatechol. **-chinaldin** n naphthoquinaldine. **-chinhydron** n naphthoquinhydrone. **-chinolin°** n naphthoquinoline. **-chinon** n naphthoquinone. **-cumarin** n naphthocoumarine
Naphthoe·aldehyd m naphthaldehyde, naphthoic aldehyde. **-säure** f naphthoic acid
Naphtho·hydrochinon n naphthohydroquinone
Naphthol·artikel m (Dye.) naphthol style
Naphtholat n (-e) naphtholate
Naphthol·blau n naphthol blue
naphtholieren vt naphtholize, naphtholate
Naphthol·natrium n sodium naphthoxide. **n–sauer** a naphtholsulfonate of . . . **N–schwarz** n naphthol black
Naphth·sultam n naph(tho)sultam. **-sulton** n naph(tho)sulton
Naphthyl·amin n naphthylamine. **-amin·sulfosäure** f naphthylaminesulfonic acid. **-blau** n naphthyl blue
Naphthyridin n (-e) naphthyridine
Nappa·leder n napa leather
Narbe f (-n) 1 scar; pockmark, pit (incl. Metll.). 2 (Bot.) stigma. 3 (Agric.) topsoil, sod. **narben 1** vt grain (leather). **2** vi (form a) scar. **Narben** m (-) (Lthr.) grain (surface)
Narben· scar; (Lthr.) grain: **-bildung** f scarring, cicatrization; (Metll.) pitting. **-(binde)gewebe** n scar tissue. **-leder** n grain leather. **-membran** f grain (or hyaline) layer. **-öl** n grain oil. **pressen** vt graining. **-schicht** f grain layer. **-schleim** m (Biol.) stigmatic mucus. **-schmiere** f

grain dubbing. **-seite** f grain side. **-spalt** m grain split. **-spaltschafleder** n skiver
narbig a 1 scarred. 2 pitted. 3 grained. 4 embossed (paper). Cf NARBE
Narbung f (-en) 1 scarring. 2 grain(ing). 3 pitting. 4 embossed pattern
Narde f (-n) (Bot.) (spike)nard. **Narden·bartgras** n lemon grass. **-öl** n spikenard oil. **-salbe** f spikenard ointment
Narkose f (-n) 1 anesthesia. 2 narcosis. **-äther** m ether for anesthesia. **Narkotikum** n (.. ika) 1 anesthetic. 2 narcotic. **Narkotin** n narcotine. **narkotisch** a anesthetic; narcotic. **narkotisieren** vt anesthetize. **Narkotisierung** f (-en) anesthetization
Narr m (-en, -en) fool. **narren·sicher** a foolproof. **närrisch** a foolish; mad, crazy; merry; odd
Narzisse f (-n) narcissus
nasc. abbv = **nascierend** a nascent
Nase f (-n) 1 nose; proboscis; beak; snout. 2 spout, nozzle. 3 projection; cam, tappet, stud, lug, tab. **näseln 1** vt say through one's nose, say with a nasal twang. **2** vi sound nasal, have a nasal twang. **näselnd** p.a nasal, twangy
Nasen· nasal, naso-, rhino-: **-bein** n nasal bone. **-bluten** n nosebleed, epistaxis. **-loch** n, **-öffnung** f nostril, naris. **-rachenraum** m nasopharynx. **-(rachen)reizstoff** m nose (and throat) irritant, sternutator. **-schlacke** f (Metll.) tuyère (twyer) slag. **-schleim** m nasal mucus. **-schleimhaut** f nasal mucous membrane. **-spray** m nasal spray. **-spülwasser** n nasal rinse. **-tropfen** mpl nose drops. **-wege** mpl nasal passages
Nas·horn n rhinoceros
naß a (nässer, nässest..) wet, moist, damp: **auf nassem Wege** by wet process, the wet way
Naß·abscheider m scrubber. **-auftragen** n wet application. **-behandlung** f wet treatment. **-beize** f 1 liquid (or soak) disinfection. 2 = **-beizmittel** n liquid (or soak) disinfectant. **-betrieb** m wet operation. **-binder** m wet (or liquid) binder. **-bleiche** f wet bleach. **-dampf** m wet steam, moist vapor. **-dekatur** f (Tex.) wet decatizing

Nässe *f* wetness, moisture, dampness, humidity—**bei N.** when wet, on wet roads

naß-echt *a* wet-fast, moistureproof. **N-echtheit** *f* wet-fastness. **-element** *n* (*Elec.*) wet cell (*or* battery)

Nässe-gehalt *m* moisture content

nässen I *vt* 1 wet, moisten, soak. **II** *vi* 2 exude moisture, weep: **n-de Flechte** *f* weeping eczema. 3 wet, urinate. 4: **es näßt** it drizzles. **III: sich n.** get wet, become moist

Naß-entwicklung *f* wet developing

nässer wetter, etc: *compar of* NASS

Naß-fäule *f* wet rot. **n-fest** *a* moisture-resistant; strong when wet. **N-festigkeit** *f* moisture-resistance; wet-strength. **-fettung** *f* (*Lthr.*) wet fat-liquoring. **-gas** *n* wet (natural) gas. **-gehalt** *m* moisture content. **-gehaltmesser** *m* hygrometer. **-gießverfahren** *n* (*Plastics*) wet casting process. **-guß** *m* green mold casting. **-gußformerei** *f* green sand molding. **-kalk** *m* milk of lime. **n-kalt** *a* cold and wet, raw (weather); clammy. **-kühl** *a* chilly and damp. **-lichtecht** *a* light-fast when wet. **N-löscher** *m* water-filled fire extinguisher. **-mahlen** *n* wet grinding (*or* milling). **n-mechanisch** *a* wet-mill, wet-mechanical. **N-metallurgie** *f* hydrometallurgy. **-mühle** *f* wet mill. **-probe** *f* wet test

Näß-probe *f* moisture test (*or* determination)

Naß-reiniger *m* scrubber. **-reinigung** *f* wet purification (*or* cleaning); (*Gases*) scrubbing. **-schlamm** *m* (*Sewage*) wet sludge. **-schliff** *m* wet grinding. **-spinnverfahren** *n* (*Plastics*) wet-spinning process. **-thermometer** *n* wet-bulb thermometer. **-verfahren** *n* wet process. **n—-vermahlen** *vt* wet-grind. **N-vermahlung** *f* wet grinding. **-wäsche** *f* 1 wet washing. 2 wet wash. **n—-waschen*** *vt* wet wash, launder. **N-wäscherei** *f* wet-wash laundry

nasz. *abbv* = **naszierend** *a* nascent

Nativ-präparat *n* (*Micros.*) untreated specimen

Natracetessigester *m* sodioacetoacetic acid

Natrium *n* sodium; (*Pharm.*): **N. bromatum** sodium bromide; **N. chloratum** sodium chloride; **N. chloricum** sodium chlorate; **N. jodatum** sodium iodide. **-alaun** *m* sodium alum. **-amid** *n* sodium amide, sodamide. **-arsen(i)at** *n* sodium arsenate. **-äthyl** *n* ethylsodium. **-äthylat** *n* sodium ethylate. **-azetat** *n* sodium acetate. **-benzoat** *n* sodium benzoate. **-bikarbonat** *n* sodium bicarbonate. **-bisulfat** *n* sodium bisulfate. **-bisulfatkuchen** *m* niter cake. **-bleichlauge** *f* sodium bleach liquor (sodium hypochlorite solution). **-borat** *n* sodium borate, borax. **-brechweinstein** *m* sodium antimonyl tartrate. **-bromid** *n* sodium bromide. **-chlorid** *n* sodium chloride. **-dampf** *m* sodium vapor. **-dampflampe** *f* sodium vapor lamp. **-draht** *m* sodium wire. **-flamme** *f* sodium flame. **-fluorid** *n* sodium fluoride. **-formiat** *n* sodium formate. **-gehalt** *m* sodium content. **-goldchlorid** *n* sodium chloroaurate. **-hydrat, -hydroxid** *n* sodium hydroxide. **-hyperjodat** *n* sodium periodate. **-hyperoxid** *n* sodium peroxide. **-jodat** *n* sodium iodate. **-jodid** *n* sodium iodide. **-kaliumtartrat** *n* sodium potassium tartrate. **-karbonat** *n* sodium carbonate. **-kieselfluorid** *n* sodium fluosilicate. **-leuchte** *f*, **-licht** *n* sodium light. **-löffel** *m* sodium spoon. **-metall** *n* metallic sodium. **-methyl** *n* methylsodium. **-nitrat** *n* sodium nitrate. **-oxid** *n* sodium oxide. **-oxidhydrat** *n* sodium hydroxide. **-phosphat** *n* sodium phosphate. **-platinchlorid** *n* sodium chloroplatinate. **-presse** *f* sodium press. **-rhodanid** *n* sodium thiocyanate. **-salz** *n* sodium salt. **-seife** *f* sodium soap. **-silikofluorid** *n* sodium fluosilicate. **-sulfat** *n* sodium sulfate. **-sulfatkuchen** *m* salt cake. **-sulfhydrat** *n* sodium hydrosulfide. **-sulfozyanid** *n* sodium thiocyanate. **-superoxidbleiche** *f* sodium peroxide bleach. **-verbindung** *f* sodium compound. **-wasserstoff** *m* sodium hydride. **-wolframat** *n* sodium tungstate. **-zange** *f* sodium tongs. **-zinnchlorid** *n* sodium chlorostannate. **-zyanamid** *n* sodium cyanamide. **-zyanid** *n* sodium cyanide

Natro·kalzit n gaylussite. **-lith** m natrolite
Natron n **I** (genl.) soda; (specif, oft) sodium
oxide, sodium hydroxide, sodium carbonate, sodium bicarbonate, sodium nitrate.
II sodium = NATRIUM, now obsol exc in
old-style names such as **kohlensaures
N.** sodium carbonate. **III** (Min.) natron
($Na_2 CO_3 \cdot 10H_2O$). **-ablauge** f 1 spent
soda lye. 2 (Paper) black liquor. **-alaun** m
soda alum; (Min.) mendozite. **-ätzlauge** f
caustic soda solution. **-cellulose** f soda
cellulose; (Paper) soda pulp. **-feldspat** m
soda feldspar; (Min.) albite. **-glas** n soda
glass. **-glimmer** m soda mica, (Min.) paragonite. **n-haltig** a containing soda. **N-hydratlösung** f sodium hydroxide solution. **-hyperoxid** n sodium peroxide.
-kalkglas n soda lime glass. **-kalkrohr**
n, **-röhre** f soda lime tube. **-lauge** f soda
lye, caustic soda solution
natronlauge·echt a fast to caustic soda.
-kochecht a fast to boiling soda lye
Natron·leber f soda liver of sulfur, sodium
sulfide mixture. **-metall** n, metallic sodium. **-mikroklin** n. **-orthoklas** m
(Min.) anorthoclase. **-pastille** f (Pharm.)
sodium bicarbonate tablet. **-präparat** n
soda preparation. **-reihe** f (Petrog.) soda
series. **-salpeter** m Chile saltpeter, sodium nitrate. **-salz** n sodium salt —
prismatisches N. (Min.) thermonatrite.
-schwefelleber f soda liver of sulfur, sodium sulfide. **-see** m soda lake. **-seife** f
soda soap, hard soap. **-stoff** m (Paper)
soda pulp. **-wasserglas** n soda waterglass, sodium silicate. **-weinstein** m sodium (potassium) tartrate. **-zellstoff** m
(Paper) soda pulp. **-zellstoffabrikation** f
soda pulp manufacture
Natter f (-n) adder, viper. **--wurz(el)** f
bistort, adderwort
Natur f (-en) nature; (in compds oft) natural: **n--ähnlich** a nature-like, realistic
Naturalien pl 1 natural produce. 2 natural
history specimens. 3 material assets
Natur·benzin n natural gasoline. **-bleiche**
f sun (or grass) bleach. **n--braun** a (esp re
Wood Pulp) natural (or nature-) brown.
N-dünger m natural (or organic) fertilizer. **-erscheinung** f natural phenomenon. **-erzeugnis** n natural product.

-farbe f natural color. **n--farben** a in its
natural color; (Lthr.) fair. **N--farbfilm** m
natural-color film. **n--farbig** a in its natural color; (Lthr.) fair. **N--farbstoff** m
natural dye(stuff). **-fehler** m natural defect. **-fett** n natural fat (or grease), (specif) suint. **n--forschend** a naturalscience. **N--forscher** m naturalist; natural scientist. **-forschung** f (research in)
natural science. **-gas** n natural gas. **-gasolin** n natural gasoline. **n--gemäß** a natural, according to nature. **-genarbt** a naturally grained. **N--gerbstoff** m natural
tanning agent. **-geschichte** f natural history. **n--geschichtlich** a natural-history.
-gesetz n law of nature. **n--getreu** a true
to nature. **N--gewalt** f force of nature. **n--groß** a natural-size. **-grün** a natural
green. **N--gummi** n 1 natural gum. 2 natural rubber. **n--hart** a 1 naturally hard. 2
(Steel) self-hardening. **N--härte** f natural
hardness (or Metll. temper). **-harz** n natural resin. **-heilkunde** f nature healing.
-heilung f natural (or spontaneous) cure,
nature healing. **-historiker** m naturalist. **-holz** n natural wood. **n--identisch** a
natural, "organic" (usu food ingredients).
N--kautschuk m or n natural (India, unprocessed) rubber. **-körper** m natural
substance (matter, body). **-kraft** f natural
force (or power). **-kunde** f natural history; nature study; biology. **-latex** m natural latex. **-lehre** f general science; (elementary) physics and chemistry
natürlich a natural; innate; adv oft of
course
Natur·nässe f natural moisture. **-papier** n
stuff-colored paper; unbleached art paper. **-produkt** n natural product, produce. **-reich** n kingdom of nature. **-rostung** f natural rusting (or corrosion).
-schätze mpl natural resources. **-schutz**
m nature conservation, nature preserve
protection. **-schutzgebiet** n nature preserve (or sanctuary). **-schutzpark** m nature sanctuary, national park. **-seide** f
natural silk. **-stein** m natural (or native)
stone. **-stoff** m natural substance. **n--treu** a true to nature. **N--trieb** m instinct. **-uranreaktor** m natural uranium
reactor. **-versuch** m field trial. **n--w.** abbv

(naturwissenschaftlich) scientific. **N -wachs** n natural (or native) wax. **-wasser** n natural water. **n–widrig** a unnatural, abnormal. **N–wiss.** abbv = **-wissenschaft** f natural science (as opposed to social science and philosophy). **-wissenschaftler** m natural scientist. **n–wissenschaftlich** a (natural-) scientific, (natural-) science. **-wüchsig** a spontaneous, original. **N–zement** m natural (or Roman) cement. **-zustand** m natural state

Nauli·gummi n nauli gum

nautisch a nautical

Nd. abbv (Niederschlag) precipitate

ND abbv (Niederdruck) low pressure

Ndd. abbv (Niederschläge) precipitates

ndl. abbv (niederländisch) Dutch

Ndl. 1 (Nadel) needle. 2 (Niederlage) warehouse. 3 (Niederländisch) Dutch (language)

NE abbv (Nichteisen) nonferrous (metal)

Neapel·gelb n Naples yellow. **-rot** n Naples red

Nebel M (-) 1 fog, mist. 2 haze. 3 smoke-(screen). 4 nebula. **-apparat** m atomizer, sprayer; smoke generator. **n–artig** a foglike, foggy; mist-like, misty. **N–bildung** f fog (mist, smoke) formation. **-bläser** m crop duster. **-bombe** f smoke bomb. **-büchse** f smoke generator. **-decke** f smokescreen, smoke cover (or blanket). **-dunst** m mist, haze. **-düse** f atomizing (or spray) nozzle. **-entwicklung** f smoke generation; evolution of smoke (fog, mist). **-fleck** m nebula. **-gas** n smoke (or cloud) gas. **-gerät** n smoke generator. **-geschoß** n smoke projectile. **-gewehrgranate** f smoke rifle grenade. **-granate** f smoke shell grenade. **n–grau** a misty-gray

nebelhaft a foggy; hazy; nebulous; vague

Nebel·handgranate f smoke hand grenade

nebelig a foggy, etc = NEBLIG

Nebel·kammer f cloud chamber. **-kasten** m smoke box (or generator). **-kerze** f smoke candle. **-mittel** n smoke agent; smoke equipment. **-munition** f smoke munitions

nebeln vi 1: **es nebelt** it is foggy (or misty). 2 lay a smokescreen. 3 (aerosol-)spray (with pesticides, etc)

Nebel·patrone f smoke cartridge. **-satz** m fumigator. **-säure** f fuming sulfuric acid. **-scheinwerfer** m fog (head)light. **-schießen** n smoke-shell firing. **-schleier** m smokescreen. **-spur** f 1 cloud chamber track. 2 trace of mist (or fog). **-spuraufnahme** f cloud chamber photograph. **-stoff** m smokescreening agent. **-topf** m smoke pot. **-tornister** m backpack smoke apparatus. **-trommel** f smoke drum. **-tröpfchen** n fog (or mist) droplet. **-wand** f 1 wall of fog. 2 smokescreen. **-werfer** m smoke-shell mortar. **-wolke** f smoke cloud. **-wurfgranate** f mortar smoke shell. **-zerstäuber** m smoke sprayer. **-zerstreuung** f fog dispersion. **-zone** f smokescreen zone. **-zündmittel** n firing device for smoke container

neben I prep (dat or accus) 1 next to, next-door to, beside, alongside (of). 2 in addition to, aside from. 3 (dat) along with, accompanied by: **n. ihnen her** (moving) along with (or beside) them. 4 compared to. II pfx 5 adjacent. 6 additional. 7 side, secondary, minor, auxiliary. 8 (Elec.) shunt.: **N–achse** f 1 (Cryst.) secondary axis. 2 (Math.) minor (or conjugate) axis. **-ader** f collateral blood vessel. **n–an** adv nextdoor, close by. **N–anlage** f branch plant. **-anschluß** m extension (telephone). **-antrieb** m auxiliary drive. **-apparat** m auxiliary apparatus, accessory. **-arbeit** f 1 extra (or spare-time) work (or job); sideline. 2 secondary operation. **-bau** m adjacent building. **-bedingung** f supplementary condition (or factor). **n–bei** adv 1 by the way, incidentally. 2 in addition; on the side. **N–bestandteil** m minor constituent, secondary ingredient. **-betrieb** m 1 secondary operation. 2 subsidiary establishment. 3 neighboring establishment. **-beweis** m additional (or accessory) proof. **-bindung** f secondary bond. **-blatt** n (Bot.) bract. **-blättchen** n (Bot.) stipule. **-buhler** m rival. **-drüse** f accessory gland. **-eigenschaft** f secondary property

nebeneinander I *adv* **1** side by side, next to each other. **2** together. **3** simultaneously, concurrently.—**n. bestehen** coexist. **II** *sep pfx* side by side (*also implying comparison*), (*Elec.*) in parallel. **Nebeneinander** *n* **1** juxtaposition. **2** existence. **3** simultaneousness. **n—geschaltet** *p.a* (in) parallel; synchronized *cf* N–SCHALTEN. **-gestellt** *p.a* side by side *cf* N–STELLEN. **N–leben** *n* coexistence

nebeneinander·legen *vt* put (*or* lay) side by side; juxtapose. **·liegen*** *vi* (h *or* s) lie (*or* be located) side by side, be contiguous. **-liegend** *p.a* contiguous, adjacent. **·schalten** *vt* connect in parallel *cf* N–GESCHALTET. **·stehen*** *vi* (h *or* s) stand side by side. **-stehend** *p.a* adjacent. **·stellen** *vt* **1** place side by side, juxtapose. **2** compare. *Cf* N–GESTELLT. **N–stellung** *f* juxtaposition; comparison

Neben·einteilung *f* subdivision. **-ergebnis** *n* secondary (*or* incidental) result. **-erscheinung** *f* attendant phenomenon. **-erzeugnis** *n* by-product. **-erzeugnisgewinnung** *f* by-product recovery. **-fach** *n* minor (subject of study). **-farbe** *f* secondary color. **-fläche** *f* (*Cryst.*) secondary face. **-fluß** *m* tributary. **-folge** *f* indirect result. **-gang** *m* byway, side road (*or* passage); (*Min.*) lateral vein. **-gärung** *f* secondary fermentation. **-gemengteil** *m* subsidiary part of a mixture. **-geräusch** *n* (*Elec.*) background noise, static. **-geschmack** *m* tang, aftertaste. **-gestein** *n* (*Min.*) country (*or* native) rock, gangue. **-gewebe** *n* (*Biol.*) accessory tissue. **-gewinnung** *f* by-product recovery. **-gruppe** *f* subgroup. **n–her I** *adv* **1** incidentally, by the way. **2** in addition, besides. **II** *sep pfx* (*with verbs of motion*) along with, alongside (of). **-hergehend** *p.a* incidental; secondary, minor. **-hin** *adv* casually, incidentally. **N–hoden** *m* (*Anat.*) epididymis. **-kern** *m* (*Biol.*) paranucleus. **-kette** *f* side chain. **-klasse** *f* subclass: (*Math.*) coset. **-kosten** *pl* extra (*or* incidental) expenses, extras. **-leiter** *m* (*Elec.*) branch conductor. **-leitung** *f* (*usu Elec.*) bypass, branch line. **-linie** *f* branch line, secondary line. **-luft** *f* admixed (*or*

secondary) air. **-magen** *m* auxiliary stomach. **-niere** *f* adrenal gland. **-nierenrinde** *f* adrenal cortex

neben·ordnen *vt* coordinate

Neben·organ *n* (*Anat.*) secondary organ. **-produkt** *n* by-product.

Nebenprodukten·anlage *f* by-product plant. **-gewinnung** *f* by-product recovery. **-kokerei** *f* by-product coking (*or* coke plant). **-ofen** *m* by-product oven

Neben·quantenzahl *f* secondary (*or* azimuthal) quantum number. **-raum** *m* (smaller) adjoining room; side room. **-reaktion** *f* side (*or* secondary) reaction. **-rechnung** *f* auxiliary calculation. **-reihe** *f* secondary (*or* subsidiary) series. **-rohr** *n*, **-röhre** *f* side (*or* branch) tube (*or* pipe). **-rolle** *f* minor role (*or* part). **-sache** *f* secondary (*or* minor) matter, nonessential. **n–sächlich** *a* nonessential, minor, secondary; *adv oft* casually, incidentally; **das N—e = N–sächlichkeit** *f* **1** (*pl:* -en) incidentals, minor matters. **2** (*sg*) unimportance, triviality

neben·schalten *vt* (*Elec.*) connect in parallel

Neben·schilddrüse *f* parathyroid gland

neben·schließen* *vt* (*Elec.*) shunt. **N–schluß** *m* (*Elec., Med.*) shunt(ing)

Nebenschluß· shunt: **-schaltung** *f* shunt connection. **-strom** *m* shunt current. **-widerstand** *m* shunt resistance

Neben·serie *f* **1** secondary (*or* subsidiary) series. **2** (*Spect.*): **erste N.** diffuse series; **zweite N.** sharp series. **-setzung** *f* juxtaposition. **-sonne** *f* mock sun, parhelion. **-spirale** *f* secondary coil. **n–ständig** *a* collateral, accessory. **-stehend** *a* accompanying (text, illustration); marginal (note); *adv oft* in the margin, on the facing page: **n. abgebildet** (shown) in the accompanying illustration. **N–strom** *m* **1** (*Elec.*) induction current; *in compds oft* bypass, shunt. **2** tributary; feeder. **-teil** *m* accessory part. **-titel** *m* subtitle. **-tür** *f* side door. **-typus** *m* secondary type. **-umstand** *m* minor detail, incidental circumstance. **-valenz** *f* secondary (*or* Werner) valence. **-widerstand** *m* shunt (*or* parallel) resistance. **-winkel** *m* secondary angle. **-wirkung** f side-effect. **-zeit** f unproductive (or leisure) time. **-zelle** f secondary cell. **-zweck** m secondary (or incidental) purpose. **-zweig** m lateral branch; sideline

neblig a foggy, misty; hazy; nebulous cf NEBEL

nebst prep (dat) together (or along) with; beside, in addition to

Necessaire n (-s) case; kit

Necin·säure f necinic acid

Negativ·bild n negative (image). **n–elektrisch** a electronegative. **N–farbauszug** f (Phot.) separation negative. **Negativität** f negativity. **Negativ·lack** m (Phot.) varnish for negatives

Neger m (-) black man, Negro. **Negerin** f (-nen) black woman

negieren vt negate

nehmen* (nahm, genommen; nimmt) vt 1 (genl) take; (+ **für**) take (for); (+ d.i.o, oft) (a) take (away) (from), (b) free (from), relieve (of): **zu sich n.** take (e.g. medicine), eat, drink; **sich etwas n.** help oneself to sthg. 2 accept. 3 handle, deal with. 4 oft in idioms with noun: transl depends on noun; cf ANGRIFF, ANSPRUCH, PLATZ etc

neigbar a inclinable cf NEIGEN

Neige f (-n) 1 incline, tilt: **auf der N.** tilted. 2 decline, close, end; **zur** (or **auf die**) **N. gehen** come to an end, draw to a close. 3 (sg & pl) lees, dregs, last drop(s): **bis zur N.** to the last drop. **neigen** I vt & sich n. 1 (genl) incline. 2 bend; bow; nod. 3 tilt, tip; deflect (as: needle). II vi (usu + **zu**) 4 be inclined, lean (to, toward). 5 be susceptible (to). 6 be fond (of). III: **sich n.** 7 slope (down), drop. 8 run down, come to (its) end. Cf GENEIGT. **Neigung** f (-en) (oft + **zu**) I (genl) inclination (to, toward). II (specif) 1 bending; nod; bow. 2 slope, grade. 3 tendency, disposition (to, toward). 4 fondness (for). 5 (Engg.) (screw) pitch. 6 (Geol.) dip

Neigungs·ebene f inclined plane. **-messer** m (in)clinometer, gradiometer. **-winkel** m angle of inclination

nein intj no. **Nein** n no, nay, negative (answer)

Nekrolog m (-e) obituary

Nekrose f (-n) necrosis

Nelke f (-n) 1 carnation. 2 clove

Nelken·blätteröl n clove leaf oil. **-gewächse** npl Caryophyllaceae. **-kassie** f clove cassia (or cinnamon). **-knospenöl** n clove bud oil. **-öl** n oil of cloves **-pfeffer** m allspice, pimento. **-pfefferwasser** n (Pharm.) pimento water. **-rinde** f clove cassia. **-säure** f eugenol, caryophyllic acid. **-stein** m (Min.) iolite. **-stielöl** n clove stem oil. **-wurzel** f avens (or clove) root. **-wurz(el)öl** n clove root oil. **-zimt** m clove cinnamon (or cassia)

NE-Metall n nonferrous metal cf NE

Nenn· nominal, rated: **-betrag** m face value. **-drehzahl** f rated rpm (or speed). **-druck** m nominal pressure

nennen* (nannte, genannt; sbjc nennte) 1 vt call, name; mention. 2: **sich n.** (oft) be called, be named. **nennens·wert** a worth mentioning, noteworthy; appreciable. **Nenner** m (-) denominator: **auf einen N. bringen** reduce to a common denominator

Nenn· nominal, rated: **-gewicht** n nominal weight. **-kraft** f rated (horse)power. **-last** f rated (or nominal) load. **-leistung** f rated output. **-spannung** f rated stress; (Elec.) rated (or nominal) voltage. **-strom** m (Elec.) rated current

nennte would name, etc: sbjc of NENNEN

Nennung f (-en) naming, calling; mention(ing)

Nenn·weite f nominal width. **-wert** m face (par, nominal) value. **-wort** n noun

Neodym n neodymium

Neon·(glimm)lampe f, **-licht** n neon lamp (or light). **-rohr** n, **-röhre** f neon tube. **-verflüssigungsmaschine** f neon liquefier

Neosilikat n (-e) orthosilicate

neozoisch a (Geol.) Neozoic, Cenozoic

Nephelin m (-e) (Min.) nepheline, nephelite

Nephrit m (-e) jade

Nernst·faden m Nernst filament. **-stift** m Nernst element

Neroli-öl n Neroli (or orange-blossom) oil

Nerv m (en) 1 nerve. 2 vein, rib (of leaves, insect wings). 3 (pl) nerve (= courage, self-control). **Nervatur** f (-en) venation

Nerven· nerve, nervous, neur(o)-: **-entzün-**

dung *f* neuritis. **-faser** *f* nerve fiber. **-gas** *n* nerve gas. **-gewebe** *n* nerve tissue. **-kitt** *m* (*Anat.*) neuroglia. **-knoten** *m* ganglion. **n–krank** *a* neurotic. **N–krieg** *m* war of nerves. **-kunde, -lehre** *f* neurology. **-masse** *f* nerve (*or* neural) substance. **-reiz** *m* nervous stimulus. **-stoff** *m* nerve (*or* neural) substance. **-system, -werk** *n* nervous system. **-zelle** *f* nerve cell. **-zusammenbruch** *m* nervous breakdown

nervig *a* 1 sinewy, strong. 2 irksome. 3 (*Bot.*) ribbed, veined. **nervlich** *a* nervous (strain, etc); neural

Nervon·säure *f* nervonic acid

nervös *a* nervous. **Nervosität** *f* nervousness

Nerv·wert *m* (*Rubber*) degree of snap (*or* recovery)

Nerz *m* (-e) mink

Nessel I *f* (-n) nettle. II *m* (*Tex.*) unbleached cotton cloth; cheesecloth. **--ausschlag** *m,* **-fieber** *n* nettle rash, hives. **-gewächse** *npl* nettles, Urticaceae. **-kapsel** *f* nematocyst. **-tiere** *npl* (*Zool.*) cnidaria, nematophores

Nest *n* (-er) 1 (*usu*) nest. 2 bed. 3 hideout. 4 (*Ores*) pocket. 5 (*Tex.*) knot, tangle

Netor·säure *f* netoric acid

nett *a* 1 neat, tidy. 2 nice, pleasant; kind. 3 (*Sugar*) pure white

netto *adv & pfx* (*Com.*) net: **n. zehn Kilo** 10 kilos net (weight). **N--ertrag** *m* net yield. **-gewicht** *n* net weight

Netz *n* (-e) 1 (*usu*) net. 2 gauze. 3 open mesh bag. 4 web. 5 (*Opt. Instrum.*) reticle. 6 (*Med.*) reticulum; plexus; omentum. 7 grid, lattice. 8 network, system (*of transp., power, supply lines*); (*Elec.*) (main) power line(s)

Netz· 1 net, reticular, omental, etc; *refers to* NETZ. 2 wetting, steeping; *refers to* NETZEN: **-adern** *fpl* netted veins (*or* cracks), check(er)ing. **n–ähnlich** *a* netlike, reticular; plexiform; grid-like. **N–anschluß** *m* (*Elec.*) connection (*or* outlet) to the main power line. **n–artig** *a* netlike, etc = ·ÄHNLICH. **N–bad** *n* (*Dye.*) wetting(-out) bath. **-beize** *f* (*Turkey red Dyeing*) oil mordant. **-bindung** *f* covalence; reticular linkage. **-ebene** *f*

(*Cryst.*) lattice plane. **-ebenenabstand** *m* (*Cryst.*) interplanar spacing. **-elektrode** *f* wire gauze electrode, grid. **-empfänger** *m* (*Radio*) network receiver

netzen *vt* wet, moisten; steep

Netz·fähigkeit *f* wetting property. **-fasergewebe** *n* reticulated tissue. **-faß** *n* steeping vat (*or* tub). **-ferment** *n* omental enzyme. **-flotte** *f* (*Dye.*) wetting-out liquor. **n–förmig** *a* net-shaped, reticular. **N–haut** *f* retina

Netzhaut· retinal: **-bild** *n* retinal image. **-eindruck** *m* retinal impression. **-stäbchen** *n* retinal rod. **-zäpfchen** *n* retinal cone

Netz· net, reticular; wetting, steeping: **-kasten** *m* immersion tank. **-katalysator** *m* grid catalyst. **-kessel** *m* steeping vat. **-kraft** *f* wetting power. **-magen** *m* (*Zool.*) second stomach. **-mittel** *n* wetting agent. **-öl** *n* wetting oil. **-plan** *m,* **-plantechnik** *f* project planning and control. **-polymer** *n* cross-linked (*or* reticular) polymer; **räumliches N.** cross-linked high polymer. **-punkt** *m* 1 junction (in a network). 2 (*Polymers*) cross link. **-schalter** *m* (*Elec.*) power line switch. **-schaltung** *f* (*Elec.*) delta connection. **-schnur** *f* 1 (*Elec.*) plug-in cord. 2 fishnet cord. **-schutz** *m* protective gauze (*or* grid). **-schwefel** *m* wettable sulfur. **-silikat** *n* layer (*or* sheet) silicate. **-spannung** *f* line voltage. **-ständer** *m* (*Paper*) steeping tub. **-störung** *f* (*Elec.*) power failure. **-strom** *m* line current. **-struktur** *f* reticular structure. **-tuch** *n* open-mesh fabric

Netzung *f* moistening, wetting; steeping *cf* NETZEN

netz· net, etc; wetting: **-vermascht** *a* interconnected. **N–vermögen** *n* wetting power. **-ware** *f* netting. **-wärme** *f* heat of wetting. **-werk** *n* 1 network. 2 netting, meshwork

netzwerk·artig *a* network-like, reticular. **N–struktur** *f* (*esp Metll.*) reticular (*or* cellular) structure. **-wandler** *m* (*Cryst.*) network modifier

Netz·wirkung *f* netting action (*or* effect)

neu I *a* 1 new: **wie n.** like (*or* as good as) new. 2 fresh. 3 modern. 4 recent. *Cf* NEUER, NEUESTE. II (*as noun*): **N—es** *n* new

ER, NEUESTE. **II** (*as noun*): **N—es** *n* new things, news: **nichts N—es** nothing new; **das Alte und das N—e** the old and the new. **III** *adv oft* **5** *also in phr.*: **aufs n—e, von n—em** anew, afresh; **seit n—em** since recently. **6** (+ *verb, oft*) re-: **n. geprüft** retested, re-examined *cf* NEU-GEPRÜFT. **IV** *pfx* **7** new, neo-. **8** (*esp* + *pp*) newly, recently *cf* NEUGEPRÜFT. **9** (+ *noun derived from verb, oft*) re-: *cf* NEU-PRÜFUNG. **10** (*esp* + *language*) modern *cf* NEUHOCHDEUTSCH

Neu·aktivierung *f* reactivation. **n–artig** *a* new kind of, new-style, novel. **-aufgelegt** *a* republished; newly edited (*or* published). **N–auflage** *f* new edition. **n–aufgestellt** *a* newly set up. **N–aufstellung** *f* new set-up; reorganization. **-bau** *m* new structure; reconstruction. **n–bearbeitet** *a* (newly *or* recently) revised. **N–bearbeitung** *f* revision. **-berechnung** *f* recalculation. **-bildung** *f* **1** new formation (compound, phase, etc). **2** reconstruction, re-formation. **3** (*Med.*) neoplasm. **-blau** *n* new blue (starch-indigo mixture). **-braun** *n* new brown (copper prussiate). **-coccin** *n* (*Dyes*) new coccin, neococcin. **-druck** *m* new printing; reprint. **-eichung** *f* recalibration, restandardization. **-einrichtung** *f* **1** new installation. **2** new institution, new arrangement. **3** refurnishing. **n–entstanden** *a* newly arisen (*or* formed), recent

neuer *a* (*compar of* NEU) **1** newer; fresher, more recent. **2** fairly new (fresh, recent): **in n—er Zeit** in (fairly) recent times; **seit n—em** since fairly recently. **3** modern: **n—e Sprachen** modern languages. **neuerdings** *adv* **1** recently, lately. **2** again, anew

Neuerer *m* (-) innovator

neuerlich *a* **1** recent; *adv oft* lately. **2** renewed, repeated; *adv oft* again

Neuerung *f* (-en) innovation

Neu·erzeugung *f* **1** new production. **2** reproduction

neueste 1 *a* (*superl of* NEU) newest; latest, most recent. **2** (*as noun*): **das N—** the latest (thing, news, etc). **neuestens** *adv* (quite *or* most) recently

Neufundland *n* (*Geog.*) Newfoundland

neu·gebacken *a* **1** freshly baked, ovenfresh. **2** brand-new. **-gebildet** *a* newly formed. **-geboren** *a* newborn. **N–gelb** *n* new yellow; massicot. **n–geprüft** *a* newly tested. **N–gestaltung** *f* reshaping, reformation; reorganization

Neugier(de) *f* curiosity, inquisitiveness. **neugierig** *a* curious, inquisitive—**wir sind n. auf . . .** we wonder about . . .; **wir sind n., ob . . .** we wonder whether . . .

Neu·gold *n* Mannheim gold. **-grad** *m* (*Math.*) new (*or* centesimal) degree (1/400th of a circle). **-gradteilung** *f* centesimal graduation. **-grün** *n* new (*or* Malachite) green

Neuheit *f* (-en) **1** newness, freshness. **2** novelty. **neuheits·schädlich** *a* (*Patents*) anticipatory

Neuhochdeutsch *n* Modern (High) German

Neuigkeit *f* (-en) piece of news, news item; (*esp pl*) news

Neu·jahr *n* New Year('s Day). **-kristallisation** *f* recrystallization

neulich *adv* recently, lately

Neuling *m* (-e) **1** novice. **2** newcomer

Neu·messing *n* new brass. **-metall** *n* virgin metal. **n–modisch** *a* newfangled

neun *num* nine. **Neun** *f* (-en) (figure) nine. **n–eckig** *a* nonagonal, nine-cornered

neunerlei *a* (*no endgs*) nine kinds of, of nine kinds

neun·fach, -fältig *a* ninefold. **-flächig** *a* nine-sided (*or* -faced). **-gliedrig** *a* nine-membered. **-mal** *adv* nine times. **-malig** *a* nine-time

neunt *a* ninth. **Neuntel** *n* (-) (*Frac.*) ninth. **neuntens** *adv* in the ninth place, as the ninth item

neun·wertig *a* nonavalent, nine-valent

neunzehn *num* nineteen. **neunzehnte** *a* nineteenth

neunzig *num* ninety. **neunzigste** *a* ninetieth

Neu·ordnung *f* reorganization. **-prüfung** *f* **1** new test (*or* examination). **2** retesting, re-examination

Neurer *m* (-) innovator = NEUERER

Neurose *f* (-n) neurosis

Neu·rot *n* new red. **-rotölung** *f* (*Dye.*) new

cess. **-schöpfung** *f* 1 new creation. 2 re-creation. **-seeland** *n* (*Geog.*) New Zealand. **-silber** *n* nickel (*or* German) silver. **-sprache** *f* modern language

neuste *a* newest, etc = NEUESTE

Neu-steinzeit *f* New Stone Age, Neolithic

neutr. *abbv* (neutral, neutralisiert) neut. (neutral, neutralized)

neutral-färben *vt* dye neutral (*or* sweet)

Neutral-fett *n* neutral fat. **-grau** *n* neutral gray

Neutralisat *n* (-e) neutralized material; material to be neutralized

Neutralisations-wärme *f* heat of neutralization. **-zahl** *f* neutralization number

neutralisieren *vt/i* neutralize. **Neutralisierung** *f* (-en) neutralization

Neutralisierungs-wärme *f* heat of neutralization

Neutralität *f* neutrality

Neutral-körper *m* neutral body (*or* substance). **-öl** *n* neutral oil. **-punkt** *m* neutral point. **-rot** *n* neutral red. **-salzion** *n* neutral salt ion. **-salzwirkung** *f* neutral salt effect

Neutronen-beschuß *m* neutron bombardment. **-beugung** *f* neutron diffraction. **-einfang** *m* neutron capture. **-fluß** *m* neutron flux. **-welle** *f* neutron wave

Neu-weiß *n* permanent white (barium sulfate)

Neuwied-blau *n*, **Neuwieder Blau** Neuwied Blue (copper hydroxide carbonate pigment). **Neuwieder Grün, Neuwiedgrün** *n* Neuwied (*or* Paris) green (copper arsenite pigment)

Neuyork *n* (*Geog.*) New York

Neu-zeit *f* modern times. **neuzeitlich** *a* modern, present-day

NF *abbv* 1 (neue Folge) new series (of a publication). 2 (neue Fassung) new version. 3 (Niederfrequenz) low freq. **n.F.** *abbv* (neue Folge) = NF 1

Nichin *n* niquine

nicht I *adv* 1 not: **wenn** (*or* **wo**) **n.** if not; **n. daß** . . . not that . . . 2 (+ *compar oft*) no: **n. größer als** no bigger than. 3: **n. mehr** no longer, not . . . any longer (*or* any more): **das gilt n. mehr** that no longer applies, that does not apply any more. 4 (*with implied verb*) do (*or* did) not: **alle andern zerschmolzen, nur dieses n.** all the others melted, only this one did not. 5 (*not transl*) e.g.: **und was sie n. alles behaupten** and all the things (*or* and who knows what else) they claim. II *pfx* non-, un-, in-, etc: **-abbaubar** *a* nondegradable. **-absetzend** *a* nonsettling. **N-achtung** *f* disregard; disrespect. **n-amtlich** *a* unofficial. **-angreifbar** *a* noncorrodible. **-angreifend** *a* noncorrosive. **-anteilig** *a* (*Electrons*) unshared. **N-auflösung** *f* nondissolution. **n-aushaltend** *a* non-lasting, non-enduring; discontinuous. **-backend** *a* (*Coal*) noncaking, non-coking. **N-beachtung** *f* disregard; nonobservance, noncompliance. **n-beansprucht** *a* 1 unclaimed. 2 not loaded, unstressed. 3 (*Electron.*) inert (pair). **N-beobachtung** *f* nonobservance. **-berücksichtigung** *f* neglect, failure to consider. **n-bevorzugt** *a* nonpreferential, random. **-bindig** *a* noncohesive, loose (soil). **-brennbar** *a* noncombustible. **N-chemiker** *m* non-chemist. **n-daltonid, -daltonisch** *a* non-Daltonian, nonstoichiometric. **-dissoziiert** *a* non-dissociated. **-drehend** *a* nonturning, nonrotating

Nichte *f* (-n) niece

nicht- non-, un-, in-, dis- etc: **-eben** *a* nonplanar. **N-eignung** *f* unsuitability. **n-eisenhaltig** *a* nonferrous **N-eisenmetall** *n* nonferrous metal. **-elektrolyt** *m* nonelectrolyte. **n-entartet** *a* nondegenerate. **N-entartung** *f* nondegeneration. **-erscheinung** *f* nonappearance. **-erz** *n* non-ore deposit. **n-euklidisch** *a* non-Euclidian. **n-existenzfähig** *a* nonviable, incapable of existence. **-explosiv** *a* nonexplosive. **N-fachleute** *pl of* **-fachmann** *m*, **-fachmännin** *f* nonprofessional, layman. **n-faserig** *a* nonfibrous. **-flüchtig** *a* nonvolatile. **-fortlaufend** *a* discontinuous. **-ganzzahlig** *a* (*Math.*) nonintegral. **-gasförmig** *a* nongaseous. **N-gebrauch** *m* disuse, non-use; **bei N.** when not in use. **n-geordnet** *a* unorganized. **N-gerbstoff** *m* nontan(nin). **n-gleichmäßig** *a* nonuniform, nonhomogeneous. **-haltbar** *a* unstable, non-

durable, nonfast. **-häufig** *a* infrequent
nichtig *a* 1 trivial, insignificant; empty. 2
invalid, void *cf* NULL. **Nichtigkeit** *f* (-en)
triviality, insignificance; invalidity,
nullity
Nichtigkeits-erklärung *f* nullification,
annulment. **-klage** *f* (*Patents*) invalidity
plea (*or* suit)
Nicht- non-, un-, in-, dis-, etc: **-ion** *n* non-
ion. **N–ionen-** (*in compds oft*) non-ionic.
n–ionenbildend *a* nonionogenic, non-
ionizing. **-ionisierbar** *a* non-ionizable.
-ionisiert *a* non-ionized. **-ionogen** *a* nonionogenic, non-ionizing. **-kalkhaltig** *a*
noncalcareous, containing no calcium.
N–kämpfer *m* noncombatant. **-kar-
bonathärte** *f* (*Water*) permanent (*or* non-
carbonate) hardness. **n–klebend** *a* non-
adhesive. **-kohärent** *a* noncoherent.
-kondensiert *a* uncondensed. **-kon-
tinuierlich** *a* discontinous. **-kristalli-
nisch** *a* noncrystalline. **-lebend** *a* nonliv-
ing. **-leitend** *a* nonconducting. **N–leiter**
m nonconductor. **n–leuchtend** *a* nonluminous. **-linear** *a* nonlinear. **-lokali-
siert** *a* unlocalized. **-lohnend** *a* unre-
munerative, unrewarding. **N–löser** *m*
nonsolvent. **n–löslich** *a* insoluble. **N–
metall** *n* nonmetal. **n–metallisch** *a* nonmetallic. **-mischbar** *a* immiscible. **N–
mischbarkeit** *f* immiscibility. **n–normal** *a* abnormal. **-oxidierbar** *a* nonox-
idizable. **-oxidierend** *a* nonoxidizing.
-pathogen *a* nonpathogenic. **-perio-
disch** *a* aperiodic. **-polar** *a* nonpolar;
n—e Bindung homopolar (*or* covalent)
bond. **-plattiert** *a* unplated, unclad. **N–
raucher** *m* non-smoker; non-smoking
car (compartment, section). **n–
reduzierbar** *a* irreducible. **-rela-
tivistisch** *a* nonrelativistic. **-re-
produzierbar** *a* nonreproducible. **-ro-
stend** *a* nonrusting, rustless, stainless.
-rußend *a* sootless
nichts *indef pron* 1 nothing, not . . . anything: **sie ahnten n.** they did not suspect
anything (*or* suspect a thing); **n. als** nothing but; **n. mehr** nothing (*or* no) more,
not . . . any more; **n. Neues** nothing new;
n. anderes nothing else; **und weiter n.**

and nothing more; **n. weniger als** anything but (!). 2 no, not . . . any: **es
schadet n.** it does no harm; **es nutzt n.** it
is of no use; **sich in n. unterscheiden**
differ in no way, not differ in any way *cf*
ANGEHEN, MACHEN, NUTZEN. 3: **mir n. dir
n.** casually, for no apparent reason. *Cf*
SAGEN, WERT. **Nichts** *n* 1 nothingness,
mere nothing, nonentity. 2 void, thin air.
3 ruin
Nicht-sättigung *f* nonsaturation. **n–
schmelzbar** *a* nonfusible
nichts-destoweniger *adv-cjnc* nevertheless
nicht-selbständig *a* dependent
nichts-sagend *a* vacuous; meaningless;
noncommittal *cf* SAGEN. **-würdig** *a*
worthless; contemptible
nicht- non-, un-, a-, dis- etc: **-starr** *a* nonrigid, deformable. **-stöchiometrisch** *a*
nonstoichiometric. **-substituiert** *a* non-
substituted. **-symmetrisch** *a* asymmetrical, unsymmetrical. **-trocknend** *a* non-
drying. **N–übereinstimmung** *f* non-
agreement, noncorrespondence. **n–umkehrbar** *a* irreversible. **-umwandelbar**
a nonconvertible. **-vergasbar** *a* nongasi-
fiable, nonvolatile. **-verseifbar** *a* non-
saponifiable. **N–vorhandensein** *n* absence, nonexistence. **-wärmeleiter** *m*
nonconductor of heat. **n–wässerig** *a*
nonaqueous. **-zerfallbar** *a* nondisin-
tegrating, nonfissionable. **-zerstörend** *a*
nondestructive. **N–zucker** *m* nonsugar
Nickel-antimonglanz *m* (*Min.*) ullmannite. **n–arm** *a* low-nickel, poor in nickel
(content). **N–armierung** *f* nickel fitting
(*or* mounting). **-arsenglanz** *m* (*Min.*)
gersdorffite. **-arsenik** *n* nickel arsenide.
-arsenikglanz, -arsen(ik)kies *m* (*Min.*)
gersdorffite. **n–artig** *a* nickel-like. **N–
azetat** *n* nickel acetate. **-bad** *n* nickel
bath. **-beize** *f* nickel mordant. **-blech** *n*
nickel sheet (*or* plate). **-blüte** *f* nickel
bloom, (*Min.*) annabergite. **-bromür** *n*
nickelous bromide. **-chlorür** *n* nickelous
chloride. **-chromstahl** *m* chrome-nickel
steel. **-cyanür** *n* nickelous cyanide.
-drahtnetz *n* nickel wire gauze. **-eisen** *n*
ferronickel. **-erz** *n* nickel ore. **-fahlerz** *n*

nickelian tetrahedrite, (*Min.*) malinowskite, studerite. **-flußeisen** *n* nickel steel. **n–führend** *a* nickel-bearing. **N–gefäß** *n* nickel vessel. **-gehalt** *m* nickel content. **-gelb** *n* nickel yellow. **-gewinnung** *f* nickel recovery. **-glanz** *m* nickel glance, (*Min.*) gersdorffite. **-granalien** *pl* granulated nickel; nickel shot. **n–haltig** *a* nickel-bearing, nickeliferous. **N–hydroxid** *n* nickel hydroxide. **-hydroxydul** *n* nickelous hydroxide

Nickeli· nickeli-, nickelic: **-cyanwasserstoffsäure** *f* nickelicyanic acid

Nickelin *n* 1 (*Min.*) niccolite. 2 (*Metll.*) nickeline, copper-nickel-zinc alloy (for resistance wire)

Nickeli·oxid *n* nickelic oxide. **-verbindung** *f* nickelic compound

Nickel·jodid *n* nickel iodide. **-kies** *m* (*Min.*) millerite. **-kohlenoxid** *n* nickel carbonyl. **-kühler** *m* nickel cooler (or condenser). **-lauge** *f* (*Elec.*) nickel electrolyte. **-legierung** *f* nickel alloy. **-metall** *n* metallic nickel. **-münze** *f* nickel coin. **-niederschlag** *m* nickel precipitate. **-ocker** *m* nickel ocher, (*Min.*) annabergite

Nickelo· nickelo-, nickelous: **-cyanwasserstoffsäure** *f* nickelocyanic acid. **-verbindung** *f* nickelous compound

Nickel·oxid *n* nickel(ic) oxide. **-oxidsalz** *n* nickelic salt. **-oxydul** *n* nickelous oxide

Nickeloxydul·hydrat *n* nickelous hydroxide. **-salz** *n* nickelous salt. **-verbindung** *f* nickelous compound

Nickel·papier *n* nickel foil. **n–plattiert** *a* nickel-plated. **-sammler** *m* nickel-alkali cell (or storage battery). **-schwamm** *m* nickel sponge. **-smaragd** *m* emerald nickel, (*Min.*) zaratite. **-speise** *f* nickel speiss, (*Min.*) maucherite

Nickelspieß·glanz *m*, **-glanzerz** *n*, **-glaserz** *n* (*Min.*) ullmannite

Nickel·stahl *m* nickel steel. **-stein** *m* (*Metll.*) nickel matte. **-sulfür** *n* nickelous sulfide. **-tiegel** *m* nickel crucible. **-überzug** *m* nickel plating. **-vitriol** *n* nickel sulfate. **-zusatz** *m* addition of nickel. **-zyanid** *n* nickel(ic) cyanide. **-zyanür** *n* nickelous cyanide

nicken *vi* 1 nod. 2 nap. **Nicker** *m* (-) 1 nod. 2 nap.

Nicolsche(s) Prisma Nicol prism

nie *adv* never: **noch n.** never yet (or before); **n. mehr** (or **wieder**) never again; **fast n.** hardly ever

nieder I *a* low, lower; base, inferior; minor. **II** *adv* & *sep pfx* down: **auf und n.** up and down

Niederbewegung° *f* downward motion

nieder·blasen* *vt* blow down; blow out (a furnace)

nieder·brechen* *vt*, *vi*(s) break down

Niederdruck° *m* low pressure. **--dampf** *m* low-pressure steam

nieder·drücken *vt* press down; depress; weigh down

Niederdruck·kessel *m* low-pressure boiler. **-polyäthylen** *n* high-density polyethylene. **-schlauch** *m* low-pressure tubing (or hose)

nieder·fallen* *vi*(s) fall down

niederfrequent *a* low-frequency. **Niederfrequenz°** *f* low frequency. **--strom** *m* low-frequency current

niedergegangen gone down, etc: *pp of* NIEDERGEHEN

Niedergang *m* 1 (down)fall, descent (*incl Metll.*), drop. 2 decline. 3 (*Aero.*) landing; splashdown. 4 (*Engg.*) downstroke, downtravel. 5 setting, sunset

niedergebrochen broken down: *pp of* NIEDERBRECHEN

nieder·gehen* *vi*(s) 1 go down, descend, drop. 2 decline. 3 fall, come down (*as: rain*). 4 (*Aero.*) land; splash down. 5 (*Astron.*) set

niedergerissen torn down, etc: *pp of* NIEDERREISSEN

niedergeschlagen 1 *p.a* depressed, despondent. 2 precipitated, etc: *pp of* NIEDERSCHLAGEN. **Niedergeschlagenheit** *f* depression, despondency

niedergeschmolzen melted down: *pp of* NIEDERSCHMELZEN

niedergeschrieben written down: *pp of* NIEDERSCHREIBEN

niedergesunken 1 *p.a* sunken. 2 sunk down: *pp of* NIEDERSINKEN

nieder·halten* *vt* 1 hold down. 2 contain,

keep under control. **3** suppress

Niederholz° *n* undergrowth, underbrush

nieder·kommen* *vi*(s) (*oft* + **mit**) give birth (to). **Niederkunft** *f* (-e) childbirth, delivery

Niederlage° *f* **1** branch (office, store). **2** warehouse, depot. **3** defeat

Niederlande *npl* (*Geog*) Netherlands. **niederländisch** *a* Dutch

nieder·lassen* I *vt* **1** let down, lower. **2** drop. **II: sich n. 3** lower (*or* let oneself) down, drop. **4** sit down. **5** settle (down); establish oneself, take up residence. **Niederlassung°** *f* **1** settlement, colony. **2** settling, establishment of residence (of an office, a business). **3** branch (office, store, plant, etc)

nieder·legen* I *vt* **1** put (*or* lay) down, deposit. **2** tear down. **3** give up, close (business, etc), resign (from). **4: die Arbeit n.** walk out, go on strike. **II: sich n.** lie down. **Niederlegung°** *f* putting down, depositing. **2** demolition. **3** closing; resignation. **4: N. der Arbeit** walkout, strike

niedermolekular *a* low-molecular (weight)

Niederquarz *n* low-temperature quartz, alpha-quartz

nieder·reißen* *vt* **1** carry down, precipitate. **2** tear (*or* pull) down, demolish

Niederschlag° *m* **1** (*Chem.*) precipitation; precipitate, deposit; sediment; condensation. **2** (*Meteor.*) (*oft pl*) precipitation, rainfall. **3** (radioactive) fallout. **4: seinen N. finden** be expressed, be reflected. **5** (*in compds usu Chem.*) precipitate: **·-arbeit** *f* precipitation process, = NIEDERSCHLAGS-ARBEIT. **n–arm** *a* poor in precipitate

niederschlagbar *a* precipitable. **Niederschlagbarkeit** *f* precipitability

nieder·schlagen* I *vt* **1** precipitate; deposit; condense. **2** beat (knock, strike) down, defeat. **3** put down, suppress; quash. **4** depress, discourage. **5** cast down, lower (eyes). **6** turn (*or* fold) down. *Cf* NIEDERGESCHLAGEN. **II** *vi*(s) **7** settle, be deposited. **8** hit the ground. **III: sich n. 9** precipitate, be deposited; settle; condense. **10** be expressed (*or* reflected). **niederschlagend** *p.a* precipitant; sedative

Niederschlag· (*Chem.*) precipitation:

-gefäß *n* precipitating vessel. **-menge** *f* amount of precipitate. **-messung** *f* measurement of precipitate. **-mittel** *n* precipitant. **-raum** *m* precipitation chamber

Niederschlags· precipitation (*Chem.* or *Meteor.*): **-apparat** *m* precipitation apparatus, precipitator. **-arbeit** *f* (*genl*) precipitation process; (*specif Metll.*) iron reduction process (liberation of metals by means of iron). **n–arm** *a* low-rainfall (area, season). **N–bildung** *f* formation of a precipitate. **n–frei** *a* dry, without rain(fall). **N–kessel** *m* precipitation pan. **-menge** *f* amoung of precipitation (*or* rainfall). **-mittel** *n* precipitant. **-ziffer** *f* precipitation number

Niederschlagung· *f* **1** (*Chem.*) precipitation. **2** defeat; suppression, quashing

Niederschlag· (*Chem.*) precipitation: **-verfahren** *n* precipitation process. **-vorrichtung** *f* precipitation device, precipitator. **-wärme** *f* heat of precipitation. **-wasser** *n* precipitated (*or* condensed) water. **-zeit** *f* precipitation time

nieder·schmelzen* *vt, vi*(s) melt down

nieder·schreiben* *vt* write down, put (down) in writing. **Niederschrift°** *f* **1** writing (down). **2** written version (copy, record); minutes

nieder·sinken* *vi*(s) sink (down), fall *cf* NIEDERGESUNKEN

Niederspannung* *f* low tension; (*Gases*) low pressure; (*Elec.*) low voltage

Niedertracht° *f* **1** malice, meanness. **2** low-down trick. **niederträchtig** *a* mean, vile; malicious

nieder·transformieren *vt* (*Elec.*) step down

Niederung *f* (-en) **1** lowland area, *pl oft* lowlands. **2** *pl:* **N—en des Lebens** darker (*or* seamier) sides of life

niedervoltig *a* low-voltage

Niederwald° *m* undergrowth, underbrush

niederwärts *adv* downward

Nied·nagel *m* hangnail

niedr. *abbv* = **niedrig** *a* **1** low: **n. geschätzt** at a low estimate, conservatively estimated. **2** vulgar. **3** inferior. **4** humble. **·-gekohlt** *a* low-carbon. **Niedrigkeit** *f* lowness; vulgarity; inferiority; humbleness

niedrig·legiert *a* low-alloyed. **-molekular** *a* low-molecular (weight). **-prozentig** *a* low-percentage. **-schmelzend** *a* low-melting. **-siedend** *a* low-boiling. **N–sieder** *m* low-boiling substance, low boiler. **n–siliziert** *a* low-silicon

Niedrigst· lowest, minimum: **-gehalt** *m* minimum content

niedrig·viskos *a* low-viscosity

niemals *adv* never = NIE

niemand *indef pron* (**-es** *genit*, **-em** *dat*, **-en** *accus*) nobody, no one, not . . . anybody (*or* anyone)

Niere *f* (-n) 1 kidney. 2 nodule, reniform concretion. 3 (*Ores*) bunch, pocket

Nieren· kidney, renal, nephro-: **-baum** *m* cashew (tree). **-drüse** *f* renal gland. **-entzündung** *f* nephritis. **-erz** *n* kidney ore. **n–förmig** *a* kidney-shaped. **N–grieß** *m* renal gravel. **-haut** *f* renal capsule. **-mittel** *n* kidney remedy. **-stein** *m* I (*genl*) kidney stone. II (*specif*): 1 (*Min.*) nephrite, spherulite. 2 (*Med.*) renal calculus

nierig *a* (*Min.*) nodular, kidney-shaped; *adv* oft in pockets

Niese·fieber *n* hayfever

nieseln *vi* drizzle

Niese·mittel *n* sternutator, sneeze-inducing agent

niesen *vi* sneeze. **Niesen·erreger** *m* sternutator; (*Mil.*) sneeze gas

Nies·gas *n* sneeze gas. **-kraut** *n* sneezewort. **-wurz(el)** *f* hellebore

Niet *m* or *n* (-e) rivet *cf* NIETE

Niete *f* (-n) 1 failure, dud, washout; **eine N. ziehen** draw a blank. 2 rivet. *Cf* NIET

Niet·eisen *n* rivet iron (*or* steel)

nieten *vt* rivet

Niet·hammer *m* riveting hammer. **-nagel** *m* 1 rivet. 2 = NIEDNAGEL hangnail

Nietung *f* (-en) 1 riveting. 2 rivet joint

Niet·verbindung *f* riveted joint

Nigella·öl *n* oil of Nigella damascena

Niger·öl *n* niger(seed) oil

Nikolsche(s) Prisma Nicol prism

nikotin·frei *a* nicotine-free. **N–gehalt** *m* nicotine content. **-säure** *f* nicotinic acid. **-vergiftung** *f* nicotine poisoning

Nil·blau *n* Nile blue. **-grün** *n* Nile green. **-pferd** *n* hippopotamus

nimmer *adv* 1 (= **nie mehr**) never again. 2 never: **nie und n.** absolutely never. **-satt** *a* insatiable, gluttonous

nimmt takes: *pres of* NEHMEN

Nim·öl *n* neem (*or* margosa) oil

Niob *n* niobium (columbium). **Niobat** *n* (-e) niobate, columbate. **Niob·säure** *f* niobic (columbic) acid. **-wasserstoff** *m* niobium (columbium) hydride

Nippel *m* (-) nipple

nippen *vi* sip, take a nip, get a taste

nirgends, nirgendwo *adv* nowhere, not . . . anywhere

Nirosta·stahl *m* (*TN*) 18/8 stainless steel

Nische *f* (-n) niche, recess

nisten *vi* build (*or* have) a nest

Nitrammit *m* or *n* ammonium nitrate

Nitranilid *n* N-nitroaniline

Nitranilin·rot *n* p-Nitraniline Red

Nitrat *n* (-e) nitrate. **-ätze** *f* (*Dye.*) nitrate discharge. **-beize** *f* nitrate mordant

Nitration *f* (-en) nitration

Nitrat·ion *n* nitrate ion. **-kunstseide** *f* nitrocellulose (*or* nitrate) rayon. **n–reduzierend** *a* nitrate-reducing. **N–seide** *f* nitrate rayon

Nitrid *n* (-e) (-ic) nitride

Nitrier·apparat *m* nitrating apparatus

nitrierbar *a* 1 nitr(ifi)able. 2 (*Metll.*) nitridable

Nitrier·baumwolle *f* nitrating cotton

nitrieren *vt* 1 nitrate, nitrify. 2 (*Metll.*) nitride

Nitrier·gemisch *n* nitrating mixture. **-härteverfahren** *n* nitriding process. **-härtung** *f* nitridation, nitride hardening. **-ofen** *m* nitriding oven (*or* furnace). **-säure** *f* nitrating acid, mixed (nitric and sulfuric) acid. **-schicht** *f* (*Metll.*) nitride case, nitrided layer. **-schleuder** *f* acid hydroextractor. **-sonderstahl** *m* special steel for nitriding. **-stahl** *m* nitrided (*or* nitriding) steel. **-topf** *m* nitrating (*or* nitriding) vessel

Nitrierung *f* (-en) nitr(ific)ation; (*Metll.*) nitridation. **Nitrierungs·grad** *m* degree of nitr(ific)ation (*or* nitridation). **-stufe** *f* stage of nitr(ific)ation (*or* nitridation)

Nitrier·wirkung *f* nitrating (nitrifying; nitriding) action (*or* effect)

nitrifizieren *vt* nitrify

nitrifizieren *vt* nitrify

Nitril-gruppe *f* nitrile group. **-gummi** *n,* **-kautschuk** *m* nitrile rubber. **-sulfon-säure** *f* nitrilosulfonic acid

Nitrit *n* (-e) nitrite. **--bildner** *m* nitrite former, nitrous bacterium. **n--frei** *a* nitrite-free

Nitro-bakterien *npl* nitrobacteria. **-baryt** *m* nitrobarite. **-benzin, -benzol** *n* nitrobenzene. **-chinon** *n* nitroquinone. **-coccussäure** *f* nitrococcic acid. **-derivat** *n* nitro derivative. **-farbstoff** *m* nitro dye. **-fettkörper** *m* aliphatic nitro compound. **-fettsäure** *f* nitro fatty acid. **-glyzerin** *n* nitroglycerine. **-gruppe** *f* nitro group. **-halogenbenzole** *npl* halonitrobenzenes. **-kohlenstoff** *m* tetranitromethane. **-körper** *m* nitro compound. **-kupfer** *n* nitrocopper. **-lack** *m* nitrocelluose lacquer. **-lampe** *f* (*Elec.*) nitrogen-filled lamp

Nitrol-säure *f* nitrolic acid. **Nitron-säure** *f* nitronic acid.

Nitro-penta *n* nitropentaerythritol = **-pentaerythrit** *n* pentaerythritol tetranitrate

Nitroprussid-natrium *n* sodium nitroprusside. **-wasserstoffsäure** *f* nitroprussic acid

Nitro-pulver *n* nitro powder

nitros *a* nitrous—**n—e Säure** nitrous vitriol *cf* NITROSE

Nitro-schwefelsäure *f* nitrosulfuric acid

Nitrose *f* nitrous vitriol, (*specif*) nitrosylsulfuric acid in sulfuric acid solution (from lead chamber process)

Nitro-seide *f* nitro-silk, nitrocellulose rayon

nitrosieren *vt* nitrosite, treat with nitrous acid, introduce a nitroso group into

Nitrosi-sulfosäure *f* nitrosisulfonic acid

Nitroso-bakterien *npl* nitroso bacteria. **-benzol** *n* nitrosobenzene. **-blau** *n* nitroso blue. **-farbstoff** *m* nitroso dye. **-gruppe** *f* nitroso group. **-kobalt-wasserstoffsäure** *f* cobaltnitrous acid. **-sulfosäure** *f* nitrososulfonic acid. **-verbindung** *f* nitroso compound

Nitro-sprengstoff *m* nitro explosive. **-stärke** *f* nitrostarch. **-sulfonsäure** *f* nitrosulfonic acid

Nitrosyl-säure *f* nitrosyl acid (hyponitrous acid). **-schwefelsäure** *f* nitrosylsulfuric acid, nitrosyl hydrogen sulfate

Nitro-toluol *n* nitrotoluene. **-verbindung** *f* nitro compound. **-weinsäure** *f* nitrotartaric acid (tartaric acid dinitrate). **-zelluloselack** *m* nitrocellulose lacquer (*or* dope)

Nitroxid° *n* (any) nitrogen oxide

Nitroxyl *n* (-e) nitroxyl hydride (NOH); nitryl ion. **--gruppe** *f* nitroxyl (*or* nitryl) group; nitronium ion

Nitroxylol *n* (-e) nitroxylene

Nitroxyl-säure *f* nitroxylic acid

Nitro-zellulose *f* nitrocellulose, cellulose nitrate

Nitrür *n* (-e) (-ous) nitride

Nitryl-säure *f* nitrylic acid (nitrous acid)

Niveau *n* (-s) **1** level; (*Atom. Phys.*) energy level. **2** (high) plane: **N. haben** be on a high (cultural) plane. **3** spirit level. **--fläche** *f* level (*or* equipotential) surface. **-flasche** *f* leveling vessel (for a gas burette), leveling bulb. **-gefäß** *n* leveling vessel. **-regler** *m* level controller (*or* regulator). **-rohr** *n,* **-röhre** *f* leveling tube. **-stufe** *f* energy level (of electrons)

Nivellement *n* (-s) leveling. **Nivellements-, Nivellier-** (*in compds*) leveling

nivellieren *vt* level; even out; grade. **Nivellier-schraube** *f* leveling screw. **Nivellierung** *f* leveling; evening out; grading

Nivellier-vorrichtung *f* leveling device. **-waage** *f* spirit level

n.J. *abbv* (nächsten Jahres) next year's

NK. *abbv* (Normalkerze) std. candle

n.l. *abbv* (nicht löslich) not sol.

n-Leiter° *m* n-type conductor; excess semiconductor

nm. *abbv* (nachmittags) PM, in the afternoon

nM. *abbv* (nächsten Monats) next month's

n.M.³ *abbv* (Normalkubikmeter) m³ (*or* cu.m.) at S.T.P.

No., N° *abbv* (Numero) no. (number)

noch I *adv* **1** yet, still: **n. nicht** not yet; **n. nicht fertig** not finished yet; **n. immer, immer n.** still; **n. immer nicht fertig** still not finished; **es siedet n.** (*or more emphat:* **n. immer**) it is still boiling; **das**

tain too; **ein n. nie dagewesener Fall** an unprecedented case *cf* NIE, NICHTS. **2** (+ *specif time*) **A** (*future*): **es muß n. heute geschehen** it has to be done no later than today. **B** (*past*): **n. heute kam es an** it arrived this very day (*or* only just today); **n. vor einem Jahr war das unbekannt** only (*or* as late as) a year ago that was unknown. **3** (+ *genl future time*) **A** some day: **wir werden n. alles mögliche erleben** we will (yet) experience all sorts of things (some day). **B** (*implies threat of danger*): **das explodiert n.** that is apt to explode. **4** other, more, besides, in addition; else: **n. ein Versuch** another (*or* one more) experiment; **n. einmal** once more, again; **n. zweimal** two more times; **n. etwas** another thing; **n. etwas?** anything else?; **was n.?** what else (*or* more)?; **wie lange n.?** how much longer?; **und dazu n.** (*or* **und n. dazu**) **falsch** and wrong besides; **arbeiten und dann (auch) n. lernen** work and then study in addition. **5** (+ *compar*) even, still: **n. größer** even (*or* still) larger. **6** (+ **so** + *quantity*) ever: **jede n. so geringe Änderung** any ever so slight change. **7** (+ *nur*) **es fehlen nur n. zwei** there are now only two (*or* are only two still) missing; **wir haben nur n. ein Exemplar** we have only one copy left. **II** *cjnc*: **weder . . . n.** neither . . . nor *cf* WEDER

nochmal *adv* (= **noch einmal**) again, once more. **nochmalig** *a* second, repeated; one more: **nach n—er Zählung** after one more count, after a recount, **bei n—er Überlegung** on second thought. **nochmals** *adv* again = NOCHMAL

Nocke *f* (-n), **Nocken** *m* (-) cam. **Nocken-welle** *f* camshaft

Nodi, (*Bot.*): **Nodien** *pl* of **Nodus** *m* node

Nomenklatur *f* (-en) nomenclature

nominell *a* nominal

Nonan·säure *f* nonanonic acid

Nonius *m* (-se *or* . . nien) vernier. **--ein-teilung** *f* vernier scale

Nonne *f* (-n) 1 mold, matrix. **2** night moth. **3** concave tile. **4** nun

nonocarbozyklisch *a* nonacarbocyclic

Nopin·säure *f* nopinic acid

Noppe *f* (-n) (*Tex.*) burl; knot, noil

Noppen·beize *f* burl dye. **n—·färben** *vt* burl-dye, noil-dye. **N–stift** *m* inking pencil. **-tinktur** *f* burl dye, burl(ing) ink

Nord I *m* 1 north. **2** north wind. **II** *pfx* north, northern: **n-amerikanisch** *a* North American. **-deutsch** *a* North German. **Norden** *m* north, northern part: **nach N.** (toward the) north; **im N. von . . .** in the northern part of . . .

Nordhäuser Schwefelsäure (or **Vitriolöl**) Nordhausen acid, fuming sulfuric acid.

nordisch *a* 1 northern. **2** Nordic, Norse (language, culture)

nördlich I *a* 1 northern; **n—e Breite** north(ern) latitude; **der n—ste Punkt** the northernmost point. **2** Arctic, (North) Polar. **3** northerly (wind, direction). **II** *adv* (+ **von**), *prep* (+ *genit*) (to the) north (of): **etwas n—er gelegen** lying somewhat more to the north

Nord·licht *n* aurora borealis. **-meer** *n* Arctic Ocean. **-ost** *m* 1 northeaster. **2** = **-osten** *m* northeast. **n–östlich** *a* northeast(ern), northeasterly *cf* NÖRDLICH. **N–pol** *m* North Pole. **-see** *f* North Sea. **n–wärts** *adv* northward. **N–west** *m* 1 northwester. **2** = **-westen** *m* northwest. **n–westlich** *a* northwest(ern), northwesterly *cf* NÖRDLICH

Norgeranium·säure *f* norgeranic acid

Norge·salpeter *m* Norway saltpeter, calcium nitrate

Nor·guajakharzsäure *f* norguaiaretic acid. **-homokampfersäure** *f* norhomocamphoric acid. **-kampfer** *m* norcamphor. **-kampfersäure** *f* norcamphoric acid

Norm *f* (-en) norm, standard

normal *a* normal, standard. **Normal** *n* (-e) international standard. **--bedingungen** *fpl* 1 normal (*or* standard) conditions (of temperature and pressure). **2** standard specifications. **-benzin** *n* gasoline free from aromatics. **-druck** *m* normal (*or* standard) pressure

Normale *f* (*adj endgs*) normal, perpendicular

Normal·einheit *f* normal (*or* standard) unit. **-element** *n* (*Elec.*) standard cell. **-entropie** *f* standard entropy

normalerweise f normally
Normal·essig m standard (or proof) vinegar. **-fall** m normal case. **-feuchtigkeit** f standard (or normal amount of) moisture. **-flüssigkeit** f normal (or standard) solution; (Dye.) test liquor. **-gewicht** n standard weight. **n—·glühen** vt (Metll.) normalize. **N–glühen** n normalizing; controlled tempering (or annealing). **-größe** f normal (or standard) size (or magnitude)
Normalien npl standards; standard rules
normalisieren vt standardize, (incl Metll.) normalize. **Normalisierung** f (-en) standardization.
Normalität f normality
Normal·kalomelelektrode f normal calomel electrode. **-kerze** f standard candle. **-kupfer** n (Elec.) standard copper. **-leistung** f normal (or rated) output. **-lösung** f normal (or standard) solution; (Dye.) test liquor. **-maß** n standard measure(ment) (dimension, size, gauge). **-null** f mean sea level. **-potential** n (Elec.) standard electrode potential. **-probe** f standard sample. **-profil** n standard cross section. **-säure** f normal (or standard) acid. **-schliff** m standard-taper ground-glass joint. **-seifenlösung** f standard soap solution. **-spannung** f (Elec.) normal voltage. **-sprit** m proof alcohol. **-stärke** f (genl) normal (or standard) strength; (specif) proof strength (of alcohol). **-stellung** f normal (or standard) position; zero position. **-temperatur** f standard temperature. **-uhr** f standard clock. **-verbraucher** m normal (ordinary, average) consumer. **-verteilung** f normal distribution. **-weingeist** m proof alcohol. **-wert** m normal (or standard) value. **-widerstand** m normal (or standard) resistance. **-zeit** f standard time. **-zusammensetzung** f standard composition. **-zustand** m normal (or standard) state (or condition)
Normativ n (government-set) standard requirement, official norm (formerly in East Germany)
Norm·blatt n standard specification sheet. **-druck** m standard pressure
normen vt standardize, set norms for

Normen·aufstellung f establishment of standards (or norms). **-ausschuß** m committee on standards. **-sand** m (Cement) standard sand
norm·entsprechend a normal, standard; conforming to the norm. **N–entwurf** m draft for standard specifications
Normen·vorschrift f standard specification
norm·gebend a normative. **-gerecht** a conforming to standards (or to the norm), standard
normieren vt standardize; normalize; regulate. **Normierung** f (-en) standardization; normalization; regulation
normig a normative
Norm·maß n standard size (dimension, measure). **-prüfung** f standard test. **n–recht** a standard. **N–schliff** m standard ground joint. **-teil** m standard part. **-temperatur** f standard temperature
Normung f (-en) standardization; normalization
norm·widrig a abnormal, nonstandard. **N–zahl** f standard (or preferred) number. **-zustand** m standard condition
Norpenald·säure f norpenaldic acid
Norpin·säure f norpinic acid
Norwegen n (Geog.) Norway. **norwegisch** a Norwegian. **Norw. P.** abbv (norwegisches Patent) Norw. Pat.
Nörz m (-e) mink = Nerz
Nosean n (Min.) noselite
not adv & pred a: **n. sein, n. tun** be necessary; **n. werden** become necessary
Not I f (⁼e) 1 need: **zur N.** if need be, in the worst case; **wenn N. am Mann ist** when help is desperately needed. 2 necessity. 3 distress, trouble: **in N—·en** in distress, in serious trouble; **ohne N.** without trouble, easily; (seine liebe) **N. haben** (mit) have no end of trouble (with); **mit knapper N., mit Mühe und N.** just barely. 4 hardship, misery, privation: **einen in N. bringen** cause someone hardship; **in N. geraten** fall on hard times. 5 (at the end of compds) e.g. **Wasser·not** water shortage. **II** pfx 5 emergency cf Not-ausgang. 6 spare, reserve: **Not·reifen** spare tire
Nota f (-s) (Com.) memorandum; bill, invoice

Not·absperrhahn *m* emergency stopcock

Notar *m* (-e) notary (public)

Not·ausgang *m* emergency exit. **-auslaß** *m* emergency outlet. **-behelf** *m* makeshift; (last) resort. **n–dürftig** *a* scanty, makeshift, *adv* oft in a makeshift manner.

Note *f* (-n) **1** note (*incl musical*); (*pl* oft) (sheet) music. **2** (*Fin.*) bill (paper money). **3** (*School*) mark, grade

Not·fall *m* (case of) emergency. **n–falls** *adv* if necessary, in case of emergency. **-gar** *a* (*Lthr.*) undertanned. **-gedrungen** *a* under pressure of necessity, forced by circumstances. **N–gemeinschaft** *f:* **N. der deutschen Wissenschaft** Emergency Association for German Science. **-hilfe** *f* emergency aid (*or* service)

notieren **I** *vt* **1** take (*or* note) down. **2** note, take note of. **3** quote (stock prices). **II** *vi* (*of stocks*) stand at

nötig *a* necessary; **n. haben** need; **n. brauchen** need urgently. **nötigen** *vt* (+ **zu**) urge (to . . .); force, make it necessary for (someone to . . .): **sich n. lassen** wait to be urged; **sich genötigt sehen** (**zu**) be forced (to). **nötigenfalls** *adv* if necessary. **Nötigkeit** *f* necessity. **Nötigung** *f* (-en) urging; (+ **zu**) need, compulsion (to); duress

Notiz *f* (-en) **1** note, memorandum: (**sich**) **N–en machen** take notes. **2: N. nehmen von . . .** take notice of. **--blatt** *n* (*name used by newspapers or notice sections of newspapers*). **-block** *m* note (*or* memo) pad. **-buch** *n* notebook, memo book. **-zettel** *m* note, memo (on a slip of paper)

Not·lage *f* emergency, distress, calamity. **n–leidend** *a* needy, distressed. **N–mittel** *n* expedient, makeshift. **-stand** *m* state of emergency. **-verband** *m* temporary (*or* first-aid) bandage (*or* dressing). **-wehr** *f* self-defense

notwendig *a* necessary. **notwendigerweise** *adv* necessarily. **Notwendigkeit** *f* (-en) necessity

Novität *f* (-en) novelty

Novojodin *n* novoiodine

N.P. *abbv* (Norwegisches Patent) Norw. Pat.

Nr(n) *abbv* (Nummer[n]) no(s). (numbers)

Nro *abbv* (Numero) no. (number)

Ns. *abbv* (Normschliff) std. ground joint

NT *abbv* (Normaltemperatur) std. temp.

n-te *a:* **bis in den n-ten Grad** to the nth degree

Ntf. *abbv* (Naturforscher) scientist

Nu *n:* **im Nu** in an instant, in a split second

Nuance *f* (-n) nuance, shade; subtlety

Nuancen·abstufung *a* gradation of shades. **-abweichung** *f* deviation in shade. **n–reich** *a* rich in nuances, highly subtle. **N–unterschied** *m* difference in shade

Nuancier·blau *n* tinting blue

nuancieren *vt* **1** nuance, shade (off); tint. **2** modulate; (*Perfumes*) modify. **3** grasp in all its nuances. **Nuancier·farbstoff** *m* shading (toning, tinting) dye. **Nuancierung** *f* (-en) nuancing, shading; tinting; modulation; modification

Nubuk·leder *n* nubuck leather

nüchtern *a* **1** sober. **2: auf n—en Magen** on an empty stomach. **3** level-headed; matter-of-fact. **4** insipid. **5** plain. **Nüchternheit** *f* sobriety; level-headedness; matter-of-factness; insipidity; plainness

Nuclein·säure *f* nucleic acid

Nudel *f* (-n) **1** noodle. **2** fattening ball (for geese, etc). **nudeln** *vt* fatten, force-feed

Nuklein·säure *f* nucleic acid. **Nukleoproteid** *n* nucleoprotein

null *a* (*no endgs*) **1** zero; nothing. **2: n. Uhr** midnight. **3: n. und nichtig** null and void. **Null** *f* (-en) **1** zero. **2** nothing, nil. **--ablesung** *f* zero reading. **-achse** *f* neutral axis

Nullage *f* (= **Null·lage**) zero position. **Null·effekt** *m* zero effect; background noise. **Nulleiter** *m* (= **Null·leiter**) (*Elec.*) neutral conductor (*or* line); return wire

nullen **1** *vt* neutralize. **2** *vi* ground a neutral

Null·grad *m* zero (on a degree scale). **Null·linie** *f* (= **Null·linie**) zero (null, vacuum) line; (*Spect.*) band origin. **Null·instrument** *n* null instrument. **-methode** *f* zero method. **-niveau** *n* zero level

Nullode *f* (-n) electrodeless discharge tube

Null·punkt *m* zero (point); neutral point; origin. **-punktbestimmung** *f* zero point determination. **-punktsenergie** *f* zero point energy. **-retentionsvolumen** *n* (*Gas Chromatography*) gas hold-up. **-setzen** *n* setting equal to zero. **-spannung** *f* zero tension, (*Elec.*) zero voltage. **-stelle, -stellung** *f* zero position; (*Spect.*) band origin. **-strich** *m* zero mark

nullte *a* (*Math.*) zero(th): **n. Spalte** zero column

Nullung *f* (-en) neutralizing; (*Elec.*) grounded neutral connection; zero voltage

Null-vektor *m* zero vector. **-verfahren** *n* zero (*or* null) method. **-versuch** *m* blank experiment (*or* test). **-wert** *m* zero (value). **n–wertig** *a* zero-valent, nonvalent. **N–zweig** *m* zero branch

numerieren *vt* number. **Numerierung** *f* (-en) numbering; (e) numeration, **Numerierungs-art** *f* numbering system. **numerisch** *a* numerical. **Numero** *n* (-s) (*usu* + *specif no.,* e.g.): **N. fünf** number five. **Numerus** *m* (..ri) 1 number—**N. clausus** numerically restricted admission (to university). 2 antilogarithm

Nummer *f* (-n) **I** (*genl*) number. **II** (*specif*) 1 size. 2 issue, edition. 3 (*School*) grade, mark. **nummern** *vt* number

Nummern·folge *f* numerical order. **-scheibe** *f* (*Tlph., Clocks*) dial

nun **I** *adv* 1 now, (*with past, oft*) then. 2: **n. einmal** (*implies inevitability*): **es ist n. einmal so** that's the way it is (and nothing can be done about it). **II** *adv-cjnc* and now, so now. **III** *cjnc* now that. **IV** *intj* well. **-mehr** *adv* from now on; up to now; now (implying: after all that has happened). **-mehrig** *a* present, now existing

nur *adv* 1 (*usu*) only: **nicht n., sondern auch** not only, but also; **wenn n.** if only. 2 (*less formal & esp* + *expressed or implied imperat*) just: **es fragt sich n., ob . . .** it is just a question of whether . . .; **man sei n. vorsichtig** just be careful; **n. nicht zu langsam** just not (*i.e.* just don't proceed) too slowly. 3 (+ *mehr*): **n. mehr zwei** only two left (*or* remaining). 4 (*after interrog in clause*): **was man n. versucht . . .** whatever you try. 5 at all, in any way: **soviel wie n. möglich** as much as at all possible; **jede n. nützliche Maßnahme** any in any way useful (*or* any ever so use-

ful) measure. **6: n. so: es schmilzt n. so** it just melts away

Nürnberg *n* (*Geog.*) Nuremberg: **N—er Violett** Nuremberg violet (manganese phosphate pigment)

Nuß *f* (Nüsse) (*genl*) nut; (*specif*) walnut. **-baum** *m* (wal)nut tree; walnut (wood). **-beize** *f* (*Paper*) soluble brown. **n–braun** *a* nut-brown, hazel. **N–grießkohle** *f*, **-grus** *m* nutty slack (coal). **-kern** *m* nut kernel. **-kernmehl** *n* nut meal. **-knacker** *m* nutcracker. **-kohle** *f* nut coal. **-körnung** *f* nut size. **-öl** *n* (wal)nut oil. **-schale** *f* nutshell

Nüster *f* (-n) nostril

Nut(e) *f* (..ten) groove, slot—**N. und Zapfen** mortise and tenon. **nuten** *vt* groove, slot. **Nut·rolle** *f* grooved pulley

Nutsch·apparat *m* suction filter (apparatus). **Nutsche** *f* (-n) suction filter, Büchner funnel. **nutschen** *vt/i* suction-filter, filter by suction

Nutschen·becher *m* suction filter cup. **-filter** *n* suction filter, Büchner funnel. **-sieb** *n* suction filter sieve. **Nutsch(en)·trichter** *m* suction (*esp* Büchner) funnel

Nutt·harz *n* acaroid resin (*or* gum)

Nutz· useful, practical: **-anwendung** *f* practical application, utilization. **-arbeit** *f* useful work

nutzbar *a* 1 usable, useful. 2 profitable, productive. 3 effective. 4: (sich) **etwas n. machen** utilize sthg, take advantage of sthg; **eine Kraft n. machen** harness a power. **Nutzbarkeit** *f* usefulness; profitability, productivity; effectiveness. **Nutzbarmachung** *f* utilization, harnessing

nutz·bringend *a* profitable, advantageous. **N–dauer** *f* useful (*or* service) life

nutze, nütze *pred a:* **zu etwas n. sein** be useful (*or* of use) for sthg; **zu nichts n.** useless

Nutz·effekt *m* useful effect, efficiency

nutzen, nützen 1 *vt* make use of; take advantage of; utilize, exploit. 2 *vi* be of use: **was nützt das?** of what use is that?; **das nützt etwas (nichts, viel)** that is of some (no, much) use; **wem nutzt das?** to whom is that of use?, who profits by that?

Nutzen *m* 1 use; profit, advantage: **von N.**

sein be of use, be profitable; **etwas mit N. lesen** read sthg to one's advantage; **N. ziehen (aus)** gain an advantage (from). **2** yield

Nutz· useful, practical, effective, commercial: **-energie** *f* effective energy. **-fahrzeug** *n* commercial vehicle. **-fläche** *f* usable area (*or* surface). **-garten** *m* **1** vegetable garden. **2** orchard. **-gas** *n* (*Aero.*) impellent. **-gewächs** *n* useful plant. **-gras** *n* feed grass. **-holz** *n* commercial timber; lumber. **-kapazität** *f* useful capacity. **-ladung, -last** *f* payload; loading capacity. **-leistung** *f* effective output, efficiency; mechanical power

nützlich *a* useful. **Nützlichkeit** *f* usefulness

nutzlos *a* useless. **Nutzlosigkeit** *f* uselessness

Nutz· useful, practical, effective: **-nießer** *m* beneficiary; profiteer. **-nießung** *f* profitable use: **die N. von etwas haben** enjoy the profitable use of sthg. **-pferdestärke** *f* effective horsepower. **-pflanze** *f* useful plant, food (*or* feed) plant. **-schicht** *f* (*Ceram.*) wearing layer. **-strom** *m* (*Elec.*) useful current. **-tier** *n* (economically) useful animal

Nutzung *f* (-en) **1** use, utilization. **2** profitable use, exploitation. **3** yield, produce; revenue

Nutzungs·dauer *f* useful life. **-recht** *n* right of (profitable) use

Nutz·vieh *n* (domestic) farm animals; cattle. **-wert** *m* economic value. **-wirkung** *f* useful effect, efficiency

n.W. *abbv* (nächster Woche) of next week

Nw. *abbv* **1** (Nachrichtenwesen) news media. **2** (Nachwachsen) (*Ceram.*) expansion

Nyltest *TN* a nylon fabric

NZ *abbv* **1** (Neutralisationszahl) neutralization no. **2** (Neuzeit) modern era (*or* times). **3** (Normalzeit) standard time

O

o. *abbv* **1** (oben) above. **2** (oder) or. **3** (ohne) without. **4** (ORDENTLICH) regular, etc. **5** (ordinär) ordinary

ö. *abbv* (östlich) eastern, etc

O. *abbv* (Ost, Osten) East

o.a. *abbv* **1** (oben angeführt) mentioned above. **2** (oder andere) or others

o.ä. *abbv* (oder ähnliches) or the like

Oase *f* (-n) oasis

ob I *cnjc* **1** if, whether: **ob es zerfällt oder nicht** whether it decomposes or not; **als ob** as if. **2** (*in questions*) I wonder if, the question is whether: **ob es** (**wohl**) **zerfällt?** the question is whether it will decompose. **II** *prep* (+ *genit*) on the basis of, on account of. **III** *pfx* (*implies meanings of* OBER *as pfx*) *cf* OBDACH, OBMANN, OBER **II**

Obacht° *f* care, attention = ACHT **1**

ÖBB *f* (*no endgs*) *abbv* (Österreichische Bundesbahn) Austrian Federal Railroad

Obdach° *n* shelter. **obdachlos** *a* homeless

oben *adv* **1** up, above, on (*or* at the) top; upstairs: **o. auf dem Dach** on the roof, **hier o.** up here; **wie o. erwähnt** as mentioned above (*or* previously); **von o. bis unten** from top to bottom; **von o. nach unten kehren** turn upside down. **2** (*on packages*) this side up. **3: nach o.** up(ward0, to the top, upstairs. **4** (*implies higher authority*): **Anordnungen von o.** orders from above (from the boss, the management). **-an** *adv* at the top, in first place. **-auf** *adv* **1** on top. **2** in the pink of condition. **-aufschwimmend** *a* supernatant. **-drein** *adv* (*usu* + **noch**) besides, in addition; at that, to boot. **-erwähnt**, **-genannt** *a* above-mentioned, aforesaid. **-hin** *adv* casually, offhand; perfunctorily. **-liegend** *a* lying on top, located above; overhead. **-stehend** *a* above-mentioned; **das O—e** the above; *adv oft* above

ober I *a* upper, top; higher, superior *cf* OBERSTE. **II** *pfx* upper, top; head, chief; super, epi-; higher, superior, senior. **III** *prep* (+ *dat*) above

Ober·arm *m* upper arm. **-aufsicht** *f*: **die O. haben** supervise, be the (chief) officer in charge. **-bau** *m* superstructure; aboveground construction; overhead structure. **-bauch** *m* epigastrium; (*in compds*) epigastric. **-beamte** *m*, **-beamtin** *f* higher official. **-boden** *m* topsoil. **-dampf** *m* (*Gas*) down-run steam. **-druck** *m* downward pressure, pressure from above. **-fläche** *f* surface; area

oberflächen· surface: **-aktiv** *a* surface-active. **O—aktivität** *f* surface activity. **-ausführung** *f* surface finish. **-bau** *m* surface structure. **o-behandelt** *a* surface-treated. **O—behandlung** *f* surface treatment. **-chemie** *f* surface chemistry. **-druck** *m* surface pressure. **-einheit** *f* unit of area. **-energie** *f* surface energy. **-entfaltung** *f* surface development. **-entwicklung** *f* surface developing. **-erscheinung** *f* surface phenomenon. **-farbe** *f* surface color. **-färbung** *f* **1** surface coloring. **2** surface dyeing (*or* staining). **-fehler** *m* surface defect. **-glanz** *m* surface luster. **-güte** *f* surface quality (*or* finish). **-härte** *f* surface (*or* skin) hardness. **O—härten** *n*, **-härtung** *f* case-hardening; surface hardness; **o—inaktiv** *a* surface-inactive. **O—kondensator** *m* surface condenser. **-kultur** *f* (*Agric.*) surface cultivation. **-ladung** *f* (*Elec.*) surface charge. **-leimung** *f* (*Paper*) surface sizing. **-lösung** *f* surface solution. **-niederschlag** *m* surface precipitate. **-oxidschicht** *f* surface oxide layer (*or* film). **-riß** *m* surface crack. **-schicht** *f* surface layer. **-spannung** *f* surface tension. **-verbindung** *f* surface compound (*or* combination). **-vered(e)lung** *f* surface finishing. **-wärme** *f* surface heat (*or* energy). **-wasser** *n* surface water. **-widerstand** *m* surface resistance (resistivity, impedance). **o—wirksam** *a* surface-active. **O—wirkung** *f* surface action (*or* effect)

oberflächlich _a_ superficial. **Oberflächlichkeit** _f_ superficiality
ober·gärig _a_ top-fermenting, top-fermented: **o—e Hefe** top yeast. **O—gärung** _f_ top fermentation. **-gewalt** _f_ sovereignty, supremacy. **-grund** _m_ surface soil, topsoil. **-gruppe** _f_ (_Cryst._) supergroup. **ohalb** _prep_ (_genit_) above, over. **O—hand** _f_ 1 back of the hand. 2 upper hand. **-haupt** _n_ head, chief. **-haus** _n_ Upper House. **-haut** _f_ epidermis; (_in compds oft_) epidermal. **-häutchen** _n_ cuticle. **-hefe** _f_ top yeast. **-heizwert** _m_ gross calorific value. **oirdisch** _a_ above-ground; overhead; aerial. **O—kante** _f_ upper (_or_ top) edge. **-keim** _m_ (_Zool._) ectoderm. **-kiefer** _m_ upper jaw. **-kolbenpresse** _f_ downstroke press. **-körper** _m_ upper torso. **-land** _n_ upland(s), highlands. **-leder** _n_ upper leather; uppers (of shoes). **-leitung** _f_ 1 chief (_or_ top) management. 2 overhead wire(s). **-licht** _n_ light from above; skylight. **-lippe** _f_ upper lip. **-niere** _f_ suprarenal gland
Obers _n_ (dairy) cream
Ober·schalseife _f_ top-layer soap. **-schenkel** _m_ thigh. **-schicht** _f_ upper stratum (_or_ layer). **-schule** _f_ secondary school. **-schwingung** _f_ overtone, harmonic (vibration). **-seite** _f_ upper side; (_Tex._) right side. **-spannung** _f_ (_Elec._) high tension
Oberst _m_ (-en, -en) colonel
oberste _a_ (_superl of_ OBER) uppermost, topmost; highest; supreme. **Oberste** (_adj endgs_) **I** _m, f_ chief, boss. **II** _n_ uppermost part, top—**das O. zuunterst kehren** turn everything upside down
Ober·stufe _f_ advanced level; upper grade (_or_ stage). **-tasse** _f_ cup. **-teig** _m_ (_Brew._) upper dough; extractives (from spent grains). **-teil** _m_ upper part, top. **-ton** _m_ overtone, harmonic. **-wasser** _n_ 1 headwater. 2 overshot water. 3: **O. bekommen** get the upper hand. **-welle** _f_ higher harmonic (wave). **-zahn** _m_ upper tooth
obgleich _subord cnjc_ although
Obhut _f_ guardianship, protection, care: **in O. nehmen** take under one's care
obig _a_ above(-mentioned), foregoing
Objekt· (_Opt., Micros._) object: **-abstand** _m_ working (_or_ object) distance. **-glas** _n_ slide,

mount. **-halter** _m_ specimen (_or_ slide) holder, stage
Objektiv· (_Opt._) objective, object glass (_or_ lens): **-glas** _n_ objective, object lens. **-linse** _f_ objective lens. **-wechsler** _m_ revolving nosepiece
Objekt· (_Opt., Micros._) object: **-sucher** _m_ object finder. **-tisch** _m_ stage, stand. **-träger** _m_ 1 microscope slide, specimen holder. 2 stage, stand
Oblate _f_ (-n) wafer. **Oblaten·papier** _n_ wafer (_or_ rice) paper
Obleute chiefs, etc: _pl of_ OBMANN
ob·liegen* (_also insep in simple tenses:_ obliegt, etc; _pp_ obgelegen) _vi_ (h _or_ s) (_dat_) 1 be incumbent (on). 2 apply oneself (to). **obliegend** _p.a_ incumbent. **Obliegenheit** _f_ obligation, duty
obligat _a_ 1 inevitable, indispensable. 2 = **obligatorisch** _a_ obligatory, required
Obmann° _m_ (_pl oft_ Obleute) chief, chairman; starter; shop steward; spokesman; umpire; (jury) foreman
obschon _subord cnjc_ although
Observatorium _n_ (..rien) observatory
Obsorge° _f_ care, guardianship
Obst _n_ fruit (_incl._ berries). **-art** _f_ type of fruit. **o—artig** _a_ fruit-like, fruity. **O—bau** _m_ fruit growing. **-bauer** _m_ fruit grower. **-baum** _m_ fruit tree. **-blüte** _f_ fruit blossom (time). **-branntwein** _m_ fruit brandy. **-essig** _m_ fruit vinegar. **-garten** _m_ orchard. **-kern** _m_ fruit kernel (pit, stone). **-konserve** _f_ fruit preserves, preserved fruit. **-mark** _n_ fruit pulp. **-most** _m_ fruit juice. **-mus** _n_ (fruit) jam, marmalade. **-paprika** _m_ fruit capsicum. **-saft** _m_ fruit juice. **-schwemme** _f_ fruit glut. **-trester** _mpl_ fruit marc. **-wein** _m_ fruit wine, cider. **-weinbereitung** _f_ fruit wine manufacture. **-zucker** _m_ fruit sugar, levulose
Obtusat·säure _f_ obtusatic acid
ob·walten (_also insep in simple tenses:_ obwaltet, _etc;_ _pp_ obgewaltet) _vi_ exist, prevail
obwohl, obzwar _subord cnjc_ although
Occ . . . _cf_ OKK . . ., OKZ
Ocher _m or_ n **ocher** = OCKER
Ochras _m_ black salt, melted ashes (crude potash _or_ soda from ashes)
Ochs(e) _m_ (.. sen, .. sen) ox, bull

Ochsen·auge n 1 oxeye daisy. 2 bull's eye (window). 3 fried egg (sunny side up). **-blut** n oxblood. **-fleisch** n beef. **-frosch** m bull frog. **-galle** f ox gall, ox bile. **-gallenseife** f bile soap. **-klauenfett, -klauenöl** n neat's-foot oil. **-leder** n oxhide, neat's leather. **-schwanz** m oxtail. **-zunge** f 1 beef tongue. 2 (Bot.) oxtongue, bugloss, borage

Ocker m ocher. **o--ähnlich, -artig** a ocherlike, ocherous. **O--braun** n ocher brown. **-farbe** f ocher (as a pigment). **o--farben, -farbig** a ocher-colored. **O--gelb** n ocher yellow. **o--haltig** a ocherous, containing ocher

ockerig a ocherous, ochery

oct . . . cf OKT . . .

od. abbv (oder) or

O.D. abbv (optisches Drehungsvermögen) optical rotatory power

od. dgl. abbv (oder dergleichen) or the like

öde a 1 desolate, barren. 2 tedious, boring. **Öde** f (-n) 1 wasteland, desert. 2 desolation. 3 barrenness. 4 boredom. **--land** n fallow land

Ödem n (-e) edema

oder cnjc or

Odermenning m (-e) (Bot.) agrimony

o. dgl. abbv (oder dergleichen) or the like

Öd·land n fallow (or barren) land; wasteland

Oe . . . oft used in place of Ö . . .

Oe.P. abbv (österreichisches Patent) Austrian patent

Öfchen n (-) little stove (oven, furnace)

Ofen m (⁅) oven, furnace; stove; kiln; heater. **--auskleidung** f oven (kiln, furnace) lining. **-block** m battery of ovens (furnaces, kilns). **-bruch** m tutty, crude zinc oxide. **-darren** n kiln-drying. **-einsatz** m furnace (oven, kiln) charge. **-emaille** f stove enamel. **-farbe** f graphite, black lead. **-fluß** m electrolysis bath. **-futter** n furnace (or kiln) lining. **-galmei** m tutty, crude zinc oxide. **-gang** m furnace run, oven (or kiln) operation; (Metll.) descent of the (blast-furnace) charge. **-gewölbe** n arched furnace roof. **-gicht** f 1 furnace throat. 2 furnace charge. **-gut** n furnace (oven, kiln) charge, material (to be) treated in furnace, etc. **-kachel** m (Dutch) stove tile. **-kanone** f (Cement) kiln gun. **-lack** m stove lacquer (or polish). **-lackfarbe** f stove lacquer pigment. **-putzmittel** n stove polish. **-reise** f furnace campaign. **-rohr** n stovepipe. **-röhre** f oven (of a kitchen stove). **-ruß** m furnace (or oven) soot. **-sack** m (Metll.) furnace sow, salamander. **-schlacke** f furnace slag. **-schwamm** m tutty, crude zinc oxide. **-schwarz** n, **-schwärze** f black lead, stove polish. **o--trocken** a kiln-dried, oven-dried. **-trocknend** a kiln-drying, oven-drying, stove-drying. **O--trocknung** f kiln (oven, stove) drying. **-tür** f oven door; fire door of furnace (or kiln); (Ceram.) wicket. **-verkokung** f coking in ovens. **-ziegel** m firebrick; stove tile. **-zug** m furnace (oven, kiln) draft

offen I a 1 (usu) open; **die Apotheke hat abends o.** the pharmacy is open evenings. 2 public, open to all. 3 frank, sincere: **o. gestanden** (to put it) frankly. 4 uncertain, unsettled (matter, question); unsolved (problem). 5 blank (check, etc). 6 on tap, on draft; loose (milk, etc). 7 live (steam). II sep pfx open; undecided

offenbar a 1 obvious. 2 evident. 3 (publicly) known. **offenbaren** (pp offenbart or geoffenbart) 1 vt reveal, disclose. 2: **sich o.** be revealed, become evident. **Offenbarung** f (-en) revelation, disclosure

offen·halten* 1 vt keep open. 2 vi stay open

Offenheit f openness, frankness, sincerity

offen·kettig a open-chain, **-kundig** a 1 obvious. 2 evident. 3 notorious, well-known

offen·lassen* vt 1 leave open. 2 leave undecided

offen·sichtlich a 1 obvious. 2 evident

offensiv a offensive (= aggressive)

offen·stehen* vi (h or s) 1 be (or stand) open. 2 be (left) undecided. 3 (Fin.) be due

öffentlich a public; adv oft in public. **Öffentlichkeit** f 1 (the) public, public view, the open: **etwas an die Ö. bringen** bring sthg into the open, make sthg public. 2 public character (of an undertaking)

offerieren vt, **Offerte** f (-n) offer

offiziell a official. **Offizier** m (-e) officer

Offizin f (-en) 1 prescription pharmacy; pharmaceutical laboratory. 2 printing shop. **offizinell** a officinal

öffnen *vt* & **sich ö.** open; *vi* open the door.
öffnend *p.a* (*oft*) aperient. **Öffner** *m* (-) 1
opener. 2 (*Paper*) pulper, kneader. **Öff-**
nung *f* (-en) opening; aperture, orifice;
slit, slot; gap
Öffnungs‑mittel *n* (*Med.*) aperient. **-weite**
f width of the opening. **-winkel** *m* angle
of aperture
Offset‑druck *m* offset printing. **-farbe** *f*
offset(-printing) ink
oft *adv* often. **öfter** *adv* (*compar of* OFT) 1
more often. 2: ö., des ö—en = öfters *adv*
fairly often. **öftest: am ö—en** *adv* (*su-*
perl of OFT) most often
oft‑malig *a* frequent, repeated. **-mals** *adv*
often
ohmsch (*Elec.*) ohmic—**das O—e Gesetz**
Ohm's Law
ohne I *prep* (+ *accus*) without, with no; not
counting *cf* WEITER, WENN. **II** *cnjc* 1 (+ **zu**)
e.g. **o. etwas zu erfahren** without finding out anything. 2 (+ **daß**) e.g. **o. daß**
das Gas entweicht without the gas es-
caping. **‑dies** *adv* without that, anyway.
-gleichen *adv* & *pred* a without equal.
O—haltflug *m* nonstop flight. **o—hin** *adv*
anyway, in any case
Ohnmacht *f* 1 unconsciousness, faint: **in O.**
fallen faint. 2 powerlessness. **ohnmäch-**
tig *a* 1 unconscious, in a faint. 2 powerless
Ohr *n* (-en) ear; (*in compds oft*) auricular,
aural, oto-
Öhr *n* (-e) 1 eye (of a needle, for hook fas-
teners), eyelet, (button) loop. 2 ear, handle (of cups, etc). 3 bog iron ore
Öhrchen *n* (-) 1 little ear; auricle *cf* OHR. 2
eyelet *cf* ÖHR
Ohr‑clip *m* (-s) 1 earphones, headphones. 2
pl oft clip-on earrings
Ohren‑ ear, auricular, oto-: **-arzt** *m* ear specialist, otologist. **-schmalz** *n* ear wax,
cerumen. **-schmerzen** *mpl* earache.
-stein *m*, **-steinchen** *n* otolith. **-stöpsel**
m earstopper
Ohr‑ ear, auricular, oto-: **-finger** *m* little
finger. **o—förmig** *a* auriform. **O-mus-**
chel *f* (*Anat.*) auricle; outer ear, conch of
the ear. **-speicheldrüse** *f* parotid gland.
-stein *m* otolith. **-trommel** *f* eardrum.
-trompete *f* (*Anat.*) Eustachian tube.
-wachs *n* earwax, cerumen. **-wasser** *n*

endolymph. **-wurm** *m* earwig
Oiazin *n* pyridazine
Oker *m* ocher = OCKER
okkludieren *vt* occlude
Ökologie *f* ecology. **ökologisch** *a*
ecological
Ökonom *m* (-en, -en) 1 economist. 2 agri-
culturist. **Ökonomie** *f* economy. **Ökono-**
mik *f* economics. **ökonomisch** *a* economic, economical; financial
Ökosystem° *n* ecosystem
Oktaeder *n* octahedron. **‑gitter** *n* (*Cryst.*)
octahedral lattice. **oktaedrisch** *a* octahedral. **Oktaedrit** *m* (*Min.*) octahedrite
Oktan‑wert *m*, **-zahl** *f* octane number
Oktett‑regel *f* octet rule
okto‑ *pfx* octa-, octo-: **-carbozyklisch** *a* oc-
tacarbocyclic. **O—naphthen** *n* octanaph-
thene
Oktoxid *n* octoxide
oktyl‑ *pfx* octyl-
Okular *n* (-e) eyepiece, ocular. **‑muschel** *f*
(*Micros.*) eyepiece cup
Okulation *f* (-en) (*Bot.*) budding. **okulie-**
ren *vt* graft, bud, inoculate
okzidental(isch) *a* occidental, western
‑ol *Chem sfx* 1 (*Alcohols, Phenols*) -ol. 2 (*in*
some hydrocarbons) -ene: **Benzol** benzene. 3 (*other compounds*) -ole: **Pyrrol**
pyrrole
Öl *n* (-e) oil. **‑abdichtung** *f* oil seal. **-ablaß**
m oil drain(ing). **-abscheider** *m* oil separator; (*Brew.*) oil trap. **-anstrich** *m* 1
painting in oil (colors). 2 coat of oil (*or* oil
paint). **-anstrichfarbe** *f* oil painting col-
or; oil paint. **öl‑arm** *a* oil-deficient, poor
in oil. **-artig** *a* oily, oleaginous.
Öl‑aufnahme *f* oil absorption. **-aus-**
scheider *m* oil separator (*or* extractor).
-ausstrich *m* coat of oil (paint). **-avivage**
f (*Dye.*) brightening with oil. **-bad** *n* 1 oil
bath. 2 (*Turkey Red Dye.*) green liquor.
-basis *f* oil base; (*Dye.*) fatty base. **-baum**
m olive tree—**falscher Ö.** oleaster.
-baumgummi, **-baumharz** *n* (gum) elemi. **-beere** *f* olive. **-behälter** *m* oil container (reservoir, tank). **-beize** *f* 1 oil mordant. 2 oil stain (for wood). **öl‑beständig**
a oil-resistant. **-bildend** *a* oil-forming,
olefiant (*esp* gas). **Öl‑bindevermögen** *n*

oil-binding property. **-blase** f oil-boiling
kettle. **-blau** n smalt, oil (or Saxon) blue,
indigo copper, covellite. **-bleiche** f,
-bleichen n oil-bleaching. **-bodensatz** m
oil sediment (or foot). **-bohnermasse** f oil
floor polish. **-bohrinsel** f oil drilling platform (at sea). **-bohrung** f 1 oil drilling. 2
oil well. **-bombe** f oil bomb. **-bremse** f
(oil) hydraulic brake. **-brenner** m oil
burner. **-brunnen** m oil well. **-büchse** f
(Mach.) oil cup; oil can. **-buntlack** m colored oil enamel. **-creme** f oil cream, oil
preparation. **-dampf** m oil vapor; oil
smoke. **-dampfer** m oil tanker.
-dämpfung f oil damping. **öl-dicht** a
oiltight. **Öl-draß** m oil dregs. **-druck** m 1
oil pressure. 2 oleography; oleograph.
-druckmesser m oil-pressure gauge.
-drucktapete f sanitary wallpaper.
-dunst m oil vapor (or fumes).
öl-durchtränkt a oil-saturated
Oleanol·säure f oleanolic acid
Oleat n (-e) oleate
öl·echt a oil-fast, oil-insoluble. **Öl·echtheit**
f fastness to oil
Oliban(um) n olibanum, frankincense
Olefin·alkohol m olefin(ic) alcohol. **-ha-
loid** n olefin halide. **olefinisch** a olefinic.
Olefin·keton n olefin(ic) ketone
Olein·säure f oleic acid. **-(säure)seife** f
olein (or red-oil) soap. **-schmälze** f (Tex.)
olein softener (or emulsion)
ölen vt/i oil, grease, lubricate
Oleokreosot n (Pharm.) creosote oleate
Oleosol·farbe f oil-soluble dye
Öler m (-) oiler; oilcan, oil cup. **--glas** n glass
oil cup
Öl·ersatz m oil substitute. **-fabrik** f oil refinery. **-fänger** m oil catcher (collector,
trap). **-farbe** f oil color; oil paint. **-far-
benanstrich** m 1 painting in oil colors. 2
coat of oil paint. **-farbendruck** m 1
oleography. 2 oleograph. **-faß** n oil barrel,
oil drum. **-feld** n oil field. **öl·fest** a oil-
resistant. **Öl·feuerung** f oil firing; oil
furnace; oil heating. **-film** m oil film. **-fir-
nis** m oil varnish. **-firnisbaum** m 1
candlenut tree (Aleurites cordata). 2
Tungchou tree (Eleococea vernicia).
-fläschchen n (Micros.) immersion oil
bottle. **-fleck** m oil spot, oil stain. **öl·frei** a
oil-free. **-führend** a oil-bearing.
Öl·füllung f oil filling. **öl·fundig** a: **die**

Bohrung ist ö. the drilling has struck
oil. **-gar** a (Lthr.) oil-tanned. **Öl·gas** n oil
gas. **-gasteer** m oil-gas tar. **-gefäß** n oil
vessel (can, tank). **öl·gefeuert, -geheizt** a
oil-fired, oil-heated. **Öl·gehalt** m oil content. **-geläger** n oil dregs. **-gelb** n (Dyes)
oil yellow. **-gemälde** n oil painting.
-geschmack m oily taste. **öl·getränkt** n
oil-impregnated. **Öl·gewinnung** f oil extraction (or production). **öl·glänzend** a
oil-glossy, oily-lustered. **Öl·glanzlack** m
oil-gloss paint. **-grün** n oil green. **-grun-
dierung** f oil priming; oil primer, oil bottom. **-hahn** m oil cock (or tap). **öl·haltig** a
oil-bearing, oleiferous. **Öl·handel** m oil
trade. **-härtung** f oil hardening (or hydrogenation). **-härtungsanlage** f oil-
hardening plant. **-härtungsstahl** m oil-
hardening steel. **-harz** n oleoresin. **-haut**
f 1 oil film. 2 (Tex.) oilskin. **-häutchen** n
oil pellicle, oil film. **-hefen** fpl oil dregs,
oil sediment. **-heizung** f oil heat(ing).
öl·höffig a promising rich oil deposits
ölig a oily. **Öligkeit** f oiliness
Oligo·klas m oligoclase. **-merisation** f
oligomerization. **-zän** n (Geol.) Oligocene
öl·imprägniert a oil-impregnated. **Öl·in-
dustrie** f oil industry. **-immersion** f (Micros.) oil immersion. **-isolation** f oil
insulation
oliv a olive (green). **Olive** f (-n) (Bot. &
Mach.) olive
oliven·artig a olive-like, (Anat.) olivary.
O--baum m olive tree. **o--braun** n olive
brown. **O--farbe** f olive color. **o--farben,
-farbig** a olive-colored. **-förmig** a olive-
shaped. (Anat.) olivary. **O--gelb** n olive
yellow. **-grün** n olive green. **-kern** m olive
pit; (Anat.) olivary nucleus. **-kernöl** n
olive kernel oil. **-nachöl** n low-grade olive
oil. **-öl** n olive oil. **-ölfettsäure** f fatty acid
of olive oil. **-ölschmierseife** f olive-oil
soft soap. **-ölseife** f olive-oil soap, (genuine) Castile soap
oliv·grau a olive gray (or drab)
Öl·kalk m oil-bearing limestone. **-kammer**
f oil chamber. **-kännchen** n, **-kanne** f oilcan. **-karburierung** f oil carburizing.
-kautschuk m factice, rubber substitute.
-kelter f oil press. **-kessel** m oil-fired
boiler. **-kitt** m putty. **-kohle** f oil carbon;

carbon deposit (or residue). **-körper** m (Biol.) oily substance. **-kreidestift** m oil crayon. **-krem** f oil cream. **-krise** f oil crisis. **-kruste** f oil crust. **-kuchen** m oil cake. **-kugel** f oil drop, oil globule. **-lache** f oil slick. **-lack** m oil varnish, oil paint. **-lampe** f oil lamp. **-leder** n chamois. **-leinen** n oiled linen. **-leitung** f oil (pipe)line. **öl·los** a oilless. **Öl·lösevermögen** n oil-dissolving power. **öl·löslich** a oil-soluble. **Öl·malerei** f oil painting. **-menge** f amount of oil. **-messer** m oleometer, oil gauge. **-meßstab** m oil dipstick. **-milch** f oil emulsion. **-mischung** f oil mixture. **-mühle** f oil mill. **-nuß** f oil(-bearing) nut, butternut. **-nußbaum** m butternut tree. **-nußfett** n oilnut fat. **-ofen** m oil stove; oil-fired furnace. **-packpapier** n oiled wrapping paper. **-palme** f palm oil tree. **-papier** n oiled paper. **-pauspapier** n oiled tracing paper. **-pergament** n oiled parchment. **-pest** f oil spill(s), oil pollution. **-pflanze** f oil-yielding plant. **-präparat** n oil preparation, oil polish. **-probe** f oil sample; oil test. **-prüfer** m oil tester. **-pumpe** f oil pump. **-qualm** m oil smoke. **-quelle** f source of oil; oil well. **-raffinerie** f oil refinery. **-rauch** m oil smoke. **öl·reich(haltig)** a richly oil-bearing, rich in oil. **Öl·reinigung** f oil purification. **-rettich** m common radish. **-rückstand** m oil residue. **-ruß** m lampblack. **-saat** f oil (-bearing) seeds. **-samen** m oil seed, (specif) rape seed, linseed. **-sand** m oil sand. **-satz** m oil sediment, oil dregs. **öl·sauer** a oleate of. **Öl·säure** f oleic acid. **-säurereihe** f oleic acid series. **-säureseife** f olein (or oleic acid) soap. **-schalter** m (Elec.) oil switch. **-schicht** f layer of oil, oil film; oil-bearing stratum. **-schiefer** m oil shale. **-schlagen** n oil pressing. **-schlamm** m oil sludge. **-schmierung** f oil lubrication. **-schuhkrem** f oil shoe polish. **-schwarz** n oil black, lampblack. **-schwemmverfahren** n oil flotation. **-seide** f oiled silk. **-seife** f oil(-based) soap, (specif) Castile or Venetian soap. **-sieb** n oil strainer. **-siederei** f oil refinery. **-sodaseife** f oil(-based) soap, (specif) Castile or Venetian soap. **-sorte** f kind (grade, quality) of oil. **-spachtel** m

or f oil filler. **-spaltgas** n gas from oil gasification. **-sperre** f oil embargo. **-spur** f trace of spoil oil. **-stand** m oil level. **-staub** m oil spray. **-stein** m oil stone. **-stoff** m 1 (Chem.) olein. 2 (Tex.) oilcloth, oiled cloth. **-sud** m oil boiling. **-sylvinsäure** f oleosylvic acid. **-trester** mpl oil marc, oil press residue. **-tropfen** m drop of oil. **-tubenfarbe** f oil color in a tube. **-tuch** n oilcloth. **-überzug** m coating of oil. **-umlauf** m oil circulation

Ölung f (-en) oiling, lubrication
Öl·verbrauch m oil consumption. **-vergütung** f (Metll.) heat treatment using oil. **-waage** f oil hydrometer, oleometer. **-wachsbeize** f oil-wax stain. **-wanne** f oil pan (trough, sump). **-wasser** n oily water. **-wechsel** m oil change. **-weide** f oleaster. **-werk** n oil works (mill, factory). **-zelle** f oil cell. **-zelluloselack** m oil cellulose lacquer. **-zeug** n oilskin(s); oilcloth. **-zucker** m (Pharm.) eleosaccharum; glycerine, glycerol. **-zufluß** m oil inflow (or feed). **-zusatz** m addition of oil

Ombré·färbung f ombré (or shaded) dyeing
Önanth· (o)enanth(ic): **-äther** m enanthic ether. **-säure** f enanthic acid. **Önan·thyl·säure** f enanthylic acid
Önidin n (o)enidin. **Önin** n (o)enin
Oolith m (-e) oolite. **oolithisch** a oolitic
Ö.P. abbv (österreichisches Patent) Austrian patent
opak a opaque
Opal·artig a opaline, opal-like. **O–ausrüstung** f (Tex.) opal finish. **-blau** n opal blue. **-druck** m (Tex.) opal(ine) print
opaleszieren vi opalesce. **opaleszierend** p.a opalescent. **Opaleszenz** f opalescence.
Opal·farbe f opal color. **-firnis** m opal varnish. **-glanz** m opaline luster, opalescence. **o–glänzend** a opalescent
opalisieren 1 vt opalize, render opalescent. 2 vi opalesce. **opalisierend** p.a opalescent
Opal·mutter f opal matrix. **o–schillernd** a opalescent
Opazität f (-en) opacity
Oper f (-n) opera
operieren vt/i operate (on)
Operment n (-e) orpiment

Opfer *n* (-) **1** sacrifice; offering. **2** victim.
opfern *vt/i* sacrifice
Ophelia-säure *f* ophelic acid
ophitisch *a* (*Petrog.*) ophitic
Opian-säure *f* opianic acid
opium-haltig *a* containing opium, opiated.
O—handel *m* opium trade. **-säure** *f*
meconic acid. **-tinktur** *f* laudanum
opsonisch *a* opsonic
opt.-akt. *abbv* (OPTISCH-AKTIV) optically
active
optieren 1 *vt* focus. **2** *vi* opt. **Optierung** *f*
(-en) **1** focussing. **2** opting; option
Optik *f* **1** optics. **2** optical equipment. **Op-
tiker** *m* (-) optician, optical instrument
maker
optimal *a* optimum. **optimieren** *vt*
optimize
optisch *a* optical. **o.-aktiv** *a* optically active. **o.-inaktiv** *a* optically inactive
Orange-farbe *f* orange color (*or* dye). **o—
farben, -farbig** *a* orange(-colored). **O—
gelb** *n* orange (color). **-lack** *m* orange lac
(*or* lake). **-mennige** *f* orange minium (*or*
lead)
Orangen-blüte *f* orange blossom. **-essenz** *f*
orange (peel) oil. **-essig** *m* orange vinegar. **-farbe** *f* orange (color). **-frucht** *f* **1**
orange. **2** (*Bot.*) citrus fruit, hesperidium.
-gelb *n* orange (color). **-öl** *n* orange oil.
-samenöl *n* orange seed oil. **-schale** *f* orange peel.
orange-rot *a* orange red. **O—samenöl** *n* orange seed oil.
orangieren *vt* (*Tex.*) treat with milk of lime
Orasthin *n* oxytocin
Orbit *m* (-s) (space) orbit. **Orbita** *f* (..tae *or*
..ten) orbit of the eye, eye socket
Orchidee *f* (-n) orchid
Ord. *abbv* (ORDNUNG) order
Orden *m* (-) order (= decoration *or* society)
ordentlich *a* **1** orderly, tidy; (*adv oft*) in
(good) order. **2** thorough. **3** regular, decent, proper; full(-ranking). **4** real, sound;
considerable. **Ordentlichkeit** *f* orderliness, tidiness; (good) order; thoroughness
ordern *vt* order (for delivery)
ordinär *a* **1** ordinary, common. **2** mediocre,
inferior. **3** vulgar
Ordinarius *m* (..rien) full professor
Ordinaten-achse *f* axis of ordinates.

-strecke *f* ordinate distance
Ordination *f* (-en) **1** ordination. **2** office
hours; office consultation. **ordinieren I**
vt **1** ordain. **2** (*Med.*) prescribe. **II** *vi* have
office hours. **Ordinierung** *f* (-en) **1** ordination. **2** prescription
ordnen I *vt* **1** put in order, straighten out. **2**
arrange, organize; classify. **II: sich o.** be
arranged (organized, classified); line up.
Cf GEORDNET. **Ordner** *m* (-) **1** arranger,
organizer. **2** marshall, monitor. **3** file
(folder). **Ordnung** *f* (-en) (*usu.*) order (=
arrangement, organization, classification); orderliness, good order; proper sequence; social system; routine; rank;
group, subgroup; **in O. in** order, under
control, all right; **in O. bringen** put in
order, organize, = ORDNEN **1, 2**
ordnungs-gemäß, -mäßig *a* orderly, methodical; (*adv oft*) duly, in (due) order. **O—
verfahren** *n* method of organizing
(data); data processing. **-zahl** *f* **1** ordinal
number. **2** atomic number. **3** (*Spect.*) order number
org. *abbv* (organisch) organic
Organ *n* (-e) **1** (*esp Biol.*) organ. **2** voice. **3**
mouthpiece. **4** receptive sense, under-
standing. *Cf* ORGEL. **--brei** *m* (*Biol.*) organ
(tissue) pulp. **-eiweiß** *n* organ protein
Organiker *m* (-) organic chemist.
organisatorisch *a* organizing, organiza-
tional. **organisch** *a* organic: **o.-che-
misch** organic chemical. **organisieren**
vt/i organize. **Organismus** *m* (..smen)
organism
organo-leptisch *a* organoleptic, sense-
perception(al). **O-metall** *n* organometallic compound
Organ-teil *m* part of an organ
Orgel *f* (-n) (*Music*) organ. **--metall** *n* organ-
pipe metal
orientalisch *a* oriental, eastern. **orien-
tieren 1** *vt* orient. **2: sich o.** orient
oneself, get one's bearings. **Orien-
tierung** *f* (-en) orientation
Original-abfüllung *f*, **-abzug** *m* bottling
(*or on labels:* bottled) by the wine grower.
-ausgabe *f* first (*or* original) edition. **o—
groß** *a* full-size(d). **O—größe** *f* original
size. **-hüttenmetall** *n* virgin metal
Originalität *f* (en) **1** originality. **2** unique

character. **3** eccentricity

originell *a* **1** original. **2** ingenious. **3** peculiar, odd

Orkan *m* (-e) hurricane

Orlean *m* (*Dyes*) orlean, annatto

orogenetisch *a* (*Geol.*) orogen(et)ic

Orseille *f* (*Dyes*), **--flechte** *f* (*Bot.*) archil

Orsellin·säure *f* orsellinic acid. **Orsell·säure** *f* orsellic (diorsellinic, lecanoric) acid

Ort *m* **I** (-e) place, site, location; locality; town: **an O. und Stelle** in its proper place; **höherer O.** higher authority; **vor O.** on the spot. **II** (-̈er) (*Math.*) locus; (*Astron.*) position. **III** (*also n*) (-̈er) (shoemaker's) awl. **--beton** *m* in-situ concrete. **orten 1** *vt* locate, find, spot. **2** *vi* take bearings

Ortho·achse *f* (*Cryst.*) ortho axis. **-ameisensäure** *f* orthoformic acid. **-antimonigsäure** *f* orthoantimonious acid. **-antimonsäure** *f* orthoantimonic acid. **o--arm** *a* low-ortho. **O--arsenigsäure** *f* orthoarsenious acid. **-arsensäure** *f* orthoarsenic acid. **-borsäure** *f* orthoboric acid. **-chinon** *n* orthoquinone. **o--chromatisch** *a* orthochromatic. **O--cymol** *n* orthocymene. **-essigsäure** *f* orthoacetic acid.

orthogonal *a* orthogonal. **orthogonalisieren** *vt* orthogonalize. **Orthogonalität** *f* orthogonality

Ortho·kieselsäure *f* orthosilicic acid. **-klas** *m* orthoclase. **-kohlensäure** *f* orthocarbonic acid. **-phosphorsäure** *f* orthophosphoric acid. **-phthalsäure** *f* orthophthalic acid. **o--rhombisch** *a* orthorhombic. **O--salpetersäure** *f* orthonitric acid. **o--salpetrig** *a* orthonitrous. **O--säure** *f* ortho acid. **o--ständig** *a* in the ortho position. **O--stellung** *f* ortho position. **-tellursäure** *f* orthotelluric acid. **o--tomisch** *a* orthotomic. **-typ** *a* (*Cryst.*) orthorhombic. **O--verbindung** *f* ortho compound. **-wasserstoff** *m* orthohydrogen. **-xylol** *n* orthoxylene. **-zimtsäure** *f* orthocinnamic acid

Ort·isomerie *f* position isomerism

örtlich *a* local; regional; (*Med. oft*) topical.

Örtlichkeit *f* (-en) **1** locality. **2** lavatory, facilities

Orton·kegel *m* (*Ceram.*) Orton cone

Orts· local: **-beschreibung** *f* topography. **-bestimmung** *f* orientation, determination of position. **o--beweglich** *a* mobile. **O--bewegung** *f* locomotion

Ortschaft *f* (-en) place, town, village

orts·fest *a* stationary, fixed. **O--funktion** *f* position function. **-isomerie** *f* position isomerism. **-koordinate** *f* spatial coordinate

Ort·stein *m* (*Min.*) bog iron ore, hardpan

orts·üblich *a* locally customary. **O--veränderung** *f*, **-wechsel** *m* **1** change of place, relocation, migration. **2** displacement. **3** locomotion. **-zahl** *f* position number (*or* index). **-zeit** *f* local time

Ortung *f* (-en) orientation; direction (location, position) finding

Oschakk·pflanze *f* oshac (*or* ammoniac) plant

Öse *f* (-n) **1** eye (of needle; for hook). **2** loop; (*specif*) platinum-wire loop. **3** eyelet, ring

Ösen·blatt *n* (*Mach.*) tongue, lip. **-haken** *m* eyehook. **-schraube** *f* eyescrew, eyebolt

Osmiam·säure *f* osmiamic acid

osmig *a* osmious

Osmium-chlorwasserstoffsäure *f* chloroosmic acid (H_2OsO_6 *or* H_3OsO_6). **o--haltig** *a* osmium-bearing, containing osmium. **O--legierung** *f* osmium alloy. **-oxid** *n* osmium oxide (OsO_2). **-oxydul** *n* osmious oxide. **-salmiak** *m* ammonium chloroosmate. **-säure** *f* osmic acid, osmium tetroxide. **-verbindung** *f* osmium compound

Osmose *f* (-n) osmosis. **osmosieren** *vt* osmose. **osmotisch** *a* osmotic

Ost I *m* **1** east. **2** east wind. **II** *pfx* east, eastern

o--ständig *a* ortho, etc = ORTHOSTÄNDIG

ost·deutsch *a* East German. **O--deutschland** *n* East Germany

Osten *m* east, eastern part: **nach O.** (toward the) east; **im O. von** in the eastern part of

Oster· Easter: **-ei** *n* Easter egg

Osterluzei *f* (-en) (*Bot.*) aristolochia, birthwort

Ostern *n* Easter

österr. *abbv* (österreichisch) Aust. (Austrian)

Österreich n (Geog.) Austria. **Österreicher** m (-), **Österreicherin** f (-nen), **österreichisch** a Austrian

ost·europäisch a East European. **O-gebiete** npl Eastern territories (of pre-WW2 Germany, incl East Germany). **-indien** n (East) India. **o-indisch** a East Indian

östlich I a 1 eastern: **der ö—ste Punkt** the easternmost point; **ö—e Länge** east longitude. 2 easterly (wind, direction). **II** adv (+ **von**), prep (+ genit) (to the) east (of): **etwas ö—er gelegen** located somewhat more to the east

Ost·mark f 1 (East German) mark. 2 Eastern borderland

Ostritz·wurzel f masterwort (or its rhizome)

Östrogen n estrogen. **ö—·verdächtig** a suspected of containing estrogen

Ost·see f (Geog.) Baltic Sea

Ostwaldsch a Ostwald (dilution law, etc). **Ostwald·verfahren** n Ostwald-Brauer process

ost·wärts adv eastward. **O–zone** f Eastern Zone (later East Germany)

Osumilith m (Min.) osumilite

Oszillations·quantumzahl f vibrational quantum number

oszillieren vi oscillate

Oszilloskop n (-en) oscilloscope

Otter I m (-) otter. **II** f (-n) adder. **··fell** n otter skin. **-gift** n adder venom

Otto m short for **Otto·kraftstoff** m Otto engine gasoline. **-motor** m Otto engine

Ö.U.P. abbv (österreich-ungarisches Patent) Austro-Hungarian patent (pre-WW1)

o.V. abbv (ohne Verzögerung) without delay

oval·förmig a oval(-shaped)

Ovarium n (..rien) ovary

Ox. abbv (Oxidation) oxidation

Oxalat n (-e) oxalate

Oxal·äther m oxalic ether (ethyl oxalate). **-essigester** m oxaloacetic ester (ethyl oxalacetate). **-essigsäure** f oxaloacetic acid. **-ester** m oxalic ester (ethyl oxalate)

Oxalin n (Min.) imidazole, glyoxaline

Oxalit n (-e) (Min.) humboldtine, oxalite

Oxalkyl· hydroxyalkyl-

Oxal·salz n oxalate. **o-sauer** a oxalate of.

O-säure f oxalic acid

Oxalsäure·äthylester m ethyl oxalate. **-salz** n oxalate. **-trübung** f oxalate turbidity

Oxalur·säure f oxaluric acid

Oxamid· (or **Oxamin·**) **-säure** f oxamic acid

Oxäthyl· hydroxyethyl-

Oxhämoglobin n oxyhemoglobin

Oxhoft n (-e) (Meas.) hogshead

Oxld n (-e) oxide cf **Oxyd**

oxidabel a oxidizable

oxid·artig a oxide-like

Oxidase f (-n) oxidase

Oxidations· cf **Oxyd**: **-artikel** m (Tex.) oxidation style. **-ätze** f (Tex.) oxidation discharge. **-bad** n oxidizing bath. **-beständigkeit** f resistance to oxidation. **-bestreben** n tendency to oxidize. **-bleiche** f, **-bleichen** n, **-bleichung** f oxidizing (specif peroxide) bleach; oxidative (specif peroxide) bleaching. **-braun** n oxidation brown. **o–empfindlich** a sensitive to oxidation. **-fähig** a capable of oxidation, oxidizable. **O–farbe** f (Dye.) oxidation color (produced on fabrics). **-flamme** f oxidizing flame. **-geschwindigkeit** f rate of oxidation. **-grad** m degree of oxidation. **-mittel** n oxidizing agent. **-ofen** m oxidizing oven (or furnace). **-schlacke** f (Metll.) oxidizing slag. **-schutz** m protection against oxidation. **-schwarz** n oxidation (specif aniline) black. **-stoffwechsel** m oxidation metabolism. **-stufe** f stage (or degree) of oxidation, oxidation number. **o–verzögernd** a antioxidant, oxidation-retarding. **O–vorgang** m oxidation process. **-wärme** f heat of oxidation. **-wirkung** f oxidizing action (or effect). **-zahl, -ziffer** f oxidation number (or index)

oxidativ a oxidative, oxidizing cf **Oxyd**

Oxid·beschlag m oxide coating (or film). **o–bildend** a oxide-forming. **O–chlorid** n oxychloride. **-einschluß** m oxide inclusion. **-faden** m (Elec.) oxide-coated filament. **-firnis** m oxide varnish (thickened drying oil). **o–haltig** a oxidic, containing oxide(s). **O–haut** f, **-häutchen** n oxide film. **-hydrat** n hydrated oxide, -ic hydroxide

oxidierbar a oxidizable. **Oxidierbarkeit** f

oxidizability cf OXYD

oxidieren vt/i & **sich o.** oxidize cf OXYD.

oxidierend p.a oxidizing; under oxidizing conditions

Oxidier·mittel n, **-stoff** m oxidizing agent

Oxidierung f (-en) oxidation cf OXYD

Oxidimetrie f oxidimetry cf OXYD

oxidisch a oxidic, oxygenic; oxide-like; higher-valent cf OXYD

Oxid·oxydul n mixed (-ous & -ic) oxide = OXYDULOXID. **o–reich** a rich in oxide(s), high-oxide. **O–rot** n Turkey red, iron oxide red pigment. **-salz** n (-ic) salt. **-schicht** f oxide layer (or film). **o–überzogen** a oxide-coated. **O–überzug** m oxide coating

Oxim n (-e) oxime

Oxin n (-e) oxine

Oxm. abbv (Oxidationsmittel) oxidizing agent

Oxo. pfx (Org. Chem.) oxo-, oft equiv. to keto-: **-säure** f oxo acid (keto or aldehydic acid)

Oxtriazin n (-e) oxatriazine. **Oxtriazol** n (-e) oxatriazole

Oxy· pfx 1 (usu Org. Chem., chiefly indicating hydroxyl group) hydroxy(l), e.g. **Oxyapfelsäure** f hydroxymalic acid. 2 (with ketonic groups) keto-, oxo-. 3 (denoting ring oxygen) oxa-: **-aldehyd** m hydroxylaldehyde. **-ammoniak** n hydroxylamine. **-azoverbindung** f hydroxyazo compound. **-benzol** n phenol. **-bernsteinsäure** f hydroxysuccinic acid (malic acid). **-biazol** n oxadiazole. **-bitumen** n oxidized bitumen. **-carbonsäure** f hydroxycarboxylic acid. **-chinolin** n hydroxyquinoline. **-chinon** n hydroxyquinone. **-chlorid** n oxychloride. **-chlorkupfer** n copper oxychloride. **-cyan** n oxycyanogen

Oxyd n (-e) (usu -ic) oxide (Traditional spelling still widespread in Chem. literature, now being replaced by OXID; the y-spelling may also occur in derivative words such as OXYDATION, OXYDIEREN, HYDROXYD)

Oxydul n (-e) (lower or -ous) oxide cf OXYD. **-eisen** n ferrous iron. **-hydrat** n hydrated -ous oxide, -ous hydroxide

oxydulisch a lower-valent, -ous

Oxydul-oxid n mixed oxide (in which metal has both higher and lower valence), ous-ic oxide. **-salz** n -ous salt

Oxy·essigsäure f hydroxyacetic acid. **-fettsäure** f hydroxy fatty acid. **-fluorid** n oxyfluoride

oxygenieren vt oxygenate, oxygenize.

Oxygenierung f (-en) oxygenation

Oxy·gruppe f hydroxy group. **-keton** n hydroxyketone, -ketol. **-ketoncarbonsäure** f hydroxyketocarboxylic acid. **-methylgruppe** f hydroxymethyl group. **-salz** n oxysalt, hydroxy salt. **-säure** f oxy acid (= hydroxy acid or oxygen-containing acid). **-schwefelsäure** f (hydr)oxysulfuric acid. **-toluol** n hydroxytoluene. **-verbindung** f oxy (or hydroxy) compound. **-zellulose** f hydroxycellulose. **-zyanid** n (hydr)oxycyanide

O.Z. abbv 1 (Ordnungszahl) atomic no. 2 (Oktanzahl) octane no.

Ozäna f (..nen) (Med.) ozena

Ozean m (-e) ocean. **Ozeanien** n (Geog.) Oceania. **ozeanisch** a oceanic

o. Zers. abbv (ohne Zersetzung) without decomposition

Ozo·benzol n ozobenzene. **-kerit** n ozocerite

Ozon n ozone. **-abbau** m ozonolysis. **-bildung** f ozone formation. **-bleiche** f ozone bleach(ing). **o–erzeugend** a ozone-producing. **O–gehalt** m ozone content. **o–haltig** a containing ozone, ozoniferous.

Ozonisator m (-en), **Ozoniseur** m (-e) ozonizer. **ozonisieren** vt ozonize.

Ozonisierung f (-en) ozonization

Ozon·loch n hole in the ozone layer. **-messer** m ozonometer. **-messung** f ozonometry. **-(reagens)papier** n ozone (test) paper. **o–reich** a rich in ozone, high-ozone. **-sauer** a ozonide of. **O–sauerstoff** m ozonized oxygen, oxygen in the form of ozone. **-schicht** f ozone layer, ozonosphere. **-spaltung** f ozonolysis. **-zerstörung** f destruction of ozone

P

p. *abbv* (pro) per

p- *abbv, pfx* (PARA-)

p.a. *abbv* 1 (*Chemicals:* pro analysi) for analysis, analytical grade. 2 (pro anno) per annum, per year

p.A. *abbv* (per Adresse) c/o (care of)

PÄ *abbv* (Petroleumäther) petroleum ether

paar I *indef pron* (*no endgs*) few: **ein p. Stunden** a few hours; **alle p. Tage** every few days. II *a* 1 (*Bot., Zool.*) paired. 2 (*Math.*): **p. oder unpaar** even or odd

Paar *n* (-e) pair, couple; (*Quant: pl* -): **zwei P. Schuhe** two pairs of shoes

paaren I *vt* 1 pair, couple. 2 match. 3 conjugate. II: **sich p.** 4 mate. 5 pair off

paarig *a* paired; *adv* in pairs

Paarling *m* (-e) 1 (*Chem.*) conjugated compound. 2 (*Biol.*) allelomorph, one of a pair; (*pl oft*) homologous chromosomes

paarmal *adv* (*usu*): **ein p.** a few times

Paarung *f* (-en) pairing, coupling; mating; conjugation. **Paarungs-zeit** *f* mating season

paar-weise *adv* in pairs

Pacht *f* 1 lease: **in P. geben** rent out (on lease); **in P. nehmen** take a lease on. 2 leasehold. 3 (amount of) rent. **pachten** *vt* rent, take a lease on. **Pächter** *m* (-) lessee, tenant (farmer)

Pack I *m* (-) pack, stack; bale, bundle. II *n* riffraff. **Päckchen** *n* (-) packet, little pack (bale, bundle). **Päckchen-farben** *fpl* packet dyes. **packen** I *vt/i* pack. II *vt* 1 seize, grab. 2 come over. 3 captivate. **Packen** *m* (-) pack, stack, etc = PACK I. **Packerei** *f* (-en) 1 packing. 2 packing department

Packfong *n* paktong, packfong (copper-nickel-zinc alloy)

Pack-hahn *m* packed cock (*or* valve), gland valve. **-haus** *n*, **-hof** *m* warehouse; packing department. **-kosten** *fpl* packing charges. **-lack** *m* sealing wax for packing. **-leinen** *n*, **-leinwand** *f* packing cloth, packcloth. **-papier** *n* wrapping (*or* pack-ing) paper. **-pappe** *f* packing (card)board. **-schnur** *f* packing twine (string, cord). **-seidenpapier** *n* packing tissue (paper). **-stoff** *m* packing (material). **-tuch** *n* packing cloth

Packung *f* (-en) 1 packing. 2 pack(et), package; wrapping, wrapper; box; bale, bundle. 3 pack, compress. 4 gasket. 5 (*Cryst.*) structure: **dichte P.** close-packed structure. *Cf* PACKEN *v*

Packungs-anteil *m* (*Nucl. Phys.*) packing fraction. **-art** *f* 1 type of packing, packaging. 2 (*Cryst.*) structure. **-dichte** *f* packing density. **-effekt** *m* packing effect. **-erscheinung** *f* packing phenomenon

Pack-zeug *n* packing material(s)

Padde *f* (-n) 1 frog; toad. 2 (*Med.*) (cattle) bloat

PAe. *abbv* (Petroleumäther) petroleum ether

paginieren *vt* paginate, page

Paket *n* (-e) package, parcel

paketieren *vt* 1 package. 2 (*Metll.*) briquet. 3 (*Iron*) pile. **Paketier-schweißstahl** *m* refined iron, merchant bar. **Paketie-rung** *f* packaging; briquetting; piling

Paket-stahl *m* refined iron, merchant bar. **p–weise** *adv* by the package

Pako-wolle *f* alpaca

Paläolithikum *n* Paleolithic (Age). **Paläontologie** *f* paleontology. **Paläozän** *n* Paleocene. **Paläozoikum** *n*, **paläozoisch** *a* Paleozoic (Age)

pälen *vt* (*Lthr.*) unhair (by scraping)

Palisander *m*, **-holz** *n* rosewood

Pallad-gold *n* palladium-gold (alloy)

Palladi- palladic: **-chlorwasserstoff-säure** *f* chloropalladic acid

Palladium-asbest *m* palladium (*or* pal-ladized) asbestos. **-bromür** *n* palladous bromide. **-chlorid** *n* palladic chloride. **-chlorür** *n* palladous chloride. **-chlor-wasserstoff** *m* chloropalladic acid. **-erz** *n* palladium ore. **-gehalt** *m* palladium content. **-jodür** *n* palladous iodide.

-**legierung** *f* palladium alloy. -**mohr** *m* palladium black. -**oxid** *n* palladium (*specif* palladic) oxide. -**oxydul** *n* palladous oxide. -**oxydulnitrat** *n* palladous nitrate. -**oxydulsalz** *n* palladous salt. -**reihe** *f* palladium series. -**salz** *n* palladium salt. -**schwamm** *m* palladium sponge. -**schwarz** *n* palladium black. -**wasserstoff** *m* palladium hydride

Pallado- palladous: -**chlorid** *n* palladous chloride. -**chlorwasserstoffsäure** *f* chloropalladous acid. -**hydroxid** *n* palladous hydroxide

Palmarosa-öl *n* 1 palmarosa (*or* Turkish geranium) oil. 2 (*inaccurate usage*) gingergrass oil

Palmatin-seife *f* palmatin soap

Palm-butter *f* palm oil

Palme *f* (-n) palm (tree)

Palmen-öl *n* palm oil. -**stärke** *f* palm starch, sago. -**wachs** *n* palm wax

Palmetto-öl *n* oil of saw palmetto

Palm-fett *n* palm oil (*or* butter). -**frucht** *f* palm fruit. -**honig** *m* palm honey

Palmin *n* palmitin. -**säure** *f* palmitic acid

Palmitin-säure *f* palmitic acid. -**seife** *f* palmitin soap

Palmitolein-säure *f* palmitoleic acid

Palm-kern *m* palm kernel (*or* nut). -**kernöl** *n* palm-kernel oil. -**kern(öl)seife** *f* palm(-kernel) oil soap. -**lilie** *f* yucca. -**mehl** *n* sago. -**nuß** *f* palm nut (*or* kernel); coconut. -**öl** *n* palm oil. -**stärke** *f* palm starch, sago. -**wachs** *n* palm wax. -**zucker** *m* palm sugar; jaggery

palpieren *vt* palpate

Pampelmuse *f* (-n) grapefruit; shaddock

PAN. *abbv* (Polyacrylnitril) polyacrylonitrile

Panama-rinde *f* Panama (*or* quillai) bark

Panaschierung *f* (-en) variegation

Panazee *f* (-n) panacea

panchromatisch *a* (*Phot.*) panchromatic

Paneel *n* (-e) panel; wainscot(ting)

Panell-öl *n* olive kernel oil

panieren *vt* bread (meat, etc)

Pankreas *n*, -**drüse** *f* pancreas. -**saft** *m* pancreatic juice. **pankreatisch** *a* pancreatic

Panne *f* (-n) 1 breakdown, failure, (*specif*) car trouble. 2 (*Tex.*) panne

panschen *vt/i* water (down), adulterate

Pansen *m* (-) rumen, first stomach (of ruminants); paunch

Pantoffel *m* (-) slipper. **pantoffeln** *vt* (*Lthr.*) board, raise the grain. **Pantoffel-tierchen** *n* slipper animalcule

Pantothen-säure *f* pantothenic acid

pantschen *vt/i* adulterate = PANSCHEN

Panzer *m* (-) 1 armor (plate). 2 (*Mach.*) casing, steel jacket. 3 (*Zool.*) shell, shield. 4 (*Mil.*) tank. -**abwehrrakete** *f* antitank rocket. -**blech** *n* armor plate. -**bombe** *f* armor-piercing bomb. -**brandgranate** *f* armor-piercing incendiary shell. **p-brechend** *a* armor-piercing. **P-draht** *m* armored wire. **p-durchschlagend** *a* armor-piercing. **P-echse** *f* crocodile. -**geschoß** *n* armor-piercing shell. -**glas** *n* bulletproof (*or* armored) glass. -**granate** *f* armor-piercing grenade (*or* shell). -**holz** *n* metal-clad wood. -**kampfwagen** *m* (*Mil.*) tank. -**kopfgranate** *f* armor-piercing shell. -**kraftwagen** *m* armored car. -**mine** *f* (*Mil.*) antitank mine

panzern *vt* armor, armor-plate

Panzer-platte *f* armor plate. -**schlauch** *m* armored hose. -**sprenggeschoß** *n* armor-piercing projectile. -**stahl** *m* armor plate. -**tier** *n* armadillo

Panzerung *f* (-en) 1 armoring, armor-plating; metal-shielding. 2 armor, armor plate; metal casing, shield; shell

Panzer-wagen *m* armored car, tank

Päonidin *n* peonidin. **Päonie** *f* (-n) peony.

Päonin *n* peonin. **Päonol** *n* peonol

Papagei *m* (-en) parrot. **Papageien-krankheit** *f* psittacosis. **papagei-grün** *a* parrot-green

Papel *f* (-n) (*Med.*) papule, pimple

Papier *n* (-e) paper. -**abfall**, -**abgang** *m* paper waste. **p-artig** *a* paper-like, papery. **P-band** *n* paper tape. -**blatt** *n*, -**bogen** *m* sheet of paper. -**brei** *m* paper pulp. -**ebene** *f* plane of the paper

papieren *a* 1 paper(y). 2 bookish

Papier-fabrik *f* paper mill. -**fabrikant** *m* paper maker. -**fabrikation** *f* paper manufacture. -**farbe** *f* paper color. -**fläche** *f* surface of the paper. -**garn** *n* paper yarn (*or* twine). -**handel** *m* paper (*or* stationery) trade. -**handlung** *f* stationery store.

-holz n pulpwood. **-jod** n test paper solution (of iodine and potassium iodide). **p-kaschiert** a paper-covered (or -backed) (metal foil, etc). **P-kaschierung** f paper coating. **-kohle** f paper coal, lignite. **-korb** m waste-paper basket. **-krieg** m paper chase; red tape. **-leim** m paper size (or sizing). **-macher** m paper . maker. **-masse** f 1 paper pulp. 2 papier-mâché. **-mühle** f paper mill. **-pappe** f paperboard, cardboard. **-pergament** n parchment paper. **-prüfung** f paper testing. **-sack** m paper bag (or sack). **-scheibe** f paper disc. **-schere** f paper shears (or scissors). **-schneidemaschine** f paper cutter (or cutting machine). **-sorte** f type (or grade) of paper. **-stoff** m paper pulp. **-stoffbrei** m paper pulp slurry. **-streifen** m strip of paper. **-teig** m papier-mâché. **-trommel** f paper drum. **-tüte** f paper bag. **-überseite** f (Paper) felt side. **-währung** f paper money (or currency). **-ware(n)** f(pl) stationery. **-wolle** f shredded paper. **-zeichen** n watermark. **-zeug** n paper (or wood) pulp, stuff

Papille f (-n) papilla; nipple. **Papillen-** (in compds) papillary

papinianischer (or **Papinscher**) **Topf** Papin's digester, autoclave

Papp m 1 mush; goo. 2 paste. 3 (in compds, implies) cardboard, pasteboard: **-band** m pasteboard volume, book bound in boards, hard paperback. **-bogen** m sheet of cardboard. **-deckel** m cardboard, pasteboard. **-dose** f cardboard box. **-druck** m paste resist printing

Pappe f (-n) 1 cardboard, pasteboard; (Books): **in P. gebunden** bound in boards; **nicht von P.** (fig) not made of cardboard, no pushover; **geformte P.** millboard. 2 (in compds oft) board: **Asphaltpappe** asphalt board. 3 (= PAPP) paste; mush

Pappel f (-n) 1 poplar. 2 mallow. **-art** f variety of poplar (or mallow). **-holz** n poplar (wood). **-kraut** n mallow

pappen 1 vt paste, stick. 2 vi cake, stick (together)

Pappen- cardboard, pasteboard: **-art** f (type of) cardboard. **-deckel** m cardboard, pasteboard. **-fabrik** f (paper)board mill.

-guß m cast pressboard (or millboard). **-leim** m pasteboard glue. **-stiel** m trifle

Papp-hülse f pasteboard case

pappig a pasty, doughy, sticky

Papp-karton m cardboard box, carton. **-masse** f papier-mâché. **-reserve** f (Tex.) paste resist. **-schachtel** f cardboard box. **-schirm** m cardboard screen. **-teller** m paper plate. **-tüte** f heavy paper bag. **-waren** fpl cardboard goods. **-weiß** n (Tex.) pigment white

Paprika-schote f (bell) pepper

Papyrographie f paper chromatography

Papyrolin n cloth-centered paper

Papyrus-staude f papyrus plant

Paraban-säure f parabanic acid

Parabel f (-n) 1 parabola. 2 parable. **p-förmig, parabolisch** a parabolic. **Parabol-spiegel** m parabolic mirror

Para-chinon n paraquinone

Paracon-säure f paraconic acid

Para-cyan n paracyanogen. **-cymol** n paracymene

Parade-beispiel n (+ **für**) classic (or perfect) example (of)

Paradeis-apfel m, **Paradeiser** m (-) tomato. **Paradeis-mark** n tomato paste

Paradies n (-e) paradise. **-apfel** m tomato. **-feige** f banana. **-holz** n agalloch, aloes wood. **-körner** npl grains of paradise. **-nuß** f sapucaia nut. **-vogel** m bird of paradise

paradox a paradoxical

Paraffin-anstrich m paraffin coating. **-bad** n paraffin bath. **-durchtränkung** f impregnation with paraffin. **-einbettung** f imbedding in paraffin. **-gatsch** m paraffin-wax cake, slack wax. **p-haltig** a containing paraffin, paraffinic

paraffinieren vt (coat, impregnate with) paraffin. **paraffinisch** a paraffin(ic)

Paraffin-kaschierung f wax backing. **-kerze** f paraffin candle. **-kohlenwasserstoff** m paraffin hydrocarbon. **-lack** m paraffin varnish. **-leimung** f paraffin sizing. **-öldestillat** n paraffin distillate. **-reihe** f paraffin (or alkane) series. **-salbe** f petrolatum; paraffin ointment. **-säure** f paraffinic (saturated aliphatic) acid. **-schuppen** fpl paraffin scale. **-spaltung** f paraffin cracking.

-**tränkung** f impregnation with paraffin.
-**wachs** n paraffin wax
Para·fuchsin n para fuchsine, chloride of pararosaniline base
Paragenese° f paragenesis
Para·gummi n Pará rubber
Para·kampfersäure f paracamphoric acid
Para·kautschuk m Pará rubber
Para·kresol n paracresol
Parallele f (-n) parallel. **Parallelismus** m (..ismen), **Parallelität** f (-en) parallelism
Parallel·probe f parallel sample (or test).
-**schaltung** f (Elec.) parallel connection (or hook-up). **p–steng(e)lig** a (Cryst.) parallel-columnar. **P–versuch** m parallel experiment, duplicate determination.
p–wandig a parallel-walled (or -sided)
Paralysator m (-en) anticatalyst. **Paralyse** f (-n) paralysis. **paralysieren** vt paralyze
para·magnetisch a paramagnetic
Paramidophenol n para-aminophenol
Para·milchsäure f paralactic (or dextrolactic) acid
paramorph a paramorphic, paramorphous
Para·nuß f Brazil nut
Para·phosphorsäure f pyrophosphoric acid. -**phthalsäure** f paraphthalic acid
parasitär a parasitic. **Parasiten·kunde** f parasitology. **p–tötend** a parasiticidal. **parasitisch** a parasitic
para·ständig a in the para position. **P–stellung** f para position
parat a prepared, ready
Para·verbindung f para compound. -**wasserstoff** m parahydrogen. -**weinsäure** f paratartaric (or racemic) acid. -**xylol** n paraxylene. **p–zentrisch** a paracentric. **P–zustand** m para state. -**zyan** n paracyanogen
Pardel·katze f ocelot
Parellin·säure f parellinic acid
Parell·säure f parellic acid
Parenchym n (-e) parenchyma. **parenchymatisch** a parenchymatous
Parenthese f (-n) parenthesis, parenthetic expression; **in P.** parenthetically
Parfum n (-s) **Parfüm** n (-e) perfume. **Parfümerie** f (-n) 1 perfume industry. 2 perfume (or cosmetics) shop (or department). 3 pl perfumes. --**seife** f perfumed soap.

Parfümeur m (-e) perfumer. **Parfüm·grundstoff** m perfume base. **parfümieren** vt perfume. **Parfümör** m (-e) perfumer
pari adv, **Pari** n par: **auf** (or **zu**) **p.** at par; **(al) p. stehen** be equal, be at par
paribeschweren, pari/erschweren vt (Silk) par-weight, par-load
Parill·säure f parillic acid
Parinin·säure f parininic acid
Parin·säure f parinic acid
Pari·seide f par silk
Pariser a (no endgs) or pfx: **P. Blau, Pariserblau** Paris blue; **P. Grün** n Paris green. --**lack** m carmine lake. -**rot** n colcothar, Paris red, red lead. -**weiß** n Paris white, whiting
Parität f (-en) parity
Pari·wert m par value
parker(isiere)n vt parkerize. **Parker·verfahren** n Parker process (of phosphate rust-proofing)
parkesieren vt (Metll.) subject to Parke's process. **Parke·verfahren** n Parke's process (for desilvering lead)
Parkett·kosmetik f (parquet) floor waxing. -**kosmetikerin** f floor waxer (= person). -**pflege** f floor waxing. -**pflegerin** f floor waxer (= person). -**wachs** n floor wax
Park·platz m 1 parking lot. 2 parking space. -**uhr** f parking meter
Paroli: P. bieten (dat.) stand up (to), counter
Parrot·kohle f parrot (or cannel) coal
Partei f (-en) 1 (political, legal) party (incl. person, client, subscriber, tenant); interested party. 2 side (in a contest): **P. nehmen** take sides. **parteiisch** a partial, partisan. **parteilich** a 1 (of the) party. 2 partisan, factious. **parteilos** a non-party; impartial, independent. **Partei·nahme** f partiality, partisanship
Parterre n (-s) ground floor
Partial·bruch m (Math.) partial fraction. -**druck** m partial pressure. -**valenz** f partial valence
partiär a partial; adv oft by parts
Partie f (-n) 1 part. 2 batch, lot, parcel. 3 game. 4 outing, picnic. 5 company, group. 6 match. **partiell** a partial; adv oft partly.

partien·weise *adv* in lots (batches, parcels)

Partikel *f* (-n) particle. **Partikelchen** *n* (-) small particle

Parvolin *n* 2,3,4,5-tetramethylpyridine

Parzelle *f* (-n) lot, parcel (of land). **parzellieren** *vt* divide into parcels (or lots)

PAS. *abbv* (Paraaminosalicylsäure) para-aminosalicylic acid

Paß *m* (Pässe) 1 (*usu*) pass. 2 passport. 3 amble

Passagier *m* (-e) passenger

Passat *m* (-e), **--wind** *m* trade wind

passen I *vi* 1 be suitable (fitting, proper). 2 (+ *dat*, + **auf**, **in**) fit. 3 (+ *dat*) suit. 4 (+ **zu**) go (with), match; fit. 5 (*Games*) pass. 6 (+ **auf**) be on the lookout (for). II *vt* 7 (+ **in**) fit (sthg) in(to). III: **sich p.** be fitting (or appropriate). **passend** *p.a* fitting; suitable, proper, appropriate; matching

Paß·farbe *f* (*Dye.*) illuminating color. **-glas** *n* graduated glass

passieren I *vt* 1 pass (by, over, through) a place); traverse (a road); cross (rivers, boundaries). 2 strain, pass through a sieve; (*Dye.*) liquor. 3 let pass (through), let through. II *vi(s)* 4 pass, go by (or through). 5 (*oft* + *dat*) happen (to)

Passier·maschine *f* 1 straining (or sieving) machine; strainer, collander; (*Dye.*) liquoring machine. 2 meat grinder, food mill. **-schein** *m* pass, permit

passiv *a* passive; inert. **passivieren** *vt* passivate (metals), render inert. **Passivierung** *f* passivation. **Passivität** *f* passivity

Paß·rohr *n* pipe fitting. **-sitz** *m* snug (or tight) fit. **-stift** *m* dowel. **-stück** *n* adapter. **-teil** *m* fitting (matching, mating) part. **-toleranz** *f* (*Mach.*) fit tolerance

Passung *f* (-en) (machine) fit. **Passungs·rost** *m* fretting corrosion

Pasta, Paste *f* (..ten) paste, pulp. **Paste·kathode** *f* pasted cathode

Pastell *n* (-e) pastel (drawing). **--blau** *n* pastel blue. **-farbe** *f* pastel color. **p--farbig** *a* pastel-colored. **P--papier** *n* pastel paper. **-stift** *m* pastel crayon

pasten·artig *a* pasty. **P--farbe** *f* (*Paint*) paste color. **p--förmig** *a* pasty. **P--harz** *n*

paste resin. **-konsistenz** *f* pasty consistency

Pastete *f* (-n) pastry, pie, pâté

Pasteur *m* (-e) pasteurizer. **pasteurisieren** *vt* pasteurize

Paste·zementieren *n* (*Metll.*) cementation using a paste

pastieren *vt* (make into a) paste

pastig *a* pasty

Pastille *f* (-n) pastille; cough drop; tablet. **Pastillen·presse** *f* tablet press.

Pastinak *m* (-e), **Pastinake** *f* (-n) parsnip

Past·milch *f* pasteurized milk

pastös *a* pasty

Patent *n* (-e) patent: **zum P. angemeldet** patent applied for. **--amt** *n* patent office. **-anmeldung** *f* patent application. **-anspruch** *m* patent claim. **-anwalt** *m* patent attorney. **-beschreibung** *f* patent description (or specifications). **-blatt** *n* patent gazette. **-blau** *n* patent blue. **-brief** *m* license, permit. **-bruch** *m* patent infringement. **-dauer** *f* life of the patent. **-einspruch** *m* objection to granting of a patent. **p--fähig** *a* patentable. **P--frage** *f* patent problem; a matter of a patent. **-gelb** *n* patent yellow. **-gesetz** *n* patent law. **-grün** *n* patent green. **-gummi** *n* cut sheet rubber

patentierbar *a* patentable. **patentieren** *vt* patent (*incl Metll.*)

Patent·inhaber *m* patentee, patent holder. **-kali** *n* potassium magnesium sulfate. **-klage** *f* patent (law)suit. **-kohle** *f* briquette. **-lösung** *f* ingenious (or ideal) solution. **-recht** *n* 1 patent law. 2 patent rights. **-rot** *n* patent red (mercuric sulfide). **-salz** *n* ammonium antimony fluoride. **-schrift** *f* patent (document); patent specification. **-schutz** *m* protection by patent. **-schutzgesetz** *n* patent law. **-streit** *m* patent contest (or dispute). **-träger** *m* patentee, patent holder. **-verletzung** *f* patent infringement. **-zement** *m* Roman cement

Paternoster·aufzug *m* continuous (or non-stop) elevator. **-erbse** *f* Indian (or wild) licorice. **-werk** *n* endless bucket conveyor; noria

pathogen *a* pathogenic

pathologisch *a* pathological

patinieren *vt* patinate

Patrize *f* (-n) male (half of) die (*or* mold), die punch

Patrone *f* (-n) 1 cartridge. 2 pattern, stencil; mold; (*Tex.*) design. 3 shell, case, thimble

Patronen·hülse *f* (empty) cartridge case. **-papier** *n* cartridge paper. **-zylinder** *m* cartridge case

Patrouille *f* (-n), **patrouillieren** *vt/i* patrol

Patsche *f* (-n) 1 mud, slush; (*fig*) fix, mess. **patschen** 1 *vi*(h) clap; splash. 2 *vi*(s) slosh (through mud, etc). **patsch·naß** *a* soaking wet

Patschulen *n* patchoulene. **Patschuli·öl** *n* oil of patchouli

pattinsonieren *vt* pattinsonize

Pauke *f* (-n) 1 (kettle)drum. 2 (*Anat.*) tympanum. **pauken** 1 *vt/i* cram (=study). 2 *vi* beat the drum. **Pauken·** (*in compds, oft Anat.*) tympanic: **-fell** *n* 1 drumhead. 2 (*Anat.*) tympanic membrane

Pauli-Verbot *n* Pauli's Exclusion Principle

Pauly·seide *f* Pauly silk, cuprammonium rayon

pauschal 1 *a* & *oft in compds* lump-sum, all-inclusive; very general. 2 *adv* all included, all expenses paid; wholesale, as a whole. **Pauschale** *f* (-n) *or* *n* (..lien) lump sum, flat charge; all-inclusive rate

pauschen *vt* 1 puff up. 2 (*Metll.*) refine

Pauscht *m* (-e) (*Paper*) post (of wet sheets)

Pause *f* (-n) 1 tracing; blueprint. 2 pause, break; recess; intermission: **P. machen** take a break, rest, pause. **pausen** *vt* trace, make a tracing of. **pausieren** *vi* pause, take a break, stop over. **Paus·leinen** *n*, **-leinwand** *f* tracing cloth. **-papier** *n* tracing paper. **-zeichnung** *f* tracing

Pavian *m* (-e) baboon

PC. *n abbu* (Polyvinylchlorid) P.V.C.

·pctg. *abbu, sfx* (-procentig) -percent

Pe-Ce-Faser *f*, *TN* P.V.C. fiber *cf* PC

Pech *n* (-e) 1 pitch; asphalt. 2 hard luck. **-·art** *f* type of pitch. **p–artig**, *a* pitchy, bituminous. **P–blende** *f* pitchblende. **-draht** *m* (shoemaker's) tarred thread. **-eisenerz** *n* pitch-like iron ore (= triplite,

pitticoke, compact limonite)

pecheln 1 *vt/i* extract pitch (from). 2 *vt* (coat with) pitch, tar. 3 *vi* smell of pitch. **pechen** *vt* (coat with) pitch

Pech·erde *f* bituminous earth. **-erz** *n* 1 pitchblende. 2 pitch-like iron ore = PECHEISENERZ. **p–finster** *a* pitch-dark. **P–gang** *m* asphalt rock. **-geschmack** *m* pitchy taste. **-glanz** *m* pitchy luster. **-granat** *m* colophonite (pitchy-looking andradite). **p–haltig** *a* containing pitch, bituminous, asphaltic. **P–harz** *n* pitch resin

pechig *a* pitchy

Pech·kiefer *f* pitch pine. **-kohle** *f* pitch (*or* bituminous) coal, jet. **-koks** *m* pitch coke. **-kupfer** *n* (*Min.*) chrysocolla. **-öl** *n* tar oil. **-pflaster** *n* asphalt pavement. **-rückstand** *m* pitch residue. **p–schwarz** *a* pitch-black. **P–stein** *m* pitchstone. **-steinkohle** *f* pitch coal, jet. **-tanne** *f* pitch pine. **-torf** *m* pitch (*or* black) peat. **-uran** *n* pitchblende, uranite

pedial *a* (*Cryst.*) asymmetric

Pegel *m* (-) 1 water gauge. 2 level

pegmatitisch *a* (*Petrog.*) pegmatitic

peignieren *vt* comb (fibers)

Peil· sounding, direction-finding: **-anlage** *f* sounding (*or* direction-finding) device. **-bombe** *f* pilot bomb

peilen *vt/i* 1 sound, find the depth. 2 take a bearing (of), find the direction (of). **Peiler** *m* (-) direction finder. **Peilung** *f* (-en) sounding; direction finding

Pein *f* pain, agony, torment. **peinigen** *vt* torment, torture; afflict

peinlich *a* 1 embarrassing. 2 painstaking, meticulous. 3 penal

Peitsche *f* (-n), **peitschen** *vt/i* whip. **peitschen·förmig** *a* whip-shaped, flagelliform. **P—schlagsyndrom** *n* whiplash syndrome

Pekan·nuß *f* pecan

pektin·artig *a* pectin-like. **pektinig, pektinisch** *a* pectinous, pectinic

Pektin·körper *m* pectic substance. **-säure** *f* pectic acid. **-stoff** *m* pectic substance. **-zucker** *m* arabinose

Pektolith *m* (*Min.*) pectolite

pelagisch *a* pelagic

Pelargonium·öl n geranium oil. **Pel·argon·säure** f pelargonic acid
Péligot·rohr n, **-röhre** f Péligot tube
Pelle f (-n) skin. **pellen** 1 vt skin. 2 vt & sich p. peel
Peloteur m (-e), **Peloteuse** f (-n) (Soap) plodder
Pelz m (-e) 1 fur. 2 fur garment. 3 pelt, skin. 4 (Tex.) fleece, nap. **p-·artig** a furry, fur-like; cottony; nappy. **P-gerbung** f fur-skin tanning. **pelzig** a furry; cottony; nappy
Pelz·lustrierung f fur lustering. **-tier** n fur-bearing animal. **-ware** f, **-werk** n furs
Penaldin·säure f penaldinic acid. **Penald·säure** f penaldic acid
Pendel n (-) pendulum; (in compds oft) swing; shuttle; commuter. **--lager** n (Mach.) swing bearing. **pendeln** vi (h or s) 1 swing, oscillate. 2 run back and forth. 3 shuttle, commute. **Pendelung** f (-en) oscillation, swinging; shuttling; commuting. **Pendler** m (-) commuter
Penetrameter n (-), **Penetrations·messer** m penetrometer. **penetrieren** vt penetrate
Penetron n (-en) meson
penibel a 1 exacting (task). 2 fastidious, finicky
Penicillen·säure f penicillenic acid. **Penicilloin·säure** penicilloic acid
Penillo·säure f penilloic acid. **Penill·säure** f penillic acid
Pensée f (-s) pansy. **p-·farbig** a pansy-colored
Pension f (-en) 1 pension. 2 (room and) board. 3 boarding home. **pensioniert** a retired, pensioned
Pensum n (.. sa) 1 assignment, homework. 2 curriculum
Penta·bromphosphor m phosphorus pentabromide. **p-carbocyclisch** a pentacarbocyclic. **P-chlorphosphor** m phosphorus pentachloride. **-dekansäure** f pentadecanoic acid. **-erythrit** n pentaerythritol. **-thionsäure** f pentathionic acid. **p-zyklisch** a pentacyclic
Penthiazol n 1,3,2H- or 1,3,4H-thiazene
Penthiophen n 1,2- or 1,4-thiopyran
Pentin·säure n pentinoic acid

Pentit n pentitol
Pentosurie f pentosuria
Pentrinit n pentaerythritol tetranitrate nitroglycerine
Pepsin·drüse f peptic gland. **p—haltig** a containing pepsin
Peptisator m (-en) peptizer, peptizing agent. **peptisch** a peptic. **peptisieren** vt peptize. **Peptisierung** f peptization
peptolytisch a peptolytic
pepton·erzeugend a peptogenic. **P-fleischbrühe** f peptone beef broth
peptonisieren vt peptonize
per I prep per; by, via; as of. II pfx (Chem.) per(oxy)-, super: **Per·acidität** f hyperacidity. **-ameisensäure** f per(oxy)formic acid. **-benzoesäure** f per(oxy)benzoic acid. **-borsäure** f per(oxy)boric acid. **-bromsäure** f per(oxy)bromic acid. **-buttersäure** f per(oxy)butyric acid. **-carbonsäure** f per(oxy)carboxylic acid
Percha·gummi n gutta percha
Perchlor·äthan n perchloroethane, hexachloroethane. **-äthylen** n perchloroethylene, tetrachloroethylene. **-methan** n perchloromethane, carbon tetrachloride. **-säure** f per(oxy)chloric acid
Perchrom·säure f perchromic acid
Pereira·rinde f pereira (bark)
perennierend a perennial
Per·essigsäure f per(oxy)acetic acid. **-ferricyanwasserstoffsäure** f perferricyanic (or hexacyanoferric) acid
perforieren vt perforate
perfluoriniert a perfluorinated
Pergament n (-e) parchment. **p-·ähnlich** a parchment-like. **pergamentieren** vt parchmentize. **Pergamentierung** f parchmentization
Pergament·leder n vellum. **-papier** n 1 parchment paper. 2 wax paper. **-schlauch** m parchment tubing
Pergamyn n glassine (paper)
perhydrieren vt perhydrogenate
Peridot m (-e) (Min.) olivine, peridote
Perihel n (-e) (Astron.) perihelion
Perikard n (-e) (Anat.) pericardium
Periklas m (-e) periclase, periclasite
Perilla·aldehyd m perillaldehyde. **-alkohol** m perillic alcohol. **-öl** n perilla oil. **-säure** f perillic acid

Periode f (-n) period, (*Elec. oft*) cycle.
Perioden·system n periodic system.
-zahl f number of cycles, frequency.
periodisch a periodic(al); (*Math. oft*) e.g.
p—er Dezimalbruch recurring decimal. **Periodizität** f (-en) periodicity
peripher(isch) a peripheral
Peritektikum n (..ika), **peritektisch** a peritectic
Per·jodat n periodate. **-jodsäure** f periodic acid. **-kohlensäure** f percarbonic acid
Perkussions·ladung f detonator charge. **-zünder** m percussion fuse. **-zündhütchen** n percussion cap. **-zündung** f percussion priming
perl·artig a pearly; bead-like; nacreous. **P–asche** f pearl ash. **-auster** f pearl oyster. **-boot** n nautilus. **-druck** m (*Tex.*) yarn printing
Perle f (-n) pearl; bead. **perlen** vi(h *or* s) bubble, sparkle; form pearls (*or* beads); ripple (like pearls)
perlen·artig a pearly; bead-like; nacreous
perlend p.a bubbly, effervescent cf PERLEN
Perlen·glanz m pearly luster. **p–grau** a pearl-gray. **P–mosaik** n mosaic of beads. **-muschel** f pearl oyster (*or* mussel). **-probe** f bead test. **-schnur** f string of beads (*or* pearls)
Perl·erz n pearl ore. **p–farben** a pearl-colored. **P–glanz** m pearly luster. **-glimmer** m (*Min.*) margarite. **-gras** n pearl grass. **p–grau** a pearl-gray; zinc-gray. **P–graupen** fpl pearl barley. **-haileder** n shagreen. **-huhn** n guinea hen
perlig a pearly
Perlit m (-e) 1 (*Petrog.*) perlite. 2 (*Metll.*) pearlite. **p—ähnlich** a (*Metll.*) pearlitoid, pearloid. **P–insel** f (*Metll.*) pearlite area. **perlitisch** a (*Min.*) perlitic; (*Metll.*) pearlitic. **perlitisieren** vt (*Metll.*) pearlitize, render pearlitic
Perl·katalysator m bead catalyst. **-kohle** f pea coal. **-koks** m coke breeze. **-leim** m pearl glue. **-moos** n pearl moss, carrageen. **-muschel** f pearl oyster (*or* mussel). **-mutter** f mother-of-pearl, nacre
perlmutter·artig a like mother-of-pearl, nacreous. **P–blech** n crystallized tinplate. **-glanz** m mother-of-pearl (*or*

nacreous) luster. **p–glänzend** a having a mother-of-pearl luster. **P–papier** n nacreous paper
Perl·polymerisation f bead polymerization. **-reaktion** f bead reaction. **-rohr** n, **-röhre** f bead tube, tube filled with glass beads. **-sago** m pearl sago. **-salz** n microcosmic salt, sodium ammonium phosphate. **-samen** m seed pearl. **-schicht** f nacreous layer (of shells). **-schnur** f 1 string of beads (*or* pearls). 2 row of droplets. **-seide** f embroidery silk. **-spat** m pearl spar, pearly dolomite. **-stein** m (*Min.*) perlite, adularia. **-sucht** f bovine tuberculosis. **-weiß** n pearl white. **-zwiebel** f pearl onion
Perm n (*Geol.*) Permian (Age)
Permanent·gelb n permanent yellow. **-grün** n permanent green (chromium oxide pigment). **-rot** n permanent red, red toner. **-violett** n permanent violet (manganese phosphate pigment). **-weiß** n permanent white, barium sulfate pigment
Permanenz·satz m permanence principle
Permanganat·lösung f permanganate solution. **Permangan·säure** f permanganic acid
Permeabilitäts·änderung f change in permeability. **-messer** m permeameter
permisch a (*Geol.*) Permian
Perm·legierung f high-permeability alloy
permutieren vt 1 (*Math.*) permute, permutate. 2 (*Chem.*) treat with permutite
Permutit·verfahren n permutite process
Perm·zeit f (*Geol.*) Permian Age
Perna TN short for Perchloronaphtalen(e)
Pernambuk·holz n Pernambuco wood. **-kautschuk** m Pernambuco rubber, mangabeira
perniziös a pernicious
Perowskit n (-e) (*Min.*) perovskite
peroxidieren vt peroxidize cf OXYD
Peroxo·diphosphorsäure f peroxodiphosphoric acid. **-hydrat** n hydrogen peroxide addition compound. **-salz** n peroxo salt. **-säure** f peroxo acid
Perpendikel m or n (-) 1 perpendicular. 2 pendulum
Perpetuum mobile n (*Latin*) perpetual motion
Per·phthalsäure f perphthalic acid.

-rheniumsäure f perrhenic acid.
-ruthen(ium)säure f perruthenic acid.
-salz n per(oxy)salt
Persaner·stahl m Brescian steel
Per·säure f per(oxy)acid. **-schwefelsäure** f per(oxy)sulfuric acid
Perseit n (-e) perseitol
Persenning f (-e or -en) tarpaulin
Perser m (-) Persian (person, rug, cat, horse). **Persianer** m (-) Persian lamb coat. **Persien** n (Geog.) Persia
Persimmon(holz) n persimmon (wood).
Persimone f (-n) persimmon
persisch a Persian: **p-e Erde, p-es Rot, Persisch·rot** n Persian red, coral red
Person f (-en) 1 person, pl oft people. 2 identity. 3 actor, character, pl oft cast. **Personal** n personnel, employees; (in compds oft) personal. **Personalien** npl personal data **p-los** a unmanned. **personell** a (of the) personnel
Personen· person(al); passenger: **-fahrzeug** n passenger vehicle. **p-gebunden** a individual(-use), for personal use
personifizieren vt personify
persönlich a personal, individual. **Persönlichkeit** f (-en) personality; individual
Perspektiv n (-e) field glasses, spyglass. **Perspektive** f (-n) perspective
Per·stoff m diphosgene, superpalite (ClCO$_2$CCl$_3$). **-sulfomolybdänsäure** f perthiomolybdic (or thiopermolybdic) acid. **-sulfozyansäure** f perthiocyanic (or persulfocyanic) acid. **-technetiumsäure** f pertechnetic acid. **-thiokohlensäure** f perthiocarbonic acid
peruanisch a Peruvian. **Peru·balsamöl** n oil of Peru balsam
Perücken·baum m sumach tree
Peru·gummi n Peruvian gum (or rubber). **-rinde** f cinchona (or Peruvian) bark. **-salpeter** m Peruvian saltpeter. **-silber** n nickel silver, Cu-Ni-Ag-Zn alloy
Per·verbindung f per (or peroxy) compound
Perylen·chinon n perylenequinone
Perzent n (-e), **·perzentig** a percent = PRO-ZENT, ·PROZENTIG
Pest f plague, pestilence. **p-·artig** a pestilential, contagious

Pestilenz·kraut n (Bot.) goat's rue
Petersil m, **Petersilie** f parsley. **petersilien·ähnlich** a parsley-like. **P-öl** n parsley oil. **-samen** m parsley seed
Peters·kraut n (Bot.) wall pellitory
petiotisieren vt petiotize (wines)
Petitgrain·bergamottöl n oil of petitgrain bergamot, bergamot leaf oil. **-bigaradeöl** n oil of petitgrain bigarade, bitter orange leaf oil. **-mandarinöl** n oil of petitgrain mandarin, mandarin leaf oil. **-öl** n oil of petitgrain, orange leaf oil. **-paraguayöl** n oil of petitgrain Paraguay
Petri·schale f, **Petrisches Schälchen** n Petri dish
Petrochemie° f petrochemistry, petroleum chemicals industry. **petrochemisch** a petrochemical
Petrographie f petrography. **petrographisch** a petrographic(al)
Petrol n petroleum. **-äther** m petroleum ether, naphtha, light petroleum. **-benzin** n gasoline. **-chemie** f petrochemistry. **p-chemisch** a petrochemical
Petroleum·äther m petroleum ether, naphtha, light petroleum. **-behälter** m petroleum container (or tank). **-benzin** n gasoline. **-brunnen** m oil well. **-dampf** m petroleum vapor. **-destillationsgefäß** n petroleum still. **-essenz** f gasoline. **-geruch** m petroleum odor. **p-haltig** a oil-bearing, containing petroleum. **P-handel** m petroleum trade, oil business. **-heizung** f oil heating. **-kocher** m oil stove. **-lager** n oil deposit. **-lampe** f oil lamp. **-pech** n petroleum pitch. **-prober**, **-prüfer** m petroleum tester. **-quelle** f oil well. **-seifenbrühe** f (Agric.) kerosene emulsion
Petrol·koks m petroleum coke. **-pech** n petroleum pitch (or tar), bitumen. **-säure** f petrolic acid
Petroselin·säure f petroselinic acid
Petschaft n (-e) seal, signet
Petsche f (-n) drying room; drying rack
Pf. abbv 1 (PFENNIG). 2 (PFUND) lb.
Pfad m (-e) path
Pfaffen·hütchen n (Bot.) wahoo, spindletree
Pfahl m (-e) 1 (fence) pale, post; stake, pole. 2 (Engg.) pile. 3 prop. **pfählen** vt 1 prop

(up), stake. **2** impale

Pfahl-gründung f pile foundation. **-werk** n **1** paling, pale fence. **2** piling. **-wurm** m (Zool.) marine borer. **-wurzel** f taproot

Pfalz f **1** (Geog.) Palatinate. **2** (-en) palace

Pfand n (-er) deposit, security; pledge. **pfänden** vt seize (as security)

Pfanne f (-n) **I** (genl, usu) pan. **II** (specif) **1** frying pan. **2** (Anat.) socket, acetabulum. **3** (Brew.) copper. **4** (Metll.) ladle. **5** (Ceram.) pantile, roof tile. **6** (Mach.) bearing, bush(ing)

Pfannen-gericht n pan-fried dish (or food). **-probe** f (Metll.) ladle test. **-stein** m pan (or boiler) scale. **-stiehl** m panhandle. **-wagen** m (Metll.) ladle car. **-ziegel** m pantile

Pfann-kuchen m pancake

Pfau m (-e or -en) peacock. **pfau(en)-blau** a peacock blue

Pfd. abbv (PFUND) lb.

Pfeffer m pepper. **pf--artig** a peppery, pepper-like. **Pf--gurke** f gherkin

pfefferig a peppery

Pfeffer-korn n peppercorn. **-kraut** n **1** savory. **2** peppergrass. **3** stonecrop. **-kuchen** m gingerbread. **-minz I** m peppermint brandy. **II** n peppermint (drop, candy). **-minze** f (Bot.) peppermint

Pfefferminz-geruch m peppermint odor. **-kampfer** m menthol. **-öl** n peppermint oil

Pfeffer-nuß f spice cookie. **-öl** n pepper oil. **-schote** f bell pepper. **-stein** m peperino

Pfeife f (-n) **1** pipe. **2** fife. **3** whistle. **4** (Metll.) blowhole. **pfeifen*** (pfiff, gepfiffen) vt/i whistle, pipe

Pfeifen-erde f pipe clay. **-rohr** n pipestem. **-stein** m pipestone. **-ton** m **I** pipe clay. **II** whistle, whistling sound

Pfeif-patrone f (Mil.) whistling cartridge

Pfeil m (-e) **1** arrow, dart. **2** (Curves) camber

Pfeiler m (-) pillar, column, post; pier

pfeil- arrow: **-förmig** a arrow-shaped. **-gerade** a straight as an arrow. **Pf--gift** n arrow poison, inee. **-höhe** f **1** height, rise (of an arch, a meniscus). **2** deflection (of a spring). **-rad** n (Mach.) herringbone gear. **pf--schnell** a swift as an arrow. **Pf--verzahnung** f double helical gear. **-wurzelmehl** n arrowroot starch.

-zeichen n (directional) arrow (on signs)

Pfennig m (-e) penny, (specif) pfennig ($^1/_{100}$ of a mark). **--kraut** n (Bot.) pennycress

Pferch m (-e) (sheep) fold, corral. **pferchen** vt cram, pack; pen (up)

Pferd n (-e) horse

Pferde- horse: **-bohne** f horse bean, broad bean. **-dung, -dünger** m horse manure. **-fett** n horse grease. **-fleisch** n horsemeat, horseflesh. **-fußöl** n horse's-foot oil. **-haar** n horsehair. **-harnsäure** f hippuric acid. **-kammfett** n horse grease. **-kraft** f horsepower. **-kraftstunde** f horsepower hour. **-milch** f mare's milk. **-minze** f horsemint. **-mist** m horse manure. **-serum** n horse serum. **-stärke** f horsepower. **-stunde** f horsepower hour

-pferdig sfx -horsepower

Pfg. abbv (PFENNIG)

pfiff whistled: past of PFEIFEN

Pfiff m (-e) **1** whistle (sound). **2** trick, knack. **3** style, flair. **4** pinch (of salt, etc). **pfiffig** a clever

Pfingsten n or pl Pentecost, Whitsun. **Pfingst-rose** f peony

Pfirsich m (-e) peach. **--blüte** f peach blossom. **-branntwein** m peach brandy. **-farbe** f peach (color). **pf--farben** a peach-colored. **Pf--holz** n peachwood. **-kern** m peach kernel. **-kernschwarz** n peach black

Pfl. abbv (Pflanze) plant

Pflänzchen n little plant, seedling

Pflanze f (-n), **pflanzen** vt plant

Pflanzen- plant, vegetable: **-abfälle** mpl plant remains, vegetable waste. **-alkali** n vegetable alkali, plant potash = **-alkaloid** n plant (or vegetable) alkaloid. **-art** f type (or species) of plant. **pf--artig** a plant-like, vegetable. **Pf--asche** f plant ash(es). **-aufguß** m plant infusion. **-auszug** m plant (or vegetable) extract. **-base** f vegetable base, plant alkaloid. **-bau** m plant (or vegetable) growing; horticulture; agriculture. **-beschreibung** f phytography, description of plants. **-bestandteil** m plant (or vegetable) constituent. **-butter** f vegetable butter. **-chemie** f phytochemistry, plant chemistry. **pf--chemisch** a phytochemical. **Pf--decke** f plant (or crop) cover, vegetation.

-**dekokt** n plant decoction. -**eiweiß** n vegetable albumin (or protein). -**erde** f vegetable mold, humus. -**ernährung** f plant nutrition. -**erzeugnis** n plant (or vegetable) product. -**farbe** f vegetable color. -**farbstoff** m plant pigment, vegetable dye. -**faser** f vegetable (or plant) fiber. -**faserstoff** m vegetable fiber (gluten, cellulose). -**fett** n vegetable fat (or shortening). -**fettseife** f vegetable oil soap. -**fibrin** n vegetable fibrin (gluten, cellulose). -**forscher** m botanist. **pf-fressend** a herbivorous. **Pf–fresser** m herbivore. -**gallert** n vegetable gelatin, pectin. -**gattung** f genus of plants. -**gerbung** f vegetable tanning. -**gift** n plant poison. -**grün** n chlorophyll. -**gummi** n plant gum; natural rubber (or resin). -**haar** n 1 plant hair, vegetable horsehair. 2 palm fiber. -**harz** n vegetable resin. -**kasein** n, -**käsestoff** m vegetable casein, legumin. -**keim** m plant germ, embryo. -**kleber** m gluten. -**kohle** f vegetable charcoal. -**kost** f vegetable diet. -**krankheit** f plant disease. -**kunde** f botany. -**laugensalz** n potash. -**laus** f plant louse, aphis. -**leben** n plant life, vegetation. -**lehre** f botany. -**leim** m gliadin, vegetable glue. -**margarine** f vegetable margarine. -**nahrung** f plant food. -**nucleinsäure** f vegetable nucleic acid. -**öl** n vegetable oil. -**pech** n vegetable pitch. **pf–reich** a rich in plant life, lush. **Pf–reich** n vegetable (or plant) kingdom, flora. -**reste** mpl plant remains (or fossils), vegetable residue. -**rot** n carthamin. -**saft** m 1 sap. 2 vegetable juice. -**salz** n vegetable salt. -**säure** f vegetable acid. -**schädling** m plant pest. -**schleim** m mucilage. -**schutz** m plant (or crop) protection. -**schutzmittel** n plant protective (agent), (specif) pesticide. -**schwarz** n vegetable black. -**seide** f vegetable silk. -**stoff** m vegetable matter; plant substance. -**talg** m vegetable tallow, Japan wax. -**tier** n zoophyte. -**wachs** n vegetable wax. -**wachstum** n plant growth. -**welt** f flora, plant (or vegetable) kingdom. -**wolle** f vegetable wool. -**wuchs** m vegetation. -**zelle** f plant (or vegetable) cell. -**zellenstoff** m cellulose. -**zucht** f plant cultivation (growing, breeding)

Pflanz·erde f compost. -**garten** m nursery. -**kartoffel** f seed potato

pflanzlich a plant, vegetable

Pflaster n (-) 1 (Med.) plaster; adhesive tape. 2 pavement; floor. 3 place, town. -**käfer** m Spanish fly, blister beetle. **pflastern** vt 1 pave. 2 bandage with plaster (or tape). **Pflaster·stein** m paving stone (or brick). **Pflasterung** f (-en) 1 paving. 2 pavement

Pflaster·werkstoff m paving material. -**ziegel** m paving brick

Pflatsch m (-e) 1 wet spot; puddle. 2 sudden downpour. 3 (Tex.) fancy style. -**druck** m slop padding. **pflatschen** 1 vt (Tex.) pad. 2 vi splash

Pflatsch·farbe f (Tex.) padding color (or liquor). -**färbung** f slop pad dyeing. -**maschine** f padding machine

Pflaume f (-n) 1 plum. 2 prune. 3 = **Pflaumen·baum** m plum tree. **pf–blau** a plum-blue. **Pf–branntwein** m plum brandy. **pf–farben**, -**farbig** a plum-colored. **Pf–gummi** n plum gum. -**kernöl** n plum-kernel oil. -**sieder** m plum distiller

Pflege f 1 care, nursing. 2 cultivation. 3 servicing; maintenance, preservation. 4 (in compds, esp) foster, nursing (e.g. home). **pf–·leicht** a easy to take care of, easily maintained. **Pf–mittel** n cosmetic, toilet preparation

pflegen vt 1 care for, take care of. 2 nurse. 3 groom. 4 cultivate, foster. **II** v aux (+ **zu** + inf) be accustomed to (doing), usually (do): **sie pflegten es zu tun** they usually did it, they used to do it; **wie es zu geschehen pflegt** as (it) usually happens. Cf GEPFLEGT. **Pfleger** m (-), **Pflegerin** f (-nen) 1 nurse, nursing (or hospital) attendant. 2 guardian; foster parent. 3 curator, trustee. 4 (esp in compds) service person: **Raumpflegerin** f cleaning woman. **pfleglich** a careful

Pflicht f (-en) duty; obligation. **pf–·mäßig**, -**schuldig** a dutiful; due

Pflock m (-e) 1 stake, post. 2 plug, (dowel) pin; tampon. 3 peg. 4 (Med.) embolus

Pflücke f picking (time), harvest. **pflücken** vt pluck, pick; gather

Pflug *m* (÷e), **pflügen** *vt/i* plow. **Pflug-schar** *f* plowshare; (*Anat.*) vomer
Pfort·ader *f* portal vein
Pforte *f* (-n) **1** gate(way), entrance (door). **2** orifice. **3** porthole. **Pförtner** *m* (-) doorkeeper, doorman; (*Anat.*) pylorus
Pfosten *m* (-) post, pillar
Pfote *f* (-n) paw
Pfriem *m* (-e) awl. **Pfriemen·gras** *n* feather (*or* esparto) grass. **-kraut** *n* (*Bot.*) broom; (*Pharm.*) scoparius
Pfropf *m* (-e), **Pfropfen** *m* (-) **1** stopper, cork. **2** plug, tampon. **3** wad. **4** (*Med.*) clot, thrombus. **5** (*Bot.*) graft. **pfropfen** *vt* **1** cork, stop(per), plug (up). **2** cram, stuff. **3** (*Bot.*) graft, (*Med.*) implant. **Pfropfen·zieher** *m* corkscrew. **Pfropf·polymerisat** *n* graft polymer. **-reis** *n* graft, scion. **Pfropfung** *f* (-en) **1** corking, plugging. **2** graft(ing), graft polymerization; (*Med.*) implantation. **Pfropf·wachs** *n* grafting wax
Pfuhl *m* (-e) (muddy) pool, puddle, slough
Pfund *n* (-e) pound (= 500 grams) (*Quant: pl* -)
pfuschen *vi* bungle, do a sloppy job
Pfütze *f* (-n) puddle
pH *abbv* **1** (*Chem.*) pH. **2** (pro Hundert) percent
phagendänisches Wasser (*Pharm.*) yellow mercurial lotion
Phagozyt *n* (-e) phagocyte. **phagozytär, phagozytisch** *a* phagocytic. **Phagozytose** *f* phagocytosis
Phänokristall° *m* phenocryst, inset
Phänologie *f* phenology
Phänomen *n* (-e) phenomenon. **phänomenologisch** *a* phenomenological
Phänotyp° *m* phenotype
Phantasie *f* (-n) imagination; fancy, fantasy. **-artikel** *m* (*Tex.*) fancy style. **-leder** *n* fancy leather. **-papier** *n* fancy paper
phantasieren *vi* **1** fantasize, daydream. **2** rave, ramble (deliriously). **phantasie·voll** *a* highly imaginative. **Phantast** *m* (-en, -en) dreamer, visionary. **phantastisch** *a* fantastic; fanciful
Phäo- pheo.,
Pharao·schlange *f* Pharaoh's serpent
Pharbitin·säure *f* pharbitinic acid
Pharmakolith *m* (*Min.*) pharmacolite

Pharmakolog *m* (-en, -en) pharmacologist. **Pharmakologie** *f* pharmacology. **pharmakologisch** *a* pharmacological
Pharmakopöe *f* pharmacopeia
Pharmazeut *m* (-en, -en) pharmacist, druggist. **Pharmazeutik** *f* pharamceutics. **Pharmazeutikum** *n* (..ika) **pharmazeutisch** *a* pharmaceutical: **ph-es Papier** Seidlitz (*or* blue) paper
Pharmazie *f* (science of) pharmacy
Phasen·änderung *f* phase change. **-geschwindigkeit** *f* phase velocity. **-gesetz** *n* phase law (*or* rule). **ph–gleich** *a* of like phase, *adv* in (the same) phase. **Ph-gleichgewicht** *n* phase equilibrium. **-gleichheit** *f* phase coincidence. **-grenze** *f* phase boundary (*or* interface). **-integral** *n* phase integral. **-lehre** *f* phase theory. **-maß** *n* phase constant. **-messer** *m* phasemeter. **-raum** *m* phase space. **-regel** *f* phase rule. **-spannung** *f* (*Elec.*) phase voltage. **-umkehrung** *f* (*Colloids*) phase inversion. **-umwandlung** *f* phase transformation; change of state. **-unterschied** *m* phase difference. **-vertauschung** *f* phase reversal. **-welle** *f* phase (electron, de Broglie) wave. **-winkel** *m* phase angle. **-zahl** *f* number of phases
Phaseomannit *n* phaseomannitol, phaseomannite
·phasig *sfx* -phase(d)
Phasotropie *f* (-n) phasotropy, phasotropism
Phenacetur·säure *f* phenaceturic acid
Phenakit *m* (-e) phenacite
Phenalin *n* (-e) phenalene
Phenanthren·chinon *n* phenanthrenequinone. **Phenanthroe·säure** *f* phenanthroic acid
phenäthyl· phenylethyl
Phen·carboxonium *n* xanthylium. **-carbthionium** *n* thiaxanthylium
Phenochinon° *n* phenoquinone
Phenol· **1** phenol(ic). **2** (+ *metal, usu*) phenoxide, phenolate: **-aluminium** *n* aluminum phenoxide (*or* phenolate). **ph–artig** *a* phenol-like, phenoloid
Phenolat *n* (-e) phenolate, phenoxide
Phenol· **1** phenol(ic). **2** (+ *metal, usu*) phenolate, phenoxide: **-äther** *m* phenol ether. **-carbonsäure** *f* phenolcarboxylic acid.

-gruppe f phenol group. **ph–haltig** a phenolic, containing phenol. **Ph–harz** n phenol(ic) resin. **-kalium** n potassium phenoxide (or phenolate). **-kalzium** n calcium phenoxide (or phenolate). **ph–löslich** n phenol-soluble. **Ph–lösung** f phenol solution. **-natrium** n sodium phenolate (or phenoxide). **-öl** n carbolic acid. **-quecksilber** n mercury phenolate (or phenoxide). **-rot** n phenol red. **-säure** f phenol(ic) (or carbolic) acid. **-schwefelsäure** f phenolsulfuric acid. **-sulfo(n)säure** f phenolsulfonic acid. **-verbindung** f phenol(ic) compound. **-verkochung** f decomposition of a diazonium salt to yield phenol. **-wismut** n or m bismuth phenolate (or phenoxide)

Phenoplast m (-e) phenolic plastic (or resin)

Phen·säure f phenic acid, phenol

Phenyl·arsenchlorür n dichlorophenylarsine. **-äther** m phenyl ether. **-borchlorid** n dichlorophenylborine. **-braun** n phenyl brown. **-diazoniumsalz** n benzenediazonium salt

Phenylen·blau n phenylene blue

Phenyl·essigäther m phenyl acetate. **-essigsäure** f phenylacetic acid. **-fettsäure** f phenylated fatty acid

phenylieren vt phenylate

Phenyl·jodidchlorid n phenyliodochloride. **-milchsäure** f phenyllactic acid. **-säure** f phenol, carbolic acid. **-schwefelsäure** f phenylsulfuric acid. **-senföl** n phenyl mustard oil. **-siliciumchlorid** n trichlorophenylsilane. **-verbindung** f phenyl compound. **-wasserstoff** m phenyl hydride, benzene

Ph. g. abbv (Pharmakopoeia germanica) German pharmacopeia

philan(is)ieren vt philanize (mercerize by a special process)

philodien a dienophilic

Philosoph m (-en, -en) philosopher. **Philosophen·wolle** f philosopher's wool (sublimed zinc oxide). **philosophisch** a philosophical

Phiole f (-n) vial, phial

Phlegma n 1 phlegm, mucus. 2 (Chem.) distillate. 3 lethargy, apathy. 4 equanimity.

phlegmatisch a phlegmatic. **phlegmatisieren** vt (Explo.) desensitize

phlobaphen·lösend a dissolving phlobaphenes

phlogistisch a phlogistic

Phloion·säure f phloionic acid

Phlorchinyl n phloroquinyl

Phloro·glucid, -gluzid n phloroglucidol. **-gluzin** n phloroglucinol. **-gluzit** n phloroglucitol

Phokän·säure f phocenic (or valeric) acid

Phönizin n (Min.) ph(o)enichroite

Phorese·zelle f electrophoresis cell

Phosgen n phosgene, carbonyl chloride

Phosphat n (-e) phosphate. **-dünger** m phosphate fertilizer. **ph–führend** a phosphate-bearing. **-verseucht** a phosphate-polluted

phosphatieren vt phosphatize, phosphate

phosphatisch a phosphatic

Phosphat·puffer m phosphate buffer. **-schmelze** f phosphate bath

phosphenylig a phosphenylous

Phosphim·säure f phosphimic acid

phosphinig a phosphinous. **Ph–-säure** f phosphinous acid

Phosphin·oxid n phosphine oxide. **-säure** f phosphinic acid

Phosphor m (-e) 1 phosphorus. 2 (Electron.) phosphor. **ph–-arm** a low-phosphorus. **-artig** a phosphorus-like, phosphorous. **Ph–äther** m phosphoric ether, ethyl phosphate. **-basis** f phosphorus base. **-bestimmung** f phosphorus determination. **-blei** n lead phosphide; (Min.) pyromorphite. **-bombe** f phosphorus bomb. **-brandgranate** f phosphorus incendiary shell. **-brei** m phosphorus paste. **-bromid** n phosphorus pentabromide. **-bromür** n phosphorus tribromide. **-bronze** f phosphor bronze. **-calcium** n calcium phosphide. **-chlorid** n phosphorus pentachloride. **-chlorür** n phosphorus trichloride. **-chromit** n (Min.) vaquelinite. **-dampf** m phosphorus vapor (or fumes). **-dünger** m phosphate fertilizer. **-eisen** n 1 (Chem.) iron (or ferrous) phosphide. 2 (Metll.) ferrophosphorus. **-eisensinter** m (Min.) diadochite

phosphoren vt phosphorize, phosphorate
Phosphoreszenz f (-en) phosphorescence.
ph--erzeugend a phosphorogenic.
phosphoreszieren vi phosphoresce.
phosphoreszierend p.a phosphorescent.
Phosphoreszierung f phosphorescence
Phosphor·fleischsäure f phosphocarnic acid. **ph–frei** a nonphosphorous, phosphorus-free. **Ph–gehalt** m phosphorus content. **-geruch** m phosphorus odor. **-geschoß** n phosphorus bullet, incendiary shell. **-glas** n metaphosphoric acid. **-gruppe** f phosphorus group. **-guano** m phosphatic guano. **ph–haltig** a containing phosphorus, phosphorated, phosphatic. **Ph–handgranate** f phosphorus grenade
phosphorig a phosphorous. **--sauer** a phosphite of. **Ph–säureanhydrid** n phosphorous (acid) anhydride
phosphorisch a phosphoric. **phosphorisieren** vt phosphorize, phosphorate
Phosphor·jodid n phosphoric iodide. **-jodür** n phosphorous iodide. **-kalk** m crude calcium phosphide (containing phosphate). **-kalzium** n calcium phosphide. **-kanister** m phosphorus bomb. **-kerzchen** n wax match, vesta. **-kupfer** n 1 (Chem.) copper phosphide. 2 (Metll.) phosphor copper. 3 (Min.) libethenite, pseudomalachite. **-löffel** m phosphorus (deflagrating) spoon. **-mangan** n phosphor manganese. **-masse** f phosphorus paste (or composition). **-metall** n metallic phosphide. **-molybdänsäure** f phosphomolybdic acid. **-natrium** n sodium phosphide. **-nebelgranate** f phosphorus smoke shell. **-nitrildichlorid** n phosphorus nitride dichloride. **-öl** n (Pharm.) phosphorated oil. **-oxid** n phosphoric oxide, phosphorus pentoxide. **-oxydul** n phosphorous oxide, phosphorus trioxide. **-pentoxid** n phosphorus pentoxide. **-proteid, protein** n phosphorprotein. **-roheisen** n phosphoric pig iron. **-salz** n 1 (genl) phosphate. 2 (specif) microcosmic salt, sodium ammonium hydrogen phosphate. 3 (Min.) stercorite. **-salzperle** f microcosmic bead. **ph–sauer** a phos-

phate of. **Ph–säure** f phosphoric acid: glasige Ph. metaphosphoric acid. **-säureanhydrid** n phosphoric anhydride, phosphorus pentoxide. **-schlamm** m phosphorus mud (or sludge). **-stahl** m phosphorus steel. **-stange** f phosphorus stick. **-stickstoff** m (tri)phosphorus (penta)nitride. **-sulfid** n phosphorus sulfide. **-sulfochlorid** n thiophosphoryl chloride
Phosphorung f phosphor(iz)ation
Phosphor·verbindung f phosphorus compound. **-vergiftung** f phosphorus poisoning. **-wasserstoff** m hydrogen phosphide, phosphorus hydride. **-wasserstoffgas** n phosphine. **-wolframsäure** f phosphotungstic acid
Phosphorylen n olefinic phosphorane (Wittig reagent)
Phosphor·zink n zinc phosphide. **-zinn** n 1 (Chem.) tin phosphide. 2 (Metll.) phosphor-tin alloy. **-zündholz** n phosphorus match
Phospho·säure f phosphonic acid. **-wolframat** n phosphotungstate
Photo·anregung f photoexcitation. **-apparat** m camera. **-aufnahme** f photograph. **-bakterien** npl photobacteria. **-chemie** f photochemistry. **ph–chemisch** a photochemical. **Ph–chlorierung** f photochlorination. **-elektrizität** f photoelectricity. **ph–gen** a photogenic. **Ph–gramm** n photogram. **-grammetrie** f photogrammetry
Photograph m (-en, -en) photographer. **Photographie** f 1 photograph. 2 photography. **photographieren** vt photograph. **photographisch** a photographic
Photo·katalyse f photocatalysis. **-kopie** f, **photokopieren** vt photocopy, photostat
Photo·leitfähigkeit f photoconductivity
Photolyse f (-n) photolysis
Photometer n photometer. **Photometrie** f photometry. **photometrieren** vt measure photometrically. **photometrisch** a photometric
Photonen·absorption f photon absorption
Photo·oximierung f photooximation. **-papier** n photographic paper. **-phorese** f photophoresis. **-physik** f photophysics. **-platte** f photographic plate. **-präparat** n

photographic preparation. **-sphäre** *f* photosphere. **-strom** *m* photoelectric current, photocurrent. **-synthese** *f* photosynthesis

Phototropie *f* phototropy, phototropism. **phototropisch** *a* phototropic

Photo·vervielfacher *m* (*Elec.*) photomultiplier. **ph–voltaisch** *a* photovoltaic. **Ph–zelle** *f* photoelectric cell, photocell

Phrenosin·säure *f* phrenosic acid

Phthal· *obsol. spelling for:* **Phthal·**

Phthal· phthalic, phthalo-: **-amidsäure**, **-aminsäure** *f* phthalamic acid. **-azin** *n* phthalazine

Phthalein·farbstoff *m* phthalein dye

Phthalid *n* (-e) phthalide

Phthal·monopersäure *f* monoperphthalic acid

Phthalon·säure *f* phthalonic acid

Phthalsäure° *f* phthalic acid. **-anhydrid** *n* phthalic anhydride. **-harzlack** *m* phthalic acid resin varnish

Phthion·säure *f* phthionic acid

pH-Wert *m* pH value

Phyko·phäin *n* phycophein. **·myzetene** *npl* phycomycetes. **-zyan** *n* phycocyanin

Phyllo·hämin *n* phyllohemin. **-silikat** *n* layer (*or* sheet) silicate

Physcia·säure *f* physcic acid, physcione

Physik *f* 1 physics. 2 (*Dye.*) tin composition (solution of tin in aqua regia)

physikalisch *a* physical: **ph.-chemisch** *a* physicochemical; **ph.-metallurgisch** *a* physicometallurgical

Physik·bad *n* (*Dye.*) tin composition

Physiker *m* (-) physicist

Physiko·chemiker *m* physical chemist. **ph–chemisch** *a* physicochemical

Physik·salz *n* (*Dye.*) red spirit

Physiolog *m* (-en, -en) physiologist. **Physiologie** *f* (-n) physiology. **physiologisch** *a* physiological

physisch *a* physical

Physod·säure *f* physodic acid

Phytin·säure *f* phytinic acid

Phyto·chemie *f* phytochemistry. **ph–chemisch** *a* phytochemical. **Ph–sterin** *n* phytosterol

Phyzit *n* (-e) erythritol

Piazin *n* (-e) pyrazine

Picen·chinon *n* picenequinone

Pick·apparat *m* (*Brew.*) pitching machine

pichen *vt* (smear with) pitch

Pich·pech *n* common pitch. **-wachs** *n* propolis

Pick *m* (-e) 1 peck, stab. 2 grudge. 3 glue, paste

Picke *f* (-n) pickaxe

Pickel *m* (-) 1 (ice) pick; pickaxe. 2 pickle (brine), pickling = **Pökel**. 3 pimple, pustule. **-bad** *n* pickle (bath). **-bildung** *f* (*Metll.*) pitting. **-brühe** *f* pickle

pickeln *vi* pickle, soak in a pickle

Pickel·säure *f* pickling acid

picken I *vt/i* peck (at); pick up (*or* off). II *vi* 1 beat, pound. 2 stick, be sticky

Picolin·säure *f* picolinic acid

piepsen *vt/i* beep, peep. **Piepser** *m* (-) beeper. **Piepton** *m* (*Instrum.*) beep, tone

Piezo·chemie *f* piezochemistry. **p–elektrisch** *a* piezoelectric. **P–elektrizität** *f* piezoelectricity. **-kristall** *m* piezoelectric crystal

Pigment·anreibung *f* 1 pigment grinding. 2 pigment paste

pigmentarisch *a* pigmentary

Pigment·bakterien *npl* chromatographic bacteria. **-bildung** *f* (*Biol.*) pigment formation, chromogenesis. **-bindemittel** *n* pigment binder. **-druck** *m* pigment printing. **-farbe** *f* pigment color. **-farbstoff** *m* pigment dye. **p–frei** *a* nonpigmented. **-haltig** *a* containing pigment, pigmented

pigmentieren 1 *vt* pigment. 2 *vi* & **sich p.** become pigmented

Pigment·klotzverfahren *n* pigment padding. **-lack** *m* pigmented paint. **p–los** *a* nonpigmented. **-mattiert** *a* pigment-dulled. **P–papier** *n* carbon paper

pikant *a* 1 piquant, spicy, hot: **p–e Soße** hot sauce. 2 racy, risqué. **pikieren** *vt* 1 transplant; prick out (seedlings). 2 padstitch. 3 spice. 4 pique, needle

Piknometer *n* (-) pycnometer

Pikramin·säure *f* picramic acid

Pikrat *n* (-e) picrate. **-ion** *n* picrate ion

Pikrin·pulver *n* picric powder. **-säure** *f* picric acid

Pikrit *n* (-e) (*Geol.*) picrite

Pikro·toxininsäure f picrotoxininic acid.
-toxinsäure f picrotoxinic acid
Filé m, **Pilée** f, **··zucker** m crushed sugar
Pilier·anlage f (Soap) milling plant. **pilieren** vt mill (soap). **Pilier·maschine** f (soap) milling machine (or miller)
Pike f: **von der P. auf lernen** learn from the bottom up
Pille f (-n) pill, pellet
Pillen·dreher m 1 tumblebug. 2 pillroller (= druggist). **-farn** m pillwort. **-glas** n pill bottle (or vial). **-schachtel** f pillbox
Pilop·säure f pilopic acid
Pils n = **Pilsener** n (-) Pilsen (or Pilsner) beer
Pilsenit n (Min.) wehrlite
Pilz m (-e) 1 fungus. 2 mushroom. 3 mold. 4 (Mil.) pillbox. **p—·ähnlich, -artig** a fungoid; mushroom-like. **P–art** f species of fungus (or mushroom). **-bildung** f fungoid growth, fungus formation
Pilz(en)·entwicklung f fungus development
pilz·feind(lich) a mold-resistant, rotproof. **-förmig** a fungiform; mushroom-shaped. **P–holz** n decayed timber
pilzig a fungous, fungoid
Pilz·isolator m (Elec.) petticoat insulator. **-krankheit** f fungus disease. **-kunde** f mycology. **-maischverfahren** n (Alcohols) amylo process of saccharification. **-samen** m spawn of fungus; mycelium. **-säure** f fungic acid. **-stoff** m fungin. **p—tötend** a fungicidal. **P–vergiftung** f mushroom (or fungus) poisoning. **-vertilgungsmittel** n fungicide. **-wucherung** f (uncontrolled) fungus growth
Pimar·säure f pimaric acid
Pimel·keton n pimelic ketone (cyclohexane). **Pimel(in)säure** f pimelic acid
Piment n pimento, allspice. **··öl** n pimento (or allspice) oil. **-pfeffer** m pimento, allspice. **-rum** m bay rum
Pimpel·stein m (Copper) pimple (or blister) metal
Pimpernelle f (-n) burnet saxifrage
Pimper·nuß f pistachio nut
Pimpinelle f (-n) burnet saxifrage
Pim·stein m pumice stone

Pina·faser f piña fiber
Pinakolin n (-e) pinacolone. **··alkohol** m pinacolyl alcohol
Pinakon n (-e) pinacol. **··bildung** f pinacol formation
Pinal·drüse f pineal gland
Pincette f (-n) tweezers; forceps
Pineal·drüse f, **-organ** n pineal gland (or body)
Piney·talg m piney tallow
Pinguin m (-e) penguin
Pinie f (-n) pine (esp Pinus pinea); pine kernel, piñon (seed)
Pinien·kern m pine kernel, piñon. **-kiefer** f white pine. **-talg** m piney tallow. **-zapfen** m pine cone
Pinin·säure f pininic acid
Pinit n pinitol, pinite (D-inositol monomethyl ether)
pinken vt 1 treat with pink salt. 2 (Silk) load with tin (i.e. stannic chloride). 3 hammer
Pink·salz n (Tex.) pink salt, ammonium hexachlorostannate; double chloride of zinc and ammonium; potassium sodium tartrate. **-salzbad** n pink-salt bath
Pinne f (-n) 1 pin, peg, tack. 2 peen (of a hammer). 3 quill feather. **pinnen** vt pin, peg, tack
Pinokampfer·säure f pinocamphoric acid
Pinon·säure f pinonic acid
Pinophan·säure f pinophanic acid
Pin·säure f pinic acid
Pinsel m (-) 1 (painter's, incl artist's) brush. 2 (cosmetic) pencil. 3 tuft (of hair). **p—·förmig** a brush-shaped, brushlike. **pinseln** vt/i paint; pencil. **Pinsel·schimmel** m Penicillium, brush mold
Pinus·harz n pine resin
Pinzette f (-n) tweezers; forceps
Pion n (-en) pi meson
Pionier m (-e) 1 pioneer. 2 army engineer. **··sprengmittel** n engineering explosive
Piotin n soapstone
Pipecolin·säure f pipecolic acid
Piperin·säure f piperinic acid
Piperitolen·säure f piperitolenic acid
Piperonyl·säure f piperonylic acid
Pipette f (-n) pipette; medicine dropper
Pipetten·etagere f pipette rack (or stand).

-fläschchen *n*, **-flasche** *f* dropping bottle with a pipette. **-gestell** *n*, **-ständer** *m* pipette stand

pipettieren *vt/i* pipette, transfer (*or* measure) with a pipette

Pisang·wachs *n* pisang wax

Pistazie *f* (-n) pistachio (nut)

Pistazien·grün *n* pistachio green. **-öl** *n* pistachio oil

Piste *f* (-n) runway; track, course

Pistill *n* (-e) **1** pestle. **2** pistil

Pistole *f* (-n) pistol, (hand) gun; torpedo fuse. **Pistolen·gebläse** *n* hand-operated blast

Pita·hanf, Pite·hanf *m* pita (hemp)

Pitot·rohr *n*, **-röhre** *f*, **Pitotsches Rohr** Pitot tube

Pivalon *n* 2,2,4,4-tetramethylpentan-3-one

Pkt. *abbv* (Punkt) pt. (point)

PKW, Pkw *abbv* (Personenkraftwagen) (passenger) motor car

placentar *a* placental

placieren *vt* **1** place. **2** aim (projectiles). **3** invest (capital). *Cf* PLAZIEREN. **placiert** *p.a* well-placed, well-aimed

placken I *vt* **1** plague, pester. **2** patch. **3** stick, paste. **4** beat down, flatten. **5** card (wool). **II: sich p.** slave, toil. **Plackerei** *f* (-en) drudgery; trouble

Plage *f* (-n) **1** torment; ordeal. **2** nuisance, bother, trouble. **3** plague, affliction. **plagen 1** *vt* plague, torment; pester. **2: sich p.** (*usu* + **mit**) toil, slave (over); be plagued (by)

Plagge *f* (-n) (piece of) turf, sod

plagiedrisch *a* (*Cryst.*) plagihedral

Plagioklas *m* (-e) (*Min.*) plagioclase

Plakat *n* (-e) poster, placard. **--farbe** *f* poster color. **plakativ** *a* poster-like, boldly simple. **Plakat·papier** *n* poster paper

Plakodin *n* (*Min.*) maucherite

plan *a* flat; level; plane; smooth

Plan *m* (-̈e) **1** plan. **2** (city) street map. **3** diagram. **4** schedule, (time)table. **5** scene. **6** (*Painting*) ground. *Cf* PLANE

Plane *f* (-n) canvas cover, tarpaulin *cf* PLAN

planen *vt/i* plan

Planen·stoff *m* canvas, awning cloth; tarpaulin

planetar(isch) *a* planetary

Planeten·bahn *f* planetary orbit. **-getriebe** *n* planetary gear(ing). **-mischer** *m* planetary motion mixer. **-system** *n* planetary system

Plan·fläche *f* plane surface. **p–gemäß** *a* **1** planned. **2** systematic. **3** *adv* as planned, according to plan. **P–gitter** *n* (*Opt.*) plane grating

Planier·bank *f* spinning (*or* smoothing) lathe

planieren *vt* **1** level, grade; bulldoze. **2** (*Metll.*) planish, spin (on a lathe). **3** (*Paper*) size, glue

Planier·löffel *m* skimmer. **-maschine** *f* grader; bulldozer. **-masse** *f* (*Paper*) size. **-presse** *f* (*Paper*) size press. **-raupe** *f* bulldozer

Planierung *f* (-en) leveling, grading; bulldozing; planishing; (paper) sizing

Planier·wasser *n* glue water, size

Planimetrie *f* plane geometry. **planimetrisch** *a* planimetric

Planke *f* (-n) **1** plank, (thick) board. **2** board fence

plan·konkav *a* (*Opt.*) planoconcave. **-konvex** *a* (*Opt.*) planoconvex. **-los** *a* aimless, haphazard; random, *adv oft* at random. **-mäßig** *a* **1** planned. **2** systematic. **3** *adv* as planned, according to plan. **-parallel** *a* plane-parallel. **P–rätter** *m* gyratory screen. **-rost** *n* flat grate

planschen *vi* splash

Plan·schleifen *n*, **-schliff** *m* flat grinding. **-sichter** *m* sieve, plansifter. **-sieb** *n* flat sieve (*or* screen). **-spiegel** *m* plane mirror. **p–symmetrisch** *a* planisymmetric(al)

Plantage *f* (-n) plantation. **Plantagen·kautschuk** *m* plantation rubber

Planung *f* (-en) planning; plan, design

Plasma *n* (..men) plasma; protoplasm. **--physik** *f* plasma physics

Plasmolyse *f* (-n) plasmolysis. **plasmolytisch** *a* plasmolytic

Plast *m* (-e) plastic. **--art** *f* type of plastic. **-beton** *m* plastic concrete (containing a polymer as an additive)

Plaste plastics: *pl of* PLAST

plastifizieren *vt* plasticize, plastify. **Plas-**

tifizier·mittel n plasticizer
Plastik I n (-en or -s) plastic. **II** f (-en) 1 plastic art. 2 (piece of) sculpture. 3 plastic surgery. 4 (Opt.) stereo(scopic) effect. **III** sfx -plasty: **Nasenplastik** rhinoplasty
Plastikator m (-en) plasticizer; (Rubber) softener
Plastilin n, **Plastilina** f modeling clay; plasticene
Plastimeter° n plastometer
plastisch a 1 plastic. 2 three-dimensional; stereoscopic. 3 graphic, vivid
plastizieren vt plasticize. **Plastizierer** m (-) plasticizer
Plastizimeter° n plastimeter
Plastizität f plasticity. **Plastizitäts·messer** m plastometer
Plast·mörtel m plastic mortar
Plastometrie f plastometry
Platane f (-n) plane (tree), sycamore
Platiak n platinum ammine
Platin n platinum. **-abfall** m platinum waste (or residue). **-ammonchlorid** n platinum ammonium chloride, ammonium chloroplatinate. **p–artig** a platinum-like, platinoid. **P–asbest** m platinized asbestos
Platinat n (-e) platinate. **p–·beschichtet** a platinate-coated
Platin·bad n platinum bath. **-bariumzyanür** n barium cyanoplatinite (or cyanoplatinate). **-blase** f platinum still. **-blech** n platinum sheet; sheet platinum. **-chlorid** n platinic chloride; (as sfx) chloroplatinate. **-chlorür** n platinous chloride. **-chlorwasserstoff** m hydrogen chloroplatinate. **-chlorwasserstoffsäure** f chloroplatinic acid. **-cyanür** n platinous cyanide, tetracyanoplatinate. **-cyanürwasserstoff** m hydrogen cyanoplatinite. **-cyanürwasserstoffsäure** f cyanoplatinous acid. **-draht** m platinum wire. **-drahtöse** f platinum wire loop. **-dreieck** n platinum triangle. **-druck** m platinotype
Platine f (-n) 1 (Metll.) rolled sheet or bar; mill (sheet, plate) bar. 2 (Metalwork) blank. 3 (Tex.) hook, lifter; sinker
Platin·element n platinum thermocouple. **-ersatz** m platinum substitute. **-erz** n

platinum ore. **-fluorwasserstoff** m hydrogen fluoroplatinate. **-fluorwasserstoffsäure** f fluoroplatinic acid. **-folie** f platinum foil. **-gefäß** n platinum vessel. **-gerät** n, **gerätschaft** f platinum ware (utensil, apparatus). **p–haltig** a containing platinum. **P–hydroxid** n platinic hydroxide. **-hydroxydul** n platinous hydroxide
Platini· platinic: **-chlorid** n platinic chloride. **-chlorwasserstoff** m hydrogen chloroplatinate. **-chlorwasserstoffsäure** f chloroplatinic acid. **-cyanwasserstoff** m hydrogen cyanoplatinate. **-cyanwasserstoffsäure** f cyanoplatinic acid
platinieren vt platinize, platinum-plate. **Platinierung** f (-en) platinization, platinum plating
Platini· platinic: **-rhodanwasserstoffsäure** f thiocyanoplatinic acid. **-salz** n platinic salt. **-selencyanwasserstoffsäure** f selenocyanoplatinic acid. **-verbindung** f platinic compound. **-zyanwasserstoff** m hydrogen cyanoplatinate See also -CYANWASSERSTOFFSÄURE
Platin·kegel m platinum cone. **-kohle** f platinized charcoal. **-konus** m platinum cone. **-legierung** f platinum alloy. **-löffel** m platinum spoon. **-magnesiumzyanür** n magnesium cyanoplatinite. **-metall** n platinum (-group) metal. **-mohr** m platinum black. **-nadel** f platinum needle. **-natriumchlorür** n sodium chloroplatinite. **-netz** n platinum grid. **-oberflächenoxid** n surface platinum oxide.
Platino· platinous: **-chlorid** n platinous choride. **-chlorwasserstoff** m hydrogen chloroplatinite. **-chlorwasserstoffsäure** f chloroplatinous acid, hydrogen tetrachloroplatinite. **-cyanwasserstoffsäure** f cyanoplatinous acid. **-rhodanwasserstoffsäure** f thiocyanoplatinous acid
Platin·öse f platinum wire loop
Platino·verbindung f platinous compound
Platin·oxid n platinum (specif platinic) oxide. **-oxidverbindung** f platinic compound. **-oxydul** n platinous oxide. **-oxy-**

dulverbindung f platinous compound
Platino·zyanwasserstoff m hydrogen chloroplatinite *See also* -CYANWASSERSTOFFSÄURE
platin·plattiert a platinum-plated. **P– plattierung** f platinum plating. **-reihe** f platinum series. **-rhodium** n platinum-rhodium alloy. **-rohr** n, **-röhre** f platinum tube. **-rückstand** m platinum residue. **-salmiak** m ammonium hexachloroplatinate. **-salz** n platinum salt. **-sand** m sand for cleaning platinum. **-säure** f platinic acid. **-schale** f platinum dish. **-schiffchen** n platinum boat. **-schwamm** m platinum sponge. **-schwarz** n platinum black. **-spatel** m platinum spatula. **-spirale** f platinum coil. **-spritze** f platinum tip (*or* point). **-stern** m platinum star. **-sulfür** n platinous sulfide. **-tiegel** m platinum crucible. **-tonbad** n (*Phot.*) platinum toning bath. **-verbindung** f platinum compound. **-zyanür** n platinous cyanide. **-zyanürwasserstoffsäure** f cyanoplatinous acid

plätschern vi splash, patter

platt a 1 flat; level. 2 superficial, trivial. 3 plain, frank; sheer, out-and-out. 4 *pred* a flabbergasted. 5 *adv* Low German. *Cf* PLATTE II

Plätt· ironing **-brett** n ironing board

Plättchen n (-) platelet, lamella scale

platt·drücken vt flatten, press flat

Platte I f (-n) 1 plate. 2 sheet, lamina; planchet; panel. 3 slab, flagstone; tile. 4 disc, phonograph record. 5 tray, platter. 6 hot plate. 7 table top; leaf (of a table). 8 plateau. 9 (*Bact.*) plate culture. II m (*adj endgs*) flat (tire)

Plätte f (-n) 1 flat(tened) wire. 2 = **Plätt· eisen** n (flat)iron, electric iron. **plätten** 1 vt flatten, level (out); laminate; pave. 2 vt/i iron, press

Platten·band n plate (apron, slat) conveyor. **-bauweise** f prefabricated slab construction. **-dekatur** f hydraulic press finish. **-druck** m plate printing, stereotypography. **-flasche** f, **fläschchen** n (*Bact.*) flat culture flask. **p–förmig** a plate-like, lamellar, lamelliform; laminated. **P–gewebe** n lamellar tissue. **-glimmer** m sheet mica. **-gummi** n plate

(*or* sheet) rubber. **-kalk** m slab limestone. **-kamera** f plate camera. **-kautschuk** m plate (*or* sheet) rubber. **-kolonne** f (*Distil.*) plate column. **-kondensator** m (*Elec.*) plate condensor. **-kultur** f plate (*or* Petri dish) culture. **-kupfer** n sheet copper. **-leistung** f (*Distil.*) plate efficiency. **-plaketsystem** n Berkeley grid system. **-presse** f platen press. **-pulver** n (*Explo.*) rolled (*or* flaked) powder. **-schale** f flat dish, Petri dish. **-spieler** m record player. **-turm** m plate column (*or* tower). **-zählung** f (*Bact.*) plate count. **-zink** n slab zinc. **-zucker** m slab sugar

Platt·erbse f meadow pea, vetch

platterdings adv downright, absolutely; bluntly

Platt·fisch m flatfish, fluke

platt·gedrückt p.a flattened, flat-pressed

Plattheit f (-en) flatness; triviality; platitude; plainness, frankness cf PLATT

Plattier·blech n cladding plate

plattieren vt plate; (*Tex.*) plait. **plattiert** a plated; (*Metll.*) clad (steel). **Plattierung** f (-en) plating; (*Metll.*) cladding

Plattier·verfahren n plating (*or* cladding) process. **-werkstoff** m plating (*or* cladding) material

plattig a lamellar, laminated; (*Cryst.*) plate-like, lath-shaped, bladed

Plattine f (-n) (*Mach.*) plate; (*Metll.*) mill bar

Platt·wurm m flatworm, platyhelminth

Platz m (⸚e) 1 place (*also* = spot, location; rank; town, locality): **am P.** (a) here, in this place, (b) in (the right) place; **nicht** (*or* **fehl**) **am P.** out of place. 2 (playing) field, court, course, grounds. 3 plaza, square. 4 storage place, yard; building lot, site. 5 room, space: **P. machen** (*or* **schaffen**) make room; **P. greifen** spread, gain ground. 6 seat: **P. nehmen** take a seat, sit down. 7 position, post (*also* = job)

Plätzchen n (-) 1 little place (spot, etc) cf PLATZ 1. 2 cookie. 3 (cough) drop; piece of hard candy

platzen vi(s) 1 blow up, explode; burst, rupture. 2 go wrong, fail

plätzen vi bang, pop

Platz·mangel m lack of space. **-nummer** f position (*or* serial) number; atomic number. **-patrone** f blank cartridge. **-probe** f bursting (or rupture) test. **-quecksilber** n fulminating mercury. **-regen** m downpour, (sudden) heavy shower. **-scheibe** f bursting disc. **p–sparend** a space-saving. **P–wechsel** m 1 change of place, transposition; migration. 2 (Electrons) exchange of place(s)

Platzwechsel·integral n transposition integral. **-reaktion** f double decomposition, metathesis; replacement reaction

plazieren vt place cf PLACIEREN. **Plazierung** f (-en) placement, placing

Pleistozän n (Geol.) Pleistocene

pleite pred a bankrupt. **Pleite** f (-n) 1 bankruptcy: **P. machen** go bankrupt. 2 failure

P-Leiter m p-type (or defect) semiconductor

Plejaden pl Pleiades

Pleochroismus m (Cryst.) pleochroism. **pleochroitisch** a pleochro(it)ic

pleomorph a 1 (Biol.) pleomorphic. 2 (Cryst.) polymorphic

Pleuel·stange f connecting rod

pleuritisch a pleuritic

Pliozän n (Geol.) Pliocene

Plissée (-s) plissé, pleating

Plombe f (-n) 1 (tooth) filling. 2 lead seal. **plombieren** vt 1 fill (a tooth). 2 (lead-)seal. **Plombier·gold** n dental gold. **Plombierung** f (-en) 1 filling (of tooth). 2 (lead-)sealing

plötzlich a sudden, abrupt. **Plötzlichkeit** f suddenness, abruptness

Plücker·rohr n Plücker gas-discharge tube

Plumbagin n plumbago, graphite

Plumbi· plumbic, plumbi-, lead: **-chlorwasserstoffsäure** f hydrogen hexachloroplumbate, hexachloroplumbic acid. **-oxid** n plumbic oxide. **-salz** n plumbic salt. **-verbindung** f plumbic compound

Plumbo· plumbous, plumbo-, lead: **-chlorwasserstoffsäure** f hydrogen hexachloroplumbite, hexachloroplumbous acid. **-oxid** n plumbous oxide. **-salz** n plumbous salt. **-verbindung** f plumbous compound

plump a clumsy; crude, blatant

Plunscher m (-) plunger

Plus n (-) plus; surplus

Plüsch m plush. **--leder** n ooze (or velvet) leather

Plusin·glanz m (Min.) argyrodite

Plus·platte f (Elec.) positive plate. **-pol** m positive pole. **-zeichen** n plus sign. **-zucker** m plus (or dextrorotatory) sugar; raffinose

plutonisch a (Geol.) plutonic. **Plutonit** n (Geol.) plutonic rock

p.m. abbv 1 (pro mille) per thousand. 2 (pro mense) per month. 3 (post meridiem) PM (after noon)

Pneu m (-s) 1 (pneumatic) tire. 2 pneumothorax

Pneumatik I m (-en) (pneumatic) tire. **II** f (-en) 1 (science of) pneumatics. 2 (Mach.) pneumatic system (or action). **pneumatisch** a pneumatic

pneumatolytisch a (Petrog.) pneumatolytic

pneumonisch a pneumonic

pochen I vt 1 (Min.) crush, stamp (ores). **II** vi 2 knock, rap. 3 beat, pound. 4 (+ **auf**) insist (on); brag (about)

Poch·erz n stamping (or milling) ore. **-gänge** mpl poor ore, halvans. **-gestein** n (Min.) stamp rock. **-mehl** n pulverized ore. **-mühle** f stamping mill

Pochote·öl n pockwood oil

Poch·satz, **-schlamm**, **-schlich** m ore slime, residual stamp-mill sludge. **-stempel** m (Min.) (ore) stamp. **-stempelreihe** f stamp battery. **-trübe** f (Metll.) stamp pulp. **-werk** n stamping mill

Pocke f (-n) pock; pl pox, (specif) smallpox

Pocken·holz n lignum vitae, guaiacum wood. **-impfung** f smallpox vaccination. **-lymphe** f vaccine lymph. **-narbe** f 1 pockmark. 2 (Metll.) pit corrosion. **p–narbig** a pockmarked. **P–wurzel** f china root

Pock·holz n lignum vitae, guaiacum wood, pockwood

Podest n (-e) pedestal, base

Podophyll·säure f podophyllic acid

Pokal m (-e) 1 goblet. 2 (prize) cup

Pökel m (-) pickle, pickling (or corning) liquid. **Pökelei** f (-en) meat curing house, pickling (or corning) plant

Pökel·faß n pickling tub (or vat); salting tub. **-fleisch** n pickled (cured, corned)

meat. **-hering** *m* pickled herring. **-kufe** *f* pickling tub (*or* vat)

pökeln *vt* pickle; corn, cure (meat)

Pökel·trog *m* pickling trough. **-zunge** *f* corned tongue

Pol *m* (-e) 1 (*Geom., Phys., Elec., Geog.*) pole; (*in compds usu*) polar. 2 (*Tex.*) pile, nap. **-achse** *f* polar axis. **-anziehung** *f* polar attraction

Polar·eis *n* polar ice. **-gegend** *f* polar region

Polarisations·apparat *m* polarization apparatus. **-ebene** *f* plane of polarization. **-einrichtung** *f* (*Micros.*) polarizing attachment. **-erscheinung** *f* polarization phenomenon. **-farbe** *f* polarization color. **-prisma** *n* polarizer. **-richtung** *f* direction of polarization. **-rohr** *n*, **-röhre** *f* polarization tube. **-spannung** *f* (*Elec.*) polarization voltage. **-strom** *m* (*Elec.*) polarization current. **-verhalten** *n* polarization behavior. **-vorrichtung** *f* (*Micros.*) polarizing attachment. **-winkel** *m* angle of polarization

Polarisator *m* (-en) polarizer. **P.-Nicol** *m* polarizing Nicol prism

polarisch *a* polar

polarisierbar *a* polarizable. **Polarisierbarkeit** *f* polarizability. **polarisieren** *vt* polarize. **Polarisierung** *f* (-en) polarization

Polarität *f* (-en) polarity

Polar·kreis *m* polar circle. **-licht** *n* aurora polaris. **-meer** *n* polar sea

Polarographie *f* polarography. **polarographieren** *vt* polarograph

Polar·röhre *f* electrode tube. **-stern** *m* polestar. **-winkel** *m* polar angle

Pol·bildung *f* pole formation, polarization. **-dreck** *m* (*Metll.*) dross, skimmings

Polder *m* (-) reclaimed coastal marshland

Pole I *m* (-n, -n) Pole. **II** *f* (-n) (*Tex.*) pile, nap

Pol·eck *n*, **-ecke** *f* (*Cryst.*) summit

Polei *m* (-e) *or* *f* (-en) (*Bot.*) pennyroyal. **-öl** *n* pennyroyal oil

Polemik *f* (-en) polemic

polen *vt* 1 (*Elec.*) polarize, render polar. 2 (*Metll.: Copper Refining*) pole

Polen I *n* polarizing; poling *cf* POLEN. **vt. II** *n* (*Geog.*) Poland. **III** *pl of* POLE I & II

pölen *vt* (*Lthr.*) unhair

Pol·ende *f* electrode; pole

Polenske·zahl *f* (*Soap*) Polenske value

Pol·höhe *f* altitude of the pole, latitude

Polianit *m* (-e) pyrolusite

Police *f* (-n) (insurance) policy

polierbar *a* polishable. **polieren** *vt* polish

polier·fähig *a* readily polishable. **P-flüssigkeit** *f* (liquid) polish. **-grün** *n* green rouge, chrome green. **-kalk** *m* polishing chalk. **-lack** *m* polishing lacquer. **-masse** *f* polishing composition, (solid) polish. **-mittel** *n* polishing agent, polish. **-öl** *n* polishing oil. **-papier** *n* polishing paper, sandpaper. **-paste** *f* polishing paste. **-pulver** *n* polishing powder. **-rot** *n* polishing rouge, colcothar, crocus. **-scheibe** *f* polishing wheel (*or* disc). **-schiefer** *m* polishing slate, tripoli. **-schleifen** *n* fine-grinding (of surfaces). **-stahl** *m* (steel) burnishing tool. **-staub** *m* polishing dust. **-stein** *m* polishing stone. **-tinte** *f* polishing ink. **-trommel** *f* polishing drum. **-wachs** *n* polishing wax

polig *a* polar

Poliklinik *f* (-en) outpatient department, clinical dispensary. **poliklinisch** *a* outpatient, *adv oft* as an outpatient, in an outpatient department

Polin *f* (-nen) Pole, Polish woman

Politik *f* (-en) policy; politics. **Politiker** *m* (-) politician. **politisch** *a* political. **Politologie** *f* political science

Politur *f* (-en) 1 (French) polish; shellac varnish. 2 polish, gloss. 3 polishing. **p-fähig** *a* polishable. **P-lack** *m* polishing varnish (shellac, lacquer). **-masse** *f* polishing paste. **-messer** *m* glossometer. **-öl** *n* polishing oil

Polizei *f* police. **polizeilich** *a* (of the) police, *adv* by the police

Pol·klemme *f* (*Elec.*) (pole) terminal, binding post. **-körper** *m*, **-körperchen** *n* (*Biol.*) polar body (*or* cell)

Pollen·korn *n* pollen grain. **-übertragung** *f* pollen transfer, pollination

pol·los *a* poleless, without poles

polnisch *a* Polish

Pol·papier *n* (*Elec.*) pole paper, polarity test paper. **-platte** *f* (*Elec.*) pole plate (*or* piece). **-reagenzpapier** n polarity test paper. **-schuh** m pole piece. **-stärke** f pole strength

Polster n (-) cushion; pad; padding. **polstern** vt pad, stuff; upholster

Pol·stern m polestar

Polsterung f (-en) **1** padding, upholstering. **2** pad, upholstery

Pol·strahl m (Math.) radius vector. **-strahlung** f polar radiation. **-suchpapier** n pole (-finding) paper

Polung f (-en) polarization, making polar; (Metll.) poling cf POLEN

pol·wärts adv poleward, toward the pole. **Pol·wechsel** m change (or reversal) of poles; (Elec.) alternation of polarity. **-wechsler** m (Elec.) pole-reversing switch, polarity changer, chopper

Poly·addition f addition polymerization. **-additionsprodukt, -addukt** n addition polymer. **-äthylen** n polyethylene. **-chroismus** m polychroism, pleochroism. **p–chromatisch** a polychromatic, polychrome. **P–chromsäure** f polychromic acid. **p–cyclisch** a polycyclic. **P-eder** n polyhedron. **p–edrisch** a polyhedral. **P–esterharz** n polyester resin. **p–gonisch** a polygonic. **-heteroatomig** a polyheteroatomic. **P–hyperjodat** n polyperiodate. **-kieselsäure** f polysilicic acid. **-kondensat** n condensation polymer. **-kondensation** f condensation polymerization. **-kondensationsprodukt** n condensation polymer. **-kras** m (Min.) polycrase

polymer a polymeric. **Polymere** n (adj endgs) polymer

polymer·einheitlich a uniformly polymerized. **-homolog** a homologous-polymeric. **P–homolog** n homopolymer

Polymerie f polymerism. **Polymerisat** n (-e) polymerizate. **polymerisch** a polymeric. **polymensierbar** a polymerizable. **Polymerisierbarkeit** f polymerizability. **polymerisieren** vt/i polymerize. **Polymerisierung** f (-en) polymerization.

Polymolybdän·säure f polymolybdic acid

polymorph a polymorphous, polymorphic. **Polymorphie** f polymorphism, allotropy.

polymorphisch a polymorphous, polymorphic

polynär a multicomponent

Polynom n (-e), **polynomisch** a polynomial

Poly·oxyverbindung f polyoxy compound. **-plaste** mpl plastics. **-porsäure** f polyporic acid. **-salz** n poly salt. **-säure** f poly acid. **-schwefelwasserstoff** m hydrogen persulfide. **-siliciumsäure** f polysilicic acid. **-styrol** n polystyrene. **-technikum** n polytechnic(al) institute, institute of technology. **p–therm** a polythermal. **P–thionsäure** f polythionic acid. **-vinylazetat** n polyvinyl acetate. **-viol** n polyvinyl alcohol. **-zimtsäure** f polycinnamic acid. **p–zyklisch** a polycyclic

Pol·zelle f polar cell

Pomeranze f (-n) (bitter or Seville) orange **pomeranzen·artig** a orange-like. **P–bitter** n hesperidin. **-blütenöl** n orange flower oil, neroli. **-branntwein** m orange brandy. **p–gelb** a orange(-colored). **P–likör** n curaçao. **-(schalen)öl** n orange (peel) oil

pompejanisch a Pompeian. **P–rot** n Pompeian red

Pompelmuse f (-n) grapefruit; shaddock

ponderomotorisch a (Phys.) ponderomotive

Popanz m (-e) **1** bugaboo. **2** puppet

Popeline f (Tex.) poplin

popularisieren vt popularize

Porcellan n (-e) porcelain, china See PORZELLAN

Pore f (-n) pore; interstice, void

Poren·anteil m proportion of pores (interstices, voids). **-beton** m cellular concrete. **p–frei** a pore-free, nonporous. **-füllend** a pore-filling. **P–füller** m pore filler, primer. **-gefüge** n pore structure. **-gehalt** m pore content. **-gips** m (lightweight) cellular gypsum plaster. **-größe** f pore size. **-größenverteilung** f pore-size distribution. **-raum** m pore space. **-silikat** n cellular silicate. **-sinter** m lightweight aggregate from shale or clay. **-volumen** n pore volume, porosity. **-weg** m pore passage, passage through a filter. **-weite** f pore width (or diameter)

porig a porous. **Porigkeit** f porosity

Poroplaste *mpl* foamed plastics
porös *a* porous; permeable. **Porosität** *f* porosity; permeability
Porositäts·grad *m* degree of porosity. **-wert** *m* porosity value
Porphyr *m* (-e) porphyry. **p--·ähnlich, -artig** *a* porphyritic. **P–felsen** *m*, **-gestein** *n* porphyritic rock
porphyrisch *a* porphyritic
Porree *m* (-s) leek
Porris·säure *f* purreic acid (euxanthic acid)
Porsch, Porst *m* (-e) wild rosemary, marsh tea
Porter·würze *f* (*Brew.*) porter wort
Portier *m* (-s) doorman, doorkeeper
Portion *f* (-en) portion, serving, helping; dose
Portions·beutel *m* single-dose pack (of antacid powder, etc); single-serving package. **p–weise** *adv* in portions, by the portion; batchwise, stepwise
Portland·kalk *m* Portland limestone
Porto *n* (-s) postage. **p--·frei** *a* postage-free, postpaid
Portugal·öl *n* oil of sweet orange: **Neroli-P.** sweet orange blossom oil. **-wasser** *n* laurel water. **portugiesisch** *a* Portuguese
Porzellan *n* (-e) porcelain, china. **p--·artig** *a* porcelaneous. **P–becher** *m* porcelain beaker. **-brei** *m* porcelain slip. **-brennofen** *m* porcelain kiln. **-dampfschale** *f* porcelain evaporating dish. **-einsatz** *m* porcelain insert (shelf, tray) (*e.g.* in a desiccator). **-erde** *f* porcelain (*or* china) clay, kaolin. **-fabrikation** *f* porcelain manufacture. **-gefäß** *n* porcelain vessel. **-geschirr** *n* china (ware). **-glasur** *f* porcelain glaze. **-griff** *m* porcelain handle. **-isolator** *m* porcelain insulator. **-jaspis** *m* porcelain jasper, porcelanite. **-kachel** *f* porcelain tile. **-kitt** *m* porcelain cement. **-kugel** *f* porcelain ball. **-kugelmühle** *f* porcelain ball mill. **-löffel** *m* porcelain spoon. **-malerei** *f* china painting. **-masse** *f* porcelain paste. **-mörser** *m* porcelain mortar. **-mörtel** *m* pozzuolana mortar. **-nutsche** *f* porcelain suction filter, Buchner filter. **-ofen** *m* porcelain kiln. **-platte** *f* porcelain plate. **-rohr** *n*, **-röhre** *f*

porcelain tube. **-schale** *f* porcelain dish (*or* cup). **-schiffchen** *n* porcelain boat. **-spat** *m* (*Min.*) scapolite. **-spatel** *m* porcelain spatula. **-stab** *m* porcelain rod (*or* bar). **-tiegel** *m* porcelain crucible. **-ton** *m* porcelain clay, kaolin. **-tonumschlag** *m* (*Pharm.*) cataplasm of kaolin. **-trichter** *m* porcelain funnel. **-wanne** *f* porcelain trough. **-waren** *fpl* porcelain (ware), china(ware)
Posaune *f* (-n) trombone; (*fig*) trumpet
poschieren *vt* poach (eggs)
Pose *f* (-n) 1 quill. 2 pose
positiv·elektrisch *a* electropositive
Positiv·pause *f* (*Phot.*) positive print
Post *f* (-en) 1 mail; postal service, post office. 2 postal bus (service)
Postament *n* (-e) pedestal
Post·amt *n* post office. **-bote** *m* mailman, mail carrier
Posten *m* (-) 1 post, position (*incl* job). 2 (duty) station; guard (duty); sentry; (strike) picket: **auf dem P. (a)** on one's guard, (**b**) up to par. 3 lot (of goods); (*Metll.*) batch; (*Glass*) piece, lump. 4 (bookkeeping) entry, item; amount
Post·fach *n* post office box. **p–frei** *a* postage-free. **-karte** *f* 1 postcard. 2 postal map. **p–lagernd** *a, adv* general delivery. **P–leitzahl** *f* postal zone number. **-papier** *n* letter paper, stationery. **-stempel** *m* postmark
postum *a* posthumous
post·wendend *adv* 1 by return mail. 2 immediately. **P–zeichen** *n* postmark
Potential·abfall *m* drop in potential. **-berg** *m* potential barrier. **-gefälle** *n* drop in potential, potential gradient. **-hügel** *m* potential barrier. **-mulde** *f* potential (energy) trough. **-rand** *m* (*Elec.*), **-schwelle** *f* potential barrier. **-sprung** *m* (*Elec.*) jump in potential, potential difference. **-topf** *m* potential well. **-unterschied** *m* difference in potential. **-wall** *m* potential barrier
potentiell *a* potential
potentiieren *vt* render potent, potentize
potentiometrisch *a* potentiometric
Potenz *f* (-en) 1 (*Math.*) power, exponent. 2 potency. **--·gefälle** *n* potential gradient.

-gesetz n law of exponents

potenzieren vt 1 strengthen, intensify. 2 (*Math.*) raise to a higher power: **8 mit 4 p.** raise 8 to the fourth power. 3 (*Pharm.*) potentiate (a drug)

Potenz·reihe f (*Math.*) power (*or* exponential) series. **-zentrum** n (*Math.*) radical center

Poterie f (-n) pottery. **·-guß** m heavy earthenware

Pott m (-e) pot. **·-asche** f potash

Pottasche(n)·fluß m crude potash (from ashes). **-lauge** f potash lye. **-lösung** f potash solution. **-siederei** f potash factory

pott·echt a (*Dye.*) fast to potting, potting-resistant. **P—erde** f potter's clay. **-fisch** m sperm whale. **-fischöl** n, **-fischtran** m sperm oil

potting·echt a (*Tex.*) fast to potting = POTTECHT

Pott·lot n graphite, black lead

Poussiere f (*Zinc*) blue powder

poussieren vt 1 push, promote. 2 butter up

Pozz(u)olan n, **·-erde** f pozzuolana

Prä n: **das P. haben** have priority. **prä·** pfx pre-

Pracht f splendor, magnificence—**eine wahre P.** a real treat. **·-ausgabe** f deluxe edition. **-exemplar** n magnificent specimen, showpiece

prächtig, prachtvoll a splendid, magnificent

Präci . . . cf PRÄZI . . .

pract . . . cf PRAKT . . .

Prädikat n 1 (exam) grade, mark; rating—**mit P. bestehen** pass with distinction. 2 title, attribute. 3 predicate

prädisponieren vt predispose

Prädissoziation° f predissociation

Präexistenz° f preexistence

prägbar a stampable

Präge·anstalt f (currency) mint. **-form** f (stamping) mold, matrix. **-kalander** m (*Tex.*) embossing calender

prägen I vt 1 mint, coin (money, words). 2 stamp, (im)press; emboss. 3 mark; form, mold. **II: sich p.** impress itself, leave its mark, leave an impression

Präge·papier n embossed (*or* embossing) paper

Präglobulin n preglobulin

prägnant a concise, terse, succinct; adv oft significantly. **Prägnanz** f conciseness, etc

Prägung f (-en) 1 minting, coinage. 2 stamping, embossing. 3 stamp, character; imprint

prähistorisch a prehistoric

prahlen vi 1 boast, brag. 2 (*Colors*) be loud

Prahm m (-e) barge, lighter

Präionisation° f pre-ionization

Präjudiz n (-e *or* -ien) prejudice; (legal) precedent. **präjudizieren** vt prejudice, prejudge

präkambrisch a (*Geol.*) pre-Cambrian

prakt. abbv (PRAKTISCH) practical, etc

Praktik f (-en) 1 practice, procedure. 2 pl oft tricks, short cuts; machinations. **Praktikant** m (-en, -en), **Praktikantin** f (-nen) 1 laboratory assistant. 2 laboratory student. 3 practitioner. **Praktiker** m (-) 1 experienced practitioner, old hand. 2 (*Med.*) general practitioner. **Praktikum** n (. . ika *or* . . iken) 1 course with practical training in the field. 2 laboratory course. **Praktikums·aufgabe** f practical exercise

praktisch a 1 practical; applied. 2: **p—er Arzt** general practitioner. 3 adv practically, for all practical purposes; in practice: **p. geschult** trained in practice; **p. durchführen** put into practice

praktizieren I vt 1 put into practice; apply, use. 2 maneuver (sthg into or out of). II vi (*Med.*) practice; have office hours

prall a 1 round and firm, bulging. 2 chubby, plump. 3 taut, tight(-fitting). 4 elastic, springy. 5 blazing (sun). **Prall** m (-e) impact; shock, crash; rebound; reflection. **·-blech** n deflecting plate, baffle. **-brecher** m impact crusher. **-elastizität** f rebound resilience. **-elektrode** f reflecting electrode

prallen vi I (h) 1 beat down (*as:* hot sun). II (s) 2 (**gegen**) crash (into), collide (with), strike. 3 bounce, rebound. 4 be reflected

prall·gefüllt a jam-packed, filled to bursting

Prall·heit f 1 firm roundness, bulge. 2 plumpness. 3 tautness; tight fit. 4 elasticity, springiness. *cf* PRALL

Prallicht *n* (= **Prall·licht**) reflected light

Prall·kraft *f* elasticity, resiliency. **-mühle** *f* impact mill. **-platte** *f* baffle, deflecting plate. **-wand** *f* baffle. **-winkel** *m* angle of reflection. **-zerkleinerung** *f* impact crushing

Prämie *f* (-n) **1** premium (*incl* insurance). **2** prize, award. **3** bonus. **4** bounty; reward

prangen *vi* be resplendant, shine (forth)

Pranke *f* (-n) paw; claw, clutch

Präp. *abbv* = **Präparat** *n* (-e) **1** (*Chem.*, *Med.*) preparation. **2** (*esp Biol.*, *Med*) (instructional) model; specimen (for dissection). **3** (*Micros.*) slide. **4** stuffed animal

Präparaten·glas *n* preparation glass. **-kunde** *f* knowledge concerning preparations. **-röhrchen** *n* preparation (*or* specimen) tube. **-schachtel** *f* (*Micros.*) slide box

präparativ *a* preparative, preparational.

Präparier·arbeit *f* dissecting work. **präparieren** *vt* prepare (specimens, etc); dissect

Präparier·lupe *f* (*Biol.*) dissecting lens (*or* magnifier). **-nadel** *f* dissecting needle. **-salz** *n* (*Dye.*) preparing salt (sodium stannate)

präpariert *p.a* (*Phot.*): **p—es** Papier sensitized paper. *See also* PRÄPARIEREN

Präparier·tisch *m* dissecting table

Prasem, Prasen·stein *m* prase (green quartz)

präsentieren *vt* present, offer

Praseodym *n* praseodymium. **Praseokobalt·salz** *n* praseocobaltic salt

Präservativ *n* (-e) contraceptive sheath, condom

Präserven *fpl* preserves. **präservieren** *vt* preserve. **Präservierung** *f* (-en) preservation. **Präservierungs·mittel** *n* preservative

prasseln *vi* **1** crackle, crepitate (*esp as:* fire). **2** patter (*as:* rain); hail, pelt (down)

Pratze *f* (-n) **1** paw; claw, clutch. **2** (*Mach.*) arm, lug, bracket

Präventiv·impfung *f* preventive inoculation

praxi: in p. in practice

Praxis *f* (.. xen) **1** practice. **2** doctor's office (hours). **p—·bezogen** *a* practice-related.

-nah *a* practical, real-life; practice-oriented; *adv oft* under practical (*or* real-life) conditions. **-reif** *a* operational, ready for practical use. **P—reife** *f* operational stage, practical applicability. **-versuch** *m* practical experiment. **p—wirksam** *a* effective in practice

präzedieren, präzessieren *vi* precess. **Präzession** *f* (-en) precession

Präzipitat *n* (e) precipitate. **Präzipitations·wärme** *f* heat of precipitation. **Präzipitier·bottich** *m* precipitating vat. **präzipitieren** *vt* precipitate. **Präzipitin** *n* (-e) precipitin

präzis *a* precise. **präzisieren** *vt* define (more) precisely. **Präzision** *f* precision

Präzisions·arbeit *f* precision work. **-gewicht** *n* precision weight. **-messung** *f* precision measurement. **-waage** *f* precision balance

Prehnit·säure *f* prehnitic acid

Preis *m* (-e) **1** price. **2** prize. **3** praise

Preisel·beere *f* mountain cranberry

preisen* (pries, gepriesen) *vt* praise, laud

Preisgabe° *f* surrender; betrayal. **preis·geben*** *vt* **1** surrender, give up (*or* away); betray (secrets, etc). **2** (*d.i.o*) expose (to danger, etc)

Preis·lage *f* price range

Preißel·beere *f* mountain cranberry

preis·wert *a* reasonable, moderately priced

Prell·block *m* stop (block). **-bock** *m* (track-end) bumper

prellen I *vt* **1** (*usu* + *dat rflx* **sich**) bump, bruise (e.g. one's knee). **2** (**um**) cheat (out of)—**die Zeche p.** beat one's check. **II** *vt*, *vi*(s) **3** bounce. **III** *vi*(s) **4** strike, bump. **5** dash. **IV: sich p.** bump into sthg, get bruised

Prell·kraft *f* resiliency. **-platte** *f* baffle (plate). **-stein** *m* protective abutment. **-stoß** *m* rebound

Prellung *f* (-en) bruise, contusion

Prephen·säure *f* prephenic acid

Preslit *m* (-e) (*Min.*) tsumebite

pressant *a* pressing, urgent

Preß·artikel *m* (*Rubber*) press-cured article. **-automat** *m* automatic (pelletizing, tableting) press

preßbar *a* compressible. **Preßbarkeit** *f* compressibility

Preß·bernstein *m* amb(e)roid. **-beutel** *m* pressing bag, filter bag (for presses). **-blaseverfahren** *n* (*Glass*) press-and-blow process. **-blei** *n* (*Metll.*) inferior lead (from Carinthian process after adding charcoal). **-bolzen** *m* (*Metll.*) billet. **-bolzenofen** *f m* billet-heating furnace. **-dorn** *m* mandrel. **-druck** *m* pressing pressure, (*esp Plastics*) molding pressure

Presse *f* (-n) **1** press (*both Mach. & news medium*). **2** (*Tex. oft*) calender. **3** gloss, luster

pressen I *vt & sich p.* **1** press, squeeze. **II** *vt* **2** compress. **3** cram. **4** (*Tex.*) calender. **5** (*Metll.*) extrude. **Presser** *m* (-) presser; pressman; compressor

Preß·fehler *m* molding defect. **-filter** *n* pressure filter. **-flüssigkeit** *f* pressure fluid. **-form** *f* pressure (*or* compression) mold. **-gas** *n* compressed gas. **-glas** *n* pressed glassware. **-glimmer** *m* pressed mica. **-guß** *m* pressure die casting(s); injection molding. **-hefe** *f* compressed yeast. **-holz** *n* compressed wood

pressieren *vi* be urgent. **pressiert** *p.a* in a hurry.

Pression *f* (-en) pressure

Preß·kohle *f* pressed coal (*or* charcoal), coal (*or* charcoal) briquettes. **-kopf** *m* head cheese. **-körper** *m* pressed article (*or* object). **-kraft** *f* compressive force. **-kuchen** *m* filter cake; oil (*or* press) cake. **-lauge** *f* expressed liquor

Preßling *m* (-e) **1** pressed article, (*specif*) briquette, pellet, molding, (*Metll.*) compact. **2** (*usu pl; Sugar*) expressed beet pulp

Preß·luft *f* compressed air

Preßluft·hammer *m* jackhammer, compressed-air hammer. **-leitung** *f* compressed-air line. **-maschine** *f* air compresser. **-schlauch** *m* compressed-air hose. **-zerstäuber** *m* compressed-air atomizer

Preß·masse *f* **1** molded mass. **2** (*esp Plastics*) molding composition. **-matrize** *f* extrusion die. **-mischung** *f* molding mixture. **-most** *m* expressed fruit juice, must. **-mühle** *f* pressing mill. **-naht** *f* burr, seam (on molded articles). **-öl** *n* **1** oil obtained by pressure. **2** oil under pressure,

force-feed (*or* hydraulic oil). **-öler** *m* force-feed (*or* pressure) oiler. **-ölschmierung** *f* force-feed (*or* pressure) lubrication. **-passung** *f* force fit. **-pastille** *f* pressed tablet (*or* pellet). **-platte** *f* pressed plate, (*specif*) phonograph record

preßpolieren *vt* (*pp* preßpoliert) burnish

Preß·pulver *n* molding powder. **-pumpe** *f* pressure pump. **-rückstand** *m* expressed residue; extrusion discard. **-sack** *m* **1** filter bag (for presses). **2** head cheese. **-saft** *m* press juice. **-schmierung** *f* force-feed lubrication. **-schraube** *f* pressure (*or* thrust) screw. **-schweißung** *f* pressure welding. **-sitz** *m* force fit, press fit. **-span** *m* pressboard, chipboard. **-stahl** *m* pressed steel. **-stein** *m* briquette. **-stempel** *m* press ram (*or* plunger). **-stoff** *m* **1** pressed material. **2** molding composition, plastic. **-stück** *n* pressed (*or* molded) piece. **-talg** *m* pressed tallow. **-teil** *m* pressed (*or* molded) part. **-temperatur** *f* compression temperature. **-torf** *m* pressed peat. **-tuch** *n* filter cloth (for presses)

Pressung *f* (-en) pressing, squeezing; compression; cramming; calendering; compression molding *cf* PRESSEN

Preß·verfahren *n* pressing (*or* molding) process. **-vulkanisation** *f* (*Rubber*) press cure. **-walze** *f* press roll. **-wasserakkumulator** *m* hydraulic accumulator. **-werk** *n* extrusion plant. **-würfel** *m* (*Sugar*) pressed cube. **-zeit** *f* compression time. **-ziegel** *m* pressed brick. **-zucker** *m* compressed sugar

Preußen *n* (*Geog.*) Prussia. **preußisch** *a* Prussian

Preußisch·blau *n* Prussian blue. **-braun** *n* Prussian brown

prickeln I *vt* tickle, titillate. **II** *vi* **1** tingle, prickle; be itching (to do sthg). **2** bubble; sparkle

Priem·tabak *m* chewing tobacco

pries praised: *past of* PREISEN

prim *abbv* (primär) primary

prima *a* (*no endgs*) prime, fine, first-rate, A-one. **Prima·** *pfx* prime, quality, refined

primär *a* primary. **P--akt** *m* primary act. **-element** *n* (*Elec.*) primary cell. **-fleck** *m* primary spot. **-gefüge** *n* primary structure (*or* texture). **-kreis** *m* primary circuit. **-linie** *f* primary line. **-spannung** *f* (*Elec.*) primary voltage. **-strahl** *m* (*Opt.*) primary ray. **-strahlung** *f* primary radiation. **-strom** *m* primary current. **-vorgang** *m* primary process

Prima·soda *f* refined soda. **-sprit** *m* refined spirits, ethyl alcohol

Primat *n* primacy, preeminence

Prima·ware *f* prime (*or* quality) goods, first-rate article

Primel *f* (-n) primrose; primula. **-krankheit** *f* primrose dermatitis. **-kratzstoff** *m* unpleasant tasting substance in primrose root

Primordial·gefäß *n* (*Bot.*) protoxylem element

Prim·zahl *f* prime number

Prinz *m* (-en, -en) prince

Prinzip *m* (-ien *or* -e) principle. **prinzipiell** *a* fundamental, basic; *adv* & **im P—en** in (*or* as a matter of) principle. **Prinzipien·frage** *f* matter of principle. **Prinzipium** *n* (..pien) principle. **Prinzip·skizze** *f* diagrammatic sketch

Priorität *f* (-en) 1 priority. 2 *pl* preferred stocks

Prioritäts·beleg *m* certificate of priority. **-recht** *n* right of priority

Prise *f* (-n) 1 pinch (of salt, snuff, etc). 2 prize (capture, *usu* ship)

Prisma *n* (.. men) prism. **p—·ähnlich**, **-artig** *a* prism-like, prismatoid. **-förmig**, **prismatisch** *a* prism-shaped, prismatic. **prismatoid·förmig** *a* prismatic, prism-shaped

Prismen prisms; *pl of* PRISMA. **-fläche** *f* prism face. **-formel** *f* prism formula. **p—förmig** *a* prism-shaped, prismatic. **-glas** *n* 1 prism glass. 2 prism binoculars. **-kante** *f* prismatic edge. **-spektrum** *n* prismatic spectrum

Pritsche *f* (-n) 1 plank bed; bunk (bed). 2 (wooden) platform, (flat)bed (*esp* of a truck). 3 (*Alum*) washing floor. 4 (*Dye.*) stillage (to support goods while drying). 5 bat, racquet; slapstick. **Pritschen·wagen** *m* platform (*or* flatbed) truck

Privat·dozent *m* licensed university lecturer (receiving student tuition fees but no salary). **-wirtschaft** *f* private enterprise

pro 1 *prep* per. 2 *adv* pro: **p. und kontra** pro and con

pro anal. *abbv* (pro Analyse) for analysis

probat *a* proven, effective

Probe *f* (-n) 1 test, trial: **auf die P. stellen** put to the test; **auf P. anstellen** hire on probation. 2 sample, specimen; assay. 3 rehearsal. **-abdruck**, **-abzug** *m* (*Phot.*, *Print*) proof. **-bogen** *m* proof (sheet). **-brand** *m* fire test. **-brühe** *f* test bath, dye test. **-druck** *m* (printer's) proof. **-entnahme** *f* sampling. **-essig** *m* proof vinegar. **-exemplar** *n* test specimen (*or* copy). **-fahrt** *f* road test, trial run. **-färbung** *f* test dyeing. **-fläche** *f* test (*or* sample) area. **-fläschchen** *n*, **-flasche** *f* sampling bottle; specimen bottle. **-flug** *m* test flight. **-flüssigkeit** *f* test liquid (*or* liquor). **p—gemäß** *a* according to the sample. **P—gewicht** *n* test (sample, assay, standard) weight. **-glas** *n* test tube; sample (*or* specimen) jar (tube, glass). **-gold** *n* standard gold. **-gut** *n* sample (material), specimen. **-hahn** *m* try cock, sample (*or* specimen) tap; bleeder valve. **p—halber** *adv* as a test. **-haltig** *a* proof, standard; genuine. **P—kolben** *m* test (*or* sample) flask. **-korn** *n* assay button. **-körper** *m* test body (specimen, sample). **-lauf** *m* test (*or* trial) run. **-löffel** *m* assay spoon. **-lösung** *f* test (*or* standard) solution. **-machen** *n* testing, assaying. **-maß** *n* standard measure. **p—mäßig** *a* according to the sample (*or* pattern). **P—münze** *f* proof (*or* standard) coin. **-muster** *n* test sample (*or* pattern)

proben 1 *vt* test, try out. 2 *vt/i* rehearse

Proben· *See also* PROBE·

Probe·nadel *f* touch needle. **-nahme** *f* sample, sampling. **-nahmerohr** *n* sampling (*or* thief) tube. **-nehmen** *n* sampling. **-nehmer** *m* sampler

Proben·glas *n* specimen glass (*or* tube). **-stecher** *m* sampler, etc = PROBE·STECHER. **-teiler** *m* sample splitter

Probe·nummer *f* sample (*or* specimen) issue

Proben·vorbereitung *f* sample preparation

Probe·objekt *n* specimen, test piece (*or* object). **-ofen** *m* assay furnace (*or* oven). **-papier** *n* test paper, paper to be tested

Prober *m* (-) tester; assayer

Probe·ring *m* (*Engg.*) proof (*or* proving) ring. **-rohr, -röhrchen** *n*, **-röhre** *f* 1 (*Chem.*) test tube. 2 (*Mach.*) trial pipe. **-säure** *f* standard acid; test acid. **-schachtel** *f* sample box. **-scherbe** *f*, **-scherben** *m* (*Metll.*) cupel; (*Ceram.*) trial piece. **-sendung** *f* sample shipment; sample sent on approval. **-silber** *n* standard silver. **-spiritus** *m* proof alcohol. **-stab** *m* test (*or* trial) rod. **-stange** *f* test rod (stick, bar). **-stecher** *m* sampler (inserted into materials); (*specif*) thief tube (for liquids), proof stick (for sugar). **-stein** *m* touchstone; test (*or* specimen) stone. **-stoff** *m* sample material. **-stück** *n* sample, specimen; test piece. **-tiegel** *m* assay crucible; cupel. **-waage** *f* assay balance. **-weingeist** *m* proof alcohol. **p–weise** *a* as (or by way of) a test; on a trial basis, on probation; on approval. **P–würfel** *m* test cube. **-ziehen** *n* sampling. **-zieher** *m* sampler. **-zinn** *n* standard tin. **-zylinder** *m* test jar (*or* tube)

Prob·hahn *m* try cock, gauge cock

Probier·blei *n* test (*or* assay) lead. **-brühe** *f* (*Dye.*) test bath; dye test

probieren I *vt/i* 1 try, attempt; (+ *ob*) try and see (if), test (whether). 2 rehearse. II *vt* 3 sample, assay; taste. 4 try on (garment, etc). **Probierer** *m* (-) tester; assayer; analyst

Probier· *usu* = PROBE·: **-gefäß** *n* testing (*or* assaying) vessel. **-geräte** *npl*, **-gerätschaften** *fpl* assaying apparatus. **-gewicht** *n* assay weight. **-glas** *n* sample (*or* specimen) jar; test tube. **-glätte** *f* test litharge. **-gold** *n* standard gold. **-hahn** *m* try cock, gauge valve. **-kluft** *f* assayer's tongs. **-korn** *n* assay button. **-kunde**, **-kunst** *f* assaying. **-laboratorium** *n* assay laboratory. **-löffel** *m* assay spoon. **-metall** *n* test metal. **-methode** *f* testing (*or* experimental) method. **-nadel** *f* touch needle. **-ofen** *m* assay furnace. **-papier** *n* test paper. **-rohr** *n*, **-röhre** *f* test tube. **-röhrengestell** *n* test tube rack. **-scherbe** *f*, **-scherben** *m* cupel; (*Ceram.*) trial piece. **-stein** *m* touchstone. **-tiegel** *m* assay crucible; cupel. **-tüte** *f* crucible. **-ventil** *n* test (*or* gauge) valve. **-verfahren** *n* testing (*or* assaying) process. **-waage** *f*

assaying balance. **-zange** *f* assayer's tongs

Problematik *f* problems (connected with . . .)

Problem·kreis *m* problem area. **p–los** *a* problem-free, trouble-free

Proc. *abbv* 1 (Procent) percent *cf* PROZENT. 2 (Proceß) process *cf* PROZESS

Prod. *abbv* (Produkt) product

Produkt *n* (-e) (*incl Math.*) (+ **aus**) product (of)

Produktions·fähigkeit *f* productivity. **-leistung** *f* productive capacity. **-rückstand** *m* residual waste

Produktivität *f* productivity

Produzent *m* (-en, -en) producer, grower; manufacturer. **produzieren** *vt* produce

Professorat *n* (-e) professorship. **professorisch** *a* professional. **Professorschaft, Professur** *f* (-en) professorship

Profi *m* (-s) pro(fessional)

Profil *n* (-e) 1 profile. 2 (*Metll.*) section. 3 (tire) tread. 4 personality. **-draht** *m* sectional wire. **-drahtsieb** *n* wedge-wire screen. **-eisen** *n* sectional iron (*or* steel)

profilieren *vt* profile; shape; streamline

Profil·material *n* section material. **-messer** *n* profile cutter. **-prüfer** *m* profilometer. **-teil** *m* profiled part

Prognose *f* (-n) prognosis, prediction; forecast

Programm *n* (-e) 1 program; schedule. 2 (*Cmptrs*) software. 3 leaflet. **programmäßig** *a* according to plan, on schedule. **programmieren** *vt* program. **Programmiersprache** *f* computer (programming) language. **Programmierung** *f* (-en) programming

projektieren, projizieren *vt* project

Prokurist *m* (-en, -en) (*Com.*) vice president

Promille *n* (-) 1 per thousand. 2 unit of alcohol content of blood (for car drivers). **-grenze** *f* legal limit of blood alcohol content. **-sünder** *m* 1 driver with excessive blood alcohol. 2 petty offender. **promillig** *a* per (*or* by the) thousand

promovieren 1 *vt* award the doctor's degree to. 2 *vi* receive the doctor's degree

Propan·säure *f* propionic (*or* propanoic) acid

prophylaktisch *a* prophylactic. **Pro-**

phylaxe *f* (-n) prophylaxis
Propiol·säure *f* propiolic acid
propion·sauer *a* propionate of. **P—·säure** *f* propionic acid
Propioper·säure *f* per(oxy)propionic acid
Proportionalität *f* proportionality. **proportionieren** *vt* proportion
Propylitisierung *f* propylitization
Propyl·wasserstoff *m* propyl hydride, propane
Prospekt *m* (-e) 1 prospectus, brochure. 2 backdrop. 3 view, prospect
prosthetisch *a* prosthetic
Protaktinium *n* protoactinium
Protalbin·säure *f* protalbinic acid
protein·haltig *a* protein-bearing, containing protein. **P—körper** *m* protein (substance). **-säure** *f* proteic acid. **-silber** *n* silver proteinate. **-stoff** *m* protein (substance)
Proteinurie *f* albuminuria
Proteolyse *f* proteolysis. **proteolytisch** *a* proteolytic
Prothese *f* (-n) prosthesis; denture(s)
Proto·cetrarsäure *f* protocetraric acid. **-cocasäure** *f* protococaic acid
protogen *a* protogenic
Protokatechu·säure *f* protocatechuic acid
Protokoll *n* (-e) 1 protocol. 2 minutes (of meetings), record(s). 3 traffic ticket
Protonen·austausch *m* proton exchange. **-beschleunigung** *f* proton acceleration. **-donator, -geber** *m* proton donor. **-geschoß** *n* proton projectile. **-nehmer** *m* proton acceptor. **-stoß,** *m* proton collision. **-streuung** *f* proton scattering. **-wanderung** *f* proton migration
protophil *a* protophilic, proton-accepting
Protoplasma·bewegung *f* protoplasmic motion. **-körper** *m* protoplasmic body. **-strahl** *m* protoplasmic ray
Protoxid° *n* protoxide
Protozoen *npl* protozoa. **protozoisch** *a* protozoic, protozoan
Provence·öl *n,* **Provencer Öl** (Provence) olive oil
Provenienz *f* (-en) provenience, origin
Provenzer Öl *n* olive oil (from Provence)
Proviant *m* provisions (*usu for a trip*)
Provision *f* (-en) commission (on sales)
Provisor *m* (-en) manager of a pharmacy

provisorisch *a* provisional, temporary
·proz. *abbv* (·prozentig), **Proz.** *abbv* (Prozent) percent
Prozedere *n* (-) procedure
Prozedur *f* (-en) (tedious) procedure, ordeal
Prozent *n* (-e) 1 percent. 2 *pl* percentage, discount, rebate. **·anteil** *m* percentage composition; percentage, share. **-gehalt** *m* (**an**) percentage, percent content (of) **·prozentig** *sfx* e.g. dreip—e Lösung 3% solution
prozentisch *a* percentage, *adv* percentagewise
Prozent·satz *m* percentage. **-teilung** *f* percentage scale
prozentual, prozentuell *a* percentage, *adv* percentagewise
Prozent·zahl *f* percentage
Prozeß *m* (. . esse) 1 (*incl Chem.*) process. 2 lawsuit, trial—**kurzen P. machen (mit)** make short work (of). 3 (*Med.*) lesion. **prozessieren** *vi* litigate. **Prozeß·rechner** *m* program control computer
Prüf· testing: **-anstalt** *f* testing laboratory. **-arbeit** *f* testing, assaying; checking
prüfbar *a* testable, assayable, etc *cf* PRÜFEN
Prüf·bericht *m* test report. **-druck** *m* test(ing) pressure
prüfen *vt* 1 (*oft* + **auf**) test (for); assay; taste. 2 examine. 3 investigate, check; inspect. 4 try, subject to trials. **prüfend** *p.a* searching, scrutinizing. **Prüfer** *m* (-) 1 tester, assayer; taster; examiner; checker, inspector. 2 meter
Prüf· test(ing): **-ergebnis** *n* test result. **-feld** *n* proving ground; test(ing) floor. **-gerät** *n* testing apparatus (*or* equipment). **-glas** *n* test tube (*or* glass). **-hahn** *m* test valve. **-kelch** *m* test (*or* reaction) glass. **-körper** *m* test specimen (*or* substance)
Prüfling *m* (-e) 1 test specimen (sample, piece). 2 examinee, candidate
Prüflsgg. *abbv* (Prüflösungen) test solutions
Prüf· test(ing): **-maschine** *f* test(ing) machine. **-methodik** *f* test(ing) method (*or* procedure). **-mittel** *n* testing agent, test

reagent. **-nadel** f testing needle. **-rohr** n, **-röhre** f test(ing) tube. **-scheibe** f test disk. **-sender** m (Elec.) signal generator. **-sieb** n test sieve. **-stand** m test floor (block, rig). **-stein** m touchstone. **-stelle** f testing center (or station); inspection (or assay) office. **-stöpsel** m test plug. **-streifen** m test strip. **-stück** n test piece (or specimen)

Prüfung f (-en) 1 examination, test; assay: **eine P. ablegen** take a test. 2 testing, assaying; tasting; investigation, checking, inspection cf PRÜFEN. 3 trial, ordeal

Prüfungs·ergebnis n test result. **-methode** f testing method. **-mittel** n test(ing) agent. **-schein** m inspection (examination, testing) certificate. **-schrift** f (university) degree thesis. **-stein** m touchstone. **-stück** n test piece (or sample). **-zeugnis** n inspection (examination, testing) certificate; diploma

Prüf·verfahren n testing process. **-zeichen** n test mark

Prügel m (-) 1 cudgel, club. 2 pl beating

Prunell(en)·salz n sal prunelle (fused potassium nitrate)

Prunk m splendor, pomp, ostentation

PS abbv (Pferdestärke) hp. (horsepower): **PS_e** (effektive PS) effective hp.; **PS_h** hp-hr.; **PS_i** (indizierte PS) indicated hp.

Pseudo·harnsäure f pseudouric acid. **-katalysator** m pseudocatalyst. **-katalyse** f pseudocatalysis. **ps–katalytisch** a pseudocatalytic. **Ps–lösung** f pseudosolution

Pseudomerie f tautomerism

pseudomorph a pseudomorphic. **Pseudomorphie** f pseudomorphism. **Pseudomorphose** f (-n) 1 pseudomorphosis. 2 pseudomorph

Pseudo·säure f pseudoacid. **-schwefelzyan** n pseudothiocyanogen. **ps–wissenschaftlich** a pseudoscientific

Psorom·säure f psoromic acid

Pst. abbv (Pferdestärke) hp (horsepower)

p-ständig a para-, in the para position

Psychiater m (-) psychiatrist

psychisch a psychic

Psylla·alkohol m psyllic alcohol. **-säure** f psyllic acid. **-stearylsäure** f psyllastearic acid

Pteroin·säure f pteroic acid

Publikum n 1 (the) public. 2 audience

publizieren vt publish

Puddel·arbeiter m puddler. **-bett** n puddling furnace bed. **-eisen** n puddled iron. **-luppe** f puddle ball. **-maschine** f puddler

puddeln vt (Metll.) puddle

Puddel·ofen m puddling furnace. **-prozeß** m puddling process. **-roheisen** n forge pig iron. **-schlacke** f puddling slag, puddle cinder. **-sohle** f puddling furnace bed. **-spiegel** m specular forge pig iron. **-spitze** f puddler's paddle. **-stab** m puddle bar. **-stahl** m puddled steel. **-verfahren** n puddling process. **-walze** f puddle roll. **-werk** n puddling mill (plant, works)

Pudding·stein m pudding stone, conglomerate

Pudel m (-) poodle—**des P–s Kern** the crux of the matter

Puder m (-) (cosmetic) powder. **-email** n dry-process enamel

puderig a powdery. **pudern** vt powder.

Puder·zucker m powdered sugar

pudrig a powdery

Puff m (-e or ⸗e) 1 puff (incl Genetics); (in compds oft) puffed. 2 pop, bang. 3 blow, thump; nudge, poke. 4 (upholstered) linen hamper. **-bohne** f broad bean, Italian bean

puffen 1 vt/i puff (up, out). 2 vt thump, nudge, poke. 3 vi pop, bang

Puffer m (-) 1 (incl Chem.) buffer; damper. 2 cushion, pad. 3 bumper. 4 (potato) pancake. **-lösung** f buffer solution

puffern vt buffer, damp; cushion, pad. **Puffer·salz** n buffer salt. **Pufferung** f (-n) buffering, damping; cushioning, padding. **Pufferungs·vermögen** n buffering power (or capacity)

Puffer·wert m buffer value. **-wirkung** f buffering (damping, cushioning) action (or effect)

Puff·mais n popcorn. **-reis** m puffed rice

Pukall·masse f, **Pukallsche Masse** Pukall mass (or composition) (porous porcelain for cells)

Pulegen·säure f pulegenic acid

Pulpe f (-n) pulp. **p–·haltig** a pulpy (fruit)

Puls m (-e) pulse. **-ader** f artery. **-aderblut** n arterial blood

pulsen *vi* pulsate

Puls·glas *n*, **-hammer** *m* cryophorus, water hammer

pulsieren *vi* pulsate. **Pulsieren** *n* pulsation. **Puls·schlag** *m* pulse (beat). **Pulsung** *f* (-en) pulsation. **Puls·zahl** *f* pulse rate

Pult *n* (-e) desk, lectern. **--feuerung** *f* firing on stepped grate bars. **-ofen** *m* (*Metll.*) back-flame hearth

Pulver *n* powder, (*specif*) gunpowder. **-alitierung** *f* aluminizing with a powder, powder-calorizing. **p--artig** *a* powdery, pulverulent. **P--aufnahme** *f* (*Cryst.*) powder photograph. **-band** *n* (*Explo.*) powder strand. **-blättchen** *n* (*Explo.*) powder flake (*or* grain). **-brennzündung** *f* powder-train ignition. **-dampf** *m* powder (*or* gun) smoke. **-diagramm** *n* powder pattern. **-fabrik** *f* powder mill (*or* factory). **-fabrikation** *f* powder manufacture. **-faß** *n* powder keg. **-flasche** *f* (wide-mouth) powder flask (*or* bottle). **-form** *f* powder form. **p--förmig** *a* powdery, powdered, pulverulent. **P--förmigkeit** *f* powdered form, pulverulence. **-gas** *n* gas from burning powder. **-glas** *n* (wide-mouth) powder bottle (*or* jar). **-holz** *n* wood for gunpowder

pulverig *a* powdery, pulverulent

Pulverisator *m* (-en) pulverizer

pulverisierbar *a* pulverizable. **pulverisieren** *vt* pulverize. **Pulverisier·mühle** *f* pulverizing mill. **Pulverisierung** *f* (-en) pulverization

Pulver·kaffee *m* instant coffee. **-kammer** *f* powder magazine. **-kapsel** *f* powder capsule; small powder scoop. **-korn** *n* grain of powder. **-kuchen** *m* (*Gunpowder*) press cake. **-ladung** *f* powder charge. **-masse** *f* powder composition. **-mehl** *n* mealed powder. **p--metallurgisch** *adv* by powder metallurgy. **P--methode** *f* (*Cryst.*) powder method. **-mischverfahren** *n* (*Plastics*) powder mixing (process). **-mörser** *m* powder mortar. **-mühle** *f* powder mill

pulvern *vt* pulverize, powder

Pulver·partikel *f* powder particle. **-preß·kuchen** *m* (*Gunpowder*) press cake. **-preßling** *m* (piece of) compressed

powder. **-probe** *f* 1 powder test. 2 powder particle. **-probiermörser** *m* small powder(-testing) mortar. **-punkt** *m* powder(-coated) contact. **-punktauftrag** *m* spot powder application. **-rakete** *f* powder rocket. **-rauch** *m* powder smoke. **-satz** *m* powder composition; powder train; powder pellet. **-schlag** *m* (*Pyro.*) cracker, petard. **-seele** *f* (*Explo.*) powder core. **-sprengstoff** *m* powder (*or* low) explosive. **-staub** *m* powder dust. **-streuanlage** *f* powder scattering unit. **-streifen** *m* (*Explo.*) powder strand. **-tonne** *f* powder keg. **-trichter** *m* powder funnel. **p--trocken** *a* powder-dry

Pulverung *f* (-en) powdering, pulverization

Pulver·vertreibladung *f* propellant powder charge. **-zementieren** *n* (*Metll.*) cementation using a powder. **-zerstäuber** *m* dusting apparatus. **-zucker** *m* powdered sugar. **-zufuhr** *f* powder supply

Pulvin·säure *f* pulvinic acid

pulvrig *a* powdery

Pumpe *f* (-n), **pumpen** *vt/i* pump

Pumpen·benzin *n* pump gasoline. **-kolben** *m* pump piston. **-kühlung** *f* forced-circulation cooling

Punkt *m* (-e) 1 (*usu*) point; (*oft*) tip, dot, speck, spot: **toter P.** low point, dead point (spot, center); **charakteristischer P. der Mischbarkeit** critical solution temperature. 2 period *cf* KOMMA. 3 e.g.: (**um**) **P.** (or **p.**) **fünf Uhr** at 5 o'clock sharp (*or* on the dot). **Pünktchen** *n* (-) dot, speck, spot. **punkten** 1 *vt* dot; spot-weld. 2 *vi* score (points)

Punkt·fehlordnung *f* (*Cryst.*) point defect. **p--förmig** *a* point (-shaped), point-like, punctiform; *adv oft* in the form of (isolated) points (*or* specks): **p--er Brennpunkt** point focus; **p--e Anfressung** pitting (corrosion). **-frei** *a* (*Dye.*) free from spots or specks, spotless. **P--gitter** *n* (*Cryst.*) point lattice. **-gruppe** *f* (*Cryst.*) point group

punktieren *vt* 1 dot, point, mark (out) with dots; punctuate. 2 (*Med.*) puncture, tap. **Punktion** *f* (-en) (*Med.*) puncture, puncturing, tap(ping). **Punktions·flüssigkeit** *f* puncture exudate

Punkt·kathode f crater (or point source) cathode. **-ladung** f (Elec.) point charge. **-lage** f point position

pünktlich a 1 punctual, adv & pred a oft on time—**p. um acht Uhr** on the stroke of eight o'clock. 2 precise, exact. 3 conscientious. **Pünktlichkeit** f punctuality; preciseness, accuracy; conscientiousness

Punkt·lichtquelle f point source of light. **-linie** f dotted line. **-mechanik** f point (or particle) mechanics. **p–mechanisch** a point- (or particle-)mechanical. **P–reihe** f series of points, row of dots. **-schreiber** m chart recorder (giving a dotted trace). **-schweißung** f spot welding. **-strich** m semicolon. **p-symmetrisch** a point-symmetrical

punktuell a (Opt.) point-focal

Punktur f (-en) puncture. **p–·sicher** a puncture-proof

punkt·weise adv point by point

Punsch m (-e or ¨e) punch (= the drink)

Punze f(-n), **Punzen** m (-) 1 (Mach.) punch; engraving chisel, embossing hammer. 2 hallmark

Pupille f (-n) (Anat.) pupil. **Pupillen·erweiterung** f dilation of the pupil

Puppe f (-n) 1 pupa, chrysalis, cocoon. 2 doll; puppet, dummy. 3 (Agric.) shock (of grain). 4 (Rubber) roll of uncured rubber. **Puppen·kokon** n cocoon

pur a pure; sheer; (Drinks) straight, neat

Purgans n (.. anzien), **Purgativ** n (-e) laxative. **purgieren** I vt 1 purge; eliminate (impurities from the system). 2 give a laxative to. 3 (Silk) boil off with soap. II vi act as a laxative

Purgier·harz n (Bot.) scammony. **-kassie** f purging cassia. **-kernöl** n scammony seed oil. **-korn** n purging grain (or seed): **kleines P.** croton seed; **großes P.** castor bean. **-kraut** n purgative herb, (specif) hedge hyssop. **-lein(en)** n purging flax. **-mittel** n purgative, laxative. **-nuß** f nux vomica. **-paradiesapfel** m colocynth. **-pille** f laxative pill. **-salz** n laxative salt. **-wurzel** f 1 jalap. 2 medicinal rhubarb. 3 (Bot.) scammony

purifizieren vt purify

Purin·basen fpl purine bases. **-gruppe** f purine group

Purpur m purple, (more accurately) crimson. **--bakterien** npl purple bacteria. **p–blau** a purplish blue, indigo purple

Purpureo·(kobalt)verbindung f purpureo(cobaltic) compound

Purpur·erz n purple ore, blue billy, burnt pyrites. **-farbe** f purple (or crimson) color. **p–farben, -farbig** a purple- (or crimson-) colored. **P–färbung** f purple (or crimson) coloring. **-holz** n violet wood. **-karmin** m purple carmine, murexide. **-lack** m purple lake, (specif) madder purple. **-muschel** f purple shell

purpurn a purple, crimson cf PURPUR

purpur·rot a purplish red, crimson. **-sauer** a purpurate of. **P–säure** f purpuric acid. **-schwefelsäure** f sulfopurpuric acid

purzeln vi(s) tumble

Purzement·leim m neat cement paste

Puste f breath, wind. **--blume** f dandelion

Pustel f (-n) pustule, pimple. **p–·ähnlich, -artig** a pustular

pusten 1 vt/i blow. 2 vi pant, gasp

Puste·rohr n blowpipe; blowgun

Pust·lampe f blowtorch. **-licht** n (Phot.) flash(light) (powder blown into a flame). **-probe** f blow (or bubble) test. **-rohr** n blowpipe; blowgun. **-span** m skimmer

Pute f (-n) turkey hen. **Puter** m (-) turkey (cock)

Putz m 1 (wall) plaster, roughcast. 2 finery; trim(mings), (gaudy) ornament(s); millinery. **--baumwolle** f cotton waste (for cleaning). **putzen** 1 vt plaster, roughcast (walls, etc). 2 vt/i clean, polish; wipe; blow (nose). 3 vt & **sich p.** dress up; trim (oneself). **Putzer** m (-) 1 plasterer. 2 cleaner. **Putzerei** f (-en) 1 (Metll.) dressing (or fettling) shop. 2 (dry) cleaner's. 3 cleaning

Putz·gips m anhydrous gypsum plaster. **-kalk** m 1 plastering lime; stucco. 2 polishing chalk. **-lage** f coat of plaster. **-lappen** m cleaning (or polishing) rag. **-leder** n chamois (for polishing). **-macherei** f millinery (shop). **-maschine** f cleaning machine. **-mittel** n cleaning (or polishing) agent; cleanser; polish. **-öl** n polishing (or cleaning) oil. **-präparat** n polish(ing preparation). **-pulver** n cleaning (or polishing) powder; cleanser. **-sand** m

(abrasive) cleaning sand. **-schicht** f coat
of plaster. **-stein** m cleaning (or polishing) stone. **-tisch** m dressing table;
(Ceram.) cleaning table. **-trommel** f tumbling barrel. **-tuch** n cleaning (or polishing) cloth. **-wasser** n dilute scouring acid.
-wirkung f cleaning action (or effect).
-wolle f(Mach.) (cleaning) waste. **-zeug** n
cleaning materials (or utensils)
Puzzolane f, **Puzzolan-erde** f pozzuolana.
-zement n pozzuolana cement
Puzzuolan-erde f pozzuolana
Py. abbv (Pyridin) pyridine
Pyämie f (-n) pyemia. **pyämisch** a pyemic
pyogen a pyogenic
Pyorrhöe f pyorrhea
pyramiden-förmig a pyramidal. **P–**
oktaeder n (Cryst.) trisoctahedron.
-stumpf m frustum of a pyramid.
-würfel m tetrahexahedron
Pyrenäen pl (Geog.) Pyrenees
Pyren-chinon n pyrenequinone
Pyrit-abbrände mpl pyrites cinders. **p–**
artig a pyritic. **P–erz** n pyritic ore. **p–**
haltig a pyritiferous
pyritisch a pyritic. **pyritisieren** vt pyri-
tize. **Pyritisierung** f pyritization
Pyrit-(röst)ofen m pyrites oven (or
burner). **-schmelzen** n pyritic smelting.
-schwefel m pyritic sulfur
Pyro-antimonsäure f pyroantimonic acid.
-arsensäure f pyroarsenic acid.
-catechin n catechol. **p–chemisch** a

pyrochemical. **P–chinin** n pyroquinine.
-chromat n dichromate. **p–elektrisch** a
pyroelectric. **P–elekrizität** f pyro-
electricity. **-gallolgerbstoff** m pyrogallol
tannin. **-gallussäure** f pyrogallic acid,
pyrogallol
pyrogen a pyrogenic
Pyrokatechin n (pyro)catechol. **--gerb-**
stoff m pyrocatechol tannin
pyroklastisch a pyroclastic
Pyro-lösung f(Photo.) pyro(gallol) solution
Pyrolyse f (-n) pyrolysis. **pyrolysieren** vt
pyrolyze. **pyrolytisch** a pyrolytic
pyrometrisch a pyrometric
Pyro-perjodsäure f pyroperiodic acid
Pyrophor m (-e) pyrophorus. **pyro-**
phor(isch) a pyrophoric
pyro-phosphorig a pyrophosphorous.
-phosphorsauer a pyrophosphate of. **P–**
phosphorsäure f pyrophosphoric acid.
-säure f pyro acid. **-schleimsäure** f
pyromucic (or furoic) acid. **-schwe-**
felsäure f pyrosulfuric acid. **p–schwe-**
flig a pyrosulfurous. **P-synthese** f pyro-
synthesis. **-technik** f pyrotechnics. **p–**
technisch a pryotechnical. **P–**
traubensäure f pyroracemic (or pyruvic)
acid. **-weinsäure** f pyrotartaric acid
Pyrrol-blau n pyrrole blue
Pyruvin-säure f pyruvic acid
PZ. abbv (Portlandzement) Portland
cement

Q

q *abbv* (Quadrat) sq (square): **qcm** sq. cm.; **qdm** sq. dm. (decimeter); **qkm** sq. km.; **qm** sq. m.; **qmm** sq. mm.

QS *abbv* 1 (Quecksilbersäule) mercury column. 2 (Quecksilberstand) mercury level

quabbelig *a* 1 flabby, jelly-like. 2 boggy. **quabbeln** *vt* 1 quiver, jiggle. 2 be flabby. 3 be boggy

Quacksalber *m* (-) quack, charlatan

Quader *m* (-) 1 parallelepiped. 2 = **--stein** *m* ashlar; freestone

Quadrat *n* (e) square; (*Math. oft*) square(d): **5 Q.** (= 5²) five squared. **--centimeter** *n* or *m* square centimeter. **-dezimeter** *n* or *m* square decimeter. **q--förmig** *a* square, quadratic. **Q--fuß** *m* square foot

quadratisch *a* quadratic; square; (*Cryst.*) tetragonal

Quadrat·kilometer *n* or *m* square kilometer. **-meter** *n* or *m* square meter. **-millimeter** *n* or *m* square millimeter. **-pyramide** *f* square pyramid. **-wurzel** *f* square root. **-zahl** *f* square (number). **-zentimeter** *n* or *m* square centimeter. **-zoll** *m* square inch

quadrieren *vt* square

Quadroxid° *n* quadroxide, tetroxide

Quadrupol·kraft *f* quadrupole force. **-moment** *n* quadrupole moment

Qual *f* (-en) torment, agony. **quälen I** *vt* 1 mill; crush, disaggregate. 2 torment, torture; pester. **II: sich q.** 3 suffer. 4 struggle. 5 worry

qualifizieren *vt* & **sich q.** qualify (oneself); *vt* (*oft*) judge. **Qualifizierung** *f* (-en) qualification

Qualität *f* (-en) quality

Qualitäts· (high-)quality, high-grade: **-abweichung** *f* deviation in quality. **-guß** *m* high-grade cast iron. **-leder** *n* high-grade leather. **-stahl** *m* high-grade steel

Qualle *f* (-n) jellyfish. **quallig** *a* jellylike

Qualm *m* dense smoke. **qualmen** *vi* smoke, puff (out dense smoke). **qualmig** *a*

smoky, vaporous. **Qualm·wasser** *n* seepage water

qualvoll *a* agonizing, tormenting, *cf* QUAL

Quant *n* (-en) quantum. **quanteln** *vt* quantize. **Quantelung** *f* (-en) quantization

Quanten 1 quanta: *pl of* QUANT & QUANTUM. 2 *as pfx* quantum: **·ausbeute** *f* quantum yield. **-bahn** *f* quantum orbit. **-bedingung** *f* quantum condition. **-chemie** *f* quantum chemistry. **-empfindlichkeit** *f* quantum sensitivity. **-gewicht** *n* quantum weight

quantenhaft *a* quantum-like, quantized; *adv oft* by quanta

Quanten·hypothese *f* quantum hypothesis. **q--mäßig** *a* as regards quanta, quantumwise. **Q--mechanik** *f* quantum mechanics. **q--mechanisch** *a* quantum-mechanical. **-physik** *f* quantum physics. **-sprung** *m* quantum jump. **q--theoretisch** *a* quantum-theoretical. **Q--theorie** *f* quantum theory. **-zahl** *f* quantum number. **-zustand** *m* quantum state

·quantig *sfx* -quantum

quantifizieren *vt* quantify

quantisieren *vt* quantize. **Quantisierung** *f* (-en) quantization

Quantität *f* (-en) quantity. **Quantitäts·bestimmung** *f* quantitative determination

Quantum *n* (.. ten) 1 quantum. 2 quantity. 3 share, portion. **-ausbeute** *f* quantum yield. **-gewicht** *n* quantum weight. **-übergang** *m* quantum transition (*or* jump). **-zustand** *m* quantum state

Quarantäne *f* (-n) quarantine

Quargel *m* (-n) (round) sour-milk curd cheese

Quark I *m* 1 curds, cottage cheese. 2 nonsense, rubbish. **II** *n* (-s) (*Phys.*) quark. **quarkig** *a* curdy

Quart *n* (*Books*) quarto

Quartal *n* (-e) quarter (of a year). **-säufer** *m* periodic drunk(ard). **q--weise** *a*, *adv* quarterly

quartär *a*, **Quartär** *n* quaternary

Quart·band *m* quarto volume

Quartier *n* (-e) place to stay; lodgings; quarters

Quartierung *f* (-en) quartation; quartering

quart·mäßig *adv* by quartation (*or* quartering). **Q–scheidung** *f* separation by quartering (*or* quartation)

Quarz *n* (-e) quartz. **q–ähnlich, -artig** *a* quartzlike, quartzous. **Q–balken** *m* quartz beam. **-dreieck** *n* quartz (*or* silica) triangle. **-druse** *f* (*Min.*) quartzdruse, crystallized quartz. **q–elektrisch** *a* piezoelectric. **Q–faden** *m* quartz thread (*or* filament). **-fels** *m* quartz rock; (*Min.*) quartzite. **-fenster** *n* quartz window. **q–frei** *a* free from quartz. **Q–gang** *m* quartz vein. **-gefäß** *n* quartz vessel. **-gerät** *n* quartz apparatus. **q–gesteuert** *a* (*Elec.*) quartz(-crystal)-controlled. **Q–glas** *n* quartz glass, fused silica. **-glimmerfels** *m* quartz mica rock. **-gut** *n* quartz (*or* fused silica) ware. **q–haltig** *a* quartz-bearing, quartziferous

quarzig *a* quartzy, quartzose

Quarzit *n* quartzite. **--gestein** *n* quartzite rock

Quarz·keil *m* (*Opt.*) quartz wedge. **-kies** *m* flint. **-kiesel** *m* quartz gravel. **-kristall** *m* 1 (*Min.*) rock crystal. 2 (*Elec.*) quartz crystal. **-lager** *n* quartz deposit. **-linse** *f* quartz lens. **-mehl, -pulver** *n* quartz (*or* silica) powder. **-rohr** *n*, **-röhre** *f* quartz tube. **-sand** *m* quartz (*or* silica) sand. **-schamottestein** *m* silica brick. **-scheibe** *f* quartz (*or* crystal) plate. **-schiefer** *m* quartz schist (*or* shale). **-sinter** *m* silica sinter. **-stäbchen** *n* quartz rod. **-uhr** *f* quartz crystal timepiece (clock, watch). **-wolle** *f* quartz (*or* silica) wool. **-ziegel** *m* quartz (silica, Dinas) brick

quasi 1 *adv* virtually, as good as. 2 *pfx* quasi-: **-elastisch** *a* quasi-elastic. **-unendlich** *a* quasi-infinite

Quassia·holz, Quassien·holz *n* quassia (wood)

Quast *m* (-e) (housepainter's, paperhanger's) brush. **Quaste** *f* (-n) 1 tassel. 2 tuft. 3 brush = QUAST

quaternär *a* quaternary. **Q–·stahl** *m* quaternary steel

Quatsch *m* nonsense

Quebrachit *n* quebrachitol. **Quebracho·gerbsäure** *f* quebracho tannic acid. **-rinde** *f* quebracho bark

Quecke *f* (-n) couch grass—**rote Q.** (*Pharm.*) carex root

Quecksilber *n* mercury, quicksilver. **q–·ähnlich, -artig** *a* mercury-like, mercurial. **Q–arsenitoxid** *n* mercuric arsenite. **-äthyl** *n* diethylmercury. **-azetatoxydul** *n* mercurous acetate. **-bad** *n* mercury bath. **-beize** *f* 1 (*Dye.*) mercury mordant. 2 (*Med.*) mercury disinfectant. **-bogen** *m* mercury arc. **-branderz** *n* (*Min.*) idrialite, inflammable hepatic cinnabar. **-bromid** *n* mercuric bromide. **-bromür** *n* mercurous bromide. **-chlorid** *n* mercuric chloride. **-chlorür** *n* mercurous chloride. **-chromatoxid** *n* mercuric chromate. **-chromatoxydul** *n* mercurous chromate. **-cyanid** *n* mercuric cyanide. **-cyanür** *n* mercurous cyanide. **-cyanwasserstoffsäure** *f* cyanomercuric acid. **-dampf** *m* mercury vapor

Quecksilberdampf·gleichrichter *m* (*Elec.*) mercury vapor rectifier. **-lampe** *f* mercury vapor lamp. **-strahlpumpe** *f* mercury diffusion pump. **-turbine** *f* mercury turbine

Quecksilber·druck *m* mercury pressure. **-elektrode** *f* mercury electrode. **-erz** *n* mercury ore. **-faden** *m* mercury thread. **-fahlerz** *n* tetrahedrite containing mercury. **-falle** *f* mercury vapor trap. **-füllung** *f* mercury filling. **-gefäß** *n* mercury vessel (*or* bulb). **-gehalt** *m* mercury content. **-halogen** *n* mercury halide. **q–haltig** *a* containing mercury, mercurial. **Q–hochdruckentladungsrohr** *n* high-pressure mercury discharge tube. **-hochdrucklampe** *f* high-pressure mercury vapor lamp. **-hochvakuumpumpe** *f* mercury vapor vacuum pump. **-hornerz** *n* horn quicksilver (native calomel). **-hüttenwesen** *n* mercury metallurgy. **-jodid** *n* mercuric iodide. **-jodür** *n* mercurous iodide. **-kapsel** *f* bulb (of a mercury thermometer). **-kippschalter** *m* mercury switch. **-kugel** *f* mercury bulb. **-kuppe** *f* mercury meniscus. **-lebererz** *n* hepatic cinnabar. **-legierung** *f* amalgam. **-licht** *n* mercury vapor light. **-luftmesser** *m* mercury barometer. **-luftpumpe** *f* mercury air pump. **-masse** *f* mercury pool. **-methyl** *n* dimethylmercury. **-mittel** *n* (*Pharm.*) mercurial. **-mohr** *m* ethiops mineral (black mercuric sulfide); (*Min.*) metacinnabarite

quecksilbern *a* (of) mercury, mercurial

Quecksilber·niederdruckbrenner *m* low-pressure mercury lamp. **-nitratoxid** *n* mercuric nitrate. **-nitratoxydul** *n* mercurous nitrate. **-ofen** *m* mercury oven (*or* furnace). **-oxid** *n* mercuric oxide. **-oxidnitrat** *n* mercuric nitrate. **-oxidsalbe** *f* mercuric oxide ointment. **-oxidsalz** *n* mercuric salt. **-oxydul** *n* mercurous oxide. **-oxydulsalz** *n* mercurous salt. **-pflaster** *n* (*Pharm.*) mercurial plaster. **-pille** *f* blue pill. **-präparat** *n* mercurial preparation. **-präzipitat** *n* (*Pharm.*) mercury precipitate. **-quarzlampe** *f* quartz mercury vapor lamp. **-rhodanid** *n* mercuric thiocyanate. **-rhodanür** *n* mercurous thiocyanate. **-ruß** *m* mercurial soot, stupp. **-salbe** *f* mercurial ointment. **-salpeter** *m* mercuric nitrate. **-salz** *n* mercury salt. **-säule** *f* column of mercury. **-schalter** *m* mercury switch. **-schliff** *m* mercury ground-glass joint. **-spat** *m* (*Min.*) horn quicksilver, calomel. **-spiegel** *m* 1 mercury mirror. 2 mercury surface. **-stand** *m* mercury level. **-sublimat** *n* corrosive sublimate (mercuric chloride). **-sulfatoxydul** *n* mercurous sulfate. **-sulfid** *n* mercuric sulfide, cinnabar. **-sulfozyanid** *n* mercuric thiocyanate. **-sulfür** *n* mercurous sulfide. **-tropfelektrode** *f* dropping mercury electrode. **-verbindung** *f* mercury compound. **-verfahren** *n* mercury process. **-vergiftung** *f* mercury poisoning. **-vitriol** *n* mercuric sulfate. **-waage** *f* mercury level. **-wanne** *f* mercury trough. **-zyanid** *n* mercuric cyanide. **-zyanür** *n* mercurous cyanide

quellbar *a* capable of swelling (*or* expanding). **Quellbarkeit** *f* swelling (*or* expanding) capacity

Quell·bottich *m* steeping vat (*or* tub). **-brunnen** *m* spring well, fountain

Quelle *f* (-n) 1 source (also of information,

supply). 2 spring. 3 oil well

quellen I *vt* 1 soak, steep. 2 swell (grains, etc). II* (quoll, gequollen; quillt) *vi*(s) 3 spring, burst forth. 4 well up, flow, pour, billow. 5 pop, bulge. 6 swell (up)

Quellen·forschung *f* research into sources. **q–frei** *a* 1 source-free. 2 solenoidal. **-mäßig** *a* according to the best sources, authentic. **Q–nachweis** *m* citation of sources. **q–reich** *a* abounding in springs. **Q–verzeichnis** *n* bibliography, index of sources

Queller *m* (-) (*Bot.*) glasswort, saltwort

quell·fähig *a* capable of swelling, etc = QUELLBAR. **Q–fluß** *m* headstream. **-gas** *n* gas from springs. **-grad** *m* degree of swelling. **q–klar** *a* clear as spring water. **Q–kuppe** *f* (*Geol.*) volcanic plug. **-mittel** *n* swelling agent. **q–reif** *a* sufficiently steeped (*or* soaked). **Q–reife** *f* sufficient steeping. **-salz** *n* spring (*or* well) salt. **-sand** *m* quicksand. **-satzsäure** *f* apocrenic acid. **-säure** *f* crenic acid. **-schicht** *f* 1 source layer. 2 swelling layer. **-sole** *f* 1 spring (*or* well) brine. 2 mother liquor. **-stock** *m* steeping tank (*or* cistern). **-substanz** *f* substance that swells

Quellung *f* (-en) 1 soaking, steeping. 2 swelling. 3 springing, welling, flowing. 4 imbibation, expansion; tumefaction. Cf QUELLEN

quellungs·fördernd *a* promoting swelling. **Q–kolloid** *n* swelling colloid. **-vermögen** *n* swelling power. **-wärme** *f* heat of swelling (*or* tumefaction). **-zustand** *m* swelling state

Quell·vermögen *n* swelling power. **-wasser** *n* spring (*or* well) water. **-widerstand** *m* resistance to swelling. **-zement** *m* expanding cement

Quendel *m* (-) wild thyme

quer I *a* 1 (*also as pfx*) cross(wise), transverse. 2 diagonal. 3 strange, odd. II *adv* 4 (*esp* + **durch, über**) straight (through, across). 5 (+ **zu**) at right angles (to). 6 diagonally, at an angle *cf* KREUZ. III *sep pfx* wrong, amiss. **Q–·achse** *f* transverse axis. **-arm** *m* crossarm, crossbar. **-balken** *m* crossbeam, crossbar. **-bewegung** *f* transverse motion

Quercit *n* (-e) quercitol

Quercitron·rinde f quercitron bark
quer·durch adv straight through. **Q–durchmesser** m transverse diameter.
-durchschnitt m cross section
Quere f 1: **in der Q., der Q. nach** crosswise, transversely. 2: **einem in die Q. kommen** (a) cross a psn's path, (b) upset a psn's plans. 3: **die Kreuz und Quer(e)** helter skelter, in all directions cf QUER
Quer·ebene f lateral plane. **-effekt** m transverse effect. **-falte** f transverse fold; horizontal (or crosswise) furrow; cross pleat. **-faser** f transverse fiber; cross grain. **-feld** n transverse field
quer·gehen* vi(s) go wrong
quer·gestreift a horizontally striped; transversely striated. **-gewellt** a cross-corrugated
quer·kommen* vi(s) (dat) upset (or cross) a psn's plans
Quer·kraft f transverse force; shear(ing force). **q–laufend** a transverse. **Q–leiste** f crosspiece
quer·liegen* vi be wrong (or mistaken)
Quer·profil n cross section. **-richtung** f transverse direction. **—in der Q.** crosswise, transversely. **-riß** m 1 transverse crack (or Min. fracture). 2 cross section. **-schlag** m (Min.) crosscut.
quer·schleifen* vt grind (wood) across the grain
Quer·schneiden n cross cutting. **-schnitt** m cross section
Querschnitts·ebene f cross-sectional plane. **-fläche** f cross-sectional area. **q–gelähmt** a paraplegic. **Q–serie** f series of cross sections. **-zeichnung** f sectional drawing
Quer·schwingung f transverse vibration. **-spalte** f transverse split (or fissure). **-spießglanz** m jamesonite. **-streifen** m 1 crossbar, cross line. 2 horizontal stripe. **-streifung** f cross striping, transverse striation. **-strich** m 1 horizontal line (or stroke); crossbar. 2 dash. **-strom** m 1 cross current. 2 (Elec.) wattless current. **-stück** n crosspiece. **-teilung** f crosswise division. **-verbindung** f 1 (Chem.) cross-linkage. 2 cross connection. **q–**

verlaufend a running transversely. **Q–wand** f 1 crosswall, partition. 2 diaphragm. **-wandsporenrost** m (Bot.) brand, phragmidium. **-weg** m crossroad; shortcut. **-welle** f transverse wave
Querzit n (-e) quercitol
Querzitron·rinde f quercitron bark
Quer·zusammenziehung f lateral contraction
Quetsche f (-n) 1 squeezer; wringer; crusher, press. 2 tight spot, pinch
quetschen vt 1 squeeze, press; pinch. 2 squash; mash. 3 bruise. **Quetscher** m (-) squeezer (incl Dye.); pincher; masher
Quetsch·flasche f squeeze bottle. **-fuß** m (Elec.) pinched base (of a lamp). **-grenze** f compression limit (or yield point). **-hahn** m pinchcock, pinch clamp. **-kartoffeln** fpl mashed potatoes. **-klemme** f pinch clamp. **-mühle** f bruising (or crushing) mill. **-sand** m crushed sand.
Quetschung f (-en) 1 squeezing; pinching. 2 squashing, mashing. 3 bruising. 4 bruise, contusion
Quetsch·walze f crushing roll, squeeze roller; (Grain) flaking roll. **-werk** n crushing mill, crusher; squeezer. **-wirkung** f squeezing action (or effect). **-wunde** f bruise, contusion
Quick·arbeit f amalgamation (with mercury). **-beutel** m amalgamating skin. **-brei** m amalgam
quicken vt amalgamate (with mercury)
Quick·erz n mercury ore. **-gold** n gold amalgam. **-metall** n amalgamated metal. **-mühle** f amalgamating mill. **-wasser** n (Plating) quickening liquid (mercuric salt solution)
quieken vi squeal
quietschen vi squeak, squeal
Quillaja·rinde f quillai(a) bark. **-säure** f quillaic salt
quillt swells, etc: pres of QUELLEN
Quirl m (-e) 1 wooden twister (or mixer). 2 (Bot.) whorl. 3 fidgety person. **quirlen** vt/i twirl (to mix)
quitt pred a 1 even, square. 2 finished, through. 3 (+ genit) rid (of)
Quitte f (-n) quince
Quitten·äther m, **-essenz** f quince essence.

-kern *m* quince seed. **-öl** *n* amylacetate, banana (*or* quince) oil. **-samenöl** *n* quince seed oil. **-schleim** *m* quince mucilage

quittieren *vt* 1 sign a receipt for; receipt (a bill). 2: **den Dienst q.** quit one's job, re-sign. 3 counter (a remark)

Quittung *f* (-en) 1 receipt. 2 retort, re-joinder. 3 repayment (to get even)

quoll swelled, etc: *past of* QUELLEN

Quote *f* (-n) quota; ratio, proportion

R

r. *abbv* **1** (rechtsdrehend) dextrorotatory. **2** (rund) approx., in round numbers

R *abbv* **1** Réaumur. **2** (Rabatt) discount

R. *abbv* **1** (Referat) report, abstract. **2** ring, cyclic. **3** (Radikal) radical

Rab. *abbv* = **Rabatt** *m* (-e) discount; rebate

Rabatte *f* (-n) border (flower) bed *cf* RABATT

Rabe *m* (-n, -n) raven; crow. **raben·schwarz** *a* jet-black

rabiat *a* rabid; furious

rac. *abbv* (racemisch) racemic

racem *a* racemic. **Racemat** *n* (-e) racemate. **Racemie** *f* racemism. **racemisch** *a* racemic. **racemisieren** *vt* racemize. **Racemisierung** *f* (-en) racemization

Racem·körper *m* racemic substance (*or* compound). **-säure** *f* racemic-tartaric acid. **-verbindung** *f* racemic compound

Rache *f* revenge, vengeance

Rachen *m* (-) **1** throat, pharynx. **2** (wide-open) mouth, jaws

rächen *vt* & **sich r.** avenge, revenge

Rachen·katarrh *m* pharyngitis. **-lehre** *f* (inside) calipers. **-reizstoff** *m* throat irritant

Rachitis *f* rickets. **rachitisch** *a* rachitic

Rad *n* (-er) **1** wheel. **2** gear. **3** bicycle. **-·achse** *f* wheel axle

Radar·anlage *f* radar installation (*or* unit)

Rad·arm *m* wheel spoke

Radar·schirm *m* radar screen

Rad·bewegung *f* rotary motion. **-drehung** *f* rotation; torsion

Rade *f* (-n) (*Bot.*) cockle

Rädel·erz *n* wheel ore, bournonite

radeln *vi*(s) cycle, pedal

rädeln *vt* cut (*or* mark out) with a roulette; knurl, mill

Rädels·führer *m* ringleader

Räder·antrieb *m* gear drive. **-fahrzeug** *n* wheeled vehicle

rädern *vt* **1** furnish with wheels. **2** exhaust, tire out. **3** sift, screen

Räder·tier(chen) *n* rotifer. **-werk** *n* gears,

gearing; works, clockwork; machinery

rad·fahren* *vi*(s) cycle, ride a bicycle. **Rad·fahrer** *m* (-) cyclist

rad·förmig *a* wheel-shaped, radial, rotiform

Radi *m* (-) radish

Radial·schlag *m* (*Engg.*) lateral (*or* side) play. **-schnitt** *m* radial section. **r–strahlig** *a* (*Cryst.*) radiating. **-symmetrisch** *a* radially symmetric

Radiant *m* (-en, -en) (*Math.*) radian

Radien *mpl* radii

radieren *vt/i* **1** etch. **2** erase

Radier·firnis *m* etching varnish. **-grund** *m* etching ground. **-gummi** *m* (rubber) eraser. **-kunst** *f* (art of) etching. **-messer** *n* erasing knife. **-nadel** *f* etching needle

Radierung *f* (-en) etching; erasure, erasing

Radies/chen *n* radish. **-·bombe** *f* smoke grenade

radiieren *vi* radiate

Radikal·essig *m* radical vinegar, glacial acetic acid. **-kettenreaktion** *f* radical chain reaction

Radio·aktivität *f* radioactivity. **-blei** *n* radio lead. **-chemie** *f* radiochemistry. **-element** *n* radioactive element

radiogen *a* radiogenic

Radio·indikator *m* radioactive tracer

Radiolyse *f* (-n) radiolysis

radio·markiert *a* isotape-labeled. **R–tellur** *n* radiotellurium (= polonium). **-thor** *n* radiothorium

Radium·behälter *m* radium container. **-belag** *m* radium coating. **-erz** *n* radium ore. **-jodid** *n* radium iodide. **-leuchtmasse** *f* radium-based luminous paint. **-salz** *n* radium salt. **-strahlen** *mpl* radium rays. **-uhr** *f* radium clock

radizieren *vt/i* (*Math.*) extract the root (of). **Radizierung** *f* (-en) root extraction

Rad·kranz *m* wheel rim. **-linie** *f* cycloid. **-reifen** *m* tire. **-schlauch** *m* inner tube. **-schuh** *m* brake. **-speiche** *f* spoke. **-welle**

f 1 gear shaft. **2** paddle wheel shaft.
-zahn *m* gear (tooth), cog
Rafaelit *m* (*Min.*) paralaureonite
raff. *abbv* (raffiniert) ref. (refined)
Raff·eisen *n* scrap iron
raffen *vt* 1 gather (up). **2** (*esp* + **an sich**) snatch, grab. **2** condense (sthg written) *cf* GERAFFT
Raff·gas *n* refinery gas
Raffinade *f* (-n) refined sugar. **--zink** *n* refined spelter. **-zucker** *m* refined sugar
Raffinal *n* refined aluminum (99.99% pure)
Raffinat *n* (-e) (*Petrol.*) refined product; (*Essential Oils*) isolate. **--blei** *n* refined lead
Raffination *f* (-en) refining
Raffinations·abfall *m* refinery waste. **-anlage** *f* refining plant. **-ertrag** *m* (*Sugar*) rendement. **-fettsäure** *f* acid oil, refined fatty acid. **-ofen** *m* refining furnace. **-öl** *n* refining oil. **-verfahren** *n*, **-vorgang** *m* refining process. **-wert** *m* (*Sugar*) rendement
Raffinat·kupfer *n* refined copper. **-silber** *n* refined silver
Raffinerie *f* (-n) refinery. **--gas** *n* refinery gas
Raffinesse *f* (-n) refinement, fine point
Raffineur *m* (-e) (*Mach., Paper*) refiner
Raffinier·anlage *f* refinery, refining unit
raffinieren *vt* refine. *See also* RAFFINIERT
Raffinier·feuer *n*, **-herd** *m*, **-ofen** *n* refining furnace. **-schlacke** *f* refinery slag. **-stahl** *m* refined steel
raffiniert *a* 1 refined. **2** cunning, slick; ingenious. **Raffiniertheit** *f* (-en) refinement; slickness, finesse; ingeniousness
Raffinierung *f* (-en) refining; refinement
Raffinierungs·abfall *m* refinery waste
Raffinier·verfahren *n* refining process
ragen *vi* tower; project, extend (upward)
Rahm *m* 1 (dairy) cream. **2** crust (forming on top); soot, dirt. **r--ähnlich, -artig** *a* creamy, cream-like. **R--eis** *n* ice cream
rahmen I (*cf* RAHM) 1 *vt:* **Milch r.** skim cream off milk. **2** *vi* form cream. **II** (*cf* RAHMEN) **3** *vt* frame; mount (slides)
Rahmen *m* (-) 1 frame; framework; welt (of a shoe). **2** bounds, scope: **aus dem R. fal-**

len go out of bounds, be out of the ordinary—**im R.** (+ *genit*) within the framework (the bounds, the scope) (of). **--achse** *f* (*Opt.*) collimating axis. **-bedingungen** *fpl* overall conditions; defined limits. **-filterpresse** *f* frame filter-press. **-leder** *n* welting leather
Rahm·erz *n* (*Min.*) foaming variety of wad. **-farbe** *f* cream (color). **r--farben, -farbig** *a* cream-colored
rahmig *a* creamy
Rahm·käse *m* creamy cheese. **-messer** *m* creamometer. **-reifer** *m* cream aging (*or* souring) vat. **-reifung** *f* cream aging (*or* souring). **-säuerungskultur** *f* butter culture (*or* culturing)
Rain *m* (-e) grassy border strip (between cultivated lots). **--farn** *m*, **-farnkraut** *n* (*Bot.*) tansy. **-farnöl** *n* tansy oil. **-weide** *f* privet
Rakel *f* (-n) (*Tex.*) doctor blade. **--appretur** *f* doctor finish. **-lineal** *n* doctor rule. **-messer** *n* doctor (blade). **-stärkemaschine** *f* starching machine with doctor, backfilling starcher. **-streifen** *m* doctor streak
Rakete *f* (-n) rocket
Raketen·antrieb *m* rocket propulsion. **-bombe** *f* rocket bomb. **r--getrieben** *a* rocket-propelled. **R--leuchtgeschoß** *n* rocket flare missile. **-satz** *m* rocket composition (*or* attachment). **-sonde** *f* high-altitude rocket; (*Meteor.*) rocket sonde. **-stufe** *f* rocket stage. **-technik** *f* rocketry, rocket technology. **-triebstoff** *m* rocket fuel. **-triebwerk** *n* rocket engine. **-vortrieb** *m* rocket propulsion. **-werfer** *m*, **-wurfmaschine** *f* rocket launcher. **-zeichen** *n* rocket signal
Ramal·säure *f* obtusatic acid
Ramie *f* ramie, China grass. **--faser** *f* ramie fiber
ramifizieren *vt/i* ramify. **Ramifizierung** *f* (-en) ramification
Ramm·bär *m* pile driver. **-block** *m* rammer; pile driver. **-bock** *m* battering ram
Ramme *f* (-n) (battering) ram, rammer; pile driver. **rammen** *vt* ram, collide with
Ramm·vorrichtung *f* ramming device
Rampe *f* (-n) 1 ramp. **2** (stage) footlights
ramponieren *vt* batter, mar, damage

Ramsch *m* 1 junk. 2: **im R.** in job lots, dirt cheap. **--waren** *fpl* job-lot (*or* second-hand) goods; junk

Rand *m* (-̈er) 1 edge, border; margin; periphery: **am R-e bemerken** note marginally, remark incidentally. 2 brink. 3 boundary. 4 flange, lip. 5 circular stain, (dark) ring.—**außer R. und Band** out of hand (*or* control); **zu R-e kommen mit** . . . get . . . done, manage

Rand· marginal, peripheral: **-anmerkung** *f* marginal note. **-bedingung** *f* boundary (*or* marginal) condition. **-bemerkung** *f* marginal note; incidental remark

rändeln *vt* knurl, mill (edge, coin)

randen *vt* border, edge

Rand·entkohlung *f* marginal decarbonization

Ränder·email *n* beading enamel

rändern *vt* 1 border, edge. 2 mill = RÄNDELN

Ränder·scheibe *f* edge cutter

Rand·erscheinung *f* marginal phenomenon. **-fassung** *f* rim. **-feuerpatrone** *f* rimfire cartridge. **-fläche** *f* margin(al area); (*Cryst.*) lateral face. **-gängigkeit** *f* bypassing at the edges (*as:* liquid in packed column). **-gärung** *f* (*Brew.*) rim fermentation. **-gebiet** *n* border (peripheral, marginal) area. **r-gehärtet** *a* case-hardened. **R-härtung** *f* case-hardening. **-kante** *f* (*Cryst.*) lateral edge. **-leder** *n* welting leather

randlich *a* marginal, peripheral

Rand·lochkarte *f* edge-notched data card. **r-los** *a* rimless, borderless, unmargined. **R-note, -notiz** *f* marginal note. **-schärfe** *f* sharpness of the edge (*or* border). **-schicht** *f* marginal (*or* peripheral) layer (*or* zone). **-schubspannung** *f* edge-shearing stress. **r-ständig** *a* marginal, peripheral. **R-stein** *m* curb (stone). **-störung** *f* edge effect. **-strahl** *m* marginal (*or* peripheral) ray

Randung *f* (-en) edge, border; edging, bordering

rand·voll *a* brim-full. **-wärts** *adv* toward the edge (border, margin). **R-wert** *m* boundary value. **-winkel** *m* contact angle. **-wirkung** *f* edge (*or* wall) effect.

-zone *f* marginal (peripheral, border) zone

rang wrestled: *past of* RINGEN

Rang *m* (-̈e) 1 rank, status, grade, class: **ersten R-es** first-class (*or* -rate). 2 tier, gallery. **--folge** *f* ordered series; order (of precedence)

rangieren 1 *vt* switch, classify (railroad cars). 2 *vi* rank

Rang·ordnung *f* ranking, grading, order of precedence; hierarchy

Ranke *f* (-n) tendril, vine. **ranken** 1 *vi* send out shoots (vines, tendrils). 2: **sich r.** climb (like a vine); (*esp* + **um**) twine, coil (around). **Ranken·gewächs** *n* climbing (*or* creeping) plant, vine

rann flowed, etc: *past of* RINNEN

rannte ran: *past of* RENNEN

Ranzen *m* (-) 1 knapsack. 2 satchel. 3 belly; rumen

Ranzidität *f* rancidity. **ranzig** *a* rancid: **r. werden** turn rancid. **Ranzigkeit** *f* rancidity, rancidness

Raphia *f* (. .phien) raffia (palm)

Rapid·stahl *m* high-speed steel

Rapp *m* (-e) (*Bot.*) cole(seed), rape(seed) *cf* RAPPE

Rappe 1 *m* (-n, -n) black horse. 2 *f* (-n) grater;(*Vet.*) staggers. *Cf* RAPPEN

Rappel 1 *m* (-) fit (of madness). 2 *f* (-n) grater; (*Vet.*) staggers

Rappen *m* (-) 1 black horse. 2 (Swiss) centime. *Cf* RAPP, RAPPE

Rapport *m* (-e) 1 report. 2 (*Psych.*) rapport. 3 (*Tex.*) repeated pattern. **rapportieren** *vt/i* report

Rapputz *m* (-e) roughcast (of plaster)

Raps *m* (-e) (*Bot.*) rape(seed), cole(seed), colza—**rotblühender R.** garden rocket. **--kuchen** *m* rapeseed cake. **-mehl** *n* rapeseed meal. **-öl** *n* rape(seed) oil, colza oil. **-saat** *f*, **-samen** *m* rapeseed, colza

rasant *a* 1 flat (trajectory). 2 lightning-fast; *adv oft* at breakneck speed. 3 dashing; excellent, outstanding. **Rasanz** *f* 1 flatness (of a trajectory). 2 (breakneck) speed. 3 (*Cars, oft*) high performance. 4 dash, verve

rasch *a* fast, quick, rapid—(*Pyro.*) **r-er Satz** meal-powder (*or* fuse) composition

Rasch·binder *m* quick-setting cement
rascheln *vi* rustle, crackle
raschest 1 *a* fastest, etc: *superl of* RASCH. **2** *adv* as fast (*or* soon) as possible
Raschheit *f* rapidity, speed *cf* RASCH
Rasch·kochverfahren *n* quick-cook process. **r–trocknend** *a* quick-drying
rasen *vi*(s) **1** rave, rage. **2** speed, dash
Rasen *m* (-) lawn, grass. **–asche** *f* turf ashes. **-bleiche** *f* sun bleaching (on the grass)
rasend *p.a* **1** mad, raging. **2** breakneck, dashing. *Cf* RASEN
Rasen·eisenerz *n*, **-eisenstein** *m*, **-erz** *n* bog iron ore, limonite. **-mäher** *m* lawnmower
Rasier·apparat *m* (safety *or* electric) razor
rasieren 1 *vt & sich* r. shave. **2** *vt* level, raze
Rasier·klinge *f* razor blade. **-krem** *f* shaving cream. **-messer** *n* (straight) razor. **-seife** *f* shaving soap. **-stein** *m* styptic pencil
Raspel *f* (-n) **1** (*Tools*) rasp. **2** grater. **3** shred, flake. **raspeln** *vt/i* rasp, grate, scrape
Rasse *f* (-n) **1** race, breed, pedigree. **2** (*in compds oft*) thoroughbred, pedigreed
Rassel *f* (-n) rattle; noisemaker. **rasseln** *vi* (h *or* s) rattle
rassisch *a* racial
Rast *f* (-en) **1** rest. **2** break, pause. **3** (*Mach.*) notch; catch, detent. **4** (*Metll.*) bosh
Raste *f* (-n) notch, indentation
rasten *vi* **1** (h) rest, take a break. **2** (s) (*Mach.*) engage (in a notch, etc)
Raster *m or n* (-) **1** screen. **2** grid. **3** graph paper. **4** engraving roll
Rast·gärung *f* (*Brew.*) arrested (*or* slow and incomplete) fermentation. **-klinke** *f* catch, detent, latch. **r–los** *a* restless; ceaseless, relentless. **R–mälzerei** *f* arrest malting. **-tag** *m* day of rest
Rasur *f* (-en) **1** shave. **2** erasure
Rat *m* (–e) **1** (piece of) advice. **2** remedy, way out. **3** consultation, deliberation: **zu R–e ziehen** consult. **4** council. **5** council member
rät advises, etc: *pres of* RATEN
Ratanhia(·wurzel) *f* rhatany (root)
Rate *f* (-n) **1** rate. **2** installment

raten* (riet, geraten; rät) **I** *vt* **1** (d.i.o) recommend, advise. **2** guess (a solution); solve (a puzzle by guesswork). **II** *vi* **3** (*dat*) advise—**sich r. lassen** take advice. **4** consult, confer, deliberate. **5** (+ **auf**) guess (that the answer is). *Cf* GERATEN
Raten·kauf *m* installment purchase. **r–weise** *adv* in installments
Rat·geber *m* **1** adviser, counselor. **2** guide(book). **-haus** *n* town hall, city hall
rationalisieren *vt* rationalize; streamline. **rationell** *a* rational; efficient, streamlined; economical
rationieren *vt* **1** ration. **2** design efficiently (*or* for efficiency)
ratlos *a* helpless, at a loss for advice
ratsam *a* advisable, prudent. **Ratsamkeit** *f* advisability, prudence
Ratsche *f* (-n) **1** ratchet. **2** rattle, noisemaker
Rat·schlag *m* (piece of) advice, recommendation. **ratschlagen** *vi* deliberate, confer
Rat·schluß *m* decision; decree
Rätsel *n* (-) riddle, puzzle, enigma, mystery. **rätselhaft** *a* enigmatic, mysterious. **rätseln** *vi* puzzle, rack one's brains. **Rätsel·raten** *n* (riddle-)guessing, speculation
Ratte *f* (-n) rat
rätten *vi* screen, strain, riddle
Ratten·fraß *m* (*esp Agric.*) rat damage. **-gift** *n* rat poison. **-könig** *m* confused tangle. **-plage** *f* rat infestation. **-pulver** *n* rat powder, rat poison. **-schwanz** *m* **1** rattail (file). **2** pigtail. **3** endless chain (of problems, etc). **-tod** *m* rat poison. **-vertilgungsmittel** *n* rat killer, rat poison
Rätter *m* (-) screen, riddle; grate, grating
Ratze *f* (-n) rat. **r–kahl** *a* completely bare
Raub *m* **1** robbery; kidnaping. **2** loot, booty. **3** prey, victim; (*in compds oft*) predatory
Raub·bau *m* abusive exploitation (of land); soil exhaustion
rauben *vt/i* rob, steal; **einem etwas r.** rob sbdy of sthg, steal sthg from sbdy
räuberhaft, räuberisch *a* rapacious; predatory
Raub·tier *n* predatory animal, predator
rauch. *abbv* (rauchend) fum. (fuming)
Rauch *m* smoke; vapor; fumes. **–achat** *m*

smoky agate. **r–artig** *a* smoky. **R–bahngeschoß** *n* tracer projectile. **-ball** *m* 1 smoke bomb. 2 float light. **-behälter** *m* smokebox. **-belästigung** *f* smoke nuisance. **-bildung** *f* formation of smoke. **-bombe** *f* smoke bomb; float light. **r–dicht** *a* smoke-tight, smokeproof. **R–dichte, -dichtigkeit** *f* 1 smoke-tightness. 2 density of smoke. **-dose** *f* (*Mil.*) smoke float. **-düse** *f* smoke nozzle **rauchen** 1 *vt/i* smoke: **es raucht** there is smoke. 2 *vi* fume

Rauch·entwickler *m* smoke generator. **-entwicklung** *f* smoke generation

Raucher *m* (-) smoker (= person, compartment, etc)

Räucher· smoked, smoking; fumigating; aromatic: **-aal** *m* smoked eel

Raucher·bein *n* (*Med.*) gangrenous leg (from excessive smoking)

Räucherer *m* (-) 1 (*Meat, etc*) smoker, curer. 2 fumigator. 3 perfumer

Räucher·essenz *f* aromatic essence. **-essig** *m* aromatic vinegar. **-fisch** *m* smoked fish. **-fleisch** *n* smoked meat (*esp* pork) **räucherig** *a* smoky; dingy

Räucher·kerzchen *n*, **-kerze** *f* fumigating candle. **-lachs** *m* smoked salmon. **-lampe** *f* vaporizing lamp. **-mittel** *n* 1 fumigant. 2 meat-curing smoke

räuchern I *vt* 1 smoke (food; wood); cure (meat). 2 fumigate. 3 perfume. II *vi* burn incense

Räucher·papier *n* fumigating paper. **-pulver** *n* fumigating powder. **-schinken** *m* smoked ham. **-speck** *m* smoked bacon

Räucherung *f* 1 smoke-curing. 2 fumigation. 3 perfuming. 4 incense-burning. *Cf* RÄUCHERN

Räucher·waren *fpl* smoked goods (*esp* meat). **-werk** *n* perfume(s); incense. **-wurst** *f* smoked sausage

Rauch· smoke: **-erzeuger** *m* smoke generator (*or* producer). **-erzeugung** *f* smoke production; (*Agric.*) smudging. **-fahne** *f* smoke streamer. **-fang** *m* 1 chimney, flue. 2 smokestack. 3 hood (for smoke, fumes). **r–farben, -farbig** *a* smoke-colored, smoke-gray. **R–färbung** *f* smoke coloring. **-fleisch** *n* smoked meat (*esp* pork). **r–frei** *a* smoke-free, smokeless. **-gar** *a*

smoke-cured. **R–gas** *n* flue gas, chimney gas

Rauchgas·analyse *f* flue-gas analysis. **-vorwärmer** *m* flue-gas preheater

Rauch· smoke: **-gehalt** *m* smoke content. **-gerät** *n* smoke apparatus. **-gerbung** *f* smoke tanning. **-geschmack** *m* smoky taste, smoked flavor. **r–geschwängert** *a* smoke-filled. **-geschwärzt** *a* smoke-blackened. **R–glas** *n* smoke-tinted glass. **-granate** *f* smoke shell. **r–grau** *a* smoke-gray. **R–helm** *m* smoke helmet

rauchig *a* 1 smoky; fuming. 2 smoke-stained. 3 hoarse, husky (voice)

Rauch· smoke: **-kalk** *m* magnesian limestone. **-kammer** *f* 1 smoke box. 2 combustion chamber. 3 (meat-)smokeroom, smokehouse. **-kanal** *m* (smoke) flue. **-kasten** *m* smokebox. **-kern** *m* smoke nucleus. **-kerze** *f* smoke candle. **-körper** *m* smoke filler, smoke charge. **-ladung** *f* smoke charge. **r–los** *a* smokeless. **R–malz** *n* smoke-dried malt. **-meldepatrone** *f* smoke signal shell (*or* cartridge). **-ofen** *m* smudge pot; smoke generator. **-patrone** *f* smoke signal cartridge. **-pilz** *m* (nuclear) mushroom cloud. **-pulver** *n* fumigating powder. **-quarz** *n* smoky quartz. **-rohr** *n*, **röhre** *f* smoke flue, smokestack; fire tube. **-rohrkessel** *m* fire-tube boiler. **-satz** *m* (*Explo.*) smoke mixture (*or* composition). **-schaden** *m* smoke damage, damage caused by fumes. **-schieber** *m* smoke damper. **-schirm, -schleier** *m* smoke-screen. **-schriftmasse** *f* (*Aero.*) sky-writing composition. **r–schwach** *a* low-smoke. **-schwarz** *a* smoky black, soot-black. **R–schwimmer** *m* (*Mil.*) floating smoke pot. **-signalgerät** *n* smoke signal apparatus. **-speck** *m* smoked bacon. **-spurgeschoß** *n* tracer bullet (*or* projectile). **-stärke** *f* smoke density. **-tabak** *m* smoking tobacco. **-topas** *m* smoky topaz (*or* quartz). **-verbot** *n* no-smoking rule (*or* ordinance); (*as a sign*) No Smoking. **-verbrennung** *f* smoke combustion (*or* consumption). **-verbrennungs-einrichtung** *f* smoke consumer. **-verdünnung** *f* smoke dilution. **-vergiftung** *f* smoke poisoning. **-verhütung** *f* smoke prevention. **-verminderung** *f* smoke abatement. **-verzehrung** *f* smoke consumption. **-vorhang** *m* smoke screen, curtain of smoke. **-wand** *f* wall of smoke, smokescreen. **-waren** *fpl* 1 tobacco products; smoking accessories. 2 furs. **-werk** *n* incense. **-wolke** *f* cloud of smoke. **-wurst** *f* smoked sausage. **-zeichen** *n* smoke signal. **-zug** *m* smoke flue (*or* draft)

Räude *f* mange, scabies; scab, scurf.

räudig *a* mangy, scabious, scabby

Raufe *f* (-n) 1 (*Tex., Mach.*) hackle, hatchel. 2 (feeding) crib. **raufen** 1 *vt* hackle; pluck. 2 *vi & sich r.* scuffle, brawl. **Rauf-** **wolle** *f* plucked wool

rauh *a* 1 rough, course. 2 harsh, raw. 3 rude. 4 hoarse. 5 sore (throat); coated (tongue); hairy (side of hide). 6 raw, unprocessed. 7: **r-e Mengen** huge quantities

Rauh·artikel *m* (*Tex.*) raised style

Rauheit *f* roughness, coarseness; harshness; rawness; rudeness; hoarseness *cf* RAUH

Rauheits·beiwert *m* coefficient of roughness

rauhen I *vt* 1 roughen. 2 (*Tex.*) raise, nap. **II** *vi & sich r.* molt

Rauh·frost *m* hoarfrost. **-futter** *n* roughage. **r-gar** *a* (*Lthr.*) dressed with the hair on

Rauhheit, Rauhigkeit *f* roughness, etc = RAUHEIT

Rauh·kalk *m* (*Min.*) dolomite. **r-narbig** *a* (Lthr.) harsh-grained. **R-reif** *m* hoarfrost. **-reserve** *f* (*Tex.*) resist against raising

Rauke *f* (-n) hedge mustard

Raum *m* (⁼e) 1 room, chamber. 2 room, space (*incl* cosmic); volume. 3 area, zone, sector. 4 (*esp in pl*) open space, expanse. 5 (ship's) hold. 6 scope, opportunity (for action). **-analyse** *f* volumetric analysis (of gases). **-anzug** *m* space suit

Räum·asche *f* (*Zinc*) retort residue

Raum·beanspruchung *f* space requirement(s). **r-beständig** *a* constant-volume; sound (lime, etc.). **R-** **beständigkeit** *f* constancy of volume; soundness (of lime, etc.). **-bestimmung** *f* determination of volume. **-bewetterung** *f* ventilation; air conditioning. **-bild** *n* 1

space design. 2 stereoscopic image. 3 (man's) conception of cosmic space. **-bildentfernungsmesser** *m* stereoscopic range finder. **-chemie** *f* stereochemistry. **r-chemisch** *a* stereochemical. **R-dichte** *f* volume(tric) density. **-einheit** *f* unit of space (*or* volume). **-element** *n* spatial unit

räumen *vt* 1 vacate, leave; evacuate. 2 clear (a space, etc), empty, clean out—**das Feld r.** (+ *dat*) make way (for). 3 (re)move, clear (away): **etwas aus dem Weg r.** clear sthg out of the way, get rid of sthg. 4 (*Mach.*) ream. **Räumer** *m* (-) 1 reamer. 2 bulldozer

Raum·ersparnis *f* space saving: **wegen R.** to save space. **-fähre** *f* space shuttle. **-fahrer** *m* space traveler, astronaut. **-fahrt** *f* space flight. **-fahrzeug** *n* spacecraft. **r-fest** *a* stationary, fixed in space. **R-film** *m* stereoscopic (*or* three-dimensional) film. **-filter** *n* (*Elec.*) wave guide. **-flug** *m* space flight. **-formel** *f* spatial formula. **-forschung** *f* space research. **-frage** *f* matter of (available) space. **-fuß** *m* cubic foot. **-gebilde** *n* space diagram; three-dimensional structure. **-gehalt** *m* volume (*or* spatial) content. **-geometrie** *f* solid geometry.

Räum·gerät *n* mine-sweeping device

Raum·gewicht *n* weight per unit volume; (bulk) density. **-gitter** *n* (*Cryst.*) space lattice. **-glas** *n* stereoscope. **-größe** *f* room size; spatial extent. **-gruppe** *f* (*Cryst.*) space group. **-heizung** *f* room (*or* space) heating. **-hundertstel** *n* percent by volume

räumig *a* roomy, spacious. **Räumigkeit** *f* roominess, spaciousness, specific volume

Raum·inhalt *m* volume, cubic content, capacity. **r-isomer** *a* stereoisomeric. **R-isomerie** *f* stereoisomerism. **-kapsel** *f* space capsule. **-klima** *n* interior atmosphere (*or* environment). **-kosmetik** *f* house cleaning. **-krümmung** *f* space curvature. **-kurve** *f* solid curve; skew curve. **-labor** *n* space lab. **-ladung** *f* (*Elec.*) space charge. **-lehre** *f* geometry; space science

räumlich *a* 1 spatial, space, three-dimensional. 2 cubic; solid (geometry). 3 stereoscopic. 4 steric. 5 (*adv oft*) as regards space—**r. beengt** cramped for space; **r. getrennt** in separate rooms (*or* quarters). **Räumlichkeit** *f* (-en) **1** spatiality, spatial extension; three-dimensionality. **2** specific volume. **3** (*pl*) rooms, premises

Raum·loch *n* vent. **-mangel** *m* lack of space. **-maß** *n* **1** cubic measure. **2** measure of capacity, dimensions. **-menge** *f* amount of space, volume. **-meter** *n* or *m* **1** cubic meter. **2** cubic meter of stacked wood. **-modell** *n* space model. **-ordnung** *f* area planning. **-orientierung** *f* orientation in space. **-pendelschiff** *n* space shuttle. **-pflegerin** *f* cleaning woman. **-quantelung** *f* spatial quantization. **-richtung** *f* direction in space. **-schiff** *n* space ship. **-schiffahrt** *f* space travel, astronautics. **-sonde** *f* space probe. **-strahl** *m* cosmic ray. **-strahlenstoß** *m* burst of cosmic radiation. **-strahlung** *f* cosmic radiation. **-teil** *m* part by volume. **-temp.** *abbv* = **-temperatur** *f* room temperature. **-ton** *m* stereo(phonic) sound

Räumung *f* (en) evacuation; clearance; cleaning (out); removal; reaming *cf* RÄUMEN

Räumungs·verkauf *f* clearance sale

Raum·veränderung *f* change in volume. **-verhältnisse** *npl* space conditions; volume relationships. **-verminderung** *f* decrease in volume. **-verschwendung** *f* waste of space. **-verteilung** *f* space distribution. **-wärme** *f* room temperature. **-welle** *f* space (*or* sky) wave. **-winkel** *m* solid angle. **-zeitalter** *n* space age. **r–zentriert** *a* (*Cryst.*) body-centered

raunen *vt/i* whisper, murmur

Raupe *f* (-n) (*Zool. & Mach.*) caterpillar

Raupen·antrieb *m* caterpillar drive. **-fraß** *m* caterpillar damage. **-leim** *m* (anti-caterpillar) banding grease (for trees). **-schlepper** *m* caterpillar tractor. **-seide** *f* natural silk

Rausch *m* (ᵉe) **1** drunkenness, intoxication—**einen R. haben** be drunk. **2** narcosis, high. **3** euphoria, frenzy. **4** (*Med.*) short-term generasl anesthesia. **5** (*Min.*) pounded ore. **6** (*in compds, oft*) noise, static *Cf* RAUSCHEN. **r–-arm** *a* low-noise. **R–beere** *f* cranberry, crowberry

rauschen *vi* **1** (h) rustle; rush, roar: **es rauscht** there is a rustling noise, there is static. **2** (s) go sweeping (*or* rushing) (in, out, etc). **Rauschen** *n* rustling (rushing, roaring) noise, rustle, rush, roar; static. **rauschend** *p.a* (*oft*) noisy, thunderous. **Rauscher** *m* (-) sparkling (*or* bubbling) new wine

Rausch·gelb *n* orpiment; realgar. **-gift** *n* narcotic, drug. **r–giftsüchtig** *a* drug-addicted. **R–giftsüchtige** *m, f* (*adj endgs*) drug addict. **-gold** *n* Dutch gold, tinsel, brass foil. **-leder** *n* chamois (leather). **-messung** *f* (*Electron.*) noise measurement. **-mittel** *n* intoxicant, narcotic. **-pegel** *m* noise level. **-rot** *n* realgar. **-silber** *n* imitation silver foil, silver tinsel. **-spannung** *f* (*Elec.*) noise voltage

Raute *f* (-n) **1** (*Bot.*) rue. **2** diamond (-shaped figure *or* design), lozenge, rhombus. **3** facet (of a gem)

rauten·artig *a* **1** rue-like. **2** rhomboid. **R–flach** *m* rhombohedron. **-fläche** *f* rhombus. **-flächner** *m* rhombohedron. **r–förmig** *a* rhombic, diamond-shaped. **R–gewächse** *npl* (*Bot.*) Rutaceae. **-glas** *n* faceted glass. **-öl** *n* oil of rue. **-spat** *m* (*Min.*) rhomb spar, dolomite

Ray·gras *n* ryegrass

Rayon *m* (-s) area, district; department

raz. *abbv* = **razem** *a* racemic = RACEM

Razzia *f* (. . ien) raid

rd. *abbv* (rund) round; approx.

Reagens *n* (. . ntien) = **Reagenz** *f* (-ien) reagent. **--farbe** *f* test color. **-flasche** *f* reagent bottle. **-glas** *n* test tube

Reagenzglas·bürste *f* test tube brush. **-gestell** *n* test tube rack **-halter** *m* test tube holder. **-stichkultur** *f* (*Bact.*) test tube slab culture

Reagenzien·flasche *f* reagent bottle. **-kasten** *m* reagent box (*or* case). **-raum** *m* reagent space (*or* room)

Reagenz·kelch *m* test glass (*or* cup). **-lösung** *f* test solution. **-mittel** *n* reagent. **-papier** *n* test paper. **-rohr, -röhrchen** *n*, **-röhre** *f* test tube

reagieren *vi* (+ **auf, mit**) react (to, with), respond (to). **reagierend** *p.a* (*oft*) reactive

Reagier· = REAGENZ·. **Reagierglas·** = REAGENZGLAS·

Reaktanz f reactance. **Reaktion** f (-en) reaction: **in R. treten** react, enter into a reaction

Reaktions-ablauf m course of the reaction. **-arbeit** f (Chem.) maximum work of reaction, free energy change; (Metll.) reaction process. **-bahn** f reaction path. **r-bedingt** a reaction-dependent. **R–bedingung** f reaction condition. **-bereitschaft** f reactivity. **-druck** m reaction pressure. **r–fähig** a reactive, capable of reacting; responsive. **R–fähigkeit** f reactivity, reaction capacity; responsiveness. **-flüssigkeit** f reaction liquid. **-formel** f reaction formula (or equation). **-fortgang** m progress of a reaction. **r–freudig** a reactive, responsive. **R–freudigkeit** f reactivity, responsiveness. **-gefäß** n reaction vessel. **-gemisch** n reaction mixture. **-geschehen** n occurrence of a reaction. **-geschwindigkeit** f reaction velocity (or rate). **-gleichung** f reaction equation. **-grenze** f limit of reaction. **-hemmung** f inhibition of a reaction. **-kammer** f reaction chamber. **-kette** f reaction chain. **-kinetik** f reaction kinetics. **r–kinetisch** a kinetic. **R–klebstoff** m two-component adhesive. **-konstante** f reaction constant, velocity coefficient. **-kurve** f reaction curve

reaktionslos a reactionless, inert. **Reaktionslosigkeit** f absence of reaction, inertness

Reaktions-masse f reaction mass. **-mischung** f reaction mixture. **-mittel** n reagent. **-ordnung** f order of reaction. **-ort** m locus (or area) of the reaction. **-papier** n test paper. **-raum** m reaction area (or space). **-rohr** n, **-röhre** f reaction tube. **-stufe** f phase (or step) of the reaction. **-teilnehmer** m reactant. **r–träge** f slow to react. **R–trägheit** f slowness to react. **-turm** m reaction tower. **-verhältnis** n reaction ratio (or proportion). **-verlauf** m course of the reaction. **-vermögen** n reactivity. **-versuch** m reaction test. **-wärme** f heat of reaction. **-weise** f manner (or type) of reaction. **-zeit** f reaction time

Reaktor·kern m reactor core

reaktiv a reactive; **r–er Antrieb** reaction propulsion. **Reaktiv** n (-e) reagent. **reak-**

tivieren vt reactivate. **Reaktivierung** f (-en) reactivation. **Reaktivität** f reactivity

real a 1 real, actual. 2 practical. 3 material. 4: **r–e Seide** thrown silk. **--denkend** a objective

Realien npl 1 actual facts, realities. 2 humanities and natural sciences. 3 expert knowledge

Realin-seide f second-grade silk

realisierbar a realizable, feasible. **realisieren** I vt 1 realize, accomplish. 2 realize (= become aware of). 3 liquidate, convert to cash. **II: sich r.** be realized, become reality. **Realisierung** f (-en) 1 realization, accomplishment. 2 realization, awareness. 3 liquidation

Realität f (-en) reality

Real-lexikon n encyclopedia. **-schule** f secondary school (stressing modern languages and natural sciences) cf GYMNASIUM. **-seide** f first-grade silk. **-wissenschaft** f practical (or applied) science

Reb-bau m viticulture, wine growing

Rebe f (-n) (grape) vine; tendril

Reben-ruß m vine (vegetable, Frankfort) black. **-saft** m wine, juice of the grape. **-schwarz** n vine (vegetable, Frankfort) black

Reb-huhn n partridge. **-laus** f vine louse, phylloxera

rebromieren vt rebrominate

Reb-schwarz n vine black = REBENSCHWARZ. **-stock** m (grape) vine

Recept n (-e) prescription, recipe = REZEPT

rechen vt rake. **Rechen** m (-) 1 rake. 2 grid, screen, grate. 3 coat rack.

Rechen· arithmetic, calculating: **-anlage** f computer. **-automat** n calculator, (Electron.) computer. **-bild** n nomogram. **-fehler** m miscalculation, mistake in arithmetic. **-gerät** n calculator; computer. **-hilfsmittel** n aid in calculation. **-kunst** f arithmetic. **-maschine** f 1 calculating machine; (Electron.) calculator; computer. 2 adding machine. 3 abacus

Rechenschaft f account(ing): **zur R. ziehen** call to account; **R. fordern** demand an accounting; **R. geben** (or

ablegen) (**über**) account (for)
Rechen· calculating, arithmetic: **-scheibe**
f circular slide rule. **-schieber, -stab** *m*
slide rule. **-tafel** *f* mathematical table.
-technik *f* 1 technique of calculation. 2
computer technology. **-uhr** *f* circular slide
rule. **-verfahren** *n* arithmetical (*or*
mathematical) process. **-zentrum** *n* computer center

rechnen I *vt/i* calculate, figure (out), reckon. **II** *vt* 1 estimate. 2 (+ **unter, zu**), count,
number (among). 3 charge (a price). **III** *vi*
(+ **auf, mit**) count (on), reckon (with).
Rechnen *n* arithmetic. **Rechner** *m* (-) 1
computer. 2 arithmetician. **rechner·
gesteuert** *a* computer-controlled. **rech·
nerisch** *a* arithmetical, mathematical
Rechnung *f* (-en) 1 calculation, arithmetic:
die R. geht auf (**a**) the calculation comes
out even, (**b**) things turn out as planned. 2
bill, check: **einem etwas in R. stellen**
bill (*or* charge) someone for sthg. 3 expense(s): **auf seine R. kommen** (re)cover
one's expenses. 4 account: **auf R. kaufen**
buy on account; **in R. setzen** (*or* **ziehen**)
(+ *accus*), **R. tragen** (+ *dat*) take into
account

Rechnungs·art *f* mathematical operation.
-betrag *m* amount of the account (*or* bill).
-führer *m* bookkeeper, accountant. **-jahr**
n fiscal year. **r–mäßig** *a* mathematical,
according to calculation. **R–prüfer** *m* auditor. **-tafel** *f* mathematical table

recht *a* 1 right(-hand). 2 right(-wing). 3
(*Geom.*): **r–er Winkel** right angle. 4
right, correct; decent, proper: **alles was r.
ist** what's right is right. 5 *pred* (+ *dat*): **es
ist ihnen r.** it suits them, it's all right
with them; **ihnen ist nicht r.** they don't
feel well; **einem etwas r. machen** do
sthg to sbdy's satisfaction. 6 *usu adv*
really, quite; **erst r.** all the more (so). 7
adv right(ly): **gerade r. kommen** come
at just the right moment; **es geschieht
ihnen r.** it serves them right; **r. haben** be
right; **r. bekommen** turn out to be right;
r. behalten be proved right. *Cf* RECHTE;
LINK; FALSCH, UNRECHT

Recht *n* (-e) 1 law, justice. 2 right(s), (one's)
due: **alle R–e vorbehalten** all rights reserved; **von R–s wegen** by rights; **mit R.**

rightly; **zu R. bestehen** be valid. 3 (*in
compds* or *as pfx, oft*) legal, rect-, ortho-:
-bild *n* (*Phot.*) positive

Rechte (*adj endgs*) **I** *f* right (hand, side, party). **II** *n* the right thing—**etwas R–s**
something worthwhile; **nach dem R–n
sehen** look after things

Recht·eck *n* rectangle. **r–eckig** *a* rectangular. **R–eckwelle** *f* rectangular
wave

rechten *vi* (+ **mit**) dispute, litigate (with)
rechtens *adv* by rights, rightly
rechter·hand, -seits *adv* on the right-hand
side

rechtfertigen 1 *vt* justify; exonerate. 2:
sich r. be justified; be exonerated.
Rechtfertigung *f* (-en) justification;
exoneration

recht·gläubig *a* orthodox. **-haberisch** *a*
dogmatic, opinionated. **-laufig** *a* running
to the right; clockwise; in direct motion
rechtlich *a* 1 legal. 2 honest, just
rechtlos *a* 1 rightless, having no rights. 2
lawless

recht·mäßig *a* rightful, legitimate, legal
rechts 1 *adv* (*oft* + **von**) & *prep* (+ *genit*) to
(*or* on) the right (of), right: **von r.** (**her**)
from the right; **nach r.** to(ward) the right.
2 *pred a* right-handed. *Cf* LINKS. 3 *pfx*
right(-hand), right-wing, dextro.; law,
legal

Rechts·anwalt *m* lawyer, attorney. **-beistand** *m* legal counsel. **r–beständig** *a* legal, valid

rechtschaffen *a* honest, upright
Recht·schreibung *f* (standard) spelling,
orthography

rechts· right, dextro.; law legal: **-drehend**
a dextrorotatory. **R–drehung** *f* dextrorotation, right-hand(ed) polarization.
-gang *m* 1 right-hand rotation. 2 legal
procedure. **r–gängig** *a* right-hand(ed)
(screw, etc); clockwise. **R–gelehrte** *m, f*
(*adj endgs*) law scholar, jurist. **-gewinde**
n right-hand (screw) thread. **r–gültig** *a*
legally valid. **R–gültigkeit** *f* legal validity. **-handel** *m* lawsuit. **r–händig** *a*
right-hand(ed). **-herum** *adv* (around) to
the right, clockwise. **-kräftig** *a* legally
binding, valid. **R–lauf** *m* clockwise motion, motion to the right. **r–laufend,**

-läufig *a* right-hand, to the right, clockwise. **R–lehre** *f* jurisprudence. **-milchsäure** dextrolactic acid. **-mittel** *n* legal remedy (redress, measure). **-pflege** *f* administration of justice. **-polarisation** *f* right-hand polarization. **-quarz** *n* right-handed quartz. **-säure** *f* dextro acid. **-schraube** *f* right-hand screw. **-schutz** *m* legal protection (as by patents). **-spruch** *m* (legal) decision, judgment. **-weg** *m* legal proceedings *cf* WEG 1. **-weinsäure** *f* dextrotartaric acid. **r–wendig** *a* right-handed

recht·winklig *a* right-angled; rectangular. **-zeitig** *a* 1 punctual; *adv oft* at the right time, on time, in time. 2 *adv* early enough

Recinit *m* (*Coal*) resinite

reciprok *a* reciprocal

Reck *n* (-e) horizontal bar

Reck·alterung *f* (*Metll.*) strain aging

Recke *f* (-n) 1 rack. 2 stretching device

recken 1 *vt* & *sich* r. stretch. 2 *vt* (*Metll.*) draw; shingle

Reck·probe *f* elongation (*or* stretching) test. **-spannung** *f* tensile stress (*or* strain)

Reckung *f* (-en) stretching; drawing; shingling *cf* RECKEN

Reck·walzen *fpl*, **-werk** *n* (*Metll.*) finishing rolls

Red. *abbv* (Reduktion) red. (reduction)

Redakteur *m* (-e) editor. **Redaktion** *f* (-en) 1 editing. 2 editorial office (department, staff)

Rede *f* (-n) 1 speech. 2 talk, conversation, discussion: **die R. kam auf . . .** the discussion turned to . . .; **davon ist (nicht) die R.** that is (not) the point (of the discussion); **in R. stehen** be under discussion, be in question; **nicht der R. wert** not worth talking about; **keine R.** nothing of the sort, no way, not a chance. 3 account: **zur R. stellen** call to account. 4 rumor

reden *vt/i* talk, speak

Redens·art *f* phrase, expression, idiom; term

redestillieren *vt* redistill

Rede·weise *f* manner of speaking, style. **-wendung** *f* phrase, expression, idiom

redigieren *vt* edit, revise

redlich *a* honest; sincere

Redm. *abbv* (Reduktionsmittel) reducing agent

Redner *m* (-) speaker, orator. **--bühne** *f* (lecture) platform

reducier . . . See REDUZIER . . .

Reduktions·arbeit *f* reduction process. **-ätze** *f* (*Tex.*) reduction discharge. **-bad** *n* reducing bath. **r–beständig** *a* reduction-resistant. **R–beständigkeit** *f* resistance to reduction. **r–fähig** *a* reducible, capable of reduction. **R–fähigkeit** *f* reducibility. **-ferment** *n* reductase, reducing ferment. **-flamme** *f* reducing flame. **-geschwindigkeit** *f* rate of reduction. **-hütte** *f* (*Metll.*) reduction plant. **-kohle** *f* reducing carbon (*or* coal). **-kraft** *f* reducing power. **-mittel** *n* reducing agent. **-muffe** *f* (*Mach.*) reducing sleeve (*or* coupling). **-ofen** *m* reducing furnace. **-ort** *m* reduction region. **-rohr** *n*, **-röhre** *f* reduction tube. **-schlacke** *f* reducing slag. **-tiegel** *m* reduction crucible. **-ventil** *n* reducing valve. **-verhältnis** *n* reduction ratio. **-vermögen** *n* reducing power. **-vorgang** *m* reduction process. **-wärme** *f* heat of reduction. **-wirkung** *f* reducing action (*or* effect)

reduktiv *a* reductive, by reduction

reduktometrisch *a* reductometric

Reduktor *m* (-en) (Jones) reductor

reduzierbar *a* reducible. **Reduzierbarkeit** *f* reducibility

reduzieren 1 *vt* reduce. 2: **sich** r. be reduced, diminish. **reduzierend** *p.a*, *as adv oft* under reducing conditions

reduzier·fähig *a* reducible. **R–salz** *n* reducing salt. **-stück** *n* reducing piece, reducer

Reduzierung *f* (-en) reduction, reducing

Reduzier·ventil *n* reducing valve

reell *a* 1 real (*incl Math.*) genuine. 2 decent, respectable, fair; honest

Ref. *abbv* 1 (Referat) report, etc. 2 (Referent) speaker, etc

Referat *n* (-e) 1 report, abstract; review, evaluation. 2 department (of a public authority). **Referent** *m* (-en, -en), **Referentin** *f* (-nen) 1 speaker, writer (presenting a report or abstract), abstractor; reviewer. 2 department head (of a public authority).

Referenz f (-en) reference
Referier·dienst m abstracting (reporting, reviewing) service
referieren vt/i present a report (or an abstract), report (on)
reflektieren I vt reflect. **II** vi 1 (+ **auf**) have one's eye (on). **2** (+ **über**) reflect (on), think over
Reflektier·strahl m reflected ray
Reflektions· (in compds) = REFLEXIONS·
Reflex m (-e) reflex; reflection. **-bewegung** f reflex motion (or action). **r–frei** a nonreflecting. **R–galvanometer** n reflecting galvanometer
Reflexion f (-en) reflection
Reflexions·ebene f plane of reflection; (Cryst.) mirror (or reflection) plane. **r–frei** a nonreflecting. **R–gitter** n reflection grating. **-goniometer** n reflecting goniometer. **-kugel** f (Cryst.) sphere of reflection. **-linie** f (Cryst.) reflection line. **-punkt** m (Cryst.) reflection point. **-stab** m (Cryst.) reflection line. **-vermögen** n reflecting power. **-winkel** m angle of reflection
Reflex·projektion f (Cryst.) interference projection. **-wirkung** f reflex action
Reform·haus n health food center. **reformieren** vt reform
Refr. abbv (Refraktion) refraction
refraktär a refractory
Refraktions·winkel m angle of refraction. **-zahl** f refractive index
refraktometrisch a refractometric(al)
Refraktor m (-en) 1 (Opt.) refractor. 2 (Ceram.) refractory. 3 (Astron.) refractory telescope
refraktorisch a refractory
Regal n (-e) shelf, rack, case with shelves
Reg.-Bez abbv (Regierungsbezirk) adm. dist. (administrative district)
rege a active, lively; agile
Regel f (-n) 1 rule: **in der R.** as a rule. 2 (Med.) menstruation, period
regelatieren vt regelate, refreeze
regelbar a adjustable, controllable. **Regelbarkeit** f adjustability, controllability
Regel·einrichtung f controlling device, control mechanism. **r–fähig** a adjustable, controllable. **R–fall** m normal case;

im R. as a rule. **-größe** f regular (or standard) size; (Cybern.) controlled variable. **-hahn** m regulating cock, control valve. **-klappe** f control valve. **-leistung** f normal output. **r—los** a irregular, disorderly; random. **-mäßig** a regular, normal. **R–mäßigkeit** f regularity, normality. **-metall** n standard metal
regeln vt 1 regulate, control, put in order cf GEREGELT. 2 arrange, adjust. 3 take care of, attend to. 4 settle (a dispute, etc)
Regel·organ n controlling device. **r–recht** a 1 real, genuine; regular. 2 sheer, downright. 3 according to rules (or regulations). 4 adv oft thoroughly, properly. **R–schraube** f adjusting screw. **-stab** m control rod. **-technik** f control technology
Regelung f (-en) 1 regulation, control. 2 arrangement, adjustment. 3 attending to . . . 4 settlement. Cf REGELN
Regel·widerstand m (Elec.) variable resistance; rheostat. **r–widrig** a contrary to rules; irregular, abnormal. **R–zyklus** m menstrual period
regen 1 vt move. **2: sich r.** move, stir; be aroused, arise
Regen m (-) rain; (usu fig) hail, shower
regen·arm a low-rainfall (area). **R–bogen** m rainbow
Regenbogen·farbe f rainbow color. **-haut** f (Anat.) iris
regen·dicht a rainproof, showerproof. **R–dusche** f rain shower, downpour
Regenerat n (-e) reclaimed (or regenerated) material; (specif) reclaimed rubber. **-gummi** n reclaimed rubber
Regenerativ·feuerung f regenerative firing (or heating). **-ofen** m regenerative furnace, recuperative oven
Regenerat·mischung f (Rubber) reclaim mix
regenerierbar a regenerable, reclaimable
regenerieren vt regenerate, reclaim.
Regenerierung f (-en) regeneration, reclaiming
Regenerier·vorrichtung f regenerative device
Regen·fall m (heavy) rainfall, rainstorm. **-fang** m cistern. **-faß** n rain barrel. **r–frei** a rainless, dry. **-glatt** a rain-slick. **R–guß** m downpour, shower. **-haut** f plastic

slicker (*or* raincoat). **r–los** *a* rainless, dry. **R–mantel** *m* raincoat. **-maß** *n*, **-messer** *m* rain gauge. **r–naß** *a* rain-drenched, rain-soaked. **-reich** *a* high-rainfall (area), highly rainy. **R–rinne** *f* rain (*or* eave) gutter. **-schauer** *m* shower. **-schirm** *m* umbrella. **-schutz** *m* rain shelter, protection against rain. **-tropfen** *m* raindrop. **r–unempfindlich** *a* insensitive to rain, rainproof. **R–wald** *m* rain forest. **-wasser** *n* rainwater. **-wetter** *n* rainy (*or* wet) weather. **-wurm** *m* earthworm, angleworm. **-zeit** *f* rainy season

Ragie *f* (-n) **1** direction; management, control: **in R. nehmen** take control of. **2** government monopoly (e.g. of tobacco)

regieren 1 *vt/i* rule, govern. **2** *vt* manage. **3** *vi* reign; predominate. **Regierung** *f* (-en) government, administration

Regie·verschluß *m:* **unter R.** in bond

Register *n* (-) **1** register, index, record. **2** (organ) stop: **alle R. aufziehen** pull out all the stops. **-band** *m* index volume. **-karte** *f* index card

Registratur *f* (-en) registry; files, filing department, record office

Registrier·apparat *m* registering (recording, indexing) device

registrieren *vt* register, record; file; index

Registrier·kasse *f* cash register

Registrierung *f* registration, recording; filing; indexing

Registrier·vorrichtung *f* recording device. **-waage** *f* recording balance

Regler *m* (-) **1** (*Person or Mach.*) regulator, governor, controller *cf* REGELN. **2** (*Flot.*) conditioner. **-(bügel)eisen** *n* temperature-control iron. **-knopf** *m* control knob

Reglisse *f* (-n) **1** licorice. **2** marshmallow paste

Reglung *f* (-en) regulation, etc = REGELUNG

regnen *vt/i* rain, pour down. **Regner** *m* (-) sprinkler. **regnerisch** *a* rainy

Regreß *m* (. . esse) redress, recourse

regsam *a* active, alert; agile

Reg·stoff *m* ferment

regulär *a* regular; (*Cryst. oft*) cubic. **Regularität** *f* (-en) regularity

Regulations·vorgang *m* regulatory process

regulierbar *a* adjustable, controllable

regulieren 1 *vt* regulate, adjust, control; repair; settle. **2: sich r. (nach)** adjust itself (to)

Regulier·hahn *m* regulating (*or* control) cock. **-schraube** *f* adjusting screw

Regulierung *f* (-en) regulation, adjustment, control; repair; settlement *cf* REGULIEREN

Regulierungs·vorrichtung *f* regulating (*or* control) device

Regulier·widerstand *m* adjustable resistance; rheostat

regulinisch *a* (*Metll.*) reguline

Regulus·ofen *m* regulus furnace; reduction crucible (for antimony)

Regung *f* (-en) **1** motion, movement. **2** impulse. **3** (inner) stirring, emotion. **4** (*esp pl*) effort(s). *Cf* REGEN

regungs·los *a* motionless; impassive

Reh *n* (-e) (roe) deer; venison. **-braun** *n* **1** velvety brown (iron pigment). **2** fawn (color). **r–braun, -farben, -farbig** *a* fawn (-colored). **-leder** *n* doeskin, buckskin. **-posten** *m* buckshot. **-wild** *n* roe deer

Reib·ahle *f* reamer

Reibe *f* (-n) grater, rasp

reib·echt *a* fast to rubbing, abrasion-resistant. **R–echtheit** *f* fastness to rubbing, resistance to abrasion. **-eisen** *n* grater, rasp

Reibe·mühle *f* grinder, grinding mill

reiben* (rieb, gerieben) *vt/i* rub; abrade; chafe; rasp, grate; grind, triturate; scour

Reibe·pulver *n* abrasive powder

Reiber *m* (-) grater; rubber, brayer, pestle

Reiberei *f* (-en) (*usu pl & fig*) friction

reib·fest *a* abrasion-resistant. **R–festigkeit** *f* resistance to abrasion. **-fläche** *f* rubbing (*or* rubbed) surface; striking surface (on matchboxes). **-gut** *n* material to be ground. **-kasten** *m* grinder, grinding mill. **-keule** *f* pestle. **-löten** *n* tinning. **-mühle** *f* grinding mill. **-oxidation** *f* frictional oxidation. **-papier** *n* friction paper. **-probe** *f* rubbing (grinding, friction) test; grinding sample. **-rad** *n* friction wheel. **-satz** *m* friction composition, (*specif*) fulminate. **-schale** *f* (grinding) mortar

Reibsel *n* (-) (*usu pl*) grindings, gratings

Reib·stein *m* grindstone. **-stelle** *f* rubbed area, friction mark. **-stuhl** *m* grinding mill. **r–unecht** *a* not rub-resistant

Reibung *f* (-en) 1 friction. 2 rubbing; abrasion; chafing; rasping; grating; grinding; trituration; scouring *cf* REIBEN

Reibungs·antrieb *m* (*Mach.*) friction drive. **-beiwert** *m* coefficient of friction. **-elektrizität** *f* frictional electricity. **r–fest** *a* abrasion-resistant. **R–festigkeit** *f* resistance to abrasion. **R–fläche** *f* friction (*or* abrasive) surface. **r–frei** *a* frictionless. **R–koeffizient** *m* coefficient of friction. **-kraft** *f* frictional force. **r–los** *a* frictionless, smooth. **R–messer** *m* friction meter, tribometer. **-messung** *f* measurement of friction. **-probe** *f* friction (*or* rubbing) test. **-punkt** *m* point of friction. **-rad** *n* friction wheel. **-wärme** *f* frictional heat. **-wert** *m* coefficient of friction. **-widerstand** *m* frictional resistance. **-zahl, -ziffer** *f* coefficient of friction

Reib·verzinnen *n* wipe-tinning. **-zündholz** *n* friction match. **-zündkerze** *f* (*Explo.*) friction primer

reich *a* 1 (*oft* + **an**) rich (in), wealthy. 2 abundant. 3 generous. 4 (*as sfx, oft*) -ful, -ous: **farbenreich** colorful; **silberreich** rich in silver

Reich *n* (-e) 1 kingdom, empire. 2 rule. 3 realm, domain

Reich·blei *n* rich (*specif* argentiferous) lead

reichen I *vt* 1 (*d.i.o*) hand, pass; serve, offer. **II** *vi* 2 reach, extend, come (up to). 3 suffice, do, be (*or* go far) enough. 4 (+ **mit**) make do (with). *Cf* GEREICHEN

Reich·frischen *n* (*Copper*) enriching. **-gas** *n* rich gas. **r–haltig** *a* rich, substantial; extensive. **R–haltigkeit** *f* richness, substantial content

Reichheit *f* richness, abundance

reichlich *a* 1 ample, abundant, plentiful: **r. gerechnet** amply (*or* generously) reckoned. 2 oversize, larger than necessary. 3 (*a & adv* + *Quant.*) a good, more than: **eine r–e Stunde** a good hour, more than an hour. 4 (*adv oft*) quite, good and: **ein r. schwieriges Verfahren** quite a difficult process; **r. spät** quite (*or* good and) late. 5 *adv* plenty of: **r. (viel)**

Zeit plenty of time. **Reichlichkeit** *f* ampleness, abundance

Reichs· royal, imperial, rational, state: **-amt** *n* national (state, government) office (*or* bureau). **-bahn** *f* national (*or* state) railroad (*pre-WW2 & in DDR*)

Reich· rich: **-schaum** *m* (*Metll.*) (silver-rich) zinc crust (formed in Parkes process). **-schlacke** *f* rich slag. **-schmelzen** *n* smelting of precious metals

Reichs·metall *n* delta metal (bronze of variable composition). **-patent** *n* national (state, government) patent

Reichtum *m* (¨er) wealth, abundance

Reich·weite *f* (-n) reach; range: **in R.** within reach; **außer R.** out of reach (*or* range). **-werk** *n* (*Metll.*) refining, cupellation

reif *a* ripe, mature; (+ **für, zu** *oft*) ready (for)

Reif *m* **I** (hoar)frost. **II** (-e) = REIFEN ring, etc

Reife *f* (-n) 1 ripeness, maturity; readiness; age (of beer, etc). 2 secondary school final examinations = ABITUR, REIFEPRÜFUNG. **–grad** *m* degree of maturity (*or* ripeness)

Reif·ei *n* mature ovum

reifeln *vt* groove, channel; rifle

Reife·merkmal *n* indicator of ripeness

reifen 1 *vt & vi*(s) ripen, mature. 2 *vi*(h): **es reift** there is a frost

Reifen *m* (-) ring, hoop; tire; collar; bracelet

reifen·artig *a* ring-like; hoop-like. **R–bau** *m* tire design (*or* construction). **-mischung** *f* (*Rubber*) tire stock. **-panne** *f* tire trouble; flat tire, puncture. **-profil** *n* tire tread. **-prüfer** *m* tire gauge. **-regenerat** *n* (*Rubber*) tire reclaim. **-schlauch** *m* inner tube

Reife·prozeß *m* ripening (maturation, aging) process. **-prüfung** *f* (secondary school) comprehensive final examination = ABITUR. **-stadium** *n* stage of maturity. **-teilung** *f* maturation division, meiosis. **-vorgang** *m* ripening (*or* maturing) process. **-zeit** *f* time (*or* age) of maturity; puberty. **-zeugnis** *n* secondary school diploma *cf* ABITUR. **-zustand** *m* ripe (*or* mature) state

Reif·holz *n* 1 heartwood. 2 hoop wood

reiflich *a* mature;—**r. überlegen** think over carefully

Reif·punkt *m* frost point

Reifung *f* ripening, maturation

Reifungs·körper *m* ripening agent, accelerator; (*Photo.*) sensitizer

Reihe *f* (-n) **1** series; group, number. **2** row, line. **3** rank. **4** order, succession, turn: **nach der R., der R. nach** in succession, in turn, one after the other; **an der R. sein, an die R. kommen** have one's turn, be next; **außer der R.** out of (one's) turn; **aus der R. kommen** get out of line, get off the track, get upset

reihen I *vt* **1** arrange in a row, line up. **2** (*Elec.*) connect in series. **3** string (beads). **4** (+ **zu**) add to the ranks (of). **5** baste, tack. **II: sich r. 6** form a row, line up. **7** (+ **an**) follow; join the ranks (of). **8** (+ **um**) gather (around)

Reihen·anordnung *f* (*Elec.*) series connection. **-aufnahme** *f* **1** (*Phot.*) sequence shot. **2** (*pl*) serial radiographs (*or* X-ray photos). **-entwicklung** *f* (*Math.*) series development. **-fertigung** *f* assembly-line production. **-folge** *f* succession, order, sequence. **-formel** *f* series formula. **-haus** *n* row house. **-herstellung** *f* assembly-line production. **-saat** *f* drill sowing. **-schaltung** *f* (*Elec.*) series connection. **-schluß** *m* series winding, (*in compds usu*) series-wound. **-untersuchung** *f* mass (medical) examination. **r—weise** *adv* in rows, in lines; by turns; in large numbers

Reiher *m* (-) heron. **--gras** *n* esparto grass

reih·um *adv* in turn, in succession; all around, all along the line.—**r. gehen** make the rounds; **r. gehen lassen** pass around

Reihung *f* lining up, arrangement in rows; connecting in series, etc *cf* REIHEN *vt*

rein *a* **1** pure, genuine; nothing but, sheer; *adv oft* downright *cf* REINST. **2** net (profit, etc). **3** correct, accent-free (pronunciation). **4** clean, clear: **im r—en** in the clear, **ins r—e bringen** clear up, straighten out, settle; **mit einem ins r—e kommen** reach an understanding with a psn.; **ins r—e schreiben** make a clean copy (of)

Rein·asbest *m* pure (*or* free) asbestos. **-benzol** *n* rectified benzene. **r—blau** *a* pure blue, sky-blue. **R—blauton** *m* (*Dye.*)

sky-blue shade. **-darstellung** *f* preparation in pure form; purification

Reinecke·salz *n* Reinecke's salt, ammonium tetracyanatodiammine chromate

Reinedaude *f* greengage

Rein·element *n* pure (single-isotope) element. **-ertrag** *m* net yield (*or* profit). **-gas** *n* purified gas. **r—gelb** *a* pure yellow. **R—gewicht** *n* net weight. **-gewinn** *m* net profit (*or* gain). **-gewinnung** *f* purification. **-gummi** *n* (*Rubber*) pure gum stock

Reinheit *f* purity, genuineness; cleanness; clarity *cf* REIN

Reinheits·grad *m* degree of purity. **-probe** *f* test for purity

reinigen *vt* **1** (*genl*) purify; (*specif*) refine (metals), rectify (spirits), clarify (liquids). **2** clean, cleanse: **chemisch r.** dry-clean. **3** scrub; scour (silk). **4** purge, expurgate. **5** disinfect. **reinigend** *p.a* (*oft*) detergent; depurative; purgative. **Reiniger** *m* (-) purifier; cleaner, cleanser, etc

Reiniger·masse *f* cleansing material. **-wasser** *n* cleansing water (*or* liquid)

Reinigung *f* (-en) **1** purification; refining; rectification; clarification. **2** cleaning, cleansing. **3: (chemische) R.** (a) dry cleaning, (b) dry cleaner's. **4** scrubbing, scouring. **5** purging, expurgation. **6** disinfecting. **7** (*Med.*): **(monatliche) R.** menstruation. *Cf* REINIGEN

Reinigungs·anlage *f* purification plant. **-anstalt** *f*: **(chemische) R.** dry-cleaning establishment, dry-cleaner's. **-apparat** *m* purifying (rectifying, cleaning, cleansing) apparatus, purifier, rectifier. **-bad** *n* purifying (cleaning, clearing) bath. **-bassin** *n* purifying tank (*or* basin). **-behälter** *m* filtering tank. **-benzin** *n* gasoline for cleaning. **-creme** *f* cleansing cream. **-feuer** *n* refining fire. **-flasche** *f* wash bottle; gas washing bottle. **-kraft** *f* purifying (cleaning, cleansing) power. **-krem** *f* cleansing cream. **-masse** *f* purifying (*or* cleansing) material. **-mittel** *n* **1** purifying agent. **2** cleansing agent. **3** cleaner, cleaning fluid. **4** detergent. **5** purgative. **-möller** *m* (*Metll.*) purifying charge. **-verfahren** *n* purifying (refining, cleansing, disinfecting) process. **-vermögen** *n* purifying (cleaning, cleansing) power. **-verschluß** m cleaning door (hatch, trap). **-verstärker** m dry-cleaning detergent. **-wirkung** f purifying (cleansing, cleaning) action (or effect)

Rein·kohle f pure (clean, cleaned) coal; solid carbon. **-koks** m pure (or clean) coke. **-kultivierung** f (Bact.) pure cultivation, pure culture preparation. **-kultur** f (Bact.) pure culture.

reinlich a 1 clean. 2 neat, tidy. 3 clean, clear (distinction). **Reinlichkeit** f cleanliness, neatness, tidiness; clarity

rein·machen 1 vt clean. 2 vi clean up, clean house

Rein·öl n pure oil; clean oil. **-schrift** f clean copy. **-seide** f pure silk

reinst I a 1 superl of REIN purest, cleanest, etc. 2 pure, sheer. II pfx very pure, high-purity: **R—aluminium** n high-purity aluminum. **-kohle** f super (or high-grade) coal

Rein·toluol n pure toluene. **-wasser** n pure water. **r—weg** adv completely. **R—weiß** n pure (or clear) white. **-wichte** f absolute density, specific gravity. **-wismut** n pure bismuth. **r—wollen** a (Tex.) pure-wool, all-wool. **R—xylol** n rectified xylol, mixture of xylenes. **-zellulose** f pure cellulose. **-zucht** f (Bact.) pure culture (or strain. **-zuchthefe** f pure culture of yeast. **-züchtung** f obtaining (of) a pure culture

Reis I m rice. II n (-er) twig, sprig, shoot; (lit & fig) scion

Reis·branntwein m rice brandy (or wine)

Reise f (-n) 1 trip, journey, (pl oft) travels; (in compds oft) travel(ing), traveler's. 2 (Engg.) run, life; (Furnaces) campaign.

reisen vi(s) travel. **Reisende** m, f (adj endgs) traveler, tourist; traveling salesperson

Reis·faser f rice fiber. **-futtermehl** n rice flour, ground rice

Reisig n brushwood, twigs (esp for kindling)

Reis·kleie f rice bran. **-körper** m, **-körperchen** n (Med.) rice body. **-mehl** n rice flour. **-öl** n rice oil. **-papier** n rice paper. **-puder** m (cosmetic) rice powder

Reiß·ahle f scratch-awl, scribe(r). **-belastung** f breaking load. **-baumwolle**

f reclaimed (or re-usable) cotton (from rags). **-brett** n drawing board

Reis·schleim m rice gruel; (Pharm.) rice water

Reiß·dreieck n draftsman's triangle. **-ebene** f fracture plane

reißen* (riß, gerissen) I vt 1 rip, tear; break, rupture. 2 pull, snatch; sweep (along). 3 crush (malt); rough-grind (glass). 4 (+ an sich) take over, seize. 5 (of animals) kill (as prey). 6: **es reißt einen, zu . . .** one feels a strong urge to . . . 7 draw, trace. II vi(h) 8 tug, tear (at). 9 (+ dat): **es reißt einem** one has rheumatic pains. III vi(s) 10 tear, rip, come apart; crack, break, snap, rupture. IV: **sich r.** 11 scratch oneself. 12 (+ **um**) scramble, vie (for). Cf GERISSEN. **Reißen** n rheumatism. **reißend** p.a (oft) 1 raging (stream), ripping (current). 2 rapid, racing; (oft as adv) (sell) like hot cakes. 3 rheumatic, racking (pain). 4 rapacious (animals)

Reiß·feder f ruling (or drawing) pen. **-festigkeit** f resistance to tearing (or breaking); tensile strength. **-gelb** n orpiment. **-kohle** f sketching charcoal. **-korn** n (Lthr.) artificial grain. **-lack** m crackle lacquer. **-länge** f breaking length (of fibers). **-last** f breaking load. **-nadel** f scriber. **-nagel** m thumbtack. **-prüfgerät** n tear-testing device. **-scheibe** f bursting disc. **-schiene** f T-square

Reis·stärke f rice starch

Reiß·verschluß m zipper, slide fastener. **-wolf** m (paper or rag) shredder; teaser; willow(er). **-wolle** f reclaimed (or re-usable) wool, shoddy. **-zahn** m carnassial tooth, fang. **-zeug** n drafting instruments. **-zirkel** m drafting compass. **-zwecke** f thumbtack

Reit· riding. **reiten*** (ritt, geritten) vt & vi(s) ride (on horseback, etc). **reitend** p.a (oft) mounted. **Reiter** m (-) (lit. & Mach.) rider; tab (on index card); pointer, indicator. **-gewicht** n rider (= weight)

Reiz m (-e) 1 stimulus. 2 irritation, irritant. 3 charm, attraction, appeal. **-antwort** f reaction (to stimuli, irritations)

reizbar a 1 sensitive (to stimuli). 2 irritable. **Reizbarkeit** f 1 sensitivity (to stimuli). 2 irritability; reaction (to irritation)

Reiz·beantwortung f response to stimuli. **-bewegung** f tropism

reizen vt 1 stimulate, act on. 2 attract, arouse; (oft + **zu**) move, tempt (to). 3 irritate, annoy, anger, provoke. **reizend** p.a (oft) charming, attractive, nice; irritant

Reiz·gas n irritant gas; (specif) tear (or sneeze) gas. **-geschoß** n irritant gas projectile. **-gift** n irritant poison

Reizker m (-) agaric; lactarlus

Reiz·kerze f (Mil.) irritant candle. **-körper** m (genl) irritant (substance); (specif, esp Mil.) lacrimator. **r–lindernd** a (Pharm.) demulcent

reizlos a 1 nonirritant. 2 insipid, bland. 3 devoid of all charm, unattractive. **Reiz·losigkeit** f 1 nonirritating nature. 2 insipidity, blandness. 3 lack of charm, unattractiveness

Reiz·mittel n, **-stoff** m 1 irritant; stimulant; (Med.) adjuvant. 2 inducement. **-topf** m (Mil.) irritant candle

Reizung f (-en) stimulation; attraction; arousal; temptation; irritation cf REIZEN

reiz·voll a 1 tempting. 2 attractive, graceful. **R–wirkung** f irritating (or stimulating) effect. **-wort** n trigger (or buzz) word

rekapitulieren vt recapitulate

rekarbon(at)isieren vt recarbonate

Reklamation f (-en) 1 complaint, protest. 2 claim

Reklame f (-n) 1 advertising, publicity—**R. machen (für)** advertise, publicize. 2 advertisement, commercial

reklamieren I vt 1 complain about. 2 claim. II vi 3 protest; object; complain. 4 make inquiries (about overdue deliveries, etc)

Rekombinations·wärme f heat of recombination. **rekombinieren** vt/i recombine

rekristallisieren vt recrystallize

Rektifikations·apparat m rectifying apparatus. **-becken** n rectifying vessel. **-kolonne** f rectifying column

Rektifizier·apparat m rectifying apparatus. **-boden** m fractionating column plate

rektifizieren vt rectify

Rektifizier·kolonne, -säule f rectifying column

Rektifizierung f (-en) rectification

Relais n (-) relay

relativieren I vt 1 relate. 2 modify, weaken, cancel out (esp in the light of other data). II: **sich r.** be related; be modified (weakened, etc)

relativistisch a relativistic. **Relativität** f relativity

Relief·druck m relief printing

Remission f (-en) 1 remission. 2 (Opt.) diffuse reflection. **remittieren** vt/i 1 remit, return. 2 (Opt.) reflect

Rench·gneis m (Petrog.) paragneiss

Rendement n (-s) (esp Tex.) yield

Rendita f (Tex.) extent of weighting (fabrics)

Rendite f (-n) yield, return (on investment)

Reneklode f (-n) greengage

Renn·arbeit f (Iron) direct-process extraction. **-eisen** n direct-process iron

rennen* (rannte, gerannt; sbjc rennte) 1 vi(s) run, race. 2 vt run; (Metll.): **Eisen r.** make malleable iron directly from the ore. **Rennen** n (-) running, racing; race

Renn·feuer n bloomery hearth, (direct-process) smelting furnace. **-feuereisen** n direct-process malleable iron. **-feuerschlacke** f (Iron) direct-process slag. **-herd** m bloomery hearth

Rennin n rennet

Renn·schlacke f (Iron) direct-process slag. **-stahl** m direct-process (or natural) steel

rennte would run, etc: sbjc of RENNEN

Renn·tier n reindeer. **-verfahren** n (Iron) direct process

Renommee n (-s) reputation, renown. **renommieren** vi brag, boast. **renommiert** p.a renowned

renovieren vt renovate

rentabel a profitable. **Rentabilität** f profitability

Rente f (-n) 1 pension, annuity. 2 investment income. 3 pl oft bonds

rentieren: sich r. be profitable (or worthwhile), pay

Reparatur f (-en) repair: **in R.** closed for repairs; **etwas in R. geben** have sthg repaired. **r–-freundlich** a easily repaired. **R–werkstatt** f repair shop

reparieren vt repair

Repassier·bad n (Silk) second boiling-off bath. **repassieren** vt 1 (Tex.) mend. 2 (Dye.) repeat the treatment of. 3 (Silk) boil off a second time. 4 (*Metll.*) give a smooth surface by cold processing. 5 check, recheck

Repertorium *n* (. . rien) 1 register, index; reference work; (*oft in titles*) compendium. 2 repertory

repetieren 1 *vt/i* repeat; study (by rote). 2 *vi* repeat a class (in school). **Repetier-gut** *n* recirculated material

Repetitorium *n* (. . rien) 1 (*School*) review course. 2 review manual (*or* book)

Replik *f* (-en) 1 reply. 2 replica

Repräsentant *m* (-en, -en), **Repräsentantin** *f* (-nen) representative (*incl* salesperson). **Repräsentation** *f* (-en) 1 representation. 2 display of status, making a good impression. **repräsentativ** *a* 1 representative. 2 impressive, displaying status; distinguished

Repräsentativ-erhebung *f* statistical sampling. **-umfrage** *f* statistical sampling of opinion

repräsentieren 1 *vt* represent. 2 *vi* make a good impression, display one's status, show off

Reprise *f* (-n) 1 reprise; revival; rerun. 2 (economic) recovery

Reprivatisierung *f* (-en) denationalization, reprivatization (of industry, etc)

Reproduktions-kraft *f* reproductive power

reproduzierbar *a* reproducible. **reproduzieren** *vt* reproduce

Reps *m* rape(seed) = RAPS

Reptil-leder *n* reptile leather

resch *a* crisp, crusty

Reseda *f* (-s) mignonette—**gelbe R.** dyer's rocket. **-grün** *n* chrome green. **-samenöl** *n* mignonette oil

Resede *f* (-n) mignonette

Reservage *f* (-n) (*Tex.*) resist, reserve. **-artikel** *m* resist style. **-druck** *m* resist printing. **-papp** *m* resist paste

Reserve *f* (-n) 1 reserve; (*in compds oft*) spare, backup. 2 (*Dye., Tex.*) resist; (*Tex., oft*) resist paste, (*in compds*) = RESERVAGE e.g.: **-artikel** *m* (*Tex.*) resist style. **-druckfarbe** *f* resist printing paste. **-mittel** *n* reserve; resist. **-muster** *n* reference pattern. **-papp** *m* (*Tex.*) resist paste. **-reifen** *m* spare tire. **-stärke** *f* 1 reserve starch. 2 reserve strength. **-stoff** *m* re-

serve(s), reserve material. **-teil** *m* spare part

reservierbar *a* reservable; (*Tex.*) resistible. **Reservierbarkeit** *f* reservability; (*Tex.*) resistibility

reservieren *vt* 1 reserve. 2 (*Tex.*) treat with a reserve (*or* resist). **reserviert** *p.a* (*Tex.*) resist(-printed)

Reservierungs-artikel *m* (*Tex.*) resist-printed goods; resist style. **-mittel** *n* (*Tex.*) resist, reserve. **-wirkung** *f* (*Tex.*) resist effect

residuell *a* residual. **Residuum** *n* (. . duen) residue

Resinat *n* (-e) resinate, rosinate. **-firnis** *m* resinate varnish (*or* boiled oil)

resinogen *a* resinogenous, resiniferous

Resinoid *n* (-e) 1 oleoresin. 2 (*Perfumes*) resinoid

Resonanz-linie *f* (*Spect.*) resonance line. **-schwingung** *f* resonance vibration, covibration. **-spannung** *f* resonance potential. **-spitze** *f* resonance peak. **-zustand** *m* state of resonance

resorbieren *vt* resorb, reabsorb. **Resorbierung** *f* (-en) resorption, reabsorption

Resorcin, Resorzin *n* resorcin(ol). **-blau** *n* resorcin blue. **-gelb** *n* resorcin yellow

resp *abbv* = **respektive** *cnjc* or (as the case may be), respectively, or rather = BEZIEHUNGSWEISE, BZW

Respirations-nahrungsmittel *n* respiratory food. **respiratorisch** *a* respiratory

Rest *m* (-e) 1 (*Chem.*) residue, (*in compds oft*) residual; radical, group. 2 rest, remainder. 3 trace, vestige. 4 (*pl*) remnants, remains; leftovers, scraps; dregs. 5: **den R. geben** (+ *dat*) finish off. **-affinität** *f* residual affinity. **-alkalinität** *f* residual alkalinity

restaurieren *vt* restore

Rest- remaining, residual: **-betrag** *m* remaining amount, remainder. **-bier** *n* (*Brew.*) waste beer. **-druck** *m* residual pressure. **-feuchtigkeit** *f* residual moisture. **-flüssigkeit** *f* residual liquid. **-gas** *n* residual gas. **-glied** *n* (*Math., Logic*) remainder (*or* number); remainder (of an infinite series). **-härte** *f* (*Water*) residual (*or* permanent) hardness

restieren *vi* remain, be left over

Rest- remaining, residual: **-ion** *n* residual

ion. **-keim** m residual nucleus. **-kohle** f residual coal (or carbon), carbonized residue. **-kraft** f residual force (or power); (magnetic) remanence. **-ladung** f (Elec.) residual charge. **-lauge** f residual liquor (lye, liquid)

restlich a remaining

restlos a complete, total; without residue

Rest· remaining, residual. **-lösung** f residual solution. **-luft** f residual air. **-magnetismus** m residual magnetism, remanence. **-müll** m discarded leftovers. **-plasma** n residual plasma (or electrons). **-säure** f residual acid. **-schmelze** f (Metll.) residual melt (or heat); (Petrog.) residual magma. **-schmutz** m (Tex.) soil residue. **-spannung** f residual stress (or tension). **-stickstoff** m residual nitrogen. **-stickstoffkörper** m residual nitrogenous substance. **-strahl** m residual ray. **-strom** m residual current. **-valenz** f residual valence. **-widerstand** m residual resistance

resublimieren vt resublime

Resultante f (-n) resultant

Resultat n (-e) result, outcome. **resultatlos** a fruitless. **resultieren** vi (aus) result (from). **resultierend** p.a (oft), **Resultierende** f (adj endgs) resultant

Resümee n (-s) resumé, summary. **resümieren** vt summarize

retardieren vt retard

Retorte f (-n) retort

Retorten·bauch m belly of the retort. **-gestell** n retort stand. **-hals** m neck of the retort. **-halter** m retort holder (or stand). **-haus** n retort house. **-helm** m retort head (or helm). **-kammer** f retort chamber. **-kind** n test-tube baby. **-kitt** m retort cement (or lute). **-kohle** f retort (or gas) carbon; (Analysis) charcoal block. **-koks** m gas coke. **-mündung** f mouth of the retort. **-ofen** m retort (furnace). **-paraffin** n (Petrol.) still wax. **-produkt** n laboratory product. **-reihe** f series (or bank) of retorts. **-rückstand** m retort residue. **-vergasung** f gasification in retorts. **-verkokung** f retort coking. **-vorstoß** m 1 adapter. 2 (Zinc) condenser (of clay or iron)

Retouche f (-n) retouching. Cf RETUSCHE, etc

Retour· (in compds) return(ing)

Retourn·öl n recovered oil

Retour·schlacke f (Metll.) return slag

retten 1 vt rescue, save; salvage. 2: **sich r.** save oneself, escape. **rettend** p.a (oft) fortunate; of refuge—**ein r-er Ausweg** a way out, an avenue of escape. **Retter** m (-) rescuer, savior

Rettich m (-e) radish. **-öl** n oil of radish

Rettung f (-en) 1 rescue, salvation; salvage. 2 ambulance (or life-saving) service

Rettungs·arbeit f rescue work (or operation). **-dienst** m life-saving (emergency, ambulance) service. **-gürtel** m life belt (or preserver). **r–los** a hopeless, beyond rescue. **R–rakete** f recovery rocket. **-wagen** m ambulance

Retusche f (-n) (Phot.) retouching. **retuschieren** vt retouch

Retuschier·essenz f retouching varnish. **-farbe** f retouching color. **-lack** m retouching varnish. **-mittel** n retouching agent

Reue f remorse. **reuen** vt cause remorse (to); **es reut uns** we regret

Reu·geld n forfeit

Reuse f (-n) fish trap (or net); cage; grid

reuten vt uproot (trees); clear (land)

reverberieren vt/i reverberate. **Reverberier·feuer** n, **-ofen** m reverberatory furnace

Reversibilität f reversibility. **reversierbar** a reversible. **reversieren** vt reverse

revidieren vt 1 check, inspect; audit. 2 revise

Revier n (-e) district, area; precinct; territory; department

Revision f (-en) 1 check, inspection; audit. 2 revision

Revisions·spachtel m or f glazing putty

Revolver· (usu) revolving: **-presse** f scandal sheet(s)

Reynoldsche Zahl Reynold's number

rez. abbv (reziprok) recip. (reciprocal)

Rezensent m (-en, -en) reviewer, critic. **Rezension** f (-en) review (of books, etc)

rezent a recent

Rezept n (-e) 1 prescription. 2 recipe. **-buch** n recipe book. **-formel** f prescription formula. **r–frei** a requiring no prescription; adv without prescription

rezeptieren vt 1 (Med.) dispense; prescribe.

2 give the formula for
Rezeptor *m* (-en) receptor. **Rezeptoren-**
·seitenkette *f* receptor sidechain
rezept·pflichtig *a* requiring a prescription
Rezeptur *f* (-en) 1 dispensing (of prescriptions). 2 prescription department (of a
pharmacy). 3 prescription formula
Rezidiv *n* (-e) (*Med.*) relapse, recurrence
Rezipient *m* (-en, -en) 1 receiving vessel (*or*
tank). 2 recipient
reziprok *a,* **Reziproke** *f* (-n) reciprocal.
Reziprozität *f* reciprocity
rezyklieren *vt/i* recycle. **Rezyklierung** *f*
(-en) recycling
RGT-Regel *f* *abbv* (*R*eaktions-*g*eschwin-
digkeits-*T*emperatur-Regel) reaction-
velocity-temperature rule
Rhabarber *m* rhubarb—(*Pharm.*) **fal-**
scher R. root of European rhubarb.
·gelb *n* chrysophanic acid. **-gerbsäure** *f*
rheotannic acid. **-säure** *f* chrysophanic
acid
Rhachitis *f* rickets. **rhachitisch** *a* rha-
chitic
Rhamnit *n* rhamnitol
Rhamnon·säure *f* rhamnonic acid
rheinisch, rheinländisch *a* Rhenish,
Rhineland
Rhein·säure *f* rhenic (*or* chrysophanic)
acid. **-wein** *m* Rhine wine
Rheuma *n,* **Rheumatismus** *m* rheuma-
tism
Rhigolen *n* petroleum ether
Rhizocarpin·säure *f* rhizocarpinic acid.
Rhizocarp·säure *f* rhizocarpic acid
Rhizonin·säure *f* rhizoninic acid. **Rhizon-**
·säure *f* rhizonic acid
Rhld. *abbv* (Rheinland) Rhineland
Rhöadin *n* rheadine
Rhodan *n* thiocyanogen; (*in compds*) thio-
cyanate. **·aluminium** *n* aluminum thio-
cyanate. **-ammon(ium)** *n* ammonium
thiocyanate. **-ammonlösung** *f* am-
monium thiocyanate solution
Rhodanat *n* (-e) thiocyanate. **··lösung** *f*
thiocyanate solution
Rhodan·baryum *n* barium thocyanate.
-eisen(rot) *n* iron (*esp* ferric) thiocyanate
Rhodanid *n* (-e) (-ic) thiocyanate. **··zahl** *f*
(*Fats & Oils*) thiocyanogen number
rhodanieren *vt* (treat with) thiocyanate.

Rhodanierung *f* (-en) thiocyanation
Rhodan·ion *n* thianocyanogen ion
rhodanisieren *vt* (treat with) thiocyanate
Rhodan·kali(um) *n* potassium thiocyanate. **-kalzium** *n* calcium thiocyanate.
-kupfer *n* cupric thiocyanate. **-lösung** *f*
thiocyanate solution. **-metall** *n* metallic
thiocyanate. **-methyl** *n* methyl thiocyanate. **-natrium** *n* sodium thiocyanate.
-nickel *n* nickel thiocyanate. **-quecksil-**
ber *n* mercury thiocyanate. **-salz** *n* thiocyanate. **-säure** *f* thiocyan(at)o acid.
-tonerde *f* aluminum thiocyanate
Rhodanür *n* (-ous) thiocyanate
Rhodan·verbindung *f* thiocyanogen compound. **-wasserstoff** *m* 1 hydrogen thiocyanate. 2 = **-wasserstoffsäure** *f* thiocyanic acid. **-zahl** *f* (*Fats & Oils*)
thiocyanogen number. **-zinn(oxid)** *n*
stannic thiocyanate. **-zinnoxydul** *n* stannous thiocyanate.
Rhodiak *n* rhodiumammine
Rhodina·säure *f* rhodinic acid
Rhodium·chlorwasserstoffsäure *f*
chlororhodic acid. **-metall** *n* metallic rhodium. **-salz** *n* rhodium salt. **-schwarz** *n*
rhodium black. **-überzug** *m* rhodium
coating (*or* plating). **-verbindung** *f* rhodium compound
Rhodizon·säure *f* rhodizonic acid
Rhodochrom·salz *n* rhodochromium salt
Rhodocladon·säure *f* rhodocladonic acid
Rhodo·salz *n* rhodo salt
Rhomben·dodekaeder *n* rhombic dodecahedron. **r–förmig** *a* rhombic, rhombus-
shaped
rhombisch *a* rhombic
Rhombo·eder *n* rhombohedron. **rhom-**
boedrisch *a* rhombohedral
rhomboidisch *a* rhomboid(al)
Rhus·lack *m* rhus varnish (from Japanese
trees)
rhythmisch *a* rhythmic. **Rhythmus** *m*
(. . men) rhythm
R.-I.-Bezfg. *abbv* (Ring-Index-Beziffe-
rung) ring-index numbering
Ribon·säure *f* ribonic acid
Ribonuklein·säure *f* ribonucleic acid
Richt·analyse *f* guide (*or* approximate)
analysis. **-antenne** *f* directional antenna.
-beispiel *n* model, guide, example to fol-

low. **-blei** n plumb line (or bob)

Richte f straight line—**in die R. bringen** straighten (out)

Richt·effekt m directing effect. **-eigenschaft** f directional property

richten I vt **1** straighten (out), level. **2** set (correctly), adjust, align. **3** aim, direct; address (questions, comments to . . .). **4** set up, erect. **5** repair, fix. **6** top out (a building). **7** execute. **II** vt/i **8** judge, pass sentence (on). **9** prepare. **III: sich r. 10** aim, be aimed, be directed (or oriented). **11** (usu + **nach**) go, be guided (by), conform (to). **12: sich in die Höhe r.** rise (up). **13** spruce up. Cf GERICHTET, ZUGRUNDE

Richt·energie f directional energy

Richter m (-), **Richterin** f (-nen) judge **Richter·Skala** f (Seismology) Richter scale

Richt·funk m beam radio, directional transmission. **-glas** n (Opt.) collimator

richtig a **1** right, correct. **2** real, genuine; proper; regular. **--gehend** a **1** accurate. **2** real, regular

Richtigkeit f correctness, rightness

richtig·phasig a of the right phase

richtig·stellen vt correct, rectify. **Richtigstellung°** f correction, rectification

Richt·kraft f directive force. **-leiter** m (Elec.) rectifier. **-linie** f guideline; (pl oft) instructions. **-maß** n standard (of measure), yardstick. **-platte** f surface plate; plate for hammering sheet metal flat. **-preis** m standard (or recommended) price. **-satz** m standard rate; rule, regulation. **-scheit** n straightedge, level. **-schnur** f **1** guideline (lit & fig). **2** plumb line. **-schraube** f adjusting screw. **-strahl** m directional beam. **-strahler** m beam antenna. **-strom** m (Elec.) rectified current

Richtung f (-en) **1** direction; course, line— **(in) R. Hamburg** toward Hamburg. **2** trend, tendency. **3** regard, respect

richtung·gebend a **1** directive, direction-indicating. **2** trend-setting

Richtungs·änderung f change in direction. **-bewegung** f directional motion, tropism. **r--los** a directionless, aimless. **R--quantelung** f directional (or space) quantization

richtung·weisend a **1** directive, direction-indicating. **2** trend-setting

Richt·wert m **1** coefficient. **2** standard value. **3** approximate value. **-wirkung** f directional effect (of an antenna). **-zahl** f **1** coefficient. **2** index; governing figure

Ricinus·öl n dehydrated castor oil

Ricinol·säure f ricinoleic acid

Ricinus m (- or -se) **1** castor oil plant. **2** = **--öl** n castor oil. **-ölsäure** f ricinoleic acid. **-ölsulfosäure** f castor-oil sulfonic acid. **-samen** m castor bean (or seed). **-säure** f ricinic (or ricinoleic) acid. **-seife** f castor-oil soap

Ricke f (-n) (Zool.) doe

rieb rubbed etc: past of REIBEN

riechbar a detectable by smell

Riech·bein n (Anat.) ethmoid (bone)

riechen* (roch, gerochen) vt & vi (oft + **nach**) smell (of); vi + adv, oft: e.g. **stark r.** have a strong odor. **riechend** p.a (oft) odorous, redolent; malodorous

Riech·essig m aromatic vinegar. **-fläschchen** n smelling (-salt) bottle. **-haut** f olfactory membrane. **-kissen** n sachet, perfumed pad. **-mittel** n perfume. **-nerv** m olfactory nerve. **-organ** n olfactory organ. **-probe** f smelling test (or sample). **-salz** n smelling salts. **-spur** f scent. **-stoff** m odoriferous (or aromatic) substance, perfume. **-streifen** m perfumed paper. **-topf** m (Mil.) irritant gas generator. **-wasser** n perfumed water. **-werkzeug** n olfactory organ. **-würfel** m (Mil.) chemical capsule

Ried n (-e) **1** reed. **2** bog, marsh. **3** (vineyard) slope. **--gras** n reeds, sedge. **-moor** n reed moor

rief called: past of RUFEN

Riefe f (-n) groove, furrow; striation. **riefe(l)n** vt groove, furrow; rifle; striate; knurl, mill. **Rief(e)lung** f grooving, furrowing; rifling; striation; knurling, milling. **riefig** a grooved, furrowed, etc. **Riefung** f (-en) grooving, etc = RIEFELUNG

Riegel m (-) **1** (slide) bolt; latch. **2** (cross)bar. **3** bar (of soap, chocolate, etc). **4** beam; (fence) rail. **riegeln** vt bolt, bar, lock. **Riegel·schneiden** n (Soap) barring

Riemen m (-) **1** (leather) strap; belt (incl

Mach.). **2** oar. **--antrieb** *m* (*Mach.*) belt drive. **-fett** *n* belt dressing. **-kernstück** *n* (*Lthr.*) strap butt. **-kitt** *m* belt (*or* strap) cement. **-leder** *n* belt (*or* strap) leather, belting. **-scheibe** *f* belt pulley. **-schmiere** *f*, **-schmiermittel** *n* belt lubricant. **-trieb** *m* belt drive (*or* transmission)

Ries *n* (-e) (*Paper*) ream

Riese *m* (-n, -n) giant; colossus

Riesel·feld *n* field irrigated with sewer water. **-höhe** *f* packing depth (of a trickling tower). **-jauche** *f* liquid sewage (as fertilizer). **-kolonne** *f* trickling column. **-kühler** *f* spray (*or* trickling) cooler

rieseln I *vi* **1** (*usu* s) trickle, run. **2** (h) drizzle; ripple, drip. **II** *vt* sprinkle, spray, irrigate

Riesel·regen *m* drizzle. **-turm** *m* trickling (*or* spray) tower. **-verhalten** *n* flow characteristics (of powders). **-wasser** *n* irrigation (spray, trickling) water

riesen·groß *a* gigantic, colossal, tremendous. **R--größe** *f* gigantic (colossal, huge) size

riesenhaft *a* gigantic, colossal, huge

Riesen· giant, gigantic, tremendous, colossal: **-konzern** *m* giant corporation. **-schlange** *f* giant snake, (*specif*) python, boa constrictor, anaconda. **-zelle** *f* giant cell

riesig *a* gigantic, huge, tremendous; (*adv oft*) awfully

riet advised, etc: *past of* RATEN

Riet *n* (-e) (weaving) reed, comb

Riff *n* (-e) reef, (underwater) ledge

Riffel *f* (-n) **1** groove, corrugation. **2** (*Tex.*) rib; ripple, comb. **--blech** *n* corrugated sheet metal. **-glas** *n* corrugated (fluted, channeled) glass. **-muster** *n* ribbed pattern

riffeln *vt* groove, corrugate, flute; rib; ripple (flax)

Riffel·trichter *m* ribbed (*or* fluted) funnel. **-walze** *f* fluted (*or* corrugated) roller

rigoros *a* rigorous. **Rigorosum** *n* (. . sa *or* . . sen) oral examinations for the doctor's degree

Rille *f* (-n) groove, furrow, rill; striation.

rillen *vt* groove, furrow, flute; drill-plow *cf* GERILLT

rillen·förmig *a* ribbed, grooved, furrowed;

r--e Abzehrung corrosive graining. **R--pflug** *m* drill plow. **-scheibe** *f* grooved pulley, sheave

Rimesse *f* (-n) remittance, draft

Rind *n* (-er) **1** ox, steer, cow; *pl usu* (head of) cattle. **2** beef. **--box** *n* chrome-tanned oxhide (for shoes). **-brühe** *f* beef broth

Rinde *f* (-n) bark, rind, crust; cortex

Rinden· bark, cortical: **-auszug** *m* bark extract. **-borke** *f* outer bark. **-brand** *m* bark scorch. **-farbstoff** *m* phlobaphene. **r--förmig** *a* bark-like, rindy, corticiform. **R--gerbstoff** *m* bark tannin. **-grau** *n* (*Anat.*) gray matter (of the cortex). **-haut** *f* (*Bot.*) periderm. **-mühle** *f* bark(-grinding) mill. **-pore** *f* (*Bot.*) lenticel. **-säure** *f* corticinic acid. **-schicht** *f* cortical layer. **-span** *m* bark chip (*or* paring). **-substanz** *f* cortical substance

Rinder· beef, cattle, bovine: **-blut** *n* oxblood. **-bouillon** *f* beef broth. **-braten** *m* roast beef. **-bremse** *f* gadfly, horsefly. **-fett** *n* beef fat (*or* suet). **-galle** *f* ox gall. **-klauenöl** *n* neat's-foot oil. **-markfett** *n* beef marrow fat. **-pest** *f* cattle plague. **-pökelfleisch** *n* corned beef. **-talg** *m* beef tallow (*or* suet). **-tuberkulose** *f* bovine tuberculosis. **-zucht** *f* cattle breeding

Rind·fleisch *n* beef. **-fleischbrühe** *f* beef broth. **-haut** *f* cowhide, oxhide

rindig *a* bark-covered; rindy, bark-like

Rind·leder *n*, **r--ledern** *a* cowhide, oxhide

Rinds· *See also* RINDER: **-blase** *f* ox bladder. **-fett** *n* beef fat (*or* suet). **-haut** *f* cowhide, oxhide. **-klauenfett** *n* neat's-foot oil. **-leder** *n*, **r--ledern** *a* cowhide, oxhide

Rind·spalte *f* split cowhide

Rinds·talg *m* beef tallow (*or* fat)

Rind·vachette *f* light cowhide. **-vieh** *n* **1** cattle. **2** stupid fool

Ring *m* (-e) (*usu*) ring, (*specif*) circle, cycle; coil, hoop, loop; (*in compds oft, esp Org.* *Chem.*) cyclic

ring·ähnlich *a* ring-like, circular, cyclic; annular, areolar. **R--alkohol** *m* cyclic alcohol. **-amin** *n* cyclic amine. **r--artig** *a* ring-like, etc = R--ÄHNLICH. **R--aufspaltung** *f* ring cleavage. **-bildung** *f* ring formation. **-brenner** *m* (*Gas*) ring burner. **-düse** *f* ring nozzle

Ringel *m* (-) ring, ringlet, circlet. **--blume** *f*

marigold, calendula. **-erz** n ring (or cascade) ore. **-flechte** f ringworm

ringelig a ring-like, annular; curled, coiled

ringeln 1 vt & **sich r.** curl, coil (up). 2 vt girdle (trees)

ringen* (rang, gerungen) 1 vt/i wrestle; wring. 2 vi struggle. **Ringer** m (-) wrestler

Ringer·lösung f (Physiol.) Ringer's solution

Ring· annular, cyclic: **-erweiterung** f ring extension. **-faser** f annular fiber. **r–förmig** a annular, ring-shaped; cyclic. **R–gasbrenner** m ring (gas) burner. **-gefüge** n ring structure. **-geschoß** n (Mil.) laminated shell. **-glied** n ring member. **-gliederzahl** n number of members in the ring. **-granate** f (Mil.) laminated shell. **-keton** n cyclic ketone. **-nut** f circular groove (or slot). **-ofen** m (Ceram.) ring (or annular) kiln, Hofmann kiln. **-öffnung** f ring opening. **-platte** f ring (or annular) plate. **-pulver** n (Explo.) annular powder. **-röhre** f toroidal tube, doughnut

rings adv (all) around

Ring· annular, cyclic: **-schicht** f annular layer. **-schließen** n, **-schließung** f ring formation (or closure), cyclization. **-schlitz** m circular slot. **-schluß** m ring formation (or closure), cyclization

rings·herum adv all around, on all sides

Ring·spaltung f ring cleavage. **-spannung** f ring tension. **-sprengung** f ring cleavage (or fission)

rings·um(her) adv all around, on all sides

ring·ungesättigt a cyclically unsaturated, containing an unsaturated ring. **R–untersuchung** f cooperative investigation. **-verbindung** f 1 (Chem.) cyclic (or ring) compound. 2 (Mach.) ring (or thimble) joint; ring fastening. **-verengung** f ring contraction. **-verkleinerung** f reduction in ring size. **-weite** f ring size. **-wulst** m (Geom.) torus, anchor ring. **-wurm** m ringed worm, annelid

Rinmann·grün n Rinmann's green

Rinne f (-n) channel, furrow; gutter; trough

rinnen* (rann, geronnen) vi 1 (s) run, flow; pour. 2 (h) leak. Cf GERINNEN

Rinnen·schlacke f (Metll.) spout slag

Rinn·harz n resin. **-stein** m 1 gutter. 2 sink

Rippe f (-n) 1 (Anat., Bot., & fig) rib. 2 strip, bar. 3 fin., flange. Cf GERIPPT

Rippen· ribbed; costal: **-fell** n (costal) pleura. **-fellentzündung** f pleurisy. **-glas** n ribbed glass. **-haut** f (costal) pleura. **-heizkörper** m ribbed (or finned) radiator. **-rohr** n ribbed (or finned) pipe (or tube). **-samt** m corduroy. **-speer** m cured ribs of pork. **-trichter** m fluted funnel. **r–versehen** a (Mach.) ribbed, finned

Rips m (-e) (Tex.) rep

Risigallum n realgar

Risiko n (.. ken) risk. **riskant** a risky. **riskieren** vt/i risk

Rispe f (-n) (Bot.) panicle; wild oat. **Rispen·gras** n meadow grass

riß tore, etc: past of REISSEN

Riß m (Risse) 1 ripping, tearing; breaking, rupture. 2 rip, tear; break; crack, split, fissure; scratch—**Risse bekommen** crack, get scratched, get fissured. 3 rift, breach. 4 (engineer's) drawing, plan. Cf REISSEN

Ris·säure f risic acid

Riß·bildung f formation of cracks, fissuring. **r–empfindlich** a liable to tear (crack, split, etc). **-fest** a crack-resistant, tear-resistant. **R–festigkeit** f resistance to cracking (tearing, splitting, etc). **r–frei** a free from cracks, etc, flawless

rissig a ripped, torn; cracked, fissured, split. **Rissigkeit** f ripped (torn, cracked etc) condition. **Rissig·werden** n cracking, fissuring

Riß·wesen n graphic representation. **-wunde** f laceration

ritt rode: past of REITEN

Ritt m (-e) (horseback) ride. **Ritter** m (-) knight

Rittersporn° m larkspur, delphinium. **-öl** n larkspur oil

rittlings adv astride

Ritz m (-e) 1 scratch. 2 = **Ritze** f (-n) crack, chink, fissure

Ritzel m (-) (Mach.) pinion

ritzen vt scratch; engrave, grave. **Ritzer** m (-) scratch (wound)

Ritzhärte° f scratch hardness. **-probe** f scratch hardness test. **-prüfer** m abrasive hardness tester, sclerometer. **-ver-**

fahren n scratch hardness test. **-zahl** f (scratch) hardness number

Ritz-probe f scratch test (or sample). **-versuch** m scratch test

Rizin n Ricin. *For related words see* RICIN . . .

Rk. abbv (Reaktion) reaction. *Pl:* **Rkk.**

rm abbv (Raummeter) cu. m. (cubic meter)

Robbe f (-n) (Zool.) seal; pinniped

Robben-leder n seal leather. **-öl** n, **-tran** m seal oil

roboten vi slave, toil. **Roboter** m (-) robot; drudge, automaton

Rocell-säure f rocellic acid

roch smelled: *past of* RIECHEN

Roche m (-n) (Fish) ray

Rochelle-salz n Rochelle salt (sodium potassium tartrate)

röcheln vi wheeze, breathe noisily

Rochen m (-) (Fish) ray

Rock m (-̈e) 1 skirt. 2 (men's) coat. 3 robe

roden vt 1 uproot, clear away. 2 dig up (potatoes, etc). 3 clear (land). 4 make arable. **Rodung** f (-en) 1 uprooting; digging (up); clearing. 2 cleared area

Rogen m (-) (fish) roe(s), spawn. **-stein** m oölite. **r-steinartig** a oölitic.

Roggen m rye. **-brot** n rye bread. **-kleber** m rye gluten. **-kleie** f rye bran. **-mehl** n rye flour. **-mutter** f ergot. **-öl** n rye oil. **-schrot** n coarse rye (for pumpernickel). **-stärke** f rye starch. **-stroh** n rye straw

roh m 1 raw. 2 unprocessed, crude. 3 rough; rude, coarse. 4 brutal. 5 gross (amount, etc)—**im R-en** in the raw (or rough)

Roh-alkohol m crude alcohol. **-analyse** f rough analysis. **-antimon** n crude antimony. **-arbeit** f (Metll.) ore smelting. **-asbest** m crude asbestos. **-aufbereitung** f preliminary preparation. **-baumwolle** f raw cotton. **-benzol** n crude benzene, benzol. **-bilanz** f rough balance. **-blei** n crude lead. **-blende** f (Min.) crude blende. **-block** m (Metll.) (raw) ingot. **-boden** m virgin soil. **-bramme** f slap ingot. **-brand** m (Ceram.) first firing. **-braunkohle** f crude lignite. **-brom** n crude bromine. **-dichte** f bulk (or apparent) density. **-eisen** n crude (or pig) iron

Roheisen-gans f pig of iron. **-guß** m pig-iron casting. **-mischer** m (Iron) hot-

metal mixer. **-pfanne** f hot-metal ladle. **-verfahren** n (Iron) pig process

Roheit f (-en) 1 crudity. 2 brutality. 3 brutal act

Roh- raw, crude, rough, gross: **-ertag** m gross yield; gross returns (or earnings). **-erz** n raw ore. **-erzeugnis** n raw product. **-faser** f crude fiber. **-faserbestimmung** f crude fiber determination. **-feinkohle** f raw fine coal. **-fell** n pelt; (Rubber) rough sheet of masticated rubber. **-fett** n crude (or raw) fat. **-film** m (Phot.) 1 blank film. 2 negative (film). **-filter** n coarse filter. **-formel** f empirical formula. **-frischen** n (Metll.) first refining. **-frischperiode** f (Iron) boiling stage; (Bessemer process) boil. **-frucht** f (genl) raw produce; (Brew.) unmalted grain. **-gang** m (Metll.) irregular operation (or working); cold working. **r-gar** a partly refined. **R-gas** n crude gas. **r-gegossen** a crude-cast. **R-geschmack** m raw taste. **-gewebe** n unbleached fabric. **-gewicht** n gross weight. **-gewinn** m gross profit(s). **-gift** n crude poison (or toxin). **-gips** m gypsum rock. **-glas** n crude (or rough) glass. **-glasur** f rough glaze. **-gold** n gold bullion. **-gummi** n 1 crude (or unvulcanized) rubber. 2 crude gum. **-gummimilch** f (rubber) latex. **-haut** f rawhide. **-humus** m virgin humus. **-kalk** m limestone (for lime burning). **-kautschuk** m or n crude (raw, unvulcanized) rubber. **-kohle** f rough (or run-of-the-mine) coal. **-koks** m crude (or impure) coke. **-kost** f raw food (diet). **-kresol** n crude cresol. **-kultur** f 1 (Biol.) impure culture. 2 (Agric.) cultivation of new land. **-kupfer** n crude copper. **-leder** n crust (or rough-tanned) leather. **-leinöl** n raw linseed oil

Rohling m (-e) 1 blank, unworked piece. 2 partly fabricated article; rough forging; rough wood. 3 brute

Roh-maß n rough size (quantity, measurement). **-material** n raw material. **-mehl** n (Cement) raw mix. **-messing** n crude brass. **-metall** n crude metal. **-naphtha** f crude naphtha. **-ofen** m ore furnace. **-öl** n crude oil. **-pappe** f raw paperboard. **-paraffin** n crude paraffin. **-petroleum** n crude oil (or petroleum). **-phosphat** n

(*Agric.*) rock phosphate. **-probe** *f* crude sample. **-produkt** *n* raw (*or* crude) product. **-protein** *n* raw (*or* crude) protein

Rohr *n* (-e) 1 pipe, tube; duct, canal; flue. 2 reed; cane, rattan, bamboo. 3 (gun) barrel. 4 (pipe)stem. 5 (baking) oven. **--ansatz** *m* tube (*or* pipe) attachment. **r–artig** *a* tube-like, reed-like, reedy. **R–bogen** *m* bend, elbow (in pipe or tube). **-bruch** *m* pipe burst. **-brunnen** *m* artesian well. **-bürste** *f* (test-)tube brush; flue brush

Röhrchen *n* (-) 1 small tube; tubule; capillary tube. 2 (drinking) straw. 3 (organ) reed

Röhre *f* (-n) 1 pipe, tube; duct, canal. 2 (electronic) tube: **Braunsche R.** cathode-ray tube. 3 (baking) oven

Röhren·anlage *f* tubing, piping. **r–artig** *a* tubular. **R–bürste** *f* (test-)tube brush. **-destillationsofen** *m* pipe still. **r–förmig** *a* tubular. **R–halter** *m* tube (*or* pipe) holder. **-kassie** *f* purging cassia. **-kessel** *m* tubular boiler. **-klemme** *f* tube (*or* pipe) clamp. **-kühler** *m* tubular radiator (condenser, cooler). **-libelle** *f* spirit level, air level. **-lot** *n* pipe solder. **-manna** *f* flake manna. **-nudeln** *fpl* macaroni. **-ofen** *m* tube furnace; pipe still. **-pulver** *n* (*Explo.*) perforated powder. **-streifen** *m* tube strip. **-struktur** *f* tubular structure. **-substanz** *f* (*Anat.*) medullary substance. **-system** *n* 1 piping, system of pipes. 2 (*Sugar*) calandria. **-träger** *m* tube (*or* pipe) support. **-voltmeter** *n* tube voltmeter. **-wachs** *n* petroleum ceresin. **-werk** *n* 1 tubing, piping. 2 tube-rolling mill. **-wischer** *m* tube brush. **-wulst** *m* tubular tore (*or* doughnut). **-zelle** *f* tubular cell; (*Bot.*) tracheid

rohr·förmig *a* tubular. **R–füllung** *f* tube filling, column packing. **-geschwür** *n* fistula

Röhricht *n* (-e) reed patch, reeds

röhrig *a* 1 tubular. 2 reedy

Rohr·kessel *m* tubular boiler. **-kopf** *m* (*Petrol.*) casing head. **-kopfbenzin, -kopfgasolin** *n* casing-head gasoline. **-krümmung** *f* tube (*or* pipe) bend(ing). **-leger** *m* pipefitter. **-leitung** *f* 1 pipe, tube; pipeline. 2 (system of) pipes. **-muffe** *f* pipe sleeve (*or* joint). **-mühle** *f*

tube mill. **-netz** *n* (network of) pipes (*or* tubes) **-palme** *f* rattan. **-post** *f* pneumatic mail carrier system. **-pulver** *n* tubular powder. **-richtmaschine** *f* pipe (*or* tube) straightening machine. **-saft** *m* (sugar) cane juice. **-schelle** *f* 1 pipe clamp. 2 sleeve. **-schellenverbindung** *f* pipe sleeve joint. **-schlange** *f* coiled (*or* spiral) tube. **-seele** *f* bore of a pipe (*or* tube). **-spirale** *f* coiled (*or* spiral) tube. **-stock** *m* cane; bamboo stick. **-stuhl** *m* cane chair. **-stutzen** *m* pipe sleeve; pipe muzzle. **-sumpf** *m* reed swamp. **-verbindung** *f* pipe (*or* tube) joint. **-verschraubung** *f* (threaded) pipe joint. **-walzwerk** *n* tube rolling mill. **-wand(ung)** *f* pipe (*or* tube) wall. **-weite** *f* bore of a tube (*or* pipe). **-werk** *n* 1 tubing, piping. 2 tube mill. **-zange** *f* pipe wrench. **-ziehbank** *f* tube drawing bench. **-zucker** *m* cane sugar, sucrose

Rohrzucker·saft *m* (sugar) cane sap (syrup, juice). **-verbindung** *f* cane sugar compound, sucrate

Roh· raw, crude, coarse: **-saft** *m* crude juice (sap, syrup). **-salpeter** *m* crude saltpeter. **-säure** *f* crude acid. **-schlacke** *f* raw (*or* ore) slag. **-schliff** *m* (*Paper*) semipulp. **-schmelzen** *n* raw (*or* ore) smelting. **-schwefel** *m* crude (*or* native) sulfur. **-seide** *f* raw silk. **-sieb** *n* coarse sieve (screen, filter). **-silber** *n* crude silver, silver bullion. **-soda** *f* crude soda; (Leblanc process) black ash. **-sole** *f* raw (*or* crude) brine. **-spaltbenzin** *n* cracked gasoline. **-spießglanzerz** *n* (*Min.*) kermesite. **-spiritus** *m* raw (*or* crude) spirits. **-stahl** *m* raw (crude, natural) steel. **-stahlblock** *m* ingot steel. **-stahleisen** *n* pig iron for steel making. **-stein** *m* (*Copper*) coarse metal. **-steinschlacke** *f* coarse-metal slag. **-stoff** *m* raw material. **-stück** *n* blank (to be worked or processed). **-stückkohle** *f* run-of-the-mine coal. **-sulfat** *n* crude sulfate. **-teer** *m* crude (*or* raw) tar. **-ton** *m* 1 unburned clay. 2 beige, ecru. **-übersetzung** *f* rough translation. **-wachs** *n* crude wax. **-wasser** *n* raw (*or* untreated) water. **r–weiß** *a* (*Tex.*) raw, gray, unbleached. **R–wichte** *f* apparent (*or* bulk) density. **-wismut** *n* crude bismuth. **-wolle** f raw wool. **-zink** n crude zinc, spelter. **-zucker** m raw (or unrefined) sugar, muscovado. **-zustand** m raw (crude, unrefined) state

Rolladen m (= **Roll-laden**) roller (steel) shutter

Roll-bahn f 1 roller conveyor. 2 (airport) runway. **-bombe** f rotary autoclave

Rolle f (-n) 1 roll (of paper, etc); coil, reel, spool; scroll. 2 roller; wringer; caster; pulley; calender. 3 loop. 4 role, part: **das spielt keine R.** that plays no part, that doesn't matter

rollen 1 vt, vi (h & s), **sich r.** roll, rotate, curl. 2 vt wring; calender

rollen-artig a cylindrical. **R–förderer** m roller (rack) conveyor, rollerway. **r–förmig** a cylindrical. **R–kufe** f roller vat. **-lager** n roller bearing. **-zug** m block and tackle

Roller m (-) 1 roller (incl canary). 2 calender. 3 breaker (= wave). 4 motor scooter

Roll-faß n tumbling barrel, mixing drum. **-feld** n (airport) runway(s). **-flasche** f narrow-necked cylindrical bottle. **-gang** m roller conveyor

rollig a loose, crumbly (esp soil)

Roll-kalander m 1 (Tex., etc) roller calender. 2 (Paper) supercalender. **-maß** n (coiled) tape measure

Rollo n (-s) roller shade (or blind)

Roll-reibung f rolling friction. **-stein** m boulder. **-stuhl** m wheelchair. **-tisch** m gurney; trolley table. **-wagen** m dolly; rolling stretcher; platform truck. **-widerstand** m resistance to rolling; rolling friction

Roman m (-e) novel; fiction. **-kalk** m, **-zement** m Roman cement

Römer-kerze f Roman candle

römisch a Roman; **r–es Kamillenöl, R.-Kamillenöl** Roman (or English) chamomile oil; **r–er Kümmel** cumin; **R.-Kümmelöl** cumin oil; **r–e Minze** spearmint; **r–er Salat** romaine lettuce

rommeln vt tumble (in a tumbling barrel)

Ronde f (-n) 1 (Metal) round blank. 2 disc. 3 (Type) ronde. 4 patrol. **Ronden-schere** f disc shears

Rongalit-ätze f (Dye.) rongalite discharge

röntgen vt (pp geröntgt, first t silent) x-ray.

Röntgen n (-) roentgen (as a unit)

röntgen-amorph a (Cryst.) amorphous to X-rays. **R–analyse** f X-ray analysis. **-anlage** f X-ray laboratory. **-aufnahme** f X-ray photograph. **-beugung** f X-ray diffraction. **-bild** n X-ray photograph. **-bremsstrahlung** f X-radiation due to checking or impact. **-durchleuchtung** f (X-ray) transillumination, radioscopy, fluoroscopy. **-einheit** f roentgen (as a unit). **-einrichtung** f X-ray installation (or equipment). **-fluoreszenz** f X-ray fluorescence

röntgenisieren vt x-ray

Röntgen-kunde, -lehre f roentgenology. **-licht** n X-radiation, X-rays. **-messer** m roentgenometer

Röntgenogramm n (-e) X-ray diffraction diagram

Röntgen-prüfung f X-ray testing (or examination). **-röhre** f X-ray tube. **-spektralanalyse** f X-ray spectrum analysis. **-spektrum** n X-ray spectrum. **-strahl** m X-ray. **-strahlbündel** n X-ray beam. **r–strahlenundurchlässig** a impervious to X-rays, radiopaque. **-strahlung** f X-radiation. **-streuung** f X-ray scattering. **-untersuchung** f X-ray examination (or investigation)

rosa a (no endgs), **-farben, -farbig** a pink, rose(-colored). **R–färbung** f pink coloring (or coloration). **r–rot** a pink

rösch a 1 crisp, hard-baked (or -roasted). 2 brittle. 3 (Paper pulp) free. **Rösche** f (-n) 1 crispness; brittleness. 2 (Min.) watercourse. **röschen** vt 1 age, cure. 2 dig (a trench or canal). **Rösch-erz, -gewächs** n brittle silver ore, stephanite. **Röschheit** f crispness; brittleness

Rose f (-n) 1 rose. 2 (Med.) erysipelas

Rosen-essenz f attar of roses. **-farbe** f rose color. **r–farben, -farbig** a rose-colored, rosy. **R–geruch** m smell of roses, rose fragrance. **-gewächse** npl Rosaceae. **-holz** n rosewood. **-honig** m (Pharm.) honey of rose. **-kohl** m Brussels sprouts. **-konserve** f (Pharm.) confection of rose. **-kranz** m rosary; rose garland. **r–kranzförmig** a rosary-like, moniliform. **R–kupfer** n rose copper; (Min.) cuprite.

-**lack** m rose lake. -**lorbeer** m oleander; mountain rose. -**öl** n attar of roses, rose oil. -**quarz** n rose quartz. **r–rot** a rosered. **R–schwamm** m bedegar. -**spat** m rhodochrosite. -**stahl** m rose steel. -**stock** m rosebush. -**wasser** n rose water. -**zucht** f rose growing

Roseokobalt·salz n roseocobaltic (or pentammineaquocobalt) salt

Rosesches Metall Rose's metal. **Rose·tiegel** m Rose crucible

rosetten·artig a rosette-like. **R–bahn** f rosette orbit (or path). -**herd** m (Copper) refining hearth. -**kupfer** n 1 (Metll.) rosette copper. 2 (Min.) cuprite

rosettieren vi make rosette copper. **Rosettier·herd** m (copper) refining hearth

rosieren vt dye (sthg) pink (or rose). **Rosier·salz** n rose salt, tin composition (solution of tin in aqua regia)

rosig a rosy

Rosine f (-n) raisin

Rosmarin m rosemary. --**öl** n rosemary oil

Rosol·säure f rosolic acid

Roß n (Rosse or Rösser) horse, steed. --**bremse** f horsefly. -**fenchel** m (Pharm.) water fennel, horsebane. -**haar** n horsehair. -**haut** f horsehide. -**huf** m (Pharm.) coltsfoot. -**kammfett** n horse grease. -**kastanie** f horse chestnut. -**leder** n horsehide leather. -**schild** n (Lthr.) horse butt. -**schwefel** m horsehide brimstone (impure grayish sulfur). -**wurzel** f carline thistle (or root)

Rost m I 1 (Chem. & Bot.) rust; mildew. II (-e) 2 (oven) grate, grill, grid(iron); grating; screen. 3 (Metll.) roasting charge

Röst·abgang m weight loss on roasting cf RÖSTEN

Rost·anfressung f corrosion (of metals). -**angriff** m corrosive action (on metals)

Röst·anlage f roasting plant cf RÖSTEN

Rost·ansatz m rust deposit (or coating)

Röst·arbeit f roasting (process) cf RÖSTEN

Rost·behandlung f rust treatment, rustproofing. **r–beständig** a rust-resistant, corrosion-resistant. **R–beständigkeit** f resistance to rust (or corrosion)

Röst·betriebsdauer f roasting time. -**bett** n roasting bed. Cf RÖSTEN

Rost·bildner m rusting (or corroding)

agent. -**bildung** f rust formation

Röst·bitter n assamar. -**blende** f roasting blende. Cf RÖSTEN

Rost·brand m rust, smut, mildew. -**braten** m grilled steak. **r–braun** a rust brown

Röste f (-n) 1 roasting grid (or device), roaster. 2 roasting (process); (Metll.) calcining. 3 (Tex.: Flax) retting; (Place) rettery

rost·empfindlich a rust-sensitive. **R–empfindlichkeit** f sensitivity to rust

rosten vi (h,s) rust, get rusty

rösten vt 1 roast; calcine; torrefy. 2 (Tex.) ret, steep (flax, hemp). 3 fry; grill. 4 toast

Rost·entfernung f rust removal

Röster m (-) roaster; toaster; calciner. **Rösterei** f (-en) coffee roasting plant

Röst·erz n roasted (or calcined) ore. -**erzeugnis** n product of roasting (or calcining)

Rost·farbe f rust color. **r–farben, -farbig** a rust-colored. -**fest** a rust-resistant. **R–feuer** n grate fire. -**fläche** f grate surface. -**fleck** m rust spot. -**fleckenwasser** n rust stain remover. **r–fleckig** a rustspotted, rust-stained. **R–fraß** m (metal) corrosion. **r–frei** a rust-free; stainless (esp steel). **R–freiheit** f immunity to rust (or corrosion)

röst·frisch a freshly roasted. **R–gas** n gas from roasting

rost·gelb a rust-yellow, rusty yellow. **R–gelb** n iron buff (hydrous iron oxide)

Röst·gummi n dextrin. -**gut** n material to be roasted; roasted material

rost·hindernd a rust-preventing, anticorrosive

Röst·hütte f (Metll.) roasting plant

rostig a rusty; (of water) chalybeate. **Rostigkeit** f, **Rostig·sein** n rustiness. -**werden** n rusting; (Bot.) russeting

Röst·kaffee m roasted coffee. -**kartoffeln** fpl fried potatoes

Rost·kitt m iron-rust cement

Röst·kufe f retting vat. -**malz** n roasted (or black) malt

Rost·mittel n 1 rust preventive, antirust agent. 2 rusting (or corroding) agent. -**neigung** f tendency to rust

Röst·ofen m roasting oven (or furnace); calcining kiln; smelting furnace

Rost·öl n slushing oil. **-pilz** m rust fungus

Röst·posten m roasting charge. **-probe** f calcination assay (or test). **-produkt** n product of roasting (or calcining). **-prozeß** m roasting process; calcination. **-reaktionsarbeit** f (Min.) sulfatizing process. **-reduktionsarbeit** f (Ores) roast reduction process (oxidative roasting followed by reduction with coke)

rost·rein a rust-free

Röst·rohr n roasting (or open) tube (for qualitative tests)

rost·rot a rust-red, rust-colored

Röst·rückstand m residue from roasting. **-scherben** m roasting dish, scorifier

Rost·schicht f layer of rust; (Metll.) layer of raw matte

Röst·schlacke f slag from roasting. **-schmelzen** n (Metll.) roasting and smelting

Rost·schutz m protection against rust; rust prevention; anti-rusting agent. **r–schützend** a rust preventing

Rostschutz·farbe f antirust paint. **-grund** m antirust primer. **-lack** m antirust lacquer (or varnish). **-mittel** n antirust agent, rust preventive. **-öl** n slushing oil. **-verfahren** n rustproofing process. **-wirkung** f rust-preventive action (or effect)

rost·sicher a rustproof, stainless. **R–sicherheit** f immunity to rust(ing); stainlessness. **-stab** m grate bar

Röst·staub m dust of roasted ore

Rostung f rusting

Röstung f(-en) roasting; calcining; retting; frying; grilling; toasting. Cf RÖSTEN

Rostungs·versuch m rusting (or corrosion) test

Röst·verfahren n roasting process

rost·verhütend a rust-preventing, antirust, anticorrosive

Röst·verlust m loss on roasting. **-vorrichtung** f roasting device. **-wasser** n steeping water

Rost·widerstand m resistance to rusting

Röst·zuschlag m flux for roasting. **-zwiebeln** fpl fried onions

rot a red: **r–es Präzipitat** red precipitate (red mercuric oxide). **Rot** n red (color);

rouge cf BERLINER. **--algen** fpl red algae

Rota·messer m rotameter, gas meter

Rotang m (-s) rattan. **--palme** f rattan palm

Rotations·achse f axis of rotation. **-bewegung** f rotational (or rotary) motion. **-druck** m rotary printing. **-energie** f rotational energy. **-filter** n rotary drum filter. **-fläche** f surface of revolution. **r–frei** a rotation-free. **R–freiheit** f rotational freedom. **-körper** m solid of revolution. **-mischmaschine** f rotary mixer. **-niveau** n rotational (energy) level. **-pumpe** f rotary pump. **-quantenzahl** f rotational quantum number. **-schwingung** f rotation vibration. **-sinn** m direction of rotation. **r–symmetrisch** a axially symmetrical. **R–übergang** m (Spect.) rotational transition. **-zähler** m revolution counter. **-zustand** m rotational state (or energy level)

rotatorisch a rotational

Rot·auszug m (Color Phot.) red record. **-beize** f(Dye.) red mordant (or liquor). **r–blau** a reddish-blue. **R–bleierz** n, **-bleispat** m red lead ore, crocoite. **r–braun** a reddish-brown. **R–bruch** m (Metll.) red-shortness. **r–brüchig** a redshort (metal); rotten (wood). **R–buche** f red (or copper) beech. **r–bunt** a red and white

Röte f 1 redness, red (color, glow). 2 blush, flush. 3 (Bot., Dye.) madder

Rot·eiche f red oak

Roteisen·erz n red iron ore, (Min.) hematite. **-ocker** m red iron ocher, (Min.) earthy hematite. **-stein** m red iron ore, (Min.) hematite, bloodstone

Rötel m(-) 1 red dye, red ocher, reddle. 2 red chalk. **--erde** f red ocher, (Min.) Adamic earth. **Röteln** pl German measles, rubella

Rötel·stift m (stick of red) chalk. **-ton** m brownish-red tone

rot·empfindlich a (Phot.) red-sensitive

röten 1 vt & sich r. redden. 2: sich r. blush. **rötend** p.a (oft) rubefacient

Rot·erde f red earth. **-erle** f(Bot.) red alder. **-esche** f red ash (tree). **r–farbig** a redcolored. **R–färbung** f red color(ation), red dyeing. **-fäule** f red (or heart) rot. **-feuer**

n (Pyro.) red fire. **r–gar** *a* tanned (to a russet color). **-gebrannt** *a* red-roasted, red-burned. **-gefärbt** *a* red-dyed, dyed red. **-gelb** *a* reddish-yellow, orange

rot·gerben *vt* bark-tan

Rot·gerber *m* bark tanner. **-gerberei** *f* bark tanning; bark tannery. **-gleßer** *m* brazier. **-gießerei** *f* braziery. **-gilderz** *n* red silver ore. **-glas** *n* red glassy arsenic monosulfide

rot·glühen *vt* heat red-hot, heat to redness. **rot·glühend** *a* red-hot

Rot·glühhärte *f* *(Metll.)* red-hardness. **-glühhitze, -glut** *f* red heat. **-gluthärte** *f* *(Metll.)* red-hardness. **-gold** *n* red gold (gold-copper alloy). **-gülden(erz), -gül-tigerz** *n* red silver ore—*(Min.):* **dunkles R.** pyrargyrite; **fahles R.** miargyrite; **lichtes R.** proustite. **-guß** *m* red casting brass, gunmetal. **-heizen** *n* heating to redness. **-hitze** *f* red heat. **-holz** *n* red-wood, Brazil wood

rotieren *vi* rotate. **rotierend** *p.a (oft)* rotary. **Rotier·ofen** *m* rotating furnace (or kiln)

Rot·kali *n* red prussiate of potash, potassium ferricyanide. **-kiefer** *f* red pine. **-klee** *m* red clover. **-kohl** *m* red cabbage. **-kohle** *f* red charcoal. **-kraut** *n* red cabbage. **-kupfer(erz)** *n* red copper ore, *(Min.)* cuprite. **-lauf** *m* erysipelas

rötlich *a* reddish. **-braun** *a* reddish-brown. **-gelb** *a* reddish-yellow, orange

Rot·liegende *n (adj endgs) (Geol.)* lower Permian sandstones and shales (in Germany). **r–machend** *p.a* rubefacient. **R–messing** *n* red brass. **-nickelkies** *m (Min.)* niccolite, kupfernickel. **-ocker** *m* red ocher. **-öl** *n* red oil. **-oxid** *n* red iron oxide. **-rauschgelb** *n* realgar. **-reserve** *f (Dye.)* red resist. **-salz** *n* crude sodium acetate (from pyroligneous acid). **-säure** *f* erythric acid (erythrin). **-schönung** *f (Wine)* red-clearing (treatment with iron oxide to remove arsenic). **-silber(erz)** *n* red silver ore. **-spat** *m (Min.)* rhodonite. **-spießglanz** *m*, **-spießglanzerz** *n (Min.)* kermesite. **-stein** *m* 1 red ocher, reddle. 2 *(Min.)* rhodonite. 3 red brick (or tile). **-stich** *m* red cast (or tinge). **r–stichig** *a*

red-tinged. **R–stift** *m* red pencil (or crayon). **-tanne** *f* Norway spruce

Rotte *f (-n)* 1 *(Flax)* retting. 2 troop, gang, mob

rotten, rötten *vt* ret (flax)

Rot·tombak *m* red tombac. **-ton** *m (Color)* red shade

Rötung *f* reddening; red tint, redness

Rot·verschiebung *f (Spect.)* red shift. **r–verschoben** *a* red-shifted. **-warm** *a* red-hot. **R–warmhärte** *f* red-hardness. **-warmhitze** *f* red heat. **-wein** *m* red wine. **-weinfarbe** *f* wine red (color), claret

Rotz *m* 1 glanders. 2 *(vulgar)* snot

Rotz·inkerz *n* red zinc ore, *(Min.)* zincite. **Rotz·krankheit** *f* glanders. **-nase** *f (Paint)* run

Rouleau *n (-s)* window shade. **Rouleaux-druck** *m (Tex.)* roller printing

Roulette *f(-n)* 1 roulette. 2 *(Mach.)* horizontal ball(-and-race) mill. **·küpe** *f* continuous dyeing vat

Routine *f(-n)* 1 routine. 2 experience, practical skill. **routiniert** *a* practiced, skilled; wily, crafty

R·P. *abbv* (Reichs-Patent) Government Patent (of pre-1945 German Reich)

R·Säure *f* R-acid

R.T. *abbv* (Raumtemperatur) room temp.

Rtl. *abbv* (Raumteil) part by vol.

Rübe *f (-n) (Bot., Agric.)* 1 *(genl)* turnip, beet; *(specif):* **gelbe R.** carrot; **rote R.** (red) beet; **schwedische R.** rutabaga; **weiße R.** sugar beet. 2 rape

Rubean·wasserstoff *m* rubeanic acid

Rübel·bronze *f* Rubel bronze

Rüben· turnip, beet: **-aaskäfer** *m* beet carrion beetle. **-asche** *f* beet ashes. **-beizung** *f* beet seed disinfection. **-brei** *m* beet pulp. **-breiapparat** *m* (sugar) beet pulper. **-breimaschine** *f* beet rasp. **-brennerei** *f* beet molassses distillery. **-essig** *m* beet vinegar. **-fliege** *f* marigold (or beet) fly. **r–förmig** *a* turnip-shaped. **R–harzsäure** *f* resin acid of beets, beet sapogenin. **-kohl** *m* turnip, rutabaga. **-melasse** *f* beet molasses. **-müdigkeit** *f (Agric.)* beet sickness. **-pottasche** *f* potash from beet molasses. **-pulpe** *f* beet pulp. **-rohzucker** *m* raw beet sugar. **-rost** *m (Agric.)* beet rust.

-saft *m* beet juice; black molasses. **-samen** *m* beet (turnip, carrot, rape) seed. **-schlempe** *f* beet vinasse. **-schnitzel** *npl* beet slices (*or* chips), cassettes; dried beet pulp. **-sirup** *m* beet sirup. **-spiritus** *m* beet spirit(s) (from beet molasses). **-zucker** *m* beet sugar. **-zünsler** *m* meadow moth, webworm

Ruberythrin·säure *f* ruberythric acid

Rubidium·alaun *m* rubidium alum. **-jodid** *n* rubidium iodide. **-oxidhydrat** *n* rubidium hydroxide. **-platinchlorid** *n* rubidium chloroplatinate

Rubin *m* (-e) ruby. **--balas** *m* (*Min.*) ruby spinel. **-blende** *f* (*Min.*) pyrargyrite. **-farbe** *f* ruby color. **r--farben, -farbig** *a* ruby-colored. **R--fluß** *m*, **-glas** *n* ruby glass. **-glimmer** *m* (*Min.*) lepidocrocite. **-granat** *m* (*Min.*) rock ruby. **r--rot** *a* ruby-red. **R--schwefel** *m* ruby sulfur (realgar). **-spinell** *m* (*Min.*) ruby spinel

Rüb·öl *n* rape(seed) (*or* colza) oil

Rüböl·faktis *n* rape oil factice. **-kuchen** *m* rapeseed cake

Rubrik *f* (-en) 1 column (of print, etc). 2 heading. 3 rubric, class, category. **rubrizieren** *vt* 1 file, classify (under a heading). 2 arrange in columns. 3 tabulate. 4 provide with headings

Rüb·samen *m* turnip (beet, carrot, rape) seed *cf* RÜBE. **-samenöl** *n* rapeseed (*or* colza) oil

Rübsen *m* (-) rapeseed. **--öl** *n* rapeseed (*or* colza) oil

ruchbar *a:* **r. werden** become known

Ruch·gras *n* (*Bot.*) sweet vernal grass

ruchlos *a* vicious, wicked

Ruck *m* (-e) quick (*or* sudden) movement, jerk, jolt, tug; shift: **in einem R.** in one quick movement; **mit einem R.** suddenly; **einer Sache einen R. geben** nudge a matter forward

rück· *pfx* 1 back, rear, reverse, return, retro-, re-. 2 (*with verbs; these occur only in inf & pp forms*) return, retro-, re-, back *cf* ZURÜCK. **R--ansicht** *f* rear view

ruck·artig *a* jerky, jolting; *adv* oft in fits and starts

Rück·bau *m* 1 dismantling. 2 (*Min.*) retreating system. **-bewegung** *f* back

(-ward) motion; return travel; receding motion

rück·bilden: sich r. (*inf & pp only*) 1 re(tro)gress; degenerate, atrophy; abate. 2 recover; be re-formed, revert. **R--bildung** *f* re(tro)gression; degeneration, atrophy; abatement; recovery; reversion, re-formation. **-blende** *f* flashback. **-blick** *m* backward glance, retrospect, review. **-destillation** *f* redistillation. **r--diffundieren** *vt* (*inf & pp only*) rediffuse, diffuse back. **R--diffusion** *f* back diffusion. **-druck** *m* back pressure

rucken *vi* jerk, jolt; (+ **an**) tug (at)

rücken 1 *vt, vi*(h) (+ **an**) move, shift. 2 *vi*(s) move (over), shift; (+ **mit**) move (a chesspiece); (*Mil.*) march

Rücken *m* (-) (*Anat. & fig*) back; loin; (*fig oft*) spine, ridge; (*in compds oft*) dorsal, spinal: **-ansicht** *f* rear view. **-lage** *f* dorsal (*or* supine) position. **-mark** *n* spinal cord. **-säule** *f* spinal column. **-schmerz** *m* back pain, backache. **-seite** *f* back, rear, reverse (side). **-stärkung** *f* moral support. **-stück** *n* back piece; sirloin

Rück·entionisierung *f* (*Water*) reverse deionization. **-entwicklung** *f* 1 retrogression, etc. 2 recovery, etc = RÜCKBILDUNG

Rücken·weh *n* lumbago, backache. **-wind** *m* tailwind. **-wirbel** *m* dorsal vertebra

Rück·erinnerung *f* reminiscence. **-erstattung** *f* return, repayment, restitution. **-fahrt** *f* return (trip). **-fall** *m* relapse, reversion. **-fallfieber** *n* relapsing fever. **r--feinerung** *f* recovery, regeneration. **r--fließend** *a* back(ward)-flowing, reflux. **R--fluß** *m* back (*or* return) flow, reflux. **-fluß(luft)kühler** *m* reflux (air) condenser. **-führung** *f* return; recovery; feedback. **-gabe** *f* return, giving back. **-gang** *m* 1 decline; recession. 2 (*Mach.*) back stroke. **r--gängig** *a* 1 declining, receding, retrogressive. 2 (*Mach.*) cancel, annul, rescind. **R--gewinnung** *f* recovery, reclamation. **-glühung** *f* (*Metll.*) drawing, tempering. **-grat** *n* spine, backbone; *in compd:* **R--grats·** *oft* spinal. **-griff** *m* (+ **auf**) resorting, recourse (to): **der R. auf die Kohle** the recourse to coal (for energy). **-gut** *n* residue; recycled material.

-halt *m* 1 backing, support. 2 reserve, restraint. **r–haltlos** *a* unreserved, *adv* without restraint. **R–infektion** *f* reinfection. **-kehr** *f* return, homecoming

Rückkehr·moment *n* restoring moment (*or* torque). **-punkt** *m* 1 point of return. 2 (*Math.*) cusp

Rück·kohlung *f* recarbonization, recarburization; recharring. **-kopplung** *f* (*Elec.*) feedback; regenerative circuit. **-kristallisation** *f* recrystallization

Rückkühl·anlage *f* recooling plant. **-turm** *m* recooling tower. **-wasser** *n* condensing water

Rück·kunft *f* return, homecoming. **-kuppelung** *f* (*Elec.*) reaction, feedback—**mit R.** heterodyne. **-lage** *f* reserves. **-lauf** *m* 1 (*incl Distil.*) reflux. 2 (*Mach.*) return motion (travel, stroke). 3 rewind. 4 (*Selection Dials*) homing. 5 (*Oscilloscopes*) retrace, flyback. 6 (*Mil.*) recoil. 7 (*Astron.*) retrograde motion. 8 tailings, machine rejects. **r–laufend** *a* returning; retrogressive, retrograde; declining, recessive. **R–laufgut** *n* recycled material. **r–läufig** *a* retrogressive, retrograde; declining, recessive

Rücklauf·kondensator, -kühler, -verdichter *m* reflux condenser. **-verhältnis** *n* reflux ratio

Rück·leiter *m* (*Elec.*) return conductor (*or* wire). **-leitung** *f* return line; recycling. **r–leuchtend** *a* reflecting. **R–leuchtwirkung** *f* reflecting effect. **-licht** *n* taillight

rücklings *adv* backward; from behind

Rück·mischung *f* backmixing. **-oxidation** *f* reoxidation. **-phosphorung** *f* (*Metll.*) rephosphorization. **-porto** *n* return postage. **-prall** *m* rebound, recoil. **-prallelastizität** *f* resilience. **-reaktion** *f* back (*or* reverse) reaction. **-saugung** *f* back (*or* reverse) suction, sucking back. **-schau** *f* review; retrospect. **-schlag** *m* 1 reverse, setback; relapse. 2 atavism. 3 return stroke; rebound, recoil. 4 reaction; reverberation. 5 striking back (of gas flames)

Rückschlag·hemmung *f* prevention of (*or* device to prevent) striking back *cf* Rückschlag 5. **-ventil** *n* check valve

Rück·schluß *m* 1 (*oft* + **auf**) conclusion, inference (as to). 2 (*Elec.*) ground. **r–schreitend** *a* retrogressive. **R–schritt** *m* step backward; retrogression; relapse; recession. **-schwefelung** *f* resulfurization. **-seite** *f* reverse (side), back; wrong side. **-sendung** *f* return (by mail)

Rücksicht *f* (-en) consideration, regard, respect: **aus R. auf . . .** out of consideration for . . .; **ohne R. auf . . .** without regard for (*or* consideration of); **R. nehmen auf . . .** have consideration for, take into consideration . . .; **mit R. auf . . .** with respect to . . .; **aus wirtschaftlichen R–en** for economic reasons. **rücksichtlich** *prep* (+ *genit*) considering, in consideration of. **Rücksicht·nahme** *f* consideration: **unter R. auf . . .** considering . . ., with consideration of . . . **rücksichts·los** *a* inconsiderate, callous, ruthless. **-voll** *a* considerate

Rück·sprache *f* consultation, discussion: **R. halten** (*or* **nehmen**) **mit . . .** consult (with) . . . **-sprung** *m* rebound. **-sprunghärte** *f* rebound hardness

Rück·stand *m* 1 residue, sediment; refuse, dregs. 2 (*usu pl*) arrears, backlog: **Rückstände haben, im R. sein** be in arrears, be behind; **in R. geraten** fall behind. **-standbrenner** *m* residue burner. **r–ständig** *a* 1 residual. 2 in arrears; overdue. 3 antiquated. 4 backward

Rückstands·erzeugnis *n* residual product. **-gewicht** *n* weight of residue. **-untersuchung** *f* (*Pesticides*) residue test

Rück·stein *m* (*Metll.*) back stone; crucible bottom. **rück·stellen** *vt* (*inf & pp only*) set back (timepiece, etc); defer, restore; backspace

Rückstell·feder *f* restoring spring. **-kraft** *f* restoring force, resiliency

Rück·stellung *f* resettling; restoration; deferment; reserve(s). **-stoß** *m* 1 recoil, rebound. 2 reaction. 3 repulsion. 4 back (*or* return) stroke. 5 jet propulsion

Rückstoß·atom *n* recoil atom. **-bahn** *f* recoil path (*or* track). **-elektron** *n* recoil electron

Rück·strahl *m* reflected ray. **rück·**

strahlen *vt/i* (*inf & pp only*) reflect. R–strahler *m* reflector. -strahlung *f* reflection. -strahlungsmesser *m* reflectometer. -strahlverfahren *n* reflection method. -streuung *f* back-scattering (of rays). -strom *m* reflux, return flow; (*incl Elec.*) return (*or* reverse) current. -stromgrad *m* reflux ratio, degree of reflux. -strömung *f* backflow, reflux. -stromventil *n* non-return valve. -sturz *m* backward fall (*or* rush). -taste *f* backspace key; rewind key. -titration *f* back-titration. rück·titrieren *vt/i* (*inf & pp only*) back-titrate. -treibend *a* restoring (force, etc). R–tritt *m* 1 withdrawal. 2 resignation; retirement. 3 retrogression. 4 backpedaling. 5 = -trittbremse *f* backpedaling (*or* coaster) brake. -umwandlung, -verwandlung *f* reconversion, retransformation. -wand *f* back wall. r–wärtig *a* rear, back (door, etc). -wärts *adv* 1 backward. 2 (in the) rear, in back: nach/von r. toward/from the rear Rückwärts· backward, reverse, retro-: -bewegung *f* backward (reverse, retrograde) motion; retrogression. -drehung *f* reverse rotation. -gang *m* reverse (gear); reverse motion. rückwärts·gehen* *vi*(s) deteriorate, go downhill. R–substitution *f* reversal of substitution (replacement of a substituent by hydrogen)
Rück·weg *m* (the) way back, return (path) *cf* HINWEG
ruck·weise *adv* in fits and starts, intermittently *cf* RUCK
rück·werfend *a* reflecting. rück·wirken *vi* (*inf & pp only*) react, have repercussions. -wirkend *p.a* retroactive, retrospective; reactive. R–wirkung *f* reaction, repercussion; reactance; retroactive effect. -wurf *m* throwing back; reflection. -zahlung *f* repayment; redemption. -zug *m* retreat, withdrawal. -zugfeder *f* return spring. -zündung *f* arc(ing)-back, backfiring
Rüde *m* (-n, -n) (male) wolf (fox, dog)
Rudel *n* (-) 1 herd, pack; crowd, gang. 2 stirring pole; oar
Ruder *n* (-) 1 oar, paddle. 2 rudder, helm; (*Aero.*) controls. 3 (*Zool.*) webbed foot.

rudern *vt/i* row, paddle
Ruf *m* (-e) 1 call (*incl Tlph*); cry, shout. 2 reputation. 3 telephone number. rufen* (rief, gerufen) *vt/i* call; *vi* call out, cry (out). rufen·lassen* *vt* send for
Ruf·frequenz *f* audiofrequency. -nummer *f* telephone number
Rufigallus·säure *f* rufigallic acid
Ruh· (= RUHE·): -bütte *f* (*Brew.*) storage vat, stock tub
Ruhe *f* 1 rest. 2 pause, break. 3 peace (and quiet); stillness, silence—R. geben be quiet; in R. lassen leave alone, not bother. --energie *f* rest energy. -lage *f* position of rest. r–los *a* restless. R–masse *f* rest mass
ruhen *vi* 1 rest. 2 be idle, be halted, be at a standstill. 3 lie fallow. 4 (+ auf) rest, be based (*or* supported) (on); (+ in, *oft*) lie (in). ruhend *p.a* still; stationary, motionless; latent; stagnant, static
Ruhe·pause *f* (rest) break. -pulver *n* sedative powder. -punkt *m* point of rest; pause; fulcrum; center of gravity. -schüttung *f* (*Fluidization*) fixed bed. -spannung *f* (*Elec.*) steady potential; voltage on open circuit. -stadium *n* resting stage; encystment (of insects). -stand *m* retirement. -stellung *f* rest position. -stoffwechsel *m* resting metabolism. -strom *m* (*Elec.*) steady (continuous, closed-circuit) current. -system *n* static system. -umsatz *m* (*Physiol.*) resting metabolism. -winkel *m* angle of repose. -zustand *m* state of rest (*or* equilibrium)
ruhig *a* 1 (*genl*) still. 2 at rest, stationary, static. 3 quiet, silent. 4 calm, tranquil, peaceful. 5 even, smooth, unruffled, unbroken: r–er Glanz smooth luster; r–e Färbung solid dyeing. 6 steady. 7 (*Molten Metal*) dead. 8 (*adv oft*) safely, without worry (*or* hesitation). ruhig·stellen *vt* (*Med.*) immobilize, put in a cast
Ruhm *m* fame, renown, glory
Ruh·masse *f* rest mass, static mass
rühmen 1 *vt* praise, laud. 2: sich r. be proud; boast. rühmlich *a* praiseworthy, glorious
Ruhr *f* 1 dysentery. 2 (*Geog.*) Ruhr (river, industrial region)
Rühr· stirring, *cf* RÜHREN: -apparat *m* stir-

ring apparatus, agitator, (Ceram.) blunger. **-arm** m stirring arm, stirrer, agitator. **-äscher** m lime pit with agitator. **-aufsatz** m stirring attachment

Ruhr·bazillen mpl dysentery bacilli

Rühr· stirring, cf RÜHREN: **-blei** n (Metll.) first lead removed in the Carinthian process. **-druckgefäß** n stirred pressure vessel. **-ei** n scrambled egg(s). **-eisen** n iron stirrer (poker, mixer). **-einrichtung** f agitating (stirring, mixing) device

rühren I vt & sich r. 1 move, stir. II vt 2 agitate, mix. 3 touch. 4 beat (a drum); play (a harp). 5 crutch (soap). III vi 6 (+ an) touch (on). 7 (+ von) be the result (or the work) (of)—**das rührt daher, daß.** . that is due to the fact that . . . IV: sich r. 8 be active, bestir oneself. 9: nichts rührt sich nothing is happening. **Rührer** m (-) stirrer, agitator, mixer

Rühr·faß n churn; (Metll.) dolly tub. **-form** f form (or shape) of stirrer. **-frischen** n (Metll.) puddling

Ruhr·gebiet n (Geog.) Ruhr district (coal & heavy industry)

Rühr·geschwindigkeit f stirring velocity. **-häkchen** n small stirring hook. **-haken** m stirring instrument, (specif) rake, rabble, poker, mixer. **-holz** n wooden stirrer, stirring stick, paddle

rührig a busy, active, energetic, go-getting

Rühr·kessel m stirrer vessel; (Soap) crutching pan. **-krücke** f stirring crutch, rake. **-kühler** m condenser with a stirrer. **-laugung** f stir-leaching, stir-buckling. **-löffel** m stirring ladle. **-maschine** f stirring machine, mechanical agitator (or mixer). **-milch** f buttermilk

Ruhr·mittel n remedy for dysentery

Rühr·motor m stirring (stirrer, agitator) motor. **-ofen** m (Metll.) rabble furnace

Ruhr·rinde f Simarouba bark

Rühr·schaufel f stirring paddle (or blade). **-scheit** n (wooden) stirrer, paddle. **-schüssel** f mixing bowl. **-spatel** m stirring paddle. **-stab** m stirring rod, rabble. **-stäbchen** n stirring bar; magnetic stirrer. **-stange** f stirring rod (or pole). **-stativ** n stirring stand. **-tank** m agitating (or mixing) tank

Rührung f emotion

Rühr·versuch m stirring test. **-vorrichtung** f stirring (or agitating) device. **-wärme** f heat of agitation. **-werk** n stirrer, agitator, stirring unit; (Ceram.) blunger

Ruhr·wurzel f tormentil root; ipecac

Rühr·zeit f stirring time, period of agitation

Ruinen·marmor m marble from ruins. **ruinieren** vt ruin

Ruja·holz n Venetian sumac wood, young fustic

Rüll·öl n cameline oil

Rumänien n Rumania

Rum·äther m rum ether (ethyl butyrate or ethyl formate). **-brennerei** f rum distillery. **-essenz** f rum essence (ethyl butyrate or ethyl formate). **-fabrik** f rum distillery

Rumpel·faß n tumbling barrel. **rumpeln** vi (h,s) rumble

Rumpf m (-̈e) trunk, body, torso; carcass; (Atom, Ion) core; (Aero.) fuselage; (Ship) hull. **-elektron** n inner (core, nonvalence) electron. **-fläche** f (Geol.) peneplain. **-isomerie** f (Nucl. Phys.) core isomerism. **-wirkung** f (atomic) core effect

Rum·sprit m double rum

rund I a 1 round; rotund. 2 whole, full; well-rounded, perfect, complete. 3 round (number). II adv 4 roundly, fully, completely. 5 (usu + Quant.) approximately, about. 6 esp in phr: r. heraus bluntly, frankly. 7 (+ um) (all) around: r. um die Uhr around the clock

Rund n 1 round (form), circle, sphere. 2: im R. in the surrounding area

Rund·blick m all-around view, panorama. **-brecher** m rotary crusher. **-brenner** m ring burner

Runde f (-n) 1 round; beat, patrol; lap: über die R–n kommen (mit) cope (with). 2 circle, group. 3: in der R. in the surrounding area

Rande f roundness, rotundity

runden 1 vt make round; complete, perfect. 2 vt & sich r. round out. 3: sich r. be completed, be perfected

rund·erhaben a convex. **-erneuert** a retreaded (tire). **R–faser** f round fiber. **-feile** f round file. **-filter** n round (or circular) filter. **-flasche** f round (or Florence) flask. **-frage** *f* inquiry, survey, questionnaire

Rundfunk° *m* radio. **--gerät** *n* radio set. **-sender** *m* radio broadcaster; radio station. **-sendung, -übertragung** *f* radio program; radio broadcasting

Rund·gang *m* round, tour, patrol; circuit

Rundheit *f* roundness, rotundity

rund·heraus *adv* bluntly, frankly. **-hohl** *a* concave. **R-holz** *n* logs, round wood. **-kolben** *m* round-bottomed flask. **r–körnig** *a* round-grained, globular. **R–kupfer** *n* copper billets. **-lauf** *m* rotation; circulation

rundlich *a* roundish, rounded; plump

Rund·ofen *m* round (*or* beehive) kiln. **-profil** *n* round section; curved profile. **-reise** *f* circular tour, circuit. **-schau** *f* review. **-schreiben** *n* circular (letter). **-sichtgerät** *n* (*Radar*) panoramic scanner. **-stahl** *m* round steel bar (*or* rod). **-stange** *f* round bar (*or* rod)

rund·um *adv* all around

Rundung *f* (-en) rounding; curve, arc; roundness

rund·weg *adv* bluntly, frankly. **R–wert** *m* round (*or* approximate) number

Runge *f* (-n) stake, stanchion

Runkel *f* (-n) feed turnip. **--fliege** *f* marigold (*or* beet) fly. **-rübe** *f* feed beet (*or* turnip)

Runkelrüben·melasse *f* beet molasses. **-zucker** *m* beet sugar

Runzel *f* (-n) wrinkle, crease, fold. **--bildung** *f* wrinkling, creasing. **runzelig** *a* wrinkled, creased. **Runzel·lack** *m* wrinkle varnish (*or* finish). **runzeln** *vt* wrinkle, crease. **runzlig** *a* wrinkled, creased

Ruperts·metall *n* Prince Rupert's metal

rupfen *vt* pull up (*or* out), pluck. **Rupf·festigkeit** *f* resistance to plucking (*or* pulling)

Ruprechts·kraut *n* herb Robert

Rusa·öl *f* ginger grass oil

Rüsche *f* (-n) ruche, frill(ing)

Ruß *m* (-e) **1** soot, carbon black, lampblack. **2** (*Agric.*) loose smut, rust. **--abscheidung** *f* depositing of soot. **-ansatz** *m* soot deposit (*or* covering). **r–artig** *a* soot-like, fuliginous. **R–bildung** *f* soot formation.

-braun *n* bister. **-brennerei** *f* lampblack manufacturing (*or* factory). **-dampf** *m* sooty vapor

Russe *m* (-n, -n) Russian

Rüssel *m* (-) trunk, proboscis (of elephants); snout

rußen *vt/i* smoke, soot; blacken. **rußend** *p.a* sooty, smoky

Ruß·farbe *f* bister. **-fleck(en)** *m* soot stain. **-flocke** *f* soot flake. **-hütte** *f* lampblack factory

rußig *a* sooty, fuliginous

Russin *f* (-nen) Russian (woman). **russisch** *a* Russian: **r–es Grün** Russian green; **r–es Glas** Muscovy glass, mica. **R–-grün** *n* Russian green

Ruß·kobalt *m* asbolite. **-kohle** *f* sooty coal, mineral charcoal; (*Geol.*) fusain. **-kreide** *f* black chalk

Rußland *n* Russia

ruß·schwarz *a* soot-black. **R–schwarz** *n* lampblack, carbon black. **-vorlage** *f* soot collector

rüsten **I** *vt* **1** propose, set up; rig. **2** scaffold (a building). **II** *vi* & **sich r.** (*usu* + **zu**) prepare (for), arm (oneself) (for)

Rüster *f* (-n) elm (tree). **--rinde** *f* elm bark

rüstig *a* active; robust, vigorous. **Rüstigkeit** *f* vigor, robustness

Rüstung *f* (-en) **1** armament, arms, military arsenal. **2** scaffolding. **3** armor

Rüst·zeug *n* **1** tools, implements, equipment (for a job). **2** (professional) know-how. **3** parts for a scaffolding

Rute *f* (-n) **1** twig, switch. **2** rod (*incl Meas.*, fishing, divining); wand. **4** (*Zool.*) brush, (bushy) tail; penis. **4** (*fig*) burden. **ruten·förmig** *a* rod-shaped, wand-shaped, virgate

Ruthen *n* ruthenium. **Rutheniak** *n* ruthenium ammine compound

Ruthenium·oxid *n* ruthenium oxide (*or specif* sesquioxide). **-oxydul** *n* ruthenium (mon)oxide. **-säure** *f* ruthenic acid

Ruthen·rot *n* ruthenium red

Rutsch *m* (-e) **1** skip, slide, skid. **2** landslide. **3** short trip. **4: in einem R.** in one go. **Rutsche** *f* (-n) chute, slide. **rutschen** *vi(s)* slip, skid, slide

rutsch·fest *a* nonskid, nonslip. **R–fläche** *f* (*Geol.*) slickenslide

rutschig *a* slippery

Rutsch·kupplung *f* friction coupling (*or* clutch). **r–los** *a* nonskid, nonslip. **R–pulver** *n* talcum powder. **r–sicher** *a* nonskid, anti-skid, nonslip

Rutschung *f* (-en) slip(ping), slide, sliding, skid(ding)

Rüttel·beton *m* vibrated concrete. **-bewegung** *f* shaking motion. **-gewicht** *n* bulk density (of loose materials after shaking)

rütteln 1 *vt/i* shake, jolt. **2** *vt* (*Agric.*) winnow; (*Concrete*) vibrate. **3** *vi* (*Birds*) hover. **4** *vi* (+ **an**) rattle at, shake, try to open (*or* topple)—**daran ist nicht zu r.** that must not be touched (*or* interfered with)

Rüttel·sieb *n* vibrating sieve. **-stoß** *m* jolt, jar. **-tisch** *m* vibrating table

Rüttler *m* (-) shaker, vibrator

Rüttlung *f* shaking, jolting, winnowing, vibration *cf* RÜTTELN

RV. *abbv* (Reduktionsvermögen) reducing power

S

s *abbv* **1** (Sekunde) second (of time). **2** (sächlich) neuter

s. *abbv* **1** (siehe) see. **2** (symmetrisch) symmetrical. **3** (schwer) with difficulty. **4** (sehr) very

S *abbv* **1** (Seite) p. (page). **2** (Säure) acid. **3** (Sankt) St. (Saint)

s.a. *abbv* (siehe auch) see also

Sa. *abbv* **1** (Summa) sum. **2** (Samstag) Sat. (Saturday)

S.-A. *abbv* (Sonder-Abdruck) (special) reprint, separate

Saal *m* (Säle) (assembly, meeting, lecture) hall, room

Saat *f* (-en) **1** sowing. **2** seed(s). **3** crops. **--beet** *n* seedbed. **-beizmittel** *n* seed disinfectant

Saaten·stand *m* state of the crops. **-standbericht** *m* crop report

Saat·feld *n* grain field. **-gut** *n* seed (grain). **-gutbeize** *f* **1** seed disinfection. **2** seed steep. **-gutkrankheit** *f* seed-borne disease. **-hefe** *f* seed (*or* starter) yeast. **-korn** *n* seed grain. **-pflanze** *f* crop plant. **-schote** *f* seed pod. **-wechsel** *m* crop rotation. **-zucht** *f* seed growing

Sabadill·samen *m* sabadilla seed. **-(samen)öl** *n* sabadilla oil

Säbel *m* (-) saber, sword. **--fisch** *m* swordfish. **s--förmig** *a* saber-shaped, ensiform. **S–kolben** *m* Anschütz distillation flask

Sabinen·säure *f* sabinenic acid

Sabiner·baum *m* savin

Sabinin·säure *f* sabinic acid

saccharimetrisch *a* saccharimetric. **saccharin·haltig** *a* saccharin-bearing. **S– säure** *f* saccharinic acid. **saccharoidisch** *a* saccharoid(al)

Sach·angabe *f* factual information, particulars *cf* SACHE

Sacharin *n* saccharin(e)

Sach·bearbeiter *m* (**für**) official in charge (of), expert (on). **s–bezogen, -dienlich** *a* relevant, pertinent

Sache *f* (-n) **1** (*usu pl*) thing(s), object(s), article(s); (*may also imply:* clothes, food, drink(s), possessions, property): **so eine S.** (*oft*) one of those things. **2** (*usu sg*) matter, affair; (legal) case: **das ist ihre S.** that is their affair; **in S—n ...** in matters (*or* in the case) of. . . **3** cause: **gemeinsame S. machen** make common cause. **4** subject, point: **zur S. kommen** get to the point; **bei der S. bleiben** stick to the point; **das gehört nicht** (*or* **tut nichts**) **zur S.** that is beside the point, that is irrelevant. **5** job, (line of) work: **seine S. verstehen** know one's job; (**das ist**) **S.!** (that's a) good job! **6** (*pl only*) kilometers per hour

Sach·gebiet *n* field (of endeavor), subject. **s–gemäß, -gerecht** *a* appropriate; relevant. **S–index** *m* subject index. **-katalog** *m* subject catalog. **-kenner** *m* expert, authority (on a subject). **-kenntnis, -kunde** *f* expertise, expert knowledge. **s–kundig** *a* expert. **S–kundige** *m,f* (*adj endgs*) expert, specialist. **-lage** *f* situation, state of affairs

sachlich *a* **1** relevant. **2** material. **3** functional, practical. **4** objective; pragmatic. **5** matter-of-fact, business-like. **6** *adv oft* essentially; by subjects. **Sachlichkeit** *f* relevance; practicality; objectivity; pragmatism; matter-of-factness

sächlich *a* neuter

Sach·register *n* subject index. **-schaden** *m* property damage

Sachsen *n* (*Geog.*) Saxony. **sächsisch** *a* Saxon: **s–es Blau** = **S--blau** Saxon blue, smalt; (*appx*) Prussian blue

sacht(e) *a* **1** gentle, easy. **2** soft. **3** gradual. **4** (*Silk*) scoured

Sach·verhalt *m* situation, state of affairs; circumstance(s). **s–verständig** *a* expert. **S–verständige** *m,f* (*adj endgs*) expert, specialist, authority (on a subject). **-verzeichnis** *n* subject index. **-walter** *m* spokesman; counsel, lawyer. **-wörterbuch** *n* encyclopedia. **-zwang** *m* real

necessity, given reality
Sack *m* (⁻e) 1 bag, sack; pouch. 2 pocket. 3
sac, cyst. **-band** *n* drawstring (of a bag *or*
sack)
Säckel *m* (-) purse, money bag
sacken 1 *vt* bag, put up in bags. 2 *vi*(s) sag,
sink, give way
Sack·filter *n* bag (*or* sack) filter. **s—förmig**
a sack-like, bag-like, pouch-like. **S—garn**
n twine. **-gasse** *f* dead end, blind alley.
-gerbung *f* (*Lthr.*) bag tannage. **-leinen**
n, **-leinwand** *f* sackcloth, burlap. **-pack-
maschine** *f* bagger, bag-filling machine
Sackung *f* (-en) 1 bagging. 2 sagging, sinking. *Cf* SACKEN
Sadebaum° *m* savin. **-öl** *n* savin oil
säen *vt/i* sow (the seeds of). **Säer** *m* (-) sower
Saffian *m*, **-leder** *n* morocco (leather)
Saf(f)lor *m* (-e) safflower. **-blüte** *f* safflower blossom. **-gelb** *n* safflower (dye).
-öl *n* safflower oil. **-rot** *n* safflower red,
cardamin
Safran *m* (-e) saffron. **s—ähnlich** *a*
saffron-like, saffrony. **S—farbe** *f* saffron
color. **s—farben, -farbig** *a* saffron-
colored. **S—farbstoff** *m* crocetin. **s—gelb**
a saffron (yellow). **-haltig** *a* saffron-
bearing. **S—öl** *n* saffron oil
Saft *m* (⁻e) juice; sap; fluid; gravy; syrup;
liquor. **-blau** *n* sap blue. **-braun** *n* sap
brown. **-fluß** *m* flow of juice (sap, fluid,
etc). **s—frisch** *a* green (wood). **S—futter** *n*
moisture-rich feed. **-grün** *n* sap green
saftig *a* 1 juicy, succulent; sappy; (*Mettl.*)
wet. 2 hefty, substantial, pithy. **Saf-
tigkeit** *f* juiciness, succulence; sappiness,
etc
Saft·kanal *m* lymph(atic) canal. **s—los** *a*
sapless, juiceless; dry; insipid. **S—
pflanze** *f* succulent plant. **-raum** *m* (*Bot.*)
cell cavity, vacuole. **s—reich** *a* succulent,
rich in juice (sap, moisture). **S—röhre** *f*
lymphatic vessel (*or* duct). **-zelle** *f*
lymph(atic) cell
Sagapen·gummi *n* sagapenum. **-harz** *n*
sagapenum resin
Sage *f* (-n) legend, myth; rumor
Säge *f* (-n) saw. **s—artig** *a* saw-like, serrated. **-gezähnt** *a* sawtoothed, serrate.
S—mehl *n* sawdust. **s—mehlartig** *a*
sawdust-like. **S—mühle** *f* saw mill

sagen *vtli* 1 say, tell; express, state. 2 (*esp* +
wollen) mean: **das will s.** that means;
das hat nichts zu s that doesn't matter
(*or* mean anything); **sage und schreibe**
no less (than). **Sagen** *n:* **das S. haben**
give the orders, have the last word
sägen vt saw, cut
Sägen·gewinde *n* (*Engg.*) buttress screw
thread
sagenhaft *a* legendary; fabulous, unbelievable
Säge·späne *mpl*, **-staub** *m* sawdust.
-wespe *f* sawfly. **-zahn** *m* sawtooth; (*pl
oft*) indentations
Sago·mehl *n* sago flour. **-stärke** *f* sago
(starch)
Sagrada·rinde *f* (*Pharm.*) cascara sagrada
sah saw: *past of* SEHEN
Sahlband° *n* selvage; wall (of a lode)
Sahne *f* (dairy) cream. **sahnen** *vt* skim (the
cream off). **Sahnen·käse** *m* cream
cheese. **sahnig** *a* creamy
Saidschützer Salz Epsom salt
saiger *a* (*Min.*) perpendicular = SEIGER.
saigern *vt* (*Mettl.*) liquate, etc = SEIGERN.
See also SEIGER.
Saison *f* (-s) season. **saisonal, saison-
·bedingt** *a* seasonal. **Saison·farbe** *f* season color, fashion shade. **s—gebunden** *a*
seasonal. **-gerecht** *a* appropriate for the
season. **S—schwankung** *f* seasonal
fluctuation
Saite *f* (-n) (musical) string, catgut —
andere S—en aufziehen sing a different tune; **neue S—n anschlagen** strike
a new chord
Saiten·draht *m* music wire. **-galvanome-
ter** *n* string galvanometer
säkular *a* secular
Salär *n* (-e) salary
Salat *m* (-e) 1 salad. 2 lettuce. 3 mess.
-kopf *m* head of lettuce. **-machen** *n*
(*Brew.*) doughing-in and keeping the
mash very cold. **-öl** *n* salad oil. **-sauce,
-soße** *f* salad dressing
Salband° *n* selvage; wall (of a lode)
Salbe *f* (-n) ointment, salve
Salbei *m,f* (*Spices*) sage. **s—grün** *a* sage
green. **S—öl** *n* sage oil: **spanisches S.** oil
of Spanish sage; **Muskateller S.** oil of
clary sage

salben *vt* 1 rub with ointment. 2 embalm. 3 annoint

salben·artig *a* ointment-like; unctuous. **S–grundlage** *f* ointment base. **-spatel** *m* salve (*or* ointment) spatula

salbig *a* salve-like; unctuous

Saldo *m* (-s *or* . . di *or* . . den) (*Com.*) balance

Säle rooms, halls: *pl of* SAAL

Salep·schleim *m* salep mucilage. **-wurzel** *f* salep (root)

Salicoyl·säure *f* salicylic acid

salicyliert *a* salicylated. **salicylig** *a* salicyloxy

salicyl·sauer *a* salicylate of. **S–säure** *f* salicylic acid. **-streupulver** *n* salicylated talc. **-sulfonsäure** *f* sulfosalicylic acid. **-talg** *m* salicylated tallow

Salicylur·säure *f* salicyluric acid

Salicyl·watte *f* salicylic cotton

Saline *f* (-n) salt works, saltern. **Salinen·wasser** *n* saline water. **salinisch** *a* saline

salisch *a* (*Geol., Petrog.*) salic

salivieren *vt/i* salivate

Salizin *n* salicin

Salizyl. . . = SALICYL. . .

Salm *m* (-e) 1 salmon. 2 rigmarole, long story

Salmiak *m* sal ammoniac, ammonium chloride. **-element** *n* (*Elec.*) Leclanché (*or* sal ammoniac) cell. **-geist** *m* aqueous ammonia, ammonia solution. **-kristall** *m* crystallized sal ammoniac. **-lakritze** *f* sal ammoniac-licorice tablet. **-lösung** *f* ammonium chloride solution. **-salz** *n* 1 sal ammoniac. 2 sal volatile (ammonium carbonate). **-spiritus** *m* ammonia (gas)

Salm·rot *n* salmon red

Salniter *m* saltpeter, potassium nitrate

Salomons·siegel *n* Solomon's seal

Salpeter *m* 1 saltpeter, niter. 2 (any) nitrate: **prismatischer S.** potassium nitrate; **flammender S.** ammonium nitrate. *Cf* CHILISALPETER

salpeter·artig *a* saltpeter-like, nitrous. **S–äther** *m* nitric ether (ethyl nitrate). **-bakterien** *npl* nitrogen-fixing bacteria. **-bildung** *f* nitrification. **-blumen** *fpl* niter efflorescence. **-dampf** *m* nitrous fumes. **-druse** *f* crystallized saltpeter. **-dunst** *m* nitrous fumes. **-erde** *f* nitrous

earth. **-erzeugung** *f* niter production; nitrification. **s–essigsauer** *a* nitrate and acetate of. **S–fraß** *m* corrosion by niter, damage caused by saltpeter. **-gas** *n* 1 nitrous oxide. 2 nitric oxide. **-geist** *m* spirit of niter (=nitric acid). **-grube** *f* saltpeter mine. **-gütemesser** *m* nitrometer. **-hafen** *m* niter pot. **s–haltig** *a* containing saltpeter, nitrous: **s–haltiger Höllenstein** (*Pharm.*) mitigated silver nitrate. **S–hütte** *f* saltpeter (*or* niter) works

salpeterig *a* nitrous = SALPETRIG

Salpeter·kalk *m* calcium nitrate (containing lime). **-luft** *f* 1 nitrogen. 2 nitrous air, nitric oxide. **-messer** *m* nitrometer. **-milchsäure** *f* lactic acid nitrate. **-naphtha** *f* ethyl nitrite. **-papier** *n* niter (*or* nitrous) paper. **-pflanze** *f* nitrophyte. **-plantage** *f* saltpeter plantation. **-probe** *f* 1 saltpeter (*or* nitrate) test. 2 saltpeter sample. **-salz** *n* nitrate salt. **s–salzsauer** *a* nitromuriate of, nitrochloride of. **S–salzsäure** *f* nitrohydrochloric (*or* nitromuriatic) acid, aqua regia. **s–sauer** *a* nitrate of: **saltpetersaures Natron** sodium nitrate; **s—es Zinnoxid** *n* stannic nitrate; **s—es Zinnoxydul** *n* stannous nitrate

Salpetersäure° *f* nitric acid. **-anhydrid** *n* nitric anhydride, nitrogen pentoxide. **-äther** *m* nitric ether, ethyl nitrate. **-bad** *n* nitric acid bath. **-dampf** *m* nitric acid vapor (*or* fumes). **s–haltig** *a* containing nitric acid. **S–salz** *n* salt of nitric acid, nitrate

Salpeter·schaum *m* wall saltpeter, calcium nitrate efflorescence. **s–schwefelsauer** *a* nitrosulfate of. **S–schwefelsäure** *f* nitrosulfuric acid; nitrosylsulfuric acid, nitrosyl hydrogen sulfate; chamber crystals. **-siederei** *f* 1 saltpeter works. 2 saltpeter manufacture. **-stärkemehl** *n* nitrated starch. **-stoff** *m* nitrogen. **-strauch** *m* niter bush

Salpeterung *f* (-en) nitrification

Salpeter·verbindung *f* nitrate. **-waage** *f* nitrometer

salpetrig *a* nitrous: **s—e Säure** *f* nitrous acid; **s—e Schwefelsäure** *f* nitrosylsulfuric acid. **S–äther** *m* ethyl nitrite. **s–sauer** *a* nitrous, nitrite of: **s–saures Natrium** sodium nitrite

Salpetrigsäure° *f* nitrous acid. **--anhydrid** *n* nitrous anhydride, nitrogen trioxide. **-äther** *m* nitrous ether, ethyl nitrite
salpetrisch *a* nitrous
Saltation *f* (-en) mutation
Salve *f* (-n) salvo, volley
Sal·weide *f* sallow, broad-leaved willow
Salz *n* (-e) salt; *specif oft* -ate, -ite, e.g. **schwefelsaures S.** sulfate *cf* ·SAUER *sfx*. **--ablagerung** *f* salt deposit. **s–ähnlich** *a* salt-like, saline. **S–anteil** *m* salt content, percentage of salt. **s–arm** *a* low salt. **-artig** *a* salt-like, saline, haloid. **S–äther** *m* muriatic ether (*former name of ethyl chloride*). **-ausblühung** *f* efflorescence of salt. **-bad** *n* salt bath. **-badofen** *m* salt bath furnace. **-bedarf** *m* salt requirement. **-beize** *f* (*Lthr.*) salt dressing. **-bergwerk** *n* salt mine. **s–bildend** *a* salt-forming. **S–bildner** *m* salt-forming substance, halogen. **-bildung** *f* salt formation, salification. **s–bildungsfähig** *a* salifiable. **-bildungsunfähig** *a* nonsalifiable. **S–blumen** *fpl* efflorescence of salt. **-boden** *m* saline soil. **-brücke** *f* salt bridge. **-brühe** *f* brine, pickle. **-brunnen** *m* saline spring
Salzburger Vitriol *n* mixed (iron and copper) vitriol
Salz·chemie *f* chemistry of salts, halochemistry. **-decke** *f* salt-impregnated quilted cover (to absorb toxic gases). **-dom** *m* (*Geol.*) salt dome. **-drüse** *f* salt-secreting gland
salzen *vt* (*pp usu* GESALZEN) salt; season
salz·erzeugend *a* salt-producing. **S–erzeugung** *f* salt production, salification. **s–fähig** *a* salifiable. **S–farbe** *f* metallic (*or* basic) dye; dye salt. **-faß** *n* 1 salt (pickling) barrel. 2 salt cellar. **-fehler** *m* salt error. **-fleisch** *n* salt meat. **-fluß** *m* saline flux; salt rheum, catarrh. **s–förmig** *a* saliniform. **-frei** *a* salt-free, no-salt. **-führend** *a* salt-bearing. **S–garten** *m* salt garden (marsh, meadow). **-gehalt** *m* salt content, salinity. **-geist** *m* spirit of salt, hydrochloric acid — **Libavius' rauchender S.** fuming liquor of Libavius (stannic chloride). **-gemisch** *n* salt mixture. **-geschmack** *m* salty taste. **-gestein** *n* 1 rock salt. 2 (*Min.*) evaporite, hydrogenetic rock. **s–getränkt** *a* salt-

impregnated; (*Coal*) mineralized. **S–gewinnung** *f* extraction (production, manufacture) of salt. **-glasur** *f* salt glaze. **-grube** *f* salt mine (*or* pit). **-gurke** *f* pickled cucumber (*or* gherkin). **s–haltig** *a* salt-bearing, saliferous. **-haut** *f* film (*or* crust) of salt
salzig *a* salty, saline. **Salzigkeit** *f* saltiness, salinity
Salz·ion *n* salt ion. **-isomerie** *f* salt isomerism. **-kartoffeln** *fpl* boiled potatoes. **-korn** *n* grain of salt. **-kraut** *n* saltwort. **-krücke** *f* salt stirrer. **-kuchen** *m* salt cake. **-kupfererz** *n* (*Min.*) atacamite. **-lager** *n*, **-lagerstätte** *f* salt bed (layer, deposit). **-lake**, **-lauge** *f* brine, pickle. **-laugung** *f* brine leaching. **-lecke** *f* salt licks. **s–los** *a* saltless, salt-free. **-löser** *m* salt dissolver, brine mixer. **-lösung** *f* salt solution. **-mandel** *f* salted almond. **-marsch** *f* salt marsh. **-messer** *m* sali(no)meter. **-messung** *f* sali(no)metry. **-mutterlauge** *f* 1 saline mother liquor. 2 (*Salt Works*) bittern. **-nebelversuch** *m* (*Metll.*) salt spray test. **-niederschlag** *m* salt (*or* saline) deposit. **-papier** *n* (*Photo.*) salted paper. **-pfanne** *f* salt (*or* brine) pan. **-pfannenstein** *m* (*Salt Works*) pan scale. **-pflanze** *f* (*Bot.*) halophyte. **-probe** *f* 1 salt test. 2 salt sample. **-quelle** *f* salt (*or* saline) spring; salt well. **s–reich** *a* rich in salt, high-salt. **S–rinde** *f* salt crust. **-rückstand** *m* salt residue. **-salpetersäure** *f* aqua regia. **s–sauer** *a* 1 (*with metals*) chloride of. 2 (*with organic bases*) hydrochloride of. **S–säure** *f* hydrochloric acid. **-säuregas** *n* hydrogen chloride gas. **-säurenebel** *m* hydrochloric acid fumes. **-schicht** *f* layer of salt. **-schmelze** *f* salt melt, fused salt. **-see** *m* salt lake. **-siedepfanne** *f* salt pan. **-sieder** *m* salt boiler, salt maker. **-siederei** *f* 1 salt making. 2 salt works. **-sole** *f* 1 brine, salt water. 2 salt spring. **-speck** *m* bacon. **-sprühversuch** *m* (*Metll.*) salt spray test. **-stein** *m* 1 rock salt. 2 boiler scale. **-stock** *m* (*Nucl.*) salt dome (for waste storage). **-stoffwechsel** *m* salt metabolism. **-ton** *m* salt (*or* saliferous) clay. **-waage** *f* brine gauge, salimeter. **-wasser** *n* salt water; brine. **s–wasserecht** *a* fast to salt water.

-wasserwaage f brine gauge, salimeter. **-werk** n salt works. **-wirkung** f effect (or action) of salt. **-zusatz** m addition of salt

Sä·maschine f sowing machine

Sambesi·schwarz n Zambezi black

Same(n) m (. . men) 1 seed; semen, sperm; (in compds oft) seminal, spermatic. 2 grain, kernel. 3 progeny

Samen·anlage f ovule. **-bau** m seed planting. **-behandlung** f seed treatment. **-beizung** f seed disinfection. **-blättchen** n cotyledon. **-drüse** f spermatic gland, testicle. **-eiweiß** n (Bot.) endosperm. **-faden** m spermatozoon. **-fluß** m spermatorrhea. **-flüssigkeit** f seminal fluid. **s—führend** a seminiferous, seed-bearing. **S—gewächs** n seedling. **-haar** n (Bot.) seed hair, coma. **-handlung** f seed store. **-hefe** f seed yeast. **-keim** m germ, embryo. **-kern** m seed kernel; (Bot.) endosperm; (Physiol.) spermatic nucleus. **-korn** n (single) seed, grain. **-lappen** m seed lobe, cotyledon. **-leiter** m spermatic duct, vas deferens. **-öl** n (rape)seed oil. **-pflanze** f seed plant, spermatophyte, phanerogam. **-probe** f 1 seed test. 2 seed sample, **-saft** m seminal fluid. **-schale** f seed coat. **-staub** m pollen. **-tierchen** n spermatozoon. **-übertragung** f insemination. **-zelle** f seminal (or sperm cell). **-zucker** m quercitol, quercite. **-zwiebel** f 1 seed bulb. 2 seed onion

sämig a creamy, viscous

sämisch a, **-gar** a oil-tanned, chamois-dressed; (Lthr.) soft; **s—es Leder** wash leather. **S—gerbung** f chamois dressing. **-leder** n chamois

Sämling m (-e) seedling

Sammel· collecting, collective: **-aktion** f collection drive (or campaign). **-art** f collective species. **-band** m volume of collected works on a subject; omnibus volume. **-batterie** f storage battery. **-becken** n collecting basin (vessel, cistern, pond). **-begriff** m collective concept (or term), generic term. **-behälter** m storage tank (or hopper); receiver, collector. **-bottich** m collecting vat (tub, reservoir); (Brew.) starting tub. **-faktor** m collective factor. **-flasche** f collecting flask, receiver. **-gefäß** n 1 collecting vessel. 2

(Tin) listing pot. **-gespräch** n (Tlph.) conference call. **-glas** n 1 preparation (or specimen) tube. 2 (Opt.) converging lens. **-kristallisation** f coarsening of crystalline precipitates by collection of small particles. **-leitung** f 1 (Mach.) manifold; (lubrication) header. 2 (Elec.) omnibus circuit. **-linse** f (Opt.) converging (or convex) lens. **-literatur** f reference literature

sammeln 1 vt, vi, **sich s.** collect, gather. 2: **sich s.** collect one's thoughts; (re)gain one's composure; (Opt.) concentrate, focus

Sammel· collective, collecting: **-name** m collective name, generic term. **-probe** f cumulative sample. **-profil** n (Geol.) composite section. **-raum** m receiver, receptacle. **-referat** n collective report (or review). **-rohr** n collecting tube (pipe; wire, cable). **-schiene** f (Elec.) bus bar. **-sirup** m syrup from spilt sugar. **-spiegel** m concave mirror. **-stelle** f collecting (converging, gathering, assembly) point; depot; dump. **-trichter** m collecting funnel. **-vorrichtung** f collecting device, collector. **-werk** n compilation, collection of articles on a theme. **-zelle** f collective cell

Sammet: archaic form of SAMT velvet

Sammler m (-) collector, gatherer; compiler; (Elec.) storage battery. **-säure** f battery acid. **Sammlung** f (-en) collection, gathering; compilation; composure, collectedness

Sammlungs·glas n 1 specimen (or display) glass. 2 converging lens

Samstag m Saturday

samt 1 prep (+dat) along (or together) with. 2 adv: **s. und sonders** all and sundry, all without exception

Samt m (-e) velvet: **s—artig** a velvety; (Tex.) suede. **S—blende** f (Min.) goethite, limonite. **-blume** f marigold; amaranth. **-braun** n velvet brown (iron pigment). **-eisenerz** n goethite

samten a velvet, velvety

Samt·erz n (Min.) cyanotrichite. **s—glänzend** a velvety. **S—leder** n hide suede

sämtlich 1 a all; complete. 2 adv all, without exception

Samt·manchester m velveteen, cotton vel-

vet. **-schwarz** n ivory black. **s—weich** a velvet(y)-soft
Sandarach, Sandarak m 1 (Min.) realgar. 2 (Pharm.) sandarac. **--gummi, -harz** n sandarac (resin)
sand·artig a sand-like, sandy. **S—bad** n sand bath. **-badschale** f sand bath dish. **-beerbaum** m arbutus (tree). **-beere** f arbutus (berry). **-beerenöl** n arbutus oil. **-bestrahlung** f sandblasting. **-boden** m sandy soil. **-büchsenbaum** m sand-box tree. **-dorn** m sea buckthorn
Sandelholz° n sandalwood. **--öl** n sandalwood (or santal) oil
sandeln vt spray sand on; sandblast (wood)
Sandel-öl n sandalwood (or santal) oil. **-rot** n santalin
Sand·farbe f sand(y) color. **s—farben, -farbig** a sand-colored, sandy. **-form** f sand mold. **s—führend** a sand-bearing, sand-conveying. **S—gießerei** f sand casting. **-grieß** m coarse sand, fine gravel. **-gummi** n sandarac gum (or resin). **-guß** m sand casting. **-hafer** m lyme grass. **s—haltig** a containing sand, sandy
sandig a sandy, arenaceous
Sand·kohle f sand coal (forming a powdery coke), non-coking coal. **-korn** n grain of sand. **-körnung** f sand grading. **-mergel** m sandy marl. **-papier** n sandpaper. **s—reich** a sandy. **S—riedgras** n sea sedge, carex. **-sack** m sandbag. **-schicht** f layer of sand. **-schiefer** m sandy shale, schistous sandstone. **-segge** f sea sedge, carex. **-stein** m sandstone. **-strahl** m sand jet
sand·strahlen vt (inf & pp only) sandblast.
Sand·strahlgebläse n sandblasting equipment
sandte sent: past of SENDEN
Sand·ton m sandy clay. **-traube** f bearberry. **-uhr** f hourglass. **-uhrstruktur** f hourglass structure. **-wüste** f sandy desert. **-zucker** m raw (or brown) ground sugar
sanforisieren vt sanforize
sanft a gentle, mild, soft. **sänftigen** 1 vt soothe. **2: sich s.** abate, let up
sang sang: past of SINGEN
Sangajol n (Paint) mineral spirits
sanieren 1 vt clean up, make hygienic; rehabilitate; reorganize (on a sound finan-

cial basis). **2: sich s.** get back on a sound basis. **Sanierung** f (-en) clean-up; rehabilitation; (financial) reorganization.
Sanierungs·mittel n (Med.) prophylactic
Sanikel m (-) sanicle
sanitär, sanitarisch a sanitary, hygienic; **s—er Dienst** ambulance service. **Sanität** f public health (field). **Sanitäter** m (-) medic (esp Mil.); ambulance attendant
Sanitäts· health, hygienic, sanitary, medical: **-anstalt** f medical center. **-auto** n ambulance. **-behörde** f public health authority, board of health. **-bericht** m medical report. **-flugzeug** n ambulance plane. **-geschirr** n sanitary ware. **-kasten** m first-aid kit. **-keramik** f sanitary (ceramic) ware. **-kraftwagen** m ambulance. **-pflege** f hygiene, sanitation. **-polizei** f sanitary police, health department inspectors. **s—polizeilich** a of the sanitary police. **S—stelle** f health center; first-aid station. **-wagen** m ambulance. **-wesen** n sanitation, public hygiene; sanitary engineering
sank sank: past of SINKEN
sann thought, etc: past of SINNEN
Santelholz° n sandalwood = SANDELHOLZ
santonig a santonous. **Santonin·säure** f santoninic acid. **Santon·säure** f santonic acid
Sapan·holz n sapanwood, Brazil wood
Saphir m (-e) sapphire **—grüner S.** green corundum; **roter S.** oriental ruby. **--blau** n sapphire blue. **saphiren** a sapphire. **Saphir·spat** m (Min.) cyanite
Sapiet·säure f sapietic acid
Saponifikat·glyzerin n glycerine from saponification
Saponit n (-e) (Min.) steatite, soapstone
Saprämie f (Med.) sapremia
saprogen a saprogenic, putrefactive
Sapropelit·teer m sapropelite tar. **Sapropel·kohle** f sapropelic coal, sapropelite
saprophytisch a saprophytic
Sapukaja·öl n sapucaine (or monkey-pot) oil
Sardelle f (-n) anchovy. **Sardellen·butter** f anchovy paste
Sarder m (-) (Min.) sard
Sardinen·öl n sardine oil
Sarg m (⁻e) coffin, casket

Sarkin n sarcine, hypoxanthine
Sarkom n (-e) sarcoma. **sarkomatös** a sarcomatous
s.a.S. abbv (siehe auch Seite) see also page. . .
Sasa·palme f nipa palm
saß sat, etc: past of SITZEN
Sassafras·öl n sassafras oil
Sassaparille f (-n) sarsaparilla
Sassolin n sassolite, boric acid
Satinage f (-n) glazing, glaze finish
Satin·appretur f satin (or glazed) finish
satinieren vt satin, glaze, gloss; (Paper) (super)calender
Satin·papier n glazed paper
Sativin·säure f sativic acid
satt a 1 satiated, well-fed. 2 full: **s—e 200 PS** a full 200 HP. 3 (esp Colors) deep, saturated. 4 pred a: **sich s. essen** (**trinken**) eat (drink) one's fill; **Nudeln machen s.** noodles are filling; **sie haben** (or **sind**) **es s.** they have had enough of it, they are tired of it (or fed up with it); **etwas s. bekommen** (or **werden**) get tired of sthg. **--blau** a deep blue. **S—dampf** m saturated steam
Satte f (-n) dish, pan
Sattel m (-) saddle; bridge (of nose): **im S.** in the saddle, secure, firmly established. **--azeotrop** n azeotrope with minimum boiling point. **-faß** n rider cask. **s—fest** a firm in the saddle, secure; knowledgeable, well-versed (in a subject). **-förmig** a saddle-shaped. **S—füllkörper** m saddle packing (for fractionating columns). **-gang** m (Min.) saddle reef
satteln 1 vt saddle. 2: **sich s.** prepare oneself
Sattel·schlange f saddle-shaped coil. **-schlepper** m semitrailer tractor (truck)
satt·gelb a deep yellow
sättigbar a saturable. **sättigen** vt 1 saturate. 2 impregnate. 3 satiate, satisfy. **Sättiger** m (-) saturator. **Sättigung** f (-en) 1 saturation. 2 impregnation. 3 satiation. 4 satiety, satedness. 5 depth (of a color)
Sättigungs·apparat m saturator. **-druck** m saturation pressure. **s—fähig** a saturable. **S—fähigkeit** f saturation capacity. **-gefäß** n saturating vessel. **-grad** m degree of saturation. **-grenze** f saturation limit. **-kapazität** f saturation capacity. **-mittel** n saturating agent, saturant. **-punkt** m saturation point. **-strom** m (Elec.) saturation current. **-verhältnis** n (Air) relative humidity. **-wert** m saturation value, valence

Sattler m (-) saddler. **-leder** n saddler's leather. **-pech** n saddler's pitch (or wax)
sattsam adv sufficiently, abundantly
Saturateur m (-e) saturator
Saturations·gefäß n saturation vessel. **-saft** m (Sugar) saturation (or carbonization) juice. **-scheidung** f (Sugar) purification by saturation (separation of impurities with the lime when the syrup is saturated with carbon dioxide). **-schlamm** m (Sugar) carbonization sediment, saturation mud (or scum)
Saturei f, **--kraut** n savory. **-öl** n savory oil
saturieren vt saturate; (Sugar) carbonate
Saturn·rot n Saturn red, minimum. **-zinnober** m Paris red, red lead
Satz m (-e) 1 sediment, grounds: deposit; (Brew.) yeast. 2 set (of utensils); litter (of young), fry (of fish); battery (of furnaces). 3 batch; (Metll.) (furnace) charge. 4 (Mach.) assembly, unit. 5 rate; quota. 6 typesetting, composition. 7 sentence, clause. 8 theorem, proposition, tenet. 9 statement, dictum. 10 leap, bound, jump; stride. **--betrieb** m batch operation. **s—bildend** a deposit-forming, sediment-forming. **S—brauen** n brewing with cold malt extract. **-fehler** m misprint. **-krücke** f (Brew.) yeast rouser. **-mehl** n fecula, starch flour. **-schale** f settling dish
Satzung f (-en) 1 statute, regulation; article (of law). 2 pl oft constitution
satz·weise adv 1 intermittently. 2 by the set, in sets (or batches), batchwise
Sau f (-e) 1 sow (incl Metll.); hog. 2 drying kiln. 3 ink blot. Cf SCHWEIN
sauber a clean; neat, tidy; nice. **sauber·halten*** vt keep clean. **Sauberkeit** f cleanness, cleanliness; neatness, tidiness. **säuberlich** adv neatly; carefully, meticulously. **sauber·machen, säubern** 1 vt clean, cleanse; (+von) rid (sthg) (of. . .). 2 vi clean up, clean house
Sau·bohne f feed bean, (specif): (a) vetch,

(b) soybean. **-bohnenöl** n soybean oil.
-brot n Jerusalem artichoke
sauer I a (with endgs: **saur** ...) 1 acid(ic);
sour; tart: **s. machen** acidify; **s. einlegen**
pickle. 2 peeved, annoyed. 3 tough, difficult (work, life); adv oft the hard way. 4
gone wrong, out of order: **s. werden** go
wrong, get out of order. **II** sfx (Chem.: designates the salt of the corresponding acid)
e.g. **Essigsäure** acetic acid: **essigsaures
Natron** sodium acetate, **essigsaures
Salz** acetate Cf SÄURE
Sauer·ampfer m sorrel, sour dock. **-bad** n
(Bleaching) sour (bath)
säuerbar a acidifiable cf SÄUERN
säuer·beständig a acid-resistant; (Colors)
fast to acids
Sauer·blei n lead chromate. **-bleiche** f sour
bleaching. **-boden** m acid soil. **-brühe** f
sour (or acid) liquor. **-brunnen** m acidulated spring (water), mineral (aerated,
carbonated) spring (or spring water). **s–
chromsauer** a dichromate of. **S–dorn** m
barberry. **-dornbitter** m barberine.
-eisen n (Metll.) iron oxide. **-futter** n
(en)silage, **s–haltig** a acidiferous. **S–honig** m oxymel. **-kasten** m (Tex.) souring
tank
Sauerkeit f acidity; sourness
Sauer·kirsche f sour (or morello) cherry.
-klee m clover (or wood) sorrel. **-kleesalz**
n sorrel salt (potassium hydrogen oxalate). **-kleesäure** f oxalic acid. **-kohl** m,
-kraut n sauerkraut
säuerlich a acidulous, sourish: **s. machen**
acidulate. **Säuerlichkeit** f slight acidity
(or sourness)
säuerlich·stechend a turning sourish,
with a touch of acidity. **-süß** a sourishsweet
Säuerling m (-e) 1 acidulous (or heavily
carbonated) mineral spring (or water). 2
sour wine. 3 sour milk cheese. 4 clover
sorrel
Sauer·machen n acidification. **s–
machend** a acidifying. **S-milch** f
(cultured) sour milk, curdled milk
säuern I vt 1 acidify, sour. 2 leaven. **II** vi(s)
(turn) sour
sauer·phosphorsauer a acid phosphate
of. **S–quelle** f acidulated (or carbonated)

mineral spring. **-rahm** m sour cream.
-rahmbutter f ripened cream butter. **s–
reagierend** a acid-reacting. **S–regen** m
acid rain. **-salz** n acid salt. **s–schwefelsauer** a acid sulfate of. **-schweflig-
sauer** a acid sulfite of
Sauerst. abbv (Sauerstoff) oxyg. (oxygen)
Sauerstoff m oxygen. **--abgabe** f evolution
of oxygen. **-alterung** f aging with oxygen. **s–angereichert** a oxygen-enriched.
-arm a poor in oxygen, low-oxygen;
oxygen-starved. **S–äther** m acetaldehyde. **-atmung** f oxygen breathing;
(Biol.) aerobic respiration. **-atom** n oxygen atom. **-aufnahme** f oxygen absorption. **-bedürfnis** n oxygen requirement.
-bildung f formation of oxygen. **-bleiche**
f oxidation bleach(ing). **-bleichmittel** n
oxygen-releasing bleaching agent.
-bombe f oxygen cylinder. **-brücke** f oxygen bridge. **-druck** m oxygen pressure.
-entwicklung f evolution of oxygen. **s–
entziehend** a deoxidizing. **S–entzug** m
deoxidation, removal of oxygen.
-erzeuger m oxygen generator. **-fänger**
m oxygen absorber. **-flasche** f oxygen cylinder. **s–frei** = SAUERSTOFFREI
Sauerstoff-Frischen n (Steel) oxygen
lance process
Sauerstoff·gebläse n oxygen blowpipe.
-gehalt m oxygen content. **-gerät** n oxygen apparatus; respirator. **s–haltig** a
oxygenic, oxygenated. **S–ion** n anion.
-mangel m lack of oxygen, oxygen deficiency. **-maske** f oxygen mask. **-menge** f
amount of oxygen. **-messer** m eudiometer. **-ort** m oxidation region. **-pol** m oxygen pole, anode. **s–reich** a oxygen-rich,
high-oxygen
sauerstoffrei a (=sauerstoff·frei) oxy-
gen-free
Sauerstoff·salz n oxysalt, salt of an oxyacid. **-sättigung** f saturation with oxygen. **-säure** f oxyacid, oxygen acid.
-schlauch m oxygen tubing. **-spritzver-
fahren** n (Metll.) oxygen lance (or jet)
process. **-strom** m current of oxygen.
-träger m oxygen carrier. **-überschuß** m
excess of oxygen. **-überträger** m oxygen
carrier. **-verbindung** f oxygen compound. **-verwandtschaft** f affinity for

oxygen. **-zelt** *n* oxygen tent. **-zufuhr** *f* oxygen supply

sauer-süß *a* sour-sweet. S**–teig** *m* sourdough, leaven; yeast

Säuerung *f* (-en) acidification; souring; leavening

säuerungs-fähig *a* acidifiable; capable of souring. S**–grad** *m* degree of acidity. **-mittel** *n* acidifying agent

Sauer-wasser *n* 1 sour (water), dilute acid solution. 2 acidulous (carbonated, sparkling) water. **-wein** *m* sour wine; verjuice. s**–wein(stein)sauer** *a* acid tartrate of, bitartrate of. S**–werden** *n* souring; (*Liquors*) acetification. s**–werdend** *a* souring, acescent. S**–zone** *f* acid zone. s**–zugestellt** *a* (*Metll.*) acidlined

saufen* (soff, gesoffen; säuft) *vt/i* drink, guzzle (*applies to animals; derogatory for humans*). **Säufer** *m* (-) drunkard

Saug- suction, absorption, aspirating: **-ader** *f* lymphatic vessel. **-aderdrüse** *f* lymphatic gland. **-apparat** *m* suction apparatus, aspirator. **-druck** *m* suction pressure. **-düse** *f* suction nozzle; Venturi tube

saugen* (*reg or* sog, gesogen) I *vt/i* 1 suck, suction. 2 (*reg*) vacuum(-clean). II *vt* 3 (*oft* + **in sich**) soak up, drink in, absorb. 4 siphon. 5 (+**aus**, *oft*) draw, derive (from). III: **sich s.** 6 (+**voll**) become saturated (with), soak (*or* draw) up all of. 7 (+**in**) e.g. **die Flüssigkeit saugt sich in den Stoff** the liquid saturates (soaks into, is absorbed by) the material

säugen *vt* suckle, nurse, breast-feed

Sauger *m* (-) 1 (*Mach.*) suction apparatus; aspirator; exhauster. 2 (*Zool., esp Fish*) sucker. 3 (*Zool.*) suckling. 4 nipple (of a baby-feeding bottle); pacifier

Säuger *m* (-) 1 (*Zool., esp Fish*) sucker. 2 =**Säuge-tier** *n* mammal

saug-fähig *a* absorptive; hygroscopic. S**–fähigkeit** *f* absorptive capacity, absorptivity. **-festigkeit** *f* resistance to suction, suction strength. **-filter** *n* suction filter. **-flasche** *f* suction bottle, filter flask. **-gas** *n* suction (producer, power, aspirated) gas. **-gefäß** *n* suction (*or* absorbent) vessel. **-geschwindigkeit** *f* suction

velocity. **-glas** *n* 1 suction bottle. 2 breast pump. **-haar** *n* absorbent hair. **-hahn** *m* suction cock. **-heber** *m* siphon. **-höhe** *f* suction head, suction (*or* absorption) height; lift (of a suction pump). **-kasten** *m* suction (*or* vacuum) box. **-kolben** *m* 1 suction (*or* filter) flask. 2 valve piston. **-kopf** *m* 1 (*Pump*) mushroom strainer. 2 (*Metll.*) (Steel) crop end; (*Foun.*) shrink bob. **-korb** *m* suction strainer. **-leitung** *f* suction pipe (piping, line); vacuum line

Säugling *m* (-e) (breast-feeding) baby, infant. **Säuglings-nahrung** *f* baby food

Saug-luft *f* 1 vacuum, suction. 2 inflow, indraft. 3 intake air. **-luftanlage** *f* vacuum equipment (*or* plant). **-lüfter** *m* exhauster. **-luftkessel** *m* vacuum vessel. **-luftpumpe** *f* vacuum pump. **-maschine** *f* exhauster; aspirator. **-napf** *m*, **-näpfchen** *n* suction cup (*or* disc). **-papier** *n* absorbent paper. **-pipette** *f* suction pipette. **-pumpe** *f* suction pump. **-rohr** *n*, **-röhre** *f* suction pipe (*or* tube); siphon; Venturi tube. **-röhrchen** *n* pipette. **-schiefer** *m* absorbent shale (*or* schist). **-schlauch** *m* suction tubing (*or* hose). **-stäbchen** *n* filter stick, micro-immersion filter. **-strahlpumpe** *f* jet suction pump. **-stutzen** *m* 1 suction cylinder (fitted with perforated bottom for filtering). 2 vacuum pump intake. 3 exhaust connection. **-tiegel** *m* suction crucible. **-trichter** *m* pipe (in a casing). **-trockner** *m* suction drier

Saugung *f* suction, sucking; absorption; siphoning

Saug-ventil *n* suction valve. **-vorrichtung** *f* suction device (*or* arrangement). **-watte** *f* absorbent cotton. **-widerstand** *m* resistance to suction. **-wirkung** *f* suction effect, sucking action. **-wurm** *m* trematode, parasitic flatworm. **-zone** *f* suction (*or* absorption) zone. **-zug** *m* suction (*or* induced) draft. **-zuglüfter** *m* exhaust fan

Säulchen *n* (-) little column (pillar, post; pile; prism) *Cf* SÄULE

Säule *f* (-n) 1 column, pillar, post. 2 (*Elec.*) pile. 3 (*Cryst.*) prism. 4 (gasoline) pump

Sau-leder *n* pigskin

Säulen- columnar; (*Cryst.*) prismatic: **-achse** *f* prismatic axis. s**–artig** *a* col-

umnar; prismatic. **S–bohrmaschine** *f*
drill press. **-chromatographie** *f* column
chromatography. **s–förmig** *a* columnar,
column-shaped. **S–zelle** *f* columnar cell
säulig *a* columnar, columned
Saum *m* (-̈e) 1 hem. 2 seam. 3 edge, border,
margin; strip. 4 (*Tinplate*) list
säumen I *vt* 1 hem. 2 edge, border. 3 trim,
square. **II** *vi* 4 tarry, delay. 5 hesitate
säumig *a* 1 tardy, dilatory. 2 overdue. 3
negligent
Saum·pfanne *f* (*Tinplate*) list pot. **-riff** *n*
fringing reef. **-spiegel** *m* gray spiegeleisen. **-topf** *m* (*Tinplate*) list pot
saur (*+adj endgs*) sour, etc *Cf* SAUER
Säure *f* (-n) 1 acid (*as sfx usu designates* -ic
acid: cf SCHWEFELSÄURE). 2 acidity, sourness. **~abfall** *m* acid sludge. **-amid** *n* acid
amide. **-angriff** *m* attack by acid, acid corrosion. **-anhydrid** *n* acid anhydride.
-anzug *m* acid-proof suit (*or* clothing). **s–
artig** *a* acid-like. **S–äther** *m* ester. **-aus-
tausch** *m* acid exchange. **-austauscher**
m acid-exchange material. **-avivage** *f*
(*Dye.*) brightening with acid. **-azid** *n* acid
azide. **-bad** *n* acid bath. **-ballon** *m* acid
carboy. **-basengleichgewicht** *n* acid-
base equilibrium. **-batterie** *f* lead-acid
battery. **-behälter** *m* acid container. **-be-
handlung** *f* acid treatment. **-beize** *f*
(*Lthr.*) sour. **s–beständig** *a* acid-
resistant; (*Colors*) fast to acids. **S–
beständigkeit** *f* resistance to acids. **-be-
standteil** *m* acidic constituent. **-bestim-
mung** *f* determination of acid. **-beton** *m*
acid-resistant concrete. **s–bildend** *a*
acid-forming, acidifying. **S–bildner** *m*
acid former, acidifier. **-bildung** *f* acid formation, acidification. **s–bindend** *a* acid-
binding, acid-bonding. **S–bindungs-
vermögen** *n* power to combine with
acids, acid capacity. **-bottich** *m* acid vat.
-braun *n* acid brown. **-bromid** *n* acid bromide. **-chlorid** *n* acid chloride. **-dampf** *m*
acid vapor (*or* fumes). **-dehydrase** *f* acid
dehydrase. **-dichte** *f* acid density (*or* specific gravity). **-druckbehälter** *m* acid
blow case (*or* pressure container).
-druckbirne *f* acid egg. **-dunst** *m* acid
vapor (*or* fumes). **s–echt** *a* fast to acid. **S–
echtheit** *f* fastness to acid. **-echtviolett** *n*

(*Dyes*) Fast Acid Violet. **-eiweiß** *n* add
protein. **s–empfindlich** *a* sensitive to
acid. **S–empfindlichkeit** *f* sensitivity to
acid. **-ester** *m* ester. **s–fähig** *a* acidifiable. **S–farbstoff** *m* acid dye. **s–fest** *a*
acid-resistant (-proof, -fast). **S–flasche** *f*
acid bottle. **-flotte** *f* acid bath. **-fluorid** *n*
acid fluoride. **s–frei** *a* acid-free, non-
acid(ic). **S–gehalt** *m* acid content. **-gelb** *n*
acid yellow. **-grad** *m* (degree of) acidity.
-grün *n* acid green. **-grundlage** *f* acidifiable base. **-haloid** *n* acid halide. **s–
haltig** *a* acid-containing. **S–harz** *n*
(*Petrol.*) acid sludge. **-heber** *m* acid siphon. **-herstellung** *f* acid manufacture.
-hydrat *n* acid hydrate. **-imid** *n* acid imide.
-ion *n* acid ion, anion. **-kasein** *n* acid casein. **-kitt** *m* acid-proof cement (putty, adhesive). **s–kochecht** *a* fast to boiling
acid. **S–kochechtheit** *f* fastness to boiling acid. **s–löslich** *a* acid-soluble. **S–löslichkeit** *f* acid solubility. **-lösung** *f* acid
solution. **-maschine** *f* souring machine.
-menge *f* amount of acid. **-messer** *m*
acidimeter; hydrometer. **-meßkunst,
-messung** *f* acidimetry
Säuren·kupe *f* acid vat
Säure·politur *f* acid polishing. **-probe** *f* 1
acid test. 2 acid sample. **-pumpe** *f* acid
pump. **-radikal** *n* acid radical. **-re-
generat** *n* (*Rubber*) acid reclaim. **s–
reich** *a* acid-rich, high-acid(ity). **S–rest**
m acid residue. **-schlacke** *f* (*Metll.*) acid
slag. **-schlamm** *m* acid sludge.
-schlauch *m* acid hose. **-schleuder** *f* acid
hydroextractor. **-schutz** *m* acid protection. **-schwarz** *n* acid black. **-schwel-
lung** *f* (*Lthr.*) acid plumping. **-schwemm-
verfahren** *n* flotation process. **-spal-
tung** *f* acid cleavage (*or* hydrolysis). **-spat**
m crude calcium fluoride. **-stand** *m* acid
level. **-ständer** *m* acid cistern. **-teer** *m*
acid tar. **-titer** *m* acid titer, degree of
acidity. **-trog** *m* acid trough. **-turm**
m acid tower. **-überschuß** *m* excess of
acid. **s–unlöslich** *a* acid-insoluble.
S–unlöslichkeit *f* acid insolubility.
-ventil *n* acid valve. **-verteiler** *m* acid
distributor. **s–walkecht** *a* fast to acid
milling. **S–wechsel** *m* change in acidity.
-widerstandsfähigkeit *f* resistance to

acid. **s–widerstehend** *a* acid-resistant, fast to acid. **-widrig** *a* antacid. **S– wirkung** *f* acid action. **-zahl** *f* acid number. **-zufuhr** *f* addition of acid

Säurung *f* (-en) acidification, etc =SÄUERUNG

säuseln 1 *vt/i* whisper, murmur. 2 *vi* rustle

sausen *vi* 1 (h) buzz, hum. 2 (h,s) rush, whiz

SBB *pl abbv* (Schweizerische Bundesbahnen) Swiss Fed. RRs.

Sblp. *abbv* (Sublimationspunkt) sublimation pt.

Sc... *See also* **Sk...**

Schabe *f* (-n) I (*Zool.*) 1 cockroach. 2 moth. II (*Tools*) scraper

Schäbe *f* chaff (of flax, hemp)

schab·echt *a* scrape-resistant. **Schab· eisen** *n* scraping iron (*or* tool)

Schabe·fleisch *n* chopped meat. **-messer** *n* scraping knife; (*Lthr.*) fleshing knife

schaben *vt/i* scrape, grate; chop, mince; plane

Schaben·gift *n* roach poison, insecticide. **-pulver** *n* roach powder, *usu* white arsenic

Schaber *m* (-) scraper, grater. **-klinge** *f*, **-messer** *n* doctor (blade). **-walze** *f* doctor roll

Schabin *n* (-e), **Schabine** *f* (-n) gold-leaf parings, goldbeater's waste

Schablone *f* (-n) 1 pattern, stencil; template, jig. 2 (*fig*) conventional routine, stereotype; imposed standard

Schablonen·blech *n* sheet metal for stencils. **-druck** *m* screen printing

schablonenhaft *a* pattern-like; stereotyped, routine, hackneyed; mechanical

Schablonen·lack *m* screen varnish. **sch– mäßig** *a* pattern-like; stereotyped, etc =SCHABLONENHAFT. **Sch–papier** *n* stencil paper. **-wesen** *n* hackneyed routine

schablon(is)ieren *vt* stencil, copy; (*fig.*) stereotype, force into stereotyped patterns (*or* routines)

Schabsel *n* (-) (*usu pl*) shaving(s), scraping(s)

Schach *n* chess; check: **in Sch. halten** hold in check; **Sch. bieten** (+*dat.*) defy. **-brett** *n* chessboard. **sch–brettartig** *a* checkered

Schacht *m* (-̈e) 1 shaft (*esp Min.*), tunnel. 2 (*Geog.*) gorge. 3 (*Metll.*) stack. **-ausbau** *m* (*Min.*) shaft lining. **-deckel** *m* manhole cover

Schachtel *f* (-n) box, carton. **-halm** *m* (*Bot.*) horsetail, equisetum. **-halmsäure** *f* equisetic acid. **-kraut** *n* (*Bot.*) horsetail

schachten *vi* dig (a pit), sink a shaft

Schacht·ofen *m* shaft furnace; vertical kiln. **-speicher** *m* 1 (*Metll.*) storage bin. 2 (*Agric.*) silo. **-trockner** *m* tunnel drier

Schach·zug *m* chess move; clever maneuver

schade *pred a* (*oft* + **um**) too bad, a pity (for); **zu sch.** (**für**) too good (*or* valuable) (for)

Schädel *m* (-) cranium, skull; (*in cmpds, oft*) cranial: **-bruch** *m* skull fracture. **-höhle** *f* cranial cavity

schaden *vi* (+*dat*) harm, do harm, be detrimental (to); **es schadet nichts** it does no harm, it doesn't matter

Schaden *m* (-̈) 1 damage, harm; defect; injury, disability: **zu Sch. kommen** be harmed (injured, etc). 2 (financial) loss. **-ersatz** *m* compensation, damages

Schad·gas *n* harmful (pollutant, toxic) gas

schadhaft *a* damaged; defective, faulty; injured

schädigen *vt* 1 damage, impair; injure. 2 cause financial loss (to). **Schädigung** *f* (-en) damaging; damage; impairment; injury

schädlich *a* 1 harmful, damaging; noxious; injurious. 2 (*Mach.*): **sch—er Raum** dead space. **Schädlichkeit** *f* harmfulness; noxiousness; injuriousness. **Schädlich- keits·grad** *m* degree of harmfulness, etc. **Schädlich·machung** *f* rendering noxious, contamination

Schädling *m* (-e) pest, harmful animal (plant, substance), noxious substance

Schädlings·befall *m* pest infestation. **-bekämpfung** *f* pest control, control of noxious substances. **-bekämpfungsmit- tel** *n* pesticide

schadlos *a*: **sch. halten** (**für**) compensate (for)

Schadstoff° *m* harmful (*or* noxious) substance, pollutant. **sch–·arm** *a* low-pollutant, low-emission. **Sch–ausstoß** *m* toxic emission. **sch–beladen** *a* pollutant-laden, high-pollutant. **Sch–belastung** *f* (+*genit*) pollutant damage (to)

Schad·wirkung *f* harmful (*or* toxic) effect **Schaf** *n* (-e) sheep. **–·bein** *n* sheep bone; bone ash. **-bock** *m* ram. **-darmsaite** *f* catgut

Schaff *n* (-e) tub, vat

Schaf·fell *n* sheepskin

schaffen I* (schuf, geschaffen) *vt* create, establish: **wie geschaffen für. . .** as if created (*or* made) (for); **Ordnung sch.** create order, straighten things out. **II*** (*reg or* schuf, geschaffen) *vt* provide for, bring about (a remedy, help, clarity, etc); (*oft* + *dat* **sich**) get (sthg for oneself). **III** (schaffte, geschafft) **A** *vt* **1** make it (=be successful); manage, cope with; get (sthg) done; catch (a bus, etc). **2** get (sthg to a place). **3** wear out (*usu* a psn). **4** assign (a job to a psn). **B** *vi* **5** work, labor —**sich an etwas zu sch. machen** busy oneself with sthg; **das macht ihnen zu sch.** that causes them trouble. **6** do: **das hat nichts damit zu sch.** that has nothing to do with it. **C: sich sch.** make it, do a fine job. **Schaffen** *n* **1** creative activity, creativity. **2** work, activity. **schaffend** *p.a* **1** creative, productive. **2** working (people)

Schaffens·kraft *f* creative power

Schaf·fleisch *n* mutton

Schäffler *m* (-) cooper, barrelmaker

Schaffner *m* (-) **Schaffnerin** *f* (-nen) conductor, ticket collector

Schaffung *f* creation, establishment *Cf* SCHAFFEN **I**

Schaf·garbe *f* (*Bot.*) yarrow, milfoil. **-haut** *f* **1** sheepskin. **2** =**-häutchen** *n* amnion. **-käse** *m* ewe's-milk cheese. **-leder** *n* sheepskin. **-milch** *f* ewe's milk. **-pelz** *m* fleece. **-pergament** *n* sheepskin parchment. **-schmiere** *f* sheep dip. **-schweiß** *m* suint, (wool) yolk

Schafs·darm *m* sheep gut. **-klauenöl** *n* sheep's foot oil

Schaf·spalt *m* (*Lthr.*) skiver

Schaft *m* (-̈e) **1** (*genl*) shaft; shank. **2** (*specif*) (tool) handle, grip; (gun) stock; (boot) leg; (tree) trunk; (flower) stem. **3** bookshelf, bookcase

Schaf·talg *m* mutton tallow. **-waschmittel** *n* sheep dip. **-wasser** *n* amniotic fluid. **-wolle** *f* sheep's wool

Schagrin *n* shagreen (leather)

Schakal *m* (-e) jackal

schal *a* stale, flat; insipid; vapid

Schal *m* (-e) shawl, scarf

Schälchen *n* (-) little dish (*or* cup); capsule; cupel *cf* SCHALE

Schale *f* (-n) **1** (*usu, incl Electrons*) shell; (*specif*) peel, skin, husk, rind. **2** dish, bowl; cup; pan (of a balance). **3** (*Foun.*) chill

schälen I *vt* peel, skin, shell; strip (the bark off); decorticate. **II: sich sch. 1** peel (off), come off. **2** shed (its skin). **3** (+**aus**) appear (from out of), slip (out of)

Schalen·bau *m* shell structure. **-blende** *f* (*Min.*) fibrous sphalerite, wurtzite. **-entwicklung** *f* (*Phot.*) tray (*or* dish) developing. **sch–förmig** *a* shell-like, shell-shaped; dished, dish-shaped, bowl-shaped. **Sch–frucht** *f* caryopsis. **-guß** *m* chill casting

Schalenguß·eisen *n* chill-cast iron. **-form** *f* (*Foun.*) chill. **-kern** *m* chill core

schalen·hart *a* chilled. **Sch–haut** *f* (*Anat.*) chorion. **-lack** *m* shellac. **-lederhaut** *f* chorion. **-obst** *n* hard-shell dry fruit, nuts. **-tier** *n* shellfish. **-träger** *m* dish support, tripod

Schalheit *f* staleness, insipidity *cf* SCHAL

Schäl·holz *n* veneer

schalig *a* scaly, foliated; crusted; (hard-)shelled

Schall *m* (-̈e) sound; peal, resonance; noise; (*in compds, oft*) acoustic: **-abwehr** *f* noise abatement. **-aufnahmegerät** *n* sound recording device. **-boden** *m* sounding board. **sch–dämpfend** *a* sound-absorbing (*or* -deadening); muffling. **Sch–dämpfer** *m* sound absorber (*or* deadener); silencer; muffler. **sch–dicht** *a* soundproof. **Sch–dose** *f* sound box

Schallehre *f* (=**Schall·lehre**) acoustics, study of sound. **Schalleiter** *m* (=**Schall·leiter**) sound conductor

schallen *vi* ring (out), resound, echo; **es**

schallt there is an echo

Schall· sound: **-färbung** f tone color, timbre. **-geber** m source of sound. **-geschwindigkeit** f 1 sound velocity. 2 sonic speed. **-grenze** f sound barrier. **-lehre** f cf SCHALLEHRE. **-leiter** m cf SCHALLEITER. **-mauer** f sound barrier. **-messer** m phonometer. **-pegel** m noise level. **-platte** f (phonograph) record. **-quelle** f sound source. **sch–schluckend** a sound-absorbing. **Sch–schutz** m soundproofing. **-schwingung** f sound vibration. **sch–sicher** a soundproof. **Sch–stärke** f intensity of sound. **-technik** f acoustics; sonics. **sch–technisch** a acoustic; sonic. **-tot** a nonresonant, acoustically dead. **Sch–trichter** m horn, trumpet, megaphone. **-verstärkung** f sound amplification. **sch–weich** a sound-absorbing. **Sch–welle** f sound wave. **sch–zuleitend** a sound-conducting

Schäl·maschine f shelling (or husking) machine, peeler. **-mühle** f husking mill

Schalotte f (-n) shallot

Schalt· switch(ing), circuit: **-anlage** f switchgear, switching installation. **-bild** n circuit diagram. **-brett** n switchboard; control (or instrument) panel. **-dose** f switchbox

schalten I vt 1 set (a device) (at), turn (to). 2 operate, control (a device). 3 connect, wire; cut in, interpolate. 4 engage (clutch); shift into (a gear). II vi 5 switch. 6 shift (gears). 7 catch on, get the idea. 8: **sch. (und walten)** be in control, take charge (of things)

Schalter m (-) 1 (Elec.). switch. 2 (store) counter; (ticket) window

Schalt·getriebe n control gear, gear shift; gearbox. **-hebel** m 1 switch (lever), control lever, pl oft controls. 2 shift stick

Schal·tier n (Zool.) crustacean

Schalt·jahr n leap year. **-kupplung** f clutch. **-leistung** f switching capacity. **-plan** m circuit diagram. **-pult** n control panel (or desk). **-schlüssel** m switch key. **-skizze** f circuit diagram. **-stand** m control station. **-stück** n 1 insert. 2 (Elec.) contact (piece). **-tafel** f switchboard; instrument panel. **-uhr** f time switch

Schaltung f (-en) 1 switching. 2 circuit:

gedruckte Sch. printed circuit. 3 circuit diagram. 4 gear shift. Cf SCHALTEN

Schalt·werk n 1 controls. 2 switchboard. 3 gear shift. **-zentrale** f central control station

Schalung f (-en) (concrete-pouring) form

Schälung f (-en) peeling, skinning, shelling; decortication; desquamation; bark stripping; skin shedding cf SCHÄLEN

Scham f 1 shame. 2 pudenda. 3 (in cmpds oft) pubic: **-bein** n pubic bone, pubis

schämen vr: **sich sch.** be (or feel) ashamed

Schamotte f (-n) (Ceram.) chamotte, deadburned fireclay. **--mörtel** m fireclay mortar. **-retorte** f fireclay retort. **-stein** m esp: **halbsaurer Sch.** fireclay brick, firebrick. **-steingut** n fireclay ware. **-tiegel** m chamotte crucible. **-ton** m fireclay. **-ziegel** m chamotte brick, firebrick

schamponieren, schampuen vt/i. **Schampun** n (-s) shampoo

schandbar a disgraceful, infamous

Schande f shame, disgrace. **schänden** vt disgrace, dishonor, discredit; desecrate

Schank·bier n draft beer

Schanker m (-) chancre

Schanze f (-n) 1 trench, entrenchment. — **etwas in die Sch. schlagen** risk sthg. 2 skijump. **schanzen** I vt dig (trenches); build (entrenchments). II vi 1 dig in; entrench oneself. 2 drudge, toil

Schappe f (-n) 1 (Tex.) spun silk (from silk waste). 2 (Min.) (helical) drill

Schar f (-en) 1 crowd, troop, horde, herd; (Math.) family, group; system. 2 plowshare

Scharb·brett n chopping board. **schärben** vt chop. **Scharb·messer** n chopping knife, cleaver

Scharbocks·kraut n (Bot.) figwort

scharen 1 vt (+um sich) gather (around oneself). 2: **sich sch. (um)** gather, crowd (around)

scharen·weise adv in crowds (herds, hordes) cf SCHAR

scharf (-er, -ste) 1 (genl) sharp, keen: **sch. ins Auge fassen** keep a sharp eye on; **sch. sein (auf)** be keen (on). 2 corrosive. 3 spicy, pungent. 4 strident. 5 precise, acute, distinct. 6 strict, rigorous. 7 intense, vigorous. 8 adv at a fast pace. 9

adv: **sch. trocknen** dry at high temperatures. **10** live (shot); *adv* with live ammunition

Scharf·blick *m* sharp eye, acuity, perspicacity

Schärfe *f* (-n) **1** sharpness, keenness; corrosiveness; pungency; stridence; preciseness, acuteness, distinctness; strictness, rigor; intensity, vigor. **2** *pl oft* sharp criticism; sharp edges. *Cf* SCHARF

scharf·eckig *a* sharp-cornered. **Sch–einstellung** *f* sharp focus(ing)

schärfen *vt* **1** sharpen, intensify; strengthen (solutions). **2** whet, grind. **3** prime (bombs, mines)

Schärfen·fläche *f* (*Opt.*) surface of distinct vision. **-tiefe** *f* (*Opt.*) depth of focus

Scharf·feuer *n* (*Ceram.*) hard (*or* sharp) fire. **-feuerfarbe** *f* (*Ceram.*) hard-fire color. **sch–getrocknet** *a* dried at high temperature. **-kantig** *a* sharp-edged; acute-angled. **-konturig** *a* sharp-outlined. **-körnig** *a* sharp-grained. **Sch–manganerz** *n* hausmannite. **sch–salzig** *a* highly salty. **-sauer** *a* strongly acid; very sour. **-schmeckend** *a* sharp-tasting, spicy, tart; pungent, acrid. **-schweflig** *a* strongly sulfurous. **Sch–sicht** *f* sharp eye, perspicacity. **sch–sichtig** *a* perspicacious, sharp-eyed. **Sch–sinn** *m* acumen, astuteness. **sch–sinnig** *a* sharp-witted, astute; shrewd. **-winklig** *a* sharp-angled, acute-angled

Scharlach *m* (-e) **1** scarlet (color, dye). **2** (*Med.*) scarlet fever. **3** (*Bot.*) scarlet runner. **-beere** *f* kermes (berry). **scharlachen** *a* scarlet

Scharlach·farbe *f* scarlet color (*or* dye). **sch–farben, -farbig** *a* scarlet. **Sch–fieber** *n* scarlet fever, scarlatina. **sch–rot** *a* scarlet, bright red. **Sch–rot** *n* scarlet (red); cochineal. **-wurm** *m* cochineal insect

Schar·mittelwert *m* group mean value

Scharnier *n* (-e) hinge; (hinged) joint. **-ventil** *n* clack (*or* flap) valve

Schärpe *f* (-n) **1** scarf; sash. **2** (*Med.*) sling

Scharre *f* (-n) scraper; rake. **scharren 1** *vt/i* scrape (up); rake; bury. **2** *vi* scratch, make a scraping noise; shuffle; paw (the ground)

Scharr·harz *n* scraped resin. **-werk** *n* scraping device, scraper

Scharte *f* (-n) **1** notch, nick; fissure: **die sch. auswetzen** (*usu fig*) repair the damage. **2** (*Bot.*) sawwort. **schartig** *a* notched; jagged

Scharung *f* (-en) gathering; grouping, assembly

schatten *vi* cast a shadow; (provide) shade. **Schatten** *m* (-) shadow, shade. **-bild** *n* silhouette. **schattend** *a* shady, shade-providing

Schatten·färbung *f* shadow-dyeing. **-riß** *m* silhouette (drawing). **sch–trocken** *a* dried in the shade

schattieren *vt* shade; tint. **Schattierung** *f* (-en) shading; shade, tint; tinge

schattig *a* shady. **Schattung** *f* (-en) shading; shade

Schatz *m* (-e) **1** treasure; rich store (of. . .), abundance. **2** (natural) resource. **-amt** *n* treasury

schätzbar *a* capable of being estimated, ratable, assessable *cf* SCHÄTZEN

schätzen I *vt* **1** estimate; rate, value. **2** assess (for taxation). **3** appreciate; respect. **4** (+*clause*) guess, reckon. **II: sich sch.** (+*a, usu* **glücklich**) consider oneself (lucky)

Schatz·kammer *f* treasury; (*fig*) storehouse. **-meister** *m* treasurer

Schätzung *f* (-en) estimate, rating, valuation; assessment; appreciation, respect *cf* SCHÄTZEN

schätzungs·weise *adv* approximately, at a rough estimate. **Sch–wert** *m* estimated value

Schau *f* (-en) **1** exhibition, display, show: **zur Sch. stellen** (*or* **tragen**) make a display of, show off. **2** point of view. **3** view (of an area, etc.). **-bild** *n* diagram, graph. **sch–bildlich** *a* diagrammatic, graphic

Schauder *m* (-) shudder; chill, shiver; horror. **schauderhaft** *a* awful, frightful. **schaudern** *vt/i* shudder, shiver, tremble: **es schaudert einen** it makes one shudder (etc)

schauen I *vt* **1** look at, see; watch. **II** *vi* **2** (*usu*) look; (+**auf, oft**) attend (to); (+**nach**) look for, look after. **3** be surprised. **4** peek. **5** (+**daß**) see (to it) (that).

Note: With *sep pfxs* **schauen** *is usu synonymous with* SEHEN, *e.g.* **an·schauen** = **an·sehen.**

Schauer I *m* (-) **1** (*Meteor.*) shower. **2** =SCHAUDER shudder; etc. **II** *m* (-), *f* (-n) shed. **schauerlich** *a* chilling, bloodcurdling. **schauern** *vi* **1** shower, rain. **2** shudder, tremble

Schaufel *f* (-n) **1** shovel, scoop. **2** dustpan. **3** (*usu Mach.*) paddle, blade, vane; (dredging *or* conveyor) bucket. **schaufeln** *vt/i* shovel, dig

Schaufel·rad *n* paddle wheel. **-rührer** *m* paddle-stirrer

Schau·fenster *n* show (*or* display) window. **-glas** *n* **1** display (specimen, sample) glass. **2** sight glass, inspection window. **3** (*Opt.*) eyepiece, ocular. **4** (*Mach.*) glass gauge (oil gauge, etc)

Schaufler *m* (-) shoveler, digger

Schaukel *f* (-n) swing; seesaw. **schaukeln** *vt/i* swing, rock

Schau·linie *f* line, curve (on a graph). **-loch** *n* peephole; inspection window. **sch·lustig** *a* curious

Schaum *m* (-e) **1** foam, froth; suds; lather; bubbles. **2** (*Brew.*) head. **3** (*Baking*) meringue. **4** scum. **sch-·artig** *a* foamy, frothy. **Sch–bad** *n* foam (*or* bubble) bath. **-berg** *m* mound of foam. **sch–bekämpfend** *a* anti-foam, anti-froth. **Sch–beton** *m* aerated concrete. **-bier** *n* foaming beer. **-bildner** *m* foaming agent. **-bildung** *f* formation of foam (froth, bubbles). **-blase** *f* bubble. **-brecher** *m* foam breaker, froth killer

schäumen 1 *vt* skim (the foam off). **2** *vi* foam, froth; fizz, bubble, effervesce; (*Wine*) sparkle; (*Soap*) lather

Schaum·entwässerung *f* foam drainage

Schäumer *m* (-) foaming agent

Schaum·erde *f* (*Min.*) aphrite, calcite. **-erz** *n* soft (*or* foamy) wad. **sch–fähig** *a* foaming; lathering. **Sch–feuerlöscher** *m* foam extinguisher. **-fleck** *m* froth spot, foam stain. **-flocke** *f* small wad of foam. **-gärung** *f* frothy fermentation. **-gebäck** *n* meringue(s). **sch–gebremst** *a* lowsudsing (detergent). **Sch–gemisch** *n* froth mixture. **-gips** *m* foliated gypsum. **-glas** *n* foam glass. **-gold** *n* Dutch metal, tinsel. **-gummi** *n* foam rubber

schaumhaft *a* foamy, frothy

Schaum·hahn *m* scum valve. **-haken** *m* skimmer. **sch–haltig** *a* foam-stable; (*Beer*) head-retaining. **Sch–haltigkeit** *f* foam stability; (*Beer*) head retention. **-haube** *f* (*Brew.*) head. **-hemmungsmittel** *n* foam suppressor (*or* inhibitor)

schaumig *a* **1** foamy, frothy —**sch. rühren** (*or* **schlagen**) stir (*or* beat) to a foam. **2** (*Soap*) lathery. **3** porous. **Schaumigkeit** *f* foaminess, frothiness

Schaum·kalk *m* (*Min.*) sparry aphrite, aragonite. **-kelle** *f* skimming ladle. **-korb** *m* (*Sugar*) scum basket. **-kraft** *f* foaming (sudsing, lathering) power. **-kraut** *n* (*Bot.*) (*usu*) bitteres Sch. cardamine. **-kunststoff** *m* expanded plastic. **-lamelle** *f* foam film. **-löffel** *m* skimming ladle. **sch–los** *a* foamless, frothless; (*Beer*) flat. **Sch–löscher** *m*, **-löschgerät** *n* foam extinguisher. **-löschmittel** *n* foam fire-extinguishing agent. **-mittel** *n* foaming agent. **-öl** *n* defoamer, anti-froth oil. **-polystyrol** *n* polystyrene foam. **-rand** *m* scum line. **-reiniger** *m*, **reinigungsmittel** *n* foaming cleanser. **-rohr** *n*, **-röhre** *f* **1** foam tube. **2** scum pipe. **-schlacke** *f* foam slag, slag pumice. **-schwärze** *f* finely powdered animal charcoal. **-schwimmaufbereitung** *f* (*Ores*) froth flotation. **-seife** *f* lathering soap. **-spat** *m* (*Min.*) laumontite (hardfoliated calcite), analeite. **-stand** *m* (*Brew.*) head. **-stoff** *m* **1** spongy (*or* foamed) plastic; cellular plastic. **2** foaming agent. **-ton** *m* fuller's earth. **-verbesserer** *m* (*Dterg.*) suds booster, foam improver. **-verhinderung** *f* foam prevention. **-verhinderungsöl** *n* anti-froth oil. **-verhüter** *m*, **-verhütungsmittel** *n* antifoaming agent. **sch–verträglich** *a* foamcompatible (fire extinguishers). **Sch–wein** *m* sparkling wine. **-wert** *m* (*Dterg.*) lather value

Schau·öffnung *f* peephole, inspection hole. **-platz** *m* scene, setting

schaurig *a* blood-curdling, horrifying *cf* SCHAUER 2

Schau·sammlung *f* display collection. **-seite** *f* (*Tex.*) right side. **-spiel** *n* stage play, drama. **-spieler** *m* actor. **-spielerin** *f* actress. **-stellung** *f* exhibition. **-tafel** *f* diagram, chart. **-versuch** *m* lecture (*or* demonstration) experiment. **-zeichen** *n* visual sign (*or* signal)

Scheck *m* (-s) (*Bank*) check

schecken *vt* spot, dapple, mottle. **scheckig** *a* spotted, dappled, mottled

Scheelat *n* (-e) (*Min.*) tungstate

Scheel·bleierz *n*, **-bleispat** *m* (*Min.*) scheeletite, stolzite

Scheelesche *a:* Sch—s Grün Scheele's green; Sch—s Süß glycerol, glycerine

scheelisieren *vt* (*Wine*) scheelize (treat with glycerine)

scheel·sauer *a* tungstate of. **Sch—säure** *f* tungstic acid. **-spat** *m* scheelite

Scheffel *m* (-) (*usu fig*) bushel, whole lot

Scheibchen *n* (-) little disk (slice, pane) *cf* SCHEIBE

Scheibe *f* (-n) 1 disk, disc. 2 slice. 3 pane (of glass); windshield. 4 dial. 5 (*Mach.*) plate; gasket; washer; pulley; (potter's) wheel. 6 (*Opt.*) filter. 7 cake (of wax). 8 honeycomb. 9 (shooting) target

scheiben·ähnlich, **-artig** *a* disc-like, discoid(al). **Sch—blei** *n* window lead. **-eisen** *n* pig iron in disks. **-filter** *n* disk filter. **sch—förmig** *a* disk-shaped, discoid. **-glas** *n* window glass. **-holländer** *m* (*Paper*) disk beater. **-honig** *m* comb honey. **-kupfer** *n* rose (*or* rosette) copper. **-lack** *m* shellac. **-mühle** *f* disk mill. **-reißen** *n* (*Metll.*) conversion into disks (*or* rosettes). **-wachs** *n* cake wax. **-waschanlage** *f* windshield washer unit. **sch—weise** *adv* in slices, by the slice. **Sch—wischer** *m* windshield wiper. **-zelle** *f* disk-shaped (*or* discoid) cell

scheidbar *a* separable; analyzable *cf* SCHEIDEN. **Scheidbarkeit** *f* separability; analyzability

Scheide (-n) 1 sheath. 2 vagina. 3 boundary, borderline. 4 (*in compds, oft*) separated, separating: **-anstalt** *f* 1 refinery. 2 assay office. **-bad** *n* separating (*or* refining) bath. **-bank** *f* sorting table. **-bock** *m* retort stand. **-bürette** *f* separating burette. **-erz** *n* picked (*or* screened) ore. **sch—fähig** *a* separable, refinable. **Sch—flüssigkeit** *f* separating (*or* parting) liquid. **-gang** *m* low-grade ore (after removal of high-grade ore). **-gefäß** *n* partitioning vessel. **-gold** *n* gold purified by

partitioning. **-grenze** *f* borderline. **-gut** *n* material to be separated; (*specif oft*) scrap containing precious metal. **-kalk** *m* (*Sugar*) defecating lime. **-kapelle** *f* cupel. **-kolben** *m* separating flask. **-kuchen** *m* (*Metll.*) liquation disk. **-kunst** *f* analytical chemistry. **-linie** *f* borderline, dividing line. **-mauer** *f* partition wall. **-mehl** *n* dust of picked ore. **-mittel** *n* separating (*or* partitioning) agent. **-münze** *f* small copper coin; token

scheiden* (schied, geschieden) **I** *vt & sich sch.* 1 separate, part; divide. **II** *vt* 2 analyze. 3 decompose. 4 pick, sort (ore). 5 refine; (*Liquid*) clarify; (*Tar*) extract; (*Sugar*) defecate, lime. 6 divorce. 7 distinguish, tell apart. **III:** *sich sch.* be (*or* get) divorced

Scheiden· sheath; vaginal: **-abstrich** *m* vaginal smear. **sch—artig** *a* sheath-like; vaginal. **Sch—schleim** *m* vaginal mucus. **-tampon** *m* vaginal tampon

Scheide· separating, part(ition)ing; (*Sugar*) defecation: **-ofen** *m* parting (*or* almond) furnace. **-pfanne** *f* (*Sugar*) defecating pan, clarifier. **-punkt** *m* point of separation (*or* divergence)

Scheider *m* (-) separator; ore sorter; metal refiner

Scheid·erz *n* separated ore

Scheide· separating; (*Sugar*) defecation: **-saft** *m* (*Sugar*) defecated juice. **-saturation** *f* (*Sugar*) defecosaturation. **-schlamm** *m* (*Sugar*) defecation sludge, presscake. **-sieb** *n* separating (*or* partitioning) sieve. **-silber** *n* parting silver. **-trichter** *m* separating (*or* separatory) funnel. **-verfahren** *n*, **-vorgang** *m* separating (*or* refining) process. **-vorrichtung** *f* parting (separating, screening, sorting) device. **-wand** *f* 1 partition (wall); barrier. 2 septum, diaphragm. **-wasser** *n* (*formerly*) nitric acid (for parting). **-weg** *m* crossroads, fork

Scheidung *f* (-en) separation, part(ition)ing, division; (chemical) analysis; refining; (*Sugar*) defecation; etc, *cf* SCHEIDEN. **Scheidungs·mittel** *n* separating (*or* partitioning) agent

Scheimigkeit *f* ropiness

Schein *m* (-e) 1 shine, light, glow, gleam. 2 glimmer; tinge; (*Oils*) bloom. 3 look(s), (external) appearance. **3** document; certificate, check, coupon, voucher, ticket, receipt; bill, note (paper money). **4** illusion; sham, make-believe, pretense; (*oft in compds*) apparent, ostensible, pseudo-, false, feigned

scheinbar *a* apparent, seeming; ostensible, pretended, feigned

Schein·dasein *n* sham (*or* shadow) existence

scheinen* (schien, geschienen) *vi* **1** shine. **2** seem, appear

Schein· *cf* SCHEIN 4: **-farbe** *f* accidental color. **-fuß** *m* pseudopodium. **-gold** *n* imitation gold. **-grund** *m* apparent (*or* pretended) reason, pretext. **-leistung** *f* apparent output. **-lösung** *f* apparent solution. **-mine** *f* (*Mil.*) dummy mine. **sch-tot** *a* apparently dead. **Sch–vergiftung** *f* (*Mil.*) false contamination. **-verne-belung** *f* (*Mil.*) dummy smokescreen. **-werfer** *m* headlight; floodlight; spotlight; searchlight. **-widerstand** *m* (*Elec.*) impedance

Scheit *n* (-e *or* -er) log; stick; piece of wood

Scheitel *m* (-) **1** apex, vertex; summit, peak; zenith. **2** (*Math.*) origin (of coordinates). **3** top (of the head); head; hair; (hair) part. **5** (*in compds.*) maximum, peak; (*Anat.*) parietal: **-ausschlag** *m* amplitude. **-bein** *n* parietal bone. **-linie** *f* vertical line. **-punkt** *m* vertex; zenith. **sch–recht** *a* vertical. **Sch–spannung** *f* peak voltage. **-wert** *m* maximum value. **-winkel** *m* vertex angle

scheitern *vi* (s) **1** founder, run aground. **2** (*oft* + **an**) fail, go wrong, miscarry (because of). **Scheitern** *n* failure

Schelfe *f* (-n) husk, shell; pod. **schelfern** *vi* & **sich sch.** peel, scale (off) (*as:* skin)

Schellack *m* shellac. **-bindung** *f* shellack bond(ing). **-firnis** *m* shellac varnish

schellackieren *vt* shellac

Schellack·lösung *f* shellac solution. **-politur** *f* shellac (*or* French) polish. **-steife** *f* shellac stiffening. **-wachs** *n* shellac wax

Schellan *n* shellan (a synthetic resin)

Schelle *f* (-n) **1** (little) bell. **2** clamp, clip; shackle, *pl oft* handcuffs. **schellen** *vi* ring (the bell): **es schellt** the bell is ringing. **Schellen·nietung** *f* snap riveting

Schell·fisch *m* haddock. **-harz** *n* white

rosin. **-kraut** *n* celandine, swallowwort

Schellol·säure *f* shellolic acid

Schema *n* (.. men *or* -ta) scheme, plan, diagram; pattern, schedule, routine; model. **-bild** *n* **1** diagram. **2** flow sheet. **schematisch** *a* schematic, diagrammatic; mechanical, routine. **schematisieren** *vt/i* sketch, schematize

Schemel *m* (-) (foot)stool

Schemen *m or n* (-) phantom, shadow. **schemenhaft** *a* shadowy, phantom

Schenk·bier *n* draft beer. **Schenke** *f* (-n) tavern, pub

Schenkel *m* (-) **1** thigh, femur. **2** shank; side piece; (*incl Geom.*) leg. **3** handle (one of a pair). **4** (*in compds. oft*) femoral, crural: **-bein** *n* femur, thigh bone. **-rohr** *n*, **-röhre** *f* bent tube, V-tube; elbow tube (*or* pipe)

schenken *vt* **1** give (as a gift), present. **2** pour. **3** excuse from. **4: sich etwas sch.** do without sthg

scherardisieren *vt* sherardize

Scherbe *f* (-n) **1** (*usu pl*) shard(s), fragment(s), broken piece(s); *pl oft* broken pieces (china, pottery). **2** crock; (flower) pot; cupel. **3** (*Ceram.*) body. **4** scorifier. *Cf* SCHERBEN

Scherbel·krautwurzel *f* (*Bot.*) asarum, hazelwort

Scherben *m* (-) broken piece; pot, etc =SCHERBE. **-haufen** *m* shambles, heap of fragments. **-kobalt, -stein** *m* native arsenic

Schere *f* (-n) **1** (pair of) scissors (shears, tinsnips, wire cutters). **2** (*Zool.*) claw, chela. **3** (*Mach.*) (lathe) adjustment plate

scheren I* (schor, geschoren) *vt* **1** shear, cut, clip; trim; prune. **2** mow. **3** (*Tex.*) warp. **II** (*usu* scherte, geschert) **A** *vt* **4** concern, bother. **B: sich sch.** **5** (**um**) care (about). **6** (+*place phr*) go, get (to. . .)

Scheren·bindung *f* chelate linkage, chelation. **-schleifer** *m* scissors grinder. **-schnabel** *m* (*Zool.*) scissorbill. **-sprei-zen** *fpl* lazy tongs. **-stahl** *m* shear steel. **-zange** *f* (pair of) cutting forceps; wire cutters; (*Metll.*) shingling tongs

Scher·festigkeit *f* shearing strength

Scherflein *n* mite, bit: **sein Sch. bei-tragen** do (*or* contribute) one's bit

Scher·kraft *f* shearing force (*or* stress).

-spannung *f* shear stress

Scherung *f* 1 shear. 2 shearing, clipping, mowing; warping. *Cf* SCHEREN I

Scherungs·festigkeit *f* shearing strength. **-winkel** *m* angle of shear

Scher·versuch *m* shearing test. **-wirkung** *f* shearing action (*or* effect)

Scherz *m* (-e), **scherzen** *vi* joke, jest. **scherzhaft** *a* jocular; humorous

scheu *a* shy, bashful, timid. **Scheu** *f* (*oft* + **vor**) shyness, bashfulness, timidity (toward)

Scheuche *f* (-n) scarecrow. **scheuchen** *vt* scare away

scheuen 1 *vt* shun, avoid, shy away from. 2: **sich sch.** (**zu** + *inf*) hesitate (to. . .), shrink (from)

Scheuer *f* (-n) barn; granary

Scheuer· scouring, scrubbing: **-bürste** *f* scrub brush. **sch-fest** *a* scourproof, abrasion-resistant. **Sch-mittel** *n* scouring agent

scheuern 1 *vt* scrub, scour; rub, chafe. 2 *vi* scrape, scratch (noisily). 3: **sich sch.** (**an**) rub (against), scratch oneself (on)

Scheuer· scouring, scrubbing: **-pulver** *n* scouring powder. **-trommel** *f* scouring (*or* tumbling) barrel. **-wirkung** *f* scouring action (*or* effect). **-wulst** *m* scouring pad

Scheune *f* (-n) barn; granary

Schi·butter *f* shea butter

Schicht *f* (-en) 1 layer, stratum; bed, course. 2 coating, film; lamina. 3 (*Phot.*) emulsion. 4 (furnace) charge; batch. 5 (work) shift. **--boden** *m* (*Metll.*) mixing place. **-dicke** *f* thickness of a layer (*or* stratum). **-ebene** *f* layer (bedding, stratification) plane

schichten *vt* 1 layer, arrange in layers (*or* beds); stratify. 2 laminate. 3 pile, stack

Schichten·abstand *m* interlayer distance (*or* spacing). **-aufbau** *m* layered (*or* stratified) structure. **-aufrichtung** *f* tilt (of strata). **-bau** *m* stratification (=-AUF-BAU). **-bildung** *f* stratification, lamination. **-folge** *f* series of strata (*or* layers). **-fuge** *f* bedding plane; stratum joint. **-git-ter** *n* lattice layer, stratified lattice. **-glas** *n* laminated (shatterproof) glass. **-gruppe** *f* (*Geol.*) group of strata, formation. **-hochpolymer** *n* laminated high polymer sheet. **-kohle** *f* stratified coal.

-polymer *n* laminated polymer. **sch-weise** *a* stratified, laminated; (*adv oft*) in layers, in strata, in piles

Schicht·fläche *f* surface of a layer (*or* stratum); bedding plane. **-folge** *f* series of strata (*or* layers). **-folie** *f* laminating sheet. **-fuge** *f* bedding plane; stratum joint. **-gefüge** *n* layered (*or* stratified) structure. **-gestein** *n* stratified rock. **-git-ter** *n* (*Cryst.*) layer lattice. **-glas** *n* layered (shatterproof, safety) glass. **-glied** *n* layer, stratum. **-holz** *n* 1 plywood. 2 stacked wood, cordwood

schichtig *a* layered, stratified; *adv oft* in layers (*or* strata)

Schicht·korrosion *n* layer corrosion. **-kristall** *m* layer crystal. **-linie** *f* (*Cryst.*) layer line. **-pappe** *f* laminated cardboard. **-pressen** *n* laminate molding. **-preß-stoff** *m* molded laminate, laminated plastic. **-seite** *f* (*Phot.*) emulsion side. **-silikat** *n* layered silicate. **-spaltung** *f* delamination. **-stärke** *f* layer thickness. **-stoff** *m* laminate(d material). **-stoffharz** *n* laminated resin. **-stoffplatte** *f* laminated plate. **-textur** *f* stratified structure

Schichtung *f* (-en) arrangement in layers (*or* beds); stratification, lamination; piling, stacking *Cf* SCHICHTEN

Schichtungs·ebene *f* plane of stratification, bedding plane

Schicht·verband *m* laminar bonding. **-wasser** *n* ground water. **sch-weise** *a* laminated, stratified; *adv oft* in layers (strata, beds, piles)

schick *a* 1 chic, elegant. 2 fine, neat. **Schick** *m* 1 style, fashion. 2 elegance. 3 neatness. 4 skill

schicken I *vt* send. II *vi* (+**nach**) send (for). III: **sich sch.** 1 work out right. 2 be fitting, be (socially) acceptable. 3 happen, come about. 4 hurry. 6 (+**in**) adapt oneself (to). *Cf* GESCHICKT

schicklich *a* fitting, proper, (socially) acceptable

Schick·probe *f* Schick test

Schicksal *n* (-e) fate, destiny

Schicksche Reaktion (*Med.*) Schick reaction

Schickung *f* stroke of fate

Schiebe· sliding: **-dach** *n* sliding roof. **-hülse** *f* sliding sleeve

schieben* (schob, geschoben) **I** *vt* **1** push, shove; (+**von sich**) push aside, reject. **2** (+**auf**) put (the blame on . .), blame (on). **3** postpone, put off. **4** manage (to do). **II** *vt/i* **5** (*oft* + **mit**) racketeer, (wheel and) deal (in). **III: sich sch. 6** push (one's way) (ahead). **7** move, shift (out of place).
Schieber *m* (-) **1** (*Mach.*) slide(r), slide bar (or bolt); slide valve; damper; carriage. **2** bedpan. **3** racketeer, black market dealer; (drug) pusher
Schieber-kasten *m* (*Mach.*) slide box, valve chest. **-lineal** *n* slide rule
Schieber-rohr *n* sliding tube (or sleeve)
Schieber-ventil *n* slide valve
Schiebe-sitz *m* (*Engg.*) push fit, force fit. **-tür** *f* sliding door
schieb-fest *a* nonslip. **Sch–karren** *m* wheelbarrow. **-kraft** *f* (*Phys., Engg.*) thrust(ing power). **-lehre** *f* **1** slide gauge. **2** vernier calipers
Schiebung *f* (-en) **1** maneuver, manipulation. **2** racketeering, wheeling and dealing; (drug) pushing. **3** (*Metals*) glide. **4** (*Geol.*) shear(ing); (*Cryst.*) slippage. *See also* SCHIEBEN
Schiebungs-fläche *f* **1** (*Geol.*) shearing plane. **2** (*Cryst.*) slip plane
schied separated, etc: *past of* SCHEIDEN
Schieds-analyse *f* arbitration (or umpire) analysis. **-gericht** *n* arbitration court. **-mann, -richter** *m* arbiter, arbitrator; referee, umpire. **-spruch** *m* arbitration award; referee's decision. **-versuch** *m* referee test
schief *a* **1** leaning, inclined, oblique; skewed; *adv oft* askew. **2** crooked, uneven, lopsided, unsymmetrical. **3** sidelong (glance). **4** wry (facial expression). **5** wrong, adverse; *adv oft* awry, amiss. **--achsig** *a* oblique-axial. **Sch–agar-kultur** *f* sloped-agar culture
Schiefe *f* **1** obliqueness; inclination; skewness. **2** inclined plane. **3** crookedness; lopsidedness. **4** wryness
Schief-einfall *m* (*Phys.*) oblique incidence
Schiefer *m* **1** slate; (*Geol.*) schist; shale. **2** (*Metll.*) flaw (in iron). **3** splinter
schiefer-ähnlich, -artig *a* slate-like, slaty, schistous. **-blau** *a* slate-blue. **Sch–boden** *m* slaty (or schistous) soil. **-bruch**

m **1** slate quarry. **2** slaty fracture. **-farbe** *f* slate color. **sch–farben, -farbig** *a* slate-colored. **Sch–gips** *m* foliated gypsum. **sch–grau** *a* slate-gray. **-haltig** *a* slaty, schistous, containing slate (*or* shale)
schieferig *a* **1** slaty, schistous. **2** scaly, foliated, flaky. **Schieferigkeit** *f* **1** slatiness, schistosity. **2** scaliness, flakiness
Schiefer-kohle *f* slaty (*or* splint) coal. **-letten** *m* clay shale. **-mehl** *n* slate flour, shale meal, ground shale. **-mergel** *m* slaty marl
schiefern 1 *vi* & *sich sch.* scale off, exfoliate. **2** *vi* (*of wood*) splinter
Schiefer-öl *n* shale oil. **-paraffin** *n* (*Petrol.*) shale wax. **-platte** *f* slab (*or* sheet) of slate. **sch–schwarz** *a* slate-black. **Sch–spat** *m* slate spar, lamellar calcite. **-stein** *m* slate. **-stift** *m* slate pencil, stylus. **-talk** *m* talc slate, indurated talc. **-teer** *m* shale tar. **-ton** *m* clay slate (*or* shale), slate clay; (*Min.*) dickite
Schieferung *f* (-en) **1** scaling off, exfoliation. **2** splintering (of wood). **3** (*Geol.*) cleavage, schistosity
schiefer-weiß *a* flake-white
schief-gehen* *vi*(s) go wrong
Schiefheit *f* **1** obliqueness; inclination; slope. **2** crookedness, lopsidedness. *Usu* = SCHIEFE
schief-liegend *a* inclined, sloping, oblique
schiefrig *a* slaty; scaly, etc =SCHIEFERIG
schief-winklig *a* **1** oblique-angled; skew. **2** (*Cryst.*) clinographic
schien seemed, etc: *past of* SCHEINEN
Schien-bein *n* shinbone, tibia
Schiene *f* (-n) **1** (railroad) rail. **2** channel, track, guide bar, skid. **3** (metal) bar; (wooden) slat; straightedge. **4** (*Elec.*) busbar. **5** band, hoop, rim. **6** (*Med.*) splint
Schienen-bus *m* (self-propelled) railcar. **-fahrzeug** *n* rail vehicle. **-weg** *m* railway, railroad. **-weite** *f* track gauge (of a railroad)
schier 1 *a* sheer, pure. **2** *adv* downright, outright
Schierling *m* (*Herbs*) hemlock
Schierlings-kraut *n* hemlock, conium. **-saft** *m* hemlock juice. **-tanne** *f* hemlock (spruce)
Schieß-arbeit *f* shooting, blasting. **-baum-**

wolle f guncotton; collodion cotton, pyroxylin

schießen* (schoß, geschossen) **1** vt/i shoot (at), fire; blast; (Dye.) flush on. **2** vi(s) shoot, dash, dart; sprout; gush, spurt; rush (as: blood to the head): **in die Höhe sch. (a)** shoot up, grow tall; **(b)** go flying up.

Schieß·lehre f ballistics. **-ofen** m bomb oven, Carius tube furnace. **-pulver** n gunpowder. **-rohr** n, **-röhre** f Carius tube. **-stoff** m powder (for shooting), propellant. **-wolle** f guncotton. **-wollpulver** n guncotton powder

Schiff n (-e) **1** ship, vessel. **2** (Tex.) shuttle. **3** nave. **Schiffahrt** f shipping, navigation.

schiffbar a navigable

Schiff·bau m shipbuilding. **-bruch** m shipwreck

Schiffchen n (-) **1** little ship, boat. **2** (Tex.) shuttle. **3** (Bot.) keel

Schiffs·bedarf m naval stores

Schiffsche Base Schiff base

Schiffs·leim m marine glue. **-pech** n common black pitch. **-raum** m ship's hold. **-teer** m ship's (or wood) tar

Schild I m (-e) shield, coat of arms —**etwas im Sch. führen** have sthg secretly planned. **II** n (-er) **1** sign(board); label, tag; (name)plate; badge; ticket. **2** (Hides) butt. **3** turtle shell. **4** visor. **Schildchen** n (-) **1** little sign (label, tag, shield). **2** (Bot.) scutellum

Schild·drüse f thyroid gland. **Schild-drüsen·essenz** f, **-extrakt** m thyroid extract

Schilder·blau n (Dye.) pencil blue

schildern vt describe; depict, portray. **Schilderung** f (-en) description; depiction, portrayal

schild·förmig a shield-shaped, scutiform. **Sch—käfer** m tortoise beetle. **-knorpel** m thyroid cartilage. **-kröte** f turtle, tortoise. **-laus** f cochineal (or coccid) insect. **-patt** n tortoise shell. **sch—patten** a tortoise-shell. **Sch—wache** f sentinel, sentry

Schilf n reed(s), rush(es)

schilferig a scaly. **schilfern** vi scale off, exfoliate

Schilf·glanz m, **-glaserz** n freieslebenite.

-rohr n reed(s), rush(es). **-torf** m phragmites peat

Schiller m iridescence, play of colors; shot color. **--farbe** f changeable color. **sch—farbig** a iridescent, metallic-color. **Sch—glanz** m iridescent (or colored metallic) luster

schillerig a iridescent

schillern vi iridesce, show a play of colors: **blau sch.** give a blue iridescence. **schillernd** p.a iridescent

Schiller·quarz n (Min.) cat's-eye. **-seide** f changeable (or shot) silk. **-spat** m (Min.) schiller spar, altered enstatite; bastite, diallage. **-stein** m schiller spar. **-stoff** m iridescent substance. **-taft** m (Tex.) shot (or watered) taffeta. **-wein** m mixed red and white wine

Schimmel m (-) **1** (blue) mold; mildew. **2** white (or gray) horse

schimmel·artig a mold-like, moldy. **Sch—bildung** f mold formation. **-farbe** f mold color. **-fleck** m mold (or mildew) stain. **sch—fleckig** a spotted with mold. **Sch—geruch** m moldy (or musty) smell. **sch—grau** a moldy gray

schimmelig a moldy, musty

schimmeln vi get moldy (or mildewed)

Schimmel·pilz m mold fungus. **-rasen** m coating of mold

Schimmer m **1** shimmer, glimmer. **2** glitter, gleam; luster. **3** (Oil) bloom. **4** touch, tinge; (slight) trace. **5** (faint) idea, notion (of a matter). **schimmern** vi shimmer, glimmer, glow; (+color) have a tinge of

Schimpanse m (-n, -n) chimpanzee

Schindel f (-n) **1** shingle; clapboard. **2** (Med.) splint

schinden* (schund, geschunden) **I** vt **1** drive (a person) like a slave; harass; mistreat. **2** avoid paying. **3** exploit. **4** (orig. meaning, now obsol) skin, flay. **II: sich sch.** slave, toil. **Schinder** m slavedriver, rawhider

Schinken m (-) ham

Schinnen fpl dandruff

Schinus·öl n oil of Schinus molle

Schippe f (-n) **1** shovel, scoop, spade (incl Cards). **2** dustpan. **3** grimace. **schippen** vt shovel, scoop

schipperig a (Dye.) skittery, mixtury

Schiras *m* (-) Shiraz (woolen rug)

Schirbel *m* (-) (*Metll.*) bloom; stamp (of metal)

Schirm *m* (-e) **1** screen (both as shield and as scope), picture screen. **2** (lamp)shade. **3** visor (of cap). **4** umbrella, parasol. **5** parachute. **6** shield; protector. **7** (*Bot.*) umbel. **-bild** *n* screen image. **-blütler** *m* umbelliferous plant

schirmen *vt* (+**vor**) shield, protect, screen (from)

Schirm·gitter *n* (*Radio*) screen grid. **-gitterröhre** *f* screen-grid tube. **-pflanze** *f* umbelliferous plant. **-wirkung** *f* shielding (or screening) effect

Schirokko *m* (-s) scirocco (hot desert wind)

schlabbern I *vt* **1** slop up, slurp (food). **II** *vi* **2** overflow. **3** slobber, slop. **4** jabber

Schlabber·rohr *n* overflow pipe (or tube). **-ventil** *n* overflow (or check) valve.

Schlacht *f* (-en) battle

schlachten *vt/i* slaughter, butcher, kill. **Schlächter** *m* (-) butcher; slaughterer

Schlacht·feld *n* battlefield. **-fett** *n* (butcher's) fat. **-haus** *n*, **-hof** *m* slaughterhouse, (meat) packing house. **-haustalg** *m* packing-house tallow. **-schiff** *n* battleship. **-tier** *n* animal for slaughtering. **-vieh** *n* animals for slaughtering

Schlack *m* (-e) **1** niter deposit. **2** pulp, mash. **3** sleet (rain and snow)

Schlacke *f* (-n) **1** (*Metll. & Geol.*) slag. **2** waste matter, (undigested) residue; dross. **3** cinder(s), scoria; clinker. **sch-bildend** *a* slag-forming

schlacken *vi* (form) slag

Schlacken·abscheider *m* slag separator (or skimmer). **-absonderung** *f* slag separation. **-abstichloch** *n* (*Med.*) slag notch. **sch-ähnlich**, **-artig** *a* slag-like, scoriaceous. **Sch-auge** *n* cinder notch (or tap). **-beton** *m* slag (or cinder) concrete. **sch-bildend** *a* slag-forming. **Sch-bildner** *m* slag former, flux. **-bildung** *f* slag formation, scorification. **-binder** *m* cement from ground slag and hydrated lime. **-blei** *n* slaggy lead. **-einschluß** *m* slag inclusion. **-form** *f* **1** cinder block. **2** cinder notch. **sch-frei** *a* slag-free, cinder-free. **Sch-frischen** *n* (*Metll.*) pig boiling. **-gang** *m* slag duct, cinder fall. **-grube** *f* slag pit. **-halde** *f* slag heap (pile, dump). **sch-haltig** *a* containing slag. **Sch-herd** *m* slag hearth (or furnace). **-kammer** *f* slag chamber. **-klotz**, **-klumpen** *m* clinker, lump of slag. **-kobalt** *m* (*Min.*) safflorite. **-kost** *f* diet containing unassimilable material. **-kuchen** *m* cake of slag, clinker. **-lava** *f* slag-like lava. **-loch** *n* cinder notch (or tap), slag hole. **-mehl** *n* ground slag, Thomas meal. **-ofen** *m* slag furnace. **-pfanne** *f* slag ladle. **-puddeln** *n* pig boiling. **sch-reich** *a* high-slag, high-cinder, rich in slag (or cinders). **-rein** *a* slag-free, cinder-free. **Sch-rösten** *n* roasting of slag. **-sand** *m* granulated blast-furnace slag. **-scherbe** *f*, **-scherben** *m* scorifier. **-spieß** *m* slag iron. **-spur** *f* cinder notch (or trap), slag hole. **-staub** *m* coal dust. **-stein** *m* slag block (or brick). **-stich** *m*, **-stichloch** *n* cinder notch, slag hole. **-wolle** *f* slag (or mineral) wool. **-zacken** *m* cinder plate, front plate. **-zement** *m* slag cement. **-ziegel(stein)** *m* slag brick. **-zinn** *n* block tin (extracted from slag)

schlackig *a* slaggy, cindery, drossy; clinkery; scoriaceous

Schlaf *m* sleep: —**Sch. haben** be sleepy. **-arznei** *f* sleeping drug, soporific. **sch-bedürftig** *a* sleepy, drowsy. **-befördernd** *a* sleep-inducing

Schläfe *f* (-n) (*Anat.*) temple

schlafen* (schlief, geschlafen; schläft) sleep; lie dormant

Schläfen·bein *n* (*Anat.*) temporal bone

schlaff *a* loose, limp, flabby; weak. **Schlaffheit** *f* looseness, limpness, flabbiness; weakness

Schlaf·krankheit *f* sleeping sickness. **sch-los** *a* sleepless. **Sch-losigkeit** *f* sleeplessness, insomnia. **-mittel** *n* sleeping drug, soporific. **-mohn** *m* opium poppy

schläfrig *a* sleepy

schläft sleeps: *pres of* schlafen

Schlaf·tabletten *fpl* sleeping pills. **-trunk** *m* sleeping potion. **-wagen** *m* sleeping car. **-wandeln** *n* sleepwalking, somnambulism. **-zimmer** *n* bedroom

Schlag *m* (-e) **1** blow, stroke; impact: **mit einem Sch.** at one blow, instantaneously;

Sch. auf Sch. in rapid succession. **2** hit (*also* = **s'** **ss**). **3** (electric) shock. **4** beat (of heart). **ɔ** (*Med.*) (apoplectic) stroke. **6** thud, bang, crash. **7** cutting (down) (of trees). **8** (forest) clearing. **9** (planted) strip, field. **10** song, call (of birds). **11** helping (of food). **12** (pigeon) cote. **13** breed, stock; kind, type. **14** wobble (of record player). **-ader** *f* artery. **-anfall** *m* (apoplectic) stroke. **sch-artig** *a* sudden, instantaneous. **Sch-besen** *m* (eggbeating) whisk. **-biegefestigkeit** *f* impact bending strength. **-bohrer** *m* percussion drill. **-drehversuch** *m* torsion impact test. **-druck** *m* impact compression

Schlägel *m* (-) (*Min.*) mallet, hammer
Schlag·elastizität *f* impact elasticity
Schlägel·mühle *f* hammer mill
Schlage·lot *n* hard solder
Schlag·empfindlichkeit *f* impact (*or* percussion) sensitivity

schlagen* (schlug, geschlagen; schlägt) **I** *vt* **1** hit, strike; beat, whip. **2** knock, pound; nail. **3** break, smash. **4** cut, chop (trees). **5** defeat. **6** churn (butter). **7** press (oil). **8** play, strum. **9** wrap. **10** fold. **11** throw; put. **12** add (on). **13** (*in expr with nouns*) get, take, make do: **Kapital sch.** (**aus**) get (*or* make) a profit (from); **Falten sch.** make (*or* form) folds; **Wurzeln sch.** take root; **Alarm sch.** sound the alarm; (*For others, look under the noun*) *cf* BUCH, etc. **II** *vt/i* **14** strike (the hour). **III** *vi* (h) **15** (*usu* + **mit**) beat, flap, jerk; slam. **16** sing, call (*as:* birds). **17: das schlägt in ihr Fach** that belongs (*or* is) in their line (of work). **18** (+**nach**) take (after) (=resemble). **IV** *vi* **19** leap (*as:* flames); (*as:* smoke). **20** fall (and hit): **mit dem Kopf gegen den Tisch sch.** hit one's head on the table. **V: sich sch. 21** beat one's way. **22** move, veer (in some direction); (*esp* + **zu**) join (a group, a cause). **23** (*recip*) scuffle, brawl. **24: das schlägt sich auf die Lunge** that affects (*or* settles on) the lungs. **schlagend** *p.a* (*oft*) persuasive, convincing. **2** effective, telling. **3** (*Min.*): **sch—e Wetter** firedamp. **Schlager** *m* (-) hit (song). **Schläger** *m* (-) **1** (egg) beater, whisk. **2**

bat; paddle. **3** sword, rapier
Schlag·festigkeit *f* impact strength, shock resistance. **-figur** *f* percussion figure. **-fläche** *f* striking (*or* impact surface). **-gold** *n* gold leaf, leaf gold. **-härte** *f* impact hardness. **-kraft** *f* striking (*or* impact) force. **-kreuzmühle** *f* cross-beater mill, disintegrator. **-ladung** *f* (*Explo.*) booster charge. **-licht** *n* **1** strong light, glare. **2** highlight; spotlight. **-lot** *n* hard solder. **-mühle** *f* hammer mill; beetling mill, beetle. **-obers** *n* whipped cream. **-patrone** *f* (*Explo.*) priming cartridge. **-pressen** *n* impact molding. **-probe** *f* percussion (*or* impact) test. **-pulver** *n* fulminating powder; (*Pharm.*) antapoplectic powder. **-rahm** *m* whipped cream. **-saat** *f* hempseed. **-sahne** *f* whipped cream. **-schrauber** *m* percussion screwdriver. **-sieb** *n* vibrating screen, precipitating sieve. **-siebprobe** *f* shatter test. **-stauchversuch** *m* impact compression test. **-stiftmühle** *f* stamping mill

schlägt strikes, etc: *pres of* SCHLAGEN
Schlag·versuch *m* impact (*or* percussion) test. **-wasser** *n* bilge water. **-weite** *f* (*Elec.*) spark gap, sparking distance. **-werk** *n* **1** (*Mach.*) rammer; impact testing machine. **2** signal bell. **3** (*Clocks*) striking mechanism. **-wetter** *npl* firedamp, methane
schlagwetter·frei *a* firedamp-free. **-geschützt** *a* (*Min.*) flameproof. **Sch—grube** *f* fire mine. **sch—sicher** *a* firedamp-proof, flameproof. **-zündfähig** *a* ignitable by firedamp
Schlag·widerstand *m* impact resistance. **-wort** *n* **1** catchword, slogan. **2** cue. **-zähigkeit** *f* impact strength. **-zeile** *f* headline. **-zünder** *m* percussion fuse. **-zündhütchen** *n* percussion cap
Schlamm *m* (-̈e) **1** mud. **2** sludge. **3** silt. **4** slurry. **5** (*Ceram.*) slip
Schlämm· elutriation, washing: **-analyse** *f* elutriation (*or* sedimentation) analysis. **-apparat** *m* elutriation (*or* washing) apparatus. **-arbeit** *f* elutriation process
schlamm·artig *a* mud-like, sludge-like; muddy, slimy. **Sch—bad** *n* mud bath
schlämmbar *a* washable (by elutriation)
Schlamm·belebung *f* activation of sludge.

-boden m muddy (or boggy) soil. **-ein-dicker** m 1 (Ores) pulp thickener. 2 (Sewage) sludge densifier

schlämmen vt 1 elutriate, wash (powdered substances); levigate; slurry. 2 clear (of mud); dredge. 3 (Min.) buddle (ores). Cf SCHLEMMEN

Schlämm·faß n washing tub (for precipitates); (Min.) dolly tub

Schlamm·faulraum m sludge digestion tank

Schlämm·flasche f elutriation flask

Schlamm·flüssigkeit f mud-laden liquid

Schlämm·gefäß n elutriatring vessel (or reservoir). **-glas** n elutriating glass

Schlamm·grube f mud (sludge, slime) pit. **sch–haltig** a muddy, sludgy, containing mud (or sludge)

Schlämm·herd m (Metll.) slime pit (or table)

schlammig a muddy, sludgy, slimy

Schlammischapparat m mud (or slime) mixer (=**Schlamm·mischapparat**)

Schlamm·kalk m sludge lime

Schlämm·kelch m elutriation vessel

Schlamm·kohle f coal slime (or mud), mud coal

Schlämm·kohle f washed coal. **kreide** f prepared chalk; whiting

Schlamm·packung f mud pack. **-pfänn-chen** n (Salt) scum pan

Schlämm·pipette f sedimentation pipette

Schlamm·pumpe f sludge (or slurry) pump. **-saft** m (Sugar) carbonation (or scum) juice. **-scheider** m slime separator

Schlamm· (or **Schlämm·**) **-schlich** m washed ore slime

Schlamm·stein m mudstone

Schlämm·trichter m elutriating funnel

Schlamm·trübe f (Metll.) slime pulp

Schlämmung f (-en) elutriation, washing; levigation; slurrying; dredging; buddling; cf SCHLÄMMEN

Schlämm·verfahren n elutriation (washing, buddling, sedimentation) process. **-vorrichtung** f elutriating (or washing) apparatus

Schlamm·vulkan m mud volcano (or spring). **-wasser** n muddy (sludgy, slimy) water; (Sewage) sludge liquor

Schlämm·werk n (ore) washing mill.

-zylinder m elutriating cylinder

schlampig a sloppy, lax, disorderly

schlang wound, etc: past of SCHLINGEN

Schlange f (-n) 1 snake, serpent. 2 coil; (flexible) hose. 3 (Mach.) worm. 4 line (of waiting people or cars): **Sch. stehen** stand in line

schlängeln vr: **sich sch.** wind, meander: **sich sch–d** p.a (oft) sinuous, serpentine. Cf GESCHLÄNGELT

schlangen·artig a snake-like, serpentine. **Sch–biß** m snakebite. **-gift** n snake venom. **-holz** n snakewood. **-kühler** m coil (or spiral) condenser; tubular cooler. **-rohr** n, **-röhre** f coil(ed tube), spiral tube. **-stein** m (Min.) ophite, serpentine. **-wurzel** f snakeroot

schlank a slim, slender. **Schlankheit** f slimness, slenderness

Schlankheits·bad n slenderizing bath. **-grad** m, **-zahl** f slenderness ratio

schlankweg adv 1 flatly, bluntly. 2 downright, outright

schlapp a 1 slack, limp, flabby. 2 listless, languid

Schlapper·milch f curdled milk

schlau a cunning, crafty, sly; clever —**sch. werden (aus)** grasp, understand

Schlauch m (-e) 1 tube, hose. 2 (Tires) inner tube. 3 (wine)skin. 4 (Min.) ore pipe, chimney. 5 drunkard; glutton

schlauch·artig a tube-like, tubular. **Sch–boot** n inflatable raft

schlauchen I vt 1 hose (esp beer into barrels), siphon. 2 wear out. **II** vi drink heavily. 3 sponge (on others)

Schlauch·filter n tube filter. **-folie** f tubular film. **-klemme** f tube clamp, hose clip. **-leitung** f 1 hose line. 2 (Elec.) rubber-sheathed cable. **Sch–pilz** m asomycete. **-reifen** m tube tire. **-sicherung** f hose protection. **-stück** n tube attachment; hose coupling. **-tiere** npl coelenterata. **-verbindung** f hose connection (or coupling)

Schlaufe f (-n) (Transp.) (standees') strap, loop

schlecht 1a bad, poor(-quality), ill cf GEHEN 10; adv oft not (very) well —**ihnen ist (ihnen wird) sch.** they are (getting) sick. 2 (as pfx on adj oft) ill-: **schlecht·beraten**

ill-advised. **schlechter** *a* (*compar of* SCHLECHT) worse, poorer

schlechterdings *adv* downright, absolutely; virtually, just about

schlechteste *a* (*superl of* SCHLECHT) worst, poorest

schlecht·hin, -weg *adv* **1** per se, par excellence. **2** downright, absolutely

schlecken 1 *vt* lap up. **2** *vi* (*usu* + **an**) lap (at), feast (on)

Schlegel *m* (-) **1** mallet, (wooden) hammer, beetle. **2** (bell) clapper. **3** drumstrick. **4** thigh (bone); leg (of lamb, veal)

Schleh·dorn *m*, **Schlehe** *f* (-n) blackthorn; sloe (plum). **Schlehen-likör** *n* sloe gin

Schlei *m* (-e) tench (fish)

schleichen* (schlich, geschlichen) *vi* (s) & **sich sch.** sneak, steal, slink. **schleichend** *p.a* sneaky, stealthy; insidious; creeping (illness, inflation)

Schleich·gut *n* smuggled (*or* black-market) goods. **-handel** *m* black market, shady (*or* illicit) deal(ing); smuggling. **-werbung** *f* subliminal advertising

Schleie *f* (-n) tench (fish)

Schleier *m* (-) **1** veil. **2** haze; cloudiness, turbidity; (smoke)screen. **3** (*Phot.*) fog. **4** (*Tex.*) lawn. **--bildung** *f* haze formation, fogging; blushing, surface film formation. **sch--frei** *a* haze-free, fog-free

schleierhaft *a* veiled, mysterious

schleierig *a* hazy, foggy, cloudy

schleiern *vt* veil, cloud; (*Phot.*) fog

Schleif·apparat *m* grinder

schleifbar *a* grindable, polishable, etc. Cf SCHLEIFEN II

Schleif·drahtwiderstand *m* slide-wire resistor

Schleife *f* (-n) **1** loop. **2** noose; bow (knot). **3** slide. **4** (wood-hauling) sledge. **5** sled. **6** dragnet

schleif·echt *a* (*Lthr.*) fast to buffing. **Sch--emaillelack** *m* (*Varnish*) dull finish

schleifen I (*reg*) **1** *vt/i* drag. **2** *vt* tear down. **3** *vi* (+**an**) rub (on, against); (*Mach.*: clutch) slip. **II*** (schliff, geschliffen) *vt* **4** sharpen, whet. **5** grind; cut (glass, gems). **6** polish. **7** buff (leather). **8** settle and fit (soap)

Schleifer *m* (-) **1** sharpener; grinder, cutter; polisher. **2** (*Elec.*) slide contact =

SCHLEIFSCHUH. **Schleiferei** *f* (-en) **1** grinder's shop; grinding (trade). **2** (*Paper*) pulp mill; pulp manufacture

schleif·fähig *a* grindable, etc *cf* SCHLEIFEN II. **Sch--fläche** *f* wearing surface. **-flüssigkeit** *f* (*Varnish*) rubdown liquid. **-güte** *f* abrasive temper. **-holz** *n* pulpwood. **-kontakt** *m* (*Elec.*) sliding contact. **-korn** *n* abrasive grain. **-körper** *m* **1** (*genl*) abrasive material. **2** (*specif*) grinding wheel. **-lack** *m* rubbing (flash, flatting) varnish; sanding lacquer. **-lackwaren** *fpl* polished lacquer goods. **-leder** *n* buffed leather. **-leinen** *n* abrasive cloth. **-material, mittel** *n* abrasive. **-öl** *n* grinding oil. **-papier** *n* abrasive paper. **-paste** *f* abrasive (e.g. emery) paste; (*Varnish*) flatting paste. **-polieren** *n* fine grinding (of surfaces). **-pulver** *n* grinding (*or* polishing) powder. **-rad** *n* grinding (*or* polishing) wheel. **-ring** *m* slip ring. **-rohstoff** *m* abrasive raw material. **-scheibe** *f* grinding disk (*or* wheel). **-schuh** *m* (*Elec.*) (sliding) pick-up shoe

Schleifsel *npl* grindings

Schleif·staub *m* grindings, swarf. **-stein** *m* grindstone; whetstone. **-stoff** *m* grinding material, abrasive; (*Paper*) ground pulp

Schleifung *f* (-en) dragging; demolition; sharpening; grinding; polishing; buffing *cf* SCHLEIFEN

Schleif·wachs *n* (*Varnish*) flatting paste. **-wirkung** *f* abrasive action, abrasion; attrition

Schleim *m* **1** slime. **2** mucus. **3** (*Bot.*) mucilage. **4** cooked cereal, porridge. **--absonderung** *f* mucous secretion. **sch--artig** *a* slimy, mucous; mucoid; glutinous. **Sch--bakterien** *npl* slime bacteria, Myxobacteriaceae. **-beutel** *m* mucous sac, bursa mucosa. **-beutelentzündung** *f* bursitis. **sch--bildend** *a* slime-forming; (*Physiol.*) muciparous. **Sch--bildung** *f* slime (mucus, mucilage) formation. **-blatt** *n* mucous layer. **-drüse** *f* mucous gland. **-gärung** *f* viscous fermentation. **sch--gebend** *a* glutinous, mucilaginous. **Sch--harz** *n* gum resin. **-haut** *f* mucous membrane

schleimig *a* slimy, mucilaginous; mucous: **sch--e Gärung** mucous (*or* viscous) fermentation. **Schleimigkeit** f sliminess; mucosity

Schleim·membran f mucous membrane. **-pilz** m slime fungus (or mold). **sch–sauer** a mucate of. **Sch–säure** f mucic acid. **-schicht** f (genl) layer of slime, mucous layer; (specif) Malpighian layer (of the epidermis). **-stoff** m slimy substance, mucilage; mucin. **sch–stoffartig** a mucin-like, mucoid. **Sch–tier** n mollusk. **-zellulose** f mucocellulose. **-zucker** m levulose

Schleiße f (-n) splint(er); quill. **schleißen*** (schliß, geschlissen) vt strip (feathers); slit, split (wood) cf VERSCHLEISSEN

Schleiß·wirkung f abrasive action

Schlemm· see SCHLÄMM·. **schlemmen** vt/i feast (on) cf SCHLÄMMEN

Schlempe f (-n) 1 distiller's wash, spent wash. 2 (Brew.) pot ale, malt residue; (as feed) swill, slops. 3 (Sugar) vinasse. 4 (Cement) grout. **--asche, -kohle** f saline, crude potash from beet vinasse. **-ofen** m spent-wash furnace. **-verdampfer** m vinasse (or mash) evaporator

schlenkern 1 vt fling. 2 vt/i (h) (oft + mit) swing, dangle. 3 vi (s) skid, swerve; stroll

Schlepp·dampfer m tugboat. **Schleppe** f (-n) 1 sledge; ground leveler. 2 canal tow. 3 (Paper) felt board. **schleppen** 1 vt/i drag, trail. 2 vt tow; lug. 3 vi (+an) & **sich sch.** (+mit) lug, struggle (with a heavy load). 4: **sich sch.** drag on; drag oneself. **schleppend** p.a sluggish, lagging, halting. **Schlepper** m (-) 1 tractor. 2 tugboat. 3 hauler

Schlepp·kraft f tractive force. **-mühle** f drag(-stone) mill. **-netz** n dragnet. **-tau** n towrope

Schleppung f (-en) 1 dragging; towing. 2 time lag

Schlepp·zeiger m trailing (or friction) pointer

Schlesien n Silesia (now part of Poland). **schlesisch** a Silesian

Schleuder f (-n) 1 centrifuge; (in compds oft) centrifugal. 2 sling, catapult. **--förderer** m centrifugal conveyor. **-gebläse** n centrifugal blower. **-guß** m centrifugal casting. **-honig** m extracted honey. **-korb** m centrifuge (or hydroextractor) basket.

-kraft f centrifugal force. **-maschine** f 1 centrifugal machine; (specif) separator, (hydro)extractor. 2 catapult. **-mühle** f centrifugal mill

schleudern I vt 1 fling, catapult. 2 centrifuge; separate, (hydro)extract. 3 spin-dry. 4 sell below cost, dump (on the market). II vi (s) skid, swerve

Schleuder·pumpe f centrifugal pump. **sch–sicher** a anti-skid, nonskid. **Sch–sitz** m ejection seat. **-sortierer** m centrifugal sorter. **-trommel** f centrifugal drum. **-ware** f goods dumped on the market, catchpenny merchandise. **-wirkung** f centrifugal action (or effort)

schleunig a prompt, swift, speedy, immediate. **schleunigst** adv as fast as possible, right away

Schleuse f (-n) 1 floodgate; sluice(gate). 2 (canal) lock. 3 sewer. 4 aseptic chamber; air lock. **schleusen** vt 1 lock (a ship through a canal). 2 pilot, channel (through various stages). **Schleusen·gas** n sewer gas

schlich sneaked, etc: past of SCHLEICHEN

Schlich m (-e) 1 (Metll.) slime, slick; grinder's dust. 2 (usu pl) trick(s), artifice(s). **--arbeit** f smelting of slimes

schlicht a 1 simple, plain. 2 smooth. 3 (Files) fine. **Sch–·arbeit** f smooth-finishing (work)

Schlichte f (-n) 1 (Tex., Paper) size. 2 (Foun.) blackwash. 3 (Plaster) skim coat

schlicht·echt a sizing-resistant

schlichte·haltig a containing size. **Sch–mittel** n sizing agent, size

Schlicht·emulsion f sizing emulsion

schlichten vt 1 smooth, plane; (Metals) planish; (Lthr.) sleek. 2 arrange in order. 3 settle, mediate **—sch—d eingreifen** intervene as a mediator. 4 (Tex., Paper) size. 5 (Plaster) skim. 6 (Foun.) blackwash (molds)

Schlicht·feile f smoothing file. **-flüssigkeit** f (Tex.) size. **-leim** m sizing, size. **-masse** f sizing paste. **-mittel** n sizing agent. **-öl** n sizing oil. **-walze** f finishing roll

Schlick m (-e) 1 mud, slick, ooze, silt. 2 (Metll.) schlich, ore slime(s). 3 (Mach.) (grinding) sludge. **schlicken** vi fill up with mud (ooze, silt)

Schlicker *m* (-) 1 (*Ceram.*) slip, slop. 2 (*Metll.*) dross. **-gießen** *n* slip casting.

schlickern 1 *vt* (*Metll.*) dross. 2 *vi* (s) slip, slide

Schliech *m* (-e) (*Metll.*) slime(s), schlich

schlief slept: *past of* SCHLAFEN

schliefen* (schloff, geschloffen) *vi*(s) slip, creep; slink

Schliere *f* (-n) (*usu pl*) schlieren, streaks. **-verfahren** *n* (*Phot.*) schlieren process.

schlierig *a* 1 streaky, striated. 2 slippery

schließbar *a* closable, lockable

Schließe *f* (-n) 1 buckle; clasp. 2 fastener, fastening (pin *or* bolt). 3 (*Locks*) catch, snap

schließen* (schloß, geschlossen) I *vt, vi, sich sch.* 1 close, shut; lock; seal; button up. II *vt/i* 2 close, end, conclude. III *vt* 3 (+**an**) fasten (on, to); add (to). 4 (*usu* + **in**) enclose, embrace (in); (+**in sich**) include, contain; involve. 5 make, enter into, conclude (agreements, etc). 6 (*oft* + **aus**) conclude, infer (from); (*also* + **auf**) draw conclusions (as to). IV: *sich sch.* 7 (+**in**) lock oneself (in, into). 8 (+**an**) follow. *Cf* GESCHLOSSEN

Schließ-fach *n* lock (strong, safe deposit, post office) box

schließlich *adv* 1 finally. 2 ultimately, in the end. 3 after all

Schließ-rohr *n*, **-röhre** *f* sealed tube

schliff ground, etc: *past of* SCHLEIFEN II

Schliff *m* (-e) 1 grinding, sharpening, polishing —**der letzte Sch.** the finishing touches. 2 ground (*or* polished) surface; smooth spot. 3 grind, polish (*also* = refinement), finish; cut (of a gem). 4 flair. 5 grindings, (ground) pulp. 6 ground (glass) joint. **-apparat** *m* apparatus with ground-glass joints. **-bild** *n* microphotograph; (*Geol.*) micrograph, microsection. **-kolben** *m* flask with ground-glass stopper. **-rohr** *n*, **-röhre** *f* tube with a ground-glass joint. **-stelle** *f* ground (*or* polished) spot, grinding. **-stopfen** *m* ground-glass stopper. **-stück** *n* ground piece. **-verbindung** *f* ground-glass joint

schlimm *a* 1 (*genl*) bad; sore (finger, etc); evil, wicked (person, plans). 2 *compar:* **schlimmer** worse. 3 *superl:* **schlimmste**

worst. **schlimmstenfalls** *adv* in the worst case

Schlinge *f* (-n) 1 loop; sling. 2 slipknot; noose. 3 snare, trap. 4 mesh. 5 tendril

schlingen* (schlang, geschlungen) I *vt* 1 sling, fling. 2 wrap, wind. 3 braid, twine. 4 tie. II *vt/i* swallow; bolt (one's food). III: *sich sch.* wind, coil, twine

Schling-pflanze *f* climbing plant

Schlippesches Salz Schlippe's salt, sodium thioantimonate

schliß split, etc: *past of* SCHLEISSEN

Schlitten *m* (-) 1 sled, sledge. 2 (*Mach.*) slide, sliding carriage. 3 slide rail

Schlitt-schuh *m* (ice) skate: **Sch. laufen** skate

Schlitz *m* (-e) 1 slit, slot. 2 crack, fissure; cleft. **-aufsatz** *m* wing top (of a burner). **-brenner** *m* batswing burner. **-düsenauftragmaschine** *f* (*Plastics*) air-knife coater

schlitzen *vt* slit

schlitz-förmig *a* slit-shaped, slit-like

schloff slipped, etc: *past of* SCHLIEFEN

schloß closed, etc: *past of* SCHLIESSEN

Schloß *n* (Schlösser) 1 lock; clasp. 2 castle

Schloße *f* (-n) hailstone. **schloßen** *vi* hail

Schlosser *m* (-) locksmith; fitter, mechanic. **Schlosserei** *f* (-en) 1 locksmith's shop. 2 metalworking shop. 3 locksmithing. 4 light engineering

Schlot *m* (-e) 1 chimney (*incl Geol.*), (smoke)stack, flue. 2 soil pipe

Schlotte *f* (-n) 1 (*Min., Geol.*) cavity, sink. 2 (*Bot.*) hollow stalk

Schlotter *m* (*Salt*) sediment from boiling

schlotterig *a* 1 shaky, wobbly. 2 loose, flapping. **schlottern** *vi* 1 shake, wobble. 2 be loose(-fitting), flap

Schlucht *f* (-en) ravine, gorge

schluchzen *vi* sob

Schluck *m* (-e) sip, mouthful, drink (of liquid). **-beiwert** *m* coefficient of absorption, absorptivity. **schlucken** *vt/i* swallow; absorb. **schluckend** *p.a* (*oft*) absorbent

Schlucker-zahl *f* absorption coefficient

Schluck-stoff *m* absorbent material

Schluckung *f* absorption; swallowing

Schluff *m* (-e) 1 silt. 2 (low-grade) potter's clay

schlug struck, etc: *past of* SCHLAGEN

Schlummer *m* slumber. **schlummern** *vi* slumber; be dormant. **schlummernd** *p.a* (*oft*) dormant

Schlund *m* (-̈e) 1 throat, pharynx; (*in compds, usu*) pharyngeal. 2 chasm; pit, maw. **-̈kopf** *m* upper pharynx. **-rohr** *n,* **röhre** *f* esophagus

Schlupf *m* (-̈e) 1 hatching. 2 (*Mach.*) (wheel) slippage. 3 (*Cmptrs.*) slack. 4 refuge. 5 gap, loophole. **schlüpfen, schlupfen** *vi* (s) slip

schlupf·frei *a* nonslip, nonskid. **Sch-loch** *n* loophole

schlüpfrig *a* slippery; lubricious; lewd — **sch. machen** lubricate. **Schlüpfrigkeit** *f* slipperiness; oiliness (of lubricants); lubriciousness; lewdness

Schlüpfung, Schlupfung *f* slip(ping), slippage

Schlupf·winkel *m* hideout, haunt

schlürfen I *vt* 1 sip noisily. 2 drink in. **II** *vi*(h) 3 drink noisily. 4 (*Pumps*) suck air. **III** *vi*(s) shuffle (along)

Schluß *m* (Schlüsse) 1 closing, shutting; closing time. 2 end, close, conclusion: **Sch. machen (mit)** put an end (to), stop, quit; **zum Sch.** finally; **es ist Sch. damit** that's the end of that. 3 conclusion, inference: **Schlüsse ziehen** draw conclusions. 4 (*Tex.*) closeness, body. 5 (*Elec.*) connection; short circuit. 6 seal: **der Deckel hat einen guten Sch.** the cover closes (*or* seals) tightly. 7 (*in compds usu*) final: **-bilanz** *f* final balance. **-deckel** *m* (tight-fitting) cover (*or* lid)

Schlüssel *m* (-) 1 key. 2 wrench. 3 code. 4 ratio (of a distribution system). **-̈bein** *n* collar bone, clavicle. **-blume** *f* primrose. **-bund** *m* bunch of keys. **-formel** *f* key formula. **-loch** *n* keyhole. **-nummer** *f* key number

Schlüsselung *f* (-en) coding

Schluß·ergebnis *n* final result. **-folge-(rung)** *f* conclusion, deduction. **-härten** *n,* **-härtung** *f* final hardening

schlüssig *a* 1 conclusive, convincing. 2 e.g. **sie sind (sie werden) sich sch. über. . .** they have made up (they are making up) their minds about. . .

Schluß·prüfer *m* (*Elec.*) short-circuit test-

er. **-prüfung** *f* final examination. **-punkt** *m* final period. **-stein** *m* keystone. **-strich** *m* (*Paint*) final coat. **-weise** *f* manner of reasoning

Schlutte *f* (-n) alkekengi (ground cherry)

schm. *abbv* 1 (schmelzend) melting. 2 (schmilzt) melts

schmächtig *a* 1 frail, delicate. 2 slight, slim

Schmack *m* (-e) (tanning) sumac. **schmacken** *vt* (treat with) sumac. **schmack·gar** *a* sumac-dressed

schmackhaft *a* tasty, savory; palatable

schmackieren *vt* (treat with) sumac

schmal *a* 1 narrow; slim. 2 scanty

schmälern *vt* 1 narrow; reduce, diminish. 2 downgrade; detract from

Schmal·film *m* narrow film (*usu* 8 mm). **-leder** *n* tanned skins, *esp* calfskin for uppers. **-seite** *f* narrow side, edge. **-spur** *f* 1 narrow gauge. 2 (*compounded with field of study*) e.g.: **Schmalspurchemiker** *m* student taking a minor in chemistry

Schmalt *m* (-e) enamel. **-̈blau** *n* = **Schmalte** *f* (-n) smalt (blue). **schmalten** *vt* enamel

Schmalz *n* (-e) 1 lard, grease; melted fat; drippings. 2 schmaltz. **sch-̈artig** *a* lardaceous, lardy. **Sch-̈butter** *f* melted butter

Schmälze *f* (-n) (*Tex.*) spinning oil; lubricant

schmalzen *vt* prepare (food) with lard; deep-fry

schmälzen *vt* grease; (*Tex.*) oil

schmalzig *a* 1 lardy; greasy. 2 schmaltzy

Schmälz·masse *f* (*Tex.*) softener. **-mittel** *n* (*Tex.*) oiling agent

Schmälz·öl *n* lard (oleo, wool) oil

Schmälz·vorgang *m* (*Tex.*) oiling process

Schmant *m* 1 (dairy) cream. 2 dirt, grime

schmarotzen *vi* live as a parasite; sponge. **schmarotzend** *p.a* (*oft*) parasitic. **Schmarotzer** *m* (-) parasite; (*in compds usu*) parasitic. **schmarotzerisch** *a* parasitic **Schmarotzer·tier** *n* parasitic animal

Schmauch *m* (thick) smoke. **schmauchen** *vt/i* smoke, puff away (at)

schmecken 1 *vt/i* (*oft* + *nach*) taste (of). 2 *vi* taste good: **das schmeckt ihnen** they like (the taste of) that; **sie lassen es sich sch.** they are relishing it

Schmeck·zelle f taste cell

schmeichelhaft a flattering, complimentary. **schmeicheln** 1 vi (+dat) flatter — **sie sch. sich, viel zu wissen** they like to think they know a lot. 2: **sich sch.** (+in) steal (one's way into)

schmeißen* (schmiß, geschmissen) 1 vt & vi (mit) throw, chuck, fling. 2 vt pull off (a job). 3 vi (Insects) defecate; lay eggs. **Schmeißen** n (oft) warping, distortion

Schmeiß·fliege f blowfly, bluebottle

Schmelz m 1 enamel, glaze. 2 (Zinc) blue powder. 3 mellowness, soft glow; warm tone (of colors, voice). Cf SCHMELZE. **-anlage** f smelter(y), foundry; melting plant. **-arbeit** f 1 smelting. 2 enameling; enameled work. **-arbeiter** m 1 smelter, founder. 2 enameler. **sch-artig** a enamel-like. **Sch-ausdehnung** f expansion on melting. **-bad** n melting bath; molten bath

schmelzbar a fusible. **Schmelzbarkeit** f fusibility

Schmelz·basalt m cast basalt. **-behälter** m melting pot; smelting crucible. **-blau** n smalt. **-butter** f melted butter. **-diagramm** n melting-point diagram. **-draht** m (Elec.) fuse wire. **-druck** m melting pressure. **-druckkurve** f fusion curve, melting-point-pressure curve

Schmelze f (-n) 1 melting, fusion; (Metll.) smelting. 2 melt, (batch of) molten material; fused mass; (Soda) ball; (Petrog.) magma. 3 smeltery. 4 (Spinning) mill oil. 5 thaw. Cf SCHMELZ

Schmelz·einsatz m (Elec.) fuse. **-elektrolyse** f electrolysis of fused melt

schmelzen* (schmolz, geschmolzen; schmilzt) 1 vt, vi(s) melt, fuse; (Metll.) smelt. 2 vt: **teilweise sch.** frit. 3 vi(s) deliquesce. **Schmelzen** n (oft) fusion

Schmelz·enthalpie f melt enthalpy; heat of fusion

Schmelzer m (-) melter; founder; smelter

Schmelz·erde f fusible earth

Schmelzerei f (-en) 1 (s)melting. 2 smeltery, foundry

Schmelz·erz n smelting ore. **-farbe** f vitrifiable pigment; enamel (or majolica) col-

or. **-feuer** n (Iron) refinery. **-fluß** m 1 fused (or molten) mass, melt. 2 fusion. 3 enamel

Schmelzfluß·bad n fused-salt bath. **-elektrolyse** f electrolysis of fused melt, electrolytic smelting. **-elektrolyt** m fused electrolyte. **sch-elektrolytisch** a by electrolytic smelting

schmelz·flüssig a 1 fusible. 2 fused, molten —**sch. gegossen** fusion-cast. **Sch-gefäß** n melting pot; smelting hearth. **sch-geschweißt** a fusion-welded. **Sch-glas** n fusion glass; enamel. **-glasur** f enamel; fusible glaze. **-grad** m melting point. **-gut** n fusible material, smelter charge. **-hafen** m melting pot. **-herd** m smelting hearth; (Soda) front hearth of a black-hearth furnace. **-hitze** f melting heat. **-hütte** f foundry, smeltery

schmelzig a fusible

Schmelz·käse m processed cheese (spread). **-kegel** m (Ceram.) fusible (or Seger) cone. **-kessel** m crucible, melting vessel; (Soda) caustic pan; (Furnaces) slag-tap furnace. **-kirsche** f ignition pellet. **-kitt** m fusible cement. **-koks** m metallurgical (or smelter) coke. **-kontakt** m fused catalyst. **-körper** m ceramic pyrometer. **-kuchen** m fusion residue. **-kunst** f (art of) enameling; smelting. **-kurve** f fusion (or melting-point) curve. **-lampe** f enameler's (or glass-blower's) lamp. **-legierung** f fusible alloy. **-linie** f fusion (or melting-point) curve. **-löffel** m melting ladle. **sch-los** a unenameled, unglazed. **Sch-lösung** f molten solution, melt. **-malerei** f enamel painting. **-mittel** n flux. **-ofen** m (s)melting furnace. **-öl** n lard (oleo, wool) oil. **-p.** abbv (Schmelzpunkt) melt. pt. **-patrone** f (Elec.) fuse cartridge. **-perle** f blowpipe bead; enamel bead, bugle. **-pfanne** f melting pan. **-pfropfen** m fusible plug. **-post** f smelting charge, post. **-prozeß** m (s)melting process. **-pulpe** f (Anat.) enamel pulp. **-pulver** n (powdered) flux. **-punkt** m melting point. **-punktbestimmung** f melting-point determination

Schmelzpunktbestimmungs·rohr n, **-röhre** f melting-point tube

Schmelz·raum m (Metll.) hearth. **-raupe** f

fused-quartz spiral. **-reaktion** f fusion reaction. **-rinne** f (*Metll.*) melting channel. **-röhrchen** n melting tube; blowpipe. **-salz** n emulsifying salt. **-satz** m (*Anal.*) fusion mixture. **-schleuder** f melting centrifuge. **-schweißung** f fusion welding. **-sicherung** f (*Elec.*) fuse. **-soda** f crude sodium carbonate. **-stahl** m German (natural, furnace) steel. **-stein** m (*Min.*) mizzonite, dipyre. **-stopfen, -stöpsel** m fusible plug, fuse plug. **-temperatur** f fusing temperature. **-tiegel** m crucible, melting pot. **-tiegeldeckel** m crucible cover; (*Metll.*) tile. **-tiegelhalter** m crucible holder (*or* support). **-topf** m melting pot. **-tröpfchen** n blowpipe bead

Schmelzung f (-en) fusion, melting; smelting cf SCHMELZEN. **Schmelzungs·punkt** m melting point

Schmelz·verfahren n melting (fusion, smelting) process. **-verhalten** n behavior on melting, fusibility. **-viskosität** f melt viscosity. **-wanne** f melting tank; tank furnace. **-wärme** f 1 (latent) heat of fusion. 2 (*Metll.*) (s)melting heat. **-wasser** n water from melting ice or snow. **-werk** n 1 smeltery, foundry. 2 enameled work. **sch–würdig** a smeltable, ready for smelting. **Sch–zeit** f melting (*or* fusion) time; (*Glass*) journey. **-zement** m alumina cement. **-zeug** n smelting tools (*or* equipment). **-zone** f fusion (*or* smelting) zone

Schmer m or n fat, grease; suet

Schmergel m emery cf SCHMIRGEL

Schmerz m (-en) (*oft pl*) pain(s), ache(s); grief cf KOPFSCHMERZEN, ZAHNSCHMERZEN. **-anfall** m attack of pain. **sch–betäubend** a pain-relieving, analgesic. **-empfindlich** a sensitive to pain

schmerzen vt/i pain, hurt; vt (*oft*) grieve; vi (*oft*) ache

schmerz·erregend a pain-causing. **-frei** a pain-free, painless. **Sch–gefühl** n sensation of pain

schmerzhaft, schmerzlich a painful, sore

schmerzlos a painless. **Schmerzlosigkeit** f painlessness

Schmerz·mittel n pain reliever. **-schwelle** f pain threshold. **sch–stillend** a pain-relieving

Schmetterling m (-e) butterfly

Schmetterlings·blütler m papilionaceous (*or* fabaceous) plant. **-brenner** m batswing burner. **-ventil** n butterfly valve

schmettern 1 vt hurl; slam; bellow (orders, etc) 2 vi(h) blare; clang. 3 vi(s) slam, crash

Schmied m (-e) 1 (black)smith. 2: **Sch. seines Glücks** architect of one's fortune. Cf SCHMIEDE

schmiedbar a malleable, forgeable. SCHMIEDBARKEIT f malleability, forgeability

Schmiede f (-n) smithy, forge cf SCHMIED. **-eisen** n wrought iron; forging steel. **sch–eisern** a wrought-iron. **Sch–feuer** n forge. **-herd** m forge (hearth). **-kohle** f forge coal. **-legierung** f wrought alloy. **-metall** n malleable (*or* forgeable) metal

schmieden vt 1 forge (iron, etc). 2 shape, fashion, make. Cf GESCHMIEDET

Schmiede·presse f forging press. **-schlacke** f forge cinder. **-sinter** m forge scale. **-stahl** m wrought steel. **-stück** n forging

Schmiege f (-n) 1 bevel. 2 (adjustable) protractor

schmiegen I vt 1 bevel. 2 nestle; rest. **II: sich sch. 3** (*usu* + **an**) snuggle (up to), nestle (close to); fit snugly (against), cling (to), hug. 4 (*Geom.*) osculate

schmiegsam a 1 supple, limber. 2 flexible, pliable. **Schmiegsamkeit** f 1 suppleness. 2 flexibility, pliability

Schmiegungs·ebene f (*Geom.*) osculating plane

Schmier·apparat m lubricator. **-büchse** f grease (*or* oil) cup, lubricator

Schmiere f (-n) 1 lubricant, grease, oil. 2 ointment. 3 shoe polish. 4 patch of grease, greasy area. 5 slush, mire; dirt, grime; goo. 6 dip (for animals). 7 spread (on bread)

schmieren I vt 1 lubricate, grease, oil. 2 smear, spread; rub (ointment on. . .); butter (bread). II vt/i scribble, scrawl; daub. III vi: **diese Feder schmiert** this pen smudges

schmier·fähig a 1 good as a lubricant. 2 spreadable (on bread, etc). **Sch–fähigkeit** f 1 good lubricating quality. 2

spreadability. **-fett** n lubricating grease. **-film** m 1 lubricating film. 2 greasy surface (on roads). **-fleck** m grease spot. **-flüssigkeit** f lubricating fluid. **-hahn** m lubricating (or grease) cock

schmierig a 1 greasy, smeary, oily; glutinous; viscous. 2 grimy. 3 slovenly. 4 lewd. **Schmierigkeit** f greasiness, smeariness, etc

Schmier·kalk m fat lime. **-kanne** f oil can. **-käse** m (spreadable) cottage cheese. **-leder** n curried (or oil-dressed) leather. **-masse** f (Mach.) lubricating paste. **-material, -mittel** n 1 lubricant. 2 ointment; liniment. **-öl** n lubricating oil. **-öldestillat** n (Petrol.) lube distillate. **-ölvorrichtung** f lubricator. **-seife** f soft soap. **-stoff** m 1 lubricant. 2 ointment; liniment

Schmierung f (-en) lubrication; etc cf SCHMIEREN

Schmier·vorrichtung f lubricating device, lubricator. **-wert** m lubricating value. **-wirkung** f lubricating effect (or action). **-wolle** f greasy wool. **-zapfenöl** n journal oil

schmilzt melts: pres of SCHMELZEN

Schmink·bohne f kidney bean

Schminke f facial cosmetics, makeup. **schminken** vt & sich sch. make up, put on makeup

Schmink·mittel n cosmetic. **-pulver** n makeup powder, powdered cosmetic. **-rot** n rouge. **-topf** m makeup jar. **-weiß** n flake (or pearl) white (basic nitrate or oxychloride of bismuth)

Schmirgel m emery. **-leinen** n, **-leinwand** f emery cloth. **schmirgeln** vt rub with emery

Schmirgel·papier n emery paper. **-paste** f emery paste. **-pulver** n emery powder. **-scheibe** f emery wheel. **-staub** m emery dust

schmiß threw, etc: past of SCHMEISSEN

Schmiß m (Schmisse) 1 cut, gash; scar. 2 pep, dash

schmissig a peppy, lively

schmitzen vt dye (leather); dress (cloth)

schmolz melted: past of SCHMELZEN

schmoren 1 vt pot-roast, braise. 2 vt/i stew. 3 vi swelter; (Elec.) arc; (Radio) fry

Schmor·braten m pot roast. **-stelle** f

(Elec.) point of arcing

Schmp(t). abbv (Schmelzpunkt) melt. pt.

schmuck a trim, cute; dapper

Schmuck m 1 ornament, decoration. 2 jewelry. 3 finery. 4: im Sch. (+genit) in the splendor (of)

schmücken vt adorn, decorate, trim

Schmuck·stein m gemstone. **-waren** fpl jewelry. **-wirkung** f decorative effect

Schmutz m dirt, muck, grime; smut. **--ansatz** m deposit of dirt. **-decke** f thin upper layer of sand in a filter bed. **-farbe** f dull (dingy, dirty) color. **-fleck** m dirt spot, soil, stain. **-haltfestigkeit** f soil adherence

schmutzig a dirty, soiled, grimy; smutty. **--rot** n dull-red, dingy-red

schmutz·lösend a dirt-dissolving. **Sch--lösungsmittel** n dirt solvent. **-papier** n waste paper. **-träger** m (Dterg.) anti-redisposition agent. **-tragevermögen** n (Dterg.) soil-suspending power. **-ventil** n mud valve. **-wasser** n waste (or dirty) water; sewage. **-wolle** f wool in the yolk

Schnabel m (-) 1 beak, bill. 2 (fig) nose, nozzle; spout. **sch--förmig** a beak-shaped, beaked. **Sch--tier** n duckbill, platypus

Schnake f (-n) midge, gnat

Schnalle f (-n) 1 buckle, clasp. 2 (shoe) strap. 3 door handle. **schnallen** vt buckle, strap (on)

Schnapp·deckel m snap lid (or cover)

schnappen 1 vt catch, nab. 2 vt/i(h) snap (up), snatch (at) **—(nach)** Luft sch. gasp for air. 3 vi(s) snap shut

Schnaps m (-e) liquor; brandy. **--brenner** m distiller

schnarchen vi snore; (Mach.) snort. **Schnarch·ventil** n poppet valve

Schnarre f (-n) 1 rattle. 2 (Elec.) buzzer. **schnarren** vi rasp; buzz

schnattern vi cackle; chatter; jabber

schnauben* (reg or schnob, geschnoben) vi snort

schnaufen vi wheeze, puff, pant

Schnauze f (-n) 1 snout, (animal's) nose; mouth. 2 (fig) nozzle; spout

Schnecke f (-n) 1 snail, slug. 2 spiral, helix; volute. 3 (Mach.) worm; screw conveyor. 4 (Anat.) cochlea. 5 spiral stairway.

Schneckel f (-n) conical spiral

Schnecken·antrieb m worm drive. **sch-artig** a snail-like; spiral, helical. **Sch-feder** f coiled (or spiral) spring. **-förderer** m screw conveyor. **sch-förmig** a snail-shaped; spiral, helical. **Sch-gang** m 1 winding alley; spiral walk. 2 snail's pace. **-gehäuse** n 1 snail shell. 2 cochlea. **-gewinde** n worm (screw) thread; helix, spiral. **-klee** m snail clover. **-linie** f spiral, helix. **-presse** f screw press (or extruder); expeller. **-rad** n worm wheel (or gear). **-tempo** n snail's pace

Schnee m 1 snow. 2 whipped egg white; meringue

schnee·artig a snow-like. **Sch-flocke** f snowflake. **-gestöber** n snow flurry. **-gips** m snowy gypsum. **-glöckchen** n snowdrop. **-grenze** f snow line. **-wehe** f snowdrift. **sch-weiß** a snow-white. **Sch-weiß** n snow (or zinc) white (pigment)

schneidbar a readily cut, sectile

Schneid·brenner m cutting torch

Schneide f (-n) (cutting or knife) edge; ridge. **-brenner** m cutting torch. **-brett** n cutting (carving, chopping) board. **-diamant** m cutting diamond. **-kante** f cutting edge. **-kessel** m (Cellulose) masticator. **-maschine** f cutting machine. **-mühle** f saw mill

schneiden* (schnitt, geschnitten) **I** vt/i 1 cut; carve; clip; mow. **II** vt 2 trim; prune. 3 edit (film, tape). **III** vi bite, chill (as: cold, wind). **IV** vt & **sich sch.** (recip) intersect, cross. **V: sich sch.** cut oneself; **sich in den Finger sch.** cut one's finger

Schneiden·berührung f knife-edge contact. **-lager** n knife-edge bearing

schneidend p.a cutting, etc cf SCHNEIDEN; piercing (sound); sharp (pain)

Schneider m (-) 1 tailor, dressmaker. 2 (at end of compd oft) cutter: **Glas·schneider** glass cutter. **Schneiderin** f (-nen) dressmaker

Schneider·kreide f tailor's chalk, soapstone

Schneide·werkzeug n cutting tool

Schneid·fähigkeit f cutting quality. **-flamme** f cutting flame. **-geschwindigkeit** f cutting speed. **sch-haltig** a edge-holding (tools). **Sch-härte** f cutting hardness

schneidig a 1 sharp, keen. 2 soft, easily cut (rock). 3 spirited, bold; sporty

Schneid·kante f cutting edge. **-legierung** f cutting-tool alloy. **-maschine** f cutting machine; slicer. **-metall** n cutting metal (or alloy). **-öl** n cutting oil. **-vermögen** n cutting power

schneien vi snow

Schneise f (-n) cleared lane (through woods or bushes); firebreak

schnell a fast (incl adv), quick, rapid; high-speed: **sch—er Brüter** fast breeder; **auf die sch—e** in a hurry. **--abbindend** a quick-setting. **-abtastend** a fast-scanning. **Sch-analyse** f rapid analysis. **-arbeitsstahl** m high-speed steel

schnellaufend a high-speed, rapid. **Schnelläufer** m 1 high-speed mill (or machine). 2 (as pfx) high-speed

Schnell·auflöser m rapid dissolver

Schnellauf·stahl m high-speed steel

Schnell·bad n quick bath. **-beizverfahren** n rapid disinfection process (for seeds). **-bestimmung** f rapid determination. **-binder** m 1 quick-setting cement. 2 accelerator (for cement). **-bleiche** f quick (chemical) bleach(ing). **-dämpfer** m (Print) rapid ager. **-drehstahl** m high-speed (or rapid-machining) steel

Schnelle f 1 speed. 2 rapids (in river)

schnellebig a short-lifespan; fast-paced. **Schnelleser°** m (Cmptrs) high-speed reader

schnellen 1 vt fling, send flying. 2 vi(s) & **sich sch.** dart, dash, shoot (usu up). 3 vi(h) (+mit) snap

Schnell·entladung f rapid discharge. **-essig** m quick vinegar. **-essigbereitung** f, **essigverfahren** n quick-vinegar process. **-färbung** a (Micros.) quick staining. **-feder** f (Mach.) spring. **-feuer** n (Mil.) rapid fire. **-filter** m rapid (or quick-run) filter. **sch-fliegend** a fast-flying, high-velocity. **Sch-fluß** m quick flux. **sch-flüssig** a easily fusible, fast-fusing. **Sch-gang** m rapid motion, high speed. **-gärung** f quick (or accelerated) fermentation. **-gefrierverfahren** n accelerated

freeze-dry process. **sch–gehend** *a* fast-
(-moving). **Sch–gerben** *n* rapid tanning.
sch–härtend *a* quick-hardening
Schnelligkeit *f* speed, velocity, rapidity
Schnell·imbiß *m* fast food (restaurant).
-**kocher, -kochtopf** *m* autoclave, pressure cooker. -**kraft** *f* springiness, resilience, elasticity. **sch–kräftig** *a* resilient, elastic. **Sch–kurs(us)** *m*
intensive (*or* crash) course
Schnell·l . . . : *correctly spelled* **Schnel·l**
and alphabetically listed accordingly: cf
SCHNELLEBIG etc
Schnell·methode *f* rapid method
Schnellot° *n* soft solder; fusible metal
Schnell·photographie *f* instant(aneous)
photograph(y). -**preßmasse** *f* (*Plastics*)
fast-curing molding compound. -**rechner**
m electronic calculator. -**regler** *m* quick-
acting regulator. -**reinigung** *f* fast-
service dry cleaning. -**reparaturwerk-
stätte** *f* fast-service repair shop. -**röste**
quick (*or* chemical) retting. -**säurer** *m*
acetifier (in quick-vinegar process).
-**schalter** *m* quick-break switch. -**schnitt-
stahl** *m* high-speed steel. -**schwarz** *n*
rapid black (dye). -**spaltung** *f* (*Nucl.*) fast
fission. -**stahl** *m* high-speed steel
schnellstens *adv* as fast as possible.
schnellst·möglich *a* fastest possible;
adv as fast as possible
Schnell·straße *f* expressway. -**temperguß**
m short-cycle malleable iron. -**temper-
verfahren** *n* short-cycle annealing.
-**trockenfarbe** *f* quick-drying paint (col-
or, ink). **sch–trocknend** *a* quick-drying.
Sch–verband *m* first-aid bandage; adhesive tape. -**verfahren** *n* rapid process.
-**waage** *f* steelyard. -**wäscherei** *f* laun-
dromat. -**waschmittel** *n* fast-action detergent. **sch–wüchsig** *a* fast-growing.
Sch–zug *m* express (train). -**zünder** *m*
instantaneous fuse
Schnepfe *f* (-n) (*Zool.*) snipe
Schneppe *f* (-n) spout; nozzle
schneuzen *v*: **sich (die Nase) sch.** blow
one's nose. **Schneuz·tuch** *n*
handkerchief
Schnippel, Schnipsel *m or n* (-) scrap,
shred; chip. **schnipseln** *vt* cut up, snip at

schnitt cut: *past of* SCHNEIDEN
Schnitt *m* (-e) 1 cut, incision; (surgical) section. 2 cutting, pruning, trimming, mow-
ing. 3 (cut) edge. 4 split, separation. 5
intersection. 6 cross-section; average,
mean: **im Sch.** on the average. 7 slice. 8
(*Agric.*) crop. 9 woodcut, carving. 10
clothing pattern. 11 form, shape, style. *Cf*
SCHNITTE. --**band** *n* (*Micros.*) serial section, ribbon. **sch–bearbeitbar** *a* machinable. **Sch–bohnen** *fpl* (cut) string
beans. -**brenner** *m* batswing (*or* fishtail)
burner. -**dicke** *f* slice (*or* section) thickness
Schnitte *f* (-n) 1 slice; slice of bread:
belegte Sch. open sandwich. 2 cutlet;
steak. *Cf* SCHNITT
Schnitt·ebene *f* cutting (*or* intersecting)
plane. -**fänger** *m* (*Micros.*) section lifter.
-**färbung** *f* (*Micros.*) section staining.
sch–fest *a* firm, easy-slicing (fruit, etc.)
Sch–fläche *f* section, sectional plane.
-**geschwindigkeit** *f* cutting speed. **sch–
haltig** *a* edge-holding (knife). **Sch–holz**
n plankwood, lumber
schnittig 1 ready for cutting (*or* mowing). 2
sporty, racy; streamlined
Schnitt·kante *f* cutting edge. -**käse** *m* easy-
slicing cheese. -**länge** *f* length of the cut
(*or* section). -**lauch** *m* chives. -**linie** *f* line
of intersection. -**präparat** *n* section preparation. -**punkt** *m* point of intersection,
interface. **sch–reif** *a* ready for cutting (*or*
mowing). **Sch–salat** *m* lettuce. -**stelle** *f*
point of (inter)section. -**wachs** *n* cobbler's
wax. -**ware** *f* 1 yard goods. 2 sawed timber, lumber. -**werkzeug** *n* cutting tool.
-**wunde** *f* cut, gash (=wound). -**zeichnung** *f* sectional drawing, cross-section
Schnitz *m* (-e) cut piece, slice, sliver. --**arbeit** *f* wood carving
Schnitzel *n* (-) 1 cut piece: shred, scrap,
slice; paring, clipping, shaving. 2 (*Meat*)
cutlet. 3 (*Sugar Beets*) cossette —
ausgelaugte Sch. beet pulp. --**maschine**
f slicing machine; shredding machine (*esp*
for beets). -**masse** *f* diced plastic, molded
macerate. -**messer** *n* slicing (*or* carving)
knife; chip cutter
schnitzeln 1 *vt* cut up; clip. 2 *vt/i* carve,

whittle. **3** *vi* snip away (at)

Schnitzel·presse *f* (*Sugar*) pulp press

schnitzen 1 *vt* carve. **2** *vi* (+**an**) whittle (at)

Schnitzer *m* (-) **1** (wood) carver. **2** blunder

schnob snorted: *past of* SCHNAUBEN

Schnorchel *m* (-) snorkel

schnüffeln *vi* sniff; sniffle; snoop

Schnupfen *m* (-) (*Med.*) (head) cold, catarrh

Schnupf·tabak *m* snuff. **-tuch** *n* handkerchief

Schnur *f* (⁼e) cord (*incl Elec.*) string; filament. **schnüren 1** *vt* tie; lace (up). **2** *vt/i* be too tight (on)

Schnür·kolben *m* narrow-neck flask. **-senkel** *m* shoestring. **-senkelsand** *m* shoestring sand

Schnurre *f* (-n) **1** rattle. **2** funny story. **schnurren** *vi* whir, buzz; purr

Schnur·scheibe *f* pulley (wheel)

schob shoved, etc: *past of* SCHIEBEN

Schober *m* (-) (*Hay*) stack, rick; barn, shed. **schobern** *vt* stack, pile (hay, etc)

Schock (-e) **I** *n* **1** (*Meas.*) sixty, five dozen. **2** (*appx*) lots, dozens. **II** *m* (nervous) shock. **--einwirkung** *f:* **unter Sch. sein** be in shock. **-tisch** *m* vibrating (*or* jolting) table. **sch—weise** *f* by the dozen, in bunches

Schokolade *f* (-n) chocolate. **schokoladen·braun** *a* chocolate-brown. **-farbig** *a* chocolate(-colored). **Sch—tafel** *f* chocolate bar

Scholle *f* (-n) **I 1** clod, lump. **2** (one's native) soil. **II** (*Zool.*) summer flounder, plaice

Schöll·kraut *n* celandine

schon *adv* **1** already; (*in questions*) yet; (+*time expr*) no later than, as early as: **sch. im 18. Jahrhundert** already in (*or* as early as) the 18th century; **siedet es sch.?** is it boiling yet?; **sch. jetzt** right now. **2** even: **sch. die Ägypter wußten das:** even the Egyptians knew that; **sch. früher** even before this. **3** mere, very, no(thing) more than: **sch. der Gedanke daran** the mere (*or* very) thought of it. **4** (*usu* + *fut, implying confidence, assurance*): **es wird sch. gelingen** it will succeed all right; **das hat sch. seinen Grund** (you may be sure) there is a good reason for that. **5** (+*verb, oft*) almost, be

ready to: **sie dachten sch., es wäre zu spät** they almost believed (*or* were ready to believe) it was too late. **6** (*action still in progress; oft* + **seit**): **sie studieren sch. ein Jahr** (*or* **sch. seit einem Jahr**) dort they have been studying there for a year. **7: sch. einmal** (once) before, (*in questions*) ever. **8** (*implies impatience, annoyance*): **sch. wieder** again; **sch. längst** long ago; **wer braucht das sch.?** who needs that? **9** (*implies less than total agreement*): **das kommt sch. vor, aber. . .** that does happen, but. . . **10: wenn . . . sch.** if only

schön I *a* **1** beautiful, fine. **2** nice; neat, tidy (*incl* sum); (*Quant., oft*): **ein sch—es Stück Arbeit** (quite) a good bit of work. **3: sch. und gut** well and good. **II** *adv, oft* **4: sch. langsam** nice and slow; **sch. teuer** quite (*or* really) expensive; **ganz sch. sparen** really economize. **5** (*implies: as is proper*): **sch. der Reihe nach** one by one, all in proper order. **III** *intj* OK, (all) right

Schöne *f* **1** fining, isinglass. **2** beauty

schonen *vt* spare; treat gently, take care of

schönen *vt* **1** beautify. **2** (*esp Wine*) clarify, fine. **3** (*Colors*) brighten. **4** (*Tex.*) top. **5** (*Food*) improve

schonend *p.a* sparing, gentle; *adv oft* with care *cf* SCHONEN

Schön·färber *m* garment dyer. **-färberei** *f* garment dyeing (*or* dyer's). **-grün** *n* Paris green

Schönheit *f* (-en) beauty

Schönheits·fehler *m* blemish. **-mittel** *n* cosmetic, beauty aid. **-pflege** *f* beauty care. **-wasser** *n* liquid cosmetic

Schon·kost *f* bland diet

Schön·seite *f* right side (of fabrics)

Schonung *f* (-en) **1** protection; care; gentle (*or* sparing) treatment: **ohne Sch.** unsparingly. **2** convalescence; rest. **3** forest preserve

Schönung *f* beautification, etc *cf* SCHÖNEN

Schönungs·farbstoff *m* brightening dye

schonungslos *a* unsparing, pitiless

Schönungs·mittel *n* fining (agent); (*Dye.*) brightening agent

Schopf *m* (⁼e) **1** (head of) hair; tuft; (*fig*) forelock. **2** tree top

Schöpf·brunnen m draw well. **-bütte** f (Paper) pulp (or stuff) vat. **-eimer** m (dipping) bucket

schopfen vt top, crop

schöpfen I vt 1 scoop, bail, ladle, dip: **den Behälter leer sch.** empty the container (by ladling). **2** (oft + **aus**) draw, gather (strength, hope, etc) (from): **Atem** (or **Luft**) **sch.** catch one's breath; **Verdacht sch.** get suspicious. **3** (Glass) gather. **4** (Paper) dip, mold. **5** create; invent. **II** vi (usu + **aus**) draw (on) (experience, etc).

Schöpfer m (-) **1** scoop, ladle, dipper; drawer (of water, etc). **2** creator, originator; creative genius; (in compds oft) creative: **-geist** m creative spirit

schöpferisch a creative

Schöpf·gefäß n scoop, ladle, dipper; (dipping) bucket. **-herd** m casting crucible (of a furnace). **-kelle** f, **-löffel** m ladle. **-papier** n handmade paper. **-probe** f 1 drawn (dipped, ladled) sample. **2** ladle (or cup) test. **-rad** n bucket wheel. **-rahmen** m (Paper) deckle edge. **-schraube** f tap screw

Schöpfung f (-en) creation

Schöpf·werk n bucket elevator (or conveyor); water engine

Schoppen m (-) mug, ¼ liter. **-bier** n tap (or draft) beer. **-wein** m open wine

Schöps m (-e) (Zool.) wether; (Meat) mutton

Schöpsen·fleisch n mutton. **-talg** m mutton tallow, (Pharm.) prepared suet

schor sheared: past of SCHEREN

Schorf m (-e) **1** (Med.) scab, crust. **2** (incl Bot.) scurf. **schorfig** a scabby; scurfy

Schörl m (Min.) schorl, black tourmaline —**roter Sch.** rutile

Schornstein° m (smoke)stack, chimney. **-gas** n flue gas

schoß shot, etc: past of SCHIESSEN

Schoß m **I** (̈e) **1** lap. **2** womb. **3** coattail. **4** skirt. **II** (Schösse) sprout, shoot, sprig

schossen I vt (Bot.) shoot up; run to seed. **II** shot: past pl of SCHIESSEN cf SCHOSS

Schoß·gerinne n wooden trough

Schößling m (-e) sprig, shoot; sapling

Schote f (-n) **1** pod, shell. **2** pea. **3** bell pepper

Schoten·dorn m (Bot.) acacia, black locust. **-gewächse** npl leguminous plants. **-pfeffer** m red pepper, paprika; capsicum. **-pflanze** f leguminous plant

Schotte m (-n, -n) Scot. **Schotten** m (-) (Tex.) tartan

Schotter m gravel; crushed rock; macadam. **-boden** m gravelly soil

Schottin f (-nen) Scotswoman. **schottisch** a Scottish, Scotch

Schott·kolben m Schott flask

Schottland n Scotland

Sch. P. abbv (Schmelzpunkt) melt. pt., m.p.

schr. abbv (schriftlich) written, in writing

Sch.-R. abbv (Paper) Schopper-Riegler degree of freeness (of pulp)

schraffen, schraffieren vt/i shade, crosshatch. **Schraffur** f (-en) shading, crosshatching

schräg a **1** oblique, diagonal, slanting; adv oft on a slant. **2** inclined, sloping. **3** beveled. **4** (Print.) italic

Schräg·agar m (Bact.) agar slant. **-druck** m italics

Schräge f (-n) **1** slant; diagonal. **2** slope, incline. **3** bevel. **schrägen** vt slant; incline; bevel

Schräg·kante f bevel (edge), chamfer. **sch–liegend** a slanting; sloping. **Sch–linie** f slant(ing) line, diagonal. **-retortenofen** m inclined-retort furnace. **-rohr** n, **-röhre** f slanting (inclined, diagonal) tube (or pipe). **-rolle** f tapered roller. **sch–schichtig** a obliquely layered. **Sch–schnitt** m oblique cut (or section); (Ceram.) ellipse. **sch–stellbar** a inclinable

schräg·stellen vt incline, tilt; set diagonally

Schräg·strahlen mpl (Phys.) skew rays. **sch–über** adv aslant; diagonally opposite

Schrägung f (-en) **1** slanting; inclining; beveling cf SCHRÄGEN. **2** slant; incline; bevel, chamfer

schräg·winklig a oblique(-angled)

Schram m (̈e) (Min.) cut, kerf. **schrämen** vt cut, shear

Schramme f (-n) scratch, abrasion; scar.

schrammen vt scratch, scrape; scar.

schrammig *a* scratched; scarred

Schrank *m* (-e) closet; cupboard, cabinet; wardrobe

Schranke *f* (-n) **1** barrier, (crossing) gate; bar (*incl. Law*) **—in die Sch—n treten (für)** go to bat (for). **2** (*usu pl*) limit(s), bounds: **sich in Sch—n halten** control oneself

schränken *vt* **1** fold, cross (arms). **2** set (sawteeth)

schrankenlos *a* **1** boundless. **2** unbridled. **3** unguarded (crossing)

Schrape *f* (-n) scraper

Schrapnell *n* (-s) shrapnel

schrappen *vt* scrape. **Schrapper** *m* (-) scraper

Schräubchen *n* (-) small screw

Schraub·deckel *m* screw lid (*or* cover)

Schraube *f* (-n) screw (*incl* propeller); bolt. **schrauben 1** *vt/i* screw. **2** *vt* (+**höher, niedriger**) turn (up, down), raise, lower. **3: sich in die Höhe sch.** spiral (*or* circle) upward

Schrauben·achse *f* (*Cryst.*) screw axis. **sch–artig** *a* screw-shaped, spiral, helical. **Sch–bakterie** *f* spirillum. **-bohrer** *m* screw tap; twist drill. **-bolzen** *m* (screw) bolt. **-drehmaschine** *f* screw-cutting lathe. **-feder** *f* helical (*or* coil) spring. **sch–förmig** *a* screw-shaped, spiral, helical. **Sch–gang** *m* pitch of a screw thread. **-gewinde** *f* screw thread. **-kappe** *f* screw cap. **-klammer, -klemme** *f* screw clamp (*or* clip). **-kühler** *m* helical condenser (*or* cooler). **-lehre** *f* screw thread caliper, micrometer gauge. **-linie** *f* helical line, spiral, helix. **-mutter** *f* (screw) nut. **-presse** *f* screw press. **-quetschhahn** *m* screw pinchcock. **-rohr** *n*, **-röhre** *f* spiral (*or* helical) tube. **-rührwerk** *n* screw stirrer. **-schlüssel** *m* (screw, nut, monkey) wrench. **-schnecke** *f* helix. **-zieher** *m* screwdriver

Schraub·fassung *f* screw base (of light bulbs, etc.). **-glas** *n* screw-top jar

schraubig *a* spiraled, screwed; *adv* in a spiral

Schraub·knecht *m* screw clamp. **-lehre** *f* micrometer caliper. **-stock** *m* vise

Schraubung *f* screwing, screw motion

Schraubungs·komponente *f* (*Cryst.*) screw component. **-sinn** *m* (*Cryst.*) sense of screw

Schraub·verschluß *m* screw cap (top, plug). **-zwinge** *f* screw clamp, C-clamp

Schreck *m* fright, scare; fear. **schrecken 1** *vt* frighten, terrify, scare; startle; chill. **2** *vi*(s) start up (in fright). **Schrecken** *m* (-) fright; terror, horror

Schreck·ladung *f* booby trap

schrecklich *a* frightful; terrible, awful

Schreck·mine *f* booby-trap mine. **-wirkung** *f* chilling action (*or* effect)

Schrei *m* (-e) **1** cry, shout, shriek. **2** (*Metll.*) bloom, lump

Schreib·anzeiger *m* (*Instrum.*) recording indicator. **-art** *f* **1** style of writing. **2** spelling. **-barometer** *n* barograph. **-blatt** *n* record sheet. **-block** *m* writing pad

schreiben* (schrieb, geschrieben) *vt/i* write; (*Instrum.*) record. **Schreiben** *n* (-) (*oft*) note, letter. **Schreiber** *m* (-) **1** clerk, secretary. **2** writer. **3** recorder; recording instrument

Schreib·feder *f* (writing *or* recording) pen; stylus. **-fehler** *m* error in writing; clerical error. **-fläche** *f* **1** writing surface. **2** (*Instrum.*) record sheet. **-gerät** *n* recording instrument. **-kies** *m* (*Min.*) marcasite. **-maschine** *f* typewriter. **(mit der) Sch. schreiben** type(write). **sch–maschinengeschrieben** *a* typewritten. **-materialien** *npl* writing materials, stationery. **-papier** *n* writing paper. **-röhrchen** *n* recording siphon. **-stift** *m* stylus. **-tinte** *f* writing ink. **-tisch** *m* desk. **-trommel** *f* (*Instrum.*) recording drum

Schreibung *f* (-en) spelling (of a word)

Schreib·unterlage *f* something to put under the paper when writing. **-waren** *fpl* stationery. **-weise** *f* **1** writing style. **2** spelling. **-werk** *n* (*Instrum.*) recording mechanism

schreien* (schrie, geschrieen) *vt/i* cry (out), shout, scream. **Schreien** *n* (*of tin*) cry. **schreiend** *p.a* (*oft*) loud, gaudy, glaring

Schreiner *m* (-) carpenter, cabinet maker. **schreinern 1** *vt* (*Tex.*) schreinerize. **2** *vi* do carpentry

schreiten* (schritt, geschritten) *vi*(s) **1** stride, step. **2** (+**zu**) proceed, pass on (to)

schrie cried: *past of* SCHREIEN

schrieb wrote: *past of* SCHREIBEN

Schrift *f* (-en) **1** writing, characters, script. **2** type (face), font. **3** document, publication; article, theme; (*pl oft*) writings, works: **gesammelte Sch—en** collected works. **--absatz** *m* paragraph

Schriften·nachweis *m*, **-verzeichnis** *n* list of references, bibliography

Schrift·erz *n* (*Min.*) sylvanite. **-flasche** *f* (permanently) labeled (reagent) bottle. **-führer** *m* secretary. **-gießer** *m* type founder. **-gießerei** *f* type foundry. **-gießermetall** *n* type metal. **-gold** *n* (*Min.*) sylvanite. **-granit** *m* graphic granite. **sch–granitisch** *a* graphic, granophyric. **Sch–guß** *m* type founding, type metal. **-gut** *n* literature (on a subject). **-jaspis** *m* jasper opal. **-leiter** *m* editor

schriftlich *a* written, (*esp adv*) in writing

Schrift·malen *n*, **-malerei** *f* sign painting; lettering. **-material** *n* written (*or* recorded) material, literature, documentation; bibliography. **-metall** *n* type metal. **-mutter** *f* type mold, matrix. **-setzer** *m* typesetter, compositor. **-stein** *m* graphic granite. **-steller** *m*, **-stellerin** *f* (professional) writer, author. **sch–stellerisch** *a* literary; *adv* as a writer. **Sch–stück** *n* document. **-tellur** *m* graphic tellurium (sylvanite). **-zeichen** *n* (written) symbol, character

Schrifttum *n* literature (on a subject). **Schrifttums·angabe** *f* bibliographical reference. **Schrifttum·verzeichnis** *n* list of references, bibliography

Schrift·wechsel *m* correspondence. **-zeichen** *n* letter, character. **-zeug** *n* type metal. **-zug** *m* **1** stroke (of writing). **2** signature. **3** *pl* handwriting

schritt strode, etc: *past of* SCHREITEN

Schritt *m* (-e) **1** stride, step; pace. **2** gait. **--macher** *m* pacemaker. **sch–weise** *a*, *adv* stepwise, step by step

schroff *a* **1** steep, sheer. **2** abrupt. **3** rough

schröpfen *vt* (*Med.*) bleed, cup, scarify

Schrot *m or n* **1** chip, block. **2** grits, cracked wheat (*or* grain). **3** (lead) shot. **4** plumb bob. **5** alloy weight (of coins): **Sch. und Korn** weight and fineness. **6** (*Brew.*) grist, bruised malt. **7** (*Tex.*) selvage. **--brot** *n* whole-grain bread. **-effekt** *m* (*Phys.*) (small) shot effect

schroten *vt* **1** cut into pieces; chip. **2** rough-grind, crush, bruise; granulate. **3** (*Metll.*) rough-plane; trim, size (blanks). **4** (*Food*) gobble. **5** (*Casks*) shoot . . . into the cellar

Schröter *m* (-) **1** scrip metal crusher. **2** stag beetle

Schrot·fabrik *f* shot factory. **-gewehr** *n* shotgun. **-kleie** *f* coarse bran. **-korn** *n* **1** grain of shot. **2** bruised (*or* coarse) grain. **-kugel** *f* (single) grain of shot

Schrötling *m* (-e) (*Coin Minting*) blank, planchet

Schrot·mehl *n* coarse meal, grits. **-metall** *n* shot metal. **-mühle** *f* grist (bruising, malt) mill. **-speck** *m* lean bacon

Schrott *m* scrap metal (*esp* iron). **--haufen** *m* scrap heap. **-martinieren** *n* (*Iron*) pig-and-scrap process. **-platz** *m* scrap heap (*or* yard). **-roheisenverfahren** *n* (*Iron*) pig-and-scrap process. **-schmelze** *f* (*Metll.*) scrap heat. **-verfahren** *n* (*Iron*) pig-and-scrap process

schrubben *vt* scrub, scour. **Schrubber** *m* (-) scrubber; scrub brush

Schrüh·brand *m* (*Ceram.*) biscuit-baking (*or* -firing). **schrühen** *vt* biscuit-bake (*or* -fire)

Schrulle *f* (-n) whim, odd notion

schrumpeln *vi* (s) shrivel, shrink

schrumpf·beständig, **-echt** *a* shrink-resistant

schrumpfen *vi* (s) shrink, contract, shrivel. **schrumpfend** *p.a* (*oft*) astringent

Schrumpf·farbe *f* (*Paint*) crackle finish. **-grenze** *f* shrinkage limit

schrumpfig *a* shriveled, wrinkled

Schrumpf·leder *n* shrunk leather. **-maß** *n* (degree of) shrinkage (*or* contraction). **-niere** *f* shrunken kidney; atrophic cirrhosis. **-riß** *m* shrinkage crack. **-sitz** *m* shrink fit

Schrumpfung *f* (-en) shrinkage, contraction; shriveling. **Schrumpfungs·geschwindigkeit** *f* rate of shrinkage

Schrund *m* (ë) , **Schrunde** *f* (-n) crack; cleft, fissure; (glacial) crevasse; (skin) chap. **schrundig** *a* cracked up, fissured; chapped

schruppen *vt* rough-machine

Schub *m* (ë) **1** push, shove; thrust. **2** shear. **3** load; batch, lot. **--beanspruchung** *f* shear stress. **-bewehrung** f shear reinforcement. **-bruch** m shear fracture (or failure). **-fach** n (e.g. desk) drawer. **-fenster** n sash window. **-festigkeit** f shear(ing) strength. **-karren** m wheelbarrow. **-kraft** f thrust; shear(ing) force. **-kurbel** f (Mach.) crank. **-lade** f (e.g. desk) drawer. **-lehre** f slide gauge. **-modul** n shear modulus. **-riegel** m (sliding) bolt. **-spannung** f shearing stress. **-stange** f connecting rod. **-vektor** m thrust vector. **-viskosität** f shear viscosity. **sch—weise** adv 1 by shoves, by thrusts. 2 in batches; gradually

schuf created: past of SCHAFFEN I, II
schuften vi slave, toil
Schuh m (-e) shoe. **-band** n shoestring, shoelace. **-draht** m twine; pitched thread. **-krem** f shoe polish. **-leder** n shoe leather. **-macher** m shoemaker. **-putzer** m shoeshine man. **-putzmittel** n shoe polish, shoe cleaner. **-riemen** m shoelace; shoe strap. **-schwärze** f shoe blacking. **-sohle** f shoe sole. **-wachs** n shoe polish. **-werk** n shoes, footgear. **-wichse** f shoe polish
Schuko m (-s) (Elec.) =SCHUTZKONTAKT protective contact. **--stecker** m safety plug
Schul· school: **-abgang** m graduation. **-ausgabe** f school edition. **-beispiel** n object lesson, classical example. **-bildung** f (school) education. **-buch** n (school) textbook
schuld pred a: **sch. sein** (or **haben**) **an** be to blame for; **sch. geben** (+dat) blame. **Schuld** f (-en) 1 debt. 2 guilt; fault, blame: **die Sch. haben** (or **tragen**) be to blame. 3 offense, wrong. **--buch** n ledger, accounts. **-forderung** f claim, demand (for debt payment)
schulden vt owe
schuldig a 1 (**an**) guilty (of); to blame, responsible (for). 2 pred in debt: **sch. sein** (+dat) (oft) owe; **sch. bleiben** (+d.i.o) fail to give (or pay). 3 proper, due (respect, etc). **Schuldige** m, f (adj endgs) guilty party, culprit
Schuldner m (-) debtor
Schule f (-n) school cf GESCHULT —**Sch. machen** gain a following. **Schüler** m (-),

Schülerin f (-nen) pupil, student (below university level cf STUDENT); disciple. **Schul·geld** n tuition. **schulisch** a (of) school, adv oft in school
Schulter f (-n) shoulder. **--blatt** n shoulder blade, scapula. **-gurt** m shoulder (seat) belt
Schulung f schooling, training
Schumann·gebiet n (Spect.) Schumann region. **-platte** f (Phot.) Schumann (or gelatin-free) plate
schund harassed, etc: past of SCHINDEN
Schund m trash, rubbish
Schüppchen n (-) little scale, flake
Schüppe f (n) scale, flake; pl oft scurf, dandruff of SCHUPPEN
Schuppe f (-n) scoop, etc = SCHIPPE 1, 2
schuppen 1 vt scale (off), peel, desquamate. 2: **sich sch.** scale (off), peel, desquamate. Cf GESCHUPPT
Schuppen m (-) shed, barn; hangar cf SCHUPPE
schuppen·artig a scaly, flaky. **Sch·bildung** f scaling, flaking. **-flechte** f psoriasis. **sch·förmig** a scaly, flaky. **Sch—glanz** m (Min.) franckeite. **-glätte** f flake litharge. **-graphit** m flaky graphite. **-paraffin** n (Petrol.) scale wax. **-stein** m (Min.) lepidolite. **-tier** n scaly anteater. **sch—weise** adv in scales, in flakes
schüpperig a (Dye.) skittery
schuppig a scaly, flaky; dandruffy; squamous. **Schuppigkeit** f scaliness, etc
schüpprig a (Dye.) skittery
Schur f (-en) 1 (sheep) shearing; mowing, trimming. 2 fleece, clip
Schürbel m (-) (Metll.) bloom, etc = SCHIRBEL
Schür·eisen n (fire) poker
schüren vt poke (fire); stir up. **Schürer** m (-) poker
Schurf m (-̈e) 1 (Min.) borehole; pit. 2 prospecting. 3 scratch, abrasion. **Schürf·arbeit** f prospecting
schürfen I vt 1 dig (ore, etc). 2 open (a mine). 3 scrape, scratch. II vt/i 4 prospect (for), explore. III vi 5 (**nach**) dig (for). 6 make a scraping noise
Schürf·graben m (Min.) test pit. **-wunde** f (Med.) abrasion
Schür·haken m (fire) poker. **-loch** n stoke (or fire) hole

Schurre *f* (-n) slide, chute

Schur·wolle *f* virgin wool

Schurz *m* (-e), **Schürze** *f* (-n) apron. **schürzen I** *vt* **1** tie (knots). **2** gather (tuck, curl) up. **II: sich sch. 3** be tied. **4** curl

Schuß *m* (Schüsse) **1** (gun)shot; shooting: **zum Sch. kommen** get a chance to shoot; **es kam ihnen in den Sch.** it came at just the right moment for them. **2** round (of ammunition). **3** (gunshot) wound. **4** gun barrel. **5** blast(ing). **6** (*Bot.*) shoot: **einen Sch. machen** (or **tun**) shoot up (in height). **7** shot (of liquor); dash, touch (of. . .) **8** batch, charge. **9** (*Tex.*) weft. **10** (*Mach.*) ring, collar. **11: im Sch. sein** be in good shape; **in Sch. bringen** get (sthg, sbdy) into shape. **12: (im) Sch. fahren** go full speed, dash

Schüssel *f* (-n) bowl, dish; pan. **sch-·för·mig** *a* bowl- (or dish-)shaped. **Sch-zinn** *n* pewter

schuß·fest *a* bulletproof. **Sch-schweißung** *f* shot welding. **sch-sicher** *a* bulletproof, shellproof. **Sch-waffe** *f* firearm. **sch-weise** *a* in shots (jerks, blasts, batches). **Sch-weite** *f* (firing) range. **-wunde** *f* gunshot (or bullet) wound

Schuster *m* (-) shoemaker. **-·pech** *n* shoemaker's wax

Schute *f* (-n) scow; barge, lighter

Schutt *m* rubble; refuse, rubbish, scrap; (*esp. Geol.*) debris, scree

Schütt·beton *m* poured (-in-situ) concrete. **-boden** *m* granary

Schutt·breccie *f* (*Geol.*) debris (or scree) breccia

Schütt·dichte *f* bulk density, apparent density. **-eigenschaft** *f* bulk property (of a material)

Schüttel·apparat *m* shaking apparatus. **-aufgabe** *f* vibrating feeder, shaker. **-be·wegung** *f* shaking motion. **-faß** *n* shaking drum. **-flasche** *f* shaking bottle. **-frost** *m* chill, shivers. **-glas** *n* shaking glass (jar, tube). **-maschine** *f* shaking machine, agitator

schütteln 1 *vt, vi* & **sich sch.** shake. **2** *vt* agitate. **3** *vi* vibrate, oscillate

Schüttel·rinne *f* shaker trough. **-sieb** *n*

shaking screen. **-sortierer** *m* oscillating separator. **-tisch** *m* shaking (or rocking) table. **-trichter** *m* separatory funnel. **-vorrichtung** *f* shaking device. **-werk** *n* shaking mechanism. **-zahl** *f* shaking coefficient. **-zylinder** *m* (stoppered) shaking cylinder

schütten I *vt/i* **1** pour. **II** *vt* **2** dump (bulk goods) *cf* GESCHÜTTET. **3** spill. **III** *vi* (+**gut**) give a good yield

schütter *a* sparse; meager; faint

Schütter·gebiet *n* (*Geol.*) region of disturbance

schüttern 1 *vt/i* shake (up). **2** *vi* tremble, vibrate

Schütt·gelb *n* Dutch pink. **-gewicht** *n* bulk weight (or density). **-gut** *n* bulk goods; loose material; packing. **-gutunterhalt** *m* packing support

Schutt·halde *f* (slope covered with) rock debris; talus, scree. **-haufen** *m* heap of rubble (rubbish, debris)

Schütt·loch *n* feed hole

Schüttrichter *m* (=**Schütt·trichter**) **1** feed hopper. **2** discharge hopper

Schütt·röstofen *m* continuous roasting furnace

Schüttung *f* (-en) **1** pouring; dumping, heaping *cf* SCHÜTTEN. **2** dumped heap; heaped lead; layer, fill, ballast. **3** dam, dike. **4** yield (of a spring). **5** (*Brew.*) extract-yielding materials

Schütt·volumen *n* bulk volume; (*Plastics*) pourability, bulk factor. **-winkel** *m* angle of repose

Schutz *m* (*usu* + **vor, gegen**) **1** protection (from, against); guardianship. **2** shield, screen. **3** shelter, refuge. **4** prevention. **5** defense: **in Sch. nehmen** defend. **6** patronage

Schütz I *n* (-e) **1** (*Elec.*) relay. **2** flood (or sluice) gate. **II** *m* (-en, -en) marksman, etc = SCHÜTZE

Schutz·anstrich *m* protective coating (or coat of paint), priming (coat). **-anwen·dung** *f* protective use, **-anzug** *m* protective clothing. **-beize** *f* (*Tex.*) resist, reserve. **-beizendruck** *m* resist style (or printing). **-blech** *n* safety shield; fender. **-brille** *f* protective goggles. **-decke** *f* protective cover(ing)

Schütze I *m* (-n, -n) **1** marksman, rifleman. **2** (*Tex.*) shuttle. **III** *f* (-n) **3** (*Elec.*) relay. **4** flood (*or* sluice) gate. *Cf* SCHÜTZ

schützen 1 *vt/i* (*oft* + **vor, gegen**) protect, (safe)guard (from, against); shelter (from); defend (against). **2** *vi* provide shelter. **schützend** *p.a* (*oft*) protective

Schützen-grabenkrieg *m* trench warfare

Schutz protective, safety: **-farbe** *f* protective (*or* anticorrosive) paint. **-färbung** *f* protective coloration. **-firnis** *m* protective varnish. **-gas** *n* inert gas. **-gasglühen** *n* bright annealing. **-gasschleier** *m* inert gas shield. **-geländer** *n* guard rail(ing). **-gitter** *n* screen grid. **-glas** *n* protective (*or* safety) glass. **-gläser** *npl* safety glasses (*or* goggles). **-glocke** *f* bell jar. **-handschuh** *m* protective glove. **-helm** *m* safety helmet. **-hülle** *f* protective covering (casing, sheath). **-impfung** *f* preventive inoculation, vaccination. **-kappe** *f* protective cap. **-kleidung** *f* protective clothing. **-kolloid** *n* protective colloid. **sch-kolloidal** *a* protective-colloid. **Sch-kontakt** *m* (*Elec.*) protective contact. **-kraft** *f* protective power. **-lack** *m* protective lacquer (*or* varnish). **-leder** *n* protective leather apron

Schützling *m* (-e) protegé, charge

schutzlos *a* unprotected, defenseless

Schutz protective, safety: **-lösung** *f* antifreeze solution. **-mann** *m* policeman. **-mantel** *m* protective jacket; windbreaker. **-marke** *f* trademark. **-maske** *f* safety (*or* gas) mask. **-masse** *f* (*Tex.*) resist. **-maßnahme, -maßregel** *f* protective measure, precaution. **-mittel** *n* prophylactic, preventive (*or* protective) agent; pesticide; (*Tex.*) resist. **-muffe** *f* protective sleeve. **-panzer** *m* protective armor; safety shield. **-papp** *m* (*Tex.*) resist paste, reserve. **-pockengift** *n* vaccine virus. **-polizei** *f* police. **-raum** *m* protective space; air-raid shelter. **-rechte** *npl* patent (*or* trademark) rights. **-ring** *m* guard ring. **-rohr** *n* protective tube (*or* pipe). **-salbe** *f* protective ointment. **-salzlösung** *f* protective salt solution. **-schicht** *f* protective layer (*or* coating). **-stoff** *m* **1** protective material (*or* substance). **2** (*Biol.*) antibody; (*Med.*) vaccine.

-streifen *m* (*Glass*) colored stripe used as a trademark. **-trichter** *m* protective funnel. **-überzug** *m* protective coating (*or* cover). **-ummantelung** *f* protective sheathing (*or* casing). **-umschlag** *m* protective envelope; dust cover (of books). **-vorrichtung** *f* protective (*or* safety) device. **-wall** *m* safety (*or* protective) wall. **-wand** *f* protective wall (partition, screen). **-wert** *m* protective value. **-widerstand** *m* (*Elec.*) protective resistance. **-wirkung** *f* protective action (*or* effect). **-zahl** *f* (*Colloids*) gold number

schw. *abbv* (schwach) wk. (weak)

Schw. *abbv* (Schweizer) Sw. (Swiss)

Schwabbel *m* (-) swab; mop. **-lack** *m* swab varnish (*or* lacquer). **schwabbeln 1** *vt* buff. **2** *vi* jiggle; jabber; slosh (over). **Schwabbel-scheibe** *f* buffing wheel. **Schwabber** *m* (-) swab; mop

Schwabe I *f* (-n) cockroach. **II** *m* (-n, -n) Swabian. **Schwaben** *n* (*Geog.*) Swabia (southwestern Germany). **Schwäbin** *f* (-nen), **schwäbisch** *a* Swabian

schwach (¨er, ¨ste) weak; feeble; frail; dilute; slight; sparse; faint: **sch—e Seite** weak point; **sch. basisch** slightly basic; **sch. machen, sch. werden** weaken; **sch. kochen** boil gently. **Sch--ammoniakwasser** *n* weak ammonia water. **schbackend** *a* (*Coal*) weakly caking

Schwäche *f* (-n) **1** weakness, etc *cf* SCHWACH. **2** defect, flaw

schwächen 1 *vt/i* weaken; diminish. **2** *vt* debilitate; dilute; (*Tex.*) tender; (*Colors*, *oft*) tone down

schwach-erhitzt *a* gently heated

schwach-färben *vt* tint, tinge

Schwach-gas *n* lean gas. **sch-legiert** *a* (*Metll.*) low-alloy

schwächlich *a* weak; frail, delicate; sickly **Schwächling** *m* (-e) weakling

schwach-sauer, -säuerlich *a* weakly acid. **-siedend** *a* gently boiling. **-sinnig** *a* feebleminded. **Sch-strom** *m* (*Elec.*) weak current

Schwächung *f* (-en) weakening; diminution, etc *cf* SCHWÄCHEN

Schwächungs-gesetz *n* (*Spect.*) extinction law. **-koeffizient** *m* **1** (*Spect.*) extinction coefficient. **2** (*X-rays*) absorption coefficient. **-mittel** *n* depressant

schwach·wandig *a* thin-walled. **-wirk-sam** *a* feebly active

Schwaden *m* (-) **1** (*usu pl*) fumes, gas clouds, noxious vapors; (*Min.*) choke damp (CO_2); **feuriger Sch.** fire damp, methane. **2** (*Agric.*) swath; sweet grass. **--fang** *m* hood, ventilator

Schwa(h)l *m* (*Iron*) rich finery cinder, slag. **--arbeit** *f* (*Iron*) single refining; slag washing. **-boden** *m* slag bed (*or* bottom)

Schwalbe *f* (-n) (*Zool.*) swallow

Schwalben·nest *n* **1** (*lit*) swallow's nest. **2** (*Slag*) honeycomb. **-schwanz** *m* swallowtail; (*Joints, etc*) dovetail. **-wurz(el)** *f* (*Bot.*) swallowwort; (*Pharm.*) vincetoxicum

Schwal·boden *m* slag bed (*or* bottom)

Schwalch *m* (-e) **1** furnace flue. **2** smoke

Schwall *m* swell, surge; flood, torrent. **--wasser** *n* flood water

schwamm swam: *past of* SCHWIMMEN

Schwamm *m* (-̈e) **1** sponge. **2** mushroom; (*incl Med.*) fungus; dry rot. **3** tinder. **4** (*Zinc*) tutty. **sch-·artig** *a* spongy; fungous. **Sch--filter** *n* sponge filter. **sch-förmig** *a* spongy, porous. **Sch--gewebe** *n* spongy tissue. **-gift** *n* mushroom poison, muscarine. **-gummi** *n* sponge rubber. **-holz** *n* spongy (*or* dry-rotted) wood

schwammig *a* **1** spongy, porous; (*Paper*) bibulous. **2** fungous, fungoid

Schwamm·kupfer *n* spongy copper. **-kürbis** *m* loofah, sponge gourd. **-säure** *f* boletic (fungic, fumaric) acid. **-stoff** *m* fungin. **-tod** *m* fungicide. **-zucker** *m* mannitol

Schwan *m* (-̈e) swan

schwand vanished, etc: *past of* SCHWINDEN

Schwand *m* **1** shrinkage, etc = SCHWUND. **2** loss, ullage

Schwanenhals· (*in compds*) gooseneck

schwang swung, etc: *past of* SCHWINGEN

Schwang *m:* **im Sch.** in fashion, current; **in Sch. kommen** become current

schwanger *a* pregnant. **Schwangere** *f* (*adj endgs*) pregnant woman, expectant mother. **schwängern** *vt* make pregnant; impregnate, saturate. **Schwanger-schaft** *f* (-en) pregnancy

Schwangerschafts·abbruch *m* termina-

tion of pregnancy, abortion. **-nachweis** *m* pregnancy test. **-verhinderung, -ver-hütung** *f* contraception

Schwängerung *f* impregnation; saturation

schwank *a* **1** pliable. **2** slender. **3** loose. **4** wavering, unsteady

schwanken I *vi* (h) **1** swing, wave; sway. **2** rock, shake. **3** vary, fluctuate. **4** oscillate. **5** waver; falter. **II** *vi*(s) stagger (to a place). **schwankend** *p.a* (*oft*) variable; unsteady, uncertain. **Schwankung** *f* (-en) **1** swinging, etc. **2** variation, fluctuation

Schwankungs·grenze *f* limit of variation. **-maß** *n* standard deviation. **-weite** *f* range of variability

Schwanz *m* (-̈e) **1** tail; (*in compds oft*) caudal. **2** (tail) end. **3** line (of people). **--bande** *f* (*Spect.*) tail band. **-flosse** *f* tail (*or* caudal) fin. **-hahn** *m* stopcock (with an outlet through the end of the key). **-kügelchen** *n* bulb drawn out to a narrow opening. **-pfeffer** *m* cubebs. **-stern** *m* comet. **-steuer** *n* rudder. **-wirbel** *m* caudal vertebra

schwappen *vi* slosh (over) (*as:* liquid in vessels); surge, rise

Schwäre *f* (-n) abscess, ulcer. **schwären** *vi* fester, suppurate

Schwarm *m* (-̈e) **1** swarm, cluster, flock; (*Bact.*) colony. **2** fancy, passion; idol. **--bildung** *f* formation of swarms (clusters, colonies). **schwärmen** *vi* **1** swarm, flock, cluster. **2** roam, wander, migrate. **3** dream, rave; (+**für**) be enthusiastic (about). **Schwärmer** *m* (-) **1** dreamer, enthusiast; visionary. **2** (*Zool.*) sphinx (moth). **3** serpent (firecracker). **Schwarm·ion** *n* exchangeable ion

Schwarte *f* (-n) **1** rind, (thick) skin. **2** crust, covering; (tree) bark. **3** scalp. **4** plank (with bark on one edge). **Schwarte(n)-·magen** *m* head cheese

schwarz (-̈er, -̈este) **1** *a* black; dark-haired; swarthy; gloomy: **sch-es Öl** (*Petrol.*) black oil, blackstrap; **sch-es Wasser** (*Pharm.*) black mercurial lotion; **Glas sch. anlaufen lassen** smoke glass; **sch. liegen** (*Beer*) be settled, be clear. **2** *a, adv, sep pfx* illegal(ly), smuggled, black-market. **Schwarz** n black (color & dye); soot, lampblack, etc; carbon

Schwarz·beere f huckleberry; melastoma. **-beize** f (Dye.) black (or iron) liquor. **sch–blau** a midnight-blue, blue-black; black and blue. **Sch–blech** n black plate, untinned iron plate. **-blei** n black lead, graphite. **-bleierz** n black lead spar, carboniferous cerrusite. **sch–braun** a very dark brown, brownish black; tawny. **Sch–braunstein** m psilomelane. **-brenner** m moonshiner. **-brot** n black (or dark rye) bread. **sch–brüchig** a (Metll.) blackshort. **Sch–druck** m printing in black

Schwarze (adj endgs) **I** m, f **1** black-haired person. **2** black person. **3** cleric, clergyman. **II** n: **ins Sch. treffen** hit the bull's eye

Schwärze f **1** blackness, darkness. **2** blacking, black; soot. **3** printer's ink; India ink. **4** (Foun.) black wash. **5** (Bot.) smut

Schwarz·eisen n high-silicon pig iron

Schwärze·messung f (Opt.) densitometry

schwärzen 1 vt/i blacken, darken. **2** blackfinish; (Metll.) black; (Print.) ink

Schwarz·erde f black earth (or soil); (Geol.) chernozem. **-erle** f (Bot.) alder. **-erz** n (Min.) tetrahedrite, stephanite

schwarz·fahren* vi(s) beat the fare; drive without a license

Schwarz·farbe f black color. **-färber** m dyer in black. **sch–farbig** a black(-colored). **Sch–färbung** f black coloration, blackening; dyeing (in) black. **-fäule** f black rot. **-föhre** f Austrian pine. **sch–gar** a black-tanned. **-gebrannt** a (Metll.) kishy. **-gelb** a very dark yellow, tawny

schwarz·glühen vt black-anneal

schwarz·grau a very dark gray. **-grün** a very dark green. **Sch–gültigerz** n (Min.) stephanite, tetrahedrite, polybasite. **-guß** m black malleable cast iron. **-handel** m black market (dealing). **sch–heiß** a black-hot

schwarz·hören vi **1** listen in on an unlicensed radio. **2** attend lectures without paying tuition

schwarz·kalk m **1** lump hydraulic lime. **2** crude calcium acetate. **Sch–kerntemperguß** m blackheart malleable cast iron. **-kohle** f black coal (or char-

coal). **-kreide** f black chalk. **-kugelthermometer** n black-bulb thermometer. **-kümmel** m nutmeg flower — türkischer Sch. fennel flower. **-kümmelöl** n fennel-flower oil. **-kupfer** n **1** black (or slag) copper. **2** = **-kupfererz** n (Min.) melaconite, tenorite. **-lack** m black varnish. **-lauge** f black liquor

schwärzlich a blackish

Schwarz·manganerz n hausmannite. **-marktpreis** m black market price. **-mehl** n dark (esp rye) flour. **-öl** n black (specif dark linseed) oil. **-pech** n black (or common) pitch. **-pulver** n black (gun)powder. **sch–rot** a very dark red. **Sch–schmelz** m (Ceram.) black enamel. **-schmelze** f (Paper) black ash. **-seher** m **1** pessimist, crepe-hanger. **2** person watching unlicensed TV. **-sender** m illegal transmitter. **-senföl** n black-mustard oil. **-silbererz** n, **silberglanz** m black silver, (Min.) stephanite. **-spießglanzerz** n (Min.) bournonite. **-strahlen** mpl blackbody radiation. **-strahler** m black-body radiator. **-strahlung** f black-body radiation

Schwärzung f (-en) blackening, darkening; (Phot.) density

Schwärzungs·kurve f (Phot.) characteristic (exposure-density) curve. **-messer** m optical densitometer

Schwarz·vitriol n black vitriol, impure ferrous sulfate. **-wald** m (Geog.) Black Forest. **sch–weiß** a black-and-white. **Sch–werden** n blackening. **-wurz(el)** f (Bot.) comfrey; (Pharm.) symphytum — amerikanische Sch. baneberry root. **-zinkerz** n (Min.) franklinite

Schwatzit m (Min.) mercurian tetrahedrite

Schwebe f: **in der Sch.** uncertain, undecided; **sich in der Sch. halten** hover

Schwebe·bahn f **1** suspended monorail. **2** aerial (or cable) tramway. **-bett** n suspended (or fluidized) bed. **-fähigkeit** f floating (or suspension) power; buoyancy. **-flora** f (Biol.) phytoplankton. **-führung** f (Soil) silt content. **-körper** m **1** suspended substance (or material); floating body. **2** (Instrum.) rotameter. **-methode** f suspension (or flotation) method

schweben *vi* (h, s) 1 hover, float (in the air). 2 hang; be suspended (*incl Chem.*). 3 (*Sound*) linger on. 4 be undecided, be pending. 5 be (in a state, a situation): **in Gefahr sch.** be in danger. **schwebend** *p.a* (*oft*) suspended; undecided, pending **Schwebe‐staub** *m* airborne dust. **-stoff** *m* suspended substance (*or pl oft* matter). **-teilchen** *n* suspended particle. **-trockner** *m* flash drier. **-vergasung** *f* fluidized gasification. **-zustand** *m* 1 suspended state, suspension. 2 state of uncertainty **Schwebstoff°** *m* 1 suspended substance. 2 (*Mil.*) nonpersistent chemical agent. **-filter** *n* filter for suspended matter **Schwebung** *f* (-en) 1 hovering, etc *cf* SCHWEBEN. 2 (*Phys.*) beat; surge. **Schwebungs‐frequenz** *f* beat frequency **Schweden** *n* (*Geog.*) Sweden; (*in compds oft*) Swedish. **schwedisch** *a* Swedish: **sch—es Grün** Swedish (*or* Scheele's) green. **Sch—leder** *n* suede **Schwed. P.** *abbv* (schwedisches Patent) Swed. Pat.

Schwefel *m* sulfur; (*in compds oft*) sulfuric, sulfide of, sulfo-, thio-: **-abdruck** *m* sulfur print. **-alkali** *n* alkali sulfide. **-alkohol** *m* 1 thiol. 2 (*formerly*) carbon disulfide. **-ammon(ium)** *n* ammonium sulfide. **-antimon** *n* antimony sulfide. **-antimonblei** *n* antimony lead sulfide; (*Min.*) boulangerite. **sch—antimonig** *a* thioantimonous. **-antimonsauer** *a* thioantimonate of. **Sch—antimonsäure** *f* thioantimonic acid. **sch—arm** *a* lowsulfur. **Sch—arsen** *n* arsenic sulfide. **sch—arsenig** *a* thioarsenious **Schwefelarsenik** *n* arsenic sulfide. **-säure** *f* thioarsenic acid. **-verbindung** *f* arsenic sulfide; sulfarsenide **Schwefel‐arsensäure** *f* thioarsenic acid. **-art** *f* type of sulfur. **sch—artig** *a* sulfurous. **Sch—äther** *m* sulfuric (*or* ethyl) ether. **-aufnahme** *f* (*Metll.*) sulfur pickup. **-ausscheidung** *f* separation of sulfur. **-ausschlag** *m* sulfur bloom. **-bad** *n* 1 sulfur bath. 2 sulfur springs (resort). **-bakterien** *npl* sulfur bacteria **schwefelbar** *a* sulfurizable **Schwefel‐barium** *n* barium sulfide. **-bestimmung** *f* sulfur determination. **-bin-**

dung *f* sulfur bonding. **-blau** *n* sulfur blue. **-blausäure** *f* thiocyanic acid. **-blei** *n* lead sulfide. **-bleiche** *f* sulfur bleach. **-blumen, -blüten** *fpl* flowers of sulfur. **-brennofen** *m* sulfur kiln. **-bromid** *n* sulfur dibromide. **-bromür** *n* sulfur monobromide. **-brot** *n* loaf of sulfur. **-cadmium** *n* cadmium sulfide. **-calcium** *n* calcium sulfide. **-chlorid** *n* sulfur dichloride. **-chlorür** *n* sulfur monochloride. **-chrom** *n* chromium sulfide **Schwefelcyan** *n* thiocyanogen, cyanogen sulfide; (*in compds usu*) thiocyanate of.: **-ammonium** *n* ammonium thiocyanate. **-kali(um)** *n* potassium thiocyanate. **-metall** *n* metallic thiocyanate. **-säure** *f* thiocyanic acid. **sch—wasserstoffsauer** *a* thiocyanate of. **Sch—wasserstoffsaüre** *f* thiocyanic acid **Schwefel‐dampf** *m* sulfur vapor (*or* fumes). **-dichlorid** *n* sulfur dichloride. **-dioxid** *n* sulfur dioxide. **-dunst** *m* sulfurous vapor. **sch—echt** *a* (*Dye.*) fast to stoving (*or* to sulfurous acid). **Sch—einschlag** *m* 1 (*Casks*) sulfuring. 2 sulfur match. **-eisen** *n* ferrous sulfide. **-entfernung** *f* desulfurization. **-erde** *f* sulfurous earth. **-erz** *n* sulfur ore. **-faden** *m* sulfured wick, sulfur match. **-farbe** *f* 1 sulfur color (*or* dye). 2 (*Wool*) stoved shade. **sch—farben, -farbig** *a* sulfurcolored. **Sch—farbstoff** *m* sulfur dye. **sch—fest** *a* sulfur tolerant (*or* resistant). **Sch—festigkeit** *f* immunity to sulfur poisoning. **-form** *f* brimstone mold. **sch—frei** *a* sulfur-free. **Sch—gallium** *n* gallium sulfide. **-gang** *m* sulfur vein (*or* lode). **-gehalt** *m* sulfur content. **sch—gelb** *a* sulfur-yellow. **Sch—gerbung** *f* sulfur tanning. **-germanium** *n* germanium sulfide. **-geruch** *m* sulfur odor. **sch—gesäuert** *a* sulfurated, treated with sulfuric acid. **Sch-gold** *n* gold sulfide. **-grube** *f* sulfur mine. **-halogen** *n* sulfur halide. **sch—haltig** *a* sulfurous, sulfurbearing. **Sch—harnstoff** *m* thiourea. **-holz, -hölzchen** *n* sulfur match. **-hütte** *f* sulfur refinery **schwefelig** *a* sulfurous = SCHWEFLIG **Schwefel‐indigo** *n* thioindigo. **-indium** *n* indium sulfide. **-jodür** *n* sulfur monoiodide. **-kadmium** n cadmium sulfide. **-kalium** n potassium sulfide. **-kalk** m (*Agric.*) lime-sulfur. **-kalkbrühe** f lime-sulfur spray (*or* wash). **-kalzium** n calcium sulfide. **-kammer** f sulfur chamber, sulfuring room. **-karbolsäure** f thiophenol. **-kastenbleiche** f (*Dye.*) stoving. **-kies** m iron pyrites —**gemeiner Sch.** pyrite; **prismatischer Sch.** (*Min.*) marcasite. **-kobalt** m cobalt sulfide; (*Min.*) linnaeite. **-kohle** f high-sulfur coal. **-kohlensäure** f thiocarbonic acid. **-kohlenstoff** m carbon disulfide. **-kolben** m sulfur distilling retort. **-korn, -körnchen** n sulfur granule. **-kuchen** m cake of sulfur. **-kupfer** n copper sulfide. **-kupferoxydul** n cuprous sulfide. **-latwerge** f (*Pharm.*) confection of sulfur. **-läuterofen** m sulfur refining furnace. **-leber** f liver of sulfur, hepar, potassium sulfide. **-leinöl** n (*Pharm.*) balsam of sulfur. **-magnesium** n magnesium sulfide. **-mangan** n manganese sulfide; (*Min.*) alabandite. **-mehl** n sublimed sulfur. **-metall** n metallic sulfide. **-milch** f milk of sulfur, precipitated sulfur. **-molybdän** n molybdenum sulfide

schwefeln vt 1 treat (bleach, fumigate) with sulfur, sulfurize, sulfurate. 2 (*Rubber*) vulcanize. 3 (*Tex.*) stove. 4 (*Hops*) cure. *Cf* GESCHWEFELT

Schwefel·natrium, -natron n sodium sulfide. **-nickel** n nickel sulfide. **-niederschlag** m sulfur precipitate. **-ofen** m sulfur burner. **-öl** n dimethyl sulfide. **-oxid** n (any) oxide of sulfur. **-phosphor** m (any) sulfide of phosphorus. **-pocke** f (*Metll.*) sulfur pockmark. **-probe** f sulfur sample; sulfur test. **-pulver** n powdered sulfur. **-quecksilber** n 1 (any) sulfide of mercury. 2 (*Min.*) cinnabar. **-quelle** f 1 sulfur spring. 2 source of sulfur. **-räucherung** f sulfur fumigation. **-regen** m (volcanic) sulfur rain. **sch–reich** a high-sulfur. **Sch–rubin** m ruby sulfur, realgar. **-salbe** f sulfur ointment. **-salz** n sulfur (*or* thio) salt. **sch–sauer** a sulfate of: **sch–er Kalk** calcium sulfate **Schwefelsäure** f sulfuric acid. **–anhydrid** n sulfuric anhydride, sulfur trioxide. **-ballon** m sulfuric acid carboy. **-bestim-**

mung f sulfuric acid determination. **-chlorhydrin** n chlorosulfonic acid. **-fabrik** f sulfuric acid plant (*or* works). **-fabrikation** f sulfuric acid manufacture. **-kammer** f sulfuric acid chamber. **-salz** n sulfate

Schwefel·schlacke f sulfur dross. **-schwarz** n sulfur black. **-seife** f sulfur soap. **-selen** n selenium sulfide; (*Min.*) selensulfur. **-silber** n silver sulfide; (*Min.*) argentite. **-silicium** n silicon sulfide. **-spießglanz** m, **-spießglanzerz** n (*Min.*) stibnite. **-stange** f roll (of) sulfur. **-stickstoff** m nitrogen sulfide. **-stück** n piece of sulfur. **-tetrachlorid** n sulfur tetrachloride. **-thallium** n thallium sulfide. **-tonerde** f aluminum sulfide. **-tonung** f (*Phot.*) sulfide toning. **-trioxid** n sulfur trioxide

Schwefelung f sulfurization, sulfuration; vulcanization, etc cf SCHWEFELN. **Schwefelungs-mittel** n sulfurizing agent

Schwefel·verbindung f sulfur compound. **-wasser** n sulfur water

Schwefelwasserstoff m, **-gas** n hydrogen sulfide (gas). **-rest** m mercapto (*or* sulfhydryl) group. **-säure** f hydrosulfuric acid. **-strom** m current of hydrogen sulfide. **-verbindung** f hydrosulfide. **-wasser** n hydrogen sulfide water

Schwefel·weinsäure f ethylsulfuric acid. **-werk** n sulfur refinery. **-wismut** n bismuth sulfide. **-wurz** f brimstonewort. **-wurzel** f brimstonewort (*or* peucedanum) root. **-zink** n zinc sulfide. **-zinkweiß** n lithopone (*or* zinc sulfide) pigment. **-zinn** n stannic (*or* tin) sulfide. **-zyan** n thiocyanogen = SCHWEFELCYAN **schweflig** a sulfurous: **sch—e Säure** sulfurous acid. **–sauer** a sulfite of

Schwefligsäure f sulfurous acid. **–anhydrid** n sulfurous anhydride, sulfur dioxide. **-gas** n sulfur dioxide. **-wasser** n aqueous sulfurous acid solution

Schweflung f sulfurization, etc = SCHWEFELUNG

Schweif m (-e) 1 tail. 2 (sweeping) curve. 3 (*Tex.*) warp. **schweifen I** vt 1 curve, arch. 2 scallop; bevel; (*Tex.*) warp. **II** vi(s) roam, wander; sweep

Schweif·haar n horsehair. **-säge** f fret (*or*

compass) saw. **-stern** *m* comet

Schweifung *f* (-en) **1** curving, etc *cf* SCHWEIFEN. **2** curve; scallop

schweigen* (schwieg, geschwiegen) *vi* be silent, say nothing. **Schweigen** *n* silence. **schweigend** *p.a* silent. **schweigsam** *a* silent; secretive, uncommunicative

Schwein *n* (-e) **1** pig, hog; swine. **2** pork. **3** good luck. **--brot** *n* (*Bot.*) sow bread

Schweinchen *n* (-) **1** weighing bottle with feet. **2** little pig

Schweine- hog, pig, pork; *oft* = SCHWEINS-: **-eisen** *n* pig iron. **-fett** *n* hog fat, lard. **-fleisch** *n* pork. **-futter** *n* pig swill, hog-wash. **-pest** *f* hog cholera

Schweinerei *f* (-en) **1** filthy mess. **2** dirty trick; nasty affair. **3** obscenity

Schweine-schmalz *n*, **-schmer** *m* lard. **-zucht** *f* hog raising

Schweinfurter Grün Schweinfurt (*or* Paris) green

Schweins- hog, pig, pork; *oft* = SCHWEINE-: **-gummi** *n* hog gum. **-haut** *f* pigskin, hogskin. **-kopfsülze** *f* head cheese. **-leder** *n* pigskin, hogskin. **-wurst** *f* pork sausage

Schweiß I *m* **1** sweat, perspiration. **2** steam, fog (on windows, etc). **3** (*Wool*) suint, yolk. II (*in compds oft, cf* SCHWEISSEN) weld(ing): **-abgabe** *f* perspiration. **-apparat** *m* welding apparatus. **-arbeit** *f* welding. **-asche** *f* suint ash, raw potash. **-bad** *n* **1** (*Wool*) degreasing bath. **2** steam (*or* Turkish) bath.

schweißbar *a* weldable, welding. **Schweißbarkeit** *f* weldability

schweiß- sweat; welding: **-befördernd** *a* sudorific, sweat-inducing. **Sch-bogen** *m* welding arc. **-brenner** *m* welding torch. **-brille** *f* welder's goggles. **-drüse** *f* sweat gland. **sch--echt** *a* perspiration-resistant, sweatfast. **Sch-eisen** *n* weld iron; wrought (*or* puddled) iron

schweißen 1 *vt/i* weld. **2** *vt* heat-seal, bond by heating. **3** *vi* begin to melt; (*Liquids*) leak. **Schweißer** *m* (-) welder

schweiß- sweat; welding: **-erregend** *a* sudorific, sweat-inducing. **Sch-fehler** *m* welding defect. **-flüssigkeit** *f* sweat. **-gang** *m* sweat duct. **-gehalt** *m* (*Wool*) suint content. **sch--gewaschen** *a* (*Wool*)

washed in the grease

schweiß-härten *vt* (*Metll.*) weld-harden

Schweiß- sweat; welding: **-härtung** *f* weld-hardening. **-hitze** *f* welding heat. **-hund** *m* bloodhound. **-loch** *n* sweat pore. **-metall** *n* welding metal; wrought iron. **-mittel** *n* **1** sudorific, sweat-inducing agent. **2** welding flux. **-naht** *f* weld (seam). **-ofen** *m* **1** welding furnace. **2** reheating furnace. **-paste** *f* welding paste, flux

schweiß-plattieren *vt* (*Metll.*) hard-surface; coat (*or* deposit) by welding

Schweiß- sweat; welding: **-prozeß** *m* welding process. **-pulver** *n* **1** welding powder (*or* flux). **2** sudorific powder. **-schlacke** *f* welding cinder. **-schmiedeeisen** *n* weld iron. **-stab** *m* welding rod. **-stahl** *m* weld(ing) steel. **-stelle** *f* weld; thermocouple junction. **sch--treibend** *a* sudorific, sweat-inducing

Schweißung *f* (-en) weld; welding

Schweiß-verfahren *n* welding process. **-verbindung** *f* welded joint. **-wachs** *n* wax from suint, yolk wax. **-walzen** *fpl* roughing rolls. **sch--warm** *a* welding-hot. **Sch-wärme** *f* welding heat. **-wasser** *n* water of condensation. **-wolle** *f* wool in the yolk (*or* grease), wool containing suint

Schweiz *f* Switzerland. **Schweizer** *m* (-) or *a*, **Schweizerin** *f* (-nen), **schweizerisch** *a* Swiss. **Schweizer-käse** *m* Swiss (*esp* Gruyère) cheese

Schwel- smoldering; low-temperature carbonization: **-anlage** *f* low-temperature carbonization plant

schwelbar *a* suitable for low-temperature carbonization

Schwel-benzin *n* low-temperature crude light oil. **-braunkohle** *f* low-temperature pyrolizing lignite

schwelchen *vt* (*Brew.*) wither, air-dry (malt). **Schwelch-malz** *n* withered (*or* air-dried) malt

Schwelen 1 *vt* distil (*or* carbonize) at low temperature. **2** *vt/i* burn slowly. **3** *vi* smolder. **Schweler** *m* (-) low-temperature carbonizer. **Schwelerei** *f* (-en) low temperature carbonization process (*or* plant)

Schwel-gas *n* producer gas, low-temperature carbonization gas

schwelken vt (Brew.) wither, air-dry (malt)

Schwel·kerze f (Mil.) smoke candle

Schwelk·malz n withered (or air-dried) malt

Schwel·kohle f coal (usu lignite) suitable for low-temperature carbonization. **-koks** m low-temperature coke

Schwell· swelling; erectile: **-äscher** m (Lthr.) fresh (or white) lime

schwellbar a capable of swelling; (Anat.) erectile

Schwell·beize f (Lthr.) swelling (or plumping) liquor; bran drench

Schwelle f (-n) 1 threshold; sill. 2 (railroad) crosstie; crossbar. 3 (Geol.) uplift, swell. 4 (Phys.) (energy) barrier; (Radar) gate

schwellen I (reg) vt 1 swell, distend. 2 (Lthr.) plump. **II*** vi(s) (schwoll, geschwollen; schwillt) 3 swell, puff up. 4 (+über) spill (over), overflow. **schwellend** p.a (oft) full; lush, luxuriant; (Physiol.) tumescent

Schwellen·wert m threshold value; (Phot.) exposure factor

Schwell·farbe f (Lthr.) plumping liquor. **-gewebe** n (Anat.) erectile tissue

schwellig a threshold, liminal

Schwell·körper m (Anat.) corpus cavernosum. **-kraft** f (Lthr.) plumping power. **-mittel** n (Lthr.) plumping agent

Schwellung f (-en) swelling; (Geol.) swell; (Lthr.) plumping

Schwell·wert m threshold value

Schwel·ofen m low-temperature carbonizing furnace. **-raum** m carbonizing chamber. **-retorte** f low-temperature distillation retort. **-teer** m low-temperature carbonization tar

Schwelung f (-en) low-temperature carbonization; smoldering, slow burning

Schwel·vorgang m low-temperature carbonization process. **-wasser** n aqueous liquid from the low-temperature carbonization process. **-werk** n low-temperature carbonization plant

Schwemm·boden m alluvial soil

Schwemme f (-n) 1 watering place (for animals); trough. 2 glut. 3 raft. 4 timber floating; flume. **schwemmen** vt 1 water (animals); flush, irrigate. 2 wash (up), deposit. 3 float (logs, etc). 4 soak (hides). 5 rinse

Schwemm·gebilde n (Geol.) alluvial formation. **-kohle** f (Coal) dewatered fines. **-land** n alluvial land. **-rinne** f flume. **-sand** m alluvial sand. **-stein** m 1 porous brick from clay and gravel; porous concrete block. 2 lightweight brick. 3 pumice stone. **-verfahren** n flotation process. **-wasser** n wash water; flushing water

Schwengel m (-) 1 (pump) handle; crank. 2 bell clapper. 3 pendulum

schwenkbar a pivoted, swivel(ing)

schwenk·echt a rinse-resistant

schwenken I vt 1 swing, wave, brandish. 2 rinse. 3 toss in grease, stir-fry. **III** vt, vi(s) pivot, swivel. **III** vi(s) swing, turn; change sides

Schwenk· swivel, pivoted: **-hahn** m swivel tap (faucet, cock). **-kran** m swivel crane. **-rohr** n swing pipe

Schwenkung f (-en) 1 swinging; rinsing etc. cf SCHWENKEN. 2 deviation

schwer I a 1 heavy: **sch—es Weinöl** heavy oil of wine; **10 Kilo sch.** weighing ten kilos; **das wiegt sch.** that carries weight. 2 severe, serious, grave. 3 lethargic. 4 lots of; great, large. **II** a & sep pfx 5 difficult, hard; troublesome, laborious cf SCHWERFALLEN. 6 adv (usu) with difficulty; (+adj in **-bar**, **-lich**) e.g.: **sch. losbär** (or **löslich**) not readily soluble, difficult to dissolve. 7: **sie haben es sch., sie tun sich sch.** they have a hard time; **das macht ihnen sch. zu schaffen** that causes them a lot of trouble

Schwer·beanspruchungsmotor m heavy-duty motor. **-benzin** n heavy gasoline (B.P. over 100°C). **-benzol** n heavy benzol (B.P. 160°C+); solvent naphtha. **-beton** n dense (or heavy-aggregate) concrete. **-betrieb** m heavy operation (or duty). **-bleierz** n (Min.) plattnerite. **sch—blütig** a lethargic. **-brennbar** a combustion-resistant, not readily combustible. **Sch—brennbarkeit** f resistance to combustion, difficult combustibility. **-chemikalien** npl heavy chemicals. **sch—durchlässig** a not readily permeable

Schwere f 1 (esp Phys.) gravity. 2 heaviness; weight; body (of wine). 3 seriousness, severity. 4 difficulty. Cf SCHWER. **-beschleunigung** f acceleration due to gravity. **-feld** n gravitational field

schwerelos *a* weightless. **Schwerelosigkeit** *f* weightlessness

Schwere·messer *m* barometer; baroscope; gravimeter. **-messung** *f* gravimetry, gravity measurement

schwer·entflammbar *a* flame-resistant. **Sch–erde** *f* heavy earth, baryta

schwerer·löslich *a* more (*or* fairly) difficult to dissolve

schwer·fallen* *vi*(s) (*dat*) be difficult (for), cause trouble

schwer·fällig *a* 1 clumsy, heavy-handed. 2 ponderous, unwieldy. **-flüchtig** *a* not readily volatile. **Sch–flüchtigkeit** *f* low volatility. **sch–flüssig** *a* 1 refractory, not readily fusible. **Sch–flüssigkeit** *f* 1 heavy liquid. 2 (*Ores*) heavy (*or* dense) medium. 3 low fusibility, refractoriness. **-flüssigkeitsverfahren** *n* (*Ores*) heavy-media separation. **-frucht** *f* heavy grain (*usu* wheat, rye). **sch–gefrierbar** *a* not readily freezable, low-freezing. **Sch–gewicht** *n* main stress, burden (of an argument, etc)

schwer·halten* *vi* be difficult

schwerhörig *a* hearing-impaired, hard of hearing. **Schwerhörigkeit** *f* defective hearing

Schwer·industrie *f* heavy industry. **-ion** *n* heavy ion

Schwerkraft° *f* (force of) gravity; gravitation. **–beschleunigung** *f* acceleration due to gravity. **-feld** *n* gravitational field

Schwer·kraftstoff *m* heavy (*usu* diesel) fuel

Schwerkraft·wirkung *f* gravitational effect

Schwer·kristall *n* (*Glass*) heavy crystal

schwerl. *abbv* (schwerlöslich) not readily soluble

Schwer·leder *n* heavy leather; sole leather

schwerlich *adv* hardly, scarcely

schwer·löslich *a* not readily soluble, difficult to dissolve. **Sch–löslichkeit** *f* difficult solubility. **-metall** *n* heavy metal. **-metallsalz** *n* salt of a heavy metal. **-mut** *m* melancholy. **-naphta** *n* heavy naphtha. **-öl** *n* heavy oil

Schwerpunkt° *m* 1 center of gravity (*or* of mass). 2 focal (*or* key) point; main stress; (*in compds oft*) key: **-betrieb** *m* key operation. **-bildung** *f* concentration. **-indu-**strie *f* key industry. **sch–mäßig** *a* by key points; *adv oft* chiefly, mainly

Schwerpunkts·satz *m* principle of the center of gravity

Schwerpunkt·verlagerung *f* shift of the center of gravity

schwer·rostend *a* rust-resistant, rustless. **-schmelzbar** *a* not readily fusible, refractory. **-schmelzend** *a* fusing with difficulty. **Sch–schwarz** *n* (*Dye.*) weighted black. **sch–siedend** *a* high-boiling. **Sch–spat** *m* (*Min.*) heavy spar, barite

Schwerst·beton *m* high-density concrete

Schwer·stein *m* (*Min.*) scheelite. **-stoff** *m* (*Ores*) heavy medium

Schwert *n* (-er) sword

Schwer·tantalerz *n* (*Min.*) tantalite

Schwert·kolben *m* Anschütz distillation flask. **-länge** *f* blade length (of a saw). **-lilie** *f* (*Bot.*) iris, fleur-de-lis

Schwer·trübesortierung *f* heavy media separation. **-uranerz** *n* (*Min.*) uraninite. **sch–verdaulich** *a* indigestible, hard to digest. **-verständlich** *a* abstruse, difficult to understand. **Sch–wasser** *n* heavy water, deuterium oxide. **sch–wiegend** *a* weighty, serious

Schwester *f*(-n) 1 sister. 2 nurse. 3 (*in compds oft*) affiliated: **-firma** *f* affiliated company

schwieg was silent: *past of* SCHWEIGEN

Schwiele *f* (-n) 1 callus, callosity. 2 welt. **schwielig** *a* callous, calloused; welted

schwierig *a* difficult, troublesome, hard; *adv usu* with difficulty. **Schwierigkeit** *f* (-en) difficulty, *pl oft* trouble

schwillt swells, etc: *pres of* SCHWELLEN

Schwimm· floating: **-äscher** *m* floating lime. **-aufbereitung** *f* (*Ores*) flotation. **-auftrieb** *m* buoyancy. **-bad** *n* swimming pool. **-badeseife** *f* floating bath soap. **-blase** *f* air bladder, sound. **-decke** *f* surface scum

schwimmen* (schwamm, geschwommen) 1 *vt, vi*(s) swim; float. 2 *vi*(s) drift; be (all) at sea; be flooded; skid. **schwimmend** *p.a* (*oft*) vague, indistinct. **Schwimmer** *m* (-) 1 swimmer. 2 float; floater. 3 pontoon

Schwimmer·probe *f* float test. **-ventil** *n* float valve

Schwimmethode *f* (=**Schwimm·methode**) flotation method

schwimm·fähig *a* buoyant, floatable. **Sch–fähigkeit** *f* buoyancy, floatability. **-gerät** *n* (*Ores*) flotation apparatus. **-gerste** *f* (*Brew.*) float barley, skimmings. **-haut** *f* web (membrane), (*Anat.*) tela. **-kiesel** *m* floatstone. **-körper** *m* swimming (*or* floating) object; float. **-kraft** *f* buoyancy. **-methode** = SCHWIMMETHODE. **-sand** *m* quicksand. **-schlamm** *m* scum. **-seife** *f* floating soap. **-stein** *m* floatstone. **-stoff** *m* floating material. **-verfahren** *n* flotation method. **-vermögen** *n* floating power, buoyancy. **-weste** *f* life jacket (vest, preserver). **-ziegel** *m* floating brick

Schwindel *m* (-) **I** dizziness, vertigo. **II** swindle, fraud; lie; deal, affair. **schwindelhaft** *a* 1 dizzying. 2 fraudulent, bogus

Schwindel·korn *n* cubeb; coriander seed

schwindeln 1 *vt* make up (a story). 2 *vt/i* swindle; cheat. 3 *vi* feel dizzy. **schwindelnd** *p.a* (*oft*) dizzy, vertiginous; dizzying

schwinden* (schwand, geschwunden) *vi*(s) 1 shrink, contract. 2 disappear. 3 dwindle, wane, fade (away); atrophy

schwindlig *a* dizzy, giddy

Schwind·lunker *m* shrinkage cavity. **-maß** *n* (amount of) shrinkage, contraction. **-riß** *m* shrinkage crack. **-sucht** *f* (*Med.*) consumption (TB)

Schwindung *f* (-en) shrinkage, contraction; disappearance; dwindling, waning; fading (away); atrophy *cf* SCHWINDEN

Schwindungs·fähigkeit *f* shrinking capacity. **-loch** *n* shrinkage cavity. **-riß** *m* shrinkage crack. **-zugabe** *f* contraction allowance

Schwinge *f* (-n) 1 (*Mach.*) rocker arm. 2 (*Agric.*) winnow, fan; swingle. 3 (*Zool.*) wing, pinion

Schwingel *m* (-), **-gras** *n* fescue (grass); bluegrass

schwingen* (schwang, geschwungen) 1 *vt/i* & **sich sch.** swing; **sich in die Höhe sch.** soar. 2 *vt/i* wave, rock. 3 *vt* wield; centrifuge; winnow, swingle. 4 *vi* vibrate, oscillate; ring (*as:* sound). **schwingend** *p.a* (*oft*) vibratory, oscillatory. **Schwinger** *m* (-) 1 vibrator, oscillator. 2 (*Elec.*) piezoelectric crystal

schwing·fähig *a* capable of vibration (*or* oscillation). **Sch–festigkeit** *f* fatigue strength. **-frequenz** *f* oscillation (*or* vibration) frequency. **-kraftmühle** *f* vibratory mill. **-kreis** *m* oscillating (*or* *Elec.* resonant) circuit. **-kristall** *m* crystal oscillator. **-mahlung** *f* vibratory grinding. **-mischen** *n* vibratory mixing. **-mühle** *f* vibratory mill. **-neigung** *f* oscillating tendency. **-ofen** *m* tilting furnace. **-quarz** *n* quartz resonator, piezoelectric crystal. **-rohr** *n* swing pipe. **-röhre** *f* oscillatory valve. **-sieb** *n* rocking (*or* vibrating) screen (*or* sieve). **-speiser** *m* vibratory feeder. **-tisch** *m* vibrating (*or* rocking) table. **-tür** *f* swinging door

Schwingung *f* (-en) 1 vibration, oscillation. 2 swinging; waving, rocking etc *cf* SCHWINGEN. 3 (sweeping) curve

Schwingungs·achse *f* axis of oscillation. **-bewegung** *f* vibratory (*or* oscillatory) motion. **-bogen** *m* arc of oscillation. **-bruch** *m* break caused by vibration. **-dauer** *f* period of oscillation (*or* vibration). **-drehimpuls** *m* vibrational angular momentum. **-ebene** *f* 1 plane of vibration. 2 (*Opt.*) plane of polarization. **-energie** *f* vibrational energy. **sch–fähig** *a* capable of vibration (*or* oscillation). **Sch–festigkeit** *f* vibration (*or* fatigue) strength. **sch–frei** *a* vibrationless, non-oscillating. **Sch–freiheitsgrad** *m* degree of vibrational freedom. **-frequenz** *f* oscillatory (*or* vibratory) frequency. **-gleichung** *f* vibration (*or* wave) equation. **-knoten** *m* node. **-kreis** *m* (*Elec.*) oscillatory circuit. **-methode** *f* swing method (of weighing). **-niveau** *n* vibrational (energy) level. **-quant(um)** *n* vibrational quantum. **-schreiber** *m* oscillograph. **-spektrum** *n* vibrational spectrum. **-übergang** *m* (*Spect.*) vibrational transition. **-wärme** *f* vibrational molar heat capacity. **-weite** *f* amplitude (of vibration). **-welle** *f* vibrational wave, undulation. **-zahl** *f* vibration number (*or* frequency). **-zeit** *f* oscillation period (*or* time). **-zustand** *m* state of vibration (*or* oscillation)

Schwing·weite *f* amplitude of vibration. **-zahl** *f* vibration number (*or* frequency)

schwirren 1 *vt* centrifuge. **2** *vi* (h,s) whiz, whir, buzz

Schwitz·anlage *f* (*Lthr.*) sweating plant. **-bad** *n* steam bath

Schwitze *f*(-n) **1** (*Lthr.*) sweating: **Häute in die Sch. bringen** sweat hides. **2** roux

schwitzen 1 *vt/i* sweat, perspire. **2** *vt* fry, brown. **3** *vi* steam up, fog (*as:* windows)

Schwitz·kammer *f* sweating room (*or* chamber); steam room. **-mittel** *n* sudorific, diaphoretic. **-öl** *n* (*Petrol.*) foots oil. **-pulver** *n* diaphoretic (*or* sweat-inducing) powder). **-röste** *f* steam retting. **-verfahren** *n* sweating process. **-wasser** *n* condensed moisture; sweat, perspiration. **-wasserkorrosion** *f* condensed-moisture corrosion

Schwöde *f* (-n) (*Lthr.*) liming room (plant, grounds); liming. **--brei** *m* lime cream (*or* paste). **-faß** *n* lime vat. **-grube** *f* lime pit. **-masse** *f* lime cream

schwöden *vt* lime (hides), paint with lime

Schwöd· (=SCHWÖDE): **-grube** *f* lime pit. **-wasser** *n* lime water

schwoll swelled: *past of* SCHWELLEN II *vi*

schwor swore: *past of* SCHWÖREN

schwören (schwor, geschworen) *vt/i* swear, vow, take an oath (on)

schwül *a* sultry; humid; tense. **Schwüle** *f* sultriness, etc

schwulstig *a* puffy, thick

Schwund *m* **1** shrinkage, contraction. **2** diminution. **3** atrophy. **4** disappearance. **5** (*Radio*) fading. **6** loss; wastage, leakage. **--ausgleich** *m* (*Radio*) automatic volume control. **-spannung** *f* shrinkage stress

Schwung *m* (-e) **1** swing, (swinging) motion: **in vollem Sch.** in full swing; **in Sch. setzen** set in motion; **in Sch. sein** be running smoothly, be in stride; **in einem Sch.** in one motion, without a break. **3** tempo. **4** vitality, verve, dash; style; momentum. **5** push, impetus. **6** arch, (swinging) curve. **7** batch, group. **8** (*in compds oft*) vibration, oscillation. **-bewegung** *f* vibratory motion. **-gewicht** *n* pendulum

schwunghaft *a* swinging, sweeping; lively; booming (business)

Schwung·kraft *f* **1** (*Phys.*) centrifugal force. **2** momentum. **3** swing, vivacity, energy. **sch-los** *a* dry, lifeless, insipid. **Sch–maschine** *f* centrifuge. **-rad** flywheel; balance wheel. **sch–voll** *a* vivacious; sweeping, boldly curved

Schwur *m* (–e) oath, vow

Schwz. P. *abbv* (schweizerisches Patent) Sw. pat.

sci . . : *see* **szi . . ,** e.g.: SZINTILLIEREN

Scopolin·säure *f* scopolic acid

Scyllit *n* scyllitol

sd. *abbv* (siedend, siedet) boiling, boils. **Sd.** *abbv* (Siedepunkt) boiling point

s.d. *abbv* **1** (siehe dort) see there, which see. **2** (siehe das (den, die) . .) see the . .

Sdp. *abbv* (Siedepunkt) boiling point

SE *abbv* (Siemens-Einheit) Siemens unit

Sebacin·säure *f* sebacic acid

Sebacon *n* cyclononanone

Sebat *n* (-e) sebacate

Sebazin·säure *f* sebacic acid

secernieren *vt* secrete

sechs *num* six; (*as pfx, usu*) hex(a)-. **Sechs** *f* (-en) (number, figure) six

sechs·atomig *a* hexatomic. **S–eck** *n* hexagon. **s–eckig** *a* hexagonal

Sechser *m* (-) (number, figure) six. **--gruppe** *f* group of six, 6-group

sechserlei *a* (*no endgs*) six kinds of, of six kinds

sechs·fach, -fältig *a* sixfold, sextuple. **S–flach** *m* hexahedron; cube. **s–flächig** *a* hexahedral. **S–flächner** *m* hexahedron. **s–gliedrig** *a* **1** six-membered, six-part. **2** (*Cryst.*) hexagonal. **-jährig** *a* six-year (-old). **S–kant** *m*, *n* hexagon; (*in compds*) =S–kantig, hexagonal, six-edged

Sechskant·mutter *f* hexagon(al) nut. **-schraube** *f* hexagon(-head) bolt. **-stange** *f* hexagonal bar

sechs·mal *adv* six times. **s–monatig** *a* six-month(-old). **-monatlich** *a* semiannual. **S–ring** *m* six-membered ring. **s–säurig** *a* hexacid. **-seitig** *a* six-sided, hexagonal; six-page

sechst *a* sixth **—zu s.** in a group of six

sechstel *num a* (*no endgs*), **Sechstel** *n* (-) (*Frac.*) sixth. **sechstens** *adv* sixthly, in the sixth place

sechs·wertig *a* **1** six-valent, hexavalent. **2** (*Alcohols*) hexahydric. **S–wertigkeit** *f* hexavalence. **s–winklig** *a* six-angled,

hexangular. **-zählig** a 1 sixfold, sextuple.
2 (*Ligands*) sexidentate
sechzehn num sixteen. **sechzehnte** a sixteenth. **Sechzehntel** n (-) (*Frac.*) sixteenth
sechzig num sixty. **Sechziger** m (-) 1 (number, figure) sixty. 2 sixty-year-old (person). **sechzigste** a sixtieth. **Sechzigstel** n (-) (*Frac.*) sixtieth
Secret n (-e) secretion = SEKRET
Sedanol-säure f sedanolic acid
Sedan-schwarz n sedan black
Sedativ-salz n (*formerly*) boric acid
Sedativum n (. . ven) sedative
sedimentär a sedimentary. **Sedimentations-potential** n Dorn effect. **Sediment-gestein** n sedimentary rock. **sedimentieren** vi deposit sediment, settle out. **Sedimentierung** f (-en) sedimentation
Sedoheptit n sedoheptitol
See I m (-n) lake, pond. II f(-n) 1 sea, ocean; seashore. 2 (*in compds, oft*) marine, maritime, nautical: **-ablagerung** f sea (or lake) deposit. **-algen** fpl marine algae. **-asphalt** m marine asphalt. **-bad** n seaside resort; sea bath, dip in the sea. **-band** n edible dulse (an alga)
See-Erz n lake ore (limonite containing phosphate)
See-fahrt f 1 sea voyage. 2 navigation. **-fisch** m salt-water fish. **-flugzeug** n seaplane. **-funk** m marine radio. **-gewächs** n marine plant. **-gras** n seaweed, kelp. **s-grün** a sea-green. **S-güter** npl ocean freight, cargo. **-hafen** m seaport. **-höhe** f height above sea level. **-hund** m (*Zool.*) seal. **-hundsöl** n, **-hundstran** m seal oil. **-igel** m sea urchin. **-kabel** n submarine cable. **-kohl** m sea kale. **-krebs** m crawfish, lobster. **-kreide** f calcareous mud
Seele f(-n) 1 soul, heart; inner being; mind. 2 (*Cables*) core. 3 (*Tubes, etc*) bore. 4 (*Metll.*) shaft (of blast furnaces)
Seelen-messer m inside calipers, bore gauge. **-ruhe** f inner tranquillity
See-licht n marine phosphorescence
seelisch a psychic, mental; emotional
See-luft f sea air. **-meile** f nautical mile.

-moos n sea moss, carrageen. **-pflanze** f marine plant. **-regulierung** f (ocean) flood control. **-rose** f water lily. **-salz** n sea salt. **-sand** m sea sand. **-schlick** m sea ooze. **-seide** f sea silk (from algae); byssus silk. **-stern** m starfish. **-stichlingtran** m stickleback oil. **-tang** m seaweed, sea tang, kelp. **-tier** n marine animal. **s-wärtig** a by sea, maritime. **-wärts** adv seaward, (out) to sea
Seewasser n sea water. **s-echt** a fast to sea water, sea-water-resistant. **S-seife** f marine (or sea-water) soap
See-weg m sea route —**auf dem S.** by sea. **-zunge** f (*Fish*) sole
Segel n (-) sail; (*Anat.*) velum. **-flieger** m glider, sail plane. **-flug** m gliding, sailplaning. **-flugzeug** n glider, sailplane
segeln vt, vi(s) sail
Segel-tuch n sailcloth, canvas
Segen m (-) 1 (*usu*) blessing. 2 benediction. 3: **zum S.** (+*genit*) for the benefit of. . . 4 yield, abundance
Seger-kegel m Seger cone **-porzellan** n Seger porcelain. **Segerscher Kegel** Seger cone
Segge f (-n) sedge; rush. **Seggen-torf** m sedge (or carex) peat
segment-förmig a segmental. **Segmentierung** f(-en) segmentation
segnen vt bless cf ZEITLICH
Seh- sight, visual, optic: **-achse** f optical axis
Sehe f(-n) pupil (of the eye); (sense of) sight
sehen* (sah, gesehen; sieht) vt/i see; vi look —**sich s. lassen** show up, appear, be visible, look good
sehens-wert a worth seeing. **S-würdigkeit** f (interesting) sight
Seh-feld n field of vision. **-kraft** f eyesight. **-lehre** f optics. **-linse** f(*Anat.*) crystalline lens
Sehne f(-n) 1 (*Math.*) chord. 2 (*Anat.*) sinew, tendon. 3 cord, (bow)string; (*esp Metll.*) fiber
Sehn-eisen n fibrous iron
sehnen vr: **sich s. (nach)** long, yearn (for)
Sehnen-band n ligament. **-scheide** f synovial sheath. **-schmiere** f synovial fluid
Sehne-puddeln n puddling of fibrous iron

Seh·nerv *m* optic nerve
sehnig *a* 1 (*Anat.*) tendinous, sinewy. 2 (*Metll.*) fibrous
Sehnsucht *f* (ё) longing, yearning
Seh·purpur *m* visual purple
sehr *adv* very; (very) much, greatly, highly
Seh·rohr *n* periscope; telescope. **-schärfe** *f* sharpness of vision. **-stäbchen** *n* retinal rod. **-weise** *f* viewpoint, way of looking at things. **-weite** *f* visual range. **-zelle** *f* visual cell
sei *v* (*pl* **seien**) *sbjc of* SEIN **I:** 1 be, let (*or* may) be: **s. es groß oder klein** be it large or small; **s. dem wie es s., wie dem auch s.** be that as it may; **es s. betont, daß. . .** let it be emphasized that . . . ; **es s. denn daß. . .** unless. . . 2 (*in indir quot*) is (*pl* are) (supposed to be); (*preceded by past tense, usu*) was (*pl* were): **sie dachten, es s. Sauerstoff** they thought it was oxygen
seicht *a* shallow. **Seichtheit, Seichtigkeit** *f* (-en) shallowness
Seide *f* (-n) 1 silk. 2 (*Bot.*) dodder, love vine
Seidel *n* (-) (beer) stein, (*appx*) pint. **-bast** *m* (*Bot.*) common daphne, mezereon
seiden *a* silk, silken
Seiden· silk: **-abfall** *m* silk waste. **s–ähnlich, -artig** *a* silk-like, silky. **S–asbest** *m* silky asbestos. **-bast** *m* tussah silk; sericin. **-bau** *m* silk culture, sericulture. **-beschwerung, -erschwerung** *f* silk weighting. **-fabrik** *f* silk mill. **-faden** *m* silk thread (fiber, filament). **-faserstoff** *m*, **-fibr(o)in** *n* silk fibroin. **-flor** *m* silk gauze. **-florsieb** *n* silk gauze sieve. **-garn** *n* silk yarn. **-glanz** *m* silky luster. **s–glänzend** *a* silky. **S–grün** *n* silk green. **-holz** *n* satinwood. **-kreppapier** *n* crepe paper. **-leim** *m* silk glue, sericin. **s–matt** *a* satin-finished. **S–papier** *n* tissue paper. **-raupe** *f* silkworm. **-raupenzucht** *f* silkworm raising, sericulture. **-schrei** *m* scroop of silk. **-spinner** *m* silk moth. **-stoff** *m* silk cloth. **-substanz** *f* fibroin. **s–weich** *a* silky(-soft), soft as silk. **S–wurm** *m* silkworm. **-zwirn** *m* silk thread
seidig *a* silky, silken
Seidlitz·pulver *n* Seidlitz powder. **-salz** *n* Epsom salt
seien be, were: *pl of* SEI. **seiend** being, ex-

isting: *pres p of* SEIN **I**
Seife *f* (-n) 1 soap: **grüne** (or **schwarze**) **S.** soft soap; **kaltgerührte S.** half-boiled soap; **geschliffene S.** settled and fitted soap; **gestreckte S.** liquored soap. 2 (*Min.*) alluvial ore, placer
seif·echt *a* fast to soaping, soap-resistant
seifen 1 *vt* soap, lather. 2 *vt/i* wash (ore), buddle
Seifen· soap: **-abfälle** *mpl* soap waste (*or* scraps). **-ansatz** *m* soap stock. **s–artig** *a* soapy, saponaceous. **S–asche** *f* soap ashes. **-avivage** *f* (*Tex.*) brightening with soap. **-bad** *n* soap bath. **-balsam** *m* (camphorated) soap liniment, opodeldoc. **-baum** *m* soapbark tree, quillai. **-baumrinde** *f* soapbark, quillai bark. **-bereitung** *f* soap making. **-bildung** *f* formation of soap, saponification. **-blase** *f* soap bubble. **-brühe** *f* soap suds. **s–echt** *a* fast to soaping. **S–erde** *f* fuller's earth; marl; saponaceous clay. **-ersatz** *m*, **-ersatzmittel** *n* soap substitute. **-erz** *n* alluvial ore. **-fabrik** *f* soap factory (*or* works). **-fabrikant** *m* soap manufacturer. **-fabrikation** *f* soap manufacture. **-farbe** *f* soap color (*or* dye). **-flocken** *fpl* soap flakes. **-form** *f* soap frame. **-füllung** *f* soap filler. **-gold** *n* placer gold. **s–haltig** *a* soapy, containing soap. **S–industrie** *f* soap industry. **-kessel** *m* soap boiler. **s–kochecht** *a* fast to boiling with soap. **S–kocher** *m* soap boiler. **-kraut** *n* soap plant (*or* weed). **-krem** *f* soap cream. **-lauge** *f* soap (*or* lye) solution, soap suds. **-leim** *m* 1 soap glue (*or* paste). 2 (*Paper*) soap size. **-lösung** *f* soap solution. **-napf** *m* soap dish. **-pflaster** *n* soap plaster. **-platte** *f* soap slab. **-probe** *f* 1 soap test. 2 soap sample. **-pulver** *n* soap powder. **-riegel** *m* bar of soap. **-rinde** *f* soap bark. **-rückstand** *m* soap residue. **-schabsel** *npl* soap scraps. **-schale** *f* soap dish. **-schaum** *m* lather. **-schmiere** *f* (*Lthr.*) soap stuff(ing). **-sieder** *m* soap maker (*or* boiler). **-siederasche** *f* soap ashes. **-siederei** *f* 1 soap works. 2 soap making. **-siederlauge** *f* soap boiler's lye. **-späne** *mpl* soap chips. **-spender** *m* soap dispenser. **-spiritus** *m* alcoholic soap solution.

-stein *m* **1** soapstone, (*Min.*) steatite. **2** caustic soda. **-stoff** *m* saponin. **-tafel** *f* slab of soap. **-täfelchen** *n* cake of soap. **-teig** *m* soap paste. **-ton** *m* fuller's earth. **-wäsche** *f* (*Tex.*) soap wash. **-wasser** *n* soap suds, soapy water. **-wurzel** *f* (*Bot.*) soaproot, soapwort. **-zäpfchen** *n* soap suppository. **-zinn** *n* stream tin

seifig *a* soapy

Seif·tuch *n* washcloth

seiger *a* perpendicular, vertical *cf* SAIGER

Seiger·arbeit *f* (*Metll.*) liquation (process). **-blei** *n* liquation lead. **-dörner** *mpl* liquation dross. **s—gerade** *a* perpendicular. **S—herd** *m* liquation hearth. **-hütte** *f* liquation plant, refinery

seigern **1** *vt* (*Metll.*) liquate. **2** *vi* segregate

Seiger·ofen *m* (*Metll.*) liquation furnace. **-pfanne** *f* liquation pan. **-riß** *m* vertical section. **-rückstand** *m* liquation residue. **-schlacke** *f* liquation slag. **-stück** *n* liquation cake

Seigerung *f* (-en) (*Metll.*) liquation; segregation

Seiger·werk *n* (*Metll.*) liquation plant

Seignette·salz *n* Seignette (*or* Rochelle) salt (sodium potassium tartrate)

Seih·brühe *f* strained liquid (syrup, broth)

Seihe *f* (-n) **1** strainer; filter. **2** residue, stock; (*Brew.*) spent malt. **-beutel** *m* filter bag. **-boden** *m* strainer (*or* perforated) bottom. **-faß** *n* filtering cask (*or* tub). **-gefäß** *n* straining (*or* filtering) vessel. **-löffel** *m* straining (*or* perforated) ladle

seihen *vt* strain, sieve, percolate; filter

Seihe·papier *n* filter paper

Seiher *m* (-) strainer, colander; filter

Seihe·rahmen *m* filtering frame. **-sack** *m* filter (*or* straining) bag. **-stein** *m* filtering stone. **-tuch** *n* straining (*or* filtering) cloth. **-vermögen** *n* straining (*or* filtering) power

Seih· = SEIHE·: **-gefäß** *n* straining (*or* filtering) vessel

Seil *n* (-e) **1** rope, line. **2** cable; wire, tightrope. **Seiler** *m* (-) rope maker. **-waren** *fpl* cordage

Seil·faser *f* rope (*or* cordage) fiber. **-öl** *n* cordage oil. **-schmiere** *f* rope grease (*or* lubricant)

Seim *m* (-e) **1** (strained) honey. **2** glutinous liquid; (*Bot.*) mucilage. **seimen 1** *vt* strain (honey). **2** *vi* yield a viscous liquid.

seimig *a* glutinous, mucilaginous

sein I *v** (war, gewesen; ist, *pl* sind; *sbjc* sei, *pl* seien) **A** *vi*(s) **1** be, exist: **es ist an ihnen, das zu tun** it is up to them to do that; **lassen wir es gut s.** let's let it be, let's forget it. **2** (+**zu** + *inf*) be (+*passive inf; oft* = can, should, must be): **das ist zu beachten** this is to be (*or* should be) noted; **es war nichts zu sehen** there was nothing to be seen. **3** (+*pp used as pred a*) be, have been: **es ist erreicht** it is (*or* it has been) attained. **4** (+*dat & pred a*) be, feel: **ihnen ist kalt** they are (*or* feel) cold. **B** *v aux* (+*pp, forms perfect tenses*): **es ist geschmolzen** it melted, it has melted; **es war (es wäre) geschmolzen** it had (it would have) melted. II *possess a & pron* (*with endgs*) his, its, one's

Sein *n* being, existence

seiner *pron* (*genit of* ER, ES) (of) him, (of) it: **sie waren s. gewahr** they were aware of it (*or* of him). **·-seits** *adv* on his (*or* its) part; as for him (*or* it). **-zeit** *adv* at that time

seinesgleichen *pron* his (its, one's) equal

seinetwegen *adv* for his (*or* its) sake; on his (*or* its) account

seinige *pron* (*adj endgs*) his, its (own)

sein·lassen* *vt* let (sthg) be, drop (a matter), stop (doing sthg)

Seismik *f* seismology, **seismisch** *a* seismic

seit I *prep* (*dat*) *cf* SCHON 6: **1** since. **2** for: **das ist (schon) s. Jahren** so that has been so for years. II *cnjc* (ever) since. **seitdem** I *adv* since then, ever since. II *cnjc* (ever) since

Seite *f* (-n) **1** side; aspect: **zur** (*or* **auf die**) **S. schaffen** put aside; **einem zur S. stehen** stand beside someone; **an die S. stellen** (+*dat*) put beside, compare (to); **einem an die S. treten** come to someone's aid; **nach allen S—n** in all directions *cf* BEI-SEITE. **2** face (of a solid). **3:** **schwache (starke) S.** weak (strong) point. **4** e.g. **von zuverlässiger S.** from a reliable source. 5 page (of a book)

seiten: von s. (+*genit*) e.g. **von s. der Fabrikanten** on the part of the manufacturers

Seiten· side, lateral; page: **-achse** *f* lateral axis. **-angabe** *f* page (number) reference. **-anmerkung** *f* marginal note. **-ansatz** *m* side attachment (arm, tube, etc). **-ansicht** *f* side view. **-arm** *m* lateral arm (tube, branch, etc). **-bewegung** *f* lateral motion. **-destillat** *n* side stream (*or* cut). **-druck** *m* lateral pressure. **-eingang** *m* side entrance. **-fläche** *f* lateral face; facet. **s—gleich** *a* alike on both sides, reversible. **S—gummi** *n* sidewall (of a rubber tire). **-isomerie** *f* chain isomerism. **-kante** *f* lateral edge. **-kette** *f* side chain. **-kettenisomerie** *f* side-chain isomerism. **s—kettenständig** *a* attached to the side chain. **S—kraft** *f* component (*or* secondary) force. **-lage** *f* lateral position. **-länge** *f* lateral length, length of a side. **-rand** *m* margin. **-riß** *m* side elevation. **-rohr** *n*, **-röhre** *f* side (*or* branch) tube, branch pipe

seitens *prep* (+*genit*) on the part (of) *cf* SEITEN

Seiten· side, lateral; page: **-sprung** *m* 1 leap to the side. 2 side trip. 3 digression. 4 escapade. **s—ständig** *a* lateral. **S—stechen** *n* stitch (*or* pain) in the side. **-strahlung** *f* lateral (*or* secondary) radiation. **-strom** *m* side stream. **-stück** *n* sidepiece; counterpart. **-trogfilter** *n* side-feed filter. **-tür** *f* side door. **-wand** *f* side wall; side plate. **-zahl** *f* 1 page number. 2 number of pages

seither *adv* 1 since then, ever since. 2 up to now

·seitig *sfx* -sided: **einseitig** one-sided

seitlich 1 *a* lateral, side; collateral. 2 *adv oft* on (*or* to) the side. 3 *prep* (+*genit*) beside, alongside

seitwärts *adv* sideways, sideward, to the side, aside

sek *abbv* (Sekunde) sec., second (of time)

sek. *abbv* (sekundär) secondary

Sek. *abbv* (Sekunde) sec (secant)

Sekante *f* (-n) (*Geom.*) secant

Sekret *n* (-e) secretion

Sekretär *m* (-e) **Sekretärin** *f* (-nen) secretary

sekretieren *vt* secrete. **sekretions·hemmend** *a* secretion-inhibiting. **sekretorisch** *a* secretory

Sekret·präparat *n* preparation from a secretion. **-stoff** *m* secreted substance

Sekt *m* (-e) champagne; dry sack

Sektion *f* (-en) 1 section. 2 (*Med.*) (dis)section; autopsy, postmortem

Sektions·befund *m* postmortem results. **-chef** *m* section(al) chief (*or* head). **-tisch** *m* dissecting (*or* autopsy) table

sekundär *a* secondary. **S—farbe** *f* secondary color. **-kreis** *m* (*Elec.*) secondary circuit. **-reaktion** *f* secondary reaction. **-strahlung** *f* secondary radiation. **-strom** *m* secondary current

Sekunde *f* (-n) second (of time). **Sekunden·schnelle** *f* split-second speed: **in S.** at split-second speed. **-uhr** *f* watch (*or* clock) with a second-sweep

sekundlich *a* per second, every second

selbe *a* same *cf* DERSELBE

selber *pron* oneself, etc = SELBST 1

selbst 1 *pron* (*no endgs*) oneself, *specif* itself, themselves, etc (*esp emphat with* sich): **das Gas s.** the gas itself; **sie irrten sich s.** they were mistaken themselves; **von s.** by itself, of its own accord. 2 *adv* even: **s. wenn.** even when (*or* if) . . . ; **s. die Gase** even the gases. 3 (*in compds*) self-, auto-, automatic, spontaneous. **Selbst** *n* (the) self, ego

selbst·abdichtend *a* self-sealing. **S—achtung** *f* self-respect

selb·ständig *a* independent, self-supporting —(*Elec.*) **s—e Entladung** self-sustained discharge. **Selbständigkeit** *f* independence

Selbst·ansteckung *f* self-infection. **s—auftauend** *a* self-defrosting. **S—bedienung** *f* self-service. **-befruchtung** *f* self-fertilization. **-beherrschung** *f* self-control. **-bestäubung** *f* self-pollination. **s—bewußt** *a* self-confident. **S—bewußtsein** *n* self-confidence. **-binder** *m* (*Varnish*) self-binder. **-biographie** *f* autobiography. **s—dichtend** *a* self-sealing. **S—diffusion** *f* self-diffusion, autodiffusion. **s–emulgierend** a self-emulsifying. **-entzündbar**, **-entzündlich** a spontaneously inflammable. **S–entzündlichkeit** f spontaneous inflammability. **-entzündung** f spontaneous ignition. **s–erhaltend** a self-sustaining. **S–erhaltung** f self-preservation. **-erhitzung** f self-heating. **s–erregend** a self-exciting. **S–erwärmung** f spontaneous heating. **-erzeugung** f spontaneous generation, autogenesis. **-farbe** f self color, solid color. **s–farbig** a self-colored, single-colored. **S–gang** m automatic action. **-gärung** f spontaneous fermentation. **s–gebacken** a home-baked, home-made. **-gefällig** a smug, self-satisfied; conceited. **S–gefühl** n self-confidence. **s–gehend** a 1 automatic. 2 (Metll.) self-fluxing. **-gemacht** a 1 self-made. 2 home-made. **-genügsam** a self-sufficient. **S–gift** n autotoxin. **s–haftend** a self-adhesive, self-bonding. **-härtend** a self-hardening. **S–härter** m (Steel) self-hardener. **-heilung** f 1 self-healing. 2 (Cement) autogenous healing. **s–hemmend** a 1 self-inhibiting. 2 irreversible. **-induktivität** f self-inductance. **-kante** f selvage, list. **-klebefolie** f contact paper. **s–klebend** a self-adhesive. **S–kosten** pl prime (or production) cost. **-kühlung** f self-cooling, natural cooling. **s–leuchtend** a luminous. **S–leuchter** m phosphorescent substance, phosphor. **s–los** a selfless, unselfish. **S–löschung** f (Lime) spontaneous slaking. **-lötung** f autogenic soldering. **-mord** m suicide. **-oxidation** f auto-oxidation, self-oxidation. **-polymerisat** n autopolymer. **s–polymerisiert** a autopolymerized. **S–polymerisierung** f autopolymerization. **s–redend** a obvious, etc = SELBSTVERSTÄNDLICH. **-registrierend** a self-recording. **-regelnd** a self-regulating. **S–reglung** f self-regulation; automatic control. **-reinigung** f self-purification. **s–schmierend** a self-lubricating. **S–schmierung** f self-lubrication. **s–schreibend** a self-recording. **S–schreiber** m recording instrument. **-schutz** m self-protection; self-defense. **-schutzmittel** n prophylactic. **s–sicher** a self-

assured (or -confident). **S–spaltung** f self-fission. **-steuerung** f 1 (Biol.) self-regulation. 2 (Mach.) automatic control. **-strahlung** f self-emission (of rays), spontaneous radiation. **s–tätig** a 1 automatic. 2 (Flour) self-rising. **-tauend** a self-defrosting. **S–teilung** f spontaneous division. **s–tonend** a (Phot.) self-toning. **S–umkehr** f (Spect.) self-reversal. **-unterbrecher** m (Elec.) automatic interrupter. **-verbrennung** f spontaneous combustion. **-verdauung** f autodigestion; autolysis. **-vergiftung** f autointoxication. **-versorger** m self-supporting farmer (who grows his own food). **s–verständlich** a 1 obvious; natural. 2 adv usu (as a matter) of course. **S–verständlichkeit** f (-en) matter of course. **-verwirklichung** f self-fulfilment. **s–vulkanisierend** a self-vulcanizing. **s–wirkend** a self-acting, automatic. **S–zersetzung** f spontaneous decomposition; autolysis. **-zerstörung** f self-destruction. **-zeugung** f spontaneous generation, abiogenesis. **s–zündend** a self-igniting, pyrophoric. **S–zünder** m pyrophorus, self-igniter. **-zündung** f self-ignition. **-zweck** m end in itself

Selbung f automatization, automation

selchen vt smoke, cure (meat). **Selchfleisch** n smoked (or cured) meat

selektieren vt select. **Selektivität** f selectivity

Selen n selenium; (in compds oft) selenic, seleno-, selenide of: **-ammonium** n ammonium selenide

Selenat n (-e) selenate

Selen-äthyl n ethyl selenide. **-blei** n lead selenide, (Min.) clausthalite

Selenblei-silber n (Min.) naumannite. **-spat** m (Min.) kerstenite. **-wismutglanz** m (Min.) galenobismuthite, weibullite

Selen-bromür n selenium monobromide. **-brücke** f selenium bridge. **-chlorid** n selenium tetrachloride. **-chlorür** n selenium monochloride. **-cyanid** n selenium cyanide. **-cyankalium** n potassium selenocyanate. **-cyansäure** f selenocyanic acid. **-dioxid** n selenium dioxide. **-eisen** n ferrous selenide. **-erz** n selenium ore. **-gitter** n selenium lattice.

-gleichrichter *m* selenium rectifier.
-halogen *n* selenium halide. **s–haltig** *a* containing selenium, seleniferous. **S–harnstoff** *m* selenourea. **-hydrat** *n* hydroselenide

Selenid *n* (-e) (*usu* -ic) selenide

selenig *a* selenious: **s–e Säure** selenious acid. **--sauer** *a* selenite of

Selenigsäure *f* selenious acid. **--anhydrid** *n* selenious anhydride (SeO$_2$)

Selen·kupfer *n* copper selenide, (*Min.*) berzelianite. **-kupfersilber** *n* copper silver selenide; (*Min.*) eucairite. **-metall** **n** selenium metal, metallic selenide

Selenocyan *n* selenocyanogen

Selen·oxid *n* selenium oxide. **-quecksilber** *n* mercury selenide, (*Min.*) tiemannite. **-quecksilberblei** *n* (*Min.*) lehrbachite. **-rubin** *m* selenium ruby glass. **-salz** *n* selenide. **s–sauer** *a* selenate of **S–säure** *f* selenic acid. **-säureanhydrid** *n* selenic anhydride, selenium trioxide. **-schlamm** *m* selenium mud (*or* slime, from sulfuric acid plants). **-schwefel** *m* selenium sulfide, (*Min.*) selensulfur. **-kohlenstoff** *m* carbon sulfide selenide (CSSe). **-silber** *n* silver selenide; (*Min.*) = **-silberglanz** *m* naumannite. **-sperrschichtphotozelle** *f* selenium barrier-layer photovoltaic cell

Selenür *n* (low-valent, *usu* -ous) selenide

Selen·wasserstoff *m* hydrogen selenide. **-wasserstoffsäure** *f* hydroselenic acid. **-wismut** *m* bismuth selenide; (*Min.*) = **-wismutglanz** *m* guanajuatite. **-zelle** *f* selenium cell. **-zyanid** *n* selenium cyanide. **-zyankalium** *n* potassium selenocyanate

Self·bogen *m* (*Elec.*) flashover

selig *a* 1 blessed. 2 happy. 3 late, deceased

seligieren *vt* select

Sellerie *m, f* celery (root). **--blätteröl** *n* celery leaf oil. **-knolle** *f* celery root

selten 1 *a* (*incl Chem*) rare; unusual. 2 *adv* oft seldom, exceptionally. **S--erde** *f* rare earth. **Seltenheit** *f* (-en) rarity

Selters·wasser *n* Seltzer water

seltsam *a* strange, odd. **seltsamerweise** *adv* strangely enough. **Seltsamkeit** *f* (-en) strangeness, oddity

Semichinon *n* semiquinone

Semichrom·gerbung *f* semichrome tanning. **-leder** *n* semichrome leather

semicyclisch *a* semicyclic

Semidin·umlagerung *f* semidine rearrangement

semimer *a* semimeric

Seminar *n* (-e) 1 seminar. 2 (University) department, institute. 3 seminary

Semmel *f* (-n) (baked) roll. **--mehl** *n* 1 flour for rolls. 2 ground rolls, roll crumbs

Sende·anlage *f* transmitter. **-folge** *f* series of broadcasts. **-gerät** *n* transmitting device, transmitter

senden* 1 *vt* (*reg:* sendete, gesendet *or* *: sandte, gesandt) send; remit. 2 *vt/i* (*reg*) transmit, broadcast. **Sender** *m* (-) 1 sender, consignor. 2 transmitter; broadcasting (radio, TV) station

Sende·plan *m* broadcasting schedule. **-termin** *m* scheduled time, showing (of a TV program)

Sendung *f* (-en) 1 sending; remitting. 2 shipment, consignment; piece of mail; remittance. 3 transmission; broadcast(ing); program. 5 mission, calling

Senegal·gummi *n* Senegal gum

Senega·wurzel *f* senega root

Senf *m* mustard. **s--farbig** *a* mustard-colored (*or* -yellow). **S–gas** *n* mustard gas. **-geist** *m* (volatile) oil of mustard. **-korn** *n* mustard seed. **-mehl** *n* ground (*or* dry) mustard. **-öl** *n* mustard oil; (allyl) isothiocyanate. **-ölessigsäure** *f* 2,4-thiazoledione. **-pflaster** *n* mustard plaster. **-pulver** *n* powdered (*or* dry) mustard. **-samen** *m* mustard seed. **-säure** *f* sinapic acid

Seng·apparat *m* (*Tex.*) singeing machine

sengen 1 *vt* singe, scorch. 2 *vi* scorch, be scorching (hot); get singed (*or* scorched)

senkbar *a* sinkable, depressable

Senk·blei *n* plumb bob, plummet; sounding lead; sinker. **-boden** *m* (*Brew.*) false bottom, strainer. **-bombe** *f* depth charge

Senke *f* (-n) 1 depression, low spot; hollow; valley. 2 cesspool

Senkel *m* (-) 1 plummet. 2 shoe lace

senken 1 *vt* lower, drop, depress; sink (a shaft); (*Mach.*) stamp (with a die). 2: **sich s.** drop, fall, sink; settle (*incl Chem.*); give way, sag; slope. **Senker** *m* (-) 1 (*Bot.*)

layer. **2** (*Mach.*) countersink. **3** (*Min.,
Fishing*) sinker

Senk·grube *f* cesspool; sump; catch basin.
-körper *m* sinker, bob. **-küpe** *f* (*Tex.*) dipping frame. **-loch** *n* sinkhole; catch basin;
drain. **-lot** *n* plumb bob, plummet. **-nadel**
f needle probe. **-pumpe** *f* submersible
pump. **s–recht** *a* vertical, perpendicular.
S–rechte *f* (*adj endgs*) vertical line, perpendicular. **-schnur** *f* plumb (*or* sounding) line. **-schraube** *f* countersunk screw.
-spindel *f* hydrometer

Senkung *f* (*-en*) **1** lowering, dropping, etc *cf*
SENKEN. **2** dip, hollow, depression. **3** drop,
decline, decrease, reduction. **3** slope, incline. **4** (*Med.*) prolapse

Senk·waage *f* **1** hydrometer. **2** plumb rule.
-zylinder *m* separating cylinder

Sennes·baum *m* cassia, senna tree. **-blätter** *npl* senna leaves. **-strauch** *m* senna,
coffeeweed

Sense *f* (*-n*) scythe

sensibel *a* **1** sensory. **2** sensitive. **Sensibilisator** *m* (*-en*) sensitizer. **sensibilisieren**
vt sensitize. **Sensibilisierung** *f*
sensitization

Sensibilisierungs·farbstoff *m* sensitizing
dye. **-mittel** *n* sensitizing agent

Sensibilität *f* (*-en*) sensitiveness,
sensitivity

Sensitometer *n* (-) (*Phot.*) actinometer

Separat·(ab)druck *m* separate impression; reprint

separieren *vt* & **sich s.** separate; *vt oft*
centrifuge

Separier·trichter *m* separatory funnel

Sepia, Sepie *f* (. . pien) **1** (*Zool.*) cuttlefish.
2 (*Zool.* & *Dye*) sepia. **sepia·braun** *a*
sepia

Septikämie *f* septicemia

Septikopyämie *f* septic pyemia

septisch *a* septic

Seren serums, sera; *pl of* **Serum** *n*

Serie *f* (*-n*) **1** series. **2** mass production: **in S.
gehen** go into mass production

Serien· serial, assembly-line: **-artikel** *m*
mass-produced article. **-bad** *n* series bath
(*or* tank). **-fabrikation, -fertigung** *f*
assembly-line manufacture, mass production. **s–fremd** *a* of another (*or* of dif

ferent) series. **S–grenze** *f* series limit;
(*Spect.*) convergence limit. **s–mäßig 1** *a*
series; mass-production. **2** *adv* in (a) series; by mass-production methods. **S–
nummer** *f* serial number. **-schaltung** *f*
(*Elec.*) series connection. **-schnitt** *m* serial section. **-spektrum** *n* series spectrum. **-wagen** *m* assembly-line car

Serizin *n* sericin. **Serizitisierung** *f* (*Petrog.*) sericitization

serologisch *a* serological. **serös** *a* serous

Serpentine *f* (*-n*) **1** serpentine (*or* winding)
road. **2** hairpin turn. **serpentinieren** *vi*
wind, meander

Serpentin·schiefer *m* serpentine schist

Serum·eiweiß *n* serum albumin (*or* protein). **-krankheit** *f* serum sickness

Serviette *f* (*-n*) (table) napkin

Servo·antrieb *m* servo drive. **-lenkung** *f*
power steering

Sesam *m* sesame; (*in Anat. compds*) sesamoid: **-knorpel** *m* sesamoid cartilage.
-korn *n* sesame seed. **-kuchen** *m* sesame
(oil) cake. **-öl** *n* sesame oil. **-same(n)** *n*
sesame seed

Sesquioxid *n* sesquioxide

Sessel *m* (-) armchair, (easy) chair; seat.
-·bahn *f* chair lift. **-form** *f* chair form (of
cyclohexane)

seßhaft *a* **1** resident. **2** settled, sedentary:
sich s. machen, s. werden settle. **3**
(*Mil.: War Gases*) persistent. **Seßhaftigkeit** *f* resident status; sedentariness;
(*Mil.*) persistency

Setz·arbeit *f* settling (process); (*Ore*) jigging. **-bottich** *m* settling vat (*or* tank). **-ei**
n fried egg

setzen I *vt* **1** put, set (*incl* type)—*Idiomatic
wordings with nouns should be looked up
under the noun: cf* BEWEGUNG, KRAFT,
etc. **2** set (*or* put) up; erect. **3** seat. **4** write,
put down. **5** plant. **6** lay (bricks). **7** (*Ores*)
jig, sieve. **8** (*Animals*) give birth to. **9** assume, suppose. **II** *vt/i* **10** (+**auf**) stake,
bet (on). **III** *vi* (h,s) **11** jump, leap; (+**über**,
oft) cross. **IV: sich s. 12** sit down; alight.
13 put oneself (in a place or situation), get
(in): **sich in Gang s.** get into motion,
start moving. **14** settle, sink, subside; be
deposited, precipitate. **15** set, harden (*as:*

concrete, glue). *cf* GESETZT; SITZEN. **Set-zenlassen** *n* allowing to set (*or* settle).

Setzer *m* (-) typesetter, compositor

Setz·fehler *m* misprint, typographical error. **-kasten** *m* **1** settling tank. **2** type case

Setzling *m* (-e) **1** slip, layer. **2** seedling, sapling. **3** fry, spawn

Setz·maschine *f* (*Ores*) jig. **-phiole** *f* (flat-bottomed) vial. **-sieb** *n* jig(ging) screen (*or* sieve). **-zapfen** *m* suppository

Seuche *f* (-n) contagious disease; epidemic, pestilence

seuchen·artig *a* contagious; infectious; epidemic. **-fest** *a* immune

seuchenhaft *a* contagious, infectious; epidemic

seufzen *vi*, **Seufzer** *m* (-) sigh

Seven·baum *m*, **-kraut** *n* (*Bot.*) savin. **-krautöl** *n* savin oil

Sexual·hormon *n* sex hormone

Sexualität *f* sexuality

Sexual·zelle *f* (*Biol.*) germ cell

sexuell *a* sexual

sezernieren *vt* secrete

sezieren *vt* dissect

Sezier·lupe *f* dissecting lens. **-messer** *n* scalpel, dissecting knife

S-förmig *a* S-shaped

s.g. *abbv* (sogenannt) so-called

S.G. *abbv* (spezifisches Gewicht) sp. gr. (specific gravity)

Sg. *abbv* (Streckung) stretching, spread

sh *abbv* (*Math.*) sinh (hyperbolic sine)

sherardisieren *vt* (*Metll.*) sherardize

Shikimi·säure *f* shikimic acid

Shikimol *n* safrole

Shore·härte *f* Shore hardness

siamesisch *a* Siamese

Sibirien *n* Siberia. **sibirisch** *a* Siberian

Sibirisch·gelb *n* Siberian yellow (iron chromate pigment)

sich I *rflx pron* oneself, himself, herself, itself, themselves, yourself. **1** (*transl oft not literal*): **sie stellen es neben s.** they put it beside them(selves); **das Element an s.** the element itself (*or* in essence); **an und für s.** in its essentials, per se; **von s. aus** on one's own initiative; **wieder zu s. kommen** regain one's senses. **2** (*Forms rflx verbs; transl rarely literal, oft passive,*

oft makes vt *intransitive in meaning. Look up the verb and check the division headed* **sich**): **s. befassen mit.** . . concern oneself (*or* be concerned) with. . . *cf* BE-WEGEN, VERBINDEN, etc. **3**: **s. lassen** + *inf* can be + *pp cf* LASSEN IV. **4** *dat rflx pron, usu indir obj, with* vt (*transl usu not literal; look under the* vt *division of the verb*): **s. etwas verschaffen** get oneself sthg; **s. etwas vorstellen** imagine sthg. **II** *recip pron* each other, one another (*oft unexpressed in English*): **die Linien schneiden s. im Punkt A** the lines intersect (each other) at point A

Sichel *f* (-n) **1** sickle. **2** (moon's) crescent. **s--förmig** *a* crescent-shaped

sicher I *a* **1** (*usu* + *vor*) secure (against), safe (from). **2** sure, certain; reliable; definite; *adv oft* for sure, we can be sure that. . . **3** experienced, practiced (hand). **4** (*pred*) well versed, sure of one's ground (in a subject). **5** self-assured. **II** *sep pfx cf* SICHERGEHEN, SICHERSTELLEN

sicher·gehen* *vi*(s) play (it) safe, be on the safe side

Sicherheit *f* (-en) **1** security; safety: **in S. bringen** rescue, get to a safe place. **2** certainty; reliability. **3** assurance; guarantee. **4** self-assurance. **5** experience, practice

Sicherheits·brennstoff *m* safe-burning fuel. **-farbe** *f* **1** warning color. **2** (*Mil.*) safety paint. **-flasche** *f* safety flask. **-glas** *n* safety glass. **-grad** *m* degree of safety; safety factor. **s--halber** *adv* for safety's sake. **S--koeffizient** *m* safety factor. **-lampe** *f* safety lamp. **-maßnahme, maßregel** *f* safety measure, precaution. **-nadel** *f* safety pin. **-papier** *n* safety paper. **-rohr** *n*, **-röhre** *f* safety tube. **-schalter** *m* (*Elec.*) safety switch. **-schloß** *n* safety lock. **-sprengstoff** *m* safety (*or* self-detonating) explosive. **-technik** *f* (technology of) safety and security. **-trichterrohr** *n* safety tube. **-ventil** *n* safety valve. **-vorrichtung** *f* safety device. **-waschflasche** *f* safety wash bottle. **-wasserbad** *n* safety water bath. **-wert** *m*, **-zahl** *f* safety factor. **-zünder** *m* safety fuse. **-zündholz** *n* safety match

sicherlich *adv* certainly, etc *cf* SICHER 2

sichern *vt* 1 (*usu* + **vor**) safeguard, secure (against). 2 insure, guarantee. 3 lock; set at "safety", put the safety on. 4 cover (against attack)

sicher·stellen *vt* 1 (*usu* + **vor**) safeguard, secure, keep safe (from, against). 2 put in safe keeping. 3 verify, establish as certain. 4 guarantee. **Sicherstellung** *f* safeguarding; verification; guarantee

Sicherung *f* (-en) 1 safeguarding, safekeeping. 2 guarantee. 3 safety device (catch, lock, cutout); (*Elec.*) fuse

Sicherungs·hebel *m* safety catch. **-körper** *m* fuse box (*or* holder). **-patrone** *f* fuse cartridge. **-stöpsel** *m* safety (*or* fuse) plug

Sicht *f* (-en) 1 view, sight —**auf lange S.** on a long-term basis. 2 visibility. 3 viewpoint

sichtbar *a* visible; obvious. **Sichtbarkeit** *f* visibility; obviousness

Sichtbar·machung *f* visualization; demonstration. **-werden** *n* revelation, appearance

Sicht·beton *m* fair-faced concrete

sichten *vt* 1 (*lit & fig*) sift, screen; (*Flour*) bolt. 2 sight. **Sichter** *m* (-) sifter, sorter. **Sichter·mühle** *f* separator mill

Sicht·kalk *m* pulverized lime. **-karte** *f* (*Transp.*) pass. **-kartei** *f* visible-card index

sichtlich *a* visible, obvious; distinct

Sicht·messer *m* visibility meter. **-schmierapparat** *m* sight-feed lubricator

Sichtung *f* (-en) sifting, screening; sighting

Sicht·weite *f* range of visibility

sicilianisch *a* Sicilian *cf* SIZILIEN

Sicke *f* (-n) seam, edge; crimp, corrugation. **sicken** *vt* 1 crimp, bead; flange. 2 swage

Sicker·kühlung *f* trickle cooling. **-laugung** *f* leaching; lixiviation

sickern *vi* trickle; seep, percolate; (*incl* light) filter

Sicker·tank *m* percolating tank. **-verlust** *m* loss by seepage. **-wasser** *n* ground water; percolating (*or* infiltration) water

Siderin·gelb *n* siderin yellow

siderophil *a* (*Petrog.*) siderophilic

Sidotsche Blende Sidot blende, crystalline zinc sulfide (phosphor)

sie *pron* 1 *sg* she; it; her. 2 *pl* they; them. **Sie** *pron* you

Sieb *n* (-e) sieve, screen, strainer; riddle, bolter; (ray) filter. **--analyse** *f* screen analysis. **s–artig** *a* sieve-like, perforated, full of holes. **S–ascheanalyse** *f* ash size analysis. **-aufsatz** *m* gauze top (on a burner). **-bein** *n* ethmoid bone. **-boden** *m* perforated (*or* sieve) bottom, grid; sieve plate (of a fractionating column). **-brille** *f* gauze-frame goggles. **-draht** *m* sieve (strainer, screen) wire. **-druck** *m* screen printing. **-durchfall**, **-durchgang** *m* screen product, sieved material **-einsatz** *m* strainer (insert)

sieben I *num* seven; (*as pfx, usu*) hepta-. II *vt/i* sift, pass through a sieve; screen; riddle, bolt; (*Rays*) filter. **Sieben I** *f* (-) (number, figure) seven. II *n* sifting, screening, etc *cf* SIEBEN II

sieben· seven, hepta-: **-atomig** *a* heptatomic. **S–bürgen** *n* (*Geog.*) Transylvania. **-eck** *n* heptagon. **s–eckig** *a* heptagonal

Siebener *m* (-) (number, figure) seven

siebenerlei *a* (*no endgs*) seven kinds of

sieben·fach *a* sevenfold. **-flächig** *a* heptahedral. **-gliedrig** *a* seven-membered, seven-part. **-jährig** *a* seven-year(-old). **-mal** *a* seven times. **S–ring** *a* seven-membered ring

siebent *a* seventh —**zu s.** in a group of seven

seibentel *num a* (*no endgs*) **Siebentel** *n* (-) (*Frac.*) seventh. **siebentens** *adv* seventhly, in the 7th place

sieben·wertig *a* heptavalent, septivalent. **S–wertigkeit** *f* heptavalence, septivalence

Sieberei *f* (-en) 1 sifting, screening. 2 sifting (*or* screening) plant

Sieb· sieve, etc: **-feine** *n* (*adj endgs*) fine screenings, sieve undersize. **-feinheit** *f* sieve fineness, mesh. **-fläche** *f* screen area. **s–förmig** *a* sieve-shaped, cribriform. **S–gewebe** *n* straining (*or* screening) cloth. **-grobe** *n* (*adj endgs*) coarse screenings, screen oversize. **-klassierung** *f* screen sizing (*or* classifying).

-koks *m* screened coke. **-korb** *m* wire basket; centrifuge basket. **-kurve** *f* screen-analysis curve. **-leistung** *f* screen(ing) output; screening capacity. **-linie** *f* grading curve (of screenings). **-maschine** *f* sifting (*or* screening) machine; sifter; bolter; winnower. **-mehl** *n* sifted flour; coarse flour. **-mittel** *n* filtering medium. **-mühle** *f* (flour) bolting mill. **-parenchym** *n* (*Bot.*) phloem parenchyma. **-platte** *f* sieve (*or* filter) plate. **-probe** *f* sieve (*or* screening) test. **-rest** *m* sieve residue. **-rost** *m* sieve (*or* mesh) grating. **-rückstand** *m* screening residue. **-satz** *m* set of sieves. **-schale** *f* perforated dish. **-schaufel** *f* perforated scoop. **-schleuder** *f* centrifugal sifter

Siebsel *n* (-) (*usu pl*) screenings, siftings
Sieb·skala *f* mesh gauge. **-staub** *m* siftings. **-struktur** *f* (*Petrog.*) sieve texture

siebt *a* seventh = SIEBENT

siebtel *num* *a* (*no endgs*), **Siebtel** *n* (-) (*Frac.*) seventh. **siebtens** *adv* seventhly = SIEBENTENS

Sieb·trichter *m* filter (*or* straining) funnel. **-trommel** *f* rotary screen. **-tuch** *n* sifting (*or* straining) cloth; bolting cloth

Siebung *f* (-en) sifting, screening, etc = SIEBEN II

Sieb·walze *f* (*Paper*) dandy roll(er). **-wasser** *n* (*Paper*) pulp (*or* back) water. **-weite** *f* mesh width (*or* size). **-wuchtrinne** *f* vibrating screen feeder

siebzehn *num* seventeen. **siebzehnte** *a* seventeenth. **Siebzehntel** *n* (-) (*Frac.*) seventeenth

siebzig *num* seventy. **Siebziger** *m* (-) 1 (number, figure) seventy. 2 seventy-year-old (person). **siebzigste** *a* seventieth. **Siebzigstel** *n* (-) (*Frac.*) seventieth

Sieb·zylinder *m* cylindrical sieve, sieve drum

siech *a* sick(ly), ailing. **Siechtum** *n* chronic ailment, ill health

sied. *abbv* (siedend) boiling

Siede *f* (-n) boiled feed, mash

Siede· boiling, distilling: **-abfälle** *mpl* waste (scum, sediment) from boiling. **-analyse** *f* analysis by fractional distillation; determination of distillation range. **-apparat** *m* boiling apparatus. **-beginn**

m beginning of boiling; initial boiling point. **-bereich** *m* boiling range. **-blech** *n* boiling plate. **-calorien** *fpl* latent heat of vaporization in calories. **-ende** *n* end of boiling; final boiling point. **-erleichterer** *m* boiling aid (*or* accelerator). **-fläche** *f* boiling surface. **-flüssigkeit** *f* boiling liquid; (*Metll.*) blanching liquid. **-gefäß** *n* 1 boiling vessel, boiler. 2 distillation flask. **-grad** *m* boiling point. **-grenze** *f* boiling limit (*or* range). **-haus** *n* boiler room; distillery. **-hitze** *f* 1 boiling heat. 2 sweltering heat. 3 seething rage. **-kapillare** *f* (*Distil.*) air-leak tube. **-kessel** *m* 1 (*genl*) boiling vessel, kettle. 2 (*Soap*) soap kettle. **-kolben** *m* boiling (*or* distilling) flask. **-kühlung** *f* vapor cooling. **-kurve** *f* boiling-point curve. **-lauge** *f* boiling lye (*or* liquor). **-linie** *f* boiling-point curve

siedeln *vt/i* settle (as a dweller)

Siede·maximum *n* maximum boiling point. **-messer** *m* boiling-point instrument. **-minimum** *n* minimum boiling point

sieden* (*reg or* sott, gesotten) 1 *vt/i* boil, distill. 2 *vt* (*Sugar*) refine; (*Silver*) blanch; (*Beer*) brew. 3 *vi* seethe

siedend·heiß *a* boiling-rot; sweltering; seething

Siedenlassen *n* allowing (*or* causing) to boil

Siede·ofen *m* (*Metll.*) blanching furnace. **-p.** *abbv* (Siedepunkt) b.p. (boiling point). **-pfanne** *f* boiling pan. **-punkt** *m* boiling point. **-punktbestimmung** *f* boiling-point determination. **-punktserhöhung** *f* elevation of the boiling point

Sieder *m* (-) boiler; (*Sugar*) refiner. **Siederei** *f* (-en) boiler room (*or* unit), distillery; (*Sugar*) refinery; (*Soap*) pan room

Siede·rohr *n*, **-röhre** *f* boiling (*or* distilling) tube; boiler tube. **-salz** *n* (common, solar) salt (from evaporation). **-sole** *f* brine. **-stab** *m* boiling stick. **-stein** *m*, **steinchen** *n* boiling stone, pottery shard (as a boiling aid). **-temperatur** *f* boiling temperature (*or* point). **-trennung** *f* distillation; fractionation. **-verzug** *m* delay in boiling; superheating. **-zeit** *f* boiling time (*or* period)

Sied·kessel *m* boiling vessel, kettle.

-kolben m boiling flask

Siedlung f (-en) settlement; (suburban) housing development; (*in place names, oft*) estates

Sieg m (-e) victory

Siegel n (-) seal (=closure). **--erde** f red ocher, hemnian earth. **-lack** m sealing wax

siegeln vt seal

Siegel-ring m signet ring. **-stein** m (*Min.*) magnetite. **-wachs** n (soft) sealing wax

siegen vi win, be victorious, triumph. **Sieger** m (-) victor, winner

siehe see: *imperat of* SEHEN

sieht sees, etc: *pres of* SEHEN

Siel n (-e) 1 floodgate. 2 sewer; drain, culvert

Siemensit-stein m (electrically fused) chrome-magnesite brick

Siemens-Martin-Ofen m open-hearth furnace. **S-M-verfahren** n Siemens-Martin (*or* open-hearth) process

Sien(n)a-erde f sienna

Sigel n (-) 1 abbreviation. 2 =**-zeichen** n logo(gram), symbol

Signal-bombe f signal flare. **-emaillelack** m signal enamel. **-farbe** f signal color; signal paint. **-feuer** n flare; signal fire, beacon. **-geber** m indicator (board); flagman. **-geschoß** n, **-granate** f (*Mil.*) star shell

signalisieren vt signal, announce

Signal-lampenöl n signal (lamp) oil. **-mittel** n signaling device (*or* material). **-patrone** f signal cartridge. **-rakete** f signal rocket. **s-rot** a bright-red. **S-stern** m star signal, Very light

Signatur f(-en) 1 signature (*incl Print., Paper*); autograph. 2 (library) call number. 3 serial number. 4 map symbol. 5 sign, mark, stamp, brand

signieren vt/i 1 initial; autograph. 2 designate; mark, stamp, brand

Signier-farbe f marking color (*or* stain). **-färbung** f fugitive staining. **-tinte** f marking ink

Sikkativ n (-e) siccative, quick-drier. **--firnis** m siccative varnish. **-mittel** n siccative, drier. **-pulver** n drying powder, (*specif*) manganese borate

Silbe f(-n) syllable

Silber n silver. **--antimonglanz** m (*Min.*) miargyrite. **s--arm** a low-silver. **--artig** a silvery, silver-like. **S--ätzstein** m lunar caustic (silver nitrate). **-azetat** n silver acetate. **-bad** n silver bath. **-baum** m silver tree. **-belag** m, **-belegung** f silver coating. **-bergwerk** n silver mine. **-beschlag** m silver fitting (mounting, trim). **-besteck** n (single set of) silverware. **-blatt** n silver leaf (*or* foil). **-blech** n thin silver plate, sheet silver. **-blei** n crude lead containing silver. **-blende** f (*Min.*) pyrargyrite, proustite, silver galena. **-blick** m silver blick, brightening (*or* fulguration) of silver. **-brennen** n silver refining. **-brennherd** m silver-refining hearth. **-bromid** n silver bromide. **-bromür** n silver subbromide. **-chlorid** n silver chloride. **-chlorür** n silver subchloride. **-cyankalium** n potassium cyanoargentate. **-distel** f (Bot.) carline thistle. **-doublé** n rolled silver. **-draht** m silver wire (*or* thread). **-erz** n silver ore. **-essigsalz** n silver acetate. **-faden** m silver thread (filament, strand). **-fahlerz** n silver-bearing tetrahedrite. **-fällung** f precipitation of silver. **-farbe** f silver color (*or* paint). **s-farben, -farbig** a silver-colored, silvery. **S-fischchen** f(*Zool.*) silverfish. **-fluorür** n silver subfluoride. **-folie** f silver foil. **-fuchs** m (*Zool.*) silver fox. **s-führend** a silver-bearing. **S-gare** f silver refining. **-gehalt** m silver content. **-geld** n silver money. **-geschirr** n silver plate, silverware. **-gewinnung** f extraction of silver. **-glanz** m 1 silvery luster. 2 (*Min.*) silver glance, argentite. **s-glänzend** a silvery-bright. **S-glas(erz)** n (*Min.*) argentite. **-glätte** f (*Min.*) silver litharge. **-glimmer** m (common) mica, muscovite. **-gold** n silver-bearing gold, electrum. **s-grau** a silver-gray. **S-grube** f silver mine. **-halogen(id)** n silver halide. **s-haltig** a containing silver, silver-bearing. **-hell** a silver-bright; silver-toned. **S-hornerz** n, **-hornspat** m horn silver, cerargyrite. **-hütte** f silver foundry

silberig a silvery

Silber-jodid n silver iodide. **-jodür** n silver subiodide. **-kies** m (*Min.*) argentopyrite, sternbergite. **-korn** n grain (bead, particle) of silver. **-körper** m (*Colloids*) silver body, silver ion. **-kupferglanz** m (*Min.*) stromeyerite. **-lack** m silver enamel (or varnish). **-lauge** f silver solution (silver hydroxide or moist silver oxide). **-laugerei** f silver-leaching plant. **-legierung** f silver alloy. **-lot** n silver solution. **-metall** n metallic silver. **-münze** f silver coin

silbern a (made of) silver

Silber·niederschlag m silver precipitate. **-nitrat** n silver nitrate. **-oxid** n silver oxide

Silberoxid·ammoniak n fulminating silver. **-salz** n silver oxysalt. **-verbindung** f silver oxide compound, argentate

Silber·oxydul n silver suboxide, **-papier** n silver(ed) paper. **s–plattiert** a silverplated. **S–plattierung** f silver plating. **-probe** f 1 silver test. 2 silver assay. **-protein** n (*Pharm.*) silver protein. **s–reich** a high-silver, rich in silver content. **S–rhodanid** n silver thiocyanate. **-salbe** f (*Pharm.*) colloidal silver ointment. **-salpeter** m silver nitrate. **-salz** n silver salt. **-sau** f silver ingot. **-schaum** m thin silver leaf, foliated silver. **-scheidung** f 1 separation of silver. 2 silver refining. **-schein** m silvery luster. **-schicht** f layer of silver, silver film (or coating)., **-schlaglot** n silver solder. **-schmied** m silversmith. **-schwärze** f earthy argentite. **s–schweflig** a argentosulfurous. **S–seife** f silver soap. **-spat** m (*Min.*) cerargyrite. **-spatel** m silver spatula. **-spiegel** m silvered mirror. **-spießglanz** m antimonial silver; (*Min.*) dyscrasite. **-stahl** m silver (or carbon tool) steel, Stub's steel. **-stoff** m (*Tex.*) silver brocade. **-stück** n piece of silver; silver coin. **-tanne** f silver fir. **-tiegel** m silver crucible. **-überzug** m silver coating. **-vitriol** n silver sulfate. **-waren** n pl silverware, silverware. **-wasser** n nitric acid (for dissolving silver). **-weiß** n silver white; white lead. **-wismutglanz** m (*Min.*) matildite. **-zeug** n silverware, silver utensils. **-zyanid** n silver cyanide. **-zyankalium** n potassium cyanoargentate

silbrig a silvery

Silica·stein m silica brick

Silicat· (*see also* SILIKAT): **-farbe** f ceramic color (or pigment). **-gestein** n silicate rock. **-modul** n (*Cement*) silica modulus

silicieren vt silicify, siliconize

Silicium n silicon; (*in compds oft*) silicide, silico-: **-ameisensäure** f silicoformic acid. **s–arm** a low-silicon. **S–äthan** n silicoethane (methylsilane or disilane). **-bromid** n silicon bromide. **-bronze** f silicon bronze. **-chlorid** n silicon chloride. **-chloroform** n silicochloroform, trichlorosilane. **-eisen** n 1 iron silicide. 2 (*Metll.*) ferrosilicon. **-eisenguß** m iron containing 15–18% silicon. **-fluorid** n 1 silicon tetrafluoride. 2 =**-fluorverbindung** f fluosilicate

Siliciumfluorwasserstoff m, **--säure** f fluosilicic acid. **-salz** n fluosilicate

silicium-frei a silicon-free. **S–gehalt** m silicon content. **s–haltig** a siliceous, silicon-bearing. **S–jodid** n silicon iodide. **-karbid** n, **-kohlenstoff** m silicon carbide. **-kupfer** n copper silicide; (*Metll.*) cuprosilicon. **-legierung** f silicon alloy. **-magnesium** n magnesium silicide. **-metall** n metallic silicide. **-methan** n silicomethane, silicon tetrahydride, silane. **-oxalsäure** f silicooxalic acid. **-oxid** n silicon dioxide. **-oxidhydrat** n hydrated oxide of silicon, silicic acid. **s–reich** a high-silicon. **S–spiegel** m siliceous ferromanganese, ferrosilicomanganese, silicospiegel. **-stahl** m silicon steel. **-tetrahydrür** n silicon tetrahydride, silane. **-verbindung** f silicon compound. **-wasserstoff** m hydrogen silicide, silicon hydride, silane

Silicon n silicone (*See also* SILIKON·). **-fett** n silicone grease

Silicowolframat n silicotungstate

Silier·mittel n ensiling agent

Silifikation f (-en) silification. **silifiziert** a silicified, siliceous

Silika·gel n silica gel. **-stein** m silica brick

Silikat n (-e) silicate (*See also* SILICAT·). **--emaille** f silicate enamel. **-gestein** n silicate rock. **s–haltig** a containing silicate(s)

silikatisch a silicate-like

Silikat·schmelzfluß m silicate melt.

-schmelzlösung f silicate melt solution.
-stein m silica brick. **-technik** f silicate technology
Silika·ziegel m silica brick
Siliko·ameisensäure f silicoformic acid.
-äthan n silicoethane. **-essigsäure** f silicoacetic acid
silikogen a silicogenic, silicon-producing
Silikon n silicone (See also SILICON·).
·-gummidichtung f silicon resin packing
Silikose f silicosis
Siliko·spiegel m ferrosilicomanganese, etc = SILIZIUMSPIEGEL
Silit n silicon carbide
Silizid n (-e) silicide
silizieren vi silicify, siliconize
Silizium n silicon (Compds listed under SILICIUM)
Silo·futter n silage. **-zelle** f silo bin
Silundum n silicon carbide
silurisch a (Geol.) Silurian
Simili m or n (-s) imitation. **·-diamant** m imitation diamond. **-stein** m imitation gemstone
Simmer·ring m sealing ring, oil seal
simpel a simple, plain; obvious; ordinary; simple-minded. **simplifizieren** vt simplify
Sims m or n (-e) (window) sill; ledge; shelf; cornice, molding
Simse f (-n) (Bot.) club rush, bulrush
Simulations·anlage f simulator (unit)
simulieren I vt 1 feign; simulate. II vi 2 malinger. 3 (+über) ponder (over)
simultan a simultaneous
Sinapin·säure f sinapic acid
sind (pres pl of SEIN) 1 are. 2 (+pp) e.g: **s. verschwunden** have disappeared cf SEIN I B
singen* (sang, gesungen) vt/i sing
Singrün n (Bot.) periwinkle, myrtle
singulär a unique, singular. **Singularität** f uniqueness, singularity
Singulett n (-e) (Spect.) singlet
Singulosilikat° n monosilicate, ferrous orthosilicate; (Min.) fayalite
Sinkalin n (-e) choline
sinken* (sank, gesunken) vi(s) 1 sink, drop, fall. 2 diminish. 3 wane. 4 sag. Cf GESUNKEN
Sink·geschwindigkeit f settling velocity.

-körper m sinker. **-stoff** m 1 sediment, settlings. 2 deposit. 3 suspended matter
Sinn m (-e) 1 (usu) sense; incl: meaning, point, purpose, intent; (+für, oft) feeling (for): **bei S—en** in control of one's senses; **S. für das Wesentliche** a feeling (or sense) for the essentials; **es hat keinen S.** it makes no sense. 2 mind, incl: opinion, attitude: **von S—en** out of one's mind; **etwas im S. haben** have sthg in mind, be planning (or scheming) sthg; **ihnen steht der S. nach. . .** their minds are set on . . . ; **es geht ihnen nicht aus dem S.** they can't get it out of their minds; **es kam uns in den S.** it came into our minds; **anderen S—es werden** change one's mind (or opinion); **das will ihnen nicht in den S.** they can't grasp that; **ein Versuch nach unserem S.** an experiment after our own hearts. 3 desire. 4 (Math.) sign. **·-bild** n symbol. **s—bildlich** a symbolic
sinnen* (sann, gesonnen) 1 vi (+über) ponder (over), speculate (on). 2 vi (+auf) & vt be planning (or scheming); have in mind. Cf GESINNT, GESONNEN. **Sinnen** n deep thought, meditation
Sinnen· sensory; sensual: **-prüfung** f organoleptic (or specif taste) test(ing). **-reiz** m sensory stimulus, sensual stimulation
sinn·entstellend a sense-distorting
Sinnen·welt f real (sense-perceived) world
Sinnes· sense, sensory; mind: **-art** f mentality; way of thinking. **-organ** n sense (or sensory) organ. **-täuschung** f sensory illusion; hallucination. **-wahrnehmung** f sense perception. **-werkzeug** n sense (or sensory) organ
sinn·fällig a obvious. **S—gedicht** n epigram. **s—gemäß** a 1 rational, logical. 2 incl adv according to the general sense. **-getreu** a faithful (translation, etc). **-gleich** a synonymous
Sinngrün n (Bot.) periwinkle, myrtle
sinnig a 1 thoughtful. 2 sensible. 3 ingenious
sinnlich a 1 sensory. 2 sensual; sensuous
sinnlos a 1 senseless. 2 meaningless. 3 mindless
Sinn·pflanze f sensitive plant; mimosa. **s—**

reich *a* 1 sensible. 2 ingenious. 3 witty.
S–spruch *m* aphorism, maxim; motto.
s–verwandt *a* synonymous. **-voll** *a*
meaningful, significant; sensible. **-widrig** *a* meaningless, abstract
sinopisch *a:* **s—e Erde** (*Min.*) sinopite
Sinter *m* (-) sinter; slag, scale; iron dross;
(*Geol.*) calc sinter. **--dolomit** *m* deadburnt dolomite. **-erzeugnis** *n* sintered
product, agglomerate. **s–fähig** *a* sinterable. **S–glas** *n* sintered glass. **-gut** *n* sintered material. **-kalk** *m* calc sinter. **-karbid** *n* sintered carbide. **-kathode** *f*
powder cathode. **-kohle** *f* sinter(ing) (*or*
non-coking) coal. **-kuchen** *m* sinter cake,
agglomerate. **-metallurgie** *f* powder
metallurgy
sintern 1 *vt/i* sinter. 2 *vt* frit. 3 *vi* trickle,
drip; form deposits (clinker, slag);
(*Ceram.*) vitrify
Sinter·ofen *m* sintering furnace. **-prozeß**
m (*Metll.*) slag process. **-quarz** *n* siliceous
sinter. **-röstung** *f* (*Metall.*) sinter (*or*
blast) roasting. **-schlacke** *f* clinker.
-stein *m* sintered brick. **-tonerde** *f* sintered alumina
Sinterung *f* sintering, etc *Cf* SINTERN
Sinterungs·hitze *f* sintering heat
Sinter·vorgang *m* sintering process;
(*Metll.*) slag process. **-wasser** *n* mineral-bearing water for sinter formation
Sinus *m* (-) 1 (*Math.*) sine. 2 (*Anat.*) sinus.
s--ähnlich *a* sinusoidal. **S–feld** *n* sinusoidal field. **s–förmig** *a* sinus-shaped,
sinusoidal. **S–kurve** *f* sine curve. **s–kurvenartig** *a* sinusoidal. **S–linie** *f* sine
curve. **-welle** *f* sine wave
siphonieren *vt* siphon
Sippe *f* (-n) clan, tribe; family; kindred.
Sippschaft *f* (-en) 1 kinship. 2 clan, etc =
SIPPE
Sirene *f* (-n) siren
Sirup *m* (-e) 1 syrup. 2 molasses. **s--artig** *a*
syrupy. **S–dichte** *f* syrupy consistency.
s–dick *a* syrupy; thick as molasses.
-haltig *a* containing syrup
sirupös *a* syrupy
Sirup·pfanne *f* syrup pan
Sisal·hanf *m* sisal (hemp)
sistieren *vt* inhibit; stop; interrupt; suspend

Sitte *f* (-n) 1 custom; usage, practice. 2 manners. 3 morals; morality
Sitten·lehre *f* ethics
Sittich *m* (-e) parakeet
sittlich *a* moral. **Sittlichkeit** *f* morality,
morals
Sitz *m* (-e) 1 seat. 2 site, location. 3 headquarters. 4 fit. 5 sitting, session
Sitz.-Ber. *abbv* (SITZUNGSBERICHT) proceedings
sitzen* (saß, gesessen) *vi* (h,s) 1 sit, be
seated. 2 be (in a place, position, situation); be located. 3 be stuck, be caught. 4
be settled, live. 5 be in session, meet. 6 fit,
be in place, be on right. 7 hit home. 8: **s.
bleiben** remain seated, remain in place.
sitzen·bleiben* *vi*(s) be left behind, be
passed over. **sitzend** *p.a* (*oft*) sedentary;
(*Bot.*) sessile. **sitzen·lassen*** *vt* leave in
the lurch, pass over
Sitz·fläche *f* (*Mach.*) seating. **-gurt** *m* seat
belt. **-streik** *m* sit-down strike
Sitzung *f* (-en) sitting; meeting, session
Sitzungs·bericht *m* report on the meeting,
proceedings
Sizilien *n* (*Geog.*) Sicily. **sizilianisch** *a*
Sicilian
SK *abbv* (Segerkegel) seger cone
Skala *f* (.. len) 1 scale. 2 dial. 3 range. **--lineal** *n* linear scale, graduated rule(r)
skalar *a* scalar
Skala·scheibe *f* graduated dial
Skalen·ablesung *f* scale reading. **-aräometer** *n* graduated hydrometer. **-einteilung** *f* scale graduation. **-intervall** *m*
scale interval (*or* division)
Skalenoeder *n* (-) scalenohedron. **skalenoedrisch** *a* scalenohedral
Skalen·rohr *n* scale tube, graduated tube.
-scheibe *f* dial (face). **-teil** *m*, **-teilung** *f*
scale division (*or* graduation)
Skammonia· (*or* **Skammonien·)harz,
Skammonium** *n* scammony (resin)
skandinavisch *a* Scandinavian
Skapolith *m* (-e) scapolite
Skarifikator *m* (-en) scarifier
SKE *abbv* (Steinkohleneinheit) bituminous unit
Skelett *n* (-e) skeleton. **--schwingung** *f*
(*Spect.*) skeletal mode (of vibration).
-substanz *f* (*Dterg.*) builder

Skepsis *f* skepticism

Skier skis: *pl of* **Ski** *m*

Skizze *f* (-n), **skizzieren** *vt* sketch

Sklave *m* (-n, -n), **Sklavin** *f* (-nen) slave. **sklavisch** *a* slavish

Sklerenchym *n* (-en) (*Bot.*) sclerenchyma

Sklerose *f* (-n) sclerosis. **sklerosieren** *vi* become indurated, harden

Skleroskop·härte *f* (*Metll.*) scleroscope hardness

skontieren *vt*, **Skonto** *m or n* (-s) discount

Skorbut *m* scurvy

Skorie *f* scoria, dross, slag

Skrofel *f* scrofula, **skrofulös** *a* scrofulous

Skrubber *m* (-) scrubber (in gas treatment)

Slab·kautschuk *m* slab rubber

Sliwowitz *m* (Serbian) plum brandy

s.l.l. *abbv* (sehr leicht löslich) very readily soluble

Slowakei *f* (*Geog.*) Slovakia

Sm. *abbv* (Schmelzpunkt) melting pt.

Smalte *f* (-n) smalt. **Smaltin** *n* (*Min.*) smaltite

Smaragd *m* (-e), **smaragden** *a* emerald
smaragd·farben *a* emerald(-colored). **-grün** *a* emerald(-green). **S–grün** *n* Guignet's (hydrated chromium oxide) green. **-malachit** *m* (*Min.*) euchroite. **-spat** *m* (*Min.*) amazonite, green feldspar

Smekal·sprung *m* Smekal transition

smektisch *a* smectic

Smirgel *m* emery = SCHMIRGEL

SMK *abbv* (Sekunden-Meter-Kerze) meter-candle-second

Smp. *abbv* (Schmelzpunkt) melting pt.

SM-Stahl *m* Siemens-Martin (*or* open-hearth) steel

Smyrna·tragant *m* Smyrna gum

s.n.F. *abbv* (siehe nebenstehende Formel) see adjoining formula

so I *adv* 1 (+*adj, adv*) so, this, that: **so klein** so (this, that) small; **man kann es nicht berühren, so heiß ist es** it can't be touched, that's how hot it is. 2 (*in comparisons*) as: **so schnell wie** (*or* **als**) **möglich** as fast as possible; **noch einmal so weit** once again as far *cf* NOCH 6; **das ist so gut wie unbrauchbar** that's as good as (*or* that's practically) useless. 3 so, thus; (in) this (*or* that) way, like this (*or* that): **so geht es** so it goes, that's the way (*or* that's how) it goes; **so oder so** this way or that way, one way or another; **es sieht nur so aus** it only looks that way; **so wie es im Labor geschieht** the way it's done in the lab; **die Sache ist so** the matter is like this (*or* is as follows); **und so fort** (*or* **weiter**) and so forth (*or* on) *cf* USF., USW. 4 (+**daß** *in next clause*) so (*or* in such a way) that. 5 (+**als ob**): **so tun, als ob. . .** act as if. . . 6 (+*noun*) e.g.: **das sind so Pflanzen, die. . .** those are a kind of plants that. . . 7 about, approximately: **so hundert Exemplare** about a hundred specimens *cf* UM 8. 8 as it is, anyway: **es ist so schon gefährlich** it's dangerous as it is. 9 (*oft rhetorical, no specif transl*): **so mancher Versuch** many an experiment. II *cnjc* 10 (*with if-clause*) then: **steigt A, so sinkt B** if A rises, (then) B drops. 11 (*after* **kaum**) when, but: **kaum kam es an die Luft, so zerfiel es** hardly was it exposed to air when (*or* but) it disintegrated. 12 (+*adj, adv, oft* + **auch**) no matter how: **so vorsichtig man auch vorgeht** no matter how carefully one proceeds. 13: **so daß** so (that) *cf* 4. III (*in special combinations*) 14: **so ein** such (a); what (a) . . . ! : **so eins** one like that, one of those. 15: **so etwas** such a thing, something like that: **oder so etwas Ähnliches** or something similar; **so etwas Schwieriges** something so difficult, such a difficult thing. 16: **so lange** (+**bis** *in next clause*) *not transl:* **es muß so lange gerührt werden, bis es dick wird** it has to be stirred until it thickens. 17: **nur so** just: A (*implies sthg going on easily or unchecked*): **es siedet nur so** it just boils on (*or* away). B (*implies: for no particular reason*): **sie fragten nur so** they were just asking. 18: **so** (z.B.) **bei. . .** as (for example) in the case of. . . IV *pfx* (*on adv, forming cnjc*) as . . . as *cf* SOBALD, SOLANGE. *See also* UM 12

s.o. *abbv* (siehe oben) see above

sobald *cnjc* as soon as

Sobrerol *n* pinol hydrate

Sockel *m* (-) pedestal, base; foundation; molding (at floor level); (*Elec.*) contact base (of a bulb, etc)

socken vi 1 crystallize out. 2 (*Metals*) contract. 3 run fast

Sod m (-e) 1 boiling (water). 2 heartburn. Cf SODE

Soda 1 f or n soda (sodium carbonate). 2 n soda (water). **s-alkalisch** a alkaline with soda. **S-asche** f soda ash. **-auszug** m soda extract(ion). **-fabrik** f soda (or alkali) factory. **-fabrikation** f soda manufacture. **s-fest** a sodaproof, soda-resistant. **S-gehalt** m soda content. **s-haltig** a containing soda. **S-kalkglas** n soda lime glass. **-kochchlorbleiche** f boiling soda-chlorine bleach. **s-kochecht** a fast to boiling soda, kier-boiling-resistant. **S-kristalle** mpl soda crystals, sal soda. **-küpe** f soda vat. **-lauge** f soda lye

Sodalith m (*Min.*) sodalite

sodalkalisch a alkaline with soda

Soda-lösung f soda solution. **-mehl** n powdered sodium carbonate monohydrate. **-menge** f amount of soda

sodann adv then

Soda-ofen m soda (or black-ash) furnace. **-rückstände** mpl soda residues, tank waste (in the Leblanc process). **-salz** n soda salt (or ash), sodium carbonate. **-schmelze** f black ash. **-see** m soda lake. **seife** f soda soap

sodaß subord cnjc so (that)

Soda-stein m caustic soda. **-wasser** n soda water

Sod-brennen n heartburn

Sode f (-n) 1 salt works; salt making. 2 sod. Cf SOD

Soden-brot n St. John's bread, carob bean

Söderberg-ofen m Soederberg furnace

sodieren vt treat (or wash) with soda

so-eben adv just (now), a moment ago

sof. abbv 1 (sofern) if. 2 (sofort, sofortig) immediate(ly)

sofern cnjc if, provided **—s. nicht** unless

soff drank: past of SAUFEN

sofort adv immediately, at once, right away. **sofortig** a immediate, instant

Sofort-entscheid m on-the-spot decision. **-maßnahme** f immediate step (or measure). **-wirkung** f 1 immediate effect. 2 (*Pesticides*) knock-down action

Sog sucked, etc: past of SAUGEN

sog m suction; current, undertow; whirlpool

sog. abbv (sogenannt) so-called

sogar adv even

sogenannt a so-called; self-styled, would-be

soggen 1 vt precipitate in crystalline form, salt down. 2 vi crystallize out

Sogge(n)-pfanne f crystallizing pan

Sogge-salz n common salt

sogleich adv immediately, right away

sohin adv, cnjc so, consequently

Sohl-band n (*Min.*) matrix, gangue. **-druck** m base pressure

Sohle f (-n) 1 sole (of foot or shoe). 2 insole. 3 bottom; floor (of mine or valley); base. 4 hearth

Sohlen-leder n sole leather

Sohl-fläche f bottom surface

söhlig a horizontal, level

Sohl-leder n sole leather

Sohn m (-e) son

Soja f (..jen). **-bohne** f soybean. **-bohnen-keimling** m (soy)bean sprout. **-kuchen** m soybean cake. **-mehl** n soybean flour. **-öl** n soybean oil. **-soße** f soy sauce

Sol n (-e) (*Chem.*) sol

Sol- brine, saline, salt-water cf SOLBAD

solange cnjc as long as cf SO IV

solarisieren vt/i solarize

Solar-öl n solar oil

Sol-bad n brine (or salt-water) bath; saline spa. **-behälter** m brine cistern. **-bohr-loch** n, **-brunnen** m salt well

solch 1 a, pron such: der Vorschlag als s**—er** the suggestion as such. 2 pron: ein s**—es** one like that; andere s**—e** others like that; s**—e mit Flecken** those with spots; s**—e, die leicht zergehen** those that melt easily

solcherart 1 a (*no endgs*) such. 2 = **solchergestalt** adv in such a way

solcherlei a (*no endgs*) of such a kind, that kind of

solchermaßen adv in such a way

Sold m pay; employ

Soldat m (-en, -en) soldier

Sole f (-n) 1 brine, salt water. 2 saline spring. 3 bittern. **-bereiter** m brine mixer. **-eindampfer** m brine concentrator. **-erzeuger** m brine mixer. **-kühler** m brine cooler. **-messer** *m* brinometer, salinometer

Solenhofener *a* (*no endgs*): **S. Platte (Schiefer, Stein)** Solenhofen (*or* lithographic) stone

Sole·pumpe *f* brine pump. **-salz** *n* common salt (from brine)

Sol·faß *n* brine pump

solid(e) *a* solid; sound, substantial; reliable

Solin·glas *n* crown glass

soll is supposed to, etc: *pres of* SOLLEN

Soll I *n* 1 debit *cf* HABEN. 2 (work) quota; (production) target. **II** *n* (-e) (*Geol.*) water-filled crater

sollen* (*pres sg* soll) *v aux* **I** (+*inf*), (*pp* sollen) 1 be to: **es soll gezeigt werden. . .** it is to be demonstrated . . . ; **sie wußten nicht, was noch kommen sollte** they didn't know what was yet to come. **2** be supposed (intended, expected) to: **das soll nicht heißen, daß. . .** that is not intended (*or* supposed) to mean, that. . . **3** (*inf in English*): **man sagte ihnen, sie sollten es versuchen** they were told to try it; **sie wußten nicht, was sie tun sollten** they didn't know what to do. **4** let: **hier soll betont werden, daß. . .** let it be stressed here that . . . ; **ABC soll ein Dreieck darstellen** let ABC represent a triangle. **5** (*expresses determination*): **das soll nicht wieder geschehen** that will not happen again. **6** (*expresses rumor*): **es soll schon heute möglich sein, zu. . .** it is already said (*or* supposed) to be possible today to. . . **7** *sbjc* **sollte** should, ought to: **jeder Chemiker sollte das wissen** every chemist should know that; **sollte es regnen, so. . .** should it rain, then. . . **8: hätte sollen** (+*inf*) e.g.: **sie hätten warten s.** they should have waited. **II** (*verb* **tun** *or verb of motion left implied; pp* sollen *or* gesollt): **was soll das?** what is that (intended) for?; **das soll ins Labor** that is (supposed) to go into the lab

Soll· nominal, rated, theoretical: **frequenz** *f* nominal frequency. **-gewicht** *n* theoretical (*or* target) weight. **Soll-Leistung** *f* rated output. **Soll·maß** *n* specified size, theoretical dimension

Soll·lösung *f* sol dispersion

Soll·wert *m* theoretical (ideal, target) val-

ue. **-zahl** *f* theoretical (ideal, target) figure

Solorin·säure *f* solorinic acid

Solor·säure *f* soloric acid

Solper *m* (-) brine. **-fleisch** *n* salt-cured meat

Sol·pfanne *f* brine pan. **-quelle** *f* brine spring, salt well. **-salz** *n* spring (well, brine) salt. **-spindel** *f* brine gauge. **-teilchen** *n* sol particle

Solubilisator *m* (-en) solubilizer. **solubilisieren** *vt* solubilize

Solvatation *f* (-en) solvation. **solvatisieren** *vt* solvate

Solvay·verfahren *n* Solvay process

Solvens *n* (. ʾ. ntien *or* . . nzien) solvent; (*Pharm.*) expectorant

solvieren *vt* dissolve. **solvierend** *p.a* (*oft*) solvent

Sol·waage *f* brine gauge, salimeter. **-wasser** *n* brine, salt water

somit *adv* so, therefore, consequently

Sommer *m* (-) summer. **-benzin** *n* summer-grade gasoline. **-frische** *f* 1 summer resort. **2** summer vacation. **-gerste** *f* spring barley. **-getreide** *n* spring wheat; summer cereal (*or* grain). **s–grün** *a* deciduous. **S–hitze** *f* summer heat

sommerlich *a* summer(y), summertime; *adv oft* as in summer

Sommer·sprossen *fpl* freckles. **-weizen** *m* spring wheat. **-wurz** *f* broom rape

sommers *adv* summers, in summer

sonach *adv/cnjc* so, accordingly, consequently

Sonde *f* (-n) 1 probe, sound, sonde; (*Med.*) bougie. 2 plumb line. 3 riser (vertical pipe). 4 sounding probe; soil penetrometer. 5 (*Geol., Min.*) well, boring, pit

Sonder· special, separate: **-abdruck** *m* separate (impression), reprint. **-ausgabe** *f* special (*or* separate) edition

sonderbar *a* strange, peculiar, odd. **sonderbarerweise** *adv* strangely enough

Sonder· special, separate: **-blatt** *n* extra (special, separate) page. **-druck** *m* special printing. **-einheit** *f* special unit. **-fall** *m* special case; exception. **s–gleichen** *adv* eg: **ein Erfolg s.** an unequaled success. **S–gußeisen** *n* special cast iron. **-heft** *n* special number (*or* issue) (of a publication). **-interessen** *npl* special interests

sonderlich *a* special, particular; *adv oft* especially

Sonder·methode *f* special method. **-müll** *m* refuse (*or* waste) for special treatment

sondern I *vt* 1 separate; sort out. 2: **sich s.** split (apart), separate (oneself). II *cnjc* but (on the contrary), but rather: **nicht nur . . . , s. auch. . .** not only . . . , but also. . .

sonders *adv:* **samt und s.** each and every one *cf* SAMT

Sonder·schicht *f* special (work) shift. **-schrift** *f* 1 specialized writing (*or* notation). 2 special article (paper, treatise). **-sprache** *f* specialized language, jargon. **-stahl** *m* special steel. **-stellung** *f* separate (*or* special) position

Sonderung *f* (-en) separation; sorting (out) *cf* SONDERN I

Sonder·verfahren *n* special process. **-verkauf** *m* (bargain) sale. **-werk** *n* special work (article, treatise). **-zweck** *m* special purpose

Sondier·ballon *m* sounding (*or* weather) balloon

sondieren *vt* sound, probe; explore

Sondierungs·gespräch *n* exploratory talk

sonisch *a* sonic

Sonn·abend *m* Saturday = SAMSTAG

Sonne *f* (-n) 1 sun, sunlight: **an der S.** in the sun(light). 2 electric heater. 3 ultraviolet lamp. 4 reflector. **sonnen** 1 *vt* sun, expose to the sun's rays. 2: **sich s.** sun oneself, bask in the sunshine

Sonnen· sun, solar, helio-: **-aufgang** *m* sunrise. **-auge** *n* (*Min.*) adularia, moonstone. **-bahn** *f* ecliptic. **-batterie** *f* solar battery. **-belichtung** *f* exposure to the sun, insolation. **s—beschienen** *a* sunlit. **-beständig** *a* sun-resistant, fast to sunlight. **S—bestrahlung** *f* solar irradiation. **-bleiche** *f* sun bleach. **-blende** *f* sun shade (*or* screen); visor. **-blume** *f* sunflower

Sonnenblumen·öl *n* sunflower oil. **-same(n)** *m* sunflower seed

Sonnen· sun, solar, helio-: **-brand** *m* 1 sunburn. 2 blazing sun. **-bräune** *f* suntan. **-brenner** *m* floodlight; lamp cluster.

-brille *f* sunglasses. **-distel** *f* carline thistle. **-energie** *f* solar energy. **-finsternis** *f* solar eclipse. **-fleck** *m* sunspot. **-gas** *n* helium. **-geflecht** *n* solar plexus. **s—gereift** *a* sun-ripened. **-gelb** *a* sun-yellow. **-getrocknet** *a* sun-dried. **S—heizung** *f* solar heating. **-hitze** *f* heat of the sun. **-kraftwerk** *n* solar energy installation, solar power plant. **-licht** *n* sunlight. **-ofen** *m* solar energy installation. **-öl** *n* suntan oil. **-schein** *m* sunshine. **-schirm** *m* sunshade, parasol. **-spektrum** *n* solar spectrum. **-stäubchenphänomen** *n* Tyndall effect. **-stein** *m* sunstone, aventurine feldspar. **-stich** *m* sunstroke. **-strahl** *m* solar ray, sunbeam. **-strahlung** *f* solar radiation. **-system** *n* solar system. **-tau** *m* (*Bot.*) sundew. **-tierchen** *n* heliozoon. **s—trocken** *a* sun-dried. **S—uhr** *f* sundial. **-untergang** *m* sunset. **s—verbrannt** *a* sunburned. **S—wärme** *f* sun's heat. **-wende** *f* 1 solstice. 2 (*Bot.*) heliotrope. **-wind** *m* solar wind. **-zeit** *f* solar time

sonnig *a* sunny

Sonntag *m* Sunday. **sonntags** *adv* (on) Sundays

sonn·trocken *a* sun-dried

sonor *a* sonorous

sonst I *adv* 1 else: **was s.?** what else?; **s. jemand** someone else; **s. nichts** nothing else. 2 usual(ly), always: **besser als s.** better than usual(ly); **es ist noch wie s.** it is still as it always was (*or* as it used to be). 3: **auch s.: es ist auch s. gefährlich** it is just dangerous in any case. II *adv & cnjc* or else, otherwise: **s. verdirbt es** or else it will spoil. **sonstig** *a* 1 other possible, any other, some other (such): **s—e Eigenschaften** any other (such) properties; **sein s—es Verhalten** its other possible behavior, its behavior in other respects. 2 usual

sonst·was *pron* any (*or* some) other thing. **-wer** *pron* any (*or* some) other person. **-wie** *adv* (in) any (*or* some) other way. **-wo** *adv* anywhere (*or* somewhere) else

so·oft *subord cnjc* as often as, whenever

Sorbens *n* (. . ntien) sorbent

Sorbett *m or n* (-e) sherbet

sorbieren *vt* (ad)sorb. **sorbierend** *p.a* sorbent, sorptive

Sorbin·aldehyd *m* sorbaldehyde. **-säure** *f* sorbic acid

Sorbit *n* 1 (*Org. Chem.*) sorbitol. 2 (*Metll.*) sorbite. **sorbitisch** *a* sorbitic

Sorge *f* (-n) 1 worry, concern; anxiety; trouble: **einem S—n machen** cause someone worry; **sich S—n machen (um)** worry (about); **in S. sein** be worried; **außer S. sein** have nothing to worry about. 2 care: **S. tragen (für)** care (for), take care (of), look (after); **S. tragen, daß. . .** see to it, that. . .

sorgen 1 *vi* (+**für**) provide (for) *cf* GESORGT; (+**dafür, daß**) see to it (that). 2: **sich s. (um)** worry, be concerned (about); **dafür wird gesorgt** that is provided for

sorgen·frei, -los *a* carefree. **-schwer, -voll** *a* careworn, troubled

Sorgfalt *f* 1 care, attention to detail, meticulousness. 2 trouble, pains (in doing a job). **sorgfältig** *a* careful, painstaking, meticulous

sorglich *a* careful, etc = SORGSAM

sorglos *a* carefree

sorgsam *a* 1 careful, painstaking. 2 solicitous, attentive. 3 cautious

Sorptions·kapazität *f* (*Soil*) exchange (*or* absorption) capacity. **-komplex** *m* (*Soil*) absorption complex

Sorptiv *n* (-e) sorbate

Sorrel·salz *n* sorrel salt, potassium hydrogen oxalate

Sorte *f* (-n) 1 sort, kind, type; brand; quality, grade. 2 variety, species, breed, strain. 3 (*pl only*) foreign currency

sorten *vt* sort, grade, etc = SORTIEREN

Sorten·kurs *m* foreign exchange rate. **s—rein** *a* pure-bred, pure-variety, pure-strain. **S—schutz** *m* protection of new plant varieties. **-verzeichis** *n* index of (registered) plant varieties

Sortier·apparat *m* sorting (*or* grading) apparatus

sortieren *vt* sort (out); assort; classify; grade, size. **Sortierer** *m* (-) sorter, grader

Sortier·maschine *f* sorting machine, sorter. **-sieb** *n* sorting (*or* grading) sieve. **-trommel** *f* sorting drum

sortiert *p.a* (*oft*) choice, select. **Sortierung** *f* (-en) sorting, etc. *Cf* SORTIEREN

Sortiment *n* (-e) 1 assortment. 2 retail book trade

sosehr *cnjc* (as) much as, no matter how much

soso *adv* so-so

Soße *f* (-n) sauce; gravy; (salad) dressing

sott boiled: *past of* SIEDEN

soundso *adv* so and so, such and such: **s. viele** so and so many

Souple·schwarz *n* (*Dye.*) souple black. **-seide** *f* souple silk

souplieren *vt* souple (silk)

soviel 1 *adv* & *indef pron* so much, this much: **s. ist gewiß** this much is certain; **s. wie** as much as; **das heißt s. wie. . .** that amounts to saying. . . 2 *cnjc* (as) much as, as far as

soweit 1 *adv* 1 so far; **es kam s., daß. . .** things got to the point where. . . 2 on the whole. 3 (+**sein**) ready, at the (right) point: **wir sind noch nicht s.** we are not ready yet, we have not reached that point yet; **wenn es s. ist** when that point is reached, when the time comes. 4: **s. wie** as far as. II *cnjc* as far as, to the extent that

sowenig 1 *indef pron* so little, such a small amount. 2 *cnjc* (as) little as

sowie *cnjc* 1 as well as. 2 as soon as

sowieso *adv* anyway, anyhow, in any case

sowjetisch *a* Soviet. **Sowjet·union** *f* Soviet Union

sowohl *cnjc* e.g: **s. das eine wie** (or **als**) (**auch**) **das andere** both the one and the other, the one as well as the other

Soya·bohne *f* soybean *cf* SOJA·

So.Z. *abbv* (Sodazahl) soda number (*or* value)

sozial *a* social. **S—·wissenschaft** *f* social science. **Soziologie** *f* sociology

Sozol·säure *f* sozolic (*or* *o*-phenosulfonic) acid

sozusagen *adv* so to speak, as it were

Spachtel *m* (-) *or* *f* (-n) 1 spatula, (*Med.*) tongue depressor. 2 trowel, putty knife; paint scraper. 3 putty, filler, spackle. **··grund** *m* putty, filler; (*Paint*) foundation surfacer. **-kitt** *m* putty, filler. **-masse** *f*, **-material** *n* filler, putty; (*Paint*) foundation surfacer. **-messer** *n* putty knife

spachteln *vt/i* smooth, fill (with a spatula), spackle, putty; scrape

Spagat *m* (-e) string, cord

spagyrisch *a* spagyric(al)

spähen *vi* peer; (*usu* + **nach**) look (out), watch, scout, reconnoiter (for)

spakig *a* mold-stained; rotten

Spalier *n* (-e) **1** trellis. **2** row; lane between two rows (*oft implies honor guard*). **-obst** *n* trellis fruit

Spalt *m* (-e) **1** crack, chink, fissure, cleft; gap. **2** (*incl Opt., Spect.*) slit. **3** (*incl Lthr.*) split. **4** (*fig*) chasm; rift, schism. *Cf* SPALTE. **-anlage** *f* cracking plant (or unit). **-ausleuchtung** *f* slit illumination

spaltbar *a* cleavable, fissionable; scissile. **Spaltbarkeit** *f* cleavability, fissionability. **Spaltbarkeits·richtung** *f* (*Min.*) direction of cleavage

Spalt·benzin *n* cracked gasoline. **-bild** *n* (*Opt.*) slit image. **-bildung** *f* splitting; fissure formation. **-blende** *f* (*Opt.*) slit diaphragm. **-breite** *f* slit (or aperture) width. **sp—brüchig** *a* easily cleaved. **Sp—brüchigkeit** *f* cleavage brittleness. **-bruchstück** *n* fission fragment. **-destillation** *f* cracking (distillation)

Spalte *f* (-n) **1** (=SPALT) crack, etc; slit; split. **2** column (of type). **3** slice

Spalt·ebene *f* (*Cryst.*) cleavage plane (or face). **-eisen** *n* cleaver

spalten **I** *vt* **1** split, crack (*incl Petrol.*); cleave, fissure. **2** (*Opt.*) resolve. **II: sich sp.** split (up); decompose; dissociate; (*Nucl.*) undergo fission

Spalten·gang *m* (*Geol.*) fissure vein. **sp—reich** *a* considerably fissured

spalt·erbig *a* heterozygous

Spalt·festigkeit *f* resistance to splitting; (*Plastics*) interlaminar strength. **-fläche** *f* cleavage surface (or plane). **-frucht** *f* dehiscent fruit. **-funkenzünder** *m* (*Explo.*) jump-spark cap. **-gas** *n* cracked gas. **-glimmer** *m* sheet mica. **-glühzünder** *m* (*Explo.*) high-tension cap

spaltig **1** *a* split, fissured, cracked. **2** *sfx* -column: **ein dreispaltiger Artikel** *a* three-column article

Spalt·katalysator *m* cracking catalyst. **-kollimation** *f* slit collimation. **-körper** *m* cleavage substance (or product). **-korrosion** *f* crevice (or crack) corrosion. **-lampe** *f* (*Opt.*) slit lamp. **-leder** *n* (*Lthr.*) skivers, splits. **-material** *n* (*Nucl.*) fissionable material, reactor fuel. **-mündung** *f* (*Bot.*) stoma, stomate. **-neutron** *n* fission neutron. **-öffnung** *f* (*Bot.*) stoma, stomate. **-pflanze** *f* schizophyte. **-pilz** *m* fission fungus, schizomycete; splitting mold. **-pilzgärung** *f* schizomycetous fermentation. **-produkt** *n* cleavage (or fission) product; fragment. **-produktausbeute** *f* chain fission yield. **-riß** *m* cleavage crack (or fissure). **-rohr** *n* (*Spect.*) slit tube, collimator. **-stoff** *m* (*Nucl.*) fissionable material, reactor fuel. **-stück** *n* segment, fragment, splinter; cleavage product; (*Nucl.*) fission fragment

Spaltung *f* (-en) splitting, cracking, cleavage, fission; decomposition, dissociation; scission; (*Opt.*) resolution; schism *cf* SPALTEN

Spaltungs·arbeit *f* **1** (*Chem.*) dissociation energy. **2** (*Nucl.*) fission energy. **-ausbeute** *f* fission yield. **-einfang** *m* fission capture. **-fläche** *f* cleavage plane. **-gärung** *f* cleavage fermentation. **-gestein** *n* differentiated dike rock. **-kette** *f* fission chain. **-kristall** *m* cleavage crystal. **-probe** *f* (*Min.*) cleavage test; (*Rock*) spalling test; (*Wood*) splitting test. **-produkt** *n* cleavage (decomposition, fission) product. **-prozeß** *m* cleavage (or fission) process. **-richtung** *f* direction of cleavage. **-stück** *n* fragment, segment; cleavage product; (*Nucl.*) fission fragment. **-verzug** *m* delayed fission. **-vorgang** *m* cleavage (or fission) process. **-wärme** *f* heat of decomposition (dissociation, fission)

Spalt·versuch *m* (*Min.*) cleavage test. **-winkel** *m* (*Cryst.*) cleavage angle. **-zünder** *m* high-tension detonator

Span *m* (-e) **1** chip, splint(er), sliver, shred. **2** *pl* sawdust; shavings, filings, etc. **3: aus einem Sp.** of one piece

span·abhebend *a* metal-removing, cutting, machining: **sp—e Werkzeuge** cutting tools; **sp. bearbeiten** machine

Spanen *vt* cut, machine (metal)

Span·faß *n* =**Spänfaß** *n* (*Brew.*) chip cask. **-ferkel** *n* suckling pig

Spange *f* (-n) **1** clasp, buckle, bobby pin. **2** dental brace. **3** (*Mach.*) stay bolt. **4** bangle; brooch; (award) pin. **5** (shoe) strap

span·gebend *a* machining, etc = SPANAB-HEBEND

Spangel·eisen *n* crystalline pig iron
spang(e)lig *a* spangled, glistening; crystalline
Span·grün *n* verdigris. **-holz** *n* matchwood, chipwood
Spanien *n* Spain. **spanisch** *a* Spanish —**sp**—**e Fliegen** Spanish flies, cantharides; **sp**—**er Pfeffer** red pepper, capsicum; **sp**—**es Rohr** Spanish reed, rattan; **sp**—**e Wand** *f* folding screen
Spanisch·braun *n* Spanish brown. **-fliegen·** (*in compds*) cantharides. **-weiß** *n* flake white
spann spun: *past of* SPINNEN
Spann *m* (-e) **1** instep *cf* SPANNE. **2** (*in compds*) tension, stretching *cf* SPANNEN
spannbar *a* tensile, extensible, ductile. **Spannbarkeit** *f* extensibility, ductility
Spann·beton *m* prestressed concrete. **-draht** *m* tension wire; guy wire
Spanne *f* (-n) **1** (time) period, (brief) interval. **2** span; stretch, distance; difference. **3** price (*or* profit) margin. *cf* SPANN
Spann·eisen *n* tendon (in prestressed concrete)
spannen I *vt* **1** tighten, tense, pull taut; stretch: **hoch sp.** stretch to the limit, pitch high. **2** (+**um**) clamp, wind (around). **3** (+**in**) insert, put (in); mount (in, on). **4: Dampf sp.** increase the tension of steam, superheat steam. **5** cock (a gun). **6** fascinate, thrill. **7** notice, catch on to. **II** *vt/i* **8** span, stretch, measure. **III** *vi* **9** be (*or* fit) tight. **10** be swollen. **11** (*usu* + **auf**) be in suspense (about), listen (watch, await) eagerly. **III: sich sp. 12** tighten, become tense (taut, tight); tense up (ready to take action). **13** (+**um**) be wrapped, be clamped (around), pinch. **14** stretch, extend. *Cf* GESPANNT. **spannend** *p.a* (*oft*) suspenseful. **Spanner** *m* (-) **1** (*Mech.*) (stretching) frame, press, rack, vise; shoe tree; hanger, (*Tex.*) tenter. **2** (*Zool.*) geometrid (moth). **3** (*Anat.*) tensor
Spann·feder *f* tension spring. **-futter** *n* chuck (of a vise). **-knorpel** *m* thyroid cartilage. **-kraft** *f* **1** elasticity; (*also fig*) resiliency, buoyancy. **2** tension, expanding force. **3** expansibility, extensibility. **4** (*Mech.*) clamping power. **5** (*Anat.*) (mus-

cle) tone. **sp·kräftig** *a* elastic, resilient. **Sp·lack** *m* stiffening varnish, dope. **-muskel** *m* tensor muscle. **-rahmen** *m* (*Tex.*) tenter, stretching frame
Spannung *f* (-en) **1** tension. **2** (*Phys.*) stress, strain. **3** (*Gas*) pressure. **4** (*Elec.*) potential, voltage. **5** suspense. **6** bulge, swell(ing). **7** tautness, tenseness
spannung·führend *a* (*Elec.*) charged, live, hot
Spannungs·abfall *m* (*Elec.*) voltage drop. **-dehnungsdiagramm** *n* stress-strain diagram. **-festigkeit** *f* dielectric strength. **sp·frei** *a* tension-free, strain-free. **Sp·freiglühen** *n* (*Metll.*) stress-relief anneal. **-grad** *m* degree of tension. **-koeffizient** *m* pressure (*or* thermal-expansion) coefficient. **-kompensator** *m* (*Elec.*) potentiometer. **-korrosion** *f* stress corrosion. **sp·los** *a* tension-free, strain-free; (*Elec.*) dead. **Sp·messer** *m* strain gauge, tensiometer; (*Elec.*) voltmeter. **-optik** *f* photoelasticity. **sp·optisch** *a* photoelastic. **Sp·prüfer** *m* strain tester. **-regler** *m* voltage regulator. **-reihe** *f* electromotive (*or* electrochemical) series. **-riß** *m* stress crack. **-schreiber** *m* voltage recorder. **-teiler** *m* voltage divider. **-theorie** *f* (*Org. Chem.*) strain theory. **-unterschied** *m* **1** difference in stress. **2** (*Elec.*) voltage (*or* potential) difference. **-verfestigung** *f* (*Metll.*) strain hardening. **-vergleichsrohr** *n* voltage reference tube. **-vervielfacher** *m* voltage multiplier. **-verzerrungsbeziehungen** *fpl* stress-strain relations. **-wandler** *m* voltage transformer. **-wechsel** *m* stress (*or* voltage) reversal. **-zeiger** *m* **1** tension indicator. **2** (*Elec.*) voltmeter. **-zustand** *m* state of tension
Spann·weite *f* span; range, distance. **-zeug** *n* clamp
Span·versuch *m* glowing-splint test
Spar·beize *f* (*Metll.*) pickling bath with an inhibitor. **-beizzusatz** *m* (*Metll.*) pickling inhibitor. **-beton** *m* lean concrete. **-brenner** *m* **1** economial burner. **2** pilot burner, gas burner with a pilot flame
sparen 1 *vt* save, spare. **2** *vt/i* (*oft* + **mit**) economize, save (on). **3: sich sp.** spare oneself, save one's strength. **sparend** *p.a*

(esp as adv) thriftily, economically

Spar·flämmchen *n*, **-flamme** *f* pilot flame

Spargel *m* (-) asparagus. **--kohl** *m* broccoli. **-stein** *m* (*Min.*) stone, variety of apatite. **-stoff** *m* asparagine

Spar·gemisch *n* lean (fuel) mixture. **-kalk** *m* flooring plaster, anhydrous calcium sulfate. **-kapsel** *f* (*Ceram.*) economy (or space-saving) sagger. **-kasse** *f* savings bank. **-lampe** *f* small (or economy) lamp

spärlich *a* sparse, meager, scant; weak; few

Spar·maßnahme *f* economy measure. **-metall** *n* scarce metal. **-mittel** *n* (*Biochem.*) protein sparer; sparing substance

Sparren *m* (-) 1 rafter. 2 scantling. 3 spar

sparsam *a* 1 economical, thrifty: **sp. umgehen (mit)** be economical (with). 2 sparing. 3 meager, scant; sparse; few. **Sparsamkeit** *f* economy, thrift

Spar·stoff *m* scarce (or rationed) material. **sp--stoffarm** *a* low in scarce materials

Sparte *f* (-n) 1 branch; department. 2 subject

Spart(o)·gras *n* esparto (grass)

Sparto·papier *n* esparto paper

Spar·vorrichtung *f* economy device. **-werkstoff** *m* scarce (or critical) material

Spasmus *m* (. . men) spasm. **spasmisch** *a* spastic. **spasmodisch** *a* spasmodic. **spastisch** *a* spastic

Spat *m* (-e or ⁻e) (*Min.*) spar

spät *a* late

spat·artig *a* sparry, spathic

spät·blühend *a* late-blooming (or -flowering)

Spat·eisen *n*, **-eisenstein** *m* spathic (or sparry) iron ore, (*Min.*) siderite, chalybite

Spatel *m* (-) 1 spatula; trowel. 2 scraper. **sp--förmig** *a* spatulate, spatula-shaped **-messer** *n* steel spatula. **spateln** *vt* smooth (with a spatula)

Spaten *m* (-) spade

später *a* later. **--hin** *adv* later (on)

Spat·erz *n* spathic ore

spätestens *adv* at the latest

Spat·fluß *m* (*Min.*) fluorite, fluorspar. **sp--förmig** *a* spathic

Spät·folge *f* delayed consequence (e.g. of radiation exposure). **-gärung** *f* delayed fermentation

spatig *a* 1 (*Min.*) spathic, spathose. 2 (*Vet.*) spavined

Spat·lack *m* barytes lake

Spätling *m* (-e) 1 latecomer. 2 late fruit

Spät·obst *n* late fruit. **sp--reif** *a* latematuring, late-ripening

Spat·sand *m* silica sand. **-säure** *f* hydrofluoric acid

Spät·schaden *m* delayed damage (injury, trauma). **-sommer** *m* late summer; Indian summer

Spat·stein *m* (*Min.*) selenite

Spät·wirkung *f* tardy (or delayed) action, after-effect

Spatz *m* (-en) sparrow

Spät·zündung *f* retarded ignition; (*fig*) delayed reaction

spazieren *vi*(s) walk, stroll

spec. *abbv* 1 (specifisch) spec. (specific). 2 (species) species

Specht *m* (-e) woodpecker

speci. . . *See* SPEZI. . .

Speck *m* 1 lard; fat; blubber; bacon. 2 (*in compds, oft*) fatty, lardaceous, amyloid

speck·ähnlich, -artig *a* fatty, lardaceous

Speck·glanz *m* greasy luster. **-haut** *f* (*Physiol.*) buffy coat

speckig *a* 1 lardy, fatty, greasy. 2 lardaceous; amyloid. 3 (*Bread*) heavy. 4 (*Barley*) flinty

Speck·kalk *m* fat (or white) lime. **-kienholz** *n* resinous pine wood. **-öl** *n* lard oil. **-schwarte** *f* bacon rind. **-stein** *m* (*Min.*) talc, soapstone, steatite. **sp--steinartig** *a* steatitic. **Sp--stoff** *m*, **-substanz** *f* lardaceous (or amyloid) substance. **-torf** *m* pitch peat. **-tran** *m* train oil

Specularit *m* specular hematite

spedieren *vt* forward, ship. **Spedition** *f* (-en) forwarding

Speer *m* (-e) spear; javelin. **--kies** *m* (*Min.*) spear pyrites, marcasite

Speiche *f* (-n) spoke; rib; (*Anat.*) radius

Speichel *m* saliva. **--absonderung** *f* salivation, secretion of saliva. **sp--befördernd** *a* sialagogic; salivationpromoting. **Sp--drüse** *f* salivary gland. **sp--echt** *a* saliva-resistant, fast to saliva. **Sp--fluß** *m* flow of saliva, salivation. **-flüssigkeit** *f* saliva. **-kasten** *m* moisture (*or* saliva) chamber. **-mittel** *n* sialagogue, salivation promoter. **-saft** *m* saliva. **-stoff** *m* ptyalin

Speicher *m* (-) 1 storehouse, warehouse. 2 granary; silo. 3 storeroom; attic, loft. 4 reservoir; magazine. 5 (*Cmptrs*) data bank, memory bank. **-batterie** *f* storage battery. **-gestein** *n* (*Petrol.*) reservoir rock. **-gewebe** *n* storage tissue, endosperm. **-kapazität** *f* storage capacity

speichern *vt* store (up), accumulate; hoard; computerize. **Speicherung** *f* (-en) storage, storing, accumulation; hoarding; computerization

Speicher-werk *n* 1 (*genl*) storage unit. 2 (*Cmptrs.*) memory, data bank. 3 (*Engg*) (hydraulic) accumulation station

speien* (spie, gespieen) *vt/i* spit; spew, belch; vomit

Speik-öl *n* oil of Celtic nard, valerian oil

Speis *m* 1 mortar. 2 (*Metll.*) arsenide. Cf SPEISE

Speise I *f* (-n) 1 food, nourishment. 2 (prepared) dish. 3 dessert. 4 pantry. 5 (*Metll.*) speiss, bell metal. 6 metal. II (*in compds*) food, alimentary; feed, supply; edible: **-apparat** *m* (*Mech.*) feed apparatus. **-bestandteil** *m* food constituent. **-brei** *m* chyme. **-eis** *n* ice cream; sherbet. **-essig** *m* table vinegar. **-fett** *n* edible (*or* nutrient) fat; cooking fat (*or* grease). **-flotte** *f* replenishing liquor. **-gang** *m* alimentary canal. **-gas** *n* feed gas. **-gefäß** *n* feed vessel (*or* tank). **sp-gelb** *a* Speiss-yellow, pale bronze. **Sp-hahn** *m* feed cock. **-kanal** *m* alimentary canal. **-karte** *f* menu, bill of fare. **-kartoffel** *f* food potato. **-kobalt** *m* (*Min.*) smaltite. **-leinöl** *n* edible linseed oil. **-leitung** *f* feed (*or* supply) pipe (line, wire). **-masse** *f* ration

speisen 1 *vt* feed, supply; fill; fuel; stock. 2 *vi* dine, have dinner

Speisen-träger *m* food carrier; vacuum vessel

Speise-öl *n* edible oil; salad oil. **-ordnung** *f* diet. **-pilz** *m* edible fungus (*or* mushroom). **-pumpe** *f* feed pump. **-punkt** *m* input terminal. **-quark** *m* cottage cheese

Speiser *m* (-) feeder

Speise-rest *m* food residue; (*pl oft*) leftovers. **-rohr** *n* feed (*or* supply) pipe.

-röhre *f* esophagus. **-rübe** *f* garden turnip. **-saal** *m* dining room (*or* hall). **-saft** *m* chyle. **-salz** *n* table salt. **-schnecke** *f* feed screw. **-sirup** *m* table syrup. **-trichter** *m* supply funnel, feed hopper. **-vorrichtung** *f* feeder, feeding (*or* supply) device. **-walze** *f* feed roll. **-wasser** *n* feed water. **-zimmer** *n* dining room. **-zucker** *m* table sugar. **-zwiebel** *f* (cooking) onion

speisig *a* cobalt-containing

Speis-kobalt *m* (*Min.*) smaltite

Speisung *f* (-en) feeding, supply(ing); filling; fueling, stocking; charging; (power) supply *cf* SPEISEN

Spektral-analyse *f* spectrum analysis. **sp-analytisch** *a* spectroscopic, spectrometric; *adv oft* by spectrum analysis. **Sp-apparat** *m* spectroscopic apparatus, spectroscope, spectrometer. **-beobachtung** *f* spectroscopic observation. **-bereich, -bezirk** *m* spectral region. **-farbe** *f* spectrum color. **-gebiet** *n*, **-gegend** *f* spectral region. **-lampe** *f* spectrum lamp. **-linie** *f* spectrum line. **-photometer** *n* spectrophotometer. **-photometrie** *f* spectrophotometry. **sp-photometrisch** *a* spectrophotometric. **Sp-probe** *f* spectrum test. **-rohr** *n*, **-röhre** *f* spectrum tube. **-tafel** *f* spectral chart. **-verteilung** *f* spectral distribution

Spektren spectra *pl of* **Spektrum** *n*

Spektrochemie *f* spectrochemistry. **Spektrochemiker°** *m* spectrochemist. **spektrochemisch** *a* spectrochemical

Spektrogramm *n* (-e) spectrogram. **spektrographisch** *a* spectrographic

Spektroskopie *f* spectroscopy. **Spektroskopiker** *m* (-) spectroscopist. **spektroskopisch** *a* spectroscopic

Spelz *m* (-e) spelt. **Spelze** *f* (-n) (*Bot.*) bract; husk (of grain); glume; awn, beard. **Spelz-mehl** *n* spelt flour

Spende *f* (-n) donation, contribution. **spenden** *vt* donate, contribute; provide. **Spender** *m* (-), **Spenderin** *f* (-nen) donor, contributor; dispenser; source

Spengler *m* (-) 1 plumber. 2 mechanic. 3 sheetmetal worker

Sper-beere *f* service berry

Sperber *m* (-) sparrow hawk

Sperling *m* (-e) sparrow. **Sperlings-schna-**

bel *m* (*Min.*) separable tin, needle-shaped cassiterite

Sperma *n* (. . men *or* . . mata) sperm. **--kern** *m* sperm nucleus

Spermazet(i) *n* (. . ten) spermaceti. **--öl** *n* sperm oil

Spermie *f* (-n) spermatozoon

Sperm·öl *n* sperm oil

Sperrad *n* (=**Sperr·rad**) ratchet (*or* cog) wheel

sperr·angelweit *adv:* **sp. offen** wide open. **Sp--druck** *m* (letter-)spaced type

Sperre *f* (-n) 1 closing, closure; blocking. 2 block, obstruction. 3 gate, turnstile; barrier. 4 stop, lock, locking device; pawl, arrest; baffle; seal; (*Radio*) wave trap. 5 ban, prohibition, embargo; freeze. **sperren I** *vt* 1 close (off), seal; stop (up), stopper; insulate. 2 lock (up); turn the lock. 3 block, obstruct. 4 stop, arrest (motion). 5 cut off, shut off. 6 restrict, freeze. 7 ban, prohibit. 8 letter-space (words). **II** *vi* stick, be stuck, refuse to move. **III: sich sp.** (+**zu** + *inf*) resist, balk (at) (doing. . .).

Sperrer *m* (-) lock; block; trap

Sperr·feder *f* lock (*or* click) spring. **-flüssigkeit** *f* sealing liquid (to prevent escape of gas). **-gebiet** *n* restricted (*or* forbidden) zone. **-gut** *n* bulk goods. **-hahn** *m* stopcock. **-haken** *m* (*Mach.*) catch, pawl, ratchet, detent. **-holz** *n* plywood. **-holzkleber** *m* plywood adhesive; veneer glue

sperrig *a* bulky, unwieldy, cumbersome; spread out

Sperr·klinke *f* (*Mach.*) catch, pawl, ratchet, detent. **-kreis** *m* (*Radio*) wave trap (circuit). **-metall** *n* ply metal. **-rad** *n* cog wheel = SPERRAD. **-schicht** *f* 1 waterproofing layer. 2 (*Elec.*) barrier layer. **-schichtenzelle** *f* solid photovoltaic cell

Sperrung *f* (-en) 1 closing; seal(ing); stopping. 2 locking. 3 blocking, obstruction. 4 stoppage. 5 cutting (*or* shutting) off, cut-off, shut-off. 6 freeze, freezing. 7 ban, prohibition, embargo. 8 photometer opening, iris. *Cf* SPERREN

Sperr·ventil *n* check (*or* shut-off) valve. **-vorrichtung** *f* closing (locking, shut-off) device; barrier; catch, stop. **-waffe** *f* defensive weapon. **-wasser** *n* sealing water

Spesen *pl* expenses; charges

spez. *abbv* 1 (spezifisch) spec. (specific): **spez. Gew.** spec. gravity. 2 (speziell) special

spezial *a* special. **Sp--arzt** *m* (medical) specialist. **-ausbildung** *f* specialized training. **-benzin** *n* special gasoline. **-fach** *n* specialty, field of specialization. **-fall** *m* special case. **-gußeisen** *n* special cast iron

spezialisieren *vr:* **sich sp.** (**auf, für**) specialize (in)

Spezialist *m* (-en, -en) specialist. **Spezialität** *f* (-en) specialty

Spezial·kenntnisse *fpl* specialized knowledge. **-öl** *n* special oil. **-reagens** *n* special reagent. **-stahl** *m* special steel

speziell 1 *a* special; specialized. 2 *adv oft* especially; separately

Spezies *f* (-) 1 species. 2 (*Metll.*) specie. 3 (*Pharm.*) simple, herb. 4 (*Math.*): **die vier Sp.** the four basic operations of arithmetic

Spezifikum *n* (. . ka) (*Pharm.*) specific

spezifisch *a* specific. **Spezifität** *f* specificity

spezifizieren *vt* specify, itemize

sp.G(ew). *abbv* (spezifisches Gewicht) sp. grav. (specific gravity)

Sphäre *f* (-n) sphere; domain

Sphärimeter° *n* spherometer

sphärisch *a* spherical

Sphärokristall° *m* spherical crystal. **Sphärolith°** *m* (-e) (*Cryst.*) spherulite. **Sphärosiderit** *n* (*Min.*) spherosiderite

Sphärulith *m* (-e) spherulite

sphenoidisch *a* sphenoid(al)

spicken *vt* 1 lard, stud. 2 oil (wood). 3 smoke (food). **Spick·öl** *n* (*Wool*) spinning (*or* spike) oil

spie spewed: *past of* SPEIEN

Spiegel *m* (-) 1 mirror. 2 glass; polished surface. 3 surface (*esp* of liquids); level (*oft in compds, cf* ALKOHOLSPIEGEL, WASSERSPIEGEL). 4 (*Med., Opt.*) speculum. 5 reflector. 6 bull's eye. 7 specular cast iron, =SPIEGELEISEN. 8 (*Lthr.*) butt. 9 model, paragon

Spiegel· mirror, specular, reflection: **-ablesung** *f* mirror reading. **-beleg** *m*, **-belegung** *f* mirror coating, silvering

Spiegelbild° *n* mirror image: **Bild und Sp.** object and mirror image. **sp-·isomer** *a* enantiomorphic. **Sp–isomerie** *f* optical isomerism

spiegel·bildlich *a* enantiomorphic, mirror-image(d). **-blank** *a* 1 highly polished. 2 spick and span. **Sp-bronze** *f* speculum metal. **-ebene** *f* mirror (*or* reflection) plane. **-ei** *n* fried egg (sunny side). **-eisen** *n* specular cast iron, metallic(-looking) hematite. **-eisenerz** *n* specular iron ore, specular hematite. **-faser** *f* medullary ray. **-fernrohr** *n* reflecting telescope. **-floß** *n* spiegeleisen. **-folie** *f* 1 (mirror) silvering, mirror foil. 2 tinfoil. **-galvanometer** *n* reflecting galvanometer. **-glanz** *m* 1 reflecting surface. 2 (*Min.*) wehrlite. **-glas** *n* 1 plate glass. 2 mirror glass. **sp–glatt** *n* smooth as glass, mirror-smooth. **-gleich** *a* mirror-symmetrical. **Sp–gleichheit** *f* mirror symmetry. **-höhe** *f* liquid level (*or* height)

spiegelig *a* specular, mirror-like; mirror-smooth

Spiegel·kamera *f* reflex camera. **-kern** *m* (*Nucl.*) mirror nucleus. **-metall** *n* speculum (*or* specular) metal

spiegeln 1 *vt* mirror, reflect. 2 *vi* shine (like a mirror); be spick and span, sparkle. 3: **sich sp.** be reflected

Spiegel·prisma *n* reflecting prism. **-schirm** *m* reflector. **-skala** *f* mirror scale

Spiegelung *f* (-en) 1 reflection. 2 mirage

Spiegelungs·moment *n* (*Nucl.*) parity

Spiegel·welle *f* reflected wave

Spiegler *m* (-) reflector

spieglig *a* mirror-smooth, etc = SPIEGELIG. **Spieglung** *f* (-en) reflection, etc = SPIEGELUNG

Spieke *f* (-n) (*Bot.*) spike lavender. **Spiek-öl** *n* oil of spike

Spiel *n* (-e) 1 play (*incl Mach.* & *most fig meanings*); playing: **freies Sp. lassen** (+ *dat*) give free play; **das Sp. der Natur** the free play of nature; **wechselseitiges Sp.** interplay; **wie im Sp.** like child's play; **ein Sp. des Zufalls** a whim of fate. 2 action, motion (of machine or body parts); stroke (of a piston). 3 game (*incl* set, equipment); deck, pack (of cards) — **wie das Sp. steht** what the score is;

leichtes Sp. haben have an easy time of it. 4 gamble, gambling — **auf dem Sp. stehen** be at stake; **aufs Sp. setzen** stake, risk. 5 acting, performance. 6 matter, affair: **aus dem Sp. bleiben** keep (*or* be kept) out of the matter; **seine Hände im Sp. haben** have a hand in the affair; **das ist (auch) mit im Sp.** that is also involved (in the matter); **sich ins Sp. mischen** interfere. **–art** *f* (*esp Biol.*) variety, variant; sport. **-automat** *m* slot machine; pin-ball machine, video-game machine

spielen I *vt/i* 1 play (game, instrument, role); act, perform. **II** *vt* 2 show (a film). 3 feign. **III** *vi* 4 (*Mach.*) work; have play. 5 (*Gems*) sparkle. 6 (*Colors*) (+ **in**) shade over (into), have a . . . tinge. 7 (+ **mit**) toy, trifle (with). 8: **das Radio spielt** the radio is on. 9 take place (in some setting, location). **IV: sp. lassen** 10 display. 11 (*Radio, TV*) play, turn on. **spielend** *p.a* (*oft*) easy, effortless

Spiel·raum *m* leeway; (*Mach. oft*) tolerance, backlash. **-waren** *fpl* toys. **-warenlack** *m* varnish for toys. **-zeug** *n* plaything, toy

Spieß *m* (-e) 1 spear, lance. 2 (roasting) spit; skewer. 3 (*Cryst.*) long needle. **spießen** *vt* 1 spear, skewer, impale. 2 (+ **in**) stick (a knife, etc) (into)

spieß·förmig *a* spear-shaped, lance-like

Spießglanz *m* 1 antimony. 2 (*Min.*) stibnite. **sp-·artig** *a* antimonial. **Sp–asche** *f* antimony ash. **-bleierz** *n* (*Min.*) bournonite. **-blende** *f* (*Min.*) antimony blende, kermesite. **-blumen** *fpl* flowers of antimony, crystalline antimony sesquioxide. **-butter** *f* antimony butter (*former name of* antimony trichloride). **-erz** *n* 1 antimony ore. 2 (*specif*) **graues Sp.** stibnite; **schwarzes Sp.** bournonite; **weißes Sp.** valentinite. **-fahlerz** *n* tetrahedrite. **sp–haltig** *a* antimonial, containing antimony. **Sp–kermes** *m* kermesite, kermes mineral. **-könig** *m* regulus of antimony. **-leber** *f* liver of antimony, hepar antimonii. **-metall** *n* antimony metal. **-mittel** *n* antimonial (remedy). **-mohr** *m* (*Pharm.*) æthiops antimonialis, sulfides of antimony and

mercury. **-ocker** m antimony ocher.
-oxid n antimony trioxide. **-säure** f antimonic acid. **-schwefel** m antimony sulfide. **-silber** n (Min.) dyscrasite. **-wein** m antimonial wine. **-weinstein** m (Pharm.) tartared antimony. **-weiß** n antimony white, antimony trioxide (pigment). **-zinnober** m (Min.) kermesite

Spießglas n antimony (for compds see also SPIESSGLANZ-). **-erz** n (Min.) stibnite. **-weiß** n antimony white = SPIESSGLANZWEISS

spießig a 1 (Cryst.) spear-like, lanceolate, in long needles. 2 (Lthr.) bodily tanned. 3 (Metll.) brittle. 4 narrow-minded

Spieß-kant m diamond. **-kantkaliber** m (Metll.) diamond pass. **-kobalt** m smaltite. **-stäbchen** n (Bact.) spearshaped rod bacterium

Spik·blüten fpl spike lavender flowers. **-öl** n spike (or lavender) oil

Spill m (-e) winch, capstan

Spin·abhängigkeit f dependence on spin

Spinasterin n spinasterol

Spinat m (-e) spinach

Spin·bahn f spin orbit

Spind m or n (-e) locker

Spindel f (-n) 1 (usu) spindle; (Tex.) bobbin. 2 (Mach.) axle, pivot, arbor; pin; feed screw, worm. 3 (Bot.) stalk. 4 hydrometer. 5 (Pharm.) euonymus. **-baum** m spindle tree, euonymus. **-baumöl** n spindlewood oil. **sp–dürr** a skinny, spindly. **Sp–faser** f spindle fiber. **sp–förmig** a spindle-shaped, fusiform

spindeln I vt 1 test with a hydrometer. 2 wind on a spindle. II vi (Bot.) spindle, grow stalks

Spindel·öl n spindle oil. **-presse** f screw press. **-ventil** n spindle valve. **-waage** f hydrometer. **-zelle** f fusiform cell

Spin·drehimpuls m spin angular momentum

Spinell m (-e) spinel

Spin·glied n (Math.) spin term. **-momentdichte** f spin momentum density

Spinn·abfall m spinning waste

spinnbar a spinnable

Spinn·drüse f web gland (of spiders); silk gland (of silk caterpillars). **-düse** f (Tex.) spinneret

Spinne f (-n) spider

spinnen* (spann, gesponnen) I vt/i 1 spin. II vt 2 (Tobacco) twist. 3 contrive (a scheme). III vi 4 purr. 5 talk drivel

Spinnen·gewebe n spider web, cobweb. **-tier** n arachnid

Spinnerei f (-en) 1 spinning. 2 spinning (or textile) mill. 3 nonsense, drivel

spinn·fähig a suitable for spinning. **Sp–faser** f textile (spinning) fiber, staple rayon. **-fett** n (Tex.) mill oil. **-flüssigkeit** f spinning solution (or dope). **-gewebe** n spider web, cobweb. **-kerze** f filter candle. **-kuchen** m (Rayon) cake. **-lösung, -masse** f (esp Rayon) spinning solution. **sp–mattiert** a delustered (during spinning). **Sp–milbe** f spider mite, red spider. **-öl** n spinning oil. **-reife** f ripeness for spinning, ammonium chloride index. **-schmalz** n spinning lubricant. **-stoff** m spinning material, textile fiber. **-webe** n spider web, cobweb. **sp–webenartig** a cobweb-like; arachnoid

Spinnweb·faden m spider thread; very fine thread. **-wolle** f spinning wool

spinös a spiny

Spin·richtung f direction of spin, spin orientation

Spiräa, Spiräe f (.. räen) spirea. **Spiräa·öl** n spirea oil

Spiral·bohrer m (Mach.) twist drill

Spirale f (-n) 1 spiral, helix, coil. 2 spiral condenser

Spiral·feder f spiral (or coil) spring. **sp–förmig** a spiral, helical

spiralig a spiral

Spiral·kühler m spiral (or coil) condenser (or cooler). **-linie** f spiral (line). **-nebel** m spiral nebula. **-rohr** n, **-röhre** f spiral tube (or pipe), worm. **-stellung** f spiral arrangement

Spirille f (-n) (Bact.) spirillum. **Spirillose** f (-n) spirillosis

Spirituosen npl (alcoholic) spirits, liquors

Spiritus m (-se) (ethyl or grain) alcohol, distilled spirits. **sp–artig** a spirituous, alcoholic. **Sp–beize** f spirit mordant (or strain). **-blau** n spirit blue. **-brenner** m 1 distiller. 2 alcohol burner. **-brennerei** f distillery. **-dampf** m alcohol vapor. **-fabrik** f distillery. **-faß** n liquor barrel.

-gehalt *m* alcohol(ic) content. **-geruch** *m* alcoholic odor. **-industrie** *f* distilling industry. **-lack** *m* spirit varnish. **-lampe** *f* spirit (*or* alcohol) lamp. **sp–löslich** *a* alcohol-soluble. **Sp–messer** *m* alcoholometer. **-mischung** *f* alcoholic mixture. **-pumpe** *f* alcohol pump. **-waage** *f* 1 alcoholometer. 2 spirit level

Spirochäte *f* (-n) spirochete. **Spirochätose** *f* spirochetosis

spiroylig *a* spiroylous: **sp—e Säure** *f* spiroylous acid (*former name for* salicylaldehyde). **Spiroyl·säure** *f* *former name of* salicylic acid

Spital *n* (¨er) hospital, infirmary

spitz *a* 1 pointy; pointed: **sp. zulaufen** taper, come to a point. 2 acute (angle). 3 (*Cryst.*) acicular. 4 snide. 5 haggard

Spitz *m* (-e) (*Metll.*) paddle *cf* SPITZE. **-becher** *m*, **-becherglas** *n* sedimentary glass. **-beutel** *m* triangular filter bag

Spitze *f* (-n) 1 point; tip; cusp: **einem Problem die Sp. nehmen** (*or* **abbrechen**) take the sharp point off a problem; **die Sp. bieten** (+*dat*) resist; **die Sp. umdrehen** turn the tables. 2 peak, top; apex, vertex; crest **—auf die Sp. treiben** drive to extremes. 3 peak performance, tops. 4 maximum. 5 leading position; leader: **an der Sp. des Instituts stehen** be the head of the institute. 6 (*Mach.*) center (of a machine tool). 7 jibe. 8 lace

spitzen I *vt* 1 point, sharpen. 2 prick up (ears). **II** *vi* peek out, sprout. **III** *vi* & *sich* **sp.** (*usu* + **auf**) be on the lookout (for), look forward (to)

Spitzen· point, peak, top; lace: **-belastung** *f* peak load. **-durchmesser** *m* overall diameter. **-elektrode** *f* point (*or* needle) electrode. **-entladung** *f* (*Elec.*) point discharge. **-glanz** *m* top gloss. **-glas** *n* reticulated glass. **-kontakt** *m* point contact. **-leistung** *f* peak (*or* maximum) output, peak (*or* record) performance. **-papier** *n* lace paper. **-reiter** *m* leader, number one; peak performer. **-strom** *m* (*Elec.*) peak current. **-wert** *m* peak (top, maximum) value. **-zähler** *m* needle counter. **-zeit** *f* peak load hour(s); rush hour(s). **-zirkel** *m* compass, dividers

spitz·findig *a* 1 shrewd, crafty. 2 subtle; so-

phistic, hairsplitting. **Spitzfindigkeit** *f* (-en) 1 shrewdness, craftiness. 2 subtlety, sophistry, hairsplitting (distinction)

Spitz·geschoß *n* pointed bullet, conical shell. **-glas** *n* tapering (*or* conical) glass; sedimentation glass. **-glasmalm** *m* (*Min.*) stibnite. **-hacke** *f* pickax, pick. **-haufen** *m* (*Brew.*) couch

spitzig *a* 1 pointy; acute, etc = SPITZ. 2 (*Colors*) mixtury, not uniform

spitz·kantig *a* acute-angled. **Sp–keimer** *m* (*Bot.*) monocotyledon. **-kolben** *m* 1 taper-necked flask. 2 pointed soldering iron. **-lutte** *f* cone separator. **-malz** *n* chit malt. **-maschine** *f* sharpener. **-maus** *f* (*Zool.*) shrew. **-name** *m* nickname. **-pocken** *fpl* chicken pox. **-röhrchen** *n* centrifuge tube. **-trichter** *m* separatory funnel, Squibb funnel. **sp–winklig** *a* acute-angled. **-zulaufend** *a* tapering, tapered

Spl. *abbv* = Supplement (to a book)

Spleen *m* (**auf**) obsession (with), hankering (for)

Spleiß *m* (-e) 1 splice. 2 = **Spleiße** *f* (-n) splinter, chip. **spleißen*** (spliß, gesplissen) *vt* 1 (*also reg*) splice. 2 split, splinter, chip

Splint *m* (-e) 1 sapwood, alburnum. 2 (*Mach.*) cotter pin. **sp–·frei** *a* sapless. **Sp–holz** *n* sapwood. **-holzkäfer** *m* Lyctus borer. **-kohle** *f* splint coal

spliß spliced, etc: *past of* SPLEISSEN

Spliß *m* (Splisse) 1 splice. 2 splinter, chip

Splitter *m* (-) splinter, chip, fragment. **sp–·bindend** *a* shatterproof. **Sp–bombe** *f* fragmentation bomb. **sp–frei** *a* 1 non-splintering. 2 shatterproof **—sp—es Glas** safety glass. 3 (*Paper*) shive-free. **Sp–granate** *f* fragmentation shell

splitterig *a* 1 splintery; easily splintered; brittle, fragile. 2 (*Paper*) shivy

Splitter·kohle *f* splint coal

splittern *vi* (h,s) splinter, split (up)

Splitter·schutzbrille *f* protective goggles. **sp–sicher** *a* splinterproof *cf* SPLITTERFREI

Splitterung *f* splintering, splitting

Splitter·wirkung *f* splinter (*or* fragmentation) effect

spönne would spin: *sbjc of* SPINNEN

spontan *a* spontaneous. **Spontaneität** *f*

spontaneity, spontaneousness

Spontan·spaltung *f* (*Nucl.*) spontaneous fission. **-verdampfung** *f* flash distillation

Spor *m* (-e) 1 mold, mildew. 2 spur, etc = Sporn

Spore 1 *f* (-n) (*Biol.*) spore. 2 *pl of* Spor

sporen *vi* dry up, become moldy

Sporen 1 spores: *pl of* Spore. 2 spurs: *pl of* Sporn

Sporen·behälter *m* (*Bot.*) sporangium. **sp–bildend** *a* sporogenic. **Sp–bildung** *f* sporulation, sporogenesis. **-färbung** *f* spore-stain(ing). **sp–haltig** *a* spore-bearing. **Sp–kapsel** *f* sporogonium. **-pflanze** *f* sporophyte. **-stäbchen** *n* (*Bact.*) endospore-producing rod. **-tierchen** *n* sporazoon. **sp–tötend** *a* sporicidal. **Sp–träger** *m* spore case, sporophore

Spor·fleck *m* mold stain (*or* spot)

sporig *a* moldy, mildewed

Sporn *m* (Sporen) spur (*also fig*), spike, spine

Sport·medizin *f* sports medicine

sportlich *a* sport(s); athletic; sporty

Spott *m* derision, ridicule, mockery. **spotten** *vi* (**über**) make fun (of), ridicule

sprach spoke, etc: *past of* Sprechen

Sprache *f* (-n) 1 language. 2 speech, talk. 3 discussion: **zur Sp. kommen** come up for discussion; **zur Sp. bringen** bring up (for discussion); **die Sp. auf etwas bringen** bring the discussion round to a topic

Sprach·eigenheit *f* idiom. **-forschung** *f* philology, linguistic research. **-gebrauch** *m* linguistic usage. **-lehre** *f* grammar; grammar textbook

sprachlich *a* linguistic, (as regards) language

sprachlos *a* speechless

Sprach·rohr *n* 1 megaphone. 2 mouthpiece. **-schatz** *m* vocabulary; thesaurus. **sp–widrig** *a* ungrammatical. **Sp–wissenschaft** *f* linguistics

sprang jumped, etc: *past of* Springen

spratze(l)n *vi* sp(l)utter; (*Distil.*) bump

Spratz·kupfer *m* copper rain

Sprech·anlage *f* intercom

sprechen* (sprach, gesprochen; spricht) I *vt/i* 1 speak, talk —**alles spricht dafür,**

daß everything favors the fact that. II *vt* 2 speak to, see: **ist der Chef zu sp.?** is the boss available? 3 say. 4 pronounce. **sprechend** *p.a* (*oft*) revealing, telling — **sp. ähnlich sein** (*+dat*) be the very image (of). **Sprecher** *m* (-) speaker, spokesman

Sprech·stunde *f* office hour. **-tag** *m* business day (*esp* of an office). **-zimmer** *n* office, consultation room

spreiten *vt* spread (out). **Spreitungs·koeffizient** *m* spreading coefficient

Spreize *f* (-n) strut, stay; sprag; spreader. **spreizen** 1 *vt* spread (out, apart), splay. 2: **sich sp.** resist, play coy; strut

Spreiz·schwingung *f* (*Spect.*) bending vibration. **-silikat** *n* (*Cement*) expansive silicate

Spreizung *f* (-en) spreading, splaying

Spreng· sprinkling; blasting, explosive: **-apparat** *m* sprinkler, sparger. **-arbeit** *f* blasting. **-bohrloch** *n* blast hole. **-bombe** *f* high-explosive bomb. **-brandbombe** *f* incendiary-demolition bomb. **-eisen** *n* (*Glass*) cracking ring

sprengen I *vt* 1: (**in die Luft**) **sp.** blow up, blast. 2 break (open), break down; rupture. 3 break up, raid (a meeting), disperse (a crowd). 4 sprinkle, water; spray. II *vi*(s) gallop, dash. **Sprenger** *m* (-) 1 sprinkler. 2 blaster

spreng·fähig *a* explosive

Spreng· explosive, blasting; sprinkling: **-falle** *f* booby trap. **-flüssigkeit** *f* explosive liquid. **-füllung** *f* explosive charge. **-gelatine** *f* explosive gelatine. **-geschoß** *n* explosive shell (*or* projectile). **-granate** *f* high-explosive grenade (*or* shell). **-kammer** *f* demoliton chamber. **-kapsel** *f* detonator, blasting cap; primer, booster charge. **-kohle** *f* (*Glass*) cracking coal. **-kopf** *m* warhead. **-kraft** *f* explosive force (*or* power). **sp–kräftig** *a* high-explosive. **Sp–ladung** *f* explosive charge. **-loch** *n* blast (*or* fire) hole. **-luft** *f* liquid-air explosive. **-masse** *f* explosive charge. **-mittel** *n* explosive. **-niet** *n,* **-niete** *f* explosive rivet. **-öl** *n* polyalcohol nitrate, (*specif*) nitroglycerine. **-ölpulver** *n* nitroglycerine powder. **-patrone** *f* explosive cartridge (*or* shell). **-pulver** *n* blasting powder.

-punkt *m* bursting point. **-regen** *m* fine rain, drizzle. **-ring** *m* snap ring. **-salpeter** *m* nitrate explosive. **-satz** *m* bursting (blasting, ignition) charge. **-schlag** *m* explosion. **-schnur** *f* fuse. **-schuß** *m* shot, blast. **-stoff** *m* explosive

Sprengstoff·füllung See SPRENGSTOFFÜLLUNG. **-kammer** *f* explosive magazine. **-ladung** *f* = **Sprengstofffüllung** *f* explosive charge. **Sprengstoff·wesen** *n* (field of) explosives

Spreng·stück *n* shell fragment. **-technik** *f* explosives technology (*or* manufacture). **-trichter** *m* 1 explosion crater. 2 spray head, sprinkler (rose)

Sprengung *f* (-en) blasting, explosion; breaking, rupture; dispersion; sprinkling *cf* SPRENGEN

Spreng·wagen *m* (street) sprinkler (truck). **-wirkung** *f* explosive effect (*or* action). **-wolke** *f* blast cloud, mushroom cloud. **-zünder** *m* detonating fuse

Sprenkel *m* (-) speck, spot. **sprenkeln** *vt* speckle, mottle; sprinkle. **sprenklig** *a* speckled

Spreu *f* chaff; leftovers, scraps. **--stein** *m* (*Min.*) natrolite

spricht speaks, etc: *pres of* SPRECHEN

Sprich·wort *n* proverb, saying. **sp--wörtlich** *a* proverbial

Spriegel *m* (-) 1 (butcher's) gambrel, meat hook. 2 hoop, stake (for covered wagon)

Sprieße *f* (-n) crossbar, brace; rung (of a ladder)

sprießen* (sproß, gesprossen) *vi*(s) sprout, bud; germinate

Spring·brunnen *m* fountain

springen* (sprang, gesprungen) *vi*(s) 1 jump, leap. 2 bounce. 3 run, dash. 4 pop (out, off), blow (up) **—alle Minen sp. lassen** (*fig*) pull out all the stops. 5 (*Liquids, esp Water*) well up, spring forth; spout, gush. 6 crack, break (open); snap. **springend** *p.a* (*oft*) salient (point)

Spring·feder *f* (elastic) spring. **-feder-waage** *f* spring balance. **-flut** *f* (spring) flood; spring tide. **-gurke** *f* squirting cucumber. **-gurkenextrakt** *m* elaterium. **sp--hart** *a* brittle. **Sp--kolben** *m* Bologna flask. **-kraft** *f* elasticity, springiness;

(power of) recoil. **sp--kräftig** *a* springy, elastic. **Sp--kraut** *f* (*Bot.*) touch-me-not. **-maus** *f* (*Zool.*) jerboa. **-mine** *f* bounding (*or* antipersonnel) mine. **-quelle** *f* geyser; fountain, wellspring. **-stift** *m* (*Fuel Testing*) bouncing pin

Sprit *m* 1 (ethyl) alcohol, spirits. 2 gas(oline). 3 lighter fluid. **--beize** *f* spirit stain. **-blau** *n* spirit (*or* aniline) blue. **-drucken** *n* (*Tex.*) spirit printing. **sp--echt** *a* alcohol-resistant, alcohol-insoluble. **Sp--essig** *m* spirit (*or* triple) vinegar. **-farbe** *f* (*Tex.*) spirit color. **-gelb** *n* spirit yellow. **sp--haltig** *a* spirituous, alcoholic; (*Wine*) fortified. **Sp--lack** *m* spirit varnish (*or* lacquer). **sp--lackecht** *a* spirit-lacquer-resistant. **-löslich** *a* spirit- (*or* alcohol-)soluble. **-unlöslich** *a* spirit- (*or* alcohol-)insoluble

Spritz- spray: **-abfall** *m* overspray. **-alitieren** *n* spray-aluminizing (*or* -calorizing). **-anlage** *f* spraying unit. **-apparat** *m* sprayer. **-beton** *m* sprayed concrete. **-bewurf** *m* roughcast, rough plastering. **-brenner** *m* spray burner. **-dose** *f* spray can (*or* container). **-druck** *m* (*Tex.*) spray printing. **-düse** *f* spray (*or* injection) nozzle; steam injector

Spritze *f* (-n) 1 syringe; injector. 2 sprayer; spray gun; grease gun; irrigator. 3 nozzle, spray head. 4 injection, shot; (fig) boost. 5 fire engine

Spritz·email *n* spray enamel

spritzen I *vt/i* 1 spray; squirt, spurt; spatter, splash. II *vt* 2 inject. 3 (*Plastics*) extrude, injection-mold. 4 squirt seltzer water into (wine, etc). III *vi*(h) drizzle. IV *vi*(s) run, dash

Spritzer *m* (-) 1 squirt, splash; dash (of liquid). 2 spot, stain. 3 spray painter

spritz·fähig *a* sprayable, injectable

Spritz- spray: **-farbe** *f* spray paint (*or* color). **-färbung** *f* spray dyeing (*or* coloring). **-flasche** *f* wash bottle. **-fleck** *m* spray (*or* spatter) stain. **-flüssigkeit** *f* spray liquid. **-form** *f* injection die (*or* mold). **-gummi** *n* sealing compound (for tin cans). **-gurke** *f* squirting cucumber. **-guß** *m* die casting; injection molding

Spritzguß·masse *f* injection-molding composition. **-verfahren** *n* extrusion (*or* injection-molding) process

Spritz·kanne *f* watering can. **-kopf** *m* spray head (*or* nozzle). **-kork** *m* sprinkler stopper. **-kranz** *m* (*Brew.*) sparger. **-lack** *m* spray varnish (*or* lacquer)

spritzlackieren *vt* spray-varnish

Spritzling *m* (-e) injection-molded article

Spritz·maschine *f* **1** sprayer, spray gun. **2** extruder. **-masse** *f* spraying (extrusion, injection-molding) composition. **-mittel** *n* injected remedy, spray remedy. **-nudeln** *fpl* vermicelli. **-pistole** *f* spray gun. **-pressen, preßverfahren** *n* (*Plastics*) transfer molding. **-probe** *f* (*Dyes*) blow test. **-pulver** *n* injection-molding powder; dispersible powder. **-regen** *m* drizzle. **-rohr** *n*, **-röhre** *f* **1** syringe. **2** spray pipe. **3** wash-bottle tube. **-spachtel** *f* primer surfacer, spray filler. **-verfahren** *n* spray process. **-vergaser** *m* spray carbureter. **-vorrichtung** *f* spraying device, sprayer. **-wasser** *n* spray

sprock *a* brittle, friable

Spröd·bruch *m* brittle fracture

spröde *a* **1** brittle, (*Metals, oft*) short. **2** friable. **3** rough; dry; cracked, chapped. **4** plain. **5** reserved, standoffish

Spröd·glanzerz, -glaserz *n* brittle silver ore, (*Min.*) stephanite, polybasite. **-glimmer** *m* brittle mica

Sprödheit, Sprödigkeit *f* brittleness, shortness; friability; roughness, dryness; reserve, standoffishness *cf* SPRÖDE

Spröd·metall *n* brittle metal; metalloid

sproß sprouted, etc: *past of* SPRIESSEN

Sproß *m* (. . osse) scion, descendant; sprig *cf* SPROSSE. **--achse** *f* (*Bot.*) stem

Sprosse *f* (-n) **1** scion, sprig. **2** descendant. **3** rung; spoke, crossbar. **4** (window) sash. **5** (**Brüsseler**) **Sp—n** Brussels sprouts. *Cf* SPROSS

sprossen *vi* sprout, bud, germinate = SPRIESSEN

Sprossen·bier *n* spruce beer. **-extrakt** *m* essence of spruce. **-fichte** *f* spruce fir, (true) spruce. **-kohl** *m* Brussels sprouts; broccoli. **-tanne** *f* hemlock spruce

Sproß·keimung *f* sprouting, budding

Sprößling *m* (-e) scion, descendant

Sproß·pilz *m* gemmiparous (budding, yeast) fungus

Sprossung *f* sprouting, budding, germination

Sprotte *f* (-n) (*Fish*) sprat. **Sprotten·tran** *m* sprat oil

Spruch *m* (-̈e) **1** saying, aphorism. **2** slogan; motto. **3** (*Radio*) message, signal. **4** (*Law*) sentence. **5** =**Sprüchlein** *n*: **sein(en) Sp. aufsagen** speak one's piece

Sprudel *m* (-) **1** (sparkling) mineral water. **2** bubbling mineral spring. **--bad** *n* effervescent (*or* bubble) bath. **-bohrung** *f* (*Petrol.*) gusher

sprudeln *vi* bubble (up); spout, gush. **sprudeind** *p.a* (*oft*) bubbly, effervescent

Sprudel·platte *f* bubble tray. **-salz** *n* mineral salt, Karlsbad salt. **-stein** *m* hot-spring deposit

Sprüh· spray: **-apparat** *m* spraying (*or* sprinkling) apparatus. **-dose** *f* spray can (*or* container). **-druck** *m* spray pressure. **-düse** *f* spray nozzle, atomizer. **-elektrode** *f* ionizing electrode

sprühen I *vt/i* spray. **II** *vt* **2** vaporize, atomize. **3** spout (flame, sparks, words). **III** *vi* (h,s) **4** (*oft* + **von, vor**) bubble (up, over) (with), overflow (with). **5** flow, gush; fly (*as:* sparks). **6** scintillate, sparkle, glitter. **7** sprinkle, drizzle. **sprühend** *p.a* (*oft*) bubbly, effervescent

Sprüh·entladung *f* spray discharge. **sp—fest** *a* sprayproof. **Sp—flüssigkeit** *f* spray (liquid). **-gerät** *n* sprayer, atomizer. **sp—getrocknet** *a* spray-dried. **Sp—ion** *n* spray ion. **-kautschuk** *m/n* sprayed (*or* spray-dried) rubber. **-kupfer** *n* copper rain. **-nebel** *m* (atomized) mist. **-regen** *m* drizzle. **-trockner** *m* spray drier. **-trocknung** *f* spray drying. **-turm** *m* spray tower

Sprung *m* (-̈e) **1** crack, fissure; (*Geol.*) fault. **2** (*in curves*) break, discontinuity; (*Spect.*) transition. **3** leap, jump, hurdle (*incl fig*); dive; (*Waves*) hop; (*Biol.*) mutation. **4** change of topic. **5: auf dem Sp. (a)** on the go, **(b)** in a hurry, **(c)** on one's way; **auf dem Sp. stehen, zu. . .** be about (to). **6** short hop, stone's throw. **7: auf einen Sp.** (e.g. **vorbeikommen**) (e.g. come by) for a

moment. **8** venture. **9** trick, scheme.
-bildung *f* crack formation, fissuring.
-brett *n* springboard. **-feder** *f* (elastic)
spring

sprunghaft *a* **1** sudden, abrupt. **2** erratic,
volatile. **3** (*adv oft*) by leaps and bounds

Sprung·höhe *f* (*Geol.*) throw (of a fault).
-punkt *m* transition point. **-spektrum** *n*
transition spectrum. **-temperatur** *f* transition temperature. **-variation** *f* (*Biol.*)
mutation. **-wahrscheinlichkeit** *f* transition probability. **sp–weise** *adv* intermittently; abruptly, by leaps and jumps. **Sp–
weite** *f* **1** range; length of a jump. **2** (*Geol.*)
shift (of a fault). **-welle** *f* surge. **-wel-
lenprobe** *f* surge pressure test. **-zeit** *f*
(*Electrons; Spect.*) transition time

Spucke *f* saliva, spittle. **spucken 1** *vt/i*
spit, spew; throw up; (*Metll.*) slop. **2** *vi*
sputter

Spül·bad *n* **1** rinsing bath; rinse (of a washing machine). **2** (*Phot.*) acid bath.
-becken *n* **1** sink. **2** toilet bowl. **-bottich**
m rinsing tub

Spule *f* (-n) **1** spool, reel, bobbin. **2** (*Elec.*)
coil, solenoid. **3** quill

Spüle *f* (-n) **1** sink unit, washtubs. **2** toilet
flush (tank)

spulen 1 *vt/i* wind, reel, spool. **2: sich sp.**
wind

spülen I *vt/i* **1** rinse (out). **2** wash (the
dishes, etc). **3** (*Med.*) irrigate. **4** (*Mach.*)
scavenge. **5** flush (the toilet). **II** *vt* wash,
sweep (e.g. onto the shore)

Spulen·antrieb *m* solenoid control. **-gal-
vanometer** *n* moving-coil galvanometer.
-lack *m* **1** (*Elec.*) insulating varnish. **2**
(*Tex.*) bobbin finish. **-öl** *n* spooling oil.
-wechsel *m* change of reel (*or* tape).
-wirkung *f* (*Elec.*) coil winding. **-wider-
stand** *m* (*Elec.*) coil resistance

Spüler *m* (-) **1** rinser, washer. **2** dishwasher.
3 flush (lever, handle, button). **Spülerei** *f*
(-en) **1** rinsing, washing. **2** rinsing plant
(*or* unit)

Spül·flüssigkeit *f* rinse (liquid). **-gang** *m*
rinse cycle (of a washing machine). **-gas** *n*
rinsing (purge, scavenging) gas. **-gasver-**

fahren *n* (*Coke*) carbonization with recir-
culation. **-gefäß** *n* rinse vessel. **-gut** *n* **1**
flushed (rinsed, washed) material. **2**
flushing material

Spülicht *n* (*Distil.*) spent wash

Spül·kasten *m* flush box, flushometer.
-klosett *n* flush toilet. **-kufe** *f* rinsing
vat. **-leitung** *f* flush pipe; scavenging
duct. **-luft** *f* scavenging air. **-maschine** *f*
dishwasher. **-mittel** *n* **1** detergent. **2**
mouthwash. **3** (chemical, cosmetic) rinse.
4 fabric softener. **5** irrigating fluid. **6** disinfectant. **-mittelrest** *m* detergent (rinse,
disinfectant) residue

Spul-öl *n* (*Tex.*) spooling oil

Spül-öl *n* flushing oil. **-schwelung** *f* low-
temperature carbonization with gas re-
circulation. **-stein** *m* sink. **-topf** *m* rinsing pot (*or* jar). **-trog** *m* rinsing trough

Spülung *f* (-en) **1** rinsing, washing; irrigation; scavenging; flushing *cf* SPÜLEN. **2**
(toilet) flush. **3** drilling mud

Spül·wasser *n* wash (*or* rinse) water

Spul·wurm *m* roundworm

Spund *m* (-e *or* ⁻e) **1** bung; plug, stopper. :
Sp. und Nute tongue & groove. **spunden**
vt **1** bung (kegs). **2** barrel, cask (wines,
etc). **3** tongue-and-groove (boards)

Spund·gärung *f* bunghole fermentation.
-loch *n* bunghole. **sp–voll** *a* brim-full

Spur *f* (-en) **1** track(s); footprint; rut. **2** trail;
scent: **auf die Sp. kommen** (+ *dat*) pick
up the trail (*or* scent) (of). **3** trace, vestige,
remnant. **4** clue. **5** bit, touch —**keine Sp.**
not in the least. **6** (highway) lane; (sound
tape) track. **7** (*esp Metll.*) channel, groove.
8 (railroad track) gauge. **--arbeit** *f* (*Metll.*)
concentration

spürbar *a* **1** palpable; noticeable; clearly
felt. **2** detectable, traceable. **3** considerable

Spur-element *n* trace element

spuren 1 *vt* (*Metll.*) concentrate. **2** *vi* hold
the road

spüren I *vt* **1** feel: **etwas zu sp. bekommen** get to feel (*or* experience) sthg, get a
taste of sthg. **2** sense; get a hint of: **einen
etwas sp. lassen** give someone a hint (*or*
a clue) of sthg. **3** catch the scent of (prey);
detect, notice. **II** *vi* (+**nach**) track, trail

Spuren·analyse *f* trace analysis. **-element** *n* trace element

Spuren·menge *f* trace amount. **-metall** *n* trace metal. **-stoff** *m* trace constituent. **-suche** *f* search for traces. **-verunreinigung** *f* trace impurity. **sp–weise** *adv* in (minute) traces

Spür·gasverfahren *n* tracer gas technique

Spur·geschoß *n* tracer projectile

spurlos *a* traceless, trackless; (*adv usu*) without a trace

Spür·mittel *n* (poison-)gas-detecting agent. **-papier** *n* gas-detecting paper. **-pulver** *n* gas-detecting powder

Spur·schlacke *f* (*Metll.*) concentration slag. **-stein** *m* concentration metal (*or* matte) (copper sulfide). **-weite** *f* track gauge

sputen *vr:* **sich sp.** hurry

sp.v(ol). *abbv* (spezifisches Volumen) spec. vol.

sp.W. *abbv* (spezifische Wärme) spec. heat

Squamat·säure *f* squamatic acid

SR *abbv* (Skala Réaumur) Réaumur scale

s.S. *abbv* (siehe Seite) see page

S.S. *abbv* (Schwefelwasserstoffsäure) hydrosulfuric acid, H_2S

SS. *abbv* **1** (Säuren) acids. **2** (Seiten) pp. (pages)

st. *abbv* (stark) strong

St. *abbv* **1** (Sankt) St. (Saint). **2** (Stahl) steel. **3** (Stamm) stem. **4** (Stärke) strength, etc; starch. **5** (Stück) (a)piece. **6** (Stunde) hr. (hour)

Staat *m* (-en) **I** state, country; government. **II** finery; display —**St. machen** show off

staatlich *a* state, national, government(al); public; state-owned; state-operated

Staats· state, national, government(al), public: **-amt** *n* public office. **-angehörigkeit** *f* nationality, citizenship. **-bürger** *m* citizen. **st–eigen** *a* state-owned. **St–kunde**, **-lehre** *f* political science. **-monopol** *n* state (national, government) monopoly. **-politik** *f* national policy. **-sicherheit** *f* national security. **-verfassung** *f* state (*or* national) constitution. **-wirtschaft** *f* national economy

Stab *m* (¨e) **1** rod, baton, bar, staff. **2** staff, personnel. **--bakterien** *fpl* (rod-shaped)

bacilli. **-brandbombe** *f* stick-type incendiary bomb

Stäbchen *n* (-) **1** stick, rodlet, peg; skewer; *pl oft* chopsticks. **2** (*Anat.*) (retinal) rod. **3** (*Bact.*) (rod-shaped) bacillus. **st--artig**, **-förmig** *a* rod-like, rod-shaped

Stab·eisen *n* bar iron. **st–förmig** *a* rod-shaped

stabil *a* stable; steady; durable; robust

Stabilisator *m* (-en) stabilizer. **stabilisieren** *vt* stabilize. **Stabilisierung** *f* stabilization. **Stabilisierungs·mittel** *n* stabilizing agent

Stabilität *f* (-en) stability; robustness

Stabilitäts·prüfer *m* stability tester. **-prüfung** *f* stability test

Stab·kranz *m* (*Biol.*) corona radiata. **-kraut** *n* southernwood. **-magnet** *m* bar magnet

Stabs· staff: **-arzt** *m* staff doctor; (*Mil.*) captain of the medical corps. **-chef** *m* chief of staff. **-quartier** *n* headquarters

Stab·thermometer *n* stem-graduated thermometer

stach pricked, etc: *past of* STECHEN

Stachel *m* (-n) **1** thorn, prickle; spine, spike; quill. **2** barb (*also fig*); sharp point. **3** spur, goad. **4** (insect's) sting. **5** tongue, prong (of a buckle). **--beere** *f* gooseberry. **-draht** *m* barbed wire. **-häuter** *m* echinoderm

stachelig *a* prickly, spiny, etc = STACHLIG

stachel·los *a* without thorns (spines, prickles)

Stachel·mohn *m* prickly poppy

stacheln *vt* **1** spur, sharpen (desire, etc); incite. **2** bother, disturb. **II** *vi* be bristly (rough, stratchy)

Stachel·schnecke *f* (*Zool.*) murex (snail). **-schwein** *n* porcupine. **-walzwerk** *n* (*Mach.*) toothed rollers

stachlig *a* **1** prickly, thorny; spiny, bristly. **2** barbed; pointed (remarks, etc). **3** sharp-tongued. **4** rough, scratchy

Stadel *m* (-) **1** shed; barn. **2** (*Metll.*) open kiln, stall. **--röstung** *f* stall roasting

Stadien *pl of* STADION & STADIUM

Stadion *n* (. . ien) stadium

Stadium *n* (. . ien) stage, phase

Stadt *f* (¨e) city, town; downtown (area); (*in*

compds, usu) municipal, urban. **--gas** *n* gas for urban use (coal gas mixed with carburetted water gas). **-haus** *n* 1 town house, house in the city. 2 city (*or* town) hall

städtisch *a* city, municipal, urban

Staffage *f* accessories; façade, window dressing

Staffel *f* (-n) 1 step, rung (*usu* in a gradation); level, stage. 2 echelon. 3 squad(ron); relay team

Staffelei *f* (-en) easel

Staffel-gitter *n* echelon grating

staffeln *vt* grade, scale; stagger

Staffel-summe *f* cumulative total

staffieren *vt* 1 equip; trim. 2 prepare (a dye bath)

stagnieren *vi* stagnate. **stagnierend** *p.a* (*oft*) stagnant

stahl stole: *past of* STEHLEN

Stahl *m* (-e *or* -̈e) 1 steel. 2 cutting tool; scriber. **--abfall** *m* scrap steel. **-aluminium-seil** *n* steel-reinforced aluminum cable. **st-ähnlich** *a* steel-like. **St-arbeit** *f* steelwork. **st-artig** *a* steel-like, steely; chalybeate. **St-arznei** *f* medicine containing iron, chalybeate. **-band** *n* steel band; steel tape. **-bereitung** *f* steelmaking. **-beton** *m* reinforced concrete. **-betonbau** *m* steel-concrete structure (*or* construction). **st-blau** *a* steel-blue. **St-blech** *n* steel plate; sheet steel. **-block** *m* (*Metll.*) steel ingot. **-bombe** *f* 1 steel bomb (*or* shell). 2 steel cylinder (for gas). **-bronze** *f* aluminum bronze. **-brunnen** *m* chalybeate (*or* iron-bearing) spring. **-draht** *m* steel wire. **-eisen** *n* open-hearth pig iron, steel pig

stählen *vt* 1 convert (iron) into steel. 2 steel-face; fortify. 3 harden, temper. 4 strengthen, toughen

stählern *a* (of) steel, steely

Stahl-erz *n* steel ore, (*Min.*) siderite. **-erzeugung** *f* steel production. **st-farbig** *a* steel-colored. **St-feder** *f* steel spring. **-flasche** *f* steel bottle (*or* cylinder). **-formguß** *m* steel casting; steel mold. **-frischfeuer** *n* steel finery. **-gattung** *f* grade of steel. **-gefäß** *n* steel vessel (*or* receptacle). **-gewinnung** *f* steelmaking. **-gießerei** *f* steel founding; steel

foundry. **st-grau** *a* steel-gray. **St-guß** *m* 1 steel casting. 2 cast steel. 3 toughened cast iron. 4 *in compds* **Stahlguß-** cast-steel, steel-cast. **-güte** *f* grade of steel. **-hahn** *m* steel valve. **st-hart** *a* hard as steel. **St-härtung** *f* tempering of steel. **-hütte** *f* steel mill. **-klinge** *f* steel blade. **-knüppel** *m* steel bar (*or* billet). **-kobalt** *m* smaltite. **-kohlen** *n* conversion of wrought iron steel by carbonization. **-kraut** *n* vervain. **-kugel** *f* steel ball. **-legierung** *f* steel alloy. **-mittel** *n* chalybeate, medicine (*or* mineral water) containing dissolved iron. **-mörser** *m* steel mortar. **-ofen** *n* steel furnace. **-panzer** *m* steel armor. **-perle** *f* steel bead. **-platte** *f* steel plate. **-präparat** *n* (*Pharm.*) iron preparation. **-probe** *f* 1 steel sample. 2 test of steel. **-puddeln** *n* steel puddling. **-quelle** *f* chalybeate (*or* iron-bearing) spring. **-roheisen** *n* open-hearth pig iron. **-rohr** *n* steel pipe; tubular steel. **-sand** *m* steel grit (*or* sand). **-säuerling** *m* acidulous iron water. **-schmelzen** *n* steel melting. **-schmelzofen** *m* steel-melting furnace. **-schrott** *m* steel scrap. **-späne** *mpl* steel turnings (*or* chips). **-stange** *f* steel rod (*or* bar). **-stechen** *n*, **-stecherei** *f* steel engraving. **-stein** *m* (*Min.*) siderite. **-stich** *m* steel-plate engraving. **-trommel** *f* steel drum

Stählung *f* (-en) conversion into steel; steel-facing; hardening, tempering; strengthening, toughening *cf* STÄHLEN

Stahl-walze *f* steel roller. **-ware** *f* hardware; cutlery. **-wasser** *n* chalybeate (*or* iron-bearing) water. **-wein** *m* iron wine. **-werk** *n* steel mill, steelworks. **-wolle** *f* steel wool

stak stuck, etc: *past of* STECKEN *vi*

Staket *n* (-e), **Staketen-zaun** *m* stockade, palisade (fence)

Stalaktit *m* (-e) stalactite

Stall *m* (-̈e) stable, barn; (animal) stall, pen, coop, etc. **-dünger, -mist** *m* stable manure

Stamm *m* (-̈e) 1 (tree) trunk; stem, stalk. 2 family (tree); clan, tribe; breed, strain. 3 stock; permanent staff, care (of specialists, etc). 4 (invested) principal. 5 (*in com-*

pds, oft) parent, original, primary: **-aktie** *f* share of common stock. **-ansatz** *m (Dye.)* stock mixture (liquor, paste). **-ätze** *f (Tex.)* standard (*or* stock) discharge. **-baum** *m* 1 *(Tech.)* flow sheet. 2 *(Biol.)* phylogenetic tree. 3 family tree, genealogy; pedigree. **-buch** *n* (personal) album

stammeln *vt/i* stammer

Stamm·emulsion *f* stock emulsion

stammen *vi* 1 **(aus)** come, spring, originate (from); date back (to). 2 (+**von**) be derived (from); e.g. **die Theorie stammt von Einstein** the theory originated with Einstein

Stamm·farbe *f* 1 primary color. 2 *(Dye.)* stock mixture (liquor, paste). **-flotte** *f (Dye.)* stock liquor (*or* solution)

stämmig *a* robust, sturdy, vigorous

Stamm·körper *m* parent substance (*or* body). **-küpe** *f (Dye.)* stock vat. **-leitung** *f (Elec.)* main circuit, trunk line. **-linie** *f* trunk (*or* main) line. **-lösung** *f* stock solution. **-personal** *n* permanent staff. **-reserve** *f (Tex.)* stock resist. **-substanz** *f* parent substance. **-tafel** *f* 1 *(Tech.)* flow sheet. 2 genealogical table. **-vater** *m* ancestor, progenitor. **-verdickung** *f* stock thickening. **st—verwandt** *a* kindred, cognate. **St—werk** *n (Tech.)* parent (*or* main) plant. **-würze** *f (Beer)* original wort. **-zelle** *f* parent cell

Stampf·asphalt *m* tamped asphalt. **-beton** *m* tamped concrete

Stampfe *f* (-n) *(Mach.)* stamp(er), tamper; ram(mer), hammer; punch; pestle

stampfen I *vt* 1 tamp, ram, pound; trample. 2 stamp. 3 mash; crush, pulverize. 4: **aus dem Boden st.** conjure up. II *vi* 5 stamp (one's foot), trample. 6 hammer, pound. 7 *(Ship)* pitch. 8 tramp, trudge

Stampfer *m* (-) 1 masher. 2 plunger. 3 tamper, etc = STAMPFE

Stampf·futter *n* 1 tamped lining. 2 *(Furnaces)* monolithic lining. **-gewicht** *n* (compacted) bulk density. **-haufen** *m (Paper)* batch. **-kartoffeln** *fpl* mashed potatoes. **-klotz** *m* pile driver, ram. **-masse** *f* tamping mass (*or* compound); *(Ceram.)* castable. **-mörtel** *m* tamped mortar. **-werk** *n* stamping (*or* crushing) mill. **-zucker** *m* crushed sugar

Stampiglie *f* (-n) rubber stamp; stamp pad

stand stood, etc: *past of* STEHEN

Stand *m* (⁼e) 1 standing position; standstill —**einen guten St. haben (a)** stand firmly, **(b)** (+**bei**) be in good standing (with); **einen schweren St. haben** have a difficult time. 2 *(usu)* **fester St.** (firm) foothold. 3 position, station; (taxi) stand; stand, booth (at market, fair, etc). 4 stage (of development) —**auf den neuesten St. bringen** bring up to date. 5: **St. (der Dinge)** state (of affairs). 6 *(Astron.)* position, height. 7 level (of water, temperature, prices, etc). 8 *(Instrum.)* reading. 9 condition, shape; status, position; ability: **gut im St. (halten)** (keep) in good condition; **in St. setzen** repair; **in den St. setzen, zu. .** put in a position (to), enable (to); **außer St. setzen** disable *cf* IMSTANDE, INSTAND, ZUSTAND(E). 10 *(Agric.)* stand (of grain, etc). 11 social station, class; rank, standing

Stand·anzeiger *m* level indicator

Standard·abweichung *f* standard deviation. **-einheit** *f* standard unit. **-fehler** *m* standard error

standardisieren *vt/i* standardize

Standard·lösung *f* standard solution. **-versuch** *m* standard test (*or* experiment). **-werk** *n* standard (reference *or* literary) work. **-wert** *m* standard value

Stand·bild *n* statue. **-bremse** *f* parking brake. **-bücherei** *f* reference library. **-dauer** *f* standing time, duration of standing. **-entwicklung** *f (Phot.)* tank development

Ständer *m* (-) 1 stand; pedestal, base. 2 rack. 3 post, pillar, column. 4 *(Mach., Elec.)* stator. 5 cistern, tank, vat. **-klemme** *f* clamp for a stand. **-strom** *m* stator current

stand·fähig *a* stable, firm. **-fest** *a* firm (on one's feet), stable, steady; rigid. **St—festigkeit** *f* firmness, stability; rigidity; *(Ceram.)* green strength. **-flasche** *f* flat-bottomed flask. **-gärung** *f* standing fermentation. **-gefäß** *n* 1 flat-bottomed vessel. 2 museum jar; storage vessel; stock tub. **-glas** *n* 1 gauge glass; glass cylinder. 2 reagent bottle

standhaft *a* steadfast, firm

stand·halten* vi (dat) hold one's ground (against), resist, withstand

ständig a steady, constant; fixed, established, permanent

Stand·kugel f stationary bulb. **-lehre** f statics. **-mörser** m firm-based heavy mortar. **-motor** m stationary motor. **-öl** n stand (lithographic, heated linseed) oil. **-ort** m station, home base; stand, location; habitat. **-punkt** m position (esp on a subject), standpoint; point of view. **-rohr** n standpipe, vertical pipe. **-sicherheit** f stability. **-tropfglas** n dropping bottle. **-verlust** m storage loss. **-versuch** m creep test. **-waage** f platform balance. **-zeit** f 1 durability, useful (or operational) life (of an appliance). 2 (Furnaces) run, campaign. **-zylinder** m standing cylindrical vessel; hydrometer jar

Stange f (-n) 1 pole, post. 2 rod, bar; roost; rack. 3 stalk, stick (e.g. of cinnamon); roll (of sulfur), ingot (of gold). 4 (valve) stem. —**die St. halten** (+dat) stand up (for); **bei der St. bleiben** stick to business, persevere; **von der St.** off the shelf, ready-made

Stängel·kobalt m (Min.) chloanthite

Stangen·blei n bar lead. **-bohne** f climbing (or runner) bean. **-bohrer** m hand auger. **-eisen** n bar iron; iron rod (or bar). **st-förmig** a rod-shaped, bar-shaped. **St-gold** n ingot gold. **-kali** n stick potash. **-kitt** m stick cement. **-kupfer** n bar copper; copper rod (or bar). **-lack** m stick lac. **-schörl** m tourmaline. **-schwefel** m roll (or stick) sulfur. **-seife** f bar soap. **-silber** n ingot silver. **-spat** m (Min.) columnar barite. **-stahl** m bar (or rod) steel. **-stein** m pycnite, (columnar) topaz. **-tabak** m roll (or twist) tobacco. **-wachs** n stick wax (or polish). **-zinn** n bar tin

Stanitzel n (-) (cone-shaped) paper bag

stank stank: past of STINKEN

stänkern vi 1 stink. 2 pick a fight

Stannat·lauge f stannate liquor

Stanni·azetat n stannic acetate. **-chlorid** n stannic chloride. **-chlorwasserstoff-säure** f chlorostannic acid. **-hydroxid** n stannic hydroxide. **-jodid** n stannic iodide

Stannin n (Min.) stannite

Stanniol n tinfoil. **-kapsel** f tinfoil cap. **-papier** n tinfoil; aluminum foil

Stanni·oxid n stannic oxide. **-reihe** f stannic series. **-sulfozyanid** n stannic thiocyanate. **-verbindung** f stannic compound

Stanno·azetat n stannous acetate. **-chlorid** n stannous chloride. **-chlorwasserstoffsäure** f chlorostannous acid. **-hydroxid** n stannous hydroxide. **-jodid** n stannous oxide. **-salz** n stannous salt. **-sulfid** n stannous sulfide. **-verbindung** f stannous compound

Stanz·abfall m stamping (or punching) waste

Stanze f (-n) 1 punch, die. 2 punching (or stamping) machine; matrix. **stanzen** vt stamp, punch; emboss

Stanz·maschine f punch (or stamping) press. **-porzellan** n porcelain for punching. **-presse** f punch (or stamping) press. **-teil** n stamping (or punching) part

Stapel m (-) 1 (Tex.) staple. 2 pile, stack; heap. 3 warehouse, storehouse. 4 drydock, stocks —**vom St. lassen** launch, deliver, let loose. **-farbe** f staple color. **-faser** f 1 staple fiber. 2 short-fibered rayon. **-fehler** m (Cryst.) stacking fault (or disorder). **-gemüse** n staple vegetable. **-gewebe** n pile fabric. **-glas-seide** f chopped glass strands. **-gut** n staple(s). **-länge** f staple length. **-lauf** m launch(ing)

stapeln vt stack, pile (up); stock; store. 2: **sich st.** pile up, accumulate

Stapel·platte f pallet

Stapelung f stacking, piling (up); storage; accumulation cf STAPELN

Stapel·ware f staple goods

Star m I (-e) (Zool.) starling. II (-e) (Med.): **(grauer) St.** cataract; **grüner St.** glaucoma. III (-s) star (performer)

starb died: past of STERBEN

stark a 1 (usu) strong: **seine st-e Seite** his strong point, his forte. 2 thick: **ein 5 cm st-es Brett** a board 5 cm thick. 3 stocky, stout. 4 heavy; severe, intense; considerable. 5 great; good; (adv) well: **eine st-e Stunde lang** for a good hour; **st. besucht** well attended. **-backend** a (Coal) strongly caking

Stärke f(-n) **I** starch. **II 1** strength; (*Acids*) concentration. **2** thickness; diameter. **3** stoutness, stockiness. **4** severity, intensity. **5** greatness; extent; size. **6** numerical strength. **7** strong point, forte. **-abbau** m starch degradation. **-art** f type of starch. **st-artig** a starchy, amyloid, amylaceous. **St-bildner** m (*Bot.*) leucoplast. **-bildung** f starch formation. **-blau** n starch blue. **-fabrik** f starch factory (or plant). **st-führend** a amylaceous. **St-gehalt** m starch content. **-grad** m degree of strength, intensity. **-gummi** n starch gum, dextrin. **st-haltig** a amyloid, amylaceous. **St-kleister** m starch paste. **-korn, -körnchen** n starch granule. **-leimpulver** n starch glue powder. **-lösung** f starch solution. **-mehl** n starch (flour or powder)

stärkemehl·ähnlich, -artig a starchy, amyloid, amylaceous. **-haltig** a containing starch. **St-kleister** m starch paste

Stärke· starch; strength: **-messer** m amylometer. **-messung** f measurement of strength (size, intensity). **-milch** f thin starch paste. **-mittel** n strengthening remedy; tonic, restorative

stärken I vt **1** strengthen, fortify; support. **2** feed. **3** intensify. **4** starch. **5** thicken. **II: sich st.** fortify oneself (with food or drink)

Stärke· starch; strength: **-papier** n starch paper. **-pulver** n starch powder, powdered starch. **st-reich** a rich in starch, high-starch. **St-sirup** m starch syrup, glucose. **-verhältnis** n relation as to strength (or size). **-wäsche** f laundry to be starched. **-wasser** n starch water. **-weizen** m starch wheat, emmer. **-wert** m starch (or amyloid) value. **-zucker** m starch sugar, glucose

stark·farbig a strongly colored. **-faserig** a tough-fibered. **-gas** n rich gas, (*specif*) coal gas. **st-klopfend** a strongly knock-producing. **st-sauer** a strongly acid. **St-strom** m strong current; (*Elec.*) heavy (or power) current. **-stromkabel** n (*Elec.*) power cable. **-stromleitung** f (*Elec.*) high-tension line

Stärkung f(-en) **1** strengthening, fortification; support. **2** feeding. **3** intensification.

4 starching. **5** thickening. **6** sustenance, food, drink, refreshment

Stärkungs·mittel n tonic, restorative

stark·wandig a thick- (or stout-)walled. **St-wasser** n (*Coal Gas*) strong ammonia water. **st-wirkend** a powerful, highly effective; drastic. **-wirksam** a powerful, highly active (or effective). **-zügig** a (*Paint*) long-stroke

starr a **1** rigid, stiff. **2** frozen. **3** motionless, fixed **—st.** ansehen stare at. **4** numb. **5** inflexible, obstinate. **Starre** f **1** rigidity, stiffness. **2** motionlessness. **3** numbness. **4** inflexibility, obstinacy. **starren** vi **1** (+**auf**) stare (at). **2** (**vor, von**) be thick (stiff, covered) with, be full (of). **3** be numb, be frozen (stiff). **Starr·heit** f rigidity, etc = **STARRE. Starrheits·modul** n modulus of rigidity

Starr·krampf m tetanus. **-krampfserum** n antitetanus serum. **-leinen** n, **-leinwand** f buckram. **-schmiere** f grease, solid lubricant. **st-sinnig** a stubborn, obstinate. **St-sucht** f catalepsy. **st-süchtig** a cataleptic

starten 1 vt, vi(s) start. **2** vt launch. **3** vi(s) take off, lift off; leave. **4** vi(h) start the engine

Start·hilfsrakete f booster rocket. **-plattform** f launching pad

Stase f(-n) stasis

Statik f statics. **Statiker** m (-) expert in statics; structural engineer

Station f (-en) **1** station. **2** stopover; rest (stop). **3** post. **4** (hospital) ward. **stationär** a stationary; steady-state. **stationieren** vt station

statisch a **1** (*Phys.*) static. **2** (*Engg.*) structural

statistisch a statistical

Stativ n (-e) stand, support; tripod; stage. **-lupe** f stand magnifier

statt **1** prep (+genit), cnjc (+**zu** + inf) instead of (. . ing) = ANSTATT. **2** sep pfx: cf STATTFINDEN, etc

Statt f: **an jemandes St.** in someone's place

stattdessen adv instead (of that)

Stätte f(-n) place, locale, site; abode

statt·finden* vi take place

statt·geben* vi (+dat) accede (to), comply (with)

statt·haben* *vi* take place

statthaft *a* permissible, admissible

stattlich *a* impressive, imposing; sizable

Statuen·bronze *f* statuary bronze

statuieren *vt* establish firmly —**ein Exempel st.** set a warning example

Stau *m* (-e) 1 damming, impounding: **in St. halten** dam up, impound. 2 retention, storage; stowage. 3 dam. 4 (ice, traffic) jam; congestion. 5 blockage (of ice, blood, wind). 6: **das Wasser ist im St.** the tide is turning. **-anlage** *f* reservoir (installation)

Staub *m* (-e *or* ⸚e) dust; (*Bot.*) pollen; (*in compds, oft*) powdered *cf* STAUBZUCKER. **-abscheider** *m* dust separator. **st-artig** *a* dust-like, powdery

Staub·belastung *f* dust pollution. **-beutel** *m* (*Bot.*) anthor. **-brandpilz** *m* dust fungus

Stäubchen *n* (-) tiny particle, mote

Staub·deckel *m* dust cover. **st-dicht** *a* dustproof

Stau·becken *n* reservoir

Stäube·mittel *n* dusting compound

stauben 1 *vt* (+**von**) dust, brush (off). 2 *vi* raise (*or* give off) dust, be dusty

stäuben I *vt* 1 dust (with a powder). II *vi* 2 raise dust. 3 fly up (*as:* snow), spray (*as:* water). 4 (*Bot.*) emit pollen

Staub·faden *m* (*Bot.*) stamen. **-fänger** *m* dust catcher (*or* collector). **-farbe** *f* powdered coloring material. **st-fein** *a* fine as dust. **st-feuerung** *f* firing with pulverized fuel. **-figuren** *fpl* dust figures, powder pattern. **-fließverfahren** *n* fluidization process. **st-förmig** *a* powdery, pulverulent. **-frei** *a* dust-free, dustless. **St-gefäß** *n* (*Bot.*) stamen. **-gehalt** *m* dust content. **st-haltig** *a* dust-laden, containing dust. **St-haltiggrün** *n* Paris green. **-hefe** *f* powdery (*or* nonflocculating) yeast

staubig *a* dusty; powdery

Staub·kalk *m* powdered (*or* air-slaked) lime. **-kohle** *f* powdered (*or* pulverized) coal. **-konzentrationsmeßverfahren** *n* particle concentration measuring process. **-korn** *n* dust particle

Stau·blech *n* baffle plate

Stäubling *m* (*Bot.*) puffball

Staub·luft *f* dust-laden air. **-lunge** *f* silicosis. **-maske** *f* dust mask. **-mehl** *n* flour mill dust, dustings. **-öl** *n* floor oil. **-regen** *m* drizzle, mist; spray. **-sammler** *m* dust collector. **-sand** *m* very fine sand, sand dust

staubsaugen (*pp* staubgesaugt) *vt/i* vacuum-clean

Staub·sauger *m* vacuum cleaner; dust suction device. **-schleifen** *n* honing. **-schreiber** *m* recording dust meter. **-schutzmaske** *f* dust mask. **-schwelverfahren** *n* (*Coal*) fluidized carbonization process. **st-sicher** *a* dustproof. **-sieb** *n* dust sieve (*or* sifter). **-tee** *m* tea dust. **-teilchen** *n* dust particle. **st-trocken** *a* dry as dust, bone-dry; (*Varnish*) dry enough to prevent dust adherence. **-tuch** *n* dust cloth. **-wolke** *f* cloud of dust. **-zähler** *m* dust counter. **-zucker** *m* powdered sugar. **-zyklon** *m* cyclone dust collector

Stauch·apparat *m* (*Explo.*) brisance meter. **-druck** *m* compression, crushing pressure

stauchen I *vt* 1 pound (down, flat); hammer, beat. 2 compress (by pounding); mash. 3 (*Metals*) upset; hot-press; pressure-forge; swage; rivet over; clench. 4 throw (*or* toss) down. 5 stub (a toe, etc). 6 kick. 7 (*Agric.*) stack (for drying). II *vi* jolt (*as:* a car). III: **sich st.** buckle

Stauch·probe *f* compression (*or* hammering) test

Stauchung *f* (-en) pounding, hammering; compression; mashing; upsetting, etc *cf* STAUCHEN

Stauch·zylinder *m* (*Explo.*) crusher gauge

Stau·damm *m* (earth-fill) dam

Staude *f* (-n) 1 perennial herb. 2 shrub. 3 head (of lettuce)

Stau·druck *m* 1 (*Phys.*) dynamic pressure. 2 (*Tech. & Med.*) back pressure. **-düse** *f* Pitot tube

stauen I *vt* 1 impound (water), dam (up). 2 stop, block (blood flow, etc); baffle. 3 stow (away). II: **sich st.** 4 jam (up), become congested. 5 be(come) pent up. 6 pile up, accumulate; (*Water*) rise

Stauffer·fett *n* Stauffer (*or* cup) grease

stau·frei *a* unobstructed, unrestricted. **St-körper** *m* dam; baffle, diaphragm

staunen *vi* 1 be amazed, be astonished. 2 (+**über**) marvel (at). **Staunen** *n* astonishment, amazement. **staunend** *p.a* amazed, astonished. **staunens·wert** *a* astonishing, amazing

Staupe *f* distemper

Stau·punkt *m* stagnation point; (*Liquid Flow*) flooding point. **-quelle** *f* tidal spring. **-rand** *m* orifice plate. **-rohr** *n*, **-röhre** *f* Pitot tube. **-scheibe** *f* baffle plate; diaphragm. **-see** *m* artificial lake (behind a dam). **-strahl** *m* ramjet

Stauung *f* (-en) 1 impounding; damming. 2 stoppage, blockage. 3 stowage. 4 jam; congestion; pile-up, accumulation; rise (of water). 5 (blood) clot. *Cf* STAUEN

Stau·verband *m* tourniquet. **-wand** *f* baffle plate. **-wasser** *n* 1 impounded water. 2 backwater, stagnant water. **-wirkung** *f* damming (*or* baffle) effect

std. *abbv* (stündig) hourly; (*oft as sfx*) -hour

Std. *abbv* 1 (Standard). 2 =**Stde.** (Stunde) hr. (hour)

stdg. *abbv, oft as sfx* (stündig) hourly, -hour

stdl. *abbv* (stündlich) hourly, per hour

Stdn. *abbv* (Stunden) hrs. (hours)

Stearin·aldehyd *m* stearic aldehyde. **-kerze** *f*, **licht** *n* stearin candle. **-öl** *n* oleic acid. **-pech** *n* stearin pitch. **st–sauer** *a* stearate of. **St–säure** *f* stearic acid. **-seife** *f* (common) stearin soap

steatinisch *a* tallowy

Stech·apfel *m* 1 (*Bot.*) thorn apple, Jimson weed. 2 (*Pharm.*) stramonium. **-apfelöl** *n* datura oil

stechen* (stach, gestochen; sticht) **I** *vt/i* 1 sting, bite. 2 stick, stab, jab. 3 prick, punch. 4 engrave. 5 dig (up) (with a spade). 6 cause a stabbing pain. **II** *vt* 7 spear, skewer. 8 pierce, puncture; (*Med.*) lance. 9 tap (kegs, etc). 10 turn (malt). **III** *vi* 11: **die Sonne sticht** the sun is blazing (hot). 12 (*Colors*) e.g.: **es sticht ins Grüne** it shades over into (*or* has a tinge of) green. *Cf* GESTOCHEN. **Stechen** *n* (*oft*) stitch, stabbing pain. **stechend** *p.a* (*oft*) pungent. **Stecher** *m* (-) 1 proof stick. 2 spear; skewer. 3 engraver. 4 hair trigger

Stech·heber *m* plunging siphon, thief tube; pipette. **-kolben** *m* pipette. **-kunst** *f* engraving. **-mücke** *f* mosquito. **-palme** *f*

holly, ilex. **-palmenbitter** *n* ilicin. **-pipette** *f* pipette. **-probe** *f* touchstone test. **-stock** *m* (*Dye.*) prodding stick. **-zirkel** *m* dividers

Steck·dose *f* (*Elec.*) (plug) socket, receptacle. **-einheit** *f* plug-in unit

stecken I *vt* 1 stick, insert, put. 2 pin (up). 3 set (limits, goals). **II** *vi* (*past oft** stak) 4 be stuck, be set (rooted, implanted, ingrained), lie *cf* ANFANG. 5 be (hidden) (in a body, a situation). 6 be dressed (in)

stecken·bleiben* *vi*(s) be (*or* get) stuck **stecken·lassen*** *vt* 1 leave (sticking, stuck). 2 leave in the lurch; let down

Stecken·pferd *n* hobby

Stecker *m* (-) (*Elec.*) plug. **-büchse** *f* plug socket, outlet

Steck·kontakt *m* (*Elec.*) socket, outlet; plug

Steckling *m* (-e) (*Bot.*) shoot, layer, cutting

Steck·lot *n* grain spelter (for muffle brazing). **-nadel** *f* (common) pin. **-rübe** *f* rutabaga. **-schlüssel** *m* socket wrench. **-zwiebel** *f* (*Bot.*) bulb for planting

Steffen·abwasser *n* (*Sugar*) Steffen waste water. **-schnitzel** *npl* Steffen sugar pulp

Steg *m* (-e) 1 footpath. 2 footbridge; catwalk; gangplank. 3 bridge, crosspiece, nosepiece; bar; web. 4 (*Molding*) gate. **-reif** *m* stirrup —**aus dem St.** on the spur of the moment, extemporaneous(ly)

Steh·bild *n* still (picture). **-bolzen** *m* stay bolt. **-bütte** *f* stock tub

stehen* (stand, gestanden) **I** *vi* (h *or* s) 1 stand; be (in an upright position). 2 be located (parked, stationed). 3 (*oft* + **um**) be (in a condition). 4 (*refers to writing, print*) say, be: **es steht hier** it says here; **wie es geschrieben steht** as it is written. 5 exist, be (still) standing. 6 be at rest, stand still; (*esp of timepieces*) have stopped. 7 (+**vor**) face. 8 (+**für**) **A** guarantee; **B** stand (for), represent. 9 (+**zu**) **A** have an opinion (on); **B** stand (by). 10: **es steht bei (ihnen**, *etc*) it is up to (them, etc). *Cf* SINN 2. **II** *vi* & **sich (gut) st. (mit)** be on (good) terms (with). *Cf* GESTANDEN. **Stehen** *n* halt, standstill

stehen·bleiben* *vi*(s) 1 stop, halt. 2 stall, break down (*as:* motor). 3 stay, remain standing. 4 be left (behind). **stehenblei-**

bend *p.a* stationary, inactive; remaining
stehend *p.a* 1 standing, etc *cf* STEHEN. 2 stationary. 3 upright, vertical. 4 permanent. 5 stagnant (water). 6 standard, commonly used

stehen·lassen* *vt* 1 let stand. 2 leave standing. 3 leave behind, forget

Steh·kolben *m* flat-bottomed flask. **-lampe** *f* floor lamp

stehlen* (stahl, gestohlen; stiehlt) 1 *vt/i* steal. 2 *vt:* **Zeit st.** waste time. 3: **sich st.** sneak, steal (into a place)

Stehl·stange *f* thief rod

Steh·zeit *f* 1 standing time. 2 (*Plastics*) molding time

Steiermark *n* (*Geog.*) Styria (Austrian province)

steif *a* stiff (*also fig*); rigid. **Steife** *f* (-n) 1 stiffness, rigidity. 2 stiffening (agent), (*specif*) starch, size, glue. 3 (diagonal) brace, strut. 4 consistency (*esp* of concrete). **steifen** *vt* stiffen; starch; brace

Steifheit *f* (*Tech. oft*) **Steifigkeit** *f* stiffness, rigidity. **Steifigkeits·zahl** *f* coefficient of rigidity

Steif·leinen *n,* **-leinwand** *f* buckram. **-macher** *m* stiffener

Steifung *f* (-en) stiffening; starching, sizing. **Steifungs·mittel** *n* stiffening (agent)

Steig *m* (-e) (foot)path, trail. **--brunnen** *m* artesian well

Steige *f* (-n) 1 ladder; steps. 2 crate. 3 path = STEIG

steigen* (stieg, gestiegen) I *vi*(s) 1 climb. 2 rise. 3 step (up). 4 get (on, off a vehicle). II *vt* go up, climb (stairs, etc)

steigern 1 *vt* raise, heighten, increase. 2: **sich st.** rise, increase. 3: **st—der Stahl** rimming steel

Steige·rohr *n* vertical pipe, riser

Steigerung *f* (-en) raising, rise; increase

Steig·höhe *f* 1 height of ascent, rise; elevation. 2 capillary rise. 3 (*Mach.*) pitch (of a screw). **-höhemethode** *f* sedimentation-equilibrium method (for determining particle size). **-leitung** *f* vertical (*or* ascending) pipe; (*Elec.*) vertical line (*or* wire), rising main. **-rad** *n* escape wheel. **-raum** *m* (*Brew.*) unfilled space above the wort. **-rohr** *n,* **-röhre** *f* vertical (*or* ascending) pipe; riser

Steigung *f* (-en) 1 (up)grade, gradient, incline, slope, hill. 2 rise, ascent, climb. 3 (*Mach.*) pitch (of a screw)

steil *a* steep, precipitous

Steilbrust·flasche *f* wide-neck(ed) bottle

Steilheit *f* steepness, (steep) slope; (*Phot.*) contrast

Steil·schrauber *m* helicopter

Stein *m* (-e) 1 (*genl*) stone (*incl Med., jewels, fruit kernel*): **St. der Weisen** philosopher's stone. 2 rock. 3 brick. 4 (*Metll.*) matte. 5 (*Med. also*) calculus. **—St—e und Erden** minerals

Stein·abfälle *mpl* stone chips, spalls. **st—ähnlich** *a* stone-like, stony. **St—alaun** *m* rock alum. **st—alt** *a* old as the hills. **St—arbeit** *f* 1 metal smelting. 2 stonework. **st—artig** *a* stone-like, stony. **St—auflösungsmittel** *n* solvent for body calculi, e.g. gallstones. **-ausleser** *m* stone separator. **-bock** *m* (*Zool.*) ibex. **-boden** *m* 1 rocky soil. 2 stone floor. **-brand** *m* (*Bot.*) bunt, smut. **-brech** *m* (*Bot.*) saxifrage. **-brecher** *m,* **brechmaschine** *f* rock crusher. **-bruch** *m* quarry. **-bühlergelb** *n* barium yellow (barium chromate). **-butter** *f* rock butter. **-druck** *m* 1 lithograph. 2 lithography

Steindruck·farbe *f* lithographic ink. **-kalkstein** *m* lithographic limestone

steinern *a* (of) stone, stony

Stein·fänger *m* stone catcher (*or* remover). **-farbe** *f* stone color. **st—farbig** *a* stone-colored. **St—flachs** *m* mountain flax, amianthus. **-flasche** *f* stoneware bottle. **-frucht** *f* stone fruit, drupe. **-glättung** *f* (*Paper*) flint glazing. **-grau** *n* stone gray, gray-slate pigment. **-grieß** *m* gravel. **-grün** *n* terre verte (glauconite *or* celadonite pigment). **-gut** *n* earthenware, *usu* white ware (with white absorbent body and soft glaze)

Steingut·geschirr *n* earthenware (*or* white-ware) (articles). **-ton** *m* earthenware clay

stein·hart *a* hard as stone. **St—holz** *n* xylolith, flooring composition (magnesia cement mixed with filler)

steinig *a* stony, rocky

Stein·kern *m* stone (in fruit; in burnt lime). **-kitt** *m* stone (*or* mastic) cement. **-klee** *m* melilot; white clover. **-kohle** *f* (mineral *or* pit) coal (as distinct from lignite); (*specif*) hard coal (bituminous *or* anthracite). **-kohleeinheit** *f* hard coal unit (of energy) **Steinkohlen·asche** *f* coal ashes. **-benzin** *n* commercial benzene, benzol. **-bergwerk** *n* coal mine. **-entgasung** *f* distillation of coal. **-flöz** *n* coal seam. **-gas** *n* coal gas. **-grube** *f* coal mine. **-kampfer** *m* naphthalene. **-klein** *n* slack, culm. **-leuchtgas** *n* coal gas for illumination. **-öl** *n* coal oil. **-pech** *n* coal-tar pitch. **-schicht** *f* coal seam. **-schiefer** *m* coal-bearing shale. **-schlacke** *f* coal cinders. **-schwelteer** *m* low-temperature coal tar. **-schwelung** *f* low-temperature carbonization of coal. **-staub** *m* coal dust. **-teer** *n* coal tar

Steinkohlenteer·benzin *n* commercial benzene, benzol. **-blase** *f* coal-tar still. **-essenz** *f* first light oil. **-farbe** *f* coal-tar dye. **-kampfer** *m* naphthalene. **-öl** *n* coal oil. **-pech** *n* coal-tar pitch. **-präparat** *n* coal-tar preparation (*or* product)

Steinkohlen·verkohlung, -verkokung *f* coal coking. **-zeit** *f* Coal (*or* Carboniferous) Age

Stein·körper *m* (*Bot.*) stone cell, cell cluster. **-krankheit** *f* gall (kidney, bladder) stones, calculosis. **-kraut** *n* (*usu*) stonecrop. **-krug** *m* stone jar, earthenware jug. **-lager** *n* (*Mach.*) jeweled bearing. **-malz** *n* vitreous malt. **-mantel** *m* lithosphere. **-mark** *n* (*Min.*) lithomarge, kaolinite, halloysite. **-mauer** *f* stone wall; brick wall. **-mehl** *n* stone powder. **-meißel** *m* stonecarver's chisel. **-metz** *m* (-en, -en) stonemason, stonecutter. **-mörtel** *m* 1 hard mortar. 2 concrete. 3 Portland cement. **-obst** *n* stone fruit. **-öl** *n* petroleum. **st-ölhaltig** *a* oil-bearing. **St-pappe** *f* 1 roofing paper (fabric, board). 2 fireproof pasteboard. **-pech** *n* stone pitch, hard pitch (*or* asphalt). **-pilz** *m* boletus, mushroom. **-platte** *f* 1 stone slab. 2 flagstone. **-porzellan** *n* hard porcelain. **st-reich** *a* fabulously rich. **St-reich** *n* mineral kingdom. **-rösten** *n* roasting of the matte. **-salz** *n* rock salt. **-salzlager** *n* rock salt bed. **-säure** *f* lithic (*or* uric) acid. **-schlag** *m* 1 crushed stone (*or* rock). 2 rock slide. **-schmelzen** *n* (*Metll.*) matte smelting. **-schutt** *m* 1 rubble, detritus. 2

ballast, macadam. **-staublunge** *f* silicosis. **-stopfen** *m* stoneware stopper (for carboys). **-waren** *fpl* stoneware. **-wurf** *m* stone's throw. **-zeit** *f*, **-zeitalter** *n* Stone Age. **-zelle** *f* (*Bot.*) stone cell. **-zement** *m* concrete. **-zeug** *n* (*Ceram.*) stoneware

Steinzeug·gefäß *n* stoneware vessel. **-rohr** *n* stoneware pipe

steirisch *a* Styrian *cf* STEIERMARK — (*Metll.*) **st—e Arbeit** single refining

Steiß *m* (-e) rump, buttocks. **--bein** *n* coccyx

Stell· set-, control, adjusting, regulating

Stellage *f* (-n) stand, rack

stellbar *a* adjustable; movable. **Stellbarkeit** *f* adjustability; movability

Stell·bottich *m* 1 (*Brew.*) fermenting vat. 2 (*Dye.*) settling vat

Stelle *f* (-n) 1 place, spot, point: **auf der St.** on the spot, immediately; **von der St. bringen** move, budge (from its place); **von der St. kommen** move ahead, make headway; **zur St.** on hand, ready *cf* ORT I. **2 an St.** (+*von or* + *genit*) in place (of), instead (of) *cf* ANSTELLE. **3: an erster St.** in first place (*or* position), (first and) foremost. **4** passage (from a written work). **5** position; job. **6** authority, agency, office

stellen I *vt* **1** put, place, set; park (a car). **2** stand (sth) (upright). **3** stop, bring to a halt. **4** supply, make available. **5** make up, account for (a percentage of a total). **6** set, adjust, turn (on, up, down) (signals, clocks, radios, etc.). **7** set (conditions, etc.). **8** address, direct (a request, etc). **9** make (motions, diagnoses). **10** shade, blend (colors). **II: sich st. 11** place oneself. **12** (go and) stand. **13** (+*dat*) appear (for), report, make oneself available (to). **14** (+*adj*) pretend to be. **15** (+**zu**) take a stand, have an opinion (on a subject), feel (about a matter). **16** (+*dat*) confront, face (as: problems, challenges). **17** (+**auf**) come to, cost. *Cf* GESTELLT

stellen·weise *adv* here and there, sporadically, in places. **St—wert** *m* (*Math.*) place value. **-zahl** *f* 1 position (*or specif* atomic) number. 2 number of digits (*or* places). 3 index

Stell·farbstoff *m* shading dye. **-feder** *f* regulating spring. **-hahn** *m* regulating (*or* set) valve. **-hebel** *m* control (*or* switch)

lever. **-hefe** *f* pitching yeast

-stellig *sfx* (*Math.*) -place, -digit: **dreistellige Zahl** three-place number

Stellit *n* (*Metll.*) stellite. **stellitieren** *vt* stellitize

Stell·klappe *f* regulating valve. **-macher** *m* cartwright, wheelwright. **-marke** *f* index (mark); register line. **-mittel** *n* 1 diluent, extender. 2 adulterant. 3 standardizing agent. **-motor** *m* servomotor, booster motor. **-mutter** *f* (*Mach.*) adjusting nut. **-öl** *n* (*Petrol.*) neutral oil. **-pult** *n* control desk (*or* panel). **-rad** *n* regulating wheel, regulator. **-schlüssel** *m* monkey wrench. **-schraube** *f* set screw. **-uhr** *f* timer

Stellung *f* (-en) 1 (*genl.*) position. 2 location; placement, arrangement. 3 stand, opinion, view (on a matter): **St. nehmen (zu)** take a stand, express an opinion (on). 4 job; employment. 5 rank, social station. **--nahme** *f* opinion, critical comment

Stellungs·isomerie *f* position (*or* structural) isomerism

Stell·vertreter *m* deputy, representative; substitute. **-vertretung** *f* substitution; proxy. **-vorrichtung** *f* adjusting (*or* regulating) device. **-werk** *n* switch tower, interlocking

Stemm·eisen *n* 1 crowbar. 2 heavy chisel

Stemmeißel *m* (=**Stemm·meißel**) mortise chisel

stemmen I *vt* 1 lean, press; brace. 2 lift. 3 mortise. 4 brake, check, calk. 5 fell (trees). **II: sich st.** 6 (+**auf**) brace oneself, lean, press (against). 7 (+**gegen**) resist (firmly)

Stempel *m* (-) 1 (rubber) stamp; postmark; brand; trademark. 2 stamper; punch; die. 3 pestle. 4 (*Bot.*) pistil. 5 (*Min.*) prop. **--farbe** *f* stamping ink. **-kissen** *n* stamp pad. **-marke** *f* stamp

stempeln 1 *vt/i* stamp, postmark, brand; punch. 2 *vt* prop

Stempel·zeichen *n* stamp, mark, brand

Stengel *m* (-) 1 (*Bot.*) stalk, stem. 2 (*Cryst.*) column. **--älchen** *n* stem nematode. **-faser** *f* stem fiber. **-gewebe** *n* (*Bot.*) stem tissue

stengelig *a* 1 (*Bot.*) stalked. 2 (*Cryst.*) columnar. 3 (*Metll.*) spiky

Stengel·kohle *f* columnar coal. **-koks** *m* fingery coke

Stephans·körner *npl* stavesacre seeds

Stepp·decke *f* quilt

steppen *vt/i* quilt; stitch

Steppen·landschaft *f* steppe landscape

sterben* (starb, gestorben; stirbt) *vi*(s) die

sterblich *a* mortal. **Sterblichkeit** *f* mortality

Stercul·säure *f* sterculic acid

Stereo·chemie *f* stereochemistry. **st-chemisch** *a* stereochemical. **-isomer** *a* stereoisomeric. **St–isomerie** *f* stereoisomerism. **-metrie** *f* solid geometry. **st-metrisch** *a* stereometric **St–phonie** *f* stereo, stereophonic sound. **st–skopisch** *a* stereoscopic. **St–typiepapier** *n* stereotyping paper

stereotypieren *vt/i* stereotype

steril *a* sterile

Sterilisations·verfahren *n* sterilization process. **Sterilisator** *m* (-en) sterilizer

Sterilisier·apparat *m* sterilizing apparatus

sterilisieren *vt/i* sterilize. **sterilisier·fähig** *a* safely sterilizable. **Sterilisierung** *f* (-en) sterilization

Sterilität *f* sterility

Sterin *n* sterol. **sterisch** *a* steric, spatial

Stern *m* (-e) 1 (*usu*) star. 2 asterisk. **--anis** *m* star anise. **st–artig** *a* star-like, stellar. **St–bild** *n* 1 constellation. 2 sign of the zodiac

Sternchen *n* (-) 1 asterisk. 2 little star

stern·förmig *a* star-shaped, stellate. **St-kunde** *f* astronomy. **-leuchtkugel**, **-leuchtpatrone** *f* star shell. **-motor** *m* radial engine. **-nebel** *m* stellar nebula. **-physik** *f* astrophysics. **-rakete** *f* star rocket. **-rohr** *n* telescope. **-rubin** *m* star ruby. **-saphir** *m* star sapphire. **-schlacke** *f* (*Metll.*) antimony flux. **-schnuppe** *f* shooting star, meteor. **-signal** *n* (*Mil.*) signal flare. **-tier** *n* starfish. **-warte** *f* observatory. **-zeit** *f* sidereal time

stet, stetig *a* constant, steady, continual; stable; (*Math.*) continuous. **Stetigkeit** *f* constancy, steadiness; stability; continuity. **stets** *adv* always

Steuer I *f* (-n) tax, duty. **II** *n* (-) steering wheel; helm, rudder; controls. **--amt** *n*

revenue (*or* tax) office (*or* bureau); custom house

steuerbar *a* 1 taxable. 2 steerable, controllable, manageable

Steuer·elektrode *f* control electrode. **st–frei** *a* tax-free; duty-free. **St–gitter** *n* (*Elec.*) control grid. **-hebel** *m* control lever. **-marke** *f* revenue stamp

steuern I 1 *vt/i* drive (a car); steer; control, regulate; guide. 2 *vi* (h) (+*dat*) check; remedy. 3 *vi*(s) drive, head (for), navigate (to, toward a place). **II** *vi* pay taxes

Steuer·pult *n* control desk. **-rad** *n* steering wheel. **-stand** *m* control desk (*or* panel), controls

Steuerung *f* (-en) **I** 1 driving; steering; control(ling), regulation, guidance; navigation. 2 check(ing), remedy(ing). 3 (*Mach.*) steering (distribution, valve) gear; control system, drive **II** payment of taxes

Steuerungs·stab *m* control rod

Steuer·werk *n* controls

sthenosieren *vt* (*Tex.*) sthenosize

Stibiat *n* (-e) antimonate

Stich *m* (-e) 1 prick, jab, puncture; stab wound; (knife) thrust. 2 sting, bite. 3 stitch. 4 pang, stabbing pain. 5 pinch (of salt, etc). 6 engraving. 7 pass(age) (through a machine). 8 (*Metll.*) topping, tapped metal, tap hole. 9 (*Colors*) (+**ins**) tinge (of). 10 (*Food*) taint; **es hat einen St.** it has turned (sour, bad, etc). **11: St. halten** stand the test, hold water. **12: im St. lassen** let down, leave holding the bag. 13 (*in compds, oft*) implies: **A** testing or sampling *cf* STICHPROBE; **B** deadline: *cf* STICHTAG; **C** tie-breaking (game, vote, etc); **D** sthg fastened or connected at one end only *cf* STICHBAHN

Stich·auge *n* tap hole. **-bahn** *f* dead-end railroad *cf* STICH 13 D. **-eisen** *n* tapping bar

Stichel *m* (-) graver, (engraver's) burin. **sticheln** *vi* 1 stitch, sew. 2 jibe, needle

stich·fest *a* puncture-proof. **St–flamme** *f* 1 fine-pointed flame. 2 pilot jet. **st–haltig** *a* valid, sound *cf* STICH 11

stichig 1 *a* (*Food*) tainted, gone sour (*or* bad). 2 *sfx* (*Colors*) e.g. **blaustichig** blue-tinged, bluish

Stich·kultur *f* stab culture. **-loch** *n* tap-

hole. **-pfropf** *m* tap-hole plug. **-probe** *f* 1 spot check; random sample (*or* sampling). 2 (*Metll.*) assay of tapped metal. 3 (*Brew.*) pricking test. **-säge** *f* keyhole saw

sticht pricks, etc: *pres of* STECHEN

Stich·tag *m* target date; appointed day; deadline. **-wein** *m* sample wine. **-wort** *n* 1 catchword. 2 cue. 3 key word, entry (in a dictionary)

Stick· suffocating; nitrogen: **-dampf, -dunst** *m* choke damp; suffocating vapor. **-dioxid** *n* nitrogen dioxide

sticken *vt/i* stitch, sew, embroider. **stickend** *p.a* (*oft*) suffocating, stifling *cf* ERSTICKEN

Stickerei *f* (-en) embroidery

Stick·gas *n* 1 (*genl*) suffocating gas. 2 (*specif*) **A** nitrogen; **B** carbon dioxide

stickig *a* suffocating, stifling, stuffy

Stick·kohlenstoff *m* carbon nitride. **-luft** *f* 1 close (stuffy, stifling) air. 2 nitrogen. **-oxid** *n* nitric oxide. **-oxidentbindung** *f* liberation of nitric oxide. **-oxydul** *n* nitrous oxide. **-stoff** *m* nitrogen

Stickstoff· nitrogen: **-ammonium** *n* ammonium nitride (*or* azide). **st–arm** *a* nitrogen-poor, low-nitrogen. **St–aufnahme** *f* absorption of nitrogen. **-ausscheidung** *f* elimination of nitrogen. **-base** *f* nitrogen base. **-benzoyl** *n* benzoyl nitride (*or* azide). **-bestimmung** *f* determination of nitrogen. **st–bindend** *a* nitrogen-fixing. **St–binder** *m* nitrogen-fixing agent, azotobacter. **-bindung** *f* 1 (*Biol., Bact.*) nitrogen fixation. 2 (*Chem.*) nitrogen bond. **-bor** *n* boron nitride. **-brücke** *f* nitrogen bridge. **-cyantitan** *n* titanium cyanonitride. **-dämpfe** *mpl* nitrous vapors (*or* fumes). **-dioxid** *n* nitrogen dioxide. **-dünger** *m* nitrogenous fertilizer. **-eisen** *n* iron nitride. **-entbindung** *f* liberation of nitrogen. **st–frei** *a* *See* STICKSTOFFREI. **St–gabe** *f* (*Agric.*) nitrogen supply. **-gas** *n* nitrogen gas. **-gehalt** *m* nitrogen content. **st–gehärtet** *a* nitrided (steel). **St–gleichgewicht** *n* nitrogen equilibrium. **-halogen** *n* nitrogen halide. **st–haltig** *a* nitrogenous, containing nitrogen. **St–kalk** *m* (crude) calcium cyanamide. **-kalomel** *n* mercurous azide. **-kalzium** *n* calcium nitride.

-kohlenoxid n carbonyl nitride (or azide), carbodiazide. **-lithium** n lithium nitride. **-lost** n nitrogen mustard gas. **-magnesium** n magnesium nitride. **-metall** n metallic nitride. **-monoxid** n nitric oxide. **-natrium** n sodium nitride. **-oxid** n nitrogen oxide, specif nitrogen monoxide. **-oxydul** n nitrous oxide. **-pentoxid** n dinitrogen pentoxide. **-peroxid** n nitrogen dioxide. **-persäure** f pernitric acid. **-quecksilber** n mercury nitride. **-quecksilberoxydul** n mercurous azide

stickstoffrei a (=**stickstoff·frei**) nitrogen-free, non-nitrogenous

stickstoff·reich a nitrogen-rich, high-nitrogen

Stickstoff·sammler m nitrogen-fixing (or -storing) plant, leguminous plant. **-säure** f nitrogenous (specif hydrazoic) acid. **-sesquioxid** n nitrogen trioxide. **-silber** n silver nitride; fulminating silver. **-silicium, -silizid** n silicon nitride. **-suboxid** n nitrous oxide. **-tetroxid** n nitrogen tetroxide. **-titan** n titanium nitride. **-trioxid** n nitrogen trioxide. **-überschuß** m excess of nitride. **-vanadium** n vanadium nitride. **-verbindung** f nitrogen compound. **-wasserstoff** m hydrogen trinitride, nitrogen hydride

stickstoffwasserstoff·sauer a azide (or hydrazoate) of. **St—säure** f hydrazoic acid. **-phenylester** m phenyl azide (or hydrazoate), azidobenzene

Stickstoff·werk n nitrogen plant. **-zyan-titan** n titanium cyanonitride

Stick·wasserstoffsäure f hydrazoic acid. **-wetter** npl (Min.) choke damp, nitrous fumes

stieben* (stob, gestoben) vi (h,s) fly (about), scatter

Stief· step . . . : e.g. **-mutter** f stepmother

Stiefel m (-) 1 boot. 2 barrel, body (of a pump)

Stief·mütterchen n (Bot.) pansy. **st—mütterlich** a: **st. behandeln** treat like a stepchild

stieg climbed, etc: past of STEIGEN

Stiege f (-n) 1 stairway, stairs; ladder. 2 crate, fruit basket

Stieglitz m (-e) (Zool.) goldfinch

stiehlt steals: pres of STEHLEN

Stiel m (-e) 1 stalk, stem. 2 handle, shaft.

--bonbon n lollypop. **-granate** f stick hand grenade. **-pfanne** f long-handled frying pan. **-pfeffer** m cubeb(s). **-zelle** f stalk cell

stier a 1 fixed, vacant (star). 2 broke

Stier m (-e) (Zool.) bull, ox

stieren vi 1 stare, glare. 2 be in heat

stieß pushed, etc: past of STOSSEN

Stift m (-e) 1 pin, peg, stud. 2 brad, nail, spike. 3 dowel. 4 tag (on end of shoelace). 5 pencil, crayon; stick (deodorant, lipstick, etc). 6 stump. 7 (philanthropic) foundation; endowed institution. **--draht** m wire for making nails

stiften vt 1 donate, treat to. 2 found, establish; endow. 3 cause, bring about

Stifter m (-) 1 donor. 2 founder

Stift·farbe f 1 color of a pencil (or crayon). 2 colored pencil (or crayon). 3 pastel. **-schraube** f stud bolt

Stiftung f (-en) 1 donation. 2 foundation; endowment

Stigmasterin n stigmasterol

Stil m (-e) style

Stilben·chinon n stilbenequinone

still(e) I a 1 quiet, peaceful —**das Stille Meer** the Pacific Ocean. 2 still, silent. 3 secret. 4 stagnant. II adv: **im stillen** on the quiet, secretly, inwardly. III sep pfx (spelled **still-** before verb forms beginning with l) cf STILLEGEN, etc

Stille f 1 stillness, quiet(ness), peacefulness. 2 silence. 3: **in der** (or **aller**) **St.** on the quiet, in secret

stillegen (=**still·legen**) vt close down, abandon; immobilize, paralyze

stillen I vt 1 still, satisfy (a desire); pacify, appease. 2 stop (blood, pain, etc). II vt/i breast-feed

still·halten* vi 1 hold (or keep) still. 2 look on (or suffer) in silence

stilliegen* (=**still·liegen**) vi (h,s) be shut down, be out of service

Still·mittel n sedative

still·schweigen* vi keep quiet, say nothing. **Stillschweigen** n silence. **still·schweigend** p.a silent; tacit, implicit

Still·stand m standstill, halt; stoppage; stagnation; (Med. oft) arrest, stasis

still·stehen* vi (h,s) 1 stand still. 2 stop, halt. 3 be shut down, be out of service (or operation). 4 stagnate. **stillstehend** p.a *(oft)* inactive; stagnant; stationary

Stillung *f* (-en) stilling, satisfaction, pacification; stoppage; breast-feeding

Stillungs‧mittel *n* sedative

Stimm‧ voice, vocal; tuning; votes, voting: **-band** *n* vocal chord

Stimme *f* (-n) **1** voice. **2** (musical) part. **3** vote

stimmen I *vt* **1** tune. **2** (+*adj*) put in a ... mood. **II** *vi* **3** be right (correct, in order), come out right. **4** vote. **5** (+**auf, zu**) fit in, tally (with)

Stimmen‧mehrheit *f* majority (vote)

Stimm‧gabel *f* tuning fork. **-recht** *n* right to vote, suffrage

Stimmung *f* (-en) **1** mood, humor; atmosphere. **2** congeniality. **3** public opinion (*or* attitude). **4** morale. **5** tuning; pitch

Stimulans *n* (. . anzien *or* . . antia) stimulant. **stimulieren** *vt* stimulate. **Stimulierung** *f* (-en) stimulation

Stink‧ stinking, fetid: **-asant** *m* asafetida. **-äscher** *m* (*Lthr.*) rotten lime. **-baum** *m* stavewood. **-drüse** *f* stink gland

stinken* (stank, gestunken) *vi* stink, reek

Stink‧farbe *f* old weak tan liquor. **-fluß-(spat)** *m* fetid fluorspar, bituminous fluorite. **-harz** *n* asafetida. **-kalk** *m* anthraconite, bituminous limestone, fetid calcite. **-kohle** *f* fetid coal. **-mergel** *m* fetid (*or* bituminous) marl. **-morchel** *f* stinkhorn, carrion fungus. **-öl** *n* fetid (animal) oil. **-quarz** *n* fetid (*or* bituminous) quartz. **-raum** *m* gas chamber. **-raumprobe** *f* gas chamber test. **-schiefer** *m* fetid (*or* bituminous) shale. **-spat** *m* fetid fluorspar (*or* fluorite). **-stein** *m* stinkstone, *specif* anthraconite. **-tier** *n* skunk. **-topf** *m* stinkpot

Stipendium *n* (. . ien) scholarship (award)

Stippe *f* (-n) **1** speck, spot. **2** gravy, sauce

stippen *vt* dip, dunk; stick (a finger in, etc)

Stippen‧frei *a* spot-free, speck-free

stirbt dies: *pres of* STERBEN

Stirn(e) *f* (. . nen) forehead, brow; front

Stirn‧ front(al): **-ansicht** *f* front view. **-bein** *n* frontal bone. **-fläche** *f* face, front. **-getriebe** *n* spur gear. **-rad** *n* spur wheel. **-seite** *f* front (side), façade. **-welle** *f* wave front, impact wave

Stirr‧holz *n* (wooden) stirrer, mixer

stob scattered, etc: *past of* STIEBEN

stöbern I *vt/i* clean up. **II** *vi* **1** poke (*or* rummage) around. **2: es** (*or* **der Schnee**) **stöbert** the snow is blowing around

Stoch‧eisen *n* poker, stoker

Stocher *m* (-) **1** poker. **2** toothpick. **stochern** *vi* poke around (with a sharp instrument), pick (at)

Stöchiometrie *f* stoichiometry. **stöchiometrisch** *a* stoichiometric(al)

Stock *m* **I** (̈e) **1** stick; rod. **2** cane. **3** post; pole. **4** tree stump; trunk; stem. **5** potted plant; vine. **6** block (of wood). **7** mountain mass. **8** main part, body. **9** (*Brick*) clamp. **10** (*Brew.*) vat, back. **II** (- *or* -werke) floor, story (of a building); deck. **III** (-s) stock (of goods)

stock‧ (*implies:* totally): **-blind** *a* stone-blind. **-dunkel** *a* pitch-dark

stocken *vi* **1** get stuck, falter; (*Motors*) stall. **2** stop (short), halt. **3** slacken, stagnate. **4** congeal, curdle, thicken. **5** turn moldy, mildew. **stockend** *p.a* (*oft*) sluggish, stagnant

Stock‧ende *n* butt end. **-erz** *n* ore in lumps. **st‧finster** *a* pitch-dark. **St‧fisch** *m* stockfish, *esp* dried cod. **-fischlebertran** *m* cod-liver oil. **-fleck** *m* spot of mold, mildew stain. **st‧fleckig** *a* moldy, mold-spotted. **St‧holz** *n* stump wood

stockig *a* **1** musty; mold-spotted, mildewed. **2** obstinate, unyielding

‧stöckig *sfx* -story, -deck *cf* STOCK **II** & ZWEISTÖCKIG

Stock‧lack *m* stick-lac. **-lacksäure** *f* laccaic acid. **-probe** *f* (*Oils*) pour test. **-punkt** *m* (*esp Oils*) pour (*or* solidification) point. **-punkterniedriger** *m* pourpoint depressant. **-schlacke** *f* shingling slag

Stockung *f* (-en) **1** faltering, stalling; stoppage, halt; stagnation. **2** slowdown. **3** traffic jam; delay, tie-up. **3** coagulation. **4** (*Med.*) arrest, stasis

Stock‧werk *n* **1** floor, story (of a building). **2** (*Min.*) stockwork

Stoff *m* (-e) **1** material (*incl Tex.*); stuff (*may also imply* alcohol *or* narcotics); (*Paper, oft*) pulp. **2** (*Chem.*) substance. **3** (*Phys.*) matter. **4** (*Tex.*) fabric, cloth. **5** subject (matter). **‧abfall** *m* waste, scrap(s), refuse. **-ableitung** *f* (*Biol.*) translocation

Stoffänger m (*Paper*) saveall (=**Stoff·fänger**)

Stoff·ansatz m (*Biol.*) anabolism. **-aufnahme** f absorption (*or* intake) of material. **-austausch** m exchange of material. **-beutel** m cloth bag. **-bildung** f formation of a substance. **-brei** m (*Paper*) pulp slurry. **-bütte** f (*Paper*) stuff chest. **-dichte** f (*Paper*) pulp consistency. **-druck** m calico printing. **-filter** n, **-filterung** f See STOFFFILTER, STOFFFILTERUNG. **-flußbild** n See STOFFLUSSBILD. **-gattung** f kind of material, class of substances. **-gebiet** n subject matter. **-gemisch** n mixture of substances. **-haushalt** m (*Biol.*) metabolism

Stoffilter m cloth filter (=**Stoff·filter**).

Stoffilterung f (-en) cloth filtration

Stoff·kette f (*Kinetics*) material chain. **-kufe** f (*Paper*) stuff vat. **-leimung** f (*Paper*) pulp (*or* engine) sizing

stofflich a material

stoff·los a immaterial, unsubstantial

Stofflußbild n flow sheet

Stoff·mahlung f (*Paper*) pulp beating. **-menge** f amount of material, quantity of (a) substance. **-mischung** f (*Paper*) pulp mixture. **-mühle** f (*Paper*) stuff engine, hollander. **-patent** n patent on a material (*or* substance). **-probe** f cloth sample, swatch. **-rahmen** m filter disk. **-sammlung** f collection of material (*or* data) (on a subject). **-teilchen** n particle of matter. **-übergang** m mass transfer. **-übergangszahl** f mass transfer coefficient. **-umsatz** m change of substance, *specif* metabolism. **-verbindung** f (*Patents*) composition of material. **-verwandtschaft** f chemical affinity. **-wasser** n (*Paper*) pulp process effluent. **-wechsel** m metabolism

Stoffwechsel· metabolism, metabolic: **-analyse** f metabolism analysis. **-beschleunigung** n metabolic equilibrium. **-gefälle** n metabolic gradient. **-gleichgewicht** n metabolic equilibrium. **-größe** f metabolic rate. **-krankheit** f metabolic disorder (*or* disease). **-produkt** n metabolic product. **-störung** f metabolic disturbance. **-vorgang** m metabolic process

Stoff·wert m coefficient for a particular substance. **-zahl** f number of substances. **-zustand** m state of aggregation

stollen vt (*Lthr.*) stake

Stollen m (-) 1 (*Min.*) tunnel, gallery, drift. 2 post, prop. 3 (loaf-shaped) fruit cake

stolpern vi(s) stumble, trip; blunder

stolz a 1 proud. 2 imposing, splendid. **Stolz** m pride. 2 arrogance

Stomachal·mittel n, **Stomachikum** n (..ka) stomachic, stomach remedy

Stopf·buchse, -büchse f (*Mach.*) stuffing box

stopfen vt 1 stuff, cram; plug (up), stop (up). 2 darn, mend. II vi (*of Food*). 3 be filling. 4 cause constipation. 5 gorge oneself. **Stopfen** m (-) stopper, cork, plug.

stopfend p.a (*oft*) constipating; astringent, styptic

Stopf·garn n mending yarn. **-mittel** n (*Med.*) astringent, styptic. **-nadel** f darning needle. **-werg** n oakum

Stopp m (-s) 1 stop, halt, interruption. 2 ban; freeze. 3 hitchhiking

Stoppel m (-) 1 stopper, cork; plug. 2 stubble. **-zieher** m corkscrew

stoppen 1 vt/i stop, halt. 2 vt time with a stopwatch

Stopp·uhr f stopwatch

Stöpsel m (-) stopper, cork; plug. **-flasche** f stoppered bottle (*or* flask). **-glas** n stoppered glass. **-hahn** m stopcock. **-kasten** m (*Elec.*) resistance box

stöpseln vt stopper, cork, plug; plug in

Stöpsel-sicherung f (*Elec.*) plug fuse

Stör m (-e) sturgeon

Stör· static, interference: **st–anfällig** a susceptible to static (interference, "bugs")

Storax·harz n storax (resin)

Storch m (-̈e) stork. **-schnabel** m 1 (draftsman's) pantograph. 2 (*Bot.*) =**-schnabelkraut** n cranesbill, geranium

stören I vt 1 disturb, bother, inconvenience. 2 put out of commission, damage, impair. II vt/i 3 interfere (with), (*Radio, oft*) jam. 4 interrupt. III vi intrude, be (*or* get) in the way. **störend** p.a (*oft*) troublesome, inconvenient

Stör·fall m (nuclear) accident (in a power plant). **st–frei** a trouble-free, static-free.

St–geräusch *n* static; background noise.
-ion *n* interfering ion
stornieren *vt* cancel
Stör·öl *n* sturgeon oil
störrig, störrisch *a* troublesome, refractory, obstinate
Stör·schall *m* noise, static. **-spiegel** *m* noise level. **-stellenhalbleiter** *m* defect semiconductor. **-strahlung** *f* interfering radiation
Störung *f* (-en) 1 disturbance, trouble; perturbation. 2 interruption, interference. 3 (*Radio*) static; jamming. 4 impairment, defect, disorder; breakdown, malfunction. 5 turbulence. 6 (*Geol.*) deformation. 7 traffic jam, delay, tie-up
Störungs·energie *f* perturbation energy. **st–frei** *a* trouble-free, undisturbed. **St–gleichung** *f* perturbation equation. **-methode** *f* (*Math.*) perturbation method. **-stelle** *f* trouble spot, point of interference. **-sucher** *m* trouble shooter
Störzel *m* mother liquor from sugar candy
Stoß *m* (ˉe) 1 push, thrust (*esp* with pointed instrument). 2 blow, stroke; kick. 3 impact; shock. 4 (*Elec.*) surge; pulse. 5 impulse; beat (of the heart); gust (of wind); puff (of breath); blast (of a horn). 6 seam; (butt) joint; rail joint. 7 stack, heap, pile; file (of papers). 8 (*Min.*) side (wall) (of the shaft). 9 (*Med.*) massive dose
stoß·artig *a* intermittent, in fits and starts
Stoß·butter *f* farm butter. **st–dämpfend** *a* shock-absorbing, cushioning. **St–dämpfer** *m* shock absorber; bumper. **-dämpfung** *f* shock absorption, cushioning. **-dauer** *f* duration of impact. **-dichte** *f* collision density
Stößel *m* (-) 1 pestle. 2 ram. 3 tappet
Stoß·elastizität *f* rebound resilience. **st–empfindlich** *a* sensitive to shock
stoßen* (stieß, gestoßen; stößt) I *vt* 1 push, shove; kick. 2 strike, thrust; stick (into); butt. 3 crush, pulverize, pound; ram. 4 come over, affect. 5: **sich** (*dat*) **eine Wunde st.** get hurt by running into sthg. II *vi*(h) 6 (**an**) border (on), abut, adjoin. 7 jolt (*as:* car on road). 8 (*Boiling liquids*) bump, knock. 9 recoil. III *vi*(s) 10 (+**an, auf**) come (upon), run (into); (+**gegen**) bump (against). 11 (**auf**) lead (to). 12 (**auf**) pounce (on). 13 (**zu**) join. IV: **sich st.** 14 (**an**) bump (into). 15 (**an**) be disturbed, be bothered (by)
Stoß·energie *f* impact energy
Stößer *m* (-) 1 pestle. 2 rammer, pounder
Stoß·fänger *m* 1 shock absorber. 2 pressure equalizer. **st–fest** *a* shockproof, shock-resistant. **St–festigkeit** *f* shock resistance, impact strength. **st–frei** *a* shock-free, smooth. **St–glanz** *m* (*Lthr.*) friction glaze. **-glühung** *f* flash annealing. **-ionisation** *f* impact ionization. **-kette** *f* chain of collisions. **-kraft** *f* impact force; percussive power; (*Phys.*) impulse load. **-mahlung** *f* impact grinding. **-mine** *f* (*Mil.*) contact mine. **-naht** *f* butt joint. **-ofen** *m* pusher furnace. **-punkt** *m* point of impact. **-querschnitt** *m* collision area (cross-section, diameter); shock area. **st–reizbar** *a* shock-sensitive. **St–schweißung** *f* butt-wielding. **-spannung** *f* surge voltage. **-stange** *f* (car) bumper
stößt pushes, etc: *pres of* STOSSEN
Stoß·teilchen *n* bombarding particle. **-therapie** *f* massive dose therapy. **-verbreitung** *f* (*Spect.*) impact expansion, collision broadening. **-versuch** *m* impact test. **-waage** *f* ballistic pendulum. **st–weise** *adv* 1 in fits and starts, jerkily. 2 by impact. 3 intermittently, pulsatingly. 4 in batches. **St–welle** *f* shock (*or* impact) wave. **-widerstand** *m* impact (*or* shock) resistance. **-zahl** *f* number of collisions, impact number. **-zahn** *m* tusk. **-zeit** *f* 1 rush hour. 2 impact time. **-zünder** *m* percussion fuse
stottern *vi* stutter; (*Motor*) splutter, miss
Str. *abbv* (Straße) St. (street)
Strafe *f* (-n) punishment, penalty; fine. **strafen** *vt* punish; fine. **strafend** *p.a* (*oft*) penal
straff *a* 1 taut, tight. 2 rigorous, firm. 3 straight. 4 austere. 5 concise. **straffen** *vt* pull taut, tighten; firm (up); straighten; make concise. **Straffheit** *f* tautness, tightness; rigor, firmness; straightness; austerity; conciseness
Straf·geld *n* fine *cf* STRAFE
sträflich *a* criminal, unpardonable
straflos *a* unpunished; without penalty
Straf·recht *n* criminal law

Strahl *m* (-en) 1 ray; beam. 2 stream, jet (of gas, liquid). 3 flash (of lightning). 4 (*Math.*) straight line. **~antrieb** *m* jet propulsion. **-apparat** *m* jet apparatus, *esp* steam-jet injector. **-asbest** *m* plumose asbestos. **-baryt** *m* radiated barite, Bologna stone. **-blende** *f* (*Min.*) sphalerite. **-brenner** *m* (*Gas*) radiant burner. **-düse** *f* jet nozzle

strahlen I *vi* 1 radiate, emit rays. 2 beam, shine, gleam. **II** *vt* transmit (radio, TV signals)

Strahlen- ray, radiation; (*Anat.*) ciliary: **-art** *f* kind of rays, type of radiation. **st-artig** *a* ray-like radiating. **St-behandlung** *f* radiation treatment (*or* therapy). **-belastung** *f* exposure to radiation; (*Phys.*) radiant load. **-biologie** *f* radiation biology. **st-brechend** *a* refractive. **St-brechung** *f* refraction. **-bündel, -büschel** *n* pencil of lines (*or* rays); beam, **-chemie** *f* radiation chemistry

strahlend *p.a* (*oft*) radiant, radioactive; brilliant *cf* STRAHLEN

Strahlen-dosis *f* dose of radiation. **st-empfindlich** *a* sensitive to radiation, radiosensitive. **St-erz** *n* (*Min.*) clinoclasite. **-figur** *f* radiating figure. **-filter** *m* ray filter. **st-förmig** *a* radiate(d). **St-gang** *m* beam, light path, path of rays. **-glimmer** *m* striated mica. **-härte** *f* hardness of radiation. **-kegel** *m* cone of rays. **-krankheit** *f* radiation sickness. **-kunde** *f* radiology. **-kupfer** *n* (*Min.*) clinoclasite. **-messer** *m* radiometer; actinometer. **-optik** *f* geometrical optics. **st-optisch** *a* regarding geometrical optics. **St-parenchym** *n* (*Bot.*) medullary ray. **-pilz** *m* actinomyces. **-schaden** *m* radiation damage. **-schutz** *m* radiation safeguard; radiation shield(ing). **-sonne** *f* (*Biol.*) astrosphere. **-stein** *m* (*Min.*) actinolite; dufrenite; amianthus. **-therapie** *f* radiotherapy. **-tierchen** *npl* (*Zool.*) radiolaria. **-tod** *m* radiation death. **st-verseucht** *a* radiation-sick; radiation-contaminated. **St-verseuchung** *f* radiation sickness (*or* contamination). **-werfen** *n* (ir)radiation. **-wirkung** *f* radiation effect

Strahler *m* (-) radiator; beacon
Strahl-erz *n* (*Min.*) clinoclasite. **st-förmig** *a* ray-like. **St-gips** *m* fibrous gypsum
strahlig *a* radiated, radiant, in rays; radiosymmetrical; fibrous
Strahl-keil *m* (*Min.*) belemnite. **-kies** *m* (*Min.*) marcasite. **-kondensator** *m* jet condenser. **-körper** *m* radiating body, radiator. **-motor** *m* jet engine. **-pumpe** *f* jet pump, injector. **-punkt** *m* radiant (*or* radiating) point. **-quarz** *n* fibrous quartz. **-schörl** *m* radiated tourmaline. **-stein** *m* (*Min.*) actinolite, dufrenite; amianthus. **-strom** *m* jet stream. **-triebwerk** *n* jet-propulsion unit; jet engine. **-übergangswahrscheinlichkeit** *f* radiative transition probability

Strahlung *f* (-en) radiation; radiance
strahlungs-aktiv *a* radioactive. **St-aktivierung** *f* radioactivity. **-druck** *m* radiation pressure. **-einfang** *m* (*Nucl.*) radiative capture. **-empfänger** *m* radiation absorber. **-energie** *f* radiant energy. **-erkrankung** *f* (case of) radiation sickness. **-fläche** *f* emitting (*or* radiating) surface. **-fluß** *m* radiation flux. **-gürtel** *m* (Van Allen) radiation belt. **-intensität** *f* intensity of radiation. **st-los** *a* radiation-free, radiationless; without radiation. **St-messer** *m* radiation meter. **-meßgerät** *n* radiation detector, (*specif*) Geiger counter. **-schaden** *m* radiation damage. **-sprung** *m* radiative transition. **-stoß** *m* radiation impulse. **-übergang** *m* radiative transition. **-verbrennung** *f* radiation burn. **-vermögen** *n* radiating power. **st-verseucht** *a* radiation-contaminated. **St-wärme** *f* heat of radiation; radiant heat. **-wirkung** *f* radiation effect. **-zone** *f* 1 (*genl*) radiation zone. 2 (*specif*, =GÜRTEL) radiation belt

Strahl-vortrieb *m* jet propulsion. **-wärme** *f* heat of radiation; radiation heat. **-wäsche** *f* washing with a jet. **-zeolith** *m* (*Min.*) stilbite

Strähne *f* (-n) 1 hank, skein; strand (of hair). 2 stream (of liquid, gas); beam (of light). 3 phase, stage

Stramin *m* (-e) (*Tex.*) lining (*or* embroidering) canvas; open-mesh linen

stramm *a* 1 tight(-fitting), taut. 2 snappy (rubber). 3 strict. 4 strenuous (work). 5 strapping, well-built, buxom. **Stramm·heit** *f* tightness, tautness, etc. **stramm·ziehen*** *vt* pull taut, tighten

Strand *m* (-e *or* ¨e) beach, (sea)shore. **··ablagerung** *f* (*Geol.*) coastal (*or* shore) deposit. **-gut** *n* flotsam. **-nelke** *f* (*Bot.*) sea lavender. **-pflanze** *f* seaside (*or* littoral) plant. **-terrasse** *f* (*Geol.*) raised (*or* elevated) beach

Strang *m* (¨e) 1 rope; halter. 2 skein, hank. 3 strand, thread (*esp fig*). 4 (*Anat.*) cord, chord. 5 line (of pipe, track). 6 (*Paper*) web. 7 (*Metll.*) slab, billet. **··färberei** *f* hank (*or* rope) dyeing. **st–farbig** *a* dyed in the yarn. **St–garn** *n* skein (*or* hank) yarn. **st–gepreßt** *a* extruded. **St–gewebe** *n* (*Bot.*) vascular tissue. **-guß** *m* continuous (*or* direct) casting (process). **-presse** *f* 1 (*Ceram.*) wire-cutting press. 2 (*Metals*) extrusion press. 3 (*Soap*) plodder. **-pressen** *n* (*Metals*) extrusion

Strangpreß·erzeugnis *n* extrusion product. **-form** *f* extrusion die

Strang·preßling *m* extruded article

Strapaze *f* (-n) hardship, struggle, ordeal. **strapazieren** 1 *vt* treat roughly, be rough on. 2: **sich st.** (over)exert oneself. **strapaziös** *a* strenuous, exhausting

Straß *m* (. . asse) imitation gem; paste, strass

Straße *f* (-n) 1 road, highway. 2 (city) street. 3 sea lane. 4 (*Geog.*) strait. 5 (*Mach.*) production line. 6 (*Metll.*) rolling mill. *Cf* STRASS

Straßen·bahn *f* streetcar (*or* trolley) (line). **-bau** *m* road construction. **-bauöl** *n* road oil. **-belag** *m* road surfacing (*or* pavement). **-beleuchtung** *f* street lighting. **-decke** *f* street (*or* road) surface. **-kehricht** *m* street sweepings; road scrapings. **-laterne, -leuchte** *f* street lamp, lamppost. **-material** *n* road material. **-pflaster** *n* sheet (*or* road) pavement. **-teer** *m* road (*or* paving) tar. **-unfall** *m* highway accident. **-verkehr** *m* street (*or* highway) traffic

stratifizieren *vt* stratify

sträuben I *vt* 1 ruffle (*esp.* feathers). II:

sich st. 2 bristle, stand on end. 3 resist, refuse

sträubig *a* 1 rebellious. 2 (*Wood*) rough, coarse

Strauch *m* (¨er) shrub, bush

strauch·artig *a* shrub-like, shrubby, frutescent

straucheln *vi*(s) stumble

Strauch·holz *n* brushwood, underbrush

Strauß *m* I (¨e) 1 bunch of flowers, bouquet; (*Bot.*) thyrsus. 2 ostrich. II (-e) fight; showdown

Strebe *f* (-n) strut, brace, buttress

streben 1 *vi* (h) (+**nach**) strive (for). 2 *vi*(s) (*oft* + **nach**) push (*or* press) on (toward); tend (toward); head (for) —**in die Höhe st.** rise, soar. **Streben** *n* striving; tendency

strebsam *a* ambitious, industrious

streckbar *a* stretchable, extensible; (*Metll.*) ductile. **Streckbarkeit** *f* extensibility, ductility

Strecke *f* (-n) 1 (*genl*) stretch (implying distance or time). 2 distance, way; route; (*Tlph.* or *Transp.*) line, track. 3 while. 4 (*Math.*) line, segment. 5 (*Min.*) drift, gallery; road. 6 (*Metll.*) drawing. 7 passage (in a book)

strecken 1 (*genl*) *vt* & **sich st.** stretch, extend. 2 *vt* flatten, spread; draw (metals, thread); roll (metal, glass); liquor (soap). *Cf* GESTRECKT

Strecken·spektrum *n* continuous spectrum. **-teilchen** *n* (*Math.*) linear element. **st–weise** *adv* in stretches, here and there

Strecker *m* (-) 1 (*Anat.*) extensor (muscle). 2 (*Glass*) flattener

Streck·festigkeit *f* stretching strength; elongation resistance. **-formverfahren** *n* (*Plastics*) vacuum forming process. **-grenze** *f* 1 elastic limit. 2 yield point, proof stress. **-metall** *n* expanded metal. **-mittel** *n* 1 diluting agent, extender, filler. 2 (*Soap*) runnings. **-muskel** *m* (*Anat.*) extensor muscle. **-ofen** *m* (*Glass*) flattening (*or* annealing) furnace. **-pech** *n* coaltar pitch (with additives). **-probe** *f* tensile test. **-rahmen** *m* stretching frame. **-stahl** *m* rolled steel

Streckung *f* (-en) stretching, extension;

protraction, expansion; dilution.
Streckungs·mittel *n* 1 diluting agent,
etc. 2 (*Soap*) runnings. =STRECKMITTEL
Streck·verfahren *n* (*Plastics*) drape forming process. **-werk** *n* 1 stretching machine. 2 (*Metll.*) rolling mill
Streich *m* (-e) 1 stoke, blow. 2 prank, caper
Streich·artikel *m* coated article
streichbar *a* 1 plastic. 2 brushable; spreadable; (*Paint*) workable
Streich·beize *f* (*Lthr.*) brush pickle.
-bürste *f* paint brush; (*paper*) sizing
brush
Streiche *f* (-n) spatula
streichen* (strich, gestrichen) I *vt* 1 stroke,
caress; run (one's hand) over; brush. 2
spread, smear, apply. 3 paint; (*esp Paper*)
coat. 4 strop (razors). 5 strike (matches);
scrape (skins); card (wool); whet. 6
(*Ceram.*) mold. 7 strike (*or* cross) out, delete, cancel; erase. 8 skim off (foam); level
off (a cupful, etc) *cf* GESTRICHEN II *vi*(s) 9
roam; (*Birds*) migrate; (*Wind, Gas*)
sweep. 10 stretch, extend. 11 (+**um**)
sneak, prowl (around). 12 (*Min., Geol.*)
strike, run (in a certain direction);
zutage st. crop out
streich·fähig *a* spreadable; (*Paint*) brushable. **St–fähigkeit** *f* spreadability;
brushability. **-farbe** *f* brush-on color,
paint; (*Paper*) staining color. **st–fertig** *a*
(*Paint*) ready for brushing, ready-mixed.
St–fläche *f* striking surface (for
matches). **-garn** *n* carded yarn. **-holz,**
-hölzchen *n* (friction) match. **-instru-**
ment *n* stringed instrument. **-kappe** *f*
(*Explo.*) friction cap. **-käse** *m* cheese
spread. **-kasten** *m* (*Dye.*) color tub.
-kraut *n* dyer's rocket. **-lack** *m* brushing
lacquer. **-leder** *n* strop. **-maschine** *f* coating machine. **-masse** *f* 1 (*genl*) coating
composition. 2 (*Matches*) friction composition. **-messer** *n* spatula. **-mischung** *f*
coating mixture. **-mittel** *n* paint, varnish.
-muster *n* stained-paper pattern. **-ofen**
m reverberatory furnace. **-papier** *n*
coated paper. **-paste** *f* coating paste.
-richtung *f* (*Min., Geol.*) strike line.
-stein *m* touchstone; hone. **-torf** *m*
pressed (*or* molded) peat
Streichung *f* (-en) 1 stroking, brushing,

spreading, painting, etc *cf* STREICHEN 1–
6. 2 cut, deletion; cancelation
Streich·wolle *f* carded wool
Streif *m* (-e) stripe, etc = STREIFEN
Streif·band *n* (mailing) wrapper
Streife *f* (-n) patrol; raid
streifen I *vt/i* 1 brush (against), graze:
(*Opt.*) **st—der Einfall** grazing incidence.
II *vt* 2 touch on (a topic). 3 brush off. 4
stripe, streak; striate. 5 slip (on, off,
over); strip (off). III *vi* (h) (+**an**) border
(on). IV *vi*(s) ramble; prowl; patrol
Streifen *m* (-) 1 stripe, streak, band,
stria(tion), line. 2 strip (*incl Film*). 3 welt.
4 vein (in marble, etc). **-gefüge** *n* banded
structure. **-kohle** *f* banded coal. **-spek-**
trum *n* band spectrum. **st—weise** *adv* in
strips (stripes, streaks)
streifig *a* streaked, striped, banded;
striated
Streif·schuß *m* grazing shot
Streifung *f* (-en) striping, streaking,
striation
Streik *m* (-s) strike, walkout. **streiken** *vi*
(go on) strike, refuse (to participate)
Streit *m* (-e) quarrel, argument, dispute;
fight **—im St. liegen** be in conflict.
streiten* (stritt, gestritten) *vi* & **sich st.**
quarrel, argue, dispute; fight
Streit·fall *m* dispute; case (in court). **-frage**
f controversy; question at issue; moot
point
streitig *a* 1 disputed; disputable. 2: **einem**
etwas st. machen dispute someone's
right to sthg. **Streitigkeit** *f* (-en) dispute,
argument
Streit·kräfte *fpl* combat forces. **st–süchtig**
a quarrelsome, argumentative
streng(e) *a* 1 strict; rigorous, harsh, severe.
2 stern. 3 pungent. **Strenge** *f* strictness,
rigor, harshness, severity; sternness;
pungency
streng·flüssig *a* viscous, refractory; difficultly fusible. **St–flüssigkeit** *f* viscosity,
refractoriness; difficult fusibility. **st–**
genommen *adv* strictly speaking, in a
strict sense. **St–lot** *n* hard solder
strengstens *adv* (most) strictly, severely
Streu *f* (-en) litter (for animals)
Streu· scattering; dusting; sprinkling;
powdered: **-blau** *n* powder blue. **-brand-**

bombe f scatter bomb. **-büchse** f, **-dose** f shaker (for salt, etc), sprinkle-top can. **-düse** f spray nozzle

streuen I vt **1** sprinkle, strew; dust. **2** vt/i scatter (incl Opt.); spread; spray; sand, salt (the streets, etc)

Streu·faktor m (Cryst.) scattering factor. **-glas** n powdered glass. **-gut** n sprinkling (or spreading) material (e.g. sand for icy roads, straw for stalls); spray; dust. **-kalk** m agricultural lime. **-körper** m scattering body. **-kupfer** n copper rain. **-licht** n scattered light. **-linie** f (Spect.) scattered line. **-puder** m dusting powder. **-pulver** n dusting (specif insect, lycopodium, bleaching) powder. **-salz** n shaker salt. **-sand** m spreadable (or sprinkling) sand (for icy roads, etc)

Streusel n crumbly top (for cakes, buns)

Streu·spannung f stray voltage. **-spektrum** n scattered spectrum. **-strahlung** f scattered (or stray) radiation

Streuung f **1** sprinkling, strewing, dusting; scattering, spreading; spraying; sanding; salting cf STREUEN. **2** (Elec.) leakage. **3** deviation, variation

Streu·vermögen n **1** scattering power. **2** (Electroplating) throwing power. **-wachs** n sprinkling wax. **-welle** f scattered wave. **-wert** m erratic value; amount of scattering. **-winkel** m angle of scattering. **-würze** f powdered (mixed) condiment. **-zinn** n tin dust. **-zucker** m powdered sugar

strich stroked, etc: past of STREICHEN

Strich m (-e) **1** (pen, pencil, brush) stroke. **2** line (esp Math, under a column of figures): **unterm St. bleibt** the bottom line is; **einen St. unter etwas ziehen** put an end to sthg. **3** (graduation, calibration) mark; (compass) point; (Math.) prime('); (incl Morse Code) dash. **4** streak, stria; grain (of wood, etc). **5** (Explo.) train (of powder). **6** batch; brood, flock. **7** (Tex.) nap, pile. **8** strip (of land, city street). **9** deletion, cut. **10** (Paint) coat. **-ätzung** f line engraving. **-einteilung** f (graduated) scale

Strichelchen n (-) little stroke (line, streak) cf STRICH

stricheln vt **1** mark with a dotted line (more

accurately a line of short dashes). **2** crosshatch. **3** streak

Strich·farbe f (Min.) streak (color). **-formulierung** f bond formula (writing). **-führung** f brushwork. **-gitter** n (Spect.) simple line grating. **-kreuz** n cross hairs (or wires). **-kultur** f streak culture. **-lage** f (Tex.) nap. **-marke** f locating mark, reference line. **-platte** f **1** ruled plate; graduated dial. **2** (Min.) streak plate. **-punkt** m semicolon. **st–punktiert** a marked with a dot-and-dash line. **St–regen** m localized rain (or shower). **-vogel** m migratory bird. **st–weise** a local(ized); adv oft here and there. **-zahl** f (Paint) number of coats. **-zeichnung** f line drawing

Strick m (-e) rope **—wenn alle St—e rei-ßen** if worst comes to worst

stricken vt knit; net, reticulate

strick·förmig a rope-like; (Anat.) restiform. **St–lava** f ropy lava. **-waren** fpl knitted goods

Strippe f (-n) shoestring; strap; telephone (cord)

stritt argued, etc: past of STREITEN

strittig a disputed, in dispute

Stroh n straw. **-blume** f (Bot.) everlasting. **-dach** n thatched roof. **st–farben, -farbig** a straw-colored. **St–feile** f coarse (or rough) file. **-flachs** m raw flax. **st–gelb** a straw-yellow. **St–häcksel** m, n chopped straw, chaff. **-halm** m (single) straw. **-hülse** f straw case (or envelope)

strohig a strawy, like straw

Stroh·kessel, -kocher m (Paper) straw boiler. **-papier** n straw paper. **-pappe** f strawboard. **-ring** m straw ring. **-stein** m (Min.) carpholite. **-stoff** m (Paper) straw stuff (or pulp). **-wein** m straw wine. **-zellstoff** m (Paper) straw pulp. **-zeug** n (Paper) straw stuff (or pulp)

Strom m (-e) **1** stream; flow; torrent **—in Strömen regnen** rain in torrents, pour. **2** current (incl Elec.); electricity. **-ableitung** f (Elec.) shunt. **-abnahme** f (Elec.) **1** drop (or fall) in current. **2** current drain. **3** current collection. **-abnehmer** m (Elec.) **1** current collector. **2** (commutator) brush. **3** consumer (of electricity). **-abweichung** f current variation. **-anzeiger** m current indicator. **-art** f (Elec.) type of current.

-aufnahme f (*Elec.*) charging (rate). **-ausfall** m power outage, blackout. **-bahn** f path of the current; flow path. **-bild** n flow sheet. **-dichtigkeit** f current density. **st–durchflossen** a current-carrying, live. **St–durchgang** m passage of the current. **-einheit** f unit of current

strömen vi(s) stream, flow; (*Rain*) pour

Strom-entnahme f (*Elec.*) current consumption. **-erzeuger** m (*Elec.*) generator, dynamo. **-erzeugung** f (*Elec.*) current generation. **-feld** n (*Elec.*) current field. **st–führend** a (*Elec.*) current-carrying, live. **St–geschwindigkeit** f velocity of the current. **-indikator** m current indicator, ammeter. **-induktion** f induction of currents, electromagnetic induction. **-kreis** m (*Elec.*) circuit cf AUFDRUCKEN. **-kreisunterbrecher** m circuit breaker. **-lauf** m flow of current. **-leistung** f power wattage. **st–leitend** a (*Elec.*) conducting. **St–leiter** m (electric) conductor. **-leitung** f electric conduction. **-linie** f streamline. **st–linienförmig** a streamlined

stromlos a (*Elec.*) dead, without current.

Stromlosigkeit f absence of current

Strom-menge f (*Elec.*) amount of current, current strength. **-messer** m 1 flowmeter, current meter. 2 (*Elec.*) ammeter. **-netz** n electric, power network (*or* system). **-quelle** f source of current. **-regler** m flow (*or* current) regulator. **-richter** m (*Elec.*) converter. **-richtung** f direction of current; (*Elec. oft*) polarity. **-richtungsanzeiger** m (*Elec.*) polarity indicator. **-schiene** f (*Elec.*) 1 bus bar. 2 third rail. **-schlag** m electric shock. **-schluß** m circuit closing. **-schlüssel** m (*Elec.*) key, switch. **-schreiber** m current recorder. **-schwankung** f current fluctuation. **-spannung** f (electric) current potential, voltage. **-speicher** m (*Elec.*) accumulator, storage battery. **-spule** f (*Elec.*) current coil, solenoid. **-stärke** f current strength (*or* intensity), (*Elec. oft*) amperage. **-stärkemesser** m ammeter, galvanometer. **-störung** f (*Elec.*) power outage, blackout. **-stoß** m current pulse, surge of current. **-tor** n (*Elec.*) thyratron. **-trenner** m (*Elec.*) switch. **-umkehrer** m

(*Elec.*) commutator; reversing switch. **-umkehrung** f (*Elec.*) reversal of current

Strömung f (-en) 1 streaming, flowing cf STRÖMEN. 2 current, stream, flow. 3 (*Gases*) transpiration. 4 (magnetic) flux. 5 tendency, trend

Strömungs-doppelbrechung f streaming birefringence. **-geschwindigkeit** f flow (*or* current) velocity. **st–gleich** a isokinetic. **-günstig** a streamlined. **St–mechanik** f fluid dynamics. **-messer** m flowmeter, current meter

Strom-unterbrecher m circuit breaker; interrupter; cutout. **-verbrauch** m consumption of electricity. **-verbraucher** m consumer of electricity. **-verdrängung** f current displacement. **-verlust** m loss of current. **-wandler** m (*Elec.*) current transformer. **-wechsel** m (*Elec.*) alternation. **-wechsler** m (*Elec.*) commutator. **-weg** m path of the current. **-wender** m (*Elec.*) current reverser, (*specif*) commutator. **-zeiger** m current indicator; (*Elec. oft*) ammeter. **-zelle** f electric cell. **-zinn** n stream tin. **-zuführung** f (*Elec.*) power supply

Strontian m, **-erde** f strontia, strontium oxide. **st–haltig** a strontia-bearing, containing strontia. **St–salpeter** m strontium nitrate. **-salz** n strontium salt. **-verfahren** n (*Sugar*) strontia process. **-wasser** n strontia water, strontium hydroxide solution. **-weiß** n strontium white. **-zucker** m strontium sucrate

Strontium-gehalt m strontium content. **-jodid** n strontium iodide. **-oxyhydrat** n strontium hydroxide. **-salpeter** m strontium nitrate. **-salz** n strontium salt. **-wasser** n strontia water. **-wasserstoff** m strontium hydride

Strophanth-säure f strophanthic acid. **Strophant(h)in-säure** f strophanthinic acid

Strosse f (-n) (*Min.*) (tunnel) floor; level

strotzen vi (+von, vor) be brimming (*or* bursting) (with); be thick (with dirt)

Strudel m (-) 1 whirlpool, eddy, vortex. 2 strudel. **strudellos** a noneddying, irrotational

Struktur f (-en) 1 structure; (*in compds oft*)

structural. 2 (*Tex.*, *Petrog.*) texture. **-än-derung** *f* change in structure (*or* texture). **-bestandteil** *m* structural constituent. **-bestimmung** *f* determination of structure. **-beweis** *m* proof of structure. **-chemie** *f* structural chemistry

strukturell *a* structural

struktur·empfindlich *a* (*Cryst.*) structure-sensitive. **St–farbe** *f* structural color. **-festigkeit** *f* (*esp Plastics*) tearing strength, tear resistance. **-formel** *f* structural formula. **st–identisch** *a* structurally identical

strukturiert *p.a* structured

Struktur·isomerie *f* structural isomerism. **-lehre** *f* structure theory. **st–los** *a* unstructured. **St–veränderung** *f* change in structure (*or* texture). **-viskosität** *f* intrinsic viscosity

Strumpf *m* (⁻e) 1 stocking; sock; *pl oft* hosiery. 2 incandescent gas mantle. **--band** *n*, **-halter** *m* garter. **-waren** *fpl* hosiery

Strunk *m* (-e *or* ⁻e) 1 stump, stub. 2 stalk, stem, trunk

struppig *a* bristly; shaggy

Strychninol·säure *f* strychninolic acid

Strychninon·säure *f* strychninonic acid

Stubbe *f* (-n), **Stubben** *m* (-) stump

Stube *f* (-n) room; parlor

Stuck *m* stucco; plaster (of Paris)

Stück *m* (-e) 1 piece, unit; part; (*Soap*) cake; (*Sugar*) lump: **zwei Mark das St.** two marks apiece; **in einem St. tun** do without a break. 2 (*Quant.*) bit (of); head (of cattle); *oft no transl:* **wieviele Gläser?—20 St.** how many glasses?—20 (of them). 3 (short) distance, way. 4 coin; denomination. 5 (*pl*) securities, stocks, etc. 6 (*Metll.*) bloom. 7 trick, caper. 8 (art) work; (stage) play. 9: **in allen St—en** in every respect. 10: **große St—e halten (auf)** think highly (of). **--arbeit** *f* piecework

Stückchen *n* (-) little piece, bit, particle. **--zucker** *m* lump sugar

stückeln *vt* 1 cut into pieces. 2 piece (things) together, patch

Stücken·zucker *m* lump sugar

Stück·erz *n* lump ore. **-färbekufe** *f* piece-dyeing vat. **-färber** *m* piece dyer. **-färberei** *f* piece dyeing. **st–farbig** *a* dyed in

the piece. **St–färbung** *f* 1 piece dyeing. 2 (*Micros.*) tissue staining (in toto). **-form** *f* lump form. **-gewicht** *n* individual (*or* unit) weight

Stuck·gips *m* quick-settling plaster (of Paris), hemihydrate gypsum plaster

Stück·größe *f* lump size, piece size. **-gut** *n* 1 gun metal. 2 piece goods. 3 parcel freight. **-holz** *n* stick wood, wood in pieces

stückig *a* in pieces, in lumps

Stück·kalk *m* lump lime. **-kohle** *f* lump coal. **-koks** *m* lump cake. **-lohn** *m* piece wages. **-metall** *n* gun metal

Stuck·mörtel *m* stucco; badigeon

Stück·ofen *m* batch furnace. **-ofenstahl** *m* (*Steel*) bloom. **-preis** *m* price per piece. **-schlacke** *f* lump slag. **-seife** *f* bar soap. **-waren** *fpl* piece goods. **-wäsche** *f* piece washing (*or* scouring). **st–weise** *adv* by the piece, piecewise; (by) retail

Student *m* (-en, -en), **Studentin** *f* (-nen) (university-level *or* graduate) student

Studie *f* (-n) study. **Studien** studies: *pl of* STUDIE & STUDIUM

studieren *vt/i* study; *vi* attend a university. **Studierende** *m,f* (*adj endgs*) (university-level) student. **studiert** *p.a* (*oft*) university-educated, learned. **Studierte** *m,f* (*adj endgs*) scholar, university graduate

Studium *n* (. . ien) study; studies, university education

Stufe *f* (-n) 1 (*genl.*) step; (*pl oft*) stairs. 2 stage, phase, gradation. 3 rung, level (*esp implying rank*). 4 piece of ore; (*pl*) poorest mercury ore

stufen·artig *a* 1 step-like, stepped. 2 terraced. 3 graded, gradual, stepwise

Stufen·erz *n* raw ore; lump ore. **-folge** *f* succession of steps (*or* stages); gradation. **st–förmig** *a* 1 step-like, stepped. 2 terraced. 3 graded, gradual. **St–gang** *m* succession (of steps), gradation. **-gesetz** *n* law of stages. **-gitter** *n* (*Spect.*) echelon grating. **-heizung** *f* 1 heating in stages. 2 (*Rubber*) step-up cure. **-keil** *m* step wedge. **-leiter** *f* 1 (step)ladder. 2 succession, gradation, scale. **-linse** *f* (*Spect.*) echelon lens. **st–los** *a* stepless, continuous, infinitely variable. **St–photometer** *n* step photometer. **-prisma** *an* echelon prism. **-prozeß** *m* stepwise process. **-rakete** f multistage rocket. **-reaktion** f stepwise reaction. **-regel** f (Ostwald's) rule of stages, law of successive reactions. **-schalter** m multistage switch. **-trocknen** n graduated-temperature drying. **st–weise** a stepwise, graduated, gradual; adv oft step by step, in stages. **St–zahl** f number of stages

stufig a stepwise; stepped, terraced; graduated

Stufung f (-en) stepwise arrangement, gradation

Stuhl m (-̈e) 1 chair, seat, stool. 2 (Med.) stool: **St. haben** have a bowel movement. 3 (at end of compd nouns, oft) loom cf WEBSTUHL

stuhl· fecal, etc: **-befördernd** a laxative. **St–gang** m bowel movement, defecation. **-zäpfchen** n suppository. **-züchtung** f (Bact.) stool (or fecal) culture

Stukkateur m (-e) stucco worker, plasterer. **Stukkatur** f (-en) stucco work; plastering

stülpen vt 1 put over (as a cover). 2 turn (inside) out. 3 turn upside down. 4 turn up (cuffs, etc)

stumm a mute; silent

Stummel m (-) stub, stump; (cigar) butt

Stummheit f muteness; silence

stumpf a 1 blunt. 2 dull (incl colors, surfaces). 3 (Geom.) obtuse (angle); truncated (cone). 4 stubby, snub(bed). 5 apathetic, insentive, obtuse

Stumpf m (-̈e) stump, stub; (Geom.) frustum

stumpf·eckig a blunt-cornered; obtuse-angled. **-kantig** a blunt-edged, dull. **St–kegel** m truncated cone. **-schweißung** f butt welding. **-sinn** m 1 stupidity, denseness. 2 apathy. 3 tedium. **st–sinnig** a stupid; dull; apathetic; tedious. **St–stoß** m butt joint. **st–winklig** a obtuse-angled

Stunde f (-n) 1 hour. 2 lesson; (class) period. 3 time

stünde would stand, etc: sbjc of STEHEN

stunden vt give an extension (or more time to pay)

Stunden·geschwindigkeit f hourly rate, speed per hour. **-glas** n hourglass. **-kilometer** n kilometer per hour. **st–lang** 1 a

hours of (waiting, etc). 2 adv for hours. **St–leistung** f hourly output. **-plan** m timetable; schedule. **-zeiger** m hour hand

·stündig a sfx -hour

stündlich a hourly

Stupp n mercurial soot, stupp

Stuppea·säure f stuppe(a)ic acid

Stupp·fett n greasy mixture of hydrocarbons obtained in refining mercurial soot

stürbe would die, etc: sbjc of sterben

Sturm m (-̈e) 1 storm. 2 rush; (Mil.) attack; storming: **im St. nehmen** take by storm; **im St. angreifen, St. laufen gegen** attack, storm (a position). 3 **St. läuten** sound the alarm. **stürmen** 1 vt/i storm. 2 vi be stormy

sturm·fest a stormproof. **St–hut** m (Bot.) monkshood, aconite

stürmisch a stormy; wild

Sturz m (-̈e) 1 fall, plunge. 2 downfall; collapse; overthrow. 3 drop. 4 (Meteor.) sudden change (for the worse). 5 (Iron) slab, plate. 6 (Min.) dumping ground. 7 glass bell (as a cover). **--acker** m newly plowed field. **-blech** n (black) sheet iron. **-bomben** n dive bombing

Stürze f (-n) lid, cover

stürzen I vt 1 fling (hurl, cast) down. 2 drop. 3 overthrow; upset; turn over (and empty). 4 (Math.) transpose. 5 plow (for the first time). II vi(s) 6 fall, drop, plunge. 7 dash, rush. 8 stream, pour (out of). III: **sich st.** 9 fling oneself (e.g. into one's work). 10 make a mad dash (for)

Sturz·festigkeit f shattering strength. **-flamme** f reverberatory flame. **-guß** m downhill casting. **-güter** npl bulk goods

Stürz·mühle f tumbling mill

Sturz·probe f slump test; (Coke) shatter test. **-rinne** f chute

Stürzung f (-en) flinging, hurling; dropping; overthrow; overturning; (Math.) transposition; first plowing cf STÜRZEN I

Sturz·walzen n pack rolling

Stute f (-n) mare. **Stuten·milch** f mare's milk

Stuto. abbv (Stundentonnen) metric tons per hour

Stütz·apparat m supporting apparatus; brace. **-balken** m supporting beam

Stütze f (-n) (supporting) pillar; support, brace, prop; mainstay; aid

stutzen 1 vt crop, clip, cut short, trim; cut. **2** vi stop short; be puzzled

Stutzen m (-) **1** short length of pipe. **2** tubulure

stützen I vt **1** hold (up), support. **2** (+**auf**) base (on). **II: sich st.** (usu + **auf**). **3** lean, rest (on). **4** be based (on). **5** use (or quote) for support

Stütz·fläche f supporting surface. **-gewebe** n supporting tissue

stutzig a: **1: st. werden** stop short, be puzzled. **2: st. machen** bring up short, puzzle

Stütz·mittel n supporting medium. **-punkt** m **1** (point of) support; fulcrum; pivot; bearing surface. **2** (usu Mil.) base. **-stoff** m, **-substanz** f supporting substance

Stutz·uhr f mantle clock

Stütz·walze f backing roll

Stycerin n phenylglycerol

Styphnin·säure f styphnic acid

Styrol n styrene

Styron n cinnamyl alcohol

s.u. abbv (siehe unten) see below

Suakin·gummi n Suakin (or Tolca) gum

subaerisch a subaerial

subatomar a subatomic

Subchlorür n subchloride

Suberin·säure f suberic acid

Subhaloid° n subhalide

subkutan a subcutaneous

subl. abbv (sublimiert) sublimed

Sublimat n (-e) (genl) sublimate; (specif) corrosive sublimate, mercuric chloride

Sublimations·probe f sublimation test (or sample). **-punkt** m sublimation point. **-wärme** f heat of sublimation

Sublimat·lösung f sublimate solution

sublimierbar a sublimable. **Sublimierbarkeit** f sublimability. **sublimier·echt** a nonsubliming. **sublimieren** vt/i sublime; sublimate

Sublimier·gefäß n sublimation vessel. **-ofen** m subliming furnace. **-topf** m subliming pot, aludel

Sublimierung f (-en) sublimation

Subl.P. abbv (Sublimationspunkt) sublimation point

Suboxid° n suboxide

Subphosphor·säure f hypophosphoric acid

Subst. abbv (Substanz) substance

Substantivität f substantivity

Substanz f (-en) substance, matter, material. **-fleck** m (Dye.) dye stain. **-menge** f amount of substance. **-muster** n sample in kind. **-verlust** m loss of material. **-wechsel** m change of substance

Substituent m (-en, -en) substituent

substituierbar a replaceable. **substituieren** vt/i substitute; (+**durch**) replace (by, with). **Substituierung** f (-en) substitution, replacement

Substitutions·isomerie f structural isomerism. **-produkt** n substitution product

Substrat n (-e) **1** substratum; (esp Biochem.) substrate. **2** foundation, basis

subsumieren vt subsume, include

subtil a subtle. **Subtilität** f (-en) subtlety

subtrahieren vt/i subtract

Subvention f (-en) subsidy. **subventionieren** vt subsidize

Such· searching, tracing: **-aktion** f organized search. **-bohrung** f exploratory boring (or drilling)

Suche f (usu + **nach**) search, hunt, quest (for); **auf der S. nach. . .** in search of, on the lookout for)

suchen 1 vt; vi(s) (**nach**) search, look (for), seek. **2** vi (h) (+**zu** + inf) seek, try (to). Cf GESUCHT

Sucher m (-) **1** seeker, searcher. **2** (Instrum.) probe; detector, tracer. **3** (Phot.) viewfinder

Such·geschoß n tracer projectile. **-lauf** m (Cmptrs.) search function. **-licht** n searchlight. **-spindel** f exploring spindle

Sucht f (-e) (usu + **nach**) **1** greed; craze, mania (for). **2** addition (to). **süchtig** a (+**nach**) addicted (to) —**das macht s.** that is addictive. **Süchtige** m, f (adj endgs) addict

Sud m (-e) **1** stock, broth. **2** decoction, extract. **3** brew. **4** (Dye.) mordant. **5** (Sugar) strike

Süd I m **1** south. **2** south wind. **II** pfx south, southern: **S–afrika** South Africa; **s–a·merikanisch** South American

Sudan·braun *n* Sudan brown. **-rot** *n* Sudan red

Süden *m* south, southern part: **nach S.** (toward the) south; **im S. von** in the southern part of. . .

Süd·frankreich *n* Southern France. **-frucht** *f* tropical fruit

Sud·haus *n* brew(ing) house

südlich I *a* 1 southern: **s—e Breite** south(ern) latitude; **der s—ste Punkt** the southernmost point. 2 Antarctic, South Polar. 3 southerly (wind, direction). **II** *adv* (+**von**), *prep* (+*genit*) (to the) south (of): **etwas s—er gelegen** located somewhat more to the south

Süd·licht *n* aurora australis. **-ost** *m* 1 southeaster. 2 =**-osten** *m* southeast. **s—östlich** *a* southeast(ern), southeasterly *cf* SÜDLICH

Sud·maische *f* (*Sugar*) crystallizer

Süd·pol *m* South Pole

Sud·salz *n* boiled salt (from boiled-down brine). **-seifenbad** *n* (*Dye.*) broken soap bath, broken suds

Süd·tirol *m* (*Geog.*) South Tyrol (*now part of Italy*). **s—wärts** *adv* southward. **S—wein** *m* southern (*Eur.*) wine

Sud·werk *n* brewing plant. **-wesen** *n* brewing

Süd·west *m* 1 southwest wind. 2 =**-westen** *m* southwest. **s—westlich** *a* southwest(ern), southwesterly *cf* SÜDLICH

süffig *a* (*esp Wines*) pleasant to drink

suggerieren *vt/i* suggest

Suhle *f* (-n) muddy pool, (hog) wallow, slough

Sukzession *f* (-en) succession. **sukzessiv** *a* successive

Sukzin·dialdehyd *m* succinaldehyde. **Sukzinimid** *n* succinimide. **Sukzinyl·säure** *f* succinic acid

Sulf· sulf(o)-, thio-: **amidharz** *n* sulfamide resin. **-amidsäure, -aminsäure** *f* sulfamic acid

Sulfan *n* (-e) hydrogen sulfide

Sulfanil·säure *f* sulfanilic acid. **-säure·amid** *n* sulfanilamide

Sulfansulfon·säure *f* thiosulfuric acid

sulfantimonig *a* thioantimonious, thioantimonite. **Sulfantimon·säure** *f* thioantimonic acid

sulfarsenig *a* thioarsenious, thioarsenite. **Sulfarsen·säure** *f* thioarsenic acid

Sulfat *n* (-e) sulfate. **-ablauge** *f* sulfate waste liquor. **-anlage** *f* sulfate installation

Sulfatation *f* sulfatization

sulfat·haltig *a* containing sulfate, sulfatic. **S—härte** *f* (*Water*) sulfate hardness. **-hüttenzement** *m* supersulfated cement

sulfatieren *vt* sulfate, sulfatize. **Sulfatierung** *f* sulfatization

Sulfat·ion *n* sulfate ion

sulfatisch *a* sulfatic

sulfatisieren *vt* sulfatize. **Sulfatisierung** *f* sulfatization. **Sulfatisierungs·mittel** *n* sulfatizing agent

Sulfat·kessel *m* (*Paper*) sulfate boiler. **-kocher** *m* (*Paper*) sulfate digester. **-papier** *n* sulfate (*or* kraft) paper. **-rest** *m* sulfate residue (*or* radical). **-schmelze** *f* crude recovered sodium carbonate. **-schwefel** *m* sulfate sulfur. **-see** *m* sulfate lake. **-soda** *f* sulfate (*or* Leblanc) soda. **-stoff** *m* (*Paper*) sulfate pulp. **-terpentinöl** *n* sulfate wood turpentine. **-verfahren** *n* sulfate process. **-zellstoff** *m* sulfate cellulose (*or* pulp)

Sulf·aurat *n* thioaurate. **-carbaminsäure** *f* thiocarbamic acid. **-carbanil** *n* thiocarbanil (phenyl isothiocyanate)

Sulfen·säure *f* sulfenic acid

Sulf·hydrat, -hydrid *n* hydrosulfide. **-hydrit** *n* hydrosulfite (dithionite)

Sulfid *n* (-e) sulfide, *esp* -ic sulfide *cf* SULFÜR. **-äscher** *m* (*Lthr.*) sulfide lime. **-erz** *n* sulfide ore

sulfidieren *vt* 1 sulfidize. 2 (*Viscose*) xanthate, treat with carbon disulfide. **Sulfidierung** *f* (-en) sulfidization; treatment with carbon disulfide

sulfidisch *a* sulfidic, containing sulfides

Sulfid·schwefel *m* sulfide sulfur

sulfieren *vt* sulfonate. **Sulfierung** *f* (-en) sulfonation

Sulfigran *n* sodium sulfide

Sulfin *n* sulfonium, sulfine. **-farbe,** *f,* **-farbstoff** *m* sulfur dye. **-salz** *n* sulfonium (*or* sulfine) salt. **-säure** *f* sulfuric acid

Sulfit *n* (-e) sulfite. **-ablauge** *f* (*Paper*) sulfite waste liquor. **-ätze** *f* (*Tex.*) sulfite dis-

charge. **-auslauger** m (*Paper*) sulfite digester. **-brei** m (*Paper*) sulfite pulp
sulfitieren vt (*Tex., Food*) sulfite. **Sulfitierung** f (-en) sulfitation
Sulfit·kocher m (*Paper*) sulfite digester. **-kochung** f (*Paper*) sulfite cooking. **-lauge** f sulfite liquor. **-laugenturm** m sulfite tower. **-pappe** f sulfite paperboard. **-reserve** f (*Tex.*) sulfite resist (*or* reserve). **-sprit** m alcohol from sulfite liquor. **-stoff** m (*Paper*) sulfite pulp. **-terpentinöl** n sulfite turpentine. **-turm** m sulfite tower. **-verfahren** n sulfite process. **-zellstoff** m sulfite cellulose (*or* pulp). **-zellstoffabrikation** f sulfite cellulose manufacture. **-zellulose** f sulfite cellulose (*or* pulp)
Sulf·kohlensäure f sulfocarbonic (*or* trithiocarbonic) acid
Sulfo· sulfo-, thio-: **-azetat** n sulfoacetate. **-base, -basis** f sulfur base. **-carbonat** n thiocarbonate. **-carbonsäure** f sulfocarboxylic acid. **-cyan** n thiocyanogen, sulfocyanogen; (*for derivatives see* SULFOZYAN·). **-cyanid** n thiocyanate. **-fettsäure** f alkylsulfonic acid. **-gruppe** f sulfo (*or* sulfonic acid) group. **-harnstoff** m thiourea. **-hydrat** n hydrosulfide. **-karbolsäure** f thiophenol. **-karbonsäure, -kohlensäure** f trithiocarbonic acid
Sulfolyse f (-n) sulfolysis
Sulfo·monopersäure f monoperoxysulfuric (*or* Caro's) acid
Sulfon·carbonsäure f sulfonecarboxylic acid
sulfonierbar a capable of being sulfonated. **sulfonieren** vt sulfonate. **Sulfonierung** f sulfonization
Sulfon·säure f sulfonic acid. **-säurecarbonsäure** f sulfocarboxylic acid
Sulfo·ölsäure f sulfooleic acid. **-persäure** f per(oxy)sulfuric acid. **-ricinat** n Turkey-red oil. **-salz** n thio (*or* sulfo) salt. **s-sauer** a sulfonate of the. **S–säure** f sulfo (*or* sulfonic) acid, thio acid. **-stannat** n thiostannate. **-stibiat** n thioantimonate. **-verbindung** f sulfo compound. **-zyan** n thiocyanogen, sulfocyanogen. **-zyanat** n thiocyanate
Sulfozyan·eisen n ferric thiocyanate
Sulfo·zyanid n sulfocyanide, thiocyanate

Sulfozyan·kalium n potassium thiocyanate. **s–sauer** a thiocyanate of. **S–säure** f thiocyanic acid. **-verbindung** f thiocyanate
Sulfo·thiokohlensäure f thiolthionocarbonic acid
Sulfür n (-e) (-ous) sulfide
sulfurieren vt sulfonate, sulfonize
Sulfur·öl n sulfocarbon oil (olive oil extracted from marc)
Sulfür·schwefel m sulfide sulfur
Sulze, Sülze f (-n) 1 brine, souse. 2 salt lick. 3 meat in aspic. **Sulz·fleisch** n pickled meat, souse. **sulzig, sülzig** a gelatinous. **Sulz·wurst** f head cheese, brawn
Sumach·abkochung f sumac decoction. **-behandlung** f (*Lthr.*) sumacking. **s–gar** a sumac-tanned. **S–gerbung** f sumac tanning
sumachieren vt (*Lthr.*) sumac
Sumbul·öl n oil of sumbul. **-wurzel** f sumbul (*or* musk) root
Summand m (-en, -en) (*Math.*) term of an addition
summarisch a summary
summa summarum adv all in all, when everything is added up
Summations·gift n cumulative poison
Sümmchen n (-) (tidy) little sum
Summe f (-n) sum, total; amount: **die S. ziehen** (von) sum up, total up
summen 1 vt/i hum. 2 vi buzz; sing
Summen·formel f empirical (*or* summation) formula. **-gleichung** f summation equation. **-regel** f rule of sums. **-satz** m principle (*or* law) of sums. **-wirkung** f combined action (*or* effect)
Summer m (-) buzzer
summieren 1 vt sum up, add. 2: **sich s.** add up, accumulate
Sumpf m (⸚e) 1 swamp, marsh. 2 (*Tech.*) sump, pit. 3 pool (of mercury). 4 wave absorbent. 5 base (of a fractionating column). **-boden** m marshy (*or* swampy) ground. **-dotterblume** f (*Bot.*) cowslip; marsh marigold. **-eisenstein** m bog iron ore
sümpfen vt 1 (*Min.*) drain. 2 (*Ceram.*) knead. 3 (*Clay*) sour
Sumpf·erz n bog ore. **-gas** n marsh gas, methane

sumpfig *a* swampy, marshy

Sumpf·kalk *m* slaked lime, pit lume. **-land** *n* marshland. **-luft** *f* marsh gas. **-moos** *n* swamp moss, sphagnum. **-nelke** *f* (*Bot.*) purple avens. **-öl** *n* sump oil. **-pflanze** *f* marsh plant. **-phase** *f* semisolid (*or* liquid) phase. **-porsch, -porst** *m* marsh tea. **-silge** *f* marsh parsley. **-verfahren** *n* liquid-phase process (in hydrogenation)

Sund *m* (-e) (*Geog.*) sound, strait

Sünde *f* (-n) sin, offense

Sundtit *m* (-e) (*Min.*) andorite

Super 1 *m* (-) (*Radio*) superheterodyne receiver. **2** *n* (-) super (gasoline). **-azidität** *f* hyperacidity

superfiziell *a* superficial

Super·legierung *f* permeability alloy. **-leitfähigkeit** *f* superconductivity. **-markt** *m* supermarket. **-oxid** *n* peroxide, superoxide

Superoxid·bleiche *f* peroxide bleach. **s-oxidecht** *a* peroxide-resistant. **S–hydrat** *n* hydrated peroxide

Super·phosphatschlempe *f* superphosphate-molasses-residue fertilizer

superponieren *vt* superpose

supersaturieren *vt* supersaturate

Suppe *f* (-n) soup

Suppen·kraut *n* 1 pot herbs. 2 soup greens. **-würze** *f* soup seasoning

supra·fluid *a* superfluid. **-leitend** *a* superconducting. **S–leitelektron** *n* superconduction electron. **-leiter** *m* superconductor. **-leitfähigkeit** *f* superconductivity. **-sterin** *n* suprasterol

surren *vi* buzz, whiz, hum

Surrogat *n* (-e) substitute, replacement. **-stoff** *m* (*genl*) substitute (material); (*specif: Paper*) pulp substitute

suspendieren *vt* suspend. **Suspendierung** *f* (-en) suspension

Suspensions·kolloid *n* suspensoid sol. **-mittel** *n* dispersing agent. **-polymerisat** *n* granular (*or* suspension) polymer

süß *a* 1 sweet; sugary, honeyed. 2 (*Water*) fresh. **Süß·bier** *n* sweet beer. **Süße** *f* sweetness. **süßen** *vt/i* sweeten. **Süß·erde** *f* beryllia, glucina (beryllium oxide). **-holz** *n* licorice

Süßholz·saft *m* licorice extract. **-zucker** *m* glycyrrhizic acid

Süßigkeit *f* (-en) 1 sweetness. 2 sweet(meat), confection (candy, cookie, etc)

süßlich *a* sweetish; sentimental. **-riechend** *a* sweet-smelling

Süß·mandelöl *n* oil of sweet almonds. **-milchkäse** *m* cream cheese. **-most** *m* 1 pure fruit juice, *esp* grape juice. 2 sweet cider. **-rahm** *m* sweet (dairy) cream. **s-sauer, -säuerlich** *a* bittersweet, sourishsweet. **S–speise** *f* dessert. **-stoff** *m* sweetener, dulcifier; (*specif*) saccharin. **-waren** *fpl* sweets, confectionery. **-wasser** *n* fresh water. **-wasserablagerung** *f* freshwater deposit. **-wein** *m* sweet wine

Susz. *abbv* = **Suszeptibilität** *f* (-en) susceptibility

s.W. *abbv* (spezifische Wärme) specific heat

Sweetchrombeize° *f* neutral chrome mordant

swl. *abbv* 1 (sehr wenig löslich) very difficultly soluble. 2 (schwerlöslich) sparingly soluble

s.w.u. *abbv* (siehe weiter unten) see below

Sylvan·erz *n* (*Min.*) sylvanite

Sylvester·abend *m* New Year's Eve

Sylvin *m* (*Min.*) sylvite. **-säure** *f* sylvic (*or* abietic) acid

Symbasis *f* (. . asen) agreement, correlation; (*Biol.*) symbasis. **symbat** *a* correlative

Symbiose *f* (-n) symbiosis

Symbolik *f* (-en) symbolism; notation. **symbolisch** *a* symbolic. **symbolisieren** *vt* symbolize

Symmetrie·achse *f* axis of symmetry. **-ebene** *f* plane of symmetry. **-grad** *m* degree of symmetry

symmetrieren *vt* symmetrize, balance

Symmetrie·stab *m* (*Cryst.*) symmetry line. **-zahl** *f* symmetry factor. **-zentrum** *n* center of symmetry

symmetrisch *a* symmetrical. **Symmetrisierung** *f* (-en) symmetrization

sympathetisch *a* sympathetic

sympathisch *a* likable, nice

sympathisieren *vi* sympathize

Synärese *f* (-n) syneresis

synchron *a* synchronous

Synchron·getriebe *n* synchromesh gear

Synchronisator *m* (-en) synchronizer

synchronisieren *vt* synchronize

synchronistisch *a* synchronous
Synchron·motor *m* synchronous motor
Syndikat *n* (-e) syndicate, combine, trust
Synergismus *m* (.. men) synergism, mutual action. **synergistisch** *a* synergistic
syngenetisch *a* syngenetic
Synthese *f* (-n) synthesis. **·gas** *n* synthesis
gas. **-gummi** *n*, **-kautschuk** *m* synthetic
rubber. **-ofen** *m* synthesis oven
synthetisch *a* synthetic. **synthetisieren** *vt*
synthesize. **Synthetisierung** *f* (-en) syn-
thesizing, synthesis
Syringa·aldehyd *m* syringic aldehyde.
-säure *f* syringic acid
Syrup *m* (-e) (*for compds see* SIRUP)
System *n* (-e) system; (atomic) plan.

·analytiker *m* systems analyst
Systematik *f* systematics, systematology;
(*Bot., Zool.*) taxonomy. **systematisch** *a*
systematic
System·berater *m* systems consultant.
-zahl *f* system number; file number.
-zwang *m* system-imposed limitation of
options
s.Z. *abbv* (seinerzeit) at that time, in its (*or*
his) time
SZ, S.Z. *abbv* 1 (Säurezahl) acid number. 2
(Sommerzeit) daylight-saving time
Szene *f* (-n) scene; stage: **in Sz. setzen** put
on, arrange, organize
Szintillations·zähler *m* scintillation
counter. **szintillieren** *vi* scintillate

T

(See Note under **Th. . .**)

t *abbv* 1 (Temperatur) temp. 2 (Tonne) metric ton

t. *abbv* (täglich) daily

T. *abbv* 1 (Tag, Tage) day(s). 2 (Tausend) thousand. 3 (Teil, Teile) part(s)

Tab. *abbv* (Tabelle) table

Tabak *m* (-e) tobacco. **-asche** *f* tobacco ash(es). **-auszug** *m* tobacco extract. **-bau** *m* tobacco cultivation. **-beize** *f* sauce (for tobacco). **-blatt** *n* tobacco leaf. **t-braun** *a* tobacco-brown. **T-brühe** *f* sauce (for tobacco). **-dampf** *m* tobacco smoke. **-fabrik** *f* tobacco factory. **-kampfer** *m* nicotinian. **-rauch** *m* tobacco smoke. **-saft** *m* tobacco juice

Tabaks- =Tabak-: **-asche** *f* tobacco ash = Tabaksche

Tabaschir *m* (-) tabasheer

tabellarisch *a* tabular; *adv* oft in tabular form. **tabellarisieren** *vt* tabulate

Tabelle *f* (-n) table, tabulation; graph, chart

tabellen-förmig *a* tabular

tabellieren *vt* tabulate. **Tabellierung** *f* (-en) tabulation

Tablett *n* (-e) tray

Tablette *f* (-n) tablet, pill

Tabletten-form *f* tablet form. **-maschine** *f* tablet press

tablettieren *vt* press into tablets

Tablettier-maschine *f* tablet press

Tachymeter *n* (-) tachometer

Tachysterin *n* (-e) tachysterol

Tadel *m* (-) 1 censure, reproach, blame. 2 defect, imperfection. **tadel-frei**, **-los** *a* faultless; blameless. **tadeln** 1 *vt* reproach, criticize, blame. 2 *vt/i* find fault (with)

Tafel *f* (-n) 1 sign. 2 blackboard. 3 bulletin board. 4A panel, flat piece; plate (*incl Cryst.*); lamina, lamella. B (*specif*) bar, tablet (of chocolate), sheet (of metal);

pane (of glass). 5 (banquet) table. 6 (*Books*) table, index; plate. **t-artig** *a* tabular; laminar, lamellar. **T-berg** *m* flattened curve. **-bier** *n* table beer. **-blechschere** *f* guillotine shears. **-blei** *n* sheet lead. **-butter** *f* table butter

Täfelchen *n* little table; tablet; platelet

Tafel-farbe *f* (*Tex.*) local (*or* topical) color. **-fett** *n* (*Lthr.*) table grease (*or* stuff). **t-förmig** *a* tabular; laminar, lamellar. **T-geschirr** *n* table china, tableware. **-glas** *n* 1 sheet (*or* plate) glass. 2 table glass(ware)

tafelig *a* tabular

Tafel-kreide *f* blackboard chalk. **-lack** *m* shellac. **-land** *n* tableland, plateau. **-leim** *m* glue in flat pieces. **-messing** *n* sheet brass

tafeln *vi* dine, (have a) banquet

täfeln *vt* panel, wainscot; inlay; floor

Tafel-öl *n* salad oil, (*specif*) olive oil. **-paraffin** *n* cake (*or* block) paraffin. **-quarz** *n* tabular quartz. **-salz** *n* table salt. **-schiefer** *m* slate in slabs; (*specif*) roofing slate, blackboard slate. **-schmiere** *f* table grease. **-spat** *m* tabular spar, (*Min.*) wollastonite

Täfelung *f* (-en) paneling, wainscoting

Tafel-waage *f* 1 counter scales. 2 platform scales. **-wasser** *n* bottled water, tonic (*or* soda) water, chaser; mineral water. **-wein** *m* table wine. **t-weise** *adv* in tables; in tabular form

Taffet, **Taft** *m* (-e) (*Tex.*) taffeta. **Taft-papier** *n* satin paper

Tag *m* (-e) 1 day: **dreimal am T.** three times a day; **acht T-e** a week. 2 day(time): **am** (or **bei**) **T.** in the daytime *cf* ALL. 3 (day)light: **am hellichten T.** in broad daylight; **an den T. kommen** come to light; **an den T. bringen** bring to light; **an den T. legen** show, display; **am T.**

liegen be clear, be obvious. *Cf* ZUTAGE. **4** (*Min.*): **über T.** above ground; **über T. bringen** bring to the surface; **unter T.** underground. **-bau** *m* open-pit mining, surface mining. **-blindheit** *f* day blindness. **-buch** *n* diary, journal

Tage-(berg)bau *m* open-pit (*or* strip) mine (*or* mining). **-blatt** *n* (*in names of newspapers*) daily. **-blindheit** *f* day blindness. **-buch** *n* diary, journal. **-lang** *a* a whole day of, whole days of; *adv* for days (on end). **T-lohn** *m* day's wages, day's pay. **-löhner** *m* day laborer

tagen *vi* **1** (*oft* + *dat*) dawn (on). **2** meet, hold a meeting

Tages- day('s), daily; (*Min.*) surface, openpit: **-bedarf** *m* daily requirement. **-erz** *n* open-pit ore. **-frage** *f* question of the day, immediate problem. **-leistung** *f* daily output. **-leuchtfarbe** *f* fluorescent paint. **-licht** *n* **1** daylight, light of day *cf* TAG **3**. **2** (*Min*) surface *cf* TAG **4**. **-oberfläche** *f* surface of the ground. **-ordnung** *f* agenda, day's business; order of the day (*esp fig*). **-preis** *m* current price. **-wasser** *n* surface water. **-zeit** *f* time of day; daytime

-tägig *sfx* -day('s)

täglich *a* daily

tags *adv:* **t. darauf** on the next day. **-über** *adv* in the daytime, during the day

tagtäglich *adv* day after day

Tagundnacht-gleiche *f* equinox

Tagung *f* (-en) meeting, conference, convention

Taifun *m* (-e) typhoon

Taille *f* (-n) waist(line)

Takt *m* (-e) **1** beat, rhythm; regular interval: **im T.** on the beat, at regular intervals, in step; **außer T.** out of step, off the beat; **T. halten** keep time. **2** (*Mach.*) stroke, cycle. **3** tact. **-fertigung** *f* assembly-line production. **t-mäßig** *a* timed, measured, rhythmic. **T-straße** *f* assembly line

Tal *n* (-̈er) valley —**zu T.** downhill

Taler *m* (-) thaler (*former German coin*)

Talerkürbis-öl *n* telfairia oil

Tal-fahrt *f* (*lit*) downhill (*or* downstream) run; (*fig, esp Econ.*) downward trend, decline, drop

Talg *m* (-e) **1** tallow, solid fat; suet. **2** (*Anat.*)

sebum. **t-ähnlich** *a* tallow-like; sebaceous. **T-art** *f* type of tallow (*or* solid fat). **t-artig** *a* tallowy; sebaceous. **T-baum** *m* tallow tree. **-drüse** *f* sebaceous gland. **-einbrennung** *f* (*Lthr.*) hot stuffing

talgen **1** *vt* tallow, grease. **2** *vi* yield tallow

Talg-fett *n* stearin. **t-gebend** *a* yielding tallow (*or* fat); sebiferous. **T-grieben** *fpl* tallow cracklings, greaves. **t-haltig** *a* containing tallow; sebaceous

talgig *a* tallowy, adipose; sebaceous

Talg-kernseife *f* tallow soap. **-kerze** *f*, **-licht** *n* tallow candle. **-öl** *n* tallow oil. **-säure** *f* stearic acid. **-schmelzen** *n* rendering of tallow (*or* suet). **-seife** *f* tallow soap. **-stein** *m* (*Min.*) soapstone, steatite. **-stoff** *m* stearin. **-zelle** *f* sebaceous cell

Talit *n* talitol

Talk *m* talc(um). **t-artig** *a* talc-like. **T-eisenerz** *n* magnesite. **-erde** *f* (*Min.*) magnesia, magnesite. **-erdealaun** *m* (*Min.*) magnesium alum. **-erdehydrat** *n* (*Min.*) brucite. **-glimmer** *m* mica. **-hydrat** *n* (*Min.*) brucite

talkig *a* talc-like, talcose

Talk-pulver *n* talcum powder. **-schiefer** *m* talc schist, slaty talc. **-spat** *m* (*Min.*) magnesite. **-spinell** *m* (*Min.*) spinel. **-stein** *m* soapstone, steatite

Talkum *n* talcum. **-puder** *m* talcum powder

Tall-öl *n* tall oil, talloel. **Tallöl-säure** *f* talloleic acid

Talmi *n* **1** copper-zinc-gold alloy. **2** sham, fake. **-gold** *n* tombac, gold-like brass

Tal-mulde *f* valley basin, hollow

Talon *m* (-s) coupon, stub. **-säure** *f* talonic acid

Taloschleim-säure *f* talomucic acid

Tal-sohle *f* valley floor. **-sperre** *f* dam. **t-wärts** *adv* downhill. **T-wert** *m* minimum (on a curve)

Tamarinden-baum *m* tamarind tree. **-mus** *n* tamarind pulp. **-samenöl** *n* tamarind seed oil

Tambour *m* (-e) drum; drummer

Tang *m* (-e) seaweed. **-asche** *f* seaweed ashes, kelp

Tangens *m* (-), **Tangente** *f* (-n) tangent

Tangenten-bussole *f* tangent compass (*or* galvanometer)

tangieren *vt* touch, be tangent to
Tangkallak·fett *n* tangcallac fat
Tang·säure *f* tangic acid. **-soda** *f* kelp
tanken 1 *vt* get (some gasoline, air, etc). 2 *vi* get gas, fill up, refuel
Tank·farbe *f* tank coating. **-holz** *n* wood for producer gas. **-säule** *f* gas(oline) pump. **-schiff** *n* (oil) tanker. **-stelle** *f* gas station. **-wagen** *m* tank car; tank truck
Tanne *f* (-n), **Tannen·baum** *m* fir (tree); pine (tree). **t–baumartig** *a* dendritic. **T–baumkristall** *m* (*Cryst.*) dendrite. **-harz** *n* fir (*or* pine) resin. **-harzsäure** *f* fir-resin (*or specif* abietic) acid. **-holz** *n* fir (*or* pine) wood. **-holzstoff** *m* fir pulp. **-nadel** *f* fir (*or* pine) needle. **-nadelöl** *n* fir-needle oil. **-wald** *m* fir (*or* pine) forest. **-zapfen** *m* fir (*or* pine) cone. **-zapfenöl** *n* fir-cone (*or* pine-cone) oil
tannieren *vt* tan: (*Dye.*) mordant with tannic acid
Tannin *n* tannin, tannic acid. **--ätzartikel** *m* (*Tex.*) tannic acid resist style. **-bleisalbe** *f* tannate of lead ointment. **-buntätzdruck** *m* colored tannin discharge printing. **-druck** *m* tannic acid print. **-druckfarbe** *f* tannic acid printing color. **-farblack** *m* tannic acid color lake. **-lösung** *f* tannin solution. **-reserve** *f* tannic acid resist. **-salbe** *f* tannin ointment. **-säure** *f* tannic acid. **-stoff** *m* tannin, tannic acid
Tantal *n* tantalum. **--erz** *n* tantalum ore, (*Min.*) =**prismatisches T.** tantalite; **hemiprismatisches T.** columbite, niobite
tantalig *a* tantalous
tantalisch *a* tantalic, (of) tantalum
Tantal·lampe *f* tantalum lamp. **-oxid** *n* tantalum pentoxide. **-säure** *f* tantalic acid. **-verbindung** *f* tantalum compound
Tantieme *f* (-n) share (in profits), royalty
Tanz *m* (-̈e) dance, ball. **tanzen** *vt/i* dance
Tapet *n:* **aufs T. bringen** bring up *cf* SPRACHE 3
Tapete *f* (-n) 1 wallpaper. 2 tapestry. 3 (*Anat.*) tapetum
Tapeten·druck *m* wallpaper printing. **-teigfarbe** *f* wallpaper pulp (*or* paste) color. **-wechsel** *m* change of surroundings
tapezieren *vt/i* (wall)paper. **Tapezierer** *m* (-) paper hanger. **Tapezierung** *f* (-en)

(wall)papering. **Tapezier·verfahren** *n* (*Plastics*) layup method
tapfer *a* brave, valiant
Tapioka·mehl *n* tapioca starch
tappen *vi* (*oft* + **nach**) grope (one's way), grope (for)
täppisch *a* clumsy, awkward
Tara *f* (. . ren) (*Com.*) tare. **--fläschchen** *f* small tare flask. **-gewicht** *n* tare weight
Tarantel *f* (-n) tarantula
Taren tares: *pl of* TARA
Tarier·becher *m* tare cup. **tarieren** *vt/i* tare
Tarier·granat *m* garnet for taring. **-schrot** *m* tare shot. **-stück** *n* tare. **-waage** *f* tare balance
Tarif *m* (-e) 1 tariff, rate. 2 standard, scale (wage). **--satz** *m* tariff rate
tarnen *vt*, **Tarnung** *f* (-en) camouflage. **Tarnungs·farbe** *f* camouflage paint
Tarn·wort *n* code word
Tartar·beefsteak *n* raw chopped beef
tartarisieren *vt* tartarize
Tartron·säure *f* tartronic acid
Tasche *f* (-n) 1 pocket. 2 briefcase, attaché case. 3 pouch; handbag; shopping bag; purse. 4 pin, chamber
Taschen·ausgabe *f* pocket edition. **-buch** *n* 1 pocket(-size) book. 2 manual, handbook. 3 memo book, little black book. **-filter** *n* bag filter. **-format** *n* pocket size. **-krebs** *m* common crab. **-lampe** *f* pocket flashlight. **-lupe** *f* pocket lens (*or* magnifying glass). **-messer** *n* pocket knife. **-rechner** *m* pocket calculator. **-spiel** *n* jugglery, sleight of hand. **-tuch** *n* handkerchief. **-uhr** *f* pocket watch. **-wörterbuch** *n* pocket dictionary
Täschner·leder *n* fancy leather
Tasmanien *n* (*Geog.*) Tasmania
Tasse *f* (-n) cup; cup and saucer
tassen·fertig *a* instant (coffee, etc)
Tastatur *f* (-en) keyboard
tastbar *a* tangible, palpable
Taste *f* (-n) (*Instrum.*) key; (push)button
tasten I *vt* 1 feel, touch. 2 (*Med.*) palpate. 3 (*Tgph.*) key. II *vi* (*usu* + **nach**) grope (for). III: **sich t.** grope one's way
Tasten· key(board), push-button, digital: **-brett** *n* keyboard. **-druck** *m* push of (*or* on) a button. **t–gesteuert** *a* key-operated,

push-button, digital. **T–instrument** n key-punch (push-button, digital) instrument. **-telefon** n touch (or push-button) telephone

Taster m (-) 1 (Instrum.) (punch, tap, sending) key; (push)button. 2 calipers. 3 (Zool.) antenna, feeler. **--zirkel** m gauging calipers

Tast·gefühl n sense of touch. **-organ** n tactile organ. **-sinn** m sense of touch. **-spitze** f probe. **-versuch** m tentative (or exploratory) experiment. **-zirkel** m calipers

tat did, etc: past of TUN

Tat f (-en) 1 deed, act, thing to do —**in der T.** indeed, in fact. 2 action: **in die T. umsetzen** put into action. 3 feat, achievement. 4 offense, crime. **--bestand** m facts (of the case)

Täter m (-) doer, perpetrator

tätig a active, busy —**t. sein** (oft) be employed; **t. sein als** act as, work as. **tätigen** vt carry on, effect; conclude (deals, etc); put into operation. **Tätigkeit** f (-en) activity; occupation; action; service, operation: **in T. sein** be active, be in service (or operation); **in T. setzen** put into action (service, operation); **in T. treten** go into operation (or service); **außer T. setzen** take out of service, retire, cancel

Tat·kraft f energy. **tatkräftig** a energetic

tätlich a physical; violent

Tato. abbv (Tagestonnen) tons per day

tätowieren vt/i tattoo

Tat·sache f fact

Tatsachen·bericht m factual report; documentary. **-material** n, **-stoff** m factual material (or data), evidence

tatsächlich a real, actual, adv oft in fact, indeed

Tatze f (-n) 1 paw. 2 (Mach.) cam

Tau I m 1 dew; serein. 2 (in compds, oft) thaw cf TAUWETTER. II n (-e) (tow) rope, hawser

taub a 1 deaf. 2 numb. 3 empty, hollow, barren, sterile

Taube I f (-n) pigeon, dove. II m,f (adj endgs) deaf person

tauben·grau a dove-gray, dove-colored

Taub·glas n frosted (or ground) glass

Taubheit f deafness; numbness; barrenness

Tau·bildung f formation of dew

Taub·kohle f anthracite. **t–stumm** a deaf-mute

Tauch· dipping, immersion: **-alitieren** n aluminizing by dipping, calorizing. **-anlage** f dipping (or steeping) machine (or unit). **-artikel** mpl (Rubber) dipped goods. **-bad** n dip; immersion bath. **-bahn** f (Electrons) dip (or penetrating) orbit. **-batterie** f immersion battery. **-beize** f 1 disinfection by immersion. 2 disinfecting steep (or dip). **-boot** n submarine. **-brenner** m immersion heater. **-element** n (Elec.) immersion cell. **-email** n, **-emaillelack** m dipping enamel

tauchen I vt immerse; steep; coat (by dipping). II vt, vi (h,s) dip; submerge; plunge. III vi (h,s) 1 dive; sink. 2 (+**aus**) rise, appear, emerge (from)

Taucher m (-) diver; diving bird. **--kolben** m (Mach.) plunger. **-krankheit** f decompression sickness

Tauch·farbe f dipping color. **-färbung** f dip dyeing (or coloring); (Paper) stuffing, calender staining. **-filter** n immersion filter. **-flüssigkeit** f dipping liquid (or fluid). **-härten** n, **-härtung** f dip (or hotbath) hardening. **-heizkörper** m immersion heater. **-kammer** f diving chamber. **-kolben** m (Mach.) plunger. **-korn** n (Brew.) sinker. **-körper** m float. **-küpe** f dipping vat. **-lack** m dipping lacquer (varnish, paint). **-löten** n dip soldering, dip brazing. **-mikrotom** n immersion microtome. **-mischung** f 1 dip mixing. 2 dipping mixture. **-patentierung** f (Wire) patenting by heating and then immersion in a lead or salt bath. **-pumpe** f submersible pump. **-sieder** m immersion heater. **-stange** f plunger. **-verfahren** n 1 (genl) immersion process. 2 (Dye.) tub-dip process. **-versuch** m immersion (or diving) test. **-verzinnen** n dip tinning. **-waage** f hydrometer. **-wärmer** m immersion heater. **-zeit** f immersion time. **-zylinder** m plunge cylinder, plunger (of a calorimeter)

tauen 1 vt (Lthr.) taw. 2 vt/i thaw, melt. 3 vi: **es taut** dew is falling

Taufe f (-n) baptism —**aus der T. heben** bring into being, found. **taufen** vt baptize

Tau·fläche f (*Phase Diagrams*) liquid surface

taugen vi 1 be suitable, be of use. 2 (+**etwas, nichts,** *etc*) be worth

tauglich a fit, able, suitable, qualified. **Tauglichkeit** f fitness, suitability

tauig a dewy

Tau·linie f (*Phase Diagrams*) liquid curve

Taumel m 1 giddy feeling. 2 staggering, reeling. 3 intoxication, rapture, frenzy. 4 dizzying succession (of . . .). **··be-wegung** f (*Mach.*) tumbler motion. **-korn** n (*Bot.*) darnel, rye grass. **-mischer** m tumbler mixer

taumeln vi (h/s) 1 stagger, reel, feel giddy. 2 (*Mach., Space*) tumble

Tau·messer m drosometer. **-punkt** m dewpoint, dew point

Taurochol·salz n taurocholate. **-säure** f taurocholic acid

Tau·röste f dew retting

Tausch m exchange, barter

tauschbar a exchangeable

tauschen vt/i exchange, trade, barter

täuschen 1 vt deceive, delude; disappoint. 2 vi be deceptive. 3: **sich t.** deceive (*or* delude) oneself; be mistaken. **täuschend** p.a deceptive, misleading; striking (resemblance)

Tauscher m (-) exchanger

Täuschung f (-en) 1 deception, deceit, delusion. 2 illusion. 3 misconception, error

Tausch·wert m exchange (*or* market) value. **-zersetzung** f double decomposition

tausend num one (*or* a) thousand; *in compds oft* milli. **tausenderlei** a (*no endgs*) a thousand kinds of

Tausend·fuß, -füßler m millipede. **-güldenkraut** n (*Bot.*) lesser centaury. **t-mal** adv a thousand times. **T–schön** n daisy

tausendste a thousandth. **Tausendstel** n (-) (*Frac.*) thousandth

tautomer a tautomeric. **Tautomerie** f tautomerism

Tau·topf m condensing pot; steam trap. **-wetter** n thaw

Taxator m (-en) assessor, appraiser

Taxe f (-n) 1 assessed value; assessment. 2 (fixed) price, fee. 3 taxi (cab)

taxieren vt 1 (*usu* + **auf**) assess, estimate

(at). 2 rate, judge, appraise

Taxus m, **··hecke** f (*Bot.*) yew (hedge)

Tb, Tbc abbv (Tuberkulose) TB

Technik f (-en) 1 technology, engineering; industry. 2 engineering school, institute of technology. 3 technique, operating procedure; practice, skill. 4 mechanics, operation (of a machine). **Techniker** m (-), **Technikerin** f (-nen) 1 technologist, engineer. 2 technician, expert. **Technikum** n (. . iken) institute of technology, (postgraduate) engineering school

technisch a 1 technological, (of) engineering, high-tech: **t—e Hochschule** engineering school = TECHNIKUM. 2 technical, practical. 3 (*usu Substances*) commercial. **technisch-chemisch** a technochemical, (of) chemical engineering

technisieren vt mechanize

Technologe m (-n, -n) technologist. **Technologie** f (-n) technology. **technologisch** a technological

Tectochinon n (-e) tectoquinone (2-methylanthraquinone)

Tee m (-s) tea. **··kraut** n tea (*or* infusion) herb. **-löffel** m teaspoon. **-öl** n tea oil

Teer m (-e) tar. **··abscheider** m tar separator. **t–artig** a tarry, tar-like. **T–asphalt** m tar asphalt, coal-tar pitch. **-aus-scheider** m tar separator. **-ausschlag** m exudation of tar. **-band** n tarred tape. **-baum** m Scotch pine. **-belag** m tar surface (for roads etc). **-benzin** n benzene, benzol. **-bestandteil** m constituent of tar. **-bildung** f formation of tar. **-bitter** n picamar. **-brenner** m tar burner, tar maker. **-brennerei** f tar factory. **-dampf, -dunst** m tar fumes, tar gas. **-destillat** n tar distillate

teeren 1 vt tar. 2 vi tar up, form tarry matter

Teer·fabrikation f tar production. **-farbe** f, **-farbstoff** m coal-tar color (*or* dye), aniline dye. **-fettöl** n fatty tar oil (higher-boiling fraction of anthracene oil). **-feuerung** f tar furnace. **t-frei** a tar-free. **T–gas** n tar gas. **-gehalt** m tar content. **t–haltig** a tarry, containing tar. **T–hefe** f tar dregs

teerig a tarry, tarred

Teer·kessel m tar kettle. **-kocherei** f tar

boiling (plant). **-leinwand** f tarpaulin. **-nebel** m tar mist. **-öl** n coal-tar oil

Tee·rose f tea rose

Teer·papier n tar paper. **-pappe** f tar board. **-pech** n tar pitch, cutback pitch. **prüfer** m tar tester. **t–reich** a rich in tar, high-tar. **T–rückstand** m tar(ry) residue. **-salbe** f tar ointment. **-satz** m tar sediment. **-säure** f acid from coal tar; phenol. **-scheider** m tar extractor (or separator). **-schwelapparat** m coal tar distilling apparatus. **-schwelerei** f 1 tar distillation (or manufacture). 2 tar distillery (or factory). **-seife** f tar soap. **-substanz** f tar substance; tarry matter. **-tröpfchen** n tar droplet (or globule). **-tonne** f tar barrel (or drum). **-tuch** n tarpaulin. **-überzug** m tar coating. **-vorlage** f 1 (Coal gas) hydraulic main (or pipe). 2 (Coke ovens) collecting main (or pipe). **-wäsche** f (Gas) tar extraction. **-wäscher** m tar washer (or scrubber). **-wasser** n (Gas) ammoniacal liquor. **-werg** n tarred oakum (or tow). **-zahl** f tar number

Tegel m blue-green marl

Teich m (-e) pond

Teig m (-e) 1 (pastry) dough. 2 pulp. **t–artig** a doughy, pasty. **T–farbe** f pastel color. **-form** f doughy form

teigig a 1 doughy. 2 soggy. 3 mealy (fruit)

Teig·konsistenz f doughy consistency. **-schaber** m dough scraper. **-ware** f 1 article (e.g. dye) in paste form. 2 (usu pl) pasta (noodles, macaroni, etc)

Teil m,n (-e) 1 (usu m) part, portion, share; extent: **zum T.** partly, in part: **zu einem beachtlichen T.** to a considerable extent. 2 (esp Mach., usu n) component, part. 3 (in compds) partial, fractional, component: **-bande** f component band

teilbar a divisible: **der Satz ist nicht t.** the set can not be broken up. **Teilbarkeit** f divisibility

Teilchen n (-) particle; corpuscle. **··bahn** f particle orbit. **-beschleunigung** f particle acceleration. **-größe** f particle size. **-ladung** f particle charge. **-verteilung** f particle distribution. **-zähler** m particle counter

Teil·druck m partial pressure

teilen I vt 1 divide, break up (into parts). 2

(with numerical pfx) divide into . . . parts cf VIERTEILEN. 3 (usu + mit) share (with). II: sich t. 4 divide, separate, part (company). 5 break (or split) up, branch, go different ways. 6 (+in + accus) share, get a share (of)

Teiler m (-) divider; (Math.) divisor, factor. **t–fremd** a (Math.) prime

Teil·erhebung f sampling. **-fehler** m partial error. **-gabe** f addition in batches; batch feeding. **-gebiet** n (specialized) branch (of a field of study)

teil·haben* vi (an) share, participate (in). **Teil·haber** m partner. **-haberschaft** f partnership

teilhaft(ig) pred a: **t. werden** (genit) share (in); acquire, get

·teilig sfx -part, -partite: **dreiteilig** three-part, tripartite

Teil·kraft f (Mech.) component force. **-kreis** m divided (or graduated) circle. **-maschine** f dividing machine, distributing engine. **-menge** f subset, aliquot (part)

Teilnahme f 1 participation; cooperation. 2 complicity. 3 sympathy. 4 interest. **teilnahmslos** a indifferent, apathetic

teil·nehmen* vi (+an) take part, participate (in); sympathize (with). **Teilnehmer** m (-) participant; subscriber

Teil·niveau n partial level. **-platte** f graduated plate (or scale). **-problem** n part of the problem, subproblem. **-reaktion** f partial reaction

teils adv partly, in part

Teil·strich m graduation (mark). **-strom** m divided current; part of the current. **-stück** n constituent part, component, portion

Teilung f (-en) division, partition, distribution; sharing; separation; graduation, scale; break-up, split-up; fork (in the road), bifurcation; (Mach.) pitch. Cf TEILEN

Teilungs·analyse f partition analysis. **-bruch** m (Math.) partial fraction. **-ebene** f plane of division. **-fläche** f (Geol.) division plane. **-gesetz** n law of partition. **-koeffizient** m partition (or distribution) coefficient. **-zahl** f distribution number, partition ratio; dividend.

-zeichen n 1 division sign. 2 hyphen.
-zustand m state of (sub)division
Teil·verflüssigung f partial liquefaction.
-vorgang m partial process. **t–w.** abbv =
t–weise I a 1 partial. II adv 2 partially,
partly, in part. 3 in some cases. **T–zahl** f 1
number of divisions. 2 quotient.
-zahlung f partial payment; installment.
-zeit f part-time. **-zeitarbeit** f part time
work; flextime
Teïn n theine, caffeine in tea
Teint m (-s) complexion
Tek·holz n teakwood
Teklu·brenner m Teclu burner
Tektit n (Min.) tectite
tektonisch a tectonic, structural
Tektosilikat° n three-dimensional network
silicate
Telefon n (-e) telephone. **Telefonat** n (-e)
telephone conversation. **telefonieren**
vt/i telephone; vi (+mit) have a telephone
conversation (with). **telefonisch** a telephonic; adv by telephone. See also TELE-
PHON. . . (Both spellings are in current
use.)
telegrafieren, telegraphieren vt/i telegraph. **telegrafisch, telegraphisch** a
telegraphic, adv by telegraph
Teleobjektiv° n telephoto lens
telephonieren, telephonisch, etc = TELE-
FONIEREN, TELEFONISCH
Telephor·säure f telephoric acid
Telespiel° n video game
teleskop·artig a telescope-like, telescopic
Teller m (-) 1 plate, platter, dish. 2 disk,
disc; (esp Valves) seat. 3 palm (of the
hand). 4 (Phonograph) turntable. 5 (Mil.)
antitank mine. **t–fertig** a (Food) ready-
to-serve. **T–fuß** m disk-shaped base.
-kühlung f plate cooling. **-mine** f (Mil.)
antitank (or teller) mine. **-trockner** m
disk drier. **-ventil** n disk valve. **-wäscher**
m plate scrubber; dishwasher. **-zinn** n
plate pewter
Tellur n tellurium; (in compds also) telluric, telluride of: **-alkyl** n alkyl telluride. **-blei** n 1 lead telluride. 2 (Min.)
altaite. **-cyansäure** f tellurocyanic acid.
-diäthyl n diethyl telluride. **-eisen** n iron
telluride; native iron. **-erz** n tellurium
ore. **t–führend** a telluriferous,

tellurium-bearing. **T–glanz** m 1 tellurium glance. 2 (Min.) nagyagite. **-gold**
n gold telluride. **-goldsilber** n 1 gold silver telluride. 2 (Min.) sylvanite, petzite.
-halogen n tellurium halide. **t–haltig** a
containing tellurium
tellurig a tellurous. **T–·säureanhydrid** n
tellurous anhydride (or dioxide)
tellurisch a telluric
Tellur·kohlenstoff m carbon telluride.
-metall n metallic tellurium. **-natrium** n
sodium telluride. **-nickel** n 1 nickel telluride. 2 (Min.) melonite. **-ocker** m
(Min.) tellurite. **-oxid** n tellurium oxide.
-salz n tellurium salt, tellurate. **t–sauer**
a tellurate of. **T–säure** f telluric acid.
-schwefelkohlenstoff m carbon sulfide
telluride (CSTe). **-silber** n 1 silver telluride. 2 (Min.) hessite. **-silberblei** n lead
silver telluride, (Min.) sylvanite. **-silber-
blende** f (Min.) sylvanite, stutzite
Tellurür n (-ous) telluride
Tellur·verbindung f tellurium compound.
-vorlegierung f tellurium prealloy.
-wasserstoff m hydrogen telluride.
-wasserstoffsäure f hydrotelluric acid.
-wismut n 1 bismuth telluride. 2 (Min.)
wehrlite
Temp. abbv (Temperatur) temp.
Tempel m (-) temple
Tempera·farbe f tempera color
Temperatur·abfall m drop in temperature. **t–abhängig** a temperature-depen-
dent. **T–abhängigkeit** f temperature dependence. **-änderung** f change in
temperature. **-ausgleich** m temperature
equalization. **-beobachtung** f temperature observation. **-bereich** m temperature range. **t–beständig** a temperature-
stable, unaffected by temperature
changes. **T–dehnung** f thermal expansion. **-einfluß** m effect (or influence) of
temperature. **-erhöhung** f rise in temperature. **-gefälle** n temperature gradient. **-grad** m degree of temperature.
-grenze f temperature limit. **-leit-
fähigkeit** f thermal conductivity. **-mes-
ser** m thermometer, pyrometer. **-mes-
sung** f temperature measurement. **-mit-
tel** n mean temperature. **-regelung** f
temperature regulation. **-regler** m ther-

mostat. **-schreiber** m recording thermometer, temperature recorder. **-schwankung** f fluctuation in temperature. **-spanne** f temperature interval. **-steigerung** f rise in temperature. **-stoß** m temperature shock. **-strahlung** f thermactinic radiation. **-überführung** f heat transfer. **-umwandlung** f temperature inversion. **t–unabhängig** a temperature-independent. **T–veränderung** f change in temperature. **-wechsel** m change (or variation) in temperature. **-wechselbeständigkeit** f thermal shock resistance. **-wechsler** m heat exchanger. **-zahl** f temperature coefficient. **-zunahme** f temperature increase

Temper·eisen n malleable iron. **-erz** n ore for malleable iron. **-farbe** f temper(ing) color. **-guß** m 1 malleable casting, temper cast. 2 =**-gußeisen** n malleable cast iron

Temperier·bad n tempering bath

temperieren vt temper, moderate; bring to a certain temperature. **temperiert** p.a (oft) having a certain temperature

Temper·kohle f temper (or graphitic) carbon. **-kohlebildung** f formation of temper carbon, graphitization. **-mittel** n (Metll.) packing, annealing (or tempering) material

tempern vt malleablize; temper, anneal, age-harden

Temper·ofen m annealing furnace. **-roheisen** n malleable pig iron. **-schrott** m (Metll.) malleable scrap. **-stahlguß** m malleable cast iron = TEMPERGUSS. **-topf** m annealing pot

Temperung f (-en) tempering, etc cf TEMPERN

Tempi paces, tempos: pl of TEMPO

Tempo n (-s or . . pi) pace, rate (of speed), tempo

temporär n temporary

Tenakel n (-) 1 filtering rack. 2 (printer's) copyholder

Tenazität f tenacity

Tendenz f (-en) tendency, trend. **tendieren** vi 1 (+**nach, zu**) tend (to be), incline (to, toward). 2 (esp Meteor.) (+adv) tend to be, show a . . . trend

Tenne f (-n) (threshing) floor. **-mälzerei** f (Beer) floor-matting (area)

Teppich m (-e) carpet, rug. **-kosmetik** f carpet cleaning

Teracon·säure f teraconic acid

Terbin·erde f 1 terbia, terbium oxide. 2 (pl -n): terbium earths (Tb, Gd, Eu, etc)

Terebin·säure f terebic acid

Terephthal·säure f terephthalic acid

Teresantal·säure f teresantalic acid

Term·folge f sequence of terms

Termin m (-e) appointed time (or day); date of maturity; deadline, time limit. **t–gerecht** a, adv on time, at the appointed time

Terminus m (.. ni) (technical) term, expression

Term·schema n (Spect.) energy level diagram. **-verschiebung** f term shift. **-wert** m (Spect.) term value

ternär a ternary. **T–stahl** m ternary (or simple alloy) steel

Terne·blech n terneplate

Terpen·chemie f terpene chemistry. **t–frei** a terpene-free. **T–gruppe** f terpene group. **-kohlenwasserstoff** m terpene hydrocarbon

Terpentin n (-e) turpentine. **-alkohol** m spirit (or oil) of turpentine. **-art** f type of turpentine. **t–artig** a turpentine-like, terebinthic. **T–ersatz** m turpentine substitute. **-firnis** m turpentine varnish. **-geist** m spirit of turpentine. **t–haltig** a containing turpentine. **T–harz** n turpentine resin. **-öl** n (oil of) turpentine; rosin oil

Terpentinöl·ersatz m turpentine substitute. **-firnis** m turpentine varnish. **-seife** f turpentine-oil soap

Terpentin·pech n turpentine pitch. **-salbe** f turpentine ointment. **-spiritus** m spirit of turpentine

Terpenyl·säure f terpenylic acid

Terpin n terpinol. **-hydrat** n terpinol hydrate

Terrasse f (-n) terrace

Terrine f (-n) tureen

terrestrisch a terrestrial

tert. abbv = **tertiär** a tertiary

Tesla·transformator m Tesla coil

Tesseral·kies m skutterudite; smaltite. **-system** n (Cryst.) isometric system

Test m (-e) test; cupel. **-asche** f bone ash.

-benzin *n* (*Paint*) mineral spirits, white spirit

testen *vt/i* test

Test·gift *n* test poison (*or* toxin)

Testikel·hormon *n* testicle hormone

Test·körper *m* test body, test piece. **-mischung** *f* test mixture. **-platte** *f* test plate. **-stopp** *m* (nuclear) test ban. **-versuch** *m* laboratory test

tetanisieren *vt* tetanize. **Tetanus·impfung** *f* tetanus inoculation

tetartoedrisch *a* (*Cryst.*) tetartohedral

Tetra *n* (*esp Com.*) carbon tet = TETRACHLORKOHLENSTOFF

Tetra·borsäure *f* tetraboric acid. **-bromkohlenstoff** *m* carbon tetrabromide. **-chlorkohlenstoff** *m* carbon tetrachloride. **-chlorzinn** *n* tin tetrachloride. **-eder** *n* tetrohedron. **t–edrisch** *a* tetrahedral. **T–edrit** *n* tetrahedrite. **-fluorkohlenstoff** *m* carbon tetrafluoride. **-gyre** *f* (-n) (*Cryst.*) tetragonal rotary axis. **t–gyrisch** *a* (*Cryst.*) having a fourfold rotation axis; *as adv usu* tetragonal-: **t. pedial** tetragonal-pyramidal. **-gyroidisch** *a* (*Cryst.*) having a tetrad alternating axis—**t. pedial** tetragonal-diphenoidal. **T–hexaeder** *n* tetrahexahedron. **-jodkohlenstoff** *m* carbon tetraiodide. **-kontan** *n* tetracontane. **-kosan** *n* tetracosane. **t–mer** *a* tetrameric. **T–oxid** *n* tetroxide. **-thionsäure** *f* tetrathionic acid. **-vanadinsäure** tetravanadic acid. **-zyklin** *n* tetracycline. **t–zyklisch** *a* tetracyclic

Tetrin·säure *f* tetric acid

Tetrit *n* (-e) tetritol

Tetrol·säure *f* tetrolic acid

Tetron·säure *f* tetronic acid

Tetroxid° *n* tetroxide

teuer *a* (*with endgs usu:* **teur . . ;** *compar* **teurer,** *superl* **teuerst . .**) dear (all meanings); costly, expensive —**wie t. ist es?** what is the price of it?

Teuerung *f* (-en) general price rise

Teufe *f* (-n) (*Min.*) mining

Teufel *m* (-) devil

Teufels·dreck *m* asafetida. **-kirsche** *f* belladonna. **-wurz** *f* aconite

teufen *vt* (*Min.*) sink (a shaft)

teure, teurer etc dear(er), etc, *See* TEUER

Teurung *f* (-en) price rise = TEUERUNG

Teutsch·schwarz *n* ivory black

Text·abbildung *f* illustration in the text

textil·chemisch *a* textile-chemical. **T–faser** *f* textile fiber. **-gewerbe** *n* textile industry

Textilien *fpl* textiles

Textil·seife *f* textile soap. **-vered(e)lung** *f* textile finishing (*or* dressing). **-waren** *fpl* textile goods

Textur *f* (-en) (*usu*) texture; (*Petrog. etc*) structure

Tfl. *abbv* (TAFEL) table; plate

T-förmig *a* T-shaped

tg *abbv* (Tangens, Tangente) tan (tangent)

TGA *abbv* (thermogravimetrische Analyse) thermogravimetric analysis

Th . . : *Most words of native German origin now spelled with initial* **T** *were formerly—up to the early 20th century—spelled with initial* **Th,** e.g. **Thal** (*now* **Tal**), **Theil** (*now* **Teil**), *etc. Since that time the th-spelling has been retained only in words of foreign origin. This involves mainly Greek words in which the original* θ *is represented by* **th: Thema, Theorie,** *etc*

T.H. *abbv* (Technische Hochschule) Institute of Technology *cf* TECHNISCH 1

Thalleio·chin(in) *n* thalleioquin. **-chinolin** *n* thalleioquinoline

Thalli· thallic: **-chlorat** *n* thallic chlorate. **-chlorid** *n* thallic chloride. **-ion** *n* thallic ion. **-oxid** *n* thallic oxide. **-salz** *n* thallic salt. **-sulfat** *n* thallic sulfate

Thallium·alaun *m* thallium alum. **-bromür** *n* thallous bromide. **-chlorid** *n* thallic chloride. **-chlorür** *n* thallous chloride. **-fluorür** *n* thallous fluoride. **-hydroxid** *n* thallic hydroxide. **-hydroxydul** *n* thallous hydroxide. **-jodür** *n* thallous iodide. **-oxid** *n* thallic (*or* thallium) oxide. **-oxydul** *n* thallous oxide. **-sulfür** *n* thallous sulfide. **-verbindung** *f* thallium compound

Thalli·verbindung *f* thallic compound

Thallo· thallous: **-bromid** *n* thallous bromide. **-chlorat** *n* thallous chlorate. **-chlorid** *n* thallous chloride. **-fluorid** *n* thallous fluoride. **-ion** *n* thallous ion. **-jodat** *n* thallous iodate. **-jodid** *n* thallous iodide. **-salz** *n* thallous salt. **-sul-**

fat *n* thallous sulfate. **-verbindung** *f* thallous compound

Thapsia·harz *n* thapsia resin. **-säure** *f* thapsic acid

Theke *f* (-n) (serving) counter; bar

Thema *n* (-ta *or* . . men) subject, topic; theme

Thenards·blau *n* Thénard's blue (cobalt aluminate)

Thenhydroxam·säure *f* thenohydroxamic acid

Theoretiker *m* (-) theorist. **theoretisch** *a* theoretical. **Theorie** *f* (-n) theory

Therapeut *m* (-en, -en) therapist. **Therapeutik** *f* (-en) therapeutics. **therapeutisch** *a* therapeutic. **Therapie** *f* (-n) therapy

thermalisieren *vt* thermalize. **Thermalisierung** *f* (-en) (*esp Nucl.*) thermalization, slowing down

Thermal·quelle *f*, **Therme** *f* (-n) thermal (hot, warm) springs

Thermik *f* 1 heat (as a subject of study). 2 (-en) thermal current

thermionisch *a* thermionic

thermisch *a* thermal. **thermisch-optisch** *a* thermooptical

Thermit·(brand)bombe *f* thermite (incendiary) bomb. **-füllung, -ladung** *f* thermite filling (*or* charge). **-schweißung** *f* thermite weld(ing). **-verfahren** *n* thermite process

Thermo·analyse *f* thermal analysis. **th–analytisch** *a* thermoanalytic(al); *adv oft* by (*or* of) thermal analysis. **Th–chemie** *f* thermochemistry. **-chemiker** *m* thermochemist. **th–chemisch** *a* thermochemical. **-chrom** *a* thermochromic. **Th–chromie** *f* thermochromism. **-diffusion** *f* thermal diffusion. **th–dynamisch** *a* thermodynamic. **-elastisch** *a* thermoelastic. **-elektrisch** *a* thermoelectric. **Th–elektrizität** *f* thermoelectricity. **-element** *n* thermocouple. **-farbe** *f* thermocolor. **th–fixierecht** *a* resistant to thermosetting. **Th–fixierung** *f* thermosetting, heat setting. **-kauter** *m* thermocautery. **-kette** *f* thermocouple. **-kleber** *m* thermoplastic adhesive. **-kraft** *f* thermoelectric force (*or* power)

Thermolyse *f* (-n) thermolysis

Thermometer·kugel *f* thermometer bulb. **-röhre** *f* thermometer tube. **-stand** *m* thermometer reading

thermometrisch *a* thermometric

Thermo·paar *n* thermocouple

thermophil *a* thermophile. **Thermophil** *m, n* (-e) thermistor

Thermoplast *m* (-e) thermoplastic (resin *or* polymer). **thermoplastisch** *a* thermoplastic

Thermo·regler *m* thermoregulator, heat regulator. **th–resistent** *a* heat-stable, heat-resistant. **Th–säule** *f* thermopile

Thermos·flasche *f* thermos bottle

Thermo·spannung *f* thermal EMF, thermoelectric potential. **-strom** *m* thermoelectric current

Thermotaxie *f* thermotaxis

Thermo·waage *f* thermobalance

These *f* (-n) thesis (to be proved)

Thespasia·öl *n* mallow oil

Thiacet·säure *f* thioacetic acid

Thienon *n* dithienyl ketone

thioacetylieren *vt* thioacetylate

thioacylieren *vt* thioacylate

Thio·alkohol *m* thioalcohol, mercaptan. **-ameisensäure** *f* thioformic acid. **-antimonsäure** *f* thioantimonic acid. **-arsenigsäure** *f* thioarsenious acid. **-arsensäure** *f* thioarsenic acid. **-äther** *m* thio ether. **-carbaminsäure** *f* thiocarbamic acid

Thiochin. . . thioquin. . . e.g. **Thiochinol** *n* thioquinol

Thiocyan *n* thiocyanogen. **--kalium** *n* potassium thiocyanate. **th–sauer** *a* thiocyanate of. **Th–säure** *f* thiocyanic acid. **-verbindung** *f* thiocyanate

Thio·essigsäure *f* thioacetic acid

Thiogen·farbe *f* thiogene dye

Thio·germaniumsäure *f* thiogermanic acid. **-harnstoff** *m* thiourea. **-indigoscharlach** *m* thioindigo scarlet. **-kohlensäure** *f* trithiocarbonic acid

Thiol·säure *f* thiolic acid

Thio·milchsäure *f* thiolactic acid

Thion·carbaminsäure *f* thionocarbamic acid. **-carbonsäure** *f* thionocarbonic acid, carbothionic acid. **-farbe** *f* thion dye

Thionin·lösung *f* thionine solution

Thion·kohlensäure *f* thionocarbonic acid.

-kohlenthiolsäure *f* thiolthionocarbonic acid. -säure *f* thionic acid
Thionur·säure *f* thionuric acid
Thiophenin *n* aminothiophene
Thio·phosphorsäure *f* thiophosphoric acid. -phtalid *n* thiophthalide
Thiophten *n* thiophthene
Thio·salz *n* thio salt. -säure *f* thio acid. th–schwefelsauer *a* thiosulfate of. . . Th–schwefelsäure *f* thiosulfuric acid. th–schweflig *a* thiosulfurous. Th–sinamin *n* N-allylthiourea. -sulfosäure *f* thiosulfonic acid. -verbindung *f* thio compound. -zinnsäure *f* thiostannic acid. -zyan *n* thiocyanogen *See* THIOCYAN· *for compds*
thixotrop *a* thixotropic. Thixotropie *f* thixotropy
Thomas·birne *f* (*Metll.*) basic converter. -eisen *n* Thomas (*or* basic) iron, basic Bessemer steel. -flußeisen *n*, -flußstahl *m* Thomas low-carbon steel. -mehl *n* Thomas meal (ground basic slag fertilizer). -roheisen *n* Thomas (*or* basic Bessemer) pig iron. -schlacke *f* basic (*or* Thomas) slag. -schlackenmehl *n* Thomas meal (*or* slag fertilizer). -stahl *m* Thomas (*or* basic Bessemer) steel. -verfahren *n* Thomas (*or* basic Bessemer) process
Thor *n* thorium. --erde *f* thoria, thorium oxide
thorieren *vt* thoriate
Thorium·blei *n* thorium lead (208 isotope of Pb). th–haltig *a* thoriated, containing thorium. Th–heizfaden *m* thoriated heating filament. -salz *n* thorium salt. -verbindung *f* thorium compound
Thor·oxid *n* thorium oxide
Threït *n* erythritol
Thrombose *f* (-n) thrombosis
Thuja·öl *n* thuja oil
Thun *m* (-e), --fisch *m* tuna (fish)
Thüringen *n* (*Geog.*) Thuringia. thüringisch *a* Thuringian
Thymian *m* (-e) (*Bot.*) thyme. --kampfer *m* thymol. -öl *n* oil of thyme
Thymo·chinhydron *n* thymoquinhydrone. -chinon *n* thymoquinone. -hydrochinon *n* thymohydroquinone. -nucleinsäure *f* thymonucleic acid

Thymus·drüse *f* thymus gland
Thyrojodin *n* thyroiodine, iodothyrin
tickern *vt/i* sputter (out)
tief *a* deep (*incl Colors*), profound; low; extreme *cf* TIEFST. Tief *n* (-e) 1 low point; deep area, depression. 2 (*Meteor.*) low. --ätzprobe *f* deep-etch test (*or* sample). -ätzung *f* deep etching. -bau *m* 1 underground engineering. 2 underground structure. 3 (*Min.*) deep mining. -bauge-winnung *f* underground mining. t–blau *a* deep blue. T–bohrung *f* 1 deep drilling. 2 deep borehole. -bohrzement *m* oil-well cement. t–braun *a* deep brown. T–brunnen *m* deep well. -druck *m* 1 low pressure. 2 intaglio printing. -druck-farbe *f* intaglio printing ink. -druck-gebiet *n* (*Meteor.*) low-pressure area
Tiefe *f* (-n) 1 depth; profundity. 2 (person's) innermost being
Tief·ebene *f* lowland. t–eingreifend *a* penetrating, thoroughgoing. -eingewur-zelt *a* deep-rooted
tiefen 1 *vt* deepen; (*Metll.*) deep-draw, cup. 2 *vt/i* sound (depth)
Tiefen·gestein *n* plutonic rock. -lehre *f*, -messer *m* depth gauge. -messung *f* depth measurement, sounding. -schärfe *f* (*Opt., Phot.*) depth of focus. -wirkung *f* 1 deep-felt effect. 2 depth of penetration. 3 three-dimensional effect. 4 (*Plating*) throwing power
tiefer·siedend *a* lower-boiling
tief·gefrieren* *vt* deep-freeze
tief·gefrostet *a* deep-frozen. -gehend *a* penetrating, thoroughgoing. -gekühlt *a* deep-frozen. -gelb *a* deep yellow. -grei-fend *a* penetrating, thoroughgoing; fundamental. -grün *a* deep green. -gründig *a* 1 deep-seated, fundamental. 2 penetrating. T–gründigkeit *f* deep-seated-ness, profundity, depth
tief·kühlen *vt* deep-freeze
Tief·kühler *m* deep-freeze (unit), freezer. -kühlfach *n* deep-freeze compartment. -kühlung *f* deep-freezing. -land *n* low-land. t–liegend *a* low-lying; deep-seated. T–moortorf *m* bottom peat. -ofen *m* (*Metll.*) soaking pit. -punkt *m* low (*or* minimum) point, nadir. t–reichend *a* penetrating; far-reaching. -rot *a* deep

red. **-rund** a concave. **T–schliff** m intaglio. **t–schmelzend** a low-melting. **-schürfend** a penetrating. **-schwarz** a deep (or jet) black. **T–seeschlamm** m deep-sea ooze. **-seetauchboot** n bathyscaph. **t–siedend** a low-boiling. **-sinnig** a 1 profound. 2 pensive. 3 melancholy

tiefst a superl of TIEF **—im t—en** etc in the depths of

Tief·stand m low level, minimum, nadir. **-stanzen** n (Metll.) deep-stamping

tief·stellen vt lower; set at "low"

Tiefst·stand m lowest level, minimum. **-temperatur** f lowest (or minimum) temperature. **-wert** m lowest (or minimum) value

Tief·teer m low-temperature tar

Tieftemperatur·forschung f cryogenics. **-teer** m low-temperature tar. **-vergasung, -verkokung** f low-temperature carbonization

Tiefung f (-en) 1 deepening, sounding. 2 (Metll.) deep-drawing, cupping. 3 low spot, depression. cf TIEFEN

Tiefungs·versuch m (Metll.) deep-drawing test. **-wert** m cupping value

tief·wurzelnd a deep-rooted

tief·ziehen* vt (Metll.) deep-draw, cup, dish

Tiefzieh·verfahren n deep-drawing process

Tiegel m (-) 1 crucible, melting pot. 2 pan. 3 cosmetic jar. 4 platen press. **--boden** m crucible bottom. **-brenner** m crucible maker. **-brennofen** m crucible oven. **-deckel** m crucible lid. **-einsatz** m crucible charge. **-flußstahl** m crucible (cast) steel. **-form** f crucible mold. **-formerei** f crucible molding. **-futter** n crucible lining. **-gießerei** f, **-guß** m casting in crucibles. **-gußstahl** m crucible cast steel. **-gußstahlgießen** n casting of crucible steel. **-hohlform** f crucible mold. **-inhalt** m crucible content. **-koks** m crucible coke. **-masse** f refractory material. **-ofen** m crucible furnace. **-probe** f crucible test. **-rand** m crucible rim. **-ring** m crucible ring. **-scherben** fpl crucible shards, broken pieces of a crucible. **-schmelzofen** m crucible melting furnace. **-schmelzverfahren** n (Metll.) crucible

process. **-stahl** m crucible steel. **-ton** m crucible clay. **-trockner** m crucible drier. **-untersatz** m crucible stand. **-verkokung** f crucible coking. **-zange** f crucible tongs

Tiek·holz n teakwood

Tier n (-e) animal; beast. **--art** f species of animal. **-arzneikunde** f veterinary medicine. **-arzneimittel** f veterinary remedy. **-arzt** m veterinarian, veterinary surgeon. **t–ärztlich** a veterinary. **T–blase** f animal bladder. **-chemie** f animal chemistry

Tierchen n (-) animalcule, tiny organism; little animal

Tier·faser f animal fiber. **-fett** n animal fat. **-fibrin** n (animal) fibrin. **-futter** n animal feed. **-garten** m zoo, zoological garden. **-gattung** f genus of animals. **-gift** n animal poison, venom. **-heilkunde** f veterinary medicine

tierisch a animal; bestial, brutish

Tier·keim m animal embryo. **-kenner** m expert on animals; zoologist. **-kohle** f animal charcoal. **-körper** m animal body; carcass. **-kreis** m zodiac. **-kunde, lehre** f zoology. **-leim** m animal glue (or size). **-mehl** n animal meal, tankage. **-milch** f animal milk. **-öl** n animal oil, (specif) bone oil. **-pflanze** f zoophyte. **-reich** n animal kingdom. **-schutz** m protection of animals. **-schutzgebiet** n animal (esp game) preserve. **-schutzverein** m society for the prevention of cruelty to animals. **-schwarz** n animal (specif bone) black. **-seuche** f livestock epidemic. **-stoff** m animal substance. **-versuch** m experiment on animals. **-welt** f animal kingdom, fauna. **-zelle** f animal cell. **-zucht** f animal (esp livestock) breeding

Tiftik·wolle f mohair; Spanish moss

Tiger·auge n tiger eye. **-erz** n (Min.) stephanite. **-sandstein** n mottled sandstone

Tiglin·aldehyd m tigl(in)ic aldehyde. **-säure** f tigl(in)ic acid

tilgbar a 1 eradicable. 2 redeemable

tilgen vt 1 erase, eradicate; exterminate, wipe out. 2 delete, strike out. 3 cancel, annul. 4 pay up (debts, loans); redeem. 5 atone for. **Tilgung** f (-en) eradication, extermination, etc

tingibel *a* (*Micros.*) stainable. **tingieren** *vt* stain, tinge. **Tinktion** *f* (-en) staining, tingeing

tinktoriell *a* tinctorial

Tinktur *f* (-en) tincture

Tinte *f* (-n) ink; tint

tinten·artig *a* inky, ink-like. **T–beutel** *m* ink sac. **-fabrikant** *m* ink manufacturer. **-faß** *n* inkwell. **-fisch** *m* cuttlefish, squid. **-fischschwarz** *n* sepia (as a pigment). **-fleck** *m* ink stain (spot, blot). **-gummi** *m* ink eraser. **-klecks** *m* ink blot. **-kuli** *m* ballpoint pen. **-pille** *f* ink tablet. **stein** *m* inkstone. **-stift** *m* indelible pencil. **-wein** *m* tent, deep-red wine

tintig *a* inky

tippen 1 *vt/i* type(write). 2 *vt* prime (the carburetor). 3 *vi* (+**an, auf, in, gegen**) tap, touch (lightly), touch (on a subject) — **daran ist nicht zu t.** that's untouchable, that's beyond question. 4 *vi* (+**auf**) guess (at). **Tipp·fehler** *m* typing error

Tirolit *n* (*Min.*) tyrolite

Tisch *m* (-e) 1 table; dinner, meal(time). 2 (*Micros.*) stage. 3 (work)bench. **-bier** *n* table beer

Tischchen *n* (-) little table; tablet

Tisch·decke *f* tablecloth. **-gerät** *n* 1 table model (of an appliance). 2 =**-geschirr** *n* tableware. **-leim** *m* furniture glue

Tischler *m* (-) carpenter, joiner. **Tischlerei** *f* (-en) 1 carpentry. 2 carpenter shop

Tischler·leim *m* carpenter's glue

Tisch·platte *f* 1 tabletop. 2 (*Micros.*) stage plate. **-tuch** *n* tablecloth. **-wein** *m* table wine

Titan 1 *n* (*Chem.*) titanium. 2 *m* (-en, -en) titan. **-chlorid** *n* titanium tetrachloride. **-eisen** *n* 1 (*Chem.*) ferrous titanate. 2 (*Metll.*) titaniferous (*or* titanium-bearing) iron (ore)

Titaneisen·erz *n* titanium-bearing iron ore; (*Min.*) ilmenite. **-sand** *m* titaniferous iron sand (a form of ilmenite). **-stein** *m* titanic iron ore, (*Min.*) ilmenite

Titan·erz *n* titanium ore. **-fluorwasserstoffsäure**, **-flußsäure** *f* fluotitanic acid. **t–führend** *a* titaniferous, titanium-bearing. **T–gehalt** *m* titanium content. **-halogen** *n* titanium halide. **t–haltig** *a* containing titanium, titaniferous

titanig *a* titanous

titanisch *a* titanic

Titan–kaliumfluorid *n* potassium fluotitanate. **-kalk** *m* futile. **-karbid** *n* titanium carbide. **-metall** *n* titanium metal. **-nitrid** *n* titanium nitride

Titanofluorwasserstoffsäure *f* fluotitanous acid

Titan·oxid *n* titanium oxide. **t–reich** *a* high-titanium. **T–salz** *n* titanium salt. **-sand** *m* titanium-bearing sand. **t–sauer** *a* titanate of. . . **T–säure** *f* titanic acid. **-säureanhydrid** *n* titanic anhydride, titanium dioxide. **-schwefelsäure** *f* titanosulfuric acid. **-stahl** *m* titanium steel. **-stickstoff** *m* titanium nitride. **-sulfat** *n* titanium sulfate. **-verbindung** *f* titanium compound. **-weiß** *n* titanium white

Titel *m* (-) title. **-bild** *n* frontispiece; cover picture. **-blatt** *n* title page

Titer *m* (-) 1 (*Chem.*) titer. 2 (*Tex.*) count. **-apparat** *m* titrating apparatus. **-flüssigkeit**, **-lösung** *f* standard solution

titern *vt/i* titrate

Titer·stellung *f* standardization. **-substanz** *f* standardizing agent, titrant

Titrage *f* (-n) titration

Titrations·flüssigkeit *f* titration solution. **-verfahren** *n* titration method. **-wert** *m* titration value

Titrier·analyse *f* volumetric analysis, titration. **-apparat** *m* titrating apparatus

titrierbar *a* titratable

Titrier·becher *m* titrating beaker

titrieren *vt/i* titrate

Titrier·flasche *f* titration flask. **-flüssigkeit** *f* titrating (*or* standard) solution. **-geräte** *npl* titrating apparatus. **-lösung** *f* titrating (*or* standard) solution. **-säure** *f* titrating (*or* standard) acid

Titrierung *f* (-en) titration

Titrier·vorrichtung *f* titrating device (*or* apparatus). **-waage** *f* (*Tex.*) testing balance

titrimetrisch *a* titrimetric

tl. *abbv* (teilweise löslich) partly soluble

Tl., *pl* **Tle., Tln.** *abbv* (Teil, Teile, Teilen) part(s)

Toast·automat *m* electric toaster. **toasten** *vt/i* toast. **Toast·röster** *m* toaster

toben *vi* 1 rage, storm. 2 romp, frolic

Tob·sucht f delirium, frenzy

Tochter f (⸚) 1 daughter. 2 (in compds, oft) affiliated, subsidiary

Tod m (-e) death

tod·bringend p.a deadly, fatal, lethal

Todes·dosis f lethal dose. **-fall** m (case of) death, casualty. **-gabe** f lethal dose. **-gefahr** f mortal danger. **-kampf** m death struggle (or agony). **-stoß** m death blow. **-strafe** f death penalty. **-strahl** m death ray. **-ursache** f cause of death. **-wunde** f mortal wound

tödlich a deadly, mortal, fatal, lethal

tod·krank a mortally ill. **-müde** a dead tired. **-sicher** a dead sure (or certain)

Toilette f (-n) 1 toilet, lavatory, rest room. 2 formal attire. 3: **T. machen** get dressed (and do one's hair)

Toiletten·artikel m toilet article. **-papier** n toilet paper. **-seife** f toilet soap. **-wasser** n toilet water

Tokaier(·wein) m Tokay (wine)

Toleranz f (-en) (incl Mach.) tolerance. **-grenze** f tolerance limit

tolerieren vt tolerate. **Tolerierung** f (-en) tolerance, toleration

toll a 1 mad, insane, crazy. 2 wild, madcap; far-out

Toll·beere, -kirsche f belladonna, deadly nightshade. **-kraut** n belladonna; stramonium. **t-kühn** a foolhardy, reckless. **T-wut** f rabies, hydrophobia. **-gift** n rabies virus. **t-wütig** a rabid

Tolu·balsamöl n oil of Tolu balsam. **-chinon** n toluquinone

Toluidin·blau n toluidine blue

Toluol n (Chem.) toluene; (Com.) toluol. **-regler** m toluene regulator. **-sulfonsäure** f toluenesulfonic acid. **-süß** n saccharin. **-trockenschrank** m toluene drying closet. **t-vergällt** a denatured with toluene

Tolu·sirup m sirup of tolu. **-tinktur** f tincture of tolu

Toluylen·rot n toluylene red

Toluyl·säure f toluic acid

Tomate f (-n) tomato

Tombak m tombac, red brass (copper-zinc alloy)

Ton m I (-e) clay. II (⸚e) 1 tone (incl Color); sound; note; audio. 2 tone of voice. 3 accent(uation). 4 (Color, oft) shade. 5: **der**

gute T. good manners. **-abnehmer** m audio pick-up, tone arm. **t-angebend** a fashionable, trend-setting. **T-art** f 1 type of clay. 2 (musical) key. **t-artig** a clayey, argillaceous. **T-aufschluß** m decomposition of clay. **-aufzeichnung** f sound recording. **-bad** n (Phot.) toning bath. **-band** n recording tape. **-bandgerät** n tape recorder. **-behälter** m stoneware container. **-beize** f (Dye.) red liquor. **-beschlag** m coating of clay. **-bildnerei** f ceramics. **-bindemittel** n clay bond(ing) agent. **-boden** m clay soil. **-brei** m (Ceram.) clay slip. **-decke** f clay cover. **-dinasstein** m Dinas brick. **-dreieck** n (pipe) clay triangle

toneisen·haltig a argilloferruginous. **T-stein** m clay ironstone

tonen vt 1 (Phot.) tone. 2 (Paper) tint

tönen I vt 1 tone, tint. II vi 2 sound, ring. 3 resound, echo. 4 sound off

Tonerde f 1 alumina, aluminum oxide; argillaceous earth. 2 (in compds, oft) aluminum, aluminate; **essigsaure T.** aluminum acetate, (Pharm.) Burow's solution. **-beize** f (Dye.) alum mordant, red liquor. **-gehalt** m alumina content. **t-haltig** a aluminiferous, containing alumina. **T-hydrat** n aluminum hydroxide. **-kali** n potassium aluminate. **-lack** m alumina lake. **-metall** n aluminum. **-modul** m (Cement) alumina modulus. **-natron** n sodium aluminate. **-phosphat** n aluminum phosphate; (Min.) wavellite. **-präparat** n alumina preparation. **t-reich** a alumina-rich, high-alumina. **T-salz** n aluminum salt. **-stein** m alumina brick: **-sulfat** n aluminum sulfate. **-verbindung** f alumina compound, (specif) aluminate. **-zement** m or n aluminous (or high-alumina) cement

tönern a 1 clay(ey), argillaceous. 2 earthen(ware)

Ton·farbe f tone color, timbre. **t-farbig** a clay-colored. **T-film** m sound film (or movie). **-filter** n 1 clay filter. 2 audio filter. **-fixierbad** n (Phot.) toning and fixing bath. **-fixiersalz** n (Phot.) toning and fixing salt. **-frequenz** f audio frequency. **-gefäß** n clay (or earthen) vessel. **-geschirr** n pottery, earthenware. **-gips** m argillaceous gypsum. **t-gleich** a alike in tone (*or* shade). **T–glimmerschiefer** *m* (*Min.*) phyllite. **-gut** *n* clayware, earthenware. **t–haltig** *a* containing clay, argillaceous. **T–höhe** *f* pitch (of a sound)

tonig *a* clayey, argillaceous

Tonikum *n* (. . ika) (*Pharm.*) tonic

Ton-industrie *f* clay industry. **-industrielle** *m,f* (*adj endgs*) clayworker. **-ingenieur** *m* sound (audio, acoustics) engineer

tonisch *a* (*Pharm. & Music*) tonic

Tonka-bohne *f* tonka bean. **-bohnenkampfer** *m* coumarin

Ton-kalk *m* argillaceous limestone, argillocalcite. **-kegel** *m* clay cone. **-kerze** *f* clay filter candle. **-kitt** *m* clay cement (*or* glue). **-konserve** *f* sound recording, recorded tape. **-kopf** *m* recording head. **-krug** *m* clay pitcher, earthenware jug. **-kühlschlange** *f* earthenware cooling coil. **-kunst** *f* music. **-lager** *n* clay bed (*or* stratum). **-lehre** *f* acoustics. **-leiter** *f* (musical) scale. **t–los** *a* 1 soundless. 2 unstressed. 3 (*Elec.*) unmodulated. **T–masse** *f* (*Ceram.*) paste. **-mehl** *n* ground clay, clay meal. **-mergel** *m* clay marl. **-mörtel** *m* (fire)clay mortar. **-muffel** *f* clay muffle. **-mühle** *f* clay (*or* pug) mill

Tönnchen *n* (-) small cask, keg *Cf* TONNE 2

Tonne *f* (-n) 1 (metric) ton (1000 kg). 2 barrel, tun, drum

tonnen-förmig *a* barrel-shaped. **-weise** *adv* by the ton; by the barrel

Ton-papier *n* tinted paper. **-pfeife** *f* clay pipe. **-platte** *f* 1 clay plate. 2 disc recording, record. **-prüfung** *f* clay testing. **-reiniger** *m* (*Ceram.*) stone separator. **-retorte** *f* clay retort. **t–richtig** *a* (*Opt.*) orthochromatic. **T–rohr** *n*, **-röhre** *f* clay pipe (*or* tube). **-rohrkrümmer** *m* bent clay pipe, clay elbow. **-salz** *n* (*Phot.*) toning salt. **-sand** *m* argillaceous sand. **-sandstein** *m* argillaceous sandstone. **-scherbe** *f* clay shard, pottery fragment. **-schicht** *f* layer (*or* stratum) of clay. **-schiefer** *m* clay slate, (*Min.*) argillite. **-schlamm** *m* clayey mud; clay slide; (*Ceram.*) claywash. **-schneider** *m* 1 pug mill. 2 sound film cutter. **-seife** *f* aluminous soap. **-speise** *f* tempered clay; clay mortar. **-spur** *f* sound track; groove (of a disc). **-stein** *m* clay stone. **-steingut**

n (*Ceram.*) high-clay white ware. **-streifen** *m* sound track. **-substanz** *f* (*Min.*); (*Ceram.*) clay body. **-teller** *m* unglazed clay plate (*or* dish), porous plate. **-tiegel** *m* clay crucible. **-topf** *m* clay pot, earthenware jug (*or* jar)

Tönung *f* (-en) toning, tone; tint(ing) *cf* TÖNEN

Ton-verschiebung *f* change in tone (*or* shade). **-verstärker** *m* sound amplifier. **-waren** *fpl* pottery, earthenware. **-wasser** *n* clay wash. **-zelle** *f* (-n) unglazed clay cell. **-zeug** *n*, **-zeugwaren** *fpl* (vitreous) clayware, crockery. **-ziegel** *m* clay tile. **-zuschlag** *m* (*Metll.*) aluminous flux

Topas *m* (-e) topaz. **t–gelb** *n* topaz-yellow. **T–schörlit** *n* (*Min.*) pycnite (columnar topaz)

Topf *m* (-̈e) 1 pot. 2 (sauce)pan. 3 crock, jar; mug. **Töpfchen** *n* (-) little pot (pan, jar, etc)

Topf-deckel *m* pot lid

Topfen *m* cottage (*or* pot) cheese

Töpfer *m* (-) potter. **Töpferei** *f* (-en) 1 pottery, ceramics. 2 potter's shop

Töpfer-erde *f* potter's clay. **-erz** *n* potter's ore, alquifou. **-farbe** *f* pottery (*or* ceramic) color. **-geschirr**, **-gut** *n* pottery, earthenware, crockery

töpfern *a* (of) clay, earthen

Töpfer-ofen *m* potter's kiln. **-scheibe** *f* potter's wheel. **-ton** *m* potter's clay. **-ware** *f* pottery, earthenware, crockery

Topf-gießerei *f* crucible steel making. **-glasur** *f* pottery (*or* earthenware) glaze, alquifou. **-glühverfahren** *n* pot annealing. **-lappen** *m* pot holder. **-mühle** *f* barrel mill. **-ofen** *m* pit furnace. **-pflanze** *f* potted plant. **-rösten** *n* pot roasting. **-scheibe** *f* potter's wheel. **-scherbe** *f* potsherd, pottery fragment. **-stein** *m* potstone, soapstone. **-system** *n* (*Dye.*) pot system. **-versuch** *m* pot experiment. **-zeit** *f* pot life

Topinambur *m* (-s) or *f* (-en) Jerusalem artichoke

topisch, *a* topical, local

Tor I *n* (-e) gate(way), portal; goal. **II** *m* (-en, -en) fool

tordierend *p.a* torsional, twisting

Torf *m* peat. **t–·artig** *a* peaty, peat-like. **T-asche** *f* peat ashes. **-boden** *m* peat soil. **-bruch** *m* peat bog. **-eisenerz** *n* bog iron ore. **-erde** *f* peaty soil, peat mold. **-faser** *f* peat fiber. **-gas** *n* peat gas. **-geruch** *m* peaty odor. **-geschmack** *m* peaty taste (*or* flavor). **-gewinnung** *f* peat cutting. **t–haltig** *a* peaty, containing peat

torfig *a* peaty

Torf·kohle *f* peat charcoal. **-koks** *m* peat coke. **-lager** *n* peat bed (*or* bog). **-masse** *f* peat. **-mehl** *n* powdered peat. **-moor** *n* peat moor (*or* bog). **-moos** *n* sphagnum (*or* peat) moss. **-müll** *m* ground peat, peat dust. **-rauchgeschmack** *m* peat smoke flavor. **-staub** *m* peat dust. **-stechen** *n*, **-stich** *m* peat cutting. **-streu** *f* peat litter. **-teer** *m* peat tar. **-verkohlung** *f* peat charring. **-watte** *f* peat wadding

Tor·gummi *n* Bessora gum

Torheit *f* (-en) foolishness, foolish thing (to do)

töricht *a* foolish, silly

torkeln *vi* (h,s) stagger, reel, totter

torkretieren *vt* plaster pneumatically

Torkret·putz *m* (pneumatic) cement-sand plastering, gunite coating. **-sand** *m* plastering, (*Min.*) gunite

Tornister *m* (-) backpack, knapsack

torpedieren *vt* torpedo; (*Petrol.*) shoot

torquieren *vt* twist

Torr *n* (-) (*Meas.: Pressure*) torr

Torsions·faden *m* torsion filament (*or* wire). **-festigkeit** *f* torsional strength, torque. **-kraft** *f* torsional force. **-schwingung** *f* (*Spect.*) twisting vibration, torsional oscillation. **-spannung** *f* torsional stress. **-waage** *f* torsion balance. **-winkel** *m* angle of torsion

Torte *f* (-n) (layer) cake

Torus·fläche *f* (*Math.*) torus

tosen *vi* rage, storm

tot 1 *a* dead: **t—er Punkt** dead point (*or* center); **t—er Gang** lost motion, backlash. 2 *sep pfx* dead, to death; (*Tech.: implies overdoing: cf* TOTMAHLEN): **tot·schießen** shoot dead, kill; **Zeit totschlagen** kill time; **tot·schweigen** hush up (a matter)

total *a* total; (*esp as pfx*) general, overall

Totalität *f* (-en) totality

Total·schaden *m:* **T. haben** be totally ruined, be totaled. **-vergasung** *f* total (*or* complete) gasification. **-verlust** *m* total loss, dead loss

tot·brennen* *vt* overburn (gypsum, etc), dead-burn

Tote *m,f* (*adj endgs*) dead person

töten *vt* kill; deaden; (*Colors*) soften

Toten·blume *f* marigold. **-farbe** *f* livid color. **-fleck** *m* (*Med.*) livid spot. **-kopf** *m* 1 death's head, skull. 2 (*Pigment*) Venetian red, colcothar. **-starre** *f* rigor mortis. **-stille** *f* dead silence. **-uhr** *f* (*Zool.*) death-watch beetle. **-verbrennung** *f* cremation

Töter *m* (-) 1 killer, murderer. 2 cigarette extinguisher

tot·gar *a* (*Metll.*) overrefined. **-geboren** *a* stillborn. **-gebrannt** *p.a* dead-burned *cf* TOTBRENNEN. **-gegerbt** *a* overtanned. **-gekocht** *a* overboiled, dead-boiled. **-gemahlen** *p.a* overground *cf* TOT-MAHLEN. **T–gerbung** *f* (*Lthr.*) overtanning, casehardening. **t–gewalzt** *a* overrolled. **T–last** *f* dead weight

tot·mahlen *vt* overgrind, overmill (to the point of breaking down the structure); (*Paper*) overbeat, dead-grind. *Cf* TOTGE-MAHLEN

tot·pressen *vt* dead-press

Tot·punkt *m* dead point, dead center. **t–reif** *a* dead ripe

tot·rösten *vt* dead-roast, dead-burn

tot·schlagen* *vt* 1 slay, kill *cf* TOT 2. 2 ruin, wreck. 3 defeat

tot·spülen *vt* (*Petrol.*) muddle off, water off

Tötung *f* (-en) killing, murder; deadening *cf* TÖTEN

Tot·volumen *n* dead (*or* unused) space

tot·walzen *vt* (*Rubber*) overmasticate, kill

Tot·zeit *n* dead (*or* nonproductive) time

touchieren *vt* (*Metll.*) touch up = TU-SCHIEREN

Tour *f* (-en) 1 tour. 2 (*Mach.*) revolution, turn: **auf hohen T—en** at high speeds. 3 way, manner. 4 trick

Touren·zahl *f* number of rpm (*or* revolutions). **-zähler** *m* rpm counter; speedometer

Tourill *m* (-s) 1 (*Distil.*) stoneware receiver. 2 carboy. = TURILL

Tournant·öl *n* rancid olive oil

Tournesol *m* litmus

Toxämie *f* (-n) toxemia

toxikologisch *a* toxicological

Toxikum *n* (. . ika) toxic substance

toxin·bindend *a* toxin-binding

toxisch *a* toxic

Toxizität *f* (-en) toxicity

TpM, T.p.M. *abbv* (Touren pro Minute) rpm

Trabant *m* (-en, -en) satellite, (*Spect. oft*) attendant line

Tracht *f* (-en) 1 (*Agric.*) crop, yield. 2 (*Zool.*) litter (of young). 3 (*Bees*) swarming time. 4 (*Bot.*) honey (*or* nectar) flow. 5 costume, garb

trachten *vi* 1 (+**nach**) strive (for), aspire (to). 2 (+**zu** + *inf*) try, strive (to)

trächtig *a* pregnant. **Trächtigkeit** *f* pregnancy

traf met, etc: *past of* TREFFEN

Trafo *m* (-s) (*Elec.*) transformer, =TRANS-FORMATOR

Tragant *m* (-e) tragacanth. **-gummi** *n* gum tragacanth. **-schleim** *m* tragacanth mucilage. **-stoff** *m* bassorin

Trag·bahre *f* stretcher. **-band** *n* (arm) sling; suspensory; (suspension) strap; cable suspender

tragbar *a* 1 portable. 2 bearable; tolerable, acceptable. 3 wearable, fit for wearing. *Cf* TRAGEN

Trage *f* (-n) stretcher; litter; barrow

träge *a* inert, inactive; sluggish, languid

Trage·büchse *f* carrying case

trag·echt *a* wear-resistant

tragen* (trug, getragen; trägt) **I** *vt* 1 carry; (*esp fig*) bear. 2 wear. 3 bear, endure, tolerate. 4 harbor, have (doubts, etc); (*in idioms with nouns*): *cf* RECHNUNG 4, SCHAU 1, SORGE 2. **II** *vt/i* 5 yield, bring in (a good, a poor yield), bear fruit. **III** *vi* 6 carry, reach (*as:* missiles, etc). 7 (*Liquids*) be buoyant. 8 (*esp Farm Animals*) be pregnant. **9: schwer zu t. haben an** have a hard time carrying. **IV: sich t.** 10 wear (well, badly). 11 pay (its way). **12: sich mit dem Gedanken** (or **der Absicht**) **t., zu. . .** be thinking of (doing); **sich mit der Hoffnung t., daß. . .** be hoping that. . . **tragend** *p.a* 1 load-bearing. **2: weit t.** long-range; (voice) that carries. 3

basic, fundamental (idea, role, etc). 4 pregnant

Träger *m* (-) 1 bearer, carrier. 2 support, supporting column (beam, girder, etc). 3 shoulder strap. 4 representative. 5 vehicle (*fig*). 6 porter. 7 wearer. **-gestein** *n* (*Petrol.*) reservoir rock. **-gas** *n* carrier gas. **-metall** *n* (*Metll.*) cementing material, binder; bonding (*or* support) metal. **-rakete** *f* carrier rocket. **-stoff** *m* carrier (substance). **-strom** *m* (*Elec.*) carrier current. **-welle** *f* (*Elec., Radio, etc*) carrier wave

trag·fähig *a* 1 capable of bearing a load. 2 buoyant. 3 productive, 4 acceptable. **Tragfähigkeit** *f* 1 load bearing capacity, bearing strength. 2 buoyancy. 3 productiveness

Trägheit *f* 1 inertia. 2 laziness, sluggishness

Trägheits·kraft *f* force of inertia. **-mittelpunkt** *m* center of inertia. **-moment** *n* moment of inertia

Trag·klemme *f* suspension clamp. **-kraft** *f* carrying (*or* lifting) capacity. **-lager** *n* journal bearing. **-last** *f* load, burden

Tragödie *f* (-n) tragedy

Trag·pfeiler *m* supporting pillar. **-riemen** *m* carrying strap. **-seil** *n* supporting (*or* suspension) cable

Träg·sicherung *f* slow-blow fuse

Trag·stein *m* keystone

trägt carries, etc: *pres of* TRAGEN

Trag·tier *n* beast of burden, pack animal. **-vermögen** *n* carrying (*or* supporting) power; buoyancy. **-weite** *f* range; significance. **-zapfenreibung** *f* journal-bearing friction. **-zeit** *f* gestation period

Trakt *m* (-e) 1 tract (of land, etc). 2 wing (of a building). 3 stretch (of road, etc)

Traktat *m,n* (-e) 1 (scientific) treatise. 2 (religious) tract

traktieren *vt* 1 treat. 2 torment, mistreat

Tralje *f* (-n) bar, slat (of a trellis, fence, etc)

Tran *m* 1 train (*or* fish) oil; (*with pfx*) oil *cf* LEBERTRAN. 2 blubber. **3: im T. sein** be high (on alcohol or drugs). **-brennerei** *f* train oil works, tryworks = TRANSIEDEREI

tranchieren *vt* carve (meat)

Träne *f* (-n) tear

tranen vt oil cf TRAN

tränen vi (Eyes) tear, water

Tränen· tear, lacrimal: **-drüse** f lacrimal gland. **t–erregend** a tear-producing, lacrimatory. **T–feuchtigkeit** f lacrimal fluid. **-fließen** n, **-fluß** m flow of tears. **-flüssigkeit** f lacrimal fluid. **-gang** m tear duct. **-gas** n tear gas. **-reiz** m lacrimatory irritation. **-stoff** m tear gas. **-wasser** n lacrimal fluid, tears

Tran·fettsäure f fish oil fatty acid. **-füllung** f (Lthr.) train stuffing. **-gerbung** f fish-oil tannage. **-geruch** m odor of train oil. **t–haltig** a containing train oil, oily

tranig a 1 fish-oil-like, oily. 2 sluggish

trank drank: past of TRINKEN

Trank m ("-e) drink, beverage; potion

Tränk·bad n impregnating bath

Tränkchen n (-) sip, nip (usu of medicine)

Tränke f (-n) watering place (esp for animals)

tränken vt 1 water, give (esp animals) sthg to drink. 2 breast-feed. 3 saturate, soak, drench; impregnate, steep

Tränk·gefäß n watering vessel (for animals). **-harz** n impregnating resin. **-kessel** m steeping (or impregnating) vessel. **-masse** f impregnating material. **-rinne** f watering trough. **-stoff** m impregnating agent, steep

Tränkung f (-en) watering; breast-feeding; saturation, soaking; impregnation, steeping cf TRÄNKEN

Tränkungs·mittel n, **-stoff** m steeping (or impregnating) agent, steep

Tran·lampe f (fish-)oil lamp. **-leder** n leather dressed with train oil. **-schmiere** f (Lthr.) daubing. **-seife** f train-oil soap

Transformator m (-en) (Elec.) transformer. **transformieren** vt 1 transform. 2 (Elec.) step up, step down

Tran·siederei f (-en) tryworks, blubber rendering plant

transistorisieren vt transistorize

Trans·körper m trans substance

transkristallisieren vt transcrystallize

Translations·energie f translational energy. **-geschwindigkeit** f translational velocity. **-gruppe** f (Cryst.) translational group

translatorisch a translational

Transparent n (-e) (Phot.) slide, transparency

Transparenz f transparency. **--messer** m diaphanometer

Tran·speck m blubber

Transpiration f perspiration; (Bot.) transpiration. **transpirieren** vi perspire; (Bot.) transpire

Transplantat n (-e) 1 (Med.) organ transplant. 2 (Bot.) graft. **transplantieren** vt transplant; graft

transponieren vt transpose

Transport m (-e) 1 transportation. 2 shipment. 3 convoy. 4 (Phys.) convection. **transportabel** a transportable; portable

Transport·band n conveyor belt. **-behälter** m (freight or cargo) container. **-beton** m ready-mixed concrete

Transporteur m (-e) 1 shipper, forwarder. 2 conveyor. 3 (Math.) protractor

transport·fähig a transportable. **T–flasche** f carboy, (cargo) cylinder. **-gefäß** n (freight or cargo) container

transportieren vt transport, ship

Trans·stellung f trans position. **-uran** n transuranium element

Transversal·schwingung f transverse vibration (or oscillation)

Tran·trester mpl blubber residue(s)

Trapez n (-e) 1 trapeze. 2 (Geom.) trapezoid. **--gewinde** n (Mach.) Acme screw thread

Trapezoeder n (-) trapezohedron. **trapezoedrisch** a trapezohedral

Trapp m (-e) trap (rock) cf TRAPPE. **t–artig** a trap-like, trappean

Trappe 1 m (-n,-n) (Zool.) bustard. 2 f (-n) footstep, footprint; track. Cf TRAPP

trappeln vi (h/s) patter, pad. **trappen** vi (h/s) tramp

Traß·beton m trass concrete

Trasse f (-n) route, routing, location line

trassieren vt trace, mark out

Traß·mörtel m trass mortar

trat trod, stepped, etc: past of TRETEN

Tratte f (-n) (Com.) draft

Traube f (-n) 1 grape. 2 bunch of grapes. 3 cluster, bunch. 4 (Bot.) raceme

Trauben·abfall m grape husks, marc. **t–ähnlich**, **-artig** a grape-like; clustered;

racemose. **T–beere** f grape. **-blei** n (Min.) mimetite, pyromorphite. **-dicksaft** m concentrated grape juice. **-essig** m grape vinegar. **t–förmig** a grape-like; racemose; (esp Cryst.) botryoidal. **T–gärung** f fermentation of grapes. **-kamm** m grape stalk. **-kern** m grape seed. **-kernöl** n grapeseed oil. **-kokkus** m staphylococcus. **-kraut** n sea wormwood —**mexikanisches T.** Mexican tea. **-lese** f vintage, grape harvest. **-most** m grape must. **-öl** n grape oil. **-presse** f wine press. **-saft** m grape juice; wine, **t–sauer** a racemate of. **T–säure** f racemic (or DL-tartaric) acid. **-süßmost** m unfermented grape juice. **-trester** mpl grape marc (or pomace). **-vitriol** n copperas. **-zeit** f vintage. **-zucker** m grape sugar, dextrose, glucose. **-zuckeragar** m glucose agar. **-lösung** f glucose solution

traubig a grape-like; grape-laden; (Cryst.) botryoidal

trauen 1 vt marry. 2 vi (+dat) trust, believe. **3: sich t.** dare to go; (+zu + inf) dare, venture, have the courage (to . . .). Cf GETRAUEN

Trauer f sorrow; mourning. **--birke** f weeping birch. **-buche** f weeping beech

trauern vi (+um) mourn, grieve (for)

Trauer·spiel n tragedy. **-weide** f weeping willow

Traufe f(-n) 1 eaves. 2 (rain) gutter; trough. 3 rainwater. **träufeln** 1 vt put drops of . . . in (or on). 2 vi (h/s) drip, trickle

Trauf·rinne f rain gutter = TRAUFE 2. **-wasser** n 1 dripping (or trickling) water. 2 rainwater

traulich a 1 intimate, dear old. 2 cozy, snug

Traum m (¨e) dream

traumatisch a traumatic

träumen vt/i dream, daydream. **Träumerei** f (-en) dreaming; reverie, daydream. **träumerisch** a dreamy

traurig a sad; sorry, miserable

Treber mpl 1 (genl) spent residue. 2 (Brew.) spent malt, brewer's grains. 3 (Wine) marc. **--branntwein** m marc brandy. **-wasser** n (Brew.) grains water. **-wein** m afterwine

Trecker m (-) tractor

Treff m 1 (-e) hit, blow. 2 (-s) meeting, appointment; meeting place

treffen* (traf, getroffen; trifft) **I** vt 1 meet; find. 2 hit, strike (esp a target). 3 (of suspicion, guilt) fall on, point to. 4 hurt, insult. 5 make, conclude (decisions, agreements, preparations), take (measures, etc). **6: es gut (schlecht) t.** hit it right (wrong), strike it lucky (unlucky). **II** vi (h) 7 hit home, hit the mark. **III** vi (s) (+auf) run (into), meet; discover. **IV: sich t.** 8 (+mit) meet, have a meeting (with). **9: es trifft sich** it turns out, it happens. **V: sich t.** (recip) meet (each other). **Treffen** n (-) 1 meeting; appointment. 2 meet, match, contest; skirmish. **treffend** p.a (oft) 1 apt, appropriate. 2 striking. 3 well-captured (likeness). Cf GETROFFEN. **Treffer** m (-) 1 hit, impact. 2 (lottery) prize, win. **--zahl** f number of hits

trefflich a excellent, outstanding

Treff·punkt m 1 point of impact. 2 meeting point. **t–sicher** a sure-fire, accurate. **T–wahrscheinlichkeit** f collision probability

Treib·alkohol m fuel alcohol. **-arbeit** f 1 (Metll.) cupellation. 2 embossed work. **-asche** f cupel ashes, bone ash. **-brühe** f (Tech.) old liquor. **-druck** m (incl Coal) swelling pressure. **-eis** n drift ice. **-eisen** n white pig iron

treiben* (trieb, getrieben) **I** vt 1 drive, propel; impel. 2 raise (dough, hides); stimulate; (Bot.) force (plants). 3 (Bot.) put out (shoots, buds, roots). 4 (Metals) emboss, chase. 5 carry on, pursue, engage in (occupations, activities); study (a subject). 6 (Metll.) extract, refine, cupel. **7: es t.** (+adv) carry on (wildly, etc); (+mit) treat (well, badly). **II** vi (h/s) 8 drift. **III** vi (h) ferment, (esp Yeast, Dough) rise; (Cement) blow. 10 promote perspiration; be diuretic. 11 (Bot.) sprout, bud. Cf GETRIEBEN. **Treiben** n activity; carryings-on. **treibend** p.a (oft) sudorific; diuretic

Treibe·ofen m (Metll.) refining (or cupelling) furnace

Treiber m (-) 1 driver, drover. 2 slave driver (esp fig). 3 (Tex.) picker (on a loom). 4 (Tech.) driving (chasing, embossing) tool. 5 (Paint) active material

Treib·gas n 1 motor fuel gas. 2 propellant

(for aerosols). **-haus** *n* hothouse, greenhouse. **-hauseffekt** *m* greenhouse effect. **-herd** *m* refining hearth, cupellation furnace. **-holz** *n* driftwood. **-kraft** *f* motive power; propelling force. **-ladung** *f* propelling charge. **-mine** *f* floating (*or* drifting) mine. **-mittel** *n* 1 motor fuel. 2 (*Explo.*, *Aerosols*) propellant. 3 expanding (*or* foaming) agent. 4 leavening agent. 5 (*Med.*) purgative, laxative. 6 (*Plastics*) porosity-promoting agent. **-ofen** *m* refining (*or* cupelling) furnace. **-öl** *n* motor (*or* fuel) oil. **-pflanze** *f* forced (hothouse, greenhouse) plant. **-prozeß** *m* cupellation. **-rad** *n* driving wheel; flywheel. **-riemen** *m* drive belt. **-riß** *m* expansion crack. **-sand** *m* quicksand. **-satz** *m* (rocket) propelling composition. **-scherben** *m* cupel. **-schwefel** *m* native sulfur. **-sitz** *m* (*Mach.*) driving fit. **-sprengstoff** *m* (*Explo.*) propellant. **-stange** *f* connecting rod

Treibstoff° *m* 1 (motor) fuel. 2 propellant. 3 expanding (*or* foaming) agent. **--alkoholgemisch** *n* alcohol motor-fuel mixture, gasohol. **-lager** *n* fuel depot. **-tank** *m* fuel tank. **-verhältnis** *n* fuel ratio

Treib-strahl *m* propulsive jet. **-verfahren** *n* (*Metll.*) refining, cupellation. **-welle** *f* drive shaft

trennbar *a* separable, divisible; detachable

Trenn-düse *f* separation nozzle

trennen I *vt* & **sich t.** 1 separate, part, divide. II *vt* 2 break, sever, disconnect. 3 rip up (seams). III *vi* 4 (*Radio*) have high selectivity. 5 cause disunity. IV: **sich t.** (+**von**, *oft*) leave, give up. *Cf* GETRENNT.

Trenner *m* (-) separator, divider

Trenn-faktor *m* separation factor. **-festigkeit** *f* 1 resistance to separation. 2 rupture strength. 3 adhesion. **-fläche** *f* 1 (*Cryst.*) cleavage plane. 2 (*Liquids*) interface. **-flüssigkeit** *f* 1 partition solvent; (*Chromatography*) liquid phase. **-leistung** *f* resolution power, separating efficiency. **-linie** *f* 1 dividing line, borderline. 2 perforated line (for tearing). **-messer** *n* 1 switch-blade knife. 2 (seam-)ripping tool. **-mittel** *n* (*Plastics*) release agent. **-rohr** *n* separating column; (*Isotopes*) Clusius column, thermal diffusion tube.

-schalter *m* disconnecting switch, circuit breaker. **t--scharf** *a* (*Radio*) selective. **T--schärfe** *f* sharpness of separation; (*Radio*) selectivity. **-schicht** *f* separating layer; (*Bot.*) absciss layer. **-schleuder** *f* centrifugal separator. **-technik** *f* separation technology

Trennung *f* (-en) separation, division; severing, disconnection; ripping up *cf* TRENNEN

Trennungs-fläche *f* surface of separation; (*Cryst.*) cleavage plane. **-flüssigkeit** *f* separation (*or* partition) liquid. **-gang** *m* separation procedure, course of the separation. **-grad** *m* degree of separation. **-linie** *f* dividing line. **-methode** *f* method of separation. **-mittel** *n* means of separation, separating medium. **-schicht** *f* separating layer; (*Bot.*) absciss layer. **-strich** *m* 1 line of separation. 2 hyphen. **-verfahren** *n*, **-vorgang** *m* separation process. **-vermögen** *n* separating (*or* resolution) power. **-vorrichtung** *f* separating device, separation. **-wand** *f* dividing wall, partition; (*Tech.*, *Opt.*, *Anat.*) diaphragm, (*Anat.*) septum. **-wärme** *f* heat of separation. **-zeichen** *n* hyphen

Trenn-verfahren *n* separation process. **-wand** *f* partition, etc = TRENNUNGSWAND. **-wirksamkeit** *f* separation efficiency. **-zelle** *f* separating cell

Treppe *f* (-n) stairway, stairs; flight (of steps)

Treppen-absatz *m* stairway landing. **t--förmig** *a* stair-shaped, arranged in steps. **T--geländer** *n* banister. **-haus** *n* stairwell. **-rost** *m* step grate. **-stufe** *f* (stairway) step. **t--weise** *adv* stepwise, by stages (*or* degrees)

Tresor *m* (-e) safe; strong box; (safe-deposit) vault. **--stahl** *m* steel for safes

Trespe *f* brome grass

Trester *mpl* (*genl*) residue; (*specif*) marc, skins, husks, etc. **--branntwein** *m* brandy made from grape marc. **-essig** *m* vinegar from grape marc. **-kuchen** *m* grape cake. **-wein** *m* piquette, wine made from grape marc

Tret-eimer *m* pedal refuse can

treten* (trat, getreten; tritt) I *vt* 1 step (on), tread; trample. 2 kick; mistreat. 3 beat (a path). **II** *vi*(s) **4** step, come, move; (+**in**) enter (into); (*in idioms with nouns and adverbs: cf* ERSCHEINUNG, KRAFT, REAK-TION, ZUTAGE, etc). **III** *vi*(h) (+**nach, gegen**) kick (at)

Tret·gebläse *n* foot bellows (*or* blower). **-hebel** *m* treadle, pedal. **-mine** *f* (*Mil.*) antipersonnel (*or* tread) mine. **-mühle** *f* treadmill

treu *a* **1** true, faithful; loyal; genuine. **2** (*Radio*) high-fidelity. **Treue** *f* loyalty; fidelity; accuracy, genuineness. **treulich** *adv* faithfully, truly

Triakis·oktaeder *n* triakisoctahedron. **-tetraeder** *n* triakistetrahedron

Triakontan *n* triacontane

Triamido· triamido-, triamino-

Trias *f* (*Geol.*) Triassic

Tri·äthanolamin *n* triethanolamine. **-aze-tat** *n* triacetete. **-azojodid** *n* tirazoiodide. **t–basisch** *a* tribasic

Tribolumineszenz *f* triboluminescence

Tribüne *f* (-n) **1** platform, podium. **2** stand(s)

Tricarbonsäure *f* tricarboxylic acid

Trichine *f* (-n) (*Zool.*) trichina. **Tri-chinen·schau** *f* meat inspection for trichinosis

Tri·chinoyl *n* triquinoyl, cyclohexahex-aone. **-chloroessigsäure** *f* trichloroace-tic acid. **t–chroitisch** *a* trichroic. **T–chromsäure** *f* trichromic acid

Trichter *m* (-) **1** funnel, (hollow) cone. **2** horn; megaphone. **3** (feed) hopper. **4** cra-ter. **5** (*Foun.*) gate. **6** (*Anat.*) infun-dibulum. **7** idea. **8** right way. **t–·artig** *a* funnel-like, infundibular

Trichterchen *n* (-) little funnel; little hopper

Trichter·einlage *f* filter cone. **-falte** *f* fun-nel fold. **t–förmig** *a* funnel-shaped, in-fundibular. **T–gestell** *n* funnel stand. **-hals** *m* funnel neck. **-halter** *m* funnel holder (*or* stand). **-kolben** *m* funnel flask. **-lunker** *m* (*Metll.*) pipe, cavity. **-mühle** *f* funnel (cone, hopper) mill

trichtern *vt* funnel, pour through a funnel

Trichter·rohr *n*, **-röhre** *f* funnel tube (*or* pipe), thistle tube. **-stativ** *n* funnel stand. **-stiel** *m* funnel stem. **-wandungen** *fpl* funnel walls

Tricyan *n* tricyanogen. **-chlorid** *n* tri-cyanyl (*or* cyanuryl) chloride. **-säure** *f* cyanuric acid

Tridestillat° *n* triple distillate; (*in compds, usu*) triple-distilled

trieb drove, etc: *past of* TREIBEN

Trieb *m* (-e) **1** instinct, impulse, urge; (+**zu**) tendency (to, toward); (*Psych.*) drive. **2** driving force. **3** (*Phys.*) momentum. **4** (*Biol.*) germinating power. **5** (*Bot.*) shoot, sprout. **6** (*Mach.*) gear drive, gearing. **7** (*Dough*) rise. **-·achse** *f* drive shaft. **-feder** *f* mainspring; (*fig*) motive. **-gas** *n* **1** (mo-tor) fuel gas. **2** (*Aerosols*) propellant. **-kraft** *f* **1** driving force; motive power. **2** germinating (*or* leavening) agent. **3** mo-tive, impetus. **-malz** *n* leavening malt. **-mittel** *n* **1** motor fuel. **2** (*Explo., Aero-sols*) propellant. **3** expanding (*or* foam-ing) agent. **4** leavening agent. **5** (*Med.*) purgative, laxative. **6** (*Plastics*) porosity-promoting agent. **-rad** *n* driving wheel (*or* gear). **-rinde** *f* (*Bot.*) cambium. **-salz** *n* leavening salt. **-sand** *m* quicksand. **-scheibe** *f* driving pulley. **-schraube** *f* (*Micros.*) coarse adjustment. **-stahl** *m* pi-nion steel. **-stoff** *m* motor fuel; pro-pellant. **-wagen** *m* (self-propelled) rail-car. **2** motive power, (*specif*): (rail) motor car, locomotive, (road) tractor. **-welle** *f* drive shaft. **-werk** *n* **1** driving mecha-nism, drive. **2** motor, engine. **3** works; clockwork. **4** power plant. **5** transmission. **-wurzel** *f* main root, taproot.

Tri·eder° *n* trihedron

triefen* (*reg or* troff, getroffen) **1** *vi*(h) be dripping (*or* soaking) wet; **t—d naß** drip-ping wet. **2** *vi* (h/s) drip, trickle; ooze, (*Nose*) run, (*Eyes*) water

Triester 1 *m* (-) (*Chem.*) tri-ester. **2** *m* (-) & *a* (*no endgs*) Triestine, (of *or* from) Trieste (Italy)

Trieur *m* (-e) rotary sifter; sorter

trifft meets, etc: *pres of* TREFFEN

Trift *f* (-en) **1** ocean current. **2** drift(ing) (of timber in rivers), log driving. **3** cattle driving (to pasture). **4** cattle path (*or* trail). **5** pasture (*esp* in mtns). **triften** *vt* float, drift (logs). **Trift·holz** *n* driftwood

triftig *a* **1** weighty, cogent, valid. **2** adrift

Trift·rohr *n*, **-röhre** *f* klystron

trigonometrisch *a* trigonometric

Trigyre *f* (-n) (*Cryst.*) trigonal (*or* triad) ro-

tary axis. **trigyrisch** a trigonal
Tri·jodid n triiodide. **-kaliumphosphat** n tripotassium phosphate
triklin(isch) a triclinic
Trikosan n (-e) tricosane
Trikot (-s) 1 m (Tex.) jersey, knitted fabric. 2 n (athletic) jersey, T-shirt. **Trikotage** f (-n) knitted goods; knitted garment
Trikot·färberei f dyeing of knitted goods
Trilit n TNT, trinitrotoluene
Triller m (-) trill, warble, quaver
Trilliarde f (-n) sextillion, 10^{21}
Trillion f (-en) quintillion, 10^{18}
Trimellit(h)·säure f trimellitic (or benzene-1,2,4-tricarboxylic) acid
trimer a trimeric. **Trimer** n (-e) trimer. **trimerisieren** vt/i trimerize
Trimesin·säure f trimesic (or benzene-1,3,5-tricarboxylic) acid
tri·metrisch a trimetric. **-molekular** a termolecular. **T–molybdänsaure** f trimolybdic acid
trimorph a trimorphic, trimorphous. **Trimorphie** f (-n) trimorphisms
Tri·natriumphosphat n trisodium phosphate. **-nitrobenzolat** n trinitrobenzene complex
trinkbar a drinkable, potable
Trink·becher m drinking cup (or mug). **-branntwein** m potable spirits. **-brunnen** m drinking fountain
trinken* (trank, getrunken) vt/i drink (in, up), imbibe; soak up
trink·fertig a ready to drink. **T–gefäß** n drinking vessel. **-glas** n drinking glass. **-wasser** n drinking water
Trioden·röhre f triode tube
Trioxy· trihydroxy-, trioxy- Cf Oxy·
Trioxid° n trioxide
Tripel (-) 1 m (Min.) tripoli. 2 n triplet; (in compds oft) triple. **t–artig** a tripoline. **T–erde** f tripoli powder, rottenstone. **-glanz** m (Min.) bournonite. **-phosphat** n triple phosphate. **-punkt** m triple point. **-salz** n triple salt. **-stein** m tripoli stone
Triphenyl·siliziumchlorid n triphenyl silicon chloride, chlorotriphenylsilane
Triphylin m (Min.) triphylite
Triplett n (-e) (Spect.) triplet. **-zustand** m triplet state
triplieren vt triple, treble
Trippel (-) 1 m tripoli. 2 n triplet. = TRIPEL

Tripper m (-) gonorrhea, clap. **-gift** n gonorrheal virus
Tri-Reinigung f trichloroethylene cleaning
Tri·sauerstoff m ozone. **-sulfaminsäure** f trisulfamic acid. **-sulfan** n hydrogen trisulfide
Tritan n (-e) triphenylmethane. **Tritanol** n (-e) triphenylmethanol
tri·tetraedrisch a (Cryst.) tritetrahedral. **T–thionsäure** f trithionic acid
Tritol n 2,4,6-trinitrotoluene, TNT
tritt steps, treads, etc: pres of TRETEN
Tritt m (-e) 1 pace; gait; walk; (marching) step: **T. halten** keep pace, keep in step. **2** (audible) footstep. **3** footprint, track. **4** kick. **5** step(s) (on a vehicle). **6** stepstool. **7** treadle, pedal. **t–fest** a hard-wearing, durable (floor, linoleum, etc). **T–gebläse** f foot bellows (or blower). **-leiter** f stepladder. **-wechsel** m change of pace
tritylieren vt tritylate
triumphieren vi triumph; gloat (over a victory)
Trivial·bezeichnung f common (or popular) designation
Trivialität f (-en) triviality
Trivial·name m common (or popular) name
Tri·zyan n tricyanogen = TRICYAN
trocken 1 a dry: **auf t–em Wege** in the dry way; **t–e Destillation** dry (or destructive) distillation; **auf dem t–en sitzen** be left stranded; **im t–en sein** be safe, be out of trouble cf TROCKENE 2. 2 sep pfx cf TROCKENLEGEN etc. 3 (in compds, oft) drying: **T–anlage** f drying plant. **-apparat** m drier; desiccator. **-aufhängeboden** m drying loft. **-aufschluß** m dry-sinter process. **-batterie** f dry battery (or cell). **-beerwein** m straw wine. **-beize** f (Seeds) dry disinfection; dry disinfectant. **-bestimmung** f dry determination. **-blech** n (metal) drying plate (or sheet). **-bleiche** f dry bleaching. **-boden** m drying loft. **-brett** n drying board. **-brikett** n dry-pressed briquette. **-chlor** n (Bleaching) dry-chemicking. **-creme** f foundaton cream. **-dampf** m dry steam. **-darre** f drying kiln. **-dauer** f drying time
trocken·dekatieren vt (Tex.) steamdecatize
Trocken·dekatur f steam (or dry) decatiz-

ing. **-destillation** f dry (or destructive) distillation. **-dock** n dry dock. **-drehofen** m dry-process rotary kiln

Trockene 1 f dryness. **2** n (adj endings) dry spot (or place): **auf dem T—n stehen** be in a dry spot; **im T—n sitzen** (or **sein**) be out of the rain cf TROCKEN 1

Trocken- dry(ing), dehydrated. **-ei** n powdered egg(s). **-eis** n dry ice, carbon dioxide snow. **-element** n dry cell. **-entgasung** f dry distillation. **-entstauber** m dry dust remover. **-extrakt** m dry extract. **-farbe** f dry color, pigment; pastel color. **-fäule** f dry rot. **t-fest** a resistant to dryness (or drying). **T-festigkeit** f **1** resistance to dryness (or drying); dry strength. **2** (Ceram.) green strength. **-filter** n (Opt.) dry filter. **-filz** m drying felt. **-firnis** m siccative varnish; drying oil. **-flasche** f drying flask. **-frucht** f dried fruit. **-füllung** f dry filling; (Elec.) solid electrolyte. **-futter** n dry feed. **-fütterung** f dry (or stall) feeding. **-gas** n dry gas: (Petrol.) gas free from condensable hydrocarbons. **-gebiet** n arid area. **-gefäß** n drying vessel. **-gefrieren** n dehydrofreezing. **-gehalt** m dry (or solids) content. **-gehaltsbestimmung** f determination of dry content. **-geldecke** f dry-gel coating. **-gemüse** n dried (or dehydrated) vegetable(s). **-gerüst** n drying rack (frame, stand). **-geschwindigkeit** f drying speed (or rate). **-gestell** n drying stand (frame, rack), drain board. **-gewicht** n dry weight. **-gießverfahren** n (Plastics) dry casting process. **-glas** n drying glass. **-gleichrichter** m dry-disk (or metal oxide) rectifier. **-gummi** n dry gum; dry rubber. **-guß** m dry sand casting. **-gut** n material to be dried. **-haube** f hair drier, drying hood. **-haus** n drying house. **t-häutig** a dry-skinned. **T-hefe** f dry yeast

Trockenheit f dryness, aridity; drought.

Trockenheits-grad m degree of dryness

Trocken-horde f drying tray. **-kalk** m dry lime. **-kammer** f drying chamber. **-kanal** m drying tunnel. **-kasten** m drying closet, drier. **-kartoffeln** fpl dehydrated potatoes. **-kohle** f dry coal (low in volatile matter). **-kupplung** f (Mach.) dry clutch

trocken-legen vt drain (a swamp)

Trocken- dry, drying, dried, dehydrated: **-legung** f drainage, draining (of swamps, etc). **t-liebend** a (Bot.) xerophilous. **T-löscher** m dry-ice (fire) extinguisher. **-malerei** f pastel painting, crayon drawing. **-maschine** f drier. **-maß** n dry measure. **-masse** f dry matter, solids. **-milch** f dried (or powdered) milk. **-mittel** n drying agent, (de)siccative. **-nährboden** m (Bact.) dehydrated nutrient (or culture) medium. **-obst** n dried fruit. **-ofen** m drying oven (kiln, stove). **-öl** n drying (or siccative) oil. **-pflanze** f (Bot.) xerophyte. **-pistole** f drying pistol, blow drier. **-plasma** n dried plasma. **-platte** f drying plate; (Phot.) dry plate. **-platz** m drying area

trocken-polieren vt dry-polish

Trocken-polierung f dry polishing. **-präparat** n dry preparation. **-preßverfahren** n (Ceram.) dry pressing. **-probe** f dry test (or assay). **-puddeln** n (Metll.) dry puddling. **-pulver** n drying powder. **-quark** m solid skim-milk cheese. **-rahmen** m drying frame, tenter. **-rasierer** m dry shaver. **-raum** m drying space (or room). **-raumgewicht** n dry bulk density

trocken-reiben* vt rub dry

trocken-reinigen vt dry-clean

Trocken-reinigung f dry cleaning (shop). **-riß** m crack due to drying. **-rohr** n, **-röhre** f drying tube. **-rückstand** m dry residue. **-schälchen** n drying capsule. **-schale** f drying dish. **-scheibe** f drying plate. **-schlamm** m (Sewage) dry sludge. **-schleuder** f centrifugal (or spin) drier

trocken-schleudern vt spin-dry

Trocken- dry, dried, drying, dehydrated: **-schliff** m dry grind(ing). **-schrank** m drying closet (cabinet, chamber). **-schwindung** f, **schwund** m drying shrinkage. **-ständer** m drying stand (or rack). **-stoff** m **1** dry substance **2** drying agent, drier, siccative. **-stofflösung** f liquid drier. **-stube** f drying room (or stove). **-substanz** f **1** dry substance (matter, material). **2** drying agent. **-teller** m drying plate (or dish). **-temperatur** f drying temperature. **-thermometer** n dry-bulb thermometer. -torf *m* dry (*or* dried) peat. -treber *pl* (*Brew.*) dried grains. -trommel *f* drying drum, rotary drier. -tuch *n* (drying) towel. -tunnel *m* tunnel drier. -turm *m* drying tower; (*Lthr.*) turret drier. -verfahren *n* dry(ing) process. -verlust *m* loss on drying. -vorgang *m* drying process. -vorrichtung *f* drying apparatus (*or* device). -walze *f* drying roll(er). -wäsche *f* 1 dry cleaning. 2 (*Metll.*) dry scrubbing. -wirkung *f* drying effect (*or* action). -zeit *f* 1 drying time. 2 dry season; dry spell, drought. -zylinder *m* drying cylinder

trockne *a* dry = **trockene**, etc *cf* TROCKEN

Trockne *f* dryness = TROCKENE 1

trocknen 1 *vt/i* dry. 2 *vt* dessicate, dehydrate; season (wood). trocknend *p.a* (*oft*) (de)siccative. Trockner *m* (-) drier. Trocknerei *f* (-en) drying plant. Trocknung *f* (-en) drying; dessication, dehydration; (*Wood*) seasoning

Trocknungs·anlage *f* drying plant. -beschleuniger *m* drying accelerator. -grad *m* degree of drying. -gut *n* material to be dried. -mittel *n* drying agent (*or* medium), desiccant. -ofen *m* drying oven (kiln, stove). -vorlage *f* pre-drying device (*or* attachment)

troff dripped, etc: *past of* TRIEFEN

trog deceived, etc: *past of* TRÜGEN

Trog *m* (¨e) trough; tub vat; tank; pan; hod. --apparat *m*, -batterie *f* (*Elec.*) trough battery, galvanic trough. t–förmig *a* trough-shaped. T–stecher *m* (*Sugar*) stirrer

T-Rohr *n* T-tube, T-pipe

Trombe *f* (-n) whirlwind; waterspout; sandspout

Trommel *f* (-n) 1 (*incl Mach.*) drum; tumbler. 2 barrel, cylinder. 3 spool, reel. 4 (*Anat.*) tympanum. 5 coffee roaster. 6 drum sieve. --darre *f* drum kiln. -fell *n* 1 drumhead. 2 (*Anat.*) eardrum, tympanic membrane. -filter *n* drum filter. -haut *f* 1 drumhead. 2 eardrum = -FELL. -mälzerei *f* drum malting. -mischer *m* drum mixer. -mühle *f* drum (*or* tumbling) mill; (*Ceram.*) Alsing cylinder

trommeln *vt/i* drum, beat (the drum); screen (with a drum sieve); tumble; (*Ceram.*) rattle

Trommel·ofen *m* rotary oven (kiln, furnace); cylindrical roaster. -plattierung *f* barrel plating. -probe *f* (*Ceram.*) rattler test. -rad *n* (*Mach.*) tympan, drum wheel. -sieb *n* rotary screen, drum sieve. -trockner *m* drum (*or* rotary) drier. -versuch *m* (*Ceram.*) rattler test. -waschmaschine *f* tumbler-type washing machine

Trompete *f* (-n) trumpet; (*Anat.*) tube

Trompeten·baum *m* calabash tree

Trona *f*, --salz *n* trona salt, natural sodium sesquiborate

troostitisch *a* (*Metll.*) troostitic

Tropäolin *n* (*Dye.*) tropeolin

Tropa·säure *f* tropaic acid

Tropen *pl* tropics. --ausführung *f* (*Instrum.*) tropicalized form, tropicalization —in T. tropicalized. t–beständig, -fest *a* stable under tropical conditions. T–festigkeit *f* resistance to tropical conditions. -festmachung *f* tropicalization. -fieber *n* tropical fever. -frucht *f* tropical fruit. -gewächs *n* tropical plant (*or* vegetation). -kraftstoff *m* tropical motor fuel. -krankheit *f* tropical disease. -medizin *f* tropical medicine. -pflanze *f* tropical plant

tropfbar *a* capable of forming drops, liquid. --flüssig *a* liquid. Tropfbarkeit *f* liquidity, ability to form drops

Tropf·behälter *m* drip pan (cup, catcher). -bernstein *m* liquid amber. -blech *n* drip (*or* drain) pan (*or* tray). -brett *n* drainboard

Tröpfchen *n* (-) droplet, globule. --kultur *f* (*Bact.*) hanging-drop culture

Tropf·düse *f* dripping (*or* dropping) nozzle. -elektrode *f* dropping electrode

Tröpfel·fett *n* (grease) drippings

tröpfeln I *vt* 1 put drops of . . . on (*or* in), (*Med.*) dispense in drops. II *vi* (h/s) 2 fall in drops, drip. 3 (*Rain*) drizzle, sprinkle

Tröpfel·pfanne *f* drip pan, dripper. -werk *n* (*Salt*) graduation (*or* drying) house

tropfen 1 *vt* dispense in drops, etc. 2 *vi* (h/s) drip, etc = TRÖPFELN

Tropfen *m* (-) 1 drop (of liquid); wine. 2 (*Glass*) gob, bead. --bildung *f* drop formation. -fänger *m* (*Distil.*) drip catcher, splash head. -flasche *f* drop bottle. t–förmig *a* drop-shaped. T–glas *n* dropping glass (*or* tube). **-messer** *m* drop counter; stalagmometer, stactometer; burette. **-mixtur** *f* (*Med.*) drops. **-probe** *f* drop (*or* spot) test. **-registrierapparat** *m* drop recorder. **-wasser** *n* drip water. **t-weise** *a*/*adv* dropwise, *adv oft* in drops, drop by drop. **T–zähler** *m* drop counter; stalagmometer, stactometer

Tropfer *m* (-) (*Med.*) dropper

Tropf- drop, dropping: **-fallpumpe** *f*(*Phys.*) Sprengel pump. **-fett** *n* (grease) drippings. **-fläschchen** *n* dropping vial. **-flasche** *f* dropping (*or* dropper) bottle. **-gefäß** *n* dropper, dropping vessel. **-gewichtsmethode** *f* drop-weight method. **-glas, -gläschen** *n* dropping pipette (*or* bottle). **-hahn** *m* dropping stopcock. **-harz** *n* drops of resin. **-kante** *f*(*Tinplate*) list. **-kasten** *m* (*Paper*) save-all. **-körper** *m* (*Sewage*) trickling filter. **t–naß** *a* dripping wet. **T–öl** *n* dinitrotoluene. **-öler** *m* dropping oil feeder. **-pfanne** *f* drip pan. **-pipette** *f* dropping pipette. **-probe** *f* dropping test; pour test. **-punkt** *m* drip (*or* drop) point, drop-forming temperature. **-rinne** *f* gutter; drip molding. **-rohr** *n* dropping tube. **-schale** *f* drip pan (*or* tray). **-schmierung** *f* drip oil lubrication. **-schwefel** *m* drop sulfur. **-stein** *m* dripstone; (*specif*) stalactite, stalagmite. **t–steinartig** *a* dripstone-like, stalactitic, stalagmitic. **T–transfusion** *f* drip transfusion. **-trichter** *m* dropping funnel. **-wasser** *n* drip(ping) water; *pl oft* (*Med.*) drops. **-zink** *n* drop zinc. **-zinn** *n* drop (*or* granulated) tin. **-zündpunkt** *m* drop ignition temperature

Trophoplast *m* (-e) plastid

Tropika *f* tropical malaria

Tropin·säure *f* tropinic acid

tropisch *a* tropical

Troposphäre° *f* troposphere

Trost *m* consolation, solace. **trösten** 1 *vt* console. 2: **sich t.** console oneself, find solace

Trott *m* (-e) 1 trot. 2 routine. *Cf* TROTTE

Trotte *f* (-n) wine press, bruising mill *cf* TROTT

Trottoir *n* (-e *or* -s) sidewalk

Trotyl *n* TNT, trinitrotoluene

trotz *prep* (*genit, dat*) in spite of

Trotz *m* defiance: **dem zum T.** in spite (*or* in defiance) of that; **T. bieten** (+*dat*) defy

trotzdem 1 *adv* in spite of that, nevertheless. 2 *cnjc* even though

trotzen *vi* sulk; be obstinate; (+*dat*) defy

trotzig *a* defiant; obstinate; sulky

Troutonscher Quotient Trouton's constant

Trp. *abbv* (Tropfpunkt) drip point

trüb *a* turbid, etc = TRÜBE

Trub *m* 1 sediment, dregs. 2 (*Wine*) cloudiness, turbidity. 3 (*Beer*) wort break *cf* TRUBE, TRÜBE. **-bier** *n* beer from the sediment bag

Trube *f* (-n) slurry *cf* TRUB, TRÜBE

trübe *a* 1 turbid, cloudy; opaque: **im t—n fischen** fish in troubled waters. 2 dull, tarnished. 3 gloomy, dim. 4 (*Wine*) thick. **Trübe** *f* 1 turbidity, cloudiness; opacity; dullness, tarnish; gloom, dimness; thickness. 2 turbid liquid; slime, sludge, mud; pulp; (*Metll.*) drass. **-dichte** *f* (*Metll.*) pulp density

trüben 1 *vt* cloud; dull, dim; tarnish; (*Glass*) opacify. 2 *vt* & **sich t.** darken. 3: **sich t.** grow cloudy (dull, dim), become tarnished, turn opaque. *Cf* GETRÜBT

Trüben·entfernung *f* clarification

Trübe·punkt *m* (*Petrol.*) cloud point. **-verdicker** *m* (*Metll.*) pulp thickener

Trüb·glas *n* opaque glass

Trübheit *f* (-en) turbidity, etc = TRÜBE 1

Trub·sack *m* (*Brew.*) sediment (*or* filter) bag. **-stoff** *m* suspended matter, sediment

Trübung *f* 1 clouding, cloudiness; turbidity; opacity. 2 dulling, dimming; tarnish(ing). 3 (*Plastics, Beer*) haze

Trübungs·grad *m* degree of turbidity. **-messer** *m* turbidimeter; nephelometer; opacimeter. **-mittel** *n* opacifying agent. **-punkt** *m* turbidity (*or* cloud) point. **-schleier** *m* haze. **-stoff** *m* turbidifier, turbidity-causing substance

Trüb·würze *f* (*Beer*) first wort

trudeln *vt/i* roll, trundle; (tail)spin

Trüffel *f* (-n) truffle

trug carried, etc: *past of* TRAGEN

Trug *m* deceit, fraud; delusion. **-bild** *n* phantom, illusion. **trügen*** (trog, getrogen) 1 *vt* deceive. 2 *vi* be deceptive. **trügerisch** *a* deceptive, deceitful; false.

Trug·schluß *m* fallacy, false conclusion
Truhe *f* (-n) 1 (storage) chest. 2 *(Radio, TV)* console
Trum *m* (-e *or* ⁼mer) 1 *(Geol.)* apophysis. 2 *(Min.)* shaft compartment. 3 *(Mach.)* upper (*or* lower) half of a drive belt. **Trumm** *m, n* (⁼er) 1 chunk, hulk. 2 =**Trum 2, 3**. 3 strand; tail end, stump —**in einem T.** without a break. 4 *(Min.)* narrow vein (*or* lode). 5 strip (*esp* of film). *See also* **Trümmer**
Trümmer *pl* fragments, pieces; ruins; rubble, debris: **in T. gehen** fall to pieces; **in T. legen** reduce to rubble; **in T. schlagen** smash to pieces. **-absatz** *m (Geol.)* clastic sediment. **-achat** *m* brecciated agate. **t−förmig** *a (Cryst.)* clastic. **T−gestein** *n* rubble, debris, loose rock; breccia. **-haufen** *m* heap of ruins, shambles. **-masse** *f* debris, detritus
Trumpf *m* (⁼e) trump (card)
Trunk *m* (⁼e) 1 drink. 2 drinking, alcoholism —**im T.** under the influence of alcohol. **trunken** *a* drunk(en), intoxicated. **Trunk·sucht** *f* alcoholism. **t−süchtig** *a* alcoholic
Trupp *m* (-s) gang, crew; squad, detachment. **Truppe** *f* (-n) 1 troop(s), unit; *pl (Mil.)* troops, forces. 2 troupe
Trut·hahn *m* turkey (cock). **-henne** *f* turkey hen. **-huhn** *n* turkey
Truxil·säure *f* truxillic acid
Truxin·säure *f* truxinic acid
Trypan·blau *n* trypan blue. **-rot** *n* trypan red
tryptisch *a* tryptic
tschechisch *a* Czech. **Tschechoslowakei** *f* Czechoslovakia. **tschechoslowakisch** *a* Czechoslovakian
Ts-Seide *f* (*short for* Titerschwankungsseide) nub yarn
T-Stoff *m* (*short for* Tränenstoff) 1 tear gas, lacrimator. 2 concentrated hydrogen peroxide fuel
T-Stück *n* T-piece
Tuba·säure *f* tubaic acid
Tuben·creme *f* tube cream. **-emaille** *f* enamel for tubes. **-krem** *f* tube cream. **-lack** *m* tube lacquer (*or* varnish)
Tuberkel *f* (-n) tubercle; (*in compds usu*) tubercular. **tuberkulös** *a* tubercular.

Tuberkulose *f* tuberculosis. **Tuberkulostatikum** *n* (. . ika) tuberculostatic agent
Tuberosen·öl *n* tuberose flower oil
Tubulator *m* (-en) tubulure. **tubuliert** *a* tubulated. **tubulös** *a* tubular. **Tubulus** *m* (. . li) tubule
Tubus *m* 1 (. . bi) *(Med.)* tube; (X-Rays) cone. 2 (. . ben *or* -se) neck (of a flask); *(Opt.)* tube. **-aufsatz** *m* tube attachment. **-aufzug** *m*, **-röhre** *f (Micros.)* drawtube. **-schlitten** *m* tube slide. **-träger** *m* tube support
Tuch *n* 1 (-e) cloth, fabric, material; *pl oft* textiles. 2 (⁼er) cloth (item), (*specif*) wash (dust, table) cloth; scarf; shawl; (hand)kerchief. **t−·artig** *a* cloth-like. **T−fabrik** *f* cloth (*or* textile) mill. **-färberei** *f* cloth dyeing. **-fühlung** *f* close touch (*or* contact). **-handel** *m* cloth trade. **-rot** *n* cloth red
tüchtig *a* 1 capable, efficient. 2 thorough. 3 hardworking, industrious. 4 skilled. 5 real, sound, good and proper; *adv oft* with a vengeance. **Tüchtigkeit** *f* capability, efficiency; thoroughness; industriousness; skill; soundness
Tücke *f* (-n) malice, spite(fulness): **die T. des Objekts** the spitefulness of inanimate objects
tuckern *vi* tick, chug, (go) putt-putt
tückisch *a* malicious, spiteful; insidious, malignant (illness)
Tuff *m* (-e) tuff, tufa. **t−·artig** *a* tufaceous. **T−erde** *f* tufaceous earth. **-kalk** *m* tufaceous limestone. **-stein** *m* tufa. **t−steinartig** *a* tufaceous. **T−wacke** *f* decomposed trap rock
Tugend *f* (-en) virtue. **tugendhaft** *a* virtuous
Tüll *m (Tex.)* tulle *cf* **Tülle**
Tülle *f* (-n) 1 spout; nozzle. 2 socket. *Cf* **Tüll**
Tüll·zelle *f (Bot.)* tylosis
Tulpe *f* (-n) 1 tulip. 2 thistle tube
Tulpen·baum *m* tulip tree. **-zucht** *f* tulip growing. **-zwiebel** *f* tulip bulb
tumeszieren *vi* tumefy, swell
tummeln 1 *vt* exercise (a horse, etc). 2: **sich t.** romp; hurry, hustle
Tummel·platz *m* playground; arena, scene of action

Tummler _m_ (-) tumbler (=drinking glass)

Tümmler _m_ (-) (_Zool._) 1 dolphin, porpoise. 2 tumbler (pigeon)

Tümpel _m_ (-) 1 pool, puddle; sump. 2 (_Metll._) tymp. **--stein** _m_ (_Furnaces_) tympstone, tymp arch

Tumult _m_ (-e) commotion, row, riot. **tumultuarisch** _a_ tumultuous

tun* (tat, getan; tut) **I** _vt_ **1** do. **2** (_in idioms according to noun object_) make, take, have, perform: **einen Fall t.** have (_or_ take) a fall; _cf_ LEID, NOT, WEH, KUNDTUN. **3** put; **dazu t.** add. **4** (+**es**) do, be enough: **das tut es** that will do _cf_ GETAN. **5** (+**es**) e.g. **er hat es mit der Leber zu t.** he has trouble with his liver; **wir haben es zu t. mit. . .** we are dealing with. . . **6** (+**etwas, nichts**) matter, concern: **was tut es?** what does it matter? **II** _vi_ **7** act, pretend to be: **so t., als ob. . .** act as if. . . **8: sie haben zu t.** they are busy. **9: zu t. haben (mit)** have to do, have dealings (with). **10: gut daran t., zu. . .** do well by . . ing. **III: sich t.:** be stirring, be happening. **IV: es ist ihnen zu t. mit. . .** they are concerned with. . . **Tun** _n_ doings, action(s)

Tünche _f_ (-n) whitewash, limewash; distemper; plaster. **tünchen** _vt_ whitewash; plaster

Tünch·farbe _f_ plastering color; pigment for whitewash. **-kalk** _m_ lime for whitewashing. **-schicht** _f_ finishing coat (of plaster). **-werk** _n_ whitewashing

Tunell _n_ (-e) tunnel

Tung·baum _m_ candlenut tree. **-öl** _n_ tung oil. **-stein** _m_ (_Min._) scheelite. **-steinsäure** _f_ tungstic acid

Tunke _f_ (-n) 1 sauce. 2 trouble, predicament. **tunken** _vt/i_ dip, dunk; steep. **Tunk·verfahren** _n_ dipping (_or_ soaking) process

tunlich _a_ feasible, possible; advisable. **tunlichst** _adv_ if at all possible. **Tunlichkeit** _f_ feasibility, possibility; advisability

Tunnel·ofen _m_ tunnel kiln (_or_ furnace)

Tupf _m_ (-e _or_ -en) dot, etc = TUPFEN

Tüpfel _m,n_ (-) dot, spot, speck. **--analyse** _f_ spot (_or_ drop) analysis. **-gewebe** _n_ pitted tissue

tüpfelig _a_ speckled, spotted, dotted; pitted

Tüpfel·kolorimeter _n_ spot colorimeter. **-methode** _f_ spot (_or_ drop) method

tüpfeln _vt_ 1 dot, spot, speckle, stipple. 2 (_Anal._) test by the spot method

Tüpfel·papier _n_ spot-test paper. **-platte** _f_ spot plate. **-probe** _f_ spot (_or_ drop) test. **-reaktion** _f_ spot (_or_ drop) reaction. **-zelle** _f_ (_Bot._) pitted cell

tupfen 1 _vt_ dab, pat; dot, spot. 2 _vi_ (+**an, auf**) touch, tap (on)

Tupfen _m_ (-) (polka)dot, spot, speck

Tupfer _m_ (-) 1 dot, spot, speck. 2 tap, (light) touch. 3 cotton wad, swab, tampon

Tupf·probe _f_ spot (_or_ drop) test. **-reaktion** _f_ spot (_or_ drop) reaction

Tür _f_ (-en) door, doorway —**das steht vor der T.** that is at the door, that is coming soon

Turbidität _f_ turbidity. **Turbiditäts-messer** _m_ turbidimeter

Turbinen·antrieb _m_ turbine drive. **-schaufel** _f_ turbine blade

turbinieren _vt_ centrifuge

Turbinen·triebwerk _n_ turbojet engine

Turbo·brenner _m_ turboburner. **-gebläse** _n_ turboblower. **-grillenboden** _m_ turbogrid plate. **-lader** _m_ turbocharger. **-läufer** _m_ turborotor. **-mischer** _m_ turbomixer. **-verdichter** _m_ turbocompressor. **-zerstäuber** _m_ turbopulverizer

Turbulenz _f_ (-en) turbulence

Turill _m_ (-e) 1 carboy. 2 tourie (Woulfe bottle used in distilling acids); stoneware receiver

Türke _m_ (-n, -n) Turk. **Türkei** _f_ (_Geog._) Turkey. **Türken** _m_ (Indian) corn, maize. **Türkin** _f_ (-nen) Turkish woman

türkis _a_, **Türkis** _m_ (-e) turquoise. **--blau** _a_ turquoise blue

türkisch _a_ Turkish; **t—e Bohne** scarlet runner; **t—er Weizen** maize, corn. **T-rot** _n_ Turkey red

Türkischrot·artikel _m_ (_Tex._) Turkey red style. **-bleiche** _f_ Turkey-red (_or_ madder) bleaching. **-öl** _n_ Turkey-red oil

Türkis·grün _n_ turquoise green, Turkish (_or_ Rinmann's) green

Turm _m_ (ᵈe) tower; turret

Turmalin _n_ (-e) tourmaline. **--zange** _f_ tourmaline tongs

Turm·drehkran *m* rotary tower crane

türmen 1 *vt* & **sich** T. stack up, pile up. 2: **sich t.** tower

turm·förmig *a* tower-like, turreted. **-hoch** *a* sky-high, towering —**t.stehen (über)** tower (over, above). **T–lauge** *f* (*Paper*) tower liquor

Turnbull·blau *n* Turnbull blue

turnen *vi* do gymnastics. **Turner** *m* (-) gymnast

Turn·halle *f* gymnasium

Turnus *m* (-se) 1 cycle, (period of) rotation. 2 (work) shift. 3 (first, second, etc) sitting, session, shift

Turpet·harz *n* turpeth resin

Tusch *m* (-e) fanfare, flourish *cf* TUSCHE

Tusch·blau *n* water-color blue

Tusche *f* (-n) 1 drawing ink: **chinesische T.** India ink. 2 mascara. *Cf* TUSCH

tuschen *vt/i* draw in India ink; paint in water colors; (put on) mascara

Tusche·verfahren *n* (*Micros.*) India ink method, negative staining

Tusch·farbe *f* water color

tuschieren *vt* 1 paint with India ink or water color; ink in. 2 (*Metals*) touch up

Tuschier·platte *f* (*Engg.*) surface plate

Tusch·manier *f* aquatint. **-pinsel** *m* ink (*or* water-color) brush. **-zeichnung** *f* India ink sketch

Tussah·seide *f* tussore silk

tut does, etc: *pres of* TUN

Tüte *f* (-n) 1 assay crucible. 2 (*Glass*) glass cylinder. 3 (*Metll.*) prolong. 4 paper (*or* plastic) bag. 5 cone (for ice cream)

tuten *vi* toot, honk —**es tutet** there is a dial tone

tüten·formig *a* (paper) bag-shaped; conical. **T–papier** *n* bag paper

Tutia *f* (*Zinc*) tutty

Tutol *n* TNT, trinitrotoluene

tw. *abbv* (teilweise) partly, partially

T.W. *abbv* (Teile Wasser) parts of water

TWB *abbv* (Temperaturwechselbeständigkeit) (*Ceram.*) thermal shock resistance

T-Wert *m* (*Soil*) exchange capacity

Twitchell·verfahren *n* Twitchell process

Tyndall·kegel *m* Tyndall cone. **Tyndall-Licht** *n* Tyndall effect

Typ *m* (-en) 1 type, kind, sort. 2 prototype, standard. 3 character, individual

Type *f* (-n) 1 type (*incl* print). 2 odd character

typen *vt* standardize; (classify by) type

Typen·druck *m* type printing. **-metall** *n* type metal. **-molekül** *n* type molecule. **-muster** *n* standard sample. **-theorie** *f* type theory

Typ·färbung *f* standard dyeing

typhös *a* typhoid, typhous

Typhus·gift *n* typhus virus

typisch *a* typical

typisieren *vt* 1 standardize. 2 identify (*or* classify) as to type. 3 typefy

Typ·lösung *f* standard (*or* reference) solution

Typung *f* (-en) standardization, classification by type

Typus *m* (. . pen) type (*incl* person); prototype, very model (of)

U

u *abbv* (Undichtigkeitsgrad) degree of porosity

u. *abbv* 1 (und) &. 2 (unter) under, among. 3 (unten) below

U *abbv* (Umdrehung) rev. (revolution) *cf* UpM

u.a. *abbv* 1 (unter anderem) among other things. 2 (unter anderen) among others. 3 (und andere) and others, et al.

u.ä. *abbv* (und ähnliche) and the like

u.a.a.O. *abbv* 1 (und an anderen Orten) and elsewhere. 2 (und am angeführten Ort) et l.c. (and in the place quoted)

u.a.m. *abbv* 1 (und andere mehr) and others. 2 (und anderes mehr) and the like, and so forth

u.ä.m *abbv* (und ähnliches mehr) and (more of) the like

u.a.O. *abbv* 1 (und andere Orte) and other places. 2 (unter anderen Orten) among other places

u.a.s. *abbv* (und andere solche) and others of this type

üb. *abbv* (ÜBER) over, etc

U-Bahn *f* subway, underground (transit line)

übel *a* (*with endgs:* **übl. .** ; *compar* ÜBLER, *superl* ÜBELSTE) 1 bad, evil. 2 (*pred*): **ü. dran** badly off. 3 vile; dirty. 4 (*pred + dat*) nauseous, sick: **ihnen ist (ihnen wird) ü.** they are (getting) sick. 5 (*as pfx oft*) ill-: **übelberaten** *a* ill-advised

Übel *n* (-) 1 evil (*esp* = bad state of affairs). 2 disaster, calamity. **--befinden** *n* indisposition, sick-feeling

Übelkeit *f* (-en) nausea, sickness

übel·nehmen* *vt* take offense at; (*d.i.o*) hold (sthg) against (sbdy)

übel·riechend *p.a* ill- (*or* vile-)smelling, malodorous. **Ü--stand** *m* evil, abuse, bad situation

übelste *a* (*superl of* ÜBEL) worst, etc

üben I *vt* 1 use, exercise, show, have (discretion, caution, etc). 2 *vt/i* practice, exercise; train. III: **sich ü.** (+in) school oneself

(in), practice *cf* GEÜBT

über I *prep* **A** (*dat, accus*) 1 over, above. **B** (*accus*) 2 across. 3 via, by way of. 4: **ü.** (. . . **hinaus**) beyond. 5 (*Lapse of time*) in (a year, etc); **heute ü. acht Tage** a week from today. 6 (*Clock Time*) past (noon, one o'clock, etc). 7 about, concerning (a topic). 8 after, upon: **Versuche ü. Versuche** experiments after experiments. 9 (*Quant.*) over, more than. **C** 10 (*dat*) with, while: **ü. dem Lärm konnte man nichts hören** with all the noise nothing could be heard; **ü. dem Lesen** while reading. II *adv, pred a* 11 during: **den Sommer ü.** during the summer. 12: **ü. sein** (+*dat*) be superior (to), outmatch. 13: **es ist (es wird) ihen ü.** they are (getting) tired of it; **sie sind die Sache ü.** they are tired of the matter. 14: **ü. und ü.** all over, altogether, totally. III *pfx* over-, super-, supra-, hyper-, overly; (*esp Chem. Nomcl.*) per-. 16 (*Verbs*) (*sep & insep*) over; excessive. 17 (*insep*) out-. (*implies outdoing*)

überaktiv *a* hyperactive

überall *adv* everywhere, in all areas; universally —**ü. dort . . . , wo** wherever

überaltert *a* superannuated, outdated

Überangebot° *n* oversupply, surplus

überanstrengen (*insep*) 1 *vt* overexert, (over)strain. 2: **sich ü.** overexert (strain, overtax) oneself

überantworten (*insep*) *vt* (*d.i.o*) turn over, entrust; deliver, consign (to)

überarbeiten (*insep*) 1 *vt* revise; touch up. 2: **sich ü.** overwork oneself

überäschern (*insep*) *vt* (*Lthr.*) overlime

überaß overate: *past of* ÜBERESSEN

überätzen (*insep*) *vt* overetch

überaus *adv* extremely, exceedingly

überbasisch *a* superbasic, hyperbasic

Überbau° *m* superstructure

überbeizen (*insep*) *vt* (*Lthr.*) overbate

überbelichten (*insep*) *vt/i* (*Phot.*) overexpose. **Überbelichtung** *f* overexposure

Überbesserung° *f* overcorrection

überbetrieblich *a* non-company

überbieten* *(insep) vt* outbid, outdo; exceed

überblasen* *(insep) vt* overblow

überblättern* *(insep) vt* **1** skip over, miss (a page). **2** page through (a book)

über-bleiben* *(insep) vi(s)* be left (over). **Überbleibsel** *n* (-) leftover, remnant, residue

Überbleiche° *f* overbleaching. **überbleichen** *(insep) vt* overbleach

Überblick° *m* **1** overview, general view. **2** overall grasp (of a subject). **3** survey, synopsis. **4** view. **überblicken** *(insep) vt* **1** look over, overlook (an area). **2** grasp, have an overall grasp of (a subject)

überblies overblew: *past of* ÜBERBLASEN

überbor-sauer *a* perborate of. **Ü–säure** *f* perboric acid

überbot(en) outdid (outdone), etc: *past (& pp)* of ÜBERBIETEN

überbracht(e) delivered, etc: *pp (& past) of* ÜBERBRINGEN

überbrannt(e) overburned, etc: *pp (& past) of* ÜBERBRENNEN

überbrausen *(insep) vi(s)* fizz (or foam) over

überbrennen* *(insep) vt* overburn

überbringen* *vt* deliver, transmit. **Überbringer** *m* (-) deliverer; messenger, bearer

überbrom-sauer *a* perbromate of. **Ü–säure** *f* perbromic acid

überbrücken *(insep) vt* **1** bridge, span. **2** *(Elec. oft)* bypass, jump. **3** reconcile

überchlor-sauer *a* perchlorate of. **Ü–säure** *f* perchloric acid

überchrom-sauer *a* perchromate of. **Ü–säure** *f* perchromic acid

überdacht 1 *p.a* roofed over, covered = *pp of* überdachen. **2** thought over, etc: *pp of* ÜBERDENKEN

überdampfen *(insep) vt* distil

überdauern *(insep) vt* outlast, survive

überdecken *(insep) vt* **1** cover (over), overlay. **2** overlap. **3** mask, veil

überdem *adv* besides, in addition

überdenken* *(insep) vt* think over, consider

über-destillieren *vt* distil over

überdies *adv* **1** besides, moreover. **2** anyway, in any case

überdimensional *a* oversize(d), huge

Überdosis° *f* overdose

überdrehen *(insep) vt* overwind, strip; overspeed

Überdruck° *m* **1** excess pressure; *(Phys.)* **atmosphärischer Ü.** atmospheric pressure. **2** overprint(ing), surcharge. **3** *(Tex.)* cover printing. **-artikel** *m* *(Tex.)* cover-print style. **-behälter** *m* pressure tank. **ü–echt** *a* *(Tex.)* fast to overprinting

überdrucken *(insep) vt* overprint, surcharge

Überdruck-farbe *f* *(Tex.)* cover-print paste. **-firnis** *m* overprint varnish. **-hahn** *m* (high-)pressure valve. **-kabine** *f* pressurized cabin. **-papier** *n* transfer paper. **-pumpe** *f* booster pump. **-regler** *m* excess-pressure regulator. **-ventil** *n* pressure relief valve. **-windkanal** *m* compressed-air tunnel

Überdruß *m* surfeit: **bis zum Ü.** to the point of boredom, ad nauseam. **überdrüssig** *a* *(genit)* tired (of), bored (with)

überdurchschnittlich *a* above-average

übereck *adv* crosswise, diagonally

übereilt *a* precipitate, premature. **Übereilung** *f* excessive haste

überein- *sep pfx (implies agreement; see verbs)*

übereinander *adv & sep pfx* one over (or above) the other; about each other

übereinander-fallen* *vi(s),* **-greifen*** *vi* overlap. **-lagern** *vt* superpose. **Ü–lagerung** *f* superposition. **ü—-legen** *vt* pile (or stack) up. **-liegend** *a* superjacent. **-schichten** *vt* stack up in layers. **Ü–schweißung°** *f* lap weld(ing). **ü—-stehen** *vi* stand one above the other; overlap. **-stellen** *vt* place one on top of the other, stack up. **-werfen*** *vt* throw in a heap

überein-kommen* *vi(s)* **(mit, über)** reach an agreement (with, on). **Ü–kommen** *n* (-), **Übereinkunft** *f* (¨-e) agreement

überein-stimmen 1 *vi(s)* **(mit)** agree, harmonize (with); correspond (to); coincide (with). **2** *vt (Dye.)* dye to pattern. **-stimmend** *a (oft)* concurrent; unanimous. **Ü–stimmung** *f* (-en) agreement, harmony; correspondence. **ü—-treffen*** *vi* agree = ÜBEREINKOMMEN

übereist a ice-covered, ice-coated

überelastisch a hyperelastic; above the elastic limit

überempfindlich a hypersensitive. **Überempfindlichkeit** f hypersensitivity

überendlich a transfinite

überentwickeln (insep) vt overdevelop. **Überentwicklung** f overdevelopment; (Biol.) hypertrophy

übererfüllen (insep) vt exceed (a norm, etc)

übererregen (insep) vt overexcite

Übererzeugung f overproduction

überessen* (insep) vr: **sich u.** overeat

übereutektisch a hypereutectic. **übereutektoid** a hypereutectoid

überexponiert a overexposed. **Überexposition** f overexposure

Überf. abbv (Überführung) conv. (conversion)

über/fahren* I (sep) vt, vi(s) ferry across. II (insep) vt 1 run over (or down) (with a car). 2 run through (a stop signal). 3 pass, overshoot (a mark). 4 overpass (another road). 5 walk over, take advantage of

Überfahrt° f crossing (by boat)

Überfall° m 1 (surprise) attack, assault; raid; invasion; onslaught. 2 (Engg.) overfall

über/fallen* I (sep) vi(s) 1 fall over. 2 lap over. II (insep) vt 3 attack, assault; raid; invade. 4 overcome, come upon

überfällig a overdue

Überfall-rohr n, **-röhre** f overflow pipe. **-wagen** m squad car. **-wasser** n overflow

überfangen* (insep) vt (Glass) case, flash, plate. **Überfang-glas** n (Opt.) flashed glass. **-zapfen** m (Glass) flash cone

Überfärbe-artikel m (Tex.) cross-dyed style. **ü–echt** a fast to cross-dyeing

überfärben (insep) vt 1 overcolor, overdye, overstain. 2 cross-dye. 3 redye, top; repaint. 4 fill up, pad (cotton, warp). 5 dye on a mordant. **Überfärbung°** f overcoloring, overdyeing, etc

überfaulen (insep) vi(s) overferment

überfein a superfine

überfetten (insep) vt 1 (Soap) superfat. 2 (Lthr.) overstuff. **Überfettung°** f superfatting; overstuffing. **Überfettungsmittel** n superfatting agent

überfeuern (insep) vt overfire, overheat

überfiel attacked, etc: past of ÜBERFALLEN II (insep) & oft I (sep)

überfing flashed, etc: past of ÜBERFANGEN

überfirnissen (insep) vt varnish over

überfliegen* (insep) vt 1 fly over (or across), overfly. 2 read (or skim) through, glance over. 3 pass over (momentarily)

über/fließen* I (sep) 1 vi(s) flow (or run) over, overflow; brim; (Colors): **ineinander ü.** interflow, blend, shade over into each other. 2 **Überfließen** n: **zum Ü. voll** brimful, full to overflowing. II (insep) vt flood (fields, etc)

überflog(en) flew over (flown over), etc: past (& pp) of ÜBERFLIEGEN

überfloß, überflossen flooded, etc: past & pp of ÜBERFLIESSEN II (insep) & oft past of I (sep)

überflügeln (insep) vt outstrip, outdo

Überfluß m 1 (+**an**) excess, surplus; superfluity; abundance (of). 2: **zum Ü.** to top it all. 3 overflow. **überflüssig** a 1 superfluous, (over)abundant. 2 (Water) overflowing; waste. 3 (Phys.) superfluid. **Überflüssigkeit** f superfluity, overabundance

überfluten (insep) vt flood, inundate

überfordert a swamped (with demand); beyond one's depth

überformen vt (Ceram.) mold on an inside mold

überfraß overate: past of ÜBERFRESSEN

überfressen* (insep) vr: **sich ü.** (esp Animals) overeat, stuff oneself

über-frieren* vi(s) freeze over

überfuhr ran over: past of ÜBERFAHREN II (insep) & oft I (sep)

überführbar a convertible, etc cf ÜBERFÜHREN

über/führen (sep & insep) vt 1 change over, convert, transform. 2 transport, transfer, move, take. **Überführung°** f 1 conversion, transformation. 2 transportation, transfer(ing), moving. 3 (Engg.) overpass

Überführungs-zahl f transference (or transport) number. **-zeit** f conversion period

Überfülle° f overabundance, profusion. **überfüllt** a overcrowded; overstocked; overloaded

Überfunktion *f* (*Med.*) overfunction(ing)

Überfütterung *f* overfeeding

übergab surrendered: *past of* ÜBERGEBEN

Übergabe° *f* 1 handing over; delivery. 2 surrender; capitulation. 3 bequest. 4 opening (to the public). *Cf* ÜBERGEBEN

Übergang° *m* 1 transition; change (over). 2 crossing; passage (across). 3 (*Colors*) blending, shading over

übergangen disregarded, etc: *pp of* ÜBERGEHEN II

Übergangs-bogen *m* 1 transition curve. 2 reducing elbow. **-eisen** *n* off-grade iron. **-element** *n* transition element. **-epoche** *f* transition period. **-erscheinung** *f* transition(al) phenomenon. **-farbe** *f* transition color. **-gebiet** *n* transitional area (*or* region). **-lösung** *f* temporary solution. **-periode** *f* transition period. **-punkt** *m* transition point. **-rohr** *n*, **-röhre** *f* reducing tube (*or* pipe). **-schicht** *f* transition layer. **-stadium** *n* transition(al) stage. **-stahl** *m* borderline steel. **-stecker** *m* (*Elec.*) adapter plug. **-stelle** *f* point of transition. **-stück** *n* 1 (*genl*) transition piece, adapter. 2 (*Glass*) reduction or expansion adapter. **-stufe** *f* transitional step (*or* stage). **-temperatur** *f* transition temperature. **-wahrscheinlichkeit** *f* transition probability. **-zeit** *f* 1 transition time. 2 change of season. **-zustand** *m* transitional state

übergar *a* 1 overfermented; (*esp Food*) overdone. 2 (*Metll.*) overrefined; (*Copper*) dry; (*Steel*) overblown; (*Furnace*) too hot

über/gären* I (*sep*) *vi*(s) run over in fermenting. II (*insep*) *vi* (h) overferment

übergeben *s* (*insep*) I *vt* 1 hand (*or* turn) over; surrender; entrust. 2 open (to the public). II: **sich ü.** 3 surrender. 4 vomit

übergeblieben 1 (been) left (over): *pp of* ÜBERBLEIBEN. 2 *p.a* leftover, remaining

übergeflossen flooded: *pp of* ÜBERFLIESSEN

übergefroren frozen over: *pp of* ÜBERFRIEREN

übergegoren run over in fermenting: *pp of* ÜBERGÄREN I

übergegossen poured over: *pp of* ÜBERGIESSEN I

übergegriffen shifted, etc: *pp of* ÜBERGREIFEN

über/gehen* I (*sep*) *vi*(s) 1 (*in*) change over (into); (**ineinander**) merge, fuse, blend (into each other). 2: **in jemandes Besitz ü.** come (*or* pass over) into sbdy's possession. 3 (**zu**) proceed, go on (to another matter). 4 change sides, desert. 5 overflow. II (*insep*) *vt* pass over, disregard, omit

übergehoben lifted over: *pp of* ÜBERHEBEN A (*sep*)

Übergehung *f* (-en) disregard, omission *cf* ÜBERGEHEN II

übergekrochen crept over: *pp of* ÜBERKRIECHEN

übergelagert *p.a* (*Geol.*) overlying

Übergemengteil *m* minor constituent

übergenau *a* (overly) meticulous, hypercritical

übergenug *adv* more than enough

übergesprungen jumped over: *pp of* ÜBERSPRINGEN I

übergessen overeaten; *pp of* ÜBERESSEN (*insep*)

übergestanden protruded: *pp of* ÜBERSTEHEN I

übergetrieben distilled, etc: *pp of* ÜBERTREIBEN I

Übergewicht° *n* 1 excessive weight: **Ü. haben** be overweight. 2: **Ü. bekommen** lose its balance, topple over. 3 superiority; predominance: **das Ü. haben (bekommen, behaupten,** etc) predominate (become predominant; maintain predominance). 4 preponderance

übergewogen (been) overweight: *pp of* ÜBERWIEGEN I

über/gießen* I (*sep*) *vt* 1 pour (sthg) over. 2 pour over (into another vessel); transfuse. 3 spill. II (*insep*) *vt* 4 cover (by pouring); **etwas mit Wasser ü.** cover sthg with water, pour water over sthg. 5 water, sprinkle; (*Cooking*) baste; (*Med.*) irrigate, douche. 6 flood (*also* with light)

überging omitted, etc: *past of* ÜBERGEHEN II (*insep*) & *oft* I (*sep*)

übergipsen (*insep*) *vt* plaster, parget

überglasen (*insep*) *vt* overglaze, glaze (over). **Überglasung** *f* overglazing, overglaze. **Überglasur** *f* overglaze. **Überglasur-farbe** *f* overglaze (*or* onglaze) color

übergolden (*insep*) *vt* gild

übergor(en) overfermented, etc: *past* (*& pp*) *of* ÜBERGÄREN **II** (*insep*) *& oft past of* **I** (*sep*)

übergoß, übergossen watered, etc: *past & pp of* ÜBERGIESSEN **II** (*insep*) *& oft* **I** (*sep*)

über·greifen *vi* (+*auf*) **1** shift, spread (to). **2** encroach (on). **3:** (**ineinander**) **ü.** overlap. **übergreifend** *p.a* (*oft*) predominant

Übergriff° *m* **1** spread. **2** overstepping (of one's authority). **3** encroachment. **4** incursion, trespass

übergroß *a* oversize, outsize, huge; **der ü—e Teil** the vast majority. **Übergröße°** *f* **1** outsize, extra large size. **2** vastness

Überguß° *m* **1** sthg poured over, sauce, dressing; basting. **2** spill. **3** (*Med.*) douche. **4** (*Sugar*) crust, icing. *Cf* ÜBERGIESSEN

über·haben* *vt* **1** be tired of. **2** have left over; **nichts ü.** (**für**) have no interest (in). **3** be in charge of

Überhandnahme *f* spread, increased prevalence. **überhand·nehmen*** *vi* become widespread, gain prevalence

über/hängen* **1** (*sep*) *vi* hang over, project. **2** (*insep*) *vt* (*pp oft* **überhangen**) (+*mit*) (over)hang, cover, drape (with)

überhäufen (*insep*) *vt* (+*mit*) overload, shower, swamp (with); overstock, glut

überhaupt *adv* **1** (*esp with neg*) at all: **warum ü.?** why at all?; **ü. nichts** nothing at all. **2** generally, in general: **die Wissenschaft ü.** science in general (*or as a whole*). **3** anyway, indeed

über/heben* **A** (*sep*) *vt* lift over (to another place). **B** (*insep*) **I** *vt:* **einen einer Sache ü.** relieve sbdy of sthg. **II: sich ü. 1** injure oneself by lifting. **2** be presumptuous; (+*über*) consider oneself superior (to) *Cf* ÜBERHOBEN **1**

überheblich *a* presumptuous, arrogant

überheizen (*insep*) *vt* overheat

Überhitze *f* excess (*or* waste) heat; superheat. **überhitzen** (*insep*) *vt* overheat; (*Steam*) superheat. **Überhitzer** *m* (-) superheater. **Überhitzer-dampf** *m* superheated steam. **überhitzt** *p.a* (*oft*) overexcited, overwrought. **Überhitzung** *f* overheating; superheating

überhob relieved, etc: *past of* ÜBERHEBEN **B** (*insep*) *& oft* **A** (*sep*)

überhoben 1 *p.a* presumptuous, arrogant. **2** *pp of* ÜBERHEBEN **B**

überhoch (*no endgs*), **überhohe, überhöht** *a* excessive

überholen (*insep*) **1** *vt/i* overtake, pass; outstrip; supersede. **2** *vt* overhaul. **überholt** *p.a* (*oft*) passé, outdated. **Überholungs-gebiet** *n* (*Electrons*) overtake (*or* bunching) range

Überholz *n* resin-impregnated wood

überhören (*insep*) *vt* **1** not hear, miss (sthg said). **2** ignore, pretend not to hear. **Überhör-frequenz** *f* supersonic (*or* ultrasonic) frequency

über·impfen *vt* (**von, auf**) inoculate (from, to). **Überimpf-pipette** *f* transfer pipette

überirdisch *a* superterrestrial, heavenly

Überjodid° *n* periodide. **Überjod·säure** *f* periodic acid

überkalken (*insep*) *vt* **1** overlime. **2** whitewash

überkam overcame, etc: *past of* ÜBERKOMMEN

überkalten (*insep*) *vt* supercool, undercool. **Überkaltung** *f* (-en) supercooling, undercooling

Überkieselung *f* silification

über·kippen *vi*(s) **1** tip over. **2** (*Voice*) break

Überkiste° *f* outside box (*or* case)

überklotzen (*insep*) *vt* (*Tex.*) slop-pad

über/kochen I (*sep*) *vi*(s) boil over. **II** (*insep*) *vt* recook, reboil; overboil

überkohlen·sauer *a* per(oxy)carbonate of. **Ü—säure** *f* per(oxy)carbonic acid

Überkohlung° *f* (*Metll.*) supercarbonization

überkommen* (*insep*) **I** *vt* **1** overcome, come over. **2** inherit; get, receive. **II** *vi*(s) (+*dat* *or* + **auf**) come down, be handed down (to). **III** *p.a* (*oft*) traditional

überkompensieren *vt* overcompensate

Überkorn° *n* oversize grain

Überkorrektur° *f* overcorrection

überkreuzen (*insep*): **sich ü.** intersect, cross

über·kriechen* *vi*(s) creep over

überkritisch *a* **1** hypercritical. **2** (*Temp.*, *Speed*, *etc*) supercritical, above-critical

Überkrümmung° *f* excessive curvature

überkrusten (*insep*) **1** *vt* incrust. **2: sich ü.** form a crust. **Überkrustung** *f* (-en) incrustation

überkühlen (*insep*) *vt* overcool, supercool

überlackierbar *a* subject to overvarnishing. **überlackier·echt** *a* overvarnishresistant. **überlackieren** (*insep*) *vt* overvarnish, overlacquer

über/laden* **I** (*sep*) *vt/i* transship, shift (*or* transfer) (a load). **II** (*insep*) *vt* overload, (*Elec.*) overcharge; (*as p.a oft*) cluttered

überlagern (*insep*) **1** *vt* cover (with a layer of), overlay; superimpose; (*Elec., Radio, TV*) superhet(erodyne). **2: sich ü.** overlap. *See also* ÜBERGELAGERT. **Überlagerung** *f* (-en) covering, overlay(ing); superposition; superheterodyning; overlapping. **Überlagerungs·empfang** *m* superhet(erodyne) reception

Überland·kraftwerk *n* district power station. **-leitung** *f* (cross-country) transmission line. **-telefon** *n* long-distance telephone. **-zentrale** *f* district power station

überlappen (*insep*) *vt* & **sich ü.** overlap. **Überlapp·stoß** *m* lap joint

überlassen* (*insep*) **I** *vt* **1** (*d.i.o*) let (sbdy) have (sthg). **2: einem etwas ü.** leave (sthg to sbdy). **3** expose (to sthg). **4** give up, relinquish; yield; make over (to). **II: sich ü.** (*dat*) indulge (in); yield, abandon oneself (to). **Überlassung** *f* leaving, exposure (to); relinquishment, yielding

Überlast° *f* overload, excess load. **Überlastbarkeit** *f* overload capacity. **überlasten** (*insep*) *vt* overload, overburden, overtax. **Überlastung** *f* (-en) overloading, etc

Überlauf° *m* **1** overflow, spillway. **2** net profit. **über/laufen*** **I** (*sep*) *vi*(s) **1** run over, overflow. **2** desert (to the other side). **II** (*insep*) **3** *vt* overcome, come over. **4** overrun. **5** (*as p.a oft*) overcrowded (profession, etc)

Überlauf·gefäß *n* overflow vessel. **-pipette** *f* overflow pipette. **-rohr** *n*, **-stutzen** *m* overflow tube (*or* pipe). **-ventil** *n* overflow valve

überleben (*insep*) **1** *vt* outlive, survive, live through. **2: sich ü.** become obsolete, outlive its usefulness. **Überlebende** *m,f* (*adj endgs*) survivor

über/legen **I** (*sep*) *vt* lay (*or* put) sthg over sthg. **II** (*insep*) *vt/i* think (things) over; *vt* (*usu* + *dat rflx pron*): **sich etwas ü.** think sthg over, reflect on, consider sthg.

überlegen *p.a* (*usu* + *dat*) superior (to); self-confident, serene; *adv oft* with superior skill. **Überlegenheit** *f* superiority. **überlegt** *p.a* decisive, deliberate; well-thought-out. **Überlegung** *f* (-en) consideration, reflection, deliberation

über·leiten **I** *vt* **1** lead (conduct, pass) over. **2** introduce (sthg new). **3** (*Blood*) transfuse. **II** *vi* lead (into, up to)

überleitfähig *a* superconductive. **Überleitfähigkeit** *f* superconductivity

Überleitung° *f* lead-in, transition: **ohne Ü.** abruptly

überlichten (*insep*) *vt* (*Phot.*) overexpose

Überlicht·geschwindigkeit *f* velocity greater than that of light

Überlichtung° *f* overexposure

überlief overran, etc: *past of* ÜBERLAUFEN **II** (*insep*) & *oft* **I** (*sep*)

überliefern (*insep*) *vt* hand down (to posterity). **Überlieferung°** *f* (*usu* literary) tradition

überließ let have, left: *past of* ÜBERLASSEN

überlöst *a* overgrown (malt)

überlud overloaded, etc: *past of* ÜBERLADEN **II** (*insep*) & *oft* **I** (*sep*)

überm *contrac* = **über dem** *cf* ÜBER **I**

Übermacht° *f* **1** superior power (force, strength); superiority, upper hand. **2** predominance. **3** majority

übermangan·sauer *a* permanganate of. **Ü–säure** *f* permanganic acid

Übermaß° *n* **1** (*oft* + **an**) excess (of): **im Ü. ausgeben** spend to excess, overspend. **2** (*usu Tex.*) oversize. **3** (*Press Fits*) allowance. **übermäßig** *a* excessive

übermastiziert *a* (*Rubber*) overmasticated, dead-rolled

Übermatrix° *f* (*Math.*) matrix of matrices

übermenschlich *a* superhuman

Übermikrometer° *n* ultramicrometer

Übermikroskop° *n* electron microscope.

übermikroskopisch *a* ultramicroscopic

übermitteln (*insep*) *vt* transmit, deliver (messages), communicate; (*d.i.o*): **einem etwas ü.** inform sbdy about sthg. **Übermittlung** *f* (-en) 1 transmission, delivery, communication. 2 message

übermorgen *adv* day after tomorrow

Übermut *m* high spirits, exuberance. **übermütig** *a* spirited, exuberant, frisky

übernächst *a* (the one) after next

übernahm took over, etc: *past of* ÜBERNEHMEN

Übernahme *f* takeover; receipt; acceptance; adoption; assumption, taking on *cf* ÜBERNEHMEN

übernähren (*insep*) *vt* overfeed

übernatürlich *a* supernatural

übernehmen* (*insep*) **I** *vt* 1 take over. 2 receive; take; accept. 3 adopt (for use). 4 assume, take on (duties, etc). 5: **es ü., zu** (+*inf*) undertake (to do). **II: sich u.** undertake too much, overextend (*or* overexert) oneself; overspend; (+**beim** + *inf as noun*) overdo, over-(+**verb**)

übernommen taken over, etc: *pp of* ÜBERNEHMEN

übernormal *a* supernormal, above-normal

über-ordnen *vt* give priority to

Überosmium·säure *f* perosmic acid

Überoxid° *n* peroxide. **überoxidieren** (*insep*) *vt* peroxidize; overoxidize. **Überoxidierung°** *f* peroxidation; overoxidation

Überpflanze° *f* (*Bot.*) epiphyte. **über/pflanzen I** (*sep*) *vt* (*Bot. & Med.*) transplant; (*Bot.*) graft. **II** (*insep*) *vt* (**mit**) plant (an area) (with)

Überpflatschung *f* (*Tex.*) padding

überphosphor·sauer *a* perphosphate of. **Ü—säure** *f* perphosphoric (*or specif* peroxydiphosphoric) acid

überpolen (*insep*) *vt* overpole

Überprobe *f*, **··weingeist** *m* overproof alcohol

Überprodukt° *n* by-product; residual product

Überproduktion *f* overproduction

überprüfen (*insep*) *vt* check, verify, examine, investigate; (*oft* + **auf**) check (for). **Überprüfung°** *f* checking, verification, examination, investigation, review

über·quellen* *vi*(s) bubble (*or* boil) over. **überquellend** *p.a* (*oft*) ebullient

überquer *adv* crosswise. **überqueren** (*insep*) *vt* cross, go across

überraffinieren (*insep*) *vt* overrefine

über/ragen I (*sep*) *vi* jut out, protrude. **II** (*insep*) 1 *vt* tower over. 2 outdo, surpass. **überragend** *p.a* (*usu*) overriding; outstanding

überraschen (*insep*) 1 *vt* surprise, astonish. 2 *vi* be astonishing. **Überraschung** *f* (-en) surprise

überreden (*insep*) *vt* persuade, talk into (doing)

Überreduktion° *f* overreduction

überreichen (*insep*) *vt* hand over (in, out), deliver; present; send

überreichlich *a* (super)abundant

überreif *a* overripe, hypermature

überreizen (*insep*) *vt* overexcite; irritate

Überrest° *m* residue, remnant, vestige; (*pl oft*) remains, leftovers

Überrhenium·säure *f* perrhenic acid

überrieseln (*insep*) *vt* irrigate

Überröste *f* overroasting; overretting

überrosten (*insep*) *vi* become rust-covered. **überrostet** *p.a* rust-covered

überrösten (*insep*) *vt* 1 overroast, burn. 2 overret (flax, hemp)

überrot *a* infrared

Überruthenium·säure *f* perruthenic acid

übers *contrac.* = **über das** *cf* ÜBER **I**

übers. *abbv* (übersetzt) translated

übersäen (*insep*) *vt* sow; strew (an area) *cf* ÜBERSÄT

übersah overlooked, etc: *past of* ÜBERSEHEN

Übersalpeter·säure *f* pernitric acid

übersandt(e) sent, etc: *pp* (& *past*) of ÜBERSENDEN

übersät *p.a* (*oft*) studded, speckled; covered; littered *cf* ÜBERSÄEN

übersättigen (*insep*) *vt* supersaturate; surfeit. **Übersättigung°** *f* supersaturation; satiety

übersauer° *a* overly acid (*or* sour); (*Salts*) containing more than two equivalents of acid to one of base, e.g.: **ü—es oxalsaures Kali** potassium tetroxalate. **übersäuern** (*insep*) *vt* overacidify; peroxidize, overoxidize. **Übersäuerung** *f* overacidification; peroxidation

Überschall° *m* ultrasound; (*Phys.*) supersonics: **-flug** *m* supersonic flight. **-welle** *f* supersonic wave. **-zelle** *f* supersonic light relay (*or* valve)

überschärfen (*insep*) *vt* oversharpen; (*Solutions*) overstrengthen

überschatten (*insep*) *vt* overshadow; shade

überschätzen (*insep*) *vt* overrated, overestimate

über·schäumen *vi*(s) 1 foam (froth, bubble) over. 2 (*Beer, oft*) gush. 3 (*Soap*) prime over

Überschein° *m* reflected color, overtone

über/schichten (*sep & insep*) *vt* cover with a layer of (e.g. another liquid); arrange in layers

Überschiebung *f* (-en) (*Geol.*) overthrust; (*Math.*) transvection

Überschlag° *m* 1 estimate, rough calculation. 2 covering, coating. 3 (*Elec.*) flashover. **über/schlagen*** I (*sep*) 1 *vi*(s) (*Elec.*) arc, flash over. 2 change suddenly, jump. II (*insep*) A *vt* 3 skip, omit. 4 guess, estimate (roughly); go over in one's mind. 5 coat, cover. B: **sich ü.** 6 roll (*or* turn) over (in falling). 7 (*Voice*) break, crack. 8 become coated. 9 follow (each other) in rapid succession. III *as p.a* (*oft*) rough (estimate), roughly estimated

Überschlags·länge *f* (*Explo.*) propagation distance

Überschlag·spannung *f* (*Elec.*) flashover voltage

Überschlags·probe *f* (*Explo.*) gap test. **-rechnung** *f* rough calculation, estimate

überschlug skipped, etc: *past of* ÜBERSCHLAGEN II (*insep*) & *oft* I (*sep*)

überschmelzen* (*insep*) *vt* superfuse; supercool; enamel. **überschmilzt** *pres*, **überschmolz(en)** *past (& pp)*

überschmieren (*insep*) *vt* (**mit**) smear, daub (with)

überschneiden* (*insep*) *vr*: **sich ü.** intersect; overlap; conflict; coincide. **Überschneidung** *f* (-en) intersection; overlapping, conflict

überschnitt(en) intersected, etc: *past (& pp) of* ÜBERSCHNEIDEN

überschoben *a* (*Geol.*) overthrust

überschreiben* (*insep*) *vt* head, caption,

title; address (letters, etc)

überschreiten* (*insep*) *vt* cross; pass, go beyond; exceed. **Überschreitung** *f* (-en) crossing, passing; exceeding (of a limit or norm)

überschrieb(en) headed, etc: *past (& pp) of* ÜBERSCHREIBEN

Überschrift° *f* heading; title; caption

überschritt(en) crossed, etc: *past (& pp) of* ÜBERSCHREITEN

Überschuhe *mpl* overshoes, rubbers

Überschuß° *m* 1 profit. 2 (+**an**) surplus, excess (of). 3 (*Com.*) balance. **überschüssig** *a* surplus, excess

überschwänzen (*insep*) *vt* (*Beer*) sparge

Überschwefel·blei *n* lead persulfide. **ü–sauer** *a* persulfate of. **Ü–säure** *f* per(oxydi)sulfuric acid

überschwellen* (*insep*) *vt* (*Lthr.*) overplump

überschwemmen (*insep*) *vt* flood, inundate. **Überschwemmung** *f* (-en) flood, inundation

überschwenglich *a* exuberant; excessive

überschwer *a* overly heavy; overly difficult

überschwillt overplumps: *pres of* ÜBERSCHWELLEN

überschwimmend *a* supernatant

überschwall(en) overplumped: *past (& pp) of* ÜBERSCHWELLEN

Übersee (*no endgs*): **in Ü., nach Ü.** overseas; **aus Ü.** from overseas. **-handel** *m* overseas trade. **überseeisch** *a* overseas

übersehbar *a* visible at a glance, comprehensible; foreseeable

übersehen* (*insep*) *vt* 1 overlook, have a view of. 2 grasp, estimate (a situation). 3 overlook, miss, omit. 4 disregard. **Übersehen** *n* (*oft*) oversight, mistake

übersenden* (*insep*) *vt* send, mail, ship

über·seifen *vi*(s) (*Soap*) boil over

über/setzen I (*sep*) 1 *vt*, *vi*(s) ferry across; *vi* cross over. 2 *vt* (*Dye.*) top. II (*insep*) 3 *vt* translate; transcribe; transpose; transmit. **Übersetzung** *f* (-en) 1 translation; transcription; transposition. 2 gear (*or* transmission) ratio, mechanical advantage

Übersicht° *f* 1 comprehensive grasp (of the situation). 2 overview, overall view. 3 tabulation list. **4** outline, summary, resumé. **5** supervision. **übersichtlich** *a* **1** visible at a glance. **2** easily grasped, clear and comprehensive. **Übersichtlichkeit** *f* quick and easy comprehension, visibility at a glance

Übersichts·bericht *m* review article. **-bild** *n* general view; overall picture. **-karte** *f* general map. **-spektrum** *n* general spectrum. **-tabelle** *f* tabular summary

über/siedeln (*sep & insep*) *vi*(s) move, resettle

über·sieden* *vi*(s) boil over. **Übersieden** *n* boiling over; (*Brew.*) extra brew

übersieht overlooks: *pres of* ÜBERSEHEN

übersilbern (*insep*) *vt* silver, silver-plate

übersinnlich *a* metaphysical, abstract, supersensible

Übersinterung *f* (-en) incrustation

überspann spun, etc: *past of* ÜBERSPINNEN

überspannen (*insep*) *vt* **1** span. **2** cover. **3** pull too taut, overstretch, (over)strain. **4** (*Steam*) superheat. **überspannt** *p.a* (*oft*) eccentric; exaggerated. **Überspannung°** *f* **1** spanning. **2** covering. **3** overstretching, excessive strain. **4** (*Elec.*) overvoltage

überspinnen* (*insep*) *vt* spin a web over

übersponnen spun, etc: *pp of* ÜBERSPINNEN

übersprang skipped, etc: *past of* ÜBERSPRINGEN **II** (*insep*) & *oft* **I** (*sep*)

über/springen* I (*sep*) *vi*(s) **1** jump (over); (*Elec.*) arc, flash over. **2** (+**auf**) (*esp Med.*) shift, spread (to). **II** (*insep*) *vt* **3** skip, omit. **4** clear (a height, in jumping)

über/spritzen 1 (*sep*) *vi*(s) spurt (spatter, spray) over. **2** (*insep*) *vt* spray over (an area); overspray

Überspritz·farbe *f* overspraying color

über·sprudeln *vi*(s) bubble (*or* gush) over. **übersprudelnd** *p.a* (*oft*) exuberant, bubbly

übersprungen skipped, etc: *pp of* ÜBERSPRINGEN **II**

überspülen (*insep*) *vt* drench, wash over; irrigate

Überstand° *m* **1** projection, excess length. **2** excess; residue

überstand(en) withstood, etc: *past (& pp) of* ÜBERSTEHEN

überständig *a* **1** stale, flat. **2** overripe; overdue for harvesting. **3** (*Ores*) weath-

ered. **4** outdated. **5** leftover, residual

überstark *a* overly strong (*or* intense), excessive

über/stehen* I (*sep*) **1** *vi* jut out, protrude. **II** (*insep*) *vt* **2** (with)stand, weather, survive. **3** recover from. **4: er hat es überstanden** he has passed away (=died). **überstehend** *p.a* (*oft*) overlying; (*Liquids*) supernatant; leftover, residual

über/steigen* I (*sep*) *vi*(s) **1** climb (*or* step) over (*or* across). **2** (*Water, etc*) (*usu* + **über**) overflow, rise above (riverbank, rim, etc). **II** (*insep*) *vt* **3** cross, climb over (a mountain, etc). **4** exceed. **Übersteiger°** *m* overflow (device); siphon. **Übersteig·gefäß** *n* overflow vessel

Übersteuerung° *f* (*Instrum.*) hunting

überstieg(en) exceeded, etc: *past (& pp) of* ÜBERSTEIGEN **II** (*insep*) & *oft past of* **I** (*sep*)

überstrahlen (*insep*) *vt* **1** illuminate, irradiate. **2** overradiate. **3** outshine. **Überstrahlung°** *f* irradiation; overradiation, overexposure to radiation

überstreichen* (*insep*) *vt* **1** paint over, coat. **2** pass (*or* sweep) over (an area)

überstreuen (*insep*) *vt* (**mit**) cover, strew (with)

überstrich(en) coated, etc: *past (& pp) of* ÜBERSTREICHEN

Überstrom° *m* (*Elec.*) flow over, overflow

über·strömen *vi*(s) slow over, overflow

Überstruktur° *f* (*Cryst.*) overstructure. **--gitter** *n*, **-phase** *f* superlattice

überstumpf *a* (*Angles*) reflex

Überstunden *fpl* (hours of) overtime

über/stürzen I (*sep*) *vt*, *vi*(s) overturn. **II** (*insep*) **1** *vt* rush into (an activity). **2: sich ü.** rush into things, act rashly. **3: sich ü.** (*recip*) follow (each other) in rapid succession. **überstürzt** *p.a* (*oft*) rash, hasty, precipitate

Übersud° *m* boiled-over material; distillate

Übersulfid° *n* persulfide

übertag(e) *adv* on the surface. **Ü--arbeit** *f* (*Min.*) surface mining

übertäuben (*insep*) *vt* **1** drown out (sound). **2** cover up (an odor). **3** deaden, dull (a pain). **4** suppress (emotions)

überteuern (*insep*) *vt* overcharge for

übertönen (*insep*) *vt* drown out (sounds); cover up (odors)

übertragbar *a* 1 transferable. 2 applicable (to other cases). 3 infectious

übertragen* *(insep)* **I** *vt* *(usu* + **auf)** 1 transfer (to); carry (oxygen); propagate (waves); transfuse (blood). 2 *(Accounting)* carry over. 3 apply (to another case). 4 (re)assign. 5 copy, transcribe. 6 translate, transpose. 7 broadcast, transmit. 8 record, put on tape. 9 *(Med.)* communicate, pass on (a disease). **II: sich ü. (auf)** *(esp Med)* spread (to), be infectious. **III** *as p.a* *(oft)* figurative. **Übertrager** *m* (-) 1 *(Tlph.)* audio-frequency transformer. 2 *(Elec.)* repeating coil. **Überträger°** *m* 1 *(Chem., Med.)* carrier, transporter. 2 transcriber. 3 transmitter. 4 broadcaster.

Übertragung *f* (-en) 1 transfer(ence); carrying; propagation; transfusion. 2 carryover. 3 application. 4 assignment. 5 transcription. 6 translation; transposition. 7 broadcast, (radio *or* TV) program, show; transmission. 8 recording. 7 infection; spread (of disease). 8 convection

Übertragungs-copolymer *n* graft polymer. **-frequenz** *f* transmission frequency. **-kolonne** *f* transfer column. **-körper** *m* *(Explo.)* induced-detonation charge. **-ladung** *f* *(Explo.)* propagation *(or* primer) charge

übertraf outdid, etc: *past of* ÜBERTREFFEN

übertrat violated, etc: *past of* ÜBERTRETEN **II** *(insep)* & *oft* **I** *(sep)*

übertreffen* *(insep)* *vt* outdo, excel; exceed, go beyond. **übertreffend** *p.a:* **alles ü.** unparalleled

über/treiben* **I** *(sep)* 1 *vt* drive over; distil; sublimate. 2 *vi(s)* overflow. **II** *(insep)* 3 *vt* overdo. 4 *vt/i* exaggerate. **Übertreibkühler** *m* efflux condenser. **Übertreibung** *f* (-en) exaggeration, etc

über/treten* **I** *(sep)* *vi(s)* 1 *(River)* overflow its banks. 2 (+**in**) move (into). 3 (+**zu**) go *(or* change) over (to another group). **II** *(insep)* *vt* 4 transgress, overstep, violate. 5 cross (a boundary)

übertrieb(en) exaggerated: *past (& pp) of* ÜBERTREIBEN. **Übertriebenheit** *f* (-en) exaggeration; extravagance

übertrifft outdoes, etc: *pres of* ÜBERTREFFEN

übertritt violates, etc: *pres of* ÜBERTRETEN **II** *(insep)* & *oft* **I** *(sep)*

Übertritt° *m* changeover, conversion; crossing *cf* ÜBERTRETEN **I**

übertrocknen *(insep)* *vt* overdry

übertroffen outdone, etc: *pp of* ÜBERTREFFEN

übertrug transferred, etc: *past of* ÜBERTRAGEN

übertünchen *(insep)* *vt* whitewash; plaster; gloss over

überverdichten *(insep)* *vt* supercompress, supercharge. **Überverdichter°** *m* supercharger

Übervergrößerung° *f* overmagnification; supplementary magnification

Übervölkerung *f* overpopulation

übervorteilen *(insep)* *vt* take (unfair) advantage of

Übervulkanisation° *f* overvulcanization

überwachen *(insep)* *vt* 1 watch, observe. 2 monitor. 3 supervise, control

überwachsen *(insep)* *vt* 1 outgrow. 2 overgrow. **Überwachsung** *f* overgrowth

Überwachung *f* observation; monitoring, surveillance; supervision, control

über/wallen I *(sep)* *vi(s)* boil over. **II** *(insep)* *vt* overcome, come over, strike

überwältigen *(insep)* *vt* overpower, overwhelm

überwand overcame: *past of* ÜBERWINDEN

überweiden *(insep)* *vt* overpasture, overgraze

überweisen* *(insep)* *vt* (+*dat* or + **an**) remit, transfer; refer (to)

über/wiegen* I *(sep)* *vi* be overweight. **II** *(insep)* 1 *vt* outweigh. 2 *vi* predominate. **überwiegend** *p.a* predominant

überwies(en) remitted: *past (&* **pp)** *of* ÜBERWEISEN

überwinden* *(insep)* 1 *vt* overcome. **2: sich ü. (zu** + *inf)* bring oneself (to do). **Überwindung°** *f* 1 overcoming, conquest, elimination. 2 conscious effort. 3 reluctance

überwintern *(insep)* 1 *vi* spend the winter; hibernate. 2 *vt* keep (e.g. plants) through the winter

überwog(en) predominated, etc: *past (& pp) of* ÜBERWIEGEN **II** *(insep)* & *oft past of* **I** *(sep)*

überwuchern *(insep)* *vt/i* overgrow, grow wild. **Überwucherung°** *f* overgrowth; *(Biol.)* hypertrophy

Überwucht f overwhelming weight (or force); imbalance

überwunden I overcome: *pp of* ÜBER-WINDEN. II *p.a* 2 outmoded, outdated. 3: **sich ü. geben** admit defeat

Überwurf° m 1 roughcast. 2 wrap, robe. 3 throw blanket (or spread). 4 hasp. **-mutter** f lock nut, coupling nut. **-ring** m screw collar

Überzahl° f 1 large number. 2 majority. 3 superior numbers (or forces). 4 excess, surplus. **überzählig** a surplus, excess; superfluous, odd, spare

überzeugen (*insep*) 1 *vt* convince, persuade. 2 *vi* be convincing. **überzeugt** *p.a* (*oft*) confirmed: **ü—er Selbermacher** confirmed do-it-yourselfer. **Überzeugung** f conviction, persuasion. **Überzeugungs-kraft** f persuasive power

überziehen* (*insep*) I *vt* 1 cover, coat. 2 plate (metal). 3 frost, ice (a cake). 4 overdraw (one's account). 5 overwind (a watch). II *vi* 6 run overtime. III: **sich ü.** 7 become coated (or covered). 8 become cloudy (or overcast). **Überzieher** m (-) 1 sheath, (*specif*) condom. 2 topcoat

überzog(en) covered, etc: *past* (*& pp*) *of* ÜBERZIEHEN. **überzogen** *p.a* (*oft*) 1 overdone, excessive, exaggerated, 2 cloudy, overcast. 3 (*Pills*) sugar-coated

überzuckern (*insep*) *vt* sugar-coat

Überzug° m 1 cover(ing), coat(ing). 2 crust, incrustation. 3 plating. 4 skin, film. 5 lining. 6 icing. 7 (pillow) case, etc. *Cf* ÜBERZIEHEN

Überzugs-lack m coating varnish; organic coating. **-masse** f coating compound. **-metall** n plating (or cladding) metal. **-schicht** f coating film (or layer)

Übf. *abbv* (Überführung) conversion, etc

Übg. *abbv* (Übung) exercise

üble a bad, evil, etc = ÜBEL. **übler** a (*compar of* ÜBEL) worse, etc

üblich a usual, customary. **üblicherweise** *adv* customarily, ordinarily

U-Boot n (*short for* UNTERSEEBOOT) submarine

übrig I a 1 remaining, rest of; leftover. 2 other; else: **alles ü—e** everything else. 3 further, additional. 4: **ein ü—es tun** do sthg more, do its bit. 5 *pred* a left (over):

was ist noch ü? what is still left? II *adv* 6: **im ü—en** otherwise, as for the rest. 7 (+**haben**) A have left (over); B: **etwas ü. haben für** care for, be interested in; **nichts ü. haben für** have no interest in *cf* WÜNSCHEN. III *sep pfx: see verbs following*

übrig-bleiben* *vi*(s) be left (over), remain: **es bleibt (ihnen) nichts übrig, als . . . zu** (+*inf*) there is nothing left (or else) (for them) to do but to. . .

übrigens *adv* as for the rest; by the way

übriggeblieben *p.a* leftover, remaining; residual *cf* ÜBRIGBLEIBEN

übrig-lassen* *vt* leave (over): **das läßt (nichts) zu wünschen übrig** that leaves sthg (leaves nothing) to be desired

übsch. *abbv* (überschüssig) excess

Übung f (-en) 1 exercise. 2 practice, drill. 3 training, experience. 4 recitation (class)

Übungs-beispiel n practice example. **-bombe** f practice bomb. **-ladung** f practice charge. **-lager** n training camp. **-reizstoff** m (*Mil.*) irritant training agent. **-riechstoff** m (*Mil.*) odor-detection training agent. **-stück** n exercise

Uchatius-stahl m Uchatius (or direct) steel

u. dergl. *abbv* (und dergleichen) and the like

u.d.f. *abbv* (und die folgenden) and those following

u. dgl. *abbv* (und dergleichen) and the like

u. dgl. m. *abbv* (und dergleichen mehr) and more of the same

u.d.M. *abbv* (unter dem Mikroskop) under the microscope

UdSSR *abbv* USSR

Ue. . . *oft used in place of* **Ü. . .**

u.E. *abbv* (unseres Erachtens) in our opinion

u.e.a. *abbv* (und einige andere) and some others

U-Eisen n channel iron

u.f. *abbv* (und folgende) et seq., and following

UF *abbv* (Unterfamilie) subfamily

Ufer n (-) (river)bank, shore; seashore, beach. **uferlos** a limitless, infinite: **ins u—e gehen (a)** go on endlessly; (b) lead nowhere

u. ff., uff. *abbv* (und folgende) et seq., and following

U-förmig *a* U-shaped

Uhr *f* (-en) **1** timepiece; watch, clock. **2** meter. **3** o'clock: **ein U.** one o'clock; **zwanzig U. dreißig** eight thirty PM; **um wieviel U.?** at what time?

Uhren·öl *n* clock oil. **-vergleich** *m* time check, watch synchronization

Uhr·feder *f* watch (*or* clock) spring. **-glas** *n* **1** watch crystal; clock glass. **2** beaker cover. **-macheröl** *n* watchmaker's oil. **-werk** *n* clockwork. **-zeiger** *m* watch (*or* clock) hand

Uhrzeiger·gegensinn *m* counterclockwise direction. **-richtung** *f* clockwise direction: **in U.** clockwise **-sinn** *m* clockwise direction: **im U.** clockwise

Uhrzeit *f* (clock) time: **nach der U. fragen** ask the time

UKW *abbv* (ULTRAKURZWELLE) ultrashort wave

ul. *abbv* (unlöslich) insol. (insoluble)

Ulme *f* (-n) elm. **Ulmen·rinde** *f* elm bark

Ulmin·säure *f* ulmic acid

ultra·basisch *a* ultrabasic. **U—beschleuniger** *m* ultra-accelerator. **u—filtrieren** (*insep*) *vt* ultrafilter. **U—hochfrequenz** *f* ultrahigh frequency, UHF. **-kurzwelle** *f* ultrashort wave. **-lampe** *f* ultraviolet lamp. **-marin** *n* ultramarine (blue), (*Min.*) lazurite; **gelbes U.** barium chromate pigment. **u—mikroskopisch** *a* ultramicroscopic. **U—rot** *n* infrared

Ultrarot·absorptionsschreiber *m* recording infrared absorptiometer. **u—durchlässig** *a* infrared-transmitting, diathermanous. **U—sperre** *f* infrared block. **u—undurchlässig** *a* infrared-absorbing, athermanous

Ultraschall *m* **1** ultrasound. **2** supersonics. **3** supersonant. **4** (*in compds*) supersonic: **-welle** *f* supersonic wave. **-zelle** *f* supersonic light valve

Ultra·strahl *m* cosmic ray. **-strahlung** *f* cosmic radiation, ultraradiation. **-violett** *n* ultraviolet (light). **-waage** *f* ultrabalance. **-wasser** *n* optically empty water

um I *prep* (*accus*) **1** around, about. **2** (*Time, appx*) (at) about; (*clock time, exact*) at: **um ein Uhr** at one o'clock. **3** for (*also with*

verbs of request: *cf* BITTEN); at (a price). **4** (*Quant.*) by (*or no transl.*): **um 10 cm zu kurz** too short by 10 cm, 10 cm too short. **5** after: **Jahr um Jahr** year after year; **eine Woche um die andere** every other week. **6** (*Transl. varies with verb, e.g.*): **A** (deal) with, (be a matter) of *cf* GEHEN 11, HANDELN III. **B** (deprive) of *cf* BRINGEN 7, KOMMEN 7; **um etwas betrügen** cheat out of sthg. **7**: **um** (*genit*) **willen** for the sake of. **II** *adv* **8** (*appx*): (**so**) **um die zehn Gramm** (**herum**) about (*or* somewhere around) ten grams. **9** over, finished, up; **die Zeit ist um** the time is up. **10**: **um und um** all over, all around, completely. **III** *cnjc* **11** (+**zu** + *inf*) **A** (in order) to. **B** only to. **12**: **um so** (+*compar*) e.g.: **um so besser** all (*or* so much) the better; **um so mehr, als. . .** the more so, since. . . **IV** *pfx* **A** (+*zu*) **13** around, about. **14** over (*usu implies knocking over, cf* UMSTÜRZEN). **15** re-, trans-, over: (*implies change or doing over, cf* UMBAUEN, UMSCHREIBEN). **B** (*insep*) **16** (*usu implies embracing, surrounding with, enclosing in, circling around, circumventing: cf* UMFAHREN II, UMFASSEN)

um·aminieren *vt* transaminate. **Um·aminierung** *f* (-en) transamination

um·ändern *vt* convert, alter, change (around). **Umänderung°** *f* conversion, alteration, change

um·arbeiten *vt* **1** redo, recast, remodel. **2** rewrite, revise. **3** adapt, modify. **Um·arbeitung** *f* (-en) recasting, remodeling; rewriting, revision; adaptation, modification

Umbau° *m* **1** rebuilding, reconstruction, renovation. **2** reshaping. **um/bauen** *vt* **I** (*sep*) **1** rebuild, reconstruct, renovate. **2** reshape, transform. **II** (*insep*) **3** (+**mit**) surround (with), enclose (in) (a structure)

Umbell·säure *f* umbellic acid

umbenannt(e) renamed: *pp* (& *past*) *of* UMBENENNEN

um·benennen* *vt* rename, redesignate

Umber *m*, **-erde** *f* umber

um·bezeichnen *vt* redesignate, rename

um·biegen* **1** *vt* bend, fold over; distort. **2** *vi*(s) bend, turn (*or* double) back. **3**: **sich u.** curl up

um·bilden (*oft* + **in, zu**) 1 *vt* reshape, remodel; transform; reorganize (into). **2: sich u.** be transformed (into). **Umbildung°** *f* reshaping, remodeling; transformation; reorganization

Umbilicar(in)·säure *f* umbilicaric acid

um·blasen* *vt* blow over (*or* down)

um·blicken *vr:* **sich u.** look around

um/bördeln *vt* **I** (*sep*) turn (*or* fold) over (the edge). **II** (*insep*) (provide with a) border; edge; flange

Umbra(·erde) *f* umber

um·brennen* *vt* reburn, refire

um·bringen* *vt* kill, destroy

Umbruch° *m* 1 freshly plowed soil. 2 upheaval, revolution; radical change. 3 (*fig*) new territory

um·definieren *vt* redefine

um·denken* *vi* reorient one's views. **Umdenkungs·prozeß** *m* process of rethinking

Umdestillation° *f* redistillation. **um·destillieren** *vt/i* redistill, rectify

um·deuten *vt* reinterpret, rethink. **Umdeutung°** *f* reinterpretation

umdrehbar *a* 1 reversible. 2 revolving

um·drehen 1 *vt* & **sich u.** turn (around, over), rotate, revolve; twist; reverse. 2 *vi*(s) turn (*or* double) back. **Umdrehung** *f* (-en) turn, rotation, revolution

Umdrehungs·achse *f* axis of rotation (*or* revolution). **-bewegung** *f* rotary (*or* rotatory) motion. **-geschwindigkeit** *f* rate (*or* speed) of rotation, rotary velocity. **-kegel** *m* cone of revolution. **-punkt** *m* center of revolution (*or* rotation). **-richtung** *f* direction of rotation (*or* revolution). **-zahl** *f* number of revolutions, r.p.m. **-zähler** *m* revolution (*or* rpm) counter

Umdruck° *m* reprint(ing). **um·drucken** *vt* reprint, print over

Umdruck·farbe *f* reprinting ink. **-papier** *n* transfer (*or* offset printing) paper

umeinander *adv* & *sep pfx* around (*or* about) one another

um·estern *vt* transesterify. **Umesterung** *f* (-en) transesterification, ester interchange

um·fallen* *vi*(s) fall over (*or* down), drop

um·fällen *vt* reprecipitate

Umfang *m* (ⁿe) 1 circumference, girth; periphery; perimeter. 2 extent, range, scope; scale. 3 size

umfangen* (*insep*) *vt* 1 embrace; encompass, take in; surround. 2 hug, clasp

umfänglich *a* extensive

umfang·reich *a* extensive; voluminous, spacious; ample

Umfangs·geschwindigkeit *f* peripheral (*or* circumferential) velocity. **-kraft** *f* circumferential force

um·färben *vt* redye, dye a different color

umfassen (*insep*) *vt* 1 embrace (*lit* & *fig*). 2 enclose, surround. 3 contain, hold; include, comprise; cover (a range of). **umfassend** *p.a* extensive; comprehensive, full, complete

umfing surrounded, etc: *past of* UMFANGEN

umflechten* (*insep*) *vt* surround with woven (*or* braided) material. **umflicht** *pres.* **umflocht(en)** *past* (& *pp*)

umformbar *a* deformable; plastic. **umformen** *vt* reshape, remodel. 2 (*incl Elec.*) transform, convert. 3 deform. **Umformer** *m* (-) (*incl Elec.*) transformer, (rotary) converter; (*Steam*) desuperheater. **Umformung°** *f* reshaping, remodeling; conversion; (*incl Math.*) transformation; plastic deformation

Umfrage° *f* inquiry; opinion poll, survey: **U. halten** = **um·fragen** *vi* make inquiries, conduct a survey

um·füllen *vt* pour into another container; refill; decant; (*Brew.*) rerack. **Umfüllung°** *f* transferring (of liquids); refilling; decantation

um·funktionieren *vt* convert, remodel; reprogram

umgab surrounded: *past of* UMGEBEN

Umgang *m* 1 (*usu* + **mit**) contact, relations, dealings (with). 2 (+**mit**) handling, operation (of machines, materials). 3 company (=people). 4 rotation

umgangen circumvented: *pp of* UMGEHEN **II** (*insep*)

umgänglich *a* sociable, outgoing

umgeben* (*insep*) *vt* surround

umgebogen bent: *pp of* UMBIEGEN

umgebracht killed: *pp of* UMBRINGEN

umgebrannt refired: *pp of* UMBRENNEN

Umgebung *f* (-en) surroundings, environment; neighborhood; circle (of friends, followers)

Umgebungs·flüssigkeit f enveloping liquid. **-temperatur** f ambient temperature

umgedacht reoriented one's views: *pp of* UMDENKEN

umgegangen circulated: *pp of* UMGEHEN I

Umgegend° f surrounding area, vicinity

umgegossen decanted: *pp of* UMGIESSEN I

um/gehen* I (*sep*) 1 go round, circulate. 2 turn, rotate. 3 (make a) detour. 4 (+**mit**) associate, deal (with); (*esp Mach.*) run, handle, operate. **5: mit einem Plan u.** be considering a plan; **mit dem Gedanken u.** (+**zu** + *inf*) be thinking (of doing). II (*insep*) get around, circumvent, bypass. **umgehend** *p.a* 1 immediate. 2 *adv* oft without delay; by return mail. **Umgehung** f (-en) 1 cicumvention, avoidance. 2 detour, bypass

umgekehrt I *pp of* UMKEHREN. II *p.a* 1 reverse, opposite; (*esp Math.*) inverse. 2 (*as adv*) the other way around, vice versa; on the other hand, conversely

umgekommen died: *pp of* UMKOMMEN

umgelaufen circulated: *pp of* UMLAUFEN I

umgelegt 1 *pp of* UMLEGEN. 2 *p.a:* **u—er Hals** ring neck

umgerechnet *p.adv:* **das sind u. 3 Mio.** that is the equivalent of 3 million *cf* UMRECHNEN

umgerissen torn down: *pp of* UMREISSEN I

umgeschlagen cut down: *pp of* UMSCHLAGEN I

umgeschlungen wrapped around: *pp of* UMSCHLINGEN 1

umgeschmolzen remelted: *pp of* UMSCHMELZEN

umgeschrieben rewritten: *pp of* UMSCHREIBEN

umgesehen looked around: *pp of* umsehen

umgesotten distilled: *pp of* UMSIEDEN

um·gestalten *vt* reshape, remodel; alter; rearrange. **Umgestaltung°** f reshaping; transformation, metamorphosis; rearrangement

umgestochen dug up: *pp of* UMSTECHEN

umgetan put on: *pp of* UMTUN

umgewandt turned over: *pp of* UMWENDEN

umgezogen changed, moved: *pp of* UMZIEHEN

umgibt surrounds: *pres of* UMGEBEN

um/gießen* I (*sep*) *vt* 1 decant, pour into another container. 2 (*Metll.*) recast. 3 spill. II (*insep*) *vt* 4 surround with liquid. 5 (*Metll.*) grout

umging circumvented: *pp of* UMGEHEN II

umgoß, umgossen grouted, etc: *past & pp of* UMGIESSEN II (*insep*) & *oft past of* I (*sep*)

umgrenzen (*insep*) *vt* 1 limit, bound (on all sides). 2 enclose, embrace, include. 3 define precisely. **Umgrenzung** f (-en) limitation, bounding; boundary; enclosure; precise definition

um·größern *vt* change in size

um·gruppieren *vt* regroup, rearrange

Umguß° m decantation, transfer (by pouring); transfusion; recast(ing)

umher *adv & sep pfx* around, about = HERUM

umherziehend *p.a* itinerant

umhin·kommen*, umhin·können* *vi:* **man kommt** (or **kann**) **nicht umhin, es zu tun** one can not help doing it

umhüllen (*insep*) *vt* (*oft* + **mit**) wrap (up), envelop, shroud (in); cover, surround (with); (*Cryst.*) occlude. **Umhüllende** f (*adj endgs*) (*Math.*) envelope. **Umhüllung** f (-en) 1 wrapping, enveloping, shrouding, covering, surrounding. 2 wrap(per), envelope, shroud, cover; occlusion

U/Min *abbv* (Umdrehungen pro Minute) r.p.m.

U. Mk. *abbv* (unter dem Mikroskop) under the microscope

Umkehr f 1 turning around (*or* back), turnabout. 2 reversal; inversion. 3 sudden change, upturn. **umkehrbar** *a* reversible. **um·kehren** 1 *vt, vi*(s), **sich u.** turn around (over, back, upside down, inside out); reverse, invert. 2 *vi*(s) change, do an about-face. *Cf* UMGEKEHRT

Umkehr·entwicklung f (*Phot.*) reversal development. **-erscheinung** f reversal phenomenon. **-film** m reversal film. **-funktion** f inverse function. **-linse** f (*Opt.*) erector lens. **-spektroskop** n reversion spectroscopy

Umkehrung° f reversal, inversion

Umkehr·verfahren n reversal process

um·kippen 1 *vt, vi*(s) turn (*or* tip) over. 2

vi(s) keel over, pass out; do an about-face
umklappbar *a* folding, collapsible (table, etc)
um·klappen 1 *vt* turn (fold, flip) over, fold up (*or* down) (hinged shelf, etc.). **2** *vi*(s) keel over, collapse. **Umklappung** *f* (-en) turning (folding, flipping) over (up, down); (*Nucl.*) flip
um·kleiden 1 (*sep*) *vt* & **sich u.** change (clothes). **2** (*insep*) *vt* (*usu* + *mit*) clothe (in); cover, line, jacket (with)
um·kochen *vt* (re)boil
um·kommen° *vi*(s) die, perish; spoil
Umkreis° *m* **1** surrounding area, neighborhood. **2: im U. von** within a radius of. **3** (*Math.*) circumscribed circle. **4** (*esp Geog.*) circumference; circuit, orbit. **5** circle (of acquaintances). **umkreisen** (*insep*) *vt* circle (revolve, orbit) around; encircle. **Umkreis·geschwindigkeit** *f* peripheral velocity
umkrempeln *vt* turn inside out (*or* upside down), upset
Umkristallisation° *f* recrystallization. **um·kristallisieren** *vt* (dissolve and) recrystallize. **Umkristallisierung°** *f* (dissolving and) recrystallization
um·krücken *vt* rake, rabble; (*Brew.*) mash
um·laden° *vt* **1** reload, transship; load into another vehicle. **2** (*Elec.*) reverse the charge. **Umladung°** *f* reloading; (*Elec.*) reversal of the charge, reversal of sign. **Umladungs·potential** *n* redox potential
Umlagen *fpl* overhead expenses
um·lagern 1 (*sep*) *vt* rearrange, transpose; move to another storage place. **2** (*insep*) *vt* surround, beseige. **Umlagerung°** *f* rearrangement, transposition; re-storing; surrounding, besieging
Umlauf° *m* **1** orbit; circuit, lap. **2** circulation. **3** revolution, rotation. **4** cycle, recycling. **5** memorandum, circular. **--bahn** *f* orbit
um·laufen° **I** (*sep*) **1** *vt* run down. **2** *vi*(s) detour; circulate; revolve, rotate. **II** (*insep*) *vt* circle, orbit. **umlaufend** *p.a* (*oft*) rotary, circular
Umlauf·gebläse *n* centrifugal blower. **-geschwindigkeit** *f* speed of rotation; orbital velocity. **-kühlung** *f* (closed-circuit)

circulation cooling; (*Space*) regenerative cooling. **-motor** *m* rotary motor (*or* engine). **-pumpe** *f* rotary pump; circulation pump. **-richtung** *f* direction of rotation. **-rührwerk** *n* rotary agitator. **-schmierung** *f* closed-circuit lubrication. **-sinn** *m* direction of rotation. **-speicher** *m* (*Cmptrs.*) circulating memory. **-trocknung** *f* drying with circulating air. **-ventil** *n* bypass valve. **-zahl** *f* number of revolutions, r.p.m. **-zeit** *f* **1** period of rotation. **2** (*Astron.*) period (of revolution). **3** (*Space*) orbital period
umlegbar *a* folding, hinged; reversible
um·legen I *vt* **1** wrap around. **2** fold down, flip over. **3** knock over. **4** lay flat. **5** do in, finish off. **6** turn (malt). **II: sich u. 7** bend (*or* tip) over; be laid flat. **8** shift. *Cf* UMGELEGT
um·leiten *vt* divert, detour, reroute. **Umleitung°** *f* diversion, detour, rerouting
um·lenken 1 *vt/i* turn (a vehicle) around. **2: sich u.** change direction; change one's ways. **Umlenkung°** *f* turn(around), change of direction
um·lernen 1 *vt* relearn, learn over (*or* anew). **2** *vi* reorient oneself; learn a new trade
umlief circled, etc: *past of* UMLAUFEN **II** (*insep*) & *oft of* **I** (*sep*)
umliegend *a* surrounding
um·lösen *vt* recrystallize (without filtering)
Umluft *f* ambient air; circulating air. **--ofen** *m* air-circulating oven
Ummagnetisierung *f* magnetic hysteresis
um·manteln (*insep*) *vt* sheathe, jacket, encase. **Ummantelung°** *f* sheath(ing), jacket(ing)
um·modeln *vt* remodel, transform
Umnetzung *f* (-en) phase inversion; preferential wetting
um·ordnen *vt* rearrange, reorganize
um·packen 1 (*sep*) *vt* repack. **2** (*insep*) *vt* pack all around
um·pflanzen 1 (*sep*) *vt* transplant. **2** (*insep*) *vt* plant all around
um·polen *vt/i* reverse the polarity (of). **Umpolung** *f* (-en) polarity reversal

um·programmieren *vt* reprogram

um·pumpen *vt* recycle, recirculate (by pumping)

Umrahmung *f* (-en) framing; frame; (*Print.*) box

umranden, umrändern (*insep*) *vt* 1 border, rim. 2 (mark with a) circle. **umrändert** *p.a* (*oft*) dark-rimmed

um·rechnen *vt* recalculate, convert *cf* UM-GERECHNET. **Umrechnung°** *f* recalculation, conversion

Umrechnungs·faktor *m* conversion factor. **-kurs** *m* rate of exchange. **-tabelle, -tafel** *f* conversion table

um/reißen* 1 (*sep*) *vt* tear down. 2 (*insep*) *vt* outline, sketch *cf* UMRISSEN

umringen (*insep*) *vt* encircle, surround

umriß outlined, etc: *past of* UMREISSEN II (*insep*) & *oft of* I (*sep*)

Umriß° *f* outline, contour: **Chemie in Umrissen** outline of chemistry; **in großen Umrissen** in broad outline(s)

umrissen 1 outlined: *pp of* UMREISSEN II. 2 *p.a* (*oft*) defined, clear-cut

um·rühren *vt* stir (around); agitate; (*Iron*) puddle; (*Copper*) pole. **Umrühr·stab** *m* stirring rod, stirrer

ums *contrac* = **um das** *cf* UM I

um·satteln 1 *vt* resaddle. 2 *vi* change careers; change, switch

Umsatz° *m* 1 (*Chem.*) conversion; **doppelter U.** double decomposition. 2 (*Biol., Med.*) (energy) transformation, metabolism. 3 (*Com.*) turnover, sales. **--produkt** *n* conversion (decomposition, metabolism) product

um·schalten 1 *vt/i* change (over), switch. 2 *vt* (*Elec.*) commutate. 3 *vi* shift gears; (+**auf**) adapt (to). **Umschalter°** *m* changeover (*or* reversing) switch; commutator; shift key

Umschalt·feuerung *f* regenerative furnace. **-hebel** *m* changeover switch

Umschau *f* 1 (written, printed) review, survey. 2 looking around, reconnoitering: **U. halten (nach)** watch (for)

um·schaufeln *vt* turn over (with a shovel); stir (with a paddle)

um·schichten *vt* reshuffle, regroup; restratify

umschichtig *adv* by turns, in (alternate) shifts, alternately

Umschlag° *m* 1 envelope; (book) jacket; wrapper. 2 (*Med.*) compress. 3 conversion; (sudden, radical) change, shift. 4 reloading, transshipment. 5 (*Com.*) turnover, sale. 6 cuff; hem

um·schlagen* I *vt* 1 cut (*or* chop) down. 2 turn up (*or* down), fold over. 3 put (*or* throw) on (a garment). 4 reload, transship. II *vi*(s) 5 change, switch, be converted. 6 turn sour, spoil; decompose. 7 (*Cement*) set. 8 overturn, tip over. **Umschlagen** *n:* (*Cement*) **U. der Bindezeit** development of flash set

Umschlag·kosten *pl* handling (reloading, transshipment) charges. **-papier** *n* wrapping paper

Umschlags·gebiet *n* color change interval (of indicators). **-punkt** *m* 1 (*incl Chem.*) transition (*or* in version) point. 2 (*Anal.*) end point

Umschlag(s)·zahl *f* (*Anal.*) titer, titration value

umschlang clasped, etc: *past of* UM-SCHLINGEN 2 (*insep*) & *oft of* 1 (*sep*)

umschließen* (*insep*) *vt* surround; embrace; include

um/schlingen* 1 (*sep*) *vt* wrap around. 2 (*insep*) *vt* clasp, embrace

umschloß, umschlossen surrounded: *past* & *pp of* UMSCHLIESSEN

umschlungen clasped, etc: *pp of* UM-SCHLINGEN 2

Umschmelz·betrieb *m* remelting foundry. **-eisen** *n* remelt iron

um·schmelzen* *vt* remelt, recast; convert

Umschmelz·metall *n* remelt metal. **-ofen** *m* remelting furnace

Umschmelzung° *f* remelting, recasting; conversion

um/schreiben* I (*sep*) 1 *vt* rewrite; transcribe. 2 reassign, relabel, reregister. II (*insep*) *vt* 3 define. 4 paraphrase. 5 (*Math.*) circumscribe. **umschrieben** 1 *pp.* 2 *p.a* (*Med. oft*) localized

Umschrift° *f* transcription, rewritten version

um·schulen *vt* retrain, reeducate. **Umschulungs·kurs** *m* retraining course

um·schütteln *vt* shake up, agitate

um·schütten *vt* 1 (knock over and) spill. 2 decant

Umschweife *mpl* digressions: **ohne U.** straightforwardly, without beating around the bush

um·schwenken 1 *vt* turn over, rotate. 2 *vi*(s) swing (*or* veer) around; switch, about-face

umschwirren (*insep*) *vt* buzz (*or* whirl) around

Umschwung· *m* 1 rotation, revolution. 2 radical change, turnabout, switch

um·sehen*: sich u. look around (for, after); look back: **im U.** in a flash, in no time

umseitig *a, adv* on the other side (of the page), overleaf

umsetzbar *a* 1 convertible, transformable. 2 marketable, salable

um·setzen I *vt* 1 (+**in**) convert, transform, change *cf* TAT 2. 2 cause to react. 3 (*Food*) assimilate. 4 transpose. 5 transplant. 6 reset. 7 relocate. 8 sell, turn over. **II: sich u.** 9 be converted (*or* transformed). 10 react. **Umsetzer°** *m* converter. **Umsetzung** *f*(-en) 1 conversion, transformation; reaction. **2: doppelte U.** double decomposition. 3 assimilation. 4 transposition. 5 transplantation. 6 resetting. 7 relocation. *Cf* UMSATZ. **Umsetzungs·geschwindigkeit** *f* rate of reaction, speed of conversion

Umsichgreifen *n* spread, rampancy *cf* GREIFEN 9

Umsicht *f* circumspection, discretion. **umsichtig** *a* circumspect, discreet

um·siedeln *vt, vi*(s) move, resettle, relocate

um·sieden* *vt/i* distill

umsomehr *adv* all (*or* so much) the more

umsonst *adv* 1 free, gratis. 2 in vain, futile; (*as intj*) it's no use. **3: nicht u.** not without good reason

umsoweniger *adv* all (*or* so much) the less

umspann ensnared, etc: *past of* UMSPINNEN

um·spannen *vt* **I** (*sep*) 1 (*Elec.*) transform (current). 2 (*Mach.*) reset. **II** (*insep*) 3 span; embrace, include. **Umspanner** *m* (-) (*Elec.*) transformer. **Umspannung°** *f* transformation; resetting; spanning; inclusion

umspinnen* (*insep*) *vt* weave a web

around; ensnare; cover by lapping. **umsponnen** *pp*

um·spulen *vt* rewind, wind on another reel

umspülen (*insep*) *vt* wash, flow around; rinse

Umstand° *m* 1 circumstance, factor; **unter U.̈-en** under certain circumstances, on occasion. 2 (*pl*) bother, ceremony: **ohne U.̈-e** without ceremony, without going to any trouble. **umständehalber** *adv* owing to circumstances. **umständlich** *a* 1 troublesome, involved. 2 digressive, long-winded. 3 fussy, pedantic. 4 *adv oft* in a roundabout way

um·stechen* *vt* 1 (*Agric.*) dig up, turn. 2 reengrave

umstehend *p.a* 1 surrounding. 2 overleaf, on the other side (of the page)

umstellbar *a* transposable; adjustable; reversible; convertible *cf* UMSTELLEN

um·stellen I (*sep*) **A** *vt* 1 rearrange, shift around; transpose. 2 (*esp Mach.*) reset, (re)adjust. 3 reverse, invert. 4 convert. **B: sich u.** 5 change, shift; be converted. **II** (*insep*) surround. **Umstellung°** *f* rearrangement; transposition; resetting; (re)adjustment; reversal, inversion; conversion. **Umstell·werkstoff** *m* substitute (material)

umsteuerbar *a* (*Elec., Mach.*) reversible

um·steuern *vt* (*Elec., Mach.*) reverse

um·stimmen *vt* 1 retune. 2 change (sbdy's) mind. **Umstimmungs·mittel** *n* alterative

um·stoßen* *vt* 1 knock over, upset; overturn. 2 reverse. 3 overrule

umstrahlen (*insep*) *vt* irradiate, bathe in light. **Umstrahlung°** *f* irradiation

umstritten *p.a* disputed, controversial

um·strukturieren *vt* restructure

um·stülpen *vt* turn over, invert; turn inside out

Umsturz° *m* overthrow, upheaval; coup. **um·stürzen** 1 *vt* upset, overthrow, topple. 2 *vi*(s) fall over, topple

Umsud° *m* distillate

Umtausch° *m*, **um·tauschen** *vt* (**gegen**) exchange (for)

Umtrieb° *m* 1 (*Min.*) bypass. 2 bustle, activity. 3 (*usu pl*) machinations

um·tun* 1 *vt* put (*or* throw) on (a garment).

2: sich u. (*oft* + **nach**) make an effort (to find), look around (for)

Umwälz·becken *n* spiral flow tank

um·wälzen I *vt* 1 roll over, overturn; upset. 2 revolutionize. 3 circulate, recycle. **II: sich u.** roll over, rotate. **umwälzend** *p.a* (*oft*) revolutionary

Umwälz·gas *n* circulating (*or* recycle) gas. **-mischer** *m* circulating mixer. **-pumpe** *f* circulation pump

Umwälzung *f* (-en) 1 overturning, upsetting. 2 upheaval. 3 circulation, (re)cycling. 4 rotation. *Cf* UMWÄLZEN

umwandelbar *a* transformable, convertible

um·wandeln 1 *vt* transform, convert; metabolize. **2: sich u.** be transformed (converted, metabolized). **Umwandler°** *m* converter; transducer **Umwandlung°** *f* transformation, conversion, metamorphosis; transmutation

Umwandlungs·artikel *m* (*Tex.*) conversion style. **u–fähig** *a* transformable, convertible. **U–geschwindigkeit** *f* transformation (*or* conversion) rate. **-gestein** *n* metamorphic rock. **-kurve** *f* transition curve. **-produkt** *n* transformation (*or* conversion) product. **-pseudomorphose** *f* alteration pseudomorphism. **-punkt** *m* transition (*or* transformation) point. **-reihe** *f* (*Nucl.*) disintegration series. **-spannung** *f* (*Elec.*) transformation potential. **-tabelle** *f* conversion table. **-temperatur** *f* 1 (*Metall.*) critical temperature. 2 (*Phys.*) transmutation point. **-wärme** *f* heat of transformation; (*Metll. oft*) critical heat

Umweg° *m* detour, roundabout way: **auf U—en** in a roundabout way; **ohne U—e** directly

Umwelt *f* environment. **--angst** *f* fear (*or* concern) for the environment. **u–bedingt** *a* envrionmental. **U–bewußte** *m, f* (*adj endgs*) environmentalist. **-einfluß** *m* environmental impact (*or* influence). **u–feindlich** *a* environmentally harmful. **-freundlich** *a* environmentally safe (*or* beneficial). **-gefährlich** *a* environmentally hazardous. **U–gift** *n* environmental pollutant (*or* toxin). **U–risiko** *n* environmental hazard.

-schädiger *m* pollutant, enemy of the environment. **u–schädlich** *a* environmentally harmful. **-schonend** *a* environmentally protective. **U–schonung** *f*, **-schutz** *m* environmental protection. **-schützer** *m* environmentalist. **-verschmutzer** *m* environmental pollutant (*or* polluter). **-verschmutzung** *f* environmental pollution (*or* contamination). **u–verträglich** *a* environmentally compatible (*or* tolerable). **U–verträglichkeitsstudie** *f* environmental impact study. **-wirkung** *f* environmental impact. **-zerstörung** *f* destruction of the environment

um·wenden* **I** *vt* & **sich u.** turn (over, around, inside out). 2 *vi* reverse, invert. 3 *vi* (s) turn (around *or* back)

um·werten *vt* revalue. **Umwertung°** *f* revaluation

um/wickeln *vt* 1 (*sep*) rewind. 2 (*insep*) wrap, surround, envelop. **Umwick(e)lung°** *f* wrapping, covering, casing; winding

umwölkt *a* cloudy, overcast

um/ziehen* **I** (*sep*) 1 *vt* & **sich u.** change (clothes). 2 *vi* (s) move, change residence. **II** (*insep*) 3 *vt* surround, enclose. **4: sich u.** cloud over

umzog(en): surrounded, etc: *past* (& *pp*) of UMZIEHEN **II** (*insep*) & *oft past of* **I** (*sep*)

Umzug° *m* 1 move, change of residence; moving. 2 migration, wandering. 3 procession

umzüngeln (*insep*) *vt* (*Flames*) leap (*or* lick) around

un. . . *pfx* 1 (*negative*) un-, in-, non-, mis-, dis-. **2: Un. . .** *oft implies large size, cf* UNZAHL

unabänderlich *a* immutable, unchangeable

unabhängig *a* independent. **Unabhängigkeit** *f* independence

unabkömmlich *a* indispensable

unablässig *a* incessant, uninterrupted

unabsehbar *a* immeasurable, unbounded

unabsichtlich *a* unintentional

unabweisbar, unabweislich *a* 1 irrefusable. 2 irrefutable. 3 imperative

unabwendbar *a* inescapable, inevitable

unachtsam *a* inattentive, heedless

unähnlich *a* (*oft* + *dat*) dissimilar (to), different (from), unlike *cf* ÄHNLICH

unaktinisch *a* nonactinic

unaktuell *a* untimely, outdated

unanfechtbar *a* incontestable, indisputable

unangebracht *a* unsuitable, inappropriate

unangegriffen *a* unaffected, unattacked

unangelassen *a* untempered, unannealed

unangemessen *a* 1 disproportionate. 2 excessive. 3 inadequate

unangenehm *a* unpleasant, disagreeable

unangeregt *a* unexcited, in the normal state

unangreifbar *a* attack-proof, untouchable; incorrodible

Unannehmlichkeit *f* (-en) unpleasantness; inconvenience, annoyance

unansehnlich *a* 1 unattractive, nondescript, shabby. 2 not very sizable

unanständig *a* indecent

unanstößig *a* inoffensive

unantastbar *a* inviolable, sacrosanct

unanwendbar *a* inapplicable, unsuitable

unappetitlich *a* unappetizing, unsavory

unatembar *a* unbreathable, irrespirable

unauffällig *a* inconspicuous

unauffindbar *a* untraceable, impossible to find

unaufgefordert *a* unasked, spontaneous; *adv oft* of one's own accord

unaufgelöst *a* 1 undissolved. 2 unresolved

unaufhaltsam *a* unstoppable, inexorable

unaufhörlich *a* incessant, ceaseless

unauflösbar, unauflöslich *a* in(dis)soluble. **Unauflösbarkeit, Unauflöslichkeit** *f* in(dis)solubility

unaufmerksam *a* inattentive

unaufschiebbar *a* undeferable, urgent

unausbleiblich *a* inevitable

unausdehnbar *a* unexpandable, inextensible; (*Metll.*) inductile

unausdenkbar *a* inconceivable, unimaginable

unausführbar *a* infeasible, unworkable

unausgefällt *a* unprecipitated

unausgeglichen *a* 1 uncompensated, unbalanced. 2 unstable. **Unausgeglichenheit** *f* uncompensated state; instability

unausgemacht *a* undecided, indefinite

unausgesetzt *a* uninterrupted. *See also* AUSSETZEN

unauslöschbar, unauslöschlich *a* indelible; inextinguishable, unquenchable

unausrottbar *a* ineradicable

unaussprechlich *a* inexpressible, unspeakable

unausstehlich *a* intolerable, insufferable

unaustilgbar *a* ineradicable

unauswaschbar *a* impossible to wash out

unausweichlich *a* unavoidable, inescapable

unbändig *a* unrestrained, unruly

unbeabsichtigt *a* unintentional

unbeachtet *a* unnoticed: **u. lassen** ignore, disregard

unbeanstandet *a* unopposed, unchallenged

unbearbeitet *a* 1 unworked, untreated, unfinished. 2 raw, unprocessed. 3 (*Lthr.*) undressed. 4 (*Agric.*) untilled

unbedeckt *a* uncovered

unbedenklich *a* 1 unobjectionable. 2 unhesitating; (*adv oft*) without misgivings. 3 unscrupulous

unbedeutend *a* insignificant, minor

unbedingt *a* 1 absolute, unconditional. 2 unconditioned (reflex). 3 (*adv oft*) definitely, by all means; urgently

unbeeinflußt *a* uninfluenced, unaffected

unbefangen *a* 1 unprejudiced, open-minded. 2 uninhibited; unaffected

unbefriedigend *a* unsatisfactory. **unbefriedigt** *a* unsatisfied

unbefugt *a* unauthorized: **U—en ist der Eintritt verboten** no trespassing

unbegleitet *a* (*usu* + **von**) unaccompanied (by)

unbegreiflich *a* incomprehensible. **unbegreiflicherweise** *adv* for some unknown reason

unbegrenzt *a* unlimited, boundless

unbegründet *a* unfounded, baseless

unbehaart *a* hairless

Unbehagen *n* uneasiness, discomfort. **unbehaglich** *a* uncomfortable, uneasy

unbehandelt *a* untreated

unbehindert *a* unhindered, unimpeded

unbeholfen *a* awkward, heavy-handed

unbekannt a unknown. **Unbekannte** f (adj endgs) (Math.) unknown (quantity)
unbelebt a 1 inanimate. 2 deserted
unbelichtet a (esp Phot.) unexposed
unbeliebt a unpopular
unbemannt a unmanned
unbemerkbar a unnoticeable. **unbemerkt** a unnoticed
unbenannt a 1 unnamed, anonymous. 2 (Math.) indeterminate. 3 (Anat.) innominate
unbenutzbar a unusable, unserviceable.
unbenutzt a unused
unbeobachtet a unobserved
unbepflanzt a unplanted
unbequem a uncomfortable; inconvenient. **Unbequemlichkeit** f (-en) discomfort; inconvenience
unberechenbar a 1 incalculable. 2 unpredictable; unreliable
unberechtigt a 1 unjustified. 2 unauthorized; pirated
unberücksichtigt a unconsidered, disregarded
unberuhigt a (Metll.) unkilled, (Steel) rimmed
unberührt a untouched; unmoved
unbeschadet prep (+genit) in spite of, regardless of
unbeschädigt a undamaged, uninjured
unbeschäftigt a unemployed; idle, not busy
unbeschränkt a unlimited, unrestricted
unbeschreiblich a indescribable
unbeschwert a 1 unweighted. 2 unworried
unbesetzt a unoccupied, vacant
unbeständig a unstable, changeable, erratic. **Unbeständigkeit°** f instability
unbestellbar a 1 undeliverable. 2 untillable
unbestimmt a indefinite, indeterminate; uncertain. **Unbestimmtheit°** f indefiniteness; uncertainty. **Unbestimmtheits‧relation** f (Heisenberg's) uncertainty principle
unbestreitbar a indisputable
unbestritten a undisputed
unbeteiligt a 1 indifferent. 2 (+an) uninvolved (in)
unbeugsam a inflexible, unyielding

unbewaffnet a 1 unarmed. 2 unaided: **das u—e Auge** the naked eye
unbeweglich a immovable, fixed
unbewiesen a unproved, undemonstrated
unbewußt a 1 unconscious, unaware: **es ist ihnen u.** they are unaware of it. 2 instinctive, involuntary. **Unbewußtheit** f unconsciousness
unbezogen a 1 unrelated. 2 uncovered. 3 unoccupied
unbiegsam a inflexible, unyielding
unbildsam a nonplastic, inflexible
unbrauchbar a unusable, useless
unbrennbar a incombustible. **U—‧machung** f fireproofing
und cnjc and
undefinierbar a undefinable
Undefiniertheit f indefiniteness
undehnbar a inextensible; nonductile
Undekan n undecane. **‧‧säure** f undecanoic acid. **Undekylen‧säure** f undecenoic acid. **Undekyl‧säure** f undecanoic acid
undenkbar a unthinkable, inconceivable
undenklich a: **seit u—en Zeiten** since time immemorial
undeutlich a indistinct, vague
undicht a leaky, imperfectly sealed, porous. **Undichtheit, Undichtigkeit** f (-en) leakiness, porosity; leak, imperfect seal. **Undichtigkeitsgrad** m degree of porosity (or perviousness)
Unding° n absurdity; monstrosity
undissoziiert a undissociated
undulatorisch a undulatory. **undulieren** vi undulate. **undulös** a undulous, undulatory
undurchdringlich a impenetrable, impermeable, impervious
undurchforscht a unexplored
undurchführbar a infeasible, unworkable
undurchgängig a impenetrable, impermeable
undurchlässig a (+für) impermeable (to)
undurchscheinend a nontranslucent
undurchsichtig a nontransparent, opaque. **Undurchsichtigkeit** f nontransparency, opacity
uneben a uneven, rough. **Unebenheit** f (-en) unevenness

unecht *a* 1 untrue, nongenuine. 2 false, artificial; forged, imitation. 3 (*Colors*) nonfast, fugitive. 4 improper (fractions). **Unechtheit** *f* falsity, artificiality; nonfastness, fugitiveness

unedel *a* base (*esp Metals*); low-minded

unegal *a* unequal. **Unegalität** *a* (-en) inequality

unehrlich *a* dishonest

uneigentlich *a* figurative; (*Math.*) improper; (*adv oft*) not in a strict sense

uneinheitlich *a* nonuniform, inhomogeneous; varied, mixed

unelastisch *a* inelastic

unelektrisch *a* nonelectrified, uncharged

unempfänglich *a* (+**für**) unreceptive, unsusceptible (to)

unempfindlich *a* 1 (+**gegen**) insensitive, immune, unsusceptible; unreactive (to). 2 (*Tex.*) soil-resistant

unendlich *a* infinite; endless, unending; immense: **u. viel** an immense amount of. **Unendliche** *n* (*adj endgs*) infinity: **bis ins U.** to infinity, ad infinitum. **Unendlichkeit** *f* infinity; endlessness

unentbehrlich *a* indispensable, essential

unentdeckt *a* undiscovered; undisclosed

unentflammbar *a* non(in)flammable

unentgeltlich *a* gratuitous, free; (*adv oft*) gratis, free (of charge)

unentschieden *a* undecided; tied (score)

unentschlossen *a* undecided; indecisive

unentwickelt *a* undeveloped; (*Math.*) implicit

unentwirrbar *a* inextricable

unentzündbar, unentzündlich *a* non(in)-flammable

unerbittlich *a* inexorable, implacable

unerfahren *a* inexperienced

unerforschlich *a* unfathomable

unerforscht *a* unexplored, uninvestigated

unergiebig *a* unproductive

unerheblich *a* unimportant; inconsiderable

unerhört *a* unheard-of; outrageous

unerkannt *a* unrecognized. **unerkennbar** *a* unrecognizable

unerklärlich *a* inexplicable, unexplainable. **unerklärt** *a* unexplained

unerläßlich *a* indispensable; obligatory

unerlaubt *a* forbidden, unpermitted; *adv*

oft without permission

unerledigt *a* unaccomplished, unattended-to

unermeßlich *a* immeasurable, immense

unermittelt *a* unascertained

unermüdlich *a* untiring, indefatigable

uneröffnet *a* unopened

unerprobt *a* untested, untried

unerregt *a* unexcited

unerreichbar *a* unattainable, unreachable

unerschlossen *a* 1 unexplored. 2 undeveloped

unerschöpflich *a* inexhaustible

unersetzbar, unersetzlich *a* irreplaceable; irreparable

unerträglich *a* intolerable, unbearable

unerwähnt *a* unmentioned

unerwartet *a* unexpected

unerwünscht *a* unwanted; undesirable

unexplodierbar *a* inexplosive

unfähig *a* 1 incompetent. 2 (+**zu** + *inf*) unable (to), incapable (of). **Unfähigkeit** *f* incompetence; inability, incapability

Unfall° *m* accident, mishap. **-auto** *n* ambulance. **u–sicher** *a* accident-proof. **-trächtig** *a* 1 hazardous. 2 accident-prone. **U–wagen** *m* ambulance

unfaßbar, unfaßlich *a* incomprehensible

unfehlbar *a* unfallible, unfailing; (*adv oft*) without fail

unfern *adv* (+**von**), *prep* (+*genit*) not far (from)

unfertig *a* unfinished; immature

unfiltrierbar *a* unfilterable

unfrei *a* 1 unfree, dependent; obligated; constrained, (*Mach.*) under constraint. 2 (*Equilibrium*) invariant. 3 inhibited, self-conscious

unfreiwillig *a* 1 involuntary. 2 unintentional

unfruchtbar *a* 1 unfruitful, infertile, sterile: **u. machen** sterilize. 2 fruitless

Unfug *m* 1 mischief. 2 disorder. 3 nonsense

unfühlbar *a* impalpable, intangible

.. ung *sfx* (*forms nouns from verbs*): **abändern** modify, **Abänderung** *f* modification

unganz *a* unsound, defective. **Ungänze** *f* (-n) flaw, defect

ungar *a* 1 (*Food*) undone, underdone. 2

(Soil) unfriable. 3 *(Lthr.)* undertanned
Ungar *m* (-n, -n), **Ungarin** *f* (-nen), **unga-**
risch *a* Hungarian. **Ungarn** *n* Hungary
ungeachtet *prep* (+*genit*) regardless (of)
ungeahnt *a* unexpected, unsuspected
ungealtert *a* unaged
ungeändert *a* unchanged
ungebildet *a* uneducated, untutored
ungebleicht *a* unbleached
ungebrannt *a* 1 *(genl)* unburnt. 2 *(Coffee)*
unroasted. 3 *(Ceram.)* green, unfired
ungebräuchlich *a* unusual, uncommon
ungebührlich *a* improper, undue
ungebunden *a* unbound, unrestrained;
(Chem.) free, uncombined
ungedämpft *a* undamped, unsubdued
Ungeduld *f* impatience. **ungeduldig** *a*
impatient
ungedüngt *a* *(Agric.)* unfertilized
ungeeignet *a* unsuited, unsuitable
ungefähr *a* approximate; *(adv oft)* about;
von u. by chance
ungefährlich *a* safe, harmless
ungefärbt *a* undyed, uncolored; unstained
ungrefrierbar *a* unfreezable, nonfreezing
ungefügig *a* unwieldy, unmanageable
ungefüllt *a* unfilled
ungegerbt *a* untanned
ungegoren *a* unfermented
ungehärtet *a* unhardened, untempered
ungeheißen *adv* unasked, of one's own
accord
ungeheizt *a* unheated; unfired
ungehemmt *a* unrestrained
ungeheuer *a* huge, tremendous. **Unge-**
heuer *n* (-) monster, monstrosity
ungehörig *a* undue, improper
ungeklärt *a* unclarified
ungekocht *a* uncooked, unboiled
ungekürzt *a* unabridged; unabbreviated
ungel. *abbv* (ungelöst) undissolved
ungeladen *a* unloaded, light; uninvited
ungeläutert *a* unclarified, unpurified
ungelegen *a* inconvenient, inopportune: **u.**
kommen come at the wrong moment
ungelehrt *a* untaught, untutored
ungeleimt *a* unglued; *(Paper)* unsized
ungelernt *a* unskilled, unschooled
ungelöscht *a* unquenched; unslaked
ungelöst *a* undissolved; unsolved
ungemein *a* uncommon, unusual

ungemessen *a* unmeasured; immeasurable
ungemindert *a* undiminished
ungemischt *a* unmixed; unalloyed
ungemünzt *a* uncoined
ungemütlich *a* unpleasant, uncongenial
ungenannt *a* unnamed; anonymous;
(Anat.) innominate
ungenau *a* inaccurate, inexact. **Un-**
genauigkeit *f* (-en) inaccuracy, inexact-
ness, lack of definition
ungeneigt *a* disinclined
ungenießbar *a* unpalatable, unappetizing
ungenügend *a* 1 insufficient, inadequate.
2 unsatisfactory
ungenügsam *a* insatiable
ungenutzt *a* unused
ungeöffnet *a* unopened
ungeordnet *a* disordered, disorganized
ungepaart *a* unpaired
ungepflegt *a* uncared-for, neglected
ungeprüft *a* untried, untested
ungequantelt *a* unquantized
ungerade *a* 1 *(Numbers)* odd. 2 crooked,
curved
ungerad·wertig *a* odd-valued, odd-
valence(d). **-zahlig** *a* odd-numbered
ungerechnet *a* 1 uncounted. 2 *adv, prep*
(+*genit*) not counting, excluding
ungereinigt *a* unclean(s)ed; unpurified
ungerinnbar *a* uncoagulable
ungern *adv* unwillingly, reluctantly: **u.**
haben dislike, hate; **u. tun** dislike doing,
hate to do
ungeröstet *a* unroasted
ungerufen *a* uncalled; *adv* without being
called
unges. *abbv* (ungesättigt) unsat. (unsaturated)
ungesalzen *a* unsalted; insipid
ungesätt. *abbv* = **ungesättigt** *a* unsaturated; unsatiated
ungesäubert *a* uncleaned
ungesäuert *a* 1 non-acidified. 2 unleav-
ened
ungesäumt 1 *a* seamless. 2 *adv* immediately
ungeschält *a* unpeeled, unhusked
ungeschehen *a* undone: **u. machen** undo
ungeschichtet *a* unstratified
Ungeschicklichkeit *f* awkwardness. **un-**

geschickt *a* unskilled; awkward; inconvenient

ungeschlechtlich *a* asexual

ungeschliffen *a* unground, unpolished; unsharpened; (*Gems*) uncut

ungeschlossen *a* unclosed, open, incomplete

ungeschmolzen *a* unmelted, unfused

ungeschult *a* unschooled, untrained

ungeschützt *a* unprotected, unguarded

ungesichert *a* unsecured, unsafeguarded

ungesiebt *a* unsifted, unscreened

ungespalten *a* unsplit, uncleaved

ungespannt *a* unstrained, relaxed, slack

ungestalt *a* misshapen, deformed. **ungestaltet** *a* unformed, shapeless

ungestört *a* undisturbed, untroubled

ungesucht *a* 1 unsought. 2 unaffected, natural

ungesund *a* unhealthy

ungetan *a* undone

ungeteilt *a* undivided; ungraduated

ungetempert *a* 1 (*Metll.*) unannealed. 2 (*Synth.*) unstoved after bake

Ungetüm *n* (-e) monster; monstrosity

ungeübt *a* unpracticed, inexperienced

ungewaschen *a* unwashed; foul; (*Wool*) in the grease

ungewiß *a* unsure, uncertain; vague; dubious: **im ungewissen sein (über)** be uncertain about; **im ungewissen lassen** leave in the dark. **Ungewißheit** *f* (-en) uncertainty. **Ungewißheits·prinzip** *n* uncertainty principle

ungewöhnlich *a* unusual; abnormal

ungewohnt *a* unusual; unfamiliar; (+*accus, oft*) unaccustomed

ungezählt *a* countless, innumerable; uncounted

ungezähnt *a* untoothed; unserrated; (*Bot.*) edentate

Ungeziefer *n* vermin; insect; pest. **··bekämpfung** *f* vermin (*or* pest) control. **·vertilgung** *f* extermination (of vermin)

ungezwungen *a* unforced; casual

ungiftig *a* nonpoisonous, nontoxic. **Ungiftigkeit** *f* nontoxicity

unglasiert *a* unglazed, unfrosted

Unglaube(n)° *m* incredulity, disbelief. **unglaubhaft** *a* unbelievable, incredible. **ungläubig** *a* incredulous. **unglaublich**

a unbelievable, incredible

ungleich *a* 1 unequal. 2 unlike, different. 3 (*adv oft*) incomparably, (by) far. **··artig** *a* dissimilar, unlike, different; heterogeneous. **U–artigkeit** *f* dissimilarity; heterogeneity. **u–förmig** *a* nonuniform, different(ly shaped); uneven. **U–gewicht** *n* disequilibrium, imbalance

Ungleichheit° *f* inequality; dissimilarity; unevenness

ungleich·mäßig *a* uneven, irregular. **U–mäßigkeit°** *f* unevenness, irregularity. **u–namig** *a* 1 (*Elec.: Poles*) unlike, opposite. 2 (*Math.*) unlike (fractions). **-seitig** *a* scalene (triangle). **-stoffig** *a* inhomogeneous. **U–stoffigkeit** *f* inhomogeneity. **u–teilig** *a* 1 inhomogeneous. 2 (*Colloids*) polydisperse

Ungleichung° *f* inequality

ungleich·wertig *a* unequal-valence, of unequal valence; unequal-valued

Unglück° *n* 1 misfortune, bad luck: **zum U.** unfortunately. 2 accident; disaster. **unglücklich** *a* 1 unlucky, unfortunate. 2 unhappy. **unglücklicherweise** *adv* unfortunately. **Unglücks·fall** *m* accident; casualty; misfortune

ung. P. *abbv* (ungarisches Patent) Hungarian patent

ungreifbar *a* impalpable

ungültig *a* invalid, void, inapplicable

Ungunst *f* (-en) disfavor, unfavorableness *cf* ZUUNGUNSTEN. **ungünstig** *a* unfavorable

ungut *a* 1 uncanny. 2 unfavorable

unhaltbar *a* 1 nondurable. 2 untenable

unhaltig *a* (*Min.*) containing no metal. **Unhaltige(s)** *n* (*adj endgs*) waste, tailings

unhämmerbar *a* unmalleable

unhandlich *a* unhandy, unwieldy

Unheil *n* disaster; harm, mischief

unheilbar *a* incurable, irremediable

unheilsam *a* unwholesome

unheimlich *a* 1 uncanny, weird. 2 tremendous

unheizbar *a* unheatable

unhomogen *a* unhomogeneous. **Unhomogenität** *f* inhomogeneity

unhörbar *a* inaudible

uni *a* solid-color, unicolor(ed). **U–·artikel** *m* (*Tex.*) solid style

unieren *vt* unite
Uni·farbe *f* (*Dye.*) self color, solid (*or* uniform) color. **uni·färben** *vt* dye a self (*or* solid) color, dye solid. **uni·farbig** *a* self-colored, solid-color
unifizieren *vt* 1 unify. 2 standardize
unikal *a* unique, one-of-a-kind. **Unikum** *n* (. . ika) rarity, unique phenomenon
Union·gerbung *f* union tanning
Uni·stückware *f* solid-color piece goods
unitarisch *a* unitary; Unitarian
Uni·ton *m* solid color (*or* shade)
Universal·arznei *f* universal remedy, cure-all. **-gelenk** *n* universal joint. **-mittel** *n* cure-all, panacea
Universität *f* (-en) university
unk. *abbv* (unkorrigiert) uncorrected
Unke *f* (-n) toad
unkenntlich *a* unrecognizable, undiscernible
Unkenntnis *f* ignorance
unklar *a* 1 unclear, indistinct; vague; indefinite: **im u—en lassen** leave indefinite, leave in the dark. 2 turbid
unkondensierbar *a* uncondensable
unkontrollierbar *a* uncontrollable
unkorrigiert *a* uncorrected
unkorrodierbar *a* noncorrodable
Unkosten *pl* expenses, charges. **-konto** *n* expense account
Unkraut *n* weed(s). **-(bekämpfungs)-mittel** *n*, **-vertilger** *m* weed killer
unkristallinisch *a* noncrystalline, amorphous. **unkristallisierbar** *a* uncrystallizable
unkundig *a* (+*genit*) ignorant (of)
unl. *abbv* (unlöslich) insol. (insoluble)
unlängst *adv* recently, not long ago
unlauter *a* 1 impure. 2 dishonest, unethical
unlegiert *a* unalloyed
unlesbar *a* 1 unreadable. 2 = **unleserlich** *a* illegible
unleugbar *a* undeniable, indisputable
unlieb *a* 1 unfriendly, hostile. 2 (*Pred* + *dat*) unwelcome, unpleasant (to, for)
unliebsam *a* unpleasant, disagreeable
unlösbar *a* 1 insoluble. 2 indissoluble. 3 unsolvable. 4 undetachable
unlöschbar *a* unquenchable, unslakable, inextinguishable
unlöslich *a* insoluble. **Unlösliche(s)** *n* (*adj*

endgs) insoluble matter. **Unlöslichkeit** *a* insolubility
Unlust *f* displeasure, dislike, aversion. **unlustig** *a* reluctant
unmagnetisch *a* nonmagnetic
Unmaß° *n* excess, excessive amount
Unmasse° *f* enormous amount, heaps, loads (of)
unmaßgeblich *a* unauthoritative; humble
unmäßig *a* 1 immoderate. 2 inordinate; excessive; enormous
Unmenge° *f* enormous amount, heaps, loads (of)
unmenschlich *a* inhuman; inhumane
unmerklich *a* unnoticeable, imperceptible
unmeßbar *a* immeasurable
unmischbar *a* unmixable, immiscible. **Unmischbarkeit** *f* immiscibility
unmittelbar *a* immediate, direct
unmöbliert *a* unfurnished
unmodern *a* old-fashioned, outmoded
unmöglich *a* impossible; (*adv oft*) not . . . possibly. **Unmöglichkeit°** *f* impossibility
unmündig *a* underage. **Unmündige** *m, f* (*adj endgs*) minor. **Unmündigkeit** *f* minority in age, youth
unnachahmlich *a* inimitable
unnachgiebig *a* unyielding, inflexible
unnahbar *a* unapproachable, inaccessible
unnatürlich *a* unnatural
unnennbar *a* unutterable, inexpressible
unnötig *a* unnecessary, needless
unnütz *a* 1 useless, unprofitable. 2 unnecessary. 3 (*pred & adv oft*) in vain. **unnützerweise** *adv* needlessly
unokular *a* monocular
unordentlich *a* disorderly, untidy, messy. **Unordnung°** *f* disorder; mess, confusion: **in U. sein** (*oft*) be upset; **in U. bringen** upset, disorganize
unorganisch *a* inorganic
unoxidierbar *a* unoxidizable
unpaar *a* odd(-numbered). **unpaarig** *a* unpaired; azygous. **unpaar·wertig** *a* of odd valence
unparteiisch *a* impartial. **Unparteiische** *m, f* (*adj endgs*) referee, arbiter. **Unparteilichkeit** *f* impartiality
unpaß 1 *adv* at the wrong time. 2 *pred a* indisposed

unpassend *a* unfitting, inappropriate
unpäßlich *a* indisposed, unwell.
Unpäßlichkeit *f* (-en) indisposition
unperiodisch *a* aperiodic, nonperiodic
unplastisch *a* nonplastic
unpolar *a* nonpolar. **unpolarisierbar** *a* unpolarizable, nonpolarizing
unpraktisch *a* impractical
unpräzis *a* imprecise
unpreßbar *a* incompressible. **Unpreßbarkeit** *f* incompressibility
Unrat *m* 1 dirt; garbage, trash. 2 (*Metll.*) dross
unratsam *a* inadvisable, ill-advised
unrecht 1 *a* wrong, incorrect; (*adv oft*) inopportunely. 2 improper. 3: **u. haben** be wrong (=mistaken); **u. tun** (+*dat*) (do) wrong; **u. bekommen** be contradicted, turn out to be wrong; **u. geben** contradict, show to be wrong. **Unrecht** *n* wrong, injustice: **mit** (or **zu**) **U.** wrongly
unrechtmäßig *a* illegal; wrongful
unregelmäßig *a* irregular; anomalous, abnormal. **Unregelmäßigkeit** *f* (-en) irregularity; anomaly, abnormality
unreif *a* unripe, immature. **Unreife** *f* unripeness, immaturity
unrein *a* impure; unclean, soiled; **ins u—e schreiben** make a rough copy of. **Unreinheit** *f* (-en) impurity. **unreinlich** *a* uncleanly, slovenly. **Unreinlichkeit** *f* uncleanliness, slovenliness
unreizbar *a* nonirritable, insensitive
unrentabel *a* unprofitable
unrettbar *a* irretrievable; beyond rescue
unrichtig *a* incorrect, erroneous, wrong; false
Unruh(e) *f* (.. hen) 1 uneasiness. 2 restlessness. 3 disturbance, commotion; unrest; riot. 5 disturbing factor. 6 feverish activity. 7 fluctuation, irregularity. 8 (*Watches*) balance spring. **Unruhe·stifter** *m* troublemaker. **unruhig** *a* 1 uneasy. 2 restless. 3 restive. 4 disturbed, troubled. 5 fluctuating, irregular. 6 (*Steel*) effervescent, unkilled
uns *pron* (to) us, (to) ourselves
uns. *abbu* (unsymmetrisch) unsymmetrical
unsachgemäß *a* 1 inappropriate. 2 inadequate
unsäglich *a* unutterable, unspeakable
unsanft *a* rough, harsh

unsauber *a* 1 impure. 2 dirty, messy, untidy. 3 inexact, inaccurate. 4 shady, questionable
unschädlich *a* harmless: **u. machen** render harmless, destroy; (*incl Chem.*) neutralize
unscharf *a* blurred, fuzzy; inexact. **Unschärfe** *f* lack of sharpness, fuzziness; uncertainty. **·beziehung, ·relation** *f* uncertainty principle
unschätzbar *a* invaluable, inestimable
unscheidbar *a* inseparable, undecomposable
unscheinbar *a* inconspicuous, nondescript; (*Colors*) dull, pale
Unschlitt *m* or *n* tallow. **·kerze** *f* tallow candle. **·seife** *f* tallow soap
unschlüssig *a* undecided, indecisive, irresolute; inconclusive
unschmackhaft *a* insipid, unpalatable
unschmelzbar *a* infusible, nonmelting. **Unschmelzbarkeit** *f* infusibility
Unschuld *f* innocence. **unschuldig** *a* innocent
unschweißbar *a* unweldable
unschwer *adv* without difficulty, easily
unser 1 (*possess, with endgs*) our. 2 *pron* (*genit of* WIR) (of) us
unsererseits *adv* on our part, in our turn
unser(e)t·halben, ·wegen *adv* on our account (*or* behalf), for our sake
unserig *pron:* **der (die, das) u—e** ours, our own: **das u—e tun** do our part
unsicher *a* 1 uncertain; unsafe; insecure; (*Mach.*) unsteady. **Unsicherheit** *f* (-en) uncertainty; insecurity. **Unsicherheits·prinzip** *n* uncertainty principle
unsichtbar *a* invisible
Unsinn *m* nonsense. **unsinnig** *a* nonsensical, absurd; irrational
unspaltbar *a* uncleavable
unsr. . . = unser. . .
unstabil *a* unstable
unstarr *a* nonrigid
unstatthaft *a* inadmissible, forbidden
unstet *a* 1 restless, unsettled. 2 = UNSTETIG 1
unstetig *a* 1 unstable, unsteady. 2 variable, irregular. 3 (*Math.*) discontinuous. **Unstetigkeit** *f* instability, unsteadiness; variability, irregularity; discontinuousness
Unstimmigkeit *f* (-en) 1 discrepancy. 2 dis-

agreement, difference of opinion

unstreckbar *a* nonextensible, nonductile

unstreitig *a* indisputable, incontrovertible

unstudiert *a* 1 unlettered. 2 unsophisticated. 3 not university-educated, having no academic degree

unsulfiert *a* unsulfonated

Unsumme° *f* vast (*or* enormous) sum

Unsymmetrie *f* asymmetry. **unsymmetrisch** *a* asymmetrical unsymmetrical

unt. *abbv* (unter) under, etc

untätig *a* inactive, idle; inert; dormant. **Untätigkeit** *f* inactivity, idleness; inertness, inertia; dormancy

untauglich *a* unfit, unqualified; useless

unteilbar *a* indivisible

unteilhaftig *a* nonparticipating, uninvolved

unten *adv* down, below, underneath, at the bottom; downstairs: **ganz u.** at the very bottom; **von u.** from below; **nach u.** down(ward), below, downstairs; **da u.** down there; **weiter u.** further down *cf* OBEN. **-erwähnt**, **-genannt** *a* mentioned below, following. **-stehend** *a* (printed) below, following

unter I *prep* **A** (*dat or accus*) 1 under, below (*incl Quant.*) beneath *cf* VERSTEHEN. 2 among; **mitten u.** (in)to the midst of: *cf.* U.A. **B** (*dat*) 3 during. 4 (*esp + verbal nouns*) with, accompanied by: **u. Bezugnahme auf** with reference to. II *a* lower, bottom *cf* UNTERST. III *pfx* **C** (*nouns, adj*) 5 (*esp Chem.*) hypo-. 6 sub-, infra-, under-; lower, inferior, partial. **D** (*Verbs*) 7 (*sep & insep*) under(neath), down, sub-. 8 (*insep, oft, implies cessation, omission*) *cf* UNTERBLEIBEN, UNTERLASSEN

Unterabschnitt° *m* subsection

Unterabteilung° *f* subdivision

Unterarm° *m* forearm

Unterart° *f* subspecies, variety

Unteraugenhöhlen·ader *f* infraorbital vessel

Unterausschuß° *m* subcommittee

unterband stopped, etc: *past of* UNTERBINDEN

Unterbau° *m* foundation, substructure; chassis; roadbed; (*fig*) basis

Unterbauch° *m* hypogastrium

Unterbeamte° *m* **Unterbeamtin°** *f* subordinate official, minor clerk

Unterbegriff° *m* sub-concept

unterbelichten (*insep*) *vt* (*Phot.*) underexpose. **Unterbelichtung** *f* underexposure

unterbewußt *a*, **Unterbewußtsein** *n* subconscious

Unterbilanz° *f* deficit

unterbinden* (*insep*) *vt* 1 forestall. 2 prohibit. 3 stop. 4 (*Med.*) tie off, ligate. **Unterbindung°** *f* 1 forestalling, prohibition, etc. 2 ligature

unterbleiben* (*insep*) *vi*(s) go (*or* be left) undone, not happen, be dropped. **unterblieb(en)** *past* (*& pp*)

Unterboden° *m* subsoil

unterbrach interrupted: *past of* UNTERBRECHEN

unterbrannt(e) underburned: *pp* (*& past*) *of* UNTERBRENNEN

unterbrechen* (*insep*) *vt* interrupt, break, discontinue; disconnect. **Unterbrecher** *m* (-) interrupter, circuit breaker. **Unterbrechung** *f* (-en) interruption, break. **Unterbrechungs·strom** *m* interrupted current

unterbreiten (*insep*) *vt* present, submit

unterbrennen* (*insep*) *vt* underburn

unterbricht interrupts: *pres of* UNTERBRECHEN

unter·bringen* *vt* accommodate, find a place for; stow, store; get (sthg) accepted

unterbrochen interrupted: *pp of* UNTERBROCHEN

unterbromig *a* hypobromous: **u—e Säure** hypobromous acid. **unterbromigsauer** *a* hypobromite of

unterbunden forestalled, etc: *pp of* UNTERBINDEN

unterchlorig *a* hypochlorous: **u—e Säure** hypochlorous acid. **unterchlorig·sauer** *a* hypochlorite of. **U—säure** *f* hypochlorous acid

Unterchlor·säure *f* hypochlorous acid, chlorine dioxide

Unterdampf° *m* (*Producer Gas*) up-run steam

unterdessen *adv* meanwhile

Unterdruck° *m* reduced (*or* low) pressure; (*Med.*) hypotension, low blood pressure. **unterdrücken** (*insep*) *vt* suppress, put down; oppress

Unterdruck·kammer *f* low-pressure (*or*

decompression) chamber. **-kessel** *m* low-pressure boiler (*or* tank). **-messer** *m* vacuometer, vacuum gauge

Unterdrückung *f* suppression; oppression

unterdurchschnittlich *a* subaverage

untere *a* lower, bottom = UNTER **II**, *cf* UNTERST

untereinander 1 *adv & sep pfx* one below the other; among themselves (ourselves, yourselves); mutually. **2** *adv* in confusion. **3** *sep pfx* (*oft*) inter-

Untereinheit° *f* subunit

Untereinteilung° *f* subdivision; subclassification

unterentwickelt *p.a* underdeveloped. **Unterentwicklung** *f* underdevelopment

unterernährt *p.a* undernourished. **Unterernährung** *f* malnutrition

unteressigsauer *a* subacetate of

untereutektisch *a* hypoeutectic. **untereutektoid** *a* hypoeutectoid

unterexponiert *a* underexposed

Unterfamilie *f* subfamily

unterfangen* (*insep*) **1** *vt* prop up. **2: sich u.** dare; (*esp + genit*) venture. **Unterfangen** *n* (-) venture

unterfeuern (*insep*) *vt* underfire. **Unterfeuerung** *f* undergrate firing

unterfing ventured: *past of* UNTERFANGEN

Unterfläche° *f* undersurface, base

Unterflur-garage *f* underground garage. **-motor** *m* underfloor engine

Unterform° *f* subvariety, subspecies

Unterfutter° *n* lining

Untergang° *m* **1** (*Astron.*) setting. **2** foundering; sinking. **3** decline, (down)fall, ruin

untergärig *a* (*Beer*) bottom-fermented. **Untergärung** *f* bottom fermentation

Untergattung° *f* subgenus

untergeben *p.a* (*dat*) subordinate (to)

untergebracht accommodated: *pp of* UNTERBRINGEN

untergegangen set, sunk: *pp of* UNTERGEHEN

unter-gehen* *vi*(s) **1** (*Astron.*) set. **2** sink, go down. **3** decline; perish; become extinct. **4** be lost, be swallowed up, disappear

untergekrochen crawled underneath: *pp of* UNTERKRIECHEN

untergeordnet *p.a* subordinate. *See also* UNTERORDNEN **1**

untergeschoben pushed underneath: *pp of* UNTERSCHIEBEN **1**

Untergeschoß° *n* **1** basement. **2** ground floor

untergestanden taken shelter: *pp of* UNTERSTEHEN **1**

Untergestell° *n* underframe, chassis

untergesunken sunk: *pp of* UNTERSINKEN

Untergewicht° *n* underweight; short weight

untergezogen run underneath: *pp of* UNTERZIEHEN **I**

Unterglasur° *f* underglaze. **-farbe** *f* underglaze color

untergliedern (*insep*) *vt* subdivide, break down (into parts). **Untergliederung°** *f* subdivision, breakdown

untergraben* (*insep*) *vt* undermine

untergrub undermined: *past of* UNTERGRABEN

Untergrund I *m* (-̈e) **1** subsoil, ground; seabed. **2** (*Geol.*) substratum; bedrock. **3** (*Paint*) undercoat. **4** (*Dye.*) bottom. **5** (*Tex.*) bottom print. **6** underground (movement, world). **7** background. **II** *f* = **-bahn** *f* subway (line). **-farbe** *f* (*Paint*) undercoat; (*Dye.*) bottom color. **-strahlung** *f* background radiation

Untergruppe° *f* subgroup; (*Mach.*) subassembly

Unterguß° *m* (*Phot.*) substratum

unterhalb 1 *prep* (+*genit*) below, under. **2** *adv* underneath, below

unterhalogenig *a* hypohal(ogen)ous

Unterhalt *m* **1** upkeep, maintenance. **2** support. **3** livelihood, living

unter/halten* I (*sep*) **1** *vt* hold underneath. **II** (*insep*) **A** *vt* **1** support, maintain. **2** keep (sthg) going. **3** entertain, amuse. **B: sich u. 4** have a good time. **5** have a conversation. **Unterhalts-kosten** *pl* maintenance costs. **Unterhaltung°** *f* **1** maintenance. **2** support. **3** entertainment. **4** conversation

unterhandeln (*insep*) *vi* negotiate. **Unterhändler°** *m* negotiator, mediator. **Unterhandlung°** *f* negotiation

Unterhaut° *f* **1** hypodermis, subcutis. **2** (*in compds*) subcutaneous: **-gewebe** *n* subcutaneous tissue

Unterhefe° *f* bottom yeast
unterhielt supported, etc: *past of* UNTERHALTEN II (*insep*) & *oft* I (*sep*)
Unterhirn *n* (*Anat.*) subencephalon
unterhöhlen (*insep*) *vt* undermine
Unterholz *n* underbrush
unterirdisch *a* underground, subterranean
unterjochen (*insep*) *vt* subjugate
unterjodig *a* hypoiodous. **-sauer** *a* hypoiodite of **U-säure** *f* hypoiodous acid
Unterkiefer° *m* 1 lower jaw. 2 (*in compds*) submaxillary, mandibular
Unterkolben· presse *f* upstroke press
unter·kommen* *vi*(s) 1 find shelter; find employment. 2 (*dat*) run across one's path. **Unterkommen** *n* 1 place to stay, shelter. 2 job
Unterkorn° *n* undersize grain; *pl oft* fine particles (*or* fractions), fines
Unterkorrektur° *f* undercorrection, undercompensation
Unterkreide *f* (*Geol.*) Lower Cretaceous
unter·kriechen* *vi*(s) 1 crawl underneath. 2 find shelter
unter·kriegen *vt* get (sbdy) down, get the best of
unterkritisch *a* subcritical
unterkühlen (*insep*) *vt* supercool, undercool. **Unterkühlung** *f* supercooling, undercooling
Unterkunft *f* (ːe) accommodations, place to stay
unterlag succumbed: *past of* UNTERLIEGEN
Unterlage° *f* 1 substratum, subsoil. 2 basis, foundation; bed. 3 something to rest one's work on, pad, backing (sheet); base, stand. 4 undercoat; lining. 5 parent stock. 6 document. 7 (*usu pl*) supporting evidence, proof; literature; details, particulars. 8 (*pl*) notes, data, records. **-gestein** *n* underlying rock
unterlagert *a* underlying, subjacent
Unterlags·ring *m* supporting ring
Unterlaß *m:* **ohne U.** without letup, incessantly. **unterlassen*** (*insep*) *vt* 1 avoid (doing), neglect (to do); not do, refrain from. 2 fail to do. 3 desist from, give up, stop. **Unterlassung** *f* (-en) avoidance, neglect, failure (to do), nonfeasance; omission, cessation

Unterlauf° *m* lower reaches (of a river)
unterlaufen* (*insep*) **I** *vi*(s) (*dat*): **etwas unterläuft einem** sthg comes across one's path; **ein Fehler unterlief ihnen** a mistake crept into their work. **II** *p.a* (*oft*) extravasated, bloodshot
Unterlauge° *f* (*Soap*) underlye, spent lye
Unterleder *n* bottom (*or* sole) leather
unter/legen I *vt* (*usu* + *d.i.o*) 1 (*sep* & *insep*) attribute (to), read (into). 2 (*sep*) put (sthg) underneath. 3 (*insep*) underlay; line. **II** succumbed: *pp of* UNTERLIEGEN; (*as p.a*) inferior
Unterleg·ring *m* supporting ring. **-scheibe** *f* supporting disk; washer
Unterleib *m* abdomen, belly. **Unterleibs·abdominal**. **-entzündung** *f* peritonitis
unterlief crept, etc: *past of* UNTERLAUFEN
unterliegen* (*insep*) (*dat*) *vi* **I** (h) 1 be subject(ed) (to), be undergoing: **das unterliegt keinem Zweifel** there is no doubt about that. 2 underlie. **II** (s) succumb, lose (to), be defeated (by). **unterliegend** *p.a* (*oft*) subject; defeated; subjacent. *Cf* UNTERLEGEN II
unterließ avoided, etc: *past of* UNTERLASSEN
unterlöst *p.a* (*Malt*) insufficiently grown
unterm *contrac* = **unter dem** *cf* UNTER I
untermauern (*insep*) *vt* support; substantiate
untermeerisch *a* undersea, submarine
unter/mengen *vt* (*sep* + **zwischen, unter**; *insep* + **mit**) intermix (with)
Untermikron·größe *f* submicron size
unterminieren (*insep*) *vt* undermine
unter/mischen *vt* (*sep* & *insep*) intermix *cf* UNTERMENGEN
unternahm undertook, etc: *past of* UNTERNEHMEN
unternehmen* (*insep*) *vt* undertake, do. **Unternehmen** *n* (-) enterprise, business (concern); undertaking, venture, project. **unternehmend** *p.a* (*oft*) enterprising. **Unternehmer** *m* (-) entrepreneur, businessman; employer. **Unternehmung** *f* (-en) enterprise, etc = UNTERNEHMEN
unternimmt undertakes, **unternommen** undertaken: *pres & pp of* UNTERNEHMEN
unternormal *a* subnormal
unter·ordnen *vt* subordinate *cf* UN-

TERGEORDNET. **Unterordnung°** *f* subordination; subdivision, (*Biol.*) suborder

Unterpfand° *n* pledge, security

unterphosphorig *a* hypophosphorous. **--sauer** *a* hypophosphate of

unterphosphor·sauer *a* hypophosphate of. **U—säure** *f* hypophosphoric acid

Unterprobe° *f*, **--weingeist** *m* underproof alcohol

Unterredung *f* (-en) discussion; conference; interview

Unterricht *m* instruction, teaching; lesson(s), class. **unterrichten** (*insep*) **I** *vt* 1 instruct, teach. 2 inform. **II: sich u.** 3 teach oneself. 4 get information, inform oneself

Unterrichts·anstalt *f* educational institution, school. **-stunde** *f* (45 or 50 minute) period (of instruction), class; lesson

Unterrinde *f* underbark, undercrust

unters *contrac* = **unter das** *cf* UNTER **I**

Unters. *abbv* (Untersuchung) investigation, examination

untersagen (*insep*) *vt* forbid, prohibit

Untersalpeter·säure *f* nitrogen peroxide (formerly hyponitric acid)

untersalpetrig *a* hyponitrous. **--sauer** *a* hyponitrite of

untersättigen (*insep*) *vt* undersaturate

Untersatz° *m* 1 pad, mat; coaster. 2 saucer. 3 stand, rest. 4 chassis, undercarriage. 5 assumption, minor premise

Unterschall *m* (*Phys.*) subsonics. **--geschwindigkeit** *f* subsonic speed

unterschätzen (*insep*) *vt* underrate, undervalue

unterscheidbar *a* distinguishable; discernible

unterscheiden* (*insep*) 1 *vt* distinguish, differentiate; discern. **2: sich u.** be distinguished, differ. **Unterscheidung°** *f* distinction; differentiation; (sub)division

Unterscheidungs·markierung *f* distinctive marking. **-merkmal** *n* distinguishing feature

Unterschicht° *f* sublayer, substratum

unter/schieben* 1 (*sep*) *vt* push underneath. 2 (*insep*) *vt* (*d.i.o*) blame (on), attribute (to); read (into); substitute; forge

unterschied differentiated: *past of* UNTERSCHEIDEN

Unterschied *m* (-e) difference, distinction: **zum U. von** as distinct from

unterschieden differentiated: *pp of* UNTERSCHEIDEN

unterschiedlich *a* 1 different; differing, variable; discriminatory. 2 *adv* variably, variously; **u. viele** different (*or* varying) amounts of. **Unterschiedlichkeit** *f* (-en) variability

unterschiedslos *a & esp adv* without distinction

unterschlagen* (*insep*) *vt* 1 suppress (*esp* facts), keep (sthg) secret; omit; ignore. 2 embezzle. **unterschlug** *past*

unterschneiden* (*insep*) *vt* 1 undercut. 2 intersect. **unterschnitt(en)** *past* (& *pp*)

unterschob(en) substituted, etc: *past* (& *pp*) *of* UNTERSCHIEBEN **2**

unterschreiben* (*insep*) *vt/i* sign; underwrite

unterschreiten* (*insep*) *vi*(s) fall below, stay under (a limit)

unterschrieb(en) signed, etc: *past* (& *pp*) *of* UNTERSCHREIBEN

Unterschrift° *f* signature; caption

unterschritt(en) fell (fallen) below: *past* (& *pp*) *of* UNTERSCHREITEN

unterschüssig *a* deficient, insufficient

unterschweflig *a* hyposulfurous: **u—e Säure** hyposulfurous (*or* dithionous) acid. **--sauer** *a* hyposulfite (*or* dithionite) of

unterschwellig *a* subliminal

Unterschwingung *f* subharmonic oscillation

Untersee·boot *n* submarine

unterseeisch *a* submarine, undersea

Unterseite° *f* underside, bottom; (*Tex.*) wrong side

unter/setzen I (*sep*) *vt* 1 put underneath. **II** (*insep*) *vt* 2 mix, intermingle. 3 gear down (a transmission). **Untersetzer°** *m* pad, coaster, mat; stand. **Untersetz·scherbe** *f* saucer, dish (to put underneath); crucible stand. **untersetzt** *p.a* 1 squat, stocky. 2 reduced. 3 intermingled. **Untersetzung°** *f* (-en) gear reduction; intermingling

Untersieb·bereich *m* sub-sieve range

unter·sinken* *vi*(s) sink, go down

Untersippe° *f* subspecies

Unterspannung° *f* 1 (*Elec.*) undervoltage. 2 (*Metll.*) lower stress limit

unterst *a superl of* UNTER(E) lowest, undermost *cf* ZUUNTERST

unterstand was subordinate, etc: *past of* UNTERSTEHEN II (*insep*) & *oft of* I (*sep*)

Unterstand° *m* shelter

unterstanden (been) subordinate: *pp of* UNTERSTEHEN II

Unterstation° *f* substation

unter-stecken *vt* stick underneath; immerse, submerge

unter/stehen* I (*sep*) **1** *vi* (h *or* s) take shelter. **II** (*insep*) **2** *vi* (+*dat*) be subordinate, be responsible (to), be under the control (of); come under. **3: sich u.** (**zu**) dare (to)

unter/stellen I (*sep*) **1** *vt* put underneath; stow, park. **2: sich u.** take shelter. **II** (*insep*) *vt* **3** (+**daß**) assume; insinuate, imply (that). **4** (*d.i.o*) impute (to); subordinate (to), put under the control (of)

Unterstock° *m* (*Brew.*) underback

unterstreichen* (*insep*) *vt* underline, underscore. **unterstrich(en)** *past* (*& pp*)

Unterstufe° *f* **1** substage. **2** lower grades (of secondary school)

unterstützen (*insep*) *vt* support; subsidize. **Unterstützung** *f* (-en) support; subsidy

untersuchbar *a* capable of being investigated (examined, inspected)

untersuchen (*insep*) *vt* investigate, study; examine; inspect. **Untersuchung** *f* (-en) investigation, study; examination; inspection

Untersuchungs-arbeit *f* research work, study. **-ausschuß** *m* investigating committee. **-chemiker** *m* research chemist. **-ergebnis** *n* result of the investigation, finding(s). **-material** *n* material for (*or* under) investigation. **-methode** *f* method of investigating, research method. **-mittel** *n* means of investigation (examination, research); indicator. **-raum** *m* laboratory; examination room. **-richtung** *f* line (*or* direction) of investigation. **-verfahren** *n* investigative procedure, research (*or* test) procedure

untertage *adv* underground. **U--arbeit** *f* underground work; mining

Untertag-vergasung *f* subterranean (*or* subsurface) gasification

Untertasse° *f* saucer

untertauchbar *a* immersible, submersible

unter-tauchen 1 *vt* immerse, dip. **2** *vi*(s) dive, submerge. **Untertauchung** *f* immersion, submersion

Unterteig° *m* (*Brew.*) underdough

Unterteil° *n* lower part, bottom, base

unterteilbar *a* divisible. **unterteilen** (*insep*) *vt* subdivide; partition; (sub)classify. **Unterteilung°** *f* subdivision; partitioning; (sub)classification, breakdown

Untertemperatur° *f* subnormal temperature

Untertitel° *m* subtitle

Unterton° *m* undertone

Untervernetzung° *f* deficient crosslinking (of polymers)

Unterwald *m* underbrush

Unterwalze *f* lower (roll)er, bottom roll(er)

unterwandern (*insep*) *vt* infiltrate

unterwarf subjugated: *past of* UNTERWERFEN

unterwärts *adv* downward; underneath

Unterwäsche *f* underwear

Unterwasser- underwater: **-anstrich** *m* underwater (antifouling) paint. **-aufnahme** *f* (*Phot.*) underwater shot. **-bombe** *f* depth bomb. **-fahrzeug** *n* underwater vehicle. **-farbe** *f* underwater paint; (*Ships*) bottom paint

unterwegs *adv* **1** on the way, en route. **2** on the move, underway. **3** away, traveling. **4** knowledgeable, on one's toes

unterweichen (*insep*) *vt* understeep

unterweisen* (*insep*) *vt* instruct. **Unterweisung°** *f* instruction

unterwerfen* (*insep*) **1** *vt* subjugate; (*d.i.o*) subject (to). **2: sich u.** (+*dat*) surrender, submit (to). *Cf* UNTERWORFEN

unterwertig *a* low-value, inferior, substandard

unterwies(en) instructed: *past* (*& pp*) *of* UNTERWEISEN

unterwirft subjects: *pres of* UNTERWERFEN

unterworfen 1 *p.a* (*usu* + *dat*) subject (to). **2** subjected: *pp of* UNTERWERFEN

unterwühlen (*insep*) *vt* undermine

unterzeichnen (*insep*) *vt* sign. **unterzeichnend** *p.a* (*oft*) signatory. **Unterzeichnete** *m, f* (*adj endgs*) undersigned. **Unterzeichnung°** *f* signature; signing

Unterzeug *n* underwear

unter/ziehen* I (*sep*) *vt* **1** put on underneath. **2** run (e.g. a wire) underneath. **3** (*Cooking*) fold in. II (*insep*) **4** *vt* (*d.i.o*) subject (to). **5: sich u.** (+*dat*) submit (to), undergo. **unterzog(en)** subjected: *past* (& *pp*) *of* **II**, *oft past of* **I**

Unterzug *m* underdraft

Unterzungen-gegend *f* sublingual (*or* hypoglossal) region

untief *a* shallow. **Untiefe°** *f* **1** shallowness. **2** shoal. **3** (*fig*) abyss, bottomless depth

untilgbar *a* indelible; irredeemable

untrennbar *a* inseparable; undetachable. **Untrennbarkeit** *f* inseparability; undetachability

untrinkbar *a* undrinkable, unfit to drink

untrüglich *a* unerring, infallible; unmistakable

untüchtig *a* inept, inefficient

untunlich *a* infeasible, impractical

untypisch *a* atypical

unt. Zers. *abbv* (unter Zersetzung) with decomposition

unübersehbar *a* **1** indeterminate, incalculable; boundless. **2** unmistakable

unübersichtlich *a* not giving a clear overview, vague; unpredictable

unübersteiglich *a* insurmountable

unübertragbar *a* nontransferable; untransmittable

unübertrefflich *a* unsurpassable, unexcelled

unübertroffen *a* unsurpassed, unexcelled

unüberwindlich *a* insuperable, invincible

unumgänglich *a* unavoidable, indispensable

unumschränkt *a* unlimited, unrestricted

unumstößlich *a* irrefutable, indisputable

unumwunden *a* frank, blunt; *adv oft* without hesitation

ununterbrochen *a* uninterrupted, unbroken

ununterscheidbar *a* indistinguishable

ununtersucht *a* uninvestigated, unresearched; unexamined

unv. *abbv* (unveröffentlicht) unpubl.

unveränderlich *a* unchangeable, invariable

unverändert *a* unchanged

unverantwortlich *a* irresponsible

unverarbeitet *a* unprocessed; unwrought, not made up

unveräußerlich *a* inalienable

unverbesserlich *a* incorrigible; unimprovable

unverbindlich *a* non-binding; *adv usu* without obligation

unverbleit *a* unleaded

unverblümt *a* plain, unvarnished

unverbrannt *a* unburned

unverbrennbar, unverbrennlich *a* incombustible; slow-burning. **Unverbrennbarkeit** *f* incombustibility

unverbraucht *a* unused; unconsumed

unverbunden *a* **1** uncombined, unbound; unconnected. **2** unobligated

unvebürgt *a* unwarranted, unconfirmed

unverdaulich *a* indigestible. **unverdaut** *a* undigested

unverdichtbar *a* uncondensable, incompressible. **Unverdichtbarkeit** *f* incondensability, incompressibility

unverdient *a* undeserved

unverdorben *a* unspoiled, unadulterated

unverdrossen *a* unwearied; undeterred

unverdünnt *a* undiluted

unvereinbar *a* incompatible, irreconcilable

unverestert *a* unesterified

unverfälscht *a* unadulterated; unfalsified

unverfaulbar *a* unputrefiable, imputrescible

unverflüchtigt *a* unvolatilized

unvergänglich *a* imperishable

unvergärbar *a* unfermentable

unvergasbar *a* ungasifiable

unvergeßlich *a* unforgettable

unverglast *a* unvitrified

unvergleichbar, unvergleichlich *a* incomparable

unvergrünlich *a* ungreenable

unverhältnismäßig *a* disproportionate; (*adv oft*) excessively, far too

unverhofft *a* unhoped-for, unexpected

unverhohlen *a* unconcealed

unverholzt *a* unlignified

unverhüttbar *a* unsmeltable

unverkäuflich *a* unsalable, unmarketable

unverkennbar *a* unmistakable

unverkittet *a* uncemented, unluted

Unverkokte(s) *n* (*adj endgs*) (*Coke*) black ends, green coke
unverkürzt *a* unabridged, uncurtailed
unverletzlich *a* inviolable
unverletzt *a* unimpaired, uninjured
unvermeidbar, unvermeidlich *a* unavoidable
unvermengt *a* unmixed
unvermerkt *a* unperceived, imperceptible
unvermindert *a* undiminished
unvermischbar *a* immiscible
unvermischt *a* unmixed
unvermittelt *a* sudden, unexpected
Unvermögen *n* inability, incapacity, impotence
unvermutet *a* unexpected
Unvernunft *f* nonsense, absurdity. **unvernünftig** *a* foolish; senseless, irrational
unveröffentlicht *a* unpublished
unverpackt *a* unpacked, unpackaged, in bulk
unverrichtet *a* unaccomplished. **unverrichteter-dinge** *adv* without accomplishing one's purpose
unverrostbar *a* nonrusting, stainless
unverschnitten *a* uncut; unblended, unadulterated
unverschuldet *a* 1 undeserved. 2 debt-free
unverschwelt *a* uncarbonized, unpyrolyzed
unversehens *adv* unexpectedly, suddenly
unversehrt *a* unharmed, unscathed, undamaged
unverseifbar *a* unsaponifiable: **U—e(s)** *n* (*adj endgs*) unsaponifiable matter
unversiegbar *a* inexhaustible
unversorgt *a* unsupplied, unprovided-for
Unverstand *m* imprudence, lack of judgment
unverständig *a* unwise, imprudent, lacking common sense
unverständlich *a* unintelligible; incomprehensible
unversteuert *a* untaxed, tax-free; *adv oft* duty unpaid
unversucht *a* untried
unvertilgbar *a* ineradicable, indelible
unverträglich *a* 1 incompatible, inconsistent. 2 undigestible. 3 unsociable.
Unverträglichkeit *f* 1 incompatibility,

inconsistency. 2 undigestibility. 3 unsociableness
unverwandt *a* 1 fixed, steady; intent. 2 unrelated
unverwechselbar *a* unmistakable
unverweilt *adv* without delay
unverweslich *a* imputrescible, undecaying
unverwittert *a* unweathered
unverwüstlich *a* indestructible, durable
unverzeihlich *a* unpardonable
unverzerrt *a* undistorted, undeformed
unverzichtbar *a* indispensable
unverzollt *a* duty-free; *adv oft* duty unpaid
unverzüglich *a* immediate, prompt; *adv oft* without delay
unverzweigt *a* unbranched
unvollendet *a* incomplete, unfinished
unvollkommen *a* imperfect. **Unvolkommenheit** *f* (-en) imperfection
unvollständig *a* incomplete. **Unvollständigkeit** *f* incompleteness
unvorbereitet *a* unprepared
unvoreingenommen *a* unprejudiced
unvorhergesehen *a* unforeseen
unvorsätzlich *a* unintentional, unpremeditated
unvorsichtig *a* incautious, careless
unvorteilhaft *a* disadvantageous, unprofitable
unwägbar *a* unweighable, imponderable
unwahr *a* untrue
unwahrscheinlich *a* improbable. **Unwahrscheinlichkeit°** *f* improbability
unwandelbar *a* unchangeable, invariable
unwegsam *a* 1 impassable. 2 trackless
unweigerlich *a* undeniable
unweit *prep* (*genit*), *adv* (+**von**) not far (from)
Unwesen *n* bad state of affairs; **sein U. treiben** carry on one's mischief
unwesentlich *a* nonessential, negligible
Unwetter *n* bad weather, (violent) storm
unwichtig *a* unimportant
unwiderlegbar, unwiderleglich *a* irrefutable
unwiderruflich *a* irrevocable
unwiderstehlich *a* irresistible
unwiederbringlich *a* irretrievable
unwillkürlich *a* involuntary

unwirksam *a* ineffective. **Unwirksam-keit** *f* ineffectiveness
unwirtschaftlich *a* uneconomical
unwissend *a* uninformed, ignorant
unwissenschaftlich *a* unscientific
Unwucht *f* imbalance, disequilibrium
Unzahl *f* (+*genit or* + **von**) vast number (of)
unzählbar, unzählig *a* innumerable, countless. **unzähligemal** *adv* countless times
Unze *f* (-n) ounce
Unzeit *f*: **zur U.** at the wrong time
unzeitgemäß *a* 1 outmoded, out of date. 2 unseasonable
unzerbrechlich *a* unbreakable
unzerfreßbar *a* incorrodible, noncorroding
unzerlegbar *a* undecomposable; indivisible. **unzerlegt** *a* undecomposed
unzerreißbar *a* untearable
unzersetzbar *a* undecomposable
unzerstörbar *a* indestructible. **Unzerstörbarkeit** *f* indestructibility. **unzerstört** *a* undisturbed
unzertrennbar, unzertrennlich *a* inseparable
unziehbar *a* inductile
unzufrieden *a* dissatisfied. **Unzufriedenheit** *f* dissatisfaction
unzugänglich *a* inaccessible
unzugerichtet *a* unfinished; (*Lthr.*) rough-tanned
unzukömmlich *a* 1 inadequate. 2 unjustified
unzulänglich *a* inadequate, insufficient. **Unzulänglichkeit** *f* (-en) inadequacy, insufficiency
unzulässig *a* inadmissible, unpermitted
unzureichend *a* insufficient
unzusammendrückbar *a* incompressible. **Unzusammendrückbarkeit** *f* incompressibility
unzustellbar *a* undeliverable
unzuträglich *a* 1 unhealthy, unwholesome. 2 (+*dat*) detrimental (to)
unzutreffend *a* 1 inapplicable. 2 unfounded
unzuverlässig *a* unreliable. **Unzuverlässigkeit** *f* unreliability
unzweckmäßig *a* unsuitable, inexpedient

unzweideutig *a* unequivocal, unambiguous
unzweifelhaft *a* undoubted, indubitable
UP *abbv* (ungesättigte Polyester) unsat. polyesters
U.P. *abbv* (ungarisches Patent) Hungarian patent
Upas·baum *m* upas (tree), antiar. **-gift** *n* upas poison
UpM *abbv* (Umdrehungen *or* Umlaufungen pro Minute) rpm
üppig *a* 1 luxuriant, lush. 2 abundant. 3 exuberant. **Üppigkeit** *f* luxuriance, lushness; abundance; exuberance
ur- *pfx* 1 original, primitive, primeval, elemental, proto-. 2 great-: **Urenkel** great-grandchildren. 3 genuine; standard, master. 4 extremely *cf* URALT
-ür *sfx* (*Chem.*) -ous (*indicating lower valence of preceeding element; see* BROMÜR, CHLORÜR, etc)
Urahn *m* (-en, -en) original ancestor, greatgrandparent
Uralitisierung *f* (*Min.*) uralitization
uralt *a* age-old, ancient
Urämie *f* (-n) uremia
Uran *n* uranium. **--atom** *n* uranium atom. **-bergbau** *m* uranium mining. **-blei** *n* uranium lead. **-brenner** *m* uranium pile. **-carbid** *n* uranium carbide. **-dioxidpulver** *n* uranium dioxide powder. **-erz** *n* uranium ore
Uranfang° *m* origin, very beginning
Uran·gehalt *m* uranium content. **-gelb** *n* uranium yellow. **-glas** *n* uranium glass. **-glimmer** *m* (*Min.*) torbernite. **-grün** *n* (*Min.*) uranochalcite. **-gummi** *n* (*Min.*) gummite. **u--haltig** *a* containing uranium, uraniferous
Urani· uranic
uranig *a* uranous
Urani·nitrat *n* uranic nitrate. **-oxid** *n* uranic oxide
Uranit *n* (*Min.*) autunite
Urani·verbindung *f* uranic (*or* uranyl) compound
Uranlage° *f* original disposition
Uran·meiler *m* uranium pile. **-metall** *n* uranium metal. **-nitrat** *n* uranium nitrate
Urano· uranous, uranoso-

Uran·ocker *m* uranium ocher; (*Min.*) uraconite, uranopilite

Urano·hydroxid *n* uranous hydroxide. **-reihe** *f* uranous series. **-salz** *n* uranous salt. **-uranat** *n* uranous uranate. **-verbindung** *f* uranous compound

Uran·oxid *n* uranic (*or* uranium) trioxide. **-oxidhydrat** *n* uranium hydroxide. **-oxidoxydul** *n* uranoso-uranic oxide. **-oxidrot** *n* uranium oxide red. **-oxydul** *n* uranous oxide. **-oxyduloxid** *n* uranoso-uranic oxide. **-oxydulsalz** *n* uranous salt. **-pechblende** *f*, **-pecherz** *n* pitchblende. **-phosphat** *n* uranium phosphate. **-rot** *n* uranium red. **-salz** *n* uranium salt. **u–sauer** *a* uranate of. **U–säure** *f* uranic acid. **-spaltung** *f* uranium fission. **-strahlen** *mpl* uranium rays. **-tonbad** *n* (*Phot.*) uranium toning bath. **-tönung** *f* uranium toning. **u–uranig** *a* uranoso-uranic. **U–verbindung** *f* uranium compound. **-verstärker** *m* uranium intensifier. **-vitriol** *n* (*Min.*) johannite

Urat *n* (-e) urate

Uratom° *n* primordial atom

urbar *a* arable, tillable. **urbaren** *vt* cultivate, till; reclaim. **Urbar·machung** *f* cultivation, tilling; reclamation

Urbaustein *m* primary building material, element

Urbeginn° *m* very beginning, prime origin

Urbestandteil *m* ultimate constituent

Urbewohner° *m* original inhabitant; aborigine

Urbild° *n* prototype, original model; paragon

Urboden° *m* virgin soil

Urdestillation° *f* low-temperature distillation

ureigen *a* **1** original, innate. **2** very own

Urein *n* imidazolid-2-one

Ureltern *pl* ancestors

Urfarbe° *f* primary color

Urfels° *m* primitive rock

Urform° *f* original form, prototype

Urgebirge *n* primitive (*or* primeval) mountains; (*Geol.*) primary rocks

Urgeschichte *f* earliest history, prehistory. **urgeschichtlich** *a* prehistoric

Urgestein *n* primitive (*or* parent) rock; (*Ceram.*) native kaolinic rock (yielding kaolin by elutriation)

Urgewalt° *f* primitive (*or* elemental) force

Urgewicht° *n* standard weight

Urgranit *m* primitive granite

Urgroßeltern *pl* greatgrandparents

Urheber° *m* originator, founder, creator; author. **–recht** *n* copyright. **Urheberschaft** *f* authorship

urig *a* primitive, elemental, down-to-earth

Urin·absatz *m* urinary sediment. **-geist** *m* ammonia

urinieren *vi* urinate

Urin·küpe *f* (*Wool*) urine vat. **-stein** *m* urinary calculus

Urinstinkt° *m* primitive instinct

urin·treibend *a* diuretic

Urkalk *m* primitive limestone

Urknall *m* (*Astron.*) Big Bang

Urkoks *m* low-temperature coke, semicoke

Urkraft° *f* elemental force (*or* power). **urkräftig** *a* very powerful; hearty

Urkunde° *f* document, record; diploma; certificate

Urkunden·beweis *m* documentary evidence. **-papier** *n* document paper

urkundlich *a* documentary

Urlaub *m* (-e) leave (of absence), furlough, vacation: **U. machen** take a vacation

Urläuter *m* (*Lthr.*) sod oil, degras

Urlehre° *f* master gauge

Urlösung° *f* original solution

Urmaß° *n* standard measure; master gauge

Urmaterie *f* prime matter; basic (*or* original) material; element

Urmensch° *m* primitive man

Urmeter° *n* standard meter

Urmuster° *n* prototype, standard

Urne *f* (-n) urn; ballot box

Urocanin·säure *f* urocanic acid

U-Rohr° *n*, **U-Röhre°** *f* U-tube

Uron·säure *f* uronic acid

Uroxan·säure *f* uroxanic acid

urplötzlich *a* very sudden, totally unexpected

Urpreis° *m* original (*or* manufacturer's) price

Urquell° *m* wellspring, fountainhead, origin

Urreaktion° *f* reaction step (*or* stage)

Ursache° *f* cause; reason; motive. **ursächlich** *a* causal, causative

Ursäure *f* acidic ureide (e.g. barbituric *or* oxaluric acid)

Urschleim *m* protoplasm

Urschrift° *f* original (manuscript); first draft

Ursol·säure *f* ursolic acid

urspr. *abbv* (ursprünglich) orig.

Ursprung° *m* origin, source; **ursprünglich** *a* original, primary

Urstoff° *m* primary substance; initial material. **·-lehre** *f* atomism, atomic theory. **urstofflich** *a* elementary. **Urstoff ·teilchen** *n* primoridal particle

Ursubstanz° *f* primary substance (*or* matter)

Urteil° *n* 1 judgment; (legal) decision, verdict; sentence. 2 opinion, view

Urteilchen° *n* primary (*or* primordial) particle

urteilen *vi* 1 judge, form an opinion. 2 decide; pass judgment; pass sentence

Urtier°, Urtierchen° *n* protozoon

Urtinktur° *f* mother tincture

Urtiter *m* original titer, titrimetric standard. **·-lösung** *f* primary standard solution. **·substanz** *f* standard titrimetric substance; (*Volumetric Anal.*) primary standard

urtümlich *a* original, innate; genuine

Urverkokung *f* low-temperature coking (*or* carbonization)

Urverschwelung *f* low-temperature distillation (*or* carbonization)

Urwald° *m* virgin (*or* primeval) forest

Urwasser° *n* primordial water

urwellen *vt* (*Metll.*) double

Urwelt *f* primeval world

urwüchsig *a* 1 original, native; elemental. 2 earthy, down-to-earth

Urzeit° *f* primitive times

Urzelle° *f* primitive cell; ovum

Urzeugung° *f* abiogenesis, spontaneous generation

Urzustand° *m* original (*or* primitive) state

usf. *abbv* (und so fort) etc, and so forth

Usnetin·säure *f* usnetinic acid. **Usnet ·säure** *f* usnetic acid. **Usnin·säure** *f* usnic acid

U-Stahl° *m* channel-section steel

usw. *abbv* (und so weiter) etc, and so forth

Utensilien *npl* utensils, implements

u.U. *abbv* (unter Umständen) under certain circumstances

u.ü.V. *abbv* (unter dem üblichen Vorbehalt) with the usual reservations

UV *abbv* 1 (Ultraviolett) ultraviolet. 2 (Unverseifbares) unsaponifiable matter

u.v.a.(m.) *abbv* (und viele andere *or* andere mehr) and many others

Uvin·säure *f* pyrotritaric (*or* uvic) acid

Uviol·licht *n* ultraviolet light

Uvitin·säure *f* uvitic acid. **Uvitonin·säure** *f* uvitonic acid

u.W. *abbv* (unseres Wissens) as far as we know, to our knowledge

Uwarowit *m* uvarovite

u.Z., u. Zers. *abbv* (unter Zersetzung) with decomposition

u.zw. *abbv* (und zwar) and specifically, etc *cf* ZWAR

V

v. *abbv* 1 (variabel) var. 2 (vide) see. 3 (vom, von) of, from, by (the). 4 (vor) before. 5 (vorig) former, preceding, last. 6 (vormals) formerly. 7 (vormittags) AM

V *abbv* 1 (Vakuum) vac. 2 (Verband) assn., org., soc. 3 (Verbindung) comp(d). 4 (Volt). 5 (Volumen) vol.

V. *abbv* 1 (Verbindung) comp(d). 2 (Verein) soc., assn., union. 3 (Verlag) pub. (publishers). 4 (Vorkommen) occurrence, presence

vaccinieren *vt/i* vaccinate

Vache·leder *n* sole leather, neat's leather

Vachetten *fpl* split hide

Vacu·blitz *m* photoflash (lamp)

Vacuolen·wand *f* tonoplast

vag *a* vague

vagabundieren *vi* (s) wander, stray. **vagabundierend** *p.a* (*oft*) vagrant, stray, roving

vage *a* vague. **Vagheit** *f* (-en) vagueness

Vagus·stoff *m* vagus substance (acetylcholine *or* thiamine)

Vak. *abbv* (Vakuum) vac. (vacuum)

Vakanz *f* (-en) 1 vacancy. 2 (school) vacation

Vakua vacuums: *pl of* **Vakuum**

Vak. Eks. *abbv* (Vakuumexsikkator) vacuum desiccator

Vakuo·messer *m* vacuum gauge

Vakuum·apparat *m* 1 vacuum apparatus. 2 (*Sugar*) vacuum pan. **-birne** *f* vacuum bulb. **v–dicht** *a* vacuum-tight, airtight. **V–dichtung** *f* vacuum seal(ing) (*or* packing). **-exsikkator** *m* vacuum desiccator. **-fett** *n* vacuum grease. **-förderer** *m* vacuum conveyor. **-formen** *n* vacuum forming. **-gärung** *f* vacuum fermentation. **-glocke** *f* vacuum bell jar. **-glühlampe** *f* vacuum incandescent lamp. **-hahn** *m* vacuum tap

vakuumieren *vt* evacuate

Vakuum·kitt *m* vacuum cement. **-kochapparat** *m* vacuum pan. **-kolben** *m* vacuum flask. **-kühler** *m* vacuum cooler. **-licht-**

bogen *m* (*Elec.*) vacuum arc. **-mantelgefäß** *n* vacuum-jacketed (*or* Dewar) vessel. **-messer** *m*, **-meter** *n* vacuum gauge, vacuometer. **-öl** *n* vacuum oil. **-packung** *f*(*Food*) vacuum pack. **-pfanne** *f* vacuum pan. **-prüfrohr** *n*, **-prüfröhre** *f* vacuum testing tube. **-pumpe** *f* vacuum pump, **-rohr** *n*, **-röhre** *f* vacuum tube. **-röhrenverstärker** *m* thermionic amplifier. **-schlauch** *m* vacuum hose (*or* tubing). **-strangpresse** *f* deaeration press. **-technik** *f* vacuum technology. **-teer** *m* vacuum tar. **-trockenofen** *m* vacuum drying oven. **-trockner** *m* vacuum drier. **-verdampfer** *m* vacuum evaporator. **-verdampfung** *f* vacuum evaporation. **-verfahren** *n* vacuum process. **v–verformt** *a* vacuum-formed. **V–verformung** *f* vacuum formation. **v–verpackt** *a* vacuum-packed. **V–verpackung** *f* vacuum pack. **-zelle** *f* vacuum cell

Vakzine *f*(-n) vaccine. **vakzinieren** *vt/i* vaccinate

Val *m or* *n* (-) equivalent weight, gram equivalent

Valenz *f* (-en) valence. **-betätigung** *f* valence activity. **-bindungsmethode** *f* valence bond method. **-einheit** *f* valence unit. **v–gesteuert** *a* valence-controlled. **V–kraft** *f* valence (power *or* force). **-lehre** *f* valence concept. **-ordnung** *f* valence orientation. **-richtung** *f* valence direction. **-schwingung** *f* valence vibration (*or* frequency), stretching frequency. **-schwingungsfrequenz** *f* bond-stretching frequency. **-strich** *m* valence bond. **-stufe** *f* valence stage. **-winkel** *m* valence bond angle. **-zahl** *f* valence (number). **-zustand** *m* valence state

Valerianat *n* (-e) valerate

Valerian·öl *n* valerian oil. **-säure** *f* valeric acid

Valeriansäure·amyläther *n* amyl valerate. **-salz** *n* valerate, valeric acid salt

validieren *vt* validate. **Validität** *f* (-en) validity

Valin *n* (-e) 1 valine, alpha-aminoisovaleric acid. 2 vellum

Valone(a) *f* (*Tanning*) valonia

Valuations·tabelle *f* table of values

Valuta *f* (..ten) (*Com.*) 1 value. 2 (monetary) standard, rate. 3 foreign currency. **-·aufschlag** *m* increased value (*or* price) due to rate of exchange

Vanadat·beize *f* (*Dye.*) vanadate mordant

Vanadblei·erz *n*, **-spat** *m* (*Min.*) vanadinite

Vanadi· vanadic

vanadig *a* vanadous

Vanadin *n* vanadium

Vanadinat *n* (-e) vanadate

Vanadin·beize *f* vanadium mordant. **-bleierz** *n*, **-bleispat** *m* (*Min.*) vanadinite. **-chlorid** *n* vanadium chloride. **-eisen** *n* ferrovanadium. **v–enthaltend**, **-haltig** *a* vanadium-containing, vanadiferous. **V–salz** *n* vanadium salt. **v–sauer** *a* vanadate of. **V–säure** *f* vanadic acid. **-säureanhydrid** *n* vanadic anhydride, vanadium (pent)oxide, V_2O_5. **-spat** *m* (*Min.*) vanadinate. **-stahl** *m* vanadium steel. **-stickstoff** *m* vanadium nitride. **-sulfat** *n* vanadium sulfate. **-verbindung** *f* vanadium compound

Vanadium·chlorid *n* vanadium chloride, VCl_3. **-chlorür** *n* vanadous chloride, VCl_2. **-eisen** *n* ferrovanadium. **v–enthaltend** *a* vanadiferous. **V–säure** *f* vanadic acid. **-stahl** *m* vanadium steel

Vanado· vanadous: **-sulfat** *n* vanadous sulfate

Vanad·schwarz *n* vanadium black. **-stahl** *m* vanadium steel

Vandura·seide *f* gelatin silk

Vanille *f* vanilla

Vanillen·kampfer *m* vanillin. **-pflanze** *f* vanilla plant

Vanille(n)·schote *f* vanilla bean (*or* pod)

variabel *a* variable. **Variabilität** *f* variability. **Variabilitäts·index** *m* standard deviation

Variante *f* (-n) variant

Variations·ableitung *f* variation derivative. **-breite** *f* range (*or* amplitude) of variation. **v–fähig** *a* variable. **V–fähigkeit** *f* variability. **-kurve** *f* frequency curve. **-rechnung** *f* variation calculus. **-reihe** *f* frequency distribution. **-weite** *f* extent of variation. **-wellenfunktion** *f* variation wave function

Varietät *f* (-en) variety

variieren *vt/i* vary. **Variierung** *f* (-en) variation

Vaselin *n* petroleum jelly, petrolatum. **-·öl** *n* paraffin oil

vaskularisieren *vt* vascularize

Vater *m* (⁻) father, sire. **-·element** *n* parent element. **-land** *n* native country

väterlich *a* paternal, fatherly

Vatten·kies *m* (*Min.*) pyrrhotite; marcasite

vb *abbv* 1 (verbessert) improved. 2 (verbunden) combined, connected, bonded

Vb. *abbv* 1 (Verband) soc., assn. 2 (Verbindung) comp(d).

VB *abbv* (Valenzband) valence bond

Vbb. *abbv* (Verbindungen) compounds

Vbd. *abbv* 1 (Verband) soc., assn. 2 (Verbindung) comp(d). 3 (Verbund) assn., combine

Vbdg. *abbv* (Verbindung) comp(d).

vbl. *abbv* 1 (variabel) var. 2 (verbindlich) binding

v. Chr. *abbv* (vor Christus), **v. Chr. G.** *abbv* (vor Christi Geburt) B.C. (before Christ)

VDCh *abbv* (Verband Deutscher Chemiker) Assn. of German Chemists

VD *abbv* (Verband Deutscher Drogisten) Assn. of German Druggists

VDE *abbv* (Verband Deutscher Elektrotechniker) Assn. of German Electrical Engineers

VDEh *abbv* (Verband Deutscher Eisenhüttenleute) Assn. of German Metallurgists

VDI *abbv* (Verein Deutscher Ingenieure) Assn. of German Engineers

v.d.l. *abbv* (vor dem Lötrohr) B.B. (before the blowpipe)

VEB *abbv* (DDR) (Volkseigener Betrieb) state-owned operation

Vegetabilien *pl* 1 vegetables. 2 herbal drugs. **vegetabilisch** *a* vegetable

Vegetarier *m* (-), **vegetarisch** *a* vegetarian

Vegetations·kasten *m* plant incubator

vegetieren *vi* vegetate

Vehikel *n* (-) 1 vehicle (*incl fig*). 2 jalopy

Veilchen *n* (-) violet—**dreifarbiges V.**

pansy. **v-·blau, -farben** *a* violet (-colored). **V-holz** *n* violet wood. **-keton** *n* ionone. **-öl** *n* violet oil. **-stein** *m* iolite. **-stock** *m* violet plant. **-wurz(el)** *f* orris root. **-wurzelöl** *n* oil of orris root

Veits·bohne *f* kidney (*or*) French bean. **-tanz** *m* St. Vitus dance, chorea

Vektoren·gleichung *f* vector equation

Vektor·gerüst *n* vector diagram. **-größe** *f* vector quantity

vektoriell *a* vectorial

Vektor·rechnung *f* vector calculus. **-zusammensetzung** *f* vector addition

Velin *n* vellum. **v-·artig** *a* vellum-like. **V-form** *f* (*Paper*) wove mold. **-papier** *n* vellum (*or* wove) paper

Velour·leder *n* suede

Velours *n* (*Tex. oft*) velvet; suede. **-tapete** *f* flock wallpaper. **-teppich** *m* velvet-pile carpet

veloutieren *vt* velvetize; (*Paper*) flock

Velozität *f* (-en) velocity

Vene *f* (-n) vein

Venedig *n* Venice

Venen· venous: **-blut** *n* venous blood. **-entzündung** *f* phlebitis. **-häutchen** *n* choroid membrane

venenös *a* poisonous, toxic

Venen·stein *m* vein stone, phlebolite

venerisch *a* venereal

Venezianer *m* (-), *a* Venetian. **-rot** *n* Venetian red. **venezianisch** *a* Venetian

venös *a* venous

Ventil *n* (-e) valve; outlet, vent. **v-·artig** *a* valvular

Ventilator *m* (-en) fan; blower; ventilator

Ventil·deckel *m* valve lid. **-eimer** *m*, **-eimerchen** *n* valved container (for viscous liquids). **-einstellung** *f* valve positioning (adjusting, setting). **-gehäuse** *n* valve casing (*or* housing)

ventilieren *vt* ventilate; air

Ventil·kolben *m* valve piston. **-küken** *n* valve plug. **-öl** *n* valve oil. **-schlauch** *m* valve tubing. **-sitz** *m* valve seat. **-stahl** *m* valve steel. **-stellungsanzeiger** *m* valve position indicator. **-stopfen** *m* valve stopper. **-zelle** *f* (*Elec.*) valve cell

Venturi·messer *m* Venturi meter. **-rohr** *n* Venturi tube. **venturisch** *a* (of) Venturi

ver. *abbv* (vereinigt) united.

Ver. *abbv* (Verein) assn., soc.

ver.. *insep pfx, implying:* **1** forth, away: **vertreiben** drive away, dispel. **2** *perfection, completion of an action:* **brauchen** use, **verbrauchen** use up, consume. **3** *reversal of meaning:* **kaufen** buy, **verkaufen** sell. **4** *error* (*usu rflx*) *or excess:* **sich verrechnen** miscalculate; **versalzen** oversalt. **5** *making or turning into, or simply converting nouns or adjectives into verbs:* **Ursache** cause: **verursachen** cause; **allgemein** general, **verallgemeinern** generalize; **Käse** cheese, **verkäsen** become cheesy; (*with metals, usu implies plating*): **verchromen** chromium-plate

verabfolgen *vt* administer, give (medicine, treatment); serve, dispense

verabreden 1 *vt* arrange, set, agree on (a date, appointment). **2: sich v.** make an appointment. **verabredet** *p.a* appointed (time); **v. sein** have an appointment. **Verabredung** *f* (-en) date, appointment; agreement

verabreichen *vt* dispense, etc = VERABFOLGEN

verabschieden I *vt* **1** discharge, dismiss. **2** say goodby to. **3** pass, enact (a law); adopt (a plan). **II: sich v. (von)** take leave (of), say goodby (to)

verabsolutieren *vt* universalize, make generally valid

verachten *vt* despise, scorn. **Verachtung** *f* contempt, disdain

verallgemeinern *vt* generalize. **Verallgemeinerung** *f* (-en) generalization

veralten *vi*(s) grow old (*or* obsolete). **veraltet** *p.a* obsolete, antiquated, out of date

veraluminieren *vt* aluminize, aluminum-plate

veränderlich *a* variable, changeable; unsteady; **V—e** *f* (*adj endgs*) variable (number). **Veränderlichkeit** *f* variability, changeability; unsteadiness

verändern 1 *vt* & **sich v.** change, vary. **2** *vt* (*Ores*) concentrate. **Veränderung°** *f* change, variation

verankern *vt* anchor; establish; incorporate

veranlagt *p.a* inclined, prone, predisposed. **Veranlagung** *f* (-en) predisposition, tendency; talent

veranlassen *vt* (*reg*) **1** arrange; (*esp* +**daß**) see to it (that). **2** (+**zu**) induce, cause (to do), make (do). **veranlaßt** *p.a:*sich v. sehen (or fühlen) feel compelled. **Veranlassung** *f* (-en) **1** arranging; inducement, cause; occasion: das gibt V. zu . . . that gives rise to . . . ; auf V. von . . . at the instigation of . . .

veranschaulichen *vt* illustrate; demonstrate

veranschlagen *vt* (*reg*) (**auf**) estimate, value (at). **Veranschlagung** *f* (-en) estimate

veranstalten *vt* arrange, organize; plan. **Veranstaltung** *f* (-en) **1** arrangement, organization; planning. **2** event, function, affair (meeting, etc); performance

verantworten 1 *vt* answer for, be responsible for. **2:** sich v. account for oneself

verantwortlich *a* responsible, answerable. **Verantwortlichkeit** *f* responsibility

Verantwortung *f* (-en) responsibility: zur V. ziehen call to account. **verantwortungslos** *a* irresponsible

verarbeitbar *a* workable, machinable; processable. **Verarbeitbarkeit** *f* workability, machinability; processability

verarbeiten *vt* **1** (**zu**) make (into). **2** machine, finish. **3** work, process. **4** (*Med. & fig*) digest. **Verarbeitung** *f* machining, finishing; working, processing; digesting

verarbeitungs·fähig *a* processable, etc = VERARBEITBAR. **V—zahl** *f* (*Brew.*) evaluation number

verarmen 1 *vt* weaken; reduce in quality; impoverish. **2:** sich v. weaken, become impoverished (or poor)

verarten *vt, vi*(s) degenerate

veraschen *vt* ash, incinerate. **Veraschung** *f* incineration

Veraschungs·grundkurve *f* (*Coal*) instantaneous ash curve. **-probe** *f* incineration test. **-schale** *f* ignition crucible

verästelt *p.a* branched, ramified; (*Metll.*) feathery. **Verästelung** *f* ramification

veräthern *vt, vi*(s) etherify. **Verätherung** *f* etherification

veratmen *vt* breathe up, use up by breathing; (*Yeast*) consume in respiration

Veratrin·säure *f* veratric acid

Veratrum·aldehyd *m* veratraldehyde. **-alkohol** *m* veratric alcohol. **-säure** *f* veratric acid

verätzen *vt* **1** corrode. **2** cauterize

verausgaben I *vt* **1** spend; use up, exhaust. **2** issue. **II:** sich v. **3** overspend. **4** exhaust oneself

verb. *abbv* **1** (verbessert) improved; corrected. **2** (verbunden) combined, connected, bonded

Verb. *abbv* **1** (Verband) assn., soc. **2** (Verbindung) comp(d).

verband combined: *past of* VERBINDEN

Verband *m* (⁼e) **1** bandage; bandaging. **2** association, society; group; (*Mil.*) unit, formation. **3** bond, binding; joint; ligature; fastening

Verband(s)·kasten *m* first-aid kit. **-päckchen** *n* first-aid packet. **-stoff** *m* **1** bandaging material. **2** bonding (binding, fastening) material. **-watte** *f* absorbent cotton. **-zeug** *n* first-aid kit, bandaging materials

verbarg hid: *past of* VERBERGEN

verbauen *vt* **1** block, obstruct. **2** use up in construction

verbaumwollen *vt* cottonize

Verbb. *abbv* (Verbindungen) compounds

verbeißen* I *vt* **1:** damage by biting. **2:** sich etwas v. suppress sthg. **II:** sich v. (**in**) **3** sink one's teeth (in). **4** become obsessed (with) *cf* VERBISSEN

verbeizen *vt* (*Lthr.*) overbate

Verbene *f* (-n) verbena, vervain

verbergen* 1 *vt & sich v.* hide, conceal (oneself). **2** *vt* harbor; keep secret. *Cf.* VERBORGEN

verbessern *vt* improve; correct, revise. **Verbesserung** *f* (-en) improvement; correction, revision

verbesserungs·bedürftig *a* in need of improvement (or revision). **V-mittel, -produkt** *n* corrective, ameliorative. **-vorschlag** *m* suggestion for improvement. **-wert** *m* correction factor

verbeulen *vt* **1** batter. **2** dent. **Verbeulung** *f* (-en) dent

verbiegen* 1 *vt* bend (out of shape), distort, twist. **2:** sich v. warp, buckle, be bent.

verbieten* 1 *vt* forbid, prohibit. **2:** sich v. be impossible, be prohibitive. *Cf* VERBOTEN

verbilligen 1 *vt* cheapen, reduce in price. **2:** sich v. come down in price

verbinden* I *vt & sich v.* **1** combine, unite.

II *vt* **2** (*usu* +**mit**) connect, link, join (with); tie (to). **3** bandage. **III:sich v. 4** be combined, be associated. **5** band together, form an association (alliance, partnership). *Cf* VERBUNDEN

verbindlich *a* **1** binding (agreement, etc); obligatory. **2** friendly, obliging. **Verbindlichkeit** *f* (-en) **1** binding nature, obligatoriness. **2** civility, amiability. **3** obligation, liability

Verbindung *f* (-en) **1** (*Chem.*) compound. **2** combination. **3** (*incl Tlph.*) connection; link(age). **4** contact, touch: **in V. bleiben** stay in touch. **5** association, alliance; (student) fraternity. **6** bandaging. —**eine V. eingehen** combine, form a compound, enter into combination (an association, alliance, etc.); **in V. mit** in connection (conjunction, association) with; **in V. bringen mit** link (connect, associate) with; **in V. stehen** be connected, be linked, be in touch (*or* contact); **in V. setzen mit** connect with, put in touch with

Verbindungs- (*usu*) connecting: **-bildung** *f* compound formation. **-draht** *m* connecting wire. **-faden** *m* connecting fiber. **-fähigkeit** *f* combining ability. **-form** *f* form of combination; combining form. **-gang** *m* connecting passage. **-gewebe** *n* connective tissue. **-gewicht** *n* combining weight. **-gleichung** *f* combining equation. **-glied** *n* connecting link. **-hahn** *m* connecting stopcock. **-kitt** *m* cement for joints. **-klammer** *f* **1** brace. **2** (*Elec.*) clip splicing clamp. **-klasse** *f* class of compounds. **-klemme** *f* **1** (*Elec.*) terminal, binding post. **2** (*Lab*) fastener. **3** (*Heat Meas.*) couple connector block. **-kraft** *f* combining power. **-leitung** *f* (*Elec., Tlph., Tgph.*) connecting line (*or* circuit); (*Engg.*) connecting conduit. **-leute** mediators: *pl of* -MANN. **-linie** *f* connecting (*or* communication) line. **-mann** *m* mediator; intermediary. **-mittel** *n* **1** means of communication. **2** binding agent. **-molekül** *n* molecule of a compound. **-punkt** *m* junction (point). **-rohr** *n*, **-röhre** *f* connecting pipe (*or* tube). **-schlauch** *m* connecting tube (*or* hose). **-schliff** *m* ground joint. **-schnur** *f* connecting (*or* patch) cord. **-stange** *f* connecting rod. **-stelle** *f* junction (point). **-streben** *n* (chemical)

affinity. **-strich** *m* **1** valence bond. **2** hyphen. **-stück** *n* connecting piece; coupling, link; adapter. **-stufe** *f* stage of combination. **-verhältnis** *n* combining proportion. **-volumen** *n* combining volume. **-wärme** *f* heat of combination. **-weg** *m* **1** line of communication. **2** connecting path (*or* road). **-wert** *m* combining value. **-zeichen** *n* hyphen

verbiß suppressed, etc: *past of* VERBEISSEN

verbissen 1 *p.a* stubborn; dogged; sullen. **2** suppressed, etc: *pp of* VERBEISSEN

verblasen* *vt* **1** (*Glass*) blow; use up in blowing. **2** (*Colors*) dilute, shade off. **3** (*Metll.*) refine; (*Bessemer Process*) convert, blow

Verblase-ofen *m* blast furnace. **-röstung** *f* (*Metll.*) blast roasting. **-röstverfahren** *n* air blast roasting process. **-wind** *m* blast of air

verblassen *vi*(s) pale, fade

Verblechung *f* sheet-metal paneling (*or* sheathing)

Verbleib *m* **1** whereabouts. **2** stay. **verbleiben*** *vi*(s) **1** stay, remain. **2** (+**bei**) stick (to) (a plan, etc). **3: es verbleibt dabei** it will be left at that. **4** (+*dat*) be left (over) (for, to)

verbleichen* (verblich, verblichen) *vi*(s) pale, fade away; expire

verbleien *vt* coat (line, treat) with lead. **verbleit** *p.a* leaded. **Verbleiung** *f* **1** leading; coating (lining, treatment) with lead. **2** lead lining

verblenden *vt* **1** blind, dazzle. **2** dim, black out. **3** delude, mislead. **4** face, cover (a wall with tile, etc). **Verblender** *m* (-) (*Ceram.*) face brick

verblich(en) paled: *past* (& *pp*) *of* VERBLEICHEN

verblieb(en) stayed, etc: *past* (& *pp*) *of* VERBLEIBEN

verblies blew, etc: *past of* VERBLASEN

verblüffen *vt* astonish, amaze, bowl over. **verblüfft** *p.a* (*oft*) dumbfounded, stunned

verblühen *vi*(s) **1** fade, wither. **2** cease blooming

verbluten *vi*(s) bleed to death

verbog(en) distorted: *past* (& *pp*) *of* VERBIEGEN

verborgen I *vt* lend (*or* loan) out. **II** *pp of* VERBERGEN & *p.a:* **v. halten** keep con-

cealed; **im V—en bleiben** stay hidden
Verbot n (-e) prohibition
verbot(en) prohibited: *past (& pp) of* VER-
BIETEN; *(pp oft in signs)* no, don't:
Rauchen v. no smoking; **es ist v.,**
zu . . . don't . . .
verbr. *abbv* 1 (verbrannt) burnt. 2 (ver-
braucht) consumed, used
verbracht(e) spent, etc: *pp (& past) of*
VERBRINGEN
verbrämen *vt* border, edge; trim; veil, gloss
over
verbrannt(e) burned: *pp (& past) of*
VERBRENNEN
Verbrauch m (+an) use, consumption (of
goods, etc); exhaustion, depletion. **ver-**
brauchen 1 *vt* use (up), consume; spend,
exhaust; wear out. **2: sich v.** wear oneself
out, be(come) exhausted
Verbraucher m (-) consumer. **v—·**
freundlich a user-friendly. **-unfreund-**
lich a user-unfriendly
verbrauchs-fertig a ready for consumption. **V—gegenstand** m commodity, consumer goods item. **-güter** npl consumer
goods. **-industrie** f consumer goods industry. **-messer** m (use) meter (for gas,
etc.) **-stelle** f place of consumption (*or*
use). **-steuer** f consumer (*or* excise) tax.
-stoff m article of consumption, commodity. **-wirtschaft** f consumer (goods)
economy. **-zähler** m meter (for gas, etc)
Verbrauch(s)·zucker m table sugar
verbrausen *vi(s)* cease fermenting;
subside
Verbrechen n (-) crime. **Verbrecher** m (-)
criminal
verbreiten I *vt* 1 spread, disseminate. 2 circulate, (*esp* +**daß**) broadcast, spread the
news (that). **II: sich v.** spread, circulate,
go round; (+**über**, *oft*) expand (on a topic).
Cf VERBREITET
verbreitern *vt & sich v.* widen, broaden,
spread. **Verbreiterung** f (-en) widening,
broadening, spread
verbreitet a 1: **stark (weithin, überall) v.**
widespread, widely used (*or* known). 2
adv (in weather reports) in many areas. *Cf*
VERBREITEN. **Verbreitung** f spreading,
dissemination; dispersal, dispersion; circulation; range

Verbreitungs·gebiet n dispersion area.
-grenzen fpl dispersion (dissemination,
circulation) limits. **-mittel** n means of
dispersion (dissemination, circulation)
verbrennbar a combustible. **Verbrenn-**
barkeit f combustibility
verbrennen* 1 *vt, vi(s)* burn (up). 2 *vt* cremate; scorch; scald. **3: sich v.** be burned
(cremated, scorched, scalded)
verbrennlich a combustible, (in)flamm-
able. **Verbrennlichkeit** f combustibility,
(in)flammability
Verbrennung f (-en) combustion, burning;
cremation; incineration
Verbrennungs·analyse f combustion
analysis; analysis by combustion.
-anlage f incinerator (unit). **-bombe** f
combustion bomb, bomb calorimeter. **-e-**
nergie f combustion energy. **-ergebnis** n
combustion product. **-gas** n flue gas, combustion gas. **-geschwindigkeit** f rate of
combustion. **-glas** n combustion glass.
-hilfsstoff m combustion aid. **-intensität**
f intensity of combustion. **-institut** n cre-
matorium. **-kammer** f combustion chamber. **-kapillare** f capillary combustion
tube. **-kolben** m combustion flask.
-kraftmaschine f internal combustion
engine. **-lehre** f theory of combustion.
-löffel m combustion spoon. **-luft** f air for
combustion. **-maschine** f, **-motor** m internal combustion engine. **-narbe** f burn
scar. **-ofen** m 1 combustion furnace. 2 incinerator. 3 crematory furnace. **-probe** f
combustion (*or* ignition) test. **-produkt** n
combustion product. **-raum** m combustion chamber. **-rohr** n, **-röhre** f combustion tube. **-rückstand** m combustion
(*or* ignition) residue. **-schälchen,**
-schiffchen n combustion capsule (*or*
boat). **-versuch** m combustion experiment; attempt to burn. **-vorgang** m combustion process. **-wärme** f heat of combustion. **-wasser** n water of combustion.
-wert m combustion (*or* calorific) value.
-wirkungsgrad m combustion efficiency. **-zone** f zone of combustion
verbrieft p.a 1 documented, confirmed in
writing. 2 chartered; bonded
verbringen* *vt* 1 spend, pass (time). 2
squander. 3 take (to a place)

verbrühen *vt* scald

Verbund° *m* **1** (*Com.*) combine, integrated system (*or* service). **2** (*Engg.*) compound (*or* composite) system, interlocking (e.g. electric power) system; (*oft in compds:*) **-bauweise** *f* compound (sandwich, interlocking) construction. **-block** *m* (*Metll.*) compound ingot

verbunden I combined, etc.: *pp of* VER-BINDEN. **II** *p.a* **1** obliged. **2**: **v. sein mit** involve. **3** (*Tlph.*) **falsch v.** (sein) (have a) wrong number

Verbundenheit *f* bonds, ties; connection; solidarity

verbündet *p.a* allied

Verbund- compound, composite, interlocking: **-glas** *n* laminated (*or* compound) glass. **-guß** *m* compound casting. **-kern** *m* compound nucleus. **-koksofen** *m* twin generator coke oven. **-maschine** *f* compound machine. **-material** *n* laminated (compound, sandwich) material; interweave. **-netz** *n* (*Elec.*) integrated power grid. **-platte** *f* sandwich board (*or* panel.). **-preßstoff** *m* composite plastic. **-stahl** *m* compound steel. **-stoff** *m* compounding material. **-stück** *n* coupling, connector. **-verfahren** *n* composite method; (*Metll.*) duplexing process. **-werkstoff** *m* **1** compound plastic. **2** bimetal, clad metal; composite alloy. **-wirkung** *f* compound (*or* mutual) action (*or* effect)

verbürgen *vt* guarantee; confirm. **verbürgt** *p.a* (*oft*) authentic

verchloren *vt* chlorinate. **Verchlorung** *f* (-en) chlorination

verchromen *vt* chrome (*or* chromium) (-plate). **Verchromung** *f* chrome (*or* chromium)-plating

verd. *abbv* (verdünnt) dil., dilute(d). **Verd.** *abbv* (Verdünnung) dilution

Verdacht *m* suspicion: **in V.** under suspicion, suspected; **außer V.** beyond suspicion

verdacht(e) held against: *pp* (*& past*) of VERDENKEN

verdächtig *a* suspicious. **verdächtigen** *vt* suspect. **Verdächtigte** *m.f* (*adj endgs*) suspect

verdammen *vt* damn; condemn, doom

verdämmen *vt* dam, block; tamp; choke off

Verdampf-apparat *m* evaporator, vaporizer; carburetor

verdampfbar *a* vaporizable, volatile. **Verdampfbarkeit** *f* vaporizability, volatility

Verdampf-becken *n* evaporating basin (*or* dish)

verdampfen *vt, vi*(s) evaporate, vaporize, concentrate by evaporation. **Verdampfer** *m* (-) evaporator, vaporizer; carburetor. **-kolonne** *f* evaporating column (*or* tower)

Verdampf-oberfläche *f* evaporating surface. **-pfanne** *f* evaporating pan. **-schale** *f* evaporating dish. **-turm** *m* evaporating tower

Verdampfung *f* (-en) evaporation, vaporization: **V. durch Entspannung** flash evaporation; **mehrfache V.** multiple-effect evporation

Verdampfungs-anlage *f* evaporation plant (*or* unit). **-druck** *m* evaporation pressure. **-ente** *f* duck-shaped evaporating vessel. **v-fähig** *a* capable of evaporating; volatile. **V-fähigkeit** *f* evaporative capacity; volatility. **-geschwindigkeit** *f* rate of evaporation. **-kolben** *m* evaporating flask. **-kühlung** *f* evaporative cooling. **-kurve** *f* vaporization curve. **-messer** *m* evaporimeter. **-pfanne** *f* evaporating (*or* concentrating) pan. **-punkt** *m* vaporization point. **-röstung** *f* volatilization roasting. **-rückstand** *m* residue on evaporation. **-schale** *f* evaporating dish. **-verlust** *m* evaporation loss, loss on volatilization. **-vermögen** *n* evaporating power. **-wärme** *f* heat of vaporization. **-wert** *m*, **-zahl** *f*, **-ziffer** *f* coefficient of vaporization

verdanken *vt*: **einem etwas v.** owe (*or* be indebted to) sbdy (for) sthg; **es ist diesem Umstand zu v., daß . .** it is due to this fact that . . .

verdarb spoiled: *past of* VERDERBEN

verdauen 1 *vi/i* digest. **2: sich v.** be digested—**sich schwer v.** be hard to digest

verdaulich *a* digestbile. **Verdaulichkeit** *f* digestibility

Verdauung *f* digestion

Verdauungs-apparat *m* digestive apparatus (*or* system). **v-befördernd** *a* digestive, promoting digestion. **V-beschwer-**

den *fpl* digestive complaints, indigestion.
-dauer *f* digestion period. **-drüse** *f* digestive gland. **-eingeweide** *n* digestive tract. **v–fähig** *a* digestible. **V–fähigkeit** *f* digestibility. **-ferment** *n* pepsin. **-flüssigkeit** *f* digestive fluid, gastric juice. **v–fördernd** *a* digestive, digestion-promoting. **V–kanal** *m* alimentary canal, digestive tract. **-mittel** *n* digestive remedy (*or* aid). **-ofen** *m* digesting oven. **-organ** *n* digestive organ. **-rohr** *n* alimentary canal. **-saft** *m* gastric juice. **-schwäche** *f* dyspepsia. **-schwierigkeiten** *fpl* digestive problems. **-stoff** *m* pepsin. **-störung** *f* indigestion. **-trakt** *m* digestive tract. **-vorgang** *m* digestive process. **-weg** *m* 1 digestive tract. 2: **auf dem V.** by way of digestion. **-werkzeug** *n* digestive organ

Verdeck *n* (-e) 1 cover, tarpaulin. 2 roof; (*Cars*) top. 3 awning. 4 (ship's) deck; upper level (of a vehicle). **verdecken** *vt* cover(up); conceal, mask

verdenken* *vt:* **einem etwas v.** hold sthg against sbdy

Verderb *m* decay, spoilage; deterioration; waste *cf* GEDEIH. **verderben*** (verdarb, verdorben; verdirbt) **I** *vt* 1 ruin, spoil, damage. 2: **es v. mit** fall out with. **II** *vi*(s) spoil, go bad, decay, rot; deteriorate. *Cf* VERDORBEN. **Verderben** *n* ruin, doom

verderblich *a* 1 perishable. 2 ruinous, pernicious

Verderbnis *f* decay; spoilage; corruption

verderbt *a* corrupt

verdeutlichen *vt* clarify, elucidate; (+**daß**) make it clear (that)

verdeutschen *vt* translate into (plain) German

verdichtbar *a* condensable; compressible. **Verdichtbarkeit** *f* condensability; compressibility

verdichten 1 *vt* & **sich v.** condense; concentrate; increase. 2 *vt* compress, pack; seal. 3: **sich v.** be compressed (packed, sealed). **Verdichter** *m* (-) condenser; compressor. **Verdichtung** *f* (-en) condensation; compression; concentration; increase; packing, seal; compaction

Verdichtungs·apparat *m* condensing apparatus, condenser. **-druck** *m* compres-

sion pressure. **-grad** *m* degree of condensation (compression, concentration). **-hub** *m* compression stroke. **-ring** *m* packing ring, gasket. **-stoßwelle** *f* (*Explo.*) compression wave. **-verhältnis** *n* compression ratio. **-wärme** *f* heat of compression (*or* condensation). **-welle** *f* compression wave. **-zündung** *f* compression ignition

verdicken 1 *vt* & **sich v.** thicken, concentrate, coagulate, curdle. 2 *vt* inspissate. 3: **sich v.** jell, become viscous. **Verdicker** *m* (-) thickener. **Verdickung** *f* 1 thickening, concentration, coagulation, curdling. 2 (*Tex.*) paste

Verdickungs·mittel *n* thickener

verdienen 1 *vt/i* earn. 2 *vt* deserve, 3 *vi* earn (*or* make) money, have an income. *Cf* VERDIENT. **Verdiener** *m* (-) breadwinner, wage earner

Verdienst I *m* (-e) wage(s), earnings, income; profit, gain. **II** *n* (-e) accomplishment; merit, credit; (*pl oft*) services: **ihre V—e um . . .** their services to (*or* on behalf of) . . .

verdienstlich *a* deserving, meritorious

verdient *p.a* deserving—**sich v. machen um . . .** render services to (*or* on behalf of) . . .

verdienter·maßen *adv* deservedly

verdirbt spoils: *pres of* VERDERBEN

verdolmetschen *vt* translate; interpret

verdoppeln *vt* & **sich v.** (re)double. **Verdopplung** *f* (-en) doubling

verdorben 1 *p.a* bad, foul, rotten; corrupt. 2 spoiled, etc: *pp of* VERDERBEN. **Verdorbenheit** *f* rottenness; corruptness

verdorren *vi*(s) dry (up), wither

verdrahten *vt* wire (up). **Verdrahtung** *f* (-en) wiring. **Verdrahtungs·plan** *m* wiring diagram

verdrängen *vt* displace; replace; drive out, repel; repress. **Verdränger·pumpe** *f* displacement pump. **Verdrängung** *f* (-en) displacement; replacement; ousting; repression

Verdrängungs·chromatographie *f* carrier displacement chromatography. **-messer** *m* displacement meter. **-mittel** *n* displacing agent. **-pseudomorphose** *f* (*Cryst.*) substitution-pseudomorphism

verdreckt *a* filthy; polluted

verdrehen *vt* twist, wrench, distort; crane (neck); roll (eyes) *cf* VERDREHT

Verdreh·festigkeit *f* torsion strength. **-schwingung** *f* torsional vibration

verdreht *p.a* (*oft*) warped; deranged *cf* VERDREHEN

Verdrehung *f* (-en) twisting, wrenching; (dis)torsion

Verdrehungs·festigkeit *f* torsion strength. **-kraft** *f* torsional force. **-waage** *f* torsion balance. **-winkel** *m* angle of torsion

verdreifachen *vt* triple. **Verdreifachung** *f* (-en) tripling

verdrießen* (verdroß, verdrossen) *vt* annoy, anger; trouble. **verdrießlich** *a* 1 angered; troubled; sullen. 2 annoying

verdrillen *vt* twist. **Verdrillung** *f* (-en) twisting, torsion

verdrosseln *vt* throttle, choke

verdroß, verdrossen 1 angered, etc: *past & pp* of VERDRIESSEN. 2: **verdrossen** *p.a* (*oft*) sullen

verdrucken *vt* 1 misprint. 2 use up in printing

verdrücken *vt* crush, crumple; wrinkle

Verdruß *m* annoyance, trouble

verdübeln *vt* dowel

verduften *vi*(s) 1 evaporate; (*specif*) lose its flavor (*or* aroma). 2 get lost

verdunkeln 1 *vt* & **sich v.** darken. 2 *vt* black out. **Verdunklung** *f* (-en) darkening; blackout

verdünnbar *a* dilutable; rarefiable

verdünnen 1 *vt* dilute, thin; (*Gases*) rarefy; (*Brew.*) attenuate; (*Colors*) temper. 2: **sich v.** be(come) diluted; thin out; be rarefied (attenuated, tempered); taper (off). **verdünnt** *p.a* (*oft*) dilute, thin, rare. **Verdünnung** *f* (-en) dilution, thinning; rarefaction; attenuation; tempering

Verdünnungs·gesetz *n* dilution law. **-grad** *m* degree of dilution. **-mittel** *n* diluting agent; diluent; thinner; extender; attenuant; solvent. **-wärme** *f* heat of dilution

verdunsten *vt*, *vi*(s) evaporate, vaporize. **Verdunster** *m*(-) evaporator. **Verdunstung** *f* (-en) evaporation, vaporizer

Verdunstungs·gefäß *n* evaporating vessel; evaporimeter. **-geschwindigkeit** *f* rate of evaporation. **-kälte** *f* latent heat of evaporation; cold due to evaporation. **-kühlung** *f* evaporative cooling. **-messer** *m* evaporimeter; atmometer. **-verlust** *m* evaporative loss. **-wärme** *f* heat of evaporation

verdursten *vi*(s) die of thirst

verdüstert *p.a* dark(ened), gloomy

Verdüsungs·verfahren *n* (*Plastics*) nozzle process

verdutzt *p.a* nonplussed

veredeln I *vt* 1 improve; purify; refine; enrich. 2 (*Metals*) plate with a nobler metal; (*Corrosion*) shift toward a nobler potential. 3 (*Bot.*) cultivate; graft. 4 (*Tex.*) finish, dress, process; (*Yarn*) throw. 5 (*Paper*) convert. 6 ennoble. II: **sich v.** improve; be purified (refined, enriched, plated, etc). **Vered(e)lung** *f* (-en) improvement; purification, etc

Vered(e)lungs·bad *n* (*Tex.*) processing bath. **-mittel** *n* (*Tex.*) processing agent. **-produkt** *n* processed (refined, finished) product. **-stoff** *m* improving additive

verehren *vt* 1 honor. 2 (*d.i.o*) present sthg to sbdy. **verehrt** *p.a*: **sehr v—e Herren!** Dear Sirs, Gentlemen

vereidigen *vt* swear in, put under oath

Verein *m* (-e) 1 club, society, association. 2: **im V. mit** in conjunction with

vereinbar *a* reconcilable, compatible. **vereinbaren** *vt* 1 agree on. 2 reconcile: **sich v. lassen** be compatible. **Vereinbarkeit** *f* compatibility. **Vereinbarung** *f* (-en) agreement; reconciliation: **laut V.** as agreed on

vereinen *vt* & **sich v.** unite, combine; (*Colors*) blend: **die Vereinten Nationen** the United Nations

vereinfachen *vt* simplify; (*Math.*) reduce. **Vereinfachung** *f* (-en) simplification; reduction

vereinheitlichen *vt* make uniform, standardize, unify. **Vereinheitlichung** *f* (-en) standardization, unification

vereinigen I *vt* & **sich v.** 1 united, combine, join: **die Vereinigten Staaten** the United States. 2 (*Colors*) blend. II *vt* consolidate, unify. **Vereinigung** *f* (-en) 1 union, combination; consolidation, unification. 2 organization, association, society. 3 (*Colors*) blending

vereinzelt *p.a* isolated, solitary; sporadic; (*Meteor.*) scattered (showers, etc)

vereisen 1 *vt, vi*(s) freeze (up, over), ice up. **Vereisung** *f* (-en) freezing (up), icing (up); (*Geol.*) glaciation. **Vereisungs-schutzflüssigkeit** *f* anti-icing liquid

vereiteln *vt* thwart, frustrate, foil, nullify

vereitern *vi*(s) suppurate, fester. **Vereiterung** *f* suppuration

verenden *vi*(s) die, perish

verenge(r)n *vt* & **sich v.** narrow, contract, constrict. **Vereng(er)ung f** (-en) narrowing, contraction, constriction

vererben I *vt* 1 bequeath. 2 transmit, pass on. **II: sich v.** be hereditary; (+*auf*) be transmitted (to). **vererblich** *a*, **vererbt** *p.a* hereditary. **Vererbung** *f* (-en) inheritance; heredity; transmission

Vererbungs-forscher *m* geneticist. **-forschung** *f* genetic research. **-gesetz** *n* law of heredity. **-lehre** *f* genetics. **-substanz** *f* (*Biol.*) idioplasm

vererden *vt, vi*(s) oxidize

vererzbar *a* mineralizable

vererzen *vt* mineralize; encase in ore. **Vererzung** *f* mineralization. **Vererzungs-mittel** *n* mineralizer

veresterbar *a* esterifiable. **verestern** *vt* esterify. **Veresterung** *f* (-en) esterification

verewigen *vt* perpetuate; immortalize. **verewigt** *p.a* (*oft*) deceased, late

Verf. abbv 1 (Verfahren) proc. (process). 2 (Verfasser) author

verfahren* I *vt* 1 use up (time, fuel) in driving. 2 (*Min.*): **eine Schicht v.** work a shift. II *vi*(s) 3 proceed, act; (+*mit*) deal (with), manage, treat. **III: sich v.** 4 take the wrong road. 5 get stuck. IV *p.a* (*oft*) muddled, hopelessly entangled. **Verfahren** *n* 1 process, method, procedure, technique *cf* VORGANG. 2 (way of) dealing. 3 legal proceedings

Verfahrens-chemie *f* process chemistry. **-forschung** *f* operational research. **-ingenieur** *m* process (or chemical) engineer. **-produkt** *n* product of the process. **-steuerung** *f* process control. **-technik** *f* process technology. **v–technisch** *a* regarding process technology. **V–zeitschalter** *m* process timer

Verfahrungs-weise *f* manner of proceeding

Verfall *m* 1 decay, deterioration; decline; dissolution, collapse; **im V. sein** be decaying, be on the decline; **in V. geraten** decline, deteriorate; **in V. bringen** ruin. 2 expiration (of validity), lapse: **bei V.** on expiration. when due; **bis V.** till due. **-datum** *n* expiration date

verfallen* I *vi*(s) 1 decay, deteriorate, go to ruin, degenerate, decline. 2 lose validity, expire, lapse. 3 (+*in*) fall, lapse (into a condition). 4 (+*dat*) become dependent (on), become addicted (to), become a victim (of); **dem Staat v.** become the property of the state. 5 (+*auf*) hit (up)on, think of. II *p.a:* **v. sein** (+*dat*) be addicted (to), be a victim (of)

verfälschen *vt* adulterate, debase; falsify, forge, counterfeit, **Verfälscher** *m* (-) adulterator; falsifier, forger, counterfeiter. **Verfälschung** *f* (-en) adulteration; falsification; forgery, counterfeiting

Verfälschungs-mittel *n* adulterant

verfangen*1 *vi* work, be effective. **2: sich v.** get caught (or entangled)

verfänglich *a* risky; insidious; incriminating

verfärben I *vt* 1 stain, discolor. 2 use up in dyeing. **II: sich v.** change color; discolor, face. **Verfärbung** *f* (-en) discoloration, fading

verfassen *vt* write, compose. **Verfasser** *m* (-), **Verfasserin** *f* (-nen) author, writer. **Verfassung** *f* (-en) 1 state (of mind), condition. 2 constitution

Verfassungs-änderung *f* (constitutional) amendment. **v–gemäß** *a* constitutional. **-widrig** *a* unconstitutional

verfaulbar *a* putrescible

verfaulen *vi*(s) decay, rot, putrefy. **verfault** *p.a* (*oft*) rotten. **Verfaulung** *f* putrefaction, rotting, decay

verfechten* *vt* advocate, stand up for. **Verfechter** *m* (-) advocate, champion

verfehlen 1 *vt* miss; fall short of: **seine Wirkung v.** fall short of (or fail to have) its effect. 2 *vi* (+*zu* +*inf*) fail, neglect (to do). **verfehlt** *p.a* misdirected, unsuccessful

verfeinern 1 *vt* refine; improve, sophisticate. **2: sich v.** improve, be refined (or sophisticated). **verfeinert** *p.a* (*oft*) finely divided. **Verfeinerung** *f* (-en) improvement, refinement, sophistication

verfertigen *v1* make, produce, manufacture; compose. **Verfertigung** *f* (-en) 1 making, production. 2 make, manufacture; composition

verfestigen I *vt & sich v.* 1 harden, solidify. 2 stiffen. **II** *vt* 3 make firm; fasten. 4 strengthen. 5 consolidate. 6 (*Metll.*) cold-work; strain-harden. **II: sich. v.** 7 become firm. 8 be strengthened. 9 be consolidated. **Verfestigung** *f* (-en) hardening, solidification; stiffening; fastening; strengthening; consolidation; (*Metll.*) cold-working, strain-hardening

Verfestigungs-zeit *f* setting (*or* solidifying) time

verfettet *p.a* fatty, adipose. **Verfettung** *f* fatty degeneration

verfeuern *vt* burn (up) (as fuel); fire, use up (ammunition)

verficht advocates: *pres of* VERFECHTEN

verfiel decayed, etc: *past of* VERFALLEN

verfilmen *vt* (make into a) film

verfilzen 1 *vt* mat; (*Wool*) felt. **2: sich v.** become matted; (*Polymer chains & fig*) become entangled. **Verfilzung** *f* (-en) matting; felting; entanglement

Verfilzungs-fähigkeit *f* felting (*or* matting) property

verfing worked, etc: *past of* VERFANGEN

verfinstern 1 *vt & sich v.* darken. 2 *vt* obscure, cloud; eclipse. **3: sich v.** get cloudy; be eclipsed. **Verfinsterung** *f* (-en) darkening; clouding; eclipse; obscuration

verfitzen *vt & sich v.* tangle (up)

verflachen 1 *vt, vi*(s) *& sich v.* flatten, level (out). 2 *vt* make shallow. 3 *vi*(s) *& sich v.* become shallow; taper off, peter out. **verflacht** *p.a* (*oft*) shallow, superficial; level. **Verflachung** *f* flattening; levelling; decline

verflechten* 1 *vt & sich v.* interweave, interlace. 2 *vt* (+**in**) entangle (in). **3: sich v.** (+**in**) become entangled (in). **Verflechtung** *f* (-en) interweaving, interlacing; entanglement, involvement; complexity. *Cf* VERFLOCHTEN

verflicht interlaces: *pres of* VERFLECHTEN

verfliegen* 1 *vi*(s) volatilize, fade away; fly (by), fleet (*as:* time). **2: sich v.** fly off

course. **verfliegend** *p.a* volatile; evanescent

verfließen* *vi*(s) 1 (*Colors*) blend, merge; run. 2 (*Time*) pass, go by, elapse. *Cf* VERFLOSSEN

verflocht(en) 1 interwove(n): *past* (*& pp*) *of* VERFLECHTEN. **2: verflochten** *p.a* (*oft*) complex, intricate

verflog(en) flew, (flown): *past* (*& pp*) *of* VERFLIEGEN

verfloß, verflossen 1 elapsed, etc: *past & pp of* VERFLIESSEN. **2: verflossen** *p.a* past, last; late

verflüchtigbar *a* volatilizable

verflüchtigen *vt & sich v.* volatilize, evaporate. **Verflüchtigung** *f* volatilization

Verflüchtigungs-fähigkeit *f* volatility. **-probe** *f* volatility test. **-verlust** *m* loss by volatilization

verflüssigen 1 *vt & sich v.* liquefy; fuse; (*Metll.*) thin, (*vt*) dilute. **2: sich v.** be diluted. **Verflüssiger** *m* (-) liquefier; condenser. **Verflüssigung** *f* liquefaction, fluidization; fusion; condensation; dilution

Verflüssigungs-druck *m* liquefying pressure. **-mittel** *n* liquefacient; (*Metll.*) thinning agent; (*Ceram.*) deflocculant

verfocht(en) advocated: *past* (*& pp*) *of* VERFECHTEN

Verfolg *m:* **im V.** (+*genit*) in the course (of), pursuant (to)

verfolgen *vt* follow (up), pursue; prosecute; persecute. **Verfolgung** *f* (-en) pursuit; prosecution; persecution

verformbar *a* (de)formable; moldable, plastic, workable. **Verformbarkeit** *f* (de)formability; moldability, plasticity, workability

verformen 1 *vt & sich v.* form. 2 *vt* deform. **3: sich v.** be(come) deformed; change (its) shape. **2 Verformung** *f* (-en) formation; deformation: **bleibende V.** permanent set (*or* deformation)

Verformungs-fähigkeit *f* moldability, etc = VERFORMBARKEIT. **-rest** *m* permanent set

verfrischen *vt* (*Metll.*) refine

verfrüht *a* premature

verfügbar *a* available, at one's disposal. **Verfügbarkeit** *f* availability

verfügen 1 *vt* decree, order. 2 *vi* (über)

have available, have at one's disposal; be provided with. **Verfügung** f (-en) **1** order, decree. **2** disposal, disposition: **zur V. stehen** (+dat) be at one's disposal, be available (to); **zur V. stellen** (+dat) put at (one's) disposal, make available (to)

verfuhr proceeded, etc: *past of* VERFAHREN

verführen vt mislead, seduce

verfüllen vt fill (from another vessel); overfill

verfünffachen vt & **sich v.** quintuple

verfuttern vt use up (for food *or* feed)

verfüttern vt **1** = VERFUTTERN. **2** (+an) feed (to). **3** overfeed

Verfutterung f using up (for food *or* feed)

Verfütterung f **1** = VERFUTTERUNG. **2** feeding. **3** overfeeding

Verfütterungs·versuch m feeding experiment

vergab awarded, etc: *past of* VERGEBEN

Vergabe f award(ing), allocation, assignment; subcontracting *cf* VERGEBEN **2**

vergällen vt **1** denature, methylate. **2** embitter. **Vergällung** f denaturing

Vergällungs·mittel n denaturant

vergalt repaid, etc: *past of* VERGELTEN

vergangen 1 gone (by): *pp of* VERGEHEN. **2** p.a past, bygone; last. **Vergangenheit** f (the) past

vergänglich a transitory; perishable, unstable. **Vergänglichkeit** f transitoriness; perishability, instability

vergärbar a fermentable; attenuable. **Vergärbarkeit** f fermentability

vergären* **1** vt, vi(s) ferment. **2** vt attenuate (wort). **3** vi(h) cease fermenting. **Vergärung** f (-en) fermentation; attenuation

Vergärungs·fähigkeit f fermentability. **-grad** m degree of fermentation (*or* attenuation). **-messer** m zymometer

vergasbar a gasifiable, vaporizable

vergasen vt **1** gasify, vaporize. **2** (incl Mil.) gas. **3** fumigate. **4** carburet

Vergaser m (-) gasifier, vaporizer; carburetor. **·graphit** m retort graphite. **-kraftstoff** m gasoline

vergaß forgot: *past of* VERGESSEN

Vergasung f (-en) gasification, vaporization; gassing; fumigation; carburetion

Vergasungs·gas n manufactured gas. **-mittel** n gasifying agent. **-öl** n oil for gasification

vergeben* **I** vt **1** give away. **2** award, allot, assign; place (an order); subcontract. **3** promise. **II** vt/i forgive. **III** p.a: **schon v. sein** have a previous appointment. **vergebens** adv in vain

vergeblich a futile, vain

Vergebung f (-en) giving away; award, allotment, assignment; placement (of an order), promise; forgiveness, pardon *cf* VERGEBEN

vergegenwärtigen vt **1** (d.i.o) bring home (to). **2: sich etwas v.** visualize sthg

vergehen*I vi(s) **1** pass, go by; elapse. **2** wear off, subside, expire; (+vor) die (of); (+dat): **es vergeht ihnen** they get over it, they lose it. **3** perish, vanish. **4** be drowned out. **II: sich v. 5** (+gegen) violate. **6** (+an) assault

vergeilen vi(s) (Bot.) etiolate

vergelben vi(s) turn yellow; etiolate

vergelten* vt (d.i.o) repay; pay back, reward (for). **Vergeltung** f (-en) repayment; reward

vergesellschaften 1 vt nationalize, put under public ownership; (Com.) incorporate. **2: sich v.** (Biol.) associate. **Vergesellschaftung** f (-en) nationalization, public ownership; incorporation; (Biol.) association

vergessen* (vergaß, vergessen; vergißt) vt; vi (+auf) forget (about). **Vergessenheit** f oblivion

vergeßlich a forgetful. **Vergeßlichkeit** f forgetfulness

vergeuden vt squander, waste

vergewaltigen vt do violence to; rape

vergewissern vr: **sich v.** (oft +genit or +über) make sure (of, about). **Vergewisserung** f (-en) assurance

vergibt awards, etc: *pres of* VERGEBEN

vergießbar a castable; pourable. **Vergießbarkeit** f castability; pourability

vergießen* vt **1** cast, pour (out). **2** spill, shed. **3** fill up

Vergieß·temperatur f casting (or pouring) temperature

Vergiftbarkeit f susceptibility to poisoning

vergiften vt poison; contaminate. **Vergiftung** f (-en) poisoning; contamination

Vergiftungs·erscheinung f symptom of poisoning

vergilben *vi* (s) (turn) yellow; etiolate. **Vergilbung** *f* yellowing

vergilt repays, etc: *pres of* VERGELTEN

vergipsen *vt* plaster (up)

vergißt forgets: *pres of* VERGESSEN

vergittern *vt* screen, cover with a grate (a lattice, bars). **Vergitterung** *f* (-en) 1 screening, grating, latticing. 2 screen; grate, lattice; bars

vergl. *abbv* (vergleiche) cf, compare. **Vergl.** *abbv* (Vergleich) comparison

verglasbar *a* vitrifiable

verglasen 1 *vt, vi*(s) & *sich* v. vitrify. 2 *vt* glaze; cover with glass, glass in. **verglast** *p.a* (*oft*) vitreous; glassy; glass-covered. **Verglasung** *f* (-en) 1 vitrification; glazing. 2 glaze. 3 glasswork

Vergleich *m* (-e) 1 comparison: **im V. mit** compared to. 2 compromise, settlement

vergleichbar *a* comparable. **Vergleichbarkeit** *f* comparability

vergleichen* 1 *vt* compare. **2: sich** v. reach a settlement, come to terms. **vergleichend** *p.a* (*oft*) comparative

vergleichlich *a* comparable

vergleich·los *a* incomparable, matchless.

V–prisma *n* comparison prism

Vergleichs· comparison, comparative, standard, reference: **-ausfärbung** *f* comparative dyeing. **-bestimmung** *f* standard determination. **-elektrode** *f* reference electrode. **-fähigkeit** *f* comparability. **-feld** *n* (*Photometry*) matching field. **-flüssigkeit** *f* comparison liquid. **-kraftstoff** *m* standard motor fuel. **-küvette** *f* comparison cuvette. **-linie** *f* standard (*or* comparison) line. **-lösung** *f* standard solution. **v–mäßig** *a* stipulated, agreed on; *adv* as agreed on. **V–maßstab** *m* standard of comparison. **-messung** *f* comparative measurement. **-pflanze** *f* control plant. **-präparat** *n* comparison preparation. **-substanz** *f* standard, comparison substance. **-test** *m* comparison test. **-tier** *n* control animal. **-verfahren** *n* 1 comparison method. 2 arbitration. **-versuch** *m* 1 comparative experiment; comparison test. 2 attempt at compromise. **-weg** *m*: **auf dem V.** by compromise. **v–weise** *adv* 1 comparatively. 2 by way of compromise. **V–wert** *m* comparative value. **-zahl** *f* comparative figure. **-zweck** *m*

purpose of comparison

Vergleichung *f* (-en) comparison, comparing

Vergleich·unterlage *f* basis for comparison

verglich(en) compared: *past* (& *pp*) *of* VERGLEICHEN

verglimmen* (verglomm, verglommen) *vi*(s) die down, fade away

Verglüh·brand *m* (*Ceram.*) biscuit baking

verglühen I *vt* 1 (*Ceram.*) biscuit-fire, bake. 2 anneal poorly. **II** *vi*(s) cease glowing, smolder out. **Verglühen** *n* biscuit-firing; faulty annealing. **Verglüh·ofen** *m* (*Ceram.*) biscuit kiln. **Verglühung** *f* (-en) biscuit-firing; faulty annealing

vergnügen 1 *vt* delight, amuse. **2: sich** v. have fun, enjoy oneself. **Vergnügen** *n* (-) pleasure, fun, enjoyment. **vergnüglich** *a* enjoyable, entertaining, **vergnügt** *p.a* happy, cheerful, enjoyable. **Vergnügung** *f* (-en) entertainment; party; (*pl oft*) fun

vergolden *vt* gild; gold-plate

Vergolder·masse *f* gilding. **-wachs** *n* gilder's wax

Vergoldung *f* (-en) gilding; gold-plating

Vergoldungs·wachs *n* gilder's wax. **-wasser** *n* quickening liquid

vergolten repaid, etc.: *pp of* VERGELTEN

vergönnen *vt* grant, let have

vergor(en) fermented: *past* (& *pp*) *of* VERGÄREN

vergoß, vergossen cast etc: *past* & *pp of* VERGIESSEN

vergr. *abbv* (vergrößert) magnified. **Vergr.** *abbv* (Vergrößerung) magnification

vergraben* 1 *vt* bury; cache. **2: sich** v. bury oneself; hide away; become entrenched

Vergrämungs·mittel *n* repellent

vergrauen *vi*(s) (turn) gray

vergreifen* *vr*: **sich** v. 1 make a mistake (*or* a wrong move), strike the wrong note (key, etc). 2 (+**an**) molest, assault. *Cf* VERGRIFFEN

Vergreisenung *f* (*Petrog.*) greisenization

vergriff(en) 1 made a mistake, etc: *past* (& *pp*) *of* VERGREIFEN. **2: vergriffen** *p.a* out of print

vergröbern 1 *vt* coarsen. **2: sich** v. grow coarser. **Vergröberung** *f* coarsening

vergrößern 1 *vt* enlarge, magnify. 2 *vt* &

sich v. increase. 3: **sich v.** grow (larger).
Vergrößerung f (-en) enlargement,
magnification; increase
Vergrößerungs-apparat m (Phot.) en-
larger. **-glas** n magnifying glass. **-kraft** f
magnifying power. **-linse** f magnifying
lens
vergrub buried: past of VERGRABEN
vergrünen vi(s) 1 turn green. 2 lose its
green color, fade. **vergrünlich** a (Dye.)
greenable
vergünstigt p.a reduced (in price). **Ver-
günstigung** f (-en) 1 (price, rate) reduction. 2 privilege
Verguß° m casting, pouring; spilling, shedding; filling up cf VERGIESSEN. **-harz** n
casting resin. **-masse** f (pourable) sealing
compound, filler
vergütbar a 1 improvable; (Metll.) heat-
treatable, temperable. 2 recompensable.
Vergütbarkeit f improvability; temperability, tempering quality
vergüten vt 1 make good; (+d.i.o) (re)pay
(or refund) sthg to sbdy, compensate sbdy
for sthg. 2 improve, (Metll.) heat-treat,
temper-harden, quench and temper;
(Opt.) coat, lumenize. **Vergüte-ofen** m
quenching and tempering furnace. **Ver-
güterei** f (Metll.) heat-treating plant.
Vergütung f (-en) refund(ing); compensation. 2 improvement; (Metll.) temper-
hardening, heat-treatment; (Opt.) coating. **Vergütungs-stahl** m heat-treated
(or heat-treatable) steel
Verh. abbv 1 (Verhalten) behavior. 2 (Verhältnis) ratio, proportion. 3 (Verhandlungen) negotiations
verhaften vt arrest, take into custody. **ver-
haftet** p.a (oft) (+dat or +mit) rooted (in)
verhagern vi(s) grow lean. **verhagert** p.a
lean; emaciated
verhaken 1 vt hook. 2: **sich v.** get hooked. 3
vt & **sich v.** (+ineinander) interlock
verhalf helped: past of VERHELFEN
verhallen vi(s) (Sound) die away—
ungehört v. go unheard
verhalten* I vt 1 hold back, suppress; retard (step). 2 hold shut. II: **sich v.** 3 behave, act; keep (still); assume (or maintain) an (e.g. skeptical) attitude; **die
Sache verhält sich so** the matter is like

this; **wie verhält es sich damit?** how
does the matter stand? 4 (+zu) be related
(to); (+wie, oft) stand in the ratio (of): **A
verhält sich zu B wie 3 zu 5** A is to B as 3
is to 5. III p.a suppressed, restrained, subdued; bated (breath); (adv oft) with restraint. **Verhalten** n 1 behavior; reaction. 2 retention (of urine). 3 suppression.
Verhaltenheit f restraint, reserve
Verhaltens-weise f mode of behavior.
-wissenschaften fpl behavioral sciences
Verhältnis n (-se) 1 ratio; proportion; rela-
tion(ship): **im V. zu** in proportion (or relation) to; **im V. stehen** be in proportion, be
related, be in a ratio. 2 (pl) conditions,
circumstances. **-anzeiger** m (Math.) exponent. **-formel** f empirical formula. **v-
gleich** a proportional. **V-gleichheit** f
proportion. **v-mäßig** a proportionate;
relative. **-widrig** a disproportionate. **V-
zahl** f coefficient, numerical ratio—**konstante V.** constant of proportionality
Verhaltung f (Med.) retention (of urine)
verhandeln vt/i 1 negotiate, debate. 2 conduct (a hearing, trial). **Verhandlung°** f
negotiation; (pl oft) proceedings
verhangen p.a covered; cloudy, overcast
verhängen vt 1 hang, drape; curtain off. 2
impose; decree, proclaim; inflict. Cf
VERHANGEN
Verhängnis n (-se) disaster, doom. **v-voll**
a disastrous, fatal
verharren vi 1 remain (in a position). 2
(+auf, bei) persist (in), adhere (to)
verhärten vt, vi(s) harden. **Verhärtung** f
(-en) hardening
verharzen 1 vt, vi(s) resinify. 2 vi(s) gum;
(X-ray Phot.) blur. **Verharzung** f (-en)
resinification; gumming; blurring
verhauen vt cut up, hack, prune; beat up
verhehlen vt conceal, camouflage; keep secret cf VERHOHLEN
verheilen vi (s) heal (up)
verheimlichen vt keep secret, conceal
verheiraten 1 vt give in marriage. 2: **sich
v.** (mit) get married (to), marry
verheißen* vt promise. **verheißungs-voll**
a promising
verhelfen* vi (+dat, +zu) help (sbdy) to get
(or achieve)
verhielt held back, etc: past of VERHALTEN

verhilft helps: *pres of* VERHELFEN

verhindern *vt* prevent, hinder; detain. **verhindert** *p.a* (*oft*) unable to be present (at a meeting, etc). **Verhinderung** *f* (-en) prevention, hindrance

Verhinderungs·fall *m:* **im V.** in case anything interferes. **-grund** *m* reason for not being present, previous engagement. **-mittel** *n* preventive

verhohlen *a* hidden; camouflaged; clandestine *cf* VERHEHLEN

verholfen helped: *pp of* VERHELFEN

verholzen *vi*(s) lignify, become woody. **verholzt** *p.a* (*oft*) woody. **Verholzung** *f* lignification

Verhör *n* (-e) hearing; interrogation—**ins V. nehmen** interrogate. **verhören** 1 *vt* question, interrogate. 2 *vt* & **sich v.** mishear, misunderstand

verhornen *vi*(s) cornify, keratinize, become calloused. **Verhornung** *f* (-en) cornification, keratinization, formation of calluses

verhüllen *vt* cover up; veil, mask, disguise. **verhüllend** *p.a* (*oft*) euphemistic. **Verhüllung** *f* (-en) covering; veiling; mask(ing), disguise

verhundertfachen *vt.* & **sich v.** increase a hundredfold

verhungern *vi*(s) starve. **Verhungerung** *f* starvation

verhunzen *vt* botch up, mess up, ruin

verhüten *vt* prevent, avert. **verhütend** *p.a* (*oft*) preventive, prophylactic

verhüttbar *a* (*Ores*) smeltable

verhütten *vt* (*Metll.*) smelt, work. **Verhüttung** *f* smelting

Verhütung *f* prevention, prophylaxis

Verhütungs·maßregel *f* preventive measure. **-mittel** *n* preventive; (*Med.*) prophylactic

verifizieren *vt* verify. **Verifizierung** *f* (-en) verification

verimpfen *vt* 1 transmit (by inoculation *or* contagion). 2 (*Med.*) inoculate. 3 (*Bot.*) graft. 4 (*Cryst.*) seed

verirren *vr:* **sich v.** stray, go astray. **verirrt** *p.a* (*oft*) lost; stray. **Verirrung** *f* (-en) aberration; error, mistake

verjagen *vt* chase away, dispel

verjähren *vi*(s) become invalid, lapse. **verjährt** *p.a.* invalid, outdated. **Verjährung** *f* (expiration due to the) statute of limitations

verjaucht *p.a.* putrefied, sanious. **Verjauchung** *f* putrefaction, sanies

verjüngen I *vt* 1 rejuvenate; regenerate. 2 restock. 3 reduce; constrict. II: **sich v.** 4 be(come) rejuvenated. 5 be reduced, diminish. 6 taper, narrow, constrict. **Verjüngung** *f* (-en) 1 rejuvenation, regeneration. 2 reduction, diminution. 3 tapering, narrowing, constriction

verk. *abbv* (verkürzt) abbrev. (abbreviated)

Verkabelung f (-en) (*Elec.*) wiring

verkadmen, verkadminieren *vt* cadmium-plate

verkalkbar *a* calcifiable

verkalken *vi*(s) 1 calcify; harden. 2 suffer from arteriosclerosis. 3 turn senile. **verkalkt** *p.a* (*oft*) sclerotic; senile. **Verkalkung** *f* (-en) calcification; hardening, sclerosis; senility

verkam degenerated: *past of* VERKOMMEN

verkannt(e) 1 misunderstood, etc: *pp* (& *past*) *of* VERKENNEN. 2: **verkannt** *p.a* (*oft*) unrecognized, unappreciated

verkanten *vt* tilt, cant, turn on its edge

verkappen *vt* mask, disguise

verkapseln 1 *vt* & **sich v.** encapsulate. 2: **sich v.** become encysted. **Verkapselung** f (-en) encapsulation, encystment

verkäsen *vi*(s) become cheesy (*or* caseous). **Verkäsung** *f* caseation

verkauen 1 *vt* chew up. 2: **sich v.** be chewed up

Verkauf *m* sale(s); selling: **verkaufen** *vt/i* & **sich v.** sell. **Verkäufer** *m* (-), **Verkäuferin** *f* (-nen) salesperson. **verkäuflich** *a* salable, marketable; for sale, available

verkaufs·fähig *a* salable, marketable. **V–kraft** *f* salesperson

Verkehr *m* 1 traffic. 2 transportation. 3 circulation: **in den V. bringen** put into circulation. 4 contact, relations, dealings. 5 intercourse (*incl* sexual). 6 business; trade, trading; commerce

verkehren I *vt* 1 reverse *cf* VERKEHRT. 2 (+**in**) convert (into). II *vi* (h,s) have contact (*or* relations), associate; do business; (+**bei**) visit regularly. III: **sich v.** (+**in**) turn, be converted (into)

Verkehrs·ampel *f* traffic light. **v–sicher** *a*
1 safe for commerce: **v–er Sprengstoff**
safety explosive. 2 roadworthy (car). **V–**
stau *m.* **-stauung** *f* traffic jam. **-zeichen**
n traffic sign

verkehrt *p.a* (*oft*) 1 (turned) backward(s),
upside-down. 2 wrong, (*adv oft*) the wrong
way. 3 opposite (direction). *See also* VER-
KEHREN. **Verkehrtheit** *f* (-en) wrongness;
folly

Verkehrung *f* (-en) reversal

verkennen* *vt* 1 misunderstand. 2 misjudge; underrate. 3 (*oft* + **daß**) fail to recognize (that), overlook (the fact that):
nicht zu v. not to be overlooked, not easily missed *cf* VERKANNT. **Verkennung** *f*
(-en) misjudgment; overlooking

Verkernung *f* (-en) change from sapwood
to heartwood

verketten 1 *vt* link up, interlink; form into
a chain. **2: sich v.** be (inter)linked. **Ver-**
kettung *f* (-en) linkage, (inter)linking,
chain formation. **Verkettungs·fähig-**
keit *f* linking (*or* chain-forming) capacity

verkienen *vt* saturate with resin

verkieseln *vt* silicify. **Verkieselung** *f*
silicification

verkitten *vt* cement, lute, seal. **Verkittung**
f (-en) cementing (together); sealing

Verkittungs·fähigkeit *f* cementing properly. **-mittel** *n* cement, binder, adhesive;
(*Coal*) agglutinant

verkl. *abbv* (verkleinert) reduced

verklammern I *vt* 1 clamp (together). 2
(*Med.*) clip together. 3 (*Print.*) brace.
II: sich v. interlock; cling (together)

verklang died out, etc: *past of* VERKLINGEN

verkleben I *vt* 1 paste over (up, down); seal.
2 gum up. 3 use up in pasting. **II** *vi* 4 get
sticky. 5 agglutinate; coagulate, clot. 6
(*Med.*) form adhesions. **verklebt** *p.a* (*oft*)
pasty, sticky, gummy, closed, sealed. **Ver-**
klebung *f* (-en) pasting; sealing; gumming up; agglutination, coagulation,
clotting; (*Med.*) adhesion. **Ver-**
klebungs·stoff *m* agglutinative substance

verkleiden *vt* 1 face; line; (en)case. 2 disguise, mask

verkleinern I *vt* 1 make smaller, scale

down, reduce. 2 belittle, disparage.
II: sich v. 3 get smaller, diminish,
shrink. 4 retrench. **Verkleinerung** *f*
(-en) reduction, diminution; belittlement,
disparagement; retrenchment

verkleistern *vt* paste up; clog; make pasty;
turn into a paste. **verkleistert** *p.a* pasty.
Verkleisterung *f* (-en) pasting up, clog-
ging; conversion into a paste; glutiniza-
tion, gelatinization

verklemmen *vr:sich v.* jam, get stuck. **ver-**
klemmt *p.a* (*oft*) tense; inhibited

verklingen* *vi*(s) fade away, die out; wear
off; (*Waves, oft*) decay

verklumpen *vi* cake, form lumps

verklungen 1 died out, etc: *pp of* VER-
KLINGEN. 2 *p.a* bygone

verknallen 1 *vt, vi*(s) detonate. 2 *vt* fire off
(one's ammunition)

verknappen 1 *vt* cut short, shorten. 2 *vi*(s)
& sich v. become scarce. **Verknappung** f
(-en) shortage, scarcity

verkneten *vt* knead (together)

verknistern *vt* decrepitate. **Verkniste-**
rung *f* decrepitation. **Verkniste-**
rungs·wasser *n* water of decrepitation

verknittern *vt, vi*(s) crease, crumple

verknöchern *vi*(s) ossify, grow rigid. **Ver-**
knöcherung *f* ossification

verknorpeln *vi*(s) become cartilaginous.
verknorpelt *p.a* cartilaginous, gristly.
Verknorpelung *f* chondrification

verknoten 1 *vt* knot, tie up. **2: sich v.** get
knotted

verknüpfbar *a* combinable, linkable

verknüpfen 1 *vt* tie (up *or* together).2 *vt* &
sich v. combine; link (up), connect; associate. 3: **sich v.** knot up. **verknüpft** *p.a*
(*oft*) involved. **Verknüpfung** *f* (-en) combination; linkage, connection, association

verkobalten *vt* plate with cobalt. **Ver-**
kobaltung *f* cobalt plating

verkochbar *a* (*Varnish*) compatible with
drying on boiling

verkochen I *vt* 1 boil down, concentrate. 2
(*Sugar*) pan-boil. 3 (*Oils*) heat-treat. **II** *vt,*
vi(s) overboil, spoil by boiling

verkohlen *vt, vi*(s) char, carbonize. **Ver-**
kohlung *f* charring, carbonization

verkokbar *a* (*Coal*) cooking. **verkoken** *vt*

coke.**Verkokung** f coking
Verkokungs·endtemperatur f final coking temperature. **v–fähig** a capable of coking. **V–grad** m (*Coal*) carbonization index. **-kammer** f coking chamber. **-ofen** m coke oven. **-probe** f coking test (*or* sample). **-vorgang** m coking process. **-zeit** f coking time

verkommen* vi(s) break down; deteriorate, degenerate. **Verkommenheit** f deterioration, degenerateness

verkoppeln vt couple (up), combine, join

verkorken vt 1 cork. 2 (*Bot.*) suberize. **Verkorkung** f (-en) 1 corking. 2 suberization

verkörnen vt granulate

verkörpern 1 vt embody, personify. **2:** sich v. be embodied, be personified. **Verkörperung** f (-en) embodiment, personification

verkracken vt (*Petrol.*) crack. **Verkrackung** f cracking

verkraften vt 1 manage, cope with; stand. 2 (*Mach., Mil.*) motorize; mechanize

verkrampft p.a 1 cramped. 2 tense, inhibited

verkreiden vt calcify. **Verkreidung** f calcification

verkrümmen vt, vi(s) & sich v. bend, curve, warp. **verkrümmt** p.a (*oft*) crooked. **Verkrümmung** f curvature; crookedness; warp(ing)

verkrüppelt p.a crippled, disabled

verkrusten vi(s) form a crust, become incrusted. **Verkrustung** f (-en) incrustation

verkühlen: sich v. catch cold

verkümmern vi(s) become stunted; waste (*or* wither) away, atrophy; be worn down. **verkümmert** p.a stunted; atrophied; worn down. **Verkümmerung** f stuntedness; atrophy; deterioration

verkünden, verkündigen vt announce, proclaim; predict, forebode. **Verkünd(ig)ung** f (-en) announcement, proclamation; prophecy

verküpbar a (*Dye.*) vattable. **Verküpbarkeit** f vatting property

verküpen vt (*Dye.*) vat, reduce for dyeing

verkupfern vt copper-plate, coat with copper. **Verkupferung** f coppering, copperplating

verkuppeln vt couple

Verküpung f (*Dye.*) vatting, reduction for dyeing. **Verküpungs·dauer** f vatting time

verkürzen 1 vt shorten, curtail, cut short; abbreviate; reduce. 2 vt & sich v. (*Muscles*) contract, retract. **3: sich v.** be shortened, be reduced. **verkürzt** p.a **1:v. arbeiten** work short(er) hours. **2** adv for short, in shortened form. **Verkürzung** f (-en) shortening, curtailment; abbreviation; reduction; contraction, retraction

verl. abbv (verlängert) extended. **Verl.** abbv 1 (Verlag, Verleger) publisher. 2 (Verlängerung) extension

verlacken vt 1 vanish, lacquer. 2 (*Dye.*) convert into a lake

Verlackungs·verfahren n lacking process (*or* procedure). **-vorschrift** f lacking recipe

verladen* vt load (onto a vehicle); ship

Verlag m (-e) publisher, publishing house

verlagern 1 vt & sich v. move, shift. 2 vt displace. **3: sich v.** be displaced. **Verlagerung** f (-en) shift(ing), displacement

Verlags·buchhändler m publisher and bookseller. **-buchhandlung** f publishing and bookselling company. **-recht** n 1 publishing laws. 2 copyright

verlangen 1 vt demand; require; ask for. 2 vi (+**nach**) ask (for); long (for). **Verlangen** n 1 desire, longing. 2 demand

verlängern I vt 1 lengthen, prolong, extend. 2 (*Math.*) produce (a line). 3 renew (license, etc.). 4 dilute. 5 replenish (dye liquor). 6 stretch (food) II: **sich v.** be lengthened (prolonged, extended, renewed, stretched). **Verlängerung** f (-en) lengthening, extension; dilution; renewal; (*Math.*) production

Verlängerungsschnur f (*Elec.*) extension cord. **-stück** n extension piece

verlangsamen 1 vt & sich v. slow down, decelerate. 2 vt retard. **3: sich v.** be retarded. **Verlangsamung** f slowdown, deceleration, retardation

verlas misread, etc: *past* of VERLESEN

Verlaß m reliability: **darauf ist kein V.** that can not be relied on

verlassen * I vt 1 leave (*esp* a place); abandon, desert. II: **sich v.** (**auf**) rely, depend (on). III p.a (*oft*) forsaken. **Verlassenheit**

f forlornness, desertedness

verläßlich *a* reliable, dependable. **Verläßlichkeit** *f* reliability, dependability

Verlauf° *m* 1 course, run. 2 (*esp Paint*) flow. 3 trend. 4 outcome. **verlaufen* I** *vi*(s) 1 run, proceed, go; take its course. 2 (*Time*) pass, elapse. 3 end (up), come out. 4 melt; (*Colors*) run; (*Paint*) flow. **II: sich v.** 5 lose one's way. 6 scatter, disperse. **III** *vi*(s) & **sich v.** peter out, disappear, get lost. **IV** *p.a* (*oft*) lost, stray

verlautbaren *vt* 1 announce, make known: **v. lassen** let it be known. 2 express. **Verlautbarung** f (-en) announcement.

verlauten *vi* be reported: **wie amtlich verlautet** according to official reports; **v. lassen** let it be known

verleben *vt* spend (time); live on (money). **verlebt** *p.a* (*usu*) played out, dissipated, decrepit

verlegen A. I *vt* 1 mislay, misplace. 2 move, shift; relocate. 3 postpone, put off. 4 (*usu* + *dat rflx*) block, obstruct. 5 install, lay, put down (pipes, floors). 6 place (an event in time or a location). 7 publish. 8 advance (money). **II: sich v. (auf)** shift (one's interest), change (to); resort (to). **B. III** *a* embarrassed; (+*um*) at a loss (for), short (of). **Verlegenheit** *f* (-en) embarrassment; predicament

Verleger *m* (-) publisher *cf* VERLAG

Verlegung *f* (-en) 1 shift, transfer, relocation. 2 postponement. 3 installation. *Cf* VERLEGEN I

Verleih *m* (-e) 1 lending, renting (for hire), distribution. 2 distributor, renting (lending, for-hire) service. **verleihen*** *vt* (+*d.i.o*) 1 lend (to). 2 rent out (to). 3 bestow, confer (on), award (to). 4 give, impart (to). **Verleiher** *m* (-) lender; distributor. **Verleihung** *f* lending, renting, distribution; bestowing, conferring, awarding

verleimen *vt* cement, glue

verleiten *vt* lead; mislead, induce; entice: **sich v. lassen** be (easily) misled (induced, etc)

verlernen *vt* unlearn, forget; (+*inf as noun*) forget how to

verlesen* I *vt* 1 read (aloud) (publicly). 2

sort (out), pick (over). **II: sich v.** misread sth

verletzbar *a* vulnerable, easily injured; touchy

verletzen 1 *vt* & **sich v.** e.g.: **sich den Arm** (or **sich am Arm**) **v.** injure (or hurt) one's arm. 2 *vt* injure; insult, offend; violate. **verletzlich** *a* vulnerable, easily injured. **Verletzung** *f* (-en) injury; insult; violation

verleugnen *vt* deny; disavow; betray

verlief ran, etc: *past of* VERLAUFEN

verlieh(en) imparted, etc: *past* (& *pp*) *of* VERLEIHEN

verlieren* (verlor, verloren) 1 *vt/i* lose: **an Farbe v.** lose color. 2: **sich v.** be (or get) lost; lose one's way; fade out, disappear. *Cf* VERLOREN, VERLORENGEHEN

verließ left, etc: *past of* VERLASSEN

verlischt goes out: *pres of* VERLÖSCHEN II

verlocken 1 *vt* entice; tempt. 2 *vi* be enticing

verlohnen *vi* & **sich v.** be worthwhile, pay off

verlor(en) 1 lost: *past* (& *pp*) *of* VERLIEREN. 2: **verloren A** *p.a* lost, forlorn; lost: **v. geben** give up for lost; **v—er Schuß** random shot; **v—er Kopf** (*Metll.*) feedhead, top end. **B** *sep pfx*: **verloren-gehen** *vi*(s) be lost, get lost; fade away, disappear

verloschen 1 gone out: *pp of* VERLÖSCHEN. 2 *p.a* (*oft*) dead

verlöschen I *vt* (*reg*) 1 put out, extinguish. 2 (*Lime*) overburn. **II** *vi**(s) go out, be extinguished; fade, die (out), disappear

verlösen *vt* dissolve. **Verlösung°** *f* dissolution

verlöten *vt* solder (together)

verlud loaded, etc: *past of* VERLADEN

verlüften *vt* air, ventilate

Verlust *m* (-e) loss: **in V. geraten** be lost. **v—-arm** *a* low-loss. **V—faktor** *m* (*Elec.*) loss factor

verlustig *adv* & *prep* (+*genit*) deprived (of): **v. gehen, v. werden** lose

Verlust·leistung *f* (*Elec.*) power loss. **v—los** *a* lossless, nondissipative; (*adv oft*) without loss (or waste). **V—quelle** *f* source of loss. **-winkel** *m* (*Elec.*) phase angle (difference)

verm. *abbv* **1** (vermehrt) increased. **2** (vermindert) decreased

Vermächtnis *n* (-se) **1** legacy, bequest. **2** will

vermag is able to, can: *pres of* VERMÖGEN

vermahlen *vt* **1** grind, mill. **2** misgrind

vermählen *vt & sich v.* marry

vermanteln *vt* cover, sheathe; case, jacket

vermarkten *vt* market; put on the market. **Vermarktung** *f* marketing; sale (on the market)

vermaschen *vt* mesh, link

vermaß measured, etc: *past of* VERMESSEN

vermehren *vt & sich v.* increase; enlarge; multiply, propagate, proliferate. **Vermehrung** *f* (-en) increase; enlargement; multiplication, propagation, proliferation

vermeidbar *a* avoidable

vermeiden* *vt* avoid. **Vermeidung** *f* avoidance

vermeinen *vt/i* suppose, imagine. **vermeintlich** *a* supposed, imagined; (*adv. oft*) supposedly, presumably

vermelden *vt* report, announce

vermengen *vt & sich v.* mix (up), blend. **Vermengung** *f* (-en) mixing, mixture, blend(ing)

Vermerk *n* (-e) note, notation; memo; entry (on books, etc). **vermerken** *vt* note, take note of; take (*or* mark) down—**übel v.** take amiss, take offense at

vermessen* I *vt* **1** measure; survey; (*Space*) track. **II: sich v. 2** mismeasure. **3** (+**zu** + *inf*) venture, presume (to). **III** *p.a* presumptuous, bold. **Vermessenheit** *f* presumptuousness. **Vermesser** *m* (-) surveyor

vermessingen *vt* brass(-plate). **Vermessingung** *f* (-en) brassing, brassplating

Vermessung° *f* **1** measurement; surveying; (*space*) tracking. **2** (*Ships*) tonnage measurement

vermied(en) avoided: *past* (*& pp*) *of* VERMEIDEN

vermieten *vt* rent, hire (out): **zu v.** for rent

vermilchen *vt* emulsify

Vermillon *n* vermilion

vermindern *vt & sich v.* lessen, diminish,

decrease, reduce. **Verminderung** *f* (-en) lessening, diminution, decrease, reduction

vermischbar *a* miscible

vermischen *vt & sich v.* mix, blend. **vermischt** *p.a* (*oft*) miscellaneous. **Vermischung°** *f* mixing, mixture; blending

vermissen *vt* miss, be unable to find

vermißt I missed: *pp of* VERMISSEN; (*as p.a oft*) missing. **II** *pres of* VERMISSEN (misses) **&** VERMESSEN (measures, etc)

vermitteln I *vt* **1** (*d.i.o*) get, procure (sthg for sbdy). **2** (+**an**) place (sbdy) as an employee (with). **3** arrange, bring about. **4** provide; impart. **II** *vt/i* mediate; *vi also* intercede. **vermittelnd** *p.adv* (*oft*) as a mediator

vermittels(t) *prep* (+*genit*) by means (of)

Vermittler *m* (-) mediator. **Vermittlung** *f* (-en) **1** procurement. **2** (job) placement. **3** arrangement. **4** provision (of services); imparting (of information). **5** mediation; intercession. **6** agency, referral service. **7** service(s) (as a mediator, etc). **8** (*Tlph.*) exchange. **Vermittlungs-stelle** *f* referral (placement, procurement) office

vermocht(e) been able to (*pp*), (could: *past*) *of* VERMÖGEN

vermodern *vi*(s) rot, molder, decay. **vermodert** *p.a.* rotten, moldy, decayed. **Vermoderung** *f* rot(ting), moldering, decay

vermöge *prep* (+*genit*) by virtue, because (of)

vermögen* (*pp* vermocht) **1** *v aux* (+**zu** + *inf*) be able (to); (*pres oft*) can; (*past & sbjc usu*) could. **2** *vt* be able to do, manage. **Vermögen** *n* (-) **I** (*also oft in compds*) ability, power, capacity: **nach bestem V.** to the best of one's ability; **Leit-vermögen** conducting power, conductivity. **II** fortune (=wealth); assets, capital; property. **vermögend** *p.a* propertied; wealthy

vermuten *vt/i* presume, suppose, guess, conjecture; suspect. **vermutlich** *a* presumable; supposed; suspected; (*adv oft*) one supposes, we suppose, etc. **Vermutung** *f* (-en) presumption, supposition, guess, conjecture; suspicion; expectation

vernachlässigbar *a* negligible

vernachlässigen *vt* neglect; disregard.
Vernachlässigung *f* (-en) neglect; negligence

vernähen *vt* sew up, stitch; suture; fasten

vernahm perceived, etc: *past of* VERNEHMEN

vernarben *vi*(s) form a scar, cicatrize. **Vernarbungs-gewebe** *n* scar tissue

vernässen *vt* waterlog

vernebeln *vt* 1 atomize, convert into a mist (*or* fog). 2 (be)fog, cloud; screen. **Vernebler** *m* (-) atomizer. **Verneblung** *f* (-en) atomization; fogging; screening; smokescreen. **Verneblungs-apparat** *m* 1 atomizer. 2 (*Mil.*) smoke generator

vernehmen* *vt* 1 perceive, make out. 2 hear, find out. 3 question, interrogate. 4: **v. lassen, daß** intimate that . . . **Vernehmen** *n* 1 perception. 2 report: **dem V. nach** according to reports. 3: **sich ins. V. setzen mit** get in touch with

vernehmlich *a* perceptible; audible; distinct

verneinen *vt* 1 answer in the negative; deny. 2 reject. 3 contradict. **verneinend** *p.a* negative

vernetzen *vt* interlace, net(work), hook up; cross-link. **Vernetzer** *m* (-) (*Polymers*) cross-linking agent. **Vernetzung** *f* (-en) 1 (*incl Polymers*) cross-linkage. 2 network(ing)

Vernetzungs-dichte *f* cross-linking density. **-grad** *m* degree of cross-linking. **-kleber** *m* cross-linking adhesive. **-mittel** *n* cross-linking agent

vernichten *vt* destroy, annihilate. **vernichtend** *p.a* (*oft*) destructive, ruinous. **Vernichtung** *f* (-en) destruction, annihilation. **Vernichtungs-bombe** *f* demolition bomb

vernickeln *vt* nickel; nickel-plate. **Vernickelung** *f* (-en) nickeling; nickel-plating

vernieten *vt* rivet

vernimmt perceives, etc: *pres of* VERNEHMEN

vernommen perceived, etc: *pp of* VERNEHMEN

Vernunft *f* reason; good sense; intellect, intellectual capacity. **vernünftig** *a* reason-

able, rational; sensible

veröden I *vt* 1 lay waste, obliterate. 2 (*Med.*) deaden (nerves, etc). II *vi*(s) 3 become barren (*or* desolate). 4 atrophy, die off. **verödet** *p.a* desolate, deserted

veröff. *abbv* (veröffentlicht) published. **Veröff.** *abbv* (Veröffentlichung) publication

veröffentlichen *vt* publish, make public. **Veröffentlichung** *f* (-en) publication; announcement

verölen 1 *vt* oil. 2 *vi* (s) get oily

Veroneser· (of) Verona: **-erde** *f* Verona earth

verordnen *vt* 1 (*Med.*) prescribe. 2 order, decree. **Verordnung°** *f* 1 prescription. 2 order, decree; regulation, directive

verpacken *vt* pack (up), package; wrap up. **Verpack·flasche** *f* packing bottle. **Verpackung** *f* (-en) packing, wrapping; package, wrapper

verpassen *vt* miss, let slip (by)

verpechen *vt* treat with pitch

verpesten *vt* pollute, contaminate; poison

verpflanzen *vt* (*Bot. & Med.*) transplant. **Verpflanzung°** *f* transplantation; parallel displacement

verpflegen *vt* 1 supply with food, provision; feed. 2 (*Mach.*) maintain, service. **Verpflegung** *f* 1 feeding, provisioning. 2 food supply, provisions. 3 (*Mach.*) maintenance, service

verpflichten 1 *vt* obligate, bind. 2: **sich v.** commit (*or* obligate) oneself. **Verpflichtung** *f* (-en) obligation; commitment

verpfuschen *vt* bungle, botch up

verpichen *vt* coat (seal, treat) with pitch. **Verpichung** *f* coating (sealing, treatment) with pitch; pitch formation

verpillen *vt* pelletize

verplanen I *vt* 1 budget. 2 plan, schedule. 3 cover with a tarpaulin. II: **sich v.** misplan, miscalculate

verplatinieren *vt* platinize. **Verplatinierung** *f* platinization

verpönt *a* forbidden, taboo

verpuddeln *vt* puddle

verpuffen *vi*(s) 1 be expelled as exhaust. 2 deflagrate. 3 explode, detonate, fulminate. 4 go up in smoke, fizzle out; fall flat.

Verpuffung f (-en) 1 blowout; explosion, detonation. 2 deflagration. 3 fizzling out
Verpuffungs-apparat m explosion apparatus. **-motor** m internal combustion engine. **-probe** f deflagration test. **-röhre** f explosion test
Verputz m 1 plaster (work) (on walls, etc); roughcast. 2 finishing coat. 3 (Metll.) dressing. **verputzen** vt 1 plaster, roughcast. 2 finish. 3 dress
verqualmt p.a smoke-filled
verquellen* vi(s) swell (up); warp; flow away
verquer a & adv 1 strange; crooked; crosswise. 2: **v. kommen** come at the wrong time; **v. gehen** go wrong; **v. nehmen** take offense at
verquicken vt amalgamate
verquillt swells: pres of VERQUELLEN
verquoll(en) swelled, (swollen) past (& pp) of VERQUELLEN
verrann ran off, etc: past of VERRINNEN
Verrat m betrayal; disclosure; treason. **verraten*** vt betray; divulge, reveal; show
verrauchen 1 vt smoke up; fill with smoke. 2 vi(s) evaporate, escape (esp as smoke or fumes); blow over. **verraucht** p.a smoke-filled
verräuchern vt fill (or blacken) with smoke. **verräuchert** p.a smoke-filled; smoke-blackened
verrechnen I vt 1 reckon up, settle up; (+**mit**) balance (against). 2 (+d.i.o) credit (to sbdy's acct). II: **sich v.** miscalculate, make a mistake in figuring
verregnet p.a rainy, wet, rain-soaked
verreiben* vt 1 grind fine, triturate. 2 rub on (or in); rub out (spots)
verreist p.a away (usu. on business), traveling
verrenken 1 vt twist, contort 2 (+dat rflx): **sich ein Glied v.** twist (sprain, dislocate) a limb
verrichten vt carry out, do, perform cf NOTDURFT. **Verrichtung°**f 1 performance. 2 chore, task; duty; function
verrieb(en) ground, etc: past (& pp) of VERREIBEN
verriegeln vt bolt, bar, lock

verrieseln 1 vt irrigate. 2 vi(s) trickle away
verriet betrayed: past of VERRATEN
verringern vt & **sich v.** diminish, decrease, reduce, lessen. **Verringerung** f (-en) diminution, decrease, reduction, lessening
verrinnen* vi(s) 1 run (or trickle) off. 2 peter out, disappear. 3 (Time) pass, go by
Verrohrung f (-en) (well) casing; (shaft) tubing
verrosten vi(s) rust. **verrostet** p.a rusty. **Verrostung** f rusting
verrottbar a (bio)degradable
verrotten vi(s) rot, decay; be (bio)degradable; go to ruin. **verrottet** p.a (oft) rotten, ruined. **Verrottung** f rotting, decay; (bio)degradation
verrucht a infamous, nefarious
verrücken 1 vt & **sich v.** move (or shift) out of place. 2 vt displace; derange. 3: **sich v.** be displaced. **verrückt** p.a crazy, mad. **Verrückung** f (-en) shift, displacement; (Geol.) slip, derangement
verrühren vt stir, mix; (Copper) flap
verrußen vt, vi(s) soot up. **verrussen** vt russianize. **verrußt** p.a I sooty. II russianized
Vers. abbv 1 (Versuch) experiment. 2 (Versammlung) meeting
versacken vi(s) sink; give way; get stuck
versagen 1 vt (d.i.o) deny (sbdy sthg); (+dat rflx): **sich etwas v.** deny oneself sthg. 2 vi (oft + dat) fail, give out, break down; misfire cf. DIENST. 3: **sich v.** make an appointment. **Versagen** n failure, breakdown. **Versager** m (-) failure, dud; breakdown; misfire. **versagt** p.a:**v. sein** have a previous appointment
versah provided, etc: past of VERSEHEN
versalzen vt oversalt; mess up, spoil
versammeln vt & **sich v.** gather, assemble. **Versammlung°**f gathering, assembly; meeting, convention
Versand m shipping; mailing; exportation. **--abteilung** f shipping (or mailing) department. **-bier** n export beer
versanden vi(s) 1 silt up, fill up with sand. 2 blow over
versand·fähig a fit for shipment. **V–faß** n shipping keg (or barrel). **-geschäft** n ex-

port (*or* mail-order) business. **-kiste** *f* shipping case (*or* crate). **-schachtel** *f* shipping carton

versandt(e) mailed: *pp* (*& past*) *of* VERSENDEN

versank sank: *past of* VERSINKEN

Versatz *m* 1 displacement, shift. 2 misalignment. 3 (*Min.*) stowing, packing; gobbing. 4 (*Ceram.*) batch. 5 (*Lthr.*) layer, lay-away 6 pawning. *Cf* VERSETZEN. **--grube** *f* (*Lthr.*) lay-away vat

versauern *vi*(s) 1 turn sour. 2 go to seed

versäuern *vt* acidify, make sour. **Versäuerung** *f* acidification, souring

versäumen *vt* 1 miss; lose, waste (time). 2 fail to do, leave undone

verschachtelt *p.a* nested; interlocked; interlaced

verschaffen *vt* (*d.i.o*) procure, get (for sbdy); **sich etwas v.** get oneself sthg

verschalen *vt* board (up); encase, jacket, lag

verschärfen *vt & sich v.* sharpen, intensify; *vt* (*oft*) aggravate. **Verschärfung** *f* (-en) sharpening, intensification; aggravation

verschäumen *vi* cease foaming, foam away

verscheiden* *vi*(s) die, expire

verschenken *vt* 1 give away as a gift. 2 pour, serve (drinks)

verschicken *vt* 1 send out, ship; mail. 2 deport. **Verschickung** *f* (-en) shipping, mailing; deportation

verschiebbar *a* 1 sliding, adjustable. 2 movable, displaceable

verschieben* I *vt* 1 move, shift; switch; displace. 2 postpone, defer. 3 sell illicitly. **II: sich v.** move out of place, shift. 5 be postponed. **Verschiebung** *f* (-en) 1 shift, switching; (*inc Geol.*) displacement. 2 postponement, deferment. 3 (*Elec.*) lag. 4 illicit (*or* black-market) sale

Verschiebungs- displacement: -gesetz *n* displacement law. **-polarisation** *f* induced (*or* distorted) polarization. **-satz** *m* displacement principle. **-strom** *m* displacement current

verschied died: *past of* VERSCHEIDEN

verschieden I *a* different; varied, various

cf VERSCHIEDENES. **II** *p.a* deceased = *pp of* VERSCHEIDEN

verschieden-artig *a* various, varied; diverse; disparate; heterogeneous

verschiedenerlei 1 *a* (*no endgs*) various different kinds of, a variety of. 2 *pron* a variety of things

Verschiedenes *n* (*adj endgs*) (*esp as heading*) miscellaneous

verschieden-fach *a* a variety of; *adv oft* in various ways. **-farbig** *a* of different colors, many-colored. **-gestaltig** *a* of various shapes, heteromorphic, polymorphic

Verschiedenheit *f* difference; variety, diversity

verschiedenste *a* (*superl of* VERSCHIEDEN I) most varied; the greatest variety of

verschiedentlich *adv* at various times; repeatedly, several times

verschiedenwertig *a* of different (*or* various) valences

verschießen* I *vt* 1 shoot, fire (off) (ammunition). 2 shade off (colors). **II** *vi*(s) fade. **III: sich v.** miss the target

verschiffen *vt* ship, export. **Verschiffung** *f* (-en) shipping, shipment

verschimmeln *vi*(s) mold, get moldy. **verschimmelt** *p.a* moldy

verschlacken 1 *vt* scorify. 2 *vi*(s) to be reduced to slag (*or* scoria). **Verschlackung** *f* scorification; slagging

Verschlackungs-beständigkeit *f* resistance to slagging. **v-fähig** *a* slaggable; fluxable. **V-fähigkeit** *f* slaggability; fluxing property. **-probe** *f* scorification assay. **-verhältnis** *n* slagging ratio. **-vermögen** *n* slaggability, fluxing power. **-zustand** *m* state of scorification

verschlafen* I *vt* sleep off; sleep through; oversleep. **II** *p.a* sleepy, sleepy-eyed

Verschlag° *m* shed; shack; bin, compartment; crate

verschlagen* I *vt* 1 board up, partition off. 2 line. 3 (*esp Cooking*) blend, mix. 4 drive off course; cast, send (into a place). 5 take (e.g. one's breath) away. **II** *vi* 6 (*dat*) avail, be of use (to). **III: sich v.** 7 (*dat*) fail, break down. **IV** *p.a* 8 lukewarm, tepid. 9 cunning, wily. **Verschlagenheit**

f cunning, shrewdness

verschlammen *vi*(s), **verschlämmen** *vt* silt up, clog with mud

verschlang swallowed, etc: *past of* VER-SCHLINGEN

verschlechtern 1 *vt* & **sich v.** worsen. **2** *vt* aggravate, impair. **3: sich v.** deteriorate. **Verschlechterung** *f* worsening; aggravation; impairment; deterioration

verschleiern *vt* **1** veil, cover up, conceal. **2** cloud. **verschleiert** *p.a* (*oft*) cloudy, hazy

verschleifen* *vt* grind (down *or* off); (*Soap*) fit

verschleimen *vt, vi*(s) choke up with mucus; (*Guns*) foul. **verschleimt** *p.a* (mucus-)clogged, stuffy. **Verschleimung** *f* choking up (with mucus); stuffiness; (*Guns*) fouling

Verschleiß *m* **1** wear (and tear), deterioration; (*Com.*) depreciation—**eingeplanter V.** planned obsolescence. **2** retail sale. **~angriff** *m* abrasion, wear

verschleißen* 1 *vt, vi*(s) wear out, deteriorate; erode. **2** sell retail

verschleiß·fest *a* wear-resistant. **V–festigkeit** *f* wear resistance, wearing quality. **-fläche** *f* (*Mach.*) wearing surface. **v–frei** *a* wear-free, without wear. **V–teil** *n* part subject to wear

verschleppen *vt* **1** carry off; abduct; deport. **2** protract, delay. **3** spread (disease)

verschlief overslept, etc: *past of* VER-SCHLAFEN

verschliff(en) ground down: *past* (& *pp*) *of* VERSCHLEIFEN

verschließbar *a* closable, lockable

verschließen* I *vt* **1** close (up), lock up; stopper; seal, lute; occlude. **2** lock away. **3** hide, conceal. **II: sich v. 4** lock oneself (in a place). **5** become withdrawn; (+*dat or* +*vor*) close one's mind (to), ignore. *Cf* VERSCHLOSSEN. **Verschließung** *f* closing up, locking, stoppering, etc

verschlimmern *vt* & **sich v.** worsen = VERSCHLECHTERN

verschlingen* I *vt* **1** intertwine, entangle. **2** swallow, devour. **II: sich v.** intertwine, be(come) entangled. *Cf* VERSCHLUNGEN

verschliß, verschlissen wore out, worn out: *past* & *pp of* VERSCHLEISSEN

verschloß closed up, etc: *past of* VER-SCHLIESSEN

verschlossen 1 closed up, etc: *pp of* VER-SCHLIESSEN. **2** *p.a* withdrawn, aloof

verschlucken 1 *vt* swallow (up *or* down); suppress. **2: sich v.** choke, get food in one's windpipe

verschlug boarded up, etc: *past of* VER-SCHLAGEN

verschlungen 1 swallowed, etc: *pp of* VER-SCHLINGEN. **2** *p.a* entangled; winding, convoluted, tortuous

Verschluß° *m* **I** (= VERSCHLIESSUNG) closing, locking, etc *cf* VERSCHLIESSEN. **II 1** closure; lock: **unter V.** under lock and key. **2** clasp, catch, snap; fastener. **3** stopper, plug. **4** seal, bond. **5** cover, lid, cap. **6** (*Med.*) occlusion; thrombus

verschlüsseln *vt* encode, encipher

verschmälern *vt* & **sich v.** narrow (down); thin (out); attenuate

Verschmauchung *f* (*Ceram.*) smoking, spoiling by condensation while preheating

verschmelzen* 1 *vt* smelt; alloy; solder; synthesize. **2** *vt, vi*(s) melt (together), fuse, blend, merge. **3** *vi*(s) coalesce. **Verschmelzung** *f* (s)melting; alloy(ing); soldering; fusing, blending, merger; coalescence; (*Glass*): **innere V.** internal seal

Verschmelzungs·körper *m* alloy constituent. **-wärme** *f* heat of fusion

verschmiedbar *a* forgeable

verschmieren I *vt* **1** smear, smudge; stain, soil; mess up. **2** daub, scribble all over. **3** fill up, plaster up (cracks), lute. **II** *vi*(s) clog up; fog, blur, glaze over

verschmilzt melts, etc: *pres of* VER-SCHMELZEN

verschmolz(en) melted, (molten): *past* (& *pp*) *of* VERSCHMELZEN

verschmutzen I *vt* dirty, soil, (be)foul; pollute, contaminate. **2** *vi*(s) get dirty (soiled, etc). **Verschmutzung** *f* soil(ing), (be)fouling; pollution, contamination

verschneiden* *vt* **1** cut (to size), trim, tailor; clip, prune. **2** cut up, cut the wrong way. **3** castrate, geld. **4** dilute, reduce. **5** blend, mix; cut, adulterate

verschneit *p.a* snowbound; snow-covered

verschnitt clipped, etc: *past of* VER-SCHNEIDEN

Verschnitt *m* (-e) 1 cutting, trimming, clipping etc *cf* VERSCHNEIDEN. 2 cuttings, trimmings; chips, scraps; remnants. 3 industrial waste. 4 blend. 5 adulterated liquor (tobacco, etc); rotgut. 6 (*Tex.*) reduction (of print pastes). **--ansatz** *m* (*Tex.*) reduction paste. **-bitumen** *n* cut-back bitumen

verschnitten clipped, etc: *pp of* VER-SCHNEIDEN

verschnitt·fähig *a* dilutable; cuttable. **V–fähigkeit** *f* cuttability; (*Lacquers*) dilution value. **-mittel** *n* cutting agent; (*Lacquers*, etc) diluent, extender, filler. **-wein** *m* blended wine; adulterated wine

verschnupft *a:* **v. sein** have a cold

verschob(en) shifted: *past* (*& pp*) *of* VERSCHIEBEN

verschollen *p.a* disappeared; presumed dead

verschonen *vt* spare; **einen mit etwas v.** spare sbdy sthg *cf* VERSCHONT

verschönern *vt* enhance, beautify; embellish

verschont *p.a:* **v. bleiben (von)** be free (of), be unbothered (by), stay immune (to) *cf* VERSCHONEN

verschoß, verschossen faded, etc: *past & pp of* VERSCHIESSEN

verschottern *vt* gravel (up), fill (*or* cover) with gravel

verschränken *vt* cross, fold; interlace

verschrauben *vt* 1 screw together; screw up tight 2 screw the wrong way. **Verschraubung** *f* (-en) screwing (together, etc). 2 screw cap; screw joint

verschreiben* I *vt* 1 use up (ink, paper) in writing. 2 prescribe (for). 3 bequeath, transfer (to). **II: sich v.** 4 make a slip of the pen. 5 (+*dat*) dedicate (*or* commit) oneself (to). **Verschreibung** *f* (-en) 1 (*usu Pharm.*) prescription. 2 bequest, transfer, assignment

verschrieb(en) prescribed: *past* (*& pp*) *of* VERSCHREIBEN

verschroben *p.a* distorted; eccentric

verschroten *vt* rough-grind (grain)

verschrotten *vt* scrap (*esp* metal)

verschrumpeln, verschrumpfen *vi*(s) shrivel up

verschuldet *p.a* indebted, in debt; encumbered

verschütten *vt* 1 spill; shed. 2 fill up, cover (with debris, etc). 3 bury

Verschwächung *f* weakening

verschwamm blurred: *past of* VERSCHWIMMEN

verschwand disappeared: *past of* VERSCHWINDEN

verschweigen* *vt* keep secret *cf* VERSCHWIEGEN

verschweißen *vt* weld, bond; (*Plastics*) heat-seal

verschwelen 1 *vt* carbonize. 2 *vi*(s) smolder

verschwellen* *vi*(s) swell (up), swell shut. **Verschwellung*** *f* swelling

Verschwelung° *f* low-temperature carbonization

verschwenden *vt* waste, squander. **verschwenderisch** *a* 1 wasteful. 2 lavish. **Verschwendung** *f* waste, squandering

verschwenken *vt* tilt, incline

verschwieg(en) 1 kept secret: *past* (*& pp*) *of* VERSCHWEIGEN. **2: verschwiegen** *p.a* secretive; discreet; secluded. **Verschwiegenheit** *f* secretiveness, secrecy; silence; discretion

verschwillt swells up; *pres of* VERSCHWELLEN

verschwimmen* *vi*(s) blur, become indistinct (*or* hazy); **ineinander v.** blend, merge, fuse. *Cf.* VERSCHWOMMEN

verschwinden* *vi*(s) disappear, vanish. **Verschwinden** *n* disappearance. **verschwindend** *p.a* (*as adv*) e.g. **v. klein** infinitesimally small; **v. wenig** an infinitesimal amount (of)

verschwoll(en) swelled up, swollen: *past* (*& pp*) *of* VERSCHWELLEN

verschwommen 1 blurred: *pp of* VERSCHWIMMEN. 2 *p.a* indistinct, hazy; out of focus

verschwunden 1 disappeared; *pp of* VERSCHWINDEN. 2 *p.a* missing

versehen* I *vt* 1 (+**mit**) provide, supply (with); fit, equip (with). 2 be in charge of; look after; perform (the duties of). 3 overlook, neglect. 4 misread. 5 (+*dat rflx*):

sich es v. be aware of it. **II: sich v. 5** make a mistake, slip up. **6** (+**mit**) provide oneself (with). **Versehen** *n* oversight; error: **aus V.** = **versehentlich** *adv* by mistake, inadvertently

versehren *vt* disable, injure. **Versehrte** *m, f* (*adj endgs*) disabled (*or* handicapped) person

verseifbar *a* saponifiable. **Verseifbare(s)** *n* (*adj endgs*) saponifiable matter. **Verseifbarkeit** *f* saponifiability

verseifen *vt, vi*(s) saponify. **Verseifung** *f* saponification

Verseifungs·farbe *f* soap color. **-faß** *n* saponifying run. **v–fest** *a* saponification-resistant. **V–mittel** *n* saponifying agent. **-zahl** *f* saponification number

verselben *vt* automate

versenden* *vt* send off, dispatch, mail, ship. **Versender** *m* (-) sender, shipper

versengen *vt* singe, scorch; parch

versenkbar *a* retractable, disappearing; sinkable

versenken I *vt* **1** sink, submerge. **2** lower. **3** (*Mach.*) countersink; retract. **II: sich v.** immerse oneself, become absorbed (in an activity). **Versenk·grube** *f* (*Lthr.*) tan pit. **Versenkung** *f* (-en) sinking; submersion; absorption; subsidence

versetzbar *a* mixable; movable; permutable, etc *cf* VERSETZEN

versetzen I *vt* **1** (*usu* + **mit**) mix, alloy, dilute, flavor, spice (with). **2** (**mit**) fit, equip. **3** (*Lthr.*) handle; layer. **4** move, shift, transfer; transpose; transplant; (*Math.*) permute; (*Cryst., Geol.*) dislocate; (*School*) promote. **5** put, place. **6** obstruct; tamp. **7** (+**in** + *verbal noun*) start, cause (*implies causing an activity to start; transl. depends on noun*) e.g.: **in Schwingung v.** cause to vibrate, start (sthg) vibrating; **in Erregung v.** excite. **8** pawn. **9** sell. **10: einem etwas v.** give (*or* administer) sthg to sbdy. **II** *vt/i* **11** reply, retort. **III: sich v. 12** put oneself (in a state, a position). **13** move, shift. **14** get stopped up, be obstructed

Versetz·grube *f* (*Lthr.*) tan pit

Versetzung *f* (-en) **1** mixing, alloying; dilution; flavoring, spicing. **2** equipping. **3**

(*Lthr.*) layer. **4** moving, shifting; transfer(ing); transposition; transplantation; (*Math.*) permutation; (*Cryst., Geol.*) dislocation; (*School*) promotion. **5** putting, placement. **6** obstruction; tamping. **7** administering. *Cf* VERSATZ, VERSETZEN

verseuchen *vt* infect; contaminate, poison; infest. **Verseuchung** *f* infection; contamination, poisoning; infestation. **Verseuchungs·stoff** *m* contaminant

versichern I *vt* **1** insure. **2** affirm, declare. **3: einem etwas v.** assure sbdy of sthg. **II** *vi* (+*dat*, + **daß**) assure (sbdy that . . .). **III: sich v.** insure oneself; (+**daß**, +*genit*) make sure (that, of), assure oneself (of). **Versicherung** *f* (-en) **1** insurance; insurance company; insurance policy. **2** assurance. **3** affirmation, declaration. **Versicherungs·schein** *m* insurance policy

versickern *vi*(s) **1** seep (ooze, trickle) away; percolate. **2** peter out; ebb; wear off. **Versickerung** *f* seepage, percolation

versiegeln *vt* seal (up); reinforce

versiegen *vi*(s) dry up; dwindle, ebb; peter out

versieht provides, etc: *pres of* VERSEHEN

versiert *p.a* well-versed; experienced, skilled

versilbern *vt* silver(-plate). **Versilberung** *f* silvering, silver-plating. **Versilberungs·bad** *n* silver bath

versinken* *vi*(s) **1** sink, go down. **2** bog down. **3** disappear. **4** (+**in**) become absorbed, be lost (in). *Cf* VERSUNKEN.

versinnbildlichen 1 *vt* symbolize. **2: sich v.** be symbolized

versintern *vt* sinter. **Versinterung** *f* (-en) sintering

versöhnen *vt* reconcile; appease, conciliate

versorgen *vt* **1** (**mit**) provide, supply (with). **2** take care of, provide for; maintain. **versorgt** *p.a* well taken care of. **Versorgung** *f* (-en) supply, provision(ing); maintenance, support; care

Versorgungs·gang *m* supply channel. **-leitung** *f* utility pipe (duct, cable). **-staat** *m* welfare state. **-struktur** *f* supply infrastructure

verspannen *vt* brace, stay; guy

Verspannungs-faktor *m* (*Explo.*) loading factor

Verspanung *f* 1 machining. 2 removal of excess metal (e.g. shavings, borings)

verspäten 1 *vt* delay, defer. 2: **sich v.** be late, be delayed. **verspätet** *p.a & adv* late. **Verspätung** *f* (-en) delay, deferment; lateness, late arrival—**10 Minuten V. haben** be 10 minutes late

versperren *vt* 1 close off; bar, bolt. 2 block. 3 lock up

verspiegeln *vt* coat with metal, metalize

versprach promised: *past of* VERSPRECHEN

versprechen* 1 *vt/i* promise. 2 *vt:* **sich etwas v.** (**von**) expect (*or* hope for) sthg (from). 3: **sich v.** make a slip of the tongue. **Versprechen** *n* (-) promise

versprengen *vt* 1 scatter, disperse. 2 sprinkle

verspricht promises: *pres of* VERSPRECHEN

verspritzen *vt* 1 spray, spatter. 2 (*esp Metll.*) spill. 3 (*Metll.*) injection-mold

versprochen promised; *pp of* VERSPRECHEN

verspröden *vi*(s) become brittle, embrittle. **Versprödung** *f* embrittlement

versprühen I *vt/i* 1 spray (out). II *vi* 2 scintillate. 3 dissipate

verspüren *vt* perceive, feel; experience

Verss. *abbv* (Versuche) experiments

verstach blended, etc: *past of* VERSTECHEN

verstaatlichen *vt* nationalize, put under government control. **Verstaatlichung** *f* nationalization

verstählen *vt* steel-face

verstand understood: *past of* VERSTEHEN

Verstand *m* mind, brains, wits; senses; common sense, judgment; intelligence

verstanden understood: *pp of* VERSTEHEN

verständig *a* sensible; intelligent

verständigen I *vt* 1 inform, advise, notify. II: **sich v.** 2 (+**mit**, +**auf**) agree, come to terms (with sbdy, on sthg). 3 (+**mit**) communicate (with). **Verständigung** *f* (-en) 1 notification. 2 agreement. 3 communication. 4 (*Tlph.*) audibility, reception

verständlich *a* 1 understandable; intelligible, clear: **es ist schwer v., wie** it is difficult to understand how; **sich v. machen** make oneself understood; **einem etwas v. machen** make sthg clear to sbdy. 2 audible. **verständlicherweise** *adv* understandably. **Verständlichkeit** *f* clarity, intelligibility; audibility

Verständnis *n* understanding, comprehension; appreciation: **mit V.** understandingly; **kein V. haben für** . . . not understand (*or* appreciate) . . . ; **nach heutigem V.** by today's standards. **v--los** *a* uncomprehending, unappreciative, unsympathetic. **-voll** *a* understanding, appreciative, sympathetic

verstärken I *vt* 1 strengthen, reinforce, fortify. 2 concentrate. 3 increase, intensify, amplify. II: **sich v.** 4 grow stronger, be reinforced. 5 be concentrated. 6 increase, be intensified. *Cf* VERSTÄRKT

Verstärker *m* (-) intensifier; amplifier; activator; reinforcing agent, booster; (*Catalysts*) promoter. **--pumpe** *f* booster pump. **-röhre** *f* amplifying tube. **-säule** *f* rectifying apparatus, rectifying section of a fractionating tower

verstärkt *p.a* (*oft*) increased; *adv oft* on a larger scale, to an increased extent. *Cf* VERSTÄRKEN

Verstärkung *f* (-en) strengthening, reinforcement, etc *cf* VERSTÄRKEN

Verstärkungs-kolonne *f* concentrating column. **-mittel** *n* reinforcing (*or* fortifying) agent. **-verhältnis** *n* enrichment ratio

verstatten *vt* permit, allow

verstauben *vi*(s) get dusty

verstäuben I *vt* 1 atomize, spray. 2 cover with dust. 3 pulverize, convert to dust. II *vi*(s) 4 scatter (rise, disperse) as (a cloud of) dust. 5 be atomized, spray. **Verstäuber** *m* (-) atomizer, sprayer

verstaubt *p.a* dusty, dust-covered *cf* VERSTAUBEN

verstauchen *vt:* **sich** (e.g. **den Fuß**) **v.** sprain, wrench (one's foot, etc). **Verstauchung** *f* (-en) sprain

verstechen* *vt* 1 blend; adulterate. 2 stitch up; patch. 3 barter

Versteck *m* (-e) hiding place. **verstecken** 1 *vt & sich v.* (**vor**) hide (from). 2 *vt* conceal. **versteckt** *p.a* hidden, secret; furtive, veiled; latent: **sich v. halten** stay in hiding

verstehen* I *vt/i* 1 understand. II *vt* 2 (*esp*

+ **als**) understand, interpret (as). **3** know (how): **sie v. etwas vom Destillieren** they know sthg about distillation; **sie v. (es), damit umzugehen** they know how to deal with it. **4** (+ **unter**) mean (by): **unter Sublimierung versteht man . . .** by sublimation we mean . . . **III: sich v. 5** (*oft* + **von selbst**) be obvious, be self-evident, go without saying. **6: es versteht sich als . . .** it is to be understood (*or* interpreted) as . . . **7** (+ **mit**) get along (with). **8** (+ **auf**) know a lot (about), be an expert (at). **9** (+ **zu**) agree (to)

versteifen 1 *vt* reinforce, brace. **2** *vt, vi*(s) & **sich v.** stiffen. **3: sich v. (auf)** insist (on). **Versteifer** *m* (-) stiffener, stiffening agent. **Versteifung** *f* (-en) stiffening; reinforcement; brace, strut

versteigen* *vr:* **sich v. (zu)** presume, go so far as (to)

versteigern *vt* sell at auction, auction off

versteinern *vt, vi*(s) & **sich v.** petrify, fossilize. **Versteinerung** *f* (-en) **1** petrification, fossilization. **2** fossil

Versteinerungs·kunde *f* paleontology. **-mittel** *n* (*Min.*) mineralizing agent

Versteinung *f* (-en) petrification

verstellbar *a* adjustable

verstellen I *vt* **1** block, obstruct. **2** misplace. **3** shift, rearrange; transpose. **4** adjust, set. **5** disguise. **II: sich v.** dissimulate, pretend. **verstellt** *p.a* (*oft*) feigned, fictitious. **Verstellung°** *f* obstruction; misplacement; rearrangement; transposition; adjustment; disguise, dissimulation, pretense

Verstell·vorrichtung *f* adjusting device. **-weg** *m* displacement, travel

verstemmen *vt* calk

versteuerbar *a* taxable, dutiable. **versteuern** *vt* pay tax (*or* duty) on. **versteuert** *p.a* tax-paid; duty-paid

versticht blends, etc: *pres of* VERSTECHEN

versticken *vt* nitrogenize, nitride

Verstick·stahl *m* nitrided steel

verstochen blended, etc: *pp of* VERSTECHEN

verstocken *vi* be(come) moldy. **verstockt** *p.a* (*usu*) obdurate; hardened

verstopfen I *vt* **1** clog (up), stuff, congest; choke; constipate. **2** obstruct. **II** *vi* cause constipation. **Verstopfung** *f* (-en) clog-

ging, congestion; choking; constipation; obstruction

verstopfungs·lösend *a* laxative; decongestive. **V—mittel** *n* astringent

verstöpseln *vt* stopper

verstorben *p.a* deceased, late

verstören *vt* disturb, upset, distress

Verstoß° *m* offense; violation, infraction, **verstoßen*** **I** *vt* **1** repudiate. **2** expel. **II** *vi* (+ **gegen**) offend, infringe (against), violate; be an offense (against), be a violation (of)

verstrammen *vt* tighten; strengthen. **Verstrammungs·mittel** *n* strengthening agent

verstreben *vt* prop, brace, strut; guy

verstreichen* **I** *vt* **1** fill (cracks, etc); spackle, plaster. **2** (*Tex.*) level. **3** spread on, rub on. **II** *vi*(s) (*Time*) pass, elapse, expire

verstreuen *vt* scatter; spill

verstrich(en) plastered, etc: *past* (& *pp*) *of* VERSTREICHEN

verstricken *vt* entangle; involve

verstromen *vt* convert (in)to electric current

verstümmeln *vt* mutilate, mangle

verstummen *vi*(s) turn silent, become uncommunicative; subside, die away

Versuch *m* (-e) **1** experiment; test. **2** attempt, trial. **versuchen I** *vt* **1** try, attempt: **es wurde versucht, zu . . .** an attempt was made to . . . **2** test. **3** taste, sample. **4** tempt. **II** *vi* **5** experiment. **6** (+ **ob**) try and see (if, whether)

Versuchs· experimental test, trial: **-anlage** *f* test (pilot, experimental) plant (*or* unit). **-anordnung** *f* experimental procedure; arrangement of the experiment. **-anstalt** *f* experimental station; research institute. **-ballon** *m* trial balloon. **-bedingung** *f* experimental (*or* test) condition. **-bohrung** *f* test boring; test well. **-dauer** *f* duration of the experiment, testing time. **-durchführung** *f* experimental procedure. **-ergebnis** *n* experimental (*or* test) result. **-feld** *n* experimental area, proving ground. **-flamme** *f* test flame. **-folge** *f* series of experiments (*or* tests). **-gefäß** *n* test flask, experimental vessel. **-gegenstand** *m* experimental (*or* test) object. **-gelände** *n* test range (*or* area). **-grundlage** *f* experimental basis. **-gut** *n* experimental farm. **-kaninchen** *n* test rabbit; (*fig*) guinea pig. **-körper** *m* experimental object, test piece. **-laboratorium** *n* test (experimental, research) laboratory. **-ladung** *f* (*Explo.*) test charge. **-lauf** *m* test (*or* trial) run. **v-mäßig** *a* experimental. **V-material** *n* experimental material. **-muster** *n* test sample. **-objekt** *n* test object. **-person** *f* test subject. **-raum** *m* laboratory; research space. **-reihe** *f* series of tests (*or* experiments). **-rohr** *n,* **-röhre** *f* experimental tube. **-serie** *f* series of tests. **-stab** *m* test rod (*or* bar). **-stadium** *n* experimental stage. **-stand** *m* test stand (*or* bench); test rig. **-station, -stelle** *f* experimental (*or* research) station. **-strecke** *f* test track. **-stück** *n* test piece, **-tier** *n* experimental (*or* test) animal. **-träger** *m* test stand. **-vorschrift** *f* test directions. **v-weise** *adv* experimentally, by way of experiment; on trial, on approval. **V-wert** *m* experimental value. **-wesen** *n* research. **-zweck** *m* experimental purpose

Versuch·tier *n* experimental animal

Versuchung *f* (-en) temptation **—in V. kommen** (or **sein**) be tempted

versumpfen *vi*(s) turn marshy; degenerate; bog down. **versumpft** *p.a* marshy, boggy

versunken I sunk(en); *pp of* VERSINKEN. **II** *p.a* 1 lost, perished. 2 absorbed, wrapped up (in). **Versunkenheit** *f* oblivion

versüßen *vt* sweeten; edulcorate, purify by washing. **versüßt** *p.a:* **v—er Salpetergeist** spirit of nitrous ether; **v—er Salzgeist** sweet spirit of salt (alcoholic solution of ethyl chloride). **Versüßung** *f* sweetening; edulcoration. **Versüßungs·mittel** *n* sweetening agent, sweetener

vertagen *vt* (**auf**) adjourn, postpone (till)

vertan, vertat wasted, etc: *pp & past of* VERTUN

vertauschbar *a* interchangeable; exchangeable; (*Math.*) permutable. **vertauschen** *vt* (**mit**) exchange (for); mistake (for); permute. **Vertauschung** *f*

(-en) exchange, interchange; permutation

vertausendfachen *vt* increase a thousandfold

verte! (*Latin*) turn the page

verteeren *vt* tar

verteidigen *vt & sich v.* defend (oneself). **Verteidigung** *f* (-en) defense

Verteidigungs·minister *m* minister (*or* US secretary) of defense. **-waffe** *f* defensive weapon

verteilen 1 *vt* (**auf**) distribute (over), divide. **2** *vt & sich v.* spread, disperse; dissolve. **3: sich v.** be distributed

Verteiler *m* (-) **1** (*incl Mach.*) distributor. **2** traffic circle. **-leitung** *f* distribution line (*or* main). **-netz** *n* distribution network (*or* system). **-säule** *f* (*Chromatography*) partitioning column. **-stelle** *f* **1** distribution point. **2** retail store

Verteil·kasten *m* distribution box. **-trichter** *m* distributing funnel (*or* hopper), distributor

Verteilung *f* (-en) distribution, division; partition; dispersion, dissolution

Verteilungs·chromatographie *f* partition chromatography. **-gesetz** *n* law of distribution (*or* partition). **-grad** *m* degree of dispersion (*or* distribution). **-koeffizient** *m* coefficient of distribution (*or* partition). **-kurve** *f* distribution curve. **-mittel** *n* distributing agent. **-netz** *n* distribution network. **-rohr** *n,* **-röhre** *f* **1** distributing tube (*or* pipe). **2** (*Blast Furnace*) blast main. **-satz** *m* principle (*or* law) of distribution (*or* partition). **-stand** *m* state of (sub)division. **-stelle** *f* distribution point. **-zahl** *f* distribution number. **-zustand** *m* state of (sub)division

verteuern 1 *vt* raise (prices, taxes, etc); raise the price of. **2: sich v.** rise (in price), become more expensive. **Verteu(e)rung** *f* (-en) rise (in prices, etc)

vertiefen I *vt* **1** deepen; lower. **2** intensify. **II: sich v. 3** deepen, become deeper. **4** (+**in**) delve (into), be deeply absorbed (*or* involved) (in). **Vertiefung** *f* (-en) **1** deepening, lowering; intensification; deep absorption, involvement. **2** depression, low spot; cavity, hollow, dent; recess

vertiert *p.a* brutal, bestial

Vertikal·achse *f* vertical axis. **-bewegung** *f* vertical motion

Vertikale *f* (*adj endgs*) perpendicular, vertical line

vertilgbar *a* eradicable, exterminable

vertilgen *vt* destroy; wipe out, eradicate, exterminate. **Vertilgungs·mittel** *n* eradicator, exterminating agent

vertippen *vt & sich v.* mistype

vertorft *a* peaty. **Vertorfung** *f* peat formation, conversion into peat

Vertrag *m* (-̈e) treaty; contract; agreement

vertragbar *a* bearable, tolerable

vertragen* I *vt* 1 stand, endure, tolerate, II: **sich v.** (mit) 2 get along (with). 3 be reconciled, make up (with). 4 be compatible, (*or* consistent), go well (with)

vertraglich *a* contractual; (*adv oft*) by contract

verträglich *a* 1 peaceable, amicable. 2 compatible, consistent. 3 digestible. **Verträglichkeit** *f* amicability; compatibility

vertrags·mäßig *a* contractual, etc = VERTRAGLICH

vertrat represented: *past of* VERTRETEN

vertrauen *vt/i* (*dat* or **auf**) trust (in). **Vertrauen** *n* (**zu**) trust, confidence (in). **vertrauens·würdig** *a* trustworthy. **vertraulich** *a* confidential; intimate. **vertraut** *p.a* intimate, familiar: (**sich**) **v. machen** (**mit**) familiarize (oneself) (with). **Vertraute** *m, f* (*adj endgs*) confidant(e)

vertreiben* *vt* 1 drive out, chase away, expel; evict. 2 displace. 3 banish, dispel. 4 cure. 5 (*Colors*) scumble, soften. 6: **sich die Zeit v.** (mit) pass the time (with), kill time (..ing). 7 (*Com.*) sell, retail; market, distribute. **Vertreibung** *f* (-en) expulsion, eviction; displacement; banishment, dispelling; cure; softening (of colors); passing, spending (of time); sale, retailing, marketing distribution

vertretbar *a* 1 tenable. 2 justifiable, warrantable. 3 tolerable. 4 capable of being delegated. **Vertretbarkeit** *f* 1 tenability. 2 justifiability, warrantability. 3 tolerability. 4 possibility of being delegated. 5: **isomorphe V.** isomorphous substitution

vertreten* I *vt* 1 substitute for, replace. 2 represent, stand for. 3 advocate, stand up for (a cause, etc). 4 take responsibility for; justify, warrant; safeguard, look after (interests). 5 wear out (*or* down) (shoes). 6 tread down, trample. 7: **sich die Beine v.** A sprain one's legs, B stretch (*or* exercise) one's legs. 8: **einem den Weg v.** bar sbdy's way. II *p.a:* **v. sein** be represented; be in evidence. **Vertreter** *m* (-) 1 representative, delegate; replacement, substitute. 2 advocate, supporter. 3 attorney. **Vertretung** *f* (-en) 1 replacement. 2 representation. 3 advocacy, support. 4 justification. 5 delegation, representative body

vertrieb expelled; *past of* VERTREIBEN

Vertrieb *m* 1 sale, marketing. 2 sales department

vertrieben 1 expelled, etc: *pp of* VERTREIBEN. 2: **Vertriebene** *m, f* (*adj endgs*) displaced person, refugee

vertritt represents: *pres of* VERTRETEN

vertrocknen *vi*(s) dry up; go stale. **Vertrocknung** *f* (-en) drying up, desiccation

vertrug endured, etc: *past of* VERTRAGEN

vertun* I *vt* 1 waste, squander. 2 miss, pass up (a chance). II: **sich v.** make a mistake

verunedeln *vt* degrade, debase; shift to a lower potential

verunglücken *vi*(s) come to grief; have an accident

verunkrauten *vi*(s) be overgrown with weeds

verunreinigen *vt* contaminate, pollute. **Verunreinigung** *f* (-en) contamination, pollution; impurify

verunstalten *vt* disfigure, deform

verunzieren *vt* deface, disfigure

verursachen *vt* cause

verurteilen *vt* condemn, doom; sentence

vervielfachen *vt & sich v.* multiply

vervielfältigen 1 *vt & sich v.* multiply, increase. 2 mimeograph, duplicate. 3 (*Film*) micrograph. **Vervielfältigung** *f* (-en) multiplication, increase; mimeographing, duplication; micrographing. **Vervielfältigungs·apparat** *m* mimeograph (machine), duplicator; micrograph (machine)

vervierfachen *vt & sich v.* quadruple

vervollkommnen *vt & sich v.* perfect, improve (oneself). **Vervollkommnung** *f* (-en) perfection

vervollständigen 1 *vt* complete, round out. **2: sich v.** be completed

verwachsen* I *vt* **1** outgrow (clothes). **II** *vi*(s) **2** grow together, grow into one; intergrow; coalesce. **3** heal (up). **4** be(come) overgrown. **5** (**mit**) become involved (with), develop close ties (to). **II** *p.a* outgrown; intergrown; deformed. **Verwachsung** *f* intergrowth, coalescence; overgrowth; involvement; close ties. **Verwachsungs-fläche** *f* (*Cryst.*) composition plane

verwägen* *vt* determine the weight, weigh

verwählen *vr: sich v.* (*Tlph.*) misdial

Verwahr *m* custody, safekeeping: **in V. geben** put in safekeeping. **vewahren I** *vt* **1** keep (in a safe place, in custody), guard. **2** save for later. **3** secure, lock (doors). **II: sich v.** (**gegen**) protest (against), resist

verwahrlosen 1 *vt* neglect. **2** *vi*(s) go to ruin, degenerate

Verwahrsam *m* custody, safekeeping

Verwahrung *f* (safe)keeping, (safe)guarding; custody

verwalten *vt* administer, manage, supervise. **Verwaltung** *f* (-en) administration, etc

verwalzen *vt* (*Metll.*) roll; reduce (by rolling)

verwand got over: *past of* VERWINDEN

verwandelbar *a* transformable, convertible. **Verwandelbarkeit** *f* transformability, convertibility

verwandeln 1 *vt* transform, convert. **2: sich v.** be transformed, be converted. **Verwandlung** *f* (-en) transformation, conversion; metamorphosis

verwandt(e) 1 used: *pp* (& *past*) of VERWENDEN. **2: verwandt** *p.a* related, kindred, akin; cognate. **Verwandte** *m, f* (*adj endgs*) relative, family member. **Verwandtschaft** *f* (-en) **1** affinity; relation(ship), kinship. **2** relatives, family **Verwandtschafts-einheit** *f* unit of affinity. **-lehre** *f* doctrine of affinity

verwarf rejected: *past of* VERWERFEN

verwärmen *vt* absorb heat. **Verwärmung** *f* heat absorption

verwarnen *vt* warn, caution; fine on the spot

verwaschen *p.a* washed out, faded; watery; vague, blurred, indistinct; **v—es Gut** (*Min.*) middlings. **Verwaschenhelt** *f* washed-out (*or* faded) appearance; vagueness, indistinctness

verwässern *vt* water (down), thin, dilute. **verwässert** *p.a* watered-down; watery

verweben* 1 *vt* interweave. **2: sich v.** be interwoven, intermingle; (**+in**) get involved (in)

verwechseln *vt* (*usu* **+ mit**) interchange, mix up, confuse (one thing with another). **Verwechslung** *f* (-en) interchanging, mix-up, confusion, mistaken identity

verwegen *a* bold, daring

verwehen 1 *vt* blow away; scatter, disperse; cover with drifts. **2** *vi*(s) blow over; dissipate, subside; drift away

verwehren *vt* (*d.i.o*) refuse sbdy sthg, block sbdy's way to sthg; keep sbdy from doing sthg. **verwehrt** *p.a:* **v. bleiben** (*dat*) be (*or* remain) inaccessible (*or* unattainable) (to)

Verwehung *f* (-en) **1** blowing away, scattering, dispersion. **2** snowdrift, sanddrift. *Cf.* VERWEHEN

verweigern *vt* refuse (to give); withhold (from). **Verweigerung** *f* (-en) refusal; withholding

verweilen *vi* stay, tarry

Verweil-weg *n* tarrying path. **-zeit** *f* duration, length of stay; dwell time

Verweis *m* (-e) **1** (cross) reference. **2** reproach, reprimand

verweisen* I *vt/i* **1** refer (the reader to another page, etc). **II** *vt* **2** reproach, reprimand. **3** expel, order to leave; banish. **4** (**+auf**) order (to go) (to a place). **5: einen in seine Schranken v.** put someone in his place; **zur Ordnung v.** call to order. **III** *vi* (**+auf**) point out. **Verweisung** *f* (-en) reproach, reprimand; expulsion, banishment; referral

verwelken *vi*(s) wither, wilt, droop

verwendbar *a* usable, serviceable; applicable; available

verwenden* I *vt* **1** use. **II: sich v. 2** (**für**) devote oneself (to); intercede (for). **Verwendung°** *f* **1** use: **V. finden** be of use, be used. **2** intercession

Verwendungs·bereich *m* range of use. **v–fähig** *a* usable, applicable. **V–fähigkeit** *f* usability, applicability. **-gebiet** *n* field of application. **-möglichkeit** *f* possible use. **-stoffwechsel** *m* catabolism. **-zweck** *m* intended use

verwerfen* **I** *vt* **1** reject, turn down; discard; cancel. **2** misplace. **3** (*Min.*) dislocate. **II** *vi* (*Animals*) miscarry. **III:** **sich v.** warp; (*Geol.*) fault. *Cf* VERWORFEN. **Verwerfung** *f* (-en) rejection; misplacement; warping; (*Min.*) dislocation, slip; (*Geol.*) fault(ing)

Verwerfungs·fläche *f* (*Geol.*) fault plane. **-lette** *f* (*Min.*) gouge, floucan

verwertbar *a* **1** utilizable. **2** (*Com.*) exploitable, marketable. **Verwertbarkeit** *f* utilizability; exploitability, marketability

verwerten *vt* **1** make use of, utilize; exploit. **2** commercialize, market. **3:** **sich gut v. lassen** find a ready market. **Verwertung** *f* (-en) utilization; exploitation; commercialization, marketing. **Verwertungs·anlage** *f* waste (*or* garbage) recycling plant

verwesen *vi* (s) rot, decay, decompose. **verweslich** *a* liable to decay, perishable. **Verwesung** *f* (-en) decomposition, decay, rot(ting)

Verwesungs·pilz *m* (*Bot.*) saprophyte. **-prozeß** *m* process of decay

verwickelt *p.a* involved, complicated, intricate. **Verwicklung** *f* (-en) involvement, complication, company, intricacy

verwiegen* **1** *vt* weigh (out). **2:** **sich v.** make a mistake in weighing. **Verwiegung** *f* (-en) weighing

verwies(en) referred, etc: *past* (& *pp*) *of* VERWEISEN

verwichen *a* past, bygone, former

verwinden* *vt* **1** (*Mach.*) twist, distort. **2** get over, recover from. **Verwindung** *f* (-en) **1** distortion; twisting; torsion. **2** recovery (from). **verwindungs·steif** *a* torsion-resistant

verwirft rejects, etc: *pres of* VERWERFEN

verwirklichen **1** *vt* realize, accomplish, achieve. **2:** **sich v.** materialize, become reality. **Verwirklichung** *f* (-en) realiza-

tion, accomplishment, achievement, materialization

verwirren *vt* confuse, tangle up; perplex *cf* VERWORREN. **Verwirrung** *f* (-en) confusion, tangle, mix-up; perplexity

verwischen **1** *vt* smudge, blur; blot out, obliterate. **2:** **sich v.** be smudged (blurred, etc)

verwittern *vi* (s) weather away, erode; effloresce; (*Lime*) air-slake. **Verwitterung** *f* weathering, erosion; efflorescence

Verwitterungs· (*usu Geol.*) residual: **-boden** *m*, **-krume** *f* residual soil. **-kruste** *f* weathered crust. **-produkt** *n* weathering residue. **-schutt** *m* detritus. **-ton** *m* residual clay

verwoben interwoven: *pp of* VERWEBEN

verwogen weighed: *pp of* VERWÄGEN & VERWIEGEN

verworfen rejected, etc: *pp of* VERWERFEN

verworren *a* confused, tangled; complex *cf* VERWIRREN. **--faserig** *a* reticulate-fibrous

verwuchs outgrew, etc: *past of* VERWACHSEN

verwunden **I** *vt* wound, injure. **II** twisted; recovered, etc: *pp of* VERWINDEN

verwunderlich *a* surprising

verwundern **1** *vt* surprise, astonish. **2** *vi* be surprising. **3:** **sich v.** (**über**) wonder, be surprised (at). **Verwunderung** *f* surprise, amazement: **in V. setzen** astonish, amaze

Verwurf· *m* (*Geol.*) fault

verwurzeln **1** *vt* root, provide with roots. **2** *vi* (s) take root, become rooted

verwüsten *vt* lay waste, devastate. **Verwüstung** *f* (-en) devastation

verzählen *vr:* **sich v.** lose count, miscount

verzahnen *vt* & **sich v.** mesh, dovetail, interlock. **Verzahnung** *f* (-en) **1** gear teeth. **2** meshing (of gears); dovetailing, interlocking

verzapfen *vt* **1** join, mortise. **2** have (*esp* beer) on tap, serve. **3** bring out (ideas, etc)

verzehnfachen *vt* & **sich v.** increase tenfold

verzehren **I** *vt* **1** consume, eat up, drink up. **2** spend, use up; sap one's strength. **II:** **sich v.** (**in, vor**) be consumed (with).

verzehrt *p.a* (*oft*) drained, exhausted, wasted (away)

verzeichnen *vt* 1 note (down), record, register: **keine Änderung war zu v.** no change was to be noted. 2 index, catalog; inventory. 3 (*incl Phot., Opt.*) misdraw, distort; misrepresent. **Verzeichnis** *n* (-se) index, list; table; catalog; inventory. **Verzeichnung** *f* noting, recording, etc *cf* VERZEICHNEN

verzeihen* *vt* (*d.i.o*), *vi* (+*dat*) pardon, excuse, forgive

verzerren 1 *vt* distort, twist. 2: **sich v.** be distorted. **Verzerrung** *f* (-en) distortion; (*Stat.*) bias. **verzerrungs-frei** *a* distortion-free

verzetteln I *vt* 1 catalog (on slips of paper). 2 fritter away, dissipate. **II: sich v.** be dissipated, go to waste; get bogged down in minor matters

Verzicht *m* (-e) (*usu* + **auf**) relinquishment, renunciation (of); doing without: **V. leisten** (or **üben**) (**auf**) relinquish, etc = **verzichten** *vi* (**auf**) give up, relinquish, do without, waive. **Verzicht-leistung** *f* relinquishment; waiver

verziehen* **I** *v* **A** *vt* 1 distort. 2 thin out (plants). **B** *vi*(s) 3 move, change residence. **C: sich v.** 4 twist, curl, buckle, warp; be distorted. 5 disperse, dissolve, dissipate; blow over. 6 disappear, vanish. **II** pardoned: *pp of* VERZEIHEN

verzieren *vt* decorate; embellish. **Verzierung** *f* (-en) decoration, ornament, embellishment; flourish

verzinken *vt* 1 coat with zinc, galvanize. **Verzinkerei** *f* (-en) galvanizing unit. **Verzinkung** *f* (-en) galvanizing, galvanization

Verzinkungs-anlage *f* galvanizing plant. **-bad** *n* galvanizing bath

verzinnen *vt* tin, tinplate; line with tin: **verzinntes Eisenblech** tinplate, "tin". **Verzinnerei** *f* (-en) tinning plant. **Verzinnung** *f* tinning, tinplating

Verzinnungs-farbe *f* tinning paint. **-paste** *f* tinning paste

verzinsen 1 *vt* pay interest on. 2: **sich v.** bear interest

verzog(en) moved, etc: *past* (& *pp*) *of* VERZIEHEN

Verzögerer *m* (-) (*genl*) retarder, retarding agent; (*Phot.*) restrainer; (*Varnish*) reducer; (*Synthetics*) inhibitor. **verzögern** 1 *vt* retard, delay, postpone, protract. **2: sich v.** be delayed, be long in coming. **Verzögerung** *f* (-en) delay, retardation, postponement; time lag

Verzögerungs-bad *n* (*Phot.*) restraining bath. **-manöver** *n* delaying tactics, foot-dragging. **-mittel** *n* retarder, etc = VERZÖGERER. **-satz** *m* (*Explo.*) delay composition. **-zünder** *m* delayed-action fuse

verzollbar *a* dutiable. **verzollen** *vt* pay duty on. **verzollt** *p.a* duty-paid

verzuckern *vt* saccharify; sugar(coat); candy; (*Cakes*) ice. **Verzuckerung** *f* saccharification; sugaring, sugarcoating; candying; icing

Verzug *m* 1 delay; arrears: **in V. geraten** get into arrears, fall behind. **2: im V. sein** (**a**) be in arrears, (**b**) be approaching, be ahead. 3 moving, change of address. 4 (*Mech.*) distraction, warpage. 5 sheathing, lagging; cribbing

Verzugs-zünder *m* delay fuse

Verzunderung *f* (*Mettl.*) scaling. **verzunderungs-beständig** *a* scale-resistant

verzweifeln *vi* (**an**) despair, lose hope (of). **verzweifelt** *p.a* desperate. **Verzweiflung** *f* desperation

verzweigen *vr*: **sich v.** branch (out), ramify. **Verzweigung** *f* (-en) branching, ramification. **Verzweigungs-punkt** *m* branch(ing) point

verzwickt *a* complex, involved, intricate

verzwillingt *p.a* twinned. **Verzwillingung** *f* (-en) twinning

verzwittern *vt* hybridize. **Verzwitterung** *f* (-en) hybridization

Vesuvian *m* (*Min.*) vesuvianite, idiocrase

Vesuvin *n* Bismarck brown

Vetter *m* (-n) cousin

vexieren *vt* tease; mislead

Vf. *abbv* (Verfasser) author; *pl:* **Vff.** authors

V-förmig *a* V-shaped

vgl. *abbv* (vergleiche) cf (compare), see: **vgl. a.** see also

vH, v.H. *abbv* (vom Hundert) percent

Vhdl(g) *abbv* (Verhandlungen) transactions, proceedings

Vh.Z. *abbv* (Verhältniszahl) proportional no.

Vibrations-rinne *f* vibrating trough

vibrieren *vt* vibrate, tremble, quiver

Vibro-tisch *m* vibrating table

Viburnit *n* viburnitol

Victoria-blau *n* Victoria Blue

Vidal-schwarz *n* Vidal black

Vieh *n* 1 cattle; livestock, farm animals: **zehn Stück V.** ten head of livestock. 2 animal, beast; brute. **-arzt** *m* veterinarian. **-dünger** *m* stable manure. **-futter** *n* cattle (*or* livestock) feed, fodder. **-salz** *n* cattle salt. **-seuche** *f* 1 (*genl*) livestock disease. 2 (*specif*) cattle plague. **-waschmittel** *n* livestock dip. **-zucht** *f* livestock (*or* cattle) raising

viel 1 *indef pron, adv* (*no endgs*), *a* (*with endgs*) much, a lot (of), a great deal (of), a large quantity (*or* number) (of); *pl* **viele** (*oft*) many: **v. kleiner** much smaller; **so v. ist gewiß** this much is certain; **v. zuviel** much too much; **das v—e Öl** the large amount of oil, all that oil; **gleich v. Wasser** just as much water, an equal amount of water; **unendlich v—e Verbindungen** an infinite number of compounds; **ziemlich v.** quite a lot; **v. anderes, v—es andere** much more, many other things *cf* SOVIEL, WIEVIEL, ZUVIEL. 2 *adv oft* very: **sich v. anders verhalten** behave very differently. 2 (*as pfx, oft*) poly-, multi-: **-artig** *a* varied, diverse. **-atomig** *a* polyatomic. **-benutzt** *a* much-used. **-deutig** *a* 1 ambiguous. 2 (*Math.*) many-valued. **V–eck** *n* polygon. **v–eckig** *a* polygonal

vielen-orts *adv* in many places

vielerlei *a* (*no endgs*) a variety of, many different kinds of

vieler-orts *adv* in many places

vielfach 1 *a* multiple, varied; frequent, repeated. 2 *adv* widely, variously; often; in many ways (*or* cases): **v. diskutiert** widely discussed, **v. verwendet** widely (*or* variously) used. 3: **V—e(s)** *n* multiple; large amount. **Vielfachheit** *f* multiplicity

Vielfalt *f* (*oft* + **an**) variety, multiplicity, diversity (of, in the way of)

vielfältig *a* varied, various, a variety of;

frequent, repeated; *oft* = VIELFACH.

Vielfältigkeit *f* variety, multiplicity

viel-farbig *a* many-colored, variegated; polychromatic. **V–flach** *m* polyhedron. **v–flächig** *a* polyhedral. **V–flächner** *m* polyhedron. **v–förmig** *a* multiform, polymorphous. **-gelesen** *a* much-read. **-genannt** *a* oft-mentioned. **-gestaltig** *a* many-shaped, multiform, polymorphous. **-gliedrig** *a* many-membered; (*Math.*) polynomial

Vielheit *f* multiplicity; multitude, large number

viel-jährig *a* many years of, lasting many years. **-kantig** *a* many-edged; polygonal. **-kernig** *a* multinuclear; many-grained

vielleicht *adv* perhaps, maybe; possibly; (*with verb oft*) may: **das ist v. wahr** that may be true

Vielling *m* (-e) (*Cryst.*) multiple twin

Viel-linienspektrum *n* many-line spectrum. **v–linig** *a* many-line, multilinear. **-malig** *a* repeated; frequent

vielmehr 1 *adv* rather. 2 *cnjc* but rather, but on the contrary; and what is more

viel-phasig *a* polyphase, multiphase. **-polig** *a* multipolar. **-sagend** *a* significant, meaningful, expressive. **-säurig** *a* polyacid. **-schichtig** *a* multilayered, multiply, stratified

vielseitig *a* 1 versatile. 2 many-sided, multifaceted; varied. 3 (*Geom.*) polyhedral. 4 (*adv oft*) widely, in many ways. **Vielseitigkeit** *f* 1 versatility. 2 many-sidedness, variety

viel-stufig *a* multi-stage. **-teilig** *a* of many parts, multipartite; (*Math.*) polynomial. **-verheißend, -versprechend** *a* (highly) promising. **V–wegehahn** *m* multiway (*or* four-way) stopcock. **v–wertig** *a* polyvalent, multivalent. **V–wertigkeit** *f* polyvalence, multivalence. **-zahl** *f* multitude, large number, multiplicity. **v–zellig** *a* multicellular. **V–zweck-koffer** *m* multi-purpose trunk

vier 1 *num* four. 2 (*as pfx oft*) quadri-, tetra-, quadruple. **Vier** *f* (-en) (figure) four

vier-atomig *a* tetratomic. **-basisch** *a* tetrabasic, quadribasic. **-drittel** *a* (*Salts*) containing four units of base to three of acid: **v. kohlensaures Natrium** trona

$(Na_2Co_3 \cdot 2NaHCO_3 \cdot 3H_2O)$

Viereck° *n* quandrangle, (*in popular usage usu*) square. **viereckig** *a* four-cornered, quadrangular; square

Vierer *m* (-) (number) four, figure four

viererlei *a* (*no endgs*) four kinds (of)

Vierer·vektor *m* four-component vector

vierfach 1 *a* fourfold, quadruple. 2: **V— e(s)** *n* quadruple, four times

vierfältig *a* fourfold, quadruple

Vier·farbendruck *m* four-color printing. **-flach, -flächner** *m* tetrahedron. **v— flächig** *a* four-faced, tetrahedral. **V–fuß** *m* four-footed stand. **v–gliedrig** *a* four-membered; (*Cryst.*) tetragonal; (*Math.*) quadrinomial. **-jährig** *a* four-year (-old); quadrennial

Vierkant *m* (-e) square. **vierkantig** *a* four-cornered; square. **Vierkant·stöpsel** *m* square-headed stopper

vier·komponentig *a* four-component. **-phasig** *a* four-phase. **V–pol** *m* quadripole. **v–polig** *a* four-pole, quadripolar. **-quantig** *a* four-quantum. **V— radantrieb** *m*, **v–radgetrieben** *a* four-wheel drive. **-reihig** *a* four-series, four-row. **V–ring** *m* four-membered ring. **v— säurig** *a* tetracid. **V–seit** m quadrilateral. **v–seitig** *a* four-sided, quadrilateral; square. **v–stellig** *a* four-place. **V–stoffsystem** *n* quaternary system, four-component system. **v–stufig** *a* four-stage. **V–stundenlack** *m* four-hour varnish

viert *a* 1 fourth. 2 **:zu v.** in a group of four

Viertakt *m* (*Mach.*) four-stroke cycle. **--maschine** *f*, **-motor** *m* four-stroke engine

vierteilen *vt* quarter. **vierteilig** *a* fourpart. **Vierteilung°** *f* quartering

viertel *a* (*no endgs*) fourth, quarter. **Viertel** *n* (-) 1 (*Math.*) quarter, fourth. 2 (*Wine, etc*) one-quarter-liter glass. 3 district, neighborhood

Viertel·drehung *f* quarter turn. **-flächner** *m* (*Cryst.*) tetartohedron. **v–geleimt** *a* (*Paper*) quarter-sized. **V–jahr** *n* quarter, three months. **-jahresschrift** *f* quarterly (review, journal). **v–jährig** *a* three-month. **jährlich** *a* quarterly. **V–kreis** *m* quarter circle, quadrant

vierteln *vt* quarter, cut into quarters; (*Sampling*) reduce by quartering

Viertel·stunde *f* 1 quarter of an hour, fifteen minutes. 2 little while. 3 short walk. **v–stündig** *a* fifteen-minute. **-stündlich** *a* every fifteen minutes. **V–wellenlänge** *f* quarter wavelength

viertens *adv* fourthly, in the fourth place

Vierundzwanzig·flächner *m* icositetrahedron

vier·wandig *a* four-walled. **V— weg(e)hahn** *m* four-way stopcock. **v— wertig** *a* quadrivalent, tetravalent. **V— wertigkeit** *f* quadrivalence, tetravalence. **v–zählig** *a* 1 quadruple, fourfold. 2 (*Ligands*) quadridentate

vierzehn *num* fourteen—**v. Tage** two weeks. **vierzehnt** *a* fourteenth

vierzig *num* forty. **vierzigst** *a* fortieth

Vigogne·garn *n* yarn spun chiefly from cotton waste

Vigoureux·garn *n* yarn spun from printed tops

Viktoria·grün *n* Victoria green; malachite green

Vinyl·benzol *n* vinyl benzene. **-cyanür** *n* vinyl cyanide, acrylonitrile. **-essigsäure** *f* vinylacetic acid. **-harz** *n* vinyl resin

vinylieren *vt* vinylate

vinylog *a* vinylogous

Viole *f* (-n) (*Bot.*) violet

Violen·wurzel *f* orris root

Violett·schwarz *n* violet black. **v–stichig** *a* violet-tinged

Violur·säure *f* violuric acid

Viren viruses: *pl of* **Virus**

virginisch *a* Virginian

Virial·satz *m* virial principle

virtuell *a* virtual

Virus·forschung *f* virus research. **-krankheit** *f* virus disease

Viset·holz *n* fustet, young fustic

Visier *n* (-e) (gun)sight

visieren I *vt/i* 1 take aim (at). **II** *vt* 2 gauge. 3 stamp (with a visa)

Visier·erz *n*, **-graupen** *fpl* (*Min.*) twinned form of cassiterite. **-lupe** *f* magnifying sight lens

Visiten·karte *f* calling (*or* name) card

visitieren *vt* visit, inspect; search

Viskogramm *n* (-e) (*Oil*) viscosity chart

viskos, viskös *a* viscous

Viskose·fibranne *f* viscose rayon (staple). **-(kunst)seide** *f* viscose rayon. **-ver-**

fahren n viscose process. **-zellwolle** f viscose rayon (staple)
viskosimetrisch a viscosimetric
Viskosität f (-en) viscosity
Viskositäts·messer m visco(si)meter. **-zahl** f viscosity number (or coefficient)
visuell a visual
Vital·färbung f vital (or intra-vitam) staining
Vitalität f vitality
vitamin-arm a low-vitamin, vitamin-poor. **V–bedarf** m vitamin requirement. **-gehalt** m vitamin content
vitamin(is)ieren vt vitaminize, enrich with vitamins. **vitamin(is)iert** p.a vitamin-enriched
Vitamin·mangel m vitamin deficiency. **v–reich** a vitamin-rich
vitrifizieren vt vitrify
Vitrine f (-n) (glass) showcase; display window
vitriol·artig a vitriolic. **V–äther** m vitriolic ether. **-blei** n lead vitriol (or sulfate). **-bleierz** n, **-bleispat** m (Min.) anglesite. **-erz** n vitriolic ore. **-fabrik** f sulfuric acid plant. **-flasche** f sulfuric acid carboy. **-gelb** n (Min.) jarosite. **v–haltig** a vitriolic, containing vitriol. **V–hütte** f sulfuric acid plant
vitriolisch a vitriolic
vitriolisieren vt vitriolate
Vitriol·kies m (Min.) marcasite. **-küpe** f blue vat, copperas vat. **-ocker** m (Min.) glockerite. **-öl** n oil of vitriol, sulfuric acid. **v–sauer** a sulfuric, sulfate of. **V–säure** f vitriol, sulfuric acid. **-schiefer** m pyritic shale (or schist), alum shale. **-siederei** f sulfuric acid works
vitrophyrisch a (Petrog.) vitrophyric
Vits·bohne f kidney bean
Vize·kanzler m vice-chancellor. **-präsident** m vice-president
vizinal a neighboring; (as pfx) local, town-
v. J(r). abbv 1 (vom Jahr) of the year. 2 (vorigen Jahres) of last year
vlämisch a Flemish
Vlies n (-e) fleece. **–stoff** m woolen (fabric)
Vlissingen n (Geog.) Flushing
vm. abbv (vormittags) in the morning, AM
v.M. abbv (vorigen Monats) of last month
v.o. abbv (von oben) from above, from the top

Vogel m (⁓) **-amber** m spermaceti. **-beerbaum** m mountain ash; service tree. **-beere** f serviceberry, rowanberry. **-beeröl** n mountain ash oil. **-dünger** m bird droppings, guano. **-dunst** m fine bird shot, dust shot. **-ei** n bird's egg. **-fraß** m crop damage by birds. **-kirsche** f wild cherry. **-kunde** f ornithology. **-leim** m birdlime, bird glue. **-mist** m bird droppings. **-mistbeize** f bird dung bate. **-perspektive** f bird's-eye view. **-zug** m bird migration
Vokabel f (-n) 1 vocabulary word. 2 pl = **Vokabular** n (-e) vocabulary
Vol. abbv (Volumen) vol. (volume). **Vol.-Gew.** abbv (Volumengewicht) vol. wt. (volume weight)
Volk n (⁓er) 1 people, nation, ethnic group. 2 (common) people. 3 crowd, bunch; folk(s). 4 masses. 5 swarm, flock
Völker·kunde f ethnology. **-recht** n international law
volk·reich a populous
Volks· popular, public, national, (esp DDR) people's: **-brauch** m popular usage. **v–eigen** a (DDR) nationalized, state-owned. **V–hochschule** f adult-education school. **-kunde** f folklore. **-meinung** f popular opinion. **-schule** f public elementary school
volkstümlich a popular, traditional, folk
Volks·wirtschaft f political (or national) economy. **-zählung** f census
voll I a 1 (usu) full. 2 whole, complete—**für v. nehmen** take seriously. 3 crowded—**gedrängt v.** jam-packed. 4 rich; solid—**ins v–e greifen, aus dem v–en schöpfen** draw on rich resources. 5 (Lthr.) plump, compact. 6 (Price, oft) gross. **II** adv 7 (oft) in full. **III** pfx 8 full, fully; whole, etc cf **I**. 9 (sep) full, to capacity. 10 (insep): implies completion, fulfilment. **IV** prep (accus) full of cf VOLLER. **V** sfx (oft) -ful, -ous: **wundervoll** wonderful, wondrous
volladen vt = **voll·laden*** load up, fill to capacity
Voll·analyse f complete analysis. **-appretur** f full finish
Vollast f = **Voll·last** full load
vollauf adv 1 fully. 2 plentifully; plenty (of)

Voll·ausschlag m (*Instrum.*) full-scale deflection. **-automat°** m fully automatic device, robot. **v–automatisch** a fully automatic. **V–bau** m solid construction (*or* structure). **-bleiche** f full bleach. **v–beschäftigt** a employed full-time. **V–beschäftigung** f full(-time) employment. **v–besetzt** a fully occupied, filled to capacity. **V–bier** n whole beer, beer with a high original wort. **v–blütig** a 1 full-blooded, thoroughbred. 2 plethoric

vollbracht(e) achieved, etc: *pp (& past) of* VOLLBRINGEN

vollbringen* (*insep*) vt achieve, accomplish, carry out

Voll·brot n whole meal bread. **-chromleder** n full chrome leather. **-dampf** m full steam pressure: **mit V.** (**voraus**) full-steam (ahead). **-draht** m solid wire. **-druck** m 1 full pressure. 2 (*Tex.*) full print. **-düngemittel** n, **-dünger** m complete (*or* compound) fertilizer

Volleder n (= **Voll·leder**) n full (unsplit, grain) leather

vollenden (*insep*) vt complete, round out, perfect. **vollendet** p.a (*oft*) perfect

vollends adv 1 fully, completely, wholly. 2 all the more (so)

Vollendung° f completion, perfection

Voll·entsalzung f complete desalination. **v–entwickelt** a fully developed

voller prep (+*genit*) full of

Voll·farbe f full color. **v–fett** a (*Cheese*) full-fat, whole-cream. **v–feuer** n full fire (*or* heat). **v–flächig** a (*Cryst.*) holohedral. **V–flächner** m (*Cryst.*) holohedron

vollführen (*insep*) vt carry out, effect, accomplish. **Vollführung** f carrying out, accomplishment

voll·füllen vt fill up, fill to capacity

Voll·gehalt m (*usu Coins*) full weight and value. **v–gesogen** p.a saturated cf VOLLSAUGEN

voll·gießen*vt pour full, fill to capacity

voll·gültig a fully valid. **V–gummi** n solid rubber. **v–haltig** a full-value

Vollheit f fullness

völlig a full, complete; (*oft adv*) fully, etc. **Völligkeit** f fullness, completeness

voll·jährig a of age. **Volljährigkeit** f majority

vollkommen a complete, perfect. **Vollkommenheit** f completeness, perfection

Voll·kornbrot n whole-grain bread. **-kornmehl** n whole-grain flour. **v–körnig** a whole-grain(ed). **V–kraft** f full vigor; peak, prime. **v–kristallin(isch)** a holocrystalline. **voll·l . . .** see **voll . . . V–macht** f full power; power of attorney. **-mehl** n whole-grain flour. **-milch** f whole milk. **v–mundig** a (*Beer*, etc) full-(bodied).**V–mundigkeit** f (*Beer*, etc) body, fullness

voll·packen vt pack (*or* cram) full, load up

Voll·pappe f millboard

voll·pfropfen vt cram full, stuff to capacity

Voll·pipette f volumetric (transfer, delivery) pipette. **-reife** f full maturity (*or* ripeness). **-reifen** m solid (rubber) tire. **-rindleder** n unsplit leather. **-salz** n iodized salt

voll·saugen* vr: **sich v.** be(come) saturated; (*Insect*) suck itself full

Voll·sitzung f full meeting, plenary session

vollständig a complete, total; (*Math.*) integral, whole. **Vollständigkeit** f completeness, wholeness

Voll·stange f solid bar. **-stein** m solid (unperforated) brick

vollstrecken (*insep*) vt carry out, execute, enforce

voll·synchronisiert a fully synchronized. **-synthetisch** a fully synthetic. **-transistorisiert** a fully transistorized. **V–ton** m full tone (*or* shade). **-treffer** m direct hit. **-versatz** m (*Min.*) solid packing. **v–wertig** a 1 full-value, fully valid. 2 full, complete. 3 fully qualified. 4 (*Food*) highly nutritious. 5 adv oft with all due respect. **-wichtig** a (*usu Metal*) full-weight. **-zählig** a 1 full, complete, full-strength, full-count. 2 adv (*oft*) in full number. **V–ziegel** m solid brick (*or* tile)

vollziehen* (*insep*) 1 vt carry out, execute, accomplish; enforce; sign (a document); consummate. 2: **sich v.** take place, come about

vollzog(en) carried out, etc: *past (& pp) of* VOLLZIEHEN

Vollzug m execution, accomplishment, completion; enforcement; consummation cf VOLLZIEHEN

vol.T. abbv (Volumenteil) part by volume

voltaisch a voltaic, galvanic. **Voltameter** n voltmeter

voltasche Säule Voltaic pile

Volt·messer m voltmeter. **-spannung** f (*Elec.*) voltage

Volum·abnahme f decrease in volume. **-änderung** f change in volume. **-dichte** f density by volume. **-einheit** f unit of volume

Volumen n (..mina) volume; *in compds* = VOLUM·; **-anzeiger** m volume indicator. **-arbeit** f work of expansion (*or* contraction). **-schwund** m volume shrinkage

Volumetrie f volumetric analysis. **volumetrisch** a volumetric

Volum·gesetz n law of volumes. **-gewicht** n volume weight, weight per unit volume

Volumina volumes: *pl of* VOLUMEN

voluminös a voluminous. **Voluminosität** f bulkiness

Volum·messer m volumeter. **-minderung** f decrease in volume. **-prozent** n percent by volume. **-prozentgehalt** m volume-percent content. **v–prozentig** a percent by volume. **V–teil** m part by volume. **-verhältnis** n proportion by volume. **-verlust** m loss in volume. **-vermehrung** f increase in volume. **-verminderung** f decrease in volume. **-zunahme** f, **-zuwachs** m increase in volume

vom *contrac* = **von dem**: **v. Hundert** percent

Vomhundert·gehalt m percentage content. **-satz** m percentage

vomieren vi vomit

von prep (*dat*) **1** of, from; off: **v. nun an** (or **ab**) from now on; **v. dort her** (or **aus**) (starting) from there *cf* AUS. **2** out of: **zwei v. fünf** two out of five. **3** (+*adv of place, oft,* e.g.): **v. vorne** (**a**) from the front, (**b**) in front. **4** (+*aus,* e.g.): **v. ihnen aus** (**a**) from them, (**b**) on their initiative, on their part. **5** (*oft* + **auf**) from, since: **v. Kind auf** since childhood *cf* JEHER, VORNHEREIN. **6** of, about (a matter, a topic). **7** (*esp* + *passive or author's name*) by. **8** (*as part of family name*), e.g.: **v. Schmidt**: *orig. a token of nobility. Cf* HAND, KLEIN, LEBEN, NEU **5**, SEITEN, SELBST, SINN, WEGEN, WEIT

voneinander adv of (from, by, about) one another; apart *cf* VON I

vonnöten adv necessary

vonstatten gehen proceed, go (ahead)

vor I prep A (*dat or accus*) **1** before, in front of, ahead of: *cf* AUGE. **2**: **v. sich gehen** proceed, go ahead. B (*dat*) **3** (*Time*) ago: **v. acht Tagen** a week ago *cf* KURZ **2**. **4**: **v. allem, v. allen Dingen** above all, especially. **5** (+*City*) outside. **6** in the presence of. **7**: **Respekt v.** . . . respect (esteem, regard) for . . . **8** (*Fear,* e.g.): **Angst haben v.** . . . be afraid of . . . **9** with: **v. Schmutz starren** be stiff with dirt; **v. Angst zittern** tremble with fear. **10** (*Clock Time*): **viertel v. acht** quarter to (or off) eight. **II** adv **11** forward, ahead. **12**: **nach wie v.** as before, always. **III** pfx A (*nouns & verbs*) **13** pre·, preliminary: *cf* VORARBEIT, VORWÄRMEN. B (*sep, with verbs*) **14** forward, ahead; before, in front. **15** (*implies presentation, esp to an audience*) *cf* VORLEGEN, VORLESEN. **16** (*implies reproachful confrontation*) *cf* VORHALTEN, VORWERFEN

vorab adv **1** to begin with, by way of introduction. **2** above all, especially

Vorabend° m eve

Vorahnung° f premonition

Voralarm° m early warning

voran adv & sep pfx (on) ahead, forward

voran·bringen* vt expedite, advance, move (sthg) forward

vorangegangen p.a earlier, previous, preceding *cf* VORANGEHEN

voran·gehen* vi(s) **1** progress. **2** (*oft* + *dat*) go (on) ahead (of), precede. **3** (*usu* + **mit**) set an example (by, with). **vorangehend** p.a preceding, foregoing: **aus dem V– en ersieht man** . . . from the foregoing we see . . . *Cf* VORANGEGANGEN

vorangestellt p.a. put in ahead, preceding *cf* VORANSTELLEN

voran·kommen* vi(s) advance, progress

voran·schicken vt **1** send ahead. **2** mention beforehand, (*esp* + *d.i.o*) put (in) ahead (of), let precede

Voranschlag° m (provisional) estimate; estimated budget

voran·stellen vt **1** begin with, put in front. **2** (+*d.i.o*) put (in) ahead (of)

Voranstrich° m (*Paint*) first (or priming) coat

voran·treiben* vt expedite, speed up

Voranzeige° *f* preliminary announcement; advance notice

Vorappretur° *f* preliminary finishing (*or* processing)

Vorarbeit° *f* preliminary work (study, project, etc), preparation. **vor·arbeiten 1** *vt* pretreat; prepare. **2** *vi* (+*dat*) lay the groundwork (for). **3: sich v.** work one's way ahead. **Vorarbeiter°** *m* foreman. **Vorarbeiterin°** *f* forelady

vorauf *adv* & *sep pfx* ahead, before. **vorauf·gehen*** *vi*(s) (*oft* + *dat*) go (on) ahead (of), precede

voraus I *adv* **1** (*dat*) ahead (of). **2: im v.** beforehand, in advance. **II** *sep pfx* ahead, in advance, pre-, fore-

voraus·berechnen *vt* precalculate. **Vorausberechnung°** *f* advance calculation, precalculation

voraus·bezahlen *vt* pay in advance, prepay

Vorausblick° *m* foresight. **voraus·blicken** *vi* look ahead (*or* forward). **vorausblickend** *p.a* forward-looking

vorausgegangen *p.a* earlier, previous, preceding *cf* VORAUSGEHEN

voraus·gehen* *vi*(s) (*oft* + *dat*) go ahead (of), precede. **vorausgehend** *p.a* preceding, foregoing

vorausgenommen anticipated: *pp of* VORAUSNEHMEN

vorausgesetzt *p.a* (*usu* + **daß**) assuming, provided (that) *cf* VORAUSSETZEN

voraus·haben* *vt* e.g.: **sie haben uns das voraus** they have the better of us in that

voraus·nehmen* *vt* anticipate, forestall

Voraussage° *f* prediction; forecast. **voraus·sagen** *vt* predict, foretell; forecast

vorausschauend *p.a* forward-looking

voraus·schicken *vt* **1** send ahead. **2** mention beforehand. **3** (+*d.i.o*) put (in) ahead (of), let precede

voraussehbar *a* foreseeable. **voraus·sehen*** *vt* foresee, predict

voraus·setzen *vt* assume, presuppose; take for granted; require, demand *cf* VORAUSGESETZT. **Voraussetzung** *f* (-en) **1** assumption, presupposition. **2** premise; prerequisite, (pre)condition, stipulation. **3** qualification. **voraussetzungs·los** *a* unconditional

Voraussicht° *f* foresight—**aller V. nach** in

all probability. **voraussichtlich** *a* foreseeable, probable; *adv oft* presumably

vor·bauen I *vt* **1** (*d.i.o.*) build in front (of). **2** build in advance (*esp* models). **II** *vi* **3** provide for the future. **4** (*dat*) forestall, head off, prevent. **Vorbauung** *f* forestalling; prevention; (*Med.*) prophylaxis. **Vorbauungs·mittel** *n* (*Med.*) prophylactic

Vorbearbeitung *f* **1** rough-work(ing). **2** preliminary processing (*or* treatment)

vorbedacht *a* premeditated, aforethought. **Vorbedacht** *m* premeditation: **mit (in, aus) V.** deliberately, premeditatedly

Vorbedingung° *f* precondition, prerequisite

Vorbehalt *m* (-e) (mental) reservation; prejudice; proviso. **vor·behalten* I** *vt*: **sich etwas v.** reserve sthg (for oneself). **II** *p.a* (rights) reserved —**es bleibt ihnen v., zu** . . . it remains for them to . . . **vorbehaltlich** *prep* (*genit*) reserving, subject to. **vorbehaltlos** *a* unconditional; *adv oft* without reservation(s)

vor·behandeln *vt* pretreat, condition; (*Med.*) give pre-operative treatment. **Vorbehandlung°** *f* pretreatment, (pre)conditioning; (*Med.*) pre-operative treatment; prefabrication

vorbei 1 *adv* over, gone, past. **2** *adv* & *sep pfx* (*esp* + **an, bei**) past, by (a place); beyond: **am Haus v.** past the house; **es ist zehn Uhr v.** it is past (*or* after) ten o'clock

vorbei·fahren* *vi*(s) (+**an, bei**) drive (ride, travel, move) past (*or* by)

vorbei·fliegen* *vi*(s) (+**an, bei**) fly by (*or* past)

vorbei·führen 1 *vt/i* lead past (*or* by). **2** *vt* pass (e.g., a gas over a substance)

vorbei·gehen* *vi*(s) **1** (*usu* + **an, bei**) go (by, past), pass (by); evade, avoid; miss. **2** (+**bei**) drop in (at). **Vorbeigehen** *n: im V.* in passing

vorbei·gleiten* *vi*(s) glide by (*or* past): **aneinander v.** glide (*or* move) past (*or* over) one another

vorbei·lassen* *vt* let by, let pass

vorbei·sehen* *vt* (+**an**) ignore

vorbei·streichen* *vi*(s) (+**an**) sweep by (*or* past)

vorbei·strömen *vi*(s) (+**an**) flow, stream by (*or* past)

vorbei·streifen *vi*(s) (+**an**) brush, graze

Vorbeize° f (Dye.) bottom (or preliminary) mordant. **vor·beizen** vt (Dye.) mordant previously; (Metll.) black-pickle

vorbei·ziehen* (an, bei) 1 vt pull by (or past). 2 vi(s) move by (or past)

vor·belichten vt pre-expose, pre-illumination. **Verbelichtung°** f pre-exposure; pre-illumination

Vorbemerkung° f preliminary remark, prefatory note

Vorbenutzung° f prior use

Vorberechnung° f preliminary calculation

vor·bereiten vt & sich v. (auf, für) prepare, get ready (for). **vorbereitend** p.a preparatory. **Vorbereitung°** 1 preparation: **V—en treffen** make preparations

Vorbereitungs· preparatory: **-arbeiten** fpl preparatory work (or studies). **-küche** f pre-cooking kitchen

Vorbericht° m preliminary report

vorbeschleunlgen vt preaccelerate

Vorbesprechung° f preliminary discussion; preparatory conference

vor·beugen 1 vt & sich v. bend forward. 2 vi take preventive measures; (+dat) guard (against), prevent, avoid. **vor·beugend** p.a preventive, prophylactic; (adv oft) as a precaution. **Vorbeugung** f precaution; preventive, prophylactic

Vorbeugungs·maßnahme f preventive (prophylactic, precautionary) measure. **-mittel** n preventive, prophylactic

Vorbild° n 1 model, example (to follow).2 prototype; standard

vor·bilden vt 1 preform. 2 pretrain, give preliminary training (or education) to

vorbildlich a exemplary, model; adv in an exemplary way

Vorbildung° f 1 preformation. 2 preparatory training (or education), educational background

vor·binden* vt tie on (an apron, etc)

Vorblatt° n (Bot.) bactreole

Vorbleiche° f prebleach

Vorblick° m 1 preview. 2: im V. auf in view of

Vorblock° m (Metll.) bloom

Vorbote° m herald, harbinger; early sign (or symptom)

Vorbrand° m preliminary firing; (Ceram.) biscuit firing

Vorbrecher° m primary crusher

vor·bringen* vt 1 bring up, present (a matter); allege, claim. 2 bring (or move) up, bring forward

vor·datieren vt predate, antedate

vor·decken vt (Dye.) ground

vordem adv before that

vor·demonstrieren vt demonstrate (to spectators)

vorder a & pfx front, anterior; pfx oft fore-: **V—ansicht** f front view. **-arm** m forearm. **-asien** n (Geog.) Asia Minor, Near East. **-ende** n front end. **-glied** n 1 (Math.) antecedent. 2 (Anat.) front limb

Vordergrund° m 1 foreground; forefront, leading position. 2 fore: **in den V. treten** (**in den V. rücken**) come to (to bring to) the fore

vorderhand adv for the present, for the time being; meanwhile

Vorder·herd m (Metll.) forehearth. **-hirn** n (Anat.) forebrain. **-rad** n front wheel. **-satz** m antecedent, premise. **-schicht** f front layer. **-seite** f front, face. **v—seitig** a front(al)

vorderst a furthest front, foremost, first

Vorder·teil m front (part). **-wand** f front wall; (Blast Furance) breast. **-würze** f (Beer) first wort

vor·drängen 1 vt push to the fore. 2: sich v. push (one's way) forward, press forward

vor·dringen* vi(s) push on, advance, make headway. **Vordringen** n advance, progress. **vordringlich** a urgent: **v. behandeln** give priority to

Vordruck° m 1 (Print.) first impression, proof. 2 form, blank (to fill out). 3 (Tex.: Calico) first (or bottom) printing. 4 (Wine) first pressing, new wine

vor·drucken vt print in front, prefix; print in advance; preprint

Vordruck·reserve f (Tex.) preprinted resist

vor·eilen vi(s) 1 rush ahead. 2 (Elec.: Phase) advance, lead. 3 (Metll.) slip forward. **voreilig** a rash, ill-considered, hasty. **Voreilung°** f 1 (Elec.: Phase) advance, lead. 2 (Metll.) forward slip. **Voreilungs·winkel** m angle of lead

voreinander adv one before the other; in each other's presence

voreingenommen a prejudiced. **Vorein-**

genommenheit *f* (-en) prejudice
vorein·stellen *vt* preset
voreiszeitlich *a* (*Geol.*) preglacial
Voreltern *pl* ancestors
vor·enthalten *vt* (*d.i.o*) withhold (from).
Vorenthaltung *f* (-en) withholding
Vorerhitzer° *m* (-) preheater. **Vorerhitzung** *f* preheating
vorerst *adv* for the time being, for the present
vorerwähnt *a* aforementioned
Vorerwärmung *f* preheating
Vorerzeugnis° *n* crude (*or* semifinished) product
Vorfabrikation° *f* prefabrication. **vor·fabrizieren** *vt* prefabricate
Vorfahr *m* (-en, -en) ancestor
Vorfahrt° *f* right of way, priority —**V. gewähren** yield
Vorfall° *m* **1** incident, event, occurrence. **2** (*Med.*) prolapsus
vor·fallen* *vi*(s) **1** happen. **2** (*Med.*) prolapse
vor·färben *vt* precolor; predye, ground, bottom. **Vorfärbung°** *f* precoloring; predyeing, etc
Vorfaul·becken *n* (*Sewage*) primary digestion tank
Vorfechter *m* (-) champion, pioneer
vor·fertigen *vt* prefabricate
Vorfeile° *f* bastard file
Vorfilter° *m* first (*or* coarse) filter. **vor·filtern** *vt* prefilter, pass through the first filter
vor·finden* **1** *vt* find (present), come upon. **2: sich v.** be present, be forthcoming
Vorflut° *f* **1** rising tide. **2** drainage ditch
vor·formen *vt* preform, make up in preliminary form. **Vorformling** *m* (-e) (*Plastics*) preform
Vorfrischen *n* preliminary refining
vor·führen *vt* **1** present. **2** show, display; demonstrate (appliances); project (films, slides). **Vorführung°** *f* presentation, performance; show; display; demonstration, projection
Vorgabe° *f* requirement; specification; cue
Vorgang° *m* **1** process, course (of the action); NOTE: *stresses spontaneous action in progress, while* VERFAHREN, *also transl* process, *refers to a planned method or pro*-

cedure. **2** (chemical) reaction. **3** event, occurrence; (*esp pl*) proceedings, goings-on. **4** file, records; previous correspondence. **5** (*Cmptrs.*) process. **6** (*Distil.*) first runnings. **Vorgänger** *m* (-) predecessor; earlier model
vorgängig *a* **1** preliminary; (*adv oft*) for the time being. **2** foregoing, previous; (*esp* + *dat*) preceding
Vorgärung° *f* preliminary fermentation
vor·geben* *vt* (*oft* + *d.i.o*) **1** persuade, get (sbdy) to believe; pretend, allege, claim; profess. **2** hand (to), pass up forward (to). **3** specify (rules, goals). **4** give (odds, headstart, handicap). **5** give in advance
Vorgebirge° *n* **1** headland; foothills. **2** cape, promontory
vorgeblich *a* alleged; would-be, so-called
vorgebracht brought up: *pp of* VOR-BRINGEN
vorgeburtlich *a* prenatal
vorgedrungen pushed on: *pp of* VORDRINGEN
vorgefaßt *a* preconceived; fixed, set
vorgefertigt *p.a* prefabricated
vorgeformt *p.a* preformed
vorgefrischt *a* semirefined; semipurified
Vorgefühl° *n* anticipation; presentiment
vorgefunden found, etc: *pp of* VORFINDEN
vorgegangen gone on, etc: *pp of* VORGEHEN
vorgegriffen anticipated, etc: *pp of* VORGREIFEN
vor·gehen* *vi*(s) **1** go up front, move ahead. **2** (+ *dat*) go ahead (of), precede. **3** proceed, take steps; act, take action. **4** (*Clocks*) be fast. **5** happen, occur, go on. **6** take precedence, have priority. **Vorgehen** *n* procedure, (course of) action; advance
vorgelagert *a* (*oft* + *dat*) extending out in front (of)
Vorgelege *f* countershaft; reduction gearing; back gear mechanism
vorgemischt *a* premixed; ready-mixed
vorgenannt *a* aforementioned
vorgenommen undertaken: *pp of* VORNEHMEN
vorgequollen precooked: *pp of* VORQUELLEN
vor·gerben *vt/i* pretan. **Vorgerbung°** *f* pretanning
vorgerieben preground: *pp of* VORREIBEN

vorgerückt *p.a* advanced, late (hour) *cf* VORRÜCKEN

Vorgeschichte *f* 1 prehistoric times. 2 prehistory, previous history, case history. **vorgeschichtlich** *a* prehistoric

vorgeschaltet *p.a* (*Elec.*): **v—er Widerstand** series resistor *cf* VORSCHALTEN

Vorgeschmack *m* foretaste

vorgeschmolzen premelted: *pp of* VORSCHMELZEN

vorgeschoben pushed ahead: *pp of* VORSCHIEBEN

vorgeschossen advanced, etc: *pp of* VORSCHIESSEN

vorgeschrieben 1 prescribed, etc: *pp of* VORSCHREIBEN. 2 *p.a* (*oft*) required

vorgeschritten advanced, etc: *pp of* VORSCHREITEN

vorgesetzt *p.a* superior (in rank) *cf* VORSETZEN. **Vorgesetzte** *m,f* (*adj endgs*) superior (officer), boss

vorgespannt *a*: **v—er Beton** prestressed concrete

vorgesprungen leaped out, etc: *pp of* VORSPRINGEN

vorgestanden stuck out, etc: *pp of* VORSTEHEN

vorgestochen stood out, etc: *pp of* VORSTECHEN

vorgestern *adv* day before yesterday, two days ago—**von v.** outdated. **vorgestrig** *a* day before yesterday's

vorgetrieben propelled: *pp of* VORTREIBEN

vorgewiesen shown: *pp of* VORWEISEN

vorgewogen predominated: *pp of* VORWIEGEN

vorgeworfen reproached, etc: *pp of* VORWERFEN

vorgezogen preferred, etc: *pp of* VORZIEHEN

Vorglühen *n* preliminary heating; (*Metll.*) annealing. **Vorglüh-ofen** *m* (*Metll.*) annealing oven (*or* furnace); (*Ceram.*) biscuit kiln

vor-greifen* *vi* act prematurely; (+*dat*) anticipate, forestall. **vorgreifend** *p.a* anticipatory

Vorgriff° *m* (+*auf*) anticipation (of)

vor-grundieren *vt* (*Dye.*) ground

vor-haben* *vt* 1 be planning, intend to do. 2 be busy with. 3 have on (in front). 4 call to account. **Vorhaben** *n* (-) intention, plan, project

Vorhalle° *f* entrance hall, foyer, lobby

vor-halten* I *vt*: **einem etwas v.** 1 hold sthg in front of sbdy, point sthg at sbdy. 2 reproach sbdy for sthg. 3 (+**als Muster**) hold up as an example. II *vi* hold out, last; be enough. **Vorhaltung°** *f* reproach, remonstration: **einem V—en machen wegen** . . . reproach sbdy for . . .

vorhanden *a* present; existing; available. **V—-sein** *n* presence; existence; availability

Vorhang° *m* curtain, shade; drop

Vorhänge-schloß *n* padlock

Vorhaut° *f* (*Anat.*) foreskin, prepuce

vor-heizen *vt* preheat

vorher *adv & sep pfx* previously, before, in advance; *pfx* (*oft*) pre-

vorher-bedenken* *vt* premeditate

vorher-bestimmen *vt* predetermine

Vorherd° *m* forehearth

vorhergegangen 1 preceded: *pp of* VORHERGEHEN. 2 *p.a* preceding, previous

vorher-gehen* *vi*(s) go before, precede. **vorhergehend** *p.a* (*oft*) previous, prior; foregoing

vorherig *a* previous, preceding

vor-herrschen *vi* predominate. **vorherrschend** *p.a* predominant

Vorhersage *f* (-n) prediction, forecast. **vorher-sagen** *vt* predict, forecast

vorhersehbar *a* foreseeable, predictable. **vorher-sehen*** *vt* foresee, predict

vorher-wissen* *vt* know in advance, anticipate

vorhin *adv* just now, just a while ago

vorhinein: im v. *adv* in advance, from the start

Vorhut *f* vanguard

vorig *a* 1 last, past (month, etc). 2 preceding; previous, former

vor-ionisieren *vt* pre-ionize

Vorjahr *n* preceding (*or* last) year, year before. **vorjährig** *a* last year's, the preceding year's

Vork. *abbv* (Vorkommen) occurrence, etc

vor-kalkulieren *vt* precalculate

Vorkammer° *f* 1 (*incl Mach.*) antechamber; precombustion chamber. 2 (*Anat.*) auricle

Vorkäsen *n* (*Cheese*) pretreatment of curd

Vorkehrung f (-en) arrangement, precautionary measure: **V—en treffen** take precautionary measures, make arrangements

Vorkenntnisse fpl previous (or preparatory) knowledge; rudiments

Vorklär·becken n preliminary sedimentation tank

Vorklotzung f (Tex.) pad ground

vor·kommen* vi(s) **1** (esp Chem., re Elements, etc) occur, be found. **2** happen, take place. **3** appear, show up. **4** (+ dat) happen (to sbdy); seem, look, appear (to sbdy). **5** come out (from). **6** come forward. **Vorkommen** n (-) occurrence, incidence: deposit, bed, vein (of ore, etc); habitat; presence, existence. **Vorkommnis** n (-se) incident, event

Vorkondensator° m preliminary condenser

Vorkonzentrat n master batch

Vorkost° f first course, appetizer

Vorkracken n, **Vorkrackung** f (Petrol.) primary cracking

Vorkriegs·zeit f pre-war era

vor·kühlen vt precool. **Vorkühler** m precooler, primary cooler. **Vorkühlung°** f precooling

Vorkultur° f preliminary culture

vor·küpen vt bottom with a vat dye

Vorlack° m priming lacquer, primer

Vorlage° f **1** (genl) sthg to put in front; (specif): trap, absorption bulb; (Distil.) receiver; (Zinc) condenser; (Metll., Coke Ovens) collecting main, off-take; (Coal Gas) hydraulic main; (Explo.) antiflash bag. **2** model, prototype; pattern. **3** subject (of discussion). **4** proposal, (legislative) bill. **5** presentation, submission (of evidence, documents, etc). **6** helping (of food). **·-flüssigkeit** f distillate, liquid in the receiver

Vorlaß m (Distil.) first runnings, etc = VORLAUF

vor·lassen* vt **1** let in, admit. **2** give precedence to. **3** let by, let pass

Vorlauf° m **1** (Distil.) first runnings; (specif): (Whiskey) foreshot; (Coal Tar) first light oil. **2** (Mach.) forward travel (stroke, speed)

Vorläufer° m forerunner, precursor; harbinger; (mountain) spur

vorläufig a preliminary; tentative, temporary; adv (oft) for the time being

Vorlaugung° f preliminary leaching

vor·legen vt (usu + d.i.o) **1** put (place, set) before; put on (lock, chain); slide (a bolt) over; apply. **2** present, submit; offer. **3** put, ask (a question) (to, of). **Vorleger** m (-) rug, mat

Vorlegierung° f key (or hardener) alloy, prealloy

Vorlese f (-n) early vintage

vor·lesen* **1** vt (oft + d.i.o) read out (or aloud) (to sbdy). **2** vt/i (über) lecture (on). **Vorlesung** f (-en) **1** (university professor's) lecture. **2** reading aloud

vorletzt a next to (the) last, (year, week, etc) before last; penultimate

Vorlicht° n priming illumination

Vorlickerung° f (Lthr.) preliminary fatliquoring

Vorliebe° f preference, special fondness

vor·liegen* vi (oft + dat) **1** lie before; be present, be available; (Results, etc) be known—**es liegt ihnen ein Bericht vor** they have a report before them (or in their hands); **hier liegt ein Irrtum vor** there is a mistake here. **2** exist. **3** (Locks, Bolts, etc) be on. **Vorliegen** n presence; existence. **vorliegend** p.a present, available; existing

vorm contrac = **vor dem** cf VOR

vorm. abbv **1** (vormals) form. (formerly). **2** (vormittags) A.M.

Vormacht(stellung) f dominant position

vor·mahlen vt pregrind. **Vormahlen** n pregrinding, preliminary grinding

Vormaisch·apparat m foremashing apparatus. **-bottich** m foremashing vat

vormaischen vt (Brew.) foremash. **Vormaischer** m (-) foremasher, premasher

vormalig a former, previous. **vormals** adv formerly

Vormann° m **1** foreman. **2** person in front

Vormaterial° n initial material

vor·merken vt make a note of, put sbdy's name down for a reservation (for an appointment, on a waiting list). **Vormerkung** f note, memorandum; notation on a reservation (appointment, waiting list)

Vormilch f colostrum, foremilk

Vormischung° f 1 premixing. 2 (*Rubber*) master batch

Vormittag° m forenoon, morning. **vormittags** adv mornings, in the forenoon

Vormontage° f preassembly. **vor·montieren** vt preassemble

Vormund m (-e or ⁻er) guardian

vorn adv in front, up front, at the beginning: **nach v.** toward the front, forward; **von v. bis hinten** from front to back; **von v. und hinten** front and back; **von v.** from the beginning; **wieder von v. anfangen** start all over again; **v. im Buch** at the front (or beginning) of the book

Vornahme f taking up, undertaking; performing, carrying out cf VORNEHMEN

Vorname° m first (or given) name

vorne adv in front, etc = VORN

vornehm a 1 distinguished, refined, high-class. 2 aristocratic. 3 (*Luster*) subdued

vor·nehmen* I vt 1 carry out, perform; undertake, take up, get busy on. 2: **sich etwas v.** resolve to do, plan sthg. II: **sich v.** (**zu . . .**) resolve, make up one's mind (to . . .)

vornehmlich adv mainly, chiefly, especially

vornehmst a 1 most important, main, principal. 2 superl of VORNEHM

Vornetz·bad n (*Tex.*) wetting-out bath. **vor·netzen** vt wet out

vornherein: **von v.** from the start (or beginning)

Vorniere° f head kidney, pronephros. **Vor·nieren·gang** m Wolffian duct

Vornorm° f tentative standard

vorn·über adv forward, head first

vorn·weg adv in front, ahead; (head) first

Vorort° m suburb; (in compds oft) suburban

Voroxidation° f previous oxidation

Vorpolieren n preliminary (or rough) polishing

Vorpolymerisat n (-e) initial polymerizate

Vorposten° m outpost

Vorpräparation° f previous (or preliminary) preparation (or treatment)

Vorpreßling° m (*Plastics*) preform

Vorprobe° f preliminary test

Vorprodukt° n 1 initial (primary, crude) product, intermediate. 2 crude benzol.

-·kühler m (*Coal Tar*) crude benzol condenser

Vorprüfung° f previous examination; preliminary examination (or test)

Vorpumpe f auxiliary pump; (*Brew.*) wort pump, circulator

vor·quellen* vi(s) 1 presoak, presteep. 2 gush out; bulge

Vorr. abbv (Vorrichtung) device

Vorraffination° f prerefining; (*Lead*) softening, improving

vor·ragen vi extend (or stick) out; stand out

Vorrang m priority, precedence; superiority

Vorrat° m (**an**) supply, stock; reserve (of): **auf V. haben** have in stock. **-·flasche** f stock bottle. **-gefäß** n stock vessel; reservoir. **vorrätig** a in stock, on hand

Vorrats·behälter m storage container (tank, bin, etc). **-bottich** m storage vat. **-eiweiß** n supply (or circulating) protein. **-faß** n storage vat (barrel, tank). **-flasche** f stock bottle. **-gefäß** n supply vessel, reservoir. **-kammer** f pantry; stock room, storeroom. **-lösung** f stock solution. **-raum** m stock room, storeroom. **-schädling** m pest infecting stored products, storage pest. **-tank** m storage tank. **-teil** m spare part

Vorreaktor° m preliminary reactor, prereactor

vor·rechnen vt 1 calculate, figure out (in advance). 2 (+ d.i.o) show how to calculate. **Vorrechnung°** f (advance) calculation; demonstration of calculation

Vorrecht° n privilege, prerogative

Vorrede° f introduction, preface; preamble

Vorreduktion° f prereduction; previous reduction. **Vorreduktions·ansatz** m (*Tex.*) prereduction paste

vor·reduzieren vt prereduce

vor·reiben* vt pregrind; rough-grind

vor·reifen vt age (in advance)

vor·reinigen vt preclean, prepurify. **Vorreinigung°** f precleaning, prepurification; preliminary cleaning (or washing); (*Gas*) primary cleaning

Vorreiter° m forerunner, pioneer

vorrelativistisch a (*Phys.*) prerelativistic

vor·richten vt prepare. **Vorrichtung°** f device, contrivance, appliance

Vorröste *f* preretting; = **Vorrösten** *n*, **Vorröstung°** *f* preliminary roasting, preroasting. **Vorröst·ofen** *m* preroasting furnace

vor·rücken *vt & vi*(s) move forward (*or* ahead); advance, progress

vors *contrac* = **vor das** *cf* VOR

Vorsatz° *m* **1** resolve, resolution; intent, intention: **einen V. fassen** make a resolution; **mit V.** = VORSÄTZLICH *as adv.* **2** (*Books*) endpaper. **vorsätzlich** *a* intentional, deliberate, premeditated

Vorsatz·linse *f* front (*or* amplifying) lens. **-papier** *n* endpaper

vor·schalten *vt* **1** (*Elec.*) connect in series; (*d.i.o*) connect into the circuit ahead of. **2** (*Tech.*) add, superpose

Vorschalt·widerstand *m* series (*or* rheostatic) resistance

vor·schärfen *vt* presharpen; prestrengthen

Vorschau *f* (-en) (**auf**) preview (of)

Vorschein *m*: **zum V. kommen** appear, show up, come to light; **zum V. bringen** reveal, bring to light

vor·schieben* **I** *vt* **1** push forward (*or* ahead), advance. **2** (*Mach.*) feed. **3** (*Bolt*) slide (*or* push) over. **4** use as an excuse. **II**: **sich v.** push one's way forward, press ahead

vor·schießen* **I** *vt* **1** (*d.i.o*) advance (money) (to). **2** (*Brew.*) circulate (wort). **II** *vi*(s) shoot (dash, dart) forward

Vorschlag° *m* **1** suggestion, proposal, proposition. **2** (*Metll.*) fusion, flux. **3** (*Book*) blank space at top of first page. **vor·schlagen*** *vt* suggest, propose; offer

Vorschlichten *n* rough-polishing

Vorschliff° *m* rough grind(ing)

Vorschmack *m* (-̈e) foretaste; predominant taste

vor·schmecken *vi* (*of a flavor*) predominate

vor·schmelzen* *vt* premelt, prefuse. **Vorschmelz·ofen** *m* premelting furnace

vorschnell *a* hasty, rash

vor·schreiben* *vt* (*usu* + *d.i.o*) **1** prescribe (for), dictate, direct, give directions; specify. **2** write a rough draft of. *Cf* VORGESCHRIEBEN

vor·schreiten* *vi*(s) stride forward, advance

Vorschrift° *f* directive, decree, dictate; rule, regulation, order; instructions, directions; specification(s): **nach V.** as directed, as specified, according to regulations; **Dienst nach V.** rule-book slowdown of service. **vorschrifts·mäßig** *a* prescribed, specified, directed; regulation; (*adv oft*) as directed (specified, etc), according to regulations

Vorschub *m* **1** (*Mach.*) feed(ing), advance. **2**: **V. leisten** (*dat*) foster, promote, encourage

Vorschule° *f* **1** pre-elementary school; preschool education. **2** schooling, experience, training

Vorschuß° *m* advance (of money)

vor·schützen *vt* use as an excuse; claim to have

vor·schweben *vi*: **es schwebt einem vor** one has it in mind, one dreams of it

vor·sehen* **1** *vt* plan, provide (for); designate (for a purpose). **2** *vi* peek out, show (*esp* from under sthg). **3**: **sich v.** (*oft* + *vor*) be careful (of), watch out (for). **Vorsehung** *f* providence

vor·setzen **1** *vt & sich v.* move up (*or* forward). **2** *vt* (*oft* + *d.i.o*) put in front (of), set before; prefix; serve, dish up

Vorsicht *f* **1** caution, care; (*as exclam*) watch out, beware!—**aus V., zur V.** to play safe. **2** foresight. **vorsichtig** *a* cautious, careful. **vorsichts·halber** *adv* for safety's sake, to play safe

Vorsichts·maßnahme, -maßregel *f* precautionary measure: **V—n treffen** take precautions

Vorsieden *n* preboiling

Vorsilbe *f* (-n) prefix

vor·sintern *vt* presinter, semisinter

Vorsitz° *m* chair(manship). **vor·sitzen*** *vi* (*dat*) preside (at, over). **Vorsitzende** *m, f* (*adj endgs*) chairman, president, presiding officer

Vorsorge *f* precaution; provision (for eventualities): **V. tragen** (*or* **treffen**) take precautions, provide (for). **vorsorgend** *p.a*, **vorsorglich** *a* precautionary; *adv usu* as a precaution

Vorsortierung° *f* presorting
Vorspannung° *f* 1 (*Metll.*) initial stress, prestress *cf* VORGESPANNT. 2 (*Elec.*) bias potential
Vorspektrum° *n* preliminary spectrum
vor·spiegeln *vt:* **einem etwas v.** delude sbdy into believing sthg. **Vorspiegelung** *f:* **unter V. falscher Tatsachen** under false pretenses
Vorspiel° *n* prelude, prologue
vor·springen* *vi(s)* 1 leap out; jump ahead. 2 jut out, protrude, be prominent. **vorspringend** *p.a* jutting, prominent; (*Colors*) glaring
Vorsprung° *m* 1 projection, projecting part; salient; promontory. 2 (*Med., Anat.*) process; spur; protuberance. 3 (*oft* + **vor**) head start, lead (on); advantage (over)
Vorstabilisator° *m* prestabilizer
Vorstadium° *n* early (*or* preliminary) stage
Vorstadt° *f* suburb(s)
Vorstand° *m* directorate, management, executive committee; manager, director, head. **Vorstands·vorsitzende** *m, f* chairman of the board; president
vor·stechen* 1 *vt* bradawl (holes). 2 *vi* stand out; stick out
vor·stecken *vt* 1 pin on (in front). 2 prefix. 3 move forward. 4 stick (*or* push) out (*or* forward)
vor·stehen* *vi* 1 stick (*or* jut) out; stand out. 2 (+ *dat*) precede, stand before. 3 (+ *dat*) preside (over), direct, administer. **vorstehend** *p.a* 1 preceding, foregoing: **im v—en** in the foregoing, above. 2 jutting, prominent
Vorsteher·drüse *f* prostate gland
vorstellbar *a* imaginable, conceivable
vor·stellen I *vt* 1 (*d.i.o*) introduce, present (to); point out (to). 2 represent; play (the part of). 3: **sich etwas v.** imagine sthg, conceive of sthg; (+ **unter**) understand (by). 4 move up (forward); stick out. 5 put in front. II: **sich v.** (+ *dat*) introduce oneself (to); have an interview (with)
Vorstellung° *f* 1 introduction, presentation; interview. 2 performance. 3 representation. 4 conception, idea, mental image, notion: **sich eine V. machen von**

have a conception (notion, etc) of. 4 imagination, mind. 5 *pl* remonstrations; objections
Vorstellungs·fähigkeit *f,* **-kraft** *f,* **-vermögen** *n* imaginative (*or* conceptual) capacity
Vorstoß° *m* 1 (*Distil.*) adapter. 2 edge, edging (strip); lap (of tiles); (*Tex.*) piping. 3 forward thrust (*or* push), drive, sally, advance; offensive. **vor·stoßen*** 1 *vt* push forward; thrust out. 2 *vi(s)* advance, venture forward; protrude
Vorstudie° *f* preliminary study (of a matter)
Vorstudium° *n* preparatory (course of) study
Vorstufe° *f* early (*or* preliminary) stage; prestage; (+ **zu**) first step (to, toward)
Vorsud *m* first boiling
Vortag° *m* previous day, day before
vor·täuschen *vt* simulate, fake, pretend to have
Vorteil° *m* advantage. **vorteilhaft** *a* favorable, advantageous, beneficial
Vortiegel° *m* 1 (*Lead*) outer basin, lead pot. 2 (*Tin*) forehearth
Vortrag *m* (∺e) 1 (public) lecture. 2 recitation, recital; report. 3 (manner of) performance, playing, delivery. 4 (*Com.*) carryover; balance
vor·tragen* I *vt* 1 present, perform. 2 carry up forward. 3 (*Com.*) transfer, bring forward. II *vi* speak, lecture; perform, play, recite; report. **vortragend** *p.a* acting (chairman, etc)
Vortrags·tisch *m* lecture table
vortrefflich *a* excellent, outstanding
vor·treiben* 1 *vt* drive forward, propel. 2 *vi(s)* advance, progress
vor·treten* *vi(s)* step forward, advance; stand out; jut out, protrude
Vortrieb° *m* 1 (*Min.*) drifting; rate of advance. 2 (*Phys.*) propulsion, propulsive thrust
Vortritt° *m* priority, precedence: **den V. haben** have priority, go first
vor·trocknen *vt/i* predry. **Vortrockner** *m* predrier; (*Paper*) receiving drier
vorüber 1 *adv & sep pfx* past, by. 2 *adv* over, done, gone (by)

vorüber·gehen* *vi*(s) (+**an, bei**) go (by, past), pass (by); avoid, bypass. **vorüber·gehend** *p.a* temporary; *adv oft* for the time being

Voruntersuchung° *f* preliminary investigation

Vorurteil° *n* prejudice, bias

Vorvakuum° *n* preliminary vacuum

Vorverbrennung° *f* precombustion

Vorverdampfer° *m* pre-evaporator. **vorverdampft** *a* pre-evaporated

Vorverdauung *f* predigestion

vor·verdichten *vt* precompress; supercharge

Vorverfahren° *n* preliminary procedure (*or* process)

Vorveröffentlichung° *f* prior publication

vor·verpacken *vt* prepack

Vorverstärker° *m* pre-amplifier

Vorversuch° *m* preliminary experiment (*or* test)

Vorvulkanisation° *f* prevulcanizing. **vor·vulkanisieren** *vt* prevulcanize

Vorwachs *n* bee glue, propolis

Vorwahl·nummer *f* (*Tlph.*) area code

vor·walken *vt* (*Tex.*) scour

vor·walten *vi* prevail, predominate

vor·walzen *vt* (*Metll.*) rough down (blooms)

Vorwalz·platte *f* (*Metll.*) hot-rolled plate. **-werk** *n* (*Metll.*) roughing mill (*or* rolls)

Vorwand *m* (-̈e) pretext, excuse

Vorwärm·apparat *m* preheating (*or* forewarming) apparatus

vor·wärmen *vt* preheat, forewarm. **Vorwärmer°** *m* preheater, forewarmer

Vorwärme·rohr *n* preheating tube (*or* pipe)

Vorwärm·ofen *m* preheating furnace; annealing oven. **-schrank** *m* preheating cabinet

Vorwärmung *f* preheating, forewarming

Vorwärm·zone *f* zone of preparatory heating

Vorwarnung° *f* early warning

vorwärts *adv & sep pfx* foward, (on)ward, ahead

vorwärts·bringen* *vt* promote, make progress with

vorwärts·entwickeln: sich v. progress, advance

Vorwärts·gang *m* forward movement, progress. **vorwärts·gehen*** *vi*(s) go (move, get) ahead (*or* forward), advance, progress

vorwärts·kommen* *vi*(s) get ahead, make headway

vorwärts·treiben* *vt* drive forward, propel

vorwärtsweisend *p.a* progressive

Vorwäsche *f*, **Vorwaschen** *n* prewash, preliminary washing

Vorwegnahme *f* anticipation. **vorweg·nehmen*** *vt* anticipate

Vorwein *m* (*Wine*) first runnings

vor·weisen* *vt* **1** show, present (*esp* for inspection). **2: v. können** have (sthg) to show

Vorwelt° *f* prehistoric world. **vorweltlich** *a* prehistoric

vor·werfen* *vt* **1** throw forward. **2** (*d.i.o*) throw (in front of). **3** (*d.i.o*) reproach (for, with)

Vorwiderstand° *m* (*Elec.*) series resistance

vor·wiegen* *vi* predominate. **vorwiegend** *p.a* predominant

Vorwissen *n* foreknowledge

Vorwort *n* (-e) foreword, preface, prologue

Vorwurf° *m* reproach, rebuke; blame

Vorzahl° *f* coefficient

Vorzählung° *f* enumeration, counting out

Vorzeichen° *n* **1** (early) indication, symptom; omen. **2** (*Math.*) sign. **3** (*Cmptrs.*) polarity symbol. **-umkehr** *f*, **-wechsel** *m* (*Math.*) change of sign

vor·zeigen *vt* show, present (*esp* for inspection)

Vorzeit *f* prehistoric times

vorzeitig *a* early, premature, untimely

Vorzerkleinerung *f* preliminary crushing

Vorzerlegung° *f* **1** preliminary decomposition. **2** crude fractionation

vor·ziehen* **I** *vt* **1** prefer. **2** advance the date of. **3** draw (curtain). **4** pull forward (*or* out). **II** *vi*(s) move forward

Vorzimmer° *n* anteroom; outer office, reception desk. **-dame** *f* receptionist

Vorzug° *m* **1** preference; priority, precedence. **2** advantage; asset, merit, virtue. **3** privilege. **4** superiority. **vorzüglich** *a* **1** preferable. **2** excellent, outstanding, first-class. **3** *adv oft* mainly, chiefly

Vorzugs· preferred, preferential: **-aktien**

fpl preferred stocks. **-milch** *f* certified milk. **-richtung** *f* preferred (*or* ruling) direction. **-stellung** *f* preferential position. **v–weise** *adv* 1 preferably, by preference. 2 especially, pre-eminently

Vorzündung° *f* pre-ignition

V.St.A. *abbv* (Vereinigte Staaten von Amerika) USA (United States of America)

vT, v.T. *abbv* (vom Tausend) per thousand

VT *abbv* (Volumenteil) part by volume

v.u. *abbv* (von unten) from below (*or* underneath), from the bottom

Vulkan *m* (-e) volcano. **-asbest** *m* vulcanized asbestos. **-artikel** *m* (*Dye.*) vulcanizing style. **-fiber** *f* vulcanized fiber. **-gas** *n* volcanic gas. **-glas** *n* volcanic (*or* tempered) glass

Vulkanisat *n* (-e) (*Rubber*) vulcanizate

Vulkanisations-agens *n* vulcanizing agent. **-artikel** *m* (*Dye.*) vulcanizing style. **-bereich** *m* (*Rubber*) curing range. **-beschleuniger** *m* vulcanization accelera-

tor. **-einsatz** *m* (*Rubber*) start of the cure. **v–hemmend** *a* vulcanization-retarding. **V–mittel** *n* vulcanizing agent. **-probe** *f* vulcanization test (*or* sample). **-verzögerung** *f* retardation of vulcanization

Vulkanisator *m* (-en) vulcanizer

vulkanisch *a* volcanic

vulkanisier-echt *a* vulcanization-resistant

vulkanisieren *vt* vulcanize; (*Rubber*) cure; (*Tires*) recap. **Vulkanisierer** *m* (-) vulcanizer

Vulkanisier-ofen *m* vulcanizing (*or* curing) oven. **-presse** *f* vulcanizing press. **-trog** *m* (*Rubber*) curing trough

Vulkanisierung *f* (-en) vulcanization

Vulkanismus *m* (*Geol.*) volcanism

Vulkanit *m* volcanic rock, (*Min.*) volcanite

Vulkan-öl *n* mineral lubricating oil

Vulpin-säure *f* vulpinic acid

VZ *abbv* (Verseifungszahl) saponification number

v.Zw. *abbv* (vor Zeitenwende) B.C.

W

w. *abbv* (warm) warm, hot

W *abbv* **1** (Watt). **2** (*Elec.:* Wechselstrom) AC. **3** (Wert) value, valence. **4** (Wolfram) tungsten

W. *abbv* **1** (Wasser) water. **2** (Wert) value, valence. **3** (Widerstand) resistance

Waag·balken *m* balance beam

Waage *f* (-n) **1** balance, scales; weighing machine (*or* platform); **sich** (*or* **einander**) **die W. halten** counterbalance (each other). **2** hydrometer. **3** (carpenter's) level. **4** (*Astron.*) Libra. **--arm** *m* arm of the balance. **-balken** *m* balance beam. **-gehäuse** *n* balance case. **-haus** *n* public scales. **-kasten** *m* balance case. **w–recht** *a* horizontal. **W–schale** *f* balance pan (*or* tray). **-zunge** *f* balance needle (*or* pointer)

waag·recht *a* horizontal. **W–schale** *f* balance pan (*or* tray)

Waare *f* (-n) goods: *see* WARE

Wabe *f* (-n) honeycomb. **w--artig** *a* honeycombed, alveolar. **W–honig** *m* comb honey. **-linse** *f* multicellular lens. **-theorie** *f* alveolar theory (of protoplasm)

wach 1 *a* awake, waking; wide awake; alert: **w. werden** wake up. **2** *sep pfx* awake

Wache *f* (-n) **1** watch, guard; guard duty. **2** vigil. **3** guardhouse; police station. *Cf* WACHT

wachen *vi* be awake; watch. **wachend** *p.a* (*oft*) awake. **wach·halten*** *vt* keep (sbdy, sthg) awake

Wach·mittel *n* antisoporific

Wacholder *m* juniper (tree, brandy). **--beere** *f* juniper berry. **-beeröl** *n* oil of juniper berries. **-branntwein, -geist** *m* (Holland) gin, Jenever. **-harz** *n* juniper resin, sandarac. **-öl** *n* juniper oil— **brenzliches W.** oil of cade. **-spiritus** *m* (*Pharm.*) spirit of juniper. **-teer** *m* juniper tar, oil of cade

wach·rufen* *vt* awaken, rouse; revive

Wachs *n* (-e) wax; (*in compds also*) waxen: **--abdruck** *m* wax impression. **w–**

ähnlich *a* wax-like, waxy. **W–alaun** *m* crystallized alum

wachsam *a* watchful, vigilant

Wachs· wax(en): **-appretur** *f* wax finish. **-art** *f* type of wax. **w–artig** *a* waxy, wax-like. **W–aufnahme** *f* wax recording. **-ausscheidung** *f* secretion of wax. **-ausschmelzverfahren** *n* (*Casting*) lost-wax process. **-baum** *m* wax myrtle, bayberry tree. **-beize** *f* wax stain. **-bildnerei** *f* modeling in wax. **-bleiche** *f*, **-bleichen** *n* wax bleaching. **-boden** *m* wax cake. **-bohnermasse** *f* polishing wax. **-bottich** *m* (*Alum*) roching cask. **-draht** *m* wax-insulated wire. **-drüse** *f* ceruminous gland

wachsen* I (wuchs, gewachsen; wächst) *vi*(s) **1** grow, increase; (*incl Moon*) wax. **2** rise (up). **3** crystallize. **4** (*Malting*) sprout. **5** (*Lime*) swell. *Cf* GEWACHSEN. **II** (*reg*) *vt* wax, spread wax on

wächsern *a* wax, waxen

Wachs· wax(en): **-farbe** *f* wax color; encaustic paint. **w–farben, -farbig** *a* wax-colored. **W–firnis** *m* wax varnish. **-form** *f* wax mold. **-gagel** *m, f* (*Bot.*) wax myrtle, candleberry. **-gehalt** *m* wax content. **w–gelb** *a* wax-yellow. **W–glanz** *m* waxy luster. **w–glänzend** *a* wax-lustered. **W–handel** *m* wax trade. **-harzreserve** *f* (*Tex.*) wax-resin resist

wachsig *a* waxy; (*Med.*) amyloid (liver)

Wachs·kaschierung *f* wax backing. **-kerze** *f* wax candle. **-kitt** *m* wax cement; luting wax. **-kohle** *f* paraffin coal. **-kuchen** *m* cake of wax. **-leim** *m* wax glue. **-leinen** *m*, **-leinwand** *f* oilcloth. **-machen** *n* roching (*or* crystallization) of alum. **-maisstärke** *f* waxy cornstarch. **-malerei** *f* encaustic (*or* wax) painting, cerography. **-malkreide** *f* crayon. **-masse** *f* wax composition. **-myrte** *f* (*Bot.*) wax myrtle. **-öl** *n* wax oil. **-palme** *f* (*Bot.*) wax palm. **-papier** *n* wax(ed) paper. **-paste** *f* wax paste; cerate. **-pauspapier** *n* waxed tracing paper. **-perle** *f* wax bead (*or* pearl). **-pflaster** *n* wax plaster; cerate. **-präparat** n preparation in wax. **-raum** m growing space cf WACHSEN I. **-reserve** f (Tex.) wax resist. **-salbe** f wax-base ointment; cerate. **-scheibe** f cake of wax, wax disk. **-schmelze** f wax-melting unit. **-seife** f wax soap. **-sonde** f (Med.) wax bougie. **-stock** m wax taper (or candle)

wächst grows: pres of WACHSEN I

Wachs·tafel f wax tablet. **-taffet** m oiled (or waxed) silk. **-tuch** n oilcloth. **-tuchlack** m oilcloth varnish

Wachstum n growth; increase, accretion— **eigenes W.** home-grown product. **w-·hemmend** a growth-inhibiting

Wachstums·branche f growth industry. **w–fähig** a viable. **-fördernd** a growth-stimulating. **W–geschwindigkeit** f rate of growth (increase, accretion). **w–hemmend** a growth-inhibiting. **W–kurve** f growth curve. **-mittel** n growth stimulant. **-reaktion** f propagation reaction. **-ruhe** f dormancy. **-schicht** f (Bot.) growth layer, cambrium. **-vorgang** m growth process. **-zentrum** n center of growth, nucleus

Wachs·überzug m wax coating. **w–weich** a soft as wax. **W–zündholz** n wax match, vesta

Wacht f (-en) guard: **auf der W.** on guard; **W. haben** be on guard duty, stand guard = WACHE

Wachtel f (-n) quail

Wächter m (-) watchman, guard; custodian

wackelig a shaky, rickety; wobbly, loose

Wackel·kontakt m (Elec.) loose contact (or wire)

wackeln vi totter, wobble; (oft + **an**) shake, rock (sthg)

Wackenrodersche Flüssigkeit Wackenroder's solution

wacker a valiant; honest, upright

wacklig a shaky, etc = WACKELIG

Wade f (-n) calf (of the leg). **Waden·bein** n fibula

Wad·erz n (Min.) wad

w.a.f. abbv (wasser- und aschefrei) d.a.f. (dry ash free)

Waffe f (-n) **1** weapon, pl oft arms. **2** (Mil: in compds.) branch of service, force cf LUFTWAFFE

Waffel f (-n) waffle; wafer

Waffen·dienst m military service. **-fabrik** f arms factory. **-fett** n gun (or rifle) grease (or oil). **-haus** n arsenal, armory. **w–los** a unarmed. **W–stillstand** m armistice, truce. **-technik** f weaponry

waffnen vt arm

Wag· = WAAG·, cf WAAGBALKEN etc

wägbar a weighable; ponderable. **Wägbarkeit** f weighability; ponderability

Wage f (-n) balance, etc: now spelled WAAGE

Wäge·bühne f weighing platform. **-bürette** f weighing burette. **-fläschchen** n weighing bottle. **-garnitur** f weighing set. **-gegengewicht** n counterweight. **-glas, -gläschen** n weighing glass (or bottle)

wagen 1 vt risk, venture; stake. **2** vi (+**zu**) dare, venture (to). **3: sich w.** venture, risk going (somewhere)

wägen* (wog, gewogen) vt weigh; ponder

Wagen m (- or ⸚) wagon; car; (Typewriter) carriage; (Astron.) Dipper. **-fett** n axle grease. **-heber** m car jack. **-·lack** m automobile (or wagon) varnish. **-länge** f car length. **-last** f carload; wagon (or truck) load. **-öl** n car (or wagon) oil. **-schmiere** f axle grease. **-spur** f wheel (or tire) track

Wäge·pipette f weighing pipette

Wäger m (-) weigher

Wäge·röhrchen n, **-röhre** f weighing tube. **-schale** f weighing pan (dish, tray). **-schiffchen** n weighing boat. **-substanz** f substance to be weighed. **-vorrichtung** f weighing device. **-zimmer** n weighing (or balance) room

Waggon m (-s) wagon, car = WAGEN

Wagnis n (-se) risk, gamble

Wägung f (-en) weighing

Wahl f (-en) choice, selection; option; election

wählen 1 vt/i choose, select; (Tlph.) dial. **2** vt elect. **3** vi vote. **Wähler** m (-) **1** chooser, selector; voter. **2** (Tlph) dial; (TV) channel selector. **wählerisch** a choosy, discriminating; fastidious. **Wähler·scheibe** f (Tlph.) dial

Wahl·fach n elective (subject). **w–frei** a elective, optional. **-los** a indiscriminate; nonselective, random; adv oft at random. **W–recht** n right to vote, franchise

Wähl·schalter *m* selector switch. **-scheibe** *f* preselection dial; (*Tlph.*) dial; (*TV*) channel selector

Wahl·spruch *m* motto, slogan. **-verwandtschaft** *f* elective affinity—**einfache W.** simple substitution; **doppelte W.** double decomposition. **w—weise** *adv* by choice, alternatively, at will

Wahn *m* illusion, delusion; mania. **--bild** *n* illusion, phantom; hallucination

wähnen 1 *vi* imagine, have the delusion. **2: sich w.** (+*adv*) imagine oneself (to be)

Wahn·sinn *m* madness, insanity. **w—sinnig** *a* mad, insane. **W—vorstellung** *f* delusion

wahr I *a* 1 true; real, genuine: **das W—e** the truth; **w. machen** make come true; **w. werden** come true. **2** veritable: **eine w—e Flut von** ... a veritable flood of ... **II** *sep pfx: cf* WAHRNEHMEN, WAHRSAGEN, ETC

wahren *vt* maintain, preserve; guard; uphold, observe. *See also* GEWAHREN

währen *vi* last, endure, continue

während 1 *prep* (+*genit*) during, in the course of. **2** *subord cnjc* while, whereas. **3** *p.a* lasting *cf* WÄHREN. **--dem, -dessen** *adv* meanwhile

wahrgenommen perceived, etc: *pp of* WAHRNEHMEN

wahr·haben *vt:* **etwas nicht w. wollen** refuse to admit sthg

wahr·haft *a* true, real, genuine = **wahrhaftig** *a; usu adv* really, truly

Wahrheit *f* (-en) truth; reality

wahrlich *adv* really, truly; certainly

wahr·machen *vt* make come true *cf* WAHR

wahrnehmbar *a* perceptible, noticeable. **Wahrnehmbarkeit** *f* perceptibility, noticeability

wahr·nehmen* 1 *vt* perceive, notice. **2** observe (and carry out). **3** look out for (interests). **4** make use of, assert (rights, etc). **Wahrnehmung** *f* (-en) perception; observation; attention (to); assertion. **Wahrnehmungs·fähigkeit** *f*, **-vermögen** *n* perceptiveness

wahr·sagen 1 *vt/i* predict. **2** *vi* (+*dat*) tell (sbdy's) fortune

wahrscheinlich *a* probable, likely; plausible. **Wahrscheinlichkeit** *f* (-en) probability, likelihood; plausibility

Wahrscheinlichkeits·dichte *f* probability density. **-gesetz** *n* probability law. **-kurve** *f* probability curve. **-rechnung** *f* probability calculus; calculation of probabilities. **-wert** *m* probable value

Wahr·spruch *m* verdict; judgment

Wahrung *f* maintenance, preservation; safeguarding; upholding, observance *cf* WAHREN

Währung *f* (-en) currency. **Währungs·kurs** *m* rate of (currency) exchange

Wahr·zeichen *n* landmark; symbol, emblem

Waid *m* (-e) (dyer's) woad (blue dye). **--blau** *n* woad (blue). **-färber** *m* woad dyer. **-küpe** *f* woad vat. **-küpenschwarz** *n* woaded logwood black

Wal *m* (-e) whale

Wald *m* (¨er) forest, wood(s). **--ahorn** *m* sycamore. **-bau** *m* forestry. **-baum** *m* forest tree. **-boden** *m* forest soil. **-brand** *m* forest fire

Waldensche Umkehrung Walden inversion

Wald·gewächs *n* forest growth (*or* plants). **-humus** *m* forest humus, leaf mold

waldig *a* wooded, woody

Wald·kirsche *f* wild cherry. **-malve** *f* mallow. **-meister** *m* woodruff. **-rebe** *f* clematis. **-sterben** *n* dying forest(s), death of the forest(s) (due to acid rain)

Waldung *f* (-en) woodland, forest

Wald·wolle *f* pine (-needle) wool. **-wollöl** *n* pine-needle oil

Walfisch *m* whale. **--öl** *n* whale oil. **-speck** *m* whale blubber. **-tran** *m* whale (*or* train) oil

Walk·brühe *f* (*Tex.*) milling liquor

Walke *f* (*Tex.*) fulling, milling; fulling machine (*or* mill)

walk·echt *a* fulling-resistant, fast to fulling (*or* milling)

walken *vt* 1 (*Tex.*) full, mill; felt. 2 (*Lthr.*) beat, drum, mill. 3 (*Tires*) flex. 4 (*Dough*) roll. **Walker** *m* (-) 1 fuller; felter. 2 rolling pin. 3 June bug

Walk·erde *f* fuller's earth

Walker·distel *f* fuller's teasel

Walkerei *f* (-en) 1 (*Tex.*) fulling mill. 2 (*Lthr.*) drumming shop

Walker·erde *f* fuller's earth. **-seife** *f* fuller's (*or* milling) soap. **-ton** *m* fuller's earth

Walk· (*Tex.*) fulling, milling: **-fähigkeit** *f* fulling (milling, felting) property. **-faß** *n* (*Lthr.*) drum, tumbler. **-fett** *n* 1 (*Tex.*) fulling (*or* milling) fat. 2 (*Lthr.*) dressing grease. **-flüssigkeit** *f* milling liquor. **-mittel** *n* fulling (*or* milling) agent. **-seife** *f* fuller's soap. **-ton** *m* fuller's earth

Wall *m* (⁻e) bulwark, rampart; embankment, bank; wall

Wallach *m* (-e) gelding

wallen *vi* 1 boil (up), bubble. 2 roll, heave, billow. 3 flow, wave, undulate

Walloonen·arbeit *f* Walloon process. **-eisen** *n* Walloon iron. **-frischen** *n*, **-schmiede** *f* Walloon process

Wall·stein *m* (*Blast Furnace*) dam, damstone. **-steinplatte** *f* dam plate

Wallung *f* (-en) 1 (rolling) boil; boiling, ebullition. 2 rolling, heaving, billowing. 3 undulation

Walnuß° *f* walnut. **-baum** *m* walnut tree— **grauer W.** butternut tree; **weißer W.** hickory tree. **w–groß** *a* walnut-sized. **W– größe** *f* walnut size, size of a walnut. **-öl** *n* walnut oil

Wal·rat *n* spermaceti. **-ratöl** *n* sperm oil. **-roß** *n* walrus

Wälsch·korn *n* (Indian) corn, maize

walten *vi* 1 (*oft* + **über**) govern, manage; take charge, be in charge (of); be at work, prevail—**hier waltet ein Mißver- ständnis** there is a misunderstanding here

Wal·tier *n* cetacean. **-tran** *m* whale oil

walzbar *a* rollable *cf* WALZEN

Walz·blech *n* rolled plate, rolled sheet (of) metal. **-blei** *n* sheet lead. **-draht** *m* rod wire; wire rod

Walze *f* (-n) 1 (*Mach., Metll.*) roller, roll. 2 (*Mach., Math.*) cylinder; spindle; (*incl Cmptrs*) drum

Walz·eisen *n* rolled (*or* drawn) iron; pin; axle

walzen *vt* roll, mill

wälzen I *vt* 1 roll over (*or* around), turn over: **w–de Reibung** rolling friction. **2: von sich w.** shake off, get rid of. 3 (+ **auf**) fob off (on, onto). **II: sich w.** roll (around), revolve; writhe, wallow

Walzen·apparat *m* (*Dye.*) rolling frame. **-brecher** *m* roll crusher. **-bürste** *f* revolving brush. **-drehzahl** *f* roll(er) velocity (*or* rpm). **-druck** *m* 1 roller pressure. 2 cylinder printing. **w–förmig** *a* cylindrical. **-getrocknet** *a* drum-dried. **W–glas** *n* cylinder glass. **-glättwerk** *n* (*Paper*) calender. **-gußeisen** *n* chilled rolled iron. **-kessel** *m* cylindrical boiler. **-lager** *n* roller bearing. **-mühle** *f* roller mill, roll crusher. **-satz** *m* set of roll(er)s, roll train. **-sinter** *m* mill scale. **-spalt** *m* gap between roll(er)s. **-straße, -strecke** *f* roll train. **-stuhl** *m* rolling mill. **-trockner** *m* roller (*or* drum) drier. **-vorschub** *m* (*Mach.*) roller train

Walzer *m* (-) waltz

Wälzer *m* (-) 1 roller. 2 big tome

Walz·erz *n* ore for crushing. **-erzeugnis** *n* rolled product. **-fehler** *m* rolling (*or* milling) defect. **-fell** *n* (*Rubber*) freshly rolled sheet. **-gas** *n* circulating gas. **-gut** *n* material for rolling (milling, roll-crushing). **-gutwerkstoff** *m* industrial rolling material. **-haut** *f* mill scale (*or* cinder), rolling skin. **-hitze** *f* (*Metll.*) rolling heat

walzig *a* cylindrical, roller-shaped

Walz·kupfer *n* sheet copper. **-messing** *n* sheet (*or* rolled) brass. **-normale** *f* (*adj endgs*) normal to the plane of rolling. **-öl** *n* (*Steel*) palm oil (for rolling). **-plattieren** *n* (*Metll.*) cladding. **-produkt** *n* rolled product, sheet, plate. **-reibung** *f* rolling friction. **-schlacke** *f*, **-sinter** *m* mill scale (*or* cinder). **-spalt** *m* gap between rollers. **-stahl** *m* rolled steel. **-straße, -strecke** *f* roll train

Walzung *f* (-en) rolling, milling

Wälz·verfahren *n* rotary process

Walz·werk *n* rolling mill, roll(er) train; crushing mill. **-zink** *n* rolled (*or* sheet) zinc. **-zinn** *n* rolled (*or* sheet) tin. **-zunder** *m* (*Metll.*) mill scale (*or* cinder). **-zustand** *m* rolled state

wand coiled, etc: *past of* WINDEN

Wand *f* (⁻e) 1 (interior) wall; partition; (*Bot., Anat.*) septum. **2: spanische W.** folding screen. 3 side, cheek, panel; baffle, screen. 4 cliff. 5 (*Min.*) (large) lump of ore. 6 (*Meteor.*) cloudbank. **-behänge** *mpl* wall hangings. **-bekleidung** *f* wall covering. **-bewurf** *m* wall plaster(ing). **-dicke** *f* thickness of the wall

Wände *fpl* 1 walls, etc. 2 lump ore. *Cf* WAND

Wand·einfluß *m* wall effect

Wandel *m* 1 change, transformation. 2 course (of events, etc). 3: **Handel und W.** trade and traffic. **wandelbar** *a* changeable; transient. **Wandelbarkeit** *f* changeability; transiency

Wandel·halle *f* 1 lobby, foyer. 2 pump room (of a spa). **w–los** *a* unchanging, changeless

wandeln 1 *vt & sich* w. change, vary. 2 *vi*(s) stroll, walk; **handeln und w.** trade and traffic

Wandel·skala *f* sliding scale. **-stern** *m* planet·

Wander· migratory, traveling: **-block** *m* (*Geol.*) erratic block. **-feldröhre** *f* (*Electron.*) traveling wave tube

wandern *vi*(s) wander, migrate; creep, drift; diffuse; go, travel; hike, walk. **wandernd** *p.a* (*oft*) migratory

Wander·niere *f* (*Anat.*) floating kidney. **-rost** *m* traveling grate. **-sand** *m* shifting sand. **-schicht** *f* movable fluidizing bed. **-stärke** *f* translocatory starch. **-tier** *n* migratory animal. **-tisch** *m* conveyor table

Wanderung *f* (-en) wandering, migration; creep(ing), drift; diffusion; travel(ing); walking trip, hike

Wanderungs·geschwindigkeit *f* migratory velocity; ionic mobility (*or* conductance). **-sinn** *m* (*Ions*) direction of migration. **-zahl** *f* transport number

Wander·vogel *m* migratory bird. **-welle** *f* traveling wave, transient oscillation. **-zelle** *f* migratory cell

Wand·farbe *f* 1 indoor (wall) paint. 2 color of the wall. **w–fest** *a* attached to the wall. **W–gestell** *n* wall rack. **-katalyse** *f* wall catalysis, wall effect

Wandler *m* (-) 1 (*Elec.*) transformer; converter; transducer. 2 modifier. 3 traveler

Wandlung *f* (-en) 1 change, variation; metamorphosis. 2 travel, stroll(ing). *Cf* WANDELN. **wandlungs·fähig** *a* adaptable, versatile

Wand·malerei *f* wall painting; mural, fresco. **-montage** *f* wall mounting. **-platte** *f* wall tile. **-putz** *m* plaster(ing). **-reaktion**

f wall reaction. **w–ständig** *a* 1 (*Bot.*) parietal. 2 (*Med.*) marginal. **W–stärke** *f* thickness of the wall. **-tafel** *f* 1 blackboard. 2 wall chart

wandte turned, etc: *past of* WENDEN

Wand·tisch *m* console table. **-uhr** *f* wall clock

Wandung *f* (-en) (*esp Anat., Bot.*) wall, partition. **Wandungs·schicht** *f* (*Biol.*) parietal layer

Wand(ungs)·zelle *f* (*Biol.*) parietal cell

Wange *f* (-n) cheek; side (*or* end) piece

Wangen·bein *n* cheekbone. **-drüse** *f* buccal gland

wankelhaft *a* changeable, unsteady

Wankel·motor *m* Wankel (*or* rotary piston) engine

wanken *vi* stagger, reel; rock; waver, weaken

wann *interrog adv & cnjc* when *cf* DANN

Wanne *f* (-n) 1 bath, tub; trough; pan, vat, tank. 2 hollow (= valley)

Wannen·bad *n* tub bath. **-färberei** *f* openvat dyeing. **-form** *f* boat form. **-ofen** *m* (*Glass*) tank furnace

Wanze *f* (-n) 1 (*Zool.*) bedbug. 2 (*Electron.*) bug. 3 paper clip. **Wanzen·kraut** *f* (*Bot.*) black cohosh, marsh tea

war was: *past of* SEIN *vi; cf* WÄRE, WAREN

WaR *abbv* (Wassermannsche Reaktion) Wassermann reaction

warb advertised, etc: *past of* WERBEN

Ware *f* (-n) 1 article (of trade), commodity; merchandise, wares, goods. 2 *pl* articles, commodities

wäre *sbjc of* SEIN 1 was, were: **man dachte, es w. neu** they thought it was new; **wenn das so w.** if that were so. 2 would be: **das w. unmöglich** that would be impossible

waren were: *past pl of* SEIN *vi*

Waren·begleitschein *m* invoice. **-haus** *n* department store. **-lager** *n* 1 merchandise inventory. 2 warehouse. **-muster** *n*, **-probe** *f* sample (of goods). **-prüfung** *f* product testing. **-stempel** *m*, **-zeichen** *n* trademark

warf threw: *past of* WERFEN

warm *a* (wärmer, wärmst . . .) warm; hot: **ihnen ist w.** they are (*or* feel) warm; **w. essen** eat a hot meal; **w. machen** warm up; **w. waschen** wash in hot water;

(*Mach.*) **w. laufen** heat up
warm·ab·binden* *vt* (*Synth.*) hot-set
Wärm·aushärtung *f* (*Metll.*) hardening at
higher than room temperature
Warm·bad *n* 1 warm (*or* hot) bath. 2 warm
(hot, thermal) spring(s). **-badhärten** *n*
hot-bath hardening, martempering.
-band *n* (*Metll.*) hot-rolled strip. **-bear-
beitung** *f* hot-working. **-behandlung** *f*
heat treatment. **-biegeprobe** *f* hot bending test. **w-bildsam** *a* thermoplastic;
forgeable. **W-bildsamkeit** *f* thermoplas-
ticity; forgeability. **-blasen** *n* hot blast;
(*Metll.*) silicon blow; (*Water Gas*) air blow.
-blüter *m* warm-blooded (*or* homeothermal) animal. **w-blütig** *a* warm-blooded,
homeothermal. **-brüchig** *a* (*Metll.*) hot-
short, brittle when hot. **W-brunnen** *m*
warm (hot, thermal) spring. **w-dehnbar**
a (*Metll.*) hot-ductile

Wärme *f* 1 warmth. 2 (*Phys.*) heat. 3 ardor.
4 (*in compds usu*) thermal, thermo-,
calori-: **-abbau** *m* thermal decomposition. **-abgabe** *f* heat emission (*or* loss).
w-abgebend *a* exothermic. **W-ableiter**
m heat sink, heat dissipating unit. **-ab-
strahlung** *f* heat loss by radiation. **-al-
terung** *f* heat aging. **-änderung** *f* temperature change. **-äquivalent** *n* thermal
equivalent. **-arbeitswert** *m* mechanical
equivalent of heat. **-aufnahme** *f* absorption of heat. **-aufspeicherung** *f* heat
storage. **-ausdehnung** *f* thermal expansion. **-ausgleich** *m* 1 (*Phys.*) temperature
compensation. 2 (*Metll.*) soaking, **-aus-
nutzung** *f* thermal efficiency. **-aus-
strahlung** *f* radiation of heat. **-aus-
tausch** *m* heat exchange. **-austauscher**
m heat exchanger. **-bedarf** *m* heat requirement. **-behandlung** *f* heat treatment. **-belastung** *f* thermal stress. **w-
beständig** *a* heat-resistant; thermostable; (of) constant temperature. **W-
beständigkeit** *f* resistance to heat. **-be-
wegung** *f* thermal motion (*or* agitation).
-bilanz *f* heat balance. **-bildner** *m* heat
producer. **w-bildsam** *a* thermoplastic.
W-bildung *f* heat production. **-bindung**
f heat absorption. **-dämmstoff** *m* heat-
insulating material. **-dämmung** *f* heat
insulation; (*Windows*) thermal breaking.

w-durchlassend, -durchlässig *a* diathermic; heat-conducting. **W-
durchsatz** *m* rate of heat transfer. **-dy-
namik** *f* thermodynamics. **-einfluß** *m* 1
influence of heat. 2 heat influx. **-einheit** *f*
(*genl*) thermal unit; (*specif*) kilocalorie.
-einsparung *f* heat economy. **w-elek-
trisch** *a* thermoelectric. **W-elektrizität**
f thermoelectricity. **w-empfindlich** *a*
heat-sensitive. **-empfindung** *f* sensitivity to heat. **-energie** *f* thermal energy. **-entbindung** *f* disengagement of
heat. **-entwicklung** *f* evolution of heat.
-entziehung *f* withdrawal of heat. **w-
erzeugend** *a* heat-producing. **W-
erzeuger** *m* heat producer. **-erzeugung** *f*
generation of heat. **w-fest** *a* heat-
resistant; thermostable. **W-festigkeit** *f*
heat resistance; thermostability, high-
temperature strength. **-fluß** *m* heat flow,
thermal flux; heat transfer. **-fühler** *m*
thermostat. **-funktion** *n* thermal
function—**Gibbssche W.** heat content,
enthalpy. **w-gebend** *a* heat-yielding, exothermic; (*Windows*) thermal-break. **W-
gefälle** *n* temperature gradient. **-gehalt**
m heat content, enthalpy. **-gewicht** *n* entropy. **-gewitter** *n* heat thunderstorm.
w-gleich *a* isothermal. **W-gleiche** *f* isotherm. **-gleichgewicht** *n* thermal equilibrium. **-grad** *m* degree of heat, degree
above zero. **-gradmesser** *m* thermometer. **-gradschreiber** *m* temperature recorder. **-größe** *f* specific heat. **-haushalt**
m heat economy (*or* balance). **-impuls** *m*
heat transfer. **-inhalt** *m* heat content, enthalpy. **-ion** *n* thermion. **-isolator** *m* thermal insulator. **-isolierstoff** *m* heat-
insulating material. **-isolierung** *f* thermal insulation. **-kapazität** *f* heat
capacity

Wärmekraft·lehre *f* thermodynamics.
-maschine *f* heat engine. **-werk** *n* thermal power station.

Wärme·lehre *f* thermodynamics, science
(*or* theory) of heat. **w-leitend** *a* heat-
conducting. **W-leiter** *m* heat conductor.
-leitfähigkeit *f* thermal conductivity.
-leitung *f* conduction of heat. **-leitver-
mögen** *n*, **-leitzahl** *f* thermal conductivity. **w-liebend** *a* (*Bact.*) thermophilic.

-**liefernd** a heat-providing; exothermal. **W‐mechanik** f thermodynamics. **w‐mechanisch** a thermodynamic. **W‐menge** f amount (or quantity) of heat. -**mengenmesser** m calorimeter. -**mengenmessung** f calorimetry; **-messer** m calorimeter; thermometer; pyrometer. -**messung** f calorimetry; thermometry; pyrometry

wärmen 1 vt warm, heat. 2 vi provide heat. 3: **sich w.** warm oneself; get warm

Wärme· heat, thermal, thermo-: -**ohm** n unit of thermal resistivity. -**platte** f heating (or hot) plate. -**quantum** n heat quantum, quantity of heat. -**quelle** f source of heat

Wärmer m (-) heater, stove

Wärme· heat, thermal, thermo-: **regelung** f heat regulation (or control). -**regler** m thermostat, thermoregulator. -**rückgewinn** m heat recovery. -**rückstrahlung** f heat reflection (by the earth). -**sammler** m heat accumulator. -**satz** m law of thermodynamics. -**schrank** m warming cabinet. -**schutz** m, -**schutzmittel** n heat insulation, thermal insulator, lagging. -**schwankung** f heat fluctuation. -**schwingung** f heat vibration. **w‐sicher** a heatproof. **W‐speicher** m thermal storage device; heat accumulator, regenerator. -**speicherung** f heat storage. -**spektrum** n thermal spectrum. -**standfestigkeit** f resistance to thermal distortion. -**stau** m, -**stauung** f heat accumulation, hyperthermia. -**stich** m fever-producing puncture. -**stoß** m heat impulse. -**strahl** m heat ray. -**strahlung** f heat radiation, radiant heat. -**strömung** f heat convection. -**summe** f heat sum: **Satz der konstanten W‐n** law of constant heat summation. -**tauscher** m heat exchanger. -**technik** f heat technology. **w‐technisch** a regarding heat technology. **W‐theorie** f theory of heat. -**tisch** m warming table. -**tod** m heat (or entropy) death (of the universe). -**tönung** f heat effect (of a reaction); (Metll.) heat tone. -**träger** m heat carrier. -**transport** m, -**übergabe** f, -**übergang** m,

-**übertragung** f heat transfer. **w‐unbeständig** a thermolabile, heat-unstable. -**undurchlässig** a athermanous, impervious to heat. **W‐verbrauch** m heat consumption. **w‐verbrauchend** a heat-consuming, endothermic. **W‐vergangenheit** f prior heat treatment. -**vergütung** f (Metll.) heat treatment. -**verlust** m heat loss. -**vermögen** n heat capacity. -**vorgang** m thermal process (or phenomenon). -**wert** m heat value. **w‐widerstehend** a heat-resistant. **W‐wirkung** f thermal effect. -**wirkungsgrad** m thermal efficiency. -**wirtschaft** f heat economy. -**zahl** f temperature coefficient. -**zähler** m heat meter. **w‐verzehrend** a heat-consuming, endothermic. **W‐zerreißversuch** m heat tensile test. -**zufuhr, zuführung** f heat supply, addition of heat. -**zunahme** f heat increase. -**zustand** m thermal condition (or state). -**zustandsgröße** f entropy

warm·fest a heat-resistant, thermostable; (Steel) high-temperature. **W‐festigkeit** f heat resistance, thermostability; (Steel) resistance to high-temperature deformation

Wärm·flasche f hot-water bottle

warm·gepreßt a hot-pressed. -**gewalzt** a hot-rolled. -**gezogen** a hot-drawn

Warmhalte·kanne f thermos bottle (or jug). -**ofen** m (Metll.) holding furnace. -**packung** f thermos package, vacuum pack(aging) (to keep contents warm). -**platte** f hot plate

Warm·halter m oven, hot plate (to keep things warm). -**haus** n hothouse

warm·kleben vt heat-bond, heat-seal

Warm·kleber m hot-setting adhesive. -**lack** m lacquer applied hot

warm·laufen* vi(s) (Mach.) heat up, run hot

Warm·luft f hot air. -**luftheizung** f hot-air heat(ing)

Wärm·ofen m (re)heating furnace (oven, stove). -**pfanne** f warming pan. -**platte** f hot plate

Warm·preßstahl m hot-pressing steel. -**probe** f hot test. -**riß** m (Metll.) heating crack. **w–rissig** *a (Metll.)* hot-short. **W–rissigkeit** *f* hot-shortness; heat checking. **w–spröde** *a* hot-short. **W–sprödigkeit** *f* hot-shortness

Wärmung *f* (-en) warming, heating

warm·vergüten *vt* heat-treat, temper

Warm·versprödung *f* thermal embrittlement. **-vulkanisation** *f (Rubber)* hot cure

warm·walzen *vt* hot-roll

Warmwalz·platte *f* hot-rolled plate. **-riß** *m* hot-rolling crack. **-verfahren** *n* hot-rolling (*or* hot-pack) process. **-werk** *n* hot-rolling mill

Warmwasser· hot-water: **-anlage** *f* hot-water unit (*or* installation). **-bad** *n* hot-water bath. **-speicher** *m* hot-water (storage) tank

Warm·wind *m (Metll.)* hot blast

warm·ziehen* *vt (Metll.)* hot-draw (wire)

Warn·anlage *f* warning (*or* alarm) unit; security system. **-dienst** *m* warning (*or* alarm) service

warnen *vt/i* (**vor**) warn (of, against). **Warner** *m* (-) **1** warning voice, warner: **2** *(Mach.)* = **Warn·gerät** *n* monitor, alarm (device). **Warn·signal** *n* warning signal, alarm. **Warnung** *f* (-en) warning. **Warn·vorrichtung** *f* alarm (device)

Warte *f* (-n) **1** observatory, lookout. **2** point of view. **3** level, plane

warten 1 *vt (Mach.)* maintain, service; operate. **2** *vi* (**auf** + *accus*) wait (for): **auf sich w. lassen** keep one waiting

Wärter *m* (-), **Wärterin** *f* (-nen) caretaker, custodian; warden; keeper; attendant

Warte·zeit *f* waiting period

·wärts *sfx* ·ward(s) *cf* VORWÄRTS, etc

Wartung *f (Mach.)* maintenance, service; operation

wartungs·arm *a* low-maintenance, maintenance-free. **W–aufwand** *m* maintenance (effort). **-dienst** *m* maintenance service. **w–frei** *a* maintenance-free; unattended, automatic. **W–intervall** *n* servicing interval

warum *interrog adv & cnjc* why

Warze *f* (-n) **1** *(Anat.)* wart; nipple; pustule; excrescence; nodule, tubercle. **2** *(Mach., etc)* knob, boss

warzen·ähnlich, -artig *a* wart-like, mammillary, *(Biol.)* papillary. **W–blech** *n* relief-patterned sheet (metal). **w–förmig** *a* wart-shaped, mammillary, *(Biol.)* papillary. **W–kraut** *n* marigold. **-mittel** *n* wart remedy (*or* remover). **-schwein** *n* wart hog

warzig *a* warty, wart-covered

was I *interrog* **1** *pron* what; **w. auch, w. immer** whatever. **2: w. für (ein) A** *(interrog)* what kind of?; **B** *(exclam)* what (a)! **II** *relat pron* **3** what: **was sie sagen, ist wahr** what they say is true. **4** which *(referring to a preceding clause or idea)*: **es sublimiert, was man nicht erwartet hätte** it sublimes, which would not have been expected. **5** *(after indef pron* **etwas, nichts, alles,** etc) that (*or* omitted in English): **nichts, w. sie tun** nothing (that) they do. **6** *(indef)* e.g.: **w. diese Reaktion betrifft** (or **anbelangt**) as regards this reaction. **III** *indef pron* something = ETWAS

WAS *abbv* (waschaktive Substanzen) active detergents, active ingredients in a detergent

wasch· washing: **-aktiv** *a* surface-active, detergent. **W–anlage** *f* washing plant; flotation tank. **-anstalt** *f* laundry. **-apparat** *m* washing apparatus (*or* machine); scrubber; chemical purifier. **-aufsatz** *m* washing attachment. **-automat** *m* (automatic) washing machine. **-bad** *n* washing bath (*or* liquor)

waschbar *a* washable; *(Colors)* fast

Wasch·bär *m* raccoon

Waschbarkeit *f* washability; *(Colors)* fastness

Wasch· washing: **-becken** *n* wash basin (*or* bowl); sink. **-behälter** *m* washing tank. **-benzin** *n* gasoline for cleaning. **-benzol** *n* dry-cleaning benzene. **w–beständig** *a* launderproof, washproof. **W–beständigkeit** *f* resistance to laundering. **-blau** *n* bluing; laundry blue. **-bottich** *m* washing vat; washtub. **-brett** *n* washboard. **-bürste** *f* washing brush, scrub brush. **-bütte** *f* washtub; purification tank

Wäsche *f* (-n) **1** wash, laundry (*incl* clothes

to be washed). **2** washing, laundering. **3** washday. **4** underwear

wasch-echt *a* **1** washfast, launderproof; shrinkproof; (*Colors*) fast. **2** genuine, simon-pure

waschen* (wusch, gewaschen; wäscht) **1** *vt* wash, launder; pan (gold). **2** *vi*(s) wash (away), spill. **3:** **sich w.** wash (up); stand the acid test. **Wäscher** *m* (-) (*Mach.*) washer, scrubber

Wasch-erde *f* fuller's earth

Wäscherei *f* (-en) laundry

Wäsche-tinte *f* laundry marking ink. **-zeichen** *n* laundry mark

Wasch- wash(ing): **-faß** *n* washing vat, washtub. **-flasche** *f* wash bottle. **-flotte, -flüssigkeit** *f* washing liquid (liquor, bath). **-gefäß** *n* washing vessel. **-gelegenheit** *f* washroom, lavatory. **-gold** *n* placer gold. **-gut** *n* material to be washed. **-hilfsmittel** *n* detergent auxiliary. **-holländer** *m* washing engine. **-kraft** *f* detergent power. **-kristall** *m* washing soda (crystals); sodium carbonate decahydrate. **-lauge** *f* washing liquor; detergent solution; suds. **-leder** *n* wash leather, chamois. **-lösung** *f* washing solution. **-maschine** *f* washing machine, washer. **-mittel** *n* washing agent, detergent; (*Med.*) lotion. **-öl** *n* (*Gas*) wash(ing) oil, absorption washer. **-probe** *f* **1** washing test. **2** washing sample; (*Min.*) assay (of washed *or* buddled ore). **-pulver** *n* washing powder; powdered detergent. **-raum** *m* washroom, lavatory. **-seife** *f* washing soap. **-soda** *f* washing soda

wäscht washes: *pres of* WASCHEN

Wasch-trog *m* washtub; (*Min.*) washing trough. **-trommel** *f* washing drum (*or* cylinder); drum washer. **-turm** *m* scrubbing tower, scrubber

Wasch- und Reinigungsanstalt *f* laundry and dry cleaning shop. **-symbole** *npl* laundering and dry-cleaning instructions (in symbols, on garment labels)

Waschung *f* (-en) **1** (*genl*) washing. **2** (*Med.*) lavage, ablution

Wasch-verfahren *n*. **-vorgang** *m* washing process. **-vorrichtung** *f* washing apparatus (*or* device). **-wanne** *f* washtub.

-wasser *n* wash water; dishwater; windshield washing fluid; (*Cosmet.*) lotion. **-wirkung** *f* cleansing (*or* detergent) action (*or* effect). **-wurzel** *f* soapwort. **-zink** *n* wash zinc (New Jersey process). **-zinn** *n* stream tin. **-zyklon** *m* hydrocyclone

Wasser *n* (-n) **1** water (*incl* lotion, toilet water, perfume, mineral water, body fluid, edema, liquor). **2** waterway. **3** (*in compds oft*) watery, aqueous, hydro-, hydraulic: **-abfluß** *m* water outlet. **-abgabe** *f* elimination (*or* output) of (body) water; sweating, perspiration. **w-abhaltend** *a* waterproof. **W-ablaß** *m* drainage; drain. **-ableitung** *f* drainage. **-abscheider** *m* water separator. **-abscheidung** *f* separation (secretion, excretion) of water. **-abspaltung** *f* elimination of water, dehydration. **w-abstoßend, -abweisend** *a* water-repellent, hydrophobic. **-ähnlich** *a* water-like, watery. **W-analyse** *f* water analysis. **-anlagerung** *f* hydration, addition of water. **w-anziehend** *a* hygroscopic. **-arm** *a* low-humidity, poorly irrigated, arid. **-artig** *a* watery, aqueous. **W-aufbereitung** *f* water treatment (*or* purification). **-aufnahme** *f* absorption of water. **w-aufnahmefähig** *a* capable of absorbing water. **-aufsaugend** *a* water-absorbing. **W-ausscheidung** *f* separation (secretion, excretion) of water. **-ausspülung** *f* rinsing with water. **-austritt** *m* elimination (*or* loss) of water. **-auszug** *m* aqueous extract. **-bad** *n* water bath. **-balg** *m* (*Med.*) serous cyst. **-ballon** *m* water carboy. **-barometer** *n* hydrobarometer. **-bau** *m* **1** hydraulic engineering. **2** hydraulic installation, waterworks. **-bedarf** *m* water requirement. **w-begierig** *a* readily absorbing water, hygroscopic. **W-behälter** *m* (water) tank (cistern, reservoir); (*Brew.*) water back. **-behandlung** *f* water treatment. **-beize** *f* water stain. **-berieselung** *f* water spraying, spray irrigation. **w-beständig** *a* **1** waterproof, water-resistant. **2** stable in water. **W-bestimmung** *f* water determination. **w-bewohnend** *a* (*Biol.*) aquatic. **W-bildung** *f* formation of water. **w-bindend** *a* hygroscopic; hydraulic. **W-bindung** *f* combination

with water. **-bindungsvermögen** *n* water-binding (*or* -retaining) power. **-blase** *f* 1 bubble. 2 (*Med.*) vesicle, blister. 3 water storage (*or* heating) vessel. **w-blau** *a* water (*or* sea) blue. **W-blau** *n* water blue. **-blei** *n* (*Min.*) molybdenite; graphite. **-bleiocker** *m* (*Min.*) molybdic ocher, molybdite. **-bleisäure** *f* molybdic acid. **-bombe** *f* depth bomb (*or* charge). **-bruch** *m* hydrocele. **-dampf** *m* water vapor; steam
Wasserdampf-bad *n* steam (*or* vapor) bath. **-destillation** *f* steam distillation. **w-dicht** *a* steamproof, resistant to water vapor. **W-entwickler, -erzeuger** *m* (steam) boiler. **-gehalt** *m* steam (*or* water vapor) content
Wasser· water(y), aqueous, hydro-, hydraulic: **-deckfarbe** *f* water pigment color. **w-dicht** *a* waterproof, watertight. **W-dichte, -dichtheit, -dichtigkeit** *f* watertightness, imperviousness to water. **-dichtungsmittel** *n* waterproofing agent. **-dost** *m* dyer's weed. **-druck** *m* water (*or* hydraulic) pressure. **-drucklehre** *f* hydrodynamics. **-dunst** *m* water vapor. **w-durchlässig** *a* water-permeable. **W-durchlässigkeit** *f* permeability to water. **w-echt** *a* waterproof, water-resistant. **W-echtmachungsmittel** *n* waterproofing (*or* water-resisting) agent. **w-empfindlich** *a* water-sensitive. **W-enteisenung** *f* removal of iron from water. **-enthärter** *m* water softener. **-enthärtung** *f* water softening. **w-entziehend** *a* dehydrating; hygroscopic. **W-entzieher** *m* desiccator, dehumidifier; (*Tex.*) hydroextractor. **-entziehung** *f* desiccation, dehumidification, dehydration; (*Tex.*) hydroextraction. **-entziehungsmittel** *n* dehydrating agent. **-entzug** *m* dehydration, withdrawal of water. **-erguß** *m* edema. **-erhitzer** *m* water heater. **-fahrzeug** *n* watercraft. **-farbe** *f* 1 water color. 2 color of the water. **-faß** *n* water barrel (*or* cask), tub. **-feinkalk** *m* semihydraulic ground quicklime. **-fenchel** *m* water fennel. **w-fest** *a* waterproof, watertight. **W-fe-**

-stigkeit *f* waterproofness, watertightness. **-fläche** *f* 1 water level. 2 surface of the water. 3 sheet of water. **-flasche** *f* water bottle. **-fleck** *m* water stain. **-flut** *f* flood, inundation. **w-förmig** *a* waterlike, watery. **-fräßig** *a* water-absorbent; spongy. **-frei** *a* anhydrous; water-free, nonaqueous; dehydrated. **-führend** *a* water-bearing. **W-gang** *m* 1 waterway. 2 aqueduct. 3 drain. **-gas** *n* water gas. **-gasteer** *m* water-gas tar. **-gebläse** *n* water blast. **-gefäß** *n* water container (*or* vessel). **-geflügel** *n* waterfowl. **-gehalt** *m* water (moisture, aqueous) content. **w-gekühlt** *a* water-cooled. **W-geschwulst** *f* edema, hygroma. **-gewächs** *n* aquatic plant. **-gier** *f* hygroscopicity. **w-gierig** *a* hygroscopic; hydrophilic. **W-glanz** *m* moiré. **-glas** *n* 1 (*Chem.*) water glass, sodium silicate. 2 water (drinking) glass. 3 water gauge. **-glaskitt** *m* water-glass cement. **w-gleich** *a* 1 water-like, watery. 2 level. **w-grün** *n* water green (finely ground green verditer)
wasserhaft *a* aqueous
Wasser·hahn *m* water tap (*or* faucet). **w-haltend** *a* 1 water-retaining. 2 = **-haltig** *a* hydrous, hydrated, aqueous; water-containing. **W-haltevermögen** *n* capacity to retain water. **-haltung** *f* (*Min.*) water drainage. **w-hart** *a* (water-)impervious; (*Ceram.*) air-dried, half-dry. **W-härte** *f* water hardness. **-härten** *n*, **-härtung** *f* water hardening. **-härtungsstahl** *m* water-hardening steel. **-harz** *n* Burgundy (*or* white) pitch. **-haushalt** *m* water economy, water distribution (in an area). **-haut** *f* 1 water film. 2 (*Anat.*) hyaloid membrane. 3 (*Zool.*) ammion. **-heilkunde** *f* hydropany. **-heizung** *f* hot-water heating. **w-hell** *a* water-clear, water-white. **W-höhe** *f* water level, height (*or* depth) of water
wässerig *a* 1 watery, aqueous, hydrous. 2 dilute(d), watered. 3 (*Med.*) serous. **Wässerigkeit** *f* wateriness, aqueousness, etc
Wasser· watery, aqueous, hydro-, hydraulic: **-kalk** *m* hydraulic (*or* water) lime. **-kante** *f* (North German) seacoast. **-kasten** *m* water tank (*or* container).

-kessel *m* (*Engg.*) boiler, water tank.
-kies *m* (*Min.*) marcasite. **-kitt** *m*
hydraulic cement. **w–klar** *a* water-clear,
water-white. **W–kläranlage** *f* water purification plant. **-klee** *m* buck (*or* bog)
bean, marsh trefoil. **-klosett** *n* rest room,
(*specif*) flush toilet. **-kopf** *m* hydrocephalus, water on the brain. **-kraft** *f* wa-
terpower (source). **-kraftlehre** *f* hydrodynamics. **-kraftwerk** *n* hydroelectric
power plant. **-kristall** *m* rock crystal.
-kühler *m* water cooler. **-kühlkasten** *m*
water block. **-kühlung** *f* water cooling.
-kunst *f* 1 hydraulics. 2 (ornamental)
fountain (*usu* with special effects). **w–l.**
abbv (w–löslich) water-soluble. **-lässig** *a*
leaky. **W–lauf** *m* stream, watercourse.
w–leer *a* water-free, anhydrous; arid.
W–leitung *f* water pipe(s) (piping, main,
supply); water tap (*or* faucet); (*Engg. &
Anat.*) aqueduct
Wasserleitungs·rohr *n*, **-röhre** *f* water
(supply) pipe. **-wasser** *n* tap (*or* city)
water
Wasser·linie *f* water line. **w–los** *a* water-
less, anhydrous. **-löslich** *a* water-soluble.
W–löslichkeit *f* water solubility. **w–
löslichmachend** *a* water-solubilizing.
W–luftpumpe *f* water vacuum pump.
-maische *f* aqueous infusion; mash.
-malerei *f* water-color painting. **-mangel**
m 1 water shortage. 2 (*Med.*) dehydration. **-mantel** *m* water jacket. **-marke** *f*
high-water mark. **-maß** *n* water gauge.
-masse *f* 1 body of water. 2 *pl* masses (*or*
deluge) of water. **-melone** *f* watermelon.
-menge *f* amount of water. **-messer** *m*
hydrometer; water meter (*or* gauge).
-meßkunst *f* hydrometry. **-moos** *n* alga;
seaweed. **-mörtel** *m* hydraulic mortar
wassern *vi*(s) touch down on water, splash
down
wässern 1 *vt* water (down); hydrate; irrigate. 2 *vt/i* water; soak
Wasser· water(y), aqueous, hydro-, hydraulic: **-nabel** *m* (*Bot.*) marsh pen-
nywort. **-niederschlag** *m* deposit of
moisture. **-not** *f* water shortage; drought
cf WASSERSNOT. **-opal** *m* water opal,
hyalite; hydrophane. **-papier** *n* water-
leaf. **-parfüm** *n* aqueous perfume. **w–**

paß horizontal, (water-)level. **W–
pflanze** *f* aquatic plant, hydrophyte.
-phase *f* aqueous phase. **-presse** *f*
hydraulic press. **-probe** *f* 1 water sample.
2 water test. **-prüfer** *m* water tester.
-prüfung *f* water testing. **-pumpe** *f* water pump. **-rad** *n* water wheel. **-radiolyse**
f water radiolysis. **w–reich** *a* 1 of high
water content. 2 (*Lakes, Rivers*) water-
rich, abounding in water. **W–reich** *n*
aquatic kingdom. **-reinigung** *f* water purification. **-reinigungsmittel** *n* water
purifier. **-rest** *m* 1 residue of water. 2
(*Chem.*) hydroxyl group. **-rohr** *n*, **-röhre** *f*
water pipe (*or* tube). **-rohrkessel**, **-röh-
renkessel** *m* water-tube boiler.. **-röste** *f*
water retting. **-rübe** *f* turnip. **w–satt** *a*
water-saturated. **-saugend** *a* water-
absorbing. **W–säule** *f* water column.
-scheide *f* watershed, divide. **w–scheu** *a*
water-dreading; hydrophobia. **W–scheu**
f fear of water; hydrophobia. **-schierling**
m (*Bot.*) water hemlock. **-schlag** *m*
(*Engg.*) water hammer. **-schlange** *f*
(*Zool.*) water snake. **-schlauch** *m* 1 water
hose; (flexible) water tubing. 2 (*Bot.*)
bladderwort. **-schluß** *m* water seal (*or*
trap). **-schwein** *n* water hog, capybara.
-schwere *f* specific gravity of water.
-senf *m* dyer's weed. **-siedemesser** *m*
hypsometer
Wassers·not *f* flood disaster *cf* WASSERNOT
Wasser· water(y), aqueous, hydro-,
hydraulic: **-speicher** *m* water reservoir.
-speisung *f* water feed (*or* supply).
-spiegel *m* water level, surface of the water. **-spritze** *f* 1 syringe. 2 water
sprinkler. **-spülung** *f* 1 water flushing;
rinsing. 2 (*Min.*) water injection. 3 flushing cistern. **-stand** *m* 1 water level,
height of water. 2 constant water level
device. **-standsglas** *n* gauge glass.
-standshahn *m* gauge cock. **-stein** *m* 1
scale, fur (from water). 2 whetstone.
-steinansatz *m* (water) scale deposit
Wasserstoff *m* 1 hydrogen. 2 (*in compds: as
pfx with metals, cf* WASSERSTOFFKALIUM;
as sfx, cf ÄTHYLWASSERSTOFF) hydride.
w–ähnlich *a* resembling hydrogen.
-arm *a* low-hydrogen. **W–bindung**,
-brücke *f* hydrogen bond. **-brückenbin-**

dung f hydrogen bond(ing). **-elektrode** f hydrogen electrode. **w–enthaltend** a hydrogen-containing. **W–entwickler** m hydrogen generator. **-entwicklung** f evolution of hydrogen. **w–entziehend** a dehydrogenating. **W–entziehung** f dehydrogenation, hydrogen abstraction. **-erzeuger** m hydrogen generator. **-exponent** m hydrogen ion exponent, pH. **-flamme** f hydrogen flame. **-flasche** f hydrogen cylinder. **w–frei** cf WASSERSTOFFREI. **W–gas** n hydrogen gas. **-gehalt** m hydrogen content. **w–haltig** a hydrogen-bearing; hydrogenous, hydrogenated. **W–hyperoxid** n hydrogen peroxide. **-ion** n hydrogen ion. **-ionenkonzentration** f hydrogen ion concentration, pH. **-kalium** n potassium hydride. **-knallgas** n (hydrogen-oxygen) detonating gas. **-krankheit** f (Metll.) hydrogen embrittlement. **-lichtbogen** m hydrogen arc. **-lötung** f hydrogen soldering. **-palladium** n palladium hydride. **-peroxid** n hydrogen peroxide

wasserstoffrei a hydrogen-free = **wasserstoff·frei**

wasserstoff·reich a rich in hydrogen, high-hydrogen. **W–salz** n hydrogen salt. **w–sauer** a of a hydracid. **W–säure** f hydracid, acid containing hydrogen. **-strom** m hydrogen current. **-sulfid** n hydrogen sulfide. **-superoxid** n hydrogen peroxide. **-verbindung** f hydrogen compound. **-versprödung** f hydrogen embrittlement. **-wertigkeit** f hydrogen valence. **-zahl** f hydrogen-ion concentration, pH. **-zündmaschine** f hydrogen lamp, Döbereiner's lamp

Wasserstrahl° m water jet. **–blasvorrichtung** f water-pump blower. **-gebläse** n water-jet blower. **-(luft)pumpe** f water-jet vacuum pump

Wasser· water(y), aqueous, hydro-, hydraulic: **-straße** f waterway; (ship) channel. **-strom** m stream of water; water current. **-stückkalk** m semi-hydraulic lump (quick)lime. **-sturz** m waterfall. **-sucht** f dropsy. **w–süchtig** a dropsical, edematous. **-süffig** a water-absorbent, spongy. **W–suppe** f watery (oatmeal or farina) gruel. **-talk** m (Min.) brucite.

-teilchen n water particle. **-tiefe** f 1 depth of the water. **2** (ship's) draft. **-tier** n aquatic animal. **w–treibend** a (Med.) diuretic, hydragog. **W–trockenschrank** m water-jacketed drying closet. **-trog** m water trough. **-trommelgebläse** n waterdrum blast, trompe. **-tröpfchen** n droplet of water. **-tropfen** m drop of water. **-uhr** f water meter. **-umlauf** m water circulation. **w–undurchlässig** a impervious to water, watertight. **W–undurchlässigkeit** f imperviousness to water, watertightness

Wässerung f (-en) watering; hydration; irrigation; soaking cf WÄSSERN

wasser· water(y), aqueous, hydro-, hydraulic: **-unlöslich** a water-insoluble. **W–unlöslichkeit** f water-insolubility. **-untersuchung** f water research (or analysis). **-verbrauch** m water consumption. **-verdampfung** f evaporation of water. **-verdrängung** f displacement of water. **-verdunstung** f evaporation of water. **-vergoldung** f water gilding. **-vergüten** n, **-vergütung** f (Metll.) heat treatment using water. **-verlust** m loss of water. **-vermögen** n (Ceram.) water retentivity. **-verschluß** m water seal. **-verschmutzung** f water pollution. **-versorgung** f water supply. **-verträglichkeit** f compatibility with water. **-verunreinigung** f 1 water pollution (or contamination). 2 impurity in water. **-vogel** m aquatic bird, waterfowl. **-vollentsalzung** f demineralization (or desalination) of water. **-vorlage** f (Coal Gas) hydraulic main. **-vorrat** m water supply (or reserves). **-waage** f (Instrum.) water (or spirit) level; hydrostatic balance. **-wanne** f water trough (or bath); pneumatic trough. **-weg** m waterway: **auf dem W.** by water. **-welle** f (Cosmet.) hair set. **-werk** n 1 waterworks. 2 hydraulic engine. **-wert** m (Heat) water equivalent. **-wirtschaft** f water economy. **w–wirtschaftlich** a water-saving. **W–zähler** m water meter. **-zeichen** n (Paper) watermark. **-zement** m hydraulic cement. **-zersetzung** f decomposition of water. **w–ziehend** a water-attracting, hydrophilic. **W–zufluß** m, **-zufuhr** f wa-

ter supply. **w–zügig** *a* water-absorbent.
W–zusatz *m* addition of water
wäßrig *a* watery, aqueous, hydrous =
WÄSSERIG
waten *vi* wade
Waterkant *f* seacoast (of northern Germany)
Watt *n* 1 (-) (*Elec.*) watt. 2 (-en) tideland
Watte *f* (absorbent) cotton wadding, cotton pad(ding); glass wool. **-bausch** *m* cotton pad (ball, swab). **-filter** *n* cotton strainer. **-pfropf** *m* cotton plug (*or* wad). **-schicht** *f* layer of cotton wadding. **-verschluß** *m* plug of cotton wadding
wattieren *vt* pad (with cotton); wad
Watt· (*Elec.*): **-leistung** *f* power in watts: **w–los** *a* wattless. **W–messer** *m* wattmeter. **-sekunde** *f* watt second. **-stunde** *f* watt hour. **-zahl** *f* number of watts; wattage
Wau *m* (*Bot.*) weld, wold, dyer's weed. **--gelb** *n* luteolin
WC *abbv* (*English: water closet*) toilet *cf* WASSERKLOSETT
W.D.D. *abbv* (Wasserdampfdurchlässigkeit) water vapor permeability
WE *abbv* (Wärmeeinheit) thermal unit
Web·art *f* type of weave (*or* weaving). **-automat** *m* automatic loom
weben* (*reg or* wob, gewoben) *vt/i* weave
Weber *m* (-) weaver. **--distel** *f* (*Bot.*) fuller's teasel
Weberei *f* (-en) 1 weaving. 2 weaving (*or* textile) mill. 3 weaving material. 4 texture, tissue
Weberin *f* (-nen) weaver
Weber·glas *n* 1 web glass. 2 thread counter. **-zettel** *m* warp
Web·kante *f* selvedge. **-kunst** *f* art of weaving
Webnerit *m* (*Min.*) andorite
Web·schützen *m* shuttle. **-stuhl** *m* loom. **-ware** *f* woven goods, fabric
Wechsel *m* (-) 1 change; alternation, fluctuation; rotation. 2 exchange, interchange. 3 (*Engg.*) (pipe) junction, joint. 4 (*Com.*) draft, bill of exchange. 5 (*in compds oft*) inter-, mutual, reciprocal; alternating
wechsel·artig *a* mutual, reciprocal. **W–bad** *n* alternating hot and cold baths; (*fig*) ups and downs

wechselbar *a* changeable
Wechsel·beanspruchung *f* alternating stress. **-bewegung** *f* reciprocating motion. **-beziehung** *f*, **-bezug** *m* interrelation, correlation. **-fall** *m* alternative; (*pl usu*) ups and downs, vicissitudes. **-farbig** *a* color-changing, iridescent. **W–feld** *n* (*Elec.*) alternating field. **-festigkeit** *f* fatigue strength, resistance to periodic stress. **-fieber** *n* intermittent fever. **-folge** *f* alternation. **-geld** *n* (small) change. **-gespräch** *n* dialog. **-kontakt** *m* (*Elec.*) make and break. **-kraft** *f* variable (*or* periodic) force. **-kurs** *m* rate of (currency) exchange
wechsellagern *vi* be interbedded, be interstratified. **Wechsel·lagerung** *f* interstratification; (*Concrete*) alternating curing
wechseln 1 *vt/i* change, vary. 2 *vt* exchange. 3 *vi* (h,s) come and go, alternate, fluctuate. **wechselnd** *p.a* (*oft*) variable, intermittent
Wechsel·rede *f* dialog. **-richter** *m* (*Elec.*) inverter; vibrator. **-satz** *m* exchange principle. **-schalter** *m* (*Elec.*) two-way switch. **w–seitig** *a* reciprocal, mutual. **W–seitigkeit** *f* reciprocity. **-spannung** *f* 1 alternation of stress. 2 (*Elec.*) AC voltage. **-spiel** *n* interplay; alternation, fluctuation. **-sprechanlage** *f* intercom. **w–ständig** *a* alternate. **W–stein** *m* glazed tile (*or* brick). **-strom** *m* (*Elec.*) alternating current. **-stromwiderstand** *m* impedance. **-stube** *f* currency exchange office. **-tauchversuch** *m* alternate immersion test (for corrosion). **-ventil** *n* change-over valve. **-verhältnis** *n* reciprocal relation (*or* proportion). **w–voll** *a* varied, eventful. **-weise** *adv* 1 mutually, reciprocally. 2 interchangeably. 3 alternately. **W–winkel** *mpl* alternate angles. **-wirkung** *f* reciprocal action (*or* effect); interaction. **-wirtschaft** *f* crop rotation. **-zahl** *f* number of changes (*or* alternations); cycle number; frequency. **-zersetzung** *f* double decomposition
Wechsler *m* (-) 1 (*usu Mach.*) changer. 2 money changer. **Wechslung** *f* (-en) change, variation; exchange; alternation, fluctuation *cf* WECHSELN

Wecke *f* (-n) (*Bread*) roll; small loaf

wecken *vt* wake (up), awaken; arouse

Wecken *m* (-) roll; small loaf = WECKE

Wecker *m* (-), **Weck·uhr** *f* alarm (clock)

Wedel *m* (-) 1 (feather) duster, whisk. 2 (palm) frond. **wedeln** *vi* (**mit**) wave, wag

weder *cnjc:* w. . . . **noch** neither . . . nor

weg I *adv* 1 away, off, gone: **Hände w.** hands off. 2 (*after adv phr, implies rapid, direct action*): **in einem w.** without a break; **vom Fleck w.** on the spot, right away. 3: **über . . . w.** over; **sie sind darüber w.** they have gotten over it. 4 **von . . . w.** away from, off. **II** *sep pfx* away, off

Weg *m* (-e) (*dat sg oft* **W—e**) 1 way (to go), route, road; path, walk, trail; passage; channel: **auf dem W(—e)** on the way; **den W. weisen** point the way; **auf halbem W.** halfway; **sich auf den W. machen** be on one's way; **auf bestem W., zu . . .** well on the way to . . . ; **aus dem W. gehen** (+*dat*) get out of (sbdy's) way, steer clear of (sthg); **im W. stehen** (+*dat.*) stand (*or* be) in (sbdy's) way, be an obstacle (to); **dem steht nichts im W.** there is no obstacle to that; **etwas kommt** (*or* **läuft**) **einem über den W.** sthg crosses one's path, sthg happens to one; **einen anderen W. einschlagen** take a different route; **den Rechtsweg beschreiten** go (*or* take) the legal route; **auf amtlichem W—e** through official channels; **das hat noch gute W—e** there is still time for that. 2 distance; (*Phys.*) displacement: **ein W. von 10 km** a distance of 10 km; **der durchlaufene W.** the distance (*or* route) traveled. 3 errand: **W—e machen** run errands. 4 way (to do), manner, means, method, process: **Mittel und W—e** ways and means; **auf nassem W.** by the wet process; **auf dem W. der Verdauung, auf dem Verdauungsweg** by way of digestion, by the digestive process. **--abkürzung** *f* short cut

weg·ätzen *vt* remove by caustics, etch away

weg·begeben* *vr:* **sich w.** go away, withdraw

weg·beizen *vt* remove by caustics, etch away

weg·bekommen* *vt* 1 get rid of, get (a stain) out; (+**von**) get (sthg) away (from). 2 incur (unpleasant consequences). 3 (*oft* + *clause*) figure out, realize

Weg·bereiter *m* precursor, pioneer

weg·bleiben* *vi*(s) 1 stay away. 2 be omitted; drop out

weg·brennen* *vt, vi*(s) burn away (*or* off)

weg·bringen* *vt* 1 take away, get (sthg) away (out, off), get rid of. 2 get (sthg) on its way

weg·denken* *vt:* **sich etwas w.** imagine sthg to be absent

weg·diffundieren *vi*(s) diffuse away

weg·diskutieren *vt* argue (sthg) away

Wege·bau *m* road construction. **-dorn** *m* (*Bot.*) purging buckthorn

wegen 1 *prep* (*usu* + *genit*) on account of, because of: **w. des Wetters, des Wetters w.** because of the weather. 2 *adv:* **von . . . w.** by virtue of: **von Berufs w.** by virtue of one's profession, professionally *cf* RECHT 2

Wegerich *m* (-e) (*Bot.*) plantain

Weg·fall *m* dropping out, absence: **in W. bringen** eliminate, omit; **in W. kommen** drop out, be omitted, cease to exist. **weg·fallen*** *vi*(s) drop out (of the picture), no longer exist; be(come) unnecessary

Weg·filtern *n* filtering out (*or* off). **-gang°** *m* departure

weggeblieben stayed away: *pp of* WEG-BLEIBEN

weggebracht taken away: *pp of* WEG-BRINGEN

weggebrannt burned off: *pp of* WEG-BRENNEN

weggedacht imagined as absent: *pp of* WEGDENKEN

weggegangen gone away: *pp of* WEGGEHEN

weggegossen poured off: *pp of* WEG-GIESSEN

weg·gehen* *vi*(s) 1 (**von**) go away (from), leave, depart. 2 come out (*or* off) (*as:* spot, stain); stop, disappear. 3 (**über**) ignore

weggehoben lifted off: *pp of* WEGHEBEN

weggeholfen helped cope: *pp of* WEG-HELFEN

weggenommen taken away: *pp of* WEG-NEHMEN

weggerieben rubbed off: *pp of* WEGREIBEN

weggerissen torn away: *pp of* WEGREISSEN

weggeschmolzen melted away: *pp of* WEGSCHMELZEN

weggesogen suctioned off: *pp of* WEGSAUGEN

weggestrichen stricken out: *pp of* WEGSTREICHEN

weggetan put away: *pp of* WEGTUN

weggewesen been away: *pp of* WEGSEIN

weggeworfen thrown away: *pp of* WEGWERFEN

weggezogen pulled away, etc: *pp of* WEGZIEHEN

weg·gießen* *vt* pour off (*or* out)

weg·glühen *vt* drive off by ignition

weg·haben* *vt* 1 have out of the way, get rid of. 2 catch (a cold, etc). 3 catch on to, get the knack of

weg·halten* 1 *vt* keep at a distance. 2 *vt & sich w.* keep away

weg·heben* *vt* lift off (*or* away)

weg·helfen* *vi* (*dat*) 1 help get away. 2 (**über**) help cope (with), help (over)

weg·kochen *vt, vi*(s) boil away (*or* off)

weg·kommen* *vi*(s) 1 get away, escape; come off (*or* out) (well, badly). 2 be lost (stolen, removed, torn down). 3 (**über**) get (over)

weg·kratzen *vt* scratch off (out, away)

Weg·länge *f* distance, length of the path: **mittlere freie W.** mean free path

weg·lassen* *vt* 1 omit, leave out. 2 let go (*or* get) away

weg·legen *vt* put away (*or* aside), lay aside

wegleitend *a* efferent

weg·machen 1 *vt* remove, get rid of. 2: **sich w.** get lost, disappear

Wegnahme *f* taking away, removal, deprivation; seizure; (*incl Physiol.*) elimination *cf* WEGNEHMEN

weg·nehmen* *vt* (*d.i.o*) take away, remove, seize (from), deprive (of)—**Gas. w.** take one's foot off the accelerator, coast

weg·oxidieren *vt* oxidize away (*or* off)

weg·radieren *vt* erase, scratch out

weg·rasieren *vt* 1 shave off. 2 level, raze

weg·räumen *vt* clear away, remove

weg·reiben* *vt* rub off (*or* away)

weg·reißen* *vt* tear (rip, scratch) away

wegsam *a* pervious, penetrable, passable.
Wegsamkeit *f* perviousness, penetrability, passability

weg·saugen *vt* suction off, remove by suction

weg·schaben *vt* scrape off (*or* away)

weg·schaffen *vt* get rid of, remove, eliminate

weg·schmelzen* *vt, vi*(s) melt away (*or* off)

weg·schmirgeln *vt* remove by abrasion, grind off

weg·schütten *vt* pour out (off, away)

weg·sehen* *vi* look away (aside, the other way); (**über**) ignore

weg·sein* *vi* be away, be gone *cf* WEG

weg·setzen I *vt* 1 move away (aside, to another seat). II *vi*(s) 2 (**über**) leap, jump (over). III: **sich w.** 3 change one's seat, move. 4 (**über**) ignore, disregard

weg·sickern *vi*(s) ooze (*or* trickle) away

weg·sieden* *vt* boil away (*or* off)

weg·spülen *vt* rinse (wash, flush) away

weg·stellen *vt* put away (*or* aside)

Weg·strecke *f* distance, stretch (of the road)

weg·streichen* *vt* cross (*or* strike) out; brush away

weg·tun* *vt* put away (*or* aside)

Weg·warte *f* (*Bot.*) (wild) chicory

wegwaschbar *a* removable by washing.
weg·waschen* *vt* wash off (*or* away)

wegweisend *pred a:* **w. sein** point the way.
Wegweiser *m* (-) 1 road sign. 2 guide(book), directory

weg·werfen* *vt* throw away. **wegwerfend** *p.a* disparaging, deprecating. **Wegwerf·gesellschaft** *f* throw-away society (*or* world)

weg·wischen *vt* wipe away (*or* off); brush aside

weg·ziehen* 1 *vt* (+*d.i.o*) pull away (*or* out) (from). 2 *vi*(s) move away, change residence; migrate

weh 1 *a* sore, painful. 2 *adv:* **w. sein** (+*dat*) feel grief. 3: **w. tun** (*oft* + *dat*) hurt

Weh(e) *n* 1 grief, woe; ache, pain. 2 (*pl:* **Wehen**) labor pains; (*fig*) throes

Wehe *f* (-n) (snow) drift; (sand) dune

wehen 1 *vt* (*Wind*) blow. 2 *vi* blow, waft; be windy; wave (in the wind)

Wehen·mittel *n* ecbolic

Wehr I *f* defense: **sich zur W. setzen** defend oneself. II *n* (-e) weir, dam

Wehr·drüse *f* defensive scent gland

wehren I *vt* 1 (+*d.i.o*) prevent (from doing). **II** *vi* 2 (*dat*) resist, fight; prevent. **III: sich w.** 3 defend oneself, resist. 4 (+**gegen**, *oft*) reject; (+**zu** + *inf*) refuse (to)

wehrlos *a* defenseless

Weib *n* (-er) woman; wife (*now usu derogatory*). **Weibchen** *n* (-) (*Animals*) female. **weiblich** *a* female; feminine; womanly

weich *a* 1 soft, tender; weak: **w. werden** soften, weaken. 2 gentle; smooth. 3 (*Lthr.*) limp

Weich·bild *n* metropolitan area. **-blei** *n* soft (*or* refined) lead. **w—bleibend** *a* remaining soft, non-hardening. **W—bottich** *m* steeping tub (*or* vat). **-brand** *m* soft (*or* place) brick. **-braunkohle** *f* soft (*or* slack) lignite. **-braunstein** *m* (*Min.*) pyrolusite. **-bütte** *f* (*Brew.*) steep(ing) tank (vat, cistern). **-dauer** *f* steeping (*or* soaking) time

Weiche *f* (-n) **I** (*cf* WEICH) 1 softness. 2 (*Anat.*) side, flank. **II** (*cf* WEICHEN) 3 (*Brew.*) steep; (*Lthr.*) soaking pit. 4 (railroad) switch: **die W—n stellen (für)** set the course (for), initiate

Weich·eisen *n* soft iron. **-eisenkies** *m* (*Min.*) marcasite

weichen I (*reg*) *vt*, *vi*(s) soak, steep; soften. **II*** (wich, gewichen) *vi*(s) (*usu* + *dat, oft* + **vor**) yield, give way (to); (+**von**) move, budge (from); go, disappear (from), drain (out of)

Weichen·gegend *f* (*Anat.*) groin

Weich·faß *n* steeping tub

weich·feuern *vt* (*Puddling*) melt down

Weich·fleckigkeit *f* (*Metll.*) liability to soft spots. **-floß** *n* (*Metll.*) porous white pig. **w—-gekocht, -gesotten** *a* 1 boiled. 2 (*Eggs*) soft-boiled. **-gestellt** *a* (*Plastics*) plasticized. **W—gewächs** *n* (*Min.*) argentite. **w—gewalzt** *a* soft-rolled

weich·glühen *vt* soft-anneal

Weich·glühofen *m* annealing furnace. **-glühung** *f* soft annealing. **-grube** *f* soaking pit. **-gummi** *n* soft rubber. **-guß** *m* malleable (cast) iron. **-haltungsmittel** *n* softening agent, plasticizer. **-harz** *n* soft resin, oleoresin

Weichheit *f* softness, tenderness

Weich·holz *n* softwood. **-käse** *m* soft cheese. **-kautschuk** *m* soft rubber. **-kies**

m (*Min.*) marcasite. **-kohle** *f* soft coal. **-kopal** *m* soft copal. **-kufe** *f* steeping (*or* soaking) vat (*or* tub). **-kupfer** *n* soft copper. **-leder** *n* soft leather

weichlich *a* soft, tender; weak; delicate

Weich·lot *n* soft solder. **weich·löten** *vt* soft-solder

weich·machen *vt* soften, plasticize

Weich·machen *n* softening, plasticizing. **-macher** *m* (*genl*) softener; (*Synth.*) plasticizer

weichmacher·haltig *a* plasticized, containing a plasticizer. **W—wanderung** *f* plasticizer migration

Weichmachungs·mittel *n* plasticizer, softener

Weich·mangan(erz) *n* (*Min.*) pyrolusite. **-metall** *n* soft metal. **-paraffin** *n* soft paraffin. **-pech** *n* soft pitch. **-porzellan** *n* soft(-paste) porcelain; bone china. **-ruß** *m* soft black. **w—schalig** *a* (*Fruit*) soft-skinned; (*Zool.*) soft-shell(ed)

Weichsel·kirsche *f* mahaleb; morello

Weich·stahl *m* mild (*or* soft) steel, soft iron. **-steingut** *n* (*Ceram.*) calcareous whiteware. **-stelle** *f* soft spot. **-stock** *m* (*incl Brew.*) steep(ing) tank (*or* tub), cistern. **-teil** *m* soft part. **-tier** *n* mollusk. **-wasser** *n* 1 soft water. 2 soaking (*or* steeping) water (*or* liquor). **-werden** *n* softening

Weide *f* (-n) 1 willow (tree). 2 pasture

weiden 1 *vt* put (animals) out to pasture. 2 *vi* graze. 3: **sich w. (an)** feast one's eyes (on), gloat (over)

Weiden·bitter *n* salicin. **-geflecht** *n* wickerwork. **-gewächse** *npl* willow family, salicaceae. **-kohle** *f* willow charcoal

weidlich *adv* thoroughly, fully: **w. ausnutzen** take full advantage of

Weife *f* (-n) (*Tex.*) reel. **weifen** *vt/i* reel, wind

weigern *vr*: **sich w. (zu)** refuse, decline (to). **Weigerung** *f* (-en) refusal

Weih *m* (-e) (*Zool.*) kite, harrier *cf* WEIHE

Weihe *f* 1 consecration, dedication; solemnity. 2 (-n) (*Zool.*) kite, harrier. *Cf* WEIH

Weiher *m* (-) (fish) pond; oyster bed

Weihnachten *n* Christmas

Weihnachts·wurzel *f* (*Bot.*) black hellebore

Weihrauch° *m* (frank)incense. **--harz** *n* incense resin, (*specif*) frankincense

weil *subord cnjc* because
Weilchen *n* (-) little while
Weile *f* while, short time: **damit hat es gute W.** there is plenty of time for that
weilen *vi* stay, linger, spend some time
Weiler *m* (-) hamlet
Wein *m* (-e) 1 wine. 2 vine; grapes; **W. lesen** pick grapes. 3: **Wilder W.** Virginia creeper
wein·ähnlich *a* wine-like, vinaceous. **W-art** *f* type (*or* variety) of wine. **w-artig** *a* vinous, winy. **W-bau** *m* viticulture, wine growing. **-bauer** *m* viticulturist, wine grower. **-beere** *f* grape. **-beeröl** *n* oil of wine, grapeseed oil; enanthic ether. **-berg** *m* vineyard. **-blau** *n* wine blue (enocyanin). **-blume** *f* 1 bouquet, aroma (of wine). 2 enanthic ether (as artificial flavoring). **-blüte** *f* vine blossom. **-brand** *m* brandy *cf* BRANNTWEIN. **-brandverschnitt** *m* blended brandy. **-branntwein** *m* brandy made from wine
weinen *vi* cry, weep
Wein·ernte *f* vintage. **w-erzeugend** *a* wine-producing. **W-essig** *m* wine vinegar. **-fabrik** *f* winery. **-farbe** *f* wine color. **w-farben, -farbig** *a* wine-colored. **W-farbstoff** *m* coloring matter of wine. **-faß** *n* wine cask (*or* barrel). **-flasche** *f* wine bottle. **w-gar** *a* fermented—**w-e Maische** (*Distil.*) wash. **W-garten** *m* vineyard. **-gärung** *f* vinous fermentation. **-gegend** *f* wine district. **-gehalt** *m* wine content, vinosity
Weingeist *m* (ethyl *or* rectified) alcohol, spirits of wine; **versüßter W.** spirits of nitrous ether. **w-·artig** *a* alcoholic. **W-firnis** *m* spirit varnish. **w-haltig** *a* alcoholic, alcohol-containing
weingeistig *a* alcoholic; **w-es Ammoniak** spirits of ammonia
Weingeist·lack *m* spirit varnish. **-lampe** *f* alcohol (*or* spirit) lamp. **-messer** *m* alcoholometer
wein·gelb *a* wine yellow. **-haltig** *a* containing wine. **W-handel** *m* wine trade. **-hefe** *f* 1 wine yeast. 2 wine lees (*or* sediment). **-hefeöl** *n* grapeseed oil, oil of cognac
Weinhold·gefäß *n* vacuum (*or* Dewar) flask
weinig *a* winy, vinous

Wein·jahr *n* vintage, wine year. **-kamm** *m* grape pomace (*or* pulp). **-kellerei** *f* winery. **-kelter** *f* winepress. **-kernöl** *n* grapeseed oil. **-krankheit** *f* vine disease; fault (*or* defect) in wine. **-lese** *f* vintage. **-leser** *m* vintager. **-messer** *m* vinometer, enometer. **-most** *m* grape must (*or* juice). **-öl** *n* oil of wine (*or* cognac), enanthic ether. **-probe** *f* wine sample (*or* sampling). **-prüfung** *f* wine testing. **-ranke** *f* tendril, vine branch. **-raute** *f* (*Bot.*) (common) rue. **-rebe** *f* grapevine. **-rebenschwarz** *n* Frankfort black. **w-rot** *a* wine-red, claret. **-sauer** *a* tartrate of. **-säuerlich** *a* sourish. **W-säure** *f* 1 tartaric acid. 2 acidity of wine. **-schönung** *f* wine fining. **-sprit** *m* spirits of wine, ethyl alcohol. **-stärkemesser** *m* wine hydrometer, enometer
Weinstein *m* tartar: **roher W.** tartar, wine stone, argol; **gereinigter W.** purified tartar, cream of tartar. **w-·artig** *a* tartarlike, tartareous. **W-bildung** *f* tartar formation, tartarization. **-ersatz** *m* (*Dye.*) tartar substitute, (*specif*) sodium hydrogen sulfate. **-kohle** *f* black flux. **-präparat** *n* (*Dye.*) sodium hydrogen sulfate. **-rahm** *m* cream of tartar. **-salz** *n* tartar salt, potassium carbonate. **w-sauer** *a* tartate of. **W-säure** *f* tartaric acid
Wein·stock *m* grapevine. **-traube** *f* grape; bunch of grapes. **-trester** *pl* grape pomace (husks, skins). **-untersuchung** *f* examination (*or* investigation) of wine. **-verfälschung** *f* adulteration of wine. **-waage** *f* vinometer, enometer, wine gauge
weise *a* wise, sage
Weise I *f* (-n) 1 way, manner, fashion: **auf diese W.** in this way (*or* manner); **in der W., daß . . .** in such a way that . . . ; **die Art und W.** the way, the manner *cf* ART, *cf* ·WEISE. 2 air, melody, tune. II *m, f* (*adj endgs*) wise man (*or* woman), sage—**der Stein der W—n** the philosophers' stone
·weise *sfx* 1 (*on adj*) -ly, in a . . . way: **nötigerweise** necessarily. 2 (*on noun*) -wise, by way of: **schrittweise** stepwise, step by step; **probeweise** by way of a test. 3 (*on quantity words*) by the . . . : **dutzendweise** by the dozen

Weisel *m* (-) queen bee

weisen* (wies, gewiesen) **I** *vt* **1** instruct (*usu* to go . . .), direct (to a place). **2** (+**aus, von, von sich**) reject, repudiate. **3** (*usu* + **zu**) warn (to be . . .). **4** (*d.i.o*) point out (to), show. **II** *vi* point (in a direction), indicate. **Weiser** *m* (-) pointer, (clock) hand *cf* WEGWEISER

Weisheit *f* (-en) wisdom, sagacity

weiß I *v* know(s): *pres of* WISSEN. **II** *a* white; blank; clean; (*in combinations & compds oft*) leuco-, albus, incandescent: **w. werden** turn white, (*Varnish*) blush, chalk; **w. ätzbar** (*Dye.*) dischargeable to white; **w—e Glut** white heat; **w—er Kupferstein** (*Copper*) white metal; **w—es Vitriol** white vitriol, zinc sulfate; **w—er Fluß** leucorrhea; **w—e Magnesia** magnesia alba; **w—es Nichts** nihil album, zinc oxide; **w. sieden** blanch; **w—er Amber** spermaceti; **w—es Eisenblech** tinplate; **w—er Elektronenstrahl** heterogeneous beam of electrons; **w. gerben** taw. **Weiß** *n* white (color, dye): **Berliner W.** Berlin white *cf* WEISSE

weis-sagen *vt/i* predict, prophesy

Weiß-anlaufen *n.* (*Varnish*) blushing, chalking. **-äscher** *m* (*Lthr.*) white (*or* fresh) lime. **-ätze, -ätzung** *f* (*Tex.*) white discharge. **-bad** *n* whitening bath; (*Turkey-red Dye.*) white liquor bath. **-band** *n* tinplate strip. **-baumöl** *n* cajuput oil. **-bier** *n* pale beer (containing lactic acid), weiss beer. **-birke** *f* white birch. **-birkenöl** *n* oil of birch buds. **w—blau** *a* whitish-blue

Weißblech° *n* tinplate. **-dose** *f* tin can, tin box. **-waren** *fpl* tinware

Weiß-blei *n* tin. **-bleiche** *f* full bleach(ing). **-bleierz** *n* white lead ore, cerussite. **-blütigkeit** *f* leukemia, leucocythemia. **-boden** *m* white ground

weiß-brennen* *vt* calcine at white heat. **Weiß-brot** *n* white bread. **-brühe** *f* (*Lthr.*) tawing liquor, sod oil. **-buche** *f* (*Bot.*) hornbeam. **-dorn** *m* hawthorn

Weiße I *f* **1** whiteness; pallor. **2** (*adj endgs*): **Berliner W.** weiss beer = WEISSBIER. **II** *f, m* (*adj endgs*) white person, Caucasian

Weiß-eisen *n* white iron. **-emaille** *f* white enamel

weißen *vt* **1** whiten; paint white. **2** bleach. **3** whitewash. **4** (*Iron*) refine

Weiß-erde *f* white earth (*or* clay), terra alba, gypsum. **-erz** *n* arsenopyrite, krennerite. **-farbe** *f* white color (dye, paint, pigment)

weiß-färben *vt* bleach (and blue); color (*or* paint) white

Weiß-färber *m* bleacher. **-fäule** *f* (*Bot.*) white rot. **-feinkalk** *m* ground white quicklime. **-feuer** *n* white fire. **-föhre** *f* Scotch pine. **w—gar** *a* (*Lthr.*) tawed, alum-tanned. **W—gehalt** *m* (*Colors*) brightness. **w—gelb** *a* pale yellow

Weißgerbe-degras *n* (*Lthr.*) sod oil. **weiß-gerben** *vt* taw

Weißgerber° *m* (*Lthr.*) tawer. **-degras** *n* sod oil. **Weißgerberei°** *f* **1** tawing. **2** tawery. = WEISSGERBUNG. **Weißgerber-fett** *n* sod oil. **Weißgerbung°** *f* **1** tawing, alum tanning. **2** tawery, alum tannery

Weiß-glas *n* flint glass

weiß-glühen *vt* raise to white heat. **W—glühen** *n* (heating to) incandescence. **w—glühend** *a* white-hot, incandescent. **W—glühhitze, -glut** *f* white heat, incandescence. **-gold** *n* white gold. **-golderz** *n* sylvanite. **-grad** *m* degree of whiteness. **w—grau** *a* pale (*or* light) gray. **W—güldenerz, -güldigerz** *n* argentiferous tetrahedrite, white silver ore—**dunkles W.** (*Min.*) freieslebenite. **-guß** *m* **1** white metal. **2** white malleable cast iron. **-hitze** *f* white heat. **-kalk** *m* **1** pyrolignite of lime, crude calcium acetate. **2** white (fat, non-hydraulic) lime. **-kernguß** *m* white-heart malleable cast iron. **w—kernig** *a* (*Iron*) white-heat. **W—kerntemperguß** *m* white-heart malleable cast iron. **-kies** *m* (*Min.*) arsenopyrite. **-klee** *m* white clover

weiß-kochen *vt* degum (silk)

Weiß-kohl *m, -kraut* *n* (common) white cabbage. **-kupfer** *n* **1** white copper, nickel silver. **2** native copper arsenide. **-kupfererz** *n* (*Min.*) cubanite, impure marcasite

weißl. *abbv* (weißlich) whitish

Weiß-lauge *f* white liquor. **-leder** *n* white (*or* tawed) leather. **w—leuchtend** *a* glowing (*or* shiny) white

weißlich *a* whitish. **--grau** *a* whitish-gray

Weiß·lot *n* soft (*or* tin) solder

weiß·machen *vt* bleach; whiten. **W–macher** *m* (*Dterg.*) whitener

Weiß·mehl *n* white (*or* wheat) flour. **-messing** *n* white brass. **-metall** *n* white metal. **-nickelerz** *n*, **-nickelkies** *m* (*Min.*) white nickel ore, chloanthite, rammelsbergite. **-ofen** *m* (*Metll.*) refining furnace. **-papp** *m* (*Tex.*) white resist. **-produkt** *n* white product; (*Petrol.*) refined product. **-reserve** *f* (*Tex.*) white resist. **w–rot** *a* whitish (*or* pale) red. **W–ruß** *m* (*Synth.*) fillers. **-schliff** *m* white ground wood; (*Paper*) white mechanical pulp. **-siedekessel** *f* blanching copper (kettle). **-siedelauge** *f* blanching liquor

weiß·sieden *vt* blanch

Weiß·spießglanz *m*, **-spießglanzerz** *n* white antimony, (*Min.*) valentinite. **-stahl** *m* white pig iron. **-stein** *m* whitestone, granulite; (*Copper*) white metal. **-strahl** *m*, **-strahleisen** *n* white pig iron resembling spiegeleisen. **w–strahlig** *a* white-radiated. **W–stuck** *m* white stucco. **-stückkalk** *m* white lump quicklime. **-sud** *m* (*Silver*) blanching (solution). **-sylvanerz** *n* (*Min.*) sylvanite. **-tanne** *f* silver fir (tree). **-tellur** *n* sylvanite. **-töner** *m* (*Bleaching*) fluorescent whitening agent. **-tönung** *f* 1 white tint. 2 fluorescent whitening. **-trockner** *m* (*Paint*) white drier. **-tünchen** *n* whitewashing. **-verdünnung** *f*, **-verschnitt** *m* dilution with white. **-vitriol** *n* white vitriol, zinc sulfate. **-waren** *fpl* white goods, dry goods, linens. **w–warm** *a* white-hot. **W–wäsche** *f* white laundry goods, whites *cf* WÄSCHE. **-wein** *m* white wine. **-werden** *n* whitening, bleaching (= turning white); (*Varnish, Lacquer*) blushing, chalking. **-wurz(el)** *f* (*Bot.*) Solomon's seal. **-zeug** *n* (*Tex.*) white goods, whites. **-zucker** *m* white sugar

Weisung *f* (-en) (*sg & pl*) instructions, directions, order(s): **laut W. des Chefs** by orders of the boss, **weisungs·gemäß** *adv* as per instructions, as directed

weit I *a* 1 far, distant; long: *adv oft* a long way, (a)way: **w. und breit** far and wide; **bei w—em** by far; **von w—em** from afar, from a distance; **w. entfernt von** far (*or* a long way) from, far removed from; **es w. bringen**, go far, make good progress; **sie gingen so w., zu . . .** they went so far as to . . . ; **nicht mehr w.** not much further; **ein w—er Weg** a long way; **sie haben es noch w.** they still have a long way to go; **w. oben (unten, draußen)** away up (down, out), far above (below, out); **es ist so w.** the time has come; **es ist so w., daß** we have reached the point where; **damit ist es nicht w. her** that does not amount to much. 2 (+*exact distance*) e.g.: **10 km w. laufen** run a distance of 10 km. 3 wide, broad; roomy; wide open, dilated; (*Clothes*) loose-fitting, baggy. 4 *adv* (+*compar*) much, far (better, etc). *Cf* WEITE, WEITER, WEITESTE. **II.** *sfx* e.g. **meilenweit** for miles

weit·aus *adv* (+*compar, superl*) by far (better, best, etc). **-ausgebreitet** *a* widespread, extended. **W–blick** *m* farsightedness, vision. **W–blickend** *a* farsighted

Weite I *f* (-n) 1 distance, length. 2 width, diameter: **lichte W.** inside diameter. 3 breadth, extent; spread, expanse; spaciousness. **II** *n* (*adj endgs*): **das W. suchen** run away

weiten *vt & sich w.* widen, broaden; stretch; dilate; expand

weiter I *a, adv* (*compar of* WEIT) 1 farther, further, more distant. 2 wider, roomier, looser. 3 more, additional, other; continued, further; *adv oft* else, the rest, besides, in addition; **alles W—e** anything else, all the rest; **w—e zehn Tage** another ten days; **bis auf W—es** until further notice, pending further details; **ohne w—es** without hesitation (*or* further ado); **im w—en** in the following remarks (*also* = :) **des w—en** besides, in addition, furthermore; **und so w.** and so forth. **II** *pfx* further, continued *cf* WEITER-ENTWICKLUNG. **III** *adv & sep pfx* on, continue to: **sich w. daran beteiligen** continue to participate in it; **weiterschreiben** write on, go (*or* keep) on writing, continue to write

weiter·behandeln vt continue to treat. **W—behandlung** f further (or continued) treatment

Weiterbildung° f continuing education, further (vocational) training, new-career training

weiter·bringen* vt move (sthg) ahead, foster, advance

weiter·denken* 1 vt/i follow up (an idea). 2 vi go on thinking; think things through

weitere a further, etc of WEITER I

weiter·entwickeln vt & **sich w.** continue to develop, develop further. **W—entwicklung°** f further development

weiter·fahren* 1 vt continue driving (a car). 2 vi(s) drive (ride, travel, move) on, continue on one's way. **W—fahrt** f continued trip, rest of the trip

weiter·färben vt continue dyeing

weiter·führen vt move (sthg) forward, go on with sthg

weiter·geben* vt (oft + **an**) pass on (to)

weiter·gehen* vi(s) walk on; go on, continue (on one's way)

weiter·gerben vt retan (hides)

weiterhin I adv 1 in the future. 2 besides, beyond that. **II** cnjc furthermore, moreover

weiter·kommen* vi(s) get ahead, make headway

weiter·können* vi be able to continue

weiter·leben vi live on, go on living

weiter·leiten vt pass on, lead on, transmit

weiter·machen vi continue, go on

weiter·oxidieren vt/i continue oxidizing

weiter·reagieren vi react further, continue reacting

weiters adv besides, etc = WEITERHIN I

weiter·schicken vt send on, forward. **weiter·senden*** I vt = WEITERSCHICKEN. 2 vi go on broadcasting

weiter·treiben* vt continue, carry on (with)

Weiterung f (-en) (usu pl) (later) complication(s), unpleasant consequence(s)

weiter·verarbeiten vt continue treating (or processing); treat subsequently. **W—verarbeitung°** f continued (or subsequent) processing (or treatment). **-verwendung°** f further use

weiter·wissen* vi (usu neg): **nicht w.** not know how to go on (or what to do next)

weitest a (superl of WEIT) furthest; widest; loosest, etc. **--gehend** a furthest-reaching cf WEITGEHEND

weit·gehend, -greifend a extensive, far-reaching, sweeping; adv oft to a great extent. **-her** adv from far away. **-hergeholt** a farfetched. **W—halsflasche** f wideneck(ed) bottle. **w—halsig** a wideneck(ed). **W—halskolben** m wideneck(ed) flask. **w—hin** adv 1 a long way, far into the distance. 2 to a great extent. **-laufig** a 1 distant, remote. 2 lengthy, long-drawn-out; adv oft at great length. 3 spacious, extensive. 4 rambling. **-lochig** a wide-holed; large-pored. **-maschig** a wide-mesh(ed), coarse-mesh(ed). **-mündig** a wide-mouth(ed). **-ragend** a far-reaching. **-räumig** a roomy, spacious. **-reichend** a far-reaching. **-schweifig** a wide-ranging; digressive, long-winded. **-sichtig** a far-sighted. **-tragend** a far-reaching; long-range. **-umfassend** a extensive, encompassing

Weitung f (-en) 1 dilation. 2 widening, broadening. 3 wide part (of a bottle, passage, road, etc)

weit·verbreitet a widespread; wide-circulation (periodical). **-verzweigt** a widely ramified. **-winklig** a wide-angle(d)

Weizen m (-) wheat; **türkischer W.** maize, (Indian) corn. **--brot** n wheat bread. **-grieß** m wheat grits, semolina. **-grießkleie** f semolina with bran. **-keim** m wheat germ. **-kleie** f wheat bran. **-kleienbeize** f(Lthr.) bran drench. **-malz** n wheat malt. **-mehl** n wheat flour. **-schrot** m course-ground wheat. **-stärke** f, **-stärkemehl** n wheat starch. **-stärkekleister** m wheat starch paste. **-stroh** m wheat straw

welch pron **I** (in exclam, no endgs) what (a) . . . ! **II** (adj endgs) 1 (interrog) which (one), what; **w—e auch** whichever (ones), whatever (ones). 2 (relat) which, that. 3 (indef) some, any: **haben sie w—e?** do they have any (or some)?

welcher·art adv of what(ever) kind

welcherlei *adv* whatever (kind of)

Weldon·schlamm *m* weldon mud

welk *a* 1 withered, wilted. 2 limp, flaccid. **W–·boden** *m* (*Brew.*) withering floor. **welken** 1 *vt/i* wither, wilt. 2 *vt* (*Brew.*) air-dry; (*Lthr.*) sam, sammy. **Welk·malz** *n* withered (*or* air-dried) malt

Well·asbest *m* corrugated sheet (of asbestos). **-baum** *m* (*Mach.*) arbor, shaft, axle; (*Engg.*) crab, winch. **-blech** *n* corrugated sheet (of) iron, corrugated tinplate

Welle *f* (-n) 1 (*incl Phys.*) wave. 2 (*esp Mach.*) shaft; axle, arbor; roller. 3 bundle of sticks. 4 (*Glass*) stria

wellen I *vt* 1 wave, curl. 2 corrugate. 3 roll (out flat). 4 boil. **II: sich w.** be(come) wavy, curl; undulate. *cf* GEWELLT

Wellen· wave:-**·änderung** *f* frequency shift. **-anzeiger** *m* wave detector. **w–artig** *a* wavy, wave-like, undulatory. **W–ausbreitung** *f* wave propagation. **-bereich** *m* wave band. **-berg** *m* wave crest. **-bewegung** *f* wave motion, undulation. **-bild** *n* wave form; oscillogram. **-brecher** *m* breakwater. **-bündel** *n* wave pocket. **-bündelung** *f* beaming. **-filter** *n* wave filter. **-fläche** *f* wave surface. **w–förmig** *a* 1 undulatory, wavy. 2 corrugated. **W–front** *f* wave front (*or* head). **-führer** *m* wave guide. **-funktion** *f* wave function. **-gipfel** *m* wave crest (*or* peak). **-gleichung** *f* wave equation. **-kalk** *m* Triassic limestone. **-kamm** *m* wave crest. **-kopf** *m* wave head (*or* front). **-lager** *n* (*Mach.*) (shaft *or* axle) bearing. **-länge** *f* wavelength. **-leiter** *m* wave guide. **-leitung** *f* 1 (*Phys.*) wave guide. 2 (*Mach.*) transmission. **-linie** *f* wavy line. **-mechanik** *f* wave mechanics. **w–mechanisch** *a* wave-mechanical. **W–messer** *m* wavemeter, ondometer, cymometer. **-optik** *f* wave optics. **w–optisch** *a* wave-optical. **W–paket** *n* wave packet. **-papier** *n* corrugated paper. **-rückstrahlung** *f* wave reflection. **-sauger, -schlucker** *m* wave trap. **-schreiber** *m* oscillograph. **-schwingung** *f* undulation. **-sittich** *m* zebra parakeet; budgie. **-spannung** *f* wave potential. **-stirn** *f* wave face (*or* front). **-strom** *m* (*Elec.*) pulsating current. **-stromlichtbogen** *m* pulsating current arc. **-tal** *n* wave trough.

-theorie *f* wave theory. **-weite** *f* amplitude. **-zahl** *f* (*Spect.*) wave number. **-zug** *m* wave train

Weller *m* loam and straw (used as wall filler)

Well·fleisch *n* boiled pork

wellig *a* wavy, undulating. **Welligkeit** *f* waviness

Well·(pack)papier *n* a corrugated (wrapping) paper. **-pappe** *f* corrugated cardboard. **-platte** *f* corrugated sheet metal. **-rad** *n* wheel and axle. **-rohr** *n* corrugated pipe (*or* tube)

Wellung *f* (-en) waving, undulation, curling; corrugation

Well·wurzel *f* (*Pharm.*) symphytum

Welpe *m* (-n, -n) whelp; pup, cub

Welsch·kohl *m* Savoy cabbage. **-korn** *n* (Indian) corn, maize. **-kraut** *n* Savoy cabbage

Welt *f* (en) world: **alle W., die ganze W.** (a) all the world, the whole world, (b) everybody; **auf der ganzen W.** in the whole world; **aus der W. schaffen** do away with. **-·all** *n* universe. **-alter** *n* (historical) age, eon. **-anschauung** *f* philosophy (of life), outlook. **-auge** *n* (*Min.*) hydrophane. **-ausstellung** *f* world's fair. **w–berühmt** *a* world-famous. **W–beschreibung** *f* cosmography. **-bewegung** *f* worldwide movement. **-bild** *n* conception of the world. **-bund** *m* world federation. **w–bürgerlich** *a* cosmopolitan

Welten·raum *m* (interstellar) space

Welt·erzeugung *f* world production (*or* output). **w–fremd** *a* withdrawn (from the world), unworldly. **W–gegend** *f* region of the world. **-geltung** *f* worldwide recognition. **-geschehen** *n* world events. **-geschichte** *f* history of the world. **-gesundheitsorganisation** *f* World Health Organization. **-gürtel** *m* zone (of the earth). **-handel** *m* world trade. **-herrschaft** *f* mastery of the world. **-hungerhilfe** *f* worldwide aid for the hungry. **-jahreserzeugung** *f* annual world production. **-karte** *f* map of the world. **-klima** *n* world's climate. **-körper** *m* heavenly (*or* cosmic) body. **-krieg** *m* world war. **-kugel** *f* globe. **-lage** *f* world situation

weltlich *a* worldly, mundane; secular

Welt·macht *f* world power. **-markt** *m* world market. **-meer** *n* ocean. **-postverein** *m* (International) Postal Union. **-raum** *m* (cosmic) space

Weltraum· space: **-fahrer** *m* astronaut, space traveler. **-fahrt** *f* space travel. **-fahrzeug** *n* space ship (*or* craft). **-forschung** *f* space research (*or* exploration). **-kapsel** *f* space capsule. **-rakete** *f* space rocket. **-schiff** *n* space ship. **-station** *f* space station. **-strahl** *m* cosmic ray.

Welt·spitzenleistung *f* world's record. **-sprache** *f* 1 universal language. 2 language of worldwide importance. **-stadt** *f* metropolis. **-teil** *m* 1 part of the world. 2 continent. **-untergang** *m* end (*or* destruction) of the world. **-verbesserer** *m* utopian world reformer. **-vorrat** *m* world supply. **w—weit** *a* & *adv* worldwide; *adv oft* on a worldwide scale. **W—wirtschaft** *f* world economy

wem *pron* (*dat of* WER) 1 (*interrog*) (to) who(m)? 2 (*relat*) (to) who(m)ever

wen *pron* (*accus of* WER) 1 (*interrog*) who(m)? 2 (*relat*) who(m)ever

Wende *f* (-n) 1 turning point. 2 turn of events, change, (*specif*) reunification of Germany in 1990. 3: **W. des Jahrhunderts** turn of the century. **··kreis** *m* 1 (*Mech.*) turning circle. 2 (*Geog.*) Tropic (of Cancer *or* Capricorn)

Wendel *f* (-n) coil; spiral, helix. **··antenne** *f* helical antenna. **-bohrer** *m* twist drill. **-fläche** *f* helicoid surface. **wendeln** *vi* coil, spiral. **Wendel·treppe** *f* spiral stairway

wenden* (*reg or* wandte, gewandt) **I** *vt* 1 turn (over, around, inside out), reverse. 2 dip, dredge (in flour, etc). 3 (**an**) apply (to); (**auf**) direct (attention) (to), spend (time) (on). **II: sich w.** 4 turn (over, around). 5 (**an**) be directed (*or* addressed) (to); turn (to) (for aid, advice, etc); apply (to), consult. 6 (**zu**) take a turn (e.g. for the worse). *Cf* GEWANDT

Wende·punkt *m* 1 turning point. 2 solstice. 3 (*Math.*) point of inflection

Wender *m* (-) 1 (*Mach.*) turner, rotator, rotor: 2 (*Metll.*) manipulator. 3 (*Elec.*) reverser

Wende·schalter *m* (*Elec.*) reversing switch. **-stelle** *f* turning place

wendig *a* 1 maneuverable. 2 resourceful. 3 agile, nimble. **Wendigkeit** *f* 1 maneuverability. 2 resourcefulness. 3 agility, nimbleness

Wendung *f* (-en) 1 turn: **eine W. machen** make (*or* take) a turn. 2 turn of events; change. 3 (turn of) phrase, expression idiom

wenig **I** *pron* (*indef*) 1 *sg* (*usu no endgs*) little, a small quantity (amount, number) of, not much: **w. Salz** little (*or* not much) salt; **w. Neues** little that is new: **das w—e Salz** the small amount of salt. **2: ein w.** a little (bit of); **ein klein w.** a tiny bit (of). 3 *pl* (*usu with adj endgs*) (only a) few, not many, a small number of: **in w—en Fällen** in few (*or* a small number of) cases; **einige w—e** (only a) few. **II** *adv* 5 not much, not very, only slightly, to a very slight extent; not often. **6: ein w.** a little (bit): **ein w. mehr** a little more. *Cf* WENIGER, WENIGSTE, WENIGSTENS

weniger *pron/adv* (*compar of* WENIG) (*sg*) less, (*pl*) fewer; (*Math., oft*) minus: **viel w. Säure** much less acid; **nicht w. schnell** no less fast; **um so w.** so much the less; **zehn w. acht** ten minus eight —**nichts w. als** anything but *cf* NICHTSDESTOWENIGER

Wenigkeit *f* trifle; slight amount (quantity, number)

wenig·löslich *a* poorly (*or* only slightly) soluble

wenigste *a* (*superl of* WENIG) 1 *pron* (*adj endgs*), (*sg*) least, (*pl*) fewest. 2 *adv:* **am w—n** least, to the least extent

wenigstens *adv* at least

wenn *cnjc* 1 if; (*oft* + *doch* or *nur*) if only; (+ **auch, gleich, schon**) even if, even though: **ohne Wenn und Aber** without ifs or buts. 2 when, whenever

wenn·gleich, -schon *cnjc* even if, even though

wer *pron* 1 (*interrog*) who; **w. von ihnen** which one of them. 2 (*relat*) whoever, anyone who. 3 (*indef*) = JEMAND someone, somebody. *Cf* WEM, WEN, WESSEN

Werbe·abteilung *f* advertising (*or* publicity) department. **-mittel** *n* advertising (or publicity) medium

werben* (warb, geworben; wirbt) 1 *vt* woo, solicit; recruit; court. 2 *vi* (**um**), *vt* try to win (or gain). 3 *vi* (**für**) advertise, publicize, promote

Werbe·schrift *f* prospectus, publicity (or advertising) publication. **-sendung** *f* (radio, TV) commercial. **-wesen** *n* (field of) publicity, advertising

werblich *a* advertising, commercial

Werbung *f* (-en) 1 advertising, publicity. 2 advertising department. 3 recruitment

Werde·gang *m* 1 development. 2 history, (past) career. 3 (*Engg.*) process of manufacture

werden* (wurde, geworden; wird) **I** *vi* (s) 1 (+*pred noun* or *adj* or +*zu*) become, get (to be), grow; (*esp* + **zu**) turn (into). 2 (*Profession*) be, become (a chemist, etc). 3 (**aus, mit**) come, become (of); **aus A. wird B** A becomes (or turns into) B. 4 develop, turn out—**das** (or **daraus**) **wird nichts** nothing will come of that. 5 make progress. 6 happen. 7 (*dat*) begin to feel: **es wird ihnen warm** they are beginning to feel warm. **II** *v aux* 8 (*pres* + *inf*): **A** will (*forms future tense*); (*sbjc* + *inf*) would: **sie w. versuchen** they will try; **sie behaupteten, es werde** (or **würde**) **bald enden** they claimed it would end soon. **B** (*oft* + **wohl:** *probability in pres*, e.g.): **sie w. wohl krank sein** they are probably (or they must be) ill. 9 (*pres* + *pp* + **haben** or **sein**): **A** will have: **sie w. es bis morgen gelesen haben** they will have read it by tomorrow. **B** (*oft* + **wohl:** *probability in past*, e.g.): **sie w. wohl krank gewesen sein** they were probably (or they must have been) ill. 10 *sbjc*: **würde** (+*inf*) would (*forms conditionals*) *cf* WÜRDE. **III** *v aux* (s) (*all tenses* + *pp*) be, get (*forms passive voice*): **dann wird es mit Wasser vermischt und erhitzt** then it is mixed with water and heated *cf* WORDEN. **Werden** *n* growth, development: **im W.** in the course of development, in the making; **W. und Wachsen** growth and development. **werdend** *p.a* (*oft*) developing, budding, incipient: **w—e Mutter** expectant mother

werfen* (warf, geworfen; wirft) **I** *vt* 1 throw, fling, pitch, cast; drop (bombs), form (bubbles) **—Falten w.** get creased. **2: aufs Papier w.** jot down (on paper). **3: Junge w.** have cubs (pups, etc). **II** *vi* 4 (**mit**) throw (sthg). 5 (+**um sich**) throw (things) around; flaunt. **III: sich w.** 6 throw (or fling) oneself; dive; toss and turn. 7 (+**auf, oft**) jump, pounce (on), rush (at). 8 (*Wood*) warp; (*Pavement*) buckle. **Werfer** *m* (-) thrower; pitcher; launcher

Werft I *m* (-e) (*Tex.*) weft. **II** *f* (-en) shipyard; repair hangar

Werg *n* tow, (flax or hemp) waste, oakum

Werk *n* (-e) 1 (*usu*) work: **am W. sein** be at work; **sich ans W. machen, ans W.** (or **zu W—e**) **gehen** get to work, start working; **ins W. setzen** set in motion. 2 work of art, opus; achievement, deed; (*sg & pl*) works. 3 (industrial) works, plant, factory. 4 (*Mach.*) works, mechanism. 5 (*Metll.*) pig of raw lead. 6 (*Salt*) brine evaporated at one time. 7 (*Paper*) stuff. 8 (*Glass*) frit. **·-anlage** *f* plant layout; work equipment, (*pl oft*) industrial plants. **-arzt** *m* factory doctor. **-bank** *f* workbench. **-blei** *n* raw lead (containing silver). **-bottich** *m*, **-bütte** *f* (*Paper*) stuff vat. **-direktor** *m* plant manager. **w—eigen** *a* factory-owned, company-owned. **-fremd** *a* not factory-related, outside. **W—führer** *m* plant manager. **-halle** *f* workshop, shop floor. **-holz** *n* timber. **-holzkäfer** *m* deathwatch beetle. **-leiter** *m* plant manager. **-leitung** *f* plant management. **-loch** *n* manhole. **-meister** *m* 1 shop foreman. 2 master mechanic. **-norm** *f* standard specification. **-nummer** *f* serial number. **-probe** *f* (*Metll.*) metal sample; cast specimen

Werks· (= WERK·) **-abgabepreis** *m* factory (or ex-works) price

Werk·seide *f* floss silk. **-silber** *n* silver extracted from lead ore

Werks·leitung *f* plant management

Werk·spionage *f* industrial espionage. **-statt, -stätte** *f* 1 (*work*) shop; machine shop. 2 studio, workroom. 3 laboratory. **-stein** *m* freestone, quarry stone. **-stelle** *f* workplace; shop, studio, factory, etc. **-stoff** *m* material used in manufacturing, industrial (construction, engineering) material

Werkstofforschung f materials research

Werkstoff·kunde f science of materials. **-prüfung** f materials testing

Werk·stück n 1 work (being processed in a shop), article (piece, material) to be worked on. 2 manufactured article. **-stufe** f stage of work (or operation). **-stuhl** m loom

Werks·vertreter m factory salesman

Werk·tag m work(ing) day, weekday. **w-täglich** a workday. **w-tags** adv (on) weekdays. **-tätig** a working, employed. **W-tätige** m, f (adj endgs) working person (or pl people). **-zeug** n tool, instrument, (piece of) apparatus

Werkzeug·kasten m tool box (or kit); instrument case. **-maschine** f machine tool. **-schlosser** m toolmaker. **-stahl** m tool steel

Werk·zink n raw zinc. **-zinn** n raw tin

Wermut m 1 (Bot.) wormwood. 2 vermouth. **-·bitter** m absinthin. **-öl** n wormwood (or absinth) oil. **-schnaps, -wein** m vermouth

wert 1 pred a (+genit or accus) worth: **wieviel ist es w.?** how much is it worth?; **nichts w.** worth nothing, worthless, no good; **der Mühe w.** worth the trouble, worthwhile cf REDE. 2 (esp in polite phrases) worthy, esteemed, dear

Wert m (-e) 1 (incl Math.) value. 2 valence. 3 postage stamp. 4 (pl) variables. 5 (pl) (Com.) stocks, securities. 6 importance, quality, merit: **W. legen auf etwas** consider (sthg) important. **-·angabe** f declaration of value. **-ansatz** m assay, estimate of value. **-bestimmung** f evaluation, determination of value (valence, purity, quality, active constituent). **-brief** m registered letter. **-einheit** f unit of value

werten vt value, appreciate, esteem

Werte·paar n pair of values

Wert·ersatz m equivalent; compensation. **-gegenstand** m valuable item, pl valuables. **-gut** n article(s) of value

-wertig sfx -valent. **Wertigkeit** f (-en) valence

Wertigkeits·einheit f unit of valence. **-for-**

mel f valence (or linkage) formula. **-stufe** f valence (stage)

wert·los a worthless, no good; useless. **-mäßig** adv 1 in value, according to value. 2 ad valorem. **Wert·messer** m standard of value. **-papier** n (share of) stock, bond, note, etc; pl securities. **-sachen** fpl valuables. **w-schaffend** a productive. **W-schätzung** f appreciation. **-stoff** m valuable substance. **-stufe** f rating (on a scale of values)

Wertung f (-en) (e)valuation, rating; judgment

Wert·verhältnis n relative value. **w-voll** a valuable

Wesen n (-) 1 essence, (inner) nature, character. 2 (person's) manner. 3 ado, carrying on, fuss: **sein W. treiben** carry on, be in action; **viel W-s machen (von, um)** make a big thing (or production) of; **kein W. daraus machen** make nothing of it. 4 organism, (living) being, creature; soul (= person). 5 (as sfx, designates a field of activity:) **Gesundheitswesen** (field of) health care

wesenhaft a 1 essential, intrinsic. 2 real, substantial

Wesenheit f (-en) 1 real nature, inner essence. 2 reality; being, existence. 3 pl (oft) entities

wesenlos a nonmaterial, incorporeal; shadowy

wesens·eigen a (oft + dat) characteristic (of). **-fremd** a (oft + dat) uncharacteristic (of), alien (to). **-gleich** a essentially alike, identical in nature. **W-gleichheit** f identity. **w-verschieden** a essentially different. **-verwandt** a essentially similar, related in essence. **W-zug** m essential trait, characteristic

wesentlich a 1 essential, significant, important; main, basic: **im w-en** essentially, mainly, basically. 2 considerable, appreciable; oft as adv + compar, e.g.: **w. älter** considerably older. **Wesentliche** n (adj endgs): **das W.** the essential point, the essentials; **nichts W-es** nothing essential (important, etc)

weshalb 1 adv & cnjc (interrog) why. 2 cnjc which is why, and that is why

Wespe f (-n) wasp, hornet. **Wespen·bein** n

(*Anat.*) sphenoid bone. **-nest** *n* wasp's nest

wessen *pron: genit of* WER & WAS whose, of what

West I *m* 1 west. 2 west wind. **II** *pfx* west, western: **w—deutsch** West German. *Cf* WESTEN

Weste *f* (-n) vest —**reine W.** clean slate

Westen *m* west, western part: **nach W.** (toward the) west; **im W. von** in the western part of

Westen·tasche *f* vest pocket. **-taschenformat** *n* vest-pocket size

West·europa *n* Western Europe. **w—europäisch** *a* West European

Westfalen *n* Westphalia. **westfälisch** *a* Westphalian

west·indisch *a* West Indian

westlich I *a* 1 western; **w—e Länge** west(ern) longitude; **der w—ste Punkt** the westernmost point. 2 westerly (wind, direction). **II** *adv* (+**von**), *prep* (+*genit*) (to the) west (of): **etwas w—er gelegen** lying somewhat more to the west

west·wärts *adv* westward

weswegen 1 *adv & cnjc* (*interrog*) why. 2 *cnjc* which is why, and that is why

Wet·schiefer *m* (*Min.*) novaculite

wett 1 *pred a:* **w. sein** be quits (*or* through), be even (*or* square). 2 *sep pfx: cf* WETTMACHEN

Wett·bewerb *m* competition, contest. **-bewerber** *m* competitor. **w—bewerbsfähig** *a* competitive

Wette *f* (-n) 1 bet, wager. 2: **um die W.: a** like mad, to beat the band, **b** (+**rennen, fahren,** etc) (run a) race

Wett·eifer *m* competitive spirit, competition, rivalry. **wetteifern** (*pp* gewetteifert) *vi* (**um**) vie, compete (for)

wetten *vt/i* bet, wager

Wetter *n* (-) 1 weather. 2 (bad) weather, (thunder)storm. 3 *pl* (*Min.*): **schlagende W.** firedamp. **-bericht** *m* weather report. **w—beständig** *a* weatherproof, weather-resistant. **W—beständigkeit** *f* resistance to weather. **-dienst** *m* weather (*or* meteorological) service. **-dienststelle** *f* weather service center. **-dynamit** *n* (*Min.*) permissible dynamite. **w—echt** *a* weather-resistant. **W—eintritt** *m* (*Min.*) air inlet. **-fahne** *f* weather vane. **w—fest** *a* weath-

erproof. **W—führung** *f* (*Min.*) ventilation. **-fulminit** *n* (*Explo.*) wetterfulminite. **-glas** *n* barometer. **-hahn** *m* weather vane. **-karte** *f* weather chart. **-kunde** *f* meteorology. **-leitung** *f* (*Min.*) air channel

wetterleuchten *vi:* **es wetterleuchtet** heat lightning is flashing. **Wetterleuchten** *n* heat (*or* sheet) lighting

Wetter·maschine *f* (*Min.*) ventilating engine

wettern *vi:* **es wettert** 1 a storm is coming up. 2 it (*or* the weather) is stormy

Wetter·probe *f* (*Min.*) air sample, air test. **-prognose** *f* weather forecast. **-prüfung** *f* (*Min.*) air testing. **-rakete** *f*, **-satellit** *m* weather satellite. **-schaden** *m* weather (*or* storm) damage. **w—sicher** *a* 1 weatherproof. 2 (*Min.*) dampproof, flameproof. **W—sprengstoff** *m* (*Min.*) permissible explosive, **-stein** *m* (*Min.*) belemnite. **-strom** *m* (*Min.*) air current. **-sturz** *m* sudden drop in temperature. **-verhältnisse** *npl* weather conditions. **-voraussage** *f* weather forecast. **-zug** *m* (*Min.*) ventilation, draft. **-zünder** *m* (*Min.*) electric fuse

Wett·kampf *m* contest, competition; match, meet, etc. **-lauf** *m* race

wett·machen *vt* make up (for), make good

Wett·rennen *n* race. **-rüsten** *n* arms race. **-spiel** *n* match, (championship) game. **-streit** *m* contest

wetzen *vt* whet, sharpen; rub, scrape

Wetz·schiefer *m* (*Petrog.*) novaculite. **-stein** *m* whetstone, hone; grindstone

wf. *abbv* (wasserfrei) anhydrous, moisture-free

wich yielded, etc: *past of* WEICHEN II

Wichs·bürste *f* shoeshine brush

Wichse *f* (-n) (shoe) polish; polishing wax. **wichsen** *vt* polish, shine; wax

Wichs·kalbleder *n* waxed calf leather

Wichte *f* (-n) density; specific gravity. **-analyse** *f* float-and-sink analysis

wichtig *a* important; serious, grave: **das W—e dabei** the important thing about it; **w. nehmen** take seriously; **sich w. nehmen (machen, haben), w. tun** act important (*or* grave). **Wichtigkeit** *f* (-en) 1 importance: **ohne W.** of no importance.

2 gravity, air of importance. **3** important matter

Wicke f (-n) vetch

Wickel m (-) **1** ball (of yarn, etc). **2** reel, spool; bundle. **3** wrapper. **4** (Med.) compress, pack. **5** (hair) curler, roller. **6** (cigar) filler. **-binde** f roll bandage. **-draht** m binding wire. **-körper** m (Tex.) wound package

wickeln I vt **1** wrap; bandage. **2** wind, coil. **3** curl (hair). **4** roll (cigars). **5** (+ **aus, von**) unwrap (from, out of); (+ **von**) unwind (from, off). **II:** sich w. wind, coil, curl. **Wickelung** f (-en) winding, etc = WICKLUNG

Wicken·stroh m vetch straw

Wickler m (-) **1** wrapper. **2** winder; roller. **3** (hair) curler. **4** (Zool.) bell moth

Wicklung f (-en) **1** (incl Elec.) winding. **2** wrapping, casing. **3** bandaging

Widder m (-) **1** (Zool., Mach.) ram. **2** (Astron.) Aries

widdern vt (Brew.) turn (malt)

wider I prep (accus) against, contrary to — **das Für und W.** the pros and cons. **II** pfx **1** (insep) contra-, counter-, anti- (implies opposition). **2** (usu sep) back, re- (implies reflection)

Widerdruck° m counterpressure, reaction

widerfahren* (insep) vi (dat) happen, be done (to), befall. **widerfuhr** (past)

widergesetzlich a illegal

widerhaarig a cross-grained; perverse; unruly

Widerhaken° m barb (of an arrowhead)

Widerhall m echo, reverberation. **wider/hallen** (sep & insep) vi echo, resound

Widerhalt m support

Widerlager° n (Engg.) abutment

widerlegbar a refutable. **widerlegen** (insep) vt refute. **Widerlegung** f (-en) refutation

widerlich a repulsive, repugnant

widernatürlich a unnatural, preternatural

widerraten* (insep) vt (+ d.i.o) advise (sbdy) against

widerrechtlich a illegal, unlawful

Widerrede° f contradiction, objection

widerrief revoked: past of WIDERRUFEN

widerriet advised against: past of WIDERRATEN

Widerruf° m recall, revocation; retraction. **widerrufen*** (insep) vt recall, revoke; retract

Widersacher m (-) adversary, opponent

Widerschall° m echo

Widerschein° m reflection

Widerschlag° m rebound; return stroke

widersetzen (insep) vr: sich w. (dat) resist, oppose

widersinnig a **1** reversed, of opposite sense (or direction). **2** anticlinal. **3** absurd, nonsensical

widerspenstig a refractory, unruly

wider/spiegeln (sep & insep) **1** vt reflect, mirror. **2:** sich w. be reflected (or mirrored)

Widerspiel° n reverse, opposite, contrary

widersprach contradicted: past of widersprechen

widersprechen* (insep) vi (dat) contradict. **widersprechend** p.a. contradictory

widerspricht contradicts: pres of widersprechen

widersprochen contradicted: pp of widersprechen

Widerspruch° m contradiction; inconsistency

widerspruchs·los a without contradiction, unquestioned; adv oft without question (or protest). **-voll** a contradictory; inconsistent, incompatible

Widerstand° m resistance; obstacle; (Cmptrs.) impedance

widerstand(en) resisted: past (& pp) of WIDERSTEHEN

Widerstands·brücke f (Elec.) resistance bridge. **-büchse** f resistance box. **-draht** m (Elec.) resistance wire. **-einheit** f resistance unit. **w—fähig** a resistant; durable; refractory. **W—fähigkeit** f ability to resist, resistivity; durability. **-gefäß** n conductivity cell. **-heizung** f resistance heating. **-kasten** m resistance box. **-kraft** f power of resistance. **-legierung** f resistance alloy. **w—los** a unresisting; adv oft without resistance. **W—messer** m ohmmeter, resistance meter. **-messung** f measurement of resistance. **-moment** n moment of resistance. **-ofen** *m* resistance furnace. **-satz** *m* (*Elec.*) resistance set (*or* box). **-schaltung** *f* rheostat. **-spule** *f* (*Elec.*) resistance coil. **-thermometer** *n* resistance thermometer. **-vermögen** *n* power of resistance. **-verschiebung** *f* shift (*or* variation) in resistance. **-wert** *m* coefficient of resistance. **-zelle** *f* photoconductive cell

widerstehen* (*insep*) *vi* (*dat*) 1 resist, withstand. 2 be repugnant (to)

Widerstoß° *m* countershock

wider/strahlen (*sep & insep*) 1 *vt* reradiate, reflect. 2 *vi* be reflected

widerstreben (*insep*) *vi* (*dat*) 1 resist, oppose. 2 be repugnant (to). **widerstrebend** *p.a* 1 resistant. 2 contrary, conflicting. 3 repugnant. 4 *adv aft* reluctantly

Widerstreit° *m* (**von**) conflict, clash (between); antagonism. **widerstreiten*** (*insep*) *vi* (*dat*) conflict (with), be contrary (to). **widerstreitend** *p.a.* conflicting, contrary

widerstritt(en) conflicted: *past* (*& pp*) *of* WIDERSTREITEN

widerwärtig *a* repulsive, repugnant; disagreeable

Widerwille° *m* (**gegen**) aversion, antipathy (to). **widerwillig** *a* unwilling, reluctant

widmen 1 *vt* (+ *d.i.o*) dedicate (to). **2: sich w.** (*dat.*) devote oneself (to). **Widmung** *f* (-en) dedication; devotion

widrig I *a* 1 adverse, unfavorable. 2 repellent, repulsive. II *sfx* anti-, -resistant: **fäulniswidrig** antiseptic. **widrigenfalls** *adv* otherwise, failing that (*or* which). **Widrigkeit** *f* (-en) 1 adversity, unfavorableness. 2 repulsiveness

wie I *adv* 1 (*interrog & exclam*) how: **w. kommt es, daß** . . . ? how is it that . . . ? **w. bitte?** what did you say? *cf* HEISSEN. 2 (+**auch**) however, no matter how: **w. schwer es auch sein mag** however difficult it may be. II *cnjc* 3 (such) as, like: **so klein w.** . . . as small as . . . ; **Elemente w. Argon, Neon usw.** elements like (*or* such as) argon, neon, etc.; **ebenso w. wir** just like us, just as we do; **(so) w. man es im Labor macht** as (*or* the way) one does it in the lab. 4 as if: **w. gelähmt** as if paralyzed. 5 (= **als**) A than: **größer w.**

dieses larger than this one. B: **nichts w. . . .** nothing but . . . *cf* ALS 1, 3. 6 (*oft* + **auch**) as well as: **hier w. (auch) dort** here as well as there *cf* SOWOHL. 7 (*subord., verb at end*) when, as: **w. sie das sahen** when they saw that. 8: **w. wenn** as if. *Cf* WIEFERN, WIESO, WIEVIEL

Wiebel *m* (-) weevil

Wied *m* woad

wieder I *adv* 1 again, back: **nie w.** never again; **w. ins Labor gehen** go back to the lab; **hin und w.** now and again; **immer w.** again and again; **nichts und w. nichts** nothing at all; **w. andere** and still others *cf* SCHON 8. 2 in return. 3 e.g.:**sie w.** they on their part. II *pfx* 4 *usu sep* (*except in verb* WIEDERHOLEN I) again, back, re-. 5 (+ *other sep pfx, note separation*): **baut w. auf** rebuilds: *pres of* WIEDER·AUF·BAUEN

Wiederabdruck° *m* reprint. WIEDER·AB-DRUCKEN *vt* reprint

wieder·ab·scheiden* *vt* reprecipitate, reseparate

wieder·ab·teilen *vt* redivide, subdivide

Wieder·anfang° *m* recommencement, fresh beginning. **wieder·an·fangen*** *vt/i* recommence, begin afresh

wieder·an·knüpfen *vt* resume

Wiederanmeldung° *f* (*Patents*) reapplication

Wiederanreicherung° *f* reconcentration

wieder·an·wärmen *vt* reheat

wieder·an·wenden* *vt* re-use, reapply

wieder·auf·arbeiten *vt* recondition, redo; recycle. **Wiederaufarbeitungs·anlage** *f* (*Nucl.*) fuel recycling plant

Wiederaufbau° *m* reconstruction. **wieder·auf·bauen** *vt* reconstruct, build up again

Wiederaufbereitung° *f* recycling; repreparation

wieder·auf·finden* *vt* rediscover

wieder·auf·füllen *vt* refill

wiederaufladbar *a* rechargeable. **wieder·auf·laden*** *vt* (*Elec.*) recharge

wieder·auf·lösen *vt & sich w.* redissolve

Wiederaufnahme° *f* resumption; reabsorption; readoption; readmission; re-recording. **wieder·auf·nehmen*** *vt* 1 resume, take up again. 2 reabsorb. 3 readopt, readmit. 4 rerecord

Wiederaufrüstung° *f* rearmament

wieder·auf·saugen* *vt* reabsorb

wieder·aus·fällen *vt* reprecipitate, redeposit

wieder·aus·kristallisieren *vt/i* recrystallize, crystallize out again

wieder·aus·richten *vt* realign, readjust

wieder·aus·strahlen *vt/i* reradiate

wieder·beleben *vt* revive, resuscitate; reactivate; (*Sugar*) reburn, char. **Wiederbelebung°** *f* revival, resuscitation; reactivation. **Wiederbelebungs·mittel** *n* (*Med.*) restorative

wieder·belüften *vt* re-aerate, reventilate

Wiederbenutzung° *f* re-use, reutilization

wieder·beschicken *vt* reload, recharge

wieder·bilden *vt* re-form, form again

Wiederbrauchbarmachen *n* regeneration

Wieder·druck *m* reprint(ing)

wieder·ein·nehmen* *vt* recapture

wieder·ein·schmelzen* *vt, vi*(s) remelt

wieder·erhitzen *vt* reheat

wieder·erkennen* *vt* recognize

wieder·erscheinen* *vi*(s) reappear

wieder·erwärmen *vt* rewarm, reheat

Wiedererstattung° *f* restitution, repayment

wieder·erzeugen *vt* reproduce, regenerate

wieder·fällen *vt* reprecipitate

wieder·färben *vt* recolor, redye

Wiedergabe° *f* 1 reproduction, rendition; playback. 2 return, giving back. **wiedergebbar** *a* reproducible; translatable; returnable. **wieder·geben*** *vt* 1 reproduce, render; translate; play back. 2 give back, return

Wiedergebrauch° *m* re-use

Wiedergeburt° *f* rebirth, renascence

wiedergewinnbar *a* recoverable. **wieder·gewinnen*** *vt* regain, recover; win back. **Wiedergewinnung°** *f* recovery. **wiedergewonnen** recovered, etc: *pp*

wieder·gut·machen *vt* make good (for), make up (for). **Wiedergutmachung** *f* making good, restitution, compensation

Wiederhall° *m* echo, reverberation; resonance

wieder·her·stellen *vt* restore; reconstitute. **Wiederherstellung°** *f* restoration; reconstitution

wieder·hervor·bringen* *vt* reproduce

wiederholbar *a* repeatable, reproducible

wieder/holen I (*insep*) 1 *vt* repeat; review. **2: sich w.** recur. **II** (*sep*) *vt* get (*or* fetch) back, recover. **wiederholt** *p.a* repeated; *adv* repeatedly. **Wiederholung** *f* (-en) 1 repetition; review. 2 recurrence. **Wiederholungs·fall** *m:* **im W.** in case of recurrence

Wiederinbetriebnahme *f* resumption, reopening

Wiederinkrafttreten *n* reinstatement, restoration

Wiederinstandsetzung *f* repair, reconditioning

wieder·käuen 1 *vt* ruminate, rechew; rehash. 2 *vi* chew its cud. **Wiederkäuer** *m* (-) (*Zool.*) ruminant

Wiederkehr° *f* recurrence; return; anniversary. **wieder·kehren** *vi*(s) recur, return, come back. **Wiederkehr·spannung** *f* (*Elec.*) recovery voltage

wieder·kommen* *vi*(s) come back, return

Wiederkondensation *f* recondensation. **wieder·kondensieren** *vt, vi*(s) recondense

wieder·kristallisieren *vt* recrystallize. **Wiederkristallisierung°** *f* recrystallization

Wiederkunft *f* (¨e) return *cf* WIEDERKOMMEN

Wiederoxidation° *f* reoxidation

Wiederschein° *m* reflection

wieder·schmelzen* *vt, vi*(s) remelt, refuse

wieder·sehen* *vt* & **sich w.** (*recip*) see (each other) again, meet again. **Wiedersehen** *n* reunion—**auf W.** see you again, goodbye

Wiederstoß° *m* countershock

Wiederstrahl° *m* reflected ray. **wieder·strahlen** *vt* reflect

wiederum *adv* 1 again. 2 in (its) turn. 3 on the other hand

wieder·vereinigen *vt* reunite, reunify; recombine; (*Rays*, etc) refocus. **Wiedervereinigung°** *f* reunion, reunification; recombination; refocusing

Wiedervergeltung *f* retaliation

Wiederverkauf° *m* resale. **wieder·verkaufen** *vt* resell

wieder·verstärken *vt* reamplify, strengthen again; replenish

wieder·verwandeln *vt* reconvert

wiederverwendbar *a* re-usable, **wieder-verwenden*** *vt* re-use. **Wiederverwen-dung°** *f* re-use

wieder-verwerten *vt* recycle. **Wiederver-wertung°** *f* recycling

Wiederwässerung° *f* rewatering, resoaking, rehydration

Wiederzündung° *f* reignition

Wiege *f* (-n) 1 cradle; rocker. 2 (*in compds*): weighing; chopping *cf* WIEGEN: **-behälter** *m* weighing container, weigh-bin. **-brett** *n* chopping board. **-gläschen** *n* weighing glass. **-maschine** *f* mincing machine, chopper. **-messer** *n* chopping (*or* rocker) knife

wiegen I* (wog, gewogen) 1 *vt/i* weigh. 2 *vi* have (*or* carry) weight. *Cf* GEWOGEN. II (*reg: pp* gewiegt) 3 *vt; vi* (+**mit**) rock, nod. 4 *vt* chop (*esp* with a rocker knife). 5 *vt:* **einen** (*or* **sich**) **in Sicherheit w.** lull sbdy (*or* oneself) into a false sense of security. 6: **sich w.** rock, sway, swing. *Cf* GEWIEGT

Wiege-schale *f* scale pan. **-vorrichtung** *f* weighing device

Wiegung *f* (-en) weighing; rocking; chopping *cf* WIEGEN

Wieke *f* (-n) (*Med.*) pledget, (surgical) tent

Wien *n* Vienna. **Wiener 1** *m* (-) Viennese = 2 *a* Vienna: **W. Ätzpulver** Vienna paste (*or* caustic); **W. grün** Vienna green; **W. Kalk** Vienna lime (*or* white); **W. Lack** Vienna lake; **W. Metall** Vienna metal (white copper-antimony alloy). **wiene-risch** *a* Viennese

wies directed, etc: *past of* WEISEN

Wiese *f* (-n) meadow

Wiesel *n* (-) weasel

Wiesen-erz *n* meadow (*or* bog iron) ore. **-flachs** *m* purging flax. **-kalk** *m* freshwater limestone; chalky soil. **-klee** *m* red clover. **-knöterich** *m* (*Bot.*) bistort. **-lein** *m* purging flax. **-torf** *m* meadow peat

wieso *adv* (*interrog*) how (so)?, how come?, how does it come about that . . . ?

wieviel *adv* (*interrog*) 1 how much. 2 (= **w—e**) how many. **--mal** *adv* how many times, how often. **wievielt** *a:* **der w—e** which one (in numerical order): **zum w—en Mal** how many times is it now; **der w—e ist heute?** what day of the month is today?

wiewohl *subord cnjc* although.

wild *a* 1 wild; savage, unruly: **w—er Stahl** wild steel; **w—es Fleisch** proud flesh; **w—es Gestein** nonmetalliferous stone; **w—er Wein** Virginia creeper. 2 furious: **w. machen** infuriate. 3 (**auf**) mad, wild (about). 4 fierce. 5 (*Nucl. Phys.*) spurious (rays). 6 unauthorized, illegal, plagiarized

Wild *n* (*Zool.*) game; (*Meat*) venison. **--bach** *m* mountain torrent. **-braten** *m* roast venison. **-bret** *n* game; venison = WILD

Wilde *m, f* (*adj endgs*) savage, wild person

Wild-ente *f* wild duck. **-gehege** *f* game preserve. **-geschmack** *m* gamy taste. **w—gewachsen** *a* grown wild. **W—hefe** *f* wild yeast. **-kautschuk** *m* wild (*or* native) rubber. **-kirschenöl** *n* oil of wild cherry. **-kirschenrinde** *f* wild cherry bark. **-leder** *n* buckskin, doeskin; suede

Wildnis *f* (-se) wilderness

Wild-schutzgebiet *n* game preserve. **-schwein** *n* boar. **-werden** *n* (*Brew.*) gushing

will want(s): *pres of* WOLLEN

Wille *m* (*genit* Willens; *dat, accus* Willen) 1 will, volition; intent(ion): **wider W—n** against one's will. 2 consent: **ohne ihren W—n** without their consent. 3 way: **seinen W—en behalten** have (it) one's way

willen: **um** (*genit*) **w.** for the sake of *cf* UM 7

Willen *m* = WILLE

willenlos *a* will-less, irresolute

Willens-kraft *f* will power

willentlich *a* intentional, deliberate

willfahren (*insep*) *vi* (*reg, pp* willfahrt) (+*dat*) comply (with). **willfährig** *a* compliant, accommodating

willig *a* willing. **Willigkeit** *f* willingness

willkommen *a, adv* welcome

Willkür *f* arbitrary action, arbitrariness; despotism. **willkürlich** *a* 1 arbitrary. 2 random, *adv oft* at random. 3 voluntary. **Willkürlichkeit** *f* arbitrariness; randomness; voluntariness

Wilson-kammer *f* Wilson cloud chamber. **-nebelspur** *f* Wilson cloud track

wimmeln *vi* (**von**) swarm, teem, be alive (with)

Wimper f (-n) eyelash; cilium. **wimperig** a ciliated, ciliary. **Wimper·infusorien** npl (*Zool.*) Ciliata. **Wimpern·tusche** f mascara. **Wimper·tierchen** n ciliary animal. **-zelle** f ciliated cell

Wind m (-e) wind; (*Metll.*) (air) blast. **--bläser** m blower, blast apparatus. **-darm** m (*Anat.*) colon. **w–dicht** a airtight. **W–druck** m 1 wind pressure. 2 (*Metll.*) blast pressure. **-düse** f blast nozzle, tuyere

Winde I f (-n) 1 winch; reel. 2 hoist. 3 (*Mach.*) worm. 4 (*Bot.*) bindweed. 5 (*Glass*) stria(tion). II winds: pl of WIND

winden* (wand, gewunden) I vt 1 wind, reel, roll; wrap. 2 hoist (up). 3 (*usu* + **aus**) wrest (from). II: **sich w.** wind, coil; twist; squirm, wriggle, writhe. Cf GEWUNDEN

Wind·erhitzer m (hot-blast) air heater, Cowper stove; regenerator. **-erhitzung** f blast heating. **-erhitzungsapparat** m regenerator, etc = -ERHITZER. **-erzeuger** m blast apparatus; blower, fan. **-fahne** f wind vane. **-flügel** m (ventilator) fan; fan blade. **-form** f (*Metll.*) tuyere, twyer. **-frischapparat** m (*Metll.*) converter. **-frischen** n (*Metll.*) converting, purifying by air blast. **-frischverfahren** n (*Metll.*) converter (or Bessemer) process. **-geschwindigkeit** f wind velocity. **-geschwindigkeitsmesser** m anemometer, wind gauge. **-geschwulst** f emphysema. **w–getragen** a wind-borne. **W–hose** f 1 whirlwind, tornado. 2 waterspout

windig a 1 windy, drafty. 2 flimsy, shaky; unreliable

Wind·kanal m wind tunnel. **-kasten** m (*Metll.*) twyer (or blast) box. **-kessel** m blast pressure tank. **-klappe** f air valve. **-leitung** f (*Metll.*) blast main, main blast pipe. **-licht** n hurricane lamp. **-messer** m, **meßgerät** n anemometer; blast gauge. **-mühle** f windmill. **-ofen** m air (wind, draft) furnace. **-pocken** fpl chicken pox. **-pressung** f (*Metll.*) blast pressure. **-richtung** f wind direction. **-rohr** n (*Metll.*) blast (or twyer) pipe. **-rose** f compass card. **-röstverfahren** n (*Metll.*) air blast roasting (process); converter process. **-sack** m wind sock. **-sammler** m

(compressed-)air tank. **-schatten** m lee, leeward side. **-scheibe** f windshield. **-scheibenwischer** m windshield wiper. **w–schief** a (wind-)warped, skewed; leaning. **-schnittig** a streamlined. **W–schutz** m 1 windbreak. 2 = **-schutzscheibe** f windshield. **-seite** f windward (or weather) side. **-sichter, -sortierer** m 1 air separator; aspirator. 2 pneumatic dust remover. **-stärke** f force of the wind. **-stärketabelle** f Beaufort scale. **w–still** a, **W–stille** f calm. **-stoß** m gust of wind. **-strom** m, **-strömung** f air (or blast) current. **w–treibend** a (*Med.*) carminative. **-trocken** a wind-dried; air-dry. **W–trocknung** f wind (or blast) drying. **-trocknungsverfahren** n dry-blast process

Windung f (-en) 1 (*incl Elec.*) winding; coil, spiral. 2 twist, turn; bend, curve; convolution. 3 torsion. 4 screw thread. **Windungs-zahl** f (*Screw*) number of turns

Wind·wehe f snowdrift. **-werk** n winch, hoisting gear. **-zufuhr, -zuführung** f air (or blast) supply. **-zug** m air current, draft

Wink m (-e) 1 signal, sign; wave (of the hand), beckoning gesture; nod. 2 hint, suggestion

Winkel m (-) 1 angle, corner. 2 nook, recess; hideaway; remote outpost. 3 (*Geog.*) panhandle. 4 try square. **-abstand** m angular distance. **-abweichung** f angular deviation. **-beschleunigung** f angular acceleration. **-bewegung** f angular motion. **-eisen** n angle iron. **w–förmig** a angular. **W–frequenz** f angular frequency. **-funktion** f angular (or trigonometric) function. **-geschwindigkeit** f angular velocity. **-halbierende** f (*adj endgs*) angle bisector

winkelig a angular cf WINKLIG

Winkel·konstanz f (*Cryst.*) constancy of interfacial angles. **-linie** f diagonal. **-maß** n angular measurement; goniometer. **-messer** m protractor; goniometer. **-meßprisma** n gonioprism. **-spreizung** f angular spread (or separation). **-stück** n angle piece, elbow. **-thermometer** n bent-tube thermometer

Winkelung f (-en) angularity cf GEWINKELT

Winkel·verschiebung f angular displacement. **-verteilung** *f* angular distribution. **-zug** *m* evasion, dodge; trick. **w–zügig** *a* evasive; tricky, devious

winken 1 *vt* signal to come (*or* go). **2** *vi* (*dat*) wave, beckon, signal (to); be in store (for), await

winklig *a* angular, crooked, bent; (*as sfx, oft*) -angle(d). = WINKELIG. **Winklung** *f* (-en) angularity = WINKELUNG

Wink·zeichen *n* semaphore signal

winter·fest *a* winterproof; (*Bot.*) hardy. **W–garten** *m* conservatory. **-getreide** *n* winter grain. **-grün** *n* (oil of) wintergreen. **w–hart** *a* (*Bot.*) hardy. **W–kohl** *m* kale, winter cabbage. **-kresse** *f* wintercress

winterlich *a* wintry; (of) winter

wintern *vi* get (*or* grow) wintry

Winter·ruhe *f* hibernation

winters *adv* winters, in winter

Winter·saat *f* winter crop. **-saateule** *f* winter corn moth. **-schlaf** *m* hibernation. **-spritzung** *f* (*Agric.*) dormant spray

Winters·rinde *f* winter's bark. **s–über** *adv* during the winter

Winterung *f* (-en) winter crop

Winter·weizen *m* winter wheat

Winzer *m* (-) vintner, wine grower

winzig *a* tiny, minute:**w. klein** tiny little

Wipfel *m* (-) (tree) top

Wipla·metall *n* stainless steel for dentures

Wippe *f* (-n) 1 seesaw. 2 (*Mach.*) rocker plate. 3 (*Metll.*) tilting table. 4 balancer, counterpoise. 5 (*Elec.*) tumbler switch.

wippen *vi* 1 rock, teeter, seesaw, bob. 2 flap, flip

wir *pron* we; (*preceded by verb, oft*) let us: **versuchen w.** let us try

Wirbel *m* (-) 1 whirl, vortex, eddy; whirlwind. 2 whorl, spiral. 3 (*Anat.*) vertebra; crown (of the head): **vom W. bis zur Zehe** from top to toe. 4 peg, plug; swivel. 5 crank. 6 fuss, stir, confusion; excitement. 7 vertigo, intoxication. 8 (*Birds*) warbling; (*Drums*) roll. **··bein** *n* vertebra. **-bett** *n* fluidized bed. **-bewegung** *f* vortex motion, eddying. **-brenner** *m* vortex burner. **-diffusion** *f* eddy diffusion. **w–frei** *a* irrotational; turbulence-free

wirbelig *a* whirling; wild; giddy, dizzy. **Wirbeligkeit** *f* 1 wildness; giddiness. 2 vorticity

Wirbel·knochen *m* vertebra. **-lehre** *f*

vortex theory. **w–los** *a* 1 invertebrate. 2 vortex-free, nonvortical

wirbeln I *vt/i* 1 whirl, spin, twirl. **II** *vt* 2 fluidize. **III** *vi* 3 eddy. 4 (*Drum*) (beat a) roll. *Cf* GEWIRBELT

Wirbel·punkt *m* fluidizing point. **-säule** *f* vertebral (*or* spinal) column. **-schicht** *f* fluidized bed. **-schichtofen** *m* fluidized bed oven. **-sichter** *m* vortex separator. **-strom** *m* 1 whirlpool; turbulent flow. 2 (*Elec.*) eddy current. **-strombrenner** *m* (*Gas*) swirling-flow burner. **-strömung** *f* turbulent flow. **-sturm** *m* tornado; cyclone. **-tier** *n* vertebrate

Wirbelung *f* (-en) 1 whirling, spinning, eddying; vortex motion. 2 fluidizing

Wirbel·wind *m* whirlwind. **-wirkung** *f* whirling effect

wirbt advertises, etc: *pres of* WERBEN

wird becomes; will; is: *pres of* WERDEN

wirft throws: *pres of* WERFEN

Wirk. *abbv* (Wirkung) effect

wirken I *vt* 1 achieve, accomplish. 2 make; knit; knead; (*Salt*) boil. **II** *vi* 3 (*usu* + **als**) act, be active (as). 4 be effective, take effect; (*esp* + **auf**) act, work (on). 5 (+*adv*) have an . . . effect: **beruhigend w.** have a calming effect. 6 (+*adv*) give the impression of being, appear. 7 (+**dahin, daß** . . .) see to it that . . . **Wirken** *n* (-) work; activity; effect. **wirkend** *p.a* effective, efficient, active; **w—es Mittel** agent

Wirk·gruppe *f* active group. **-leistung** *f* effective output

wirklich *a* real, true, genuine. **Wirklichkeit** *f* (-en) reality

wirklichkeits·fern, -fremd *a* unrealistic. **-nah** *a* realistic; state-of-the-art

wirksam *a* 1 (*esp Chem.*) active. 2 effective. 3 professionally active. **Wirksamkeit** *f* 1 effectiveness. 2 activity; (*Acids*) strength. 3 operation: **in W. setzen** put into operation, throw into gear; **außer W. setzen** put out of operation, throw out of gear

Wirk·spannung *f* (*Elec.*) active voltage. **-stoff** *m* active substance, effective ingredient; (*specif oft*) vitamin, hormone, biocatalyst. **-strom** *m* (*Elec.*) active (*or* effective) current

Wirkung *f* (-en) effect, action, effectiveness: **zur W. bringen** effect, put into effect; **mit W. vom 10. Juni** effective June 10th; **W. und Gegenwirkung** action and reaction. **2** influence. **3** impression

Wirkungs-art f mode of operation, kind of action. **-bereich** m effective range. **-bombe** f high-capacity (or demolition) bomb. **-dauer** f duration of the effect, (Chem.) persistence. **w–fähig** a effective, efficient; active, capable of acting. **W–fähigkeit** f effectiveness, efficiency, activity. **-faktor** m effective (or responsible) factor; (Fertilizers) plant growth efficiency factor. **-gehalt** m active content. **-geschwindigkeit** f rapidity of action. **-grad** m efficiency, (degree of) effectiveness. **-gruppe** f (Chem.) active group. **-kraft** f effective force; efficiency, efficacy. **-kreis** m range (or sphere) of activity (or effectiveness), scope

wirkungslos a ineffectual, inefficient; inactive, inert. **Wirkungslosigkeit** f ineffectiveness, inefficiency; inactivity, inertness

Wirkungs-möglichkeit f possible effect-(iveness). **-ort** m point (or site) of effectiveness. **-quant(um)** n quantum of action; Planck's constant. **-querschnitt** m 1 effective cross-section. **2** (Nucl. Phys.) capture cross-section. **-radius** m effective (or active) radius; (Bombs) efficient range. **-sphäre** f sphere of action (or effectiveness). **-stätte** f point (or site of effectiveness. **-variable** f (adj endgs) action variable. **-vermögen** n power of action, effective power. **w–voll** a 1 effective, efficient. **2** impressive, striking. **W–weise** f (mode of) action, way (a thing) works; mode of operation. **-wert** m effective value, efficacy; (Chem.: Acids, etc) strength

Wirk-waren fpl knitted goods, knitwear. **-zeit** f reaction time

wirr a confused, tangled, mixed up. **Wirre** f (-n) 1 confusion. **2** pl disturbances, disorder(s). **Wirr-faser** f tangled fiber

Wirrnis f (-se), **Wirrsal** n (-e) confusion, tangle; chaos

Wirr-seide f silk waste

Wirrwarr m confusion, confused tangle

Wirsing(-kohl) m savoy (cabbage)

Wirt m (-e) **1** (incl Biol.) host. **2** innkeeper. **3** landlord

Wirtel m (-) whorl. **wirtelig** a whorled

Wirtin f (-nen) **1** landlady. **2** inkeeper('s wife)

Wirt-kristall m oikocryst

Wirtschaft f (-en) **1** economy, economic system; (trade and) industry. **2** management; affairs. **3** household; housekeeping: **die W. führen** keep house, manage the business. **4** small farm. **5** (country) inn (with restaurant). **6** mess; confused tangle of affairs; fuss, bother, trouble. **wirtschaften I** vt **1: zugrunde** (or **in Grund und Boden) w.** mismanage completely, run . . . into the ground. **II** vi **2** manage the place (or one's money), run things; keep house. **3** (mit) squander, waste. **4** work, carry on, be busy. **Wirtschafter** m (–), **Wirtschafterin** f (-nen) **1** manager. **2** housekeeper. **3** = **Wirtschaftler** m (-) economist. **wirtschaftlich** a **1** economic; financial. **2** economical, thrifty. **3** profitable. **Wirtschaftlichkeit** f economy; good management; profitability

Wirtschafts- economic; industrial, business, commercial:**-betrieb** m **1** business enterprise; industrial unit. **2** industrial (business, farm, household) management. **-chemie** f **1** chemical plant management. **2** industrial chemistry. **-gebäude** n **1** industrial building. **2** farm building. **-gemeinschaft** f: **Europäische W.** European Economic Community (or Common Market). **-jahr** n business (fiscal, budget) year, farm year. **-krieg** m economic warfare. **-krise** f economic crisis. **-lage** f economic situation. **-lehre** f economics. **-wunder** n (West German postwar) economic miracle

Wirts- host: **-haus** n tavern, inn. **-körper** m host body. **-organismus** m host (organism) (for parasites, etc). **-pflanze** f host plant. **-tier** n host animal. **-zelle** f host cell

Wisch m (-e) **1** scrap (or slip) of paper; note. **2** rag. **3** bundle, tuft; mop. **w–echt** a wipe-resistant

wischen 1 vt wipe, mop. **2** vi (s) (**über**) whisk; brush (by accident). **Wischer** m (-) **1** wiper; windshield wiper; swab, sponge.

2 duster. 3 eraser. 4 blow, stroke. 5 track, trace

Wisch·gold n gold leaf. **-gummi** m rubber wiper. **-kontakt** m (Elec.) wiping (or self-cleaning) contact. **-lappen** m, **-tuch** n wiping (or cleaning) cloth, dustcloth; dish cloth

Wisent m (-e) (Zool.) (European) bison

Wismut n bismuth. **wismut·artig** a bismuthal. **Wismutat** n (-e) bismuthate

Wismut·bleierz n (Min.) schapbachite, matildite. **-blende** f (Min.) bismuth blende, eulytite. **-blüte** f bismuth ocher, (Min.) bismite. **-bromid** n bismuth bromide. **-butter** f butter of bismuth, bismuth trichloride. **-chlorid** n bismuth (tri)chloride

wismuten vt/i solder with bismuth (solder)

Wismut·erz n bismuth ore. **-gehalt** m bismuth content. **-glanz** m bismuth glance, bismuthinite; **prismatischer W.** aikinite. **-glätte** f bismuth litharge (bismuth oxide). **-gold** n (Min.) maldonite. **w–haltig** a bismuthiferous. **W–jodid** n bismuth iodide. **-kupfererz** n (Min.) emplectite; klaprothite; wittichenite. **-legierung** f bismuth alloy. **-lot** n bismuth solder. **-metall** n bismuth metal. **-nickelglanz** m bismuthian ullmanite. **-nickel(kobalt)kies** m (Min.) grünauite. **-niederschlag** m bismuth precipitate, (usu) bismuth oxynitrate. **-ocker** m bismuth ocher, bismite. **-oxid** n bismuth (tri)oxide. **-oxychlorid** n bismuth oxychloride. **-oxyjodid** n bismuth oxyiodide. **-raffination** f bismuth refining. **-salz** n bismuth salt. **w–sauer** a bismuthate of. **W–säure** f bismuthic acid, hydrogen bismuthate. **-säureanhydrid** n bismuthic anhydride, dibismuth pentoxide. **-schwamm** m bismuth sponge. **-silber** n (Min.) chilenite; schapbachite. **-spat** m (Min.) bismutite. **-subjodid** n bismuth subiodide. **-sulfid** n bismuth sulfide. **-tellur** n (Min.) tetradymite, telluric bismuth. **-verbindung** f bismuth compound. **-wasserstoff** m bismuth hydride. **-weiß** n bismuth white

wiss. abbv (wissenschaftlich) scientific-(ally)

Wiß·begier(de) f (intellectual) curiosity. **w–begierig** a inquisitive

wissen* (wußte, gewußt; weiß) I vt/i 1 know: **einem etwas zu w. geben** let sbdy know sthg; **nicht daß wir wüßten** not that we know (of); **soviel wir w.** as far as we know; **noch w.** (oft) remember. II vt 2 (usu + adj) know to be. 3 (+ zu + inf) know how, be able (to); manage (to) **—zu schätzen w.** appreciate. 4 (+adj) see (sthg accomplished, etc). III vi (+um, von) know (of, about). **Wissen** n 1 (range of) knowledge: **unseres W—s** to the best of our knowledge. 2 knowledge, awareness: **ohne ihr W.** without their knowledge (or awareness); **mit W.** knowingly; **W. um . . .** awareness of . . . 3 technical knowledge, know-how. 4 judgment: **gegen unser besseres W.** against our better judgment

Wissenschaft f (-en) 1 science. 2 scientific world. **Wissenschaftler** m (-), **Wissenschaftlerin** f (-nen) scientist. **wissenschaftlich** a scientific. **Wissenschaftlichkeit** f scientific nature

Wissenschafts·rat m (West German) Science Advisory Council. **-zweig** m branch of science

Wissens·gebiet n field of knowledge, discipline. **-stand** m level (or stage) of knowledge (on a subject). **w–wert** a worth knowing. **W–zweig** m branch of knowledge

wissentlich a knowing, conscious

wittern vt 1 sniff out, catch the scent of. 2 sense, detect. **Witterung** f 1 sense of smell, nose. 2 scent, trail **—W. bekommen** get wind of. 3 weather

witterungs·bedingt a weather-dependent. **W–bedingungen** fpl weather conditions. **w–beständig** a weather-resistant; (Steel) stainless, rust-resisting. **W–beständigkeit** f resistance to weathering. **-einfluß** m atmospheric influence. **-kunde, -lehre** f meteorology. **-schaden** m weather damage. **-umschlag** m (drastic) change in the weather. **-verhältnisse** npl atmospheric (or weather) conditions

Wittsche Scheibe Witt plate

Witz m (-e) 1 joke; prank. **—das ist der ganze W.** that is the whole point. 2 wit(s).

witzig *a* humorous, witty; comical

w.l. *abbv* (wenig löslich) difficultly (*or* poorly) soluble

wlösl. *abbv* (wasserlöslich) water-soluble

wo I *adv* **1** (*interrog*) where? **2** (*indef*) where; (*Time*) when, that; **zur Zeit, w. es geschah** at the time (when *or* that) it happened. **3** (*indef*) somewhere = IRGENDWO. **II** *cnjc* **4: w. . . . doch** since, because. **5: w. nicht** if not *cf* WOMÖGLICH. **III** *pfx* (*on preps*) prep + which (*depending on regular and any idiomatic uses of the prep*) *cf* WOFÜR, WOR·

w.o. *abbv* (wie oben) as above

wo·anders *adv* elsewhere, somewhere (*or* anywhere) else

wob wove: *past of* WEBEN

wobei *adv* **I** (*interrog*) where, at what point; on what occasion, in what connection. **II** (*relat*) where; while; and while (*or* in) doing so; in which (case), on which occasion; in connection with which. *Cf* BEI, WO **III,** WOFÜR

Woche *f* (-n) week; (*pl oft*) confinement (in childbirth)

Wochen· week, weekly; childbirth, maternity: **-bett** *n* childbed. **-bettfieber** *n* puerperal fever. **-binde** *f* sanitary napkin (*or* pad). **-ende** *n* week end. **-fluß** *m* (*Med.*) lochia. **w–lang 1** *a* weeks of. **2** *adv* for weeks. **w–schrift** *f* weekly (publication). **-tag** *m* weekday. **w–tags** *adv* (on weekdays

wöchentlich *a, adv* weekly

·wöchig *sfx* -week

Wöchnerin *f* (-nen) woman in childbirth

wodurch *adv* **I** (*interrog*) how, by what means, through where (*or* what). **II** (*relat*) whereby, by (means of) which, because of which, through which. *Cf* DURCH, WO **III,** WOFÜR

wofür *adv* **I** (*interrog*) **1** what . . . for, for what: **w. gebraucht man das?** what is that used for? **2** (*idiomat*) e.g.: **wir wissen, w. sie sich interessieren** we know what they are interested in. **II** (*relat*) **3** for which: **die Forschungen, w. er den Preis bekam** the research for which he got the prize, the research he got the prize for. **4** (*idiomat*) e.g.: **alles, w. wir uns interessieren** everything we

are interested in. *Cf* FÜR, WO **III**

wog weighed: *past of* WÄGEN & WIEGEN

Woge *f* (-n) (sea) wave, billow

wogegen *adv* **I** (*interrog*) against what. **II** (*relat*) against which; compared to which, while; in return for which. *Cf* GEGEN, WO **III,** WOFÜR

wogig *a* billowing, surging

woher *adv* **I** (*interrog*) where . . . from — **w. wissen sie das?** how do they know that? **II** (*relat*) from where, whence, (place) from which

wohin *adv* (*interrog*) where(to), where . . . to. **II** (*relat*) where, to which place. **III** (*indef*) (to) someplace, somewhere

wohingegen *cnjc* while, whereas

wohinter *adv* **1** (*interrog*) what . . . behind, behind what. **2** (*relat*) behind which. *Cf* HINTER, WO **III,** WOFÜR

wohl I *a* **1** well, healthy; happy, contented. **II** *adv* **2** well. **3: w. oder übel** like it or not. **4** probably, presumably:**das ist w. wahr** that is probably true, that must be true *cf* WERDEN **8 B, 9 B. III** *cnjc* **5 w. . . . aber** to be sure . . . but: **w. klein, aber nicht billig** small, to be sure, but not cheap. **6** (*after neg*) **w. aber** e.g.: **damals nicht, w. aber jetzt** not then, but definitely now

Wohl *n* good, welfare, benefit

wohl· well:**-angebracht** *a* well-timed, appropriate. **-ausgebildet** *a* well-developed. **-bedacht** *a* well-considered. **W–befinden** *n* health, well-being. **w–begründet** *a* well-founded. **W–behagen** *n* feeling of comfort. **w–behalten** *a* safe and sound, in good condition. **-bekannt** *a* well-known. **-beleibt** *a* corpulent. **-belesen** *a* well-read. **-beraten** *a* well-advised. **-beschaffen** *a* in good condition. **-besetzt** *a* well-filled (position, etc). **-bewandert** *a* well-versed. **-definiert** *a* well-defined. **-durchdacht** *a* well-thought-out, well-devised. **-erfahren** *a* experienced. **W–ergehen** *n* welfare, well-being. **w–erhalten** *a* well-preserved, in good condition. **-erwogen** *a* well-considered. **W–fahrt** *f* welfare; charity. **w–gebaut** *a* well-built. **-gebildet** *a* **1** well-formed. **2** well-bred, well-educated. **W–gefallen** *n* satisfaction; pleasure, delight. **w-gefällig** *a* pleasant, agreeable. **-gehärtet** *a* well-hardened; well-tempered. **-gelitten** *a* well-liked, popular. **-gemeint** *a* well-meant. **-gemerkt** *a* (*usu as imper*) note carefully! **-gemut** *a* merry, cheerful. **-genährt** *a* well-fed, well-nourished. **-geordnet** *a* well-arranged, well-organized. **-geraten** *a* well-turned-out, perfect; well-bred. **W-geruch** *m* agreeable odor, fragrance; perfume. **-geschmack** *m* agreeable taste (*or* flavor). **w-gesinnt** *a* well-disposed, benevolent. **-gestaltet** *a* well-shaped, well-formed. **-geübt** *a* practiced, skilled. **-gewogen** *a* well-disposed. **-habend** *a* well-to-do, well-off

wohlig *a* comfortable, cozy, contented

Wohl- well: **-klang** *m* melodious sound, euphony, harmony. **w-klingend** *a* melodious; harmonious. **W-laut** *m* melodiousness, euphony; harmony. **-leben** *n* good living, life of luxury. **w-redend** *a* well-spoken, eloquent. **-riechend** *a* fragrant, aromatic. **-schmeckend** *a* good-tasting, tasty. **W-sein** *n* wealth, well-being. **-stand** *m* prosperity, affluence. **-tat** *f* good deed, act of charity. **-täter** *m* benefactor. **w-tätig** *a* charitable; beneficial. **W-tätigkeit** *f* charity; benefit, beneficence. **w-tuend** *a* pleasant, comforting; beneficial

wohl-tun* *vi* (*dat*) do good, do well; help

wohl- well: **-überlegt** *a* well-considered. **-unterrichtet** *a* well-informed. **-verdient** *a* well-deserved. **W-verleih** *m* (*Bot.*) arnica. **-verleihwurzel** *f* arnica root. **w-verschlossen** *a* well-closed, well-sealed. **-versehen** *a* well-provided. **-weislich** *adv* wisely, prudently

wohl-wollen *vi* (*dat*) be well-disposed (to). **Wohl-wollen** *n* good will, benevolence. **w-wollend** *a* well-disposed, benevolent

wohnbar *a* habitable

wohnen *vi* live, dwell, reside *cf* GEWOHNT

wohnhaft *a* living, residing (at an address)

Wohn-haus *n* dwelling, residential building

wohnlich *a* livable; cozy, comfortable

Wohn-ort *m* address, (place of) residence. **-raum** *m* 1 residential room (*or* space). 2 living room, **-sitz** *m* domicile, permanent

address. **-stätte** *f* dwelling, habitation. **-stube** *f* living room

Wohnung *f* (-en) 1 home, residential unit, place to live. 2 apartment

Wohn-wagen *m* (house) trailer, caravan. **-zimmer** *n* living room

wölben *vt* & *sich w.* arch, vault, curve. **Wölbung** *f* (-en) vault; curve, curvature; camber. **wölbungs-frei** *a* free of curvature, flat

Wolf *m* (-e) 1 (*Zool.*) wolf. 2 (*Iron*) lump, ball, bloom. 3 (*Mach.*): **A** (*Tex.*) willow(er); **B** meat grinder; **C** shredder, devil. 4 (*Med.*) chafing, abrasion, intertrigo. **Wölfin** *f* (-nen) she-wolf

Wolfram *n* 1 tungsten. 2 (*Min.*) wolframite. **Wolframat** *n* (-e) tungstate, wolframate **Wolfram-blau** *n* wolfram (tungsten, mineral) blue. **-bleierz** *n* (*Min.*) stolzite. **-bogenlampe** *f* tungsten arc lamp. **-bronze** *f* tungsten bronze. **-draht** *m* tungsten wire. **-erz** *n* tungsten ore. **-faden** *m* tungsten filament. **-gelb** *n* yellow tungsten bronze. **w-haltig** *a* containing tungsten, tungstic

Wolframid *n* (-e) tungstide

Wolframit *n* (*Min.*) wolframite

Wolfram-karbid *n* tungsten carbide. **-lampe** *f* tungsten lamp. **-metall** *n* metallic tungsten. **-ocker** *m* tungstic ocher, (*Min.*) tungstite. **-oxid** *n* tungsten oxide. **-punktlampe** *f* point-source tungsten arc lamp. **-salz** *n* tungsten salt. **w-sauer** *a* tungstate of. **W-säure** *f* tungstic (*or* wolframic) acid

Wolframsäure-anhydrid *n* tungstic anhydride. **-salz** *n* tungstate

Wolfram-stahl *m* tungsten steel. **-stickstoff** *m* tungsten nitride. **-sulfid** *n* tungsten sulfide. **-trioxid** *n* tungsten trioxide

Wolfs-bohne *f* lupine. **-kirsche** *f* belladonna. **-milch** *f* (*Bot.*) wolf's milk, spurge, euphorbia. **w-milchartig** *a* euphorbiaceous. **W-(milchsamen)öl** *n* spurge oil. **-stahl** *m* natural (*or* bloom) steel. **-wurz** *f* baneberry

Wolke *f* (-n) 1 cloud. 2 (*Insects*) swarm. 3 (*Gems*) flaw

Wolken-achat *m* clouded agate. **-angriff** *m* cloud gas attack. **w-artig** *a* cloud-like, cloudy. **W-bildung** *f* cloud formation.

-schicht *f* cloud layer (*or* stratum)

wolkig *a* 1 cloudy, overcast. 2 (*Chem.*) flocculent. 3 (*Gems*) milky

Woll· wool: **-abfälle** *mpl*, **-abgang** *m* wool waste (*or* scraps), waste wool. **w–artig** *a* woolly, wool-like. **W–asche** *f* wool ashes

Wollaus *f* (= **Woll·laus**) aphid

Woll· wool: **-baum** *m* cotton tree *cf* BAUM-WOLLE. **-blumen** *fpl* (*Pharm.*) mullen flowers. **-decke** *f* woolen blanket. **-druck** *m* wool printing

Wolle *f* 1 wool; (woolen) yarn: **in der W. färben** dye in the wool (*or* in grain). 2 (*Bot.*) down. 3 hair, (animal's) coat. 4 (*Metll.*) shavings

wollen A *v* (*pres sg* will, *otherwise reg*) **I** *vt/i* (*pp* gewollt) 1 want, wish; (+ **daß**): **sie w., daß wir** ... they want us to ... 2 (*sbjc* + *clause*) wish (that ...): **wir wollten, es wäre wahr** we wish it were true. 3 (would) like: **wie man will** as one likes (*or* pleases). **II** *v aux* (+*inf*) (*pp* wollen) 4 want (to): **sie w. es sehen** they want to see it; **ohne zu w.** unintentionally. 5 (*inf of motion left implied*): **sie w. hinaus** they want (to get) out. 6 will (*in non-future meanings; past* would): **der Motor wollte nicht anspringen** the engine wouldn't start; **der Nagel will nicht ins Holz** the nail won't go into the wood *cf* 5; **wir w. annehmen** we will assume, let us assume. 7 shall: **w. wir annehmen** ... ? shall we assume ... ? 8 (+ *pp* + **haben** *or* **sein**) claim: **sie w. es bewiesen haben** they claim to have proved it. 9 (+ *pp* + **sein**) need: **das will geplant sein** that needs to be planned. 10 (*past, usu* + **eben** *or* **gerade**) was about to —**was wir sagen wollten** as we were saying. 11 (+ **sagen, heißen**) mean: **das will nichts heißen** that doesn't mean anything; **was w. sie damit sagen?** what do they mean by that? 12 (*usu sbjc*) may: **es koste, was es wolle** whatever it may cost. **III** *vt/i* & *aux* (+ **lieber**) prefer *cf* GERN 1. **B** *a* woolen

Woll· wool: **-entfettung, -entschweißung** *f* degreasing of wool. **-färber** *m* wool dyer. **w–farbig** *a* dyed in the wool. **W–farbstoff** *m* wool dye. **-faser** *f* wool fiber. **-fett** *n* wool fat (*or* grease), lanolin. **-fett-**

säure *f* wool fat acid. **-filz** *m* wool felt. **-garn** *n* woolen yarn. **-gras** *n* cotton grass. **-grün** *n* wool green

wollig *a* woolly; woolen

Woll· wool: **-kraut** *n* mullen. **-öl** *n* wool oil. **-pulver** *n* wool flock (*or* powder). **-schmiere** *f* wool yolk, suint. **-schwarz** *n* wool black

Wollschweiß° *m* wool yolk, suint. **--asche** *f* suint potash. **-fett** *n* wool grease. **-küpe** *f* suint vat. **-salz** *n* suint salt

Woll· wool: **-seide** *f* poplin. **-spicköl** *n* wool (lubricating) oil. **-staub** *m* wool flock (*or* powder). **-stoff** *m* woolen (fabric). **-veredelung** *f* wool processing. **-wachs** *n* wool wax. **-wäsche** *f* 1 wool scouring. 2 woolen underwear. **-weberei** *f* wool weaving (mill). **-weste** *f* woolen vest; sleeveless sweater

womit *adv* **I** (*interrog*) with (*or* by) what, what ... with, how: **w. können wir dienen?** how can we help you? **II** (*relat*) with (*or* by) which. *Cf* WO III, WOFÜR, MIT

womöglich *adv* possibly, if possible: **es ist w. schon fertig** it may already be finished

wonach *adv* **I** (*interrog*) after (for, about, by) what, whereby. **II** (*relat*) after (for, about, by, according to) which, whereafter, whereby: **die Vorschriften, w. wir uns richten** the regulations we are guided by. *Cf* WO III, WOFÜR, NACH

Wonne *f* (-n) bliss, delight, joy

Wood-öl *n* wood oil. **Woodsche Legierung** Wood's metal

Wootz-stahl *m* wootz (steel)

wor· *pfx* (*on preps*) = WO III. where ..., what?, which? *cf* WORAN, WORIN, WOR-ÜBER etc

woran *adv* **I** (*interrog*) on (to, by, of) what, whereof, how: **w. erkennt man es?** how (*or* whereby) does one recognize it? **II** (*relat*) on (to, by, of) which, whereby, whereof. *Cf* WO III, WOFÜR, AN

worauf *adv* **I** (*interrog*) on what, whereon. **II** (*relat*) on which, whereon; whereupon. *Cf* WO III, WOFÜR, AUF

woraus *adv* **I** (*interrog*) from (*or* out of) what, wherefrom. **II** (*relat*) from (*or* out of), which, wherefrom. *Cf* WO III, WOFÜR, AUS

worden been *pp of* WERDEN *when used in passive constructions:* **es ist gefunden w.** it has been (*or* it was) found

worein *adv* **I** (*interrog*) in(to) what, wherein. **II** (*relat*) in(to) which, wherein. *Cf* wo **III**, EIN *pfx*

worfeln *vt* winnow, fan

worin *adv* **I** (*interrog*) in what, wherein: **w. leigt ihr Vorteil?** wherein does their advantage lie? **II** (*relat*) in which, wherein. *Cf* wo **III**, WOFÜR, IN

Wort *n* (-e *or* ⁻er) 1 word: **mit anderen W—en** in other words. 2 saying, expression; utterance. **··ableitung** *f* etymology. **w—brüchig** *a:* **w. werden** break one's word

Wörter·buch *n* dictionary

Wort·folge *f* word order. **-fügung** *f* wording, construction, syntax. **-führer** *m* spokesman. **w—getreu** *a* faithful (translation). **W—kunde** *f* etymology. **-laut** *m* wording; text

wörtlich *a* literal; *adv* (*oft*) verbatim

Wort·register *n* word index, vocabulary. **w—reich** *a* wordy, verbose. **W—schatz** *m* vocabulary. **-stellung** *f* word order. **-wahl** *f* choice of words. **-zeichen** *n* logo(gram); trade mark

worüber *adv* **I** (*interrog*) over (above, across, about) what, what . . . about. **II** (*relat*) about which. *Cf* wo **III**, WOFÜR, ÜBER

worum *adv* **I** (*interrog*) around (about, for) what, what . . . about: **sie wissen, w. es sich handelt** they know what it is about. **II** (*relat*) around (about, for) which. *Cf* wo **III**, WOFÜR, UM

worunter *adv* **I** (*interrog*) under what, what . . . under. **II** (*relat*) under (*or* among) which. *Cf* wo **III**, WOFÜR, UNTER

Woulfische Flasche Woulfe flask

wovon *adv* **I** (*interrog*) what . . . of (about, from), whereof, wherefrom, whence: **sie wissen, w. sie sprechen** they know what they are talking about, they know whereof they speak. **II** (*relat*) of (about, from) which. *Cf* wo **III**, WOFÜR, VON

wovor *adv* **I** (*interrog*) before (*or* in front of) what, what . . . of (about, for): **w. haben sie Angst?** what are they afraid of? **II** (*relat*) before (*or* in front of) which, of (about, for) which: **etwas, w. man Respekt haben muß** something one must

have respect for. *Cf* wo **III**, WOFÜR, VOR

wozu *adv* **I** (*interrog*) what . . . for, to what purpose: **w. das?** why that?, what's that for? **II** (*relat*) for which (*or* idiomat): **der Plan, w. sie uns raten** the plan they advise us to follow. *Cf* wo **III**, WOFÜR, ZU

wrack *a* wrecked; smashed. **Wrack** *n* (-e) wreck. **··gut** *n* wreckage

wrang wrung: *past of* WRINGEN

Wrasen *m* (-) steam; hot fumes

wringen* (wrang, gewrungen) wring. **Wring·maschine** *f* wringer

Wrkg. *abbv* (Wirkung) effect, action

Wruke *f* (-n) kohlrabi

Ws *abbv* (*Elec.*) 1 (Wattsekunde) watts per sec. 2 (Wechselstrom) AC

Ws. *abbv* 1 (Wassersäule) water gauge (pressure). 2 (Werkzeugstahl) tool steel

wss. *abbv* (wässerig) aqueous, hydrous

Wssb. *abbv* (Wasserbad) water bath

Wucher *m* usury, excessive interest

wuchern *vi* (h, s) 1 proliferate, grow (*or* be) rampant. 2 practice usury; (+ **mit**) profiteer (with). **wuchernd** *p.a* rampant. **Wucherung** *f* 1 proliferation, uncontrolled growth. 2 (*Med.*) swelling, tumor

wuchs grew: *past of* WACHSEN

Wuchs *m* (⁻e) 1 growth, development. 2 build, physique, stature. **··stoff** *m* growth hormone

Wucht *f* (-en) 1 weight. 2 pressure. 3 force. 4 brunt, impact. 5 (*Phys.*) kinetic energy, momentum. 6 (*Quant.*) a whole lot (of), heaps (of). **wuchtig** *a* 1 massive, ponderous. 2 powerful. **Wucht·rinne** *f* vibrating feeder

wühlen **I** *vt* 1 burrow, dig. **II** *vi* 2 rummage; wallow. 3 bore; agitate. 4 toil, slave

Wühl·maus *f* (*Zool.*) vole

Wulst *m, f* (-e *or* ⁻e) 1 pad, roll; torus. 2 swelling, lump, bulge. 3 bulb. 4 padded (*or* rolled) edge; bead. **wulstig** *a* 1 thick; fleshy. 2 padded, stuffed. 3 swollen, tumid. **Wulst·naht** *f* reinforced weld

wund *a* sore; wounded; chafed. **W—·balsam** *m* salve, ointment (for wounds). **-behandlung** *f* surgery

Wunde *f* (-n) wound, lesion; sore

Wunder *n* (-) wonder, marvel, miracle. **wunderbar** *a* wonderful, marvelous; miraculous

Wunder·baum *m* 1 castor oil plant. 2 locust

tree. **-baumöl** m castor oil. **-erde** f litharge. **-erscheinung** f miraculous phenomenon. **-kerze** f sparkler. **-kind** n prodigy

wunderlich a strange, odd. **Wunderlichkeit** f (-en) strangeness; oddity, eccentricity

wundern 1 vt surprise. **2: sich w. (über)** wonder, be surprised (at)

wunder-nehmen* vt surprise, amaze

Wunder-pfeffer m allspice, **-salz** n Glauber's salt, sal mirabile. **w-schön** a beautiful, exquisite. **-voll** a wonderful. **W-waffe** f miracle weapon. **-wasser** n aqua mirabilis

Wund-fieber n traumatic fever. **w-gelegen** a bedsore. **W-klee** m kidney vetch. **-kraut** n vulnerary herb. **-pflaster** n adhesive tape. **-pulver** n vulnerary powder. **-saft** m wound exudate. **-salbe** f (healing) ointment. **-schwamm** m surgeon's agaric; surgical sponge. **-stein** m copper aluminate. **-streusalz** n vulnerary powder. **-wasser** n vulnerary lotion

Wunsch m (⁻e) wish; request. **-bild** n ideal, dream vision. **-denken** n wishful thinking

Wünschel-rute f divining (or dowsing) rod

wünschen vt/i wish (for): **etwas zu w. übrig lassen** leave sthg to be desired

wünschens-wert a desirable

würbe would advertise: sbjc of WERBEN

wurde became, etc: past of WERDEN

würde 1 would: sbjc of WERDEN. **2 = w. werden** would be (get, become): **weil es sonst zu kalt w.** because otherwise it would get too cold; **Energie, die sonstwo besser eingesetzt w.** energy that would better be applied elsewhere

Würde f (-n) **1** dignity. **2** rank, title; honor. **würde-los** a undignified. **-voll** a dignified

würdig a **1** dignified. **2** (+genit) worthy (of). **3** adv, e.g.: **sich w. anreihen an** be worthy of joining ranks with

würdigen vt appreciate, honor; (+genit) consider worthy (of). **Würdigung** f (-en) **1** appreciation, honor. **2** appraisal

Wurf m (⁻e) **1** throw, cast, pitch —**alles auf einen W. setzen** put all one's eggs in one basket. **2** strike, hit: **der große W.** the big hit, the master stroke. **3** litter (of pups,

etc). **4** sketch, outline; projection. **-bahn** f trajectory

würfe would throw: sbjc of WERFEN

Würfel m (-) **1** cube; (Sugar) lump. **2** die, (pl) dice. **-alaun** m cubic alum. **-brikett** n cubical briquet. **-diagonale** f (Cryst., Math.) cube diagonal. **-eck n, -ecke** f corner of a cube, cubic apex. **-erz** n cube ore, (Min.) pharmacosiderite. **-festigkeit** f cubic stability to crushing. **-fläche** f cube face. **w-förmig** a cubic(al), cubiform; diced. **W-gambir** n cube gambier. **-gips** m anydrite. **-gitter** n cubic lattice

würfelig a cubic(al), cubed; diced: **w. schneiden** dice

Würfel-inhalt m cubic content. **-kante** f cube edge. **-kohle** f lump coal, cobbles

würfeln 1 vt cube, dice; checker. **2** vi shoot dice. Cf GEWÜRFELT

Würfel-nickel n cube(d) nickel. **-pulver** n (Explo.) cube-cut powder. **-salpeter** m cube salpeter (or sodium nitrate). **-schiefer** m clay slate, (Min.) argillite. **-spaltbarkeit** f (Cryst.) cubic cleavage. **-spat** m (Min.) anhydrite, cubic spar. **-stein** m (Min.) boracite. **-werk** n checkerwork. **-zahl** f (Math.) cube. **-zeolith** m (Min.) analcite, chabasite. **-zucker** m lump sugar, cube sugar

Wurf-geschoß n missile, projectile. **-granate** f mortar shell. **-granatzünder** m mortar shell fuse. **-körper** m projectile. **-kraft** f throwing (or projectile) force. **-ladung** f propellent charge. **-lehre** f ballistics

würflig a cubed, diced, etc = WÜRFELIG

Wurf-linie f trajectory, line of projection. **-maschine** f **1** catapult; missile launcher, **2** (Agric.) winnowing machine. **-mine** f trench mortar shell (or bomb). **-rauchkörper** m smoke bomb. **-schaufel** f shovel, scoop. **-weite** f projectile (mortar, bomb, missile) range

Würgel-pumpe f rotary pump

würgen 1 vt strangle. **2** vt/i choke. **3** vi gag, retch; (+an, oft) struggle, toil (over)

Wurm m (⁻er) worm; maggot. **w-abtreibend** a vermifuge, anthelmintic. **-artig** a vermicular, worm-like. **W-arznei** f anthelmintic, worming medicine. **-farn** m male fern. **w-förmig** a worm-shaped, vermiform. **W-fortsatz** m (Anat.) vermiform appendix. **-getriebe** *n* worm gear(ing)

wurmig *a* wormy

Wurm·kraut *n* 1 (*genl*) anthelmintic herb. 2 (*specif*) tansy; **amerikanisches W.** pinkroot. **-loch** *n* wormhole. **-mehl** *n* worm(hole) dust. **-mittel** *n* vermicide, anthelmintic. **-moos** *n* worm (*or* Corsican) moss. **-pulver** *n* worm (*or* vermicidal) powder. **-rad** *n* (*Mach.*) worm wheel (*or* gear). **-rinde** *f* worm bark. **-samen** *m* wormseed. **-samenöl** *n* wormseed oil, oil of chenopodium. **w-stichig** *a* worm-eaten. **W-tang** *m* worm moss. **w-treibend, -vertilgend, -vertreibend, -widrig** *a* anthelmintic

Wurst *f* (ẅe) **I** sausage; *specif:*1 (*small sizes usu whole*) hot dog, frankfurter, knockwurst, etc. 2 (*large sizes usu sliced as cold cuts*) bologna, salami, liverwurst, etc. **II** (cylindrical) pad(ding), roll. **··darm** *m* sausage casing. **-fleisch** *n* sausage meat (*or* filling). **w-förmig** *a* sausage-shaped; toroidal. **W-gift** *n* sausage poison. **-hülle** *f* sausage casing. **-kraut** *n* sausage herb, *specif* savory, marjoram. **-kunstpapier** *n* synthetic sausage casing. **-masse** *f* sausage filling. **-stein** *m* pudding stone. **-vergiftung** *f* sausage poisoning

Würt. *abbv* Württemberg (*area in SW Germany*)

·wurz *f* (*in compds only*) **1** = WURZEL root. **2** (*Bot., Brew.*) wort

Würze *f* (-n) **1** spice; condiment, seasoning. **2** (*Brew.*) wort, malt liquor. **··brechen** *n* (*Brew.*) breaking of the wort. **-filter** *n* wort filter. **-kochen** *n* wort boiling. **-kühler** *m* wort cooler

Wurzel *f* (-n) root: **W. fassen** take root; (*Math.*) **W. ziehen** extract the (square) root

Würzelchen *n* (-) rootlet; radical

Wurzel· root: **-faser** *f* root fiber. **-fäule** *f* root rot. **-frucht** *f* root vegetable. **-früchtler** *m* (*Bot.*) rhizocarp. **-füß(l)er** *m* rhizopod. **-gemüse** *n* root vegetable(s). **-gerbstoff** *m* root tannin. **-gesetz** *n* square root law. **-gewächs** *n* root plant

(*or* crop). **-gummi** *n* root rubber. **-haar** *n* (*Bot.*) root hair. **-haube** *f* (*Bot.*) root cap. **-kautschuk** *m* root rubber. **-keim** *m* (*Bot.*) radicle. **-knollen** *m* tuber, bulb. **-knoten** *m* node. **w-los** *a* rootless

wurzeln *vi* be rooted, have roots

Wurzel· root: **-ranke** *f* runner, sucker, (*Bot.*) solon. **-schößling** *m* root sucker, runner. **-stock** *m* rhizome, rootstock. **-werk** *n* root system, roots. **-zahl** *f* (*Math.*) root of a number. **-zeichen** *n* (*Math.*) radical (*or* root) sign

würzen *vt* spice, season *cf* GEWÜRZT

Würze·pfanne *f* (*Beer*) copper. **-siede-pfanne** *f* wort boiler, brewing copper. **-vorwärmer** *m* wort preheater

Würz·geruch *m* spicy odor. **-geschmack** *m* spicy taste

würzig *a* spicy, aromatic

Wurzit *m* (*Min.*) wur(t)zite. **··gitter** *n* wurtzite lattice

würz·los *a* unspiced, unseasoned, flat

Würz·mittel *n* condiment. **-nägelein** *n*, **-nelke** *f* clove. **-stoff** *m* aromatic essence. **-wein** *m* spiced wine

wusch washed: *past of* WASCHEN

wußte knew: *past of* WISSEN

Wust *m* tangled mass, jumble; mess; rubbish, trash

wüst *a* **1** waste, desert(ed), desolate. **2** wild, messy; chaotic

Wüste *f* (-n) desert, waste(land), wilderness

wüsten *vi* (**mit**) waste, squander

Wüsten·bewohner *m* desert dweller. **Wüstenei** *f* (-en) wild (*or* chaotic) mess. **Wüsten·lack** *m* (*Geol.*) desert varnish

Wut *f* **1** rage, fury: **W. haben** (**auf**) be furious (with), rage (against). **2** mania. **3** (*Med.*) rabies

wüten *vi* **1** rage, be furious. **2** wreak havoc. **wütend** *p.a* furious, enraged

Wut·gift *n* rabies virus

wütig *a* furious, raging; rabid

Wut·krankheit, -seuche *f* rabies

Wutz(·stahl) *m* wootz (steel)

Wz. *abbv* (Warenzeichen) trade mark

W-Z-Faktor *m*, **WZW** *abbv* (Wasser-Zement-Wert) water-cement ratio

X

X-Achse *f* x-axis

Xanthan·wasserstoff *m* xanthan hydride (perthiocyanic acid)

xanthisch *a* xanthic

Xanthogenat *n* (cellulose) xanthate. **··kunstseide** *f* viscose rayon

xanthogen·sauer *a* xanthate of. **X–säure** *f* xanth(ogen)ic acid, cellulose xanthic acid

Xantho·kobaltchlorid *n* xanthocobaltic chloride. **-proteinsäure** *f* xanthoproteic acid

x-beliebig *a* any (desired)

X-E. *abbv* = **X-Einheit** *f* X-ray unit (of wave length)

x-fach *adv* any number of times. **x-förmig** *a* X-shaped

Xenon·hochdrucklampe *f* xenon high-pressure arc lamp

Xeres(·wein) *m* sherry

Xerographie *f* xerography

Xeron·säure *f* xeronic acid

x-geschnitten *a* x-cut, cross-cut

Ximenyl·säure *f* ximenylic acid

Ximenyn·säure *f* ximenynic acid

x-mal *adv* any number of times, x times

X-Stoß *m* double V joint

X-Strahlen *mpl* X-rays, roentgen rays

x-te *a* nth

Xylidin·rot *n* xylidine red

Xylit *n* xylitol

Xylo·chinol *n* xyloquinol. **-chinon** *n* xyloquinone

Xylol *n* xylene

Xylon·säure *f* xylonic acid

Xylorcin *n* xylorcin(ol)

Xylyl·säure *f* xylic acid

X-Zeit *f* zero hour

Y

Y-Achse *f* y-axis
Yacht-farbe *f* yacht paint
Yam(s)-wurzel *f* yam
Yangona-säure *f* yangonic acid
Yerba-strauch *m* maté, Paraguay tea
y-förmig *a* Y-shaped
y-geschnitten *a* y-cut
Y-Legierung *f*, Y-Metall *n* Y-alloy, Y-metal
Yohimboa-säure *f* yohimbic acid
Ypsilon *n* (-s) 1 (*Ger. letter*) Y. 2 (*Greek letter*) upsilon
Y-Rohr *n*, Y-Röhre *f* Y-tube
Ysop *m* hyssop. --öl *n* oil of hyssop

Ytterbin *n*, --erde *f* ytterbia, ytterbium earth
Ytter-erde *f* yttria, yttrium earth -phosphorsaure Y. xerotime. -flußspat *m* (*Min.*) yttrocerite. y-haltig *a* yttriferous, containing yttrium
Ytter-oxid *n* yttrium oxide. -salz *n* yttrium salt. -spat *m* (*Min.*) xenotime
Yttrium-erde *f* yttrium earth, yttria (yttrium oxide)
yttrium-haltig *a* yttriferous, containing yttrium
Yukka *f* (-s) yucca

Z

z. *abbv* (zu, zum, zur) to (the) etc, *cf* ZU *prep*

Z *abbv* **1** (Zähigkeit) toughness. **2** (Zeitschrift) Journal

Z. *abbv* **1** (Zahl) no. (number). **2** (Zeile) line. **3** (Zeit) time. **4** (Zeitschrift) Journal. **5** (Ziffer) no., fig. **6** (Zoll) in. (inch). **7** (Zone) zone

za. *abbv* (zirka) approx.

Z-Achse *f* Z axis

Zachäus·ol, Zachunöl *n* zachun (oil), bito oil (from *Balanites aegyptiaca*)

Zacke *f* (-n) **1** (sharp) point; (jagged) peak. **2** prong, spike. **4** (*Combs*, etc) tooth. **3** (*Radar*) pip. **4** *pl oft* jagged edge; scalloping, pinking. **5** (*Metll.*) plate. **6** twig

Zackel·wolle *f* refuse wool

zacken *vt* tooth, make jagged; scallop, serrate *cf* GEZACKT

Zacken *m* (-) **1** (sharp) point; peak; prong, etc = ZACKE **1–6**. **2** (*in compds oft*) jagged, scalloped, serrated, toothed: **-blatt** *n* scalloped leaf. **-rolle** *f* toothed roller

zackig *a* jagged; scalloped, serrated, notched; (*Fracture*) hackly

Zaffer *m* (-) zaffer (impure cobalt oxide)

Zaffetika *f* asafetida

Zagel·eisen *n* slab iron

zagen *vi* hesitate, be uncertain, be timid

Zaggel *m* (-), *f* (-n) (*Metll.*) billet

zäh *a* **1** tough (*incl Food*). **2** viscous, viscid, stringy. **3** tenacious. **Zähe** *f* toughness; viscosity; tenacity. **-grad** *m* (degree of) viscosity

Zäh·eisen *n* toughened iron

Zäheit *f* toughness; viscosity; tenacity *cf* ZÄH

zäh·fest *a* tenacious, tough. **Z–festigkeit** *f* tenacity, toughness. **-fluß** *m* viscosity. **z–flüssig** *a* viscous; difficultly fusible, refractory. **Z–flüssigkeit** *f* viscosity; difficult fusibility, refractoriness.

-flüssigkeitsmesser *m* visco(si)meter.

z–gepolt *a* (*Metll.: Copper*) tough-pitch.

-hart *a* tough

Zähigkeit *f* **1** toughness; viscosity; tenacity = ZÄHEIT. **2** ductility

Zähigkeits·einheit *f* unit of viscosity (tenacity, ductility). **-grad** *m* degree of viscosity (tenacity, ductility). **-maß** *n* measure of viscosity (tenacity, ductility). **-messer** *m* viscosimeter. **-modul** *n* (*Ceram.*) rigidity modulus. **-probe** *f* test (or sample to test) for viscosity (tenacity, ductility). **-zahl** *f* viscosity index

Zäh·kupfer *n* tough-pitch copper

Zahl *f* (-en) number, numeral; numerical value: **zehn an der Z.** ten in number; **es gehört zur Z. der Edelgase** it belongs among (*or* is one of) the inert gases

Zähl·apparat *m* counter, counting device

zahlbar *a* payable, due

zählbar *a* countable, enumerable; computable

zahlen *vt/i* pay

zählen I *vt/i* count, number —**das Kind zählt fünf Jahre** the child is five years old. **II** *vt* **2** (+**zu**): **wir z. sie zu den Halogenen** we count them among (*or* under) the halogens. **III** *vi* count (= be of importance). **4** (+**auf**) count, rely (on). **5** (+**nach**) be numbered (in), run (into) (e.g. the hundreds). **6** (+**zu**) e.g.: **sie z. zu den Edelgasen** they belong to (they are some of, they are counted among) the inert gases *cf* ZAHL

Zahlen·angaben *fpl* numerical data. **-anzeiger** *m* digital indicator. **-faktor** *m* numerical factor. **-folge** *f* numerical order. **-größe** *f* numerical quantity. **-index** *m* numerical index (*or* suffix). **-koeffizient** *m* numerical coefficient. **z–mäßig** *a* numerical; quantitative. **Z–**

853

material n numerical data. **-reihe** f number (or numerical) series. **-tafel** f table of figures, numerical table. **-theorie** f number theory. **-verhältnis** n numerical ratio. **-wert** m numerical value

Zähler m (-) 1 (Mach.) counter, meter; (Nucl.) Geiger counter. 2 (Math.) numerator. **-öl** n counter (or meter) oil

Zähl-flasche f counting bottle. **-gerät** n counter, (rate) meter

-zählig sfx -fold, -tuple: **fünfzählig** fivefold, quintuple

Zähligkeit f multiplicity

Zähl-kammer f counting chamber

zahllos a countless, innumerable; adv oft in large number(s)

zahlr. abbv = ZAHLREICH a numerous

Zähl-rohr n counting tube, tube counter

Zahl-tag m payday

Zahlung f (-en) payment

Zählung f (-en) count(ing), enumeration; census; computation

zahlungs-fähig a (Com.) solvent. **Z—fähig-keit** f solvency. **-schein** m receipt. **z—unfähig** a insolvent

Zähl-vorrichtung f, **-werk** n counting device, counter

Zahl-wert m numerical value. **-wort** n numeral (word). **-zeichen** n numerical symbol, numeral, figure

zahm a tame; domestic (esp animal); mild, gentle. **zähmen** 1 vt tame, domesticate; restrain, control. **2: sich z.** become tame, be domesticated; restrain (or control) oneself

Zahn m (-̈e) 1 (Anat.) tooth, tusk. 2 (Mach.) cog, gear tooth —**Z. und Trieb** rack and pinion. 3 (in compds) tooth(ed), dental, odonto-; notched; cog, gear: **z—ähnlich** a toothlike, (Anat.) odontoid. **Z—arzt** m, **-ärztin** f dentist; dental surgeon. **z—ärztlich** a dentist's, dental. **Z—bein** n dentine. **-blei** n (Bot.) leadwort. **-bürste** f toothbrush

zähneln vt cut teeth into, serrate; indent. **Zähnelung** f toothing, serration, indentation

Zahn-email n tooth enamel

zahnen 1 vt cut teeth into, serrate; cog (a wheel) cf GEZAHNT. 2 vl teethe, cut one's baby teeth

Zahn-ersatz m false tooth, dental prosthesis

zäh-nervig a (Plastics) snappy

Zahn-fäule f tooth decay, (dental) caries. **-fisch** m red snapper. **-fleisch** n gum(s). **-formel** f dental formula. **z—förmig** a tooth-shaped, dentiform. **Z—füllung** f tooth filling. **-geschwür** n abscess; gum boil. **-gewebe** n dental tissue. **-gummi** n dental rubber. **-heilkunde** f dentistry

zahnig a toothy; toothed, dentate; indented; jagged

Zahn-keim m dental pulp. **-kitt** m dental cement. **-knorpel** m dental cartilage. **-kranz** m (Mach.) toothed wheel (or rim), gear ring. **-krone** f crown (of a tooth). **z—los** a toothless, (Zool.) edentate. **Z—mark** n dental pulp, pulpa. **-pasta**, **-paste** f toothpaste. **-pastenspender** m toothpaste pump. **-pflege** f dental care. **-plombe** f tooth filling. **-prothese** f false tooth; denture. **-pulpa** f dental pulp(a). **-pulver** n tooth powder. **-putzglas** n toothbrush glass. **-rad** n gear, toothed wheel

Zahnrad-bahn f rack railway. **-fett** n gear grease. **-getriebe** n gear transmission (or mechanism). **-pumpe** f gear pump

Zahn-reinigungsmittel n dentifrice. **-schmelz** m dental enamel. **-schmerzen** mpl toothache. **-stange** f (Mach.) toothed rack. **-stein** m tartar (on the teeth). **-stocher** m toothpick. **-substanz** f tooth substance, dentine. **-techniker** m dental technician. **-trieb** m rack and pinion drive

Zahnung f (-en) 1 (Anat.) teething, dentition. 2 toothed (or jagged) edge, serration. 3 (Saws, Gears) type of teeth, tooth system

Zahn-wasser n tooth wash, dental lotion. **-wechsel** m second dentition. **-weh** n toothache. **-wehholz** n (Bot.) prickly ash. **-weinstein** m (dental) tartar. **-werk** n gearing. **-zement** m dental cement. **-wurzel** f root (of a tooth)

zäh-polen vt (Metll.) toughen by poling. **-schlackig** a forming tough slag. **-schleimig** a mucous. **Z—stoff** m highly viscous material

Zain m (-e) (Metll.) bar, pig, ingot. **zainen** vt

make into ingots (*or* bars); stretch, draw out (iron). **Zain-form** *f* ingot mold

Zängchen *n* (-) tweezers, small forceps

Zange *f* (-n) (pair of) tongs (pliers, pincers, nippers); forceps

Zänge-arbeit *f* (*Metll.*) shingling. **zängen** *vt* (*Metll.*) shingle

zangen-förmig *a* pincer-shaped. **Z–geburt** *f* forceps delivery; (*fig*) tough job, ordeal. **-griff** *m* handle of tongs (pliers, etc)

Zank *m* quarrel(ing). **zanken 1** *vi* (*mit*) scold. **2: sich z.** quarrel

Zäpfchen *n* (-) **1** small plug (tap, cone, etc) *cf* ZAPFEN. **2** (*Anat.*) uvula; (*in compds*) uvular. **3** (*Med.*) suppository

zapfen I *vt* **1** tap (kegs, etc). **2** join by mortise and tenon, dovetail. **II** *vi* fill up (at a gas station)

Zapfen *m* (-) **1** peg, pin, plug, stud; tenon *cf* NUTE. **2** tap, bung, spigot. **3** (axle) journal, trunnion. **4** (*Bot.*) cone. **5** cork, stopper. **z–-artig** *a* plug-like; cone-like. **Z–baum** *m* conifer. **z–förmig** *a* peg-shaped; cone-shaped; conical; (*Bot.*) strobiliform. **-korn** *n* ergot. **-lager** *n* journal bearing. **-loch** *n* peg (*or* pivot) hole; (*Kegs*) tap (*or* bung) hole. **z–tragend** *a* coniferous. **Z–träger** *m* conifer. **-wein** *m* leaked wine. **-zieher** *m* corkscrew

Zapf-hahn *m* tap, faucet; drain valve. **-kolophonium** *n* gum rosin. **-loch** *n* **1** (*Kegs*) tap (*or* bung) hole. **2** hole for a peg, pivot hole. **-säule** *f* gasoline pump. **-stelle** *f* filling (*or* gas) station. **2** (water) tap connection. **3** (*Elec.*) charging station; outlet

Zapfung *f* (-en) tapping; joining by mortise and tenon

zaponieren *vt* varnish with cellulose ester. **Zapon-lack** *m* cellulose ester (acetate, nitrate) lacquer, zapon varnish (pyroxylin and amyl acetate). **-verdünnung** *f* zapon thinner

zappeln *vi* thrash around, kick; fidget

Zarge *f* (-n) sidewall (e.g. of a tin can); frame (*esp* of door, window); rim (of chair seats, etc)

zart *a* **1** tender, soft. **2** delicate, sensitive. **3** gentle, mild. **4** faint, (*esp Colors*) pale. **Zartheit** *f* tenderness; sensitivity; gentleness. **Zart-macher** *m* (meat) tenderizer. **z–rosa** *a* pale pink

Zaser *f* (-n) fiber, filament. **Zäserchen** *n* (-) fibril, (fine) filament. **zaserig** *a* fibrous; filamentous

Zäsium *n* cesium = CÄSIUM

Zaspel *f* (-n) (*Yarn*) skein, hank

Zauber *m* (-) **1** magic. **2** magic spell, charm. **3** nonsense; fuss. **4: fauler Z.** swindle, fraud. **Zauberei** *f* (-en) magic (trick), sorcery. **Zauberer** *m* (-) magician, sorcerer. **Zauber-formel** *f* magic formula. **zauberhaft** *a* magic(al), enchanting

Zauber-hasel *m* witch hazel. **-kreis** *m* magic circle

zaubern 1 *vt* conjure (up). **2** *vi* work (*or* do) magic

Zauber-nuß *f* witch hazel. **-wurzel** *f* mandrake

zaudern *vi* hesitate, waver; temporize

Zaum *m* (ᐨe) bridle **—im Z. halten = zäumen** *vt* restrain, hold in check

Zaun *m* (ᐨe) fence **—lebender Z.** hedge; **vom Z. brechen** unleash, launch. **--draht** *m* fence wire. **-gitter** *n* lattice fence. **-könig** *m* wren. **-latte** *f* (fence) picket. **-pfahl** *m* fence post**—ein Wink mit dem Z.** a broad hint. **-rebe** *f* Virginia creeper. **-rübe** *f* bryony. **-rübenöl** *n* bryony oil

z.B. *abbv* (zum Beispiel) e.g. (for example)

z.b.V. *abbv* (zur besonderen Verwendung) for special use

Z-Drehung *f* (*Tex.*) Z-twist

z.E. *abbv* (zum Exempel) e.g. (for example)

Zebra-streifen *m* **1** zebra stripe. **2** *pl oft* marked pedestrian crosswalk, white lines

Zeche *f* (-n) **1** (*Min.*) (coal) mine; mining company. **2** (tavern) bill

Zechen-kohle *f* mine coal. **-kokerei** *f* coke oven plant (at the mine). **-koks** *m* (*Metll.*) furnace coke. **-teer** *m* coke tar

Zechstein *m* (*Geol.*) Upper Permian. **--kalk** *m* Permian limestone

Zecke *f* (-n) (*Zool.*) tick

Zeder *f* (-n) cedar. **zedern** *a* (made of) cedar

Zedern-harz *n* cedar resin. **-holz** *n* cedar(wood). **-holzöl** *n* cedarwood oil. **-öl** *n* cedar oil

Zedernuß-öl *n* cedar nut oil

zedieren *vt* cede; transfer, assign

Zedrat-öl, Zedro-öl *n* citron oil

Zeh *m* (-en), **Zehe** *f* (-n) **1** toe. **2** (*Garlic*) clove. **3** (*Ginger*) root, stick. **Zehen-spitze** *f* tip of the toe, tiptoe

zehn *num* ten; (*as pfx oft*) deca-, deci-: **-basisch** *a* decabasic. **Z—eck** *n* decagon. **z—eckig** *a* decagonal, ten-cornered

Zehner *m* (-) (number) ten, tenner; (*in compds oft*) decimal

zehnerlei *a* (*no endgs*) ten kinds of

Zehner-logarithmus *m* logarithm to the base ten. **-potenz** *f* (*Math.*) tenth power. **-stein** *m* hollow concrete block. **-stelle** *f* decimal place. **-system** *n* decimal system. **-waage** *f* decimal balance

zehn-fach *a* tenfold. **-fachnormal** *a* decanormal, 10 *N*. **-fältig** *a* tenfold. **Z—flach, -flächner** *m* decahedron. **-füßer** *m* (*Zool.*) decapod. **z—jährig** *a* ten-year(-old). **-jährlich** *a* decennial, every ten years. **-mal** *adv* ten times. **-malig** *a* ten-time. **-prozentig** *a* ten-percent

zehnt: zu z. in a group of ten *cf* ZEHNTE

zehn-tägig *a* ten-day. **-tausend** *num* ten thousand

zehnte *a* tenth *cf* ZEHNT

Zehntel *n* (-) (*Frac.*) tenth. **-grad** *m* tenth of a degree. **-lösung** *f* tenth-normal solution. **z—normal** *a* tenth-normal. **Z—silberlösung** *f* tenth-normal silver solution

zehntens *adv* tenthly, in the tenth place

zehren *vi* **1** cause loss of weight. **2** (**an**) weaken, wear down. **3** (**von**) live, survive; reminisce (on). **zehrend** *p.a* (*oft*) consuming. **Zehrung** *f* (-en) **1** provisions (for a trip). **2** living expenses. **3** loss (of weight). **4** lubrication **5** (*Explo.*) priming

Zeichen *n* (-) **1** sign, symbol; logo. **2** mark, marker; punctuation mark. **3** signal, indication, cue. **4** token. **5** symptom. **6** (*in compds, also*) drawing, drafting *cf* ZEICHNEN: **-brett** *n* drawing board. **-büro** *n* drafting room. **-ebene** *f* plane of the (drawing) paper. **-erklärung** *f* (*Maps*) key to symbols. **-farbe** *f* drawing (*or* marking) color. **-kohle** *f* drawing charcoal. **-kreide** *f* drawing (*or* marking) chalk. **-papier** *n* drawing paper. **-pult** *n* drafting table. **z—rechtlich** *adv:* **z. geschützt** (*Trade Marks*) registered. **Z—saal** *m* drafting room; art studio. **-set-**

zung *f* punctuation. **-sprache** *f* symbolic notation; sign language. **-stab** *m* drawing rule(r), draftsman's scale. **-stift** *m* drawing pencil (*or* crayon). **-tinte** *f* drawing (*or* marking) ink. **-vorlage** *f* drawing copy (*or* model)

zeichnen I *vt/i* **1** draw, sketch. **2** sign (one's name). **II** *vt* **3** plot, map. **4** mark (with a sign, symbol, initials). **III** *vi* (*Com.*) (**für**) subscribe (to). **Zeichnen** *n* drawing, drafting; sketching, etc. **Zeichner** *m* (-) **1** drawer, draftsman, graphic artist. **2** signer, subscriber. **zeichnerisch** *a* drawing, draftsman's, graphic. **Zeichnung** *f* (-en) **1** drawing, sketch, design; plan; illustration, figure. **2** marking, mark. **3** signing; subscription. **4** (*Wood*) grain

Zeige-finger *m* index finger, forefinger

zeigen I *vt* **1** (*usu* + *d.i.o*) show; display; present. **2** (*Timepieces, Meters*) indicate, read. **II** *vi* **3** (**auf**) point (to), point out. **III: sich z.** **4** show up; appear, become evident. **5** (+*pred a*) show that one is. **6** attract attention, show off

Zeiger *m* (-) **1** (*usu Instrum.*) pointer, needle, indicator; (clock) hand. **2** (*Math.*) index, vector. **3** index finger. **4** presenter, bearer. **-ablesung** *f* needle (pointer, meter) reading. **-ausschlag** *m* needle (*or* pointer) deflection. **-diagramm** *n* vector diagram. **-galvanometer** *n* needle galvanometer. **-hebel** *m* pointing lever, pointer, **-platte, -scheibe** *f* dial (plate *or* face). **-thermometer** *n* dial thermometer

Zeige-stock *m* pointer

Zeile *f* (-n) **1** line (*usu* of print, etc); (*TV*) scan line. **2** row. **3** streak, band

Zeilen-abstand *m* line spacing. **-folge** *f* series (*or* sequence) of lines. **-frequenz** *f* (*TV*) line frequency. **-struktur** *f* banded structure. **z—weise** *adv* line by line, by the line

Zein I *n* (*Chem.*) zein. **II** *m* (-e) ingot = ZAIN

Zeisig *m* (-e) (*Zool.*) siskin, greenfinch. **z—gelb** *a* siskin-yellow, light greenish yellow. **-grün** *a* siskin-green, light yellowish green

zeit *prep:* **z. ihres Lebens** in their lifetime

Zeit *f* (-en) **1** time: **zur Z.** at the time, at present; **das hat noch Zeit** that can wait. **2** (period) of time, while; season. **3** age, era = ZEITALTER. 4 (verb) tense. **-ab-häbhängigkeit** _f_ time dependence. **-ablenkung** _f_ (_Elec._) time base. **-abschnitt** _m_ period (of time). **-abstand** _m_ interval. **-alter** _n_ age, era, (one's) time. **-angabe** _f_ date (and time). **-ansage** _f_ (_Radio_) announcement of the time. **-aufnahme** _f_ (_Phot._) time exposure. **-aufwand** _m_ expenditure of time. **z–aufwendig** _a_ time-consuming. **-bedingt** _a_ 1 caused by (the) time(s). 2 seasonal. **-begrenzt** _a_ limited in time. **Z–bombe** _f_ time bomb. **-dauer** _f_ duration, length of time. **-druck** _m_ pressure of time. **-einheit** _f_ unit of time. **-einteilung** _f_ division (_or_ organization) of (the available) time, timetable. **z–ersparend** _a_ time-saving. **Z–ersparung** _f_ saving of time. **-faktor** _m_ time factor. **-festigkeit** _f_ prolonged strength (_or_ stability). **-folge** _f_ chronological order. **-frage** _f_ 1 topic of the day. 2 question (_or_ matter) of time. **-geist** _m_ spirit of the age. **z–gemäß** _a_ 1 in keeping with the times, contemporary. 2 modern, up to date. 3 seasonable. **Z–genosse** _m_, **-genossin** _f_, **z–genössisch** _a_ contemporary. **-geordnet** _a_ chronological. **Z–gewinn** _m_ time gain. **-gleichung** _f_ equation of time. **-grenze** _f_ time limit. **-härtung** _f_ age hardening. **z–her** _adv_ ever since

zeitig _a_ 1 early, timely. 2 ripe, mature
zeitigen 1 _vt_ bring about, produce, lead to. 2 _vi_ ripen, mature. **Zeitigung** _f_ (-en) production; ripening, maturation
Zeit·lang _f_: **eine Z.** for some time, for a while. **-läufte** _pl_ current trend(s). **z–lebens** _adv_ all one's life
zeitlich _a_ 1 temporal, as regards time. 2 chronological. 3 transitory; temporary — **das Z—e segnen** pass away, die
zeitlos _a_ timeless; ageless, not affected by time
Zeit·lose _f_ (-n) (_Bot._) autumn crocus, meadow saffron. **-lupe** _f_ slow motion. **-lupenwiederholung** _f_ slow-motion replay. **-mangel** _m_ lack of time. **-maß** _n_ 1 tempo, (_Music_) time. 2 unit of time. **-maßstab** _m_ time scale. **-messer** _m_ timepiece; chronometer; metronome. **-messung** _f_ timekeeping; chronometry. **-mittel** _n_

time average. **z–nahe** _a_ topical, of current interest. **Z–not** _f_: **in Z.** pressed for time. **-plan** _m_ schedule, timetable. **-punkt** _m_ moment, time, point (in time), juncture. **-raffer** _m_ 1 time lapse camera (_or_ photography). 2 fast motion. **z–raubend** _a_ time-consuming. **Z–raum** _m_ space (_or_ period) of time, time interval. **-rechnung** _f_ 1 time reckoning, chronology. 2 era: **im 18. Jht. unserer Z.** in the 18th century of our era (= A.D.). **-schalter** _m_ (_Elec._) time switch. **-schnur** _f_ time fuse.

Zeitschr. abbv (ZEITSCHRIFT)
Zeit·schreiber _m_ chronograph. **-schrift** _f_ periodical, journal, review. **-schriftenliteratur** _f_ periodical literature. **-schriftenliteratur** _f_ periodical literature. **-spanne** _f_ period (_or_ stretch) of time. **z–sparend** _a_ time-saving. **Z–standversuch** _m_ (_Engg._) creep test. **-tafel** _f_ chronological table. **z–unabhängig** _a_ independent of time
Zeitung _f_ (-en) (news)paper
Zeitungs·ausgabe _f_ edition of a newspaper, **-ausschnitt** _m_ newspaper clipping. **-beilage** _f_ supplement (to a newspaper). **-bericht** _m_ newspaper report. **-druckfarbe** _f_ newsprint color. **-papier** _n_ 1 newsprint. 2 (old) newspaper(s). **-wesen** _n_ newspaper business, journalism
Zeit·verhältnis _n_ time relation. **-verlauf** _m_ lapse (_or_ course) of time. **-verlust** _m_ loss of time. **-verschiebung** _f_ time shift; time lag. **-verschwendung** _f_ waste of time. **-vertreib** _m_ pastime. **z–weilig** _a_ 1 temporary; _adv oft_ for a while, for the time being. 2 periodic, intermittent; occasional. **-weise** _adv_ 1 for a while, for the time being. 2 now and then, occasionally. **Z–wort** _n_ verb. **-zeichen** _n_ time symbol. **-zünder** _m_ time(-delay) fuse, delay igniter. **-zündung** _f_ delayed ignition. **-zwischenraum** _m_ time interval. **-zyklus** _m_ time cycle
zelebrieren _vt_ celebrate
zell· cell: **-ähnlich** _a_ cell-like, celloid, cytoid. **Z–beständigkeit** _f_ cellular stability. **-bestandteil** _m_ cell constituent. **-bildung** _f_ cell formation. **Zellchen** _n_ (-) cellule, small cell. **Zell·dehydrase** _f_ cellular dehydrogenase

Zelle f (-n) **1** (incl Biol., Elec.) cell. **2** booth
Zelleib° m (=**Zell-leib**) cell body
zellen· cell, cellular: **-ähnlich** a cell-like, cellular. **Z–art** f type of cell. **z–artig** a cell-like, cellular. **Z–aufbau** m cell(ular) structure. **-beton** m cellular concrete. **z–bildend** a cell-forming. **Z–bildung** f cell formation. **-faser** f cell fiber. **-faserstoff** m cellulose. **-filter** n revolving filter. **-flüssigkeit** f cell fluid. **z–förmig** a cell-shaped, cellular. **-frei** a cell-free, non-cellular, adv oft outside the cell. **Z–gehalt** m cell content. **-gewebe** n cellular tissue. **z–haltig** a containing cells, cellular. **Z–haut** f cell membrane. **-inhalt** m cell content. **-keim** m cell germ. **-kern** m cell nucleus. **-kies** m cellular pyrites. **-körper** m **1** cellular body. **2** body of a cell. **-lehre** f cytology. **-protoplasma** n cell protoplasm, cytoplasm. **-rad** n bucket wheel. **-saft** m cellular fluid. **-schale** f cell membrane. **-schalter** m (Elec.) battery switch. **-spannung** f cell voltage. **-stoff** m cellulose. **-struktur** f cellular structure. **-teilung** f cell division. **-tiefofen** m soaking-pit furnace. **-wand, -wandung** f cell wall. **-wucherung** f cell proliferation. **-zwischensubstanz** f intercellular substance
Zell· cell, cellular: **-faden** m cell filament. **-faser** f cellular fiber. **-faserstoff** m cellulose. **z–förmig** a cellular. **-frei** a cell-free, noncellular; adv oft outside the cell. **Z–gewebe** n cellular tissue, parenchyma. **-glas** n cellophane, transparent cellulose film. **-gummi** n foam (or sponge) rubber. **-horn** n celluloid
zellig 1 a & esp sfx cellular, celled: **einzellig** unicellular, one-celled. **2** vesicular; honeycombed
Zell· cell, cellular: **-kautschuk** m or n foam (or sponge) rubber. **-kern** m cell nucleus. **-kernteilung** f (Biol.) nuclear division. **-kies** m (Min.) marcasite. **-körper** m cell body. **-masse** f cellular substance. **-membran** f cellular membrane. **-mund** m cytostome
Zelloidin·einbettung f celloidin embedding. **-papier** n celloidin (paper)
Zellon n cellulose acetate. **--lack** m

cellulose acetate lacquer
Zellophan n (-e) cellophane
Zell· cell, cellular: **-pech** n cell(ulose) pitch (from evaporation of sulfite liquor); (Paper) lignin sulfonate. **-plasma** n cell plasm, protoplasm. **-reste** mpl cell debris. **-saft** m cell fluid. **-stoff** m **1** cellulose. **2** (Paper) pulp
Zellstof· (= **Zellstoff** when prefixed to nouns beginning with **F**): **-fabrik** f cellulose factory; (Paper) pulp mill. **-fabrikation** f cellulose manufacture; (Paper) pulp making. **-faser** f cellulose fiber. **-flocken** fpl. loose (paper) pulp
Zellstoff· cellulose; (Paper) pulp: **-ausbeute** f (Paper) yield of pulp. **-chemie** f cellulose chemistry. **-fabrik, -fabrikation, -faser, -flocken:** see ZELLSTOF· above. **-garn** n cellulose yarn. **-glashaut** f transparent cellulose film. **-holz** n pulpwood. **-karton** m paperboard, pulpboard. **-kocher** m cellulose digester. **-pappe** f paperboard, pulpboard. **-prüfung** f cellulose (or specif pulp) testing. **-schleifer** m pulp grinder. **-schleim** m cellulose pulp slurry. **-seide** f rayon. **-trockner** m pulp drier. **-tuch** n cellulose tissue (paper). **-verbindung** f cellulose compound. **-watte** f cellulose wadding (or cotton)
Zell-stoffwechsel m cell metabolism. **-teilung** f cell division
Zelluloid n celluloid. **--lack** m celluloid varnish
Zellulose f cellulose (For compds. see CELLULOSE·)
Zell·vermehrung f cell growth. **-verschmelzung** f cell fusion. **-wachstum** n cell growth. **-wand** f cell wall. **-wolle** f rayon staple; artificial wool. **-wollgarn** n rayon staple yarn
Zelt n (-e) tent; pavilion; canopy. **--bahn** f tent canvas; (specif) tent square
Zeltchen n **1** little tent. **2** pill, tablet
Zelt·dach n awning; tent top. **-plane** f tarpaulin. **-stoff** m canvas, tent cloth
Zement m or n (-e) cement. **z--artig** a cement-like
Zementation f (-en) **1** (Metll.) cementation. **2** (Geol.) reduction
Zement·beton m cement concrete. **-brei** m

cement slurry. **-brennofen** *m* cement kiln (*or* furnace). **-drehofen** *m* rotary cement kiln. **z-echt** *a* (*Colors*) cement-stable. **Z-fabrik** *f* cement factory. **-gerbstahl** *m* shear steel. **-hydrat** *n* hydrated cement

zementieren *vt* 1 cement. 2 (*Metll.*) subject to cementation, carburize

Zementier-faß *n* precipitation vat. **-mittel** *n* cementing agent. **-ofen** *m* cementation furnace

Zementierung *f* (-en) cementation; carburization

Zementierungs-ofen *m* cementing furnace. **-pulver** *n* cementing powder

Zementier-verfahren *n* cementation process

Zementit *n* cementite

Zement-kalk *m* hydraulic lime. **-kalkmörtel** *m* lime-cement mortar. **-kalkstein** *m* hydraulic limestone. **-kohle** *f* cementation carbon. **-kufe, -küpe** *f* cement vat. **-kupfer** *n* cement (*or* precipitated) copper. **-leim** *m* cement paste. **-mastix** *m* mastic cement. **-metall** *n* metal precipitated by cementation. **-milch** *f* thin cement mortar, grout. **-mörtel** *m* cement mortar. **-mörtelbrei** *m* grout. **-ofen** *m* cement kiln. **-prüfung** *f* cement testing. **-pulver** *n* cement(ing) powder. **-putz** *m* cement plaster. **-rohschlamm** *m* raw cement slurry. **-schlacke** *f* cement clinker. **-silber** *n* precipitated silver. **-spritzverfahren** *n* cement-grout spraying process. **-stahl** *n* cementation (*or* cemented steel. **-stein** *m* cement stone, hardened cement paste. **-tiegel** *m* cement(ation) crucible. **z-verkleidet** *a* cement-lined. **Z-wasser** *n* cement(ing) water. **-werk** *n* cement plant (*or* works)

Zenit *m* *or* *n* zenith

zensieren *vt* 1 mark, grade (exams, etc). 2 censor

Zensur *f* (-en) 1 (*School*) grade, mark. 2 censorship. 3 censor's office

Zentigramm *n* (-) centigram. **Zentiliter** *n* (-) centiliter. **Zentimeter** *n* (-) centimeter

Zentner *m* (-) (metric) hundredweight = 50 kg (=100 *Eur.* lbs, 110 *Am.* lbs)

zentral *a* central. **Z-·amt** *n* central (*or* main) office, headquarters. **-blatt** *n* (*esp*

in names of periodicals): **Z. für** General Journal of . . .

Zentrale *f* (-n) 1 center, central office, headquarters; central (railroad *or* bus) terminal, central (power *or* control station). 2 (*Math.*) median line; line joining centers

Zentral-formel *f* (*Benzene*) centric formula. **z-geheizt** *a* centrally heated. **Z-gruppe** *f* central group. **-heizung** *f* central heating

zentralisieren *vt* centralize

Zentral-körperchen *n* (*Biol.*) centrosome. **-kraftsystem** *n* central force system. **-lager** *n* central storehouse (*or* storage place). **-nervensystem** *n* central nervous system. **-stelle** *f* center, central (*or* main) office, headquarters = ZENTRALE. **-strahl** *m* central (*or* principal) ray. **z-symmetrisch** *a* centrosymmetric. **Z-verein** *m* central association (*or* organization). **-wert** *m* (*Stat.*) median. **z-winkelständig** *a* axial

Zentren centers: *pl of* ZENTRUM

zentrieren *vt & sich z.* center

Zentrifugal-abscheider *m* (centrifugal) separator. **-gebläse** *n* centrifugal blower (*or* fan). **-kraft** *f* centrifugal force. **-sichter** *m* 1 centrifugal separator. 2 (*Flour*) bolting mill. **-stoffmühle** *f* (*Paper*) centrifugal beater

Zentrifuge *f* (-n), **zentrifugieren** *vt/i* centrifuge

zentrisch *a* centric, central; concentric. **Zentrizität** *f* (-en) centricity

Zentrum *n* (. . . tren) 1 center. 2 downtown (area)

Zeolith *m* (-e *or* -en) zeolite —**schwarzer Z.** (*Min.*) gadolinite. **zeolithisch** *a* zeolitic

Zer *n* cerium (For compds., etc see CER·)

zer.. *insep pfx* apart, to (*or* in) pieces, dis- (*implies disintegration, breaking up; destruction; dispersion*)

zerarbeiten 1 *vt* crush, work to bits; overwork. 2: *sich z.* work oneself to death

Zerasin *n* cerasin

Zerat *n* (-e) cerate

zerätzen *vt* destroy with caustics

zerbeißen* *vt* bite (*or* crunch) to pieces

zerbiß, zerbissen bit(ten) to pieces: *past*

(& *pp*) *of* ZERBEISSEN

zerbomben *vt* bomb to bits, destroy by bombing

zerbrach broke up, etc: *past of* ZERBRECHEN

zerbrechen* I *vt* 1 smash. 2 (*Emulsions*) break down. II *vt, vi*(s) break up (apart, to pieces), shatter

zerbrechlich *a* breakable, fragile; brittle. **Zerbrechlichkeit** *f* fragility, brittleness

Zerbrechungs·festigkeit *f* breaking strength

zerbricht breaks up, etc: *pres of* ZERBRECHEN

zerbrochen broken up, etc: *pp of* ZERBRECHEN

zerbröckeln *vt, vi*(s) crumble, break up. **zerbröckelnd** *p.a* (*oft*) friable. **Zerbröck(e)lung** *f* crumbling

zerbröseln *vt/i* crumble, disintegrate

Zerdehnung° *f* radial distortion

Zerdrehung° *f* rotational distortion, torsion, twist

zerdrückbar *a* friable, brittle, easily crushed. **zerdrücken** *vt* crush, crumple; squash. **Zerdrück(ungs)·fähigkeit** *f* crush strength

Zerealien *npl* cereals

zerebral *a* cerebral

Zer·eisen *n* pyrophoric iron-cerium alloy, "flint"

Zeremonie *f* (-n) ceremony. **zeremoniell** *a* ceremonial

Zeresin *n* ceresin

zerfahren I* *vt* rut, tear up (roads, by driving). II *p.a.* confused; absent-minded; disorganized. **Zerfahrenheit** *f* confused state, absent-mindedness

Zerfall° *m* decomposition, disintegration; dissociation; decay; ruin, collapse. **zerfallbar** *a* decomposable, dissociable; apt to disintegrate (collapse, etc) *cf* ZERFALLEN I. **Zerfallbarkeit** *f* decomposability, dissociability, disintegrability, etc.

Zerfall·elektron *n* decay electron (resulting from disintegration) *See also* ZERFALLS·

zerfallen I* *vi*(s) 1 decompose, disintegrate, dissociate; decay. 2 fall apart, go to pieces, collapse; go to ruin. 3 divide, be divided, break down (into parts). II *p.a*

(**mit**) dissatisfied, at odds (with)

Zerfalls·elektron *n* decay electron = ZERFALLELEKTRON. **-energie** *f* (*Nucl.*) decay (*or* disintegration) energy. **-erscheinung** *f* symptom of decay. **-geschwindigkeit** *f* rate of decomposition (*or* disintegration). **-grad** *m* degree of dissociation (*or* decomposition). **-grenze** *f* limit of dissociation (*or* decomposition). **-konstante** *f* dissociation (disintegration, radioactive decay) constant. **-kurve** *f* decay (*or* disintegration) curve. **-leuchten** *n* decomposition luminescence; (*Spect.*) continuous emission with dissociation. **-produkt** *n* decomposition (disintegration, dissociation) product. **-prozeß** *m* decomposition (*or* dissociation) process. **-reihe** *f* disintegration (*or* decay) series. **-wärme** *f* heat of decomposition (*or* dissociation). **-zeit** *f* disintegration (*or* radioactive decay) time.

Zerfaserer *m* (-) 1 (*Wood*) chip crusher. 2 (*Paper*) stuff grinder, pulper, kneader; (*Waste Paper*) perfecter. 3 (*Rags*) unraveling machine. **zerfasern** *vt* & **sich z.** break up into fibers, (un)ravel, fray

zerfetzen *vt* shred, tear into rags; maul, slash. **zerfetzt** *p.a* (*oft*) tattered, ragged

zerfiel decomposed, etc: *past of* ZERFALLEN

zerfl. *abbv* (zerfließlich) deliquescent

zerfleischen *vt* maul, mangle, butcher; lacerate

zerfließbar *a* deliquescent. **Zerfließbarkeit** *f* deliquescence

zerfließen* *vi*(s) deliquesce, melt, dissolve; (*Colors*) run. **zerfließend** *p.a* deliquescent

zerfließlich *a* deliquescent. **Zerfließlichkeit** *f* deliquescence

Zerfließung *f* deliquescence, dissolution

zerfloß, zerflossen deliquesced: *past & pp of* ZERFLIESSEN

Zer·fluor *m* cerium fluoride

zerfraß corroded: *past of* ZERFRESSEN

zerfressen I* *vt* corrode, chew up, eat away (at). II *p.a* (*oft*) worm-eaten, moth-eaten **zerfressend** *p.a* corrosive. **Zerfressung** *f* (-en) corrosion

zerfrieren* *vi*(s) freeze and break up, freeze to pieces

zerfrißt corrodes: *pres of* ZERFRESSEN

zerfror(en) froze, etc: *past (& pp) of* ZERFRIEREN

zergangen melted, etc: *pp of* ZERGEHEN

zergehen* *vi*(s) melt, dissolve; dwindle away; deliquesce

zerging melted, etc: *past of* ZERGEHEN

zergliedern *vt* dissect, dismember; break up (*or* down); analyze. **Zergliederung** *f* (-en) dissection, dismemberment; breakdown, analysis. **Zergliederungs·kunde, -kunst** *f* anatomy

zerhacken *vt* chop up, mince, hack to pieces. **Zerhacker** *m* (-) chopper; (*Elec., oft*) vibrator

zerhauen *vt* chop up, cut to pieces

Zerin *n* cerin. **Zerit** *n* (-e) cerite. **·-erde** *f* cerite earth. **Zerium** *m* cerium

Zerkleinerer *m* (-) comminutor, crusher, pulverizer. **zerkleinern** *vt* crush, pulverize, break into small pieces, comminute. **Zerkleinerung** *f* (-en) crushing, pulverization, comminution

Zerkleinerungs·grad *m* degree of comminution (*or* pulverization); reduction ratio. **-maschine** *f* crusher, pulverizer, comminutor

zerklopfen *vt* pound to pieces, crush

zerklüftet *p.a* fractured, fissured; cloven, split. **Zerklüftung** *f* (-en) fissuring, cleavage, splitting

Zerknall° *m* explosion, detonation. **zerknallbar** *a* explosive. **zerknallen** *vt, vi*(s) explode, detonate. **Zerknall·stoß** *m* shock of the explosion

zerknicken *vt* crack, snap (by bending)

zerknistern *vt* decrepitate

zerknittern, zerknüllen *vt* crumple

zerkochen *vt* boil till it falls apart; (*Paper*) pulp

zerkörnen *vt* granulate

zerkratzen *vt* scratch (up), scrape

zerkrümeln 1 *vt* reduce to crumbs, pulverize. **2: sich z.** crumble, fall apart into crumbs

zerlassen* *vt* melt; render, clarify (fat)

zerlegbar *a* decomposable, dissociable; dissectible, easily taken apart, divisible; dispersible. **Zerlegbarkeit** *f* decomposability, etc.

zerlegen *vt* decompose, dissociate; disassemble, dismantle, take apart; divide,

split up; analyze, break down, resolve; (*Opt.*) disperse, diffract; (*Elec.*) scan. **Zerlegung** *f* (-en) decomposition, dissociation; disassembly; division; analysis, breakdown, resolution; (*Opt.*) dispersion, diffraction; (*Elec.*) scanning. **Zerlegungs·wärme** *f* heat of decomposition (*or* dissociation)

zerließ melted: *past of* ZERLASSEN

zerlöchern *vt* perforate, puncture holes in. **zerlöchert** *p.a* (*oft*) full of holes

zermahlbar *a* grindable, crushable. **zermahlen** *vt* grind up, grind fine, pulverize, triturate

zermalmen *vt* crush, grind up. **Zermalmung** *f* crushing, grinding up

Zer·metall *n* cerium metal

zermürben *vt* wear down, grind down; (*esp Tex.*) cause to rot. **Zermürb· ·erscheinung** *f* fatigue. **Zermürbung** *f* wearing down, attrition; rotting (of fabrics)

Zermürbungs·versuch *m* fatigue test. **-widerstand** *m* fatigue resistance

zernagen *vt* chew (*or* gnaw) to pieces; erode

Zer·nitrat *n* cerium nitrate

zerpflücken *vt* pull (*or* pick) apart

Zerplatz·druck *m* bursting pressure. **zerplatzen** *vi*(s) burst, explode

Zerplatz·probe *f* bursting test. **-widerstand** *m* resistance to bursting

zerpulvern *v.* pulverize, powder

zerquetschen *vt* crush, squash

zerrann melted, etc: *past of* ZERRINNEN

zerrannt(e) refined: *pp (& past) of* ZERRENNEN

Zerr·bild *n* distorted picture, caricature

zerreibbar *a* friable, triturable. **zerreiben*** *vt* grind up, pulverize, triturate. **zerreiblich** *a* friable, triturable. **Zerreiblichkeit** *f* friability. **Zerreibung°** *f* pulverization, trituration

zerreißbar *a* tearable. **zerreißen*** *vt, vi*(s) tear up, tear apart; rupture

Zerreiß·festigkeit *f* tensile strength, resistance to tearing. **-grenze** *f* breaking point. **-last** *f* breaking load. **-maschine** *f* tensile test machine. **-probe** *f* **1** tearing (breaking, tensile strength) test. **2** sample for tensile test. **-versuch** *m* tensile test

zerren *vt, vi*(+**an**) tug, pull, tear (at)

Zerrenn·boden *m* (*Metll.*) slag bottom.
zerrennen* *vt* refine. **Zerrenner** *m* (-)
refiner

Zerrenn·feuer *n* refining fire. **-herd** *m* refining hearth

zerrieb(en) ground up: *past* (& *pp*) *of*
ZERREIBEN

zerrieseln *vi*(s) (*Slag*) disintegrate

zerrinnen* *vi*(s) melt (away), dwindle away

zerriß tore up, **zerrissen** torn up: *past* & *pp*
of ZERREISSEN

zerronnen melted away: *pp* of ZERRINNEN

zerrühren *vt* mix by stirring, stir to dissolve (*or* to break up)

Zerrung *f* (-en) strain, overstretching; tensile stress

zerrütten *vt* break (down), shatter; disorganize, throw into confusion *cf*
NERVENZERRÜTTUNG

zers. *abbv* **1** (zersetzbar) decomposable. **2**
(zersetzend) decomposing. **3** (zersetzt) decomposed; decomposes. **Zers.** *abbv* (Zersetzung) decomp. (decomposition)

zerschellen *vi*(s) shatter, break into pieces

zerschlagen I* 1 *vt* smash, break (pound,
hammer) to pieces; destroy, annihilate. **2:**
sich z. break, break down, fail. **II** *p.a* exhausted, knocked out. **Zerschlagenheit** *f*
exhaustion

zerschlug smashed, etc: *past of* ZERSCHLA-
GEN

zerschmelzen* *vt, vi*(s) melt away:
zerschmilzt *pres;* **zerschmolz(en)** *past*
(& *pp*)

zerschmettern *vt* smash, shatter

zerschneiden* *vt* cut up, cut to pieces;
carve up, dissect: **zerschnitt(en)** *past*
(& *pp*)

zerschnitzeln *vt* cut into little pieces, shred

zersetzbar *a* decomposable. **Zer-
setzbarkeit** *f* decomposability

zersetzen *vt* & **sich z.** decompose, disintegrate. **Zersetzer** *m* (-) decomposer, decomposing agent

zersetzlich *a* decomposable, unstable. **Zer-
setzlichkeit** *f* decomposability, instability

Zersetzung *f* (-en) decomposition, disintegration

Zersetzungs·destillation *f* destructive dis-

tillation. **-druck** *m* dissociation pressure.
-erscheinung *f* symptom of decomposition. **-erzeugnis** *n* decomposition product. **z-fähig** *a* decomposable. **Z-gas** *n*
gaseous decomposition product. **-gefäß** *n*
decomposition vessel. **-geruch** *m* odor of
decomposition. **-kolben** *m* decomposition
(*or* reaction) flask. **-kunst** *f* analysis.
-mittel *n* decomposing agent. **-pfanne** *f*
decomposing pan. **-produkt** *n* decomposition product. **-prozeß** *m* decomposition process. **-punkt** *m* decomposition
point. **-wärme** *f* heat of decomposition.
-widerstand *m* electrolytic resistance

zerspalten *vt* & **sich z.** split (up, apart)

zerspanbar *a* machinable. **Zerspan-
barkeit** *f* machinability. **zerspanen** *vt*
machine, cut metal from; chip off.
Zerspanung *f* (-en) machining.
Zerspanungs·eigenschaft *f* machinability, cutting property

zersplittern *vt, vi*(s) splinter, break into
splinters, fragment. **Zersplitterung** *f*
splintering, fragmentation; dispersal

zersprang cracked, etc: *past of* ZERSPRIN-
GEN

zersprengen *vt* blow up, shatter; disperse.
zersprengend *p.a* explosive

zerspringen* *vi*(s) crack (*or* burst) apart
(*or* into pieces), explode

zersprühen *vt* atomize, spray (in all
directions)

zersprungen cracked, etc.: *pp of* ZER-
SPRINGEN

zerstampfen *vt* pound to pieces; grind
(with a mortar); trample; break
(emulsions)

zerstäubbar *a* atomizable, sprayable; dispersible. **Zerstäubbarkeit** *f* atom-
izability; dispersibility

zerstäuben I *vt* **1** atomize, spray. **2** reduce
to dust, pulverize. **3** (scatter as) dust. **II** *vt,
vi*(s) **4** disperse. **5** (*Mercury*) flour. **III** *vi*(s)
6 turn (*or* crumble) into dust. **7** (*Electrodes*) sputter

Zerstäuber *m* (-) spray(er), atomizer.
-brenner *m* atomizing burner. **-rohr** *n*
atomizer tube

Zerstäubung *f* **1** atomization, spraying. **2**
pulverization. **3** scattering as dust, dusting. **4** dispersion. **5** (*Mercury*) flouring. **6**

crumbling into dust. **7** (*Cathodes*) sputtering

Zerstäubungs·apparat *m* atomizer. **-düse** *f* spray (*or* atomizing) nozzle. **-mittel** *n* atomizing agent. **-trockner** *m* atomizing (*or* spray) drier. **-trocknung** *f* spray drying. **-verfahren** *n* atomizing (*or* spraying) process

zerstieben* *vi*(s) scatter (as dust); spray, disperse; vanish

zerstieß crushed: *past of* ZERSTOSSEN

zerstob(en) scattered; *past* (& *pp*) *of* ZERSTIEBEN

zerstochen *p.a* pin-pricked

zerstörbar *a* destructible. **Zerstörbarkeit** *f* destructibility

zerstören *vt* destroy, ruin. **zerstörend** *p.a* (*oft*) destructive, ruinous. **Zerstörer** *m* (-) (*incl* naval) destroyer. **zerstörerisch** *a* destructive. **Zerstör·ladung** *f* demolition charge. **Zerstörung** *f* (-en) destruction, ruin

Zerstörungs·bombe *f* demolition bomb. **z–frei** *a* nondestructive

zerstoßen* *vt* pound to pieces; crush, pulverize, mash

Zerstrahlung *f* (-en) annihilation (by) radiation

zerstreuen I *vt* & *sich* z. **1** scatter, disperse. **2** dissipate, diffuse. **II** *vt* **3** disseminate. **4** distract; divert, entertain. **III:** *sich* z. get some diversion, relax. **zerstreuend** *p.a:* z—e Linse dispersion lens. **zerstreut** *p.a oft* **1** diffuse. **2** distracted, absent-minded. **Zerstreutheit** *f* **1** dispersion. **2** diffuseness. **3** distraction, absent-mindedness. **zerstreut·porig** *a* diffuse-porous. **Zerstreuung** *f* (-en) **1** scattering, (*incl* Opt.) dispersion; (*Opt. oft*) divergence. **2** dissipation, diffusion. **3** dissemination. **4** distraction; diversion, entertainment; relaxation. **5** absent-mindedness

Zerstreuungs·linse *f* dispersion (*or* diverging) lens. **-vermögen** *n* (*Opt.*) dispersive power. **-wirkung** *f* (*Electrons, etc*) scattering

zerstückeln 1 *vt* cut (chop, break) into pieces, dismember. **2: sich** z. fall (*or* break) into pieces, crumble; (*Ores*) spall. **zerstückelt** *p.a* (*oft*) fragmented.

Zerstückelung *f* (-en) fragmentation, dismemberment; crumbling; spalling

zerteilbar *a* divisible. **zerteilen** *vt* & *sich* z. divide, split up, separate. **Zerteilung** *f* (-en) division, separation, breakup, breakdown

Zerteilungs·grad *m* degree of division, fineness. **-kraft** *f* (*Explo.*) fragmentation power. **-mittel** *n* (*Med.*) resolvent

zertrat trampled: *past of* ZERTRETEN

zertrennen *vt* separate, rip (tear, take) apart, tear the seams of. **Zertrennung** *f* (-en) separation, ripping, tearing (the seams)

zertreten* *vt* **1** trample, crush (under foot); stamp out (*esp* fire). **2** wear out (shoes). **zertritt** *pres*

zertrümmerbar *a* demolishable, disruptable; (*Nucl.*) fissionable. **zertrümmern** *vt* smash, shatter, wreck, demolish. **Zertrümmerung** *f* (-en) smashing, shattering, wrecking, demolition

Zerussit *m* cerussite

Zervelat·wurst *f* saveloy

Zerwürfnis *n* (-se) falling-out, disagreement; split, rift

zerzupfen *vt* pick (*or* pluck) to pieces

zessieren *vi* cease

Zession *f* (-en) cession, transfer, assignment *cf* ZEDIEREN. **Zessionar** *m* (-e) assignee, transferee

Zettel *m* (-) **1** slip (*or* scrap) of paper. **2** note. **3** ballot. **4** file card. **5** label. **6** receipt, check, stub, voucher. **7** (*Tex.*) warp, chain. **-kartei** *f* card index file. **-katalog** *m* card catalog

zetteln I *vt* **1** file (in a card catalog). **II** *vi* **2** do careless paperwork. **3** (*Tex.*) warp

Zeug *n* (*oft used in genit* Z–s *implying an indef Quant.*) **1** tools, equipment, gear, rigging, harness, etc. **2** (*Tex.*) cloth, fabric, material. **3** clothes, *esp* underwear. **4** things (*oft* = **1, 3**). **5** junk, trash, rubbish; nonsense, drivel. **6** (*genl*) stuff; (*specif*) mortar, (baking) dough, (*Paper*) pulp, (*Brew.*) yeast, (*Metll.*) metal, ore, (*Fireworks*) composition; *also oft* = **1, 2, 5.** — **das Z. dazu haben** have what it takes; **arbeiten, was das Z. hält** work with all one's might. **-bütte** *f* (*Paper*) stuff chest, paper vat. **-druck** *m* cloth (*or* calico) printing. **-drucker** m cloth (or calico) printer. **-druckerei** f calico printing (factory)

Zeuge m (-n, -n) witness

Zeuge·linie f generating line, generatrix

zeugen I vt **1** beget, produce, procreate; generate. **II** vi **2** testify, bear witness. **3** (von) testify (to), give evidence (of)

Zeugen· witness: **-beweis** m (witness's) evidence

Zeug·fabrik f textile (or specif woolen) mill. **-fänger** m (Paper) stuff catcher. **-färberei** f textile dyeing (plant). **-festigkeit** f (Paper) pulp strength. **-geben** n (Brew.) adding the yeast, pitching the wort. **-handel** m textile trade. **-haus** n arsenal, armory

Zeugin f (-nen) witness

Zeug·kasten m **1** tool chest. **2** (Paper) stuff chest. **-lumpen** mpl (Paper) cloth rags

Zeugnis n (-se) **1** report (card). **2** mark, grade. **3** expert opinion, (specif) doctor's statement. **4** proof, evidence. **5** certificate, diploma; credentials. **6** reference, testimonial. **7** testimony: **Z. ablegen** testify, make a deposition

Zeug·schmied m toolsmith. **-schmiede** f tool works. **-sichter** m (Paper) pulp strainer

Zeugung f (-en) procreation, (re)production, generation, breeding

Zeugungs·flüssigkeit f seminal fluid. **-kraft** f procreative power; virility. **-mittel** n procreative agent; aphrodisiac. **-organ** n genital organ. **-stoff** m semen

Zeug·wanne f (Brew.) yeast tub

ZF abbv (Zwischenfrequenz) (Elec.) intermediate frequency

z.H. abbv **1** (zur Hälfte) (by) one half. **2** (zu Händen) for the attention of

Zibebe f (-n) **1** muscatel raisin. **2** (Pharm.) cubeb

Zibet m civet. **-katze** f civet (cat). **-ratte** f muskrat

Zibeton n civetone

Zichorie f (-n) chicory. **Zichorien·wurzel** f chicory root

Zickel n (-) (Zool.) kid. **-fell, -leder** n kid (leather), kidskin

Zicklein n (-) (Zool.) kid

Zickzack m zigzag: **im Z.** zigzag, in a zigzag path. **-bewegung** f zigzag motion. **z–förmig** a zigzag, staggered. **Z–linie** f zigzag line

Zider m (-) cider cf CIDER·

Ziege f (-n) (genl) goat, (specif) she-goat

Ziegel m (-) **1** brick. **2** (roofing) tile. **3** briquet. **-arbeiter** m brick (or tile) maker. **-backstein** m (burnt) brick. **-brand** m **1** brick (or tile) burning. **2** batch of bricks (or tiles). **-brennen** n brick (or tile) burning. **-brenner** m brick (or tile) burner (or maker). **-brennerei** f **1** brick (or tile) burning. **2** brick (or tile) factory. **-brennofen** m brick (or tile) kiln. **-dach** n tile roof

Ziegelei f (-en) brickyard, brick (or tile) works (or kiln). **-ton** m brick clay

Ziegel· brick, tile: **-erde** f brick clay (or earth). **-erz** n tile ore; (Min.) cuprite. **-fachwerk** n brick checkerwork. **-farbe** f brick color, brick red. **z–farben, -farbig** a brick-colored, brick-red. **-förmig** a brick-shaped, in brick form. **Z–hütte** f brick (or tile) works (or kiln). **-mauerung** f brick masonry. **-mehl** n brick dust. **-ofen** m brick (or tile) kiln. **z–rot** a brick-red. **Z–sorte** f grade (or sort) of brick (or tile). **-stein** m brick. **-ton** m brick (or tile) clay

Ziegen· goat, goat's: **-bock** m he-goat. **-butter** f goat's butter. **-fell** n goatskin, buckskin. **-käse** m goat's(-milk) cheese. **-lab** n goat rennet. **-leder** n goatskin, kidskin. **-milch** f goat's milk. **-peter** m (Med.) mumps. **-stein** m bezoar. **-talg** m goat tallow

Ziegler m (-) brickmaker, tile maker

Zieh·arm m (pull) handle, crank. **-bank** f (wire-)drawing bench

ziehbar a ductile. **Ziehbarkeit** f ductility

Zieh·brunnen m draw-well. **-eisen** n (wire-)drawing block (or die)

ziehen* (zog, gezogen) I vt **1** pull, haul; pull (or take) out; draw, derive; extrude. **2: auf sich z.** attract. **3** (esp in idioms with nouns) draw, make, etc: **Folgerungen (Parallele, Vergleiche) z.** draw conclusions (parallels, comparisons) cf SCHLUSS **3; Kerzen z.** dip (or mold) candles; **Linien (Draht) z.** draw lines (wire) cf **5; Gräben z.** dig ditches; **Gewehre z.** rifle guns;

Wasser z. leak *cf* BETRACHT, BILANZ, BLASE, ERWÄGUNG, FADEN I 1, FLASCHE 1, LÄNGE, MITLEIDENSCHAFT, NUTZEN, RAT 3, RECHENSCHAFT, VERANTWORTUNG, ZWEIFEL. **4: es zieht einen** one feels drawn (*usu* to a place). **5** extract (drugs, *Math.* roots). **6** take off (hat, etc). **7** thread (beads), string (violins, etc); leave (tracks, ruts). **8** (*Grain*) couch, floor. **9** put up, erect (walls, partitions, washline); lay, install (wires, pipes). **10** grow, raise, breed *cf* GROSSZIEHEN. **11: nach sich z.** involve, bring on. **II** *vi*(h) **12** (**an**) pull, tug (at, on); puff, take a drag (on a cigar, pipe, etc). **13** (*Mach.*) run, work. **14** (*Chimney, etc*) draw. **15** steep. **16** work, have an effect; be well received. **17** (*Colors*) e.g.: **ins Blaue z.** incline to blue, have a blue tinge. **18: es zieht** there is a draft. **19: es zieht einem** (**in**) one feels a twinge (*or* pain) (**in**) *cf* ZIEHEND. **III** *vi*(s) **20** move, travel, go; march; roam, migrate (*esp* Birds). **IV: sich z. 21** stretch, extend, run. **22** (*oft* **+ in die Länge**) drag on and on. **23** (*Wood*) warp. **Ziehen** *n* **1** pulling, drawing; draft, traction. **2** (*Liquids*) ropiness. **3** move. **4** twinge, pain. **ziehend** *p.a* (*oft*): **z—e Schmerzen** twinge, sharp (*usu* rheumatic *or* arthritic) pain. **Ziehenlassen** *n* steeping, infusion (as of tea); allowing (a vessel) to fill up

Zieh·fähigkeit *f* (*Metll.*) ductility, drawing capacity. **-feder** *f* drafting pen. **-fett** *n* wire-drawing grease. **-frosch** *m* (pair of) pliers. **-grad** *m* degree of ductility. **-haken** *m* drag (*or* hauling) hook. **-hitze** *f* wire-drawing heat. **-kraft** *f* **1** pulling power, tractive force. **2** publicity value, attractiveness (to customers, etc). **-matrize** *f* wire-drawing die. **-öl** *n* wire-drawing oil. **-pappe** *f* molded paperboard. **-presse** *f* wire-drawing (*or* extruding) press. **-probe** *f* **1** wire-drawing (*or* extruding) sample. **2** tensile test. **-schnur** *f* pull cord; drawstring. **-seife** *f* wire-drawing soap. **-spachtel** *f* glazing filler. **-stahl** *m* wire-drawing steel. **-stein** *m* (wire-)drawing die

Ziehung *f* (-en) **1** drawing (*esp* of lots). **2** pulling; draft, traction

Zieh·vermögen *n* (*Dye.*) affinity. **-wert** *m*

(*Metll.*) ductility value

Ziel *n* (-e) **1** goal; target, mark; finish line: **das Z. verfehlen** miss the mark. **2** destination. **3** aim, objective: **Z. dieser Versuche ist** ... the aim of these experiments is ... **4** due date, credit deadline, term. **--bereich** *m* target area. **z—bewußt** *a* single-minded, purposeful, (clearly) goal-oriented

zielen *vi* (**auf**) aim, take aim; drive (at); strive (for). **zielend** *p.a:* **z—es Zeitwort** transitive verb

Ziel·fernrohr *n* telescopic sight. **z—gerichtet** *a* goal-oriented. **Z—linie** *f* **1** goal (*or* finish) line. **2** line of sight. **Z—los** *a* aimless; random, haphazard. **Z—objekt** *n* target. **-punkt** *m* target, mark; goal. **-scheibe** *f* target. **-setzung** *f* goal-setting; aim, objective. **z—sicher** *a* unerring. **Z—sprache** *f* target language (in translating). **z—strebig** *a* single-minded, purposeful, goal-oriented. **Z—sucher** *m* homing device, (*specif*) radar, sonar

ziemen *vi* (*dat*) **& sich z.** be fitting, be suitable (for) *cf* GEZIEMEN

ziemlich 1 *a* considerable, appreciable. **2** *adv* fairly, rather, more or less: **z. hoch** fairly (*or* pretty) high; **so z. dasselbe** pretty much the same

Zienst *m* (-e) (*Bot.*) betony

Zier *f* ornament, adornment; (*in compds, oft*) ornamental, decorative. **Zierat** *m* (-e) ornament, decoration. **Zierde** *f* (-n) **1** ornament, adornment. **2** credit: **eine Z. des Vereins** a credit to the association. **zieren 1** *vt* adorn, embellish; be an asset to. **2: sich z.** act coy *cf* GEZIERT. **Zierlack** *m* decorative varnish. **zierlich** *a* fragile, delicate; dainty. **Zierpflanze** *f* ornamental plant

Ziffer *f* (-n) **1** number, numeral, figure. **2** numbered section (*or* paragraph); item, clause (in a document). **3** cipher, code. **--blatt** *n* (clock, meter) dial, clock face

ziffern·mäßig *a* numerical, digital. **Z—rechenmaschine** *f* digital computer. **-uhr** *f* digital clock (*or* watch)

zig *num* umpteen, zillion, thousands of

Zigarette *f* (-n) cigarette

Zigaretten·rauch *m* cigarette smoke. **-stummel** *m* cigarette butt

Zigarre f (-n) 1 cigar. 2 bowling-out
zigarren·förmig a cigar-shaped. **Z–rauch**
m cigar smoke. **-stummel** m cigar butt
zig·fach a lots and lots, thousands (of). **-mal**
adv a million (or zillion) times. **-tausend**
num thousands and thousands of cf ZIG
Zimmer n (-) 1 room; cabin; chamber. 2 (in
compds, oft) indoor; carpentry cf ZIM-
MERN: **-antenne** f indoor antenna. **-ar-
beit** f carpentry. **-decke** f ceiling.
-einrichtung f (room) furniture
Zimmerer m (-) carpenter
Zimmer·holz n lumber. **-leute** pl of **-mann**
m carpenter
zimmern 1 vt make, put together; shingle
(roofs). 2 vi do carpentry
Zimmer·temperatur f room temperature.
-thermometer m/n indoor thermometer
Zimmerung f (-en) (Min.) timberwork
Zimmer·wärme f room temperature
zimolisch a cimolitic = CIMOLISCH
Zimt m (-e) 1 cinnamon: **chinesischer Z.**
cassia bark; **weißer Z.** canella (bark). 2
junk, trash; nonsense. **-·aldehyd** m cin-
namaldehyde. **-alkohol** m cinnamyl alcohol. **-baum** m cinnamon tree. **-blät-
teröl** n oil of cinnamon leaves. **-blüte** f
cinnamon flower, cassia bud. **-blütenöl** n
oil of cassia. **z–braun** a cinnamon brown.
Z–carbonsäure f carboxycinnamic acid.
z–farben, -farbig a cinnamon-colored.
Z–granat m (Min.) cinnamon stone, essonite. **-kaneel** m canella (bark). **-kassia**
f cassia (bark). **-kassienöl** n oil of cassia.
-öl n oil of cinnamon—**chinesisches Z.**
cassia oil. **-rinde** f cinnamon bark.
-röhrchen n cinnamon stick. **z–sauer** a
cinnamate of. **Z–säure** f cinnamic acid.
-stange f cinnamon stick. **-stein** m (Min.)
cinnamon stone, essonite. **-wasser** n cinnamon water
Zinder m (-) cinder
Zineb n zineb, zinc diethyldithiocarbamate
Zink n zinc; spelter. **z–·artig** a zinc-like. **Z–
asche** f zinc ash; zinc dross (or oxide)
Zinkat n (-e) zincate
Zink·äthyl n diethylzinc. **-ätze** f zinc-
etching bath (or solution). **-ätzung** f zinc
etching. **-auflage** f zinc coating (or layer).
-azetat n zinc acetate. **-bad** n zinc bath.
-becher m zinc case (of a dry cell). **-beize**

f zinc mordant. **-beschlag** m zinc
sheathing. **-blech** n zinc plate, sheet (of)
zinc. **-blende** f (Min.) zinc blende,
sphalerite. **-blumen** fpl 1 flowers of zinc,
zinc oxide. 2 zinc spangles. **-blüte** f zinc
bloom, hydrozincite. **-butter** f butter of
zinc, zinc chloride. **-chromgelb** n zinc
chrome, zinc yellow. **-dampf** m zinc vapor
(or fumes). **-destillierofen** m zinc (distillation) furnace. **-draht** m zinc wire.
-druck m zinc printing. **-druckguß** m
pressed zinc casting
Zinke f (-n) 1 prong, tine; tooth (of a comb,
etc); spike. 2 dovetail, tenon. 3 proboscis.
Cf ZINKEN
Zinkeisen·erz n (Min.) franklinite. **-spat**
m (Min.) ferriferous smithsonite. **-stein**
m (Min.) franklinite
Zink·elektrolysenanlage f electrolytic
zinc plant
Zinken m (-) prong, etc = ZINKE 1, 2
Zink·entsilberung f desilvering of zinc by
the Parkes process. **-entsilberungsan-
lage** f zinc-desilvering plant (or unit)
Zinkerei f (-en) zinc plant (or works)
Zink·erz n zinc ore **—rotes Z.** zincite.
-fahlerz n (Min.) tennantite. **-farbe** f
zinc paint, **-feile** f, **-feilspäne** mpl zinc
filings. **-folie** f zinc foil. **-formiat** n zinc
formiate. **z–führend** a zinc-bearing. **Z–
gehalt** m zinc content. **-gekrätz** n zinc
dross (or oxide). **-gelb** n zinc yellow (zinc
chromate). **-gewinnung** f extraction of
zinc, zinc production. **-glas(erz)** n (Min.)
hemimorphite; silicious calamine; zinc
glance. **-granalien** pl granulated zinc.
-grau n zinc gray. **-grün** n zinc green. **z–
haltig** a containing zinc, zinciferous. **Z–
harz** n (Min.) zinc blende, sphalerite.
-hütte f zinc works, zinc smelter, **-hüt-
tenrauch** m zinc smelter fumes
zinkig a pronged, tined; toothed; jagged
Zinkit n (-e) zincite
Zink·jodid n zinc iodide. **-kalk** m zinc ash
(or dross), zinc oxide. **-kalkküpe** f zinc-
lime vat. **-kasten** m zinc case (or tank).
-kiesel m, **-kieselerz** n silicious
calamine; hemimorphite. **-kitt** m zinc cement. **-kohlenbatterie** f zinc-carbon battery. **-legierung** f zinc alloy. **-lösung** f
zinc solution. **-mehl** n zinc powder (or

dust). **-muffel** *f* zinc muffle, zinc distilling retort. **-nebel** *m* zinc smoke mixture. **-ofen** *m* zinc furnace. **-ofenbruch** *m* tutty, cadmia, crude zinc oxide. **z– organisch** *a* organozinc. **Z–oxid** *n* zinc oxide —**rotes Z.** zincite. **-oxidnitrat** *n* zinc nitrate. **-pecherz** *n* sphalerite. **-pol** *m* (Cells) zinc pole, cathode. **-pulver** *n* zinc powder. **-raffinerie** *f* zinc refinery. **-rauch** *m* zinc fumes. **-salbe** *f* zinc (oxide) ointment. **-salz** *n* zinc salt. **z–sauer** *a* zincate of. **Z–schale** *f* zinc dish (or basin). **-schaum** *m* (Metll.) zinc scum. **-schlicker** *m* zinc dross. **-schnitzel** *npl* zinc shavings (or filings). **-schwamm** *m* tutty, cadmia, crude zinc oxide. **-span** *m* zinc shavings (or chips). **-spat** *m* zinc spar, smithsonite. **-stab** *m* zinc rod. **-staub** *m* zinc dust; (Metll.) blue dust. **-staubätze** *f* (Tex.) zinc dust discharge. **-überzug** *m* zinc coating. **-verbindung** *f* zinc compound. **-vitriol** *n* zinc (or white) vitriol ($ZnSO_4 \cdot 7H_2O$); (Min.) goslarite. **-wanne** *f* zinc tub (or bath). **-weiß** *n* zinc white. **-werk** *n* zinc works. **-wolle** *f* flowers of zinc; mossy zinc metal. **-zyanid** *n* zinc cyanide

Zinn *n* tin; tinware, pewter; (in Chem. compds oft) stannic, stannous: **-abstrich** *m* tin scum (or dross). **-ader** *f* tin vein (or lode). **-after** *m* tin (ore) refuse. **z–ähnlich** *a* tin-like. **Z–ammoniumchlorid** *n* ammonium chlorostannate. **z–arm** *a* poor (or low) in tin, low-tin. **-artig** *a* tinlike, tinny. **Z–asche** *f* tin ashes, stannic oxide

Zinnatron *n* (= **Zinn·natron**) sodium stannate

Zinn·ätzfarbe *f* (Tex.) tin discharge paste. **-bad** *n* tin bath. **-beize** *f* tin mordant. **-beizendruck** *m* tin mordant printing. **-bergwerk** *n* tin mine. **-blatt** *n* tinfoil. **-blech** *n* tinplate, sheet (of) tin. **-bleilegierung** *f* tin-lead alloy. **-bleilot** *n* tin-lead solder. **-bromid** *n* stannic bromide. **-bromür** *n* stannous bromide. **z– bromwasserstoffsauer** *a* bromostannate of. **Z–bromwasserstoffsäure** *f* bromostannic acid. **-bronze** *f* tin bronze. **-butter** *f* butter of tin, stannic chloride. **-charge** *f* (Tex.) tin weighting. **-chloram-**

monium *n* ammonium chlorostannate, (Dye.) pink salt. **-chlorid** *n* stannic chloride. **-chloridchlorwasserstoffsäure** *f* hydrogen hexachlorostannate. **-chlorür** *n* stannous chloride. **z–chlorwasserstoffsauer** *a* chlorostannate of. **Z–chlorwasserstoffsäure** *f* chlorostannic acid. **-diphenylchlorid** *n* diphenyltin chloride. **-draht** *m* tin wire

Zinne *f* (-n) pinnacle, battlement

zinnen I *vt* tin, tinplate. **II** *a* = **zinnern** (made of) tin, pewter

Zinn· tin, stannic, stannous: **-erschwerung** *f* (Tex.) tin weighting. **-erz** *n* tin ore, (Min.) cassiterite. **-erzseife** *f* tin placer deposit. **-farbe** *f* (Dye.) tin-mordant color. **-feilicht** *n,* **-feilspäne** *mpl* tin filings. **-flammofen** *m* reverberatory tin furnace. **-folie** *f* tinfoil. **z–führend** *a* tin-bearing, stanniferous. **Z–gehalt** *m* tin content. **-gekrätz** *n* tin refuse (or sweepings). **-gerät** *n* tin utensil. **-geschirr** *n* tinware, pewter. **-geschrei** *n* tin cry. **-gießer** *m* tin founder. **-gießerei** *f* tin foundry. **z–glasiert** *a* tin-glazed. **Z–glasur** *n* tin glaze (or glazing). **-granalien** *pl* granulated tin. **-graupen** *fpl* twinned cassiterite crystals. **-grube** *f* tin mine. **-grundierung** *f* (Dye.) bottoming (or grounding) with tin. **z–haltig** *a* containing tin, stanniferous. **Z–hütte** *f* tin smeltery (or works). **-hydrat** *n* tin hydroxide. **-hydroxid** *n* stannic hydroxide. **-hydroxydul** *n* stannous hydroxide. **-jodid** *n* stannic iodide. **-jodür** *n* stannous iodide. **-kalk** *m* stannic oxide. **-kapsel** *f* tin capsule (or case); tin cap. **-kies** *m* tin pyrites, (Min.) stannite. **-knirschen** *n* tin cry. **-krätze** *f* tin dross (or waste). **-kreischen** *n* tin cry. **-lagerstätte** *f* tin deposit. **-legierung** *f* tin alloy. **-lösung** *f* tin solution; (Dye.) tin spirit. **-lot** *n,* **-löte** *f* tin (or soft) solder

Zinnober *m* **1** (red) cinnabar, (Pigment) vermilion. **2** fuss; mess. **--ersatz** *m* vermilion substitute. **-farbe** *f* vermilion (color). **z–farben, -farbig** *a* vermilion. **Z–grün** *n* cinnabar green, chromium oxide pigment. **-rot** *n* vermilion. **-salz** *n* vermilion substitute. **-spat** *m* crystallized cinnabar

Zinn· tin; stannic, stannous: **-ofen** *m* tin furnace. **-oxid** *n* stannic oxide; (*in compds with other elements in archaic terminology*) stannic. **-oxidsalz** *n* stannic salt. **-oxydul** *n* stannous oxide; (*in compds with other elements in archaic terminology*) stannous. **-oxydulsalz** *n* stannous salt. **-pest** *f* tin plague. **-pfanne** *f* (*Tinning*) tin pot. **-platte** *f* tinplate, sheet of tin(plate). **z–plattiert** *a* tin-plated. **Z–probe** *f* 1 tin sample. 2 tin test (*or* assay). **-puder** *m* powdery tin dross. **-pulver** *n* powdered tin. **-raffination** *f* tin refining. **z–reich** *a* tin-rich, high-tin. **Z–rohr** *n*, **-röhre** *f* tin pipe. **-rückstände** *mpl* tin residue(s). **-salmiak** *m/n* ammonium hexachlorostannate, pink salt. **-salz** *n* 1 tin salt. 2 (*Dye. usu*) tin crystals, stannous chloride. **-salzätze** *f* (*Dye.*) tin crystal discharge. **-sand** *m* (*Min.*) grain tin, alluvial cassiterite. **z–sauer** *a* stannate of. **Z–saum** *m* tin strip (list, edging); selvedge. **-säure** *f* stannic acid. **-säureanhydrid** *n* stannic anhydride. **-schlich** *m* (*Ores*) tin slimes, fine tin. **-schrei** *m* tin cry. **-seife** *f* (*Min.*) stream tin. **-soda** *f* sodium stannate. **-staub** *m* tin dust. **-stein** *m* tinstone, cassiterite. **-sulfid** *n* stannic sulfide. **-sulfocyanid** *n* stannic thiocyanate. **-sulfür** *n* stannous sulfide. **-verbindung** *f* tin compound. **-waren** *fpl* tinware, pewter. **-wasserstoff** *m* tin hydride, stannate. **-weiß** *n* tin white. **-wolle** *f* mossy tin

Zins *m* (-en) 1 (*Com.*) interest. 2 rent. **Zinses·zins** *m* compound interest

Zins·fuß *m* interest rate. **z–pflichtig** *a* subject to interest charges; liable to pay rent. **Z–satz** *m* interest rate

Zipfel *m* (-) 1 corner (of a cloth); flap, lobe. 2 end, tip; point. 3 (*Geog.*) spit (of land), panhandle. **-bildung** *f* (*Tensile Testing*) earing. **zipfelig** *a* pointy, pointed; peaked

Zipp·verschluß *m* zipper

Zirbe(l) *f* (-n) Swiss pine

Zirbel·drüse *f* (*Anat.*) pineal body, ephiphysis. **-fichte, -kiefer** *f* Swiss pine. **-nuß** *f* cedar nut

zirka *adv* circa, about, approximately

zirkassisch *a* Circassian

Zirkel *m* (-) 1 (*Instrum.*) compass(es). 2 (*Geom. & fig*) circle. 3 (*in compds. oft*) circular. **-schluß** *m* vicious circle, circular reasoning

Zirkon *n* (-e) 1 (*Min.*) zircon. 2 (*esp in compds*) zirconium: **-alkoxide** *npl* zirconium alkoxides. **Zirkonat** *n* (-e) zirconate

Zirkon· zircon(ium): **-brenner** *m* zirconium (incandescent) filament. **-erde** *f* zirconia, zirconium oxide. **-glas** *n* zirconium glass

Zirkonium·karbid *n* zirconium carbide. **-stahl** *m* zirconium steel. **-verbindung** *f* zirconium compound

Zirkon· zircon(ium): **-lampe** *f* zirconium lamp. **-licht** *n* zircon light. **-oxid** *n* zirconium oxide. **-präparat** *n* zirconium preparation. **-säure** *f* zirconic acid. **-stahl** *m* zirconium steel

Zirkulante *f* (-n) (*Math.*) circulant. **zirkular, zirkulär** *a* circular

Zirkulations·störung *f* (*Med.*) circulatory disorder. **-trocknung** *f* (*Gas, Air*) circulation drying

Zirkulator *m* (-en) (*Elec.*) circulator

zirkulieren *vi* circulate. **Zirkulier·gefäß** *n* circulating vessel

Zirrhose *f* cirrhosis

Zisch *m* (-e) hiss. **zischen** 1 *vt/i* hiss. 2 *vi*(s) whiz, zip. **Zisch·hahn** *m* compression tap

Ziselier·arbeit *f* chasing, chiseled work. **ziselieren** *vt/i* engrave, chase, chisel. **Ziselier·werkzeug** *n* engraving tool

Zissoide *f* (-n) cissoid

Zisterne *f* (-n) cistern, tank. **Zisternen·wagen** *m* tank car; tank truck

Zitat *n* (-e) quotation. **zitieren** *vt* 1 quote. 2 call; cite, subpena

Zitrat *n* (-e) citrate. **z–löslich** *a* citrate-soluble

Zitra·weinsäure *f* citratartaric acid

Zitrin *n* (-e) citrine

Zitronat *n* (-e) candied lemon peel

Zitron·bartgras *n* lemon grass

Zitrone *f* (-n) lemon; citron

Zitronell· citronella: **-öl** *n* citronella oil. **-säure** *f* citronellic acid

Zitronella·säure *f* citronellic acid

Zitronellan·baum *m* lime tree

Zitronellol *n* (-e) citronellol

Zitronell·öl *n* citronella oil. **-säure** *f* citronellic acid

Zitronen·äther *m* citric ester. **-farbe** *f*

lemon (*or* citron) color. **z–farben, -far-
big** *a* lemon-colored, lemon-yellow. **-gelb**
a lemon-yellow, citrine. **Z–gras** *n* lemon
grass. **-kern** *m* lemon seed. **-kernöl** *n*
lemon-seed oil. **z–löslich** *a* citrate-
soluble. **Z–melisse** *f* lemon balm, balm
mint. **-öl** *n* lemon oil. **-presse** *f* lemon
squeezer. **-saft** *m* lemon juice. **-salz** *n* citrate. **z–sauer** *a* citrate of. **Z–säure** *f*
citric acid. **-schale** *f* lemon peel. **-wasser**
n lemonade

zitron·sauer *a* citrate of. **Z–säure** *f* citric
acid

Zitrulle *f* (-n) watermelon

Zitrus·öl *n* citrus oil, shaddock oil

Zitter·aal *m* electric eel. **-bewegung** *f*
trembling, variable motion; (*Light*) twin-
kling. **-elektrode** *f* vibrating electrode.
-fisch *m* sting ray. **-gras** *n* quaking grass

zitterig *a* tremulous, shaky, jittery

zittern *vi* tremble, shake, quiver, vibrate

Zitter·pappel *f* trembling aspen. **-sieb** *n*
vibrating screen. **-wurzel** *f* zedoary

zittrig *a* tremulous, etc = ZITTERIG

Zitwer *m* (-) **1** zedoary. **2** aconite. **3:**
(deutscher) Z. sweet flag. **4** ginger.
-kraut *n* tarragon. **-samen** *m* wormseed,
santonica; zedoary seed. **-wurzel** *f* zedoary (root)

Zitz *m* (*Tex.*) chintz

Zitze *f* (-n) (*Anat.*) nipple, teat, dug

zitzen·förmig *a* mamillary, mastoid. **Z–
tier** *n* mammal

zivil *a* **1** civil. **2** civilian. **3** normal, everyday.
4 (*Price*) reasonable. **Zivil** *n* **1** civilian
clothing; plainclothes. **2: für Z.** for
civilian use. **3: in Z.** in civilian life; in
civilian garb

zivilisieren *vt* civilize

Zivil·leben *n* civilian life. **-recht** *n* civil law

zl. *abbv* (ziemlich löslich) fairly soluble

Zle. *abbv* (Zeile) line

zll. *abbv* (ziemlich leicht löslich) fairly
soluble

Z-Mittel *n* (*Polymers*) Z-average (molecular
weight)

Zobel *m* (-) sable

Zober *m* (-) tub

zog pulled, etc: *past of* ZIEHEN

zögern *vi* hesitate, delay. **zögernd** *p.a.* hes-
itant. **Zögerung** *f* (-en) hesitation, delay

Zögling *m* (-e) pupil

Zölestin *m* (-e) (*Min.*) celestite

Zoll *m* **I** (ˉe) **1** (customs) duty. **2** customs
(authorities). **3** toll (*fig*). **II** (-) inch. **-amt**
n custom house

zollbar *a* dutiable

Zoll·beamte *m* (*adj endgs*) customs officer.
-bestimmungen *fpl* customs regulations

zollen *vt* (+*d.i.o*) show, accord, pay (sbdy
respect, etc)

zoll·frei *a* duty-free. **Z–freiheit** *f* exemption from customs duty. **-gebühr** *f* (customs) duty. **-haus** *n* custom house. **-kon-
trolle** *f* customs check. **-maß** *n*
measurement in inches

Zöllner *m* (-) customs official

zoll·pflichtig *a* dutiable. **Z–revision** *f* customs inspection. **-satz** *m* tariff rate.
-spesen *pl* customs charges. **-stock** *m*
ruler (calibrated in inches), yardstick.
-verein *m* customs (*or* tariff) union. **-ver-
schluß** *m* bond, customs seal

zonal *a* zonal; Eastern-Zone *cf* ZONE 2.

zonar *a* zonal. **Zonar·struktur** *f* (*Cryst.*)
zonal structure

Zone *f* (-n) **1** (*genl*) zone, region. **2** (*specif*)
Eastern Zone, East Germany (when it
was the Soviet-occupied zone)

Zonen·achse *f* (*Cryst.*) zone axis. **z–förmig**
a zonal. **Z–grenze** *f* zone boundary (*oft
implies:*) Eastern Zone border *cf* ZONE 2.
-richtung *f* (*Cryst.*) zone axis.
-schmelzen *n* zone refining

Zoochemie *f* zoochemistry. **zoochemisch**
a zoochemical

Zoologe *m* (-n, -n) zoologist. **Zoologie** *f* zoology. **Zoologin** *f* (-nen) zoologist. **zoo-
logisch** *a* zoological

Zoomarin·säure *f* hexadec-9-enoic acid

Zootechnik *f* zootechnology

Zootin·salz *n* Chile saltpeter

Zopf *m* (ˉe) **1** braid, pigtail. **2** twist bread. **3**
quaint old custom

Zorn *m* anger, wrath. **zornig** *a* angry,
furious

Zotte(l) *f* (-n) **1** tuft of shaggy (unkempt,
matted) hair. **2** (*Anat.*) villus. **zottelig** *a*
disheveled, shaggy, unkempt; matted.
Zottel·wolle *f* shaggy wool

Zotten·haut *f* (*Anat.*) chorion

zottig *a* shaggy, etc = ZOTTELIG

Zp. *abbv* (Zersetzungspunkt) decomp. pt.

z.T. *abbv* (zum Teil) in part, partly

Ztg. *abbv* (Zeitung) newspaper

Ztr. *abbv* (Zentner) metric hundredweight

Ztrbl. *abbv* (Zentralblatt) journal

Zts(chr). *abbv* (Zeitschrift) periodical, journal

zu I *prep* (*dat*) to, toward *cf* BIS III 3; KOMMEN 1 E, 2, 3G, 6; SICH I 1. **2** at (a time, price): **zu Ostern** at Easter *cf* ANFANG, ENDE 1, HAUS 1. 3 as: **es dient zum Vorbild** it serves as a model. **4** for (*implying purpose*): **zum Vergleich** for comparison. **5** (+Mal) e.g.: **zum letzten Mal, zum letztenmal** for the last time. **6** by: **zu Lande** by land; **zu Fuß** on foot. **7** (*numbers in pl*): **zu Dutzenden (Hunderten, Tausenden)** by the dozen (hundreds, thousands). **8** for, to make: **Stoff zu einer Jacke** material for (*or* to make) a jacket. **9** into, to, to form (*implies becoming, converting*): **zu Eis werden** turn to ice; **zu Asche verbrennen** burn to ashes; **die Elemente verbinden sich zu . . .** the elements combine into (*or* to form) . . . ; **zum Präsidenten erwählen** elect president. **10** (*Quant.*): **zu 10%** by 10%, to the extent of 10%. **11** to go with: **Kaffee zum Kuchen** coffee to go with the cake. **12** in addition to, on top of: **zu allem andern** on top of everything else; **zu den Akten legen** put in the files. **13** (*in Titles, oft*) on, concerning (a topic). **14** (+ *ordinal num*): **zu dritt** in a group of three; **sie sind zu viert** there are four of them. **15** (+ *ordinal num*): **zum ersten** first of all, firstly = ERSTENS. **16** (*esp* + *inf as noun*) to, for: **Zeit zum Nachdenken** time to reflect, time for reflection; **zur Aushilfe** to help out. **17** (+ *psn, implying place*): **zu ihnen** to their place (house, office, etc), to see (*or* visit) them. **18** (*in combination with advs*): **zur Tür hin(über)** (over) toward the door; **zum Fenster hinaus** out the window. **II** *prep* (+ *verb*) **19** (+ *inf*) to, of: **sie versuchten es auszurechnen** they tried to figure it out; **die Möglichkeit, es zu tun** the chance to do it, the possibility of doing it. **20** (+ *pres p*) to be (+ *pp*): **das zu gewinnende Element** the element to be obtained. **21** (+ *inf after v.* sein) to be:

das ist zu vermeiden that is to be avoided *cf* SEIN I A 1, 2. **III** *adv* **22** (*with prep or other adv*) toward: **dem Ausgang zu** toward the exit; **nach Westen zu** toward the west *cf* AB 1. **23** too (= excessively): **zu groß** too big *cf* GAR, *cf* ZUVIEL, ZUWENIG. **24** closed, shut; (turned) off: **die Tür ist zu** the door is closed; **der Gashahn ist zu** the gas tap is (turned) off. **IV** *pfx* **25**: *Many former prep + noun phrases now fused into advs*: *cf* ZUGRUNDE, (*orig* **zu Grunde**), ZUSTANDE, etc. **V** *sep pfx* **26** (*usu* + **auf**) to, toward, up to *cf* ZUGEHEN 1. **27** shut, closed *cf* ZUMACHEN; (*also implies covering*) *cf* ZUDECKEN. **28** (*implies preparing for specif use*) *cf* ZUSCHNEIDEN. **29** (*implies addition, increase*) *cf* ZUGIESSEN 1, ZUNEHMEN

zuallererst *adv* first of all, first and foremost. **zuallerletzt** *adv* last of all

Zubehör *n* accessories, fittings; (*Mach. oft*) equipment. **--teil** *m* accessory

zu·bekommen* *vt* 1 get as an extra (*or* as a bonus). **2** get (a door, etc) closed

zubenannt *a* surnamed, called

Zuber *m* (-) tub

zu·bereiten *vt* prepare (*esp* food); dress, finish; (*Med.*) dispense; make up, fill (prescriptions). **Zubereitung** *f* (-en) preparation; dressing; finishing, etc

zu·bessern *vt* improve; (*Bath*) feed up, strengthen. **Zubesserung°** *f* improvement; feeding addition, food supplement

zu·billigen *vt* (*d.i.o*) grant, concede

zu·binden* *vt* tie up; bind up, bandage

zu·brennen* *vt* 1 roast, calcine. **2** cauterize. **3** close by heating

zu·bringen* *vt* 1 spend (time). **2** (*d.i.o*) take (sthg to sbdy), deliver; feed, transfer (material to be processed). **3** get (e.g. a door) closed. **Zubringer** *m* (-) **1** (*Mach.*) feeder, conveyor. **2** (*Elec.*) transmission line. **3** approach (*or* access) road. **4** feeder (bus, van), transfer bus; (*oft as pfx*): **-dienst** *m* transfer (*or* feeder) service

Zubrühen *n* (*Brew.*) addition of boiling water in mashing

Zubuße *f* (-n) **1** allowance. **2** contribution. **3** additional payment

Zucht *f* (-en) **1** breeding, raising, growing, cultivation; training. **2** (*Animals*) breed;

(*Bacteria, etc*) strain. **3** breeding farm; nursery. **4** discipline, order: **in Z. haben** have under control. **5** (good) breeding, propriety. **6** (*as pfx*) breeding, stud *cf* ZUCHTTIER. **züchten 1** *vt/i* breed. **2** *vt* grow, raise, cultivate. **Züchter** *m* (-) breeder, grower

Zucht-stier *m* (*Zool.*) bull. **-tier** *n* breeding animal, stud

Züchtung *f* (-en) breeding, growing, raising, cultivation; culture *cf* ZÜCHTEN

Zucht-wahl *f* (*Biol.*) selection

Zuck-anzeige *f* ballistic (kick, flash) reading (*or* indication)

zucken I *vi* **1** twitch; flinch, wince; kick. **2** flash, dart. **3** (+**an**) tug (at). **II** *vt* shrug (shoulders). **zuckend** *p.a* convulsive, spasmodic

Zucker *m* sugar. **-abbau** *m* sugar degradation. **z-ähnlich** *a* sugary, saccharoid. **Z-ahorn** *m* sugar maple. **-art** *f* variety (*or* type) of sugar; *pl oft* sugars. **z-artig** *a* sugary, saccharine, saccharoid. **Z-ausbeute** *f* sugar yield. **-bäcker** *m* confectioner, pastry chef. **-baryt** *n* barium sucrate. **-beschwerung** *f* weighting with sugar. **-bestimmung** *f* sugar determination. **-bildung** *f* sugar formation, saccharification; (*Physiol.*) glycogenesis. **-branntwein** *m* sugared alcohol. **-brot** *n* **1** sugarloaf. **2** (piece of) pastry. **-busch** *m* sugar bush, honey flower. **-couleur** *f* caramel. **-dicksaft** *m* syrup; molasses. **-dose** *f* sugar bowl. **-erbse** *f* sweet pea. **-erde** *f* (*Sugar Refining*) animal charcoal, clay. **-ersatz** *m* sugar substitute. **-fabrik** *f* sugar refinery. **-fabrikation** *f* sugar manufacture. **-farbe** *f* confectionery color. **-form** *f* sugar mold. **-gärung** *f* sugar (*or* saccharine) fermentation. **-gast** *m* sugar mite. **-gehalt** *m* sugar content. **-gehaltmesser** *m* saccharimeter. **-geist** *m* rum. **-geschmack** *m* sugary taste. **-gewinnung** *f* sugar extraction (*or* manufacture). **-glasur** *f*, **-guß** *m* icing, frosting; sugar coating. **z-haltig** *a* sugar, containing sugar, saccharated. **Z-harnen** *n* glycosuria. **-harnruhr** *f* diabetes mellitus. **-hirse** *f* sorghum. **-honig** *m* molasses, treacle; crystallized honey. **-hut** *m* sugarloaf

zuckerig *a* sugary, saccharine

Zuckerin *n* saccharin

Zucker-industrie *f* sugar industry. **-inversion** *f* inversion of a sugar (sucrose). **-kalk** *m* calcium sucrate, sugar-lime. **-kand(is)** *m* sugar candy —**brauner Z.** caramel. **-katalyse** *f* catalysis of sugar reaction. **-kessel** *m* sugar kettle (*or* boiler). **-kiefer** *f* sugar pine. **-kohle** *f* charcoal from sugar. **-korn** *n* grain of sugar. **z-krank** *a* diabetic. **Z-krankheit** *f* diabetes. **-küpe** *f* (*Wool*) glucose vat. **-lauge** *f* (*Sugar*) limewater. **-lösung** *f* sugar solution; syrup. **-mais** *m* sweet corn. **-mehl** *n* powdered sugar. **-melasse** *f* (beet) molasses. **-melone** *f* sweet melon. **-messer** *m* saccharimeter. **-meßkunst, -messung** *f* saccharimetry. **-mühle** *f* sugar mill

zuckern *vt* sugar, sweeten; sugarcoat; ice, frost (cakes)

Zucker-palme *f* sugar (*or* gomuti) palm. **-pflanzung, -plantage** *f* sugar plantation. **-probe** *f* **1** sugar sample. **2** sugar test. **-raffinade** *f* refined sugar. **-raffinerie** *f* sugar refinery. **-rohr** *n* sugar cane—**chinesisches Z.** sorghum. **-röhrchen** *n* sugar tube (small funnel for sugar determinations)

Zuckerrohr-faser *f* sugarcane (*or* bagasse) fiber. **-melasse** *f* cane molasses. **-rückstände** *mpl* bagasse. **-saft** *m* cane juice. **-wachs** *n* sugarcane wax

Zucker-rose *f* red rose. **-rübe** *f* sugar beet

Zuckerrüben-essig *m* sugar-beet vinegar. **-melasse** *f* beet molasses. **-saft** *m* sugarbeet juice, beet syrup. **-schnitzel** *n* beet chip, cossette. **-zucker** *m* beet sugar

Zucker-ruhr *f* diabetes mellitus. **-saft** *m* saccharine juice. **-satz** *m* molasses. **z-sauer** *a* saccharate of. **Z-säure** *f* saccharic acid. **-schaum** *m* powdered animal charcoal. **-schlamm** *m* (*Beet Sugar*) beet scum. **-schleuder** *f* sugar centrifuge. **-schotenbaum** *m* honey locust (tree). **-sieden** *n* sugar boiling (*or* refining). **-siederei** *f* sugar refinery. **-sirup** *m* (sugar) syrup, molasses. **-spaltung** *f* sugar cleavage. **-spiegel** *m* blood-sugar level. **-stein** *m* (*Min.*) granular albite. **-strontian** *m* strontium sucrate. **z-süß** *a* sugary, sweet as sugar, saccharine. **Z–umwandlung** *f* sugar conversion (*or* metabolism). **-untersuchung** *f* examination (*or* investigation) of sugar. **-verbindung** *f* sugar compound, saccharate, sucrate. **-waren** *fpl* confectionery, sweets. **-watte** *f* cotton candy. **-werk** *n* confectionery, sweets. **-würfel** *m* sugar cube (*or* lump). **-zange** *f* sugar tongs

zuckrig *a* sugary = ZUCKERIG

Zuckung *f* (-en) twitch(ing); flinch(ing); convulsion; flash(ing), darting; tug(ging) *cf* ZUCKEN

zu-decken *vt* cover (up), put a lid on

zudem *adv* besides, in addition

zu-denken* *vt* (*d.i.o*) intend, earmark (for)

Zudrang *m* influx, inrush; rush, run (on a place)

zu-drehen *vt* 1 turn off, shut off. 2 (*d.i.o*) turn (sthg toward sbdy)

zudringlich *a* obtrusive, pushy

zu-drücken *vt* press shut, close —**ein Auge z.** look the other way

zu-eignen *vt* 1 (*d.i.o*) dedicate (to). **2: sich etwas z.** appropriate (take, usurp) sthg (for oneself)

zueinander *adv & sep pfx* to (toward, at) each other

zuende *adv*: **z. gehen** (come to an) end *cf* ZU IV

zuerkannt awarded: *pp of* ZUERKENNEN

zu-erkennen* *vt* (*d.i.o*) allot, award (to)

zuerst *adv* first; at first

zu-erteilen *vt* (*d.i.o*) allot, award (to)

Zufahrt° *f* approach (*or* access) road, driveway

Zufall° *m* 1 coincidence. 2 chance, accident; luck: **durch Z.** by chance. 3 (random) incident

zu-fallen* *vi*(s) 1 close, fall shut: **selbst z—d** self-closing. 2 (*dat*) come, accrue (to), devolve (on); fall toward

zufällig *a* 1 accidental, chance, random; (*as adv* oft) by chance, by accident, happen to (+*v*): **z. bemerkten wir** by chance we noticed, we happened to notice. 2 (*oft* + *dat*) incidental (to). **zufälligerweise** *adv* by chance = ZUFÄLLIG *as adv.* **Zufälligkeits-fehler** *m* accidental (*or* random) error

Zufalls- chance, random: **-auswahl** *f* random selection. **z—bedingt** *a* accidental,

chance, random. **Z–ergebnis** *n* chance result. **-fehler** *m* random error. **-folge** *f* random series. **-gesetz** *n* law of chance (*or* probability). **-kurve** *f* probability curve. **-moment** *n* chance factor. **-ordnung** *f* random order. **-orientierung** *f* random orientation. **-treffer** *m* lucky (*or* random) hit (shot, strike, draw). **-verteilung** *f* random distribution. **-wert** *m* chance (*or* random) value

zu-fassen *vi* 1 reach out and grab; hold fast. 2 pitch in, get to work

zu-fließen* *vi*(s) (*dat*) 1 flow, run (toward, into). 2 be awarded, accrue, come (easily) (to)

Zuflucht° *f* 1 refuge; shelter; asylum. **2: letzte Z.** last resort. **3: Z. nehmen** (**zu**) resort, have recourse (to). **Zufluchts-ort** *m* (place of) refuge

Zufluß° *m* 1 inflow, influx, supply (of liquid, gas, air). 2 inlet. 3 (*Geog.*) tributary. **--behälter** *m* feed tank (*or* vessel). **-rohr** *n* feed (*or* supply) pipe

zufolge *prep* 1 (*genit*) owing to, as a result of. 2 (*dat*) e.g.: **dem Unglück z.** as a result of the accident; **diesen Erscheinungen z.** on the strength of these phenomena. *Cf.* ZU IV

zufrieden *a* satisfied, content(ed). **zu-frieden-geben*** *vr*: **sich z.** (**mit**) content oneself, be satisfied, make do (with). **Zufriedenheit** *f* satisfaction, contentment. **zufrieden-stellen** *vt* satisfy. **-stellend** *p.a* satisfactory. **Z–stellung** *f* satisfaction

zu-frieren* *vi*(s) freeze up (*or* over)

zu-fügen *vt* (*d.i.o*) 1 add (to). 2 inflict (on)

Zufuhr *f* (-en) 1 supply, delivery. 2 supplies, provisions. 3 importation. 4 influx, intake; (*Mach.*) feed. 5 (*Med.*) administration (of medicines). **zu-führen I** *vt* (*usu* + *d.i.o*) 1 take, bring, lead (to). 2 supply, feed, deliver (to). **II** *vi* 3 (**auf**) lead (to). **Zuführer** *m* (-) 1 supplier. 2 (*Mach.*) feed, conveyor. **Zuführung°** *f* 1 supply; (*Mach.*) feed; conveying. 2 importation. 3 inlet. 4 (*Elec.*) lead(-in) wire

Zuführungs-draht *m* conductor (*or* lead) wire. **-rohr** *n* feed (*or* supply) pipe. **-vor-richtung** *f* feeding device. **-walze** *f* feed roll(er)

Zug *m* (⁻) 1 (railroad) train. 2 pull(ing), tug(ging); (bell) cord; (organ) stop; drawstring; (*Micros.*) draw tube; (*Phys., Engg.*) traction, tension; (*in compds usu*) tractive, tensile. 3 (*Board Games*) move: **er ist am** (*or* **kommt zum**) **Z.** it is his move (*or* his turn). 4 breath, puff; draft, sip–**in vollen Z.⁻en genießen** enjoy fully. 5 draft (of air); draw (of a chimney); suction (of a pump). 6 draw, attention; (*Phys.*) attractive force. 7 flue; groove; *pl oft* rifling (in a gun). 8 movement, motion; migration, journey, trip; direction (of motion); course: **im Z—e** (+ *genit*) in the course (of), within the scope (of); **in einem Z—e** in one motion (*or* effort), without a break; **in Z. bringen** set in motion; **in Z. kommen** start moving: **im** (**besten**) **Z. sein** be (well) on the way, be on the move. 9 tendency, trend: **der Z. der Zeit** the trend of the times. 10 feature (of face, handwriting, etc); trait, characteristic; line, stroke; *pl oft* handwriting: **in großen Z.⁻en** in bold strokes; **Z. um Z.** stroke by stroke, one after another. 11 row (of houses); line, column (of marchers); procession; march. 12 group, platoon, flock (of migrating birds); (*Fish*) school, catch. 13 order: **Z. in etwas bringen** establish order in a matter, get sthg organized; **etwas gut im Z—e haben** have sthg well in hand. 14 (*Dye.*) dip

Zugabe° *f* 1 bonus, extra; supplement, filler: **als Z.** as a bonus (*or* an extra). 2 encore. 3 (*Engg., Mach.*) tolerance, allowance. **⁻material** *n* added material, filler

Zugang° *m* 1 admittance, access: **Z. finden** (**zu**) gain access (to). 2 approach (road, driveway). 3 entrance, gateway. 4 affinity: **sie haben keinen Z. zur Wissenschaft** they have no affinity for science. 5 (**an**) increase, influx (of customers, funds, supplies, etc); accessions; (*Fin.*) receipts. 6 newcomer(s), new customer (patient, member), new arrival, new merchandise

zugänglich *a* 1 accessible, approachable. 2: **einer Sache z. sein** have an affinity for a thing. **Zugänglichkeit** *f* accessibility

Zug·beanspruchung *f* tensile strength. **-dauer** *f* (*Dye.*) duration of the dip
Zug-Dehnungskurve *f* stress-strain curve. **Zug-Druckversuch** *m* tensile-compression test
zu·geben* *vt* 1 add (*oft as* a bonus, an encore), throw in. 2 admit, confess. *Cf.* ZUGEGEBEN
zugebracht delivered, etc: *pp of* ZUBRINGEN
zugebrannt roasted, etc: *pp of* ZUBRENNEN
zugebunden tied up: *pp of* ZUBINDEN
zugedacht intended: *pp of* ZUDENKEN
zugeflossen flowed, etc: *pp of* ZUFLIESSEN
zugefroren frozen up: *pp of* ZUFRIEREN
zugegangen gone on: *pp of* ZUGEHEN
zugegeben 1 added, etc: *pp of* ZUGEBEN. 2 *p. adv* = **zugegebenermaßen** admittedly
zugegen *pred a* present (at a place)
zugegossen poured on, etc: *pp of* ZUGIESSEN
zugegriffen seized, etc: *pp of* ZUGREIFEN
zu·gehen* *vi(s)* 1 (**auf**) go (up to, toward), get close (to), approach. 2: **spitz z.** come to a point. 3 hurry, hustle. 4 (*dat*) be on its way (to). 5: **einem etwas z. lassen** send sthg to sbdy. 6 close, shut. 7 go (on), proceed, happen
zu·gehören *vi* (*dat*) belong (to), fit in (with).
zugehörig *a* (*dat*) 1 appropriate (to), fitting. 2 associated, affiliated (with); accompanying. **Zugehörigkeit** *f* 1 appropriateness. 2 affiliation, association
Zügel *m* (-) rein: **Z. anlegen** (*dat*) tighten the reins (on), rein in. **zügellos** *a* unbridled, uncontrolled. **zügeln** 1 *vt* control, restrain. 2: **sich z.** control (*or* restrain) oneself
zugenommen increased, etc: *pp of* ZUNEHMEN
zugesandt sent: *pp of* ZUSENDEN
zugeschliffen ground: *pp of* ZUSCHLEIFEN
zugeschmolzen heat-sealed: *pp of* ZUSCHMELZEN
zugeschnitten cut to fit: *pp of* ZUSCHNEIDEN
zugeschoben pushed shut: *pp of* ZUSCHIEBEN
zugeschossen shot: *pp of* ZUSCHIESSEN
zugeschrieben ascribed: *pp of* ZUSCHREIBEN
zu·gesellen *vr:* **sich z.** (*dat*) join

zugespitzt *p.a* 1 pointy, sharp. 2 (**auf**) aimed (at) *cf* ZUSPITZEN

zugesprochen attributed: *pp of* ZU-SPRECHEN

zugestanden 1 admitted: *pp of* ZU-GESTEHEN. 2 entitled: *pp of* ZUSTEHEN.

zugestandener·maßen *adv* admittedly

Zugeständnis *n* (-se) admission; acknowledgement, concession

zu·gestehen* *vt* (*d.i.o*) admit; grant, concede (to): **man muß ihnen z., daß** . . . one must admit that they . . .

zugetan 1 *p.a* (*dat*) attached, devoted (to), fond (of). 2 closed, etc: *pp of* ZUTUN

zugetroffen been true, etc: *pp of* ZU-TREFFEN

zugewandt turned toward: *pp of* ZU-WENDEN

zugezogen contracted, etc: *pp of* ZUZIEHEN

Zug·feder *f* tension spring. **-festigkeit** *f* tensile strenth. **-fisch** *m* migratory fish. **z–frei** *a* 1 draft-free, wind-sheltered. 2 tension-free

zu·gießen* *vt* 1 add, pour on, pour in. 2 seal (*or* close up) by casting (*or* pouring in a liquid)

zugig *a* drafty, windy

zügig *a* 1 swift, brisk. 2 smooth, steady, unbroken. 3 ductile, tensile. 4 (*Lthr.*) flexible, pliant

Zugigkeit *f* draftiness

Zügigkeit *f* 1 swiftness, briskness. 2 smoothness, smooth (*or* steady) flow. 3 ductility. 4 flexibility, pliability; elasticity

zu·gipsen *vt* plaster up (a hole)

Zug·kanal *m* flue. **-klappe** *f* flue damper. **-kraft** *f* 1 tractive (*or* tensile) force. 2 drawing power, appeal, attractiveness. **z–kräftig** *a* attractive, appealing, popular. **Z–kurve** *f* stress-strain curve

zugl. *abbv* (zugleich) sim. (simultaneously)

zugleich *adv* at the same time, simultaneously; also **—z. billig und praktisch** both cheap and practical

Zug·leistung *f* tractive output (*or* power). **-linie** *f* (*Geom.*) tractrix, trajectory. **-loch** *n* air hole, air vent. **-luft** *f* draft(s), air current. **-maschine** *f* 1 tractor. 2 prime mover. **-messer** 1 *m* tensometer; draft gauge. 2 *n* draw knife. **-mittel** *n* 1 tension

medium. 2 tractor. 3 (*Med.*) vesicant. 4 drawing card, attraction. **-modul** *n* (Young's) modulus of extension. **-muffel** *f* (*Ceram.*) continuous muffle. **-netz** *n* dragnet. **-ofen** *m* blast (*or* draft) furnace. **-pflaster** *n* blistering plaster, vesicatory. **-probe** *f* tensile test

zu·greifen* *vi* 1 reach out and grasp, help onself, jump at the opportunity. 2 take action, intervene. 3 pitch in, lend a hand

Zugriff° *m* 1 grip, grasp. 2 action, intervention. 3 (*Cmptrs*) access. **Zugriffs·zeit** *f* (*Cmptrs.*) access time

Zug·rohr *n*, **-röhre** *f* air pipe (*or* vent)

zugrunde *adv* 1: **z. gehen** (*oft* + **an**) be destroyed (*or* ruined) (by), die (of). 2: **z. legen** (*d.i.o*) use as a basis (for). 3: **z. liegen** (*dat*) serve as the basis (for). 4: **z. richten** ruin, destroy. 5: **z. wirtschaften** ruin (by mismanagement). *Cf* ZU **IV**

Zugrunde·gehen *n* ruin, destruction. **-legung** *f*: **unter Z.** (*genit*) using . . . as a basis. **z—liegend** *p.a* underlying, basic

Zug·salbe *f* (*Pharm.*) resin cerate (vesicant). **-spannung** *f* tensile stress. **-standfestigkeit** *f* resistance to sustained stretch. **-stange** *f* tie bar, tow bar, drawbar. **-tier** *n* draft animal

zugunsten *prep* (*genit;* + *dat when it follows the noun*) in favor of, for the sake (*or* benefit) (of) *cf* ZU **IV**

Zuguß° *m* 1 infusion. 2 addition of liquid; added liquid

zugute *adv* 1: **z. kommen** (*dat*) work to the advantage (of), benefit; **z. kommen lassen** (*d.i.o*) give the benefit (of). 2: **z. halten** (*d.i.o*) make allowances for (sbdy's) . . . , give credit for (doing). 3: **sich z. halten** (*or* **tun**) (**auf**) pride oneself (on). 4: **z. machen** (*Ores*) work (up)

zuguterletzt *adv* last of all, finally *cf* ZU **IV**

Zug·verformung *f* tensile deformation. **-versuch** *m* tensile test. **-vogel** *m* migratory bird. **-wagen** *m* tractor. **-winde** *f* pulley. **-wirkung** *f* pulling effect, pull, tension

zu·haben* 1 *vt* have (doors, windows) closed. 2 *vi* (*Store, etc*) be closed

zu·halten* 1 *vt* hold (*or* keep) shut. 2 *vi* (**auf**) head, make (for). **Zuhaltung** *f* (-en) (*Locks*) tumbler

zuhanden 1 *prep* (*genit*) (*Letters*) (for the) attention (of). **2** *adv* available, on hand — **einem z. kommen** come into one's hands. *Cf* ZU IV

Zuhause *n* home *cf* HAUS

zu·heilen *vi*(s) heal up

Zuhilfenahme: unter Z. (*genit*) with the aid (of)

zuhinterst *adv* at the very back, last of all

zuhöchst *adv* **1** at the very top. **2** extremely, highly

zu·hören *vi* (*dat*) listen (to). **Zuhörer°** *m*, **Zuhörerin** *f* listener, *pl oft* audience. **Zuhörer·raum** *m* auditorium. **Zuhörerschaft** *f* (-en) (listening) audience

zuinnerst *adv* deep inside; most deeply

zu·kitten *vt* cement up, cement shut

zu·klappen *vt* shut, close, snap (*or* clap) shut

zu·kleben, zu·kleistern *vt* paste up (*or* together), seal (*or* close up) with paste

zu·kommen* *vi*(s) **1** (**auf**) come (toward), approach; come (upon), confront. **2** (*dat*) be coming (to), be in store (for); befit: **es kommt ihnen zu** it befits them, they deserve it, they have it coming to them. **3** (*dat*) come (to, into sbdy's hands): **es kam ihnen zu, daß** . . . they got word that . . . **4: einem etwas z. lassen** send sthg (*or* have sthg sent) to sbdy, bestow sthg on sbdy

zu·korken *vt* cork up, stopper up

Zukunft *f* future. **zukünftig** *a* future, coming; next. **Zukünftige** *m*, *f* (*adj endgs*) fiancé(e), husband- (wife)-to-be. **zukunft·sicher** *a* guaranteed (for the future)

Zukunfts·prognose *f* prediction for the future. **z–trächtig** *a* promising (for the future). **zukunft(s)·weisend** *a* future-oriented, forward-looking, momentous, promising

Zulage° *f* **1** supplement; bonus. **2** additional pay, raise. **3** allowance

zu·langen *vi* **1** help oneself. **2** be sufficient, go far enough. **zulänglich** *a* sufficient, adequate, satisfactory. **Zulänglichkeit** *f* sufficiency, adequacy

Zulaß *m* (. . lässe) admission, access; (*Mach.*) tolerance. **zu·lassen*** *vt* **1** (**zu**) admit, give access (to), let in, let (at); turn

on (gas, steam, etc). **2** allow, permit, tolerate; license, authorize. **3** leave closed. **zulässig** *a* admissible, permissible. **Zulässigkeit** *f* admissibility, permissibility. **Zulassung** *f* admittance, admission, access; license, registration. **Zulassungs·papiere** *npl*, **-schein** *m* license, registration (card), permit

Zulauf *m* **1** supply, intake, influx, feed. **2** addition, admixture. **3: Z. haben** be popular, have (*or* attract) a clientele (an audience, followers, adherents). **zu·laufen*** *vi*(s) **1** (**auf**) run (toward, up to), lead (to) **—spitz z.** come to a point. (*dat*) flock (to, to join). **3** (*dat*) stray in (to). **4** (*Liquids*) run into: **Wasser z. lassen** run in (let in, add) some water

Zulauf·flotte *f* feeding (*or* replenishing) liquor. **-gefäß** *n* receptacle, receiver. **-temperatur** *f* inlet temperature

zu·legen I *vt* **1** add, put on (more), pay an additional . . . **2** (**zu**) contribute (to). **3** (*dat rflx*): **sich etwas z.** acquire (procure, get onself) sthg. **4** cover up (*or* over). **II** *vt/i* lose (money on a deal). **III** *vi* step on it, rev it up

zu·legieren *vt* (*d.i.o*) e.g.: **dem Kupfer etwas Zink z.** alloy some zinc with the copper

zu·leimen *vt* glue (cement, paste) up (*or* shut)

zu·leiten *vt* (*d.i.o*) supply, pass on (to), lead, conduct (to, into); let in (*usu* liquid). **Zuleitung°** *f* **1** supplying, conduction, leading in; letting in, admission. **2** supply pipe (tube, channel, duct), feed (*or* inlet) pipe; (*Elec.*) lead wire, power supply wiring

Zuleitungs·rohr *n*, **-röhre** *f* supply (feed, inlet) pipe (*or* tube). **-schnur** *f* (*Elec.*) extension (*or* plug-in) cord

zu·lernen *vt* learn more of

zuletzt *adv* (at) last, finally **—bis z.** up to the end

Zuliefer·betrieb *m* supplier, subcontractor. **-industrie** *f* supply (*or* subcontracting) industry

zu·liefern *vt* (*d.i.o*) supply (to). **Zuliefer·teil** *m* part supplied by a subcontractor. **Zulieferung°** *f* supply(ing)

zu·löten *vt* solder up, seal by soldering

zum *contrac* = **zu dem** *cf* ZU **I**
zu·machen 1 *vt* close, shut; button up. 2 *vi*
hurry up
zumal 1 *adv* especially. 2 *cnjc* especially
since
zumeist *adv* for the most part; generally
zu·messen* *vt* (*d.i.o*) measure out, allot
(to); attribute, ascribe (to)
zumindest *adv* at least
zu·mischen *vt* (*d.i.o*) add (to), mix in, admix
Zumisch·pulver *n* additive (*or* admixed)
powder; (*Dynamite*) dope. **-stoff** *m* additive, admixture
Zumischung° *f* admixture
zumutbar *a* reasonable—**das ist ihnen
nicht z.** that can not be expected of them
zumute *pred a* (+*dat*, +*adv*) e.g.: **ihnen ist
unbehaglich z.** they feel uncomfortable
cf ZU **IV**
zu·muten *vt* (*d.i.o*): **einem etwas z.** attribute sthg to sbdy; expect sthg of sbdy.
Zumutung *f* (-en) sthg expected (of sbdy);
imposition, unreasonable demand (*or*
thing to expect)
zunächst I *adv* 1 (at) first. 2 for the present. **II** *prep* (*dat*) nearest, next, closest
(to)
zu·nageln *vt* nail up, nail shut
zu·nähen *vt* sew up; suture, close up
Zunahme *f* (-n) 1 increase, growth, rise. 2
weight gain
Zuname° *m* 1 family name. 2 nickname
Zünd· ignition, lighting, priming:**-akt** *m* 1
ignition reaction. 2 (*Kinetics*) chain-
initiating reaction. **-apparat** *m* 1 ignition apparatus, igniter, magneto. 2 priming apparatus, primer
zündbar *a* (in)flammable, ignitable.
Zündbarkeit *f* (in)flammability, ignitability
Zünd·blättchen *n* (toy pistol) cap. **-draht**
m ignition wire. **-einrichtung** *f* ignition
device. **-elektrode** *f* ignition electrode,
igniter. **-empfindlichkeit** *f* ignition
sensibility
zünden 1 *vt* set fire to, kindle. 2 *vt/i* ignite.
3 *vi* catch fire; inspire enthusiasm. **zündend** *p.a* (*oft*) 1 incendiary. 2 inspiring,
rousing
Zunder *m* 1 tinder. 2 (*Metll.*) flaky oxide
layer, forge scale

Zünder *m* (-) lighter, igniter; detonator;
fuse
zunder·beständig *a* (*Metll.*) non-scaling.
Z—beständigkeit *f* non-scaling property.
-bildung *f* (forge) scale formation. **-erz** *n*
(*Min.*) impure jamesonite (*or* stibnite). **z—
fest** *a* 1 tinderproof. 2 (*Metll.*) non-
scaling. **-frei** *a* free from (forge) scale. **Z—
freiglühen** *n* scale-free annealing
Zünder·füllmasse *f* fuse composition
Zunder·papier *n* touch paper
Zünder·satz *m* fuse composition
Zunderung *f* (*Metll.*) scaling
zünd· ignition, lighting, priming: **-fähig** *a*
(in)flammable, ignitable. **Z—fähigkeit** *f*
(in)flammability, ignitability. **-flämm-
chen** *n*, **-flamme** *f* igniting flame, pilot
flame. **z—freudig** *a* easily (*or* highly)
flammable. **Z—freudigkeit** *f* (high) flammability. **-funke(n)** *m* ignition spark.
-gas *n* fuse gas. **-gerät** *n* ignition device;
lighter; primer. **-geschwindigkeit** *f* ignition velocity. **-holz**, **-hölzchen** *n* match
Zündholz· match: **-draht** *m* match splint.
-fabrik *f* match factory. **-masse** *f* match
composition. **-papier** *n* match paper.
-satz *m* match composition
Zünd· ignition, lighting, priming:
-hütchen *n* percussion (detonating,
blasting) cap, detonator. **-hütchensatz** *m*
priming composition. **-kabel** *n* 1 (*Cars*)
ignition cable. 2 (*Mil.*) firing wire. **-kam-
mer** *f* ignition chamber. **-kapsel** *f* percussion cap, detonator = **-HÜTCHEN**. **-kerze** *f*
spark plug. **-kerzenspannung** *f* spark
plug voltage. **-kirsche** *f* ignition pellet:
Goldschmidtsche Z. pellet of aluminum powder and barium peroxide.
-ladung *f* priming charge. **-lücke** *f* non-
ignition zone. **-magnet** *m* magneto.
-maschine *f* (*Explo.*) blasting (*or* priming) machine; exploder. **-masse** *f* igniting
composition, ignition mixture. **-metall** *n*
(in)flammable metal; flint. **-mittel** *n* igniting (primary, detonating) agent,
primer. **-papier** *n* ignition (*or* touch) paper. **-patrone** *f* ignition (*or* priming)
cartridge. **-pille** *f* pellet primer, priming
drop. **-pulver** *n* priming powder. **-punkt**
m ignition point (*or* temperature). **-satz**
m priming composition, detonator.

-schloß *n* ignition lock. **-schlüssel** *m* ignition key. **-schnur** *f* (*Explo.*) fuse, primacord. **-schwamm** *m* tinder, touchwood, punk. **-spannung** *f* 1 (*Cars*) ignition voltage. 2 (*Explo.*) firing voltage. 3 (*Elec. Engg.*) striking potential. 4 (*Glow Tubes*) glow potential. **-spule** *f* ignition coil. **-stelle** *f* place of ignition. **-stoff** *m* 1 (in)flammable material. 2 igniting agent, primer

Zündung *f* (-en) ignition (*incl Cars*); primer; fuse

Zündungs·schlüssel *m* ignition key. **-temperatur** *f* ignition temperature

Zünd·vorrichtung *f* ignition device, igniter. **-waren** *fpl* (in)flammable goods. **-wärme** *f* heat of ignition. **-willigkeit** *f* 1 flammability. 2 (*Fuel*) ignition quality; cetane number

zu·nehmen* 1 *vi* (an) grow, increase (in); wax, get stronger; (*esp Days*) get longer. 2 *vt/i:* (**Gewicht**) z. gain (*or* put on) weight. 3 *vt* (+ *Quant.*) increase by . . .

zu·neigen 1 *vt & sich z.* (*dat*) incline, bend (to, toward): **sich dem Ende z.** approach its end, draw to a close. **2: sich z.** (*dat*) grow fond (of). 3 *vi & sich z.* (*dat*) be inclined, tend (toward). **Zuneigung°** *f* 1 inclination. 2 affection, attachment—**Z. fassen** (**zu**) grow fond (of), take a liking (to)

Zunft *f* (-̈) 1 clique, group. 2 (*Hist.*) guild, corporation. **zünftig** *a* 1 trained, professional. 2 fine, great; sound

Zunge *f* (-n) 1 (*Anat., Language*) tongue. 2 pointer, indicator, needle. 3 blade, reed. 4 (railroad) switch point. 5 (*Geog.*) spit of land

züngeln *vi* (*Flames*) shoot, dart, lick (at). **züngelnd** *p.a* (*oft*) lambent

Zungen· lingual, glosso-, hyoid: **-bein** *n* hyoid bone. **z—förmig** *a* tongue-shaped. **Z—spitze** *f* tip of the tongue. **-zäpfchen** *n* epiglottis

zunichte *pred a* ruined, destroyed: **z. machen** ruin, wreck, destroy; **z. werden** go to ruin, be destroyed, come to nothing

zunutze *adv:* **sich etwas z. machen** take advantage of sthg *cf* ZU IV

zuoberst *adv* on top, at the top, uppermost

zu·ordnen *vt* (*d.i.o*) assign (to), (*Cmptrs.*)

allocate (to). **Zuordnung°** *f* assignment, allocation, correspondence

zupaß *adv* (*dat*): **es kommt ihnen z.** it comes at just the right moment for them

zupfen 1 *vt* pull up (weeds). 2 *vt/i* (**an**) pluck, pick (at), tease. 3 *vi* (**an**) tug (at)

Zupf·leinwand *f* lint. **-nadel** *f* dissecting needle. **-präparat** *n* teased-out preparation

zu·propfen *vt* cork (up), stopper (up)

Zupf·seide *f* silk ravelings. **-wolle** *f* picked wool, wool pickings

zu·pichen *vt* pitch (up), cover (*or* fill) with pitch

zu·pressen *vt* press shut, constrict

zur *contrac* = **zu der** *cf* ZU I

zu·raten* *vi:* **einem z. zu . . .** advise sbdy to . . .

Zürbel·kiefer *f* (*Bot.*) Swiss pine

zu·rechnen *vt* (*d.i.o*) 1 count, number (among), consider (part of); ascribe, attribute (to). 2 add on (to), include (in). **Zurechnung** *f:* **unter Z. aller Kosten** all expenses included. **zurechnungs·fähig** *a* accountable, responsible

zurecht· *sep pfx* 1 (*implies making sthg for a special use*). 2 (*implies straightening, rectifying*): **Z—bestehen** *n* validity. **z—bringen*** *vt* set (*or* put) right, arrange (properly). **-finden** *vr:* **sich z.** find one's way around, get along. **-gebracht** *pp* of ZURECHTBRINGEN. **-gefunden** *pp* of ZURECHTFINDEN. **-geschnitten** *pp* of ZURECHTSCHNEIDEN. **-gewiesen** *pp* of ZURECHTWEISEN. **-hämmern** *vt* hammer into shape. **-kommen*** *vi*(s) 1 come out (all) right. 2 get along, manage (*esp Fin.*); (+ **mit**) get along (with). 3 arrive on time. **-legen** *vt* (*oft* + *d.i.o* or *dat rflx* **sich**) 1 lay out, get (sthg) ready (for sbdy, oneself). 2 plan, prepare (for). **3: sich etwas z.** (*oft*) explain (*or* account for) sthg. **-machen** I *vt* (*d.i.o*) 1 think up, devise. 2 doctor, manipulate. II *vt & sich z.* 1 get ready, prepare. 3 tidy up. **-rücken** *vt* straighten out. **-schneiden*** *vt* cut to fit (to size, to pattern), trim. **-setzen** 1 *vt* straighten out. **2: sich z.** settle (oneself, properly, on a seat). **-stellen** I *vt* 1 straighten out, set straight. 2 prepare, fix. 3 set out (ready for use). II: **sich z.** straighten up, set oneself straight.

·**weisen*** vt 1 censure, reprimand; reproach, rebuke. 2 show the way, give directions to. **Z—weisung°** f censure, reprimand, rebuke. ·**zimmern** vt (usu + dat rflx) rig up (for oneself), slap (hammer, patch) together

zu·reden vi (dat) persuade, coax, encourage. **Zureden** n persuasion, encouragement

zu·reichen 1 vi do, suffice, be enough, go far enough. 2 vt (d.i.o) hand (up to). **zureichend** p.a sufficient, adequate, satisfactory

Zuricht·deckfarbe f (Lthr.) coating color **zu·richten** vt 1 cut, trim (to shape); prepare, fix; (Lthr., Metll.) dress; (Tex.) finish, size; (Baking) leaven; (Air) condition. 2 patch together, rig up. 3 (usu ironic): **schön z.** make a fine mess of. 4: **übel z.** maul, beat up

Zuricht·masse f (Tex.) sizing material. -**mittel** n (Lthr.) dressing agent

Zurichtung° f cutting, trimming (to shape); preparation; dressing; finishing, sizing; leavening cf ZURICHTEN

Zurschau·stellung f display, exhibition **zurück** 1 adv back; backward; behind, in arrears; behind the times. 2 sep pfx back, behind, re-

zurück·befördern vt 1 move (carry, take) back; return. 2 demote

zurück·begeben* vr: **sich z.** go back, return

zurück·behalten* vt (oft + dat rflx) hold back, keep, retain, withhold (for oneself)

zurück·bekommen* vt get back, recover

zurück·bilden vr: **sich z.** regain its original shape, recover, regress; (Med.) remit, go into remission

zurück·bleiben* vi(s) 1 stay behind, lag behind; be retarded. 2 remain, be left; survive. 3 (hinter) fall short (of), be inferior (to). **zurückbleibend** p.a remaining, left (over), residual

zurück·blicken vi look back. **zurück·blickend** p. adv (oft) in retrospect

zurück·bringen* vt (d.i.o) bring back, take back (to), restore; (Math.) reduce

zurück·datieren 1 vt mark with an earlier date; determine the date (of a past event, etc). 2 vi (bis) date back (to)

zurück·drängen vt push (drive, force) back, repel; repress, suppress; restrain

zurück·drehen vt turn back

zu·rücken vt, vi(s) (auf) move, push (toward)

zurück·erhalten* vt get back, recover

zurück·fahren* I vt 1 take back (in a vehicle), drive back. II vi(s) 2 go (ride, drive, move, travel) back. 3 shrink back, recoil, flinch

zurück·fallen* vi(s) 1 fall (or drop) back; relapse; revert. 2 (Rays) be reflected. 3 (auf) reflect (unfavorably) (on)

zurück·federn vi(s) spring back, recoil

zurück·finden vi & sich z. find one's way back

zurück·fließen* vi(s) flow back, reflux

zurück·fluten vi(s) stream back; go (or come) flooding back; ebb, recede

zurück·fordern vt demand back, reclaim

zurückführbar a (auf) traceable, reducible (to). **zurück·führen** I vt 1 take (or lead) back. 2 (auf) trace back; attribute, ascribe (to). 3 (auf) reduce (to). II vi lead (or go) back

zurück·geben* vt (d.i.o) give back, return (to)

zurück/geblieben 1 p.a retarded, underdeveloped. 2 remained, etc: pp of ZURÜCKBLEIBEN. -**gebracht** brought back: pp of ZURÜCKBRINGEN. -**gefunden** found one's way back: pp of ZURÜCKFINDEN. -**gegangen** gone back: pp of ZURÜCKGEHEN. -**gegriffen** reached back: pp of ZURÜCKGREIFEN

zurück·gehen* vi(s) 1 go back; retreat. 2 recede, ebb. 3 diminish, drop (off), decrease, decline

zurück·gelangen vi(s) get back (to a place), arrive back, return

zurück/gelegen been behind: pp of ZURÜCKLIEGEN. -**genommen** taken back: pp of ZURÜCKNEHMEN. -**geschnitten** cut back: pp of ZURÜCKSCHNEIDEN. -**gesogen** sucked back: pp of ZURÜCKSAUGEN. -**gesprungen** sprung back: pp of ZURÜCKSPRINGEN. -**gestanden** stood back: pp of ZURÜCKSTEHEN. -**gestiegen** climbed back: pp of ZURÜCKSTEIGEN. -**gesunken** sunk(en) back: pp of ZURÜCKSINKEN. -**getan** put back: pp of

ZURÜCKTUN. **-getrieben** repelled: *pp of* ZURÜCKTREIBEN. **-gewandt** turned back: *pp of* ZURÜCKWENDEN. **-gewichen** receded: *pp of* ZURÜCKWEICHEN. **-gewiesen** rejected: *pp of* ZURÜCKWEISEN

zurückgewinnbar *a* recoverable. **zurück·gewinnen*** *vt* win back, regain, recover. **Zurückgewinnung°** *f* recovery, regaining

zurück/gewonnen regained: *pp of* ZURÜCKGEWINNEN. **-geworfen** rejected: *pp of* ZURÜCKWERFEN. **-gezogen** 1 *p.a* secluded, solitary. 2 withdrawn, etc: *pp of* ZURÜCKZIEHEN. **Z–gezogenheit** *f* seclusion, solitude

zurück·greifen* *vi* (**auf**) reach (*or* go) back (to), fall back (on)

zurück·halten* I *vt* 1 hold back, keep, detain. 2 repress, suppress. 3 withhold. 4 (*oft* + **von**) keep, restrain (from doing); stop. II *vi* 5 be reserved. 6 (**mit**) hold back, withhold. III: **sich z.** 7 control (*or* restrain) oneself. 8 (**von**) refrain (from doing). 9 be reserved. **zurückhaltend** *p.a* reserved, shy, restrained; withdrawn; (**mit**) sparing (with). **Zurückhaltung** *f* reserve, shyness; restraint

zurück·kehren *vi*(s) return, come (*or* go) back; revert

zurück·kommen* *vi*(s) 1 (**auf**) come back, return (to). 2 (**mit**) fall behind (with work, etc)

Zurückkunft *f* (⁻) return, coming back

zurück·lassen* *vt* leave behind, abandon. **Zurücklassung** *f* (-en) leaving, abandonment: **unter Z. von . . .** leaving behind . . .

zurück·laufen* *vi*(s) 1 run back; walk back. 2 flow back. 3 recede, ebb

zurück·legen *vt* 1 put (*or* lay) back. 2 bend back. 3 (*oft* + *dat* or + **für**) put (*or* lay) by, put aside (for). 4 cover, travel (a distance)

zurück·liegen* *vi* (h *or* s) 1: **das liegt weit zurück** that is a long way back, that was a long time ago. 2 be (a distance) behind

Zurücknahme *f* (-n) taking back, withdrawal; retraction

zurück·nehmen* *vt* 1 take back. 2 withdraw, retract. 3 cancel. 4 mark down (in price)

Zurückprall *m* rebound, recoil. **zurück-**

·prallen *vi*(s) 1 bounce back, rebound. 2 recoil, shrink back. 3 be reflected, reverberate

zurück·rufen* *vt/i* call back, recall

zurück·saugen* *vt* suck back, draw back (by suction)

zurück·schalten *vt/i* shift (gears) back (*or* down)

zurück·schlagen* I *vt* 1 hit (*or* strike) back. 2 beat back, repel. 3 fold back, move back, open. II *vi* 4 backfire; (*Burners*) strike (*or* flash) back

zurück·schneiden* *vt* cut back, prune, lop (off)

zurück·schnellen *vi*(s) snap (*or* spring) back, recoil

zurück·setzen I *vt* 1 set (put, move) back; demote; defer. 2: **im Preis z.** reduce in price. 3 slight, overlook, discriminate against. II *vt/i* reverse, shift into reverse. III: **sich z.** sit back down; sit down further back

zurück·sinken* *vi*(s) sink (fall, drop) back

zurück·spiegeln 1 *vt* reflect. 2: **sich z.** be reflected

zurück·springen* *vi*(s) spring (jump, leap) back, rebound, recoil; (*Angles*) re-enter

zurück·spulen *vt* rewind

zurück·spülen *vt* rinse (*or* wash) back

zurück·stehen* *vi* (h *or* s) 1 stand back, be set (*or* located) further back. 2 (**hinter**) be left (behind), be inferior (to). 3 be left out, miss out. **zurückstehend** *p.a* (*oft*) inferior

zurück·steigen* *vi*(s) rise (step, climb) back (up)

zurück·stellen *vt* 1 put (set, move) back. 2 (*d.i.o*) put (*or* lay) aside, reserve (for). 3 (*d.i.o*) give (*or* send) back, remit (to). 4 postpone, defer. **Zurückstellung°** *f* putting (setting, moving, giving, sending) back; putting aside, reserving; return; postponement, deferment

zurück·stoßen* *vt* push back; repel, repulse. **Zurückstoßung** *f* (-en) repulsion, pushing back. **Zurückstoßungs·kraft** *f* repulsive force

zurück·strahlen 1 *vt* reflect. 2 *vi* be reflected

zurück·stufen *vt* demote, downgrade

zurück·titrieren *vt/i* back-titrate. **Zu-**

rücktitrierung° *f* return (*or* back) titration

zurück·treiben* *vt* drive back, repel, repulse

zurück·treten* *vi*(s) 1 step back, recede. 2 retire, withdraw; resign. 3 (**hinter**) yield (to), be inferior (to)

zurück·tun* *vt* put back

zurück·übersetzen *vt* retranslate

zurück·verfolgen *vt* trace back, retrace

zurück·versetzen *vt* put back, retransfer; demote

zurück·verwandeln 1 *vt* retransform, reconvert. 2 *vt* & **sich z.** change back. 3: **sich z.** be reconverted

zurück·weichen* *vi*(s) 1 recede. 2 (**vor**) yield (to); retreat (before)

zurück·weisen* *vt* reject, refuse admission to. **Zurückweisung** *f* (-en) rejection, refusal of admission

zurück·werfen* *vt* throw back; repel; reflect

zurück·wirken *vi* (**auf**) react, retroact (on)

zurück·zahlen *vt/i* pay back, repay; refund. **Zurückzahlung°** *f* repayment; refund

zurück·ziehen* 1 *vt* & **sich z.** pull (*or* draw) back, withdraw. 2 *vt* retract. 3 *vi*(s) move (*or* migrate) back. 4: **sich z.** retire; retreat; (+ **von**, *oft*) avoid

Zuruf° *m* shout; (*specif*) cheer, catcall

zu·runden *vt* round off (to fit)

zu·rüsten *vt* prepare, equip

Zurverfügungstellung *f* making available *cf* VERFÜGUNG

zurzeit *adv* at present; for the present

zus. *abbv* (zusammen) together

Zus. *abbv* 1 (Zusammensetzung) comp. (compound). 2 (Zusatz) addition

Zusage° *f* 1 consent, agreement. 2 promise, pledge. 3 acceptance. **zu·sagen I** *vt* 1 (*d.i.o*) promise (sbdy sthg). **II** *vi* 2 accept (an invitation), give an affirmative answer. 3: **es sagt einem zu** it suits (pleases, appeals to, agrees with) one. **zusagend** *p.a* affirmative; agreeable, suitable, pleasing

zusammen I *adv* & *sep pfx* together; at the same time. **II** *sep pfx oft* co-, com-, con-; *can also imply:* 1 (*with vt*) *collecting to form a whole, cf* ZUSAMMENBRINGEN. 2

(*with vt*) *amateurish or improvised construction, cf* ZUSAMMENBASTELN. 3 (*with vt*) *destruction* & (*with vi*) *collapse or compression to a smaller volume, cf* ZUSAMMENBRECHEN, ZUSSAMMENFAHREN, ZUSAMMENFASSEN. 4 (*with vi*) *recoiling cf* ZUSAMMENFAHREN

Zusammen·arbeit *f* cooperation, collaboration. **z–·arbeiten** *vi* cooperate, collaborate

zusammen·backen* *vi* (h *or* s) stick together, cake, agglutinate

zusammen·ballen 1 *vt* roll into a ball; amass. 2: **sich z.** accumulate, agglomerate. **Zusammenballung** *f* amassing, accumulation, agglomeration, concentration

zusammen·basteln *vt* rig up, throw together

Zusammen·bau *m* (-e) assembly (from parts). **z–·bauen** *vt* assemble; put (*or* piece together, rig up

zusammen·binden* *vt* tie (*or* bind) together. **·brechen*** *vi*(s) break down, collapse; (*Traffic*) come to a halt. **·brennen*** *vt*, *vi*(s) burn down (to a heap of ashes). **·bringen*** *vt* 1 bring (get, gather) together, amass, accumulate. 2 put together, assemble. 3 (**mit**) introduce (to); reconcile (with)

Zusammen·bruch° *m* collapse, breakdown

zusammen·drängen 1 *vt* compress, condense, concentrate. 2 *vt* & **sich z.** crowd together. 3: **sich z.** be compressed (condensed, concentrated). *Cf* ZUSAMMENGEDRÄNGT

zusammen·drehen 1 *vt* & **sich z.** twist (*or* twine) together. 2: **sich z.** become entwined (*or* entangled). **Z–·drehung°** *f* twisting, torsion; entwining, entanglement

zusammen·drückbar *a* compressible. **Z–·drückbarkeit** *f* compressibility; (*Plastics*) compression set

zusammen·drücken *vt* compress, compact, crush. **Z–·drücker** *m* (-) compressor. **·drückung** *f* compression. **·drückungsmesser** *m* compressometer

zusammen·fahren* I *vt* 1 wreck, smash up (a car). **II** *vi*(s) 2 (*oft* + **mit**) collide (with). 3 shrink back, recoil, flinch

Zusammen·fall m coincidence, coalescence; conflict, clash. **z--·fallen*** vi(s) **1** (oft + mit) coincide, coalesce, conflict, clash (with); converge (with). **2** collapse, cave in. **3** weaken, deteriorate

zusammen·falten vt fold up

zusammen·fassen vt **1** summarize, express concisely, condense. **2** combine, put (or bring) together; assemble, gather, collect; compile. **3** (**unter**) classify (under). **-fassend** p.a synoptic; (as adv) to sum up, by way of a summary. **Z--fassung°** f **1** summary, synopsis, condensed version. **2** combination; assembly, assemblage; gathering, collection; compilation

zusammen·flicken vt patch up (or together)

zusammen·fließen* vi(s) flow (or run) together, merge. **-fließend** p.a (oft) confluent. **Z--fluß** m confluence, merging

zusammen·frieren* vi(s) **1** freeze together (or solid). **2** freeze and shrink up. **·fügen** vt & sich z. join, unite, combine, assemble. **·geben*** vt put together, join

zusammen/gebracht amassed: pp of ZUSAMMENBRINGEN. **-gebunden** tied together: pp of ZUSAMMENBINDEN. **-gedrängt** p. adv in compressed (or condensed) form. See also ZUSAMMENDRÄNGEN. **-geflossen** merged: pp of ZUSAMMENFLIESSEN. **-gefroren** frozen solid: pp of ZUSAMMENFRIEREN. **-gegossen** mixed: pp of ZUSAMMENGIESSEN

zusammen·gehen* vi(s) **1** go (or fit) together; match. **2** meet, converge. **3** shrink

zusammen·gehören vi belong together, make a good match. **-gehörig** a matching, homogeneous; homologous; correlated: **z. sein** belong together, match, correlate, correspond. **Z-gehörigkeit** f homogeneity; homologousness; correlation

zusammen/genommen 1 p. adv: **alles z.** all in all, all things considered. **2** gathered, etc: pp of ZUSAMMENNEHMEN. **-gerieben** rubbed together: pp of ZUSAMMENREIBEN. **-geschliffen** ground together: pp of ZUSAMMENSCHLEIFEN. **-geschlossen** consolidated: pp of ZUSAMMENSCHLIESSEN. **-geschmolzen** fused: pp

of ZUSAMMENSCHMELZEN. **-geschoben** telescoped: pp of ZUSAMMENSCHIEBEN. **-geschrieben** solid-spelled, written as one word: pp of ZUSAMMENSCHREIBEN. **-gesunken** caved in: pp of ZUSAMMENSINKEN. **-getan** put together: pp of ZUSAMMENTUN. **-getroffen** met: pp of ZUSAMMENTREFFEN. **-gewürfelt** p.a **1** thrown together, (hastily) assembled. **2** varied; oddly assorted. **-gezogen** contracted: pp of ZUSAMMENZIEHEN

zusammen·gießen* vt mix (liquids), pour together

Zusammen·halt m holding together, unity; cohesion, (inner) bond; consistency. **z-haltbar** a **1** cohesive; (Gases) coercible. **2** comparable. **·halten*** I vt **1** hold (or keep) together. **2** hold side by side (for comparison). II vi **3** hold together, cohere, stay intact. **4** stay (or stick) together

Zusammen·hang m (. . h⁼e) **1** connection, link. **2** (inter)relation(ship). **3** coherence; consistency. **4** context. **z--·hängen*** vi (oft + mit) be connected, be linked (with); be related (to). **-hängend** p.a **1** connected; continuous; coherent. **2** related—**damit z.** (oft) relevant. **-hangslos** a disconnected, disjointed; incoherent; adv oft out of context

zusammen·häufen vt heap up, pile up, accumulate. **·ketten** vt chain (or link) together. **·kitten** vt glue (cement, bond) together

zusammen·klappbar a collapsible, folding (e.g. chair). **·klappen 1** vt, vi(s) fold up. **2** vi(s) collapse

zusammen·kleben 1 vt/i stick together, agglutinate. **2** vt splice. **·klumpen** vi agglutinate, get lumpy. **·kommen*** vi(s) **1** get together, meet, gather; accumulate. **2** come together, converge, merge. **Z--kunft** f (. . k⁼e) meeting, gathering, assembly; conference, convention

zusammen·lagern 1 vt/i assemble, join. **2** vt arrange together; store together. **3** vi be arranged (be stored) together. **Z--lagerung°** f assembly, assemblage

zusammen·laufen* vi(s) **1** run together; meet, gather; converge, merge; blend. **2** concur. **3** coagulate

Zusammenleben n living together; mutual relationship; coexistence, (*Biol.*) symbiosis

zusammen·legbar *a* collapsible, folding (e.g. table). **·legen I** *vt* 1 fold (up). 2 combine, merge; put together, put side by side. **II** *vt/i* pool (resources). **Z–legung** *f* (-en) combination, merger, consolidation

zusammen·leimen *vt* glue together, agglutinate. **Z–leimung** *f* gluing together; agglutination

zusammen·lesen* *vt* gather, collect; compile. **·löten** *vt* solder together

Zusammenmündung *f* confluence; (*Biol.*) anastomosis

zusammen·nähen *vt* sew up, sew together; suture. **·nehmen* I** *vt* 1 combine, merge, join. 2 gather (*or* muster) up (courage, strength, etc). 3 fold up. **II: sich z.** pull oneself together; control oneself. *Cf* ZUSAMMENGENOMMEN. **·passen** *vi* go (*or* fit) together; match, harmonize; (**mit**) go, fit in (with)

Zusammen·prall° *m* clash, collision, impact. **z–·prallen** *vi*(s) (**mit**) collide, clash (with)

zusammen·preßbar *a* compressible. **·pressen** *vt* press together, compress

zusammen·raffen 1 *vt* gather, muster, pick up, collect. 2: **sich z.** pull oneself together. **·rechnen** *vt* reckon (figure, add, count) up. **·reiben*** *vt* rub (*or* grind) together. **·rollen** 1 *vt* & **sich z.** roll (coil, curl) up. 2 *vi*(s) collide. **·rücken** *vt*, *vi*(s) move (closer) together, move side by side. **·rühren** *vt* stir together, mix (by stirring); concoct. **·sacken** *vi*(s) collapse. **·scharen** *vr:* **sich z.** gather, congregate

Zusammenschau *f* (-en) (overall) view; synopsis

zusammen·schiebbar *a* telescopic, collapsible. **·schieben*** *vt* push closer together, telescope

zusammen·schleifen* *vt* grind together

zusammen·schließen* *vt* & **sich z.** combine, unite, merge, consolidate. **Z–schluß°** *m* combination, union, merger, consolidation

zusammen·schmelzen* *vt*, *vi*(s) 1 melt together (*or* into one), fuse. 2 melt down to nothing. **·schreiben*** *vt* 1 write as one word *cf* ZUSAMMENGESCHRIEBEN. 2 (hast-ily) throw together, scribble (an article, etc). **Z–schreibung°** *f* solid spelling, writing as one word. **·schrumpfen** *vi*(s) shrink down, dwindle away; contract. **·schütteln** *vt* shake to form a mixture. **·schütten** *vt* pour together, mix. **·schweißen** *vt* weld together

zusammen·setzen I *vt* 1 put together, compose, combine, assemble; compound. 2 (*Glass*) laminate. **II: sich z.** 3 combine, unite. 4 (**aus**) be composed, consist (of). 5 sit down together (*or* side by side); assemble. **Z–setzung** *f* (-en) 1 compound; combination. 2 assembly. 3 composition, make-up. 4 synthesis; putting together, combining, etc.

Zusammensetzungs·dreieck *n* composition triangle. **·verhältnis** *n* combining proportion

zusammen·sinken* *vi*(s) collapse. **·sintern** *vt/i* sinter together, agglomerate

Zusammenspiel *n* interaction, interplay; teamwork

zusammen·stecken 1 *vt* put (stick, pin) together. 2 *vi* stick (= stay) together

zusammen·stellen I *vt* 1 put together, compose, assemble, make up. 2 make out, write out, draw up (a document). 3 compile. 4 put side by side. **II: sich z.** stand side by side. **Z–stellung°** *f* 1 composition, assembly; set. 2 making out, writing out, drawing up. 3 compilation; tabulation. 4 juxtaposition

zusammen·stimmen *vi* be in agreement, be in harmony

Zusammen·stoß° *m* collision; clash. **z–·stoßen*** *vi*(s) 1 collide, clash. 2 adjoin, abut. **·stoßend** *p.a* (*oft*) adjacent

Zusammen·sturz° *m*, **z–·stürzen** *vi*(s) collapse

zusammen·tragen* *vt* gather, collect, compile. **·treffen*** *vi*(s) 1 (**mit**) meet (with), run (into); coincide (with). 2 convene. **Z–treffen** *n* (-) 1 meeting, convention. 2 coincidence. **·treten*** 1 *vi*(s) meet, come (*or* get) together; (**zu**) join, combine (to form). 2 *vt* trample underfoot. **·trocknen** *vi*(s) dry up, shrivel up. **·tun*** 1 *vt* put together. 2: **sich z.** join forces, cooperate. **·wachsen*** *vi*(s) 1 grow (back) together; heal up; (*Bones*) knit. 2 coalesce; merge.

·**wirken** *vi* 1 work together, collaborate. 2 have a combined effect. ·**zählen** *vt* add (*or* count) up

zusammen·ziehbar *a* contractible. ·**ziehen*** I *vt* 1 pull together. 2 concentrate, mass. II *vt* & sich z. contract, shrink. III *vt/i* have an astringent effect (on). IV *vi*(s) (**mit**) move in together (with). V: sich z. (*Storm*) gather, brew. -**ziehend** *p.a* astringent. **Z–ziehung** *f* (-en) contraction; concentration

zusamt *prep* (*dat*) (together) with, including

Zusatz° *m* 1 addition: **unter Z. von Wasser** (while) adding water, with the addition of water. 2 admixture, added amount (*or* substance), additive; supplement. 3 addendum, appendix, postscript; corollary. 4 (*Soldering*) filler metal. 5 (*in compds oft*) additional, added, additive, supplementary: -**element** *n* added element; (*Metll.*) alloying element. -**feld** *n* additional (*or* supplementary) field. -**gas** *n* make-up gas. -**gerät** *n* supplementary apparatus, attachment, accessory. -**glied** *n* additional (*or* supplementary) term (*or* member). -**komponente** *f* additional (*or* added) component, additive, admixture. -**kosten** *pl* additional expenses. -**lack** *m* blending varnish. -**legierung** *f* key (*or* hardening) alloy

zusätzlich *a* additional, added, extra; supplementary, auxiliary; (*as adv* oft) in addition, besides

Zusatz· additional, etc: -**linse** *f* supplementary lens. -**luft** *f* additional (*or* supplementary) air. -**lüfter** *m* (*Min.*) auxiliary fan. -**luftverdichter** *m* make-up air compressor. -**material** *n* additional material, additive, admixture. -**metall** *n* alloy. -**mittel** *n* additive, admixture; addition agent. -**patent** *n* addition to a patent. -**spannung** *f* (*Elec.*) booster voltage. -**stoff** *m* admixture, additive; filler. -**studium** *n* additional study (program). -**waschmittel** *n* supplementary detergent. -**wasser** *n* added (*or* make-up) water

zu·schalten *vt* (*dat* or **zu**) connect (to), insert (in), hook up (with); synchronize (with)

zuschanden *adv* 1: **z. machen** ruin, foil. 2: **z. werden** go to ruin, collapse. 3: **sich z. arbeiten** work oneself to death. 4: **z. fahren** *vt* wreck (a vehicle). *Cf* ZU IV

zu·schärfen *vt* sharpen, point; (*Cryst.*) bevel

zu·schauen *vi* (*dat*) watch, look on: **einem beim Arbeiten z.** watch sbdy work, look on as sbdy works. **Zuschauer** *m* (-) spectator, member of the audience; observer, onlooker; (*TV*) viewer

Zuschauer·raum *m* auditorium; house. -**tribüne** *f* (grand)stand

zu·schicken *vt* (*d.i.o*) send, ship, mail (to): **zugeschickt bekommen** receive by mail

zu·schieben* *vt* 1 slide (*or* push) shut. 2 (*d.i.o*) push (over to, toward), shift (e.g. blame to sbdy else)

zu·schießen* I *vt* (*d.i.o*) 1 shoot (a ball to, glances at). 2 contribute (to): **einem Geld z.** (*oft*) subsidize sbdy. II *vi*(s) (**auf**) rush, dash (at, to toward)

Zuschlag° *m* 1 additive, admixture; (*Metll.*) flux; (*Concrete*) aggregate. 2 surcharge, extra charge; extra fare; surtax. 3 bonus. 4 acceptance of a bid—**den Z. erteilen** award the contract

zu·schlagen* I *vt* (*oft* + *d.i.o*) 1 add (on, onto). 2 nail (hammer, slam) shut. 3 chop, hammer (to fit, into shape). 4 play, hit, serve (a ball to). 5 award (to an heir, a bidder); (*Auction*) knock down (to). II *vi* (h) strike (*or* lash) out, attack, take action. III *vi*(s) slam shut, snap shut

Zuschlag·erz *n* fluxing ore. -**gebühr** *f* surcharge, extra charge. -**kalkstein** *m* fluxing limestone

zuschläglich *prep* (*genit*) with the addition of, plus

zuschlag·pflichtig *a* liable to surcharge, etc *cf* ZUSCHLAG 2

Zuschlags· *see* ZUSCHLAG·

Zuschlag·stoff *m* additive, admixture; filler; aggregate; flux *cf* ZUSCHLAG 1

zu·schleifen* *vt* grind to shape (*or* to fit)

zu·schmelzen* 1 *vt* melt (*or* fuse) shut, seal. 2 *vi*(s) fuse shut

zu·schmieren *vt* gum up, clog; fill up, smear up

zu·schneiden* *vt* (*Tex.* & *fig*) cut out (from

a pattern); trim, tailor (to fit). **Zuschnitt** *m* cut, style, tailoring

zu·schnüren *vt* lace (*or* tie) up; constrict; strangle

zu·schrauben *vt* screw shut

zu·schreiben* I *vt* (*d.i.o*) 1 ascribe, attribute (to), hold responsible (for). 2 bequeath (to), credit, transfer (to sbdy's account). 3 add (sthg in writing). II *vi* accept (an invitation). **Zuschrift°** *f* letter, note, communication; reply (to an inquiry *or* ad)

zuschulden *adv: sich etwas z. kommen lassen* be (inadvertently) guilty of sthg *cf* ZU IV

Zuschuß° *m* subsidy; benefit, grant; contribution, allowance. **-betrieb** *m*, **-unternehmen** *n* subsidized enterprise

zu·schütten *vt* 1 fill up (a ditch, etc). 2 (*d.i.o*) pour on more

zu·sehen* *vi* 1 (*dat*) watch, look on = ZUSCHAUEN. 2 wait and see. 3 stand (idly) by and watch. 4 (+ *daß*) see to it, take care (that). **zusehends** *adv* 1 visibly. 2 notably, considerably. 3 rapidly

zu·senden* *vt* (*d.i.o*) send (ship, mail) to *cf* ZUSCHICKEN. **Zusendung°** *f* (piece of) mail, shipment, consignment

zu·setzen I *vt* 1 (*d.i.o*) add (to); mix in (with); contribute (to). **2: nichts zuzusetzen haben (a)** have no fat (*or* weight) to spare, **(b)** have no resistance (to illness). II *vi* (*dat*) harass, molest; mistreat; take a lot out of. III *vt/i* (**Geld**) **z.** lose (invested) money. **Zusetz·mittel** *n* reagent

zu·sichern *vt: einem etwas z.* guarantee sbdy sthg, assure sbdy of sthg. **Zusicherung°** *f* assurance, guarantee

Zus.-Patent *n, abbv* (Zusatzpatent) addition to a patent

zu·sperren *vt* lock (*or* close) up

zu·spitzen 1 *vt* point, sharpen (to a point). 2 *vt & sich z.* intensify. 3: *sich z.* come to a point; come to a head, become more critical. *Cf* ZUGESPITZT

zu·sprechen* I *vt* (*d.i.o*) 1 award (to); credit (sbdy) with. **2: einem** (z.B. **Mut**) **z.** offer sbdy words (e.g. of courage). **3: einem ein Telegramm z.** read sbdy a telegram by telephone. II *vi* (*dat*) 4 (+ *adv*) talk (e.g.

gently) (to). **5: dem Essen (dem Trinken) z.** do justice to the food (the drinks); **dem Alkohol z.** indulge in alcohol

Zuspruch *m* 1 words of encouragement (sympathy, etc) *cf* ZUSPRECHEN 2. **2: Z. haben** enjoy popularity; have a clientele, have followers, etc. **3: Z. finden (bei)** meet with (sbdy's) approval, be popular (with)

Zustand° *m* condition, state; phase; state of affairs

zustande *adv* **1: z. bringen** bring about, accomplish, achieve, create. **2: z. kommen** come about, be achieved, materialize. *Cf* ZU IV. **Zustande·bringen** *n* accomplishment, achievement (of sthg). **-kommen** *n* rise, achievement, materialization; occurrence

zuständig *a* responsible, authorized, having jurisdiction. **Zuständigkeit** *f* jurisdiction, authority, responsibility

Zustands·änderung *f* change of state. **-diagramm** *n* phase diagram. **-feld** *n* phase region. **-form** *f* state, form. **-formel** *f* state formula. **-gebiet** *n* phase region. **-gleichung** *f* equation of state. **-größe** *f* variable of state; variable quantity (*or* magnitude), (*specif, oft*) entropy. **-kurve** *f* curve of state. **-schaubild** *n* phase (*or* equilibrium) diagram. **-summe** *f* partial function. **-verschiebung** *f* change (*or* shift) in state

zustatten *adv: z. kommen* (*dat*) come in handy (for), be welcome (to), help

zu·stecken 1 *vt* pin together. 2 *vi* (h) slip (sthg to sbdy)

zu·stehen* *vi* (h *or* s) 1 (*dat*): **es steht ihnen nicht zu, das zu tun** it is not up to them (*or* incumbent on them) to do that. **2: es steht ihnen zu** they are entitled to it. **3: man muß ihnen z., daß . . .** one must consider in their favor that . . .

zu·stellen *vt* 1 block, close off. 2 line (a furnace). 3 (*d.i.o*) deliver (mail, etc) (to); serve a summons (on). **Zusteller** *m* (-), **Zustellerin** *f* (-nen) mail carrier; deliverer, messenger; process (*or* summons) server. **Zustellung°** *f* delivery, etc. **Zustellungs·masse** *f* (*Metll.*) lining material

zu·stimmen *vi* (*dat*) agree (with), consent

(to); approve (of). **zustimmend** *p.a* approving, etc; *as adv oft* in agreement, affirmatively. **Zustimmung°** *f* agreement, consent, approval, endorsement: **Z. finden** meet with approval

zu·stopfen *vt* **1** plug up. **2** darn, mend
zu·stöpseln *vt* stopper, cork
zu·stoßen* **I** *vt* **1** push (e.g. a door) to (*or* shut). **II** *vi*(h) **2** stab, lunge (with a knife, etc). **III** *vi*(s) **3** (**auf**) head (for). **4** (come and) join those present. **5** (*dat*) happen (to)

Zustrebe·kraft *f* centripetal force
zu·streben *vi* (*dat*) strive (to reach), tend (toward)
Zustrom *m* **1** (**an**) influx, inflow (of). **2** crowd; visitors, customers, followers: **Z. gewinnen** win followers, gain popularity. **zu·strömen** *vi*(s) (*dat*) flow in, come streaming (*or* flocking) (to, toward)

zutage *adv* **1**: **z. bringen, z. fördern** bring to light. **2**: **z. kommen, z. treten** come to light, (*Min., Geol.*) appear, crop out *cf* STREICHEN 12. **3**: **z. liegen** be obvious. *Cf* ZU **IV**

Zutat° *f* **1** (additional) ingredient, seasoning. **2** addition, addendum, extra. **3** accessory; *pl oft* furnishings, trimmings
zu·teilen *vt* (*d.i.o*) assign, allot, ration out (to); (*Mach.*) feed, supply (to). **Zuteiler** *m* (-) distributor, feeder. **Zuteilung°** *f* allotment, ration: **auf Z. erhalten** receive as one's ration. **Zuteilungs·verhältnis** *n* mixing ratio
zutiefst *adv* deeply; at the very bottom
zu·tragen* **I** *vit* **1** (*d.i.o*) take, carry, deliver (to). **2** slip (to) secretly, report (to). **II**: **sich z.** happen, take place
zuträglich *a* **1** (*dat*) conducive; beneficial, advantageous (to). **2** healthy, wholesome
zu·trauen *vt* (*d.i.o*) think (sbdy) capable of, expect (sthg) (of). **Zutrauen** *n* (**zu**) trust, confidence (in); trustworthiness. **zutraulich** *a* trusting; sociable
zu·treffen *vi* (**auf, für**) (prove to) be true (of), applicable, relevant; be appropriate
zu·treten* **I** *vi* (s) **1** (**auf**) come (step, walk) up (to), approach: **z. lassen** (*dat or* **auf**) give access (to). **II** *vt* **2** fill up (a hole) and tread it down. **3** kick (e.g. a door) shut
Zutritt *m* admittance, admission, entry, ac-

cess: **Z. verboten, kein Z.** no admittance, do not enter
zu·tröpfeln, zu·tropfen 1 *vt* (*d.i.o*), *vt* (*dat*) drop in, drip in. **2** *vt* add drops of . . . **3** *vi* (+ **lassen**) add drop by drop
zu·tun* *vt* **1** close, shut. **2** (*d.i.o*) add (to). **3** (*dat rflx*): **sich etwas z.** get (oneself) sthg. *Cf* ZUGETAN. **Zutun** *n* aid, cooperation, involvement: **ohne ihr Z.** without their being involved
zuungunsten *prep* (*genit*) to the disadvantage (of), in (sbdy's) disfavor *cf* ZU **IV**
zuunterst *adv* all the way down, at the very bottom: **das oberste z. kehren** turn everything upside down
zuverlässig *a* reliable, dependable, trustworthy; *adv oft* for sure, for certain. **Zuverlässigkeit** *f* reliability, dependability
Zuversicht *f* confidence, trust. **zuversichtlich** *a* confident. **Zuversichtlichkeit** *f* confidence
zuviel *adv/pron* too much. **Zuviel** *n* excess. **zuviele** *a/pron* too many
zuvor 1 *adv* before (hand): **tags z., am Tag z.** the day before, the preceding day. **2** *sep pfx*: *cf* ZUVORKOMMEN, ZUVORTUN
zuvorderst *adv* all the way forward, at the very front
zuvörderst *adv* first of all, to begin with
zuvor·kommen* *vi*(s) (*dat*) anticipate; forestall, head off. **zuvorkommend** *p.a* obliging, accommodating, helpful. **Zuvorkommenheit** *f* helpfulness, obliging manner
zuvor·tun* *vt*: **es einem z.** (**an**) outdo sbdy (in)
zuw. *abbv* (zuweilen) sometimes
Zuwachs *m* **1** (**an**) increase, growth (in): **auf Z. berechnet** designed for growth (*or* future expansion). **2** addition to the family, progeny
zu·wachsen* *vi*(s) **1** heal up, grow back together. **2** be(come) overgrown. **3** (*dat*) accrue, be assigned (to)
zuwege *adv* **1**: **z. bringen** bring about, accomplish. **2**: **z. kommen** (**mit**) make progress (with). **3**: **gut** (**schlecht**) **z. sein** be well (badly) off, feel well (badly). *Cf* ZU **IV**
zuweilen *adv* sometimes, at times, now and then

zu·weisen* *vt* (*d.i.o*) assign, allocate (to).
Zuweisung° *f* assignment, allocation
zu·wenden* 1 *vt & sich z.* (*dat*) turn (to,
toward). 2 *vt* (*d.i.o*) bestow (on), award (to).
3: *sich z.* (*dat*) devote oneself, direct one's
attention (to); resort (to)
zuwenig *adv* too little, *pl* too few
zu·werfen* I *vt* 1 (*d.i.o*) throw, cast (to, toward). 2 fill up, shovel full (e.g. a ditch). 3
slam (shut)
zuwider I *a, usu pred* (*dat*) distasteful, re-
volting: **das ist ihnen z.** they hate that.
II *prep* (*dat*) contrary to: **der Vernunft z.**
contrary to reason. III *sep pfx* contrary,
counter: **·handeln** *vi* (*dat*) act contrary to,
violate. **Z–handelnde** *m, f* (*adj endgs*)
violator. **-handlung** *f* violation. **z–
·laufen** *vi*(s) (*dat*) run counter to,
contradict
Zuwuchs *m* 1 increase, growth, etc =
ZUWACHS. 2 (*Math.*) increment
zuzeiten *adv* at times, sometimes *cf* ZU IV
zu·ziehen* I *vt* 1 pull (a door) to (*or* shut);
pull, draw (blinds, etc); pull tight, tight-
en. 2 (*oft* + **zur Beratung**) consult. 3 (+
dat rflx **sich**) contract (an illness), incur,
suffer (an injury). II *vi*(s) move in, immigrate; join those already settled in a
place. III: *sich z.* tighten (up)
Zuzug *m* 1 moving in, immigration. 2 new-
comers, new personnel. (adherents, set-
tlers, etc); reinforcements
zw. *abbv* 1 (zwar) to be sure. 2 (zwecks) for
the purpose of. 3 (zwischen) btw. (between)
Zwack·eisen *n* (pair of) pincers. **zwacken**
vt pinch
zwang forced: *past of* ZWINGEN
Zwang *m* 1 coercion, force; pressure, duress: **sich Z. antun** control (*or* restrain)
oneself. 2 restraint, constraint; obligation: **Prinzip des kleinsten Z—es** Le
Châtelier's principle of least restraint (*or*
of mobile equilibrium). 3 control; despotic
rule, tyranny. 4 (*incl Psych.*) compulsion
zwängen *vt & sich z.* squeeze, wedge, cram
(oneself)
Zwang·lauf *m* (*Mach., Engg.*) guided (*or*
positive) motion. **z–läufig** *a* guided (motion); (*specif*): positive (drive), geared

(transmission), force-feed (oil circulation) *cf* ZWANGSLÄUFIG. **Z–läufigkeit** *f*
guided motion. **-lauflehre** *f* kinetics,
kinematics
zwanglos *a* 1 casual, informal; unconven-
tional. 2 (published) at irregular
intervals
Zwangs·lage *f* predicament; quandary, dilemma. **z–läufig** *a* necessary, unavoidable; inevitable; legally required; *adv oft*
of necessity *cf* ZWANGLÄUFIG: *Caution:
the two words are sometimes confused
even by German writers!* **-mischer** *m* mechanical mixer. **-mischung** *f* mechanical
mixing. **-umlauf** *m* forced circulation. **z–
weise** *a* 1 inevitable. 2 compulsory, obligatory. 3 forcible; *adv oft* by force. **Z–
wirtschaft** *f* 1 government control. 2 controlled economy
zwanzig *num* twenty; *as pfx oft* icosa-. (*For
compds and derivs, e.g.* **zwanzigfach,** *cf
similar forms under* ZWEI.) **Z–·flächner**
m icosahedron, **zwanzigste** *a,* **Zwanzig-
stel** *n* (*Frac.*) twentieth
zwar *cnjc* 1: **z. . . . aber** to be sure . . . but,
of course . . . but. 2: **und z.** and in fact,
and as a matter of fact, and to be specific
Zweck *m* (-e) purpose, object; aim, end;
point: **zu diesem Z.** for this purpose; **Mittel zum Z.** means to an end; **Ziel und Z.**
aim and purpose; **das hat keinen Z.** that
is of no use, there is no point to (doing)
that. *Cf* ZWECKE. **z–·dienlich** *a* useful,
relevant, appropriate, suitable
Zwecke *f* (-n) tack; (wooden) peg *cf* ZWECK.
zwecken *vt* tack, peg
zweck·entsprechend *a* suitable, appropriate. **Z–forschung** *f* practical (*or* applied)
research. **z–gebunden** *a* earmarked
zweckhaft *a* 1 suitable, appropriate. 2
practical
zweck·los *a* useless; pointless, aimless;
futile. **-mäßig** *a* suitable, appropriate; expedient. **-mäßigerweise** *adv* suitably,
appropriately. **Z–mäßigkeit** *f* suitability,
appropriateness; expediency
zwecks *prep* (*genit*) for the purpose (of)
zweck·widrig *a* unsuitable, inappropriate;
inexpedient
zwei *num* two; (*as pfx oft*) duo-, di-, bi-, dou-

ble: **-achsig** *a* biaxial. **-armig** *a* two-armed. **-atomig** *a* diatomic. **-äugig** *a* binocular; two-eyed. **-badig** *a* two-bath. **Z—badverfahren** *n* two-bath process. **z—basisch** *a* dibasic. **-deutig** *a* ambiguous, equivocal. **Z—deutigkeit** *f* ambiguity. **z—dimensional** *a* two-dimensional. **-einhalb** *num* two and a half **zweieinhalb·fach** *a* two-and-a-half-fold. **-mal** *adv* two and a half times **zwei·einwertig** *a* bi-univalent **zweier** of two: *genit of* ZWEI. **Zweier** *m* (-) (figure, number) two **zweierlei** *a* (*no endgs*) of two different kinds. **·-wertig** *a* having two (different) valences **Zweier·schale** *f* duplet shell (*or* ring). **-stoß** *m* two-body collision **zweifach** *a* twofold, double, dual; (*as pfx*) di-, bi-: **-basisch** *a* dibasic. **-chromsauer** *a:* **z—es Kali** potassium dichromate. **-frei** *a* bivariant, having two degrees of freedom. **-ionisiert** *a* doubly ionized. **-kohlensauer** *a:* **z—es Natron** sodium bicarbonate. **-sauer** *a* (*in Acid Salts*) dihydrogen: **z—es Natriumphosphat** sodium dihydrogen phosphate. **-schwefeleisen** *n* iron disulfide. **z—schwefelsauer** *a* bisulfate of. **Z—schwefelzinn** *n* tin disulfide. **-verbindung** *f* binary compound **zwei·fädig** *a* two-threaded, bifilar. **Z—farbendruck** *m* two-color print(ing). **z—farbig** *a* two-color(ed), bi-colored, dichromatic, dichroic. **Z—farbigkeit** *f* dichroism **Zweifel** *m* (-) (**an, über**) doubt (about): **ohne Z.** without a doubt, undoubtedly; **es steht außer Z.** it is beyond doubt; **Z. hegen** (or **haben**), **ob** . . . have doubts (as to) whether . . . **über allen Z. erhaben** beyond all doubt; **etwas in Z. stellen** (or **ziehen**) question sthg, cast doubt on sthg. **zweifelhaft** *a* doubtful, dubious, questionable. **zweifellos** *a* doubtless, undoubted, *adv oft* without (*or* beyond) a doubt. **zweifeln** *vi* (**an**) have doubts (about), doubt, question **Zweifels·fall** *m* doubtful case: **im Z.** in case of doubt. **z—ohne** *adv* without a doubt, doubtless

zweifel·süchtig *a* skeptical **zwei·flächig** *a* two-surfaced, two-faced, dihedral **Zweifler** *m* (-) doubter, skeptic **Zweig** *m* (-e) branch (*incl fig*); twig, limb; (*Mountain*) spur **zwei·gängig** *a* (*Mach.*) double-thread(ed) **Zweig·betrieb** *m* branch operation (*or* establishment). **-büro** *n* branch office **zwei·gestaltig** *a* dimorphic. **Z—gestaltung** *f* dimorphism **Zweig·gesellschaft** *f* affiliate, branch establishment **zwei·gipfelig** *a* two-peaked, double-peaked **Zweig·leitung** *f* branch line (pipe, wire) **zwei·gliedrig** *a* two-membered, binary; (*Math.*) binomial **Zweig·niederlassung** *f* branch (establishment), affiliate. **-rohr** *n* branch pipe (*or* tube). **-stelle** *f* branch office. **-strom** *m* branch current. **-wissenschaft** *f* affiliated science **zwei·halsig** *a* two-necked **Zweiheit** *f* duality, dualism **zwei·höckerig** *a* two-humped, double-hump **zwei·hundert** *num* two hundred. **-hundertste** *a*, **-hundertstel** *a* (*Frac.*) two hundredth. **-hundertjährig** *a* 1 bicentennial. 2 two-hundred-year(-old) **zwei·jährig** *a* 1 two-year(-old). 2 biennial. **-keimblätterig** *a* (*Bot.*) dicotyledonous. **Z—kern·** (*in compd nouns*) = **z—kernig** *a* binuclear. **Z—körperverdampfer** *m* double-effect evaporator. **z—lebig** *a* amphibious **zwei·mal** *adv* twice; (*when ordering*) two (portions, orders) (of). **-malig** *a* two; twice-done, double. **Z—malschmelzerei** *f* (*Metll.*) Walloon (two-stage refining) process **Zwei·metall** *n* bimetal. **z—monatig** *a* two-month(-old). **-monatlich** *a* bimonthly. **Z—phasen·** (*in compd nouns*) = **z—phasig** *a* two-phase, diphase. **-polig** *a* bipolar, double-pole. **-quantig** *a* two-quantum. **Z—rad** *n* bicycle. **z—räderig** *a* two-wheel(ed). **-reihig** *a* two-row; two-series; double-breasted. **Z—richtungsbetrieb** *m* bidirectional operation. **z—säurig** *a*

(*Bases*) diacid(ic). **-schalig** *a* bivalve, bivalvular. **-schenk(e)lig** *a* two-legged, two-branch. **-schichtig** *a* two-layer; two-shift. **Z–schlangendurchfluß** *m* two-coil flow. **z–schneidig** *a* two-edged, double-edge(d). **-seitig** *a* 1 two-sided, bilateral. 2 two-page. 3 double-face, duplex. **-spitzig** *a* (*Curves*) two-peaked, two-pointed. **-sprachig** *a* bilingual. **-spurig** *a* two-track, two-lane. **-stöckig** *a* two-story, double-deck

Zweistoff·legierung *f* binary alloy. **-system** *n* two-component system

zwei·stufig *a* two-step, two-stage. **-stündig** *a* two-hour

zweit: zu z. in twos, in a twosome: **sie sind zu z.** there are two of them *cf* ZWEITE

Zweitakt·motor *m* two-cycle (*or* two-stroke) engine. **-verfahren** *n* two-stroke cycle

zweit·ältest *a* second-oldest. **-best** *a* second-best

zweite *a* second *cf* ZWEIT

zwei·teilig *a* two-part, two-piece. **Z–teilung** *f* 1 division in two, (bi)partition; bisection. 2 dichotomy. 3 bifurcation. 4 binary fission

zweitens *adv* secondly, in the second place

Zweit·haar *n* hair replacement. **z–klassig** *a* second-class. **Z–luft** *f* (*Combustion*) secondary air. **-produkt** *n* 1 second product. 2 (*Sugar*) second crop of crystals. **z–rangig** *a* second-rate, second-rank

Zweiunddreißiger·schale *f* shell of thirty-two electrons

zwei·wandig *a* two-walled, double wall(ed)

Zweiweg(e)·gleichrichter *m* full-wave rectifier. **-hahn** *m* two-way stopcock. **-ventil** *n* two-way valve

zwei·wellig *a* two-wave; (*Curves*) double-hump. **Z–welligkeit** *f* two-wave property. **z–wertig** *a* bivalent, divalent. **Z–wertigkeit** *f* bivalence, divalence. **-zählig** *a* 1 twofold, double, binary; (*Bot.*) binate. 2 (*Ligands*) bidentate, chelate. **Z–zweck·** (*in compd nouns*) dual-purpose . . .

Zwerch·fell *n* (*Anat.*) diaphragm

Zwerg *m* (-e) dwarf, midget; (*as pfx oft, esp Instrum.*) miniature. **zwerg(en)haft** *a* dwarfish, dwarf-like; diminutive

Zwerg·holunder *m* dwarf elderberry. **-welle** *f* microwave

Zwetsche, Zwetschge, Zwetschke *f* (-n) plum

zwetschgen·blau *a* plum-blue. **Z–schnaps** *m* plum brandy

Zwickel *m* (-) 1 wedge. 2 (*Tex.*) gusset. **--(kapillar)wasser** *n* toroidal capillary water

zwicken 1 *vt/i* pinch, nip. 2 *vi* be too tight.

Zwicker *m* (-) 1 (pair of) pincers. 2 pince-nez

Zwick·mühle *f* predicament, tight squeeze, dilemma. **-zange** *f* (pair of) pincers

Zwie·back *m* rusk, zwieback

Zwiebel *f* (-n) 1 onion. 2 (*Bot.*) bulb. **z–artig** *a* onion-like, onion-shaped; bulbous, alliaceous. **-förmig** *a* bulb-shaped, onion-shaped, bulbar. **Z–gewächs** *n* bulbous plant. **-linse** *f* biconvex lens. **-marmor** *m* (*Min.*) cipolin. **-öl** *n* onion oil. **-ring** *m* onion ring. **z–rot** *a* onion-red. **Z–saft** *m* onion juice. **-schale** *f* onion skin

zwie- = ZWEI-: **-fach** *a* twofold, double. **Z–gespräch** *n* dialog, conversation. **-licht** *n* twilight. **-metall** *n* bimetal, clad metal. **-natur** *f* double nature

Zwiesel *m* (-) 1 forked branch. 2 fork, bifurcation

Zwie·spalt *m* inner conflict; dilemma; discord. **z–spältig** *a* 1 split, forked, bifid. 2 disunited; discordant; conflicting. **Z–sprache** *f* dialog: **Z. halten** have a dialog. **-tracht** *f* discord, dissension; strife. **z–trächtig** *a* discordant, disunited

Zwillich *m* (-e) (*Tex.*) tick(ing), drill

Zwilling *m* (-e) twin

Zwillings· twin: **-achse** *f* (*Cryst.*) twin(ning) axis. **z–artig** *a* twin-like, *adv oft* as twins, in the form of twins. **Z–bildung** *f* twin formation, twinning. **-doppelverbindung** *f* conjugated double linkage. **-ebene, -fläche** *f* (*Cryst.*) twinning plane. **-kerne** *mpl* twin nuclei. **-kristall** *m* twin crystal. **-paar** *n* pair of twins; twin pair. **-prisma** *n* biprism. **-salz** *n* double salt

Zwinge *f* (-n) 1 C-clamp; vise. 2 ferrule, metal tip (*or* ring). 3 (*Mach.*) wing nut

zwingen* (zwang, gezwungen) I *vt* 1 force, compel: **sich gezwungen sehen (zu)** find oneself compelled (to). 2 manage, cope with. **II** *vi* (**zu**) make it necessary (*or* unavoidable) to. **III: sich z.** force oneself; control onself. **zwingend** *p.a* urgent, compelling (reasons, etc); conclusive (evidence)

zwinkern *vi* blink, wink; twinkle

Zwirn *m* (-e) twine, double yarn; thread. **--band** *n* cloth tape. **zwirnen** *vt* (*Tex.*) twist, twine; (*Silk*) throw.

Zwirn(s)·faden *m* (twisted) thread

zwischen 1 *prep* (*dat, accus*) between, among. **2** *pfx* inter-, mid-, intermediate, middle: **Z–achse** *f* tertiary axis. **-bad** *n* intermediate bath. **-behälter** *m* intermediate container; (*Petrol.*) surge tank (*or* hopper). **-bemerkung** *f* interjected remark. **-bild** *n* intermediate image. **-boden** *m* **1** intermediate bottom. **2** false (*or* dropped) ceiling. **3** diaphragm. **-dampf** *m* reheat steam. **-ding** *n* something between, intermediate stage; hybrid, cross. **z–durch** *adv* **1** in between, meanwhile. **2** here and there; now and then, occasionally. **3** through. **Z–ergebnis** *n* intermediate (*or* interim) result. **-erzeugnis** *n* intermediate product. **-fall** *m* incident, episode. **-farbe** *f* intermediate color; (*Lthr.*) intermediate pit. **-ferment** *n* intermediate enzyme. **-fläche** *f* interface. **-form** *f* intermediate form. **-frage** *f* interjected question. **-frequenz** *f* intermediate frequency. **-gefäß** *n* intermediate vessel (*or* receptacle). **-gerbung** *f* intermediate tannage. **z–geschichtet** *a* interstratified. **Z–gewebe** *n* interstitial tissue. **-gewebesubstanz** *f* intercellular substance. **-gitteratom** *n* interstitial atom. **-gitterplatz** *m* (*Cryst.*) interstitial position (*or* space). **-glasurmalerei** *f* (*Ceram.*) painting between glazes. **-glied** *n* intermediate member, connecting link. **-glühen** *n*, **-glühung** *f* (*Metll.*) intermediate anneal(ing). **-händler** *m* **1** middleman; agent, intermediary. **2** mediator. **-hirn** *n* midbrain. **-klemmungsmasse** *f* interstitial matter. **-körper** *m* intermediate body (*or* substance). **-kühler** *m* intercooler. **-kühlung** *f* intercooling. **-lage** *f* intermediate layer (*or* position).

z–·lagern *vt* intercalate; interstratify. **z–·landen** *vi*(s) make an intermediate landing. **Z–legierung** *f* intermediate alloy. **z–liegend** *a* **1** intermediate. **2** interim, intervening. **Z–lösung** *f* intermediate (*or* interim) solution. **-maß** *n* intermediate size. **-masse** *f* **1** ground substance. **2** interstitial matter. **z–molekular** *a* intermolecular. **Z–niveau** *n* intermediate level. **-optik** *f* intermediate optical device. **-norm** *f* intermediate (*or* subsidiary) standard (*or* norm). **-pause** *f* interval, intermission; break. **z–planetarisch** *a* interplanetary. **Z–produkt** *n* **1** intermediate (*or* semi-finished) product. **2** (*Ores*) middlings. **-raum** *m* **1** space between, gap; interspace, interstice. **2** interval—**in Z–en** intermittently. **-reaktion** *f* intermediate reaction. **z–·schalten** *vt* insert, interpose; connect in between; (*Elec.*) connect in series, cut in. **Z–schaltung** *f* insertion, interposition; cutting in, etc: **unter Z. eines Gleichrichters** cutting in a rectifier, with a rectifier cut in between. **-schicht** *f* intermediate layer (stratum; shift), interlayer. **-sorte** *f* intermediate grade (quality, brand). **-spiel** *n* interlude. **-spülung** *f* intermediate rinsing. **z–staatlich** *a* international. **Z–stadium** *n* intermediate stage. **z–städtisch** *a* intercity. **-ständig** *a* intermediate. **Z–stecker** *m* (*Elec.*) adapter (plug). **-stein** *m* (*Copper*) blue metal. **-stellung** *f* intermediate position, **-stück** *n* **1** intermediate piece (*or* part); insert. **2** connecting piece. **-stufe** *f* intermediate stage (*or* step). **-substanz** *f* intermediate (interstitial, ground) substance. **-summe** *f* subtotal. **-ton** *m* (*Colors*) intermediate shade (*or* tone). **-träger** *m* intermediate carrier (*or* support), intermediary. **-trocknung** *f* intermediate drying. **-verbindung** *f* intermediate compound. **-vergütung** *f* (*Metll.*) austempering. **-walzen** *n* (*Metll.*) intermediate rolling. **-wand** *f* **1** partition, bulkhead; baffle, shelf. **2** diaphragm; septum. **3** (*Paper*) midfeather (of a beater). **-wärmung** *f* intermediate heating. **-wasser** *n* capillary water. **-weite** *f* distance between, interval. **-wirbelknorpel** *m* intervertebral cartilage. **-wirt** *m* (*Biol.*) intermediate host. **-zeit** *f* interval, meantime, interim. **z–zeitlich** *adv* meanwhile. **-zelle** *f* interstitial cell. **z–zellig** *a* intercellular. **Z–zellraum** *m* intercellular space. **-zunder** *m* (*Explo.*) intermediate charge, booster. **-zustand** *m* intermediate state

Zwist *m* (-e) dissension, discord; strife

zwitschern *vt/i* chirp, twitter; warble

Zwitter *m* (-) hybrid, hermaphrodite; mongrel; bastard; (*Petrog.*) zwitter. **--ion** *n* hybrid (*or* amphoteric) ion

zwl. *abbv* (ziemlich wenig löslich) rather difficultly soluble

zwo *num* two (= ZWEI; *used mainly on Tlph.*)

zwölf *num* twelve; (*as pfx oft*) dodeca-:**Z–eck** *n* dodecagon. (*For compds & derivs not listed here or below cf similar ones with* ZWEI.)

Zwölfer·system *n* (*Math.*) duodecimal system

Zwölf· twelve, dodeca-: **-fingerdarm** *m* (*Anat.*) duodenum. **-flach** *m* dodecahedron. **z–flächig** *a* dodecahedral. **Z–flächner** *m* dodecahedron. **z–seitig** *a* twelve-sided, dodecahedral; dodecagonal

zwölfte *a,* **Zwölftel** *n* (*Frac.*) twelfth

zwote *a* second = ZWEITE, *cf* ZWO

Zyan *n* cyanogen; (*in compds, oft*) cyanic, cyanide. *See* CYAN·

Zyane *f* (-n) cornflower

zygotisch *a* zygotic

zyklisch *a* cyclical

Zyklisieren *n* cyclization

Zykloide *f* (-n) cycloid. **zykloidisch** *a* cycloidal

Zyklon *m* (-e) 1 (*Meteor.*) cyclone. 2 (*Mach.*) cyclone separator

Zyklotron *n* (-e) cyclotron

Zyklus *m* (. . klen) cycle

Zylinder *m* (-) 1 (*Math., Mach., Engg.*) cylinder. 2 (lamp) chimney. 3 top hat. **Zylinderchen** *n* small cylinder

Zylinder·dämpfer *m* cylinder steamer. **z–förmig** *a* cylinder-shaped, cylindrical **zylindern, zylindrieren** *vt* calender **Zylinder·öl** *n* cylinder oil **zylindrisch** *a* cylindrical

Zymotechnik *f* zymotechnology. **zymotechnisch** *a* zymotechnical

Zypern *n* (*Geog.*) Cyprus

Zyper·vitriol *n* blue vitriol, copper sulfate. **-wein** *m* Cyprus (*or* Cypriot) wine. **-wurz(el)** *f* (*Bot.*) galingale

Zypresse *f* (-n) cypress. **Zypressen·öl** *n* cypress oil

zyprisch *a* Cyprian, Cypriot; **z—es Vitriol** blue vitriol

Zyste *f* (-n) cyst = CYSTE

Zytolyse *f* (-n) cytolysis

Zytoplasma *n* cytoplasm

z.Z., z.Zt. *abbv* (zur Zeit) at present, for the present = ZURZEIT, *cf* ZEIT 1